FOR ALL PRACTICAL PURPOSES

Project Director **Solomon Garfunkel**
Consortium for Mathematics and Its Applications

Contributing Authors

PART I **MANAGEMENT SCIENCE**
Joseph Malkevitch
York College, The City University of New York

PART II **STATISTICS: THE SCIENCE OF DATA**
Marsha J. Davis
Eastern Connecticut State University
[*This section was previously authored by David S. Moore, Purdue University (editions 1–7), and Lawrence M. Lesser, The University of Texas at El Paso (editions 8–9).*]

PART III **VOTING AND SOCIAL CHOICE**
Alan D. Taylor | **Bruce P. Conrad** | **Michael A. Jones**
Union College | Temple University | American Mathematical Society
[*Chapter 12 was previously authored by Steven J. Brams, New York University.*]

PART IV **FAIRNESS AND GAME THEORY**
Alan D. Taylor | **Bruce P. Conrad** | **Michael A. Jones**
Union College | Temple University | American Mathematical Society
[*Chapter 15 was previously authored by Steven J. Brams, New York University.*]

PART V **THE DIGITAL REVOLUTION**
Joseph A. Gallian
University of Minnesota Duluth

PART VI **ON SIZE AND GROWTH**
Paul J. Campbell
Beloit College

PART VII **YOUR MONEY AND RESOURCES**
Paul J. Campbell
Beloit College

FOR ALL PRACTICAL PURPOSES

Mathematical Literacy in Today's World

Tenth Edition

W. H. FREEMAN & COMPANY

A Macmillan Education Imprint

Publisher:	Terri Ward
Acquisitions Editor:	Nikki Miller
Marketing Manager:	Cara LeClair
Development Editors:	Leslie Lahr and Jorge Amaral
Executive Media Editor:	Laura Judge
Assistant Media Editor:	Danielle Lindsey
Associate Editor:	Marie Dripchak
Editorial Assistant:	Victoria Garvey
Marketing Assistant:	Bailey James
Photo Editors:	Cecilia Varas and Nick Ciani
Cover Designer:	Blake Logan
Art Manager:	Matthew McAdams
Text Designer:	Marsha Cohen
Senior Project Editor:	Vivien Weiss
Production Supervisor:	Susan Wein
Illustrations:	Eli Ensor and Lisa Winden
Cover Art:	© Illustration source/Kiki Tikiriki
Composition:	MPS Limited
Printing and Binding:	RR Donnelley & Sons

Library of Congress Control Number: 2015939496

Student Edition (hardcover):
ISBN-13: 978-1-4641-2473-0
ISBN-10: 1-4641-2473-6

Student Edition (loose-leaf):
ISBN-13: 978-1-4641-2483-9
ISBN-10: 1-4641-2483-3

Instructor Comp Copy (hardcover):
ISBN-13: 978-1-4641-6633-4
ISBN-10: 1-4641-6633-1

NASTA-spec version:
ISBN-13: 978-1-4641-2480-8
ISBN-10: 1-4641-2480-9

Printed in China
Second printing

W. H. Freeman and Company
One New York Plaza Suite 4500
New York, NY 10004-1562
www.macmillanhighered.com

Brief Contents

Contents

Robert Daly/Caiaimage/
OJO+/Getty Images

Part I
MANAGEMENT SCIENCE / 2

Carl Court/Getty Images

Blend Images, Hill Street Studios/Getty Images

Yana Gayvoronskaya/ Shutterstock

Part IV FAIRNESS AND GAME THEORY / 536

Justin Ziewe/Ikon Images/SuperStock

SnapshotPhotos/Shutterstock

Part VII
YOUR MONEY AND RESOURCES / 866

CUSTOM CHAPTERS

The following chapters are available through W. H. Freeman's custom publishing (restrictions may apply):

- Sets
- Problem Solving
- Logic
- Geometry
- Counting and Probability
- Numeration Systems
- Personal Finance

Preface

To the Student

For All Practical Purposes, Tenth Edition, continues our effort to bring the excitement of contemporary mathematical thinking to the nonspecialist. In science and industry, mathematical models are the main tools for analyzing and solving problems that arise. In this book, our goal is to convey the power of mathematics by showing you the wide variety of problems that can be modeled and solved by quantitative means. An extensive package of supplements designed to make study time supremely effective complements the tenth edition. Highlights of the supplements package include the *Student Study Guide* and *Student Solutions Manual.* Between the text and the available resources, *For All Practical Purposes* offers you the tools to succeed in the course and apply your new knowledge to daily life experiences.

There are many ways to talk about why mathematics and its applications matter. You will hear expressions such as "mathematical literacy" or "quantitative literacy." They mean, essentially, that math is important. It is important because knowing it can make your life easier. In other words, it can help explain how your world works. We created this course and this book because we know that not everyone looks at mathematics in this way.

In school, you spent a great deal of time learning the tools of mathematics—how to manipulate symbols and how to solve equations. In this course, you will spend time learning about the power of mathematics, which helps us understand many different parts of our everyday lives and the world itself. We hope this exploration will give you a broader sense of the subject and why we wanted you to take a math course every year you were in school. It's "for all practical purposes" because, in a sense, you've learned to hammer nails and saw wood—and now we're going to build houses.

Enjoy!

To the Instructor

Because *For All Practical Purposes* stresses the connections between contemporary mathematics and modern society, our text must be flexible enough to accommodate new ideas in mathematics and their new applications to our daily lives. We maintain this flexibility in the tenth edition.

Our primary goal for this edition was to further improve the ease of use for instructors and students alike. An extensive supplements package is available within LaunchPad, W. H. Freeman's new online homework system. LaunchPad offers content that has been curated and organized for easy assignability in a simple but powerful interface. Assets integrated into LaunchPad include an interactive e-Book, LearningCurve, practice quizzes, exercise solutions, interactive applets, flashcards, video clips, and much more.

New to the Tenth Edition

New Algebra Review Appendix

An all-new Algebra Review Appendix offers reviews on basic concepts used in the text and includes straightforward examples, technology tips, and practice exercises. Answers to all practice exercises are also included. References to relevant sections of the Algebra Review Appendix appear throughout the text in the margins.

New Self Check Exercises

Self Check exercises have been added throughout the text, with answers at the end of each chapter, allowing students to check their understanding of new concepts as the material is being taught. Instructors can also use these exercises in class as part of their lectures.

New Examples

New and updated examples are included throughout the text to address new topics and changes to the material. Examples provide new topics for class discussion and new ways of relating to essential concepts.

New Exercises

- Exercise sets, including Skills Check questions, have been updated and refreshed.
- Over 200 Self Check exercises have been added throughout the chapters.
- The new Algebra Review Appendix includes over 200 practice exercises.
- New Chapter Review exercises have been added to the exercise sets. These exercises are not organized by section and test student understanding of the chapter material as a whole.

Part-Specific Content Changes

Part I: Management Science

- New discussion of the importance of operations research for improving health care (Chapters 1, 3, and 4).
- Revised example illustrating how to cut the costs of installing a local cable TV network (Chapter 2).
- A new case study, which closes Part I, offers a discussion of the ideas behind Alvin Roth and Lloyd Shapley's Nobel Prize–winning work about stable allocations, such as pairing hospitals and medical school graduates for residencies and pairing schoolchildren with schools they want to attend (school choice).

Part II: Statistics: The Science of Data

- Throughout Part II, more graphic displays have been added to help students visualize the material.
- Spotlight features have been added about the use of technology (TI-84 graphing calculator, Excel) for calculations, creation of graphic displays, and selection of simple random samples (Chapters 5–8).
- Expanded directions for constructing histograms and stemplots, along with examples on changing histogram class intervals and expanding stemplot stems (Chapter 5).
- Addition of an example showing computation of correlation from a formula (Chapter 6, Example 7).
- Explanation of least-squares criterion (Chapter 6).
- Addition of an example showing the dangers of extrapolation (Chapter 6, Example 10).
- Discussion of how the Pew Research Center modifies its telephone survey plans to correct for undercoverage (Spotlight 7.2).
- Expanded instructions on selecting a simple random sample (Chapter 7).
- Expanded discussion of ethics in experiments (Spotlight 7.3).
- Expanded discussion of the basic rules of probability (Chapter 8).
- Inclusion of estimation of probabilities based on survey data (Chapter 8).
- Coverage of conditional probability and Bayes' Rule (Section 8.2).

Part III: Voting and Social Choice

- Simplified coverage of the Banzhaf Power Index (Chapter 11).
- New section on the use of apportionment in presidential primaries, including an example with real data from the 2008 Democratic primary (Section 12.1).
- Improved presentation of spatial modeling by introducing discrete models first and using them to analyze the Electoral College (Chapter 12).

Part IV: Fairness and Game Theory

- Expanded coverage of the history of the apportionment of seats in the U.S. House of Representatives (Chapter 14).
- New section highlights mechanism design—designing a game to achieve a particular outcome (Section 15.4).
- New examples model a work location versus schedule decision, as well as the goalie/kicker interaction in a soccer penalty kick (Chapter 15).

Part V: The Digital Revolution

- New examples of check-digit schemes (Chapter 16).
- Expanded summary of error-detection schemes (Chapter 16).
- Expanded spotlight on history of bar codes (Chapter 16).

- New title for Chapter 17 to more accurately describe the content.
- New spotlight features, covering facts about social security numbers, enigma machines, Mavis Batey, Alan Turing, and smart cards (Chapter 17).
- Added coverage of permutation, Playfair, and Jefferson wheel ciphers (Chapter 17).
- New website and video suggestions added (Chapter 17).

Part VI: On Size and Growth

- Revised exercises that consider production of solar energy, consumption by electric cars, and units for measuring water use (Chapter 18).
- New spotlight featuring an award-winning young mathematician whose interest was prompted by rhythms that he learned from playing drums and reading Sanskrit poetry (Spotlight 19.2).
- New spotlight on the use of Fibonacci numbers in optimizing the design of a solar power plant (Spotlight 19.3).
- New examples of patterns on ancient and modern pottery (Chapter 19).
- New illustration of a Penrose pattern in San Francisco architecture, and new figures and exercises about the inflation of patterns, as well as a discussion of the Nobel Prize awarded for the discovery of quasicrystals (Chapter 20).
- New spotlight on mathematics and autism (Spotlight 20.5).

Part VII: Your Money and Resources

- Incorporation of current student loan interest rates into examples (Chapter 22).
- New section on discounted and add-on loans, with exercises about payday loans (Chapter 22).
- Example showing details of costs at a real estate closing (Chapter 22).
- New spotlight on how minimum-size catch limits manipulate fish genetics (Spotlight 23.4).
- New exercises on radioactive isotopes, including those used in medicine and those released in the meltdown of Japanese reactors (Chapter 23).

Focus on Accuracy

For this edition, we once again implemented a detailed accuracy-checking plan to sustain the quality of the exercises and solutions. To this end, we are very grateful to Dennis Evans of Concordia University–Wisconsin and Paul McCombs of Rock Valley College.

Custom Options

In addition to the extensive topics covered in the text, more traditional chapters (including Problem Solving, Sets, Logic, Geometry, Counting and Probability, Numeration Systems, and Personal Finance) are available with *For All Practical Purposes* through custom publishing. For more information, please contact your W. H. Freeman representative or go to www.macmillanhighered.com/fapp10e. Restrictions apply.

Media and Supplements

The media and supplements package for the tenth edition has been updated to reflect changes in the book. Both instructors and students will benefit from the innovative materials available to them.

Macmillan Education LaunchPad W. H. Freeman's new online homework system, **LaunchPad,** offers quality content that has been curated and organized for easy assignability in a simple but powerful interface. We have taken what we have learned from thousands of instructors and hundreds of thousands of students to create a new generation of W. H. Freeman/Macmillan technology.

Curated units. Combining a curated collection of videos, homework sets, tutorials, applets, and e-Book content, LaunchPad's interactive units give instructors a building block to use as is or as a starting point for customized learning units. A majority of exercises from the text can be assigned as online homework, including an abundance of algorithmic exercises. An entire unit's worth of work can be assigned in seconds, drastically reducing the amount of time it takes for instructors to have their course up and running.

Easily customizable. Instructors can customize the LaunchPad units by adding quizzes and other activities from our vast wealth of resources. They can also add a discussion board, a drop box, and an RSS feed, with just a few clicks. LaunchPad allows instructors to customize students' experiences as much or as little as desired.

Useful analytics. The gradebook quickly and easily allows instructors to look up performance metrics for classes, individual students, and individual assignments.

Intuitive interface and design. The student experience is simplified. Students' navigation options and expectations are clearly laid out at all times, ensuring they can never get lost in the system.

Assets integrated into LaunchPad include the following:

Interactive e-Book. Every LaunchPad e-Book comes with powerful study tools for students, video and multimedia content, and easy customization for instructors. Students can search, highlight, and bookmark, making it easier to study and access key content. And teachers can ensure that their classes get just the book they want to deliver by customizing and rearranging chapters; adding and sharing notes and discussions; and linking to quizzes, activities, and other resources.

Macmillan Education LearningCurve **LearningCurve** provides students and instructors with powerful adaptive quizzing, a game-like format, direct links to the e-Book, and instant feedback. The quizzing system features questions tailored specifically to the text, and it adapts to students' responses, providing material at different difficulty levels and topics based on student performance.

Freeman SolutionMaster **SolutionMaster** offers an easy-to-use web-based version of the instructor's solutions, allowing instructors to generate a solution file for any set of homework exercises.

Other online homework options include:

WebAssign. **WebAssign** integrates the text exercises from *For All Practical Purposes,* Tenth Edition, into a popular and trusted online homework system, making it easy to assign algorithmically generated homework and quizzes.

Student Resources

- **Student Solutions Manual** provides solutions to the odd-numbered exercises, with step-by-step solutions to select problems.
- **Student Study Guide** offers study tips and tools to help students gain a better understanding of course material, including key ideas for each section and additional examples and practice exercises.
- **Interactive applets** help students master key mathematical concepts and work exercises.
- **Math Clips** are animated whiteboard videos that illuminate key concepts in the text by showing students step-by-step solutions to selected exercises.
- **Self-quizzes, flash cards,** and other projects offer additional study help.

Instructor Resources

- **Instructor's Guide with Full Solutions** includes teaching suggestions, chapter comments, and detailed solutions to all exercises.
- **Teaching Guide for First-Time Instructors** helps instructors, adjuncts, and teaching assistants plan their course more easily and effectively. This guide also offers fresh perspectives and ideas to experienced instructors.
- **Test Bank** offers thousands of multiple-choice questions.
- **Lecture slides** offer a detailed lecture presentation of concepts covered in each chapter of *For All Practical Purposes,* Tenth Edition.
- **Clicker Questions** are available for each chapter.

Companion Website

www.macmillanhighered.com/fapp10e This open-access website provides students with access to the applets referenced throughout the text.

Acknowledgments

For All Practical Purposes continues to evolve in great part because of our many friends and colleagues who have offered suggestions, comments, and corrections. We are grateful to them all.

Michael Allen, *Glendale Community College*
Hamid Attarzadeh, *Kentucky Community and Technical College System*
Robert J. Bass, *Gardner-Webb University*
Debra D. Bryant, *Tennessee Technological University*
Annette M. Burden, *Youngstown State University*
Christine Cedzo, *Gannon University*
Gina Poore Dunn, *Lander University*
Dennis Evans, *Concordia University*
Kevin Ferland, *Bloomsburg University*
Gregory Goeckel, *Presbyterian College*
Doug Greiner, *Rogers State University*
Samuel S. Gross, *Bloomsburg University*
Patricia Humphrey, *Georgia Southern University*
Jamie L. King, *Arkansas Tech University*
Jay Malmstrom, *Oklahoma City Community College*
Paul McCombs, *Rock Valley College*
Linda McGuire, *Muhlenberg College*
Amanda Nutt, *Western Kentucky University*
Jacquelyn O'Donohoe, *Plymouth State University*
Daniel Pinzon, *Georgia Gwinnett College*
Stacy Reagan, *Caldwell Community College*
David Shannon, *Transylvania University*
Sharon Sullivan, *Catawba College*
Susan Toma, *Madonna University*
Kim Y. Ward, *Eastern Connecticut State University*
Weicheng Xuan, *Arizona Western College*

We owe our appreciation to the people at W. H. Freeman and Company who participated in the development and production of this edition. We wish especially to thank the editorial staff for their tireless efforts and support. Among them are Terri Ward, Publisher; Leslie Lahr and Jorge Amaral, Development Editors; Laura Judge, Executive Media Editor; Marie Dripchak, Associate Editor; Victoria Garvey, Editorial Assistant; Vivien Weiss, Senior Project Editor; Susan Wein, Production Supervisor; Matthew McAdams, Art Manager; Blake Logan, Cover Designer; Cecilia Varas and Nick Ciani, Photo Editors; Jennifer MacMillan, Felicia Ruocco, and Hilary Newman, Permissions; Patti Brecht and Laura Cooney, copyeditors; Edward Dionne, Project Manager, MPS Limited.

We'd also like to give special thanks to the authors of the ancillaries to accompany this tenth edition: Lauren Fern, University of Montana, Instructor's Guide with Full Solutions and Student Solutions Manual; Doug Greiner, Rogers State University, Teaching Guide for First-Time Instructors and Student Study Guide; Hee Seok Nam, Washburn University, Practice Quizzes and Test Bank; Kathy Rodgers, University of Southern Indiana, Clicker Questions and Lecture Slides.

Through ten editions, this text has been used by well over a million students. When we first suggested our new approach, we were turned down by every major (and minor) textbook publisher. Only W. H. Freeman, under the leadership of Linda Chaput and the faith of mathematics editor Jerry Lyons, was willing to take a chance. That chance has permanently changed the face of introductory undergraduate mathematics. Words cannot express the gratitude we feel for the staffs of W. H. Freeman and COMAP and for the authors past and present through these almost 30 years. To them and everyone who made our purposes practical, we offer our appreciation for an exciting and exhilarating ride.

Solomon Garfunkel, COMAP

FOR ALL PRACTICAL PURPOSES

Part I

Management Science

Getting through a typical day can be a challenge: getting to or from school or your job on time; finding a parking spot when you are late for a date; making sure you have food in the refrigerator or getting to your favorite fast-food restaurant to stay properly fed and "fueled" with coffee; making sure your body is fit by getting to the gym or exercising at home; and seeing the doctor for a regular checkup or when you are ill. While your personal life may seem complex, consider what goes on at any of the large, modern medical centers found across America, in cities and on university campuses. In a typical 24-hour period, babies will be born; people will die; a friend may need an emergency appendectomy; and a relative may need treatment for a heart attack, asthma attack, stroke, or puncture wound. And there are the more mundane things: blood tests, mammograms, garden variety X-rays, CT and MRI scans, elective surgery to remove a cyst, or a routine checkup. Medical centers need many kinds of workers to make them hum: doctors and nurses with different specialties, technicians of different kinds, workers who pay attention to the "business" aspects of the medical center—not to mention people who prepare food for the patients, clean the premises, or plow the access roads after a snowstorm.

So what does this have to do with mathematics? To deal with the emergencies and day-in and day-out demands, there must be specialists (doctors, nurses, technicians, etc.) and other workers either physically present or available by phone. The part of mathematics concerned with efficient operations of businesses and governments is called operations research (OR) or management science. The domain of OR includes resource allocation, scheduling, queues (waiting lines), inventory analysis, routing problems, and cost minimization, to mention but a few of OR's growing areas of applicability. Medical centers rely on the expertise of mathematical specialists to make their operations run smoothly day and night, rain or shine, winter or summer, so that when you or your loved ones need to use the center, its services are there for you.

Chalk up another triumph for OR! What follows will help you, too, to know about and use such tools.

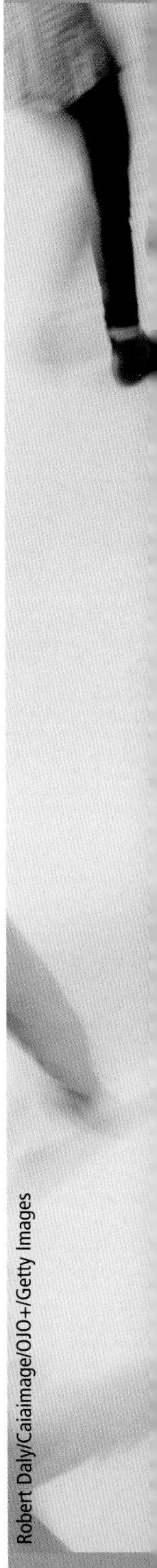

Robert Daly/Caiaimage/OJO+/Getty Images

Urban Services

1

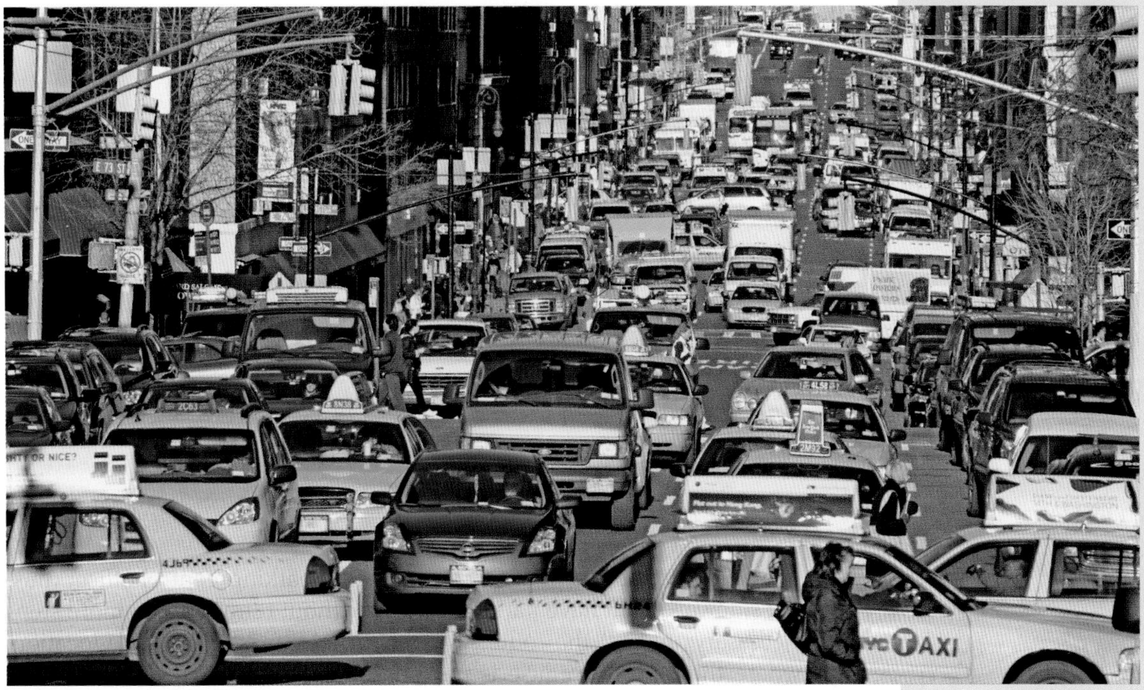

Mitchell Funk/Getty Images

The underlying theme of management science, also called **operations research (OR),** is finding the best method for solving some problem—what mathematicians call the **optimal solution.** In some cases, the goal may be to finish a job or get somewhere as quickly as possible. In other situations, the objective might be to maximize profit or minimize cost. In this chapter, our goal is to save time (and usually taxpayer money) in traversing a street network while providing services such as checking parking meters, collecting garbage or bottles for recycling, de-icing roads, inspecting for potholes or gas leaks, or delivering packages or mail.

Let's begin by assisting the parking department of a city government. Most cities and many small towns have parking meters that must be regularly checked for parking violations or emptied of coins. We will use an imaginary town to show how management science techniques can help to make parking control more efficient.

1.1 Euler Circuits

The street map in Figure 1.1 is typical of many villages and cities across the United States, with streets, residential blocks, and a town park. Our job, or that of the commissioner of parking, is to find the most efficient route for the parking-control officer, who travels on foot, to check the meters in an area. Efficient routes save money. Our map shows only a small area, allowing us to start with an easy problem. But the problem occurs on a larger scale in all cities and towns and for larger areas. The bigger the region involved, the greater the potential for cost savings.

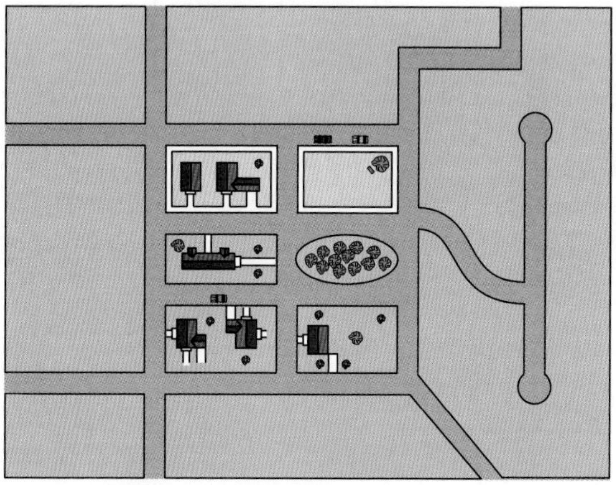

Figure 1.1 A street map for part of a town.

The commissioner has two goals in mind: (1) The parking-control officer must cover all the sidewalks that have parking meters without retracing any more steps than are necessary; and (2) the route should start and end at the same location, perhaps where the officer's vehicle is parked. To be specific, suppose there are only two blocks that have parking meters, the two lightly shaded blocks that are side by side toward the top of Figure 1.1. Suppose further that the parking-control officer must start and end at the upper left corner of the left-hand block. You might enjoy working out some routes by trial and error and evaluating their good and bad features. We are going to leave this problem for the moment and establish some concepts that will give us a better method than trial and error to deal with this kind of problem.

What Is a Graph? DEFINITION

A **graph** is a finite set of dots and connecting curved or straight links. The dots are called **vertices** (a single dot is called a **vertex**), and the links are called **edges.** Each edge must connect two different vertices. A graph can represent a city map, a social network, a system of air routes, or electrical power lines.

Path and Circuit DEFINITION

A **path** is a connected sequence of edges showing a route on the graph that starts at a vertex and ends at a vertex; a path is usually described by naming in turn the vertices visited in traversing it. A path that starts and ends at the same vertex is called a **circuit.**

EXAMPLE 1 ➔ Parts of a Graph

We can use the graph in Figure 1.2 to help explain these technical terms. The graph shown has five vertices and eight edges. The vertices represent cities, and the edges represent nonstop airline routes between them. We see that there is a nonstop flight between Berlin and Rome, but no such flight between New York and Berlin. There are several paths that describe how a person might travel with this airline from New York to Rome. The path that seems most direct is New York, London, Rome. But New York, Miami, Rome is also a path with only one "stop." Furthermore, New York, London, Berlin, Rome is a path. We can describe these three paths as *NLR*, *NMR*, *NLBR*.

Another path would be New York, Miami, London, Berlin, Rome, which can be written *NMLBR*. An example of a circuit is Miami, Rome, London, Miami. It is a circuit because the path starts and ends at the same vertex. This circuit can best be described in symbols by *MRLM*. Another example of a circuit in this graph would be *LRBL*, which is the circuit involving the cities London, Rome, Berlin, and back to London. In this chapter, we are especially interested in circuits, just as we are in real life. Most of us end our day in the same place that we start it—at home!

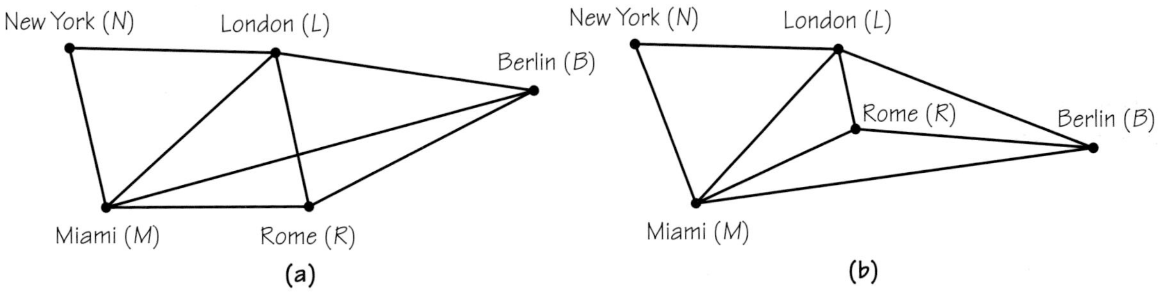

Figure 1.2 (a) The edges of the graph show nonstop routes that an airline might offer. (b) The graph in (a) redrawn without the accidental crossing.

Notice that the edges *MB* (which could also be denoted *BM*) and *RL* shown in Figure 1.2a meet at a point that has no label. Furthermore, this point does not have a dark dot. This is because this point does not represent a vertex of our graph; it does not represent a city. It arises as an "accidental" consequence of the way this diagram has been drawn. We could join *M* and *B* with a curved line segment so that the edges *LR* and *MB* do not cross, or redraw the diagram so as to avoid a crossing in this case. We will be working often in situations where graphs can be drawn without accidental crossings, and we will try to avoid such crossings when it is convenient to do so. However, there are infinitely many graphs for which—when they are drawn on a flat piece of paper—accidental crossings are unavoidable. (Figure 2.12 on page 52 is an example of such a graph.) ■

Returning to the case of parking enforcement in Figure 1.1, we can use a graph to represent the whole territory to be patrolled: Think of each street intersection as a vertex and each sidewalk that contains meters as an edge, as in Figure 1.3. Notice in Figure 1.3b that the width of the street separating the blocks is not explicitly represented; it has been shrunk to nothing. In effect, we are simplifying our problem by ignoring any distance traveled in crossing streets. In drawing graph diagrams such as those in Figure 1.3 or Figure 1.5, we usually use straight line segments to

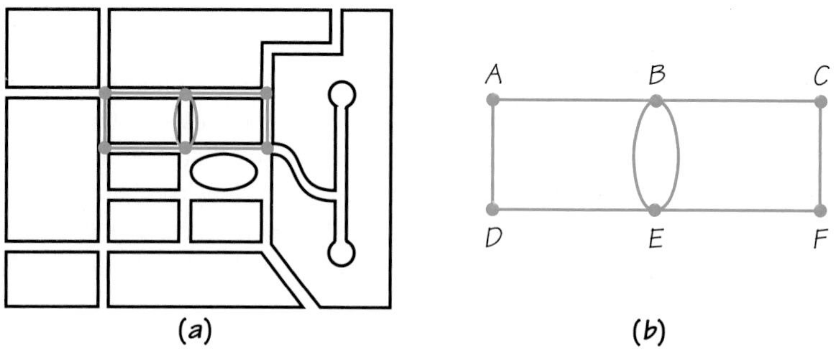

Figure 1.3 (a) A graph superimposed upon a street map. The edges show which sidewalks have parking meters. (b) The same graph enlarged.

represent edges. However, sometimes we cannot avoid the use of "curves," or we may prefer to use curved edges because they convey aspects of the original problem that we desire to emphasize.

The sequence of numbered edges in Figure 1.4a shows one circuit that covers all the meters. (Note that it is a circuit because its path returns to its starting point.) However, one edge is traversed three times. Figure 1.4b shows another solution that is better because its circuit covers every edge (sidewalk) exactly once. In Figure 1.4b, no edge is covered more than once, or *deadheaded* (a term borrowed from shipping, which means making a return trip without a load). When deadheading is required in an applied situation, such as inspecting parking meters or pothole inspection, typically time and effort is being spent but no productive work is accomplished because the productive work was done the first time the edge was covered (traversed).

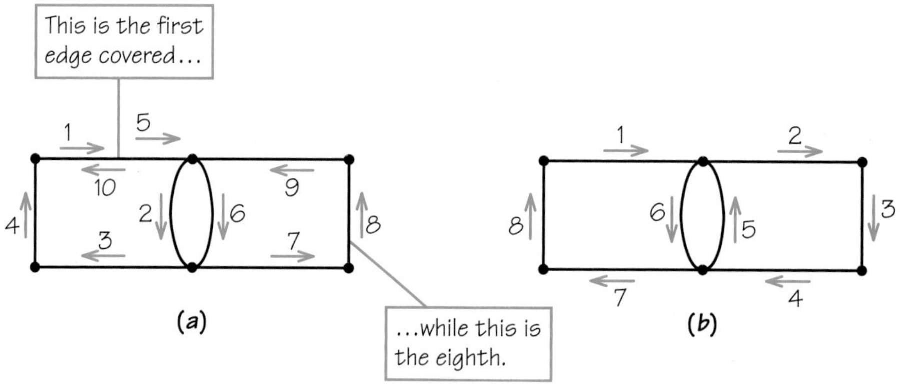

Figure 1.4 (a) A circuit and (b) an Euler circuit.

Euler Circuit DEFINITION

A circuit that covers each edge of a graph once, but not more than once, is called an **Euler circuit.**

Figure 1.4b shows an Euler circuit. These circuits get their name from the great Swiss-born mathematician Leonhard Euler (pronounced *oy' lur*), who first studied them (see Spotlight 1.1). Euler was the founder of the theory of graphs, or graph theory. One of his first discoveries was that some graphs have no Euler circuits at all.

Leonhard Euler

SPOTLIGHT 1.1

Leonhard Euler (1707–1783) was remarkable in many ways. He was extremely prolific, publishing over 500 works in his lifetime. But he wasn't devoted just to mathematics; he was a people person, too. He was extremely fond of children and had 13 of his own, of whom only five survived childhood. It is said that he often wrote difficult mathematical works with a child or two in his lap.

Human interest stories about Euler have been handed down through three centuries. He was a prodigy at doing complex mathematical calculations under less than ideal conditions, and he continued to do them even after he became totally blind later in life. His blindness diminished neither the quantity nor the quality of his output. Throughout his life, he was able to mentally calculate in a short time what would have taken ordinary mathematicians hours of pencil-and-paper work. A contemporary claimed that Euler could calculate effortlessly, "just as men breathe, as eagles sustain themselves in the air." His collected works and numerous letters to other scholars of his day are still being published.

Leonhard Euler

Euler invented the idea of a graph in 1736 when he solved a problem in "recreational mathematics." He showed that it was impossible to stroll a route visiting the seven bridges of the German town of Königsberg exactly once. Ironically, in 1752 he discovered that three-dimensional polyhedra obey the remarkable formula $V - E + F = 2$ (that is, number of vertices − number of edges + number of faces = 2) but failed to give a proof because he did not analyze the situation using graph theory methods. Sometimes even a genius can miss something.

For example, in the graph in Figure 1.5b, it would be impossible to start at one point, return to that starting vertex and cover all the edges without retracing some steps: If we try to start a circuit at the leftmost vertex, we discover that once we have left the vertex, we have "used up" the only edge meeting it. We have no way to return to our starting point except to reuse that edge. But this is not allowed in an Euler circuit. If we try to start a circuit at one of the other two vertices, we likewise can't complete it to form an Euler circuit.

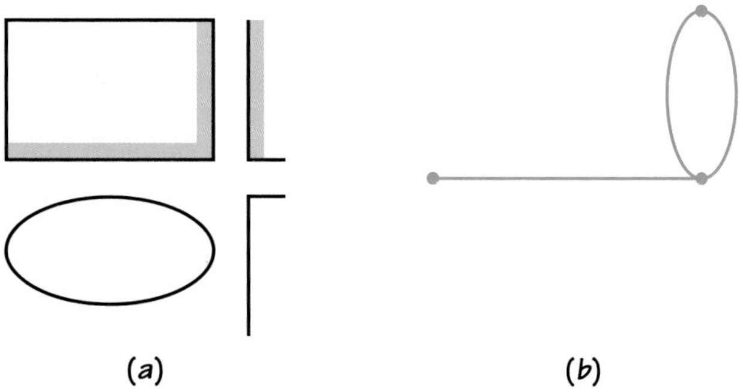

Figure 1.5 (a) The three shaded sidewalks cannot be covered by an Euler circuit. (b) The graph of the shaded sidewalks in part (a).

As mentioned in Spotlight 1.2, realistic problems of this type involve larger neighborhoods that might require the use of a computer. In addition, there may be other complications that might take us beyond the simple mathematics we want to stick to.

The Human Aspect of Problem Solving

SPOTLIGHT 1.2

Thomas Magnanti, professor of operations research and management science, is former Dean of Engineering at MIT. Here are some of his observations:

> Typically, a management science approach has several different ingredients. One is just structuring the problem—understanding that the problem is an Euler circuit problem or a related management science problem. After that, one has to develop the solution methods.
>
> But one should also recognize that you don't just push a button and get the answer. In using these underlying mathematical tools, we never want to lose sight of our common sense, understanding, intuition, and judgment. The computer provides certain kinds of insights. It deals with some of the combinatorial complexities of these problems very nicely. But a model such as an Euler circuit can never capture the full essence of a decision-making problem.
>
> Typically, when we solve the mathematical problem, we see that it doesn't quite correspond to the real problem we want to solve. So we make modifications in the underlying model. It is an interactive approach, using the best of what computers and mathematics have to offer and the best of what we, as human beings, with our own decision-making capabilities, have to offer.

Courtesy of Thomas Magnanti

Thomas Magnanti

Because we are interested in finding circuits, and Euler circuits are the most efficient ones, we want to know how to find them. If a graph has no Euler circuit, we want to develop efficient, alternative tours, those having minimum deadheading. These topics make up the rest of this chapter.

1.2 Finding Euler Circuits

Now that we know what an Euler circuit is, we are faced with two obvious questions:

1. Is there a way to tell by calculation or logical reasoning, not by trial and error, if a graph has an Euler circuit?
2. Is there a method, other than trial and error, for finding an Euler circuit when one exists and finding it quickly?

Loosely speaking, the first question lies within the concerns of mathematicians because it asks whether or not a certain problem admits a solution. Typically, the second question lies in the domain of computer science because it concerns finding the actual answer to a complex version of a problem in a short enough time to be useful.

Euler investigated these questions in 1735 by using the concepts of **valence** and **connectedness.**

Valence DEFINITION

The **valence** of a vertex in a graph is the number of edges meeting at the vertex.

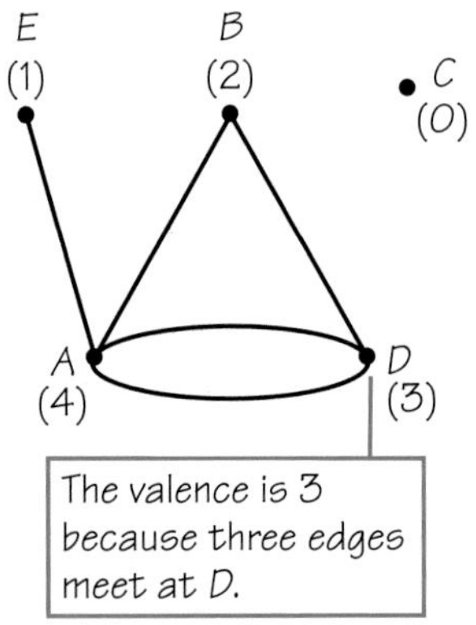

Figure 1.6 Valences of vertices.

Figure 1.6 illustrates the concept of valence, with vertex *A* having valence 4, vertex *D* having valence 3, vertex *B* having valence 2, and vertex *C* having valence

0. Vertex *E* has valence 1. Isolated vertices such as vertex *C* are an annoyance in Euler circuit theory. Because they don't occur in typical applications, we henceforth assume that our graphs have no vertices of valence 0.

Figure 1.3b has four vertices of valence 2, namely, *A, C, F,* and *D.* This graph also has two vertices, *B* and *E,* of valence 4. Notice that each vertex has a valence that is an even number. We'll soon see that this is very significant.

Connected Graph DEFINITION

A graph is said to be **connected** if for every pair of its vertices, there is at least one path connecting the two vertices.

Given a graph, if we can find even one pair of vertices not connected by a path, then we say that the graph is not connected. For example, the graphs in Figure 1.6 and Figure 1.7 are not connected because we are unable to join *A* to *C* with a path of edges. However, the graph in Figure 1.7 does consist of two "pieces" or connected components, one containing the vertices *A, B, F,* and *G,* the other containing *C, D,* and *E.* A connected graph contains a single connected component. Notice that the parking-control graph of Figure 1.3b is connected.

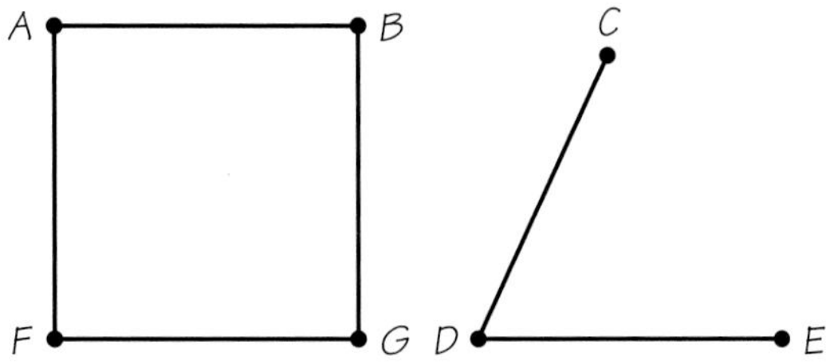

Figure 1.7 A graph that is not connected.

We can now state Euler's theorem, his simple answer to the problem of detecting when a graph *G* has an Euler circuit.

Euler Circuit Theorem THEOREM

1. If *G* is connected and has all valences even, then *G* has an Euler circuit.
2. Conversely, if *G* has an Euler circuit, then *G* must be connected and all its valences must be even numbers.

Because the parking-control graph of Figure 1.3b conforms to the connectedness and even-valence conditions, Euler's theorem tells us that it has an Euler circuit. We already have found an Euler circuit for the graph shown in Figure 1.4b by trial and error. For a very large graph, however, trial and error may take a long time. It is usually quicker to check whether the graph is connected and even-valent as a way to be sure an Euler circuit must exist rather than produce a specific Euler circuit.

Once we know there is an Euler circuit in a certain graph, how do we find it? Many people find that, after a little practice, they can find Euler circuits by trial and error, and they don't need detailed instructions on how to proceed. At this point you should see if you can develop this skill by trying to find Euler circuits in Figure 1.8a, Figure 1.9a, and Figure 1.10. In doing your experiments, draw your graph in ink and the circuit in pencil so you can erase if necessary.

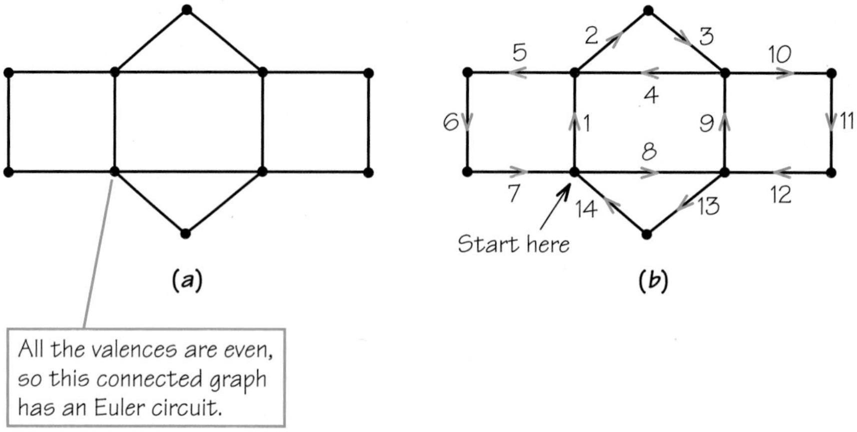

(a)
(b)
Start here
All the valences are even, so this connected graph has an Euler circuit.

Figure 1.8 (a) A connected graph having (b) an Euler circuit.

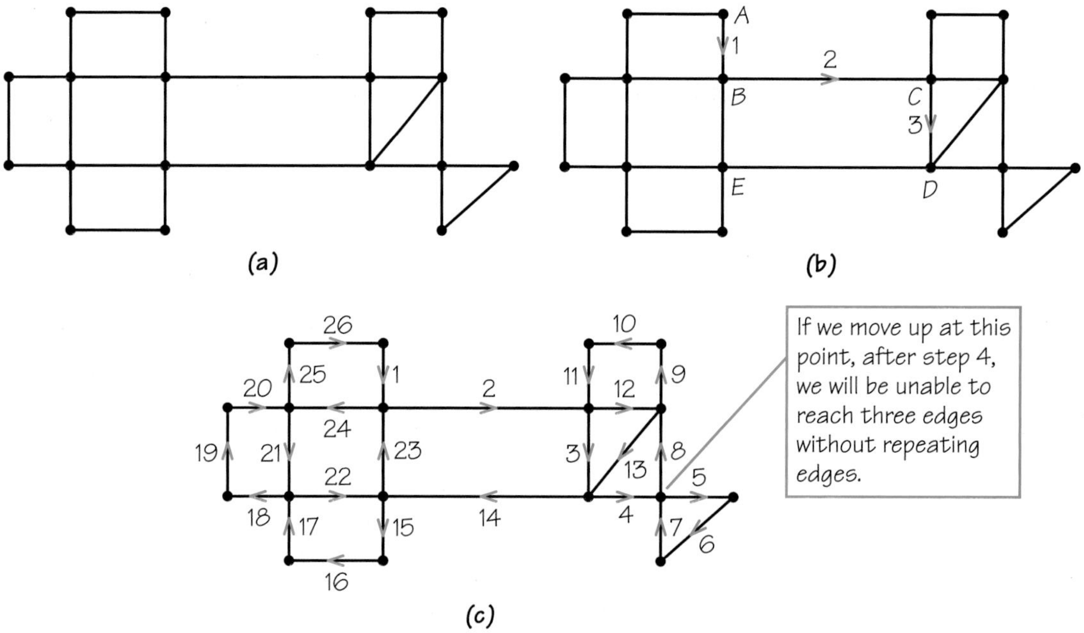

(a)
A
B
C
D
E
(b)
If we move up at this point, after step 4, we will be unable to reach three edges without repeating edges.
(c)

Figure 1.9 (a) A graph that has an Euler circuit. (b) A critical junction in finding an Euler circuit in this graph, starting from vertex *A*. (c) A description of a full Euler circuit for this graph.

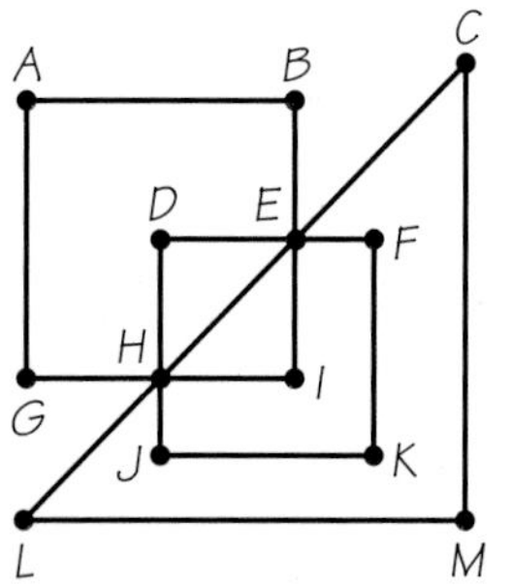

A
B
C
D
E
F
H
I
G
J
K
L
M

Figure 1.10 A graph with an Euler circuit.

If you would like more guidance on how to find an Euler circuit without trial and error, here is a method that works: Never use an edge that is the only link between two parts of the graph that still need to be covered. Figure 1.9b illustrates this. Here we have started the circuit at *A* and gotten to *D* via *B* and *C*, and we want to know what to do next. Going to *E* would be a bad idea because the uncovered part of the graph would then be disconnected into left and right portions. You will never be able to get from the left part back to the right part because you have just used the last remaining link between these parts. Therefore, you should stay on the right side and finish that before using the edge from *D* to *E*. This kind of thinking needs to be applied every time you need to choose a new edge.

Let's see how this works in Figure 1.9, starting at *A*. From vertex *A* there are two possible edges, and neither of them disconnects the unused portion of the graph. Thus, we could have gone either to the left or down. Having gone down to *B*, we now have three choices, none of which disconnects the unused part of the graph. After choosing to go from *B* to *C*, we find that any of the three choices at *C* is acceptable. Can you complete the Euler circuit? Figure 1.9c shows one of many ways to do this.

The method just described leaves many edge choices up to you. When there are many acceptable edges for your next step, you can pick one at random.

EXAMPLE 2 Finding an Euler Circuit

Check the valences of the vertices and the connectivity of the graph in Figure 1.8a to verify that the graph does have an Euler circuit. Now try to find an Euler circuit for that graph. You can start at any vertex. When you are done, compare your solution with the Euler circuit given in Figure 1.8b. If your path covers each edge exactly once and returns to its original vertex (which is a circuit), then it is an Euler circuit, even if it is not the same as the one we give.

Understanding Euler's Theorem

We start by showing that if a graph has an Euler circuit *R*, then it must have only even-valent vertices and it must be connected. Let *X* be any vertex of the graph. We will show that the edges at *X* can be paired up, and this will prove that the valence is even. Every edge at *X* is used by *R* as an outgoing edge (leaving from *X*) or an incoming edge (arriving at *X*). If the Euler circuit starts at *X*, then pair up the first edge used by *R* with the last one (when the circuit returns to *X* for the last time). In addition, each other edge at *X* that is used by the circuit as an incoming edge will be paired with the outgoing edge that is used next. Because all edges at *X* are used by the Euler circuit, none more than once, this pairs up the edges. Hence, *X* must be even-valent because we have "organized" the edges of *R* in pairs.

But what if *X* is not the start of the Euler circuit? If that is the case, do the pairing like this: The first incoming edge at *X* is paired with the outgoing one used next, the second incoming edge at *X* is paired with the outgoing one used next, and so on. For example, in Figure 1.11 at vertex *B*, we would pair up edges 2 and 3 and edges 9 and 10. At vertex *C*, we would pair up edges 4 and 5 and edges 8 and 9. Can you see how the pairings would work at *D*? How about vertex *A*?

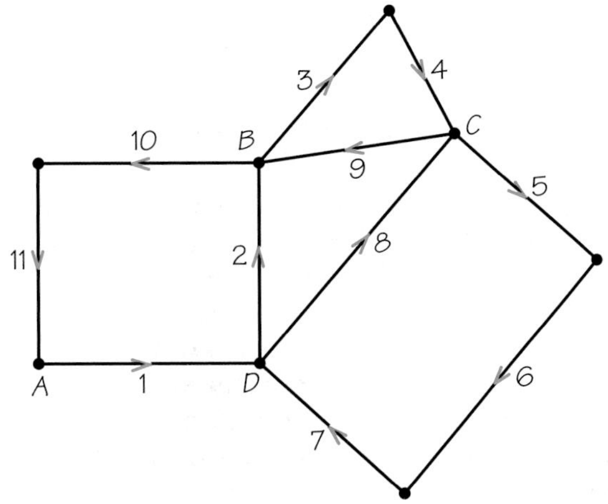

Figure 1.11 An Euler circuit starting and ending at *A*.

In studying this example, you might think it would be simpler to count the edges at a vertex to see that the valence is even. True, but our pairing method works for a graph about which we know nothing except that it has an Euler circuit.

To see that a graph with an Euler circuit is connected, note that by following the Euler circuit around we can get from any edge to any other edge (it covers them all) using a portion of the Euler circuit. Because every vertex is on an edge (there are no vertices of valence 0), we can get from any vertex to any other using a portion of the Euler circuit.

So far, this is not a complete proof of Euler's theorem. To complete the proof, we would need to prove that if a graph has all vertices even-valent and is connected, then an Euler circuit can be found for it. One way to do this is to piece together shorter circuits in the even-valent connected graph to get circuits using more edges.

Self Check 1

(a) How many vertices and edges does the graph in Figure 1.12b have?
(b) List the vertices in the graph that are even-valent (have even numbers for their valences).
(c) Is *FEABFGCDHGF* an Euler circuit for the graph? ■

1.3 Beyond Euler Circuits

Now let's see what Euler's theorem tells us about the three-block neighborhood with parking meters, represented by dots in Figure 1.12a. Figure 1.12b shows the corresponding graph. (Because we use edges to represent only sidewalks along which the officer must walk, the sidewalk with no meters is not represented by any edge in the graph.) This graph has vertices with odd valences (at vertices C and G), so Euler's theorem tells us that there is no Euler circuit for this graph.

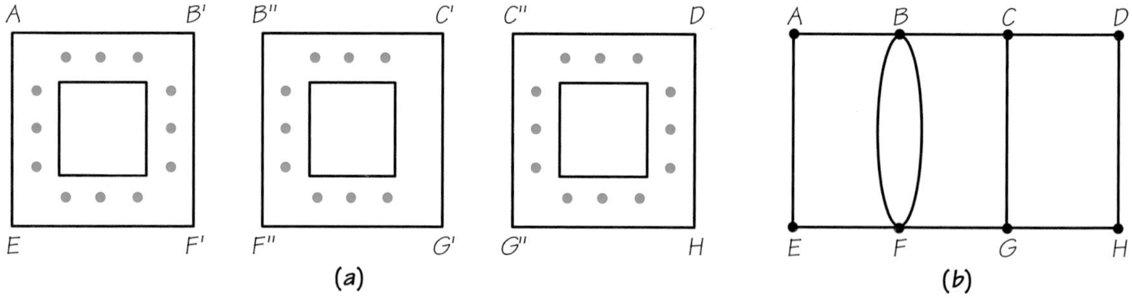

Figure 1.12 (a) A street network and (b) its graphic representation. Locations such as *B'* and *B''*, *C'* and *C''*, *F'* and *F''*, and *G'* and *G''* are merged to form the vertices *B*, *C*, *F*, and *G*. The dots shown represent parking meters.

Because we must reuse some edges in this graph to cover all edges in a circuit, for efficiency we need to keep the total length of reused edges to a minimum. Note that we are still looking for a tour that starts and ends at the same vertex. This type of problem, in which we want to minimize the length of a circuit by carefully choosing which edges to retrace, is often called the **Chinese postman problem.** (Like parking-control routes, mail routes need to be efficient.) The problem was first studied by the Chinese mathematician Meigu Guan in 1962—hence the name. The remainder of this chapter is dedicated to solving the Chinese postman problem and discussing applications beyond parking control.

Solving the Chinese Postman Problem

In a realistic Chinese postman problem, we need to consider the lengths of the sidewalks, streets, or whatever "costs" the edges have (represent) because we want to minimize the total length of the reused edges. However, to simplify things at the start, we can suppose that all edges represent the same length. (This is often called the *simplified* Chinese postman problem.) In this case, we need only count reused edges and need not add up their lengths. To solve the problem, we want to find a circuit that covers each edge and that has the minimal number of reuses of edges already covered.

To follow the procedure we are going to develop, look at the graph in Figure 1.13a, which is essentially the same graph as in Figure 1.12b. This graph has no Euler circuit, but there is a circuit that has only one reuse of an edge (*CG*), namely, *ABCDHGCGFBFEA*. Let's draw this circuit so that when edge *CG* is about to be reused, we install a new, extra, blue edge in the graph for the circuit to use. By duplicating edge *CG*, we can avoid reusing the edge. To duplicate an edge, we must add an edge that joins the two vertices that are already joined by the edge we want to duplicate. (It makes no sense to join vertices that are not already connected by an edge, because such edges would not represent sidewalk sections with meters; see Figure 1.15.)

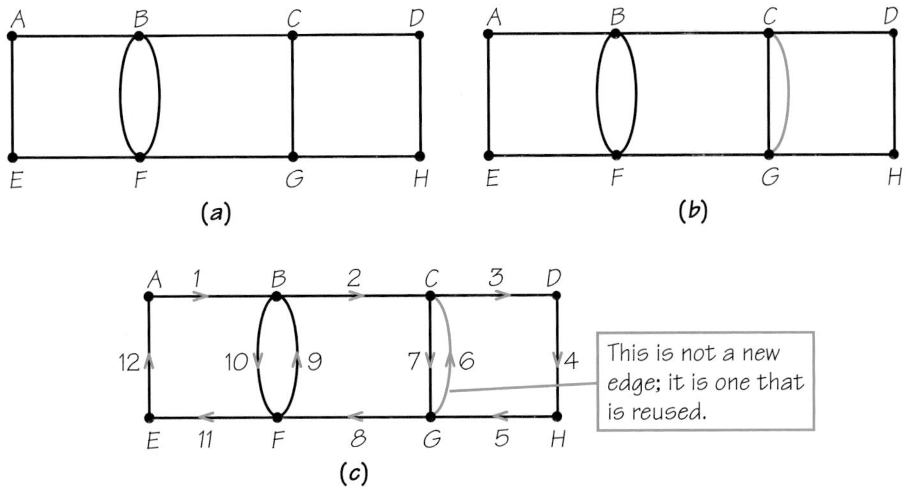

Figure 1.13 Making a circuit by reusing an edge.

We have now created the graph of Figure 1.13b. In the graphs we draw, the edges that are added will be shown in color to distinguish them from the original edges, which are shown in black. (In the graphs that you draw, you may want to use a similar scheme, perhaps dotted edges, to help you remember which edges are the originals and which are duplicated.) In the graph of Figure 1.13b, the original circuit can be traced as an Euler circuit, using the new edge when needed. The circuit is shown in Figure 1.13c. Our theory will be based on using this idea in reverse, as follows:

1. Take the given graph and add edges by duplicating existing edges, until you arrive at a graph that is connected and even-valent. Note that after a graph is *eulerized*, the new graph produced will have an Euler circuit.

2. Find an Euler circuit on the eulerized graph.
3. "Squeeze" this Euler circuit from the eulerized graph onto the original graph by reusing an edge of the original graph each time the circuit on the eulerized graph uses an added edge.

Eulerizing a Graph DEFINITION

Adding edges that duplicate existing edges to a connected graph to make all valences even is called **eulerizing** the graph.

EXAMPLE 3 Eulerizing a Graph

Suppose we want to eulerize the graph of Figure 1.14a. When we eulerize a graph, we first locate the vertices with odd valence. The graph in Figure 1.14a has two, *B* and *C*. Next, we add one end of an edge at each such vertex, matching up the new edge with an existing edge in the original graph. Figure 1.14b shows one way to eulerize the graph. Note that *B* and *C* have even valence in the second graph. After eulerization, each vertex has even valence. To see an Euler circuit on the eulerized graph in Figure 1.14c, simply follow the edges in numerical order and in the direction of the arrows, beginning and ending at vertex *A*. The final step, shown in Figure 1.14d, is to "squeeze" our Euler circuit onto the original graph. There are two reuses of previously covered edges. Notice that each reuse of an edge corresponds to an added edge.

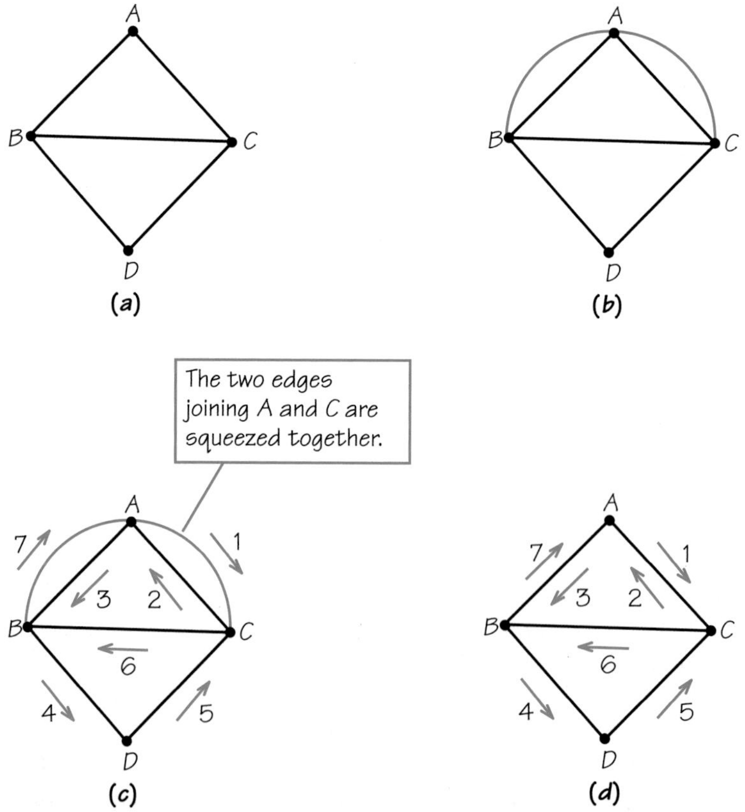

Figure 1.14 Eulerizing a graph.

In the previous example, we noticed that we could count how many reuses we needed by counting added edges. This is generally true in this type of problem: *If you add the new edges correctly, the number of reuses of edges equals the number of edges added during eulerization.*

Adding new edges correctly means adding only edges that are duplicates of existing edges. Doing this makes the rule just stated in italics always true, and so it is easy to count the needed reuses.

To see why we add only duplicate edges, examine Figure 1.15a. We need to alter the valences of vertices *X* and *Y* by adding edges so that they become even-valent. Adding one long edge from *X* to *Y* (Figure 1.15b) might seem like an attractive idea, but adding this edge is equivalent to asking a snowplow, say, to get from *X* to *Y* without moving along existing streets. At times it is necessary to traverse sections of the graph that have been previously traversed. This is the significance of the duplicated edges. Here the structure of the graph forces us to repeat some edges. We cannot get away with fewer than three repeats: the three edges *XU, UV,* and *VY* (Figure 1.15c). The duplicated edges are shown in color.

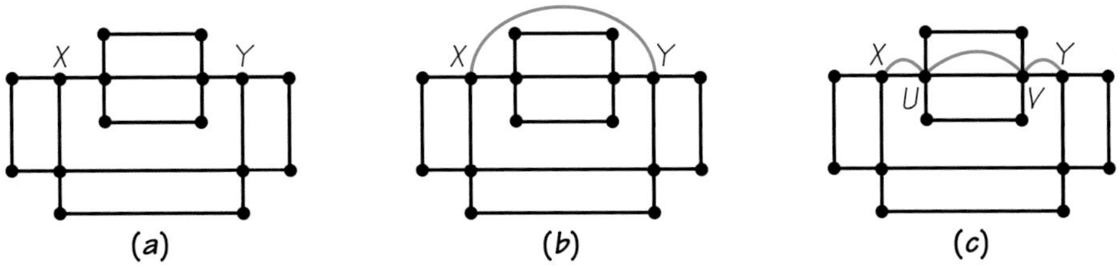

Figure 1.15 Eulerizing when the odd-valent vertices are more than one edge apart.

Now that we have learned to eulerize, the next step is to try to get a best eulerization—one with the fewest added edges. It turns out that often there are many ways to eulerize a graph. It is even possible that the smallest number of added edges can be achieved with two different eulerizations. This is the reason we use the phrase "a best eulerization" rather than "the best eulerization." Remember, we want a best eulerization because this enables us to find the circuit for the original graph that has the minimum number of reuses of edges, which in typical applications means saving time or money by avoiding retraversing edges where the productive work has already been accomplished (avoiding deadheading).

EXAMPLE 4 A Better Eulerization

In Figure 1.16a, we begin with the same graph as in Figure 1.14, but we eulerize it in a different way—by adding only one edge (see Figure 1.16b). Figure 1.16c shows an Euler circuit on the eulerized graph, and in Figure 1.14d we see how it is squeezed onto the original graph. There is only one reuse of an edge, because we added one edge during eulerization.

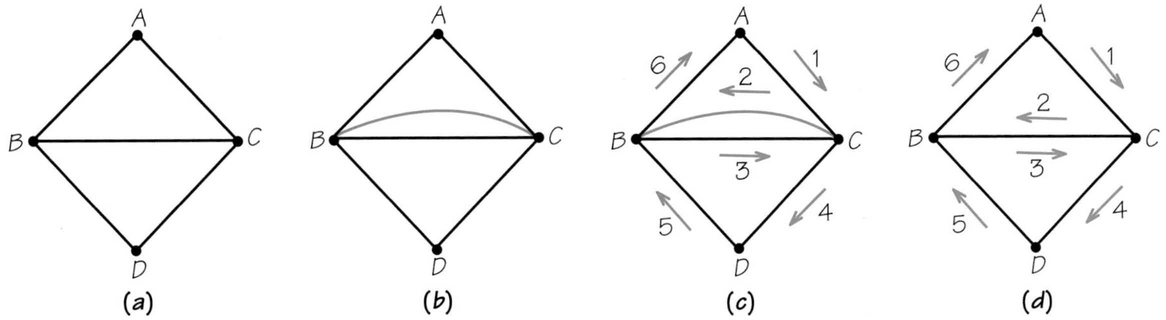

Figure 1.16 A better eulerization of Figure 1.14.

The solution in Figure 1.16 is better than the solution in Figure 1.14 because one reuse is better than two. These examples suggest the following addition to our solution procedure: Try to find an eulerization with the smallest number of added edges. This extra requirement makes the problem both more interesting and more difficult. For large graphs, a best eulerization may not be obvious. We can try out a few and pick the best among the ones we find, but there may be an even better one that our haphazard search does not turn up.

A systematic procedure for finding a best eulerization does exist, but the process is complicated. There is a particularly easy technique for eulerizing a special category of networks often found in residential neighborhoods. A street network is called *rectangular* when composed of a series of rectangular blocks that form a large rectangle a certain number of blocks high and a certain number of blocks wide.

Examples of rectangular street networks (a 3-by-3, a 3-by-4, and a 4-by-4) are shown in Figure 1.17. The graph on the right in each pair shows a best eulerization for the rectangular street network on the left. There appear to be three different eulerization patterns, depending upon whether the rectangle height and width in the original graph are odd or even numbers. In Figure 1.17a, both lengths are 3, both odd; in Figure 1.17b, one length is odd (3) and one is even (4); in Figure 1.17c, both lengths are 4, an even number.

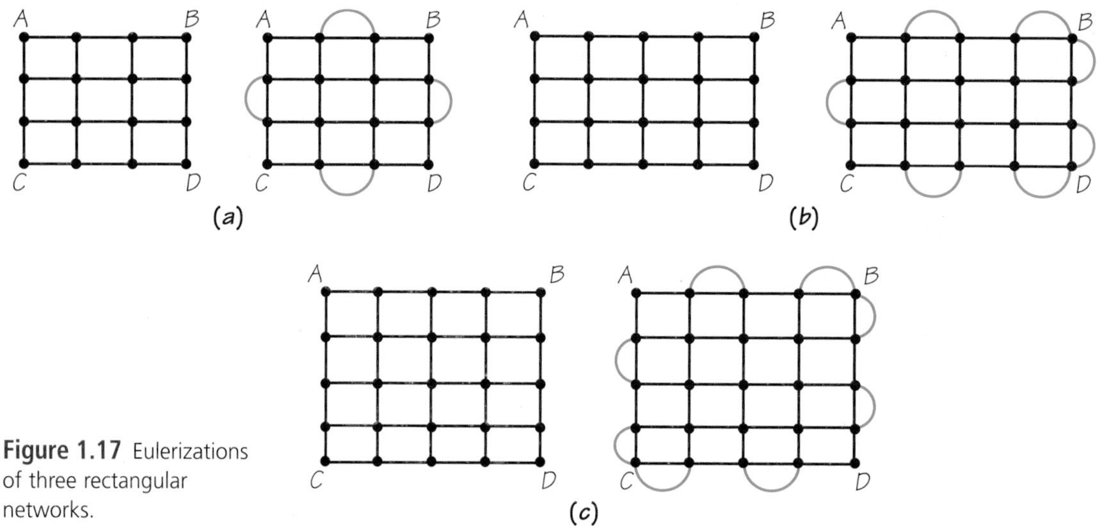

Figure 1.17 Eulerizations of three rectangular networks.

Although the patterns appear different, one technique can be used to create all of them. This technique can be thought of as involving an "edge walker" who walks around the outer boundary of the large rectangle in some direction, say, clockwise. He starts at any corner, say, the upper-left corner. As he goes around, he adds edges by the following rules. When he comes to an odd-valent vertex, he links it to the next vertex with an added edge. This next vertex now becomes either even or odd. If it became even, he skips it and continues around, looking for an odd vertex. If it became odd (this could happen only at a corner of the big rectangle), the edge walker links it to the next vertex and then checks this vertex to see whether it is even or odd. Each of the three parts of Figure 1.17 has been eulerized by this method.

In a street network that is not rectangular, the eulerization process is started by locating all the vertices with odd valence and then pairing these vertices with each other and finding the length of the shortest path between each pair. We look for the shortest paths because each edge on the connecting paths will be duplicated. The idea is to choose the pairings cleverly so that the sum of the lengths of those paths is the smallest it can be. With a little practice, most people can find a best or nearly best eulerization using this idea, together with trial and error and some ingenuity.

Finding Good Eulerizations

Suppose we want an optimal procedure for eulerizing a graph. What theoretical ideas and methods could we use to build such a tool?

One building block we could use is a method for finding the shortest path between two given vertices of a graph. For example, let us focus on vertices X and Y in Figure 1.18a. Both have odd valence. We can connect them with a pattern of duplicate edges, as in Figure 1.18b. The cost of this is the length of the path we duplicated from X to Y. A shorter path from X to Y, such as the one shown in Figure 1.18c, would be better. Fortunately, the *shortest-path problem* has been well studied, and we have many good procedures for solving it exactly, even in large, complex graphs.

But there is more to eulerizing the graph in Figure 1.18a than dealing with X and Y. Notice that we have odd valences at Z and W. Should we connect X and Y with a path, and then connect Z and W, as in Figure 1.18d? Or should we connect X to Z and Y to W, as in Figure 1.18e? Another alternative is to use connections X to W and Y to Z, as in Figure 1.18f. It turns out that the alternatives in both Figures 1.18e and 1.18f are preferable to the one in Figure 1.18d because they involve seven added edges, whereas Figure 1.18d uses nine.

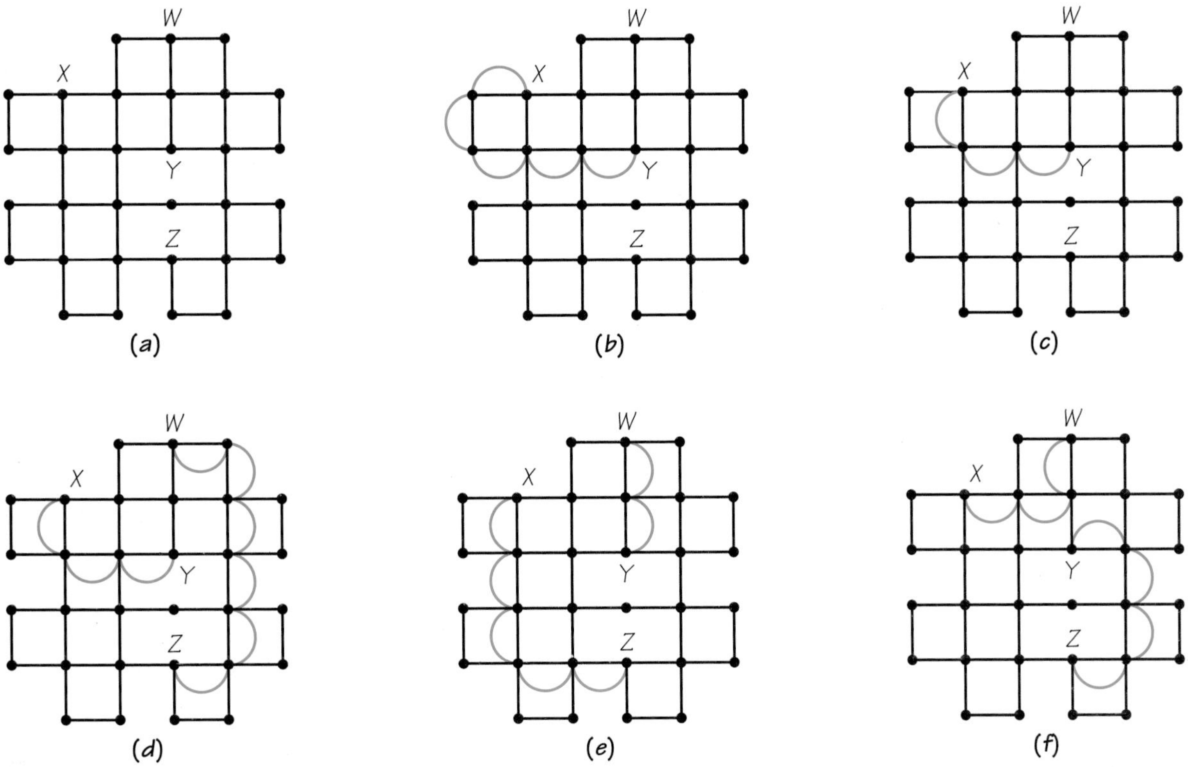

Figure 1.18 Choosing among eulerizations.

We know there is a simple way to test whether a connected graph has an Euler circuit: Check to see if the graph is even-valent. Is there a very easy way to compute the number of edges in a best eulerization of a graph? Unfortunately, the answer is "no." However, there is a simple observation that often saves a lot of work. Suppose we count the number of odd-valent vertices in a graph. This number must always be an even number. When we duplicate an existing edge, we can never change more than two odd-valent vertices to even-valent vertices. Thus, in a best eulerization of a graph, the number of edges that must be duplicated is at least the number of odd-valent vertices divided by two. If, for example, we have a graph with 10 odd-valent vertices and we find an eulerization with five added edges, there may be other eulerizations that also have five duplicated edges, but there can be no eulerization with fewer than five duplicated edges.

Remember that when an unweighted graph is eulerized in an optimal way, then the total cost of traversing each edge at least once can be found by adding the total number of edges in the graph to the number of edges that are reused (duplicated). Small problems involving eulerization can be carried out by trial-and-error methods. Unfortunately, although there is an algorithm that can be applied to find the best eulerization for large problems, the details of this algorithm are quite complex. However, the procedure works quickly not only for graphs without weights but also for graphs with weights on the edges.

Self Check 2

(a) Does the graph in Figure 1.12b have an Euler circuit? (If so, write down the Euler circuit.)

(b) If the graph does not have an Euler circuit, what is the minimum number of duplicated edges required for a best possible eulerization? ■

1.4 Urban Graph Traversal Problems

Euler circuits and eulerizing have many more practical applications than just checking parking meters. Almost any time services must be delivered along streets or roads, our theory can make the job more efficient. Examples include collecting garbage, salting icy roads, plowing snow, inspecting subway or railroad tracks for flaws, collecting debris or leaves from urban curbs, mowing grass along superhighways, or reading electric meters at private houses (see Spotlight 1.3).

Each of these problems has its own special requirements that may call for modifications in the theory. For example, in the case of garbage collection, the edges of our graph represent streets, not sidewalks. If some of the streets are one-way, we need to put arrows on the corresponding edges, resulting in a directed graph, or **digraph.** The circuits we seek will have to obey these arrows. In the case of salt spreaders and snowplows, each lane of a street needs to be modeled as a directed edge, as shown in Figure 1.19. Note that the arrows on the map and digraph are not in color because these arrows denote restrictions in traversal possibilities, not parts of circuits.

Like salt spreaders, street-sweeping trucks can travel in only one lane at a time and must obey the direction of traffic. Street sweepers, however, have an additional complication: parked cars. It is very difficult to clean the

Israel Electric Company Reduces Meter-Reading Task

SPOTLIGHT 1.3

Electric meters and icy roads don't seem to have much in common, but private companies and governments can draw on a common source—mathematics—to save customers money and drivers from accidents. Thus, in Israel there was a branch of the major electric company that needed to make the job of reading its electric meters more efficient, while in Minnesota and Ohio, the counties and cities must plow and/or de-ice winter roads. Both problems start in the same place: constructing a graph theory model of the streets and roads the meter readers need to traverse and winter maintenance vehicles must travel. Next, applying the ideas about the Chinese Postman problem for efficiently providing services along the edges of a graph at least once can lead to time-saving routes for meter reading or de-icing. Because the customers of the electric company change as new houses and businesses need service and because the roads needing plowing vary with storm patterns, digitized versions of street networks are usually now maintained online; solutions to either the Chinese Postman problem or variants of it can be solved in real time. In Minnesota and Ohio, routes are typically reevaluated every winter season. In one Ohio county where nearly 200 segments of road involving about 300 miles need plowing, the number of trucks involved was reduced by 30% and the time to do the work was cut 40% when Chinese Postman ideas were used. In Israel, meter reader routes could be traversed in 40% less time and deadheading time was reduced to 1% of the time spent by using Chinese Postman algorithms.

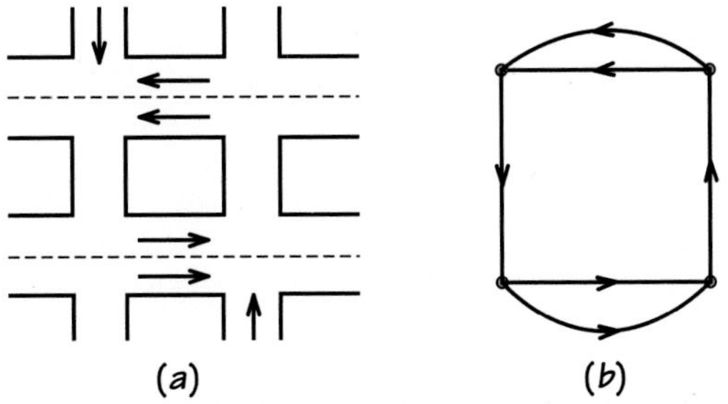

Figure 1.19 (a) Salt-spreading route, where each east–west street has two traffic lanes in the same direction, and (b) an appropriate digraph model.

street if cars are parked along the curb. Yet for overall efficiency, those who are responsible for routing street sweepers want to interfere with parking as little as possible. The common solution is to post signs specifying times when parking is prohibited. Because the parking-time factor is a constraint on street sweeping, it is important to find not only an Euler circuit, or a circuit with very few duplications, but also a circuit that visits streets when they are free of cars. Once again, mathematicians have developed techniques to handle this constraint.

Finally, because towns and cities of any size have more than one street sweeper, parking officer, garbage truck, or pothole-filling crew, a single best route may not suffice. Instead, it becomes necessary to divide the territory into multiple routes. The general goal is to find optimal solutions while taking into account traffic direction, number of lanes, parking-time restrictions, and divided routes (see Figure 1.20).

Management science makes all this possible. For example, a pilot study done in the 1970s in New York City showed that applying these techniques to street sweepers in just one district could save about $30,000 per year. With 57 sanitation districts in New York, this would amount to a savings of more than $1.5 million in a single year. This translates to over $6 million in today's dollars. In addition,

Figure 1.20
(a) Residential neighborhoods, whether they be in cities or the suburbs, require many services such as mail delivery, garbage collection, street sweeping, meter reading, or sewage systems. The mathematical techniques of operations research make it possible to provide these services as cheaply as possible. When optimal solutions to providing such services can be found, everyone is a winner. (b) Computers can be used to extract the essential information needed from photographs to solve routing problems.

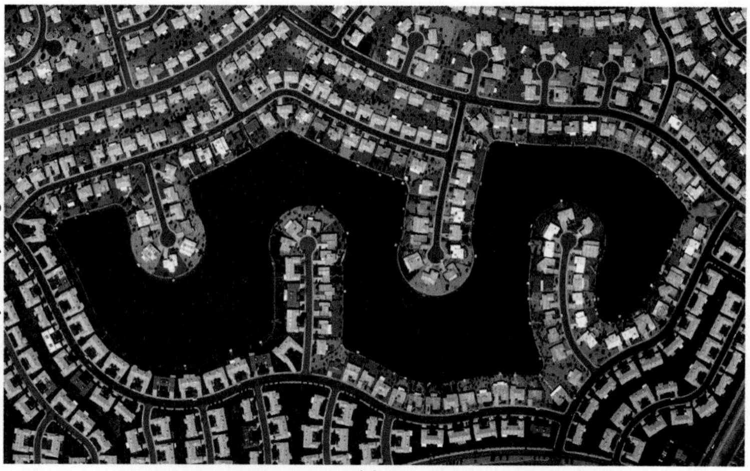

Harald Sund/Stockbyte/Getty Images

(*a*)

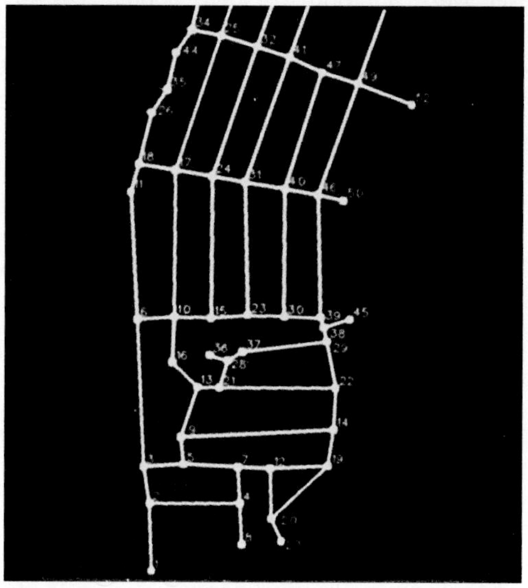

(*b*)

the same principles could be extended to garbage collection, parking control, and other services carried out on street networks.

This plan was not adopted when first proposed. Because city services take place in a political context, several other factors come into play. For example, union leaders try to protect the jobs of city workers, certain bureaucrats try to keep their departmental budgets high, and elected politicians rarely want to be accused of cutting the jobs of their constituents. Thus, political obstacles can overrule management science. As mentioned in Spotlight 1.2 on page 10, such human factors often arise when applying management science. Perhaps a more acceptable street-sweeping plan would have been devised for New York City if more attention had been paid to the human factors earlier.

Despite the complications of real-world problems, management science principles provide ways to understand these problems by using graphs as models. We can reason about the graphs and then return to the real-world problem with a workable solution. The results we get can have a lasting effect on the efficiency and economic well-being of any organization or community.

Review Vocabulary

Chinese postman problem The problem of finding a circuit on a graph that covers every edge of the graph at least once and that has the shortest possible length. (p. 14)

Circuit A path that starts and ends at the same vertex. (p. 6)

Connected graph A graph wherein it is possible to reach any vertex from any specified starting vertex by traversing edges. (p. 11)

Digraph A graph in which each edge has an arrow indicating the direction of the edge. Such directed edges are appropriate when the relationship is "one-sided" rather than symmetric (for instance, one-way streets as opposed to regular streets). (p. 20)

Edge A link joining two vertices in a graph. (p. 6)

Euler circuit A circuit that traverses each edge of a graph exactly once. (p. 8)

Eulerizing Adding new edges which duplicate existing edges to a connected graph so as to make a graph that possesses an Euler circuit. (p. 16)

Graph A mathematical structure in which points (called vertices) are used to represent things of interest and in which links (called edges) are used to connect vertices, denoting that the connected vertices have a certain relationship. (p. 6)

Management science A discipline in which mathematical methods are applied to management problems in pursuit of optimal solutions that cannot readily be obtained by common sense. (p. 5)

Operations research (OR) Another name for management science. (p. 5)

Optimal solution When a problem has various solutions that can be ranked in preference order (perhaps according to some numerical measure of "goodness"), the optimal solution is the best-ranking solution. (p. 5)

Path A connected sequence of edges in a graph. (p. 6)

Valence (of a vertex) The number of edges touching that vertex. (p. 10)

Vertex A point in a graph where one or more edges end. (p. 6)

Self Check Answers

1. (a) 8 vertices and 11 edges
(b) *A, B, D, H, F,* and *E* are even-valent.
(c) It is not an Euler circuit. One of the two edges joining *B* and *F* was not traversed; *FG* was traversed twice; edge *BC* was not traversed.

2. (a) No. Though the graph is connected, its vertices are not all even-valent.
(b) By duplicating one edge, we get a new graph, all of whose vertices are even-valent.

Skills Check

1. The accompanying graph has

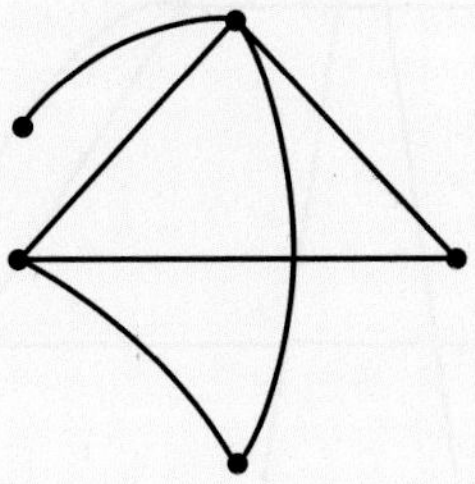

(a) four vertices and six edges.
(b) four vertices and four edges.
(c) five vertices and six edges.

2. The number of vertices in the following graph is _____, while the number of edges in this graph is __________.

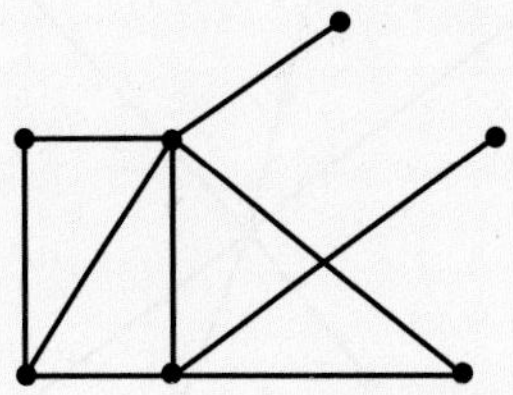

3. The accompanying graph

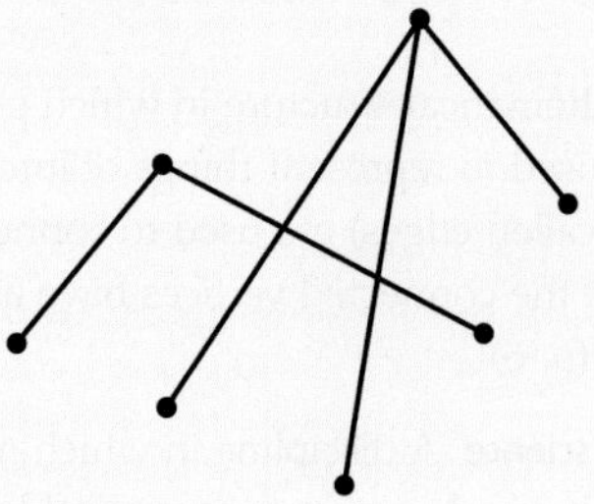

(a) is even-valent.

(b) is not connected.

(c) has six components.

4. The graph shown below is not connected because it consists of _______ parts.

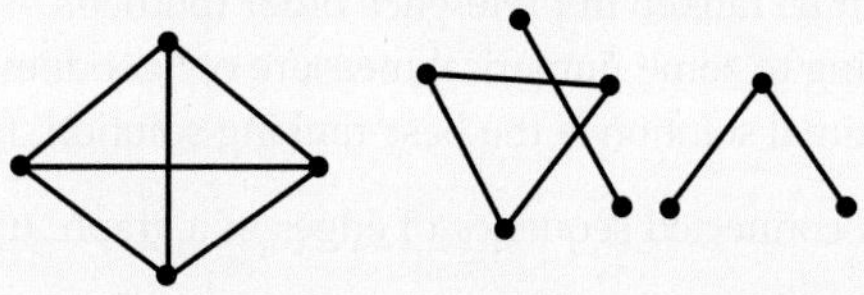

5. What is the valence of vertex *B* in the graph below?

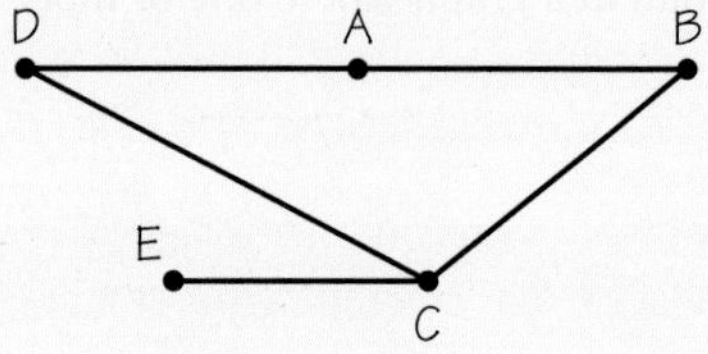

(a) 2 (b) 1 (c) 3

6. A graph has 24 edges and all the vertices of the graph have valence 3. The number of vertices of the graph must be _______ .

7. The valences of the vertices in the accompanying graph, listed in decreasing order, are

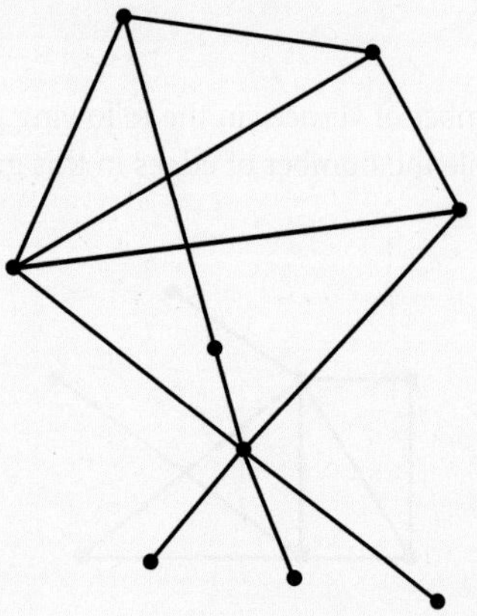

(a) 6, 4, 3, 3, 3, 2, 1, 1, 1.

(b) 1, 3, 4, 4, 5, 5.

(c) 5, 5, 4, 3, 3, 1, 1, 1.

8. The alphabetically ordered list of even-valent vertices of the graph below is ______ , ______ .

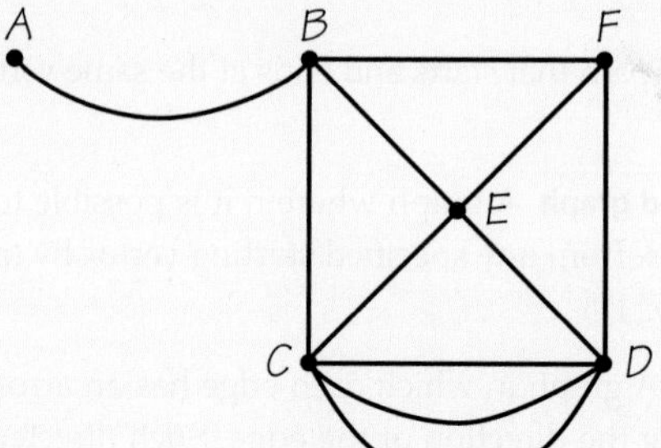

9. Which of the statements about the accompanying graph is false?

(a) It is connected.

(b) It is not a graph because it includes curved edges.

(c) It is even-valent.

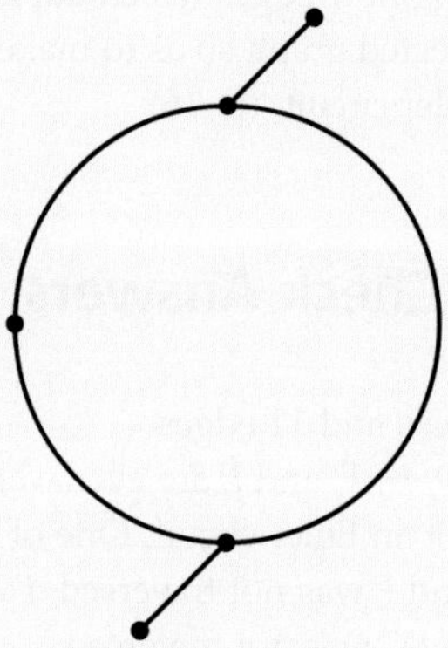

10. The accompanying graph has _______ edges and _______ vertices.

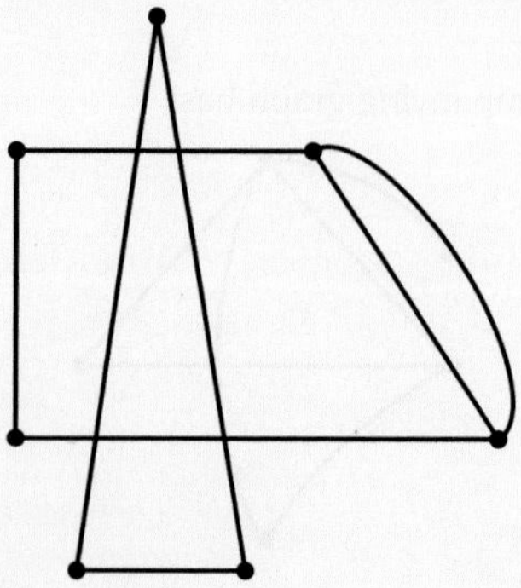

11. Which of the following statements about a *path* is true?

(a) A path always forms a circuit.

(b) A path is always connected.

(c) A path can visit any vertex only once.

12. A graph G has 10 edges, and all its vertices have the same valence. The possible valences of the vertices of G are ______, ______, ______, ______, ______.

13. It is not possible for a graph to have three vertices of valence 3 and six vertices of valence 4 because

(a) there are no graphs with exactly 11 vertices.

(b) a graph cannot have an even number of 4-valent vertices.

(c) a graph cannot have an odd number of odd-valent vertices.

14. If a graph consists of five vertices and every pair of vertices is connected by a single edge, the number of edges in the graph is exactly ________ .

15. For which of the situations below is it most desirable to find an Euler circuit or an efficient eulerization of the graph?

(a) Sweeping the sidewalks of a small town

(b) Planning a new highway

(c) Planning a parade route in Muncie, Indiana

16. For the accompanying graph, the vertex that has the largest valence is ________, and the number of edges in the graph is ________ .

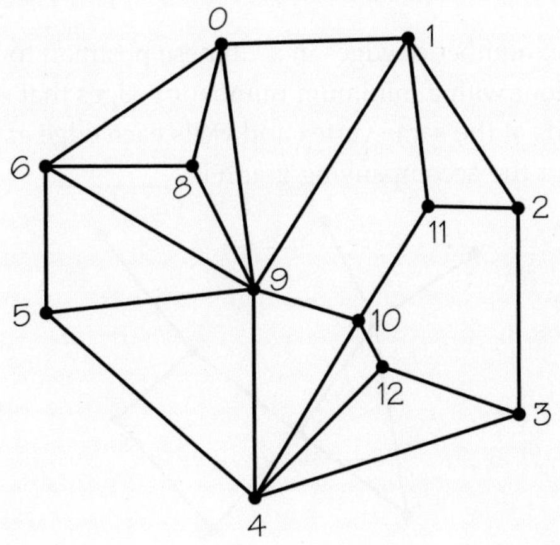

17. The following graph has no Euler circuit because it

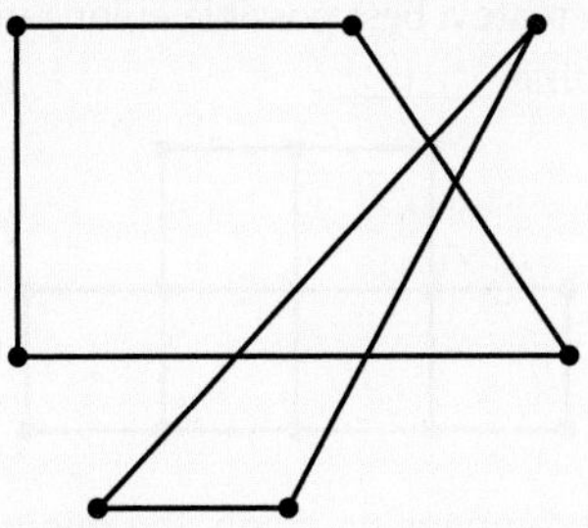

(a) is even-valent.

(b) has seven vertices.

(c) is not connected.

18. If a graph is connected and has nine vertices, the graph must have at least __________ edges.

19. Consider the path represented by the sequence of numbered edges on the graph below. Which statement is correct?

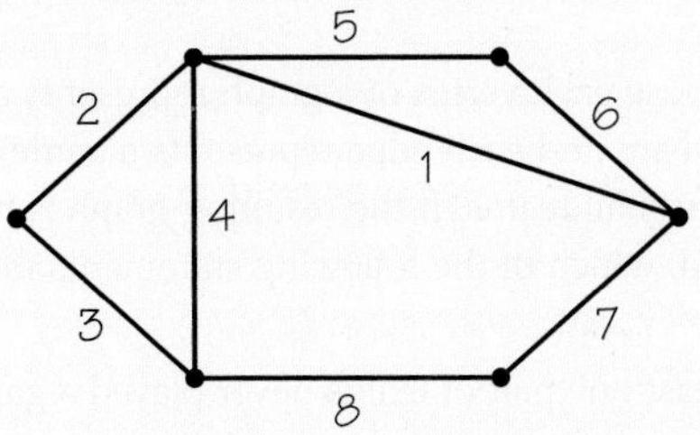

(a) The sequence of numbered edges forms an Euler circuit.

(b) The sequence of numbered edges forms a circuit but not an Euler circuit.

(c) The sequence of numbered edges traverses each edge exactly once but is not an Euler circuit.

20. The accompanying graph has no Euler circuit because it

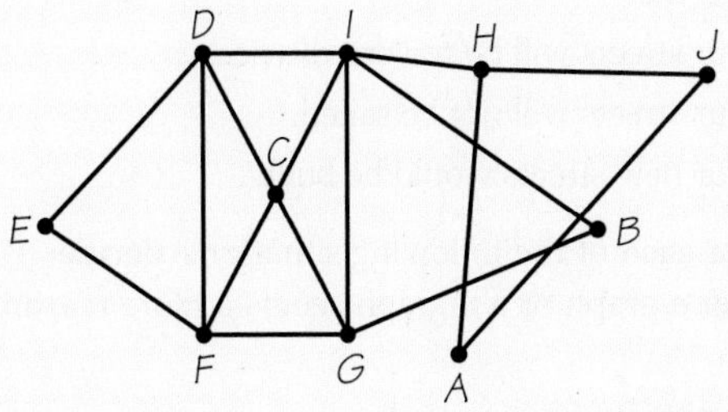

(a) is not even-valent.

(b) has 16 edges.

(c) is not connected.

21. The minimum number of edges that must be duplicated to create a best possible eulerization of the following graph is _______ .

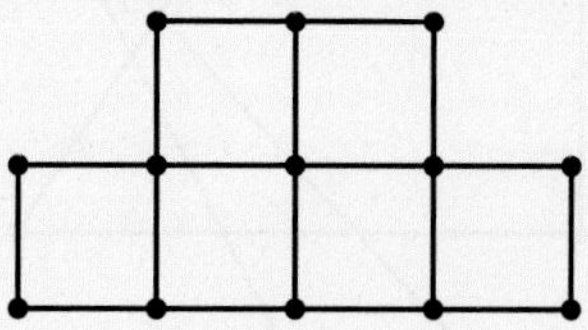

22. For the accompanying graph, the minimum *total* number of edges which constitutes a tour of the graph, starting and ending at the same vertex, and which visits each edge at least once, is _______ .

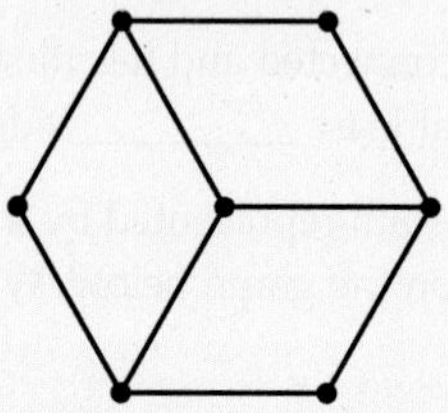

23. Suppose each vertex of a graph represents a baseball team and each edge represents a game played by two baseball teams. If the resulting graph is not connected, which of the following statements must be true?

(a) At least one pair of teams never played a game.

(b) At least one team played every other team.

(c) The teams play in distinct leagues.

24. If a graph has eight vertices of odd valence, the absolute minimum number of edges that must be added (duplicated) to eulerize the graph is ________ .

25. Suppose the edges of a graph represent streets that must be salted after a snowstorm. To eulerize the graph, four edges must be added. The real-world interpretation of this is that

(a) four streets will be traversed twice.

(b) four streets will not be salted.

(c) four new streets would be built.

26. For each of the following situations, decide whether a graph or a digraph seems a more reasonable model.

(a) A system of hiking trails: ________________

(b) A bus route map: ________________

(c) An electrical wiring plan for a home: _________

27. The smallest number of edges needed to eulerize the graph below is

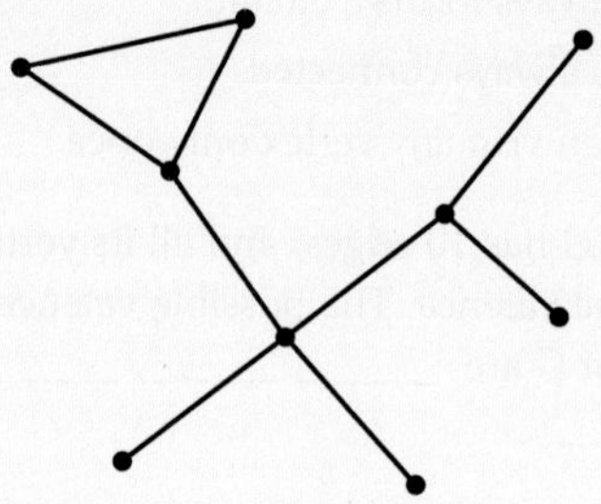

(a) 4.

(b) 6.

(c) 8.

28. If the valences of the vertices of a graph G are: 5, 4, 4, 4, 4, 3, 2, 2, and 2, the number of vertices of G is _______ and the number of edges of G is _______ .

29. The smallest number of edges that must be added to the accompanying graph for it to have an Euler circuit is

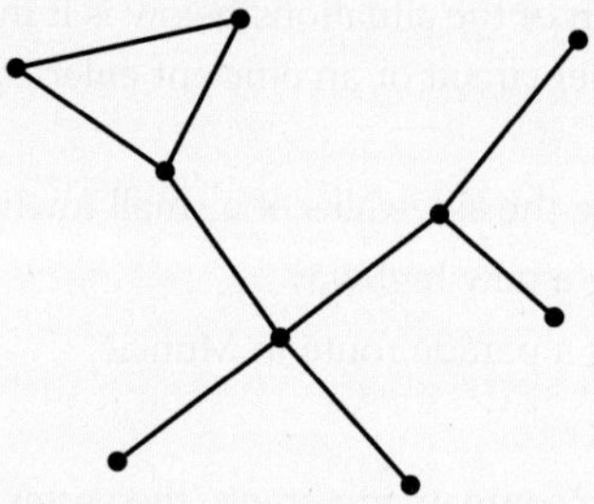

(a) 3.

(b) 6.

(c) 7.

30. The number of edges in a Chinese postman tour (i.e., a tour with a minimum number of edges that starts and ends at the same vertex and visits each edge at least once) for the accompanying graph is _________ .

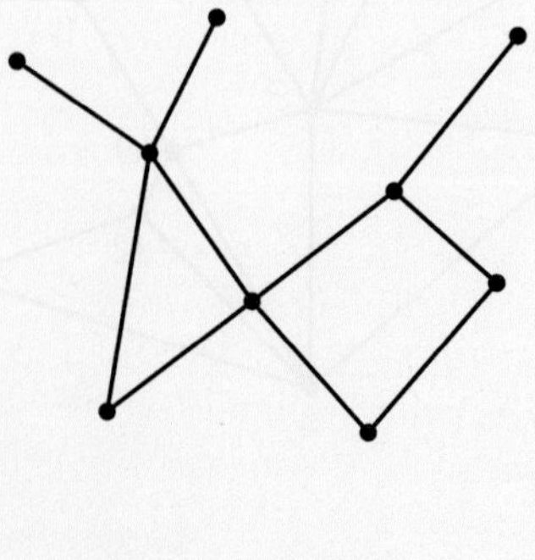

Chapter 1 Exercises

Challenge Discussion

1.1 Euler Circuits

1.2 Finding Euler Circuits

1. Refer to the accompanying graph *G* (which might represent the two-way streets of a small housing subdivision).

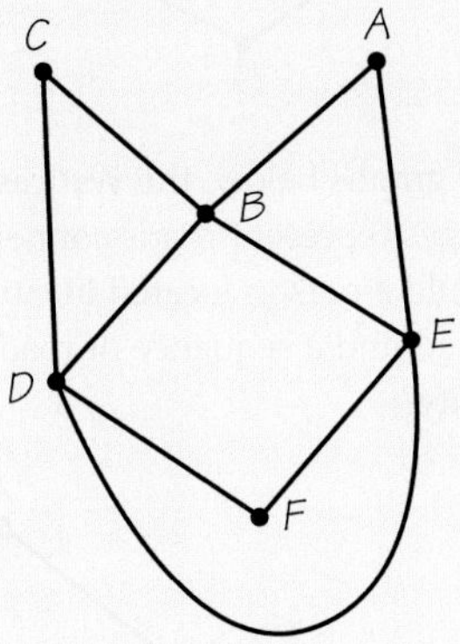

(a) Determine the number of vertices.

(b) Determine the number of edges.

2. Use the diagram for graph *G* in Exercise 1.

(a) Find two ways to get from *C* to *A* by traversing four edges.

(b) Find two ways to get from *C* to *A* by traversing three edges.

(c) Find two ways to start at *B* and return to *B*, traversing four edges.

3. (a) Locate within Figure 1.3a a section of the street network shown that indicates how the graph below could be interpreted as meters that require inspection in this urban street layout.

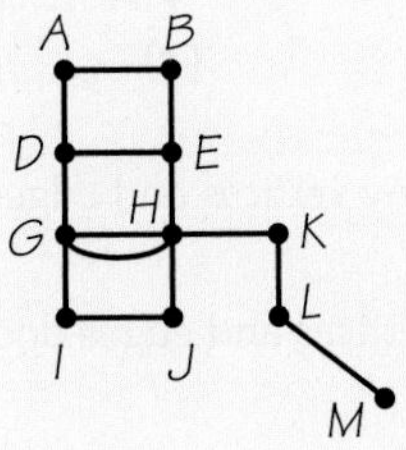

(b) Can you find a tour of the edges in the graph shown above that starts at *A* and allows for inspection of the parking meters with as few deadheaded edges as possible?

4. (a) Determine the number of vertices and edges in the accompanying graph *G*.

(b) What is the smallest number of edges that must be added to graph *G* so that the result is a connected graph?

(c) What are the valences of the vertices in *G*?

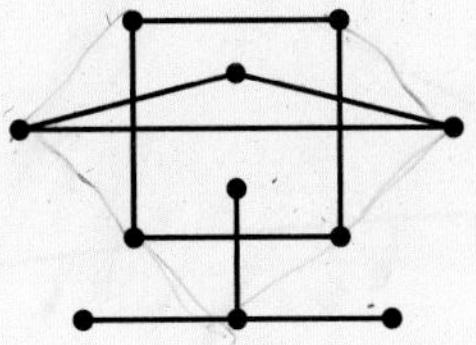

5. (a) Is the figure below a graph? Explain your answer.

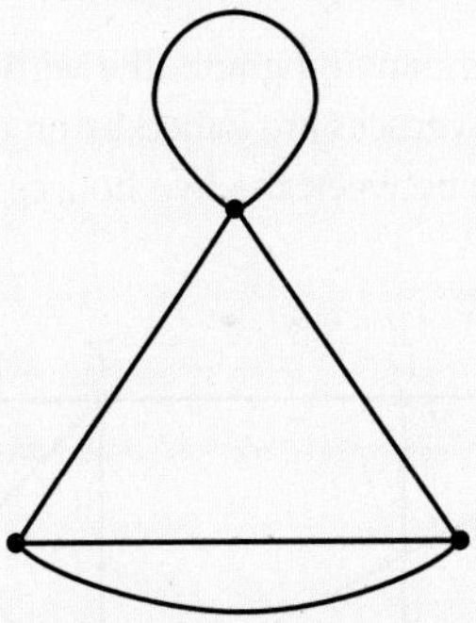

(b) The graph below has edges that "cross" at points that are not vertices of the graph. Which edges are these?

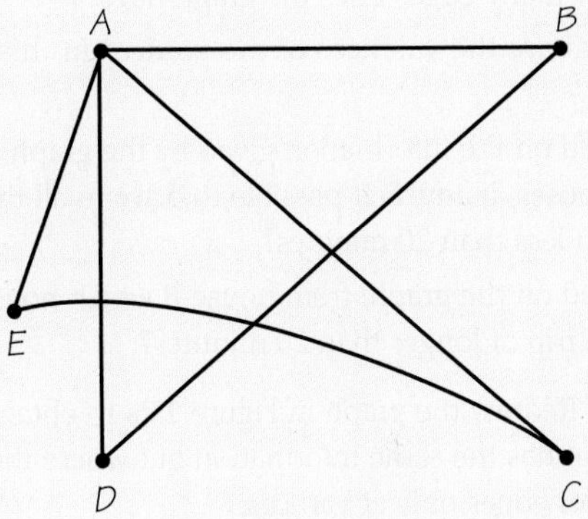

(c) How many vertices and edges are there in the preceding graph?

6. The graph below shows the stores and roads connecting them in a small shopping mall.

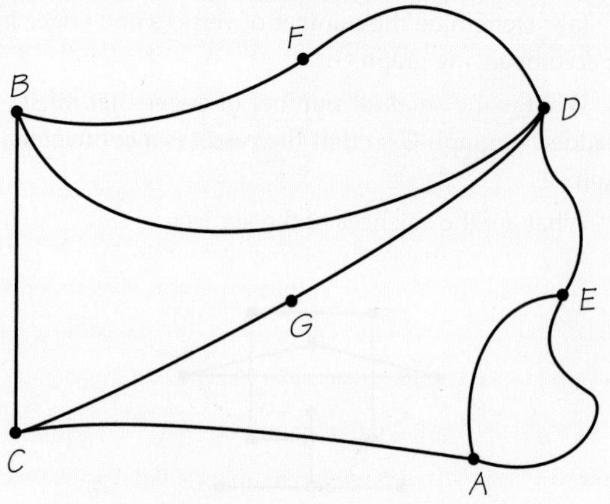

(a) How many stores does the mall have?

(b) How many roads connect up the stores in the mall?

(c) Write down a path from *C* to *F*.

(d) Write down a path from *E* to *B*.

7. In the accompanying graph, the vertices represent houses and two vertices are joined by an edge if it is possible to drive between the two houses in under 10 minutes.

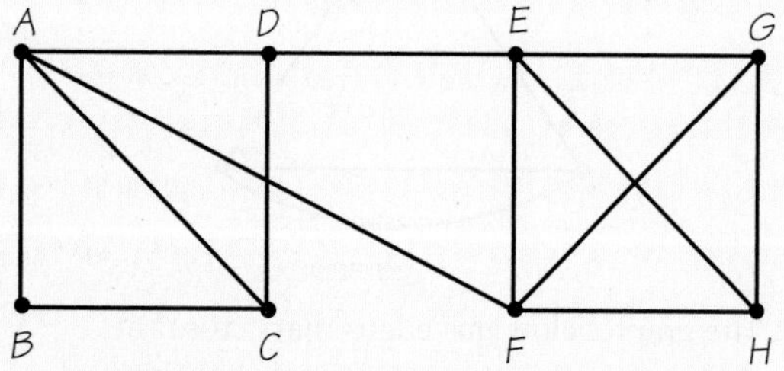

(a) How many vertices does the graph have?

(b) How many edges does the graph have?

(c) What are the valences of the vertices in this graph?

(d) Based on the information given by the graph, for which houses, if any, is it possible to drive to all the other houses in less than 20 minutes?

(e) Based on the graph, from house *B* which houses require a trip of longer than 20 minutes?

8. (a) Redraw the graph in Figure 1.2a to obtain a graph that has the same information but where the edges meet other edges only at vertices.

(b) List all the routes that start on the U.S. side of the Atlantic Ocean and cross the ocean once and immediately.

9. In the graph below, the vertices represent cities and the edges represent roads connecting them. What are the valences of the vertices in this graph? (Keep in mind that *E* is part of the graph.) What might the valence of city *E* be showing about the geography?

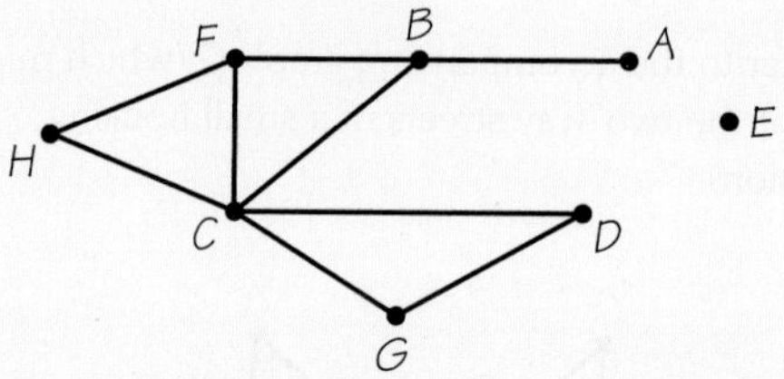

10. In the three graphs below, the vertices represent cities and the edges represent roads connecting them. In which graphs could a person located in city *A* choose any other city and then find a sequence of roads to get from *A* to that other city?

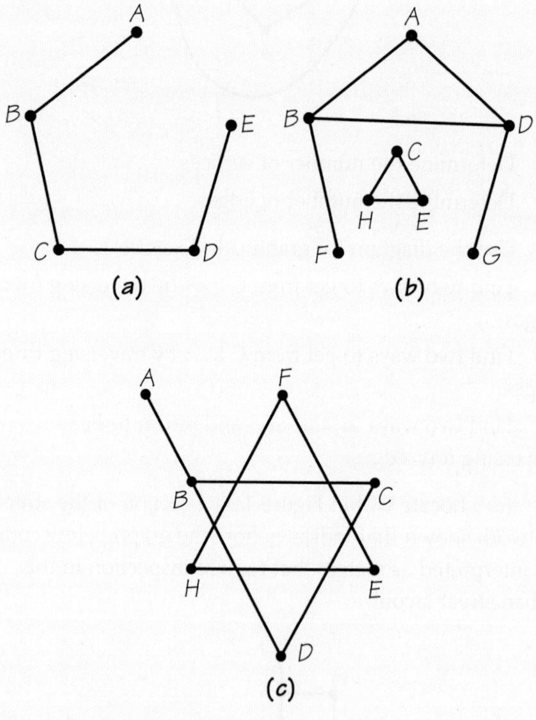

11. (a) How many vertices and edges does the graph in Figure 1.6 have?

(b) How many vertices and edges does the graph in Figure 1.7 have?

(c) How many vertices and edges does the graph in Figure 1.8a have?

12. Jack and Jill are located in Miami and want to fly to Berlin (see Figure 1.2).

(a) Find three paths for them to carry out this trip.

(b) What is the largest number of paths that can be used to carry out this trip that do not repeat a vertex (city)?

(c) Explain why it is reasonable not to want to repeat a vertex in this situation.

13. Refer to the figure in Exercise 6.

(a) Write down a circuit that includes the vertices *G* and *D* but does not start or end at either of these vertices.

(b) If two paths are considered different if they use different edges, write down

(i) two different paths from *B* to *D*.

(ii) three different paths from *C* to *F*.

(iii) a circuit that has four edges.

14. In the graphs in Figure 1.17, find the smallest possible number of edges you could remove that would disconnect the graph.

15. Draw a graph with eight vertices that is connected where each vertex has

(a) degree (valence) 2.

(b) valence 3.

(c) valence 4.

(d) Do all graphs with eight vertices having valence 2 have the same number of edges?

16. (a) Add up the numbers you get for the valences of the vertices in Figure 1.6.

(b) Add up the numbers you get for the valences of the vertices in Figure 1.7.

(c) Add up the numbers you get for the valences of the vertices in Figure 1.8a.

(d) Describe the pattern you see in your answers for parts (a) through (c).

(e) Show that the pattern describes a fact that is true for any graph. (*Hint*: How many endpoints does an edge have?)

17. In the graph in Figure 1.8a, find the smallest possible number of edges you could remove that would disconnect the graph.

18. (a) Can you draw a graph with at least two vertices for which all the vertices have different valences?

(b) Can you do this so that the valences are consecutive integers?

19. (a) Can you draw a graph that is connected and for which the valences of vertices are the consecutive integers from 2 to 6?

(b) If a graph such as the one described in part (a) exists, can it have an Euler circuit? (Explain your answer.)

20. (a) What is the number of vertices and edges in the accompanying graph *G*?

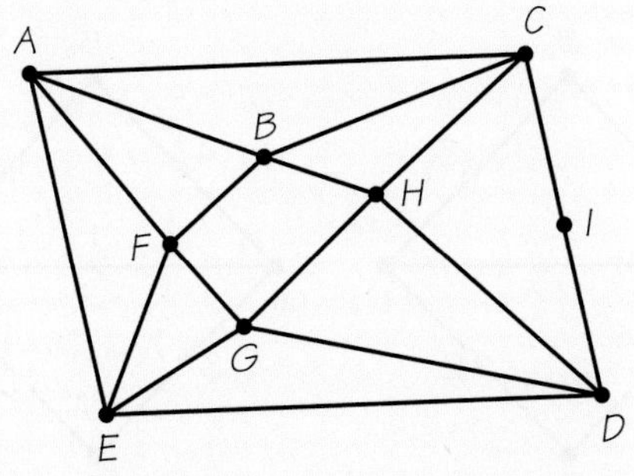

(b) Write down an Euler circuit for the accompanying graph, first by numbering the edges consecutively starting at an edge emanating from vertex *D* and then using an alternative description that lists the vertices as they are traversed by the Euler circuit.

21. Draw a graph that is not connected with eight vertices, all of whose vertices have valence 2.

22. (a) Is it possible that a street network gives rise to a disconnected graph? If so, draw such a network of blocks and streets and parking meters (in the style of Figure 1.12a). Then draw the disconnected graph it gives rise to.

(b) Draw a graph where every vertex has valence of at least 3 but where removing a single edge disconnects the graph.

(c) In what urban settings might a road network be represented by a graph that has an edge whose removal would disconnect the graph?

23. (a) Find a graph of eight vertices in which the valences of the vertices are 1, 2, 2, 2, 3, 3, 3, 4.

(b) Find a graph with the same valences as the one you found in part (a) but that is "different" from the one in part (a).

24. For some urban services provided along streets, it may matter whether the roads are one-way or two-way. Give some examples where the street directions do and do not matter for our graph model analysis.

25. A postal worker is supposed to deliver mail on all streets represented by edges in the accompanying graph by traversing each edge exactly once. The first day, the worker traverses the numbered edges in the order shown in graph (a), but the supervisor is not satisfied—why? The second day the worker follows the path indicated in graph (b), and the worker is unhappy—why? Is the original job description realistic? Why?

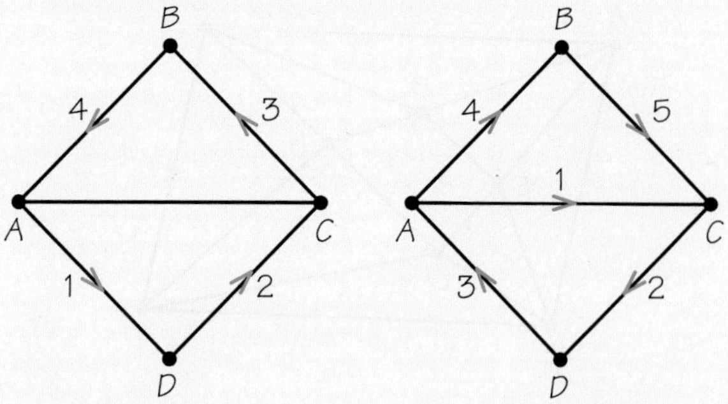

(a) (b)

26. For the street network in Exercise 25, draw the graph that would be useful for routing a snowplow. Assume that all streets are two-way, one lane in each direction, and that you need to pass down each lane separately.

27. Find an efficient route for the snowplow to follow in the graph you drew in Exercise 26.

28. (a) Give examples of services that could be performed by a vehicle that moved in the direction of traffic down either lane of a two-way street.

(b) Give examples of services that would probably require a vehicle to travel down each of the lanes of a two-way street (in the direction of traffic for that lane) to perform the service.

29. For the street network shown below, draw the graph that would be useful for finding an efficient route for checking parking meters. (*Hint:* Notice that not every sidewalk has a meter; see Figure 1.12.)

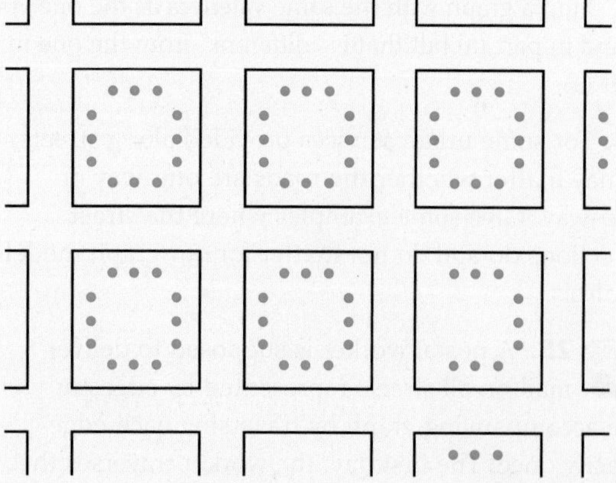

30. (a) For the street network in Exercise 29, draw a graph that would be useful for routing a garbage truck. Assume that all streets are two-way and that passing once down a street suffices to collect from both sides.

(b) Do the same problem as in part (a), using the assumption that one pass down the street suffices to collect from only one side.

31. (a) In the accompanying graph, find the largest number of paths from *A* to *C* that do not have any edges in common.

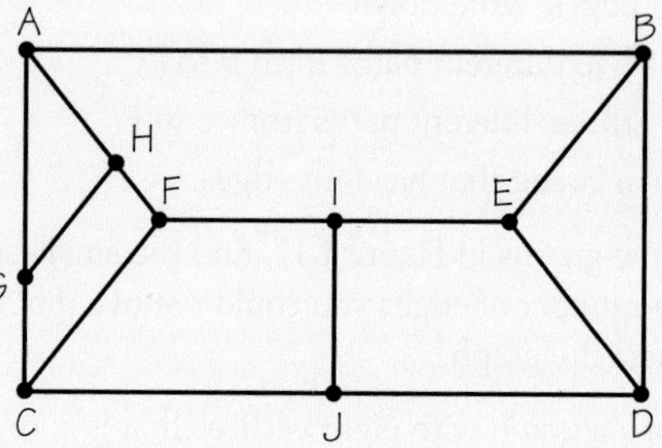

(b) Verify that the largest number of paths with no edges in common between any pair of vertices in this graph is the same.

(c) Why might we want to be able to design graphs such that we can move between two vertices of a graph using paths that have no edges in common?

32. Examine the paths represented by the numbered sequences of edges in both parts of the figure below. Determine whether each path is a circuit. If it is a circuit, determine whether it is an Euler circuit.

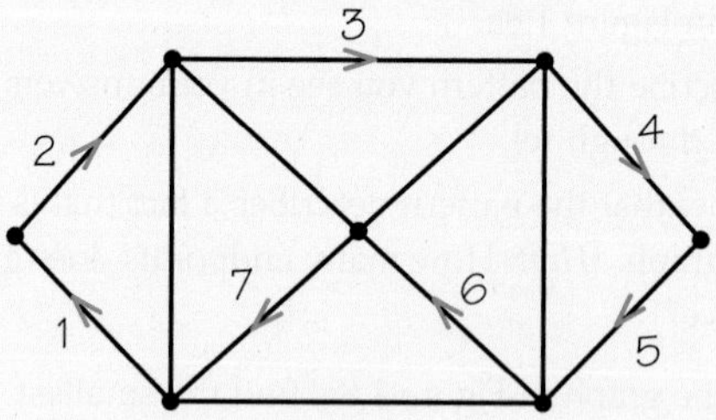

(a)

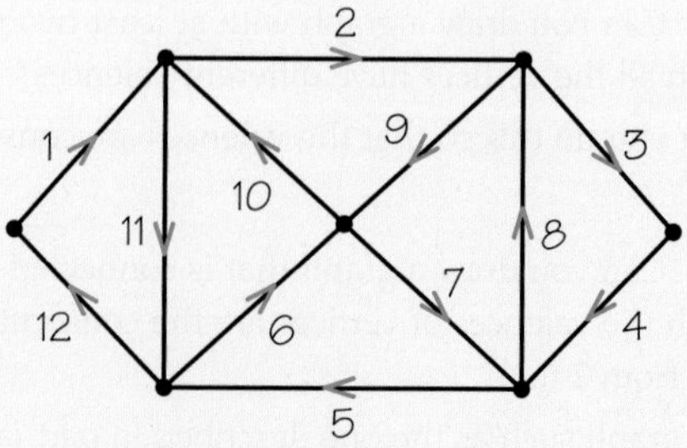

(b)

33. The company that distributes natural gas in a small town is located at vertex C in the accompanying graph G, which models the two-way streets for the town.

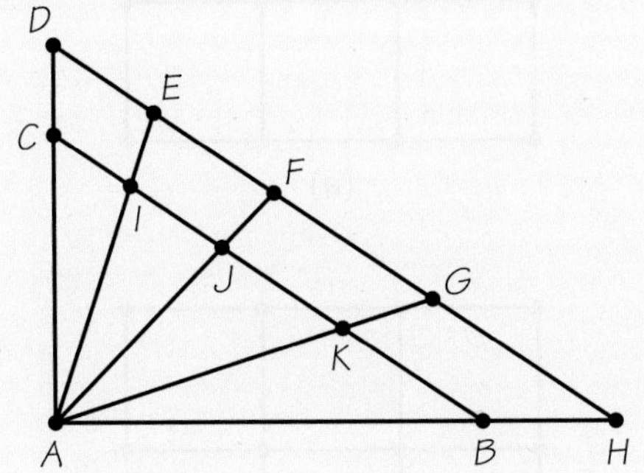

(a) Can an inspector for gas leaks travel along the street network and find a route that starts and ends at C and has no deadheads (repeated edges)?

(b) What is the minimum number of repeated edges in a shortest-length tour T, starting and ending at C, that looks for gas leaks?

(c) By duplicating existing edges in G, use tour T to modify graph G to obtain a new graph G^*, which shows that T can be interpreted as an Euler circuit in G^*.

34. In Figure 1.8b, suppose we started an Euler circuit using this sequence of edges: 14, 13, 8, 1, 4 (ignore existing arrows on the edges). What does our guideline for finding Euler circuits tell you *not* to do next?

35. In Figure 1.13c, suppose we started an Euler circuit using this sequence of edges: 6, 7, 8, 9 (ignore existing arrows on the edges). What does our guideline for finding Euler circuits tell you *not* to do next?

36. Find Euler circuits in the right-hand graphs in Figures 1.17a and 1.17b.

37. Find an Euler circuit on the graph of Figure 1.15c (including the blue edges).

38. An Euler circuit visits a four-valent vertex X, such as the one in the accompanying graph, by using the edges AX and XB consecutively, and then using CX and XD consecutively. When this happens, we say that the Euler circuit cuts through at X.

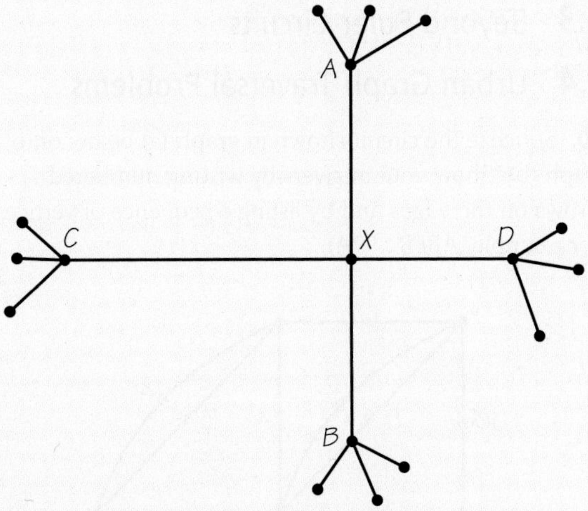

Suppose G is a four-valent graph such as that in the diagram below. Is it possible to find an Euler circuit of this graph that never cuts through any vertex? Explain why it might be desirable to find an Euler circuit of this special kind in an applied situation.

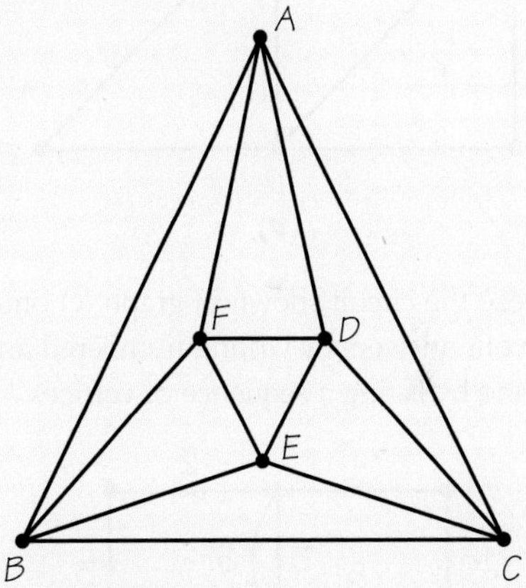

39. (a) Which vertices in the accompanying graph are odd-valent?

(b) In the accompanying graph, we see a territory for a parking-control officer that has no Euler circuit. How many sidewalks (edges) need to be omitted in order to enable us to find an Euler circuit? What effect would this have in the associated real-world situation?

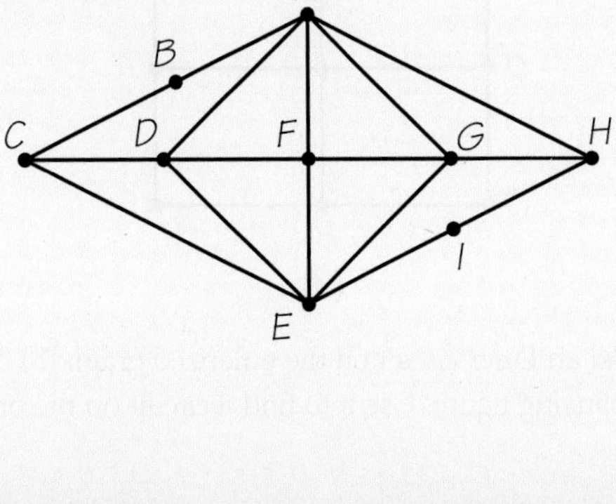

1.3 Beyond Euler Circuits

1.4 Urban Graph Traversal Problems

40. Squeeze the circuit shown in graph (a) below onto graph (b). Show your answers by writing numbered arrows on the edges and by listing a sequence of vertices (for example, *ABEB . . . A*).

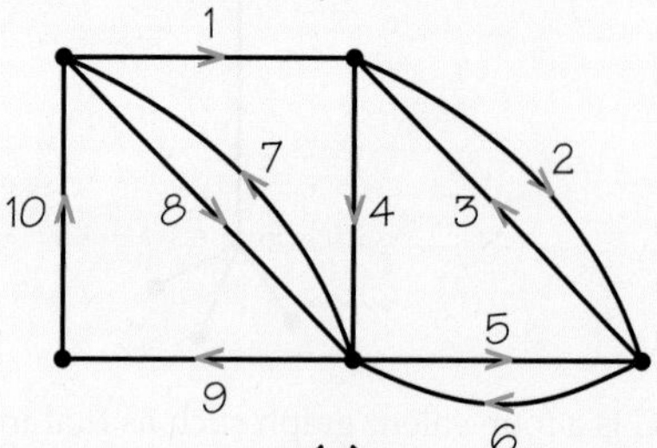

(a)

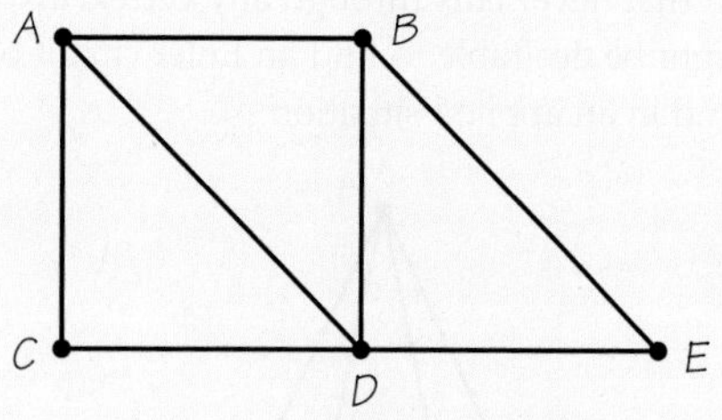

(b)

Next, squeeze the circuit shown in graph (c) onto graph (d). Show your answers by writing numbered arrows on the edges and by listing a sequence of vertices.

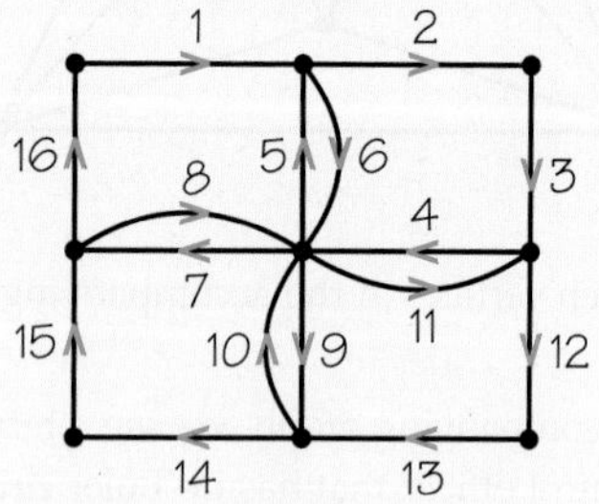

(c)

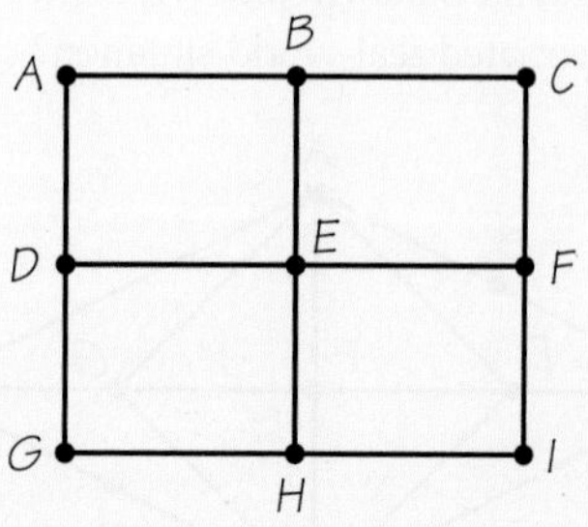

(d)

41. Find an Euler circuit on the eulerized graph (b) of the accompanying figure. Use it to find a circuit on the original graph (a) that covers all edges and reuses edges only five times. Can fewer than five reused edges be achieved?

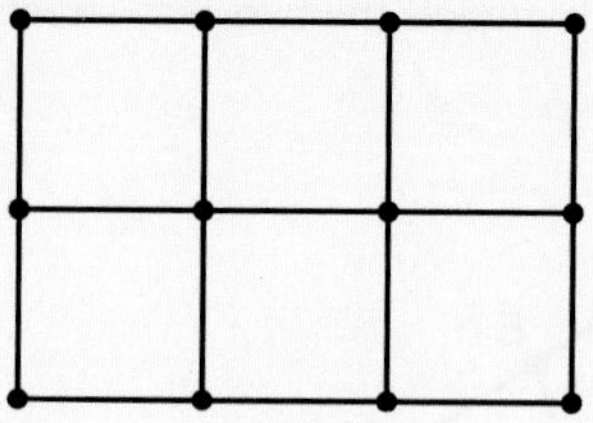

(a)

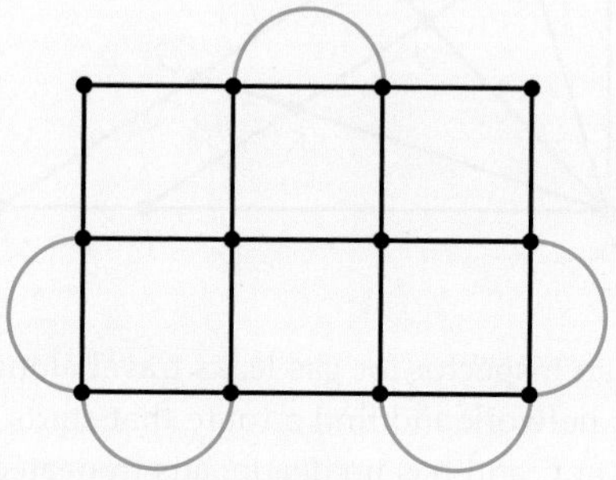

(b)

42. In the accompanying graph, add one or more edges to produce a graph that has an Euler circuit.

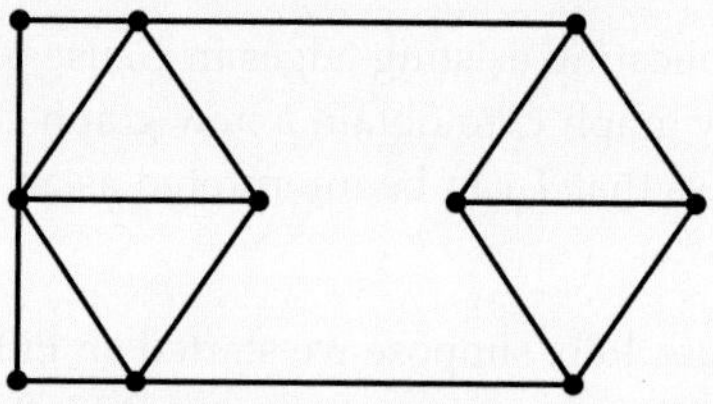

43. A college campus has a central square with sides arranged as shown by the edges in the graph below. Show how all these sidewalks can be traversed at least once in a tour that starts and ends at the same vertex.

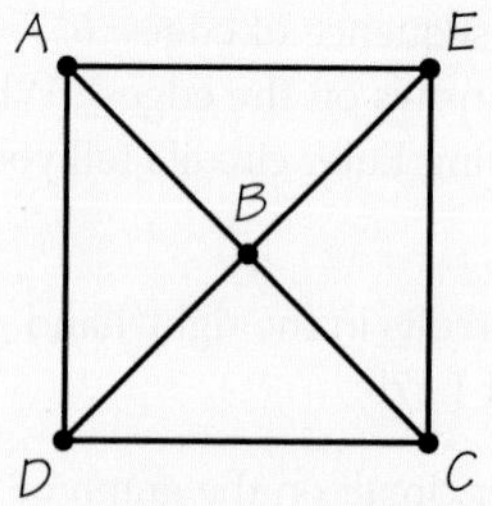

44. Find a circuit in the accompanying graph that covers every edge and has as few reuses as possible.

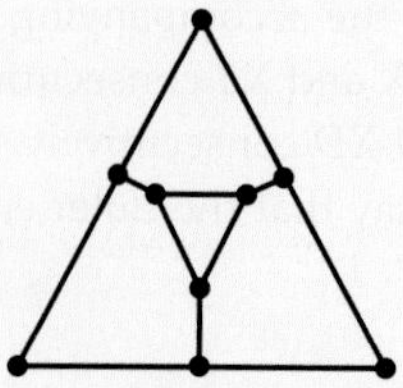

45. Eulerize the following rectangular street networks using the same patterns that would be used by the edge walker described on page 18.

(a) A 5 × 5 rectangle

(b) A 4 × 5 rectangle

(c) A 6 × 6 rectangle

(d) Can you find an eulerization with nine added edges for a 2-by-7-block rectangular street network? Can you do better than nine added edges?

46. Find good eulerizations for the accompanying graphs, using as few duplicated edges as you can. See "Finding Good Eulerizations" (page 19) for hints.

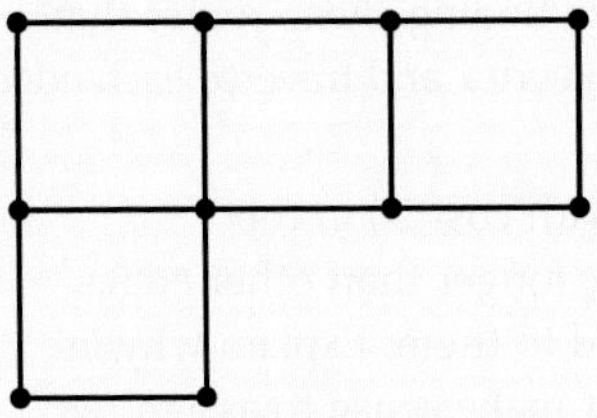

(a)

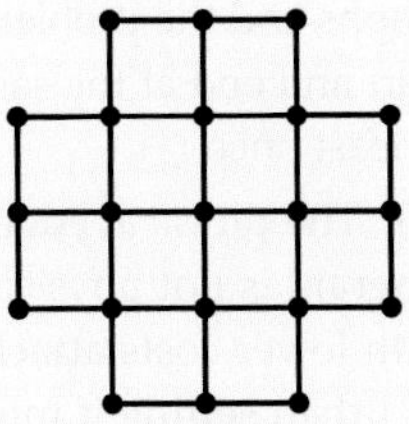

(b)

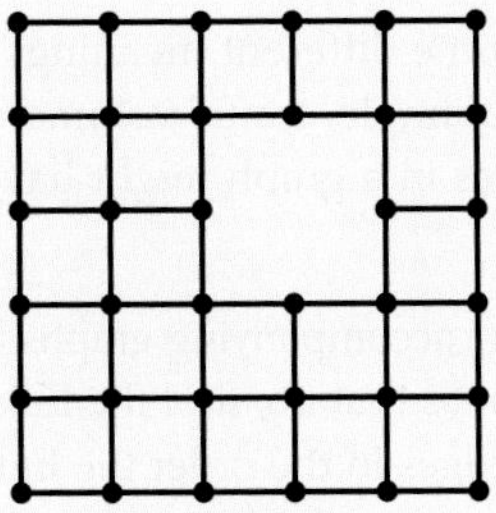

(c)

47.

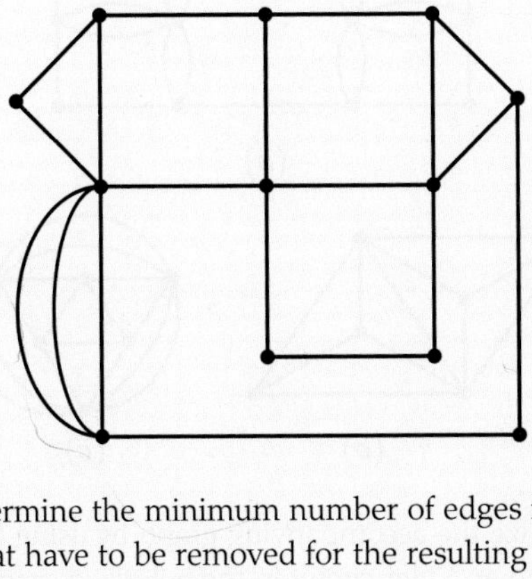

(a) Determine the minimum number of edges in the graph that have to be removed for the resulting graph to have all even-valent vertices.

(b) Does the graph you obtain in part (a) have an Euler circuit?

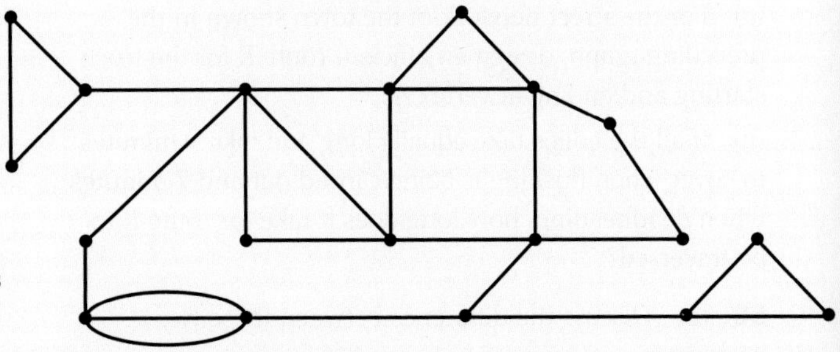

(c) Determine the minimum number of edges in the graph that have to be removed for the resulting graph to have all even-valent vertices.

(d) Does the graph you obtain in part (c) have an Euler circuit?

48. The following figure shows a river, some islands, and bridges connecting the islands and riverbanks. A charity is sponsoring a race in which entrants have to start at *A*, go over each bridge at least once, and end at *A*. Draw a graph that would be useful for finding a route that requires the least recrossing of bridges. Show what that route would be. (*Historical note:* This situation resembles the one that inspired Leonhard Euler's 1736 "recreational mathematics" problem that resulted in the first work in graph theory.)

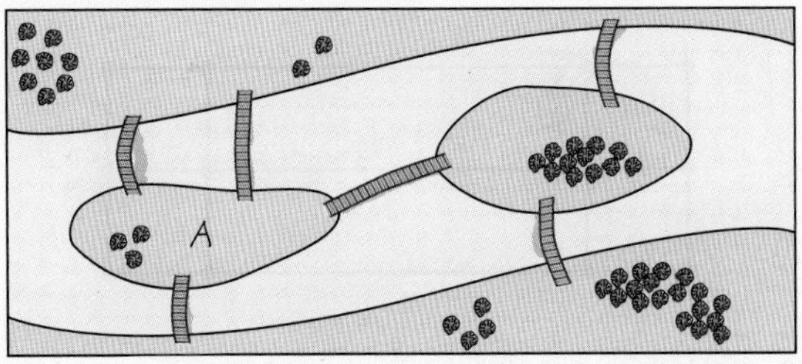

49. On Wednesdays during the summer, a small town that has only two-way streets does not allow on-street parking in the mornings. A truck that flushes water onto the streets to clean off "debris" can thus easily perform its work.

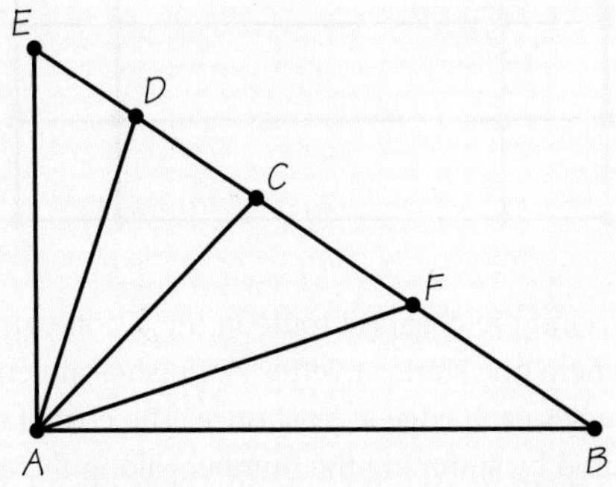

(a) For the street network of the town shown in the preceding graph, design an efficient route R for the truck starting and ending at vertex A.

(b) If all the edges take equally long and take 9 minutes to cover when flushing is being carried out and 7 minutes when deadheading, how long does it take for route R to be traversed?

50. (a) Discuss the difference between these two problems:

(i) Adding the minimum number of edges to a graph to make all its vertices even-valent

(ii) Finding the best eulerization of a connected graph

(b) In (i), must the graph that results from adding a minimum number of edges to make all the vertices even-valent have an Euler circuit?

51. Draw a graph with exactly two odd-valent vertices, which requires exactly nine edges to be duplicated in order to find the best eulerization of the graph.

52. In the following graph, the outer blocks are 1000-by-1000 feet, and the middle blocks are 1000-by-4000 feet. Find a circuit of minimum total length that covers all edges.

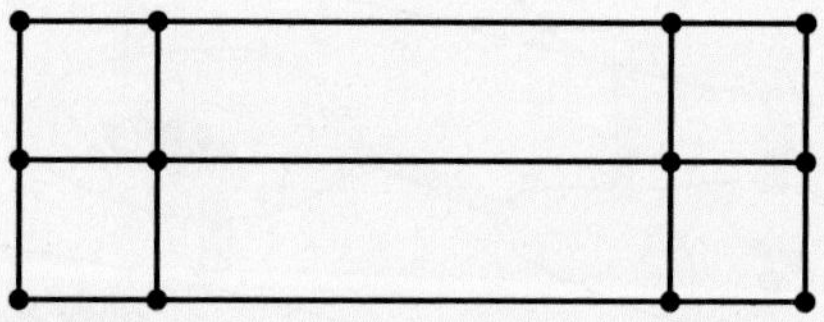

53. In the graph below, the outer blocks are 1000-by-1000 feet, and the middle blocks are 1000-by-4000 feet. Find a circuit of minimum total length that covers all edges.

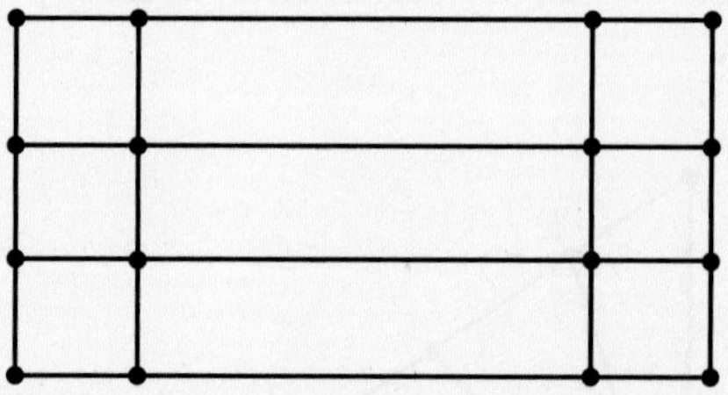

54. (a) Find the cheapest route in the accompanying graph, where one starts at vertex A, finishes at vertex A, and traverses each edge at least once. The cost of a route is computed by summing the numbers along the edges that one uses.

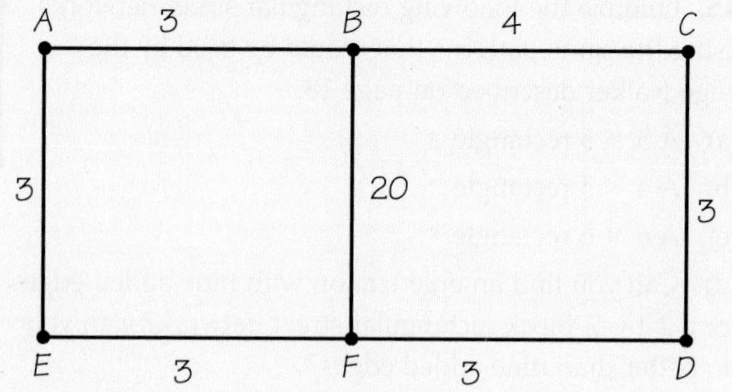

(b) How many edges are repeated in the minimal-cost route?

(c) Discuss the implications of this example for the relation between finding good eulerizations of graphs and the problem of finding cheap routes that start and end at the same vertex and traverse each edge at least once.

(d) The physical edge with cost 20 in the diagram is not physically longer than other edges with lower costs attached to them. Explain why in an urban setting it might make sense to assign two stretches of street of similar length very different "costs" for traversing them.

(e) What are some different meanings that "weights" (for example, traffic volume) potentially assigned to edges in a graph might have in an urban setting?

55. Which of the accompanying graphs have Euler circuits? In the ones that do, find the Euler circuits by numbering the edges in the order the Euler circuit uses them. For the ones that don't, explain why no Euler circuit is possible.

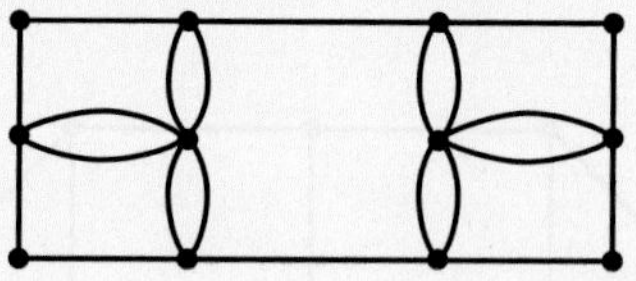

(a)

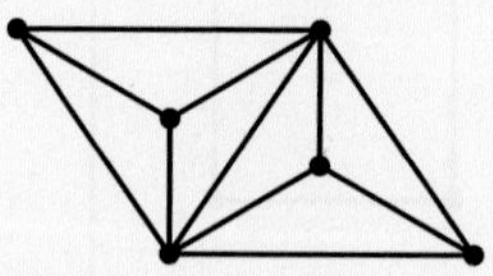

(b)

(c)

56. Eulerize the accompanying graph by using four new edges. Find an Euler circuit in the eulerized graph and use that circuit to find a circuit of the original graph that covers all edges but reuses edges only four times. How many different ways can the four edges be chosen? What is the total number of edges in the Euler circuit in the eulerized graph?

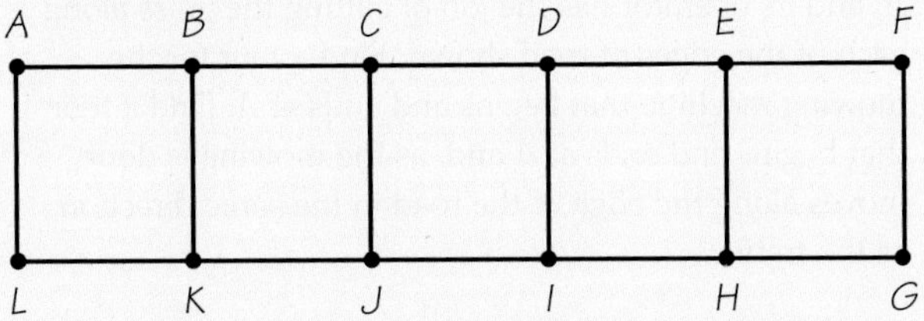

57. A graph G represents a street network to be traveled by a postal worker on foot, who must deliver mail to the houses on both sides of each street. How does one obtain from G a new graph H that represents the sidewalks that must be traversed? Does graph H always have an Euler circuit? Explain your answer.

58. In the following graph, find a circuit that covers every edge and has as few reuses as possible.

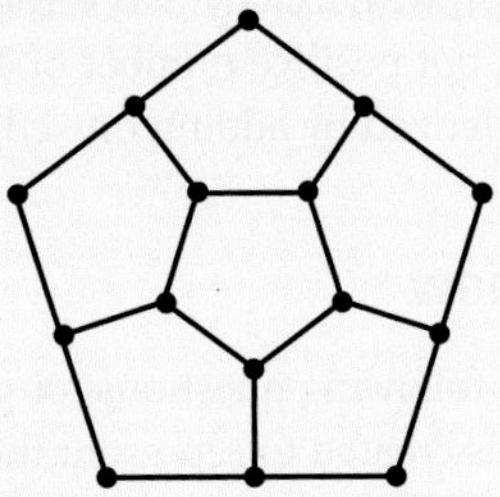

59. (a) Find the best eulerizations you can for the two accompanying graphs.

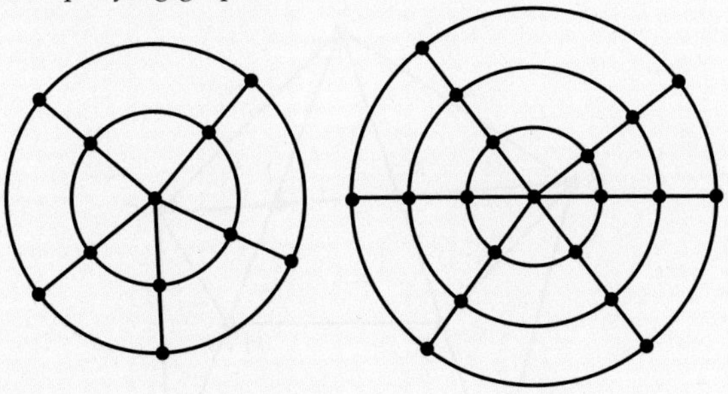

(a) (b)

(b) Graph (a) can be thought of as having five rays and two circles, and graph (b) as having six rays and three circles. Draw a graph with four rays and four circles and find the best eulerization you can for this graph.

(c) Find a "formula" involving r and s for the smallest number of edges needed to eulerize a graph of this type having r rays and s circles.

60. Suppose that for a certain connected graph, it is possible to disconnect it by removing one edge. Explain why such a graph (before the edge is removed) must have at least one vertex of odd valence. (*Hint:* Show that it cannot have an Euler circuit.)

61. (a) Can you draw a graph with six vertices where the valence of each vertex is 5?

(b) If such a graph exists, how many edges must it have?

62. Each of the accompanying graphs represent the sidewalks to be cleaned in a fancy garden (one pass over a sidewalk will clean it). Can the cleaning be done using an Euler circuit? If so, show the circuit in each graph by numbering the edges in the order the Euler circuit uses them. If not, explain why no Euler circuit is possible.

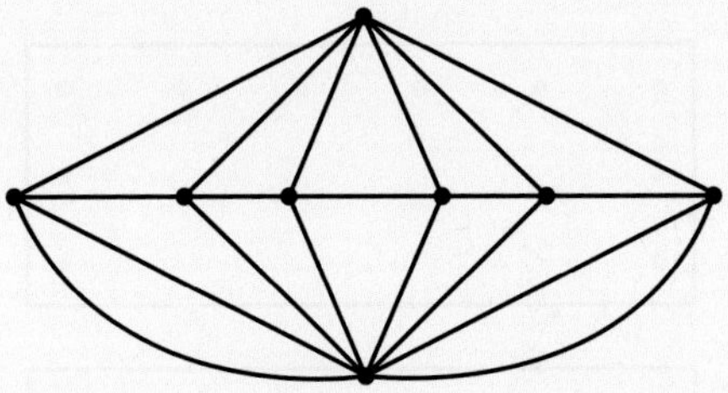

(a)

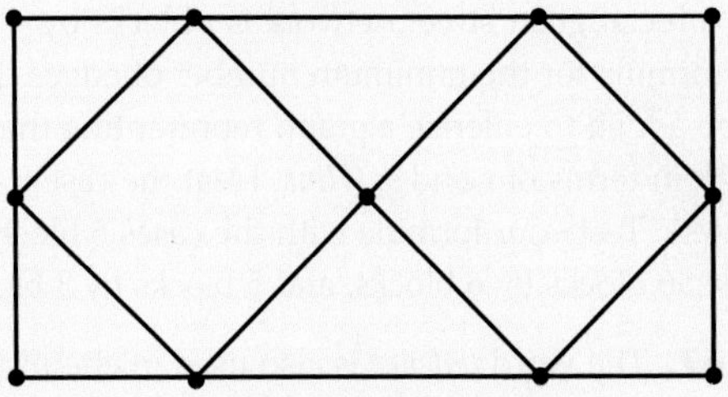

(b)

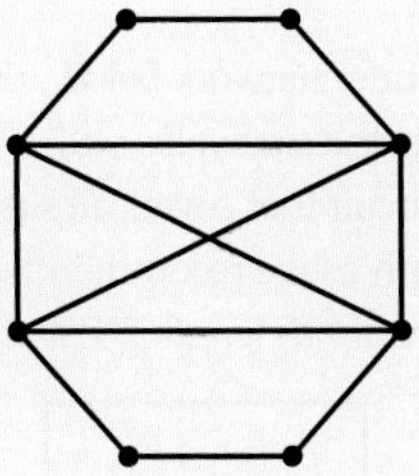

(c)

63. If an edge is added to an already existing graph, connecting two vertices already in the graph, explain why the number of vertices with odd valence has the same parity before and after. (This means that if it was even before, it is even after, while if it was odd before, it remains odd.)

64. Any graph can be built in the following fashion: Put down dots for the vertices and then add edges connecting the dots as needed. When you have put down the dots, and before any edges have been added, is the number of vertices with odd valence an even number or an odd number? What is the number of vertices with odd valence when all the edges have been added (see Exercise 63)?

65. Draw the graph for the parking-control territory shown in the figure below. Label each vertex with its valence and determine whether the graph is connected.

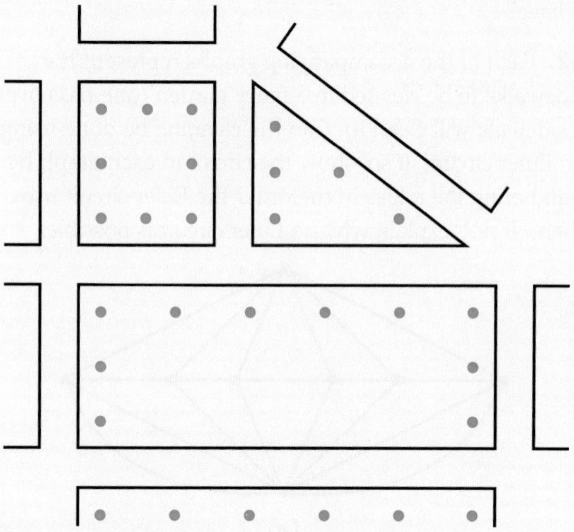

66. If a rectangular street network is r blocks by s blocks, find a formula for the minimum number of edges that must be added to eulerize a graph representing the network in terms of r and s. (*Hint:* Treat the case $r = 1$ separately. Test your formula with the cases 6 blocks by 5 blocks, 6 blocks by 6 blocks, and 5 blocks by 3 blocks.)

67. The word *valence* is also used in chemistry. Find out what it means in chemistry, and explain how this usage is similar to its use here.

68. For the street network below, draw a graph that represents the sidewalks with meters. Find the minimum-length circuit that covers all sidewalks with meters. If you drew the graph as we recommended, you would find that the shortest circuit has length 18 (it reuses every edge).

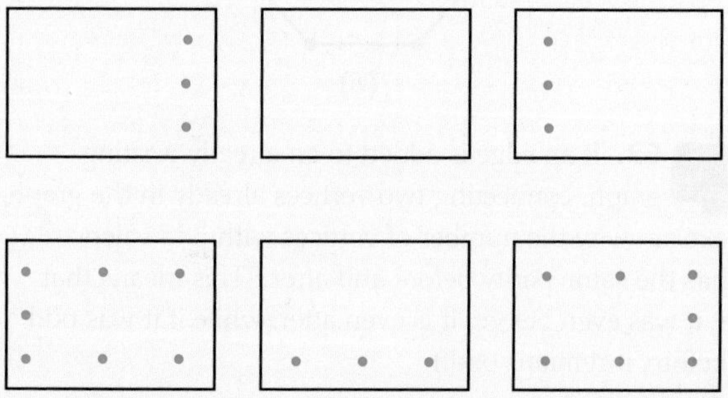

But the meter checker comes to you and says, "I don't know anything about your theories, but I have found a way to cover the sidewalks with meters using a circuit of length 10. My trick is that I don't rule out walking on sidewalks with no meters." Explain what he means and discuss whether his strategy can be used in other problems.

69. Each edge of the accompanying graph represents a two-lane highway. A grass-mowing machine is located at A, and its operator has the job of cutting the grass along each of the edges of road shown. Find a tour for the mowing machine that begins and ends at A. Find a tour that begins and ends at A and, as the mowing is done, moves along the edge of the road in the same direction as the traffic.

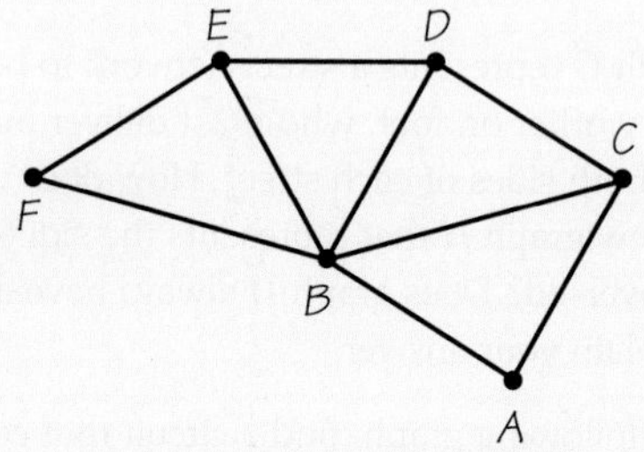

70. Draw a graph with exactly two vertices of valence 3 and a minimal but positive number of 4-valent vertices, which can be eulerized by adding one edge.

Chapter Review

71. Answer the following questions for graph G below, which was created to represent the roads of a shopping mall.

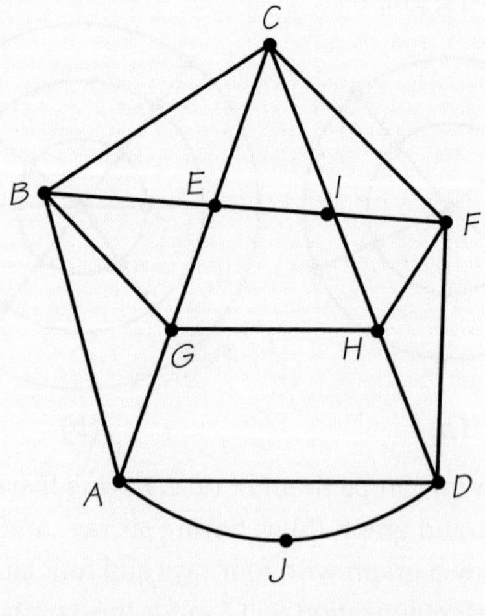

(a) Determine the number of vertices and edges of the graph.

(b) Determine the valences of the vertices A, F, and D.

(c) If the graph has an Euler circuit, write it down starting with vertex E.

72. Answer the following questions for accompanying graph G, which is used to model roads that must be inspected for potholes.

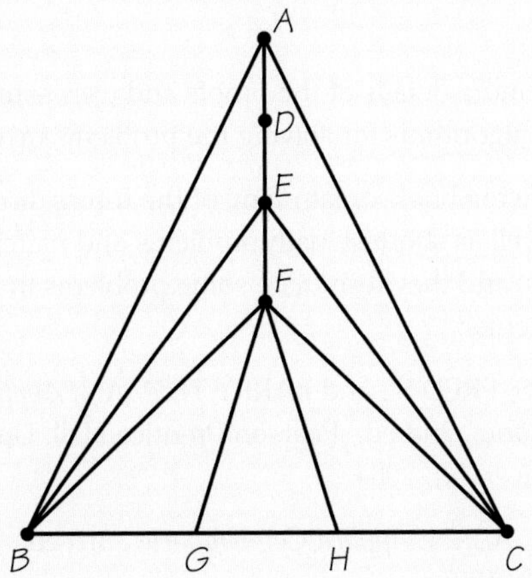

(a) How many odd-valent vertices does *G* have?

(b) If *G* has an Euler circuit, write it down.

(c) If *G* does not have an Euler circuit, what is the minimum number of edges that must be repeated to find a tour of *G* that starts and ends at *C* and traverses each edge of *G* at least once?

(d) How does your answer to part (c) relate to the answer to part (a)?

Applet Exercises

To do these exercises, go to www.macmillanhighered.com/fapp10e.

1. We learned that if a graph has exactly two vertices with odd valences, then an Euler circuit does not exist—but an Euler path does. It is also possible to produce an Euler circuit through the process of eulerization, by duplicating certain edges of the graph. But how many duplications are necessary to obtain an Euler circuit? Investigate this problem and more general related topics using the *Eulerizing a Graph* applet.

2. We know that if all the vertices have even valence, then an Euler circuit exists. Try your hand at finding such circuits in the *Euler Circuit* applet.

Writing Projects

1. Write a memo to your local department of parking control (or police department) in which you suggest that management science techniques like the ones in this chapter be used to plan routes. Assume that the person to whom you are writing is not extensively trained in mathematics but is willing to read through some technical material, provided you make it seem worth the trouble.

2. Write a letter as in Writing Project 1, but address it to the department in charge of spreading salt on roads after snowstorms.

3. If you were making a recommendation to the mayor of New York City concerning proposed new street-sweeping routes designed using the theory of this chapter, would you recommend that the changes be adopted or not? Write a memo that outlines the pros and cons as fairly as you can, and then conclude with your recommendation.

Suggested Readings

BELTRAMI, EDWARD J., *Models for Public Systems Analysis*, Academic Press, New York, 1977. This book gives a good overview of the way that operations research has provided and continues to provide new tools for solving societal problems. Among the ideas discussed are police patrol tactics, organization of emergency services, and scheduling. Some of the mathematics used is advanced.

MALKEVITCH, JOSEPH, and WALTER MEYER, *Graphs, Models, and Finite Mathematics*, Prentice-Hall, Englewood Cliffs, NJ, 1974. This introductory book includes much of the same material presented in this chapter but provides more details of the proofs and uses somewhat different algorithms for solving the problems involved.

The following books treat many of the topics discussed here as well as shortest-path problems and matching problems, and they formulate some problems in more realistic terms:

ROBERTS, FRED S., and BARRY TESMAN, *Applied Combinatorics*, 2nd ed., Pearson Prentice Hall, Upper Saddle River, NJ, 2004.

TUCKER, ALAN. *Applied Combinatorics*, 3rd ed., Wiley, New York, 1995.

Suggested Websites

www.hsor.org/what_is_or.cfm This site discusses the history of OR and some of the areas where OR is being applied. Be sure to follow the "Networks Routing" link to see applications of the Chinese postman problem.

www.geom.uiuc.edu/~doty/applications This web page provides some examples of how to apply Euler circuits.

www-history.mcs.st-andrews.ac.uk/Biographies/Euler.html This essay discusses the numerous contributions that Euler made to mathematics and provides biographical information about him.

www.ams.org/samplings/feature-column/fcarc-urban-geom This web page includes an introduction to how graph theory has provided tools for urban operations research.

Business Efficiency

2

Walter Hodges/Stone/Getty Images

In the preceding chapter, we saw that there was an easy way of telling whether a connected graph has a circuit that traverses each of the edges of a graph exactly once—for example, a route for a snowplow that covers the streets of a section of a town. However, the situation changes drastically if we make a seemingly small change in the problem: When is it possible to find a route along distinct edges of a graph that visits each *vertex* once and only once in a simple circuit? Perhaps there has been a hurricane and it is important to check whether the storm sewers at every corner in town are clogged. We want all the sewers to be checked. However, doing so typically will not require that all the streets in the town be traversed.

This problem is called the *Hamiltonian circuit problem,* and, like the Euler circuit problem, it is a graph theory problem. The Hamiltonian circuit problem has many applications—for example, the delivery of parcels, traffic light inspections, or gas meter readings. Suppose inspections or deliveries need to be made at each vertex (rather than along each edge) of a graph. An "efficient" tour of the graph would be a route that started and ended at the same vertex and passed through all the vertices without reuse, or repetition; that is, the route would be a **Hamiltonian circuit.** Such routes would be useful for inspecting for blocked sewers, for delivering mail to drop-off boxes, which hold heavy loads of mail so that urban postal carriers do not have to carry them long distances, or for delivering meals to needy elderly people.

Hamiltonian Circuit DEFINITION

A tour that starts at a vertex of a graph and visits each vertex once and only once, returning to where it started, is called a **Hamiltonian circuit.**

For example, the wiggly line in Figure 2.1a shows a circuit we can take to tour that graph, visiting each vertex once and only once. This tour can be written *ABDGIHFECA.* Note that another way of writing the same circuit would be *ACEFHIGDBA* or *EFHIGDBACE.* A different circuit visiting each vertex once and only once would be *CDBIGFEHAC* (Figure 2.1b). Do not be confused because C is written twice when we write down this list of vertices. We can think of the circuit as starting at any of its vertices, but we do start and end at the same vertex.

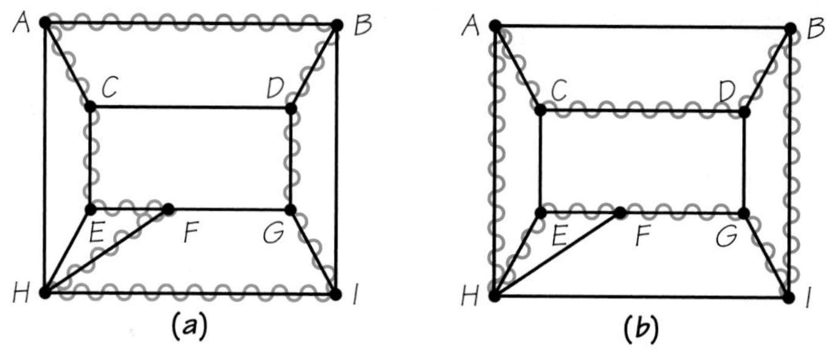

Figure 2.1 Wiggly edges illustrate Hamiltonian circuits.

2.1 Hamiltonian Circuits

The concept is named for the Irish mathematician William Rowan Hamilton (1805–1865), who was one of the first to study it. We now know that the concept was discovered somewhat earlier by Thomas Kirkman (1806–1895), a British minister with a penchant for mathematics.

The concepts of Euler and Hamiltonian circuits are similar in that both forbid reuse. An Euler circuit forbids the reuse of edges, while a Hamiltonian circuit forbids the reuse of vertices. However, it is far more difficult to determine which connected graphs possess Hamiltonian circuits than to determine which connected graphs have Euler circuits. Using the concepts of Euler circuit and Hamiltonian circuit in practical applications in the real world involves actually being able to find such circuits in large graphs. As we saw in Chapter 1, looking at the valences of vertices tells us whether a connected graph has an Euler circuit, but we have no such simple method for telling whether or not a graph has a Hamiltonian circuit.

Some special classes of graphs are known to have Hamiltonian circuits, and some special classes of graphs are known to lack them. For example, here is a method for constructing an infinite family of graphs where each graph in the family cannot have a Hamiltonian circuit. Construct a vertical column of m vertices and a parallel column of n vertices, where m is bigger than n, as shown in Figure 2.2a. The figure illustrates a typical case where $m = 4$ and $n = 2$. Now join each vertex on the left in the figure to every vertex on the right. As m and n vary, we get a family of different graphs.

No graph obtained in this manner can have a Hamiltonian circuit. If a Hamiltonian circuit existed, it would have to alternately include vertices on the left and right of the figure. This is not possible because the number of vertices on the left and right are not the same. It is unlikely that a method will ever be found to easily determine whether or not a graph has a Hamiltonian circuit based on theoretical results developed in computer science. If Hamiltonian circuits were easy to find in any graph at all, many applied problems could be solved in a less costly way.

In many urban operations research situations, "grid graphs" such as the one in Figure 2.2b are of interest. If we wanted an efficient route (circuit) to inspect traffic surveillance cameras located at urban street intersections, we would need to find a Hamiltonian circuit for the graph in Figure 2.2b. Note that in going from one vertex to another, we move from a vertex of one color to a vertex of the other color. Since colors would alternate in a Hamiltonian circuit, it follows that the number of vertices of each color would have to be the same if there is a Hamiltonian circuit in this graph. Since the number of vertices of the two colors is *not* the same, there is no Hamiltonian circuit and, hence, no fully efficient route for inspecting the traffic control cameras.

The Hamiltonian circuit problem itself has many applications. This is not unusual in mathematics. Often mathematics used to solve a particular real-world problem leads to new mathematics that suggests applications to other real-world situations. One class of problems to which we can apply Hamiltonian circuits is vacation planning.

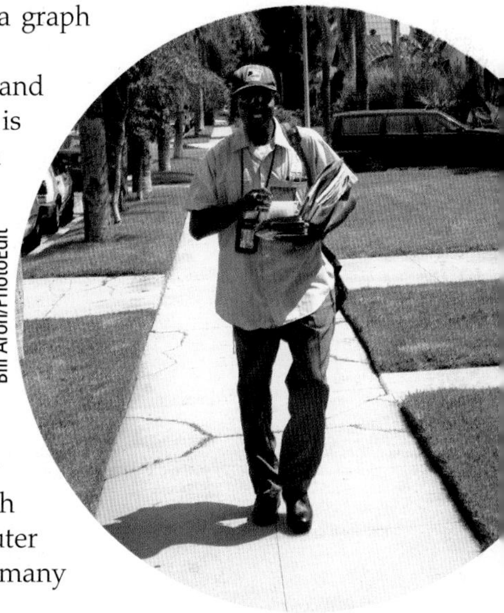

Bill Aron/PhotoEdit

It may seem that delivering mail in a cheap and timely manner should not be that hard. However, finding the best way to deliver mail over a variety of environments, rural, suburban, and urban, is very complex. How should a large geographic area be divided into smaller sections? Should each mail carrier use a truck as a "depot" to resupply small amounts of mail for delivery or should there be deposit boxes on street corners? Mathematics can be used to find answers to such questions.

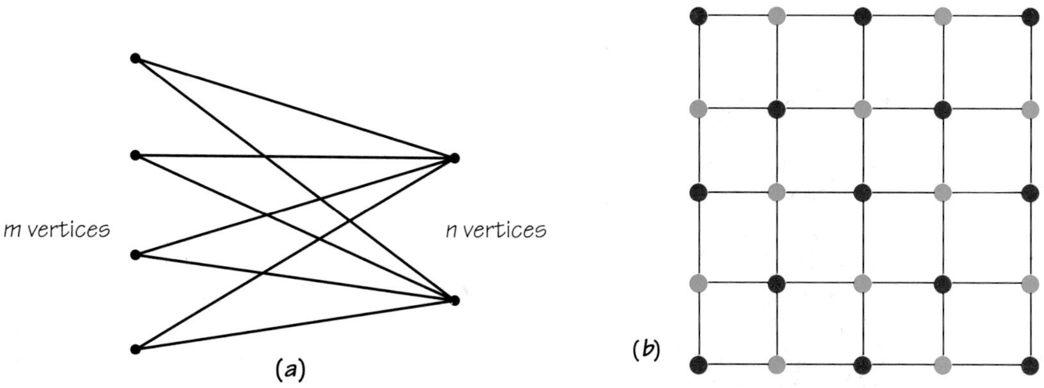

Figure 2.2 (a) An example of one graph from a family of graphs that has no Hamiltonian circuit. The number of vertices m on the left is chosen to be greater than the number of vertices n on the right. The case $m = 4$ and $n = 2$ is shown. (b) A graph used to model a portion of a city. Since the graph reflects the block structure of the city, it is known as a "grid graph."

Express mail and parcel post delivery companies need to make complicated patterns of deliveries and pick-ups. To do this, they need to know driving distances between the various geographical locations involved. Using this information, together with driving times, they can use mathematics to cut costs and to make the pick-ups and deliveries on time.

EXAMPLE 1 ➔ Vacation Planning

Let's imagine that you are a college student studying in Chicago. During spring break, you and a group of friends have decided to take a car trip to visit other friends in Minneapolis, Cleveland, and St. Louis. There are many choices as to the order of visiting the cities and returning to Chicago, but you want to design a route that minimizes the distance you have to travel. Presumably, you also want a route that cuts costs, and you know that minimizing distance will minimize the cost of gasoline for the trip. Similar problems with different complications would arise for bus, railroad, or airplane trips.

Imagine now that the Internet has provided you with the intercity driving distances between Chicago, Minneapolis, Cleveland, and St. Louis. We can construct a graph model with this information, representing each city by a vertex and the legs of the journey between the cities by edges joining the vertices. To construct the model, we add a number called a **weight** to each graph edge, as in Figure 2.3. In this example, the weights represent the distances between the cities, each of which corresponds to one of the endpoints of the edges in the graph. (In other examples the weight might represent a cost, time, satisfaction rating, or profit.) We want to find a minimum-cost tour that starts and ends in Chicago and visits each other city only once. Using our earlier terminology, what we wish to find is a **minimum-cost Hamiltonian circuit**—a Hamiltonian circuit with the lowest possible sum of the weights of its edges.

Steps 2 and 3 of the algorithm are straightforward. Thus, we need worry only about Step 1, generating all the possible Hamiltonian circuits in a systematic way. To find the Hamiltonian tours, we use the **method of trees,** as follows: Starting from Chicago, we can choose any of the three cities to visit after leaving Chicago. The first stage of the enumeration tree is shown in Figure 2.4. If Minneapolis is chosen as the first city to visit, then there are two possible cities to visit next, Cleveland and St. Louis. The possible branchings of the **tree** at this stage are shown in Figure 2.5. In this second stage, however, for each choice of first city to visit, there are two choices from this city to the second city to visit. This would lead to the diagram in Figure 2.6.

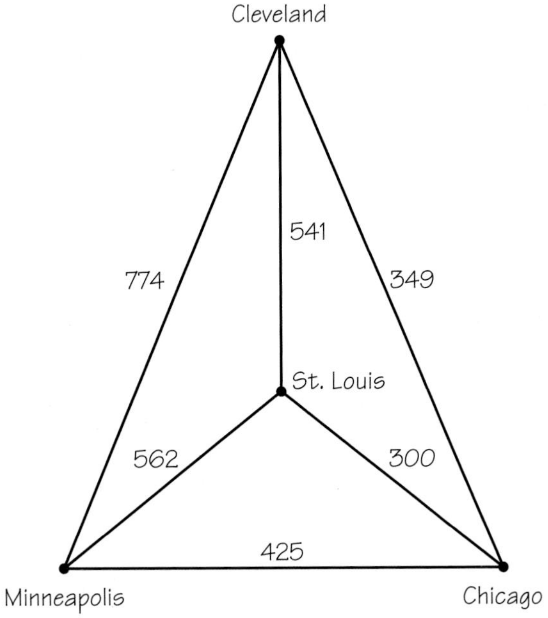

Figure 2.3 Road mileages between four cities.

Finding a Minimum-Cost Hamiltonian Circuit

PROCEDURE

How can we determine which Hamiltonian circuit has minimum cost? There is a conceptually easy **algorithm,** or a step-by-step process like a recipe in cooking, for solving this problem:

1. Generate all possible Hamiltonian tours (starting from Chicago).
2. Add up the distances on the edges of each tour.
3. Choose a tour with total distance being a minimum, that is, as small as possible.

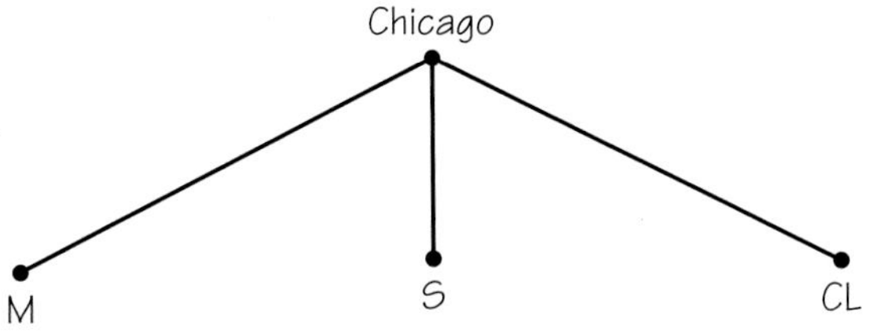

Figure 2.4 First stage in finding vacation-planning routes.

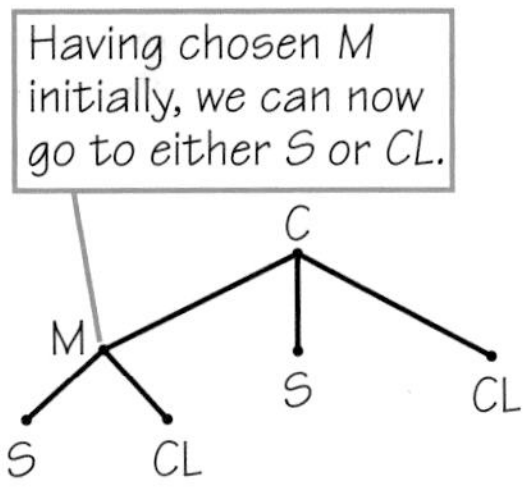

Figure 2.5 Part of the second stage in finding vacation-planning routes.

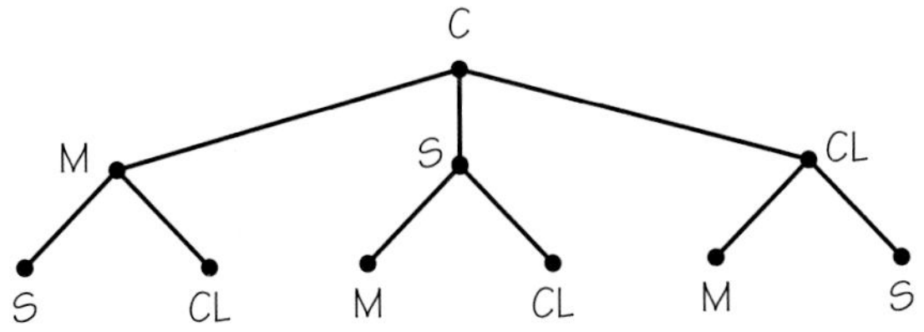

Figure 2.6 Complete second stage in finding vacation-planning routes.

Having chosen the order of the first two cities to visit, and knowing that no revisits (reuses) can occur in a Hamiltonian circuit, we have only one choice left for the next city. From this city, we return to Chicago. The complete tree diagram showing the third and fourth stages for these routes is given in Figure 2.7. Notice, however, that because we can traverse a circular tour in either of two directions, the paths shown in the tree

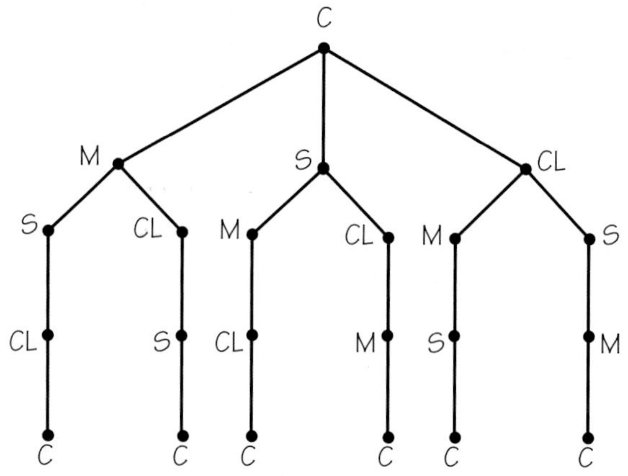

Figure 2.7 Completed enumeration of routes using the method of trees for the vacation-planning problem.

diagram of Figure 2.7 do not correspond to different Hamiltonian circuits. For example, the leftmost path (C–M–S–CL–C) and the rightmost path (C–CL–S–M–C) represent the same Hamiltonian circuit. Thus, among what appear to be six different paths in the tree diagram, only three in fact correspond to different Hamiltonian circuits. These three distinct Hamiltonian circuits are shown in Figure 2.8.

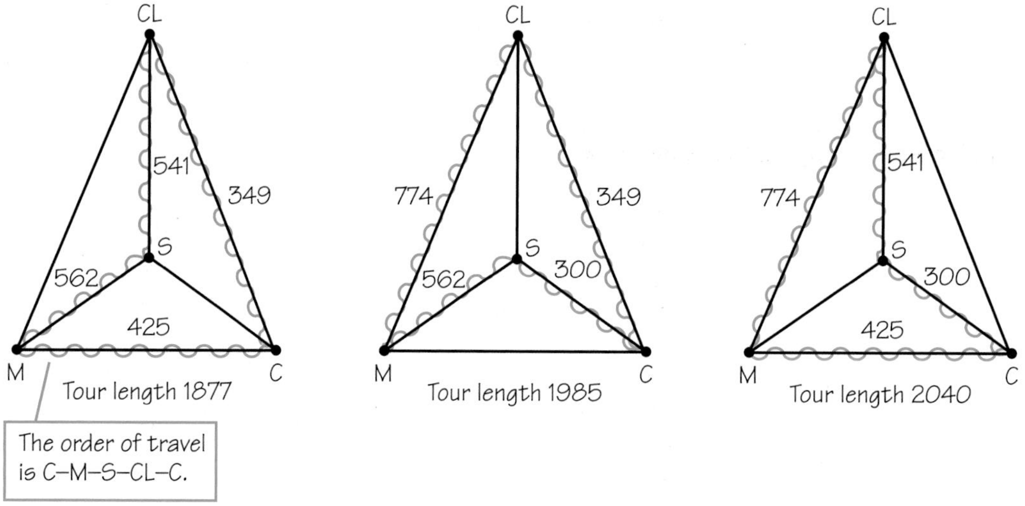

Figure 2.8 The three Hamiltonian circuits for the vacation-planning problem of Figure 2.3.

Note that in generating the Hamiltonian circuits, we disregard the distances involved. We are concerned only with the different patterns of carrying out the visits. To find the optimal, or best, route, however, we must add up the distances on the edges to get each tour's length and select a tour whose total length is smallest. Figure 2.8 shows that the optimal tour is Chicago, Minneapolis, St. Louis, Cleveland, Chicago. The length of this tour is 1877 miles, which saves 163 miles over the longest choice of tour. ■

The method of trees is not always as easy to use as our example suggests. Instead of doing our analysis for four cities, consider the general case of n cities. The graph model similar to that in Figure 2.3 would consist of a weighted graph with n vertices, with every pair of vertices joined by an edge.

Complete Graph DEFINITION

A graph is called **complete** if there is exactly one edge between each pair of vertices in the graph.

A complete graph with five vertices is illustrated in Figure 2.9. The graph in Figure 2.3 is a weighted complete graph with four vertices.

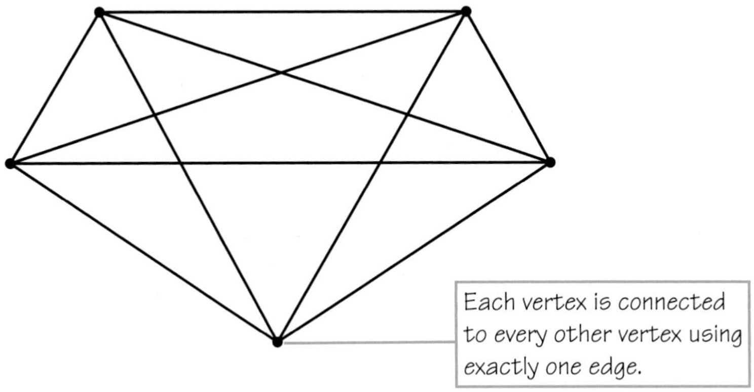

Each vertex is connected
to every other vertex using
exactly one edge.

Figure 2.9 A complete graph with five vertices. Every pair of vertices is joined by an edge.

Fundamental Principle of Counting

How many Hamiltonian circuits are in a complete graph of n vertices? We can solve this problem by using the same type of analysis that we used in the method of trees. The method of trees is a visual application of the **fundamental principle of counting,** a procedure for counting outcomes in multistage processes. Using this procedure, we can count how many patterns occur in a situation by looking at the number of ways in which the component parts can occur. For example, if Jack has 10 shirts and 4 pairs of trousers, he can wear $10 \times 4 = 40$ shirt–pants outfits. Each shirt can be worn with any of the pants. (This can be verified by drawing a tree diagram, but such a diagram is cumbersome for big numbers.)

The Fundamental Principle of Counting DEFINITION

In general, the **fundamental principle of counting** can be stated this way: If there are a ways of choosing one thing, b ways of choosing a second after the first is chosen, . . . , and z ways of choosing the last item after the earlier choices, then the total number of choice patterns is $a \times b \times c \times \ldots \times z$.

Note that each choice cannot depend on a prior choice for this count to be valid. When choosing which of the 10 shirts and 4 pants to wear, Jack does not depend on the pants he chooses or the other way around. Therefore, he can make $10 \times 4 = 40$ choices.

EXAMPLE 2 Counting

Here are some examples of how to use the fundamental principle of counting:

1. In a restaurant, there are 4 kinds of soup, 12 entrees, 6 desserts, and 3 drinks. How many different four-course meals can a patron choose? The four choices can be made in 4, 12, 6, and 3 ways, respectively. Hence, applying the fundamental principle of counting, there are $4 \times 12 \times 6 \times 3 = 864$ possible meals.
2. In a state lottery, a contestant gets to pick a four-digit number that does not contain a zero followed by an uppercase or lowercase letter. How many such sequences of digits and a letter are there? Each of the four digits can be chosen in 9 ways (i.e., 1, 2, . . . , 9), and the letter can be chosen in 52 ways (i.e., A, B, . . . , Z plus a, b, . . . , z). Hence, there are $9 \times 9 \times 9 \times 9 \times 52 = 341{,}172$ possible patterns.
3. A corporation is creating a musical logo consisting of four different ordered notes from the scale C, D, E, F, G, A, and B. How many logos are there to choose from? The first note can be chosen in 7 ways, but because reuse is not allowed, the next note can be chosen in only 6 ways. The remaining two notes can be chosen in 5 and 4 ways, respectively. Using the fundamental principle of counting, $7 \times 6 \times 5 \times 4 = 840$ musical logos are possible. If reuse of notes is allowed, $7 \times 7 \times 7 \times 7 = 2401$ logos are possible.

Let's now return to the problem of enumerating Hamiltonian circuits for the complete graph with n vertices. The city visited first after the home city can be chosen in $n - 1$ ways, the next city in $n - 2$ ways, and so on, until only one choice remains. Using the fundamental principle of counting, there are $(n - 1)! = (n - 1)(n - 2) \times \cdots \times 3 \times 2 \times 1$ routes. The exclamation mark in $(n - 1)!$ is read "factorial" and is shorthand notation for the product $(n - 1)(n - 2) \times \cdots \times 3 \times 2 \times 1$. For example, $5! = 5 \times 4 \times 3 \times 2 \times 1 = 120$.

As we saw in Figure 2.7, pairs of routes correspond to the same Hamiltonian circuit because one route can be obtained from the other by traversing the cities in reverse order. Thus, although there are $(n - 1)!$ possible routes, there are only half as many, or $(n - 1)!/2$, different Hamiltonian circuits. Now, if we have only a few cities to visit, $(n - 1)!/2$ Hamiltonian circuits can be listed and examined in a reasonable amount of time. Analysis of a six-city problem would require generation of $(6 - 1)!/2 = 5!/2 = 120/2 = 60$ tours. But for, say, 25 cities, $24!/2$ is approximately 3×10^{23}. Even if these tours could be generated at the rate of 1 million per second, it would take almost 10 billion years to generate them all. Because it would take so long to solve large vacation-planning problems using this method, it is sometimes referred to as a **brute force method** (i.e., trying all the possibilities). Computer scientists and engineers have made it possible to market faster and faster computers. However, governments and businesses need to solve larger-scale problems; say, for example, finding a Hamiltonian circuit in a graph with 10,000 vertices. If the methods one knows for solving such problems are not much better than brute force, then it's unlikely that even these faster computers can optimally solve large versions of such problems. Mathematicians and computer scientists are actively seeking procedures that will significantly improve our ability to solve large versions of important problems.

2.2 Traveling Salesman Problem

If the only benefit were saving money and time in vacation planning, the difficulty of finding a minimum-cost Hamiltonian circuit in a complete graph with n vertices for large values of n would not be of great concern. However, the problem we are discussing is one of the most common in *operations research,* the branch of mathematics concerned with getting governments and businesses to operate more efficiently. This problem is usually called the **traveling salesman problem (TSP)** because of its early formulation.

Traveling Salesman Problem (TSP) DEFINITION

The **traveling salesman problem (TSP)** involves finding the trip of minimum cost that a salesman can make to visit the cities in a sales territory once and only once (represented by a complete graph with weights on the edges), starting and ending the trip in the same city.

Many situations require in essence solving a TSP:

1. Physical records generated at automated teller machine (ATM) locations—as backup in case of failure of the electronic systems—must be picked up periodically.
2. A local ethnic fast-food restaurant must hire someone to deliver the food quickly and while it is still hot, before returning to the restaurant for the next batch of orders to be delivered.
3. A limousine service with a van located at an airport must pick up five customers and deliver them to the airport in time to catch their flights.
4. In drilling holes in a series of plates, the drill press operator (perhaps a robot) must drill the holes in a predetermined order.
5. The electric (or gas) company needs to design a route for its meter readers.
6. A minibus must pick up six day campers, deliver them to camp, and return them home later in the day.
7. The telephone company wishes to pick up the coins from its pay telephone booths. (To avoid the high cost of picking up these coins, phone companies in many countries have adopted a system that uses prepurchased phone cards to operate phones. This means that there are no coins to collect.) Due to the increased use of cell phones, fewer pay phones are available.
8. A lobster fisherman has set out traps at various locations and wishes to pick up his catch.

Perhaps surprisingly, TSPs are also solved regularly in the design of computer chips. The components must be located so that the machines involved in the assembly can insert them on the chips as efficiently as possible. Because many chips are manufactured, even a small improvement in the time needed to make a chip can save a lot of money.

The meaning of *cost* can vary from one formulation of a TSP to another. We can measure cost as distance, airplane ticket prices, time, or any other factor that is to be optimized. In many situations, the TSP arises as a subproblem of a more complicated problem. For example, a supermarket chain may have a very large number of stores to be served from a single large warehouse. If there are fewer trucks than stores, the stores must be grouped into clusters so that one truck can serve each cluster. If we then solve the TSP for every truck, we can minimize total costs for the supermarket chain. Similar vehicle-routing problems—for dial-a-ride services that transport senior citizens to activity centers, for example, or that deliver children to their schools or camps—often involve solving the TSP as a subproblem.

2.3 Helping Traveling Salesmen

Because the traveling salesman problem arises so often in situations where the associated complete graphs would be very large, we must find a faster method than the brute force method we have described. One intuitive idea is to try to visit nearby locations sooner. Recall that our goal is to find the minimum-cost vertex tour that forms a Hamiltonian circuit.

Nearest-Neighbor Algorithm PROCEDURE

Starting from the home city, first visit the nearest city, then visit the nearest city that has not already been visited. We return to the start city when no other choice is available. This approach is called the **nearest-neighbor algorithm.**

EXAMPLE 3 Applying the Nearest-Neighbor Algorithm

Applying this algorithm to the TSP in Figure 2.3 quickly leads to the tour of Chicago, St. Louis, Cleveland, Minneapolis, and Chicago, with a length of 2040 miles. Here is how this tour is determined. Because we are starting in Chicago, there is a choice of going to a city that is 425, 300, or 349 miles away. Because the smallest of these numbers is 300, we next visit St. Louis, which is the nearest neighbor of Chicago not already visited. At St. Louis, we have a choice of visiting next cities that are 541 or 562 miles away. Hence, Cleveland, which is nearer (541 miles), is visited. To complete the tour, we visit Minneapolis and return to Chicago, thereby adding 774 and 425 miles to the length of the tour. The total length of the tour is 2040 miles. ■

The nearest-neighbor algorithm is an example of a **greedy algorithm** because at each stage a best (greedy) choice, based on an appropriate criterion, is made. Unfortunately, this is not the optimal tour, which we saw was C–M–S–CL–C, for a total length of 1877 miles. Making the best choice at each stage may not yield the best "global" solution. However, even for a large TSP, one can always find a nearest-neighbor route quickly.

EXAMPLE 4 Applying the Nearest-Neighbor Algorithm Revisited

Figure 2.10 again illustrates the ease of applying the nearest-neighbor algorithm, this time to a weighted complete graph with five vertices. Starting at vertex *A*, we get the tour *ADECBA* (cost 2800) (Figure 2.10a). Note that the nearest-neighbor algorithm starting at vertex *B* yields the tour *BCADEB* (cost 3050) (Figure 2.10b).

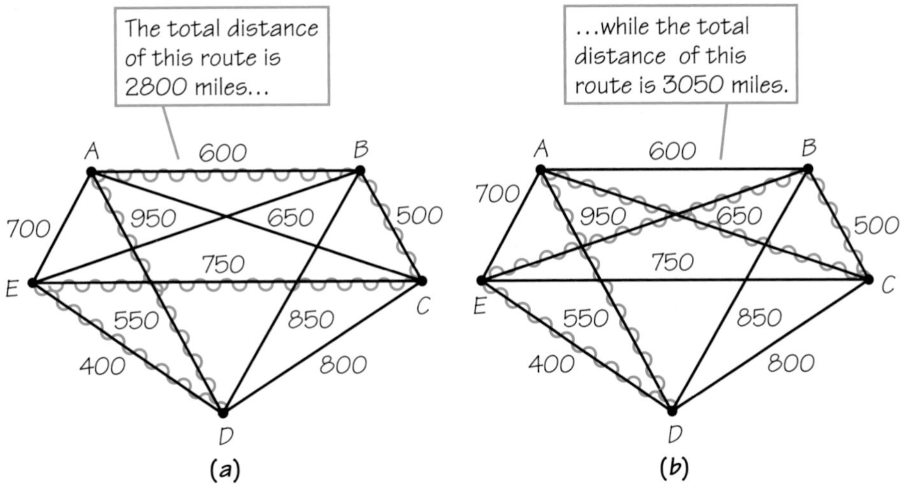

Figure 2.10 (a) A weighted complete graph with five vertices that illustrates the use of the nearest-neighbor algorithm (starting at *A*). (b) TSP tour generated by the nearest-neighbor algorithm (starting at *B*).

This example illustrates that a nearest-neighbor tour can be computed for each vertex of the complete graph being considered and that different nearest-neighbor tours can be obtained starting at different vertices. Thus, even though we may seek a tour starting at a particular vertex—say, *A* in Figure 2.10—because a Hamiltonian circuit can be thought of as starting at any of its vertices, we can just as easily apply the nearest-neighbor procedure starting at vertex *B* (rather than at *A*). The Hamiltonian circuit we get can still be thought of as beginning at vertex *A* rather than *B*. Even for complete graphs with a large number of vertices, it would still be faster to apply the nearest-neighbor algorithm for each vertex and pick the cheapest of the tours generated (though such a tour might not be optimal) than to apply the brute force method.

Self Check 1

Apply the nearest-neighbor algorithm to the graph in Figure 2.10a starting at vertex *D*. What is the cost of the associated tour?

We now consider a different approach to the TSP that might find a good solution quickly.

Sorted-Edges Algorithm — PROCEDURE

Start by sorting or arranging the edges of the complete graph in order of increasing cost (or, equivalently, arranging the intercity distances in order of increasing distance). Then at each stage select an edge that has not been previously chosen of least cost that (1) never requires that three used edges meet at a vertex (because a Hamiltonian circuit uses up exactly two edges at each vertex) and (2) never closes up a circular tour that doesn't include all the vertices. This algorithm is called the **sorted-edges algorithm.**

EXAMPLE 5 Applying the Sorted-Edges Algorithm

Applying the sorted-edges algorithm to the TSP in Figure 2.3 works as follows: First, the six weights on the edges listed in increasing order would be 300, 349, 425, 541, 562, and 774. Because the cheapest edge in this sorted list is 300, this is the first edge we select for the tour we are building. Next we add the edge with weight 349 to the tour. The next-cheapest edge would be 425, but using this edge together with those already selected would result in having three edges at a vertex (Figure 2.11a), which is not consistent with having a Hamiltonian circuit. Hence, we do not use this edge. The next-cheapest edge, 541, used together with the edges already selected, would create a circuit (see Figure 2.11b) that does not include all the vertices. Thus, this edge, too, would be skipped over. However, we are able to add the edges 562 and 774 without either creating a circuit shorter than one including all the vertices or having three edges at a vertex. Hence, the tour we arrive at is Chicago, St. Louis, Minneapolis, Cleveland, and Chicago. Again, this solution is not optimal because its length is 1985. Note that this algorithm, like the nearest-neighbor, is greedy.

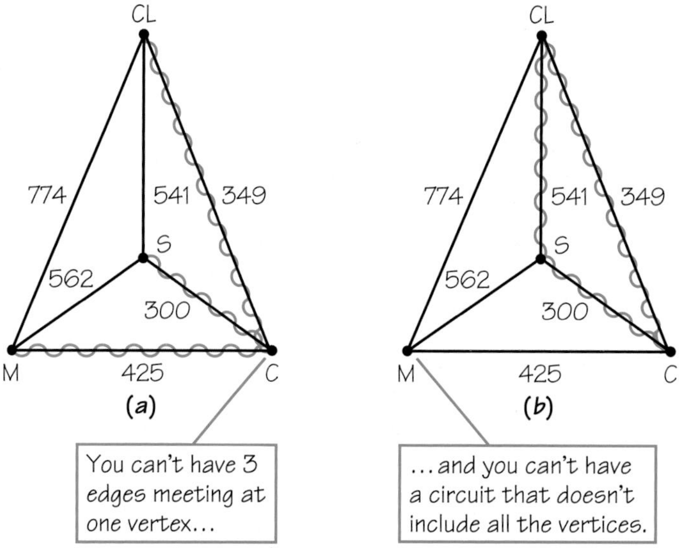

Figure 2.11 (a) When three shortest edges are added in order of increasing distance, three edges at a vertex are selected, which is not allowed as part of a Hamiltonian circuit. (b) When the edges of distances 300, 349, anu 541 are selected, a circuit that does not include all vertices results.

EXAMPLE 6 Applying the Sorted-Edges Algorithm Revisited

Although the edges selected by applying the sorted-edges method to the example in Figure 2.3 are connected to each other at every stage, this does not always happen. For example, if we apply the sorted-edges algorithm to the graph in Figure 2.10a, we build up the tour first with edge *ED* (400) and then edge *BC* (500), which do not touch. The edges that are then selected are *AD*, *AB*, and *EC*, giving the circuit *EDABCE*, which is the same as the nearest-neighbor circuit starting at vertex *A*. Note we must sort 10 numbers in order to find 5 edges.

Self Check 2

Apply the sorted-edges algorithm to the graph in Figure 2.12 on page 52. What is the cost of the associated tour?

NP-Complete Problems

SPOTLIGHT 2.1

Steven Cook, a computer scientist at the University of Toronto, showed in 1971 that certain computational problems are equivalently difficult. This class of problems, now referred to as **NP-complete problems,** has the following characteristic: If a "fast" algorithm for solving one of these problems could be found, then a fast method would exist for all these problems.

In this context, "fast" means that as the size n of the problem grows (the number of cities gives the problem size in the TSP), the amount of time needed to solve the problem grows no more rapidly than a polynomial function in n. (A polynomial function has the form $a_k n^k + a_{k-1}n^{k-1} + \cdots + a_1 n + a_0$.) On the other hand, if it could be shown that any problem in the class of NP-complete problems required an amount of time that grows faster than any polynomial (an exponential function, such as 3^n, is an example of a function that grows faster than any polynomial) as the problem size increased, then all problems in the NP-complete class would share this characteristic. The TSP, along with a wide variety of other practical problems, is known to be NP-complete. It is widely believed that large versions of these problems cannot be solved quickly. Furthermore, the security of some recent cryptographical systems relies on the hope that large NP-complete problems are actually as time consuming to solve as they appear to be. The Clay Foundation is offering a $1 million prize for determining whether NP-complete problems are truly computationally hard. The prize is still unclaimed! Researchers are also exploring whether the development of new approaches to computer design, such as quantum computing, will offer faster ways to solve very difficult problems.

Many "quick-and-dirty" methods for solving the TSP have been suggested; while some methods give an optimal solution in some cases, none of these methods guarantees an optimal solution. Surprisingly, most experts believe that no efficient method that guarantees an optimal solution for the TSP will ever be found (see Spotlight 2.1).

Recently, mathematical researchers have adopted a somewhat different strategy for dealing with TSPs. If finding a fast algorithm to generate optimal solutions for large problems is unlikely, perhaps we can show that the quick-and-dirty methods, usually called **heuristic algorithms,** come close enough to giving optimal solutions to be important for practical use. For example, suppose we could prove that the nearest-neighbor heuristic was never off by more than 25% in the worst case or by more than 15% in the average case. For a medium-sized TSP, we would then have to choose whether to spend a lot of time (or money) to find an optimal solution or instead to use a heuristic algorithm to obtain a fairly good solution. Investigators at AT&T Research have developed many remarkably good heuristic algorithms. The best-known guarantee for a heuristic algorithm for a TSP is that it yields a cost no worse than one and a half times the optimal cost. Interestingly, this heuristic algorithm involves solving a Chinese postman problem (see Chapter 1), for which a "fast" algorithm is known to exist.

Throughout our discussion of the TSP, we have concentrated on the goal of minimizing the cost (or time) of a tour that visited each of a variety of sites once and only once. However, the subtle issues that arise in specific real-world situations (or that provide a contrast between seemingly similar situations) are the things that make mathematical modeling exciting. For example, suppose the TSP situation is to pick up day campers and take them to and from the camp. The camp wants to minimize the total length of time that the bus needs to pick up the campers. The parents of the campers, however, may want to minimize the time their children spend on the bus. For some problems, the tour that minimizes the mean (average) time that a child spends on the bus may not be the same tour that minimizes the total time of the tour. (Specifically, if the bus first picks up the child who lives the farthest from the camp, and then picks up the other children, this may yield a relatively short time on the bus for the kids but a relatively long time for the tour itself.) Mathematicians return to examine these subtleties between problems at a later time, after the basic structure of the main problem itself is well understood. It is in this way that mathematics continues to grow, explore new ideas, and find new applications.

2.4 Minimum-Cost Spanning Trees

The TSP is but one of many graph theory optimization problems that have grown out of real-world problems in both government and industry. Here is another.

EXAMPLE 7 Cable TV Service

Imagine that cable TV (CT) service will be set up on one large university campus. The graph in Figure 2.12 shows the possible links that might be included in the CT network, with each edge showing the cost in hundreds of thousands of dollars to create that particular link. To send video between two locations, a direct cable link is not necessary because it is possible to send a video indirectly via another site. Thus, in Figure 2.12, sending a video from *A* to *C* could be achieved by sending the video from *A* to *B*, from *B* to *E*, and from *E* to *C*, provided the links *AB*, *BE*, and *EC* are part of the network. We assume that the cost of relaying a video, compared with the cost of the direct communication link, is so small that we can neglect this amount. The problem that concerns us,

Ariel Skelley/Blend Images/Getty Images

When students are given access to local cable TV networks, minimizing the cost of creating such networks might involve the techniques of finding a minimum-cost spanning tree.

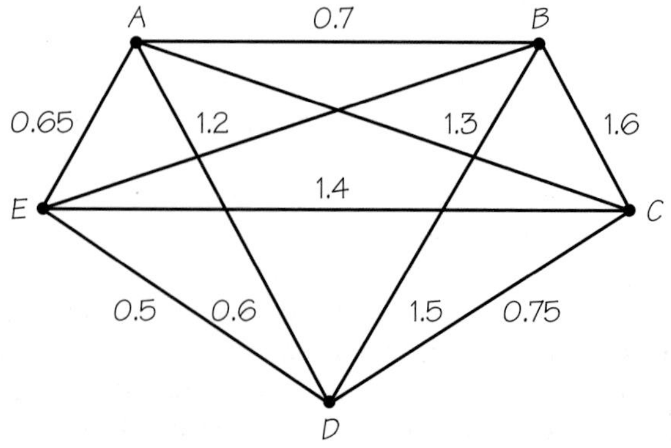

Figure 2.12 Costs of installing cable service among five locations.

therefore, is to provide service between any pair of locations in a way that minimizes the total cost of the links.

Our first guess at a solution is to put in the cheapest possible links between locations first, until all sites could send video to any other site. Such an approach is analogous to the sorted-edges method that was used to study the traveling salesman problem. In our example, if the cheapest links are added until all locations are joined, we obtain the connections shown in Figure 2.13a.

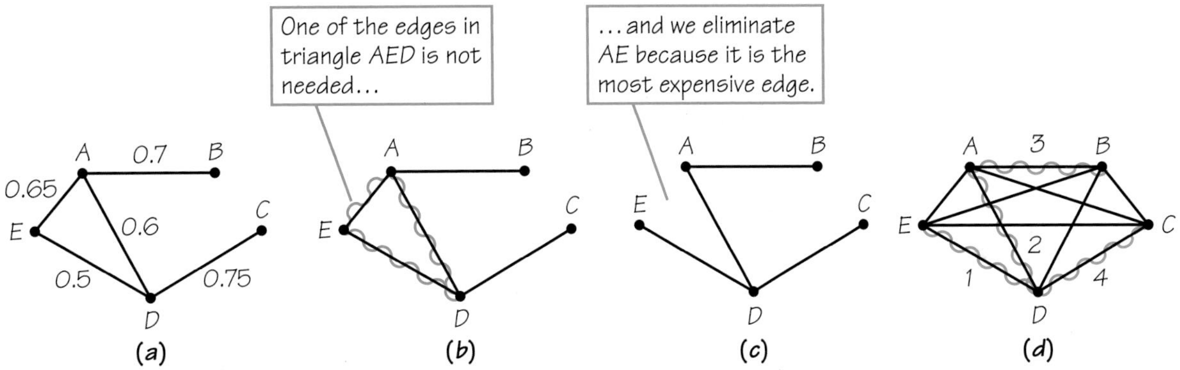

Figure 2.13 (a) Sites are linked in order of increasing cost until all locations are connected. (b) Circuit in part (a) highlighted. (c) Most expensive link in circuit in part (a) deleted. (d) Highlighted edges show, as a subgraph of the original graph, those links connecting the locations with minimum cost, obtained using Kruskal's algorithm. The numbers show the order in which Kruskal's algorithm selects the edges.

The links were added in the order *ED*, *AD*, *AE*, *AB*, *DC*. However, because this graph contains the circuit *ADEA* (wiggly edges in Figure 2.13b), it has redundant edges: We can still send videos between any pair of sites using relays after omitting the most expensive edge in the circuit—*AE*. After an edge of a circuit is deleted, a video can still be relayed among the sites of the circuit by sending signals the long way around. After *AE* is deleted, videos going from *A* to *E* can be sent via *D* (Figure 2.13c). These ideas constitute a procedure developed by Joseph Kruskal (1928–2010) in 1956.

Kruskal's Algorithm PROCEDURE

Kruskal's algorithm: Add links in order of cheapest cost so that no circuits form and so that every vertex belongs to some link added (Figure 2.13d).

In Kruskal's procedure, as in the sorted-edges method for the TSP, the edges that are added need not be connected to each other until the end. A subgraph formed in this way is a **tree;** that is, it consists of one piece and contains no circuits. It also includes all the vertices of the original graph. A subgraph that is a tree and that contains all the vertices of the original graph is called a **spanning tree** of the original graph.

To understand these concepts better, consider the graph *G* in Figure 2.14a. The wiggly edges in Figure 2.14b constitute a subgraph of *G* that is a tree (because it is connected and has no circuit), but this tree is not a spanning tree of *G* because the vertices *D* and *E* are not included. On the other hand, the wiggly edges in Figure 2.14c and 2.14d show subgraphs of *G* that include all the vertices of *G* but are not trees because the first is not connected and the second contains a circuit. Figure 2.14e shows a spanning tree of *G*; the wiggly edges are connected and contain no circuit, and every vertex of the original graph is an endpoint of some wiggly edge.

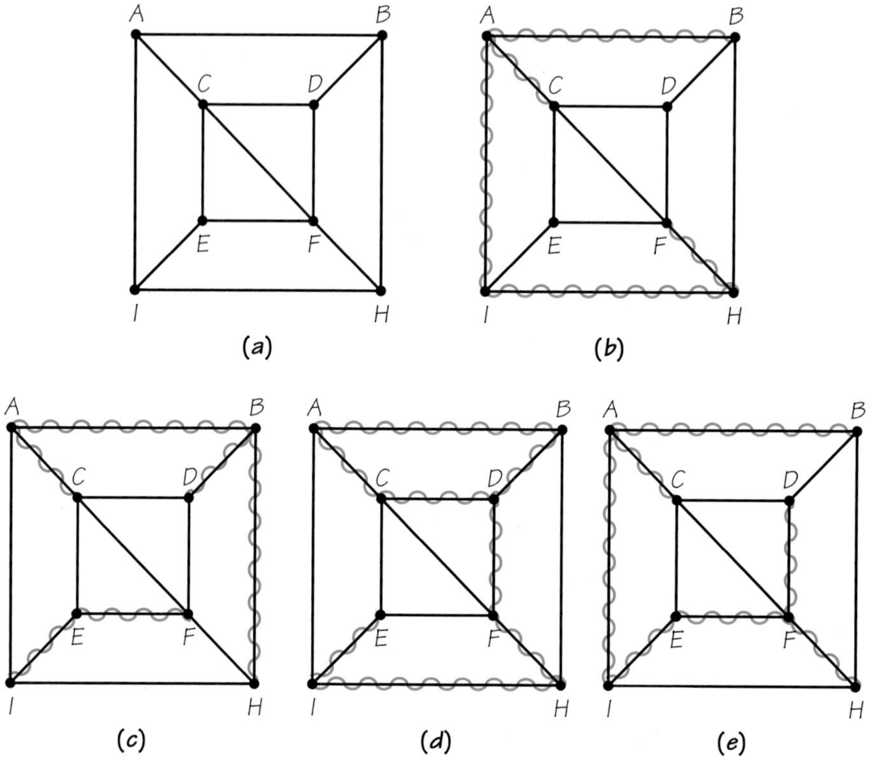

Figure 2.14 (a) A graph to help illustrate the concept of a spanning tree. (b) The wiggly edges are a tree, but not a spanning tree, because vertices *D* and *E* are not part of the tree. (c) The wiggly edges are not a tree because they are not connected. All of the vertices of the graph are, however, endpoints of wiggly edges. (d) The wiggly edges are not a tree, because they contain the edges of the circuit *BDCAB*. All the vertices of the graph are, however, endpoints of wiggly edges. (e) The wiggly edges form a tree and include all of the vertices of the graph as endpoints of wiggly edges. Thus, the wiggly edges are a spanning tree.

Self Check 3

Apply Kruskal's algorithm to find a minimum-cost spanning tree for the graph in Figure 2.3 (page 43). What is the cost associated with the tree you find?

Finding a **minimum-cost spanning tree**—that is, a spanning tree whose edge weights sum to a minimum value—solves the cable TV problem. Note that having a different goal in the cable TV problem led to a different mathematical question from that of finding a Chinese postman tour or TSP tour. This application required that we find a minimum-cost spanning tree. In Figure 2.15a, we have a graph model showing the costs of putting in roads to connect new houses in a suburban land-development project. Applying Kruskal's algorithm—adding the edges in the order of increasing cost, but avoiding the creation of a circuit—yields a minimum-cost spanning tree, indicated by the wiggly edges in Figure 2.15b. This tree is the cheapest one that makes it possible to drive

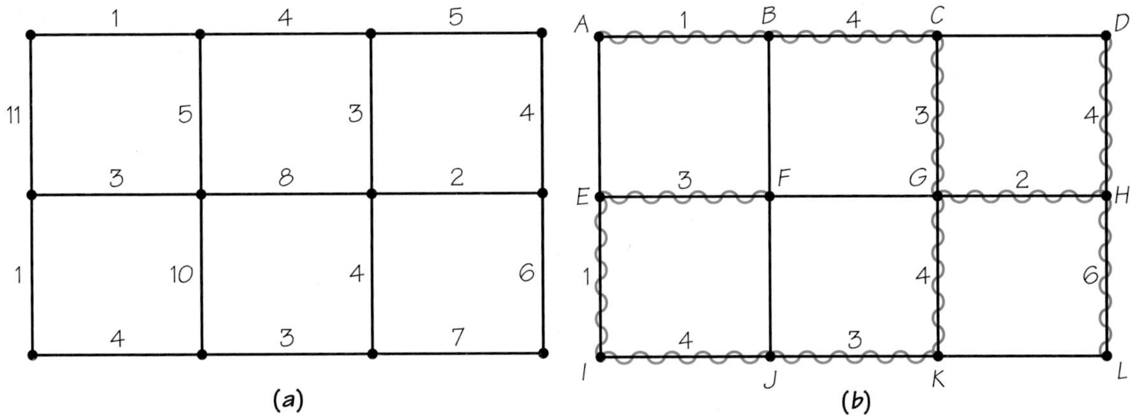

Figure 2.15 (a) A graph showing costs for construction of roads between houses. (b) Wiggly edges show a minimum-cost spanning tree for the graph in part (a).

between any pair of homes, though the driving distance between some of the homes will be relatively large because only roads corresponding to wiggly edges will be built.

Remember that the weights on the edges of the graph in Figure 2.15a represent the costs of building roads, not the driving distances between the houses. Note that Figure 2.15a is not a complete graph, one in which all possible edges are included. Edges that correspond to roads that would be economically prohibitive to build have not been shown in the graph model. Also, in Figure 2.15b, the two edges of weight 5 (shown in Figure 2.15a) do not become part of the minimum-cost spanning tree because they would create circuits with edges already chosen.

Although Kruskal's algorithm worked in our example, how do we know that the spanning tree found by this algorithm will always achieve the minimum possible cost? While this sounds very plausible, our experience with the TSP should suggest caution. Remember that for the TSP, the sorted-edges algorithm did not necessarily give an optimal solution even though it is a greedy algorithm like Kruskal's. On what basis should we have more faith in Kruskal's algorithm?

Kruskal proposed his algorithm as a way to solve a pure mathematics problem put forward by Czechoslovakian mathematician Otakar Borůvka. In mathematics, it is surprising but not uncommon to find that ideas used to solve problems with no apparent application often turn out to have many real-world uses. Kruskal's solution to the problem of finding a minimum-cost spanning tree in a graph with weights is a good example of this phenomenon.

Kruskal showed that the greedy algorithm described does yield the minimum answer, and his work led to applications of these and related ideas in designing minimum-cost computer and cable TV networks, phone connections, sewer systems, and road and railway systems. For additional discussion of operations research in the communications industry, see Spotlight 2.2. To explore how one can reconstruct full information from partial information using the tree concept, see Spotlight 2.3.

In our discussion of routing problems in graphs, we have not touched on one of the most obvious ones: finding the path between two specified, distinct vertices while keeping the sum of the weights of the edges in the path as small as possible. (Here there is no need to cover all vertices or to cover all edges.) We have seen that the weights on the edges have many possible interpretations,

AT&T Manager Explains How Long-Distance Calls Run Smoothly

SPOTLIGHT 2.2

Although long-distance calls are now routine, it takes great expertise and careful planning for a company like AT&T to handle its vast amounts of telephone traffic. Rich Wetmore was district manager of AT&T's Communications Network Operations Center in Bedminster, New Jersey. Here are his responses to questions about how AT&T handles its huge volumes of long-distance traffic and how it tracks its operations to keep things running smoothly.

How do you make sure that a customer doesn't run into a delayed signal when attempting a long-distance call?

We monitor the performance of our AT&T network by displaying data collected from all over the country on a special wallboard. The wallboard is configured to tell us if a customer's call is not going through because the network doesn't have enough capacity to handle it.

That's when we step in and take control to correct the problem. The typical control we use is to reroute the call. Instead of sending the customer's call directly to its destination, we'll route it via a third city—to someplace else in the country that has the capacity to complete the call.

It would seem that routing via another city would take longer. Is the customer aware of this process?

Routing a call via a third city is entirely transparent [imperceptible] to the customer. I'm an expert about the network, and even when I make a phone call, I have no idea how that individual call was routed. It's transparent both in terms of how far away the other person sounds and in how quickly the telephone call gets set up. With the signaling network we use, it takes milliseconds for switching systems to "talk" to each other to set up a call. So the fact that you are involved in a third switch in some distant city is something you would never know.

You want to be sure to keep costs down while supplying enough service to customers. So how do you balance company benefits with customer benefit?

In terms of making the network efficient, we want to do two things. First, we want our customers to be happy with our service and for all their calls to go through, which means we must build enough capacity in the network to allow that to happen. Second, we want to be efficient for stockholders and not spend more money than we need to for the network to be at the optimum size.

There are basically two costs in terms of building the network. There is the cost of switching systems and the cost of the circuits that connect the switching systems. Basically, you can use operations-research techniques and mathematics to determine cost trade-offs. It may make sense to build direct routing between two switching systems and use a lot of circuits, or maybe to involve three switching systems, with fewer circuits between the main two, and so on.

including time, distance, and cost. The following are some of the many possible applications:

1. Design routes to be used by an ambulance, police car, or fire engine to get to an emergency as quickly as possible.
2. Design delivery routes that minimize gasoline use.
3. Design routes to bring soldiers to the front as quickly as possible.
4. Design a route for a truck carrying nuclear or biomedical waste.
5. Help food deliverers fill burger and pizza orders quickly—and while they're still hot.

Many people find it increasingly convenient to use the Web or software installed in their cars to get driving directions and driving time estimates to a place they wish to visit. The software that provides this information relies on algorithms that compute the shortest-path route in an appropriate weighted graph, which involves distances or times. The software sometimes makes use of global positioning

Common Ancestors?

SPOTLIGHT 2.3

In the study of ancient manuscripts, different manuscripts of the same book are available, even though the original manuscript upon which they are based has been lost. Examples of this include Euclid's *Elements* and Chaucer's *Canterbury Tales*. What interests scholars is reconstructing the relationships between the manuscripts and the common ancestors of the manuscripts, even when some of the ancestors are now missing.

Similarly, perceptual psychologists may be interested in which colors people perceive as being closely related and comparing these perceptions with those of people who are color-blind. Linguists are interested in the connections between languages that seem very different today, but have some words that are similar. Finally, in studying different species, biologists are interested in determining which species are more closely related to each other, including species known only in fossil form, and constructing a "tree" of life that shows which species were ancestors of others.

Reconstructions of this kind are made possible by using graph theory, specifically using the graph theory concept of a tree. The value of the graph theory in these and many other situations lies in using the distance between pairs of vertices in the tree as a way of reflecting the closeness of the relationships that pairs of manuscripts, pairs of colors, or pairs of species have. The distance between two vertices in a tree is the sum of the weights along the one path that joins the two vertices. If there are no weights on the edges, the distance is the number of edges in the path. In some reconstruction problems, a special vertex of the tree called the *root* is singled out. This root plays the role of the original common ancestors, and distances to the root are of critical interest.

In the case of species, trees of family relatedness were traditionally constructed based on similarities of bones and physical appearance. With the discovery of molecular biology, many new avenues have been opened up. We can now draw trees of relatedness based on an organism's genetic material, DNA, or the proteins that the DNA codes for. The traditional trees based on physical traits often show different species as being more closely related than trees based on newer molecular biological approaches. These differences cause scholars to focus on how to resolve the discrepancies and thereby reach a deeper understanding of the unity of life.

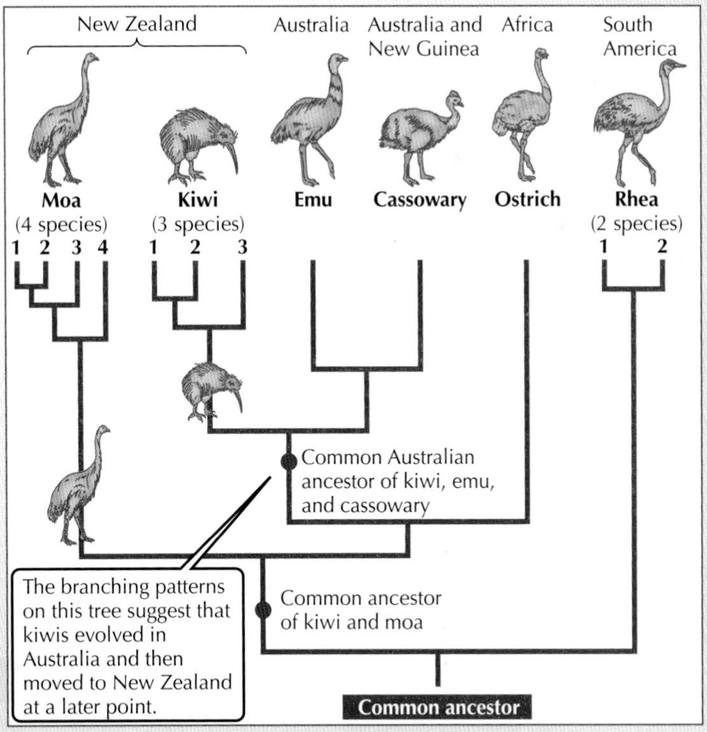

system data involving lots of mathematics—mathematics different from what we are considering here.

The need to find shortest paths seems natural. Next we investigate a situation in which finding a *longest* path is the right tool.

2.5 Critical-Path Analysis

Mathematics can confirm the obvious in certain situations while showing that our intuition is wrong in other circumstances. Our next group of applications will illustrate this point.

A characteristic of American life is its fast pace. People are interested in getting things done quickly and efficiently. This means that when you take your car in to be repaired before going to work, you want to know for sure that the repairs will be done when you pick up the car. You want the trains and the bus that take you to your doctor's appointment to run on time. When you arrive at the doctor's office, you want a technician to be free to take a blood sample and a throat culture. You want your outpatient appointment for an X-ray at the local hospital to occur on schedule. You want the X-ray to be interpreted quickly and the results reported back to your internist.

Scheduling machines and people is a big part of modern life. It is involved in running a school, a hospital, an airline, or in landing a person on Mars, and modern mathematics plays a big part in solving scheduling problems.

Part of what makes scheduling complicated is that the tasks that make up a job usually cannot be done in a random order. For example, to make Thanksgiving dinner, you must buy and prepare the turkey before putting it in the oven, and you must set the table before serving the food.

If the tasks cannot be performed in a random order, we can specify the order in an **order-requirement digraph.** The term *digraph* is short for "directed graph." A digraph is a geometrical tool similar to a graph except that each edge has an arrow on it to indicate a direction for that edge. Digraphs can be used to illustrate that traffic on a street must go in one direction or that certain tasks in a job must be completed before other tasks. A typical example of an order-requirement digraph is shown in Figure 2.16. There is a vertex in this digraph for each task. If one task must be done immediately before another, we draw a directed edge, or arrow, from the prerequisite task to the subsequent task. The numbers within the circles representing vertices are the times it takes to complete the tasks. In Figure 2.16, there is no arrow from T_1 to T_5 because task T_2 intervenes. Tasks T_7 and T_8 are independent of each other. Either task can be performed before the other, and we do not have to take into account priorities for the other tasks to schedule them. There are no directed arrows entering or leaving T_7 or T_8, and such arrows would force them to be performed in a particular order with respect to the other tasks or to each other. Also, T_1, T_7, and T_8 have no tasks that must precede them. Hence, if there are at least three processors (such as people or machines) available, tasks T_1, T_7, and T_8 can be worked on simultaneously at the start of the job.

Let's investigate a typical scheduling problem faced by a business.

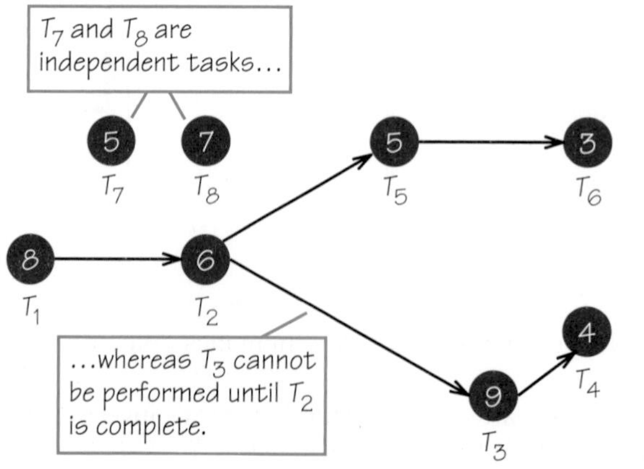

Figure 2.16 A typical order-requirement digraph.

EXAMPLE 8 Turning a Plane Around

Consider an airplane that carries both freight and passengers. The plane must have its passengers and freight unloaded and new passengers and cargo loaded before it can take off again. Also, the cabin must be cleaned before departure can occur. Thus, the job of "turning the plane around" requires completion of five tasks:

Task *A*	Unload passengers	13 minutes
Task *B*	Unload cargo	25 minutes
Task *C*	Clean cabin	15 minutes
Task *D*	Load new cargo	22 minutes
Task *E*	Load new passengers	27 minutes

Altrendo Travel/Getty Images

Turning a plane around, which involves such tasks as refueling, unloading, cabin cleanup, and then reloading cargo and passengers, entails very careful scheduling to avoid time slippage.

The order-requirement digraph for the problem of turning an airplane around is shown in Figure 2.17. The presence or absence of an edge in the order-requirement digraph depends on the analysis made as part of the modeling process for the problem. It seems natural that we need an arrow between task *A* and task *C*, because the passengers have to be unloaded before the cabin is cleaned. Other arrows may not seem natural—say, perhaps the arrow from task *B* (unload the cargo) to task *E* (load new passengers). This arrow may be due to government rules or union requirements.

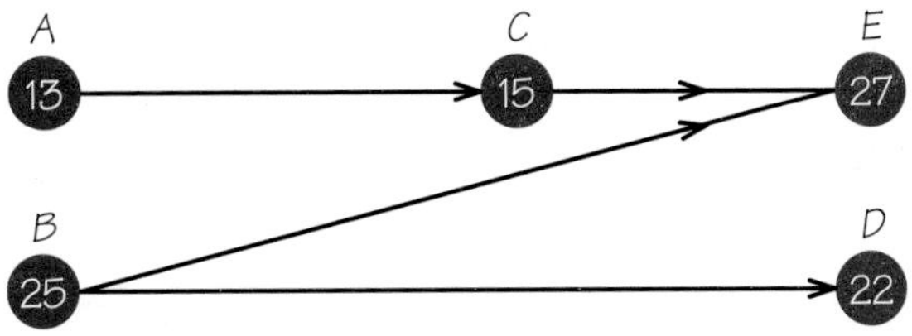

Figure 2.17 An order-requirement digraph for turning an airplane around after landing.

What matters is that the mathematics of solving the problem does not depend on the reason that the order-requirement digraph looks the way that it does. The person solving the problem constructs the order-requirement digraph and then the mathematical techniques we will develop can be applied, regardless of whether or not another business faced with a similar problem might model the problem in a different way by using a different order-requirement digraph. Because we want to find the earliest completion time, it might seem that finding the shortest path in the digraph (path *BD* with time length $25 + 22 = 47$) would solve the problem. But this approach shows the danger of ignoring the relationship between the mathematical model (the digraph) and the original problem.

The time required to complete all the tasks, *A* through *E*, must be at least as long as the time necessary to do the tasks on any particular path. Consider the path *BD*, which has length 25 + 22 = 47. Recall that here *length* of a path refers to the sum of the times of the tasks that lie along the path. Because task *B* must be done before task *D* can begin, the two tasks *B* and *D* cannot be completed before time 47. Hence, even if work on other tasks (such as *A*, *C*, and *E*) proceeds during this period, all the tasks cannot be finished before the tasks on path *BD* are finished. The same statement is true for every other path in the order-requirement digraph. Thus, the earliest completion time actually corresponds to the length of the longest path. In the airplane example, this earliest completion time is 55 (= 13 + 15 + 27) minutes, corresponding to the path *ACE*. We call *ACE* the **critical path** because the times of the tasks on this path determine the earliest completion time.

Critical Path DEFINITION

A **critical path** in an order-requirement digraph is a longest path. The length is measured in terms of summing the task times of the tasks making up the path.

Self Check 4

Determine the tasks that make up the critical path in the order-requirement digraph shown in Figure 2.16. What is the earliest completion time for the job made up of this collection of tasks?

Note that if none of these tasks could go on simultaneously, the time to complete all the tasks would be 13 + 25 + 15 + 22 + 27 = 102 minutes. However, even though some tasks may be performed simultaneously, the fact that the length of the critical path is 55 means that completion of the tasks in less than 55 minutes is not possible. Only by speeding up the times to complete the critical-path tasks themselves can a completion time less than 55 minutes be achieved.

Suppose it were desirable to speed the turnaround of the plane to less than 55 minutes. One way to do this might be to build a second jetway to help unload passengers more quickly. For example, we could unload passengers (task *A*) in 7 minutes instead of 13. However, reducing task *A* to 7 minutes does not reduce the completion time by 6 minutes, because in the new digraph (Figure 2.18) *ACE* is no longer the critical (longest) path. The longest path is now *BE*, which has a length of 52 minutes. Thus, shortening task *A* by 6 minutes results in only a 3-minute saving in completion time. This may mean that building a new jetway is uneconomical.

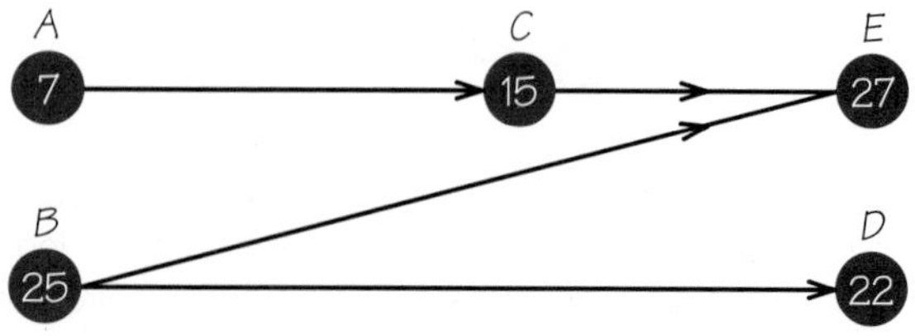

Figure 2.18 An order-requirement digraph for turning an airplane around in reduced time due to construction of a new jetway.

Note also that shortening the time to complete tasks that are not on the original critical path *ACE* will not shorten the completion time at all. Speeding up tasks on the critical path will shorten completion time of the job only up to the point where a new critical path is created. Also note that a digraph may have more than one longest path.

Not all order-requirement digraphs are as simple as the one shown in Figure 2.17. The order-requirement digraph in Figure 2.19 has 12 paths, which can be found by exhaustive search. Examples of such paths are $T_1T_2T_3$, $T_1T_5T_9$, $T_4T_5T_9$, and $T_7T_5T_3$. (Although we have not discussed them here, fast algorithms for finding longest and shortest paths in graphs are known.) The critical path is $T_7T_8T_6$ (length 21), and the earliest completion time for all nine tasks is time 21, though the actual completion time may be later than time 21, depending on the resources available to carry out the tasks. Completing the tasks by time 21 depends on having sufficient resources available so that some of the tasks can be worked on simultaneously.

These examples are typical of many scheduling problems that occur in practice (see Spotlight 2.4). Perhaps the most dramatic use of critical-path analysis is in the construction trades. No major new building project is now carried out without a critical-path analysis first being performed to ensure that the proper personnel and materials are available at the right times in order to have the project finished as quickly as possible. Many such problems are too large and complicated to be solved without the aid of computers.

The critical-path method was popularized and came into wider use as a consequence of the *Apollo* project. This project, which aimed at landing a man on

Every Moment Counts in Rigorous Airline Scheduling

SPOTLIGHT 2.4

When people think of airline scheduling, the first thing that comes to mind is how quickly a particular plane can safely reach its destination. But using ground time efficiently is just as important to an airline's timetable as the time spent in flight. Bill Rodenhizer was the manager of control operations for an airline that provided shuttle service between Boston and New York. He is considered to be an expert on airplane turnaround time, the process by which an airplane is prepared for almost immediate takeoff once it has landed. He tells us how this well-orchestrated effort works:

> Scheduling, to the airline, is just about the whole ball game. Everything is scheduled right to the minute. The whole fleet operates on a strict schedule. Each of the departments responsible for turning around an aircraft has an allotted period of time in which to perform its function. Manpower is geared to the amount of ground time scheduled for that aircraft. This would be adjusted during off-weather or bad-weather days or during heavy air-traffic delays.
>
> Most of our aircraft in Boston are scheduled for a 42- to 65-minute ground time. Boston is the end of the line, so it is a "terminating and originating station." In plain talk, that means almost every aircraft that comes in must be fully unloaded, refueled, serviced, and dispatched within roughly an hour's time.
>
> This is how the process works: In the larger aircraft, it takes passengers roughly 20 minutes to load and 20 minutes to unload. During this period, we will have completely cleaned the aircraft and unloaded the cargo, and the caterers will have taken care of the food. The ramp service may take 20 to 30 minutes to unload the baggage, mail, and cargo from underneath the plane, and it will take the same amount of time to load it up again. We double-crew those aircraft with heavier weights so that the workload will fit the time it takes passengers to load and unload upstairs.
>
> While this has been going on, the fueler has fueled the aircraft. As to repairs, most major maintenance is done during the midnight shift, when [most of our] several hundred aircraft are inactive. We all work under a very strict time frame.

New security requirements in the wake of the World Trade Center attack (9/11/2001) have increased the difficulty of adhering to timetables in operating shuttle services between East Coast cities such as New York and Boston. This makes it even more important to use analytical tools in keeping operations on schedule.

the moon within 10 years of 1960, was one of the most sophisticated projects in planning and scheduling ever attempted. The dramatic success of the project can be attributed partly to the use of critical-path ideas and the related program evaluation and review technique (PERT), which helped keep the project on schedule.

In Chapter 3, we will see how mathematical ideas drawn from outside of graph theory can be used to gain insight into scheduling problems.

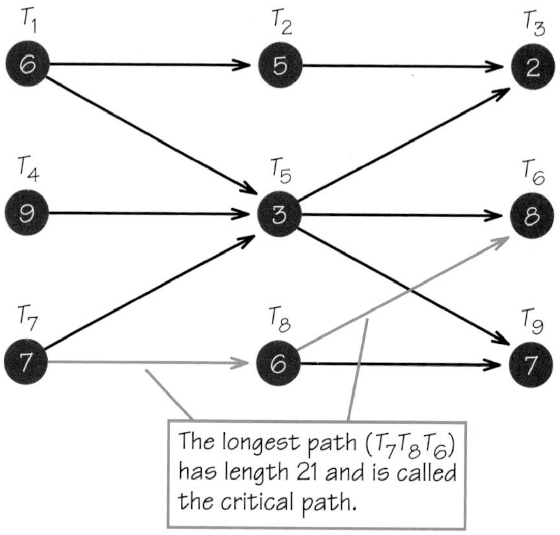

Figure 2.19 An order-requirement digraph with 12 paths, to examine how to find the length of the longest path.

Review Vocabulary

Algorithm A step-by-step description of how to solve a problem. (p. 43)

Brute force method The method that solves the traveling salesman problem (TSP) by enumerating all the Hamiltonian circuits and then selecting the one with minimum cost. (p. 46)

Complete graph A graph in which every pair of vertices is joined by an edge. (p. 45)

Critical path The longest path in an order-requirement digraph. The length of this path gives the earliest completion time for all the tasks making up the job consisting of the tasks in the digraph. (p. 60)

Fundamental principle of counting A method for counting outcomes of multistage processes. (p. 45)

Greedy algorithm An approach for solving an optimization problem, where at each stage of the algorithm the best (or cheapest) action is taken. Unfortunately, greedy algorithms do not always lead to optimal solutions. (p. 48)

Hamiltonian circuit A circuit using distinct edges of a graph that starts and ends at a particular vertex of the graph and visits each vertex once and only once. A Hamiltonian circuit can start at any one of its vertices. (p. 40)

Heuristic algorithm A method of solving an optimization problem that is "fast" but does not guarantee an optimal answer to the problem. (p. 51)

Kruskal's algorithm An algorithm developed by Joseph Kruskal (AT&T Research) that solves the minimum-cost spanning-tree problem by selecting edges in order of increasing cost, but in such a way that no edge forms a circuit with edges chosen earlier. It can be proved that this algorithm always produces an optimal solution. (p. 53)

Method of trees A visual method of carrying out the fundamental principle of counting. (p. 42)

Minimum-cost Hamiltonian circuit A Hamiltonian circuit in a graph with weights on the edges, for which the sum of the weights of the edges of the Hamiltonian circuit is as small as possible. (p. 42)

Minimum-cost spanning tree A spanning tree of a weighted connected graph having minimum cost. The cost of a tree is the sum of the weights on the edges of the tree. (p. 54)

Nearest-neighbor algorithm An algorithm for attempting to solve the TSP that begins at a "home" vertex and visits next that vertex not already visited that can be reached most cheaply. When all other vertices have been visited, the tour returns to home. This method may not give an optimal answer. (p. 48)

NP-complete problems A collection of problems, which includes the TSP, that appear to be very hard to solve quickly for an optimal solution. (p. 51)

Order-requirement digraph A directed graph that shows which tasks precede other tasks among the collection of tasks making up a job. (p. 58)

Sorted-edges algorithm An algorithm for attempting to solve the TSP where the edges added to the circuit being built up are selected in order of increasing cost, but no edge is chosen that would prevent a Hamiltonian circuit from forming. These edges must all be connected at the end, but not necessarily at earlier stages. The tour obtained may not have the lowest possible cost. (p. 49)

Spanning tree A subgraph of a connected graph that is a tree and includes all the vertices of the original graph. (p. 53)

Traveling salesman problem (TSP) The problem of finding a minimum-cost Hamiltonian circuit in a complete graph where each edge has been assigned a cost (or weight). (p. 47)

Tree A connected graph with no circuits. (pp. 42, 53)

Weight A number assigned to an edge of a graph that can be thought of as a cost, distance, or time associated with that edge. (p. 42)

Self Check Answers

1. *DEABCD*; cost = 3000.

2. The edges are added in the order *DE, DA, AB, BC,* and *CE*. One way to represent this tour is *ABCEDA*, and the cost of the tour is 4.8.

3. The edges to form the minimum-cost spanning are added in the order St. Louis–Chicago, Chicago–Cleveland, and Chicago–Minneapolis. The cost for the tree is 1074.

4. The tasks on the critical path are T_1, T_2, T_3, and T_4. The job consisting of all the tasks cannot be completed before time 27.

Skills Check

1. Which of the statements below is false for the graph *G* shown?

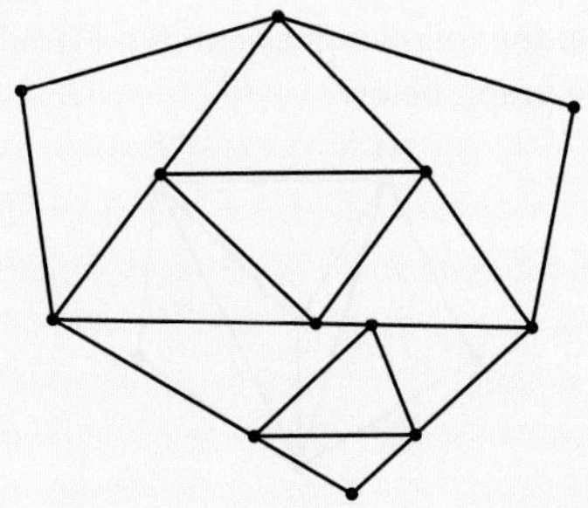

(a) *G* has an Euler circuit.

(b) *G* has no spanning tree.

(c) *G* is not even-valent.

(d) *G* has no Hamiltonian circuit.

2. The cost of the nearest-neighbor tour for the accompanying graph, starting at vertex 3, is ______________________.

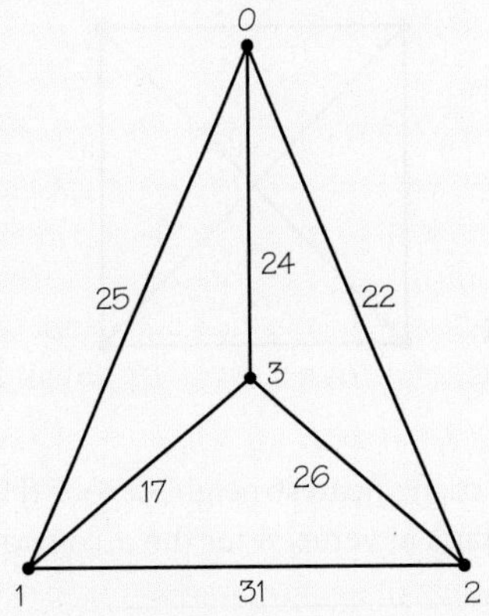

3. The cost associated with the TSP tour 0, 2, 1, 3, 0 in the accompanying graph is

(a) 48.

(b) 47.

(c) −48.

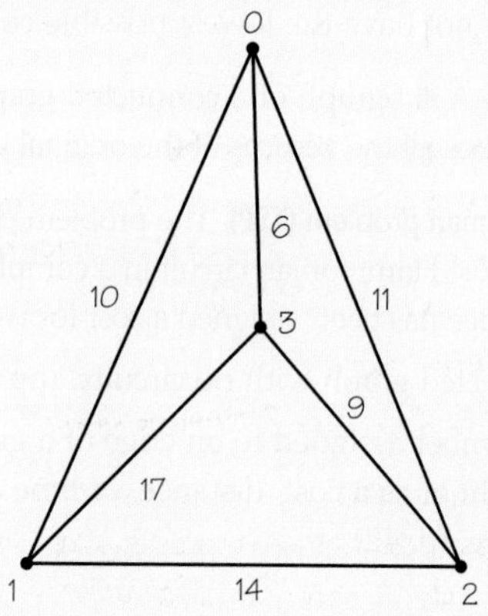

4. The difference between the cost of a nearest-neighbor tour starting at *D* and the cost of a sorted-edges tour for the accompanying graph is __________.

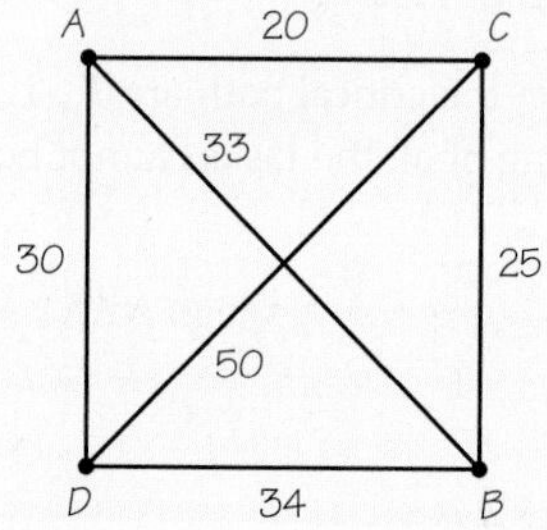

5. The tour 1, 3, 2, 4 is not a Hamiltonian circuit because

(a) it does not include all of the vertices of the graph.

(b) it is not a circuit.

(c) there are other ways to get from 1 to 4 in this graph.

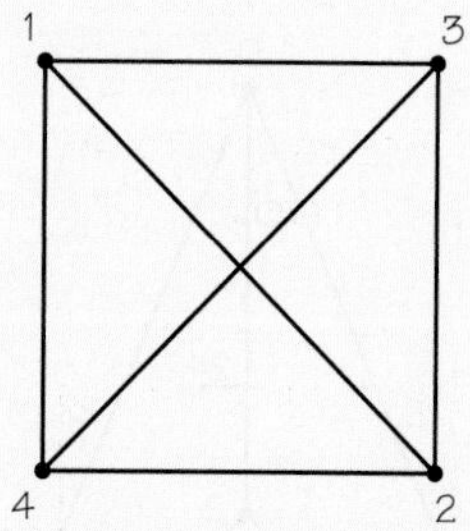

6. The cost of the nearest-neighbor tour (Hamiltonian circuit) that starts at vertex *A* for the accompanying graph is ______________________________ .

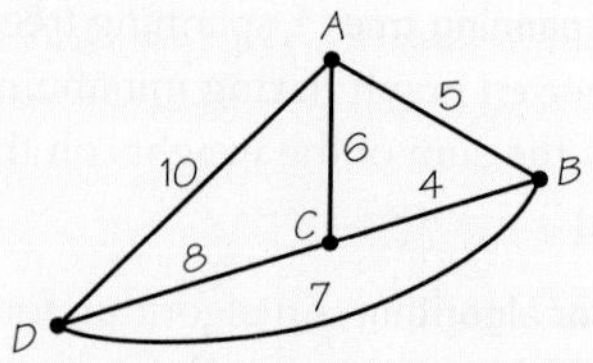

7. The accompanying graph has no Hamiltonian circuit because

(a) it is even-valent.

(b) it has an odd number of vertices.

(c) it is not connected.

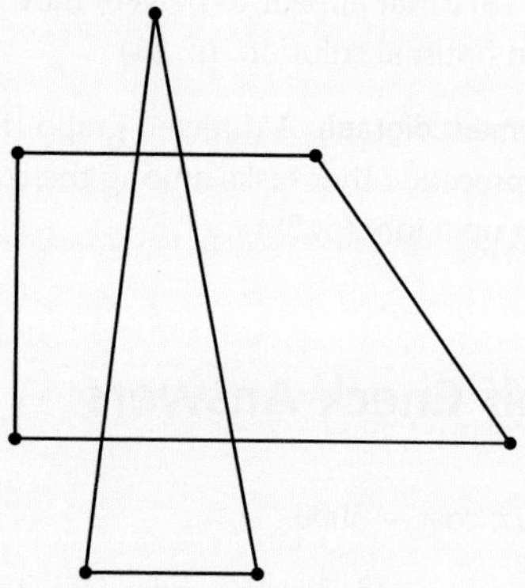

8. The cost of the sorted-edges tour (Hamiltonian circuit) for the accompanying graph is

______________________________ .

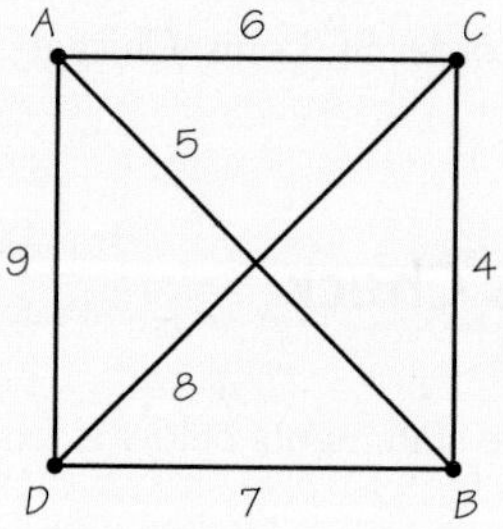

9. Which of the following describes a Hamiltonian circuit for the graph below?

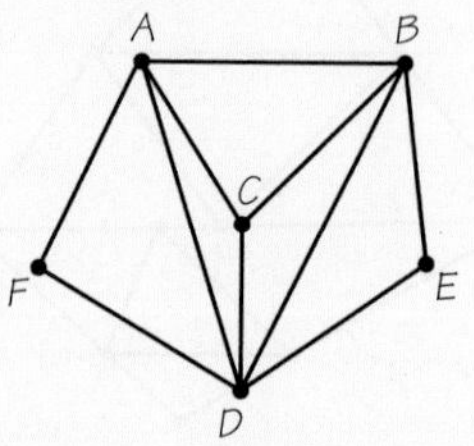

(a) *ABCDFA*

(b) *AFDCBE*

(c) *ACBEDFA*

(d) *ACEBDFA*

10. The cost of the nearest-neighbor traveling salesman tour that starts at *D* for the graph below is ______ .

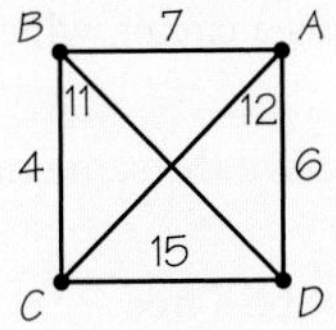

11. Suppose that after a hurricane, a van is dispatched to pick up five nurses at their homes and bring them to work at the local hospital. Which of these techniques is most likely to be useful in solving this problem?

(a) Finding an Euler circuit in a graph

(b) Finding a minimum-cost spanning tree in a graph

(c) Solving a TSP (traveling salesman problem)

12. If a graph has *E* edges and *V* vertices as well as a Hamiltonian circuit, then the number of edges in the Hamiltonian circuit is ______________ .

13. The following graph has

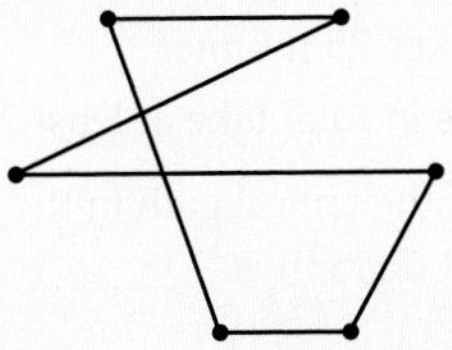

(a) no Hamiltonian circuit and no Euler circuit.

(b) an Euler circuit and a Hamiltonian circuit.

(c) no Hamiltonian circuit, but it has an Euler circuit.

14. The longest-circuit tour of vertices that arises by applying the nearest-neighbor algorithm starting at any of the sites 0, 1, 2, 3 in the graph below has a cost of ________.

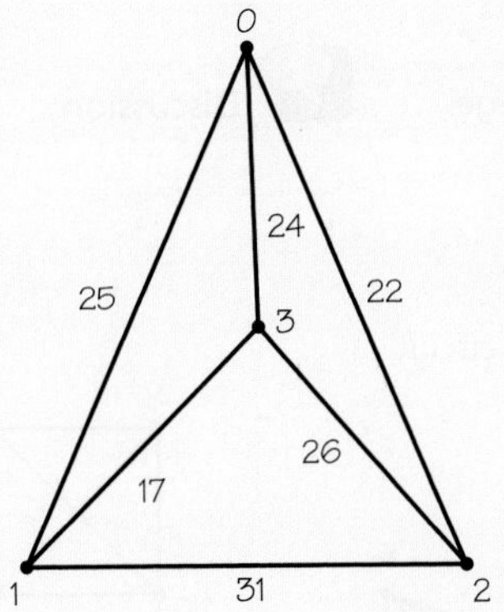

15. When the sorted-edges method and nearest-neighbor method are applied to a complete graph on seven vertices with nonnegative weights,

(a) both methods always give different answers.

(b) both methods always give the same answer but that answer may not be optimal.

(c) neither method may give an optimal answer.

16. The number of different lunches that Jules can design by selecting one of three meats, one of four salads, and one of six vegetables is exactly ____________ .

17. When the sorted-edges method is applied to the TSP where all the edges have distinct costs, which of the following *must* be true of the tour *S* obtained?

(a) The largest-weight edge cannot be part of tour *S*.

(b) The shortest-weight edge must be part of tour *S*.

(c) The length of the tour *S* must be odd.

18. If a three-character password system must begin with a lowercase letter of the English alphabet followed by two decimal digits that cannot be repeated, the number of different possible passwords is ______________ .

19. Paul has packed four ties, three shirts, and two pairs of pants for a trip. How many different outfits can he create if he never wears a tie?

(a) Fewer than 12

(b) Between 11 and 25

(c) More than 30

20. A company is designing a logo with two identical lowercase letters with a single nonzero digit between the letters. A typical possible example would be d3d. The number of such logos is ________.

21. An ice-cream shop offers 3 types of cones, 20 flavors, and 4 different toppings (crushed peanuts, crushed almonds, chocolate bits, or corn flakes). If a customer is allergic to nuts, how many different choices can she choose from?

(a) 240

(b) 120

(c) 25

22. A minimum-cost spanning tree for the weighted accompanying graph has weight ______ and ______ edges.

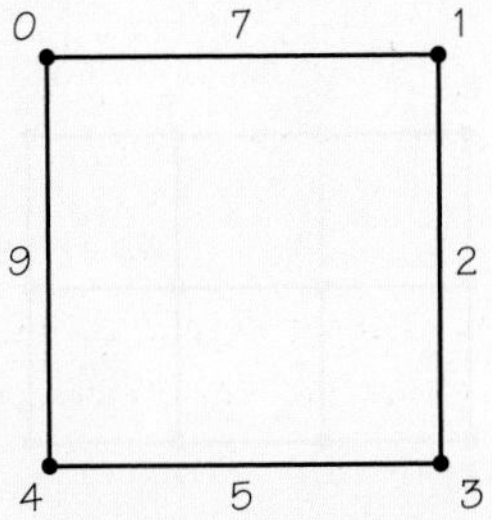

23. Assuming a graph with *E* edges and *V* vertices has a minimum-cost spanning tree *T*, which of the following statements *must* be true?

(a) The graph is connected.

(b) The tree *T* includes every minimum-cost edge.

(c) The tree *T* has exactly *V* edges.

24. When arranged in increasing order, the weights of the edges in the following graph that are not part of the minimum-cost spanning tree selected when Kruskal's algorithm is applied are ______, ______, ______.

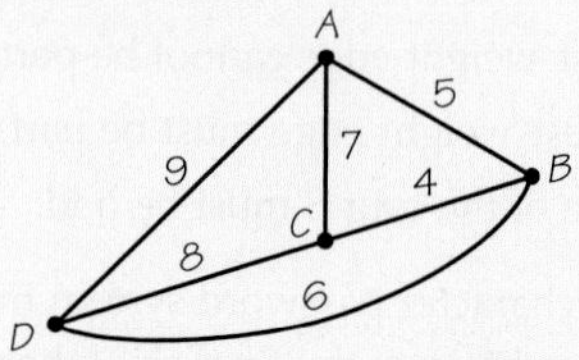

25. The cost of the last edge that Kruskal's algorithm selects to be part of a minimal-cost spanning tree for the weighted graph below is ________.

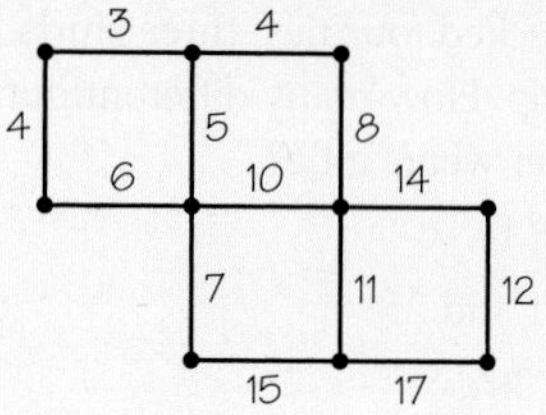

26. Assume that every edge of a graph *G* has a different cost. If Kruskal's algorithm is used to find the minimum-cost spanning tree *T* for graph *G*, which of the following statements *must* be true?

(a) Any other spanning tree for graph *G* will have more edges than *T*.

(b) Any other spanning tree for graph *G* will have a greater cost than *T*.

(c) The edge of graph *G* having greatest weight is included in *T*.

27. If a graph contains a circuit, which of the following statements is true?

(a) The graph must have the same number of vertices as edges.

(b) The graph cannot be a tree.

(c) The graph is not connected.

28. The earliest completion time (in minutes) for a job with the following order-requirement digraph is ______.

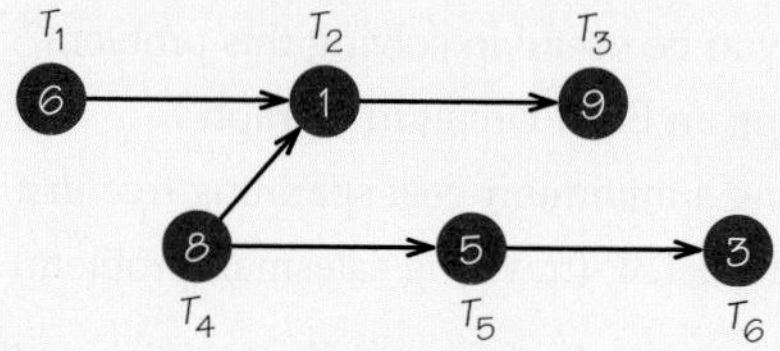

29. Assume a job has an order-requirement digraph with five tasks whose critical path is 25 minutes in length. Based on this information, what can be said about the tasks?

(a) Each task takes exactly 5 minutes.

(b) Some task takes 25 minutes.

(c) The five tasks in total take at least 25 minutes.

30. The length of the critical path in the accompanying order-requirement digraph is ______________ minutes.

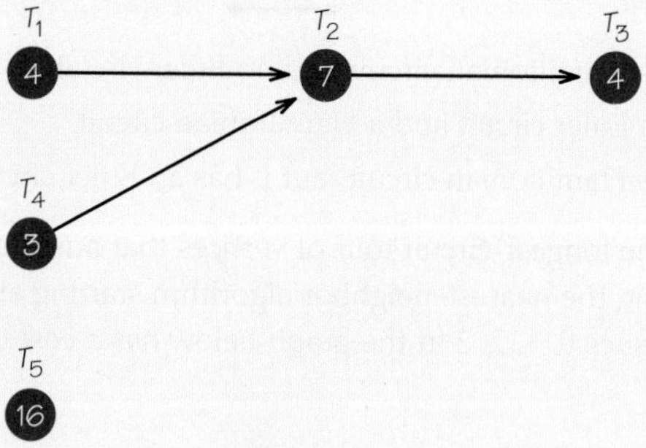

Chapter 2 Exercises

Challenge

Discussion

2.1 Hamiltonian Circuits

1. For the accompanying graphs (a) through (c), write down a Hamiltonian circuit starting at X_2.

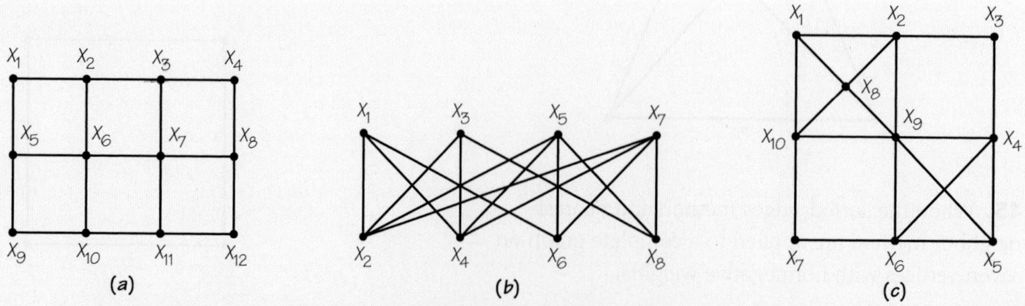

2. Construct a grid graph with m rows (m at least 3) and n columns (see Figure 2.2b for a 4-by-4 grid graph) that

(a) has a Hamiltonian circuit.

(b) does not have a Hamiltonian circuit.

3. For the accompanying graphs (a) through (c), write down, if possible, a Hamiltonian circuit starting at X_3.

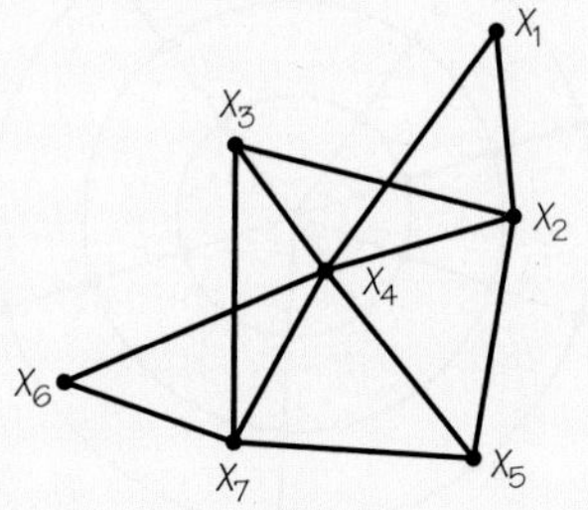

(a)

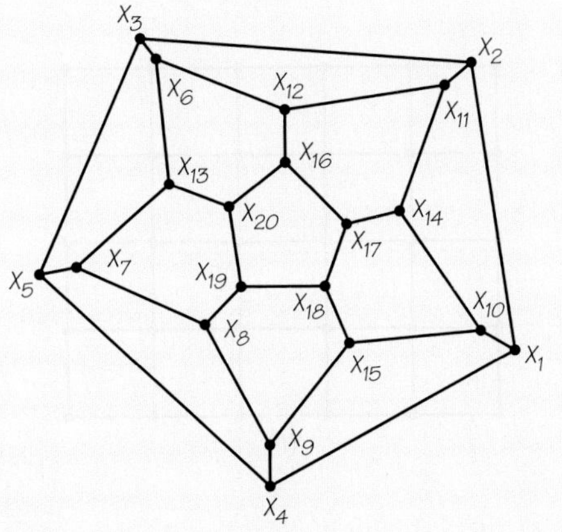

(b)

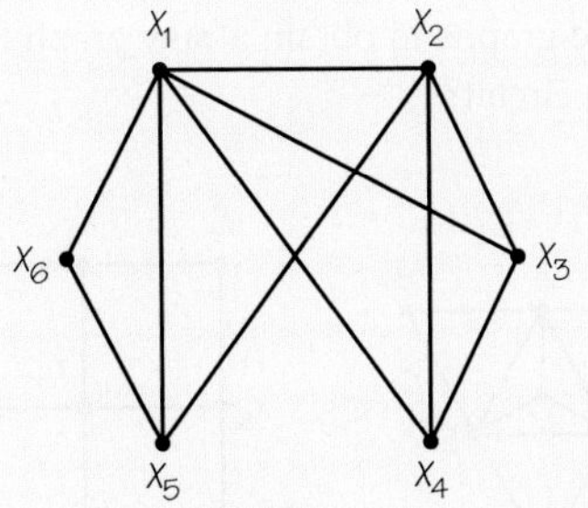

(c)

4. (a) For the graph below, write down a Hamiltonian circuit that starts at X_3.

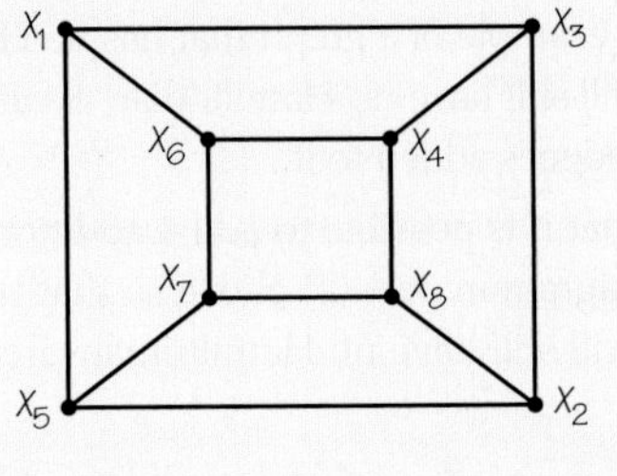

(b) How many vertices are there in the Hamiltonian circuit you found in part (a)?

(c) How many edges are there in the Hamiltonian circuit you found in part (a)?

(d) What is the largest number of edges you can remove from the graph shown in part (a) that will still allow one to find a Hamiltonian circuit in the graph after the edges are removed?

5. For the accompanying graphs (a) through (c), write a Hamiltonian circuit starting at X_3.

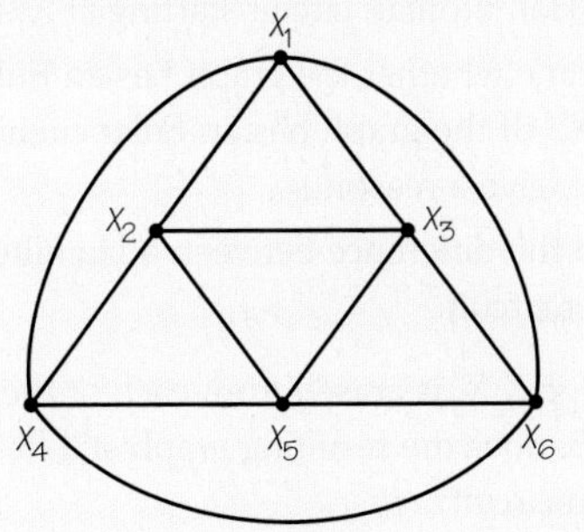

(a)

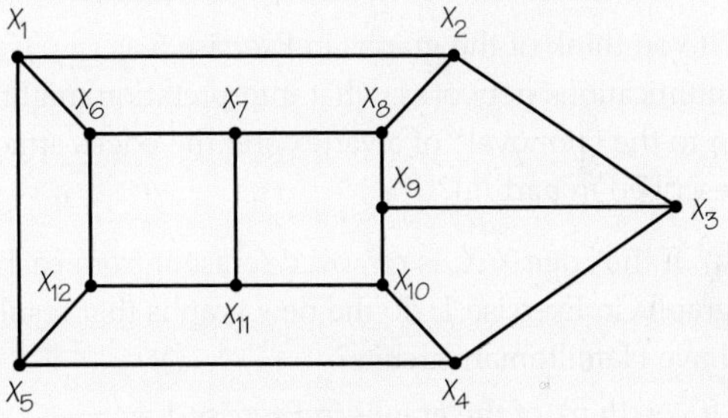

(b)

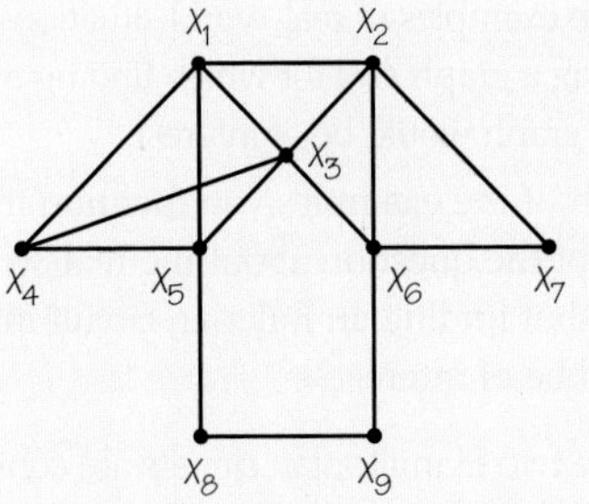

(c)

6. Refer to the accompanying graph.

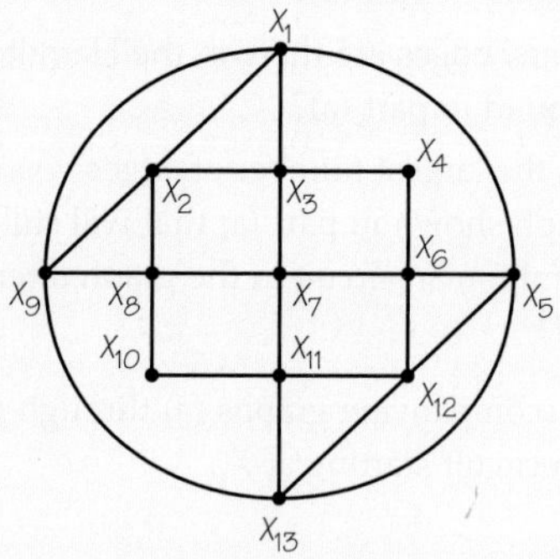

(a) Find a Hamiltonian circuit starting at X_1.

(b) Determine whether the graph has an Euler circuit starting at X_1. (If the graph has an Euler circuit, write it down; if not, give a reason.)

(c) Explain the difference between a Hamiltonian circuit and an Euler circuit.

7. If the edge X_2X_3 is erased from each of the graphs in Exercise 5, does the resulting graph still have a Hamiltonian circuit?

8. (a) If the vertex X_6 and the edges attached to X_6 are removed from the graphs in Exercise 5, do the new graphs that result still have Hamiltonian circuits?

(b) If you think of the graphs in Exercise 5 as communications networks, what interpretation might be given to the "removal" of a vertex and the edges attached as described in part (a)?

9. (a) If the edge X_6X_7 is removed (erased) from each of the graphs in Exercise 1, do the new graphs that result still have Hamiltonian circuits?

(b) If you think of the graphs in Exercise 1 as communications networks, what interpretation might be given to the "removal" of an edge as described in part (a)?

10. (a) Give examples of real-world situations that can be modeled using a graph and for which finding a Hamiltonian circuit in the graph would be of interest.

(b) For each of the examples you mention in part (a), can you adapt the question about the real-world situation involved so that finding an Eulerian circuit in the same graph would be of interest?

11. Suppose two Hamiltonian circuits are considered different if the collections of edges that they use are different. How many other Hamiltonian circuits can you find in the graph in Figure 2.1 that are different from the two discussed?

12. (a) For each of the following graphs, add wiggly edges to indicate a Hamiltonian circuit.

(b) Can you see how to construct an infinite family of graphs, each with a Hamiltonian circuit, inspired by how you constructed a Hamiltonian circuit in graph (b) of part (a)?

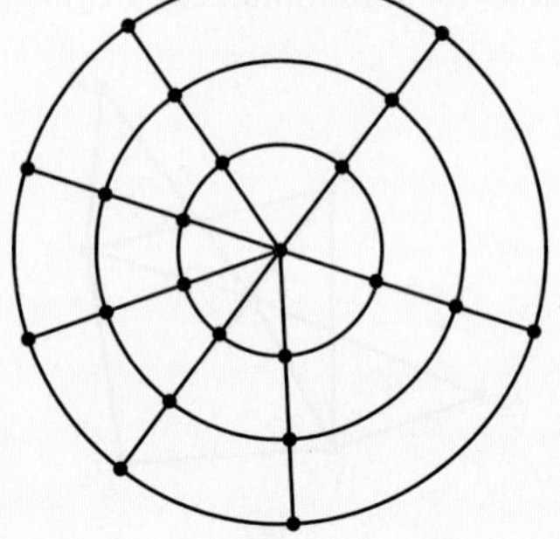

(*a*)

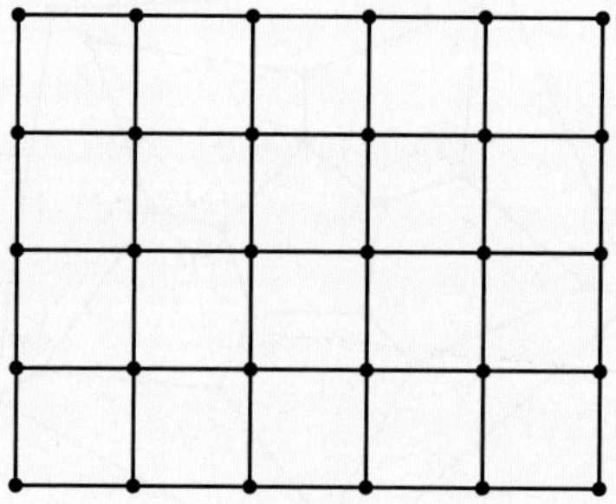

(*b*)

13. (a) Neither of the following graphs has a Hamiltonian circuit. Is it possible to add a single new edge to these graphs to obtain a new graph that has a Hamiltonian circuit?

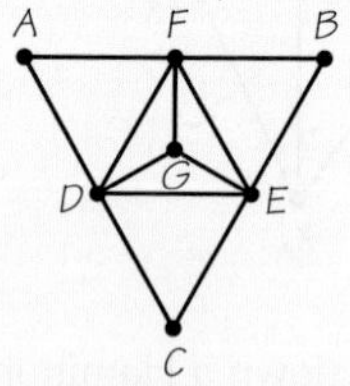

(*a*)

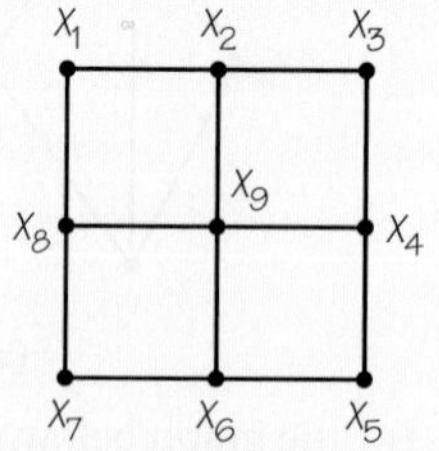

(*b*)

(b) Find an example of a graph that has no Hamiltonian circuit and will still have no Hamiltonian circuit no matter what single edge is added to it.

(c) Show that it is possible to add 4 additional edges to the graph diagram in part (b) above so that the resulting new graph will still have no Hamiltonian circuit.

14. Explain why the graph below has no Hamiltonian circuit.

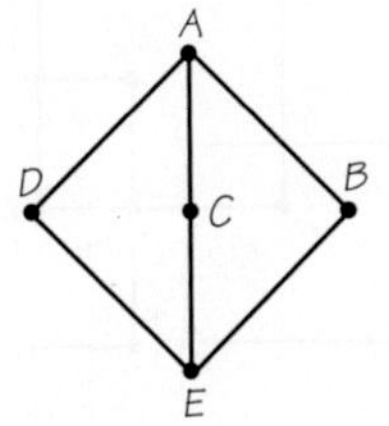

15. Use the graph shown in Exercise 14 to help you construct a connected graph for which every vertex has valence 3 and that does not have a Hamiltonian circuit.

16. Explain why the tour *CFECBADBC* is not a Hamiltonian circuit for the accompanying graph. Does this graph have a Hamiltonian circuit?

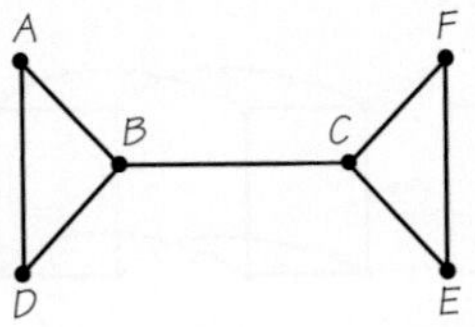

17. Do the following graphs have Hamiltonian circuits? If not, can you demonstrate why not?

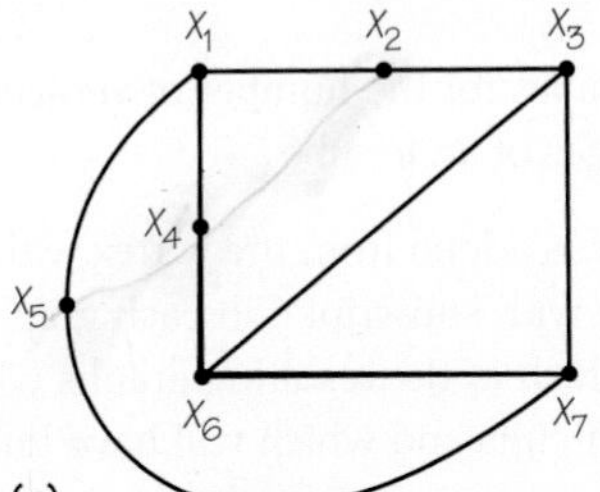

(a)

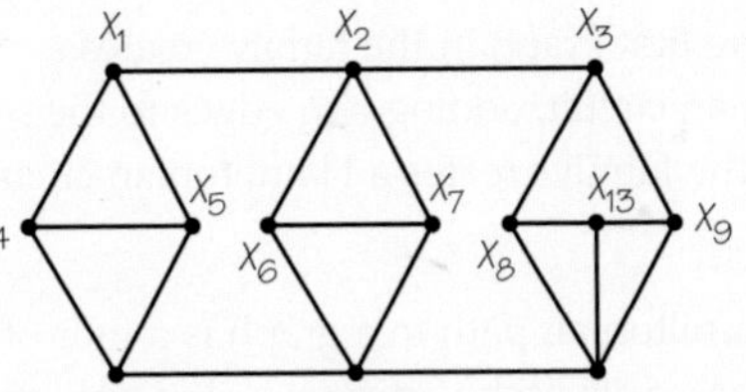

(b)

18. If an edge from X_2 to X_5 is added to each graph in Exercise 17, do the new graphs that result have a Hamiltonian circuit?

19. For each of the accompanying graphs, determine whether there is a Hamiltonian circuit.

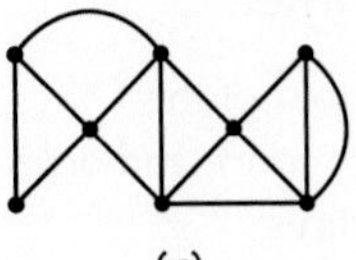

(a)

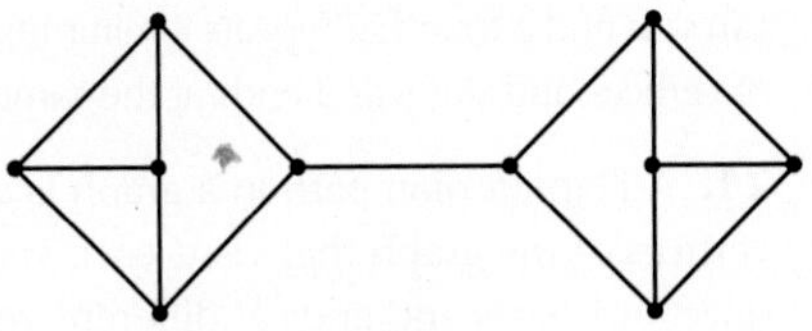

(b)

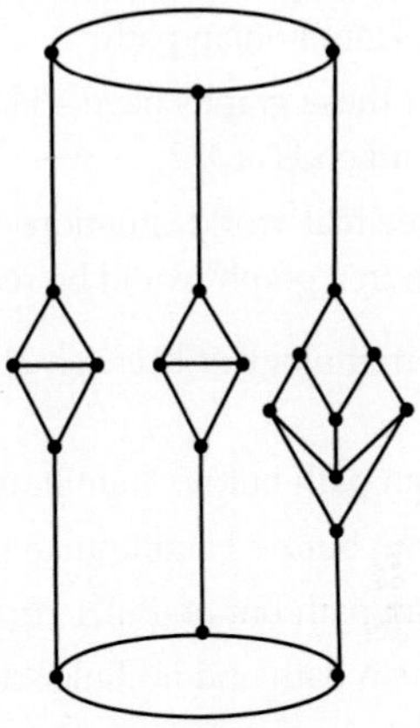

(c)

20. (a) The graph below is known as a four spokes and three concentric circles graph. What conditions on *m* and *n* guarantee that an *m* spokes and *n* concentric circles graph has a Hamiltonian circuit? (Assume $m \geq 2$, $n \geq 1$.)

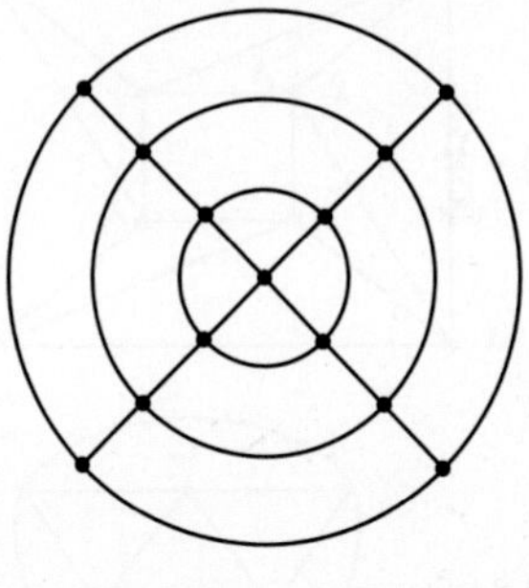

(b) The graph below is known as a 3×4 grid graph. What conditions on *m* and *n* guarantee that an $m \times n$ grid graph has a Hamiltonian circuit?

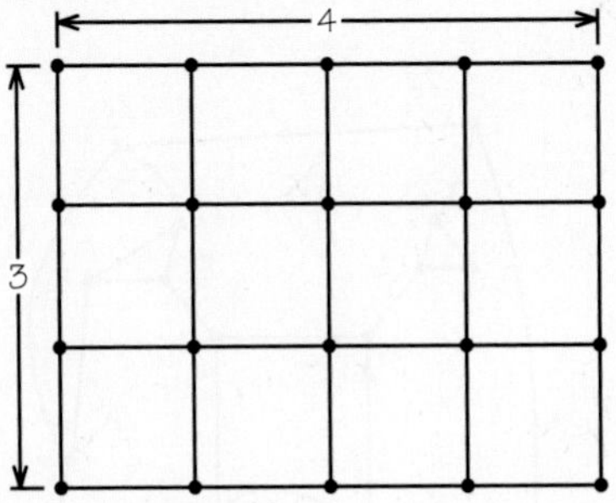

Can you think of a real-world situation in which finding a Hamiltonian circuit in an $m \times n$ grid graph would represent a solution to the problem? If an $m \times n$ grid graph has no Hamiltonian circuit, can you find a tour that repeats a minimum number of vertices and starts and ends at the same vertex?

21. A Hamiltonian path in a graph is a tour of the vertices of the graph that visits each vertex once and only once and starts and ends at different vertices.

(a) For each of the graphs shown in Exercise 17, does the graph have a Hamiltonian path?

(b) Does each of these graphs have a Hamiltonian path that starts at X_1 and ends at X_2?

(c) Describe three real-world situations where finding a Hamiltonian path in a graph would be required.

22. Using the terminology of Exercise 21, draw a graph that has

(a) a Hamiltonian path but no Hamiltonian circuit.

(b) an Euler circuit but no Hamiltonian path.

(c) a Hamiltonian path but no Euler circuit.

(d) no Hamiltonian path and no Euler circuit.

23. To practice your understanding of the concepts of Euler circuits and Hamiltonian circuits, determine for the accompanying graphs (a) through (d) whether there is an Euler circuit and/or a Hamiltonian circuit. If so, write it down.

(a)

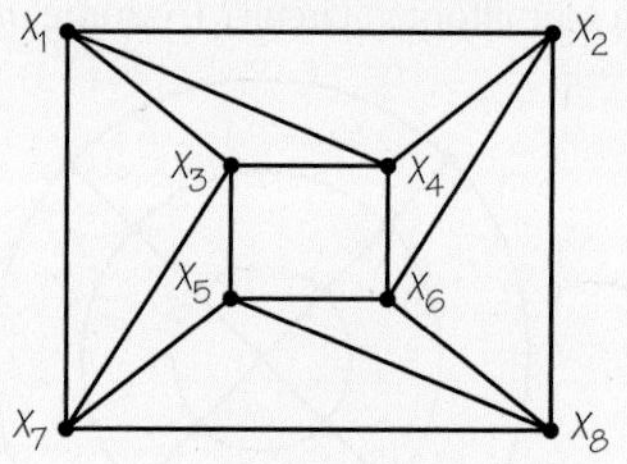

(b)

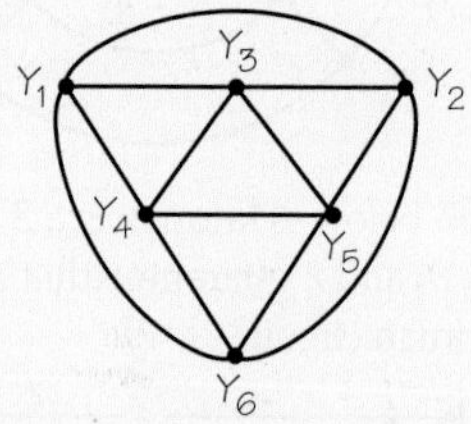

(c)

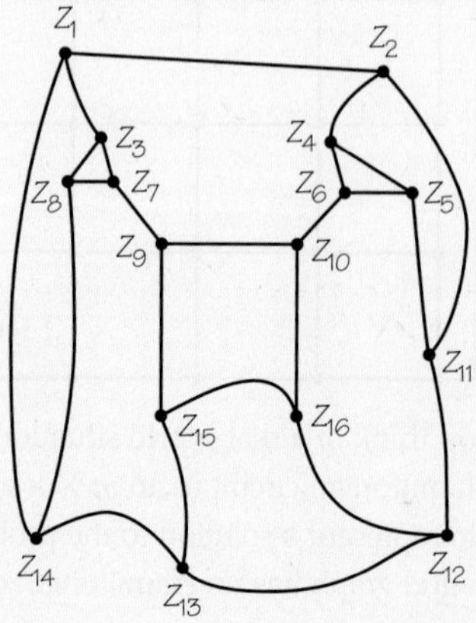

(d)

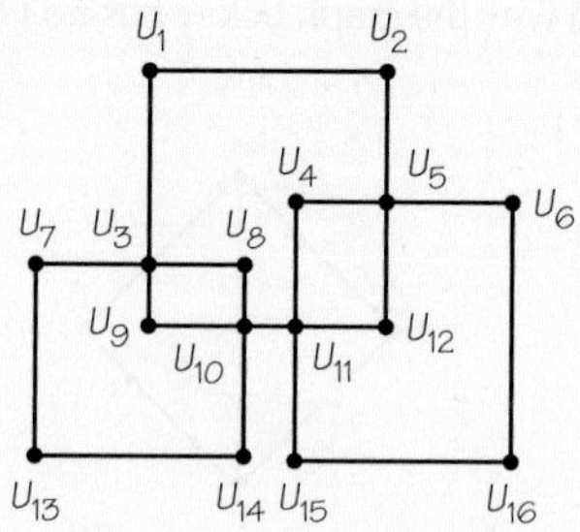

24. (a) The n-dimensional cube is obtained from two copies of an $(n - 1)$-dimensional cube by joining corresponding vertices. (The process is illustrated for the 3-cube and the 4-cube in the following figure.) Can you show that every n-cube has a Hamiltonian circuit? [*Hint:* Show that if you know how to find a Hamiltonian circuit on an $(n - 1)$-cube, then you can use two copies of this to build a Hamiltonian circuit on an n-cube.]

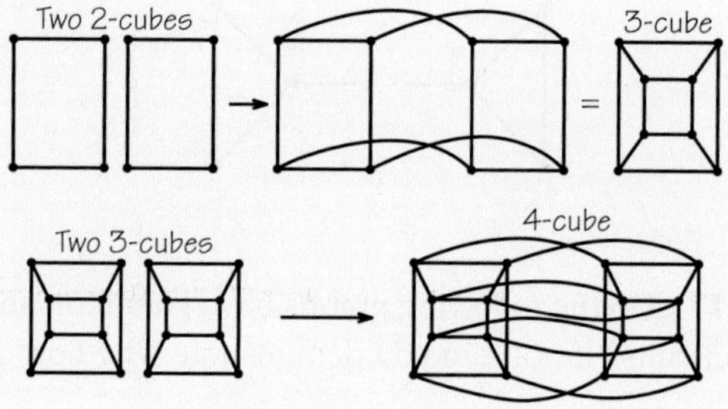

(b) Find formulas for the number of vertices and the number of edges of an n-cube.

25. If an edge is added from the vertex with subscript 3 to the vertex with subscript 6 in each graph in Exercise 23, which of the resulting graphs will have Hamiltonian circuits and which will have Euler circuits?

26. Find a family of graphs, none of which has a Hamiltonian circuit but for which adding a single edge to the first graph in the family creates a Hamiltonian circuit, adding two edges to the second graph in the family creates a Hamiltonian circuit, and so forth.

27. A Hamiltonian path in a graph is a tour of the vertices that visits each vertex once and only once and that starts and ends at different vertices.

(a) Draw an example of a graph that has no Hamiltonian path and where all the vertices are 3-valent.

(b) Draw a graph that has no Hamiltonian path but that does have an Euler circuit.

(c) By analogy with the Hamiltonian path, develop a definition of "Euler path."

28. (a) When going outside on a cold winter day, Jill can choose from four winter coats (three are red), five wool scarves (one is green), four pairs of boots, and three ski hats (two are blue). How many outfits might her friends see her in?

(b) If Jill always insists on wearing a red coat and a green scarf, how many outfits might her friends see her in?

29. The notes C, D, E, F, G, A, and B are to be used to form an ordered five-note musical logo. In how many ways can this be done under the following circumstances?

(a) No note can be repeated.

(b) Notes can be repeated.

(c) Notes can be repeated, but all the notes cannot be the same.

30. A lottery game requires a person to select an uppercase or lowercase letter followed by two different odd digits. How many choices can a customer make?

31. (a) In designing a security system for its accounts, a bank asks each customer to choose a five-digit number, all the digits to be distinct and nonzero. How many choices can a customer make?

(b) A suitcase with a liquid-crystal display allows one to unlock it with a specific combination of three capital letters that are not necessarily different. How many choices would a thief have to go through to be sure that all the possibilities had been tried? How does this compare to a "standard" combination lock?

32. To encourage her son to try new things, a mother offers to take him for a dish of ice cream with a topping once a week, for as many weeks as he does not get the same choice as on a previous occasion. If the store offers 12 flavors and 6 toppings, for how many weeks will she have to do this if her son never picks either of the two types of chocolate ice cream or the three types of nut topping that the store carries?

33. A large corporation has found that it has "outgrown" its current code system for routing interoffice mail. The current system places a code of three ordered, distinct nonzero digits on the mail. The new proposal calls for the use of two ordered capital letters. Does the new system have more code numbers than the old system? If so, how many more locations will the new system enable the company to encode over the current system?

34. Repeat Exercise 29a, except that exactly one of the notes in the musical logo must be a sharp and the note chosen to be sharped cannot appear elsewhere (e.g., BCD#AG, where D# denotes D sharp).

35. (a) In New York State, one type of license plate has three letters followed by three numbers. Suppose the digits from 0, 1, · · ·, 9 can be used, except that all three digits cannot be zero, and that any letter from A to Z (repeats allowed) can be used. How many plates are possible?

(b) Investigate what schemes for license plates are used in your state and determine how many different plates are possible.

36. A restaurant offers 6 soups, 10 entrees, and 8 desserts. How many different choices for a meal can a customer make if one selection is made from each category? If 3 of the desserts are pies and the customer will never order pie, how many meals can the customer choose?

37. In the last several years, heavily populated regions that previously had only one area code have been divided into service areas with more than one area code. What is the largest number of different phone numbers that can be served using one area code? If an area code cannot begin with a zero, how many different area codes are possible?

38. (a) A credit-card company makes it easier for customers to memorize their PIN (personal identification number) by using a four-digit PIN that consists of three different digits selected from 0, 1, 2, · · · 9, where one of the digits must be a zero, another is a nonzero digit that is repeated, and another is a digit different from these two. How many different PINs of this kind are there?

(b) How many PINs are possible if there are no restrictions on repeats of the 10 possible digits that can be used?

2.2 Traveling Salesman Problem

2.3 Helping Traveling Salesmen

39. Draw complete graphs with four, five, and six vertices. How many edges do these graphs have? Can you generalize to n vertices? How many TSP tours would these graphs have? (Tours yielding the same Hamiltonian circuit are considered the same.)

40. Calculate the values of 5!, 6!, 7!, 8!, 9!, and 10!. Then find the number of TSP tours in the complete graph with eight vertices.

41. The following table shows the mileage between four cities: Springfield, Ill. (*S*); Urbana, Ill. (*U*); Effingham, Ill. (*E*); and Indianapolis, Ind. (*I*).

	E	I	S	U
E	—	147	92	79
I	147	—	190	119
S	92	190	—	88
U	79	119	88	—

(a) Represent this information by drawing a weighted complete graph on four vertices.

(b) Use the weighted graph in part (a) to find the cost of the three distinct Hamiltonian circuits in the graph. (List them starting at *U*.)

(c) Which circuit gives the minimum cost?

(d) Would there be any difference in parts (b) and (c) if the starting vertex were at *I*?

(e) If one applies the nearest-neighbor method starting at *U*, what circuit would be obtained? Does the answer change if one applies the nearest-neighbor algorithm starting at *S*? At *E*? At *I*?

(f) If one applies the sorted-edges method, what circuit would be obtained? Does one get the optimal answer?

42. After a party at her house, Francine (*F*) has agreed to drive Mary (*M*), Rachel (*R*), and Constance (*C*) home. If the times (in minutes) to drive between her friends' homes are shown below, what route gets Francine back home the quickest?

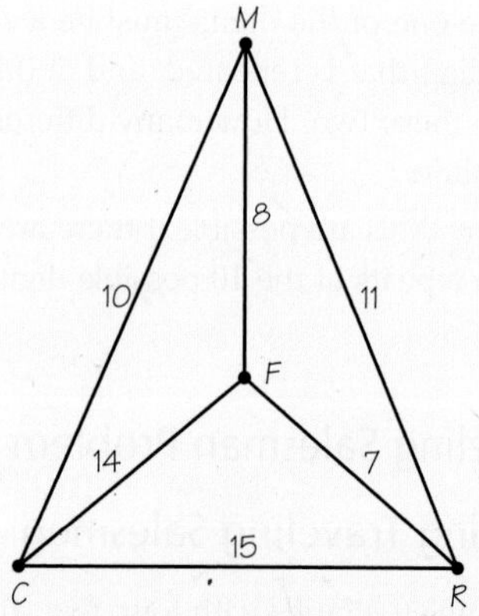

43. In Exercise 42, what route would Francine have to follow to get home as quickly as possible, assuming she promised to drive Constance home first?

44. In Exercise 42, Francine is planning to deliver her friends home and then spend the night at Rachel's house. What would her fastest route be?

45. A religious charity arranges free pickups for donated goods to encourage such donations but would like to keep its pickup costs low. The accompanying diagram shows the estimated amounts of time to get between the charity (*H*) and the locations of the three pickups scheduled for Wednesday.

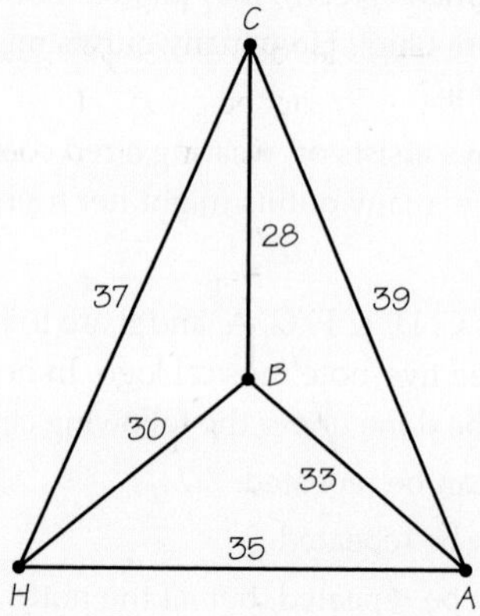

(a) What route is selected using the nearest-neighbor algorithm starting at *H*?

(b) What route is selected using the sorted-edges algorithm?

(c) Use "brute force" to determine if the solution in either (a) or (b) yields an optimal solution.

46. An organization that delivers meals to the needy elderly must deliver food to three clients and wants to keep down its costs. Typical times to get between its home site *H* and the clients are given in the accompanying graph.

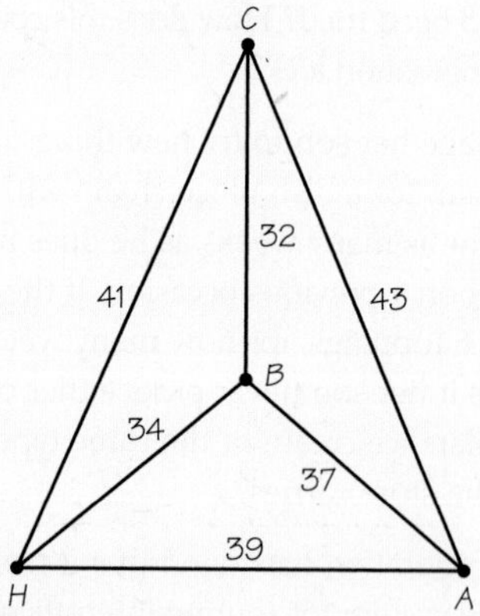

(a) What route is selected using the nearest-neighbor algorithm starting at *H*?

(b) What route is selected using the sorted-edges algorithm?

(c) Use "brute force" to determine whether the solution in either part (a) or (b) yields an optimal solution.

(d) Compare the times in the diagrams in this excercise and Exercise 45. How do they differ? Can you state a general result about solving TSPs based on what you notice?

47. Starting from the location where she moors her boat (*M*), a fisherwoman wishes to visit three areas—*A*, *B*, and *C*—where she has set fishing nets. If the times (in minutes) between the locales are given in the accompanying figure, what route to visit the three sites and return to the mooring place would be optimal?

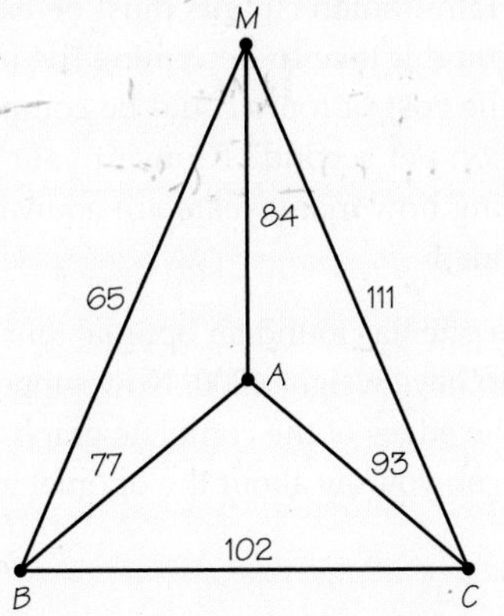

48. (a) For the two complete graphs that follow, find the costs of the nearest-neighbor tour starting at *B* and of the tour generated by the sorted-edges algorithm.

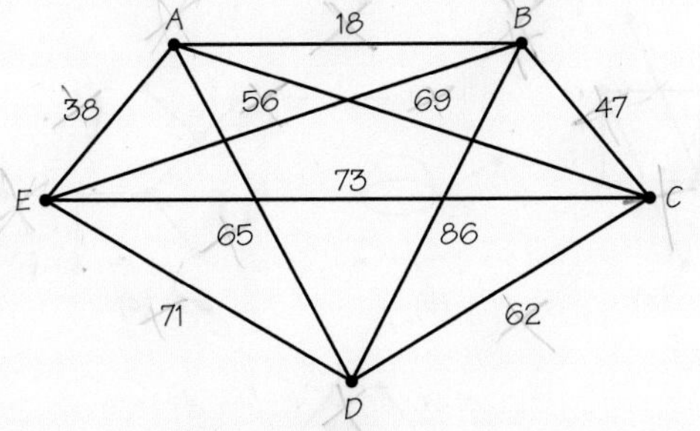

(a)

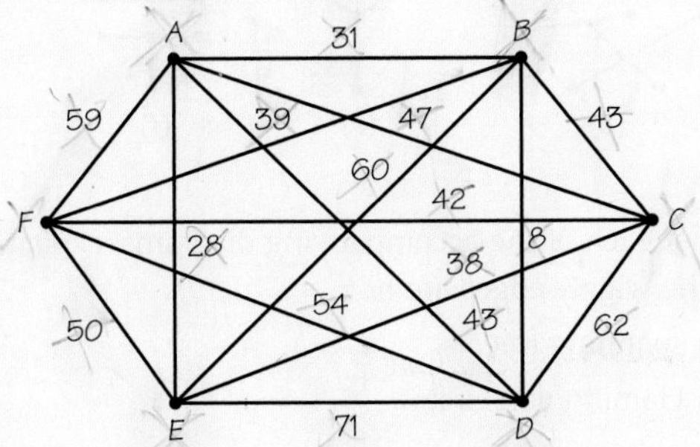

(b)

(b) How many Hamiltonian circuits would have to be examined to find a shortest route for part (a) by the brute force method?

(c) Invent an algorithm different from the sorted-edges and nearest-neighbor algorithms that is easy to apply for finding TSP solutions.

49. An airport limo must take its five passengers from the airport to different downtown hotels. Is this a traveling salesman problem, a Chinese postman problem, or an Euler circuit problem?

50. For each of the accompanying graphs with weights, apply the nearest-neighbor method (starting at vertex *A*) and the sorted-edges method to find (it is hoped) a cheap tour.

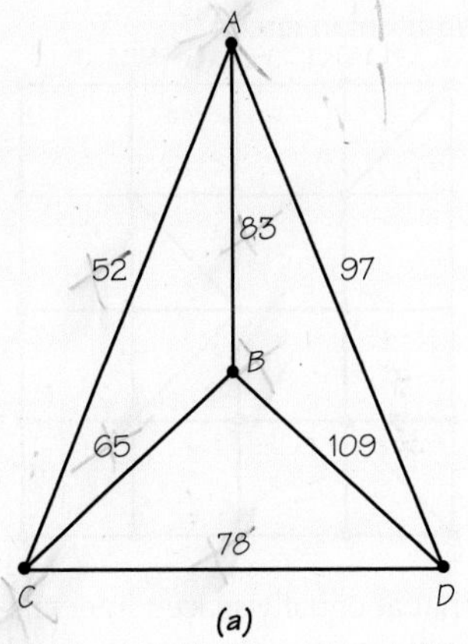

(a)

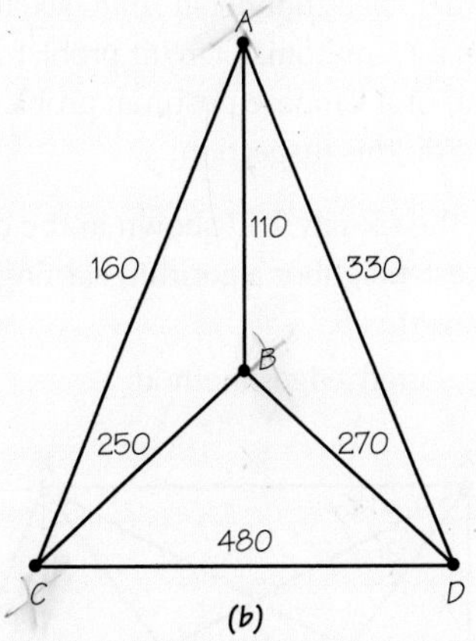

(b)

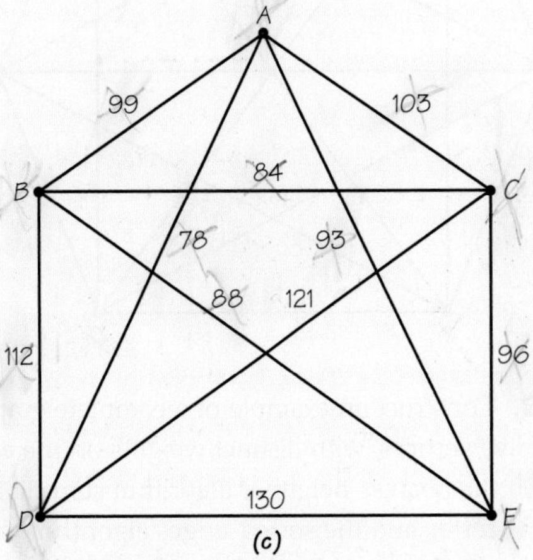

(c)

51. The accompanying figure represents a town where there is a sewer located at each corner (where two or more streets meet). After every thunderstorm, the department of public works wishes to have a truck start at its headquarters (at vertex *H*) and make an inspection of sewer drains to be sure that leaves are not clogging them. Can a route start and end at *H* that visits each corner exactly once? (Assume that all the streets are two-way streets.) Does this problem involve finding an Euler circuit or a Hamiltonian circuit?

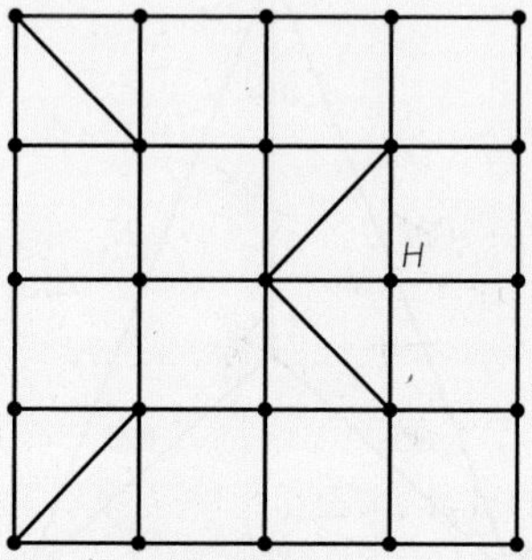

Assume that at equally spaced intervals along the blocks in this graph, there are storm sewers that must be inspected after each thunderstorm to see if they are clogged. Is this a Hamiltonian circuit problem, an Euler circuit problem, or a Chinese postman problem? Find an optimal tour to do this inspection.

52. (a) Solve the six-city TSP shown in the diagram using the nearest-neighbor algorithm starting at vertex *A* and starting at vertex *B*.

(b) Apply the sorted-edges method.

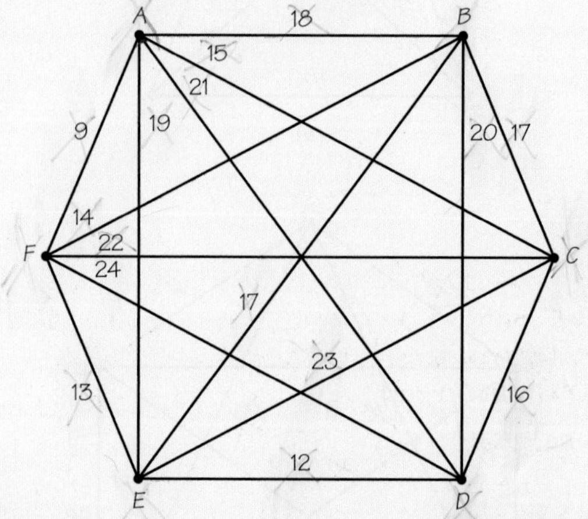

53. Construct an example of a complete graph of five vertices, with distinct weights on the edges for which the nearest-neighbor algorithm starting at a particular vertex and the sorted-edges algorithm yield different solutions for the traveling salesman problem.

Can you find a five-vertex complete graph with weights on the edges in which the optimal solution, the nearest-neighbor solution, and the sorted-edges algorithm solution are all different?

54. If the brute force method of solving a 20-city TSP is employed, use a calculator to determine how many Hamiltonian circuits must be examined. How long would it take to determine the minimum-cost tour if the cost of tours could be computed at the rate of 1 billion per second? (Convert your answer to years by seeing how many years are equivalent to a billion seconds!)

55. Suppose one has found an optimal tour for a given 10-city TSP to have weight 4200. Now suppose the weights on the edges of the complete graph are increased by 50. What can you say about the optimal tour and its weight?

2.4 Minimum-Cost Spanning Trees

56. For each graph below, explain why it is or is not a tree.

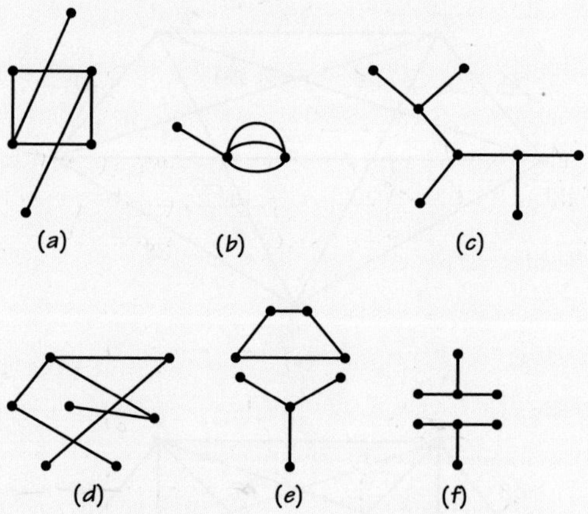

57. For each of the accompanying diagrams, explain why the wiggly edges are not

(a) a spanning tree.

(b) a Hamiltonian circuit.

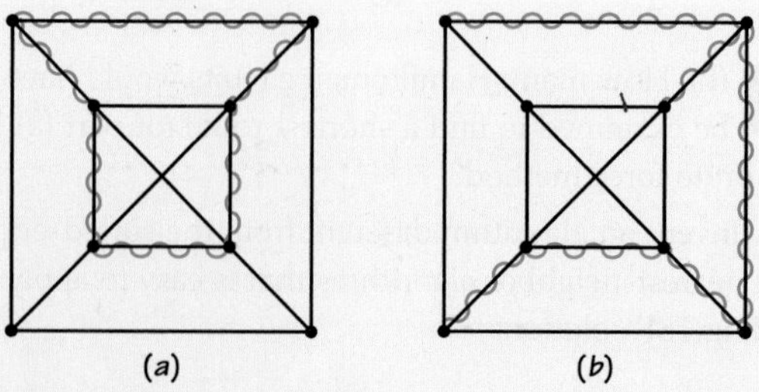

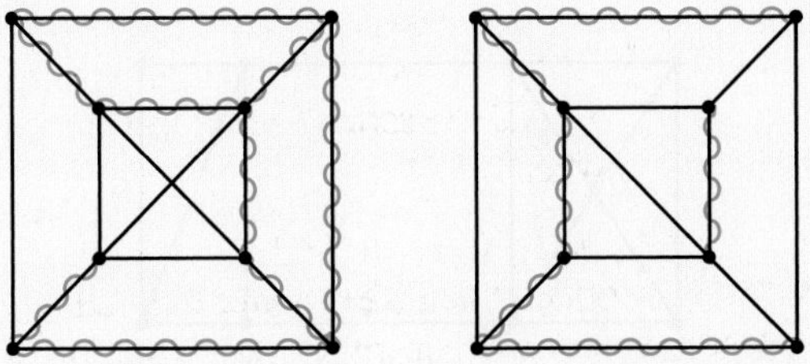
(c) (d)

58. Find all the spanning trees in the accompanying graphs.

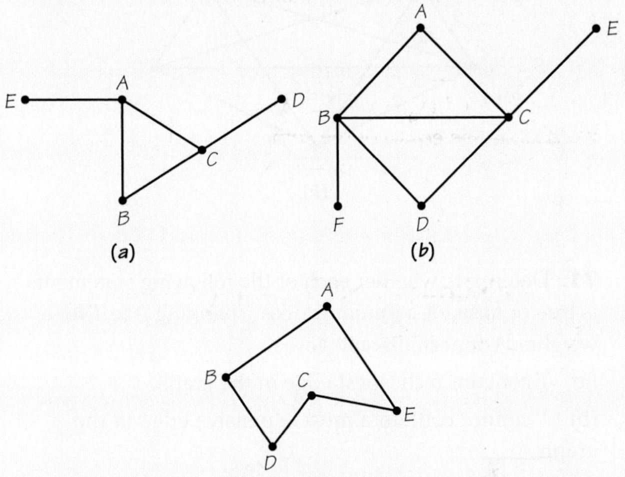

59. Use Kruskal's algorithm to find a minimum-cost spanning tree for the following graphs (a), (b), (c), and (d). In each case, what is the cost associated with the tree?

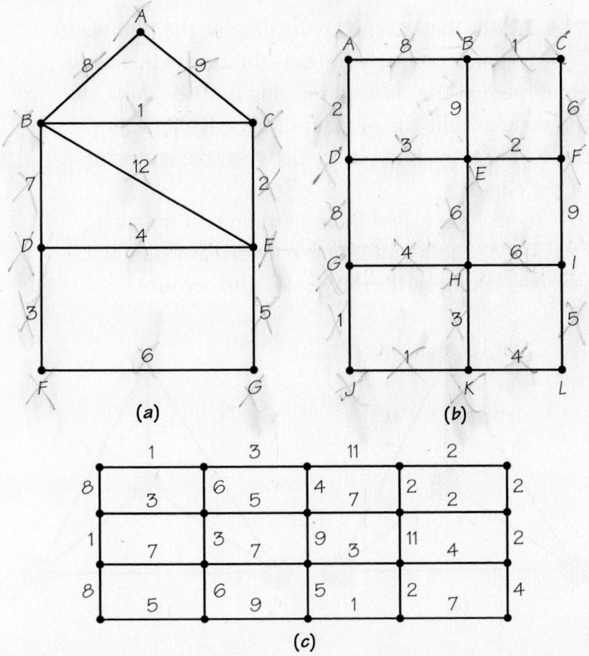

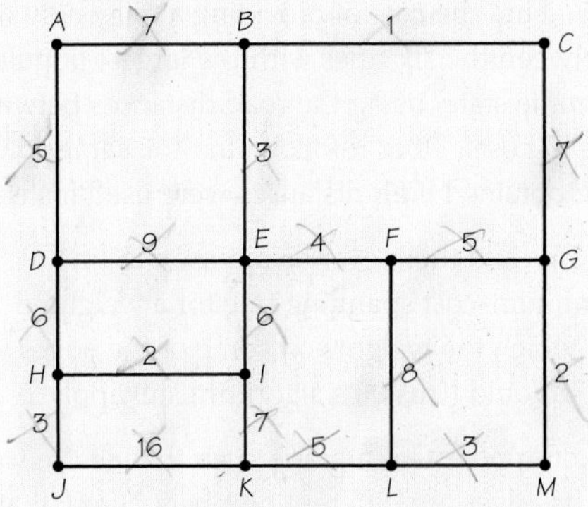

(d)

60. A connected graph G has 14 vertices. How many edges does a spanning tree of G have? How many vertices does a spanning tree of G have? What can one say about the number of edges G has?

61. A connected graph H has a spanning tree with 26 edges. How many vertices does the spanning tree have? How many vertices does H have? What can one say about the number of edges H has?

62. A large company wishes to install a pneumatic tube system that would enable small items to be sent between any of 10 locales, possibly by using relay. If the nonprohibitive costs (in \$100) are shown in the accompanying graph model, between which sites should the tube be installed to minimize the total cost?

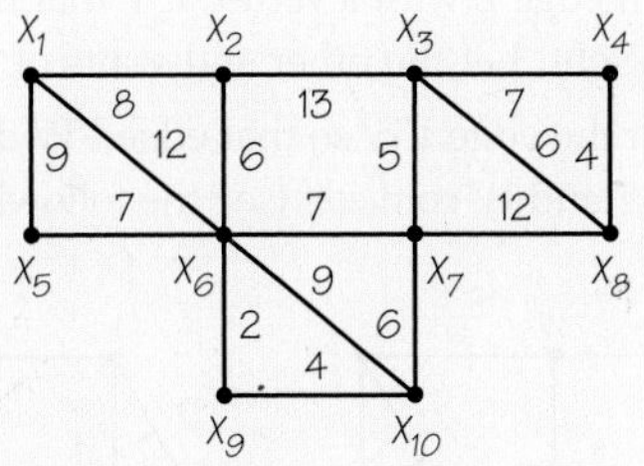

63. If the weight of each edge in Exercise 60 is increased by 3, will the tree that achieves minimum cost for the new collection of weights be the same as the one that achieves minimum cost for the original set of weights?

64. Give examples of real-world situations that can be modeled using a weighted graph and for which finding a minimum-cost spanning tree for the graph would be of interest.

65. Can Kruskal's algorithm be modified to find a maximum-weight spanning tree? Can you think of an application for finding a maximum-weight spanning tree?

66. Find the cost of providing a relay network between the six cities with the largest populations in your home state, using the road distances between the cities as costs. Does it follow that the same solution would be obtained if air distances were used instead?

67. Would there ever be a reason to find a minimum-cost spanning tree for a weighted graph in which the weights on some of the edges were negative? Would Kruskal's algorithm still apply?

68. Suppose G is a graph such that all the weights on its edges are different numbers. Show that there is a unique minimum-cost spanning tree.

69. Two spanning trees of a (weighted) graph are considered different if they use different edges. Show that the following graph has different minimum-cost spanning trees, though all these different trees have the same cost.

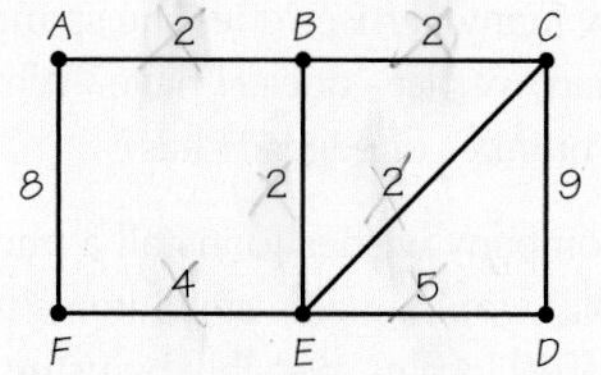

70. Let G be a graph with weights assigned to each edge. Consider the following algorithm:

(a) Pick any vertex V of G.

(b) Select an edge E with a vertex at V that has a minimum weight. Let the other endpoints of E be W.

(c) Contract the edge VW so that edge VW disappears and vertices V and W coincide (see the following figures).

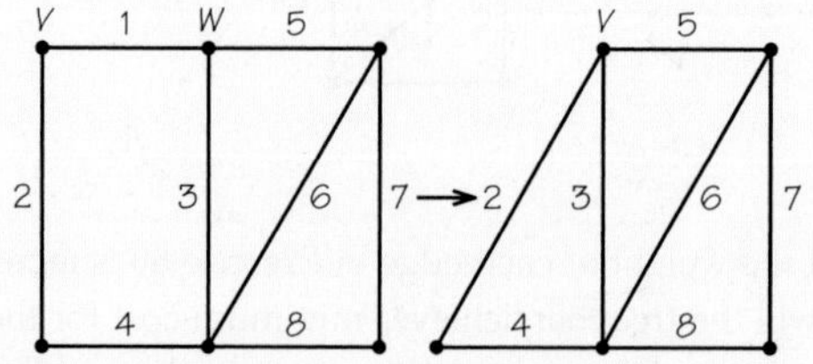

If in the new graph two or more edges join a pair of vertices, delete all but the cheapest. Continue to call the new vertex V.

(d) Repeat steps (b) and (c) until a single point is obtained. The edges selected in the course of this algorithm (called Prim's algorithm) form a minimum-cost spanning tree. Apply this algorithm to the accompanying graphs.

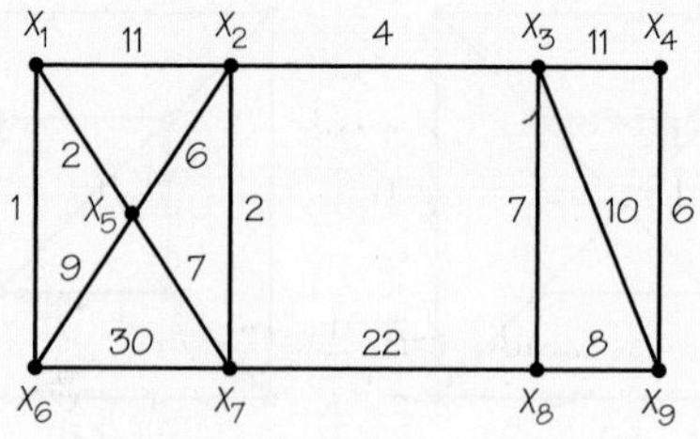

(a)

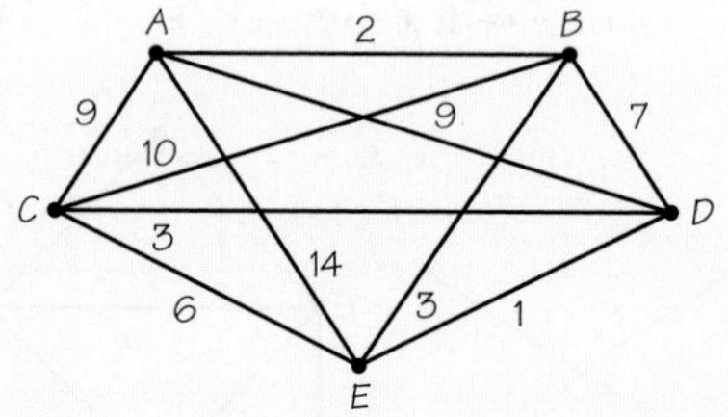

(b)

71. Determine whether each of the following statements is true or false for a minimum-cost spanning tree T for a weighted connected graph G:

(a) T contains a cheapest edge in the graph.

(b) T cannot contain a most expensive edge in the graph.

(c) T contains one fewer edge than there are vertices in G.

(d) There is some vertex in T to which all others are joined by edges.

(e) There is some vertex in T that has valence 3.

72. In the accompanying graphs, the number in the circle for each vertex is the cost of installing equipment at the vertex if relaying must be done at the vertex, while the number on an edge indicates the cost of providing service between the endpoints of the edge.

In each case, find the minimum cost (allowing relays) for sending messages between any pair of vertices, taking vertex relay costs into account.

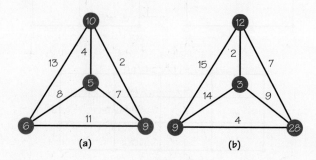

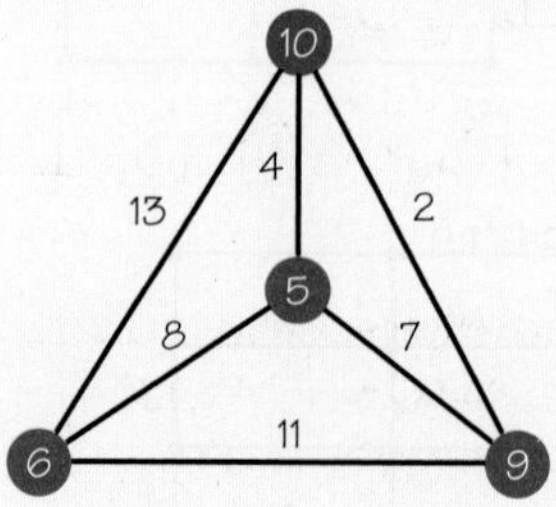

(a)

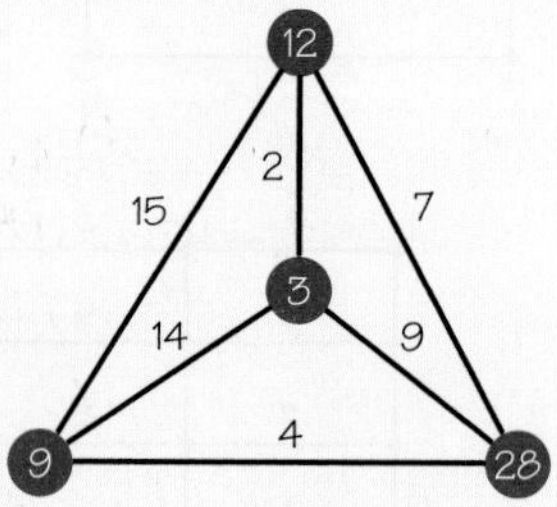

(b)

Would your answer be different if vertex relay costs were neglected? (*Warning:* Kruskal's algorithm cannot be used to answer the first question. This problem illustrates the value of having an algorithm over relying on "brute force.")

73. (a) Show that for each edge of graph J, which follows, there is a spanning tree of J that avoids that edge.

(b) For each spanning tree that you found in graph J, count the number of vertices and edges. Do you notice any pattern?

(c) For graph H, which follows, and each edge in the graph, is there a spanning tree that does not include that edge of H?

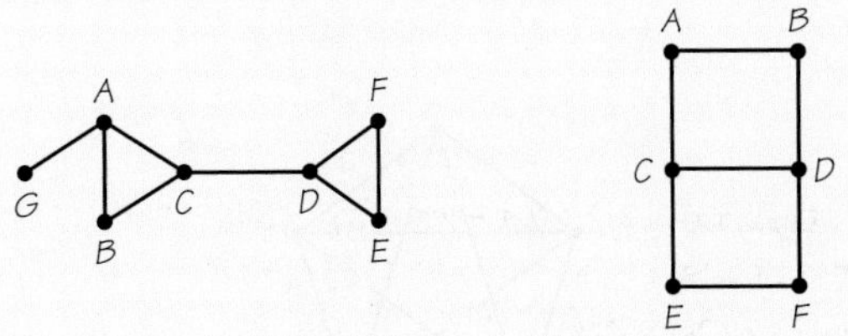

Graph H   Graph J

74. (a) The table shown gives the "closeness" or distance values between four objects. Construct a four-vertex tree with weights on its edges such that the distances between pairs of vertices of the tree (as measured by the sum of the weights on the path in the tree between these vertices) give rise to this table.

	A	B	C	D
A	0	3	10	14
B	3	0	7	11
C	10	7	0	4
D	14	11	4	0

(b) Produce several real-world contexts that might give rise to the situation described here.

75. The accompanying figure represents four objects using a tree with weights on the edges. Construct a table with four rows and four columns recording how "close" pairs of vertices in the tree are to each other. To find how close a pair of objects is, add together the weights along the path that joins these two objects.

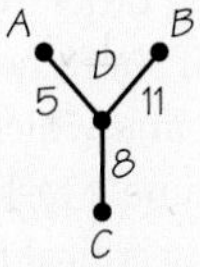

2.5 Critical-Path Analysis

76. Find the earliest completion time and critical paths for the accompanying order-requirement digraphs.

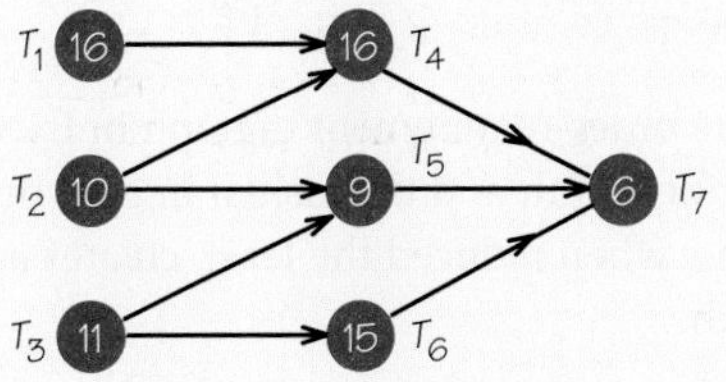

(*a*)

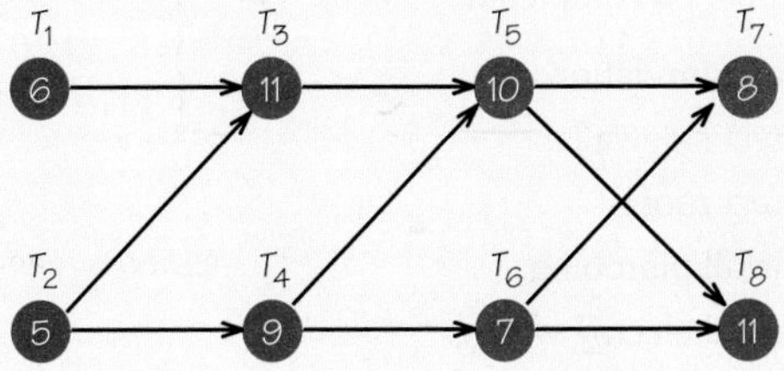

(*b*)

77. Find the earliest completion time and critical paths for the following order-requirement digraphs.

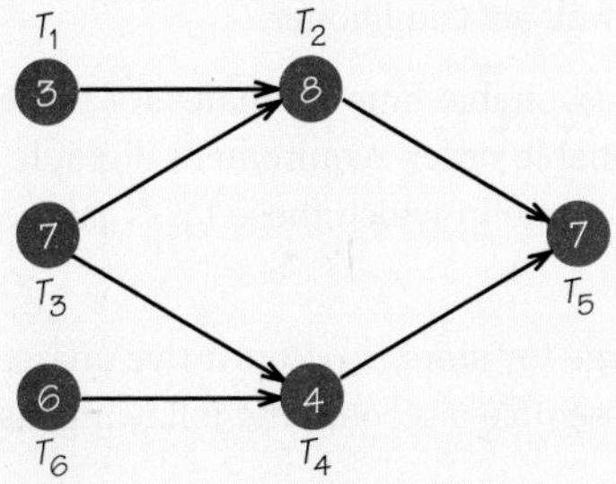

(*a*)

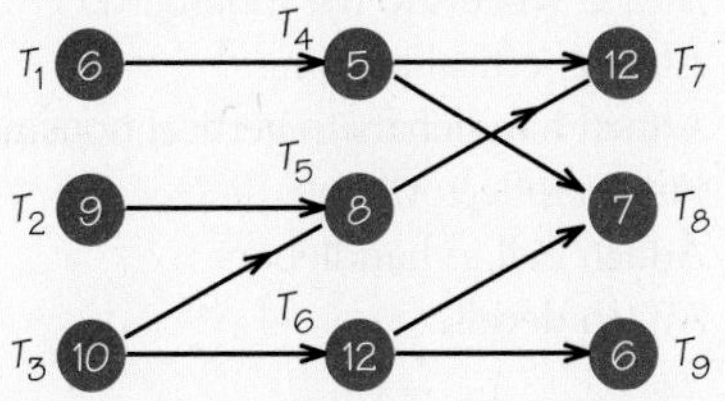

(*b*)

78. Construct an example of an order-requirement digraph with exactly three different critical paths.

79. In the accompanying order-requirement digraph, determine which tasks, if shortened, would reduce the earliest completion time and which would not. Then find the earliest completion time if task T_5 is reduced to time length 7. What is the new critical path?

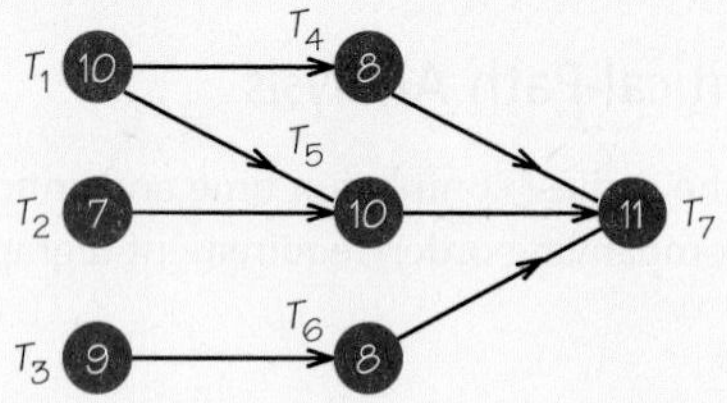

80. For the order-requirement digraph in Exercise 79, find the critical path and the task(s) in the critical path whose time, when reduced the least, creates a new critical path.

81. To build a new addition on a house, the following tasks must be completed:

(a) Lay foundation.
(b) Erect sidewalls.
(c) Erect roof.
(d) Install plumbing.
(e) Install electric wiring.
(f) Lay tile flooring.
(g) Obtain building permits.
(h) Put in door connecting new room to existing house.
(i) Install track lighting on ceiling.
(j) Install wall air-conditioner.

Construct reasonable time estimates for these tasks and a reasonable order-requirement digraph. What is the fastest time in which these tasks can be completed?

82. At a large toy store, scooters arrive unassembled in boxes. To assemble a scooter, the following tasks must be performed:

Task 1. Remove parts from the box.
Task 2. Attach wheels to the footboard.
Task 3. Attach vertical housing.
Task 4. Attach handlebars to vertical housing.
Task 5. Put on reflector tape.
Task 6. Attach bell to handlebars.
Task 7. Attach decals.
Task 8. Attach kickstand.
Task 9. Attach safety instructions to handlebars.

Give reasonable time estimates for these tasks and construct a reasonable order-requirement digraph. What is the earliest time by which these tasks can be completed?

83. Construct an order-requirement digraph with six tasks that has three critical paths of length 26.

Chapter Review

84. Use the nearest-neighbor method starting at D and the sorted-edges method to find a hopefully short way to deliver packages from the depot at D to the other four sites in the accompanying complete graph.

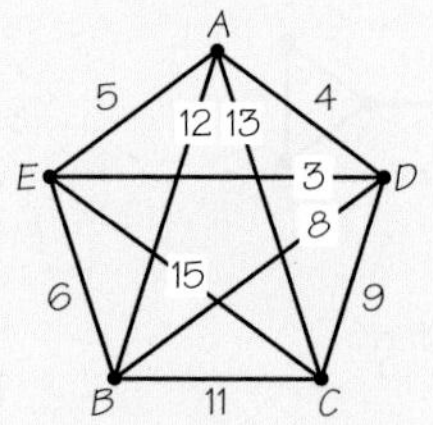

85. Mary must choose a new password where the first and last choices are possibly repeated lowercase letters; the second and third positions must be distinct capital letters; the fourth position must be a #, $, or & symbol; and the next three positions are distinct nonzero digits. In how many ways can she choose a password?

86. Does the graph G in Chapter 1, Exercise 71 (page 36), have a Hamiltonian circuit? If so, write one down that starts at H.

87. Use Kruskal's algorithm to make cable television available for the sites in the accompanying graph, assuming that video can be relayed. The costs of putting in connections are shown.

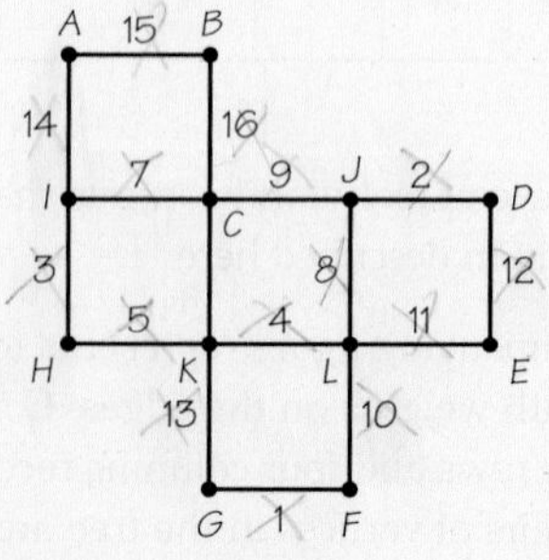

Applet Exercises

To do these exercises, go to www.macmillanhighered.com/fapp10e.

1. There is an extended version of the nearest-neighbor algorithm in which you compare the total distances of the Hamiltonian circuits produced by applying the ordinary nearest-neighbor algorithm starting at each of the vertices of the graph (rather than just a specific one). Explore the effectiveness of this algorithm using the *TSP: Nearest-Neighbor* applet.

2. Go to the *TSP: Sorted Edges* applet, where you can apply the sorted-edges algorithm to see if it solves the traveling salesman problem for the following graphs (and others):

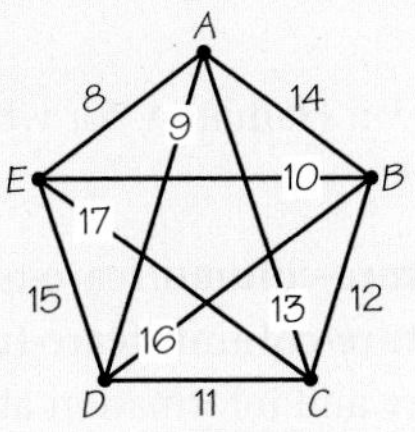

(*a*)

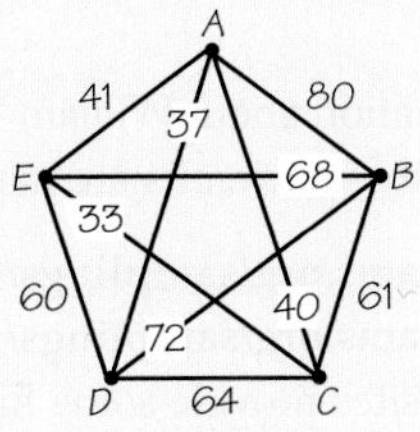

(*b*)

3. Go to the *Kruskal's Algorithm* applet, where you can apply Kruskal's algorithm to find the minimum-cost spanning trees in the following graphs (and others):

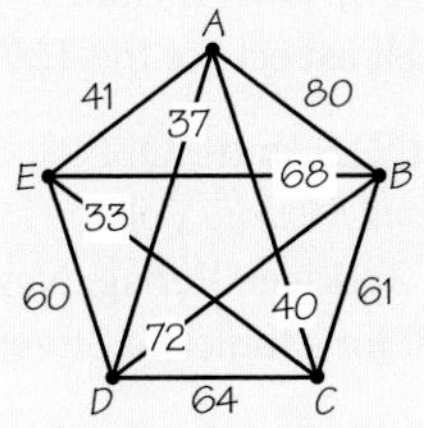

(*a*)

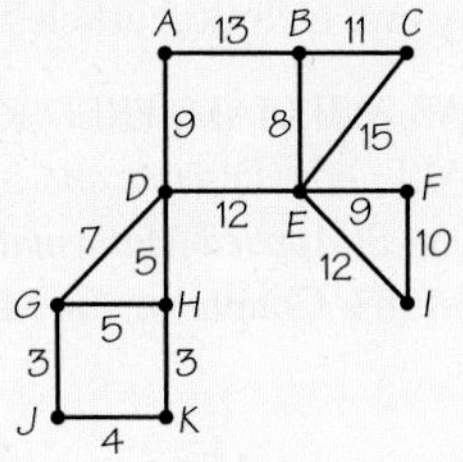

(*b*)

Writing Projects

1. Write an essay about a variety of situations in which you are personally involved for which a solution of the TSP is (perhaps implicitly) required. Explain under what circumstances it might be valuable to carry out a formal mathematical solution to such TSPs rather than use an ad hoc solution.

2. Construct an example that shows that in a situation where three day campers must be picked up and brought to camp, it may make a difference if the optimization criterion is minimizing distance traveled by the camp bus versus minimizing average time that the children spend on the bus.

3. Determine the six largest cities in the state in which you live. By consulting a road atlas (or by some other means), construct the graph that represents the road distances between your hometown and these six other cities. Now apply (a) the nearest-neighbor method, (b) the sorted-edges method, and (c) the nearest neighbor from each city, and pick the minimum tour method to solve the associated TSP. Do you have reason to believe that the answers you get might include an optimum solution among them?

Suggested Readings

BODIN, LAWRENCE. Twenty years of routing and scheduling, *Operations Research,* 38 (1990): 571–579. A survey of real-world situations where routing and scheduling were used, written by a pioneer in this area.

DOLAN, ALAN, and JOAN ALDUS. *Networks and Algorithms: An Introductory Approach,* Wiley, Chichester, UK, 1993. An excellent introduction to graph theory algorithms.

GUSFIELD, DAN. *Algorithms on Strings, Trees, and Sequences,* Cambridge University Press, New York, 1997. Details applications of graph theory in pattern recognition and reconstruction problems.

JONES, NEIL C., and PAVEL A. PEVZNER. *An Introduction to Bioinformatics Algorithms,* MIT Press, Cambridge, MA, 2004. This book contains material on how graph theory ideas, particularly those related to Hamiltonian circuits, are being used in molecular genetics and computational biology.

LAWLER, EUGENE, J. LENSTRA, RINNOY KAN, and D. SHMOYS, eds. *The Traveling Salesman Problem,* Prentice-Hall, Englewood Cliffs, NJ, 1985. Includes survey and technical articles on all aspects of the TSP.

LUCAS, WILLIAM, FRED ROBERTS, and ROBERT THRALL, eds. *Discrete and Systems Models,* Vol. 3: *Modules in Applied Mathematics,* Springer-Verlag, New York, 1983. Chapter 6, "A Model for Municipal Street Sweeping Operations," by A. Tucker and L. Bodin, describes street-sweeping and related models in detail. Other chapters detail many recent applications of mathematics.

ROBERTS, FRED S., and BARRY TESMAN. *Applied Combinatorics,* 2nd ed., Pearson Prentice Hall, Upper Saddle River, NJ, 2004. The material on network-optimization problems is excellent.

ROBERTS, FRED. *Graph Theory and Its Applications to Problems of Society,* Society for Industrial and Applied Mathematics, Philadelphia, 1978. A very readable account of how graph theory is finding a wide variety of applications.

Suggested Websites

www.math.uwaterloo.ca/tsp This site provides a detailed history and many applications of the TSP.

en.wikipedia.org/wiki/Minimum_cost_spanning_tree This web page provides basic ideas about minimum-cost spanning trees, their applications, and extensions of this idea.

www-gap.dcs.st-and.ac.uk/~history/Biographies/Hamilton.html This site provides biographical information about William Rowan Hamilton, for whom Hamiltonian circuits are named.

www.ams.org/samplings/feature-column/fcarc-tsp
www.ams.org/samplings/feature-column/fcarc-trees
These sites provide some history and information about applications of the Traveling Salesman Problem and of minimum-cost spanning trees.

Planning and Scheduling

3

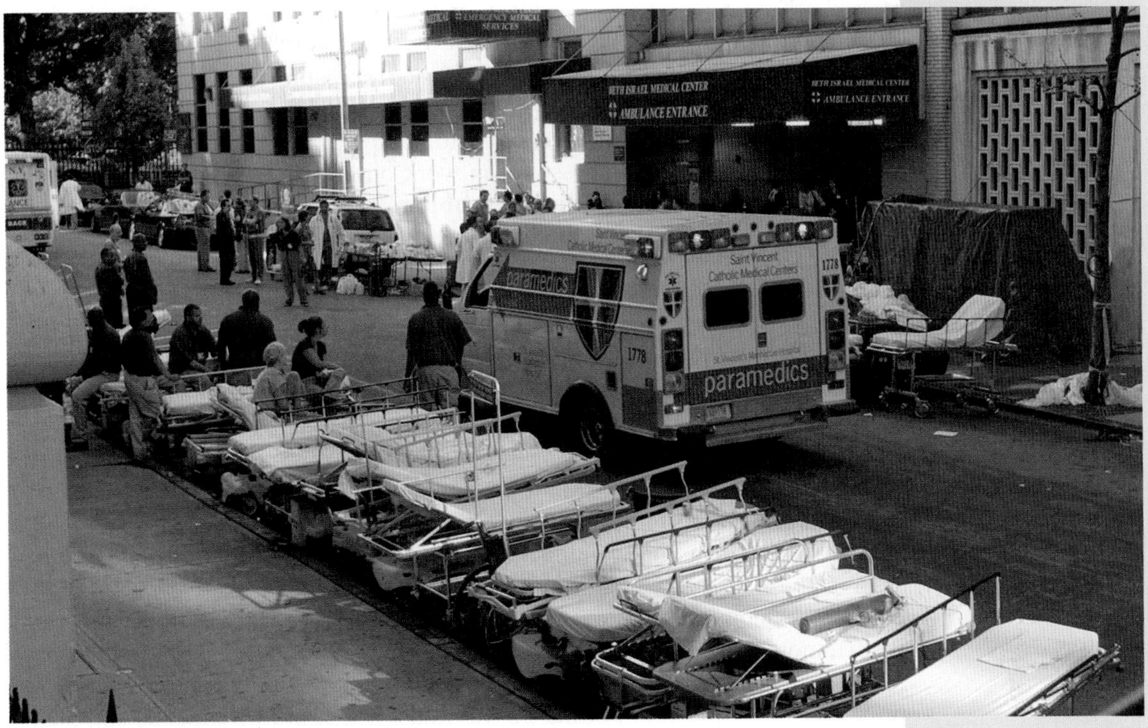

AP Photo/Amy Sancetta

3.1 **Scheduling Tasks**

3.2 **Critical-Path Schedules**

3.3 **Independent Tasks**

3.4 **Bin Packing**

3.5 **Resolving Conflict via Coloring**

In a society as complex as ours, everyday problems such as providing services efficiently and on time require accurate planning of both people and machines. Take the example of a medical center in a major city. Around-the-clock scheduling of nurses, doctors, and emergency room staff must be provided to guarantee that people with particular expertise are available during each shift. The operating rooms must be scheduled in a manner flexible enough to deal with emergencies and scheduled procedures. Equipment used for X-ray, CT, or MRI scans must be in good working order and scheduled for maximum efficiency.

Although many scheduling problems are often solved on an ad hoc basis, we can also use mathematical ideas to gain insight into the complications that arise in scheduling. The ideas we develop in this chapter have practical value in a relatively narrow range of applications, but they shed light on many characteristics of more realistic, and hence more complex, scheduling problems.

3.1 Scheduling Tasks

Assume that a certain number of identical **processors** (machines, humans, or robots) work on a series of tasks that make up a job. Associated with each task is a specified amount of time required to complete the task. For simplicity, we assume that any of the processors can work on any of the tasks. Our problem, known as the **machine-scheduling problem,** is to decide how the tasks should be scheduled so that the completion time for the tasks collectively is as early as possible.

Even with these simplifying assumptions, complications in scheduling arise. Some tasks may be more important than others and perhaps should be scheduled first. When "ties" occur, they must be resolved by special rules. As an example, suppose we are scheduling patients to be seen in a hospital emergency room staffed by one doctor. If two heavily bleeding patients arrive simultaneously, one with a bleeding thigh, the other with a bleeding arm, which patient should be treated first? Suppose the doctor treats the arm patient first, and while treatment is going on, a person in cardiac arrest or having a stroke arrives. Scheduling rules must establish appropriate priorities for cases such as these.

Another common complication arises with jobs consisting of several tasks that cannot be done in an arbitrary order. For example, if the job of putting up a new house is treated as a scheduling problem, the task of laying the foundation must precede the task of putting up the walls, which in turn must be completed before work on the roof can begin. The electrical system can be scheduled for installation later.

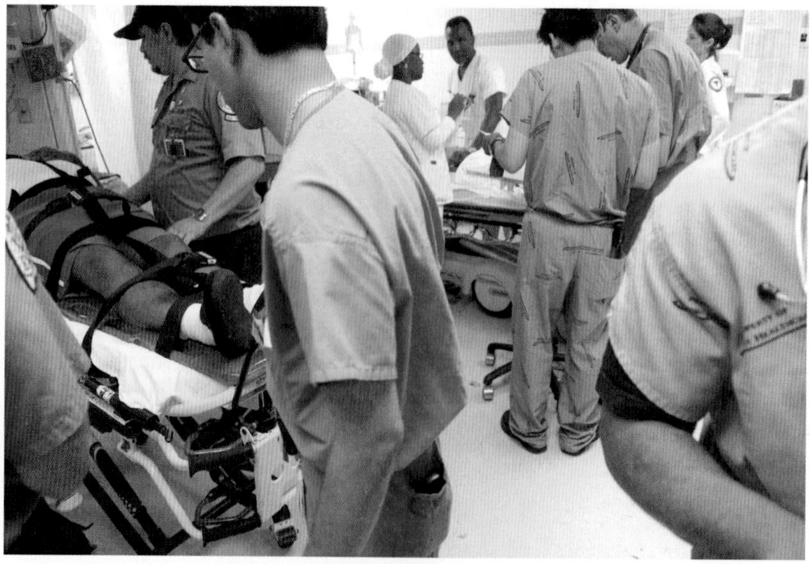

Erik S. Lesser/ZUMA Press

Hospitals are increasingly making use of mathematical techniques applied to scheduling problems. Making efficient use of one or more emergency rooms (operating rooms) requires the complicated assembly of a team of doctors, nurses, equipment, and support staff. Mathematical techniques for scheduling have made it possible to treat more patients in less time.

Assumptions and Goals

To simplify our analysis, we need to make clear and explicit assumptions:

1. If a processor starts work on a task, the work on that task will continue without interruption until the task is completed.
2. No processor stays voluntarily idle. In other words, if there is a processor free and a task available to be worked on, then that processor immediately begins work on that task.
3. The requirements for ordering the tasks are given by an order-requirement digraph. (A typical example is shown in Figure 3.1, with task times highlighted within each vertex. The ordering of the tasks imposed by the order-requirement digraph often represents constraints of physical reality. For example, you cannot fly a plane until it has taken fuel on board.)
4. The tasks are arranged in a **priority list** that is independent of the order requirements. (The priority list is a ranking of the tasks according to some criterion of "importance.")

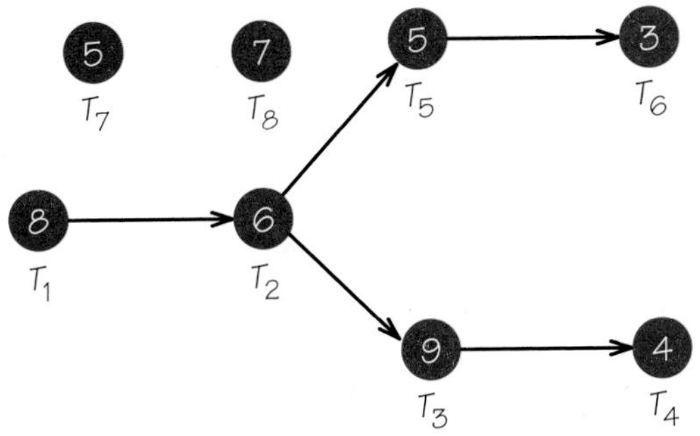

Figure 3.1 A typical order-requirement digraph. Tasks with no edges entering or leaving the vertices representing them (T_7, T_8) can be more flexibly scheduled than the other tasks.

EXAMPLE 1 Home Construction

Let's see how these assumptions might work for an example involving a home construction project. In this case, the processors are human workers with identical skills. Assumption 1 means that once a worker begins a task, the work on this task is finished without interruption. Assumption 2 means that no worker stays idle if there is some task for which the predecessors are finished. Assumption 3 requires that the ordering of the tasks be summarized in an order-requirement digraph. This digraph would code facts such as that the site must be cleared before the task of laying the foundation is begun. Assumption 4 requires that the tasks be ranked in a list from some perspective, perhaps a subjective view.

The task with highest priority rank is listed first in the list, followed left to right by the other tasks in priority rank. The priority list might be based on the size of the payments made to the construction company when a task is completed, even though these payments have no relation to the way the tasks must be done, as indicated in the order-requirement digraph. Alternatively, the priority list might reflect an attempt to find an algorithm to schedule the tasks needed to complete the whole job more quickly.

When considering a scheduling problem, there are various goals we might want to achieve. Among them are the following:

Goal 1. Minimizing the completion time of the tasks that make up the job.

Goal 2. Minimizing the total time that processors are idle.

Goal 3. Finding the minimum number of processors necessary to finish the job by a specified time.

In the context of the construction example, Goal 1 would complete the home as quickly as possible. Goal 2 would ensure that workers, who are perhaps paid by the hour, were not paid for doing nothing. One way of doing so would be to hire one fewer worker even if it means the house takes longer to finish. Goal 3 might be reasonable if the family wants the house done before the first day of school, even if they have to pay a lot more workers to get the house done by this time.

For now we will concentrate on Goal 1, finishing all the tasks that make up the job at the earliest possible time. Note, however, that optimizing for one goal may not optimize for another. Our discussion here goes beyond what was discussed in Chapter 2 (see Section 2.4) by dealing with how to assign tasks in a job to the processors that do the work. To build a new outpatient clinic involves designing a schedule for who will do what work when.

List-Processing Algorithm

The scheduling problem we have described sounds more complicated than the traveling salesman problem (TSP). Indeed, like the TSP, it is known to be NP-complete. This means that it is unlikely anyone will ever find a computationally fast algorithm that can find an optimal solution for scheduling problems involving a very large number of tasks. Thus, we will be content to seek a solution method that is computationally fast and gives only approximately optimal answers.

List-Processing Algorithm: Part I and Ready Task PROCEDURE

The algorithm we use to schedule tasks is the **list-processing algorithm.** In describing it, we will call a task **ready** at a particular time if all its predecessors as indicated in the order-requirement digraph have been completed at that time. In Figure 3.1 at time 0, the ready tasks are T_1, T_7, and T_8, while T_2 cannot be ready until 8 time units after T_1 is started. The algorithm works as follows: At a given time, assign to the lowest-numbered free processor the first task on the priority list that is *ready* at that time and that hasn't already been assigned to a processor.

In applying this algorithm, we will need to develop skill at coordinating the use of the information in the order-requirement digraph and the priority list. It will be helpful to cross out the tasks in the priority list as they are assigned to a processor to keep track of which tasks remain to be scheduled.

EXAMPLE 2 Applying the List-Processing Algorithm

Let's apply the list-processing algorithm to one possible priority list—$T_8, T_7, T_6, \ldots, T_1$—using two processors and the order-requirement digraph in Figure 3.1. The result is the schedule shown in Figure 3.2, where idle processor time (time during which a processor

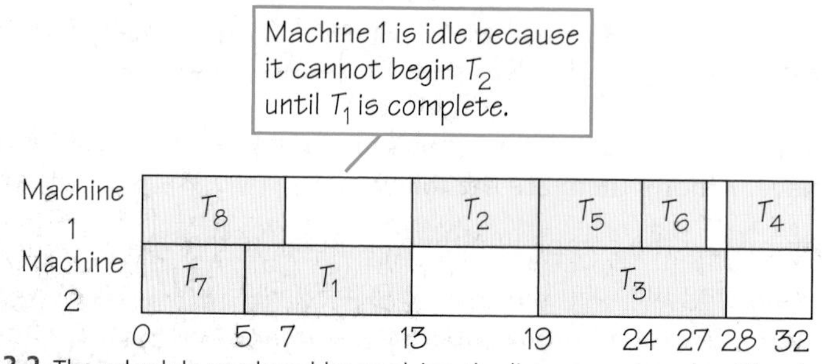

Figure 3.2 The schedule produced by applying the list-processing algorithm to the order-requirement digraph in Figure 3.1 using the list $T_8, T_7, \ldots, T_1$.

is not at work on a task) is indicated by white. How does the list-processing algorithm generate this schedule?

T_8 (task 8) is first on the priority list and ready at time 0 since it has no predecessors. It is assigned to the lowest-numbered free processor, processor 1. Task 7, next on the priority list, is also ready at time 0 and thus is assigned to processor 2. The first processor to become free is processor 2 at time 5. Recall that by assumption 1, once a processor starts work on a task, its work cannot be interrupted until the task is complete. Task 6, the next unassigned task on the list, is not ready at time 5, as can be seen by consulting Figure 3.1. The reason task 6 is not ready at time 5 is that task 5 has not been completed by time 5. In fact, at time 5, the only ready task on the list is T_1, so that task is assigned to processor 2. At time 7, processor 1 becomes free, but no task becomes ready until time 13.

Thus, processor 1 stays idle from time 7 to time 13. At this time, because T_2 is the first ready task on the list not already scheduled, it is assigned to processor 1. Processor 2, however, stays idle because no other ready task is available at this time. The remainder of the scheduling shown in Figure 3.2 is completed in this manner.

We can summarize this procedure as follows:

List-Processing Algorithm: Part II PROCEDURE

As the priority list is scanned from left to right to assign a task to a processor at a particular time, we pass over tasks that are not ready to find tasks that are ready. If no task can be assigned in this manner, we keep one or more processors idle until such time that, reading the priority list from the left, there is a ready task not already assigned. After a task is assigned to a processor, we resume scanning the priority list for unassigned tasks, starting over at the far left.

When Is a Schedule Optimal?

The schedule in Figure 3.2 has a lot of idle time, so it may not be optimal. Indeed, if we apply the list-processing algorithm for two processors to another possible priority list $T_1, \ldots, T_8$, using the digraph in Figure 3.1, the resulting schedule is that shown in Figure 3.3.

Here are the details of how we arrived at this schedule. Remember that we must coordinate the list $T_1, T_2, \ldots, T_8$ with the information in the order-requirement digraph shown in Figure 3.3a. At time 0, task T_1 is ready, so this task is assigned to processor 1. However, at time 0, tasks $T_2, T_3, \ldots, T_6$ are not ready because their predecessors are not done. For example, T_2 is not ready at time 0 because T_1, which precedes it, is not done at time 0. The first ready task on the list, reading from left to right, that is not already assigned is T_7, so task T_7 gets assigned to processor 2. Both processors are now busy until time 5, at which point processor 2 becomes available to work on another task (Figure 3.3b).

Tasks T_1 and T_7 have been assigned. Reading from left to right along the list, the first task not already assigned whose predecessors are done by time 5 is T_8, so this task is started at time 5 on processor 2; processor 2 will continue to work on this task until time 12, because the task time for this task is 7 time units. At time 8, processor 1 becomes free, and reading the list from left to right, we find that T_2 is ready (because T_1 has just been completed). Thus, T_2 is assigned to processor 1, which will stay busy on this task until time 14. At time 12, processor 2 becomes free, but the tasks that have not already been assigned from the list

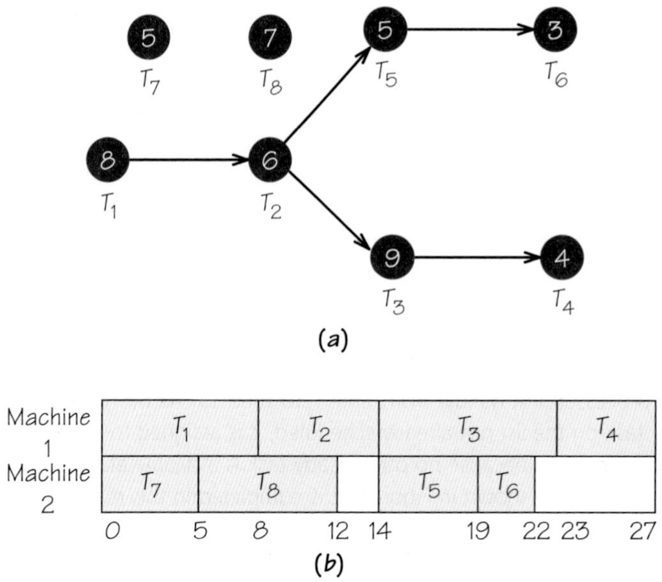

Figure 3.3 (a) A typical order-requirement digraph (repeat of Figure 3.1). (b) The schedule produced by applying the list-processing algorithm to the order-requirement digraph in Figure 3.3a using the list $T_1, T_2, \ldots, T_8$.

(T_3, T_4, T_5, T_6) are not ready, because they depend on T_2 being completed before these tasks can start. Thus, processor 2 must stay idle until time 14. At this time, T_3 and T_5 become ready. Since both processors 1 and 2 are idle at time 14, the lower numbered of the two, processor 1, gets to start on T_3 because it is the first ready task remaining to be assigned on the list scanned from left to right. Task T_5 gets assigned to processor 2 at time 14. The remaining tasks are assigned in a similar manner.

The schedule shown in Figure 3.3b is optimal because the path T_1, T_2, T_3, T_4, with length 27, is the critical path in the order-requirement digraph. As we saw in Chapter 2, the earliest completion time for the job made up of all the tasks is the length of the longest path in the order-requirement digraph. Different lists can give rise to the same or different optimal scheduling diagrams. The optimal schedule may have a completion time equal to or longer than the length of the critical path, but it cannot have a completion time shorter than the length of the critical path.

There is another way of relating optimal completion time to the completion time that is yielded by the list-processing algorithm. Suppose we add all the task times given in the order-requirement digraph and divide by the number of processors. The completion time using the list-processing algorithm must be at least as large as this number. For example, the task times for the order-requirement digraph in Figure 3.3a sum to 47. Thus, if these tasks are scheduled on two processors, the completion time is at least $\frac{47}{2} = 23.5$ (in fact, 24, because the list-processing algorithm applied to integer task times must yield an integer solution), whereas for four processors the completion time is at least $\frac{47}{4}$ (in fact, at least 12).

Why is it helpful to take the total time to do all the tasks in a job and divide this number by the number of processors? Think of each task that must be scheduled as a rectangle that is 1 unit high and t units wide, where t is the time allotted for the task. Think of the scheduling diagram with m processors as a rectangle that is m units high and whose width, W, is the completion time for the tasks. The scheduling diagram is to be filled up (packed) by the rectangles that represent the tasks. How small can W be? The area of the rectangle that represents the scheduling diagram must be at least as large as the sum of all the rectangles representing tasks that fit into it. The area of the scheduling diagram rectangle is mW. The combined areas of all the tasks, plus the area of rectangles corresponding to idle time, equal mW. Width W is smallest when the idle time is zero. Thus, W must be at least as big as the sum of all the task times divided by m.

Sometimes the estimate for completion time given by the list-processing algorithm from the length of the critical path gives a more useful value than the approach based on adding task times. Sometimes the opposite is true. For the order-requirement digraph in Figure 3.1, except for a schedule involving one processor, the critical-path estimate is superior. For some scheduling problems, both these estimates may be poor.

The number of priority lists that can be constructed if there are n tasks is $n!$ and can be computed using the fundamental principle of counting. For example, for eight tasks, $T_1, \ldots, T_8$, there are $8 \times 7 \times 6 \ldots \times 1 = 40{,}320$ possible priority lists. For different choices of the priority list, the list-processing algorithm may schedule the tasks, subject to the constraints of the order-requirement digraph, in different ways. More specifically, two different lists may yield different completion times or the same completion time, but the order in which the tasks are carried out will be different. It is also possible that two different lists produce identical ordering of the assignments of the tasks to processors and completion times. Soon we will see a method that can be used to select a list that, if we are lucky, will give a schedule with a relatively good completion time. In fact, no method is known, except for very specialized cases, of how to choose a list that can be guaranteed to produce an optimal schedule when the list algorithm is applied to it. In a nutshell, designing good schedules is difficult.

Self Check 1

Apply the list-processing algorithms to the list with the tasks in numerical order for the order-requirement digraph in Figure 3.3 with three processors. ■

Strange Happenings

The list-processing algorithm involves four factors that affect the final schedule. The answer we get depends on the following:

1. The times to carry out the tasks
2. Number of processors
3. Order-requirement digraph
4. Ordering of the tasks on the priority list

To see the interplay of these four factors, consider another scheduling problem, this time asociated with the order-requirement digraph shown in Figure 3.4. (The highlighted numbers are task time lengths.) The schedule generated by the list-processing algorithm applied to the list $T_1, T_2, \ldots, T_9$, using three processors, is given in Figure 3.5.

Treating the list $T_1, \ldots, T_9$ as fixed, how might we make the completion time earlier? Our alternatives are to pursue one or more of these strategies:

1. Reduce task times.
2. Use more processors.
3. "Loosen" the constraints by having fewer directed edges in the order-requirement digraph.

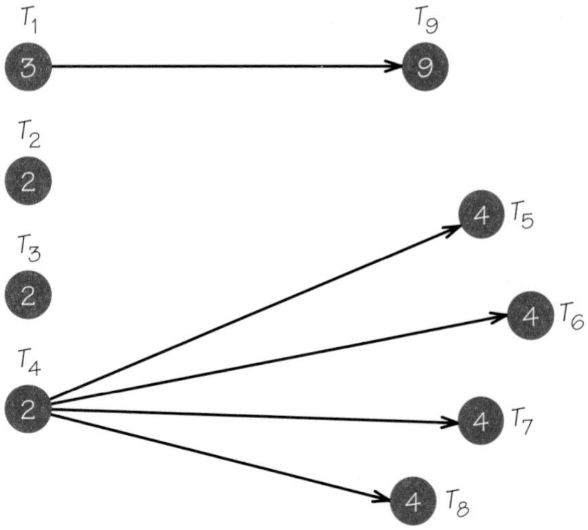

Figure 3.4 An order-requirement digraph designed to help illustrate some paradoxical behavior produced by the list-processing algorithm.

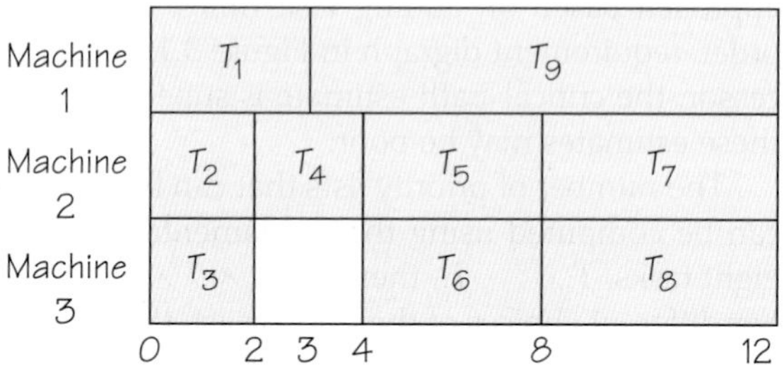

Figure 3.5 The schedule produced by applying the list-processing algorithm to the order-requirement digraph in Figure 3.4 using the list $T_1, T_2, \ldots, T_9$ with three processors.

Let's consider each alternative in turn, changing one feature of the original problem at a time, and see what happens to the resulting schedule. If we use strategy 1 and reduce the time of each task by one unit, we would expect the completion time to go down. Figure 3.6 shows the new order-requirement digraph, and Figure 3.7 shows the schedule produced for this problem, using the list-processing algorithm with three processors applied to the list $T_1, \ldots, T_9$.

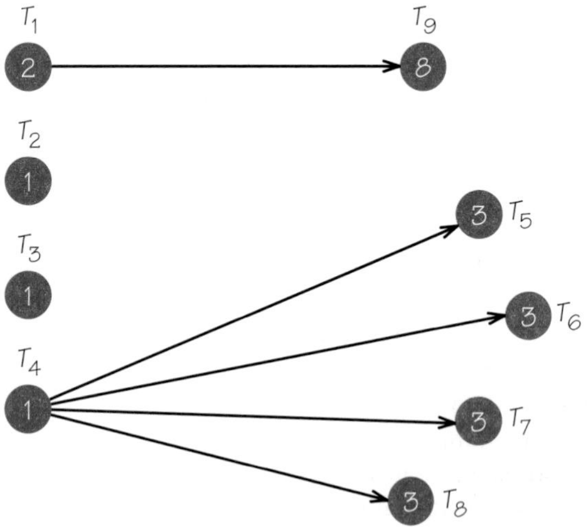

Figure 3.6 The order-requirement digraph obtained from the one in Figure 3.4 by reducing by 1 unit each of the task times shown there.

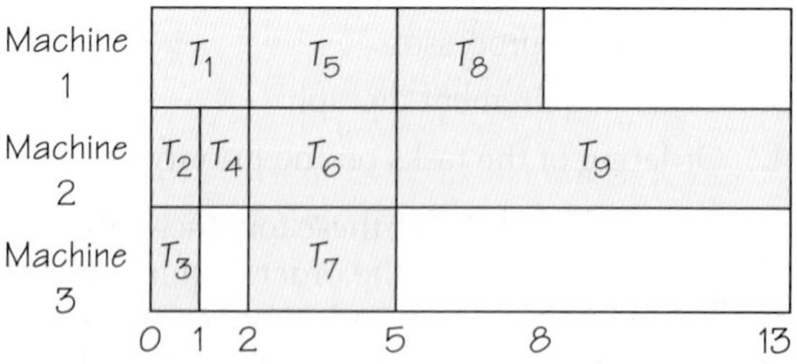

Figure 3.7 The schedule produced by applying the list-processing algorithm to the order-requirement digraph in Figure 3.6 using the list $T_1, T_2, \ldots, T_9$ with three processors.

The completion time is now 13—longer than the completion time of 12 for the case (Figure 3.5) with longer task times. This is unexpected! Let's explore further and see what happens.

Next we consider strategy 2, increasing the number of machines. Surely, this should speed matters up. When we apply the list-processing algorithm to the original graph in Figure 3.4, using the list $T_1, \ldots, T_9$ and four machines, we get the schedule shown in Figure 3.8. The completion time is now 15—an even later completion time than for the previous alteration.

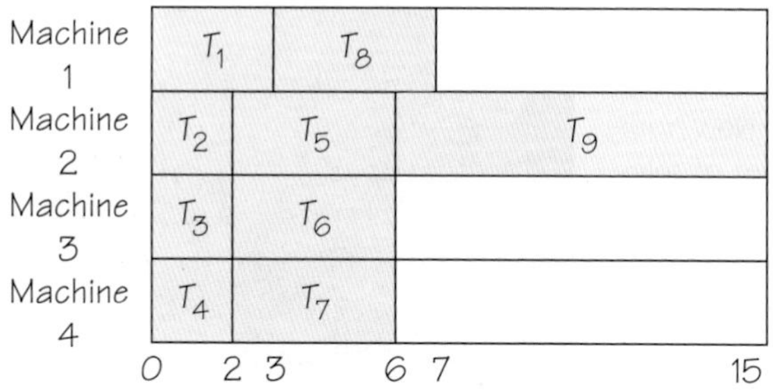

Figure 3.8 The schedule produced by applying the list-processing algorithm to the order-requirement digraph in Figure 3.4 using the list $T_1, T_2, \ldots, T_9$ with four processors.

Finally, we consider strategy 3, trying to shorten completion time by erasing all constraints (edges with arrows) in the order-requirement digraph shown in Figure 3.4. By increasing flexibility of the ordering of the tasks, we might guess we could finish our tasks more quickly. Figure 3.9 shows the schedule using the list $T_1, \ldots, T_9$—now it takes 16 units! This is the worst of our three strategies to reduce completion time.

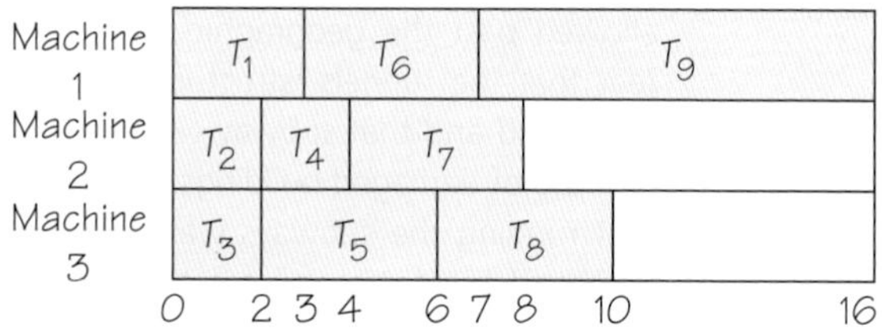

Figure 3.9 The schedule produced by applying the list-processing algorithm to the order-requirement digraph in Figure 3.4, modified by erasing all its directed edges, using the list $T_1, T_2, \ldots, T_9$ with three processors.

The failures we have encountered here are surprising at first glance, but they are typical of what can happen when a situation is too complex to analyze with naïve intuition. The value of using mathematics rather than intuition or trial and error to study scheduling and other problems is that it points out flaws that can occur in unguarded intuitive reasoning.

It is tempting to believe that we can make an adjustment in the rules for scheduling that we adopted to avoid the paradoxical behavior that has just been illustrated. Unfortunately, operations research experts have shown that there are no "simple fixes." This means that, in practice, for large scheduling problems such as those that face our medical centers and transportation system, finding the best solution to a particular scheduling problem cannot be guaranteed (see Spotlight 3.1).

3.2 Critical-Path Schedules

In our discussion so far, we have acted as though the priority list used in applying the list-processing algorithm was given to us in advance based on external considerations. Let's now consider the question of whether there is a systematic method of *choosing* a priority list that yields optimal or nearly optimal schedules.

Management Science and Disaster Recovery

SPOTLIGHT 3.1

AP World Wide Photo

The city of New York depends on a public transportation system of subways and roads to bring hundreds of thousands of people who live in the four outer boroughs (Queens, Brooklyn, the Bronx, and Staten Island) into Manhattan to work and "play." New York City (NYC) also has a communications system of telephones, radio and television stations, and computer networks. These systems speed information between New York's citizens and people outside the city and around the world. The area in southern Manhattan, in the vicinity of the World Trade Center (WTC), was a center for banking, insurance, financial markets, and domestic and international commerce. The attack on the World Trade Center on September 11, 2001, disrupted these networks and markets but did not destroy them, partly because the principles of operations research and management science were used in the design and development of these systems over a long period of time.

The diagram below shows a very simple subway (train) system between an eastern and a western terminus.

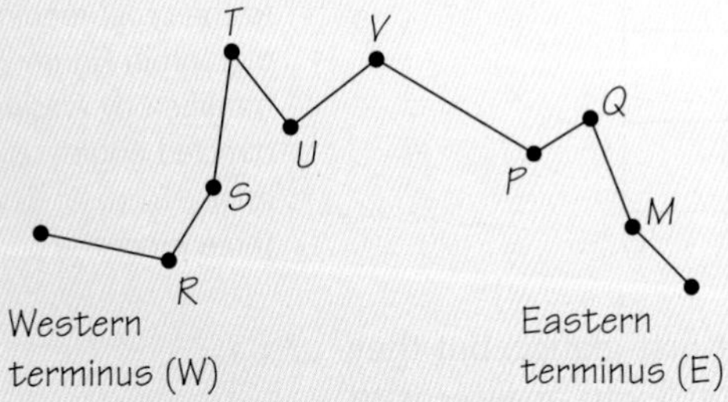

There are two tracks, each dedicated for use by westbound or eastbound trains to run between the two termini. The only place where trains can be turned around is at these termini. Simple graph theory tells us that in such a system, if a vertex is "destroyed" or out of service, or an edge is "destroyed" or out of service, the system totally breaks down. However, the simple provision that trains can be turned around at *U*, even though this is usually only one stop on the way from *W* to *E*, gives much greater flexibility to the system if there is a water main break or a gas leak and so on. Thanks to simple principles of this kind and the creation of routes that use independent lines with many transfer points, New Yorkers were able to use the subway system in a flexible way after the World Trade Center disaster. In the days right after the WTC collapsed, trains were not allowed past the geographic area near the WTC for fear that the tunnels' structural foundation had been weakened and that subway vibrations could cause the collapse of damaged buildings. After it was ascertained that running the subways was safe, both for partially damaged buildings and for the subways themselves, routes were altered several times to give rescue workers and people returning to their daily routines maximum support. One line's tunnels did collapse, and several stations had to be closed for extended periods, but due to the redundancy and flexibility of the design of the system, a remarkable amount of service was restored quickly. Recent projects to improve the infrastructure of the NYC subway system are also making it more flexible in dealing with potentially crippling events such as a huge snowfall.

Good planning and wise application of the principles of management science make it possible to minimize the effects of natural and manmade disasters.

We show how to construct a specific priority list based on this principle, to which the list-processing algorithm can then be applied.

Recall from our discussion of critical-path analysis in Chapter 2 that no matter how a schedule is constructed, the finish time cannot be earlier than the length (in terms of weight rather than number of edges) of the longest path in the order-requirement digraph. This suggests that we should try to schedule first those tasks that occur early in long paths, because they might be a bottleneck for the other tasks. This idea leads to **critical-path scheduling.**

EXAMPLE 3 ➔ Scheduling Two Processors

To illustrate this method, consider the order-requirement digraph in Figure 3.10a. Suppose we wish to schedule these tasks on two processors. Initially, there are two critical paths of length 64: T_1, T_2, T_3 and T_1, T_4, T_3. Thus, we place T_1 first on the priority list. With T_1 "gone," there is a new critical path of length 60 (T_5, T_6, T_4, T_3) that starts with T_5, so T_5 is placed second on the priority list. At this stage, with T_1 and T_5 removed, we have the residual order-requirement digraph shown in Figure 3.10b. In this diagram, there are paths of length 50 (T_2, T_3), 56 (T_6, T_4, T_3), 36 (T_6, T_4, T_7), and 24 (T_8, T_9, T_{10}). Because T_6 heads the path that is currently longest in length, it is placed third in the priority list. Once T_6 is removed from Figure 3.10b, there is a tie for which is the longest path remaining, because both T_2, T_3 and T_4, T_3 are paths of length 50.

When there is a tie between two longest paths, we place next on the priority list the lowest-numbered task heading a longest path. In the example shown here, this means that T_2 is placed next on the priority list, to be followed by T_4. Continuing in this fashion, we obtain the priority list T_1, T_5, T_6, T_2, T_4, T_3, T_8, T_9, T_7, T_{10}. Note that the order of T_7 and T_{10} was decided using the rule for breaking ties. The list-processing algorithm is now applied using this priority list and the order-requirement digraph in Figure 3.10a. We obtain the schedule in Figure 3.11.

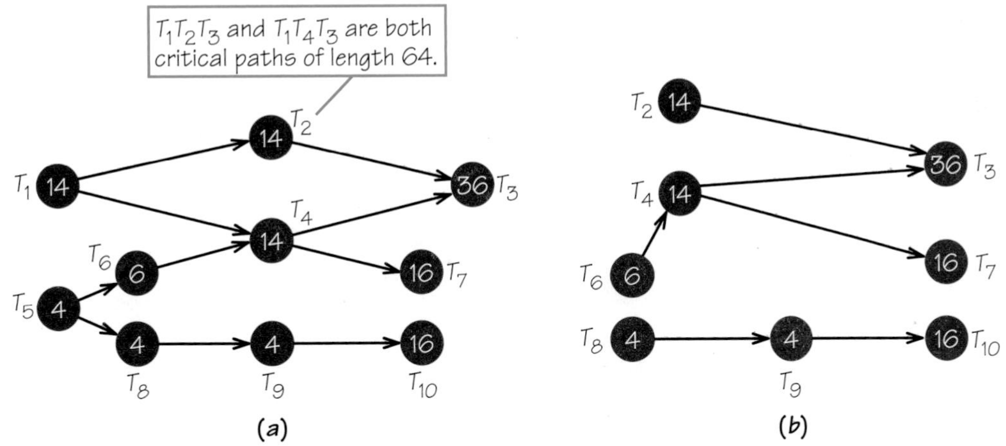

Figure 3.10 (a) An order-requirement digraph used to illustrate the critical-path scheduling method. (b) Residual order-requirement digraph after tasks T_1 and T_5 have been removed.

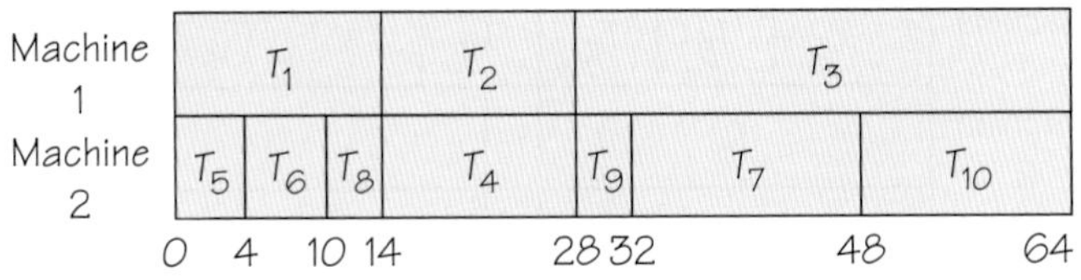

Figure 3.11 The optimal schedule produced by applying the critical-path scheduling method to the order-requirement digraph in Figure 3.10. The list used was T_1, T_5, T_6, T_2, T_4, T_3, T_8, T_9, T_7, T_{10}.

Critical-Path Scheduling

PROCEDURE

The **critical-path scheduling** algorithm applies the list-processing algorithm using the priority list L obtained as follows:

1. Find a task that heads a critical (longest) path in the order-requirement digraph. If there is a tie, choose the task with the lower number.
2. Place the task found in Step 1 next on the list L. (The first time through the process, this task will head the list.)
3. Remove the task found in Step 1 and the edges attached to it from the current order-requirement digraph, obtaining a new (modified) order-requirement digraph.
4. If there are no vertices left in the new order-requirement digraph, the procedure is complete; if there are vertices left, go to Step 1.

This procedure will terminate when all the tasks in the original order-requirement digraph have been placed on the list L.

The preceding example shows that critical-path scheduling can sometimes yield optimal solutions. Unfortunately, this algorithm does not always perform well. For example, the critical-path method employing four processors applied to the order-requirement digraph shown in Figure 3.12 yields the list T_1, T_8, T_9, T_{10}, T_{11}, T_5, T_6, T_7, T_{12}, T_2, T_3, T_4 and then the schedule in Figure 3.13. (Note that T_5, T_6, T_7 are thought of as heading paths of length 10.) In fact, there can be no worse schedule than this one. An optimal schedule is shown in Figure 3.14.

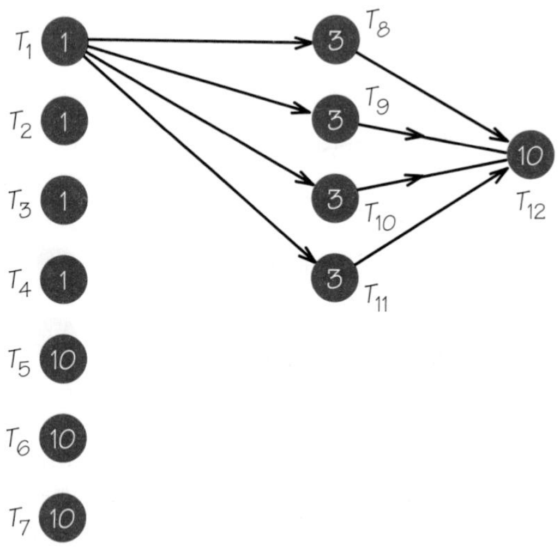

Figure 3.12 An order-requirement digraph used to illustrate how poorly the critical-path scheduling method can sometimes behave.

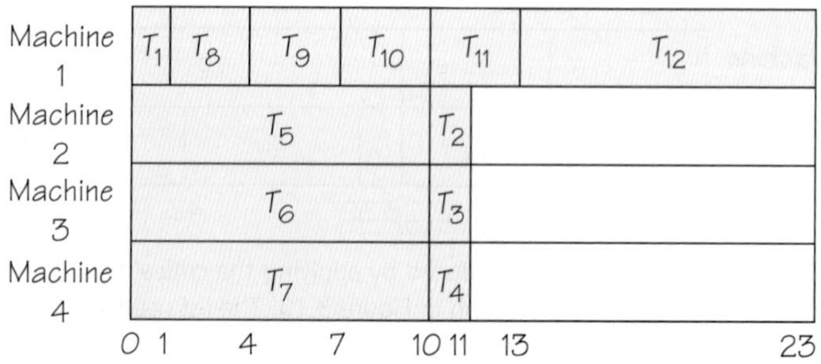

Figure 3.13 The schedule produced by applying the critical-path scheduling method to the order-requirement digraph in Figure 3.12 using four processors. The list used was T_1, T_8, T_9, T_{10}, T_{11}, T_5, T_6, T_7, T_{12}, T_2, T_3, T_4.

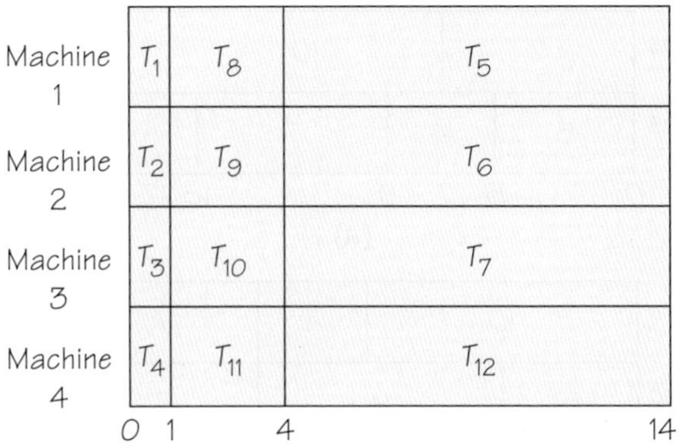

Figure 3.14 An optimal schedule for the order-requirement digraph in Figure 3.12 using four processors.

Many of the results we have examined so far are negative because we are dealing with a general class of problems that defy our using computationally efficient algorithms to find an optimal schedule. But we can close on a more positive note. Consider an arbitrary order-requirement digraph, but assume all the tasks take equal time. It turns out that we can always construct an optimal schedule using two processors in this situation. Ironically, we can choose among many algorithms to produce these optimal schedules. The algorithms are easy to understand (though not easy to prove optimal) and have all been discovered since 1969! Many people think that mathematics is a subject that is no longer alive, and that all its ideas and methods were discovered hundreds of years ago—but as we have just seen, this is not true. In fact, more new mathematics has been discovered and published in the last 30 years than during any previous 30-year period. This new mathematics typically results in new applications (better telecommunications, medical care, etc.), leading to better lives for everyone.

3.3 Independent Tasks

Mathematicians suspect that no computationally efficient algorithm for solving general scheduling problems optimally will ever be found. Owing to our limited success in designing algorithms for finding optimal schedules for general order-requirement digraphs, we will consider a special class of scheduling problems for which the order-requirement digraph has no edges. In this case, we say that the tasks are *independent* of one another, because they can be performed in any order. (No edges in the order-requirement digraph indicates that no tasks need to precede others; that is, the tasks can be done in any order.) In this section, we consider the problem of scheduling **independent tasks.**

Geometrically, we can think of the independent tasks as rectangles of height 1 whose lengths are equal to the time length of the task. Finding an optimal schedule amounts to packing the task rectangles, with no "idle time" gaps between adjacent rectangles, into a longer rectangle whose height equals the number of machines. For example, Figure 3.15 shows two different ways to schedule tasks of length 10, 4, 5, 9, 7, 7 on two machines. (For convenience, the rectangles in the case of independent tasks are labeled with their task times rather than their task numbers.) Scheduling basically means efficiently packing the task rectangles into the scheduling rectangle. There cannot be processor idle time between independent tasks because no task has to wait for any other task to be completed first. Finding the optimal answer among all possible ways to pack these rectangles is like looking for a needle in a haystack. The list-processing algorithm produces a packing, but it may not be a good one.

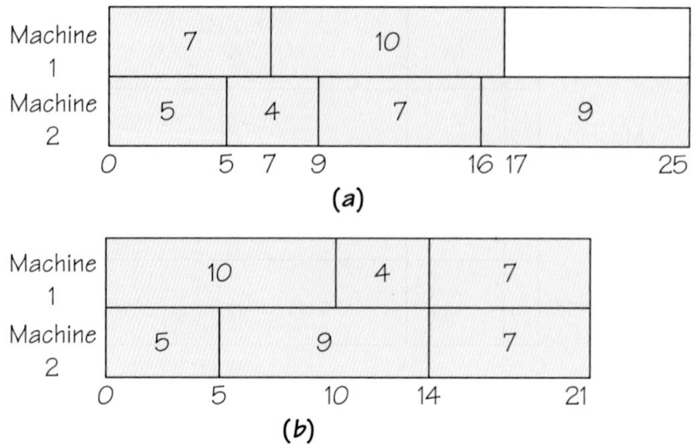

Figure 3.15 (a) A non-optimal way to schedule independent tasks of time lengths 10, 4, 5, 9, 7, 7 using two processors. (b) An optimal way to schedule independent tasks of time lengths 10, 4, 5, 9, 7, 7 using two processors.

There are two approaches we can consider. To study **average-case analysis,** we might ask: *Is the average (mean) of the completion times arrived at by using the list-processing algorithm with all the possible different lists close to the optimal possible completion time?* To study **worst-case analysis,** we might ask: *How far from optimal is a schedule obtained using the list-processing algorithm with one particular priority list?* What is being contrasted with these two points of view is that an algorithm may work well most of the time (give an answer close to optimal) even though there may be a few cases in which it performs very badly. Average-case analysis is amenable to mathematical solution but requires methods of great sophistication.

Decreasing-Time Lists

Is there some way of choosing a priority list for independent tasks that consistently yields relatively good schedules? The surprising answer is yes! The idea is that when long tasks appear toward the end of the list, they often seem to "stick out" on the right end, as in Figure 3.15a. This suggests that before one tries to schedule a collection of tasks, the tasks should be placed in a list where the longest tasks are listed first.

Decreasing-Time-List Algorithm PROCEDURE

The list-processing algorithm applied to a list of task times arranged in order of nonincreasing size is called the **decreasing-time-list algorithm.**

If we apply the **decreasing-time-list algorithm** to the set of tasks listed previously (10, 4, 5, 9, 7, 7), we obtain the times 10, 9, 7, 7, 5, 4 and the schedule (packing) shown in Figure 3.16. This packing is again optimal, but it is different from the optimal scheduling in Figure 3.15b. It is worth noting that the decreasing-time list and the list obtained by the critical-path method discussed earlier will coincide in the case of independent tasks. The decreasing-time list can also be constructed for the case in which the tasks are not independent. For general order-requirement digraphs, the decreasing-time list does not produce particularly good schedules.

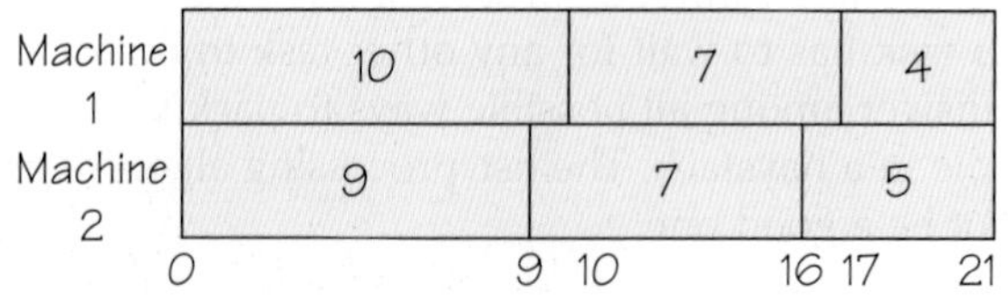

Figure 3.16 The optimal schedule resulting from applying the decreasing-time-list algorithm to a collection of independent tasks. The list used, written in terms of task times only, is 10, 9, 7, 7, 5, 4.

It is important to remember that the decreasing-time-list algorithm does not *guarantee* optimal solutions. This can be seen by scheduling the tasks with times 11, 10, 9, 6, 4 (Figure 3.17). The schedule has a completion time of 21. However, the rearranged list 9, 4, 6, 11, 10 yields the schedule in Figure 3.18, which finishes at time 20. This solution is obviously optimal because the machines finish at the same time and there is no idle time. Note that when tasks are independent, if there are m machines available, the completion time cannot be less than the sum of the task times divided by m.

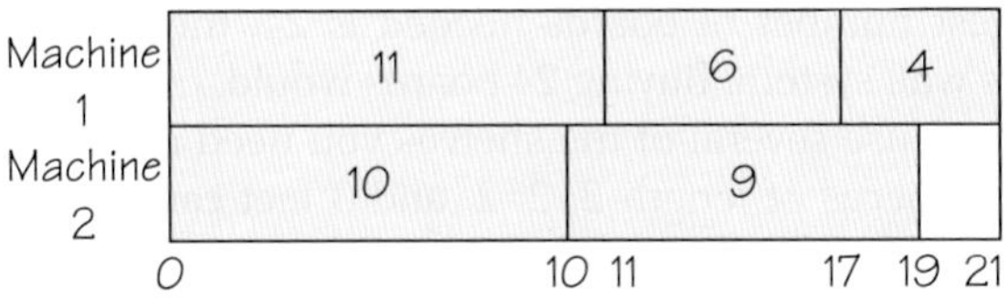

Figure 3.17 The non-optimal schedule resulting from applying the decreasing-time-list algorithm to a collection of independent tasks. The list used, written in terms of task times only, is 11, 10, 9, 6, 4.

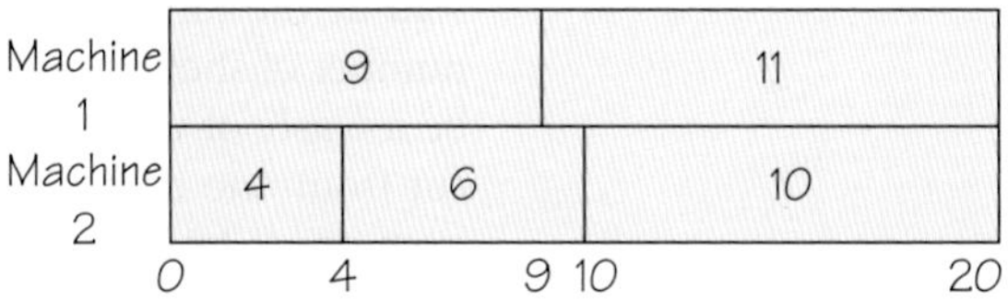

Figure 3.18 The optimal schedule resulting from applying the list-processing algorithm to a collection of independent tasks. The list used, written in terms of task times only, is 9, 4, 6, 11, 10.

Self Check 2

Apply the list-processing algorithm to the independent tasks with times shown in Figure 3.15, using the decreasing-time list with three processors. Is the resulting schedule optimal?

EXAMPLE 4 Photocopy Shop and Data Entry Problems

Imagine a photocopy shop with three photocopiers. Photocopying tasks that must be completed overnight are accepted until 5 P.M. The tasks are to be done in any manner that minimizes the finish time for all the work. Because this problem involves scheduling machines for independent tasks, the decreasing-time-list algorithm would be a good heuristic to apply.

For another example, consider a data entry pool at a large corporation or college, where individual entry tasks can be assigned to any data entry specialist. In this setting, however, the assumption that the data entry workers are identical in skill is less likely to be true. Hence, the tasks might have different times with different processors. This phenomenon, which occurs in real-world scheduling problems, violates one of the assumptions of our mathematical model.

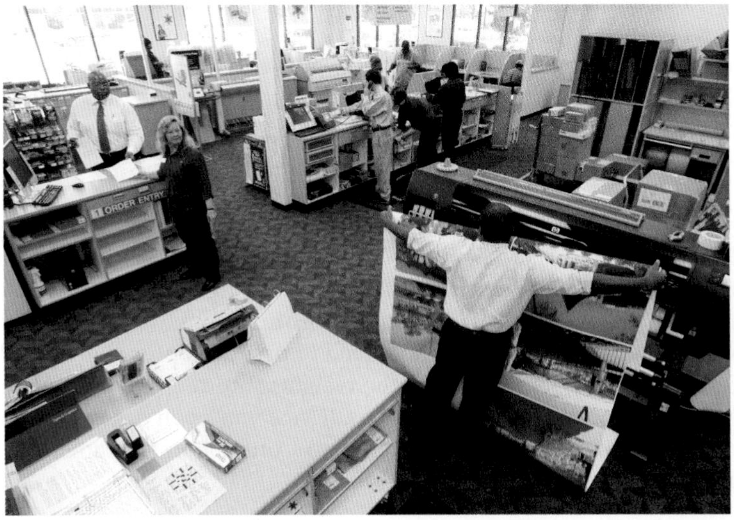

The Commercial Appeal/ZUMApress.com

A modern copy shop provides a wide array of services ranging from copying a few sheets for a "drop in" customer, to printing elaborate reports for small businesses, to publishing monographs and advertising flyers. Using mathematical scheduling techniques can save time for the customer and increase profit for the shop owner by ensuring that the many tasks are completed most efficiently.

3.4 Bin Packing

Suppose you plan to build a wall system for your books, CDs, DVDs, and stereo set. This project requires 24 wooden shelves of various lengths: 6, 6, 5, 5, 5, 4, 4, 4, 4, 2, 2, 2, 2, 3, 3, 7, 7, 5, 5, 8, 8, 4, 4, and 5 feet. The lumberyard, however, sells wood only in boards of length 9 feet. If each board costs \$8, what is the minimum cost to buy sufficient wood for this wall system?

Because all shelves required for the wall system are shorter than the boards sold at the lumberyard, the largest number of boards needed is 24, the precise number of shelves needed for the wall system. Buying 24 boards would, of course, be a waste of wood and money because several of the shelves you need could be cut from one board. For example, pieces of length 2, 2, 2, and 3 feet can be cut from one 9-foot board, assuming no loss of wood is created by the cutting process.

To make the ideas we develop more flexible, we think of the boards as bins of capacity W (9 feet in this case) into which we will pack (without overlap) n weights (in this case, lengths) whose values are $w_1, \ldots, w_n$, where each $w_i \leq W$. We wish to find the minimum number of bins into which the weights can be packed. In this formulation, the problem is known as the **bin-packing problem.**

Bin-Packing Problem DEFINITION

The **bin-packing problem** involves finding the minimum number of bins of weight capacity W into which weights $w_1, w_2, \ldots, w_n$ (each less than or equal to W) can be packed without exceeding the capacity of the bins.

At first glance, bin-packing problems may appear unrelated to the machine-scheduling problems we have been studying. However, there is a connection.

Let's suppose we want to schedule independent tasks so that each machine working on the tasks finishes its work by time W. Instead of fixing the number of machines and trying to find the earliest completion time, we must find the minimum number of machines that will guarantee completion by the fixed completion time (W). Despite this similarity between the machine-scheduling problem and the bin-packing problem, the discussion that follows will use the traditional terminology of bin packing.

By now, it should come as no surprise to learn that no one knows a fast algorithm that always picks the optimal (smallest) number of bins (boards). In fact, the bin-packing problem belongs to the class of NP-complete problems (see Spotlight 2.1 on page 51), which means that most experts think it unlikely that any fast optimal algorithm will ever be found. Relatively good algorithms for problems that come up in actual applications are known.

Bin-Packing Heuristics

We will think of the items to be packed, in any particular order, as constituting a list. In what follows we will use the list of 24 shelf lengths given for the wall system. We will consider various **heuristic algorithms,** namely, methods that can be carried out quickly but cannot be guaranteed to produce optimal results. Probably the easiest approach is simply to put the weights into the first bin until the next weight won't fit, and then start a new bin. (Once you open a new bin, don't use leftover space in an earlier, partially filled bin.) Continue in the same way until as many bins as necessary are used.

The resulting solution is shown in Figure 3.19. This algorithm, called **next fit (NF),** has the advantage of not requiring knowledge of all the weights in advance. Only the remaining space in the bin currently being packed must be remembered. The disadvantage of this heuristic is that a bin packed early on may have had room for small items that come later in the list.

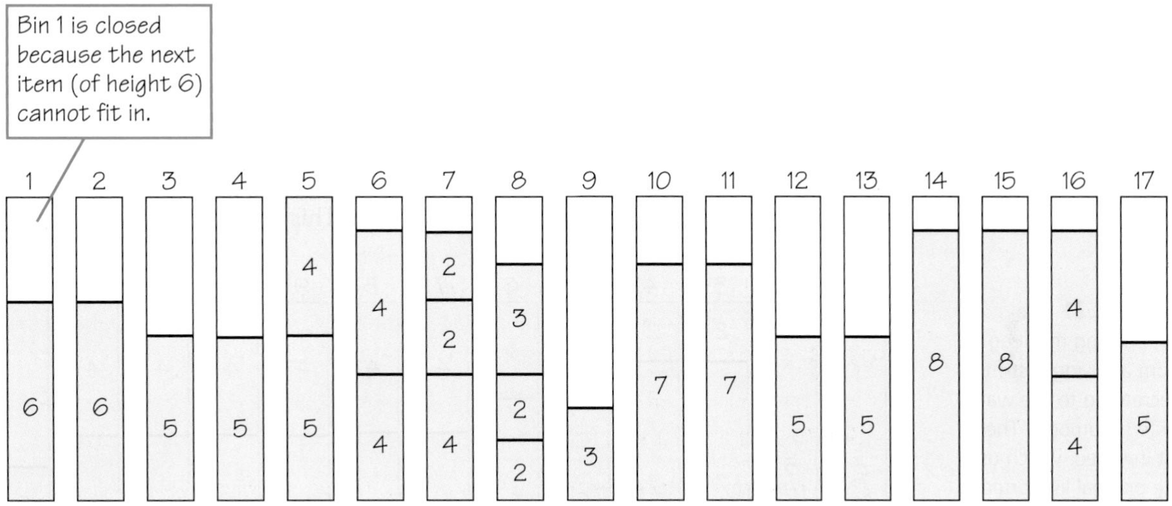

Figure 3.19 The list 6, 6, 5, 5, 5, 4, 4, 4, 4, 2, 2, 2, 2, 3, 3, 7, 7, 5, 5, 8, 8, 4, 4, 5 packed in bins using next fit.

Our wish to avoid permanently closing a bin too early suggests a different heuristic—**first fit (FF):** Put the next weight into the first bin already opened that has room for this weight. If no such bin exists, start a new bin. Note that a computer program to carry out first fit would have to keep track of how much room was left in all the previously opened bins. For the 24 wall-system shelves, the FF algorithm would generate a solution that uses only 14 bins (see Figure 3.20) instead of the 17 bins generated by the NF algorithm.

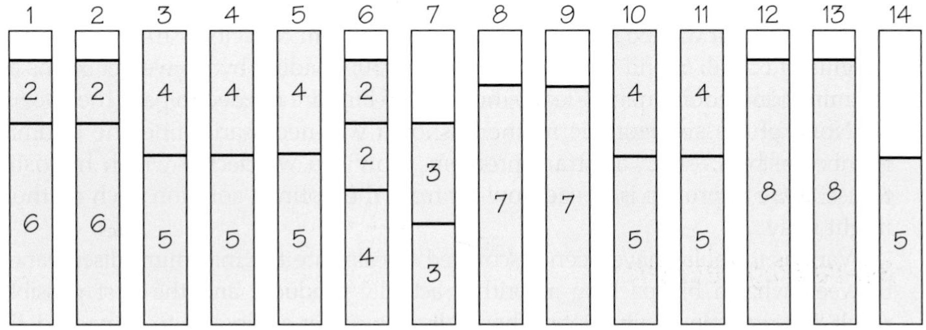

Figure 3.20 The list 6, 6, 5, 5, 5, 4, 4, 4, 4, 2, 2, 2, 2, 3, 3, 7, 7, 5, 5, 8, 8, 4, 4, 5 packed in bins using first fit. Worst fit would yield a packing that would look identical.

If we are keeping track of how much room remains in each partially filled bin, we can put the next item to be packed into the bin that currently has the most room available. This heuristic will be called **worst fit (WF).** The name *worst fit* refers to the fact that an item is packed into a bin with the most room available, that is, into which it fits "worst," rather than into a bin that will leave little room left over after it is placed in that bin ("best fit"). The solution generated by this approach looks the same as that shown in Figure 3.20. Although this heuristic also leads to 14 bins, the items are packed in a different order. For example, the first item of size 2, the tenth item in the list, is put into bin 6 in worst fit, but into bin 1 in first fit.

Decreasing-Time Heuristics

One difficulty with all three of these heuristics is that large weights that appear late in the list can't be packed efficiently. Therefore, we should first sort the items to be packed in order of decreasing size, assuming that all items are known in advance. We can then pack large items first and the smaller items into leftover spaces. This approach yields three new heuristics: **next-fit decreasing (NFD), first-fit decreasing (FFD),** and **worst-fit decreasing (WFD).** Here is the original list sorted by decreasing size: 8, 8, 7, 7, 6, 6, 5, 5, 5, 5, 5, 5, 4, 4, 4, 4, 4, 4, 3, 3, 2, 2, 2, 2. Packing using FFD order yields the solution in Figure 3.21. This solution uses only 13 bins.

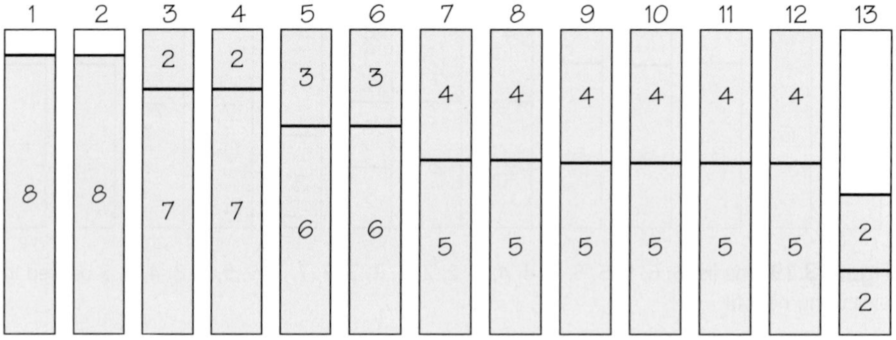

Figure 3.21 The bin packing resulting from applying first-fit decreasing to the wall-system numbers. The list involved, which uses the original list sorted in decreasing order, is 8, 8, 7, 7, 6, 6, 5, 5, 5, 5, 5, 5, 4, 4, 4, 4, 4, 4, 3, 3, 2, 2, 2, 2.

Is there any packing that uses only 12 bins? No. In Figure 3.21, there are only 2 free units (1 unit each in bins 1 and 2) of space in the first 12 bins, but 4 occupied units (two 2s) in bin 13. We could have predicted this by dividing the total length of the shelves (110) by the capacity of each bin (board): $\frac{110}{9} = 12\frac{2}{9}$. Thus, no packing could squeeze these shelves into 12 bins; there would always be at least 2 units left over for the 13th bin. (In Figure 3.21, there are 4 units in bin 13 because of the 2 wasted empty spaces in bins 1 and 2.) Even if this division created a zero remainder, there would still be no guarantee that the items could be packed to fill each bin without wasted space. For example, if the bin capacity is 10 and there are weights of 6, 6, 6, 6, and 6, the total weight is 30; dividing by 10, we get 3 bins as the minimum requirement. Clearly, however, 5 bins are needed to pack the five 6s.

None of the six heuristic methods shown will necessarily find the optimal number of bins for an arbitrary problem. How can we decide which heuristic to use? One approach is to see how far from the optimal solution each method might stray.

Various formulas have been discovered to calculate the maximum discrepancy between what a bin-packing algorithm actually produces and the best possible result. For example, in situations where a large number of bins are to be packed, FF can be off by as much as 70%, but FFD is never off by more than 22%. Of course, FFD doesn't give an answer as quickly as FF, because extra time for sorting a large collection of weights may be considerable. Also, FFD requires knowing the whole list of weights in advance, whereas FF does not. It is important to emphasize that a 22% margin of error is a worst-case figure. In many cases, FFD will perform much better. Results obtained by computer simulation indicate excellent average-case performance for this algorithm. In 2013, it was shown that if OPT (short for "optimum") denotes the fewest bins to pack a collection of weights (scaled to lie between 0 and 1), then first fit will never need more than 1.7(OPT) bins to contain the weights. Examples needing 1.7(OPT) bins were constructed.

When solving real-world problems, we always have to look at the relationship between mathematics and the real world. Thus, first-fit decreasing usually results in fewer bins than next fit, but next fit can be used even when all the weights are not known in advance. Next fit also requires much less computer storage than first fit, because once a bin is packed, it need never be looked at again.

Fine-tuning of the conditions of the actual problem often results in better practical solutions and in interesting new mathematics as well. See Spotlight 3.2 for a discussion of some of the tools mathematicians use to verify and even extend mathematical truths by raising new mathematical problems.

Self Check 3

How many bins are used when you apply the first-fit bin-packing algorithm to the weights shown in Figure 3.19, using bins of size 8? ■

Using Mathematical Tools

SPOTLIGHT 3.2

The tools of a carpenter include the saw, T square, level, and hammer. A mathematician also requires tools of the trade. Some of these tools are the proof techniques that enable verification of mathematical truths. Another set of tools consists of strategies to sharpen or extend the mathematical truths already known. For example, suppose that if *A* and *B* hold, then *C* is true. What happens if only *A* holds? Will *C* still be true? Similarly, if only *B* holds, will *C* still be true?

This type of thinking is of value because such questions will result either in more general cases where *C* holds or in examples showing that *B* alone and/or *A* alone can't imply *C*. For example, we saw that if a graph *G* is connected (hypothesis *A*) and even-valent (hypothesis *B*), then *G* has a circuit that uses each edge only once (conclusion *C*). If either hypothesis is omitted, the conclusion fails to hold. The figures illustrate this point. On the left is an even-valent but nonconnected graph; on the right, a connected graph with two odd-valent vertices. Neither graph has an Euler circuit.

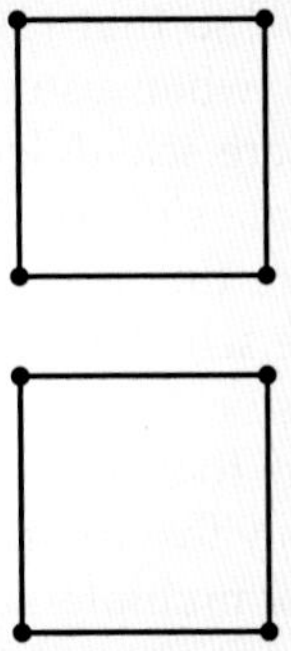

No Euler circuit
Nonconnected
Even-valent

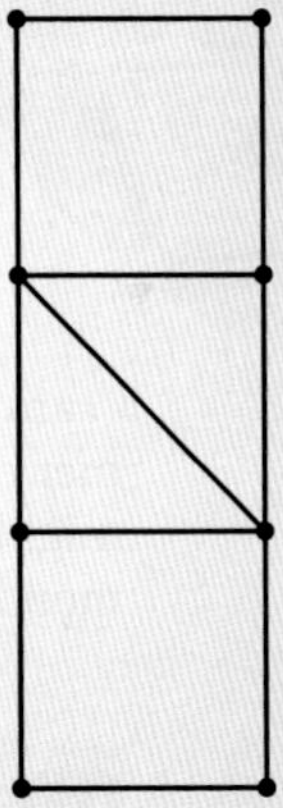

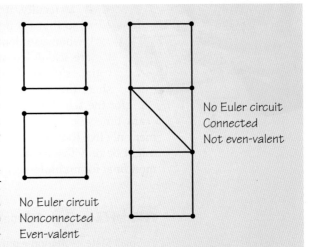

No Euler circuit
Connected
Not even-valent

Here is another way that a mathematician might approach extending mathematical knowledge. If *A* and *B* imply *C*, will *A* and *B* imply both *C* and *D*, where *D* extends the conclusion of *C*? For example, not only can we prove that a connected, even-valent (hypotheses *A* and *B*) graph has an Euler circuit, but we can also show that the first edge of the Euler circuit can be chosen arbitrarily (conclusions *C* and *D*). It turns out that being able to specify the first two edges of the Euler circuit may not always be possible. Mathematicians are trained to vary the hypotheses and conclusions of results they prove, in an attempt to clarify and sharpen the range of applicability of the results.

We have seen that machine scheduling and bin packing are probably computationally difficult to solve because they are NP-complete. A mathematician could then try to find the simplest version of a bin-packing problem that would still be NP-complete: What if the items to be packed can have only eight weights? What if the weights are only 1 and 2? Asking questions like these is part of the mathematician's craft. Such questions help to extend the domain of mathematics and hence the applications of mathematics.

3.5 Resolving Conflict via Coloring

Graph theory can be used to resolve scheduling conflicts that occur in trying to provide students access to limited database or computer resources.

Bananastock/Picturequest

In attempting to understand situations that involve scheduling, we might desire to achieve a wide variety of goals. For example, in certain types of scheduling problems, as we have seen here, we are interested in optimization issues. What is the earliest completion time for getting a collection of tasks done on two identical processors? However, in other situations, a different goal may arise. For example, in sports, consider a league of baseball teams. Each team has to play some games during the day, some at night, some at home, and some away from home. In the interests of *equity*, it may be desirable for each team to play the same number of day games and night games both at home and away against each of the other teams in the league. If, for example, team *A* plays 8 games away against team *B* and 2 games at home against *B*, then if *A* wins both home games but loses 7 out of 8 away games, it may appear that *B* had an advantage due to the way its games against *A* were scheduled.

Another goal of scheduling, other than optimization and equity, may be to prevent conflicts from occurring. We can use our knowledge of graph theory to solve some interesting scheduling problems where the goal is "conflict resolution." For example, at most colleges, final examinations must be scheduled every semester and summer session. From the point of view of students as well as faculty, it would be desirable to schedule these examinations so that (1) no two examinations are scheduled at the same time when a student is enrolled in both courses and (2) the examinations are scheduled in as "compact" a way as possible—that is, in as few time slots or days as possible. The administration of the college may share the desire for these two features and want still another property for the scheduling: (3) No more than five examinations are scheduled for any time slot. The reason for the last condition might be that during the summer only five rooms with reliable enough air conditioning are available (or there might be only five rooms large enough to hold all the students taking the common final for multiple-section courses).

EXAMPLE 5 Scheduling Examinations

Small State is offering eight courses during its summer session. The table shows with an X which pairs of courses have one or more students in common. Only two air-conditioned lecture halls are available for use at any one time. To design an efficient

	F	*M*	*H*	*P*	*E*	*I*	*S*	*C*
French (*F*)		X		X	X	X		X
Mathematics (*M*)	X				X	X		
History (*H*)						X	X	X
Philosophy (*P*)	X							X
English (*E*)	X	X				X		
Italian (*I*)	X	X	X		X		X	
Spanish (*S*)			X			X		
Chemistry (*C*)	X		X	X				

way to schedule the final examinations, we can represent the information in this table by using a graph, as shown in Figure 3.22a. In the graph, courses are represented by vertices and two courses are joined by an edge if there is any student enrolled in both courses.

We are faced with the following graph theory problem: Can we assign labels to the vertices of the graph in such a way that vertices that are joined by an edge get different labels? We think of the labels as the time slots the courses are assigned for final examinations. Traditionally, in graph theory such labels are referred to as *colors*. In this language, we seek to color the vertices of the graph so that vertices that are joined by an edge get different colors. Such a coloring is called a **vertex coloring.**

Vertex Coloring DEFINITION

The **vertex coloring** problem for a graph requires assigning each vertex of the graph a color (label) such that two vertices joined by an edge are assigned different colors.

Figure 3.22b shows one way to color the vertices of the graph so that each vertex gets a different color. Note that numbers are being used to represent the different colors. This solution is not very valuable, however, because it means that each course must be given its own time slot.

Can we improve upon eight colors? Note that *F*, *M*, *I*, *E*, and the edges that join them form a complete graph on four vertices. To have the vertices that are joined by an edge colored differently, we must assign *F*, *M*, *I*, and *E* four different colors because each of these four vertices is joined to the other three (see Figure 3.22c). Thus, any coloring of the graph in Figure 3.22a must use at least four colors. Exactly four colors will do, because the four colors used for *F*, *M*, *I*, and *E* can be used to color the remaining vertices while ensuring that no two vertices joined by an edge have the same color. The improved coloring in Figure 3.22c was found by trial and error.

Chromatic Number DEFINITION

The **chromatic number** is the minimum number of colors (labels) needed to label the vertices of a graph so that no two vertices of the graph joined by an edge get the same color.

The examination graph we have been studying has chromatic number 4; hence, we can schedule the eight examinations in four time slots without a conflict. Notice, however, that the coloring in Figure 3.22c schedules three different courses for the time slot corresponding to color 2. This means that not enough rooms with air conditioning will be available. Is there a way to recolor the graph with four colors so that each of the four colors is used only twice? Figure 3.22d shows that the answer is "yes."

Thus, we are able to schedule the eight final examinations in four time slots, using only two air-conditioned rooms, and no student will have a conflict under this schedule!

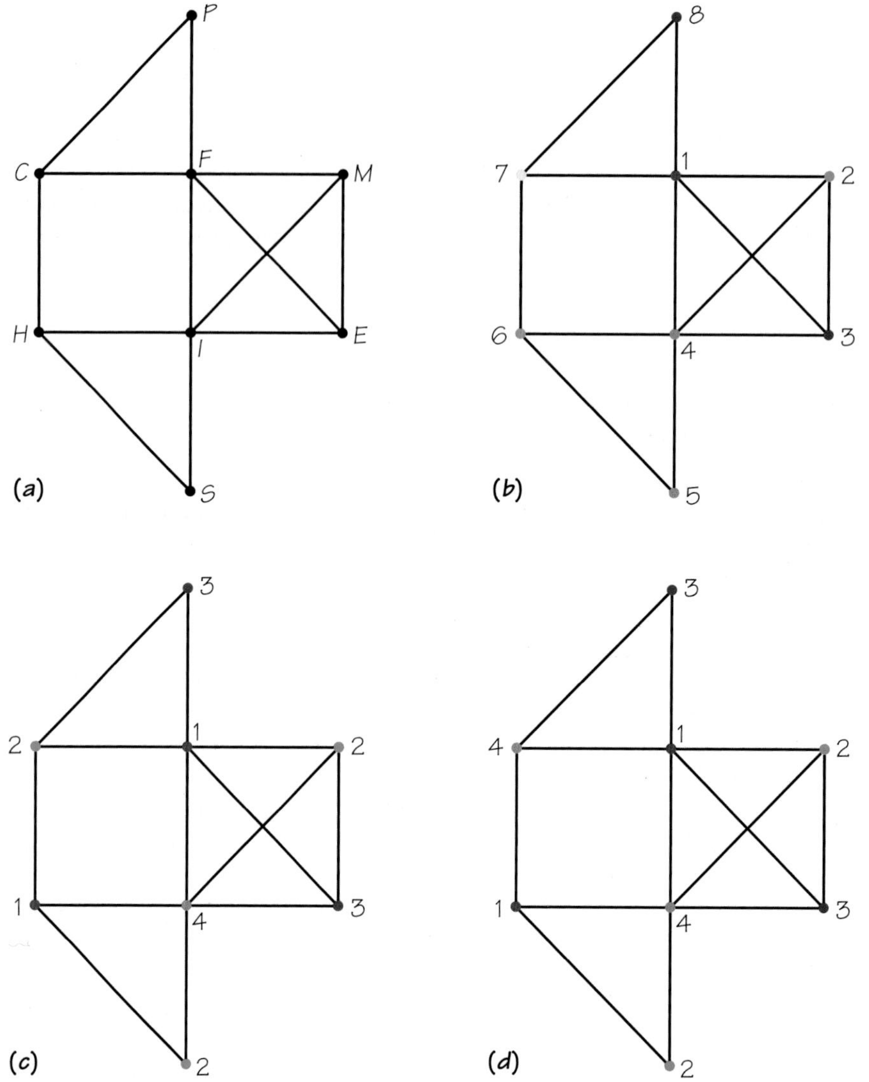

Figure 3.22 (a) A graph used to represent conflict information about courses. When two courses have a common student, an edge is drawn between the vertices that represent these courses. (b) A coloring of the scheduling graph with 8 colors, representing 8 time slots. Using this coloring would lead to a schedule where 8 time slots are used to schedule the examinations. This number is far from optimal. (c) A coloring of the scheduling graph with 4 colors. This translates into a way of scheduling the examinations during 4 time slots, and it is not possible to design a schedule with fewer time slots. However, this schedule calls for the use of three different rooms, because three examinations are scheduled during time slot 2. (d) A coloring of the scheduling graph with 4 colors. This means that the examinations can be scheduled in 4 time slots. However, because each color appears only twice, all the examinations can be scheduled in two air-conditioned rooms.

Self Check 4

What is the chromatic number of the graph shown on the right in Spotlight 3.2 on page 99? Show a coloring using the numbers 1, 2, 3, . . . as colors that achieve the chromatic number of colors.

Realistic problems in scheduling government committees, high school and university final examinations, and job interviews (see Spotlight 3.3) are usually so large that graph coloring algorithms have to be incorporated into elaborate software packages to solve them.

Mathematicians have examined many kinds of coloring problems. Many developments about coloring graphs have been an outgrowth of work on the Four Color Problem (see Spotlight 3.4). We can study problems that involve the coloring of the edges of a graph rather than its vertices. Using techniques that have emerged

Scheduling Job Interviews

SPOTLIGHT 3.3

A group of companies is coming to campus for job interviews. Different companies may want different numbers of time slots to hold their interviews. In each time slot one student can be interviewed. In the example below, all the companies have requested contiguous time slots for the interviews, but this need not be the case. Due to the fact that classes are going on at the same time, five departmental conference rooms have been made available to the companies to conduct their interviews.

The interviews will follow the school's regular hourly periods, which start at 9 A.M. and end at 4 P.M. (Companies will be scheduled for continuous interviews during lunch-hour times. Interviews cannot be scheduled beyond the end of the period that starts at 4 P.M. and ends at 5 P.M.)

	Company	Time Slot Requested
A	(Apricot Computers)	7
B	(Big Green)	1
C	(Challenge Insurance)	4, 5
D	(Daisy Printers)	7, 8
E	(Earnest Engine)	4, 5, 6
F	(Flexible Systems)	2, 3
G	(Gutter Leaders)	1, 2
H	(Halley's Combs)	6, 7
I	(Indelible Ink Corporation)	7, 8
J	(Jay's Produce)	4, 5
K	(Kelly's Detective Agency)	2, 3
L	(Large Clothes)	4, 5, 6
M	(Metropolitan TV)	1, 2
N	(Nationwide Bank)	4, 5, 6, 7

Look at the list of time blocks that the companies requested (where 1 = 9–10 A.M., . . . , 8 = 4–5 P.M.). Is it possible to accommodate all the companies that wish to do interviewing in the five rooms available while meeting their desired schedule times?

Problems of this kind seem simple enough, and you should try your hand at solving this particular one, for which a schedule does exist! However, this situation is not simple at all. The following facts are known about problems of this kind.

Fact 1. Suppose there are i interviews, p time periods, and r rooms where interviews can be scheduled. Each interviewer has specified periods during which he or she wishes to conduct interviews. Is it possible to design a schedule that meets the desired specifications? It turns out that this problem is NP-complete (see Spotlight 2.1); that is, it belongs to a large group of problems for which, among other things, the fastest known algorithms run very slowly on large-problem versions.

Fact 2. The problem just described remains NP-complete even for the case where there are only three rooms to be scheduled ($p = 3$).

The moral is *surprisingly simple:* Scheduling problems are very hard to solve.

However, the situation is not always as hopeless as it might seem. If you look at the list of time requests for the corporations, you will note again that, not surprisingly, each company has requested a contiguous block of times. It turns out that when this condition holds, it is possible to determine whether there is a feasible schedule using an algorithm that works relatively quickly.

Four Color Problem

SPOTLIGHT 3.4

Many people perceive mathematics as complex because it often uses strange notations and algebraic symbols. Thus, it may come as a surprise that a problem that is relatively easy to state and understand without complex symbolism eluded solution for about 100 years. When it was finally solved, it set off a "firestorm," with some saying that it had not truly been solved. More important, many of the ideas that have been developed in the theory of graphs were expanded or developed in the course of trying to prove this "guess."

When a graph can be drawn on a flat piece of paper so that edges meet only at vertices, we can talk about not only the vertices and edges of the graph, but also about its regions or *faces*. Such graphs are known as *plane graphs*. Two examples of plane graphs are shown in the diagram below.

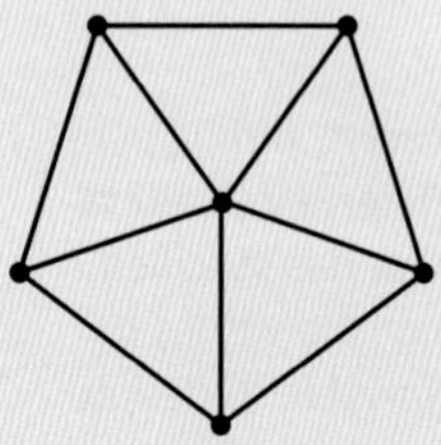

(*a*)

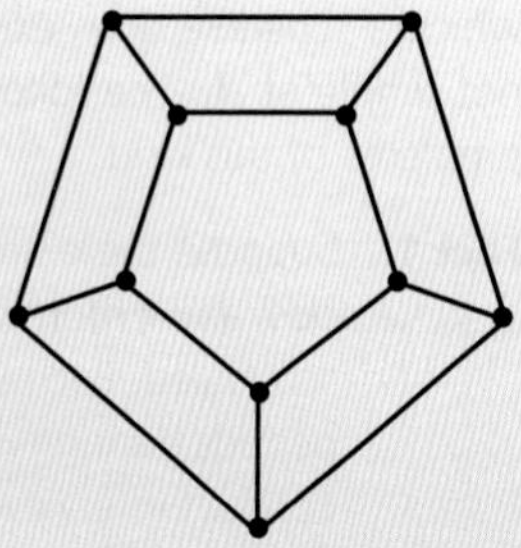

(*b*)

Graph (a) has 6 regions (the area "outside" the graph is counted as one of the 6 regions), 5 of which have 3 sides and 1 of which has 5 sides. Graph (b) has 7 regions, 2 of which have 5 sides and 5 of which have 4 sides. To count the number of sides of a region, imagine that you are a small ant and are following the edges around the region, starting at some vertex *w*. You count edges until you get back to *w*. Note that for each of these graphs, there is one *unbounded* (goes off to "infinity") region, in addition to the other regions. When you color the regions of a plane map, do not forget to assign a color to the unbounded region.

If you think of the regions of a plane graph as being distinct countries on a page that is to appear in an atlas, it would be nice if countries that share a border got different colors so that they can be distinguished. Countries that meet at a vertex, but do not share an edge representing a common border, can be given the same color. It is convenient to use the term *map* for the regions created by the drawing of a plane graph.

The following provocative question was raised in a letter (1852) from Augustus De Morgan to William Rowan Hamilton that was based on a problem posed to De Morgan by his student Fredrick Guthrie, who heard the question from his brother Francis:

Can the regions of any (plane) map always be colored with four or fewer colors?

A clever approach to proving the "Four Color Conjecture" was suggested by Alfred Kempe. Kempe's "proof" had a subtle error, which defied detection for many years, showing that proofs in mathematics really depend on the community of mathematicians to guarantee their accuracy. The British mathematician Percy Heawood discovered the error Kempe made. Heawood adapted Kempe's proof to show correctly that any map can be colored with five or fewer colors. Approximately 100 years elapsed before a proof that the Four Color Conjecture was true was found. This occurred in 1976, but there was a curious loose end: The proof found by Wolfgang Haken and Kenneth Appel required that a computer verify a large collection of "calculations," which were too numerous to be done by hand.

This proof troubled some philosophers and mathematicians, but it has been widely accepted by the mathematics community. In 1995, Neil Robertson, Daniel Sanders, Paul Seymour, and Robin Thomas found another proof. This proof, while simpler and shorter than the earlier Haken-Appel proof, also required computer calculations too numerous to be checked by "hand." Though it is possible that some new approach to the Four Color Conjecture will avoid the use of computers, this is not widely thought to be likely. However, human ingenuity sometimes surprises us!

from the study of coloring problems, problems involving such diverse contexts as scheduling government committees, using runways at airports efficiently, assigning frequencies for use by mobile pagers and cell phones, and designing timetables for public transportation have been solved—all these benefits from a problem that at first glance looks as if it belongs to recreational mathematics!

Review Vocabulary

Average-case analysis The study of the list-processing algorithm (more generally, any algorithm) from the point of view of how well it performs in all the types of problems it may be used for and seeing on average how well it does. *See also* worst-case analysis. (p. 94)

Bin-packing problem The problem of determining the minimum number of containers of capacity W into which objects of size $w_1, \ldots, w_n$ ($w_i \leq W$) can be packed. (p. 96)

Chromatic number The chromatic number of a graph G is the minimum number of colors (labels) needed in any vertex coloring of G. (p. 101)

Critical-path scheduling A heuristic algorithm for solving scheduling problems where the list-processing algorithm is applied to the priority list obtained by listing next in the priority list a task that heads a longest path in the order-requirement digraph. This task is then deleted from the order-requirement digraph, and the next task placed in the priority list is obtained by repeating the process. (pp. 91, 92)

Decreasing-time-list algorithm The heuristic algorithm that applies the list-processing algorithm to the priority list obtained by listing the tasks in decreasing order of their time length. (p. 94)

First fit (FF) A heuristic algorithm for bin packing in which the next weight to be packed is placed in the lowest-numbered bin already opened into which it will fit. If it fits in no open bin, a new bin is opened. (p. 97)

First-fit decreasing (FFD) A heuristic algorithm for bin packing where the first-fit algorithm is applied to the list of weights sorted so that they appear in decreasing order. (p. 98)

Heuristic algorithm An algorithm that is fast to carry out but that doesn't necessarily give an optimal solution to an optimization problem. (p. 96)

Independent tasks Tasks are independent when there are no edges in the order-requirement digraph. These are tasks that can be performed in any order. (p. 93)

List-processing algorithm A heuristic algorithm for assigning tasks to processors: Assign the first ready task on the priority list that has not already been assigned to the lowest-numbered processor that is not working on a task. (p. 84)

Machine-scheduling problem The problem of assigning tasks to processors so as to complete the tasks by the earliest time possible. (p. 82)

Next fit (NF) A heuristic algorithm for bin packing in which a new bin is opened if the weight to be packed next will not fit in the bin that is currently being filled; the current bin is then closed. (p. 97)

Next-fit decreasing (NFD) A heuristic algorithm for bin packing where the next-fit algorithm is applied to the list of weights sorted so that they appear in decreasing order. (p. 98)

Priority list An ordering of the collection of tasks to be scheduled for the purpose of attaining a particular scheduling goal. One such goal is minimizing completion time when the list-processing algorithm is applied. (p. 83)

Processor A person, machine, robot, operating room, or runway with time that must be scheduled. (p. 82)

Ready task A task is called ready at a particular time if its predecessors, as given by the order-requirement digraph, have been completed by that time. (p. 84)

Vertex coloring A vertex coloring of a graph G is an assignment of labels, which can be thought of as "colors," to the vertices of G so that vertices joined by an edge get different labels (colors). (p. 101)

Worst-case analysis The study of the list-processing algorithm (more generally, any algorithm) from the point of view of how well it performs on the hardest problems it may be used on. *See also* average-case analysis. (p. 94)

Worst fit (WF) A heuristic algorithm for bin packing in which the next weight to be packed is placed into the open bin with the largest amount of room remaining. If the weight fits in no open bin, a new bin is opened. (p. 97)

Worst-fit decreasing (WFD) A heuristic algorithm for bin packing where the worst-fit algorithm is applied to the list of weights sorted so that they appear in decreasing order. (p. 98)

Self Check Answers

1.

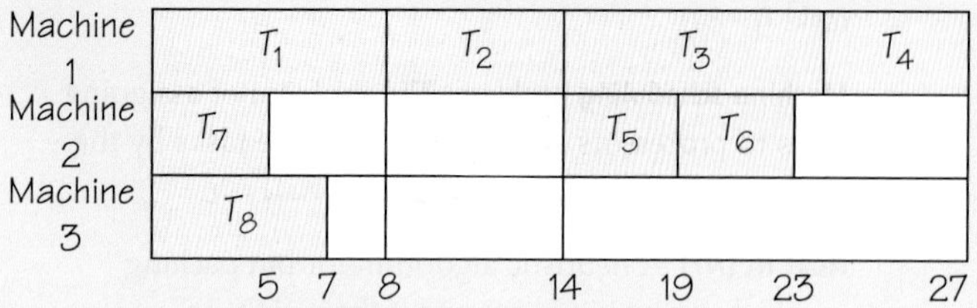

2. Using the list 10, 9, 7, 7, 5, 4, we get the result in the accompanying figure.

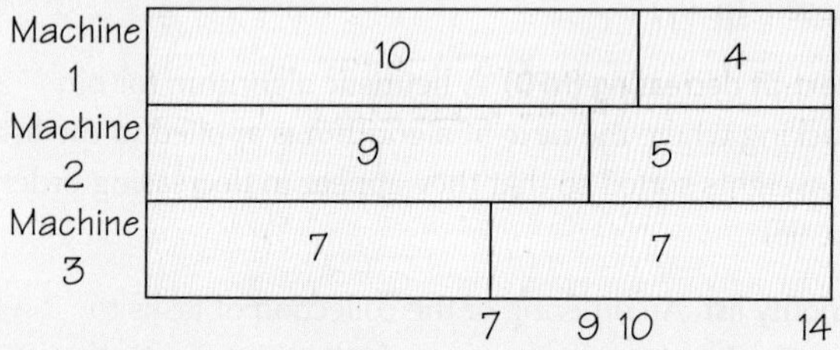

3. Fifteen bins are required, as shown in the accompanying diagram. A solution with fewer bins is not possible.

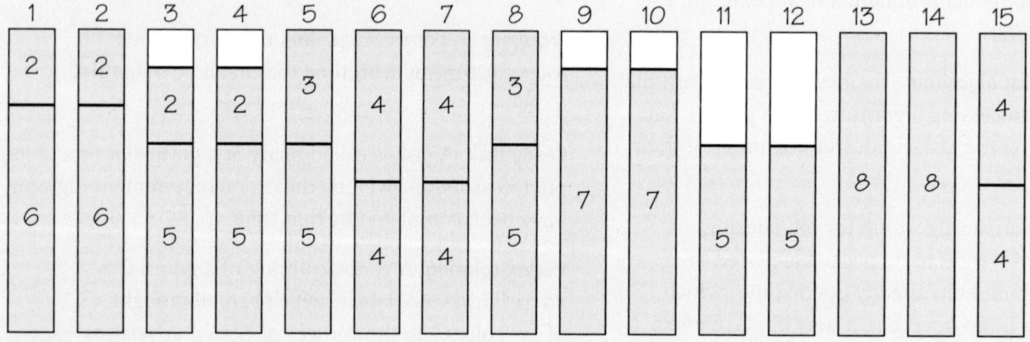

4. The chromatic number for this graph is 3, as indicated by the coloring. There are other ways to color this graph with only three colors.

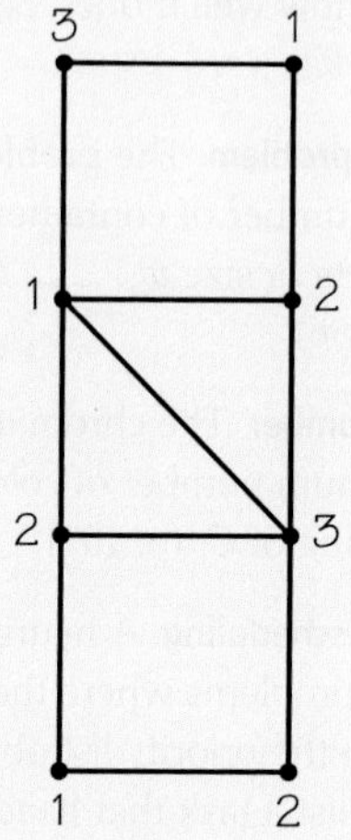

Skills Check

1. What is the minimum time required to complete 8 independent tasks with a total task time of 72 minutes on 4 machines?

(a) Less than 9 minutes

(b) Between 9 and 12 minutes

(c) More than 16 minutes

2. If the list-processing algorithm is applied to the accompanying order-requirement digraph (task time in minutes) using the list T_1, T_2, T_3, T_4, and three machines are available, the earliest completion time for all the tasks is ________.

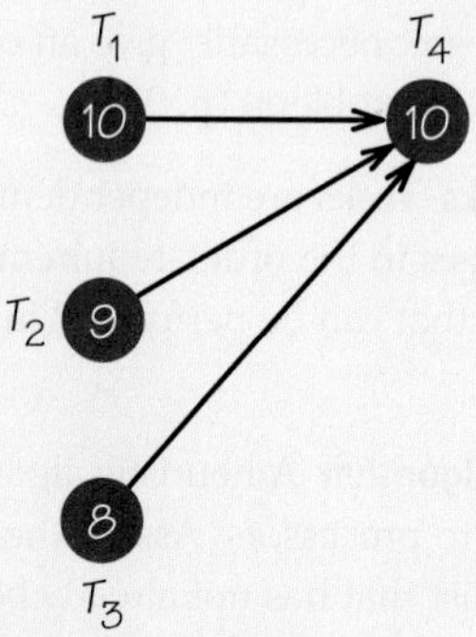

3. The following digraph cannot be an order-requirement digraph because

(a) no vertex has four edges that enter that particular vertex.

(b) all the tasks require the same time to complete.

(c) it has a directed circuit.

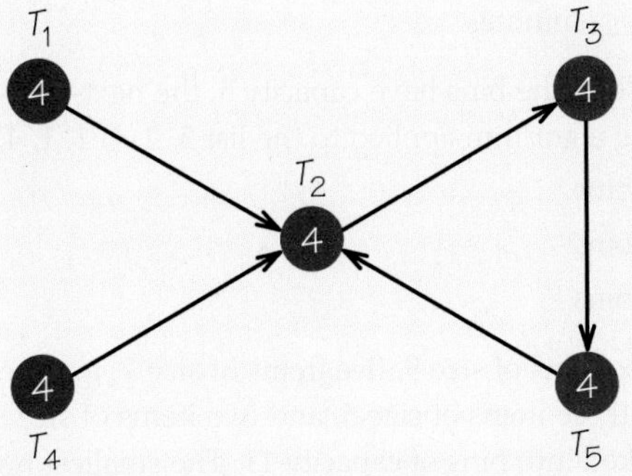

4. The shortest path and longest path, respectively, in the accompanying task-analysis digraph have, respectively, lengths

(a) 1 and 9.

(b) 4 and 15.

(c) 4 and 11.

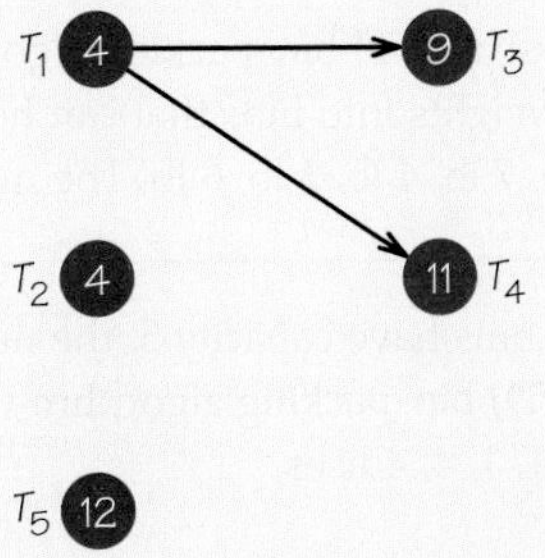

5. Given the accompanying order-requirement digraph (time in minutes) and the priority list T_6, T_5, T_4, T_3, T_2, T_1, apply the list-processing algorithm to construct a schedule using two processors. The completion time of the resulting schedule is ______________________.

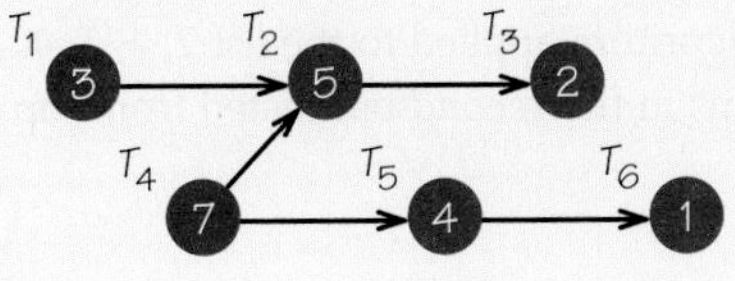

6. The subscripts for the tasks that make up a critical path for the order-requirement digraph below are ______, ______, ______.

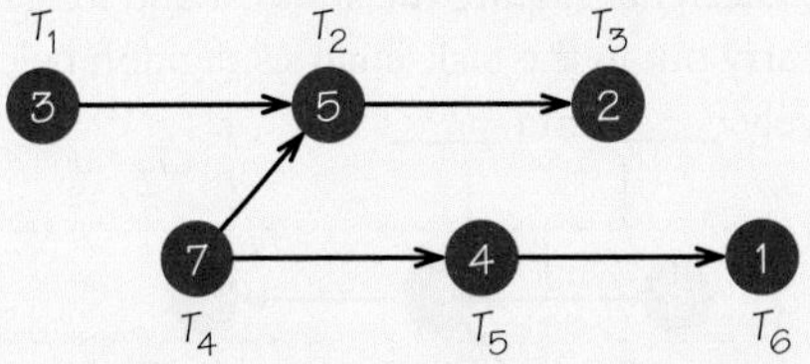

7. Suppose that a crew can complete in a minimum amount of time the job whose order-requirement digraph is shown below. If task T_2 is shortened from 5 minutes to 1 minute, then the maximum amount by which the completion time for the entire job can be shortened is ______________________________.

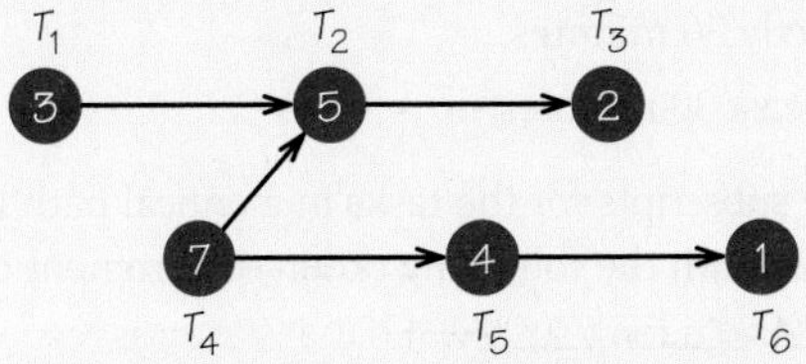

8. Suppose that independent tasks require a total of 30 minutes, while only one task takes as long as 10 minutes. If these tasks are scheduled on two machines, they

(a) can never take longer than 15 minutes to complete.

(b) might take longer than 17 minutes to complete.

(c) can always be completed within 16 minutes.

9. Which statement about the accompanying digraph is true?

(a) This digraph cannot be the order-requirement digraph for a scheduling problem because the digraph has no (directed) edges.

(b) This digraph can be the order-requirement digraph for a scheduling problem.

(c) This digraph cannot be the order-requirement digraph for a scheduling problem because it is not allowed for all the tasks to have the same time length.

10. The tasks that require the shortest and longest time to carry out in the task-analysis digraph below are, respectively, _____ and _____.

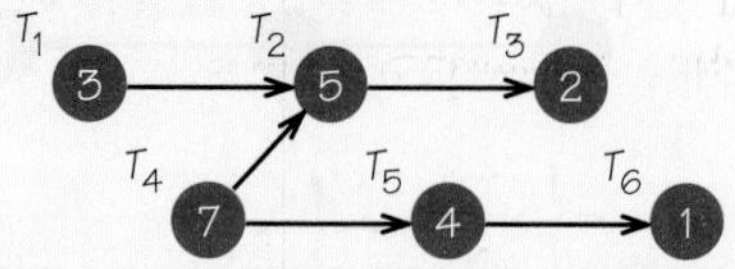

11. Assume an order-requirement digraph has a critical path with length 30 minutes. Based on this information, when the tasks are scheduled on two machines, how much time will be required?

(a) Exactly 10 minutes

(b) Exactly 30 minutes

(c) At least 30 minutes

12. The subscripts for the tasks in a critical path list associated with the following order-requirement digraph are _____, _____, _____.

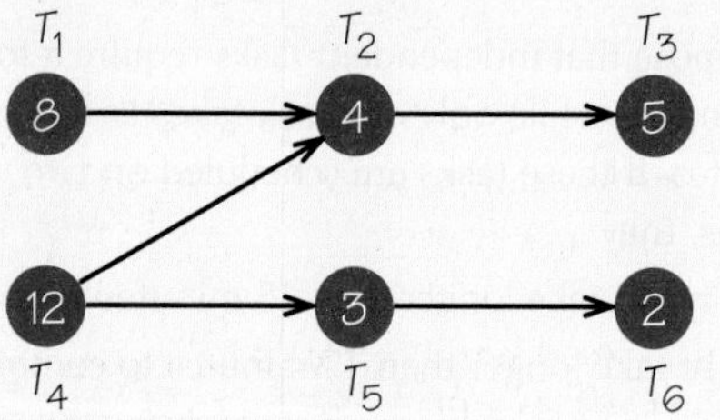

13. Assume that a job consists of six independent tasks ranging in time from 2 to 10 minutes and totaling 27 minutes. Efficiently scheduled on three machines, how much time will the job require?

(a) Exactly 10 minutes

(b) Exactly 9 minutes

(c) More than 10 minutes

14. The list-processing algorithm is used to schedule independent tasks lasting 6, 7, 4, 3, and 6 minutes on three machines, using these times as given for a list. The completion time for all the tasks will be __________.

15. A radio announcer has 10 songs of various lengths to schedule into several segments. The announcer must identify the station at least once every 15 minutes, so the segments cannot be longer than 15 minutes. This job can be solved using the

(a) list-processing algorithm for independent tasks.

(b) critical-path scheduling algorithm.

(c) first-fit algorithm for bin packing.

16. When the decreasing-time-list algorithm is used to schedule independent tasks lasting 6 minutes, 7 minutes, 4 minutes, 3 minutes, and 6 minutes, on two machines, a schedule results where the tasks are completed after ________ minutes.

17. When the bins have capacity 5, the next-fit (NF) bin-packing algorithm applied to the list 3, 2, 4, 1, 1, 4, 4 uses

(a) 7 bins.

(b) 4 bins.

(c) 5 bins.

18. Six items of size 9, five items of size 7, four items of size 6, three items of size 5, and two items of size 9 are to be packed into bins of capacity 11. The smallest number of bins that this can be accomplished with is ______.

19. Use the worst-fit-decreasing (WFD) bin-packing algorithm to pack the following weights into bins that can hold no more than 10 lb: 6 lb, 7 lb, 4 lb, 3 lb, 6 lb. How many bins are holding a full 10 lb?

(a) 2 bins

(b) 1 bin

(c) 0 bins

20. Use the first-fit (FF) bin-packing algorithm to pack the following weights into bins that can hold no more than 10 lb: 6 lb, 7 lb, 4 lb, 3 lb, 6 lb. The number of bins required is ______________________.

21. When the bins have capacity 5, the next-fit decreasing (NFD) bin-packing algorithm when applied to the list 3, 2, 4, 1, 1, 4, 4 uses

(a) 4 bins.

(b) 5 bins.

(c) 7 bins.

22. When the bins have capacity 5, the items packed in the second bin (listed from top to bottom) using the list 3, 2, 4, 1, 1, 4, 4 by the first-fit (FF) bin-packing algorithm will be ______ and ______.

23. When the bins have capacity 5, the best-fit bin-packing algorithm applied to the list 2, 4, 1, 4, 3 packs which items in the second bin (listed from top to bottom)?

(a) 1, 4

(b) 4

(c) 4, 1

24. The first-fit-decreasing (FFD) bin-packing algorithm is applied to the weight list 1, 2, 3, 4, 5, 5, 6, 8 for packing into bins of capacity 10. The item of weight 2 is packed into the bin numbered ________ when the packed bins are numbered from left to right.

25. A vertex coloring seeks to color the vertices of a graph to ensure which of the following traits?

(a) Vertices of the same color are never connected by an edge.

(b) Every edge connects vertices of the same color.

(c) Every color is used.

26. Which of the following statements is true?

(a) The vertices of a graph that can be drawn in the plane so that the edges meet only at the vertices can always be colored with at most three colors.

(b) The number of inequivalent (non-isomorphic) graphs that can be vertex-colored with exactly three colors is finite.

(c) The vertices of a graph that can be drawn in the plane so that the edges meet only at the vertices can be colored with at most four colors.

27. Assume the 8 corners of a cube represent vertices of a graph and the 12 edges of a cube represent the cube's edges. The chromatic number of this graph is ________________________________.

28. Graphs that have circuits of only even lengths have chromatic number ________.

29. The minimum number of colors needed to color the vertices of the accompanying graph is

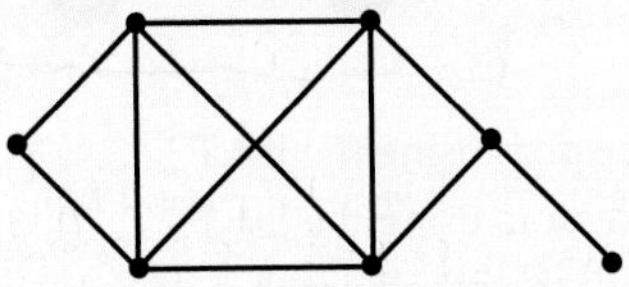

(a) 2.

(b) 4.

(c) 3.

30. A graph that has a circuit of length 3 can always be vertex colored with no fewer than _____ colors.

Chapter 3 Exercises

Challenge Discussion

3.1 Scheduling Tasks

3.2 Critical-Path Schedules

1. You and your two housemates are planning to have a party this Friday night at your apartment. Eight guests are expected, and you plan to serve a small homemade dinner. List the tasks involved in carrying out such a party and the types of processors to be used to carry out the tasks. Can any of the tasks be done simultaneously?

2. Compare and contrast the scheduling problems that arise at a

(a) fast-food restaurant.

(b) standard sit-down restaurant.

3. List as many scheduling situations as you can for these environments:

(a) Your school

(b) Bus/train terminal

(c) Medical center

(d) Police station

(e) Bookstore

(f) 24-hour drug store

(g) Firehouse

4. Jane is planning a getaway weekend at a ski resort. She plans to leave work in Manhattan at 1 P.M. and must make her way to a local airport for a 5 P.M. shuttle plane to Boston. She then hopes to get a bus to the nearby resort. Discuss the tasks that Jane must complete to be at the resort by 10 P.M. What are the different types of processors involved in getting these tasks done? Can any of these tasks be done simultaneously?

5. Schedule the six tasks of the job indicated in the accompanying order-requirement digraph on one processor, using the list-processing algorithm and the list $T_1, T_2, T_3, T_4, T_5, T_6$.

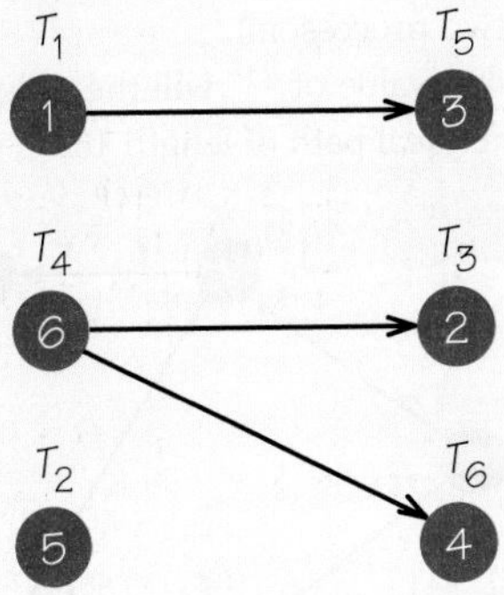

6. Use the list-processing algorithm to solve the scheduling problem in Exercise 5 on two processors.

7. Use the list-processing algorithm to schedule the tasks in the accompanying order-requirement digraph on

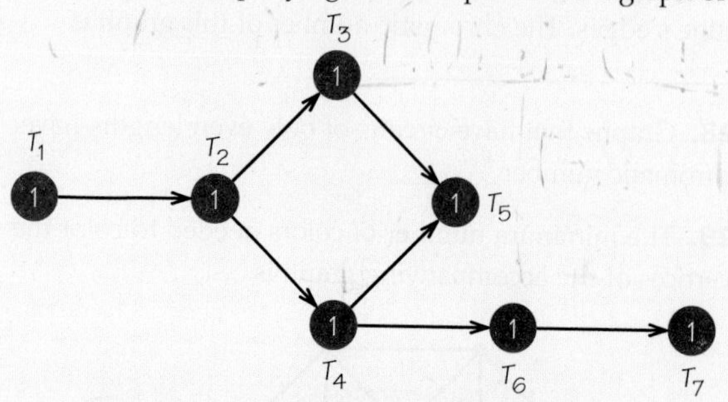

(a) two processors using the list $T_1, \ldots, T_7$.

(b) two processors using the list $T_1, T_2, T_3, T_4, T_6, T_5, T_7$.

(c) Is either of the schedules that you obtain optimal?

(d) Will adding a third processor enable the tasks to be finished earlier?

(e) Which tasks in this order-requirement digraph can be shortened and not affect the completion time of all the tasks?

8. Consider the order-requirement digraph below:

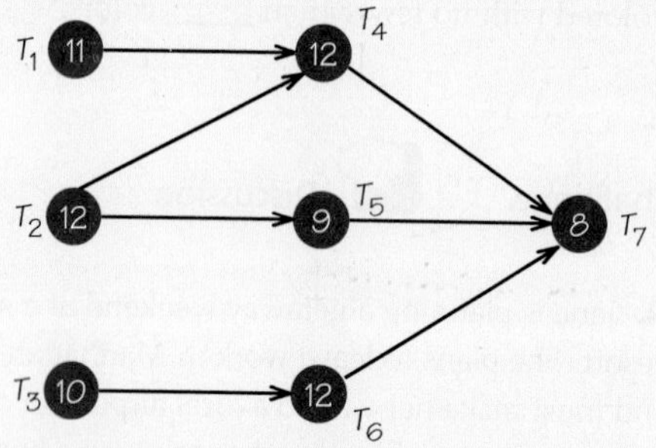

(a) Find the length of the critical path.

(b) Schedule these seven tasks on two processors using the list algorithm and the lists:

(i) $T_1, T_2, T_3, T_4, T_5, T_6, T_7$

(ii) $T_2, T_1, T_3, T_6, T_5, T_4, T_7$

(c) Does either list lead to a completion time that equals the length of the critical path?

(d) Show that no list can ever lead to a completion time equal to the length of the critical path (providing the schedule uses two processors).

9. (a) For what value of "?" will the order-requirement digraph have a critical path of length 16?

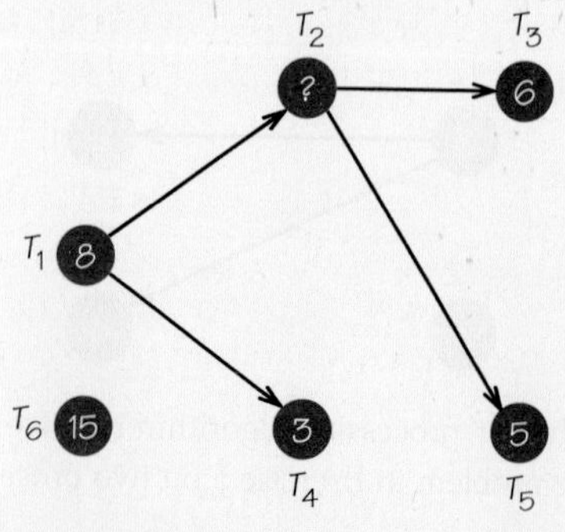

(b) If the time for T_2 is set to the answer in part (a), what is the largest amount of time that T_5 can increase so the earliest completion time for the job involving the six tasks is still 16?

10. For the accompanying order-requirement digraph, answer the following questions.

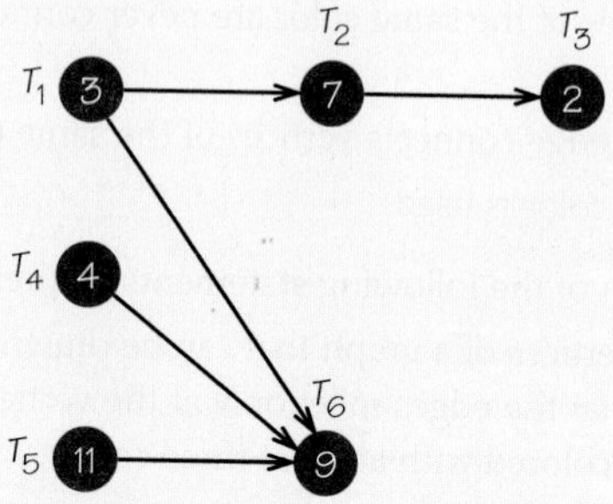

(a) Find the length of the critical path(s) and which tasks are on the critical path(s).

(b) Which task(s) (taken one at a time) would not alter the length if the time of that task were to increase by 1 time unit?

(c) With two processors, can these tasks be scheduled to finish by time 20? If so, what list L would enable you to apply the list-processing algorithm and finish by time 20 on two processors?

11. An order-requirement digraph has exactly two vertices that have no edges coming into them. Explain why, if there are three processors available, one or more of these processors must be idle some of the time.

12. (a) If the tasks subject to an order-requirement digraph are scheduled on only one machine, explain what different goals one might have in choosing to use different lists to schedule the tasks.

(b) Find the schedule for the accompanying order-requirement digraph, using the list $T_1, T_2, T_3, T_4, T_5, T_6$ on one processor.

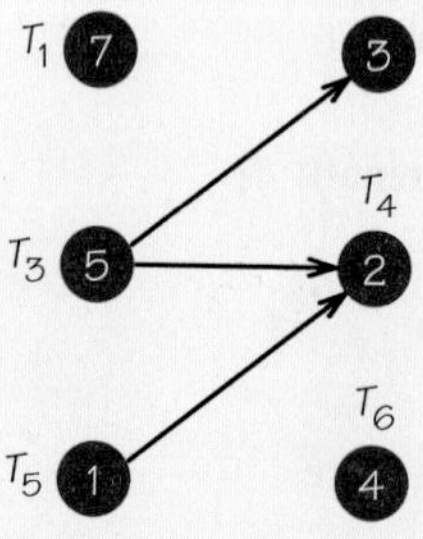

13. (a) Use the accompanying order-requirement digraph to schedule the 6 tasks $T_1, T_2, T_3, T_4, T_5, T_6$ on two processors with the priority lists:

(i) $T_1, T_2, T_3, T_4, T_5, T_6$

(ii) $T_1, T_6, T_3, T_5, T_4, T_2$

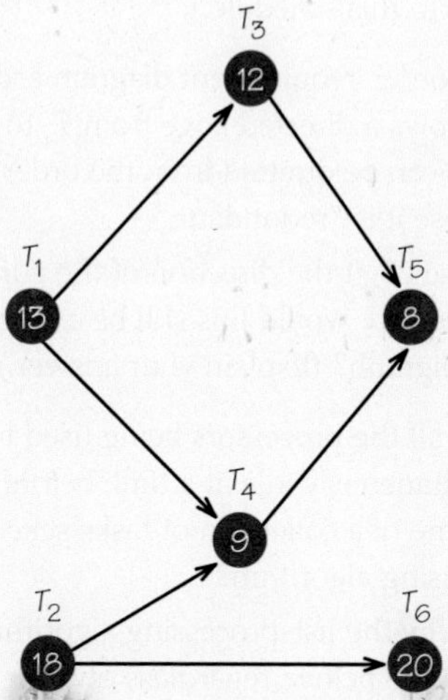

(b) Are either of the schedules produced from these lists optimal? If not, can you find a priority list that will result in an optimal schedule?

(c) Find the critical path and its length. Explain why no schedule has an earliest completion time equal to the length of the critical path.

14. (a) Repeat Exercise 13, but interchange the task times of tasks T_2 and T_6.

(b) How does the completion time for an optimum schedule for this situation compare with the optimum schedule for Exercise 13?

15. (a) If we add a new directed edge to an order-requirement digraph *D*, can the critical path in the new order-requirement digraph *D′* have longer length?

(b) If we add a new directed edge to an order-requirement digraph *D*, can the critical path in the new order-requirement digraph *D′* have shorter length?

16. (a) Discuss scheduling problems for which it is not reasonable to assume that once a processor starts a task, it will always complete that task before it works on any other task. Give examples for which this approach would be reasonable.

(b) Give an example where the assumption that all processors have identical capabilities in a scheduling situation is not realistic.

17. Can you give examples of scheduling problems for which it seems reasonable to assume that all the task times are the same?

18. Use the list-processing algorithm to schedule the tasks in the accompanying order-requirement digraph on

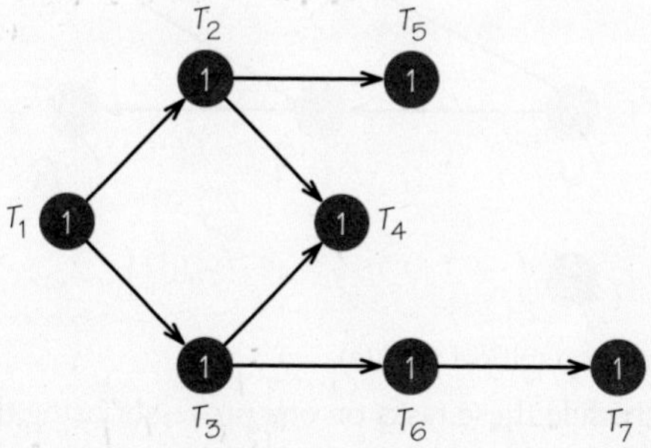

(a) two processors using the list $T_1, \ldots, T_7$.

(b) two processors using the list $T_1, T_2, T_3, T_4, T_6, T_5, T_7$.

(c) Is either of the schedules that you obtain optimal?

19. For the accompanying order-requirement digraph, apply the list-processing algorithm, using three processors for lists (a) through (c). How do the completion times obtained compare with the length of the critical path?

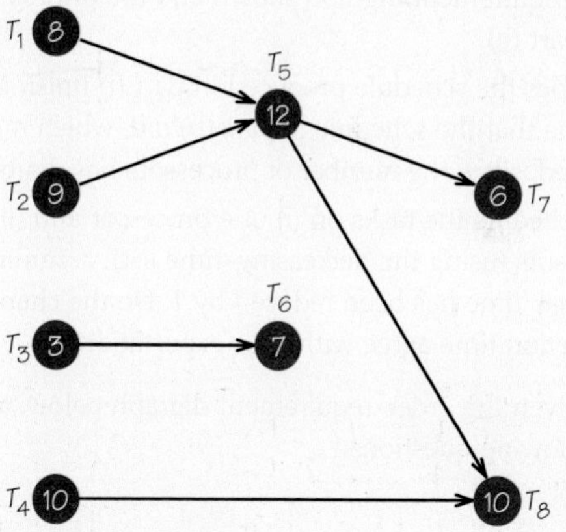

(a) $T_1, T_2, T_3, T_4, T_5, T_6, T_7, T_8$

(b) $T_1, T_3, T_5, T_7, T_2, T_4, T_6, T_8$

(c) $T_8, T_6, T_4, T_2, T_1, T_3, T_5, T_7$

20. (a) Can you find an order-requirement digraph with four tasks for which every priority list used to schedule the tasks on two machines assigns task T_4 to machine 1 at time 0?

(b) Can you choose the order-requirement digraph in part (a) so that machine 2 stays idle for all lists from time 0 to time 3?

21. Can you find a list that gives rise to the optimal schedule shown in Figure 3.14 (on page 93) for the order-requirement digraph in Figure 3.12 (on page 92)?

22. Consider the accompanying order-requirement digraph.

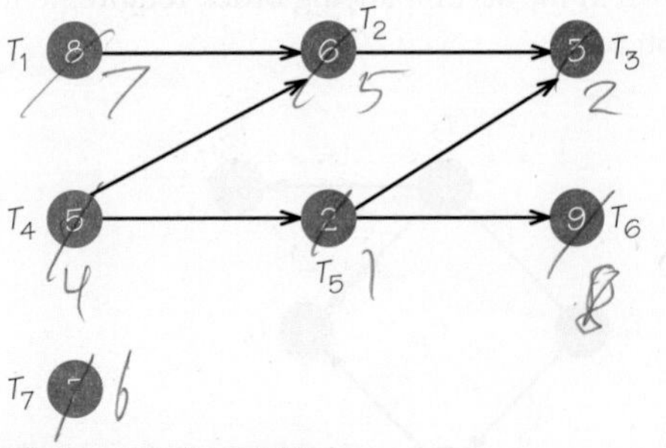

(a) Find the critical path(s).

(b) Schedule these tasks on one processor using the critical-path scheduling method.

(c) Schedule these tasks on one processor using the priority list obtained by listing the tasks in order of decreasing time.

(d) Does either of these schedules have idle time? How do their completion times compare?

(e) If two different schedules have the same completion time, what criteria can be used to say one schedule is superior to the other?

(f) Schedule these tasks on two processors using the order-requirement digraph shown and the priority list from part (b).

(g) Does the schedule produced in part (f) finish in half the time that the schedule in part (b) did, which might be expected, since the number of processors has doubled?

(h) Schedule the tasks on (i) one processor and (ii) two processors (using the decreasing-time list), assuming that each task time has been reduced by 1. Do the changes in completion time agree with your expectations?

23. Given the order-requirement digraph below, answer the following questions.

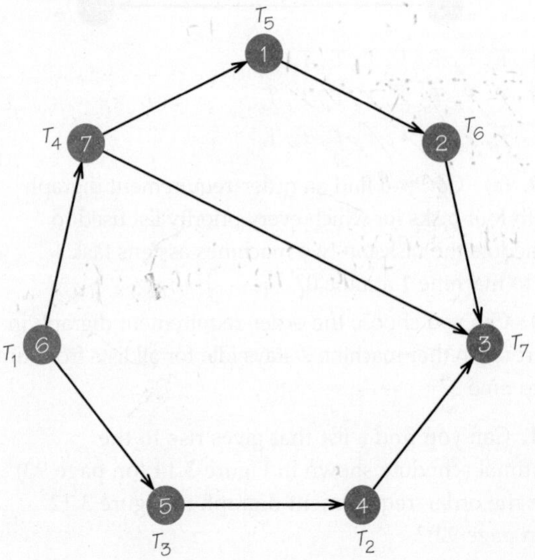

(a) Find the length of the shortest path from T_1 to T_2.

(b) What is the length of the critical path?

(c) Give a schedule that completes the tasks by the time length of the critical path on two machines, or explain why this is not possible. (If it is possible, provide a list that gives rise to this schedule.)

24. (a) The order-requirement diagram accompanying Exercise 23 shows a directed edge from T_4 to T_7. Explain why this edge can be omitted from the order-requirement digraph because it is "redundant."

(b) In Exercise 23, if the direction of the edge from T_6 to T_7 were reversed, would this still be a "legal" order-requirement digraph? (Explain your answer.)

25. (a) Can all the processors being used to schedule tasks be simultaneously idle at a time before the completion time of a collection of tasks scheduled using the list-processing algorithm?

(b) Explain why the list-processing algorithm cannot yield the schedule below, regardless of what priority list was used to schedule the tasks on the three processors.

Machine 1	T_1	T_4	T_6
Machine 2	T_2		T_7
Machine 3	T_3	T_5	

(c) Construct an order-requirement digraph and a priority list that will yield the following schedule on two processors.

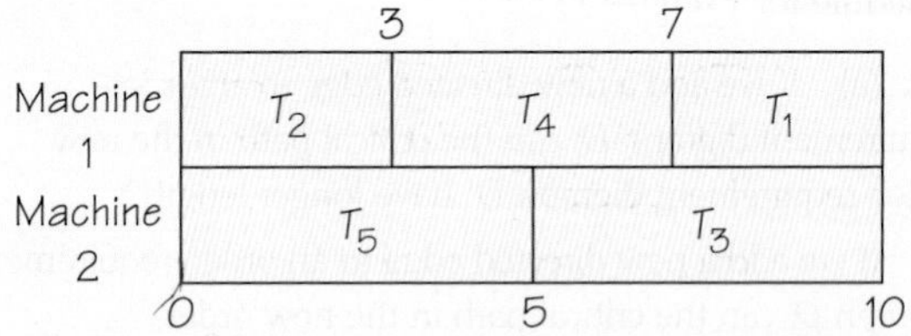

26. To prepare a meal quickly involves carrying out the tasks shown (time lengths in minutes) in the accompanying order-requirement digraph:

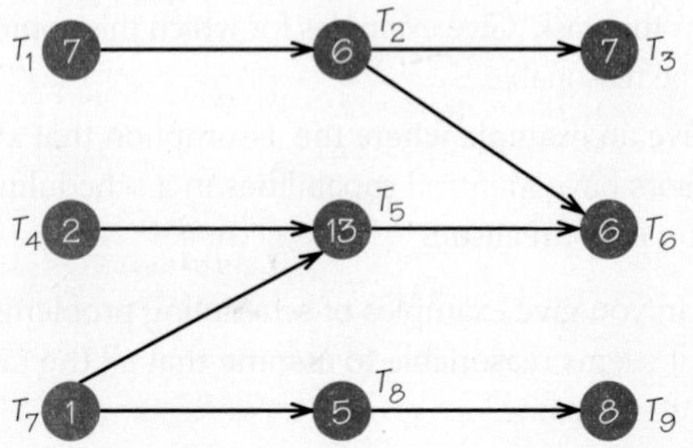

(a) If Mike prepares the meal alone, how long will it take?

(b) If Mike can talk Mary into helping him prepare the meal, how long will it take them if the tasks are scheduled using the list $T_5, T_9, T_1, T_3, T_2, T_6, T_8, T_4, T_7$ and the list-processing algorithm?

(c) If Mike can talk Mary and Jack into helping him prepare the meal, how long will it take if the tasks are scheduled using the same list as in part (b)?

(d) What would be a reasonable set of criteria for choosing a priority list in this situation?

27. (a) Making use of the order-requirement digraph below, determine at time 0 which tasks are ready.

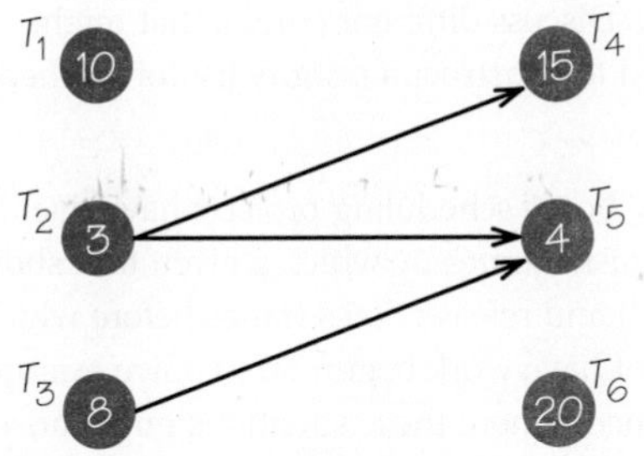

(b) What is special about tasks T_1 and T_6?

(c) What is the critical path, and what is its length?

(d) Schedule the tasks on three processors with the priority list $T_1, \ldots, T_6$.

(e) Is the schedule found in part (d) optimal?

(f) Schedule the tasks on three processors using the priority list $T_6, \ldots, T_1$.

(g) Is the schedule found in part (f) optimal?

(h) Can you find a priority list that yields an optimal schedule?

28. (a) In Exercise 27, what priority list would be used if you applied the critical-path scheduling method?

(b) Use this priority list to schedule the tasks on three processors. Is this schedule optimal?

(c) How does this schedule compare with the schedules that you found using the lists in Exercise 27?

29. Consider the following order-requirement digraph. Suppose one plans to schedule these tasks on two identical processors.

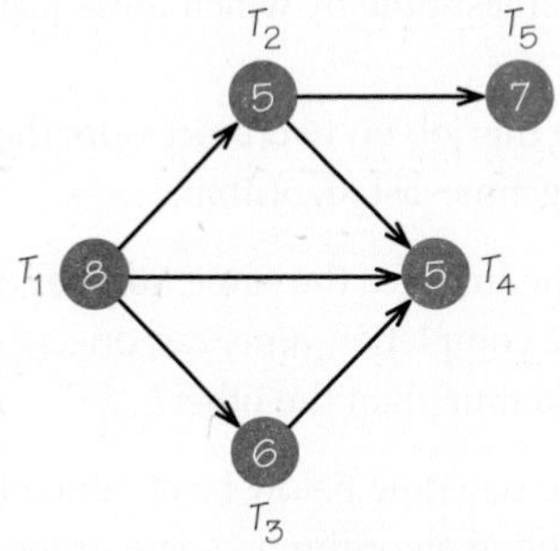

(a) How many different priority lists can be used to schedule the tasks?

(b) Can all these priority lists lead to different schedules? If not, why not?

(c) Can an optimal schedule have no idle time? Can you give two different reasons why an optimal schedule must have some idle time?

(d) Is there any list that produces a schedule where the second processor has no idle time?

30. (a) In Exercise 29, how many different lists are there that do not list T_1 first?

(b) Would it make any sense not to list T_1 first in a list?

(c) Construct a list and schedule the tasks on two processors.

(d) Can you find another list that leads to a different completion time than the schedule you found for part (c)?

(e) Find a list that leads to an optimal schedule.

31. Can you find an order-requirement digraph with four tasks for which every possible list yields exactly the same schedule?

32. Can you find an order-requirement digraph involving three tasks such that the schedule corresponding to every list is different?

33. At a large toy store, scooters arrive unassembled in boxes. To assemble a scooter, the following tasks must be performed:

Task 1. Remove parts from the box.
Task 2. Attach wheels to the footboard.
Task 3. Attach vertical housing.
Task 4. Attach handlebars to vertical housing.
Task 5. Put on reflector tape.
Task 6. Attach bell to handlebars.
Task 7. Attach decals.
Task 8. Attach kickstand.
Task 9. Attach safety instructions to handlebars.

(a) Give reasonable time estimates for these tasks and construct a reasonable order-requirement digraph.

What is the earliest time by which these tasks can be completed?

(b) Schedule this job on two processors (humans) using the decreasing-time-list algorithm.

34. If two schedules for the same number of processors have the same completion time, can one schedule have more idle time than the other?

35. Could the schedule below be obtained by applying the list-scheduling algorithm to some order-requirement digraph?

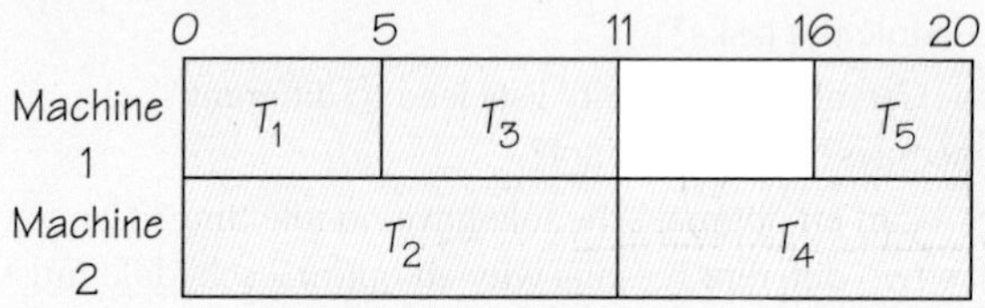

3.3 Independent Tasks

36. Could the following schedule be obtained by applying the list-scheduling algorithm to some order-requirement digraph?

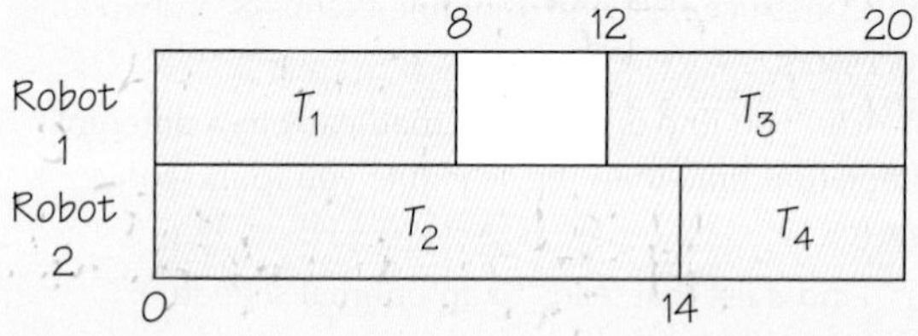

37. For the accompanying schedules, can you produce a list so that the list-processing algorithm produces the schedule shown when the tasks are independent? What are the times for each task?

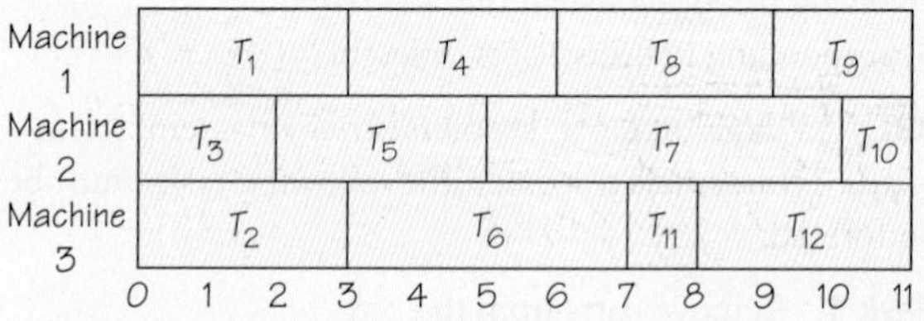

(a)

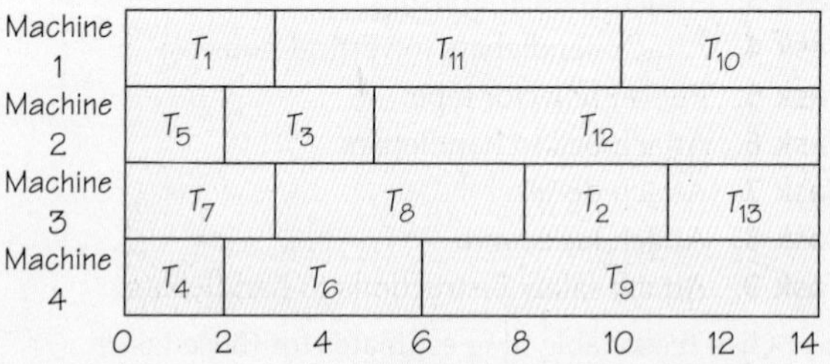

(b)

38. Once an optimal schedule has been found for independent tasks (see the diagrams in Exercise 37), usually the scheduling of the tasks can be rearranged and the same optimal time achieved.

We can, among other things, reorder the tasks done by a particular processor. Discuss criteria that might be used to implement the rearrangement process.

39. The task times of eight independent tasks T_1 to T_8 are 1, 2, 3, 4, 5, 6, 7, 8.

(a) Schedule the tasks on two processors using the lists (i) $T_1, T_2, \ldots, T_8$ and (ii) $T_8, T_7, \ldots, T_1$.

(b) Is either of the schedules you get in part (a) optimal? If not, find a list that gives an optimal schedule.

40. Repeat Exercise 39, but schedule the tasks (with the same lists) on three processors. If the schedules you get are not optimal, find a list that gives an optimal schedule.

41. Discuss different criteria that might be used to construct a priority list for a scheduling problem.

42. Some scheduling projects have due dates for tasks (times by which a given task should be completed) and release dates (times before which a task cannot have work begun on it). Give examples of circumstances where these situations might arise.

43. Using the lists you found in Exercise 37 and the task times you computed for those independent tasks, schedule the tasks for part (a) on four processors and the tasks for part (b) on five processors. Can you see why for any schedule you may produce for part (a) on four processors and part (b) on five processors, there must be some idle time for one or more processors?

44. Use the order-requirement digraph below to answer the following questions.

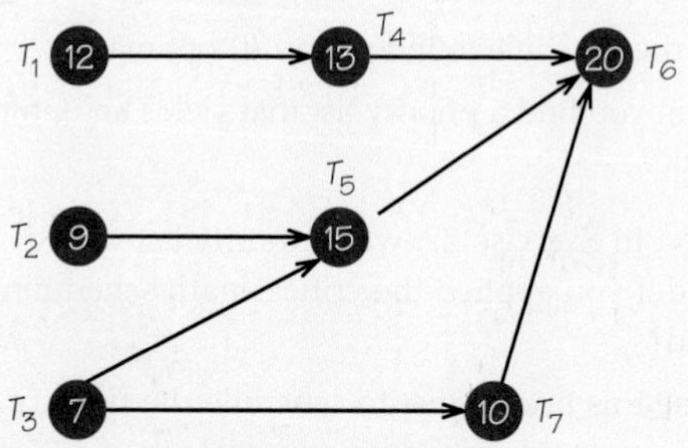

(a) Use the list-processing algorithm to schedule these seven tasks on two processors using these lists:

(i) $T_1, T_3, T_7, T_2, T_4, T_5, T_6$

(ii) $T_1, T_3, T_2, T_4, T_5, T_6, T_7$

(iii) The list obtained by listing the tasks in order of decreasing time

(b) Try to determine whether any of the resulting schedules are optimal.

(c) Schedule the tasks using the critical-path scheduling method. Try to determine whether this schedule is optimal.

45. Repeat the questions in Exercise 44 using the order-requirement digraph obtained by erasing all the (directed) edges shown there. How do the schedules you obtain compare with the ones you originally got?

46. (a) Find the completion time for independent tasks of length 8, 11, 17, 14, 16, 9, 2, 1, 18, 5, 3, 7, 6, 2, 1 on two processors, using the list-processing algorithm.

(b) Find the completion time for the tasks in part (a) on two processors, using the decreasing-time-list algorithm.

(c) Does either algorithm give rise to an optimal schedule?

(d) Repeat for tasks of lengths 19, 19, 20, 20, 1, 1, 2, 2, 3, 3, 5, 5, 11, 11, 17, 18, 18, 17, 2, 16, 16, 2.

47. Repeat parts (a)–(c) of Exercise 46 for independent tasks of lengths 19, 19, 20, 20, 1, 1, 2, 2, 3, 3, 5, 5, 11, 11, 17, 17, 18, 18, 17, 2, 16, 16, 2.

48. Suppose that independent tasks require a total of 36 minutes, but only one of the tasks takes as long as 12 minutes. If these tasks are scheduled on two machines, show by an example that the earliest completion time may be as long as 22 minutes.

49. A photocopy shop must schedule independent batches of documents to be copied. The times for the different sets of documents are (in minutes) 12, 23, 32, 13, 24, 45, 23, 23, 14, 21, 34, 53, 18, 63, 47, 25, 74, 23, 43, 43, 16, 16, 76.

(a) Construct a schedule using the list-processing algorithm on three machines.

(b) Construct a schedule using the list-processing algorithm on four machines.

(c) Repeat parts (a) and (b), but use the decreasing-time-list algorithm.

(d) Suppose union regulations require that an 8-minute rest period be allowed for any photocopy task over 45 minutes. Use the decreasing-time-list algorithm, with the preceding times modified to take into account the union requirement, to schedule the tasks on three human-operated machines.

50. Find a list that produces the following optimal schedule when the list-processing algorithm is applied to this list. (Assume that the tasks are independent.)

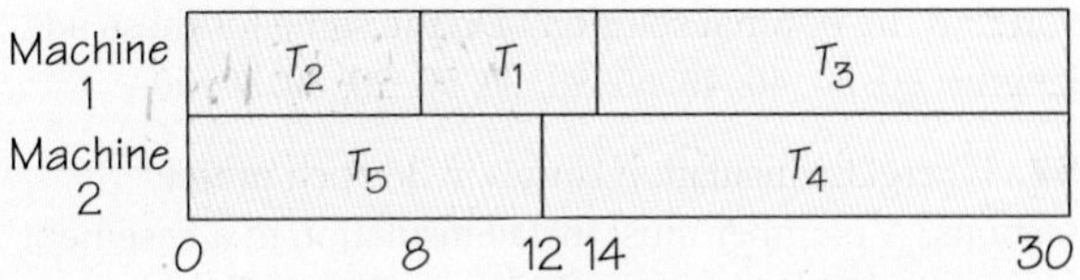

What completion time and schedule are obtained when the decreasing-time-list algorithm is applied to this list?

51. Can you think of situations other than those mentioned in the text where scheduling independent tasks on processors occurs?

52. Can you think of real-world scheduling situations in which all the tasks have the same time and are independent? Find an algorithm for solving this problem optimally. (If there are n independent tasks of time length k, when will all the tasks be finished?)

53. Show that when tasks to be scheduled are independent, the critical-path method and the decreasing-time-list method are identical.

3.4 Bin Packing

54. Two wooden wall systems are to be made of pieces of wood with lengths shown in the accompanying diagram. If wood is sold in 10-foot planks and can be cut with no waste, what number of boards would be purchased if one uses the FFD, NFD, and WFD heuristics, respectively?

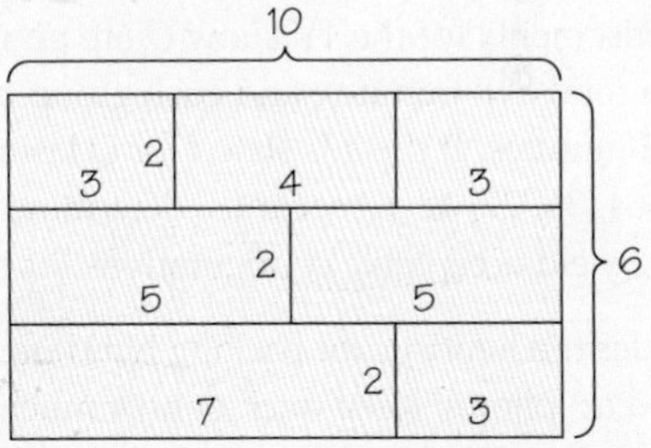

In solving this problem, does it make a difference if the 10-foot horizontal shelves and 6-foot vertical boards employ single-length pieces, compared with using pieces of boards that add up to 10- and 6-foot lengths?

55. It takes 4 seconds to photocopy one page. Manuscripts of 10, 8, 15, 24, 22, 24, 20, 14, 19, 12, 16, 30, 15, and 16 pages are to be photocopied. How many photocopy machines would be required, using the FFD algorithm, to guarantee that all manuscripts are photocopied in 2 minutes or less? Would the solution differ if WFD were used?

56. A radio station's policy allows advertising breaks of no longer than 2 minutes, 15 seconds. Using FF and FFD algorithms, determine the minimum number of breaks into which the following ads will fit (lengths given in seconds): 80, 90, 130, 50, 60, 20, 90, 30, 30, 40. Can you find the optimal solution? Do the same for these ad lengths: 60, 50, 40, 40, 60, 90, 90, 50, 20, 30, 30, 50.

57. Fiberglass insulation comes in 36-inch pre-cut sections. A plumber must install insulation in a basement on piping that is interrupted often by joints. The distances between the joints on the stretches of pipe that must be insulated are 12, 15, 16, 12, 9, 11, 15, 17, 12, 14, 17, 18, 19, 21, 31, 7, 21, 9, 23, 24, 15, 16, 12, 9, 8, 27, 22, 18 inches. How many pre-cut sections would he have to use to provide the insulation if he bases his decision on

(a) next fit?

(b) next-fit decreasing?

(c) worst fit?

(d) worst-fit decreasing?

58. The files that a company has for its employees dealing with utilities occupy 100, 120, 60, 90, 110, 45, 30, 70, 60, 50, 40, 25, 65, 25, 55, 35, 45, 60, 75, 30, 120, 100, 60, 90, 85 sectors. If, after operating systems are installed, a disk can store up to 480 sectors, determine the number of disks needed to store the utilities if each of these heuristics is used to pack the disk with files.

(a) Next fit

(b) Next-fit decreasing

(c) First fit

(d) First-fit decreasing

59. Advertisements for the TV show *Q* are permitted to last up to a total of 8 minutes, and each group of ads can last up to 2 minutes. If the ads slated for *Q* last 63, 32, 11, 19, 24, 87, 64, 36, 27, 42, 63 seconds, determine whether FF and FFD yield acceptable configurations for the ads.

60. Consider the heuristic for packing bins known as *best fit* described as follows: Keep track of how much room remains in each unfilled bin and put the next item to be packed into that bin that would leave the least room left over after the item is put into the bin. (For example, suppose that bin 4 had 6 units left, bin 7 had 5 units left, and bin 9 had 8 units left. If the next item in the list had size 5, then first fit would place this item in bin 4, worst fit would place the item in bin 9, while best fit would place the item in bin 7.) If there is a tie, place the item in the bin with the lowest number. Apply this heuristic to the list 8, 7, 1, 9, 2, 5, 7, 3, 6, 4, where the bins have capacity 10.

61. We have described two algorithms for bin packing called worst fit and best fit (see page 97 and Exercise 60). The words *best* and *worst* have connotations in English. However, the performance of algorithms depends on their merits as algorithms, not on the names we give them.

(a) On the basis of experiments you perform with the best-fit and worst-fit algorithms, which one do you think is the "better" of the two?

(b) Can you construct an example where worst fit uses fewer bins than best fit?

62. The best-fit heuristic (see Exercise 60) also has a "decreasing" version, where the list is first sorted in decreasing order. Using bins of capacity 10, apply the best-fit heuristic and its decreasing version to the following list: 6, 9, 5, 8, 3, 2, 1, 9, 2, 7, 2, 5, 4, 3, 7, 6, 2, 8, 3, 7, 1, 6, 4, 2, 5, 3, 7, 2, 5, 2, 3, 6, 2, 7, 1, 3, 5, 4, 2, 6.

63. One pianist's recording of the complete Mozart piano sonatas takes the following times (given in minutes and seconds): 13:46, 6:15, 3:29, 5:37, 7:52, 2:55, 5:00, 4:28, 4:21, 7:39, 7:55, 6:42, 4:23, 3:52, 4:21, 4:20, 5:46, 6:29, 5:34, 6:23, 6:39, 7:19, 5:54, 6:54, 2:58, 5:22, 1:42, 5:00, 1:29, 5:47, 7:30, 8:19, 4:44, 4:57, 4:09, 14:31, 3:55, 4:04, 4:01, 6:06, 6:50, 5:27, 4:28, 5:40, 2:52, 5:16, 5:34, 3:10, 7:22, 4:40, 3:08, 6:32, 4:47, 6:59, 5:38, 7:57, 3:38. If the maximum time that can be recorded on a compact disc is 70:30, can all the music be performed on four compact discs? Can all the music be performed on five compact discs?

64. In the wall-system example in the text, first fit and worst fit required equal numbers of bins (see Figure 3.20 on page 97). Can you find an example where first fit and worst fit yield different numbers of bins? Can you find an example where first fit, worst fit, and next fit yield answers with different numbers of bins?

65. A common suggestion for heuristics for the bin-packing problem with bins of capacity W involves finding weights that sum to exactly W. Discuss the pros and cons of a heuristic of this type.

66. A recording company wishes to record all the Beethoven string quartets (16 quartets, each consisting of several consecutive parts called movements) on reissued LPs. It wishes to complete the project on as few records as possible. Recording can be done on two sides as long as the movements are consecutive. Is this an example of a bin-packing problem? (Defend your answer.) If the project were to record the quartets on (standard) tape cassettes or compact discs, would your answer be different?

67. Give examples where it would be realistic to keep bins open as more items "arrive" to be packed, rather than to close a bin permanently based on some criterion.

68. Give examples where it would be unrealistic to keep bins open as more items "arrive" to be packed, rather than to close a bin permanently based on some criterion.

69. A data entry group must handle 30 (independent) tasks that will take the following amounts of time (in minutes) to type: 25, 18, 13, 19, 30, 32, 12, 36, 25, 17, 18, 26, 12, 15, 31, 18, 15, 18, 16, 19, 30, 12, 16, 15, 24, 16, 27, 18, 9, 14. Use these times as a priority list.

(a) Use the list-processing algorithm to find the completion time for scheduling tasks with four secretaries. Also, solve with five secretaries.

(b) Repeat the scheduling using the decreasing-time-list algorithm.

(c) Can you show that any of the schedules you get are optimal?

Say the typing needs to be finished in 1 hour.

(d) Use the FFD heuristic to find how many typists are needed.

(e) Repeat for the NFD and WFD heuristics.

(f) Can you show that any of the solutions you get are optimal?

70. Find the minimum number of bins necessary to pack items of size 8, 5, 3, 4, 3, 7, 8, 8, 6, 5, 3, 2, 1, 2, 1, 2, 1, 3, 5, 2, 4, 2, 6, 5, 3, 4, 2, 6, 7, 7, 8, 6, 5, 4, 6, 1, 4, 7, 5, 1, 2, 4 in bins of capacity (a) through (d) using the FF and FFD algorithms. Can you determine whether any of the packings you get are optimal?

(a) 9

(b) 10

(c) 11

(d) 12

71. Two-dimensional bin packing refers to the problem of packing rectangles of various sizes into a minimum number of $m \times n$ rectangles, with the sides of the packed rectangles parallel to those of the containing rectangle.

(a) Suggest some possible real-world applications of this problem.

(b) Devise a heuristic algorithm for this problem.

(c) Give an argument to show that the problem is at least as hard to solve as the usual bin-packing problem.

(d) If you have $1 \times m$ rectangles with total area W to be packed into a single rectangle of area $p \times q = W$, can the packing always be accomplished?

72. In what situations would packing bins of different capacities be the appropriate model for real-world situations? Suggest some possible algorithms for this type of problem.

73. Find an example of weights that, when packed into bins using first fit, use fewer bins than the number of bins used when the first-fit algorithm is applied with the first weight on the list removed.

74. Formulate "paradoxical" situations for bin packing that are analogous to those we found for scheduling processors.

3.5 Resolving Conflict via Coloring

75. Draw a connected graph with 12 vertices and 11 edges whose vertices can be colored with two colors.

76. Add a single edge to the graph you drew for Exercise 75 so that the vertices of the resulting graph cannot be colored with two colors but require three colors.

77. For each of the graphs below, answer the following questions.

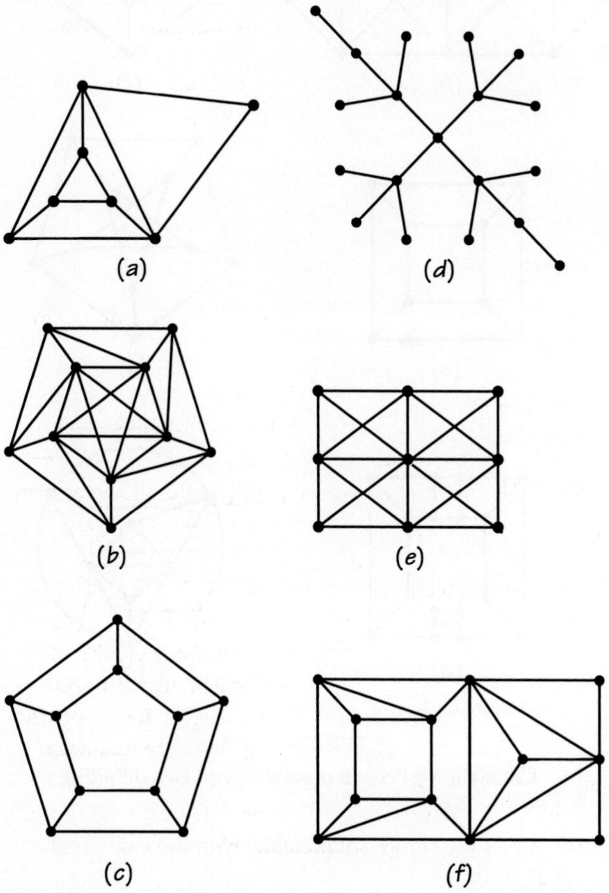

(a) Color the vertices (if possible) with three different colors.

(b) Color the vertices (if possible) with four different colors.

(c) Find the chromatic number of the graph.

78. For each of the accompanying graphs, answer the following questions.

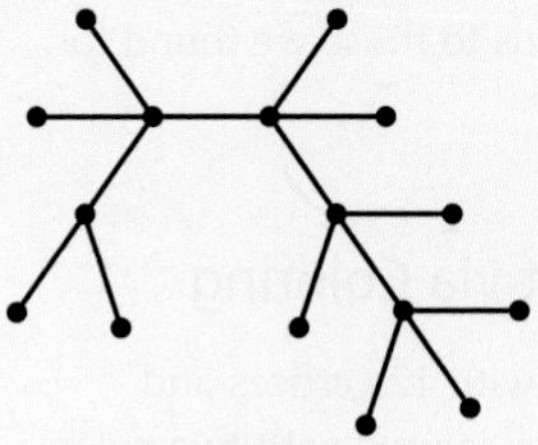

(a)

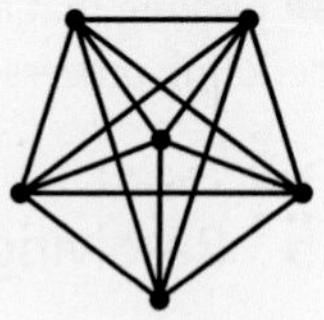

(e)

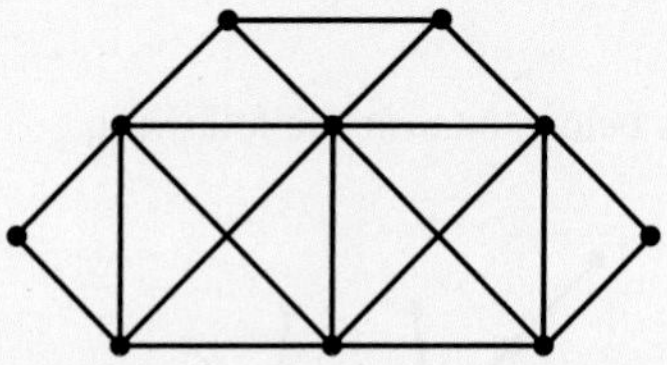

(b)

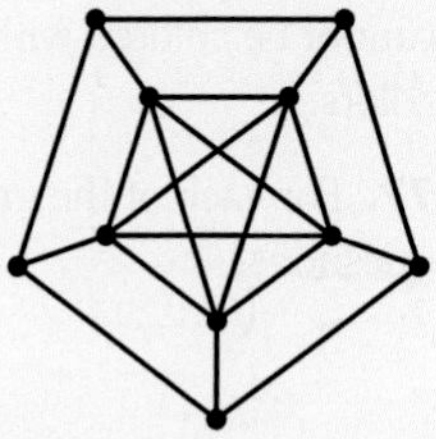

(f)

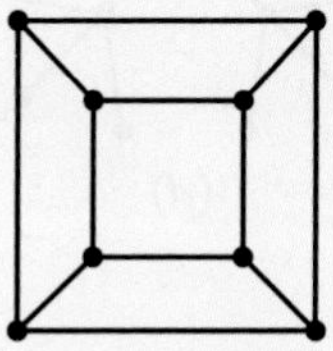

(c)

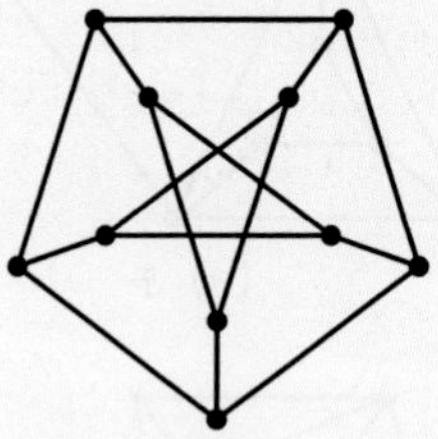

(g)

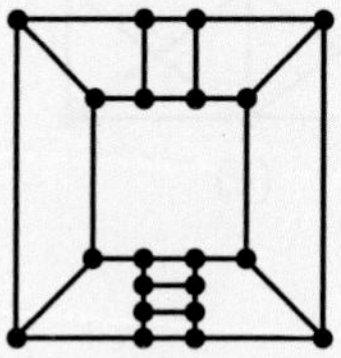

(d)

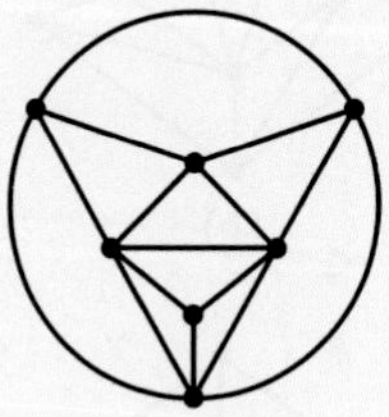

(h)

(a) Color the vertices (if possible) with two different colors.

(b) Color the vertices (if possible) with three different colors.

(c) Find the chromatic number of the graph.

79. The owner of a new pet store wishes to display tropical fish in display tanks. The accompanying table shows the incompatibilities between the species, in the sense that an X indicates that it is unwise to allow those species in the row and column that meet at the X to be in the same tank.

	A	*B*	*C*	*D*	*E*	*F*	*G*	*H*	*I*
A						X	X		X
B			X					X	
C		X			X			X	
D					X	X		X	
E			X	X			X		
F	X			X			X		X
G	X				X	X		X	X
H		X	X	X			X		
I	X					X	X		

(a) Draw an appropriate graph to represent the information in the table.

(b) What is the minimum number of tanks needed to display all the fish she wishes to sell?

(c) Display the species so that the number of species in each tank is as nearly equal as possible.

80. The managers of a zoo are planning to open a small satellite branch. The animals are to be in enclosures in which compatible animals are displayed together. The accompanying table indicates those pairs of animals that are compatible. (Thus, an X in a particular row and column means that the animals that label this row and column *can* share an enclosure.)

	A	*B*	*C*	*D*	*E*	*F*	*G*	*H*	*I*	*J*
A	X	X		X	X	X	X			
B	X	X			X	X	X		X	X
C			X		X	X	X			
D	X			X	X	X	X		X	X
E	X	X	X	X	X			X	X	
F	X	X	X	X		X	X	X	X	
G	X	X	X	X		X	X	X		
H					X	X	X	X		
I		X		X	X	X			X	
J		X		X						X

(a) Draw an appropriate graph to represent the information in the table.

(b) What is the minimum number of enclosures needed to avoid housing incompatible animals in the same enclosure?

(c) Is it possible to enclose the animals in such a way that each enclosure contains the same number of animals?

(d) Why might that be desirable? Why might this approach to grouping the animals not be ideal?

81. The nine standing committees of a state legislature are designing a schedule for when the committees can meet. The matrix shown in the following table has an X in a position where the committees corresponding to the row and column have a common member and, hence, should not be scheduled to meet at the same hour. The committees involved are Agriculture (*A*), Commerce (*C*), Consumer Affairs (*CA*), Education (*E*), Forests (*F*), Health (*H*), Justice (*J*), Labor (*L*), and Rules (*R*).

	A	***C***	***CA***	***E***	***F***	***H***	***J***	***L***	***R***
A		X	X			X			
C	X		X	X	X				
CA	X	X					X		X
E		X			X	X			
F		X		X		X	X		
H	X			X	X			X	
J			X		X			X	X
L						X	X		X
R			X				X	X	

(a) Draw a graph that will be of value in determining the minimum number of time slots the committees can meet in without any legislator having to be in two places at one time.

(b) What is the minimum number of time slots in which the committees can be scheduled without a conflict?

(c) How many different rooms are needed at any time that a committee is scheduled to meet? (Why might this issue matter?)

82. Determine the minimum number of colors, and how often each color is used, in a vertex coloring of the graphs below.

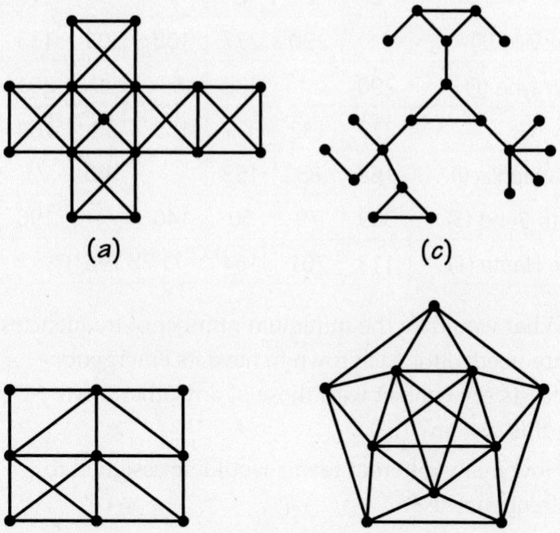
(*a*) (*c*) (*b*) (*d*)

83. The faculty–student governing council at All State College has nine standing committees (such as Curriculum, Academic Standards, Campus Life) that are designated *A*, *B*, *C*, *D*, …, *I* for convenience. The following table shows which committees have no member in common:

	A	***B***	***C***	***D***	***E***	***F***	***G***	***H***	***I***
A		X		X		X	X		X
B	X				X	X		X	X
C				X		X	X	X	X
D	X		X			X		X	
E		X					X	X	X
F	X	X	X	X					
G	X		X		X			X	
H		X	X	X	X		X		X
I	X	X	X		X			X	

(a) Draw an appropriate graph to represent the information in the table.

(b) What is the minimum number of time slots in which all the committee meetings can be scheduled?

(c) How many rooms are needed during each time slot to accommodate the committees that are scheduled to meet in that time slot?

84. When two towns are within 145 miles of each other, the frequency used by a certain type of emergency response system for the towns requires that they be on different frequencies to avoid possible interference with each other. The following table shows the mileage distances between six towns:

	E	*F*	*G*	*I*	*S*	*T*
Evansville (*E*)		290	277	168	303	133
Ft. Wayne (*F*)	290		132	83	79	201
Gary (*G*)	277	132		153	58	164
Indianapolis (*I*)	168	83	153		140	71
South Bend (*S*)	303	79	50	140		196
Terre Haute (*T*)	113	201	164	71	196	

(a) What would be the minimum number of frequencies that are needed for each town to have its emergency broadcasts not conflict with those of any other town using this system?

(b) How many different towns would be assigned to each frequency used?

85. The legislature of a city has committees devoted to the following governmental areas:

A = agriculture; P = planning; D = districting; F = finance; E = education; T = transportation; H = housing, C = courts. To schedule meetings for these committees in as few time slots as possible, consider the graph model below.

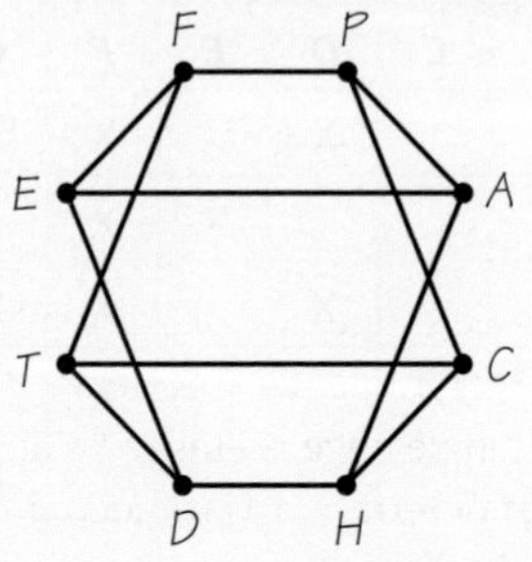

(a) Complete the entries in the table below that would display the "conflict" information represented by the graph, where an X indicates two committees that cannot meet at the same time because they have at least one common member.

	F	*E*	*H*	*P*	*T*	*A*	*D*	*C*
F								
E								
H								
P								
T								
A								
D								
C								

(b) What is the minimum number of time slots in which it is possible to schedule these committees?

(c) If there are only three rooms with the audio and video setups needed for the committees to meet, what is the minimum number of time slots during which the committees can meet?

86. A small college has a mini-session in January (between its regular semesters) in which nine classes are offered: A = Art; B = Biology, C = Chemistry; D = Diversity; E = English, F = French, G = German, H = Health, I = Italian. The accompanying table indicates those courses that have students in common.

	A	*B*	*C*	*D*	*E*	*F*	*G*	*H*	*I*
A		X	X	X					
B	X				X	X			
C	X			X			X		X
D	X		X				X		
E		X				X		X	
F		X			X			X	X
G			X	X				X	X
H					X	X	X		X
I			X			X	X	X	

(a) Draw a graph model for the information in this table.

(b) What is the minimum number of time slots in which final examinations can be scheduled for these nine courses?

(c) Can this minimum be achieved if only three rooms are large enough to hold the finals?

87. Show that the vertices of any tree can be colored with two colors.

88. Can you find a family of graphs H_n ($n \geq 1$) that require n colors to color their vertices?

89. The edge-coloring number of a graph G is the minimum number of colors needed to color the edges of G so that edges sharing a common vertex get different colors. Determine the edge-coloring number for each of the graphs in Exercise 77. Can you make a conjecture about the value of the minimum number of colors needed to color the edges of any graph?

90. Can you think of any applications that require determining the minimum number of colors needed to color the edges of a graph?

91. When a graph has been drawn on a piece of paper so that edges meet only at vertices, the graph divides the paper up into regions called *faces.* The faces include one called the "infinite" face, which surrounds the whole graph. The face-coloring number of a graph G (which can be drawn in this special way) is the minimum number of colors needed to color the faces of G so that two faces sharing an edge receive different colors. (Note that if two faces meet only at a vertex, they can be colored the same color.)

(a) Determine the minimum number of colors needed to color the faces of the accompanying graphs. In each case, remember to color the infinite face, which is labeled I (for "infinite").

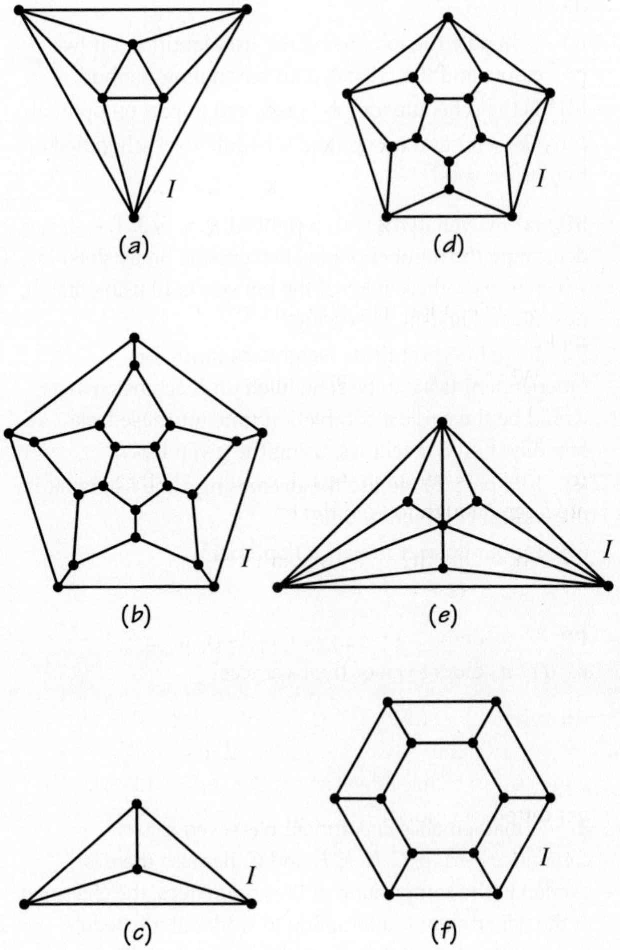

(b) Can you think of an application of the problem of coloring the faces of a graph with a minimum number of colors?

92. For each of the graphs in Exercise 78 where the graph shown has edges that meet only at vertices, verify that the Four Color Theorem holds by showing that the regions (faces) of the graph can be colored with four or fewer colors so that regions sharing an edge get different colors. (Remember to assign a color to the unbounded, so-called infinite region.)

93. A company sells herbs, each of which requires a certain level of proper watering. The accompanying graph is constructed by having one vertex for each type of herb. The vertices representing two herbs are joined by an edge if they must have different levels of watering. What is the minimum number of terrariums that the herbs can be displayed in so that herbs in the same terrarium can be watered at the same level?

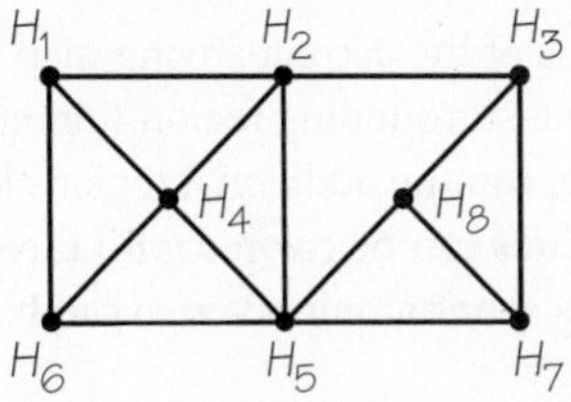

94. The company in Exercise 93 is disappointed by the minimum number of terrariums needed to display the herbs with the proper watering requirements. One company employee suggests that if the information about watering requirements is altered for a single pair of herbs (e.g., a single edge is erased from the diagram), then the number of terrariums needed will be reduced by 1. Is this true?

95. Each vertex in the graph below represents a child who attends a daycare center. An edge between two children indicates these children tend to cause problems when they are in the same play group. What is the minimum number of play groups that will ensure that no conflicts arise? Can conflict-free play groups with the same number of children in each group be formed?

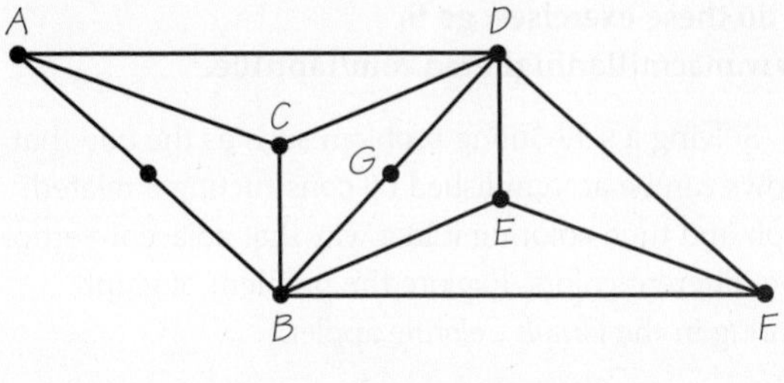

Chapter Review

96. Determine the minimum number of time slots to schedule the seven committees represented by vertices in the accompanying graph, if edges show which committees cannot meet at the same time.

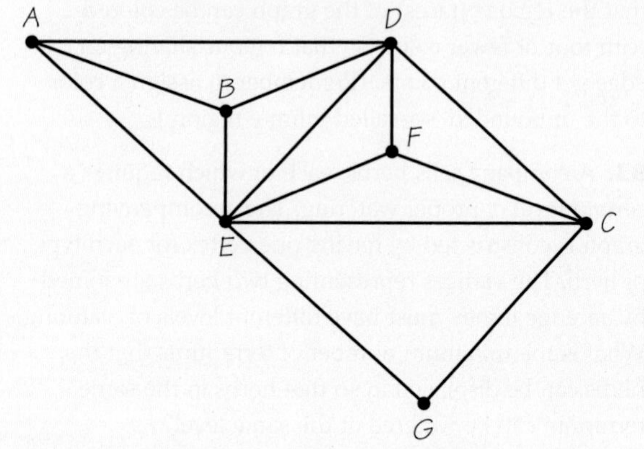

97. The regions of the accompanying map will appear in an atlas. If the surrounding region (infinite face) is to be colored blue, can the additional regions be colored so that all the regions can be colored with three or fewer colors? Regions sharing only a vertex can be colored with the same color.

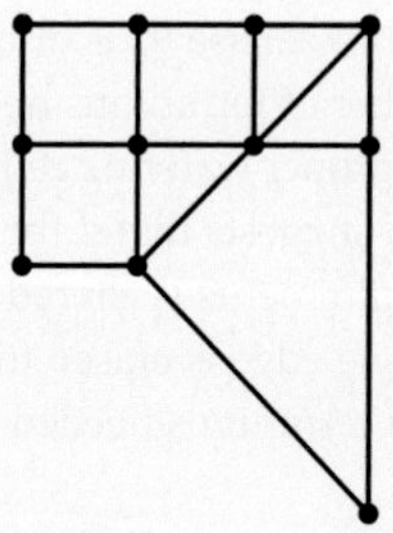

98. Given the job J (times shown in months) to publish a new edition of a book consisting of the tasks given in the order-requirement digraph D below, answer the following questions.

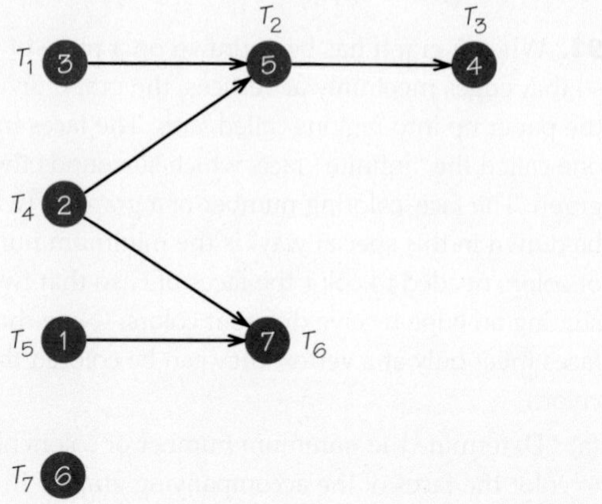

(a) Determine the length of the critical path(s) in D and list the tasks on the critical path.
(b) Explain why even with five processors, this job can't be completed by time 8, which is the total task time divided by 5.
(c) Schedule the job J using the list algorithm on two processors and the critical-path scheduling method.
(d) Is the schedule for job J produced in part (c) optimal?
(e) Can you find an optimal schedule for J scheduled on two processors?

99. (a) Given items with weights 3, 9, 6, 5, 3, 1, 7, 4, 5, 6, determine the number of bins (advertising break slots) in which to pack these items if the bin size is 10 using first fit, next fit, and first-fit decreasing.
(b) If we interpret these weights as times for independent tasks to be scheduled on machines, what would be the earliest completion time for these tasks scheduled on 3 machines, using the given list?
(c) If in part (b) we use the decreasing-time list, what is the earliest completion time?
(d) Is your answer in part (c) optimal?

Applet Exercises

To do these exercises, go to www.macmillanhighered.com/fapp10e.

1. Solving a scheduling problem such as the one that follows can be accomplished by constructing a related graph and then coloring it in a way that adjacent vertices have different colors. Explore the problem of graph coloring in the *Graph Coloring* applet.

2. A mathematics department has seven faculty committees—*A, B, C, D, E, F,* and *G*. Because there is overlap in the composition of the committees, the chairman of the department is attempting to work out a schedule that will avoid conflicts among the committees. The accompanying chart indicates the overlapping committee structure. Help the chairman arrange a schedule without conflicts in the *Scheduling* applet.

	A	B	C	D	E	F	G
A		X		X		X	
B	X		X			X	
C		X			X		X
D	X						X
E			X			X	X
F	X	X			X		
G			X	X	X		

Writing Projects

1. Scheduling is important for medical centers schools, transportation systems, police services, and fire services. Pick one of these areas and write about the different scheduling situations that come up, the types of processors, and the extent to which the assumptions of the list-processing model hold for the area you pick.

2. Compare and contrast the basic scheduling problem we investigated with the scheduling version of the bin-packing problem.

3. One of the oversimplifications made in our discussion of scheduling was that there were no "due dates" involved for the tasks making up a job. Develop an algorithm for solving a scheduling problem under the assumption that each task has a due date as well as a time length. You will probably want to decide on a penalty amount that will occur when a due date is exceeded.

4. Consider the problem of scheduling tasks on a single machine. Design different algorithms for achieving different goals. You will probably wish to assume that each task has a due date such that if the task is not finished by this date, some penalty payment must be made.

5. Discuss the role of graph colorings for scheduling committee meetings so as to avoid conflicts. Research whether or not these ideas are used in the legislature of your home state.

6. In choosing a location (vertex) for trains to turn around in the graph shown in Spotlight 3.1 (on page 90), explain why it seems to be a much better choice to use *V* as a place to allow the turnarounds, rather than at *M* or at *R*.

Suggested Readings

BRUCKER, P. *Scheduling Algorithms,* 4th ed., Springer-Verlag, Heidelberg, Germany, 2004. A detailed mathematical look at scheduling.

GRAHAM, RONALD. Combinatorial scheduling theory, in Lynn Steen (ed.), *Mathematics Today,* Springer-Verlag, New York, 1978, pp. 183–211. This essay on scheduling is one of many excellent accounts of later 20th-century developments in mathematics in this book.

GRAHAM, RONALD. The combinatorial mathematics of scheduling. *Scientific American,* March 1978, pp. 124–32. A very readable introduction to scheduling and bin packing.

JENSEN, T. R., and BJARNE TOFT. *Graph Coloring Problems,* Wiley, New York, 1995. A detailed summary of what is known about coloring problems and many questions that await answering.

LAWLER, E., et al. Sequencing and scheduling algorithms and complexity, in S. C. Graves et al. (Eds.), *Handbooks in OR and MS,* Vol. 4, Elsevier, New York, 1993, pp. 445–522. A survey of results about scheduling.

LEUNG, JOSEPH Y-T. *Handbook of Scheduling,* Chapman & Hall/CRC, Boca Raton, FL, 2004. This book offers an encyclopedic treatment of scheduling algorithms and the great variety of situations where mathematical analysis has assisted schedulers, ranging from sports to hospitals.

PARKER, R. GARY. *Deterministic Scheduling Theory,* Chapman & Hall, London, 1995. A wide-ranging look at scheduling methods and their applications.

Suggested Websites

www.ctl.ua.edu/math103/scheduling/schedmnu .htm This site provides an overview of scheduling as discussed in this chapter.

www.ams.org/samplings/feature-column/ fcarc-machines1

www.ams.org/samplings/feature-column/ fcarc-packings1

www.ams.org/samplings/feature-column/ fcarc-bins1 These web pages describe mathematical aspects of machine scheduling and bin packing and discuss the relationship between these two mathematical problems.

www.ie.bilkent.edu.tr/~ie672/docs/resources.html This web page contains links to many aspects of scheduling theory, including research on the frontier.

Linear Programming and the Transportation Problem

4

Dragan Milovanovic/Shutterstock

A manager's job often calls for making very complicated decisions. Some decisions involve planning what products the business is to make and determining what resources are needed to make these products. In the modern business world, diversification of products provides a company with stability in a climate of changing tastes and needs.

So it is not surprising that companies would produce many products, some of which share resource needs. For example, any bakery uses many resources—like butter, sugar, eggs, and flour—to make its products such as cookies, cakes, pies, and breads. Similarly, car manufacturers use many kinds of metals in the different models of cars they make, and manufacturers of gasoline use different kinds of crude oils to make their product.

Resources can include more than just raw materials. Farmland, time, machinery, and a labor force with appropriate skills are also resources. Typically, resources are limited: A farmer owns only so much land; there are only so many hours in a day; in a year of drought the wheat crop is very small; a winter freeze may damage an orange crop. Resource availability is also limited by location and competition.

Because resources are limited, management faces important questions: How should the available resources be shared among the possible products and different parts of the company's operation? One goal of management is to maximize profit. How can that determine how much of each product should be produced? There are usually so many alternative product mixes that it is impossible to evaluate them all individually. Despite this complexity, millions of dollars may ride on management's decision.

Many business and government agencies must deal with supply-and-demand problems. The general idea is that goods or services can be provided by different providers to individuals or businesses who need these goods or services. There are varying costs to the suppliers to provide different recipients with these goods or services. The goal is to find how to meet the demands from the supplies as cheaply as possible. For example, what is the cheapest way for a company with several oil refineries to provide oil distributors, in many different geographical locations, with the oil they need? Perhaps surprisingly, ideas related to linear programming have also been used to help decide whether tumors are malignant or benign.

4.1 Linear Programming and Mixture Problems: Combining Resources to Maximize Profit

Here, we learn about **linear programming,** a management science technique that helps a business allocate its resources. Whether those resources are at hand or can be purchased, the goal is to decide on a particular mix of products that will maximize profit or that will operate in ways that cut costs. For example, a large medical center needs to stock resources such as bandages, syringes, and alcohol wipes, as well as have different kinds of laborers, toilet paper, and paper towels available when needed.

Linear Programming DEFINITION

Linear programming is a tool for maximizing or minimizing a quantity, typically a profit or a cost, subject to a set of constraints.

The technique is so powerful that it is estimated that much more computer time is used solving linear programming types of problems than for any other purpose for which business managers and decision makers use computers.

Linear programming is an example of "new" mathematics. It came into being, along with many other management science techniques, during and shortly after World War II, in the 1940s. It is quite young as intellectual ideas go. Yet, during its short history, linear programming has changed the way businesses and governments make decisions, from "seat-of-the-pants" methods based on guesswork and intuition to using an algorithm based on available data and guaranteed to produce an optimal decision.

Linear programming is but one operations research tool belonging to a family of tools known as mathematical programming. Another such tool is integer programming. The difference between linear programming and integer programming is that for linear programming, the quantities being studied can take on values such as $\pi = 3.14159\ldots$ or $7\frac{1}{8}$; in integer programming, the values are confined to whole numbers such as 8, 50, or 1,102,362. Whole numbers are conceptually easier than the broader group consisting of all numbers that can be represented by decimals (1.32, 1.455555 . . .), yet integer-programming problems have proved much harder to solve.

In the discussions that follow, we often describe "relaxed" versions of integer-programming problems as linear-programming problems. For example, it would make no sense to produce 3.24 dolls to sell. So, strictly speaking, we must find an optimum whole number of dolls to produce. If we are "lucky," the linear-programming problem associated with an integer-programming problem has an integer solution. In this case, we have also found the correct answer to the integer-programming problem. Some other examples that fall into this category are discussed below.

The assembly of an automobile requires many complicated steps and processes. The use of linear-programming techniques enables the robots and humans to carry out their tasks faster and more accurately than would be possible without the use of mathematics. This makes American cars more competitive and of a higher quality than otherwise would be the case. Plus Pix/age fotostock

Linear programming has saved businesses and governments billions of dollars. Of all the management science techniques presented in this book, linear programming is far and away the most frequently used. It can be applied in a variety of situations, in addition to the one we study in this chapter. Some of the problems studied in Chapters 1, 2, and 3—for example, the traveling salesman (TSP) and scheduling problems—can be viewed as linear-programming problems. Linear programming is an excellent example of a mathematical technique useful for solving many different kinds of problems that at first do not seem to be similar problems at all. It has been suggested that without linear programming, management science would never have been born as a special branch of knowledge.

Next, we study how to use linear programming to solve a special kind of problem—a **mixture problem.** Realistic versions of such problems would be much more involved. Our discussion is designed to give you the flavor of what is actually done. Realistic examples of what follows are commonly used in the manufacture of different kinds of breads from the grain flours available, and in the making of different kinds of sausages from meats such as beef and pork.

Mixture Problem DEFINITION

In a **mixture problem,** limited resources are combined into products so that the profit from selling those products is as large as possible—a maximum.

Mixture problems are widespread because nearly every product in our economy is created by combining resources. A typical example would be how different kinds of aviation fuel are manufactured using different kinds of crude oil.

Let's analyze small versions of the kinds of problems that might confront a toy or a beverage manufacturer. Both manufacturers can sell many different products on which each company can make a profit. There could be dozens of possible products and many resources. A manufacturer must periodically look at the quantities and prices of resources and then determine which products should be produced in which quantities in order to gain the greatest, or optimum, profit (see Spotlight 4.1). This is an enormous task that usually requires a computer to solve.

What does it mean to find a solution to a linear-programming mixture problem? A solution to a mixture problem is a production policy that tells us how many units of each product to make.

Optimal Production Policy DEFINITION

An **optimal production policy** has two properties:

1. It is possible; that is, it does not violate any of the limitations under which the manufacturer operates, such as availability of resources.
2. It gives the maximum profit.

Common Features of Mixture Problems

Although our first mixture problem (Example 1) has only two products and one resource, it does contain the essential features that are common to *all* mixture problems:

- *Resources.* Definite resources are available in limited, known quantities for the time period in question. The resource in Example 1 is containers of plastic.
- *Products.* Definite products can be made by combining, or mixing, the resources. In Example 1, the products are skateboards and dolls.

Case Studies in Linear Programming

SPOTLIGHT 4.1

Linear programming is not limited to mixture problems. Here are two case studies that do not involve mixture problems, yet applying linear-programming techniques produced impressive savings:

- The Exxon Corporation spends several million dollars per day running refineries in the United States. Because running a refinery takes a lot of energy, energy-saving measures can have a large effect. Managers at Exxon's Baton Rouge plant had over 600 energy-saving projects under consideration. They couldn't implement them all because some conflicted with others, and there were so many ways of making a selection from the 600 that it was impossible to evaluate all selections individually.

  Exxon used linear programming to select an optimal configuration of about 200 projects, resulting in millions of dollars in savings.
- Edwards Lifesciences uses heart valves from pigs to produce artificial heart valves for human beings. Pig heart valves come in different sizes. Shipments of pig heart valves often contain too many of some sizes and too few of others. However, each supplier tends to ship roughly the same imbalance of valve sizes in every order, so the company can expect consistently different imbalances from the different suppliers. Thus, if they order shipments from all the suppliers, the imbalances could cancel each other out in a fairly predictable way. The amount of cancellation will depend on the sizes of the individual shipments. Unfortunately, there are too many combinations of shipment sizes to consider all combinations individually.

  Edwards Lifesciences used linear programming to figure out which combination of shipment sizes would give the best cancellation effect. This reduced the company's annual cost by $1.5 million.

- *Recipes.* A recipe for each product specifies how many units of each resource are needed to make one unit of that product. Each skateboard in Example 1 uses five units of plastic, and each doll uses two units.
- *Profits.* Each product earns a known profit per unit. (We assume that every unit produced can be sold. More complicated mathematical models, which we will not discuss here, are needed if we want to consider the possibility of items being produced but not sold.)
- *Objective.* The objective in a mixture problem is to find how much of each product to make so as to maximize the profit without exceeding any of the resource limitations.

The examples we show are not designed to be realistic. Rather, our goal is to demonstrate how ideas whose roots are in basic algebra and geometry can solve, when scaled up to realistic versions, problems that save Americans much time and money and make our government and American businesses more efficient.

EXAMPLE 1 Making Skateboards and Dolls

Patrik Giardino/Corbis

A toy manufacturer can manufacture only skateboards, only dolls, or some mixture of skateboards and dolls. Skateboards require 5 units of plastic and can be sold for a profit of $1, while dolls require 2 units of plastic and can be sold for a $0.55 profit. If 60 units of plastic are available, what numbers of skateboards and/or dolls should be manufactured for the company to maximize its profit?

Attacking this and other mixture problems requires carrying out a series of steps that determine the essence of the problem.

As a first step, we need to construct a mathematical model to take the "verbal" information that we have been given and display it in a form that makes it easier to convert into the mathematics necessary to solve the problem. This is done by making a **mixture chart** for the information we are given (see Figure 4.1).

In the rows of this chart, we display the products we want to make, and in the column of the chart, we display the resources and the profit margin information that

PRODUCTS	RESOURCE(S): Containers of Plastic 60	PROFIT
Skateboards (x units)	5	$1.00
Dolls (y units)	2	$0.55

Figure 4.1 Mixture chart for Example 1.

is available. In this case, we have two products, so we have two rows. We have one resource, which accounts for there being one column.

The other column is reserved for profit information. Since this information is used in a somewhat different way from the information about the resources, we will separate the resource column(s) from the profit column by a double bar.

Because we do not know the number of skateboards the company should make, we will use a letter x to represent the unknown number of skateboard units that the company might manufacture. Similarly, y will represent the number of dolls that the company might manufacture. We enter these letters as part of the labels of the rows of our table.

We can now enter the numbers about resources in the columns based on the information we have been given. In this case, there is one resource: plastic. Thus, for the 5 units of plastic needed for a skateboard, we record a 5 in the skateboards row and the containers of plastic column. Similarly, we enter a 2 in the second row and first column because dolls require 2 units of plastic each. Because we have 60 units of plastic available, we display this fact by placing the number 60 at the top of this column. We complete the table with the information about profit. We enter \$1 in the skateboards row and profit column and \$0.55 in the dolls row and profit column.

EXAMPLE 2 Making a Mixture Chart

Make a mixture chart to display (model) this situation: A clothing manufacturer has 60 yards of cloth available to make shirts and decorated vests. Each shirt requires 3 yards of cloth and provides a profit of \$5. Each vest requires 2 yards of cloth and provides a profit of \$3. A mixture chart is shown in Figure 4.2.

PRODUCTS	RESOURCE(S) Yards of Cloth 60	PROFIT
Shirts (x units)	3	\$5
Vests (y units)	2	\$3

Figure 4.2 Mixture chart for the clothing manufacturer.

Translating Mixture Charts into Mathematical Form

Consider again the mixture chart in Figure 4.1. What can we say about the numbers of skateboards and dolls that might be manufactured? Clearly, we cannot make negative numbers of skateboards or dolls. Because we are using the letter x to represent the number of skateboards we plan to make, we can write down the algebraic expression that $x \geq 0$. Here, we are using the standard symbol $\geq$ for "greater than or equal to."

Algebraic expressions that involve the symbol $\geq$ or its companion symbols $\leq$ (less than or equal to), $>$ (greater than), and $<$ (less than) are known as *inequalities*. We can also write down an inequality for the y number of dolls we plan to make, based on the fact that we cannot make a negative number of dolls. Thus, we must have $y \geq 0$. We will use the phrase **minimum constraints** for these two inequalities, $x \geq 0$ and $y \geq 0$, which mean that we cannot manufacture negative numbers of objects.

However, we also have only a limited number of units of plastic available. How can we represent this information? Consulting the mixture table, we see that we need 5 units of plastic for every skateboard we make. Thus, we will need $5x$ (5 times x) units of plastic for the x skateboards we make. Similarly, we will need $2y$ (2 times y) units of plastic for the y dolls we make. Hence we will need $5x + 2y$ units of plastic for the mixture of skateboards and dolls we make. We added the $5x$ and $2y$ because we need to find the total plastic used when we make a mixture of skateboards and dolls.

Reading from the table, we see that we are limited by having only 60 units of plastic. So we can express the **resource constraint** imposed by the limited number of units of plastic by writing that $5x + 2y \leq 60$. Here, we use the symbol for less than or equal to, $\leq$, to express the fact that we cannot use more than the amount of plastic we have available.

Notice that all the numbers in this inequality can be obtained from a column of the mixture chart. One of the reasons that we construct a mixture chart is that it helps speed up the conversion of the information about the problem we wish to solve into inequalities. The setup phase of a realistic linear programming problem, constructing the mathematical model of a manufacturing situation, is often the hardest and most complex step in getting an answer.

In addition to the resource inequalities (of which realistic problems will often have hundreds), there is one additional algebraic expression, this time an equality, that the mixture table allows us to create. Using the mixture table, we can compute the profit that will be produced when we manufacture different mixtures. For each skateboard, we make a profit of \$1, so if x skateboards are made, the profit is $1x$ (1 times x). For each doll made, the profit is \$0.55. So if y dolls are made, the profit is $0.55y$ (0.55 times y). Denoting by P the total profit from making x skateboards and y dolls, we get the equation

$$P = 1x + 0.55y$$

Note that unlike the situation for the resources where we got an inequality, here we get an expression for what the profit will be as we vary the numbers of skateboards and dolls manufactured. Our goal is to find which values of x and y (skateboards and dolls) make this profit as large as possible.

Algebra Review Appendix
Linear Equations in Two Variables

EXAMPLE 3 Revisiting Our Clothing Manufacturer

We can also translate the information in the mixture chart shown in Figure 4.2 into inequalities and an equation for expressing the profit in terms of how many shirts and vests are produced. Using the first column of the mixture chart and the fact that only 60 units of cloth are available, we can write

$$3x + 2y \leq 60$$

And using the last column, we get the following expression for the profit P:

$$P = 5x + 3y$$

Now that we have the information from the original problems represented in mathematical terms, we will return our attention to finding a solution to the problems.

Finding the best (largest profit) mixture of skateboards and/or dolls to make can be carried out in two phases.

1. Determine those mixtures of skateboards and/or dolls that can be manufactured subject to the limited resources that are available. This step involves finding the **feasible set** for the mixture problem.

Feasible Set or Feasible Region DEFINITION

The **feasible set,** also called the **feasible region,** for a linear-programming problem is the collection of all physically possible solution choices that can be made.

We can use a geometric diagram such as the one in Figure 4.3 to help us understand the feasible set of options that the manufacturer of skateboards and dolls has available. The geometric diagram we draw will have as many "dimensions" as there are products being manufactured. We have two products represented by the variables *x* and *y*, so we use a two-dimensional picture. Even diagrams involving three variables are hard to draw and visualize. Though these diagrams helped with developing algorithms for solving mixture problems, they are of little practical use for realistic problems.

2. Determine how to pick out, from the feasible set, the mixture (or mixtures) that gives rise to the largest profit.

Representing the Feasible Region with a Picture

After we have constructed inequalities using a mixture chart or have the inequalities that must be obeyed for a more general linear-programming problem, we can draw a helpful picture to visualize the choices to be made in solving the problem. This picture will show in a convenient way the different choices that are available in solving the linear-programming problem at hand. To get the picture that will help us, we need to draw graphs of the inequalities associated with the linear-programming problem.

To draw the graph of an inequality, let's first review how to draw the graph of the equation of a straight line. Remember that two points can be used to uniquely determine a straight line. Let's use the equation associated with the less-than-or-equal-to inequality

Algebra Review Appendix
Linear Inequalities in Two Variables

$$5x + 2y \leq 60$$

namely,

$$5x + 2y = 60$$

There are two points that are easy to find on this line: the x- and y-intercepts. When $x = 0$, this gives rise to one point on the line, and when $y = 0$, we can find another point. (See Figure 4.3.) When $x = 0$, if we substitute this value in the equation $5x + 2y = 60$, we get $5(0) + 2y = 60$. Solving this equation, we discover that $y = 30$. Similarly, if we substitute $y = 0$ in the equation $5x + 2y = 60$, we get $5x + 2(0) = 60$, from which we conclude that $x = 12$. We now have two points (0, 30) and (12, 0) that lie on the line $5x + 2y = 60$.

We are using the usual convention that when we write a pair such as (3, 10), we are describing a point that has $x = 3$ and $y = 10$. We always list the x-value (x-coordinate) first in such a pair and the y-value (y-coordinate) second.

Furthermore, when a point (x, y) is represented in a diagram, larger values of x are shown farther to the right (east) and larger values of y are shown farther up (north).

Using the two points we found on the line $5x + 2y = 60$, we can draw the graph shown in Figure 4.3a, where we have also displayed the point (3, 10). How do we know that (3, 10) is not on the line? We can see this by replacing x by 3 and y by 10 in the equation $5x + 2y = 60$, getting $5(3) + 2(10) = 15 + 20 = 35$. To have been on the line, we would have to have had the value 60. Furthermore, we know that the point (3, 10) is below the line $5x + 2y = 60$ because when $x = 3$, and we replace x by this value in the equation $5x + 2y = 60$, we get $5(3) + 2y = 60$, which means that $y = 45/2$. Since 45/2 is greater than 10, the y-value for the point (3, 10), we conclude that (3, 10) is below the line.

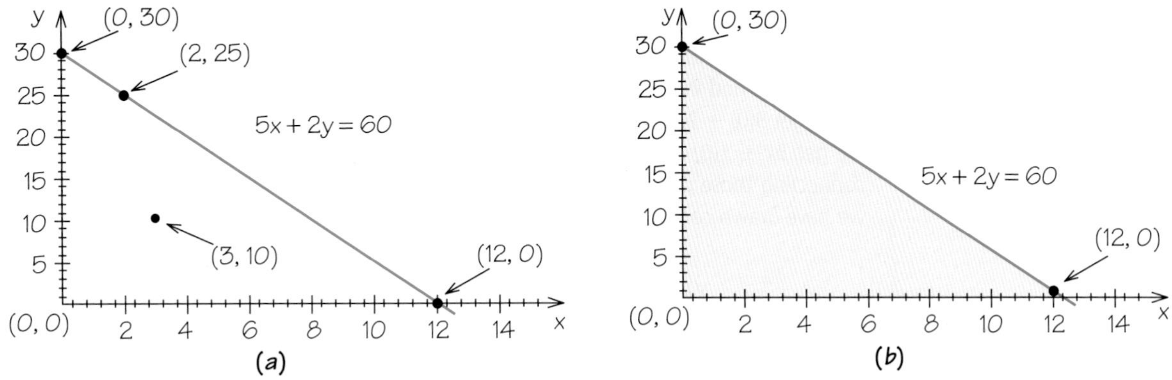

Figure 4.3 The feasible region for Example 3. (a) Graph of $5x + 2y = 60$. (b) Shading of the half-plane $5x + 2y < 60$, and where $x \geq 0$, $y \geq 0$.

Now that we know what the graph of the equation $5x + 2y = 60$ looks like, we can think through where points (x, y) that satisfy $5x + 2y < 60$ are located. The points that are either on the line $5x + 2y = 60$ or satisfy $5x + 2y < 60$ will satisfy $5x + 2y \leq 60$.

Any line, for example, $5x + 2y = 60$, divides the xy-plane into three parts: those points on the line and the points in one of two half-planes. In one of these half-planes, we have the points for which $5x + 2y < 60$, and in the other, we have the points for which $5x + 2y > 60$. How can we tell which of the two half-planes is above the line $5x + 2y = 60$ and which is below?

The key is the use of a test point (x, y) that is not on the line and whose half-planes we wish to distinguish. We saw above that (3, 10) is not on the line $5x + 2y = 60$ and is below the line. This enables us to see that the half-plane for which $5x + 2y < 60$ consists of the points below the line $5x + 2y = 60$.

To complete the drawing of the points that are feasible for the skateboard and dolls manufacturing problem, we also have to know which points satisfy the constraints that state that the number of skateboards produced x cannot be negative ($x \geq 0$) and the number of dolls produced y cannot be negative ($y \geq 0$). Each of these inequalities corresponds to a half-plane, and we can again test which of the half-planes associated with the line $x = 0$ is determined by $x \geq 0$.

This can be done using the point (3, 10) as a test point again. Because $x = 3$ is greater than 0, $x \geq 0$ determines the half-plane to the right of the line $x = 0$ (the y-axis). Similarly, using the point (3, 10), we see that because $y = 10$ is greater than 0, $y \geq 0$ determines the half-plane above the line $y = 0$ (the x-axis).

Putting this information together leads us to the conclusion that the collection of points (x, y) that meets the three inequalities involved ($x \geq 0$, $y \geq 0$, $5x + 2y \leq 60$) corresponds to the shaded region in Figure 4.3b. Remember that when graphing equations, lines with equations like $x = 2$ are vertical lines, whereas those like $y = 4$ are horizontal lines.

Note that since the minimality conditions are always present in the kind of linear-programming problems we are dealing with, the points that are feasible for these problems are always in the upper right region (quadrant) that is created by the x-axis and y-axis. Next, we draw the feasible region for the clothing manufacturing problem.

EXAMPLE 4 ➔ Drawing a Feasible Region

In the earlier clothing manufacturer example, we developed a resource constraint of $3x + 2y \leq 60$. Draw the feasible region corresponding to that resource constraint, using the reality minimums of $x \geq 0$ and $y \geq 0$.

First we find the two points where the line $3x + 2y = 60$ crosses the axes. When $x = 0$, we get $3(0) + 2y = 60$, giving $y = \frac{60}{2} = 30$, which yields the point (0, 30). For $y = 0$, we get $3x + 2(0) = 60$, or $x = \frac{60}{3} = 20$, so we have the point (20, 0). We draw the line connecting those points. Testing the point (0, 0), we find that the down side of the line we have drawn corresponds to $3x + 2y < 60$. The feasible region is shown in Figure 4.4.

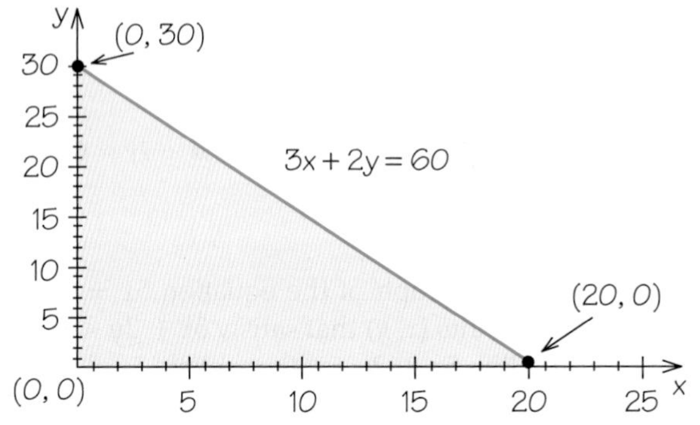

Figure 4.4 Feasible region for the clothing manufacturer.

Self Check 1

Draw the feasible region associated with constraints given by the equations $x \geq 0$, $y \geq 0$, $5x + 2y \leq 60$, and $x \leq 10$.

4.2 Finding the Optimal Production Policy

Our next step is that we still must find the *optimal production policy,* a point within the feasible region that gives a maximum profit. There are a lot of points in that region. If you consider points with only whole numbers as values for x or y, there are many points, but, in fact, either x or y or both could be some fractional number. There are so many points in this feasible region that to consider the profit at each one of them would require us to calculate profits from now until we grow very old, and still the calculations would not be done. Here is where the genius of the linear-programming technique comes in, with the **corner point principle,** which we define in terms of our mixture problems.

Corner Point Principle THEOREM

The **corner point principle** states that in a linear-programming problem, the maximum value for the profit formula always corresponds to a corner point of the feasible region.

The corner point principle is probably the most important insight into the theory of linear programming. Later in this chapter, we will explain why this principle works. The geometric nature of this principle explains the value of creating a geometric model from the data in a mixture chart.

The corner point principle gives us the following method to solve a linear-programming problem:

1. Determine the corner points of the feasible region.
2. Evaluate the profit at each corner point of the feasible region.
3. Choose the corner point with the highest profit as the production policy.

Let's look at the feasible region that we drew in Figure 4.3 on page 133. It is a triangle having three corners; namely, (0, 0), (0, 30), and (12, 0). Now all we need to do is find out which of these three points gives us the highest value for the profit formula, which in this problem is $\$1.00x + \$0.55y$. We display our calculations in Table 4.1. The maximum profit for the toy manufacturer is \$16.50, and that happens if the manufacturer makes 0 skateboards and 30 dolls. The point (0, 30) is called the *optimal production policy*.

TABLE 4.1 Calculation of the Profit Formula for Skateboards and Dolls

Corner Point	Value of the Profit Formula: $\$1.00x + \$0.55y$
(0, 0)	\$1.00(0) + \$0.55(0) = \$0.00 + \$0.00 = \$0.00
(0, 30)	\$1.00(0) + \$0.55(30) = \$0.00 + \$16.50 = \$16.50
(12, 0)	\$1.00(12) + \$0.55(0) = \$12.00 + \$0.00 = \$12.00

Optimal Production Policy THEOREM

An **optimal production policy** corresponds to a corner point of the feasible region where the profit formula has a maximum value.

EXAMPLE 5 Finding the Optimal Production Policy

Our analysis of the clothing manufacturer problem resulted in a feasible region with three corner points: (0, 0), (0, 30), and (20, 0). Which of these maximizes the profit formula, $\$5x + \$3y$, and what does that corner represent in terms of how many shirts and vests to manufacture?

The evaluation of the profit formula at the corner points is shown in Table 4.2. The maximum profit of \$100 occurs at the corner point (20, 0), which represents making 20 shirts and no vests.

TABLE 4.2 Evaluating the Profit Formula in the Clothing Example

Corner Point	Value of the Profit Formula: \$5x + \$3y
(0, 0)	\$5(0) + \$3(0) = \$0 + \$0 = \$0
(0, 30)	\$5(0) + \$3(30) = \$0 + \$90 = \$90
(20, 0)	\$5(20) + \$3(0) = \$100 + \$0 = \$100

General Shape of Feasible Regions

The shape of a feasible region for a linear-programming mixture problem has some important characteristics, without which the corner point principle would not work:

1. The feasible region is a polygon in the first quadrant, where both $x \geq 0$ and $y \geq 0$. This is because the minimum constraints require that both x and y be nonnegative.
2. The region is a polygon that has neither dents (as in Figure 4.5a) nor holes (as in Figure 4.5b). Figure 4.5c is a typical example. Such polygons are called *convex.*

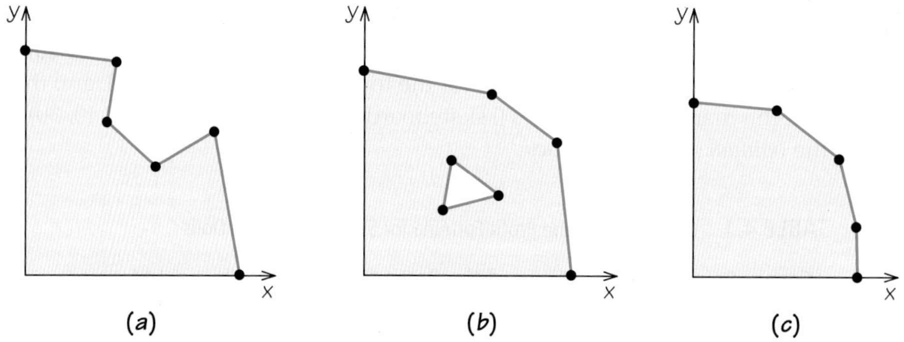

Figure 4.5 A feasible region may not have (a) dents or (b) holes. Graph (c) shows a typical feasible region.

The Role of the Profit Formula: Skateboards and Dolls

In practice, there are often different amounts of resources available in different time periods. The selling price for the products can also change. For example, if competition forces us to cut our selling price, the profit per unit can decrease. To maximize profit, it is usually necessary for a manufacturer to redo the mixture problem calculations whenever any of the numbers change.

Suppose that business conditions change, and now the profits per skateboard and doll are, respectively, \$1.05 and \$0.40. Let us keep everything else about the skateboards and dolls problem the same. The change in profits would give us a new profit formula of $\$1.05x + \$0.40y$. When we evaluate the new profit formula at the corner points, we get the results shown in Table 4.3. This time, the optimal

TABLE 4.3 A Different Profit Formula: Skateboards and Dolls

Corner Point	Value of the Profit Formula: \$1.05x + \$0.40y
(0, 0)	\$1.05(0) + \$0.40(0) = \$0.00 + \$0.00 = \$0.00
(0, 30)	\$1.05(0) + \$0.40(30) = \$0.00 + \$12.00 = \$12.00
(12, 0)	\$1.05(12) + \$0.40(0) = \$12.60 + \$0.00 = \$12.60

production policy, the point that gives the maximum value for the profit formula, is the point (12, 0). To get the maximum profit of $12.60, the toy manufacturer should now make 12 skateboards and 0 dolls.

We see from this example that the shape of the feasible region, and thus the corner points we test, are determined by the constraint inequalities. The profit formula is used to choose an optimal point from among the corner points, so it is not surprising that different profit formulas might give us different optimal production policies.

We started the exploration of skateboard and doll production with the idea that a toy manufacturer has a product line with either one to two products. But both linear-programming solutions we have found tell the manufacturer that to maximize profit, make just one product. This is probably not an acceptable result for the manufacturer, who might want to produce both products for business reasons other than profit, such as establishing brand loyalty. And it certainly would be very difficult for the manufacturer to be ready to switch back and forth between producing either skateboards or dolls every time the profit formula changed. Linear programming is a flexible enough technique that it can accommodate the desire for there to be both products in the optimal production policy. This is done by specifying that there be nonzero minimum quantities for each period.

Summary of the Pictorial Method Using a Feasible Region

Let's stop and summarize the steps we are following to find the optimal production policy in a mixture problem:

1. Read the problem carefully to identify the resources and the products.
2. Make a mixture chart showing the resources (associated with limited quantities), the products (associated with profits), the recipes for creating the products from the resources, the profit from each product, and the amount of each resource on hand. If the problem has nonzero minimums, include a column for those as well.
3. Assign an unknown quantity, x or y, to each product. Use the mixture chart to write down the resource constraints, the minimum constraints, and the profit formula.
4. Graph the line corresponding to each resource constraint and determine which side of the line is part of the feasible region. If there are nonzero minimum constraints, graph lines for them also, and determine which side of each is in the feasible region. Sketch the feasible region by finding the common points in the half-planes from all the resource constraints plus the minimum constraints. (This process is called finding the "intersection" of the half-planes.)
5. Find the coordinates of all the corner points of the feasible region. Some of these may have been calculated so that you can graph the individual lines. Proceed in order around the boundary of the feasible region. Be sure that every point you consider is really a corner point of the feasible region.
6. Evaluate the profit formula for each of the corner points. The production policy that maximizes profit is the one that gives the biggest value to the profit formula.

Two Products and Two Resources: Skateboards and Dolls

We return to the toy manufacturer now to consider two limited resources instead of one. The second limited resource will be time, the number of person-minutes available to prepare the products. Suppose that there are 360 person-minutes of labor available and that making one skateboard requires 15 person-minutes and making one doll requires 18 person-minutes. We will continue to use the original values regarding containers of plastic, the first of our two profit formulas, and to keep the problem relatively simple, we use the zero minimum constraints: $x \geq 0$ and $y \geq 0$. We need a new mixture chart. In general, we will include a column for minimums in a mixture chart only if there are any nonzero minimum constraints. In Figure 4.6, we have the mixture chart for this problem. Using the mixture chart, we can write the two resource constraints:

$$5x + 2y \leq 60 \qquad \text{for containers of plastic}$$

and

$$15x + 18y \leq 360 \qquad \text{for person-minutes}$$

We can also write the profit formula: $\$1.00x + \$0.55y$.

PRODUCTS	RESOURCE(S) Containers of Plastic 60	Person-minutes 360	PROFIT
Skateboards (x units)	5	15	$1.00
Dolls (y units)	2	18	$0.55

Figure 4.6 Mixture chart for Skateboards and Dolls (two resources).

Algebra Review Appendix
Systems of Linear Equations and Inequalities

The half-plane corresponding to the plastic resource is shown in Figure 4.7a. We now need to graph the half-plane corresponding to the time constraint. We find where the line $15x + 18y = 360$ intersects the two axes by substituting first $x = 0$ and then $y = 0$ into that equation.

The line corresponding to the time constraint contains the two points (0, 20) and (24, 0). When we insert the point (0, 0) into the inequality $15x + 18y < 360$, we get $15(0) + 18(0) < 360$, or $0 < 360$, which is true, so (0, 0) is on the side of the

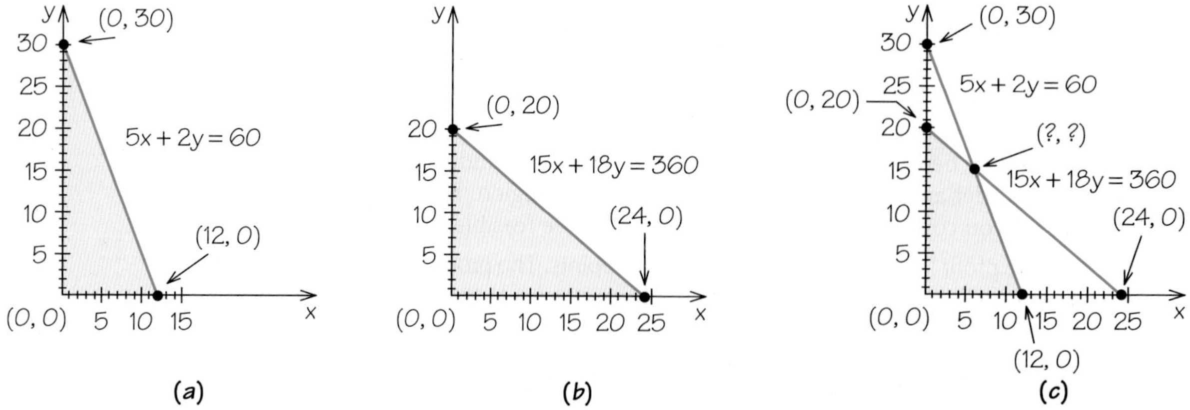

Figure 4.7 Feasible region for skateboards and dolls (two resources). (a) Half-plane for the plastic resource constraint. (b) Half-plane for the time resource constraint. (c) Intersection of the two half-planes.

line that we shade. Putting all this together, we get the half-plane in Figure 4.7b as the correct half-plane for the time resource constraint.

We are not permitted to exceed the supply of even a single resource. Therefore, the feasible region must be made up of points that are shaded twice—both in the half-plane for the plastic resource constraint, shown in Figure 4.7a, and in the half-plane for the time resource constraint in Figure 4.7b. For the procedure with several half-plane constraints, we build our feasible region by finding the intersection, or overlap, of the individual half-planes in the problem. In Figure 4.7c, we show the result of intersecting the half-planes from the two resource constraints. Because this problem has minimums that are zeroes, the shaded region in Figure 4.7c is in fact the feasible region for the problem.

The next step that we need to carry out to use the pictorial method for solving this problem is to find the corner points of the feasible region. This is carried out by doing the algebra necessary to solve two linear equations in two unknowns. This leads to the points (0, 0), (0, 20), (6, 15), and (12, 0). Three of these points have a zero value for one or more of the unknowns, so the calculations are easy.

To find the coordinates of the point (6, 15), it is necessary to solve for the point that satisfies both of the equations $5x + 2y = 60$ and $15x + 18y = 360$. To solve these two equations simultaneously, you must multiply one or both of the equations by a number to create equivalent equations, so that when added together, one of the variables will cancel out. One way to do this is to multiply the first of these equations by -3, obtaining the equation $-15x - 6y = -180$. When this is added to the second equation, the x term "drops out" and we can solve $12y = 180$, to get $y = 15$. Now it is an easy matter to substitute this value into either of the original equations to get the x value of 6.

We are ready to finish the problem. In Table 4.4 we have evaluated the profit formula at the four corner points of the feasible region. The optimal production policy for the toy manufacturer would be to make 6 skateboards and 15 dolls, for a maximum profit of $14.25.

TABLE 4.4 The Profit at the Four Corner Points

Corner Point	Value of the Profit Formula: $1.00x + $0.55y
(0, 0)	$1.00(0) + $0.55(0) = $0.00 + $0.00 = $0.00
(0, 20)	$1.00(0) + $0.55(20) = $0.00 + $11.00 = $11.00
(6, 15)	$1.00(6) + $0.55(15) = $6.00 + $8.25 = $14.25
(12, 0)	$1.00(12) + $0.55(0) = $12.00 + $0.00 = $12.00

Here is another mixture problem example of how the pictorial method using a feasible region works from start to finish.

EXAMPLE 6 Mixtures of Two Fruit Juices: Beverages

A juice manufacturer produces and sells two fruit beverages: 1 gallon of cranapple is made from 3 quarts of cranberry juice and 1 quart of apple juice; and 1 gallon of appleberry is made from 2 quarts of apple juice and 2 quarts of cranberry juice. The manufacturer makes a profit of 3 cents on a gallon of cranapple and 4 cents on a gallon of appleberry. Today, there are 200 quarts of cranberry juice and 100 quarts of apple juice available. How many gallons of cranapple and how many gallons of appleberry should be produced to obtain the highest profit without exceeding available supplies? We use zeroes as "reality minimums." The mixture chart for this problem is shown in Figure 4.8.

		RESOURCE(S)		
		Cranberry 200 quarts	Apple 100 quarts	PROFIT
PRODUCTS	**Cranapple** (x gallons)	3 quarts	1 quart	3 cents/gallon
	Appleberry (y gallons)	2 quarts	2 quarts	4 cents/gallon

Figure 4.8 A mixture chart for Example 6.

For each resource, we develop a resource constraint reflecting the fact that the manufacturer cannot use more of that resource than is available. The number of quarts of cranberry juice needed for x gallons of cranapple is $3x$. Similarly, $2y$ quarts of cranberry are needed for making y gallons of appleberry. So if the manufacturer makes x gallons of cranapple and y gallons of appleberry, then $3x + 2y$ quarts of cranberry juice will be used. Because there are only 200 quarts of cranberry available, we get the cranberry resource constraint $3x + 2y \leq 200$. Note that the numbers 3, 2, and 200 are all in the cranberry column. We get another resource constraint from the column for the apple juice resource: $1x + 2y \leq 100$. We also have these minimum constraints: $x \geq 0$ and $y \geq 0$.

Finally, we have the profit formula. Because $3x$ is the profit from making x units of cranapple and $4y$ is the profit from making y units of appleberry, we get the profit formula $3x + 4y$.

We summarize our analysis of the juice mixture problem. Maximize the profit formula, $3x + 4y$, given these constraints:

cranberry:	$3x + 2y \leq 200$
apple:	$1x + 2y \leq 100$
minimums:	$x \geq 0$ and $y \geq 0$

Remember, in a mixture problem, our job is to find a production policy (x, y), that makes all the constraints true and maximizes the profit.

Figure 4.9a shows the result of graphing the constraint associated with the cranberry resource, while Figure 4.9b shows the result of graphing the constraint associated with the apple resource, taking into account that the amounts of these resources used

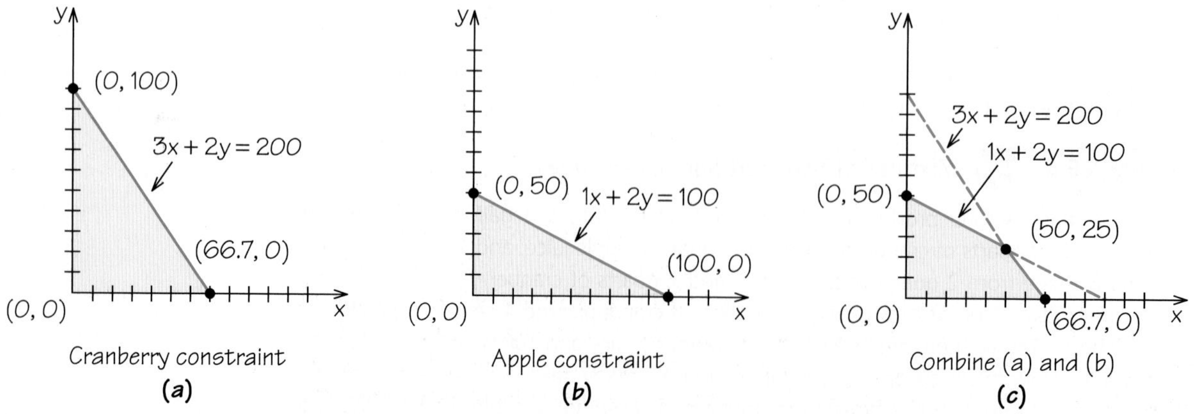

Figure 4.9 Feasible region for Example 6.

cannot be negative. When these two diagrams are superimposed, we get the diagram in Figure 4.9c. Now, to carry out the pictorial method, we need to find the profits associated with the four corner points shown. This is done in Table 4.5.

TABLE 4.5 Finding the Optimal Production Policy for Beverages

Corner Point	Value of the Profit Formula: $3x + 4y$ cents
(0, 0)	3(0) + 4(0) = 0 cents
(0, 50)	3(0) + 4(50) = 200 cents
(50, 25)	3(50) + 4(25) = 250 cents
(66.7, 0)	3(66.7) + 4(0) = 200 cents (rounded)

When we evaluate the profit formula at the four corner points, we see that the optimal production policy is to make 50 gallons of cranapple and 25 gallons of appleberry, for a profit of 250 cents.

Self Check 2

Maximize the profit function $P = 5x + 3y$ for the feasible region associated with constraints given by the equations $x \geq 0$, $y \geq 0$, $5x + 2y \leq 60$, and $x \leq 10$ (see Self Check 1 on page 134).

4.3 Why the Corner Point Principle Works

In finding solutions to our mixture problems, we have been using the corner point principle, which says that the highest profit value on a convex polygonal feasible region is always at a corner point. A feasible region has infinitely many points, making it impossible to compute the profit for each point. The corner point principle gives us a finite set of points, making the calculation possible.

You can visualize a mathematical proof of the corner point principle by imagining that each point of the plane is a tiny light bulb that is capable of lighting up. For the juice mixture example, whose feasible region is shown in Figure 4.9c, imagine what would happen if we ask this question: Will all points with profit = 360 please light up? What geometric figure do these lit-up points form?

In algebraic terms, we can restate the profit question in this way: Will all points (x, y) with $3x + 4y = 360$ please light up? As it happens, this version of the profit question is one that mathematicians learned to answer hundreds of years before linear programming was born.

The points that light up make a straight line because $3x + 4y = 360$ is the equation of a straight line. Furthermore, it is a routine matter to determine the exact position of the line. We call this line the **profit line** for 360; it is shown in Figure 4.10. For numbers other than 360, we would get different profit lines. Unfortunately, there are no points on the profit line for 360 that are feasible—that is, which lie in the feasible region. Therefore, the profit of 360 is impossible. *If the profit line corresponding to a certain profit value doesn't touch the feasible region, then that profit value isn't possible.*

Because 360 is too big, perhaps we should ask the profit line for a more modest amount, say, 160, to light up. You can see that the new profit line of 160 in Figure 4.10 is parallel to the first profit line and closer to the origin. This is no accident: All profit lines for the profit formula $3x + 4y$ have the same coefficients for x and y—namely, 3 for x and 4 for y. Because the slope of the line is determined

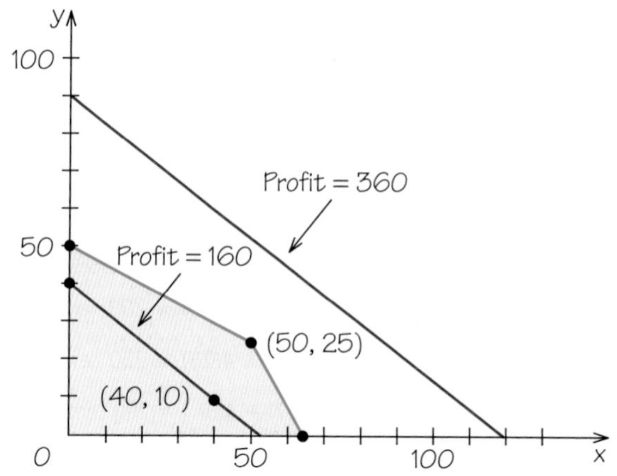

Figure 4.10 The profit line for 360 lies outside the feasible region, whereas the profit line for 160 passes through the region.

by those coefficients, they all have the same slope. Changing the profit value from 360 to 160 has the effect of changing where the line intersects the y-axis, but it does not affect the slope. These different profit lines are parallel to each other.

The most important feature of the profit line for 160 is that it has points in common with the interior of the feasible region. For example, (40, 10) is on that profit line because $3(40) + 4(10) = 160$; in addition, (40, 10) is a **feasible point.** This means that it is possible to make 40 gallons of cranapple and 10 gallons of appleberry and that if we do so, we will have a profit of 160.

Can we do better than a 160 profit? As we slowly increase our desired profit from 160 toward 360, the location of the profit line that lights up shifts smoothly upward away from the origin. So long as the line continues to cross the feasible region, we are happy to see it move away from the origin because the more it moves, the higher the profit represented by the line. We would like to stop the movement of the line at the last possible instant, while the line still has one or more points in common with the feasible region. It should be obvious that this will occur when the line is just touching the feasible region either at a corner point (Figure 4.11a) or along a line segment joining two corners (Figure 4.11b). That point or line segment corresponds to the production policy or policies with the maximum achievable profit. This is just what the corner point principle says: The maximum profit always occurs at a corner or along an edge of the feasible region.

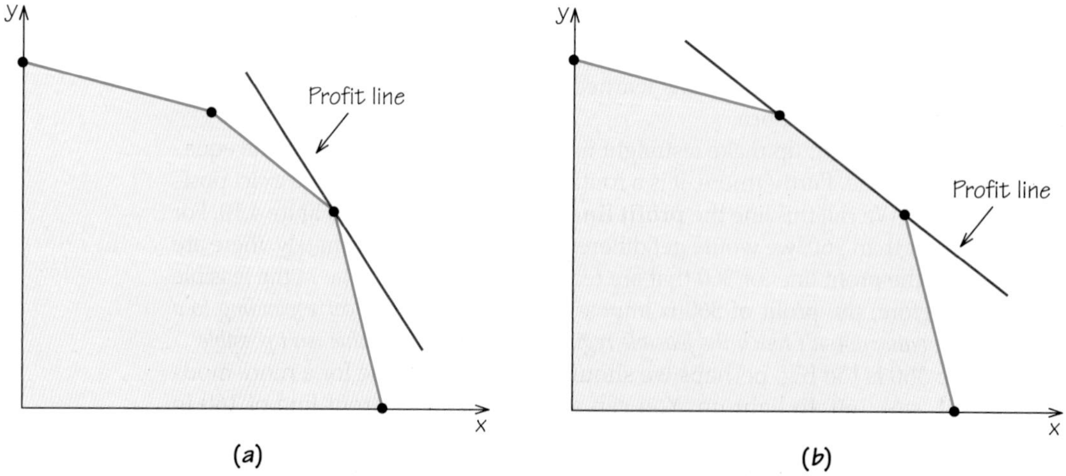

Figure 4.11 The highest profit will occur when the profit is just touching the feasible region, either (a) at the corner point or (b) along a line segment, which will include corner points.

EXAMPLE 7 Adding Nonzero Minimums: Beverages

Suppose that in Example 6, the profit for cranapple changes from 3 cents per gallon to 2 cents and the profit for appleberry changes from 4 cents per gallon to 5 cents. You can verify that this change moves the optimal production policy to the point (0, 50); no cranapple is produced. This result is not surprising: Appleberry is giving a higher profit and the policy is to produce as much of it as possible. But suppose the manufacturer wants to incorporate nonzero minimums into the linear-programming specifications so that there will always be both cranapple, *x*, and appleberry, *y*, produced. Specifically, the manufacturer decides that $x \geq 20$ and $y \geq 10$ are desirable minimums. Figure 4.12 is the mixture chart showing the new profit formula and the nonzero minimums, along with the unchanged rest of the beverage problem.

PRODUCTS	RESOURCE(S): Cranberry Juice 200 quarts	RESOURCE(S): Apple Juice 100 quarts	MINIMUMS	PROFIT
Cranapple (x gallons)	3	1	20	2 cents
Appleberry (y gallons)	2	2	10	5 cents

Figure 4.12 Mixture chart for Example 7.

The feasible region for beverages in Example 6 is shown in Figure 4.13a. The feasible region for beverages in this example is shown in Figure 4.13b. You can verify that, starting at the lower-left corner of the new feasible region and moving clockwise around its boundary, we have corner points (20, 10), (20, 40), (50, 25), and (60, 10). (One of those points was also a corner point of the old feasible region. Can you explain why?) Table 4.6 shows the evaluation of the profit formula at these corner points. For this modified problem, the optimal production policy is to produce 20 gallons of cranapple and 40 of appleberry, for a maximum profit of 240 cents.

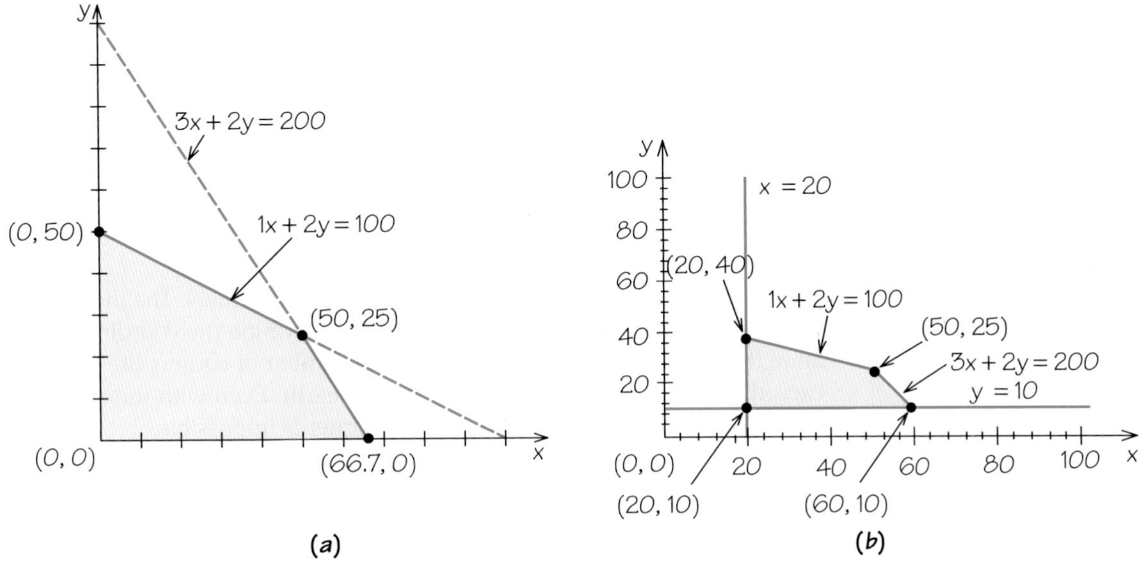

Figure 4.13 Feasible region for Examples 6 and 7. (a) Zero minimums. (b) Nonzero minimums.

TABLE 4.6 Profit Evaluation for Beverages

Corner Point	Value of the Profit Formula: $2x + 5y$
(20, 10)	2(20) + 5(10) = 40 + 50 = 90 cents
(20, 40)	2(20) + 5(40) = 40 + 200 = 240 cents
(50, 25)	2(50) + 5(25) = 100 + 125 = 225 cents
(60, 10)	2(60) + 5(10) = 120 + 50 = 170 cents

One final note about this solution concerns the resources. The point (20, 40) is on the resource constraint line for the apple juice resource, so it represents using up all the available apple juice. We can see this by inserting the apple juice resource constraint: 1(20) + 2(40) = 100 is true. However, (20, 40) is *below* the line for the cranberry juice resource, indicating that there will be *slack*, or leftover, amounts of cranberry juice. Specifically, substituting (20, 40) into the cranberry juice constraint gives $3x + 2y = 3(20) + 2(40) = 60 + 80 = 140$, which is 60 quarts less than the 200 quarts available. The slack is 60 quarts of cranberry juice. Dealing with slack can be an important consideration for manufacturers. Can you see why?

4.4 Linear Programming: Life Is Complicated

Every algorithm for solving a linear-programming problem has the following three characteristics, which hold true regardless of the number of products or the number of resources in the problem:

1. The algorithm can distinguish between "good" production policies—those in the feasible set that satisfy all the constraints—and those that violate some constraint(s) and are thus not feasible. There are usually many good points, each of which corresponds to some production policy; for example, "Make x units of product 1 and y units of product 2."
2. The algorithm uses some geometric principles—one of which is the corner point principle—to select a special subset of the feasible set.
3. The algorithm evaluates the profit formula at points in the special subset to find which corner point actually gives the maximum profit.

The various algorithms for linear programming differ in how they process the feasible set and in how quickly the algorithm finds the production policy—corner point—that gives the optimal profit.

In practical linear-programming problems, the feasible region will not be as simple as the ones we have examined here. There are two ways the feasible region can be more complex:

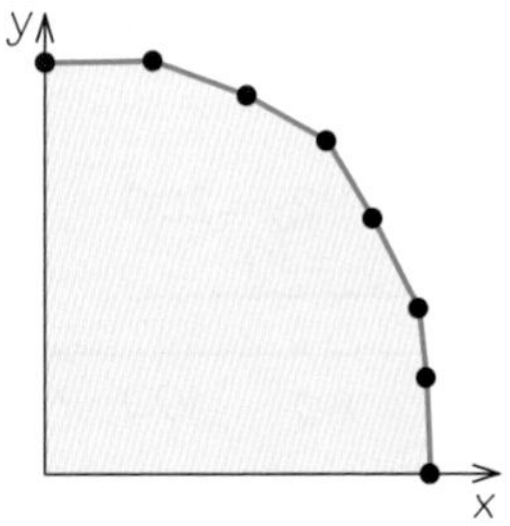

Figure 4.14 A feasible region with many corners.

1. Sometimes, as in Figure 4.14, we have a great many corners. The more corners there are, the more calculations we need to determine the coordinates of all of them and the profit at each one. The number of corners literally can exceed the number of grains of sand on the earth. Even with the fastest computer, computing the profit of every corner is impossible.
2. It is not possible to visualize the feasible region as a part of two-dimensional space when there are more than two products. Each product is represented by an unknown, and each unknown is represented by a dimension of space. If we have 50 products, we would need 50 dimensions and couldn't visualize the feasible region.

Another type of complication can occur even in simple two-dimensional regions: Corner points can have fractional coordinates, not the integer ones we see in the specially constructed problems in this text. Making 3.75 skateboards and 5.45 dolls is not possible. As mentioned earlier, integer programming, a special type of linear programming, is used when it is not realistic to use fractional answers.

We have looked at some simple mixture problems to illustrate the power of linear programming to solve optimization problems. Solving similar problems scaled up to a realistic level is widely applied by governments, businesses, and humanitarian organizations to make the world a better place that runs more smoothly.

The Simplex Method

Several methods are used for the typically large linear-programming problems solved in practice. The oldest method is the **simplex method,** which is still the most commonly used. Devised by the American mathematician George Dantzig (see Spotlight 4.2 on page 146), this ingenious mathematical invention makes it possible to find the best corner point by evaluating only a tiny fraction of all the corners. With the use of the simplex method, a problem that might be impossible to solve if each corner point had to be checked can be solved in a few minutes or even a few seconds on a typical business computer.

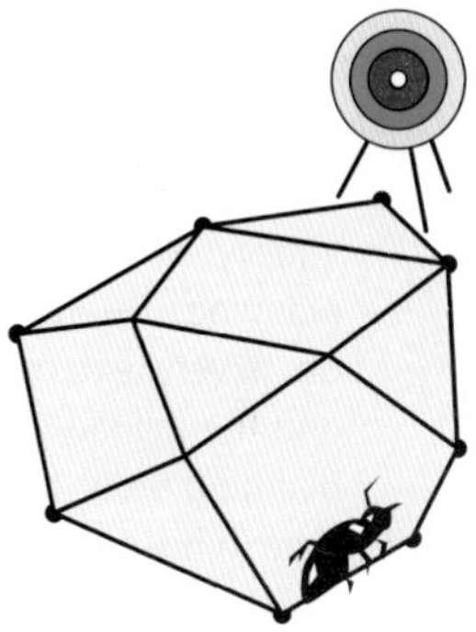

Figure 4.15 The simplex method can be compared to an ant crawling along the edges of a polyhedron, looking for the "target"—the optimal corner point.

The operation of the simplex method may be likened to the behavior of an ant crawling on the edges of a polyhedron (a solid with flat sides) looking for an optimal corner point—one that gives the highest profit (Figure 4.15). The ant cannot see where the optimal corner is. As a result, if it were to wander along the edges randomly, it might take a long time to reach that corner. The ant will do much better if it has a temperature clue to let it know it is getting warmer (closer to the optimal corner) or colder (farther from the optimal corner).

Think of the simplex method as a way of calculating these temperature hints. We begin at any corner. All neighboring corners are evaluated to see which ones are warmer and which are colder. A new corner is chosen from among the warmer ones, and the evaluation of neighbors is repeated—this time checking neighbors of the new corner. The process ends when we arrive at a corner point, all of whose neighbors are colder than it is.

Part of what the simplex method has going for it is that it works faster in practice than its worst-case behavior would lead us to believe. Although mathematicians have devised artificial cases for which the simplex method bogs down in unacceptable amounts of arithmetic, the examples arising from real applications are never like that. This may be the world's most impressive counterexample to Murphy's law, which says that if something can go wrong, it will.

Although the simplex method usually avoids visiting every corner, it may require visiting many intermediate ones as it moves from the starting corner to the optimal one. The simplex method has to search along edges on the boundary of the polyhedron. If it happens that there are a great many small edges lying between the starting corner and the optimal one, the simplex method must operate like a slow-moving bus that stops on every block.

Many computer programs are available that will use the simplex method to produce an optimal production policy if we just supply the computer with the constraint inequalities and profit formula. Simplex method programs can be found in a variety of places, including electronic spreadsheets, packages of mathematics programs designed for business applications or finite mathematics courses, and large "all-purpose" mathematics packages. A graphical solution is possible only for problems limited to two products; these special exercises involve more than two products.

The Father of Linear Programming Recalls Its Origins

SPOTLIGHT 4.2

George Dantzig, who died in 2005, spent most of his career as a professor of operations research and computer science at Stanford University. He is credited with inventing the linear-programming technique called the simplex method. Since its invention in the 1940s, the simplex method has provided solutions to linear-programming problems that have saved both industry and the military time and money. Here Dantzig talks about the background of his famous technique:

Kees Roos

George Dantzig (left), sometimes referred to as the "father" of linear programming, shown with Leonid Khachiyan (right) who developed an important new approach to solving linear-programming problems. Sadly, though much younger than Dantzig, Khachiyan died a few weeks before him.

> Initially, all the work we did had to do with military planning. During World War II, we were planning on a very extensive scale. The civilian population and the military were all performing scheduling and planning tasks, perhaps on a larger scale than at any time in history. And this was the case up until about 1950. From 1950 on, the whole emphasis shifted from military planning to practical planning for the civilian population, and industry picked it up.
>
> The first areas of industry to use linear programming were the petroleum refineries. They used it for blending gasoline. Nowadays, all of the refineries in the world (except for one) use linear-programming methods. They are one of the biggest users of it, and it's been picked up by every other industry you can think of—the forestry industry, the steel industry. You could fill up a book with all the different places it's used.
>
> The question of why linear programming wasn't invented before World War II is an interesting one. In the postwar period, various technologies just evolved that had never been there before. Computers were one example. These technologies were talked about before. You can go back in history and you'll find papers on them, but these were isolated cases that never went anywhere. . . .
>
> The problems we solve nowadays have thousands of equations, sometimes a million variables. One of the things that still amazes me is to see a program run on the computer—and to see the answer come out. If we think of the number of combinations of different solutions that we're trying to choose the best of, it's akin to the stars in the heavens. Yet we solve them in a matter of moments. This, to me, is staggering. Not that we can solve them—but that we can solve them so rapidly and efficiently.
>
> The simplex method has been used now for roughly 70 years. There has been steady work going on trying to use different versions of the simplex method, nonlinear methods, and interior methods. It has been recognized that certain classes of problems can be solved much more rapidly by special algorithms than by using the simplex method. If I were to say what my field of specialty is, it is in looking at these different methods and seeing which are more promising than others. There's a lot of promise in this—there's always something new to be looked at.

An Alternative to the Simplex Method

In 1984, Narendra Karmarkar, a mathematician working at AT&T Bell Laboratories, devised an alternative method for linear programming that finds the optimal corner point in fewer steps than the simplex algorithm by making use of search routes through the interior of the feasible region. The applications of Karmarkar's algorithm are important to a lot of industries, including telephone communications and the airlines (see Spotlight 4.3). Routing millions of long-distance calls, for

Finding Fast Algorithms Means Better Airline Service

SPOTLIGHT 4.3

Linear-programming techniques have a direct impact on the efficiency and profitability of major airlines. Thomas Cook, once director of operations research at American Airlines, made these comments concerning why optimal solutions are essential to the airline business:

> Finding an optimal solution means finding the best solution. Let's say you are trying to minimize a cost function of some kind. For example, we may want to minimize the excess costs related to scheduling crews, hotels, and other costs that are not associated with flight time. So we try to minimize that excess cost, subject to a lot of constraints, such as the amount of time a pilot can fly, how much rest time is needed, and so forth.
>
> An optimal solution, then, is either a minimum-cost solution or a maximizing solution. For example, we might want to maximize the profit associated with assigning aircrafts to the schedule; so we assign large aircraft to high-need segments and small aircraft to low-load segments.
>
> The simplex method, which was developed some 50 years ago by George Dantzig, has been very useful at American Airlines and, indeed, at a lot of large businesses. The difference between his method and Narendra Karmarkar's is speed. Finding fast solutions to linear-programming problems is also essential. With an algorithm like Karmarkar's, which is 50 to 100 times faster than the simplex method, we could do a lot of things that we couldn't do otherwise. For example, some applications could be real-time applications, as opposed to batch applications. So instead of running a job overnight and getting an answer the next morning, we could actually key in the data or access the database, generate the matrix, and come up with a solution that could be implemented a few minutes after keying in the data.
>
> A good example of this kind of application is what we call a major weather disruption. If we get a major weather disruption at one of the hubs, such as Dallas or Chicago, then a lot of flights may get canceled, which means we have a lot of crews and airplanes in the wrong places. What we need is a way to put that whole operation back together again so that the crews and airplanes are in the right places. That way, we minimize the cost of the disruption, as well as passenger inconvenience.

example, means deciding how to use the resources of long-distance landlines, repeater amplifiers, and satellite terminals to best advantage. The problem is similar to the juice company's need to find the best use of its stocks of juice to create the most profitable mix of products.

Many airlines use software based on Karmarkar's algorithm to reduce fuel costs and deal with delays caused by storms.

In the 1980s, scientists at Bell Labs applied Karmarkar's algorithm to a problem of unprecedented complexity: deciding how to economically build telephone links between cities so that calls can get from any city to any other, possibly being relayed through intermediate cities. Figure 4.16 shows a graph theory model (with color coding to show the intensity of traffic) for a portion of the Internet as it existed in the 1990s. Now, the number of ways to route Internet traffic has become unimaginably large, so picking the most efficient way to route email or other data packages is very difficult without using OR techniques.

Now that cell phones and Internet phone services are increasingly being used instead of landlines, new kinds of linear- and integer-programming problems and solutions have emerged to help provide better communications services. A typical example is using linear programming to determine the optimal locations for cell phone tower placement to achieve the best service. Integer programming is also being used to improve Wi-Fi and sensor network service.

Courtesy of Narendra Karmarkar

Indian-born Narendra Karmarkar received his Ph.D in Computer Science in the United States. In 1984, he developed an algorithm that solved many linear-programming problems more efficiently than the simplex method.

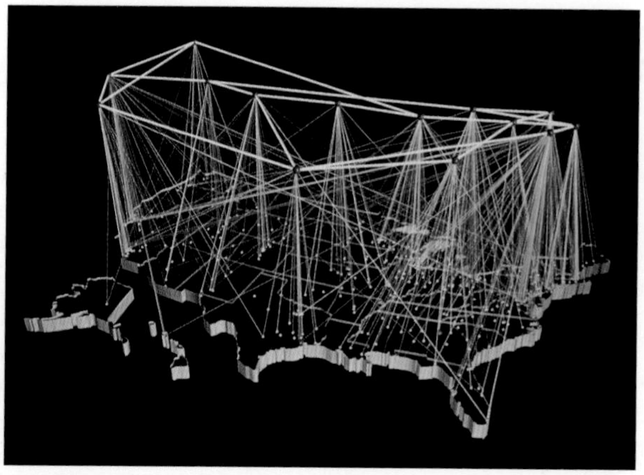

NCSA/Science Source

Figure 4.16 A map of the United States showing Internet traffic among a collection of major sites, with the volume of traffic color coded. Routing traffic (email, data packets, streaming video) over immense networks such as these can sometimes benefit from sophisticated linear-programming techniques and high-speed computers.

4.5 A Transportation Problem: Delivering Perishables

A supermarket chain gets bread deliveries from a bakery chain that does its baking in different places. Each supermarket store needs a certain number of loaves each day, and the supplier bakes, in total, enough breads to exactly meet the demands. Figure 4.17 shows the cost to ship a loaf from a particular baking location to the store involved. How many breads should be shipped from each locale to each of the stores to stay within the demands and to minimize the cost?

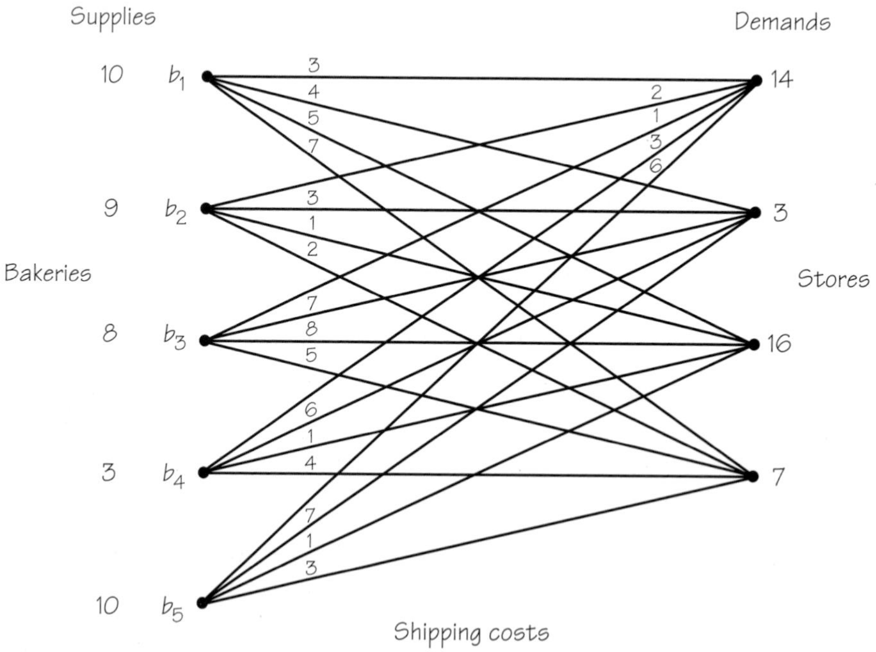

Figure 4.17 A graph theory representation of a supply-and-demand transportation problem that involves shipping breads from bakeries to stores.

Similarly, after a long holiday weekend, a car rental company will have extra cars in some cities and too few cars in other cities. It is faced with the problem of reshuffling the cars at minimal cost so that each city has the right number of cars. Companies that provide bicycles for use at specific hire/drop locations in urban areas face similar issues. Problems such as these go under the general name of

transportation problems, and they form a special class of linear-programming problems that can be solved by a specialized method.

Transportation Problem DEFINITION

A group of suppliers must meet the needs of users of these supplies. There is a cost for shipping from a particular supplier to a particular user (demander). The **transportation problem** involves minimizing the total shipping cost of meeting the required demands from the supplies available.

EXAMPLE 8 Delivering Bread

Imagine that we have three bakeries and three stores, though the ideas we develop will also solve problems where the number of stores and bakeries are not the same. The three stores require 3 dozen, 7 dozen, and 1 dozen loaves of bread, respectively, while the three bakeries can supply 8 dozen, 1 dozen, and 2 dozen loaves, respectively. The information given so far can be displayed in Figure 4.18, where the "suppliers" are represented by the rows of the table (labeled with Roman numerals) and the "demanders" are represented by the columns (labeled with Hindu-Arabic numerals).

	Stores 1	Stores 2	Stores 3	Breads baked (supply available)
Bakeries I	8	9	3	8
Bakeries II	15	1	12	1
Bakeries III	1	3	5	2
Breads needed (demands)	3	7	1	11 (Equal totals)

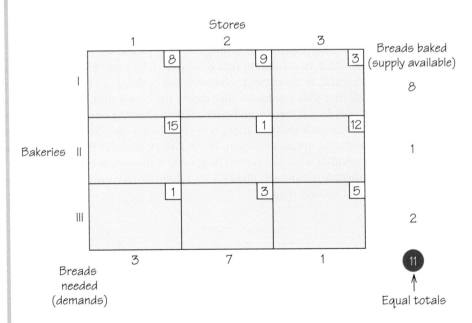

Figure 4.18 A representation of a specific problem involving meeting the demands of three stores for breads from the supplies available at three bakeries. Shipping costs between bakeries and stores are also shown.

The numbers of breads available and the numbers being required are shown on the right side and bottom of the table and will be referred to as **rim conditions.** Each entry of the table shown in Figure 4.18 is known as a cell. It is convenient to have a name for each of these cells. For example, the cell in the third row and second column will be denoted (III, 2). The first number always corresponds to a row, the second to a column. Thus cell (I, 2) refers to Bakery I and Store 2.

In deciding which bakeries should ship to which stores, it seems natural to take into account the costs of shipping a dozen breads from a particular bakery to a particular store. If Bakery I is farther from Store 2 than is Bakery II, it seems reasonable that the shipping cost for I will be higher than for II when shipping to that particular store.

However, the costs of shipping may also involve time considerations. (The distance to a store may be shorter, but it may be that this route is a very slow one.) Also, it may take extra time for a truck coming from I to park when making the next delivery.

The numbers we use in our diagrams are "aggregate" costs. The nice thing about what we are doing is that the solution method works independently of the way the costs are arrived at or computed. These costs (see Figure 4.18) are shown in the upper-right-hand corner of a cell. Thus the number 9 shown in cell (I, 2) means that it costs nine units to ship a bread from Bakery I to Store 2. Our goal will be to supply the stores with the breads they require from the supplies available at the bakeries so that the total cost of providing the breads to the stores is as small as possible (a minimum).

The tools for solving transportation problems like these were developed during World War II in conjunction with getting supplies from different ports in the United States to different ports in Europe (mostly the United Kingdom) as efficiently as possible. (The U.S. ports were like the bakeries, and the British ports were like the stores that needed the breads.)

We can think of finding a solution to a problem like this as a special kind of linear-programming problem because we can express the objective of minimizing the cost using a linear relationship. The constraints that express that the rim conditions are met can also be expressed using linear equations. However, it turns out there are algorithms that make it possible to solve problems of this kind that are rather larger than general linear-programming problems which can be solved by hand. These algorithms are intuitively appealing.

We can divide the problem of finding a solution to a transportation problem into two phases, as we did for general linear-programming problems. First, find a solution that is feasible (that is, a solution that does not violate any of the constraints of the problem). Second, if the current solution is not optimal, we move to a better one. Thus, we will first find a solution that meets the constraints and then try to find an improved solution. If there is no better solution than the one we have, under suitable circumstances we show that there is never a better one. Thus, the solution that we have found is an optimal solution. We will work our way through a simple example that is typical of what is required in general transportation problems.

Let's turn to the table shown in Figure 4.19, where certain numbers have been inserted with circles around them.

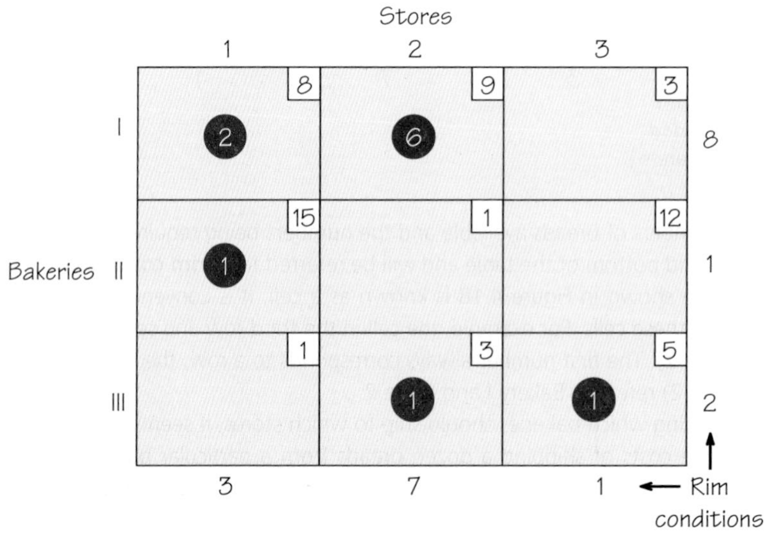

Figure 4.19 A possible solution to meeting the needs of three stores for bread from supplies available at three bakeries. The circled numbers show the amounts shipped from the bakeries to the stores.

Tableau DEFINITION

A table showing costs and rim conditions for a transportation problem is known as a **tableau.**

When we see a circled number such as the 6 in row I, column 2, this means that we plan to ship six breads to Store 2 from Bakery I. Similarly, the circled number 1 in row III and column 3 means we plan to ship one bread from Bakery III to Store 3. The cells that have no circled numbers are thought of as having zero entries; no breads are being shipped between these stores and these bakeries. Note, for example, that the row sum of the circled numbers in the first row is 8. This means that all the breads available at Bakery I are being shipped to some store.

Similarly, the fact that the circled entries in column 2 add up to 7 means that all the breads needed by Store 2 are being supplied to it. You can verify that all the row sums and column sums add to exactly the numbers that we want to ship from each bakery to each store. Note that 11 breads have been shipped by the bakeries and received by the stores. When this happens, the circled numbers are said to be a *feasible* solution to the problem.

How much will it cost to ship these amounts of breads (see Figure 4.19) to the stores? The number can be computed by multiplying the circled numbers by the cost shown in the associated cell. For example, to ship two breads from Bakery I to Store 1 costs $2(8) = 16$ because the cost associated with the cell in which the 2 appears is 8. The cost of shipping six breads from Bakery I to Store 2 is $6(9) = 54$. To get the total cost of this "shipment plan," we sum all the shipped amounts by the associated costs to get

$$2(8) + 6(9) + 1(15) + 1(3) + 1(5) = 16 + 54 + 15 + 3 + 5 = 93$$

However, at this point we do not know if there is a cheaper way to ship the breads to the stores. Notice that the number of cells with circled numbers is exactly equal to the number of rows m plus the number of columns n minus 1. This is the general pattern with transportation problems. Cells that are used for shipping are circled. On occasion, we ship a zero amount because the procedure works only when $m + n - 1$ cells are circled.

If we look at the pattern of circled numbers in the tableau in Figure 4.20, we see that there is a difficulty even though 11 breads are involved (the sum of all the circled numbers).

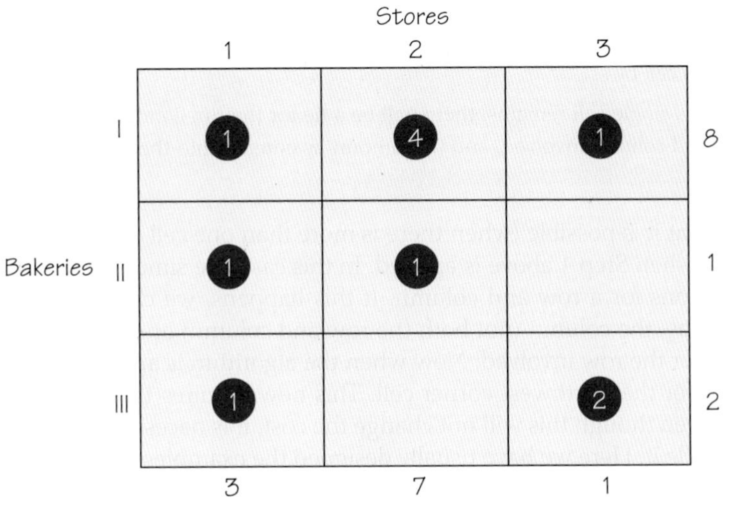

Figure 4.20 The circled numbers are not a possible solution to this transportation problem because the rim conditions are not satisfied.

The numbers in the first row add to 6, which means that at Bakery I there will be breads left over that have not been shipped. In row 2, the sum of the circled numbers is 2, but this means that something is wrong. How can Bakery II, which has a supply of only 1 bread, ship 2 breads? Furthermore, column 3 sums to 3, which means that 3 breads have been shipped to Store 3 despite the fact that it only requested 1 bread! These facts add up to the realization that this assignment of numbers to the cells violates the rules we are requiring. This proposed shipment plan also violates our rule that we are not allowed to circle more than five cells.

How can we find a solution that meets the constraints of the problem (the rim conditions)? We will show two ways to do this. The first is "fast and dirty" but typically does not find a very good solution with which to start. The second (developed in the exercises) usually gives a better "initial" solution but is a little harder to carry out.

This pair of approaches displays a common tension in problem solving: the ease of getting started but requiring more work later on, or more work at the start, which often proves to be a good investment of extra effort because less work is needed to find an optimal solution. If we know in advance the method being chosen to solve a problem, we can often find an example where this particular method does poorly. Mathematicians work hard to find methods that work well on the kinds of problems that come up in genuine applications.

Northwest Corner Rule

The easier approach involves what is called the **Northwest Corner Rule.** This rule is simple because it is based on the geometry of the table that is involved and does not even look at the costs associated with the cells in the table, which in the long run cannot be a good idea, because these costs come into play when trying to get an optimal solution.

How does the Northwest Corner Rule work? The algorithm carries out the following procedure until exactly one cell remains in the "altered tableau."

Northwest Corner Rule PROCEDURE

1. Locate that cell of the current tableau that is as far to the top and to the left as possible (that is, in the northwest corner). Ship via this cell the smaller of the two rim values (call the value s) associated with the row and column of this cell. (Indicate that this cell is being used by putting a circle around the entry in the tableau.)
2. Cross out the row or column that had rim value s and reduce the other rim value for this cell by s.
3. When a single cell remains, there will be a tie for the rim conditions of both the row and column involved, and this amount is entered into the cell and circled.

Note that it is possible (when there is more than one cell at the start) for there to be a tie when Step 1 above is applied. In this case, we simultaneously fulfill the rim conditions for a row and column. If this happens, we can always choose to cross out, say, the column (not both the row and column) and reduce to 0 the rim condition for the row involved. Now when the algorithm is applied, one has a rim value of 0 for the northwest corner cell. This now requires that 0 be shipped via that cell. Even though this will not change the cost, it is necessary to put a 0 in this cell and circle it. Here we have usually designed the examples to avoid ties so as to make it easier to get the essential ideas across.

EXAMPLE 9 Using the Northwest Corner Rule

Applying the Northwest Corner Rule to our original tableau (see Figure 4.18), we get the sequence of tableaux in Figure 4.21 as we cross out the rows or columns, where for clarity the costs associated with the cells are suppressed. The last diagram in the sequence shows the results on the original tableau, with the cost restored. Note that, at the steps in between, the costs played no role. It is a good idea to check that the circled numbers in each row and column really add up to the rim value for that row and column and that exactly $m + n - 1$ cells are filled.

We can now compute the cost of the associated solution that we have found (feasible solution), which obeys the rim conditions. As we did previously, we add up the

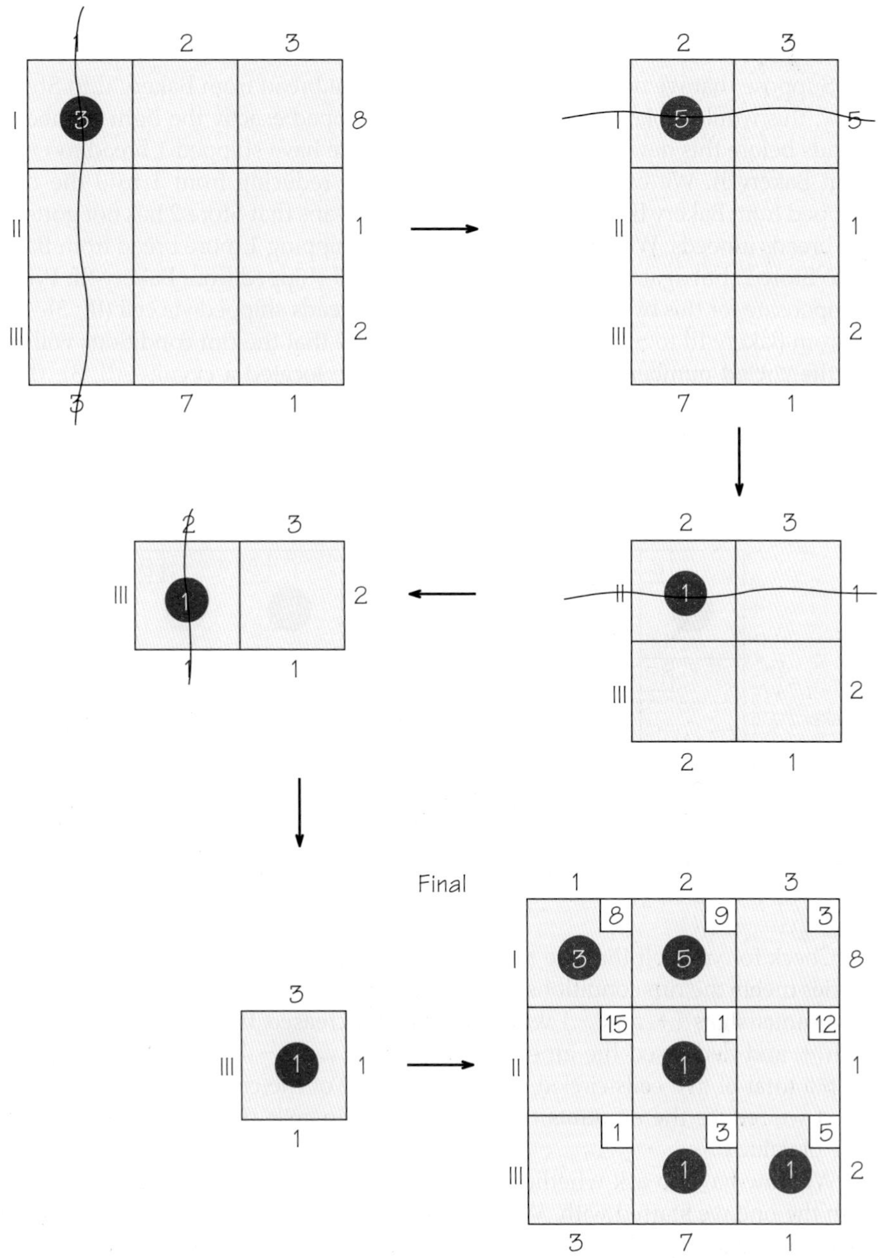

Figure 4.21 The construction of an initial solution to a transportation problem using the Northwest Corner Rule.

cost multiplied by the amount shipped for each cell with a circled entry. We get the following calculation:

$$3(8) + 5(9) + 1(1) + 1(3) + 1(5) = 78$$

This shows a cost that is smaller than the solution we found earlier. That solution involved a cost of 93. But is this solution the cheapest one? The fact that this feasible solution was found without using the costs on the cells suggests that no, it is not very likely this solution is cheapest.

Improving the Feasible Solution

The next phase of the transportation problem algorithm attempts to answer the question of how to tell if the feasible solution found by using the Northwest Corner Rule is the best. If this solution is not the best, we should be able to find a way to improve it.

Suppose that we decided to ship an additional bread from Bakery II to Store 3. Now, this would violate the fact that we had shipped exactly the right numbers of breads before this new additional shipment, so we have shipped 1 bread too many from Bakery II. We can compensate for this by reducing from 1 to 0 the bread shipped from Bakery II to Store 2. But this now means that Store 2 has not gotten all the breads it needs. We can take care of this by shipping 1 more bread from Bakery III to Store 2, but again we now have 1 extra bread shipped from Bakery III. We can compensate for this by reducing the number of breads shipped via cell (III, 3)—that is, from Bakery III to Store 3. This step will ensure that the rim conditions will hold for the circled numbers. This is because we have located a circuit—(II, 3), (II, 2), (III, 2), (III, 3), (II, 3)—where, if we increase and decrease the breads alternately going around that circuit, we maintain the rim conditions (see Figure 4.22).

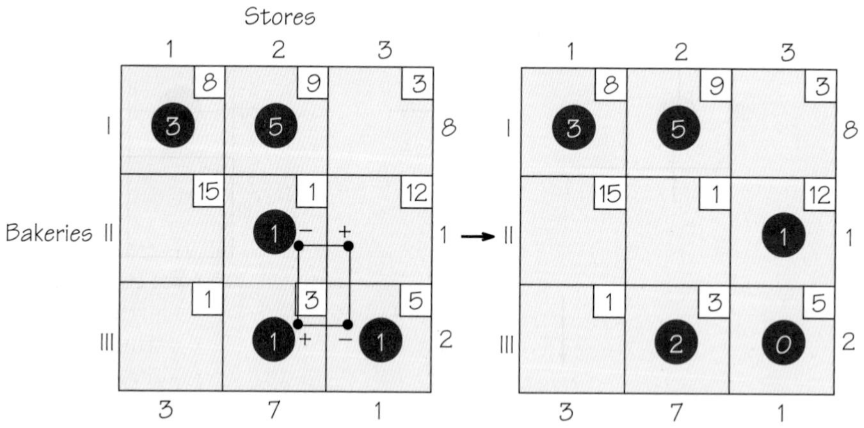

Figure 4.22 An illustration of how to take a current solution to a transportation problem and try to get an improved cheaper solution that still meets the rim conditions.

Check for yourself that the tableau on the right in Figure 4.22 with the circled entries meets the rim conditions. On the left, we show the circuit of cells with plus and minus signs (+ and −) where we have increased the amounts in the cells with + and decreased the amounts in the cells with − by 1 unit. (Note that to keep a total of five cells circled, we have set one of the circled cells to 0, because when we reduce the amounts of bread shipped in cells (II, 2) and (III, 3), we get a tie of value 0.)

We now have to ask whether this new solution is cheaper or more expensive than the one we started with. We can figure out whether this is a better or worse solution by tracking the costs of moving from the previous solution to the new one.

We went around a circuit where we increased a cost, decreased a cost (because we reduced the number of breads in that cell), increased a cost, and then decreased a cost before coming back to where we started, having traversed a circuit of length 4. The net effect of this collection of cost changes is $+12 - 1 + 3 - 5 = +9$. Thus, these changes, while producing a new feasible solution, give a more costly solution!

Perhaps increasing the amount shipped via a different circuit of cells would be better. Suppose we try the same process for cell (I, 3) (Figure 4.23)—that is, increase the shipping of breads from Bakery I to Store 3. To see if this is worthwhile, check the circuit formed by shipping more via cell (I, 3). It takes a bit of practice to find the circuit that this cell forms. In the case of cell (I, 3), the circuit we get is (I, 3), (I, 2), (III, 2), (III, 3), (I, 3). The cost of moving around this circuit is $+3 - 9 + 3 - 5 = -8$. We will refer to this number as the **indicator value** for this cell.

Indicator Value DEFINITION

The **indicator value of a cell** C (not currently a circled cell) is the cost change associated with increasing or decreasing the amounts shipped in a circuit of cells starting at C. It is computed by summing with alternating signs the costs of the cells in the circuit.

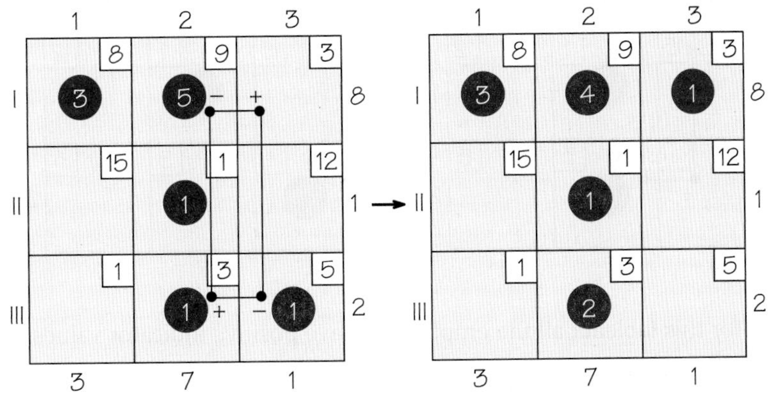

Figure 4.23 Using a cell with a negative indicator value, we can find a cheaper way of meeting the demands from the supplies.

The -8 means that we can lower the cost of shipping breads by using a different pattern of meeting the demands from the supplies. We have computed the saving for shipping one bread more via cell (I, 3), but perhaps we might be able to save even more by shipping even more breads via this cell. To determine whether we could, we look at the circuit that begins at cell (I, 3). To maintain a feasible solution, we have to increase the amounts shipped via some cells of this circuit and decrease the amounts shipped via others. Because we cannot decrease the amount shipped via any cell below zero, the minimum value of any cell that must be reduced is the maximum amount that can be shipped via cell (I, 3). In this case, it means that only one bread can be shipped via cell (I, 3), thereby lowering the cost from the previous solution by 8.

When we looked to improve the solution shown in Figure 4.21 (page 153), we have now seen that by shipping via cell (I, 3), we can get a better solution. However, there might be several cells in the solution shown in Figure 4.21 that would lead to improvement. Which one should we choose? The answer is that we should adopt a greedy point of view. If there are several cells with a negative indicator value, pick the one that is "most negative" to improve the solution.

Given a current feasible solution (one that satisfies the rim condition), we check each cell that does not have a circled number for improvement if we ship via that cell. If a cell leads to a positive indicator value with the circuit associated with it, no improvement is possible. If a cell has a negative indicator value associated with the circuit for that cell, we can get an improvement. We select as the cell to increase that cell with the largest negative indicator value. We now have a new feasible solution that is cheaper than the one we started with and can repeat our procedure just described starting from this new feasible solution.

It turns out that there was no better cell than (I, 3) (using this greedy approach) to get an improved solution. We will take the current best solution and see if we can improve it more. It turns out that for the current tableau (Figure 4.23), all the cells have a positive indicator except for cell (III, 1):

$$\text{Indicator for cell (III, 1):} \qquad +1 - 3 + 9 - 8 = -1$$

Because the minimum of the circled numbers in the cell with a negative label is 2 in cell (III, 2), we can increase by 2 the amount shipped in cell (III, 1) and get a new solution, as shown in Figure 4.24.

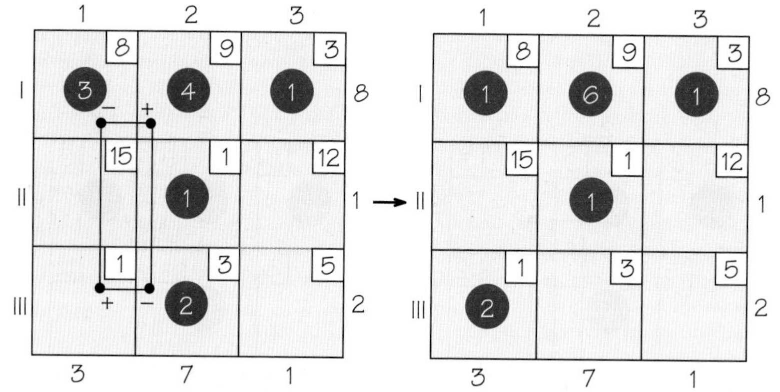

Figure 4.24 We can find an even cheaper way of meeting the demands from the supplies available using a cell with a negative indicator value.

Now, for this tableau, all the empty cells have positive indicator values.

Indicator (II, 1):	$+15 - 1 + 9 - 8 = 15$
Indicator (II, 3):	$+12 - 3 + 9 - 1 = 17$
Indicator (III, 2):	$+3 - 9 + 8 - 1 = 1$
Indicator (III, 3):	$+5 - 3 + 8 - 1 = 9$

This means that the current solution is optimal. The cost of this solution is

$$1(8) + 6(9) + 1(3) + 1(1) + 2(1) = 8 + 54 + 3 + 1 + 2 = 68$$

It turns out that if all the cells associated with a feasible solution have positive indicator values, then the solution that one has reached is optimal. (Cells with a zero indicator value show that there are other solutions that achieve the same optimal value.)

How to Recognize an Optimal Solution THEOREM

We are given a transportation problem with m suppliers and n demanders where the amount of the supplies equals the amount of demands. A collection of $m + n - 1$ circled cells is optimal (i.e., the circled cells determine a minimum cost solution) if the indicator value associated with each of the empty cells is positive. If some indicator cells are positive and some are zero, there are multiple solutions for an optimal value.

This theorem is the analog of the result for linear programming that states that if a corner point is feasible, and if no neighbor of the corner point has a better value of the objective function, then the corner point we are at is already an optimal one. Note that there may be other optimal solutions that use a different number of cells than $m + n - 1$, but we can never do any better in terms of the cheapness of a solution than what we have described above.

For those interested in the exciting fact that one piece of mathematics is often useful for other mathematics, we see an example of that here. The reason that an empty cell gives rise to a unique circuit with which we can try to improve the current solution of a transportation problem is the fact that when an edge not in a tree is added to a tree, it creates a unique circuit (see Chapter 3). Because we have m rows and n columns, a tree associated with a graph on $m + n$ vertices has $m + n - 1$ edges, exactly the number of cells we need to fill in a transportation problem!

4.6 Improving on the Current Solution

Now we will describe a method guaranteed to find an optimal solution to a transportation problem.

The Stepping Stone Method DEFINITION

The **stepping stone method** consists of taking some feasible solution of a transportation problem and improving this solution, if it is not optimal, by shipping an additional amount using a cell with a negative indicator value.

EXAMPLE 10 Applying the Stepping Stone Method

We will work out another small example to illustrate the technique of applying the Northwest Corner Rule to get an initial solution, and then improving this solution if it is not optimal. Again, we do so by computing the indicator values of the cells and improving the current solution by shipping using a cell with a negative indicator value.

We start with an initial tableau where there are two mines that can supply ore to three companies that extract ore. There are 10 units of ore being mined and the extractors need 10 units to run at full capacity. The initial tableau for the problem is displayed in Figure 4.25.

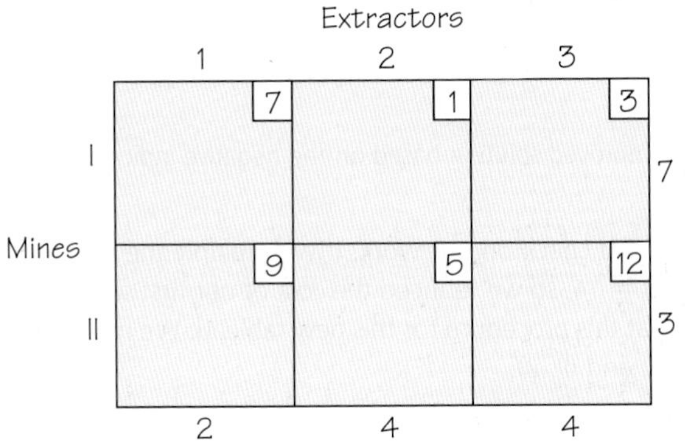

Figure 4.25 A transportation problem where two mines supply ore to three companies that extract metal from the ore. The shipping costs are indicated.

Using the Northwest Corner Rule, we find an initial feasible solution as shown in Figure 4.26. When applying the Northwest Corner Rule, we eliminate a row or column as follows: first column 1, then column 2, then row I, and we are now left with a single cell. The cost of the feasible solution shown is

$$2(7) + 4(1) + 1(3) + 3(12) = 14 + 4 + 3 + 36 = 57$$

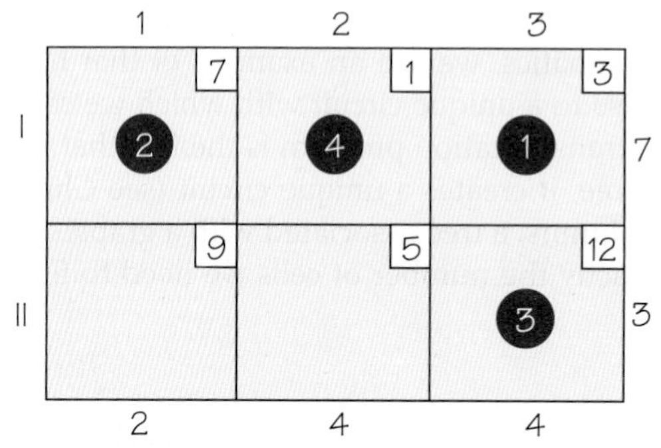

Figure 4.26 The Northwest Corner Rule has been used to find a possible way to meet the demands from the supplies for the tableau in Figure 4.25.

The two empty cells we have are (II, 1) and (II, 2). We compute the indicator value for each of these cells:

Indicator for cell (II, 1): $+9 - 12 + 3 - 7 = -7$
Indicator for cell (II, 2): $+5 - 12 + 3 - 1 = -5$

Because cell (II, 1) has a more negative indicator value, we can reduce the cost more by using that cell. Increasing by 2 (because this is the minimum of circled numbers with negative signs in the computation of the indicator) the amount of metal shipped via cell (II, 1) and cell (I, 3) and reducing by 2 the amount in cells (I, 1) and (II, 3), we obtain the new tableau in Figure 4.27. This has cost

$$4(1) + 3(3) + 2(9) + 1(12) = 4 + 9 + 18 + 12 = 43$$

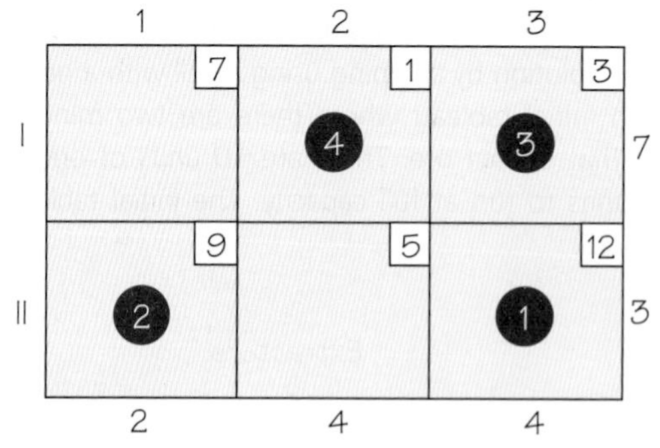

Figure 4.27 An improved solution based on the negative indicator value for cell (II, 1) in Figure 4.26.

Note that as a partial check on our work, if we multiply the indicator (−7) by 2, this is −14 and 57 − 43 = 14, so we reduced the cost of our first solution by 14, as expected.

We now repeat this procedure for this new tableau. We must compute the indicator value of cells (I, 1) and (II, 2).

Indicator for cell (I, 1): $+7 - 9 + 12 - 3 = +7$
Indicator for cell (II, 2): $+5 - 12 + 3 - 1 = -5$

Thus, it turns out that we can increase by 1 the amount shipped by (II, 2) and get the tableau in Figure 4.28.

From this tableau, we need to compute the indicator values for the cells (I, 1) and (II, 3). We obtain

Indicator for cell (I, 1): $+7 - 9 + 5 - 1 = +2$
Indicator for cell (II, 3): $+12 - 3 + 1 - 5 = +5$

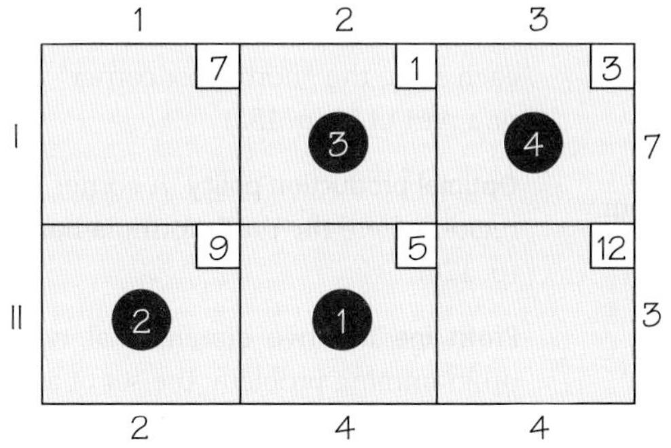

Figure 4.28 An improved solution, which turns out to be optimal, based on the negative indicator value for cell (II, 2) in Figure 4.27.

Not surprisingly, the cell (II, 2) has a positive indicator value because in the previous tableau, that cell was the one that, when we shipped less via it, enabled us to reduce the cost. The fact that both of these indicator values are positive means that the current shipping schedule is an optimal one; that is, using a shipping schedule that ships via only four cells, we cannot find any other solution with the same value.

Self Check 3

Two dairies supply three supermarket chains with the demands for sour cream as shown in the table below. Also indicated in a cell is the cost of shipping between pairs of sites.

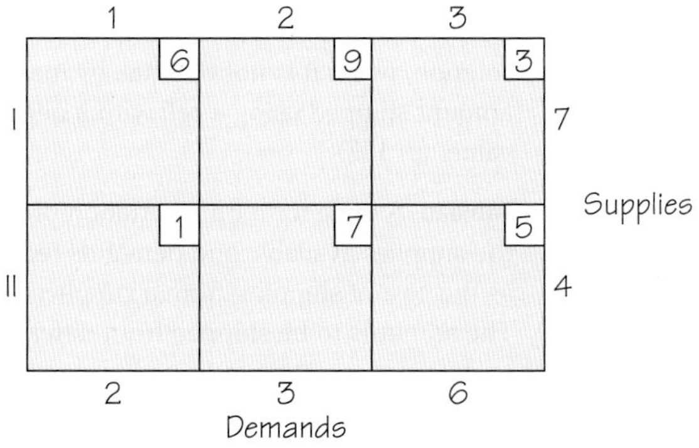

(a) Use the Northwest Corner Rule to obtain a feasible solution, and compute the cost of this feasible solution.

(b) If the feasible solution in part (a) is not optimal, find an optimal solution.

Transportation problems arise in a very large range of situations including shipping milk from dairies to supermarkets, vegetables to health food stores, and vitamins to your local drug store. The next time you sit down to breakfast, think about how many mathematics problems were solved for you to have a healthy breakfast!

Review Vocabulary

Corner point principle This principle states that there is a corner point of the feasible region that yields the optimal solution. (pp. 134, 135)

Feasible points A possible solution (but not necessarily the best one) to a linear-programming problem. With just two products, we can think of a feasible point as a point on the plane. (p. 142)

Feasible region The set of all *feasible points*—that is, possible solutions to a linear-programming problem. For problems with just two products, the feasible region is a part of the plane. Also called **feasible set.** (p. 132)

Indicator value of a cell The change in cost due to shipping an increased or decreased amount, using the cells in a transportation tableau that form a circuit consisting of circled cells together with a selected cell that is not circled. When an indicator value is negative, a cheaper solution can be found by shipping using this cell. (p. 155)

Linear programming A set of organized methods of management science used to solve problems of finding optimal solutions, while at the same time respecting certain important constraints. The mathematical formulations of the constraints in linear-programming problems are linear equations and inequalities. Mixture problems usually are solved by some type of linear programming. (p. 126)

Minimum constraint An inequality in a mixture problem that gives a minimum quantity of a product. Negative quantities can never be produced. (p. 131)

Mixture chart A table displaying the relevant data in a linear-programming mixture problem. The table has a row for each product and a column for each resource, for any nonzero minimums, and for the profit. (p. 129)

Mixture problem A problem in which a variety of resources available in limited quantities can be combined in different ways to make different products. It usually is desirable to find the way of combining the resources that produces the most profit. (p. 127)

Northwest Corner Rule A method for finding an initial, but rarely optimal, solution to a transportation problem starting from a tableau with rim conditions. The amounts to be shipped between the suppliers and demanders are indicated by circling numbers in the cells in the tableau. The number of cells circled after applying the method will equal the number of rows plus the number of columns minus 1. The method depends on locating at each stage the "northwest corner" of the original tableau or a part of it. (p. 152)

Optimal production policy A corner point of the feasible region where the profit formula has a maximum value. (p. 128)

Profit line In a two-dimensional, two-product, linear-programming problem, the set of all points that yield the same profit. (p. 141)

Resource constraint An inequality in a mixture problem that reflects the fact that no more of a resource can be used than what is available. (p. 131)

Rim conditions The supplies available (listed in a column at the right of a transportation tableau) and demands required (listed in a row at the bottom of a transportation tableau) in a transportation problem. The supplies available are usually taken to meet exactly the demands required. (p. 149)

Simplex method One of a number of algorithms for solving linear-programming problems. (p. 145)

Stepping stone method A method for solving a transportation problem that improves the current solution, when it is not optimal, by increasing the amount shipped using a cell with a negative indicator value. (p. 157)

Tableau A table for a transportation problem indicating the supplies available and demands required, as well as the cost of shipping from a supplier to a demander. The amounts to be shipped from different suppliers to different users are indicated by circled cells in the tableau. The number of such circled cells is always the number of rows plus the number of columns diminished by 1 for the tableau. (p. 151)

Transportation problem A special type of linear-programming problem where we have sources of supplies and users of, or demand for, these supplies. There is a cost to ship an item from a supplier to a demander. The goal is to minimize the total shipping cost to meet the demands from the supplies. (p. 149)

Self Check Answers

1.

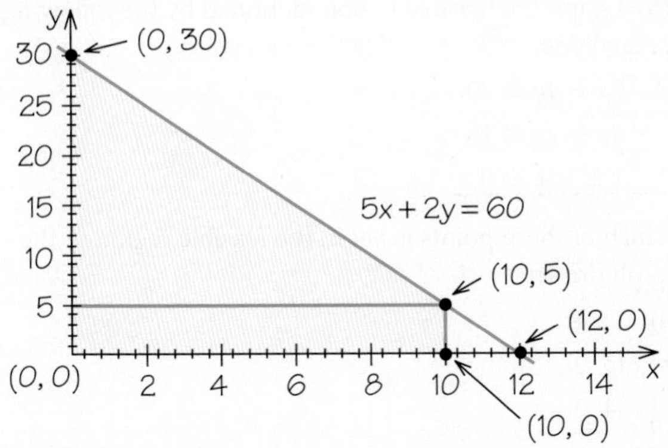

2. The maximal profit occurs at the feasible point (0, 30) and has value P equal to 90. The values of P at the other corner points are 0, 65, and 50.

3. (a) Applying the Northwest Corner Rule gives rise to the feasible solution shown here.

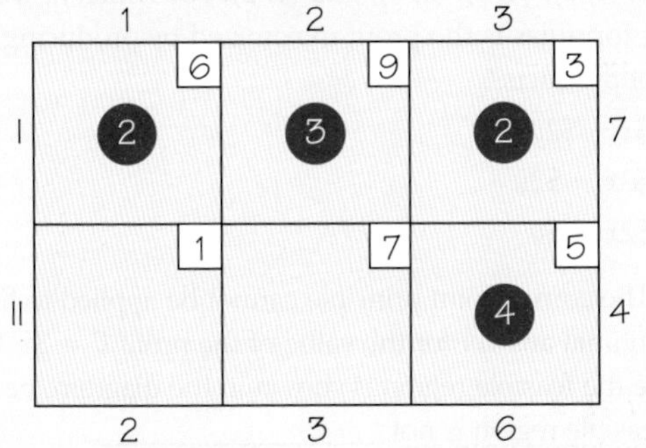

The cost of this solution is 65. The indicator value for the cell (II, 1) is −7, and the indicator value for the cell (II, 2) is −4.

(b) Consequently, this solution is not as cheap as possible. Using two iterations of the stepping stone method, we reach the following tableau, which is optimal (cost of 43).

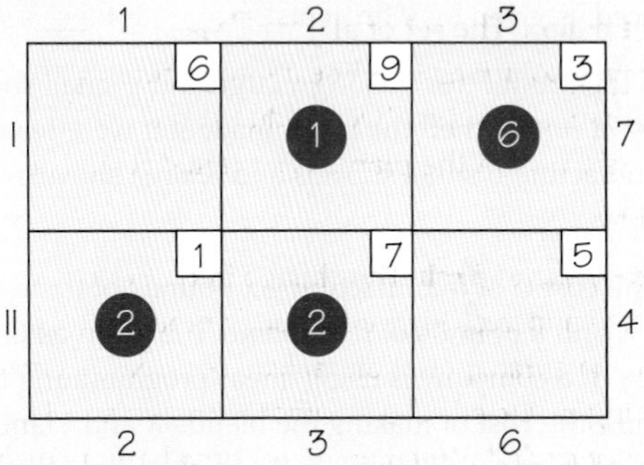

Skills Check

1. Where do the lines $3x + 5y = 26$ and $2x + 3y = 16$ intersect?

(a) At the point (2, 4)

(b) At the point (6, 2)

(c) At the point (4, 2)

2. The x-coordinate and y-coordinate of the point where the line $x = -2$ intersects the line $5y - 3x = -9$ are _____ and _____, respectively.

3. The y-coordinate of the point whose x value is 3 on the line $3x + 2y = 12$ is

(a) 3/2.

(b) 3.

(c) 2.

4. The lines $x + 3y = 12$ and $y = 3$ intersect at the point with x-coordinate _____ and y-coordinate _____.

5. The two lines $2x + 3y = 12$ and $6x + 9y = 7$

(a) intersect at (3, 2).

(b) intersect at (−3, −2).

(c) are parallel.

6. The x-coordinate and y-coordinate of the points where the line $7x + 2y = 28$ crosses the x-axis and y-axis, respectively, are _____ and _____.

7. Which of these points lie in the region $4x + 3y \geq 24$, $x \geq 0$, $y \geq 0$?

(a) Points (2, 5) and (3, 4)

(b) Points (5, 2) and (3, 4)

(c) Points (5, 2) and (2, 5)

8. The difference between the set of those pairs (x, y) that satisfy $x + 2y < 8$ and $x + 2y \leq 8$ is that the second inequality holds for points on the line _____, whereas the first inequality does not hold for points on this line.

9. A tart requires 3 oz of fruit and 2 oz of dough; a pie requires 13 oz of fruit and 7 oz of dough. There are 140 oz of fruit and 90 oz of dough available. Each tart earns 6 cents profit; each pie earns 25 cents profit. The profit formula for this situation, if x represents the numbers of pies produced and y represents the number of tarts produced, is given by P (in cents) = ____ x + ____ y.

10. If the profit P for making x large and y small shovels is $(P = 7x + 6y)$, then the profit made if it were feasible to manufacture 6 small shovels and 8 large shovels would be ________________.

11. The cost C of manufacturing x pounds of flour blend X and y pounds of flour blend Y is given as $C = 9x + 4y$. If a company is using linear programming to minimize the cost of making the blends X and Y and this happens when $x = 11$ and $y = 9$, which of the following statements must hold?

(a) $x = 11$ and $y = 9$ cannot be an interior point of the feasible region.

(b) $x = 11$ and $y = 9$ cannot be a corner point.

(c) $x = 11$ and $y = 9$ is the only point where the minimum can occur.

12. Producing a bench (x) requires 3 boards, and producing a table (y) requires 5 boards. There are 25 boards available. The resource constraint associated with this situation is _____ x + _____ $y \leq 25$.

13. For the feasible region of a linear-programming problem defined by the inequalities $x \geq 0$, $y \geq 0$, and $2x + 5y \leq 10$, which of the following pairs of points lies within the feasible region?

(a) (5, 0) and (1/2, 2)

(b) (0, 2) and (0, 0)

(c) (2, 1) and (3, 1)

14. A tart requires 3 oz of fruit and 2 oz of dough; a pie requires 13 oz of fruit and 7 oz of dough. There are 140 oz of fruit and 90 oz of dough available. Each tart earns 6 cents profit; each pie earns 25 cents profit. What are the resource inequalities of this situation?

(a) $3x + 2y \leq 140$
$13x + 7y \leq 90$
$x \geq 0,\ y \geq 0$

(b) $3x + 13y \leq 140$
$2x + 7y \leq 90$
$x \geq 0,\ y \geq 0$

(c) $3x + 2y \leq 6$
$13x + 7y \leq 25$
$x \geq 0,\ y \geq 0$

15. Graph the feasible region identified by the following inequalities:

$2x + 4y \leq 20$
$4x + 2y \leq 16$
$x \geq 0,\ y \geq 0$

Which of these points is *not* in the feasible region of the graph drawn?

(a) (2, 4)

(b) (5, 0)

(c) (4, 0)

16. Suppose the feasible region has four corners, at points (0, 0), (5, 0), (0, 3), and (3, 2). If the profit formula is $\$3x - \$2y$, the maximum value for the profit is ________________________.

17. Suppose that the feasible region has four corners, at points (0, 0), (4, 0), (0, 3), and (3, 2). For which of these profit formulas is the profit maximized by producing a mix of products?

(a) $\$x + \$2y$

(b) $\$2x - \$2y$

(c) $\$2x - \$y$

18. The corner point principle cannot be applied to find the optimal answer for the value of the profit $P = 3x + 7y$, where the feasible region is shown in the diagram, because the feasible region is not ________________.

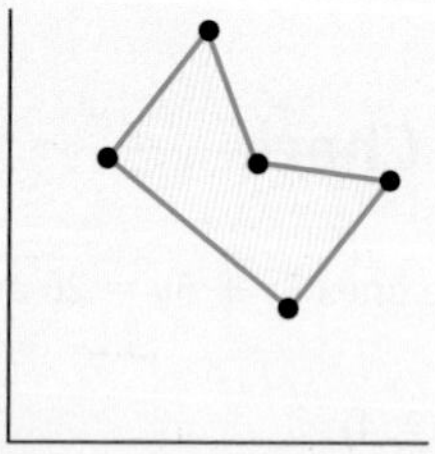

19. The shaded region in the accompanying diagram is an example of a region

(a) that is not convex.

(b) that is not bounded by straight-line segments.

(c) whose area is not bounded.

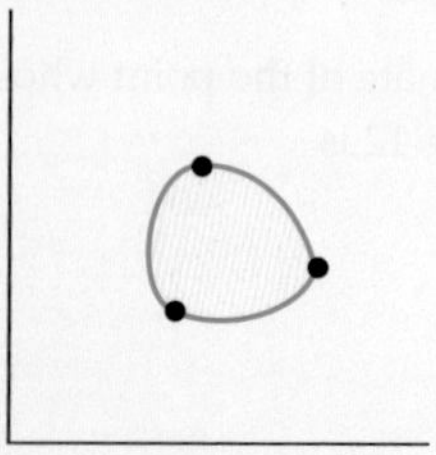

20. Suppose that the feasible region has five corners, at points (1, 1), (2, 1), (3, 2), (2, 4), and (1, 5). Which of these points is *not* in the feasible region?

(a) (1, 3)

(b) (2, 2)

(c) (0, 0)

21. Given the feasible region for a linear-programming problem defined by the inequalities $x \geq 0$, $y \geq 0$, and $2x + y \leq 12$, the feasible point where $x = y$ with the largest possible x-coordinate has x-coordinate ______ and y-coordinate ______.

22. How does the line representing the maximum feasible profit intersect the feasible region?

(a) There are no points of intersection.

(b) There is only one point of intersection.

(c) There is at least one point of intersection, and sometimes more than one point of intersection.

23. Consider the feasible region identified by the inequalities $x \geq 0$, $y \geq 0$, $3x + y \leq 10$, and $x + 2y \leq 6$. The corner point of this region, which is not (0, 0), that has x-coordinate 0 has y-coordinate ________________.

24. Suppose the feasible region has five corners, at points (1, 1), (2, 1), (3, 2), (2, 4), and (1, 5). If the profit formula is $\$5x - \$3y$, the corner point which maximizes the profit has x-coordinate ____ and y-coordinate ____.

25. Consider the feasible region for a linear-programming problem involving the inequalities $x \geq 0$, $y \geq 0$, $3x + y \leq 10$, and $x + 2y \leq 5$. The corner point for this feasible region that has no zero coordinates has x-coordinate ______ and y-coordinate ______.

26. Given the feasible region for a linear-programming problem defined by the inequalities $x \geq 0$, $y \geq 0$, and $2x + y \leq 15$, the feasible point where $x = y$ with the largest possible value for a y-coordinate has coordinates

(a) (5, 5).

(b) (3, 3).

(c) (0, 15).

27. When the Northwest Corner Rule is applied to the accompanying transportation problem tableau, the cells that remain empty are

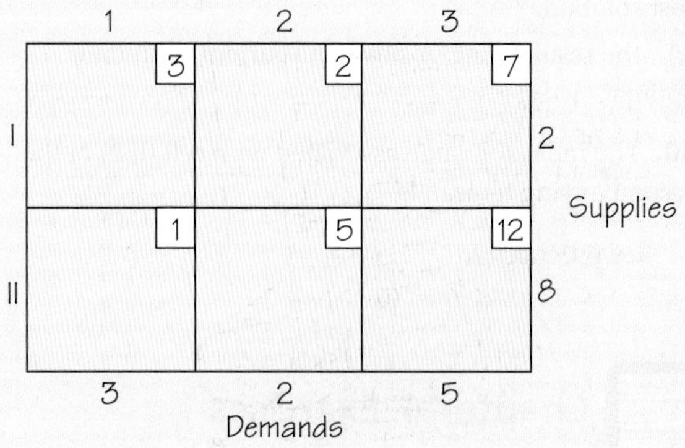

(a) cell (I, 2) and cell (I, 3).

(b) cell (II, 2) and cell (I, 2).

(c) cell (I, 2) and cell (II, 3).

28. The circled cells in the accompanying tableau give a solution that satisfies the rim conditions. The cost associated with this solution is ______________.

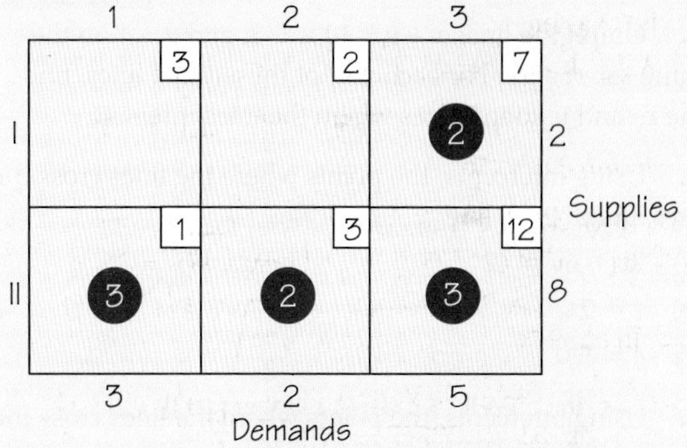

29. The circled cells in the accompanying tableau satisfy the rim conditions. When the indicator value for cell (I, 2) is computed,

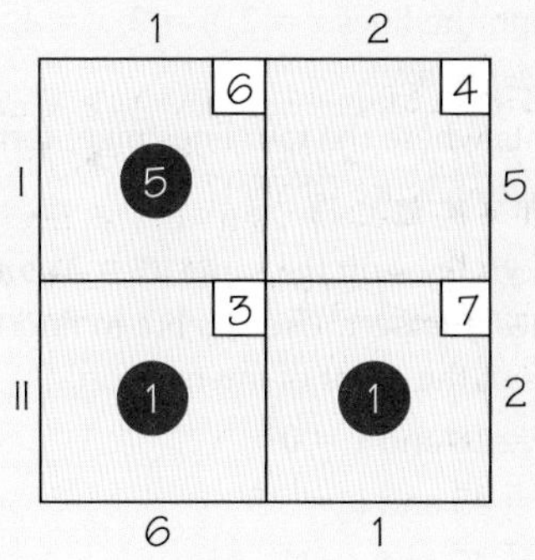

(a) the result being negative, this tableau does not give a minimal cost solution.

(b) the result being positive, this tableau has no minimal cost solution.

(c) the result being positive, the current solution is optimal.

30. The indicator value associated with cell (I, 2) of the accompanying tableau is ______________.

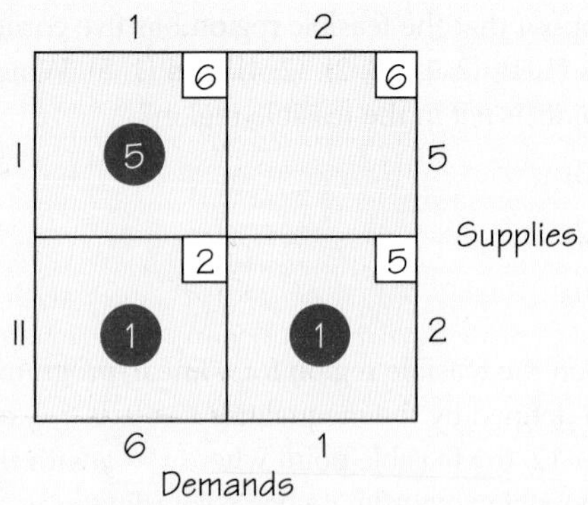

Chapter 4 Exercises

Challenge Discussion

4.1 Linear Programming and Mixture Problems: Combining Resources to Maximize Profit

1. Find the coordinates of all points where the lines $x + y = 8$, $y = 5$, and $x = 1$ intersect.

2. Find the x- and y-intercepts of the line $6y + 5x = 30$.

3. Graph the lines $x + y = 10$, $x = 3$, and $y = 4$ on the same set of axes. For each pair of these three lines, find the x- and y-coordinates where the lines intersect.

4. Using intercepts, the points where the lines cross the axes, graph each line.

(a) $7x + 6y = 42$

(b) $x = 2$

(c) $y = 5$

5. Using intercepts, the points where the lines cross the axes, graph each line.

(a) $2x + 3y = 12$

(b) $3x + 5y = 30$

(c) $4x + 3y = 24$

6. (a) Graph the lines $x = 2$, $y = 3$, $x = 4$, and $y = 7$ on the same set of axes.

(b) The points where the four lines meet form what kind of geometric figure (shape)?

7. Graph both lines on the same axes. Put a dot where the lines intersect. Use algebra to find the x- and y-coordinates of the point of intersection.

(a) $4x + 6y = 18$ and $x = 0$

(b) $5x + 3y = 45$ and $y = -5$

(c) $5x + 3y = 45$ and $x = 3$

8. Graph both lines on the same axes. Put a dot where the lines intersect. Use algebra to find the x- and y-coordinates of the point of intersection.

(a) $x = 3$ and $y = -2$

(b) $3x + 5y = 45$ and $x = -5$

(c) $5x + 3y = 45$ and $x = -3$

9. Graph both lines on the same axes. Put a dot where the lines intersect. Use algebra to find the x- and y-coordinates of the point of intersection.

(a) $x + y = 10$ and $x + 2y = 14$

(b) $y - 2x = 0$ and $x = 2$

10. Graph the line and half-plane corresponding to the inequality, a typical constraint from a mixture problem.

(a) $x \geq 6$

(b) $y \geq 4$

(c) $5x + 3y \leq 15$

(d) $4x + 5y \leq 30$

11. Graph the line and half-plane corresponding to the inequality, a typical constraint from a mixture problem.

(a) $x \geq 2$

(b) $y \geq 8$

(c) $3x + 2y \leq 18$

(d) $7x + 2y \leq 42$

For each description in Exercises 12–14, write one or more suitable resource-constraint inequalities. The unknown to use for each product is given in parentheses.

12. (a) One bridesmaid's bouquet (x) requires 2 roses, and one corsage (y) requires 4 roses. There are 28 roses available.

(b) Maintaining a large tree (x) takes 1 hr of pruning time and 30 min of shredder time; maintaining a small tree (y) takes 30 min of pruning time and 15 min of shredder time. There are 40 hr of pruning time and 2 hr of shredder time available.

13. (a) Manufacturing one package of hot dogs (x) requires 6 oz of beef, and manufacturing one package of bologna (y) requires 4 oz of beef. There are 300 oz of beef available.

(b) It takes 30 ft of 12-in. board to make one bookcase (x); it takes 72 ft of 12-in. board to make one table (y). There are 420 ft of 12-in. board available.

14. Manufacturing one salami (x) requires 12 oz of beef and 4 oz of pork. Manufacturing one bologna (y) requires 10 oz of beef and 3 oz of pork. There are 40 lb of beef and 480 oz of pork available.

In Exercises 15–20, graph the feasible region, label each line segment bounding it with the appropriate equation, and give the coordinates of every corner point.

15. $x \geq 0; y \geq 0; 2x + y \leq 10$

16. $x \geq 0; y \geq 0; x + 2y \leq 12; x + 2y \leq 8$

17. $x \geq 0; y \geq 0; 2x + 5y \leq 60$

18. $x \geq 10; y \geq 0; 3x + 5y \leq 120$

19. $x \geq 0; y \geq 4; x + y \leq 20$

20. $x \geq 2; y \geq 6; 3x + 2y \leq 30$

In Exercises 21–22, determine whether the points (2, 4) and/or (10, 6) are points of the given feasible regions of:

21. Exercises 15, 17, and 19.

22. Exercises 16, 18, and 20.

23. In the toy problem, x represents the number of skateboards and y the number of dolls. Using the version of that problem whose feasible region is presented in Figure 4.3b, with the profit formula $\$2.30x + \$3.70y$, write a sentence giving the maximum profit and describing the production policy that gives that profit.

24. In the toy problem, x represents the number of skateboards and y the number of dolls. Using the version of that problem whose feasible region is presented in Figure 4.3b, with the profit formula $\$5.50x + \$1.80y$, write a sentence giving the maximum profit and describing the production policy that gives that profit.

25. Graph both lines on the same axes. Put a dot where the lines intersect. Use algebra to find the x- and y-coordinates of the point of intersection.

(a) $5x + 4y = 22$ and $x + 2y = 8$

(b) $x + y = 7$ and $4x + 3y = 24$

In Exercises 26–29, graph the feasible region, label each line segment bounding it with the appropriate equation, and give the coordinates of every corner point.

26. $x \geq 0; y \geq 0; 3x + y \leq 9; x + y \leq 7$

27. $x \geq 0; y \geq 0; 2x + y \leq 4; 4x + 4y \leq 12$

28. $x \geq 0; y \geq 2; 5x + y \leq 14; x + 2y \leq 10$

29. $x \geq 4; y \geq 0; 5x + 4y \leq 60; x + y \leq 13$

30. Determine whether the points (4, 2) and/or (1, 3) are points of the given feasible regions of Exercises 27 and 29.

31. Determine the maximum value of P given by $P = 3x + 2y$ subject to the constraints $x \geq 0$, $y \geq 0$, $x \leq 7$, and $y \leq 5$.

32. A linear-programming problem has constraints given by $x \geq 0$, $y \geq 0$, $5x - y \leq 15$, and $4y + x \leq 24$.

(a) What are the corner points of the feasible region for this LP problem?

(b) Sketch a graph of the feasible region.

4.2 Finding the Optimal Production Policy

4.3 Why the Corner Point Principle Works

4.4 Linear Programming: Life Is Complicated

33. A linear-programming problem has constraints given by $x \geq 0$, $y \geq 0$, $x \leq 4$, and $4y + 13x \leq 60$.

(a) Which, if any, of the constraints involve vertical and horizontal lines?

(b) Sketch a graph of the feasible region.

(c) What are the corner points of the feasible region?

(d) If profit is given by the expression $P = 3x + 7y$:

(i) What is the profit associated with the point (3, 1)?

(ii) Which corner points have higher profit than (3, 1)?

34. Nuts Galore sells two spiced nut mixtures: Grade A and Grade B. Grade A requires 7 oz of peanuts for every 8 oz of almonds. Grade B requires 9 oz of peanuts for every 8 oz of almonds. There are 630 oz of peanuts and 640 oz of almonds available. Grade A makes Nuts Galore a profit of \$1.70, and Grade B makes a profit of \$2.40 per unit assembled. How many units of Grade A and Grade B nut mixtures should be made to maximize the company's profit, assuming that all units made can be sold?

35. Find the maximum value of P where $P = 3x + 2y$ subject to the constraints $x \geq 3$, $y \geq 2$, $x + y \leq 10$, and $2x + 3y \leq 24$.

36. Find the maximum value of P where $P = 3x - 2y$ subject to the constraints $x \geq 2$, $y \geq 3$, $3x + y \leq 18$, and $6x + 4y \leq 48$.

37. Find the maximum value of P where $P = 5x + 2y$ subject to the constraints $x \geq 2$, $y \geq 4$, and $x + y \leq 10$.

38. Given profit $P = 21x + 11y$ subject to the constraints $x \geq 0$, $y \geq 0$, $7x + 4y \leq 13$:

(a) Graph the feasible region.

(b) Determine a corner point where there is an optimal solution. (*Warning:* The corner point where the optimal solution occurs may not have integer values for both x and y.)

39. (a) Referring to Exercise 38, use the usual rounding rule to round the x-coordinate and the y-coordinate of the point where the optimal linear-programming solution occurs. Call the point with these coordinates Q.

(b) Determine whether Q's coordinates define a feasible point by checking them against the constraints.

(c) Evaluate the profit value P at point Q. How does the profit value compare with the point where the optimal value occurred in Exercise 38?

(d) Let R be the point with coordinates (0, 3). Is R in the feasible region? Evaluate P at point R and compare the result with the answer at Q and where the optimum linear-programming value occurred.

(e) Explain the significance of the situation here for solving maximization problems where $P = ax + by$ (with a and b known in advance) is subject to linear constraints but where the variables must be nonnegative integers rather than arbitrary nonnegative decimal numbers.

Exercises 40–51 each have several steps leading to a complete solution to a mixture problem. Practice a specific step of the solution algorithm by working out just that step for several problems. The steps are:

(a) Make a mixture chart for the problem.

(b) Using the mixture chart, write the profit formula and the resource- and minimum-constraint inequalities.

(c) Draw the feasible region for those constraints and find the coordinates of the corner points.

(d) Evaluate the profit information at the corner points to determine the production policy that best answers the question.

(e) (Requires technology) Compare your answer with the one you get from running the same problem on a simplex algorithm computer program.

40. A clothing manufacturer has 600 yd of cloth available to make shirts and decorated vests. Each shirt requires 3 yd of material and provides a profit of \$5. Each vest requires 2 yd of material and provides a profit of \$2. The manufacturer wants to guarantee that under all circumstances, there are minimums of 100 shirts and 30 vests produced. How many of each garment should be made to maximize profit? If there are no minimum quantities, how, if at all, does the optimal production policy change?

41. A car maintenance shop must decide how many oil changes and how many tune-ups can be scheduled in a typical week. The oil change takes 20 min, and the tune-up requires 100 min. The maintenance shop makes a profit of \$15 on an oil change and \$65 on a tune-up. What mix of services should the shop schedule if the typical week has 8000 min available for these two types of services? How, if at all, do the maximum profit and optimal production policy change if the shop is required to schedule at least 50 oil changes and 20 tune-ups?

42. A clerk in a bookstore has 90 min at the end of each workday to process orders received by mail or on voice mail. The store has found that a typical mail order brings in a profit of \$30 and a typical voice-mail order brings in a profit of \$40. Each mail order takes 10 min to process and each voice-mail order takes 15 min. How many of each type of order should the clerk process? How, if at all, do the maximum profit and optimal processing policy change if the clerk must process at least three mail orders and two voice-mail orders?

43. In a certain medical office, a routine office visit requires 5 min of doctors' time and a comprehensive office visit requires 25 min of doctors' time. In a typical week, there are 1800 min of doctors' time available. If the medical office clears \$30 from a routine visit and \$50 from a comprehensive visit, how many of each should be scheduled per week? How, if at all, do the maximum profit and optimal production policy change if the office is required to schedule at least 20 routine visits and 30 comprehensive ones?

44. A bakery makes 600 specialty breads—multigrain or herb—each week. Standing orders from restaurants are for 100 multigrain breads and 200 herb breads. The profit on each multigrain bread is \$8 and on herb bread, \$10. How many breads of each type should the bakery make to maximize profit? How, if at all, do the maximum profit and optimal production policy change if the bakery has no standing orders?

45. A student has decided that passing a mathematics course will, in the long run, be twice as valuable as passing any other kind of course. The student estimates that passing a typical math course will require 12 hr a week to study and do homework. The student estimates that any other course will require only 8 hr a week. The student has 48 hr available for study per week. How many of each kind of course should the student take?

(*Hint:* The profit could be viewed as 2 "value points" for passing a math course and 1 "value point" for passing any other course.) How, if at all, do the maximum value and optimal course mix change if the student decides to take at least two math courses and two other courses?

Exercises 46–49 require finding the point of intersection of two lines, each corresponding to a resource constraint.

46. The firm WebsAreUs creates and maintains websites for client companies. There are two types of websites: "Hot" sites change their layout frequently but keep their content for long times; "cool" sites keep their layout for a while but frequently change their content. Maintaining a hot site requires 1.5 hr of layout time and 1 hr for content changes. Maintaining a cool site requires 1 hr of layout time and 2 hr for content changes. Every day, WebsAreUs has 12 hr available for layout changes and 16 hr for content changes. Net profit is $50 for a set of changes on a hot site and $250 for a set of changes on a cool site. To maximize profit, how many of each type of site should WebsAreUs maintain daily? How, if at all, do the maximum profit and optimal policy change if the company must maintain at least two hot and three cool sites daily?

47. A paper recycling company uses scrap cloth and scrap paper to make two different grades of recycled paper. A single batch of grade A recycled paper is made from 25 lb of scrap cloth and 10 lb of scrap paper, whereas one batch of grade B recycled paper is made from 10 lb of scrap cloth and 20 lb of scrap paper. The company has 100 lb of scrap cloth and 120 lb of scrap paper on hand. A batch of grade A paper brings a profit of $500, whereas a batch of grade B paper brings a profit of $250. What amounts of each grade should be made? How, if at all, do the maximum profit and optimal production policy change if the company is required to produce at least one batch of each type?

Ton Koene/age fotostock

48. Jerry Wolfe has a 100-acre farm that he is dividing into one-acre plots, on each of which he builds a house. He then sells the house and land. It costs him $20,000 to build a modest house and $40,000 to build a deluxe house. He has $2,600,000 to cover these costs. The profits are $25,000 for a modest house and $60,000 for a deluxe house. How many of each type of house should he build to maximize profit? How, if at all, do the maximum profit and optimal production policy change if Wolfe is required to build at least 20 of each type of house?

49. The maximum production of a soft-drink bottling company is 5000 cartons per day. The company produces regular and diet drinks and must make at least 600 cartons of regular and 1000 cartons of diet per day. Production costs are $1.00 per carton of regular and $1.20 per carton of diet. The daily operating budget is $5400. How many cartons of each type of drink should be produced if the profit is $0.10 per regular and $0.11 per diet? How, if at all, do the maximum profit and optimal bottling policy change if the company has no minimum required production?

50. Wild Things raises pheasants and partridges to restock the woodlands and has room to raise 100 birds during the season. The cost of raising one bird is $20 per pheasant and $30 per partridge. The Wildlife Foundation pays Wild Things for the birds; the latter clears a profit of $14 per pheasant and $16 per partridge. Wild Things has $2400 available to cover costs. How many of each type of bird should they raise? How, if at all, do the maximum profit and optimal restocking policy change if Wild Things is required to raise at least 20 pheasants and 10 partridges?

51. Lights Aglow makes desk lamps and floor lamps, on which the profits are $2.65 and $4.67, respectively. The company has 1200 hr of labor and $4200 for materials each week. A desk lamp takes 0.8 hr of labor and $4 for materials; a floor lamp takes 1.0 hr of labor and $3 for materials. What production policy maximizes profit? How, if at all, do the maximum profit and optimal production policy change if Lights Aglow wants to produce at least 150 desk lamps and 200 floor lamps per week?

In Exercises 52–55, there are more than two products in the problem. Although you cannot solve these problems using the two-dimensional graphical method, you can follow these steps:

(a) Make a mixture chart for each problem.

(b) Using the mixture chart, write the resource- and minimum-constraint inequalities. Also write the profit formula.

(c) (Requires software) If you have a simplex method program available, run the program to obtain the optimal production policy.

52. A toy company makes three types of toys, each of which must be processed by three machines: a shaper, a smoother, and a painter. Each Toy A requires 1 hr in the shaper, 2 hr in the smoother, and 1 hr in the painter, and brings in a $4 profit. Each Toy B requires 2 hr in the shaper, 1 hr in the smoother, and 3 hr in the painter, and brings in a $5 profit. Each Toy C requires 3 hr in the shaper, 2 hr in the smoother, and 1 hr in the painter, and brings in a $9 profit. The shaper can work at most 50 hr per week, the smoother 40 hr, and the painter 60 hr. What production policy would maximize the toy company's profit?

53. A rustic furniture company handcrafts chairs, tables, and beds. It has three workers, Chris, Sue, and Juan. Chris can work only 80 hr per month, but Sue and Juan can each put in 200 hr. Each of these artisans has special skills. To make a chair takes 1 hr of Chris's time, 3 from Sue, and 2 from Juan. A table needs 3 hr from Chris, 5 from Sue, and 4 from Juan. A bed requires 5 hr from Chris, 4 from Sue, and 8 from Juan. Even artisans are concerned about maximizing their profit, so what product mix should the company stick with if it gets $100 profit per chair, $250 per table, and $350 per bed?

54. A candy manufacturer has 1000 lb of chocolate, 200 lb of nuts, and 100 lb of fruit in stock. The Special Mix requires 3 lb of chocolate, 1 lb each of nuts and fruit, and it brings in $10. The Regular Mix requires 4 lb of chocolate, 0.5 lb of nuts, and no fruit, and brings in $6. The Purist Mix requires 5 lb of chocolate, no nuts or fruit, and brings in $4. How many boxes of each type should be produced to maximize profit?

Arina P. Habich/Shutterstock

55. A gourmet coffee distributor has on hand 17,600 lb of African coffee, 21,120 oz of Brazilian coffee, and 12,320 oz of Colombian coffee. It sells four blends—Excellent, Southern, World, and Special—on which it makes these per-pound profits, respectively: $1.80, $1.40, $1.20, and $1.00. One pound of Excellent is 16 oz of Colombian; it is not a blend at all. One pound of Southern consists of 12 oz of Brazilian and 4 oz of Colombian. One pound of World requires 6 oz of African, 8 oz of Brazilian, and 2 oz of Colombian. One pound of Special is made up of 10 oz of African and 6 oz of Brazilian. What product mix should the gourmet coffee distributor prepare to maximize profit?

In Exercises 56 and 57, use the fact that the corner point approach can also solve minimization problems to minimize the given expression for cost C.

56. Minimize C given by $C = 7x + 8y$ over the feasible region for Exercise 35.

57. Minimize C given by $C = 5x + 11y$ over the feasible region for Exercise 36.

58. Show by example that a feasible region that has the nonnegativity constraints $x \geq 0$, $y \geq 0$, and $x + y \leq 0.5$ can have no feasible points with integer coordinates other than (0, 0).

59. Courtesy Calls makes telephone calls for businesses and charities. A profit of $0.50 is made for each business call and $0.40 for each charity call. It takes 4 min (on average) to make a business call and 6 min (on average) to make a charity call. If there are 240 min of calling time to be distributed each day, how should that time be spent so that Courtesy Calls makes a maximum profit? What changes, if any, occur in the maximum profit and optimal production policy if they must make at least 12 business and 10 charity calls every day?

60. A refinery mixes high-octane and low-octane fuels to produce regular and premium gasolines. The profits per gallon on the two gasolines are $0.30 and $0.40, respectively. One gallon of premium gasoline is produced by mixing 0.5 gal of each of the fuels. One gallon of regular gasoline is produced by mixing 0.25 gal of high octane with 0.75 gal of low octane. If there are 500 gal of high octane and 600 gal of low octane available, how many gallons of each gasoline should the refinery make? How, if at all, do the maximum profit and optimal production policy change if the refinery is required to produce at least 100 gal of each gasoline?

61. A toy manufacturer makes bikes, for a profit of $12, and wagons, for a profit of $10. To produce a bike requires 2 hr of machine time and 4 hr painting time. To produce a wagon requires 3 hr machine time and 2 hr painting time. There are 12 hr of machine time and 16 hr of painting time available per day. How many of each toy should be produced to maximize profit? How, if at all, do the maximum profit and optimal production policy change if the manufacturer must produce at least two bikes and two wagons daily?

4.5 A Transportation Problem: Delivering Perishables

4.6 Improving on the Current Solution

62. Apply the Northwest Corner Rule, thereby finding a feasible solution that obeys the rim conditions, to the following transportation problem tableaux.

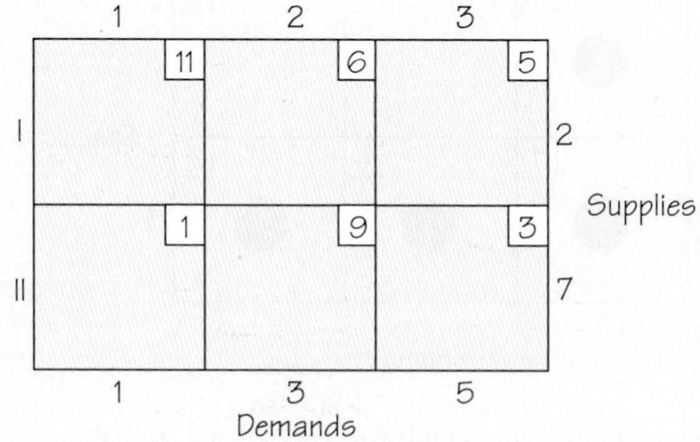

(a)

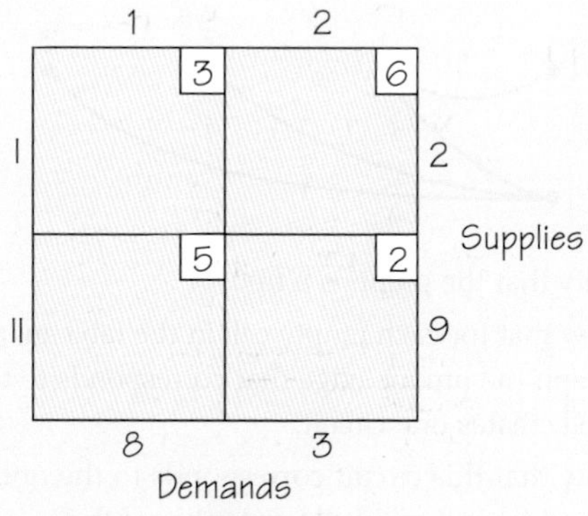

(b)

(c) For each tableau, give a possible real-world setting for the problem.

(d) For each tableau, find the cost of shipping using the cells that were circled when you used the Northwest Corner Rule.

63. The accompanying tableau represents the shipping costs and supply-and-demand constraints for supplies of purified water to be shipped to companies that resell the water to office buildings.

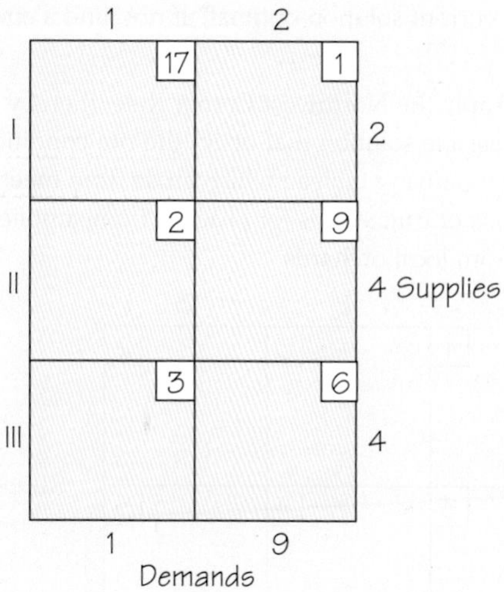

(a)

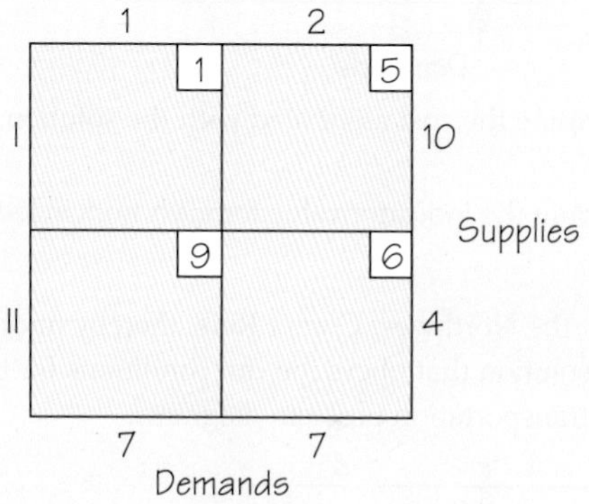

(b)

(c) For each tableau, give a possible real-world setting for the problem.

(d) For each tableau, find the cost of shipping using the cells that were circled when you applied the Northwest Corner Rule.

64. The accompanying tableau represents the shipping costs and supply-and-demand constraints for supplies of purified water to be shipped to companies that resell the water to office buildings.

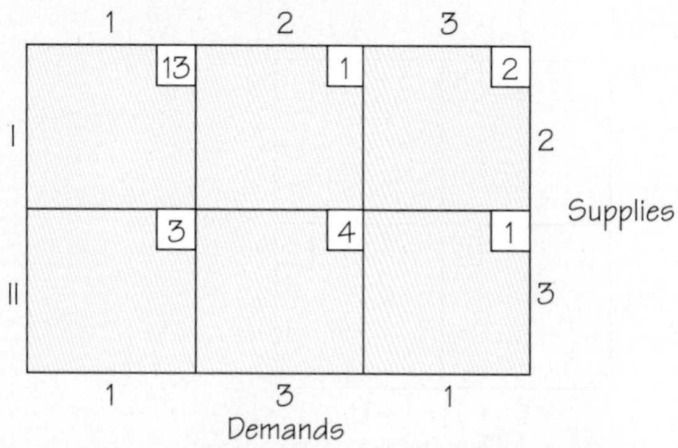

(a) Find the Northwest Corner Rule initial solution.

(b) Determine the indicator value for each noncircled cell.

(c) Is the current solution optimal? If not, find a cheaper solution.

65. (a) Apply the Northwest Corner Rule, thereby finding a feasible solution that obeys the rim conditions, to the accompanying tableau which arose from meeting the demands of fruit stands for peaches from supplies available from local orchards.

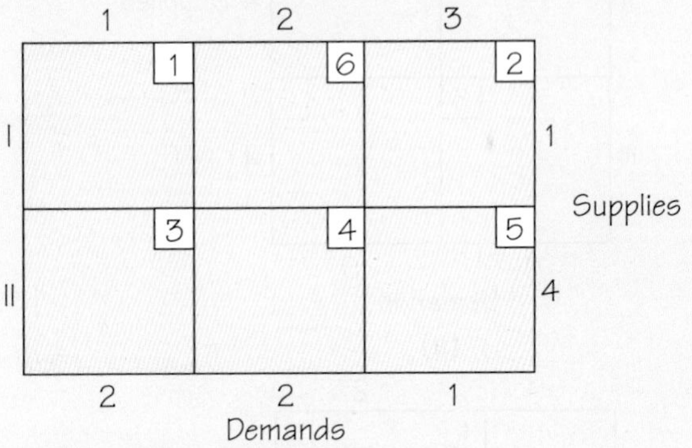

(b) Determine the cost associated with the solution that you found.

(c) Compute the indicator value for each noncircled cell.

66. Apply the Northwest Corner Rule, thereby finding a feasible solution that obeys the rim conditions for the following transportation problem diagrams.

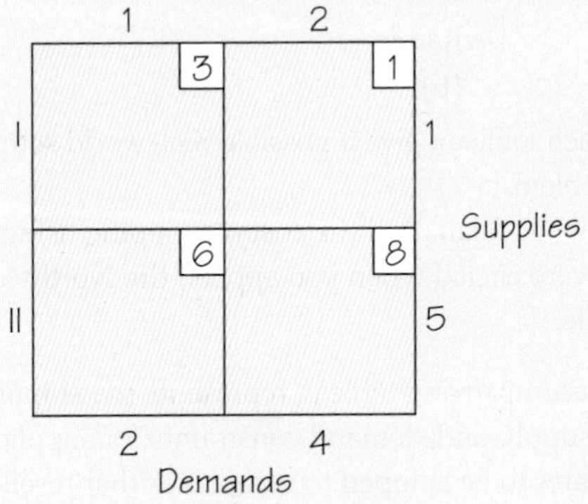

(a)

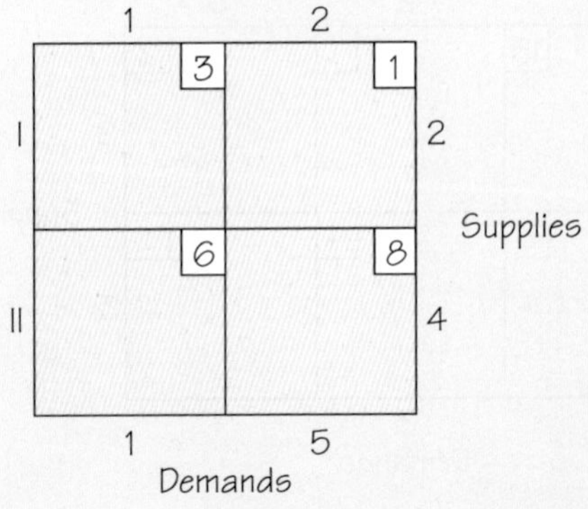

(b)

(c) For each diagram, give a possible real-world setting for the problem.

(d) For each diagram, find the cost of shipping using the cells that were circled when you applied the Northwest Corner Rule.

(e) Can you describe how the two diagrams are related?

(f) Compare the cost of the Northwest Corner feasible solutions you found for each of these two situations. Are you surprised by what you found?

67. The accompanying tableau arose by applying the Northwest Corner Rule.

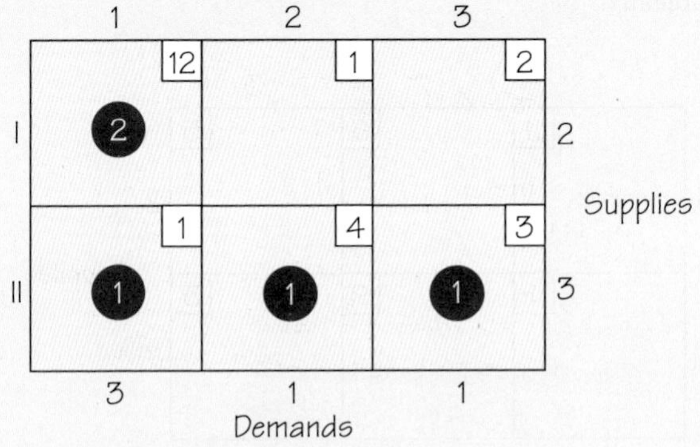

The accompanying graph was constructed so that there is one edge for each circled vertex in the tableau above.

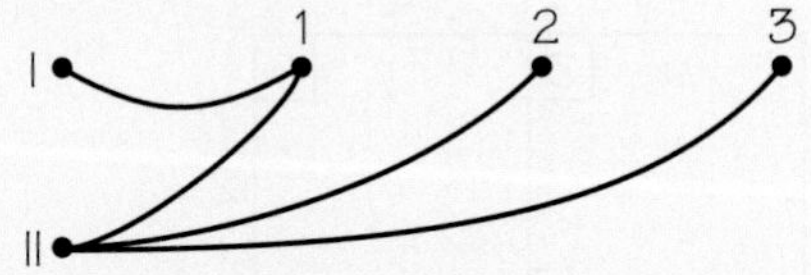

(a) Verify that the graph is a tree.

(b) Show that for each empty cell in the tableau, adding to the graph the unique edge that corresponds to the empty cell creates one circuit.

(c) Show that this circuit corresponds to the one that is used to find the indicator value of the empty cell.

68. (a) For each row of the following tableau, compute the minimum cost for that row. Now select the row R that among all the rows has the smallest row minimum. In a way similar to the Northwest Corner Rule, use the cheapest cell in row R and ship as much as possible via that cell, crossing out a row or a column, and adjust the rim conditions and repeat the process. This is known as the *minimum row entry method.* Use the minimum row entry method to find an initial solution to the following transportation problem, which shows the costs of returning rental cars from cities that have more cars than necessary to cities that have too few cars.

	1	2	3	Extra cars
I	13	1	3	3
II	2	5	4	5
Need cars	4	2	2	

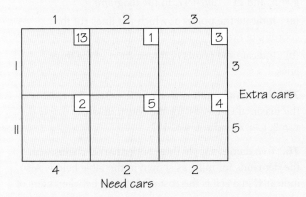

(b) Compute the cost of the solution you find using the minimum row entry method.

(c) Compare the cost found in part (b) with the cost of the initial solution obtained using the Northwest Corner Rule.

69. (a) For each of the following tableaux, find an initial solution using the Northwest Corner Rule.

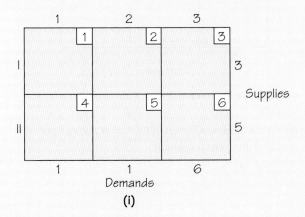

	1	2	3	Supplies
I	1	2	3	3
II	4	5	6	5
Demands	1	1	6	

(i)

	1	2	3	Supplies
I	6	5	4	3
II	3	2	1	5
Demands	1	1	6	

(ii)

	1	2	3	Supplies
I	6	5	1	5
II	2	3	4	4
Demands	2	1	6	

(iii)

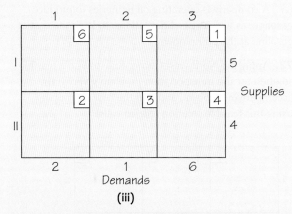

(b) If the solution you find using the Northwest Corner Rule is not optimal, then apply the stepping stone algorithm to find an optimal solution.

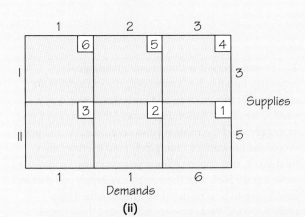

70. (a) Apply the Northwest Corner Rule to the following tableau.

	1	2	3	Supplies
I	3	1	6	5
II	5	2	3	4
III	1	7	8	1
Demands	1	2	7	

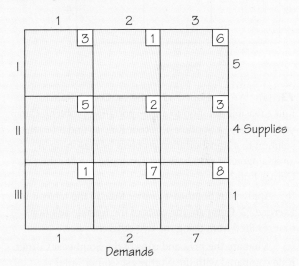

(b) Determine the cost associated with the solution you found.

(c) Compute the indicator value for each noncircled cell.

(d) Does the Northwest Corner Rule give rise to an optimal solution?

71. For each of the situations below, explain whether it seems reasonable to try to model it as a transportation problem.

(a) A supermarket chain is arranging to control costs in supplying the delivery of vegetables from its suppliers to its many branch stores.

(b) A mining company is trying to control the costs of repaving the roads that form the road network within the mine premises.

(c) A company is operating oil refineries to produce gasoline, as well as gasoline stations, to keep the cost of gasoline at the pump down.

72. Apply the Northwest Corner Rule to find a feasible solution for the transportation problem given in the accompanying tableau, and find the cost of shipping using this feasible solution (circled cells).

	1	2	3	Supplies
I	1	3	2	5
II	6	5	4	6
III	7	9	8	4
Demands	7	7	1	

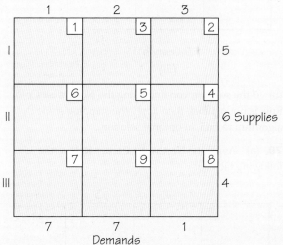

73. (a) We developed a Northwest Corner Rule to find a feasible solution to a transportation problem. Can you formulate a Southeast Corner Rule to find a feasible solution to a transportation problem by reasoning in an analogous way to how we developed the Northwest Corner Rule?

(b) Apply your Southeast Corner Rule to the tableau in Exercise 72. Compare the costs of the feasible solutions that resulted from using the two different rules.

(c) What are the pros and cons of the Southeast Corner Rule compared with the Northwest Corner Rule?

Chapter Review

74. Graph on the same set of axes the equations $x = 3$, $y = 5$, and $2x + 3y = 17$. In the diagram:

(a) Indicate the points at which the lines cut the coordinate axes.

(b) Indicate the points of intersection of the three lines.

75. If maximal profit is given by $P = 3x + 5y$, what is the maximal value of P for the constraints $x \geq 0$, $y \geq 0$, $x \geq 3$, $y \geq 7$, and $x + y \leq 11$?

76. Two dairies supply three supermarket chains with the demands for milk, as shown in the table below. Also indicated in a cell is the cost of shipping between pairs of sites.

Prices	Supermarkets 1	2	3	Supplies
I	6	9	3	7
II	1	7	5	4
Demands	2	3	6	

(a) Use the Northwest Corner Rule to obtain a feasible solution.

(b) If the feasible solution in part (a) is not an optimal one, find an optimal solution.

Writing Projects

1. Interview a local businessperson who is in charge of deciding the product mix for a business. Must this business take into consideration situations other than minimum and resource constraints? If so, what are these considerations? Find out what methods the person uses to make production policy decisions. Is linear programming used? Are other methods used? If so, what are they? Write a report of your findings, and add some of your own conclusions about the usefulness of linear programming for this business.

2. In economics, it is often useful to distinguish between a firm that has a monopoly (for example, is the only supplier of a product) and firms that supply only a small share of the market. How would the presence of a monopoly affect the relation between production and price? Would the presence of a monopoly tend to ensure the fixed-profit assumption of linear programming, or would it make it more likely that the interplay of supply and demand would have to be considered in order to have a truly realistic model? Write an essay addressing these issues.

Suggested Readings

ANDERSON, DAVID R., DENNIS J. SWEENEY, and THOMAS A. WILLIAMS. *An Introduction to Management Science: Quantitative Approaches to Decision Making,* West, St. Paul, MN, 1985. A business management text with seven chapters on linear programming.

DOLAN, ALAN, and JOAN ALDUS, *Networks and Algorithms: An Introductory Approach,* Wiley, NY, 1993. A graph theoretical approach to network optimization problems, including the transportation problem.

GASS, SAUL I. *Decision Making, Models, and Algorithms,* Krieger, Melbourne, FL, 1991. This book demonstrates how to use linear programming and related ideas to solve a variety of industrial and governmental problems.

HARDWICK, I., *Decision and Discrete Mathematics,* Albion Publishing, Chichester, England, 1996. A survey of situations that can be modeled using graphs in the area of operations research. It treats the simplex method for solving linear-programming problems and the transportation problem.

Note: Simplex software can be found in *Maple* (keyword is "simplex"), *Mathematica* (keyword is "Linear Programming"), in both *Lotus 1-2-3* and *MSExcel* via *Solver,* and in other software packages, especially those intended for quantitative mathematics courses focusing on business applications.

Suggested Websites

www.informs.org This website is maintained by the Institute for Operations Research and the Management Sciences, the main professional organization in these fields in the United States. It contains information on (and/or links to) news items about operations research and management science and employment opportunities and summer internships; it also has a student newsletter. Much of the material is written in a nontechnical style.

www.hsor.org/what_is_or.cfm?name=linear_programming This website discusses how linear programming fits into the broader subject of operations research.

www-gap.dcs.st-and.ac.uk/~history/Biographies/Dantzig_George.html This site contains biographical information about George Dantzig, who, by developing the simplex method, greatly expanded the use and applicability of linear programming.

www.neos-guide.org/content/lp-faq This site is the "frequently asked questions" section of an online newsgroup for people interested in linear programming.

en.wikipedia.org/wiki/Linear_programming This website outlines the theory of linear programming.

Improving Medical Care Using Mathematics

CASE STUDY Part 1

*This case study can be read independently of much of Chapters 1–4.

We started this section of the book with a description of some of the many complexities that are involved with running a medical center. Our goal in the prior chapters has been to try to demonstrate the way that relatively elementary mathematical ideas can be used to make our society work better. In this closing section, we will look briefly at a very recent example of this kind that makes it possible for medical centers to carry out their work better. At the same time, we will again show how elementary mathematics makes this possible; the mathematical work involved here has many applications beyond helping hospitals serve patients better. Furthermore, the work we will describe won two of the three mathematicians who developed this work, Lloyd Shapley and Alvin Roth, the Nobel Memorial Prize in Economics for 2012. It is likely that the third person involved, David Gale, would also have shared that prize had he not died in 2008.

Reed Saxon/AP Photo

Lloyd Shapley

Mandel Ngan/AFP/Getty Images

Alvin Roth

University of California, Berkeley

David Gale

Students who go to medical school complete their training to become physicians with clinical work called residencies in hospitals. Hospitals need residents for carrying out their medical missions, and students need hospital assignments to finish their clinical training to become physicians. Rather than develop the theory in detail, we will describe what Shapley and Gale discovered, and what Roth did to embellish what they had done, for the greater effectiveness of this system. The example will of necessity be small whereas the practical version is quite a bit more complex, but all of the essential ideas will be described.

Imagine there are five hospitals that need residents and five residents who need hospital residency assignments. We will assume that the students can rank the hospitals from their first choice to their least favorite (fifth choice) and that hospitals can rank the residents from 1 to 5 (1 being high and 5 being low) without ties. Hospitals might be indifferent among residents, and students might be indifferent among some hospital choices. While this more general possibility is allowed in the actual system Roth helped develop, here we make a simpler assumption. It might also be that some students would rather not have a residency at all than go to a particular hospital, and some hospitals might rather have no resident join their facility than accept a particular resident.

We will make the modeling assumption that this will not be permitted for hospitals or residents. In our example, shown below, we have generated two tables, one for the five hospitals and one for the five residents, which show each hospital's preferences for residents and each resident's preferences for the hospitals, without ties.

How can one interpret Table 1 below? Resident 3, for example, likes Hospital 1 the best and Hospital 4 the least. Resident 3's fourth choice is Hospital 5. Similarly, Hospital 2 likes resident 4 the least and Resident 1 the most. Hospital 2 likes Resident 2 second. Note that residents can have identical views about the hospitals and hospitals can have identical views about some of the residents, though that does not occur in this example.

TABLE 1 Residents Rank the Hospitals

	First	Second	Third	Fourth	Fifth
r1	h2	h3	h1	h5	h4
r2	h2	h1	h3	h5	h4
r3	h1	h2	h3	h5	h4
r4	h5	h3	h4	h2	h1
r5	h4	h3	h5	h1	h2

Improving Medical Care Using Mathematics

TABLE 2 Hospitals Rank the Students Applying for Residencies

	First	Second	Third	Fourth	Fifth
h1	r2	r1	r3	r5	r4
h2	r1	r2	r3	r5	r4
h3	r5	r1	r3	r2	r4
h4	r4	r5	r1	r2	r3
h5	r3	r4	r5	r1	r2

What is wrong with just matching any hospital with any student, as long as each of the hospitals is assigned one student and one student is assigned one hospital?

For example, what about the matching *M* (a particular way of pairing hospitals to residents), where Resident 1 (r1) is paired with Hospital 4 (h4), r2 with h5, r3 with h1, r4 with h2, and r5 with h3? Since h4 is r1's last choice, and h4 prefers r4 or r5 to r1, we see that r1 would rather be paired with h1 than h4 and that h1 would rather be paired with r1 than r3, who is its partner in the current assignment. Thus, r1 and h1 would rather be paired with each other than with their current "mates." The proposed set of assignments is not "stable" because there is a pair (in this example, r1 and h1) who would rather be together than with the "mate" to whom each is assigned in *M*. In the language of the mathematics of two-sided markets, which is what we are dealing with here, r1 and h1 are called a *blocking pair* for matching *M*. They would be happier together than with their assigned partners in *M*.

So *M* is not an ideal matching, and we have come up with one notion of what makes one matching better than some others: the idea of a stable matching. Summarizing what we learned from the example, a Hospital *h*–Resident *r* pair is called a blocking pair for matching *M* if *h* and *r* are not paired in *M* but would prefer to be paired to each other than to whom they are paired in *M*. A matching *M* is unstable if there is some blocking pair for *M*. A matching *M** is stable if there are no blocking pairs for *M**. What David Gale and Lloyd Shapely showed is that for any *m* hospitals and *m* residents with tables such as those above (strict rankings and no one would rather be unpaired than paired with someone in particular), there must always be at least one stable matching. Not only did they show this, but they showed an easy-to-understand procedure for finding, typically, at least two different such matchings (though in some cases there is only one stable matching).

While the problem being discussed has been described in terms of matching hospitals and residents, more often than not the problem is often described in terms of pairing men and women—perhaps as dancing partners at a college dance—a setting with suggestive terminology if lacking the "seriousness" of the setting we started with. So in describing the procedure, or algorithm, that Gale and Shapley developed, I will use a blend of terminology suggestive of matching men and women and/or hospitals and residents.

Gale–Shapley Deferred Acceptance Procedure (Algorithm)

Before starting the details of the procedure, here is a very brief description of the idea. In "rounds," the women (men) make proposals to the men (women). If a man (woman) gets one or more proposals, he (she) will temporarily accept the best of these proposals based on the rankings that he (she) has. Those who don't get matched in some round try again in the next round.

Think of the hospitals as setting up tables in front of a room. The process will proceed in rounds, with the residents "proposing" to the hospitals based on their preferences in Table 1 above. The algorithm is known as the deferred acceptance algorithm, and the idea is that at each stage (round), if more than one "suitor" shows up at a hospital desk, the hospital selects the highest-ranked resident from the hospital's point of view, on a "temporary" basis, and the rejected students will move on to the next round. If in a future round a resident "proposes" who is superior to the currently accepted choice, the hospital will accept temporarily the highest-ranking choice available and send other "suitors" to the next round. The procedure terminates when there is one resident located at each hospital desk; when this happens, one will have a stable matching. It is worth noting at the start that if all of the residents have different first choices of hospitals, then the procedure terminates right in the first round.

So let us carry out this process for the example above.

Round 1

Residents 1 and 2 (r1 and r2) head for the table of Hospital 2 (h2); r3 heads for the table of h1; r4 goes to

Improving Medical Care Using Mathematics *(continued)*

h5's table and r5 to h3's table. Note that no one appears at the table of h3. For the moment, h1, h4, and h5 accept r3, r5, and r4, respectively. What does h2 do? It is tickled pink to see its first choice, r1, arrive among the two people who would be pleased to come to h2, so it accepts r1 and sends r2 into the second round. So at this stage, the result of Round 1 can be expressed as follows:

	h1	h2	h3	h4	h5
Round 1	r3	r1		r5	r4

Note that r2 and h3 are unmatched.

Round 2

Resident 2 (r2) is disappointed about being rejected by his/her first choice, h2, but in Round 2 proceeds to the next highest choice on his/her list, which is h1.

Hospital 1 (h1) is pleased to have r3 at its table and the newly arrived r2. Since h1 prefers r2 to r3, it chooses r2 over r3 and sends r3 "packing"—off to the next round. Thus, at the end of Round 2 we have the following situation:

	h1	h2	h3	h4	h5
Round 2	r2	r1		r5	r4

Note that r3 and h3 are unmatched.

Round 3

At the start of Round 3, only r3 is not yet matched. So r3 goes to the next highest choice in his/her ranking after h1, namely h2.

Now, h2 has the choice of r3 or r1. From h2's perspective, r1 is better than r3, so again r1 is temporarily accepted and r3 is sent out to find another hospital. We now have the following:

	h1	h2	h3	h4	h5
Round 3	r2	r1		r5	r4

Note that r3 and h3 are unmatched.

Round 4

At this point, r3 is unmatched, so r3 goes to his next highest choice after h2, which is h3. Since h3 had no resident, h3 gladly accepts r3. At this stage, each hospital has a resident and each resident has a hospital.

	h1	h2	h3	h4	h5
Round 4	r2	r1	r3	r5	r4

The matching of h1 to r2, h2 to r1, h3 to r3, h4 to r5, and h5 to r4 is stable! The reason is that no resident has any hope of having a hospital ranked higher than the one he/she is matched with—because at every stage, residents try to form a match with the hospital highest on their list and only go on to other hospitals when they are rejected. So we cannot have any blocking pairs for this matching. In this example, at the end of all rounds except the last, we had only one resident who was unmatched; but often there will be several residents seeking a hospital in the next round.

Why can't we run this algorithm with the residents manning the tables and the hospitals proposing? The answer is we can! Here, in fewer words, are the results of doing that:

Round 1

	r1	r2	r3	r4	r5
Round 1	h2	h1	h5	h4	h3

Everyone is matched because there were different first choices by each of the hospitals. Thus, we have the matching *M* (hospitals propose) that pairs h1 to r2, h2 to r1, h3 to r5, h4 to r4, and h5 to r3.

This matching must be stable in that no hospital can do better in this process because at each stage it goes on to a lower choice only if a high choice has rejected the hospital. In fact, here, each hospital got its first choice.

We can see that the two matchings are not the same, though sometimes they can be identical. Can anything be said about these two stable matchings?

First of all, it is worth observing that although the Gale–Shapley deferred acceptance algorithm can be used to find two special stable matchings, there can be many more stable matchings. The details of an algorithm to find all stable matchings are not outlandishly complex, but they will not be discussed here.

To see the special nature of the two matchings we have found, let us look at how pleased the matched parties are with the choice they got.

Improving Medical Care Using Mathematics

Residents Propose

	h1	h2	h3	h4	h5
Rank of resident it got	1	1	3	2	2

	r1	r2	r3	r4	r5
Rank of hospital it got	1	2	3	1	1

Hospitals Propose

	h1	h2	h3	h4	h5
Rank of resident it got	1	1	1	1	1

	r1	r2	r3	r4	r5
Rank of hospital it got	1	2	4	3	2

What is going on? Among all stable matchings, the one in which the hospitals propose is the very best one from the hospitals' perspectives but the worst stable marriage from the residents' perspectives. Among all stable marriages, the one in which the residents propose is the best that the residents can do but the worst from the hospitals' perspectives. Here, best and worst are measured in terms of the ranks of the hospitals and residents for the "other side" of the market.

The National Residents Matching Program (NRMP) was a voluntary, two-sided market system to match hospitals to residents. Though it was established prior to the work of Gale and Shapley, it used an algorithm that produced stable marriages. In Great Britain, similar matching markets sometimes broke down because they did not produce stable marriages. Alvin Roth's major contribution was to deal with complications that were causing the NRMP difficulties. In particular, many married medical students wanted residencies in the same hospital or at hospitals in the same city, and this caused difficulties for the way the NRMP operated. Roth helped develop an algorithm that overcame the "couples" issue and used matching market ideas to match students with schools ("school choice").

The algorithm is also used for pairing kidneys that become available (due to an organ donor's death or a volunteer donor) to people in need of a transplant. New uses of two-sided markets are being developed regularly. For example, recently these ideas are being used to pair computers to Wi-Fi networks.

Practice

Below are the rankings of four men and four women.

Men Rank the Women

	First	Second	Third	Fourth
m1	w1	w2	w3	w4
m2	w2	w1	w4	w3
m3	w3	w4	w1	w2
m4	w4	w3	w2	w1

Women Rank the Men

	First	Second	Third	Fourth
w1	m4	m3	m2	m1
w2	m3	m4	m1	m2
w3	m2	m1	m4	m3
w4	m1	m2	m3	m4

1. Use the Gale–Shapley deferred acceptance algorithm to find the male optimal stable marriage.
2. Use the Gale–Shapley deferred acceptance algorithm to find the female optimal marriage.
3. For the male optimal stable marriage, find the rank of each woman in the matching.
4. For the female optimal stable marriage, find the rank of each man in the matching.

Note that there are in fact eight other stable marriages for this example!

Suggested Reading and Website

To learn more about the mathematical ideas discussed here, and their applications, consult the following:

ROTH, A. and M. SOTOMAYOR. *Two-Sided Matching: A Study in Game-Theoretic Modeling and Analysis,* Cambridge University Press, New York, 1992.

www.nrmp.org This web page describes the National Resident Matching Program that is used to match hospitals with residents in the United States.

Part II

Statistics: The Science of Data

Data are collected every day. Whether you know it or not, you, too, contribute to the vast amount of data collected daily. Every time you make a phone call (or send a text message), the number, location, date, and length of the call (or number of characters in the text message) are saved by the phone company. If you download a movie, the title, genre, movie rating, date and time of download get recorded. When you use a supermarket card to get store savings, the date, products you buy, amount you spend, and amount you save are stored in a data bank. At the end of each term, your university or school records your courses and grades for future reference. And that's just a small subset of the personal data that *you* generate. So data collecting is constantly going on all around you. Analysis of such data is being used for marketing, security, political advocacy, and much, much more.

And you are not the only source of data—data are collected on traffic patterns, mercury level in fish, consumer products, emergency room admissions, climate change, crops, standardized testing, and just about everything you can imagine. With so much data out there, how do we make sense of it? That's where statistics comes in. *Statistics* is the science of collecting, organizing, analyzing, and interpreting data.

Chapters 5 and 6 concern *data analysis,* the art of studying what data reveal. We learn from data by making graphs and doing calculations, guided by principles that help us decide what graphs to make, what to look for in our graphs, and what calculations are helpful based on what we see.

Sometimes we want to know more: An opinion poll or a medical study looks at only some people, but we want conclusions that apply to all voters or all patients. This is called *statistical inference* because we infer conclusions about a large group from data on a small part of the group. Chapter 7 discusses inference from beginning to end—from how to produce data when we have inference in mind to how to say just how much confidence we can have in our conclusions. Confidence, uncertainty, risk, chance—the mathematics that describes all these ideas is *probability theory,* the topic of Chapter 8. Probability is the mathematics behind statistical inference, but that's just a small part of its usefulness.

Exploring Data: Distributions

5

NASA

5.1 Displaying Distributions: Histograms

5.2 Interpreting Histograms

5.3 Displaying Distributions: Stemplots

5.4 Describing Center: Mean and Median

5.5 Describing Variability: Range and Quartiles

5.6 The Five-Number Summary and Boxplots

5.7 Describing Variability: The Standard Deviation

5.8 Normal Distributions

5.9 The 68–95–99.7 Rule for Normal Distributions

If a map showed every pothole, traffic sign, and store, it would be far too cluttered to be useful for planning a road trip. On the other hand, if a map included only a couple of reference points, it would be too easy to get lost and miss the destination. So a good map gives us just the right level of detail, calling our attention to special features and main roads.

The undigested blizzard of data we encounter in modern society can feel overwhelming, like that first type of map. But if we simply ignore data, we risk the pitfall of the second type of map. Failure to detect patterns in data in a timely manner has had serious consequences, ranging from the loss of a NASA spacecraft to large-scale misguided financial trading practices that caused billions of dollars of losses. Therefore, we need to develop good skills to "read" and appropriately summarize data so that we can navigate the terrain of information and numbers where we live and travel. Just as it helps for directions to have both numerical information (e.g., "3.2 miles on Gluckin Avenue") and visual diagrams or landmarks (e.g., "Turn right just after you pass the water tower"), it is important for data analysis to have both numerical and graphical techniques as well.

This chapter starts with an introduction to the concepts of exploring data from one quantitative variable. We will begin with a graphical technique called a *histogram,* which can serve as a middle-of-the-road approach in terms of the amount of detail it reveals about the data. Later, we explore graphical techniques that include more detail (stemplots and dotplots) and then less detail (boxplots). Along the way, we'll throw in some calculations that can help us judge the center and spread of our data. The chapter's concluding sections highlight distributions of data of a particular shape, known as normal distributions. This class of distributions is special because of its pervasiveness in statistics and its presence in the real world. Happy travels through statistics!

5.1 Displaying Distributions: Histograms

Any set of data contains information about some group of **individuals.** The information is organized in **variables.** We will briefly note some basics about data before we learn tools for exploring and summarizing.

Individuals DEFINITION

Individuals are the entities about which information (data) is collected. Individuals may be people, but they may also be groups, animals, or things.

Variable DEFINITION

A **variable** is a characteristic or trait that can take on different values for different individuals. A particular variable may be either qualitative (e.g., gender) or quantitative (e.g., age).

EXAMPLE 1 Data from a Student Questionnaire

Table 5.1 shows a partial dataset that describes the students in a statistics class. The data come from anonymous responses to a class questionnaire. Most data tables follow this format: Each row records data on one individual (in this case, one student) and each column contains values of one variable for all the individuals. This dataset appears in a *spreadsheet* program that has rows and columns ready for your use. Spreadsheets are commonly used to enter and transmit data, and spreadsheet programs also have functions for basic statistics.

TABLE 5.1 Excerpt of a Dataset Displayed in the Microsoft Excel Spreadsheet Program

	A	B	C	D	E
1	SEX	HAND	HEIGHT	STUDY	COINS
2	F	L	65	200	50
3	M	L	72	30	35
4	M	R	62	95	35
5	F	L	64	120	0
6	M	R	63	220	0
7	F	R	58	60	76
8	F	R	67	1500	215

Sheet1 Sheet2 Sheet3

The partial spreadsheet shows information on seven individuals in rows 2–8. The questionnaire consisted of five questions, which are represented by the five columns in the spreadsheet. Sex (female or male) and handedness (left-handed or right-handed) are variables that are usually described as *qualitative* or *categorical* because they categorize individuals by traits and do not take numerical values. The remaining three variables are *quantitative* or *measurement* variables because they do take numerical values. They are height (inches), time spent studying (in minutes) on a typical weeknight, and the amount of money in coins (cents) students are carrying. Our main focus in this chapter will be on variables involving quantitative or numerical data, because you probably have already had much experience with the usual ways to summarize categorical data (proportions, pie charts, and bar charts).

Knowing the context of the data—that these are student responses to a class questionnaire—helps us make sense of them. For example, one student claimed to study 1500 minutes on a typical night. We know that this is impossible! (Perhaps the student miscalculated when converting from hours to minutes or it was a typographical error.) ■

Self Check 1

Here are the first few rows of a professor's dataset at the end of a mathematics course:

	A	B	C
1	**Name**	**Major**	**Points**
2	Advani, Sura	COMM	397
3	Bartin, David	HIST	323
4	Boaz, Judah	BIOL	446
5	Chiu, Sun	PSYC	405
6	Davis, Lauren	PSYC	461

(a) Are the individuals for this dataset the students, total points, or majors?
(b) Identify each of the variables as qualitative or quantitative. ■

Statistical tools and ideas help us examine data and describe their main features. This examination is called **exploratory data analysis (EDA).** Like an

John Wilder Tukey, Champion of Exploratory Data Analysis

SPOTLIGHT 5.1

Mathematician John Wilder Tukey (1915–2000) directed Princeton University's Statistical Research Group when it was set up in 1956 and was the first chairman of Princeton's Department of Statistics after it was formed in 1965. He divided his time between Princeton and AT&T Bell Laboratories, where his work involved developing statistical methods for computers. In the field of computer science, Tukey is credited with introducing the terms *bit* (contraction for binary digit) and *software*.

Tukey led the way in the field of exploratory data analysis (EDA). EDA is an approach for data analysis that focuses heavily on graphical techniques. The goal is to tease out information from datasets in order to unlock their stories. Stemplots and boxplots (graphic displays that appear later in this chapter) became popular after the publication of Tukey's 1977 book *Exploratory Data Analysis*.

Alfred Eisenstaedt/The LIFE Picture Collection/Getty Images

explorer crossing unknown lands, we first want to describe simply what we see so that we can start to draft our map of the terrain. In this chapter and the next, we use both numbers and graphs to explore data. As we learn from Spotlight 5.1, the use of EDA to explore and analyze data was promoted by John Wilder Tukey. Here is a process for exploratory analysis of data.

Exploring Data PROCEDURE

1. Begin by examining each variable by itself. Start with one or more graphs, and then add numerical summaries of specific aspects of the data.
2. Explore possible relationships among variables, using graphical displays and then numerical summaries.

These principles also organize the material in Chapters 5 and 6. In this chapter, we look at data on a single variable. Chapter 6 moves on to relationships among two or more variables.

Data analysis begins with graphical displays of the values of a single variable. For example, universities may want information about their students' study times. Because individual study times vary so much, we are interested in the **distribution** of study times.

Distribution DEFINITION

The **distribution** of a variable gives information (as a table, graph, or formula) about how often the variable takes certain values or intervals of values.

This definition has several different manifestations. The simplest one is a "tally chart" called a **frequency distribution.**

Frequency Distribution DEFINITION

A **frequency distribution** classifies data on a single variable into non-overlapping classes or intervals (class intervals) and records how many times data values are in each class.

A frequency distribution can be displayed in a table such as this one for the variable HAND in the dataset excerpt of Table 5.1 (page 183):

Class	R	L
Frequency	4	3

Here, the classes are the possible outcomes of the qualitative variable HAND. The bar chart in Figure 5.1 is one way to represent this frequency distribution graphically.

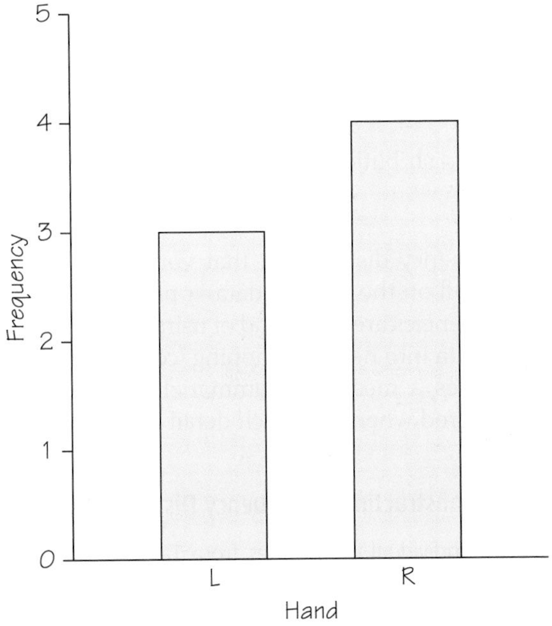

Figure 5.1 Bar chart of HAND data from Table 5.1.

The variable COINS in Table 5.1 is a quantitative variable. In Table 5.2, we construct a frequency distribution in which each outcome is its own class, similar to what was done for HAND. In addition to listing the raw frequencies or counts in each category, we have added the **relative frequency distribution,** which gives the fraction of the time that each value occurs. Since there are seven data values, we obtain the relative frequencies by dividing the frequencies by 7.

TABLE 5.2 Frequency Distribution of COINS Data from Table 5.1

Class	0	35	50	76	215
Frequency	2	2	1	1	1
Relative Frequency	$\frac{2}{7} \approx 0.29$	$\frac{2}{7} \approx 0.29$	$\frac{1}{7} \approx 0.14$	$\frac{1}{7} \approx 0.14$	$\frac{1}{7} \approx 0.14$

Relative Frequency Distribution DEFINITION

A **relative frequency distribution** classifies data on a single variable into non-overlapping classes or intervals and records what fraction (or percentage) of the data values are in each class.

Table 5.3 gives the complete dataset for the variable COINS, which would appear in column E of Table 5.1 (page 183). To save space, we have listed these data in four rows rather than a single column.

TABLE 5.3 Complete Data for COINS (Column E in Table 5.1)

50	35	35	0	0	76	215	77	62	175
189	120	54	26	145	0	0	35	47	125
55	35	78	157	225	92	85	35	59	145
137	142	62							

Self Check 2

Construct a frequency distribution for the data in Table 5.3. Use the data values as the classes.

Although the frequency distribution that you constructed for Self Check 2 contains complete detail on the COINS data, it provides little useful information since most of the frequencies are 1. Instead of using distinct data values as classes, we can partition the data into non-overlapping, consecutive intervals called **class intervals.** This provides a means of summarizing the data and often reveals patterns that are obscured when too much detail is visible.

EXAMPLE 2 **Constructing a Frequency Distribution and Histogram**

Instead of using the individual data values from Table 5.3 for the classes, we set up class intervals for the COINS data and construct a frequency distribution based on the class intervals. We then display the frequency distribution graphically in a histogram.

Constructing a Frequency Distribution

Step 1: Choose the classes. Determine an interval that is wide enough to contain all the data. Subdivide this interval into a reasonable number of class intervals of equal width. Be sure to specify the classes precisely so that each individual falls into exactly one class.

The data in Table 5.3 range from 0 to 225. So here's one way to choose the class intervals. All the data are between 0 and 250. We subdivide this interval into five class intervals of equal width:

$$0 \leq \text{COINS} < 50$$
$$50 \leq \text{COINS} < 100$$
$$100 \leq \text{COINS} < 150$$
$$150 \leq \text{COINS} < 200$$
$$200 \leq \text{COINS} < 250$$

Step 2: Setting up the table. Set up a table with three columns for the following: class interval, tally, and frequency. (Remove the tally column in the final table.)

Step 3: To complete the table, determine the frequency with which data values fall into each class interval.

Step 4: If desired, add a fourth column for relative frequency. The entries in this column should be the frequencies divided by the number of data values.

Table 5.4 shows the construction of a frequency and relative frequency table for the COINS data (Table 5.3). Since there were 33 data values, the relative frequencies were determined by dividing the frequencies by 33.

TABLE 5.4 Frequency and Relative Frequency Distribution for COINS Data

Class Interval	Tally	Frequency	Relative Frequency
$0 \le \text{COINS} < 50$	𝍸 𝍸 |	11	$11/33 \approx 0.333$
$50 \le \text{COINS} < 100$	𝍸 𝍸 |	11	$11/33 \approx 0.333$
$100 \le \text{COINS} < 150$	𝍸 |	6	$6/33 \approx 0.182$
$150 \le \text{COINS} < 200$	|||	3	$3/33 \approx 0.091$
$200 \le \text{COINS} < 250$	||	2	$2/33 \approx 0.061$

Making a Histogram

The best way to represent a frequency distribution graphically is with a histogram. Here are the steps for making a histogram.

Step 1: Draw a set of axes. On the horizontal axis, mark the boundaries of the class intervals. On the vertical axis, set up a scale appropriate for the frequencies (or relative frequencies).

Step 2: Label the horizontal axis with the name of the variable being measured and the units. Label the vertical axis as "Frequency" (or "Relative Frequency").

Step 3: Over each class interval, draw a rectangle with the interval as its base. The height of the rectangle should match the frequency (or relative frequency) of data contained in that interval.

In the case of the coin data, we draw a horizontal axis with tick marks every 50 units from 0 to 250 to mark the class intervals. On the vertical axis, we place tick marks every two units from 0 to 12 for the frequencies. We then add the rectangular bars to make the histogram shown in Figure 5.2.

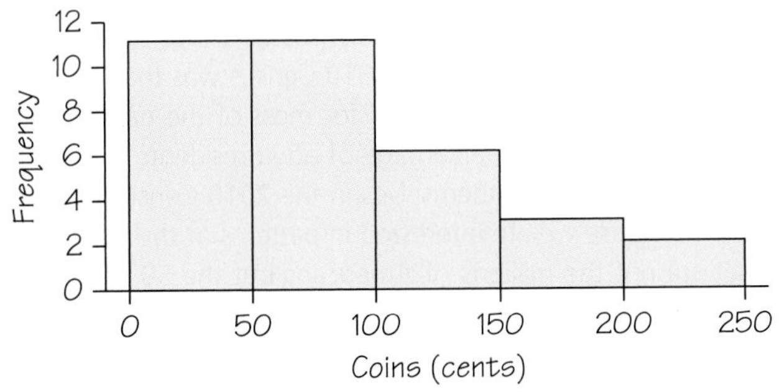

Figure 5.2 Histogram of COINS data.

Our eyes respond to the area of the bars in a histogram. Because the classes are all the same width, area is determined by height and all classes are fairly represented. While something between 5 and 20 classes works for most real-world datasets, there is no one right choice for the number of the classes in a histogram.

Histogram DEFINITION

A **histogram** is a graphical representation of a frequency distribution for a single numerical variable. Bars are drawn over each class interval on a number line. The areas of the bars are proportional to the frequencies (or relative frequencies) with which data fall into the class intervals.

Although a histogram (e.g., Figure 5.2) and a bar chart (e.g., Figure 5.1 on page 185) both use rectangular bars, they are different because they display one quantitative variable and one qualitative variable, respectively. Only a histogram's bars start from an axis that represents a numerical scale. Notice also that in a bar chart the bars are separated whereas in a histogram there is no horizontal space between the bars unless a class is empty so that the bar has height 0.

5.2 Interpreting Histograms

Making a statistical graph is not an end in itself. The purpose of the graph is to help us understand the data. After you make a graph, always ask, "What do I see?" Once you have displayed a distribution, you can see its important features as follows.

Outlier DEFINITION

In any graph of data, look for the overall pattern and for striking deviations from that pattern. You can describe the overall pattern of a distribution by its shape, center, and variability. An important kind of deviation is an **outlier,** an individual value that falls outside the overall pattern.

We now explore Example 3, which focuses on the percentage of each state's Hispanic population. The data appear in Table 5.5.

EXAMPLE 3 ➔ Describing a Distribution

Every 10 years, the Census Bureau (www.census.gov) tries to contact every household in the United States. One finding of the 2010 Census was that the Hispanic population (which is now over 50 million) accounted for most of the nation's growth in the past decade. Table 5.5 presents the percentage of adult residents (age 18 and over) in each of the 50 states who identified themselves in the 2010 Census as "Hispanic, Latino, or Spanish origin." Because we are interested in patterns at the state level, the *individuals* in this dataset are not the millions of Americans but the 50 states. The *variable* is the percentage of Hispanics in a state's adult population.

Table 5.5 contains too much detail to find patterns and trends easily. Again, we begin by grouping the data into convenient intervals (or "classes") to make a frequency distribution. Since no more than 45% of the residents of any state identified as Hispanic, we subdivide the interval from 0% to 45% into nine class intervals of width 5% and then classify the data into these intervals. The resulting frequency distribution is given in Table 5.6.

TABLE 5.5 Percent of Adult Population of Hispanic Origin, by State (2010 Census)

State	Percent	State	Percent	State	Percent
Alabama	3.2	Louisiana	4.0	Ohio	2.5
Alaska	4.7	Maine	1.0	Oklahoma	7.1
Arizona	25.0	Maryland	7.3	Oregon	9.1
Arkansas	5.0	Massachusetts	8.1	Pennsylvania	4.6
California	33.1	Michigan	3.5	Rhode Island	10.2
Colorado	17.5	Minnesota	3.7	South Carolina	4.3
Connecticut	11.6	Mississippi	2.5	South Dakota	2.1
Delaware	6.7	Missouri	2.9	Tennessee	3.8
Florida	21.1	Montana	2.3	Texas	33.6
Georgia	7.5	Nebraska	7.2	Utah	11.3
Hawaii	7.2	Nevada	22.3	Vermont	1.3
Idaho	9.0	New Hampshire	2.2	Virginia	6.9
Illinois	13.4	New Jersey	16.3	Washington	8.9
Indiana	4.8	New Mexico	42.3	West Virginia	1.0
Iowa	3.8	New York	16.2	Wisconsin	4.6
Kansas	8.4	North Carolina	6.8	Wyoming	7.5
Kentucky	2.5	North Dakota	1.5		

TABLE 5.6 Frequency Distribution for Hispanic Percentage

Class	Frequency	Class	Frequency	Class	Frequency
0.0 to 4.9	22	15.0 to 19.9	3	30.0 to 34.9	2
5.0 to 9.9	15	20.0 to 24.9	2	35.0 to 39.9	0
10.0 to 14.9	4	25.0 to 29.9	1	40.0 to 44.9	1

Now we can draw a histogram to represent the information from Table 5.6. Although the histogram contains the same information as the table, a graphic display often helps us identify patterns more easily.

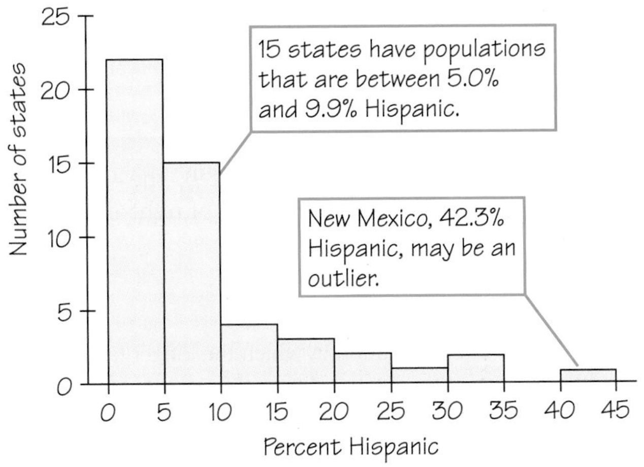

Figure 5.3 Histogram of the percentage of Hispanics among the adult residents of the states.

Next, we use all the information we have gathered so far to describe features of this dataset.

- **Shape:** The distribution has a *single peak*, which represents states in which less than 5% of adults are Hispanic. Most states have no more than 10% Hispanics, but some states have much higher percentages, so the graph trails off to the right.
- **Center:** From the frequency distribution, we know that 22 of the 50 states had a Hispanic adult population of less than 5%. The middle point for the data is somewhere between 5% and 10%. From Table 5.5 we find that about half the states have less than 7% Hispanics among their adult residents and the rest have more. So the middle of the distribution is around 7%.
- **Variability:** The data's span is from about 1% to 42% (a difference of 41%), but only six states exceed 20%.
- **Outliers:** New Mexico stands out. Whether this is an outlier or just part of the long right tail of the distribution is a matter of judgment.

Some statistical software packages "flag" outlier values using methods such as those in Exercises 35 or 62, but there is no one universal rule for calling an observation an outlier. Once you have spotted possible outliers, look for an explanation. Some outliers are due to mistakes, such as the student who studied 1500 minutes on a typical weekday (Table 5.1 on page 183), when perhaps it should have been 150 minutes. Other outliers point to the special nature of some observations—such as the high percentage of Hispanics in New Mexico, territory which was under the control of Spain and Mexico before it became part of the United States.

When you describe a distribution, concentrate on the main features. Look for major peaks, not for minor ups and downs, in the bars of the histogram. Look for clear outliers, not just for the smallest and largest observations. Look for rough *symmetry* or clear departures from it.

Distributions come in a variety of shapes. Some distributions have a shape in which the bulk of the values form a heap on one side, close to the distribution's balance point, and the rest of the values stretch out into a long tail on the other side of the balance point. This trait is known as **skewness,** and the direction of skewness may be either to the right or to the left as described in following definitions.

Right-Skewed Distribution DEFINITION

A **right-skewed distribution** is a distribution in which the longer tail of the histogram is on the right side. (Because positive numbers lie on the right side of a number line, such a distribution is also called "positively skewed.")

Both Figures 5.2 (page 187) and 5.3 (page 189), the COINS and Hispanic percentage distribution, respectively, are examples of right-skewed distributions.

Left-Skewed Distribution DEFINITION

A **left-skewed distribution** is a distribution in which the longer tail of the histogram is on the left side. (Because negative numbers lie on the left side of a number line, such a distribution is also called "negatively skewed.")

For example, an easy exam may yield a left-skewed distribution because most students will cluster together with high scores, but there are usually still a few students who perform low (possibly due to lack of attendance or effort) and give the distribution a tail stretching out to the left. Figure 5.4 illustrates this situation.

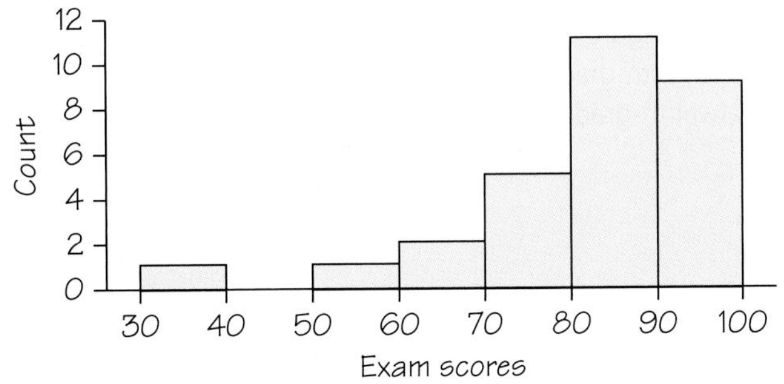

Figure 5.4 Exam scores on an easy exam.

Other distributions may have little or no skewness. For example, the distribution of heights in an adult population may look like two hills of equal size, as is the case in Figure 5.5, which depicts Brian Joiner's living histogram of Penn State students grouped by height. (See Suggested Readings for more information on Joiner's living histograms.)

Figure 5.5 Living histogram of student height.

A more common and more important shape without skewness is the bell-shaped histogram, the subject of Section 5.8. Many biological measurements (such as height, length of thigh bone, and so on) on specimens from the same species and sex have a bell shape. So do scores on most standardized tests, the subject of Example 4 (see Figure 5.6.). Distributions without much skewness where values are distributed similarly on both sides of the distribution can typically be described as **symmetric.**

Symmetric Distribution DEFINITION

A **symmetric distribution** is one in which a vertical line could be superimposed on the histogram and the left and right sides are approximate mirror images of each other.

EXAMPLE 4 Iowa Test Scores

Figure 5.6 displays the scores of all 947 seventh-grade students in the public schools of Gary, Indiana, on the vocabulary part of the Iowa Test of Basic Skills. The distribution is *single-peaked* and *symmetric*. In mathematics, the two sides of symmetric patterns are exact mirror images, but real-life data are almost never exactly symmetric. We are content to describe Figure 5.6 as symmetric. The center (half above, half below) is close to 7. This is a seventh-grade reading level. The scores range from 2.0 (second-grade level) to 12.1 (twelfth-grade level).

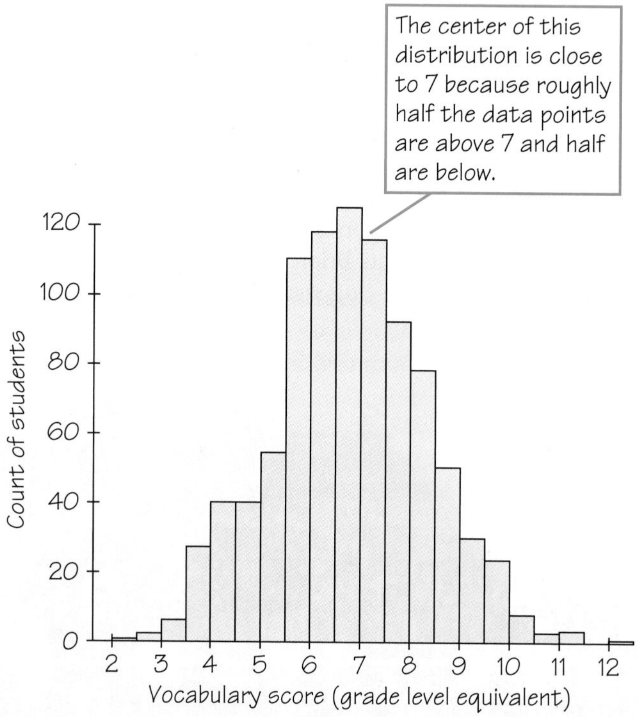

Figure 5.6 Histogram of the Iowa Test of Basic Skills vocabulary scores for all 947 seventh-grade students in Gary, Indiana, public schools.

EXAMPLE 5 College Tuition

Jeanna plans to attend college in her home state of Massachusetts. She looks up the annual tuition and fees for all 64 four-year colleges in Massachusetts (omitting art schools and other special colleges). The data for the 2014/2015 academic year are displayed in the histogram in Figure 5.7. Notice that there are three bars tied for the tallest bar, representing 12 colleges charging between \$30,000 and \$35,000, \$40,000 and \$45,000, and \$45,000 and \$50,000, respectively. As is often the case, we can't call this irregular distribution either symmetric or skewed. It does show two separate *clusters* of colleges, 12 with tuition and fees less than \$15,000 and the remaining 52 costing more than \$20,000. More generally, clusters suggest that different types of individuals are mixed in a dataset. In fact, the histogram in Figure 5.7 distinguishes 12 state colleges in Massachusetts from 52 private colleges, which charge much more.

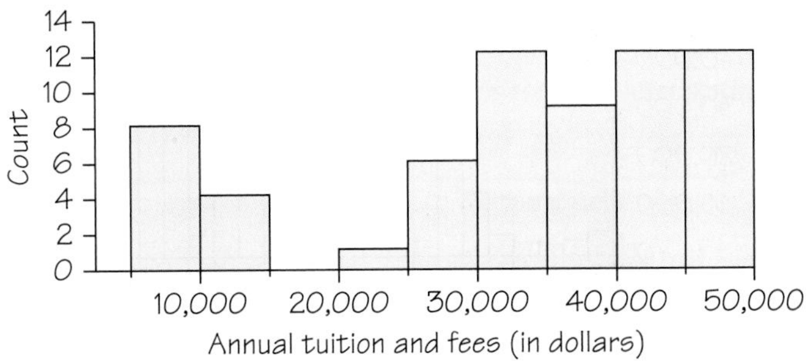

Figure 5.7 Histogram of the annual tuition and fees charged by four-year colleges in Massachusetts.

Self Check 3

Which of the statements below can be concluded from the histogram in Figure 5.7?

(a) There is at least one college with tuition and fee charges between \$7500 and \$12,500.

(b) There are no colleges with tuition and fee charges between \$12,500 and \$22,500.

(c) There are no colleges with tuition and fee charges between \$16,000 and \$18,000.

Up to this point, we have not investigated how changing the class intervals might change the look of a histogram. We return to our map analogy—too much detail can obscure patterns but too little detail can miss important information. In Example 6, we see how changing the class interval widths on histograms can change the shape of histograms and reveal new information.

EXAMPLE 6 Patterns in Traffic Density

Consider the distribution of the weekday traffic density on a portion of the Massachusetts Turnpike. The histogram shown in Figure 5.8 is based on class intervals of 4 hours. Other than showing that peak traffic density is between 4 p.m. and 8 p.m. (16 hours after midnight to 20 hours after midnight) and that traffic density is very low between 12 A.M. and 4 A.M., the histogram is not very informative.

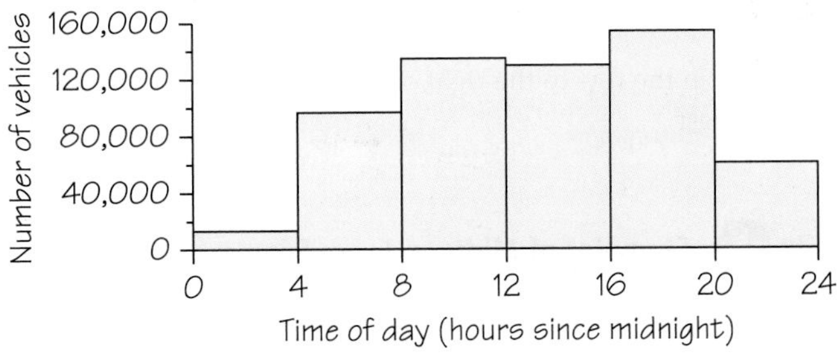

Figure 5.8 Histogram of traffic density in 4-hour intervals.

In Figure 5.9, the class interval widths have been reduced to 1 hour each. In this histogram, the increased traffic densities during morning rush hour (about 7 A.M. to 9 A.M.) and evening rush hour (about 3 P.M. to 7 P.M.) are clearly visible.

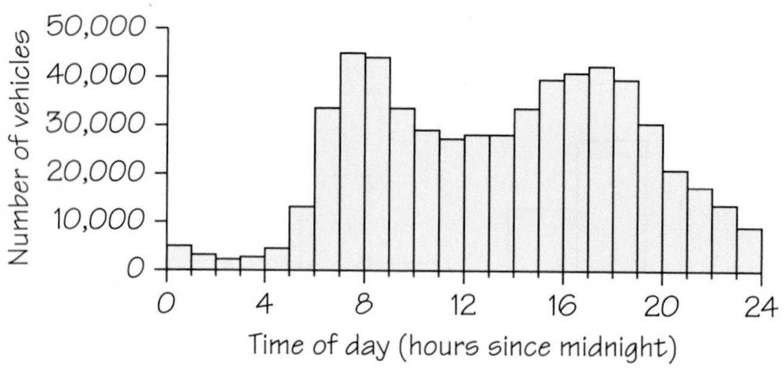

Figure 5.9 Histogram of traffic density in 1-hour intervals.

As Figures 5.8 and 5.9 illustrate, when looking for patterns in data, experiment with using different class intervals. Changing the class-interval widths may help you tease out more information from the data.

5.3 Displaying Distributions: Stemplots

Histograms are not the only way to display distributions graphically. For small datasets, a **stemplot** is quicker to make and presents more detailed information.

Stemplot DEFINITION

A **stemplot** (or **stem-and-leaf plot**) is a display of the distribution of a variable that attaches the final digits of each observation as a *leaf* on a *stem* made up of all but the final digit.

Making a Stemplot PROCEDURE

1. Separate each observation into a *stem*, which consists of all but the final (rightmost) digit, and a *leaf*, which is the final digit. (To make stems meaningful, it may be necessary to truncate or round the observed values. Tukey advocated truncating in his book *Exploratory Data Analysis.* Statistical packages are split over which approach to take.) Stems may have as many digits as needed, but each leaf contains only a single digit.
2. Write the stems in a vertical column, with the smallest at the top, and draw a vertical line at the right of this column. Include all stems, even if they are not used.
3. Write each leaf in the row to the right of its stem. Arrange the leaves from smallest to largest.

EXAMPLE 7 ➔ Stemplot of Midterm Exam Scores

The midterm exam scores of a class of 20 students are given below.

41	87	88	90	68	92	93	40	91	96
76	85	88	86	82	69	72	79	80	79

Since the exam scores range from 44 to 96, the stems are 4, 5, 6, 7, 8, and 9 (Step 1). Figure 5.10 shows how to complete Steps 2 and 3. Notice that the stem of 5 is included in the plot even though there are no data values in the 50s.

4		4	10	4	01
5		5		5	
6		6	89	6	89
7		7	6299	7	2699
8		8	7858620	8	0256788
9		9	02316	9	01236

Step 2: Write stems. Step 3: Add leaves. Step 3 (completed): Order leaves.

Figure 5.10 Constructing a stemplot.

Self Check 4

Make a stemplot of the following systolic blood pressures (in millimeters of mercury) of 10 randomly chosen adults. (Notice that to save space we have presented these data in two rows. However, if you wanted to enter these data into a spreadsheet, you would enter them into a single column.)

147	141	120	124	127
132	98	112	120	128

EXAMPLE 8 Stemplot of the Percentage of Hispanics

To make a stemplot of the percentage of Hispanics from the data in Table 5.5 (page 189), take the whole-number part of the percentage as the stem and the final digit (in this case, the tenths place) as the leaf. Figure 5.11 is the complete stemplot for the data in Table 5.5. The entries for Idaho and Oregon, 9|01, represent 9.0% and 9.1%, respectively.

If we rotate Figure 5.11 a quarter-turn counterclockwise, the stemplot would look like a histogram (of a distribution skewed to the right). Comparing the stemplot in Figure 5.11 with the histogram in Figure 5.4 (page 191) reveals the strengths and weaknesses of stemplots. The stemplot, unlike the histogram, preserves the actual value of each observation, at least in cases where the data values have not been truncated or rounded. But you can choose the classes in a histogram, whereas the classes (the stems) of a stemplot are not as flexible. Whether the large number of classes in Figure 5.11 is an improvement over Figure 5.4 is a matter of taste. To change the classes on the stemplot, we could truncate the tenths place; for example, 11.3% and 11.6% both become 11% when the tenths place is truncated. In Figure 5.12, we construct a stemplot of the truncated data. Notice that now the leaves represent 1% so that 0|1 and 1|0 represent 1% and 10%, respectively.

0	111122222222333334444445666777777788899
1	0113667
2	125
3	33
4	2

Figure 5.12 Stemplot using truncation.

There are too many leaves on the first stem in Figure 5.12, and no outliers are obvious. Just as we might zoom in on a digital map to view added detail, we can zoom in by expanding each stem into two stems, using the first stem for leaves 0, 1, 2, 3, and 4 and the second stem for leaves 5, 6, 7, 8, and 9. Figure 5.13 shows the result.

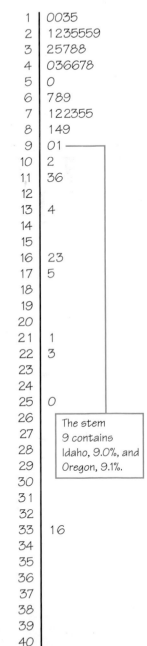

1	0035
2	1235559
3	25788
4	036678
5	0
6	789
7	122355
8	149
9	01
10	2
11	36
12	
13	4
14	
15	
16	23
17	5
18	
19	
20	
21	1
22	3
23	
24	
25	0
26	
27	
28	
29	
30	
31	
32	
33	16
34	
35	
36	
37	
38	
39	
40	
41	
42	3

The stem 9 contains Idaho, 9.0%, and Oregon, 9.1%.

Figure 5.11 Stemplot of the percentage of Hispanics among the adult residents of the U.S. states.

0	1111222222233333444444
0	566677777788899
1	0113
1	667
2	12
2	5
3	33
3	
4	2

Figure 5.13 Stemplot with expanded stem.

Now our stemplot reveals the same information as the histogram in Figure 5.3 but gives the added detail of the truncated data values.

Stemplots do not work well for large datasets, like the 947 Iowa Test scores in Figure 5.6, because some stems (like the 0 stem in Figure 5.12) must hold such a large number of leaves.

Graphical representations are good for analyzing the shape of a distribution of values. To answer precise questions about features of a dataset, such as its center, however, it helps to have numerical summaries as well. We explore this next to help us obtain a statistical "map" of our data with just the right degree of detail.

5.4 Describing Center: Mean and Median

What kind of gas mileage do you get with new cars in the Environmental Protection Agency's "midsized cars" category? Table 5.7 gives the city and highway gas mileage (from www.fueleconomy.gov) for a representative sample of 2015 midsized cars.

TABLE 5.7 Fuel Economy (Miles per Gallon) for Model Year 2015 Vehicles

Model	City Miles per Gallon	Highway Miles per Gallon
Acura RDX 2WD	20	28
Mercedes Benz AMG S 4matic (wagon)	15	21
BMW 428i	23	34
Buick Verano	21	32
Chevrolet Silverado	16	23
Ford Fusion S	22	34
Infiniti QX50	17	25
Kia Optima	30	31
Lexus GS 350	17	28
Mitsubishi Galant	26	31
Nissan Maxima	21	29
Toyota Camry	25	35
Toyota Prius	58	52

jaboo2foto/Shutterstock

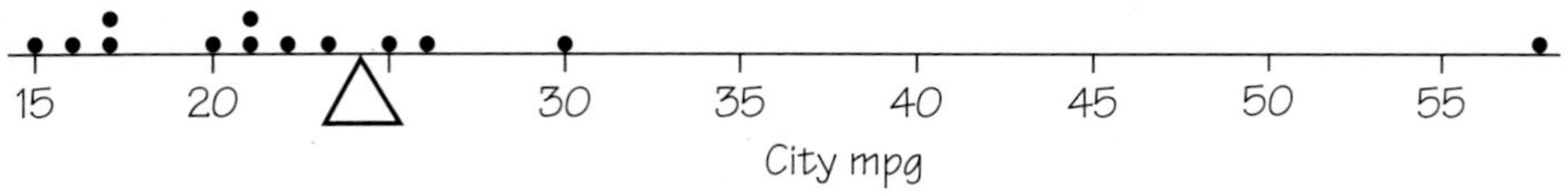

Figure 5.14 Dotplot of the city gas mileages of the sample of midsized cars. The Toyota Prius is an outlier.

We start with graphs. Figure 5.14 is a **dotplot** of the city mileages of the 13 cars in the sample of midsized cars.

Dotplot DEFINITION

A **dotplot** is a display of the distribution of a variable in which each observation is represented by a dot above a horizontal axis. If two or more data values are the same, the dots are placed directly above each other. For large datasets, each dot may represent a specified number of observations.

Numerical summaries make the comparison that we want more specific. A numerical description of a distribution begins with a measure of its center. The two most common measures of center are the **mean** and the **median.** Basically, the mean is the arithmetic "average value" and the median is the "middle value." Sometimes the **mode,** the data value that occurs most frequently, is also used as a measure of center. We need to explore the precise procedures for calculating these measures and observe how they behave differently.

One way to visualize the value of the mean of a dataset is to imagine where the fulcrum would have to be placed for its dotplot to "balance." We've placed a triangle to mark that spot in Figure 5.14. This analogy tells us that the mean is always between the largest and smallest values, and by visual inspection, we can estimate further that the balance point appears to be somewhere between 20 and 25. Let's see what exact value the formula yields.

Finding the Mean $\bar{x}$ PROCEDURE

1. Find the sum of the values.
2. Divide the sum by the number of values.

If the n observations are $x_1, x_2, \ldots, x_n$, the formula for the mean is

$$\bar{x} = \frac{x_1 + x_2 + \cdots + x_n}{n}$$

The common notation for the mean of all the x-values is a bar over the x, and $\bar{x}$ is pronounced "x-bar."

EXAMPLE 9 Calculating the Mean

The mean city mileage for the 13 midsized cars in Table 5.7 is

Algebra Review Appendix
Using Formulas

$$\bar{x} = \frac{\text{sum of the observations}}{\text{number of observations}} = \frac{x_1 + x_2 + \cdots + x_n}{n}$$

$$= \frac{20 + 15 + 23 + 21 + 16 + 22 + 17 + 30 + 17 + 26 + 21 + 25 + 58}{13}$$

$$\approx 23.9 \text{ mpg (miles per gallon)}$$

We said that the Toyota Prius may be an outlier and not belong with the other cars. If we exclude the Prius, the mean city mileage drops to $\frac{253}{12} \approx 21.1$ mpg. The single outlier adds more than 2 mpg to the mean city mileage. This illustrates an important weakness of the mean as a measure of center: *The mean is sensitive to the influence of extreme observations.* These may be outliers, but a skewed distribution that has no outliers will also pull the mean toward its long tail.

We have used the middle of a distribution as an informal measure of center. The *median* is the formal version of the middle, with a specific rule for calculation. The median M is a number where half the observations are smaller and the other half are larger.

Finding the Median M PROCEDURE

1. Arrange all observations (including any repeated values) in increasing order (from smallest to largest).
2. If the number of observations n is *odd*, the median M is the center observation in the ordered list. To find it, start at the bottom of the ordered data values and count up $(n + 1)/2$ observations. If the number of observations is *even*, the median M is the mean of the two center observations in the ordered list.

EXAMPLE 10 Calculating the Median

Since we're exploring the gas mileage cars get on the road, you might have noticed the connection that just as a median divides a road into two halves (with opposite directions of travel), a median divides a dataset into two halves! To find the median city mileage for the 2015 midsized cars, arrange the data in increasing order:

15 16 17 17 20 21 **21** 22 23 25 26 30 58

The median is the observation that is $(13 + 1)/2$, or seventh from the smallest, the bold 21. Because the dataset is small, you can find this by eye—there are six observations to the left and six to the right.

What happens if we drop the Toyota Prius? The remaining 12 gas-powered cars have the following city mileages:

15 16 17 17 20 **21** **21** 22 23 25 26 30

Because the number of observations $n = 12$ is even, there is no single center observation. [In this case, $(n + 1)/2 = 13/2 = 6.5$.] There is a center *pair* of observations—the sixth and seventh observations in the ordered list, which in this case are both 21. There are five observations to the left of the pair of 21s and five to the right. The median M is the mean of the center pair, which is $(21 + 21)/2 = 21$. ■

You see that the median resists the influence of extreme observations better than the mean does. A very high value like the Toyota Prius is simply one observation to the right of center, and, in this case, removing it did not change the median. The *Mean and Median* applet (www.macmillanhighered.com/fapp10e) is an excellent way to compare the resistance of M and x to outliers. (Try Applet Exercise 1.)

The median and mean are the most common measures of the center of a distribution. The mean and median are close together in a roughly symmetric distribution and are equal in a perfectly symmetric distribution. In a skewed distribution, the mean is generally farther out in the long tail than is the median. For example,

Which Mean Do You Mean?

SPOTLIGHT 5.2

The word "average" can be ambiguous—sometimes used to refer to the mean, sometimes the median, and so on. But even if it is clear that we are using the mean, we still need to say what unit is the basis for our averaging. Suppose a tiny school has only 4 classrooms and their class sizes are 10, 3, 4, and 3. Most people would say this school's "average (mean) class size" is (10 + 3 + 4 + 3)/4 = 5. But that's not how it seems to the students, half of whom are in a class twice that big! So we could find the mean from the student point of view by taking the mean of what each of the 20 students would report as the size of his or her class. This yields 134/20 = 6.7, a number 34% larger than 5. It turns out that the per-student mean is *always* at least as large as the per-class mean, which is good to know when viewing statistics about schools you are considering attending. (More information on this example can be found in Lawrence Lesser's article "Sizing Up Class Size," in the January 2010 *Mathematics Teacher*.)

for the right-skewed city-miles-per-gallon data in Figure 5.14, the median is 21 mpg, while the mean (marked by the triangle) is 23.9 mpg. But that's not the full story when it comes to the mean. As we see in Spotlight 5.2, sometimes the mean changes according to our point of view.

Another common numerical summary of a distribution is the *mode,* the most frequently occurring value in a dataset. Like the mean and median, it is a measure of location for a distribution. However, the mode is not necessarily a good measure of center. Consider a skewed distribution with its maximum height at one end and a long tail on the other, as shown in Figure 5.15. In this case, the mode, which marks the highest point, is considerably smaller than either the median or the mean and does not appear to be a good measure of the center of this distribution.

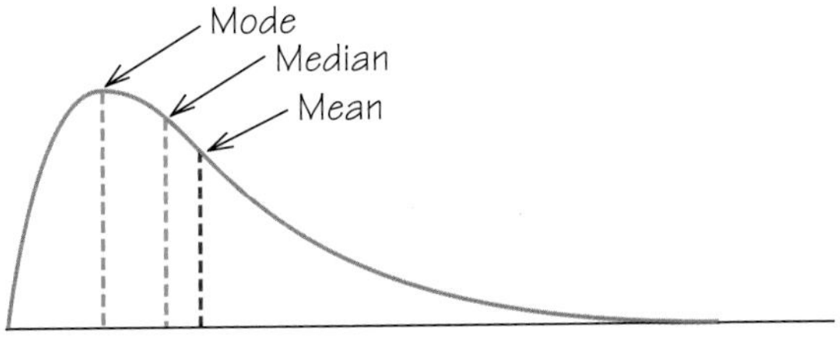

Figure 5.15 The smooth curve describes a right-skewed distribution with one mode. The mean is pulled toward the long tail more than the median is.

Mode DEFINITION

The **mode** is the most frequently occurring value in a set of numerical observations. It is possible for a dataset to have no mode, one mode, or more than one mode.

In the city-miles-per-gallon data in Table 5.7 (page 196) there are two modes, 17 and 21. Each value occurs twice in the dataset and all other values occur only once. So a distribution can have multiple modes. (The mode can also work with qualitative data: the mode for sex in Table 5.1 (page 183) is female.) We can also identify the mode(s) of a histogram. For example, Figure 5.6 (page 192) does not give the raw data of individual values for Iowa vocabulary test scores, but we can say that the modal class is grade 6.5 to 7.

Self Check 5

Find the mean, median, and mode of the systolic blood pressure data from Self Check 4 (page 195).

5.5 Describing Variability: Range and Quartiles

The mean and median provide two different measures of the center of a distribution. But a measure of center alone can be limiting. It would not be comfortable to live in a home with a mean temperature of 70°F if half the time it was 40°F and half the time it was 100°F! Two neighborhoods with a median house price of $193,000 can still be quite different if one has both mansions and modest homes and the other has little variation among houses. We are interested in the variability of house prices, as well as in their centers. *A useful numerical description of a distribution needs to consist of both a measure of center and a measure of variability.*

Jonathan Newton/The Washington Post/Getty Images

The simplest way to measure variability is with the **range,** which is the difference between the smallest and largest observations. For example, the percentages of Hispanics in the states are as low as 1% (Maine or West Virginia) and as high as 42.3% (New Mexico), so the range would be 42.3% − 1% = 41.3%. Likewise, the range of the city mileage numbers in Table 5.7 is 58 − 15 = 43 mpg. The range tells us the full span of the data, but it may be greatly affected by an outlier. Without the Toyota Prius, the preceding answer becomes 30 − 15 = 15 mpg.

Range DEFINITION

The **range** is a measure of variability of a set of observations. It is obtained by subtracting the smallest observation from the largest observation.

$$\text{range} = \text{maximum} - \text{minimum}$$

We can improve our description of variability by looking at the spread of the middle half of the data. The first and third **quartiles** delineate the middle half. At the end of the first quarter of a football game, one quarter of the game is complete. Similarly, the first quartile of a distribution or dataset is the point that exceeds one-quarter (or 25%) of the values. Q_1 is also the 25th percentile. The third quartile is the point that exceeds three-quarters (or 75%) of the values. (You usually won't hear the phrase "second quartile" because it's equivalent to something we already named: the median!) The quartiles break the dataset into four groups with equal numbers of observations. To make the idea of quartiles more exact, we need a procedure to find them.

Finding the Quartiles Q_1 and Q_3 PROCEDURE

1. Arrange all observations (including any repeated values) in increasing order.
2. Use the median to split the ordered dataset into two halves—an upper half and a lower half. (If the number of values is odd, don't include the middle observation in either half.)
3. The first quartile, Q_1, is the median of the lower half. The third quartile, Q_3, is the median of the upper half.

EXAMPLE 11 Finding Quartiles

The city mileages of the 12 gasoline-powered midsized cars, after sorting, are

15 16 17 17 20 21][21 22 23 25 26 30

Lower half — Upper half

We have indicated with brackets a split of the data into a lower half and an upper half. The first quartile is the median of the six observations in the lower half, so $Q_1 = 17$. Similarly, the third quartile is the median of the upper half: $Q_3 = 24$.

For an example with an odd number of observations, try the city mileages of all 13 midsized cars in Table 5.7 (page 196). Below are the mileages in increasing order, with the median in the center, which will be excluded to form two equal-sized groups:

15 16 17 17 20 21] 21 [22 23 25 26 30 58

Lower half — Upper half

We find the quartiles by finding the median of each half of the dataset: $Q_1 = 17$ and $Q_3 = 25.5$.

Some software packages or calculators may use a slightly different procedure to find the quartiles, so their results may be a bit different from our work here. Don't worry about this. The differences will be too small to be important.

5.6 The Five-Number Summary and Boxplots

We started by using the smallest and largest observations to indicate the variability of a distribution. These two observations tell us little about the distribution as a whole, but they give information about the tails of the distribution that is missing if we know Q_1, M, and Q_3. To get a quick summary of both center and variability, combine all five numbers.

These five numbers offer a reasonably complete description of center and variability. For the 13 midsized cars in Table 5.7, you can verify that the five-number summary for city gas mileage is

15 17 21 25.5 58

Five-Number Summary DEFINITION

The **five-number summary** of a distribution consists of the smallest observation, the first quartile, the median, the third quartile, and the largest observation, written in order from smallest to largest. The five-number summary is expressed as follows:

minimum Q_1 M Q_3 maximum

Self Check 6

Determine the five-number summary for the highway gas mileage in Table 5.7 (page 196).

A **boxplot** can visually represent both the location and the variability of datasets. To compare the cars' city fuel efficiency to highway fuel efficiency, we can create boxplots from the five-number summaries of the data in Table 5.7. Figure 5.16 shows boxplots for both city and highway gas mileages for midsized cars.

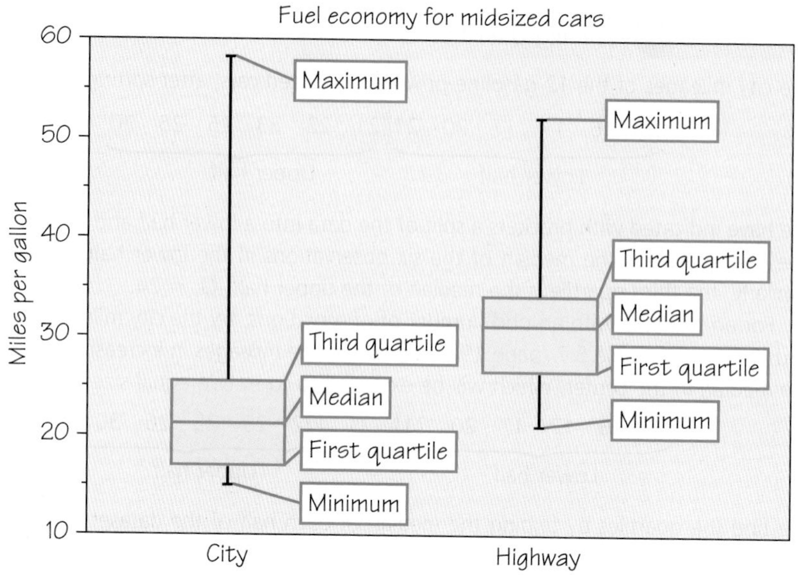

Figure 5.16 Boxplots of the highway and city gas mileages for 13 cars classified as midsized by the Environmental Protection Agency. These boxplots are drawn vertically, but it is equally correct to draw them horizontally.

Boxplot DEFINITION

A **boxplot** (or **box-and-whisker plot**) is a graph of the five-number summary.

- A central box spans the quartiles Q_1 and Q_3.
- A line somewhere inside the box marks the median M of the dataset.
- Lines extend from the box out to the smallest and largest observations.

Because boxplots show less detail than histograms or stemplots, they are best used for side-by-side comparison of more than one distribution, as in Figure 5.16. When you look at a boxplot, first locate the median, which marks the center of the distribution. Then look at the variability. The quartiles show the variability of the middle half of the data, and the extremes (the smallest and largest observations) show the variability of the entire dataset. So is there really much of a difference in gas mileages between city and highway?

From the boxplots, we see at once that highway mileages are noticeably higher than city mileages: The third quartile city mileage is less than the first quartile of highway mileage. Boxplots can also indicate a distribution's skewness. In both boxplots, the upper whiskers are longer than the lower whiskers, meaning that the upper quarter of the data is more spread out than the lower quarter. The upper whisker of the city mileage is longer and extends higher than the upper whisker of the highway mileage. That is due to the hybrid nature of Toyota's Prius (an outlier), which gets better gas mileage in the city than on the highway.

We also see that the variability of highway mileages has a somewhat different pattern than the variability of city mileages. The range of the highway mileages is represented by the length of the boxplot (maximum – minimum) and is smaller for the highway mileage data than for the city mileage data. The variability of the middle half of the data is represented by the length of the box ($Q_3 - Q_1$) and appears to be about the same for both diagrams.

Be aware that some calculators and software packages offer an alternative option for boxplots in which the lines go to the farthest values within

1.5 box-lengths of the quartiles, so they do not automatically go out to the minimum and maximum values. The advantage of this modified boxplot is that any individual values more than 1.5 box-lengths beyond either quartile can be marked as outliers. Figure 5.17 shows a modified boxplot for the city and highway gas mileages data. Only the mileage for the Prius was more than 1.5 box-lengths beyond $Q3$ and these values have been plotted as dots in the modified boxplot.

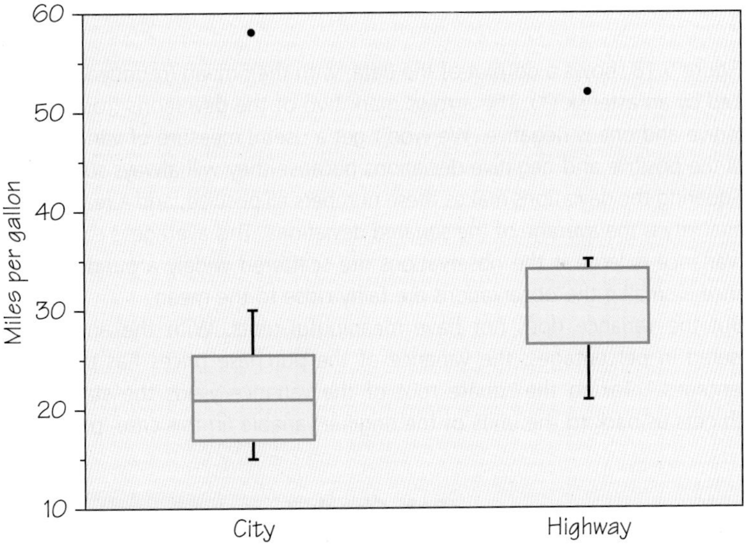

Figure 5.17 Modified boxplot for city and highway mileages of midsized cars.

5.7 Describing Variability: The Standard Deviation

Although the five-number summary is the most generally useful numerical description of a distribution, it is not the most common. That distinction belongs to the combination of the mean with the **standard deviation.** The mean, like the median, is a measure of center. The standard deviation, like the quartiles and the extremes in the five-number summary, measures variability. The standard deviation and its square, the *variance,* measure variability by looking at how far the observations are from their mean.

EXAMPLE 12 Understanding the Standard Deviation

When you buy stocks or mutual funds, you need to be aware of how to quantify and balance mean gain with the variability or risk of the investment, especially given the volatile years the market has experienced recently. Consider the PIMCO Total Return A (symbol: PTTAX), a fund that invests in intermediate-term, fixed-income securities. Here are its annual total returns for a recent 10-year period:

Neilson Barnard/Getty Images

TABLE 5.8 $PTTAX Annual Total Return

Calendar Year	2004	2005	2006	2007	2008	2009	2010	2011	2012	2013
Return (%)	4.65	2.41	3.51	8.57	4.32	13.33	8.36	3.74	9.93	−2.30

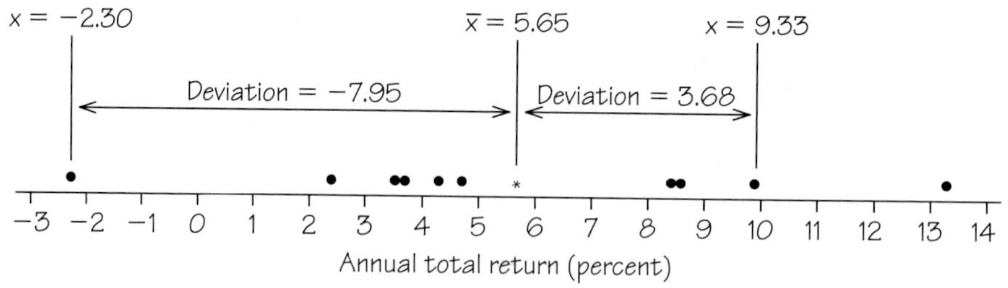

Figure 5.18 Measuring variability by looking at the deviations of observations from their mean.

Figure 5.18 shows a dotplot of the data, with their mean (rounded to two decimals) marked by an asterisk (*). The arrows mark two of the deviations from the mean: One is positive and one is negative. We won't get a useful measure of variability by totaling up all the positive and negative deviations because they will always sum to 0!

Squaring the deviations makes these numbers all positive, and a reasonable measure of variability is the average of the squared deviations. This average is called the *variance*. The variance is large if the observations are scattered widely around their mean. The variance is small if the observations are fairly close to the mean.

But the variance does not have meaningful units. With the annual return data measured in percentages, the variance of the purchase prices has units of "squared percentages." Taking the square root of the variance yields the standard deviation, which gets us back to the units of the original variable (in this case, percentages).

Self Check 7

Figure 5.18 shows the calculations for two of the deviations from the mean for the data in Table 5.8. Calculate the deviations from the mean corresponding to the other eight data values. Then verify that the sum of all 10 deviations is 0 (or close enough to 0 that any difference from 0 is due to rounding the mean to two decimals).

Standard Deviation

DEFINITION

The **standard deviation** is a kind of "standard" or average amount that observed data values deviate from their mean. More precisely, it is the square root of the mean of the squared deviations, except that the mean involves dividing by $n - 1$ instead of the usual n. (It turns out $n - 1$ makes this particular formula more accurate, but the justification is beyond the scope of this book.) In symbols, the standard deviation s of n observations $x_1, x_2, \ldots, x_n$ is

$$s = \sqrt{\frac{(x_1 - \bar{x})^2 + (x_2 - \bar{x})^2 + \cdots + (x_n - \bar{x})^2}{n - 1}}$$

Algebra Review Appendix
Powers and Roots

For simple datasets, standard deviation often can be estimated mentally by applying the first sentence of the above definition. For example, for the dataset {25, 25, 25, 30, 35, 35, 35}, we can readily see that 30 is the mean and the other numbers are each 5 units away from it. So we might assume that the standard deviation would be a value close to or equal to 5—and it is! Here are the calculations.

$$s = \sqrt{\frac{(25-30)^2+(25-30)^2+(25-30)^2+(30-30)^2+(35-30)^2+(35-30)^2+(35-30)^2}{7-1}}$$

$$= \sqrt{\frac{25+25+25+0+25+25+25}{6}} = \sqrt{\frac{150}{6}} = \sqrt{25} = 5$$

Even for more complex datasets, it is helpful to make a mental estimate first as a way to catch any errors caused by using a calculator. For the 10 return rate values in Example 12, a quick visual inspection might result in an estimate of the mean to be near 6 and a typical amount of deviation from 6 to be around 4, and certainly less than 8 (which is the approximate deviation of the most extreme data value from 6). Let's keep this estimate in mind as we do the formal calculation in Example 13.

EXAMPLE 13 Calculating the Standard Deviation

To find the standard deviation of the 10 return rates in Example 12, first find the mean.

$$\bar{x} = \frac{4.65 + 2.41 + 3.51 + 8.57 + 4.32 + 13.33 + 8.36 + 3.74 + 9.93 + (-2.30)}{10}$$

$$= 5.652\% \approx 5.65\%$$

For readability of Table 5.9, we have used the mean rounded to two decimals. You will get more accuracy if you include more decimal places for the mean throughout the process and do not round until the end.

TABLE 5.9 Step-by-Step Approach to Calculating Standard Deviation

Observations	Deviations (observation minus mean)	Squared Deviations
x_i	$x_i - \bar{x}$	$(x_i - \bar{x})^2$
4.65	4.65 − 5.65 = −1.00	$(-1.00)^2$ = 1.0000
2.41	2.41 − 5.65 = −3.24	$(-3.24)^2$ = 10.4976
3.51	3.51 − 5.65 = −2.14	$(-2.14)^2$ = 4.5796
8.57	8.57 − 5.65 = 2.92	$(2.92)^2$ = 8.5264
4.32	4.32 − 5.65 = −1.33	$(-1.33)^2$ = 1.7689
13.33	13.33 − 5.65 = 7.68	$(7.68)^2$ = 58.9824
8.36	8.36 − 5.65 = 2.71	$(2.71)^2$ = 7.3441
3.74	3.74 − 5.65 = −1.91	$(-1.91)^2$ = 3.681
9.93	9.93 − 5.65 = 4.28	$(4.28)^2$ = 18.3184
−2.30	−2.30 − 5.65 = −7.95	$(-7.95)^2$ = 63.2025
		Sum = 177.8680

The variance s^2 is the sum of the squared deviations divided by 1 less than the number of observations, so it would be $\frac{177.868}{10-1} \approx 19.763$. The standard deviation is the square root of the variance, so we obtain $s = \sqrt{19.763} \approx 4.45\%$. This value, 4.45%, can be considered small for this context, which suggests that this mutual fund happened to have a great deal of stability during a very turbulent decade. ■

Self Check 8

If the 10 observations from the fund in Example 12 still had a mean of 5.65%, but had less variability, their deviations from 5.65% would be smaller, and the standard deviation would be even smaller. To explore this dynamic, make the following changes to the data in Table 5.8: Change 13.33 to 10.33 and −2.30 to 0.7. The modified data values should have less variability because the two most extreme data values have been replaced with values closer to the mean.

(a) Verify that the mean of the modified data is the same as the mean of the original data from Table 5.8.
(b) Calculate the standard deviation for the modified data. Is this value the same as, smaller than, or larger than the standard deviation of the original data?

As you probably have noticed, calculating standard deviations using the formula given in the definition can be time consuming. In Spotlight 5.3, technology comes to the rescue!

Using Technology to Calculate Standard Deviation

SPOTLIGHT 5.3

While the formula in the definition box for standard deviation has conceptual clarity and a straightforward implementation (as shown in Table 5.9), it can be tedious to apply to large datasets with a basic-level calculator (such as a cell-phone calculator). Even with the most basic calculator, you'll get the same answer faster using the following more computationally oriented formula:

$$s = \sqrt{\frac{(x_1^2 + x_2^2 + \cdots + x_n^2) - n(\overline{x})^2}{n - 1}}$$

However, most of us have access to technology that is a bit more sophisticated than a basic-level calculator. The remainder of this spotlight is devoted to using a variety of technologies to calculate the standard deviation with a single command.

If you have a *scientific calculator,* put it into a STAT MODE if required, clear out any old data, and then enter your data one number at a time. (After each number, press your calculator's data-entry button—it may say DATA or have a symbol such as [Σ+] or [M+].) Once the data are entered, you can find the standard deviation by hitting the key labeled something like [$\sigma n - 1$], [$\sigma x n - 1$], or [s].

If you have a *graphing calculator* in the TI-83/84+ family (and you already used [STAT] → EDIT to enter one variable of quantitative data in a list, say, L1), then hit the following sequence of buttons: .

[STAT] → CALC [1] (for 1-Var Stats) [2ND] [1] (for L1) [ENTER]

You will get not only the standard deviation (*Sx*), but also other descriptive statistics, including the mean and the five-number summary! Keystrokes for other specific calculator models can be found online (see the Suggested Websites for this chapter).

If your data have been entered into a column of an Excel spreadsheet, you can calculate the mean and standard deviation as follows: To calculate the mean, in an empty cell enter =AVERAGE(and then click on the first data value and drag down to the last data value. Finish the command with) and press Enter. To calculate the standard deviation, replace =AVERAGE(with =STDEV(.

Statistical software such as JMP, Minitab, and SPSS all compute summary statistics of a dataset that include both the mean and standard deviation.

EXAMPLE 14 Calculating Standard Deviation Using a TI-84 Graphing Calculator

Next, we compare the fund from Example 12 (Table 5.8) with a different one—the Cohen & Steers Realty Shares (symbol: CSRSX), a mutual fund that invests in real estate investment trusts. Table 5.10 displays its calendar year total returns (in percentages) for the same 10-year period.

TABLE 5.10 CSRSX Percentages of Annual Total Return

2004	2005	2006	2007	2008	2009	2010	2011	2012	2013
38.48	14.88	37.13	−19.19	−34.40	32.50	27.14	6.18	15.72	3.09

To calculate the mean and standard deviations for these data, we turn to a TI-84 graphing calculator.

Step 1: Press STAT 1 (for EDIT). Enter the 10 percentages into list L1 (be sure to clear out any old data first). Here's a screen shot after entry of the last data value:

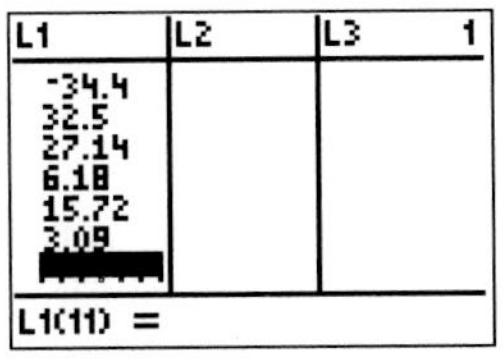

Step 2: Press STAT and enter the command for 1-var Stats.

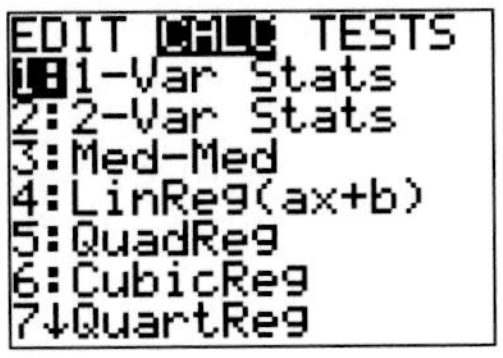

Step 3: Press ENTER to obtain the mean and standard deviation. Press the down arrow to obtain the five-number summary.

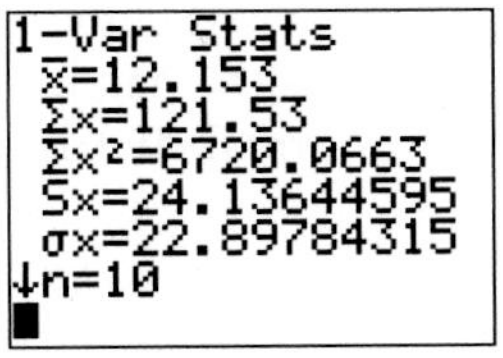

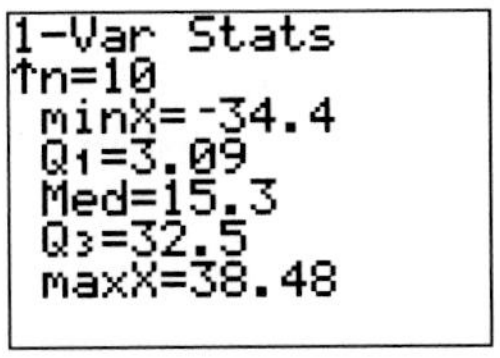

Now we are ready to compare the results for the Cohen & Steers Realty Shares (CSRSX) with the results for PIMCO Total Return A (PTTAX). On the one hand, the mean of the 10 CSRSX numbers is approximately 12.15%, which is more than double the mean from the PTTAX data. However, it comes with a tradeoff—a much higher standard deviation of approximately 24.14%. Scanning the numbers in Table 5.10, we see the dramatic lows and highs that make this fund feel like a rollercoaster ride! Knowing how to interpret these numbers is critical when making investment choices to fit your financial goals and tolerance for risk. ■

Self Check 9

Test grades of a sample of four students are given below. Determine the mean and standard deviation of the test grades. First, perform the calculations by applying the formulas for mean and standard deviation, and then check your results using your calculator's (or spreadsheet's) built-in statistical capabilities.

Test grades: 70 72 79 87

More important than the details of calculation of the standard deviation are the properties that determine the usefulness of the standard deviation:

1. s measures variability about the mean $\bar{x}$. Use s to describe the variability of a distribution only when you use $\bar{x}$ to describe the center.
2. $s = 0$ only when there is *no variability.* This happens only when all observations have the same value. (If every value is the same, every value equals the mean and thus has zero deviation from the mean!) Otherwise, $s > 0$. As the observations display more variability about their mean, s gets larger.
3. s has the same units of measurement as the original observations. For example, if you measure metabolic rates in calories, both the mean $\bar{x}$ and the standard deviation s are also in calories.
4. The use of squared deviations makes s even more sensitive than $\bar{x}$ to a few extreme observations. For example, dropping the Toyota Prius from our list of midsized cars drops the standard deviation of city mileages by nearly 60%, from approximately 11.10 mpg to 4.48 mpg without it. Distributions with outliers and strongly skewed distributions have large standard deviations. The number s does not give much helpful information about such distributions.

We now have a choice between two descriptions of the center and variability of a distribution: (1) the five-number summary or (2) $\bar{x}$ and s. Because $\bar{x}$ and s are sensitive to extreme observations, they can be misleading when a distribution is strongly skewed or has outliers. In fact, because the two sides of a skewed distribution differ in variability, no single number such as s describes the variability well. The five-number summary, with its two quartiles and two extremes, does a better job.

Although the standard deviation is widely used, it is not a natural or convenient measure of the variability of any possible distribution. The real reason for the popularity of the standard deviation is that it is the natural measure of variability for the special class of distributions called **normal distributions,** which we will discuss next.

Remember that a graph gives the best overall picture of a distribution. Numerical measures of center and variability report specific facts about a distribution, but they do not describe its entire shape—for example, numerical summaries do not disclose the presence of clusters. *Always start with a graph of your data.*

Choosing a Summary — RULE

The five-number summary is usually better than the mean and standard deviation for describing a skewed distribution or a distribution with outliers. Use $\bar{x}$ and s only for reasonably symmetric distributions that are free of outliers.

5.8 Normal Distributions

We now have a kit of graphical and numerical tools for describing distributions. What's more, we have a clear strategy for exploring data on a single numerical variable:

1. Always plot your data: Make a graph, usually a histogram, a dotplot, or a stemplot.
2. Look for the overall pattern (shape, center, variability) and for striking deviations such as outliers.
3. Calculate a numerical summary to give some description of center and variability.

Here is one more step to add to this strategy:

4. If the overall pattern of a large number of observations is so regular that we can describe it by a smooth curve, then draw that curve superimposed on the histogram.

Figure 5.6 (page 192) is a histogram of the Iowa Test vocabulary scores of 947 seventh-grade students. Like most histograms from national standardized tests, the histogram is symmetric, is single-peaked, and has a distinctive bell shape. In Figure 5.19, we draw a smooth curve through the tops of the histogram bars to describe the shape. The curve is an idealized description of the distribution. It gives a compact picture of the overall pattern of the data but ignores minor irregularities as well as any outliers. The curve in Figure 5.19 is a *normal curve.* A distribution whose shape is described by a normal curve is a *normal distribution.*

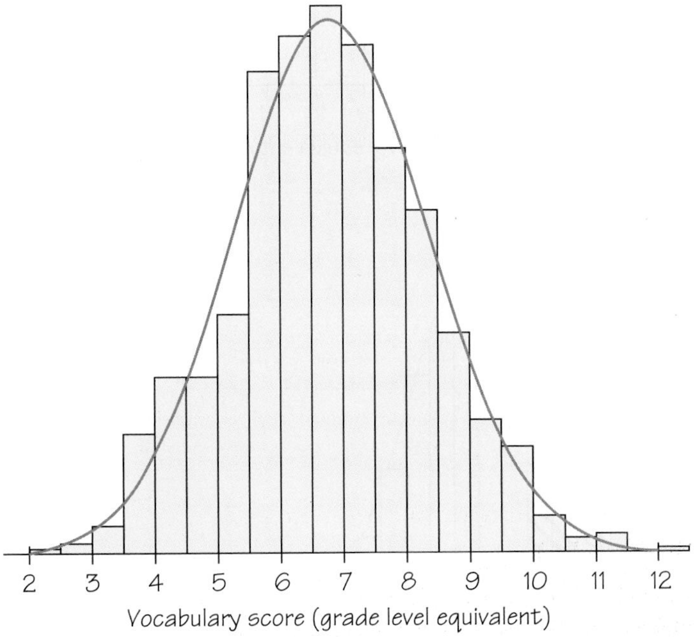

Figure 5.19 A histogram of the vocabulary scores of 947 seventh-grade students in Gary, Indiana. The smooth curve shows the overall shape of the distribution.

EXAMPLE 15 From Histogram to Density Curve

You can think of a normal curve as a smoothed-out histogram when there is symmetry and one mode. Our eyes respond to the *areas* of the bars in a histogram. The bar areas represent proportions of the observations. Figure 5.20a is a copy of Figure 5.19 with the leftmost bars shaded. The area of the shaded bars in Figure 5.20a represents the students with vocabulary scores of 6.0 or lower. This area reflects the proportion $287/947 \approx 0.30$ of Gary, Indiana, seventh graders, so 6.0 is the 30th *percentile*.

Now look at the curve drawn through the bars. In Figure 5.20b, the area under the curve to the left of 6.0 is shaded. We know that the areas of histogram bars represent proportions of all the observations, but we don't worry about the actual total area. Note that all the bars together represent 100% of the students, so we treat the total area under the normal curve as 1 for 100%, which turns the curve into a **normal density curve.** Now, areas under the density curve actually *are* proportions of the observations. The shaded area under the normal density curve in Figure 5.20b is the proportion of students with scores of 6.0 or lower. This area turns out to be 0.293, only 0.010 away from the histogram result. You see that areas under the normal density curve give quite good approximations of areas given by the histogram.

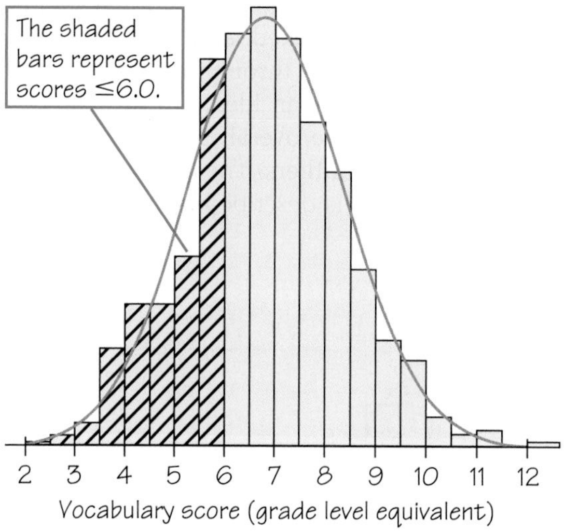

Figure 5.20a The proportion of scores less than or equal to 6.0 from the histogram is 0.30.

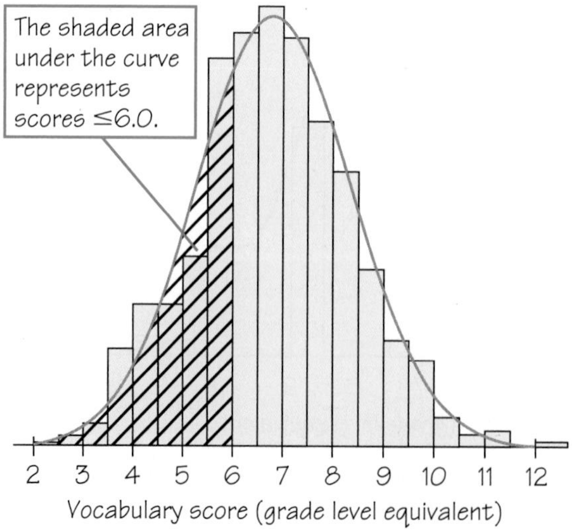

Figure 5.20b The proportion of scores less than or equal to 6.0 from the normal density curve is 0.293.

Density Curve DEFINITION

A **density curve** is a curve that

- is always on or above the horizontal axis
- has an area under the curve that is exactly 1

A density curve summarizes the overall pattern of a distribution. The area under the curve above any interval is the proportion of all observations that fall in that interval.

Normal Distribution DEFINITION

The *distribution* of a variable tells us what values the variable takes and how often it takes these values. A **normal distribution** is described by a *normal* density curve, which has a bell-shaped graph.

As is illustrated in Example 15, if a variable's distribution can be described by a normal curve, then the area under the normal density curve above any interval of values tells us what proportion of all values of the variable lie in that interval. We apply that idea in the next example.

EXAMPLE 16 Heights of American Women

The normal curve is a good approximation of the real-life distribution for a variety of biological measures (height, weight, heart rate, blood pressure, and so on), when examined for a particular species and gender. Figure 5.21 shows the heights of American women between the ages of 18 and 24. The proportion of young women who are between 60 inches (5 feet) and 65 inches tall is given by the area under the density curve between 60 and 65. This area is about 0.54, so approximately 54% of these women are between 60 and 65 inches tall.

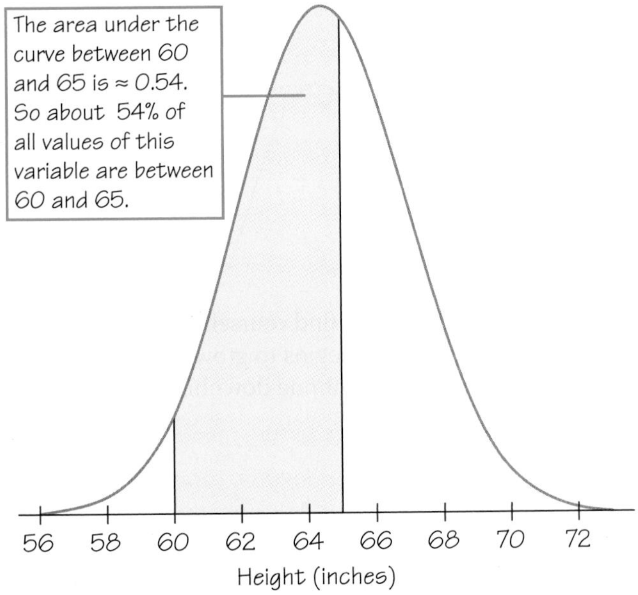
The area under the curve between 60 and 65 is ≈ 0.54. So about 54% of all values of this variable are between 60 and 65.
56
58
60
62
64
66
68
70
72
Height (inches)

Figure 5.21 Areas under a normal density curve describe a normal distribution. This normal curve describes the distribution of heights of American women.

Self Check 10

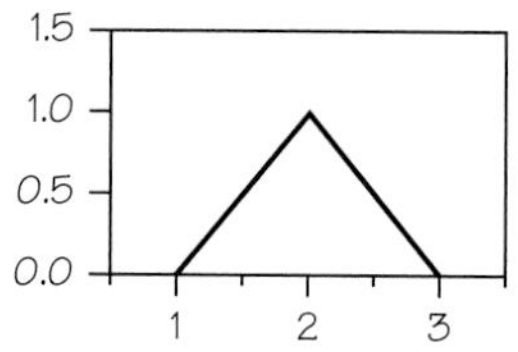

Suppose that a smooth curve drawn over a histogram has the triangular shape shown at left. This distribution, called a triangular distribution, is different from a normal distribution.

(a) Check that the area under this density curve is exactly 1.

(b) What proportion of data from this distribution would fall below 1.5? Explain your calculations.

The everyday meaning of *normal* is "typical" or "natural," and there are certainly some natural phenomena (e.g., Example 16) that are approximated well by the normal distribution. The specific form of a normal distribution and the major role it plays in statistical theory, however, are very special, not ordinary. Normal curves can be specified exactly by an equation, but we will be content with pictures. All normal curves have the same general *shape.* They are symmetric and bell-shaped, with tails that fall down rapidly from a central peak. The *center* of the normal curve is the center of the distribution in more than one way:

- It is the mean (balance point) of the distribution.
- It is also the median because half the observations (half the area under the curve) lie on each side of the center.

What about the *variability* of a normal curve? Normal curves have the special property that their variability is determined completely by a single number, the standard deviation. We have learned how to calculate the standard deviation from a set of observations. For normal distributions, the standard deviation, like the mean, can be found directly from the curve. Here's how: Imagine that you are skiing down a mountain that has the shape of a normal curve. At first, you descend at an ever-steeper angle as you go out from the peak.

Fortunately, before you find yourself going straight down, the slope now begins to grow flatter rather than steeper as you continue downhill.

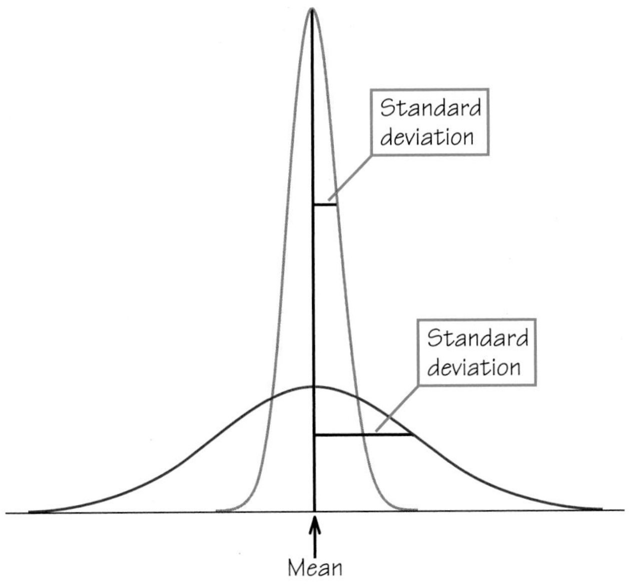

Figure 5.22a Two normal curves with the same mean but different standard deviations. The standard deviation for each curve is the distance from the center (the mean) to the change-of-curvature point on one side of the center.

The points at which the change of curvature takes place are located 1 standard deviation from the mean on either side. You can feel the change as you run your finger along a normal curve, and in that way you can find the standard deviation. Try it on the two normal curves in Figure 5.22a, which have the same means but different standard deviations. Notice that the distribution with the larger standard deviation is more spread out and has a flatter normal curve.

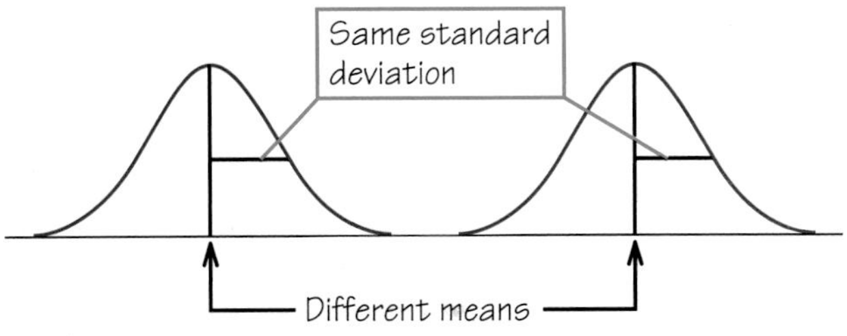

Figure 5.22b Two normal curves with the same standard deviation but different means.

Normal curves with the same standard deviation have exactly the same shape. Changing the mean just moves the center of the curve to a new location, as Figure 5.22b shows, whereas changing the standard deviation changes the variability of the curve, as Figure 5.22a shows. A normal distribution is completely determined by two numbers: the mean and the standard deviation.

Mean and Standard Deviation of a Normal Distribution DEFINITION

The **mean of a normal distribution** is at the center of symmetry of the normal curve. The **standard deviation of a normal distribution** is the distance from the center to the change-of-curvature points on either side.

Self Check 11

(a) Which of the normal density curves at right, the solid curve or the dashed curve, represents the distribution with the larger mean?

(b) Which normal density curve represents a distribution with the larger standard deviation?

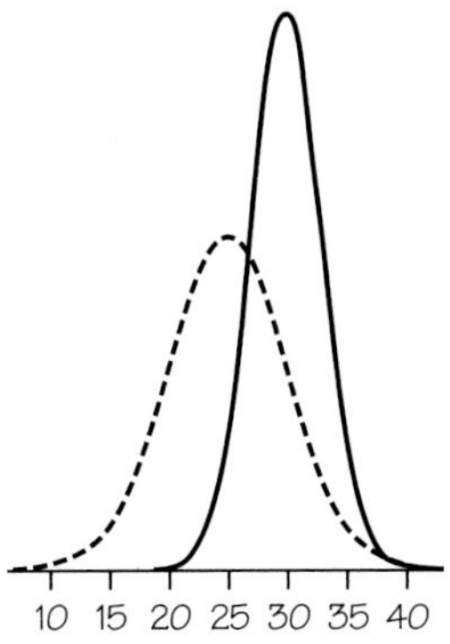

We have often used the quartiles to indicate the variability of a distribution. Because the standard deviation completely describes the variability of any normal distribution, it tells us where the quartiles are.

Quartiles of a Normal Distribution DEFINITION

The **quartiles of a normal distribution** are located about 0.67 (which is about 2/3) of a standard deviation away from the mean. In particular, the *first quartile* is located at 0.67 standard deviation below the mean, and, by symmetry, the *third quartile* is located at 0.67 standard deviation above the mean.

EXAMPLE 17 Heights of American Women: Finding the Quartiles

The distribution of heights of young American women (ages 18 to 24) is approximately normal, with a mean of 64.5 inches (that is, 5 feet 4.5 inches) and a standard deviation of 2.5 inches. Figure 5.23 shows this normal curve. The quartiles are 0.67 standard deviation, or (0.67)(2.5 inches) $\approx$ 1.7 inches away from the mean. The first quartile is 64.5 − 1.7, or 62.8 inches. The third quartile is 64.5 + 1.7, or 66.2 inches. The middle 50% of women's heights lie approximately between 62.8 inches and 66.2 inches. These numbers are exact for the normal distribution with a mean of 64.5 inches and a standard deviation of 2.5 inches, but only approximately true for the actual heights of the women because real-life distributions of biological measurements such as heights are only approximately normal.

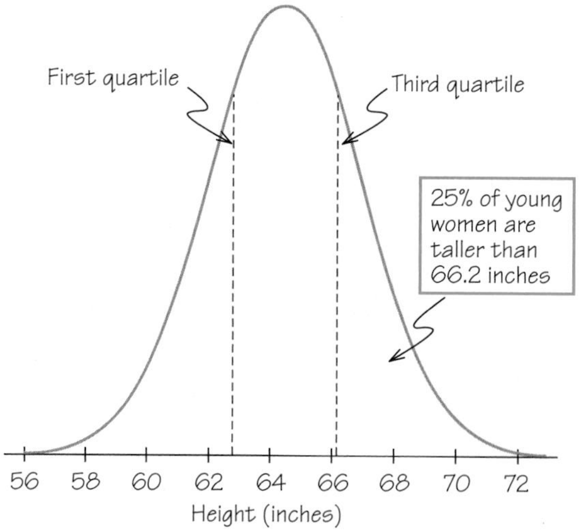

Figure 5.23 The quartiles of a normal distribution are located at 0.67 standard deviation on either side of the mean. For this normal curve, the mean is 64.5 inches and the standard deviation is 2.5 inches.

Why are normal distributions important in statistics? First, normal distributions are good models or approximations for some distributions of *real data.* Distributions that are often close to normal include scores on tests taken by many people (such as SAT exams and many psychological tests), repeated careful measurements of the same quantity, and characteristics of biological populations (such as heights of young women, yields of corn, and wingspans of a particular type of bird). Second, normal distributions are good approximations to the results of many kinds of *chance outcomes,* such as tossing a coin many times. We will return to normal curves when we study the mathematics of chance in Chapter 8.

Don't forget that many sets of data do not follow a normal distribution. Most income distributions, for example, are skewed to the right and thus are not normal. Take a look at Figure 5.24, which shows the distribution of the gross annual income in the United States for 2013. Notice that the deciles, the values that divide the distribution into ten groups of equal frequency, are marked on the graph. (The deciles are the 10th, 20th, 30th, . . . , and 90th percentiles.) The increments between deciles become wider as you go out farther into the right tail of the distribution.

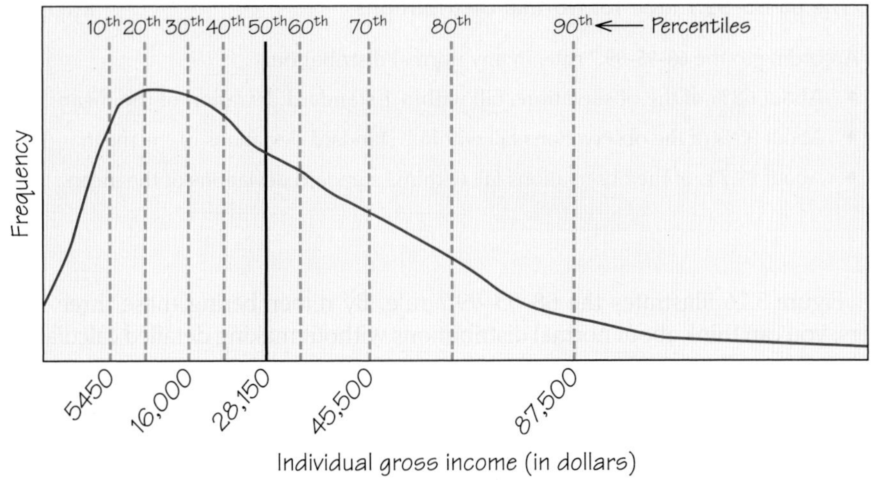

Figure 5.24 Distribution of 2013 gross annual income in the United States.

In Spotlight 5.4, we discuss another density curve, which has been fit to data using software.

Density Estimation

SPOTLIGHT 5.4

Smooth curves that describe the overall pattern of distributions of data are called *density curves.* Normal curves are one type of density curve. There are many other types used for different purposes. Clever software for "density estimation" will calculate a density curve to describe any set of observations you give it.

Figure 5.25 shows a strongly skewed distribution, the survival times of 72 guinea pigs in a medical experiment. Two graphs of the distribution are overlaid: a histogram and a density curve produced by software from the data. The histogram and density curve agree on the overall shape and on the "bumps" in the long right tail.

The density curve shows a higher single peak as a main feature of the distribution. The histogram divides the observations near the peak between two bars, thus reducing the height of the peak. Because density estimators don't depend on dividing the data into classes as histograms do, many statisticians prefer them when they need a picture of a distribution.

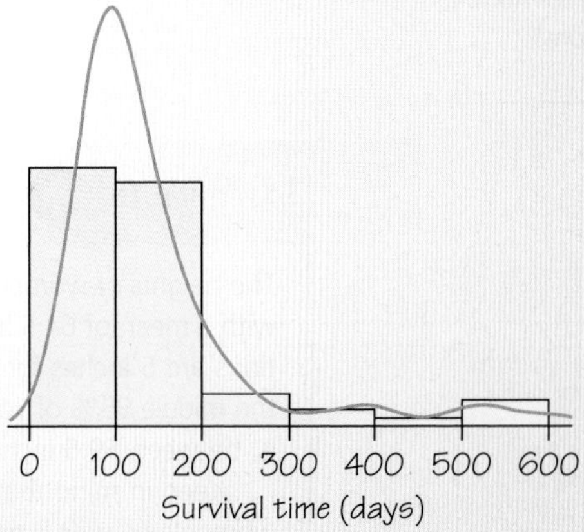

Figure 5.25 Using density estimation software to fit a smooth curve to data on the survival time of 72 guinea pigs.

5.9 The 68–95–99.7 Rule for Normal Distributions

Because any particular normal distribution is completely determined by its mean and standard deviation, it is not surprising that all normal distributions are the same in terms of what proportion of observations are within any given number of standard deviations of the mean. Here is an important rule based on this fact.

The 68–95–99.7 Rule for Normal Distributions RULE

According to the **68–95–99.7 rule,** in any normal distribution:

- About 68% of the observations fall within 1 standard deviation of the mean.
- About 95% of the observations fall within 2 standard deviations of the mean.
- About 99.7% of the observations fall within 3 standard deviations of the mean.

Figure 5.26 illustrates the 68–95–99.7 rule. By remembering these three numbers, you can think about normal distributions without making detailed calculations.

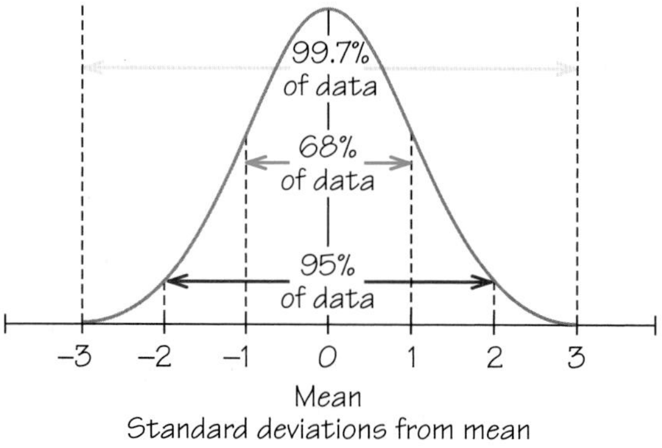

Figure 5.26 The 68–95–99.7 rule for normal distributions.

EXAMPLE 18 **Heights of American Women: Application of the 68–95–99.7 Rule**

The heights of women between the ages of 18 and 24 are distributed roughly normally, with a mean of 64.5 inches and a standard deviation of 2.5 inches. Two standard deviations are 5 inches for this distribution. The "95" part of the 68–95–99.7 rule says that the middle 95% of young women are between 64.5 − 5 and 64.5 + 5 inches tall; that is, between 59.5 inches and 69.5 inches.

Keep in mind that this is only an approximation because the distribution of young women's heights is approximately normal.

The other 5% of American women have heights outside the range from 59.5 to 69.5 inches. Because the normal distributions are symmetric, half of these women are on the tall side and half on the short side. So the tallest 2.5% of young women are taller than 69.5 inches.

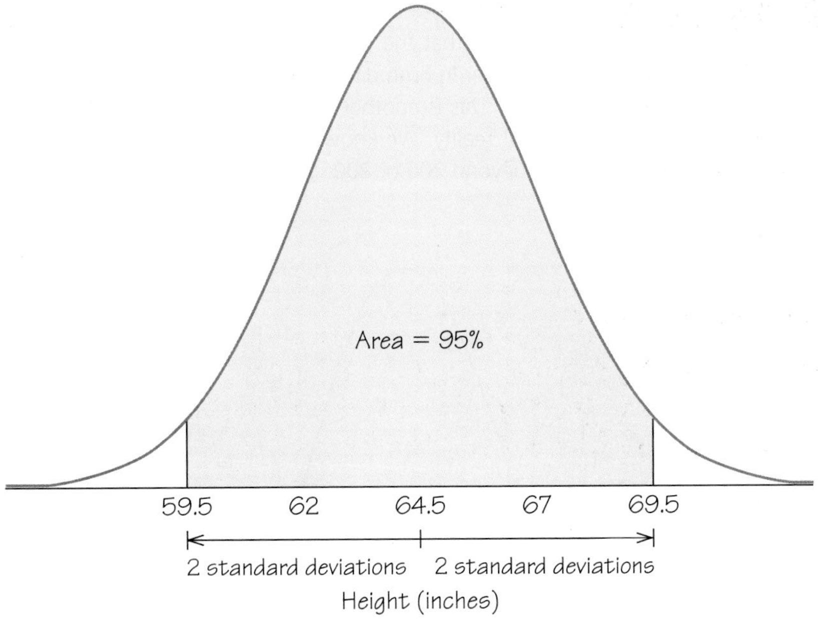

EXAMPLE 19 SAT Reasoning Test Scores

The distribution of scores on tests such as the SAT college entrance examination is close to normal. Scores on each of the three sections (math, critical reading, and writing) of the SAT are adjusted so that the mean score is about $\mu = 500$ and the standard deviation is about $\sigma = 100$. (The Greek letters μ and σ designate the population mean and standard deviation, respectively, as opposed to $\bar{x}$ and s, which designate the mean and standard deviation of sample data). This information allows us to answer many questions about SAT scores.

- *How high must a student score to fall in the top 25%?* This score is the 75th percentile or the third quartile. The third quartile is (0.67)(100) = 67 points above the mean. So scores above 567 are in the top 25% (i.e., the top quarter).
- *What percentage of scores falls between 200 and 800?* Scores of 200 and 800 are 3 standard deviations on either side of the mean—for example, $500 + 3 \times 100 = 800$. The "99.7" part of the 68–95–99.7 rule says that 99.7% of all scores lie in this interval. (In practice, the SAT makes this 100% by reporting as 200 those rare scores below 200, or as 800 those rare scores above 800.)
- *What percentage of scores is above 600?* A score of 600 is 1 standard deviation above the mean. By the "68" part of the 68–95–99.7 rule, 68% of all scores fall between 400 and 600 and 32% fall below 400 or above 600. Because normal curves are symmetric, half of this 32% are above 600. So a score above 600 places a student in the top 16% of test-takers. We can also say that a score of 600 is at the 84th percentile of all test-takers.

Sketching a normal curve and scaling the horizontal axis with the mean and points 1, 2, and 3 standard deviations from the mean can help you use the 68–95–99.7 rule.

Figure 5.27 shows the distribution of SAT scores with the areas needed to find the percentage of scores above 600. Note that the tails of Figure 5.27, like those of any bell curve, technically stretch out forever in both directions (even as the amount of faraway area becomes vanishingly small). This is another reminder that the bell curve is a very good, but not perfect, model of reality. We know that real-life SAT subtest scores are scaled so that they do not go beyond 200 or 800.

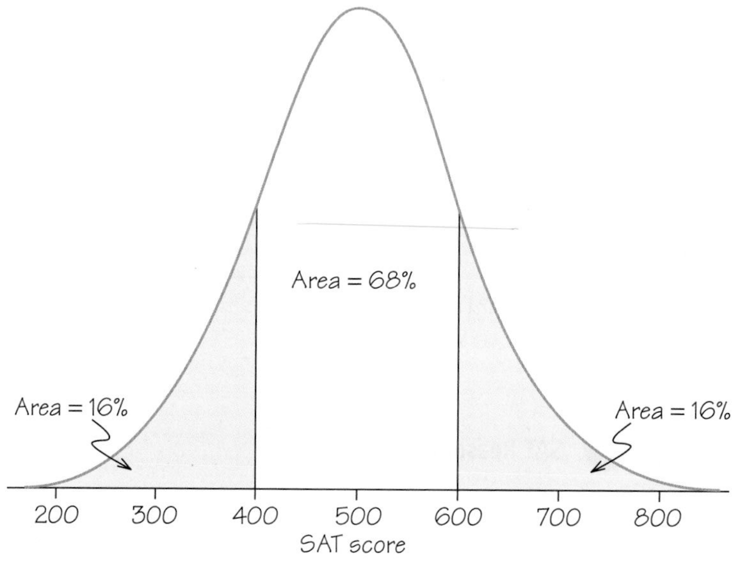

Figure 5.27 Using the 68–95–99.7 rule to find the percentage of SAT section scores that are above 600. This normal curve has a mean of 500 and a standard deviation of 100.

Self Check 12

Use the distribution of SAT scores given in Example 19 to answer the following questions.

(a) How low must a student's score be for it to be in the bottom 25%?
(b) What percentage of the SAT scores is below 300?

The 68–95–99.7 rule allows you to find selected areas under a normal curve—areas for outcomes bounded by 1, 2, or 3 standard deviations away from the mean. You can use tables, software, a graphing calculator, or the *Normal Density Curve* applet to find *any* area under a normal curve. See Applet Exercises 2 to 4 to practice using the applet to find proportions.

Review Vocabulary

Boxplot A graph of the five-number summary. A box spans the quartiles, with an interior line marking the median. Lines extend out from this box to the extreme high and low observations. A modified version of the boxplot extends the lines only to the extreme high and low observations that are within 1.5 box widths of the quartiles. Any values that are more extreme are marked (often with asterisks) as outliers. (pp. 201, 202)

Class intervals Non-overlapping, consecutive intervals into which data are classified to give an idea of the distribution. Class intervals are generally of equal width. (p. 186)

Density curve A curve that summarizes the overall pattern of a distribution. A density curve lies on or above the horizontal axis and has an area under the curve equal to 1. (p. 211)

Distribution The pattern of how often a variable takes certain values or intervals of values. (p. 184)

Dotplot A display of the distribution of a variable in which each observation (or group of a specified number of observations) is represented by a dot above a horizontal axis. Dots representing the same value are stacked vertically above that value. (p. 197)

Exploratory data analysis (EDA) The practice of using graphs and numbers to examine data for overall patterns and special features, without necessarily seeking answers to specific questions. (p. 183)

Five-number summary A summary of a distribution that gives the smallest observation, first quartile, median, third quartile, and largest observation, in that order. (p. 201)

Frequency distribution Classification of all observed values of a variable into non-overlapping classes or intervals that records how many times data values occur in each class. (pp. 184, 185)

Histogram A graph of the distribution of outcomes (often divided into classes) for a single quantitative variable. The height of each bar is the number of observations in the class of outcomes covered by the base of the bar. All classes should have the same width, and each observation must fall into exactly one class. (p. 188)

Individuals The people, animals, or things described by a dataset. (p. 182)

Left-skewed distribution A distribution in which the longer tail of the histogram is on the left side. (p. 190)

Mean The ordinary arithmetic average of a set of observations. To find the mean, add all the observations and divide the sum by the number of observations. (p. 197)

Mean of a normal distribution The balance point of a normal density curve that represents a normal distribution. The mean is at the line of symmetry of the normal curve. (p. 213)

Median The middle of a set of ordered observations. Half the observations fall below the median, and half fall above. (pp. 197, 198)

Mode The most frequently occurring value in a set of numerical observations. (pp. 197, 199)

Normal density curve Symmetric, bell-shaped curve. The center line of the normal curve is at the mean. The change-of-curvature in the bell-shaped curve occurs 1 standard deviation on either side of the mean. All density curves are scaled so that the area under the curve is 1 (p. 210).

Normal distributions A family of distributions that describe how often a variable takes its values by areas under a curve, called a normal density curve. A specific normal curve is completely described by two numbers: its mean and its standard deviation. (pp. 208, 211)

Outlier A data point that falls clearly outside the overall pattern of a set of data. (p. 188)

Quartiles The first quartile (Q_1) of a distribution is the point with one-quarter of the observations falling below it; the third quartile (Q_3) is the point with three-quarters below it. Calculate Q_1 and Q_3 by determining the median of the lower half and upper half of the ordered observations, respectively. (p. 200)

Quartiles of a normal distribution The first and third quartiles of a normal distribution are around 0.67 standard deviation below and above the mean, respectively. (p. 214)

Range The measure of variability obtained by subtracting the smallest observation from the largest observation. (p. 200)

Relative frequency distribution Classification of all observed values of a variable into non-overlapping classes or intervals that records what fraction (or percentage) of data values occur in each class. (pp. 185, 186)

Right-skewed distribution A distribution in which the longer tail of the histogram is on the right side. (p. 190)

68–95–99.7 rule In any normal distribution, approximately 68% of the observations lie within 1 standard deviation on either side of the mean, 95% lie within 2 standard deviations of the mean, and 99.7% lie within 3 standard deviations of the mean. (p. 216)

Standard deviation A measure of the variability of a distribution about its mean as center. It is the square root of the average squared deviation of the observations from their mean. (pp. 203, 204)

Standard deviation of a normal distribution The distance from the mean to the change-of-curvature point on either side of the normal density curve, which represents the distribution. (p. 213)

Stemplot A display of the distribution of a variable that attaches the final digits of the observations as leaves on stems made up of all but the final digit. (p. 194)

Symmetric distribution A distribution with a histogram, stemplot, or dotplot in which the part to the left of the median is roughly a mirror image of the part to the right of the median. (p. 191)

Variable A particular characteristic that can take on different values for different individuals. (p. 182)

Self Check Answers

1. (a) The individuals are the students.
(b) "Major" is a qualitative variable and "points" is a quantitative variable.

2.

Class	0	26	35	47	50	54	55	59
Frequency	4	1	5	1	1	1	1	1
Class	62	76	77	78	85	92	120	125
Frequency	2	1	1	1	1	1	1	1
Class	137	142	145	157	175	189	215	225
Frequency	1	1	2	1	1	1	1	1

3. Based on Figure 5.7, only part (c) is guaranteed to be true. The values \$16,000 and \$18,000 fall within the class interval from \$15,000 to \$20,000, which contains no data. For part (a), it could be the case that all data values that fell into the first class interval were less than \$7500 and all data values that fell into the second class interval were greater than \$12,500. This would make part (a) false. The interval from part (b), between \$12,500 and \$22,500, is wider than the third class interval that contains no data. It is possible that one or more of the data values that fell into the second class interval was between \$12,500 and \$15,000, which would make part (b) false.

4.

9	8
10	
11	2
12	00478
13	2
14	17

5. To find the mean: $\bar{x} = \dfrac{1249}{10} = 124.9$

To find the median, order the data from largest to smallest:

98 112 120 120 124 127 128 132 141 147

Since n is even, find the median by taking the mean of the middle two observations: median = (124 + 127)/2 = 125.5. The mode is the value that occurs most frequently. In this case, the value 120 occurs twice while all other values occur only once. Hence, the mode is 120.

6. 21, 26.5, 31, 34, 52

7. Here are all 10 deviations from the mean:

−1.00 −3.24 −2.14 2.92 −1.33 7.68 2.71 −1.91 4.28 −7.95

The sum of the deviations is 0.02, a difference from 0 that can be attributed to rounding the mean to two decimals.

8. (a) $\bar{x} \approx 5.65\%$

(b) The squared deviations for the two altered data values are $(10.33 - 5.65)^2 = 21.9024$ and $(0.7 - 5.65)^2 = 24.5025$. The new sum of squared deviations from the mean is 102.088. Hence, $s = \sqrt{102.088/9} \approx 3.4\%$. This value is smaller than the standard deviation for the original data.

9. $\bar{x} = \dfrac{70 + 72 + 79 + 87}{4} = \dfrac{308}{4} = 77$

$$s^2 = \frac{(70-77)^2 + (72-77)^2 + (79-77)^2 + (87-77)^2}{4-1}$$

$$= \frac{178}{3}; s = \sqrt{\frac{178}{3}} \approx 7.70$$

10. (a) Since this is a triangle, its area is half its base times its height: (1/2)(2)(1) = 1.

(b) To calculate the proportion, we need to calculate the area of the triangular region under the density curve to the left of 1.5 as shown below.

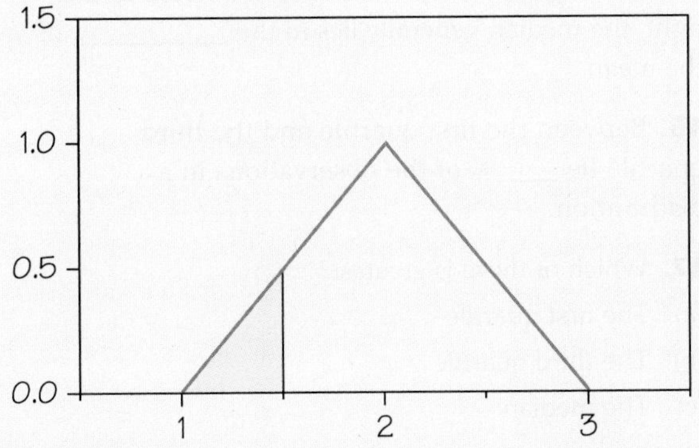

This is a triangular region with base 0.5 and height 0.5. It has area (1/2)(0.5)(0.5) = 0.125.

11. (a) The solid curve represents the distribution with the larger mean since its center of symmetry is around 30 and the dashed curve's center of symmetry is around 25.

(b) The dashed curve represents a distribution with the larger standard deviation since it is the flatter of the two density curves.

12. (a) To be in the bottom 25%, a student's score must be below the 25th percentile, which is (0.67)(100) = 67 points below the mean. So scores below 433 are in the bottom 25%.

(b) A score of 300 is 2 standard deviations below the mean. Recall that 95% of observations are within 2 standard deviations of the mean, and hence, 5/2 or 2.5% are below 300.

Skills Check

1. Here are the first rows of a dataset used for a student's research project:

Student ID	Gender	Scholarship Award	High School Percentile	Residency	First-year GPA
10256	Male	2000	64	Commuter	2.81
10260	Female	4000	72	On campus	3.62
10349	Female	2000	65	On campus	2.35
10388	Male	6000	85	On campus	3.76
10394	Female	6000	87	On campus	3.53
10422	Female	2000	59	On campus	3.01
10597	Male	4000	75	Commuter	3.26
10600	Male	4000	78	On campus	3.42

Describe the individuals in these data.

2. The number of variables in Skills Check 1 is _______. Classify each as quantitative or qualitative.

Figure 5.7 (page 193) is a histogram of the annual tuition and fee charges for 64 four-year universities/colleges in Massachusetts. Skills Checks 3, 4, and 5 are based on this histogram.

3. The number of universities/colleges with tuition and fee charges covered by the leftmost bar in the histogram is _______.

4. What is the relative frequency for the interval covered by the leftmost bar?

5. What percentage of universities/colleges had tuition and fees of at least $40,000?

6. The histogram below represents the number of job applications completed by recent college graduates before securing a job. This distribution is best described as _______-skewed.

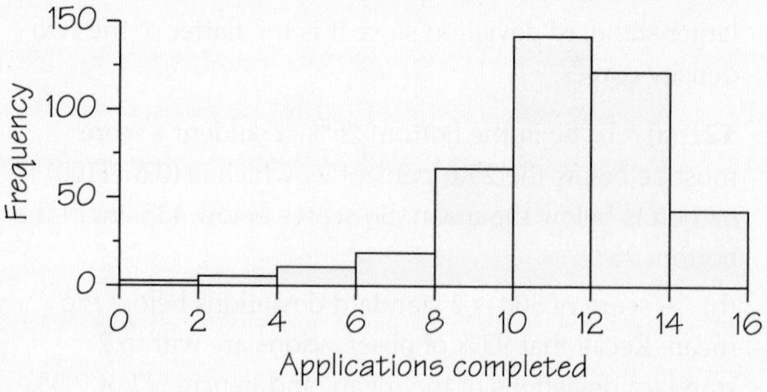

7. Brenner looks at real estate ads for houses in Sarasota, Florida. There are many houses ranging from \$200,000 to \$400,000 in price. A few houses on the coast, however, have prices of up to \$15 million. The distribution of house prices will be

(a) skewed to the left.

(b) roughly symmetric.

(c) skewed to the right.

8. Here are foot lengths (in centimeters) of 15 randomly chosen soldiers:

27.2	28.0	26.2	25.7	32.8
25.0	26.3	28.0	27.4	25.8
26.9	26.8	27.3	29.0	28.8

In a stemplot of these measurements, the largest stem is _____.

9. For Figure 5.11 (page 195), interpret the meaning of 10|2.

(a) Ten states have a Hispanic population of 2% .

(b) Two states have a Hispanic population of 10%.

(c) One state has a Hispanic population of 10.2%.

10. The final (rightmost) digit of an observation would be called a ___________ in a stemplot.

11. For Figure 5.12 (page 195), what does the stem in 4|2 represent?

(a) 4

(b) 40

(c) 2

12. The mean foot length of the 15 soldiers in Skills Check 8 is _______.

13. The median foot length of the 15 soldiers in Skills Check 8 is ________.

14. The mode of the foot length data in Skills Check 8 is ________.

15. If a single-peaked distribution is skewed to the right, the median generally lies to the __________ of the mean.

16. Between the first quartile and the third quartile lie ____% of the observations in a distribution.

17. Which of these is greatest?

(a) The first quartile

(b) The third quartile

(c) The median

18. An outlier's effect on the difference between the third quartile and the first quartile is _____ than its effect on the range.

19. In degrees Fahrenheit, a typical January day in Houston has a low of 46 and a high of 63, while in El Paso a typical January day has a low of 32 and high of 57. Which city has a larger temperature range for January?

(a) Houston, because $63 > 57$ and $46 > 32$

(b) El Paso, because $57 - 32 > 63 - 46$

(c) El Paso, because $46 - 32 > 63 - 57$

20. The first quartile of the dataset {1, 2, 3, 4, 5, 6} is _______.

21. Which of these measures is not in a five-number summary?

(a) Median

(b) Minimum

(c) Mean

22. Determine the five-number summary of the 15 soldiers' foot lengths in Skills Check 8. (Remember to list the five numbers in increasing order.)

23. The standard deviation of the 10 blood pressures in Self Check 4 (page 195) is _______. (Support your answer by showing the calculations.)

(a) 13.23

(b) 13.95

(c) 194.6

24. You have data on the weights (measured in grams) of five crackers. The correct units for the standard deviation of these weights are _______________.

25. What are all the values that a standard deviation s can possibly take?

(a) $0 \le s$

(b) $0 \le s \le 1$

(c) $-1 \le s \le 1$

26. To specify the shape of a normal distribution completely, you must give its mean and its ________________.

27. If two normal curves have the same mean but different standard deviations, the curve with the larger standard deviation will be ___________ the other curve.

(a) as tall as

(b) taller than

(c) shorter than

28. The steepest part of a normal curve is at the

(a) first and third quartiles.

(b) points at which the curvature changes.

(c) mean.

29. The scale of scores on an IQ test is approximately normal with a mean of 100 and a standard deviation of 15. The organization Mensa, which calls itself "the high-IQ society," requires an IQ score of 130 or higher for membership. What percentage of adults would qualify for membership?

(a) 95%

(b) 5%

(c) 2.5%

30. The length of human pregnancies from conception to birth varies according to a distribution that is approximately normal, with a mean of 266 days and a standard deviation of 16 days. We can expect that about _____% of all completed pregnancies are between 234 and 298 days.

Chapter 5 Exercises

Challenge Discussion

Some exercises require use of a calculator (or software or Internet applet) that will find mean and standard deviation from keyed-in data.

5.1 Displaying Distributions: Histograms

1. Table 5.11 shows a small part of a dataset that describes the fuel economy (in miles per gallon) of 2014 model motor vehicles.

(a) What are the individuals in this dataset?

(b) For each individual, what variables are given?

(c) For which of these variables would a histogram be helpful? (That is, which variables do not yield categorical data?)

2. The femur (thighbone) is the longest bone in the human body. Femur lengths (in millimeters) of 15 people are given below.

435	507	448	435	463
440	448	413	432	458
473	465	428	472	439

(a) Summarize the data on femur lengths in a frequency table. Use class intervals that start at 400 and have width 20.

(b) Add a column to your table from part (a) for the relative frequencies.

(c) Draw a histogram that represents your frequency table. (Use either frequency or relative frequency for the vertical axis.)

TABLE 5.11 Data on 2014 Model Cars

Make and Model	Vehicle Type	Transmission Type	Number of Cylinders	City mpg	Highway mpg
Mazda MX-5	Two-seater	Manual	4	22	28
Toyota Yaris	Subcompact	Automatic	4	30	36
Honda Accord	Large car	Automatic	6	21	34
Jaguar XF	Midsize car	Automatic	8	15	23

3. Eating fish contaminated with mercury can cause serious health problems. Mercury contamination from historic gold mining operations is fairly common in sediments of rivers, lakes, and reservoirs today. A study was conducted on Lake Natoma in California to determine whether the mercury concentration in fish in the lake exceeded guidelines for safe human consumption. A sample of 83 largemouth bass was collected, and the concentration of mercury from sample tissue was measured. Mercury concentration is measured in micrograms of mercury per gram or $\mu g/g$. Figure 5.28 presents a histogram of the results of the study.

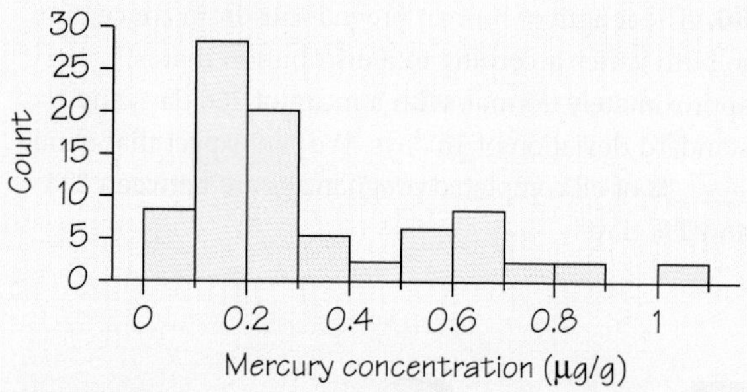

Figure 5.28 Mercury concentration in fish tissue.

(a) Which class interval contains the highest number of data values? Approximately what percentage of the fish in the sample had mercury concentrations that fell within this class interval?

(b) The primary objective of the study was to determine whether mercury concentrations in fish tissue exceeded safety guidelines for human consumption. The U.S. Environmental Protection Agency (USEPA) human-health criterion for methylmercury in fish is 0.30 $\mu g/g$. Approximately how many of the fish in the sample had mercury concentrations below the level set by the USEPA (and hence were considered safe for human consumption)?

(c) Approximately what percentage of the sample had mercury concentrations higher than the level set by the USEPA? Show how you arrived at your answer.

5.2 Interpreting Histograms

4. Figure 5.29 is a histogram of the lengths of words used in Shakespeare's plays. Because there are so many words in the plays, the vertical axis of the graph is the percentage of words that are of each length, rather than the count. In this case, the class intervals are centered at integer values, since the data consist only of counting numbers.

What is the overall shape of this distribution? What does this shape say about word lengths in Shakespeare? Do you expect other authors to have word-length distributions of the same general shape? Why?

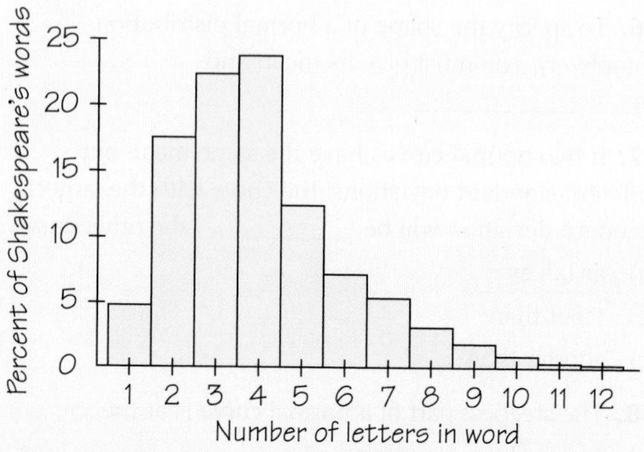

Figure 5.29 Relative frequency histogram of the lengths of words used (the percentage rather than the count that are of each length).

5. Suppose that you and your friends emptied your pockets of coins and recorded the year marked on each coin. Would you expect the histogram for the distribution of dates to be skewed to the left or right? Explain your answer and make a sketch of this histogram.

6. Make a histogram of the city gas mileages of the midsized cars in Table 5.7 (page 196). Use classes with widths of 5 mpg. Do you prefer the histogram or the dotplot in Figure 5.14 (page 197) of the same data? Why?

7. Burning fuels in power plants or motor vehicles emits carbon dioxide (CO_2), which contributes to global warming. Table 5.12 displays CO_2 emissions per person from 48 countries with populations of at least 20 million.

(a) Why do you think we choose to measure emissions per person rather than total CO_2 emissions for each country?

(b) Display the data of Table 5.12 in a histogram. Describe the shape, center, and variability of the distribution. Which countries appear to be outliers?

8. A survey of a large college class asked the following questions:

(i) Are you female or male? (In the data, male = 0, female = 1.)

TABLE 5.12 Carbon Dioxide Emissions, Metric Tons per Person

Country	CO_2	Country	CO_2	Country	CO_2	Country	CO_2
Algeria	2.3	Germany	10.0	Myanmar	0.2	South Korea	8.8
Argentina	3.9	Ghana	0.2	Nepal	0.1	Spain	6.8
Australia	17.0	India	0.9	Nigeria	0.3	Sudan	0.2
Bangladesh	0.2	Indonesia	1.2	North Korea	9.7	Tanzania	0.1
Brazil	1.8	Iran	3.8	Pakistan	0.7	Thailand	2.5
Canada	16.0	Iraq	3.6	Peru	0.8	Turkey	2.8
China	2.5	Italy	7.3	Philippines	0.9	Ukraine	7.6
Colombia	1.4	Japan	9.1	Poland	8.0	United Kingdom	9.0
Congo	0.0	Kenya	0.3	Romania	3.9	United States	19.9
Egypt	1.7	Malaysia	4.6	Russia	10.2	Uzbekistan	4.8
Ethiopia	0.0	Mexico	3.7	Saudi Arabia	11.0	Venezuela	5.1
France	6.1	Morocco	1.0	South Africa	8.1	Vietnam	0.5

(ii) Are you right-handed or left-handed? (In the data, right = 0, left = 1.)

(iii) What is your height in inches?

(iv) How many minutes do you study on a typical weeknight?

Figure 5.30 shows histograms of the student responses, in scrambled order and without scale markings. Which histogram goes with each variable? Explain your reasoning. Would the 0-1 coding scheme work for someone who is ambidextrous (or transgendered)?

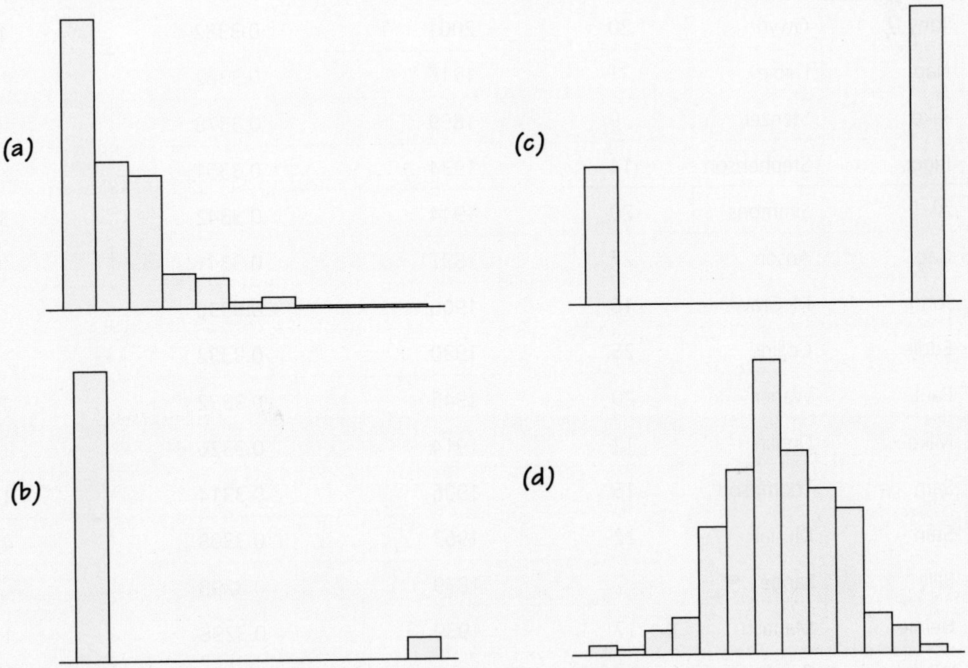

Figure 5.30 For Exercise 8, match each histogram with its variable.

TABLE 5.13 Top 100 Career Batting Averages in Baseball (for Exercises 9 and 10)

Rank	First	Last	Career Years	Last Career Year	Career Batting Average	Career Home Runs
1	Ty	Cobb	24	1928	0.3664	117
2	Rogers	Hornsby	23	1937	0.3585	301
3	Shoeless Joe	Jackson	13	1920	0.3558	54
4	Lefty	O'Doul	11	1934	0.3493	113
5	Ed	Delahanty	16	1903	0.3458	101
6	Tris	Speaker	22	1928	0.3447	117
7	Billy	Hamilton	14	1901	0.3444	40
	Ted	Williams	19	1960	0.3444	521
9	Dan	Brouthers	19	1904	0.3421	106
	Babe	Ruth	22	1935	0.3421	714
11	Dave	Orr	8	1890	0.3420	37
12	Harry	Heilmann	17	1932	0.3416	183
13	Pete	Browning	13	1984	0.3415	16
14	Willie	Keeler	19	1910	0.3413	33
15	Billy	Terry	14	1936	0.3412	154
16	Lou	Gehrig	17	1939	0.3401	493
	George	Sisler	15	1930	0.3401	102
18	Jesse	Burkett	16	1905	0.3382	75
	Tony	Gwynn	20	2001	0.3382	135
	Nap	Lajoie	21	1916	0.3382	82
21	Jake	Stenzel	9	1899	0.3378	71
22	Riggs	Stephenson	14	1934	0.3361	63
23	Al	Simmons	20	1944	0.3342	307
24	Cap	Anson	27	1897	0.3341	97
25	John	McGraw	16	1906	0.3336	13
26	Eddie	Collins	25	1930	0.3332	47
	Paul	Waner	20	1945	0.3332	113
28	Mike	Donlin	12	1914	0.3326	51
29	Sam	Thompson	15	1906	0.3314	126
30	Stan	Musial	22	1963	0.3308	475
31	Billy	Lange	7	1899	0.3298	39
	Heinie	Manush	17	1939	0.3298	110
33	Wade	Boggs	18	1999	0.3279	118

TABLE 5.13

Rank	First	Last	Career Years	Last Career Year	Career Batting Average	Career Home Runs
34	Rod	Carew	19	1985	0.3278	92
35	Honus	Wagner	21	1917	0.3276	101
36	Tip	O'Neill	10	1892	0.326	52
37	Hugh	Duffy	17	1906	0.3255	106
	Bob	Fothergill	12	1933	0.3255	36
39	Jimmie	Foxx	20	1945	0.3253	534
40	Earle	Combs	12	1935	0.3247	58
41	Joe	DiMaggio	13	1951	0.3246	361
42	Babe	Herman	13	1945	0.3245	181
43	Joe	Medwick	17	1948	0.3236	205
44	Eddie	Roush	18	1931	0.3227	68
45	Sam	Rice	20	1934	0.3223	34
46	Ross	Youngs	10	1926	0.3222	42
47	Kiki	Cuyler	18	1938	0.321	128
48	Charles	Gehringer	19	1942	0.3204	184
49	Miquel	Cabrera	12	2014+	0.3201	390
	Chuck	Klein	17	1944	0.3201	300
51	Mickey	Cochrane	13	1937	0.3196	119
	Pie	Traynor	17	1937	0.3196	58
53	Ken	Williams	14	1929	0.3192	196
54	Joe	Mauer	11	2014+	0.3186	109
55	Kirby	Puckett	12	1995	0.3181	207
56	Earl	Averill	13	1941	0.3178	238
57	Vladimir	Guerrero	16	2011	0.3176	449
	Arky	Vaughan	14	1948	0.3176	96
59	Billy	Everitt	7	1901	0.3174	11
60	Roberto	Clemente	18	1972	0.3173	240
	Joe	Harris	10	1928	0.3173	47
	Ichiro	Sizuki	14	2014+	0.3173	112
63	Albert	Pujols	14	2014+	0.3171	520
64	Chick	Hafey	13	1937	0.317	164
65	Joe	Kelley	17	1908	0.3169	65
66	Zack	Wheat	19	1927	0.3167	132
67	Roger	Connor	18	1897	0.3164	138

(*continued*)

TABLE 5.13 *(continued)*

Rank	First	Last	Career Years	Last Career Year	Career Batting Average	Career Home Runs
	Todd	Helton	17	2013	0.3164	369
	Lloyd	Waner	18	1945	0.3164	27
70	George	Van Haltren	17	1903	0.3163	69
71	Frankie	Frisch	19	1937	0.3161	105
72	Goose	Goslin	18	1938	0.3160	248
73	Lew	Fonseca	12	1933	0.3158	31
74	Bibb	Falk	12	1931	0.3145	69
75	Cecil	Travis	12	1947	0.3142	27
76	Hank	Greenberg	13	1947	0.3135	331
77	Jack	Fournier	15	1927	0.3132	136
78	Elmer	Flick	13	1910	0.313	48
79	Ed	Morgan	7	1934	0.3128	52
80	Nomar	Garciaparra	14	2009	0.3127	229
	Larry	Walker	17	2005	0.3127	383
82	Billy	Dickey	17	1946	0.3125	202
83	Dale	Mitchell	11	1956	0.3122	41
	Manny	Ramirez	19	2014+	0.3122	555
85	Jonny	Mize	15	1953	0.3121	359
	Joe	Sewell	14	1933	0.3121	49
87	Fred	Clarke	21	1915	0.3120	67
	Deacon	White	20	1890	0.3120	24
89	Bug	Holliday	10	1898	0.3119	65
90	Barney	McCosky	11	1953	0.3118	24
91	Hughie	Jennings	18	1918	0.3117	18
92	Edgar	Martinez	18	2004	0.3115	309
93	Johnny	Hodapp	9	1933	0.3114	28
	Freddie	Lindstrom	13	1936	0.3114	103
95	Bing	Miller	16	1936	0.3113	116
	Jackie	Robinson	10	1956	0.3113	137
97	Baby Doll	Jacobson	11	1927	0.3112	83
	Taffy	Wright	9	1949	0.3112	38
99	Rip	Radcliff	10	1943	0.3110	42
100	Ginger	Beaumont	12	1910	0.3108	39

Table 5.13 lists the top 100 baseball players ranked by career batting average. (These data were collected after the completion of the 2014 season.) Exercises 9 and 10 require use of the data from Table 5.13. (You will revisit these data in Chapter 6, Exercise 60.)

9. Focus on the variable "Career Home Runs" in Table 5.13.

(a) Create a frequency distribution for career home runs. Use class intervals of width 100.

(b) Draw a histogram that represents your frequency distribution from part (a).

(c) Describe the shape of the career home runs data. Identify any gaps in the data and potential outliers.

10. Focus on the variable "Career Years" in Table 5.13. (Note that the career years were based on data from 2014. Some players continued after 2014, which is noted by 2014+ in the "Last Career Year" column.)

(a) Make two histograms for career years. Use the following class intervals for your two histograms.

- Histogram 1: 5–10, 10–15, 15–20, 20–25, and 25–30.
- Histogram 2: 5–7, 7–9, 9–11, 11–13, 13–15, 15–17, 17–19, 19–21, 21–23, 23–25, 25–27, and 27–29.

If a data value falls on the boundary of a class interval, classify that data value in the interval to the right. (For example, a player with 10 career years would be counted in the interval 10–15 for histogram 1.)

(b) Describe the overall shape of each of the two histograms. Did changing the class intervals affect the shape of the distribution? Explain.

5.3 Displaying Distributions: Stemplots

11. The population of the United States is aging, though less rapidly than in other developed countries. Figure 5.31 is a stemplot of the percentage of residents aged 65 and over in the 50 states, according to the 2010 Census. The stems are whole percentages and the leaves are tenths of a percentage. (The software JMP was used to create the stemplot. Notice that this software put the low stems at the bottom of the plot and the high stems at the top of the plot.)

(a) Alaska is an outlier with the lowest percentage of older residents. Florida has the highest. What is the percentage for Florida?

(b) Ignoring Alaska, describe the shape, center, and variability of this distribution.

Stem	Leaf	Count
17	4	1
16	1	1
15	69	2
14	012344555599	12
13	02334555566778889	17
12	2233345899	10
11	49	2
10	479	3
9	0	1
8		
7	8	1

7|8 represents 7.8

Figure 5.31 For Exercise 11, stemplot of the percentages of residents aged 65 and over in the 50 states.

12. People with diabetes must monitor and control their blood glucose level. The goal is to maintain "fasting plasma glucose" between about 90 and 130 milligrams per deciliter (mg/dl). Here are the fasting plasma glucose levels for 18 diabetics enrolled in a diabetes management class, 5 months after the end of the class:

78	103	141	148	172	255
95	112	145	153	172	271
96	134	147	158	200	359

(a) Round these values to the nearest 10 and then drop the zero. For example, 141 rounds to 14 and 158 rounds to 16. Make a stemplot of the *rounded* data.

(b) Describe the main features of the distribution. Are there outliers? How well is the group as a whole achieving the goal for controlling glucose levels?

13. The Survey of Study Habits and Attitudes (SSHA) is a psychological test that evaluates college students' motivation, study habits, and attitudes toward school. A private college gives the SSHA to 18 of its incoming first-year women students. Their scores are (sorted in ascending order):

101	115	129	140	154	165
103	126	137	148	154	178
109	126	137	152	165	200

(a) Make a stemplot of these data. The overall shape of the distribution is irregular, as often happens if only a few observations are available. Are there any outliers?

(b) About where is the center of the distribution (the score with half the scores above it and half below)? What is the variability of the scores (ignoring any outliers)?

14. In 1798, the English scientist Henry Cavendish measured the density of the Earth in a careful experiment with a torsion balance. In sorted order, here are his 29 measurements of the same quantity (the density of the Earth relative to that of water) made with the same instrument. [*Source:* S. M. Stigler, Do robust estimators work with real data? *Annals of Statistics,* 5 (1977): 1055–1098.]

4.88	5.29	5.36	5.47	5.58	5.68
5.07	5.29	5.39	5.50	5.61	5.75
5.10	5.30	5.42	5.53	5.62	5.79
5.26	5.34	5.44	5.55	5.63	5.85
5.27	5.34	5.46	5.57	5.65	

(a) Make a stemplot of the data.

(b) Describe the distribution: Is it approximately symmetric or distinctly skewed? Are there gaps or outliers?

15. Here is a stemplot for the percentage of live births to unmarried mothers for each state in the United States in 2007. (*Source:* 2010 report on Centers for Disease Control website.)

```
2 | 0
2 | 56
3 | 13333344
3 | 555567777889999
4 | 0001111112223344
4 | 5677
5 | 124
```

(a) Explain how and why there are repeated stems.

(b) Describe the shape of the distribution.

5.4 Describing Center: Mean and Median

16. In Malay, the expression for the *mean* is *sama rata,* which roughly translates as "same level." To understand this cultural and conceptual connection, take some poker chips (or other equal-sized, stackable objects) and make stacks with 3, 7, and 8 chips.

(a) Explain how to redistribute chips among the stacks until they are at the same level.

(b) How does this relate to the mean?

17. Refer to the data and the stemplot in Exercise 13.

(a) Find the mean of the 18 values in Exercise 13.

(b) Your stemplot of the scores suggests that the score 200 is an outlier. Find the mean for the 17 observations that remain when you drop the outlier. [*Hint:* Can you use the work you did in part (a) to avoid calculating this new mean from scratch?]

(c) How does the outlier change the mean?

18. As of 2014, the Major League Baseball career and single-season home run records are held by Barry Bonds of the San Francisco Giants. Here are Bonds's annual home run totals from 1986 (his first year) through 2007 (his last year):

16	25	24	19	33	25	34	46
37	33	42	40	37	34	49	73
46	45	45	5	26	28		

Sandy Huffaker/Bloomberg via Getty Images

(a) Make a stemplot of the data. Are there any outliers?

(b) Find Bonds's career mean and median number of home runs. How do these change when you drop his 2001 season total of 73? What general fact about the mean and median does your result illustrate?

19. A male nursing home patient has his pulse taken every day. His pulse readings (beats per minute) over a 1-month period appear below.

72 56 56 68 78 72 70 70 60 72 68 74
76 64 70 62 74 70 72 74 72 78 76 74
72 68 70 72 68 74 70

(a) Make a stemplot of the pulse data. Expand the stem into five (for digits 0 and 1, 2 and 3, 4 and 5, 6 and 7, 8 and 9).

(b) Determine the mean, median, and mode for these data. Be sure to include units in your answers.

(c) Based on these data, which measure(s)—the mean, median, or mode—best describe(s) the man's typical pulse rate? Explain your reasoning.

20. The distribution of income in the United States is skewed to the right. According to the Census Bureau's Current Population Survey report, the mean and median incomes of American households were $51,017 and $71,274 in 2012. Explain how you can tell which of these numbers is the mean and which is the median.

21. The basic unit of census data is the household, not the person. If divorce breaks one household into two, but no individual person's income changes, how (if at all) is mean household income affected?

22. Which college football team is #1? In addition to polls of coaches and journalists, rankings from six computer programs (which have various ways to value factors such as the quality of the opponent played) determine the Bowl Championship Series (BCS) standings in major college football.

(a) At the end of the 2007 regular season, Hawaii (the only undefeated team) received these computer rankings: 12th, 8th, 14th, 10th, 8th, and 13th. The BCS formula throws out the high and low of the six computer rankings and uses the mean of the remaining four ranks. Find this mean.

(b) Why do you think the high and low values are excluded from the mean? Is your reason connected to why the median is sometimes preferred to the mean?

23. Make up an example of a small set of data for which the mean lies in the top 25% of the observations.

24. A sample of five households is selected, and the size of each household is recorded. The median size is 3 and the mode is 5. What is the mean? (*Hint:* Find the only possible dataset.)

5.5 Describing Variability: Range and Quartiles

5.6 The Five-Number Summary and Boxplots

25. The stemplot in Figure 5.31 (page 229) displays the distribution of the percentage of residents aged 65 and over in the 50 states. Stemplots help you find the five-number summary because they arrange the observations in order from smallest to largest. Give the five-number summary of this distribution.

26. In chronological order, here are the percentages of the popular vote won by each successful candidate in the last 16 presidential elections, starting in 1952:

Year	Percent	Year	Percent
1952	54.9	1984	58.8
1956	57.4	1988	53.4
1960	49.7	1992	43
1964	61.1	1996	49.2
1968	43.4	2000	47.9
1972	60.7	2004	50.7
1976	50.1	2008	52.9
1980	50.7	2012	51.1

(a) Make a stemplot of the winners' percentages.

(b) What is the median percentage of the vote won by the successful candidate in presidential elections?

(c) Call an election a landslide if the winner's percentage falls at or above the third quartile. Find the third quartile. Which elections were landslides?

(d) Find the range.

27. Figure 5.7 (page 193) is a histogram of the tuition and fees charged by the 64 four-year colleges in the state of Massachusetts for the 2014/2015 academic year. Here are those charges (in dollars), arranged in increasing order:

7519	8054	8080	8110	8157	8297	8524	8985
10,355	11,881	12,097	13,258	24,320	26,180	29,012	29,320
29,494	29,930	29,950	30,447	30,859	30,968	31,000	31,000
32,630	32,660	32,830	32,870	33,455	34,060	34,390	35,415
35,532	35,750	36,160	36,215	36,230	37,350	37,426	38,910
40,730	40,954	41,865	42,325	42,511	42,656	43,440	43,498
43,938	44,025	44,222	44,724	45,078	45,080	45,120	45,692
46,664	46,671	47,436	47,710	47,725	48,310	48,488	49,812

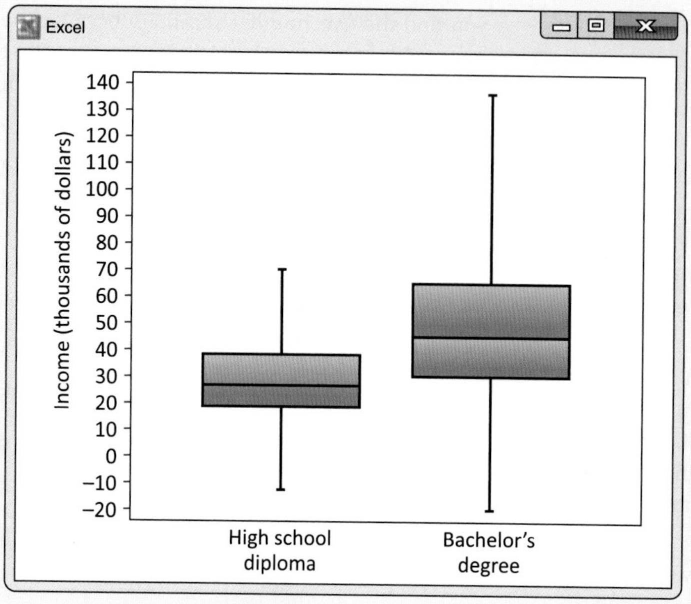

Figure 5.32 Boxplots comparing the incomes of full-time workers aged 25 to 64 years at two levels of education. The top 5% of incomes in each group are omitted.

(a) Find the five-number summary and make a boxplot.

(b) What distinctive feature of the Figure 5.7 histogram do these summaries miss? Remember that numerical summaries are not a substitute for looking at the data.

28. Find the five-number summary of Cavendish's measurements of the density of the Earth in Exercise 14 (page 230). How is the symmetry of the distribution reflected in the five-number summary?

29. Table 5.12 (page 225) gives CO_2 emissions per person for countries with populations of at least 20 million. The distribution is strongly skewed to the right. The United States and several other countries appear to be high outliers. Give the five-number summary. Explain why this summary suggests that the distribution is right-skewed.

30. Find the five-number summary of the data from Exercise 11 (Figure 5.31, page 229).

31. Figure 5.32 at the top of this page shows boxplots of the incomes of a large sample of people who have a high school diploma but no further education and another large group of people with a bachelor's degree but no higher degree. The data come from a Census Bureau survey and represent all people aged 25 to 64 in the United States. Because there are a few extremely high incomes, the boxplot leaves out the highest 5% in each group. Based on the plot, compare the distributions of income for these two levels of education. Comment on both center and variability.

32. The data that generate Figure 5.32 include the incomes of 14,959 people whose highest level of education is a bachelor's degree.

(a) What is the position of the median in the ordered list of incomes (1 to 14,959)? From the boxplot, about what is the median income of people with a bachelor's degree?

(b) What is the position of the first and third quartiles in the ordered list of incomes for these people? About what are the numerical values of Q_1 and Q_3?

33. How much oil the wells in a given field will ultimately produce is key information in deciding whether to drill more wells. Below are the estimated total amounts of oil recovered from 64 wells in the Devonian Richmond Dolomite area of the Michigan basin, in thousands of barrels. [*Source:* J. Marcus Jobe and Hutch Jobe, A statistical approach for additional infill development, *Energy Exploration and Exploitation,* 18 (2000): 89–103.]

2.0	18.5	34.6	47.6	69.5
2.5	20.1	34.6	49.4	69.8
3.0	21.3	35.1	50.4	79.5
7.1	21.7	36.6	51.9	81.1
10.1	24.9	37.0	53.2	82.2
10.3	26.9	37.7	54.2	92.2
12.0	28.3	37.9	56.4	97.7
12.1	29.1	38.6	57.4	103.1
12.9	30.5	42.7	58.8	118.2
14.7	31.4	43.4	61.4	156.5
14.8	32.5	44.5	63.1	196.0
17.6	32.9	44.9	64.9	204.9
18.0	33.7	46.4	65.6	

(a) Make a histogram and describe its main features.

(b) Find the mean and median of the amounts recovered. Explain how the relationship between the mean and the median reflects the shape of the distribution.

(c) Give the five-number summary and explain briefly how it reflects the shape of the distribution.

34. Look at the histogram of lengths of words in Shakespeare's plays shown in Figure 5.29 (page 224). The heights of the bars tell us what percentage of words have each length. (Analysis of such tendencies helps determine authorship of newly discovered manuscripts.)

(a) The median length is the length with half of all words shorter and half longer. What is the median length of the words Shakespeare used?

(b) Give the five-number summary for Shakespeare's word lengths.

35. A common criterion for identifying an outlier in a set of data is if an observation falls more than 1.5 × IQR above the third quartile or below the first quartile. (IQR stands for the interquartile range, which is the difference between the quartiles: $Q_3 - Q_1$, the width of the box in a boxplot.)

(a) Use the stemplot in Figure 5.11 (page 195) to determine a five-number summary for the data on percentage of Hispanics.

(b) Calculate IQR, $Q_1 - 1.5 \times \text{IQR}$, and $Q_3 + 1.5 \times \text{IQR}$.

(c) Any data value below $Q_1 - 1.5 \times \text{IQR}$ or above $Q_3 + 1.5 \times \text{IQR}$ should be considered an outlier. Determine the outliers, and then use Table 5.5 (page 189) to find which states are associated with the outliers.

36. Forty 6-year-olds were randomly selected from the participants in a study investigating childhood obesity. The children's weights (in kilograms) are arranged below in order from smallest to largest:

16.9	17.0	17.1	17.5	17.7	18.1	18.3	18.6	18.8	18.9
19.1	19.1	19.2	19.5	19.6	19.9	20.0	20.2	20.3	20.4
20.5	20.8	20.8	20.8	21.0	21.3	21.9	22.2	22.5	22.7
22.9	23.0	23.4	23.5	24.4	25.6	26.5	34.2	38.2	44.8

(a) Give the five-number summary of the weights of these 6-year-olds, and then draw a boxplot to represent this summary.

(b) The width of your box, $Q_3 - Q_1$, gives the interquartile range (IQR). Calculate the IQR for these data.

(c) Consider any weights that fall more than 1.5 × IQR above the third quartile or below the first quartile as outliers. Identify any outliers.

(d) Use the information from parts (a) through (c) to draw a modified boxplot of the weight data (similar to the ones shown in Figure 5.17 on page 203).

(e) Which boxplot do you think better represents these data, the one from part (a) or the modified boxplot in part (d)? Why?

5.7 Describing Variability: The Standard Deviation

37. Do you think the standard deviation of the tuition and fees of the public colleges in Massachusetts (Figure 5.7 on page 193) is likely to be bigger or smaller than the standard deviation for the private colleges? Why?

38. The level of various substances in the blood influences our health. Here are measurements of the level of phosphate in the blood of a patient, in milligrams of phosphate per deciliter of blood, made on six consecutive visits to a clinic:

5.6 5.2 4.6 4.9 5.7 6.4

(a) Find the mean.

(b) Find the standard deviation.

39. Many standard statistical methods are intended for use with distributions that are symmetric and have no outliers. These methods start with the mean and standard deviation, $\bar{x}$ and s. An example of scientific data for which standard methods should work well is Cavendish's measurements of the density of the Earth in Exercise 14 (page 230).

(a) Summarize this dataset by giving $\bar{x}$ and s.

(b) Find the median. Is the median quite close to the mean, as we expect it to be for symmetric distributions?

40. Here is a tale of two cities: Portland, Oregon, and Montreal, Canada. The average monthly precipitation (in inches) of these two cities is given in the table below.

Month	Portland	Montreal
January	5.4	2.8
February	3.9	2.6
March	3.7	2.8
April	2.5	2.9
May	2.2	2.7
June	1.5	3.3
July	0.6	3.4
August	0.9	3.6
September	1.5	3.3
October	3.1	3.0
November	5.5	3.5
December	5.9	3.4

Calculate the mean and standard deviation of the monthly average precipitation data for each city. What can you conclude about precipitation in these two cities from these means and standard deviations?

41. The mean $\bar{x}$ and standard deviation s are not generally a complete description. Datasets with different shapes can have the same mean and standard deviation.

(a) To demonstrate this fact, use your calculator (or software) to find $\bar{x}$ and s for the two small datasets below.

(b) Make a stemplot of each dataset and comment on the shape of each distribution. (Either round or truncate data values to one decimal place.)

Dataset A:	9.14	8.14	8.74	8.77
	9.26	8.10	6.13	3.10
	9.13	7.26	4.74	
Dataset B:	7.46	6.77	12.74	7.11
	7.81	8.84	6.08	5.39
	8.15	6.42	5.73	

42. "Conservationists have despaired over destruction of tropical rainforest by logging, clearing, and burning." These words begin a report on a statistical study of the effects of logging in Borneo. [*Source:* C. H. Cannon, D. R. Peart, and M. Leighton, Tree species diversity in commercially logged Bornean rainforest, *Science,* 281 (1998): 1366–1368.] Researchers compared forest plots that had never been logged (Group 1) with similar plots nearby that had been logged one year earlier (Group 2) and eight years earlier (Group 3). All plots were 0.1 hectare in area. Here are the counts of trees for plots in each group, courtesy of Charles Cannon:

Group 1:	27	22	29	21	19	33
	16	20	24	27	28	19
Group 2:	12	12	15	9	20	18
	17	14	14	2	17	19
Group 3:	18	4	22	15	18	
	19	22	12	12		

Give a complete comparison of the three distributions, using both graphs and numerical summaries. To what extent has logging affected the count of trees? The researchers used an analysis based on x and s. Explain why using this analysis is reasonably well justified.

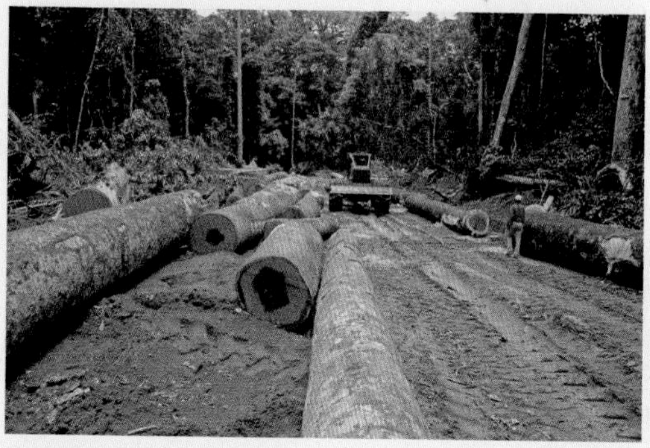
Edward Parker/Alamy

43. This is a standard deviation contest. You must choose four numbers from the whole numbers 0 to 10, with repeats allowed.

(a) Choose four numbers that have the smallest possible standard deviation.

(b) Choose four numbers that have the largest possible standard deviation.

(c) Is more than one choice possible in part (a)? Explain.

(d) Is more than one choice possible in part (b)? Explain.

44. Your data consist of observations on the ages of several subjects (measured in years) and the reaction times of these subjects (measured in seconds). In what units are each of the following descriptive statistics measured?

(a) Mean age of the subjects

(b) Standard deviation of the subjects' reaction times

(c) Variance of the subjects' reaction times

(d) Median age of the subjects

5.8 Normal Distributions

45. Figure 5.33 shows four normal density curves. Match the density curves with each of the following means, μ, and standard deviations, σ. Explain how you matched the curves to their means and standard deviations.

(a) $\mu = 4$ and $\sigma = 1$

(b) $\mu = 5$ and $\sigma = 2$

(c) $\mu = 4$ and $\sigma = 0.5$

(d) $\mu = 15$ and $\sigma = 3$

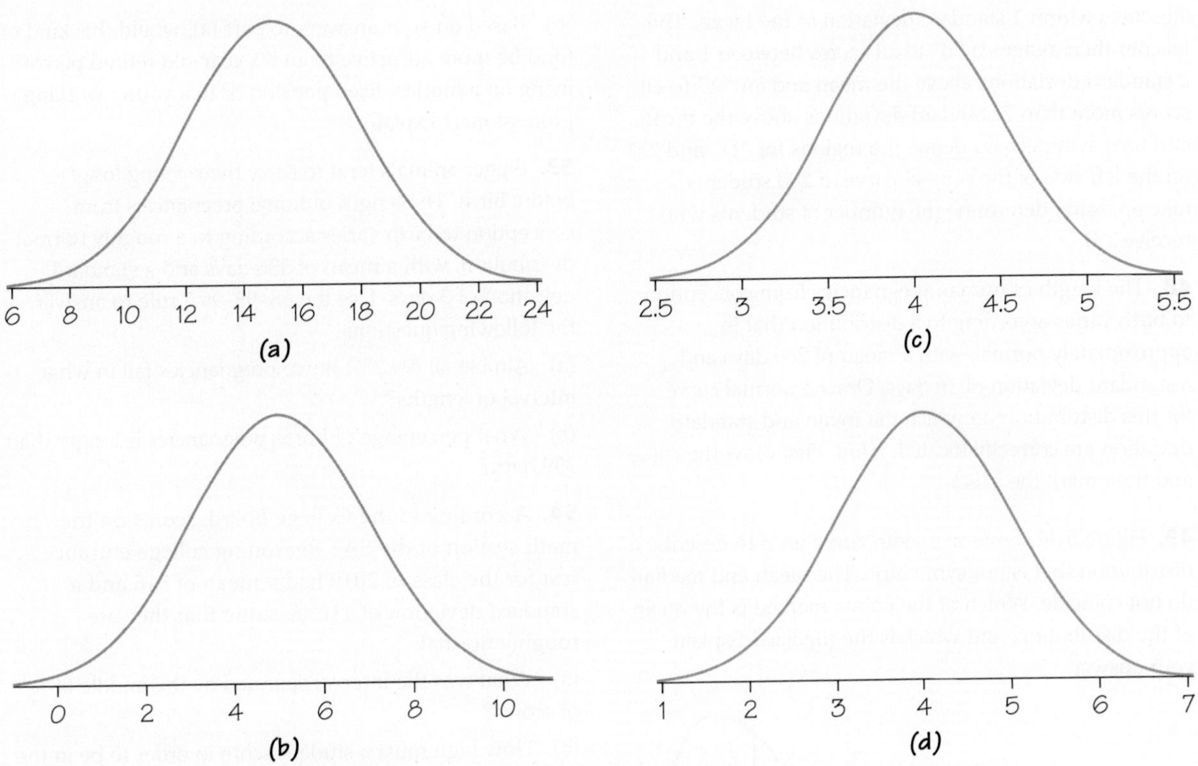

Figure 5.33 Four normal density curves.

46. Figures 5.34 and 5.35 show histograms of the height and body mass index (BMI), respectively, of 6-year-olds participating in an investigation into childhood obesity.

(a) Make a sketch of the histogram in Figure 5.34. Do you think that the height of 6-year-olds follows a normal distribution? If yes, draw a normal curve over your histogram. If no, explain why not and draw a density curve that provides a better fit.

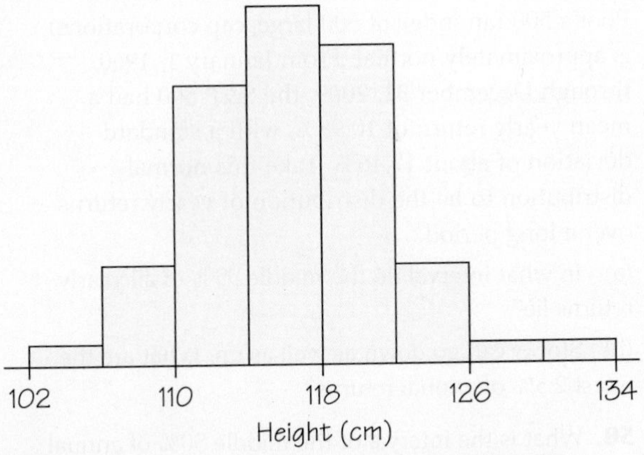

Figure 5.34 Height (in centimeters) of 6-year-olds.

(b) Make a sketch of the histogram in Figure 5.35. Do you think that the BMI of 6-year-olds follows a normal distribution? If yes, draw a normal curve over it. If no, explain why not and draw a density curve that provides a better fit.

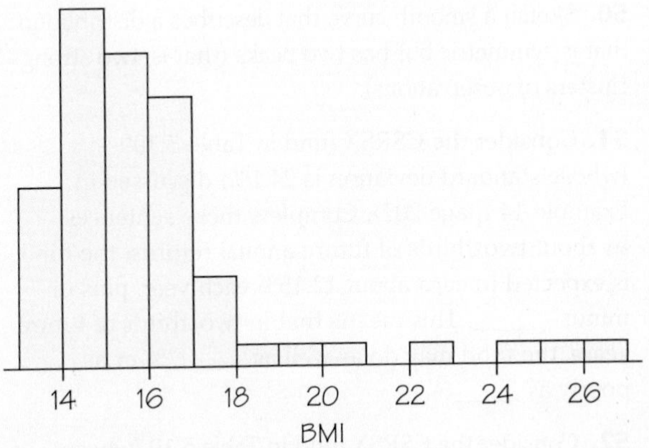

Figure 5.35 BMI of 6-year-olds.

5.9 The 68–95–99.7 Rule for Normal Distributions

47. Some teachers grade on a "(bell) curve" based on the belief that classroom test scores are normally distributed. One way of doing this is to assign a "C" to all scores within 1 standard deviation of the mean. The teacher then assigns a "B" to all scores between 1 and 2 standard deviations above the mean and an "A" to all scores more than 2 standard deviations above the mean, and uses symmetry to define the regions for "D" and "F" on the left side of the normal curve. If 200 students take an exam, determine the number of students who receive a B.

48. The length of human pregnancies from conception to birth varies according to a distribution that is approximately normal, with a mean of 266 days and a standard deviation of 16 days. Draw a normal curve for this distribution on which the mean and standard deviation are correctly located. (*Hint:* First draw the curve and then mark the axis.)

49. Figure 5.36 shows a smooth curve used to describe a distribution that is not symmetric. The mean and median do not coincide. Which of the points marked is the mean of the distribution, and which is the median? Explain your answer.

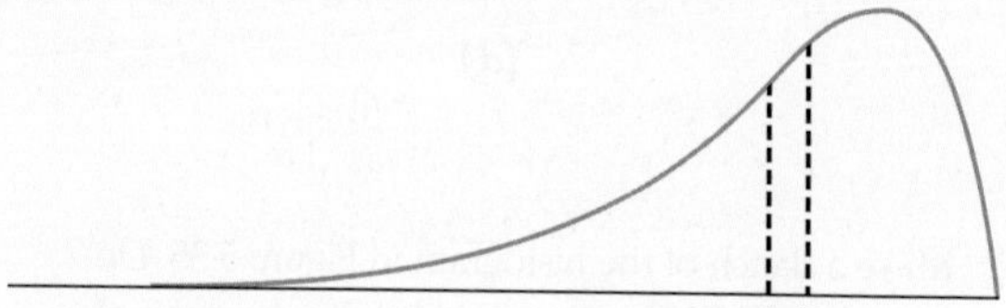

Figure 5.36 For Exercise 49, a curve describing a left-skewed distribution.

50. Sketch a smooth curve that describes a distribution that is symmetric but has two peaks (that is, two strong clusters of observations).

51. Consider the CSRSX fund in Table 5.10 (whose standard deviation is 24.1%) discussed in Example 14 (page 207). Complete these sentences: In about two-thirds of future annual returns, the fund is expected to earn about 12.15% each year, plus or minus ______. This means that in two-thirds of future years, the fund may do as well as ______% or as poorly as ______%.

52. Consider the CSRSX fund in Table 5.10 (whose standard deviation is 24.1%) discussed in Example 14 (page 207).

(a) Complete these sentences, a slight variation of which is commonly used in investment advising: In about 95% of future annual returns, the fund is expected to earn about 12.15% each year, plus or minus ______. This means that in 95% of future years, the fund may do as well as ___% or as poorly as _____%.

(b) Based on your answers to part (a), would this kind of fund be more attractive to an 80-year-old retired person living on a modest fixed pension or to a young working professional? Explain.

53. Bigger animals tend to carry their young longer before birth. The length of horse pregnancies from conception to birth varies according to a roughly normal distribution, with a mean of 336 days and a standard deviation of 3 days. Use the 68–95–99.7 rule to answer the following questions.

(a) Almost all (99.7%) horse pregnancies fall in what interval of lengths?

(b) What percentage of horse pregnancies is longer than 339 days?

54. According to the College Board, scores on the math section of the SAT Reasoning college entrance test for the class of 2010 had a mean of 516 and a standard deviation of 116. Assume that they are roughly normal.

(a) What was the interval spanned by the middle 68% of scores?

(b) How high must a student score in order to be in the top 2.5% of scores?

55. What are the quartiles of scores from the math section of the SAT Reasoning test, according to the distribution in Exercise 54?

56. The Wechsler Adult Intelligence Scale (WAIS) is the most common "IQ test." The scale of scores is set separately for each age group and is approximately normal, with a mean of 100 and a standard deviation of 15. People with WAIS scores below 70 are generally considered eligible to apply for Social Security disability benefits. By this criterion, what percentage of adults are in this IQ category?

57. The yearly rate of return on the Standard & Poor's 500 (an index of 500 large-cap corporations) is approximately normal. From January 1, 1960, through December 31, 2009, the S&P 500 had a mean yearly return of 10.98%, with a standard deviation of about 17.46%. Take this normal distribution to be the distribution of yearly returns over a long period.

(a) In what interval do the middle 95% of all yearly returns lie?

(b) Stocks can go down as well as up. What are the worst 2.5% of annual returns?

58. What is the interval of the middle 50% of annual returns on stocks, according to the distribution given in Exercise 57? (*Hint:* What two numbers mark off the middle 50% of any distribution?)

59. The concentration of the active ingredient in capsules of a prescription painkiller varies according to a normal distribution with $\mu = 10\%$ and $\sigma = 0.2\%$.

(a) What is the median concentration? Explain your answer.

(b) What interval of concentrations covers the middle 95% of all the capsules?

(c) What interval covers the middle half of all capsules?

60. Answer the following questions for the painkiller in Exercise 59.

(a) What percentage of all capsules has a concentration of the active ingredient higher than 10.4%?

(b) What percentage has a concentration higher than 10.6%?

61. One reason that normal distributions are important is that they describe how the results of an opinion poll would vary if the poll were repeated many times. About 40% of adult Americans say they are afraid to go out at night because of crime. Take many randomly chosen samples of 1050 people. The proportions of people in these samples who stay home for fear of crime will follow the normal distribution with a mean of 0.4 and a standard deviation of 0.015. Use this fact and the 68–95–99.7 rule to answer these questions.

(a) In many samples, what percentage of samples gives results above 0.4? Above 0.43?

(b) In a large number of samples, what interval contains the middle 95% of proportions of people who stay home because of crime?

62. You can compare observations from different normal distributions if you measure in standard deviations away from the mean. Scores expressed in standard deviation units are called *standard scores* (or *z-scores*), and tables and technology commands can convert *z*-scores into percentiles. A *z*-score that is more than 3 or less than −3 would definitely be considered an outlier.

(a) Scores on the ACT college entrance exam in a recent year were roughly normal, with a mean of 21.2 and a standard deviation of 4.8. Jermaine scores 27 on the ACT. Express his score in standard deviation units by calculating

$$\text{standard score} = \frac{\text{score} - \text{mean}}{\text{standard deviation}}$$

(b) Scores on the SAT Reasoning college entrance exam in the same year were roughly normal, with mean 1511 and standard deviation 194. Tonya scores 1718 on the SAT. What is her standard score?

(c) Assuming that the ACT and the SAT tests measure the same thing, did Jermaine or Tonya have the better performance?

63. The Boston Beanstalks Club is a social club for tall people. To join the club, women must be at least 5 feet 10 inches (70 inches) and men at least 6 feet 2 inches (74 inches). Both men's and women's heights are approximately normally distributed, but from different normal distributions. You can compare observations from different normal distributions if you measure in standard deviations away from the mean, which converts the observation to a *z-score.* To compute an observation's *z*-score, subtract the mean and then divide the result by the standard deviation:

$$z = \frac{\text{observation} - \text{mean}}{\text{standard deviation}}$$

(a) Assume that women's heights follow an approximately normal distribution with $\mu = 63.8$ inches and $\sigma = 4.2$ inches. How many standard deviations above the mean must a woman be in order to join the Boston Beanstalks?

(b) Assume that men's heights follow an approximately normal distribution with $\mu = 69.4$ inches and $\sigma = 4.7$ inches. How many standard deviations above the mean must a man be in order to join the Boston Beanstalks?

(c) In terms of joining the Boston Beanstalks, which is more stringent, the height requirements for women or the height requirements for men? Explain.

64. In order for men to join the Boston Beanstalks (see Exercise 63), they must be at least 6 feet 2 inches (74 inches) tall. Assume that men's heights are approximately normal with $\mu = 69.4$ inches and $\sigma = 4.7$ inches. Use the 68–95–99.7 rule to estimate the percentage of men who are eligible to join the Boston Beanstalks.

Chapter Review

Different varieties of the bright tropical flower *Heliconia* are fertilized by different species of hummingbirds. Over time, the lengths of the flowers and the form of the hummingbirds' beaks have evolved to match each other. Below are data on the lengths in millimeters of two varieties of these flowers on the island of Dominica. Exercises 65–69 use these data.

Heliconia caribaea Red				
37.40	38.07	38.87	40.66	41.93
37.78	38.10	39.16	41.47	42.01
37.87	38.20	39.63	41.69	42.18
37.97	38.23	39.78	41.90	43.09
38.01	38.79	40.57		
***Heliconia caribaea* Yellow**				
34.57	35.45	36.03	36.66	37.02
34.63	35.68	36.11	36.78	37.10
35.17	36.03	36.52	36.82	38.13

Thanks to Ethan J. Temeles of Amherst College for providing the data. His work is described in Ethan J. Temeles and W. John Kress, Adaptation in a plant–hummingbird association, *Science*, 300 (2003): 630–633.

65. Make stemplots of the lengths of each of the two varieties (red and yellow). Briefly describe the overall shape of the two distributions.

66. Find the five-number summaries of the two distributions of flower lengths. Make side-by-side boxplots to give a quick picture that compares the two distributions.

67. The biologists who collected the flower length data compared the two *Heliconia* varieties using statistical methods based on the mean and standard deviation.

(a) Find $\bar{x}$ and s for each variety.

(b) Based on Exercise 65, which distribution is more suitable for use of $\bar{x}$ and s as summaries? Why?

68. Your stemplot in Exercise 65 suggests that the distribution of lengths of yellow *Heliconia* flowers is roughly normal. Suppose that the distribution is exactly normal. Use the mean and standard deviation you found in Exercise 67 as the μ and σ of the distribution.

(a) What interval of lengths covers the middle 50% of yellow flowers?

(b) What interval of lengths covers the middle 95% of yellow flowers?

69. Continue to work with the normal distribution of lengths of yellow flowers in Exercise 68. The shortest red flower was 37.4 millimeters long. Using the 68–95–99.7 rule and the location of the quartiles in normal distributions, what can you say about the percentage of yellow flowers that are longer than 37.4 millimeters?

70. Without a calculator (or other technology), find the standard deviation of these five numbers: 0, 1, 3, 4, 12. Use the approach in the standard deviation definition box on page 204.

71. If every number in a dataset is increased by 10, which of these measures will increase: range, standard deviation, mode, mean, or median?

72. Bob is two years older than one brother and five years younger than his other brother. Find the standard deviation of the three brothers' ages.

73. If you ask a computer (or your graphing calculator) to generate "random numbers" between 0 and 1, you will get data from a uniform distribution. Figure 5.37 shows a graph of the density curve for this distribution.

(a) Check that the area under the uniform density curve is 1.

(b) What is the mean μ of this distribution?

(c) What proportion of outcomes from this distribution lie between 0.2 and 0.8?

(d) What percentage of random numbers between 0 and 1 would you expect to lie between 0 and 0.5?

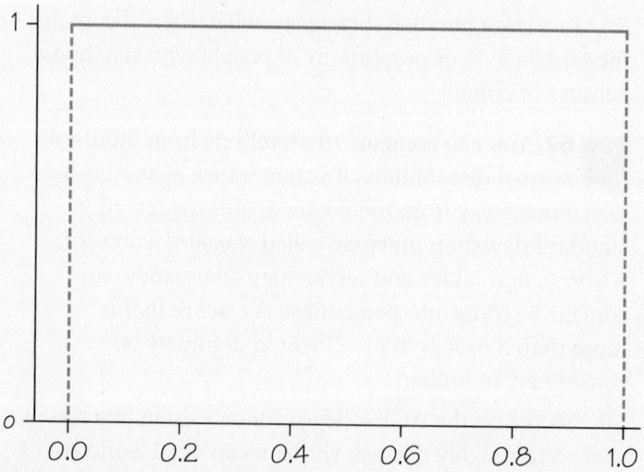

Figure 5.37 For Exercise 73, density curve for the uniform distribution.

Applet Exercises

To do these exercises, go to www.macmillanhighered.com/fapp10e.

1. The *Mean and Median* applet allows you to place observations on a line and see their mean and median visually.

(a) Place two observations on the line by clicking below it. Why does only one arrow appear?

(b) Now move the rightmost point close to the other point. (Place the cursor on the point, hold down the mouse button, and drag the point.) Add a third point that is somewhat to the right of the other two. Pull the single rightmost observation out to the right. How does the mean behave? How does the median behave? Explain briefly why each measure acts as it does.

2. In Example 19 (page 217), we used the fact that SAT section scores are close to normal and are adjusted so that the mean is close to 500 and the standard deviation is close to 100. (Actual scores in a particular year have a slightly different mean and standard deviation.) Use the *Normal Density Curve* applet with mean $\mu = 500$ and standard deviation $\sigma = 100$ to answer these questions:

(a) What proportion of SAT scores is above 640? (You may want to uncheck the 2-Tail box.)

(b) What proportion of SAT scores is between 420 and 640? (If you drag one flag across the other, the applet shows the area between the flags.)

3. Because Internet browsers have limited resolution, the *Normal Density Curve* applet can't always get exactly the values you want. Use the applet as set in the previous exercise to come close to exact answers to these questions.

(a) How high must an SAT score be to fall in the top 10% of all scores?

(b) How high must an SAT score be to fall in the top 1% of all scores?

4. The 68–95–99.7 rule for normal distributions is a useful approximation. You can use the *Normal Density Curve* applet to see how accurate the rule is. Drag one flag across the other so that the applet shows the area under the curve between the two flags.

(a) Place the flags 1 standard deviation on either side of the mean. What is the area between these two values? What does the 68–95–99.7 rule say this area is?

(b) Repeat for locations 2 and 3 standard deviations on either side of the mean. Again, compare the 68–95–99.7 rule with the area given by the applet.

Writing Projects

1. Go online and look up information about "statistical quality control" and "six-sigma." Write a paragraph or two about what you learned and how it connects to variability in general and to standard deviation and the normal curve in particular.

2. Many social issues involve data and interpreting data. For example, income inequality (roughly speaking, the gap in income between people toward the top of the income scale and people toward the bottom) has increased in the past few decades. A good place to find data is on the Census Bureau website (www.census.gov). Under People & Households (at the bottom), click on "Income" and look for the latest report on income in the United States. Select a few facts from this detailed collection of income data to describe the extent of income inequality. Write a few paragraphs based on these facts.

3. Let's produce some data and describe them to gain insight into chance behavior. The mathematics of chance is the topic of Chapter 8, but for now, we concentrate on data rather than math. You need two things: a standard six-sided die and a thumbtack with a rounded back (like a satellite dish). Toss the thumbtack 100 times (to speed things up, you could do 10 tosses of 10 tacks each) and record each outcome (pointing straight up or angled down). Also, toss the die 180 times and record each outcome (1, 2, 3, 4, 5, or 6). Use graphs and numbers to describe each set of results. Is the die roughly balanced, so that all six outcomes come up almost equally? What about the thumbtack: Is "point up" or "point down" noticeably more common?

Suggested Readings

CLEVELAND, WILLIAM S. *The Elements of Graphing Data*, rev. ed., Hobart Press, Summit, NJ, 1994. A careful study of the most effective elementary ways to present data graphically, with much sound advice on improving simple graphs.

JOINER, B. L. Living Histogram, *International Statistical Review*, 3 (1975): 339–340. This article examines how the percentage of females in groups of students can affect the shape of "living histograms" of height.

LESSER, LAWRENCE M. Critical values and transforming data: Teaching statistics with social justice, *Journal of Statistics Education* 15(1) (2007): www.amstat.org/publications/jse/v15n1/lesser.html. Resources for finding social justice data to extend Writing Project 2.

MOORE, DAVID S, WILLIAM I. NOTZ, and MICHAEL A. FLIGNER. *The Basic Practice of Statistics*, 10th ed., W. H. Freeman, New York, 2015. This text is a natural next step to learning more detail on all the material in Part II, at about the same mathematical level. The first three chapters provide a more extensive treatment of the material of Chapter 5.

ROSSMAN, ALLAN J., and BETH L. CHANCE. *Workshop Statistics: Discovery with Data*, 4th ed., Wiley, Hoboken, NJ, 2011. A different approach to basic data analysis, using hands-on activities. There are several versions, keyed to graphing calculators and to several different software packages.

Suggested Websites

www.census.gov The website of the U.S. Census Bureau is a good source of information on many topics. The latest estimates for the populations of the United States and the world are on the home page, updated regularly. See what data you can find within "American Fact Finder" or the *Statistical Abstract of the United States*. Canadians can find similar help at the website of Statistics Canada: **www.statcan.gc.ca.**

nces.ed.gov Interested in data about schools, colleges, and students? The National Center for Education Statistics is the place to look. Go to the "What's New" section.

www.learner.org/courses/againstallodds/ Want to view videos showing "statistics in action" or to learn more about statistical techniques and analyses? Take a look at *Against All Odds: Inside Statistics*. Units 2 through 8 provide a more extensive treatment of the material of Chapter 5.

www.amstat.org/sections/educ/applets.html Entire collections of statistics applets can be found here.

education.ti.com/en/us/downloads-and-activities?active=guidebooks#! Texas Instruments guidebooks are available here.

Exploring Data: Relationships

6

JGI/Jamie Grill/Blend Images/Getty Images

In Chapter 5, we analyzed data one variable at a time. In Chapter 6, we analyze data on two (or more) variables simultaneously. For example, medical studies have linked weight and the risk of heart disease. Heavier people are more likely to develop the disease than those of average weight. Data from the Bureau of Labor Statistics (BLS) indicate that there is a relationship between the educational attainment of Americans and both earnings and unemployment. In general, those with high levels of education earn more and are less likely to be unemployed, whereas the reverse is true for the least educated Americans. A report prepared for the U.S. Department of Energy finds that as vehicle footprint (a measure of vehicle size) increases, fatality risk to drivers decreases. Moreover, vehicle footprint and weight are correlated, and as vehicle weight increases, fuel efficiency decreases. All these studies focused on investigating *relationships between two variables.*

In this chapter, we explore the *relationship between two variables,* measured on the *same individuals.* If we measure the height and weight of each individual in a large group of people, we know which height goes with each weight. These data allow us to study the connections between height and weight, whereas a list of heights and a separate list of weights—two sets of single-variable data—do not provide information about the possible connections between height and weight.

In Chapter 5, we used histograms, dotplots, and boxplots to extract information from data on a single variable such as weight. In this chapter, we expand our graphical tools to include **scatterplots** in order to visually explore the relationships between two variables such as weight and height. For example, the scatterplot in Figure 6.1 shows how the weights of 4-year-olds are related to their heights.

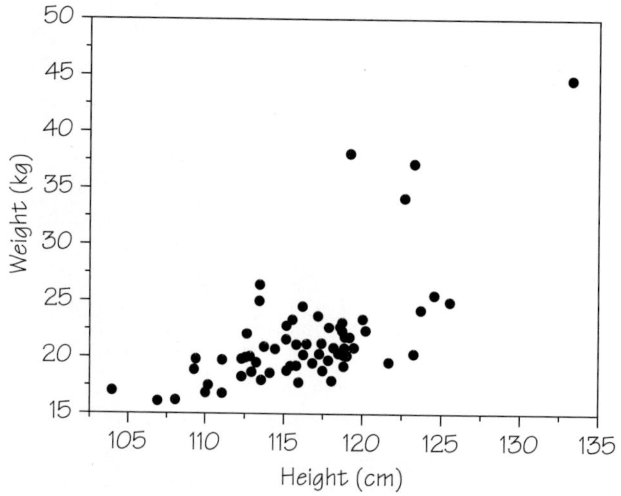

Figure 6.1 Scatterplot of heights and weights of a group of 4-year-olds.

As we delve into relationships in this chapter, we will find cases where a scatterplot shows a linear pattern. In such cases, we will summarize that pattern by drawing a line and then use the line to make predictions. In order to judge just how tightly data hug a straight line, we will calculate the correlation, a numerical measure of the strength of the linear relationship. For more precise predictions, we'll need to obtain the equation of a line that best fits the data. Then, after learning data analysis techniques for extracting information from two related variables, we'll discuss some limitations of those techniques.

6.1 Displaying Relationships: Scatterplot

The world is full of relationships between variables. For example, children's weight and height are related. As seen in Figure 6.1, taller 4-year-olds tend to be heavier than shorter 4-year-olds. In this case, neither height nor weight explains or causes the other. The two variables just go together in describing bigger or smaller 4-year-old children. Now contrast the height–weight situation with the relationship between smoking cigarettes and life expectancy. In this case, there is a great deal of evidence that smoking does explain or influence life expectancy. People who smoke more cigarettes per day tend to not live as long as those who smoke fewer. So, we call smoking an **explanatory variable** and life expectancy a **response variable.**

Response Variable DEFINITION

A **response variable** measures an outcome or result of a study.

Explanatory Variable DEFINITION

An **explanatory variable** is a variable that we think explains or causes changes in the response variables.

Even when we know two quantitative variables are related, the relationship is rarely an exact trend such as a line-shaped pattern, free of any "scatter" or deviation from that pattern. The most useful graph for displaying the relationship between two quantitative variables (whether that relationship fits a trend perfectly or not) is a scatterplot.

Scatterplot DEFINITION

A **scatterplot** is a graph of plotted points showing the relationship between two numerical variables measured on the same individuals. In the case of one explanatory and one response variable, the values of the explanatory variable are shown on the horizontal (x) axis and the values of the response variable are shown on the vertical (y) axis. The values of the explanatory and response variables for one particular individual in the dataset become the x- and y-coordinates, respectively, of a point representing that individual in the scatterplot.

EXAMPLE 1 Beer and Blood Alcohol

How well does the number of beers a student drinks predict his or her blood alcohol content (BAC)? In a study at The Ohio State University, 16 student volunteers drank a randomly assigned number of cans of beer. Thirty minutes later, a police officer measured their BAC in grams of alcohol per deciliter of blood. Throughout the United States, the legal BAC limit is 0.08. Here are the data:

TABLE 6.1 Number of Beers Consumed and Blood Alcohol Level (g/dl)

Student	**1**	**2**	**3**	**4**	**5**	**6**	**7**	**8**
Beers	5	2	9	8	3	7	3	5
BAC	0.10	0.03	0.19	0.12	0.04	0.095	0.07	0.06
Student	**9**	**10**	**11**	**12**	**13**	**14**	**15**	**16**
Beers	3	5	4	6	5	7	1	4
BAC	0.02	0.05	0.07	0.10	0.085	0.09	0.01	0.05

Monika Gruszewicz/Shutterstock

The students were equally divided between men and women and differed in weight and usual drinking habits. Because of this variation, many students don't believe that the number of drinks ingested predicts BAC well. What do the data say?

Figure 6.2 is a scatterplot of these data. Because we think that the number of beers helps explain BAC, "number of beers" is the explanatory variable and hence is put on the horizontal axis. This is also indicated by the wording used in the Figure 6.2 caption—it is common usage for the word "against" to follow the response variable and precede the explanatory variable. In terms of plotting the data values, one student (Student 2) drank 2 beers and had a BAC of 0.03. This student's point on the scatterplot is (2, 0.03), above $x = 2$ and to the right of $y = 0.03$. We have marked this point in Figure 6.2.

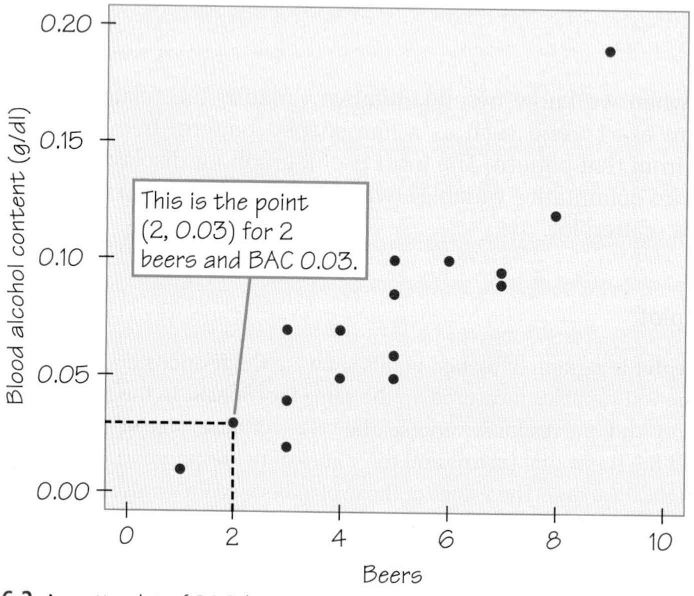

Figure 6.2 A scatterplot of BAC (response variable) against the number of beers a student drinks (explanatory variable).

Self Check 1

Table 6.2 contains the midterm and final exam scores of a sample of students in an introductory statistics course. Make a scatterplot of these data. Which variable did you use for the explanatory variable? Why?

TABLE 6.2 Midterm and Final Exam Scores

Midterm	Final
78	90
86	84
94	92
93	96
67	71
78	75
75	68
86	84

Next, we consider a procedure for examining a scatterplot.

Examining a Scatterplot PROCEDURE

1. Describe the *overall pattern* of a scatterplot by the *form, direction,* and *strength* of the relationship. (Be open to the possibility that there may be two or more trends or clusters in the same graph.)
2. Then look for any striking *deviations* from the pattern. Identify each occurrence of an **outlier**—an individual value that falls outside the overall pattern of the relationship.

In discussing the procedure for examining a scatterplot, we start with item 2, identifying outliers (even though in practice, you would begin the analysis with item 1). When dealing with a single quantitative variable (such as city gasoline mileage from Table 5.7 on page 196), potential outliers are easy to identify numerically because they are usually the minimum or maximum values in the data or they satisfy a numerical criterion such as being more than 1.5 box widths below the first quartile or above the third quartile (see Figure 5.17 on page 203 or Chapter 5, Exercise 35, on page 233). When the dataset consists of ordered pairs, however, an outlier may or may not include an extreme value in one or both coordinates. As you will see in Example 2, it is even more critical to use a graphical representation (i.e., a scatterplot) to look for deviations from the overall pattern.

EXAMPLE 2 Identification of Outliers

Consider two datasets:

- Table 6.3 gives the results of an 8-year-old competitive swimmer's first 14 races of the 50-yard butterfly. (The time for race 10 did not get recorded and hence is missing here.)
- Table 6.4 shows reading and IQ test scores for a group of fifth-grade students.

TABLE 6.3 A Swimmer's 50-Yard Butterfly Times for Her First 14 Races

Race number	1	2	3	4	5	6	7
Time (seconds)	60.81	66.11	47.32	42.69	43.40	44.82	42.67
Race number	8	9	10	11	12	13	14
Time (seconds)	45.17	41.20	missing	42.47	41.74	40.40	42.90

Purdue 9394/Getty Images/ iStock/Getty Images Plus

TABLE 6.4 Fifth-Grade Students' IQ and Reading Test Scores

IQ test score	100	102	110	115	118	123	124
Reading test score	40	65	55	70	75	95	45
IQ test score	125	126	130	135	140	143	147
Reading test score	70	85	90	75	95	85	95

Compassionate Eye Foundation/ Robert Daly/OJO Images/Getty Images

Scatterplots of these datasets appear in Figure 6.3 and Figure 6.4, respectively. Outliers have been circled. The scatterplot in Figure 6.3 shows two circled outliers—they are associated with the highest values of the response variable, time, and lowest values

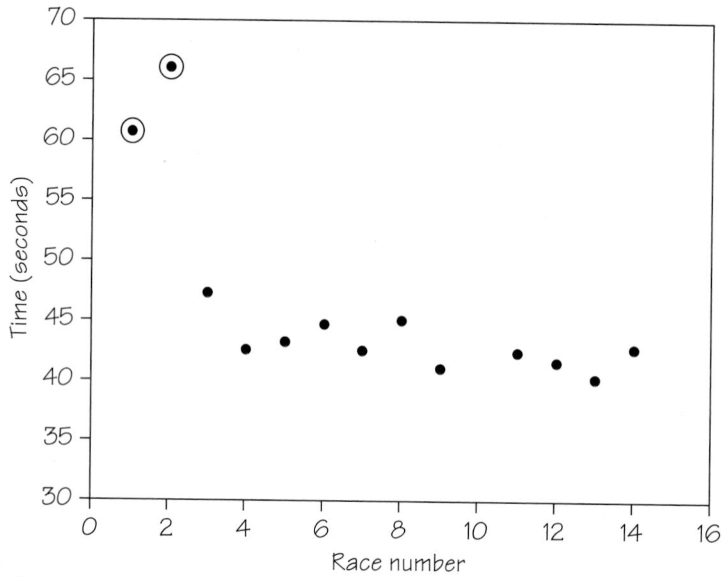

Figure 6.3 Swimmer's times for the 50-yard butterfly in consecutive races.

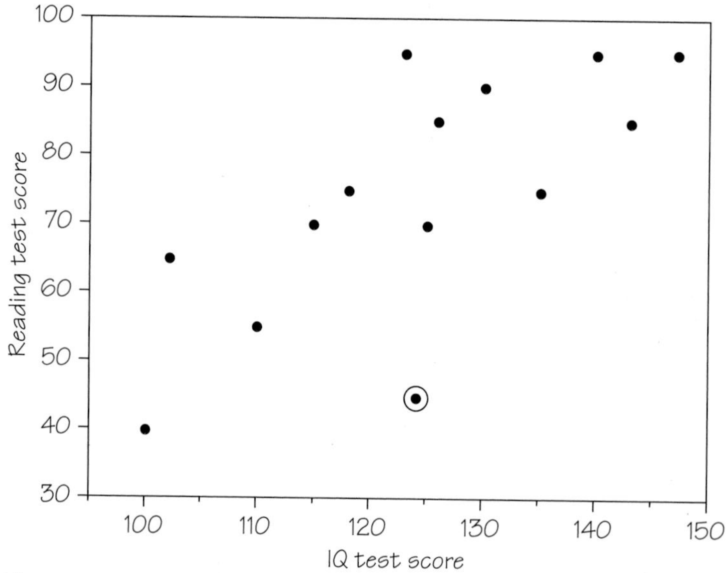

Figure 6.4 Scatterplot of reading test scores against IQ test scores.

of the explanatory variable, race number. Whenever possible, look for explanations for the presence of outliers. In this case, the swimmer had just learned the butterfly, which explains why her times in the first two races (when she was worried about getting disqualified) were unusually slow.

The outlier circled in Figure 6.4 was flagged as an outlier by a statistical program. In this case, the outlier does not correspond to the minimum or maximum values of either the response or explanatory variables. Instead, the point (124, 45) indicates a reading test score that is low in comparison to the reading test scores of other students with IQ test scores close to 124. Without additional information about this student, we don't have an explanation for the presence of this outlier.

Having dealt with item 2, outliers (deviations from an overall pattern) in the Examining a Scatterplot procedure, we return to item 1, a description of the overall pattern. The *form* of the relationship between BAC and beers consumed (Example 1, Figure 6.2) is roughly a straight-line pattern. If you look ahead a bit, Figure 6.6 (page 251) shows a line drawn through the plot to describe the overall pattern. The *direction* of the relationship is clear: As the number of beers increases, BAC also increases. We call this a **positive association** between the two variables.

Positive Association

DEFINITION

Two variables have **positive association** if their changes tend to be in the same direction. This means an increase in one variable tends to accompany an increase in the other variable. Also, a decrease in one variable tends to accompany a decrease in the other variable.

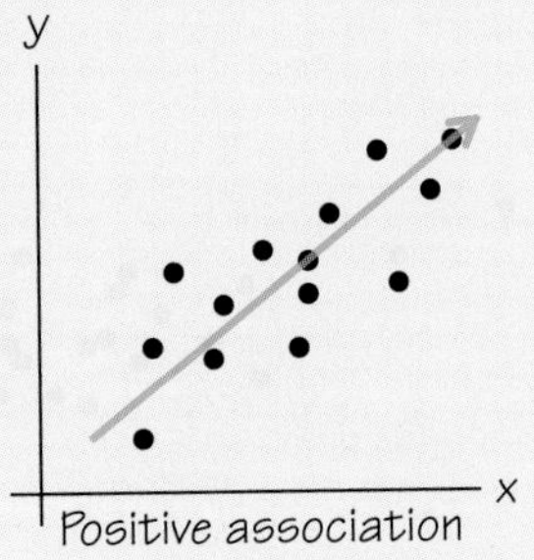

Negative Association

DEFINITION

Two variables have **negative association** if their changes tend to be in opposite directions. This means an increase in one variable tends to accompany a decrease in the other variable.

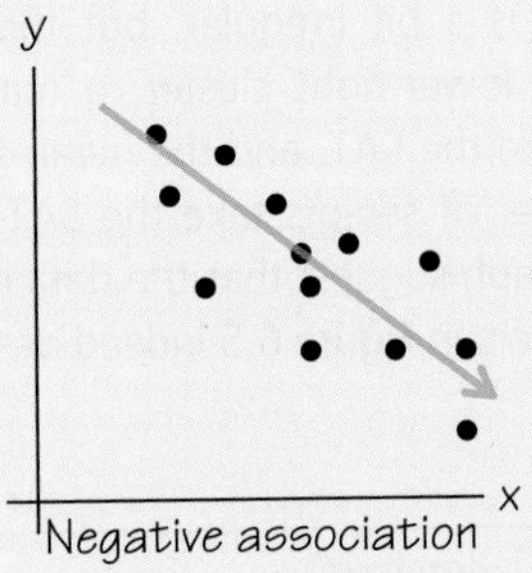

Self Check 2

Identify whether the scatterplots in Example 2, Figures 6.3 and 6.4, are examples of positive or negative association. Justify your answers.

The *strength* of a relationship describes how closely the points in a scatterplot follow a simple form such as a straight line. Figure 6.2 shows only a small amount of scatter about the straight line, so the relationship is fairly strong. (We will soon learn, in Section 6.3, a numerical measure of the strength of a straight-line relationship.)

EXAMPLE 3 SAT Mathematics Scores by State

Each year, more than 1 million high school seniors take the SAT standardized test, which has three parts: Mathematics, Critical Reading, and Writing. We sometimes see individual states rated or ranked by the average SAT scores of their high school seniors. However, this is misleading because the mean SAT score is explained largely by what percentage of a state's graduating seniors take the SAT. For example, the scatterplot in Figure 6.5 shows a negative association between the mean score on the Mathematics section and the percentage of test-takers for the class of 2013. Each dot represents a particular state.

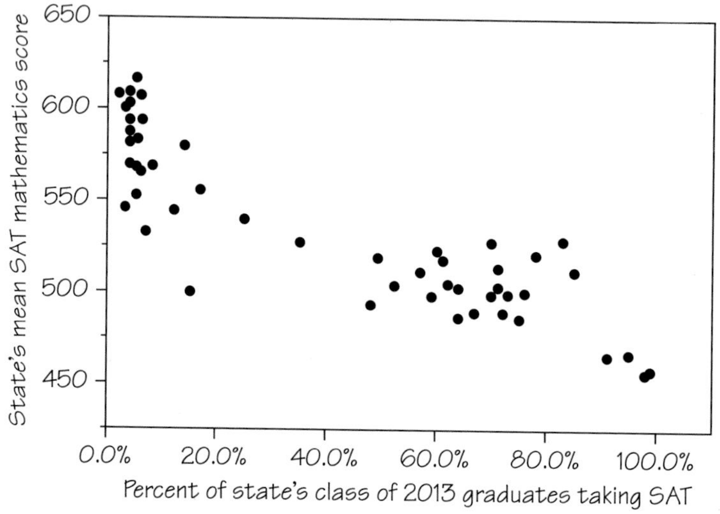

Figure 6.5 A scatterplot of states' mean SAT Mathematics scores (the response variable) against the percentage of states' class of 2013 high school graduates who take the SAT (the explanatory variable).

The *form* of Figure 6.5 is a bit irregular, but there are two distinct clusters of states. In each state in the lower-right cluster, a majority or near-majority of high school graduating seniors take the SAT, and the mean scores are low. In the upper-left cluster's states, 35% or fewer of seniors take the SAT—and these states have higher mean scores. Clusters in a graph suggest that the data describe several distinct kinds of individuals, and the two clusters in Figure 6.5 indeed describe two distinct sets of states.

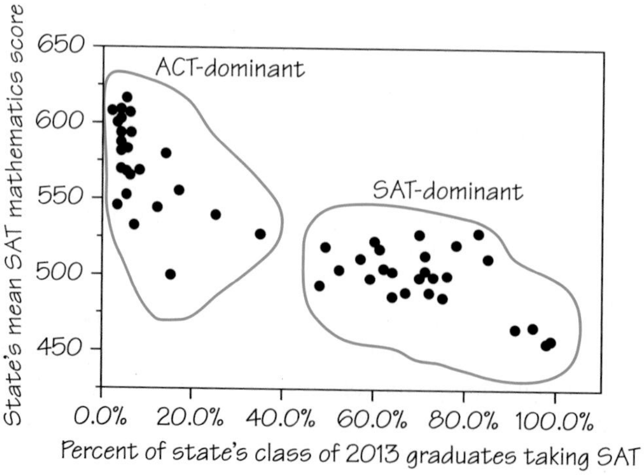

There are two common college entrance examinations, the SAT and the ACT, and each state tends to prefer one or the other. In ACT-dominant states (the left cluster in Figure 6.5, where a smaller fraction of those states' seniors take the SAT), most students who do take the SAT are applying to selective, out-of-state colleges. This select group performs well. In SAT-dominant states (the right cluster), a higher percentage of seniors take the SAT, and this broader group has a lower mean score.

The relationship in Figure 6.5 also has a clear *direction:* States in which a higher percentage of students take the SAT tend to have lower mean scores. This is true both between the clusters and within each cluster. That is, there is a *negative association* between the two variables.

There are no clear *outliers* in Figure 6.5, but each cluster does include at least one state whose mean SAT Mathematics score is lower than we would expect from the percentage of its students who take the SAT. In the cluster of ACT-dominant states, this occurs with West Virginia (WV). In the cluster of SAT-dominant states, this occurs with the District of Columbia (DC)—which is actually a federal district, not a state—and Maine (ME), Delaware (DE), and Iowa (IA).

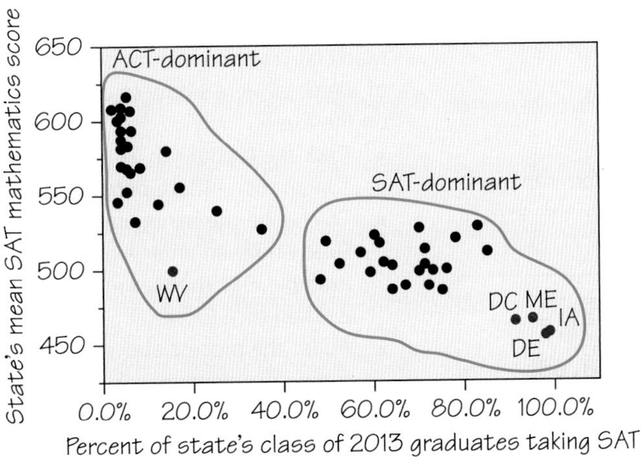

Although scatterplots can be very informative, they provide reliable information only when properly drawn to scale, which can be difficult and/or tedious to do by hand. Spotlight 6.1 provides instructions for creating a scatterplot using a graphing calculator (TI/83/84) or spreadsheet software (Excel).

Creating Scatterplots Using Technology

SPOTLIGHT 6.1

In this spotlight, we provide instructions for making a scatterplot of the IQ-Reading score data in Table 6.4 (page 245) using either a TI-83/84 graphing calculator or Excel.

Using a TI-84

Step 1. Preparation

- Turn off all STAT PLOTS: [2nd] [Y=] (for STAT PLOT) [4] [ENTER].
- Press [Y=] and erase (or turn off) any functions stored in the Y = menu.
- Press [STAT] [1]. Clear any data from lists L1 and L2.

Step 2. Entering the data and making the scatterplot:

- You should still be in the lists from Step 1. Enter the data on the explanatory variable, IQ test score, in list L1, and the response variable, Reading test score, in list L2 (be sure not to interchange these lists).

Creating Scatterplots Using Technology *(continued)*

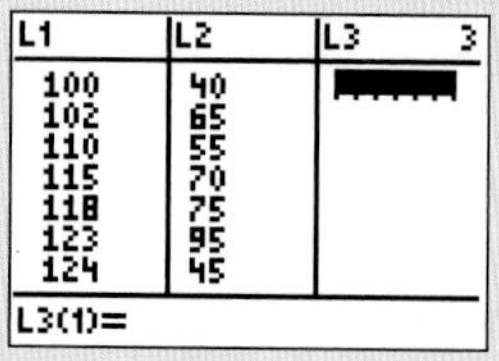

- Press [2nd] [Y=] (for STAT PLOT) [1] [ENTER] to turn on STATPLOT 1.
- For TYPE select the first scatterplot; L1 and L2 should be entered as the Xlist and Ylist, respectively; choose the mark to be used for the dots in the scatterplot.

- Press [ZOOM] [9] (for ZoomStat). You can press [WINDOW] and adjust the scaling on the axes if desired.

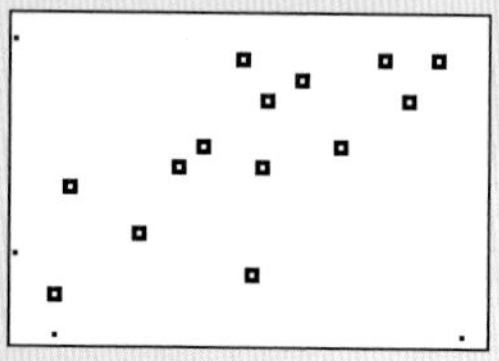

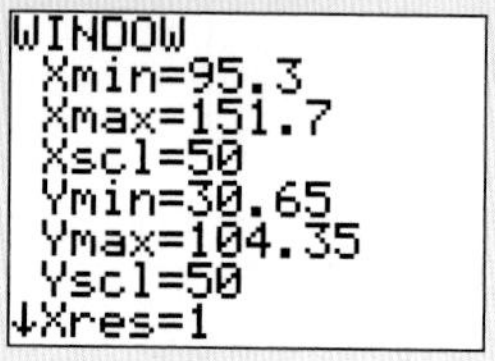

Step 3. Turn off all STAT PLOTS, as shown in Step 1.

Using Excel

Again, we use the data from Table 6.4.

- Enter the IQ test scores into column A and the Reading test scores into column B. (Don't interchange the order of the columns.)
- Highlight your data by clicking at the top of column A, dragging over to the top of column B, and then down to your last data entry in column B.
- Click the Insert tab and then the Scatter tab. Select the first scatterplot (Scatter with only Markers). You should now see a scatterplot of the data. Here's how to adjust the scaling on the axes:
  - Click the Layout tab → Axes → Primary horizontal axis → More primary horizontal axis.
  - To change the minimum, click the circle for Fixed and then enter a new minimum, for example, 90. The maximum should be fine as it is. Then click Close.
  - Adapt the process above for the vertical axis, setting its minimum value to 30.
  - If desired, click Axis Titles and add labels to the axes.

	A	B
1	IQ test score	Reading test score
2	100	40
3	102	65
4	110	55
5	115	70
6	118	75
7	123	95
8	124	45
9	125	70
10	126	85
11	130	90
12	135	75
13	140	95
14	143	85
15	147	95
16		

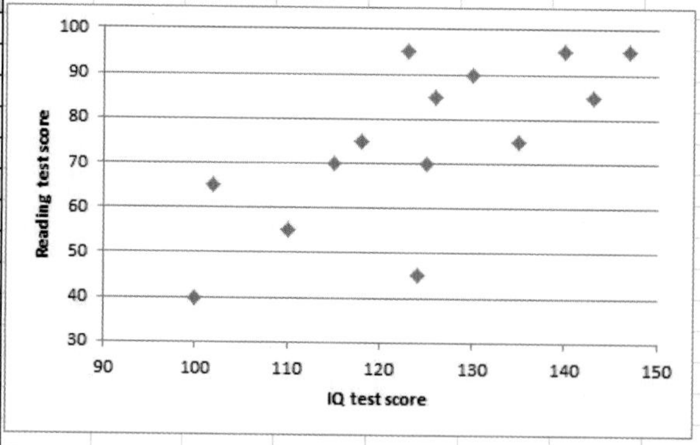

6.2 Making Predictions: Regression Line

If the overall pattern in a scatterplot is a straight-line relationship, it is useful to summarize this pattern by drawing a line on the scatterplot. A **regression line** summarizes the relationship between two variables, but only in a specific setting: when one variable helps explain or predict the other. That is, regression describes a relationship between an explanatory variable and a response variable.

Regression Line

DEFINITION

A **regression line** is a straight line that describes how a response variable y changes as an explanatory variable x changes. A regression line is often used to *predict* the value of y for a given value of x.

EXAMPLE 4 Predicting Blood Alcohol Content (BAC)

The scatterplot in Figure 6.2 shows a straight-line relationship between how many beers a student drinks and his or her BAC 30 minutes later. Figure 6.6 repeats this scatterplot and adds a regression line that summarizes the pattern of the data. We can use this line to predict BAC for a student based on the number of beers consumed.

Figure 6.6 shows the prediction in graphical form for a student who drinks 6 beers. Start at $x = 6$, go up to the line, and then head left to the y-axis. We hit the y-axis at BAC $= 0.095$. This is the BAC that corresponds to 6 beers, according to the regression line. (Recall that the legal limit for driving is 0.08.) The line represents only the overall pattern of the data, so the BAC of a randomly chosen student after 6 beers will probably not be exactly 0.095. But because the points for the 16 students in the Ohio State study are not far from the line, we expect the prediction to be reasonably accurate.

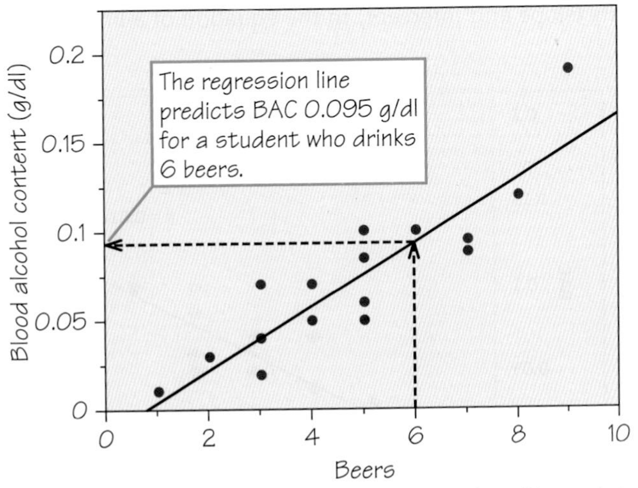

Figure 6.6 A regression line for predicting BAC from the number of beers that a student drinks.

However, for a more precise prediction, it is easier to use the *equation of the line* than to estimate the prediction from the graph. With the application of formulas that will be given in Section 6.4, the equation of the line in Figure 6.6 is

$$\text{predicted BAC} = -0.0127 + 0.01796 \times \text{beers}$$

For a student who drinks 6 beers, we have

$$\text{predicted BAC} = -0.0127 + (0.01796 \times 6) = 0.095$$

Because two points determine a unique line, you could plot a line by using its equation to determine any two particular points that lie on that line, plot those points, and then draw the line through them. For example, from the equation

$$\text{predicted BAC} = -0.0127 + 0.01796 \times \text{beers}$$

we just determined that one point is (6, 0.095). By plugging in $x = 2$, we could obtain another point. Drawing the line through those two points yields the line in Figure 6.6.

Self Check 3

The equation of the line in Figure 6.6 is

$$\text{predicted BAC} = -0.0127 + 0.01796 \times \text{beers}$$

Determine a second point on the line by plugging in $x = 2$ for the number of beers. What is that point?

Statistical software and spreadsheets as well as many calculators will give you the equation of a regression line from data that you enter. For example, Figure 6.7 shows the results for calculating the equation of the line in Figure 6.6 using a TI-84 graphing calculator and Excel. (Instructions for calculating the equation can be found in Spotlight 6.5 on page 265.)

You should know how to use a regression line even if you don't look into the details needed to calculate the line from data. First, recall some basic facts about the coefficients (slope and intercept) in the equation of a line.

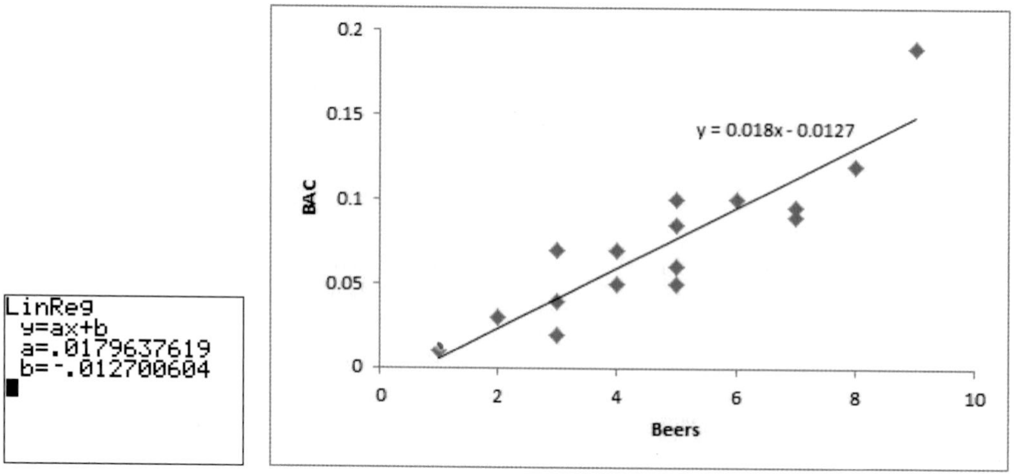

Figure 6.7 Calculation of equation of the regression line shown in Figure 6.6.

Equation of a Regression Line

DEFINITION

Suppose that y is a response variable (plotted on the vertical axis) and x is an explanatory variable (plotted on the horizontal axis). If we call $\hat{y}$ the predicted value of y, then the resulting regression line for predicting y from x has an equation of the form[1]

$$\hat{y} = mx + b$$

In this equation, m is the **slope,** which is the amount by which y changes when x increases by 1 unit. The number b is the **y-intercept,** which is the value of y when $x = 0$ (i.e., when the line intercepts the y-axis).

Algebra Review Appendix

Slope of a Line

Graphing a Line in Slope-Intercept Form

EXAMPLE 5 Interpreting the Slope of a Regression Line

The slope of the line in Example 4 is $m = 0.01796$. This says that as we move to the right along the line, predicted BAC goes up by 0.01796 for each additional beer that a student drinks. So, if a student has 3 additional beers, the BAC would increase by $3 \times 0.01796 = 0.05388$ g/dl. The slope tells us how quickly y changes as we change x, which is important for understanding the pattern in the data. The slope is positive ($m > 0$) when there is a positive association between the variables, as there is between BAC and beers consumed. It is negative when there is a negative association.

You might think that because the slope $m = 0.01796$ is small that x, beers consumed, has little influence on y, BAC. Unfortunately, the size of a slope is affected by the units in which we measure the two variables. In Table 6.1, BAC is measured in grams of alcohol per deciliter (g/dl) of blood. That is, when the number of beers consumed increases by 1, the alcohol in a deciliter of blood increases by 0.01796 grams. There are 1000 milligrams in a gram. So, if we changed the BAC units to milligrams of alcohol per deciliter (mg/dl) of blood, the slope would be 1000 times as large: $m = 17.96$. *You can't say how important a relationship is just by looking at how big the slope is.* ■

Self Check 4

A study was conducted to see whether women's total cholesterol [in milligrams per deciliter (mg/dL)] is related to age. A scatterplot of the data indicated a straight-line relationship, described by the following equation:

$$\text{predicted total cholesterol} = 140.7 + 1.64 \times \text{age}$$

Interpret the slope of the regression line in the context of this study. ■

Although it is generally useful to interpret the slope of a regression line, that is not the case for the y-intercept. Mathematically, the y-intercept is the predicted value of y when $x = 0$. The y-intercept of the regression line in Example 4 is $b = -0.0127$. Although we need the value of the intercept to draw the line, it is statistically meaningful only when x can actually take values close to zero.

[1] The letters m and b are from the slope-intercept form from algebra class, but be aware that some technologies such as the TI-84 graphing calculator or statistics books use different letters, such as $y = b + ax$ or $y = a + bx$, respectively. Always play it safe by checking that the letter used for the slope corresponds to the number multiplied by the explanatory variable.

Even then, you should think of the intercept as describing the line rather than taking it seriously as a prediction. In this situation, if a student drinks no beers, $x = 0$, then the predicted value of his or her BAC is $y = -0.0127$ g/dl, which is clearly impossible (but is at least close to 0, the value we would expect for a student's BAC if he or she consumed 0 beers).

6.3 Correlation

A scatterplot displays the form, direction, and strength of the relationship between two quantitative variables. Straight-line relations are particularly important because a straight-line pattern is quite common and is easy to interpret. We say a straight-line association is strong if the points lie close to a line and weak if they are widely scattered about a line.

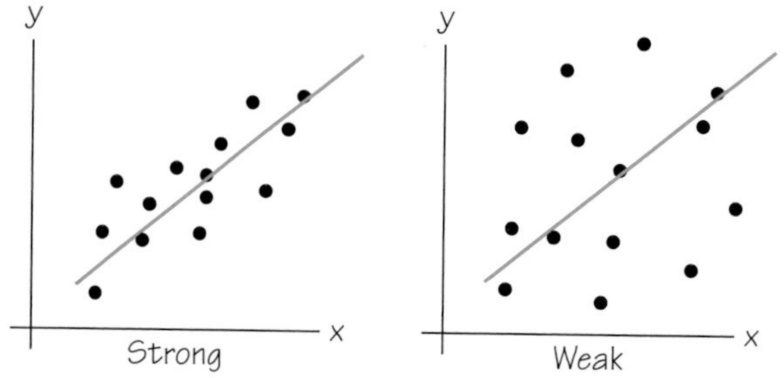

Figure 6.8 Examples of strong and weak straight-line associations.

But this language is vague and our eyes can be fooled by the choice of scaling used in the scatterplot. We need to follow our strategy for data analysis by using a numerical measure along with the graph. Correlation is the measure we use. **Correlation** is usually denoted as r, thanks to 19th-century statistician Sir Francis Galton, who was studying related ideas of ***r***egression and ***r***eversion.

Correlation DEFINITION

The **correlation** measures the direction and strength of the straight-line relationship between two quantitative variables.

The correlation r is always a number between -1 and 1, inclusive. It has the same sign as the slope of a regression line for that dataset: $r > 0$ for positive association and $r < 0$ for negative association.

Perfect correlation $r = 1$ or $r = -1$ occurs only when all points lie exactly on a straight line. As you will observe in Example 6, the correlation moves away from 1 or -1 as the straight-line relationship gets weaker.

EXAMPLE 6 Scatterplots and Correlation

The scatterplots in Figure 6.9 all involve the same scale for the horizontal and vertical axes and the same standard deviation value for the x- and y-variables. From these scatterplots, we are able to see how values of r closer to 1 or -1 correspond to stronger straight-line relationships.

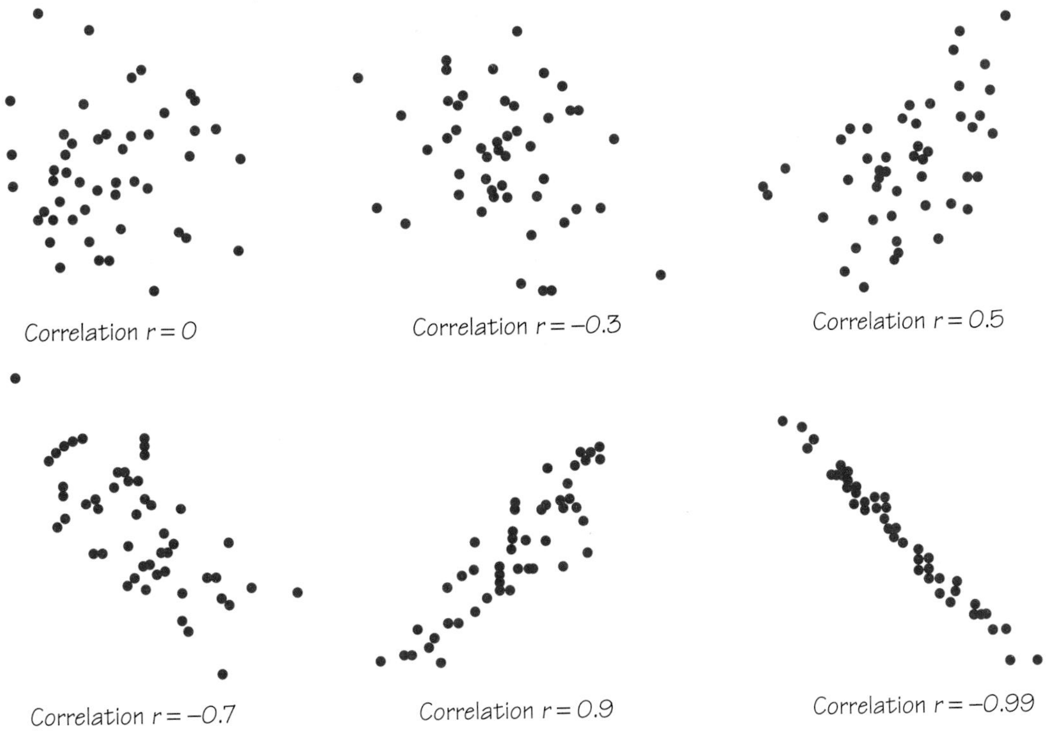

Figure 6.9 How the correlation r measures the direction and strength of straight-line association.

Figure 6.10 provides some guidelines for using correlation to classify both the direction and strength of a straight-line relationship. Using these guidelines, the relationship in Figure 6.9 corresponding to $r = -0.7$ would be classified as a moderate negative straight-line relationship.

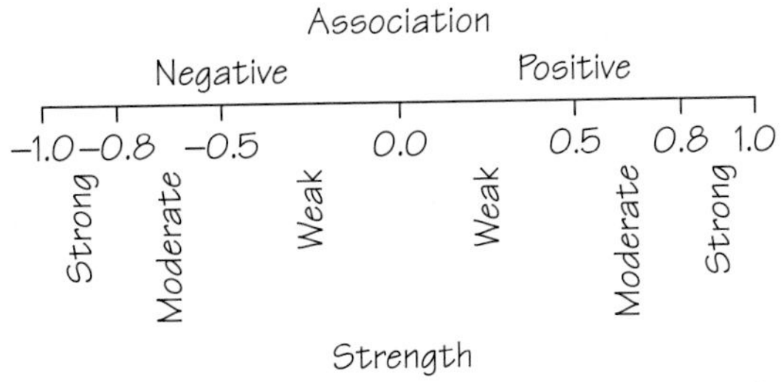

Figure 6.10 Using r to classify the direction and strength of a straight-line relationship.

Earlier, we said that Figure 6.2 (page 244) shows a fairly strong positive straight-line relationship between how many beers a student drinks and his or her BAC. The correlation between these variables is $r = 0.894$, which puts this relationship into the strong category. Figure 6.5 (page 248), despite the clusters, also shows a strong straight-line relationship between the percentage of a state's high school seniors who take the SAT exam and their mean SAT score. Here, the association is negative: Higher percentages taking the SAT go with lower mean scores. The correlation is $r = -0.877$.

Before giving a formula for computing the correlation r, here are more facts about correlation:

1. Correlation makes no sense for nonnumerical variables (such as ethnicity and occupation).

2. The correlation r measures the strength of only a straight-line relationship. It does *not* measure the strength of a relationship with a curved pattern, no matter how strong that curved relationship is. For example, the data in Figure 6.11 show a strong pattern indicating a relationship between the two variables. However, that relationship is not linear and $r \approx 0$.

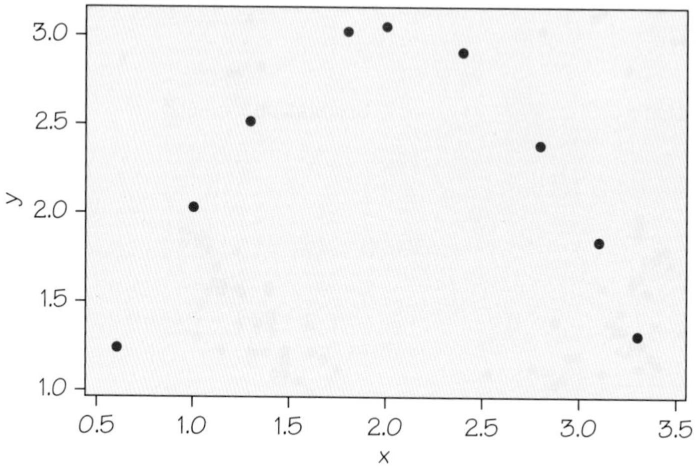

Figure 6.11 The two variables have a strong relationship, but $r \approx 0$.

3. Unlike regression, correlation makes no distinction between explanatory and response variables. It makes no difference which variable you call x and which you call y in interpreting (or calculating) a correlation.
4. The correlation r does not change when we change the units of measurement of x, y, or both. For example, measuring height in inches rather than centimeters, and weight in pounds rather than kilograms, does not change the correlation between height and weight. (This is different from the case of the slope of a regression line, which changes when the units of x and y change.) The correlation r itself has no unit of measurement; it is just a number.
5. Like the mean and standard deviation, the correlation is affected strongly by a few outlying observations. Use r with caution when outliers appear in the scatterplot, especially for small datasets. Figures 6.12a and 6.12b display two

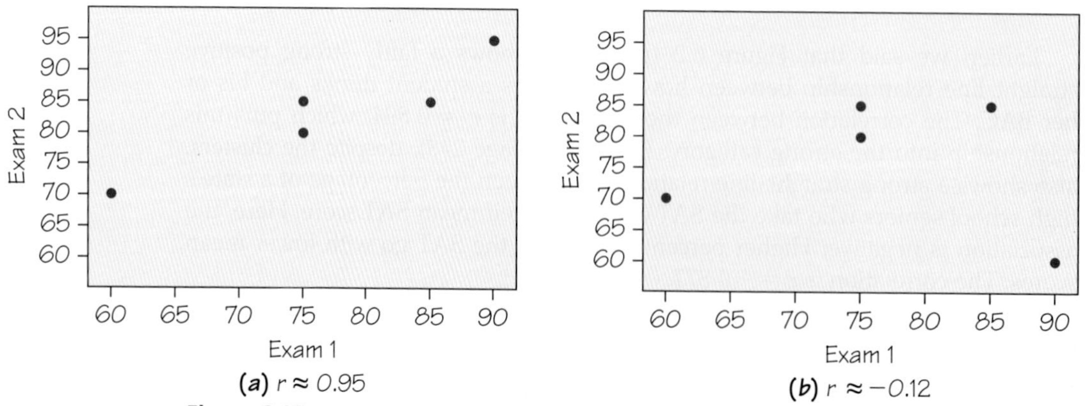

Figure 6.12 Two scatterplots that differ by only one point.

scatterplots showing the relationship between scores on two exams. The data points are identical except for one point, which happens to be an outlier in part (b). Changing that one point switches the correlation from positive to negative.

Now get ready to compute some correlations. In practice, you will use a calculator or software to find the correlation from keyed-in data, as described in Spotlight 6.2 on page 259. That's fortunate, because using the procedure box formula for correlation is quite a bit of work!

Formula for Correlation PROCEDURE

Suppose that we have data on variables x and y for n individuals. The means and standard deviations of the two variables are $\bar{x}$ and s_x for the x-values, and $\bar{y}$ and s_y for the y-values. The correlation r between x and y is

$$r = \frac{1}{n-1}\left[\left(\frac{x_1 - \bar{x}}{s_x}\right)\left(\frac{y_1 - \bar{y}}{s_y}\right) + \left(\frac{x_2 - \bar{x}}{s_x}\right)\left(\frac{y_2 - \bar{y}}{s_y}\right) + \cdots + \left(\frac{x_n - \bar{x}}{s_x}\right)\left(\frac{y_n - \bar{y}}{s_y}\right)\right]$$

Step 1. Find the mean $\bar{x}$ and standard deviation s_x of the x-values. Find the mean $\bar{y}$ and standard deviation s_y of the y-values. (Use your calculator and refer to Spotlight 5.3 on page 206.)

Step 2. Find the standardized value $(x - \bar{x})/s_x$ for each of the x-values.

Find the standardized value $(y - \bar{y})/s_y$ for each of the y-values.

Step 3. Insert your numbers from Step 2 into the formula for r. Remember that n is the number of (x, y) points or ordered pairs plotted in the scatterplot.

Algebra Review Appendix
Using Formulas

EXAMPLE 7 Calculating Correlation

Table 6.5 gives the SAT Math and Critical Reading scores for five randomly chosen students entering a college.

TABLE 6.5 SAT Math and Critical Reading Scores

Math SAT, x	Critical Reading SAT, y
610	550
440	410
550	520
520	540
420	410

Next, we walk through the steps of calculating the correlation between the SAT Math and Critical Reading scores.

Step 1. Calculate the mean and standard deviations:

Math: $\bar{x} = 508$ and $s_x = 78.5$
Critical Reading: $\bar{y} = 486$ and $s_y = 70.2$

Step 2. Calculate the standardized scores for each observation by subtracting the mean and dividing by the standard deviation:

$\left(\frac{x - \bar{x}}{s_x}\right)$	$\left(\frac{y - \bar{y}}{s_y}\right)$
$\left(\frac{610 - 508}{78.5}\right) \approx 1.299$	$\left(\frac{550 - 486}{70.2}\right) \approx 0.912$
$\left(\frac{440 - 508}{78.5}\right) \approx -0.866$	$\left(\frac{410 - 486}{70.2}\right) \approx -1.083$
$\left(\frac{550 - 508}{78.5}\right) \approx 0.535$	$\left(\frac{520 - 486}{70.2}\right) \approx 0.484$
$\left(\frac{520 - 508}{78.5}\right) \approx 0.153$	$\left(\frac{540 - 486}{70.2}\right) \approx 0.769$
$\left(\frac{420 - 508}{78.5}\right) \approx -1.121$	$\left(\frac{410 - 486}{70.2}\right) \approx -1.083$

Step 3. Substitute into the formula:

$$r = \frac{1}{5-1}[(1.299)(0.912)+(-0.866)(-1.083)+(0.535)(0.484)+(0.153)(0.769)$$
$$+(-1.121)(-1.083)]$$
$$r \approx 0.928$$

Self Check 5

Suppose the fifth student retook the SATs and earned 530 and 510 on his Math SAT and Critical Reading SAT, respectively. Replace the last row of Table 6.5 with the new results and recalculate the correlation.

The formula for correlation starts by *standardizing* each observation value (as was done in Chapter 5, Exercises 62 and 63, on page 237). That is, subtract the mean for that variable from the observation and then divide by the standard deviation. Standardizing turns each original data value into "number of standard deviations from the mean." This removes the original units and explains why r has no units and doesn't change when we change the units of x or y. The formula says that the correlation is an average of the products of the standardized x- and y-values for n individuals. Although the procedure-box formula for correlation has conceptual clarity, it can be tedious to apply even for relatively small datasets. Spotlight 6.2 shows how technology can be used to speed up the calculation of correlation.

Spotlight 6.3 on page 260 shows the connection between correlation and fitting a line to data.

Correlation Calculation

SPOTLIGHT 6.2

We return to the problem of calculating the correlation between Math and Critical Reading SATs (see Example 7, Table 6.5, on page 257). With a basic calculator, you'll get the same answer faster using this mathematically equivalent but more computationally efficient formula:

$$r = \frac{(x_1 y_1 + x_2 y_2 + \cdots + x_n y_n) - n\bar{x}\bar{y}}{(n-1)s_x s_y}$$

Here are the computations, using the means and standard deviations calculated in Example 7:

$$r = \frac{(610)(550) + (440)(410) + (550)(520) + (520)(540) + (420)(410) - 5(508)(486)}{(5-1)(78.5)(70.2)}$$

$$= \frac{20{,}460}{22{,}042.8} \approx 0.928$$

However, you probably have access to technology that is a bit more sophisticated than a basic-level calculator. The remainder of this spotlight is devoted to using a variety of technologies to calculate correlation with a single command.

If you have a *scientific calculator,* select the calculator mode to be able to do two-variable regression statistics, clear out any old data, and then enter your new *x*- and *y*-values. Once the data are entered, you can find the correlation (or regression-line slope and *y*-intercept, for that matter) by hitting the appropriate key(s). On the Internet, you can find websites that will help with keystrokes for specific models.

TI Graphing Calculators

If you have a *graphing calculator* in the TI-83/84+ family, we illustrate one procedure for calculating correlation using the data from Table 6.5, Example 7.

Step 1. Press STAT ENTER to bring up the calculator lists. Then enter the Math SATs into list L1 and the Critical Reasoning SATs into list L2.

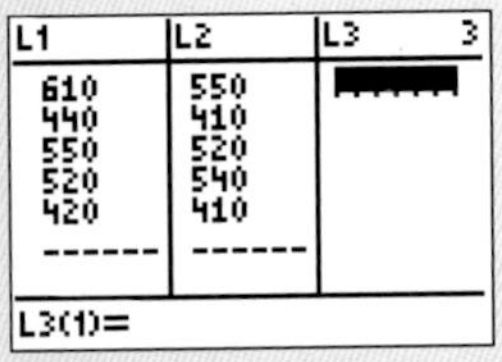

L1
L2
L3 3
610
440
550
520
420
550
410
520
540
410
L3(1)=

Step 2. Press the following sequence of keys:

- STAT → TESTS → LinRegTTest and press ENTER.
- Enter the lists where you stored the data from your independent (*x*-list) and dependent (*y*-list) variables. (To enter the lists L1 and L2, press 2ND followed by the number keys 1 and 2, respectively.)

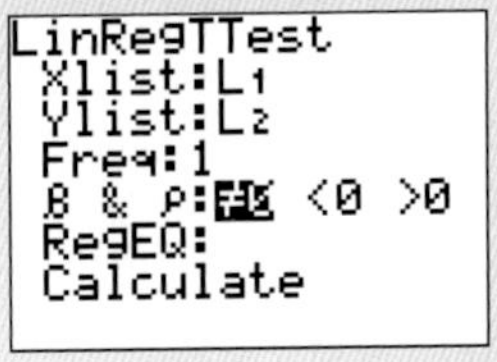

- Use the down arrow key to select Calculate and press ENTER.
- Scroll to the end of the output to see the correlation *r*.

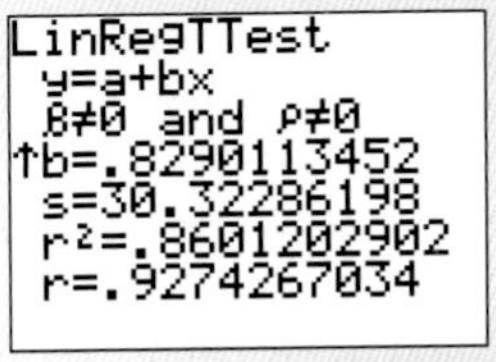

Keystrokes for other specific TI calculators can be found online in the Texas Instruments guidebooks, which are downloadable from education.ti.com/calculators/downloads/US/#Guidebooks.

Excel

Next, we show how to calculate the correlation of the data in Table 6.5 using Excel. Begin by entering the Math and Critical Reasoning SAT scores into columns A and B, respectively, of an Excel spreadsheet. Here's the procedure:

- In an empty cell, enter = CORREL(.
- Click on the first data value for your *x*-values (in this case, 610) and drag down to the last data value (420). Press the comma-key. Then click on the first data value for your *y*-values (550) and drag down to the last data value (410). Finish the command with) and press Enter.

A7 fx =CORREL(A2:A6,B2:B6)

	A	B
1	Math SAT	Critical Reading SAT
2	610	550
3	440	410
4	550	520
5	520	540
6	420	410
7	0.927426703	

Regression Toward the Mean

SPOTLIGHT 6.3

Sir Francis Galton (1822–1911) studied predicting the heights of men from the heights of their fathers. He found that tall fathers tend to have taller than average sons, but that there is a reversion or regression (i.e., going back) of a son's height toward the average height for sons. Assume the mean height of men is about 70 inches. We use the following regression equation to demonstrate this concept:

son's height = 0.516 × (father's height) + 33.73

Suppose a father is above average in height, say, 75 inches tall. Then using the regression equation, we predict the height of his son to be

son's height = 0.516 × (75) + 33.73 = 72.43

Our prediction for the son's height, 72.43 inches, is shorter than the father's height but still an above-average height.

It goes the other way, too—a very short father tends to have a son who is also shorter than average, but not quite as short as his father.

More generally, the idea that extreme measurements including some random variation are likely to be followed by measurements that are not quite as extreme is called "regression toward the mean." This dynamic shows up in many areas involving some combination of skill and luck. Students who are the very top performers on one test will tend to do a bit worse, on average, on the next test, and the very worst performers will tend to experience some improvement. An example of this idea in the world of sports is the "*Sports Illustrated* cover jinx," in which athletes appear on the magazine's cover after an "outlier performance," but their subsequent performance is usually less impressive. Singer-songwriter Christine Lavin offers another example in this couplet from her song "Attractive Stupid People":

Popperfoto/Getty Images

> but the problem is the kids won't look as good as mom or dad, and they're always slightly smarter, which drives their pretty parents mad.

The following formula is an algebraic representation of regression to the mean because so long as the correlation r is not equal to 1 or −1, the predicted standardized value of y is closer to its mean than the standardized value of x is to its mean. In other words, the $\hat{y}$-value (e.g., the son's predicted height) has a less extreme deviation than the x-value (e.g., the height of that son's father).

$$\frac{\hat{y}-\bar{y}}{s_y} = r\frac{x-\bar{x}}{s_x}$$

Self Check 6

Suppose that a father is of below-average height, say, 64 inches.

(a) Use the regression equation from Spotlight 6.3 to predict the height of his son.
(b) Explain how this result fits in with the topic of regressing toward the mean. ■

6.4 Least-Squares Regression

In Example 4 (page 251), we used the straight line given by the equation

$$\text{predicted BAC} = -0.0127 + 0.01796 \times \text{beers}$$

to predict BAC from the number of beers consumed. How did we get this particular equation? What makes it the equation of the "best-fitting" line for predicting BAC from beers consumed? Before we can answer the first question, we will first have to decide what we mean by the "best line."

For a given scatterplot, different people might draw different summarizing lines by eye. For example, Figure 6.13 shows a scatterplot of the BAC–beer

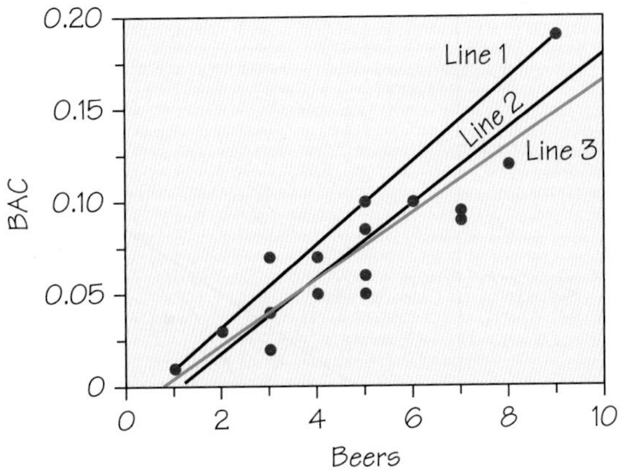

Figure 6.13 Three lines summarizing the pattern of the BAC–beer data

data along with three lines. Line 3 is the graph of the equation in the previous paragraph.

Line 1 was drawn through the points corresponding to the fewest and most beers consumed, (1, 0.01) and (9, 0.19). Line 2 was drawn through the points (3, 0.04) and (6, 0.10). So how do you determine which is the "best" line? If the criterion for the "best" line is the line passing through the most data points, Line 1 (which passes through three data points) would be the best, and Line 3 (which passes through one data point) the worst. However, Line 1 does a poor job of summarizing the overall pattern of these data. Only one data point lies above Line 1 and the rest lie on or below it. So picking a line that passes through the most data points is clearly a poor criterion for selecting a "best-fitting" line.

In looking for a new criterion for a "best" line, keep in mind that we will use this line to predict y from x. Thus, we want a line that is as close as possible to the points in the *vertical* dimension. That's because the prediction errors that we make are errors in the y-variable, which is the vertical dimension in the scatterplot.

Table 6.1 (page 243) shows that Student 12 drank 6 beers and was observed to have a BAC of 0.10. However, the regression-line equation in Example 4 showed that the predicted BAC for a student who drinks 6 beers is 0.095. These values are close, but not the same. Indeed, from Figure 6.6, we can see that the observed data point (6, 0.10) lies a bit above the line. The vertical deviation of this gap is the **residual error** (or simply **residual**), which is calculated as follows:

$$\text{residual} = \text{observed BAC} - \text{predicted BAC} = 0.10 - 0.095 = 0.005$$

Residual DEFINITION

A residual (or residual error) is the difference between an observed value of the response variable and the value predicted by the regression line. In other words, a residual is the prediction error that remains after we have chosen the regression line:

$$\text{residual} = \text{observed } y - \text{predicted } y = y - \hat{y}$$

When the observed response lies above the line (e.g., the data point for Student 1, who had 5 beers and a BAC of 0.10, lies above Line 3 in Figure 6.13), the residual is positive. And when the response lies below the line (e.g., the data point for Student 14, who had 7 beers and a BAC of 0.09, lies below Line 3), the residual is negative. Figure 6.14 shows a graphical representation of these two residuals, the vertical gaps between these data points and the line.

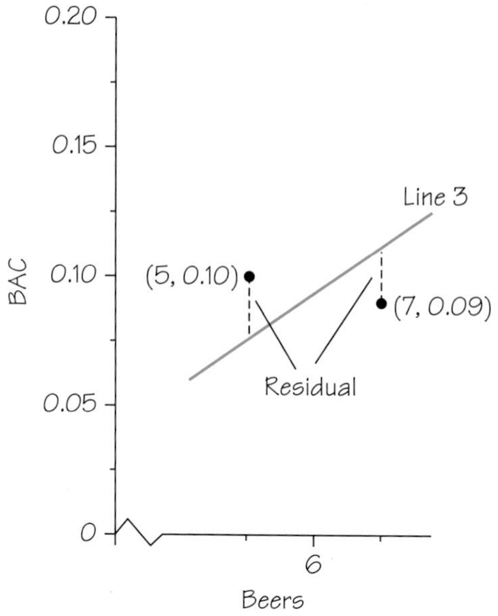

Figure 6.14 Graphical representation of two residuals.

Self Check 7

Using the regression-line equation from Example 4 (given at the start of this section), calculate the predicted BAC corresponding to 5 beers and 7 beers. Then calculate the residuals for the data values from Students 1 and 14 (these are the residuals represented graphically in Figure 6.14). Verify that one of these residuals is positive and the other negative.

Now, we return to the problem of picking a "best" line. The most common way to make the collection of residual errors for the entire dataset as small as possible is *least-squares regression.* Line 3 in Figure 6.13 is the **least-squares regression line.**

Least-Squares Regression Line DEFINITION

The **least-squares regression line** is the line that makes the sum of the squares of the residual errors, the vertical distances of the data points from the line, the least value possible.

So we now have a criterion for finding the "best-fitting" line—find the line that results in the smallest sum of the squares of the residual errors. Lines 2 and 3 in Figure 6.13 both appear to be good choices for this line. With a modest amount of algebra, we determined an equation for Line 2:

$$\text{predicted BAC} = -0.02 + 0.02 \times \text{beers}$$

Using Excel, we then calculated the predicted values, residuals, and squares of the residuals, which are shown in Table 6.6. At the bottom of the last column of Table 6.6 is the sum of the squares of the residuals, which for Line 2 is 0.00625—a fairly small value. We then did the same calculations for Line 3: predicted BAC $= -0.0127 + 0.01796 \times$ beers, and found that the sum of the squares of the residuals was even smaller, 0.00585. So, based on the least-squares criterion, Line 3 is better than Line 2.

TABLE 6.6 Using Excel to Calculate the Sum of the Residual Errors for Line 2

Beers	BAC	Predicted BAC (Line 2)	Residuals	$(\text{Residuals})^2$
5	0.1	0.08	0.02	0.000400
2	0.03	0.02	0.01	0.000100
9	0.19	0.16	0.03	0.000900
8	0.12	0.14	−0.02	0.000400
3	0.04	0.04	0	0.000000
7	0.095	0.12	−0.025	0.000625
3	0.07	0.04	0.03	0.000900
5	0.06	0.08	−0.02	0.000400
3	0.02	0.04	−0.02	0.000400
5	0.05	0.08	−0.03	0.000900
4	0.07	0.06	0.01	0.000100
6	0.1	0.1	0	0.000000
5	0.085	0.08	0.005	0.000025
7	0.09	0.12	−0.03	0.000900
1	0.01	0	0.01	0.000100
4	0.05	0.06	−0.01	0.000100
				0.006250

We can use the least-squares criterion for deciding which of two lines fits the data better. However, we still need a solution to the following mathematical problem: Starting with n observations on variables x and y, find the line that makes the sum of the squares of the vertical errors (the residuals) as small as possible. Here is the solution to this problem.

Finding the Least-Squares Regression Line PROCEDURE

1. From our data on an explanatory variable x and a response variable y for n individuals, calculate the means $\bar{x}$ and $\bar{y}$ and the standard deviations s_x and s_y of the two variables (see Chapter 5, Spotlight 5.3, on page 206.).
2. Calculate the correlation r (recall Spotlight 6.2, page 259).
3. The regression line's slope m is given by
$$m = r\frac{s_y}{s_x}$$
4. The regression line's y-intercept b is given by
$$b = \bar{y} - m\bar{x}$$
5. If we call $\hat{y}$ the predicted value of y, then the equation of the least-squares regression line for predicting y from x (we also can say from "regressing y on x") can now be stated:
$$\hat{y} = mx + b$$

This equation gives insight into the behavior of least-squares regression by showing that it is related to the means and standard deviations of the x and y observations and to the correlation between x and y. For example, it is clear that the slope m always has the same sign as the correlation r.

In practice, you don't need to calculate the means, standard deviations, and correlation first. Statistical software, spreadsheet software, and many calculators can give the slope m, intercept b, and equation of the least-squares line from keyed-in values of the variables x and y (see Spotlight 6.5 on page 266).

EXAMPLE 8 Least-Squares Regression of BAC on Number of Beers

Go back to the BAC–beer data in Table 6.1 (page 243). Use your calculator or spreadsheet software to verify the following:

- The mean and standard deviation of x, number of beers consumed, are $\bar{x} = 4.8125$ and $s_x = 2.1975$.
- The mean and standard deviation of y, BAC, are $\bar{y} = 0.07375$ and $s_y = 0.04414$.
- The correlation between the number of beers and BAC is $r = 0.8943$.

The least-squares regression line of BAC (y) on number of beers (x) has slope

$$m = r\frac{s_y}{s_x} = 0.8943 \times \frac{0.04414}{2.1975}$$
$$= 0.01796$$

and y-intercept

$$b = \bar{y} - m\bar{x} = 0.07375 - (0.01796)(4.8125)$$
$$= -0.0127$$

The equation of the least-squares line is therefore

$$\hat{y} = -0.0127 + 0.01796x$$

just as we claimed earlier.

When doing calculations like this by hand, you should carry extra decimal places in the intermediate calculations to get accurate values of the slope and intercept and not round until your final answer. Using software or a calculator with a regression function eliminates this worry.

Self Check 8

Return to the data in Table 6.5 (page 257) from Example 7.

(a) Make a scatterplot of Critical Reading SAT, y, against Math SAT, x. Does the pattern of the dots appear roughly linear?

(b) As was done for Example 8, use the formulas from the procedure to determine the equation of the least-squares regression line. (The means, standard deviations, and correlation were previously calculated in Example 7 on page 257.)

You now see that correlation and least-squares regression are connected closely. The expression $m = rs_y/s_x$ for the slope says that along the regression line, a change of 1 standard deviation in x corresponds to a change of r standard deviations in y. When the variables are correlated perfectly ($r = 1$ or $r = -1$), the change in the predicted y is the same (in standard deviation units) as the change in x. Otherwise, because $-1 \leq r \leq 1$, the change in the predicted y is less than the change in x.

Now that we've covered both correlation and regression, check out Spotlight 6.4, which discusses how colleges might use regression and correlation in their admissions process.

Regression and Correlation in Action: College Success

SPOTLIGHT 6.4

Can college success be predicted? Colleges with more applicants than spaces want to admit students who are most likely to succeed. There are many ways that we might define what success in college means, but admissions officers often focus on grades during the first year of college. There are also many choices of what variable might help admissions officers predict first-year grades: high school grade point average (GPA), number of advanced (e.g., AP) classes taken, scores on standardized tests (ACT or SAT), and so on. No one of these variables (or even all of them together) will generate a perfect prediction for an individual person because many other variables (such as work ethic or motivation) are not directly taken into account.

According to a 2013 report by the College Board, first-year college GPA has a correlation (corrected for having a range restricted by analyzing only admitted and enrolled students) of about 0.5 with any one of the three SAT section tests. (The correlations between first-year GPA and the SAT Critical Reasoning, Mathematics, and Writing tests were $r = 0.50$, $r = 0.49$, and $r = 0.54$, respectively.) This is a measure of (predictive) validity for the SAT, and it turns out that squaring this number yields the interpretation that the SAT alone explains about one-quarter of the variation of first-year college GPA. The correlation between first-year GPA and high school GPA is only slightly higher at 0.55.

Multiple regression extends regression to allow more than one explanatory variable to help explain a response variable. When the SAT and high school GPA are used together to try to predict first-year college GPA, the adjusted correlation jumps up to 0.63. The response variable of the associated regression equation can yield an "index" that admissions officers use to create a rough ordering of applicants.

Although the procedure box for calculating the equation of the least-squares regression line provides valuable insight about the line, it is extremely time consuming to compute that equation without the use of technology. Spotlight 6.5 gives instructions for using technology to calculate the regression equation.

Calculating the Equation of the Least-Squares Regression Line

SPOTLIGHT 6.5

Finally, technology comes to the rescue!

If you have a *scientific calculator,* select the calculator mode to be able to do two-variable regression statistics, clear out any old data, and then enter your new x- and y-values. Once the data are entered, you can find the regression-line slope and y-intercept by hitting the appropriate key(s). On the Internet, you can find websites that will help with the keystrokes for specific calculator models.

Calculating the Equation of the Least-Squares Regression Line *(continued)*

TI-83/84+ Family and Other TI Graphing Calculators

You can use your graphing calculator to compute the coefficients of the least-squares regression line and the correlation, make a scatterplot of the data, and overlay a graph of the least-squares regression line. We use the data on midterm and final exam scores from Table 6.2 (page 244) as an example.

Step 1. Preparation

- Turn off all STAT PLOTS: [2nd] [Y=] (for STAT PLOT) [4] [ENTER].
- Press [Y=] and erase (or turn off) any functions stored in the Y = menu.
- Turn on Diagnostics: [2nd] [0] (for CATALOG). Press [x^{-1}] (for D) to scroll to the Ds. Then use the down arrow key to scroll to DiagnosticOn. Press [ENTER] twice. (Your calculator should respond "Done.")
- Press [STAT] [1]. Clear any data from lists L1 and L2.

Step 2. Determining the equation of the least-squares line and correlation

- You should still be in the lists from Step 1. Enter the data on the explanatory variable, midterm exam scores, in list L1, and the response variable, final exam scores, in list L2 (be sure not to interchange these lists).
- To calculate the coefficients of the least-squares regression line and the correlation:
  - Press [STAT] → CALC → LinReg(ax + b), and then [ENTER].

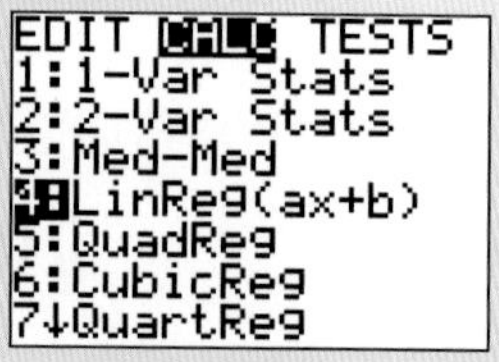

  - Complete the command as follows: [2nd] [1] (for L1) [,] [2nd] [2] (for L2) [,] VARS → Y-VARS → Function, and then press [ENTER].
  - Press [1] to store the equation as Y1 and then [ENTER].

- Read off the coefficients of the line $y = ax + b$. (Notice that the slope is designated as a and the y-intercept as b.) The correlation r appears at the bottom of the screen.

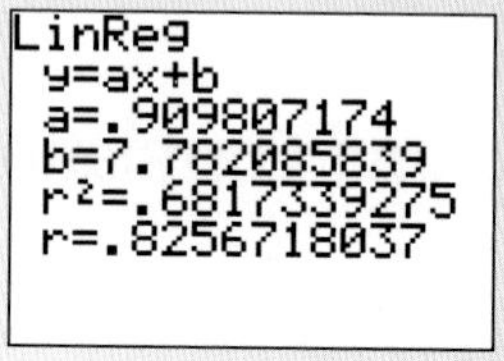

You can put the equation together by substituting the values for a and b to get $\hat{y} = 0.909807174x + 7.782085839$.

Step 3. Making a scatterplot and graphing the least-squares line

- Press [2nd] [Y=] (for STAT PLOT) [1] [ENTER] to turn on STATPLOT 1.
- For TYPE, select the first scatterplot; L1 and L2 should be entered as the Xlist and Ylist, respectively; choose the mark to be used for the dots in the scatterplot.

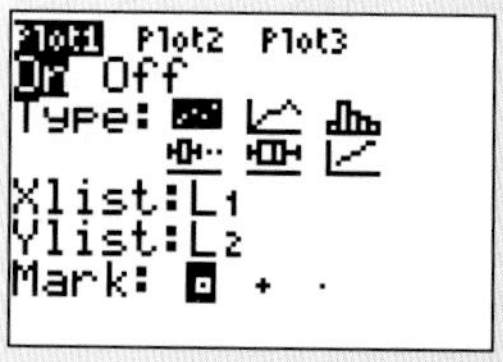

- Press [ZOOM] [9] (for ZoomStat).

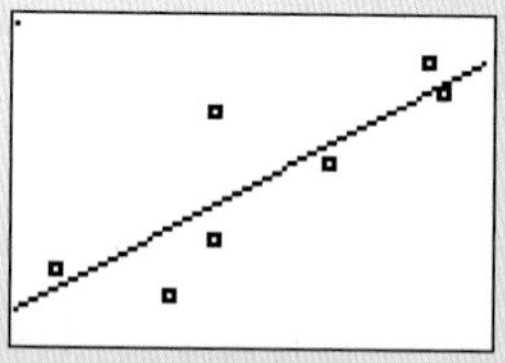

Step 4. After completing this problem, turn off all STAT PLOTS.

Keystrokes for other specific TI graphing calculators can be found online in the Texas Instruments guidebooks, which are downloadable from education.ti.com/calculators/downloads/US/#Guidebooks.

Calculating the Equation of the Least-Squares Regression Line

Excel

Again, we use the data from Table 6.2. The process will be to make a scatterplot of final exam scores against midterm exam scores and then to fit a line to the data.

- Enter the *x*-data, midterm exam scores, in column A and the *y*-data, final exam scores, in column B. (Don't interchange the order of the columns.)
- Highlight your data by clicking at the top of column A, dragging over to the top of column B, and then down to the bottom of column B.
- Click the Insert tab and then the Scatter tab. Select the first scatterplot (Scatter with only Markers). You should now see a scatterplot of the data. Adjust the scaling for the axes so that the minimum for both the *x*- and *y*-axes is 60 (see Spotlight 6.1 on page 249).
- Next, right click on a dot in the scatterplot and select Add Trendline.
- Notice that Linear is selected by default. At the bottom, click the box opposite "Display Equation on chart." Then click Close. A graph of the least-squares regression line along with its equation will be added to the scatterplot.

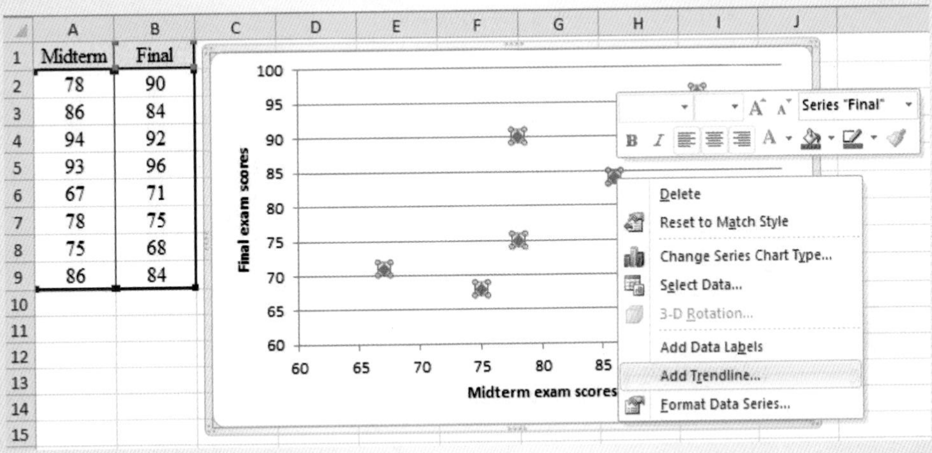

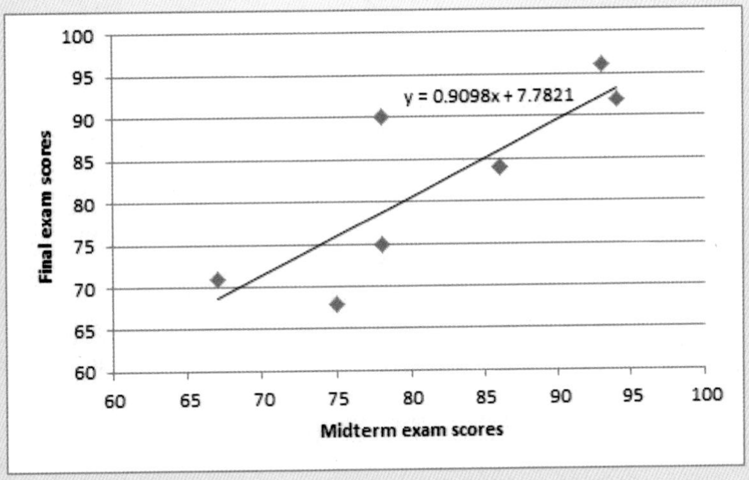

6.5 Interpreting Correlation and Regression

Correlation and regression are among the most used statistical methods for analyzing data on two or more quantitative variables. Here are a few cautions to keep in mind when you use or see these methods.

Both the correlation r and the least-squares regression line can be influenced strongly by a few outlying points. Always make a scatterplot before doing any calculations. Here is an artificial example that illustrates what can happen.

EXAMPLE 9 Beware the Outlier!

Figure 6.15 shows a scatterplot of data that have a strong positive straight-line relationship. In fact, the correlation is $r = 0.987$, close to the $r = 1$ of a perfect straight line. The line on the plot is the least-squares regression line for predicting y from x. One point is an extreme outlier in both the x- and y-directions. Let's examine the influence of this outlier.

First, suppose we omit the outlier. The correlation for the five remaining points (the cluster at the lower left) is $r = 0.523$. The outlier extends the straight-line pattern and greatly increases the correlation.

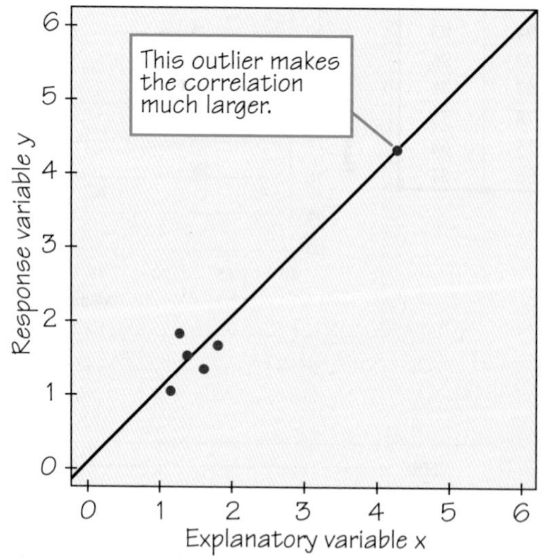

Figure 6.15 The outlier increases the correlation and fixes the location of the least-squares line.

Next, suppose we grab the outlier and pull it straight down, as in Figure 6.16. The least-squares line chases the outlier down, pivoting until it has a negative slope. This is the least-squares idea at work: The line stays close to all six points. However, in this situation its location is determined almost entirely by the one outlier. Of course, the correlation is now also negative, $r = -0.796$. Never trust a correlation or a regression line if you have not plotted the data.

One way to explore this concept is to use the *Correlation and Regression* applet. Applet Exercise 1 (page 288) asks you to animate the situation shown in Figures 6.15 and 6.16 so that you can watch r change and the regression line move as you pull the outlier down.

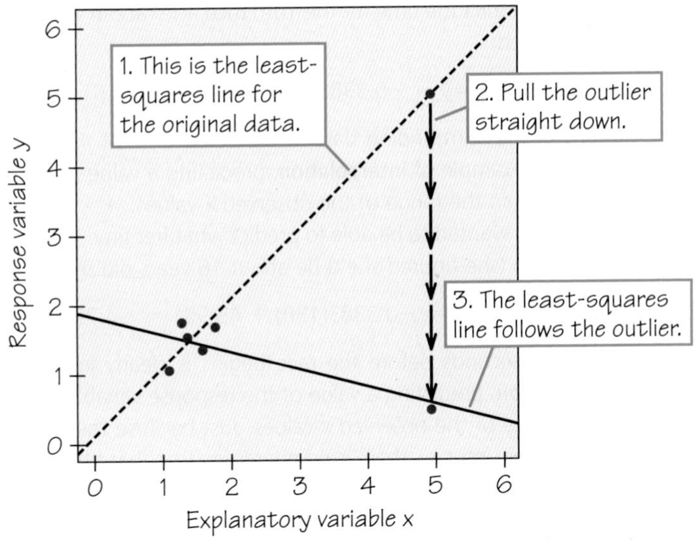

Figure 6.16 Moving the outlier unduly changes the correlation and moves the least-squares line.

Even if the correlation is moderate to strong and there are no outliers in the data that we used to find our regression line, we also must not be quick to extrapolate and make predictions well beyond the data collected. Example 10 illustrates this point.

EXAMPLE 10 Extrapolation Is Risky!

Return to the data in Table 6.3 (page 245), which gives the times for an 8-year-old competitive swimmer's 50-yard butterfly over 14 races (her time for race 10 was never recorded). A scatterplot of these data appears in Figure 6.3 (page 245). After removing the two circled outliers (corresponding to the times of the swimmer's first two races), the form of the remaining data points appears to be a straight line. Figure 6.17 shows the result of using Excel (see Spotlight 6.5 on page 265) to make a scatterplot of the data and determine the equation of the least-squares regression line.

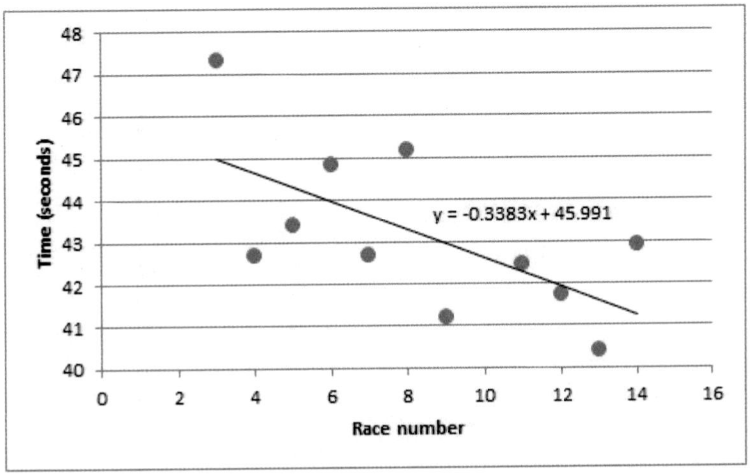

Figure 6.17 Fitting a least-squares regression line to data from a swimmer's 50-yard butterfly races.

First, we'd like to predict the time for the 10th race, the race in which the swimmer's time was never recorded:

$$\text{predicted time} = \hat{y} = -0.3383(10) + 45.991 \approx 42.61 \text{ seconds}$$

Given the pattern in the surrounding data, a predicted time of 42.61 seconds seems reasonable. This is an example of **interpolation,** predicting a value of the response variable for an x-value within the range of the observed x-values.

The swimmer really wanted to be able to predict what her time would be after many races, say, for race 150 (she figured she'd be about 16 years old by that time):

$$\text{predicted time} = \hat{y} = -0.3383(150) + 45.991 \approx -4.75 \text{ seconds}$$

Finishing a race 4.75 seconds before the race begins is clearly impossible! This is an example of **extrapolation,** predicting a value of the response variable for an x-value that lies outside of the range of the observed x-values. Just because the data fit a particular linear trend over a certain interval, there is no guarantee that that trend will continue into the future. So, avoid extrapolation—particularly for x's far from the x-values in the data.

Correlation and regression *describe* relationships. *Interpreting* relationships requires more thought. *Often the relationship between two variables is influenced strongly by other variables.* You should always think about the possible effect of other variables before you draw conclusions based on correlation or regression.

EXAMPLE 11 Money Helps SAT Scores?

The College Board, which administers the SAT, offers this information on its website about the Class of 2013 seniors who take the test (the 55% of test-takers who did not respond to this income question had a mean score of 515):

TABLE 6.7 Family Incomes and Mean Math SAT Scores

Family income (in $1000s)	Mean Math SAT score
0–20	462
20–40	482
40–60	500
60–80	511
80–100	524
100–120	536
120–140	540
140–160	548
160–200	555
Over 200	586

This information suggests a strong positive association between the test-taker's score and family income. But there's no direct mechanism that causes this association—wealthy families are not sending secret bribes to the College Board. It may simply be that children of wealthy parents are more likely to have advantages, such as well-educated role models, high expectations, access to extra tutoring or test preparation, smaller class sizes, and schools with more experienced, better qualified teachers.

Example 11 brings us to the most important caution about correlation and regression. When we study the relationship between two variables, we often hope to show that changes in the explanatory variable *cause* changes in the response variable. A strong association between two variables is not enough to draw conclusions about cause and effect. Sometimes an observed association really does reflect cause and effect. Drinking more beer does cause an increase in BAC. But in many cases, as in Example 11, a strong association is explained by other variables that influence both x and y. Here is another example.

EXAMPLE 12 Evaluation Correlation?

Grades that students earn in courses are correlated positively with the ratings that students give on anonymous end-of-course surveys administered by the university. One very simple interpretation is that instructors give easy tests with "low standards," which in turn causes students to express appreciation through high instructor ratings. But perhaps there is a third variable that drives the other two variables: A professor who is a skillful teacher and motivator may be more likely both to be rated well and to inspire high performance. Or perhaps courses that have higher grade distributions are more likely to be upper-level courses for majors in that subject, and such students would be more prepared for and favorably inclined toward the course.

EXAMPLE 13 Does Running Lead to Winning in Football?

Otto Greule Jr/Getty Images

A football broadcaster discussed how often a team wins when it runs the ball at least 30 times in a game. For the 2010 NFL regular season, the correlation between wins and number of running plays was indeed close to being moderately positive ($r = 0.48$). Could this mean that running causes winning—that all any team has to do to win more games is to run the ball more? No. In the extreme, if a team executed only running plays, the other team would simply adjust its defense to focus on and stop the run. Basically, once teams get a good lead in a game (regardless of their mix of special teams, running, and passing), they tend to start running the ball more often as a way to minimize the risk of losing the ball (pass plays are riskier) and to use up the clock faster (an incomplete pass stops the clock). And when teams fall far behind late in the game, they begin passing more often as a last chance to catch up before time runs out.

Correlations such as the one described in Example 13 are sometimes called "nonsense correlations." The correlation is real, but it is nonsense to conclude that increasing the number of running plays will cause an increase in the number of wins that season. So correlations require thoughtful interpretation, not just computation.

Association Does Not Imply Causation RULE

An association between an explanatory variable x and a response variable y, even if it is very strong, is not by itself good evidence that changes in x actually cause changes in y. Causation also requires the following: (1) Demonstrating that you have ruled out the possibility that the change in the response variable was due to any other variable besides the explanatory variable, (2) showing that the association happens under a variety of conditions, and (3) having a reasonable mechanism or model to explain how x causes changes in y.

Here is a final example in which we use a scatterplot, correlation, and a regression line to understand data.

EXAMPLE 14 What Does Growth Hormone Do in Adults?

In most species, adults stop growing but still release growth hormone from the pituitary gland to regulate metabolism. Physiologists subjected groups of adult rats to various conditions that activated muscle tissue that was either fast-twitch (as sprinters use) or slow-twitch (as distance runners use). They then measured levels of a form of growth hormone (BGH) in the blood and in pituitary tissue. Units are hundreds of nanograms per milliliter of blood and micrograms per milligram of tissue, respectively.

Blood	15.8	20.0	26.7	25.0	23.0	23.8	24.7	16.3	0.8	0.8
Tissue	38.0	36.7	27.8	28.3	34.9	34.1	33.2	32.7	38.1	39.1
Blood	0.6	10.8	37.6	41.3	39.0	57.5	84.8	82.8	28.8	16.5
Tissue	43.9	42.8	19.3	13.7	11.2	14.2	9.7	9.5	31.7	32.8

Data from G. E. McCall, K. L. Gosselink, et al., Muscle afferent-pituitary axis: A novel pathway for modulating the secretion of a pituitary growth factor, *Exercise and Sport Science Reviews, 29* (2001): 1642–1649.

Figure 6.18 is a scatterplot of these data. The plot shows a strong negative straight-line association with correlation $r = -0.90$. Here's the physiological mechanism for this association: When there is a higher BGH level in the blood, we can assume that means BGH must have been recently secreted by the pituitary gland so that less BGH now remains in pituitary tissue. The least-squares regression line is

$$\hat{y} = 41.081 - 0.43343x$$

or

$$\text{predicted pituitary BGH} = 41.081 + ((-0.43343) \times \text{blood BGH})$$

The slope $m = -0.43343$ is negative, which reflects how blood and pituitary tissue levels of BGH move in opposite directions. The y-intercept $b = 41.081$ is the estimated amount of BGH that the pituitary gland has if it does not release any into the blood.

Next, we dig deeper into the data from this experiment. Each data point represents the mean BGH levels in the blood and in pituitary tissue for a group of rats undergoing the same treatment (Figure 6.19). The two highest points of the scatterplot represent groups of rats whose slow-twitch muscles were activated, while the five lowest points on the scatterplot involved the activation of fast-twitch muscles in separate groups

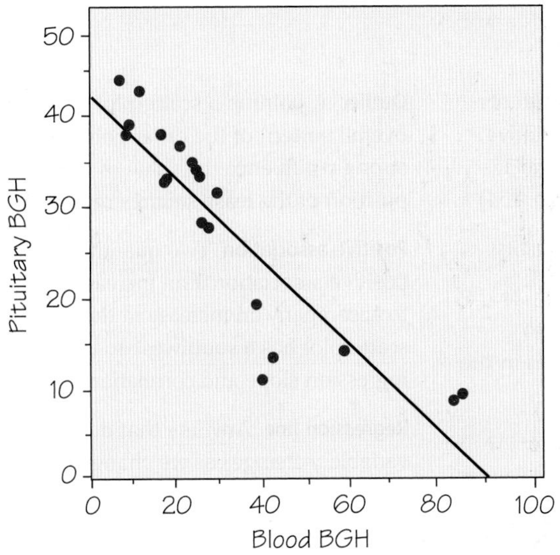

Figure 6.18 This scatterplot of BGH level in pituitary tissue versus BGH level in the blood shows a strong negative association.

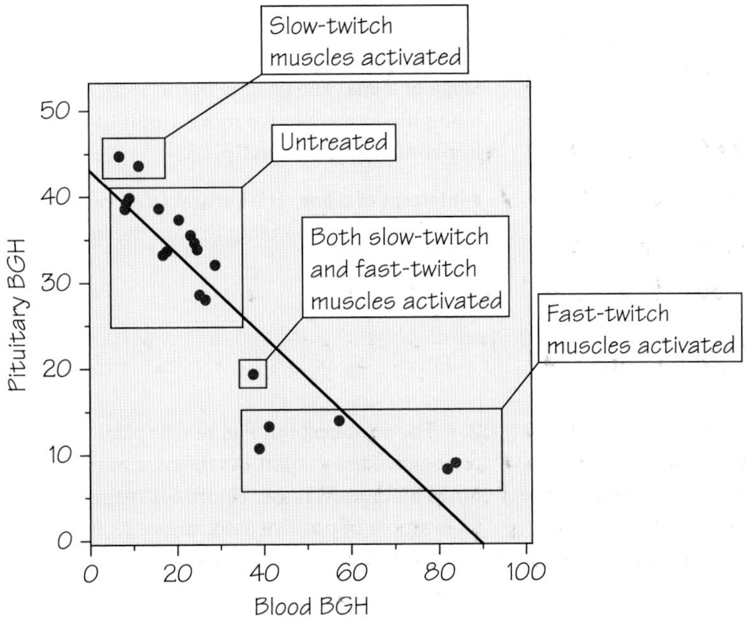

Figure 6.19 Scatterplot of BGH level in pituitary tissue versus BGH level in the blood with specific groups of rats identified by treatment.

of rats. The point (37.6, 19.3) comes from a group of rats that were exercised on a treadmill to activate fast- and slow-twitch muscles simultaneously. The remaining points represent groups that were untreated.

These data come from an *experiment* that assigned rats randomly to treatment (or no treatment) conditions. Random assignment makes us reasonably confident that slow-twitch muscle activation *causes* a decrease in BGH secretion (and hence lower levels of blood BGH) and that fast-twitch muscle activation *causes* an increase. We will discuss experiments in detail in Chapter 7.

Review Vocabulary

Correlation A measure of the direction and strength of the straight-line relationship between two quantitative variables. Correlations take values between −1 and 1, with the same sign as the regression line slope. (p. 254)

Explanatory variable A variable that attempts to justify the observed outcomes. (p. 242)

Extrapolation The use of the regression equation to predict y for values of x outside the range of values in the data used to fit the line. (p. 270)

Interpolation The use of the regression equation to predict y for values of x that lie inside the range of values in the data used to fit the line. (p. 270)

Least-squares regression line A line drawn on a scatterplot that makes the sum of the squares of the vertical distances of the data points from the line as small as possible. The regression line can be used to predict the response variable y for a given value of the explanatory variable x. (p. 262)

Negative association Two variables have negative association if an increase in one variable tends to accompany a decrease in the other variable. The scatterplot has a northwest-to-southeast pattern, and the regression slope and correlation are both negative. (p. 247)

Outlier A point in a scatterplot that lies outside the overall pattern of the other points. Outliers sometimes strongly influence the value of the correlation and the position of the least-squares regression line. (p. 245)

Positive association Two quantitative variables have positive association if an increase in one variable tends to accompany an increase in the other variable. The scatterplot has a southwest-to-northeast pattern, and the regression slope and correlation are both positive. (p. 247)

Regression line Any line that describes how a response variable y changes as we change an explanatory variable x. The most common such line is the least-squares regression line. (p. 251)

Response variable A variable that measures an outcome of a study. (p. 242)

Scatterplot A graph made by plotting ordered pairs of data to show the relationship between two quantitative variables. (p. 242)

Slope of a line The change in the vertical (y) direction along the line when we move 1 unit to the right in the horizontal (x) direction. (p. 253)

y-intercept of a line The vertical (y) coordinate of the point on the line corresponding to 0 on the horizontal (x) axis. (p. 253)

Self Check Answers

1. Midterm exam scores can be used to explain final exam scores. Hence, Midterm is the explanatory variable. The midterm grade may help explain what a student might get on the final exam. Here's a scatterplot of these data.

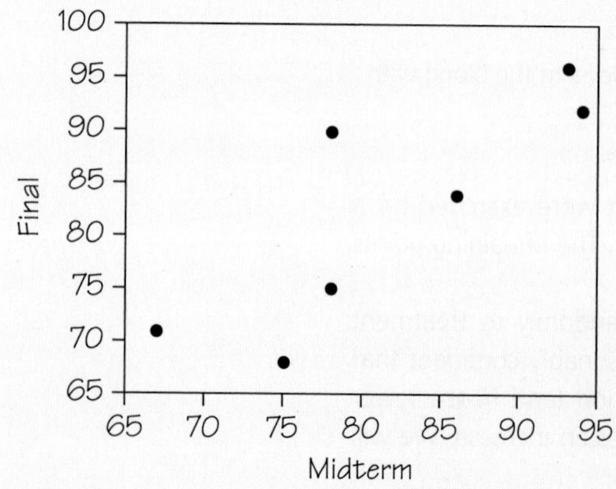

2. The scatterplot in Figure 6.3 (page 246) is an example of negative association. As the race number increases, the times tend to decrease. Figure 6.4 (page 246) provides an example of positive association. As IQ test scores increase, the reading test scores also tend to increase.

3. (2, 0.023) (The second coordinate has been rounded to three decimal places.)

4. For each additional year of age, cholesterol increases by 1.64 mg/dl.

5. Math SAT: $\bar{x} = 530$, $s_x = 61.2$

Critical Reading: $\bar{y} = 506$, $s_y = 55.9$

$$r = \frac{1}{5-1}[(1.307)(0.787) + (-1.471)(-1.717) + (0.327)(0.250) + (-0.163)(0.608) + (0)(0.072)]$$

$$r \approx 0.884$$

6. (a) Prediction for son's height = 0.516 × (64) + 33.73 ≈ 66.75 inches.

(b) This prediction indicates that the son is taller than his father but is still below average in height, which is 70 inches.

7. Student 1: BAC = −0.0127 + 0.01796(5) = 0.0771; residual = 0.10 − 0.0771 ≈ 0.02.

Student 14: BAC = −0.0127 + 0.01796(7) = 0.11302; residual = 0.09 − 0.11302 ≈ −0.02.

8. (a) TI-84-created scatterplot of Critical Reading SAT against Math SAT:

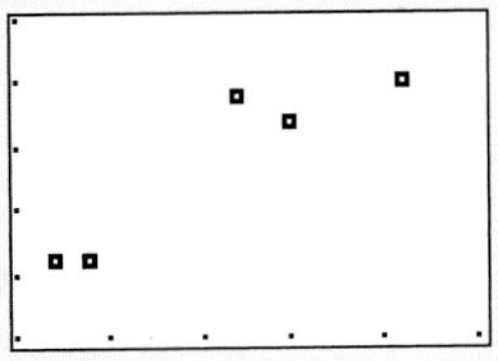

(b) $m = (0.928)\left(\frac{70.2}{78.5}\right) \approx 0.82988$; $b = 486 - 0.82988(508) \approx 64.42096$; $\hat{y} = 0.83x + 64.42$

Skills Check

1. You have data for many families about the parents' income and the years of education that their eldest child completes. When you make a scatterplot, the explanatory variable on the x-axis

(a) is the parents' income.

(b) is years of education.

(c) doesn't matter.

2. The outcome or result of a study is measured by a(n) __________ variable.

3. The data in Example 1 (page 243) consist of

(a) 16 ordered pairs of values.

(b) 2 unpaired sets of values, each of which has 16 values.

(c) 32 ordered pairs of values.

4. The explanatory variable is plotted on the ___ axis of a scatterplot.

5. If two variables have a negative association, an increase in one variable tends to accompany __________ in the other variable.

(a) an increase

(b) a decrease

(c) no change

6. You expect to see a __________ association between the parents' income and the years of education that their oldest child completes.

7. A high school teacher was curious whether the SAT Math test score was a good predictor of students' SAT Critical Reading test score. To find out, he collected SAT data from 15 graduating seniors. Figure 6.20 is a scatterplot of SAT Critical Reading test scores against the SAT Math test scores of these 15 students.

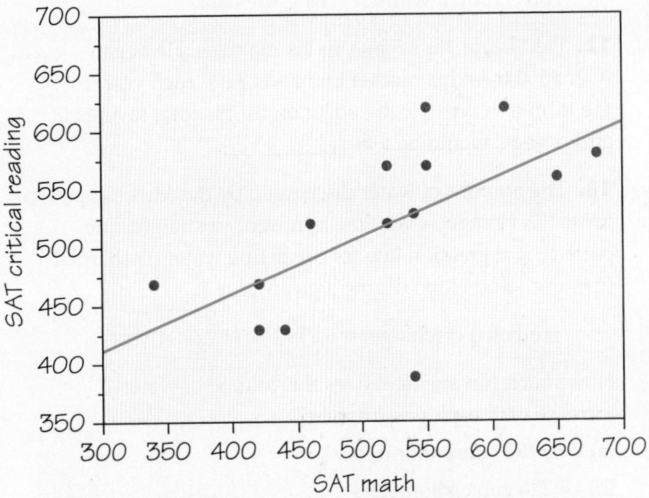

Figure 6.20 A scatterplot of the SAT Critical Reading test scores (the response variable) against SAT Math test scores (the explanatory variable) of 15 high school seniors (applies to Skills Checks 7–9).

There is one low outlier in the plot. The SAT Critical Reading and Math scores for this student are

(a) Math = 390, Critical Reading = 540.

(b) Math = 540, Critical Reading = 530.

(c) Math = 540, Critical Reading = 390.

8. The line in Figure 6.20 is a regression line for predicting the SAT Critical Reading score from the SAT Math score. If another senior from this school has an SAT Math score of 600, then which of the following choices would you predict for the student's SAT Critical Reading score?

(a) 510

(b) 560

(c) 580

9. The slope of the line in Figure 6.20 is closest to

(a) 2.

(b) 1.

(c) 1/2.

10. The points on a scatterplot lie close to the line whose equation is $y = 2 - 5x$. The slope of this line is __________.

11. Starting with a fresh bar of soap, you weigh the bar each day after you take a shower. Then you find the regression line for predicting the weight from the number of days elapsed. The slope of this line will be

(a) positive.

(b) negative.

(c) You can't tell without seeing the data.

12. Fred keeps his savings in his mattress. He began with \$500 from his mother and adds \$100 each year. In the form $y = mx + b$, the equation for his total savings y after x years would be $y =$ __________.

13. The amount of water discharged by the Mississippi River has changed over time in roughly a straight-line pattern. A regression line for predicting water discharged (in cubic kilometers) during a given year is

$$\text{predicted discharge} = -7792 + (4.226 \times \text{year})$$

How much (on average) does the volume of water increase with each passing year?

(a) −7792 cubic kilometers

(b) 4.226 cubic kilometers

(c) 7792 cubic kilometers

14. According to the regression line in Skills Check 13, the predicted Mississippi River discharge in the year 2016 will be ________ cubic kilometers.

15. You have data on the body weight x and brain weight y for many species of mammals. Body weight is given in kilograms, and brain weight is given in grams. There are 1000 grams in a kilogram. The slope of the regression line for predicting y from x is $m = 1.4$. If brain weight were given in kilograms, the slope would

(a) change to 0.0014.

(b) still be 1.4.

(c) change to 1400.

16. Suppose $y = 2x + 3$, where x and y are measured in meters. If x is re-expressed in centimeters instead, the equation becomes $y =$ __________________.

17. Given the following set of five ordered pairs, the correlation r equals what?

x	0	1	2	3	4
y	2	3	5	6	14

(a) 0.3

(b) 0.6

(c) 0.9

18. The correlation between brain weight and body weight in Exercise 15 is $r = 0.86$. If brain weight had been measured in kilograms rather than grams, the correlation would have a value of ________________.

19. The points on a scatterplot lie very close to the line whose equation is $y = 5 - 3x$. The correlation between x and y is close to which of the following values?

(a) −3

(b) −1

(c) 1

20. High coffee prices give farmers in Indonesia an incentive to cut forest in order to plant more coffee. Here are data on coffee price x (dollars per pound) and percentage y of deforestation in a national park for five years:

x	0.29	0.40	0.54	0.55	0.72
y	0.49	1.59	1.69	1.82	2.98

Using a calculator, we can determine that the correlation between x and y has a value of _________, to the nearer hundredths place.

21. Return to the data in Skills Check 20. Using a calculator, find the standard deviations of the coffee prices (x) and percentages of deforestation (y). Use these standard deviations and your answer to Skills Check 20 to calculate the slope of the least-squares regression line. Round the slope to one decimal place.

22. Using Table 6.1 (page 243) and the prediction equation in Example 4 (page 251), what is the residual error for Student 4?

(a) −0.01

(b) 0.01

(c) 0.13

23. In the least-squares method, what is it that is being squared?

(a) Each observed y value

(b) The vertical distance of each data point from the line

(c) The sum of the vertical distances of the data points from the line

24. Putting the symbol ^ (i.e., a caret or circumflex) over y indicates that this is the __________ value of y.

25. If the slope and correlation value are equal, then the standard deviation of y must equal

(a) 1.

(b) the slope.

(c) the standard deviation of x.

26. If the residual error is positive, the observed response lies ________ the line.

27. There is a strong positive correlation between the number of firefighters at a fire and the amount of damage the fire causes. The reason for this is that

(a) more firefighters cause more damage at the fire scene.

(b) bigger fires require more firefighters and also do more damage.

(c) more damage requires more firefighters to clean it up.

28. Making predictions well beyond the data collected is called ___________.

29. Make a scatterplot with the six ordered pairs in the table below. Of the first three ordered pairs listed in the table, the one that, if deleted, would cause the biggest change in the value of the correlation is

(a) (0, 0).

(b) (8, 3).

(c) (2, 4).

Ordered pair number	1	2	3	4	5	6
x	0	8	2	0	1	1
y	0	3	4	1	1	0

30. Association (correlation) does not imply that changes in x __________ changes in y.

31. A high school teacher conducted a small study on school achievement. For a group of 20 of her juniors, she recorded the number of days the students attended school in their junior year and their grade point average at the end of the year. A scatterplot of her data appears in Figure 6.21. Would you describe the relationship between GPA and Days Attended as positive, negative, or neither? Briefly explain.

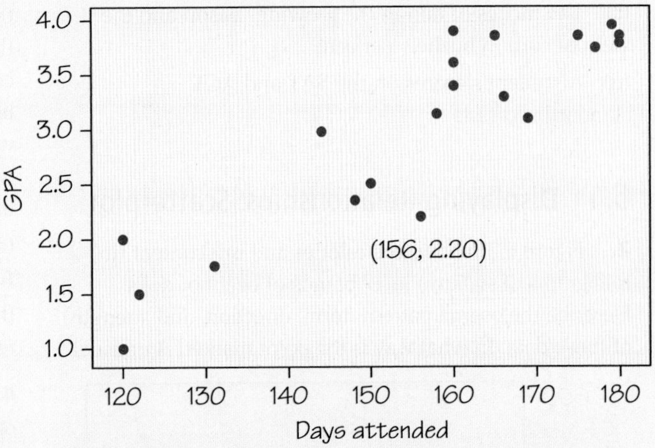

Figure 6.21 A scatterplot of high school GPA junior year against days of school attended (applies to Skills Checks 31–35).

32. The correlation between Days Attended and GPA is $r = 0.905$. Would you classify the relationship shown in Figure 6.21 as weak, moderate, or strong?

33. The teacher fit a least-squares regression line to the data displayed in Figure 6.21. Given the following summary statistics, determine the equation of this line:

Days Attended, x: $\bar{x} = 156$, $s_x = 19.87$

GPA, y: $\bar{y} = 2.989$, $s_y = 0.913$

Correlation: $r = 0.905$

34. The ordered pair (156, 2.20) is marked on the scatterplot in Figure 6.21. Use your equation of the least-squares regression line from Skills Check 33 to determine the residual error for this point.

35. Use the equation of the least-squares regression line you calculated in Skills Check 33 to predict the GPA of a student who attended school on only 60 days. Does your answer make sense? Is this an example of extrapolation or interpolation?

Chapter 6 Exercises

Challenge Discussion

Some exercises require use of a calculator (or software or Internet applet) that will find correlation and the slope and intercept of the least-squares regression line from keyed-in data.

1. In each of the following situations, is it more reasonable simply to explore the relationship between the two variables or to view one of the variables as an explanatory variable and the other as a response variable? In the latter case, which is the explanatory variable?

(a) The amount of time spent studying for a statistics exam and the grade on the exam

(b) The weight in kilograms and height in centimeters of a person

(c) The inches of rain in the growing season and the yield of corn in bushels per acre

(d) A student's scores on the SAT and ACT standardized tests

6.1 Displaying Relationships: Scatterplot

2. Figure 6.22 shows the calories and salt content (in milligrams of sodium) in 17 brands of beef hot dogs. Describe the overall pattern (form, direction, and strength) of these data. In what way is the point marked *A* unusual?

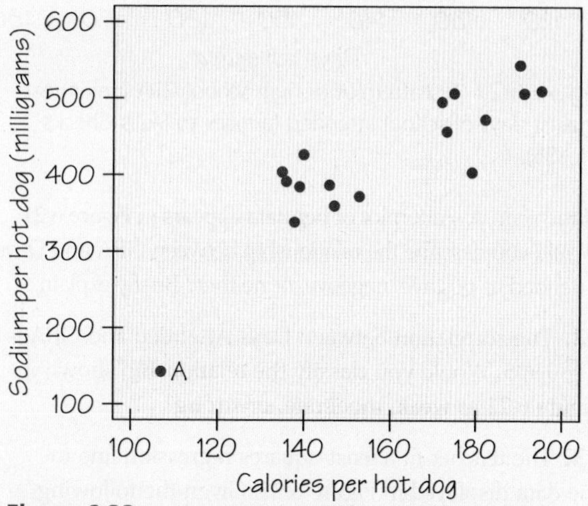

Figure 6.22 A scatterplot of sodium content versus calories in beef hot dogs, for Exercises 2 and 13.

3. Figure 6.23 is a scatterplot of data from the World Bank. The individuals are all the world's nations for which data are available. The explanatory variable is a measure of how rich a country is, which is the gross domestic product (GDP) per person. GDP is the total value of the goods and services produced in a country, converted into dollars. The response variable is life expectancy at birth. Three African nations are outliers, with lower life expectancy than usual for their GDP. A full study would ask what special circumstances explain these outliers.

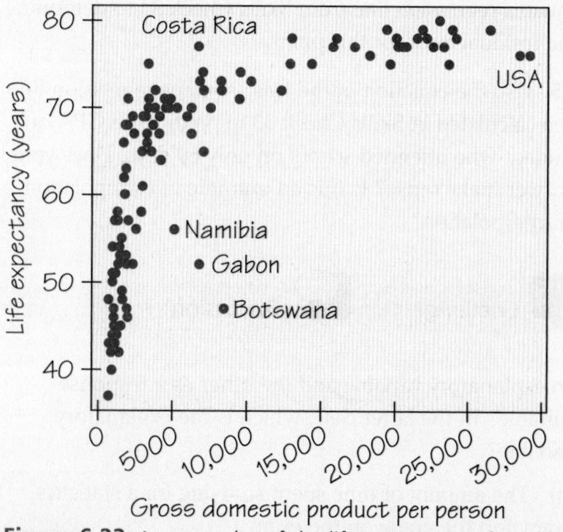

Figure 6.23 A scatterplot of the life expectancy of people in many nations against each nation's GDP per person, for Exercise 3.

(a) Describe the direction and form of the relationship. Aside from the outliers, it is moderately strong.

(b) Explain why the direction and form of this relationship make sense.

4. Global warming is due to increased concentrations of greenhouse gases such as carbon dioxide (CO_2). Here are data from the National Oceanic and Atmospheric Administration website (www.noaa.gov), where atmospheric CO_2 is measured in parts per million per unit of volume. The data below were measured at the Mauna Loa Observatory:

CO_2	315.98	324.62	336.78	352.90	368.14	387.35
Year	1959	1969	1979	1989	1999	2009

(You will revisit these data in Exercise 20.)

(a) Which is the explanatory variable?

(b) Make a scatterplot. Is the association between these variables positive or negative? Explain why you expect the relationship to have this direction.

(c) Describe the form and strength of the relationship.

5. Table 5.7 (page 196) gives the city and highway gas mileages for 13 midsized cars. Omit the hybrid car (Toyota Prius) and make a scatterplot, taking city mileage as the explanatory variable. Describe in words the form, direction, and strength of the relationship between highway mileage and city mileage. (You will revisit these data in Exercise 19.)

6. How fast do icicles grow? Here are data on two variables, Time measured in minutes and Length measured in centimeters, for one set of conditions: no wind, temperature −11°C, and water flowing over the icicle at 12 milligrams per second.

Time (minutes)	10	20	30	40	50
Length (centimeters)	0.6	1.8	2.9	4.0	5.0
Time (minutes)	60	70	80	90	100
Length (centimeters)	6.1	7.9	10.1	10.9	12.7
Time (minutes)	110	120	130	140	150
Length (centimeters)	14.4	16.6	18.1	19.9	21.0

Data from N. Maeno et al., Growth rates of icicles, *Journal of Glaciology*, *40* (1994): 319–326.

Which is the explanatory variable? Make a scatterplot. Describe in words the direction, form, and strength of the relationship.

7. How does the fuel consumption of a car change as its speed increases? Here are data for a British Ford Escort. Fuel consumption is measured in miles per gallon of gasoline used and speed is measured in miles per hour.

Speed	6.2	12.4	18.6	24.9	31.1
Fuel	11.2	18.1	23.5	29.4	33.6
Speed	37.3	43.5	49.7	55.9	62.1
Fuel	39.9	37.3	33.8	31.1	28.4
Speed	68.4	74.6	80.8	87.0	93.2
Fuel	26.0	23.8	21.8	20.0	18.3

Data from T. N. Lam, Estimating fuel consumption from engine size, *Journal of Transportation Engineering, 111* (1985): 339–357.

(a) Which is the explanatory variable?

(b) Make a scatterplot. Describe the form of the relationship. Explain why the form of the relationship makes sense.

(c) How would you describe the direction of this relationship?

(d) Is the relationship reasonably strong or quite weak? Explain your answer.

8. Give an example of two variables from everyday life that have a positive association. Give an example of two variables that have a negative association.

9. The following table shows excerpted and rounded data collected on crickets by Harvard physics professor George W. Pierce in his 1948 book *The Song of Insects*:

Chirps per 15 seconds	44	37	31	25	15
Ground temperature (in °F)	80	68	73	63	55

(a) Which is the explanatory variable?

(b) Make a scatterplot. Is the association between these variables positive or negative?

(c) Describe the form and strength of the relationship.

(d) Do the data seem consistent with a rule of thumb from *The Old Farmer's Almanac* to count the number of chirps in 14 seconds and then add 40 to get the (Fahrenheit) temperature?

10. On the NASA space shuttle, six primary O-rings were used to seal the sections of the two solid-fuel rocket motors and keep hot gases from escaping and catastrophically igniting the liquid hydrogen fuel tank. The number of O-ring erosion problems and the launch temperature (in degrees Fahrenheit) are given for 23 successful flights between April 12, 1981, and January 12, 1986:

Mission	O-Ring Incidents	°F	Mission	O-Ring Incidents	°F
1	0	66	13	0	67
2	1	70	14	2	53
3	0	69	15	0	67
4	0	68	16	0	75
5	0	67	17	0	70
6	0	72	18	0	81
7	0	73	19	0	76
8	0	70	20	0	79
9	1	57	21	2	75
10	1	63	22	0	76
11	1	70	23	1	58
12	0	78			

(a) Which is the explanatory variable?

(b) Make a scatterplot. Is the association between these variables positive or negative?

(c) Describe the form and strength of the relationship.

(d) The forecasted temperature the morning of the January 28, 1986, launch of the *Challenger* was between 26°F and 29°F. Should this have been a cause for concern or not? Why?

(e) Would a different conclusion be reached by someone whose scatterplot contained only the seven flights for which there was at least one problem? Explain.

11. Use spreadsheet software or a graphing calculator for this exercise. (Spotlight 6.1 on page 249 provides instruction for TI-83/84 graphing calculators and Excel.) The presence of mercury in fish is a health hazard, particularly for women who may become pregnant and children. Table 6.8 contains data on mercury concentration in tissue samples from 20 largemouth bass taken from Lake Natoma (California). Only fish of legal/edible size were used in this study. (Save your data and work from this exercise for use in Exercise 31.)

TABLE 6.8 Fish Length and Mercury Concentration in Fish Tissue Samples

Total Length, *x* (mm)	Mercury Concentration, *y* (μg/g wet wt.)
341	0.515
353	0.268
387	0.450
375	0.516
389	0.342
395	0.495
407	0.604
415	0.695
425	0.577
446	0.692
490	0.807
315	0.320
360	0.332
385	0.584
390	0.580
410	0.722
425	0.550
480	0.923
448	0.653
460	0.755

(a) Which is the explanatory variable and which is the response variable?

(b) Make a scatterplot of the response variable against the explanatory variable.

(c) Based on your scatterplot, as fish length increases, does mercury concentration in fish tissue tend to increase, decrease, or remain about the same? Is this an example of positive or negative association?

(d) Would you describe the form of the relationship as linear or nonlinear? Explain.

(e) Why do you think that only fish of edible/legal size were included in the dataset?

12. Use spreadsheet software or a graphing calculator for this exercise. (Spotlight 6.1 on page 249 provides instruction for TI-83/84 graphing calculators and Excel.) Satellites are one of the many tools used for predicting flash floods, heavy rainfall, and large amounts of snow. Geostationary (GEOS) satellites collect data on cloud top brightness temperatures (measured in degrees Kelvin). It turns out that colder cloud temperatures are associated with higher and thicker clouds, which in turn are associated with heavier precipitation. Data consisting of temperature and rainfall rate measured by ground radar appear in Table 6.9. Because ground radar can be limited by location and obstructions, having an alternative for predicting the rainfall rates can be useful. (Save your data and work from this exercise for use in Exercise 32.)

TABLE 6.9 Sixteen Data Pairs of (Temperature, Rain Rate) Data

Temperature (°K)	Radar Rain Rate (mm/h)	Temperature (°K)	Radar Rain Rate (mm/h)
195	150	203	44
196	150	204	39
197	150	205	39
198	118	206	35
199	109	207	38
200	95	208	31
201	63	209	20
202	66	210	24

(a) You want to use cloud top temperature to explain rainfall rate. With this in mind, make a scatterplot of the data from Table 6.9.

(b) Is the association between cloud top temperature and radar rainfall rate positive or negative?

(c) Would you describe the form of the relationship as linear?

6.2 Making Predictions: Regression Line

13. Figure 6.22 (page 278) shows the salt content (in milligrams of sodium) and calories in 17 brands of beef hot dogs. If we ignore the outlying point marked *A*, a regression line for predicting sodium from calories passes close to these two observations:

calories = 139, sodium = 386 mg
calories = 191, sodium = 506 mg

Use this fact to estimate the slope of this regression line (round your answer to two decimal places).

Algebra Review Appendix
Slope of a Line

14. Exercise 4 (page 278) gives data on CO_2 concentration (in parts per million) over time. A regression line for predicting CO_2 for a given year is

$$\text{predicted } CO_2 \text{ concentration} = 311.662 + 1.43866 \times (\text{years elapsed since 1959})$$

(a) What is the slope of this line? What does the slope say about how CO_2 is changing over time?

(b) What does the model predict the CO_2 concentration for 2006 to be? In fact, the observed value that year was 381.85. How accurate is your prediction?

15. Researchers studying acid rain measured the acidity of precipitation in a Colorado wilderness area for 150 consecutive weeks. Acidity is measured by pH, and lower pH values mean higher acidity. The acid rain researchers observed a straight-line pattern in acidity levels over time. They reported that the regression line

$$\text{predicted pH} = 5.43 - (0.0053 \times \text{weeks})$$

fit the data well. [Data from W. M. Lewis and M. C. Grant, Acid precipitation in the western United States, *Science, 207* (1980): 176–177.]

(a) Draw a graph of this line. Explain what the line says about how pH was changing over time.

(b) According to the regression line, what was the pH at the beginning of the study (weeks = 1)? At the end (weeks = 150)?

(c) What is the slope of the regression line? Explain what this slope says about the rate of change in pH.

16. A study at the University of Massachusetts, Amherst, published in the May 2007 *Journal of Marriage and Family* found that married women do about one fewer hour of housework a week for every $7500 they earn as full-time workers outside the home, regardless of their husband's income.

(a) What would be the numerical value of the slope coefficient in the regression model that predicts women's housework hours from their income? What does the sign of the slope (positive or negative) tell us about the relationship between these variables?

(b) Suppose Lynette's salary is $30,000 greater than Gabrielle's. What would you predict to be the difference in hours of housework they each do?

17. A 21-year-old college student drinks heavily at a party until his BAC is 0.15 g/dl—almost twice the legal driving limit of 0.08. He now stops drinking, and each hour his BAC falls by 0.015 g/dl.

Algebra Review Appendix
Using Formulas

(a) What is a regression-line equation that would predict his BAC from the number of hours after he stopped drinking?

(b) In how many hours would the student be able to drive legally?

(c) In how many hours would no alcohol be present in his body?

18. Suppose that the slope of the regression line of weight on height for a group of young men is $m = 1.1$ when we measure height x in centimeters and weight y in kilograms. That is, when height increases by 1 centimeter, weight increases by 1.1 kilograms. There are 1000 grams in a kilogram. If we measured weight in grams, what would the slope be?

6.3 Correlation

19. Find the correlation between the city and highway gas mileages for the 12 nonhybrid midsized cars (i.e., omit the Toyota Prius) in Table 5.7 (page 196). Explain why the value of r supports the scatterplot that you created in Exercise 5.

20. Exercise 4 (page 278) gives data on CO_2 concentration (in parts per million) over time.

(a) Use a calculator or spreadsheet software such as Excel to find the correlation r (refer to Spotlight 6.2 on page 259). Explain from looking at the scatterplot why this value of r is reasonable.

(b) Suppose that the concentration had been recorded in parts per billion instead of parts per million. For example, the value 354.16 would become 354,160. How would the value of r change?

21. Find the correlation between city and highway mileage for all 13 midsized cars in Table 5.7 (page 196), including the Toyota Prius. Compare your r with the value you found in Exercise 19. Explain why adding the Prius changes r in this direction.

22. In Example 11, Table 6.7 (page 270), the positive association between family income and SAT score is clear from the lockstep pattern, but to calculate a specific numerical value for the correlation, we need to make a simplifying assumption. For each income bracket with fixed endpoints, choose its midpoint. Now, use a calculator (or Excel) to calculate a correlation value (refer to Spotlight 6.2 on page 259).

23. Exercise 7 (page 279) gives data on gas used versus speed for a small car. Make a scatterplot, if you did not do so in Exercise 7. Calculate the correlation. Explain why r is close to 0 despite a strong relationship between speed and gas use.

24. Consider the data in Exercise 6 (page 278) on forming icicles.

(a) Find the correlation between time and icicle length.

(b) If icicle length were measured in inches rather than centimeters (*Note:* 1 inch = 2.54 cm), how would the correlation change?

25. If heterosexual women always dated men who are three years older than themselves, what would the correlation be between the ages of each man and each woman? (*Hint:* Draw a scatterplot for several ages.)

26. We want to find the correlation between

(a) the heights of fathers and the heights of their adult sons.

(b) the heights of married men and the heights of their wives.

(c) the heights of women at age 4 and their heights at age 18.

The answers (in scrambled order) are $r = 0.2$, $r = 0.5$, and $r = 0.8$. Match the r values to the variable pairings and explain your choices.

27. For each of the following pairs of variables, would you expect a substantial negative correlation, a substantial positive correlation, or a small correlation?

(a) The age of used cars and their prices

(b) The weight of new cars and their gas mileages in miles per gallon

(c) The heights and weights of adult women

(d) The heights and IQ scores of adult men

28. Each of the following statements contains a mistake. Explain what is wrong in each case.

(a) "There is a high correlation ($r = 0.89$) between the hair color of American workers and their income."

(b) "We found a high correlation ($r = 1.09$) between students' ratings of faculty teaching and ratings made by other faculty members."

(c) "The correlation between age and income was found to be $r = 0.53$ years."

29. Mutual fund reports often give correlations to describe how the prices of different investments are related. You look at the correlations between three Fidelity funds and the Standard & Poor's 500 Stock Index, which describes stocks of large U.S. companies. The three funds are Dividend Growth (stocks of large U.S. companies), Small Cap Stock (stocks of small U.S. companies), and Emerging Markets (stocks in developing countries). For a recent year, the three correlations are $r = 0.35$, $r = 0.81$, and $r = 0.98$.

(a) Which correlation goes with each fund? Explain your answer.

(b) The correlations of the three funds with the index are all positive. Does this tell you that stocks went up that year? Explain your answer.

30. *Archaeopteryx* is an extinct animal that had feathers like a bird but teeth and a long bony tail like a reptile. Only six fossil specimens are known. If the specimens belong to the same species and differ in size because some are younger than others, there should be a straight-line relationship between the lengths of a pair of bones from all individuals. An outlier from this relationship would suggest a different species. Here are data on the lengths in centimeters of the femur (a leg bone) and the humerus (a bone in the upper arm) for the five specimens that preserve both bones:

Femur length x	38	56	59	64	74
Humerus length y	41	63	70	72	84

Data from M. A. Houck et al., Allometric scaling in the earliest fossil bird, *Archaeopteryx lithographica, Science, 247* (1990): 195–198.

(a) Make a scatterplot. Do you think that all five specimens come from the same species?

(b) Find the correlation r step by step, as in the procedure box in Section 6.3 (page 257).

(c) Now use one of the methods discussed in Spotlight 6.2 (page 259) to find r and check that you get the same result as in part (b).

31. Exercise 11, Table 6.8 (page 280) presented data on fish length and mercury concentration in fish tissue.

(a) If you did not do so in Exercise 11, make a scatterplot of mercury concentration against fish length. Based on your scatterplot, would you expect the correlation between these two variables to be positive or negative? Explain.

(b) Using either your calculator or spreadsheet software such as Excel, calculate the correlation between mercury concentration and fish length (refer to Spotlight 6.2). Using the guidelines in Figure 6.10 (page 255), classify the strength of the linear relationship as strong, moderate, or weak.

32. Exercise 12, Table 6.9 (page 280) presented data on cloud top brightness temperatures (measured from a satellite) and radar rain rate.

(a) If you did not do so in Exercise 12, make a scatterplot of radar rain rate against temperature. Based on your scatterplot, would you expect the correlation between these two variables to be positive or negative? Explain.

(b) Using either your calculator or spreadsheet software such as Excel, calculate the correlation between temperature and rain rate (refer to Spotlight 6.2). Using the guidelines in Figure 6.10 (page 255), classify the strength of the linear relationship as strong, moderate, or weak.

6.4 Least-Squares Regression

33. In Exercise 5, you made a scatterplot of city and highway gas mileage for the 12 nonhybrid midsized cars (omitting the Prius) in Table 5.7 (page 196).

(a) The equation of the least-squares regression line for predicting highway mileage from city mileage is

$$\text{highway mpg} = 13.4 + 0.75 \text{ city mpg}$$

Algebra Review Appendix
Graphing a Line in Slope-Intercept Form

Redo your scatterplot of highway mileage against city mileage from Exercise 5 (omitting the Prius). Add a graph of the least-squares regression line to the plot. Be sure to show how you were able to plot the line starting with its equation.

David R. Frazier Photolibrary, Inc./Science Source

(b) Use the "up-and-across" method illustrated in Figure 6.6 (page 251) to show the predicted highway mileage of a midsized car that gets 18 mpg in the city. Approximately what is the predicted highway mileage for this car?

(c) Now use the equation of the least-squares regression line to predict the highway mileage of a car that gets 18 mpg in the city. Compare your result with your graphical estimate in part (b).

(d) Based on the scatterplot and the graph of the least-squares regression line, do you expect your prediction from part (c) to be very accurate? Why or why not?

34. In Exercise 6 (page 278), you made a scatterplot of the length of an icicle and the number of minutes that water has been flowing over the icicle.

(a) The equation of the least-squares regression line for predicting icicle length from time is

$$\text{icicle length} = -1.9 + 0.15 \times \text{time}$$

Redo your scatterplot of icicle length against time from Exercise 6. Add a graph of the least-squares regression line to the plot. Be sure to show how you were able to plot the line starting with its equation.

(b) Use the "up-and-across" method illustrated in Figure 6.6 (page 251) to show the predicted length of the icicle after 75 minutes. Approximately what is the predicted length?

(c) Now use the equation of the least-squares regression line to predict the length of the icicle after 75 minutes.

35. Exercise 7 (page 279) gives data on fuel consumption in miles per gallon (mpg) and speed in miles per hour (mph) for a small car. From these data, the least-squares regression line to predict gas performance from speed is

$$\text{predicted fuel} = 27.033 - 0.0125 \times \text{speed}$$

(a) What are the observed fuel consumption values for speeds of 6.2, 43.5, and 93.2 mph?

(b) What are the predicted fuel consumption values for speeds of 6.2, 43.5, and 93.2 mph? (Round your answers to two decimal places.)

(c) Redo your scatterplot of fuel against speed from Exercise 7. Draw the least-squares regression line on your scatterplot.

(d) Based on the answer to part (c), is this line a useful model? Why or why not?

36. Scientists are concerned that rising sea temperatures will have an adverse effect on coral growth. A small study on this issue produced the data in the table below:

Temperature (°C), x	29.7	29.9	30.2	30.2	30.5	30.7	30.9
Coral growth (mm), y	2.63	2.58	2.60	2.48	2.26	2.38	2.26

(a) Make a scatterplot of coral reef growth versus the average sea surface temperature. Do these data tend to support or refute the scientists' concern? Explain.

(b) After drawing a scatterplot, Jason thought that the line through the points (29.7, 2.63) and (30.9, 2.26) did a good job summarizing the pattern of these data. Draw Jason's line on your scatterplot from part (a). Do you agree with Jason?

(c) Verify that Jason's line can be represented by the following equation:

$$\text{coral growth} = 11.78 - 0.308 \times \text{temperature}$$

Show the calculations for obtaining the slope and y-intercept.

(d) Carol used her calculator to find the equation of something called the median-median line. The equation of her line is

$$\text{coral growth} = 11.09 - 0.285 \times \text{temperature}$$

Using the least-squares criterion (i.e., the line with the smaller sum of squares of residual errors), which line is better, Jason's or Carol's? Justify your answer. (Feel free to use spreadsheet software or operations on calculator lists to respond to this question.)

37. A random sample of femur bone lengths (in millimeters) and heights (in centimeters) from six males appears in the table below (these data are from the Forensic Data Bank at the University of Tennessee):

Person	Femur length (mm)	Height (cm)
1	520	191
2	422	160
3	522	193
4	459	170
5	447	168
6	482	178

The equation of the least-squares regression line for predicting height from femur length is

$$\text{height} = 21.48 + 0.3265 \times \text{femur length}$$

(a) Calculate the residual error for Person 2. Based on this residual, does the data point for Person 2 lie above or below the regression line?

(b) Calculate the residual error for Person 4. Based on this residual, does the data point for Person 4 lie above or below the regression line?

38. The length of the icicle in Exercise 6 (page 278) is measured in centimeters. There are 2.54 centimeters in an inch. If length were measured in inches, how would the slope of the regression line given in Exercise 34 change?

39. The mean height of American women in their early twenties is about 64.5 inches and the standard deviation is about 2.5 inches. The mean height of men the same age is about 68.5 inches, with a standard deviation of about 2.7 inches.

(a) If the correlation between the heights of married heterosexual men and their wives is about $r = 0.5$, what is the equation of the regression line of the husband's height on the wife's height in young couples?

(b) Predict the height of the husband of a married heterosexual woman who is 67 inches tall.

40. These data, from the National Oceanic and Atmospheric Administration website (www.noaa.gov), are the mean annual number of named Atlantic storms (hurricanes, tropical storms, and tropical depressions), during five-year windows ending with the year shown in the table:

Five-year period ending	2012	2007	2002	1997	1992	1987	1982	1977
Number named storms	16.2	16.2	13.6	11.0	10.4	8.2	10.0	8.8
Five-year period ending	1972	1967	1962	1957	1952	1947	1942	
Number named storms	11.2	9.2	8.8	10.6	10.4	9.4	7.4	

(a) What is the slope of the least-squares regression line of named storms against year? What is the intercept?

(b) Use the regression line to predict the mean annual number of named storms for the five-year window ending with the year 2017.

41. Use the general equation for the least-squares regression line to show that this line always passes through the point $(\bar{x}, \bar{y})$. That is, set $x = \bar{x}$ and show that the line predicts that $y = \bar{y}$.

42. Exercise 6 gives data on the growth of an icicle (page 278).

(a) Find the mean and standard deviation of the times and icicle lengths. Find the correlation between the two variables. Use these five numbers to find the equation of the regression line for predicting length from time. Verify that your result agrees with that given in Exercise 34.

(b) Use the same five numbers to find the equation of the regression line for predicting from an icicle's length the time that it has been growing. Use your line to predict the time that an icicle 15 centimeters long has been growing. *There is just one correlation between two variables, but there are two different least-squares lines, depending on which variable you choose as the response variable.*

43. Fidelity Investments, like other large mutual fund companies, offers many "sector funds" that concentrate their investments in narrow segments of the stock market. These funds often rise or fall by much more than the market as a whole. Here are the percent returns for 23 Fidelity "Select Portfolios" funds for the years 2002 (when stocks fell) and 2003 (when stocks went up).

2002 return	2003 return	2002 return	2003 return	2002 return	2003 return
−17.1	23.9	−0.7	36.9	−37.8	59.4
−6.7	14.1	−5.6	27.5	−11.5	22.9
−21.1	41.8	−26.9	26.1	−0.7	36.9
−12.8	43.9	−42.0	62.7	64.3	32.1
−18.9	31.1	−47.8	68.1	−9.6	28.7
−7.7	32.3	−50.5	71.9	−11.7	29.5
−17.2	36.5	−49.5	57.0	−2.3	19.1
−11.4	30.6	−23.4	35.0		

Do a careful statistical analysis of these data, using both graphs and whatever numerical measures you think are appropriate. Make a side-by-side comparison of the distributions of returns in 2002 and 2003 and also describe the relationship between the returns of the same funds in these two years. What are your most important findings? (The outlier is Fidelity Gold Fund.)

6.5 Interpreting Correlation and Regression

44. Here are data collected on six individuals:

x	1	2	3	4	10	10
y	1	3	3	5	1	11

(a) Make a scatterplot of the data.

(b) Use your calculator to show that the correlation is about 0.5.

(c) What feature of the data is responsible for reducing the correlation to this value despite a strong straight-line association between x and y in most of the observations?

45. Table 6.10 offers four datasets prepared by statistician Frank Anscombe to show the dangers of calculating without first plotting the data.

(a) Without making scatterplots, find the correlation and least-squares regression line for all four datasets. What do you notice? Use the regression line to predict y for $x = 10$.

(b) Make a scatterplot for each of the datasets and add the regression line to each plot.

(c) In which of the four cases would you be willing to use the regression line to describe the dependence of y on x? Explain your answer in each case.

46. Children who watch many hours of TV get lower grades in school, on average, than those who watch less TV. Explain clearly why this fact does not show that watching TV causes poor grades. In particular, suggest some other characteristics of households where children watch lots of TV that may contribute to poor grades.

47. People who use artificial sweeteners in place of sugar tend to be heavier than people who use sugar. Does this mean that artificial sweeteners cause weight gain? Give a more plausible explanation for this association.

48. "Based on an examination of 22 companies that announced large layoffs during 1994, Downs found a strong correlation between the size of the layoffs and the compensation of the CEOs" (K. Phillips, *Wealth and Democracy*, Broadway Books, New York, 2002, p. 151). Discuss why this positive correlation is probably explained by a third variable, the size of the company as measured by its number of employees.

49. "The positive correlation between health and income per capita is one of the best-known relations in international development. This correlation is commonly thought to reflect a causal link running from income to health. … Recently, however, another intriguing possibility has emerged: that the health–income

TABLE 6.10 Four Datasets for Exploring Correlation and Regression

Dataset A											
x	10	8	13	9	11	14	6	4	12	7	5
y	8.04	6.95	7.58	8.81	8.33	9.96	7.24	4.26	10.84	4.82	5.68
Dataset B											
x	10	8	13	9	11	14	6	4	12	7	5
y	9.14	8.14	8.74	8.77	9.26	8.10	6.13	3.10	9.13	7.26	4.74
Dataset C											
x	10	8	13	9	11	14	6	4	12	7	5
y	7.46	6.77	12.74	7.11	7.81	8.84	6.08	5.39	8.15	6.42	5.73
Dataset D											
x	8	8	8	8	8	8	8	8	8	8	19
y	6.58	5.76	7.71	8.84	8.47	7.04	5.25	5.56	7.91	6.89	12.50

Data from Frank J. Anscombe, Graphs in statistical analysis, *The American Statistician, 27* (1973): 17–21.

correlation is partly explained by a causal link running the other way—from health to income" [D. E. Bloom and D. Canning, The health and wealth of nations, *Science, 287* (2000): 1207–1208]. Explain how higher income in a nation can cause better health. Then explain how better health can cause higher national income. There is no simple way to determine the direction of the link.

50. The effect of an outside variable can be surprising when individuals are divided into groups. In recent years, the mean SAT score of all high school seniors has increased. But the mean SAT score has decreased for students at each level of high school grades (A, B, C, etc.). Explain how grade inflation in high school can account for this pattern. *A relationship that holds for each group within a population need not hold (and may even be in the opposite direction!) for the population as a whole.*

Chapter Review

51. Consider the dataset below:

x	4	8
y	7	12

(a) Calculate $\bar{x}$, $\bar{y}$, s_x, and s_y.

(b) Calculate the correlation r. Round your answer to two decimal places. (Why should you not be surprised at this result?)

(c) Use your results from parts (a) and (b) to find the slope and y-intercept of the least-squares regression line.

(d) In this case, did you really need to work through the procedure for finding the least-squares regression line in order to find the equation of the best-fitting line? Explain.

52. A vehicle's tire pressure can affect tire wear (in terms of the length of life of the tire). When tire pressure is low, the tire flattens out and more of its surface contacts the road. This causes more friction between the tire and the roadway and increases the amount of wear on the tire. Data collected on tire pressure and tire wear appear in the table below:

Tire Pressure (psi), x	Tire Wear (in thousands of miles driven), y
30	29.8
30	30.2
31	32.4
31	34.5
32	36.2
32	35.0
33	38.4
33	37.6

(a) Fit a least-squares regression line to these data. What is the equation?

(b) Make a scatterplot of the data with a graph of the least-squares regression line superimposed. Does the line do a reasonable job of describing the pattern in these data?

(c) Use the equation of the least-squares line to predict the tire wear corresponding to a tire pressure of 40 psi. Is this an example of interpolation or extrapolation?

(d) Four more data values shown in the table below were collected for tire pressures higher than 33 psi. Redraw your scatterplot so that it includes these additional data values. Does the least-squares regression line from part (a) do a reasonable job of describing the pattern in the complete dataset? Explain.

Tire Pressure (psi), x	Tire Wear (1000 miles), y
34	38.0
34	37.2
35	35.3
35	34.6

(e) Based on your answer to part (d), comment on the accuracy of your prediction in part (c).

53. Major recalls of toys with lead paint refocused people on the dangers of lead exposure. Below are data from research exploring the association with student achievement for blood lead levels below the "danger threshold" of 10 mcg/dl set by the Centers for Disease Control [M. L. Miranda et al., The relationship between early childhood blood lead levels and performance on end-of-grade tests, *Environmental Health Perspectives, 115* (2007): 1242–1247].

Blood lead level	1	2	3	4	5
Mean fourth-grade reading score	255.9	253.8	252.6	251.0	250.4
Blood lead level	6	7	8	9	
Mean fourth-grade reading score	249.5	248.5	247.8	249.3	

(a) What are the explanatory and response variables?

(b) Do you expect a positive or negative association between these variables? Why? Does the scatterplot support your answer?

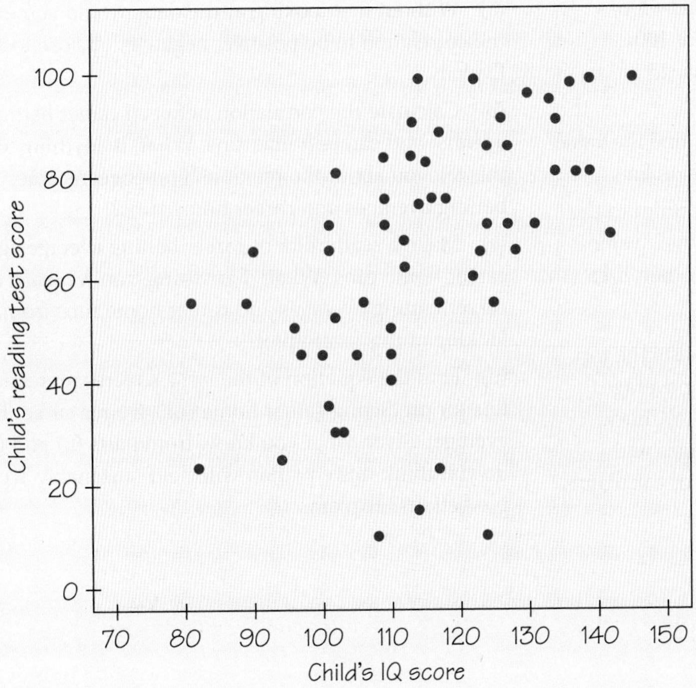

Figure 6.24 IQ and reading test scores for 60 fifth-grade children, for Exercise 54.

54. A study of reading ability in schoolchildren chose 60 fifth-grade children at random from a school. The researchers obtained the children's scores on an IQ test and on a test of reading ability. Figure 6.24 plots reading test score (response variable) against IQ score (explanatory variable).

(a) Explain why we should expect a positive association between IQ score and reading score for children in the same grade.

(b) Does the scatterplot show a positive association?

(c) Four points in a group appear to be outliers. In what way do these children's IQ and reading scores deviate from the overall pattern?

(d) Ignoring the outliers, is the form of the association between IQ score and reading score roughly a straight line? Is it very strong? Explain your answers.

55. A student wonders if tall women tend to date taller people than do short women. She measures herself, her sister, and the women in the adjoining dorm rooms. Then she measures the next person each woman dates and obtains the following data (in inches):

Heights of women (*x*)	66	64	63	65	70	65
Heights of their dates (*y*)	72	68	70	68	71	64

(a) Based on a scatterplot (with the women's heights as the explanatory variable), do you expect the correlation to be positive or negative? Near ±1 or not?

(b) Find the correlation *r* between the heights of the women and their dates.

56. In Exercise 55, you found the correlation *r* between the heights in inches of several college women and the heights in inches of the next person each woman dates.

(a) How would *r* change if all the dates were 2 inches shorter than the heights given in the table?

(b) How would *r* change if heights were measured in centimeters rather than inches? (*Note:* 1 inch = 2.54 cm.)

57. The equation of the least-squares regression line for predicting dates' heights from women's heights for the data in Exercise 55 is

predicted height of date = 41.08 + 0.42 × woman's height

(a) What is the slope of this line?

(b) Explain in simple language what the numerical value of the slope tells us about the heights of the people these women date.

(c) Use the regression line to predict the height of the next person dated by a woman who is 67 inches tall.

58. From 2000 to 2005, sales and file sharing (i.e., free downloading) intensity were tracked within seven musical genres (rock, alternative, R&B, rap/hip-hop, country, jazz, classical). The correlation between change in sales and file-sharing intensity was −0.648. Is this evidence that file sharing helps or hurts sales? Explain.

59. In issue 49 of *Stats: The Magazine for Students of Statistics*, Schuyler Huck presents a dataset of 100 ordered pairs in which 25 of them are (17, 1), 25 are (18, 2), 25 are (19, 3), and 25 are (20, 4).

(a) Without doing much formal calculation, find the value of r and the slope of the least-squares regression line.

(b) Now, suppose someone adds the 101st point to the dataset: the ordered pair (1, 20). Predict the new value of r and the slope of the regression line, and then do a calculation to see how close your answer is.

60. Return to Table 5.13 (page 226), which lists the top 100 baseball players ranked by career batting average. Consider only the top 50 players—Ty Cobb to Chuck Klein. Enter the data on career batting averages and career home runs into an Excel spreadsheet or calculator lists.

(a) Without first looking at the data, would you expect the correlation to be positive, negative, or close to zero? Explain.

(b) Calculate the correlation between career batting average and career home runs. What, if anything, does this tell you about the relationship between career batting averages and career home runs?

(c) Make a scatterplot of career batting average against career home runs. What, if anything, can we learn about career batting averages and career home runs from looking at this scatterplot?

(d) Find the equation of the least-squares regression line for predicting career home runs from a career batting average. Given what you know from parts (b) and (c), do you think this equation will yield reasonably accurate predictions? Explain.

Applet Exercises

To do these exercises, go to www.macmillanhighered.com/fapp10e.

1. In the *Correlation and Regression* applet, imitate Figure 6.16 (page 269). Click to locate five points at the lower left of the scatterplot, and then click "Show least-squares line."

(a) What is the correlation r for these five points? If necessary, move points with the mouse to get a value near $r = 0.5$, as in Example 9 (page 268).

(b) Now add an outlier at the upper right that lies exactly on the line. What is the correlation r for the six points?

(c) Use the mouse to drag the outlier down and then to the left. Watch the least-squares line follow this one point. How negative can you make the correlation r?

(d) What have you learned from parts (a) through (c) about the effect that a single outlier can have on the correlation r?

2. You are going to use the *Correlation and Regression* applet to make different scatterplots with 10 points that have a correlation close to 0.7. *Many patterns can have the same correlation. Always plot your data before you trust a correlation.*

(a) Stop after adding the first two points. What is the value of the correlation? Why does it have this value no matter where the two points are located?

(b) Make a lower-left to upper-right pattern of 10 points with a correlation of about $r = 0.7$. (You can drag points up or down to adjust r after you have 10 points.) Make a rough sketch of your scatterplot.

(c) Make another scatterplot with nine points in a vertical stack at the left of the plot. Add one point far to the right and move it until the correlation is close to 0.7. Make a rough sketch of your scatterplot.

(d) Make yet another scatterplot with 10 points in a curved pattern that starts at the lower left, rises to the right, then falls again at the far right. Adjust the points up or down until you have a smooth curve with a correlation close to 0.7. Make a rough sketch of this scatterplot as well.

(e) Based on your answers to parts (b) through (d), what can you conclude about the pattern of dots in a scatterplot if $r \approx 0.7$?

3. It isn't easy to guess the position of the least-squares line by eye (at least not without some practice). Use the *Correlation and Regression* applet to compare a line that you draw with the least-squares line. Click on the scatterplot to create a group of 15 to 20 points from the lower left to the upper right with a clear, positive straight-line pattern (with a correlation of around 0.7). Click the "Draw your own line" button. Then click on two points, one in the lower left and the other in the upper right. Move these two points so that your line is drawn through the middle of the cloud of points and appears to do a good job of summarizing the pattern in the scatterplot.

(a) You drew a line by eye through the middle of the pattern. Read off the value for Relative SS (directly below the "Draw your own line" button). This is the ratio of the sum of squares of residuals for your line and the sum of squares of the residuals for the least-squares regression line.

(b) Move your line until the Relative SS = 1. (This means that the value of Relative SS is approximately 1, or equals 1 when rounded to two decimal places.) Your line should now closely match that of the least-squares regression line. To check this out, click the "Show least-squares line" box.

(c) Repeat this exercise several times with different sets of points. Try to guess the "best" line without looking at the value of Relative SS. Does your ability to pick a line that closely matches the least-squares regression line improve with practice?

4. This time you will use the *Correlation and Regression* applet to examine the residual errors. Click Clear to remove any work done for Exercises 1–3. Click on the scatterplot to create a group of 15 to 20 points from the lower left to the upper right with a clear, positive straight-line pattern (with a correlation in the moderate range, say between 0.65 and 0.75). Click the "Show least-squares line" and "Show residuals" buttons.

(a) How can you tell from this graph which of the residuals are positive and which are negative? Do the residuals appear roughly balanced between positive and negative values?

(b) Pick a point that has an x-value somewhere in the middle of x-values of the other data points. If this point lies below the line, drag it vertically down. If this point lies above the line, drag it up. As you drag the point vertically, what happens to the size of its residual? What happens to the slope (or the tilt) of the least-squares line? Then switch the direction in which you drag the point and note the effect on the slope of the line.

(c) Return the point you were dragging in part (b) to approximately its original position. Next, click on a point that has the largest x-value. Try dragging this point vertically both in the upward and downward direction. What effect does this have on the slope of the least-squares line?

(d) Which type of outlier, one with an x-value that lies near the middle of the x-values of the other data points or one with an x-value that lies near the maximum (or minimum) of the x-values of the other data points, will have a greater influence on the slope of the least-squares line?

Writing Projects

1. Choose two variables that you think have a roughly straight-line relationship. Gather data on these variables and do a statistical analysis: Make a scatterplot, find the correlation, find the regression line (use a graphing calculator or software), and draw the line on your plot. Then write a short report on your work. Some examples of suitable pairs of variables are as follows:

(a) The height and arm span of a group of people

(b) The height and walking stride length of a group of people

(c) The price per ounce and bottle size in ounces for several brands of shampoo and several bottle sizes for each brand

(d) The weight and fuel efficiency of motor vehicles [Google "vehicle weight" for lists of weights of specific vehicles, then look up the fuel efficiency (pick among city, highway, or combined mpg) of the particular vehicles.]

2. Can regression help protect voting rights? This example is adapted from *FAPP* author Lawrence Lesser's work as a statistician for the Texas Legislative Council. To comply with the Voting Rights Act, a state cannot redraw its districts in a way that dilutes the voting strength of a protected group. Because we cannot know how individuals voted, we cannot directly measure if minority and majority persons tend to prefer different candidates. Although there are technical details and assumptions we cannot fully discuss here, you can begin to understand how voting preference might be estimated by exploring the following dataset for nine equal-sized districts, where X is the percentage of voters who are Hispanic and Y is the percentage of voters that voted for the candidate preferred by most Hispanics:

Y	14	7	19	27	37	36	53	48	65
X	12	18	24	36	42	53	68	79	86
District	1	2	3	4	5	6	7	8	9

(a) Produce a scatterplot, correlation value, and regression equation. Describe the relationship between the concentration of Hispanic population and the proportion of votes that went to the Hispanic-preferred candidate.

(b) Give a practical interpretation of the value of the slope coefficient. Give a practical interpretation of the value of Y that would be predicted when $X = 0$ and when $X = 100$.

3. Spotlight 6.3 (page 260) discussed Sir Francis Galton's study of the heights of men and their fathers. In this writing assignment, you will adapt his study to students. In order to do so, you will need to collect the following data from one or two classes:

- Have male students record their heights and the heights of their fathers.
- Have female students record their heights and the heights of their mothers.

(a) Using parent's height as the explanatory variable, make two scatterplots, one for the males and the other for the females. Find the correlations between parent's height and student's height for males and females. Compare the patterns in the two scatterplots for form, strength, and direction.

(b) Discuss the following algebraic representation of regression to the mean for males and females:

$$\frac{\hat{y} - \bar{y}}{s_y} = r\frac{x - \bar{x}}{s_x}$$

(c) Determine the equations for the least-squares regression lines for predicting student's height from parent's height. Compare and interpret the slopes of your two lines.

4. Table 5.13 (page 226) contains data on baseball players ranked in the top 100 in terms of career batting averages. Exercises 9 and 10 in Chapter 5 and Exercise 60 in Chapter 6 were based on these data. Write a report based on your analysis of these data. Feel free to include analysis of other data that can be found at www.baseball-reference.com/. (You will need to use technology—graphing calculators, spreadsheet, or statistical software—for this project.)

Suggested Readings

CLEVELAND, WILLIAM S. *The Elements of Graphing Data,* rev. ed., Hobart Press, Summit, N.J., 1994. A careful study of the most effective elementary ways to present data graphically, with much sound advice on improving graphs such as scatterplots.

LESSER, LAWRENCE M. The "Ys" and "why nots" of line of best fit, *Teaching Statistics, 21*(2) (1999): 54–55.

MOORE, DAVID S., WILLIAM I. NOTZ, and MICHAEL A. FLIGNER. *The Basic Practice of Statistics (BPS),* 7th ed., Freeman, New York, 2015. Chapters 4 and 5 of *BPS* give more extensive treatment of *FAPP* Chapter 6 material, at about the same mathematical level.

Suggested Websites

www.census.gov or **fedstats.sites.usa.gov/** The websites suggested in Chapter 5 as data sources provide information enabling us to investigate relationships as well. How is the number of medical doctors per 100,000 people in each state related to how rich a state is? To infant mortality in the state? To the cost of medical care? You can study these and many other relationships using data from government websites.

www.learner.org/courses/againstallodds Want to view videos showing applications of the material from this chapter? Find out how the deaths of Florida manatees are related to the number of powerboat registrations, or how height can be predicted from skeletal remains; or learn about the classic correlation study on genes versus the environment. You'll find these examples and more in Units 10–12 of *Against All Odds: Inside Statistics.*

www.amstat.org/publications/jse The *Journal of Statistics Education* contains many articles with interesting data and examples. Look, for example, in its archive for "Exploring Relationships in Body Dimensions" by G. Heinz et al. in the July 2003 issue. Here, you will find information about measuring body dimensions, actual data from 247 men and 260 women, and some examples of both distributions and relationships.

www.causeweb.org CAUSEweb is a searchable digital library of resources on a wide range of statistics topics offered by the Consortium for the Advancement of Undergraduate Statistics Education.

demonstrations.wolfram.com Wolfram Demonstrations Project contains a variety of interactive applets. Click "About" at the top of the page to find out how to download the free Computable Document Format (CDF) player for these applets. Then Click "Explore" to get to the topic index and select the following progression of topics: Mathematics > Statistics > Data Analysis. Check out these applets: *Correlation and Regression Explorer, Regression Toward the Mean,* and *Local Regression for Country Data.*

Data for Decisions

7

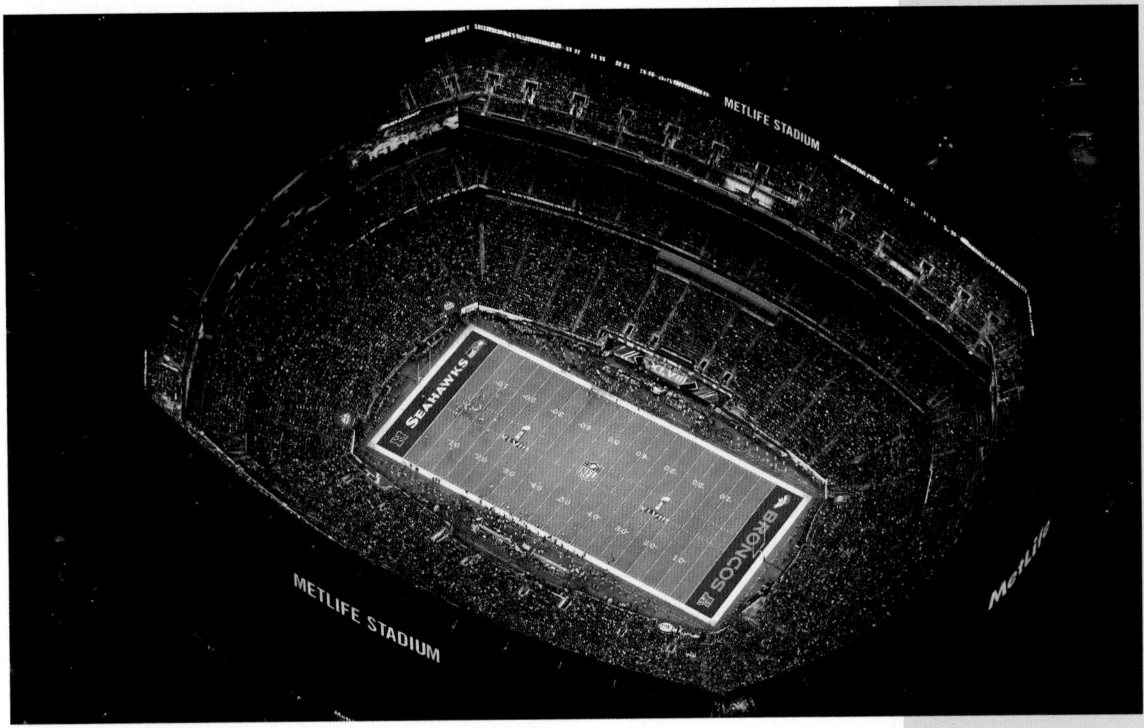

John Moore/Getty Images

The day after Super Bowl XLVIII in 2014, it was announced that the game was viewed by an estimated 111.5 million viewers, making that game the most-watched show in U.S. history. A politician looks at the latest poll to decide if a new strategy or policy will have enough support, or if he would win if an election were held that day. A doctor looks at a study in a medical journal to decide if there is strong enough evidence in a new approach to treating a disease for her to recommend it to a patient. Every day that we read a newspaper like *USA Today,* we encounter the results of some kind of poll or study that affects commerce, politics, or the growth of knowledge. Clearly, data can be used to help answer a huge variety of questions in our world.

However, we also have to think about *how data are produced* in the first place. We can analyze data using the most sophisticated techniques, but if the data are not reliable, our conclusions will be baseless. In this chapter, we explore three ways of collecting data—surveys, experiments, and observational studies—and we examine the reliability of the data produced in each of these situations. The chapter's concluding section discusses *statistical inference*, which allows us to go beyond describing a sample to making an estimate about the entire population from which our sample comes and to quantify our level of confidence when doing so.

7.1 Sampling

- A political scientist wants to know what percentage of college students consider themselves conservatives.
- An automaker hires a market research firm to learn what percentage of adults aged 18 to 35 recall seeing television advertisements for a new sport utility vehicle.
- Government economists inquire about average household income for Americans.

In all these cases, we want to gather information about a large group of individuals. Time, cost, and inconvenience preclude contacting every person, so we gather information about only part of the group in order to draw conclusions about the whole. In some cases when gathering data, the subject is destroyed. For example, when Frito Lay checks the salt level of the potato chips it produces, it tests only a small sample of chips so that the company still has product to sell. And if your doctor's appointment includes a blood test, you want only *some* of your blood removed! In these situations, it is obviously necessary to use only a sample.

Population DEFINITION

The **population** in a statistical study is the entire group of individuals about which we want information.

Sample DEFINITION

A **sample** is a part of the population from which we actually collect information that is used to draw conclusions about the whole. *Sampling* refers to the process of choosing a sample from the population.

For an example of a population and sample, let's refer back to the first bullet listed above—a political scientist wants to know what percentage of college students consider themselves conservatives. Here, the population would be all college students (or perhaps just the millions of college students in the United States), and the sample would be the small subset of students (typically between 500 and 1000) actually selected to participate in the sample.

We often draw conclusions about a whole on the basis of a sample. Everyone has sipped a spoonful of soup and judged the entire bowl on the basis of that taste. But a bowl of soup is homogeneous, so the taste of a single spoonful does represent the whole. On the other hand, a spoonful of salad dressing may be misleading because its elements may separate if the bottle has not been shaken recently. A spoonful taken from the top might be mostly oil.

MartineDee/Shutterstock

Vasily Kovalev/Shutterstock

Choosing a representative sample from a large and varied population can be difficult. The first step in a proper sample survey is to state carefully just what population we want to describe. The second step is to define exactly what we want to measure. These preliminary steps can be complicated, as the following example illustrates.

EXAMPLE 1 ➔ How Can a Survey Measure Unemployment?

The monthly unemployment rate comes from the government's Current Population Survey (CPS; www.census.gov/cps/), which involves a sample of about 50,000 households each month. To measure unemployment, we must first specify the population that we want to describe:

- Which age groups will we include?
- Will we include illegal immigrants or people in prisons?
- Should we include military personnel?

The CPS defines its population as all U.S. residents (whether citizens or not), 16 years of age and over, who are civilians and not in an institution such as a prison. The civilian unemployment rate announced in the news refers to this specific population.

The second question is more difficult: What does the term *unemployed* mean? Someone who is not looking for work—for example, a full-time student—should not be called unemployed just because he or she is not working for pay. If you are chosen for the CPS sample, the interviewer first asks whether you are available to work and whether you actually looked for work in the past four weeks. If not, you are neither employed nor unemployed; you are not in the labor force.

If you are in the labor force, the interviewer goes on to ask about employment. Any work for pay that you performed the week of the survey, whether for someone else or in your own business, qualifies you to be counted as employed. So does at least 15 hours of unpaid work in a family business. In addition, you are considered employed if you have a job but didn't work for reasons such as being on vacation or on strike.

So, an unemployment rate of 6.7% means that 6.7% of the sample was unemployed, using the exact CPS definitions of both *labor force* and *unemployed*.

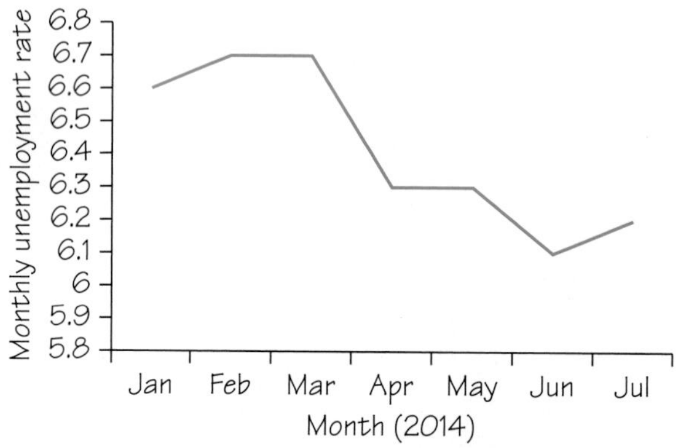

Figure 7.1 Line graph showing the monthly unemployment rates calculated from CPS data for the first seven months of 2014.

Self Check 1

A local television station conducts nightly polls of public opinion by announcing a question on the 6 o'clock news and asking viewers to call in or text their response of "yes" or "no" to the station. The results are announced on the 11 o'clock news later the same night. One such poll finds that 76% of those who called in or texted are opposed to a proposed local gun control ordinance.

(a) What do you think the population is in this situation?
(b) Do you believe this sample is representative of the population? Explain.

7.2 Bad Sampling Methods

How can we choose a sample that is truly representative of the population? The easiest way, but not the best way, to select a sample is to choose individuals close at hand. If we are interested in finding out how many people have jobs, for example, we could go to a shopping mall and ask people passing by if they are employed.

Convenience Sample DEFINITION

A **convenience sample** is a sample of individuals who are selected because they are members of a population who are easy (i.e., convenient) to reach. Usually, such a sample cannot be trusted to be representative of the population.

EXAMPLE 2 The Inconvenient Truth About Convenience Samples

Going to the mall, standing by a particular entrance, and surveying as many of the people walking through that entrance as you can seems like a fast, convenient way of finding out Americans' opinions. But people at malls tend to be more prosperous than typical Americans. They are also more likely to be teenagers or retired. The kinds of stores that are near the particular entrance you are standing by could affect the type of people you might more readily encounter. Also, when we decide which people to approach, we may tend (even unconsciously) to avoid poorly dressed or tough-looking individuals. In short, our shopping mall interviews will result in a sample that is not representative of the entire population because we underrepresent certain types of people. For that matter, we also are underrepresenting those Americans who rarely go to malls in the first place.

Bjanka Kadic/Alamy

In your classroom, your professor may try to "sample" the understanding the class has about a topic by simply calling on the nearest two students in the front row. If students who sit near the front have higher levels of preparation, interest, and engagement, the professor will overestimate how well the class as a whole understands the material. ■

In both scenarios in Example 2, the inaccuracies obtained cannot simply be explained as a sample's "bad luck." They are likely to happen every time, with the same pattern, because unscientific sampling methods have **bias.** In this context, bias refers matter-of-factly to the built-in systematic error of the procedure itself and not to any political or personal prejudice that the person conducting the poll may have.

Bias DEFINITION

The design of a statistical study has **bias** (i.e., is biased) if it systematically favors certain outcomes.

EXAMPLE 3 Are Online Polls in Line?

The American Family Association (AFA) is a conservative group that claims to stand for "traditional family values." It has often posted online poll questions on its website; just click on a response to take part. Because the respondents are people who visit this site, the poll results always support AFA's positions. Well, almost always; a recent AFA online poll asked about allowing same-sex marriage, and before long, email lists and social-network sites favored mostly by young liberals pointed to the AFA poll. Almost 850,000 people responded, and 60% of them favored the legalization of same-sex marriage. This example shows that the results of an online poll can be skewed one way or the other by particular characteristics of the people who choose to go to that website and participate in the poll.

A related example is the website www.ratemyprofessors.com, where students have chosen to evaluate and post comments on more than 1 million college and university instructors worldwide. However, focus group research indicates the ratings may not be representative because the students

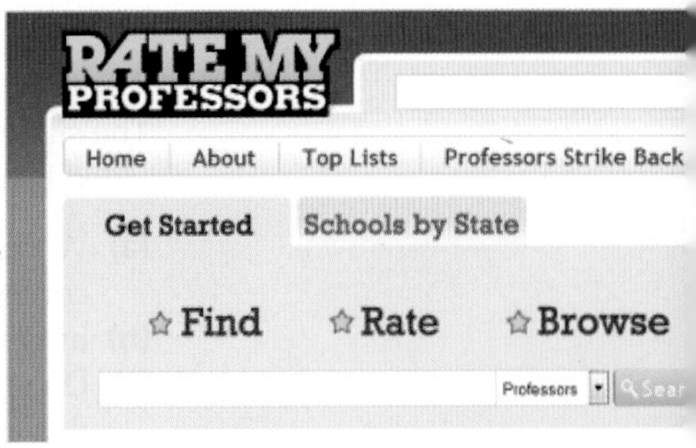

Source: Rate My Professors

most motivated to post assessments are those who believe the teacher is either extremely bad or extremely good. A different kind of problem is that this website may have no way of keeping out multiple responses from the same student—or even from the instructors themselves! It is usually better to rely on the official end-of-course evaluation data compiled by the university. ■

Online polls are now everywhere; some sites such as SurveyMonkey will even provide help in conducting your own online poll. As Example 3 illustrates, however, the results can't be trusted. People who take the trouble to write in, call in, or visit a website to respond to an open invitation are not representative of the general population. Polls like these are examples of **voluntary response sampling.**

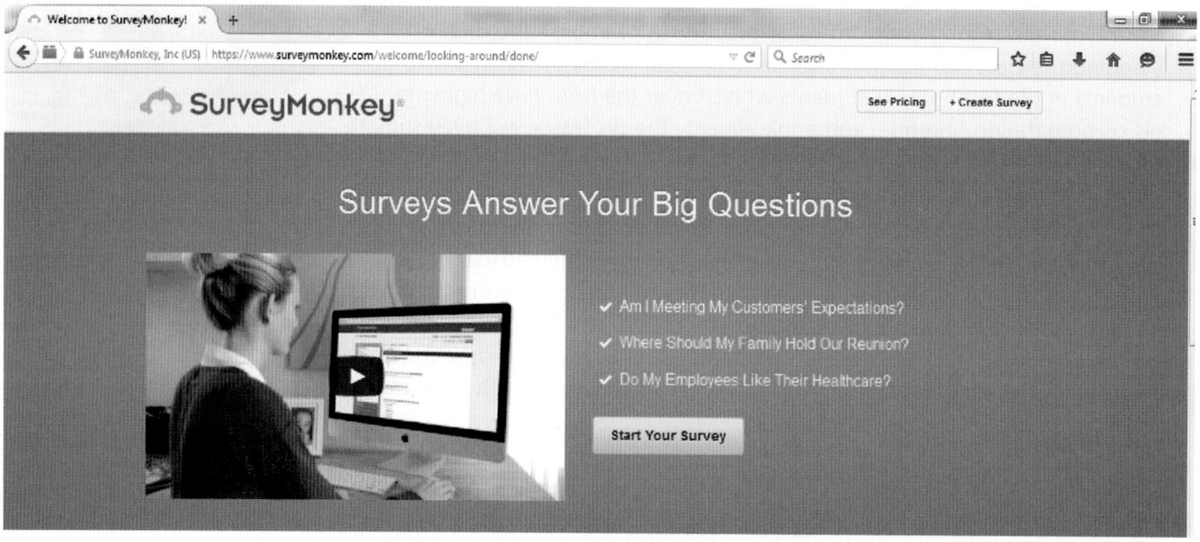

Source: SurveyMonkey

Voluntary Response Sample DEFINITION

A **voluntary response sample** consists of people who choose themselves by responding to a general appeal. Voluntary response samples are biased because people with strong opinions are more likely to respond and will therefore be overrepresented in the sample.

Self Check 2

A hospital sends out surveys to patients who visited there for routine medical tests. The survey asks questions related to the patients' satisfaction with the service provided by the hospital. The hospital analyzes the data from all responses.

(a) What is the sample? Is this an example of a voluntary response sample or a convenience sample?
(b) In what population is the hospital interested?
(c) Do you expect these survey results to accurately portray patients' satisfaction? Explain. ■

7.3 Simple Random Samples

In a voluntary response sample, people choose whether to respond. In a convenience sample, the interviewer makes the choice. In both cases, personal choice produces bias. The statistician's remedy (pioneered by such people as George Gallup in the 1930s) is to allow impersonal chance to choose the sample. A sample chosen by chance allows neither favoritism by the sampler nor self-selection by respondents. Choosing a sample by chance avoids bias by giving all individuals an equal chance to be chosen. Any individual, whether rich or poor, young or old, black or white, and so on, has the same chance to be included in the sample.

The simplest way to use chance to select a sample is to place slips of paper with the names of all individuals in the population in a hat, shake the hat vigorously, and then draw out only a few names (the sample). This is the idea behind **simple random sampling.**

Simple Random Sample DEFINITION

A **simple random sample (SRS)** of size n consists of n individuals from the population chosen in such a way that every set of n individuals has an equal chance to be in the sample actually selected.

Picturing drawing names from a hat helps us understand what an SRS is. The same picture helps us see that an SRS is a better method of choosing samples than convenience or voluntary response sampling because it doesn't favor any part of the population. But writing names on slips of paper and drawing them from a hat is a slow and inconvenient process, especially if, as in the CPS (Example 1), we must draw a sample of 50,000 participants from slips of paper representing all U.S. households. We can speed up the process by using a table of random digits. In practice, samplers use computers to do the work, but we can start doing it by hand for small samples from relatively small populations. Then we can move on to using technology to select the sample.

Table of Random Digits DEFINITION

A **table of random digits** is a list of the digits 0, 1, 2, 3, 4, 5, 6, 7, 8, 9 with these two properties:

1. Each entry in the table is equally likely to be any of the ten digits 0 through 9.
2. The entries are independent of one another. That is, knowledge of one part of the table gives no information about any other part.

James Steidl/Stock/Getty Images Plus

Table 7.1 is an excerpt from a table of random digits. (You can find longer tables on the Internet.) The digits in the table are displayed in groups of 5 to make the table easier to read, and the rows are numbered so we can refer to them, but the groups and row numbers are just for convenience. Table 7.1 in its entirety is one long string of 1000 randomly chosen digits. In order to use the table to select a sample, keep the following in mind:

- Each single entry in the table is equally likely to be any digit 0 through 9.
- Two consecutive digits from the table are equally likely to be any of the 100 possibilities 00 to 99.
- Three consecutive digits from the table are equally likely to be any of the 1000 possibilities 000 to 999.

TABLE 7.1 Random Digits

101	19223	95034	05756	28713	96409	12531	42544	82853
102	73676	47150	99400	01927	27754	42648	82425	36290
103	45467	71709	77558	00095	32863	29485	82226	90056
104	52711	38889	93074	60227	40011	85848	48767	52573
105	95592	94007	69971	91481	60779	53791	17297	59335
106	68417	35013	15529	72765	85089	57067	50211	47487
107	82739	57890	20807	47511	81676	55300	94383	14893
108	60940	72024	17868	24943	61790	90656	87964	18883
109	36009	19365	15412	39638	85453	46816	83485	41979
110	38448	48789	18338	24697	39364	42006	76688	08708
111	81486	69487	60513	09297	00412	71238	27649	39950
112	59636	88804	04634	71197	19352	73089	84898	45785
113	62568	70206	40325	03699	71080	22553	11486	11776
114	45149	32992	75730	66280	03819	56202	02938	70915
115	61041	77684	94322	24709	73698	14526	31893	32592
116	14459	26056	31424	80371	65103	62253	50490	61181
117	38167	98532	62183	70632	23417	26185	41448	75532
118	73190	32533	04470	29669	84407	90785	65956	86382
119	95857	07118	87664	92099	58806	66979	98624	84826
120	35476	55972	39421	65850	04266	35435	43742	11937
121	71487	09984	29077	14863	61683	47052	62224	51025
122	13873	81598	95052	90908	73592	75186	87136	95761
123	54580	81507	27102	56027	55892	33063	41842	81868
124	71035	09001	43367	49497	72719	96758	27611	91596
125	96746	12149	37823	71868	18442	35119	62103	39244

Here is the procedure to use the random-digit table to choose a simple random sample.

Using a Table of Random Digits

PROCEDURE

Step 1. **Label.** Give each member of the population a numerical label of the same length. Up to 10 items can be labeled with one digit: 1, 2, . . . 9, 0. Up to 100 items can be labeled with two digits: 01, 02, . . . , 99, 00. Up to 1000 items can be labeled with three digits, and so on.

Step 2. **Pick a row from the table.** Without looking at the table, pick a row (between line numbers 101 and 125, inclusive). From the row you picked, read successive groups of digits of the same length that you used as labels. Just skip over all groups of digits not used as labels or that duplicate a label already selected (unless you want to allow repeated values). If you run out of digits in a row before you have reached your sample size, continue to the next row (or repeat Step 2 and continue on another row from the table).

Step 3. **Stop.** It is generally easiest to make a relatively long list of groups of digits of the desired length and then to cross out nonlabel groups and duplicates. Stop when you have identified sufficient labels for the sample size that you need.

Step 4. **Identify the individuals in your sample.** Your sample contains the individuals corresponding to the labels you found in Steps 2 and 3.

EXAMPLE 4 Sampling Songs

Professor Lesser has all 27 songs from the album called *The Beatles One* stored on a digital media player and wants to play four randomly chosen songs to accompany his morning commute. Let's follow the four-step procedure for using the random-digit table to choose a simple random sample of size 4 from the 27-song playlist.

Step 1. Give each song a numerical label. Because two digits are needed to label the 27 songs, all the labels will have two digits. Table 7.2 lists the 27 songs with labels from 01 to 27.

TABLE 7.2 Songs from *The Beatles One* Album, Assigned to Two-Digit Labels

01 Love Me Do	10 Help!	19 Hello, Goodbye
02 From Me to You	11 Yesterday	20 Lady Madonna
03 She Loves You	12 Day Tripper	21 Hey Jude
04 I Want to Hold Your Hand	13 We Can Work It Out	22 Get Back
05 Can't Buy Me Love	14 Paperback Writer	23 The Ballad of John and Yoko
06 A Hard Day's Night	15 Yellow Submarine	24 Something
07 I Feel Fine	16 Eleanor Rigby	25 Come Together
08 Eight Days a Week	17 Penny Lane	26 Let It Be
09 Ticket to Ride	18 All You Need Is Love	27 The Long and Winding Road

Step 2. We pick line 125 from Table 7.1. We begin writing two-digit groups as we read across line 125 of the table (since our song labels have two digits). We cross out any two-digit groups that are greater than our population size of 27 and any duplicates (18 is a duplicate):

~~96~~ ~~74~~ ~~61~~ 21 ~~49~~ ~~37~~ ~~82~~ ~~37~~ 18 ~~68~~ ~~18~~ ~~44~~ 23 ~~51~~ 19 62 10 33 92 44

Step 3. We stop when we reach label 19 because we've identified 4 labels for our sample.

Step 4. Corresponding to the selected labels shaded in Step 2, our media player will play the following song sequence: "Hey Jude," "All You Need Is Love," "The Ballad of John and Yoko," and "Hello, Goodbye." ■

Self Check 3

Use the process outlined in Example 4 to select a simple random sample of 5 songs. Enter the random digits table, Table 7.1, on line 110. ■

After using the random number table to select the random sample in Self Check 3, you are probably ready to turn to technology to generate random samples. When the population from which the sample is selected is relatively small (≤144), you could use the *Simple Random Sample* applet to select your sample (see Applet Exercise 1, page 338 for directions). Spotlight 7.1 outlines the steps for using a TI-84 calculator or Excel to select the sample.

Using Technology to Select an SRS

SPOTLIGHT 7.1

Graphing calculators and spreadsheet and statistical software have random number generators. Rather than using the random digits table, you can use technology to select the random sample. Below are instructions for completing Self Check 3 using a TI-84 graphing calculator or Excel.

TI-84 Instructions

Below are instructions on using a TI-84 graphing calculator to select a sample of size 4 from the 27 songs listed in Table 7.2.

- Press [MATH] → PRB and then [5] (for randInt( ).

```
MATH NUM CPX PRB
1:rand
2:nPr
3:nCr
4:!
5:randInt(
6:randNorm(
7↓randBin(
```

- Complete the command by pressing [1] [,] [27] [)].
- Press [ENTER] repeatedly to generate a list of labels of the songs in the random sample. (If there are repeats in your sample, just skip over them.)

```
randInt(1,27)
                11
randInt(1,27)
                20
randInt(1,27)
                 2
```

On newer TI-84 Plus calculators, you can avoid the problem with repeats by using the command randInt-NoRep instead.

- Press [MATH] → PRB and then [8] (for randIntNoRep(.
- Complete the command by pressing [1] [,] [27] [)].

This command will put the entire set of 27 labels in random order. So, start at the beginning of the output list and read off the first four labels for your sample.

Using Technology to Select an SRS

Excel Instructions

In an empty cell in row 1, enter = RANDBETWEEN(1,27) and press ENTER. The label of the first song in the sample will appear. Select this cell, then click on the lower right corner of the selected cell, and drag down to row 5. The column will now contain the labels of the songs in the random sample. If there are repeats in your sample, click on the cell in row 5 and continue to drag down to randomly produce more label numbers until you have your desired sample size.

Using Rand to Select the Sample

Using randInt on the TI-84 or RANDBETWEEN on Excel to select a sample works really well when the sample size is small relative to the population size. Otherwise, eliminating duplicates becomes a time-consuming endeavor. An alternative is to use Rand, which generates an observation from the uniform distribution on the interval from 0 to 1 (see Chapter 5, Exercise 73, and Figure 5.37 on page 238).

Suppose you want to draw a random sample of 5 students from a class of 10 students:

Step 1. Enter the students' names (or label numbers) into a column of an Excel spreadsheet. Name the column Students (see Table 7.3).

Step 2. Name the next column Rand. Then in this column, click on the cell in the row opposite the top name. Enter =RAND() and then press ENTER. Click to select this cell, then click the lower right corner of the cell, hold and drag down until you have reached the last name on the list. You now have a column of randomly drawn observations from the uniform distribution.

Step 3. Highlight the columns Students and Rand. Click the Sort icon and select Custom Sort. In the "Sort by" drop-down menu, select Rand and click OK. Your results will be similar to Table 7.4.

If for some reason you need to expand your sample size, simply go to the next name or group of names on the list.

TABLE 7.3 Columns of Student Names and Random Observations from the Uniform Distribution (Your random numbers will differ from those shown in this table.)

Students	Rand
Joe	0.305127
Sally	0.130861
Kelly	0.335956
Bruce	0.525466
Marsha	0.288252
Caitlin	0.036762
George	0.084562
Jian	0.763097
Cheryl	0.75338
Ying	0.77542

TABLE 7.4 Names Randomly Scrambled by Ordering the Column of Random Observations from Smallest to Largest (The names selected for the sample of size 5 are highlighted.)

Subjects	Rand
Caitlin	0.036762
George	0.084562
Sally	0.130861
Marsha	0.288252
Joe	0.305127
Kelly	0.335956
Bruce	0.525466
Cheryl	0.75338
Jian	0.763097
Ying	0.77542

So although digital media players may have a "shuffle" option to put a playlist in a randomized order, after reading Example 4 and Spotlight 7.1, you are now familiar with a procedure for obtaining a random sample for any situation where you have an ordered list for the population.

Online polls and mall interviews produce samples. We can't trust results from these samples because they are chosen in ways that invite bias. We have more confidence in results from an SRS because it uses impersonal chance to avoid bias. So, the first question to ask about any sample is whether it was chosen at random. Opinion polls and other sample surveys carried out by people who know what they are doing use random sampling. Most national sample surveys use sampling schemes that are more complex than SRS. For example, the monthly national sample from the Current Population Survey discussed in Example 1 is pieced together from many smaller samples. However, the big idea remains the deliberate use of chance to choose the sample.

EXAMPLE 5 The Plane Truth

According to a *USA Today* Gallup poll conducted in 2010, 71% of people who had flown at least twice in the past year believed the potential loss of personal privacy from full-body scans or pat-downs was worth it as a method to prevent acts of terrorism. How much do we trust the quality of this survey? Ask first how the sample was selected. On the Gallup website, we learn that the results were based on telephone interviews with a randomly selected national sample of 3018 adults selected using random-digit-dial sampling conducted November 19–21, 2010.

It is a good start toward confidence in the poll to know the intended population, the sample size, the tight window of time (so that there is minimal influence from changes in current events), and—most importantly—random selection. In the next section, we address a few other important considerations.

David McNew/Getty Images

7.4 Cautions About Sample Surveys

Random sampling eliminates bias in the choice of a sample from a list of the population. Sample surveys of large human populations, however, require more than a good sampling design to eliminate bias.

To begin with, we need an accurate and complete list of the population. Because such a list is rarely available, most samples suffer from some degree of **undercoverage.** A sample survey of households, for example, will miss not only homeless people but also prison inmates and students in dormitories. An opinion poll conducted by random-digit-dialing of landline phones will miss the more than 40% of U.S. households without landline phones. Hispanics, African Americans, younger adults, and the poor are more likely to rely solely on cell phones; hence, their views won't be represented in the opinion poll.

Pew Research Center

SPOTLIGHT 7.2

On January 15, 2014, the Pew Research Center announced that it was changing how it surveyed Americans by telephone. Starting in 2014, 60% of interviews in its national polls would be conducted via cellphones and 40% on landline phones. This represents a change from the 50-50 split used in 2013. Although cellphone surveys are more expensive to conduct than landline surveys, Pew believes that this change will improve the quality of its surveys. For example, with the 50-50 split used in 2013, only 19% of the respondents to their surveys were under 35. By making the shift to a 60-40 split, Pew projects that 22% of the respondents will be under 35. Nationally, this group comprises 31% of the adult population. Pew will continue to make statistical adjustments to correct for the undercoverage.

Andrey Orletsky/Shutterstock

A more serious source of bias in most sample surveys is *nonresponse,* which occurs when a selected individual cannot be contacted or refuses to cooperate. Nonresponse to sample surveys often reaches 50% or more, even with careful planning and several callbacks. Because nonresponse is higher in urban areas, most sample surveys substitute other people in the same area to avoid favoring rural areas in the final sample. If the people who respond to the survey hold views different from those who are rarely at home or who refuse to answer questions, some bias remains.

Undercoverage — DEFINITION

Undercoverage occurs when some groups in the population are left out of the process of choosing the sample.

As Spotlight 7.2 shows, major research institutions, such as the Pew Research Center, take undercoverage into account when designing their polls.

Nonresponse — DEFINITION

Nonresponse occurs when an individual chosen for the sample can't be contacted or refuses to participate.

EXAMPLE 6 How Bad Is Nonresponse?

The CPS (Example 1, page 293) has the lowest nonresponse rate of any poll we know. Only about 4% of the households in the CPS sample refuse to take part, and another 3 or 4% can't be contacted. People are more likely to respond to a government survey such as the CPS, and the CPS contacts its sample in person before doing later interviews by phone. (On a related note, the national mail participation rate for the 2010 Census was 74%.)

What about polls done by the media and by market research and opinion polling firms? We don't know their rates of nonresponse because they won't say. That nondisclosure is a bad sign. The Pew Research Center imitated a telephone survey and published the results: Out of 2879 households called, 1658 were never at home, refused the interview, or would not finish it. That's a nonresponse rate of $\frac{1658}{2879} \approx 0.58$, or about 58%.

Algebra Review Appendix
Fractions, Percents, and Percentages

Self Check 4

The campus food service wants to know how students feel about their food. They hand out a survey during Friday morning breakfast between 7 AM and 9 AM. Are the survey results more likely to be affected by undercoverage or nonresponse? Explain.

Another danger is that when people do respond to a survey question, we can't always rely on them to tell the truth. For example, will students answer truthfully when asked: On how many occasions (if any) have you taken cocaine during the last 30 days? Consider another example. People know that they should take the trouble to vote. Therefore, many who didn't vote in the last election will tell a pollster that they did. Fortunately, there are strategies to help determine the level of participants' truthfulness and to improve accuracy.

EXAMPLE 7 Validating Truthfulness and Encouraging Honesty

In a major national study, researchers were concerned about the truthfulness of student answers to questions about illicit drug use. Researchers compared data among twelfth-graders' responses to questions about their own drug use, their friends' drug use, and their own exposure to drug use. In any given year, comparisons across these three measures tended to be consistent from drug to drug. Because respondents should have little reason to answer untruthfully about their friends or their general exposure to drugs, the researchers considered this consistency as evidence of the truthfulness of student responses about their own drug use.

Another strategy for encouraging honesty with sensitive topics is called *randomized response,* invented by sociologist S. L. Warner in 1965 and explored in Exercises 20 and 21 (page 331). By introducing randomness into the responses in a structured way, researchers use their knowledge of probability distributions to get reasonably accurate information about the overall group while allowing each potentially embarrassing answer to be "camouflaged." Because the interviewee knows that, for example, the researcher has no way to distinguish which "yes" answers are real and which are simply introduced by the random mechanism, they will feel safe answering honestly a question of the form "Have you ever done [some embarrassing or illegal action]?"

Finally, the *wording of questions* strongly influences the answers given to a sample survey. Confusing or leading questions can introduce strong bias, and even minor changes in wording or order can change a survey's outcome. Here are some examples.

EXAMPLE 8 Watch That Wording!

How do Americans feel about government help for the poor? Only 13% think we are spending too much on "assistance to the poor," but 44% believe we are spending too much on "welfare." How did the Scots feel about the movement to become independent from England? Well, 51% voted for "independence for Scotland," but only 34% supported "an independent Scotland separate from the United Kingdom."

It seems that *assistance to the poor* and *independence* are nice, hopeful terms, whereas *welfare* and *separate* are negative words. Other topics that have produced survey results that vary greatly depending on the wording of the questions include abortion, immigration, gay rights, gun control, and affirmative action.

Self Check 5

Consider the two questions below. Do you believe that either of these questions is designed to elicit a particular response? If so, explain how.

Q1. Do you think that smokers should have the freedom to smoke in a hotel room in which they are staying? Yes No

Q2. Given the recent reports on the dangers of second-hand smoke, do you think that companies should allow their employees to smoke in the workplace? Yes No

The design of sample surveys is a science, but this science is only part of the art of sampling. Because of nonresponse, false responses, and the difficulty of posing clear and neutral questions, you should analyze critically before fully trusting reports about complicated issues based on surveys of large human populations. Insist on knowing the exact questions asked, the rate of nonresponse, and the date and method of the survey before you trust a poll result.

7.5 Experiments

Sample surveys gather information on part of the population to make conclusions about the whole. When the goal is to describe a population, statistical sampling is the right tool to use.

Suppose, however, that we want to study the response to a stimulus, to see how one variable affects another when we change existing conditions. For example:

- Will a new mathematics curriculum improve the scores of sixth-graders on a standardized test of mathematics achievement?
- Will taking small amounts of aspirin daily reduce the risk of suffering a heart attack?
- Does a woman's smoking during pregnancy reduce the IQ of her child?

Studies that simply *observe and describe* are ineffective tools for answering these questions. **Experiments** give us clearer answers.

Experiment DEFINITION

An **experiment** deliberately imposes a *treatment* on individuals to observe their responses. The purpose of an experiment is to study whether the treatment *causes* a change in the response.

Experiments are the preferred method for examining the effect of one variable on another. By imposing the specific treatment of interest and controlling other influences, we can pin down cause and effect. A sample survey may show that two variables are related, but it cannot demonstrate that one causes the other. Statistics has something to say about how to arrange experiments, just as it suggests methods for sampling.

EXAMPLE 9 ➔ **An Uncontrolled Experiment**

A college regularly offers a review course to prepare candidates for the Graduate Management Admission Test (GMAT) required by most graduate business schools. However, one year the college decided to offer only an online version of the course. The average GMAT score of students in the online course was 10% higher than the longtime average of those who took the classroom review course. Can we conclude that the online course is more effective?

This experiment has a very simple design. A group of subjects (the students) were exposed to a treatment (the online course), and the outcome (GMAT scores) was observed. The design can be represented as

online course → observe GMAT scores

or, in general form, as

treatment → observe response ■

Most laboratory experiments use a design like that in Example 9: Apply a treatment and measure the response. In the controlled environment of the laboratory, simple designs often work well. But field experiments and experiments with human subjects have more sources of variability that can influence the outcome. With greater variability comes a greater need for statistical design, as we will see in Example 10.

A closer look at the GMAT review course showed that the students in the online review course were indeed quite different from the students who took the classroom course in past years. In particular, they were older and more likely to be employed. An online course appeals to these mature people, but we can't compare their performance with that of the undergraduates who previously dominated the course. The online course might even be less effective than the classroom version for these students. The effect of online versus in-class instruction is hopelessly mixed up with influences lurking in the background, and this entanglement is displayed in Figure 7.2.

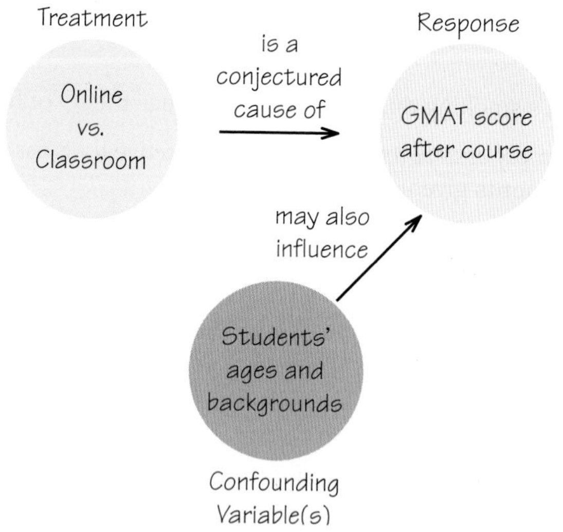

Figure 7.2 Confounding. We can't distinguish the effects of the treatment from the effects of other influences on the response.

We say that student age and background are **confounded** with whatever effect the change to online instruction may have. In everyday usage, someone who is confounded is confused or mixed up. In statistics, confounded variables have their effects mixed together so that it's hard to tell what effect is due to each variable separately.

Confounding DEFINITION

Variables are said to be **confounded** when their effects on the outcome (response or dependent variable) cannot be distinguished from each other. Such variables may or may not have been intended to be part of the study.

Self Check 6

A researcher is studying the effect of food (or lack of food) on college students' ability to perform well on math tests. In a math course, students were asked if they had eaten food within the last hour and were then given a math test. The researcher found that, on average, students who had eaten within an hour before the test did better than students who had not eaten. Explain how another variable described as "had a class right before math class" could be confounded with the variable "ate within an hour of math class."

The remedy for confounding is to do a *comparative experiment* (sometimes called a *quasiexperiment*) in which some students are taught in the classroom and other similar students take the course online. The first group is called a **control group** (because that is the mode of instruction that had been the norm). Most well-designed experiments compare two or more treatments. Of course, comparison alone isn't enough to produce results we can trust. If the treatments are given to groups that differ markedly when the experiment begins, bias will result. For example, if we allow students to choose whether they get online or classroom instruction, older employed students are likely to sign up for the online course. Personal choice will bias our results in the same way that volunteers bias the results of call-in opinion polls. The solution to the problem of bias is the same for experiments and for samples: Use impersonal chance to select the groups.

EXAMPLE 10 A Randomized Comparative Experiment

A college decides to compare the progress of 25 students taught in the classroom with that of 25 students taught the same material online. They select which students will be taught online by taking a simple random sample of size 25 from the 50 available students. The remaining 25 students form the control group. They will receive classroom instruction. The result is a **randomized comparative experiment** with two groups. Figure 7.3 outlines the design in graphical form.

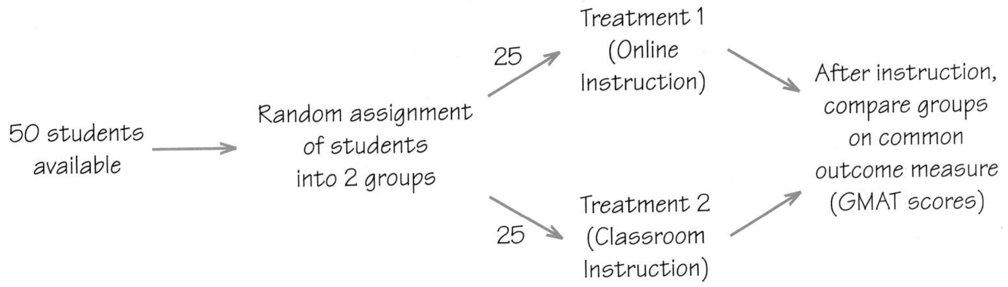

Figure 7.3 The design of a randomized comparative experiment to compare online and classroom instruction.

The selection procedure of randomly assigning students to groups is exactly the same as it is for sampling:

Step 1. *Label* the 50 students 01 to 50.

Step 2. Go to the table of random digits (Table 7.1, page 298) and read successive two-digit groups. Skip over two-digit groups greater than 50 since these values are not used as labels. Also skip over any duplicates. For example, if you begin at line 106 in Table 7.1, the first five students chosen are those labeled 41, 50, 13, 15, and 27.

Step 3. Stop when you have selected 25 labels. The students corresponding to these 25 labels form the online group.

If you prefer, you can skip Steps 2 and 3 and use technology as outlined in Spotlight 7.1 (page 300) to randomly select the 25 labels for the Treatment 1 group. Here is the output using a TI-84 graphing calculator and the randIntNoRep command to select the labels for the Treatment 1 group sample:

{26, 29, 46, 4, 25, 11, 12, 14, 41, 7, 3, 17, 22,
2, 15, 42, 39, 8, 43, 47, 21, 37, 9, 24, 20}

The GMAT experiment is *comparative* because it compares two treatments (the two instructional settings). The experiment could have even had a third treatment if there had been a "hybrid" (part-classroom, part-online) course option. It is *randomized* because the subjects are assigned to the treatments by chance. Randomization creates groups that are similar to each other before we start the experiment. Possible confounding variables act on both groups at once, so their effects tend to balance out and do not greatly affect the results of the study. The *only* difference between the groups is the online versus in-class experience. So if we see a difference in performance, it must be due to the different setting. That is the basic logic of randomized comparative experiments. This logic shows why experiments can give good evidence that the different treatments really *caused* different outcomes.

There is a fine point: The performance of the two groups will differ even if the treatments are identical. That's because the individuals assigned at random to the groups differ. It is only differences *larger than would plausibly occur just by chance* that show the effects of the treatments. The laws of probability (discussed in Section 8.1) allow statisticians to say how big of an effect is **statistically significant.** You probably understand this concept intuitively because you would not find it unusual if all the children were girls in a family with two or three children, but you probably would in a family with six or more children. We now present a convention that assigns a concrete numerical benchmark to this intuitive gut feeling.

Statistically Significant DEFINITION

An observed effect that is so large that it would rarely occur by chance is called **statistically significant.** ("Rarely" usually means less than 5% of the time.)

Although experiments provide the best way to establish cause and effect, ethical considerations, such as those discussed in Spotlight 7.3, prevent some experiments from being conducted.

Ethics in Experiments

SPOTLIGHT 7.3

The last century saw some unethical experiments; for instance, look up the painful and deadly medical experiments done by the Nazis during World War II or the sexually transmitted disease experiments conducted by the United States in Guatemala (1946–1948) and in Tuskegee, Alabama (1932–1972).

The Tuskegee experiment, conducted by the U.S. Public Health Service, studied the natural progression of untreated syphilis in African American men. The men in the study, all impoverished sharecroppers, believed they were receiving free healthcare from the U.S. government. Those who had contracted syphilis before the study were not told that they had the disease, and in 1947, when treatment with penicillin became available, they were not treated. Prompted in part by the Tuskegee syphilis study, the Belmont Report (released in 1979) proposed ethical principles and guidelines for research involving human subjects. Today, there is great emphasis on having studies approved by an Institutional Review Board (IRB) and on following ethical codes such as those outlined in the Belmont Report. Key principles of ethics codes include the following:

- Voluntary participation
- Informed consent
- The right to quit at any time
- The avoidance of unnecessary suffering or risk

Also, when randomized clinical trials accumulate enough evidence to make clear that a treatment is dangerous or is less effective than another, there is a mechanism to stop the data-gathering process before the originally scheduled end of the experiment.

For some experiments, however, giving the subject full disclosure about the true purpose of the experiment could prevent researchers from obtaining accurate data. In such cases, an IRB may give the researcher permission to (at least temporarily) withhold a piece of information from participants. Consider this example. An Internet field experiment conducted by Andrew Hanson and Michael Santas (whose results were published in the July 2014 issue of *Southern Economic Journal*) investigated discrimination against Hispanics in the U.S. housing market. Hispanics were divided into two groups: recent immigrants and those who were assimilated into American culture. Responses from (fictitious) Hispanics who were portrayed as recent immigrants got significantly fewer callbacks and invitations to showings of apartments than did otherwise identical responses from (fictitious) Whites or Hispanics who were portrayed as assimilated into American culture. While landlords may not be willing to acknowledge their prejudices, this method of research was able to yield an accurate assessment of prejudice in the rental market in metropolitan areas of the United States.

7.6 Experiments Versus Observational Studies

The first randomized comparative experiment was published in 1948—a British study of the effectiveness of streptomycin in treating tuberculosis. Randomized comparative experiments quickly became common tools of industrial, academic, and medical research. For example, federal regulations require that the safety and effectiveness of new drugs be demonstrated by randomized comparative experiments. Let's look at a medical experiment as an example.

EXAMPLE 11 St. John's Wort—Treatment for Depression?

Although prescription drugs must pass the test of randomized comparative experiments before being sold, herbs and other "natural remedies" are exempt. Because these treatments are so popular, some are now being studied more carefully. Fans of natural remedies often use extracts of the herb St. John's wort to treat depression. Is the herb safe? Does it work? The *Journal of the American Medical Association* reported a "randomized, double-blind, placebo-controlled clinical trial" in which 200 patients with major depression were assigned at random to take either herb extract or a dummy pill that looked and tasted the same. Results: The herb is safe, but "[i]n this study, St. John's wort was not effective for treatment of major depression." ■

If you read accounts of medical studies, you will often see language like "randomized, double-blind, placebo-controlled clinical trial." A clinical trial is a medical experiment with actual patients as subjects. *Randomized* and *controlled* tell us that this was a randomized comparative experiment (that's good). A *placebo* looks like a real treatment but is actually a fake pill that has no medication in it and should not actually have an effect in this study. Here, we encounter a new idea: the importance of the **placebo effect,** a special kind of confounding.

Placebo Effect DEFINITION

The **placebo effect** is the tendency of patients to respond favorably to any apparent treatment (even one that is in reality a "fake treatment") because of their expectations about the treatment.

If depressed patients given St. John's wort are compared with patients who receive no treatment, the first group gets the benefit of both the herb and the placebo effect. Any beneficial effect that St. John's wort may have is confounded with the placebo effect. To prevent confounding, it is important that some treatment be given to all subjects in any medical experiment.

Neither the subjects nor the experimenters who worked with them knew which treatment any specific subject received. Subjects might react differently if they knew they were getting "only a placebo." Knowing that a particular subject was getting "only a placebo" also could influence the health workers who interviewed and examined the subjects. Only the study's statistician knew which treatment each subject received. Because both the subjects and the health workers were "blind" to this information, this study was considered a **double-blind experiment.**

Double-Blind Experiment DEFINITION

A **double-blind experiment** is an experiment in which neither the experimental subjects nor the persons who interact with them know which treatment each subject received.

The difference between the St. John's wort groups and placebo groups was *not statistically significant;* that is, the difference was no larger than would be expected when we divide 200 depressed patients at random into two groups and do nothing else. Larger numbers of subjects would give more precise results. It's unlikely that there is exactly no difference between St. John's wort and a placebo. If the clinical trial had used 2000 patients rather than 200, it might have picked up a small effect (in either direction). The researchers believed that a group of 200 patients was sufficient to pick up any effect large enough to be medically important.

The logic of experimentation, the statistical design of experiments, and the laws that govern chance behavior combine to give compelling evidence of cause and effect. Only experimentation can produce the most convincing evidence of causation.

By way of contrast, in Example 12, we consider historical statistical evidence linking cigarette smoking to lung cancer. We can't ethically assign groups of people to smoke or not, so a direct experiment isn't possible.

EXAMPLE 12 Smoking and Health

One of the earliest studies linking smoking and lung cancer was conducted by Ernst Wynder and Evarts Graham. This study compared people with and without lung cancer, looking for differences in their backgrounds or habits. The one habit that stood out was smoking. While lung cancer was rare in nonsmokers, among patients with lung cancer, cigarette use was high. The results of Wynder and Graham's study were reported in a 1950 article in the *Journal of the American Medical Association.*

One potential problem with Wynder and Graham's study was that it relied on self-reported smoking habits of people who already had lung cancer and those who did not. People who knew they had lung cancer might be more likely to overestimate how much they smoked, whereas people who did not have lung cancer might be more likely to underestimate how much they smoked.

In January 1952, E. Cuyler Hammond and Daniel Horn, scientists working for the American Cancer Society, designed a different style of study to avoid the potential problem discussed in the previous paragraph. They recruited about 188,000 men between the ages of 50 and 69. Participants completed questionnaires, which asked if they smoked cigarettes, and if so, how often and how many (along with other questions). In November 1952, Doctors Hammond and Horn collected the first follow-up set of data. These data classified participants as alive or dead (or unknown status). For those who had died, the cause of death listed on the official death certificate was recorded.

Hammond and Horn reported their preliminary findings in a 1954 article published in the *Journal of the American Medical Association.* They found: "The death rate from lung cancer was much higher among men with a history of regular cigarette smoking than among men who never smoked regularly." They concluded their article stating that it was their opinion that a cause-and-effect relationship existed between regular smoking and lung cancer. ■

Look back at the two studies described in Example 12. The researchers could not impose treatments of smoking or not smoking on the participants in their studies. Instead, they recorded participants' smoking status and then observed associations between smoking and other data they collected from the participants. These studies are examples of **observational studies.**

Observational Study — DEFINITION

An **observational study** does not try to manipulate the environment (such as by assigning treatments to people); it simply observes the measurements of variables of interest that result from people's free choices. This kind of study is generally done when assignment of a treatment to a person is unethical (e.g., smoking while pregnant) or impossible (e.g., ethnicity).

For the Wynder and Graham study, participants belonged to one of two groups, those diagnosed with lung cancer and those not diagnosed with lung cancer. The researchers then collected data on participants' backgrounds, including medical history and smoking habits. They searched for a connection between the data on participants' pasts and their current status as having or not having lung cancer. This is an example of a **retrospective study.**

Retrospective Study DEFINITION

A **retrospective study** starts with an outcome (e.g., a group of cancer and noncancer patients) and then looks back to examine exposures to suspected risk or protective factors that might be linked to that outcome.

On the other hand, the Hammond and Horn study recruited healthy participants and gathered data about their backgrounds—including medical history and smoking habits. Then they followed them forward in time for a period of years to see which participants died and from what causes. This is an example of a **prospective study.**

Prospective Study DEFINITION

A **prospective study** starts with a group and watches for outcomes (e.g., the development of cancer or remaining cancer-free) during the study period and relates this to suspected risk or protective factors that might be linked to the outcomes.

Figure 7.4 summarizes the differences between retrospective and prospective observational studies.

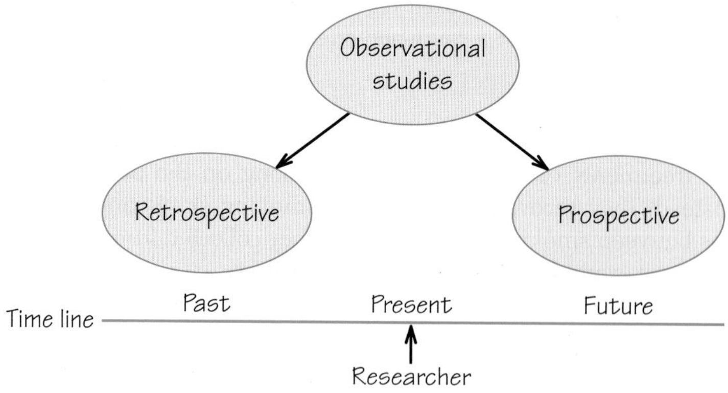

Figure 7.4 Retrospective studies start with a current outcome and look to the past to try to explain the outcome; prospective studies start with a group and look to the future for outcomes.

Self Check 7

Researchers investigating the connection between serial killers and child abuse identified a group of convicted serial killers and then found out which of them had a history of childhood abuse. Is this an example of a retrospective or prospective study? ■

Based on many studies, we can say that the connection between smoking and lung cancer is statistically significant. That is, it is far stronger than would occur by chance. We can be confident that something other than chance links smoking to cancer. But observation of samples cannot tell us *what* factors other than chance might be at work. Perhaps some other factor is involved, such as differences in diet or exposure to air pollution. Perhaps there is something in the genetic makeup of some people that predisposes them to both nicotine addiction and lung cancer. If so, we would observe a strong link even if smoking itself had no effect on the lungs.

However, the statistical evidence that points to cigarette smoking as a cause of lung cancer is about as strong as nonexperimental evidence can be. First, the connection has been observed in many studies in numerous countries. This eliminates factors peculiar to one group of people or to one specific study design. Second, there is a *dose–response relationship*: People who smoke more are more likely to get lung cancer than those who smoke less, and quitting cigarettes reduces the cancer risk. Third, specific ways in which smoking could cause cancer have been identified; cigarette smoke contains tars that have been shown by experiments to cause tumors in animals. Finally, no plausible alternative explanation is available. For example, the genetic hypothesis cannot explain the increase in lung cancer among women that occurred as more and more women became smokers. Also, the hypothesis was not supported by studies of identical twins where only one smoked.

It is very difficult and complicated to amass convincing evidence from observational studies and rule out all possible alternative explanations. This is why we have a strong preference for the more conclusive statistical evidence that we get from randomized comparative experiments when it is possible and ethical to conduct them. Despite their status as the "gold standard" of research, however, experiments can also have weaknesses. The most common of these is a contrived condition that makes it hard to say how far results may apply beyond a controlled laboratory setting.

EXAMPLE 13 ➜ Is the Experiment Realistic?

Clinical trials give medical treatments to actual patients with the condition that the treatments are supposed to help. But some experiments are less realistic in terms of how well experimental conditions align with the usual circumstances of what is of greatest interest. For example, a researcher studying stages and cycles of sleep observes patients overnight in a special "sleep lab" that can monitor their electroencephalography (EEG) waves and other data. However, individuals may sleep quite differently in a lab setting than in the natural sleeping environment of their bedroom at home.

Another type of example is that some studies on animals may be limited in how reliably their conclusions might apply to humans. Penicillin, for instance, is highly toxic to guinea pigs, but it has been a very helpful medicine for humans.

These are not statistical questions. Researchers must use their understanding of their academic domain to judge how far their results apply. Good statistical design enables us to trust results for the participants in the study at hand, but additional knowledge and judgment are needed to decide the extent to which conclusions might be generalized to other settings. ■

Figure 7.5 gives a conceptual overview of the types of designs we have covered in these last two sections.

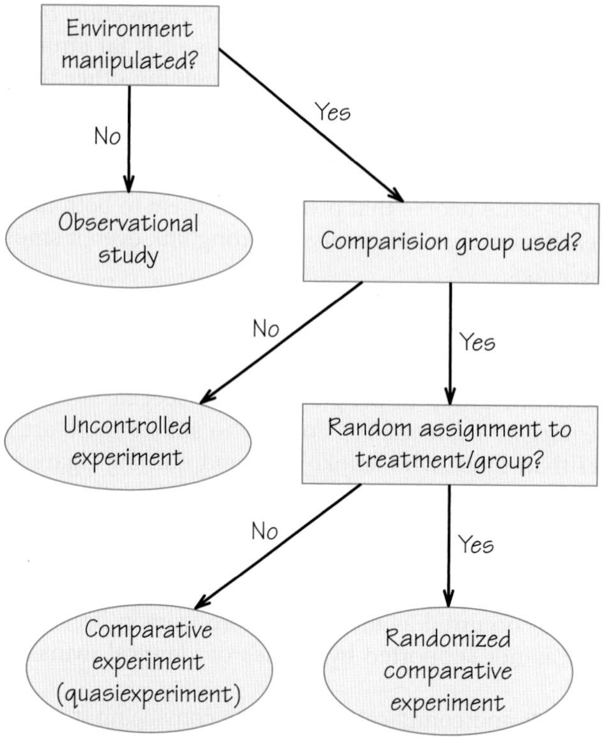

Figure 7.5 This flowchart shows how to distinguish the types of studies in Sections 7.5 and 7.6.

7.7 Inference: From Sample to Population

A *USA Today* Pew Research Center poll conducted in 2014 interviewed a random sample of 1500 adults. The result was that 60% agreed with the following statement: "Most people who want to get ahead can make it if they're willing to work hard." This applies to the 1500 people in the sample. But what is the truth about the 230 million American adults who make up the population? Because the sample was chosen at random, it's reasonable to think that these 1500 people represent the entire population fairly well. So the researchers turn the *fact* that 60% of the *sample* believe that people can get ahead if they're willing to work hard into an *estimate* that about 60% of *all* adults feel this way. That's a fundamental operation in statistics: Use a fact about a sample to estimate the truth about the whole population. We call this **statistical inference.**

Statistical Inference DEFINITION

Statistical inference refers to methods for drawing conclusions about an entire population on the basis of data from a sample.

If the selected individuals were chosen at random, we think that they fairly represent the population and inference makes sense. However, if we have data from only a convenience sample or a voluntary response sample, the data do not represent the population and we can't use them for inference. *Statistical inference works only if the data come from a random sample or randomized comparative experiment.* That's why this chapter starts with producing reliable data before moving on to inference from the data to a larger population.

To think about inference, we must keep straight whether a number describes a *sample* or a *population.*

Parameter DEFINITION

A **parameter** is a fixed (usually unknown) number that describes some characteristic of a population.

Statistic DEFINITION

A **statistic** is a number that describes some characteristic of a sample. The value of a statistic is known when we have taken a sample, but it can change from sample to sample. We often use a statistic to estimate an unknown parameter.

To avoid confusing these terms, remember Parameters are for Populations and Statistics are for Samples. We can't determine the true value of a parameter unless we examine the entire population, which isn't usually possible. However, we can estimate the unknown parameter based on information from a sample statistic.

EXAMPLE 14 Working Hard and Getting Ahead: A Sample Statistic

The actual results from the poll discussed at the start of this section were that 885 of the 1484 people who answered the question agreed with the statement "Most people if they want to get ahead can make it if they work hard." (Notice not all of the 1500 people surveyed answered this question.) The proportion of the respondents who agreed was

$$\hat{p} = \frac{855}{1484} = 0.59636 \quad \text{or approximately} \quad 60\%$$

Algebra Review Appendix
Fractions, Percents, and Percentages

The symbol $\hat{p}$ is read "p-hat." The ˆ symbol here tells us that a quantity has been estimated, just as the use of $\hat{y}$ in Chapter 6 (page 253) told us that a value was estimated by using a regression-line model. The number $\hat{p} = 0.596$ is a *statistic*. The corresponding *parameter* is the proportion (call it p) of all adult U.S. residents who would have responded "Agree" if questioned about the same statement. We don't know the value of the parameter p, so we use the statistic $\hat{p}$ to estimate it. ■

From Example 14, we have an estimate $\hat{p} = 0.596$ for the population proportion p. But how good is our estimate? If the Pew Research Center took a second random sample of 1500 adults, the new sample would have different people in it. It is almost certain that there would not be exactly 885 responses in agreement with the statement. That is, the value of the statistic $\hat{p}$ will vary from sample to sample. If the variation when we take repeat samples from the same population is too great, then we can't trust the results of any one sample.

In practice, it is too expensive to take many samples from a large population, such as all adult U.S. residents. But we can use a computer to imitate drawing many samples at random from a population that we specify. This is called *simulation.* Example 15 explores what happens when we do this.

EXAMPLE 15 What Happens in Many Samples?

We start with a scenario in which we know that the population proportion responding "Agree" to some statement is $p = 0.6$. Using computer simulation, we repeatedly take random samples, first of size 100 and later of size 1500. For each sample, we calculate the sample proportion $\hat{p}$.

Results from Samples of Size 100

$$\text{Sample 1:} \quad \hat{p} = \frac{55}{100} = 0.55$$

$$\text{Sample 2:} \quad \hat{p} = \frac{62}{100} = 0.62$$

$$\vdots \qquad \vdots \qquad \vdots$$

$$\text{Sample 1000:} \quad \hat{p} = \frac{63}{100} = 0.63$$

A histogram of the 1000 values of $\hat{p}$ from our computer-simulated data appears in Figure 7.6. This histogram gives us an idea of the shape, center, and spread of the distribution of the sample proportion $\hat{p}$ for samples of size 100 drawn from a population in which $p = 0.6$.

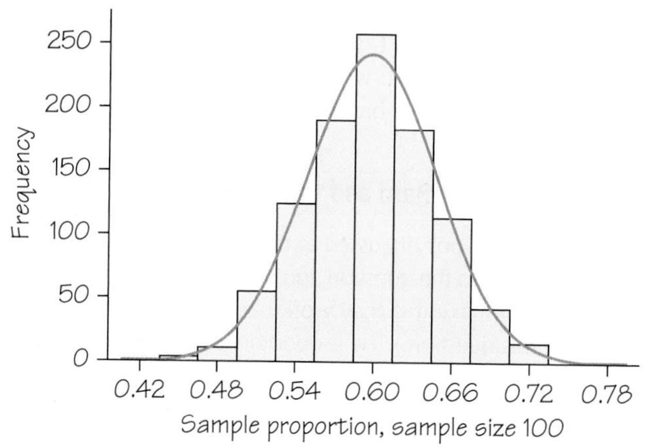

Figure 7.6 Draw 1000 SRSs of size 100 from a population with proportion $p = 0.60$ of "Agrees." The histogram shows the distribution of the 1000 sample proportions $\hat{p}$.

The *USA Today* Pew Research Center poll interviewed around 1500 people, not just 100. Again, we use computer simulation to generate 1000 samples of size 1500 and record the value of the sample proportion $\hat{p}$ for each sample. A histogram of the $\hat{p}$-values based on these 1000 samples is shown in Figure 7.7.

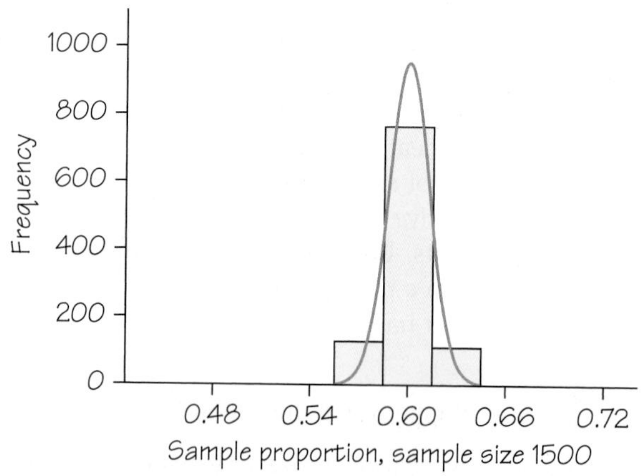

Figure 7.7 Draw 1000 SRSs of size 1500 from a population with proportion $p = 0.60$ of *Agrees*. The histogram shows the distribution of the 1000 sample proportions $\hat{p}$.

For comparative purposes, Figures 7.6 and 7.7 are drawn using the same horizontal scale. This allows us to compare what happens when we increase the size of our samples from 100 to 1500. These histograms display the **sampling distribution** of the statistic $\hat{p}$ for two sample sizes. Notice that for both situations, the histograms are centered at $p = 0.60$, the known value of the parameter. The histograms are single-peaked and roughly symmetric—what we would expect for a normal distribution (refer to Section 5.8, page 209). However, the variability is much smaller for the situation in which the sample size is 1500 compared to only 100.

Self Check 8

(a) Under the scenario described in Example 15 ($p = 0.6$), what are the possible values for $\hat{p}$ if the sample size is 2?

(b) Suppose a computer simulation is used to generate 1000 samples of size 2 from a population in which $p = 0.6$. Which of the values for $\hat{p}$ listed in part (a) do you expect to occur least frequently?

Sampling Distribution DEFINITION

The **sampling distribution** of a statistic is the distribution of values taken on by the statistic in all possible samples of the same size from the same population.

Strictly speaking, the sampling distribution is the ideal pattern that would emerge if we looked at all possible samples of the same size from our population. A distribution obtained from a fixed number of trials, like the 1000 trials in Figures 7.6 and 7.7, is only an approximation of the sampling distribution. However, the results from our simulations support a general theoretical result, which is based on probability theory (introduced in Chapter 8). We now turn to probability theory to learn the mathematical facts that lie behind the simulations. We'll use the word *success* for whatever we are counting, such as "Agree" responses in the *USA Today* Pew Research Center poll. Note that *success* does not necessarily have the positive (or negative) association it does in real life, but is simply a convenient way to identify an outcome. For example, in a cancer study, success might unfortunately signify that a person developed cancer.

Sampling Distribution of a Sample Proportion THEOREM

Choose an SRS of size n from a large population that contains population proportion p of successes. Let $\hat{p}$ be the **sample proportion** of successes, expressed as

$$\hat{p} = \frac{\text{count of successes in the sample}}{n}$$

Then:

- **Shape:** For large sample sizes ($n \geq 30$), the sampling distribution of $\hat{p}$ is *approximately normal.*
- **Center:** The *mean* of the sampling distribution of $\hat{p}$ is p.
- **Variability:** The *standard deviation* of the sampling distribution of $\hat{p}$ is

$$\sqrt{\frac{p(1-p)}{n}}$$

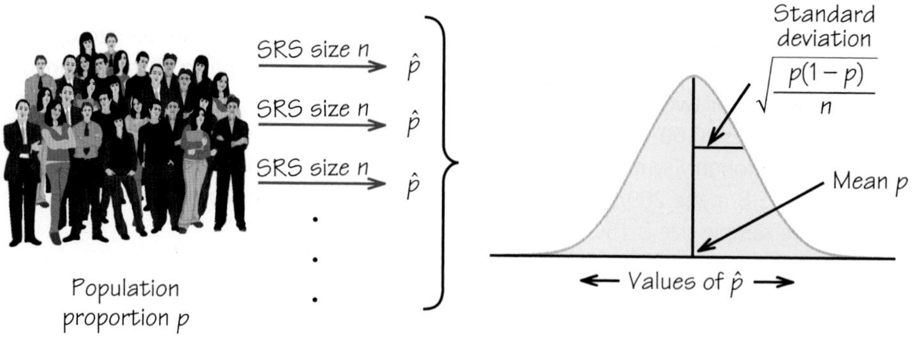

Figure 7.8 Repeat many times the process of selecting an SRS of size n from a population of which the proportion p is number of successes. The values of the sample proportion of successes $\hat{p}$ have this normal sampling distribution.

Figure 7.8 summarizes these facts in a form that reminds us that a sampling distribution describes the results of lots of samples from the same population.

EXAMPLE 16 Comparing Simulation Results to Theory

Return to the scenario in which the population proportion who would respond "Agree" to some statement is $p = 0.6$. In Example 15, we used computer simulation to generate 1000 samples of size 1500 from this population and record the sample proportion of "Agree" responses $\hat{p}$ for each sample. Figure 7.7 (page 316) shows one histogram of these $\hat{p}$-values. However, the histogram in Figure 7.9 (based on the same data) gives a better sense of the overall shape of the data. We also computed the mean and standard deviation of the $\hat{p}$-values: $\bar{x} = 0.59982$ and $s = 0.01255$.

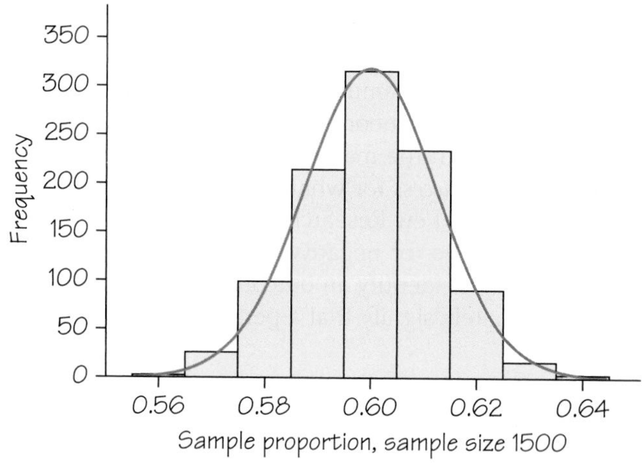

Figure 7.9 Histogram of the same data used for Figure 7.7. The horizontal scale has been changed to better show the normal shape of the data.

Next, we look at what the theorem tells us about the sampling distribution of $\hat{p}$. The distribution of $\hat{p}$ in many samples

- Is close to normal
- Has mean 0.6
- Has standard deviation 0.0126

To show our work for the last number, note that $p(1 - p)/n = (0.6)(0.4)/1500 = 0.00016$, and the square root of 0.00016 is approximately 0.0126:

$$\sqrt{\frac{p(1-p)}{n}} = \sqrt{\frac{(0.6)(0.4)}{1500}} \approx 0.0126$$

Finally, we compare the simulation results to the results from the theorem. First, the normal curve in Figure 7.9 does a reasonable job of summarizing the shape of the histogram. Second, the mean of 0.60 and standard deviation of 0.0126 from the mathematics are very close to the mean of 0.59982 and standard deviation of 0.01255 we observed in our simulation data. If the simulation used more than 1000 trials, the results would be still closer to the mathematical theory.

Self Check 9

Return to Example 16. Suppose currently that 70% of all adults in a population would reply "Agree" to some statement. Take simple random samples of 1500 adults.

(a) What is the shape of the distribution of $\hat{p}$ in many samples?
(b) What is the mean of the distribution?
(c) What is its standard deviation?
(d) How did the increase in the value of the population proportion p from 0.6 to 0.7 change the distribution of $\hat{p}$?

Look back at Figure 7.9. Notice that most of the $\hat{p}$-values lie close to the actual population proportion of $p = 0.6$. Hence, the sampling distribution shows why we can trust the results of a large random sample—a high percentage of such samples give results (values of $\hat{p}$) that are close to the truth about the population.

EXAMPLE 17 The 68–95–99.7 Rule Again

In Example 16, the population parameter, the proportion of adults who agreed with some statement, is $p = 0.6$. If we take SRSs of size 1500, the sample proportions $\hat{p}$ follow the normal distribution with a mean of 0.6 and a standard deviation of 0.0126. The "95" part of the 68–95–99.7 rule from Section 5.9 (page 216) says that 95% of all samples give a $\hat{p}$ within 2 standard deviations of the truth about the population. So in this example, 95% of all samples have $\hat{p}$ within 2×0.0126 of 0.60, that is, between 0.57 and 0.63 (rounded to two decimal places). Figure 7.10 illustrates this use of the 68–95–99.7 rule.

Algebra Review Appendix
Rounding Numbers

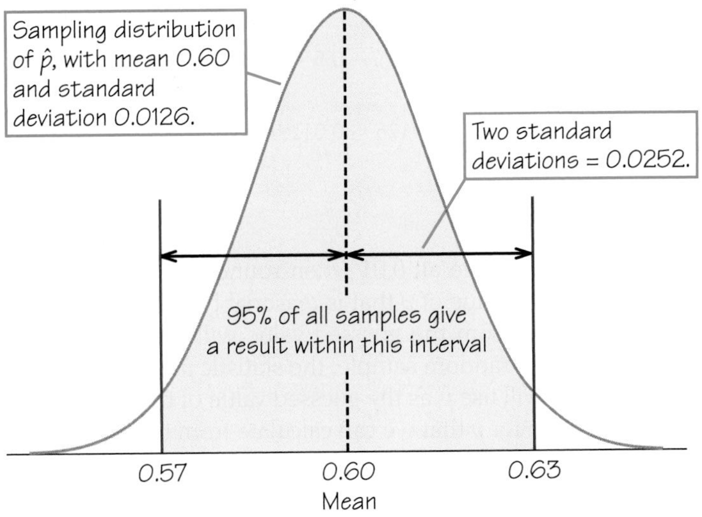

Figure 7.10 The sampling distribution of $\hat{p}$ for Example 16. By the 68–95–99.7 rule, 95% of all samples have a sample proportion $\hat{p}$ within 0.0252 of the true population proportion $p = 0.6$.

We can repeat this reasoning for any value of the parameter p and the sample size n. Using the 68–95–99.7 rule, it is always true that 95% of all samples give a sample proportion $\hat{p}$ within 2 standard deviations of the population proportion p. Now, suppose a sample is one of the 95% of all samples for which $\hat{p}$ lies within 2 standard deviations of p, as shown in Figure 7.11. Then we can turn things around and say that the interval from $\hat{p} - 2$ (standard deviation) to $\hat{p} + 2$ (standard deviation) contains p.

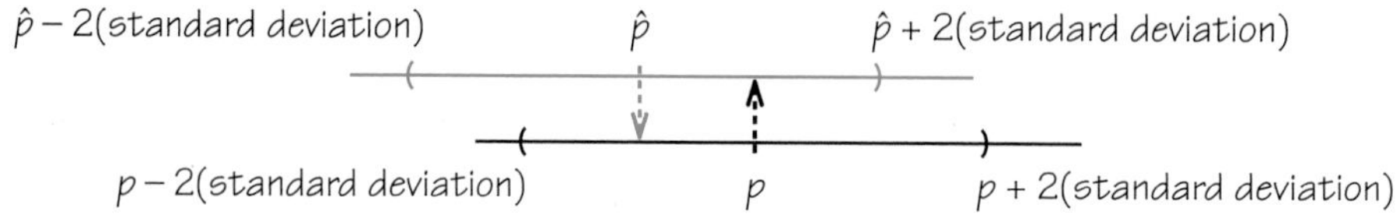

Figure 7.11 If $\hat{p}$ is within 2 standard deviations of p (lower interval), then p is within 2 standard deviations of $\hat{p}$ (upper interval).

That means 95% of all samples catch p in the interval extending 2 standard deviations on either side of $\hat{p}$, which is the interval

$$\hat{p} \pm 2\sqrt{\frac{p(1-p)}{n}}$$

This formula tells us how close the unknown parameter p lies to the observed statistic $\hat{p}$ in 95% of all samples. But there is one problem: We can't calculate the interval from the data because the standard deviation involves the population proportion p, and in practice we don't know p.

What to do? The standard deviation of the statistic $\hat{p}$ does depend on the parameter p, but it doesn't change a lot when p changes. We can go back to Example 16 and redo the calculation for other values of p when $n = 1500$. The results appear in Table 7.5. (You can fill in the value for $p = 0.7$ from your results to Self Check 9c.)

TABLE 7.5 Standard Deviation for Different Values of p

Value of p	0.4	0.5	0.6	0.7	0.8
Standard deviation: $\sqrt{\frac{p(1-p)}{n}}$	0.0126	0.0129	0.0126		0.0103

The standard deviations are all 0.01 when rounded to the hundredths place. You see that if we guess a value of p that is reasonably close to the true value, the standard deviation found from the guessed value will be about right. We know that when we take a large random sample, the statistic $\hat{p}$ is almost always close to the parameter p. So we will use $\hat{p}$ as the guessed value of the unknown p. Now we have an interval estimate for p that we can calculate from the sample data. We call it a **confidence interval.**

Confidence Interval DEFINITION

A 95% **confidence interval** is an interval obtained from the sample data by a method in which 95% of all samples will produce an interval containing the true population parameter.

Confidence Interval for Population Proportion FORMULA

Choose an SRS of size n from a large population that contains an unknown proportion p of successes. A 95% confidence interval for p is approximately

$$\hat{p} \pm 2\sqrt{\frac{\hat{p}\,(1-\hat{p})}{n}}$$

The $\pm$ sign is read "plus or minus," so, for example, 0.5 ± 0.2 yields two numbers: $0.5 - 0.2 = 0.3$ and $0.5 + 0.2 = 0.7$. This can be written as an interval: (0.3, 0.7).

This formula is only approximately correct but is quite accurate when the sample size n is large (≥ 30). Here, $\hat{p}$ is the proportion of successes, and $2\sqrt{\hat{p}\,(1-\hat{p})/n}$, the expression to the right of the $\pm$ sign, is the **margin of error.** When results of polls are reported in the news, the margin of error is commonly reported along with the percentage estimate for the population proportion.

Margin of Error DEFINITION

The **margin of error** is equal to half of the width of a confidence interval. For a 95% confidence interval, it equals about 2 standard deviations of the sampling distribution of the estimated parameter. If you conducted a very large number of polls, about 95% of the time the difference between a particular poll's result and the true value of the population parameter would be within the margin of error.

EXAMPLE 18 Americans' Concern About Climate Change

Are Americans concerned about climate change, and if so, what would they be willing to sacrifice? In 2014, the Bloomberg National Poll surveyed 1005 U.S. adults on climate change and other topics. Of those surveyed:

- 462 viewed climate change as a major threat.
- 623 indicated that they would be willing to pay more for energy if air pollution from carbon emissions could be reduced.

The sample proportion who viewed climate change as a major threat is

$$\hat{p} = \frac{462}{1005} \approx 0.4597$$

Algebra Review Appendix
Powers and Roots

A 95% confidence interval for the proportion p of all U.S. adults who view climate change as a major threat is

$$\hat{p} \pm 2\sqrt{\frac{\hat{p}(1-\hat{p})}{n}} = 0.4597 \pm 2\sqrt{\frac{(0.4597)(0.5403)}{1005}}$$

$$= 0.4597 \pm 0.0314 \quad \text{or} \quad 45.97\% \pm 3.14\%$$

$$= 0.4283 \text{ to } 0.4911 \quad \text{or} \quad 42.83\% \text{ to } 49.11\%$$

A report of these calculations might say, "The study found that approximately 46% of U.S. adults viewed climate change as a major threat. The margin of error for this result is 3.1%."

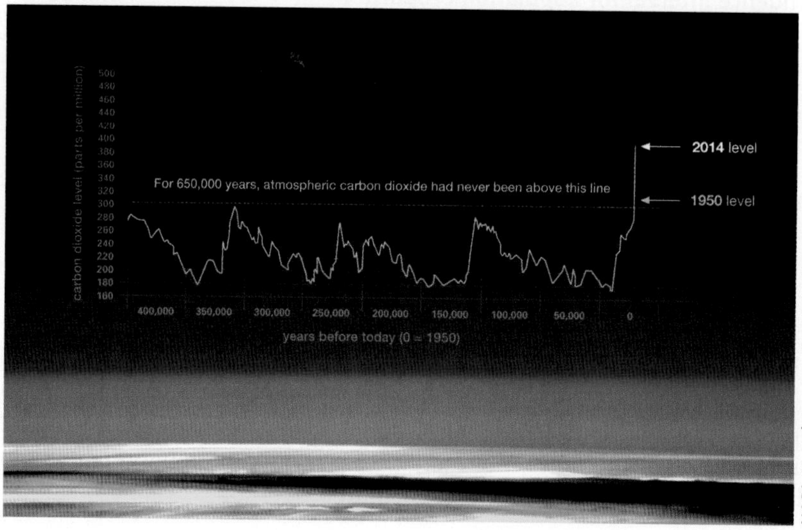

NASA; Vostok ice core data/J. R. Petit et al.; NOAA Mauna Loa CO_2 record

Self Check 10

(a) Using the information from Example 18, determine a 95% confidence interval for the percentage of U.S. adults who would be willing to pay more for energy if air pollution from carbon emissions could be reduced.

(b) Compare the margin of error on the confidence interval you calculated in part (a) to the one calculated in Example 18.

We got the interval in Example 18 by using a formula that catches the true unknown population proportion in 95% of all samples. The shorthand for this is: We are *95% confident* that the true percentage of adults in the United States who view climate change as a major threat lies between 42.83% and 49.11%.

Keep in mind that the 95% confidence level refers to the track record of using the confidence interval formula. This formula results in an interval that contains the true unknown population proportion p in 95% of all samples. However, that also means the true value of the population proportion p lies outside the calculated interval in 5% of all samples. We'll never know if our interval contains p or not. (To gain a better understanding of the meaning of confidence level, check out Applet Exercise 4, on page 339.)

The length of a confidence interval depends on how confident we want to be that the interval does capture the true parameter value. It is common to use 95% confidence, but you can ask for higher or lower confidence if you want. Our 95% confidence interval was based on the middle 95% of a normal distribution.

A 99% confidence interval requires the middle 99% of the distribution and therefore is wider (has a larger margin of error), as can be seen from Figure 7.12.

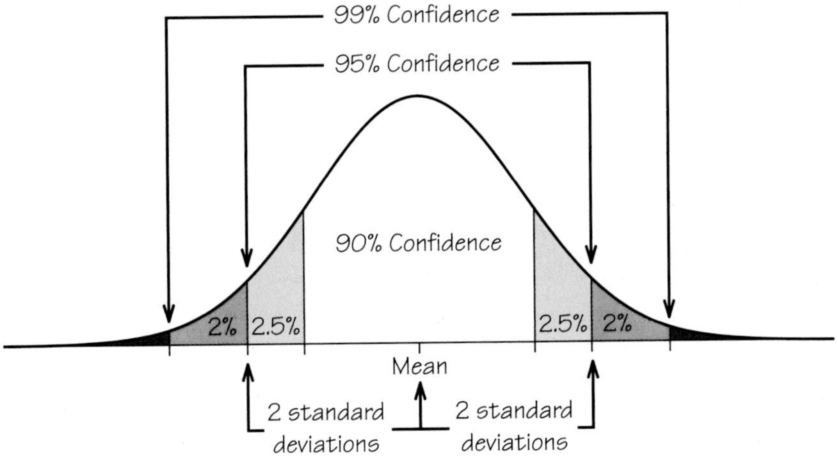

Figure 7.12 Determining margins of error for different confidence levels.

The length of a 95% confidence interval also depends on the size n of the sample. Larger samples give shorter intervals because of the $\sqrt{n}$ in the denominator of the margin of error $= \sqrt{\frac{\hat{p}(1-\hat{p})}{n}} = \frac{\sqrt{\hat{p}(1-\hat{p})}}{\sqrt{n}}$. But the interval does not depend on the size of the population. This is true so long as the population is much larger than the sample. The confidence interval in Example 18 works for a sample of 1005 from a city with 100,000 adults as well as it does for a sample of 1005 from a nation of 308 million. What matters more is how many people we interview, not what percentage of the population we contact.

EXAMPLE 19 Understanding the News

Here's what the TV news announcer says: "A new Gallup poll on American exercise habits finds that 45% of adults are not engaging in vigorous sports or physical activities. The margin of error for the poll was 3 percentage points." Plus or minus 3%, starting at 45%, is 42% to 48%. People with minimal statistics knowledge may think that the truth about the entire population must be in that interval, but now we know better!

This is the full background Gallup actually gives: "For results based on this sample, one can say with 95% confidence that the maximum error attributable to sampling and other random effects is 3 percentage points. In addition to sampling error, question wording and practical difficulties in conducting surveys can introduce error or bias into the findings of public opinion polls." That is, Gallup tells us that the margin of error works only for 95% of all its samples; "95% confidence" is shorthand for that longer fact. The news report left out the "95% confidence." In fact, *almost all margins of error in the news are for 95% confidence.* If you don't see the confidence level in a scientific poll, it's usually safe to assume 95%. ■

Gallup's mention of "question wording and practical difficulties" in Example 19 takes us back to our cautions about sample surveys in Section 7.4. *The margin of error does not address nonresponse and other practical difficulties.* The margin of error in a confidence interval comes from the sampling distribution of the statistic. The sampling distribution describes the variation of the statistic due to chance in repeated random samples. This random variation is the only source of error covered by the margin of error.

Real-life samples also suffer from undercoverage and nonresponse. Errors from these practical difficulties are usually more serious and harder to quantify than random sampling error. The actual error in sample surveys may be much larger than the announced margin of error. What is worse is that we can't say how much larger. Statistical conclusions are approximations of a complicated truth, not mathematical results that are simply true. As we will see in Spotlight 7.4, responsible polling organizations tell the public something about both the precision and limitations of their poll results.

Truth in Polling

SPOTLIGHT 7.4

College student newspapers may not have the resources to conduct polls using random sampling, so it is refreshing when the polls that it publishes from voluntary response samples are accompanied by a disclaimer, such as this one that has been used by *The Prospector,* the student newspaper of The University of Texas at El Paso:

> This poll is not scientific and reflects the opinions of only those Internet users who have chosen to participate. The results cannot be assumed to represent the opinions of Internet users in general, nor the public as a whole.

Because of this limitation, *The Prospector* simply reports the breakdown of responses given but without any margin of error, since sampling error cannot be quantified from a (voluntary response) sample that is not probability-based.

The Harris Poll accompanies its polls with the following disclaimer:

> All sample surveys and polls, whether or not they use probability sampling, are subject to multiple sources of error which are most often not possible to quantify or estimate, including sampling error, coverage error, error associated with nonresponse, error associated with question wording and response options, and post-survey weighting and adjustments. Therefore, Harris Interactive avoids the words "margin of error" as they are misleading. All that can be calculated are different possible sampling errors with different probabilities for pure, unweighted, random samples with 100 percent response rates. These are only theoretical because no published polls come close to this ideal.

ABC Review Vocabulary

Bias A systematic error that tends to cause the observations to deviate in the same direction from the truth about the population whenever a sample or experiment is repeated. (p. 295)

Confidence interval An interval of values used to estimate a population parameter with a specific level of confidence. A 95% confidence interval is an interval computed from a sample by a method that surrounds the unknown parameter 95% of the time, so when we calculate the interval for a single sample, we are 95% confident that the interval contains the unknown parameter. (pp. 320, 321)

Confounding Two variables are confounded when their effects on the outcome of a study cannot be distinguished from each other. (pp. 306, 307)

Control group A group of experimental subjects that is given a standard treatment, no treatment, or a fake treatment (such as a placebo). (p. 307)

Convenience sample A sample that consists of individuals who are easily reachable, such as people passing by on the street. A convenience sample is usually biased. (p. 294)

Double-blind experiment An experiment in which neither the experimental subjects nor the persons who interact with them know which treatment each subject received. (p. 310)

Experiment A study in which treatments are applied to people, animals, or things to observe the effect of such treatments. (p. 305)

Margin of error The number to the right of the ± sign in a 95% confidence interval and equal to half of the width of the full interval. For a 95% confidence interval, it equals about 2 standard deviations of the sampling distribution of the estimated parameter. If you conducted a very large number of polls, about 95% of the time the difference between a particular poll's result and the true value of the population parameter would be within the margin of error. (p. 321)

Nonresponse Some individuals chosen for a sample cannot be contacted or refuse to participate. (p. 303)

Observational study A study (e.g., a sample survey) that observes individuals and measures variables of interest but does not attempt to influence the responses. (p. 311)

Parameter A number that describes some characteristic of the population. In statistical inference, the goal is often to estimate an unknown parameter or make a decision about its value. (p. 315)

Placebo effect The beneficial effect of a dummy treatment (such as an inert pill in a medical experiment) on the response of subjects. (p. 310)

Population The entire group of people or things about which we want information. (p. 292)

Prospective study An observational study that follows two or more groups of subjects forward in time. (p. 312)

Randomized comparative experiment An experiment to compare two or more treatments in which people, animals, or things are assigned to treatments by chance. (p. 307)

Retrospective study An observational study that uses interviews or records to collect information about past behaviors of subjects in two or more groups. (p. 312)

Sample A part of the population that is actually observed and used to draw conclusions, or inferences, about the entire population. (p. 292)

Sample proportion The proportion $\hat{p}$ of the members of a sample having some characteristic (e.g., agreeing with an opinion poll question). The sample proportion from a simple random sample is used to estimate the corresponding proportion p in the population from which the sample was drawn. (p. 317)

Sampling distribution The distribution of values taken by a statistic when all possible random samples of the same size are drawn from the same population. The sampling distributions of sample proportions are approximately normal. (p. 317)

Simple random sample (SRS) A sample chosen by chance, so that every possible sample of the same size has an equal chance of being selected. (p. 297)

Statistic A number that describes a sample. A statistic can be calculated from the sample data alone; it does not involve any unknown parameters of the population. (p. 315)

Statistical inference Methods for drawing conclusions about an entire population on the basis of data from a sample. Confidence intervals are one type of inference method. (p. 314)

Statistically significant An observed effect is statistically significant if it is so large that it is unlikely to occur just by chance in the absence of a real effect in the population from which the data were drawn. (p. 308)

Table of random digits A table whose entries are the digits 0, 1, 2, 3, 4, 5, 6, 7, 8, 9 in a completely random order. That is, each entry is equally likely to be any of the 10 digits and no entry gives information about any other entry. (p. 297)

Undercoverage The process of choosing a sample may systematically leave out some groups in the population, such as households without landline telephones. (p. 302)

Voluntary response sample A sample of people who select themselves by responding to a general invitation to give their opinions. Such a sample is usually strongly biased. (p. 296)

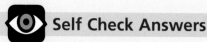

Self Check Answers

1. (a) Reasonable populations are all residents in the station's viewing area or all viewers of the 6 o'clock news. However, certain viewers who feel strongly about this question could call their friends and encourage them to vote even if they are outside of the viewing area. So, it could be all residents in the station's viewing area and their friends. (There is no clear answer to this question.)

(b) Sample answer: If the population is all residents in the station's viewing area, the sample is not representative of the population. For example, people who do not watch the 6 o'clock news will not be represented in this sample. The views of the 6 o'clock news watchers could be quite different from those who do not watch the 6 o'clock news. Also, people who feel strongly about the question are more apt to reply and will thus be overrepresented in the sample.

2. (a) The sample consists of the people who completed the survey and mailed it back to the hospital. This is an example of a voluntary response survey.

(b) The population is all people having routine medical tests done at this hospital.

(c) The data will not accurately portray patient satisfaction. Many patients will not bother to fill out the survey, particularly if they did not encounter any problems. People who had a bad experience of some type are probably more likely to complete and mail in the survey.

3. Entering Table 7.1 on line 110, here is a list of two-digit numbers:

~~38~~ ~~44~~ ~~84~~ ~~87~~ ~~89~~ 18 ~~33~~ ~~82~~ ~~46~~ ~~97~~ ~~39~~ ~~36~~ ~~44~~ 20 06 ~~76~~ ~~68~~ ~~80~~ ~~87~~ 08

Since we have not reached a sample of size 5, we move to line 111 and continue the process:

~~81~~ ~~48~~ ~~66~~ ~~94~~ ~~87~~ ~~60~~ ~~51~~ ~~30~~ ~~92~~ ~~97~~ ~~00~~ ~~41~~ 27

Our random sample of labels consists of 18, 20, 06, 08, and 27. So, the randomly selected songs will be "All You Need Is Love," "Lady Madonna," "A Hard Day's Night," "Eight Days a Week," and "The Long and Winding Road."

4. Although a few students might refuse to complete the questionnaire, the major problem affecting this survey is undercoverage. The sample will not include students who never eat breakfast or students who were out late Thursday night and didn't show for Friday breakfast. So, the views of large segments of the student population will not be represented in this survey.

5. Both questions have been designed to elicit particular responses: the first to elicit support for smoking in a hotel room, and the second to obtain support for banning smoking in the workplace. The use of the word *freedom* in Q1 makes smoking appear as a right and therefore makes it more likely that the respondent will answer "yes." The introductory phrase in Q2 focuses respondents' attention on the dangers of smoking and makes it more likely that the respondent will answer "no."

6. Students who had a class right before their math class might be more tired than those who did not. Because they had back-to-back classes, they might not have had enough time to eat before their math class. Students who did not have a class right before their math class would have had time both to eat and to study for the test before going to math class.

7. This is a retrospective study.

8. (a) In a sample of size 2, it is possible to get 0 "yes" responses, 1 "yes" response, or 2 "yes" responses. Hence, the possible values for $\hat{p}$ would be 0, 0.5, and 1.0.

(b) Since $p = 0.6$, it is least likely that both responses are "no." Hence, $\hat{p} = 0$ will occur least frequently.

9. (a) The distribution is normal and thus it is bell-shaped.

(b) The mean is 0.7.

(c) The standard deviation is $\sqrt{(0.7)(0.3)/1500} \approx 0.0118$.

(d) The distribution is still normal, but its center is shifted to the right compared to the distribution in Example 16. In addition, its standard deviation is slightly smaller.

10. (a) $\hat{p} = \dfrac{623}{1005} \approx 0.6199$

$$\hat{p} \pm 2\sqrt{\frac{\hat{p}(1-\hat{p})}{n}} = 0.6199 \pm 2\sqrt{\frac{(0.6199)(0.3801)}{1005}}$$

$$= 0.6199 \pm 0.0306 \quad \text{or} \quad 61.99\% \pm 3.06\%$$

$$= 0.5893 \text{ to } 0.6505 \quad \text{or} \quad 58.93\% \text{ to } 65.05\%$$

(b) The margins of error are close. In fact, if rounded to one decimal place, the margins of error are both 3.1%.

Skills Check

1. An opinion poll contacts 1021 adults and asks them, "Which political party do you think has better ideas for leading the country in the 21st century?" In all, 723 of the 1021 say "the Democrats." The sample in this setting is

(a) all 220 million adults in the United States.

(b) the 1021 people interviewed.

(c) the 723 people who chose the Democrats.

2. In a part-to-whole relationship, we would say the population is the __________.

3. A committee on community relations in a college town plans to survey local businesses about the importance of students as customers. From the 10,000 businesses listed in the telephone book, the committee chooses 150 businesses at random. Of these, 72 return the questionnaire mailed by the committee. The sample is

(a) all 10,000 businesses in the college town.

(b) the 150 businesses chosen.

(c) the 72 businesses that returned the questionnaire.

4. A call-in poll asks who people are planning to vote for in the next presidential election. People who believe major change is needed are likely to be represented in this poll __________ than they should be, if the goal is to get results that are representative of all voters.

5. On January 2, 2008, the *American Idol* website (www.americanidol.com) conducted an online poll that asked respondents which contestant they liked best among six former contestants. To become part of the sample, respondents simply clicked on a response. Of the 941,434 responses to this poll, 55% voted for Clay Aiken. We can conclude that

(a) most Americans preferred Clay Aiken out of those six contestants.

(b) the sample was too small a fraction of the millions of people who watched the TV show to draw any conclusion.

(c) the poll used voluntary response, so the results tell us little about the population of all adults.

6. A sample consisting of people who chose to have their opinions be part of a poll is a __________ sample.

7. If we use the range of a sample to estimate the range of a population, our estimate would likely be

(a) a bit too high.

(b) a bit too low.

(c) unbiased and right on target.

8. You are using the table of random digits to choose a simple random sample of 6 students from a class of 30 students. You label the students 01 to 30 in alphabetical order. Go to line 113 of Table 7.1 (page 298). Of the labels corresponding to the six students selected for your sample, the label that is largest is __________.

9. You must choose an SRS of 10 of the 420 retail outlets in New York that sell your company's products. How would you label this population to use Table 7.1 (page 298)?

(a) 001, 002, 003, . . . , 419, 420

(b) 000, 001, 002, . . . , 419, 420

(c) 1, 2, 3, . . . , 419, 420

10. From an alphabetical list of the 7200 salaried employees of a corporation, you label the employees 0001 to 7200. Using line 111 of Table 7.1 (page 298), choose an SRS of 5 of the 7200 employees. Of the five employees selected for your sample, the label that is the largest is __________.

11. Which of these is more likely to occur when selecting a sequence of three digits from a very large table of random digits?

(a) 123

(b) 111

(c) The above sequences are equally likely.

12. There are __________ possibilities for each digit drawn from a table of random digits.

13. A sample of households in a community is selected at random from the telephone directory. In this community, 4% of households have no telephone and another 35% have unlisted telephone numbers. The sample will certainly suffer from

(a) nonresponse.

(b) undercoverage.

(c) false responses.

14. For the survey in Skills Check 3, the nonresponse rate is __________.

15. Nonresponse is a type of

(a) coverage.

(b) sample.

(c) bias.

16. In research about the population of a county with 100,000 inhabitants, a sample of 3000 households was selected to be interviewed. For 1200 of the 3000 households that researchers attempted to contact, there was no one home willing to participate. For this survey, the nonresponse rate was ___________ .

17. A clinical trial compares an antidepression medicine with a placebo for relief of chronic headaches. There are 36 headache patients available to serve as subjects. To choose 18 patients to receive the medicine, you would

(a) assign labels 01 to 36 and use Table 7.1 to choose 18.

(b) assign labels 01 to 18 because only 18 need to be chosen.

(c) assign the first 18 who signed up to get the medicine.

18. An experiment is designed to see if a treatment is the ___________ of the response.

19. A comparative experiment

(a) does not use a treatment.

(b) has two or more groups.

(c) is statistically significant.

20. A study of cell phones and the associated risk of brain cancer looked at a group of 519 people who have brain cancer. The investigators matched each cancer patient with a person of the same sex, age, and race who did not have brain cancer, then asked about their use of cell phones. This kind of study is known as ___________.

21. Studies that follow subjects forward in time are called

(a) retrospective.

(b) prospective.

(c) double-blind.

22. A treatment consisting of a "dummy pill" that looks like real medicine (but isn't) is known as a ___________.

23. A study of religious practices among college students interviewed a sample of 125 students; 105 of the students said that they prayed at least once in a while. What is the sample proportion who said they pray?

(a) 105

(b) 84

(c) 0.84

24. Suppose that 35% of all adults in a population would say "good" and 65% would say "bad" if they were asked how they view the state of the economy. An opinion poll asks this question of an SRS of 1000 adults from the population. In repeated samples, the sample proportion $\hat{p}$ who say "good" would follow a normal distribution, with the mean having a value of ___________________.

25. Referring to Skills Check 24, the standard deviation of the distribution of the sample proportion of adults who view the economy as "good" is about

(a) 0.00023.

(b) 0.015.

(c) 0.03.

26. Referring to Skills Check 24, the standard deviation of the distribution of the sample proportion of all adults who view the economy as "bad" is ___________.

27. The sample survey in Skills Check 23 actually called 150 students, but 25 of the students refused to say whether they pray. This nonresponse could cause the survey result to be in error. The error due to nonresponse

(a) is in addition to the margin of error.

(b) is included in the margin of error.

(c) can be ignored because it isn't random.

28. To the nearer half of a percentage point, the margin of error for a 95% confidence interval is ___________ when we use the result of Skills Check 23 to estimate what percentage of all college students pray.

29. A survey of folk music fans yields this 95% confidence interval estimate of the proportion of fans who love the music of David Wilcox: 0.74 to 0.86. The estimated mean percentage of fans who love Wilcox's music must be about

(a) 95%.

(b) 86%.

(c) 80%.

30. To the nearer percentage point, the margin of error for the survey in Skills Check 29 is ___________.

Chapter 7 Exercises

Challenge Discussion

7.1 Sampling

1. A Gallup poll asked, "How would you describe your own personal weight situation right now?" Thirty-eight percent of American adults answered "very/somewhat overweight." Gallup reported that these results were based on telephone interviews of 1021 adults conducted on November 4–7, 2010.

(a) What was the population for this sample survey?

(b) What was the sample size?

2. Starting with the 2010 Census, the decennial "long form" sample was replaced with the annual American Community Survey (ACS; www.census.gov/acs/). The main part of the ACS contacts 250,000 households by mail each month, with follow-up by phone and in person if there is no response. Each household answers questions about its housing, economic, and social status. What is the population for the ACS?

3. On the Hudson Valley, New York, Patch Facebook page, readers were asked to send in stories of awful Valentine's Day gifts. The following were selected:

- Leftover chocolate (and he had eaten one!)
- Flowers purchased the day *before* Valentine's because it was cheaper to buy them the day before
- A recycled card from an ex-boyfriend with an open box of chocolates

Readers were then asked to vote on the best "worst Valentine's Day gift ever" story.

(a) Describe the population.

(b) Describe the sample.

(c) Do you think the response to this poll is representative of the views of the residents of the Hudson Valley, New York? Explain.

7.2 Bad Sampling Methods

4. You see a student standing in front of the Student Center, stopping other students now and then to ask them questions. The student says that she is collecting student opinions for a class assignment. Explain why this sampling method is almost certainly biased.

5. A member of Congress is interested in whether her constituents favor a proposed gun-control bill. Her staff reports that letters on the bill have been received from 361 constituents and that 323 of these oppose the bill. What is the population of interest? What is the sample? Is this sample likely to represent the population well? Explain your answer.

6. Highway planners made a main street in a college town one-way. Local businesses were against the change. The local newspaper invited readers to call a telephone number to record their comments. The next day, the paper reported:

> Readers overwhelmingly prefer two-way traffic flow to one-way streets. By a 6:1 ratio, callers to the newspaper's Express Yourself opinion line on Wednesday complained about the one-way streets that have been in place since May. Of the 98 comments received, all but 14 said "no" to one-way streets.

(a) What population do you think the newspaper wants information about?

(b) Is the proportion of this population who favor one-way streets almost certainly larger or smaller than the $\frac{14}{98}$ proportion in the sample? Why?

7. Your college wants to gather student opinion about a proposed student fee increase. It isn't practical to contact all students.

(a) Give an example of a way to choose a sample of students that is poor practice because it depends on voluntary response.

(b) Give an example of a bad way to choose a sample that doesn't use voluntary response.

8. Explain why each of the following samples might be biased:

(a) A large university wants to conduct a focus group on campus satisfaction. A poster was hung in the Student Union inviting students to participate (with a promise of free food!).

(b) The president of the United States wants to check his approval rating after two years in office. A sample of 1000 voters is selected from California.

7.3 Simple Random Samples

9. You have just been blessed with quadruplets (all girls). You decide to select their names using an SRS of four names from the following list of the most popular names given to American girls born in the past decade. To do this, use Table 7.1 (page 298), starting at line 122.

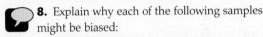

1. Emily	6. Abigail	11. Alexis	16. Brianna
2. Madison	7. Isabella	12. Sarah	17. Lauren
3. Emma	8. Samantha	13. Sophia	18. Chloe
4. Olivia	9. Elizabeth	14. Alyssa	19. Natalie
5. Hannah	10. Ashley	15. Grace	20. Kayla

10. (a) Would pulling out and lining up several dollar bills to use the eight-digit serial numbers be a reasonable substitute for Table 7.1? Explain.

(b) How about using the telephone numbers on a page of the phone book? Explain.

11. There are approximately 371 active three-digit telephone area codes covering Canada, the United States, and some Caribbean areas (more are created regularly).

You want to choose an SRS of 25 of these area codes for a study of available telephone numbers.

(a) How would you label the area codes to use Table 7.1?

(b) Use Table 7.1 (page 298), starting at line 125, to choose the first three labels of the members of this sample.

12. Each March, the Current Population Survey (CPS) is expanded to gather a wider variety of information than what is collected in the monthly reports. Suppose in one March survey, we are interested in participants whose highest level of education is a bachelor's degree and who are between the ages of 25 and 64. It turns out that 14,959 of the survey respondents fall into this category. Think of them as a population.

(a) To select an SRS of these people, how would you assign labels?

(b) Use Table 7.1 (page 298), starting at line 107, to choose the first three members of the SRS.

13. In using Table 7.1 repeatedly to choose samples, you should not always choose the same row, such as line 101. Why not?

14. Which of the following statements are true of a table of random digits and which are false? Explain your answers.

(a) There are exactly four 0s in each row of 40 digits.

(b) Each pair of digits has chance 1/100 of being 00.

(c) The digits 0000 can never appear as a group because this pattern is not random.

15. Your dog just had a litter of 5 male puppies. Picking out good names for dogs can be difficult. On the Internet, you found a list of the top 20 names for male puppies, which are shown below. You decide to randomly select 5 names from the list as names for the puppies. What names did you select? Explain how you randomly selected these names. [Use Excel or a graphing calculator and one of the techniques discussed in Spotlight 7.1 (page 300) for this exercise.]

Max	Buddy	Charlie	Rocky
Cooper	Duke	Bear	Jack
Bently	Toby	Zeus	Tucker
Diesel	Jake	Milo	Teddy
Jax	Buster	Bandit	Harley

16. The students listed below are enrolled in an elementary French course. Students are assigned to one of two smaller conversation sections at random. [Use Excel or a graphing calculator and one of the techniques discussed in Spotlight 7.1 (page 300) for this exercise.]

1. Arnold	11. Ellis	21. Martinez	31. Randall
2. Barrett	12. Fernandez	22. Moore	32. Rodriguez
3. Bartkowski	13. Flury	23. Munroe	33. Schiller
4. Burns	14. Garcia	24. Neale	34. Scott
5. Campbell	15. Hardy	25. Nguyen	35. Smith
6. Chang	16. Holmes	26. Oakley	36. Stevenson
7. Colon	17. Jones	27. Orsini	37. Swokowski
8. Davies	18. Juarez	28. Perlman	38. Taylor
9. Dodington	19. Kempthorne	29. Prizzi	39. Vuong
10. Drummond	20. Levine	30. Putnam	40. Ward

Choose a simple random sample of 20 of these students to form Section 01. Explain how you obtained the names for this section. The remaining students will be assigned to Section 02.

17. The last stage of the CPS uses a *systematic sample.* An example will illustrate the idea of a systematic sample. Suppose that we must choose 4 rooms out of the 100 rooms in a dormitory. Because 100/4 = 25, we can think of the list of 100 rooms as 4 lists of 25 rooms each. Choose 1 of the first 25 rooms at random, using Table 7.1 (page 298). The sample will contain this room and the rooms 25, 50, and 75 places down the list from it. If 13 is chosen, for example, then the systematic random sample consists of the rooms numbered 13, 38, 63, and 88.

(a) Use Table 7.1 (page 298) to choose a systematic random sample of 5 rooms from a list of 200. Enter the table at line 120.

(b) Your sample gives every room the same chance to be chosen. Explain why.

(c) Despite the answer in part (b), this sample is not an SRS. Explain why.

18. An ethics institute selected a random sample of 100 U.S. high schools and then gave an in-class survey to all students in each selected school. Of the 29,760 students surveyed, 64% have cheated on a test and 30% have stolen from a store. This type of sample is known as a *cluster sample.* Why is this sample not an SRS from the population of all U.S. high school students?

7.4 Cautions About Sample Surveys

19. An opinion poll calls 1334 randomly chosen residential telephone numbers, and then the interviewer asks to speak with an adult member of the household, inquiring, "How many movies have you watched in a movie theater in the past 12 months?"

(a) What population do you think the poll has in mind?

(b) In all, 931 people respond. What is the rate (percent) of nonresponse?

(c) Many responses to this question are likely to be inaccurate. Why?

20. Randomized response: Suppose 30 students in a class participate in a survey in which they each flip a coin and do not reveal the result. If the result is tails, the student is supposed to give an honest answer to the question "Have you ever used a fake ID?" If the result is heads, the student is supposed to say "yes" to that question, regardless of what the true answer is. Suppose the results in the class are 18 "yes" answers and 12 "no" answers.

(a) If students follow the procedure correctly, is it true that all students who answered "no" have not used a fake ID?

(b) If students follow the procedure correctly, is it true that all students who have not used a fake ID answered "no"?

(c) On average, about half of the students who have not used a fake ID flipped tails, so what is your best estimate of the true number of students who have not used a fake ID?

(d) Based on the answer to part (c), what is your estimate of the true number and proportion of students who have used a fake ID?

(e) Do we have any way to know which of the 18 "yes" answers are truthful?

21. Randomized response: Suppose 50 students in a college class participate in a survey in which they each flip a coin and do not reveal the result. If the result is tails, the student is supposed to give an honest answer to the question "Have you ever cheated on an exam in high school or in college?" If the result is heads, the student is supposed to say "yes" to that question, regardless of what the true answer is. Suppose the results in the class are 42 "yes" answers and 8 "no" answers.

(a) If students follow the procedure correctly, is it true that all students who answered "no" have not cheated on an exam either in high school or in college?

(b) If students follow the procedure correctly, is it true that all students who have not cheated on an exam either in high school or in college answered "no"?

(c) On average, about half of the students who have not cheated on an exam in high school or in college flipped tails, so what is your best estimate of the true number of students who have not cheated on an exam?

(d) Based on the answer to part (c), what is your estimate of the true number and proportion of students who have cheated on an exam either in high school or in college?

(e) Do we have any way to know which of the 42 "yes" answers are truthful?

22. Comment on each of the following as a potential sample survey question. If the question is unclear, slanted, or too complicated, restate it using better words.

(a) Which of these best represents your opinion on gun control?

(i) The government should confiscate our guns.

(ii) We have the right to keep and bear arms.

(b) In view of the escalating environmental degradation and predictions of serious resource depletion, would you favor economic incentives for recycling of resource-intensive consumer goods?

(c) More people have seen the movie *Gone with the Wind* than any other motion picture produced over the past century. Have you seen this movie?

23. The wording of questions can strongly influence the results of a sample survey. You are writing an opinion poll question about a proposed amendment to the Constitution. You can ask if people are in favor of "changing the Constitution" or "adding to the Constitution" by approving the amendment. Which of these choices of wording will likely produce a much higher percentage in favor? Why do you think this is true?

7.5 Experiments

24. As reported in *College Teaching,* in a 2006 article entitled "Humor in Pedagogy: How Ha-Ha Can Lead to Aha" (Vol. 54, Issue 1), R. L. Garner randomly assigned 117 undergraduates to "review lecture videos" on statistics research methods. The videos either did or did not have short bits of humor inserted. Students who viewed the humor-added version of the video gave significantly higher ratings in their opinion of the lesson, how well the lesson communicated information, and the quality of the instructor. Even more importantly, that same group of students also recalled and retained significantly more information on the topic.

(a) What is the explanatory variable?

(b) What is the response variable?

(c) Why is this an experiment?

(d) Why were students not initially told that the true purpose of the study was to assess the use of humor?

(e) Why do you think the study was done using a fixed-video format rather than through live teaching?

25. In a study on the attitude of gratitude, 192 undergraduates were assigned randomly to one of three clusters and asked to keep a regular report on psychological and physical indicators. One cluster was given a prompt to list things in their lives they are grateful for, another cluster's prompt was to list recent hassles, and the third cluster's prompt was to simply list events that recently had an impact on them. The "gratitude group" generally reported higher well-being. [R. A. Emmons and M. E. McCullough, Counting blessings versus burdens, *Journal of Personality and Social Psychology 84*(2) (2003):377–389.]

(a) What is the explanatory variable?

(b) What is the response variable?

(c) Why is this an experiment?

(d) Does this experiment address whether it is more reasonable to say that well-being causes gratitude or that gratitude causes well-being?

26. Will owning a video-game system hurt the academic development of young boys? You are interested in tracking time spent playing video games, time spent in academic activities, teacher-reported learning problems, and reading and writing scores four months later. Outline the design of an experiment to study the effect of video-game ownership.

Victoria Blackie/Getty Images

27. We want to investigate the following question: Will classroom programs explaining the health advantages of drinking water rather than sugary sodas reduce obesity among children aged 7 to 11 years? Because children are already grouped in school classrooms, we must randomize classes rather than individual children. An experiment assigned 15 classes to receive the program and another 14 to form a control group. After 12 months, obesity had increased in the control group and remained steady in the treatment group. Outline the design of the experiment, label the available classes, and use Table 7.1 (page 298), beginning at line 103, to carry out the random assignment.

28. A college allows students to choose either classroom or self-paced instruction in a basic mathematics course. The college wants to compare the effectiveness of self-paced and regular instruction. Someone proposes giving the same final exam to all students in both versions of the course and comparing the average score of those who took the self-paced option with the average score of students in regular sections.

(a) Explain why confounding makes the results of that study worthless.

(b) Given 30 students who are willing to use either regular or self-paced instruction, outline an experimental design to compare the two methods of instruction. Then use Table 7.1 (page 298), starting at line 108, to carry out the randomization.

29. Two second-grade teachers, Miss Earls (who is an experienced teacher) and Mrs. Gifford (who is in her second year of teaching), were really excited about a new curriculum that utilized animations to teach science. They decided to use their classrooms for an experiment. Since Miss Earls had access to computers in her class, she used the animation lessons. Mrs. Gifford covered similar material with her students using handouts followed by discussions. After students had completed the materials, they were given a test designed by Miss Earls. There were 21 students in Miss Earls's class and 29 students in Mrs. Gifford's class. Miss Earls's class scored, on average, 12 points higher on the test.

(a) Explain why confounding makes the results of these teachers' study worthless. What are some of the confounding variables?

(b) Given that the principal would allow the 50 students in the two classes to be reassigned to these two teachers for this experiment and would allow the teachers to switch rooms, outline an experimental design to compare the two methods of science instruction. Use either Table 7.1 or technology to carry out the randomization.

30. Track down a print or online copy of the Bible. Chapter 1 of the book of Daniel (especially verses 12 through 16) appears to include the first clinical trial in recorded history. Outline the design of the experiment. Is this an example of an uncontrolled experiment, a comparative experiment, or a randomized comparative experiment? Explain.

31. Will people spend less on healthcare if their health insurance requires them to pay some part of the cost themselves? An experiment on this issue asked if the percentage of medical costs that is paid by health insurance has an effect either on the amount of medical care that people use or on their health. The treatments were four insurance plans, each of which paid all medical costs above a ceiling. Below the ceiling, the plans paid 100%, 75%, 50%, or 0% of costs incurred. Outline the design of a randomized comparative experiment suitable for this study.

32. The research question for an undergraduate research project was whether hearing-impaired customers were treated differently by store clerks than non-hearing–impaired customers. There were 20 customers, 10 of whom were hearing impaired. The customers were sent in pairs into stores. The hearing-impaired pairs used sign language to communicate with each other and the non-hearing–impaired pairs entered stores speaking English to each other. The subjects consisted of 77 salesclerks in 27 stores (from 175 stores) in a large shopping mall. The response variable was the time that elapsed from when the pair entered the store and made eye contact with the salesclerk until the clerk approached and offered assistance. Describe how you would design the rest of the experiment.

33. Stores advertise price reductions to attract customers. What type of price cut is most attractive? Market researchers prepared ads for athletic shoes announcing different levels of discounts (20%, 40%, or 60%). The student subjects who read the ads were also given "inside information" about the fraction of shoes on sale (50% or 100%). Each subject then rated the attractiveness of the sale on a scale of 1 to 7.

(a) Each treatment in this experiment is a combination of values of two explanatory variables: discount level and fraction on sale. List all the treatments.

(b) Outline a randomized comparative experiment using 60 student subjects. Use Table 7.1 (page 298) at line 123 to choose the subjects for the first treatment.

34. You wish to learn if students in an English course write better essays when they are required to use computer word-processing than when they write and revise their essays by hand. There are 120 students in an English course available as subjects.

(a) Outline the design of an experiment to determine whether word-processing results in better essays.

(b) What precautions would you take in completing this experiment that do not appear in your outline of the design?

35. Eye cataracts are responsible for over 40% of blindness around the world. Can drinking tea regularly slow the growth of cataracts? We can't experiment on people, so we use rats as subjects. Researchers injected 14 young rats with a substance that causes cataracts. Half the rats also received tea extract; the other half got a placebo. The response variable was the growth of cataracts over the next six weeks. The researchers found that the tea extract did slow cataract growth in the rats.

(a) Outline the design of this experiment.

(b) Use Table 7.1 (page 298), starting at line 108, to assign rats to treatments.

36. The rats in the previous exercise were labeled 01 to 14. Unknown to the researchers, the 5 rats labeled 01 to 05 have a genetic defect that favors cataracts. If we simply put rats 01 to 07 in the tea group, the experiment would be biased against tea. We can observe how random selection works to reduce bias by keeping track of how many of these 5 rats are assigned to the tea group. Use one of the technology procedures in Spotlight 7.1 (page 300) to carry out the random assignment of 7 rats to the tea group 25 times, keeping track of how many of rats 01 to 05 are in the tea group each time. Make a histogram of the count of rats 01 to 05 assigned to tea. What is the average number in your 25 tries? Based on your results, describe how random selection works to reduce bias.

7.6 Experiments Versus Observational Studies

37. The article "Smoking, Smoking Cessation, and Risk for Type 2 Diabetes Mellitus" published in the *Annals of Internal Medicine* (January 2010) reported on a study that followed 10,892 middle-aged adults over a nine-year period. At the start of the study, none of the subjects had diabetes. Roughly 45% of the subjects were smokers. The study found that compared to those who never smoked, subjects who quit smoking had an increased risk of diabetes.

(a) Is the study described above an observational study or an experiment? Explain.

(b) Based on this study, should you conclude that quitting smoking causes diabetes? Justify your answer.

38. Healthcare providers are giving more attention to relieving the pain of cancer patients. An article in the journal *Cancer* surveyed a number of studies and concluded that controlled-release (CR) morphine tablets, which release the painkiller gradually over time, are more effective than giving standard morphine when the patient needs it. The "methods" section of the article begins: "Only those published studies that were controlled (i.e., randomized, double-blind, and comparative), repeated-dose studies with CR morphine tablets in cancer pain patients were considered for this review" [C. A. Warfield, Controlled-release morphine tablets in patients with chronic cancer pain, *Cancer, 82*(12) (1998): 2299–2306]. Explain the terms in parentheses to someone who knows nothing about medical trials.

39. Could the magnetic fields from power lines cause leukemia in children? Investigators who wanted to explore this question spent five years and $5 million comparing 638 children who had leukemia and 620 who did not. They went into the homes and actually measured the magnetic fields in the children's bedrooms, in other rooms, and at the front door. They recorded facts about nearby power lines for the family home, as well as for the mother's residence when she was pregnant. Result: They found no evidence of more than a chance connection between magnetic fields and childhood leukemia. Explain carefully why this study is not an experiment, and state what kind of study it is.

40. A typical hour of prime-time television shows three to five violent acts. Linking family interviews and police records shows a clear association between time spent watching TV as a child and later aggressive behavior.

(a) Explain why this is an observational study rather than an experiment.

(b) Suggest several variables describing a child's home life that may be confounded with how much TV he or she watches.

(c) Explain why confounding makes it difficult to conclude that more TV *causes* more aggressive behavior.

41. The Nurses' Health Study has interviewed a sample of more than 100,000 female registered nurses every two years since 1976. Beginning in 1980, the study asked questions about diet, including alcohol consumption. The researchers concluded that "light-to-moderate drinkers had a significantly lower risk of death" than either nondrinkers or heavy drinkers.

(a) Is the Nurses' Health Study an observational study or an experiment? Why?

(b) What does *significant* mean in a statistical report?

(c) Suggest some confounding variables that might explain why moderate drinkers have lower death rates than nondrinkers. (The study adjusted for these variables.)

42. The financial aid office of a university asks a sample of students about their employment and earnings. The report says that "for academic year earnings, a statistically significant difference was found between the sexes, with men earning more on the average. No significant difference was found between the earnings of black and white students." Explain both of these conclusions, for the effects of sex and of race on average earnings, in language understandable to someone who knows nothing about statistics. Do not use the words *significant* or *significance* in your answer.

43. People who eat lots of fruits and vegetables have lower rates of colon cancer than those who eat little of these foods. Fruits and vegetables are rich in antioxidants such as vitamins A, C, and E. Will taking antioxidant pills help prevent colon cancer? A clinical trial studied this question with 864 people who were at risk for colon cancer. The subjects were divided into four groups: those who took daily beta carotene (related to vitamin A), those who took daily vitamins C and E, those who took all three vitamins every day, and those who took a daily placebo. After four years, the researchers were surprised to find no significant difference in colon cancer among the groups.

(a) Outline the design of the experiment. Use your judgment in choosing the group sizes.

(b) Assign labels to the 864 subjects and use Table 7.1 (page 298), starting at line 118, to choose the first five subjects for the "beta carotene" group.

(c) The study was double-blind. What does this mean?

(d) What does "no significant difference" mean in describing the outcome of the study?

(e) Suggest some characteristics of the kind of people who eat lots of fruits and vegetables that might explain lower rates of colon cancer. The experiment suggests that these variables, rather than the antioxidants, may be responsible for the observed benefits of fruits and vegetables.

44. Dr. Megan Moreno sent a cautionary message to a randomly selected half of a sample of MySpace users (ages 18–20) whose public profiles included references to sex and substance abuse. A review of all profiles from the original sample three months later showed that those who had received the email were more likely to have removed the references from their online profiles or to have changed their profile setting to "private." Is this an experiment or observational study, and how do you know?

45. A study reported on 533,715 women at least 40 years old who were diagnosed with invasive breast cancer and reported to the National Cancer Data Base (NCDB). The study found strong evidence that patients without health insurance were more likely to have a more advanced stage of cancer (i.e., stage III or IV). Is this an experiment or observational study, and how do you know? [M. T. Halpern et al., Insurance status and stage of cancer at diagnosis among women with breast cancer, *Cancer, 110*(2) (2007): 403–411.]

7.7 Inference: From Sample to Population

46. An opinion poll uses random digit dialing equipment to select 2000 residential telephone numbers. Of these, 631 are unlisted numbers. This isn't surprising, because 35% of all residential numbers are unlisted. For each underlined number, state whether it is a parameter or a statistic.

47. In the 1980s, the Tennessee Student Teacher Achievement Ratio experiment randomly assigned more than 7000 children to regular or small classes during their first four years of school. Even though the treatment lasted only from kindergarten to third grade, there were differences (in favor of the students in the smaller classes) that were noticeable even many years later. For example, when these same children reached high school, 40.2% of Black students from the small classes took the ACT or SAT college entrance exam. Only 31.7% of Black students from the regular classes took one of these exams. For each underlined number, state whether it is a parameter or a statistic.

48. At a college in Singapore, students were randomly selected and asked to complete a Web-based survey about sexual behavior. Of those selected, 534 students completed the survey. Suppose that the population proportion of those having had sexual intercourse in the past six months was $p = 0.24$.

(a) What are the mean and standard deviation of the proportion $\hat{p}$ of the sample who have had sexual intercourse over the past six months?

(b) In what interval of values do the proportions $\hat{p}$ from 95% of all samples fall?

(c) In what interval of values do the proportions $\hat{p}$ from 99.7% of all samples fall?

49. Harley-Davidson motorcycles make up 14% of all the motorcycles registered in the United States. You plan to interview an SRS of 500 motorcycle owners.

Tony Harrington/Getty Images

(a) What is the approximate distribution of the proportion of your sample who own Harleys?

(b) In 95% of all samples like this one, the proportion of the sample who own Harleys will fall between ________ and ________. What are the missing numbers?

50. Exercise 48 asks what values the sample proportion $\hat{p}$ is likely to take when the population proportion is $p = 0.24$ and the sample size is $n = 534$. What interval covers the middle 95% of values of $\hat{p}$ when $p = 0.24$ and $n = 400$? When $n = 1600$? When $n = 6400$? What general fact about the behavior of $\hat{p}$ do your results illustrate?

51. You can use a table of random digits to *simulate* sampling from a population. Suppose that 60% of the population bought a lottery ticket in the last 12 months. We will simulate the behavior of random samples of size 40 from this population.

(a) Let each digit in the table stand for one person in this population. Digits 0 to 5 stand for people who bought a lottery ticket, and 6 to 9 stand for people who did not. Why does looking at one digit from Table 7.1 (page 298) simulate drawing one person at random from a population with 60% "yes"?

(b) Each row in Table 7.1 contains 40 digits. So the first 10 rows represent the results of 10 samples. How many digits between 0 and 5 does the top row contain? What is the percentage of "yes" responses in this sample? How many of your 10 samples overestimated the population proportion of 60%? How many underestimated it? (You could program a computer to continue this process, say, 1000 times, to produce a pattern like that in Figure 7.6 on page 316.)

52. In a random sample of students who took the SAT Reasoning college entrance exam twice, it was found that

427 of the respondents had paid for coaching courses and that the remaining 2733 had not.

(a) What is the sample proportion $\hat{p}$ of coaching among students who retake the SAT?

(b) Give a 95% confidence interval for the proportion of coaching among students who retake the SAT.

53. A Gallup poll asked each of 1785 randomly selected adults whether he or she happened to attend a house of worship in the previous seven days. Of the respondents, 750 said "yes."

(a) Determine the sample proportion $\hat{p}$ of those who answered "yes."

(b) Give a 95% confidence interval for the proportion of all adults who claim that they attended a house of worship during the week preceding the poll. (The proportion who actually attended may be lower; some people might say "yes" if they often attend, even if they didn't attend that particular week.)

54. A *CBS News* poll conducted July 29–August 4, 2014, surveyed 1344 randomly selected American adults. Of those surveyed, 726 say that their sympathies in the Middle East situation lie more with the Israelis than with the Palestinians.

(a) Give a 95% confidence interval for the proportion of all American adults whose sympathies in the Middle East situation lie more with the Israelis than with the Palestinians.

(b) The poll reported a margin of error of ±3%. Explain how your results agree with this statement.

55. A telephone survey of 880 randomly selected drivers asked, "Recalling the last 10 traffic lights you drove through, how many of them were red when you entered the intersections?" Of the 880 respondents, 171 admitted that at least one light had been red.

(a) Give a 95% confidence interval for the proportion of all drivers who ran one or more of the last 10 red lights they came across.

(b) A practical problem with this survey is that people may not give truthful answers. What is the likely direction of the bias: Do you think more or fewer than 171 of the 880 respondents really ran a red light? Why?

56. A Gallup poll conducted May 2–7, 2013, by telephone interviews (both landlines and cellular phones) of 1535 American adults found that 59% of Americans regarded gay and lesbian relations as morally acceptable.

(a) How many of the 1535 people interviewed said gay and lesbian relations were morally acceptable?

(b) Gallup indicates that the margin of error for this poll is ±3 percentage points. Explain to someone who knows nothing about statistics what "margin of error ±3 percentage points" means.

(c) Give a 95% confidence interval for this survey. Does your margin of error agree with the 3 percentage points announced by Gallup?

57. Consider the margin of error formula

$$2\sqrt{\frac{\hat{p}(1-\hat{p})}{n}}.$$

(a) For a fixed value of n, what value of $\hat{p}$ between 0 and 1 causes this formula to attain its largest possible value?

(b) Using the answer to part (a), what would be a simplified (and slightly more conservative) formula for calculating the margin of error?

58. A news article reports that in a recent Gallup poll, 78% of the sample of 1108 adults said they believe heaven exists. Only 60% said they believe there is a hell. The news article ends, "The poll's margin of sampling error was plus or minus 4 percentage points." Can we be certain that between 56% and 64% of all adults believe hell exists? Explain your answer.

59. A survey of Internet users found that males outnumbered females by nearly 2 to 1. This was a surprise because earlier surveys had put the ratio of men to women closer to 9 to 1. Later, the article about the research states that surveys were sent to 13,000 organizations and that 1468 of these responded. The survey report claims that "the margin of error is 2.8 percent, with 95% confidence."

(a) What was this survey's *response rate*? (The response rate is the percentage of the planned sample that responded.)

(b) Do you think that the small margin of error is a good measure of the accuracy of the survey's results? Explain your answer.

60. A recent Gallup telephone poll found that 68% of adult Americans favor teaching creationism along with evolution in public schools. The Gallup press release states:

> For results based on samples of this size, one can say with 95% confidence that the maximum error attributable to sampling and other random effects is plus or minus 3 percentage points.

Give one example of a source of error in the poll result that is *not* included in this margin of error.

61. The Internal Revenue Service (IRS) plans to examine an SRS of individual income tax returns from each state that were filed electronically. One variable of interest is the proportion of returns that were filed by a tax practitioner rather than by an individual taxpayer. The total number of e-filed tax returns in a state varies from 4.9 million in California to 97,000 in Vermont.

(a) Will the margin of error for estimating the proportion change from state to state if an SRS of 1000 e-filed returns is selected in each state? Explain your answer.

(b) Will the margin of error change from state to state if an SRS of 1% of all e-filed returns is selected in each state? Explain your answer.

62. Exercise 56 describes a Gallup poll that interviewed 1535 people. Suppose that you want a margin of error half as large as the one you found in that exercise. How many people must you plan to interview?

63. Though opinion polls usually make 95% confidence statements, some sample surveys use other confidence levels. The monthly unemployment rate, for example, is based on the CPS of about 50,000 households. The margin of error in the unemployment rate is announced as about ±0.15% with 90% confidence. Is the margin of error for 90% confidence larger or smaller than the margin of error for 95% confidence? Why? (*Hint:* Look again at Figure 7.12, on page 323.)

Chapter Review

64. The proportion of one's body that is fat is a key indicator of fitness. The many ways to estimate this have different margins of error (given in percentage points):

Method	Calipers pinch	Bioelectrical impedance	Body mass index calculator	Hydrostatic weighing (dunk test)
Margin of error	±3	±4	±10	±1

(a) Which of these tests is the least accurate?

(b) If the pinch test says that you have 21% body fat, what is the 95% confidence interval for this estimate?

65. Many medical trials randomly assign patients to either an active treatment or a placebo. These trials are always double-blind. Sometimes the patients can tell whether they are getting the active treatment. This defeats the purpose of blinding. Reports of medical research usually ignore this problem. Investigators looked at a random sample of 97 articles reporting on placebo-controlled randomized trials in the top five general medical journals. Only 7 of the 97 discussed the success of blinding. Give a 95% confidence interval for the proportion of all such articles that discuss the success of blinding. [Dean Fergusson et al., Turning a blind eye: The success of blinding reported in a random sample of randomised, placebo-controlled trials, *British Medical Journal, 328* (2004): 432–436.]

66. Tomeka wants to ask a sample of students at her college, "Do you think that Social Security will still be paying benefits when you retire?" She obtains the college email addresses of all 2654 students attending the college.

(a) How would you label the addresses to choose a simple random sample of 100 students?

(b) Use Table 7.1 (page 298), starting at line 103, to choose the first three labels in the sample.

(c) Tomeka emails her question to the 100 addresses in her sample. Although she has chosen an SRS, a serious practical difficulty may make it hard to draw clear conclusions from her sample. What practical difficulty do you expect Tomeka to encounter?

67. Suppose that exactly 10% of all articles in major medical journals that describe placebo-controlled randomized trials discuss the success of blinding. That is, the proportion of "successes" in the population is $p = 0.1$. What is the approximate probability that fewer than 7% of an SRS of 97 articles from this population discuss the success of blinding?

68. The ability to grow in shade may help pines found in the dry forests of Arizona resist drought. How well do these pines grow in shade? Investigators planted pine seedlings in a greenhouse in either full light or light reduced to 5% of normal by shade cloth. At the end of the study, they dried the young trees and weighed them.

(a) Explain why this study is an experiment.

(b) What are the individuals, the treatments, and the response variable in this experiment?

(c) You have 200 pine seedlings available. Outline the design that you would use for this experiment.

69. The National Children's Study (NCS), the largest and most detailed study ever on children's health in the United States, is examining environmental effects on a large sample of children (from roughly 100,000 families) from before birth to age 21 years. Learn more at www.nationalchildrensstudy.gov.

(a) Explain why this is an observational study.

(b) Is this observational study prospective or retrospective?

(c) Why couldn't this study be done as an experiment?

70. A random sample of 2454 twelfth-grade American students responded to the following question: "Taking all things together, how would you say things are these days—would you say you're happy or not too happy?" Of the responses, 2098 students selected "happy."

(a) Calculate a 95% confidence interval for the population proportion of twelfth-grade American students who are happy.

(b) Would a 97% confidence interval for the proportion of happy students be wider or narrower than the one you calculated for part (a)? Justify your answer.

71. Rasmussen Report conducted a national telephone survey of a random sample of 1000 U.S. adults from June 19 to 20, 2013. Results indicated that 63% of adults nationwide would agree with the statement "Most Americans want the government to have less power and money."

(a) Use the information from the report to calculate a 95% confidence interval for the proportion of Americans who would agree with the statement above. Restate your confidence interval in terms of percentages. What is the margin of error?

(b) The report concluded with the following statement: "The margin of error is ±3% with a 95% level of confidence." Compare this statement with the margin of error you calculated in part (a).

(c) Was a sample of size 1000 sufficiently large to guarantee that the margin of error was less than 3% even if the sample percentage had been as low as 50% or as high as 80%? Explain.

(d) How large a sample size was needed to guarantee that the margin of error was below 3% regardless of the sample proportion?

Algebra Review Appendix
Solving for One Variable in Terms of Another

Applet Exercises

To do these exercises, go to www.macmillanhighered.com/fapp10e.

1. Use the *Simple Random Sample* applet to choose the sample of songs in Example 4 (on page 299). Here's how:

- Assign labels 01 to 27 by entering 27 in the "Population 1 to" box and clicking "Reset."
- Then enter 4 in the "Select a sample of size" box and click "Sample."

Which songs from the list in Example 4 (page 299) make up your sample? Click "Reset" and choose another sample. Which songs did you choose this time? You see that random sampling gives different samples each time—what matters is that all songs have the same chance to be chosen.

2. You can use the *Simple Random Sample* applet to choose treatment groups at random for a randomized comparative experiment. In outlining the design of the experiment in Exercise 35 (page 333) to compare the growth of cataracts in rats who were given tea extract to rats who were given a placebo, you should randomly choose the subjects (rats) for the first treatment (tea extract).

(a) Use the applet to choose an SRS of 7 out of 14 to receive the first treatment. Which subjects make up this group?

(b) The applet allows you to assign subjects randomly to more than two groups. Suppose you had a total of 36 rats and you wanted to assign a different treatment to each of four 9-rat groups. After you choose the first group, the "Population Hopper" contains the 27 subjects that were not chosen, in scrambled order. Click "Sample" again to choose 9 of these remaining subjects to receive the second treatment. Do this once more to choose the third group. The 9 subjects that remain in the "Population Hopper" form the fourth group. Which of the 36 subjects will receive each of the four treatments?

3. Suppose that 60% of the population bought a lottery ticket in the last 12 months. (This is the setting for Exercise 51.) You can use the *Probability* applet to simulate the behavior of random samples of size 50 from this population. You want to take many samples from this population to observe how the sample proportion that plays the lottery varies from sample to sample. By moving the sliders, specify the "Probability of Heads" setting in the applet as 0.6 and the number of tosses as 50. This simulates an SRS of size 50 from a large population. Each head in the sample is a person who plays the lottery, and each tail is a person who does not play. By alternating between "Toss" and "Reset," you can take many samples quickly.

(a) Take 25 samples, recording the proportion in each sample that plays the lottery. (The applet gives this proportion at the top left of its display.) Make a histogram of the 25 sample proportions.

(b) Another population contains only 20% of people who play the lottery. Take 25 samples of size 50 from this population, record the number in each sample that plays, and make a histogram of the 25 sample proportions. How do the centers of your two histograms reflect the differing truths about the two populations?

4. The idea of an 80% confidence interval is that the interval captures the true parameter value in 80% of all samples. That's not high enough confidence for practical use, but 80% hits and 20% misses make it easy to see how a confidence interval behaves in repeated samples from the same population. Go to the *Confidence Interval* applet.

(a) Use the slider to set the confidence level to 80%. Click "Sample" to choose an SRS and calculate the confidence interval. Do this 10 times to simulate 10 SRSs with 10 confidence intervals. How many of the 10 intervals captured the true mean? How many missed?

(b) You see that we can't predict whether the next sample will hit or miss. The confidence level, however, tells us what percentage of responses will hit in the long run. Reset the applet and click "Sample 25" to get the confidence intervals from 25 SRSs. How many hit? (You can read the number of hits and misses under the "Sample 25" button.)

(c) Keep clicking "Sample 25" and record the percent of hits among 100, 200, 300, 400, and 500 SRSs. Even 500 samples is not truly "the long run," but we expect the percentage of hits in 500 samples to be fairly close to the confidence level of 80%.

Writing Projects

1. Go to the website of the Gallup Organization (www.gallup.com). You should be able to find a press release that you can access and read without charge. Newspapers publish short articles based on press releases. Write a news article about two paragraphs long, based on the press release.

2. Recall how Example 8 (page 304) shows how wording can affect survey results. You can explore this by doing an experiment disguised as a survey. Choose a topic, then design two questions with a key difference in wording.

Use randomization to choose which version of the question you ask each person. Don't reveal the design of the experiment to participants until after they have provided their answers. After you have collected roughly 20 or more responses to each version of the question, compare and interpret your results. If you're interested in reading about an example of such an experiment, see John Rubin's article "Weighing Anchors" in the June 1990 issue of *Omni.*

3. How would you design a double-blind experiment in which participants test which of two brands of tissue they prefer? Conduct this experiment and write up the results. How do the results compare with any claims made in advertisements for the products?

4. Choose an issue of current interest to students at your school. Prepare a short questionnaire (no more than five questions) to determine opinions on this issue. Choose a sample of about 25 students, administer your questionnaire, and write a brief description of your findings. Also write a short discussion of your experiences in designing and carrying out the survey.

Although 25 students are too few for you to be statistically confident of your results, this project centers on the practical work of a survey. You must first identify a population; if it is not possible to reach a wider student population, use students enrolled in this course. Did the subjects find your questions clear? Did you write the questions so that it was easy to tabulate the responses? At the end, did you wish that you had asked different questions?

5. Select a topic on which national polls have been conducted over a period of years. Conduct research on how Americans' views on this topic have changed over time. Some possible topics include acceptance of gay and lesbian relationships, gun control, legalization of marijuana, tobacco use/smoking, capital punishment, and the pursuit of the American Dream for a better life (the question in Example 14 on page 315).

Suggested Readings

ANDERSON-COOK, C. M., and SUNDAR DORAIRAJ. An active learning in-class demonstration of good experimental design, *Journal of Statistics Education, 9,* no. 1 (2001). See www.amstat.org/publications/jse/v9n1/anderson-cook.html. A good example of issues that arise when designing a randomized experiment. This article includes an applet that students can use to experience the experiment.

BOCK, DAVID E., PAUL F. VELLEMAN, and RICHARD D. DE VEAUX. *Stats: Modeling the World,* 4th ed., Pearson, Boston, 2015. This is another text on the topic of statistics, aimed at the mathematical level just above that of this book. The title offers a chapter (Chapter 19, "Confidence Intervals for Proportions") that uses estimating a population proportion to introduce confidence intervals.

HANSON, ANDREW, and MICHAEL SANTAS. Field experiment for discrimination against Hispanics in the U.S. rental housing market, *Southern Economic Journal, 81,* no. 1 (July 2014): 135–167. See epublications.marquette.edu/cgi/viewcontent.cgi?article=1445&context=econ_fac. Read the details of Hanson and Santas's Internet field experiment (discussed in Spotlight 6.4 on page 265), which was designed to investigate discrimination against Hispanics in the rental housing market. Pay particular attention to Section 3, "Experiment Design."

LESSER, LAWRENCE M., and ERIK NORDENHAUG. Ethical statistics and statistical ethics: Making an interdisciplinary module, *Journal of Statistical Education, 12,* no. 3 (2004). See **www.amstat.org/publications/jse/v12n3/lesser.html**. This article's discussion includes ethical issues associated with surveys, experiments, and observational studies.

MOORE, DAVID S., WILLIAM I. NOTZ, & MICHAEL A. FLIGNER. *The Basic Practice of Statistics (BPS),* 7th ed., Freeman, New York, 2015. Chapters 8 and 9 discuss samples and experiments, Chapter 14 presents the reasoning of confidence intervals, and Chapter 19 addresses confidence intervals for a population proportion.

Suggested Websites

www.ncpp.org The National Council on Public Polls has published a statement "20 Questions a Journalist Should Ask About a Poll" that makes interesting reading. The explanations expand on our cautions about sample surveys in practice. You can find similar information on the website of the American Association for Public Opinion Research, **www.aapor.org.** (Click on Resources and Education, and select "For Students." Then take a look at the "Poll & Survey FAQs.") Also, read the American Statistical Association publication *What Is a Survey?* (2nd ed.) at **www.whatisasurvey.info.**

www.census.gov/cps/ The single most important sample survey in the United States is probably the government's monthly CPS, completed by the Census Bureau on behalf of the Bureau of Labor Statistics (BLS). The CPS website contains a wealth of information.

www.learner.org/courses/againstallodds/ The online materials in Units 14, 15, 17 and 28 from *Against All Odds: Inside Statistics* are directly applicable to material discussed in this chapter. The video for Unit 14, "The Question of Causation," provides a history of the evidence leading up to the U.S. Surgeon General's 1962 report that concluded smoking causes cancer. Unit 15, "Designing Experiments," discusses two studies: an observational study on the health of coral reef ecosystems and an experiment to determine whether taking dietary supplements effects joint pain. Units 17 and 28, "Samples and Surveys" and "Inference for Proportions," expand on the content on sampling designs and inference contained in this chapter.

Probability: The Mathematics of Chance

8

PNC/Stockbyte/Getty Images

Ever wonder how gambling can be a business? A business needs predictable revenue from the service it offers, even when the service is a game of chance. Individual gamblers can never say whether a day at the casino will turn a profit or a loss, but the casino itself takes few chances. Casinos are consistently profitable, and state governments make money both from running lotteries and from selling licenses for other forms of gambling.

It is striking that an individual roll of the dice, spin of the wheel, or flip of a coin is a total unknown, but the aggregate result of thousands of chance outcomes can be known with near certainty. To illustrate, Figure 8.1 shows the outcomes of four rolls of a die. Based on Figure 8.1, it is impossible to predict the outcome of the fifth roll. Figure 8.2, however, shows a histogram of 10,000 rolls of a die—and a very predictable pattern. The frequencies are nearly equal for each possible outcome 1, 2, 3, 4, 5, and 6.

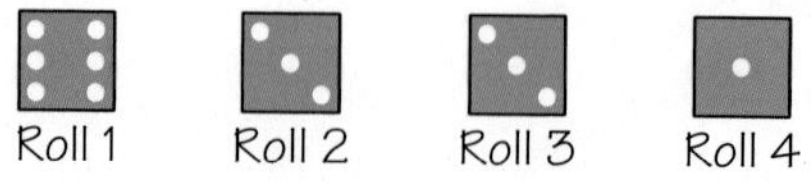

Figure 8.1 Outcomes from rolling a die four times.

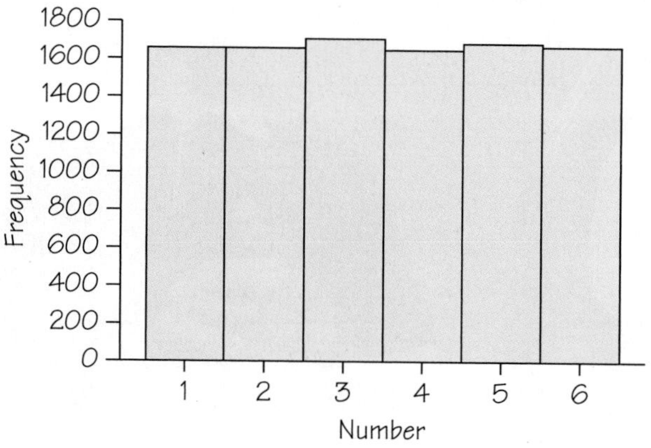

Figure 8.2 Histogram of 10,000 rolls of a die.

The casino need not load the dice, mark the cards, or alter the roulette wheel. It knows that in the long run, each dollar bet will yield its five cents or so of revenue. It is, therefore, good business to concentrate on free floor shows or inexpensive bus fares to increase the flow of dollars bet. The flow of profit will follow.

A casino is not the only business that relies on the fact that a chance outcome many times repeated is quite predictable. For example, although a life insurance company does not know *which* of its policyholders will die next year, it can predict quite accurately *how many* will die. It sets its premiums according to this knowledge, just as the casino sets its jackpots. These are just two of the many types of companies and people that rely on chance behavior being predictable in the long run.

We begin this chapter with an introduction to the nature of randomness and probability, including how to use rules to calculate numerical values of probabilities. Then we examine further ways to model and calculate probabilities by gaining tools for counting the number of possibilities and by modeling real-life phenomena with discrete and continuous models. The mean and standard deviation, first introduced in Chapter 5, will be expanded to probability models. The chapter concludes with a discussion of a central result of inferential statistics known as the *central limit theorem,* which tells us about how the sample mean behaves regardless of the shape of the population from which it was calculated.

8.1 Random Phenomena and Probability

Roll a die or choose a simple random sample (SRS) from a population. The results can't be predicted in advance. When you roll the die, you know you will get a 1, 2, 3, 4, 5, or 6. You expect each of these outcomes to be equally likely, but you don't know for certain which outcome will occur the next time you roll the die. An instructor chooses a random sample of three students from each class to put homework problems on the board. Because the selection is random, each possible size-3 sample is equally likely to be selected. Suppose that one day Josh comes to class without having done his homework. If the class is small, his chances of getting chosen are pretty good. On the other hand, if the class is large, he is not as likely to be in the selected sample. Josh won't know if he will be caught unprepared for class until the sample is actually drawn.

Rolling a die and choosing a random sample are both examples of **random** phenomena.

Random DEFINITION

A phenomenon or trial is said to be **random** if individual outcomes are uncertain but the long-term pattern of many individual outcomes is predictable.

In statistics, *random* does not mean "haphazard." Randomness is actually a kind of order, an order that emerges in the long run, over many repetitions. Take the example of tossing a coin. The result can't be predicted in advance because the result will vary from coin toss to coin toss. But there is nonetheless a regular pattern in the results, a pattern that emerges clearly only after many repetitions. This remarkable fact is the basis for the idea of **probability.**

EXAMPLE 1 Heads Up When Tossing a Coin: Long-Run Frequency Interpretation of Probability

When you toss a coin, there are only two possible outcomes: heads or tails. Figure 8.3 shows the results of tossing a coin 5000 times twice. Let's focus on Trial A, the red graph. For Trial A, the first four tosses result in tail, head, tail, tail. After four tosses, the proportion of heads is $1/4 = 0.25$. Notice that corresponding to the first 100 tosses, there is quite a bit of fluctuation in the proportions. Now, compare the amount of fluctuation about the horizontal black line for relatively few tosses, say between 1 and 100 tosses, with the amount of fluctuation corresponding to relatively many tosses, say between 2000 and 5000 tosses. Comparatively, there is very little fluctuation in the latter interval.

Next, compare the proportions of heads for Trial A (red graph) in Figure 8.3 with those plotted for Trial B (blue graph). Trial B starts with five straight heads, so the proportion of heads is 1 until the sixth toss. Notice that the proportion of tosses that produces heads for both Trials A and B is quite variable at first. Trial A starts low and Trial B starts high. As we make more and more tosses, however, the proportions of heads for

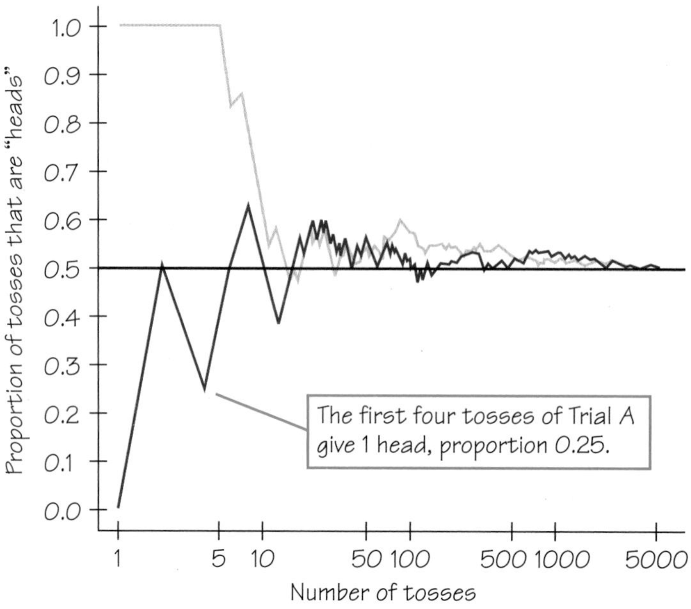

Figure 8.3 The proportion of tosses of a coin that gives heads varies as we make more tosses. Eventually, however, the proportion approaches 0.5, the probability of a head. This figure shows the results of two trials of 5000 tosses each. (The horizontal scale is transformed using logarithms to show both short- and long-term behavior.)

both trials get close to 0.5 and stay there. If we made yet a third trial at tossing the coin a great many times, the proportion of heads would again settle down to 0.5 in the long run. We say that 0.5 is the *probability* of a head. The probability 0.5 is marked by the horizontal line on the graph.

Algebra Review Appendix
Fractions, Percents, and Percentages

Probability

DEFINITION

The **probability** of any outcome of a random phenomenon is the proportion of times the outcome would occur in a very long series of repetitions. Probabilities can be expressed as decimals, percentages, or fractions.

The *Probability* applet (see Applet Exercise 1, page 399) animates Figure 8.3. It allows you to choose the probability of a coin landing on heads and simulate a specific number of tosses of a coin with that probability. Try it. You will see that the proportion of heads gradually settles down close to the probability. Equally important, you will also see that the proportion in a small or moderate number of tosses can be far from the probability. *Probability describes only what happens in the long run.* Random phenomena are irregular and unpredictable in the short run.

Dave Greenwood/Getty Images

Figure 8.4 Tacks sitting point down and point up.

We might suspect that a coin has probability 0.5 of coming up heads just because the coin has two sides. However, such suspicions are not always correct. For example, suppose we flip a tack, which can land either point up or point down, as shown in Figure 8.4. Would the probability of the tack landing point up still be 0.5? (To find out, complete Exercise 3 on page 390.) Since probability describes what happens in a great many trials,

The Problem of Points

SPOTLIGHT 8.1

In 1654, Antoine Gombaud, Chevalier de Méré, an amateur mathematician, posed the following problem, called the "Problem of Points":

> Two players agree to play rounds of a game of chance in which each has an equal chance of winning. They both contribute an equal amount of money to a prize. The first player to win a preset number of rounds gets the prize. Unfortunately, the game is interrupted before completion of all the rounds. The question is: How should the prize be divided fairly?

Mathematicians Blaise Pascal and Pierre de Fermat took up the challenge to solve this problem. This led to a series of correspondences between Pascal and Fermat, the content of which laid the foundation for the modern theory of probability.

DEA/G. Dagli Orti/ Getty Images

Blaise Pascal (1623–1662)

Photo Researchers/Getty Images

Pierre de Fermat (1601–1662)

In Exercise 13 (page 391), you can solve a simplified version of the "Problem of Points."

you will need to observe the outcomes of many flips of a tack in order to pin down this probability.

Gamblers have known for centuries that the fall of coins, cards, and dice displays clear patterns in the long run. In fact, Spotlight 8.1 presents a question about a gambling game that launched probability as a formal branch of mathematics.

The idea of probability rests on the observed fact that the average result of many thousands of chance outcomes can be known with near certainty. But a definition of probability as "long-run proportion" is vague. Who can say what the "long run" is? We can always toss the coin another 1000 times. Instead, we give a mathematical description of *how probabilities behave,* based on our understanding of long-run proportions. To see how to proceed, we return to the very simple random phenomenon of tossing a coin once.

When we toss a coin, we cannot know the outcome in advance. What do we know? We are willing to say that the outcome will be either heads or tails. We believe that each of these outcomes is equally likely, hence each has probability $\frac{1}{2}$. This description of coin tossing has two parts:

1. A list of possible outcomes
2. The probability for each outcome

We will see that this description is the basis for all the **probability models** in Section 8.4. Here is the vocabulary we use.

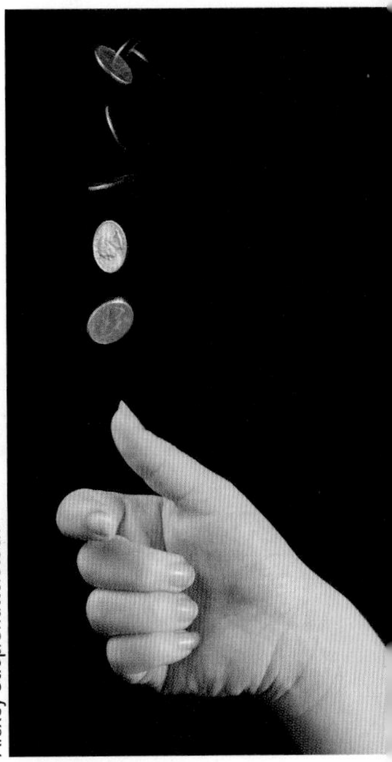

Alexey Stiop/Shutterstock

Sample Space DEFINITION

The **sample space** S of a random phenomenon is the set of all possible outcomes that cannot be broken down further into simpler components.

Event DEFINITION

An **event** is any outcome or any set of outcomes of a random phenomenon. That is, an event is a subset of the sample space. A simple event is a set of a single outcome from the sample space.

Probability Model DEFINITION

A **probability model** is a mathematical description of a random phenomenon consisting of two parts: a sample space S and a way of assigning probabilities to events.

The sample space S can be very simple or very complex. When we toss a coin once, there are two possible outcomes, heads or tails. So the sample space is $S = \{H, T\}$. If we draw a random sample of 1000 U.S. residents that are 18 years of age or over, as opinion polls often do, the sample space contains all possible choices of 1000 of the 235 million adults in the country. This S is extremely large: 2.9×10^{5803}.

EXAMPLE 2 Tossing Two Coins: The Importance of Sample Space

Probabilities can be hard to determine without detailing or diagramming the sample space. For example, E. P. Northrop notes that even the great 18th-century French mathematician Jean le Rond d'Alembert tripped on the question "In two coin tosses, what is the probability that heads will appear at least once?" Because the number of heads could be 0, 1, or 2, d'Alembert reasoned (incorrectly) that each of those possibilities would have an equal probability of $\frac{1}{3}$, and so he reached the (wrong) answer of $\frac{2}{3}$.

What went wrong? Well, {0, 1, 2} could not be the fully detailed sample space because "1 head" can happen in more than one way. For example, if you flip a dime and a penny once apiece, you could display the sample space with a *table*, such as the one in Figure 8.5.

Another way to generate these four outcomes is with the tree diagram shown in Figure 8.6. Each possible left-to-right pathway

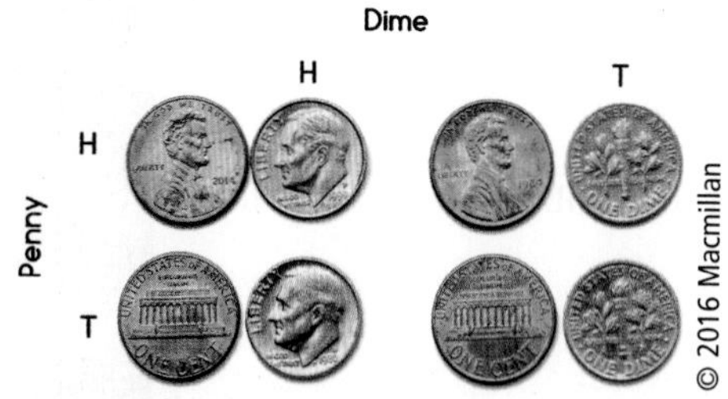

Figure 8.5 A table illustrating the outcomes of flipping two coins. As you can see, Figure 8.5 is a table with 2 rows and 2 columns, which displays $2 \times 2 = 4$ outcomes: {HH, HT, TH, TT}.

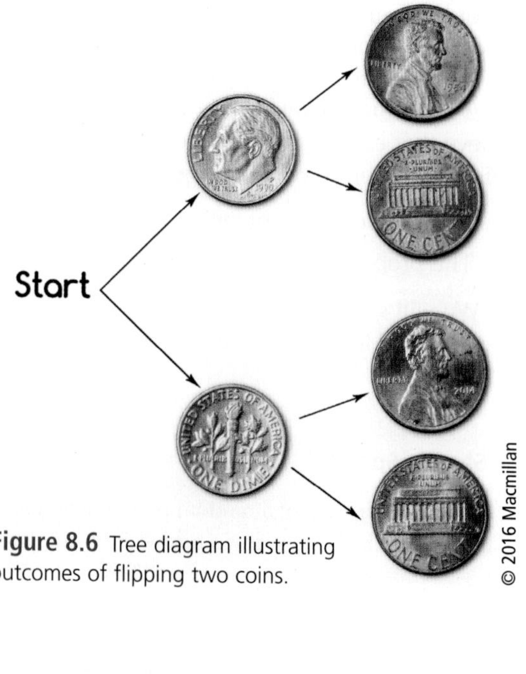

Figure 8.6 Tree diagram illustrating outcomes of flipping two coins.

through the branches generates an outcome. For example, going up (to dime "heads") and then down (to penny "tails") yields the outcome HT.

Either way, we can see that the sample space has 4, not 3, equally likely outcomes. With the table or tree diagram in view, you may already see that the correct probability of at least 1 head is not $\frac{2}{3}$, but $\frac{3}{4}$.

Self Check 1

At an intersection, a driver has the choice of turning right or left or going straight (use R for right, L for left, and S for straight). You are standing at this intersection and watch the next two cars go through. Use a tree diagram to identify the sample space for the outcomes of this situation.

EXAMPLE 3 Pair-a-Dice: Outcomes for Rolling Two Dice

Undergroundarts.co.uk/
Shutterstock

Rolling one six-sided die has an obvious sample space of six equally likely outcomes: {1, 2, 3, 4, 5, 6}. But many board games (and casino games) involve rolling two dice and noting the sum of the spots on the two sides that are facing up. We know from our experience playing games like Monopoly that the 11 possible sums (2, 3, 4, 5, 6, 7, 8, 9, 10, 11, 12) are *not* equally likely because, for example, there are many ways to get a sum of 7 but only one way to get a sum of 12.

We start by identifying the sample space in order to find the exact probabilities and patterns of the various dice sums. Because of the large number of possible outcomes, the table in Figure 8.7 is a more straightforward representation of the sample space S than a tree diagram would be. Figure 8.7 shows $6 \times 6 = 36$ possible (and equally likely) ways to roll two dice.

The longest rising diagonal of the table shows the six ways that the sum can equal 7. Therefore, it makes sense that the probability of the sum being 7 is $\frac{6}{36} = \frac{1}{6}$.

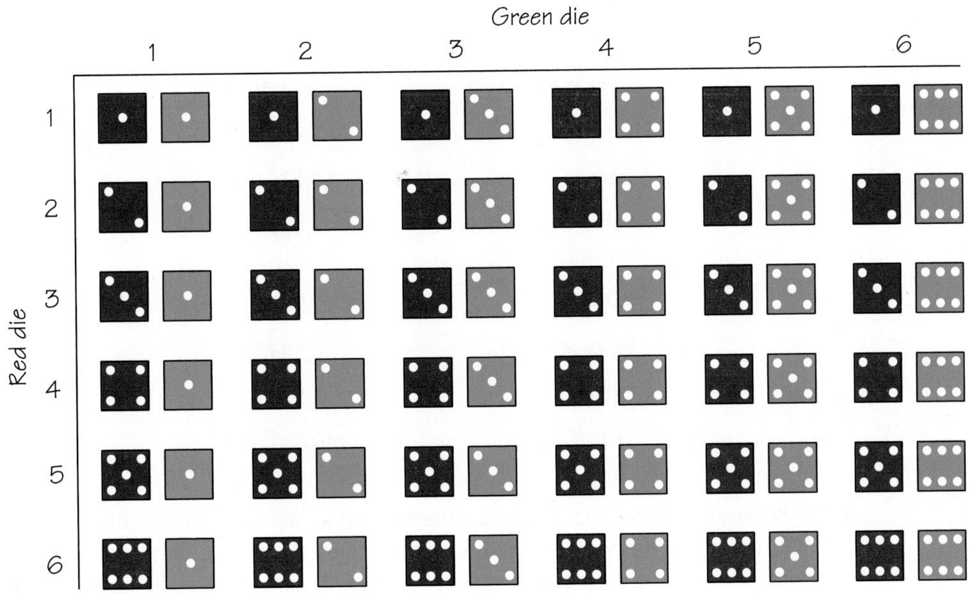

Figure 8.7 Table of the 36 outcomes for rolling two dice.

Self Check 2

Using the table in Figure 8.7, how many possibilities are there for rolling a sum of 6? What is the probability of a sum of 6?

8.2 Basic Rules of Probability

Recall from Section 8.1 that a probability model consists of two parts: the sample space and a way of assigning probabilities to events. Events are usually designated by capital letters near the beginning of the alphabet; for example, event A. The notation $P(A)$ is shorthand for "the probability that event A will occur." There are many ways to assign probabilities, so we will need some basic rules that any assignment of probabilities to events must obey. These rules follow from the idea of probability as "the long-run proportion of repetitions on which an event occurs." Here are the first two rules:

Rule 1. **Any probability is a number between 0 and 1 inclusive.** Any proportion is a number between 0 and 1 inclusive, so any probability is also a number between 0 and 1 inclusive. An event with probability 0 never occurs, an event with probability 1 always occurs, and an event with probability 0.5 occurs in half the trials in the long run.

Rule 2. **All possible outcomes together must have probability 1.** Because some outcome must occur on every trial, the sum of the probabilities for all possible (simplest) outcomes must be exactly 1.

EXAMPLE 4 Assigning Probabilities to Means of Transportation

How do people in the United States get to work? Table 8.1 shows the results of an American Community Survey by the U.S. Census Bureau.

TABLE 8.1 Survey Results from the American Community Survey, 2013

Means of Travel	Frequency
Drive alone	107,460,210
Carpool	13,675,867
Public transportation (excluding taxis)	7,053,456
Walk	3,969,058
Work at home	6,143,943
Other	2,560,426
Total	140,862,960

Because this is a U.S. Census Bureau survey, we can assume that the sample fairly represents the workers in the United States. Given the large sample size, the sample proportions should be good estimates of the probabilities for each category of transportation. Table 8.2 turns the data from the survey into a probability model for

TABLE 8.2 Probability Model for Means of Transportation to Work

Means of Travel	Proportion/Probability
Drive alone	0.763
Carpool	0.097
Public transportation (excluding taxis)	0.050
Walk	0.028
Work at home	0.044
Other	0.018
Sum	1

means of transportation to work. Notice that the probabilities are all between 0 and 1 (Rule 1) and that the sum of the probabilities is 1 (Rule 2).

Based on the probabilities in Table 8.2, a randomly selected worker is almost 8 times more likely to drive alone to work than to carpool and more than 15 times more likely to drive alone than to use pubic transportation.

When interpreting probabilities, particularly when they are small, it is helpful to make comparisons with something concrete, as we demonstrate in Spotlight 8.2.

Probability and Psychology

SPOTLIGHT 8.2

Our judgment of probability can be affected by psychological factors. Our desire to get rich quick may lead us to overestimate the tiny probability of winning the lottery. Our feeling that we are "in control" when we are driving may make us underestimate the probability of an accident. (This may be why some people prefer driving to flying, even though flying has a lower probability of death per miles traveled.)

The probability of winning (a share of) the 44-state Mega Millions jackpot is 1 in 258,890,850. This is less likely than picking out a particular sheet of printer paper from a stack 2.5 times the height of Mount Everest, or guessing a particular second from a period of about 8.2 years. Without concrete analogies, it is hard to grasp the meaning of very small probabilities, and some players may greatly overestimate their chances of winning, even if they buy lots of tickets. For example, suppose someone buys 20 \$1 Mega Millions tickets every week for 50 years. She would have spent over \$50,000, and yet her probability of winning at least one jackpot in that whole time would still be only about 1 in 5000. For comparison, the probability of dying in a car accident during a lifetime of driving is about 50 times greater than this!

Stan Honda/AFP/Getty Images

Andrew Gelman, professor of statistics and political science at Columbia University, reports that most people say they would not switch to a situation in which they had a small probability p of dying and a large probability $1 - p$ of gaining \$1000. And yet, people will not necessarily spend that much for air bags for their cars. Becoming more aware of our inconsistencies and biases can help us make better use of probability when deciding what risks to take.

Sometimes it is easier to determine the probability of an event A indirectly by finding the probability of its logical opposite—that is, the event that A does *not* happen. The special name for this event is consistent with how the word *complement* is used in other contexts: the event that together with its opposite forms a complete whole—in this context, the "whole" is the entire sample space.

Complement of an Event DEFINITION

The **complement of an event** A is the event that A does *not* occur, written as A^C. (The superscript C stands for complement. Some books use the notation $\bar{A}$ or A' or "not A.")

In the diagram in Figure 8.8, the area inside the rectangle represents the sample space. The area inside the circle represents event A, and the area inside the rectangle but outside of the circle is A^C.

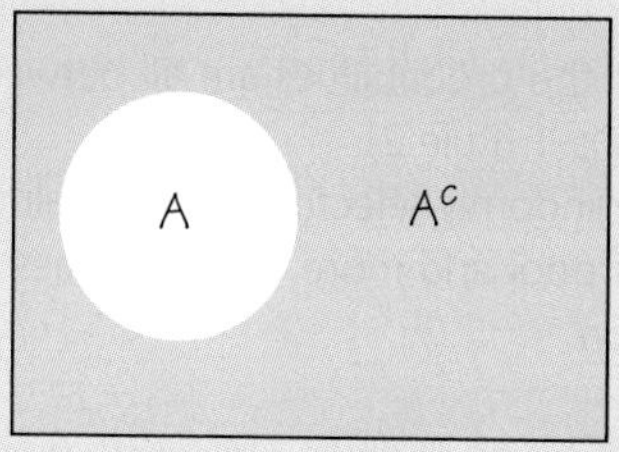

Figure 8.8 Diagram of an event and its complement.

For example, in the dice game of craps, rolling a sum of 7 on two dice (the most common roll) is an outcome that instantly loses the round once it's underway. Suppose we want to know the probability of rolling anything other than a 7. If we let event A be rolling a sum of 7, then we want $P(A^C)$, the probability of *not* rolling a sum of 7. The relationship between $P(A)$ and $P(A^C)$ is given in Rule 3, the complement rule.

Rule 3. Complement Rule. The probability that an event does not occur is 1 minus the probability that the event does occur. Continuing the discussion above, we know from Example 3 (page 347) that the probability of rolling a sum of 7 is $\frac{1}{6}$. Therefore, the probability of not rolling a sum of 7 is $1 - \frac{1}{6} = \frac{5}{6}$. This is really just another way of saying that the probability that an event occurs and the probability that it does not occur always add to 1, or 100% of the sample space. Referring back to the diagram in Figure 8.8, you can see how the complementary blue and white regions add up to fill the space inside the rectangle.

Another useful distinction to make when discussing two events is whether or not it is possible for the two events to happen simultaneously. If it is not possible, then the two events are said to be **disjoint** or **mutually exclusive.** An event and its complement are always mutually exclusive.

Disjoint Events (Mutually Exclusive Events) DEFINITION

Two events are **disjoint events** if they have no outcomes in common. Disjoint events are also called **mutually exclusive events.**

In the diagram in Figure 8.9, the two circular areas represent events A and B. Since there is no overlap in these areas, the two events are disjoint.

Rule 4, the addition rule, addresses how to assign probabilities to the event that either A or B occurs in situations where events A and B are mutually exclusive.

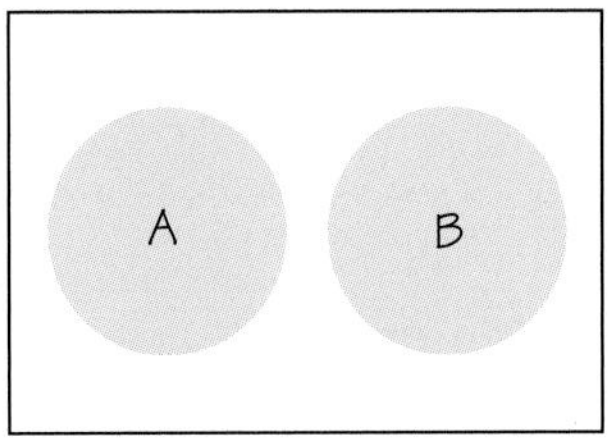

Figure 8.9 Diagram of two disjoint events.

Rule 4. Addition Rule for Disjoint Events. If two events are *disjoint*, the probability that one or the other occurs is the sum of their individual probabilities. Return to Example 4 (page 348). Suppose we want to determine the probability that a person drives to work either alone (event A) or in a carpool (event B). In other words, we want to determine $P(A \text{ or } B)$. From the survey data, we can calculate the number of workers who fell into events A or B: 107,460,210 + 13,675,867 = 121,136,077. So we estimate the probability from the sample proportion:

$$121{,}136{,}077/140{,}862{,}960 \approx 0.860.$$

Instead of estimating the "long-run" proportion from the data, we can apply the addition rule for disjoint events. Simply add the corresponding probabilities from Table 8.2:

$$P(A \text{ or } B) = P(A) + P(B) = 0.763 + 0.097 = 0.860$$

Self Check 3

Use Table 8.2 (page 349) to determine the following probabilities:

(a) The probability that a randomly selected worker in the United States either uses public transportation or walks to work.

(b) The probability that a randomly selected worker in the United States does not drive to work (either alone or in a carpool).

Sometimes we are interested in determining the probability that either A or B occurs in cases where A and B are not disjoint. We return to the situation in which a red and green die are rolled together (for the sample space, see Figure 8.7, page 347). Consider the following events:

$$A = \text{red die shows "1"}$$

$$B = \text{red and green dice add up to 7}$$

Suppose we want to determine $P(A \text{ or } B)$. This situation is depicted by the diagram in Figure 8.10. There is an overlap in these two events: The outcome that red shows "1" and green shows "6" belongs to both A and B. So we can't use Rule 4 to determine the probability that "either A or B occurs." Unlike in usual everyday usage, the mathematical use of "or" is *inclusive,* which means that the event "A or B" happens so long as at least one of the two events happens. In set theory, this is called "the union of A and B," and it includes A's and B's "separate property" (the red and blue areas in the diagram) as well as their "community property" (the area where red and blue are blended). Rule 5 will provide the adjustment needed to deal with the overlap of A and B.

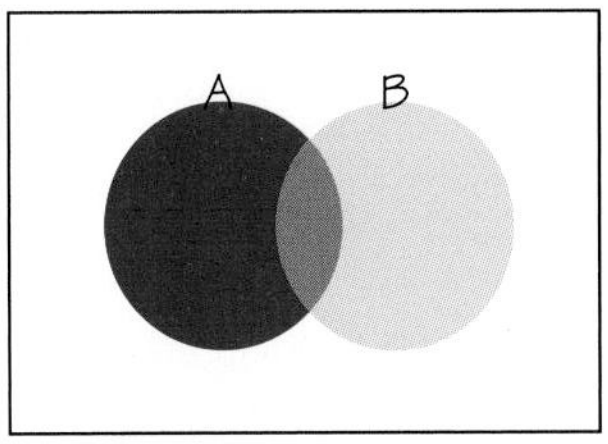

Figure 8.10 Events A and B are not disjoint.

Rule 5. General Addition Rule. The probability that one event or the other occurs is the sum of their individual probabilities minus the probability of their intersection. This general addition rule makes sense if we look at Figure 8.10. Simply adding the probabilities of the two events A and B would overshoot the answer because we would be incorrectly counting the overlap twice. The way to adjust for this is to subtract the overlap so that it is counted exactly once. Now, we return to events A and B above. Their intersection corresponds to rolling "red = 1, green = 6," which has a $\frac{1}{36}$ probability. Therefore,

$$P(A \text{ or } B) = \frac{6}{36} + \frac{6}{36} - \frac{1}{36} = \frac{11}{36}$$

We now state Rules 1 through 5 more concisely using more formal mathematical notation. As you apply these rules, remember that they are just another form of true facts about long-run proportions.

Probability Rules RULE

Rule 1. The probability $P(A)$ of any event A satisfies $0 \le P(A) \le 1$.

Rule 2. If S is the sample space in a probability model, then $P(S) = 1$.

Rule 3. The complement rule: $P(A^C) = 1 - P(A)$.

Rule 4. The addition rule for *disjoint* events: $P(A \text{ or } B) = P(A) + P(B)$.

Rule 5. The general addition rule: $P(A \text{ or } B) = P(A) + P(B) - P(A \text{ and } B)$.

In Example 5, we return to the situation of rolling two dice, which will provide an opportunity to practice applying Rules 1 through 5.

EXAMPLE 5 Probabilities for Rolling Two Dice

Figure 8.7 (page 347) displays the 36 possible outcomes of rolling two dice. For casino dice, it is reasonable to assign the same probability to each of the 36 outcomes in Figure 8.7. Because all 36 outcomes together must have probability 1 (Rule 2), each outcome must have probability $\frac{1}{36}$.

Suppose we want to determine the probability of rolling a sum of 5. Because there are four ways to roll a sum of 5, the addition rule for disjoint events (Rule 4) says that its probability is

P (roll a sum of 5) = P() + P() + P() + P()

$$= \frac{1}{36} + \frac{1}{36} + \frac{1}{36} + \frac{1}{36}$$

$$= \frac{4}{36} \approx 0.111$$

Similarly, we can find the probabilities for the other possible sums and, in this way, get the full probability model (sample space and assignment of probabilities) for rolling two dice and summing the spots on the sides facing up. The result is shown in Table 8.3.

TABLE 8.3 Probability Model for the Sum of the Spots on Two Dice

Outcome (sum of two dice)	2	3	4	5	6	7	8	9	10	11	12
Probability	$\frac{1}{36}$	$\frac{2}{36}$	$\frac{3}{36}$	$\frac{4}{36}$	$\frac{5}{36}$	$\frac{6}{36}$	$\frac{5}{36}$	$\frac{4}{36}$	$\frac{3}{36}$	$\frac{2}{36}$	$\frac{1}{36}$

The model in Table 8.3 assigns probabilities to individual outcomes. Note that Rule 2 is satisfied because all the probabilities add up to 1. To find the probability of an event, just add the probabilities of the outcomes that make up the event. For example:

$$\begin{aligned} P(\text{outcome is odd}) &= P(3) + P(5) + P(7) + P(9) + P(11) \\ &= \frac{2}{36} + \frac{4}{36} + \frac{6}{36} + \frac{4}{36} + \frac{2}{36} \\ &= \frac{18}{36} = \frac{1}{2} \end{aligned}$$

Suppose we want the probability of rolling an even number. We could find this probability by finding the sum of the following:

$$P(2) + P(4) + P(6) + P(8) + P(10) + P(12)$$

But a faster way would be to use the complement rule (Rule 3):

$$\begin{aligned} P(\text{outcome is even}) &= 1 - P(\text{outcome is odd}) \\ &= 1 - \frac{1}{2} = \frac{1}{2} \end{aligned}$$

For an example of Rule 5, let event A be "sum is odd" and event B be "sum is a multiple of 3." Suppose we want $P(A \text{ or } B)$. Earlier, we calculated $P(A) = \frac{1}{2}$. You can verify that $P(B) = \frac{1}{3}$ and $P(A \text{ and } B) = \frac{1}{6}$. Now we are ready to apply Rule 5:

$$P(A \text{ or } B) = P(A) + P(B) - P(A \text{ and } B) = \frac{1}{2} + \frac{1}{3} - \frac{1}{6} = \frac{4}{6} = \frac{2}{3}$$

When the outcomes for a probability model are numbers, such as for the model in Table 8.3, we can use a histogram to display the assignment of probabilities to the outcomes. Figure 8.11 is a **probability histogram** of the probability model

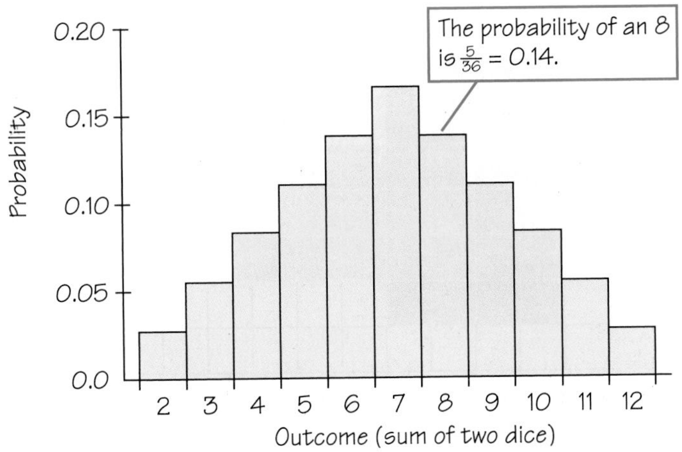

Figure 8.11 A probability histogram showing the probability model for rolling two balanced dice and counting the spots on the sides facing up.

in Table 8.3. The height of each bar shows the probability of the outcome at the base of the bar. Because the heights are probabilities, they add to 1. Think of Figure 8.11 as an idealized picture of the results of very many rolls of a pair of dice. As an idealized picture, it is perfectly symmetric about the middle bar corresponding to a sum of 7.

8.3 Rules of Probability: Independent and Dependent Events

While everyday speech might make you think that the word *independent* could also be a synonym for words like *disjoint* and *mutually exclusive,* independent events have a different meaning in statistics. Independent events do not affect each other's probability of occurrence, just as an individual's probability of being chosen in a simple random sample (SRS) (see Section 7.3 on page 297) is not affected by whether another particular individual is selected.

Independent and Dependent Events DEFINITION

Two events are **independent events** if the occurrence of one event has no influence on the probability of the occurrence of the other event. If two events are not independent, then they are **dependent events.**

The diagram in Figure 8.12 depicts two independent events, *A* and *B*, represented by rectangular-shaped areas. In this diagram, $P(A) = 1/4$ since *A* fills one-quarter of the area representing the sample space *S*. Next, assume that *B* has occurred—so the outcome lies inside *B*'s rectangle. Given this information, what is the likelihood that *A* occurs? Notice that the purple area shows the outcomes in *A* that are in *B*. This overlap fills one-fourth of *B*'s rectangle. So the probability that *A* occurs, given we know that *B* has occurred, expressed as $P(A|B)$, is still 1/4. The information about *B* did not influence how likely it was for *A* to occur. Therefore, the two events are independent.

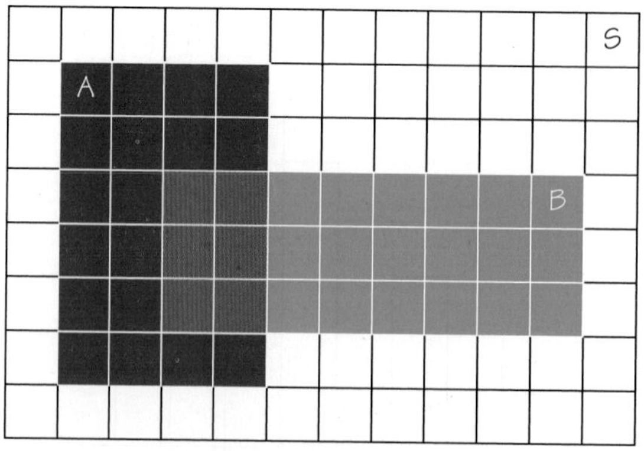

Figure 8.12 Diagram of two independent events.

EXAMPLE 6 Dependent or Independent?

For many sweepstakes, the consumer is automatically entered into the drawing after making a purchase. However, often sweepstake rules state that "no purchase is required to enter," and the consumer is given the option to enter by completing an online form or mailing in a postcard. Let event D be winning the sweepstakes and event E be making a purchase. According to the sweepstake rules, your chance of winning the contest is not affected by whether or not you make a purchase. If we use the notation $P(D|E)$ as shorthand for "the probability of winning, given a purchase is made," then according to the rules, $P(D) = P(D|E)$. In this case, events D and E are independent.

For the next example, return to Figure 8.7 (page 347) in Example 3. Consider the following events:

$$B = \text{red and green dice add up to 7}$$

$$C = \text{both red and green dice show a number less than 5}$$

Are events B and C independent or dependent? We know $P(B) = \frac{1}{6}$. But what about the probability that the sum on the dice is 7, given both dice show values less than 5—in other words, $P(B|C)$? In Figure 8.13 the outcomes in event C are outlined by a blue rectangle. Each outcome in C, of which there are 16, is equally likely and there are only two outcomes in C for which the sum of the dice is 7. Hence, $P(B|C) = \frac{2}{16} = \frac{1}{8} \neq P(B)$. (The proportion of sums of 7 in the outcomes for C differs from the proportion of sums of 7 in the sample space.) In this case, knowledge that event C has occurred makes it less likely that B occurs. Therefore, events B and C are dependent.

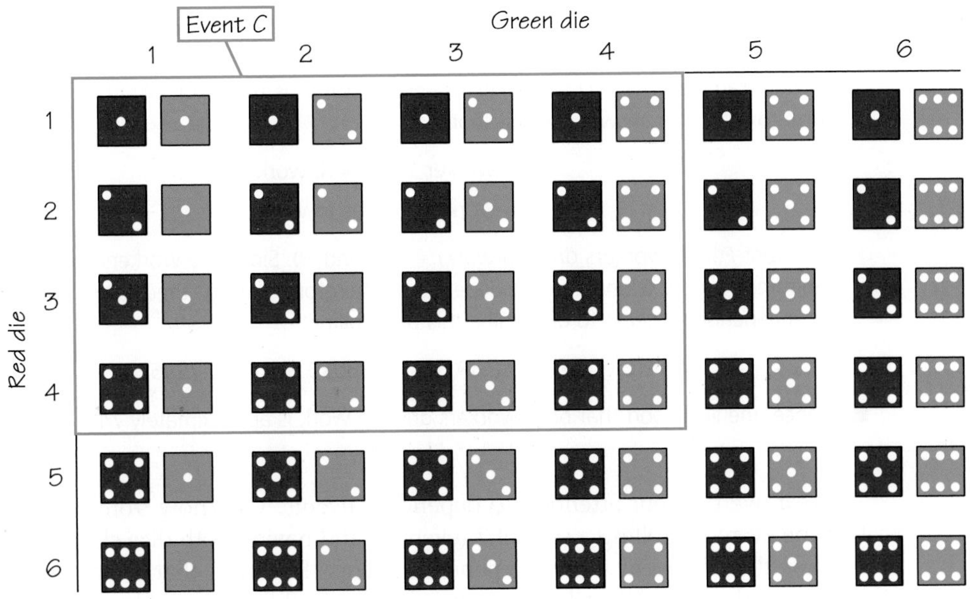

Figure 8.13 Identifying the outcomes in event C.

Self Check 4

Return to the context of rolling two dice and the events:

$$A = \text{red die shows "1"}$$
$$B = \text{red and green dice add up to 7}$$

Are A and B independent? Support your answer.

In Self Check 4, you should have found that events A and B were independent. (If not, check the Self Check answer at the end of this chapter.) Next, we learn how to find the probability that two independent events occur simultaneously. We begin with an example. Start with independent events A and B from Self Check 4. We want to determine $P(A \text{ and } B)$. Rule 6, the multiplication rule for independent events, is the rule needed for calculating this probability.

Rule 6. Multiplication Rule for Independent Events. If two events are *independent*, then the probability that one event and the other both occur is the product of their individual probabilities. In Self Check 4, you found that events A and B are independent. So we can find the probability that both A and B occur by multiplying the individual event probabilities:

$$P(A \text{ and } B) = P(A) \times P(B) = \frac{1}{6} \times \frac{1}{6} = \frac{1}{36}$$

To confirm this result, notice that the overlap of A and B is the simple event of rolling red = 1 and green = 6, which has probability $\frac{1}{36}$.

EXAMPLE 7 Probabilities for Driving to Work

In the discussion of Rule 4 (the addition rule for disjoint events, page 351), we determined that 86% of workers drive to work (either alone or in a carpool). Suppose a random sample of two U.S. workers is chosen. What is the probability that they both drive to work? To answer this question, let:

$$G = \text{worker 1 drives to work}$$
$$H = \text{worker 2 drives to work}$$

We want $P(\text{both workers drive to work}) = P(G \text{ and } H)$. Since the workers were chosen randomly, events G and H are independent. Therefore, we use the multiplication rule for independent events to determine this probability:

$$P(G \text{ and } H) = P(G)P(H) = (0.860)(0.860) = 0.7396$$

Hence, the likelihood that both workers drive to work is approximately 74%.

Next, we turn our attention to dependent events. We know from Example 6 that the events of rolling a sum of 7 (event B) and having both dice show a number less than 5 (event C) are dependent events. Knowing that C has occurred influences the probability that B occurs. The result, $P(B|C)$ is called a **conditional probability.**

Conditional Probability DEFINITION

The **conditional probability** of A given B, written as $P(A|B)$, can be computed by dividing the probability that both events occur by the probability that B occurs:

$$P(A|B) = \frac{P(A \text{ and } B)}{P(B)}$$

In cases where A and B are independent, we can apply the multiplication rule for independent events:

$$P(A|B) = \frac{P(A)\,\cancel{P(B)}}{\cancel{P(B)}} = P(A)$$

Notice, that in order to calculate $P(A|B)$, we need $P(B) > 0$.

Solving for $P(A \text{ and } B)$ in the formula for calculating a conditional probability allows us to generalize the multiplication rule so that it applies to situations where the two events are dependent. This leads to Rule 7.

Algebra Review Appendix
Solving for One Variable in Terms of Another

Rule 7. General Multiplication Rule. Given two events A and B with $P(B) > 0$, the probability that both occur is the conditional probability that A occurs given B has occurred times the probability that B occurs. Take, for example, the situation in Figure 8.14. We want $P(A \text{ and } B)$. The outcome "A and B" is represented by the overlap of A and B (purple area). It is easy to see that $P(A|B) = 1/2$ because A's area overlaps half of B's area. In addition, $P(B) = 1/4$ because B's area covers one quarter of the area in the large rectangle representing the sample space S. Applying the general multiplication rule we get $P(A \text{ and } B) = P(A|B)P(B) = 1/2 \times 1/4 = 1/8$. To confirm this result, notice that the area for "A and B" (purple area) does, in fact, cover one-eighth of the area in the large rectangle representing S.

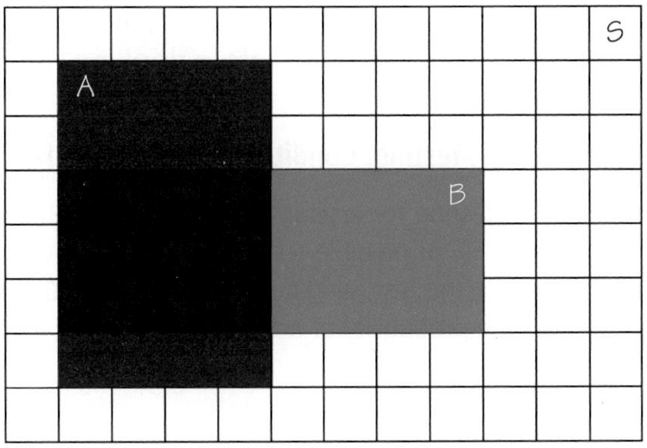

Figure 8.14 Two dependent events.

As we did for Rules 1 through 5, we now state Rules 6 and 7 more concisely using more formal mathematical notation.

Probability Rules RULE

Rule 6. The multiplication rule for *independent* events: $P(A \text{ and } B) = P(A) \times P(B)$.

Rule 7. The general multiplication rule: $P(A \text{ and } B) = P(A|B) \times P(B)$ provided $P(B) > 0$.

EXAMPLE 8 Arsenic Testing: Using the General Multiplication Rule

Tests, whether medical screening tests for diseases or testing drinking water for health-risk contaminants, are not perfect. Take, for example, testing for arsenic in drinking water. Suppose that a test for arsenic in the water supply correctly reports the presence of arsenic (positive test result) with probability 0.97, and correctly reports the absence of arsenic (negative test result) with probability 0.86. Only about 4% of the drinking water in the United States is thought to contain arsenic.

Using the information above, we want to determine the following two probabilities:

1. The probability that a randomly chosen water sample actually contains arsenic and yields a positive test result.
2. The probability that a randomly chosen water sample does not contain arsenic and yields a negative test result.

Next, we translate what we know and what we want to find into mathematical notation. Let A be the event that a water sample contains arsenic, and let A^C be the event that the sample does not contain arsenic. Let "+" represent the event that the test results come back positive for arsenic, and "−" the event that the test results come back negative.

- What we know: $P(+|A) = 0.97$, $P(-|A^c) = 0.86$, and $P(A) = 0.04$
- What we want to find: $P(A \text{ and } +)$ and $P(A^c \text{ and } -)$

Using the information from the first bullet and the general multiplication rule, we calculate probabilities 1 and 2 as follows:

$$P(A \text{ and } +) = P(+|A)\,P(A) = (0.97)(0.04) = 0.0388$$

$$P(A^c \text{ and } -) = P(-|A^c)\,P(A^c) = (0.86)(1 - 0.04) = 0.8256$$

Conditional probabilities also follow Rules 1 through 5 given in Section 8.2. In the next example, we use the complement rule (Rule 3) to find another set of probabilities associated with testing for arsenic in drinking water.

EXAMPLE 9 Arsenic Testing: Conditional Probability Models

Using the information on testing for arsenic given in Example 8, we construct a model for the conditional probability of possible test results given the water contains arsenic.

From information about the test, we know that $P(+|A) = 0.97$. Using the complement rule (Rule 3) gives the probability of a *false negative:*

$$P(-|A) = 1 - P(+|A) = 1 - 0.97 = 0.03$$

Fortunately, this probability is quite low! Table 8.4 gives the model for the conditional probability of the test results given the water sample contains arsenic.

TABLE 8.4 Conditional Probability Model

Outcome	Positive test	Negative test
Conditional probability given arsenic	0.97	0.03

Self Check 5

When the water sample does not contain arsenic, there are two possible outcomes for the test results: positive (in this case, a false positive) and negative. Make a table similar to Table 8.4 that gives the conditional probability model of the test results given the water sample does not contain arsenic. ■

In Examples 8 and 9, we have used probability rules to find a number of probabilities associated with testing water samples for arsenic. Now, suppose your water is the one randomly chosen for testing and the test comes back positive. How worried should you be that your water is contaminated by arsenic? Your level of worry should be connected to the probability that the water sample contains arsenic given the test comes back positive, $P(A|+)$. We find this probability in Example 10.

EXAMPLE 10 Arsenic Testing: How Worried Should You Be about a Positive Test Result?

We begin with a tree diagram, shown in Figure 8.15, to identify the sample space of all possible outcomes, and use the general multiplication rule (Rule 7) to assign probabilities to each outcome in the sample space. Two of these probabilities were already computed in Example 8. In addition, the tree diagram contains the conditional probability models from Example 9 and Self Check 5.

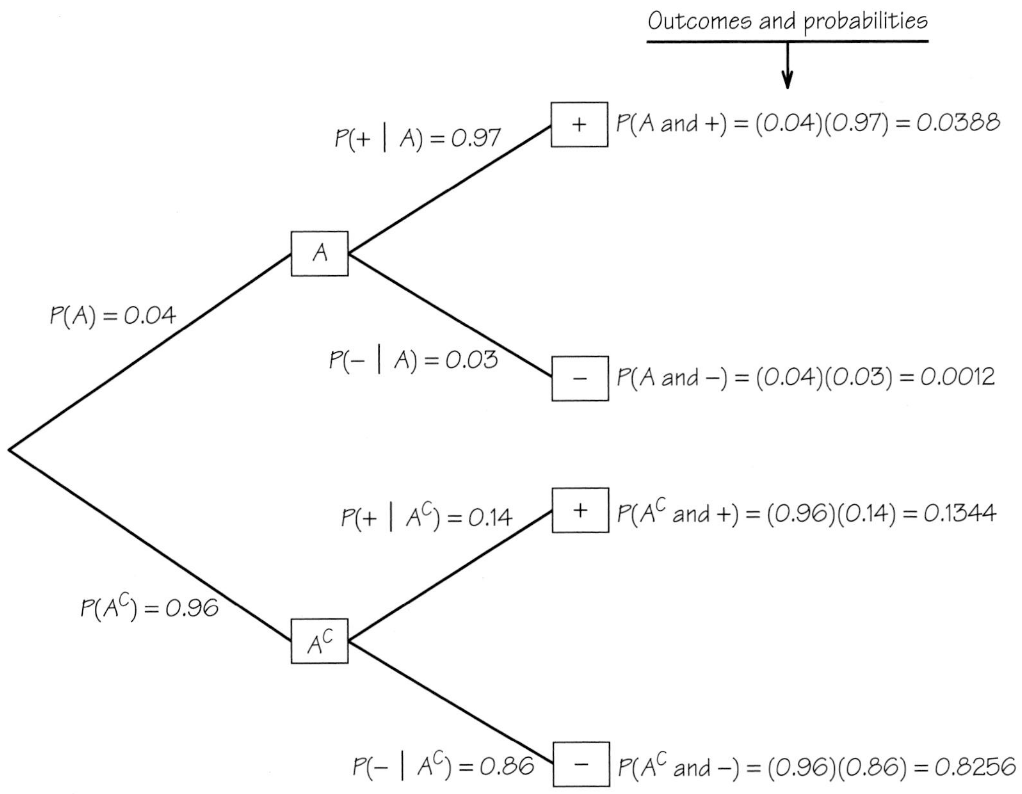

Figure 8.15 Tree diagram of sample space and calculation of probabilities.

We want $P(A|+)$, which by the definition of conditional probability we calculate as follows:

$$P(A|+) = \frac{P(A \text{ and } +)}{P(+)}$$

From the tree diagram, we know $P(A \text{ and } +) = 0.0388$. Next, we need $P(+)$. There are two ways to get a positive result: "A and $+$" or "A^C and $+$". Since these are disjoint events, we use the addition rule for disjoint events (Rule 4) to find $P(+)$:

$$P(+) = P(\text{"}A \text{ and } +\text{" or "}A^C \text{ and } +\text{"})$$

$$= P(A \text{ and } +) + P(A^C \text{ and } +) = 0.0388 + 0.1344 = 0.1732$$

Next, we substitute $P(A \text{ and } +) = 0.0388$ and $P(+) = 0.1732$ into the conditional probability formula:

$$P(A|+) = \frac{P(A \text{ and } +)}{P(+)} = \frac{0.0388}{0.1732} \approx 0.2240$$

In this situation, only around 22% of the water samples that test positive for arsenic actually contain arsenic. So, until further testing is done, you probably shouldn't be in total panic mode. In Self Check 6, you will see how this probability changes when the likelihood of having arsenic in the water supply is increased.

Self Check 6

Suppose that, due to an environmental spill in a particular region, the probability of arsenic in the water supply is thought to be 0.5. What is the probability that a randomly chosen water sample (from this region) that tests positive for arsenic actually contains arsenic? If that water sample happens to be from the water that you drink, should you be worried?

Bayes Rule Applied to Testing for a Characteristic RULE

The mathematics involved in Example 10 on testing for some characteristic A can be summarized as follows. $P(A)$ and $P(A^c)$ are called the *prior probabilities*. Given the prior probabilities and the conditional probabilities, $P(+|A)$ and $P(+|A^c)$, we can determine what are called the *posterior probabilities*, $P(A|+)$ and $P(A^c|+)$, using the following formulas:

$$P(A|+) = \frac{P(+|A)\,P(A)}{P(+|A)P(A) + P(+|A^c)P(A^c)}$$

$$P(A|-) = \frac{P(-|A)\,P(A)}{P(-|A)P(A) + P(-|A^c)P(A^c)}$$

8.4 Discrete Probability Models

In this chapter, we will work with two kinds of probability models: discrete probability models and continuous probability models. The probability models in Tables 8.3 (page 353) and 8.4 (page 358) are examples of the first kind. In both cases, the number of possible outcomes—the sums from two dice (Table 8.3) or the test results (Table 8.4)—is finite, hence the outcomes can be listed. If all the outcomes in a sample space can be put into a list, the number of outcomes is said to be *countable.* A probability model for which the sample space is countable is called a **discrete probability model.**

Discrete Probability Model DEFINITION

A **discrete probability model** is a probability model with a countable number of outcomes in its sample space.

To assign probabilities in a discrete model, list the probability of all the individual outcomes. By Rules 1 and 2, these probabilities must be numbers between 0 and 1 inclusive and must sum to 1. The probability of any event is the sum of the probabilities of the outcomes making up the event.

Up to this point, all the probability models discussed have finite sample spaces. But, as you will see in Example 11, that is not always the case.

EXAMPLE 11 Probability Model: Rolling a Pair of Dice Until You Get Doubles

According to the Donovan family's custom rules for Monopoly, if you land on the "Go to jail" square, the only way to get out of jail is to roll doubles. Let D represent rolling doubles and N represent any roll that does not result in doubles. The sample space for this situation is $S = \{D, ND, NND, NNND, \ldots\}$. In this case, the sample space contains an infinite number of outcomes, which can be put into a list. To form a discrete probability model, we need to assign probabilities to each outcome in the list. Here are calculations for some of the probabilities:

$$P(\text{first roll is doubles}) = P(D) = \frac{6}{36} = \frac{1}{6} \approx 0.167$$

$$P(\text{two rolls are needed to roll doubles}) = P(ND)$$

$$= P(N) \times P(D) \quad \text{Apply multiplication rule for independent events.}$$

$$= \left(1 - \frac{1}{6}\right) \times \left(\frac{1}{6}\right) \quad \text{Apply complement rule.}$$

$$= \frac{5}{6^2} \approx 0.139$$

$$P(\text{three rolls are needed to roll doubles}) = P(NND)$$

$$= P(N) \times P(N) \times P(D)$$

$$= \left(\frac{5}{6}\right) \times \left(\frac{5}{6}\right) \times \left(\frac{1}{6}\right) = \frac{5^2}{6^3} \approx 0.116$$

Algebra Review Appendix
Powers and Roots
Operations with Rational Numbers

Continuing this pattern, we can form a probability model in which we list the possible outcomes (number of rolls needed to get doubles) and their corresponding probabilities. Table 8.5 shows this probability model.

TABLE 8.5 Probability Model for Rolling a Pair of Dice Until You Get Doubles

Number of rolls until doubles	1	2	3	4	5	. . .	n	. . .
Probability	$\frac{1}{6}$	$\frac{5}{6^2}$	$\frac{5^2}{6^3}$	$\frac{5^3}{6^4}$	$\frac{5^4}{6^5}$	. . .	$\frac{5^{n-1}}{6^n}$	. . .

It takes a bit of work to show that the probabilities sum to 1, but they do! So, this is a valid probability model. ■

Self Check 7

We continue with the game of Monopoly and of finding probabilities associated with getting out of jail.

(a) The official rules allow a jailed player to try for doubles on three consecutive turns. If after three tries the player does not roll a double, then the player must pay $50 to get out of jail. What is the probability that the player will be able to get out of jail without paying a fine?

(b) Return to Example 11. After a Donovan family member was stuck in jail for 10 turns, the Donovans changed their rules so that jailed players get out of jail only after rolling a sum less than 7. Write a probability model for getting out of jail under the Donovans' new rule.

(c) Does the Donovans' new rule make it less likely that a player will spend a long time in jail? As part of your answer, calculate the probability that a jailed player gets out of jail within his or her first three rolls. ■

Example 12 gives another example of a discrete probability model, but this time the sample space is finite.

EXAMPLE 12 Benford's Law: One Is the Likeliest Number You'll Ever Know

Faked numbers in tax returns, invoices, or expense account claims often display patterns that aren't present in legitimate records. Some patterns, like too many round numbers, are obvious and easily avoided by a clever crook. Others are more subtle. It is a striking fact that the first (leftmost) digits of numbers in legitimate records often follow a model known as Benford's law, which is shown in Table 8.6. (Note that a first digit can't be 0).

TABLE 8.6 Probability Model Known as Benford's Law

First digit	1	2	3	4	5	6	7	8	9
Probability	0.301	0.176	0.125	0.097	0.079	0.067	0.058	0.051	0.046

You should check that the probabilities of the outcomes sum exactly to 1 to verify that this is a legitimate discrete probability model. Using this model, investigators can detect fraud by comparing the first digits in records such as invoices paid by a business with these probabilities. For example, consider the events A = "first digit is 1" and B = "first digit is 2." Applying Rule 4, the addition rule for disjoint events, to the table of probabilities yields $P(A \text{ or } B) = 0.301 + 0.176$, which is 0.477 (almost 50%). Crooks trying to "make up" the numbers probably would not make up numbers starting with 1 or 2 this often. ■

Self Check 8

You decide to fake 20 invoices. To make sure that you don't introduce any pattern into your invoice numbers, you randomly assign numbers. Use Table 7.1, the random digits table (page 298), to assign the first digits to 20 fake invoices. Enter the table on line 106 (skip any 0s).

(a) Determine the proportion of your fake invoices that have 1, 2, or 3 as the first digit in their invoice numbers.

(b) Compare your answer in part (a) to the probability of observing an invoice number with a first digit of 1, 2, or 3 based on Benford's law (Table 8.6). Do you think your fraud will be detected? ■

8.5 Equally Likely Outcomes

An SRS gives all possible samples an equal chance to be chosen. Rolling two casino dice gives all 36 outcomes the same probability. When randomness is the product of human design, it is often the case that the outcomes in the sample space are all equally likely. In this case, Rules 1 and 2 force the assignment of probabilities.

Finding Probabilities of Equally Likely Outcomes PROCEDURE

If a random phenomenon has equally likely outcomes, then the probability of event A is

$$P(A) = \frac{\text{count of outcomes in event } A}{\text{count of outcomes in sample space } S}$$

Furthermore, if another event B has a positive probability of occurring, then the probability of event A given B is

$$P(A|B) = \frac{\text{count of outcomes in event } A \text{ and } B}{\text{count of outcomes in event } B}$$

When outcomes are equally likely, we find probabilities by counting outcomes. The study of counting methods used to count those outcomes is called **combinatorics.**

Combinatorics DEFINITION

Combinatorics is the study of methods for counting.

One example of a counting method is the **fundamental principle of counting** (from Chapter 2, page 45): If there are a ways of choosing one thing, b ways of choosing a second after the first is chosen, . . . , and z ways of choosing the last item after the earlier choices, then the total number of choice sequences is $a \times b \times \cdots \times z$.

EXAMPLE 13 ➔ DNA Sequences

A strand of deoxyribonucleic acid (DNA) is a long sequence of the nucleotides adenine, cytosine, guanine, and thymine (abbreviated A, C, G, T). One helical turn of a DNA strand would contain a sequence of 10 of these acids, such as ACTGCCATGT. How many possible sequences of this length are there?

There are 4 letters that can occur in each position in the 10-letter sequence. Any of the 4 letters can be in the first position. Regardless of what is in the first position, any of the 4 letters can be in the second position, and so on. The order of the letters matters, so a sequence that begins AC will be different from one that begins CA. The number of different 10-letter sequences is more than 1 million:

$$\underbrace{4 \times 4 \times 4 \times 4 \times 4 \times 4 \times 4 \times 4 \times 4 \times 4}_{10 \text{ times}} = 4^{10} = 1{,}048{,}576$$

As big as that number is, consider that it would take a DNA sequence about 3 billion letters long to contain your entire genetic "blueprint"!

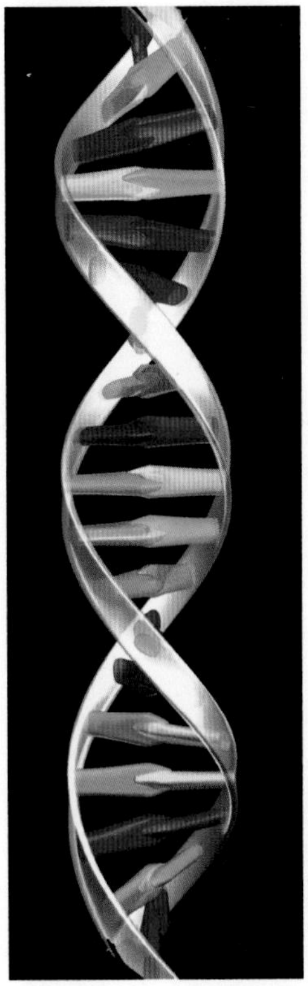
Pasieka/Science Source

Knowing the number and frequency of DNA sequences has proven important in criminal justice. When skin or bodily fluids from a crime scene are "DNA fingerprinted," the specific DNA sequences in the recovered material are extremely unlikely to be found in any suspect other than the perpetrator. The counting technique used in this example is expressed as Rule A on page 365.

EXAMPLE 14 Playing Songs

Sam Edwards/Caiaimage/Getty Images

Chapter 7, Example 4 (page 299), involved choosing a random sample of four different songs from a digital media player with a playlist of 27 different songs from Professor Lesser's *The Beatles One* album. Now we ask: How many 4-song samples are possible from a collection of 27 songs? Like DNA sequences in Example 13, order matters here. (Performers and DJs know that the same four songs can feel quite different when the songs are played in a different order.) Unlike DNA sequences, listing the same item more than once is not allowed.

Any of the 27 songs can be chosen to be played first, but only the remaining 26 songs are available to be listed as the second song, so that there are 27×26 choices for the first two songs. Any of these choices leaves 25 songs for the third position and 24 for the fourth position. Surprisingly, the number of playlists of four different songs chosen from a list of 27 songs is almost half a million!

$$27 \times 26 \times 25 \times 24 = 421{,}200$$

Now, suppose Professor Lesser's favorite song is "Let it Be." What is the probability that a randomly chosen playlist of four different songs will include this song? To answer the question, let event A be the playlists that include "Let it Be." In order to apply the procedure for finding probabilities of equally likely outcomes, we need to count the number of outcomes in A. "Let it Be" could be the first, second, third, or fourth song on a playlist. If it is the first song, then there are $26 \times 25 \times 24 = 15{,}600$ ways to complete the playlist by choosing three songs in order from the remaining 26 songs. The same is true if "Let it Be" is the second, third, or fourth song. Therefore, the number of outcomes in A is $4 \times 15{,}600 = 62{,}400$ and $P(A) = \frac{62{,}400}{421{,}200} \approx 0.148$. Thus, "Let it Be" will be included roughly 15% of the time in randomly selected playlists of four different songs.

This scenario of choosing an ordered subset of k songs from a playlist of n songs is called a **permutation.**

Self Check 9

We continue with playlists from *The Beatles One* album.

(a) How many 5-song samples are possible from a collection of 27 songs?

(b) Professor Lesser's least favorite song from the album is "The Ballad of John and Yoko." What is the probability that a randomly chosen playlist of five different songs will include this song?

Permutation DEFINITION

A **permutation** is an ordered arrangement of k items that are chosen without replacement from a collection of n items. The number of arrangements can be notated as $P(n, k)$, ${}_nP_k$, or P_k^n and has the formula

$${}_nP_k = n \times (n-1) \times \cdots \times (n-k+1)$$

which is restated as Rule B in the rule box below.

Examples 13 and 14 both involve counting the number of arrangements of distinct items. They can each be viewed as specific applications of the fundamental principle of counting, and it is easier to think your way through the counting than to memorize a recipe. Nevertheless, because these two situations occur so often, they deserve to be given their own formal recognition as Rules A and B, respectively.

Counting Ordered Collections of Distinct Items RULE

Rule A. Suppose we have a collection of n distinct items. We want to arrange k of these items in order, and the same item can appear more than once in the arrangement. The number of possible arrangements is

$$\underbrace{n \times n \times \cdots \times n}_{k \text{ times}} = n^k$$

Rule B. (Permutations) Suppose we have a collection of n distinct items. We want to arrange k of these items in order, and any item can appear no more than once in the arrangement. The number of possible arrangements is

$${}_nP_k = n \times (n-1) \times \cdots \times (n-k+1)$$

EXAMPLE 15 ➔ Four-Letter Words

Suppose you have the following four Scrabble tiles: T, S, O, and P. How many four-letter sequences can you make using all four tiles? The only way to make a four-letter sequence is to use each letter exactly once. So, there are no repeats. This is a permutation by Rule B with $n = 4$ and $k = 4$. To think through the problem (rather than simply plugging into the formula), proceed like this: Any of the four letters can be chosen first, any of the three that remain can be chosen second, and so on. The number of permutations, therefore, is

$$4 \times 3 \times 2 \times 1 = 24$$

blickwinkel/Alamy

This example shows us that the permutation of all n elements of a collection yields the product of the first n integers. This expression of factors is special enough to have its own name: **factorial** (discussed in Chapter 2, and again in Chapter 11).

Now, let's get back to the Scrabble game. It turns out that only 6 of the 24 possible sequences of 4 letters are actually words in the English language (can you figure out all 6 words?), so the probability that a permutation (chosen at random from these 4 tiles) will be an actual word is $\frac{6}{24} = \frac{1}{4}$.

Factorial DEFINITION

The **factorial** for a positive integer n equals the product of the first n positive integers. The notation for "n factorial" is $n!$:

$$n! = n \times (n-1) \times (n-2) \times \cdots \times 3 \times 2 \times 1$$

By convention, we define 0! to equal 1, which can be interpreted as saying there is one way to arrange zero items.

Self Check 10

Determine the value of 6!

Factorial notation lets us write a long string of multiplied factors very compactly. Using factorial notation, the expression for permutations in Rule B can now be rewritten as follows:

$$_nP_k = \frac{n!}{(n-k)!}$$

Permutations and factorials can be tedious to compute for large values of n, but even a scientific calculator should have a key labeled $n!$ or $x!$ If you have only a basic calculator without a factorial key, the expression $n \times (n-1) \times \cdots \times (n-k+1)$ will involve fewer multiplications than evaluating $\frac{n!}{(n-k)!}$, by first calculating $n!$ and $(n-k)!$, because it has already incorporated all the cancellations between numerator and denominator—namely, cancelling the positive integers from 1 to $n - k$.

EXAMPLE 16 Computing a Permutation: Factorial Formula or Rule B?

Suppose a club has 10 members. How many slates of president, vice president, and treasurer are possible? This is a permutation. We must select three club members and arrange them in the three elective offices. In other words, we need to compute $_{10}P_3$. We'll do this in two ways. First, we'll use the formula above and the factorial key on a calculator. Next, we'll expand the formula, cancel, and arrive at the calculation shown for Rule B. You can decide which is easier.

$$_{10}P_3 = \frac{10!}{(10-3)!} = \frac{10!}{7!} = \frac{3{,}628{,}800}{5{,}040}$$ Calculation using calculator's factorial key

$$_{10}P_3 = \frac{10!}{(10-3)!} = \frac{10!}{7!} = \frac{(10)(9)(8)(\cancel{7!})}{\cancel{7!}} = 720$$ Calculation done by hand

In Example 16, counting the number of slates for club officers is a permutation because the arrangement of people in the elected offices matters. Being listed as president is different from being listed as treasurer. However, what if the club planned to send a delegation of three of its members to a conference? In this situation, {Joe, Mia, Mizan} is the same delegation as {Mizan, Mia, Joe}. The arrangement of the names doesn't change who is representing the club at the conference. How many different delegations can be selected? This is an example of a **combination.** The 720 ways to select the three-person slate of officers from Example 16 overshoots the number of ways to select a delegation because it counts each of the 3! ways to list three different club members chosen as distinct. So the number of ways to choose the three-member delegation is $\frac{720}{3!} = 120$.

Combination DEFINITION

A **combination** is an unordered selection of k items that are chosen without replacement from a collection of n items. The number of combinations can be notated as $\binom{n}{k}$, $C(n, k)$, or ${}_nC_k$, which is often pronounced "n choose k":

$$ {}_nC_k = \frac{{}_nP_k}{k!} = \frac{n \times (n-1) \times \cdots \times (n-k+1)}{k!} = \frac{n!}{k!(n-k!)} $$

which is Rule D in the next rule box.

EXAMPLE 17 Winning the Texas Lottery?

Most states have lottery games. The Texas Lottery (www.txlottery.org) has a Lotto Texas game that involves choosing six numbers from the set of whole numbers from 1 to 54. You win (at least a share of) the jackpot so long as the collection of numbers you pick is the same collection that the lottery selects. Repetition is not allowed—that is, the same number can't be picked twice in the same drawing. Unlike permutations, order does not matter here. A person who chooses {49, 12, 18, 26, 22, 5} will receive the same payout as a person who chooses {26, 18, 12, 49, 5, 22}.

This is a perfect setup for a combination because the order of the selected numbers doesn't matter. The number of collections of lottery balls is "54 choose 6," or

$$ {}_{54}C_6 = \frac{{}_{54}P_6}{6!} = \frac{54 \times 53 \times 52 \times 51 \times 50 \times 49}{6 \times 5 \times 4 \times 3 \times 2 \times 1} = 25{,}827{,}165 $$

possible sets of numbers. Because only one of these sets of numbers will correspond to the jackpot, the probability of your ticket winning (at least a share of) the jackpot is $\frac{1}{25{,}827{,}165}$.

More generally, the scenario of choosing an unordered subset of k balls from a collection of n different balls is a combination. ■

If it's hard to remember the difference between combinations (Rule D) and permutations (Rule B), use this memory aid: "Permutations presume positions; combinations concern collections." For completeness, we also provide a formula

(Rule C) for unordered collections in which repetition *is* allowed, but an explanation is beyond the scope of this chapter and we will not use it again.

Counting Unordered Collections of Distinct Items RULE

Rule C. Suppose that we have a collection of n distinct items. We want to select k of those items with no regard to order, and any item can appear more than once in the collection. The number of possible collections is

$$\frac{(n + k - 1)!}{k!(n - 1)!}$$

Rule D. (Combinations) Suppose we have a collection of n distinct items. We want to select k of these items with no regard to order, and any item can appear no more than once in the collection. The number of possible selections is

$$_nC_k = \frac{n!}{k!(n - k)!}$$

Rules A, B, C, and D can be summarized in Table 8.7.

TABLE 8.7 Ways to Choose k Items from n Distinct Items

	Repetition is allowed	**Repetition is *not* allowed**
Order does matter	Rule A: $n \times n \times \cdots \times n = n^k$ n is multiplied by itself k times	Rule B (*permutation*): $\frac{n!}{(n-k)!} =$ $n \times (n - 1) \times \cdots \times (n - k + 1)$
Order does *not* matter	Rule C: $\frac{(n+k-1)!}{k!(n-1)!}$	Rule D (*combination*): $\frac{n!}{k!(n-k)!}$

Self Check 11

In each of the following situations, decide whether the permutation rule or combination rule should be used to determine the number of different ways the positions can be filled by the job applicants. Show the results for each situation.

(a) A university library posts an ad for three library technicians. There are 15 applicants.

(b) A hospital posts an ad for three nursing positions: a day nursing supervisor, a night nursing supervisor, and a nursing coordinator. There are 15 applicants. ■

Although it is important to have some experience computing permutations, factorials, and combinations by hand, it can be time consuming. Spotlight 8.3 shows how technology can be used to speed up the process.

Using Technology to Compute Permutations, Factorials, and Combinations

SPOTLIGHT 8.3

Graphing calculators and spreadsheets have built-in functions to compute permutations, factorials, and combinations. Below are instructions for TI-84 graphing calculators and Excel. (Not using a TI-84 calculator or Excel? Find the instructions for your particular calculator or spreadsheet on the Internet.)

TI-84 Graphing Calculators

Permutation

To calculate $_{n}P_{r}$:

- Enter the value for *n*.
- Press [MATH] → PRB and select [2] (for $_{n}P_{r}$).
- Enter the value for *r* and then press [ENTER].

We illustrate with the example $_{54}P_{6}$:

Press [5] [4] [MATH] → PRB, and then press [2] [6] [ENTER]. Here are the results.

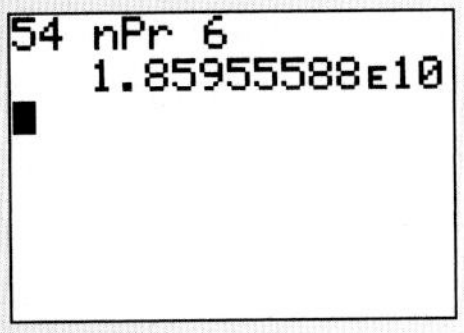

Algebra Review Appendix
Scientific Notation

Because the answer—18,595,558,800—is so large, the calculator used its version of scientific notation to report it.

Factorial

To calculate *n*!:

- Enter the value for *n*.
- Press [MATH] → PRB, select [4] (for !), and then press [ENTER].

We illustrate with the example 6!:

Press [6] [MATH] → PRB and then press [4] [ENTER].

The answer should be 720 (as you should have discovered in Self Check 10.)

Combination

To calculate $_{n}C_{r}$:

- Enter the value for *n*.
- Press [MATH] → PRB and select [3] (for $_{n}C_{r}$).
- Enter the value for *r* and then press [ENTER].

We illustrate with the example $_{54}C_{6}$:

Press [5] [4] [MATH] → PRB and then press [3] [6] [ENTER]. Here are the results.

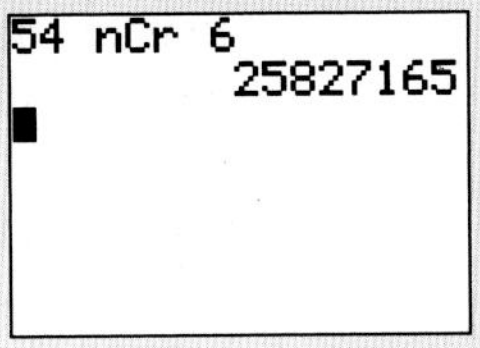

Excel

Permutation

To calculate $_{n}P_{k}$:

- Click on an empty cell.
- Key in the formula =PERMUT(the value for *n*, the value for *k*) and then press Enter

We illustrate with the example $_{54}P_{6}$:

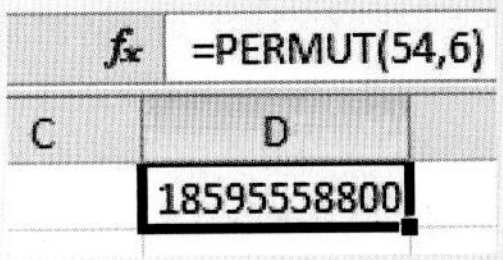

Factorial

To calculate *n*!:

- Click on an empty cell.
- Key in the formula =FACT(the value for *n*) and then press Enter.

We illustrate with the example 6!:

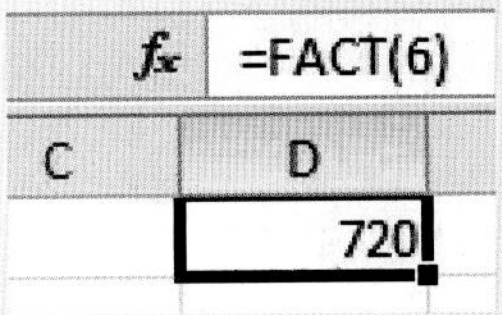

Combination

To calculate $_{n}C_{k}$:

- Click on an empty cell.
- Key in the formula =COMBIN(the value for *n*, the value for *k*) and then press Enter.

We illustrate with the example $_{54}C_{6}$:

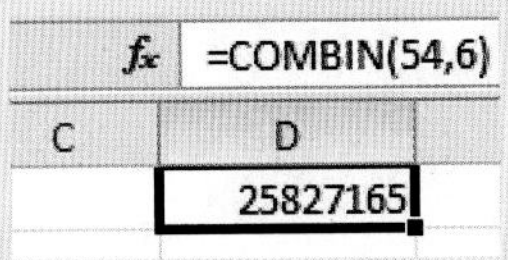

Self Check 12

Use a calculator or spreadsheet program with built-in functions for calculating permutations, factorials, and combinations to determine the following: ${}_{30}P_7$, $7!$, ${}_{30}C_7$. ■

The topic of Spotlight 8.4 is determining the likelihood that a group of people share the same birthday. In determining this likelihood, we assume a discrete uniform probability model in which all days of the year are assumed to be equally likely to be a birthday. (While this assumption is not perfectly satisfied, it turns out that any deviations from it only make the probability of a match even higher.) If we consider the continuous uniform probability model (see Chapter 5, Exercise 73 on page 238) of the day and time someone is born, then more advanced mathematics shows that only 17 people are required for there to be at least a 50% chance that at least two birth times are within 24 hours of each other! In the next two sections, we will examine continuous and discrete probability models in more detail.

Birthday Coincidences

SPOTLIGHT 8.4

If we ignore leap day (February 29), there are 365 possible birthdays a person can have. So if 366 people are gathered, there's a 100% chance that at least two people share the same birthday. Now, if only 23 people are gathered, what do you think is the probability of any birthday matches? Guess before reading further.

Now imagine these 23 people enter a room one at a time, adding their birthday to a list in the order they enter. Using $n = 365$ and $k = 23$, Rule A gives us the total number of lists of 23 birthdays: 365^{23}. Rule B gives us how many of those lists have birthdays that are all different: ${}_{365}P_{23}$. Using the rule for equally likely outcomes (assuming each day of the year is equally likely to be a randomly chosen person's birthday), we conclude that the probability of all birthdays being different is the result from Rule B divided by the result from Rule A:

$$\frac{{}_{365}P_{23}}{365^{23}} = \frac{365 \times 364 \times \cdots \times 343}{365^{23}}$$

Alternatively, we could assume independence of birthdays and use Probability Rule 6 (multiplication rule for independent events). The second person who walks in has a 364/365 chance of not matching person #1. The third person who walks in has a 363/365 chance of not matching persons #1 or #2, and so on. Verify that you get the same product by multiplying this string of fractions:

$$\frac{364}{365} \times \frac{363}{365} \times \cdots \times \frac{343}{365}$$

Either way, our final step to find the probability of getting at least one match is to subtract that answer from 1 (using Rule 3, the complement rule), and we obtain the surprisingly high value of 51%!

A rough way to make the result seem plausible (devised by Manfred Borovcnik, a researcher in probability and statistics education) is to say that with 23 people, you would expect to have about two people born in each month. In one month, the chance that one person matches someone else's birthday is about 1/30. Expecting "1/30 of a match" per month adds up to 12/30 of a match for the year, and 12/30 is not much less than 50%.

Because we underestimate the number of potential opportunities for "coincidences," we are surprised that they happen as often as they do. But as statistician Jessica Utts notes, if something has a one in a million chance of happening to any person on a given day, this rare event will happen to roughly 300 people in the United States each day!

Michael Rosenfeld/
Photographer's Choice/
Getty Images

8.6 Continuous Probability Models

When we use the table of random digits (Table 7.1, page 298) to select a digit between 1 and 9, the discrete probability model assigns probability $\frac{1}{10}$ to each of the 10 possible outcomes. Suppose we want to choose a number at random between 0 and 1, allowing *any* number between 0 and 1 as the outcome. You can do this with technology, such as by using the TI-84 calculator command sequence MATH → PRB → rand or the Microsoft Excel spreadsheet software command = RAND(). [Recall that Spotlight 7.1 on page 300 uses RAND( ) to select a random sample.] You can visualize such a random number by thinking of a spinner needle (Figure 8.16) that turns freely around its center and slowly comes to a stop. The pointer can come to rest anywhere on a circle that is marked from 0 to 1.

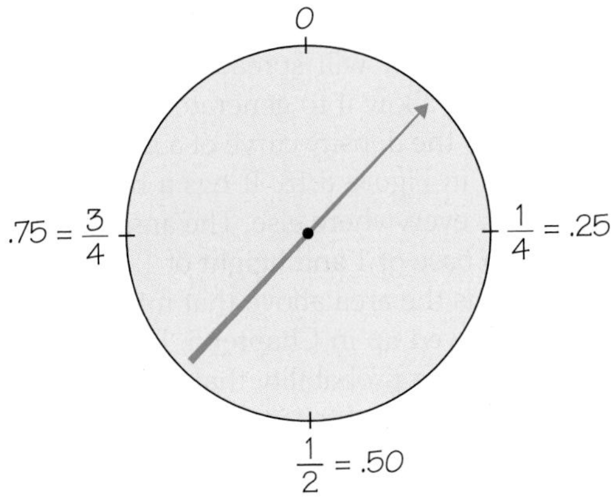

Figure 8.16 This spinner chooses a number between 0 and 1 at random. That is, it is equally likely to stop at any point on the circle.

The sample space is now an entire interval of numbers:

$$S = \{\text{all numbers } x \text{ such that } x \text{ is between 0 and 1}\}$$

How can we assign probabilities to events such as $\{0.3 \le x \le 0.7\}$? See Figure 8.17.

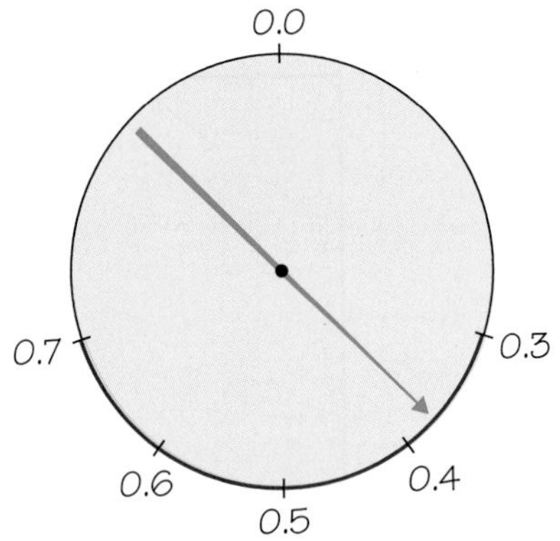

Figure 8.17 Finding the probability that the spinner will point to a number between 0.3 and 0.7.

As in the case of selecting a random digit, we would like all possible outcomes to be equally likely. But we cannot assign probabilities to each individual value of x and then sum because there are *infinitely* many possible values—too many to count or list. Instead, we use a second way of assigning probabilities directly to events—*as areas under a curve.* By Probability Rule 2, the curve must have a total area of 1 underneath it, corresponding to a total probability of 1. We call such curves **density curves,** which were introduced in Chapter 5 (see Example 15, page 210). We restate the definition here.

Density Curve DEFINITION

A **density curve** is a curve that

- is always on or above the horizontal axis
- has an area of exactly 1 underneath it

Continuous Probability Model DEFINITION

A **continuous probability model** is a probability model that assigns probabilities as areas under a density curve. The area under the curve and above any interval of values is the probability of an outcome in that interval.

The random-number generator will spread its output uniformly across the entire interval from 0 to 1 if we allow it to generate many numbers. The results of many trials are represented by the density curve of a *uniform probability model.* This density curve appears in red in Figure 8.18. It has a height of 1 over the interval from 0 to 1, and a height of 0 everywhere else. The area under the density curve is 1, the area of a square with a base of 1 and height of 1. The probability of generating a number in any interval is the area above that interval and under the density curve. (This density curve showed up in Chapter 5, Exercise 73, page 238.)

As Figure 8.18a illustrates, the probability that the random-number generator produces a number X between 0.3 and 0.7 inclusive is

$$P(0.3 \leq X \leq 0.7) = 0.4$$

because the rectangular area under the density curve above the interval from 0.3 to 0.7 is 0.4. (The area of a rectangle is the product of its height and length. The height of this density curve is 1, and the width of this interval is $0.7 - 0.3 = 0.4$. So the area is $1 \times 0.4 = 0.4$.)

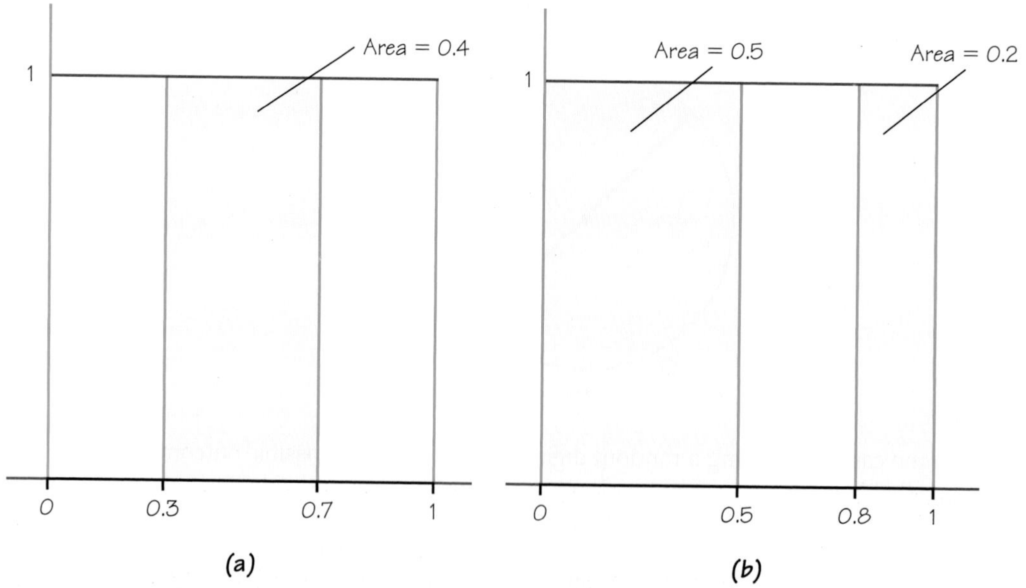

Figure 8.18 Assigning probabilities for generating a random number between 0 and 1 inclusive for the spinner of Figure 8.16. (a) The probability of an outcome between 0.3 and 0.7. (b) The probability of an outcome less than 0.5 or greater than 0.8.

Also, we can apply Probability Rule 4 (addition rule for disjoint events) to non-overlapping intervals such as

$$\begin{aligned} P(X < 0.5 \text{ or } X > 0.8) &= P(X < 0.5) + P(X > 0.8) \\ &= 0.5 + 0.2 = 0.7 \end{aligned}$$

The last event consists of two non-overlapping intervals, so the total area above the event is found by adding two areas, as illustrated by Figure 8.18b. This assignment of probabilities obeys all our rules for probability.

The probability model for a continuous random variable assigns probabilities to *intervals* of outcomes rather than to individual point outcomes. In fact, *all continuous probability models assign probability 0 to every individual outcome.* Only *intervals of values* have positive probability. To see that this is true, consider a specific outcome such as $P(X = 0.6)$ in Figure 8.18. In this example, the probability of any interval is the same as its length. The point 0.6 has no length, so this probability is 0.

EXAMPLE 18 Roundoff Error: Application of the Continuous Uniform Model

Before data values are presented, they sometimes get rounded to, say, the nearest whole number for ease of reading. For example, rounding 32.7 to 33 creates a roundoff error of $32.7 - 33 = -0.3$, and rounding 14.17 to 14 yields a roundoff error of $14.17 - 14 = 0.17$. Roundoff error can be critical to keep track of in data analysis and is one of many applications of the continuous uniform probability model. By rounding to the nearest whole number, the absolute value of the roundoff error cannot exceed $\frac{1}{2}$, and it is usually assumed that each roundoff error is equally likely to be any number between -0.5 and 0.5. Note that this example shows that so long as the total area under the density curve is 1, there is no reason the horizontal axis variable has to be between 0 and 1.

Algebra Review Appendix
Rounding Numbers

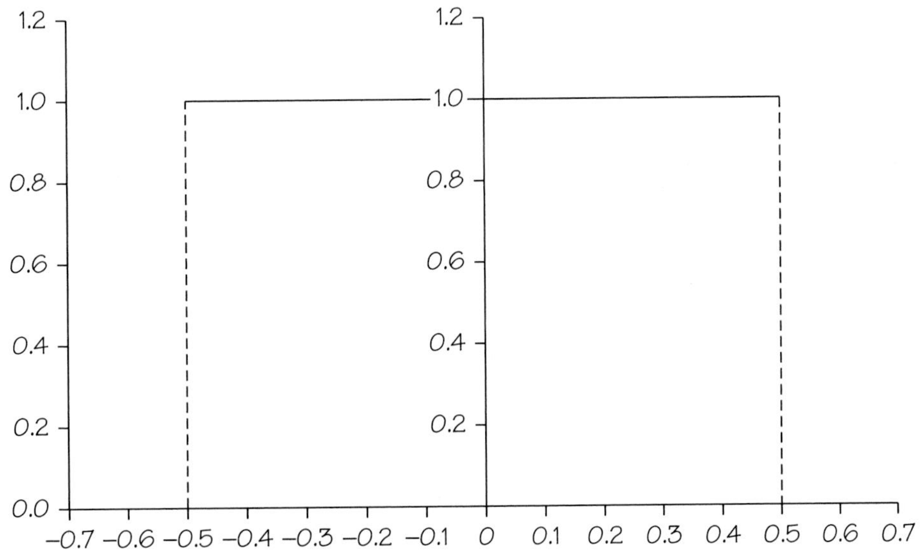

Figure 8.19 Density curve for uniform distribution on the interval from -0.5 to 0.5.

Going further, the horizontal axis variable is also not limited to an interval of length 1. For example, consider choosing a random number from the continuous interval from 1 to 3. For a rectangular area under a density curve to remain 1, a horizontal base of length $3 - 1 = 2$ would require the vertical height to be $\frac{1}{2}$.

Self Check 13

Suppose you want to pick a random number from the uniform distribution over the interval from 2 to 6.

(a) Draw the graph of a uniform density curve on the interval from 2 to 6.

(b) Find the probability that a randomly selected number X lies between 3 and 5. In other words, find $P(3 \le X \le 5)$.

The density curves that are most familiar to us are the normal curves, which were introduced in Chapter 5. Because any density curve describes an assignment of probabilities, normal distributions are *continuous probability models.* Recall the total area under a normal curve is 1. Let's revisit Example 17 from Chapter 7 (page 319), using the language of probability.

EXAMPLE 19 Areas under a Normal Curve Are Probabilities

Suppose that 60% of adults agree with this statement: "Most people who want to get ahead can make it if they're willing to work hard." All adults form a population, with population proportion $p = 0.6$. Interview an SRS of 1500 people from this population and find the proportion $\hat{p}$ of the sample who agree with the statement. We know that if we take many such samples, the statistic $\hat{p}$ will vary from sample to sample according to a normal distribution, with

$$\text{mean} = p = 0.6$$

$$\text{standard deviation} = \sqrt{\frac{p(1-p)}{n}}$$

$$= \sqrt{\frac{(0.6)(0.4)}{1500}} \approx 0.013 \text{ (rounded to three decimal places)}$$

The 68–95–99.7 rule now gives *probabilities* for the value of $\hat{p}$ from a single SRS. The probability is 0.95 that $\hat{p}$ lies between 0.574 and 0.626 (within 2 standard deviations of the mean). Figure 8.20 shows this probability as an area under the normal density curve.

All that is new is the language of probability. "Probability is 0.95" is shorthand for "95% of the time in a very large number of samples."

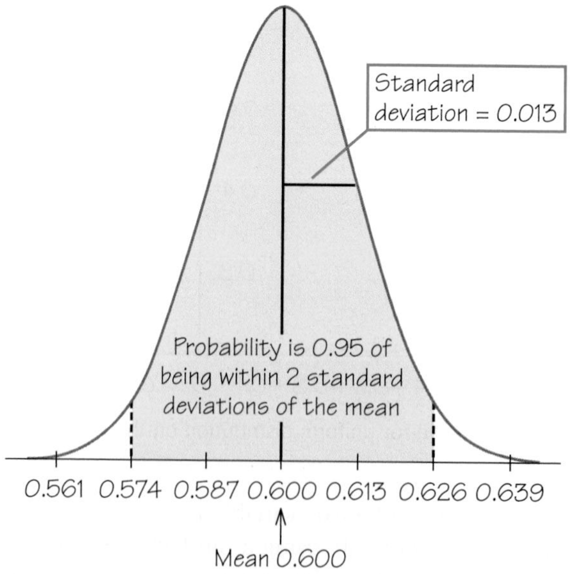

Figure 8.20 Probability shown as the area under a normal curve. The 68–95–99.7 rule gives some probabilities for normal probability models.

8.7 The Mean and Standard Deviation of a Probability Model

We return to a discrete probability model governing possible bets. Suppose that you are offered this choice of bets, each costing the same:

Bet *A* pays \$10 if you win and you have probability $\frac{1}{2}$ of winning
Bet *B* pays \$10,000 if you win and offers probability $\frac{1}{10}$ of winning

It would be foolish to decide which bet to make just on the basis of the probability of winning. How much you can win is also important. When a random phenomenon has numerical outcomes, we are concerned with their amounts as well as with their probabilities.

What will be the average payoff of our two bets in many plays? Recall that the probabilities are the long-run proportions of plays in which each outcome occurs. Bet *A* produces \$10 half the time in the long run and nothing half the time. So the average payoff should be

$$\left(\$10 \times \frac{1}{2}\right) + \left(\$0 \times \frac{1}{2}\right) = \$5$$

Bet *B*, on the other hand, pays out \$10,000 on $\frac{1}{10}$ of all bets in the long run. So bet *B*'s average payoff is

$$\left(\$10{,}000 \times \frac{1}{10}\right) + \left(\$0 \times \frac{9}{10}\right) = \$1000$$

If you can place many bets, you should certainly choose *B*. In general, to take into account values and probabilities at the same time, we can add up the values, each weighted by their respective probability, so that more likely values get more weight. Here is a procedure of the kind of "average outcome" that we used to compare the two bets.

Mean of a Discrete Probability Model PROCEDURE

Step 1. Make a table with two rows. The first row needs to list all the possible numerical outcome values in the sample space.

Step 2. In the second row of the table, list the respective probabilities of each of the outcome values from the first row of the table.

Step 3. Write (or imagine) a third row where each entry is the product of the two items in the same column from the first two rows. Now add up all the values in the third row, and you will get the mean of the discrete probability model, which we designate as μ.

We can express the above procedure with algebraic notation. If there are k possible outcome values, we can use a subscript as an index in labeling each of the k outcomes as follows: $x_1, x_2, \cdots, x_k$. If we write their respective corresponding probabilities, $p_1, p_2, \cdots, p_k$, then the mean μ of a discrete probability model can be calculated as follows:

$$\mu = x_1p_1 + x_2p_2 + \cdots + x_kp_k$$

Sometimes the mean μ of a probability model is referred to as the *expected value.*

EXAMPLE 20 Mean Family Size

The first two rows in Table 8.8 give a probability distribution for U.S. family size, x. (Although there are some families that have more than eight members, the likelihood is so small that we ignored this possibility in the discrete probability model.) The third row shows Step 3 in the procedure for calculating the mean—it contains the products of family size and corresponding probability.

TABLE 8.8 Discrete Probability Model of U.S. Family Size

x_i	1	2	3	4	5	6	7	8
p_i	0.15	0.23	0.19	0.23	0.12	0.05	0.02	0.01
$(x_i)(p_i)$	0.15	0.46	0.57	0.92	0.6	0.3	0.14	0.08

To calculate the mean, simply sum the entries in the third row:

$$\mu = 0.15 + 0.46 + 0.57 + 0.92 + 0.60 + 0.30 + 0.14 + 0.08 = 3.22$$

In Chapter 5 (page 197), we discussed the sample mean $\bar{x}$, the average of n observations that we actually have in hand. Take, for example, the data below on family size from a random sample of 30 families in the United States.

2	5	1	5	7	1	1	3	5	3
2	1	2	2	3	4	3	5	5	5
7	1	2	1	3	5	2	5	5	5

In this case, the sample mean $\bar{x}$ is computed by summing the data values and dividing by 30: $\bar{x} = 3.367$.

As shown in Example 20, the mean μ describes the probability model rather than any one collection of observations from a sample. The lowercase Greek letter *mu* (μ) is pronounced "myoo." The mean family size in Example 20 is $\mu = 3.22$. We know that it is not possible to have a family of 3.22 people. Instead, think of μ as a theoretical mean that gives the average outcome that we expect in the long run. In the case of family size, we would expect the long-run average, over many, many, many families, to be 3.22.

EXAMPLE 21 The Mean of the Probability Model for Benford's Law

In Self Check 8, you were asked to create the first digits for 20 fictitious invoice numbers. To ensure these numbers were random, you used a random digits table. That way each integer, 1 through 9, was equally likely to be chosen as a first digit. The table below shows the probability model governing your selection of first digits.

First digit	1	2	3	4	5	6	7	8	9
Probability	$\frac{1}{9}$	$\frac{1}{9}$	$\frac{1}{9}$	$\frac{1}{9}$	$\frac{1}{9}$	$\frac{1}{9}$	$\frac{1}{9}$	$\frac{1}{9}$	$\frac{1}{9}$

The mean of this model is

$$\mu = (1)\left(\frac{1}{9}\right) + (2)\left(\frac{1}{9}\right) + (3)\left(\frac{1}{9}\right) + (4)\left(\frac{1}{9}\right) + (5)\left(\frac{1}{9}\right) + (6)\left(\frac{1}{9}\right) + (7)\left(\frac{1}{9}\right) + (8)\left(\frac{1}{9}\right) + (9)\left(\frac{1}{9}\right) = 5$$

If, on the other hand, legitimate records obey Benford's law, the distribution of the first digit is

First digit	1	2	3	4	5	6	7	8	9
Probability	0.301	0.176	0.125	0.097	0.079	0.067	0.058	0.051	0.046

The mean of Benford's model is

$$\mu = (1)(0.301) + (2)(0.176) + (3)(0.125) + (4)(0.097) + (5)(0.079) + (6)(0.067) + (7)(0.058) + (8)(0.051) + (9)(0.046) \approx 3.441.$$

The comparison of means between Benford's law and random digits, $3.441 < 5$, reflects the greater probability of smaller first digits under Benford's law. Probability histograms for these two models appear in Figure 8.21. Because the histogram for random digits (Figure 8.21a) is symmetric, the mean lies at the center of symmetry. We can't determine the mean of the right-skewed Benford's law model precisely by simply looking at Figure 8.21b; calculation is needed.

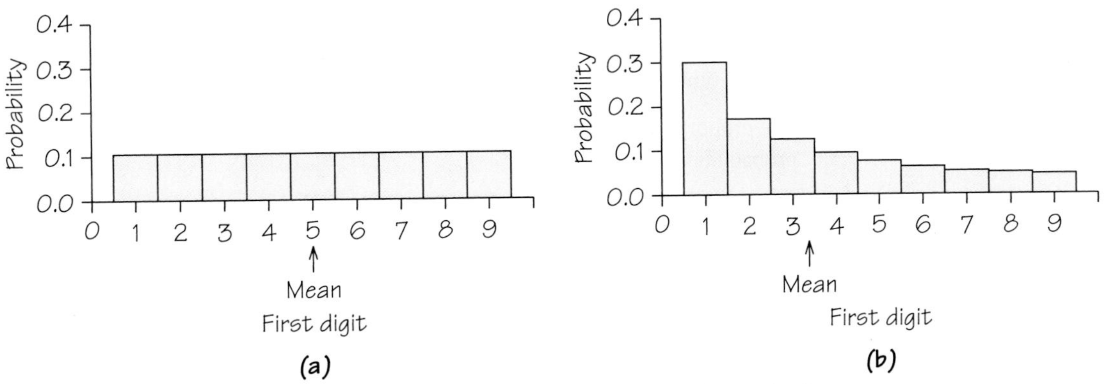

Figure 8.21 Probability histograms of two models for first digits in numerical records. The mean of each distribution is marked. (a) Digits are equally likely. (b) Digits follow Benford's law.

Self Check 14

A small business finds that the number of employees who call in sick on any given day can be described by the following probability model. Calculate the mean number of employees who call in sick on any given day.

Number who call in sick, *x*	0	1	2	3	4
Probability, *p*	0.49	0.26	0.15	0.07	0.03

What about continuous probability models? Think of the area under a density curve as being cut out of solid homogenous material. The mean μ is the point at which the shape would balance. Figure 8.22 illustrates this interpretation of the mean. The mean lies at the center of symmetric density curves, such as the uniform density in Figure 8.18 (page 372) and the normal curve in Figure 8.20 (page 374). Exact calculation of the mean of a distribution with a skewed density curve requires advanced mathematics.

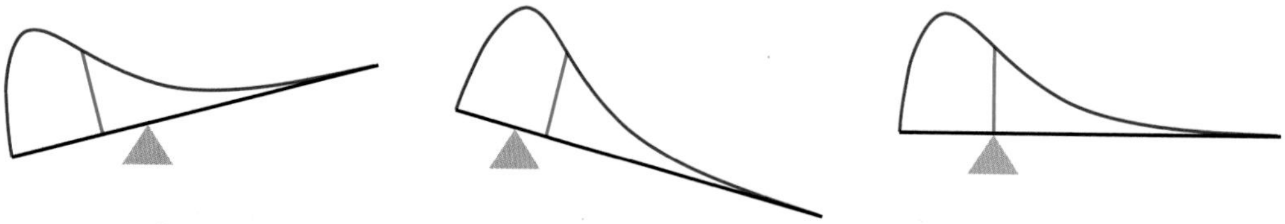

Figure 8.22 The mean of a continuous probability model is the "balance point" for the density curve.

The mean μ is an average outcome in two senses. The definition for discrete probability models says that it is the average of the possible outcomes weighted by their probabilities. More likely outcomes get more weight in the average. An important fact of probability, the **law of large numbers,** says that μ is the average outcome in another sense as well.

Law of Large Numbers — THEOREM

Observe any random phenomenon having numerical outcomes with finite mean μ. According to the **law of large numbers,** as the phenomenon is repeated a large number of times, the following occurs:

- The proportion of trials in which an outcome occurs gets closer and closer to the probability of that outcome.
- The mean $\bar{x}$ of the observed values gets closer and closer to μ.

EXAMPLE 22 The Law of Large Numbers and the Gambling Business

The law of large numbers explains why gambling can be a business. In a casino, the house (i.e., the casino) always has the upper hand. Even when the edge is very small, such as in blackjack—where according to "The Wizard of Odds," the house edge is only around 0.3%—the casino makes money. The house edge is defined as the ratio of the average loss to the initial bet. That means for every \$10 initial bet, the gambler will lose, *on average,* $0.003 \times \$10 = \$0.03$, or 3 cents per game. Over thousands and thousands of gamblers and games, that small edge starts to generate big revenues. Unlike most gamblers, casinos are playing the long game and not just hoping for a short-term payout.

The winnings (or losses) of a gambler on a few plays are highly variable or uncertain; that's why gambling is exciting. It is only *in the long run* that the mean outcome is predictable. Take, for example, roulette. An American roulette wheel has 38 slots, with numbers 1 through 36 (not in order) on alternating red and black slots and 0 and 00 on two green slots. The dealer spins the wheel and whirls a small ball in the opposite direction within the wheel. Gamblers bet on where the ball will come to rest (see Figure 8.23). One of the simplest wagers is to choose red. A bet of \$1 on red pays off an additional \$1 if the ball lands in a red slot. Otherwise, the player loses the \$1.

Julian Rovagnati/Shutterstock

nexus 7/Shutterstock

Figure 8.23 One round of blackjack and roulette. (a) A winning hand in blackjack! (b) Red wins!

Lou bets on red. He wins if the ball stops in one of the 18 red slots. He loses if it lands in one of the 20 slots that are black or green. Because casino roulette wheels are carefully balanced so that all slots are equally likely, the probability model is

Net Outcome for Gambler

	Win \$1	Lose \$1
Probability	$\frac{18}{38} = 0.474$	$\frac{20}{38} = 0.526$

The mean outcome of a single \$1 bet on red is

$$\mu = (\$1)\left(\frac{18}{38}\right) + (-\$1)\left(\frac{20}{38}\right)$$
$$= -\$\frac{2}{38} = -\$0.053 \text{ (a loss of 5.3 cents)}$$

The law of large numbers says that the mean μ is the average outcome of a very large number of individual bets. In the long run, gamblers will lose (and the casino will win) an average of 5.3 cents per bet.

So the house, unlike individual gamblers, can count on the long-run regularity described by the law of large numbers. The average winnings of the house on tens of thousands of plays will be very close to the mean of the distribution of winnings. Needless to say, gambling games have mean outcomes that guarantee the house a profit; though, as we have seen, some games such as blackjack give the house a smaller advantage than others such as keno. (According to "The Wizard of Odds," the house edge on keno is 25% to 29%!) ■

Self Check 15

On the roulette wheel described in Example 22, you can also bet on the ball landing on the numbers 1 through 12, in which case the payout is \$2 on an initial bet of \$1.

(a) What is the probability that the gambler wins? Loses?

(b) For each \$1 bet, what are the expected winnings for the house (or losses for the gambler)? How does the house edge in this game compare with the house edge in betting on red? ■

We know that the simplest description of a distribution of data requires both a measure of center and a measure of variability. The same is true for probability models. The *mean* is the average value for both a set of data and a discrete probability model. All the observations are weighted equally in finding the mean $\bar{x}$ for data, but the values are weighted by their probabilities in finding the mean μ of a probability model. The measure of variability that goes with the mean is the **standard deviation.** In Section 5.7, on page 203, we learned that the standard deviation s of data is the square root of the average squared deviation of the observations from their mean. We apply exactly the same idea to probability models, using probabilities as weights in the average. Here is the definition.

Standard Deviation of a Discrete Probability Model — DEFINITION

Suppose that the possible outcomes $x_1, x_2, \ldots, x_k$ in a sample space S are numbers, and that p_j is the probability of outcome x_j. The **standard deviation of a discrete probability model** with mean μ is denoted by the lowercase Greek letter *sigma* (σ) and is given by this formula:

$$\sigma = \sqrt{(x_1 - \mu)^2 p_1 + (x_2 - \mu)^2 p_2 + \cdots + (x_k - \mu)^2 p_k}$$

EXAMPLE 23 **Standard Deviation of the Probability Model for Benford's Law**

If the first digits in a set of records obey Benford's law, the discrete probability model is as follows:

First digit	1	2	3	4	5	6	7	8	9
Probability	0.301	0.176	0.125	0.097	0.079	0.067	0.058	0.051	0.046

We saw in Example 21 that the mean is $\mu = 3.441$. To find the standard deviation,

$$\begin{aligned}\sigma &= \sqrt{(x_1 - \mu)^2 p_1 + (x_2 - \mu)^2 p_2 + \cdots + (x_k - \mu)^2 p_k} \\ &= \sqrt{(1 - 3.441)^2 (0.301) + (2 - 3.441)^2 (0.176) + \cdots + (9 - 3.441)^2 (0.046)} \\ &= \sqrt{1.7935 + 0.3655 + \cdots + 1.4215} \\ &= \sqrt{6.061} \approx 2.46\end{aligned}$$

You can follow the same pattern to find the standard deviation of the equally likely model and show that the Benford's law model, by virtue of clustering near the left side, has less variability than the equally likely model. ■

Self Check 16

Return to the distribution of net outcomes for betting on red at the roulette wheel from Example 22 (page 378). You know that $\mu = -0.053$. Find the standard deviation for a single $1 bet on red. ■

Finding the standard deviation of a continuous probability model usually requires advanced mathematics (calculus). Section 5.8 provided the answer in one important case (see page 213): The standard deviation of a normal curve is the distance from the center (the mean) to the change-of-curvature point on either side.

8.8 The Central Limit Theorem

The key to finding a confidence interval that estimates a population proportion (Chapter 7) is the fact that the sampling distribution of a population proportion is close to normal when the sample is large. This fact is an application of one of the most important results of probability theory, the **central limit theorem.** This theorem says that the distribution of *any* random phenomenon tends to be normal if we average it over a large number of independent repetitions. The central limit theorem allows us to analyze and predict the results of chance phenomena when we average over many observations.

Central Limit Theorem THEOREM

Draw an SRS of size n from any large population with mean μ and finite standard deviation σ. Then

- The mean of the sampling distribution of $\bar{x}$ is μ.
- The standard deviation of the sampling distribution of $\bar{x}$ is $\frac{\sigma}{\sqrt{n}}$.
- The **central limit theorem** says that the sampling distribution of $\bar{x}$ is approximately normal when the sample size n is large ($n \geq 30$).

The first two parts of this statement can be proved from the definitions of the mean and the standard deviation. They are true for any sample size n. The central limit theorem is a much deeper result. Pay attention to the fact that the standard deviation of a mean decreases as the number of observations n increases. Together with the central limit theorem, this supports three general statements that help us understand a wide variety of random phenomena:

- Averages are less variable than individual observations.
- Averages are more normal than individual observations.
- Averages of large samples are less variable and more normal than averages of smaller samples.

The *Central Limit Theorem* applet (see Applet Exercise 4, page 399) enables you to watch the central limit theorem in action. You can select a distribution that is strongly skewed, not at all normal. As you increase the size of the sample, the distribution of the mean $\bar{x}$ gets closer and closer to the normal shape.

EXAMPLE 24 Heights of Young Women

Christopher Futcher/Hemera/Getty Images

The distribution of heights of young adult women is approximately normal, with mean 64.5 inches and standard deviation 2.5 inches. This normal distribution describes the population of young women. It is also the probability model for choosing one woman at random from this population and measuring her height. For example, the 68–95–99.7 rule says that the probability is 0.95 that a randomly chosen woman is between 59.5 and 69.5 inches tall.

Now choose an SRS of 25 young women at random and take the mean $\bar{x}$ of their heights. The mean $\bar{x}$ varies in repeated samples; the pattern of variation is the sampling distribution of $\bar{x}$. The sampling distribution has the same center ($\mu = 64.5$ inches) as the population of young women. The standard deviation of the sampling distribution of $\bar{x}$ is

$$\frac{\sigma}{\sqrt{n}} = \frac{2.5}{\sqrt{25}} = \frac{2.5}{5} = 0.5 \text{ inches}$$

The standard deviation σ describes the variation when we measure many *individual* women. The standard deviation $\sigma/\sqrt{n}$ of the distribution of $\bar{x}$ describes the variation in the average heights of *samples* of women when we take many samples. The average height is less variable than individual heights.

Figure 8.24 compares the two distributions: Both are normal and both have the same mean, but the average height $\bar{x}$ of 25 randomly chosen women has much less variability. For example, the 68–95–99.7 rule says that 95% of all averages $\bar{x}$ lie between 63.5 and 65.5 inches because two of $\bar{x}$'s standard deviations make 1 inch. This 2-inch span is just one-fifth as wide as the 10-inch span that catches the middle 95% of heights for individual women.

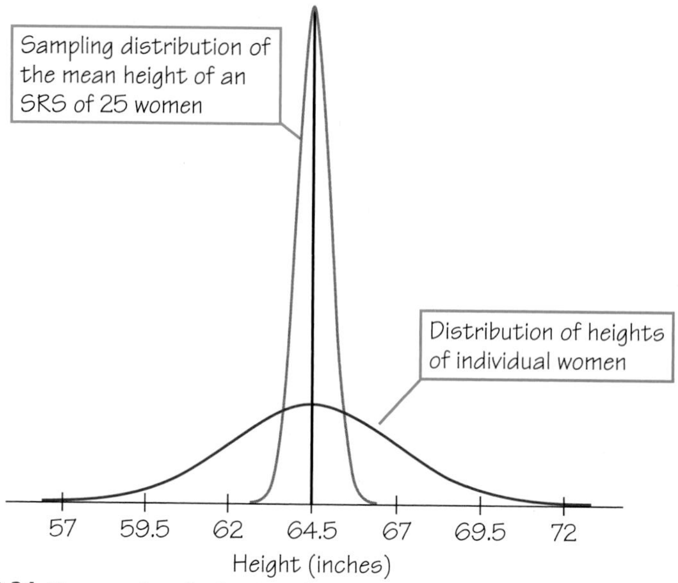

Figure 8.24 The sampling distribution of the average height of an SRS of 25 women has the same center (mean) as the distribution of individual heights but has much less variability because $\frac{2.5}{\sqrt{25}} < 2.5$.

The central limit theorem says that in large samples, the sample mean $\bar{x}$ is approximately normal. In Figure 8.24, we show a normal curve for $\bar{x}$ even though the sample size of 25 is not very large. Is that acceptable? How large a sample is needed for the central limit theorem to work depends on how far from a normal curve the model we start with is. The closer to normality we start, the quicker the distribution of the sample mean becomes normal. In fact, if individual observations follow a normal curve, the sampling distribution of $\bar{x}$ is exactly normal for any sample size. So Figure 8.24 is accurate. The central limit theorem is a striking result because as n gets large, it works for *any* model we may start with, no matter how far it is from normal—as you will see in Example 25.

EXAMPLE 25 ➔ Lou Gets Entertainment

Return to Example 22 (page 378) and to Lou, who bets on red at the roulette wheel. Figure 8.25 shows the probability model and corresponding probability histogram for Lou's favorite bet.

The probability model in Figure 8.25 is discrete, with just two possible outcomes: win $1 or lose $1. Yet the central limit theorem says that the average outcome of many bets follows a normal curve. Lou is a habitual gambler who places 50 bets of $1 on red almost every night. Because we know the probability model for a bet on red, we can simulate Lou's experience over many nights at the roulette wheel. The histogram in Figure 8.26, made from a simulation of 1000 nights, shows Lou's average winnings per bet, $\bar{x}$, from $n = 50$ bets.

As the central limit theorem says, the distribution looks normal and all we need to completely specify the normal curve shown in Figure 8.26 is its mean and standard deviation. For that we return to the distribution of outcomes for one bet on red in Figure 8.25. From Example 22 (page 378) and Self Check 16 (page 380) we know that

Outcome for Lou, x	Wins \$1	Loses \$1
x	1	−1
Probability	$\frac{18}{38} \approx 0.474$	$\frac{20}{38} \approx 0.526$

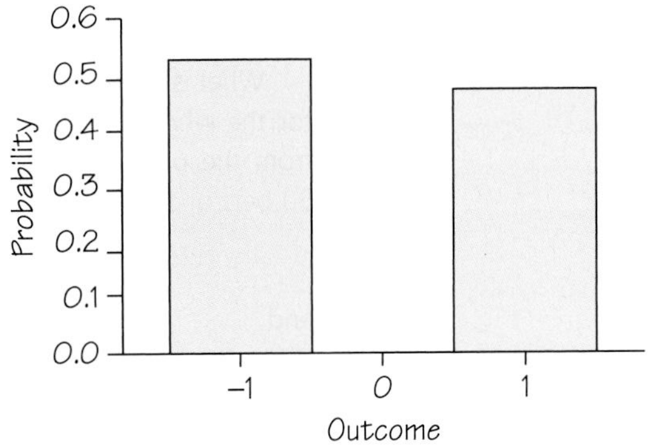

Figure 8.25 The probability model and probability histogram for betting on red at the roulette wheel.

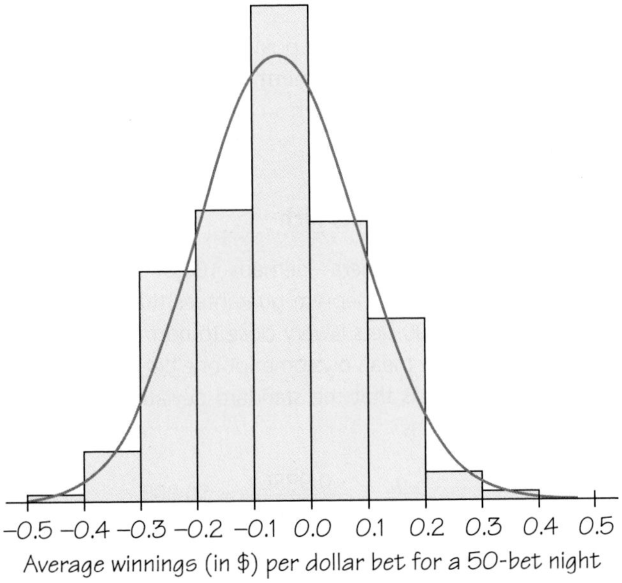

Figure 8.26 (Approximately normal) histogram of a gambler's winnings from a simulation of 1000 nights, where each night had 50 bets on red (or black) in roulette.

the mean and standard deviation are $\mu = -\$0.053$ and $\sigma = \$0.9986$, respectively. Next, we use the information from the central limit theorem to specify the mean and standard deviation of the average outcome, $\bar{x}$, from $n = 50$ bets.

$$\text{mean} = \mu \approx -\$0.053$$

$$\text{standard deviation} = \frac{\sigma}{\sqrt{n}} = \frac{\$0.9986}{\sqrt{50}} \approx \$0.141$$

Now that we have completely specified the approximate distribution of $\bar{x}$, we can apply the 99.7 part of the 68–95–99.7 rule from Section 5.9 (page 216): Almost all average nightly winnings per bet will fall within 3 standard deviations of the mean, that is, between

$$-\$0.053 - (3)(\$0.141) = -\$0.476$$

and

$$-\$0.053 + (3)(\$0.141) = \$0.370$$

This gives us a 99.7% confidence interval for Lou's mean winnings per bet, μ: between $-\$0.476$ and $\$0.370$.

What is more interesting to Lou is not average winnings per bet, but total winnings for the whole 50-bet night. To find this out, Lou can simply multiply both endpoints from the preceding 99.7% confidence interval by 50. So Lou's total winnings after 50 bets of \$1 each will almost surely fall between

$$(50)(-\$0.476) = -\$23.80$$

and

$$(50)(\$0.370) = \$18.50$$

Each night, Lou may win as much as \$18.50 or lose as much as \$23.80. Note that he will usually lose more on a bad night than he will win on a good night. Some people find gambling exciting because the outcome, even after an evening of bets, is uncertain. It is possible to beat the odds and walk away a winner. It's all a matter of luck.

The casino, however, is in a different position than individual gamblers such as Lou. It doesn't want rollercoaster excitement, just steady income. As we'll see from Example 26, that is exactly what they get.

EXAMPLE 26 The Casino Gets Rich

The casino bets with all its customers—perhaps 100,000 individual red/black roulette bets in a week. The central limit theorem guarantees that the distribution of average customer winnings on 100,000 bets is very close to normal. The mean, from the gambler's point of view, is still the mean outcome for one bet, $-\$0.053$, a loss of 5.3 cents per dollar bet. The key point is that the standard deviation is much smaller when we average over 100,000 bets. It is

$$\frac{\sigma}{\sqrt{n}} = \frac{0.9986}{\sqrt{100{,}000}} = \$0.003$$

Here is what the 99.7% confidence interval estimate of the average result looks like after 100,000 bets:

$$\begin{aligned} &\text{mean} \pm 3 \text{ standard deviations} \\ &= -\$0.053 \pm (3)(\$0.003) \\ &= -\$0.053 \pm \$0.009, \text{ which generates an interval from } -\$0.062 \text{ to } -\$0.044. \end{aligned}$$

Because the casino covers so many bets, the standard deviation of the average winnings per bet becomes very small. Not only is the mean negative, but the entire 99.7% confidence interval is also in the negative region, so the total result is virtually certain to be in the casino's favor. The gamblers' losses and the casino's winnings are almost certain to average between 4.4 and 6.2 cents for every dollar bet.

The gamblers who collectively place those 100,000 bets will lose money. The probable window of their losses is

$$(100{,}000)(-\$0.062) = -\$6200 \quad \text{to} \quad (100{,}000)(-\$0.044) = -\$4400$$

The gamblers are almost certain to lose—and the casino is almost certain to collect—between \$4400 and \$6200 on those 100,000 bets. What's more, the interval of average outcomes continues to narrow as still more bets are made. That is how a casino can make a business out of gambling.

Review Vocabulary

Addition rule (general) The probability that one event or the other occurs is the sum of their individual probabilities minus the probability of any overlap they have. (p. 351)

Central limit theorem The average of many independent random outcomes is approximately normally distributed. When we average n independent repetitions of the same random phenomenon, the resulting distribution of outcomes has mean equal to the mean outcome of a single trial and standard deviation equal to $1/\sqrt{n}$ times the standard deviation of a single trial. (p. 380)

Combination An unordered collection of k items chosen (without allowing repetition) from a set of n distinct items. (p. 367)

Combinatorics The branch of mathematics that studies how to count and choose elements. (p. 363)

Complement of an event The complement of an event A is the event "A does not occur," which is denoted A^C. (p. 350)

Complement rule The probability that an event does not occur is 1 minus the probability that the event does occur: $P(A^C) = 1 - P(A)$. (p. 350)

Conditional probability Written as $P(A|B)$, this can be computed by dividing the probability that both events occur by the probability that B occurs. (pp. 356, 357)

Continuous probability model A probability model that assigns probabilities to events as areas under a density curve. (p. 372)

Density curve A curve that is always on or above the horizontal axis and has area exactly 1 underneath it. A density curve describes a continuous probability model. (pp. 371, 372)

Discrete probability model A probability model that assigns probabilities to each of a countable number of possible outcomes. (p. 361)

Disjoint (mutually exclusive) events Events that have no outcomes in common. (p. 350)

Event A collection of possible outcomes of a random phenomenon; a subset of the sample space. (p. 346)

Factorial The product of the first n positive integers is n factorial, denoted as $n!$ (pp. 365, 366)

Fundamental principle of counting A multiplicative method for counting outcomes of multistage processes. (p. 363)

Independent events Events that do not influence each other's probability of occurring. Two events A and B are independent if $P(A) = P(A|B)$. (p. 354)

Law of large numbers As a random phenomenon is repeated many times, the mean $\bar{x}$ of the observed outcomes approaches the mean μ of the probability model. (p. 378)

Mean of a discrete probability model The average outcome of a random phenomenon with numerical values. When possible values $x_1, x_2, \ldots, x_k$ have probabilities $p_1, p_2, \ldots, p_k$, the mean is the average of the outcomes weighted by their probabilities, $\mu = x_1p_1 + x_2p_2 + \cdots + x_kp_k$. Also called *expected value*. (p. 375)

Multiplication rule for independent events If two events are independent, then the probability that one event and the other both occur is the product of their individual probabilities: $P(A \text{ and } B) = P(A) \times P(B)$, when A and B are independent events. (p. 356)

Permutation An ordered arrangement of k items chosen (without allowing repetition) from a set of n distinct items. (pp. 364, 365)

Probability A number between 0 and 1 that gives the long-run proportion of repetitions of a random phenomenon on which an event will occur. (pp. 343, 344)

Probability histogram A histogram that displays a discrete probability model when the outcomes are numerical. The height of each bar is the probability of the event at the base of the bar. (p. 353)

Probability model A sample space S together with an assignment of probabilities to events. The two main types of probability models are *discrete* and *continuous*. (pp. 345, 346)

Random A phenomenon or trial is random if it is uncertain what the next outcome will be, but each outcome nonetheless tends to occur in a fixed proportion of a very long sequence of repetitions. These long-run proportions are the probabilities of the outcomes. (p. 343)

Sample space The set of all possible (simplest) outcomes of a random phenomenon. If the outcomes in a sample space can be listed, the sample space is discrete. (p. 345)

Standard deviation of a discrete probability model A measure of the variability of a probability model. When the possible values $x_1, x_2, \ldots, x_k$ have probabilities $p_1, p_2, \ldots, p_k$, the standard deviation is the square root of the average (weighted by probabilities) of the squared deviations from the mean:

$$\sigma = \sqrt{(x_1 - \mu)^2 p_1 + (x_2 - \mu)^2 p_2 + \cdots + (x_k - \mu)^2 p_k}$$ (p. 379)

Self Check Answers

1.

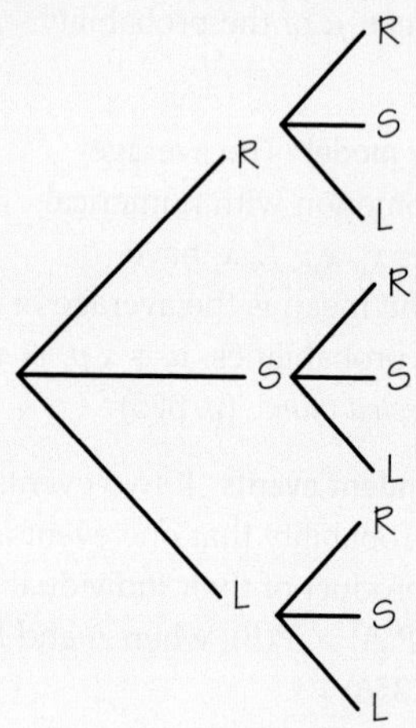

$S = \{RR, RS, RL, SR, SS, SL, LR, LS, LL\}$

2. Following the diagonal directly above the longest rising diagonal shows five possibilities for a sum of 6. The probability is 5/36.

3. (a) Let A = worker uses public transportation and B = worker walks. $P(A \text{ or } B) = P(A) + P(B) = 0.050 + 0.028 = 0.078$.

(b) Let A = worker drives to work; $P(A^c) = 1 - P(A) = 1 - 0.860 = 0.140$.

4. $P(B) = \frac{1}{6}$. Next, we find $P(B|A)$. Event A, shown in the blue rectangle, has six outcomes, each of which is equally likely. Given A has occurred, there is only one way to get a sum of 7 (red = 1 and green = 6). $P(B|A) = \frac{1}{6} = P(B)$. Therefore, A and B are independent.

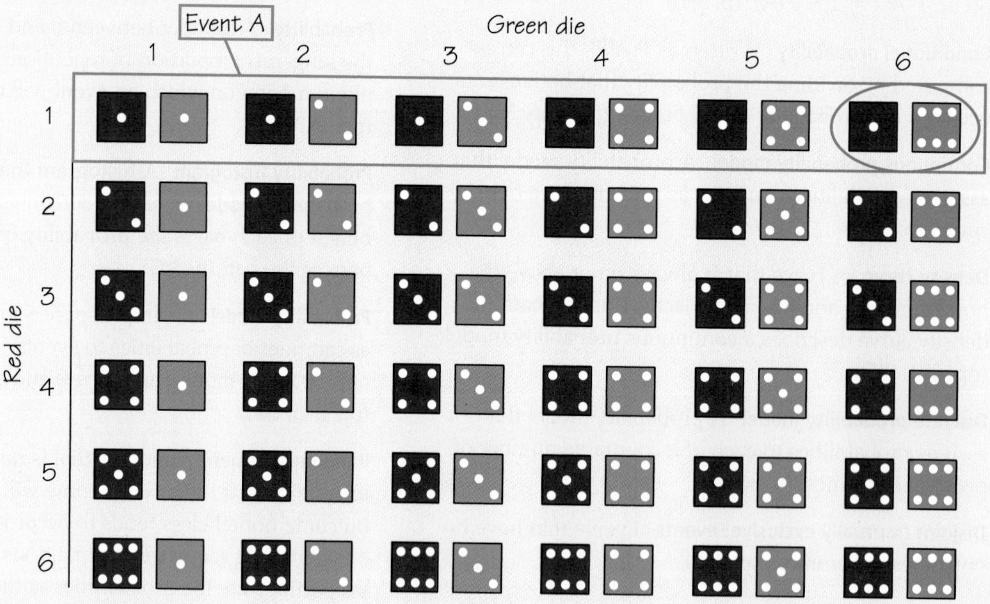

5.

Outcome	Positive test	Negative test
Conditional probability given no arsenic	1 − 0.86 = 0.14	0.86

6. $P(A \text{ and } +) = (0.5)(0.97) = 0.485$

$$P(+) = P(A \text{ and } +) + P(A^C \text{ and } +)$$

$$= 0.485 + (0.5)(0.14) = 0.555$$

$$P(A|+) = \frac{P(A \text{ and } +)}{P(+)} = \frac{0.485}{0.555} \approx 0.874$$

Should be worried; the chance that arsenic is in the water, given the test was positive, is high.

7. (a) P(rolls doubles within first three tries)

$$= \frac{1}{6} + \frac{5}{6^2} + \frac{5^2}{6^3} \approx 0.42$$

(b)

Number of rolls until sum 7	1	2	3	4	5	. . .	n	. . .
Probability	$\frac{5}{12}$	$\frac{7 \cdot 5}{12^2}$	$\frac{7^2 \cdot 5}{12^3}$	$\frac{7^3 \cdot 5}{12^4}$	$\frac{7^4 \cdot 5}{12^5}$	. . .	$\frac{7^{n-1} \cdot 5}{12^n}$	. . .

(c) This rule makes it less likely that players will spend a long time in jail. P(getting out of jail within first three rolls) $= \frac{1385}{1728} \approx 0.80$, which is nearly twice the probability under the Donovans' old rules.

8. (a) Using the table, the first digits on the invoice numbers are:

6, 8, 4, 1, 7, 3, 5, 1, 3, 1, 5, 5, 2, 9, 7, 2, 7, 6, 5, 8

The proportion of numbers assigned a first digit of 1, 2, or 3 is 0.35.

(b) The probability of observing a 1, 2, or 3, according to Benford's law is 0.602. That is quite a bit higher than the answer to part (a); hence our fraud is likely to be detected.

9. (a) There are (27)(26)(25)(24)(23) = (421,200)(23) = 9,687,600 possible five-song playlists.

(b) Let A be the playlists that include "The Ballad of John and Yoko." This song could be the first, second, third, fourth, or fifth song on a playlist. If it is the first song, then there are $26 \times 25 \times 24 \times 23 = 358{,}800$ ways to complete the playlist by choosing four songs in order from the remaining 26 songs. The same is true if "The Ballad of John and Yoko" is the second, third, fourth, or fifth song. Therefore, the number of outcomes in A is $5 \times 358{,}800 = 1{,}794{,}000$ and

$$P(A) = \frac{1{,}794{,}000}{9{,}687{,}600} \approx 0.185.$$

10. $6! = 6 \times 5 \times 4 \times 3 \times 2 \times 1 = 720$

11. (a) Combination: ${}_{15}C_3 = \frac{15!}{3!(15-3)!} = \frac{(15)(14)(13)}{(3)(2)}$
$= 455$

(b) Permutation: ${}_{15}P_3 = \frac{15!}{(15-3)!} = (15)(14)(13)$
$= 2730$

12. ${}_{30}P_7 = 10{,}260{,}432{,}000$; $7! = 5{,}040$; ${}_{30}C_7 = 2{,}035{,}800$

13. (a) The graph of the density curve (a) along with shaded region corresponding to part (b) appears below.

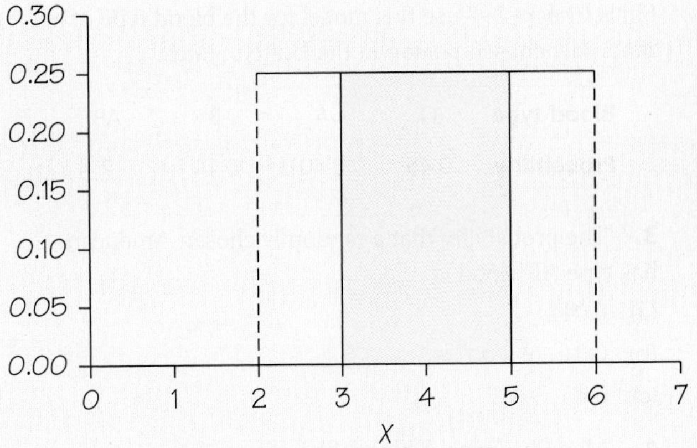

(b) $P(3 \le X \le 5) = (5-3)(0.25) = 0.5$

14. Using products from the table below:

$$\mu = 0 + 0.26 + 0.30 + 0.21 + 0.12 = 0.89$$

Number who call in sick, x	0	1	2	3	4
Probability, p	0.49	0.26	0.15	0.07	0.03
Product, $x \cdot p$	0	0.26	0.30	0.21	0.12

15. (a) $P(\text{win}) = 12/38 \approx 0.3158$; the gambler wins 31.58% of the time over many, many games. $P(\text{loss}) = 1 - P(\text{win}) \approx 0.6842$; the gambler loses 68.42% of the time over many, many games.

(b) $(-1)(0.6842) + (2)(0.3158) = -0.0526$; the house edge is exactly the same as for betting on red.

16. $\sigma = \sqrt{(1-(-0.053))^2 \frac{18}{38} + (-1-(-0.053))^2 \frac{20}{38}}$

$= \sqrt{(1.053)^2 \frac{18}{38} + (-0.947)^2 \frac{20}{38}}$

$= \sqrt{0.9972} \approx \$0.9986$

Skills Check

1. You read in a book on poker that the probability of being dealt three of a kind in a five-card poker hand is $\frac{1}{50}$. What does this mean?

(a) If you deal thousands of poker hands, the fraction of them that contain three of a kind will be very close to $\frac{1}{50}$.

(b) If you deal 50 poker hands, exactly one of them will contain three of a kind.

(c) If you deal 10,000 poker hands, exactly 200 of them will contain three of a kind.

2. If two coins are flipped and then a die is rolled, the sample space would have ________ different outcomes.

Skills Checks 3–7 use this model for the blood type of a randomly chosen person in the United States:

Blood type	O	A	B	AB
Probability	0.45	0.40	0.11	?

3. The probability that a randomly chosen American has type AB blood is

(a) 0.044.

(b) 0.04.

(c) 0.4.

4. María has type A blood. She can safely receive blood transfusions from people with blood types O or A. The probability that a randomly chosen American can donate blood to María is ________.

5. What is the probability that a randomly chosen American does not have type O blood?

(a) 0.55

(b) 0.45

(c) 0.04

6. A random sample of two Americans is selected. What is the probability that both will have type O blood?

(a) 0.25

(b) 0.90

(c) 0.2025

7. A random sample of two Americans is selected. The probability that neither will have type O blood is ________.

8. Figure 8.7 (page 347) shows the 36 possible outcomes for rolling two dice. These outcomes are equally likely. A "soft 4" is a roll of 1 on one die and 3 on the other. The probability of rolling a soft 4 is ________.

9. A discrete probability model has

(a) only two outcomes.

(b) equally likely outcomes.

(c) a countable number of outcomes.

10. According to Benford's law, the most likely first (leftmost) digit of a number from financial data is ________.

11. In a table of random digits such as Table 7.1 (page 298), each digit is equally likely to be any of 0, 1, 2, 3, 4, 5, 6, 7, 8, or 9. What is the probability that a digit in the table is a 0?

(a) 1/9

(b) 1/10

(c) 9/10

12. In a table of random digits such as Table 7.1 (page 298), each digit is equally likely to be any of 0, 1, 2, 3, 4, 5, 6, 7, 8, or 9. The probability that a digit in the table is 7 or greater is ________.

13. Toward the end of a game of Scrabble, you hold five tiles with the letters A, E, P, R, and S. In how many ways can you arrange these five letters (whether or not they form actual words)?

(a) 5

(b) $5 \times 4 \times 3 \times 2 \times 1 = 120$

(c) $5 \times 5 \times 5 \times 5 \times 5 = 3125$

14. Toward the end of a game of Scrabble, you hold the letters D, O, G, and Q. You can choose three of these four letters and arrange them in order in ________ different ways, assuming that you are not trying to form actual words.

15. A 52-card deck contains 13 cards from each of the four suits: clubs ♣, diamonds ♦, hearts ♥, and spades ♠. You deal four cards without replacement from a well-shuffled deck, so that you are equally likely to deal any four cards. What is the probability that *all* four cards are clubs?

(a) $\frac{1}{4}$, because $\frac{1}{4}$ of the cards are clubs

(b) $\frac{13}{52} \times \frac{12}{51} \times \frac{11}{50} \times \frac{10}{49} = 0.0026$

(c) $\frac{13}{52} \times \frac{12}{52} \times \frac{11}{52} \times \frac{10}{52} = 0.0023$

16. You deal four cards as in the previous exercise. The probability that you deal *no* clubs is ________.

17. Figure 5.19 (page 209) shows that the normal distribution with mean $\mu = 6.8$ and standard deviation $\sigma = 1.6$ is a good description of the Iowa Test vocabulary scores of seventh-grade students in Gary, Indiana. The probability that a randomly chosen student has a score higher than 8.4 is

(a) 0.68.

(b) 0.32.

(c) 0.16.

18. Figure 8.18 (page 372) shows the density curve of a continuous probability model for choosing a number at random between 0 and 1 inclusive. The probability that the number chosen is less than or equal to 0.4 is __________.

19. In Figure 8.18 (page 372), the probability that x is greater than 0.65 is

(a) 0.65.

(b) 0.35.

(c) 1.

20. The total area under a density curve is __________.

21. Annual returns on the more than 5000 common stocks available to investors vary a lot. In a recent year, the mean return was 8.3% and the standard deviation of returns was 28.5%. The law of large numbers says that

(a) you can get an average return higher than the mean 8.3% by investing in a large number of stocks.

(b) as you invest in more and more stocks chosen at random, your average return on these stocks gets ever closer to 8.3%.

(c) if you invest in a large number of stocks chosen at random, your average return will have approximately a normal distribution.

22. Figure 8.18 (page 372) shows the density curve of a continuous probability model for choosing a number at random between 0 and 1. The mean of this model is

(a) 0.5 because the curve is symmetric.

(b) 1 because there is area 1 under the curve.

(c) impossible to figure out at this point—this requires advanced mathematics.

23. According to the law of large numbers, as the random phenomenon is repeated a large number of times, $\bar{x}$ gets closer and closer to __________.

24. The expected value is the __________ value.

(a) median

(b) mean

(c) mode

25. The mean payoff of a $\frac{1}{10}$ chance of winning \$500 (with a $\frac{9}{10}$ chance of winning \$0) is __________.

26. The density curve of a continuous probability model would balance on the __________.

(a) mode

(b) median

(c) mean

27. Self Check 14 (page 377) gave a probability model for the number of employees of a small business who call in sick each day. Given $\mu = 0.89$, determine the standard deviation.

28. If $\mu = 25$, $\sigma = 16$, and $n = 64$, the standard deviation of the sampling distribution of $\bar{x}$ is __________.

29. Scores on the SAT Reasoning Mathematics college entrance test for the class of 2010 were roughly normal, with mean 516 and standard deviation 116. You take an SRS of 100 students and average their SAT scores. If you do this many times, the mean of the average scores that you get from all those samples would be __________.

30. Referring to Skills Check 29, the standard deviation of the average scores that you get from all those samples would be __________.

31. The number of hours that a light bulb burns before failing varies from bulb to bulb. The distribution of burnout times is strongly skewed to the right. The central limit theorem says that

(a) as we look at more and more bulbs, their average burnout times get ever closer to the mean μ for all bulbs of this type.

(b) the average burnout times of a large number of bulbs have a distribution of the same shape (strongly skewed) as the distribution for individual bulbs.

(c) the average burnout times of a large number of bulbs have a distribution that is close to normal.

Chapter 8 Exercises

Challenge Discussion

8.1 Random Phenomena and Probability

1. Identify three random phenomena that occur in your life.

2. Random phenomena can't be predicted for certain in the short term but exhibit regular patterns in the long term. Which of the data sets in parts (a) through (d) do not appear to be from the random phenomena of coin tossing? Explain.

(a) H T H T H T H T H T H T H T H T H T H T H T H T

(b) T T T T H H H H T T T T H H H H T T T T

(c) H H T H H T T H H H H T H T H H T T T T H H H H

(d) T H H T T H H T T H H T T H H T T H H T T H H T

3. (a) Hold a penny upright on its edge under your forefinger on a hard surface, and then snap it with your other forefinger so that it spins for some time before falling. Based on 50 spins, estimate the probability of heads.

(b) Toss a thumbtack with a gently curved back (similar to Figure 8.4, page 344) on a hard surface 100 times. (To speed up the process, toss 10 at a time.) How many times did it land with the point up? What is the approximate probability of landing point up?

4. Suppose there is a forecast of a 70% chance of rain. Using the "long run" interpretation of the definition of probability, if it does not rain the next day, is it appropriate to say that the forecaster was "wrong"? Explain.

5. The table of random digits (Table 7.1, page 298) was produced by a random mechanism that gives each digit probability 0.1 of being a 0. What proportion of the first five lines in the table are 0s? This proportion is an estimate of the true probability, which in this case is known to be 0.1.

In Exercises 6–8, describe a reasonable sample space *S* for the random phenomena mentioned. In some cases, you must use judgment to choose a reasonable *S*.

6. A student is randomly chosen from a large mathematics class.

(a) Ask if the subject is male or female.

(b) Ask if the student lives on or off campus.

(c) Ask how tall the student is.

(d) Ask how much money the student is carrying in change.

7. A basketball player shoots four free throws.

(a) You record the sequence of hits and misses.

(b) You record the number of shots she makes.

8. A randomly chosen subject (participant) arrives for a study of exercise and fitness.

(a) The subject is either female or male.

(b) After 10 minutes on an exercise bicycle, you ask the subject to rate his or her effort on the Rate of Perceived Exertion (RPE) scale. The RPE goes in whole-number steps from 6 (no exertion at all) to 20 (maximal exertion).

(c) You also measure the subject's maximum heart rate (beats per minute).

9. Consider flipping a dime, nickel, and penny.

(a) Why would a tree diagram be a more convenient way than a table to represent the sample space?

(b) Make a tree diagram, and then use it to write out the sample space.

(c) Use the diagram to find the probability that at least one of the three coins lands on heads.

10. At the start of class, an instructor gives a brief quiz on assigned reading. The quiz consists of two questions: a true-or-false question and a multiple-choice question with choices (a), (b), or (c). A student who hasn't done the assignment guesses at both questions. Use a table to represent the sample space of possible answers for this two-question quiz, and then list the elements in the sample space.

8.2 Basic Rules of Probability

11. Probability is a measure of how likely an event is to occur. Match one of the probabilities that follow with each statement about an event. (The probability is usually a much more exact measure of likelihood than is the verbal statement.)

0, 0.01, 0.3, 0.6, 0.99, 1

(a) This event is impossible. It can never occur.

(b) This event is certain. It will occur on every trial of the random phenomenon.

(c) This event is very unlikely, but it will occur once in a while in a long sequence of trials.

(d) This event will occur somewhat more often than not.

12. Based on data from the American Community Survey, a high percentage of U.S. workers drive alone to work (see Tables 8.1 and 8.2, pages 348 and 349). A 2009 survey asked respondents about the number of vehicles they had access to within their household. The results appear below.

TABLE 8.9 Results from a 2009 American Community Survey

Workers and Vehicles	Frequency
Workers in households with zero vehicles	5,138,372
Workers in households exceeds vehicles	12,357,761
Workers in households equals vehicles	66,463,188
Vehicles exceed workers in households	67,413,776
Total	151,370,000

(a) Assume that the sample fairly represents workers in the United States. Use the sample proportions in each category as probabilities and create a probability model for Workers and Vehicles (round probabilities to three decimal places).

(b) Verify that Rule 2 is satisfied.

(c) Find the probability that there are at least as many vehicles as there are workers in the household. Which probability rule did you use to answer this question?

(d) Find the probability that the workers in the household have access to at least one vehicle. Which probability rule did you use to answer this question?

13. In Spotlight 8.1 (page 345), we noted that a question posed to Blaise Pascal in 1654 by an amateur mathematician launched the formal study of probability. Here's a simplified version of this "Problem of Points." Suppose two players are playing a coin flip game where "heads" earns Player A one point and "tails" earns Player B one point. The winner is the first player to reach a total of four points. The game is interrupted with Player A ahead by a score of 3 to 2. Based on the sample space of possible ways that the game can be finished, what would be a fair division of the jackpot money between Players A and B?

14. Choose a young adult (aged 25 to 34 years) at random. The probability is 0.12 that the person chosen did not complete high school, 0.31 that the person has a high school diploma but no further education, and 0.29 that the person has at least a bachelor's degree.

(a) What must be the probability that a randomly chosen young adult has some education beyond high school but does not have a bachelor's degree?

(b) What is the probability that a randomly chosen young adult has at least a high school education?

15. While it is a less common way (than probability) of expressing likelihood and does not follow the five basic Probability Rules, statements of "odds" are often encountered in gambling contexts. The odds against an event A happening are the ratio $P(A^c)/P(A)$. In a horse race, suppose the odds against a particular horse winning the race are 3:2 (or "3 to 2"); in other words, $P(A^c)/P(A) = 3/2$. What is the probability that the horse wins?

Algebra Review Appendix
Solving for One Variable in Terms of Another

8.3 Rules of Probability: Independent and Dependent Events

16. The Punnett square is a diagram that biologists use to determine the probability of offspring having certain genetic makeup. Suppose that B represents the gene for brown eyes and b represents the gene for blue eyes. In genetics, capital letters refer to dominant traits, so a person receiving both B and b generally has brown eyes. This diagram shows the possibilities for the child of two Bb parents. Each parent gives the child one of its two genes with equal probability. What is the probability that this child will receive the genetic makeup for brown eyes? Discuss how this relates to Example 2 (page 346).

	Mother gives B	Mother gives b
Father gives B	BB	Bb
Father gives b	Bb	bb

Diane MacDonald/ Getty Images

17. Suppose two events A and B are mutually exclusive and both have positive probabilities of occurring. Can A and B also be independent events? Explain.

18. Given $P(A) = 0.3$, $P(B) = 0.6$, and $P(A|B) = 0.2$, determine the following probabilities.

(a) $P(A \text{ and } B)$

(b) $P(A \text{ or } B)$

19. What is the probability that Laurie rolls a pair of dice and gets an even sum in each of her first three rolls in the game of Monopoly?

20. If you are in the championship round of a tournament and you are better than your opponent, are you better off having the championship determined by playing a single game or by playing a two-out-of-three-game series? You might try intuition, simulation, or a tree diagram.

21. Return to the probability model that you created for Workers and Vehicles in Exercise 12. Suppose a random sample of two U.S. workers is chosen.

(a) What is the probability that they both live in households where the number of vehicles exceeds the number of workers in the household?

(b) What is the probability that they both live in households where the number of vehicles does not exceed the number of workers in the household?

Exercises 22–24 are based on data from the 2013 March Supplement of the U.S. Census Bureau's Current Population Survey.

22. According to the 2013 March Supplement Survey, 22.3% of U.S. households are in the Midwest. In addition, 9.6% of households earn $75,000 or more per year and are located in the Midwest. Determine the probability that a randomly selected American household earns $75,000 or more per year, given that the house is located in the Midwest.

23. According to the 2013 March Supplement Survey, 32.0% of U.S. households are in the South. In addition, 7.9% of households earn $100,000 or more per year and are located in the South. Determine the probability that a randomly selected Southern household earns $100,000 or more per year.

24. According to the 2013 March Supplement Survey, 18.8% of U.S. households are in the Northeast. In addition, 21.2% of Northeastern households earn $75,000 or more per year. What percentage of households earn $75,000 or more per year and are located in the Northeast?

25. Each year, the study *Monitoring the Future: A Continuing Study of American Youth* surveys twelfth-grade students on a wide range of topics related to behaviors, attitudes, and values. One of the survey questions asks students to identify their sex. Another question asks students to rate their intelligence compared with others their age. Assume that the sample used in this survey is representative of twelfth-grade students. Here are some of the results from the more than 13,000 students who participated in the survey: 49.8% were male; 50.2% were female; 60.3% of the females rated their intelligence as above average; 68.6% of the males rated their intelligence as above average. Since the number of participants in the survey was large, the survey percentages should be close to the probabilities for the population of twelfth-grade students.

(a) Use probability notation to express each of the percentages above as probabilities.

Find the following probabilities for parts (b) through (e), rounding answers to three decimal places.

(b) P(female and above average)

(c) P(male and above average)

(d) P(above average)

(e) P(female | above average)

(f) Suppose a twelfth-grade student is selected at random. Does knowing that the student rated his or her own intelligence as above average increase, decrease, or have no effect on the probability that the student is female? Explain.

26. The concern over drug use among teens has prompted some schools to consider mandatory drug testing of its students. However, schools should be warned that drug tests, while generally reliable, are not perfect. Sometimes a person who does use drugs gets a negative result (a false negative). Sometimes a person who does not use drugs tests positive (a false positive). Companies that produce tests for drugs usually provide information on two characteristics of their test: p, the probability that their test correctly identifies a drug user, and r, the probability that it correctly reports the absence of drugs.

Suppose that a large high school is considering a mandatory drug testing program. The particular test under consideration correctly identifies 95% of the users ($p = 0.95$) and correctly reports the absence of drugs in 90% of the nonusers ($r = 0.90$).

(a) Suppose 25% of the students are drug users. Draw a tree diagram similar to Figure 8.15 in Example 10 (page 359). What percentage of the positive test results will be from users?

(b) Suppose 2% of the students use drugs. What percentage of positive tests will be from users? What does this imply about the rest of the positive tests?

8.4 Discrete Probability Models

27. Table 8.10 gives a probability model for total household income.

TABLE 8.10 Probability Model of Total Income of U.S. Households (based on data from the March 2013 Supplement, Current Population Survey)

Total Household Income	Probability
Under $25,000	0.174
$25,000 to $49,999	0.218
$50,000 to $74,999	0.185
$75,000 to $99,999	0.140
$100,000 or over	0.283

(a) Check to see whether the probability model in Table 8.10 is legitimate. Explain what you checked.

(b) What makes the probability model in Table 8.10 a discrete probability model?

(c) What is the probability that a randomly chosen household will have a total income less than $50,000?

(d) What is the probability that a randomly chosen household will have a total income less than $100,000?

(e) Suppose two U.S. households were randomly selected. What is the probability that both households will have a total income less than $100,000?

28. Role-playing games like Dungeons & Dragons use many different types of dice. One type of die has a tetrahedral (pyramidal) shape with four triangular faces (see Figure 8.27). Each triangular face has a number (1, 2, 3, or 4) next to each of its edges. Because the top of this die is not a face but a point, the way to read it is by the number at the top of the face that is visible when the die comes to rest. Suppose that the intelligence of a character is determined by rolling this four-sided die twice and adding 1 to the sum of the results.

Brittny/Shutterstock

Figure 8.27 Rolling a pair of tetrahedral dice: red = 1 and green = 3; intelligence of character = 4 + 1 = 5, for Exercise 28.

(a) Give a probability model for the character's intelligence. (Start with a display in the style of Figure 8.7 on page 347, adapted for the outcomes of the two rolls of the four-sided die. These outcomes are equally likely.)

(b) What is the probability that the character has intelligence 7 or higher?

29. North Carolina State University posts the grade distributions for its courses online. Students in Statistics 101 in a recent semester earned 21% As, 43% Bs, 30% Cs, 5% Ds, and 1% Fs. Here is the probability model for the grade of a randomly chosen Statistics 101 student.

Grade	0 (= F)	1 (= D)	2 (= C)	3 (= B)	4 (= A)
Probability	0.01	0.05	0.30	0.43	0.21

(a) Make a probability histogram for this model. Does it have the shape of a normal distribution?

(b) What is the probability that the student got a grade of B or better?

30. How do rented housing units differ from units occupied by their owners? Here are probability models for the number of rooms for owner-occupied units and renter-occupied units, according to the Census Bureau:

# of Rooms	**1**	**2**	**3**	**4**	**5**
Owned	0.000	0.001	0.014	0.099	0.238
Rented	0.011	0.027	0.229	0.348	0.224
# of Rooms	**6**	**7**	**8**	**9**	**10**
Owned	0.266	0.178	0.107	0.050	0.047
Rented	0.105	0.035	0.012	0.004	0.005

Make probability histograms of these two models, using the same scale. What are the most important differences between the models for owner-occupied and rented housing units?

31. In each of the following situations, state whether or not the given assignment of probabilities to individual outcomes is legitimate—that is, satisfies the rules of probability. If not, give specific reasons for your answer.

(a) Choose a college student at random and record gender and enrollment status: $P(\text{full-time female}) = 0.56$, $P(\text{part-time female}) = 0.24$, $P(\text{full-time male}) = 0.44$, $P(\text{part-time male}) = 0.17$.

(b) Choose a college student at random and record the season of that student's birth: $P(\text{spring}) = 0.39$, $P(\text{summer}) = 0.28$, $P(\text{fall}) = 0$, $P(\text{winter}) = 0.33$.

32. What is the probability that a housing unit has five or more rooms? Use the models in Exercise 30 to answer this question for both owner-occupied and rented units.

33. Balanced six-sided dice with altered labels can produce interesting distributions of outcomes. Construct the probability model (sample space and assignment of probabilities for each sum) for rolling the dice that is featured in Joseph Gallian's article "Weird Dice" in the February 1995 issue of *Math Horizons*. Instead of using the regular values {1, 2, 3, 4, 5, 6}, one die has the labels 1, 2, 2, 3, 3, 4, and the other die has the labels 1, 3, 4, 5, 6, 8. How does this model compare with the model for regular dice?

8.5 Equally Likely Outcomes

34. If you play the lottery, there are two possibilities—you could either win or not win. Explain whether or not this means that you have a 1 out of 2 chance (i.e., a 50% probability) of winning.

35. A party host gives a door prize to one guest chosen at random. There are 42 men and 48 women at the party. What is the probability that the prize goes to a woman?

36. At a party a cooler is filled with cans of drinks: 12 Cokes, 6 Diet Cokes, 4 lime seltzers, and 8 lemon seltzers. You grab a can from the cooler at random.

(a) What is the probability that you grabbed either a Coke or Diet Coke?

(b) Before you look at the can, your friend says, "I didn't know you drank seltzer." Given this information, what is the probability that you have grabbed your least favorite drink, a lime seltzer?

37. Suppose you have 10 books in a stack but you only have space for five books on your bookshelf.

(a) In how many different ways can you select five books from the stack and arrange them in the empty space on your bookshelf?

(b) How many different selections of five books can you make? (Don't worry about how you will arrange the books.)

38. Abby, Boaz, Carmen, Dani, and Eduardo work in a firm's public relations office. Their employer must choose two of them to attend a conference in Paris. To avoid unfairness, the choice will be made by drawing two names from a hat. (This is an SRS of size 2.)

(a) Write down the sample space of all possible choices of two of the five names.

(b) The random drawing makes all choices equally likely. What is the probability of each choice?

(c) What is the probability that Abby is chosen?

(d) What is the probability that neither of the two men (Boaz and Eduardo) is chosen?

39. You toss a balanced coin 10 times and write down the resulting sequence of heads and tails, such as HTTTHHTHHH.

(a) How many possible outcomes are there for the 10 tosses?

(b) What is the probability that your 10-toss sequence is either all heads or all tails?

40. In the Texas Hold 'Em style of poker, play begins with each player being dealt two cards face down. From a standard 52-card deck, how many possible 2-card hands could be dealt to you?

41. A computer assigns three-character log-in IDs that may contain the digits 0 to 9 as well as the letters *a* to *z*, with repeats allowed.

(a) What is the probability that your ID contains no *x*?

(b) What is the probability that your ID contains no digits?

42. Consider a typical combination lock on a locker or briefcase.

(a) If you ask for the three numbers in the combination needed to open the lock, and they are given to you in order from smallest to largest as 3–5–8, why is this not enough information to open the lock?

(b) What would be the probability that you could open the lock with one try?

(c) Is such a combination lock accurately named, or is it really a "permutation lock"?

43. You may have heard that a monkey hitting keys at random on a typewriter keyboard for an infinite amount of time could eventually type a particular chosen text, such as the complete works of Shakespeare. Let's focus on a monkey who just types the letters *a*, *p*, and *s* in random order in three-letter sequences.

(a) How many possible three-letter "words" can the monkey type using only these letters?

(b) Which of these are words in an English dictionary?

(c) What is the probability that the word the monkey typed is in an English dictionary?

44. Mozart composed a 16-bar Viennese minuet ("Musical Dice Game") in which bars 1 through 7 each have 11 choices, bar 8 has 2, bars 9 through 15 each have 11, and bar 16 has 1. How many possible versions of this minuet are there?

45. In poker, a royal flush is a five-card hand containing (in any order) an ace, king, queen, jack, and 10, all of the same suit.

(a) How many royal flush hands are possible?

(b) What is the number of five-card hands possible from a 52-card deck?

(c) What is the probability that five cards drawn at random from a 52-card deck will yield a royal flush?

46. The King James Version of the Old Testament has its 39 books canonized in a different order than the Hebrew Bible does. What mathematical expression would yield the number of possible orders of these 39 books? Is this number larger than you expected?

47. Use technology (see Spotlight 8.3, page 369) for this exercise. A university IT department receives a shipment of 30 printers—20 are inkjet printers and 10 are laser printers. A particular technician randomly chooses 7 of the printers to process (check that it works, tag the printer, etc.). What is the probability that exactly 4 of the printers are inkjets and 3 are laser printers (we'll call this event *A*)?

8.6 Continuous Probability Models

48. Books on reserve at a university library can be checked out for at most 2 hours. The density curve for the amount of time the book is checked out is the shaded triangle shown in Figure 8.28.

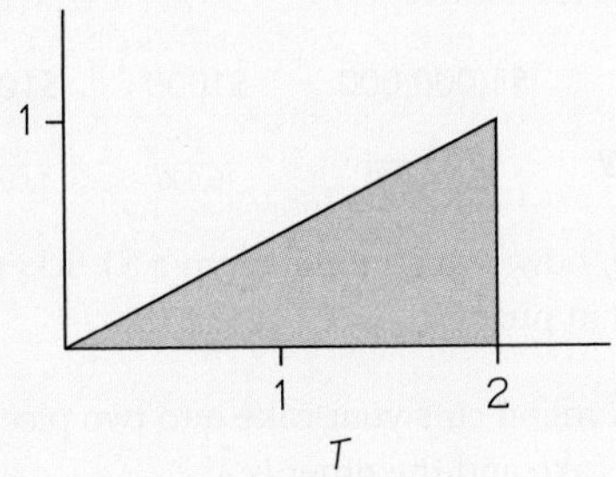

Figure 8.28 Amount of time T that a book is checked out from the reserve shelf, for Exercise 48.

(a) Explain why Figure 8.28 satisfies the definition of a density curve.

(b) What is the probability that the book is checked out for less than 1 hour?

(c) What is the probability that the book is checked out for more than 1 hour?

49. Generate two random real numbers between 0 and 1 and take their sum. The sum can take any value between 0 and 2. The density curve is the shaded triangle shown in Figure 8.29.

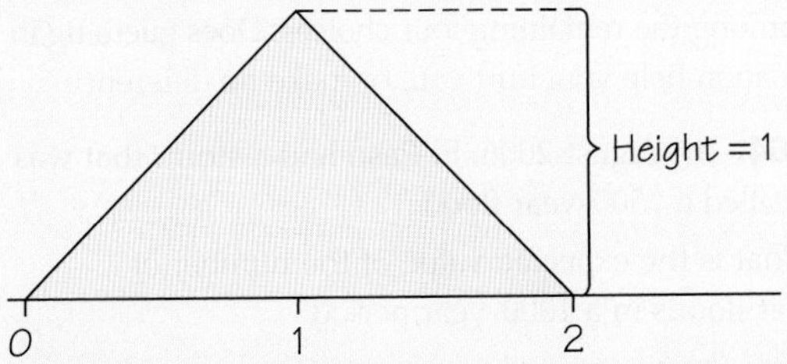

Figure 8.29 The density curve for the sum of two random numbers, for Exercise 49.

(a) Verify by geometry that the area under this curve is 1.

(b) What is the probability that the sum is less than 1? Sketch the density curve, shade the area that represents the probability, and then find that area.

(c) What is the probability that the sum is less than 0.5? Sketch the density curve, shade the area that represents the probability, and then find that area.

50. Suppose two data values are each rounded to the nearest whole number. Make a density curve for the sum of the two roundoff errors, assuming each error has a continuous uniform distribution. (Exercise 49 might help.)

51. On the TV show *The Price Is Right,* the "Range Game" involves a contestant being told that the suggested retail price of a prize lies between two numbers that are \$600 apart. The contestant has one chance to position a red window with a span of \$150 that will contain the price. On one episode, the price of a piano is between \$8900 and \$9500. If we assume a uniform continuous distribution (i.e., that all prices within the \$600 interval are equally likely), what is the probability that the contestant will be successful?

8.7 The Mean and Standard Deviation of a Probability Model

52. You have a campus errand that will take only 15 minutes. The only parking space anywhere nearby is a faculty-only space, which is checked by campus police about once every hour. If you're caught, the fine is \$25.

(a) Give the probability model for the money that you may or may not have to pay.

(b) What's the expected value of the money that you will pay for your unauthorized parking?

53. Exercise 29 gives a probability model for the grade of a randomly chosen student in Statistics 101 at North Carolina State University, using the 4-point scale. What is the mean grade in this course? What is the standard deviation of the grades?

54. In Exercise 28, you gave a probability model for the intelligence of a character in a role-playing game. What is the mean intelligence for these characters?

55. Exercise 30 gives probability models for the number of rooms in owner-occupied and rented housing units. Find the mean number of rooms for each type of housing. Make probability histograms for the two models and mark the mean on each histogram. You see that the means describe an important difference between the two models: Owner-occupied units tend to have more rooms.

56. Typographical and spelling errors can be either "nonword errors" or "word errors." A nonword error is not a real word, as when "the" is typed as "teh." A word error is a real word, but not the right word, as when "lose" is typed as "loose." When undergraduates write a 250-word essay (without checking spelling), the number of nonword errors has the following probability model:

Errors	0	1	2	3	4
Probability	0.1	0.2	0.3	0.3	0.1

The number of word errors has the following model:

Errors	0	1	2	3
Probability	0.4	0.3	0.2	0.1

(a) What is the mean number of nonword errors in an essay?

(b) What is the mean number of word errors in an essay?

(c) How does the difference between the means describe the difference between the two models?

57. Find the mean for the following probability models. Explain how you determined your answers.

(a) The continuous probability model in Exercise 49

(b) The probability model in Exercise 50

58. The idea of insurance is that we all face risks that are unlikely but carry a high cost. Think of a fire destroying your home. Insurance spreads the risk: We all pay a small amount, and the insurance policy pays a large amount to those few of us whose homes burn down. An insurance company looks at the records for millions of homeowners and sees that the mean loss from fire in a year is μ = \$250 per person. (The great majority of us have no loss, but a few lose their homes. The \$250 is the average loss.) The company plans to sell fire insurance for \$250 plus enough to cover its costs and profit. Explain clearly why it would be unwise to sell only 12 policies. Then explain why selling thousands of such policies is a safe business.

59. A company is considering offering an extended warranty on new washing machines. The company looks at past repair records and determines that there are two outcomes: an 85% probability of needing no repairs, and a 15% probability of needing a \$200 repair during the warranty period. To help the company set a price for an extended warranty on new washing machines, determine the mean outcome for this model. This would be the break-even price for a company selling the extended warranties. (The company, of course, will charge more than this in order to make a profit.)

60. An American roulette wheel has 38 slots numbered 0, 00, and 1 to 36. The ball is equally likely to come to rest in any of these slots when the wheel is spun. The slot numbers are laid out on a board on which gamblers place their bets. One column of numbers on the board contains multiples of 3—that is, 3, 6, 9, . . . , 36. Joe places a \$1 "column bet" that pays out \$3 (so he gains \$2) if any of these numbers comes up.

(a) What is the probability model for the outcome of one bet, taking into account the \$1 cost of a bet?

(b) What are the mean and standard deviation for this model?

(c) Joe has played roulette every day for years. What does the law of large numbers tell us about his results?

61. This table shows the prizes and respective probabilities for a lottery:

Net prize	\$1,000,000	\$1000	\$100	\$4
Probability	$\frac{1}{10{,}000{,}000}$	$\frac{1}{10{,}000}$	$\frac{1}{1{,}000}$	$\frac{3}{100}$

On average, how much money from a \$1 ticket comes back to you in prizes?

62. A friend cuts your cake into two pieces: one is $\frac{1}{3}$ of the cake and the other is $\frac{2}{3}$.

(a) If you flip a coin to decide which piece is yours, what is the expected value of the proportion of the original cake that you will get?

(b) More generally, what is the expected value of your share of the cake if the pieces' proportions are p and $1 - p$?

63. On five-choice questions on the SAT, you get 1 point for a correct answer and lose $\frac{1}{4}$ point for a wrong answer.

(a) Find the expected value of a completely random guess on such a question. Does guessing in this situation help you, hurt you, or make no difference?

(b) Suppose you eliminate one of the five choices as definitely not being the correct answer and then randomly guess among the remaining four choices. Does guessing in this situation help you, hurt you, or make no difference?

64. In August 2006, El Paso had a storm that was called a "500-year flood."

(a) What is the expected value of the number of 500-year floods in a 1000-year period?

(b) After the city moved to raise money to guard against such a major flood in the future, a city council representative was quoted as protesting, "But we still have 490 more years to deal with this." What false assumption was he making?

8.8 The Central Limit Theorem

65. Newly manufactured automobile radiators may have small leaks. Most have no leaks, but some have one, two, or more. The number of leaks in radiators made by one supplier has mean 0.15 and standard deviation 0.4. The distribution of the number of leaks cannot be normal because only whole-number counts are possible. The supplier ships 400 radiators per day to an auto assembly plant. Take $\bar{x}$ to be the mean number of leaks in these 400 radiators. Over several years of daily shipments, what interval of values will contain the middle 95% of the many $\bar{x}$ values?

AppStock/Shutterstock

66. The scores of eighth-grade students on the National Assessment of Educational Progress (NAEP) mathematics test in 2007 have a distribution that is approximately normal, with mean $\mu = 281$ and standard deviation $\sigma = 35$.

(a) Choose one eighth grader at random. What is the probability that his or her score is higher than 281? Higher than 316?

(b) Now choose an SRS of four eighth graders. What is the probability that their mean score is higher than 281? Higher than 316?

67. Antonio measures the alcohol content of whiskey for his Chemistry 101 lab. He actually measures the mass of 5 milliliters (ml) of whiskey—a chemical calculation—and then finds the percentage of alcohol from the mass. The standard deviation of students' measurements of mass is $\sigma = 10$ milligrams (mg). Antonio repeats the measurement three times and records the mean $\bar{x}$ of his three measurements.

(a) What is the standard deviation of Antonio's mean result?

(b) How many times must Antonio repeat the measurement to reduce the standard deviation of $\bar{x}$ to 5 mg? Explain to someone who knows nothing about statistics the advantage of reporting the average of several measurements rather than the result of a single measurement.

68. In Exercise 60, you found the mean and standard deviation of the outcome of a column bet in roulette. The central limit theorem says that the average outcome of a large number of bets has a distribution that is close to normal.

(a) What is the 99.7% confidence interval estimate (mean ± 3 standard deviations) of a gambler's average winnings after 100 bets?

(b) What is the 99.7% confidence interval estimate of a gambler's average winnings after 1000 bets?

69. Averages of several measurements are less variable than individual measurements. The true mass of the whiskey sample in Exercise 67 is 4.6 grams, or 4600 mg. Antonio's measurements have the normal distribution with mean 4600 mg and standard deviation 10 mg. In this case, the mean of his three measurements also has a normal distribution.

(a) Sketch on the same graph the two normal curves, for individual measurements and for means of three measurements. Figure 8.24 (page 382) is an example of this kind of graph.

(b) What interval of values covers the middle 95% of Antonio's individual measurements?

(c) What interval of values covers the middle 95% of averages of three measurements?

70. Exercise 28 gives the probability model for the intelligence assigned by chance to a character in a role-playing game. You found the mean intelligence of such characters in Exercise 54. Jermaine plays this character often. What interval covers (approximately) the middle 68% of average intelligence scores for 100 of Jermaine's games?

71. The scores of high school seniors on the ACT college entrance examination in 2010 were roughly normal with mean $\mu = 21.0$ and standard deviation $\sigma = 5.2$.

(a) What is the approximate probability that a single student randomly chosen from all those taking the test scores 26.2 or higher?

(b) Now take an SRS of nine students who took the test. What are the mean and standard deviation of the sample mean score $\bar{x}$ of these nine students?

(c) What is the approximate probability that the mean score $\bar{x}$ of these nine students is 26.2 or higher?

72. Although cities encourage carpooling to reduce traffic congestion, most vehicles carry only one person. For example, 70% of vehicles on the roads in the Minneapolis–St. Paul metropolitan area are occupied by just the driver. You choose 84 vehicles at random.

(a) What are the mean and standard deviation of the proportion of vehicles in your sample that carry only one person? (See Chapter 7, Sampling Distribution of a Sample Proportion, page 317.)

(b) What is the probability that more than 60% of the vehicles in your sample carry only one person?

73. Among high-performing (and low-performing) schools, there is an unrepresentatively _____ proportion of smaller schools. Explain whether you would complete this sentence with the word "high" or "low," in light of the formula for the standard deviation of the sampling distribution of the mean.

Chapter Review

74. Give either an intuitive or an algebraic argument to explain why ${}_nC_k = {}_nC_{n-k}$.

75. License plates in Florida have the form A12BCD—that is, a letter followed by two digits followed by three more letters.

(a) How many possible different license plates are there?

(b) Jerry would like a plate that ends in AAA. How many such plates are there?

(c) If license plates are issued at random from all possible plates, what is the probability that Jerry will get a plate that ends in AAA?

76. After you tell Jerry the probability that you calculated in Exercise 75, he realizes that he's unlikely to get a plate ending in AAA. So he asks you, "What's the probability I will get a plate in which all four letters are from my name?" These letters are J, E, R, and Y.

(a) Suppose that Jerry insists that the letters appear in order, so that his plate reads J*nm*ERY, where *n* and *m* stand for any number. What is the probability?

(b) Suppose Jerry allows his letters to appear in the plate in any order and also allows repeats. What is now the answer to Jerry's question?

77. Choose at random a person aged 19 to 25 years and ask, "In the past four days, how many days did you do physical exercise or work out?" Based on a large sample survey, here is a discrete probability model for the answer you will get:

# of Days	0	1	2	3	4
Probability	0.61	0.17	0.10	0.08	0.04

(a) What is the probability that the person you chose worked out either two or three days in the past four?

(b) What is the probability that the person you chose worked out at least one day in the past four?

78. Using the information in Exercise 77, what is the mean number of days that randomly chosen 19- to 25-year-olds worked out in the past four days? If you interview many people in this age group, what does the law of large numbers say about the average number of days that these people work out?

79. Use the information in Exercise 77 and your result from Exercise 78 to answer these questions.

(a) What is the standard deviation of the number of days in the past four that a randomly chosen 19- to 25-year-old has worked out?

(b) You interview 100 randomly chosen 19- to 25-year olds. You ask each how many days in the past four he or she has worked out, and you calculate the average number of days. According to the central limit theorem, there is a probability of 0.95 that your average will fall between what two values?

80. In Example 22 (page 378), we saw that a \$1 bet on red has a mean outcome of $-\$2/38 \approx -\$0.053$. It turns out not all \$1 bets in American roulette have the same mean outcome. The "five-number bet" {0, 00, 1, 2, 3} pays an additional \$6 if one of those five numbers comes up—otherwise, the player loses his \$1.

(a) Find the expected value for this five-number bet.

(b) Is this five-number bet better or worse than a bet on red?

81. Suppose you select 10 people at random. Find the probability of each event below:

(a) At least one match in the day of the week that they were born

(b) At least one match in the day of the month that they were born (assume 31 days per month)

(c) At least one match in the day of the year that they were born

82. In the 1970s, a group of children in Lyme, Connecticut, developed rheumatoid arthritis. However, it took until the 1980s for researchers to determine the cause of their disease, now known as Lyme disease—deer ticks infected with *Borrelia burgdorferi*. ELISA (enzyme-linked immunosorbent assay) is a commonly administered first test for Lyme disease. It correctly identifies patients with Lyme disease (the test is positive) 93.7% of the time and gives false positive results (the test is positive for patients who do not have Lyme disease) 6% of the time.

(a) Suppose that in a community in the Northeast, the prevalence of Lyme disease is 0.5%. Given that a person from this community tests positive for Lyme disease, what is the probability that the person actually has Lyme disease?

(b) Most people are tested only when there is a suspicion that they may have been bitten by a tick. Suppose the prevalence of Lyme disease among those tested is 10%. Given that a person tests positive for Lyme disease, what is the probability that the person actually has Lyme disease?

Applet Exercises

To do these exercises, go to www.macmillanhighered.com/fapp10e.

1. When we toss a coin, experience shows that the probability (long-term proportion) of a head is close to $\frac{1}{2}$. Suppose now that we toss the coin repeatedly until we get a head. What is the probability that the first head comes up in an odd number of tosses (1, 3, 5, and so on)? Use the *Probability* applet to estimate this probability. Set the probability of heads to 0.5. Toss coins one at a time until the first head appears. Do this 50 times (click "Reset" after each trial). What is your estimate of the probability that the first head appears on an odd toss?

2. The table of random digits (Table 7.1, page 298) was produced by a random mechanism that gives each digit probability 0.1 of being a 0.

(a) What proportion of the digits in the first row of Table 7.1 are 0s? This proportion is an estimate, based on 40 repetitions, of the true probability, which in this case is known to be 0.1.

(b) The *Probability* applet can imitate random digits. Set the probability of heads in the applet to 0.1. Check "Show true probability" to show this value on the graph. A head stands for a 0 in the random digit table and a tail stands for any other digit. Simulate 200 digits (50 at a time—don't click "Reset"). If you kept going forever, presumably you would get 10% heads. What was the percentage of heads in your 200 tosses?

3. One of the few players to have a better field goal percentage than free throw percentage, basketball star Shaquille O'Neal made about half (53%) of his free throws in his 21-year NBA career. Use the *Probability* applet to simulate 100 free throws shot independently by a player who has probability 0.53 of making each shot. (Toss 50, 50, without clicking "Reset.")

(a) What percentage of the 100 shots were made?

(b) Start the process again, this time repeating 10 tosses 10 times so that you can keep track of the individual hits and misses. Examine the sequence of hits and misses after each click on "Toss" and keep track of the longest run of shots made and the longest run of shots missed. How long were the longest runs in the 100 shots taken? (Sequences of random outcomes often show longer runs than our intuition expects.)

4. The central limit theorem is the basis for the confidence intervals that have been discussed in this chapter and in Chapter 7 (page 321). Next, you will use the *Central Limit Theorem* applet to generate individual data values from two different continuous probability models: the uniform probability model and the exponential probability model. You will find data from these distributions don't look very normal, and then you will take samples of size 30 and generate means from the samples.

(a) Go to the *Central Limit Theorem* applet. Choose "Uniform" for the distribution. Set the sample size to 1, so that you can see the results of individual data values drawn from this distribution. (The "Show normal curve" should be unchecked.) Click the "Generate samples" button. Do data from the uniform distribution appear to have the characteristic normal shape?

(b) Now change the sample size to 10. Instead of generating individual outcomes from a uniform distribution, the applet will draw many samples of size 10 and then make a histogram of the sample means, $\bar{x}$. Click the "Generate Samples" button. Do these data appear to be from a normal distribution? Check the box for "Show normal curve."

(c) This time, choose "Exponential" for the distribution. Set the sample size to 1 as you did in part (a). Click the "Generate samples" button. Describe the shape of exponential data.

(d) Now, continue with the exponential distribution but change the sample size to 10. Instead of generating individual outcomes from an exponential distribution, the applet will draw many samples of size 10 and then make a histogram of the sample means, $\bar{x}$. Click the "Generate samples" button. Do these data appear to be from a normal distribution? Check the box for "Show normal curve."

(e) Repeat part (d), but this time change the sample size to 30.

(f) Summarize the patterns you have observed in parts (a) through (e). How do these patterns relate to the central limit theorem?

Writing Projects

1. Psychologists have shown that our intuitive understanding of chance behavior is rather poor. Amos Tversky (1937–1996) was a leader in the study of how we make decisions in the face of uncertainty. In its obituary of Tversky, the *New York Times* cited the following example:

> Tversky asked subjects to choose between two public health programs that affect 600 people. One had a probability of $\frac{1}{2}$ of saving all 600 and a probability of $\frac{1}{2}$ that all 600 will die. The other was guaranteed to save exactly 400 of the 600 people. Most people chose the second program. He then offered a different choice. One program had a probability of $\frac{1}{2}$ of saving all 600 and a probability of $\frac{1}{2}$ of losing all 600, while the other would definitely lose exactly 200 lives. Most people chose the first program.

Discuss this example. What is the difference between the two choices offered? What is the mean number of people saved by the two options in each choice? What do the reactions of most subjects to these choices show about how people make decisions?

2. There are about 1×10^{44} air molecules in the atmosphere and about 2×10^{22} molecules of air in a single breath taken at rest. What is the probability that the breath that you took just now contained at least one molecule of air that was exhaled by Pythagoras in his last breath? What probability rules did you use to calculate this? What assumptions did you make, and why do you think they were reasonable?

3. We have seen that by betting on "red" in American roulette, you have a $\frac{18}{38}$ chance of winning, therefore doubling the money you bet. Suppose that you have \$5 and you want to bet until either you reach (and stop with) \$10 or you go broke. Is placing individual \$1 bets on red more, less, or equally likely to reach this goal than just placing a single \$5 bet? First, try to give an answer based on intuition, taking into account the casino's advantage.

You could also explore the following formula that gives the probability of going from h dollars to N dollars without going broke by making \$1 bets on red in American roulette:

$$\frac{1 - (20/18)^h}{1 - (20/18)^N}$$

Discuss how the strategy for maximizing the chance of reaching a financial target compares with the strategy for maximizing the length of time that your money lasts (for entertainment value).

Suggested Readings

COMAP. *Principles and Practice of Mathematics,* Springer, New York, 1997. Chapters 4 and 8 present combinatorics and probability at the next level beyond this book.

LESSER, LAWRENCE M. Take a chance by exploring the statistics in lotteries. *Statistics Teacher Network,* 65 (2004): 6–7. This article shows how lotteries can illustrate all major topics of an introductory statistics course, using a graphing calculator. The article is available online at www.amstat.org/education/stn/pdfs/STN65.pdf.

MOSTELLER, FREDERICK, ROBERT E. K. ROURKE, and GEORGE B. THOMAS. *Probability with Statistical Applications,* Addison-Wesley, Reading, MA, 1970. A rich treatment of basic probability that requires only high school algebra but is somewhat sophisticated. Although out of print, this book is a classic that deserves mention.

WAINER, HOWARD. The most dangerous equation. *American Scientist,* 95(3) (2007): 249–256. This article explores many real-world examples of the standard deviation of the sampling distribution of the sample mean.

Suggested Websites

www.shodor.org/interactivate/activities/ "Buffon's needle" is a probability problem first stated in 1777 by Count Buffon: If you drop a needle on a sheet of lined paper, what is the probability that the needle crosses one of the lines? In the simplest case, the length of the needle is the same as the distance between the lines. Some fairly advanced math shows that the answer is $2/\pi$, or about 0.637. A number of websites simulate dropping a needle many times to estimate this probability. A simulation of this activity, as well as many other probability exercises, can be found by clicking on "Probability" on this website.

www.gamblingexposed.org/gamblingexposed_708-389-1127__008.htm You may also be interested in the debate over legalized gambling. For the case against this practice, visit the National Coalition Against Legalized Gambling.

www.americangaming.org For the defense by the casino industry, visit the American Gaming Association.

illuminations.nctm.org/Lesson.aspx?id=956 For an example of a noncasino game involving probability and expected value, try playing the dice game "Skunk."

www.glennshafer.com/assets/downloads/articles/article50.pdf This website links to the article "The Early Development of Mathematical Probability" by Glen Shafer.

www.learner.org/courses/againstallodds/ The online materials in Units 19 and 22 from *Against All Odds: Inside Statistics* are directly applicable to material discussed in this chapter. Unit 19, Probability Models, provides a good review of the Probability Rules in Sections 8.2 and 8.3. To help gain a better grasp on the sampling distribution of $\bar{x}$ and the Central Limit Theorem, try viewing the video of Unit 22, Sampling Distributions. (This ties in nicely with Applet Exercise 4.)

Part III

Voting and Social Choice

Chapter 9
Social Choice: The Impossible Dream

Chapter 10
The Manipulability of Voting Systems

Chapter 11
Weighted Voting Systems

Chapter 12
Electing the President

The application of mathematics to the study of human beings—their behavior, values, interactions, conflicts, and methods of making decisions—is generally considered to be a recent revolution. Yet the study of voting and social choice, which is very much the root of this revolution, goes back several centuries.

We begin in Chapter 9 with the question of how a group of individuals, each with his or her own set of values, selects one outcome from a list of possibilities. While majority rule is a good system for deciding an election with just two candidates, it turns out that there is no perfect way of deciding an election in which there are three or more candidates.

Group decision making is often a strategic encounter, and citizens need to be aware of the difficulties that can arise when some participants have an incentive to manipulate the outcome. We turn to this issue in Chapter 10.

In Chapter 11, we consider decision-making bodies in which the individual voters or parties do not have equal power. In particular, we look at weighted voting systems in which a voter's power need not be proportional to the number of votes that he or she is entitled to cast.

In Chapter 12, we analyze not only how the Electoral College influences resource allocation in a campaign but also how polls and the positioning of candidates on a left–right continuum affect the strategies of candidates and the choices of voters.

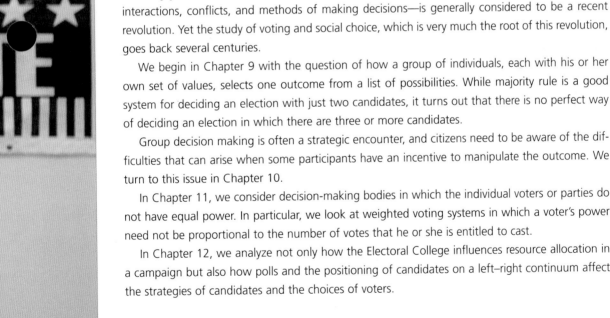

Social Choice: The Impossible Dream

9

Vallery Jean/FilmMagic/Getty Images

The basic question of social choice, of how groups can best arrive at decisions, has occupied social philosophers and political scientists for centuries. One primary example of a social-choice problem is the selection of a "good" voting system. Indeed, voting is a subject that lies at the very heart of representative government and participatory democracy.

Social-choice theory attempts to address the problem of finding good procedures that will turn individual preferences for different candidates into a single choice by the whole group. An example of such a choice would be the selection of the *winner* of an election. The goal is to find such procedures that will result in an outcome that "reflects the will of the people."

This search for good voting systems, as we shall see, is plagued by a variety of counterintuitive results and disturbing outcomes. In fact, it turns out that one can prove (mathematically) that no one will ever find a completely satisfactory voting system in an election with three or more candidates.

9.1 An Introduction to Social Choice

The elections with which we are most familiar often involve only two candidates. However, there are real-world situations in which elections must be held to choose a single winner from among three or more candidates, as in the presidential election of 2012 in which, in addition to Barack Obama and Mitt Romney, there were 25 other candidates (although these other 25 candidates garnered only a total of 2,093,848 votes, with more than half of those going to the Libertarian Party candidate Gary Johnson).

There are several methods that can be used to elect a single candidate from a choice of three or more, and we will investigate some of them in this chapter. Most of these methods use a ballot in which a voter provides a rank ordering of the candidates (without ties) that indicates the order in which he or she prefers them.

In the 2004 election, President George W. Bush was challenged by Massachusetts senator John Kerry. Bush received a slim majority of the vote. The election results have been dogged by charges of irregularities in the important swing state of Ohio.

KMazur/WireImage for Turner/Getty Images

Dave Kaup/Getty Images

Preference List Ballot DEFINITION

A ballot consisting of such a rank ordering of candidates (which we often picture as a vertical list with the most preferred candidate on top and the least preferred on the bottom) is called a **preference list ballot** because it is a statement of the preferences of the individual who is voting.

Preference list ballots allow voters to make a much clearer statement of their preferences than ballots allowing a single vote. Preference list ballots are already used in a wide range of applications, such as rating football teams and scoring track meets.

Although we do not allow ties in a preference list ballot, most voting rules of interest will, in some elections, result in a tie for the win among two or more of the candidates. In the real world, the number of voters is often so large that ties seldom occur. Nevertheless, to avoid excessive annoyances in the theory that we develop, and to simplify what we do in this chapter, we make the following assumption throughout.

The Number of Voters Assumption RULE

Throughout this chapter, we consider only elections in which there is an odd number of voters.

With this in hand, we are ready to begin our study of social choice.

9.2 Majority Rule and Condorcet's Method

When a choice is being made between two candidates, the first type of voting system to suggest itself is **majority rule:** Each voter indicates a preference for one of the two candidates, and the one with the most votes wins. With two candidates, there is no real distinction between a ballot that indicates a voter's choice for one of the two candidates and what we have called a preference list ballot. The point is that we can, for example, identify a choice for A (however indicated) with the list that has A over B and a choice for B with the list that has B over A.

Majority rule has at least three desirable properties:

1. All voters are treated equally. That is, if any two voters were to exchange (marked) ballots before submitting them, the outcome of the election would be the same.
2. Both candidates are treated equally. That is, if a new election were held and every voter were to reverse his or her vote, then the outcome of the previous election would be reversed as well.
3. It is **monotone.** That is, if some candidate X is a winner of an election, and a new election is held in which the only ballot change made is for some voter to change his or her ballot from not being a vote for X to being a vote for X, then X will remain the winner.

The desirability of these three properties, at least in the most common kinds of elections, is easy to see. For example, condition 1 reflects the (grammatically outdated) tenet "one-man, one-vote" that gained prominence from its use in a 1964 Supreme Court majority opinion. Indeed, if any voter had two or more votes, then an exchange of his or her ballot with that of a voter who only had one vote might well change the outcome of an election. If a voting system violated condition 2, then one of the candidates must have an advantage built in by the system. And as for condition 3, any system that is not monotone has the rather bizarre property that a candidate could have enough support to win, but gathering additional support might cause this same candidate to lose.

Self Check 1

Explain (in a sentence or two each) why majority rule satisfies each of the three desirable properties. ■

It is easy to devise voting systems for two candidates in which these properties fail, but each such voting system quickly reveals its undesirability. For example, condition 1 is not satisfied by a *dictatorship* (in which all ballots except that of the dictator are ignored); condition 2 is not satisfied by *imposed rule* (in which

Candidate *X* wins regardless of who votes for whom); and condition 3 is not satisfied by *minority rule* (in which the candidate with the fewest votes wins).

But maybe there are voting systems in the two-candidate case that are superior to majority rule in the sense of satisfying the three properties just listed *and* some other properties that we might also wish to have satisfied. This, however, turns out not to be the case. In 1952, Kenneth May proved the following, aptly named **May's theorem.**

May's Theorem THEOREM

Among all two-candidate voting systems that never result in a tie, majority rule is the *only* one that treats all voters equally, treats both candidates equally, and is monotone.

This is an important and elegant result. Thus, mathematical reasoning spares us the trouble of searching for a better voting system for two candidates.

But what if there are three or more candidates? Perhaps we can design a voting system for this situation that, in some way, builds on the success of majority rule in the two-candidate case. In point of fact, there does exist a voting system that arises from precisely this hope, and it is known today as **Condorcet's method.**

Our description of Condorcet's method begins with the observation that if we have a sequence of preference list ballots, then—for each pair of candidates—we can determine who the winner would have been had the election involved only these two in a one-on-one contest using majority rule.

To illustrate this notion of a one-on-one contest, consider the following preference list ballots:

Rank	Number of Voters (3)		
First	*A*	*B*	*C*
Second	*B*	*C*	*A*
Third	*C*	*A*	*B*

In this election, Candidate *A* would defeat Candidate *B* in a one-on-one contest (two votes to one), while *B* would, in turn, defeat *C* in a one-on-one contest, again by a score of 2 to 1. We'll return to this example in a moment, but we now have at hand all we need to describe Condorcet's voting system for three or more candidates.

Description of Condorcet's Method PROCEDURE

With the voting system known as Condorcet's method, a candidate is a winner precisely when he or she would, on the basis of the ballots cast, defeat every other candidate in a one-on-one contest using majority rule.

Historically, the voting system we are calling Condorcet's method dates back at least to Ramon Llull in the 13th century (see Spotlight 9.1). It was rediscovered and popularized in the 18th century by the Marquis de Condorcet (1743–1794).

Let's look at a couple of examples of elections using Condorcet's method.

The Historical Record

SPOTLIGHT 9.1

The following letter was written by Friedrich Pukelsheim of the University of Augsburg, Germany. He is imagining what Ramon Llull (1232–1316) might say if he were alive today.

Dear Editors:

It is my distinct pleasure to respond "from the beyond" to your kind invitation to set the historical record straight. I was born in 1232 on the Island of Mallorca in the Mediterranean Sea, which in your times is known as a popular tourist place. In my days it was a strong political center of that part of the world, with a population that was a mix of Christians, Jews, and Muslims. It was my dream to persuade people of the virtues of Christian belief by relying, not on force, but on reason.

Unfortunately, people did not find it easy to follow my arguments, so I was more than pleased to discover some down-to-earth applications, including an election system. My idea was to oppose every pair of candidates, one on one, and ask the electors whom of the two they would prefer—very much like a medieval jousting tournament. But how to combine the results from all the duels into a winner of the election? I first proposed electing the candidate who won the most duels, then later suggested a system of successive eliminations.

I wrote three papers on the topic, the second of which I "smuggled" into my novel *Blanquerna* in 1283. More than a century after my death, in 1428, the young German scholar Nicolaus Cusanus (1401–1464) journeyed to Paris to read my works in libraries there. He even copied out the third of my electoral writings, which I had completed on 1 July 1299 in Paris, and his manuscript is the only copy handed down to your days. Reading my papers, Cusanus was inspired to invent his own electoral system. Did he not understand mine, or just find it inadequate? Who knows?

While I had been concerned with electing Church officials, Cusanus sought a system to elect the Holy Roman Emperor. In his system, each elector assigns each candidate a rank score, with the lowest candidate getting a score of 1, the second lowest a score of 2, and the best candidate the highest score possible, that is, 10 when there are 10 candidates. The scores are totaled for each candidate and the candidate with the highest score wins. If you are a soccer player or a hockey player, you will have a good sense for one difference between our systems: Whereas I count victories, Cusanus adds up goals. Cusanus applauded himself for having invented an absolutely ingenious and novel electoral system.

Also, I advocated open voting, whereas Cusanus favored a secret ballot. He was concerned that voters might sell their votes, or that the candidates might pressure the voters. Well, that certainly happened all of the time in elections for worldly authorities! But for election to clerical office, I thought it good enough if electors took an oath to vote for the most worthy candidate and submitted themselves to the social control that comes with an open election.

Cusanus was famous in his times, as I was in mine, but fame indeed is transitory. Sure enough, my electoral system was reinvented by the Marquis de Condorcet (1743–1794), and Cusanus's system was proposed afresh by the Chevalier de Borda (1733–1799)—neither of whom, I am sure, wasted a thought on the possibility that "their" systems might already be on record. But, as my works had fallen into oblivion as had those of Cusanus, neither Condorcet nor Borda should be blamed for failing to acknowledge our priority.

My first electoral paper—actually the one that is longest and most detailed, written around 1280—was rediscovered only in 2000, filed away in the Vatican Library. How would you feel if your work attracts fresh attention after more than 700 years? Actually, I am utterly pleased that mine has resurfaced at last! The text was excavated by a mathematician interested in voting systems, Friedrich Pukelsheim of the University of Augsburg, Germany. Since the text is handwritten in Latin, handling it became an interdisciplinary project that brought together experts on medieval manuscripts, Church Latin and theology, and even computer scientists. As a result, my electoral writings are now on the Internet (in the original and in translations into English and German) at www.uni-augsburg.de/llull/.

Looking back on my lack of success in preaching peace among Christians, Jews, and Muslims, and all the writing and copying by hand of my works, I hope you can appreciate how highly I value the printed book (such as this one) and, even more, instant communication worldwide over the Internet. May that ease of communication help facilitate the religious peace that I so dearly sought.

Yours truly,
Ramon Llull (1232–1316)
Left Choir Chapel
San Francisco Cathedral
Palma de Mallorca

EXAMPLE 1 Condorcet's Method

The mathematics department is hiring a new faculty member and the five-person hiring committee has interviewed four candidates: Adam, Beth, Carol, and Dan. They have decided to use Condorcet's method on their five ballots (reproduced in the table below). Let's see who gets the offer.

	Number of Voters (5)				
Rank	**Voter 1**	**Voter 2**	**Voter 3**	**Voter 4**	**Voter 5**
First	Adam	Dan	Carol	Adam	Beth
Second	Beth	Carol	Beth	Carol	Dan
Third	Carol	Beth	Dan	Dan	Carol
Fourth	Dan	Adam	Adam	Beth	Adam

To find a winner using Condorcet's method, we begin by choosing an ordering of the candidates, which we'll take to be alphabetical for this example. Thus, we first pit Adam against each of the others in a one-on-one contest based on the ballots. In an Adam-versus-Beth election, based on these ballots, Voters 1 and 4 would vote for Adam; and Voters 2, 3, and 5 would vote for Beth. Hence, Beth would win this one-on-one contest against Adam, so we know Adam is not going to be the winner using Condorcet's method. But Beth still has a chance, so we move on to see how Beth would fare in a one-on-one contest against Carol (knowing already that Beth would defeat Adam). But here it is easy to see that Beth would lose to Carol (with Voters 2, 3, and 4 voting for Carol).

Hence, Beth is not a winner with Condorcet's method, but Carol has not yet been eliminated (and neither has Dan, but remember that we have chosen to check out the candidates in alphabetical order). First pitting Carol against Adam, we see that Carol wins 3 to 2. And pitting Carol against Dan, we see that Carol again wins by this same 3 to 2 score. This shows that Carol is the winner using Condorcet's method!

By the way, if it seems like we never gave Dan a chance, notice that our determination that Carol is the Condorcet winner means that we already know that she beats Dan one on one. So Dan can't be a winner in this election using Condorcet's method. ■

Self Check 2

Suppose that the department makes an offer to Carol, as the above vote suggests, but she refuses the offer. The department decides to again use Condorcet's method with the same ballots, but with Carol's name erased. Who gets the next offer? ■

EXAMPLE 2 Condorcet's Method

Suppose we have four candidates (*GB, AG, RN,* and *PB,* with these initials chosen for a soon-to-be-revealed reason) and the following sequence of preference list ballots, where the heading of "6" indicates that 6 of the 15 voters hold the ballot with *GB* over *AG* over *PB* over *RN,* the heading of "5" indicates that 5 of the 15 voters hold the ballot with *AG* over *RN* over *GB* over *PB,* and so on.

	Number of Voters (15)			
Rank	**6**	**5**	**3**	**1**
First	*GB*	*AG*	*RN*	*PB*
Second	*AG*	*RN*	*AG*	*GB*
Third	*PB*	*GB*	*GB*	*AG*
Fourth	*RN*	*PB*	*PB*	*RN*

We claim that *AG* is the winner in this election if we use Condorcet's method. Let's check the one-on-one scores for each possible pair of opponents:

AG versus *GB: AG* is over *GB* on 5 + 3 = 8 of the ballots, while the reverse is true on 6 + 1 = 7 of the ballots. Thus, *AG* defeats *GB* by a score of 8 to 7.

AG versus *RN: AG* is over *RN* on 6 + 5 + 1 = 12 of the ballots, while the reverse is true on 3 of the ballots. Thus, *AG* defeats *RN* by a score of 12 to 3.

AG versus *PB: AG* is over *PB* on 6 + 5 + 3 = 14 of the ballots, while the reverse is true on 1 of the ballots. Thus, *AG* defeats *PB* by a score of 14 to 1.

This shows that *AG* is the winner using Condorcet's method.

Like majority rule, Condorcet's method satisfies some very desirable properties, as we'll see later in this section. But it also has a tragic flaw, and this flaw is called **Condorcet's voting paradox.**

Condorcet's Voting Paradox THEOREM

With three or more candidates, there are elections in which Condorcet's method yields *no* winners. In particular, the following ballots (often called the "Condorcet voting paradox ballots") constitute an election in which Condorcet's method yields no winner.

Rank	**Number of Voters (3)**		
First	*A*	*B*	*C*
Second	*B*	*C*	*A*
Third	*C*	*A*	*B*

The Condorcet voting paradox ballots given above are the same ones we used earlier in illustrating the notion of a one-on-one contest. We pointed out then that *A* defeats *B* one on one and *B* defeats *C* one on one. The additional observation needed is that we also have *C* defeating *A* one on one. Thus, *A* cannot be a winner using Condorcet's method (he or she loses to *C*), *B* cannot be a winner (he or she loses to *A*), and *C* cannot be a winner (he or she loses to *B*). We will revisit Condorcet's voting paradox in Section 9.3.

Notice that because of our assumption that the number of voters is odd, Condorcet's method yields either no winner or a unique winner (see Exercise 35).

It is tempting at this point to suggest modifying Condorcet's method as we have presented it by declaring all the candidates to be tied for the win if there is no candidate who defeats each of the others one on one. The drawback to this modification is that a number of the desirable properties possessed by Condorcet's method then evaporate. We'll explore this in the upcoming exercises.

9.3 Other Voting Systems for Three or More Candidates

With three or more candidates, we find no shortage of additional procedures that suggest themselves and that seem to represent perfectly reasonable ways to choose a winner. Closer inspection, however, reveals shortcomings with all of these. We illustrate this with a consideration of several well-known procedures. Additional procedures (and additional shortcomings) can be found in the exercises.

Plurality Voting and the Condorcet Winner Criterion

In **plurality voting,** only first-place votes are considered. Thus, while we will consider plurality voting in the context of preference list ballots, a ballot here might just as well be a single vote for a single candidate. The candidate with the most votes wins, even though this may be considerably fewer than one-half the total votes cast. This is perhaps the most common system in use today. It is how the voters in Florida chose George W. Bush over Al Gore, Ralph Nader, and Patrick J. Buchanan in the presidential election of 2000.

EXAMPLE 3 Plurality Voting and the 2000 Presidential Election

On the evening of December 12, 2000, Al Gore conceded the presidential election of 2000 to George W. Bush, thus bringing to a close one of the most remarkable elections in modern times. The outcome, ultimately decided in the Electoral College, came down to whether Bush or Gore would carry Florida. With more than 6 million votes cast in Florida, the ultimate margin of victory for George W. Bush was only a few hundred votes.

There is little doubt that if the 2000 presidential election had pitted Al Gore solely against any one of the other three candidates, then Gore would have won both the election in Florida and the presidency. The point is that while most of the Buchanan supporters probably would have voted for Bush, the far more numerous Nader supporters probably would have gone largely for Gore. In fact, the illustration of Condorcet's method that we gave in Example 2 is a simplified version of this Florida election (with *GB* standing for George Bush, *AG* for Al Gore, *PB* for Patrick Buchanan, and *RN* for Ralph Nader).

Thus, although plurality voting led to Bush's winning the 2000 election in Florida (and hence the presidency), Gore was, in this example, what is called a **Condorcet winner:** He would have won the election if Condorcet's method had been used.

Ron Edmonds/Associated Press

Governor George W. Bush and Vice President Al Gore debate the issues before the 2000 election, possibly the most controversial election in U.S. history. Gore was the Condorcet winner of the election, but Bush eked out a victory that relied on the rules of the Electoral College. Many voters were suddenly put on notice that the U.S. Constitution makes the election of the president indirect—and not a pure expression of the majority's choice.

Condorcet Winner Criterion DEFINITION

A voting system is said to satisfy the **Condorcet winner criterion (CWC)** provided that, for every possible sequence of preference list ballots, either (1) there is no Condorcet winner (as is often the case) or (2) the voting system produces exactly the same winner for this election as does Condorcet's method.

The CWC is certainly a property that one would like to see satisfied. We record plurality voting's failure in this respect with the following.

The Failure of the CWC with Plurality Voting THEOREM

The Florida vote in the 2000 presidential election shows that plurality voting fails to satisfy the CWC.

Perhaps a more fundamental drawback of plurality voting is the extent to which the ballots provide no opportunity for a voter to express any preferences except for naming his or her top choice. No use is made, for example, of the fact that a candidate may be no one's first choice but everyone's close second choice.

To illustrate this point, we return to the mathematics department's five-person hiring committee that has interviewed four candidates: Adam, Beth, Carol, and Dan. They have ranked the candidates on their five ballots (reproduced in the table below), but now, let's suppose that instead of using Condorcet's method, they decide to use plurality.

EXAMPLE 4 Plurality Voting

	Number of Voters (5)				
Rank	**Voter 1**	**Voter 2**	**Voter 3**	**Voter 4**	**Voter 5**
First	Adam	Dan	Carol	Adam	Beth
Second	Beth	Carol	Beth	Carol	Dan
Third	Carol	Beth	Dan	Dan	Carol
Fourth	Dan	Adam	Adam	Beth	Adam

Clearly, Adam wins with the plurality method because he received two first-place votes and each of the others received only one. But 60% of the voters (i.e., three of the five voters) rank Adam last! ■

Finally, there is yet another shortcoming of plurality voting: There are elections in which it is to a voter's advantage to submit a ballot that misrepresents his or her true preferences.

Manipulability DEFINITION

A voting system is subject to **manipulability** (or is **manipulable**) if there are elections in which it is to a voter's advantage to submit a ballot that misrepresents his or her true preferences.

Self Check 3

Suppose there is an election in which a voter can change the outcome of an election to one he prefers *less* by submitting a disingenuous ballot. Does this mean the voting system is manipulable (according to the above definition)? ■

In the presidential election of 2000, many voters who ranked Ralph Nader or Patrick J. Buchanan over George W. Bush and Al Gore chose to vote for Bush or Gore rather than to "throw away" their vote on a candidate they believed had no chance. Condorcet's method, it turns out, is not manipulable, and this is one of its most desirable properties. We'll explore this further in the next chapter.

The Borda Count and Independence of Irrelevant Alternatives

In many elections that use preference list ballots, the goal is to arrive at a final group rank ordering of all the contestants that best expresses the desires of the electorate. The purpose is not only to determine the winner—say, the class valedictorian—but also to arrive at who finished second, third, and so on, as in the case of one's rank in the senior class. In other applications, such as an election to a hall of fame, the first few finishers each win, while the remaining nominees are also-rans.

One common mechanism for achieving this objective is to assign points to each voter's rankings and then to sum these for all voters to obtain the total points for each candidate. If there are 10 candidates, for example, then we could assign 9 points to each first-place vote for a given candidate, 8 points for each second-place vote, 7 for each third-place vote, and so forth. The candidate with the highest total number of points is the winner. Subsequent positions are assigned to those with the next-highest tallies.

Description of Rank Methods and the Borda Count PROCEDURE

A *rank method* of voting assigns points in a non-increasing manner to the ordered candidates on each voter's preference list ballot and then sums these points to arrive at a group's final ranking. The special case in which there are n candidates with each first-place vote worth $n - 1$ points, each second-place vote worth $n - 2$ points, and so on down to each last-place vote worth 0 points is known as the **Borda count.** The actual point totals are referred to as a candidate's *Borda score*.

The Borda count is named after Jean-Charles de Borda (1733–1799), who was a contemporary of Condorcet.

Rank methods other than the Borda count are common. For example, a track meet can be thought of as an "election" in which each event is a "voter" and each of the schools competing is a "candidate." If the order of finish in the 100-meter dash is school *A*, school *B*, school *C*, school *D*, then points are often awarded to each school as follows: 5 points for first place, 3 for second place, 2 for third place, and 1 for fourth place.

Sports polls often use point assignments that qualify as rank methods according to our definition. The following example provides an illustration of this.

EXAMPLE 5 Rank Methods and a Basketball Poll

In February 2014, the Associated Press issued its weekly ranking of the top 25 teams in men's college basketball, shown at right.

An interesting question is whether or not this is a ranking system. If it is, who are the candidates and how many are there? In fact, this can be regarded as a ranking system, but the number of candidates is not 25. That is, although 25 teams appeared on each ballot, at least one ballot included each of the teams listed at the bottom in the category "Others receiving votes."

For this to be regarded as a ranking system, the set of candidates must include the entire set of eligible collegiate men's basketball teams. We must also infer that each ballot lists all teams other than that voter's top 25 *below* that voter's top 25, perhaps in alphabetical order. The point assignments are then like those in the newspaper clipping, except that we also assign 0 points for a 26th place vote, 0 points for a 27th place vote, and so on. This is why our definition states that a rank method "assigns points in a *non-increasing* manner" instead of "assigns points in a *decreasing* manner."

We can use this poll to illustrate how total points are arrived at with a ranking method. With the top-ranked team, Syracuse, it's quite easy. Each first-place vote is worth 25, and Syracuse received all 65 first-place votes. This accounts for its total of $25 \times 65 = 1625$ points. But the calculation is more interesting for the second-ranked team, Arizona, and requires some speculation on our part because we don't actually have the ballots to examine. We know that there were 65 ballots (because there were exactly 65 first-place votes altogether), and we know that Arizona had no first-place votes. It stands to reason that Arizona's 1517 points must have come from a vast majority of the second-place votes, together with a few lower rankings. (We know that the team didn't receive *all* the second place-votes; otherwise, its point total would have been $24 \times 65 = 1560$.)

One possibility is that Arizona received:

0 first-place votes (at 25 points each)
40 second-place votes (at 24 points each)
15 third-place votes (at 23 points each)
4 fourth-place votes (at 22 points each)
4 fifth-place votes (at 21 points)
2 sixth-place votes (at 20 points)

The total would then be

$$0 + 960 + 345 + 88 + 84 + 40 = 1517$$

Grant Halverson/Getty Images

AP TOP 25

MEN'S BASKETBALL

The top 25 teams in The Associated Press' college basketball poll, with first-place votes in parentheses, records through Feb. 2, total points based on 25 points for a first-place vote through one point for a 25th-place vote and last week's ranking:

	Record	Pts	Prv
1. Syracuse (65)	21-0	1,625	2
2. Arizona	21-1	1,517	1
3. Florida	19-2	1,482	3
4. Wichita St.	23-0	1,447	4
5. San Diego St.	19-1	1,370	5
6. Villanova	19-2	1,252	9
7. Cincinnati	21-2	1,182	13
8. Kansas	16-5	1,141	6
9. Michigan St.	19-3	1,136	7
10. Michigan	16-5	949	10
11. Duke	17-5	940	17
12. Creighton	18-3	790	20
13. Saint Louis	20-2	728	19
14. Louisville	18-4	723	12
15. Texas	17-4	719	25
16. Iowa St.	16-4	717	16
17. Iowa	17-5	669	15
18. Kentucky	16-5	653	11
19. Oklahoma St.	16-5	420	8
20. Virginia	17-5	364	–
21. Oklahoma	17-5	361	23
22. UConn	17-4	252	–
23. Gonzaga	20-3	237	–
24. Memphis	16-5	114	22
25. Pittsburgh	18-4	110	18

Others receiving votes: Wisconsin 79, Ohio St. 45, VCU 44, SMU 15, New Mexico 12, California 9, UCLA 9, Harvard 4, George Washington 3, LSU 3, Tennessee 2, American U. 1, Southern Miss. 1.

Self Check 4

Suppose Arizona received 39 second-place votes (instead of 40) and 3 fourth-place votes (instead of 4). What must the number of third-place votes be if no other change is made and the total for Arizona is still 1517? ■

There is an easy way to calculate the Borda score of a candidate. You can count the number of occurrences of other candidate names that are below this candidate's name. For example, consider the following ballots:

Rank	Number of Voters (5)					Points
First	*A*	*A*	*A*	*B*	*B*	2
Second	*B*	*B*	*B*	*C*	*C*	1
Third	*C*	*C*	*C*	*A*	*A*	0

Because there are three candidates, each first-place vote is $n - 1$, or $3 - 1 = 2$; each second-place vote is $n - 2 = 1$; and each third-place vote is $n - 3 = 0$. If we were to calculate the Borda score of Candidate *B* algebraically, we would say that *B* has two first-place votes, worth 2 points each (a total of 4 points), and three second-place votes, worth 1 point each (a total of 3 more points). Thus, the Borda score of Candidate *B* is $4 + 3 = 7$.

But instead of calculating this Borda score algebraically, we can mentally replace each occurrence of a letter below *B* by a box, □, and simply count the boxes.

Rank	Number of Voters (5)				
First	*A*	*A*	*A*	*B*	*B*
Second	*B*	*B*	*B*	□	□
Third	□	□	□	□	□

Notice that there are seven boxes, giving us the correct value of 7 as the Borda score for Candidate *B*. Of course, you don't actually have to draw any boxes. We are just emphasizing the fact that, in the counting process, it is "spaces" that we are counting, without regard to which letter occurs in the space. A quick glance at the original ballots (without the boxes) reveals that the Borda score of Candidate *A* is 6 and the Borda score of Candidate *C* is 2. When calculating Borda scores this way, be sure that each individual ballot is listed separately, as opposed to using a single list to represent the ballots of several voters (as we often do).

The Borda count certainly seems to be a reasonable way to choose a winner from among several candidates (or to arrive at a group ranking of the candidates). It also has its shortcomings, however, one of which is the failure of a property known as **independence of irrelevant alternatives (IIA).**

Independence of Irrelevant Alternatives DEFINITION

A voting system is said to satisfy **independence of irrelevant alternatives (IIA)** if it is impossible for a Candidate X to move from nonwinner status to winner status unless at least one voter reverses the order in which he or she had X and the winning candidate ranked.

To describe this property, suppose that an election yields one candidate (call it *A*) as a winner and another candidate (call it *B*) as a nonwinner. Suppose that a new election is now held and that, although some of the voters may have changed their preference list ballots, no one who had previously ranked *A* over *B* changed his or her ballot to rank *B* over *A* now.

If this new election were to yield *B* as a winner, the new outcome would seem strange, especially because not one of the relative individual preferences for *A* over *B* had changed in *B*'s favor. The ballot changes responsible for the new outcome involve candidates *other than A* or *B*. One could argue that these other candidates ought to be "irrelevant" to the question of whether *A* is more desirable than *B* or *B* is more desirable than *A*. This inspires the name "independence of irrelevant alternatives."

Condorcet's method satisfies IIA. That is, if we have a sequence of preference list ballots that yield *A* as a Condorcet winner and *B* as a nonwinner, then *A* defeats every other candidate, and *B* in particular, in a one-on-one contest according to these ballots. If no voter reverses the order in which he or she ranked *A* and *B*, then *A* will still defeat *B* one on one, and thus *B* remains a nonwinner.

The following illustration shows that the Borda count, unlike Condorcet's method, fails to satisfy IIA. Suppose the initial five ballots are as follows:

Rank	Number of Voters (5)				
First	*A*	*A*	*A*	*C*	*C*
Second	*B*	*B*	*B*	*B*	*B*
Third	*C*	*C*	*C*	*A*	*A*

Our counting procedure shows that the Borda scores are as follows:

Borda score of *A* is 6.
Borda score of *B* is 5.
Borda score of *C* is 4.

The winner is *A* (with 6 points), and *B* is a nonwinner (with 5 points). But now suppose that the two voters on the right change their ballots by moving *C* down between *A* and *B*. The ballots then become:

Rank	Number of Voters (5)				
First	*A*	*A*	*A*	*B*	*B*
Second	*B*	*B*	*B*	*C*	*C*
Third	*C*	*C*	*C*	*A*	*A*

Our counting procedure shows that the Borda scores are as follows:

The Borda score of *A* is 6.
The Borda score of *B* is 7.
The Borda score of *C* is 2.

The Borda count therefore now yields *B* as the winner (with 7 points). Thus, *B* has gone from being a nonwinner to being a winner, even though no one changed his or her mind about whether *B* is preferred to *A*, or vice versa. Hence, the fact that *A* wins and *B* loses is *not* "independent" of where the "irrelevant" alternative *C* is ranked.

The above discussion establishes the following theorem.

The Failure of IIA with the Borda Count THEOREM

The Borda count fails to satisfy IIA.

Sequential Pairwise Voting and the Pareto Condition

In our voting-theoretic context, an **agenda** will be understood to be a listing (in some order) of the candidates. This listing is not to be confused with any of the preference list ballots, and to avoid confusion, we will present agendas as horizontal lists and continue to present preference list ballots vertically.

Description of Sequential Pairwise Voting PROCEDURE

Sequential pairwise voting starts with an agenda and pits the first candidate against the second in a one-on-one contest. The winner then moves on to confront the third candidate in the list, one on one. Losers are deleted. This process continues throughout the entire agenda, and the one remaining at the end wins.

For a given sequence of individual preference list ballots, the particular agenda chosen can greatly affect the outcome of the election, as we'll show in the next chapter. Nevertheless, we will see later in this chapter that sequential pairwise voting arises naturally in the legislative process. Notice also that because of our assumption that the number of voters is odd, there is always a unique winner with sequential pairwise voting.

EXAMPLE 6 ➔ Sequential Pairwise Voting

Assume we have four candidates and that the agenda is *A, B, C, D*. Consider the following sequence of three preference list ballots:

Rank	Number of Voters (3)		
First	*D*	*B*	*C*
Second	*C*	*D*	*A*
Third	*A*	*C*	*B*
Fourth	*B*	*A*	*D*

The first one-on-one pits *A* against *B*, and *A* wins by a score of 2 to 1 (meaning that two of the voters—the two not in the middle—prefer *A* to *B*, and one of the voters prefers *B* to *A*). Thus, *B* is eliminated and *A* moves on to confront *C*. Because *C* wins this one on one (by a score of 3 to 0), *A* is eliminated. Finally, *C* takes on *D*, and *D* wins by a score of 2 to 1. Thus, *D* is the winner.

Self Check 5

Is Candidate *D* a Condorcet winner in this example?

EXAMPLE 7 Sequential Pairwise Voting (with a different agenda)

Assume we have the same ballots as in Example 6, but now let's use sequential pairwise voting with the agenda *D, C, B, A*. The first one-on-one pits *D* against *C*, and *D* wins by a score of 2 to 1 (as only the third voter ranked *C* over *D*). Thus, *C* is eliminated and *D* moves on to confront *B*. Because *B* wins this one-on-one (by a score of 2 to 1), *D* is eliminated. Finally, *B* takes on *A*, and *A* wins by a score of 2 to 1. Thus, *A* is the winner.

Self Check 6

Is Candidate *A* a Condorcet winner in this example?

There is something very troubling about the outcome of the preceding example, especially if you are Candidate *C*. *Everyone* prefers *C* to *A*, but *A* ends up winning! This example shows that sequential pairwise voting fails to satisfy what is called the **Pareto condition.**

Pareto Condition DEFINITION

A voting system is said to satisfy the **Pareto condition** provided that in every election in which every voter prefers one candidate *X* to another candidate *Y*, the latter candidate *Y* is not among the winners.

Self Check 7

Suppose a voting system satisfies the Pareto condition and we have an election in which every voter prefers Candidate *X* to Candidate *Y*. Is Candidate *X* definitely among the winners ("yes" or "no")?

Again, the Pareto condition (named after Italian economist Vilfredo Pareto, 1848–1923) is a property we would like to see satisfied. But Example 6 (with Candidate *B* in the role of *X* and Candidate *D* in the role of *Y*) establishes the following.

The Failure of the Pareto Condition with Sequential Pairwise Voting THEOREM

Sequential pairwise voting fails to satisfy the Pareto condition.

The following sequence of three preference list ballots illustrates the Pareto condition further:

Rank	Number of Voters (3)		
First	*C*	*C*	*A*
Second	*A*	*A*	*B*
Third	*B*	*B*	*C*

Every one of the three voters prefers *A* to *B*. Hence, if we were using a voting rule that satisfies Pareto, we would conclude that *B* is *not* among the winners. However, we cannot conclude that *A* *is* among the winners. Indeed, there are very reasonable voting rules, like plurality, that satisfy the Pareto condition but would produce *C* as the unique winner using these ballots.

Runoff Systems and Monotonicity

The voting system known as the **Hare system,** which was introduced by Thomas Hare in 1861, is also known by names such as the "single transferable vote system." In 1862, John Stuart Mill described the Hare system as being "among the greatest improvements yet made in the theory and practice of government." Today, the system is used to elect public officials in Australia, Malta, the Republic of Ireland, and Northern Ireland.

Description of the Hare System PROCEDURE

The Hare system proceeds to arrive at a winner by repeatedly deleting candidates that are "least preferred" in the sense of being at the top of the fewest ballots. If a single candidate remains after all others have been eliminated, he or she alone is the winner. If two or more candidates remain and all of these remaining candidates would be eliminated in the next round (because they all have the same number of first-place votes), then these candidates are declared to be tied for the win.

EXAMPLE 8 ➜ The Hare System

Suppose we have the following sequence of preference list ballots:

Rank	Number of Voters (5)				
First	*A*	*A*	*B*	*B*	*C*
Second	*C*	*C*	*C*	*C*	*A*
Third	*B*	*B*	*A*	*A*	*B*

Candidate *C* has only 1 first-place vote (while *B* and *A* have 2 each). Thus, *C* is eliminated in round 1, and the ballots for the second round are as follows:

Rank	Number of Voters (5)				
First	*A*	*A*	*B*	*B*	*A*
Second	*B*	*B*	*A*	*A*	*B*

In the second round, *B* has only 2 first-place votes (while *A* has 3), so *B* is eliminated in round 2. Because *A* is the only candidate left, he or she is the unique winner of this election.

We now give another example of the Hare system. It is a bit more complicated than Example 8, but as with earlier examples in this section, it will reveal a serious shortcoming of a seemingly very reasonable voting system (the Hare system, in this case).

EXAMPLE 9 Shortcomings of the Hare System

Suppose we have the following sequence of preference list ballots, where, as before, the heading of "5" indicates that 5 of the 13 voters hold the ballot with *A* over *B* over *C*, the heading of "4" indicates that 4 of the 13 voters hold the ballot with *C* over *B* over *A*, and so forth.

	Number of Voters (13)			
Rank	**5**	**4**	**3**	**1**
First	*A*	*C*	*B*	*B*
Second	*B*	*B*	*C*	*A*
Third	*C*	*A*	*A*	*C*

Candidates *B* and *C* have only 4 first-place votes (while *A* has 5). Thus, *B* and *C* are eliminated in the first round, and *A* wins the election.

Now, suppose that the voter in the last column moves Candidate *A* up on his list. Let's look at the new election. Notice that, even though *A* won the last election, the only change we are making in ballots for the new election is one that is favorable to *A*. The ballots for the new election are as follows:

	Number of Voters (13)			
Rank	**5**	**4**	**3**	**1**
First	*A*	*C*	*B*	*A*
Second	*B*	*B*	*C*	*B*
Third	*C*	*A*	*A*	*C*

If we apply the Hare system again, only *B* is eliminated in round 1, as he or she has 3 first-place votes, as opposed to 4 for *C* and 6 for *A*. Thus, after this round, the ballots are as follows:

	Number of Voters (13)			
Rank	**5**	**4**	**3**	**1**
First	*A*	*C*	*C*	*A*
Second	*C*	*A*	*A*	*C*

We now have *A* on top of 6 lists and *C* on top of 7 lists. Thus, at stage 2, *A* (our previous winner!) is eliminated and *C* is the winner of this new election.

Clearly, this is once again quite counterintuitive. Alternative *A* won the original election, the only change in ballots made was one favorable to *A* (and no one else), and then *A* lost the next election. This example shows that the Hare system fails to satisfy what is called **monotonicity.**

Monotonicity DEFINITION

A voting system for three or more candidates is said to satisfy **monotonicity** provided that, for every election, if some candidate *X* is a winner and a new election is held in which the only ballot change made is for some voter to move this winning candidate *X* higher on his or her ballot (and to make no other changes), then *X* will remain a winner.

As with the CWC and the Pareto condition, monotonicity is a property we would like to see satisfied. But Example 9 (with Candidate *A* in the role of *X*) establishes the following.

The Failure of Monotonicity with the Hare System THEOREM

The Hare system fails to satisfy monotonicity.

The fact that the Hare system does not satisfy monotonicity is considered by many—and with good reason—to be a glaring defect. A 17-voter example in which only a single candidate is eliminated in the first round can also be used to show that the Hare system does not satisfy monotonicity (see Exercise 28 on page 436). For an even more glaring version of this defect, one in which alternative *A* goes from winning to losing because voters move

A from last place on their ballots to first place on their ballots, see Exercise 29 (page 436).

In spite of these drawbacks, the Hare system is used in important ways today. For example, it is essentially the method that was used to choose Rio de Janeiro as the site of the 2016 Summer Olympics. Chicago was eliminated in the first round on the basis of fewest first-place votes, then Tokyo in the second round, and Madrid in the final round.

There are other runoff systems, some more frequently used than the Hare system. One such example is the following.

Description of the Plurality Runoff Method PROCEDURE

Plurality runoff is the voting system in which there is a runoff (i.e., a new election using the same ballots) between the two candidates receiving the most first-place votes. If there are ties, then the runoff is among either those tied for the most first-place votes, or the lone candidate with the most first-place votes along with those tied for the second-most first-place votes (and plurality voting is used).

EXAMPLE 10 Plurality Runoff

The plurality runoff method is somewhat similar in spirit to the Hare system. In fact, you might wonder if they aren't just two different descriptions of the same voting system. That is, you might ask if the plurality runoff method and the Hare system always yield the same winner.

The answer is "no," however, as we now demonstrate. Consider the following sequence of preference list ballots:

	Number of Voters (13)			
Rank	**4**	**4**	**3**	**2**
First	*A*	*B*	*C*	*D*
Second	*B*	*A*	*D*	*C*
Third	*C*	*C*	*A*	*A*
Fourth	*D*	*D*	*B*	*B*

With the plurality runoff method, *A* and *B* initially tie with 4 first-place votes each, with 3 for *C* and 2 for *D*. In the runoff between *A* and *B*, the ballots are as follows:

	Number of Voters (13)			
Rank	**4**	**4**	**3**	**2**
First	*A*	*B*	*A*	*A*
Second	*B*	*A*	*B*	*B*

With the plurality runoff method, *A* is the winner, defeating *B* in the runoff by a score of 9 to 4.

On the other hand, with the Hare system, we find that the only alternative deleted in the first round is *D*, with only 2 first-place votes. With this deletion of *D*, the ballots are as follows:

	Number of Voters (13)			
Rank	**4**	**4**	**3**	**2**
First	*A*	*B*	*C*	*C*
Second	*B*	*A*	*A*	*A*
Third	*C*	*C*	*B*	*B*

A and *B* now have only 4 first-place votes compared to the 5 first-place votes that *C* has. Hence, *A* and *B* are now deleted, leaving *C* as the winner with the Hare system.

Alas, the plurality runoff method also does not satisfy monotonicity. Exercise 25 (page 435) asks you to verify this.

9.4 Insurmountable Difficulties: Arrow's Impossibility Theorem

All the voting systems for three or more candidates that we have discussed turn out to be flawed in one way or another. You may well ask at this point why we don't simply present *one* voting method for the three-candidate case that has all the desirable properties that we want to satisfy. That is, after all, exactly what we did for the two-candidate case (with majority rule filling the bill, and being the only one to do so by May's theorem).

The answer to this question is extremely important. The difficulties in the three-candidate case are not in any way tied to a few particular systems that we present in a text such as this (or that we choose to use in the real world). The fact is, there are difficulties that will be present *regardless* of what voting system is used, and this applies even to voting systems not yet discovered.

Nothing in the remarkable body of work produced by Nobel laureate Kenneth J. Arrow of Stanford University is as well known or widely acclaimed as the result known as **Arrow's impossibility theorem** (see Spotlight 9.2).

Arrow's Impossibility Theorem THEOREM

With three or more candidates and any number of voters, there does not exist—and there never will exist—a voting system that always produces a winner, satisfies the Pareto condition and IIA, and is not a dictatorship.

Arrow's impossibility theorem isn't obvious, and we won't be saying anything about the proof. But we can state and prove a much weaker result of some interest in its own right. This version is taken from the 2008 text *Mathematics and Politics* (cited in the Suggested Readings on page 438), and replaces Arrow's assumption of the Pareto condition and nondictatorship by the CWC.

Kenneth J. Arrow

SPOTLIGHT 9.2

Stanford News Service

For centuries, mathematicians have been in search of a perfect voting system. Finally, in 1951, economist Kenneth Arrow proved that finding an absolutely fair and decisive voting system is impossible. Arrow is the Joan Kenney Professor of Economics and Professor of Operations Research, Emeritus at Stanford University. In 1972, he received the Nobel Memorial Prize in Economic Science for his outstanding work in the theory of general economic equilibrium. His numerous other honors include the 1986 von Neumann Theory Prize for his fundamental contributions to the decision sciences. He has served as president of the American Economic Association, the Institute of Management Sciences, and other organizations. Dr. Arrow talks about the process by which he developed his famous impossibility theorem and his ideas on the laws that govern voting systems:

> My first interest was in the theory of corporations. In a firm with many owners, how do the owners agree when they have different opinions, for example, about the prospects of the company? I was thinking of stockholders. In the course of this, I realized that there was a paradox involved—that majority voting can lead to cycles. I then dropped that discussion because I was frustrated by it.
>
> I happened to be working with The RAND Corporation one summer about a year or two later. They were very interested in applying concepts of rationality, particularly of game theory, to military and diplomatic affairs. That summer, I felt not like an economist but instead like a general social scientist or a mathematically-oriented social scientist. There was tremendous interest in game theory, which was then new.
>
> Someone there asked me, "What does it mean in terms of national interest?" I said, "That's a very simple matter." He then asked me to write a memorandum on the subject. That memorandum led to a sharper formulation of the social-choice question, and I realized that I had been thinking of it earlier in that other context.
>
> Society must choose among a number of alternative policies. These policies may be thought of as quite comprehensive, covering a number of aspects: foreign policy, budgetary policy, or whatever. Each individual member of the society has a preference, or a set of preferences, over these alternatives. I guess you can say one alternative is better than another. These individual preferences have a property I call rationality or consistency, or more specifically, what is technically known as transitivity: If I prefer *a* to *b*, and *b* to *c*, then I prefer *a* to *c*.
>
> Imagine that society has to make these choices among a set. Each individual has a preference ordering, a ranking of these alternatives. But we really want society, in some sense, to give a ranking of these alternatives. You can always produce a ranking, but you would like it to have some properties. One is that, of course, it be responsive in some sense to the individual rankings. Another is that when you finish, you end up with a real ranking, that is, something that satisfies these consistency, or transitivity, properties. And a third condition is that when choosing between a number of alternatives, all I should take into account are the preferences of the individuals among those alternatives. If certain things are possible and some are impossible, I shouldn't ask individuals whether they care about the impossible alternatives, only the possible ones.
>
> It turns out that if you impose the conditions I just stated, there is no method of putting together the individual preferences that satisfies all of them.
>
> The whole idea of the axiomatic method was very much in the air among anybody who studied mathematics, particularly among those who studied the foundations of mathematics. The idea is that if you want to find out something, to find the properties, you say, "What would I like it to be?" [You do this] instead of trying to investigate special cases. I was really accustomed to this approach. Of course, the actual process did involve trial and error.
>
> But I went in with the idea that there was some method of handling this problem. I started

Kenneth J. Arrow *(continued)*

out with some examples. I had already discovered that these led to some problems. The next thing that was reasonable was to write down a condition that I could outlaw. I constructed another example, another method that seemed to meet that problem, and something else didn't seem very right about it. Then I had to postulate that we have some other property. I found I was having difficulty satisfying all of these properties that I thought were desirable, and it occurred to me that they couldn't be satisfied.

After having formulated three or four conditions of this kind, I kept on experimenting. Lo and behold, no matter what I did, there was nothing that would satisfy these axioms. So after a few days of this, I began to get the idea that maybe there was another kind of theorem here, namely, that there was no voting method that would satisfy all the conditions that I regarded as rational and reasonable. It was at this point that I set out to prove it. It turned out to be a matter of only a few days' work.

It should be made clear that my impossibility theorem is really a theorem [showing that] the contradictions are possible, not that they are necessary. What I claim is that given any voting procedure, there will be some possible set of preference orders for individuals that will lead to a contradiction of one of these axioms.

But you say, "Well, okay, since we can't get perfection, let's at least try to find a method that works well most of the time." Then when you do have a problem, you don't notice it as much. So my theorem is not a completely destructive or negative feature any more than the second law of thermodynamics means that people don't work on improving the efficiency of engines. We're told you'll never get 100% efficient engines. That's a fact—and a law. It doesn't mean you wouldn't like to go from 40% to 50%.

A Weak Version of Arrow's Impossibility Theorem THEOREM

With three or more candidates and an odd number of voters, there does not exist—and there never will exist—a voting system that satisfies both the CWC and IIA and that always produces at least one winner in every election.

To see why this is true, we'll handle only the case of exactly three voters. Our plan will be to assume that we have some kind of hypothetical voting system that satisfies both the CWC and IIA, and to show that when confronted by the Condorcet voting paradox ballots, it produces *no* winner.

The argument really comes in three separate, but extremely similar, pieces—one for each of the three candidates. Piece 1 argues that *A* can't be among the winners, piece 2 that *B* can't be among the winners, and piece 3 that *C* can't be among the winners. We'll do piece 1 and leave the others for you. The sequence of ballots that we are considering is the following, which we have already seen has no Condorcet winner:

Rank	Number of Voters (3)		
First	*A*	*B*	*C*
Second	*B*	*C*	*A*
Third	*C*	*A*	*B*

Our starting point, however, will be to ask what our hypothetical voting rule must do when confronted by a slightly different sequence of ballots:

Rank	Number of Voters (3)		
First	*A*	*C*	*C*
Second	*B*	*B*	*A*
Third	*C*	*A*	*B*

Here, *C* is clearly a Condorcet winner, and thus it must be the unique winner of the election contested under our hypothetical voting rule. Therefore, *C* is a winner and *A* is a nonwinner (for *this* sequence of ballots).

However, because our hypothetical voting rule satisfies IIA, we know that *A* will remain a nonwinner so long as no voter reverses his or her ordering of *A* and *C*. But to arrive at the voting paradox ballots, we can move up *B* (the candidate that is irrelevant to *A* and *C*) one slot in the second voter's list.

Thus, because of IIA, we know that *A* is a nonwinner when our voting rule is confronted by the voting paradox ballots. This is one-third of the argument. As we mentioned before, similar arguments (see Exercise 32 on page 436) show that *B* and *C* are also nonwinners when our voting rule is confronted by the voting paradox ballots.

We conclude this section with an example that yields a somewhat surprising application of Arrow's impossibility theorem in the context of what are called *social welfare functions.*

EXAMPLE 11 Organ Transplant Policies and Arrow's Impossibility Theorem

Finding an equitable procedure for determining a rank ordering of patients in need of an organ transplant is complicated: there are several criteria that should be considered in arriving at such a "priority ranking." Three such criteria are, for example, (1) the length of time that a patient has been waiting, (2) the probability of success as measured by the numbers of antigens that the patient and donor have matched, and (3) the fraction of the population unsuitable as donors for this potential recipient due to the presence of certain antibodies. A further discussion of these issues occurs in Section 13.3.

Each of the three criteria gives us a ranking (with ties) of the patients according to the more appropriate recipient of the next available organ, according to that particular criterion. Although these rankings are often determined by measurements, the use of different scales for different criteria muddies the water sufficiently so that you might want to work simply with the rankings derived from the measurements, as opposed to working directly with the measurements themselves. This is the context in which we will frame the problem.

So what does the search for a procedure to rank-order potential recipients of an organ have to do with voting? In a sense, *everything,* if looked at the right way. We can think of each criterion as a "voter" and each potential recipient as an "alternative." The procedure that we seek is what social choice theorists call a *social welfare function.* It differs from a social choice procedure in that the result of an election is not a single winner or a group tied for the win, but a listing of the alternatives—the priority ranking, in our organ-transplant situation.

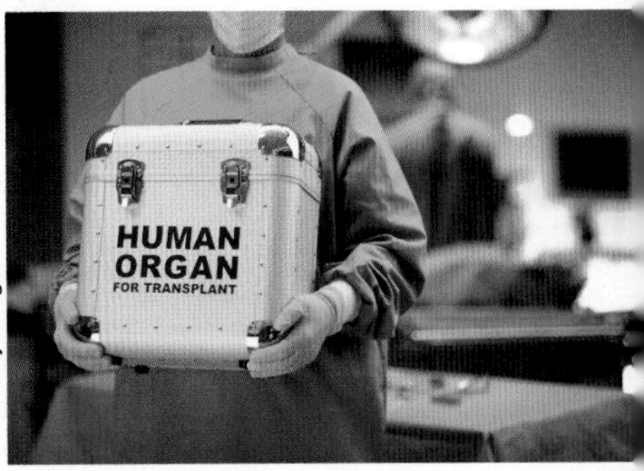

sturti/Getty Images

For a moment, let's return to the particular task of seeking a priority ranking of the potential recipients of an organ based on how they are ranked according to each of several criteria, like the three we mentioned earlier. What "reasonable" properties might we expect any such procedure to satisfy? Consider the following:

1. If one potential recipient *A* is ranked above another potential recipient *B* with respect to every single criterion, then we should expect *A* to be ranked above *B* in the priority ranking.
2. If potential recipient *A* is ranked above potential recipient *B* in the priority ranking, and there are subsequent changes in how potential recipients are ranked with respect to one or more of the criteria, then potential recipient *B* should not be ranked above potential recipient *A* in the priority ranking unless *B* has moved from being below *A* to being above *A* with respect to at least one criterion.
3. No single criterion should dominate, in the sense that one potential recipient *A*'s ranking above another potential recipient *B*'s ranking, with respect to that criterion, guarantees that *A* will be ranked above *B* in the priority ranking.

If we accept these as being required of any "reasonable" procedure, then we have a striking (and highly non-obvious) fact to report: Our task of finding a reasonable procedure is impossible! In fact, this is precisely the statement of Arrow's impossibility theorem in the context of social welfare functions: There is no social welfare function (for three or more alternatives) that satisfies Pareto (our first condition above), IIA (our second condition above), and nondictatorship (our third condition above).

9.5 A Better Approach? Approval Voting

Elections in which there are only two candidates present no problem. Majority rule is, as we have seen, an eminently successful voting system in both theory and practice. If there are three or more candidates, however, the situation changes quite dramatically. While several voting systems suggest themselves (plurality, the Borda count, sequential pairwise voting, and the Hare system), each fails to satisfy one or more desired properties (the CWC, IIA, the Pareto condition, and monotonicity). Manipulability is an ever-present problem, as we'll see in the next chapter. Moreover, when all is said and done, Arrow's impossibility theorem says that any search for an ideal voting system of the kind that we have discussed is doomed to failure.

Where does this leave us? More than intellectual issues are at stake here: More than 550,000 elected officials serve in approximately 80,000 governments in the United States. Whether it is a small academic department voting on the best senior thesis or a democratic country electing a new leader, multicandidate elections will be contested in one way or another. If there is no perfect voting system—and perhaps not even a best voting system (whatever that may mean; i.e., best in what way?)—what can we do?

Perhaps the answer is that different situations lend themselves to different voting systems, and what is required is a judicious blend of common sense with an awareness of what the mathematical theory has to say. For example, while both the Hare system and the Borda count are subject to manipulability, it seems easier to manipulate the latter. Thus, people may tend to vote more sincerely, rather than strategically, if the Hare system is used instead of the Borda count. This may be a consideration when choosing a voting system for a faculty governance system, for example.

For national political elections, there are also practical considerations. The kind of ballot that we are considering (a preference list ballot) is certainly more complicated than the ballots we now employ. There is, however, a voting system that avoids the practical difficulties caused by the type of ballot being used that has much else to commend it. It is called **approval voting.**

Description of Approval Voting PROCEDURE

Under approval voting, each voter is allowed to give one vote to as many of the candidates as he or she finds acceptable. No limit is set on the number of candidates for whom an individual can vote. Voters show disapproval of other candidates simply by not voting for them. The winner under approval voting is the candidate who receives the largest number of approval votes. This approach is also appropriate in situations where more than one candidate can win, for example, in electing new members to an exclusive society such as the National Academy of Sciences or the Baseball Hall of Fame.

EXAMPLE 12 Approval Voting

To illustrate approval voting, suppose that we have nine members of a mathematics department who are trying to choose among five finalists for an open faculty position. They decide to use approval voting—the ballots are indicated in the following table. An X indicates an approval vote. For example, Voter 1, in the first column, approves of Candidates *A*, *C*, and *D*.

	Voter (department member)								
Candidate	**1**	**2**	**3**	**4**	**5**	**6**	**7**	**8**	**9**
A	X	X		X		X	X	X	
B		X	X	X		X	X	X	X
C	X	X			X		X		
D	X		X	X	X			X	
E		X	X		X		X	X	

Counting the Xs in each row shows that six department members (1, 2, 4, 6, 7, and 8) approved of Candidate *A*, seven approved of Candidate *B*, four approved of Candidate *C*, and five approved of *D* and *E*. Thus, Candidate *B* wins, with seven approval votes.

Approval voting was proposed independently by several analysts in the 1970s. Probably the best-known official elected by approval voting today is the secretary-general of the United Nations. In the 1980s, several academic and professional societies initiated the use of approval voting. Examples include the Institute of Electrical and Electronics Engineers (IEEE), with about 400,000 members, and the National Academy of Sciences. In Eastern Europe and some former Soviet republics, approval voting has been used in the form wherein one disapproves of (instead of approving of) as many candidates as one wishes.

Is approval voting the perfect voting system? Certainly not. For example, the type of ballot that is used limits the extent to which voter preferences can be expressed. However, it is certainly a voting system with much potential, and the reader wishing to explore it in more detail can start with Brams and Fishburn's 1983 monograph, listed in the Suggested Readings on page 438.

Review Vocabulary

Agenda An ordering of the candidates to be considered, which is often used in sequential pairwise voting. (p. 418)

Approval voting A method of electing one or more candidates from a field of several in which each voter submits a ballot that indicates which candidates he or she approves of. Winning is determined by the total number of approvals that a candidate obtains. (pp. 428, 429)

Arrow's impossibility theorem Kenneth J. Arrow's discovery that any voting system can give undesirable outcomes. (p. 424)

Borda count A voting system for elections with several candidates in which points are assigned to voters' preferences; these points are summed for each candidate to determine a winner. The actual point totals are referred to as a candidate's *Borda score*. (p. 414)

Condorcet's method A voting system for elections with several candidates in which a candidate is a winner precisely when he or she would, on the basis of the ballots cast, defeat every other candidate in a one-on-one contest. (p. 408)

Condorcet winner A candidate in an election who, based on the ballots, would have defeated every other candidate in a one-on-one contest. (p. 412)

Condorcet winner criterion (CWC) A voting system satisfies the Condorcet winner criterion if, for every election in which there is a Condorcet winner, that candidate wins the election when that voting system is used. (p. 413)

Condorcet's voting paradox The observation that there are elections in which Condorcet's method yields no winner. (p. 411)

Hare system A voting system for elections with several candidates in which candidates are successively eliminated in an order based on the number of first-place votes. (p. 420)

Independence of irrelevant alternatives (IIA) A voting system satisfies independence of irrelevant alternatives if the only way a candidate (called *A*) can go from losing one election to being among the winners of a new election (with the same set of candidates and voters) is for at least one voter to reverse his or her ranking of *A* and the previous winner. (pp. 416, 417)

Manipulability A voting system is subject to manipulability (or is manipulable) if there are elections in which it is to a voter's advantage to submit a ballot that misrepresents his or her true preferences. (p. 414)

Majority rule A voting system for elections with two candidates (and an odd number of voters) in which the candidate preferred by more than half the voters is the winner. (p. 407)

May's theorem Kenneth May's discovery that, for two alternatives and an odd number of voters, majority rule is the only voting system satisfying three natural properties. (p. 408)

Monotonicity A voting system satisfies monotonicity provided that ballot changes favorable to one candidate (and not favorable to any other candidate) can never hurt that candidate. (p. 422)

Pareto condition A voting system satisfies the Pareto condition provided that every voter's ranking of one candidate higher than another precludes the possibility of this latter candidate winning. (p. 419)

Plurality runoff A voting system for elections with several candidates in which, assuming there are no ties, there is a runoff between the two candidates receiving the most first-place votes. (p. 423)

Plurality voting A voting system for elections with several candidates in which the candidate with the most first-place votes wins. (p. 412)

Preference list ballot A ballot that ranks the candidates from most preferred to least preferred, with no ties. (p. 406)

Sequential pairwise voting A voting system for elections with several candidates in which one starts with an agenda and pits the candidates against each other in one-on-one contests (based on preference list ballots), with losers being eliminated as one moves along the agenda. (p. 418)

Self Check Answers

1. Suppose the two candidates are A and B and that A has won the election using majority rule, with n votes to B's m votes where $n > m$. First, if any two voters exchange ballots, then A still receives n votes (although from a slightly different collection of voters) and B still receives m votes (again from a slightly different collection of voters). Thus, A is still the winner. Second, if every voter were to change his or her ballot, then B would receive the n votes that A previously received and vice versa. Hence, B would be the new winner with n votes. Third, if some voter who had voted for B changed his or her vote to one for A, then A's total would become $n + 1$ and B's total would become $m - 1$. Hence, A would still win.

2. Beth gets the next offer using Condorcet's method because she defeats Adam and Dan one on one by identical scores of 3 to 2.

3. No. According to the definition, a voting system is manipulable only if there is an election in which a voter gets a *more* preferred outcome (rather than a *less* preferred outcome) by submitting a disingenuous ballot.

4. Arizona would have lost 24 points (for losing a second-place vote) and 22 points (for losing a fourth-place vote), for a total loss of 46 points. Hence, if Arizona received 17 third-place votes instead of 15, it would gain back the 46 lost points and still have a total of 1517.

5. No, Candidate D is not a Condorcet winner, because D loses to B in a one-on-one contest.

6. No, Candidate A is not a Condorcet winner, because A loses to C (and, incidentally, to D as well) in a one-on-one contest.

7. No. An example is given in the text on page 419.

Skills Check

1. A preference list ballot
(a) indicates only a voter's top choice.
(b) is a rank ordering of the candidates, with no ties.
(c) will often have ties.

2. To say that a voting system treats all voters equally means that ___________.

3. To say that a voting system for two candidates treats both candidates equally means that
(a) each wins if he or she receives all the votes.
(b) if all voters reverse their ballots, the election outcome changes.
(c) if any two voters exchange ballots, the election outcome is unchanged.

4. A two-candidate voting system is monotone if ___________.

5. May's theorem says that, with an odd number of voters, among all two-candidate voting systems that never result in a tie, majority rule is the only one that
(a) treats both candidates equally.
(b) treats both candidates equally and all voters equally.
(c) treats both candidates equally and all voters equally and is monotone.

6. When a choice is being made between two candidates, the first type of voting system to suggest itself is ___________.

7. In this chapter, the "number of voters assumption" refers to the assumption that
(a) there is more than one voter.
(b) the number of voters is odd.
(c) the number of voters is even.

8. The winner with Condorcet's method is the candidate who ___________.

9. Which of the following does not satisfy exactly two of the conditions in May's theorem?
(a) A dictatorship
(b) Imposed rule
(c) Minority rule
(d) None of the above

10. The Hare system fails to satisfy ___________.

11. Suppose Condorcet's method is being used in an election in which Candidate A is ranked first on more than half of the ballots. Then Candidate A is
(a) the unique winner.
(b) among the winners, but there may be others.
(c) not necessarily among the winners.

12. The flaw in Condorcet's method is that it ___________.

13. Condorcet's voting paradox refers to the fact that

(a) people vote even though an individual vote virtually never affects the outcome of an election.

(b) the statement "This statement is false" can be neither true nor false.

(c) there are elections in which there is no winner using Condorcet's method.

14. With plurality voting, the winner is the candidate who ___________.

15. George W. Bush's defeat of Al Gore in the state of Florida in the 2000 presidential election shows that

(a) plurality voting does not satisfy the CWC.

(b) majority rule is not monotone.

(c) the Borda count does not satisfy IIA.

16. With the Borda count, the election winner is the candidate who ___________.

17. Rather than assigning points and doing arithmetic, the Borda score of a candidate can be found by

(a) scanning the ballots and counting the number of occurrences of other candidates below that one.

(b) counting the number of first-place votes and multiplying by 4.

(c) counting the number of candidates that it defeats one on one.

18. Independence of irrelevant alternatives says that a nonwinner can never switch to being a winner unless at least one voter changes his or her ballot in a way that ___________.

19. The Borda count fails to satisfy

(a) monotonicity.

(b) the Pareto condition.

(c) IIA.

20. The term *single transferrable vote system* is sometimes used to refer to the voting system in this chapter called ___________.

21. A voting system satisfies the CWC provided that, in every election,

(a) there is a Condorcet winner and this candidate is the winner of the election.

(b) if there is a Condorcet winner, then this candidate is among the winners of the election.

(c) if there is a Condorcet winner, then this candidate is the unique winner of the election.

22. Sequential pairwise voting is the voting system in which ___________.

23. Sequential pairwise voting fails to satisfy

(a) monotonicity.

(b) the Pareto condition.

(c) the CWC.

24. The voting system in which a voter can vote for as many candidates as he or she wishes to vote for is called ___________.

25. Suppose the Borda count is being used in an election in which Candidate A is ranked first on more than half of the ballots. Then Candidate A is

(a) the unique winner.

(b) among the winners, but there may be others.

(c) not necessarily among the winners.

26. A voting system is manipulable if there are elections in which ____.

27. Suppose the Hare system is being used in an election in which Candidate A is ranked first on more than half of the ballots. Then Candidate A is

(a) the unique winner.

(b) among the winners, but there may be others.

(c) not necessarily among the winners.

28. Both the Hare system and the plurality runoff method are defective in that ___________.

29. Arrow's impossibility theorem says that with three or more candidates and any number of voters, there is no voting system that

(a) is not a dictatorship.

(b) satisfies IIA and is not a dictatorship.

(c) satisfies the Pareto condition and IIA, and is not a dictatorship.

(d) always produces a winner, satisfies the Pareto condition and IIA, and is not a dictatorship.

30. The weak version of Arrow's impossibility theorem asserts that, with three or more candidates and an odd number of voters, there is no voting system that ___________.

Chapter 9 Exercises

Challenge

Discussion

9.1 An Introduction to Social Choice

9.2 Majority Rule and Condorcet's Method

1. In a few sentences, explain why minority rule (the voting procedure for two alternatives that is described on page 408) satisfies conditions 1 and 2 on page 407, but not condition 3.

2. In a few sentences, explain why imposed rule (the voting procedure for two alternatives that is described on pages 407–408) satisfies conditions 1 and 3 on page 407, but not condition 2.

3. In a few sentences, explain why a dictatorship (the voting procedure for two alternatives that is described on page 407) satisfies conditions 2 and 3 on page 407, but not condition 1.

4. Find (or invent) a voting rule for two alternatives that satisfies condition

(a) 1 on page 407, but neither 2 nor 3.

(b) 2 on page 407, but neither 1 nor 3.

(c) 3 on page 407, but neither 1 nor 2.

5. Construct a real-world example (perhaps involving yourself and two friends) where the individual preference lists for three alternatives are as in the voting paradox of Condorcet.

6. Condorcet's voting paradox shows that with three voters (or three equal-size groups of voters) and the three alternatives *A*, *B*, and *C*, it is possible to have two-thirds prefer *A* to *B*, two-thirds prefer *B* to *C*, and two-thirds prefer *C* to *A*. Find four preference lists that show that with four voters and the four alternatives *A*, *B*, *C*, and *D*, it is possible to have three-fourths prefer *A* to *B*, three-fourths prefer *B* to *C*, three-fourths prefer *C* to *D*, and three-fourths prefer *D* to *A*.

7. Generalize the result in Exercise 6 from four alternatives to *n* alternatives: $A_1, \ldots, A_n$.

8. The mathematics department is hiring a new faculty member and the five-person hiring committee has interviewed four candidates: Adam, Beth, Carol, and Dan. They have decided to use Condorcet's method on their five ballots (reproduced in the following table). Who gets the offer?

	Voter 1	Voter 2	Voter 3	Voter 4	Voter 5
First choice	Adam	Dan	Carol	Dan	Beth
Second choice	Beth	Beth	Adam	Adam	Adam
Third choice	Carol	Carol	Beth	Carol	Dan
Fourth choice	Dan	Adam	Dan	Beth	Carol

9. Suppose that votes on the five mathematics department ballots described in Exercise 8 were distributed according to the table below. Who would get the offer now?

	Voter 1	Voter 2	Voter 3	Voter 4	Voter 5
First choice	Dan	Beth	Beth	Carol	Carol
Second choice	Beth	Adam	Adam	Beth	Adam
Third choice	Adam	Dan	Dan	Dan	Dan
Fourth choice	Carol	Carol	Carol	Adam	Beth

9.3 Other Voting Systems for Three or More Candidates

10. Plurality voting is illustrated by the 1980 U.S. Senate race in New York among Alfonse D'Amato (*D*, a conservative), Elizabeth Holtzman (*H*, a liberal), and Jacob Javits (*J*, also a liberal). Reasonable estimates (based largely on exit polls) suggest that voters ranked the candidates according to the following table:

22%	23%	15%	29%	7%	4%
D	*D*	*H*	*H*	*J*	*J*
H	*J*	*D*	*J*	*H*	*D*
J	*H*	*J*	*D*	*D*	*H*

(a) Is there a Condorcet winner?

(b) Who won using plurality voting?

11. Condorcet's method can be used to create a new voting system that operates in a manner similar to the Hare system in that it involves repeatedly deleting candidates that are "least preferred." But now we use Condorcet's method to decide what *least preferred* means, and we do this in the following clever kind of way. We tip the ballots *upside-down* and we look for a Condorcet winner from these inverted ballots. Intuitively, a candidate that wins when all the ballots are reversed is "least preferred," according to the original ballots. So this new system works by tipping the ballots upside-down and then repeatedly deleting a Condorcet winner, if there is one. Use this new system to find the winner for the following ballots:

	Voter 1	Voter 2	Voter 3	Voter 4	Voter 5
First choice	A	D	C	D	B
Second choice	B	B	A	A	A
Third choice	C	C	B	C	D
Fourth choice	D	A	D	B	C

12. Use the voting system introduced in the preceding problem to find the winner for the following ballots:

	Voter 1	Voter 2	Voter 3	Voter 4	Voter 5
First choice	D	B	B	C	C
Second choice	B	A	A	B	A
Third choice	A	D	D	D	D
Fourth choice	C	C	C	A	B

13. (Everyone wins.) Consider the following set of preference lists:

	Number of Voters (9)						
Rank	**3**	**1**	**1**	**1**	**1**	**1**	**1**
First	A	A	B	B	C	C	D
Second	D	B	C	C	B	D	C
Third	B	C	D	A	D	B	B
Fourth	C	D	A	D	A	A	A

Note that the first list is held by three voters, not just one. Calculate the winner using

(a) plurality voting.

(b) the Borda count.

(c) the Hare system.

(d) sequential pairwise voting with the agenda *A, B, C, D.*

14. Consider the following set of preference lists:

	Number of Voters (7)				
Rank	**2**	**2**	**1**	**1**	**1**
First	C	D	C	B	A
Second	A	A	D	D	D
Third	B	C	A	A	B
Fourth	D	B	B	C	C

Calculate the winner using

(a) plurality voting.

(b) the Borda count.

(c) the Hare system.

(d) sequential pairwise voting with the agenda *B, D, C, A.*

15. Consider the following set of preference lists:

	Number of Voters (5)				
Rank	**1**	**1**	**1**	**1**	**1**
First	A	B	C	D	E
Second	B	C	B	C	D
Third	E	A	E	A	C
Fourth	D	D	D	E	A
Fifth	C	E	A	B	B

Calculate the winner using

(a) plurality voting.

(b) the Borda count.

(c) the Hare system.

(d) sequential pairwise voting with the agenda *A, B, C, D, E.*

16. Consider the following set of preference lists:

	Number of Voters (7)				
Rank	**2**	**2**	**1**	**1**	**1**
First	A	B	A	C	D
Second	D	D	B	B	B
Third	C	A	D	D	A
Fourth	B	C	C	A	C

Calculate the winner using
(a) plurality voting.
(b) the Borda count.
(c) the Hare system.
(d) sequential pairwise voting with the agenda *B*, *D*, *C*, *A*.

17. Consider the following set of preference lists:

Rank	Number of Voters (7)				
	2	2	1	1	1
First	C	E	C	D	A
Second	E	B	A	E	E
Third	D	D	D	A	C
Fourth	A	C	E	C	D
Fifth	B	A	B	B	B

Calculate the winner using
(a) plurality voting.
(b) the Borda count.
(c) the Hare system.
(d) sequential pairwise voting with the agenda *A*, *B*, *C*, *D*, *E*.

18. In a few sentences, explain why Condorcet's rule satisfies
(a) the Pareto condition.
(b) monotonicity.

19. In a few sentences, explain why plurality voting satisfies
(a) the Pareto condition.
(b) monotonicity.

20. In a few sentences, explain why the Borda count satisfies
(a) the Pareto condition.
(b) monotonicity.

21. In a few sentences, explain why sequential pairwise voting satisfies
(a) the CWC.
(b) monotonicity.

22. In a few sentences, explain why the Hare system satisfies the Pareto condition.

23. In a few sentences, explain why the plurality runoff method satisfies the Pareto condition.

24. Use the following ballots to show that the plurality runoff method does not satisfy the CWC:

Rank	Number of Voters (5)		
	2	2	1
First	A	B	C
Second	C	C	B
Third	B	A	A

25. Use the following ballots to show that the plurality runoff method does not satisfy monotonicity:

Rank	Number of Voters (13)				
	4	3	3	2	1
First	A	B	C	D	E
Second	B	A	A	B	D
Third	C	C	B	C	C
Fourth	D	D	D	A	B
Fifth	E	E	E	E	A

26. Consider the following two elections among Candidates *A*, *B*, and *C*:

Rank	Number of Voters (4)			
	1	1	1	1
First	A	A	B	C
Second	B	B	C	B
Third	C	C	A	A

Rank	Number of Voters (4)			
	1	1	1	1
First	A	A	B	B
Second	B	B	C	C
Third	C	C	A	A

(a) Use these two elections to show that plurality voting does not satisfy IIA.
(b) Use these two elections to show that the Hare system does not satisfy independence of irrelevant alternative.

27. Construct ballots for the alternatives *A*, *B*, and *C* to show that the Borda count does not satisfy the CWC.

28. Show that the nonmonotonicity of the Hare system can also be demonstrated by the following 17-voter, 4-alternative election. (In a number of recent books, this example is used to show the nonmonotonicity of the Hare system. The 13-voter, 3-alternative example given in the text was pointed out to us by Matt Gendron when he was an undergraduate at Union College.)

	Number of Voters (17)			
Rank	**7**	**5**	**4**	**1**
First	*A*	*C*	*B*	*D*
Second	*D*	*A*	*C*	*B*
Third	*B*	*B*	*D*	*A*
Fourth	*C*	*D*	*A*	*C*

29. The following example illustrates how badly the Hare system can fail to satisfy monotonicity. Consider the following sequence of preference lists:

	Number of Voters (21)			
Rank	**7**	**6**	**5**	**3**
First	*A*	*B*	*C*	*D*
Second	*B*	*A*	*B*	*C*
Third	*C*	*C*	*A*	*B*
Fourth	*D*	*D*	*D*	*A*

(a) Show that *A* is the unique winner if the Hare system is used.

(b) Find the winner using the Hare system in the new election, wherein the three voters on the right all move *A* from last place on their preference lists to first place on their preference lists.

30. In a few sentences, explain why, with an odd number of voters,

(a) sequential pairwise voting always yields a unique winner.

(b) we can never have exactly two winners with the Hare system.

31. Suppose there are three voters and three alternatives *A*, *B*, and *C*.

(a) If each alternative has exactly one first-place vote, what is the election outcome if the Hare system is used? What if plurality runoff is used?

(b) If an alternative has two or more first-place votes, what is the election outcome if the Hare system is used? What if plurality runoff is used?

(c) Can the Hare system and plurality runoff yield different election outcomes when there are three voters and three alternatives? Explain your answer in one sentence.

9.4 Insurmountable Difficulties: Arrow's Impossibility Theorem

32. Complete the proof of the version of Arrow's impossibility theorem from the text by showing that neither *B* nor *C* can be a winner in the situation described. (Your argument will be almost word for word the same as the proofs in the text.)

9.5 A Better Approach? Approval Voting

33. The 10 members of a board vote by approval voting on eight candidates for new positions on their board, as indicated in the following table. An X indicates an approval vote. For example, Voter 1, in the first column, approves of Candidates *A, D, E, F,* and *G*, and disapproves of *B, C,* and *H*.

	Voters									
Candidate	**1**	**2**	**3**	**4**	**5**	**6**	**7**	**8**	**9**	**10**
A	X	X	X			X	X	X		X
B		X	X	X	X	X	X	X	X	
C			X					X		
D	X	X	X	X	X		X	X	X	X
E	X		X		X		X		X	
F	X		X	X	X	X	X	X		X
G	X	X	X	X	X			X		
H		X		X		X		X		X

(a) Which candidate is chosen for the board if just one of them is to be elected?

(b) Which candidates are chosen if the top four are selected?

(c) Which candidates are elected if 80% approval is necessary and at most four are elected?

(d) Which candidates are elected if 60% approval is necessary and at most four are elected?

34. The 45 members of a school's football team vote on three nominees, *A*, *B*, and *C*, by approval voting for the award of "most improved player," as indicated in the following table. An X indicates an approval vote.

	Number of Voters (45)							
Nominee	**7**	**8**	**9**	**9**	**6**	**3**	**1**	**2**
A	X			X	X		X	
B		X		X		X	X	
C			X		X	X	X	

(a) Which nominee is selected for the award?

(b) Which nominee gets announced as runner-up for the award?

(c) Note that two of the players "abstained"; that is, approved of none of the nominees. Note also that one person approved of all three of the nominees. What would be the difference in the outcome if one were to "abstain" or "approve of everyone"?

Chapter Review

35. In a sentence or two, explain why it's impossible, with an odd number of voters, to have two distinct candidates win the same election using Condorcet's method.

36. Consider the following set of preference lists:

	Number of Voters (7)						
Rank	**1**	**1**	**1**	**1**	**1**	**1**	**1**
First	*C*	*D*	*C*	*B*	*E*	*D*	*C*
Second	*A*	*A*	*E*	*D*	*D*	*E*	*A*
Third	*E*	*E*	*D*	*A*	*A*	*A*	*E*
Fourth	*B*	*C*	*A*	*E*	*C*	*B*	*B*
Fifth	*D*	*B*	*B*	*C*	*B*	*C*	*D*

Calculate the winner using

(a) plurality voting.

(b) the Borda count.

(c) sequential pairwise voting with the agenda *A*, *B*, *C*, *D*, *E*.

(d) the Hare system.

37. An interesting variant of the Hare system was proposed by psychologist Clyde Coombs. It operates exactly as the Hare system, but instead of deleting alternatives with the fewest first-place votes, it deletes those with the most last-place votes.

(a) Use the Coombs procedure to find the winner if the ballots are as in Exercise 36.

(b) Show that for two voters and three alternatives, it is possible to have ballots that result in one candidate winning if the Coombs procedure is used and a tie between the other two if the Hare system is used.

38. In a few sentences, explain why the plurality runoff method can never elect a candidate ranked last on a majority of ballots, assuming there are no ties for first or second place in the voting.

39. Produce ballots showing that plurality voting can, in fact, elect a candidate ranked last on a majority of the ballots.

40. A voting system is said to satisfy the *majority criterion* if a candidate ranked first by a majority of the voters is always among the winners. For each of the following, either give a sentence or two explaining why the answer is "yes," or give a collection of ballots showing that the answer is "no."

(a) Does plurality voting satisfy the majority criterion?

(b) Does the Borda count satisfy the majority criterion?

(c) Does the Hare system satisfy the majority criterion?

(d) Does sequential pairwise voting satisfy the majority criterion?

41. Every voting system *P* can be used to create a new voting system P^* in the manner that we did with Condorcet's method in Exercise 11 on page 434. That is, P^* works as follows: We tip the ballots upside-down and repeatedly delete the "winners" using the voting system *P*. The last candidate (or group of candidates) to be eliminated is the winner. Describe in one sentence the voting system *P* that yields the Hare system as P^*.

Writing Projects

1. In the 2000 presidential election in Florida, the final results were as follows:

Candidates	Number of Votes	Percentage of Votes
Bush	2,911,872	49
Gore	2,910,942	49
Nader	97,419	2
Buchanan	17,472	0

Making reasonable assumptions about voters' preference schedules, give a one-page discussion of how the election might have turned out under the different voting methods discussed in this chapter.

2. Frequently in presidential campaigns, the winner of the first few primaries is given front-runner status that can lead to the nomination of his or her party. Moreover, there are often several candidates running in early primaries such as New Hampshire. In one page, consider a recent election and discuss how the nominating process might have proceeded through the campaign if approval voting had been used to decide primary winners.

Suggested Readings

BLACK, DUNCAN. *The Theory of Committees and Elections,* Kluwer, Dordrecht, The Netherlands, 1986. The historical highlights and development of voting methods in the 19th and 20th centuries are traced in this economist's volume.

BRAMS, STEVEN J., and PETER C. FISHBURN. *Approval Voting,* Birkhäuser, Boston, 1983. This volume is a research-level work on development in the recently popular (but rediscovered) method now called *approval voting.* The first chapter, however, is an elementary exposition of this voting method and its uses.

SAARI, DONALD G. *Chaotic Elections! A Mathematician Looks at Voting,* American Mathematical Society, Providence, R.I., 2001. This expository book begins with the 2000 presidential election and discusses a number of paradoxical results in voting.

SZPIRO, GEORGE G. *Numbers Rule: The Vexing Mathematics of Democracy, from Plato to the Present,* Princeton University Press, Princeton, N.J., 2010. The subtitle says it all—a very enjoyable read.

TAYLOR, ALAN D., and ALLISON M. PACELLI. *Mathematics and Politics: Strategy, Voting, Power, and Proof,* 2d ed., Springer-Verlag, New York, 2008. Chapters 1 and 7 give an expanded treatment of the topics considered here, with proofs included. This book is also intended for nonmath majors.

The Manipulability of Voting Systems

10

Lisa S./Shutterstock

People know almost by instinct that you sometimes can achieve the election result you prefer by submitting a ballot that misrepresents your actual preferences. This type of strategic voting is called **manipulation,** and a ballot that misrepresents a voter's true preferences is referred to as an **insincere** or **disingenuous ballot.**

All three of these terms—manipulation, insincere, disingenuous—are widely used in the social-choice literature, but in daily life, we use these terms pejoratively; they aren't exactly warm praise. In fact, your choice to manipulate

Popperfoto/Getty Images

Charles L. Dodgson was a mathematical lecturer at Oxford University. Dodgson, who used the pen name Lewis Carroll, wrote on mathematical topics and even manipulability. But he achieved greater fame for his satirical works. In the *Alice* books, he refers to the mathematical operations as Ambition, Distraction, Uglification, and Derision, and his characters play nonsensical, easily manipulated games.

a voting system typically is no more inherently evil than your submission of a sealed bid for a lamp at an auction at a price considerably below its actual worth. *Strategyproof*—a term with considerably less negative content—is sometimes used in place of *nonmanipulable,* but the latter is more common, so we'll stick with it here.

Historical references to the manipulability of voting systems include a comment by 19-century mathematician C. L. Dodgson (1832–1898), better known by the pseudonym Lewis Carroll, under which he wrote *Alice's Adventures in Wonderland* (1865). Dodgson commented that voters have a tendency to "adopt a principle of voting which makes it more of a game of skill than a true test of the wishes of the electors" and that it would be "better for elections to be decided according to the wishes of the majority than of those who have the most skill at the game."

But the most famous manipulability quote in the history of social choice is Jean Charles de Borda's reply to a colleague who had pointed out to him how easily the Borda count can be manipulated. "My scheme," Borda replied, "is only intended for honest men!"

Do there exist other "schemes" that need not be intended for just "honest men"? How do the voting systems we saw in Chapter 9 stack up when it comes to manipulability? And is there an impossibility result—like Arrow's theorem—lurking out there? We shall see.

10.1 An Introduction to Manipulability

Let's look at an example to illustrate how the Borda count can be manipulated.

EXAMPLE 1 Manipulating the Borda Count with Four Candidates and Two Voters

Suppose there are two voters and four candidates, and suppose the true preferences of the voters are reflected in the following ballots:

Voter 1	Voter 2
A	*B*
B	*C*
C	*A*
D	*D*

Using the Borda count with point values 3, 2, 1, 0 (or by counting the number of occurrences of other candidates below the one in question, as described in Section 9.3), we see that the Borda scores of the four candidates are as follows:

The Borda score of *A* is 4.
The Borda score of *B* is 5.
The Borda score of *C* is 3.
The Borda score of *D* is 0.

Thus, Candidate *B* wins this election. Voter 1, however, would have preferred to see Candidate *A*—her top choice, according to her true preferences—win this election rather than Candidate *B*, her second choice.

Assume that Voter 1 had known that Voter 2 planned to submit the ballot that he cast above. Could Voter 1 have secured a victory for Candidate *A* by submitting a disingenuous ballot?

The answer here, as we'll show, turns out to be "yes." The intuition is fairly transparent: Voter 1 wants to pretend that *B* is not her second choice, but her last choice. Let's see if this is enough to bring about the desired switch in winner from *B* to *A*. The new ballots and Borda scores are as follows:

Voter 1	Voter 2
A	*B*
C	*C*
D	*A*
B	*D*

The Borda score of *A* is 4.
The Borda score of *B* is 3.
The Borda score of *C* is 4.
The Borda score of *D* is 1.

Close, but not quite what we wanted: Candidates *A* and *C* now tie for the win, and we wanted the winner to be just Candidate *A*. But a moment's inspection reveals that Voter 1 can achieve this if, in addition to plunging Candidate *B* to the bottom of her ballot, she also flip-flops *C* and *D*. That is, the desired ballots (and Borda scores) that yield Candidate *A* as the sole winner are as follows:

Voter 1	Voter 2
A	*B*
D	*C*
C	*A*
B	*D*

The Borda score of *A* is 4.
The Borda score of *B* is 3.
The Borda score of *C* is 3.
The Borda score of *D* is 2.

Thus, Voter 1 can change her ballot and—with Voter 2 making no change at all—cause the election outcome to go from *B* to *A*. Moreover—and this is very important—Voter 1 prefers *A* to *B*! The reason we know that Voter 1 prefers *A* to *B* is that we are assuming the original ballots represented the voters' true preferences, and Voter 1 ranked *A* over *B* on her original ballot.

Self Check 1

Voter 2 could also change the outcome of the original election by moving *B* to the bottom of his ballot (with Voter 1 making no change in her ballot). Explain why this would be a pointless thing for Voter 2 to do.

In presenting an example of a voting system's susceptibility to manipulation, we will typically present two elections—the original one ("Election 1") in which we assume all ballots are sincere, and the one that contains a disingenuous ballot from a voter ("Election 2"). For example, if we collect the pieces of what we just did, this instance of manipulation of the Borda count could be presented succinctly as follows:

Election 1		
Rank	**Number of Voters (2)**	
First choice	*A*	*B*
Second choice	*B*	*C*
Third choice	*C*	*A*
Fourth choice	*D*	*D*

Election 2		
Rank	**Number of Voters (2)**	
First choice	*A*	*B*
Second choice	*D*	*C*
Third choice	*C*	*A*
Fourth choice	*B*	*D*

There are three aspects of manipulation taking place in this example that deserve comment.

First, there is only one voter (the voter on the left, in this example) changing his or her ballot; we call this a **unilateral change** in ballot. An example involving a unilateral change of ballot is sometimes referred to as an instance of "single-voter manipulation" to distinguish it from a situation wherein a group of voters, acting in concert, can change their ballots so that all of them prefer the new winner to the original winner. We'll see examples of group manipulation in Section 10.2.

Second, the original election produced a single winner, as did the new election held after we finished constructing Voter 1's disingenuous ballot. Thus, because we know each voter's sincere preference ranking for the candidates, we also know exactly which of the two election outcomes each voter will prefer. Ties, on the other hand, present a problem. For example, if a voter has sincere preferences that rank A over B over C over D, then it's not at all obvious whether this voter will prefer an election outcome that ties A and D to an election outcome that ties B and C or vice versa.

Third, the voter who is changing her ballot changes the election outcome to *one that she prefers*. It really is pointless to submit an insincere ballot if doing so only makes the election outcome *worse* in your opinion.

A voting system is **manipulable** if there is at least one scenario in which some voter can achieve a more preferred election outcome by unilaterally changing his or her ballot. The precise definition follows.

Manipulability DEFINITION

A voting system is said to be **manipulable** if there exist two sequences of preference list ballots and a voter (call the voter Jane) such that

1. Neither election results in a tie.
2. The only ballot change is by Jane.
3. Jane prefers—assuming that her ballot in the first election represents her true preferences—the outcome of the second election to that of the first election.

With this definition at hand, we can now turn to the study of the manipulability of some of the particular voting systems introduced in the last chapter.

EXAMPLE 2 Finding a Successful Manipulation with the Borda Count

Suppose there are two voters and four candidates, and suppose the true preferences of the voters are reflected in the following ballots:

Voter 1	Voter 2
A	*C*
B	*D*
C	*B*
D	*A*

Using the Borda count with point values 3, 2, 1, 0 (or by counting the number of occurrences of other candidates below the one in question, as described in Section 9.3), we see that the Borda scores of the four candidates are as follows:

The Borda score of *A* is 3.
The Borda score of *B* is 3.
The Borda score of *C* is 4.
The Borda score of *D* is 2.

Thus, *C* is the winner of this election using the Borda count. Voter 2 can indeed change the outcome of the election, but only in a pointless way. That is, the winning alternative *C* is her top choice, and so it is quite impossible for her to do any better than this!

Voter 1, however, is quite unhappy with *C* being the winner, and she would like to change her ballot so as to make either *A* or *B* the winner (as she prefers both of these to *C*). Let's first see if Voter 1 can change her ballot so as to make *A* the winner. Voter 1 certainly cannot increase *A*'s Borda score of 3 from the original election. But *C* had a Borda score of 4 and the most Voter 1 can do is to reduce this by 1 (by moving *C* to the bottom of her ballot). Hence, there is no way that Voter 1 can change her ballot to make *A* the unique winner.

Voter 1, however, *can* make *B* the unique winner. To do this, she first moves *B* to the top of her ballot (an obvious thing to do because she is trying to make *B* the winner). That raises *B*'s Borda score to 4. But *C* still has a Borda score of 4, so she must move *C* to the bottom of her ballot, reducing the Borda score of *C* to 3. Now we must check that neither *A* nor *D* have a Borda score of 4, but this is easily done. Hence, Voter 1 can submit a disingenuous ballot and achieve a result she prefers to that of the original election. Hence, this example also serves to show that the Borda count is manipulable. ■

Self Check 2

Show that Voter 2 can change her original ballots so that (assuming Voter 1 submits her original ballot) the winner becomes *B*. And again, why is this a pointless move on Voter 2's part? ■

10.2 Majority Rule and Condorcet's Method

Throughout this section, we assume that the number of voters is odd. In Section 9.2 (page 407), we pointed out that with two candidates, majority rule has three very desirable properties: It treats all voters equally, it treats both candidates equally, and it is *monotone,* meaning that a single voter's change in ballot from a vote for the loser to a vote for the winner has no effect on the election outcome. More strikingly, May's theorem told us that among all voting systems in the two-candidate case that never result in a tie, majority rule is the *only* one satisfying these three properties.

But let's consider for a moment what monotonicity is saying in this two-candidate case for voting systems that never yield ties. It says that if you rank *A* over *B* on your ballot, and the election winner is *B,* then the election winner will remain *B* if you switch to a ballot with *B* over *A.* But there are only two possible choices for a ballot in this two-candidate case: *B* over *A* and *A* over *B.* Monotonicity is thus saying that if you rank *A* over *B,* then no unilateral change in your ballot can make the outcome *A.* This is simply the assertion that you can't manipulate the voting system!

Thus, in the two-candidate case, nonmanipulability and monotonicity are exactly the same thing. This allows us to restate May's theorem from Section 9.2, with the word *monotonicity* replaced by *nonmanipulability.*

May's Theorem for Manipulability THEOREM

Among all two-candidate voting systems that never result in a tie, majority rule is the only one that treats all voters equally, treats both candidates equally, and is nonmanipulable.

There are examples of two-candidate voting systems that are manipulable, even though they treat all voters equally and both candidates equally. For example, the voting system that declares the winner to be the alternative with the fewest first-place votes is manipulable, as is the one that declares the winner to be whichever alternative has an odd number of first-place votes (even if that's fewer than half). Exercises 1 and 2 ask you to provide an example of voter manipulation for each of these systems.

Turning to the case of three or more candidates, we begin with Condorcet's method, as we did in Chapter 9. Condorcet's method is based on majority rule, and as we've just seen, majority rule is nonmanipulable. So the following result, as pleasing as it is, comes as no surprise.

The Nonmanipulability of Condorcet's Method THEOREM

Condorcet's method is nonmanipulable in the sense that a voter can never unilaterally change an election result from one candidate to another candidate that he or she prefers.

Let's see why Condorcet's method is nonmanipulable, regardless of the number of voters. Suppose that we have an election in which you, as one of the voters, prefer Candidate *A* to Candidate *B,* but *B* wins using Condorcet's method. We'll show that any attempt that you might make to manipulate the election so that *A* becomes the winner is doomed to failure, even if there are more than these two candidates in the election.

Because Candidate *B* is the winner, using Condorcet's method, we know that *B* defeats every other candidate in a one-on-one contest based on the ballots cast.

In particular, B defeats A in a one-on-one contest, even with your original ballot that has A over B. This means that more than half of the other voters ranked B over A, so, regardless of how you change your ballot, B will *still* defeat A in a one-on-one contest. While this need not ensure that B remains a winner with Condorcet's method, it certainly guarantees that A isn't.

Hence, you cannot unilaterally cause A to be a winner using Condorcet's method, and so your attempt at manipulation will have failed.

EXAMPLE 3 Exploiting the Condorcet Voting Paradox

We had to be careful in stating the theorem that asserts Concorcet's method is nonmanipulable because, as we've seen, elections occur in which there is no winner using Condorcet's method. With three voters and three candidates, it is possible for a voter (the one on the left in this example) to unilaterally change an election from one that yields his or her second choice as the sole winner (Candidate C in the example), to one in which there is no winner at all, as this example shows:

Election 1

Rank	**Number of Voters (3)**		
First choice	*A*	*B*	*C*
Second choice	*C*	*C*	*A*
Third choice	*B*	*A*	*B*

Election 2

Rank	**Number of Voters (3)**		
First choice	*A*	*B*	*C*
Second choice	*B*	*C*	*A*
Third choice	*C*	*A*	*B*

A voter's ability to bring about this kind of change in an election unilaterally, however, is not something that falls within the scope of our formal definition of manipulation. Nevertheless, one could argue that there are situations in which you might well prefer having an election with no outcome at all to having an election in which a candidate other than your top choice emerges as the sole winner.

Andy Sacks/Photographer's Choice/Getty Images

Paper ballots are still used in elections in many states. A lingering controversy from the 2004 presidential election is the use of electronic ballots, which do not leave physical evidence, thus making it extremely difficult to do a recount in potential disputes about the plurality or majority.

We now move on to voting systems with three or more candidates, systems that, unlike Condorcet's method, always produce at least one winner. As you might expect from the results in Chapter 9, in terms of nonmanipulability, these voting systems are not as perfect as one might hope for.

10.3 The Manipulability of Other Voting Systems for Three or More Candidates

Manipulability and the Borda Count

Example 1 showed how a single voter can manipulate an election in which the Borda count is being used. But Example 1 involved four candidates. Is there a simpler example involving only three candidates?

The answer turns out to be "no," provided that we continue to interpret the notion of a "more preferred election outcome" to be a switch from a single winner to another single winner (as opposed to a switch creating or breaking a tie). This negative answer is formalized in the following theorem.

The Nonmanipulability of the Borda Count with Exactly Three Candidates THEOREM

With exactly three candidates, the Borda count cannot be manipulated in the sense of a voter unilaterally changing an election outcome from one single winner to another single winner that he or she prefers according to that voter's ballot in the first election, which we take to be sincere preferences.

Let's see why this is true. Suppose the candidates are *A*, *B*, and *C*, and that you prefer *A* to *B*, but *B* is the election winner using the Borda count. We'll show that any attempt you make to manipulate the election by changing your ballot so that *A* emerges as the winner (using the Borda count) is doomed to failure.

Because you prefer *A* to *B*, your sincere ballot can be one of only three possibilities, corresponding to whether *C* is ranked first, second, or third. We'll consider each case in turn.

Case 1. **Your sincere ballot is *A* over *B* over *C*.** No ballot change on your part can increase *A*'s Borda score, and you can decrease *B*'s Borda score by no more than 1. Thus, at best, you can make a unilateral change that results in *A* and *B* having the same Borda score, whereas successful manipulation on your part requires that *A* have a strictly higher Borda score than *B* after your ballot change.

Case 2. **Your sincere ballot is *C* over *A* over *B*.** No ballot change on your part can decrease *B*'s Borda score, and you can increase *A*'s Borda score by no more than 1. Thus, at best, you can make a unilateral change that results in *A* and *B* having the same Borda score, whereas successful manipulation on your part requires that *A* have a strictly higher Borda score than *B* after your ballot change.

Case 3. **Your sincere ballot is *A* over *C* over *B*.** No ballot change on your part can increase *A*'s Borda score or decrease *B*'s Borda score. Thus, after your ballot change, *B* will still have a higher Borda score than *A*, so your attempt at manipulation has failed in this case as well.

So with three candidates, the Borda count is nonmanipulable. With more than three candidates, the Borda count does not fare as well, regardless of how many voters there are.

The Manipulability of the Borda Count with Four or More Candidates THEOREM

With four or more candidates (and two or more voters), the Borda count can be manipulated in the sense that there exists an election in which a voter can change the election outcome unilaterally from one single winner to another single winner that he or she prefers according to that voter's ballot in the first election, which we take to be sincere preferences.

As we've seen in Example 1, the Borda count can be manipulated in the case of four candidates and two voters. This is really half the battle, as we can modify that example to serve in any case in which the number of voters is even, as follows:

1. Any candidates in addition to *A, B, C,* and *D* can be placed below those four on every ballot.
2. The rest of the voters can be paired off with the members of each pair holding ballots that rank the candidates in exactly opposite orders (thus "canceling each other out" in terms of the Borda scores).

The following example illustrates this method of generalizing our earlier instance of manipulation of the Borda count to the case of five candidates and six voters.

EXAMPLE 4 ➔ Manipulating the Borda Count with Five Candidates and Six Voters

Consider the following two elections:

Election 1					
A	*B*	*A*	*E*	*A*	*E*
B	*C*	*B*	*D*	*B*	*D*
C	*A*	*C*	*C*	*C*	*C*
D	*D*	*D*	*B*	*D*	*B*
E	*E*	*E*	*A*	*E*	*A*

Election 2					
A	*B*	*A*	*E*	*A*	*E*
D	*C*	*B*	*D*	*B*	*D*
C	*A*	*C*	*C*	*C*	*C*
B	*D*	*D*	*B*	*D*	*B*
E	*E*	*E*	*A*	*E*	*A*

The ballots of the first two voters (in both elections) are the same as in Example 1 (the manipulation of the Borda count with four candidates and two voters), with the new candidate *E* placed at the bottom of both ballots. The last four voters contribute exactly 8 to the Borda score of each candidate, and so, taken together, they have no effect on who is the winner of the election. This is what we mean by "canceling each other out."

In the first election, as in Example 1, Candidate *B* wins. But if we take these ballots to represent true preferences, the voter on the far left prefers *A* to *B*. Moreover, that voter can achieve this better outcome—Candidate *A*—by submitting the disingenuous ballot that he or she cast in Election 2. ■

To handle the case where the number of voters is odd, we need to start with a four-candidate, three-voter example of manipulation of the Borda count. Exercise 8 (page 457) provides this. We can then modify this example to work for any odd number of voters by again adding pairs of ballots that cancel each other out exactly as we did before. Exercises 9 and 10 (page 457) fill in some of the details needed for this part of the argument and ask you to provide the necessary explanations and calculations.

Manipulability of Runoff Systems

EXAMPLE 5 ➔ Manipulability of Runoff Systems

Both the plurality runoff rule and the Hare system are manipulable. But rather than give the whole story away, we'll just present the sequences of sincere ballots in each case. Exercises 13 and 14 (page 457) ask you to figure out how the leftmost voter in each case can secure a more preferred outcome by a unilateral change of ballot.

Election 1 for the Hare System				
A	*B*	*C*	*C*	*D*
B	*A*	*B*	*B*	*B*
C	*C*	*A*	*A*	*C*
D	*D*	*D*	*D*	*A*

Election 1 for the Plurality Runoff Rule				
A	*A*	*C*	*C*	*B*
B	*B*	*A*	*A*	*C*
C	*C*	*B*	*B*	*A*

EXAMPLE 6 Manipulating Sequential Pairwise Voting

Sequential pairwise voting can also be manipulated by a single voter, even in the case of three voters and three candidates. For example, consider the following two elections with the agenda *A, B,* and *C:*

Election 1			
Rank	**Number of Voters (3)**		
First choice	*A*	*B*	*C*
Second choice	*B*	*C*	*A*
Third choice	*C*	*A*	*B*

Election 2			
Rank	**Number of Voters (3)**		
First choice	*B*	*B*	*C*
Second choice	*A*	*C*	*A*
Third choice	*C*	*A*	*B*

In Election 1, *A* defeats *B* by a score of 2 to 1, so *A* moves on to meet *C*. But *C* defeats *A* by a score of 2 to 1, so *C* is the winner in Election 1. Election 2 is the result of Voter 1 (on the left) submitting a disingenuous ballot in which he or she has elevated *B* (his or her actual second choice) to first place. It is now clear that *B* first defeats *A* by a score of 2 to 1 and then moves on to defeat *C* by this same score. Hence, *B* is the winner in Election 2. This is an instance of manipulation in which Voter 1 has secured a more preferred outcome by submitting an insincere ballot, because Voter 1 actually prefers *B* to *C* (assuming that his or her ballot in Election 1 represents his or her true preferences). This shows that sequential pairwise voting is manipulable.

Sequential Pairwise Voting and Agenda Manipulability

Thus, sequential pairwise voting can also be manipulated by a single voter, even in the case of three voters and three candidates. But there is another aspect of manipulability that arises with this particular voting system that is of even more interest, and this is something called **agenda manipulation.**

Agenda Manipulation DEFINITION

Agenda manipulation refers to the ability to control who wins an election with sequential pairwise voting by a choice of the agenda.

William H. Riker, in his book *The Art of Political Manipulation,* spoke of the possibility that "those in control of procedures can manipulate the agenda by, for example, restricting alternatives [candidates] or by arranging the order in which they are brought up." The following example provides a striking illustration of this with sequential pairwise voting.

EXAMPLE 7 Agenda Manipulation of Sequential Pairwise Voting

Suppose that we have four candidates and three voters who we know will be submitting the following preference list ballots:

Rank	Number of Voters (3)		
First choice	*A*	*C*	*B*
Second choice	*B*	*A*	*D*
Third choice	*D*	*B*	*C*
Fourth choice	*C*	*D*	*A*

Now suppose that we have agenda-setting power in the sense that we get to choose the order in which the one-on-one contests will take place. Remarkably, we can arrange for the winner to be whichever of the four candidates we want.

The intuition behind finding an agenda that will yield a certain candidate as the winner arises from the observation that candidates who appear later in the agenda are favored over candidates who appear early in the agenda. For example, if we want *A* to win, we place *A* last and look for which candidates would, in fact, defeat *A* one on one. Here, only *C* defeats *A*, and so we want to arrange for *C* to be eliminated along the way. But *B* defeats *C* one on one, so if we choose the agenda *B, C, D, A,* we have that *C* is eliminated by *B* in the first round, then *D* is eliminated by *B* in the second round, and finally *B* is eliminated by *A* in the third round, leaving *A* as the winner. Exercise 16 (page 457) asks you to find the three other agendas that will, in turn, yield *B, C,* and *D* as the winner.

Self Check 3

The agenda *B, C, D, A* yielded *A* as the winner in Example 7. There is a trivial change to this agenda that also yields *A* as the winner. What is this trivial change?

Self Check 4

In Example 7, is there an agenda in which *D* is first in the agenda, *A* is last in the agenda, and *A* is the winner?

Self Check 5

In Example 7, why is it the case that if *A* is first in the agenda, then *A* will definitely not win?

Plurality Voting and Group Manipulability

In the real world, all other voting systems pale in comparison to plurality voting in terms of the significance of the role played by disingenuous voting. "Throwing away your vote"—as some accuse Nader voters in Florida of doing in the 2000 presidential election—represents a choice, conscious or otherwise, to forgo obtaining a more desired outcome through strategic considerations.

Ironically, plurality voting, like Condorcet's method, is nonmanipulable according to the formal definition given on page 444. However, a *group* of voters, acting together, can change an election outcome into something they *all* prefer. This observation gives rise to the following definition, theorem, and explanation of why the theorem is true.

Group Manipulability DEFINITION

A voting system is **group manipulable** if there are elections in which a group of voters can change their ballots so that the new winner is preferred to the old winner by everyone in the group, assuming that the original ballots represent the true preferences of each voter in the group.

The Group Manipulability of Plurality Voting THEOREM

Plurality voting cannot be manipulated by a single individual. However, it is group manipulable.

The Green Party holds its convention. Ralph Nader ran for the presidency as a Green in the 2000 election. By doing so, he brought up many questions of social choice—some would say deliberately. Was Nader a spoiler candidate? Were Nader supporters casting sincere votes for him? Would other voters who liked his positions be hedging their bets and voting insincerely if they chose another candidate?
Mark Leffingwell/AFP/Getty

First of all, let's see why no individual can manipulate plurality voting. Suppose that you prefer A to B, but B is the winner with plurality voting. Then B has at least one more first-place vote than A. Now, because you prefer A to B, we know that B is not on top of your sincere ballot, so no ballot change that you make can subtract from B's number of first-place votes. Moreover, by moving A to the top of your ballot, you only increase A's number of first-place votes by 1. Thus, the best you can do with a unilateral change in ballots is to move A into a tie with B.

To see that plurality voting is group manipulable, we only have to look at any real-world election in which a third-party candidate acted as the "spoiler." As we've said, Ralph Nader was exactly that in the state of Florida in the 2000 presidential election. Another example occurs in Exercise 18 (page 458).

At this point, we've seen that several of our familiar voting systems for three or more candidates—the Borda count, runoff systems, sequential pairwise voting—can be manipulated. Can't we do better than this in attempting to improve on Condorcet's method? We turn to this question next.

10.4 Impossibility

Condorcet's method, as we've seen, has a number of very desirable properties, including the following four:

1. Elections (with an odd number of voters) never result in ties.
2. It satisfies the Pareto condition.
3. It is nonmanipulable.
4. It is not a dictatorship.

Unfortunately, Condorcet's voting paradox on page 411 shows that there are elections in which Condorcet's method produces no winner at all.

Can we find a voting system that satisfies all four of these properties and that, unlike Condorcet's method, always yields a winner? Several possibilities suggest themselves. For example, to avoid ties, we could modify any of our usual methods by agreeing to use a fixed ordering of the candidates to break any ties that occur. Or we could extend Condorcet's method by making the winner be the candidate with the best "win-loss record" in one-on-one contests (a method called *Copeland's rule*).

Alas, any such attempt is doomed. In the early 1970s, Allan Gibbard and Mark Satterthwaite independently proved the following remarkable result.

The Gibbard–Satterthwaite Theorem THEOREM

With three or more candidates and any number of voters, there does not exist—and there never will exist—a voting system that always produces a winner, never has ties, satisfies the Pareto condition, is nonmanipulable, and is not a dictatorship.

The **Gibbard–Satterthwaite theorem** (often called the **GS theorem** for short) is a deep result that is related in important ways to Arrow's impossibility theorem. In particular, you shouldn't find it at all obvious, and we won't be saying anything about the proof. But we can state and prove a much weaker result that is of some interest in its own right.

A Weak Version of the GS Theorem THEOREM

Any voting system for three candidates that agrees with Condorcet's method whenever there is a Condorcet winner—and that additionally produces a unique winner when confronted by the ballots in the Condorcet voting paradox—is manipulable.

Let's see why this is true. With the Condorcet voting paradox, the winner is either A or B or C. For the moment, we'll assume it is C (and leave the other two cases to you—see Exercise 24 on page 458). Consider the following two elections:

Election 1			
Rank	**Number of Voters (3)**		
First choice	*A*	*B*	*C*
Second choice	*B*	*C*	*A*
Third choice	*C*	*A*	*B*

Election 2			
Rank	**Number of Voters (3)**		
First choice	*B*	*B*	*C*
Second choice	*A*	*C*	*A*
Third choice	*C*	*A*	*B*

In Election 1, the winner is C (our assumption in this case), and in Election 2, the winner is B (because we are assuming that our voting system agrees with Condorcet's method when there is a Condorcet winner, as B is here). Notice that the voter on the left, by a unilateral change in ballot, has improved the election outcome from being his or her third choice to being his or her second choice. This is what that voter set out to do and this is the desired instance of manipulation.

But the nonintuitive nature of voting and manipulation does not end here. It also turns out that sometimes "more is less" when it comes to "voting power." We illustrate this with the so-called *chair's paradox*.

10.5 The Chair's Paradox

We conclude this chapter with an aspect of manipulability that is so counterintuitive that it is referred to as the **chair's paradox.** To illustrate the situation, we'll consider a hypothetical college in upstate New York that is trying to choose among three academic-year calendars:

- *Terms:* A term system (10 weeks–10 weeks–10 weeks)
- *Semesters:* A semester system (14 weeks–14 weeks)
- *J-Plan:* A January-plan system (12 weeks–4 weeks–12 weeks)

The trustees say that the issue will be decided by majority rule with three voters:

- Someone representing the administration
- Someone representing the student body
- Someone representing the faculty

The administration, however, is given **tie-breaking power.** That is, if each proposed calendar gets one vote, then the one the administration voted for wins. (In this presentation of the paradox, the administration is playing the role of the chair.)

The preferences are given by the following table (and notice that these preference lists exactly mirror the ballots in Condorcet's voting paradox):

	Administration	Students	Faculty
First choice	J-Plan	Terms	Semesters
Second choice	Terms	Semesters	J-Plan
Third choice	Semesters	J-Plan	Terms

We assume that everyone knows everyone else's preferences and that these are real preferences, even after taking into consideration how the other constituencies feel.

The goal now is to analyze the situation and to determine how each of the three will vote if they are all rational in the sense of being willing to vote strategically (i.e., to manipulate the system) if it's in their own best interest. This is really a game-theoretic analysis, and it's useful to borrow a couple of pieces of game-theoretic terminology.

First, a choice of which calendar to vote for is called a **strategy.** So each of the voters has three strategies at its disposal: Vote for J-Plan, vote for Terms, and vote for Semesters. The second piece of terminology arises from the observation that if everyone is rational and acting in his or her own self-interest, no one will vote for his or her least-preferred calendar. The point is that voting for either a first or second choice **weakly dominates** the strategies of voting for a third choice, in the sense that the former choices always yield outcomes that are either the same as or better than the latter.

With this, we can see that the administration's strategy of voting for its first choice (the J-Plan) weakly dominates its strategy of voting for its second choice (Terms). That is, if both students and faculty vote for Semesters, the outcome is Semesters regardless of how the administration votes, but otherwise the administration does strictly better by voting for the J-Plan rather than Terms. Hence, assuming that the administration is rational, we know that it will, in fact, vote for its top choice, the J-Plan.

Now, given that we know what the administration will do, the claim is that the faculty's strategy of voting for Semesters weakly dominates its strategy of voting for the J-Plan. That is, if the students vote for Terms, the outcome is the J-Plan regardless of whether the faculty votes for Semesters or the J-Plan. On the other hand, if the students vote for Semesters, then the faculty can secure its best outcome Semesters by voting for Semesters. Assuming that the faculty is rational, we thus know that the faculty will, like the administration, vote for its top choice, which is Semesters.

But let's see where these decisions leave the students. They know that the administration is voting for the J-Plan and that the faculty is voting for Semesters. So if they vote for Terms, then the outcome is the J-Plan—their last choice. However, if they vote for Semesters along with the faculty, then the outcome is Semesters, their second choice. There is no way that the students can secure their top choice, Terms, as the winner. If they are rational, then they will also vote for Semesters, and thus Semesters will win the election.

So why is this paradoxical? Well, the administration clearly had the most "power," but the eventual winner of the election was its least-preferred calendar! The administration would have been better off handing over the tie-breaking power to either the faculty or the students.

The Chair's Paradox THEOREM

With three voters and three candidates, the voter with tie-breaking power can, if all three voters act rationally in their own self-interest, end up with his or her least-preferred candidate as the election winner.

The chair's paradox represents only one of manipulability's first cousins, some of which involve not only the fields of mathematics and political science but psychology as well. One of the authors relates the following from his early years:

> I recall a third-grade penmanship contest in which each of us had a writing sample taped to the blackboard, and the teacher, Mrs. Levy, announced that we'd get to vote for the one we thought best, with the proviso that the voter couldn't vote for his or her own paper. She also announced that if two or more were tied, we'd have a runoff among those.
>
> I remember being torn as to which of three particular ones to vote for, all of which I thought were very good and considerably better than the rest, including my own. When the votes were counted, these three were, in fact, tied for the win, with my writing sample alone in fourth, and only one vote out of the tie.
>
> After announcing the results, Mrs. Levy went on to say that the runoff would involve not three of us, but four, as she had decided also to vote, and she was voting for me! I don't remember the final tally, or what Mrs. Levy then said to the class, or what my three classmates, all plenty smart enough to realize what had just happened, later said to me. But I do remember sitting back and smiling—absolutely sure of the outcome—as soon as she had announced her intention to vote for me.

A woman casts her ballot on Election Day, the most important day in the American civic ritual of political campaigns and elections. Although she is acting as a responsible citizen, she may also contribute to some remarkable and contradictory results: Condorcet's voting paradox and the Gibbard–Satterthwaite theorem warn us that some elections produce strange results.

AP Photo/Chris Gardner

Review Vocabulary

Agenda manipulation The ability to control who wins an election with sequential pairwise voting by a choice of the agenda—that is, a choice of the order in which the one-on-one contests will be held. (p. 448)

Chair's paradox The fact that with three voters and three candidates, the voter with tie-breaking power (the "chair") can end up with his or her least-preferred candidate as the election winner, if all three voters act rationally in their own self-interest. (pp. 452, 453)

Disingenuous ballot Any ballot that does not represent a voter's true preferences. Also called an *insincere ballot.* (p. 439)

Gibbard–Satterthwaite (GS) theorem Alan Gibbard and Mark Satterthwaite's independent discovery that every voting system for three or more alternatives and any number of voters that satisfies the Pareto condition, that always produces a unique winner, and that is not a dictatorship can be manipulated. (p. 451)

Group manipulability A voting system is group manipulable if there exists at least one election in which a group of voters can change their ballots (with the ballots of voters not in the group left unchanged) in such a way that they all prefer the winner of the new election to the winner of the old election, assuming that the original ballots represent the true preferences of these voters. (p. 450)

Manipulation A voting system is manipulable if there exists at least one election in which a voter can change his or her ballot (with the ballots of all other voters left unchanged) in such a way that he or she prefers the winner of the new election to the winner of the old election, assuming that the original ballots represent the true preferences of the voters. (p. 439)

May's theorem for manipulability Kenneth May's discovery that for two candidates and an odd number of voters, majority rule is the only voting system that treats both candidates equally, treats all voters equally, and is nonmanipulable. (p. 444)

Strategy In the chair's paradox, a choice of which candidate (calendar, in our presentation) to vote for. This is a special case of the use of the term *strategy* in general game-theoretic situations. (p. 452)

Tie-breaking power The aspect of the voting rule used in the chair's paradox that says that the winner will be whichever candidate the chair votes for if there is a tie (which happens only if each candidate gets exactly one vote). (p. 452)

Unilateral change A change (in ballot) by one voter, while every other voter keeps his or her ballot exactly as it was. (p. 442)

Weak dominance One strategy (e.g., a choice of whom to vote for) weakly dominates another if it yields an outcome that is at least as good, and sometimes better, than the other. (p. 452)

Self Check Answers

1. In the original election, Voter 2's most-preferred candidate (*B*) won. If Voter 2 were to move *B* to the bottom of his ballot, Voter 2 would then get a (much!) less preferred candidate as the winner. Doing something to get an outcome you like less is really pointless.

2. Voter 1 had a ballot with *A* over *B* over *C* over *D*. If Voter 2 were to change from having *C* over *D* over *B* over *A* to a ballot having *B* over *C* over *D* over *A*, then *B* would become the winner. Again, this is pointless because Voter 2 prefers the result of the original election (*C* being the winner) to the result obtained by her disingenuous ballot.

3. Switch the first two to get *C, B, D, A.* The trivial change of switching the order of the first two in the agenda never has an effect on which one-on-ones take place.

4. Yes. There are only two agendas with *D* first and *A* last, and *A* wins with both of them.

5. If *A* is first in the agenda, then one of two things must happen. Either *A* loses to some alternative in the agenda between *A* and *C,* or *A* defeats all of these and then loses to *C.*

Skills Check

1. A "unilateral change in ballot" refers to the fact that

(a) only one candidate's position is being altered.

(b) no communication is taking place.

(c) only one voter is changing his or her ballot.

2. The quote "My scheme is intended only for honest men!" is from ____________________.

3. If a voter has sincere preferences of *A* over *B* over *C* over *D*, then

(a) she will prefer a tie between *A* and *D* to a tie between *B* and *C*.

(b) she will prefer a tie between *B* and *C* to a tie between *A* and *D*.

(c) it's not at all clear which tie—*AD* or *BC*—she will prefer.

4. A ballot that misrepresents a voter's true preference is referred to as ____________________.

5. A ballot that does not represent a voter's true preference is often called

(a) an insincere ballot.

(b) a disingenuous ballot.

(c) either (a) or (b).

6. Suppose Voter 1 ranks *A* over *B* over *C* over *D* and Voter 2 ranks *B* over *C* over *A* over *D*. Assume the Borda count is being used, so that *B* wins. If Voter 1 knows that Voter 2 will submit his or her true preferences, then Voter 1 can secure a win for *A* by submitting the following ballot: ____________.

7. In presenting an example of a voting system's susceptibility to manipulation, we present two elections (Election 1 and Election 2). We assume that

(a) all ballots in Election 1 are sincere.

(b) all ballots in Election 2 are sincere.

(c) both (a) and (b).

8. Nonmanipulability and monotonicity are equivalent if the number of candidates is ____________________.

9. The two-candidate voting system in which the winner is the alternative (or alternatives) with the fewest first-place votes

(a) is manipulable.

(b) treats all voters and candidates equally.

(c) both (a) and (b).

10. An example of a two-candidate voting system that is not monotone is ____________________.

11. Suppose that two elections show that a voting system is manipulable. Then

(a) neither election results in a tie.

(b) the winners are the same in both elections.

(c) every voter has changed his or her ballot.

12. In the two-candidate case, nonmanipulable is equivalent to ____________________.

13. Condorcet's method

(a) can be manipulated but always produces a winner.

(b) is nonmanipulable but sometimes produces no winner.

(c) sometimes results in a tie, so manipulability is hard to assess.

14. May's theorem for manipulability says that, with an odd number of voters, among all voting systems for two candidates that never result in a tie, majority rule is the only one that is nonmanipulable and ____________.

15. With the Borda count, two ballots "cancel each other out" if

(a) they are identical.

(b) each is arrived at by turning the other one upside down.

(c) other voters also hold these same ballots.

16. The Borda count is nonmanipulable in the special case in which ____________________.

17. A 6-voter example of manipulation with the Borda count can be modified to yield a 10-voter example by

(a) adding 4 ballots that are identical to each other.

(b) adding 4 ballots that are identical to Voter 1's ballot.

(c) adding 2 pairs of ballots, with the ballots in each pair canceling each other out.

18. With any voting system that satisfies the Pareto condition, an n-voter example of manipulation with k candidates can be modified to yield an n-voter example with $k + j$ candidates by ____________________.

19. Of the Hare system and the plurality runoff method,

(a) only the Hare system is manipulable.

(b) only plurality runoff is manipulable.

(c) both are manipulable.

20. Sequential pairwise voting is susceptible to a kind of manipulation called ________________.

21. Plurality voting

(a) cannot be manipulated by a single voter.

(b) can be manipulated by a single voter.

(c) is subject to agenda manipulation.

22. Plurality voting is susceptible to a kind of manipulation called ________________________.

23. Group manipulability was discussed in connection with

(a) Condorcet's method.

(b) the Borda count.

(c) sequential pairwise voting.

(d) plurality voting.

24. One strategy weakly dominates another strategy if it yields an outcome that is ________________________.

25. Agenda manipulation was discussed in connection with

(a) Condorcet's method.

(b) the Borda count.

(c) sequential pairwise voting.

(d) plurality voting.

26. The deep result in this chapter that is related to Arrow's impossibility theorem is called the ____________.

27. The Gibbard–Satterthwaite theorem says that with three or more candidates and any number of voters, there is no voting system that

(a) is not a dictatorship.

(b) is nonmanipulable and is not a dictatorship.

(c) satisfies the Pareto condition, is nonmanipulable, and is not a dictatorship.

(d) always yields a unique winner, satisfies the Pareto condition, is nonmanipulable, and is not a dictatorship.

28. The weak version of the Gibbard–Satterthwaite theorem asserts that if we have a voting system that agrees with Condorcet's method whenever there is a Condorcet winner and that also produces a unique winner when confronted by the ballots in the Condorcet voting paradox, then the system is ________________.

29. The voters' preferences in the chair's paradox are

(a) precisely the Condorcet voting paradox ballots.

(b) all the same.

(c) dictated by the chair.

30. The chair's paradox is paradoxical because ________________.

Chapter 10 Exercises

Challenge Discussion

10.1 An Introduction to Manipulability

10.2 Majority Rule and Condorcet's Method

1. Consider the voting system for two candidates (*A* and *B*) and three voters in which the candidate with the *fewest* first-place votes wins. Produce two elections that show this voting system is manipulable.

2. Consider the voting system for two candidates (*A* and *B*) and three voters in which the candidate receiving an odd number of first-place votes wins. Produce two elections that show this voting system is manipulable.

3. Consider the voting system for two candidates (*A* and *B*) and three voters in which the candidate receiving an even number of first-place votes wins. Produce two elections that show this voting system is manipulable.

4. There are at least two voting systems for two candidates (*A* and *B*) and three voters that are nonmanipulable and that treat all voters the same (meaning that if two voters were to exchange ballots, then the election outcome would be unchanged).

(a) What does May's theorem tell us about such a voting system?

(b) In one sentence, give an example of such a voting system (i.e., produce the rule that determines which of the two candidates, *A* or *B*, wins an election).

(c) In one sentence, give another example that is different from the example you gave in part (b) in that it produces a different winner for at least one election.

5. There are at least three voting systems for two candidates (*A* and *B*) and three voters that are nonmanipulable and that treat both candidates the same (meaning that if all three voters change their ballots, then the election outcome also changes).

(a) What does May's theorem tell us about such a voting system?

(b) In one sentence, give an example of such a voting system (i.e., produce the rule that determines which of the two candidates wins an election).

(c) In one sentence, give two other examples that are different from the example you gave in part (b) in that they produce a different winner for at least one election.

10.3 The Manipulability of Other Voting Systems for Three or More Candidates

6. Consider the following election with four candidates and two voters:

B	A
C	D
A	C
D	B

Show that if the Borda count is being used, the voter on the left can manipulate the outcome (assuming the above ballot represents his or her true preferences).

7. Example 4 (page 447) showed that the Borda count is manipulable if there are five candidates and six voters. Mimic what was done there to construct an example with seven candidates and eight voters.

8. Use the following election to illustrate the manipulability of the Borda count with three voters and four candidates:

A	B	B
B	A	A
C	C	C
D	D	D

9. Show that the Borda count is manipulable if there are four candidates and five voters. (*Hint:* Start with the ballots in the previous exercise, and then add two ballots that cancel each other out.)

10. Building on the idea in the previous exercise, show that the Borda count is manipulable if there are six candidates and nine voters.

11. Assume the following ballots give the true preferences of the voters and that the Borda count is being used. Show that at least one of the voters can improve the election outcome from her point of view by a unilateral change in her ballot.

B	D	C	B
C	C	A	A
D	A	B	C
A	B	D	D

12. *Coombs's rule* is the voting system that operates like the Hare system, except that instead of deleting candidates with the *fewest* first-place votes one after another, it deletes candidates with the *most* last-place votes one after another. Use the following ballots to show that Coombs's rule is manipulable:

A	B	B	A	A
B	C	C	C	C
C	A	A	B	B

13. Use the following election to show that the Hare system is manipulable:

A	B	C	C	D
B	A	B	B	B
C	C	A	A	C
D	D	D	D	A

14. Use the following election to show that the plurality runoff rule is manipulable:

A	A	C	C	B
B	B	A	A	C
C	C	B	B	A

15. Use the following election to show that sequential pairwise voting is manipulable. (Assume that the agenda is *A*, *B*, and *C*.)

A	B	C
B	C	A
C	A	B

16. Given the ballots below, mimic what was done in Example 7 (page 449) to find an agenda for which ______ is the winner using sequential pairwise voting.

(a) *B*

(b) *C*

(c) *D*

A	C	B
B	A	D
D	B	C
C	D	A

17. Suppose we have an election in which there is a single winner, using the Hare system. In a couple of sentences, explain why we know for sure that there is at least one voter who cannot manipulate this election in the sense of making a unilateral change in his or her ballot that will yield a preferred outcome for that voter, assuming the original ballot represented his or her true preferences.

18. Suppose that we have a voting system that satisfies unanimity: If every voter ranks the same candidate first, then that candidate is the unique winner. In a few sentences, explain why it is that if the system fails to satisfy the Pareto condition, it can be manipulated by some group.

19. Assume that the following ballots give the voters' true preferences, and the Borda count is being used. Find a voter who can manipulate this election in the sense of making a unilateral change in his or her ballot that will yield a single winner that is preferred by that voter to the original winner. Explain your answer.

A	*E*	*F*	*C*	*B*
B	*A*	*D*	*D*	*C*
F	*B*	*E*	*E*	*D*
C	*F*	*A*	*F*	*F*
D	*C*	*B*	*A*	*E*
E	*D*	*C*	*B*	*A*

20. Assume that the following ballots give the voters' true preferences, and the Borda count is being used. Find all voters who *cannot* manipulate this election in the sense of making a unilateral change in their individual ballots that will yield a single winner who is preferred by that voter to the original winner. Explain your answer.

A	*E*	*F*	*C*	*B*
B	*A*	*D*	*D*	*C*
F	*B*	*E*	*E*	*D*
C	*F*	*A*	*F*	*F*
D	*C*	*B*	*A*	*E*
E	*D*	*C*	*B*	*A*

21. Alfonse D'Amato (*D*) won the 1980 U.S. Senate race in New York by defeating Elizabeth Holtzman (*H*) and Jacob Javits (*J*). Reasonable estimates (based largely on exit polls) suggest that voters ranked the candidates according to the following table:

22%	23%	15%	29%	7%	4%
D	*D*	*H*	*H*	*J*	*J*
H	*J*	*D*	*J*	*H*	*D*
J	*H*	*J*	*D*	*D*	*H*

Use these ballots to show that plurality voting is group-manipulable.

22. Consider the voting rule in which an alternative is among the winners if it receives at least one first-place vote. In one sentence, explain why this voting system is *not* manipulable.

23. Consider the voting system in which the winner is determined by the total number of first- and second-place votes, with ties broken (when possible) according to the number of first-place votes. Thus, a candidate with no first-place votes and three second-place votes would defeat a candidate with two first-place votes and no second-place votes, but a candidate with two first-place votes and three second-place votes would defeat a candidate with one first-place vote and four second-place votes. Given Election 1 below, find a change in Voter 1's ballot that shows that this voting system is manipulable.

Election 1			
Rank	**Number of Voters (3)**		
First choice	*A*	*C*	*E*
Second choice	*B*	*D*	*D*
Third choice	*C*	*A*	*A*
Fourth choice	*D*	*B*	*B*
Fifth choice	*E*	*E*	*C*

10.4 Impossibility

24. Complete the proof of the weak version of the Gibbard–Satterthwaite theorem by handling the case where the winner with the voting paradox ballots is

(a) *A*.

(b) *B*.

25. The Gibbard–Satterthwaite theorem says that the following four properties of voting systems cannot be satisfied simultaneously:

(1) Elections always have unique winners.

(2) It satisfies the Pareto condition.

(3) It is nonmanipulable.

(4) It is not a dictatorship.

Which of the four properties are satisfied by a dictatorship?

26. Which of the four properties in Exercise 25 are satisfied by an "antidictatorship," where the election winner is whichever candidate Voter 1 ranks *last* on his or her ballot?

27. Which of the four properties in Exercise 25 are satisfied if we use the plurality rule, with Voter 1's ballot utilized to break any ties that occur?

10.5 The Chair's Paradox

For Exercises 28 and 29, consider the preference lists from the chair's paradox (reproduced here) and assume that everyone knows the administration will vote for the J-Plan, but that no one knows anything about how the faculty will vote.

	Administration	Students	Faculty
First choice	J-Plan	Terms	Semesters
Second choice	Terms	Semesters	J-Plan
Third choice	Semesters	J-Plan	Terms

28. In a sentence or two, explain why the students' strategy to vote for Terms does not weakly dominate their strategy to vote for Semesters.

29. In a sentence or two, explain why the students' strategy to vote for Semesters does not weakly dominate their strategy to vote for Terms.

Chapter Review

30. With the ballots in Exercise 21, who would have won if Condorcet's method had been used instead of plurality?

31. There is a modified version of Condorcet's method called the *weak Condorcet rule:* A candidate is among the winners precisely if he or she would defeat or tie every other candidate in a one-on-one contest. Notice that with an odd number of voters, the weak Condorcet rule is identical to Condorcet's method. Use the following ballots to show that the weak Condorcet rule is manipulable:

A	*C*	*B*	*D*
B	*A*	*D*	*C*
C	*B*	*C*	*A*
D	*D*	*A*	*B*

32. *Copeland's rule* is a voting system that, like Condorcet's method, looks at one-on-one contests. It, however, takes as the election winner the candidate with the best "win-loss record." Use the following ballots to show that Copeland's rule is manipulable:

A	*C*	*A*	*D*
B	*E*	*E*	*B*
C	*D*	*D*	*E*
D	*B*	*C*	*C*
E	*A*	*B*	*A*

33. Suppose we have an election in which there is a single winner, using plurality voting. In a couple of sentences, explain why we know for sure that there is at least one voter who cannot manipulate this election in the sense of making a unilateral change in his or her ballot that will yield a preferred outcome for that voter, assuming the original ballot represented his or her true preferences.

34. Consider the voting rule in which an alternative is among the winners if it has at least two first-place votes.

(a) In one sentence, explain why this voting system is *not* manipulable.

(b) Explain why the following two elections don't contradict part (a).

Election 1				
Rank	**Number of Voters (4)**			
First choice	*B*	*A*	*A*	*C*
Second choice	*C*	*B*	*B*	*B*
Third choice	*A*	*C*	*C*	*A*

Election 2				
Rank	**Number of Voters (4)**			
First choice	*C*	*A*	*A*	*C*
Second choice	*B*	*B*	*B*	*B*
Third choice	*A*	*C*	*C*	*A*

(c) Intuitively, does it seem to you that Voter 1, on the left in part (b), has secured a better outcome by submitting a disingenuous ballot?

Writing Projects

1. In the chair's paradox, we assume that all voters act rationally. This means each voter will forgo a strategy that is weakly dominated by another strategy. While this assumption is enough to conclude that the administration will vote for its top choice, it's not actually enough to conclude that the faculty will vote for its top choice. The point is that this latter conclusion required knowing how the administration will vote. Thus, we really need to assume that the faculty *knows* the administration is rational. But now we can ask what we need to assume about what the students know to conclude that they will vote for Semesters. Answer this (with explanation) and phrase the assumption in terms of the words *knows* and *rational,* as opposed to explicitly speaking of knowing how others will vote.

2. In a paragraph or two, explain why Condorcet's method is not group manipulable.

Suggested Readings

MOULIN, HERVÉ. *The Strategy of Social Choice,* North Holland, New York, 1983. Manipulability from an economist's point of view.

RIKER, WILLIAM. *The Art of Political Manipulation,* Yale University Press, New Haven, CT, and London, 1986. Manipulability from a political scientist's point of view.

TAYLOR, ALAN. *Social Choice and the Mathematics of Manipulation,* Cambridge University Press, Cambridge, UK, 2005. Manipulability from a mathematician's point of view.

Weighted Voting Systems

11

Wiktor Dabkowski/picture-alliance/Newscom

Voting often is used to decide yes or no questions. Legislatures vote on bills, stockholders vote on resolutions presented by the board of directors of a corporation, and juries vote to acquit or convict a defendant. In this chapter, we shall concentrate on situations where there are just two alternatives, "yes" or "no." The theorem of Kenneth May quoted in Chapter 9 says that majority rule is the only system with the following properties:

1. All voters are treated equally.
2. Both alternatives are treated equally.
3. If you vote "no," and "yes" wins, then "yes" would still win if you switched your vote to "yes," provided that no other voters switched their votes.
4. A tie cannot occur unless there is an even number of voters.

There are systems in which the voters appear to be unequal in power, but actually have all the properties required by May's theorem. Any student of politics will attest that not all legislators are equally powerful. (Think of the speaker of the U.S. House of Representatives versus a freshman member, or the prime minister versus a backbencher in Parliament.) Nevertheless, the voting system actually treats the legislators equally: Each is allowed one vote. Our interest is in the voting system itself and not in the influence that some voters might acquire as a result of experience or accomplishment.

In this chapter, we shall consider voting systems that do not treat the voters equally. For example, shareholders of corporations are asked to vote on motions presented by the corporation's board of directors. Each shareholder is allotted one vote per share that he or she owns. If two shareholders own different numbers of shares, they are not treated equally as voters. The voter with the larger number of shares has the greater investment and is given at least the appearance of greater influence. In this case, the stockholders are treated unequally because they are actually unequal.

The Council of Ministers of the European Union consists of the prime ministers from each of the 28 member states. The populations of the member states range from Malta's population of half a million to Germany's, which is more than 80 million. Until the voting provisions of the Treaty of Lisbon[1] went into effect on November 1, 2014, the Council of Ministers gave the ministers from the larger states more votes as a way of enhancing their influence. Voting systems in which the voters have varying numbers of votes are called **weighted voting systems.**

We will see that the number of votes that a participant is allowed to cast does not always reflect that participant's influence. To assess a voter's power to affect the outcome of a vote, we will define and explore two mathematical models of the decision-making process, each of which leads to a numerical measure of voting power, called a **power index.**

Our *mathematical models* start with a set of assumptions designed to capture the essence of a voting system—how a system allocates decision-making power to the voters, without regard to political alliances, skillful manipulation of the system by the more able participants, etc. The first model, developed by a mathematician, Lloyd S. Shapley, and an economist, Martin Shubik, is based on the assumption that the content of a proposal to be voted upon is subject to negotiation in order to attract votes. The second, developed by an attorney, John F. Banzhaf III, assumes that there is no communication between the participants. Formally, the Banzhaf model best describes a vote in which each participant decides by a coin toss. This ignores political reality, but it does extract the essence of the way the voting system allocates power when the participants operate independently of each other.

11.1 How Weighted Voting Works

In a weighted voting system, each participant has a specified number of votes, their **voting weight.** If the voting weight of Voter A is more than that of Voter B, then A may have more power than B to influence the outcome and certainly won't have less power. (We will see that voters with different voting weights may actually have equal power.)

A weighted voting system is provided with a **quota.** If the sum of the voting weights of voters who favor a motion is greater than or equal to the quota, then "yes" wins. If the total voting weight of voters favoring the motion is less than the quota, then "no" wins.

---

[1] Writing Project 2 (page 501) explores the voting system in use under the provisions of the Treaty of Lisbon.

EXAMPLE 1 The Film Selection Committee

The Classic Film Society presents a film each month. A three-member committee—whose members are a faculty adviser, Allen, and two students, Betty and Cao—selects the films. A selection must be approved by Allen and at least one of the students.

The selection committee could make its decisions by using a weighted voting system in which Allen has a voting weight of 2, and each student has a voting weight of 1. Thus, Allen, together with one student, would have a total voting weight of 3, while the two students without Allen, or Allen by himself, would have a voting weight of 2. Therefore if we set the quota to be 3, selections will be approved if and only if the committee's criteria for approval are satisfied. ■

We have to be careful when setting a quota. For example, if the Film Selection Committee described in Example 1 had a quota more than 4, a film could not be approved even if the members were unanimously in favor. In general, the quota should not be greater than the total weight of all the voters.

The quota for the Film Selection Committee should also not be 2 (or less). With 2 as the quota, Betty and Cao could approve *Superman* while Allen could approve *The Great Dictator*. Which would be shown? In general, the quota should be greater than half of the total weight of all the voters, to avoid situations where contradictory motions can pass.

In the United States, some county legislatures use weighted voting systems (see Spotlight 11.4 on page 480). Although the framers of the Constitution intended otherwise, the U.S. Electoral College functions as a weighted voting system when electing the president. The voters are the states (see Spotlight 11.1 on page 465).

Notation for Weighted Voting Systems — DEFINITION

A weighted voting system is described by specifying the **voting weights** w_1, $w_2, \ldots, w_n$ of the participants, and the **quota,** q. The following notation is a shorthand way of making these specifications:

$$[q : w_1, w_2, \ldots, w_n]$$

For example, the weighted voting system we found for the Film Selection Committee (Example 1) would be [3 : 2, 1, 1].

Self Check 1

For the system $[q : 8, 4, 2, 1]$, make a list of all whole numbers that could be the quota, q. ■

Dictator — DEFINITION

A voter D is a **dictator** if a motion will pass if and only if D is in favor, and the votes of the other members make no difference.

Self Check 2

If one of the voters in the system $[q : 8, 4, 2, 1]$ is a dictator, what whole number or numbers could q be? ■

Dummy Voter DEFINITION

A participant in a voting system who never has an opportunity to cast a deciding vote is called a **dummy voter.**

If a voting system has a dictator, all the participants except the dictator are dummy voters, but there are situations where there are dummy voters who may not be aware that they have no power.

EXAMPLE 2 ➔ Dummy Voters

In the voting system [8 : 5, 3, 1], the weight-1 voter is a dummy because a motion will pass only if it has the support of the weight-5 and weight-3 voters, and then the additional 1 vote is not needed. For a subtler example, consider a committee with members Andy, Beth, Cathy, and Don. They use the weighted voting system [52 : 26, 26, 26, 25]. Any two of Andy, Beth, and Cathy have enough votes to pass a motion. Don plus one other member have just 51 votes, less than the quota. To pass a motion, Don would have to get two of the other members to vote with him. Because the other two members could pass the motion without Don, he is a dummy voter. ■

Self Check 3

If exactly one of the voters in the system $[q : 8, 4, 2, 1]$ is a dummy, what whole number or numbers could q be? Can q be chosen so that no voter is a dummy? ■

EXAMPLE 3 ➔ A Six-Voter System

In 1958, the Board of Supervisors of Nassau County (on Long Island, New York) had six members. Each was to represent a district in the county. Because the districts had significantly different populations, the supervisors were given different voting weights: 9, 9, 7, 3, 1, 1, with 30 votes in all. The quota was to be a simple majority, 16. In our notation, the voting system was [16 : 9, 9, 7, 3, 1, 1]. To pass a measure, two of the three supervisors with voting weight 7 or 9 would be sufficient. The other three supervisors could not cast a decisive vote, even if all three teamed with one of the higher-weight supervisors, because the combined weight of the four would still be less than the quota. If these three joined two of the higher-weight supervisors, a motion would pass, but it would also pass without their votes. In other words, these voters were dummies. This voting system was used by John F. Banzhaf III to call attention to the need for mathematical analysis of weighted voting. (See Spotlight 11.4 on page 480.) ■

Veto Power DEFINITION

A voter whose vote is necessary to pass any motion is said to have **veto power.**

Allen, the faculty adviser in the Film Selection Committee, has veto power because his vote is required in order to approve a selection. In any weighted voting system, a voter has veto power if the sum of the voting weights of all of the other voters is less than the quota. There are voting systems, such as juries in criminal trials, that require the voters to be unanimous to pass a motion. In these voting systems, each voter has veto power.

EXAMPLE 4 The Film Selection Committee Revisited

Betty has just graduated and is now working on her Master's degree. To reflect her distinction, the Film Selection Committee now uses the system [6 : 5, 3, 1]. Allen, the weight-5 voter, still has veto power because the combined weight of Betty and Cao is only 4, and the quota is 6. No motion can pass if Allen is opposed. ■

Self Check 4

If exactly two of the voters in the system [q : 8, 4, 2, 1] have veto power, what whole number or numbers could q be? Would the other voters be dummies? ■

The voters in Example 4 are not equally powerful: Allen has veto power, while neither Betty nor Cao do, yet Betty and Cao are not dummies. We can't compare power by comparing the voting weights, either. Betty will be disappointed to find that she has the same voting power as Cao. Together, they can prevent Allen from passing a motion, and either one can combine with Allen to pass a motion.

The Electoral College

SPOTLIGHT 11.1

In the United States, the Electoral College votes to elect the president and vice president. Each state is represented by electors, appointed in a manner prescribed by its legislature. The number of electors for a state is equal to the size of its delegation in Congress (senators and representatives). In addition, the District of Columbia appoints three electors. In every state, there is a general election in which voters choose a "ticket" consisting of a candidate for president and a candidate for vice president, nominated by a political party. In 2012, the Republican and Democratic parties nominated the Romney–Ryan and the Obama–Biden tickets, respectively. All but two states chose their electors by a statewide contest. For example, in California the Democratic ticket received 60% of the vote and the Republican ticket received 37%, but California's 55 electors all voted for the Obama–Biden ticket. The two exceptions are Maine and Nebraska, which use the District System to choose their electors. With the District System, two electors (corresponding to senators) go to the ticket that received a plurality in the statewide election. The remaining electors are chosen by congressional district. In 2012, Maine's four electors voted for the Democratic ticket, and Nebraska's five electors voted Republican. (In 2008, the Nebraska electors were split: Four voted Republican and one voted Democratic.) If the District System interests you, see Writing Project 3 on page 501 and Exercises 12 and 25 on pages 497 and 498, respectively.

Although the framers may have envisioned the Electoral College as a deliberative body, each elector has to vote his or her party's ticket—there is no room for negotiation. Effectively, the Electoral College functions as a weighted voting system, in which there are 56 participants: the 50 states, the District of Columbia, and five congressional districts in Maine and Nebraska.

11.2 The Shapley–Shubik Model

A mathematical model for a social structure is usually based on some simplifying assumptions. The Shapley–Shubik model for voting systems assumes that on any issue to be voted upon there is a *spectrum of opinion.* This concept is on display in the political arena as politicians are characterized as ultra-conservative, conservative, moderate, liberal, and ultra-liberal—or, in terms of colors, red, purple, and blue. There are variations; for example, a politician might be a fiscal conservative and a social liberal (or the other way around). This illustrates that various issues under consideration have different spectra of opinion.

The Shapley–Shubik model is based on **voting permutations.**

Algebra Review Appendix
Permutations

Voting Permutation DEFINITION

A **voting permutation** is an ordered list of all the voters in a voting system.

For example, the gasoline tax might be voted upon. Imagine the voters in a line, ordered by how much they think the gasoline tax should be—from a taxi driver who favors $0.00 to a bicycle commuter who favors $100.00 per gallon. The order in which the voters appear in that line would be the voting permutation associated with the gasoline tax issue.

Power Indices

SPOTLIGHT 11.2

The first widely accepted numerical index for assessing power in voting systems was the Shapley–Shubik power index, developed in 1954 by a mathematician, Lloyd S. Shapley, and an economist, Martin Shubik. A particular voter's power as measured by this index is proportional to the number of different permutations (or orderings) of the voters in which he or she has the potential to cast the pivotal vote—the vote in the permutation that first turns the permutation from losing to winning.

The Banzhaf power index was introduced in 1965 by John F. Banzhaf III, a law professor who is also well known as the founder of the anti-smoking organization Action on Smoking and Health (ASH). The Banzhaf index is the one most often cited in court rulings, perhaps because Banzhaf brought several cases to court and established precedent. A voter's Banzhaf index is the number of different possible votes in which he or she casts a critical vote—a vote in favor of a motion that is necessary for it to pass.

Photo by UC Irvine via Getty Images

Lloyd S. Shapley

Walter Bennett/The LIFE Picture Collection/Getty Images

John F. Banzhaf III

Martin Shubik/Yale University

Martin Shubik

The Shapley–Shubik model is based on two assumptions:

- Every issue to be voted upon is associated with a voting permutation.
- Every voting permutation has the same chance of being associated with an issue that may be considered.

These assumptions are not intended to be realistic. If "political reality" is considered, we can find issues where the spectrum of opinion is not one-dimensional, in defiance of the first assumption. Voter *B* might be a protégé of voter *A*, and appear next to *A* in all permutations that are associated with issues to be considered—permutations where *A* and *B* are far apart would not occur. By ignoring these political realities, the Shapley–Shubik model can focus on the way a *voting system* distributes voting power.

If a bill is being drafted to set a gasoline tax rate, it must be drawn so as to attract sufficient votes to meet the quota. Putting the voters in line according to the gasoline tax permutation, one could walk down that line, adding voting weights until the total becomes equal to or more than the quota. The voter who puts the total over the quota is the **pivotal voter.**

Pivotal Voter DEFINITION

The first voter in a voting permutation who, when joined by those coming before him or her, would have enough voting weight to win, is the **pivotal voter** in the permutation. Each voting permutation has exactly one pivotal voter.

Self Check 5

In the Film Selection Committee (see Example 4 on page 465), find a voting permutation in which Cao is the pivotal voter. ■

The gasoline tax story is repeated every time a bill is drafted for consideration by a legislature. To pass, the bill must be approved by the pivotal voter in the voting permutation on the issue it addresses. In a political campaign, there are multiple issues, and the candidates try to address their campaign messages to appeal to the pivotal voter on each issue. This may be complicated, because the pivotal voters for different issues may not agree.

Spotlight 11.1 (page 465) explains how the Electoral College is effectively a 56-voter weighted voting system. Table 11.1 displays the permutation resulting from the 2012 election for president of the United States, in which the winning Democratic ticket was Barack Obama for president and Joseph Biden for vice president, and the Republican ticket was Mitt Romney for president and Paul Ryan for vice president. The table is presented as a voting permutation of the electors in the Electoral College.

Each of the voters in the Electoral College is selected by and represents an electorate. Some, such as the Utah popular vote and that of the Nebraska third congressional district, were heavily in favor of the Republican ticket; the Florida popular vote was almost equally split between the two tickets; and still others, such as the District of Columbia popular vote, favored the Democratic ticket by a wide margin. Table 11.1 lists all voters in the Electoral College, in decreasing order by their margin in favor of the Democratic ticket. (The margin is the number

TABLE 11.1 The Permutation Resulting from the General Election for the President of the United States in 2012

State	Democrat's Margin	Electors	Running Total	State	Democrat's Margin	Electors	Running Total
DC	12.49104	3	3	FL	1.0178	29	332
HI	2.5340	4	7	NC	0.9595	15	347
VT	2.1493	3	10	NE.2	0.8646	1	348
NY	1.7992	29	39	GA	0.8533	16	364
RI	1.7791	4	43	AZ	0.8311	11	375
MD	1.7264	10	53	MO	0.8255	10	385
CA	1.6228	55	108	IN	0.8116	11	396
MA	1.6168	11	119	SC	0.8080	9	405
ME.1	1.5604	1	120	MS	0.7921	6	411
DE	1.4659	3	123	MT	0.7533	3	414
NJ	1.4362	14	137	AK	0.7447	3	417
CT	1.4256	7	144	TX	0.7239	38	455
IL	1.4141	20	164	NE.1	0.7110	1	456
ME	1.3730	2	166	LA	0.7022	8	464
WA	1.3601	12	178	SD	0.6887	3	467
OR	1.2868	7	185	ND	0.6636	3	470
NM	1.2369	5	190	TN	0.6570	11	481
MI	1.2124	16	206	KS	0.6363	6	487
ME.2	1.1929	1	207	NE	0.6359	2	489
MN	1.1711	10	217	AL	0.6336	9	498
WI	1.1488	10	227	KY	0.6249	8	506
NV	1.1463	6	233	AR	0.6089	6	512
IA	1.1258	6	239	WV	0.5705	5	517
NH	1.1202	4	243	ID	0.5055	4	521
CO	1.1163	9	252	OK	0.4976	7	528
(PA)	1.1156	20	<u>272</u>	WY	0.4053	3	531
VA	1.0819	13	285	NE.3	0.3961	1	532
OH	1.0625	18	303	UT	0.3400	6	538

of popular votes cast for the Democratic ticket divided by the number of votes cast for the Republican ticket. Other parties were ignored in this calculation.) A running total of electoral votes gives the total weight of each voter and all who came before it in the table. In listing the states and other voters in this order, we have recorded a permutation of the Electoral College participants. The pivotal voter in this particular permutation was Pennsylvania, which brought the running total from 252 to 272, thus exceeding the quota of 270. This is noted in the table by circling the code for Pennsylvania and underlining its running total.

EXAMPLE 5 Pivotal Voters in the Film Selection Committee

Allen, Betty, and Cao vote to approve film selections by using the voting system [6 : 5, 3, 1]. We observed in Example 4 that the weight-5 voter (Allen) has veto power. Betty and Cao have voting weights 3 and 1, respectively. Let's consider their voting permutations, shown in Table 11.2. (Voters are identified by initials.) Next to each voting permutation, the total weights of the first voter, the first two voters, and all three voters are shown in sequence. The first number in the sequence that equals or exceeds the quota (6) is underlined, and the corresponding pivotal voter's symbol is circled. We see that Allen is pivotal in four permutations, while Betty and Cao are each pivotal in one. Hence Allen's share of the voting permutations is $\frac{4}{6}$, while Betty and Cao each have a $\frac{1}{6}$ share—even though their voting weights differ.

TABLE 11.2 Permutations and Pivotal Voters for the Film Selection Committee

Permutations			Weights			Pivotal Voters		
A	*B*	*C*	5	8	9		Betty	
A	*C*	*B*	5	6	9			Cao
B	*A*	*C*	3	8	9	Allen		
B	*C*	*A*	3	4	9	Allen		
C	*A*	*B*	1	6	9	Allen		
C	*B*	*A*	1	4	9	Allen		

In Example 5, we found that Allen is the pivot in four permutations, while Betty and Cao are each pivot in one permutation. In effect, we determined the **Shapley–Shubik power index** for this voting system.

Shapley–Shubik Power Index

DEFINITION

The **Shapley–Shubik power index** of each voter is computed by counting the number of voting permutations in which that voter is pivotal, and dividing that number by the number of all voting permutations.

Let's see how many permutations there are. If there is one voter, there is only one permutation. Two voters would have two permutations, because one could be first, the other second. With three voters, there are three that could be first. Once the first voter is identified, there are two permutations of the remaining two—thus there are $3 \times 2 = 6$ permutations in all. The six permutations listed in Table 11.2.

Self Check 6

If there are just two voters, and neither is a dictator, what is the Shapley–Shubik power index of each?

With four voters, there are four who could be first, and for each choice of a first voter, there are six permutations of the remaining three voters. In all there are $4 \times 6 = 24$ permutations. This reasoning can be continued *ad infinitum.* With five voters, the number of permutations is $5 \times 24 = 120$.

Self Check 7

If there are seven voters, how many permutations are there? ■

Factorial DEFINITION

The number of permutations of a set of n voters is called the **factorial** of n and is denoted $n!$.

It is necessary to consider the number $0!$, the number of ways to order the empty set. By convention, $0! = 1$, meaning there is one way to do this. The factorial of a *positive* whole number n is the product all the whole numbers from 1 up to n:

$$n! = 1 \times 2 \times 3 \times 4 \times \cdots \times n$$

We saw one of the 56! permutations of the voters in the Electoral College in Table 11.1. Factorials grow rapidly in size, an instance of the combinatorial explosion. For example, 10! is more than 3.6 million, 15! is more than 1.3 trillion. Enter 56! in the Google search box—you will find it is a 75-digit number. Here's a puzzle: How many zeros are at the end of 56!?

EXAMPLE 6 The Corporation with Four Shareholders

A corporation has four shareholders, *A, B, C,* and *D,* with 49, 48, 2, and 1 shares, respectively. It uses the weighted voting system [51 : 49, 48, 2, 1].

The $4! = 24$ permutations of the shareholders are shown in Table 11.3. (The arrangement is the same as in Table 11.2, which lists permutations and finds their pivotal voters in a three-voter system.) In 10 of the permutations, *A* is the pivotal voter; *B* and *C* are each pivotal voters in 6; and *D* is the pivotal voter in 2 permutations. Therefore, the Shapley–Shubik power index for this weighted voting system is

$$\frac{10}{24}, \frac{6}{24}, \frac{6}{24}, \frac{2}{24} = \frac{5}{12}, \frac{1}{4}, \frac{1}{4}, \frac{1}{12}$$ ■

Self Check 8

If the voting system in Example 6 were changed by giving voter *D* one more vote (everything else the same), in which permutations listed in Table 11.3 would the pivot be different? Find the Shapley–Shubik power index of the altered system. ■

How to Compute the Shapley–Shubik Power Index

It is practical to calculate the Shapley–Shubik power index of a system with up to four voters by making a list of all the voting permutations and identifying the pivotal voter in each, as we have done in the previous two examples. This is the

TABLE 11.3 Permutations and Pivotal Voters for the Four-Person Corporation

Permutations				Weights				Pivot			
A	(*B*)	*C*	*D*	49	97	99	100		*B*		
A	(*B*)	*D*	*C*	49	97	98	100		*B*		
A	(*C*)	*B*	*D*	49	51	99	100			*C*	
A	(*C*)	*D*	*B*	49	51	52	100			*C*	
A	*D*	(*B*)	*C*	49	50	98	100		*B*		
A	*D*	(*C*)	*B*	49	50	52	100			*C*	
B	(*A*)	*C*	*D*	48	97	99	100	*A*			
B	(*A*)	*D*	*C*	48	97	98	100	*A*			
B	*C*	(*A*)	*D*	48	50	99	100	*A*			
B	*C*	(*D*)	*A*	48	50	51	100				*D*
B	*D*	(*A*)	*C*	48	49	98	100	*A*			
B	*D*	(*C*)	*A*	48	49	51	100			*C*	
C	(*A*)	*B*	*D*	2	51	99	100	*A*			
C	(*A*)	*D*	*B*	2	51	52	100	*A*			
C	*B*	(*A*)	*D*	2	50	99	100	*A*			
C	*B*	(*D*)	*A*	2	50	51	100				*D*
C	*D*	(*A*)	*B*	2	3	52	100	*A*			
C	*D*	(*B*)	*A*	2	3	51	100		*B*		
D	*A*	(*B*)	*C*	1	50	98	100		*B*		
D	*A*	(*C*)	*B*	1	50	52	100			*C*	
D	*B*	(*A*)	*C*	1	49	98	100	*A*			
D	*B*	(*C*)	*A*	1	49	51	100			*C*	
D	*C*	(*A*)	*B*	1	3	52	100	*A*			
D	*C*	(*B*)	*A*	1	3	51	100		*B*		
Number of Pivots								10	6	6	2

brute force way of determining the Shapley–Shubik power index. With a computer, brute force can be used to determine the Shapley–Shubik power index of somewhat larger systems, but eventually the combinatorial explosion renders the brute force method impossible to implement. The Shapley–Shubik power index of the Electoral College is shown in Spotlight 11.5 (page 484). The calculations were performed with an online calculator (see the web citation at the end of this chapter). Brute force would not work here because, as we have noted, the number of voting permutations is more than astronomical. It would be impossible to examine each permutation and identify its pivotal voter.

In special cases where almost all of the voters have the same weight, the Shapley–Shubik power index can be calculated by relying on the following two principles:

- Voters with the same voting weight have the same Shapley–Shubik power index.
- The sum of the Shapley–Shubik power indices of all the voters is 1.

EXAMPLE 7 A Nine-Person Committee

Alice is chairperson of a committee. She has 3 votes, and there are eight other members, each with 1 vote. The quota for passing a measure is a simple majority, 6 of the 11 votes. In our notation, this voting system is [6 : 3, 1, 1, 1, 1, 1, 1, 1, 1].

Each weight-1 member has the same power index. Our strategy is to compute Alice's Shapley–Shubik power index first. By subtracting her index from 1, we will get the share of power for the remaining members of the committee. Because there are eight of them, and they are equally powerful, we can find the index of each weight-1 member by dividing by 8. Thus, we'll avoid having to examine all 9! = 362,880 voting permutations.

Alice will be the pivotal voter in any voting permutation where she is in the fourth position, when her vote would bring the total voting weight in favor from 3 to 6; the fifth position, when she would increase the total weight in favor from 4 to 7; or the sixth position, when the total would increase from 5 to 8 with her vote. If she is in any other position in a permutation, another member of the committee will be the pivot.

Because Alice is pivotal in three of the nine positions of a permutation, her Shapley–Shubik index is $\frac{3}{9} = \frac{1}{3}$. The remaining $\frac{2}{3}$ of the voting power is shared equally by the eight other voters. Therefore, each has $\frac{2}{3} \div 8 = \frac{1}{12}$ of the power.

According to the Shapley–Shubik model, Alice is 4 times as powerful as a weight-1 member, although her voting weight is only 3. (Divide her Shapley–Shubik power index by that of a weight-1 voter: $\frac{1}{3} \div \frac{1}{12} = 4$.) ■

The method used to compute the Shapley–Shubik index of each voter in Alice's committee is applicable to any weighted voting system where all but one of the voters have the same voting weight. Call the voter with the different voting weight the *singular voter*. For another example, suppose that the singular voter has 1 vote, and 10 other voters have 3 votes. If the quota is 16, the singular voter will be pivotal if preceded by 5 of the other voters. There are 11 voters, so the singular voter can appear in 11 places in a permutation—but is pivotal in only one place. Therefore, the Shapley–Shubik power index of the singular voter is $\frac{1}{11}$. The other 10 voters share the remaining $\frac{10}{11}$ of the voting power, so each has Shapley–Shubik power index $\frac{1}{11}$, too. In other words, according to the Shapley–Shubik model, the voters are equally powerful.

Self Check 9

Alice has been promoted. She now has 30 votes, and her new committee has 89 one-vote members. The quota is 60. Use the fact that Alice is the singular voter in this system to find the Shapley–Shubik index of each member of this new committee. (Spotlight 11.6 on page 487 has more to say about this voting system.) ■

Some of the weight-1 members of Alice's nine-person committee have been scheming to counteract Alice's power. Bill, a weight-1 member, has convinced three of the other weight-1 members to give him their votes. Effectively, the system now has six voters: Alice has three votes, Bill has four votes, and there are four 1-vote members. The weighted voting system is [6 : 3, 4, 1, 1, 1, 1]. Zoë, a weight-1 voter who did not cede her vote to Bill, wonders if this pact will affect her power negatively.

EXAMPLE 8 Zoë's Power After the Revolt

Zoë will be the pivotal voter of a permutation if and only if the voters coming before her in the permutation have a combined weight of exactly 5. There are two kinds of voting permutations that meet this condition (Z is Zoë):

- $Y_1Y_2Y_3ZY_4Y_5$, where one of Y_1, Y_2, Y_3 is Alice, one of Y_4 or Y_5 is Bill, and the remaining three Ys are weight-1 voters
- $Y_1Y_2ZY_3Y_4Y_5$, where Y_1 or Y_2 is Bill, and one of Y_3, Y_4, Y_5 is Alice

Counting these permutations will involve the fundamental principle of counting. Let's count the voting permutations of the first type. There are three places that Alice could occupy before Zoë, and two places that Bill could occupy after Zoë. By the fundamental principle of counting, there are $3 \times 2 = 6$ ways we could position Alice and Bill in a voting permutation with Alice before Zoë and Bill after Zoë. We can count the number of permutations in each of the six groups, where Alice, Zoë, and Bill have already been positioned. The three other committee members can be ordered in $3! = 6$ ways, and put accordingly into the three open spaces. Using the fundamental principle of counting (again), we see that the number of voting permutations of the first type is $6 \times 6 = 36$. The number of permutations of the second type, where Bill comes before Zoë and Alice after, is the same. Therefore, there are $36 + 36 = 72$ voting permutations in which Zoë is pivotal. The Shapley–Shubik index of Zoë is therefore $\frac{72}{6!} = \frac{1}{10}$. Before the revolt, Zoë had $\frac{1}{12}$ of the power according to the Shapley–Shubik model, so her power has increased a bit. ■

Alice has also been wondering about the situation. Bill has only one more vote than she does—will her power be much less than his?

EXAMPLE 9 Alice's Power After the Revolt

Alice will be pivot in any permutation where the following occurs:

- She is second, with Bill first. There are $4! = 24$ permutations of this type, since the 1-vote members, who come after Alice, can be in any order.
- She is third, with Bill and a 1-vote member first. There are $2 \times 4! = 48$ of these permutations, because Bill could be in one to two positions, and the 1-vote members can be in any order.
- She is fourth, with Bill and a 1-vote member following. Again, there are 48 permutations in this category.
- She is fifth, with Bill last. There are $4! = 24$ of these permutations, because there are $4! = 24$ ways to order the four weight-1 voters who came before Alice.

All told, Alice is pivot in 144 permutations, and her Shapley–Shubik power index is $\frac{144}{6!} = \frac{1}{5}$. The revolt has reduced her power considerably. The four remaining 1-vote members of the committee control $4 \times \frac{1}{10} = \frac{2}{5}$ of the power, so the remaining $\frac{2}{5}$ of the power—twice Alice's—is Bill's. ■

Self Check 10

Zoë has decided to cede her vote to Bill, giving him 5 votes. The system is now [6 : 3, 5, 1, 1, 1]. Determine the Shapley–Shubik power indices for Alice, Bill, and the remaining weight-1 voters. ■

11.3 The Banzhaf Model

You, along with 99,999 other voters, are deciding if your city should issue bonds to build a convention center. You will vote "yes," but to pass the measure, a two-thirds majority is required. The Shapley–Shubik model would simply say that your power to influence the outcome is the same as that of any other voter, because each voter has the same chance to occupy the pivotal position (the 66,667th) in a voting permutation. Your Shapley–Shubik power index would be 1/100,000. The Banzhaf model focuses on the chance that you would cast the deciding vote. To pass, the measure must be approved by at least two-thirds—or 66,667—of the voters. You would cast the deciding vote if exactly 66,666 of the other voters were in favor of the measure. Otherwise, neither you nor anyone else will cast a deciding "yes" vote.

Coalition DEFINITION

A **coalition** is a set of participants in a voting system that might vote in favor of a measure. The empty coalition is allowed, and it represents a situation when the voters unanimously oppose a motion.

In the bond issue story, the set of voters in favor of passing the motion is a coalition. Your Banzhaf power index would be equal to the number of possible coalitions consisting of exactly 66,667 voters that you could belong to. Like the Shapley–Shubik power index, the Banzhaf power index would recognize that each voter has an equal chance to cast a decisive vote, but it counts the actual number of chances that a voter has to change the outcome.

In the voting permutation resulting from the 2012 United States presidential election, the pivotal voter was Pennsylvania (see Table 11.1 on page 468). But Pennsylvania *did not cast a deciding vote.* The Democratic ticket amassed 332 electoral votes, 62 votes more than the quota. If Pennsylvania's 20 electors had been Republican, the Democratic ticket would have won anyway, with 312 electoral votes.

The states whose margin in favor of the Democratic ticket was greater than 1.000 formed the **winning coalition** in the 2012 presidential election.

Winning Coalition and Losing Coalition DEFINITION

A set of participants in a weighted voting system whose combined voting weight is equal to or greater than the quota is called a **winning coalition.** If the set of participants has a combined voting weight less than the quota, it is a **losing coalition.**

The Banzhaf model focuses on winning coalitions, and the essential voters within them. It is important to recognize that neither the Shapley–Shubik model nor the Banzhaf model are concerned with opinions of individual voters. Alice and Bill may not get along, but in the Shapley–Shubik model the permutations in which they are in consecutive positions are counted. In the Banzhaf model, if the coalition {Alice, Bill} is winning, it counts even if Alice and Bill always vote on opposite sides. The models are concerned only with the way the voting system distributes power.

A winning coalition may include some voters who are just along for the ride. If they were to desert the coalition, it would still win. The Banzhaf model counts the voters who are *critical* to its success.

Critical Voter DEFINITION

Let W be a winning coalition, and let V be a voter who belongs to W. Then V is a **critical voter** in W if the coalition consisting of all the voters in W *except* V is a losing coalition.

EXAMPLE 10 A Criminal Trial

After the evidence is presented and the summations are complete, the jury deliberates and each juror votes "guilty" or "not guilty." We will regard these as two separate motions. Thus there is a "guilty" motion and a "not guilty" motion for each count that is before the jury. The jury must reach a unanimous verdict—there is only one winning coalition, the entire jury, and every juror is a critical voter. If neither "guilty" nor "not guilty" is approved in this system, a mistrial is declared, and the prosecution may demand a new trial. ■

In the landslide 1984 election the Republican ticket carried every state except one and had a total weight of 525, which was 255 votes more than the quota. There were no critical voters in the winning coalition, because any state could have switched to the Democratic ticket, and the Republican ticket would still have won. In the close election of 2000, the Republican ticket's winning coalition had a total weight of 271 electoral votes, one more than the quota. All voters in the Republican coalition, except for the Nebraska congressional districts, were critical.

Self Check 11

Refer to Table 11.1 on page 468 to determine if any state was a critical voter in the 2012 election. Notice that a participant's location in the voting permutation does not matter. You only need to check if the state was in the winning coalition, and if it was, did it have sufficient electoral votes to make a difference? ■

Banzhaf Power Index DEFINITION

In a weighted voting system, the **Banzhaf power index** of a voter V is the number of winning coalitions in which V is a critical member.

EXAMPLE 11 A Dictator

Consider the five-voter system [5 : 5, 1, 1, 1, 1]. We'll name the voters D and S_1, S_2, S_3, S_4, where D stands for dictator. A set of voters is a winning coalition if and only if D belongs to it. We can find the number of subsets of $\{S_1, S_2, S_3, S_4\}$ by applying the multiplication principle. Construct a subset by deciding if S_1 is in or out, two possibilities, and do the same for S_2, S_3, and S_4. The number of subsets is $2 \times 2 \times 2 \times 2 = 16$. All of the winning coalitions consist of D combined with one of these 16 subsets. In each winning coalition, D is the only critical voter. Thus, D has a Banzhaf index of 16, and each of the other participants has a Banzhaf index of 0. ■

A dummy voter is one whose vote never makes a difference. A dummy voter is never a critical voter in any winning coalition, so its Banzhaf index is 0. In Example 11, S_1, S_2, S_3, S_4 were all dummy voters. Nevertheless, their presence increased the dictator's Banzhaf power index—without them there would have been just one winning coalition, $\{D\}$, and the dictator's Banzhaf power index would have been 1.

Self Check 12

Find the Banzhaf index of each voter in the system [4 : 4, 1, 1, 1].

Should Blocking Coalitions Be Counted?

SPOTLIGHT 11.3

A *voting combination* is a list of the voters, recording how each voted. A voter is critical in a voting combination if a change in that voter's vote, with no other voter switching, will change the outcome. The definition of the Banzhaf power index that we are using counts only a voter's critical "yes" votes in a voting combination. However, Banzhaf's original definition[a] counted all voting combinations in which the voter was critical as a "yes" voter *or* as a "no" voter. Previous editions of this text used the original definition.

Let V be a voter, and suppose that V is a critical voter in b winning coalitions (so that according to the definition we are using, the Banzhaf power index of V is equal to b). Let $\mathcal{W}$ be one of the winning coalitions in which V is a critical voter, and let $\mathcal{K}$ be the set formed by combining V with all of the voters who do not belong to $\mathcal{W}$ (see the figure).

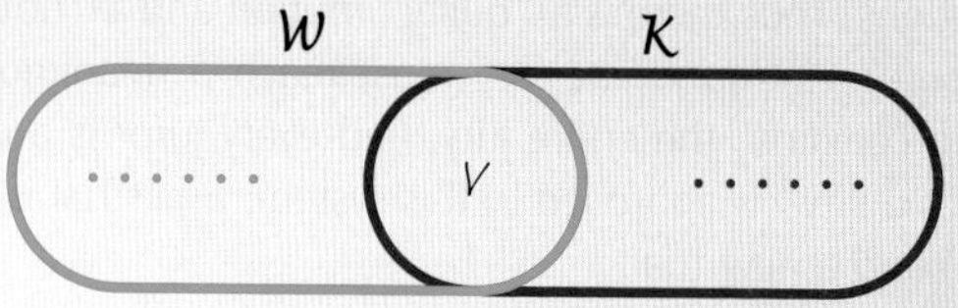

We will say that $\mathcal{K}$ is a *blocking coalition* because, by voting "no," it can prevent a motion from passing. Voter V is simultaneously a critical "yes" voter in the winning coalition $\mathcal{W}$ and, if she should change her vote, a critical "no" voter in the blocking coalition $\mathcal{K}$. Therefore, for each of the b winning combinations where V is a critical "yes" voter, there is a corresponding blocking combination—where V but no one else switches their vote—in which she is a critical "no" voter; there are b of these voting combinations, too. With the original definition, the Banzhaf index would be $2b$. Both indices yield the same distribution of voting power, but the original Banzhaf power index of each voter is double the index according to the definition that we are using.

[a]See the article "Weighted Voting Doesn't Work," cited in this chapter's Suggested Readings.

Calculating the Banzhaf Index by Brute Force

The brute force way to compute the Banzhaf power index of a weighted voting system is to start with a list of all of the winning coalitions. Determine the total weight of each coalition by adding the weights of all the voters in the coalition. The total weight will be at least the quota because it is a winning coalition, so subtract the quota to obtain the **extra votes** that the coalition has.

Extra Votes DEFINITION

The **extra votes** of a winning coalition indicate the number of votes by which its total weight exceeds the quota.

When we know all of the winning coalitions and the extra votes that each has, we can identify the critical voters in each: They are the voters who belong to the coalition, and whose voting weights are more than the extra votes. For example, the winning coalition in the 2000 presidential election had one extra vote. Nebraska belonged to that coalition and had a voting weight of 2. Therefore, Nebraska was a critical voter in the coalition. The Nebraska first congressional district also belonged to the winning coalition, but it had a voting weight of only 1, and was therefore not a critical voter. Maine also had a voting weight of 2, but since Maine voted for the Democratic ticket, Maine was not a critical voter in the Republican coalition.

Grand Coalition DEFINITION

The **grand coalition** is the coalition that includes every voter.

In every voting system, the grand coalition is a winning coalition, because the quota must be less than or equal to the total weight of all of the voters.

EXAMPLE 12 The Banzhaf Model Applied to the Film Selection Committee

Let's return to the committee of Example 4 on page 465. The members are Allen, Betty, and Cao, and they use the voting system [6 : 5, 3, 1]. Let's identify the critical voters in the grand coalition, {Allen, Betty, Cao}. If Allen leaves the coalition, what happens?

Allen	Betty	Cao	Votes	Outcome
Yes	Yes	Yes	9	Pass
↓				
No	Yes	Yes	4	Fail

Allen is a critical voter because the remaining voters form a losing coalition when he changes his vote. This is not surprising: He has veto power.

If Betty changes her vote, the remaining voters still form a winning coalition:

Allen	Betty	Cao	Votes	Outcome
Yes	Yes	Yes	9	Pass
	↓			
Yes	No	Yes	6	Pass

Betty is not a critical voter in this coalition, and you can verify for yourself that Cao is also not a critical voter.

To determine the Banzhaf power indices of Allen, Betty, and Cao, refer to Table 11.4. The table lists all winning coalitions and the extra votes in each—that's the total weight of the coalition minus the quota, 6. To obtain the list, the noncritical voters were removed, one by one, from the grand coalition. This yielded two winning coalitions, {Allen, Betty} and {Allen, Cao}. We cannot remove voters from these two coalitions without making losing coalitions, so we stop. The voters in each winning coalition with weight greater than the extra votes are the critical voters. We find Allen is critical in all three coalitions, while Betty and Cao are each critical in just one. Hence the Banzhaf index is (3, 1, 1).

TABLE 11.4 Winning Coalitions in Allen's Committee

			Critical Voters		
Coalition	**Total Weight**	**Extra Votes**	**Allen**	**Betty**	**Cao**
{Allen, Betty, Cao}	9	3	X		
{Allen, Betty}	8	2	X	X	
{Allen, Cao}	6	0	X		X

Self Check 13

In Example 12, Allen was a critical voter in the grand coalition and he had veto power. Is it possible for a voter in a weighted voting system to be a critical voter in the grand coalition and *not* have veto power?

The coalitions {Allen, Betty} and {Allen, Cao} are called **minimal winning coalitions,** because if one removes any voter the result will be a losing coalition.

Minimal Winning Coalition DEFINITION

A winning coalition is **minimal** if every voter who belongs to the coalition is critical in that coalition.

Self Check 14

Can the number of extra votes of a minimal winning coalition be greater than 0?

To make a complete list of all of the winning coalitions in a weighted voting system, start with the grand coalition. Step 1 is to form new winning coalitions by removing noncritical voters, one at a time, from the grand coalition. The number of winning coalitions formed in Step 1 is equal to the number of members of the grand coalition who are not critical voters. This is the procedure that was followed in Example 12. If some of the winning coalitions formed in Step 1 are not minimal, Step 2 is to remove noncritical voters, one at a time, from each of the winning coalitions obtained in Step 1. If some of the winning coalitions created in Step 2 are not minimal, proceed with them as in Step 2. The process ends when it cannot continue, because all of the winning coalitions formed in the last step were minimal.

You must be careful not to list any winning coalition more than once. This is illustrated in the following example.

EXAMPLE 13 A Five-Voter System

A, B, C, D, E use the weighted voting system [12 : 7, 4, 4, 3, 1]. List the winning coalitions and find their critical voters.

We will start with the grand coalition. It has a total weight of 19, so there are 7 extra votes. Since no voter has more than 7 votes, the grand coalition has no critical voters.

Step 1 is to remove members of the grand coalition one at a time and obtain five winning coalitions: {*B*, *C*, *D*, *E*} (0 extra votes, a minimal winning coalition); {*A*, *C*, *D*, *E*} and {*A*, *B*, *D*, *E*} (3 extra votes each; all voters except *D* and *E* are critical); {*A*, *B*, *C*, *E*} (4 extra votes, only *A* is critical); and {*A*, *B*, *C*, *D*} (5 extra votes, only *A* is critical).

Step 2 is to remove noncritical voters from the winning coalitions found in Step 1. We obtain {*A*, *C*, *E*}, a minimal winning coalition with 0 extra votes, by removing *D* from {*A*, *C*, *D*, *E*} or by removing *B* from {*A*, *B*, *C*, *E*}. Only count this once! By removing *E* from {*A*, *C*, *D*, *E*} or *B* from {*A*, *B*, *C*, *D*}, one gets {*A*, *C*, *D*}. In the same way, there are two ways to get the minimal winning coalitions {*A*, *B*, *E*} and {*A*, *B*, *D*} from {*A*, *B*, *D*, *E*} and one other four-member coalition listed above. Finally, the minimal winning coalition {*A*, *B*, *C*} can be obtained from {*A*, *B*, *C*, *D*} or {*A*, *B*, *C*, *E*} by removing the last voter from either one.

Table 11.5 summarizes the results and determines the Banzhaf power index of each participant. There are 11 winning coalitions in all. When compared with 120 permutations for the Shapley–Shubik index, it is manageable.

TABLE 11.5 Winning Coalitions in the Committee of Example 13

		Critical Votes				
Coalition	**Extra Votes**	***A***	***B***	***C***	***D***	***E***
{*A*, *B*, *C*, *D*, *E*}	7	0	0	0	0	0
{*B*, *C*, *D*, *E*}	0	0	1	1	1	1
{*A*, *C*, *D*, *E*}	3	1	0	1	0	0
{*A*, *B*, *D*, *E*}	3	1	1	0	0	0
{*A*, *B*, *C*, *E*}	4	1	0	0	0	0
{*A*, *B*, *C*, *D*}	6	1	0	0	0	0
{*A*, *C*, *E*}	0	1	0	1	0	1
{*A*, *C*, *D*}	2	1	0	1	1	0
{*A*, *B*, *E*}	0	1	1	0	0	1
{*A*, *B*, *D*}	2	1	1	0	1	0
{*A*, *B*, *C*}	3	1	1	1	0	0
Banzhaf power index		9	5	5	3	3

Self Check 15

Show the other ways to get the minimal winning coalitions {*A*, *B*, *E*} and {*A*, *B*, *D*}. ■

Combinations

In Alice's committee (see Example 7 on page 472), Alice has voting weight 3, and there are eight other members with voting weight 1. The quota is 6, so the voting system is [6 : 3, 1, 1, 1, 1, 1, 1, 1, 1]. In calculating the Shapley–Shubik power index, it was helpful that the members, except for Alice, were equally weighted. We would like to make use of this feature in determining the Banzhaf power index, too. There are differences, though: Unlike the Shapley–Shubik power indices, which always sum to 1, the Banzhaf power indices have no fixed sum. Therefore, one cannot determine the Banzhaf power indices of the other participants, after

A Real Mathematical Quagmire

SPOTLIGHT 11.4

The U.S. Supreme Court, in its 1962 decision of the case *Baker v. Carr,* instituted a principle that legislative districts within states should be equal in population. The decision was based on the "Equal Protection" clause of the Fourteenth Amendment to the U.S. Constitution. Dissenting, Justice Frankfurter declared that the process of equalizing populations would lead to a "mathematical quagmire."

The decision had repercussions for state and county legislative districts throughout the United States. Some district boundaries were redrawn to equalize populations. To avoid redrawing district boundaries the New Jersey legislature, the New York City Board of Estimate, Nassau County, and several other county boards of supervisors in New York State resorted to weighted voting, where the voting weight of a legislator (or supervisor) would be roughly proportional to the population being represented.

In a 1965 law review article, John F. Banzhaf III pointed out that three of the six supervisors of Nassau County, New York, were dummy voters (see Example 3 on page 464). In the same article he argued that the weighted voting system employed by the New Jersey legislature was unfair. The article inspired legal action against several elected bodies that employed weighted voting systems.

A lawsuit invalidated the voting system of the Board of Supervisors of Washington County, New York, in 1967. In its decision, the New York State Court of Appeals provided a way to fix a weighted voting system: Each supervisor's Banzhaf power index, rather than his or her voting weight, should be proportional to the population of the district that he or she represents. Borrowing Justice Frankfurter's phrase, the Court predicted that its remedy would lead to a "mathematical quagmire."

Five lawsuits filed over a period of 25 years challenged weighted voting in the Nassau County Board of Supervisors. Thus, the mathematical quagmire that the appeals court had feared actually materialized. Although Nassau County made a sincere attempt to comply with the Washington County decision, every voting system that it devised faced a new lawsuit. With conflicting expert testimony, the U.S. District Court finally ruled in 1993 that weighted voting was inherently unfair.

Banzhaf's article was aptly titled "Weighted Voting Doesn't Work." (It, and the Washington County decision, are included in the Suggested Readings at the end of this chapter. The Nassau County quagmire is the subject of Exercise 5 on page 496. You can explore what happened to the Board of Estimate in Exercises 38 and 39 on page 499).

Alice's index is known, without further calculation. A second difference has to do with permutations versus combinations. The Shapley–Shubik power index is based on counting permutations, which are different ways of ordering all of the members of a set. The Banzhaf index is based on counting winning coalitions, which are different subsets of the grand coalition. When counting subsets of a given set, we are dealing with **combinations.**

Algebra Review Appendix
Combinations

Let's see how we would determine Alice's Banzhaf power index. She would be a critical voter in any winning coalition that has less than 3 extra votes. She would need to join with at least three of the other members to meet the quota, but with less than six of the others. Thus, she needs to assemble a subset of the other eight members consisting of three, four, or five members. The number of ways to choose three members of a set of eight is given the symbol ${}_8C_3$. You may pronounce it "eight choose three."

Combinations and Combinations Symbol DEFINITION

The **combinations symbol** ${}_nC_k$ stands for the number of ways to choose k members from a set of n elements—that is, the number of **combinations** of n objects, taken k at a time.

For example, ${}_nC_0 = 1$ because there is just one way to get zero elements from a set of n, independent of n. If we are selecting one element from a set of n, there are obviously n ways to do it, so ${}_nC_1 = n$. We can determine a formula for ${}_nC_2$ by considering a toast in which each guest clinks glasses with each other guest. The number of "clinks" is ${}_nC_2$. If I'm a guest, I will clink with $n - 1$ others. Each of the n guests will do that, so multiply $n \times (n - 1)$ and divide by 2, because we've counted every "clink" twice. (For example, I clink with you, and you clink with me, but that is just one "clink.") Thus

$$ {}_nC_2 = \frac{n(n-1)}{2} $$

Alice's Banzhaf power index is ${}_8C_3 + {}_8C_4 + {}_8C_5$. Fortunately, there is a simple formula for ${}_nC_k$.

Formula 1: Basic Combinations Formula FORMULA

$$ {}_nC_k = \frac{n!}{(n-k)! \times k!} $$

Formula 1 holds, provided n and k are whole numbers and are not negative. We can apply it to calculate ${}_nC_0$ if we remember that $0! = 1$:

$$ {}_nC_0 = \frac{n!}{n! \times 0!} = 1 $$

EXAMPLE 14 ➔ Calculate ${}_8C_3$

By formula 1, ${}_8C_3 = \frac{8!}{5! \times 3!}$. Using $8! \times 40{,}320$; $3! = 6$, and $5! = 720$; we get ${}_8C_3 = \frac{40{,}320}{6 \times 720} = 56$.

This is not how we should calculate combinations, because it is more efficient to cancel before multiplying. The first cancellation is

$$ \frac{8!}{5!} = \frac{8 \times 7 \times 6 \times 5!}{5!} = 8 \times 7 \times 6 $$

Then

$$ {}_8C_3 = \frac{8!}{5!} \div 3! = \frac{8 \times 7 \times 6}{3 \times 2 \times 1} = 8 \times 7 = 56 $$

Using the cancellation trick, and $n! = n \times (n-1)!$, we get a simple way to verify ${}_nC_1 = n$.

Self Check 16

Use Formula 1 to verify that ${}_nC_2 = \frac{n \times (n-1)}{2}$.

Self Check 17

Calculate ${}_8C_4$ and ${}_8C_5$ without using a calculator.

EXAMPLE 15 Banzhaf Power in Alice's Committee

Alice's Banzhaf index has already been determined: It is ${}_8C_3 + {}_8C_4 + {}_8C_5 = 56 + 70 + 56 = 182$.

Bill, a weight-1 participant, will be a critical voter in any winning coalition with exactly 6 votes. He will need, in addition to his own vote, Alice's vote and the votes of two other participants, or, without Alice, he will need the votes of five other participants. He must get the two or the five votes from the *seven other* weight-1 participants—he can't vote twice!

There are ${}_7C_2$ winning coalitions in which Bill casts a critical vote and Alice is included, and ${}_7C_5$ winning coalitions where Bill casts a critical vote and Alice is not included. Bill's Banzhaf power index is ${}_7C_2 + {}_7C_5 = 21 + 21 = 42$. The other participants have the same index.

By the Banzhaf model, Alice is $\frac{182}{42} = 4\frac{1}{3}$ times as powerful as Bill. This is reasonably close to the Shapley–Shubik model's assessment that Alice is 4 times as powerful as Bill.

You have probably noticed that ${}_8C_3 = {}_8C_5 = 56$, and ${}_7C_2 = {}_7C_5 = 21$. If you suspect that there is symmetry here, you are right.

Formula 2: Symmetry of Combinations FORMULA

$${}_nC_k = {}_nC_{n-k}$$

We have seen that Formula 2 holds for ${}_8C_3$ and ${}_7C_2$—it would have saved a bit of time in calculating ${}_8C_5$ and ${}_7C_5$.

There are two ways to verify this formula. One is to consider a coalition of k members from a committee with n members. The k members vote "yes" and the remaining $n - k$ members vote "no." There are ${}_nC_k$ such coalitions. Now suppose everyone reverses his or her vote. If that happens, there are k "nays" and $n - k$ "yeas," corresponding to a coalition with $n - k$ members. There are ${}_nC_{n-k}$ of these coalitions. The ${}_nC_k$ coalitions of k members from a set of n can therefore be matched, one to one, with the ${}_nC_{n-k}$ coalitions of $n - k$ members from the set of n members.

Self Check 18

Use Formula 1 to verify Formula 2.

In Example 9 (page 473), we saw that, according to the Shapley–Shubik model, Alice's power was cut in half when three members of the committee agreed to cede their votes to Bill. In the next example, we will explore this situation in the context of the Banzhaf model. Writing Project 1 (page 501) provides an opportunity to explore the effect of a pact between states in the Electoral College that might have a similar effect.

EXAMPLE 16 The Revolt: Banzhaf Model

In Example 9, Alice's committee was transformed to a weighted voting system [6 : 3, 4, 1, 1, 1, 1] by a pact in which three weight-1 voters agreed to let Bill have their votes. Alice will be a critical voter in any winning coalition when she is joined by a losing coalition with weight 3, 4, or 5. This losing coalition could be Bill alone or Bill joined by one weight-1 voter, or it could be three or all four weight-1 voters. There is ${}_4C_0 = 1$ coalition consisting of Bill alone, and ${}_4C_1 = 4$ coalitions with Bill and one weight-1 voter. There are ${}_4C_3 + {}_4C_4 = {}_4C_1 + {}_4C_0 = 5$ coalitions of three or four weight-1 voters, so Alice's Banzhaf index is $1 + 4 + 5 = 10$.

Bill will be a critical voter in any winning coalition when the other voters have a total weight of 2, 3, 4, or 5. These other voters could be Alice plus up to two weight-1 voters; there are ${}_4C_0 + {}_4C_1 + {}_4C_2 = 11$ of these coalitions. Without Alice, Bill would need two or more weight-1 voters; there are ${}_4C_2 + {}_4C_3 + {}_4C_4 = {}_4C_2 + {}_4C_1 + {}_4C_0 = 11$ such coalitions, so Bill's Banzhaf index, 22, is more than twice Alice's.

A weight-1 voter, Zoë, must be joined by other voters with total weight 5 to form a winning coalition in which she is critical. That could be Alice and two of the three other weight-1 voters; there are ${}_3C_2 = 3$ of these coalitions. If Zoë is joined by Bill, exactly one of the three other weight-1 voters is needed. There are three of these coalitions too, so Zoë's Banzhaf index is 6.

One way to compare the Banzhaf indices between the two committees is to assign to each participant his or her share of the critical votes. Thus, in the committee as structured in Example 15, one participant (Alice) has 182 critical votes, and the eight weight-1 voters each have 42 critical votes. In all, the members of the committee have between them $182 + 8 \times 42 = 518$ critical votes. Alice's share of the crucial votes is $\frac{182}{518} = 35\%$, while each weight-1 voter has $\frac{42}{182} = 8\%$ of the critical votes.

After the pact, as in Example 16, Alice has 10 critical votes, Bill has 22, and four weight-1 participants each have 6, for a total of $10 + 22 + 4 \times 6 = 56$ critical votes. Now Alice has $\frac{10}{56} = 18\%$ of the critical votes, Bill has $\frac{22}{56} = 39\%$ of the critical votes, and each of the four remaining weight-1 voters has $\frac{6}{56} = 11\%$ of the critical votes.

Probability of Casting a Critical Vote

Unless a voter has veto power or is a dummy voter, it is not realistic to use a power index to measure his or her ability to influence the outcome of an election. Many factors—such as politics, for example—are not addressed by these models. Suppose that a voter, Charles, wants to know the probability that his "yes" vote will be critical. The Banzhaf model can answer this question, under the assumption that the other voters are not influenced by Charles or anyone else. We will therefore assume that each voter except Charles will toss a coin: heads for yea, tails for nay.

The probability that Charles will cast a critical vote is equal to

$$\frac{\text{number of winning coalitions in which Charles casts a critical vote}}{\text{number of winning or losing coalitions that include Charles}}$$

The numerator of this fraction is Charles's Banzhaf power index. To determine the denominator, let n be the number of voters. Each coalition that includes Charles also includes a subset of the other $(n - 1)$ participants—the ones that got heads when they tossed their coins. We can count the coalitions that include Charles by

The Electoral College: Presidential Elections of 2012, 2016, and 2020

SPOTLIGHT 11.5

The table below displays the Shapley–Shubik (SSPI) and Banzhaf (BPI) power indices of the voters in the Electoral College, as compared with the voter's weight as a percent of 538 (PCT), the total weight of all the voters. It shows that for the most part both measures of power agree closely with the actual share of power that a participant in the college has by virtue of its voting weight. There is an exception, though. California, whose voting weight is slightly more than 10% of 538, has more than its share of power by either measure. The power indices shown were calculated with the online Power Index Calculator (see page 502).

Voting Weight	Voters with This Weight	PCT (%)	SSPI (%)	BPI (%)
55	CA	10.22	11.03	11.36
38	TX	7.06	7.32	7.21
29	NY, FL	5.39	5.51	5.43
20	IL, PA	3.72	3.72	3.68
18	OH	3.35	3.33	3.30
16	GA, MI	2.97	2.95	2.93
15	NC	2.79	2.76	2.74
14	NJ	2.60	2.57	2.74
13	VA	2.42	2.38	2.37
12	WA	2.23	2.20	2.19
11	AZ, IN, MA, TN	2.04	2.01	2.01
10	MD, MN, MO, WI	1.86	1.82	1.82
9	AL, CO, SC	1.67	1.64	1.64
8	KY, LA	1.49	1.45	1.45
7	CT, OK, OR	1.30	1.27	1.27
6	AR, IA, KS, MS, NV, UT	1.12	1.09	1.09
5	NM, WV	0.93	0.90	0.91
4	HI, ID, NH, RI	0.74	0.72	0.73
3	AK, DE, DC, MT	0.56	0.54	0.55
3	ND, SD, VT, WY	0.56	0.54	0.55
2	ME, NE	0.37	0.36	0.36
1	ME, NE congressional districts	0.19	0.18	0.18

determining how many outcomes are possible if $(n - 1)$ people each toss a coin. Since each person has two possible outcomes, the answer is

$$\underbrace{2 \times 2 \times 2 \times 2 \times 2}_{(n-1)\text{ factors}} = 2^{n-1}$$

Therefore, the probability that Charles is a critical voter is equal to

$$\frac{\text{Charles's Banzhaf power index}}{2^{n-1}}$$

EXAMPLE 17 The Jury

A jury of 12 members must reach a unanimous verdict. This means that the only winning coalition is the grand coalition. The Banzhaf power index of each member of the jury is 1. Thus the probability that a juror will be a critical voter is $\frac{1}{2^{11}} = \frac{1}{2048}$. One might think that most criminal trials would end as mistrials, since it is so unlikely that the jurors would agree. But the Banzhaf model only touches on the voting system. It does not consider all of the evidence that the jurors see before they vote.

EXAMPLE 18 Majority Rule

A committee uses majority rule. What is the probability that a member of the committee will cast a critical vote in a winning coalition? Since every voter has weight 1, no one is a critical voter in a winning coalition unless the coalition has 0 extra votes; it must be a *minimal winning coalition.*

If there are three members, A, B, and C, then the minimal winning coalitions with A are $\{A, B\}$ and $\{A, C\}$. The Banzhaf power index of A is 2, so the probability that A will be critical is $\frac{2}{2^2} = 50\%$.

With four voters, the minimal winning coalitions will have three members. A would have to be joined by two of the other three members, so A's Banzhaf power index is ${}_3C_2 = 3$. The probability that A will be critical is $\frac{3}{2^3} = 37.5\%$.

If the number of voters is even, say $2n$, then a minimal winning coalition has $n + 1$ members. Voter A would be joined by n of the $2n - 1$ other members to form a minimal winning coalition. The Banzhaf power index of A would be ${}_{(2n-1)}C_n$ and the probability that A will be a critical voter is

$$\frac{{}_{(2n-1)}C_n}{2^{2n-1}}$$

With an odd number $(2n + 1)$ of voters a minimal winning coalition would still have $(n + 1)$ members. Voter A would be joined by n of the $2n$ other voters to form a minimal winning coalition, so the Banzhaf power index of A would be ${}_{(2n)}C_n$, and the probability that A is critical would be

$$\frac{{}_{(2n)}C_n}{2^{2n}}$$

Table 11.6 shows a few of these probabilities.

TABLE 11.6 Probability That a Voter Will Cast a Critical Vote in the Majority Rule Voting System (N is the number of voters; P is the probability)

N	*P*	*N*	*P*	*N*	*P*
1	100.0%	10	24.61%	1,000	2.523%
2	50.0%	11	24.61%	2,000	1.784%
3	50.0%	20	17.62%	10,000	0.564%
4	37.5%	21	17.62%	20,000	0.399%
5	37.5%	100	7.96%	100,000	0.252%
6	31.2%	101	7.96%	1,000,000	0.056%

Table 11.6 reveals some unexpected features. First, it appears that the probability that a voter will cast a critical vote when there is an even number $2n$ of voters is the same as when there are $(2n + 1)$ voters! This phenomenon can be explained by using the sum formula for combinations.

Formula 3: Sum Formula for Combinations FORMULA

$$_{n}C_{(k-1)} + {}_{n}C_{k} = {}_{(n+1)}C_{k}$$

Here's why the sum formula works: Suppose you are involved in a voting system with $(n + 1)$ voters. The number of combinations with k voters is $_{(n+1)}C_{k}$. Of these, $_{n}C_{k}$ coalitions do not include you. Any coalition that you belong to also has $(k - 1)$ of the other n voters; there are $_{n}C_{(k-1)}$ of these coalitions. Adding these numbers, we obtain the above sum formula.

By the sum formula, $_{(2n-1)}C_{(n-1)} + {}_{(2n-1)}C_{n} = {}_{(2n)}C_{n}$. By symmetry (Formula 2), $_{(2n-1)}C_{(n-1)} = {}_{(2n-1)}C_{n}$. Combining these equations, we get

$$2 \times {}_{(2n-1)}C_{n} = {}_{(2n)}C_{n}$$

The probability of casting a critical vote when there are $2n$ voters is $P = \frac{_{(2n-1)}C_{n}}{2^{2n-1}}$, and the probability of casting a critical vote in a $(2n + 1)$-voter system is also equal to

$$\frac{_{(2n)}C_{n}}{2^{2n}} = \frac{2 \times {}_{(2n-1)}C_{n}}{2 \times 2^{2n-1}} = \frac{_{(2n-1)}C_{n}}{2^{2n-1}} = P$$

The second interesting feature of Table 11.6 is that while a voter's chance of casting a critical vote decreases as the number of voters increases, the rate of decrease is slow. One would not think there would be much chance of casting a critical vote in a system with a million voters, but the chances are better than 1 in 2000. When the number of voters is large (hundreds or more), it is not practical to use the combination formula (Formula 1) to determine $_{(2n)}C_{n}$ and then divide by 2^{2n} because these numbers will be unmanageably large. Instead, use an approximation.

Formula 4: An Approximation FORMULA

$$\frac{_{2n}C_{n}}{2^{2n}} \text{ is approximately equal to } \frac{1}{\sqrt{n\pi}}.$$

In this formula, which can be derived using calculus, π is the ratio 3.14159... of the circumference of a circle to its diameter. The approximation is not too bad even if $n = 5$, and improves as n increases.

Self Check 19

Use Formula 4 to approximate the probabilities for the even values of N in the first two columns of Table 11.6. Compare your results with the probabilities given in the table (which were calculated using Formula 1, not the approximation). Don't bother with the third column because Formula 4 was used to compute those probabilities. ■

Can the Banzhaf and Shapley–Shubik Models Disagree?

SPOTLIGHT 11.6

In Self Check 9 (page 472) you calculated the Shapley–Shubik index of a 90-voter system in which Alice had 30 votes and the other 89 voters each had 1 vote apiece, where the quota was a simple majority, 60. You found that Alice had a Shapley–Shubik index of $\frac{1}{3}$, just as in the committee with voting system [6 : 3, 1, 1, 1, 1, 1, 1, 1, 1]. According to the Banzhaf model Alice had a slightly larger share of the power (about 35%) in the nine-voter system. You are not expected to calculate the Banzhaf index of the 90-member committee, because—while it is simple to do so in principle—the arithmetic is hard to manage. Alice will be a critical voter in any winning coalition when joined by between 30 and 59 of the other voters. There are

$$_{89}C_{30} + {_{89}C_{31}} + \cdots + {_{89}C_{58}} + {_{89}C_{59}}$$

Algebra Review Appendix
Scientific Notation

of these coalitions, and this is her Banzhaf index. Expressed in scientific notation, Alice's Banzhaf index is approximately 6.18×10^{26}. A weight-1 voter will be critical when joined by Alice and 29 other weight-1 voters or by 59 weight-1 voters. This voter's Banzhaf index is $_{88}C_{29} + {_{88}C_{59}} = 2 \times {_{88}C_{29}}$. This is a pretty big number too, approximately 3.03×10^{23}, or about the number of molecules in 9 milliliters of water (nearly 2 teaspoons). Alice's Banzhaf index is 2043 times as large, comparable to the number of molecules in 18.5 liters (about 5 gallons) of water. In the 90-member committee, the models differ radically. Let's use the water analogy to measure this. Alice has 18.5 liters, each of the 89 weight-1 voters has 0.009 liters. The sum of all the Banzhaf indices is $18.5 + 0.009 \times 89 = 19.3$ liters. Alice's share is $18.5 \div 19.3 = 95.8\%$ of the power according to the Banzhaf model, but 33.3% of the power in the Shapley–Shubik model.

How can we explain this difference? With the Shapley–Shubik model, Alice is a pivot if she is in one of 30 places in a permutation that has 90 places. Hence she has one-third of the power. For the Banzhaf index, suppose Alice votes "yes" and each of 89 weight-1 voters tosses a coin. What is the probability that between 29 and 59 of the tosses is heads? The average number of heads is 44.5, and it's very unlikely that the total number of heads will be more than 15 away from that value. A weight-1 voter will be critical if 29 of the other 88 weight-1 voters, and Alice, throw heads, or if 59 of the other weight-1 voters throw heads and Alice throws tails. This is improbable, because on the average there will be 44 heads when 88 coins are thrown, and it will not be often that the number of heads will be exactly 15 more or less than 44.

11.4 Voting Systems—Without Weights

Voting systems can be established without mentioning weights. In the United States, a bill becomes law when it achieves the support of a majority of the House of Representatives, a majority of the Senate, and the president; or without the president, a two-thirds majority of both houses. (This ignores filibusters, pocket vetoes, etc.) The United Nations Security Council also has a voting system that we will consider in more detail (see Example 21 on page 490).

Instead of using weights, these voting systems describe the winning coalitions directly, usually by specifying *minimal winning coalitions.* And then a winning coalition is a set of voters that contains one of the specified minimal winning coalitions. The following example is to review the definition of *minimal winning coalition* (see page 478).

EXAMPLE 19 **Minimal Winning Coalitions in the Four-Shareholder Corporation**

Example 6 (page 470) was about the Shapley–Shubik index of the corporation with four shareholders, *A*, *B*, *C*, *D*. It uses the voting system [51 : 49, 48, 2, 1], and Table 11.7 lists the winning coalitions and their extra votes.

The minimal winning coalitions are those in which each voter is a critical voter: {*A*, *B*}, {*A*, *C*}, and {*B*, *C*, *D*}. These minimal winning coalitions are displayed in Figure 11.1.

TABLE 11.7 Winning Coalitions in the Four-Shareholder Corporation

Winning Coalition	Extra Votes	Critical Voters
Grand coalition	49	None
{*A*, *B*}	46	*A*, *B*
{*A*, *B*, *C*}	48	*A*
{*A*, *B*, *D*}	47	*A*, *B*
{*A*, *C*}	0	*A*, *C*
{*A*, *C*, *D*}	1	*A*, *C*
{*B*, *C*, *D*}	0	*B*, *C*, *D*

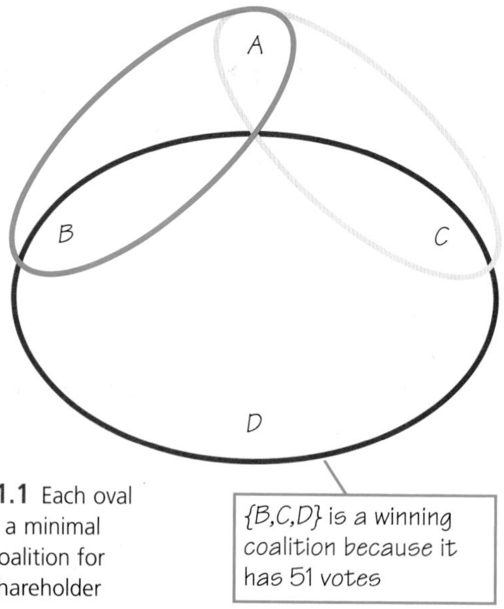
A
B
C
D
{B,C,D} is a winning coalition because it has 51 votes

Figure 11.1 Each oval surrounds a minimal winning coalition for the four-shareholder corporation.

Figure 11.1 shows that every voter belongs to at least one minimal winning coalition.

Self Check 20

What can be said of a voter who doesn't belong to any minimal winning coalition?

Every pair of minimal winning coalitions in Figure 11.1 has some overlap. This is necessary if the voting system is to be decisive, because if two winning coalitions had no overlap, one coalition could pass a motion and the other could pass a contradictory motion.

No minimal winning coalition in the figure is entirely contained in another minimal winning coalition. If there had been a minimal winning coalition, say {*A*, *B*, *C*}, that contained another minimal winning coalition, {*A*, *B*} then *C* would not be a critical voter, since {*A*, *B*} can pass a motion without *C*—hence {*A*, *B*, *C*} would not be minimal.

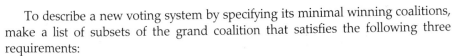

To describe a new voting system by specifying its minimal winning coalitions, make a list of subsets of the grand coalition that satisfies the following three requirements:

1. Your list can't be empty. You have to name at least one coalition; otherwise, there would be no way to approve a motion.
2. You can't have one minimal winning coalition that contains another one.
3. Every pair of coalitions in the list has to overlap.

For dictatorship with dictator *D*, there would be one minimal winning coalition, $\{D\}$. All dictatorships with the same population are the same as voting systems. While we have presented dictatorships as weighted voting systems, the weights are an unnecessary formality.

Self Check 21

Suppose a voting system has just two voters, *A* and *B* who are not dummies, and possibly some other voters who are dummies. Can $\{A\}$ and $\{B\}$ both be minimal winning coalitions? ■

Equivalent Voting Systems

DEFINITION

Two voting systems are **equivalent** if there is a way for all the voters of the first system to exchange places with the voters of the second system and preserve all winning coalitions.

The weighted voting systems $[50 : 49, 1]$ and $[4 : 3, 3]$, involving pairs of voters *A*, *B* and *C*, *D*, respectively, are equivalent because in each system, unanimous support is required to pass a measure. We could have *A* exchange places with *C*, and *B* exchange places with *D*. The grand coalition is a minimal winning coalition for each system.

Now consider two voting systems, $[2 : 2, 1]$ and $[5 : 3, 6]$, involving the same pair of voters, *A* and *B*. In the first, *A* is a dictator, while in the second, *B* dictates. By having *A* and *B* exchange places with each other, we see that the two systems are equivalent. "Equivalent" does not mean "the same." Voter *A* would tell you that the system where he is the dictator is not the same as the system where *B* is the dictator. The systems are equivalent because each has a dictator.

Every two-voter system is equivalent either to a system with a dictator or to one that requires consensus. As the number of voters increases, the number of different types of voting systems increases.

Now let's construct some voting systems.

EXAMPLE 20 Three-Voter Systems

We will make a list of all voting systems that have three participants, *A*, *B*, and *C*. No two voting systems on the list should be equivalent.

Suppose that $\{A\}$ is a minimal winning coalition. Then *A* can pass a motion unilaterally and is a dictator. *B* and *C* are dummies. Systems where *B* or *C* is dictator are equivalent to this one.

If there is no dictator, every minimal winning coalition must contain either two or all three voters. Let's consider the case in which the grand coalition is a minimal winning coalition. It is the only winning coalition, because any other winning coalition would have to be entirely contained in this coalition, which requirement 2 doesn't allow. In this voting system, a unanimous vote is required to pass a measure. We will call this system *consensus rule.*

Finally, let's suppose that there is a two-voter minimal winning coalition, {*A*, *B*}. If it is the only minimal winning coalition, then *C* is a dummy and *A* and *B* make all the arrangements. We will call this system the *clique*. Of course, the clique could be {*A*, *C*} or {*B*, *C*}, but these are equivalent systems.

There could be two 2-voter minimal winning coalitions, say {*A*, *B*} and {*A*, *C*}. Neither coalition contains the other, and there is an overlap, so all of the requirements are satisfied. In this system, *A* has veto power. We encountered this system in Example 1 (page 463)—where Allen had veto power—and we will call it the *chair veto*. There are two other voting systems equivalent to this one, where *B* or *C* is chair.

It is possible that all three two-member coalitions are minimal winning coalitions. Because there are only three voters, any two distinct two-member coalitions will overlap, so the requirements are still satisfied. This is the *majority rule* system.

Self Check 22

In general, what can we say about a voter who belongs to all of the minimal winning coalitions?

Table 11.8 lists all five of these three-voter systems. Each system can be presented as a weighted voting system, and suitable weights are given in the table. If we want to make a similar list of all types of four-voter systems, we can start by changing each three-voter system into a four-voter system, by putting a fourth voter, *D*, a dummy, into the system. Of course, *D* would not belong to any of the minimal winning coalitions. You may be interested to know that there are an additional nine 4-voter systems that don't have any dummies. Try to list as many of these systems as you can.

TABLE 11.8 Voting Systems with Three Participants

System	Minimal Winning Coalitions	Weights	Banzhaf Index
Dictator	{*A*}	[3 : 3, 1, 1]	(4, 0, 0)
Clique	{*A*, *B*}	[4 : 2, 2, 1]	(2, 2, 0)
Majority	{*A*, *B*}, {*A*, *C*}, {*B*, *C*}	[2 : 1, 1, 1]	(2, 2, 2)
Chair veto	{*A*, *B*}, {*A*, *C*}	[3 : 2, 1, 1]	(3, 1, 1)
Consensus	{*A*, *B*, *C*}	[3 : 1, 1, 1]	(1, 1, 1)

EXAMPLE 21 The United Nations Security Council

The United Nations Security Council has five permanent members—China, France, Russia, the United Kingdom, and the United States—and ten other members that serve two-year terms. To resolve a dispute not involving a member of the Security Council, 9 votes are required, including the votes of each of the permanent members. (Thus, each permanent member has veto power.) The Security Council voting system is thus specified by describing its minimal winning coalitions as consisting of the five permanent members and four other members. Exercise 40 (page 500) is an opportunity to consider this interesting voting system in depth.

EXAMPLE 22 The Scholarship Committee

A university offers scholarships on the basis of either academic excellence or financial need. Each application for a scholarship is reviewed by two professors, who rate the student academically, and two financial aid officers, who rate the applicant's need. If both professors or both financial-aid officers recommend the applicant for a scholarship, the dean of admissions decides whether to award a scholarship. Is it possible to assign weights to the professors, the financial-aid officers, and the dean to reflect this decision-making system? The answer is no.

To see why, let's focus on the minimal winning coalitions. The participants are the two professors, *A* and *B*; the financial aid officers, *E* and *F*; and the dean, *D*, who has veto power. The minimal winning coalitions (see Figure 11.2) are {*A*, *B*, *D*} and {*D*, *E*, *F*}.

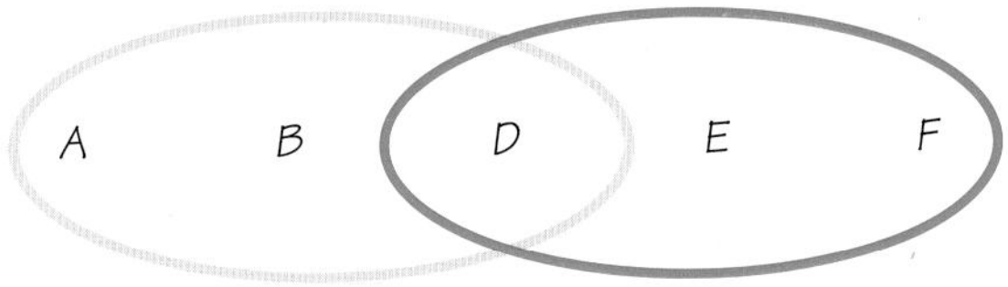

Figure 11.2 Minimal winning coalitions in the scholarship committee.

Consider the following two winning coalitions: In C_1, all except the financial-aid officer *F* favors an award; while in C_2, Professor *B* dissents. Thus

$$C_1 = \{A, B, D, E\} \text{ and } C_2 = \{A, D, E, F\}$$

In C_1, we notice that *A* is a critical voter and *E* isn't, while in C_2 the tables are turned because *E* is critical while *A* is not. If this were a weighted voting system, then in any winning coalition, the critical voters would all have greater weight than those who are not critical. Thus, *A* would have to have both more weight than *E* (because of the situation in C_1) and less weight than *E* (because of C_2), which is impossible. ■

Self Check 23

Let's change the rules of the scholarship committee: To be approved, an application must receive the votes of at least one professor and both aid officers, or both professors and one aid officer. The dean has veto power, as before. List the minimal winning coalitions. (Exercise 44 on page 500 asks about weighted voting in this version of the committee.) ■

EXAMPLE 23 Power Indices of the Scholarship Committee

The dean has veto power. Therefore, she will be the pivotal voter in any permutation where she appears last. If she is second to last in a permutation, she will still be the pivotal voter, because among the three voters before her, there will be either both professors or both financial-aid officers. In the middle position, she will be pivotal if and only if both professors or both aid officers come first. Adding this up, we have $2 \times 4! = 48$ permutations in which the dean is in fourth or fifth position. There are four permutations of the form Prof, Prof, Dean, Aid, Aid, because the professors and the aid officers can be in either order, and another four of the form Aid, Aid, Dean, Prof, Prof. The dean is not the pivotal voter when she is first or second because at least three people have to approve a scholarship. We conclude that the dean is pivotal in $48 + 4 + 4 = 56$ permutations in all. Her Shapley–Shubik power index is therefore $\frac{56}{5!} = \frac{7}{15}$. Each of the other participants is equally powerful, and they share the remaining $\frac{8}{15}$ of the power. Thus each professor and each aid officer has a Shapley–Shubik power index of $\frac{2}{15}$.

To compute the Banzhaf index, we will make a list of winning coalitions. There are seven of them:

{*A*, *B*, *D*}, {*D*, *E*, *F*}, {*B*, *D*, *E*, *F*}, {*A*, *D*, *E*, *F*}, {*A*, *B*, *D*, *F*}, {*A*, *B*, *D*, *E*}, and {*A*, *B*, *D*, *E*, *F*}

The dean has veto power, so she is a critical voter in each of them. Her Banzhaf power index is therefore 7.

Professor *A* is a critical voter in {*A*, *B*, *D*}, {*A*, *B*, *D*, *E*}, and {*A*, *B*, *D*, *F*}. His Banzhaf power index is 3. The remaining participants have the same power.

In the Shapley–Shubik model, the dean is $3\frac{1}{2}$ times as powerful as a faculty member or aid officer. In the Banzhaf model, she is only $2\frac{1}{3}$ times as powerful as one of the other members of the Scholarship Committee.

Review Vocabulary

Banzhaf power index In a voting system, the number of winning coalitions in which voter *V* is a critical voter. (p. 475)

Coalition A set of participants in a voting system that might vote in favor of a measure. The empty coalition is allowed, and it represents a situation when the voters unanimously oppose a motion. (p. 474)

Combinations and Combinations symbol The combinations symbol ${}_nC_k$ stands for the number of ways to choose *k* members from a set of *n* elements—that is, the number of combinations of *n* objects, taken *k* at a time. (p. 480)

Critical voter Let *W* be a winning coalition, and let *V* be a voter who belongs to *W*. Then *V* is a critical voter in *W* if the coalition consisting of all the voters in *W except V* is a losing coalition. (p. 475)

Dictator A voter *D* is a dictator if a motion will pass if and only if *D* is in favor and the votes of the other members make no difference. (p. 463)

Dummy voter A participant in a voting system who never has an opportunity to cast a deciding vote is called a dummy voter. (p. 464)

Equivalent voting systems Two voting systems are equivalent if there is a way for all the voters of the first system to exchange places with the voters of the second system and preserve all winning coalitions. (p. 489)

Extra votes The number of votes by which a winning coalition's total weight exceeds the quota. (p. 476)

Factorial The number of permutations of a set of *n* voters is called the factorial of *n* and is denoted *n*!. (p. 470)

Formulas involving combinations

Basic combinations formula: ${}_nC_k = \frac{n!}{(n-k)! \times k!}$ (p. 481)

Symmetry formula: ${}_nC_k = {}_nC_{n-k}$ (p. 482)

Sum formula: ${}_nC_{(k-1)} + {}_nC_k = {}_{(n+1)}C_k$ (p. 486)

Approximation formula: ${}_{2n}C_n = \frac{2^{2n}}{\sqrt{n\pi}}$, approximately (p. 486)

Grand coalition The coalition that includes every voter. (p. 477)

Losing coalition A set of participants in a weighted voting system whose combined voting weight is less than the quota; or, in an unweighted voting system, a set of voters that does not contain a minimal winning coalition. (p. 474)

Minimal winning coalition A winning coalition is minimal if every voter who belongs to the coalition is critical in that coalition. (p. 478)

Pivotal voter The first voter in a voting permutation who, when joined by those coming before him or her, would have enough voting weight to win. Each voting permutation has exactly one pivotal voter. (p. 467)

Quota In a weighted voting system, the required number of votes, or total voting weight, necessary to pass a measure. (p. 462)

Shapley–Shubik power index Among *n* voters, the number of voting permutations in which a voter is pivotal, divided by *n*! (the factorial of *n*). (p. 469)

Veto power A voter whose vote is necessary to pass any motion is said to have veto power. (p. 464)

Voting permutation An ordered list of all the voters in a voting system. (p. 466)

Weighted voting system Specifies the voting weights w_1, $w_2, \ldots, w_n$ of the participants, and the quota, q. A motion will pass if the sum of the weights of the voters in favor is at least equal to the quota. The notation $[q : w_1, w_2, \ldots, w_n]$ is used to denote a system in which there are n voters, with voting weights $w_1, w_2, \ldots, w_n$; and the quota is q. (p. 462)

Winning coalition A set of participants in a weighted voting system whose combined voting weight is greater than or equal to the quota; or, in an unweighted voting system, a set of voters that contains a minimal winning coalition. (p. 474)

Self Check Answers

1. The sum of the weights is 15; so the quota must not be greater than 15. The quota cannot be less than half the sum of the weights, so the quota is at least 8. The possible values for the quota are 8, 9, 10, 11, 12, 13, 14, and 15.

2. The dictator's voting weight must be greater than or equal to q. If there is a dictator, it has to be the weight-8 voter. The quota must be at least 8, by the result of Self Check 1. Thus if the quota is 8, the first voter is dictator. If the quota is more than 8, no voter is dictator.

3. The dummy would be the weight-1 voter. If q is an even number, this voter will make no difference. If some of the other voters get together to pass a motion, and their weights add up to a number S less than the quota, S will be even, so at least 2 more votes will be needed to reach the quota. If the quota is 8, then the weight-8 voter is dictator, so the other three voters are dummies. If the quota is 12, the weight-2 voter is also a dummy. For $q = 10$ or 14, the weight-1 voter is the only dummy. On the other hand, if the quota is odd—9, 11, 13, or 15—there are no dummy voters.

4. The weight-8 voter has veto power for any value of the quota from 8 to 15, because if she votes "no," there are at most 7 votes in favor. If the quota is 12 or more (up to 15), the weight-4 voter also has veto power; if he votes "no," there are at most 11 votes in favor. If the quota is 14 or 15, the weight-2 voter has veto power, because if she votes "no," there are at most 13 votes in favor. Thus, if the quota is 12 or 13, exactly two voters, those with weights 8 and 4, have veto power. For $q = 12$, both of the other voters are dummies; for $q = 13$, no voter is a dummy.

5. In the voting permutation Allen, Cao, Betty, the first member, Allen, has 5 votes, less than the quota. Cao and Allen together have 6 votes, which is equal to the quota, so these two voters can pass a measure. Hence Cao is the pivot.

6. The quota must be more than either voter's weight, because there is no dictator. Therefore, both voters are needed to pass a motion; so in both permutations of the voters, the second voter is pivotal. Thus, each voter is pivotal in one permutation, and each has a Shapley–Shubik index equal to $\frac{1}{2}$.

7. We have seen that with five voters there are 120 permutations. With six voters, there are $6 \times 120 = 720$ permutations. Finally, with seven voters, there are $7 \times 720 = 5040$ permutations.

8. Four permutations would acquire new pivots: *D* would be the new pivot in *ADBC* and *ADCB* (the pivots had been *B* and *C*, respectively); and *A* would be the new pivot in *DABC* and *DACB* (again, the pivots had been *B* and *C*, respectively). Thus, *B* and *C* each lost two permutations, and *A* and *D* each gained two. Now *A* is pivot in 12 permutations, while *B*, *C*, and *D* are each pivot in four. The Shapley–Shubik index of *A* is now $\frac{1}{2}$, while *B*, *C*, and *D* have Shapley–Shubik indices equal to $\frac{1}{6}$.

9. Alice will be the pivot in any permutation with between 30 and 59 weight-1 voters coming before her—in other words, if she is in position 31 up to 60. If we separate the permutations into 90 groups, according to Alice's position, she is the pivot in every permutation in 30 of the groups and not the pivot in the remaining 60. Her Shapley–Shubik power index is therefore $\frac{30}{90} = \frac{1}{3}$. The other 89 members of the committee have equal shares of the remaining $\frac{2}{3}$ of the power. Each weight-1 voter's Shapley–Shubik power index is $\frac{2}{3} \div 89 = \frac{2}{267}$.

10. Bill is one vote short of being a dictator. He will be pivot in any permutation when he is in position 2, 3, or 4. He will not be pivot in position 1, for then the pivot will be whoever is in position 2, and he will not be pivot in position 5, for then the pivot will be the occupant of position 4. Therefore, he is pivot in $\frac{3}{5}$ of the permutations. Alice's power is the same as a weight-1 voter's, so she and the three weight-1 voters share the remaining $\frac{2}{5}$ of the power. Each has a Shapley–Shubik index equal to $\frac{2}{5} \div 4 = \frac{1}{10}$.

11. No state was a critical voter. It is only necessary to see if California, which had the largest number of electoral votes in the winning coalition, was not a critical voter. The winning coalition had 332 electoral votes, 62 more than the quota. California had 55 votes, so the Democratic ticket would still win by a margin of 7 without California.

12. The weight-4 voter is a dictator and is thus a critical voter in every winning coalition. Each winning coalition consists of the dictator and a subset of the three weight-1 voters. There are $2^3 = 8$ subsets of a three-element set—hence there are 8 winning coalitions. The dictator's Banzhaf power index is therefore 8; the other voters are dummies, and their Banzhaf power indices are all 0.

13. No. Suppose that voter *V* is a critical voter in the grand coalition, and let *C* be any coalition. If *V* does not belong to *C*, then *C* is a subset of the grand coalition with *V* removed, which we know is a losing coalition. Therefore, all winning coalitions must include *V*, and hence *V* has veto power.

14. Yes. In Example 12, we saw that {Allen, Betty} was a minimal winning coalition. It has 2 extra votes. The extra votes of a minimal winning coalition must be less than the voting weight of each member of the coalition.

15. You could get {*A*, *B*, *E*} by removing *C* from {*A*, *B*, *C*, *E*}, and {*A*, *B*, *D*} by removing *C* from {*A*, *B*, *C*, *D*}.

16. $${}_nC_2 = \frac{n!}{(n-2)! \times 2!} = \frac{n \times (n-1) \times (n-2)!}{(n-2)! \times 2} = \frac{n \times (n-1)}{2}$$

17. $${}_8C_4 = \frac{8!}{4! \times 4!} = \frac{8 \times 7 \times 6 \times 5}{4 \times 3 \times 2 \times 1} = 2 \times 7 \times 1 \times 5 = 70$$
$${}_8C_5 = \frac{8!}{5! \times 3!} = \frac{8!}{3! \times 5!} = {}_8C_3 = 56$$

18. $${}_nC_{n-k} = \frac{n!}{(n-k)! \times k!} = \frac{n!}{k! \times (n-k)!} = {}_nC_k$$

19.

N	*P*	Approximation	*N*	*P*	Approximation
2	50.0%	$1/\sqrt{\pi} = 56.4\%$	10	24.61%	$1/\sqrt{5\pi} = 25.23\%$
4	37.5%	$1/\sqrt{2\pi} = 39.9\%$	20	17.62%	$1/\sqrt{10\pi} = 17.84\%$
6	31.2%	$1/\sqrt{3\pi} = 32.6\%$	100	7.96%	$1/\sqrt{50\pi} = 7.98\%$

20. Suppose *X* is a voter who belongs to no minimal winning coalition. If *X* happens to belong to some winning coalition *W*, then because *W* is winning, it contains a minimal winning coalition *M*. We know *X* does not belong to *M*, so the coalition formed by removing *X* from *W* would still be a winning coalition. This means *X* has no opportunity to cast a critical vote in any winning coalition: *X* is a dummy.

21. No, {*A*} and {*B*} do not overlap.

22. Let *X* be a voter who belongs to all of the minimal winning coalitions. If *C* is a coalition that *X* doesn't belong to, then *C* contains no minimal winning coalition and is thus a losing coalition. Therefore the vote of *X* is necessary to win; that is, *X* has veto power.

23. There are four minimal winning coalitions; each consists of the dean and three of the other four members.

Skills Check

1. In the weighted voting system [65 : 60, 30, 10],
(a) the weight-60 voter is a dictator.
(b) the weight-60 voter has veto power.
(c) the three voters are equally powerful.

2. A voting system has 200 voters, and a simple majority is needed to pass a motion. The quota for this system is __________.

3. Two daughters and a son administer a trust fund. Each daughter has 4 votes, and the son has 7 votes; the quota for passing a measure is 8.
(a) The son has veto power.
(b) The daughters are dummies.
(c) The three siblings have equal voting power.

4. How many dummies are there in the voting system [16 : 8, 8, 4, 2, 1]?

5. The voters *A*, *B*, and *C* use the voting system [3 : 2, 1, 1]. In which permutation is *C* the pivotal voter?
(a) *BCA*
(b) *ABC*
(c) *ACB*

6. Four voters, *A*, *B*, *C*, *D*, use the weighted voting system [6 : 4, 3, 2, 1]. In the permutation *DBCA*, the pivotal voter is __________.

7. If the first voter in some permutation is the pivotal voter, then
(a) that voter must be the dictator.
(b) that voter has veto power.
(c) the other voters are dummies.
(d) Answers (a), (b), and (c) are all correct.

8. The Shapley–Shubik index of the weight-1 voter in the voting system [6 : 4, 3, 2, 1] is ________.

9. The number 7! (7 factorial) is

(a) more than 1 million.

(b) between 10,000 and 1 million.

(c) less than 10,000.

10. A jury's decision in a criminal trial must be unanimous. In any permutation of the jury's members, the member who is pivotal is ________.

11. If a winning coalition is minimal, the number of extra votes is

(a) zero.

(b) less than the weight of the least powerful member of the coalition.

(c) more than the weight of the least powerful member who is opposed.

12. If a motion passes in the weighted voting system [6 : 4, 3, 2, 1] with only the weight-2 voter dissenting, then the critical voters are those with weights ________.

13. A voting system has the following winning coalitions: {*A*, *B*, *C*}, {*A*, *B*, *D*}, {*B*, *C*, *D*}, and {*A*, *B*, *C*, *D*}. Which voter is the most powerful?

(a) *A*

(b) *B*

(c) *C*

(d) *D*

14. The minimal winning coalitions of a voting system are {*A*, *B*, *C*}, {*C*, *D*, *E*}, and {*B*, *C*, *D*}. In the winning coalition {*A*, *B*, *C*, *E*}, ________ are the critical voters.

15. Referring to the voting system in Skills Check 14, which voter is not critical in the winning coalition {*A*, *C*, *D*, *E*}?

(a) *A*

(b) *B*

(c) *C*

(d) *E*

16. In the voting system described in Skills Check 14, voter *C* has ________ power.

17. If voter *X* is critical in every winning coalition, then

(a) *X* has veto power.

(b) *X* is a dictator.

(c) *X* will be needed to prevent a motion from passing.

(d) Answers (a), (b), and (c) are all correct.

18. In a system with *n* voters, *A* is a dictator. The Banzhaf power index of *A* is ________.

19. In the weighted voting system [5 : 3, 2, 1], what is the Banzhaf power index of the weight-1 voter?

(a) 2

(b) 1

(c) 0

20. Voters *A*, *B*, *C*, . . . , *K* use the voting system [19 : 8, 7, 6, 5, 4, 1, 1, 1, 1, 1, 1]. ________ is a minimal winning coalition that has 1 extra vote.

21. ${}_{15}C_7 + {}_{15}C_8 =$

(a) ${}_{16}C_7$

(b) ${}_{16}C_8$

(c) Neither (a) nor (b) is correct.

22. ${}_6C_3 =$ ________.

23. ${}_{12}C_3 =$

(a) ${}_{12}C_9$

(b) 220

(c) Answers (a) and (b) are both correct.

24. In the voting system [6 : 4, 1, 1, 1, 1, 1], the Banzhaf power index of a weight-1 voter is ________.

25. Four voters, *A*, *B*, *C*, *D*, use the weighted voting system [6 : 4, 3, 2, 1]. The Banzhaf index of B is

(a) 0.

(b) 1.

(c) 2.

(d) 3.

26. The Banzhaf power index of the weight-3 voter in the system [7 : 3, 2, 2, 2] is ________.

27. The weighted voting system [6 : 4, 3, 2, 1] is equivalent to

(a) [5 : 3, 2, 2, 1].

(b) [8 : 5, 4, 3, 2].

(c) [12 : 8, 5, 5, 2].

(d) All of the above answers are right.

28. The minimal winning coalitions of the system [7 : 3, 2, 2, 2], with voters named *A*, *B*, *C*, and *D*, are ________.

29. The minimal winning coalitions of a voting system are {*A*, *B*, *C*} and {*B*, *C*, *D*}. Which weighted voting system is equivalent to it?

(a) [5 : 1, 2, 2, 1]

(b) [6 : 1, 3, 3, 1]

(c) [6 : 1, 4 ,4, 1]

(d) The system is not equivalent to a weighted voting system.

30. If the minimal winning coalitions are {*A*, *B*} and {*B*, *C*, *D*}, ______________ has veto power.

Chapter 11 Exercises

Challenge Discussion

11.1 How Weighted Voting Works

1. How would you explain to the weight-12 voter in the weighted voting system [27 : 14, 14, 13, 12] that he is a dummy?

2. Consider a weighted voting system in which there are five voters who have weights 5, 4, 3, 2, and 1.

(a) What is the least possible quota, and what is the greatest?

(b) If the weight-5 voter has veto power, and no other voter does, what is the quota?

(c) If the weight-1 voter is a dummy, what is the quota?

3. Which voters, if any, have veto power in the weighted voting system [9 : 5, 4, 3]? Is any voter a dummy?

4. Given a voting system [q : 33, 32, 31, 4], such that exactly one of the voters has veto power, answer the following questions:

(a) The weight-__________ voter is the one with veto power.

(b) Find q.

(c) Is any voter a dummy?

(d) The voter with veto power wields more power than the others. Is there any difference in the power between the other three voters?

5. The various weighted voting systems used by the Board of Supervisors of Nassau County, New York, turned out to be the mathematical quagmire described in Spotlight 11.4 (page 480). Before the county's weighted voting was declared unconstitutional by a federal district court in 1993, it was changed several times. The weights in use since 1958 were as follows:

		Weights					
Year	**Quota**	H_1	H_2	N	B	G	L
1958	16	9	9	7	3	1	1
1964	58	31	31	21	28	2	2
1970	63	31	31	21	28	2	2
1976	71	35	35	23	32	2	3
1982	65	30	28	15	22	6	7

Here, H_1 is the presiding supervisor, always from the community of Hempstead; H_2 is the second supervisor from Hempstead; and N, B, G, and L are the supervisors from the remaining districts: North Hempstead, Oyster Bay, Glen Cove, and Long Beach.

(a) List the dummy voters in each year.

(b) Suppose that the two Hempstead supervisors always vote together. Now list the dummy voters in each year.

6. In the weighted voting system [q : 10, 8, 7, 5, 4, 4, 3], for which whole numbers q, not less than 21 and not greater than 41,

(a) does the weight-10 voter *not* have veto power?

(b) is the weight-10 voter the only one with veto power?

(c) is the weight-3 voter a dummy?

11.2 The Shapley–Shubik Model

7. Voters *A*, *B*, *C*, *D* use the weighted voting system [51 : 30, 25, 24, 21].

(a) List all permutations in which *A* is pivotal.

(b) List all permutations in which *B* is pivotal.

(c) Calculate the Shapley–Shubik power index of the system.

8. How would the Shapley–Shubik power index in Exercise 7 change if the quota were increased to

(a) 52?

(b) 55?

(c) 58?

9. Voters *A*, *B*, *C*, *D*, *E*, and *F* use the weighted voting system [7 : 3, 2, 2, 2, 2, 2].

(a) Describe the set of permutations in which Voter *A* is pivotal.

(b) What is the number of permutations that you found in part (a)?

(c) Use the answer that you have given in part (b) to determine the Shapley–Shubik power index of the system.

10. Refer to the permutation of the 2012 presidential election (see Table 11.1 on page 468). The Democratic ticket received 2,990,274 votes in Pennsylvania, to the Republican ticket's 2,680,434. Which

Brian Baer/Sacramento Bee/MCT via Getty Images

state would be the pivotal voter if, at the last minute, a television ad convinced 150,000 voters to switch from the Democratic ticket to the Republican ticket? Assume that no votes are changed outside of Pennsylvania.

11. *D* was a member of a committee until he resigned, after discovering that with the voting system in use, he was a dummy. Did the Shapley–Shubik power indices of the other committee members change as a result of *D*'s departure?

12. If a state uses the District System (described in Spotlight 11.1 on page 465) to choose its electors in a two-candidate presidential election, as Maine and Nebraska do, then some electoral permutations are impossible because the electors corresponding to the senators must not appear in a permutation before all of the electors representing the congressional districts, or after all of them. How would this affect the calculation of the Shapley–Shubik power index?

11.3 The Banzhaf Model

13. A committee has four members, *A*, *B*, *C*, and *D*. It makes decisions by majority rule.

(a) What is the quota if each member has a voting weight of 1?

(b) List all the winning coalitions in which *A* is critical.

14. *A*, *B*, *C*, *D*, *E*, *F* use the weighted voting system [15 : 10, 7, 5, 3, 1, 1]. Decide which of the following coalitions are winning and which are losing. Identify the critical voters in the winning coalitions.

(a) {*B*, *C*, *D*, *E*, *F*}

(b) {*A*, *B*, *C*}

(c) {*B*, *C*}

(d) {*A*, *D*, *E*, *F*}

15. For each of the following weighted voting systems, make a list of all winning coalitions. Identify the critical voters in each coalition, and calculate the Banzhaf power index for each voter. Give the voters the names *A*, *B*, ...

(a) [51 : 52, 48]

(b) [3 : 2, 2, 1]

(c) [8 : 5, 4, 3]

(d) [51 : 45, 43, 8, 4]

(e) [51 : 45, 43, 6, 6]

16. Four voters, *A*, *B*, *C*, and *D*, use the weighted voting system [51 : 30, 25, 24, 21]. Make a table showing all winning coalitions in the left column. In the second column, put the number of extra votes. This will enable you to identify the critical voters of the winning coalitions. List the critical voters in a third column, and use this table to determine the Banzhaf power index of each participant.

17. *This exercise is intended for a group of 2 to 4 students.* If the quota for the voting system in Exercise 16 increases, you can quickly modify the table you made by reducing the extra votes—ignore a coalition when its extra votes become negative. As the number of extra votes decreases, more of a coalition's voters will be critical—until the coalition switches from winning to losing.

For example, {*A*, *D*} is a winning coalition with 0 extra votes when the quota $q = 51$. If $q = 52$, it is a losing coalition, so *A* and *D* are no longer credited with critical votes for that coalition. On the other hand, *B* and *C* gain critical votes in {*A*, *B*, *C*} and {*A*, *C*, *D*}, respectively.

Determine the Banzhaf index for this system with the following quotas:

(a) 52
(b) 55
(c) 56
(d) 57
(e) 71
(f) 76
(g) 80

18. Calculate the following terms[2]:

(a) ${}_7C_3$

(b) ${}_{50}C_{100}$

(c) ${}_{15}C_2$

(d) ${}_{15}C_{13}$

[2] Exercise 18b is not a misprint. How many 100-element subsets does a 50-element set have?

19. Calculate the following terms.

(a) ${}_6C_3$

(b) ${}_{100}C_2$

(c) ${}_{100}C_{98}$

(d) ${}_9C_5$

20. A committee has 10 members and decides measures by weighted voting. The chairperson, Franklin, has voting weight 4; each of the 9 other members has weight 1, and the quota is 7. Determine the Shapley–Shubik and Banzhaf power indices of each member.

21. Agnes, Boris, and Carla, weight-1 voters in the committee described in Exercise 20, cede their votes to a fourth weight-1 voter, Essie. Use the following steps to recalculate the power indices.

(a) Use the notation $[q : w_1, w_2, \ldots, w_?]$ to describe the weighted voting system as it stands after the pact.

(b) Gerry, a weight-1 member of the committee, will be pivot in permutations where she is in the middle, with Essie before her and Franklin after, or the other way around. How many of these permutations are there?

(c) Determine Gerry's Shapley–Shubik power index, and then the Shapley–Shubik power index of each of the other members.

(d) Describe the winning coalitions that would have Gerry as a critical voter.

(e) Describe the winning coalitions in which Franklin is a critical voter.

(f) Determine the Banzhaf power index for each member of the committee after the pact.

22. Refer to Exercise 5 (page 496) for a brief history of weighted voting in the Nassau County Board of Supervisors. Assume that the two Hempstead supervisors always agree, so that the board is effectively a five-voter system. Determine the Banzhaf power index of this system in each year. You should be able to do this by hand. If you would like to find the index for the full system each year, you may use the Power Index Calculator (see page 502).

23. If each member of a 12-person jury in which a unanimous decision is required tosses a coin to determine his or her decision, the probability that a given juror will cast a critical vote is $\left(\frac{1}{2}\right)^{11} = \frac{1}{2048}$. In some states, civil cases are tried before a six-person jury, and the quota for a decision is 5 votes. With such a jury, what is the probability that a given juror will cast a critical vote, if each juror uses a coin toss to determine his or her vote?

24. A voting system with $2n + 1$ voters follows majority rule. That is, each voter has weight 1, and the quota is $n + 1$. How many minimal winning coalitions are there? What does this number have to do with the Banzhaf power index?

25. This exercise is about the District System for allocating votes in the Electoral College (see Spotlight 11.1 on page 465). You will need to use the Power Index Calculator for this (see page 502). Find the total percentage of power that Nebraska has in the Electoral College by adding the percentages of power of the two electors representing the state and the three individual electors who represent congressional districts, according to the Banzhaf power index. Next, consider what would happen if Nebraska changed its law so that all of its electors would be committed to vote for the ticket that won the statewide contest: There would be 53 participants in the Electoral College, and Nebraska would have a voting weight of 5. Would this change result in an increase in Nebraska's percentage of Banzhaf power, or a decrease?

11.4 Voting Systems—Without Weights

26. Consider a four-person voting system with voters *A*, *B*, *C*, and *D*. The winning coalitions are {*A*, *B*, *C*, *D*}, {*A*, *B*, *C*}, {*A*, *B*, *D*}, {*A*, *C*, *D*}, and {*A*, *B*}.

(a) List the minimal winning coalitions.

(b) A minimal blocking coalition is a set of voters that is voting against a measure, has sufficient voting weight to prevent the measure from passing, and in which the votes of each participant is essential to do so. Show that *A* has veto power and therefore that {*A*} is a minimal blocking coalition.

(c) Find another minimal blocking coalition.

(d) Determine the Banzhaf power index of each voter.

(e) Find an equivalent weighted voting system. (*Hint:* If two voters have the same Banzhaf index, give them the same weight.)

(f) Calculate the Shapley–Shubik index of each voter.

27. In Exercise 26, the term minimal blocking coalition was defined. Must minimal blocking coalitions overlap, as minimal winning coalitions do?

28. A five-member committee has a voting system in which the chairperson can pass or block any motion that she supports or opposes, provided that at least one other member is on her side. Show that this voting system is equivalent to the weighted voting system [4 : 3, 1, 1, 1, 1].

29. Find weighted voting systems that are equivalent to the following:

(a) A committee of three faculty members and the dean. To pass a measure, at least two faculty members and the dean must vote "yes."

(b) A committee of five faculty members, the dean, and the provost. To pass a measure, three faculty, the dean, and the provost must vote "yes."

30. A four-member faculty committee and a three-member administration committee vote separately on each issue. The measure passes if it receives the support of a majority of each committee. Show that this system is not equivalent to a weighted voting system.

31. Calculate the Banzhaf index of the voting system in Exercise 30. Who is more powerful according to the Banzhaf model, a faculty member or an administrator?

32. Determine the Shapley–Shubik index of the system in Exercise 30. Who is more powerful according to the Shapley–Shubik model, a faculty member or an administrator?

33. Explain why a voting system in which no voter has veto power must have at least three minimal winning coalitions.

34. How many distinct (nonequivalent) voting systems with four voters can you find? Systems that have dummies don't count. The challenge is to find all nine distinct systems and to find—if possible—weighted voting systems equivalent to each.

35. A corporation has four shareholders and a total of 100 shares. The quota for passing a measure is the votes of shareholders owning 51 or more shares. The number of shares owned by each shareholder is as follows:

Shareholder	Shares Owned
A	40
B	26
C	24
D	10

There is also an investor, *E*, who is interested in buying shares but does not own any shares at present. Sales of fractional shares are not permitted.

(a) List the winning coalitions and compute the number of extra votes for each. Make a separate list of the losing coalitions, and compute the number of votes that would be needed to make the coalition winning.

(b) How many shares can *A* sell to *B* without causing any of the winning coalitions listed in part (a) to lose or any of the losing coalitions in part (a) to win?

(c) How many shares can *A* sell to *D* without changing the sets of winning or losing coalitions?

(d) How many shares can *A* sell to *E* without changing the winning coalitions? Because *E* is now a dummy, he must remain a dummy after the trade.

36. Which of the following voting systems is equivalent to the voting system in use by the corporation in Exercise 35?

(a) [5 : 3, 2, 1, 1, 1]

(b) [5 : 3, 2, 2, 1, 0]

(c) [6 : 4, 2, 2, 2, 1]

(d) [8 : 3, 3, 2, 2, 1]

37. A nine-member committee has a chairperson and eight ordinary members. A motion can pass if and only if it has the support of the chairperson and at least two other members, or if it has the support of all eight ordinary members.

(a) Find an equivalent weighted voting system.

(b) Determine the Banzhaf power index.

(c) Determine the Shapley–Shubik power index.

(d) Compare the results of parts (b) and (c). Do the power indices agree on how power is shared in this committee?

38. The New York City Board of Estimate consists of the mayor, the comptroller, the city council president, and the presidents of each of the five boroughs. It used to employ a voting system in which the city officials each had 2 votes and the borough presidents each had 1; the quota to pass a measure was 6. This voting system was declared unconstitutional by the U.S. Supreme Court in 1989. Although the boroughs had unequal populations, they had equal representation on the board, in violation of the equal protection clause of the 14th amendment to the U.S. Constitution (*New York City Board of Estimate v. Morris*).

(a) Describe the minimal winning coalitions.

(b) Determine the Banzhaf power index.

39. A proposed weighted voting system for the New York City Board of Estimate (see Exercise 38) that is based on the populations of the boroughs is [71 : 35, 35, 35, 11.3, 7.3, 9.6, 6.0, 1.8]. Find a simpler system of weights that yields an equivalent voting system. Do you think this system would satisfy the Supreme Court's objections?

40. The voting system in use by the U.N. Security Council is described in Example 21 (page 490).

(a) Show that this voting system is equivalent to the weighted voting system in which each of the 5 permanent members has 7 votes, the 10 nonpermanent members each have 1 vote, and the quota is 39.

(b) Compute the Banzhaf power index for a permanent member and for a nonpermanent member.

(c) The Security Council originally had 5 permanent members and 6 members who served two-year terms. Each permanent member had veto power, and 6 votes were required to resolve an issue. Devise an equivalent weighted voting system and compute its Banzhaf index.

(d) Do you think that the addition of four more nonpermanent members caused the permanent members to lose significant power?

41. A committee has senior members *A*, *B*, and *C*, and junior members *D*, *E*, and *F*. Each senior member has voting weight 3, and junior members each have voting weight 1. The quota for passing a measure is 7. Find the minimal winning coalitions of the committee and determine the Banzhaf power index for members of each class.

42. A new junior member joins the system described in Exercise 41. Again, describe the minimal winning coalitions and determine the Banzhaf power index for members of each class. According to the Banzhaf model, does the presence of this new junior member increase or decrease the share of power of each junior member?

43. Use the Shapley–Shubik model to compare the share of power of a junior member in the systems described in Exercises 41 and 42.

44. Is the committee in Self Check 23 on page 491 equivalent to a weighted voting system? If so, find suitable weights.

45. An alumni committee consists of 3 rich alumni and 12 recent graduates. To pass a measure, a majority, including at least 2 of the rich alumni, must approve.

(a) Describe the minimal winning coalitions.

(b) Suppose this is a weighted voting system. Give the recent graduates each a weight of 1, and let r be the weight of each of the rich alumni. Find the total weight of each minimal winning coalition. For example, the coalition with all three rich alumni and five of the recent graduates would have weight $3r + 5$.

(c) Compare the total weight of the losing coalition consisting of all recent graduates and one rich alumna with the total weight of one of the minimal winning coalitions to show that $r > 6$.

(d) Compare the total weight of the largest losing coalition that includes all three rich alumni with the same winning coalition you used in part (c). Does any weight r that works here also satisfy the inequality in part (c)?

(e) Is this voting system equivalent to a weighted voting system?

46. Find the Banzhaf power index of each participant in the voting system of Exercise 45.

47. List the minimal winning coalitions in each of the following three-voter systems. Match each system with an equivalent system listed in Table 11.8 on page 490.

(a) [10 : 9, 5, 1]

(b) [10 : 9, 7, 3]

(c) [10 : 11, 7, 1]

(d) [10 : 6, 4, 2]

(e) [10 : 5, 4, 3]

48. In Skills Checks 14–16, we considered a five-voter system in which the minimal winning coalitions are {*A*, *B*, *C*}, {*B*, *C*, *D*}, and {*C*, *D*, *E*}.

(a) Show that this system is not equivalent to any weighted voting system.

(b) Explain why *A* and *E* have equal power and why *B* and *D* have equal power.

(c) Make a list of all of the winning coalitions and use it to determine the Banzhaf power index of each voter.

Chapter Review

49. A 5-member committee will use weighted voting. The members are assigned voting weights of 12, 7, 4, 3, and 1. They would like to set the quota so that no member has veto power and no member is a dummy voter. Help them by giving them a list of the quotas that meet their specifications. For each quota that you list, find all of the minimal winning coalitions.

50. A committee has two co-chairs and five other members. The voting weight of each co-chair is 4, the other members each have voting weight 1, and the quota is 7. Find the Banzhaf and Shapley–Shubik power indices for each member.

51. For a bill to become a federal law, it must be passed by a majority of the 435-member House of Representatives, as well as of the 101-member Senate (we count the vice president as a voting member of the

Senate); then it must be approved by the president. If the president vetoes the bill, it can still become law if a two-thirds majority of each house of Congress votes to override the veto. (The vice president does not participate in votes to override a veto, so for veto overrides, the Senate has 100 members.) Describe the minimal winning coalitions in this voting system.

52. In the system [5: 3, 1, 1, 1, 1, 1, 1] the weight-3 voter votes "yes" and each of the weight-1 voters uses a coin toss to determine his or her vote. What is the probability that the weight-3 voter has cast a critical vote?

53. Find the Shapley–Shubik power index for each voter in the system [5: 3, 1, 1, 1, 1, 1, 1].

Writing Projects

1. The most important weighted voting system in the United States is the Electoral College (see Spotlight 11.1 on page 465). An organization, National Popular Vote (NPV), is attempting to replace the Electoral College with a system in which the president is elected by popular vote. One way to accomplish this would be to amend the U.S. Constitution. NPV has what might prove to be a simpler approach, lobbying state legislatures to pass a bill that pledges their state's electoral votes to the ticket that receives the plurality of the popular vote. As of this writing, 11 states, with a total voting weight of 165 electoral votes, have passed this bill. The bill includes language that says that it will only take effect when the total voting weight of the states that have enacted it reaches 270, for then these states will make up a winning coalition.

What would happen if the states were to take effect for the next presidential election, without waiting for a winning coalition to form? The NPV website lists the states that have passed the NPV bill. Construct a new voting system in which one voter, NPV, represents all the states that have passed the NPV bill. These states have ceded their votes to NPV. The other voters are the states (or congressional districts in states with the District System) that have not passed the NPV bill, with their electors chosen as usual (see Suggested Websites on the next page). Example 9 on page 473 and Example 16 on page 483 show how a voting system can be altered if some voters combine to act together. Because the voting systems will be sizable, you will need to use the Power Index Calculator.

2. On November 1, 2014, the Council of Ministers of the European Union (E.U.) replaced its old weighted voting system with a new system, agreed to in the Treaty of Lisbon. A winning coalition now consists of the ministers from at least 55% of the countries in the E.U., comprising 65% of the population. There are currently 28 countries in the E.U., so we may say a winning coalition would have to comprise at least 16 countries that had at least 65% of the population. Show that no country is a dummy voter in this system, even though the largest country, Germany, has a population almost 200 times greater than the population of the smallest, Malta. Do you think that this system is equivalent to a weighted voting system?

Because the European Council is too large for hand calculation, consider the Nassau County Board of Supervisors in 1958, with weights and quota as shown in Example 3 on page 464. If the Board had added a requirement that a winning coalition must include at least four supervisors, show that there would be no dummy voters. Is the resulting voting system equivalent to a weighted voting system? List the winning coalitions, find the critical voters, and determine the Banzhaf and Shapley–Shubik power indices for this system.

3. Maine and Nebraska use the District System to select their electors in presidential elections. (The District System is explained in Spotlight 11.1 on page 465.) Discuss the complications involved in determining the power indices when usage of the District System is taken into account. What would happen if all states, except California, adopted the District System for choosing electors?

Suggested Readings

BANZHAF, JOHN F. III. Weighted voting doesn't work, *Rutgers Law Review*, 1965 (vol. 19), pp. 317–343. The author defines the Banzhaf index and uses it to show that the weighted voting system in use by the Nassau County Board of Supervisors was unfair.

BRAMS, STEVEN J. *Game Theory and Politics*, 2d Ed., Dover Publications, New York, NY, 2004.

Iannucci v. Board of Supervisors of Washington County. This case opened a "mathematical quagmire." You can find the text of the court's opinion with a web search engine.

TAYLOR, ALAN D. *Mathematics and Politics: Strategy, Voting Power, and Proof*, Springer-Verlag, New York, NY, 1995. Chapter 4 of this book covers weighted voting systems and their analysis using the Shapley–Shubik and Banzhaf models. It has no mathematical prerequisites, but it does include carefully written logical arguments that must be carefully read.

Suggested Websites

www.nationalpopularvote.com The website of the National Popular Vote organization. It has a current list of the states that have passed the NPV bill (see Writing Project 1 on page 501).

math.temple.edu/~conrad/Power/BPIandSSPI.html This site includes a brief discussion of the power indices that we have been exploring. It also contains a link to the *Power Index Calculator*. This online application can determine the Banzhaf and Shapley–Shubik power indices of some weighted voting systems that are too large for hand calculation, such as the U.S. Electoral College or the E.U. Council of Ministers before the Treaty of Lisbon.

Electing the President

12

Vicki Beaver/Alamy

12.1 Narrowing the Field through the Primary Process

12.2 Spatial Models for Two-Candidate Elections: Discrete Distributions

12.3 Spatial Models for Two-Candidate Elections: Continuous Distributions

12.4 Spatial Models for Multicandidate Elections

12.5 Spatial Models and the Electoral College

The American cartoonist and inventor Rube Goldberg (1883–1970) is synonymous with elaborate contraptions that solve a simple task in an overly complicated manner, often by setting off a chain reaction. The chain reaction of electing the president of the United States begins with the primaries and ends when electors cast electoral votes based on the results of the general election. Goldberg even used primaries and presidential elections as fodder for his drawings. This chapter uses mathematics to illuminate the sequence of sometimes overly complicated steps

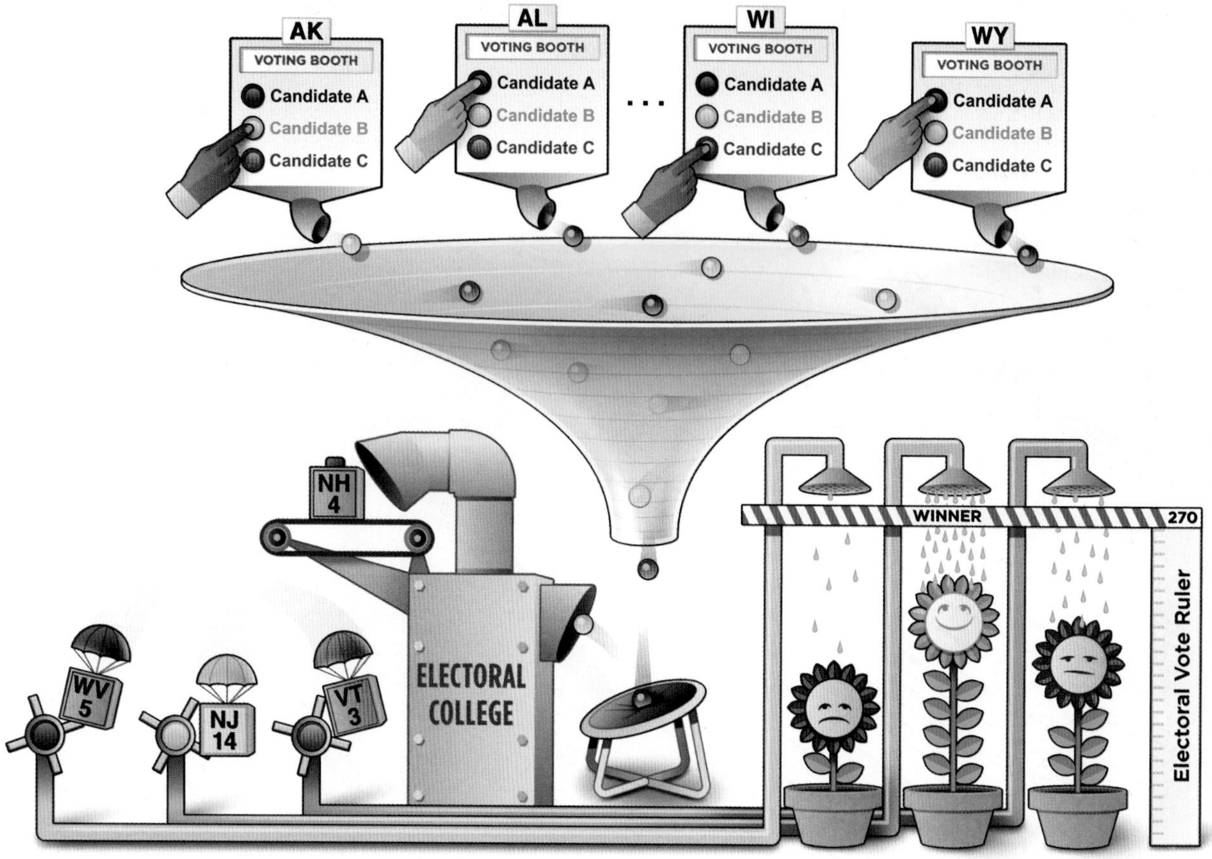

A Rube Goldberg-esque machine to invoke the complications of electing the president.

used to elect the president—from the assigning of delegates in the primaries to the strategic positioning of candidates in the general election and its relationship to the **Electoral College.**

12.1 Narrowing the Field through the Primary Process

The first phase of a presidential election year begins in January, when Democratic and Republican candidates seek their party's nomination for president by running in state caucuses and primaries. Caucuses and primaries are a series of political events used by the Democratic and Republican parties to determine their nominees for president. The first caucus of the election season is held in Iowa. Voters from Iowa's precincts assemble in a public place to discuss the candidates and to select representatives to attend county conventions. The process builds through conventions at the county, congressional district, and state levels. In comparison, primaries are statewide elections. Candidates earn delegate votes based on the number of voters who support them at the caucuses' conventions and primaries. The primary season concludes with a party convention, where a candidate becomes the party's nominee if he or she has received a majority of the delegate votes from the primary season. If no candidate has a simple majority, then the convention is called *brokered* and the party's nominee is determined through a negotiation process.

That Iowa is the first caucus of the primary season seems to have been an accident, but it is now written into the parties' rules. The winner of the Iowa caucus achieves a sort of front-runner status that translates into greater opportunities to raise funds to support his or her campaign. Other states have tried to jockey for position so that their caucus or primary appears earlier in the primary season and, hence, has greater impact. Delegate counts have been discounted when a state has pushed its primary before the Iowa caucus, as both Florida and Michigan did in the 2008 Democratic primary.

There is some variability in how delegates are awarded for the Republican caucuses and primaries. We will return to their most frequently used rule from the 2008 Republican primary after we examine how delegates are tallied in the Democratic primaries. The same Democratic Delegate Selection Rules are used for all primaries to translate votes into delegate counts. The 2008 Democratic primary was one of the closest primaries in recent U.S. history, second only to the 1976 Republican contest between Gerald Ford and Ronald Reagan that was not decided until the convention. The close 2008 race between Barack Obama and Hillary Clinton was due in part to the Democratic Party rules that assign delegates to each candidate based on his or her proportion of the popular vote. Although these rules assign delegates based on the number of votes each candidate receives, these rules prevent candidates who do not receive 15% of the popular vote from receiving any delegates. As such, these rules help narrow the field of candidates. For example, of the eight candidates who mounted national campaigns in the 2008 Democratic primary, only three were awarded delegates.

Because delegates are individuals and cannot be divided into fractions, awarding delegates proportionally requires a systematic way of rounding the fractional values to whole numbers. Methods for doing so, referred to as apportionment methods, are best known for their role in determining the number of representatives allotted to each state in the U.S. House of Representatives (see Chapter 14). The Democratic Delegate Selection Rules appear in Table 12.1. Example 1 demonstrates the five steps.

TABLE 12.1 Democratic Delegate Selection Rules (Section 13, Part D)

Step 1.	Tabulate the percentage of the vote that each presidential preference (including uncommitted status) receives in the congressional district to three decimals.
Step 2.	Retabulate the percentage of the vote to three decimals, received by each presidential preference, excluding the votes of presidential preferences whose percentage in Step 1 falls below 15%.
Step 3.	Multiply the number of delegates to be allocated by the percentage received by each presidential preference.
Step 4.	Delegates shall be allocated to each presidential preference based on the whole numbers which result from the multiplication in Step 3.
Step 5.	Remaining delegates, if any, shall be awarded in order of the highest fractional remainders in Step 3.

EXAMPLE 1 ➔ Proportionally Awarding Delegates in the Democratic Primary

The New Hampshire primary is the first primary after the Iowa caucuses. For the 2008 Democratic primary, New Hampshire was divided into two districts, each with seven delegates. Of the 21 candidates who received votes in District 2, the top five vote getters were Hillary Clinton, John Edwards, Dennis Kucinich, Barack Obama, and Bill Richardson. Their vote totals are listed in Table 12.2, and all other candidates who received votes are bundled together under "Others."

Darren McCollester/Getty Images

The columns of Table 12.2 demonstrate the steps outlined in the Democratic Delegate Selection Rules (Table 12.1). Following Step 1, the first column contains the percentage of

TABLE 12.2 Election Data for the 2008 Democratic New Hampshire Primary (District 2)

Candidates	Popular Votes	Percentage of Votes	Adjusted Votes	Adjusted Percentage	Quota	Initial Delegates	Final Delegates
Hillary Clinton	55,418	37.732	55,418	40.602	2.8421	2	3
John Edwards	25,224	17.174	25,224	18.48	1.1936	1	1
Dennis Kucinich	2,176	1.482					
Barack Obama	55,848	38.025	55,848	40.917	2.8642	2	3
Bill Richardson	7,220	4.916					
Others	985	0.671					
Totals	146,871	100	136,490	100	7	5	7

the popular vote for the candidates. For example, Clinton's percentage of the popular vote is the total number of votes she received (55,418) divided by the total number of votes (146,871); hence, her percentage is 55,418/146,871 $\approx$ 0.37743 = 37.743%. As described in Step 2, all candidates that receive less than 15% of the popular vote are eliminated; these are Kucinich, Richardson, and the candidates collected under the category "Others." The percentages of the remaining candidates—Clinton, Edwards, and Obama—are adjusted accordingly. This is achieved by discarding the votes for all candidates that failed to meet the 15% threshold. For example, Clinton's adjusted percentage becomes 40.602 by calculating 55,418/136,490 $\approx$ 40.602%.

Steps 3 to 5 are usually referred to as Hamilton's method, a method of apportionment attributed to U.S. founding father Alexander Hamilton. In Step 3, the remaining candidates' adjusted percentages are multiplied by 7, the total number of delegates to be awarded in District 2, to calculate their **quotas**—the proportion of the delegates the candidate should receive given his or her adjusted percentage of the vote. For instance, Clinton's quota of 2.8421 satisfies 40.602% $\times$ 7 = 0.40602 $\times$ 7 = 2.8421.

In Step 4, the quotas are rounded down to give an initial apportionment of 2, 1, and 2 delegates, respectively, for Clinton, Edwards, and Obama. For example, Clinton's quota of 2.8421 is rounded down to 2. As a consequence, Step 4 awards the first 5 (= 2 + 1 + 2) delegates, meaning that there are 2 delegates left to allocate. In Step 5, the remaining 2 delegates are awarded to the candidates with the largest fractional/decimal remainders. Obama has the largest decimal remainder of 2.8652 − 2 = 0.8652. Clinton's remainder is the second highest. It follows that both Obama and Clinton receive an extra delegate, rounding their delegate totals to 3 each. The final delegate counts for District 2 of the 2008 New Hampshire Democratic primary are 3 delegates each to Clinton and Obama and 1 delegate to Edwards. ■

Self Check 1

How was Obama's adjusted percentage in Table 12.2 calculated? ■

The proportional allocation of delegates in the Democratic primary is in contrast to the most frequently used rule from the 2008 Republican primary: the winner-take-all strategy. As the name implies, under this strategy the candidate that receives the most popular votes receives all of the delegates. The 2008 Democratic primary was not decided until June, whereas John McCain had wrapped up the 2008 Republican nomination on March 4. The reason for the difference in how long it takes a candidate to lock down the nomination could be that the winner-take-all strategy focuses attention on the first-place candidate. The candidates that win early primaries also receive the additional campaign funds that come with the increased media attention and popular support. The effect is that the candidate field narrows more quickly in the Republican primary than in the Democratic primary. Also, the shorter time it takes for the Republican Party to determine their presidential nominee can be viewed as advantageous, as it gives the Republican candidate more time to focus on the potential Democratic opponent for the general election in the fall.

When data are rounded, unusual behavior or a counterintuitive result may be observed. Such behavior is referred to as a *paradox* in the apportionment literature (see Chapter 14). Requiring candidates to receive 15% of the total vote to be awarded a delegate introduces surprising behavior that can affect even the non-eliminated candidates in ways that are not predictable. It may seem that eliminating candidates who receive less than 15% of the vote can only help the remaining candidates by giving them more delegate votes. The following example shows that this may not always be the case.

EXAMPLE 2 Paradoxical Behavior Under the Democratic Delegate Selection Rules

Suppose that the Democratic Delegate Selection Rules are used to allocate five delegates in a district for vote totals given in Table 12.3. The application of the rules results in Jones, Umberto, and Viktor receiving 4, 1, and 0 delegates, respectively. If the election official forgets to drop Viktor for receiving less than 15% of the vote and instead uses Steps 3 to 5 on the original/non-adjusted data, then the outcome changes, as given in Table 12.4: Jones now receives 3 delegates and Umberto receives 2. Viktor still fails to receive a single delegate. But how the five delegates are split between Jones and Umberto changes depends on whether or not Viktor's votes are discarded or not! By eliminating Viktor from consideration by the 15% rule, Umberto's delegate count decreases from 2 to 1.

TABLE 12.3 Implementing the Democratic Delegate Selection Rules for Example 2

Candidates	Popular Vote	Percentage of Votes	Remaining Votes	Adjusted Percentage	Quota	Initial Delegates	Final Delegates
Stephen Jones	6625	66.25	6625	70.479	3.524	3	4
Tracey Umberto	2775	27.75	2775	29.521	1.476	1	1
George Viktor	600	6					
Totals	10,000	100	9400	100	5	4	5

TABLE 12.4 Using Steps 3 to 5 to Award Delegates for Example 2

Candidates	Popular Vote	Percentage of Votes	Quota	Initial Delegates	Final Delegates
Stephen Jones	6625	66.25	3.313	3	3
Tracey Umberto	2775	27.75	1.387	1	2
George Viktor	600	6	0.300		
Totals	10,000	100	5	4	5

Self Check 2

How is Stephen Jones's adjusted quota determined in Table 12.3? ■

Despite the paradoxical behavior that can occur under the Democratic Delegate Selection Rules, the 15% threshold achieves its goal of narrowing the field by keeping weakly supported candidates from being allocated any delegates. For states with few delegates, a candidate with less than 15% of the vote may not receive any delegates in a district anyway, but that is not the case for states with many delegates. And even states with small districts use the same apportionment method with 15% threshold to allocate Pledged Party Leaders and Elected Officials (or PLEO) delegates and at-large delegates using the statewide vote totals for the candidates. The number of these additional delegates usually surpasses the number allocated in any one district.

The multi-stream calculation of delegate totals is part of the Rube Goldberg aspect of the primary process. As discussed, the winner-take-all policy used in the Republican primaries consolidates support more quickly for a single candidate. After the two party conventions, the Democratic and Republican nominees and any third party candidates square off in the general election. How candidates position themselves in the general election is discussed in the next section. These strategic implications are also useful in explaining how candidates compete strategically in the primary elections.

12.2 Spatial Models for Two-Candidate Elections: Discrete Distributions

Two-candidate contests are most common in the general election—but occasionally there are other significant candidates. One example is Ross Perot, who ran under the banner of the Reform Party in 1992 and garnered 19% of the popular vote. Although no minor-party candidate has ever won a presidential election, some have affected which of the two major-party nominees did win. For example, Ralph Nader impacted the 2000 presidential election, in part because many voters who voted for Nader in Florida would have voted for Al Gore otherwise and, consequently, would have tipped the election in Gore's favor. We begin with the situation in which there are just two candidates in the general election.

To model such elections, we assume that voters respond to the candidates' positions on various issues. This is not to say that other factors—such as personality, ethnicity, religion, and race—have no effect on election outcomes, but rather that issues take precedence in a voter's decision. How can a candidate's position be represented? We start by assuming that there is a single overriding issue, or set of issues, on which the candidates must take a definite stand, such as the degree of governmental intervention in the economy. We assume that the voters' attitudes on this issue or dimension can be represented along a left–right continuum, ranging from very liberal on the left (much intervention) to very conservative on the right (little intervention).

Other examples of one-dimensional issues include the amount of military buildup, the amount of foreign aid to developing countries, and so on. Geometrically, the left–right continuum is a portion of the real line. For example, foreign aid can be represented in dollars and can range from "no foreign aid" (i.e., 0 on the number line) to any positive dollar amount.

To derive conclusions about the behavior of voters from their attitudes and the positions candidates take in a campaign, some assumption is necessary about how voters decide for whom to vote. We assume that a candidate announces a **policy position**—a point on the left–right spectrum. Voters are assumed to vote for the candidate whose policy position is closest to the **voter's ideal position,** often called the **voter's ideal point**—the favorite position of the voter. For two candidates A and B with distinct policy positions $a < b$, a voter V with ideal point v votes for Candidate A if $v < (a + b)/2$ and votes for Candidate B if $v > (a + b)/2$. The voter's ideal point is equidistant from a and b if $v = (a + b)/2$. In this case, the voter is indifferent to the choice between the two candidates and could choose to vote for either one, possibly based on some other criterion used as a tiebreaker. The position $(a + b)/2$ is key and is related to geometry.

Midpoint

DEFINITION

For points a and b on the real line, the **midpoint** is $(a + b)/2$. Because a and b are numbers, $(a + b)/2$ is also called the *arithmetic average,* or *average.*

Algebra Review Appendix
Distance and Midpoint Between Two Points on a Line

For a two-candidate election, which candidate a voter votes for depends on how the voter's ideal point compares to the midpoint of the candidates' policy positions. To determine which candidate wins the election, each voter's ideal position is compared with the midpoint. For this reason, the *number* of voters who have particular attitudes along the left–right continuum is often more important than the attitude of any *individual* voter. A distribution of voters describes or counts the numbers of voters with each ideal point.

Discrete Distribution of Voters

DEFINITION

A **discrete distribution of voters** counts the number of voters whose ideal points are located at any point along the left–right continuum.

EXAMPLE 3 Determining the Winner of an Election in a Spatial Model

Suppose that there are 21 eligible voters in a tiny town who must decide between Ann and Bob for mayor. Each voter has an ideal position that describes how much of the town's taxes should be used to subsidize child care. A voter with an ideal position closer to 0 desires a higher subsidy. As the voter's ideal point increases, then the voter prefers a smaller subsidy. The 21 voters have 7 distinct ideal positions, as described by the discrete distribution given in Table 12.5. For example, there are 4 voters who have 6 as their ideal position.

TABLE 12.5 Discrete Distribution of the 21 Voters in Example 3

Ideal points	0	2	3	4	6	8	10
Number of voters	1	3	6	3	4	2	2

As part of their election platforms, Ann and Bob announce their positions on the child-care subsidy. Bob is more supportive of the child-care subsidy, having announced a position of 2 on the 0 to 10 scale. Ann is less supportive, having announced a position of 7. Each voter votes for the candidate whose announced position is closest to his or her ideal position. Figure 12.1 shows a graphical representation of the voters' ideal points, as well as the policy positions announced by Ann (red line) and Bob (blue line).

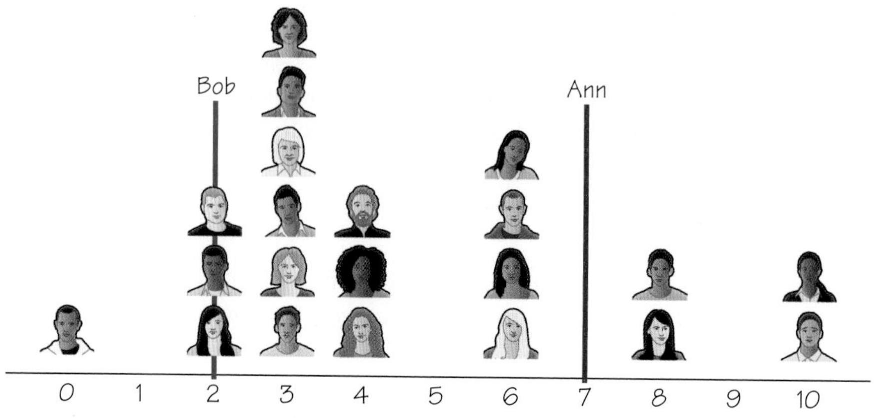

Figure 12.1 A graphical representation of the discrete distribution from Table 12.5.

Voters at positions 0, 2, 3, and 4 all vote for Bob because each of these positions is to the left of the midpoint $4.5 = (2 + 7)/2$ and therefore closer to Bob's position of 2 (the blue line in Figure 12.1) than Ann's position of 7 (the red line in Figure 12.1). For example, the distance between a voter's ideal position 4 and Bob's position 2 is $4 - 2 = 2$, whereas the distance to Ann's position is $7 - 4 = 3$. Because 2 is less than 3, a voter with ideal position 4 votes for Bob over Ann. The other voters with ideal positions at 6, 8, and 10 vote for Ann over Bob because their ideal positions are greater than the midpoint $4.5 = (2 + 7)/2$ and are closer to Ann's position of 7 than Bob's position of 2. Consequently, 13 of the 21 voters vote for Bob and 8 of the 21 voters vote for Ann. Bob wins the election.

What if Ann now thinks about how she could have changed her policy position? If she had announced a position to the right of Bob's 2, then the best she could have hoped for is that all voters with ideal positions greater than 2 voted for her. There are 17 such voters. If Ann wanted to attract these 17 voters and win the election, then she could have announced a policy position of 3 or any other policy position greater than 2 and less than 4. However, if she had announced a policy position of 4, she would have still won the election, receiving 11 of the 21 votes. In fact, the policy position 4 has a special property: it is the median of the distribution. ■

Self Check 3

If Ann had announced a policy position of 5 instead of 7, then which voters would have voted for Ann and which for Bob? Who would have won the election? ■

Self Check 4

Suppose that Ann announces policy position 4 and Bob announces policy position 3.9. Who wins the election? What if Bob announces policy position 4.1? Who wins the election? ■

Median DEFINITION

The **median** M of a distribution of voters' ideal points is a point on the horizontal axis where at least half the voters have ideal points that are equal or to the left of the point and where at least half the voters have ideal points that are equal or to the right of the point.

In practice, with an odd number of voters, finding the median requires listing all the ideal points and counting. For the distribution in Example 3, the following list contains all voters' ideal points from least to greatest:

0, 2, 2, 2, 3, 3, 3, 3, 3, 3, **4,** 4, 4, 6, 6, 6, 6, 8, 8, 10, 10

In the list, each ideal point appears as many times in the list as there are voters that share the ideal point. There are 21 numbers in the list, and the bolded 4 is the eleventh, or middle, entry in the list—this is the median of the distribution, so $M = 4$.

In Self Check 4, you determined that Ann would've won the election if Bob had announced a policy position of 3.9 or 4.1. In fact, Bob cannot defeat Ann with any policy position less than 4 because 11 of the voters have ideal points *greater than or equal to* 4. However, Bob cannot defeat Ann with any policy position greater than 4 because 11 of the voters have ideal points *less than or equal to* 4. Ann's policy position is undefeatable! The observation that 4 is an undefeatable position is generalized through the following definitions.

Maximin DEFINITION

A position is **maximin** for a candidate if there is no other position that can *guarantee* a better outcome—more votes for that candidate—whatever position the other candidate adopts.

In an election in which two candidates choose distinct policy positions and there is an odd number of voters, a candidate is guaranteed more than 50% of the total vote by selecting a position at M, regardless of what the other candidate does. Moreover, there is no other position that can guarantee the candidate with the median position more votes—and likewise for the opponent. If both candidates choose M, voters will be indifferent to their choices between the candidates on the basis of the announced policy positions alone and would presumably make their choices on other grounds. M is also *stable,* because if one candidate adopts this position, the other candidate has no incentive to choose any position other than M.

If Ann were to change her policy position in Example 3, then she should change it to $M = 4$. Thus, M is both the maximin position for each candidate (it offers a guarantee of a minimum of 50% of the votes), and if *both candidates choose* M, then these choices are in equilibrium (one candidate does worse by departing from M if the other candidate stays at M).

Equilibrium DEFINITION

A pair of positions is in **equilibrium** if, once chosen by both candidates, neither candidate has an incentive to depart from it unilaterally (i.e., by himself or herself).

More formally, we have the **median-voter theorem.**

Median-Voter Theorem THEOREM

In a two-candidate election with an odd number of voters, M is the unique equilibrium position. (The theorem is applicable if there is an even number of voters, as we show later.)

We required an odd number of voters for convenience. The same type of analysis is possible for an even number of voters. The difficulty is that if there were an even number of voters, then there may be more than one policy position that satisfies the definition of a median, as demonstrated in the following example.

EXAMPLE 4 A Discrete Distribution with an Even Number of Voters

Laurin Johnson/istockphoto/Getty Images

Suppose that a Home Owner Association of a condominium complex wants to build a swimming pool and clubhouse for its residents along a 1-mile stretch of road. Residents live in one of six buildings, and each building gets a single vote to indicate its ideal point for where the clubhouse and pool should be located. Let the line segment from 0 to 1 represent the 1-mile stretch of road on which to build the facility. Suppose the six buildings have ideal points, ordered from smallest to largest, of 0.1, 0.3, 0.3, 0.5, 0.8, and 0.9. A Building Company (ABC) and Barry's Builders are to propose different locations on the road—policy positions—for the new facility. Where should they build it?

Because there are 6 locations, there is no middle item in the list: 0.1, 0.3, 0.3, 0.5, 0.8, 0.9. That is, there is no median value in the list. In the absence of such a median position, is there still an equilibrium? It is not too difficult to show that if a discrete distribution has an even number of voters, then there is at least one equilibrium position. In fact, there may be a range of policy positions that, when paired, are in equilibrium with one another.

Suppose that ABC proposes to build at 0.4 and Barry proposes to build at 0.5. Each builder will receive half of the votes: Buildings with ideal points 0.1 and 0.3 vote for ABC, while the other buildings vote for Barry. If Barry were to change the proposed building site, notice that he could do no better than receiving 3 of the 6 votes. Indeed, this is the case for ABC, too.

This example can be generalized as follows: If ABC and Barry each announce distinct positions between 0.3 and 0.5, inclusive, then each builder will receive half the votes. This is no accident, as 0.3 and 0.5 are the two middle values in the ordered list of ideal points. Furthermore, if each builder announces a position between 0.3 and 0.5, inclusive, then neither builder will have an incentive to change. They will be in equilibrium. For this example, we call the interval [0.3, 0.5] of policy positions the *extended median* of the discrete distribution. ■

Extended Median DEFINITION

For a discrete distribution of an even number (say, $2n$) of voters' policy positions, the **extended median** is the set of positions between, and including, the nth and $(n + 1)$st values of the list of the voters' ideal points, ordered from least to greatest.

In statistics, when there are $2n$ items in an ordered list, the average of the nth and $(n + 1)$st is usually defined as the median. For a discrete distribution of an even number of ideal points, if the nth and $(n + 1)$st values are the same, then there is still just a single median; otherwise, no single value satisfies the definition of median. The policy positions that define the extended median are so-called because they extend the median-voter theorem to an even number of voters.

Self Check 5

Find the range of values that make up the extended median if the voters' ideal points are 0, 1, 1, 2, 2, 3, 3, 3, 3, and 4.

The **spatial models** considered so far can be modified to model situations in which voters prefer candidates with policy positions farther from their ideal points and to consider elections in which there is more than one important issue. Such situations are demonstrated in the next two examples.

EXAMPLE 5 Wind Farm Location: NIMBY (Not In My Backyard)

Teun van den Dries/ Shutterstock

Up to this point, we assumed that a voter would vote for a candidate whose policy position was closest to the voter's ideal position. In the following example, assume that the election is to determine the location of a wind farm off a coastline—a left–right continuum. Although many voters appreciate that a wind farm generates clean energy, voters do not want to have the wind farm ruin their views. That is, numerous voters may prefer a wind farm but don't want it to be built "in their backyard."

There may be other reasons not to support a wind farm. Indeed, a plan to install a wind farm of 170 wind turbines off the coast of Cape Cod, Massachusetts, split normally similarly spirited organizations into camps for and against the proposal. For example, the Humane Society, International Fund for Animal Welfare, and environmental lawyer Robert F. Kennedy, Jr., were against the placement of a wind farm off of Cape Cod because the wind turbines can be harmful to animals. However, the Natural Resources Defense Council, the Union of Concerned Scientists, and Greenpeace supported the wind farm.

Suppose that the 15 voters who live along the coast of a small island determine the placement of a wind farm. Like the child-care funding example (Example 3 on page 509), the voters have positions on a left–right continuum from 0 to 100. A voter's ideal point could represent his or her house and the preference that the wind farm be as far away as possible so that it does not ruin the ocean view. Hence, for this example, voters vote for the location that is farthest away from their ideal points.

Let the 15 voters live at the following positions: 10, 15, 20, 25, 40, 45, 50, 55, 55, 55, 60, 75, 80, 85, and 95, as depicted in Figure 12.2. Assume that there are two

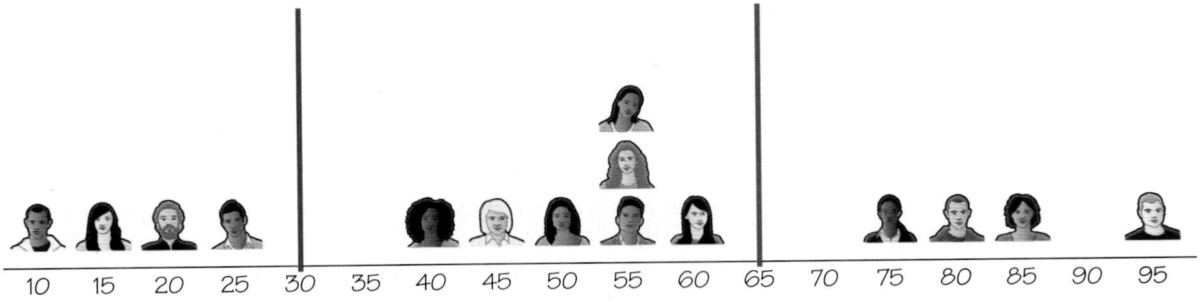

Figure 12.2 Locations of residences and proposed locations of the wind farm for Example 5.

proposals for the location of the wind farm: one at 30 and the other at 65. Voters with residences closer to position 30 vote for the wind farm to be built at 65. Similarly, voters with residences closer to position 65 vote for the wind farm to be built at 30. Even though voters now prefer the location or policy position that is farthest from their ideal points, the midpoint still plays a central role. The midpoint between 30 and 65 is 47.5 = (30 + 65)/2. Voters who reside at locations less than 47.5 are closer to location 30. These 6 voters vote for the wind farm to be built at location 65. The other 9 voters reside at locations greater than 47.5 and are closer to 65. They vote for the wind farm to be built at location 30. Because more voters are closer to location 65 than to location 30, the wind farm will be built at location 30. ■

Although some elections have a single dominant issue (such as when campaign strategist James Carville's "The economy, stupid" became a rallying cry in Bill Clinton's 1992 presidential campaign against incumbent George H. W. Bush), other elections may have more than one salient issue. Spatial models are still relevant. And the midpoint still plays an important role.

EXAMPLE 6 Two-Dimensional Spatial Voting

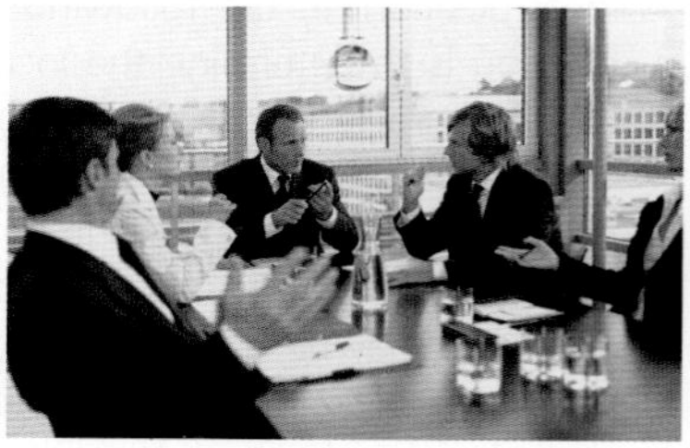

Altrendo Images/Getty Images

Five board members of a company must decide between two other board members (A and B) for the vaunted position of chairman of the board. The voters are concerned with two issues: overseas expansion and the timing of the initial public offering. The voters' ideal points and the candidates' policy positions are ordered pairs (x, y), where x indicates the position on overseas expansion (a larger value of x means a more vast expansion) and where y indicates the time until the initial public offering (a larger value of y represents a longer time until the initial public offering will be held).

Algebra Review Appendix Plotting Points in the Plane

Assume that A and B announce policy positions of (4, 2) and (8, 4), respectively. The ideal points of board members 1 through 5 are (5, 1), (4, 4), (6, 4), (9, 6), and (8, 2), respectively. The policy and ideal positions are graphed in Figure 12.3a.

Algebra Review Appendix Distance and Midpoint Between Two Points in the Plane

The midpoint of (4, 2) and (8, 4) is on the line that connects these points; this is the dashed line in Figure 12.3b. The midpoint is calculated by taking the average of each coordinate. The x-coordinate of the midpoint is $(4 + 8)/2 = 6$ and the y-coordinate of the midpoint is $(2 + 4)/2 = 3$. Equivalently, the midpoint satisfies

$$m = [(4, 2) + (8, 4)]/2 = [(4 + 8, 2 + 4)]/2 = (6, 3)$$

Just as the midpoint of two policy positions divides the left–right continuum into three regions in which (1) voters prefer one policy position, (2) voters prefer the other policy position, and (3) voters are indifferent to the choice between the two policy positions, there is also a way to use the midpoint to divide the two-dimensional policy space into three similar regions. In high school geometry, we learn how to draw a line through a point that is perpendicular to another line, using either a compass or dynamic geometry software.

The solid line in Figure 12.3c is through the midpoint m and is perpendicular to the dashed line between the policy positions of A and B. It divides the two-dimensional space so that voters with ideal points below the solid line prefer A over B, voters with ideal points above the solid line prefer B over A, and voters with ideal points on the

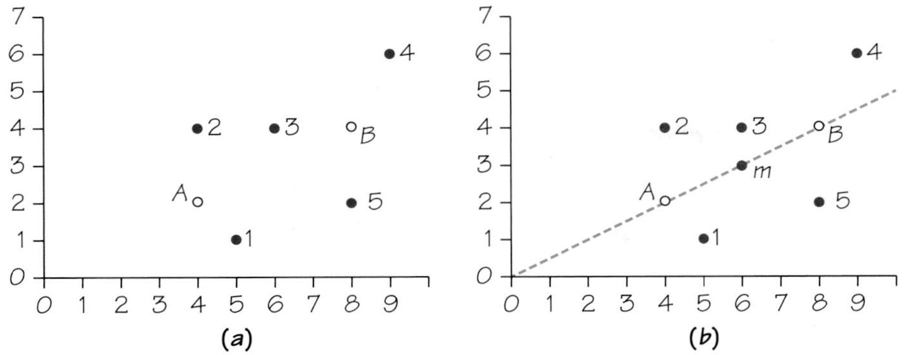

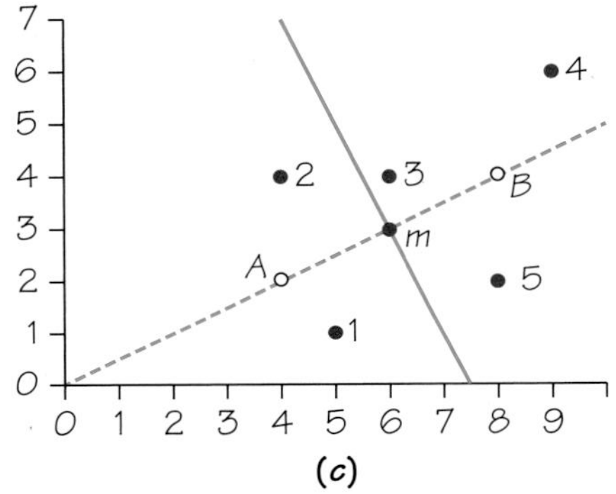

Figure 13.3 (a) Policy positions for Candidates *A* and *B*, as well as ideal positions for voting board members 1 through 5. (b) The midpoint *m* and the line connecting the policy positions of *A* and *B*. (c) The solid line divides the two-dimensional policy space into three regions. Voters with ideal points below the solid line prefer *A* over *B*, voters with ideal points above the solid line prefer *B* over *A*, and voters with ideal points on the solid line are indifferent between *A* and *B*.

solid line are indifferent between *A* and *B*. Board members 3, 4, and 5 have ideal points above the solid line and vote for Candidate *B*, while board members 1 and 2 have ideal points below the solid line and vote for Candidate *A*. Board member *B* is elected as chairman of the board.

12.3 Spatial Models for Two-Candidate Elections: Continuous Distributions

For the spatial models considered so far, there were relatively few voters and their ideal points were known. For a large number of voters, it is impractical to ask each voter to indicate his or her ideal point so that the candidates can determine a median policy position, or a policy position in the extended median. Political candidates use polls to assess the attitudes of individual voters. This information is used to approximate the opinions of the electorate on different issues. Pollsters may represent the discrete distribution of voters by a continuous curve that describes voters' attitudes on the left–right spectrum.

Continuous Voter Distribution THEOREM

A **continuous voter distribution** is a curve that gives, on the vertical axis, the number (or percentage) of voters who have attitudes at that point on the left–right continuum.

For simplicity, we picture the distribution as continuous (see Section 8.6), although in fact, because the number of voters is finite, there cannot be voters at all points along the continuum. The analysis of how to select a maximin policy position on the left–right continuum to defeat an opponent is still applicable to the continuous setting.

EXAMPLE 7 A Continuous Distribution of Voters' Attitudes

Suppose that voters' attitudes are described by the continuous distribution over the left–right spectrum given in Figure 12.4a. The red and blue lines represent the policy positions of Candidates *A* and *B*, respectively. The voters represented by the tan-shaded area vote for Candidate *A*; these are voters with ideal points less than and to the left of the midpoint between the policy positions of *A* and *B*. The voters represented by the blue-shaded area vote for Candidate *B*; these are voters with ideal points greater than and to the right of the midpoint between policy positions of *A* and *B*. Because of the symmetry of the distribution—which means there is a line in which the distribution looks the same to the left as it does to the right—it is easy to see that more voters vote for *A* than for *B* because the tan region is greater in size than the blue region. Hence, *A* would win the election.

As in the discrete model, there is a maximin position given by the median position of the distribution. For a continuous distribution, a median position covers half the area beneath the curve to the left and half the area beneath the curve to the right. Because the distribution is symmetric, the median position is the middle of the distribution. In Figure 12.4b, *B* moves to this position—and wins the election! Candidate *A* has no response to this move and can do no better than to join *B* at the median.

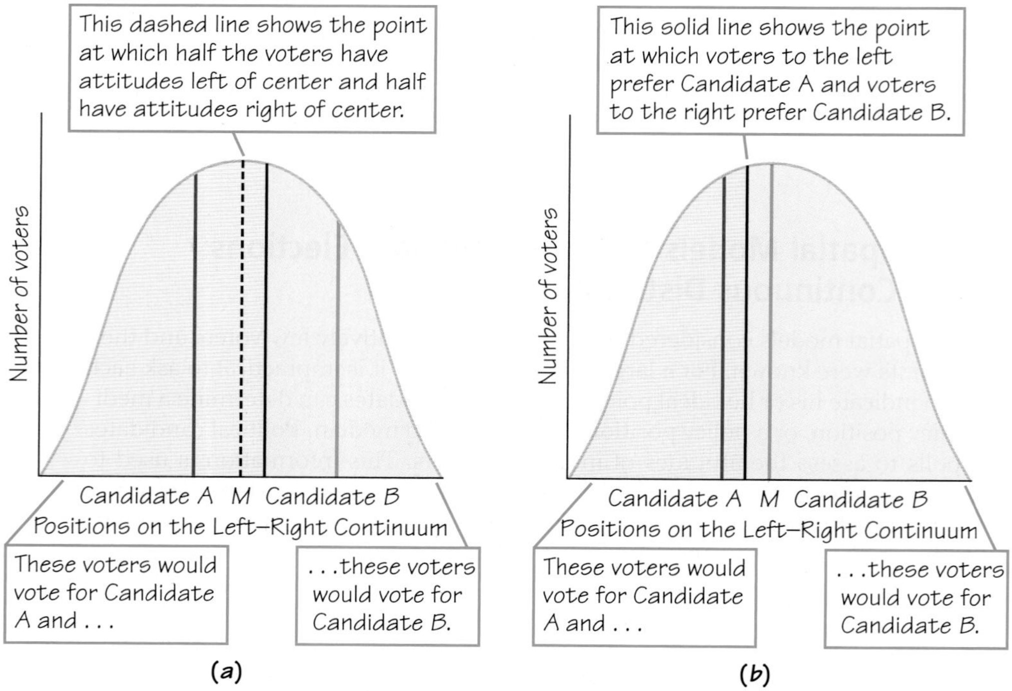

Figure 12.4 Spatial model of two-candidate voting with a continuous distribution of the voters' attitudes. (a) Candidate *A* wins the election because more voters (tan shading) vote for Candidate *A* than vote for Candidate *B* (blue shading). (b) Candidate *B* announces a policy at the median of the continuous distribution. Candidate *B* now wins the election.

Median of a Continuous Voter Distribution DEFINITION

The **median M of a continuous voter distribution** is the collection of points on the horizontal axis where half the area underneath the continuous voter distribution is to the left of M and half is to the right of M.

The median may be a single point or may consist of an interval, as could happen for discrete distributions. If the median is an interval, call it an extended median, as before. Because the median can be defined as before, the median-voter theorem is applicable *whatever* the distribution of the electorate's attitudes.

Figure 12.5a shows that one can still find the median for distributions that are not symmetric. The median is marked by M. This means that half the area under the curve is to the left of M and half the area under the curve is to the right of M. Figure 12.5b shows that the notion of an extended median may also be present in a continuous distribution. Any position between $\underline{m}$ and $\overline{m}$, inclusive, is an equilibrium, and the extended median is the interval $[\underline{m}, \overline{m}]$.

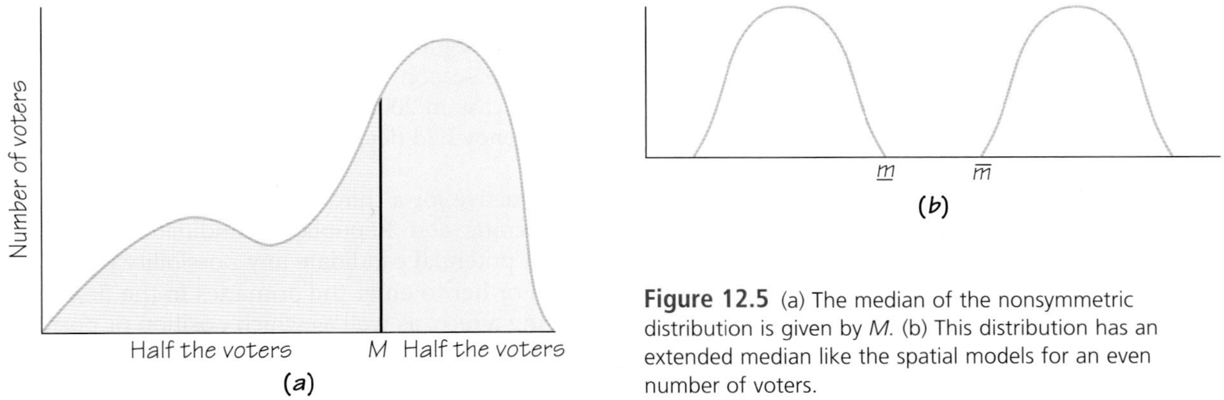

Figure 12.5 (a) The median of the nonsymmetric distribution is given by M. (b) This distribution has an extended median like the spatial models for an even number of voters.

For a discrete distribution, we have already shown that a median M is an equilibrium, which means that if both candidates select M as their policy positions, then neither candidate has an incentive to deviate from this position. The median of a continuous voter distribution is also an equilibrium; that is, if both candidates choose M, these choices are in equilibrium. Suppose there is a unique median position (as opposed to an extended median). Is there another equilibrium position or positions? There are two possibilities: (1) It is the same position for both candidates, which we call a *common position*, or (2) it is two distinct positions, one taken by each candidate. If it were a common position—suppose it is to the left of M (an analogous argument works if it is to the right)—then one candidate can always do better by moving right but staying to the left of M. This contradicts the supposition that the common position is in equilibrium. Now suppose the equilibrium were two distinct positions. Then one candidate can always do better by moving alongside the other candidate but staying closer to M. This contradicts the supposition that these two positions are in equilibrium.

Thus, in both cases, one candidate would have an incentive to depart from his or her position—holding the position of the other candidate fixed—so a nonmedian position of one or both candidates cannot be in equilibrium. Therefore, M is the only equilibrium position. With an extended median, any pair of policy positions from this set is in equilibrium. A similar argument could be used to show that if one or more of the two positions is outside of this set, then the pair could not be in equilibrium. The notion of the median of a voter distribution is the same as the notion of a median of a data sample given in Section 5.4 (page 196).

Given the stability of the median or the extended median in a two-candidate, single-issue election, is it any wonder that candidates who want to win try to avoid extreme positions? However, if there is a unique median position, and both candidates select it for their policy position, then the candidates become indistinguishable to the voters—and, in the general election, this opens up the door for a third-party candidate to enter the presidential race.

12.4 Spatial Models for Multicandidate Elections

Spatial modeling can also be used to understand the strategic behavior in multicandidate elections. Once the major parties have selected their candidates, there may be opportunities for third-party candidates to enter the race. Spatial models may also be useful in modeling behavior when there are more than two candidates in primary elections, as many candidates are likely to jump into the fray, especially in the states that go early in the season, if the incumbent president or vice president is not running (as was the case in 2008, when George Bush was unable to run for a third term and Dick Cheney had decided not to enter the Republican primary race).

Under what conditions is it attractive for a third-party candidate to enter the presidential race once the Democratic and Republican candidates have been determined? If no positions offer a potential candidate any possibility of success, then it will not be rational for him or her to enter the primaries in the first place. Therefore, the rationality of entering a race, as well as which position or positions he or she might take once in the race, are really two aspects of the same decision. However, the major party candidates are aware of the possibility of a third-party candidate entering the election. How can the major party candidates prevent a third-party challenger from entering the race?

The median-voter theorem brings the two major party candidates to the middle because the median is a maximin position. If the Democratic and Republican candidates both take positions at M, is there any room for a third candidate?

EXAMPLE 8 Entry of a Third Candidate in a Two-Candidate Race

Look at Figure 12.6, in which A and B are both at M and therefore split the vote. If a third candidate, C, enters and takes a position on either side of M (say, to the right), then voters with ideal positions to the right of the midpoint of M and C's policy position vote for Candidate C. Even though the area under the distribution to C's right may encompass less than $\frac{1}{3}$ of the total area, C may still win a plurality of votes, that is, more votes than either A or B.

To show why this is so, consider the portion of the electorate's vote that A/B will receive and the portion that C will receive. If C's area (tan) is greater than half of A/B's area (blue), C will win more votes than A or B, because C's area includes not only the votes to the right of his or her position but also some votes to the left. More precisely, C will attract voters

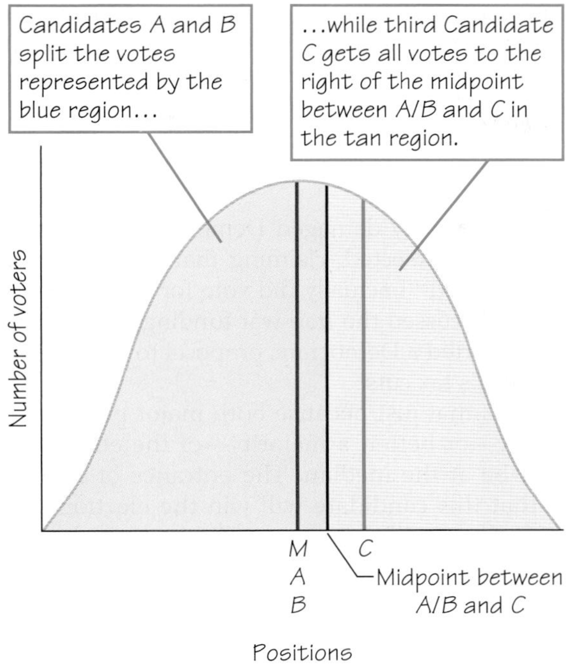

Candidates A and B split the votes represented by the blue region...
...while third Candidate C gets all votes to the right of the midpoint between A/B and C in the tan region.
Number of voters
M
A
B
C
Midpoint between A/B and C
Positions

Figure 12.6 Unimodal distribution with three candidates. Candidate C can take a position with less than $\frac{1}{3}$ of the voters to his or her right and still win if candidates A and B are at the median M and split the remainder of the vote.

up to the point midway between his or her position on the horizontal axis and that of *A/B; A* and *B* will split the votes to the left of this midway point. Because *C* picks up some votes to the left of his or her position, less than $\frac{1}{3}$ of the electorate may lie to the right and still enable *C* to win a plurality of more than $\frac{1}{3}$ of the total vote.

By similar reasoning, it is possible to show that a fourth candidate, *D,* could take a position to the left of *A/B* and further chip away at the total of the two centrists. Indeed, *D* could beat Candidate *C,* as well as *A* and *B,* by moving closer to *A/B* from the left than *C* moves from the right.

Clearly, *M* has little appeal, and is in fact quite vulnerable, to a third or fourth candidate contemplating a run against two centrists. Indeed, it is not difficult to show that *whatever* positions two candidates adopt—the same or different—at least one of these candidates will be vulnerable to a third candidate.

Self Check 6

Return to the discrete distribution of voters from Example 3. Assume that Ann and Bob both announce the same median policy position of 4. If a third candidate, Carl, announces policy position 5, then all voters with ideal points greater than or equal to 5 vote for Carl; there are 8 such voters. The other 13 voters split between Ann and Bob, say, 6 for Ann and 7 for Bob. Carl wins the election. What is the greatest whole number Carl can announce as his policy position so that he wins a plurality of the 21 votes?

In Example 8, both Candidates *A* and *B* announce policy positions at the median. Even the casual observer of presidential politics would notice that the Democratic and Republican candidates do not run on the same policy platform.

Because the major party candidates go through primary processes, their policy positions are closer to the median of the voters in their respective parties, as opposed to the median of the entire population. A candidate may try to adjust his or her position to attract more of the general population but will be unable to simply announce a completely new position. Such a sudden change or perceived change of policy position, often just before or after an election, is referred to as flip-flopping.

Indeed, critics successfully damaged Democratic presidential candidate John Kerry's chances for being elected, claiming that he had flip-flopped on issues. Famously, Kerry remarked "I actually did vote for the $87 billion, before I voted against it," when he discussed the Iraq war funding at a campaign appearance—meaning that he supported a Democratic proposal to fund the $87 billion through an elimination of Bush's tax cuts.

The implication is that just because both major party candidates are trying to attract a plurality—or better, a majority—of the electorate, the candidates' positions may not be at the median. The entrance of a third-party candidate does not assure that this candidate will win the election. There are obstacles and opportunities for a third-party candidate, which are summarized in Figure 12.7. (The reasoning behind these is explored in Exercises 34 and 35 on page 533.)

1/3-Separation Obstacle DEFINITION

For two candidates, *A* and *B*, with previously announced policy positions, the **1/3-separation obstacle** occurs when there is little room in the middle between the positions of *A* and *B*, enabling *C* to beat *A* or *B* but not both.

2/3-Separation Opportunity DEFINITION

The **2/3-separation opportunity** occurs when there is a wide separation between *A* and *B*, giving enough room in the middle for *C* to win.

The 1/3-separation obstacle theorem on the next page is stated for the case in which the continuous distribution of voters is symmetric about the median and has the candidates equidistant from the median. A similar statement for more general distributions can be made, but it requires more detailed hypotheses. For this reason, the more intuitive version is presented. The case in which Candidates *A* and *B* are both positioned at the median (as in Example 8) is not a counterexample to the theorem because Candidates *A* and *B* are required to have distinct policy positions.

When a continuous distribution of voters has a highest point or peak, it is referred to as **unimodal.** For a symmetric distribution to be unimodal, the highest peak must occur at the median. This means that more voters have their ideal points closer to the median than farther from the median.

The 2/3-separation opportunity theorem given on the next page requires the continuous distribution of voters to be symmetric about the median and to be unimodal. Again, this assumption makes the result more intuitive.

The 1/3-Separation Obstacle and the 2/3-Separation Opportunity THEOREM

The 1/3-separation obstacle. If *A* and *B* are distinct positions that are equidistant from the median of a symmetric distribution and separated from each other by no more than $\frac{1}{3}$ of the area under the curve (so that no more than $\frac{1}{3}$ of the voters lie between *A* and *B*), *C* can take no position that will displace both *A* and *B* and enable *C* to win (see below).

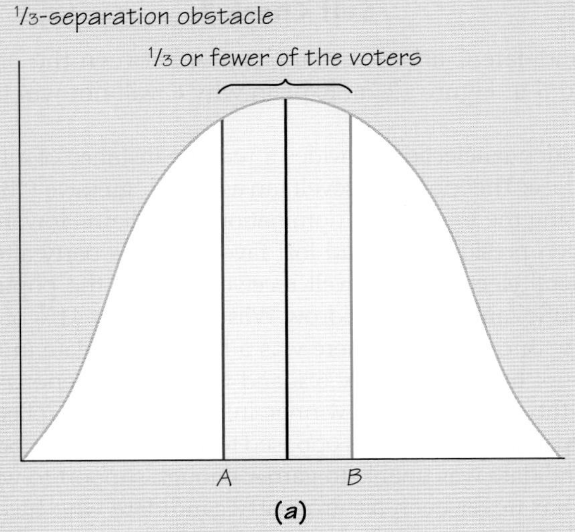

The 2/3-separation opportunity. If *A* and *B* are distinct positions that are equidistant from the median of a symmetric unimodal distribution and separated from each other by at least $\frac{2}{3}$ of the area under the curve (so that at least $\frac{2}{3}$ of the voters lie between *A* and *B*), *C* can defeat both *A* and *B* by taking a position at *M* (exactly between them, as shown below).

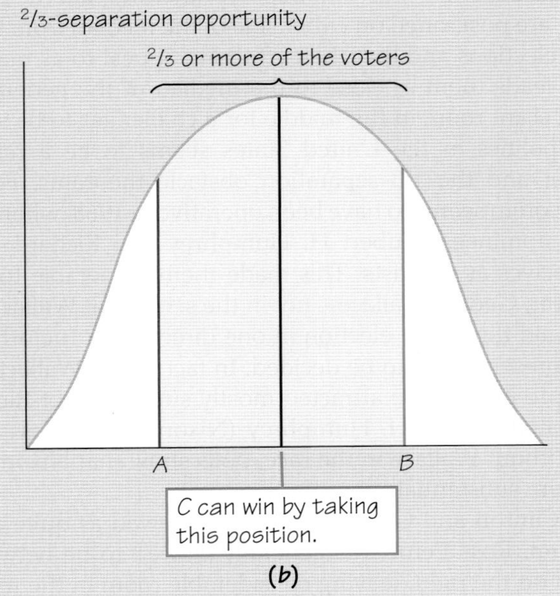

Figure 12.7 Parts (a) and (b) show both an obstacle and an opportunity for a third candidate, *C*, to enter a race.

Both portions of the theorem can be proved by considering the outcome if a Candidate *C* were to announce a policy position either to the left of *A*'s policy position, between the policy positions of *A* and *B*, or to the right of *B*'s policy position in Figure 12.7. For example, in Figure 12.7a, a Candidate *C* with policy position to the left of *A*'s policy position would capture the voters to the left of *A*'s policy position. Although Candidate *C* would receive more votes than Candidate *A*, Candidate *B* would still receive half the votes—those of all voters with ideal points to the right of the median.

Self Check 7

Suppose that Candidate *C* announces a position between the policy positions of Candidates *A* and *B* in Figure 12.7a. Explain why *C* will not win the election. ■

Topical Press Agency/Hulton Archive/ Getty Images

Theodore Roosevelt.

The 1912 presidential election provides a real-life instance of a third-party candidate entering the race: Theodore Roosevelt ran as the Progressive ("Bull Moose") Party candidate after losing the Republican nomination to William Howard Taft. (Roosevelt had previously been president but had lost favor with his party after sitting out one term.) In the general election, Roosevelt received 27% of the popular vote and Taft 24%. The Democratic candidate, Woodrow Wilson, who got 41% of the popular vote, handily defeated both candidates. There was a fourth candidate in this race, socialist Eugene V. Debs, but he received only 6% and was never a serious threat to Wilson. Wilson was also the overwhelming winner in the Electoral College. This situation illustrates a 1/3-separation obstacle, as opposed to a 2/3-separation opportunity, in that the third-party candidate was unable to garner enough support to win the election.

In fact, there has never been a third-party candidate who has won an election for U.S. president. Again, the 1912 election is notable, as it was the only election in which a third-party candidate received enough of the popular vote to defeat even one of the major-party candidates.

The stability of the two-party system in the United States may be partially explained by the fact that the two major parties, anticipating the possible entry of a third-party candidate, deliberately position themselves far enough away from the median to discourage entry on the left or right—but not so far away as to make entry in the middle advantageous. Another hypothesis is that the major-party candidates naturally are positioned on either side of the median from their positioning in their primary elections and that their efforts to appeal to as many voters in the general election leads them as close to the median of the general population as possible, leaving scant room in the middle for a challenger to be successful.

Presidential politics in the United States seems to be a reflection of both the median-voter and the 1/3-separation obstacle theorems. For example, the median-voter theorem seems to have been operative in 1968, when the Democratic and Republican nominees, Hubert H. Humphrey and Richard M. Nixon, both presented themselves as centrists. This made them vulnerable to the third-party candidate that year, George Wallace—not in the sense that Wallace could win, but rather that he could throw the election in one direction or the other, or even into the House of Representatives to be decided. In fact, while Wallace won only 14% of the popular vote in 1968, he attracted mostly supporters of Richard M. Nixon, who barely defeated Hubert H. Humphrey (Nixon won by less than 1% in the popular vote). Without Wallace in the race, polls show that Nixon's victory would have been far more substantial.

In 1992, Bill Clinton and George Bush were viewed as quite far apart on the left–right spectrum. Ross Perot was generally viewed to be between Clinton on the left and Bush on the right, leaving considerable room in the middle that Perot could better exploit than by trying to displace one of the major-party nominees on the left or right. In winning 19% of the popular vote, Perot drew almost equally from each candidate. However, he did not come close to winning, because there was not enough room between Clinton and Bush for there to be a 2/3-separation opportunity. (Clinton won decisively, with 43% of the popular vote to Bush's 38%.)

Up to now, we have looked at the spatial game that candidates play as they vie to position themselves optimally in two-candidate and multicandidate races so as to (1) receive a plurality of the vote or (2) deter new candidates from entering. We will continue to assume that candidates take positions along a left–right spectrum, but now we consider the effect of one or more of the candidates dropping out of the election.

Assume that three candidates take positions, from left to right, as follows: *A*–*B*–*C*. Clearly, if *A* or *C* drops out, their supporters mostly likely will switch to *B*, giving the centrist a boost. But what if *B* is the first to drop out? Then it is unclear whether *A*, *C*, or neither will benefit; it depends on the number of *B*'s supporters who prefer *A* next, *C* next, or neither (and hence may not vote at all). In any event, with *B* out of the race, the winner must be one of the candidates on the extremes.

The possibilities become more interesting when there are four candidates arranged from left to right as follows: *A*–*B*–*C*–*D*. If one of the extremists, *A* or *D*, drops out, then one of the two centrists, *B* or *C*, will benefit. But what if a centrist, say *C*, drops out? Does this benefit one of the extremists, or does the other centrist (*B*) benefit? At first glance, one might think that, with only one centrist remaining, he or she will surely benefit. This will not be the case, however, if most *C* supporters prefer *D* to *B*, which is certainly possible. Then the extremist *D* will benefit, which will be most upsetting to *A*'s supporters. Conceivably, *A*'s supporters might encourage *A* to withdraw so that they can throw their support to *B*, whom they definitely prefer to *D*.

Does this sound implausible? Think back to the 2000 election, in which our four hypothetical candidates are replaced by the following ordering from left to right: Nader–Gore–Bush–Buchanan. (Ralph Nader was the Green Party candidate on the left and Pat Buchanan the Reform Party candidate on the right.) Just before the election, the polls were showing that Buchanan was not much of a threat to Bush, but Nader—who ended up with 2.7% of the popular vote nationwide (Buchanan got only 0.4%)—was definitely a threat to Gore. Despite pleas from some of his supporters, Nader refused to withdraw and, consequently, gave Bush a victory in Florida and maybe a few other close states that won him the presidency.

This 2000 scenario is not the same as the previous four-candidate hypothetical scenario, in which we argued that the extremist on the right, *D*, might win if one of the centrists, *C*, dropped out. In the 2000 scenario, the extremist on the left, Nader, could have dropped out to "save" the centrist closer to him, Gore. Unfortunately for Gore, Nader not only refused to make this sacrifice but contended afterward that Gore's loss was due to Gore's own poor performance, not Nader's presence in the race. We assumed that it was rational for a candidate to enter a race when he or she has a possibility of winning the election. However, a candidate may be less concerned about representing the public and instead may be interested in promoting an agenda. Among other issues, Nader supported campaign finance reform, the environment, universal healthcare, increasing the minimum wage, and the abolition of the three-strikes laws. Despite Nader's belief that Gore was responsible for losing to Bush, Nader's role in the 2000 election was that of a **spoiler.**

Spoiler DEFINITION

A **spoiler** is a candidate who cannot win but "spoils" the election for a candidate who otherwise would win.

12.5 Spatial Models and the Electoral College

The Electoral College had a decisive effect in the 2000 presidential election, in which Al Gore received 537,000 more popular votes (0.5%) than did George W. Bush, but Bush won the electoral-vote tally by 4 votes. Thirty-six days after the election, a 5–4 Supreme Court decision blocked further vote recounts in Florida and determined that Bush won the state by a razor-thin margin of 537 votes (less than 0.01% of those cast). By capturing all 25 of Florida's electoral votes, Bush received a majority in the Electoral College and became the 43rd president of the United States.

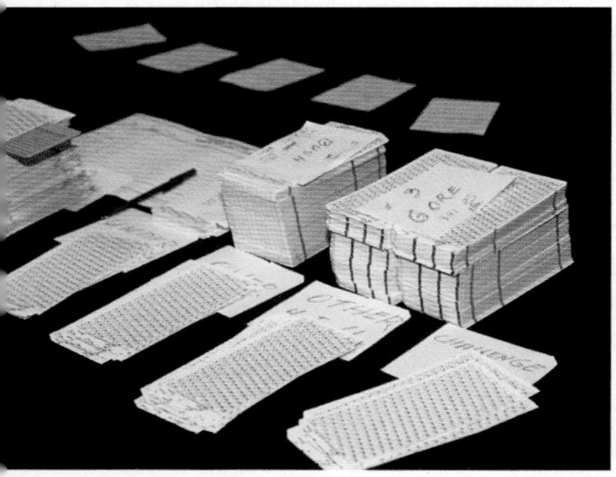

Mary Steinbacher/Alamy

Recounted ballots stacked during the 2000 U.S. presidential election.

What is the justification for the Electoral College? Its original purpose was to place the selection of a president in the hands of a body that, while its members would be chosen by the people, would be sufficiently removed from them that it could make more deliberative choices. As for its composition, each state gets 2 electoral votes for its two senators (total for all states: 100). In addition, a state receives 1 electoral vote for each of its representatives in the House of Representatives, whose numbers are based on population (see Chapter 14) and range from 1 representative for the seven smallest states to 53 representatives for the largest state, California. The House has a total of 435 representatives. The District of Columbia, like the smallest states, is given 3 electoral votes. Altogether, there are 538 electoral votes, and a candidate needs 270 to win. In 2000, George W. Bush got 271 electoral votes.

Although there is nothing in the U.S. Constitution mandating that the popular-vote winner in a state receive all its electoral votes, this has been the tradition almost from the founding of the republic. Only in Maine and Nebraska can the electoral votes be split among candidates, depending on who wins each of the two congressional districts in Maine and the three congressional districts in Nebraska. Because the statewide winner receives the two senatorial electoral votes, the closest split possible in these two states is 3–1 in Maine and 3–2 in Nebraska. In the actual election, Gore won all of Maine's 4 electoral votes, and Bush won all of Nebraska's 5 electoral votes, so winner-take-all prevailed in all 50 states and the District of Columbia.

The spatial voting models developed so far can be extended to examine some of the complexities introduced by the winner-take-all nature of the Electoral College. The models also allow for a comparison with two proposals. The first is to abolish the Electoral College and replace it by a direct popular-vote election, and the second is the **National Popular Vote law.** As of July 2015, ten states and the District of Columbia—which account for 165 electoral votes—have passed laws to give all their electoral votes to the national popular-vote winner; the law goes into effect when states with a majority of electoral votes (that is, 270) pass such a law. If enough states pass such a law, then the popular-vote winner would be guaranteed a victory in the Electoral College.

EXAMPLE 9 Spatial Modeling of an Election with Winner-Take-All Districts

Let's return to the mayoral election between Ann and Bob described in Example 3. Suppose that the tiny town is divided into three districts. One district consists of 5 voters with ideal points 2, 3, 3, 3, and 6; we denote this district by {2, 3, 3, 3, 6}. The other two districts have 7 and 9 voters, respectively; their ideal points are represented as {0, 2, 2, 3, 3, 6, 6} and {3, 4, 4, 4, 6, 8, 8, 10, 10}. Assume that the candidate who wins a majority of the votes from a district with k voters wins k district votes—akin to electoral votes. The candidate with the majority of the district votes is elected mayor.

Bob is aware of the median of the ideal points of the 21 voters and announces $M = 4$ as his policy position. Ann is shrewder. She realizes that she needs to win two of the three districts and therefore calculates the medians of each district. The median of districts {2, 3, 3, 3, 6}, {0, 2, 2, 3, 3, 6, 6}, and {3, 4, 4, 4, 6, 8, 8, 10, 10} are 3, 3, and 6, respectively. By announcing a policy position of 3, she wins a majority of the votes in two of the three districts, becoming mayor by a tally of 12 to 9 district votes. Indeed, 3 is the unique equilibrium policy position and cannot be defeated by any other policy position.

Self Check 8

Assume that a population of 9 voters has the following ideal points: 1, 2, 3, 4, 5, 6, 7, 8, and 9. Find the median of the ideal points. Assume the population is broken into the three districts: {1, 3, 5}, {2, 6, 8}, and {4, 7, 9}. Find the median for each district. If each district awards one district vote to the candidate that receives a majority of the votes in the district, then what is the equilibrium policy position?

Ann shrewdly calculated the median for each district. And it so happened that the median of the medians was the unique equilibrium strategy. Indeed, this occurred in Self Check 8, too. However, this may not happen in general. In the two previous examples, it was the case because winning any two of the three districts would ensure enough of the district votes to win the election. Because electoral votes from the Electoral College are based on the population of each state (according to the most recent census), a state with a greater population has at least as many electoral votes as a state with a smaller population. It is the size disparity that prevents the median of the medians from yielding an equilibrium strategy in general.

EXAMPLE 10 The Median of the Medians May Not Be an Equilibrium Strategy

Student Council is divided into groups *A*, *B*, and *C*, so that *A* and *B* each consist of 3 students and *C* consists of 8 students. Each student votes for a candidate based on the candidates' announced policy positions on the left–right continuum from 1 to 9. Assume that the median positions of the distribution of the voters' ideal points for groups *A*, *B*, and *C* are 5, 6, and 8, respectively. Assume that a candidate must win a majority of the "group votes" and that groups *A*, *B*, and *C* award 2, 2, and 5 group votes, respectively, to the candidate who receives a majority of the votes from the students in the group.

The median of the medians is 6 because 6 is the middle value in 5, 6, and 8. However, if a candidate announces 6 as his or her policy position, then an opponent who announces 8 would defeat the candidate. This follows because group *C* is so large in comparison with the other groups and its group votes alone are a majority of all group votes. (In Chapter 11, group *C* is referred to as a dictator, and groups *A* and *B* are referred to as dummies in the corresponding simple weighted-voting game that models this voting scenario.)

The median of each group is weighted by the number of group votes each group has to award. Groups *A*, *B*, and *C* have weights 2, 2, and 5, respectively. The **weighted median** is the median of 5, 5, 6, 6, 8, 8, 8, 8, 8. The data string 5, 5, 6, 6, 8, 8, 8, 8, 8 includes the median of each group multiple times, according to the group's weight. Thus, 5 is listed twice because group *A* has weight 2, 6 is listed twice because group *B* has weight 2, and 8 is listed five times because group *C* has weight 5. The median of this new data is 8, which is an equilibrium strategy.

Weighted Median DEFINITION

The **weighted median** of $x_1, x_2, \ldots, x_n$ with positive integer weights $w_1, w_2, \ldots, w_n$, respectively, is the median of the data $x_1, x_1, \ldots, x_1, x_2, x_2, \ldots, x_2, \ldots, x_n, x_n, \ldots, x_n$, where x_k appears in the list w_k times for each k.

When the sum $w_1 + w_2 + \cdots + w_n$ is odd, the list consists of an odd number of terms and the weighted median is a single value. When the sum is even, the weighted median may be an interval, as it is the extended median of an even number of terms.

Self Check 9

Find the weighted median of 1, 2, 3 with corresponding weights 1, 2, and 3. ■

We can now extend the median-voter theorem to model two-candidate elections under the Electoral College.

Weighted Median Voter Theorem THEOREM

In a two-candidate election with n states where, for all k, the median distribution of voters in state k is M_k and state k has E_k electoral votes, the weighted median of $M_1, M_2, \ldots, M_n$ with positive integer weights $E_1, E_2, \ldots, E_n$ is an equilibrium position.

The furor caused by the divided outcome in the 2000 presidential election, in which George W. Bush won the electoral vote and Al Gore won the popular vote, spurred efforts for reform of the election system. That the candidate with the most electoral votes may not be the same as the popular-vote winner is only one criticism of the Electoral College. Another criticism is that states have unequal power under the Electoral College, as larger states have been shown to wield more power in the winner-take-all awarding of electoral votes in the Electoral College. Under the Electoral College, candidates are less likely to court voters in states for which the election is de facto determined, just as voters are less likely to turn out to vote if one political party dominates the state. These behaviors create swing or toss-up states that garner the attention of the candidates. Indeed, Florida was the key swing state in the 2000 presidential election, deciding the election between Bush and Gore. Exercise 45 (page 534) highlights how a state with few electoral votes may have a limited impact on an election.

Besides the call to abolish the Electoral College, the immediate discussion of election reform after the 2000 election centered on making balloting more accurate and reliable and eliminating election irregularities—especially those practices that discriminate among different types of voters, such as people who do not have a state driver's license. Post-2000, there have been other initiatives, including the aforementioned National Popular Vote law that was proposed in 2006. But, for the most part, such initiatives have been focused on the state level because the U.S. Constitution grants each state the power to allocate its electoral votes.

As mentioned previously, all but two states use a winner-take-all system to allocate their electoral votes. Historically, states have used other methods, such as dividing the state into electoral districts. This is similar to how states are divided into districts for a primary, where the popular-vote winner of the district would receive all electoral votes for the district. A major criticism for this method worked its way into the English language: Gerrymandering, in which electoral districts are constructed to gain political advantage, is named for the 1812 salamander-shaped district that was created to favor the party of then Massachusetts Governor Gerry. By 1832, no states were using electoral districts, until Michigan re-introduced winner-take-all districts for the 1892 election.

After the 2008 and 2012 presidential elections, a number of states proposed changes to the method used to allocate their electoral votes. In 2010, Republicans in Pennsylvania proposed using the state's congressional districts as electoral districts. This seemed to be in response to the outcome of the 2008 presidential election, in which Barack Obama was the popular-vote winner yet won only a minority of the districts. If the change had been in effect for the 2008 election, Obama would have received 11 of Pennsylvania's 21 electoral votes, including the 2 electoral votes to the candidate who received the most statewide votes.

Republicans in Virginia, Wisconsin, Ohio, and Michigan have also considered electoral district proposals to change how the electoral votes are awarded in their states. Perhaps not surprisingly, the states that are considering these proposals are those for which electoral districts would have the biggest impact. Republicans ironically control the states in which Democrats would benefit the most from the use of electoral districts; these include Arizona, Georgia, North Carolina, and Texas.

The motivation the founding fathers had for placing the decision of who will be president in the hands of Electoral College members, rather than in the hands of voters, may no longer be justified. A simpler method of determining the president without the Rube Goldberg-esque complications of the Electoral College would be to count equally the votes of *all* voters, wherever they reside. A direct popular-vote election of a president would best accomplish this goal; but because this would require the abolishment of the Electoral College and a change to the U.S. Constitution, it seems unlikely to occur. However, enacting the National Popular Vote law would not only ensure that the electoral-vote winner is the popular-vote winner if states with a majority of electoral votes signed on, but it would also not require changing the Constitution.

Mathematics illuminates the effect of different ways of awarding delegates and electoral votes and also highlights the strategic decisions candidates make during a campaign. Furthermore, it provides a method to analyze possible changes to the Electoral College, whether proposed now or in the future.

ABC Review Vocabulary

Continuous voter distribution A curve that gives, on the vertical axis, the number (or percentage) of voters who have attitudes at that point on the left–right continuum. (p. 515)

Discrete distribution of voters A distribution in which voters are located at only certain positions along the left–right continuum. (p. 509)

Electoral College A body of 538 electors that selects the U.S. president. (p. 504)

Equilibrium A pair of policy positions from which neither candidate has an incentive to depart unilaterally. (p. 511)

Extended median The equilibrium position of two candidates when there is no median. (p. 512)

Maximin position A candidate's position in which there is no other position that can guarantee a better outcome (e.g., more votes) whatever position another candidate adopts. (p. 511)

Median The point M on the horizontal axis of a voter distribution where at least half the voters have attitudes that lie to the left and half to the right. (p. 511)

Median of a continuous voter distribution The collection of points on the horizontal axis where half the area underneath the continuous voter distribution is to the left of M and half is to the right of M. (p. 517)

Median-voter theorem In a two-candidate election with an odd number of voters, the median is the unique equilibrium position. (p. 512)

Midpoint For points a and b on the real line, the midpoint is $(a + b)/2$. (p. 509)

National Popular Vote law Proposed law that would give all the electoral votes of a state to the national popular-vote winner if states with a majority of electoral votes enact the law. (p. 524)

1/3-separation obstacle An obstacle for the entry of a third candidate, created if two previous entrants are sufficiently close together. (p. 520)

Policy position A point on a line that represents a candidate's position on an issue in the left-right political spectrum. (p. 509)

Quota The proportion of all delegates a candidate should receive, given his or her adjusted percentage of the vote. It is the adjusted percentage times the total number of delegates. (p. 506)

Spatial models The representation of candidate positions along a left–right continuum or in higher dimensions in order to determine the equilibrium or optimal positions of the candidates. (p. 513)

Spoiler A candidate who cannot win but "spoils" the election for a candidate who otherwise would win. (p. 523)

2/3-separation opportunity An opportunity for the entry of a third candidate, created if two previous entrants are sufficiently far apart. (p. 520)

Unimodal A distribution is unimodal if it has one peak. (p. 520)

Voter's ideal position (or ideal point) A point on a line that represents a voter's most preferred or ideal election outcome on an issue in the left–right political spectrum. (p. 509)

Weighted median The weighted median of $x_1, x_2, \ldots, x_n$ with positive integer weights $w_1, w_2, \ldots, w_n$, respectively, is the median of the data $x_1, x_1, \ldots, x_1, x_2, x_2, \ldots, x_2, \ldots, x_n, x_n, \ldots, x_n$, where x_k appears in the list w_k times for each k. (pp. 525, 526)

Weighted median-voter theorem In a two-candidate election under the Electoral College, if each state's median is weighted by the state's electoral votes, then the weighted median is an equilibrium position. (p. 526)

Self Check Answers

1. Obama's adjusted percentage is 55,848/136,490. The numerator 55,848 is the number of votes Obama received. The denominator 136,490 is the sum of the votes cast for candidates that receive at least 15% of the votes.

2. Stephen Jones's adjusted quota 3.524 is the adjusted percentage 70.479% times 5 delegates.

3. If Ann announced a policy position of 5 instead of 7, then the midpoint between Bob's position of 2 and Ann's position of 5 is $3.5 = (2 + 5)/3$. All voters with ideal points less than 3.5 vote for Bob. There are 10 such voters. The 11 other voters with ideal points greater than 3.5 vote for Ann. Ann wins the election.

4. If Ann announces the policy position 4 and Bob announces 3.9, then voters with ideal points greater than or equal to 4 vote for Ann. There are 11 such voters, so Ann wins the election 11 to 10. If Bob changes his policy to 4.1, then the 11 voters with ideal points less than or equal to 4 now vote for Ann. Ann still wins the election 11 to 10.

5. There are 10 numbers in the list 0, 1, 1, 2, 2, 3, 3, 3, 3, 4. The 5th and 6th numbers in the list are 2 and 3. The extended mean is the interval [2, 3].

6. If Carl announces policy position 7, then voters with ideal points 6 or greater vote for Carl. Carl receives 8 of the 13 votes and wins the election because the other 13 voters split between Ann and Bob.

7. If all voters with ideal points between the policy positions of A and B voted for C, then less than 1/3 of all the voters would vote for C. Because A and B are equidistant from the median and the distribution is symmetric, A and B split the remaining votes. Both A and B receive more votes than C.

8. The median of the 9 voters' ideal points is 5. The medians of the districts {1, 3, 5}, {2, 6, 8}, and {4, 7, 9} are 3, 6, and 7, respectively. The equilibrium position is 6.

9. The weighted median is the extended median of 1, 2, 2, 3, 3, 3; it is the interval [2, 3].

Skills Check

1. If Candidates *A, B,* and *C* receive 55, 47, and 10 votes, respectively, what percentage of the votes does each candidate receive? If the Democratic Delegate Selection Rules are used to award delegates, will one of the candidates lose his or her support? Which one and why?

2. In Skills Check 1, Candidate *C* is dropped from the election. Calculate the adjusted percentages of the popular vote for Candidates *A* and *B*.

3. If Candidate *A* receives 14,056 of the 23,903 votes and the district has 12 delegates to award, then what is Candidate *A*'s quota?

4. Candidates *A, B, C,* and *D* have quotas of 5.32, 3.92, 3.36, and 1.4, respectively. Use Hamilton's method to award the 14 delegates.

5. If the number of delegates to be awarded for the election in Skills Check 1 is 20, then how many delegates will each candidate receive? (Use Hamilton's method, but do not use the Democratic Delegate Selection Rules.)

6. If the number of delegates to be awarded for the election in Skills Check 1 is 20, then how many delegates will each candidate receive? (Use the Democratic Delegate Selection Rules.)

7. If five voters have ideal points 4, 5, 5, 6, and 9, then what is the median of the distribution of their ideal points?

8. If six voters have ideal points 4, 5, 5, 6, 9, and 10, then what is the extended median of the distribution of their ideal points?

9. Give an example of an even number of ideal points where the extended median is a single value, as opposed to an interval.

10. For a discrete distribution of voters' ideal points and an odd number of voters, can there be more than one equilibrium position? Explain.

11. Explain why, for a discrete distribution of voters' ideal points and an even number of voters, there can be more than one equilibrium. Must there be more than one equilibrium? Explain.

12. In a two-candidate election, suppose the attitudes of the voters are distributed symmetrically around the median *M*. Of the two candidates, *A* and *B*, *A* is positioned far to the left of *M*, and *B* is positioned just to the right of *M*. Which, if either, candidate will receive more votes?

(a) *A* will receive a majority of the votes.

(b) *B* will receive a majority of the votes.

(c) *A* and *B* will both receive exactly one-half of the votes.

13. In a two-candidate election, suppose that the attitudes of the voters are skewed to the left of the median *M*, so they are more spread out to the left than to the right. Assume that Candidates *A* and *B* take positions to the left and right of *M*, respectively, so that there are the same numbers of voters between their positions and *M*. Candidate ____________ will receive a majority.

14. In a two-candidate election, which of the following positions is an equilibrium position for both Candidates *A* and *B*?

(a) *A* and *B* just to the left and right of *M*

(b) *A* and *B* far to the left and right of *M*

(c) *A* and *B* both at *M*

15. Assume that *A* and *B* take distinct equidistant positions from the median of a symmetric distribution (not necessarily unimodal), with half the voters between the announced policy positions of *A* and *B*. Which of the following statements is true?

(a) The 1/3-separation obstacle applies.

(b) The 2/3-separation opportunity applies.

(c) Neither the 1/3-separation obstacle nor the 2/3-separation opportunity applies.

(d) Both the 1/3-separation obstacle and the 2/3-separation opportunity apply.

16. In a three-candidate election, suppose the attitudes of the voters are distributed symmetrically around the median *M*. Of the three candidates, *A, B,* and *C*, *A* is positioned far to the left of *M*, *B* is positioned just to the right of *M*, and *C* is positioned at *M*. Candidate ____________ will receive the most votes.

17. Suppose that Candidate *A* is situated so that exactly 1/3 of the voters lie to his left. In a two-candidate race, if Candidate *A* must remain fixed, Candidate *B*—to maximize her vote total—should situate herself

(a) just to the right of *A*.

(b) at the median *M*.

(c) to the right of *M* so that 1/3 of the voters lie to her right.

18. In a three-candidate election, if Candidates *A* and *B* are positioned at *M*, the election-winning position of Candidate *C* is ________________________.

19. In a three-candidate election, if Candidates *A* and *B* are positioned just to the left and just to the right of *M*, are there election-winning positions for Candidate *C*? What are they?

(a) At *M*

(b) Far to the left or right of *M*

(c) There is no election-winning position for Candidate *C*.

20. In a four-candidate election, if candidates are aligned in order *A–B–C–D*, Candidate ____________ benefits if *A* drops out of the race.

(a) *B*

(b) *C*

(c) *D*

21. In a five-candidate election, if candidates are aligned in order *A–B–C–D–E* and each receives about 20% of the vote, Candidate ____________ benefits most if Candidates *A*, *D*, and *E* drop out.

22. Suppose that the distribution of voters' ideal points is symmetric and that two candidates have announced positions on opposite sides of and equidistant to the median such that there is no 2/3-separation opportunity. Would a third candidate that enters with a policy position farther from the median be a spoiler?

23. What is the median of 3, 3, 4, 4, 4, 5, 6, 6, 6, 6, 6? How does this compare with the weighted median of 3, 4, 5, and 6 with associated weights of 2, 3, 1, and 5?

24. True or False: The median-voter theorem is the special case of the weighted median-voter theorem in which all weights are 1.

25. If states *A*, *B*, and *C* have median positions on a left–right spectrum of 25, 32, and 28, respectively, and the states have 3, 5, and 3 electoral votes, respectively, then what is the weighted median?

26. How does the proposed use of awarding electoral votes based on votes in electoral districts compare with how delegates are awarded in primaries?

27. The National Popular Vote law is a

(a) constitutional amendment.

(b) federal law.

(c) state law.

28. If states with a majority of electoral votes passed the National Popular Vote law, it would be impossible for a candidate for president to win the popular vote and yet lose the Electoral College because ______________________________.

Chapter 12 Exercises

Challenge   Discussion

12.1 Narrowing the Field Through the Primary Process

1. The vote totals for Ann, Bob, and Carl are 6300, 2700, and 1000, respectively. If 19 delegates are to be awarded, then how many delegates does each receive? (Assume that Hamilton's method of apportionment is used to round the quotas.)

2. Use the accompanying table and the Democratic Delegate Selection Rules to allocate 19 delegates to Ann, Bob, and Carl. Compare your answer with your answer for Exercise 1.

Candidates	Popular Vote	Percentage of Votes	Remaining Votes	Adjusted Percentage	Quota	Initial Delegates	Final Delegates
Ann	6300						
Bob	2700						
Carl	1000						
Totals	10,000						

3. For the 2004 Democratic primary, only General Clark, Governor Howard Dean, and Senators John Edwards, John Kerry, and Joe Lieberman received a significant number of votes in New Hampshire's second district. Use the Democratic Delegate Selection Rules to allocate the district's seven delegates by filling in the following table.

Candidates	Popular Vote	Percentage of Votes	Percentage after Drops	Quota	Initial Delegates	Final Delegates
Clark	14,010					
Dean	33,493					
Edwards	13,396					
Kerry	42,847					
Lieberman	1,784					
Others	8,507					

4. Calculate the delegate totals for the candidates in Exercise 3, assuming that no cutoff rule was used. Compare your answer with the answer to Exercise 3.

5. It is possible for every candidate to receive less than 15% of the popular vote. What is the minimum number of candidates for which this can happen?

6. Under winner-take-all used in the Republican primaries, it is possible for a candidate to win all the delegates from a state but still not have a majority of the popular vote. Suppose that 1001 votes are cast for the 10 candidates. What is the fewest number of votes a candidate could receive and still earn all delegates?

7. Use Hamilton's method to allocate eight delegates in both district 1 and district 2 if Candidates Anchor, Barber, and Coachman have the vote totals given in the accompanying table. Compare your answer if instead the 16 delegates were awarded based on the sum of the vote totals for the candidates.

Candidates	District 1	District 2
Anchor	55,175	47,023
Barber	45,321	23,145
Coachman	19,112	9821

8. Candidates *A*, *B*, *C*, and *D* receive 44%, 36%, 12%, and 8% of the popular vote, respectively. If Hamilton's method is used to allocate 100 delegates, how many delegates does each candidate receive? What if a 10% cutoff (instead of the 15% cutoff in the Democratic Delegate Selection Rules) is used? What if a 15% cutoff is used? Suppose that the supporters of Candidate *D* realize that *D* will be eliminated and decide to back *C* instead. How would the allocation of delegates change?

9. In Exercise 8, Candidates *C* and *D* receive 12 and 8 delegates, respectively. When all the supporters of *D* vote for *C* instead, *C* ends up with 20 delegates. Will it always be the case that, if the votes for the two candidates are combined, the sum of the delegate counts for two candidates is the same as the delegates received? Try constructing a three-candidate example in which eight delegates are awarded and the sum of the delegates awarded to *B* and to *C* is different from the number of delegates awarded to the combined Candidate *B*/*C*.

10. Suppose that Hamilton's method is used to allocate five delegates to Candidates *A*, *B*, and *C* who receive 88%, 9%, and 3% of the vote, respectively. How does your answer change if candidates who do not receive 5% of the vote are eliminated from the election?

11. Does your answer to Exercise 10 exhibit the same type of paradoxical behavior as in Example 2? Explain.

12.2 Spatial Models for Two-Candidate Elections: Discrete Distributions

Refer to the following table for Exercises 12–15.

Ideal Point	2	4	5	8	9	10	12
Number of Voters	2	2	2	5	6	4	2

12. If Candidates *A* and *B* announce policy positions of 4 and 10, respectively, then which candidate wins the election?

13. Given that Candidate *B* announces policy position 10, what is the best position that Candidate *A* can announce to maximize votes? How many votes would *A* receive?

14. What is the equilibrium policy position? Explain.

15. If Candidates *A* and *B* announce policy positions of 4 and 10, respectively, then how many voters with ideal point 2 would have to be introduced so that *A* wins the election?

16. A recycling center is to be placed on Main Street. The voters have ideal points between 0 and 10, as given in the following table. As in Example 5, each voter wants the recycling center to be as far away from his or her ideal point as possible. The two proposed sites for the recycling center are 3 and 7. Where will the recycling center be built?

Ideal Point	0	2	3	4	7	10
Number of Voters	2	5	4	5	6	7

17. Prove that if a distribution is discrete and there is no single median position, there is always an *extended median.*

18. Twenty voters have ideal points on a left–right spectrum. What is the extended median of the distribution of voters' ideal points? What is the equilibrium position if an additional voter is added to the distribution with (a) an ideal point of 0.1? (b) An ideal point of 0.4? (c) An ideal point of 0.7? What is the relationship between the equilibrium position and the new voter's ideal point as it compares with the extended median?

Ideal Point	0	0.2	0.3	0.5	0.8	1
Number of Voters	2	4	4	6	2	2

19. Consider two competitive retail businesses, such as department stores, that consider locating their stores somewhere along the main street that runs through a city. Assume that because transportation is costly, people will buy at the store closer to them. Explain how determining a store location is the same as the spatial voting model.

20. Extending Exercise 19, assume that there are 100 shoppers at each whole number position from 0 to 10. What is the equilibrium position for the two department stores?

21. Because the distribution in Exercise 20 is symmetrical, the median and equilibrium position is for the stores to locate at the center. Indeed, similar stores frequently cluster near the center of many main streets, although these stores may not be particularly convenient to people who live far from the city's center. The centralized location does not seem to be in the public interest. Wouldn't it be better if some of the same kinds of stores were near one end of the main street and some near the other? It has been argued that the "social optimum" for the location of the two stores in Exercise 20 should be 2.5 and 7.5, because then no shopper has to travel more than 2.5 from their ideal point. Would the stores still split the shoppers? Explain.

22. Extending Exercise 19, which is better for consumers: (a) to minimize the maximum distance they must travel to a store or (b) to foster price competition, which would presumably be encouraged if the two stores were located at $M = 5$?

23. For a two-dimensional policy space, there are four voters with ideal points (1, 0), (0, 1), (−1, 0), and (0, −1). Candidate A announces a policy position of (0, 0). The points are plotted in the accompanying figure. Can Candidate B do any better than to announce (0, 0) as his or her policy position? Explain.

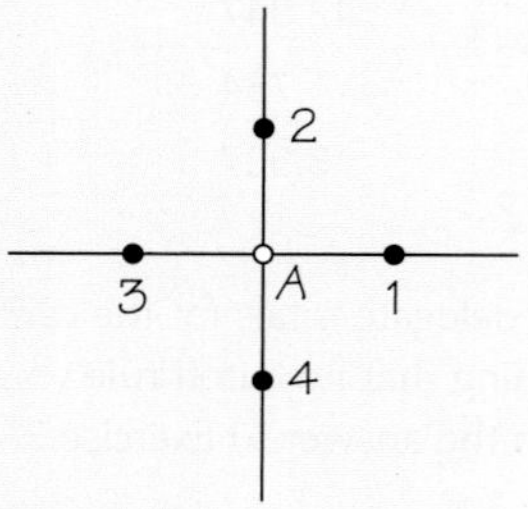

12.3 Spatial Models for Two-Candidate Elections: Continuous Distributions

24. Define an outcome to be in equilibrium if, given that one candidate chooses it, the other candidate cannot do better than to take the same position. Show that this definition is equivalent to the text's definition of being in equilibrium.

25. Suppose that the continuous distribution of voters' ideal points is the bent solid line in the accompanying figure. (*Note:* This is a continuous distribution because the triangle under the bent solid line has area 1.) Assume that A's policy position is 3 on the left–right continuum; it is denoted by the dashed line. If B announces a policy position between 0 and 1, between 1 and 3, or between 3 and 4, explain which candidate wins the election. What is the median position? Describe which voters vote for B if B announces the median.

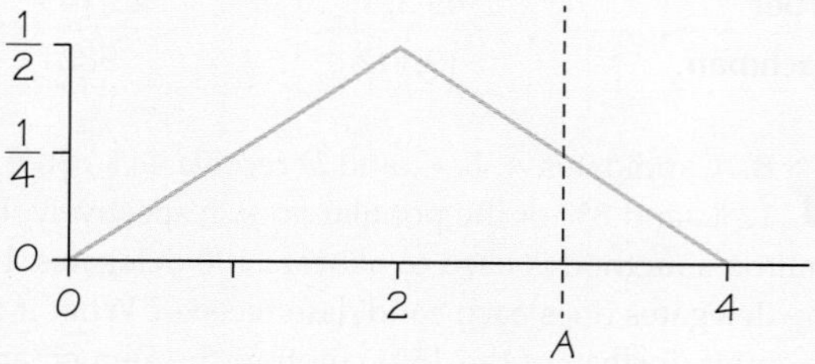

26. Consider a symmetric bimodal distribution, which has two peaks of the same height. Will taking a position at either peak be in equilibrium? Explain. (Although this question is for any such distribution, an example is given in Exercise 29.)

27. Define A's position in a two-candidate race to be *opposition-optimal* if, given that the position of B is fixed, it maximizes A's vote total. Show that A's opposition-optimal position must be adjacent to B's position and closer to M, except when B is at the median.

(Roughly speaking, being "adjacent" means being a very small distance away.)

28. Along the lines of Exercises 19–21, assume that a distribution of shoppers is given by the graph that accompanies Exercise 25. If Store *A* locates at 3 (as in the graph) and Store *B* locates at 1, then no shopper would have to travel more than 1 to go to a store. This notion of a "social optimum" was discussed in Exercise 21. Does the same idea hold for this distribution? Would it be acceptable for shoppers closer to 0 and 4 to travel farther because there are fewer of them?

29. A continuous distribution of voters' ideal points is symmetric and bimodal, as illustrated below. Determine the extended median for the distribution. If Candidate *A* announces a policy position of 2.75, then what is the greatest value that *B* can announce as a policy position and still be in equilibrium with *A*? Similarly, what is the least value that *B* can announce and still be in equilibrium?

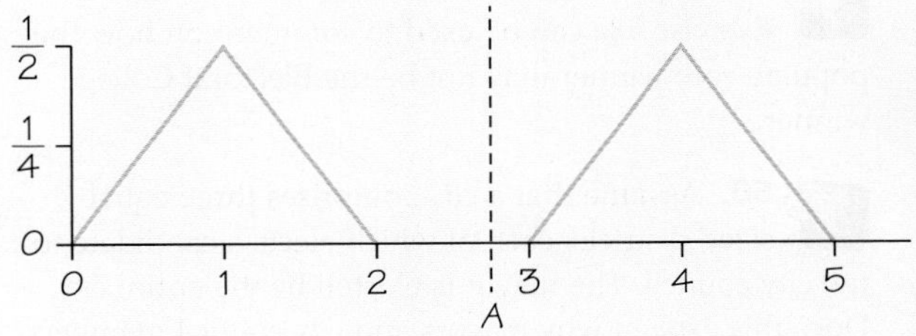

12.4 Spatial Models for Multicandidate Elections

30. Assume that there are 11 voters; one voter each has ideal point *k* for each integer $k = 0$ to 10. Assume that *A* and *B* take the *same* median position. Is there more than one position that *C* can take to maximize his or her vote total? If so, what are those positions? If not, why not? Can *C* win the election? Explain.

31. Assume that there are 10 voters; one voter each has ideal point *k* for each integer $k = 0$ to 9. Assume that *A* and *B* announce policy positions at 4.2 and 5, respectively. Is there more than one position that *C* can take to maximize his or her vote total? If so, what are those positions? If not, why not? Can *C* win the election? Explain.

32. Assume that *A* and *B* take the *same* nonmedian position. What position should *C* take to maximize his or her vote total? Is *C*'s position always a winning one?

33. Assume that *A* and *B* take *different* positions, with one possibly being at *M*. What position should *C* take to maximize his or her vote total? Is *C*'s position always a winning one?

34. Suppose that 22 voters have the ideal points given in the following table.

Ideal Point	0	1	2	3	4	5	6	7	8
Number of Voters	2	4	2	2	3	2	3	3	1

If Candidate *A* announces a policy position of 1 and Candidate *B* announces a policy position of 7, is there a 2/3-separation opportunity for a third candidate, *C*, to enter the race? Explain.

35. Suppose that two voters are added to the distribution given in Exercise 34. Both voters have ideal points at position 5. Is there a 2/3-separation opportunity for a third candidate, *C*, to enter the race? Explain.

36. If the distribution is not symmetric, but no more than $\frac{1}{6}$ of the area under the curve separates *A* (on the left) from *M*, and no more than $\frac{1}{6}$ of the area separates *B* (on the right) from *M*, is there a 1/3-separation obstacle? What if these $\frac{1}{6}$-or-less areas on the left and the right are not the same?

37. If the distribution is not unimodal, but at least $\frac{1}{3}$ of the area under the curve separates *A* (on the left) from *M*, and at least $\frac{1}{3}$ separates *B* (on the right) from *M*, is there a 2/3-separation opportunity? (*Hint:* Start by assuming that the distribution is uniform between *A* and *B*—and hence not unimodal—and that exactly $\frac{2}{3}$ of the voters lie between *A* and *B*. Can *C* always win by taking a position at *M*? If not, is there a distribution that affords *C* this opportunity?)

38. If *A* and *B* are equidistant from the median of a symmetric distribution and separated from each other by exactly $\frac{1}{2}$ of the area under the curve, under what conditions is this separation an obstacle, and under what conditions is it an opportunity? (*Hint:* Start by constructing examples of symmetric distributions in which *C* would either win or lose by taking a position at *M*.)

39. It is known that *A*, *B*, and *C* will enter an election in that order, with *A* announcing his position first, then *B*, and finally *C*. If the distribution is uniform (rectangular) over [0, 1], what position should each candidate take to maximize his or her vote total, anticipating—in the case of *A* and *B*—the entry of future candidates? [*Hint:* Start by assuming that *A* takes a position at $\frac{1}{4}$. Is *B*'s position at $\frac{3}{4}$ optimal, anticipating the entry of *C*? Or can *B* do better at some other position (perhaps by influencing *C*'s choice of a maximizing position)?]

40. Assume that the four candidates in the 2000 presidential election can be arranged from left to right as follows: Nader–Gore–Bush–Buchanan. Suppose a poll reveals Gore at 48%, Bush at 47%, Nader at 3%, and Buchanan at 2%. Would Bush be well advised to offer Buchanan a cabinet position to drop out of the race (the same way Adams offered Clay the secretary-of-state post after the 1824 election)? What if Bush knew that, after Buchanan dropped out, only half of Buchanan's supporters would switch to him, with most of the remainder not voting, except for a few who would switch to Gore?

41. Assuming the same poll results as in Exercise 40, suppose that Gore offered the same deal to Nader, knowing that only one-third of Nader supporters would switch to him and the rest would not vote. However, suppose Gore also thought that if Nader dropped out, so would Buchanan, and all Buchanan supporters would vote for Bush. Should Gore set off this chain of events?

42. Is there any evidence that the four presidential candidates in 2000 might have contemplated "deals" of the kind indicated in Exercises 40 and 41? If you cannot find any evidence, do you think this is because the candidates found such ploys unethical or because they thought they might be found out if they tried to engage in them?

43. One tactic that was considered by Nader supporters who thought their votes for Nader might kill Gore's chances in some states was to swap votes: In close states that Gore might lose if Nader supporters stuck with their candidate, these supporters would switch to Gore if Gore supporters in less contested states, where Gore would almost surely win, would switch to Nader. Thereby, the popular-vote totals for the two candidates would not change overall, but Gore would be able to win in the close states he might otherwise lose. Is this a sensible way of dealing with problems created by the Electoral College, which puts a premium on winning in large states?

12.5 Spatial Models and the Electoral College

44. Nine states have medians 2, 4, 5, 6, 8, 9, 10, 11, and 12 with associated electoral votes of 10, 8, 4, 4, 2, 3, 5, 5, and 2. Determine an equilibrium position. Is there more than one? Explain.

45. If states *A*, *B*, and *C* have median positions on a left–right spectrum of 3, 4, and 5, respectively, and the states have 25, 3, and 50 electoral votes, respectively, can the median position of state *B* ever be the weighted median? Explain. Can you infer anything about how state size impacts elections? Explain.

46. Nine registered voters have ideal points 1, 2, 3, 4, 5, 6, 7, 8, and 9. Suppose two candidates, *A* and *B*, have announced policy positions of 4 and 7, respectively. Which candidate wins the election? Candidate *B*'s strategy is to "get out the vote" by registering more voters with ideal points to the right of the median 5. How many voters must *B* register to defeat *A*?

47. In Exercise 46, Candidate *B* was trying to "get out the vote" by registering more voters who would ultimately vote for him or her. There have been attempts to increase political awareness in college-aged students, including an initiative (Rock the Vote!) in which musicians encouraged students to vote. Would Republicans or Democrats benefit the most by such efforts? Why do you think so?

48. Nine voters have ideal points 1, 2, 3, 4, 5, 6, 7, 8, and 9. Divide the voters into three districts of size 3 such that (a) the median of the medians of the districts is not the median of the nine voters and (b) the median of the medians of the districts is the median of the nine voters.

49. Explain how your partition of the voters in Exercise 48a can be used to comment on how the popular-vote winner may not be the Electoral College winner.

50. Assume that a city comprises three equal-sized districts, each of which elects a candidate to the city council. The mayor is elected by the entire city. Does this explain why mayors and city council members often disagree? Explain your answer and relate it to the districts in Exercise 48.

51. Not every state holds its primary on the same day. Suppose five primaries are held in succession and that delegates are awarded in a winner-take-all manner, in which the candidate with the most popular votes is awarded all the state's delegates. The primaries of the states 1, 2, 3, 4, and 5 occur in numerical order. If the median of state *k* is *k* and each state has the same number of electoral votes, then the median of the entire population is 3. Assume Candidates *A* and *B* announce policy positions of 2 and 3.5, respectively. Determine the outcome of each primary. Who should win the combined primary election?

The winner of early primaries receives more campaign contributions than the losers of the primaries. These funds can be used to influence the opinion of the voters. Suppose that Candidate *A* uses the money raised after winning states 1 and 2 in such a way that the electorate in states 3, 4, and 5 believe *A*'s ads in which Candidate *B* is portrayed as having a policy position of 4.1. Who wins the election now? Do you believe that money can influence an election in such a way? Explain.

52. Can there be an instance in which it is worse that the candidate with the most (popular) votes (in the country) wins a state election instead of the candidate who wins a majority of the electoral votes? How does the outcome of the 2000 presidential election and Florida's deciding role relate to the proposed National Popular Vote law? What if the law were passed in enough states?

53. Assume that the National Popular Vote law is adopted. Would the election winner be the same if there were a direct popular vote (i.e., without the Electoral College) in which the winner is the candidate with the most votes?

Writing Projects

1. The use of spatial voting models is predicated on issues being representable on a left–right spectrum. List 10 issues from recent political campaigns. Which of them are representable on a left–right spectrum? Which are discrete, and which are continuous? Some issues that may appear to be discrete, such as the availability of guns and gun control laws, may be more nuanced. Explain how such issues may also be modeled on the left–right spectrum.

2. Under the Electoral College, most states award all their electoral votes to the winner of the state's popular election. A complaint about the Electoral College is that the candidate with the most votes under the Electoral College may not be the popular-vote winner. Legislatures in some states, such as Pennsylvania, have proposed dividing their states into electoral districts and then using a winner-take-all method to award all the electoral votes of an electoral district to the popular-vote winner. Can a similar complaint occur at the state level, in which the candidate with the most electoral votes in the state is not the popular-vote winner in the state? Explain. How does this relate to the National Popular Vote law and the Electoral College?

Suggested Readings

BRAMS, STEVEN J. *The Presidential Election Game,* 2d ed., A K Peters, Wellesley, MA, 2008. Focuses on the strategic aspects of presidential elections—from primaries to conventions to general elections—and also includes an analysis of the "game" played between President Richard Nixon and the Supreme Court over the release of the Watergate tapes that led to Nixon's resignation in 1974. (Nixon is the only president to have resigned from the presidency.)

GEIST, KRISTI, MICHAEL A. JONES, and JENNIFER WILSON. Apportionment in the Democratic Primary Process, *Mathematics Teacher,* Volume 104, Number 3 (October 2010), 214–220. Applies the Democratic Delegate Selection Rules to allocate delegates for every district in the 2008 New Jersey Democratic primary, as well as to allocate the PLEO and at-large delegates. Considers the effect of using the 15% threshold, but using a different apportionment method (e.g., Jefferson's method instead of Hamilton's method; other apportionment methods are introduced in Chapter 14). It includes activity worksheets.

HINICH, MELVIN J., and MICHAEL C. MUNGER. *Analytical Politics,* Cambridge University Press, Cambridge, UK, 1997. Extends spatial modeling to more than one dimension, analyzes probabilistic voting, and introduces game-theoretic solution concepts relevant to the study of elections.

SAARI, DONALD G. *Chaotic Elections! A Mathematician Looks at Voting,* American Mathematical Society, Providence, RI, 2001. Argues that elections—in particular, the 2000 presidential election, but others as well—have chaotic features that can be understood through mathematics. The mathematics used is an unusual kind of geometry that is accessible to people with some mathematical background.

SHEPSLE, KENNETH A. *Analyzing Politics: Rationality, Behavior, and Institutions,* 2nd ed., Norton, NY, 2010. Rational strategies in voting and elections are a major component of this textbook, but it also includes sections on collective action and political institutions, such as courts and legislatures. Several case studies illustrate the theory.

Suggested Websites

www.fec.gov The website of the Federal Election Commission talks about federal elections and voting in the United States.

www.ifes.org The website of the International Foundation for Electoral Systems.

www.nationalpopularvote.com The website for the National Popular Vote bill.

Part IV

Fairness and Game Theory

Chapter 13
Fair Division

Chapter 14
Apportionment

Chapter 15
Game Theory: The Mathematics of Competition

The central thrust of the first two chapters in Part IV is the fair division of divisible and indivisible objects. Whereas a cake or a parcel of land is divisible, the representatives who are apportioned to the different states are indivisible. Sometimes, however, seemingly indivisible objects, like a car, can be shared, rendering them divisible. By contrast, Chapter 15 focuses on what rational players will choose in different strategic situations, which may be highly unfair to some.

Chapter 13 describes fair-decision schemes in which a group of individuals with different values can be assured of each receiving what he or she views as a fair share when dividing objects like cakes or the goods in an estate.

Chapter 14 discusses the apportionment problem, which is to round a set of fractions to whole numbers while preserving their sum; of course, the sum of the original fractions must be a whole number to start. Apportionment problems occur when resources must be allocated in integer quantities—for instance, when legislators allocated seats in the U.S. House of Representatives to the 50 states.

Chapter 15 introduces the mathematical field called game theory, which describes situations involving two or more decision makers having different goals. Game theory provides a collection of models to assist in the analysis of conflict and cooperation as well as strategies for resolution. Interestingly, you will find that the games covered in this chapter provide us with insights into certain social paradoxes that we routinely encounter in our daily lives.

Yana Gayvoronskaya/Shutterstock

Fair Division

13

Charles Dharapak/Associated Press

When the demands or desires of one party are in conflict with those of another—be it a divorce, a labor-management negotiation, or an international dispute—no one wants to be treated unfairly. And with 1.2 million divorces every year in the United States alone, and crises such as we've seen in the Middle East for decades, it is certainly worth asking: Can mathematics help in the search for procedures that can ensure fair and equitable resolutions of such conflicts?

13.1 The Adjusted Winner Procedure

To illustrate the **adjusted winner procedure,** we will consider an application to the multibillion-dollar world of business mergers. It turns out that one of the most elusive ingredients in the success of a merger is what deal-makers call *social issues*—how power, position, sacrifice, and status are allocated between the merging companies and their executives.

As a case in point, let's revisit the 1998 proposed merger between two giant pharmaceutical companies, Glaxo Wellcome and SmithKline Beecham. While most of the details underlying this aborted deal are still unknown to outsiders, the role of social issues is clearly underscored by reports that the companies saw nearly $19 billion of stock market value vanish in the clash of two corporate egos.

Timothy A. Clary/AFP/Getty images

Exactly what kinds of issues might bring on a "clash of two corporate egos"? While not privy to the details of the Glaxo Wellcome–SmithKline Beecham merger attempt, we can speculate as to their nature. For purposes of illustration, let's assume that the following five social issues were paramount:

1. The name that the combined company would use
2. The location of the headquarters of the combined company
3. The question of who would serve as chair of the combined company
4. The question of who would serve as CEO of the combined company
5. The question of where the necessary layoffs would come from

Each of these five social issues is known to have been a major factor in other recent proposed mergers. For example, when Chrysler merged with Daimler-Benz in 1998, the issue of the choice of a name for the combined company was described as a "standoff" before both sides finally agreed to DaimlerChrysler.

So let's assume that these were the five social issues confronting Glaxo Wellcome and SmithKline Beecham, and let's see how the adjusted winner procedure would have suggested a resolution. The starting point—and something that is quite difficult when dealing with issues (as in a negotiation) as opposed to objects (as in a divorce)—is to have each side quantify the importance it attaches to getting its own way on each of the issues.

With the adjusted winner procedure, quantification is done by having each side—independently and simultaneously—spread 100 points over the issues in a way that reflects the relative worth of each issue to that party. In our present example, let's assume that the companies allocated their 100 points as shown in Table 13.1. The adjusted winner procedure is now used to decide which side gets its way on which issues, but the procedure requires that a compromise of sorts may have to be reached on one of the issues.

TABLE 13.1 Applying the Adjusted Winner Procedure to a Merger of Two Companies

	Point Allocations	
Issue	**Glaxo Wellcome**	**SmithKline Beecham**
Name	5	10
Headquarters	25	10
Chair	35	20
CEO	15	35
Layoffs	20	25
Total	**100**	**100**

Here's how the procedure works. Suppose that we have two parties and a list of either issues to be resolved in one party's favor or the other's (as in our merger example) or objects to be awarded either to one party or to the other (as in a divorce or a two-person inheritance). To have a single word covering both issues and objects, we will often speak of "items." The adjusted winner procedure follows these basic steps:

Basic Steps in the Adjusted Winner Procedure PROCEDURE

Step 0. As described earlier, each party distributes 100 points over the items in a way that reflects each item's relative worth to that party.

Step 1. Each item on which the assigned points differ is initially given to the party that assigned it more points. Add up the total number of points each party feels that he or she has received. The party with the fewest points is now given all the items on which both parties placed the same number of points. Once again, add up the total number of points each party feels that he or she has received. The party with the most points is called the *initial winner;* the other party is called the *initial loser.*

Step 2. For each item given to the initial winner, calculate the **point ratio.**

Step 3. Start moving items from the initial winner to the initial loser in ascending order of point ratio. Stop when you get to an item whose move will cause the initial winner to have fewer points than the initial loser. This item will need to be split or shared and is thus called the *shared item.*

Step 4. Let x represent the fractional part of the shared item that will be moved from the initial winner to the initial loser. Write a formula that equates each party's total points after the sharing of this item.

Step 5. Solve the equation and state the final division of items between the two parties.

Let's demonstrate the adjusted winner procedure by continuing with our analysis of the proposed merger between Glaxo Wellcome and SmithKline

Beecham. Why the order of transfer given in Step 3 is so important will be explained later.

Step 0. Assume that Glaxo Wellcome and SmithKline Beecham have given us the point assignments shown in Table 13.1.

Step 1. Because Glaxo Wellcome has placed more points on headquarters (25) and chair (35), it is initially "given" these issues, while SmithKline Beecham is initially given name (10), CEO (35), and layoffs (25). Notice that SmithKline Beecham now has $10 + 35 + 25 = 70$ of its points, whereas Glaxo Wellcome has only $25 + 35 = 60$ of its points.

Step 2. We now calculate the point ratio of the three issues held by the initial winner, SmithKline Beecham:

Layoffs has point ratio $25/20 = 1.25$
Name has point ratio $10/5 = 2.00$
CEO has point ratio $35/15 = 2.33$

Step 3. We now start transferring issues from SmithKline Beecham to Glaxo Wellcome until the point totals of the two sides are equal. Because the layoff issue has the lowest point ratio, we start to transfer that item first. But we now see that transferring the entire layoff issue (worth 25 to SmithKline Beecham and 20 to Glaxo Wellcome) gives Glaxo Wellcome more points ($60 + 20 = 80$) than SmithKline Beecham has ($70 - 25 = 45$). Thus, the entire layoff issue cannot be transferred. Glaxo Wellcome and SmithKline Beecham will need to compromise on the issue of layoffs (since it is the shared item). But compromise may not mean meeting each other halfway. Our goal is to equalize points between the two companies, and a little algebra will tell us exactly the extent to which Glaxo Wellcome and SmithKline Beecham should get their way on the issue of layoffs.

Step 4. Let x be the fractional part of the layoff issue that will be transferred from SmithKline Beecham to Glaxo Wellcome. Because SmithKline Beecham had 25 points on the layoff issue, it loses $25x$ points; because Glaxo Wellcome had 20 points on the layoff issue, it gains $20x$ points.

Hence, the original point totals of 70 for SmithKline Beecham and 60 for Glaxo Wellcome become, after the transfer, $70 - 25x$ and $60 + 20x$, respectively. Thus, if we want a fraction x that will make SmithKline Beecham's total points equal Glaxo Wellcome's total points, then we need to solve the following equation:

Algebra Review Appendix
Linear Equations in One Variable

$$70 - 25x = 60 + 20x$$

Step 5. We use algebra to solve this equation.

$$\begin{aligned} 70 - 60 &= 20x + 25x \\ 10 &= 45x \\ 10/45 &= x \end{aligned}$$

Reducing the fraction, we see that $x = 2/9$. Inserting 2/9 back into the equation, we see that

$$70 - 25(2/9) = 60 + 20(2/9)$$

or approximately 64 points for each side. Thus, equality of points is achieved when SmithKline Beecham gives up two-ninths of what it wanted on the issue of layoffs and Glaxo Wellcome gets two-ninths of its way.

Self Check 1

Suppose SmithKline Beecham had placed 25 points on CEO and 35 points on Layoffs, with all other point assignments as in Table 13.1. SmithKline Beecham is still the initial winner and still holds the same three items: Layoffs, Name, and CEO. What are the new point ratios? What is the new order in which issues are transferred?

Having seen how the adjusted winner procedure works, we must now ask the following question: Exactly what is it about the allocation produced by this scheme that would make someone want to use it? To answer this question, we need three definitions.

Equitable DEFINITION

A fair-division procedure is said to be **equitable** if each player believes he or she received the same fractional part of the total value.

Envy-free DEFINITION

A fair-division procedure is said to be **envy-free** if each player has a strategy that can guarantee him or her a share of whatever is being divided that is, in the eyes of that player, at least as large (or at least as desirable) as that received by any other player, no matter what the other players do.

Pareto-Optimal DEFINITION

A fair-division procedure is said to be **Pareto-optimal** if it produces an allocation of the property such that no other allocation achieved by any means whatsoever can make any one player better off without making some other player worse off.

The answer to our earlier question is given by the following theorem (whose proof can be found in *Fair Division* by Brams and Taylor, listed in the Suggested Readings on page 570):

Properties of the Adjusted Winner Allocation THEOREM

For two parties, the adjusted winner procedure produces an allocation based on each player's assignment of 100 points over the items to be divided that has the following properties:

- The allocation is equitable.
- The allocation is envy-free.
- The allocation is Pareto-optimal.

Economists consider Pareto optimality (named after Vilfredo Pareto) to be an extremely important property, and the order of transfer in Step 3 on page 541 of the adjusted winner procedure is so important because it guarantees that the outcome is Pareto-optimal. The fact that the adjusted winner procedure produces an allocation that is efficient in this sense leads us to hope that it can and will play a future role in real-world dispute resolution.

13.2 The Knaster Inheritance Procedure

The adjusted winner procedure can be applied in the case of an inheritance if there are only two heirs. For *more than two heirs,* there is quite a different scheme: the **Knaster inheritance procedure,** first proposed by Bronislaw Knaster in 1945. It has a drawback, though, in that it requires the heirs to have a large amount of cash at their disposal.

EXAMPLE 1 A Four-Person Inheritance

Suppose (for the moment) that there is just one object—a house—and four heirs—Bob, Carol, Ted, and Alice. Knaster's scheme begins with each heir bidding (simultaneously and independently) on the house. Assume, for example, that the bids are as follows:

Bob	Carol	Ted	Alice
$120,000	$200,000	$140,000	$180,000

Carol, being the high bidder, is awarded the house. Her fair share, however, is only one-fourth of the $200,000 she thinks the house is worth, and so she places $150,000 (which is three-fourths of the $200,000 she bid) into a temporary "kitty."

Each of the other heirs now withdraws from the kitty his or her fair share—that is, one-fourth of his or her bid.

Bob withdraws $120,000/4 = $30,000
Ted withdraws $140,000/4 = $35,000
Alice withdraws $180,000/4 = $45,000

Thus, from the $150,000 kitty, a total of $30,000 + $35,000 + $45,000 = $110,000 is withdrawn, and each of the four heirs now feels that he or she has the equivalent of one-fourth of the estate. Moreover, there is a $40,000 surplus ($150,000 kitty − $110,000 withdrawn), which is now divided equally among the four heirs (so each receives an additional $10,000). The final settlement is as follows:

Bob	Carol	Ted	Alice
$40,000	House − $140,000	$45,000	$55,000

This illustrates Knaster's procedure for the simple case in which there is only one object. But what if our same four heirs have to divide an estate consisting of (say) a house (as before), a cabin, and a boat? There are actually two ways to handle this situation, and we'll illustrate both, assuming that our four heirs submit the following bids:

	Bob	Carol	Ted	Alice
House	$120,000	$200,000	$140,000	$180,000
Cabin	$60,000	$40,000	$90,000	$50,000
Boat	$30,000	$24,000	$20,000	$20,000

The first way to deal with the situation is simply to handle the estate one object at a time, proceeding for each object as we just did for the house. We have already settled the house. Let's handle the cabin the same way. Thus, Ted is awarded the cabin based on his high bid of $90,000. His fair share is one-fourth of this, so he places three-fourths of $90,000 (which is $67,500) into the kitty.

Bob withdraws from the kitty $60,000/4 = $15,000. Carol withdraws $40,000/4 = $10,000, and Alice withdraws $50,000/4 = $12,500. Thus, from the $67,500 kitty, a total of $15,000 + $10,000 + $12,500 = $37,500 is withdrawn. The surplus left in the kitty is $30,000, and this is again split equally ($7500 each) among the four heirs. The final settlement on the cabin is as follows:

Bob	Carol	Ted	Alice
$22,500	$17,500	Cabin − $60,000	$20,000

If we were now to do the same for the boat (we leave the details to you), the corresponding final settlement would be as follows:

Bob	Carol	Ted	Alice
Boat − $20,875	$7625	$6625	$6625

Putting the three separate analyses (house, cabin, and boat) together, we get a final settlement of

Bob: Boat + ($40,000 + $22,500 − $20,875 = $41,625)
Carol: House + (−$140,000 + $17,500 + $7625 = −$114,875)
Ted: Cabin + ($45,000 − $60,000 + $6625 = −$8375)
Alice: $55,000 + $20,000 + $6625 = $81,625

Notice that in this situation Carol gets the house but must pay $114,875 in cash (and Ted gets the cabin but must put up $8375 in cash). This cash is then disbursed to Bob and Alice. In practice, Carol's having this amount of cash available may be a real problem—the key drawback to Knaster's procedure. Nevertheless, Knaster's procedure shows again that whenever some participants have different evaluations of some objects, there is an allocation in which everyone obtains more than what they would normally consider a fair share.

EXAMPLE 2 Another Way

The second way begins by adding two rows to the chart of bids, one giving the total value of the estate to each heir (arrived at by summing the columns) and the other giving each heir's fair share (which is one-fourth the value of the estate because there are four heirs).

	Bob	Carol	Ted	Alice
House	$120,000	$200,000	$140,000	$180,000
Cabin	$60,000	$40,000	$90,000	$50,000
Boat	$30,000	$24,000	$20,000	$20,000
Total value	$210,000	$264,000	$250,000	$250,000
Fair share	$52,500	$66,000	$62,500	$62,500

Next, we give each item to the party who values it most. Bob gets the boat, Carol gets the house, and Ted gets the cabin. This is certainly not fair because Carol got the most valuable item and Alice got nothing. We fix this in the following way:

- Bob got \$30,000 in value but feels slighted since he felt his share was \$52,500. The estate gives him the difference in cash: \$52,500 − \$30,000 = \$22,500.
- Carol received more than her fair share, so we have her pay the difference to the estate: \$200,000 − \$66,000 = \$134,000.
- Ted received \$90,000 in value when he believed his fair share was only \$62,500, so we have him pay the estate the difference: \$90,000 − \$62,500 = \$27,500.
- Alice received nothing when she believed her fair share to be \$62,500, so the estate gives her \$62,500 in cash.

At this point, every party has his or her fair share. However, the estate has taken in more than it has paid out. This is called the *surplus.*

$$\begin{aligned}\text{Surplus} &= (\$134{,}000 + \$27{,}500) - (\$22{,}500 + \$62{,}500)\\ &= (\$161{,}500 - \$85{,}000) = \$76{,}500\end{aligned}$$

We now divide the surplus evenly among the parties.

$$\$76{,}500 \div 4 = \$19{,}125$$

Finally, we give an additional \$19,125 to each party, making the final division (as before) the following:

Bob gets the boat and \$41,625 (\$22,500 + \$19,125).
Carol gets the house and pays the estate \$114,875 (\$134,000 − \$19,125).
Ted gets the cabin and pays the estate \$8375 (\$27,500 − \$19,125).
Alice gets \$81,625 cash (\$62,500 + \$19,125).

We summarize Knaster's inheritance procedure as follows.

Basic Steps in Knaster's Inheritance Procedure with *n* Heirs PROCEDURE

For each object, the following steps are performed:

Step 1. The heirs—independently and simultaneously—submit monetary bids for the object.

Step 2. The high bidder is awarded the object, and he or she places all but $1/n$ of his or her bid in a kitty. So, if there are four heirs ($n = 4$), then he or she places all but one-fourth—that is, three-fourths—of his or her bid in a kitty.

Step 3. Each of the other heirs withdraws from the kitty $1/n$ of his or her bid.

Step 4. The money remaining in the kitty is divided equally among the n heirs.

13.3 Fair Division and Organ Transplant Policies

In 1984, the U.S. Congress passed the National Organ Transplant Act and established a unified transplant network known as the Organ Procurement and Transplantation Network (OPTN). One of the primary goals of the OPTN was to increase the equity in the national system of organ allocation.

Achieving an equitable system of organ allocation is complicated by factors other than demand exceeding supply. For example, should an available organ go to the patient who needs it the most or to the one for whom the likelihood of a successful transplant is greatest? Should both of these be taken into consideration, and if so, how? Questions such as these reveal the extent to which an equitable system of organ allocation is a challenging problem in fair division.

To illustrate some of the issues (and paradoxes!) arising in the search for an equitable system for organ allocation, we'll (roughly) follow Peyton Young's synopsis of the fair division procedure for kidney allocation adopted by the OPTN in the late 1980s (see Suggested Readings on page 570).

There were three (main) criteria used in arriving at a final ranking of those needing a kidney, and each potential recipient was awarded points according to a fixed method that we now describe.

- **Criterion 1: Waiting time.** A list of potential recipients was made according to how long they had been waiting for an organ. For each potential recipient, one calculates the fraction of people at or below the spot on the list he or she occupies and then awards that person a number of points equal to 10 times that fraction. So if there are five people on the list, the first (waiting the longest) gets $10 \times 1 = 10$ points, the second gets $10 \times (4/5) = 8$ points, the third gets $10 \times (3/5) = 6$ points, and so on.
- **Criterion 2: Suitability.** The donor and potential recipient each have six relevant antigens that are either matched or not matched; the likelihood of a successful transplant increases with more matches. Two points are awarded for each match.
- **Criterion 3: Disadvantage.** Each person has antibodies that rule out a certain percentage of the population as being potential donors for that person. For some, only 10% are ruled out, whereas for others it may be as high as 90%. Those in the latter category are at a serious disadvantage compared with those in the former. Thus, potential recipients are awarded one point for each 10% of the population they are "sensitized against."

To illustrate this allocation procedure, let's assume we have five potential recipients—*A, B, C, D,* and *E*—with the following characteristics:

Potential Recipient	Months Waiting	Antigens Matched	Percent Sensitized
A	5	2	10
B	4.5	2	20
C	4	0	0
D	2	3	60
E	1	6	90

According to the procedure we described, points would be allocated as follows:

Potential Recipient	Months Waiting	Antigens Matched	Percent Sensitized	Total Points
A	10	4	1	15
B	8	4	2	14
C	6	0	0	6
D	4	6	6	16
E	2	12	9	23

Thus, if one kidney became available, it would go to E (with 23 points). Presumably, if two kidneys became available at the same time, E would get one and D (with 16 points) would get the other.

But now things get interesting. Peyton Young, being well versed in the paradoxes of voting theory, fair division, and apportionment (among other things), observed the following possibility: What if two kidneys become available, but one is delayed slightly? Presumably, E gets the first one, and then we redo the chart with only A, B, C, and D. This yields the following:

Potential Recipient	Months Waiting	Antigens Matched	Percent Sensitized
A	5	2	10
B	4.5	2	20
C	4	0	0
D	2	3	60

According to the procedure that we described, points would be allocated as follows:

Potential Recipient	Months Waiting	Antigens Matched	Percent Sensitized	Total Points
A	10	4	1	15
B	7.5	4	2	13.5
C	5	0	0	5
D	2.5	6	6	14.5

Thus, A (not D!) now gets the second kidney, having 15 points to 14.5 for D. This is an example of what is called the "priority paradox." For more on this, consult Peyton Young's book, listed in the Suggested Readings.

13.4 Taking Turns

For many of us, an early lesson in fair division happens in elementary school with the choosing of sides for a spelling bee or when picking teams on the playground. In terms of importance, these pale in comparison with the issue of property settlement in a divorce. Remarkably, however, the same fair-division procedure—*taking turns*—is often used in both.

Taking turns is fairly self-explanatory. With two parties (and that's all we'll consider here), one party selects an object, the other party then selects one, the first party then selects again, and so on. But in this context, there are several interesting questions to consider.

1. How do we decide who chooses first?
2. Because choosing first is often quite an advantage, shouldn't we compensate the other party in some way, perhaps by giving him or her extra choices at the next turn?
3. Should a player always choose the object he or she most favors from those that remain, or are there strategic considerations that players should take into account?

The answer to Question 1 is often "toss a coin," but there are other possibilities—for example, the two parties could "bid" for the right to go first, as in an auction. The answer to Question 2 is less clear, but in Writing Project 2 (on page 570), we outline a discussion of the issue it raises.

Question 3, on the other hand, is remarkably interesting, and it is this one that we want to pursue. Let's look at an easy example. Suppose that Bob and Carol are getting a divorce, and their four main possessions, ranked from best to worst by each, are as follows:

	Bob's Ranking	**Carol's Ranking**
Best	Pension	House
Second best	House	Investments
Third best	Investments	Pension
Worst	Vehicles	Vehicles

If Carol knows nothing of Bob's preferences, then we can assume that she will choose sincerely—selecting at her turn whichever item she most prefers from those not yet chosen. Now, if Bob is also sincere, and if he chooses first, the items will be allocated as follows:

First turn: Bob takes the pension.
Second turn: Carol takes the house.
Third turn: Bob takes the investments.
Fourth turn: Carol is left with the vehicles.

Hence, Bob gets his first and third favorites (the pension and the investments). However, if Bob opens by choosing the house—and bypassing the pension for the moment—then the allocation will be as follows:

First turn: Bob takes the house.
Second turn: Carol takes the investments.
Third turn: Bob takes the pension.
Fourth turn: Carol is left with the vehicles.

Thus, by being insincere, Bob does better—getting his first and second favorites (the pension and the house).

Self Check 2

Suppose Bob and Carol both choose sincerely, but Carol goes first. Who gets what? ■

Self Check 3

Suppose that Carol again goes first and chooses sincerely (each time), but suppose now that Bob knows Carol's preferences (and knows that she will choose sincerely). Can Bob do better than he did in Self Check 2? ■

In general, then, what is the optimal strategy for rational players to use, assuming that both know the preferences of the other? The answer is something called the **bottom-up strategy,** discovered by the mathematicians D. A. Kohler and R. Chandrasekaran in 1969. We will illustrate it with an example.

Suppose that we have five objects—*A, B, C, D, E*—and Bob is choosing first. Suppose that Bob and Carol have the following rankings of the objects (called **preference lists** in what follows):

Bob	Carol
A	*C*
B	*E*
C	*D*
D	*A*
E	*B*

It will turn out that Bob should open with *C* (his third choice), followed by Carol's choice of *D* (skipping over *E*, for the moment). Bob will then take *A*, Carol will follow with *E*, and finally Bob will get *B*. Bob gets his first, second, and third choices without selecting his first choice first! Where does this strategy come from?

The intuition here is quite easy. Let's make two assumptions about rational players: A rational player will never willingly choose his or her least-preferred alternative, and a rational player will avoid wasting a choice on an object that he or she knows will remain available and thus can be chosen later.

With these assumptions as motivation, let's return to the preceding example and think about the mental calculation that Bob will go through in deciding what his first choice will be. Bob knows the eventual sequence of choices will fill in all the following blanks:

Bob:	______		______		______
Carol:		______		______	

Now, working mentally from right to left, Bob knows that Carol will not choose *B* because it is at the bottom of her list. Thus, he will get stuck with *B*, and so he will avoid wasting anything but his last choice on alternative *B*. Thus, Bob can pencil in alternative *B* as his last choice.

Bob:	______		______		*B*
Carol:		______		______	

Bob, placing himself momentarily in Carol's shoes, knows she will reason the same way, and thus he pencils Carol in for the bottom alternative, *E*, on his list.

Bob:	______		______		*B*
Carol:		______		*E*	

Mentally now, Bob reasons as if alternatives *B* and *E* never existed (and the choice sequence had been Bob–Carol–Bob) and continues to pencil in alternatives from right to left, with Bob working from bottom to top on Carol's preference list and Carol working from bottom to top on Bob's preference list. This yields the following sequence of choices mentally penciled in by Bob:

Bob:	*C*		*A*		*B*
Carol:		*D*		*E*	

Remember, this is just a mental calculation that Bob went through to decide upon the actual choice—in this case, *C*—with which he will open. Bob has no guarantee that Carol will, in fact, respond with *D*, so the use of this strategy involves some risk on Bob's part.

This bottom-up strategy can also be viewed as a procedure that a mediator could use to specify a division of several objects between two parties. Given the preference lists of both parties, the mediator could construct a list—exactly as we did for Bob and Carol above—and then offer this to the parties as the suggested allocation. In effect, the mediator is simultaneously playing the role of two rational parties who choose to employ optimal strategies.

13.5 Divide-and-Choose

AP Photo/Oddvar Walle Jensen, Scanpix/file

There are vast mineral resources under the seabed, all of which, one might argue, should be available to both developed and developing countries. In the absence of some kind of agreement, however, what is to prevent the developed countries from mining all of the most promising tracts before the developing countries have reached a technological level where they can begin their own mining operations? Such an agreement, called the **Convention of the Law of the Sea,** went into effect on November 16, 1994, with 161 signatories (including the United States). Also known as the *Law of the Sea Treaty,* it protects the interests of developing countries by means of the following fair-division procedure.

Whenever a developed country wants to mine a portion of the seabed, that country must propose a division of the portion into two tracts. An international mining company called the Enterprise, funded by the developed countries but representing the interests of the developing countries through the International Seabed Authority, then chooses one of the two tracts to be reserved for later use by the developing countries.

Divide-and-Choose PROCEDURE

With **divide-and-choose,** one party divides the object into two parts in any way that he desires, and the other party chooses whichever part she wants.

As a fair-division procedure, the origins of divide-and-choose go back thousands of years. The Hebrew Bible tells the story of Abram (later to be called Abraham) and Lot, who settled a dispute over land via a proposed division by Abram—"If you go north, I will go south; and if you go south, I will go north" (Gen. 13:8–9)—and a choice (of the plain of Jordan) by Lot. Divide-and-choose resurfaced later in Hesiod's book *Theogony*. The Greek gods Prometheus and Zeus had to divide a portion of meat. Prometheus began by placing the meat into two piles, and Zeus selected one.

Actually, a fair-division procedure consists of both rules and strategies, and all we have described so far are the rules of divide-and-choose. But the natural strategies here are quite obvious: The divider makes the two parts equal in his estimation, and the chooser selects whichever piece she feels is more valuable.

Rules and strategies differ from each other in the following sense: A referee could determine whether a rule is being followed, even without knowing the preferences of the players. Strategies represent choices of how players follow the rules, given their individual preferences (and any other knowledge or goals they may have).

The strategies on which we focus in our discussion of fair-division procedures are those that require no knowledge of the preferences of the other players and yet provide some kind of minimal degree of satisfaction even in the face of collusion by the other players. For example, the strategies just given for divide-and-choose guarantee each player a piece that he or she would not wish to trade for that received by the other.

There are, to be sure, other strategic considerations that might be relevant. For example, in divide-and-choose, would you rather be the divider or the chooser? The answer, given our assumptions that nothing is known of the preferences of the others, is to be the chooser. However, if you knew the preferences of your opponent (and how much she may value spite), then you might want to be the divider.

As a final comment on strategic considerations, we need only look to the origins of the well-known expression "the lion's share." It comes from one of Aesop's fables, as reported by Todd Lowry in *Archaeology of Economic Ideas* (1987, p. 130).

> It seems that a lion, a fox, and an ass participated in a joint hunt. On request, the ass divides the kill into three equal shares and invites the others to choose. Enraged, the lion eats the ass, then asks the fox to make the division. The fox piles all the kill into one great heap except for one tiny morsel. Delighted at this division, the lion asks, "Who has taught you, my very excellent fellow, the art of division?" to which the fox replies, "I learnt it from the ass, by witnessing his fate."

13.6 Cake-Division Procedures: Proportionality

The modern era of fair division in mathematics began in Poland during World War II (see Spotlight 13.1). At this time, Hugo Steinhaus asked what is, in retrospect, the obvious question: What is the "natural" generalization of divide-and-choose to three or more people? The metaphor that has been used in this context, going back at least to the English political theorist James Harrington (1611–1677), is a cake. We picture different players valuing different parts of the cake differently because of concentrations of certain flavors or the depth of frosting.

Cake-Division Procedure DEFINITION

A **cake-division procedure** for n players is a procedure that the players can use to allocate a cake among themselves (no outside arbitrators) so that each player has a strategy that will guarantee that player a piece with which he or she is "satisfied," even in the face of collusion by the others.

As we have seen, divide-and-choose is a cake-division procedure for two players, if by "satisfied" we mean either "thinks his piece is of size or value at least one-half" or "does not want to trade what she received for what anyone else received." We define the first notion here; the second notion is an example of envy-free allocations, defined in Section 13.1.

Seventy Years of Cake Cutting

SPOTLIGHT 13.1

The modern era of cake cutting began with the investigations of the Polish mathematician Hugo Steinhaus during World War II. His research, and that of dozens of others since, involved dealing with two fundamental difficulties. First, allocation schemes that work in the context of two or three players often do not generalize easily to the context of four or more players. Second, procedures that yield envy-free allocations are considerably harder to obtain than procedures that yield proportional allocations.

The mathematics inspired by these two difficulties constitutes a rather elegant corner of the large and important area of fair division. Steinhaus's investigations in the 1940s led to his observation that there is a rather natural extension of divide-and-choose to the case of three players. This is the "lone-divider method" described on page 554. Steinhaus's method was generalized to an arbitrary number of players by Harold W. Kuhn of Princeton University in 1967.

Unable to extend his procedure from three to four players, Steinhaus proposed the problem to some Polish colleagues. Two of them, Stefan Banach and Brønislaw Knaster, solved this problem in the mid-1940s by producing the "last-diminisher method" described on page 554.

In addition to the procedures devised by Banach, Knaster, and Kuhn, there are other well-known constructive procedures for obtaining a proportional allocation among four or more players. One of these is by A. M. Fink of Iowa State University and appears in Exercise 27 (on page 567).

Another constructive procedure of note, although different in flavor from the others, is the 1961 recasting by Lester E. Dubins and Edwin H. Spanier of the University of California at Berkeley of the last-diminisher method as a "moving-knife method" (illustrated in Exercise 40 on page 569). The trade-off here involves giving up the "discrete" nature of the last-diminisher method in exchange for the conceptual simplicity of the moving knife.

Although the existence of an envy-free allocation (even for four or more players) was known to Steinhaus in the 1940s, the first constructive procedure for producing an envy-free allocation among three players was not found until around 1960. At that time, John L. Selfridge of Northern Illinois University and, later but independently, John H. Conway of Princeton University found the elegant procedure presented on page 556. Although never published by either, the procedure was quickly and widely disseminated by Richard K. Guy of the University of Calgary and others. Eventually it appeared in several treatments of the problem by different authors.

In 1980, a moving-knife procedure for producing an envy-free allocation among three players was found by Walter R. Stromquist of Daniel Wagner Associates. Then another procedure, capable of being recast as a moving-knife solution of the three-player case, was found by a law professor at the University of Virginia, Saul X. Levmore, and a former student of his, Elizabeth Early Cook.

In 1992, Steven J. Brams, a political scientist at New York University, and Alan D. Taylor, a mathematician at Union College, succeeded in finding a constructive procedure for producing an envy-free allocation among four or more players. In 1994, Brams, Taylor, and William S. Zwicker (also from Union College) found a moving-knife solution to the four-person envy-free problem. No moving-knife procedure is known that will produce an envy-free allocation among five or more players.

Proportional Procedure DEFINITION

A cake-division procedure (for n players) will be called **proportional** if each player's strategy guarantees that player a piece of size or value at least $1/n$ of the whole in his or her own estimation.

It turns out that for $n = 2$, a procedure is envy-free if and only if it is proportional—that is, for $n = 2$, the two notions of fair division are exactly the same. For $n > 2$, however, all we can say is that an envy-free procedure is automatically proportional. For example, if a three-person allocation is not proportional, then one player (call him Bob) thinks that he received less than one-third.

Bob then feels that the other two are sharing more than two-thirds between them, and thus that at least one of the two (call her Carol) must have more than one-third. But then Bob will envy Carol, and so the allocation is not envy-free. Because all nonproportional allocations fail to be envy-free, it follows that if an allocation is envy-free, then it must be proportional.

Many procedures that are proportional, however, fail to be envy-free, as we shall soon show. Thus, proportional procedures are fairly easy to come by, but envy-free procedures are fairly hard to come by.

EXAMPLE 3 The Steinhaus Proportional Procedure for Three Players (Lone Divider)

Given three players—Bob, Carol, and Ted—we have Bob divide the cake into three pieces (call them *X*, *Y*, and *Z*), each of which he thinks is a size or value of exactly one-third. Let's speak of Carol as "approving of a piece" if she thinks it is of a size or value of at least one-third. Similarly, we will speak of Ted as "approving of a piece" if the same criterion applies. Notice that both Carol and Ted must approve of at least one piece.

If there are distinct pieces—say, *X* and *Y*—with Carol approving of *X* and Ted approving of *Y*, then we give the third piece, *Z*, to Bob (and, of course, *X* to Carol and *Y* to Ted), and we are done. The problem case is where both Carol and Ted approve of only one piece and it is the *same* piece.

Brand X Pictures/Punchstock

Let's assume that Carol and Ted approve of only one piece, *X*, and hence (of more importance to us) both *disapprove* of piece *Z*. Let *XY* denote the result of putting piece *X* and piece *Y* back together to form a single piece. Notice that both Carol and Ted think that *XY* is at least two-thirds of the cake because both disapprove of *Z*. Thus, we can give *Z* to Bob and let Carol and Ted use divide-and-choose on *XY*. Because half of two-thirds is one-third, both Carol and Ted are guaranteed a proportional share (as is Bob, who approved of all three pieces). ■

Self Check 4

Explain why it is impossible to have a proportional procedure for Bob, Carol, and Ted in which Bob begins by dividing the cake into three pieces *X*, *Y*, and *Z*, as in Example 3, and in which the final allocation always gives one of those pieces to Carol and another of those pieces to Ted (i.e., without any additional cutting). ■

The procedure just described, which guarantees proportional shares but is not necessarily envy-free and is sometimes called the **lone-divider method,** was discovered by Hugo Steinhaus around 1944. Unfortunately, it does not extend easily to more than three players. It was left to Steinhaus's students, Stefan Banach and Brønislaw Knaster, to devise a method for more than three players. Picking up where Steinhaus left off (and traveling in quite a different direction), they devised the proportional procedure that today is referred to as the **last-diminisher method.** Like the lone-divider method, it is proportional but not envy-free. We illustrate it for the case of four players (Bob, Carol, Ted, and Alice), and we include both the rules and the strategies that guarantee each player his or her fair share.

EXAMPLE 4 The Banach–Knaster Proportional Procedure for Four or More Players (Last Diminisher)

Bob cuts from the cake a piece that he thinks is of size one-fourth and hands it to Carol. If Carol thinks the piece handed her is larger than one-fourth, she trims it to size one-fourth in her estimation, places the trimmings back on the cake, and passes the diminished piece to Ted. If Carol thinks the piece handed her is of size at most one-fourth, she passes it unaltered to Ted.

Ted now proceeds exactly as did Carol, trimming the piece to size one-fourth if he thinks it is larger than this and passing it (diminished or unaltered) on to Alice. Alice does the same, but being the last player, she simply holds onto the piece momentarily instead of passing it to anyone.

Notice that everyone now thinks the piece is of size at most one-fourth, and the last person to trim it (or Bob, if no one trimmed it) thinks the piece is of size exactly one-fourth. Thus, the procedure now allocates this piece to the last person who trimmed it (and to Bob, if no one trimmed it).

Assume for the moment that it was Ted who trimmed the piece last, so he takes this piece and exits the game. Bob, Carol, and Alice all think that at least three-fourths of the cake is left, so they can start the process over with (say) Bob beginning by cutting a piece from what remains that he thinks is one-fourth of the original cake. Carol and Alice are both given a chance to trim it to size one-fourth in their estimation, and again, the last one to trim it takes that piece and exits the game. The two remaining players both think that at least half the cake is left, so they can use divide-and-choose to divide it between themselves and thus be assured of a piece that is of size at least one-fourth in their estimation.

Self Check 5

Describe how Example 4 needs to be modified for the case where there are five players instead of four.

13.7 Cake-Division Procedures: The Problem of Envy

Divide-and-choose has a property that neither of the last two procedures possesses: It can ensure that each player receives a piece of cake he or she considers the largest or tied for the largest. In the case of only two players, this means that each player can get what he or she perceives to be at least half the cake, no matter what the other player does. Thus, divide-and-choose is an envy-free procedure.

Steinhaus's $n = 3$ proportional procedure (the lone-divider method) is not envy-free. For example, consider the case where Carol and Ted both find one piece unacceptable (and this piece is given to Bob). Carol and Ted will not envy each other when one divides and the other chooses, but Bob may think that this is not a 50–50 split. Indeed, if Bob divided the cake initially into what he thought was three equal pieces, an unequal split of the remaining two-thirds of the cake by Carol and Ted means that Bob will prefer the larger of these two pieces to the one-third he got. Consequently, Bob will envy the person who got this larger piece.

The last-diminisher method is also not envy-free. For example, if Bob initially cuts a piece of cake of size one-fourth, and no one else trims it, then Bob receives this piece and exits the game. If Carol is the one to make the next initial cut, she may well cut a piece from the cake that she thinks is of size one-fourth but that Bob thinks is of size considerably more than one-fourth. But Bob is out of the game.

Thus, if Ted and Alice think this piece is of size less than one-fourth, then Carol receives it, and so Bob will envy Carol.

Nevertheless, there do exist cake-division procedures that are envy-free. We present one of these in what follows.

EXAMPLE 5 The Selfridge–Conway Envy-Free Procedure for Three Players

We start with a cake and three people. The point we wish to arrive at is an envy-free allocation of the entire cake among the three people in a finite number of steps. This task may seem formidable, but quite often in mathematics, an important part of solving a problem involves breaking the problem into identifiable parts. In this case, let's call our starting point *A* and the final point that we want to reach *C*. Now let's identify an appropriate in-between point *B* that makes going from *A* to *C*—via *B*—more manageable. Our in-between point *B* stands for getting a constructive procedure that gives an envy-free allocation of *part* of the cake.

Can we constructively obtain three pieces of cake, whose union may not be the whole cake, which can be given to the three people so that each thinks he or she received a piece at least tied for largest? This turns out to be quite easy with the solution given by John Selfridge and John Conway.

1. Player 1 cuts the cake into three pieces that he considers to be the same size. He hands the three pieces to Player 2.
2. Player 2 trims at most one of the three pieces to create at least a two-way tie for largest. Setting the trimmings aside, Player 2 hands the three pieces (one of which may have been trimmed) to Player 3.
3. Player 3 now chooses, from among the three pieces, one that he considers to be at least tied for largest.
4. Player 2 next chooses, from the two remaining pieces, one that she considers to be at least tied for largest, with the proviso that if she trimmed a piece in Step 2, and Player 3 did not choose this piece, then she must now choose it.
5. Player 1 receives the remaining piece.

Let's reconsider the five steps of this trimming procedure to assure ourselves that each player experiences no envy. Recall that Player 1 cuts the cake into three pieces, and Player 2 trims one of these three pieces. Now Player 3 chooses, and, as the first to choose, he certainly envies no one. Player 2 created a two-way tie for largest, and at least one of these two pieces is still available after Player 3 selects his piece. Hence, Player 2 can choose one of the tied pieces she created and will envy no one. Finally, Player 1 created a three-way tie for largest, and because of the proviso in Step 4, the trimmed piece is not the one left over. Thus, Player 1 can choose an untrimmed piece and therefore will envy no one.

So far we have gone from point *A* to point *B:* Starting with a cake and three players, we have constructively obtained (in finitely many steps) an envy-free allocation of all the cake, except the part *T* that Player 2 trimmed from one of the pieces. We will now describe how *T* can be allocated among the three players in such a way that the resulting allocation of the whole cake is envy-free. (This is the rest of the **Selfridge–Conway envy-free procedure.**)

The key observation for the $n = 3$ case is that Player 1 will not envy the player who received the trimmed piece, even if that player were to be given all of *T*. Recall that Player 1 created a three-way tie and received an untrimmed piece. The union of the trimmed piece and the trimmings yields a piece that Player 1 considers to be exactly the same size as the one he received. Thus, assume that it is Player 3 who received the trimmed piece. (It could as well be Player 2.) Player 1 will not envy Player 3, no matter how T is allocated.

The next step ensures that neither Player 2 nor Player 3 will envy another player when it comes time to allocate T. Let Player 2 cut T into three pieces she considers to be the same size. Let the players choose which of the three pieces they want in the following order: Player 3, Player 1, Player 2.

To see that this yields an envy-free allocation, notice that Player 3 envies no one, because he is choosing first. Player 1 does not envy Player 2, because he is choosing ahead of her; and Player 1 does not envy Player 3 because, as pointed out earlier, Player 1 will not envy the player who received the trimmed piece. Finally, Player 2 envies no one because she made all three pieces of T the same size.

Hence, for $n = 3$, the Selfridge–Conway procedure will give an envy-free allocation of all the cake except T, followed by an allocation of T that gives an envy-free allocation of all the cake.

A naïve attempt to generalize to $n = 4$ what we have done for $n = 3$ would proceed as follows: We would begin by having Player 1 cut the cake into four pieces he considers to be the same size. We would then have Players 2 and 3 trim some pieces (but how many?) to create ties for the largest. Finally, we would have the players choose from among the pieces—some of which would have been trimmed—in the following order: Player 4, Player 3, Player 2, Player 1.

This approach fails because Player 1 could be left in a position of envy. To understand how the approach could fail, consider how many pieces Player 3 might have to trim to create a sufficient supply of pieces tied for largest so that he is guaranteed to have one available when it is his turn to choose. Player 3 might have to trim one piece to create a two-way tie for largest. Player 2 might need to trim two pieces to create a three-way tie for largest (because if there were only a two-way tie for largest, Player 3 might further trim one of these pieces and Player 4 might choose the other). This leaves Player 1 in a possible position of envy because we could have a situation where Player 2 trims two pieces and Player 3 trims a third piece, and Player 4 then chooses the only untrimmed piece. If this happens, Player 1, by being forced to choose a trimmed piece, will definitely envy Player 4.

All is not lost, however, because there are modifications of the Selfridge–Conway procedure that will work for arbitrary n. For more on this, see *Fair Division* (cited in the Suggested Readings).

Although we have used the metaphor of cake cutting throughout our discussion of the problem of envy, the idea of successive trimming is nonetheless applicable to problems of fair division other than parceling out the last crumbs of a cake. The main practical problem in applying the trimming procedure is that many fair-division problems involve goods that cannot be divided up at all, much less trimmed in fine amounts. Such goods are said to be *indivisible.*

It is interesting to recall that when the Allies agreed in 1944 to partition Germany into sectors after World War II (Stage 1), they initially did not reach an agreement about what to do with Berlin (see the figure on the next page). Subsequently, they decided to partition Berlin itself into sectors (Stage 2), even though this city fell 110 miles within the Soviet sector. Berlin was simply too valuable a "piece" for the western Allies (Great Britain, France, and the United States) to cede to the Soviets, which suggests how, after a leftover piece is trimmed off, it can be divided subsequently under the trimming procedure.

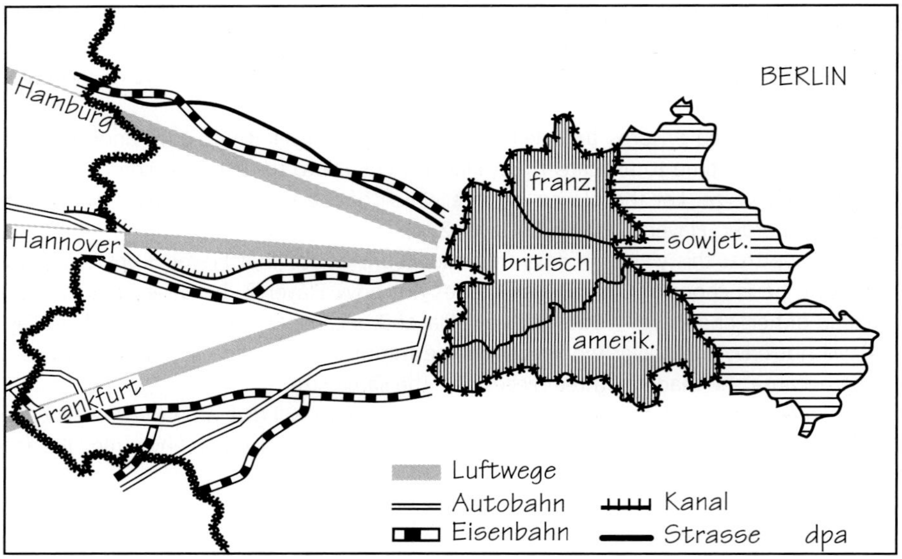

Yet, what if a large piece like Berlin is not divisible? In the settlement of an estate, this might be the house, which may be worth half the estate to the claimants. In this situation, there may be no alternative but to sell this big item and use the proceeds to make the remaining estate more liquid or, in our terms, "trimmable."

13.8 Vickrey Auctions

Among fair-division procedures, auctions are one of the oldest, dating back more than 2500 years. They are used in applications ranging from the determination of who gets to own a promising racehorse to the selection of a contractor to build a new science center at a local college. And, of course, the online auction site eBay has become a story unto itself.

But even the two examples above—an auction for a racehorse versus an auction to select a contractor—illustrate some fundamental differences between the kinds of auctions in use today. For example, is the winner the high bidder or the low bidder? With the racehorse, it's the former, whereas with the contractor, it's the latter. Are the bids oral (so everyone knows the last bid) or are they submitted in a sealed envelope (with no one knowing what anyone else bid)? Again, with the racehorse, it's the former, whereas with the contractor, it's the latter.

The subject of auctions is both large and important. Indeed, we could well devote an entire chapter (or book) to that topic alone. But we limit ourselves here to one particular kind of auction—known as a **Vickrey auction**—that is reminiscent of what is used today on eBay (and we will describe the latter momentarily as well). To avoid confusion, we will assume the high bidder wins (as opposed to the low bidder winning) in the auctions we are considering, and we will assume that ties in the bidding simply do not occur.

A Vickrey Auction PROCEDURE

In a Vickrey auction, bidders independently submit sealed bids for the object being sold. The winner is the high bidder, but he or she pays only the amount of the second-highest bid.

For example, if there are four bids of \$40, \$50, \$60, and \$80, then the fourth bidder wins the auction with his bid of \$80 but pays only \$60 for the object being auctioned off. Vickrey auctions were introduced in a famous 1961 paper by William Vickrey, a Canadian economist and Nobel laureate. Vickrey spent his career at Columbia University and died in 1996, just three days after the announcement of his Nobel Memorial Prize in Economics.

Here is how an eBay auction typically works: When a seller places an item up for auction on the eBay site, he or she indicates a minimum sale price and sets the value for the bid increments. Bidders submit their bids independent of one another. As each new bid is submitted, bids are submitted automatically on behalf of the highest bidder, one increment above the "going price." This continues until time expires and the person with the highest bid wins. In his text *Introduction to Economic Analysis,* R. Preston MacAfee explains the connection between a Vickrey auction and the system used on eBay.

> The Vickrey auction underlies the eBay outcome because when a bidder submits a bid in the eBay auction, the current "going" price is not the highest bid, but the second-highest bid, plus a bid increment. Thus, up to the granularity of the bid increment, the basic eBay auction is a Vickrey auction run over time.

Vickrey auctions are interesting because of the answer they provide to the following question: If the object being sold is worth, say, \$100 to a bidder, how much less than \$100 should he or she bid?

Intuition suggests that in any auction situation, you should bid less than what the object being sold is actually worth to you. In fact, there are mathematical results that suggest that in a standard sealed-bid auction in which the high bidder wins and pays what he or she bid, you should bid only half of what the object is worth to you if there are two bidders, two-thirds of what it is worth to you if there are three bidders, three-fourths if there are four bidders, and so forth.

Remarkably, nothing like this is true with a Vickrey auction. Indeed, with a Vickrey auction, there is a very real sense in which "honesty is the best policy."

Strategy for Bidding in a Vickrey Auction THEOREM

In a Vickrey auction, a bidder can never do better than that achieved by a bid of exactly what the object is worth to that bidder.

EXAMPLE 6 The Vickrey Auction

To see why this theorem is true, let's assume that a lamp is being auctioned off and that it is worth \$100 to a bidder named Bob. This means that Bob would prefer winning the lamp and paying less than \$100 to losing the auction, but that he would rather lose the auction than wind up paying more than \$100 for the lamp.

Let x denote the highest of the bids other than Bob's bid of \$100. Either $x < 100$, in which case Bob wins the auction and pays x dollars, or $x > 100$, in which case Bob loses the auction. We'll consider these two cases separately and show that no bid for Bob can ever do better than his (sincere!) bid of \$100.

Case 1: Bob wins the auction (so $x < 100$). Any bid by Bob greater than x yields the same outcome for Bob as does his bid of \$100. He wins the auction and gets the lamp for x dollars. So these bids are no better for Bob than his bid of \$100.

On the other hand, any bid less than x is strictly worse for Bob than that achieved with his bid of \$100 because he would lose the auction instead of getting the lamp for less than he actually thought it was worth.

Case 2: Bob loses the auction (so $x > 100$). Any bid by Bob less than x yields the same outcome for Bob as does his bid of \$100. He loses the auction. So these bids are no better for Bob than his bid of \$100. On the other hand, any bid greater than x is strictly worse for Bob than that achieved with his bid of \$100 because he would win the auction and pay more for the lamp than he actually thought it was worth, instead of just losing the auction. ∎

This completes the proof of the theorem. There is something quite satisfying in having a rigorous mathematical proof that establishes—at least in this one context—the fact that honesty is indeed the best policy.

Review Vocabulary

Adjusted winner procedure A fair-division procedure introduced by Steven Brams and Alan Taylor in 1993. It works only for two players and begins by having each player independently spread 100 points over the items to be divided so as to reflect the relative worth of each object to that player. The allocation resulting from this procedure is equitable, envy-free, and Pareto-optimal. It requires no cash from either player, but one of the objects may have to be divided or shared by the two players. (p. 540)

Bottom-up strategy A strategy under an alternating procedure in which sophisticated choices are determined by working backward. (p. 549)

Cake-division procedure A fair-division procedure that uses a cake as a metaphor. Such procedures involve finding allocations of a single object that is finely divisible, as opposed to the situation encountered with either the adjusted winner procedure or the Knaster inheritance procedure. In a cake-division procedure, each player has a strategy that will guarantee that player a piece with which he or she is "satisfied," even in the face of collusion by the others. (p. 552)

Convention of the Law of the Sea An agreement based on divide-and-choose that protects the interests of developing countries in mining operations under the sea. This is also referred to as the *Law of the Sea Treaty.* (p. 551)

Divide-and-choose A fair-division procedure for dividing an object or several objects between two players. This method produces an allocation that is both proportional and envy-free (the two being equivalent when there are only two players). (p. 551)

Envy-free A fair-division procedure is said to be envy-free if each player has a strategy that can guarantee him or her a share of whatever is being divided that is, in the eyes of that player, at least as large (or at least as desirable) as that received by any other player, no matter what the other players do. (p. 543)

Equitable A fair-division procedure like adjusted winner is said to be equitable if each player believes he or she received the same fractional part of the total value. (p. 543)

Knaster inheritance procedure A fair-division procedure for any number of parties that begins by having each player (independently) assign a dollar value (a "bid") to the item or items to be divided so as to reflect the absolute worth of each object to that player. The allocation resulting from this procedure leaves each party feeling that he or she received a dollar value at least equal to his or her fair share (and often more so). It never requires the dividing or sharing of an object, but it may require that the players have a large amount of cash on hand. (pp. 544, 546)

Last-diminisher method A cake-division procedure introduced by Stefan Banach and Brønislaw Knaster. It works for any number of players and produces an allocation that is proportional but not, in general, envy-free. (p. 554)

Lone-divider method A cake-division procedure introduced by Hugo Steinhaus. It works only for three players and produces an allocation that is proportional but not, in general, envy-free. (p. 554)

Pareto-optimal A fair-division procedure is said to be Pareto-optimal if it produces an allocation with the property that no other allocation achieved by any means whatsoever can make any one player better off without making some other player worse off. (p. 543)

Point ratio The fraction in which the numerator is the number of points one party placed on an object and the denominator is the number of points the other party placed on the object. (p. 541)

Preference lists Rankings of the items to be allocated, from best to worst, by each of the participants. (p. 550)

Proportional A fair-division procedure is said to be proportional if each of n players has a strategy that can guarantee that player a share of whatever is being divided that he or she considers to be at least $1/n$ of the whole in size or value. (p. 553)

Selfridge–Conway procedure A cake-division procedure introduced independently by John Selfridge and John Conway. It works only for three players but produces an allocation that is envy-free (as well as proportional). (p. 556)

Taking turns A fair-division procedure in which two or more parties alternate selecting objects. (p. 548)

Vickrey auction A sealed-bid auction in which the high bidder wins but pays only the amount of the second-highest bid. (p. 558)

Self Check Answers

1. Layoffs now has a point ratio of 35/20 = 1.75; Name has a point ratio of 10/5 = 2.00; and CEO has a point ratio of 25/15 = 1.67. The order of transfer is now CEO, and then Layoffs, and finally Name.

2. Carol takes the house, Bob takes the pension, Carol takes the investments, and Bob is left with the vehicles.

3. No. Instead of taking the pension when choosing second, he could take the investments or the vehicles. If Bob took the investments when choosing second, then he would wind up with the two things he values least. Perhaps, surprisingly, if Bob instead took the pension (his least preferred item) on his second choice, then he would get his most-preferred item (the pension) on his fourth choice. But this is exactly what he had before, and so he is doing no better.

4. If Carol and Ted view X as being three-quarters of the total value, then any allocation using X, Y, and Z unaltered will give one of them a piece they think is of at most one-fourth the total value, and thus it will not be a proportional allocation.

5. Simply replace "one-fourth" everywhere by "one-fifth," and give the fifth player a chance to trim the piece in the same way that the others do.

Skills Check

1. The adjusted winner procedure applies to

(a) only two-party disputes or disputes that can be recast as two-party disputes.

(b) either two-party or three-party disputes.

(c) n-party disputes for all n.

2. The starting point with the adjusted winner procedure is to have each side—independently and simultaneously—spread 100 points over the issues in such a way that it ____________________.

3. The *winner* part of the name *adjusted winner* refers to the fact that

(a) each party initially "wins" (that is, is given) each issue on which he or she places more points than the other party.

(b) objects ultimately go to whichever party bid more.

(c) it is impossible for both parties to win with any fair-division scheme.

4. The *adjusted* part of the name *adjusted winner* refers to ____________________.

5. In transferring items from one party to the other with the adjusted winner procedure,

(a) the order in which items are transferred is extremely important.

(b) only one item will need to be split or shared.

(c) the order of transfer is obtained by looking at so-called point ratios.

(d) all of the above.

6. With the adjusted winner procedure, suppose A is the party with the highest point total. Then the fraction

$$\frac{A\text{'s point value of the item}}{B\text{'s point value of the item}}$$

is called the item's ____________________.

7. With the adjusted winner procedure, the final allocation is

(a) equitable.
(b) envy-free.
(c) Pareto-optimal.
(d) all of the above.

8. With a fair-division procedure, if each player believes that he or she received the same fractional part of the total value, then the procedure is said to be ______________.

9. A fair-division procedure that produces an allocation with the property that no other allocation, achieved by any means whatsoever, can make any one player better off without making some other player worse off is said to be

(a) equitable.
(b) envy-free.
(c) Pareto-optimal.
(d) none of the above.

10. Chris and Terry must make a fair division of three objects. They assign points to the objects (as shown below) and use the adjusted winner procedure. Chris ends up with ______________.

Object	Chris	Terry
Boat	30	20
Land	50	60
Car	20	20

11. In this chapter, the number of ways to generalize Knaster's inheritance procedure from the case of a single object to the case of several objects is

(a) 0.
(b) 1.
(c) 2.
(d) 3.

12. The inheritance procedure that begins with each heir submitting a monetary bid for each object is known as ______________.

13. Chris and Terry use the Knaster inheritance procedure to divide a coin collection. Chris bids $1000 and Terry bids $800. What is the outcome?

(a) Chris gets the coins and pays Terry $200.
(b) Chris gets the coins and pays Terry $450.
(c) Chris gets the coins and pays Terry $500.

14. Four children bid on two objects (as shown in the table below). Using the Knaster inheritance procedure, Adam ends up with the ______________.

Object	Adam	Beth	Carl	Dietra
House	$80,000	$75,000	$90,000	$60,000
Car	$10,000	$12,000	$13,000	$15,000

15. With the procedure known as *taking turns*, the optimal strategy for players is called

(a) the sincere strategy.
(b) the bottom-up strategy.
(c) the top-down strategy.

16. With taking turns, we assume that a rational player will ______________.

17. Two people use the divide-and-choose procedure to divide a field. Suppose that Jeff divides and Karen chooses. Which statement is true?

(a) Karen always believes she gets more than her fair share.
(b) Karen can guarantee that she always gets at least her fair share.
(c) Karen can possibly believe she gets less than her fair share.

18. A fair-division procedure is envy-free when each player believes that ______________.

19. Using the Steinhaus procedure for three players (lone divider), what happens if there is a single portion that is the only one approved of by both nondividers?

(a) One of the other portions is given to the divider.
(b) The two nondividers flip a coin to determine who receives the approved portion.
(c) All portions are returned to the cake, and a different person serves as the new divider.

20. Using the Steinhaus procedure for three players (lone divider), if the two nondividers approve different portions, then ______________.

21. Using the Banach–Knaster procedure for three or more players (last diminisher), what happens to the first portion after each person has inspected and possibly trimmed it?

(a) The portion goes to the last person to approve the portion, whether or not it was trimmed.
(b) The portion goes to the last person to trim the portion.
(c) The portion goes to the first person to approve and not trim the portion.

22. Using the Banach–Knaster procedure for three or more players (last diminisher), the player who receives the first portion ______________.

23. Using the Banach–Knaster procedure for three or more players (last diminisher), suppose that Scott initially cuts a piece and passes it among the other people, none of whom trims it. What happens next?

(a) Scott gets this piece.

(b) The last person who is handed the piece keeps it.

(c) The piece is returned to the cake, and someone else cuts a piece.

24. Using the Banach–Knaster procedure for three or more players (last diminisher), when only two people remain, ______________.

25. For the Selfridge–Conway procedure for three players, which of the following statements is true?

(a) Each of the three players has the opportunity to trim the portions if they appear to be unfair.

(b) Each player receives a portion that he or she believes to be exactly one-third of the total.

(c) The first player may believe that the third player received more than a fair share.

26. For the Selfridge–Conway procedure for three players, the player who will definitely not receive the trimmed piece in Stage 1 is ______________.

27. An example of a proportional cake-division procedure is

(a) Steinhaus's lone-divider procedure.

(b) the Banach–Knaster last-diminisher procedure.

(c) the Selfridge–Conway procedure.

(d) all of the above.

28. In a Vickrey auction, the winner is the highest bidder, but he pays only ______________.

29. If the bids in a Vickrey auction are \$40, \$50, \$85, and \$90, the winner is the

(a) \$90 bidder and he pays \$90.

(b) \$85 bidder and she pays \$85.

(c) \$90 bidder and he pays \$85.

(d) \$85 bidder and she pays \$90.

30. The only person mentioned in this chapter in connection with two different fair-division procedures is ______________.

Chapter 13 Exercises

Challenge Discussion

13.1 The Adjusted Winner Procedure

1. The 1991 divorce of Ivana and Donald Trump was widely covered in the media. The marital assets included a 45-room mansion in Greenwich, Connecticut; the 118-room Mar-a-Lago mansion in Palm Beach, Florida; an apartment in the Trump Plaza; a 50-room Trump Tower triplex; and just over \$1 million in cash and jewelry. Assume that points are assigned as follows:

Marital Asset	Point Allocations	
	Donald's Points	Ivana's Points
Connecticut estate	10	38
Palm Beach mansion	40	20
Trump Plaza apartment	10	30
Trump Tower triplex	38	10
Cash and jewelry	2	2

Use the adjusted winner procedure to determine a fair allocation of the marital assets. (Exercise 1 courtesy of Catherine Duran.)

2. Suppose that Calvin and Hobbes discover a sunken pirate ship and must divide their loot. They assign points to the items as follows:

Object	Calvin's Points	Hobbes's Points
Cannon	10	5
Anchor	10	20
Unopened chest	15	20
Doubloon	11	14
Figurehead	20	30
Sword	15	6
Cannon ball	5	1
Wooden leg	2	1
Flag	10	2
Crow's nest	2	1

Use the adjusted winner procedure to determine a fair allocation of the loot. (Exercise 2 courtesy of Erica DeCarlo.)

3. This exercise illustrates how the adjusted winner procedure can be used to resolve disputes as well as to achieve fair allocations. Suppose that Mike and Phil are roommates in college, and they encounter serious conflicts during their first week at school. Their resident adviser decides to use the adjusted winner procedure to resolve the dispute. The issues agreed upon, and the (independently assigned) points, turn out to be the following:

Issue	Mike's Points	Phil's Points
Stereo level	4	22
Smoking rights	10	20
Room party policy	50	25
Cleanliness	6	3
Alcohol use	15	15
Phone time	1	8
Lights-out time	10	2
Visitor policy	4	5

Use the adjusted winner procedure to resolve this dispute. (Exercise 3 courtesy of Erica DeCarlo.)

4. Suppose that a labor union and management are trying to resolve a dispute that involves four issues: the base salary of the workers, the annual salary increase that workers can expect, the benefits package the workers will receive, and the amount of vacation time to which each worker will be entitled. Use the adjusted winner procedure to resolve this dispute, with the following point assignments:

Issue	Labor	Management
Base salary	30	50
Salary increases	20	40
Benefits	35	5
Vacation time	15	5

5. Mary and Fred are serving as co-chairs of the mathematics department. There are a number of time-consuming tasks that must be done by one or the other. It really doesn't matter who does which task, except that they disagree on how unpleasant particular tasks are. They decide to use the adjusted winner procedure by phrasing items as "The other co-chair will handle ________." The items and point assignments are as follows:

The other co-chair will handle:	Mary's Points	Fred's Points
Salary recommendations	11	19
Class schedules	19	9
Hiring	14	20
Department meetings	20	10
Calculus placement	21	11
External review	15	31

(a) Which tasks will Mary do?

(b) Which tasks will Fred do?

(c) Which task do they share, and who takes on more of the burden for this task?

6. Make up an example involving two people and several objects for which the adjusted winner procedure can be used, and then use the adjusted winner procedure to determine a fair division.

7. Make up an example involving two people and several issues for which the adjusted winner procedure can be used, and then use the adjusted winner procedure to determine a fair resolution of the dispute.

8. Suppose we have three items (*X*, *Y*, and *Z*) and three people (Bob, Carol, and Ted). Assume that each person spreads 100 points over the items (as in the adjusted winner procedure) to indicate the relative worth of each item to that person.

Item	Bob	Carol	Ted
X	40	30	30
Y	50	40	30
Z	10	30	40

For each of the allocations listed below, indicate the following:

(a) Whether or not it is proportional.

(b) Whether or not it is envy-free.

(c) Whether or not it is equitable.

(d) For the ones that are not Pareto-optimal, another allocation that makes one person better off without making anyone else worse off.

Allocation 1: Bob gets *Z*, Carol gets *Y*, and Ted gets *X*. (This is not Pareto-optimal.)

Allocation 2: Bob gets *Y*, Carol gets *Z*, and Ted gets *X*. (This is not Pareto-optimal.)

Allocation 3: Bob gets *X*, *Y*, and *Z*. (This is Pareto-optimal; explain why.)

Allocation 4: Bob gets *Y*, Carol gets *X*, and Ted gets *Z*. (This is Pareto-optimal.)

Allocation 5: Bob gets *X*, Carol gets *Y*, and Ted gets *Z*. (This is Pareto-optimal.)

13.2 The Knaster Inheritance Procedure

9. If John bids $28,225 and Mary bids $32,100 on their aging parents' old classic car, which they no longer drive, how would you reach a fair division?

Jeremy Woodhouse/Getty Images

10. John and Mary inherit their parents' old house and classic car. John bids $28,225 on the car and $55,900 on the house. Mary bids $32,100 on the car and $59,100 on the house. How should they arrive at a fair division?

11. Can you modify your fair-division procedure in Exercise 10 so that both John and Mary receive one of the two objects while still considering the allocation as fair?

12. Describe a fair division for three heirs, *A*, *B*, and *C*, who inherit a house in the city, a small farm, and a valuable sculpture, and who submit sealed bids (in dollars) on these objects as follows:

	A	B	C
House	$145,000	$149,999	$165,000
Farm	$135,000	$130,001	$128,000
Sculpture	$110,000	$80,000	$127,000

13.3 Fair Division and Organ Transplant Policies

13. Construct the table showing how points would be allocated among the following four potential recipients for a kidney transplant, according to the criteria in Section 13.3.

Potential Recipient	Months Waiting	Antigens Matched	Percent Sensitized
A	9	2	20
B	6	3	0
C	5	4	40
D	2	6	60

14. Does the example in Exercise 13 give rise to the same kind of paradox as in Section 13.3? Explain why or why not.

13.4 Taking Turns

15. Suppose that Bob and Carol rank a series of objects, from most preferred to least preferred, as follows:

Bob	Carol
Car	Boat
Investments	Investments
MP3 player	Car
Boat	Washer-dryer
Television	Television
Washer-dryer	MP3 player

Assume that Bob and Carol use the bottom-up strategy and that Bob gets to choose first. Determine Bob's first choice and the final allocation.

16. Repeat Exercise 15 under the assumption that Carol gets to choose first.

17. Mark and Fred have inherited a number of items from their parents' estate, with no indication of who gets what. They rank the items from most preferred to least preferred, as follows:

Mark	Fred
Truck	Boat
Tractor	Tractor
Boat	Car
Car	Truck
Tools	Motorcycle
Motorcycle	Tools

Assume that Mark and Fred use the bottom-up strategy and that Mark gets to choose first. Determine Mark's first choice and the final allocation.

18. Repeat Exercise 17 under the assumption that Fred gets to choose first.

19. Suppose that Donald and Ivana Trump decide to settle their divorce (described in Exercise 1 on page 563) by taking turns. Assume that they both use the bottom-up strategy, and Donald chooses first. Determine Donald's first choice and the final allocation. (Assume that Donald values the Connecticut estate slightly more than he values the Trump Plaza apartment.)

20. Repeat Exercise 19 under the assumption that Ivana gets to choose first.

13.5 Divide-and-Choose

21. Suppose that Bob is entitled to one-fourth of a cake and Carol is entitled to three-fourths. In a few sentences, explain how divide-and-choose can be used to achieve an allocation in which each party is guaranteed to receive at least as much as he or she is entitled to.

22. Suppose that Bob, Carol, and Ted view a cake as having 18 units of value, with each unit of value represented by a small square (as in the accompanying illustration). Suppose, however, that the players value various parts of the cake differently (or that Bob views the cake as being perfectly rectangular, whereas Carol and Ted see it as skewed in opposite ways). We represent this pictorially as follows:

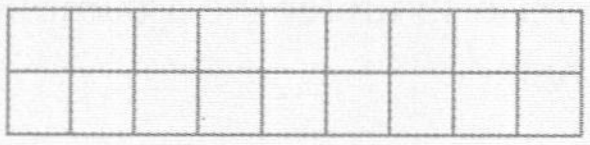

Bob

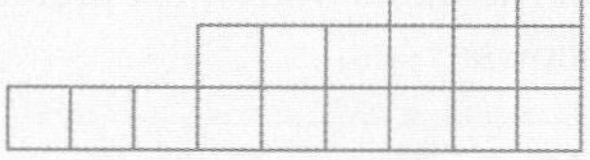

Carol

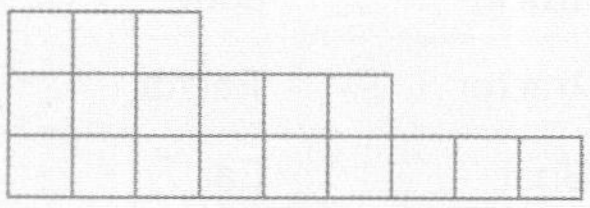

Ted

Assume that all cuts that will be made are vertical.

(a) If Bob and Carol use divide-and-choose to divide the cake between them, how large a piece will each receive (assuming they follow the suggested strategies that go with divide-and-choose and that Bob is the divider)?

(b) If Carol and Ted use divide-and-choose to divide the cake between them, how large a piece will each receive (assuming that they follow the suggested strategies that go with divide-and-choose and that Carol is the divider)?

23. Assume that Bob and Carol view the cake as in Exercise 22, but assume also that each knows how the other values the cake, and that neither is spiteful. Suppose they are to divide the cake using the rules, but not necessarily the strategies, of divide-and-choose. Is Bob better off being the divider or the chooser?

13.6 Cake-Division Procedures: Proportionality

24. Suppose that Players 1, 2, and 3 view a cake as in Exercise 22. Notice that each player views the cake as having 18 square units of area (or value). Assume that each player regards a piece as acceptable if and only if it is at least $18/3 = 6$ square units of area (his or her "fair share"). Assume also that all cuts made correspond to vertical lines.

Angela Wyant/Stone/Getty Images

(a) Provide three drawings to show how each player views a division of the cake by Player 1 into three pieces he or she considers to be the same size or value. Label the pieces *A*, *B*, and *C*.

(b) Identify two of these pieces that Player 2 finds acceptable and two that Player 3 finds acceptable.

(c) Show that a feasible assignment of fair pieces can be achieved by letting the players choose in the following order: Player 3, Player 2, Player 1. Indicate how many square units of value each player thinks he or she received. Is there any other order in which players can choose pieces (in this example) that also results in a feasible assignment?

25. Suppose that Players 1, 2, and 3 view a cake as follows:

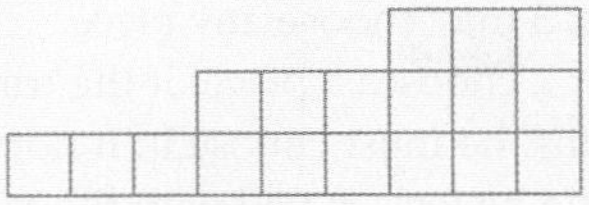

Player 1

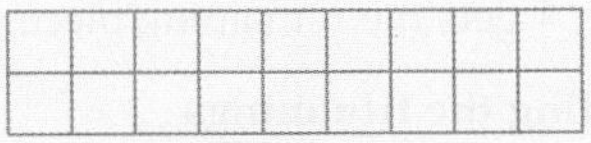

Player 2

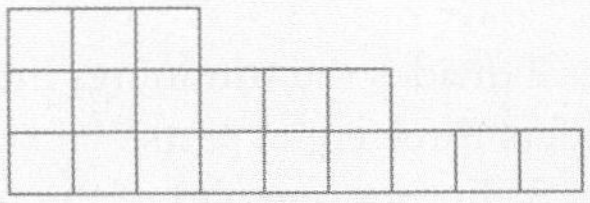

Player 3

(a) Provide three drawings to show how each player views a division of the cake by Player 1 into three pieces that he or she considers to be the same size or value. Label the pieces *A*, *B*, and *C*. (We are still assuming that all cuts correspond to vertical lines, so this will require a cut along a vertical center line of some of the squares.)

(b) Show that neither Player 2 nor Player 3 finds more than one of the three pieces acceptable (with "acceptable" defined as in Exercise 24).

(c) Identify a single piece that Player 2 and Player 3 agree is not acceptable. (There are actually two such pieces; for definiteness, find the one on the right.)

(d) Assume that Players 2 and 3 give the piece from part (c) to Player 1, and that they reassemble the rest and Players 2 and 3 divide it between themselves using divide-and-choose (with a single vertical cut). Determine what size piece each of the three players will think he or she received (i) if Player 2 divides and Player 3 chooses, and (ii) if Player 3 divides and Player 2 chooses.

26. Suppose Players 1, 2, and 3 view a cake as in Exercise 25. Illustrate the last-diminisher method (still restricting attention to vertical cuts and, in addition, assuming that the piece potentially being diminished is a piece off the left side of the cake) by following steps (a) through (f) below.

(a) Draw a picture showing the third of the cake (6 squares) that Player 1 will slice off the cake.

(b) Determine whether Player 2 will pass or further diminish this piece. If he or she would further diminish it, make a new drawing.

(c) Determine whether Player 3 will pass or further diminish this piece. If he or she would further diminish it, make a new drawing.

(d) Determine who receives the piece cut off the cake and what size or value he or she thinks it is. (Actually, we knew what size the person receiving this first piece would think it was, assuming that he or she followed the prescribed strategy. How did we know this?)

(e) Finish the last-diminisher method using divide-and-choose on what remains, with the lowest-numbered player who remains doing the dividing.

(f) Redo step (e) with the other player doing the dividing.

27. The Banach–Knaster last-diminisher method is not the only well-known cake-division procedure that yields a proportional allocation for any number of players. There is also one due to A. M. Fink (sometimes called the *lone-chooser method*). For three players (Bob, Carol, and Ted), it works as follows:

(i) Bob and Carol divide the cake into two pieces using divide-and-choose.

(ii) Bob now divides the piece he has into three parts that he considers to be the same size. Carol does the same with the piece she has.

(iii) Ted now chooses whichever of Bob's three pieces that he (Ted) thinks is largest, and Ted chooses whichever of Carol's three pieces that he thinks is largest.

(iv) Bob keeps his remaining two pieces, as does Carol.

(a) Explain why Ted thinks he is getting at least one-third of the cake.

(b) Explain why Bob and Carol each think they are receiving at least one-third of the cake.

(c) Explain why, in general, this scheme is not envy-free.

28. In A. M. Fink's procedure (described in Exercise 27), suppose that a fourth person (Alice) comes along after Bob, Carol, and Ted have already divided the cake among themselves so that each of the three thinks that he or she has a piece of size at least one-third. Mimic what was done in the three-person case to obtain an allocation among the four that is proportional. (*Hint:* Begin by having Bob, Carol, and Ted divide the pieces they have into a certain number—how many?—of equal parts.)

13.7 Cake-Division Procedures: The Problem of Envy

29. Suppose Players 1, 2, and 3 view the cake as in Exercise 25. Illustrate the envy-free procedure for $n = 3$ (yielding an allocation of part of the cake) by following steps (a) through (c) below. Again, restrict attention to vertical cuts.

(a) Provide a total of three drawings to show how each player views a division of the cake by Player 1 into three pieces he or she considers to be the same size or value. Label the pieces *A*, *B*, and *C*. (This is the same as Exercise 25a.)

(b) Redraw the picture from Player 2's view, and illustrate the trimming of piece *A* that he or she would do. Label the trimmed piece *A* and the actual trimmings *T*.

(c) Indicate which piece each player would choose (and what he or she thinks its size is) if the players choose in the following order, according to the envy-free procedure: Player 3, Player 2, Player 1.

30. There is a two-person moving-knife cake-division procedure due to A. K. Austin that leads to each player receiving a piece of cake that he or she considers to be of size exactly one-half. It begins by having one of the two players (Bob) place two knives over the cake, one of which is at the left edge and the other of which is parallel to the first and placed so that the piece between the knives (*A* in the picture below) is of size exactly one-half in Bob's estimation.

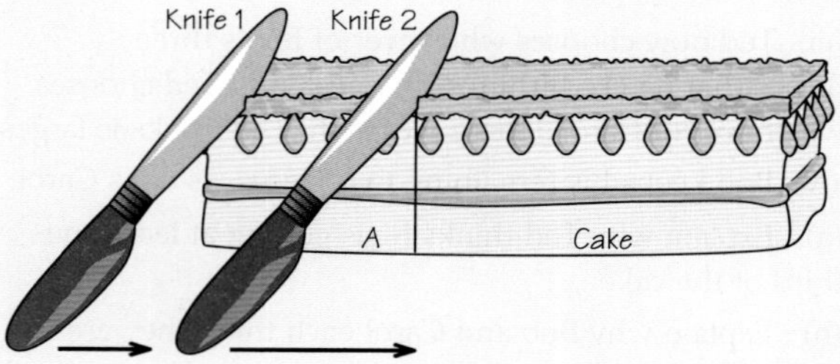

If Carol agrees that this is a 50–50 division, we are done. Otherwise, Bob starts moving both knives to the right—perhaps at different rates—so that the piece between the knives remains of size one-half in his eyes. Carol calls "stop" at the point when she also thinks the piece between the two knives is of size exactly one-half.

(a) If the knife on the right were to reach the right-hand edge, where would the knife on the left be?

(b) Explain why there definitely is a point where Carol thinks the piece between the two knives is of size exactly one-half. (*Hint:* If Carol thinks the piece is too small at the beginning, what will she think of it at the end?)

31. Here are the steps in the Selfridge–Conway procedure for three players.

Stage 1. The initial division

Step 1. Player 1 cuts the cake into, what in his view, is three equal pieces.

Step 2. Player 2, if he thinks one piece is largest, trims from that piece to create what he believes is a two-way tie for largest piece. The trimmings are set aside. If Player 2 thinks that the original split was fair, he does nothing.

Step 3. Player 3 may choose any piece.

Step 4. Player 2 chooses a piece. If the trimmed piece remains, he must choose it. If not, he chooses the one he feels is tied with the trimmed piece for largest.

Step 5. Player 1 gets the remaining piece.

Stage 2. Dividing the trimmings

Assume that Player 3 received the trimmed piece in Stage 1.

Step 6. Player 2 divides the trimmings into what he considers three equal parts.

Step 7. Player 3 chooses one part of the trimmings.

Step 8. Player 1 chooses a piece of the trimmings.

Step 9. Player 2 receives the remaining trimmings.

(a) Explain why Player 1 is envy-free after Stage 1.

(b) Explain why Player 2 is envy-free after Stage 1.

(c) Explain why Player 3 is envy-free after Stage 1.

(d) Explain why Player 1 is envy-free after Stage 2.

(e) Explain why Player 2 is envy-free after Stage 2.

(f) Explain why Player 3 is envy-free after Stage 2.

(Thanks to Michael Rosenthal for suggesting this exercise topic.)

13.8 Vickrey Auctions

32. Consider the Vickrey auction for a lamp that is worth $100 to our bidder Bob (page 559). Suppose someone suggests that Bob would always do as well with a bid of $80. Show that this is false by playing the role of another bidder who could make Bob regret a choice of bidding $80 rather than his (sincere) bid of $100. You are free to make any assumptions that you want about the bids other than Bob's and your own.

33. Consider the Vickrey auction for a lamp that is worth $100 to our bidder Bob (page 559). Suppose someone suggests that Bob would always do as well with a bid of $120. Show that this is false by playing the role of another bidder who could make Bob regret a choice of bidding $120 rather than his (sincere) bid of $100. You are free to make any assumptions you want about the bids other than Bob's and your own.

Chapter Review

34. Beth and Harvey are co-captains of their intramural softball team. As in Exercise 5 (page 564), there are a number of time-consuming tasks that must be done by one or the other. It really doesn't matter who does which task, except that they disagree on how unpleasant particular tasks are. Like Fred and Mary, Beth and Harvey decide to use the adjusted winner procedure by phrasing items as "The other co-chair will handle ________." The items and point assignments are as follows:

The other co-chair will handle:	Beth's Points	Harvey's Points
Selection of player positions	9	40
Coordination of game schedule	9	10
Reserving the field	9	11
Scheduling practices	9	12
Checking equipment	9	13
Planning the end-of-season party	55	14

(a) Which tasks will Beth do?

(b) Which tasks will Harvey do?

(c) Which task do they share, and who takes on more of the burden for this task?

35. Describe a fair division for three children, *E*, *F*, and *G*, who inherit equal shares of their parents' classic car collection and who submit sealed bids (in dollars) on these five cars as follows:

	E	*F*	*G*
Duesenberg	$18,000	$15,000	$15,000
Bentley	$18,000	$24,000	$20,000
Ferrari	$16,000	$12,000	$16,500
Pierce-Arrow	$14,000	$15,000	$13,500
Cord	$24,000	$18,000	$22,000

36. In the scheme for arriving at a priority ranking for organ transplants, how might one change the way points are assigned for "waiting time" so that the kind of paradox that arose in Section 13.3 could not occur?

37. Suppose that Mary and Fred decide to settle the question of which co-chair performs which tasks (described in Exercise 5 on page 564) by taking turns. Assume that they both use the bottom-up strategy, and Mary chooses first. Determine the final allocation of tasks.

38. Repeat Exercise 37 under the assumption that Fred chooses first.

39. If you and another person are using the divide-and-choose procedure to divide something between you, would you rather be the divider or the chooser? (Assume that neither of you knows anything about the preferences of the other.)

40. There is a moving-knife version of the Banach–Knaster procedure that is due to Dubins and Spanier. To describe it, we picture the cake as being rectangular and the procedure beginning with a referee holding a knife along the left edge, as illustrated below.

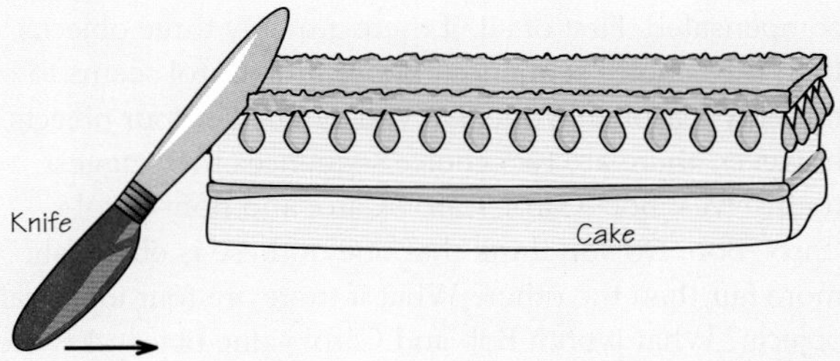

Assume, for the sake of illustration, that there are four players (Bob, Carol, Ted, and Alice). The referee starts moving the knife from left to right over the cake (keeping it parallel to the position in which it started) until one of the players (assume it is Bob) calls "cut." At this time, a cut is made, the piece to the left of the knife is given to Bob, and he exits the game. The knife starts moving again, and the process continues. The strategies are for each player to call "cut" whenever it would yield him or her a piece of size at least one-fourth.

(a) Explain why this procedure produces an allocation that is proportional.

(b) Explain why the resulting allocation is not, in general, envy-free.

(c) Explain why, if you are not the first player to call "cut," there is a strategy different from the one suggested that is never worse for you, and sometimes better.

Writing Projects

1. It turns out that there is no way to extend the adjusted winner procedure to three or more players. That is, there are point assignments by three players to three objects so that no allocation satisfies the three desired properties of equability (equal points), envy-freeness, and Pareto optimality. On the other hand, there are separate procedures that will realize any two of the three properties. Thus, tradeoffs must be made, and these may depend on the circumstances. In a few paragraphs, discuss the relative importance of the three properties and circumstances that may affect the choice of which two of the three properties one might wish to have satisfied.

2. If we use taking turns to divvy up a collection of objects between two people (Bob and Carol), then there is an obvious advantage to going first. Assume that we have decided that Bob will, in fact, choose first (say, by the toss of a coin). Let's think about how Carol might be compensated. First of all, if there are only three objects, then the "choice sequence" Bob–Carol–Carol seems to be the only reasonable one. Do you agree? For four objects, however, there are two choice sequences that suggest themselves: Bob–Carol–Carol–Carol and Bob–Carol–Carol–Bob. Do you think that one of these is obviously more fair than the other? What if there are four identical objects? What if both Bob and Carol value object A twice as much as B, and B twice as much as C, and C twice as much as D? What sequences suggest themselves for five objects? For eight objects?

In one page or less, discuss these questions. (For more on this, see *The Win–Win Solution*, in the Suggested Readings at the end of the chapter.)

3. One of the most important differences between the three-person and the n-person envy-free procedures is that the latter procedure may take more than two stages. And, of course, the more stages there are, the more cuts and trimmings may be necessary. Do you consider this a serious practical problem, or is it mainly a theoretical problem? In one paragraph, explain your reasons.

4. One often hears of the importance of "process" versus "product," the latter referring to what is achieved and the former referring to *how* it was achieved. In a couple of sentences, comment on the relevance of this to fair division as illustrated by the following rough paraphrasing of an exchange between two old friends, Ralph Kramden (played by Jackie Gleason) and Ed Norton (played by Art Carney) in the 1950s sitcom *The Honeymooners.*

> *Ralph to Ed* (as the two are sitting alone at the dinner table): I can't believe you did that.
>
> *Ed:* Did what, Ralph?
>
> *Ralph:* There were two potatoes there, and you reached right out and took the big one.
>
> *Ed:* What would you have done, Ralph?
>
> *Ralph:* Why, I'd have taken the little one.
>
> *Ed:* You got the little one, Ralph.

Suggested Readings

BRAMS, S. J., and A. D. TAYLOR. An envy-free cake division protocol. *American Mathematical Monthly,* 102 (1995): 9–18. Brams and Taylor describe in detail the finite version of their envy-free procedure for $n = 4$; in addition, they review earlier work on "protocols" (step-by-step procedures) that led up to their constructive solution of the envy-freeness problem for $n > 3$.

BRAMS, S. J., and A. D. TAYLOR. *Fair Division: From Cake-Cutting to Dispute Resolution,* Cambridge University Press, Cambridge, MA, 1996. Brams and Taylor provide a book-length treatment of the kinds of topics introduced in this chapter, as well as divide-and-choose in the political arena, moving-knife procedures for cake cutting, and fairness as it applies to different auction and election procedures.

BRAMS, S. J., and A. D. TAYLOR. *The Win-Win Solution: Guaranteeing Fair Shares to Everybody,* Norton, New York, NY, 1999. Brams and Taylor further discuss the adjusted winner procedure, as well as divide-and-choose and taking turns.

ROBERTSON, J., and W. WEBB. *Cake-Cutting Algorithms: Be Fair If You Can,* A. K. Peters, Wellesley, MA, 1998. Robertson and Webb cover a great deal of cake-cutting ground in a text that includes exercises.

YOUNG, P. *Equity in Theory and Practice,* Princeton University Press, Princeton, NJ, 1994. Contains considerably more on fair division in real-world situations, such as the organ transplant example in Section 13.3 on page 546.

Apportionment

14

clearviewstock/Shutterstock

This chapter is about how to divide a set of objects into shares. Many objects, from pieces of hard candy to seats in the U.S. House of Representatives, cannot be distributed in fractional shares, and the apportionment problem occurs when the shares are rounded. The rounded shares may add up to more or less than the whole. We are interested in situations where it is imperative that the sum of the rounded shares is equal to the whole—even if the rounding of some shares must be adjusted to achieve that goal.

For example, we will consider the dilemma of a teacher who covers classes in geometry, precalculus, and calculus. She can teach five classes. There are 52 students enrolled in geometry, 33 in precalculus, and 15 in calculus. Do the math: With 100 students in all, the classes will average 20 students each.

There ought to be 2.60 classes of geometry, 1.65 classes of precalculus, 0.75 classes of calculus—except it is ridiculous for her to teach fractional classes! How would you round these numbers to get a total of 5?

The *apportionment problem* actually gets its name from Article 1, Section 2, of the U.S. Constitution, which, as modified by the Fourteenth Amendment, requires that "[r]epresentatives shall be apportioned among the several States according to their respective numbers, counting the whole number of persons in each State. . . ." Representatives must be apportioned to states in whole numbers, which involves rounding. The framers of the Constitution gave no guidance in how to round. To be fair, an explicit method for apportionment must be chosen and agreed upon in advance by all parties. Many such methods have been proposed, and in the course of U.S. history, four methods (listed in Table 14.16 on page 597) have been employed to apportion seats in the House of Representatives. We will explore all of these methods.

You may wonder why we don't just find the best apportionment method and dismiss the inferior ones. The answer is that there is no one "best" method. When apportioning classes to the teacher as in the situation we have alluded to, one would not choose the Hill–Huntington method—currently in use to apportion seats in Congress—because there would be no way to cancel a class with a very small enrollment. If only one student signed up for calculus, 60 students enrolled in geometry, and 39 students were in precalculus, the obvious solution would be 3 geometry classes, 2 precalculus classes, and no calculus class. The Hill–Huntington method would apportion two classes each for geometry and precalculus, and one class (with one student) for calculus.

14.1 The Apportionment Problem

Winnie has convinced two friends, Louise and Tim, to buy lottery tickets. Winnie buys 18 tickets, Louise buys 4, and Tim buys 1. They are very lucky and win the top prize: a necklace with 100 diamonds! They decide to divide the diamonds in proportion to the number of tickets each bought.

Table 14.1 shows how the diamonds should be shared—if it made sense to cut one of the diamonds into three pieces! They round each share to the nearest whole number: When the fractional part is less than 0.500, they round down, and if any fractional part is greater than or equal to 0.500, they round up. Winnie's share, 78.26 diamonds, is rounded down to 78; Louise's share, 17.39 diamonds, is rounded down to 17; and Tim's 4.35 diamonds are rounded down to 4. Because all three shares were rounded down, there is one diamond left. Winnie declares that since it was her idea to participate in the lottery, that diamond is hers.

TABLE 14.1 Sharing the Diamonds

	Tickets	Shares	Diamonds
Winnie	18	$\frac{18}{23}$	$\frac{1800}{23} = 78.26$
Louise	4	$\frac{4}{23}$	$\frac{400}{23} = 17.39$
Tim	1	$\frac{1}{23}$	$\frac{100}{23} = 4.35$
Total	23	1	100

Apportionment Problem DEFINITION

Suppose we have a set of numbers whose sum is a whole number. The **apportionment problem** is to round each number so that the rounded numbers still add up to the original sum.

Winnie, Louise, and Tim have an apportionment problem because their shares of the diamonds are not whole numbers but the sum of their shares is a whole number. Simple rounding leaves one diamond undistributed. Winnie imposed an arbitrary solution to this apportionment problem: That diamond is hers!

Self Check 1

An administrative assistant decides to keep track of his activities for a 24-hour period. He worked 10 hours (600 minutes), commuted to and from work for 2 hours 20 minutes (140 minutes), and the remaining 11 hours 40 minutes (700 minutes) were spent on personal activities. What percent of his time was devoted to each activity? Express your answer in whole percentages. How would you round the numbers so that they add up to 100%? ■

Apportionment Method DEFINITION

An **apportionment method** is a procedure for solving apportionment problems without making arbitrary choices.

Because we will be discussing the apportionment of seats in the U.S. House of Representatives based on the the census of 1790, it is useful to see the relevant text from Article I, Section 2, of the Constitution.

> *Representatives . . . shall be apportioned among the several States which may be included within this Union, according to their respective Numbers, which shall be determined by adding to the whole Number of free Persons, including those bound to Service for a Term of Years, and excluding Indians not taxed, three fifths of all other Persons.* The actual Enumeration shall be made within three Years after the first Meeting of the Congress of the United States, and within every subsequent Term of ten Years, in such Manner as they shall by Law direct. The Number of Representatives shall not exceed one for every thirty Thousand, but each State shall have at Least one Representative.

The *italicized* text includes the notorious "three-fifths rule" that, for the purpose of apportionment, a state's population included three-fifths of the number of slaves residing in the state. In 1868, this wording was replaced by the part of the Fourteenth Amendment that was quoted at the beginning of this chapter. In this text, we use the term *apportionment population* to mean a state's population for apportionment purposes. Until 1868, this would be the "Numbers" specified in the Constitution as quoted above. Currently, each state's apportionment population includes those from the state who are stationed abroad for military or other United States government service.

Art Resource, NY

Howard Chandler Christy's *Scene at the Signing of the Constitution of the United States.*

The Constitution leaves to Congress the task of determining the apportionment method, subject to the following guidelines:

- Apportionment must be carried out every 10 years.
- The number of House seats awarded to a state must be "according" to its apportionment population.

- The number of representatives cannot exceed 1 per 30,000.
- Every state must be apportioned at least one representative.

There was a vigorous discussion of apportionment methods in President Washington's cabinet. Alexander Hamilton, the Secretary of the Treasury, proposed an apportionment method that will be the subject of Section 14.2. The Secretary of State, Thomas Jefferson—Hamilton's rival—proposed another method that was eventually adopted. We will see Jefferson's method in Section 14.3.

The Constitution does not specify the number of seats in the House of Representatives. It assigns to Congress the primary responsibility for all matters regarding apportionment, subject to the above guidelines. To accomplish the reapportionment of the House of Representatives, which is done following the decennial censuses, an apportionment bill must be passed by the House and the Senate, and signed into law by the president. Like any bill passed by Congress, the president has the option to veto it.

Congress interpreted the phrase, "one [seat] per thirty Thousand," to mean that the number of seats in the House cannot be more than the total apportionment population of the United States, divided by 30,000. According to the census of 1790, the total apportionment population was 3,615,920, which, divided by 30,000, yields a quotient between 120 and 121. Thus, in the first apportionment bill, passed in 1792, Congress set the House size to be 120. The apportionment that resulted is shown in Table 14.2.

TABLE 14.2 The Congressional Apportionment that George Washington Vetoed

State	Population	District Population	Apportionment
Virginia	630,560	30,027	21
Massachusetts	475,327	29,708	16
Pennsylvania	432,879	30,920	14
North Carolina	353,523	?	12
New York	331,589	?	11
Maryland	278,514	30,946	9
Connecticut	236,841	29,605	8
South Carolina	206,236	29,462	7
New Jersey	179,570	29,928	6
New Hampshire	141,822	28,364	5
Vermont	85,533	28,511	3
Georgia	70,835	35,418	2
Kentucky	68,705	34,352	2
Rhode Island	68,446	34,223	2
Delaware	55,540	27,770	2
Totals	3,615,920		120

This bill was the subject of the first presidential veto in U.S. history. (President Washington only vetoed two bills during his time in office.) The reason for the veto can be explained in terms of **district population.**

District Population DEFINITION

The population of a state, divided by the number of seats apportioned to it, is the state's **district population.**

President Washington's interpretation of "one [seat] per thirty Thousand" was that no *state* was allowed to have a district population less than 30,000—and 8 of the 15 states had district populations that violated that limit. His veto message, probably written by Jefferson, also contains a statement about fractional parts being unrelated to the state's populations. This indicates that the apportionment was determined by the Hamilton method, which we will see in Section 14.2.

Self Check 2

Fill in the blank entries in Table 14.2, and determine which state's district population was too small: New York's or North Carolina's.

Now let's see how to set up an apportionment problem. Although many apportionment problems do not involve the House of Representatives, our terminology refers to *states, populations,* and a *house size.* The first step in solving an apportionment problem is to identify the states, populations, and house size.

EXAMPLE 1 Sharing the Diamonds

In the problem of rounding shares of the diamonds so that their sum is 100, the house size is 100. The participants, Winnie, Louise, and Tim, correspond to the states, and the numbers of tickets each bought correspond to the populations of the states.

Standard Divisor DEFINITION

The **standard divisor** is the total population divided by the house size. If p is the total population, h is the house size, and s is the standard divisor, then

$$s = \frac{p}{h}$$

The standard divisor is the average district population. It is *not* the average of the district populations for each state, but the average district population for the nation as a whole.

The second step in addressing an apportionment problem is to determine the standard divisor. Continuing with our example of sharing the diamonds, the house size is 100 so the standard divisor is the total population divided by 100. The total population was the number of tickets bought, 23, and the standard divisor was $23 \div 100 = 0.23$.

Self Check 3

A total of 123 pieces of hard candy are to be divided among three families, according to the number of children in each family. The Browns have five children, the Joneses have three children, and the Robinsons have four children. Identify the states, house size, and populations for this apportionment problem. What is the standard divisor?

Quota

DEFINITION

In an apportionment problem, the **quota** for a state is the exact share that would be allocated to the state if a *whole number were not required.*

$$\text{a state's quota} = \frac{\text{that state's population}}{\text{the standard divisor}}$$

Because the standard divisor is equal to $\frac{\text{total population}}{\text{house size}}$, we can also say that

$$\text{a state's quota} = \frac{\text{that state's population}}{\text{total population}} \times \text{house size}$$

This formula emphasizes that a state's quota is its fair share of the seats in the House, but it is less efficient for calculation.

The third step in solving an apportionment problem is to calculate the quota for each state. In the diamond-sharing example, the number of tickets is interpreted as the population of a state. The quota for each participant is calculated by dividing its population by the standard divisor. If there are many states, this may entail a bit of work. With a calculator, it is helpful to store the standard divisor in memory to avoid having to enter it repeatedly. If you would like to use a spreadsheet to compute an apportionment, there are pointers in the introduction to the Writing Projects (page 619).

Self Check 4

Find the quota for each family in the problem of apportioning the hard candy in Self Check 3.

The quotas in the diamond-sharing scenario, $18 \div 0.23 = 78.26$ for Winnie, $4 \div 0.23 = 17.39$ for Louise, and $1 \div 0.23 = 4.35$ for Tim, are shown in the right column of Table 14.1.

All apportionment methods start with these three initial steps leading to the determination of the quotas. The next step, rounding the quotas to obtain whole numbers whose sum is the house size, is where the various methods differ.

EXAMPLE 2 The High School Mathematics Teacher

Jose Luis Pelaez Inc/Blend

A high school has one mathematics teacher who teaches all geometry, precalculus, and calculus classes. She has time to teach a total of five sections, and 100 students are enrolled as follows: 52 for geometry, 33 for precalculus, and 15 for calculus. How many sections of each course should be scheduled?

This is an apportionment problem because the number of sections is specified (5), and the number of sections allotted to each course must be a whole number. The three courses correspond to the states; the number of students enrolled in each course corresponds to each state's population; and the total number of sections to be taught, 5, is the house size. Thus, the populations of geometry, precalculus, and calculus are 52, 33, and 15, respectively. The total population is 100, so the standard divisor is $s = 100 \div 5 = 20$. Table 14.3 displays the calculations of the quotas.

TABLE 14.3 Calculation of the Quotas for High School Mathematics Courses

Course	Population	Quota	Rounded
Geometry	52	$52 \div 20 = 2.60$	3
Precalculus	33	$33 \div 20 = 1.65$	2
Calculus	15	$15 \div 20 = 0.75$	1
Totals	100	5.00	6

The sum of the quotas is the house size (in this case, 5). It is tempting to round each quota to the nearest whole number, as in the right column of the table, but this makes 6 sections in all—too many! The purpose of an apportionment method is to find an equitable way to round a set of numbers such as these quotas without increasing or decreasing the original sum.

EXAMPLE 3 California's Quota

The Census Bureau recorded the apportionment population of the United States—as of April 1, 2010—to be 309,183,463. There are 435 seats in the House of Representatives; therefore, the standard divisor is

$$309{,}183{,}463 \div 435 = 710{,}767$$

California's quota was determined by dividing its apportionment population, 37,341,989, by this standard divisor. Thus,

$$\text{California's quota} = 37{,}341{,}989 \div 710{,}767 = 52.538 \text{ seats}$$

This quota is slightly more than the quota that was computed with the 2000 census data, 52.447 seats. However, California's apportionment, which must be a whole number, was unchanged at 53 seats.

Self Check 5

The 2010 apportionment populations of Rhode Island and North Carolina were 1,055,247 and 9,565,781, respectively. Find their quotas. Given that the apportionments were 2 seats in the House for Rhode Island and 13 for North Carolina, find the district populations for these states.

Ideally, each state's apportionment should be close to its quota. It is unrealistic to expect that any state will be apportioned its exact quota because each apportionment is required to be a whole number and the quota is unlikely to be a whole number. In choosing an apportionment method, we must decide what we mean by the phrase "each state's apportionment should be close to its quota."

Apportionment always involves rounding, and there are many ways to round. "Rounding down" means discarding the fractional part of a number q to obtain a whole number that we will denote as $\lfloor q \rfloor$. Thus, $\lfloor 7.0001 \rfloor = 7$, $\lfloor 7 \rfloor = 7$, and $\lfloor 6.9999 \rfloor = 6$. "Rounding up" gives the next whole number, denoted as $\lceil q \rceil$. Thus, $\lceil 7.0001 \rceil = 8$, but $\lceil 7 \rceil = 7$.

No apportionment method is perfect for all occasions. Our goal is to understand how to choose a method that is appropriate for a particular apportionment problem.

The apportionment method of largest fractions is named for Alexander Hamilton.

National Portrait Gallery, Smithsonian Institution/Art Resource, NY

14.2 The Hamilton Method

In debates within Washington's cabinet about apportionment, Alexander Hamilton recognized that there would be occasions—analogous to the examples of the three friends' shares of 100 diamonds summing to 99, as in Example 1, or the high school teacher receiving an extra class to teach, as in Example 2—when the total number of seats apportioned by rounding each quota to the nearest whole number would be either less or more than the intended house size. Hamilton proposed an apportionment method that he called *largest fractions.*

Alexander Hamilton's Method and Upper and Lower Quotas DEFINITIONS

With the **Hamilton method,** each state receives either its **lower quota,** $\lfloor q \rfloor$—its quota rounded down—or its **upper quota,** $\lceil q \rceil$—obtained by rounding the quota up. The states that receive their upper quotas are those whose quotas have the largest fractional parts.

Once the quota for each state has been determined, implementing the Hamilton method is a two-step procedure.

Hamilton Method PROCEDURE

1. Tentatively assign to each state its lower quota of representatives. Each state whose quota is not a whole number loses a fraction of a seat at this stage, so the total number of seats assigned at this point will be less than the house size. This leaves additional seats to be apportioned.
2. Allot the remaining seats, one each, to the states whose quotas have the largest fractional parts, until the house is filled.

It is possible that a tie will occur, with the quotas of two states having identical fractional parts, but in practice, this rarely happens when large populations are involved.

Self Check 6

In Self Check 1, you found that the administrative assistant spent 41.67% of his time on his duties, 9.72% of his time on commuting, and 48.61% of his time on personal activities. Using the Hamilton method, round the percentages to whole numbers that add up to 100%. ■

EXAMPLE 4 The High School Teacher's Dilemma

Let us use the Hamilton method to determine how many sections of geometry, precalculus, and calculus the high school teacher in Example 2 (page 577) should teach. Table 14.3 (page 577) displays the quotas for the three subjects,. 2.60, 1.65, and 0.75, respectively. Table 14.4 shows how to obtain the apportionment by the Hamilton method.

TABLE 14.4 Apportioning High School Mathematics Classes

Course	Quota	Lower Quota	Apportionment
Geometry	2.60	2	2
Precalculus	1.65	1 ↑	2
Calculus	0.75	0 ↑	1
Totals	5	3	5

The lower quotas are the **tentative apportionments,** and their sum is 3, leaving two sections still to be apportioned. These go to precalculus and calculus, because the quotas for these courses have the largest fractional parts, 0.65 and 0.75, respectively. ■

Self Check 7

Determine the "district population" for each of the subjects in Example 4. Explain, in familiar terms, what the district populations represent in this context. ■

EXAMPLE 5 Sharing Diamonds by the Hamilton Method

In Table 14.1 (page 572), we saw that Winnie's, Louise's, and Tim's quotas for the 100 diamonds were 78.26, 17.39, and 4.35, respectively. We will round these quotas so that they sum to 100. To start, we apportion $\lfloor 78.28 \rfloor = 78$ diamonds to Winnie, $\lfloor 17.39 \rfloor = 17$ diamonds to Louise, and $\lfloor 4.35 \rfloor = 4$ diamonds to Tim. This leaves one diamond, which must go to Louise because her fraction, 0.39, is the largest. The final apportionment is 78 diamonds for Winnie, 18 diamonds for Louise, and 4 diamonds for Tim. Winnie will want to veto this apportionment. ■

EXAMPLE 6 The Vetoed Apportionment Bill by Hamilton's Method

The congressional apportionment bill that President Washington vetoed involved 15 states, and the House had 120 seats. According to the 1790 census, the U.S. population was 3,615,920. The standard divisor, $3{,}615{,}920 \div 120 = 30{,}133$, represents the population of the average congressional district.

Table 14.5 displays the apportionment, as calculated by Hamilton's method. Each quota shown in the table was calculated by dividing the state's population by this standard divisor. Adding the lower quotas, we find that their sum, 111, leaves 9 seats to be apportioned. These go to the 9 states whose quotas had the largest fractional parts.

TABLE 14.5 Calculation of the Apportionment of Seats in Congress by the Hamilton Method

State	Quota	Lower Quota	Seats Apportioned
New Jersey	5.959	5 ↑	6
Virginia	20.926	20 ↑	21
Connecticut	7.860	7 ↑	8
South Carolina	6.844	6 ↑	7
Delaware	1.843	1 ↑	2
Vermont	2.839	2 ↑	3
Massachusetts	15.774	15 ↑	16
North Carolina	11.732	11 ↑	12
New Hampshire	4.707	4 ↑	5
Pennsylvania	14.366	14	14
Georgia	2.351	2	2
Kentucky	2.280	2	2
Rhode Island	2.271	2	2
Maryland	9.243	9	9
New York	11.004	11	11
Totals	120.000	111	120

We do not know if the Hamilton method was actually used to compute the apportionment in the vetoed bill. It may be that each state's population was simply divided by 30,000 and rounded to the nearest whole number; that would yield the apportionment specified in the vetoed bill. The Hamilton method also provided the apportionment given in the bill, as you can verify by comparing Tables 14.1 (page 572) and 14.5 above.

Apportionment in the U.S. House of Representatives Before 1900

SPOTLIGHT 14.1

The apportionments following the censuses of 1790 through 1840 used a method proposed by Thomas Jefferson.

A divisor, representing the least population to be allowed per representative, was chosen. A state's apportionment was obtained by dividing its apportionment population by that divisor, and rounding the quotient down to get a whole number. Following the censuses of 1790 and 1800, the divisor was 33,000. Thereafter, the divisor increased, to 35,000 in 1810, 40,000 in 1820, and 47,700 in 1830. In 1840, Webster's method was used; that is, each state's apportionment population was divided by a divisor (70,680 for the 1840 census), and the quotient rounded to the nearest whole number. (This is the same procedure that may have been used, with 30,000 as the divisor, for the vetoed apportionment bill—although Webster, then 10 years old, was obviously not involved.)

In 1850, the legislation that provided for the census also specified, for the first time, the house size (233 for the 1852 apportionment). The Hamilton method of apportionment was also specified. The apportionment acts of 1862 through 1892 were debated and enacted after the census counts were known. From 1882 on, house sizes where the Hamilton and Webster methods gave different results were avoided, so it could be claimed that either method was used. For an account of the apportionment following the 1870 census—the first after the Civil War—see Spotlight 14.2 on page 597.

Paradoxes of the Hamilton Method

In 1881, the Census Bureau provided Congress with a table of congressional apportionments, determined by the Hamilton method, for a range of house sizes from 275 to 350 seats, based on the 1880 census. The table revealed a strange phenomenon, which is summarized in Table 14.6. Before reading Example 7, see if you detect what's wrong.

TABLE 14.6 The Alabama Paradox: Hypothetical Congressional Apportionments Based on the 1880 Census

	House Size	
State	**299**	**300**
Alabama	8	7
Illinois	18	19
Texas	9	10

EXAMPLE 7 The Apportionment of 1882

In 1880, the apportionment population of the United States was 49,373,329. When the house size was set at 299, the standard divisor was 49,373,329 ÷ 299 = 165,128. Alabama's apportionment population was 1,262,505. Thus, Alabama's quota was 1,262,505 ÷ 165,128 = 7.64561. Alabama received its upper quota, so if the house size of 299 had been chosen, Alabama would have been apportioned 8 seats. All states with quotas having fractional parts less than Alabama's were apportioned their lower quotas. The states following Alabama when comparing fractions were Illinois and Texas, with apportionment populations of 3,077,871 and 1,591,729, respectively. Their quotas were 18.63930 and 9.63936, respectively, and they were apportioned their lower quotas.

When the house size increased to 300, the standard divisor was less, 49,373,329 ÷ 300 = 164,578. Dividing the three states' apportionment populations by this smaller divisor yielded the following quotas: Alabama, 7.67117; Illinois, 18.70159; and Texas, 9.67158. One additional seat was available for apportionment. With this house size, the quota for Illinois had the largest fractional part of the three, followed by Texas; Alabama was in third place. The seat that had been Alabama's, and the 300th seat, went to Illinois and Texas; both states were apportioned their upper quotas. Alabama received its lower quota. Table 14.6 summarizes the result of these calculations. ■

The paradox was that Alabama's apportionment decreased as a result of an increase in the number of seats in the House of Representatives. Congress avoided controversy by selecting a house size of 325, but an opportunity for abuse was revealed.

Following the census of 1900, there were multiple occurrences of the Alabama paradox. The apportionment tables showed Maine's apportionment, in the words of a Maine representative, "bobbing up and down" as the house size ranged between 350 and 400. Colorado was apportioned 3 seats for each house size in this range—except with a house size of 357, Colorado would receive only 2 seats. A bill to set the house size at 357 caused an uproar; the bill was defeated, and again, another house size was chosen. (The resulting apportionment revealed *another* paradox associated with the Hamilton method, the new states paradox, which is described on page 585.)

Alabama Paradox DEFINITION

The **Alabama paradox** occurs when a state loses a seat as the result of an increase in the house size, or gains a seat due to a reduction of the house size, with no change in any state's population.

EXAMPLE 8 A Mathematics Department Meets the Alabama Paradox

A mathematics department has 30 teaching assistants (TAs) to cover recitation sections for College Algebra, Calculus I, Calculus II, Calculus III, and Contemporary Mathematics. The enrollments of these courses are given in Table 14.7. The department will use the Hamilton method to apportion the TAs to the five subjects. In this problem, the house size is 30 (the number of TAs) and the population is the number of students, 750. The states are the five courses to be offered. The standard divisor is 750 ÷ 30 = 25, which represents the average number of students per recitation section. Each quota shown in the table was determined by dividing the enrollment of the course by this divisor.

TABLE 14.7 Apportioning 30 TAs

Course	Enrollment	Quota	Lower Quota	Apportionment
College Algebra	188	7.52	7	7
Calculus I	142	5.68	5 ↑	6
Calculus II	138	5.52	5	5
Calculus III	64	2.56	2 ↑	3
Contemporary Mathematics	218	8.72	8 ↑	9
Totals	750	30.00	27	30

The lower quotas add up to 27, so the three courses whose quotas have the largest fractional parts, Calculus I and III and Contemporary Mathematics, were given their upper quotas.

After the TAs were given their teaching assignments, the graduate school authorized the department to hire an additional TA. To determine which course should get the new TA, the department had to recalculate the apportionment. With 31 TAs, the standard divisor was 750 ÷ 31 = 24.19355. The new quotas, determined by dividing each population by this new divisor, are shown in Table 14.8. Now the lower quotas add up to 28, so again three additional TAs go to the subjects whose quotas have the largest fractions. The Calculus III fraction, which had been larger than the College Algebra fraction when there were just 30 teaching assistants, has been surpassed. The new TA was placed in College Algebra, and one of the Calculus III TAs had to be reassigned to Calculus II.

TABLE 14.8 Apportioning 31 TAs

Course	Enrollment	Quota	Lower Quota	Apportionment
College Algebra	188	7.771	7 ↑	8
Calculus I	142	5.869	5 ↑	6
Calculus II	138	5.704	5 ↑	6
Calculus III	64	2.645	2	2
Contemporary Mathematics	218	9.011	9	9
Totals	750	31.000	28	31

Self Check 8

Suppose that the reverse had occurred. The department started with 31 TAs, and just before the semester began, one of the TAs was awarded a research position and released from teaching duties, so that only 30 TAs were available. Is there a paradox in this case?

The size of the House of Representatives is now fixed at 435 members by statute. Therefore, the Alabama paradox can no longer occur when apportioning seats in Congress. A second paradox, called the **population paradox,** is associated with a fixed house size.

Population Paradox DEFINITION

Two apportionments are made by the same apportionment method, based on censuses taken at different times. The house size is the same for both apportionments, but the populations for each of the states have changed. The **population paradox** occurs if there is a pair of states A and B such that when the first census results are compared with those of the second, it is found that

- A's population increased and its apportionment decreased.
- B's population decreased, but its apportionment increased.

To see an example of the population paradox, we will consider parliamentary elections. In some countries that have parliamentary systems of government, voters do not elect individual candidates; instead they vote for party lists. Each party nominates a ranked list of candidates. Votes are cast, and each party is apportioned a number of seats "proportionally" to the number of votes that its list received. If, for example, a party was apportioned 10 seats after an election, then the first 10 candidates on the party's list will be seated in the parliament.

In a parliamentary election, the house size corresponds to the number of seats in the parliament, the states correspond to the political parties, and the population of a "state" is the number of voters that selected that party. The Hamilton method is used in the parliamentary elections in Russia and several other countries. (In the context of parliamentary apportionment, the Hamilton method of largest fractions is known as the **Hare method,** after Thomas Hare, whose name also appears in Chapter 9 in connection with the Hare system for deciding multicandidate elections.)

EXAMPLE 9 Apportioning Seats in Parliament

A country has four political parties. Its parliament has 100 members, and seats are apportioned by the Hare method after each election so that the number of seats that each party is awarded is as close as possible to being proportional to the number of votes that the party receives.

An election is held, but the parties are unable to form a government, so there is a repeat election. Table 14.9 shows the results of the two elections. The three major parties—Whigs, Tories, and Liberals—all received more votes in the second election, but the Centrists received fewer. The quotas for each party, shown in Table 14.10, were determined by dividing each party's votes by the standard divisors

$$13{,}060{,}962 \div 100 = 130{,}609.62$$

for the first election, and 132,517.70 for the second election.

TABLE 14.9 Election Results

Party	First Election	Repeat Election
Whigs	5,525,381	5,657,564
Tories	3,470,152	3,507,464
Liberals	3,864,226	3,885,693
Centrists	201,203	201,049
Totals	13,060,962	13,251,770

TABLE 14.10 Quotas for the Parties

Party	First Election	Repeat Election
Whigs	42.3045	42.6929
Tories	26.5689	26.4679
Liberals	29.5861	29.3221
Centrists	1.5405	1.5171

The lower quotas for the results of the first election were 42, 26, 29, and 1, with a sum of 98; thus, the Tories and the Liberals, with the largest fractions, get extra seats. The apportionment after the first election was Whigs, 42; Tories, 27; Liberals, 30; and Centrists, 1.

For the repeat election, the lower quotas were the same, but now the largest fractions belong to the Whigs and the Centrists. Therefore, the new apportionment is Whigs, 43; Tories, 26; Liberals, 29; and Centrists, 2.

The Centrists have *gained* a seat, although they received fewer votes in the repeat election, while the Tories and the Liberals each lost a seat, even though their vote totals increased in the repeat election. This is an instance of the population paradox. This has a disturbing implication: A group of voters might have unintentionally caused the Centrist Party to gain a seat by switching their votes to the Liberal Party. ■

A third paradox of the Hamilton method of largest fractions, the **new states paradox,** was observed when Oklahoma achieved statehood in 1907. Following the 1900 census, the House of Representatives was apportioned with 386 seats by the Hamilton method. It was anticipated that Oklahoma's population would warrant 5 seats, so the house size was increased to 391 in 1907, to accommodate the new Oklahoma representatives. If the entire house had then been reapportioned by the Hamilton method, Oklahoma would still have received its 5 seats. The paradox is that the apportionment of two other states would have changed even though their populations remained the same. Maine's apportionment would increase from 3 to 4, and New York's apportionment would decrease from 38 to 37. For the details, see Exercise 14 (page 615).

14.3 Divisor Methods

The Jefferson Method

Following President Washington's veto of the first apportionment bill in 1792, Congress moved quickly to pass a new apportionment bill. Each state's apportionment population was divided by 33,000, and the resulting quotient was rounded down to get a whole number, which was the state's apportionment. Thus, Virginia, with apportionment population 630,560, received

$$\left\lfloor \frac{630{,}560}{33{,}000} \right\rfloor = \lfloor 19.1 \rfloor = 19 \text{ seats}$$

Delaware's apportionment was $\lfloor 55{,}540 \div 33{,}000 \rfloor = \lfloor 1.7 \rfloor = 1$ seat. Because the number of seats in the House was not specified in advance, the apportionment problem was avoided. (It turned out that 105 seats were apportioned.)

If a state with population p was apportioned a seats, its district population would be $\frac{p}{a}$. Since

$$a = \left\lfloor \frac{p}{33{,}000} \right\rfloor \le \frac{p}{33{,}000}$$

it follows that the district population

$$\frac{p}{a} \ge \frac{p}{\left(\frac{p}{33{,}000}\right)} = 33{,}000$$

In words, the state's district population would be at least 33,000.

Thomas Jefferson favored a method of apportionment biased in favor of states with large populations.

National Portrait Gallery, Smithsonian Institution; gift of the Regents of the Smithsonian Institution, the Thomas Jefferson Memorial Foundation, and the Enid and Crosby Kemper Foundation/Art Resource, NY

Self Check 9

Referring to Table 14.2 on page 574, determine the number of seats that would have been apportioned if the divisor had been set to be 30,000, and explain why this was the maximum house size that would conform to the requirement that each state's district population must be at least 30,000. ■

The method just described was suggested by Thomas Jefferson, and it is called the **Jefferson method.**

With this method, a divisor d is specified. Using d, rather than the standard divisor s, as the divisor, an **apportionment quotient,** rather than a quota, is obtained for each state. A state's apportionment quotient is rounded down to obtain its apportionment.

Apportionment Quotient DEFINITION

The result of dividing a state's population by a divisor is the state's **apportionment quotient.**

Jefferson Method DEFINITION

A divisor is chosen, and, using that divisor, each state's apportionment quotient is determined. The apportionment quotients are rounded down to obtain the apportionments for the states.

The Jefferson method is one of a family of apportionment methods called **divisor methods.**

Divisor Method DEFINITION

A **divisor method** of apportionment determines each state's apportionment by selecting an appropriate divisor and rounding the resulting apportionment quotients, using a specified rounding rule. Divisor methods differ in the rule used to round the apportionment quotient.

In an apportionment problem, there is a specified house size. When a divisor method is used in an apportionment problem, the challenge is to choose the divisor that achieves the intended house size. Select a divisor by trial and error, with the insight that to reduce the number of seats apportioned the divisor must be increased, and to increase the number of seats, a smaller divisor must be used.

The Jefferson Method is the divisor method associated with the rounding rule that rounds each apportionment quotient down. In this case, the divisor must be less than the standard divisor, because if the standard divisor were used, each state would be apportioned its lower quota—and we know that the lower quotas sum to less than the house size (except in the very unusual case where all the quotas turn out to be whole numbers). Let's see how apportioning the diamonds (Example 5 on page 579) works with the Jefferson method.

EXAMPLE 10 Sharing Diamonds by the Jefferson Method

Winnie, Louise, and Tim have bought 18, 4, and 1 lottery tickets, respectively, and have won 100 diamonds. The Jefferson method can be used to share the diamonds. The house size is 100 diamonds, and the total population is the 23 lottery tickets. The standard divisor is 0.23. Table 14.11 shows what happens with a few trial divisors, starting with 0.22, which turns out to be too small, as it apportions a total of 103 diamonds. We gradually increase the divisor until 100 are apportioned. For each trial divisor, there are two columns: the apportionment quotients on the left, and the same quotients rounded down on the right. The sums of the rounded apportionment quotients are shown, and we see that the divisor 0.227 gets the correct house size, 100. (There is no need to add the apportionment quotients, as their sums play no role in determining the apportionment.)

TABLE 14.11 Apportioning 100 Diamonds by the Jefferson Method

		Divisors					
Participant	Population	0.22		0.225		0.227	
Winnie	18	81.82	81	80.00	80	79.30	79
Louise	4	18.18	18	17.78	17	17.62	17
Tim	1	4.55	4	4.44	4	4.41	4
Total	23		103		101		100

The apportionment is what Winnie wanted: 79 diamonds for her, 17 for Louise, and 4 for Tim. ■

Self Check 10

Without any calculations, explain why the divisor 0.226 would also produce the apportionment of 79 diamonds for Winnie, 17 for Louise, and 4 for Tim. ■

Apportioning with the Jefferson Method[1] — PROCEDURE

1. Make a table with the "states" in the left column.
2. Choose a divisor that is slightly less than the standard divisor.
3. Divide each state's population by the chosen divisor to get its apportionment quotient. Round down to a whole number to get its tentative apportionment. Thus, if the apportionment quotient is 9.999, round it to 9.
4. Add the tentative apportionments. If the sum equals the house size, you have completed the apportionment.
5. If the total in Step 4 is less than the house size, choose a smaller divisor and repeat Steps 3 and 4. If the total in Step 4 is larger than the house size, choose a larger divisor—but not larger than the standard divisor—and repeat Steps 3 and 4.

We will implement this procedure for our high school mathematics teacher. The Jefferson divisor d in this context can be interpreted as a minimum class size because the number of classes allocated for each subject is obtained by dividing the number of students enrolled by d, and discarding any fractional section in the quotient. Thus, each section will have at least d students.

EXAMPLE 11 ➔ Apportioning Classes by the Jefferson Method

The teacher can be assigned five classes. There are 52 students enrolled in geometry, 33 in precalculus, and 15 in calculus. The calculations to determine her teaching assignment by the Jefferson method are shown in Table 14.12. We determined in Example 2 (page 577) that the standard divisor is 20, so we'll use 18 as our first trial divisor.

This time, our starting divisor was too large (only three sections were apportioned). We need to try a smaller divisor. With $d = 15$, 6 sections would be apportioned. Thus, the Jefferson divisor has to be more than 15 and less than 18. It turns out that $d = 16$

[1] For pointers on using spreadsheets to compute apportionments with divisor methods, see the Writing Projects (page 619).

TABLE 14.12 Apportioning High School Mathematics Classes

		Divisors					
Subject	**Population**	**18**		**15**		**16**	
Geometry	52	2.89	2	3.47	3	3.25	3
Precalculus	33	1.83	1	2.20	2	2.06	2
Calculus	15	0.83	0	1.00	1	0.94	0
Total	100		3		6		5

works as a divisor, resulting in 3 geometry classes and 2 precalculus classes. There will be no calculus class because the 15 students enrolled are not enough for a class when the minimum class must have at least 16 students.

Comparing the results of Example 11 and Example 4 (page 579), where the Hamilton and Jefferson apportionment methods were applied to the same apportionment problem, we see that the results differ. Examples 10 and 5 also exhibit different apportionments from another apportionment problem. Now let us apportion the TAs in the scenario we encountered in Example 8 on page 582. Will the Alabama paradox occur with the Jefferson method? Recall that with the Hamilton method, Calculus III had 3 TAs when 30 TAs were available, but only 2 TAs when another TA was hired.

EXAMPLE 12 Apportioning TAs with the Jefferson Method

In Example 8 (page 582), a mathematics department was to apportion its TAs among five courses. The populations were the numbers of students enrolled in each course, the states were the courses, and the house size was the number of TAs that were available. Here are the enrollment data: College Algebra, 188; Calculus I, 142; Calculus II, 138; Calculus III, 64; and Contemporary Mathematics, 218. For a house size of 30, the standard divisor was 25. In Table 14.13, we will follow the procedure used in the previous two examples. The apportionment quotients are not shown, just the tentative apportionments that were obtained by rounding them.

TABLE 14.13 Apportioning TAs by the Jefferson Method

		Divisors		
Subject	**Population**	**23**	**24**	**23.5**
College Algebra	188	8	7	8
Calculus I	142	6	5	6
Calculus II	138	6	5	5
Calculus III	64	2	2	2
Contemporary Math	218	9	9	9
Totals	750	31	28	30

The first trial divisor was 23, less than the standard divisor. The number of sections apportioned was 31, which is what we will need when the additional TA arrives.

Attempting to apportion 30 sections, we increase the divisor to 24, but that reduces the seats apportioned to 28. The divisor that we need is therefore between 23 and 24, so we will try 23.5, which indeed produces an apportionment of exactly 30 sections.

Self Check 11

The Mathematics Department has made a rule that all sections must have at least 24 students enrolled. Assuming that the department intends to open the maximum number of sections, how many TAs will be needed?

Table 14.13 shows that when the house size increases from 30 to 31, Calculus II gets the new TA, and no course loses a TA. In fact, the Jefferson method prevents the Alabama paradox, as the following theorem confirms.

Divisor Methods Avoid Paradoxes THEOREM

The Alabama paradox, the population paradox, and the new states paradox cannot occur with any divisor method of apportionment.

We will verify that the Alabama paradox is impossible if the Jefferson method is used.

Let d_1 be the divisor used to apportion a house with h seats. For a house size of $h + 1$, the divisor d_2 must be less than d_1. Consider a state with population p. Its apportionment with house size h will be $\left\lfloor \frac{p}{d_1} \right\rfloor$, and its apportionment with house size $h + 1$ will be $\left\lfloor \frac{p}{d_2} \right\rfloor$.

Since $d_2 < d_1$, it follows that $\frac{p}{d_2} > \frac{p}{d_1}$ and hence

$$\left\lfloor \frac{p}{d_2} \right\rfloor \geq \left\lfloor \frac{p}{d_1} \right\rfloor$$

Therefore, an increase in the house size, with no change in any state's population, cannot lead to a decrease in any state's apportionment. This shows that the Jefferson method prevents the Alabama paradox.

Because divisor methods only differ by the rounding rule used, the same argument shows that any divisor method will prevent the Alabama paradox. The other two paradoxes also do not occur with divisor methods (see Exercises 26 and 27 on page 615).

Here is another distinction between the Hamilton and Jefferson methods—and this time it is in the Hamilton method's favor. The Hamilton method always gives each state either its upper quota or its lower quota.

Self Check 12

Explain why it is that, since the Jefferson divisor must not be more than the standard divisor, the Jefferson method always apportions to each state at least its lower quota.

Although the house size was ignored when the Jefferson method was used to apportion seats in the House of Representatives, when the apportionment was complete, it was possible to compare each state's apportionment with its quota. In 1822, there was a surprise.

EXAMPLE 13 **A Problem with the 1822 Congressional Apportionment**

The 1820 census reported that New York had an apportionment population of 1,368,775. The total apportionment population of the United States was 8,969,878. Using the divisor $d = 40{,}000$, a total of 213 seats were apportioned by the Jefferson method. With the 213 as the house size, the standard divisor was $8{,}969{,}878 \div 213 = 42{,}112$. New York's quota was $1{,}368{,}775 \div 42{,}112 = 32.503$. The Hamilton method, with 213 seats in the House, would have apportioned to New York its upper quota, 33 seats. However, New York's apportionment quotient was $1{,}368{,}775 \div 40{,}000 = 34.219$. The Jefferson method rounded the apportionment quotient down and apportioned 34 seats to New York, *one more than New York's upper quota.*

Quota Condition DEFINITION

An apportionment method is said to satisfy the **quota condition** if in every situation, each state's apportionment is equal to either its lower quota or its upper quota.

The Jefferson method never gives a state fewer seats than its lower quota, but it takes only one example like the 1822 apportionment to show that a state's Jefferson apportionment may be more than its upper quota. Therefore *the Jefferson method does not satisfy the quota condition.* In fact, if the House had continued to use the Jefferson method, it would have given some state more than its upper quota in every apportionment since 1850. For example, the Jefferson apportionment of the House according to the 2010 census gives California 55 seats, although its quota is 52.54; and Texas, with quota 35.55, would get 37 seats.

Self Check 13

Use the Jefferson method to round to whole percentages:

$$97.2\% + 1.9\% + 0.9\% = 100.0\%$$

(You may use the divisor 0.98 to do this.) Is the quota condition violated by this apportionment?

The Hamilton method satisfies the quota condition. This was obvious to Congress in 1850, and that is why it switched to the Hamilton method in that year.[2]

Congress has never used an apportionment method that satisfies the quota condition and avoids the paradoxes of the Hamilton method. In the 1970s, the

[2] By 1850, Hamilton's involvement with apportionment was forgotten, and the method was called the Vinton method, honoring Congressman Samuel Vinton of Ohio (1792–1862), who rediscovered it.

mathematicians Michel L. Balinski and H. Peyton Young tried to devise such a method. Their research showed that the population paradox is possible for all apportionment methods except divisor methods. (The Hamilton method paradoxes are impossible with divisor methods.) However, *no divisor method satisfies the quota condition!* Thus, while Balinski and Young failed to meet their objective, they did prove that there is no apportionment method that satisfies the quota condition and prevents the population paradox. Their theorem is like Kenneth Arrow's impossibility theorem, which tells us that there is no completely satisfactory way to decide multicandidate elections based on voter preference schedules (see Section 9.4, page 424).

The Jefferson method favors the larger states. It is not an accident that in every example that we have considered, the "state" with the largest population fared better with the Jefferson method than it did with the Hamilton method. Winnie, not Louise, received the extra diamond when the Jefferson method was used, as compared with the Hamilton method, and in the example of the high school math teacher, there were more sections of geometry and no sections of calculus when the Jefferson method was substituted for the Hamilton method.

Let's see why the Jefferson method is biased in favor of larger states. Because the Jefferson method always rounds the apportionment quotients down, each state's apportionment quotient must be greater than its quota (otherwise, each state would get its lower quota of seats or fewer). Therefore, the divisor used to obtain these apportionment quotients must be *less* than the standard divisor. If we compare the quotas to the apportionment quotients, we will find that their ratio is equal to the ratio of the standard divisor to the Jefferson divisor:

$$\frac{\text{state's apportionment quotient}}{\text{state's quota}} = \frac{\text{standard divisor}}{\text{Jefferson divisor}}$$

Denote this ratio by R. Since the Jefferson divisor is less than the standard divisor, $R > 1$, so we will let $R = 1 + P$. Then

$$\text{state's apportionment quotient} = (1 + P) \times \text{state's quota}$$

and so the amount the state's apportionment quotient exceeds the quota is

$$\text{state's apportionment quotient} - \text{state's quota} = P \times \text{state's quota}$$

The larger the state's quota is, the more it increases when the apportionment quotient is substituted. This is an advantage to the states with large populations.

Consider the congressional apportionment of 1822. The standard divisor is $s = 42{,}112$, and the Jefferson divisor is $d = 40{,}000$. The quotient is

$$R = 1 + P = 42{,}112 \div 40{,}000 = 1.0528 = 1 + 5.28\%$$

thus $P = 5.28\%$. Now suppose that the quota for state is q. The state's apportionment quotient is obtained by increasing the state's quota by $5.28\% \times q$. A state with a large quota will get a greater raise under these circumstances than a small state will.

To see how this works with numbers, consider a state X with $q = 18.997$. The apportionment quotient is $5.28\% \times 18.997$, or 1.003 greater than q, which brings it to 20. The upper quota for X is 19, but X is awarded 20 seats! This violates the quota condition, and in fact, every state whose quota is 18.997 or more will be guaranteed to get at least its upper quota. If a state has a quota greater than $2 \times 18.997 = 37.994$, an identical calculation shows that it will receive at least its upper quota plus 1 seat. On the other hand, consider a small state whose lower quota is 1. To increase its apportionment to 2, its quota must be at least $2 \div (1 + P) = 1.8997$. Thus, a state with quota 18.997 gets more than its upper quota, and a state with quota 1.899 has to settle for its lower quota.

When parliamentary seats are being apportioned, the Jefferson method's bias favors major parties, which typically attract many votes, over less popular parties. The Jefferson method is used by many nations that prefer voters to select party lists rather than to vote for individual candidates for parliament. It is believed that this practice favors more stable government because there will be less need for a major party to be forced into a coalition with a small party.

For historical reasons, when the Jefferson method is used to apportion seats in parliament, it is called the **d'Hondt method,** after Victor d'Hondt (1841–1901). Countries that use the d'Hondt method to apportion seats in their parliaments include Argentina, Denmark, and Israel.

The procedure for apportioning seats by the d'Hondt method is not the same as the one that we have used for the Jefferson method, but it leads to the same apportionment. With the d'Hondt method, seats are apportioned one at a time until the house size is reached.

d'Hondt Method PROCEDURE

1. Each party is given a priority number, which is reduced as the party accumulates seats. A party's priority number is initially the number of votes that it received.
2. After a party has received n seats, its priority number is equal to the number of votes that it received divided by $n + 1$.
3. Award the first seat to the party that has the largest number of votes; divide that party's vote total by 2 to get its new priority number.
4. Award the second seat to the party that now has the highest priority; recalculate its priority. Continue in this way until all the seats have been distributed.

With the d'Hondt method, it is unnecessary to compute the standard divisor or the quota for any party. The d'Hondt method provides the order in which members of parliament receive their seats. The first member to be seated is the first on the list of the party with the plurality of votes, and so on. However, it achieves exactly the same result as the Jefferson method. Think of the parties' priority numbers as divisors. When a party receives the last seat (because its priority is highest), its priority number (before that seat is awarded to it) is the divisor that will produce the Jefferson apportionment. Here is why this is so: Suppose the party that is about to receive the last seat got V votes and already has n seats. Its priority number is $d = V \div (n + 1)$. Thus, $V \div d = n + 1$ (exactly—no rounding needed), so with divisor d, party P will get $n + 1$ seats. The seats awarded to the other parties were awarded with priority numbers larger than d, so the divisor d will award those seats to them as well. They will not get any additional seats, though, because their current priority numbers are less than d.

EXAMPLE 14 Using the d'Hondt Method

We will use the d'Hondt method to apportion the first 30 seats in the parliament of Example 9 (page 584), based on the data from the first election. The calculations are presented in Table 14.14, a d'Hondt table. There is a column in the table for each party. The first entry in each column is the number of votes that the party received—the party's initial priority number. Running down each column, the entries are the number of votes for the parties divided by 2, 3, 4, and so on. When a seat is awarded, the priority number for the party that receives the seat is crossed out because it has been used. The next seat goes to the party with the highest priority that has not been crossed out.

TABLE 14.14 A d'Hondt Table

Apportioned	Whigs	Tories	Liberals	Centrists
1	#1 ~~5,525,381~~	#3 ~~3,470,152~~	#2 ~~3,864,226~~	201,203
2	#4 ~~2,762,691~~	#7 ~~1,735,076~~	#5 ~~1,932,113~~	100,602
3	#6 ~~1,841,794~~	#10 ~~1,156,717~~?	#9 ~~1,288,075~~	67,068
4	#8 ~~1,381,345~~	#14 ~~867,538~~	#12 ~~966,057~~	50,301
5	#11 ~~1,105,076~~	#17 ~~694,030~~	#16 ~~772,845~~	40,241
6	#13 ~~920,897~~	#21 ~~578,359~~	#19 ~~644,038~~	33,534
7	#15 ~~789,340~~	#25 ~~495,736~~	#23 ~~552,032~~	28,743
8	#18 ~~690,673~~	#28 ~~433,769~~	#26 ~~483,028~~	25,150
9	#20 ~~613,931~~	385,572	#29 ~~429,358~~	22,356
10	#22 ~~552,538~~	347,015	386,423	20,120
11	#24 ~~502,307~~	315,468	351,293	18,291
12	#27 ~~460,448~~	289,179	322,019	16,767
13	#30 ~~425,029~~	266,935	297,248	15,477
14	394,670	247,868	276,016	14,372

At any stage of the d'Hondt process, we can see how many seats have been assigned to a party: It is the number of entries in its column that have been crossed out. If a party has been assigned n seats so far, the first remaining entry in its column is its population divided by $(n + 1)$.

As shown in the table, the Whigs get the first seat. The first priority number is marked #1 and gets crossed out because it has been used. The greatest remaining priority number is for the Liberals, who get the second seat. As seats are awarded, the priority numbers are numbered in sequence and crossed out, in decreasing order. The number of seats awarded to each party is equal to the number of numbers crossed out in its column. Of the first 30 seats, the Whigs get 13, the Tories get 8, the Liberals get 9, and the Centrists get none. ■

Self Check 14

Which party gets seat #31? ■

National Portrait Gallery, Smithsonian Institution/Art Resource, NY

Statesman and orator Daniel Webster (1782–1852), who developed a divisor method for apportioning the U.S. House of Representatives.

The Webster Method

A second divisor method was introduced in 1832 by Senator Daniel Webster of Massachusetts.

The Webster Method DEFINITION

The **Webster method** is the divisor method that rounds the apportionment quotient to the nearest whole number, rounding up when the fractional part is greater than or equal to $\frac{1}{2}$, and rounding down when the fractional part is less than $\frac{1}{2}$.

The Webster and Jefferson methods are free of the paradoxes that we have seen with the Hamilton method, but neither satisfies the quota condition. However, the Jefferson method favors the large states, while the Webster method is neutral, favoring neither the large nor the small states. Furthermore, the Webster method rarely violates the quota condition by giving a state more than its upper quota, or fewer seats than its lower quota, and would not have done so in any of the 23 congressional apportionments that have occurred so far.

The Webster Method PROCEDURE

1. Obtain the tentative apportionments by rounding each state's quota q to the nearest whole number. (Round down to $\lfloor q \rfloor$ if the fractional part of q is less than 0.5; otherwise, round up to $\lceil q \rceil$.)
2. Add the rounded quotas. If their sum is equal to the house size, the job is finished. The tentative apportionments calculated in Step 1 are the final apportionments.
3. When the rounded quotas of the states don't add up to the house size, calculate apportionment quotients, using a trial divisor as with the Jefferson method. All trial divisors must be *larger* than the standard divisor if the sum of the rounded quotas is *larger* than the house size, and *smaller* than the standard divisor if the sum of the quotas is *smaller* than the house size.
4. Round the apportionment quotients from Step 3. If their sum is more than the house size, increase the divisor; if the sum is less, decrease the divisor. If the sum is equal to the house size, the rounded apportionment quotients provide the correct apportionment.

It is possible that first use of the Webster method was before Webster invented it! The apportionment bill that was vetoed by President Washington was consistent with the Webster apportionment, with 30,000 as the divisor. The resulting apportionment was unconstitutional, but perhaps it would have not been vetoed if a large enough divisor had been used.

Let's revisit the three friends who won 100 diamonds in a lottery (see Example 10 on page 586).

EXAMPLE 15 **Sharing Diamonds by the Webster Method**

Tim and Louise have convinced Winnie to go with the Webster method because the Jefferson method is biased in her favor. The quotas are 78.26 diamonds for Winnie, 17.39 for Louise, and 4.35 for Tim; these are rounded to get the tentative shares of 78, 17, and 4 diamonds, respectively, which add up to 99. The standard divisor was 0.23, and because we need to apportion one more seat, we will have to try a smaller divisor. Our trials are shown in Table 14.15.

TABLE 14.15 Apportioning Diamonds by Webster's Method

		Divisors					
Participant	**Tickets**	**0.227**		**0.228**		**0.229**	
Winnie	18	79.2952	79	78.9474	79	78.6026	79
Louise	4	17.6211	18	17.5439	18	17.4672	17
Tim	1	4.4053	4	4.3860	4	4.3668	4
Total	23		101		101		100

The first divisor that we tried was 0.227. The apportionment quotients are in the left column under that divisor in the table, and to their right are the rounded values. Because the total number of "seats" with that divisor turned out to be 101 diamonds, we tried the divisor 0.228, with the same result. The divisor 0.229 produced our goal, 100 diamonds. We never tried any divisor greater than 0.23 because we knew from the start that 0.23 is too large. The result was Winnie's preferred apportionment: 79 diamonds for her, 17 for Louise, and 4 for Tim.

Self Check 15

You saw in Self Check 13 (page 590) that rounding the percentages in

$$97.2\% + 1.9\% + 0.9\% = 100.0\%$$

by the Jefferson method produced a violation of the quota condition. What happens with the Webster method?

EXAMPLE 16 Apportioning Classes by the Webster Method

Let us return to the case of the mathematics teacher who is to teach a total of five classes in geometry, precalculus, and calculus. The enrollments are 52 for geometry, 33 for precalculus, and 15 for calculus. With a total of 100 students enrolled, and a house size of 5, the standard divisor is 20. The quotas, determined by dividing the enrollments for the three subjects by the standard divisor, are 2.60, 1.65, and 0.75, respectively. The tentative apportionments are 3, 2, and 1; their total, 6, exceeds the house size. We therefore will try a divisor greater than 20. Using 21 as the divisor, we find that the apportionment quotients are 2.48 for geometry, 1.57 for precalculus, and 0.71 for calculus. Rounded, these quotients become 2, 2, and 1, respectively, for a total of 5 classes. Thus, geometry and precalculus are each apportioned 2 classes, and calculus gets 1 class. This is the same apportionment that we obtained with the Hamilton method.

To see why the Webster method is not biased in favor of large states or small states, we will use the same analysis that we applied to Jefferson's method. Let s be the standard divisor, let d be the divisor that is used in the Webster method, and let $R = s \div d$. Each state's apportionment quotient is equal to

$$R \times \text{the state's quota}$$

as with the Jefferson method. When $R > 1$ (this happens when the number of seats tentatively apportioned is less than the house size), the states all receive an across-the-board increase in their quotas; and just as in a company where the workers receive the same percentage raise, the larger states are favored. When $R < 1$, the reverse is true because a large number multiplied by R will decrease more than a small number would. Thus, the Webster method favors neither large nor small states when the tentative apportionment exactly fills the house; it favors small states when the tentative apportionment must be reduced, and it favors large states when the tentative apportionment must be expanded. On balance, the Webster method is neutral because it is equally likely that the tentative apportionment will be less than the house size or that it will be greater than the house size.

The Webster Method Has No Population Bias THEOREM

Among all divisor methods, the Webster method alone shows no bias with regard to state population.

When seats are apportioned to parties after an election, a method that is equivalent to the Webster method can be used. This method, called the **Sainte-Laguë method** (after André Sainte-Laguë, 1878–1950), involves a process that is analogous to the d'Hondt method. A Sainte-Laguë table is constructed with a column for each party. The number of votes received is at the top of each party's column, and below that the number of votes is divided by successive odd numbers, 3, 5, 7, ... Seats are apportioned to the parties by treating the entries in the table as a priority list. The party with the highest priority gets a seat, and then that priority number is crossed out. The next seat goes to the party with the largest remaining priority in the Sainte-Laguë table. Germany and New Zealand use the Sainte-Laguë method to apportion seats to their parliaments.

If you are curious to know why the Sainte-Laguë method produces the same apportionment as the Webster method, refer again to the party that is about to get the last seat. Let V be the number of votes that it received, and suppose that it has n seats already, before the last seat is awarded. Its priority, $V \div (2n + 1)$, is the highest remaining in the table. [The $(n + 1)$st odd number is $2n + 1$.] Multiply this priority by 2 to get a number d. Using this d as the divisor, we see that

$$V \div d = V \div \left(\frac{2 \times V}{(2n + 1)}\right) = n + \frac{1}{2}, \text{ exactly}$$

(The Vs cancel each other out, as we divide the quotients by inverting the divisor and multiplying.) This quotient can be rounded up to get $n + 1$ seats. The other parties have lower priority, and so they get no new seats with d as the divisor.

Self Check 16

Referring again to the high school teacher who was to teach five classes, where 52 students were enrolled in geometry, 33 students were enrolled in precalculus, and 15 students were enrolled in calculus, make a Sainte-Laguë table to determine the Webster apportionment. ■

So far, three of the four methods that have been used to apportion seats in the U.S. House of Representatives have been introduced. The fourth is the Hill–Huntington method, to be described in Section 14.4. This divisor method was adopted by statute following the 1940 census and is still used. In Table 14.16, the properties of the four apportionment methods are described.

TABLE 14.16 Apportionment Methods Used by Congress

	Hamilton	Jefferson	Webster	Hill–Huntington
Censuses	1850; 1860*–1900*	1790*–1830*	1840*, 1910, 1930	1940–present
Divisor method	No	Yes	Yes	Yes
Paradoxes[3]	Yes	No	No	No
Bias for	None	Large states	None	Small states
Quota condition	Satisfied	Often exceeds upper quota	Not satisfied, rarely violates	Not satisfied, rarely violates

In years marked with an asterisk (*), the house size was determined after the apportionment was calculated. In asterisk years when a divisor method was in use, the divisor was chosen and the house size was determined by the calculation. For the Hamilton method, Congress was presented with apportionments for a range of house sizes and chose the one that suited.

A Puzzling Apportionment of the U.S. House of Representatives

SPOTLIGHT 14.2

The strangest apportionment of seats in the U.S. House of Representatives followed the census of 1870. This was the first apportionment to be based on the Fourteenth Amendment to the U.S. Constitution, which redefined the apportionment populations. On February 2, 1872, President Grant signed an apportionment bill specifying 283 seats. With this house size, the Hamilton and Webster methods provided the same apportionment (see Spotlight 14.1 on page 580).

Later that year, the Congress passed a bill awarding one additional seat to each of nine states. The bill was signed by the president on May 30, bringing the house size to 292. With this house size, the Hamilton and Webster methods gave different apportionments, and the new apportionment did not agree with either of them! Circumstantial evidence indicates that the supplemental apportionment was not due to partisan considerations: the House was controlled by the Democrats, while the president and the Senate were Republican. Five of the nine states that received additional seats voted Republican, and four voted Democratic in the 1876 presidential election. Northern and southern states were among those favored with extra seats.

The supplemental apportionment may have influenced the outcome of the controversial presidential election in 1876, in which Rutherford B. Hayes lost the popular vote but defeated Samuel J. Tilden, 185 to 184 electoral votes. (Because Colorado was admitted to the Union on August 1, 1876, with one seat in the House, the House size at the time of the election was 293.) Without the extra seats from the bill on May 30, 1872, the election would have been tied with 180 electoral votes for each candidate. The House of Representatives would have decided the election, under the provision of Article II, Section 1, of the U.S. Constitution, in which each state's delegation has one vote. It is probable, but not certain, that the House would have elected Tilden, 20 states to 18.

[3] See the theorem on page 589.

14.4 Which Divisor Method Is Best?

Why was the supplemental apportionment bill of 1872 enacted? (See Spotlight 14.2 on the preceding page.) Perhaps the goal was to get what was perceived to be a fairer apportionment. Let's focus on the district populations of the states in the original apportionment bill, which agreed with the Webster and Hamilton methods. Ohio had a population of 2,665,260 and was apportioned 20 seats, so its district population was 133,263. Vermont's population was 330,551, with 2 seats, for a district population of 165,275.5. This apportionment was advantageous to Ohio, since its district population was 32,012.5 less. Suppose a seat were transferred from Ohio to Vermont. Ohio's district population would then be

$$2{,}665{,}260 \div (20 - 1) = 140{,}276.8$$

and Vermont's would be

$$330{,}551 \div (2 + 1) = 110{,}183.7$$

Now the advantage is to Vermont, by 30,093.1. Since the difference in district populations was reduced by the transfer, it would be fairer to transfer a seat from Ohio to Vermont, if one believes that the fairest apportionment is one that cannot be improved by such a transfer of seats.

Of course, it would not do to reduce Ohio's apportionment, but if one just gave an extra seat to Vermont, other inequities emerge. The solution is to give additional seats to other states. While it was not necessary to add nine seats (providing one additional seat each to Florida and Vermont would have been sufficient), the supplemented apportionment is optimal in the sense that it is impossible to reduce the difference in district populations between any two states by transferring a seat from one of the states to the other.

The **Dean method** is a divisor method that minimizes differences in district populations; thus, a Dean apportionment leads to an **equitable apportionment** in terms of district population. The method was proposed by James Dean (1776–1849) in 1832, in a letter to Senator Daniel Webster. Dean was a mathematics professor who knew Webster when he was an undergraduate at Dartmouth College. The Dean method has never been officially used to apportion seats in the House, but it may have been the source of the supplemental apportionment in 1872. To explore this interesting apportionment method, see Writing Project 2 on page 619.

Self Check 17

According to the 2010 census, the apportionment population of California was 37,341,989 and the state was apportioned 53 seats in the House of Representatives. Montana, with apportionment population 994,416, was apportioned just one seat. The district population of California was 704,565. Determine the district populations that would result if a seat were transferred from California to Montana. Would the difference be less if this transfer took place? ■

One way to compare apportionment methods is to investigate bias based on population. The Dean method is known to be biased in favor of states with small populations—but it does provide a way to minimize inequities in district population. There are other ways to measure apportionment inequities between states. For an account of two recent disputes that focused on such inequities, see Spotlight 14.3.

Legal Challenges to Apportionment

SPOTLIGHT 14.3

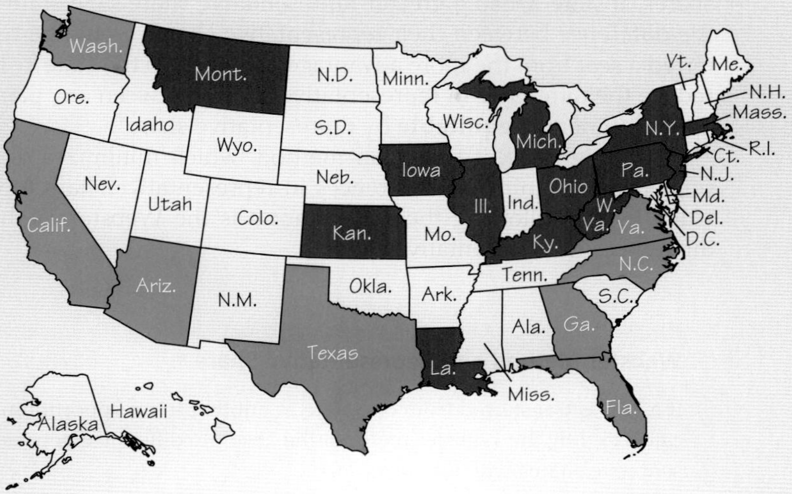

In 1991, the Census Bureau reported the apportionment of congressional seats resulting from the 1990 census. The states colored red on the above map lost representatives, and those in blue gained representatives. New York lost three representatives, and Ohio and Pennsylvania lost two apiece. Montana, whose apportionment decreased from 2 to 1, sustained the greatest percentage loss, and Montana sued to restore the lost seat. As precedents, Montana referred to the two famous cases, *Baker v. Carr* and *Wesberry v. Sanders,* in which the U.S. Supreme Court required state legislative and congressional district boundaries to be drawn so as to make district populations equal. Montana argued that the correct apportionment would be the one that met the *Baker* and *Wesberry* criterion of having district populations as nearly equal as possible, and asked the court to require the Census Bureau to recompute the apportionments using the Dean method. This would have resulted in the transfer of a congressional seat from Washington to Montana.

The Montana case proceeded concurrently with another federal lawsuit, *Massachusetts v. Mosbacher,* which asked the Federal District Court to order the apportionment to be calculated by the Webster method. If Massachusetts had won this suit, a seat would have been transferred to it from Oklahoma (see Exercise 42 on page 617). The apportionments of Montana and Washington would not have changed.

In *U.S. Department of Commerce v. Montana,* the Supreme Court unanimously rejected Montana's claim. The opinion of the court, written by Justice John Paul Stevens, pointed out that intrastate districts, which were the subject of the *Baker* and *Wesberry* cases, could be equalized in population by drawing district boundaries correctly. Because congressional districts can't cross state lines, some inequity is inevitable in congressional apportionment. The opinion conceded that there were alternative apportionment methods but concluded that the choice of apportionment method must be left to Congress.

Representative Share DEFINITION

Let a be the apportionment given to a state whose population is p. The quotient $a \div p$ is called the **representative share** for that state. It represents the share of a congressional seat given to each citizen of the state.

Ideally, every state would have the same representative share. This is impossible, just as we cannot expect each state to have the same district population. We can measure how close a given apportionment is to being ideal by using district population as the standard of comparison, or we can compare representative shares. If state X had a greater representative share than state Y, we would see if the difference between their representative shares would be reduced if we transferred a seat from the more advantaged state, X, to the state Y. If it is impossible to reduce the difference in representative shares between any pair of states, then the apportionment is **equitable** by representative share.

One might ask if the Dean method, which minimizes differences in district populations, minimizes differences in representative share too. The answer provided by the following theorem is no; it is the Webster method that minimizes differences in representative share.

Webster Method and Representative Share THEOREM

If a seat is transferred from one state to another in an apportionment that was calculated by the Webster method, the absolute difference between their representative shares will not decrease.

Here's the reasoning behind this theorem. Suppose that the divisor used in a Webster apportionment is d. The apportionment, a, given to a state with population p is the whole number nearest to $\frac{p}{d}$. The representative share is $\frac{a}{p}$, which is the closest multiple of $\frac{1}{p}$ to $\frac{1}{d}$. Hence, with the Webster apportionment, all states have representative shares as close as possible to $\frac{1}{d}$. A transfer of a seat from one state to another cannot move either states' representative shares closer to $\frac{1}{d}$, and thus not closer to each other.

For example, if seats in a parliament are apportioned by using a Sainte-Laguë table, then the apportionment is equitable from the point of view of representative share. This means that among all apportionments of parliament, this particular apportionment gives all voters—as nearly as possible—equal shares of a seat in the parliament.

EXAMPLE 17 Inequity in the 113th Congress?

The 2010 census reported the following apportionment populations: North Carolina, 9,565,781, and Rhode Island, 1,055,247. The states were apportioned 13 seats and 2 seats in Congress, respectively. Would it have been fairer to give North Carolina 14 of the 15 seats between them? We will calculate the district populations and representative shares for both apportionments and make a comparison.

With 13 seats to North Carolina, the district populations are 9,565,781 ÷ 13 = 735,829 for North Carolina and 1,055,247 ÷ 2 = 527,634 for Rhode Island. The representative share for the states are 13 ÷ 9.565781 million = 1.359 seats per million for North Carolina and 2 ÷ 1.055247 million = 1.895 seats per million for Rhode Island. With a smaller district population and a larger representative share, Rhode Island has the advantage with this apportionment. The bottom line: Rhode Island's district populations are 208,195 smaller than North Carolina's, and Rhode Island's representative share is 0.536 seats per million larger than North Carolina's.

Table 14.17 summarizes what we have calculated so far and compares the results with those that would be obtained if North Carolina had been apportioned 14 seats to Rhode Island's 1. With the 14-1 apportionment, North Carolina would have the advantage. Its district size would have been 371,977 less than Rhode Island's. This would be a greater discrepancy than exists with the 13-2 apportionment. On the other hand, North Carolina's representative share would have been 0.516 seats per million greater than Rhode Island's—a *smaller* discrepancy than exists with the 13-2 apportionment!

TABLE 14.17 Was the 2012 Apportionment Fair to North Carolina? (The units for representative share are seats per million population. The lesser differences are in red in the bottom lines.)

	Dist Pop		***Rep Share***	
State	**13-2**	**14-1**	**13-2**	**14-1**
NC	735,829	683,270	1.35901	1.46355
RI	527,634	1,055,247	1.89529	0.94765
Difference	208,195	371,977	0.53628	0.51590

Edward V. Huntington, a mathematics professor at Harvard University, pointed out that if percentage differences are compared instead of absolute differences, then district population and representative share would give identical comparisons of apportionments[4]—and he suggested a compromise.

Absolute and Percentage Difference DEFINITION

Given two positive numbers A and B, with $A > B$, the **absolute difference** is equal to $A - B$ and the **percentage difference** is equal to the quotient $\frac{A-B}{B} \times 100\%$. In computing the percentage difference between two numbers, the lesser of the two numbers goes in the denominator.

Self Check 18

The world record for the men's 100-meter race, 9.6 seconds, was set in 2009 by Usain Bolt. A century earlier, the record was 10.6 seconds, held by Knut Lindberg.

- Find the percentage difference between these records.
- Find the average speed, in meters per second, for both runners.
- Find the percentage difference in the speeds of the runners.

EXAMPLE 18 Percentage Inequity in the 113th Congress

Let's refer to Table 14.17 again and recalculate the "bottom line" as percentage differences. For example, for the percentage difference in district population with the "13-2" apportionment, we would divide the absolute difference, 208,195, by the *lesser* district

[4] You are asked to verify this statement in Exercise 44 on page 617.

population, which is Rhode Island's, 527,634. The result is 39.46%. You can calculate the remaining entries and obtain the following.

	Dist Pop		Rep Share	
	13-2	14-1	13-2	14-1
Percentage difference	39.46%	54.44%	39.46%	54.44%

You can see that the percentage differences in district population and representative share are the same, as Professor Huntington said that they would be, and that in terms of these, the "13-2" apportionment, where North Carolina gets 13 seats and Rhode Island gets 2, is preferred.

To optimize apportionment by the percentage difference criterion for equity, Professor Huntington and Joseph Hill, a statistician from the Census Bureau, designed a new divisor method. It has been used to apportion seats in the U.S. House of Representatives after each decennial census since 1940.

The Hill–Huntington Method

Like the Jefferson, Webster, and Dean methods, the **Hill–Huntington method** calculates the apportionment by rounding apportionment quotients. The only difference between the four divisor methods is in the rounding procedure.

The Hill–Huntington rounding procedure is related to the **geometric mean.**

Geometric Mean DEFINITION

The **geometric mean** of two positive numbers A and B is equal to the square root of their product, $\sqrt{A \times B}$.

Consider the rectangle $\mathcal{R}$, displayed in Figure 14.1. The area of $\mathcal{R}$ is the product of the lengths A and B, or $A \times B$. The geometric mean of A and B is equal to the length E of the edge of a square S with the same area as $\mathcal{R}$. For example, suppose $A = 4$ and $B = 9$. The area of $\mathcal{R}$ equals $4 \times 9 = 36$. The square with the same area as $\mathcal{R}$ would have edge $E = 6$, since $6^2 = 36$. In general, $E = \sqrt{A \times B}$.

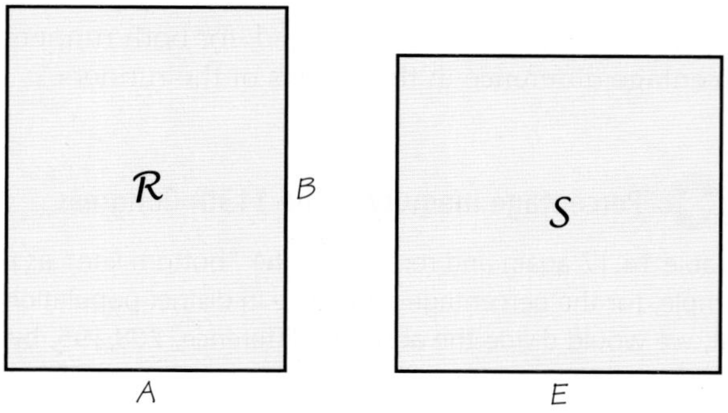

Figure 14.1 The edge of the square S is the geometric mean of the edges of the rectangle $\mathcal{R}$, because the two figures have the same area.

Self Check 19

Find the geometric mean of 5 and 20.

Hill–Huntington Rounding

DEFINITION

Given a number q that is not negative, let q^* be the geometric mean of $\lfloor q \rfloor$ and $\lceil q \rceil$; that is,

$$q^* = \sqrt{\lfloor q \rfloor \times \lceil q \rceil}$$

This q^* is called the Hill–Huntington rounding point for q. The **Hill–Huntington rounding** of a number q is equal to $\lfloor q \rfloor$ if $q < q^*$ but is equal to $\lceil q \rceil$ if $q \geq q^*$.

EXAMPLE 19 Hill–Huntington Rounding

Suppose that $q = 7.485$. Jefferson and Webster would round 7.485 down to 7. Because $\lfloor q \rfloor = 7$ and $\lceil q \rceil = 8$, the rounding point is $q^* = \sqrt{56} = 7.483$, approximately. Since $q > q^*$, Hill–Huntington rounds q up to get 8.

The procedure for calculating the Hill–Huntington apportionment is similar to the Webster procedure.

Hill–Huntington Apportionment

PROCEDURE

1. Make a table showing the quota for each state and the Hill-Huntington rounding point for each quota.
2. Calculate the Hill–Huntington rounding of each quota.
3. Add the rounded quotas from Step 2. If their sum is equal to the house size, they are the apportionment.
4. If the sum of the rounded quotas is not equal to the house size, then compute apportionment quotients, using trial divisors that are *greater* than the standard divisor if the sum of the Hill–Huntington rounded quotas is *greater* than the house size, and *smaller* than the standard divisor if the sum is *smaller* than the house size.
5. Round the apportionment quotients from Step 4 and calculate the sum. If the sum is equal to the house size, the job is finished; otherwise, repeat Step 4.

If a state's (possibly adjusted) quota q is less than 1, then $\lfloor q \rfloor = 0$. Hence $q^* = \sqrt{0 \times 1} = 0$. It follows that the Hill–Huntington rounding of any quota less than 1 is equal to 1. This proves the following theorem, which distinguishes the Hill–Huntington method from the Hamilton, Jefferson, and Webster methods. (Each of these can be modified to guarantee that each state receives at least one seat.)

No Zero Apportionments THEOREM

The Hill–Huntington method is incapable of apportioning zero seats to any state.

EXAMPLE 20 Rounding Summands

Let's use the Hill–Huntington method to round each summand in

$$0.10 + 1.43 + 2.47 = 4.00$$

Each summand is the quota, for the standard divisor is 1. The rounding points are as follows:

For numbers between	0 and 1	1 and 2	2 and 3
Approximate Hill–Huntington rounding point	0	$\sqrt{2} = 1.41$	$\sqrt{6} = 2.45$

Each summand is greater than its corresponding rounding point, so all are rounded up, resulting in a *too large* sum: $1 + 2 + 3 = 6$. We'll need to choose a divisor *larger* than 1; let's try 2. The apportionment quotients are

$$0.10 \div 2 = 0.05, \quad 1.43 \div 2 = 0.715, \quad 2.47 \div 2 = 1.235$$

The first two apportionment quotients are between 0 and 1, and so they are rounded to 1; the third is below the rounding point between 1 and 2, so it too rounds to 1. The sum, 3, is *too small.*

Let's try again with a *smaller* divisor, $d = 1.5$. The new apportionment quotients are

$$0.10 \div 1.5 = 0.067, \quad 1.43 \div 1.5 = 0.953, \quad 2.47 \div 1.5 = 1.647$$

The first two apportionment quotients are still between 0 and 1, so they round to 1. The third is greater than the rounding point between 1 and 2, so it rounds to 2. Therefore the Hill–Huntington rounding of this sum is

$$1 + 1 + 2 = 4$$

Self Check 20

Use the Hill–Huntington method to round the percentages:

$$98.1\% + 1.8\% + 0.1\% = 100.0\%$$

(You may use 1.007 as the divisor.) Does the apportionment satisfy the quota condition?

An apportionment is equitable by percentage differences (in either representative share or district population) if it is impossible to reduce the percentage difference in representative share (or district population) between any pair of states by taking a seat from one and giving it to the other. The advantage of the Hill–Huntington method is that it provides a way to determine the apportionment that is equitable by percentage differences.

Equity and the Hill–Huntington Method THEOREM

An apportionment is equitable by percentage differences if and only if it is the same as the apportionment that is produced by the Hill–Huntington method.

EXAMPLE 21 Percent Effort

Faculty members at a certain university must state the percentage of their time spent in several activities. Professor Worktorule has requisitioned five stopwatches to keep track of her activities. Table 14.18 shows, in its left columns, what she recorded over the course of one week.

TABLE 14.18 Professor Worktorule's Effort Report by the Hill–Huntington Method

Effort Category	Effort (in minutes)	Quota	Rounding Point	Tentative Apportionment
Instruction	300	8.33%	8.485	8%
Lecture prep	705	19.58%	19.494	20%
Indep. Study	31	0.86%	0.000	1%
Research	2475	68.75%	68.498	69%
Committees	89	2.47%	2.449	3%
Totals	3600	100%	—	101%

The professor is too busy to convert the data into percentages—which the university requires in whole numbers with sum 100%—so we'll do it, using the Hill–Huntington method. (She requested this method because she wanted the result to display a nonzero percentage for each of her activities.) As with any percentage apportionment problem, the house size is 100, so the standard divisor—one percentage unit—is the population, 3600, divided by 100, or 36 minutes. Table 14.18 shows in its right columns, the quotas, the rounding points, and the tentative apportionment, obtained by rounding the quotas up or down, depending on whether the quota is above the rounding point or not.

Because too many "seats" were awarded, we'll try divisors *larger* than the standard divisor, 36. The results are shown in Table 14.19. The first divisor we tried, 36.5, produced 98 "seats" too few. The left column under that divisor shows the corresponding apportionment quotients (AQ), obtained by dividing the minutes devoted to each activity by that divisor. The middle column shows the Hill–Huntington rounding points (RP) for the apportionment quotients, and the right column displays the Hill–Huntington tentative apportionment (TA). To increase the number of "seats" apportioned, the next trial divisor must be closer to 36 (but more than 36). We tried the divisor 36.2 and found that 99 "seats" were apportioned. Our third trial divisor was 36.1 (not shown in the table), which apportioned 101 "seats"—too many. The divisor we needed was therefore between 36.1 and 36.2. We set the divisor equal to 36.15, and the table shows that exactly 100 "seats" were apportioned. The right column of Table 14.19 displays the percentages that Professor Worktorule should put into her effort report.

TABLE 14.19 Trial Divisors by the Hill–Huntington Method

		Divisors								
		36.5			**36.2**			**36.15**		
Cat.	Pop	AQ	RP	TA	AQ	RP	TA	AQ	RP	TA
Inst	300	8.219	8.485	8	8.287	8.485	8	8.299	8.485	8
Prep	705	19.315	19.494	19	19.475	19.494	19	19.502	19.494	20
Ind St	31	0.849	0.000	1	0.856	0.000	1	0.858	0.000	1
Research	2475	67.808	67.498	68	68.370	68.498	68	68.465	68.498	68
Com	89	2.438	2.449	2	2.459	2.449	3	2.462	2.449	3
Totals	3600			98			99			100

EXAMPLE 22 The 435th Seat

The Webster and Hill–Huntington methods give almost the same apportionments, based on the 2010 census. The only difference is that the Webster method apportions the last seat to North Carolina, while the Hill–Huntington method gives it to Rhode Island.

The standard divisor s is the apportionment population of the United States, 309,183,463, divided by the number of seats, 435. Thus, $s = 710{,}767$. The quotas, obtained by dividing the states' populations by s, are $9{,}565{,}781 \div s = 13.458$ for North Carolina and $1{,}055{,}247 \div s = 1.485$ for Rhode Island. With the Webster method, both quotas would be rounded down and only 434 seats would be apportioned. If we reduce the divisor to 708,000, North Carolina's quotient becomes 13.51, which will be rounded up, and Rhode Island's quotient is 1.49, which will be rounded down. North Carolina will get 14 seats, and Rhode Island will get 1.

With the Hill–Huntington method, the rounding point for numbers between 13 and 14 is $q^* = \sqrt{13 \times 14} = 13.49$, which is greater than North Carolina's quota. The rounding point for numbers between 1 and 2 is $q^* = \sqrt{2} = 1.41$, which is less than Rhode Island's quota. Therefore, North Carolina's quota is rounded down to get its apportionment, 13, and Rhode Island's quota is rounded up to get its apportionment, 2, with the Hill–Huntington method.

The 435th seat was in play between Michigan and Arkansas as a result of the 1940 census, and the resulting dispute led to the permanent adoption of the Hill–Huntington method for the apportionment of seats in the House of Representatives (see Spotlight 14.4).

We have seen that the Jefferson method is biased in favor of populous states and that the Webster method is not biased in regard to state population size. It's natural to ask if the Hill–Huntington method exhibits any bias with respect to state population.

A divisor method will show bias in favor of large states when the quotas are adjusted by using a divisor that is smaller than the standard divisor. If the quotas must be adjusted downward—that is, a divisor larger than the standard divisor is used—small states are favored. Because the rounding point for the Webster method is halfway between whole numbers, it is just as likely for the divisor to be smaller than the standard divisor as it is for it to be larger.

For any positive number q, the rounding point used by the Hill–Huntington method is closer to $\lfloor q \rfloor$ than to $\lceil q \rceil$ (see Exercise 40 on page 617). This means

Mathematics and Politics: A Strange Mixture

SPOTLIGHT 14.4

Photo taken in 1961 in his study in Ithaca, NY; first appeared in the February 1961 issue of the American Statistical Association's Journal *The American Statistician*. Photographer unknown.

Walter F. Willcox

Edward V. Huntington

The first American to consider apportionment from a theoretical point of view was Walter Willcox (1861–1964), who strongly advocated the Webster method and had computed the apportionment of the 78th Congress in 1902. His arguments convinced Congress to use the Webster method again in 1912. In 1911, Joseph Hill, a statistician at the Census Bureau, proposed the Hill–Huntington method—with the endorsement of Edward V. Huntington, a mathematics professor at Harvard.

In 1920, the two methods were in competition. There were significant differences in the apportionments determined by the two methods, and the result was Washington gridlock: No apportionment bill passed during the decade, and the 1912 apportionment was retained throughout the 1920s. In preparation for the 1930 census results, the National Academy of Sciences formed a committee of distinguished mathematicians to study apportionment. In 1929, the committee endorsed the Hill–Huntington method.

The 1930 census was remarkable in that the apportionments calculated by the Webster method were the same as the Hill–Huntington apportionments. Therefore, the House of Representatives was reapportioned, but the method used could be claimed to be either one of the competing methods. The coincidence was almost repeated in the 1940 census, but there was one difference: The Hill–Huntington method gave the last seat to Arkansas, while Webster's method gave it to Michigan. At the time, Michigan was a predominantly Republican state and Arkansas was in the Democratic column. The vote on the apportionment bill split strictly along party lines, with Democrats supporting the Hill–Huntington method and Republicans voting for the Webster method. Because the Democrats had the majority, the Hill–Huntington method became law.

that a random number q is more likely to be above the rounding point, and thus rounded up to $\lceil q \rceil$, than it is to be less, and thus rounded down to $\lfloor q \rfloor$. The difference between the Webster and Hill–Huntington ways of rounding is not significant for relatively large numbers. For example, the Hill–Huntington rounding point between 50 and 51 is 50.498. Therefore, a number q between 50 and 51 will be rounded up to 51 by Hill–Huntington if it is larger than 50.498. The Webster method would round q to 51 if $q \geq 50.500$.

The differences are more significant when rounding smaller numbers. Hill–Huntington rounds all numbers between 0 and 1 up to 1; Webster rounds only the numbers between 0.500 and 1 up to 1. When the Hill–Huntington method is used for apportionment, the sum of the tentative apportionments is more likely to exceed the house size than it is to be less, especially if there are many states with small populations. Therefore, the Hill–Huntington method is likely to use a divisor larger than the standard divisor. This favors the less populous states.

In conclusion, the Webster method is the only divisor method that is unbiased regarding population size, and it minimizes differences between representative shares. Although the Webster method is capable of violating the quota condition, it is the divisor method least likely to do so.

For apportionment of seats in the U.S. House of Representatives by the Webster method, a slight modification is needed, because no state can receive a zero apportionment. The rounding point for quotas less than 1 is set to 0, rather than 0.5.

There are situations where other apportionment methods could be considered. See Exercise 46 (page 617) to explore ways to make teaching assignments.

Review Vocabulary

$\lfloor q \rfloor$ The result of rounding a number q down; for example, $\lfloor \pi \rfloor = 3$. (p. 578)

$\lceil q \rceil$ The result of rounding a number q up to the next whole number; for example, $\lceil \pi \rceil = 4$. (p. 578)

Absolute difference The result of subtracting a smaller number from a larger number. (p. 601)

Alabama paradox A state loses a representative solely because the size of the House is increased. This paradox is possible with the Hamilton method but not with divisor methods. (p. 582)

Apportionment method A systematic way of computing solutions of apportionment problems. (p. 573)

Apportionment problem Given a list of fractions whose sum is a whole number, to round each fraction in a way that preserves the original sum. (p. 573)

Apportionment quotient The result of dividing a state's quota by the divisor when apportioning by a divisor method. The apportionment quotient is rounded to obtain the state's apportionment, using the particular rounding rule associated with the chosen divisor method. (pp. 585, 586)

Dean method A divisor method of apportionment that minimizes differences in the district populations of the states. (p. 598)

d'Hondt method An apportionment method that is equivalent to the Jefferson method. It is typically used to apportion seats in parliaments to political parties in proportion to their votes. (p. 592)

District population A state's population divided by its apportionment. (p. 575)

Divisor method One of many apportionment methods in which the apportionments are determined by dividing the population of each state by a common divisor to obtain apportionment quotients. The apportionments are calculated by rounding the apportionment quotients. Divisor methods differ in their rounding rules. The Jefferson, Webster, Dean, and Hill–Huntington methods are divisor methods. (p. 586)

Equitable apportionment An apportionment is equitable—by a specified measure of inequity—if its inequity cannot be reduced by transferring a seat from one state to another state. The measures of inequity that we have considered are absolute differences in representative share (p. 599), absolute differences in district population (p. 575), and percentage differences (p. 601) in either representative share or district population. (p. 600)

Geometric mean For numbers A and B, neither of which is negative, the geometric mean is defined to be $\sqrt{A \times B}$. (p. 602)

Hamilton method An apportionment method that assigns to each state either its lower quota or its upper quota. The states that receive their upper quotas are those whose quotas have the largest fractional parts. (p. 578)

Hare method The Hamilton method applied to apportion seats to political parties in a parliament. (p. 584)

Hill–Huntington method A divisor method that minimizes percentage differences in both representative shares and district populations. This method has been used to apportion seats in the U.S. House of Representatives since 1941. (p. 602)

Hill–Huntington rounding A number q is rounded down if it is less than the geometric mean of $\lfloor q \rfloor$ and $\lceil q \rceil$; otherwise it is rounded up. (p. 603)

Jefferson method A divisor method based on rounding all fractions down. Thus, if U is the apportionment quotient of state X, the state's apportionment is $\lfloor U \rfloor$. (pp. 585, 586)

Lower quota The whole number part $\lfloor q \rfloor$ of a state's quota q. (p. 578)

New states paradox After an apportionment, a new state joins the Union. The house size is increased by the number of seats that the new state would be apportioned by the method in use. Yet when the house is reapportioned, two or more states that had participated in the original apportionment find that their apportionments have changed. (p. 585)

Percentage difference (between two positive numbers) Subtract the smaller number from the larger and express the result as a percentage of the smaller number. Thus, the percentage difference between 120 and 100 is 20%. (p. 601)

Population paradox A situation in which the house size is unchanged and the apportionment of one state, A, decreases although its population has increased, while another state, B, loses population and gains a seat. (p. 583)

Quota The quota is the quotient $V \div s$ of a state's population V divided by the standard divisor s. The quota is the number of seats a state would receive if fractional seats could be awarded. (p. 576)

Quota condition A requirement that an apportionment method should always assign to each state either its lower quota or its upper quota in every situation. The Hamilton method satisfies this condition, but no divisor method does. (p. 590)

Representative share The state's apportionment divided by its population. It is intended to represent the amount of influence a citizen of that state would have on his or her representative. (p. 599)

Sainte-Laguë method A way of calculating an apportionment that leads to the same result as the Webster method. It is typically used when apportioning seats in parliament to political parties. (p. 596)

Standard divisor The ratio $p \div h$ of the total population p to the house size h. In a congressional apportionment problem, the standard divisor represents the average district population. (p. 575)

Tentative apportionment For the Hamilton method, a state's lower quota (p. 578); for a divisor method, the appropriately rounded apportionment quotient. (p. 579)

Upper quota The result of rounding a state's quota up to a whole number $\lceil q \rceil$. (p. 578)

Webster method A divisor method of apportionment that is based on rounding fractions to the nearest whole number. The Webster method minimizes the absolute differences of representative share between states. (p. 594)

Self Check Answers

1. Convert all times to minutes, and note that 24 hours is 1440 minutes. Thus, the administrative assistant worked $\frac{600}{1440} = 41.67\%$ of the day. He commuted $\frac{140}{1440} = 9.72\%$ of the day, and spent $\frac{700}{1440} = 48.61\%$ of the day on personal activities. Rounding these percentages to their nearest whole numbers we get

$$42\% + 10\% + 49\% = 101\%$$

To make the sum 100%, one of the percentages must be rounded down. You have no guidance yet on how to do this, so there is no wrong way to choose which one.

2. North Carolina was apportioned 12 seats, so its district population is its population of 353,523 divided by 12, or 29,460, which is below the limit. New York's population, 331,589, divided by its apportionment of 11 seats, is equal to 30,144, so New York's district population does satisfy the requirement.

3. The states are the families (Browns, Joneses, and Robinsons); the house size is the number of candies to be apportioned (123); the populations are Browns (5), Joneses (3), and Robinsons (4), for a total of 12. The standard divisor is the total population divided by the house size, $\frac{12}{123} = 0.09756$.

4. The quotas are: Browns, $\frac{5}{0.09756} = 51.25$; Joneses, $\frac{3}{0.09756} = 30.75$; Robinsons, $\frac{4}{0.09756} = 41.00$.

5. Rhode Island's quota was $\frac{1{,}055{,}247}{710{,}767} = 1.48466$, and North Carolina's quota was $\frac{9{,}565{,}781}{710{,}767} = 13.45839$. The district populations are $1{,}055{,}247 \div 2 = 527{,}624$ for Rhode Island, and $9{,}565{,}781 \div 13 = 735{,}829$ for North Carolina.

6. Add the lower quotas: 41% + 9% + 48% = 98%. The fractional parts are 0.67%, 0.72%, and 0.61%. Round the first two quotas—which have the greatest fractional parts—up to get the apportionment 42% + 10% + 48% = 100%.

7. The district populations for the subjects are the average number of students in a class: 26 for geometry; 16.5 for precalculus; and 15 for calculus.

8. Yes. The house size decreased, the populations did not change, and one subject, Calculus III, received an increased apportionment.

9. The following table is the apportionment. In any apportionment in which no state receives more than one representative per 30,000 population, the maximum apportionment permitted is

$$\left\lfloor \frac{\text{state's population}}{30{,}000} \right\rfloor$$

These are the apportionments shown in the table. Thus, the maximum possible house size is 112.

State	Population	Pop/ 30,000	Apportionment
Virginia	630,560	21.02	21
Massachusetts	475,327	15.84	15
Pennsylvania	432,879	14.43	14
North Carolina	353,523	11.78	11
New York	331,589	11.05	11
Maryland	278,514	9.28	9
Connecticut	236,841	7.89	7
South Carolina	206,236	6.87	6
New Jersey	179,570	5.99	5
New Hampshire	141,822	4.73	4
Vermont	85,533	2.85	2
Georgia	70,835	2.36	2
Kentucky	68,705	2.29	2
Rhode Island	68,446	2.28	2
Delaware	55,540	1.85	1
Totals	3,615,920	120.53	112

10. Table 14.11 shows that with the divisor 0.225, the apportionment quotient for Winnie was exactly 80. The apportionment quotients for Louise and Tim were 17.78 and 4.44, respectively. The divisor 0.227 reduced the apportionment quotient for Winnie to 79.30, while Louise and Tim had apportionment quotients of 17.62 and 4.41, respectively. Thus, the divisor 0.226 would produce apportionment quotients for Winnie between 79.30 and 80, for Louise between 17.62 and 17.78, and for Tim between 4.41 and 4.44. These quotas, rounded down, would be 79, 17, and 4 diamonds, respectively.

11. The maximum number of sections for a course would be the population for that course, divided by 24, and rounded down. This would be found by the Jefferson method, using the divisor 24. Referring to Table 14.13, where that divisor was employed, we see that 28 TAs are needed.

12. A state has population p. Let s be the standard divisor and d be the Jefferson divisor. The state's apportionment quotient is $\frac{p}{d}$ and its quota is $\frac{p}{s}$. Since $d \le s$,

$$\frac{p}{s} \le \frac{p}{d}$$

It follows that rounding the apportionment quotient down results in a whole number not less than the lower quota, which we obtain by rounding the quota down.

13. The sum of the lower quotas is 98%. Divide each quota by 0.98, as suggested, to get apportionment quotients 99.18%, 1.94%, 0.92%. Rounding down we have

$$99\% + 1\% + 0\% = 100\%$$

This violates the quota condition, because the apportionment to the first percentage, 97.2%, is greater than its upper quota, 98%.

14. Compare the current priority numbers: Whigs, 394,670; Tories, 385,572; Liberals, 386,423; Centrists, 201,203. The Whigs have the highest priority and get seat #31.

15. With the Webster method, start by rounding each quota to the nearest whole number. This yields 97% + 2% + 1% = 100%, which is the Webster apportionment. There is no violation of the quota condition.

16.

Geometry	Precalculus	Calculus
#1 ~~52~~	#2 ~~33~~	#4 ~~15~~
#3 ~~17.33~~	#5 ~~11~~	5
10.4	6.6	3

Geometry and precalculus get two sections, and calculus gets one section.

17. With the current apportionment, the difference in district populations is in California's favor by 289,851. If a seat had been transferred from California to Montana, California's district population would increase to 37,341,989 ÷ 52 = 718,115. Montana's district population would be 994,416 ÷ 2 = 497,208. The difference, 220,907, is less than the difference before the transfer. Therefore, this apportionment is not equitable by absolute difference in district population. (However, the percentage difference after the transfer, $\frac{220{,}907}{497{,}208} = 44\%$, is greater than the percentage difference before the transfer, $\frac{289{,}851}{718{,}815} = 40.3\%$. This was to be expected, because the Hill–Huntington method minimizes percentage differences.)

18.

- The percentage difference in the records is $\frac{10.6 - 9.6}{9.6} =$ 10.42%.
- Bolt's average speed was $\frac{100 \text{ meters}}{9.6 \text{ seconds}} = 10.417$ meters per second. Lindberg's average speed was $\frac{100 \text{ meters}}{10.6 \text{ seconds}} =$ 9.434 meters per second.
- The percentage difference in speeds is $\frac{10.417 - 9.434}{9.434} =$ 10.42%.

The percentage difference in speeds is *exactly* equal to the percentage difference in the times taken to complete the run.

19. $\sqrt{5 \times 20} = 10$

20. By Hill–Huntington, all numbers between 0 and 1 are rounded to 1, all numbers between 1.414 and 2 are rounded to 2, and 98.1 would be rounded to 98. Thus, the rounded percentages are 98% + 2% + 1% = 101%. We have to use a divisor greater than the standard divisor (which is 1) to reduce the sum. As suggested, we'll take 1.007 as the divisor. The apportionment quotients are 98.1% ÷ 1.007 = 97.42%, 1.8% ÷ 1.007 = 1.79%, and 0.1% ÷ 1.007 = 0.099%. The rounding point for numbers between 97 and 98 is

$$\sqrt{97 \times 98} = 97.4987$$

Thus, 97.42% is rounded down to 97%, and the other two percentages are rounded up. The Hill–Huntington rounding,

$$97\% + 2\% + 1\% = 100\%$$

violates the quota condition because the first percentage is apportioned less than its lower quota.

Skills Check

1. Amy, Bill, and Connie have bought lottery tickets that won 13 identical rubies. The number of tickets each bought was: Amy, 13; Bill, 5; and Connie, 19. They wish to divide the rubies fairly. What was Connie's quota?

(a) Less than 7 rubies

(b) Between 7 and 8 rubies

(c) More than 8 rubies

2. Two calculus teachers can teach a total of 8 classes. Enrollments are as follows: Calculus I, 200; Calculus II, 100; Calculus III, 80. In this apportionment problem, the population is ____________, the standard divisor is ____________, and the quotas are ____________ for Calculus I, ____________ for Calculus II, and ____________ for Calculus III.

3. *A*, *B*, and *C* are arguing about fractions of a cent. On a project, they worked exactly 33, 34, and 35 minutes, respectively, and were paid $100. Use the Hamilton method to see who gets his upper quota (in cents!).

(a) *A*

(b) *B*

(c) *C*

4. Round each number in the sum 14.48 + 12.40 + 17.49 + 16.33 + 19.30 = 80 to a whole number.

____ + ____ + ____ + ____ + ____ = 80

(What is it about these numbers that makes this complicated?)

5. The population paradox occurs when

(a) a state's apportionment decreases because the house size increased.

(b) a state's apportionment decreases and its apportionment increases, while another state's apportionment decreases even though its population has increased.

(c) the Jefferson method is used.

6. The Alabama paradox occurred when it was noticed that Alabama would lose a seat, in apportionment by the Hamilton method, if the house size was changed from 299 to ____________.

7. If the Jefferson method is used to do the rounding in Skills Check Question 4, it will be necessary to find a divisor that is

(a) less than 1.

(b) at least 1, but less than 2.

(c) at least 2.

8. Using the Webster method, round the numbers in Skills Check 4.

____ + ____ + ____ + ____ + ____ = 80

9. When rounding the numbers in the sum 20.45 + 30.30 + 49.25 = 100 by the Jefferson method, which number gets its upper quota?

(a) 20.45

(b) 30.30

(c) 49.25

10. Seats in a parliament are apportioned by the Hare (Hamilton) method. Jane had planned to vote for party *B* but changed her mind and voted for party *A*. If her vote switch caused party *A* to lose a seat, this would be an instance of the ____________ paradox.

11. The parliament in Skills Check Question 10 will be apportioned by the d'Hondt (Jefferson) method. Now is it possible for Jane's vote switch to cause party *A* to lose a seat?

(a) Yes, because the Jefferson method does not satisfy the quota condition.

(b) No, because the Jefferson method is not susceptible to the Alabama paradox.

(c) No, because the Jefferson method is not susceptible to the population paradox.

12. Use the Jefferson method to apportion the sum 0.8 + 0.9 + 98.3 = 100 as a sum of whole numbers.

_____ + _____ + _____ = 100

13. The Jefferson method frequently

(a) gives the smallest state less than its lower quota.

(b) gives the largest state more than its upper quota.

(c) gives a state a lesser apportionment if the house size increases.

14. Use the Webster method to apportion the sum 0.8 + 0.9 + 98.3 = 100 as a sum of whole numbers.

_____ + _____ + _____ = 100

15. We want to apportion the sum 1.6 + 3.7 + 5.5 + 89.2 = 100 as a sum of whole numbers. Which method will violate the quota condition?

(a) Hamilton

(b) Jefferson

(c) Webster

16. The sum 1.6 + 2.6 + 3.6 + 4.6 + 5.6 = 18 has to be rounded as a sum of whole numbers. If the ____________ method is used, there will be a tie.

17. When rounding the numbers in the sum 20.45 + 30.30 + 49.25 = 100 by the Webster method, which number gets rounded up?

(a) 20.45

(b) 30.30

(c) 49.25

18. States *A* and *B* have populations of 1 million and 2 million, respectively. If they are apportioned 2 and 3 seats, respectively, then the absolute difference in representative share is ____________ per million.

19. If the apportionment in Skills Check 18 gave state *A* 1 seat and gave state *B* 4 seats, then

(a) the absolute difference in representative share would increase.

(b) the absolute difference in representative share would decrease.

(c) the absolute difference in representative share would be unchanged.

20. If the criterion is absolute difference in district population, the equitable apportionment of 5 seats to states *A* and *B* in Skills Check 18 is ____________ for *A* and ____________ for *B*.

21. If the initial calculations leading to the Hill–Huntington apportionment result in a sum that is too large, what happens next?

(a) Apportionment quotients must be calculated, using a divisor slightly larger than the standard divisor.

(b) Apportionment quotients must be calculated, using a divisor slightly less than the standard divisor.

(c) The largest apportionment is reduced.

(d) A different method must be used.

22. The ____________ method has been used since 1941 to apportion seats in the U.S. House of Representatives.

23. Which divisor method never apportions to a state fewer seats than its lower quota?

(a) Hill–Huntington

(b) Webster

(c) Jefferson

24. A school principal is apportioning sections of the school's mathematics classes. She wants to set a minimum section size and to adjust it so that a total of 32 sections are open. She should use the ____________ method.

25. The U.S. Constitution says that each state must get at least one representative in the House. Which apportionment method or methods automatically satisfy this requirement?

(a) Hamilton
(b) Jefferson
(c) Webster
(d) Hill–Huntington

26. Five parties, *A*, *B*, *C*, *D*, and *E*, participate in a parliamentary election. The parliament has 10 seats. The number of votes received (in thousands) were *A*, 120; *B*, 78; *C*, 50; *D*, 35; and *E*, 20. Here is a d'Hondt table.

A	B	C	D	E
120	78	50	35	20
60	39	25	17.5	10
40	26	16.67	11.67	6.67
30	19.5	12.5	8.75	5

When the seats are apportioned, the seventh seat goes to party ______________, the eighth seat goes to party ______________, the ninth goes to party ______________, and the tenth goes to party ______________.

27. The Hill–Huntington method minimizes percentage differences in

(a) district population.
(b) representative share.
(c) Both (a) and (b) are correct.

28. The divisor method that shows the least bias in favor of either large states or small states is the ________________ method.

29. A parliament has 466 seats and there are 13 parties with lists on the ballot. It is proposed that there should be a minimum number *N* of votes to qualify for a seat, and that number should be chosen so that exactly 466 seats are filled. We should point out that this idea is not new; it is the method of

(a) Hare.
(b) d'Hondt.
(c) Sainte-Laguë.

30. In the election described in Skills Check 29, party *A* received exactly 12 million of the votes, and each of the other 12 parties received exactly 1 million votes. Although Party *A* received exactly half of the votes, the number of seats that it will receive is ______________ more than half of the seats, and therefore it can form a government without a coalition partner. This will violate the ______________ condition.

Chapter 14 Exercises

Challenge Discussion

14.1 The Apportionment Problem

1. Jane has decided to track her daily expenses and finds them to be as listed in the table. Express these as percentages. If rounded to whole numbers, do the percentages add up to 100 percent?

Jane's Expenses	
Rent	$31
Food	16
Transportation	7
Gym	12
Miscellaneous	5

2. A mathematics department uses 20 teaching assistants to aid in its four-semester calculus course. The number of teaching assistants assigned to each level of the course depends on enrollment. Here are the fall enrollments:

Calculus I	500
Calculus II	100
Calculus III	350
Calculus IV	175
Total	1125

How many teaching assistants should be assigned to each level of the course?

3. Should the mathematics department in Exercise 2 revise the assignments for its TAs? Grades have been posted for the previous semester, and some students need to repeat the previous level of the course. A total of 45 students move from Calculus II to Calculus I, 41 students move from Calculus III to

Calculus II, and 12 students move from Calculus IV to Calculus III.

4. Here is a typical apportionment problem. Round the numbers in the following sum to whole numbers:

$$8.37 + 10.33 + 12.38 + 5.47 + 3.45 = 40$$

The rounded numbers must add up to 40. How would you approach this?

5. How would you round the numbers in the following sum to whole numbers? The rounded numbers must add up to 60.

$$11.63 + 9.67 + 7.62 + 14.53 + 16.55 = 60$$

14.2 The Hamilton Method

6. Use the Hamilton method to round the numbers in the following sum to whole numbers. The sum of the rounded numbers must be the same as the original sum.

$$2.64 + 1.41 + 2.01 + 0.67 + 0.62 + 0.65 = 8$$

7. Repeat Exercise 6 with the following sum:

$$0.36 + 1.59 + 0.99 + 2.33 + 2.38 + 2.35 = 10$$

8. The 37th pearl. Three friends have bought a bag guaranteed to contain 36 high-quality pearls for $14,900 at an auction. Abe contributed $5900, Beth's contribution was $7600, and Charles supplied the remaining $1400. After taking the bag to your house, they pour the 36 pearls from the bag onto the kitchen table.

(a) How many should each friend get if the Hamilton method is used to apportion the pearls according to the size of the contributions?

(b) Charles has noticed that the bag isn't empty! Another pearl comes out, so recalculate the apportionment.

(c) How do you explain the result to Charles?

9. A country has three political parties, and it allots seats in its 102-seat parliament by the Hare (Hamilton) method proportionately to the number of votes each receives. In a recent election, the Pro-UFO Party received 254,000 votes, the Anti-UFO Party got 153,000 votes, and the Who Cares Party polled 103,000 votes. Show that two of the parties are tied.

10. A small high school has one mathematics teacher who can teach a total of five sections. The subjects that she teaches, and their enrollments, are as follows: Geometry, 52; Algebra, 33; and Calculus, 12. Use the Hamilton method to apportion sections to the subjects.

11. Repeat Exercise 10 using the following enrollments: Geometry, 77; Algebra, 18; and Calculus, 20.

12. Use the Hamilton method to express the summands of the following expression as whole number percentages of the total:

$$2746 + 1725 + 1921 + 100 = 6492$$

Repeat the calculation for the following sum:

$$2814 + 1745 + 1933 + 99 = 6591$$

Do you see a paradox?

13. Abe, Beth, Charles, and David have decided to invest in rare coins. A dealer has offered to sell them a parcel containing 100 identical coins for $10,000. Each person invests all that he or she can afford, but there is not quite enough money, so Charles asks his Aunt Esther to join the group. The coins will be apportioned by the Hamilton method. Here are the amounts invested:

Investments	
Abe	$3,619
Beth	1,862
Charles	2,258
David	2,010
Esther	251
Total	$10,000

John S. Sfondilias/Shutterstock

(a) How should the coins be apportioned among the five contributors?

(b) After the coins are distributed, the dealer mentions that there will be $50 in excise tax. Everyone empties his or her wallet: Abe finds $16 more, Beth has $2, Charles has $1, and David finds $32. This adds up to $51, so

$1 is returned to Aunt Esther. The apportionment is recalculated, and one of the coins changes hands. Who has to give a coin to whom?

(c) Explain what happened.

To see how this situation works out with a different apportionment method, refer to Exercise 31 on page 616.

14. The new states paradox. The census of 1900 recorded the following apportionment populations:

Maine	694,466
New York	7,264,183
United States	74,562,608

The house size was 386. Apportionment was by the Hamilton method.

(a) Determine the quotas of New York and Maine. Given that New York was the last state to receive its upper quota, determine the numbers of seats that were apportioned to these two states.

(b) In 1907, Oklahoma became a state. Its population was stipulated to be 1,000,000 for the purpose of apportionment. Using the standard divisor that you found in part (a), determine the apportionment for Oklahoma.

(c) Add the stipulated population of Oklahoma to the 1900 population of the United States, and the number of seats that you found in part (b) to the house size before Oklahoma became a state. With these data, find out if the numbers of seats apportioned to New York and Maine changed. (New York and Maine actually retained their 1902 apportionments.)

15. A country has five political parties. Here are the numbers of votes each received in a recent election: 5,576,330; 1,387,342; 3,334,241; 7,512,860; and 310,968. Seats in its parliament are apportioned by the Hare (Hamilton) method. Calculate the apportionments for house sizes of 89, 90, and 91. Does the Alabama paradox occur?

14.3 Divisor Methods

16. Explain why the tentative Webster apportionment of a state with quota q is $\lfloor q + 0.5 \rfloor$.

17. Reapportion the classes in Exercise 11 (page 614), using the Jefferson method.

18. Reapportion the classes in Exercise 10 (page 614), using the Webster method.

19. The three friends who bought the pearls (see Exercise 8 on page 614) ask you to suggest a different apportionment method to distribute their purchase. Before answering, determine the apportionments given by the Jefferson and Webster methods for the 36- and 37-pearl house sizes, and then make your suggestion.

20. The three friends in Exercise 8 have bought a lot of 36 identical diamonds, at a total cost of $36,000; Abe's investment was $15,500, Beth's was $10,500, and Charles's was $10,000. They decided to apportion the diamonds using the Webster method, and they can't make it work out. Can you help?

21. A country has a 20-seat parliament. Seats are apportioned to parties by the d'Hondt method. The following table displays the results of a recent election. Make a d'Hondt table and determine the number of seats allocated to each party.

The Election Results	
Demopublicans	44,856
Repocrats	34,944
Greenocrats	20,004
Greenicans	19,002
Independents	9,804

22. Referring to the voting data in Exercise 21, make a Sainte-Laguë table to apportion the parliament.

Exercises 23–25 refer to the parliament in Example 9 on page 584. We will use only the data for the first election.

23. Make a Sainte-Laguë table and apportion the first 30 seats.

24. In Example 14 (page 593), we used a d'Hondt table to apportion the first 30 seats. The Centrists did not get any of the first 30 seats. Which of seats 31–100 will be the first one that the Centrists receive?

25. Which will be the first seat that the Centrists receive if the Sainte-Laguë table is used to apportion the seats?

26. Explain why the population paradox cannot occur if a divisor method of apportionment is used. (If one state receives an increased apportionment even though its population decreased, how did the divisor used in the second apportionment differ from the divisor used in the first one?)

27. Explain why the new states paradox (see page 585) cannot occur if a divisor method of apportionment is used. (Would the divisor used in the apportionment after the new state joined the union have to be different from the divisor used before?)

28. Round the following to whole percentages using the Hamilton, Jefferson, and Webster methods:

$$87.85\% + 1.26\% + 1.25\% + 1.24\% + 1.23\% + 1.22\% + 1.21\% + 1.20\% + 1.19\% + 1.18\% + 1.17\% = 100\%$$

Do any of these apportionments show a violation of the quota condition?

29. Round the following percentages to whole numbers, using the methods of Hamilton, Jefferson, and Webster.

$$92.15\% + 1.59\% + 1.58\% + 1.57\% + 1.56\% + 1.55\% = 100\%$$

Do any of these apportionments show a violation of the quota condition?

30. Use the Webster method to apportion the House of Representatives, based on the census of 1790. Choose a divisor such that each state's district population is greater than 30,000, and as many seats as possible are apportioned.

31. Recalculate the apportionment of the coins in Exercise 13 (on page 614) by the Webster method. Again, after the excise tax is paid, a coin changes hands. Who gives it to whom?

32. A country has two political parties, the Liberals and the Tories. The seats in its 99-seat parliament are apportioned to the parties according to the number of votes they receive in the election. If the Liberals receive 49% of the vote, how many seats do the Liberals get with the Hamilton (Hare) method? With the Webster (Sainte-Laguë) method? With the Jefferson (d'Hondt) method?

33. A country with a parliamentary government has two parties that capture 100% of the vote between them. Each party is awarded seats in proportion to the number of votes received.

(a) Explain why the Webster (Sainte-Laguë) and Hamilton (Hare) methods will always give the same apportionment in this two-party situation.

(b) Explain how to use the result of part (a) to show that the Alabama and population paradoxes cannot occur when the Hamilton method is used to apportion seats between two parties or states.

(c) Explain why the result of part (a) implies that the Webster method satisfies the quota condition when the seats are apportioned between two parties or states.

(d) Will the Jefferson and Hill–Huntington methods also yield the same apportionments as the Hamilton method?

14.4 Which Divisor Method Is Best?

34. A barista uses 11 grams of coffee to make an espresso and 16 grams to make a doppio. Find the percentage difference in the amounts of coffee used for the two drinks.

35. Jim is 72 inches tall and Alice is 65 inches tall. What is the percentage difference in their heights?

36. Find the Hill–Huntington rounding points for numbers between 0 and 1; between 1 and 2; between 2 and 3; and between 3 and 4.

37. A high school has one math teacher who can teach five sections. A total of 56 students have enrolled in the algebra class, 28 have signed up for geometry, and 7 students will take calculus. Use the Hill–Huntington method to decide how many sections of each course to schedule.

38. One year later, the high school described in Exercise 37 still has just one math teacher who teaches 5 sections. The enrollments are algebra, 36; geometry, 61; and calculus, 3. Apportion the classes by the Webster and Hill–Huntington methods. Which apportionment do you think the school principal would prefer?

39. In 2001, Utah sued to increase its apportionment (*Utah v. Evans*). Federal employees stationed abroad are counted in the apportionment population of the state of their residence, and Utah wanted to include in its apportionment population religious missionaries who were based in the state and serving abroad.

(a) In the apportionment based on the 2000 census, North Carolina received the last seat. Its apportionment population was 8,067,673, and it was apportioned 13 seats. Find the largest divisor that would allow North Carolina 13 seats.

(b) Utah's apportionment population (not counting missionaries) was 2,236,714, and Utah was apportioned 3 seats in the House of Representatives. Find the population that would be required for Utah to be apportioned 4 seats, with the same divisor that you found in part (a).

(c) To justify the transfer of a seat from North Carolina to Utah, by how much would Utah's population have to increase? (In the suit, Utah claimed that its residents should include 11,000 missionaries.)

40. (a) Show that for any positive numbers A and B, the geometric mean is less than the arithmetic mean,[5] except when $A = B$; in that case, the two means are equal. (*Hint:* Show that the triangle in Figure 14.2 is a right triangle.)

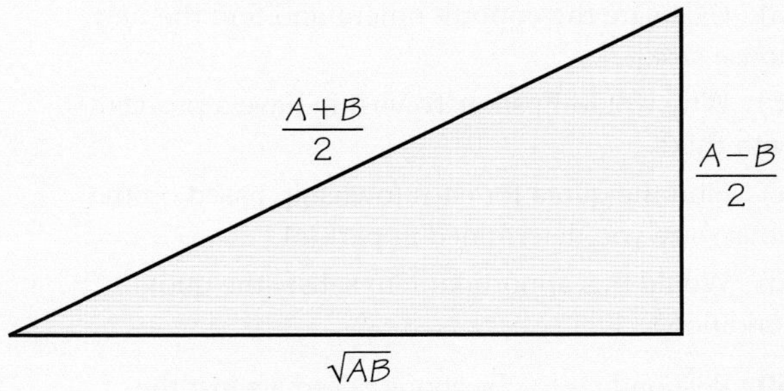

Figure 14.2 Is this a right triangle?

(b) If q is a number such that the Webster and Hill–Huntington roundings of q differ, show that q is greater than the Hill–Huntington rounding point $q^* = \sqrt{\lfloor q \rfloor \times \lceil q \rceil}$ and q is less than the Webster rounding point $\lfloor q \rfloor + \frac{1}{2}$. Conclude that, in this case, the Hill–Huntington rounding of q is equal to $\lceil q \rceil$ and the Webster rounding is equal to $\lfloor q \rfloor$.

(c) Explain why the fact established in part (b) implies that the Hill–Huntington method is more favorable to small states than the Webster method.

41. A city has three districts with populations of 100,000, 600,000, and 700,000, respectively. Its council has 20 members, and seats on the council are apportioned by the Hill–Huntington method according to the district populations. Show that there is a tie. Would a tie occur with any of the other apportionment methods that we have considered?

42. In a 1991 federal lawsuit, *Massachusetts v. Mosbacher,* Massachusetts claimed that the Hill–Huntington method of apportionment is unconstitutional because it does not reflect the "one person, one vote" principle as well as the Webster method does. Would Massachusetts have gained a seat from Oklahoma if the Webster method had been used to apportion the House of Representatives in 1991? In your calculation, use the following populations and Hill–Huntington apportionments: Massachusetts was apportioned 10 seats for a population of 6,029,051, and Oklahoma was apportioned 6 seats for a population of 3,157,604.

43. In 1822, Congressman William Lowndes of South Carolina proposed an apportionment method, which was never used. Lowndes started, as Hamilton did, by giving each state its lower quota. But where Hamilton apportions the remaining seats to the states whose quotas have the largest fractional parts—in other words, the states for which the *absolute* difference between the quota and the lower quota is greatest—Lowndes gives the extra seats to the states where the *percentage* difference is greatest, increasing the apportionments of as many states as necessary to their upper quotas to fill the House.

(a) Compared with the Hamilton method, would this method be more beneficial to states with large populations or small populations?

(b) Does the Lowndes method satisfy the quota condition?

(c) Would there be any trouble with paradoxes with the Lowndes method?

(d) Use the method to apportion the 1790 House of Representatives with a 120-seat house. The populations and quotas resulting from the 1790 census are in Table 14.5 on page 580.

44. Let the populations of states A and B be p_A and p_B, respectively. The apportionments will be a_A and a_B. Assuming that district populations for state A are larger than district populations for state B, show that the percentage difference in district populations is

$$\frac{p_A a_B - p_B a_A}{p_B a_A} \times 100\%$$

Also show that same expression is equal to the percentage difference in representative share. Hence the percentage difference in district populations is equal to the percentage difference in representative shares.

45. John Quincy Adams, the sixth president of the United States, proposed that the House of Representatives should be apportioned by a divisor method based on the rounding rule that rounds each fraction up to the next whole number.

(a) Is it likely that the initial tentative apportionment will be final?

(b) Will the divisor that produces the correct apportionment be greater than or less than the standard divisor?

(c) Does the method favor small states or large states?

(d) Can the Adams method apportion any state zero seats?

(e) The Adams method can violate the quota condition. Explain why it can never apportion to a state more than its upper quota.

46. The choice of a divisor method for apportioning classes to subjects according to

[5] The arithmetic mean of A and B is equal to $(A + B)/2$.

enrollments, as in Example 2 (page 577), depends on what the school principal considers most important.

(a) The principal wants to set a minimum class size. For example, if the minimum class size is 20, and 39 students are signed up for English III, there would be one section, because there are not enough students for two sections with enrollment of at least 20. If there were 40 students, there would be two sections. The minimum class size is adjusted so that as many sections as possible are running. What apportionment method should she use?

(b) The principal prefers to set a maximum class size. For example, if the maximum class size is 33, and 67 students are taking History I, there will be three sections because there are too many students to fit in two 33-student sections. If there were only 66 students taking History I, there would be two sections. The maximum class size is adjusted so that as many sections as possible are running. What apportionment method should she use? (*Hint:* This divisor method is not described in the text but is mentioned in one of the previous exercises.)

(c) The principal wants to cancel any class that has an enrollment of just one student. Which apportionment methods should she avoid using?

47. Let $q_1, q_2, \ldots, q_n$ be the quotas for n states in an apportionment problem, and let the apportionments assigned by some apportionment method be denoted $a_1, a_2, \ldots, a_n$. The absolute deviation for state i is defined to be $|q_i - a_i|$; it is a measure of the amount by which the state's apportionment differs from its quota. The maximum absolute deviation is the largest of these numbers. Explain why the Hamilton method always gives the least possible maximum absolute deviation.

Chapter Review

48. The Legis County Board of Supervisors has 145 seats and a total population of 115,275. The county is divided into five townships, whose populations are as follows:

Township	Population
Alpha	16,210
Beta	40,052
Gamma	8,284
Delta	48,018
Epsilon	2,711

Use the Webster method to determine the apportionment of seats on this Board, then repeat, using the Hamilton method. Are the results the same?

49. The Legis County Board of Supervisors (see Exercise 48) has too many seats for its space in the County Administration building. It has decided to use the Jefferson method of apportionment, with 3000 as divisor.

(a) Calculate the apportionment and find the new house size.

(b) Why would Epsilon Township have a problem with this?

(c) Find the quota for each township based on the house size you determined in part (a).

(d) Would this apportionment satisfy the quota condition?

50. Responding to objections raised against the apportionment scheme in Exercise 49, the Legis County Board of Supervisors has adopted the Webster method of apportionment, with 3000 as the divisor.

(a) Using the same divisor, compute the apportionment and find the house size.

(b) Find the quota for each township based on the house size you determined in part (a).

(c) Is the quota condition satisfied?

51. A county has five townships and elects a 301-seat board of supervisors, using the Webster method of apportionment. The populations of the townships are as follows:

Township	Population
A	109,050
B	55,920
C	67,770
D	61,260
E	7,500

(a) Determine the Webster apportionment.

(b) Determine the district population for each township. Call the township with the greatest district population the "disadvantaged township."

(c) To make the apportionment more equitable in terms of district population, which township should give a seat to the disadvantaged township?

52. Considering the problem of apportioning goods to people according to financial contributions, as in Exercise 8 on page 614 (allocating pearls bought at auction) or Exercise 13 on page 614 (rare coins), list the pros and cons for each of the apportionment methods listed below. Consider the appropriateness of the following standards of comparison: paradoxes, quota condition, bias favoring small or large states, district population equity, representative share equity, and percentage equity.

(a) Hamilton

(b) Jefferson

(c) Webster

(d) Dean

(e) Hill–Huntington

Writing Projects

Both writing projects require you to use a spreadsheet to compute an apportionment, based on the 2010 census. While hand calculation is theoretically possible, it is not recommended.

The first step is to download the data as an XLS file from the Census Bureau. Go to **www.census.gov/population/apportionment/**

The spreadsheet with the 2010 apportionment can be found under the "Data" tab. To modify this spreadsheet to suit the purpose of either project, you will need to put in some additional columns. Place the divisor in cell `G63`. Start with standard divisor, 710,767, which you may adjust later. Type "Divisor" in cell `A63` as a reminder of what you have entered. You can input formulas for the apportionment quotients by entering `= B12 / G$63` in cell `G12` and copying this expression to the range `G13 . . .G61` in column `G`. (The dollar sign in the expression is important; without it, when the expression is copied to cell `G13`, for example, it would be `= B13 / G64`, and a divide by zero error would be noted.)

The spreadsheet function `INT( )` can be used to put $\lfloor q \rfloor$, for each apportionment quotient q, in the next column, `H`. Thus, enter `INT(G12)` in cell `H12`, and copy this formula to the range `G13 . . . G61`. You will put the rounding point (Hill–Huntington or Dean) in column `I`. For example, for Hill–Huntington, put `SQRT (H12 * (1 + H12))` in cell `I12`, and copy the formula through `I61`.

The tentative apportionments go in column `J`. You'll use the `IF(; ;)` function. The first entry of this function is a logical expression—that is, a statement that the computer can evaluate as true or false. The second entry is evaluated if the expression is true, the third if the expression is false. A state's tentative apportionment is obtained by rounding its apportionment quotient down if it is less than the rounding point, and rounding it up otherwise. Therefore, enter

```
= IF(G12 < I12; H12; 1 + H12)
```

in cell `J12`, and copy this through cell `J61`.

The final modification is to put the sum of the tentative apportionments,

```
SUM (J12 : J61)
```

in cell `J62`. When this is done, you can experiment by adjusting the divisor in cell `G63`. When the value in this cell changes, you will immediately see the changes in the tentative apportionments and their total.

1. In the reapportionment resulting from the 2010 census, two states, New York and Ohio, lost 2 house seats, and eight other states lost 1 seat. The 12 seats were transferred to states that had experienced dramatic increases in population: Texas got 4 additional seats, Florida received 2 more, and 6 other states got one apiece. If Congress should decide to increase the statutory house size so that no state's delegation in the House would be reduced, how many seats would have been added to the House, and which states would have gotten them? (*Warning:* It may be more than 12.)

2. The Webster method was proposed in 1832 after New York received an apportionment in excess of its upper quota. Two other apportionment methods were proposed in the same year: the method of John Quincy Adams (see Exercise 45 on page 617) and the Dean method (see page 598).

Suppose that state A has population p and its tentative apportionment is a, while state B has population q and tentative apportionment b. Which of these states is most deserving of the next seat, in the sense that absolute differences in district population are minimized? (Of course, another state may be more deserving, but we are comparing states two at a time.) To see how to answer this question:

(a) Calculate the absolute difference in district populations if A gets the seat, so that its apportionment is increased to $a + 1$. Let that value be denoted d_{AB}. Repeat the calculation for the situation when B gets the seat, and get a number d_{BA}.

(b) Show that $d_{AB} - d_{BA} = 2\left(\frac{q}{b^{\#}} - \frac{p}{a^{\#}}\right)$, where $a^{\#}$ and $b^{\#}$ are the harmonic means (you may have to Google this term) of a and $a + 1$ and of b and $b + 1$, respectively.

(c) Which state is entitled to the next seat if the difference calculated in part (b) is negative?

(d) The *Dean rounding point* for numbers between a whole number n and $n + 1$ is $n^{\#}$. Explain why the divisor method that rounds an apportionment quotient r down to $\lfloor r \rfloor$ if r is less than the Dean rounding point between $\lfloor r \rfloor$ and $\lceil r \rceil$, and up to $\lceil r \rceil$ otherwise, provides the apportionment that minimizes differences in district population.

(e) Find a formula for the Dean rounding point that can be applied to find the rounding point between 0 and 1 without division by 0.

(f) Compare the Dean rounding point $n^{\#}$ with the Hill–Huntington rounding point n^* for $n = 0, 1, 2, 3, 4, 5, 6$. Which is smaller? Is the Dean method more or less biased in favor of small states? Is it possible for any state to get an apportionment of zero with the Dean method?

(g) Compute the apportionment of the House of Representatives according to the 2010 census by the Dean method. How does it differ from the Hill–Huntington apportionment shown on the spreadsheet as you downloaded it?

Suggested Readings

BALINSKI, M. L., and H. P. YOUNG. *Fair Representation: Meeting the Ideal of One Man, One Vote,* Yale University Press, New Haven, CT, 1982. In the 1970s, Balinski and Young analyzed apportionment methods in depth. Their approach was to postulate the desirable properties of an apportionment method as axioms and to deduce from the axioms which method is best. This book combines an account of the history of apportionment of the U.S. House of Representatives with the results of their research.

ERNST, LAWRENCE R. Apportionment methods for the House of Representatives and the court challenges, *Management Science,* 40 (1994): 1207–1227. Ernst, who wrote briefs for the government in both the Montana and the Massachusetts cases, reviews the apportionment problem and the arguments in favor of and against each of the divisor methods. The article includes a summary of the arguments used by both sides in the two court cases.

WILLCOX, WALTER F. Methods of apportioning seats in the House of Representatives, *Journal of the American Statistical Association,* 49 (1954): 685–695. Go to **www.jstor.org/stable/2281533**

This is an insider's view of apportionment by a Cornell professor, who computed the apportionments based on the 1900 and 1910 censuses. Willcox was a strong proponent of the Webster method and argues here that the Hill–Huntington method does not reflect the intentions of the framers of the Constitution. He refers to the Hamilton method as the *Vinton method.* Samuel Vinton was a congressman who, in 1850, reinvented the Hamilton method.

YOUNG, H. PEYTON. *Equity,* Princeton University Press, Princeton, NJ, 1994. Chapter 3 of this book covers apportionment and focuses on which apportionment method is the most equitable.

Suggested Websites

www.census.gov/population/apportionment/ The Census Bureau's apportionment website provides a summary of the history of apportionment of seats in the United States House of Representatives. It has the data for the 2010 apportionment and a video, entitled *The Amazing Apportionment Machine,* that shows how it implements the Hill–Huntington method (which it calls the "method of equal proportions"). You will be interested to see that the "Amazing Apportionment Machine" works much as a d'Hondt or Sainte-Laguë table does, by lining the states up in priority order.

nia977.wix.com/drbcap Visit Charles Biles's website to read about apportionment as it was debated in the 18th and 19th centuries. Views of what constitutes fair apportionment of seats in the House of Representatives have changed over the years.

www.bundestag.de/htdocs_e/bundestag/elections/arithmetic/ The German parliament (the Bundestag) has provided a brief but very informative English language description of its apportionment system at this website.

penguincompaniontoeu.com/additional_entries/dhondt–system/ The d'Hondt system, as it is used by several member states of the European Union to elect members of the European Parliament, is described at this website.

Game Theory: The Mathematics of Competition

15

wavebreakmedia/Shutterstock

How can mathematics help us understand bargaining tactics in labor-management disputes, military choices in international crises, threats by animals in habitat acquisition, strategic play in a soccer match, and resource-allocation decisions in political campaigns? Mathematics, and game theory in particular, can help whenever two or more individuals with different values or goals compete to try to control the course of events.

In this chapter, we will develop the necessary mathematical tools to study situations, called *games*, involving both conflict and cooperation (see Spotlight 15.1). Because the *players* in a game may be people, animals, organizations, or even countries, game theory can be used to model business, biology, politics, sports, and international relations, as well as decisions that you make in your life.

The Early History of Game Theory

SPOTLIGHT 15.1

As early as the 17th century, such outstanding scientists as Christiaan Huygens (1629–1695) and Gottfried W. Leibniz (1646–1716) proposed the creation of a discipline that would apply the scientific method to the study of human conflict and interactions. Throughout the 19th century, several leading economists created simple mathematical models to analyze particular examples of competitive encounters. These models eventually became known as games, and the mathematical subject took on the name *game theory.* In a textbook written by 2007 Nobel Prize–winning game theorist Roger Myerson, he lamented that the name "game theory" suggests that the subject focuses on the study of recreational activities. Although game theory can be used to analyze recreational games such as poker or checkers, it more broadly is used to analyze any situation in which cooperation and/or conflict is present.

The first general mathematical theorem in game theory was proved by the distinguished logician Ernst Zermelo (1871–1953) in 1912. Zermelo's result showed that a certain class of games, called *games of perfect information*—in which at each stage of play, every player is aware of all past moves (by oneself and by others), as well as all future choices that are possible—has an optimal strategy. This is an example of an *existence theorem:* It states that there must exist a best way to play such a game, but it does not describe how to play optimally. An example of such a game is chess; optimal play is too complicated to determine, but people still enjoy playing chess—unlike tic-tac-toe, another, less complex, game of perfect information.

After Zermelo, a veritable who's who of famous mathematicians and economists have worked in game theory, including mathematician F. E. Émile Borel (1871–1956), Hungarian-American mathematician John von Neumann (1903–1957), and John F. Nash, Jr. (1928–2015), who was portrayed in the movie *A Beautiful Mind* (2001). Collectively, they advanced the subject and its application by showing that more complicated classes of games also have optimal solutions.

In 2007, Myerson was awarded the Nobel Memorial Prize in Economic Sciences, along with Leonid Hurwicz and Eric S. Maskin, for their early work in designing games to achieve particular outcomes—called *mechanism design*—examined later in this chapter. Mechanism design analyzes auctions, bargaining, voting, and other procedures, especially providing incentives for players to be truthful. Spotlight 15.2 has additional information about Nobel Prizes awarded for game theory.

The French artist Georges Mathieu designed a medal for the Musée de la Monnaie in Paris in 1971 to honor game theory. It was the 17th medal to "commemorate 18 stages in the development of Western consciousness."

Roger Myerson

Tim Boyle/Bloomberg via Getty Images

Leonid Hurwicz

University of Minnesota via Getty Images

Eric S. Maskin

William Thomas Cain/Getty Images

Georges Mathieu

Keystone-France/Gamma-Keystone via Getty Images

15.1 Two-Person Total-Conflict Games: Pure Strategies

The application of game theory requires modeling some interaction as a game by determining the players, their **strategies**—a list of options or courses of action available to them—and the possible *outcomes,* which describe the consequences of their choices. We assume that the players have *preferences* for the outcomes, which are represented by numbers called *payoffs.* Game theory analyzes the **rational choice** of strategies—that is, how players select strategies to obtain preferred outcomes. In this section and the next, we start with the simplest type of two-person games: **total-conflict games,** in which what one player wins the other player loses.

EXAMPLE 1 Determining Work Schedule and Location

After moving to a new town so that Lisa can take a job as a newspaper editor, newlyweds Mark and Lisa must decide which position Mark should accept as an emergency room nurse. There are three area hospitals, each with openings for three 8-hour shifts. Concerned about saving for a down payment on a house, Mark wants to choose the shift and hospital location that pays the most money. Lisa is less concerned about finances and more concerned about their quality of life. A tougher schedule and a more active emergency room translate into a higher hourly wage, but a lower quality of life. In this way, their preferences are diametrically opposed. What is better for Mark is worse for Lisa, and vice versa.

Lisa and Mark make a table with the shifts listed along the top row and the hospitals listed along the left column (see Table 15.1). The entries in the table give the hourly wage Mark will receive for working at the hospital, given by the row, and for the shift, given by the column. The hourly wages are the payoffs to Mark for the different outcomes.

TABLE 15.1 Hourly Wage (in Dollars) for the 9 Hospital-Shift Possibilities

	Shifts		
Hospitals	**12 A.M.–8 A.M.**	**8 A.M.–4 P.M.**	**4 P.M.–12 A.M.**
Rural	23	24	22
Suburban	27	26	29
Downtown	30	23	25

Payoff Matrix DEFINITION

A **payoff matrix** (illustrated by Table 15.1) is a table whose rows and columns correspond to the strategies of the two players. For a total-conflict game, the numerical entries of the matrix give the payoffs to the row player when these strategies are chosen.

Lisa and Mark turn the table into a competitive game to determine which job Mark should accept—such a game represented by a payoff matrix is called a game *in strategic form.* Mark will select one of the three hospitals—Rural, Suburban, or Downtown—and Lisa will simultaneously choose one of the three shifts—beginning at 12 A.M., 8 A.M., or 4 P.M. Because their choices will be made simultaneously, neither knows beforehand what the other will do.

Because Lisa's preferences are diametrically opposed to Mark's preferences, Lisa's payoffs are represented by −1 times the hourly wage. For example, if the outcome is for Mark to work at the rural hospital from 12 A.M. to 8 A.M., then Mark's payoff is 23 while Lisa's payoff is −23. This is the origin of the term *zero-sum game.*

Zero-Sum Game DEFINITION

A **zero-sum game** is one in which the payoff to one player is the negative of the corresponding payoff to the other, so the sum of the payoffs to the two players is always 0.

Zero-sum games are total-conflict games in which what one player wins the other loses. But not all total-conflict games are zero-sum—in particular, the sum of the payoffs could be some constant other than 0. Nevertheless, the strategic nature of these two types of games is the same.

Continuing the game in Example 1, Mark, worried that Lisa will choose a shift that will result in lower pay, tries to determine the highest hourly wage he can guarantee by picking one of the three hospitals. For each choice of a hospital, this means considering the worst-case (lowest) hourly wage. These are the dollar amounts 22, 26, and 23, which are the respective *row minima,* indicated in the right-hand column of Table 15.2. He notes that the highest of these values is 26. By choosing the corresponding suburban hospital, Mark can guarantee himself an hourly wage of $26 per hour.

TABLE 15.2 Hourly Wage (in Dollars) from Table 15.1, with the Row Minima (Maximum Circled) and Column Maxima (Minimum Circled)

		Lisa *Shifts*			
	Hospitals	**12 A.M.–8 A.M.**	**8 A.M.–4 P.M.**	**4 P.M.–12 A.M.**	**Row Minima**
	Rural	23	24	22	22
Mark	*Suburban*	27	26	29	(26)
	Downtown	30	23	25	23
	Column Maxima	30	(26)	29	

Maximin DEFINITION

The **maximin** is the maximum value of the minimum numbers in the rows in a table. The strategy of the row player—Mark, in this case—that corresponds to the maximin is called its **maximin strategy.**

The number 26 in the right-hand column of Table 15.2, which is circled, is the **maximin.** The suburban hospital is Mark's maximin strategy.

Self Check 1

In the following table, the payoffs represent gains to the row player and losses to the column player. This makes the game a zero-sum game like the game between Mark and Lisa. Calculate the row minima to help determine which row is the maximin strategy for the game represented by the table.

9	6	2
7	8	5
4	1	3

■

Lisa likewise does a worst-case analysis and lists the highest—for her, the worst in terms of quality of life—hourly wages. These numbers, 30, 26, and 29, are the column maxima and are listed in the bottom row of Table 15.2. From Lisa's point of view, the best of these outcomes is 26. If she picks the 8 A.M. to 4 P.M. shift, then Lisa is guaranteed that Mark's workload will result in him earning an hourly wage of no more than $26 per hour.

Minimax DEFINITION

The **minimax** is the minimum value of the maximum numbers in the columns. The strategy of the column player—Lisa, in this case—that corresponds to the minimax is called its **minimax strategy.**

The number 26 at the bottom row of Table 15.2, which is circled, is the **minimax.** The 8 A.M. to 4 P.M. shift is Lisa's minimax strategy.

Self Check 2

In the following table, the payoffs represent gains to the row player and losses to column player; again, this is a zero-sum game. Calculate the column maxima to help determine which column is the minimax strategy for the game represented by the table.

9	6	2
7	8	5
4	1	3

■

To summarize the results of the game in Example 1, Mark has a strategy that will ensure an hourly wage of $26 per hour or more, and Lisa has a strategy that will ensure an hourly wage of $26 per hour or less. The hourly wage of $26 per hour is, simultaneously, the lowest amount Mark can earn at the suburban hospital and the most he can earn for working the 8 A.M. to 4 P.M. shift. In other words, the maximin and the minimax are both equal to $26 per hour for this location/schedule game.

Saddlepoint and Value DEFINITION

For a zero-sum game, when a row minimum and a column maximum are the same, the resulting payoff is called a **saddlepoint.** In this case, it is also called the **value** of the game.

To see the reason for the term **saddlepoint,** consider the saddle-shaped payoff surface of Mark's hourly wage shown in Figure 15.1. The middle point on a horse saddle (26) is simultaneously the lowest point along the spine of the horse (because 27 and 29 are higher) and the highest point between the rider's legs (because 23 and 24 are lower). The saddle structure assures Mark an hourly wage of $26 an hour, which is simultaneously the maximin and minimax. If a game has a saddlepoint, players can guarantee the saddlepoint payoff by choosing their maximin and minimax strategies.

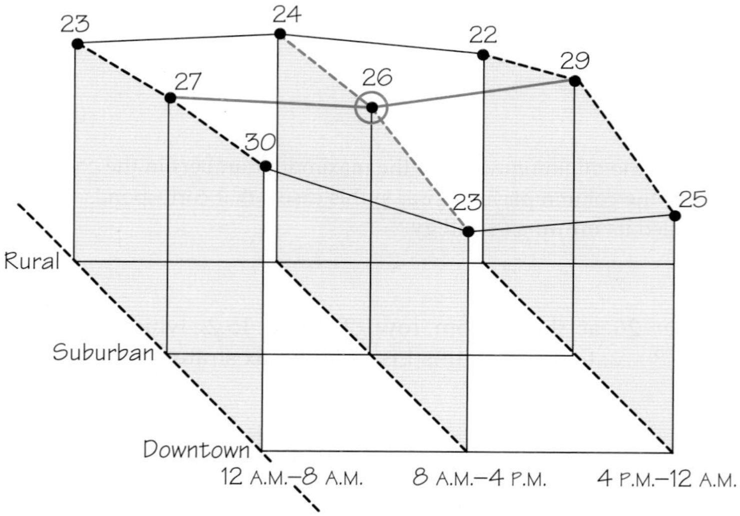

Figure 15.1 The payoff surface shows Mark's hourly wage as it depends on Mark's and Lisa's possible choices (in dollars per hour).

Self Check 3

In the following table, the payoffs represent gains to the row player and losses to the column player. What is the value for the game represented by the table?

9	6	2
7	8	5
4	1	3

■

There is no need for secrecy in a game with a saddlepoint. For the location/schedule game, Lisa has no incentive to change her selection for Mark's schedule, even if Mark were to reveal that he is going to select the suburban hospital, because $26 is the lowest hourly wage at the suburban hospital. Similarly, Mark has no incentive to change his selection of a hospital location, even if Lisa were to reveal that she is going to select the 8 A.M. to 4 P.M. shift, because $26 is the highest hourly wage for the 8 A.M. to 4 P.M. shift. Hence, the value of $26 per hour occurs when Mark picks the suburban hospital and Lisa picks the 8 A.M. to 4 P.M. time slot.

In games with saddlepoints, players' worst-case analyses lead to the best *guaranteed* outcome—in the sense that each player can ensure that he or she does not do worse than a certain amount (26 in our example) and may even do better (if the opponent deviates from a maximin or minimax strategy). Zero-sum games without saddlepoints also have a "value," as we shall see later.

Another well-known game with a saddlepoint is tic-tac-toe. Two players alternately place an X or an O, respectively, in one of the unoccupied spaces in a 3×3 grid with 9 cells. The winner is the first player to have three Xs, or three Os, in the same row, in the same column, or along a diagonal; if no player does this when all cells are filled in, the game ends in a tie.

An explicit list of all strategies for either the first- or second-moving player in tic-tac-toe is long and complicated because it specifies a complete plan for all possible contingencies that can arise. For the first-moving player, for example, a strategy might be to "put an X in the middle cell, then an X in the corner if your opponent puts an O in a noncorner cell," and so on. While young children initially find this game fun to play, before long they discover that each player can always prevent the other player from winning by forcing a tie, making the game uninteresting. Unlike the game between Mark and Lisa, the list of possible strategies in tic-tac-toe is lengthy, but only those that force a tie, it turns out, are a saddlepoint. Tic-tac-toe is an example of a combinatorial game. These two-player games have players alternating moves without the use of any randomizing elements such as a die.

EXAMPLE 2 The Restricted Location/Schedule Game

Assume that Mark and Lisa revisit the location/schedule game once Mark and Lisa have their first child. To limit the number of hours their child must spend in daycare, Mark and Lisa agree that Mark cannot work the 8 A.M. to 4 P.M. shift, because this conflicts with Lisa's work schedule. A new game is formed by eliminating the 8 A.M. to 4 P.M. shift column in Table 15.1. As before, Mark and Lisa can each do a worst-case analysis. Mark is worried about the minimum number in each row, and Lisa is concerned with the maximum number in each column. The result, along with minimum row and maximum column information, appears in Table 15.3.

TABLE 15.3 Hourly Wage (in dollars) from Table 15.1, with the Row Minima (Maximum Circled) and Column Maxima (Minimum Circled)

		Lisa *Shifts*		
	Hospitals	**12 A.M. – 8 A.M.**	**4 P.M. – 12 A.M.**	**Row Minima**
	Rural	23	22	22
Mark	*Suburban*	27	29	(27)
	Downtown	30	25	25
	Column Maxima	30	(29)	

Mark sees that his maximin is now 27, so he can guarantee an hourly wage of $27 per hour, instead of the $26 per hour from before. Lisa notices that her minimax is now 29. When the maximin and the minimax are not the same, then the game does *not* have a saddlepoint, but it does have a value (described in the next section).

If Mark plays his maximin strategy, the suburban hospital, and Lisa plays her minimax strategy, the 4 P.M. to 12 A.M. shift, then the resulting payoff is 29. However, Lisa may be motivated to gamble in this case by playing her other strategy, the 12 A.M. to 8 A.M. shift. If she switches while Mark still selects the suburban hospital, then Mark's hourly wage will decrease to $27 per hour.

If Mark thinks that Lisa will select the 12 A.M. to 8 A.M. shift, then he has an incentive to select the downtown hospital instead. The highest hourly wage that Mark can receive is by working the 12 A.M. to 8 A.M. shift at the downtown hospital.

However, if Lisa looks far enough ahead, she can respond to Mark's reasoning by playing her minimax strategy. The result is $25 per hour if Mark selects the downtown hospital and Lisa selects the 4 P.M. to 12 A.M. shift. The $25 per hour is less than the $29 per hour when Lisa plays her minimax strategy and Mark plays his maximin strategy. This means that Mark has an incentive to return to his maximin strategy.

In two-player games that have saddlepoints, like our original 3 × 3 location/schedule game and tic-tac-toe, each player can calculate the maximin and minimax strategies for both players before the game is even played. Once the solution has been determined by either mathematical analysis or practical experience (as was probably true of tic-tac-toe), there may be little interest in actually playing the game.

But this is decidedly not the case for much more complex games, like chess, whose solution has not yet been determined—and is unlikely to be determined in the foreseeable future. Even though computers are able to beat world champions, the computer's winning moves will not necessarily be optimal against those of *all* other opponents. Nevertheless, we know that chess, like tic-tac-toe, has a saddlepoint. (All games of perfect information, in which the players know each other's moves at every step, have a saddlepoint.) What we do not know is whether it yields a win for white, a win for black, or a draw.

Unlike chess, many games, such as the modified location/schedule game (the 3 row × 2 column game), do not have an outcome that can always be guaranteed. These games, which include poker, involve uncertainty and risk. In such games, one does not want to have one's strategy detected in advance because this information can be exploited by an opponent. It is no surprise, then, that poker players try to keep a "poker face," revealing nothing about their likely choices. But keeping a poker face does not tell the players what to actually do in the game, such as how many cards to ask for in draw poker.

We will show that there are optimal ways to play two-person total-conflict games without a saddlepoint so as not to reveal one's choices. But their solution is by no means as straightforward as that of games with a saddlepoint.

15.2 Two-Person Total-Conflict Games: Mixed Strategies

Most competitive games do not have a saddlepoint like the one we found in our first location-game example. Rather, as is illustrated in our modified location/schedule game—in which the maximin and minimax are not the same—players must try to keep secret their strategy choices, lest their opponent use this information to his or her advantage.

In particular, players must take care to conceal the strategy they will select until the encounter actually takes place, when it is too late for the opponent to alter his or her choice. If the game is repeated, a player will want to *vary* his or her strategy in order to surprise the opponent.

In parlor games like poker, players often use the tactic of *bluffing*. This tactic involves a player sometimes raising the stakes when he has a low hand so that opponents cannot guess whether or not his hand is high or low—and may, therefore, miscalculate whether to stay in or drop out of the game. (A player would prefer opponents to stay in when he has a high hand and drop out when he has a low hand.) In military engagements, too, secrecy and even deception are often crucial to success.

In many sporting events, a team tries to surprise or mislead its opposition. A pitcher in baseball will not signal the type of pitch he or she intends to throw in advance, mixing up pitches throughout the game to try to keep the batter off balance. A football team will vary how it approaches two-point conversions, sometimes rushing the ball and other times passing it. A kicker in a penalty shootout of a soccer match will kick the ball sometimes to the left of the goal and sometimes to the right, while the goalie will often guess one side or the other. A kicker–goalie interaction is considered in the following example.

EXAMPLE 3 A Penalty Kick Shootout

After a tie overtime period in a FIFA World Cup soccer match, penalty shootouts are used to determine which team advances to the next round. A penalty shootout consists of five kicks for each team, in alternating order. If at the end of the five kicks, the teams are still tied, then the two teams continue to alternate penalty kicks until the tie is broken—the team that scores when the other doesn't advances to the next round. Penalty shootouts have even decided the World Cup winner, as when Italy defeated France 5-3 in a shootout in 2006.

In soccer, the goal is 8 feet tall by 24 feet wide. A kick in a penalty shootout is taken at 36 feet from the goal, and the ball travels upwards of 100 miles per hour. Because of the wide goal and the distance and speed that the ball travels, the goalie typically guesses either left or right. Likewise, the kicker decides on either kicking the ball to the goalie's left or to the goalie's right. Players vary their kicking techniques; if they don't, then a well-prepared goalie can learn players' tendencies to kick to one side or the other. Indeed, Petr Cech saved all five penalty kicks in a penalty shootout by guessing the correct direction, thereby helping Chelsea Football Club of England defeat Bayern Munich of Germany to win the 2012 UEFA European Champions League. Cech had prepared by watching every penalty kick the Bayern Munich players attempted since 2007.

anekoho/Shutterstock

Because of the speed of the ball and the size of the goal, the goalie often guesses that the penalty kick will go to his right or left. This may be an educated guess based on the kicker's tendencies. In this case, the goalie guessed incorrectly.

Assume that a particular left-footed kicker can kick the ball either to the goalie's left or to the goalie's right and so has two strategies: *goalie's left* (denoted by *GL*) or *goalie's right* (*GR*). The goalie decides between diving to the left or to the right, and likewise has two strategies: *left* (denoted by *L*) and *right* (*R*). Being left-footed, the kicker's strong side is to kick to the goalie's left, resulting in a more powerful, but less accurate, shot. The kicker is more accurate when kicking the ball to the right, which results in a less forceful shot. Assume that the kicker's success rates for the different strategies by the players are given below and that both players know these rates:

- 0.2 if the kicker kicks the ball to the goalie's left (*GL*) and the goalie dives left (*L*)
- 0.9 if the kicker kicks the ball to the goalie's left (*GL*) and the goalie dives right (*R*)
- 0.95 if the kicker kicks the ball to the goalie's right (*GR*) and the goalie guesses left (*L*)
- 0.15 if the kicker kicks the ball to the goalie's right (*GR*) and the goalie guesses right (*R*)

The kicker's success rate is the number of goals scored divided by the number of attempts. Hence, 0.2 means that the kicker succeeds 20% of the time, or 2 out of every 10 shots, on average.

This game is summarized in Table 15.4. The payoffs in the table, or matrix (called a *payoff matrix*), represent the likelihood that the kicker is successful. Because the goalie wishes to minimize this likelihood, the game is zero sum. We see from the right-hand column in the table that the kicker's maximin is 0.2, which is realized when the kicker chooses *GL*. Always going with the power side, the kicker is assured of successfully scoring at least 20% of the time—hardly a successful outcome.

We see from the bottom row of the table that the goalie's minimax is 0.9, which is obtained by guessing right (*R*). The maximin and minimax are not equal, which means that the penalty shootout game does not have a saddlepoint. By varying his or her strategies, the kicker's success rate should be somewhere between 0.2 and 0.9. In a sense, the strategies are to determine how much of the difference 0.7 = 0.9 − 0.2 is split between the two players. The goalie wishes to push down the success rate as far as possible, whereas the kicker wants to raise it from 0.2 as much as possible.

TABLE 15.4 Success Rate in a Penalty Kick Shootout Game

		Goalie		**Row Minima (maximum circled)**
		L	***R***	
Kicker	***GL***	0.2	0.9	(0.2)
	GR	0.95	0.15	0.15
	Column Maxima (minimum circled)	0.95	(0.9)	

Trying to Think Ahead and Outguess One's Opponent

To see that there is no saddlepoint, let's consider the four possible outcomes separately and how the players might try to outguess one another. If the kicker chooses *GL* (the maximin strategy) and the goalie chooses *R* (the maximin strategy), then the outcome would be 0.9 and the goalie has an incentive to switch to *L*, thereby keeping the kicker at the maximin payoff of 0.2. However, the kicker can anticipate the goalie's choice of *L* and counteract by choosing *GR*, giving the kicker the highest likelihood of success with a rate of 0.95. But the goalie can think this far ahead, too, and return to the original strategy of *R*, which results in the lowest success rate of 0.15 for the kicker. From the proposed (*GR*, *R*) outcome, the kicker now wishes to switch back to the original strategy of *GL*!

This type of cyclical reasoning, in which the players anticipate one another's decisions and try to outguess one another, can go on forever, as depicted in the diagram in Figure 15.2.

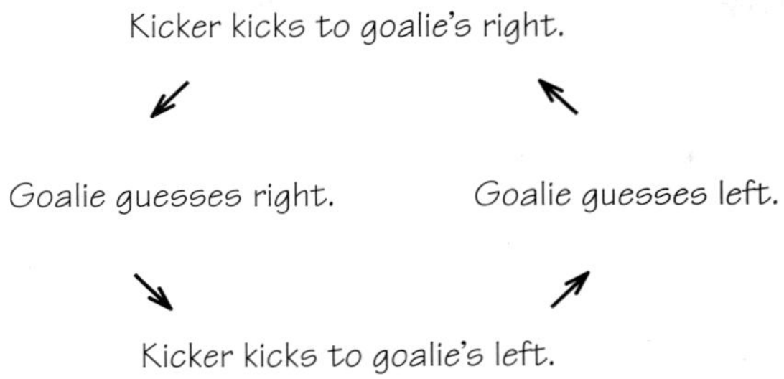
Kicker kicks to goalie's right.
Goalie guesses right.
Goalie guesses left.
Kicker kicks to goalie's left.

Figure 15.2 Anticipating one another's decisions and trying to outguess one another may lead to cyclical reasoning.

It fails to provide resolution to the players' decision problem, precisely because the players fail to vary their behaviors. Once the kicker settles on a particular side to kick the ball, then the goalie can exploit it. Similarly, once the goalie chooses left or right, then the kicker can exploit the decision.

Clearly, there is no best side to kick the ball and no best side to guess under all circumstances. Nevertheless, both the kicker and the goalie *can* do better, but not by trying to anticipate each other's choices. The answer to their problem lies in the notion of a **mixed strategy.**

A Better Idea

The play of many games requires an element of surprise, which can be realized in practice by making use of a mixed strategy, in which a player randomizes what he or she does.

Pure Strategy DEFINITION

Each of the definite courses of action that a player can choose is called a **pure strategy.**

All the choices for players—Mark and Lisa, the kicker and the goalie—that we have considered so far are **pure strategies,** in which each player in the end makes a definite choice.

Mixed Strategy DEFINITION

A **mixed strategy** is a strategy in which the course of action is randomly chosen from one of the pure strategies in the following way: Each pure strategy is assigned some probability, indicating the relative frequency with which that pure strategy will be played. The specific strategy used in any given play of the game can be selected using some appropriate randomization device (like rolling a die, flipping a coin, or using a random number generator on a computer).

Note that a pure strategy is a special case of a mixed strategy, with the probability of 1 assigned to just one pure strategy and 0 to all the rest. When a player resorts to a mixed strategy, the resulting outcome of the game is no longer predictable. (For example, if a goalie chooses to guess L and R with equal probability of 0.5 each, then the kicker cannot predict in which direction the goalie will move. This may be viewed as if in repeated interactions the goalie chooses L and R half of the time.) The payoff must be described in terms of the probabilistic notion of an **expected value** or **expected payoff,** which is the average payoff of the game if it were played many, many times.

Expected Value *E* DEFINITION

If each of the n payoffs, $s_1, s_2, \ldots, s_n$, will occur with the probabilities $p_1, p_2, \ldots, p_n$, respectively, then the average, or **expected value *E*,** is given by

$$E = p_1s_1 + p_2s_2 + \cdots + p_ns_n$$

We assume that the probabilities sum to 1 and that each probability p_i is never negative. That is, we assume that $p_1 + p_2 + \cdots + p_n = 1$, and $p_i \geq 0$ $(i = 1, 2, \ldots, n)$.

Self Check 4

Suppose a player has three strategies and plays them with probabilities p_1, p_2, and p_3. If $p_1 = 1/2$ and $p_2 = 1/3$, then what is p_3? If the three payoffs $s_1 = 4$, $s_2 = 6$, and $s_3 = -6$ occur with probabilities p_1, p_2, and p_3, then what is the expected value? ■

To see how mixed strategies and expected payoffs are used in the analysis of games, we turn to what is perhaps the simplest of all competitive games without a saddlepoint.

EXAMPLE 4 Matching Pennies

In Matching Pennies, each of two players simultaneously shows either a head H or a tail T. If the two coins match—either two heads or two tails—then the first player (Player I) receives both coins (a win of 1 for Player I). If the coins do not match, that is, if one is an H and the other is a T, then the second player (Player II) receives the two coins (a loss of 1 for Player I). These wins and losses for Player I are shown in the zero-sum payoff matrix in Table 15.5.

TABLE 15.5 Wins and Losses for Player I in Matching Pennies

		Player II	
		H	***T***
Player I	***H***	1	−1
	T	−1	1

It is fruitless for one player to attempt to outguess the other in this game. Both should instead resort to mixed strategies and use expected values to estimate their likely gains or losses.

To illustrate, assume that Player I randomly selects H half the time and T half the time. This mixed strategy can be expressed as

$$(p, 1 - p) = \left(\frac{1}{2}, \frac{1}{2}\right)$$

Note that the probabilities p of choosing H and $(1 - p)$ of choosing T do indeed sum to 1, as required; in particular, when $p = \frac{1}{2}$, $1 - p = 1 - \frac{1}{2} = \frac{1}{2}$.

Player II gains nothing by knowing that Player I is using the optimal mixed strategy $\left(\frac{1}{2}, \frac{1}{2}\right)$. However, Player I must not reveal to Player II whether H or T will be displayed *in any given play* of the game before Player II makes his or her own choice of H or T. Even without this information, if Player II knew that Player I was using a particular *nonoptimal* mixed strategy $(p_1, p_2) = (p, 1 - p)$, where $p \neq \frac{1}{2}$ (i.e., not choosing a 50-50 mixture between H and T), then Player II could take advantage of this knowledge and increase his or her average winnings over time to something greater than the value of 0 (see Exercise 18 on page 661).

This mixture can be realized in practice by the flip of a coin. Whenever Player II plays H (first column of Table 15.5), Player I's resulting expected value is

$$E_H = \frac{1}{2}(1) + \frac{1}{2}(-1) = 0$$

Similarly, whenever Player II plays T (second column), Player I's resulting expected value is

$$E_T = \frac{1}{2}(-1) + \frac{1}{2}(1) = 0$$

We will develop the tools later in this section to see that neither player can guarantee a better payoff than to choose H and T with equal likelihood (i.e., $p = \frac{1}{2}$), making this strategy optimal (see Exercise 18).

Like Mark's maximin strategy from the location/schedule game, an optimal mixed strategy in a zero-sum game for the row player provides the highest expected payoff that the row player can guarantee, regardless of the column player's strategy. Similarly, an optimal strategy for the column player is similar to Lisa's minimax strategy, as it caps the amount that the row player can make in expected value. Just as when there is a saddlepoint, these two values are equal.

Value DEFINITION

The **value** of a zero-sum game is the expected value when the row player or the column player plays an optimal strategy.

When the value of a zero-sum game can be realized only by mixed strategies, then the value generalizes the notion of a saddlepoint. Because the value of 0 in Matching Pennies is an expected value, it must be understood in a statistical sense. That is, in a given play of the game, Player I will either win 1 or lose 1. However, his or her expectation over many plays of this game is 0. The optimal mixed strategy for Player II is likewise a 50-50 mix of H and T, which also leads to an expectation of 0, making the game **fair.**

Fair Game DEFINITION

A zero-sum game is **fair** if it has a value of 0, and, consequently, it favors neither player when at least one player uses an *optimal* (mixed) strategy—one that guarantees the resulting expected payoff is the best that this player can obtain against all possible strategy choices (pure or mixed) by an opponent.

EXAMPLE 5 Another Matching Game

In this game, Players I and II can again show either heads *H* or tails *T*. When two *H*s appear, Player II pays \$5 to Player I. When two *T*s appear, Player II pays \$1 to Player I. When one *H* and one *T* are displayed, then Player II collects \$3 from Player I. The payoffs (that Player I receives from Player II) for this game are given in Table 15.6.

TABLE 15.6 Payoffs for Player I in Another Matching Game

		Player II	
		H	***T***
Player I	***H***	5	−3
	T	−3	1

A worst-case analysis, like that which solved our initial location/schedule game, is of little help here. Player I may lose \$3 whether he plays *H* or *T*, making his maximin −3. Player II can keep down her losses to \$1 by always playing *T* (and thus avoiding the loss of \$5 when two *H*s appear), so Player II's minimax is 1. However, if Player II chooses *T* and Player I knows this, then Player I will also play *T* and collect \$1 from Player II. Can Player II do better than lose \$1 in each play of the game?

Consider the situation where Player I uses a mixed strategy $(p, 1 - p)$, which involves playing *H* with probability p and playing *T* with probability $1 - p$, where $0 \leq p \leq 1$. Against Player II's pure strategy *H*, Player I's expected value is

$$E_H = (5)(p) + (-3)(1 - p) = 5p + (-3) + 3p = 8p - 3$$

Against Player II's pure strategy *T*, Player I's expected value is

$$E_T = (-3)(p) + (1)(1 - p) = -3p + 1 + -p = -4p + 1$$

Algebra Review Appendix
Graphing a Line in Slope-Intercept Form

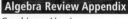

These two linear equations in the variable p are depicted in Figure 15.3. Note that the four points where these two lines intersect the two vertical lines, $p = 0$ and $p = 1$, are the four payoffs appearing in the payoff matrix.

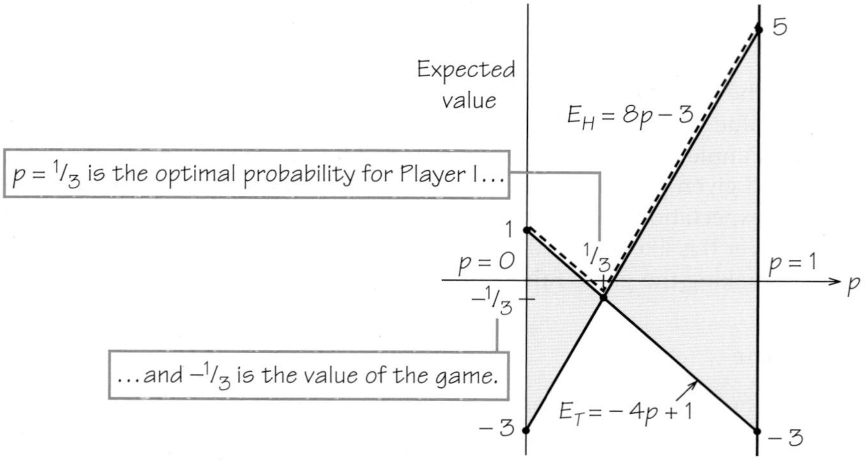

Figure 15.3 Solution to the matching game of Example 5.

The point at which the lines given by E_H and E_T intersect can be found by setting $E_H = E_T$, yielding

Algebra Review Appendix
Linear Equations in Two Variables

$$8p - 3 = -4p + 1$$
$$12p = 4$$

so $p = \frac{1}{3}$. To the left of $p = \frac{1}{3}$, $E_T > E_H$, and to the right, $E_H > E_T$; at $p = \frac{1}{3}$, $E_H = E_T$. If Player I chooses $(p_H, p_T) = (p, 1 - p) = \left(\frac{1}{3}, \frac{2}{3}\right)$, he can ensure an average payoff of

$$E_H = 8\left(\frac{1}{3}\right) - 3 = E_T = -4\left(\frac{1}{3}\right) + 1 = -\frac{1}{3}$$

regardless of what Player II does.

In other words, Player I's optimal mixed strategy is to pick H and T with probabilities $\frac{1}{3}$ and $\frac{2}{3}$, respectively, which gives Player I an expected value of $-\frac{1}{3}$ and prevents him from having a bigger average loss. As can be seen from Figure 15.3, $-\frac{1}{3}$ is the highest expected value that Player I can guarantee against *both* strategies H and T of Player II. Although T yields Player I a higher expected value for $p < \frac{1}{3}$, and H yields him a higher expected value for $p > \frac{1}{3}$, Player I's choice of $(p_H, p_T) = \left(\frac{1}{3}, \frac{2}{3}\right)$ protects him against an expected loss greater than $-\frac{1}{3}$, which neither of his pure strategies does (each may produce a maximum loss of -3). Put another way, the intersection of E_H and E_T at $p = \frac{1}{3}$ is the minimum of the function given by E_T to the left and E_H to the right (shown by the dashed line in Figure 15.3). If Player II had more than two strategies, this approach to finding a minimum that puts a floor on Player I's expected loss can be extended.

A similar calculation for Player II results in the same optimal mixed strategy $\left(\frac{1}{3}, \frac{2}{3}\right)$ and expected value $-\frac{1}{3}$. But because the payoffs for Player II are losses, $-\frac{1}{3}$ means that she gains $\frac{1}{3}$ on the average. It is a coincidence that Player I and II's optimal mixed strategies are identical; in the penalty shootout that we will return to in Example 6, this is not the case.

Therefore, this game is unfair, even though the sum of the amounts (\$6) that Player I might have to pay Player II when he loses is the same as the sum that Player II might have to pay Player I when she loses. Interestingly, Player II, who will win an average of $33\frac{1}{3}$ cents each time the game is played, is favored, even though she may have to pay more to Player I when she loses (a maximum of \$5) than Player I will ever have to pay her (a maximum of \$3).

Scoring in professional chess tournaments usually assigns a payoff of 1 for winning, 0 for losing, and $\frac{1}{2}$ to each player for a tie, making the sum of the payoffs to the two players always 1. Such games, called **constant-sum games,** can readily be converted to zero-sum games. Thus, chess could as well be scored -1 for a loss, $+1$ for a win, and 0 for a tie, making the constant 0 in this case. Although constant-sum and zero-sum games have the same strategic nature, constant-sum games are a more general class because the constant need not be 0.

The solution to Matching Pennies illustrated how the mixed strategy of $\left(\frac{1}{2}, \frac{1}{2}\right)$ guarantees each player the value of 0, but we did not give a *solution technique* for finding optimal mixed strategies. In the second matching game, we illustrated a procedure that can be applied to *every* payoff matrix in which each player has only two strategies.

We must use more complex methods, which we will not describe here, to find mixed-strategy solutions when one or both players have more than two strategies. However, one should always check first to see whether a game has a saddlepoint before employing any technique for finding optimal mixed strategies.

In our next example, which picks up on the penalty shootout between the kicker and the goalie given by the 2 × 2 payoff matrix in Table 15.4, there is no saddlepoint, as we already showed. Thus, the solution will necessarily be in mixed strategies. We now proceed to find what mix is optimal.

EXAMPLE 6 The Penalty Shootout Revisited

In Table 15.7, we add probabilities, which we explain next, to Table 15.4. The goalie should use a mixed strategy $(p_1, p_2) = (p_L, p_R) = (p, 1 - p)$, where p_L represents the probability that the goalie guesses left (i.e., dives to the left) and p_R represents the probability that the goalie guesses right. Notice that $p_L + p_R = p + (1 - p) = 1$, as required, assuring that the goalie chooses a side! The probabilities p and $1 - p$ (where $0 \leq p \leq 1$) are indicated below the game matrix and under the corresponding strategies, L and R, for the goalie. If the goalie plays a mixed strategy $(p, 1 - p)$ against the two pure strategies, GL and GR, for the kicker, then the respective expected values for the kicker are

$$E_{GL} = (0.2)p + 0.9(1 - p) = -0.7p + 0.9$$
$$E_{GR} = (0.95)p + 0.15(1 - p) = 0.8p + 0.15$$

TABLE 15.7 The Penalty Shootout with Probabilities

		Goalie		
		L	***R***	
Kicker	***GL***	0.2	0.9	q
	GR	0.95	0.15	$1 - q$
		p	$1 - p$	

As in the second matching game, the solution to this game occurs at the intersection of the two lines given by E_{GL} and E_{GR}. Setting the equations of these lines equal to each other so that $-0.7p + 0.9 = 0.8p + 0.15$ yields $0.75 = 1.5p$ or $p = 7.5/15 = 1/2$, giving $E_{GL} = E_{GR} = E = 0.55$.

Thus, the goalie should use the optimal mixed strategy by choosing L with probability $p = 1/2$ and R with probability $1 - p = 1/2$. This choice limits the success rate of the kicker to 0.55, or 55%, which is the value of the game. The value of the game is statistical in nature in that it means that if the goalie follows the prescribed strategy, then the kicker will, on average, successfully score 55% of the time. Remember that this doesn't provide information about what will happen on any one particular kick.

Assume that the kicker uses a mixed strategy $(q_1, q_2) = (q_{GL}, q_{GR}) = (q, 1 - q)$, as indicated to the right of the game matrix in Table 15.7. This means that the kicker kicks the ball to the goalie's left with probability q_{GL} and kicks the ball to the goalie's right with probability q_{GR}. This mixed strategy, when played against the goalie's pure strategies, L and R, results in the following expected values:

$$E_L = (0.2)q + 0.95(1 - q) = -0.75q + 0.95$$
$$E_R = (0.9)q + 0.15(1 - q) = 0.75q + 0.15$$

The probability q can be found by solving for q in the equation $E_L = -0.75q + 0.95 = 0.75q + 0.15 = E_R$ so that $1.5q = 0.8$. The intersection of these two lines occurs at the point $q = 8/15$, giving $E_L = E_R = E = 0.55$. The kicker's optimal mixed strategy is therefore $(q_{GL}, q_R) = (8/15, 7/15)$, to kick the ball with probability 8/15 to the goalie's left and with probability 7/15 to the goalie's right, which gives the kicker the same success rate of 0.55. ■

Algebra Review Appendix
Linear Equations in Two Variables

Self Check 5

Suppose that $E_1 = 6q - 1 = 4 - 9q = E_2$. Solve for q and find the expected value E, so $E = E_1 = E_2$. ■

We have seen that the expected payoff or expected success rate of 0.55, which is the value of the game, occurs when either the goalie selects the optimal mixed guessing strategy (1/2, 1/2) or the kicker selects the optimal mixed kicking strategy (8/15, 7/15). That there is a value for every zero-sum game when the two players are allowed to use mixed strategies is called the *fundamental theorem* of zero-sum games and is known as the **minimax theorem.**

Minimax Theorem THEOREM

The **minimax theorem** guarantees that there is a unique game value and an optimal strategy for each player so that either player alone can realize at least this value by playing this strategy, which may be pure or mixed.

The unique value in Example 6 is 0.55.

15.3 Partial-Conflict Games

Many interactions involve a delicate mix of cooperative and noncooperative behavior. In business, for example, firms in an industry cooperate to gain tax breaks even as they compete for market share. The 2×2 matrix games (two players, each with two strategies) presented so far have been total-conflict games: One player's gain is equal to the other player's loss. Although most parlor games, like chess or poker, are games of total conflict, and therefore constant-sum, most real-life games are surely not. (Elections, in which there is usually a clear-cut winner and one or more losers, probably come as close to being games of total conflict as we find in the real world.) Here, we consider two well-known **partial-conflict games** that have often been used to model real-world conflicts. In these games, even though cooperation may be possible—because the players' preferences are not diametrically opposed—players may still have reasons not to cooperate.

Partial-Conflict Games DEFINITION

A **partial-conflict game** is one in which both players can benefit by cooperation, but they may have strong incentives not to cooperate.

Variable-Sum Games DEFINITION

Games of partial conflict are **variable-sum games,** in which the sum of payoffs to the players at the different outcomes varies.

There is some mutual gain to be realized by both players if they can cooperate in partial-conflict games, but cooperation might be difficult in the absence of either good communication or trust. When these elements are lacking, players are less likely to comply with any agreement that is made. *Noncooperative games* are games in which a binding agreement cannot be enforced. Even if communication is allowed in such games, there is no assurance that a player can trust an opponent to choose a particular strategy that he or she promises to select because the selections are made simultaneously.

In fact, the self-interest of the players may lead them to make strategy choices that yield lower payoffs to both parties than they could have achieved by cooperating. Two partial-conflict games illustrate this problem. The first is so well known that it has entered into common language.

EXAMPLE 7 Prisoners' Dilemma

The **Prisoners' Dilemma** is a two-person variable-sum game. It provides a simple explanation of the forces at work behind arms races, price wars, and agreeing to environmental regulations. In these and other similar situations, the players can do better by cooperating. But there may be no compelling reasons for them to do so unless there are credible threats of retaliation for not cooperating. The name *Prisoners' Dilemma* was first given to this game by Princeton mathematician Albert W. Tucker (1905–1994) in 1950.

Before defining the formal game, we introduce it through a story.

Prisoners' Dilemma STORY

The **Prisoners' Dilemma** involves two people, each accused of the same crime, who are separated so that they cannot speak to one another (to corroborate their stories). Each prisoner has two choices: to maintain his or her innocence or to sign a confession accusing the partner of being the mastermind behind the crime. It is in each suspect's interest to confess and implicate the partner, in an effort to receive a reduced sentence. Yet if both suspects confess, they ensure a bad outcome—namely, they are both found guilty. What is good for the prisoners as a pair—to deny having committed the crime, leaving the state with insufficient evidence to convict them—is frustrated by their pursuit of their own individual interests.

The game of Prisoners' Dilemma, as we already noted, has many applications, but we will use it here to model a recurrent problem in international relations: arms races between antagonistic countries, which earlier included the superpowers of the United States and the Soviet Union but more recently have included such countries as India and Pakistan and Israel and some of its Arab neighbors. Other countries, such as Iran, may be antagonistic to more than one other country (e.g., Israel and the United States).

For simplicity, assume there are two nations, Red and Blue. Each can independently select one of two policies:

A: Arm in preparation for a possible war (noncooperation).
D: Disarm, or at least try to negotiate an arms-control agreement (cooperation).

There are four possible outcomes:

(*D*, *D*): Red and Blue disarm, which is *next best* for both because, while advantageous to each, it also entails certain risks.
(*A*, *A*): Red and Blue arm, which is *next worst* for both because they spend needlessly on arms and are comparatively no better off than at (*D*, *D*).

(*A*, *D*): Red arms and Blue disarms, which is *best for Red* and *worst for Blue* because Red gains a big edge over Blue.

(*D*, *A*): Red disarms and Blue arms, which is *worst for Red* and *best for Blue* because Blue gains a big edge over Red.

This situation can be modeled by means of the matrix in Table 15.8, which gives the possible outcomes. Here, Red's choice involves picking one of the two rows, whereas Blue's choice involves picking one of the two columns.

TABLE 15.8 The Outcomes in an Arms Race, as Modeled by the Prisoners' Dilemma

		Blue	
		A	***D***
Red	***A***	Arms race	Favors Red
	D	Favors Blue	Disarmament

We assume that the players can rank the four outcomes from best to worst, where 4 = best, 3 = next best, 2 = next worst, and 1 = worst; thus, the higher the number, the greater the payoff. The resulting game is an **ordinal game:** It indicates an ordering of outcomes from best to worst but says nothing about the *degree* to which a player prefers one outcome over another. To illustrate, if a player despises the outcome that he or she ranks 1 but sees little difference among the outcomes ranked 4, 3, and 2, the "payoff distance" between 4 and 2 will be less than that between 2 and 1, even though the numerical difference between 4 and 2 is greater.

Self Check 6

Return to the payoff matrix in Table 15.3 from the restricted location/schedule game between Mark and Lisa. Rewrite the payoff matrix using ordinal payoffs from Mark's perspective. How does this compare to rewriting the payoff matrix using ordinal payoffs from Lisa's perspective?

The ordinal payoffs to the players for choosing their strategies of *A* and *D* are shown in Table 15.9, where the first number in the pair indicates the payoff to the row player (Red), and the second number the payoff to the column player (Blue). Thus, for example, the pair (1, 4) in the second row and first column signifies a payoff of 1 (worst outcome) to Red and a payoff of 4 (best outcome) to Blue. This outcome occurs when Red unilaterally disarms while Blue continues to arm, making Blue, in a sense, the winner and Red the loser.

TABLE 15.9 Ordinal Payoffs in an Arms Race, as Modeled by the Prisoners' Dilemma

		Blue	
		A	***D***
Red	***A***	(2, 2)	(4, 1)
	D	(1, 4)	(3, 3)

Let's examine this strategic situation more closely. Should Red select Strategy *A* or *D?* There are two cases to consider, which depend on what Blue does:

- If Blue selects *A:* Red will receive a payoff of 2 for *A* and 1 for *D,* so it will choose *A.*
- If Blue selects *D:* Red will receive a payoff of 4 for *A* and 3 for *D,* so it will choose *A.*

In both cases, Red's first strategy (*A*) gives it a more desirable outcome than its second strategy (*D*). Consequently, we say that *A* is Red's **dominant strategy** because it is always advantageous for Red to choose *A* over *D.*

In the Prisoners' Dilemma, *A dominates D* for Red, so we presume that a rational Red would choose *A.* A similar argument leads Blue to choose *A* as well—that is, to pursue a policy of arming. Thus, when each nation strives to maximize its own payoffs independently, the pair is driven to the outcome (*A, A*), with payoffs of (2, 2). The better outcome for both, (*D, D*), with payoffs of (3, 3), appears unobtainable when this game is played noncooperatively.

The outcome (*A, A*), which is the product of dominant strategy choices by both players in the Prisoners' Dilemma, is called a **Nash equilibrium.**

Nash Equilibrium — DEFINITION

When no player can benefit by departing unilaterally (by itself) from its strategy associated with an outcome, the strategies of the players constitute a **Nash equilibrium.** Technically, while it is the set of strategies that defines the equilibrium, the choice of strategies leads to an outcome that we shall also refer to as the *equilibrium.*

Note that in the Prisoners' Dilemma, if either player departs from (*A, A*), the payoff for the departing player who switches to *D* drops from 2 to 1 at (*D, A*) and (*A, D*). Not only is there no benefit from departing, but there is actually a loss, with the player selecting *D* being punished with its worst payoff of 1. These losses would presumably deter each nation from moving away from the Nash equilibrium of (*A, A*), assuming that the other nation sticks to *A.*

Even if both nations agreed in advance jointly to pursue the socially beneficial outcome, (*D, D*), in which both nations disarm and receive payoff 3, the result is called *unstable:* If either nation alone reneges on the agreement and secretly arms (as North Korea did when it developed nuclear weapons), it will benefit, obtaining its best payoff of 4. Consequently, each nation would be tempted to go back on its word and select *A.* Nations with no great confidence in the trustworthiness of their opponents have good reason to try to protect themselves against the other side's defection from an agreement by arming.

Prisoners' Dilemma — DEFINITION

The **Prisoners' Dilemma** is a two-person variable-sum game in which each player has two strategies, cooperate or defect (not cooperate). Defect dominates cooperate for both players, even though the mutual-defection outcome, which is the unique Nash equilibrium in the game, is worse for both players than the mutual-cooperation outcome.

Note that if 4, 3, 2, and 1 in the Prisoners' Dilemma were not just ranks but numerical payoffs, their sum would be 2 + 2 = 4 at the mutual-defection outcome and 3 + 3 = 6 at the mutual-cooperation outcome. At the other two outcomes, the sum, 1 + 4 = 5, is still different, illustrating why the Prisoners' Dilemma is a variable-sum game.

In real life, of course, people often manage to escape the noncooperative Nash equilibrium in the Prisoners' Dilemma. Either the game is played within a larger context, in which other incentives are at work, such as cultural norms that prescribe cooperation [though this is just another way of saying that defection from (*D, D*) is not rational, rendering the game not the Prisoners' Dilemma], or the game is played on a repeated basis—it is not a one-shot affair—so players can induce cooperation by setting a pattern of rewards for cooperation and penalties for noncooperation.

In a repeated game, factors like reputation and trust may play a role. Realizing the mutual advantages of cooperation in costly arms races, players may inch toward the cooperative outcome by slowly phasing down their acquisition of weapons over time, or even destroying them. (The United States and Russia have been doing exactly this.) They may also initiate other productive measures, such as improving their communication channels, making inspection procedures more reliable, writing agreements that are truly enforceable, or imposing penalties for violators when their violations are detected (as has occurred through reconnaissance or spy satellites).

The Prisoners' Dilemma illustrates the intractable nature of certain competitive situations that blend conflict and cooperation. The standoff that results at the Nash equilibrium of (2, 2) is obviously not as good for the players as that which they could achieve by cooperating—but they risk a good deal if the other player defects.

While saddlepoints are Nash equilibria in total-conflict games, they can never be worse for *both* players than some other outcome (as in partial-conflict games like the Prisoners' Dilemma). The reason is that if one player does worse in a total-conflict or zero-sum game, the other player must do better.

The fact that the players must forsake their dominant strategies to achieve the (3, 3) cooperative outcome (see Table 15.9) makes this outcome a difficult one to sustain in one-shot play. On the other hand, assume that the players can threaten each other with a policy of tit-for-tat in repeated play: "I'll cooperate on each round unless you defect, in which case I will defect until you start cooperating again." If these threats are credible, the players may well shun their defect strategies and try to establish a pattern of cooperation in early rounds, thereby fostering the choice of (3, 3) in the future.

Let's look at one other two-person game of partial conflict, known as **Chicken,** that can also lead to troublesome outcomes.

EXAMPLE 8 Chicken

In Chicken, two drivers approach each other at high speed. Each must decide at the last minute whether to swerve to the right (to avoid crashing) or to not swerve. This scenario occurs with tractors in the movie *Footloose* (1984) and with buses in the 2011 remake, and a variation occurs when two cars approach the edge of a cliff in the classic James Dean movie *A Rebel without a Cause.* Here are the possible consequences of their actions:

1. Neither driver swerves, and the cars collide head-on, which is the worst outcome for both because they are both killed (payoff of 1).
2. Both drivers swerve—and each is mildly disgraced for "chickening out"—but they do survive, which is the second-best outcome for both (payoff of 3).

3. One of the drivers swerves and badly loses face, which is his second-worst outcome (payoff of 2), whereas the other does not swerve and is perceived as the winner, which is her best outcome (payoff of 4).

These outcomes and their associated strategies are summarized in Table 15.10.

TABLE 15.10 Payoffs in a Driver Confrontation, as Modeled by Chicken

		Driver 2	
		Swerve	**Not Swerve**
Driver 1	**Swerve**	(3, 3)	(2, 4)
	Not Swerve	(4, 2)	(1, 1)

If both drivers persist in their attempts to "win" with a payoff of 4 by not swerving, the resulting outcome will be mutual disaster, giving each driver his or her worst payoff of 1. Clearly, it is better for both drivers to back down and each obtain 3 by swerving, but neither wants to be in the position of being intimidated into swerving (payoff of 2) when the other does not (payoff of 4).

Notice that neither player in Chicken has a dominant strategy. His or her better strategy depends on what the other player does: Swerve if the other does not, don't swerve if the other player swerves, making this game's choices highly interdependent, which is characteristic of many games. The Nash equilibria in Chicken, moreover, are (4, 2) and (2, 4), suggesting that the compromise of (3, 3) will not be easy to achieve because both players will have an incentive to deviate in order to try to be the winner. This game is referred to as an anticoordination game because the Nash equilibria of (4, 2) and (2, 4) require the players to play different strategies.

Chicken DEFINITION

Chicken is a two-person variable-sum game in which each player has two strategies: to swerve to avoid a collision or not to swerve and possibly cause a collision. Neither player has a dominant strategy. The compromise outcome, in which both players swerve, and the disaster outcome, in which both players do not, are not Nash equilibria. The other two outcomes, in which one player swerves and the other does not, are Nash equilibria.

In fact, there is a third Nash equilibrium in Chicken, but it is in mixed strategies, which can be computed only if the payoffs are not ranks, as we have assumed here, but numerical values. Even if the payoffs were numerical, however, it can be shown that this equilibrium is always worse for both players than the cooperative (3, 3) outcome. Moreover, it is implausible that players would sometimes swerve and sometimes not—randomizing according to particular probabilities—in the actual play of this game, compared with either trying to win outright or reaching a compromise.

The two pure-strategy Nash equilibria in Chicken suggest that, insofar as there is a "solution" to this game, it is that one player will succeed when the other caves in to avoid the mutual-disaster outcome. But there are certainly real-life cases in which a major confrontation was defused and a compromise of sorts was achieved in a Chicken-type game. This fact suggests that the one-sided solutions given by the two pure-strategy Nash equilibria may not be the only pure-strategy solutions, especially if the players are farsighted and think about the possible untoward consequences of their actions.

International crises, labor–management disputes, and other conflicts in which escalating demands may end in wars, strikes, and other catastrophic outcomes have been modeled by the game of Chicken (see Spotlight 15.2 on page 646 for more on game theorists who have analyzed these and other games). But it can be shown that Chicken, like the Prisoners' Dilemma, is only one of the 78 essentially different 2 × 2 ordinal games in which each player can rank the four possible outcomes from best to worst.

Chicken and the Prisoners' Dilemma, however, are especially disturbing because the cooperative (3, 3) outcome in each is not a Nash equilibrium. Unlike a constant-sum game, in which the losses of one player are offset by the gains of the other, *both* players can end up doing badly—at (2, 2) in the Prisoners' Dilemma and (1, 1) in Chicken—in these variable-sum games.

What seems to be needed is a way to either change a game by providing incentives for players to follow desirable behavior, or for games to be designed to yield desirable outcomes. The latter notion is called *mechanism design* and will be considered in the next section.

15.4 Mechanism Design and Larger Games

We have shown how to compute optimal pure and mixed strategies, and the values ensured by using them, in 2 × 2 constant-sum games. In 2 × 2 variable-sum games, we focused on Nash equilibria as a solution concept in the Prisoners' Dilemma and Chicken, but we found that this notion of a stable outcome did not justify the choice of cooperative strategies in either of these games.

We turn next to a game in which there may be any number of players, each of whom can choose among an infinite number of strategies! Although this game sounds more complex, it demonstrates that it may be easy to determine a Nash equilibrium, and it may be easy for each player to determine an optimal strategy.

The game models a specific type of auction. An auction is a common way to sell an item when the seller is unsure of the going price of the item; the auction is designed to get the seller the highest price for the item. For example, Sotheby's auction house specializes in high-end contemporary art and rare items such as the British Guiana one-cent magenta stamp, considered the rarest and most valuable stamp in the world. Because these items are one of a kind, the seller has difficulties in knowing what the market value for the item is. Auctions have become more common as websites like eBay hold auctions in which bidders can participate online at any hour of the day. Stamp collectors had used the following type of auction well before Columbia University economist William Vickrey analyzed it formally in 1961.

A Vickrey Auction DEFINITION

In a **Vickrey auction,** also called the *second-price sealed-bid auction,* bidders independently submit sealed bids for an item being sold. The winner is the one who bids the highest, but he or she pays only the amount of the second-highest bid.

A Vickrey auction involves a bidding strategy that is at least as good as any other. This notion weakens the idea of a dominant strategy that was introduced when analyzing the Prisoners' Dilemma. A weakly dominant strategy is one that is at least as good as any other strategy. Flipping perspective, we define a dominated strategy.

Dominated Strategy DEFINITION

A player's strategy *S* is **dominated** by his or her strategy *T* if the outcome under *T* is always better for the player than the outcome under *S*, regardless of the decisions made by the other players.

In the location/schedule game of Example 1 with payoff matrix from Table 15.1 (page 623), Mark chooses the location of the hospital, while Lisa chooses the shift. Recall that Mark is concerned about making the highest hourly wage possible for his emergency room work. For this reason, Mark will never select the Rural hospital because he would receive a lower hourly wage than he would at the Suburban hospital, regardless of which shift Lisa selects. This is apparent because the numbers in the Suburban row of Table 15.1 are greater than the numbers in the Rural row. We say that the Rural strategy is dominated by the Suburban strategy.

Dominated strategies can be eliminated from consideration, as there will never be a situation in which a dominated strategy should be played.

EXAMPLE 9 The Vickrey Auction

Suppose that Anneliese, Binh, and Charlie bid for a stamp in a Vickrey auction. Anneliese, Binh, and Charlie value the stamp at \$300, \$200, and \$100, respectively. To see that it is always at least as good for Anneliese to bid her true valuation of the stamp, consider the two possible results of bidding \$300. To simplify the analysis, we will assume that each bidder's bid is distinct, so no two bidders bid the same amount.

- If Anneliese's bid of \$300 is the highest bid, then she wins the item but pays the second-highest bid \$*SH*, which is strictly less than \$300. To represent the Vickrey auction as a game, Anneliese's payoff in this case is given by \$300 − \$*SH*, which is a positive number because Anneliese was able to buy the stamp for less than she valued it. If she had decided to bid more than \$300, then she would still be the highest bidder and still pay the same amount \$*SH*. However, if she decides to bid less than \$300, then it is possible that her bid will be less than \$*SH*, in which case she loses the item and has a payoff of \$0. Anneliese has no incentive to raise or to lower her bid because she can do no better than a payoff of \$300 − \$*SH*, but she can do worse, receiving a payoff of \$0 if she bids too low.
- If Anneliese's bid of \$300 is not the highest bid, then she does not receive the item, and naturally does not have to pay for it—she receives a payoff of \$0. Let \$*H* be the highest bid. If she had decided to bid more than \$300, say, more than \$*H*, then she becomes the highest bidder and would have had to pay \$*H*, which is more than \$300—the value she placed on the stamp. This means that

she would have overpaid for the stamp and her payoff would be the negative amount of \$300 − \$*H*. If she bids less than \$300, she would still fail to have the highest bid and would still not have to pay anything. As before, Anneliese has no incentive to raise or to lower her bid.

Anneliese's strategy of bidding \$300 weakly dominates any other strategy she could use. A similar analysis holds for Binh and Charlie. Collectively, it is a Nash equilibrium for Anneliese, Binh, and Charlie to each bid his or her true valuation for the stamp. To see this, if Anneliese, Binh, and Charlie bid \$300, \$200, and \$100, respectively, for the stamp, then Anneliese will be the highest bidder and pay \$200 for the stamp. The payoffs for Anneliese, Binh, and Charlie are given by the triplet (\$300 − \$200 = \$100, \$0, \$0). Neither Binh nor Charlie could change his bid to receive a positive payoff. Anneliese's bid cannot affect the price she pays; it can determine only whether or not she is the highest bidder. Likewise, she cannot change her bid to do better and could only do worse by bidding too little and losing the stamp to Binh. ■

If more than one player has the highest bid, then one of the players is selected at random to receive the item and the second highest bid is equal to the highest bid. The analysis in the example extends for a Vickrey auction with any number of players. Each player has a weakly dominant strategy to bid his or her true valuation of the item. It leads to the following result.

Nash Equilibrium in a Vickrey Auction THEOREM

In a Vickrey auction, it is a Nash equilibrium for each bidder to bid the amount at which he or she values the item.

The Vickrey auction induces the bidders to reveal truthfully their values for the item being auctioned (see Chapter 13, Section 13.8, for more on Vickrey auctions). However, there are other equilbria for the Vickrey auction game. For the stamp auction, bids of \$201, \$200, and \$100 and of \$200, \$301, and \$100 by Anneliese, Binh, and Charlie, respectively, are also Nash equilibria. The second of these two equilibria is surprising because Binh can win the item! For games in which each player has a dominant strategy (like the Prisoners' Dilemma), then there is a single Nash equilibrium. It is also a Nash equilibrium for every player to use a weakly dominant strategy, but there may be more than one equilibrium if each player has a weakly dominant strategy, as demonstrated by the Vickrey auction game.

Self Check 7

Suppose Anneliese, Binh, and Charlie bid \$150, \$301, and \$100, respectively, at the stamp auction. Why does Anneliese not prefer to change her bid? Why does Binh prefer not to change his bid? Does your answer to the question about Anneliese also hold for Charlie? Answering these three questions shows that when Anneliese, Binh, and Charlie bid as described, it is a Nash equilibrium: Collectively, if no player wishes to change his or her bid, the bids form a Nash equilibrium. ■

Mechanism design is the act of designing a game to achieve a particular outcome as a Nash equilibrium. Despite the multiple equilibria for the Vickrey auction game, the Vickrey auction does give each bidder an incentive to reveal truthfully how much he or she values the item. Another example of mechanism design is the design of matching markets, described in Spotlight 15.2. Similar objectives border on the intersection of economics and psychology, in which decision problems are posed in a particular way to encourage people to make better decisions. For example, the placement of healthy food by a checkout register in a school's lunchroom results in more healthy choices by students.

The Nobel Prize in Economics

SPOTLIGHT 15.2

The Nobel Memorial Prize in Economics was awarded to three game theorists in 1994, marking the 50th anniversary of the publication of the first text on game theory (*Theory of Games and Economic Behavior*) by mathematician John von Neumann and economist Oskar Morgenstern.

- *John C. Harsanyi* (1920–2000) of the University of California, Berkeley, a Hungarian-American who emigrated from Hungary to Australia in 1950 and then to the United States in 1956. He is well known for extending game theory to the study of ethics and showing how societal institutions, each of whose members' satisfaction can be measured against that of others, choose among alternatives. His other major contribution was to give a precise definition to "incomplete information" in games in which players may be thought of as different types, and probabilities are assigned to each type.
- *John F. Nash, Jr.* (1928–2015) of Princeton University, an American mathematician who did pathbreaking work on both noncooperative game theory (the Nash equilibrium is named after him) and cooperative game theory, especially on bargaining, in which axioms or assumptions are specified and a unique solution that satisfies these axioms is derived. Nash obtained his results in the early 1950s, when he was only in his 20s, after which he became mentally ill and was unable to work. Fortunately, he made a remarkable recovery and resumed research until he was killed in a car accident.
- *Reinhard Selten* (b. 1930) of the University of Bonn, a German mathematician who proposed significant refinements in the concept of the Nash equilibrium that help to distinguish those that are most plausible in games (often there are many such equilibria, which creates a selection problem). Selten is also noted for pioneering work on developing game-theoretic models in evolutionary biology.

Jan Collsioo/AFP/Getty Images
John C. Harsanyi

Bob Strong/AFP/Getty Images
John F. Nash, Jr.

© Vario images GmbH & Co. KG/Alamy
Reinhard Selten

In 2005, mathematician Robert J. Aumann was awarded the prize for a variety of advances in cooperative and noncooperative game theory, and economist Thomas C. Schelling also was awarded the prize for his contributions to the study of conflicts involving promises, threats, and other kinds of commitments. Mathematician Lloyd Shapley and economist Alvin Roth were awarded the 2012 Nobel Prize for their application of game theory to the design of markets (see the Case Study at the end of Part I, page 174). Highly successful implementations of these types of matching markets include "The Match," in which physicians are paired up with residency programs, and school choice, in which students are enrolled in schools across Boston and New York based on different criteria.

Because the bidders could bid any amount, it was not practical to use a payoff matrix to describe the Vickrey auction game. However, for two-player games, we have used payoff matrices to describe games in strategic form. In these games, the row and column players' choices of strategies led to an outcome from which each player received a payoff. These strategy choices were assumed to be simultaneous, though mentally the players might eliminate some dominated strategies. This was the case of the schedule/location game from Example 1, as Mark would never select the Rural hospital.

In the next example, which involves a larger game, we start by assuming simultaneous choices and show what outcome would occur. Then we assume that the choices of the players need not be simultaneous, instead allowing for one player to move first. We use a **game tree** to analyze the *sequential choices* players can then make, as occurs when first you move, then I move, and so on—which are called *games in extensive form*. As we will see, the outcome in such a game may be wholly different from the outcome in a game with simultaneous choices, which raises the question of which kind of game is the most realistic model of a situation.

EXAMPLE 10 A Truel

A *truel* is like a duel, except that there are three players. Truels are depicted in several movies, including *The Good, the Bad and the Ugly* (1966), *Reservoir Dogs* (1992), and *Pulp Fiction* (1994). The photo below shows three of the characters in *The Good, the Bad and the Ugly,* about to engage in a truel—each man is armed and planning his next move.

P.E.A./The Kobal Collection at Art Resource, NY

In a truel, each player can either fire or not fire his or her gun at either of the other two players. We assume the goal of each player is, *first,* to survive and, *second,* to survive with as few other players as possible. Each player has one bullet and is a perfect shot; no communication (e.g., to pick out a common target) leading to a binding agreement with other players is allowed, making the game noncooperative. We will discuss the answers that simultaneous choices, on the one hand, and sequential choices, on the other, give to what is optimal for the players to do in the truel.

If choices are simultaneous, at the start of play, each player will fire at one of the other two players, killing that player.

Why will the players all fire at each other? Because their own survival does not depend an iota on what they do. Since they cannot affect what happens to themselves but can affect how many others survive (the fewer the better, according to the postulated secondary goal), they should all blaze away at each other. In fact, the players all have dominant strategies to shoot at each other because whether or not a player survives—we will discuss shortly the probabilities of doing so—he or she does at least as well shooting an opponent.

The game, and optimal strategies in it, would change if the players (1) were allowed more options, such as to fire in the air and thereby disarm themselves, or (2) did not have to choose simultaneously but, instead, a particular order of play were specified. Thus, if the order of play were *A,* followed by *B* and *C* choosing simultaneously, followed by any player with a bullet remaining choosing, then *A* would fire in the air, and *B* and *C* would subsequently shoot each other. (*A* is no threat to *B* or *C,* so neither of the latter will fire at *A* and waste a bullet; on the other hand, if one of *B* or *C* did not fire immediately at the other, that player would not survive to get in the last shot, so both *B* and *C* will fire at each other.) Thus, *A* will be the sole survivor.

In 1992, a modified version of this scenario was played out in late-night television programming among the three major TV broadcasting networks of the time, with ABC effectively going first with *Nightline,* its well-established news program, and CBS and NBC dueling about which host, David Letterman or Jay Leno, to choose for their entertainment shows. Regardless of their ultimate choices, ABC "won" when CBS and NBC were forced to divide the entertainment audience. In 2002, ABC, presumably to attract a younger audience than the one that watched *Nightline,* attempted unsuccessfully to hire Letterman from CBS. Ted Koppel retired in 2005, but *Nightline* continues with other hosts.

To return to the original game (all choose simultaneously), the players' strategies of all firing have two possible consequences: Either one player survives (even if two players fire at the same person, the third must fire at one of them, leaving only one survivor), or no player survives (if each player fires at a different person). In either event, there is no guarantee of survival. In fact, if each player has an equal probability of firing at one of the two other players, the probability that any player will survive is only 25%. The reason is that if the three players are *A, B,* and *C, A* will be killed if *B* fires at him or her, *C* does, or both do. The only circumstance in which *A* will survive is if *B* and *C* fire at each other, which gives *A* 1 chance in 4.

If choices are sequential, no player will fire at any other, so all will survive.

At the start of the truel, all the players are alive, which satisfies their primary goal of survival, though not their secondary goal of surviving with as few others as possible. Now assume that *A* contemplates shooting *B,* thereby reducing the number of survivors. Looking ahead, however, *A* knows that by firing first and killing *B,* he or she will be defenseless and be immediately shot by *C,* who will then be the sole survivor.

It is in *A*'s interest, therefore, not to shoot anybody at the start, and the same logic applies to each of the other players. Hence, everybody will survive, which is a happier outcome than when choices are simultaneous, in which case everyone's primary goal of survival is not satisfied—or, quantitatively speaking, satisfied only 25% of the time. ■

While sequential choices produce a "happier" outcome, do they provide a plausible model of a strategic situation that mimics what people might actually think and do in such a situation? We believe that the players in the truel, artificial as this kind of shootout may seem, would be motivated to think ahead, given the dire consequences of their actions. Therefore, they would hold their fire, knowing that if one fired first, he or she would be the next target. Indeed, this truel scenario is parodied in "Stand Off," a *Saturday Night Live* skit broadcast on November 17, 2012, in which the three characters go about a day in their lives with guns trained on one another and no one wanting to fire first.

In Figure 15.4, we show this logic somewhat more formally with a *game tree,* in which *A* has three strategies, as indicated by the three branches that sprout from *A:* not shoot ($\overline{S}$), shoot *B,* or shoot *C.* The latter two branches, in turn, give survivors *C* and *B,* respectively, two strategies: not shoot ($\overline{S}$) or shoot *A.*

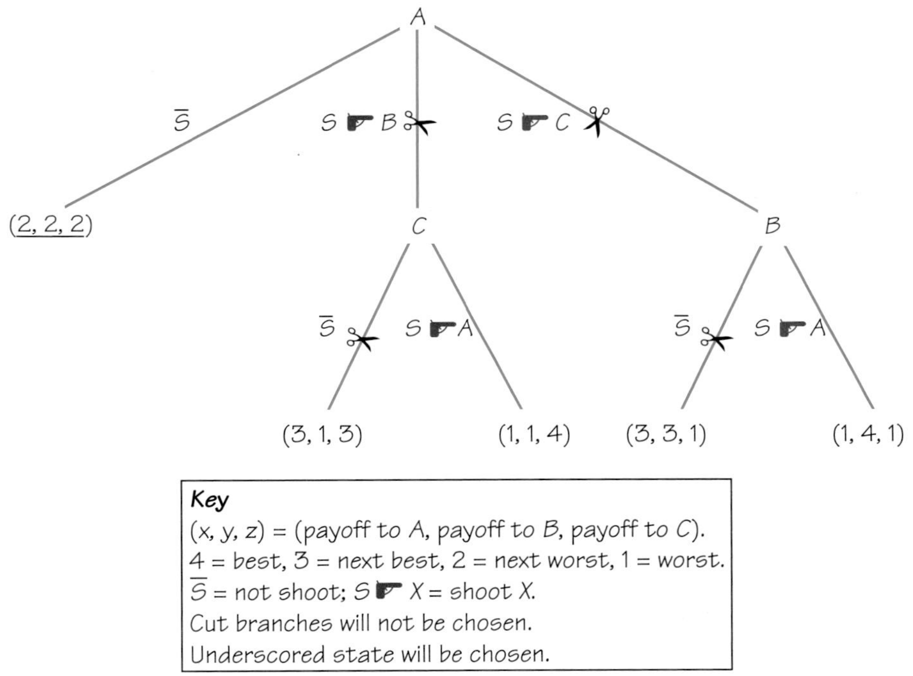

Figure 15.4 A game tree of a truel.

We assume that the players rank the outcomes as follows, which is consistent with their primary and secondary goals: 4 = best (lone survivor), 3 = next best (survivor with one other), 2 = next worst (survivor with two others), and 1 = worst (nonsurvivor). These payoffs are given for ordered triples (*A, B, C*); thus, (3, 3, 1) indicates the next-best payoffs for *A* and *B* and the worst payoff for *C.*

Note that play necessarily terminates when there is only one survivor, as is the case at (1, 1, 4) and (1, 4, 1). To keep the tree simple, we assume that play also terminates when either *A* initially chooses $\overline{S}$, or *B* or *C* subsequently chooses $\overline{S}$, giving outcomes of (2, 2, 2), (3, 3, 1), and (3, 1, 3), respectively. Of course, we could allow the two or three surviving players in the latter cases to make subsequent choices in an extended game tree, but this example is meant only to illustrate the analysis of a game tree, not to be the definitive statement on truel possibilities. (More will be explored in the exercises.)

In a game in extensive form represented by a game tree, players work backward, starting the analysis at the bottom of the tree. By "bottom" we mean where play terminates; because this is where the tree branches out, the tree looks upside down in Figure 15.4. The players then work up the tree, using backward induction. **Backward induction** is a reasoning process in which players, working backward from the last possible moves in a game, anticipate each other's rational choices.

To illustrate, because *C* prefers (1, 1, 4) to (3, 1, 3), we indicate that *C* would not choose $\overline{S}$ by "cutting" this branch with scissors. Similarly, *B* would not choose $\overline{S}$. Thus, if play got to the bottom of the tree, *C* would shoot *A* if *C* were the survivor, and *B* would shoot *A* if *B* were the survivor, following *A*'s shooting *C* or *B*, respectively.

Moving up to the next level, *A* would know that if he or she shot *B*, (1, 1, 4) would be the outcome. If he or she shot *C*, (1, 4, 1) would be the outcome, making one or the other the outcome from the bottom level. Choosing between these two outcomes and (2, 2, 2), *A* would prefer the latter, so *A* would cut the two branches, *S* would shoot *B*, and *S* would shoot *C*. Hence, *A* would choose $\overline{S}$, terminating play with nobody shooting anybody else.

This, of course, is the conclusion that we reached earlier, based on the reasoning that if *A* shot either *B* or *C*, he or she would end up dead, too. Because we could allow each player, like *A*, to choose among his or her three initial strategies in a 3 × 3 × 3 game, and subsequently make moves and countermoves from the initial state (if feasible), the foregoing analysis applies to all players.

Self Check 8

Use backward induction to determine how players *A* and *B* would make their sequential decisions for the game tree in Figure 15.5.

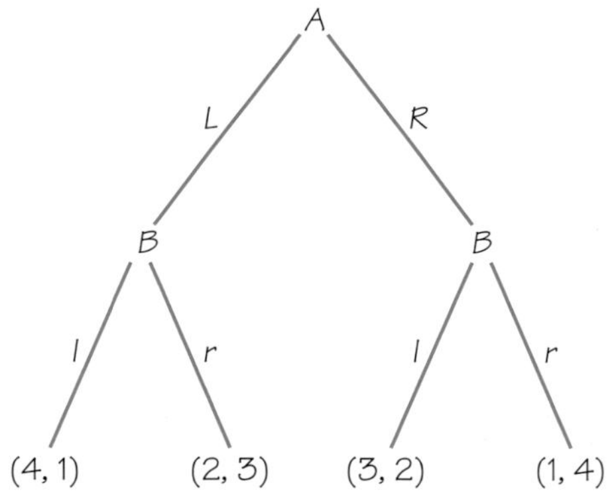

Figure 15.5 An extensive game in which player *A* first chooses between *L* and *R* and player *B* then chooses between *l* and *r*.

■

Underlying the completely different answers given by the simultaneous and sequential choices in a truel is a change in the rules of play. If play is sequential, the players do not have to fire simultaneously at the start, as assumed in a 3 × 3 × 3 strategic-form game. Rather, a player who moves first (*A* in our example)—and then the later players—would not fire, given that play continues until all bullets are expended or nobody chooses to fire.

In the extensive-form analysis, we ask of each player (it need not be *A*): Given your present situation (all alive), and the situation you anticipate will ensue if you fire first, should you do so? Because each player prefers living to the state he or she would induce by being the first to shoot (certain death), no one shoots.

This analysis suggests that truels might be more effective than duels, at least when played sequentially, in preventing the outbreak of conflict.

We will not try to develop this argument into a more general model. The main point is that a game tree allows for a look-ahead approach, whereby players compare the present state with possible future states—perhaps several steps ahead—to determine which moves to make. These choices, as we have seen, may lead to radically different outcomes compared with outcomes based on simultaneous choices.

15.5 Using Game Theory

Solving Games

For a two-player game with a payoff matrix, the first thing we ask is whether it is zero-sum (or constant-sum). If so, we check to see whether it has a saddlepoint by determining the minimum number of each row and the maximum number of each column, as we did in several earlier examples. If the maximum of the row minima (maximin) is equal to the minimum of the column maxima (minimax), then the game has a saddlepoint. The resulting value, and the corresponding pure strategies, provide a solution to the game.

This value will appear in the payoff matrix as the smallest number in its row and the largest in its column. In the 3×3 location/schedule game in Example 1 (page 623), this number was \$26 per hour, and the corresponding strategies for Mark and Lisa were to select the suburban location and the 8 A.M. to 4 P.M. shift, respectively.

Because Mark's strategy of selecting the rural hospital is dominated by selecting the suburban hospital, Mark has no reason to ever choose the rural hospital. Once Lisa knows that Mark will never select the rural hospital, then Lisa can eliminate the 4 P.M. to 12 A.M. shift, because the 8 A.M. to 4 P.M. shift is better from her perspective under both the suburban and downtown locations. Sometimes successively eliminating dominated strategies leads to a single outcome. Indeed, after Mark eliminates the rural location, Lisa also can eliminate the 12 A.M. to 8 A.M. shift. Once Lisa selects the 8 A.M. to 4 P.M. shift, then Mark chooses the suburban location over the downtown hospital. The successive elimination of dominated strategies leads to the saddlepoint of the game—and a saddlepoint is always a Nash equilibrium.

Unfortunately, the successive elimination of dominated strategies does not work to find the saddlepoint in *all* two-person zero-sum games larger than 2×2.

Recall that instead of eliminating dominated strategies in the 3×3 location/schedule game, we eliminated the 8 A.M. to 4 P.M. shift, which dominated other strategies in the successive process, to obtain the 3×2 restricted location game in Table 15.3. In this game, the rural location is still dominated, but no other eliminations are possible; hence, the game has no saddlepoint.

If a two-person zero-sum game does not have a saddlepoint—which was the case not only in the restricted location/schedule game but also for Matching Pennies, the second matching game, and the kicker/goalie duel—the solution will be in mixed strategies. To find the optimal mix in a 2×2 game, we calculate the expected value to a player from choosing its first strategy with probability p and its second with probability $1 - p$, assuming that the other player chooses its first pure strategy (yielding one expected value) and its second pure strategy (yielding another expected value).

Setting these two expected values equal to each other yields a unique value for p that gives the optimal mix, $(p, 1 - p)$, with which the player should choose its first and second strategies. Inserting the numerical solution of p back into either expected-value equation gives the value of the game, which each player can guarantee for itself, whatever strategy its opponent chooses.

Several general algorithms have been developed to find mixed-strategy solutions for large constant-sum games. This work has mostly been done in the field of linear programming, using such algorithms as the simplex method of G. B. Dantzig and the more recent method of N. K. Karmarkar (see Chapter 4).

In variable-sum games, we also begin by successively eliminating dominated strategies, if there are any. The outcomes that remain do not depend on the numerical values that we attach to them but only on their ranking from best to worst by the players.

Solving variable-sum games can be fairly demanding, as it may involve considerable calculational abilities on the part of the players. Less demanding, of course, is that players simply choose their dominant strategies or weakly dominant strategies, as is possible in the Prisoners' Dilemma and the Vickrey auction, respectively. Of course, games may not have such strategies.

In the game of Chicken, for example, neither player has a dominant (or dominated) strategy, so the game cannot be reduced. In such situations, we ascertain what outcomes are Nash equilibria. There are two (in pure strategies) in Chicken, suggesting that the only stable outcomes in this game occur when one player gives in and the other does not. In the Prisoners' Dilemma, by comparison, the choice by the players of their dominant strategies singles out the mutual-defection outcome as the unique Nash equilibrium, which is worse for both players than the cooperative outcome.

Practical Applications

The element of surprise, as captured by mixed strategies, is essential in many encounters. For example, mixed strategies are used in various inspection procedures and auditing schemes to deter potential violators. When inspection or auditing choices are made random, they are rendered unpredictable.

Police and regulatory agencies monitor certain activities to check for illegal actions. Investigators who conduct surveillance include FBI agents, customs agents, bank auditors, insurance investigators, quality-control experts, and drug testers. The National Bureau of Standards is responsible for monitoring the accuracy of measuring instruments and for maintaining reliable standards. The Nuclear Regulatory Agency demands an accounting of dangerous nuclear material as part of its safeguards program. The Internal Revenue Service attempts to identify people who cheat on taxes.

Military or intelligence services may wish to intercept a weapon hidden among many decoys or to plant a secret agent disguised to look like a respectable individual. Because it is prohibitively expensive to check the authenticity of each and every possible item or person, efficient methods must be used to check for violations. Both optimal detection and optimal concealment strategies can be modeled as a game between an inspector trying to increase the probability of detection and a violator trying to evade detection. Since the World Trade Center attack on September 11, 2001, United States government agencies have taken much stronger measures to prevent such evasion, including the formation of the Transportation Security Administration, whose agents check for contraband when passengers board flights.

Some inspection games are constant-sum: The violator "wins" when the evasion is successful and "loses" when it is not. On the other hand, cheating on arms-control agreements may well be variable-sum if both the inspector and the cheater would prefer that no cheating occur rather than cheating and public disclosure of it. The latter could be an embarrassment to both sides, especially if it undermines an arms-control agreement that both sides wanted and the cheating is minor.

We alluded earlier to the strategy of bluffing in poker, used by a player to try to keep the other player or players guessing about the true nature of one's hand. The optimal probability with which one should bluff can be calculated in a particular situation (see Exercise 21 on page 661). Besides poker, bluffing is common in many bargaining situations, whereby a player raises the stakes (e.g., labor threatens a strike in labor–management negotiations), even if it may ultimately have to back down if its "hand is called."

Jewel Samad/AFP/Getty Images

Perhaps the greatest value of game theory is the framework it provides for understanding the rational underpinnings of conflict in the world today. As a case in point, a confrontation over the budget between Democratic president Barack Obama and the Republican House of Representatives resulted in the third longest shutdown of (part of) the federal government, from October 1 to 16, 2013. The shutdown occurred due to the failure to pass legislation for the 2014 fiscal year or to pass a continuing resolution to allow the government to continue spending money. Many government workers were frustrated at not being able to do their jobs, even though they realized that they would be paid for not working given past shutdowns, not to mention the many citizens who were either greatly hurt or substantially inconvenienced by the shutdown. Viewed as a game of Chicken, in which each side wanted not only to get its way for the moment but also to influence future negotiations, this conflict may not have been as foolish as it seems at first glance.

Another example in which game theory is useful is to analyze the behavior of free riders. A **free rider** is someone who benefits from resources, goods, services, or a policy without paying the cost of the benefit. For example, if someone hops the turnstile on a subway train and gets to ride the train for free, then the other riders are paying for his or her fare. The other riders pay for the free rider by increased fares to accommodate the extra use. If there is no repercussion for jumping the turnstile, then every rider would become a free rider and there would not be enough revenue to continue subway service. Other examples of free riding behavior include citizens who fail to pay their taxes yet continue to use services provided by the government or employees who benefit from conditions negotiated by a union without joining the union.

As another broad example, consider the case in which companies collude with one another on setting prices, which is definitely not advantageous to consumers. In this case, the consumers may be thought of as a collective player whose interests are represented by the government. The government can prosecute the companies for price fixing, or the consumers themselves can file a class-action suit in a "larger" game. The government has frequently been involved in antitrust suits (e.g., against Microsoft) and in setting the rules for auctions of airwaves, in which telecommunication companies—advised by game theorists—have paid billions of dollars for the right to construct cellular phone and other networks.

All in all, game theory offers fundamental insights into conflicts at all levels, especially *seemingly* irrational features that, on second look, are often well conceived and effective.

Review Vocabulary

Backward induction A reasoning process in which players, working backward from the last possible moves in a game, anticipate each other's rational choices. (p. 650)

Chicken A two-person variable-sum game in which each player has two strategies: to swerve to avoid a collision, or not to swerve and cause a collision if the opponent has not swerved. Neither player has a dominant strategy; the compromise outcome, in which both players swerve, is not a Nash equilibrium, but the two outcomes in which one player swerves and the other does not are Nash equilibria. (pp. 641, 642)

Constant-sum game A game in which the sum of payoffs to the players at each outcome is a constant, which can be converted to a zero-sum game by an appropriate change in the payoffs to the players that does not alter the strategic nature of the game. (p. 635)

Dominant strategy A strategy that is sometimes better and never worse for a player than every other strategy, whatever strategies the other players choose. (p. 640)

Dominated strategy A strategy that is sometimes worse and never better for a player than some other strategy, whatever strategies the other players choose. (p. 644)

Expected value If each of the n possible payoffs, $s_1, s_2, \ldots, s_n$, occurs with respective probabilities $p_1, p_2, \ldots, p_n$, then the expected value E, also called the *expected payoff*, is

$$E = p_1 s_1 + p_2 s_2 + \cdots + p_n s_n$$

where $p_1 + p_2 + \cdots + p_n = 1$ and $p_i \geq 0$ $(i = 1, 2, \ldots, n)$. (p. 632)

Fair game A zero-sum game is fair when the (expected) value of the game, obtained by using optimal strategies (pure or mixed), is zero. (p. 633)

Free rider A free rider is someone who benefits from resources, goods, services, or a policy without paying the cost for the benefit. (p. 653)

Game tree A symbolic tree, based on the rules of play in a game, in which the vertices, or nodes, of the tree represent choice points, and the branches represent alternative courses of action that the players can select. (p. 647)

Maximin In a two-person zero-sum game, the largest of the minimum payoffs in each row of a payoff matrix. (p. 624)

Maximin strategy In a two-person zero-sum game, the pure strategy of the row player corresponding to the maximin in a payoff matrix. (p. 624)

Mechanism design The act of designing a game to achieve a particular outcome as a Nash equilibrium. (p. 646)

Minimax In a two-person zero-sum game, the smallest of the maximum payoffs in each column of a payoff matrix. (p. 625)

Minimax strategy In a two-person zero-sum game, the pure strategy of the column player corresponding to the minimax in a payoff matrix. (p. 625)

Minimax theorem The fundamental theorem for two-person constant-sum games, stating that there always exist optimal pure or mixed strategies that enable the two players to guarantee the value of the game. (p. 637)

Mixed strategy A strategy that involves the random choice of pure strategies, according to particular probabilities. A mixed strategy of a player is optimal if it guarantees the value of the game. (p. 631)

Nash equilibrium Strategies associated with an outcome such that no player can benefit by choosing a different strategy, given that the other players do not depart from their strategies. (p. 640)

Ordinal game A game in which the players rank the outcomes from best to worst. (p. 639)

Partial-conflict game A variable-sum game in which both players can benefit by cooperation but may have strong incentives not to cooperate. (p. 637)

Payoff matrix A rectangular array of numbers. In a two-person game, the rows and columns correspond to the strategies of the two players, and the numerical entries give the payoffs to the players when these strategies are selected. (p. 623)

Prisoners' Dilemma A two-person variable-sum game in which each player has two strategies, cooperate or defect. Defect dominates cooperate for both players, even though the mutual-defection outcome, which is the unique Nash equilibrium in the game, is worse for both players than the mutual-cooperation outcome. (pp. 638, 640)

Pure strategy A course of action a player can choose in a game that does not involve randomized choices. (p. 631)

Rational choice A choice that leads to a preferred outcome. (p. 623)

Saddlepoint In a two-person constant-sum game, the payoff that results when a row minimum and a column maximum are the same, which is the value of the game. The saddlepoint has the shape of a saddle-shaped surface and is also a Nash equilibrium. (p. 626)

Strategy One of the courses of action that a player can choose in a game; strategies are mixed or pure, depending on whether they are selected in a randomized fashion (mixed) or not (pure). (p. 623)

Total-conflict game A zero-sum or constant-sum game, in which what one player wins the other player loses. (p. 623)

Value The best outcome that both players can guarantee in a two-person zero-sum game. If there is a saddlepoint, that is the value. Otherwise, the value is the expected payoff resulting when the players choose their optimal mixed strategies. (pp. 626, 633)

Variable-sum game A game in which the sum of the payoffs to the players at the different outcomes varies. (p. 637)

Vickrey auction An auction in which bidders independently submit sealed bids for an item and the winner is the bidder who bids the highest, but he or she pays only the amount of the second-highest bid. (p. 643)

Zero-sum game A constant-sum game in which the payoff to one player is the negative of the payoff to the other player, so the sum of the payoffs to the players at each outcome is zero. (p. 624)

Self Check Answers

1. The row minima for rows 1, 2, and 3 are 2, 5, and 1, respectively. Because 5 is the maximum of the minimum values, row 2 is the associated minimax strategy.

2. The column maxima for columns 1, 2, and 3 are 9, 8, and 5, respectively. Because 5 is the minimum of the maximum values, column 3 is the associated maximin strategy.

3. The game has a saddlepoint because the maximum of the minimum row values is equal to the minimum of the maximum column values. The value of the game is 5, which is the payoff when the players follow the maximin and minimax strategies.

4. The probabilities must add to 1. Because $p_1 + p_2 + p_3 = 1/2 + 1/3 + p_3 = 1$, then $p_3 = 1/6$. The expected value $E = 3$

because $E = p_1s_1 + p_2s_2 + p_3s_3 = (1/2)4 + (1/3)6 + (1/6)(-6) = 2 + 2 + -1) = 3$.

5. Solving for q in the equation $E_1 = 6q - 1 = 4 - 9q = E_2$ yields $15q = 5$ or $q = 5/15$, which reduces to $q = 1/3$.

6. Table 15.3 is rewritten with ordinal payoffs from Mark's perspective in the payoff matrix to the left and from Lisa's perspective in the payoff matrix to the right:

2	1
4	5
6	3

5	6
3	2
1	4

For a fixed row and a fixed column, the entries of the payoff matrices sum to 7.

7. When Anneliese, Binh, and Charlie have bids of \$150, \$301, and \$100, respectively, then Binh gets the item, pays \$150 (the second highest bid), and achieves a \$200 − \$150 = \$50 payoff. Because Anneliese and Charlie don't get the item and don't pay anything, each receives a payoff of 0.

If Anneliese bids anything less than \$301, then she will not get the item; this means that she has a payoff of 0. If she bids \$301 or more, then she could receive the item and pay more than what it is worth to her. This is a negative payoff. Because she cannot increase her payoff, she has no desire to change her bid. Binh already gets the item, so bids above \$150 do not change who gets the item or how much Binh pays. If Binh bids less than \$150, then he doesn't get the item and his payoff is 0. Binh doesn't wish to risk bidding \$150 and possibly not getting the item. Binh has no incentive to change his bid. Like Anneliese, if Charlie bids more than \$301, he will win the stamp but pay too much for it. This result could occur if he bids \$301, too. For bids less than \$301, Charlie still doesn't get the item and pays nothing. Charlie has no reason to change his bid, either.

8. Player B would play r for both of his decision nodes. For the left node, B prefers r to l because r gives 3 as opposed to 1 from l. Similarly, for the right node, B prefers r to l because r gives 4 as opposed to 2 from l. Because A can look ahead and anticipates B playing r regardless of her decision, she will prefer to play L (because she gets 2 instead of 1).

Skills Check

1. In the following two-person zero-sum game, the payoffs represent gains to row Player I and losses to column Player II.

$$\begin{bmatrix} 3 & 7 & 3 \\ 8 & 6 & 1 \\ 5 & 9 & 4 \end{bmatrix}$$

What is the maximin strategy for Player I?

(a) Play the first row.

(b) Play the second row.

(c) Play the third row.

2. In Skills Check 1, the minimax strategy of Player II is to play the __________ column.

3. In the following two-person zero-sum game, the payoffs represent gains to row Player I and losses to column Player II.

$$\begin{bmatrix} 2 & 9 & 3 \\ 7 & 6 & 4 \\ 1 & 5 & 8 \end{bmatrix}$$

What is the minimax strategy for Player II?

(a) Play the first column.

(b) Play the second column.

(c) Play the third column.

4. In Skills Check 3, the maximin strategy for Player I is to play the __________ row.

5. Does the game in Skills Check 1 have a saddlepoint? What about the game in Skills Check 3?

For Skills Check Questions 6–11, consider the following three two-person zero-sum games, where the payoffs represent gains to the row Player I and losses to the column Player II.

$$\begin{bmatrix} 3 & 6 \\ 5 & 4 \end{bmatrix}$$

$$\begin{bmatrix} -1 & 3 \\ 2 & 0 \end{bmatrix}$$

$$\begin{bmatrix} 6 & 5 & 6 & 5 \\ 1 & 4 & 2 & -1 \\ 8 & 5 & 7 & 5 \\ 0 & 2 & 6 & 2 \end{bmatrix}$$

6. Do any of the above games have a saddlepoint?

7. In which two games does neither player have a dominant strategy?

(a) The first two games

(b) The last two games

(c) The first and third games

8. How many of Player I's strategies are dominated in the third game?

9. Which strategy of Player II is dominant in the third game?

(a) Strategy 1

(b) Strategy 2

(c) Strategy 4

10. Strategy ___________ of Player I is dominant in the third game.

11. The third game has

(a) no saddlepoint.

(b) one saddlepoint.

(c) more than one saddlepoint.

12. If a game has a saddlepoint, then ___________ is the value of the game.

13. In the game of Matching Pennies, Player I wins a penny if the coins match; Player II wins a penny if the coins do not match. Given this information, it can be concluded that the 2×2 matrix representing this game has

(a) two -1s and two 1s.

(b) four 1s.

(c) four -1s.

14. A mixed strategy uses randomization to ___________________________ .

15. Which of these games does not have a saddlepoint?

(a) Tic-tac-toe

(b) Chess

(c) Poker

16. A game is fair if its saddlepoint is equal to ________.

17. In the following game of kicker-versus-goalie, the kicker's likelihood of being successful in kicking certain goals is given in the game matrix. Recall that the kicker can kick to the goalie's left *GL* or the goalie's right *GR* and the goalie can dive to the left *L* or to the right *R*.

		Goalie	
		L	***R***
Kicker	***GL***	0.2	0.7
	GR	0.8	0.3

Which is the best strategy for the kicker?

(a) Always kick the ball to the goalie's left.

(b) Always kick the ball to the goalie's right.

(c) Kick the ball equally often to the goalie's left and to the goalie's right.

18. In Skills Check 17, the goalie's optimal strategy is to dive to the left with probability *p*. What is *p*?

19. In Skills Check 17, if the kicker and the goalie use their optimal mixed strategies, then, on average, the kicker successfully kicks a goal with probability _______.

20. In Skills Check 17, if the kicker uses her optimal strategy, do you know beforehand whether she will kick the ball to the goalie's left or to the goalie's right? (Yes or No)

21. In the following game of batter-versus-pitcher in baseball, the batter's batting averages are given in the game matrix. The batter tries to maximize his batting average while the pitcher tries to minimize the batter's batting average; this is a zero-sum game. The pitcher decides between throwing a fastball and a curveball. When the batter is in the batter's box, he guesses which pitch is coming (either a fastball or a curveball). If he guesses correctly, then his batting average is higher. For example, it is easier for him to hit a fastball when he correctly guesses that one is coming. However, if he guesses incorrectly, then his batting average goes down.

		Pitcher	
		Fastball	**Curveball**
Batter	**Fastball**	0.350	0.250
	Curveball	0.100	0.500

(a) If the pitcher always pitches a fastball, what should the batter do?

(b) If the pitcher always pitches a curveball, what should the batter do?

(c) If the batter always guesses fastball, what should the pitcher do?

(d) If the batter always guesses curveball, what should the pitcher do?

22. For Skills Check 21, find the optimal mixed strategy for the batter. Using this strategy, the batter will guess fastball with what probability?

23. In Skills Check 21, the pitcher's exact optimal strategy is to ______________.

24. In the following game of batter-versus-pitcher in baseball, the batter's batting averages are given in the game matrix.

		Pitcher	
		Fastball	**Curveball**
Batter	**Fastball**	0.250	0.350
	Curveball	0.200	0.500

(a) If the pitcher always pitches a fastball, what should the batter do?

(b) If the pitcher always pitches a curveball, what should the batter do?

(c) If the batter always guesses fastball, what should the pitcher do?

(d) If the batter always guesses curveball, what should the pitcher do?

25. In Skills Check 24, the pitcher's optimal strategy is to ________.

26. In Skills Check 24, given your answer to Skills Check 25, the optimal strategy for the batter is to ________.

27. Consider the following partial-conflict game with ordinal payoffs (in which a player rank-orders the outcomes 1, 2, 3, and 4).

		Player II	
		Choice *A*	**Choice *B***
Player I	**Choice *A***	(4, 4)	(1, 3)
	Choice *B*	(3, 1)	(2, 2)

Both players might select *B* because ________.

28. In Skills Check 27, what strategy constitutes a Nash equilibrium?

(a) Only when both players select *A*

(b) Only when both players select *A* or both select *B*

(c) Only when one player selects *A* and the other selects *B*

29. In the following game, Player I has the preferences of the row player in the Prisoners' Dilemma, and Player II has the preferences of the column player in Chicken.

		Player II	
		Choice *A*	**Choice *B***
Player I	**Choice *A***	(3, 3)	(1, 4)
	Choice *B*	(4, 2)	(2, 1)

Does the player with a dominant strategy benefit more than the player without one?

(a) Yes

(b) No

(c) It doesn't make any difference; both players do equally well by choosing strategies associated with the Nash-equilibrium outcome.

30. In Skills Check 29, the strategies associated with (4, 2) constitute a Nash equilibrium, but those associated with (3, 3) do not because ________.

31. A game tree is used to

(a) determine the possible strategies of a player.

(b) anticipate each other's choices through backward induction.

(c) plan a deception strategy.

32. Suppose that Chicken is played sequentially instead of simultaneously. Use backward induction to determine the equilibrium outcome for the game tree for Chicken.

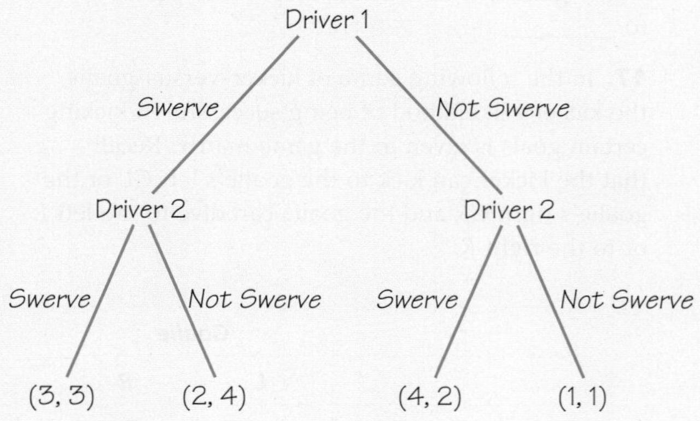

Chapter 15 Exercises

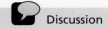

Challenge Discussion

15.1 Two-Person Total-Conflict Games: Pure Strategies

Consider the following five two-person total-conflict games, in which the payoffs represent gains to the row Player I and losses to the column Player II.

1. $\begin{bmatrix} 6 & 5 \\ 4 & 2 \end{bmatrix}$

2. $\begin{bmatrix} 0 & 3 \\ -5 & 1 \\ 1 & 6 \end{bmatrix}$

3. $\begin{bmatrix} -2 & 3 \\ 1 & -2 \end{bmatrix}$

4. $\begin{bmatrix} 13 & 11 \\ 12 & 14 \\ 10 & 11 \end{bmatrix}$

5. $\begin{bmatrix} -10 & -17 & -30 \\ -15 & -15 & -25 \\ -20 & -20 & -20 \end{bmatrix}$

(a) Which games have saddlepoints?

(b) Find the maximin strategy of Player I, the minimax strategy of Player II, and the value for the games given in part (a).

(c) List strategies in these games that the players should avoid because the resulting payoffs are worse than the payoffs for some alternative strategy.

6. If a player has two strategies in a game and one is dominated, must the other strategy be dominant? Why?

7. If a player has more than two strategies in a game and one is dominated, it is possible that no single strategy is dominant. Fill in the payoffs in the following two-person zero-sum game in which Player I can choose between Top, Middle, and Bottom and Player II can choose between Left and Right so that both Top and Middle dominate Bottom, but neither Top nor Middle is dominant.

		Player II	
		Left	**Right**
Player I	**Top**		
	Middle		
	Bottom	2	2

15.2 Two-Person Total-Conflict Games: Mixed Strategies

Determine the value of the following two games of a kicker versus a goalie. As in the chapter, the kicker can kick the ball to the goalie's left or to the goalie's right and the goalie can guess left or right. The kicker's success rates are given in the game matrix for each problem.

8.

		Goalie	
		Left	**Right**
Kicker	**Goalie's Left**	0.25	0.8
	Goalie's Right	0.7	0.25

9.

		Goalie	
		Left	**Right**
Kicker	**Goalie's Left**	0.1	0.8
	Goalie's Right	0.8	0.25

10. It is possible for a player's pure strategy to be dominated by a mixed strategy. For the following game, Player I's strategy of playing Top and Middle, each with probability 1/2, dominates the pure strategy of playing Bottom. Calculate the expected value for each of these strategies against Player II playing the pure strategies Left and Right. Use your answer to explain why Bottom is dominated by the mixed strategy.

		Player II	
		Left	**Right**
Player I	**Top**	2	8
	Middle	8	2
	Bottom	4	4

11. In the following game of batter-versus-pitcher in baseball, the batter's batting averages are given in the game matrix. The batter tries to maximize

Masakazu Watanabe/Aflo/ Getty Images

his batting average, while the pitcher tries to minimize the batter's batting average; this is a zero-sum game. The pitcher decides between throwing a fastball or a knuckleball. When the batter is in the batter's box, he guesses which pitch is coming (either a fastball or a knuckleball). If he guesses correctly, then his batting average goes up. For example, it is easier for him to hit a fastball when he correctly guesses that one is coming. However, if he guesses incorrectly, then his batting average goes down. If the pitcher and the batter follow their mixed-equilibrium strategies, what will the batter's batting average be?

		Pitcher	
		Fastball	**Knuckleball**
Batter	**Fastball**	0.400	0.200
	Knuckleball	0.200	0.250

12. A businessperson has the choice of either not cheating on his income tax or cheating and making \$1000 if not audited. If caught cheating, he will pay a fine of \$2000 in addition to the \$1000 he owes. He feels good if he does not cheat and is not audited (worth \$100). If he does not cheat and is audited, he evaluates this outcome as −\$100 (for the lost day). Viewing the game as a two-person zero-sum game between the businessperson and the tax agency, what are the optimal mixed strategies for each player and the value of the game?

13. In American football, if a team tries for a 2-point conversion, then the ball is placed on the 2-yard line and the team has one play to try to get into the end zone. The quarterback can decide to run the ball or pass it. Similarly, the other team can commit itself to defend more heavily against a run or a pass. This can be modeled as a 2 × 2 zero-sum matrix game, in which the payoffs are the probabilities of getting in the end zone and earning the 2 points. Find the value of the game.

Ezra Shaw/Getty Images

		Defense	
		Run	**Pass**
Offense	**Run**	0.4	0.65
	Pass	0.75	0.3

14. Refer to Exercise 13 to answer the following two questions.

(a) Football teams use "signs" to pass information from the coaches on the sideline to the quarterback on the field. Assume that the defensive team has stolen the offensive team's signs and knows with certainty that the Offense will Run the ball. What should the Defense do?

(b) Assume that one of the players on the Defense used to play on the opposing team. He believes that he knows their signs, but he isn't certain. He is 70% percent confident that the Offense will Pass the ball. What is the Defense's optimal play against a mixed strategy of 70% Pass and 30% Run?

15. You have the choice of either parking illegally on the street or parking in the lot and paying \$16. Parking illegally is free if the police officer is not patrolling, but you receive a \$40 parking ticket if she is. However, you are peeved when you pay to park in the lot on days when the officer does not patrol, and you are willing to assess this outcome as costing \$32 (\$16 for parking plus \$16 for your time, inconvenience, and grief). It seems reasonable to assume that the police officer ranks her preferences in the order (1) giving you a ticket, (2) not patrolling with you parked in the lot, (3) patrolling with you in the lot, and (4) not patrolling with you parked illegally.

(a) Describe this as a matrix game, assuming that you are playing a zero-sum game with the officer.

(b) Solve this matrix game for its optimal mixed strategies and its value.

(c) Discuss whether it is reasonable or not to assume that this game is zero-sum.

(d) Assuming that you play this parking game each working day of the year, how do you implement an optimal mixed strategy?

16. Describe how a pure strategy for a player in a matrix game can be considered as merely a special case of a mixed strategy.

17. (a) Describe in detail *one* pure strategy for the player who moves first in the game of tic-tac-toe. This strategy must tell how to respond to all possible moves of the other player. (*Hint:* You may wish to make use of the symmetry in the 3 × 3 grid in this game; that is, there are one "center" box, four "corner" boxes, and four "side" boxes.)

(b) Is your strategy optimal in the sense that it will guarantee the first player a tie (and possibly a win) in the game?

18. In the Matching Pennies example, consider the case where Player I favors heads H over tails T. For example, assume that Player I plays H three-fourths of the time and T only one-fourth of the time—a nonoptimal mixed strategy. What should Player II do if he knows this?

19. Assume in the matching game of Example 5 that Player II is using the nonoptimal mixed strategy $(p, 1-p) = (\frac{1}{2}, \frac{1}{2})$; that is, he is playing H and T with the same frequency. What should Player I do in this case if she knows this?

20. You plan to manufacture a new product for sale next year, and you can decide to make either a small quantity, in anticipation of a poor economy and few sales, or a large quantity, hoping for brisk sales. Your expected profits are indicated in the following table.

		Economy	
		Poor	**Good**
Quantity	**Small**	\$500,000	\$300,000
	Large	\$100,000	\$900,000

If you want to avoid risk and believe that the economy is playing an optimal mixed strategy against you in a two-person zero-sum game, then what is your optimal mixed strategy and the resulting expected value? Discuss some alternative ways to go about making your decision.

21. Consider the following poker game with two players, I and II. Each antes \$1. Each player is dealt either a high card H or a low card L, with probability 1/2. Player I then folds or bets \$1. If Player I bets, then Player II either folds, calls, or raises \$1. Finally, if II raises, Player I either folds or calls.

Most choices by the players are rather obvious, at least to anyone who has played poker: If either player holds H, that player always bets or raises if he or she gets the choice. The question remains of how often one should bluff—that is, continue to play (by calling or raising) while holding a low card in the hope that one's opponent also holds a low card.

This poker game can be represented by the following matrix game, in which the payoffs are the expected winnings for Player I (depending on the random deal) and the dominated strategies have been eliminated.

		Player II (when holding L)		
		Folds	**Calls**	**Raises**
Player I (when holding L)	**Folds initially**	−0.25	0	0.25
	Bets first, folds later	0	0	−0.25
	Bets first, calls later	−0.25	−0.25	0

(a) Are there any strategies in this matrix game that a player should avoid?

(b) Solve this game.

(c) Which player is in the more favored position?

22. Considering the scenario described in Exercise 21, should either of the players ever bluff?

23. If Person A threatens Person B but does not intend to carry out his or her threat, we say he or she is bluffing. When is such a bluff rational?

24. (a) Describe in detail *one* pure strategy for the player who moves second in the game of tic-tac-toe.
(b) Is your strategy in part (a) optimal in the sense that it will guarantee the second player a tie (and possibly a win) in the game?

25. On an overcast morning, deciding whether to carry your umbrella can be viewed as a game between yourself and nature as follows:

		Weather	
		Rain	**No rain**
You	**Carry umbrella**	Stay dry	Lug umbrella
	Leave it home	Get wet	Hands free

Let's assume that you are willing to assign the following numerical payoffs to these outcomes, and that you are also willing to make decisions on the basis of expected values (i.e., average payoffs):

(Carry umbrella, rain) $= -2$
(Carry umbrella, no rain) $= -1$
(Leave it home, rain) $= -5$
(Leave it home, no rain) $= 3$

(a) If the weather forecast says there is a 50% chance of rain, should you carry your umbrella or not? What if you believe there is a 75% chance of rain?

(b) If you are conservative and wish to protect against the worst case, what pure strategy should you pick?

(c) If you are rather paranoid and believe that nature will pick an optimal strategy in this two-person zero-sum game, what strategy should you choose?

(d) Another approach to this decision problem is to assign payoffs to represent what your *regret* will be after you know nature's decision. In this case, each such payoff is the best payoff you could have received under that state of nature, minus the corresponding payoff in the previous table.

		Weather	
		Rain	**No rain**
You	**Carry umbrella**	$0 = (-2) - (-2)$	$4 = 3 - (-1)$
	Leave it home	$3 = (-2) - (-5)$	$0 = 3 - 3$

What strategy should you select if you wish to minimize your maximum possible regret?

15.3 Partial-Conflict Games

Consider the following five two-person variable-sum games. Discuss the players' possible behavior when these games are played in a noncooperative manner (with no prior communication or agreements). The first payoff is to the row player; the second, to the column player. Are the Nash equilibria in these games sensible? Why or why not?

26.

	Player II	
Player I	(4, 4)	(1, 3)
	(3, 1)	(2, 2)

27. Battle of the sexes:

		She buys a ticket for:	
		Boxing	**Ballet**
He buys a ticket for:	**Boxing**	(4, 3)	(2, 2)
	Ballet	(1, 1)	(3, 4)

28.

	Player II	
Player I	(2, 1)	(4, 2)
	(1, 4)	(3, 3)

29.

	Player II	
Player I	(2, 4)	(4, 3)
	(1, 2)	(3, 1)

30.

	Player II	
Player I	(3, 4)	(2, 3)
	(1, 2)	(4, 1)

31. In Exercise 26, players maximize their possible gains by choosing their first strategies, but they minimize their possible losses by choosing their second strategies. Which strategy would you choose, and why?

32. Assume that two countries in an arms race assign points to all their own weapons so that the total for each is 1000. Each side can then designate weapons of the *other* side, totaling 100 points, that must be eliminated in the next year, thereby effecting a 10% reduction. Would these countries have any reason to lie about how they value their own weapons? Is this procedure practical as an arms-reduction scheme?

15.4 Mechanism Design and Larger Games

33. In Example 9 (page 644), Anneliese, Binh, and Charlie valued a stamp at \$300, \$200, and \$100, respectively. Although it is a weakly dominant strategy for each player to bid his or her true valuation of the stamp under a Vickrey auction, there are many other equilibrium bids besides (\$300, \$200, \$100). Determine which, if any, of the following bids are equilibrium bids for the stamp: (\$250, \$50, \$50), (\$150, \$100, \$50), and (\$400, \$299, \$1).

34. There are other auction formats besides the second-price, or Vickrey, auction considered in Example 9. In a first-price auction, the highest bidder gets the item and pays his or her bid for the item. Explain why it would not be an equilibrium for Anneliese, Binh, and Charlie to bid their true valuations of \$300, \$200, and \$100, respectively, for the stamp if a first-price auction were used in Example 9. (In a first-price auction, every player submits a bid at the same time and the person who bids the highest gets the item for his or her bid.)

35. Players Odd and Even play Low Person Wins, the rules of which are as follows:

(a) Odd announces an odd number between 1 and 5 (inclusive).

(b) Independently, Even announces an even number between 2 and 6 (inclusive).

(c) Whoever announces the lower number gets *twice* this number as its payoff.

(d) Whoever announces the higher number gets the *lower* number as its payoff.

What is the Nash equilibrium of this game, based on the successive elimination of dominated strategies? Is there another Nash equilibrium? What are the similarities and differences between this game and the Prisoners' Dilemma?

36. Return to the sequential model of Chicken from Skills Check 32. Use a game tree to model the situation in which Driver 2 goes first. Does the analysis change from your analysis in Skills Check 32?

37. In a sequential duel, why will the first player to act in a truel shoot in the air (if this option is allowed by the rules)? Is this choice optimal if a second player should succeed in firing in the air at the same time?

38. In a sequential truel with no firing in the air allowed, suppose *A*, who hates *B*, goes first; *B*, who hates *C*, goes second; *C*, who hates *A*, goes third. (If a player fires, he will shoot only his *antagonist*—the player he hates.)

(a) Which player is in the best position, and why?

(b) Does the outcome change if *B* hates *A* rather than *C*?

Answer these questions (1) if each player can take only one turn (if alive), and (2) if, after one round, the game continues (if there is more than one player alive) and each player can take more than one turn.

15.5 Using Game Theory

39. Find a two-person zero-sum game with a saddlepoint in which the successive elimination of dominated strategies does *not* lead to the saddlepoint. (*Hint:* Restrict yourself to 3 × 3 games. Can you construct such a game that has a saddlepoint but for which no strategies are dominated?)

40. Consider a free-rider problem in which two bus riders can choose to pay their fares or not. As long as at least one player pays his or her fare, then the bus continues to operate—though the rider who pays will have to pay twice as much for a ride. Each player ranks the outcomes from best to worst as follows:

4: Ride the bus for free and have the other player pay double.

3: Pay to ride the bus and have the other player pay his or her fare, too.

2: Pay to ride the bus and have the other player ride free.

1: The bus stops operating.

This game is modeled by the following matrix. Use it to analyze the optimal behavior. Does this game share payoffs with any other game considered in this chapter?

		Rider 2	
		Pay Fare	**Ride Free**
Rider 1	**Pay Fare**	(3, 3)	(2, 4)
	Ride Free	(4, 2)	(1, 1)

Chapter Review

41. Every zero-sum game may be written using ordinal payoffs as used for the partial-conflict games analyzed in this chapter. Return to Example 1 (page 623) and change the payoffs to ordinal payoffs—remember to include payoffs for both Mark and Lisa. Do Mark's and Lisa's decisions to choose the Suburban hospital and the 8 A.M. to 4 P.M. shift remain an equilibrium? Explain, using ordinal payoffs.

42. Would a similar approach to Exercise 41 work for the kicker–goalie game of Example 3 (page 629)? Explain.

43. Use a game tree to model the Prisoners' Dilemma in Table 15.9. Does your analysis depend on which country chooses to act first? Explain.

Applet Exercise

To do this exercise, go to www.macmillanhighered.com/fapp10e.

1. What happens to the value of a game if you or your opponent deviates from the optimal strategy? Can you exploit such deviations by your opponent to your advantage? Explore these possibilities in the *Game Theory* applet.

Writing Projects

1. *Prisoners' Dilemma* sometimes is written with the apostrophe between the *r* and the *s:* Prisoner's Dilemma. The latter placement of the apostrophe indicates that a single player faces the dilemma; the former placement indicates that all players face the dilemma. Argue both positions. Which do you think is a better representation of the problem?

2. The game-theoretic descriptions of interactions between batters and pitchers in baseball used in the exercises of this chapter are simplified. The original cover of Boston Red Sox legend Ted Williams's *The Science of Hitting* included Williams's batting average for different regions of the strike zone; for current players, this information is available on ESPN's website through their game tracker, where it is possible to follow Major League Baseball games while they are being played. Can you use batting average information to better model the batter-versus-pitcher game by including pitch location (and the batter's guess of location)?

3. Former San Francisco Giants slugger Barry Bonds has been much maligned for his use of performance-enhancing drugs. But he also was shrewd about how baseball was played. Once, before an at-bat, he described to a teammate exactly how he would be pitched; the at-bat resulted in a homerun. This suggests that the batter−pitcher interaction may be better modeled as an extensive-form game, like a truel. Pitchers often describe using pitches to set up a hitter. ESPN's website includes a game tracker feature so that users can follow a game being played in real time; this feature tracks the sequence of pitches thrown during an at-bat. Explain how to develop such an extensive-form game model of the batter–pitcher interaction. What information would you use in your model? Should the information be limited to one at-bat or for all at-bats in the game? Should it extend to previous games?

4. Consider a conflict that you had—with a parent, a boss, a girlfriend or boyfriend, or some other acquaintance—in which each person had to make a choice without being sure of what the other person would do. What strategies did you seriously consider adopting, and what options do you think the other person considered? What plausible outcomes do you think each set of strategy choices would have led to? How would you rank these outcomes from best to worst, and how do you think the other player would have ranked them? In two to three pages, analyze the resulting game, and state whether you believe you and the other person made optimal choices. If not, what interfered with your or the other person's rationality?

5. It is sometimes argued that game theory does not take account of the (irrational?) emotions of people, such as anger, jealousy, or love. What is your opinion about this question? In one to two pages, give an example, real or hypothetical, that supports your position, paying particular attention to whether the players acted consistently with or contrary to their preferences.

6. In tennis, one player often prefers to play from the baseline while her opponent prefers a serve-and-volley game (i.e., likes to come to the net). The baseline player attempts to hit passing shots. This player has a choice of hitting "down the line" or "crosscourt." The net player must often correctly guess in which direction the ball will go to cover the shot. In one to two pages, formulate this situation as a matrix game and discuss appropriate strategies for the players.

7. Quentin Tarantino's films *Reservoir Dogs* (1992) and *Pulp Fiction* (1994) both have truels, but the choices that the characters make in each film are completely different. Does the truel analysis offer any insight into why? Discuss in one to two pages.

Suggested Readings

BINMORE, KEN. *Playing for Real: A Text on Game Theory,* Oxford University Press, Oxford, U.K., 2007. A comprehensive intermediate text.

BRAMS, STEVEN J. *Game Theory and the Humanities: Bridging Two Worlds.* MIT Press, Cambridge, MA, 2011. Unusual applications of game theory to history, literature, the Bible, theology, philosophy, and law.

DIXIT, AVINASH, SUSAN SKEATH, & DAVID H. RILEY. *Games of Strategy,* 3d ed., Norton, New York, 2009. An excellent game-theory text that requires only a minimal mathematical background.

NASAR, SYLVIA. *A Beautiful Mind,* Simon & Shuster, New York, 1998. A biography of John Nash that is also a fascinating account of the early history of game theory.

In 2001, a fictionalized version of this biography was made into a movie, which received four Oscars, including Best Picture, in 2002.

OSBORNE, MARTIN J. *An Introduction to Game Theory,* Oxford University Press, New York, 2004. A fine intermediate text with several interesting applications.

POUNDSTONE, WILLIAM. *Prisoner's Dilemma,* Anchor, New York, 2003. A nontechnical look at the development of game theory and one of its most colorful characters, John Von Neumann.

WILLIAMS, J. D. *The Compleat Strategyist: Being a Primer on the Theory of Games of Strategy,* rev. ed., Dover, New York, 1986. Originally written in 1954, this text provides many zero-sum game examples solvable by the techniques introduced in this chapter.

Suggested Websites

www.economics.utoronto.ca/osborne Martin Osborne's home page (game theory).

kuznets.fas.harvard.edu/~aroth/alroth.html Alvin Roth's Game Theory and Experimental Economics page.

www.gametheory.net The most comprehensive game theory website.

www.nobelprize.org/nobel_prizes/economic-sciences/laureates Information on game theory laureates of the Nobel Memorial Prize in Economic Sciences.

Part V

The Digital Revolution

When this part of the book made its debut in the 1993 edition, it was titled "Coding Information." In the 2003 edition, the title was changed to "The Digital Revolution." Since none of the mathematics in this part had changed, why was the title changed? Here is a partial list of reasons: broadband, wireless Internet connections, Google, Facebook, Twitter, YouTube, eBay, Amazon, smart phones, text messages, Tumblr, Instagram, Snapfish, blogs, iTunes, global position satellites, e-Books, digital cameras, and online dating sites.

Little did we know in 2003 that within another decade, the digital revolution would spawn social, political, and commercial revolutions as well. In this part of the book, we examine some of the mathematics that made these revolutions possible.

Jurgen Ziewe/Ikon Images/SuperStock

Identification Numbers

16

Nick Koudis/Digital Vision/Getty Images

Identification numbers and bar codes are everywhere: retail items, books, airline tickets, checking accounts, credit cards, ZIP codes, and driver's licenses. Did you know that mathematics is involved in creating them? In this chapter, we explain how mathematics is used to assign identification numbers and bar codes that allow data to be tracked accurately, efficiently, and inexpensively.

16.1 Check Digits

Look at the 13-digit **International Standard Book Number (ISBN)** printed on the back cover of this book. The number 978-1-4641-2473-0 (978-1-4641-2488-4 for the paperback version) distinguishes this book from all others.

The last digit, 0, is there solely to detect errors that may occur when the ISBN is entered into a computer. Grocery items, credit cards, overnight mail, magazines, personal checks, travelers cheques, soft-drink cans, automobiles, and many other items that you encounter daily have identification numbers that code data, as well as a digit called a **check digit** for error detection. In this chapter, we examine some of the methods used to assign identification numbers and check digits.

Division by 9 Schemes

Let's begin by considering the U.S. Postal Service money order shown in Figure 16.1. The first 10 digits of the 11-digit number 17620289526 simply identify the money order. The last digit, 6, serves as an **error-detecting** mechanism. Let's see how this mechanism works. The 11th (last) digit of a Postal Service money order number is the remainder obtained when the sum of the first 10 digits of the number is divided by 9. In our example, the last digit is 6 because $1 + 7 + 6 + 2 + 0 + 2 + 8 + 9 + 5 + 2 = 42$ and the remainder when 42 is divided by 9 is 6. (Recall that if we divide a positive integer a by a positive integer b, there are unique integers q and r such that $a = bq + r$, where r is nonnegative and less than b. The numbers q and r are called the *quotient* and *remainder*. For example, the remainder when 42 is divided by 9 is 6 because $42 = 9 \times 4 + 6$.)

Algebra Review Appendix
Remainders

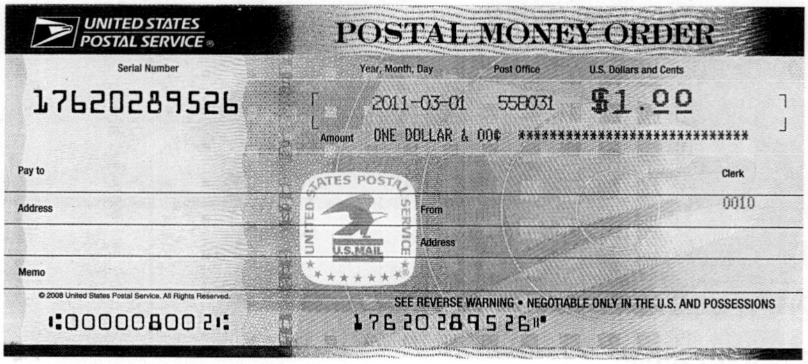

Source: U.S. Postal Service

Figure 16.1 Money order with identification number 1762028952 and appended check digit 6. The check digit is the remainder after dividing the sum of the digits by 9.

Now suppose that instead of the correct number, the number 17640289526 (an error in the fourth position) was entered into a computer that had been programmed for error detection in money orders. The machine would divide the sum of the first 10 digits of the entered number, 44, by 9 and obtain a remainder of 8. Because the last digit of the entered number is 6 rather than 8, the entered number cannot be correct. This crude method of error detection detects all errors involving a single digit except replacing a 0 with a 9, or vice versa. Because the value of a sum does not depend on the order in which the numbers are added, this method does not detect the transposition of digits such as 17260289526 instead of 17620289526. (The digits in positions 3 and 4 have been transposed.)

American Express travelers cheques, VISA travelers cheques, and Euro banknotes also use a check digit determined by division by 9. In these cases, the check digit is the smallest nonnegative integer such that the sum of the digits, including the check digit, is evenly divisible by 9.

EXAMPLE 1 ➔ The American Express Travelers Cheque

The American Express Travelers Cheque with the identification number 387505055 has the check digit 7 because $3 + 8 + 7 + 5 + 0 + 5 + 0 + 5 + 5 = 38$ and $38 + 7$ is evenly divisible by 9.

Self Check 1

If the sum of the digits of a Euro banknote serial number excluding the check digit is 68, what is the check digit?

Division by 7 Schemes

The scheme used on airline tickets and for Avis and National rental cars assigns the remainder after division by 7 of the number itself as the check digit, rather than dividing the sum of the digits by 7. For example, the check digit for the number 540047 is 4 because $540047 = 7 \times 77149 + 4$. This method will not detect the substitution of 0 for a 7, 1 for an 8, 2 for a 9, or vice versa. However, unlike the Postal Service method, it will detect transpositions of adjacent digits with the exceptions of the pairs that differ by 7, namely, 0, 7; 1, 8; and 2, 9. For example, if 5400474 were entered into a computer as 4500474 (the first two digits are transposed), the machine would determine that the check digit should be 3 because $450047 = 7 \times 64292 + 3$. Because the last digit of the entered number is not 3, the error has been detected.

One can use Google to determine easily the check digits that require division by 7 or 9. To find the Avis check digit for the number 540047, enter "540047 mod 7" in the Search box; for division by 9, enter the number followed by "mod 9." To use a calculator to find the remainder when a is divided by b, first enter $a \div b$. The integer portion of the number is the quotient q. The remainder is $r = a - bq$. Doing this for $123 \div 7$, we get 17.57, then $123 - (7 \times 17) = 4$. When a positive integer is divided by 10, the remainder is the first digit after the decimal point, but this is not true when dividing by other integers.

Universal Product Code

The scheme used on grocery products, the **Universal Product Code (UPC),** is more sophisticated. Consider the number 0 38000 00127 7 found on the bottom of a box of Kellogg's Corn Flakes. The first digit identifies a broad category of goods, the next five digits identify the manufacturer, the next five identify the product, and the last is a check digit. Suppose that this number were entered into a computer as 0 58000 00127 7 (a mistake in the second position). How would the computer recognize the mistake?

For any UPC number $a_1a_2a_3a_4a_5a_6a_7a_8a_9a_{10}a_{11}a_{12}$, the computer is programmed to carry out the following computation:

$$3a_1 + a_2 + 3a_3 + a_4 + 3a_5 + a_6 + 3a_7 + a_8 + 3a_9 + a_{10} + 3a_{11} + a_{12}$$

If the result doesn't end with a 0, the computer knows the entered number is incorrect.

For the incorrect corn flakes number, we have

$$3 \cdot 0 + 5 + 3 \cdot 8 + 0 + 3 \cdot 0 + 0 + 3 \cdot 0 + 0 + 3 \cdot 1 + 2 + 3 \cdot 7 + 7 = 62$$

Because 62 doesn't end with 0, the error is detected. Notice that had we used the correct digit 3 in the second position instead of 5, the sum would have ended in a 0 as it should. This simple scheme detects *all* single-position errors and about 89% of all other kinds of errors.

Beginning in January 2005, U.S. retailers were required to have software that could read the 12-digit UPC code used in the United States and the 13-digit **European Article Number (EAN)** code used in Europe. This change paves the way for the 13-digit EAN to become the worldwide standard. Existing UPC numbers will be converted to EAN numbers by adding an extra 0 at the beginning. The check digit for a 13-digit EAN number $a_1a_2a_3a_4a_5a_6a_7a_8a_9a_{10}a_{11}a_{12}a_{13}$ is selected so that

$$a_1 + 3a_2 + a_3 + 3a_4 + a_5 + 3a_6 + a_7 + 3a_8 + a_9 + 3a_{10} + a_{11} + 3a_{12} + a_{13}$$

ends with 0. Adding an extra 0 in the front of a UPC number does not affect the check digit. The coefficient 3 for the terms with even subscripts is called a **weight.** Because $1a_i = a_i$, we say that the terms with odd subscripts have weight 1.

Besides error detection, check-digit schemes that use weighted sums can be used to find a digit that has been corrupted in some way. Say, for example, that the packaging for a product with UPC number 1 640002 202034 was damaged or defective in such a way that the second digit was unintelligible. How would we know that it is supposed to be 6? Well, let's call it x. Then we know that the weighted sum

$$1 + 3 \cdot x + 4 + 3 \cdot 0 + 0 + 3 \cdot 0 + 2 + 3 \cdot 2 + 0 + 3 \cdot 2 + 0 + 3 \cdot 3 + 4 = 3x + 32$$

must end with 0. Because 6 is the only digit that makes this true, we know the corrupted digit is 6. This example also illustrates why the weight 3 is superior to the weight 2. While there is only one digit that makes the expression $3x + 32$ end with 0, there are two digits that make $2x + 32$ end with 0: $x = 4$ and $x = 9$.

Bank Identification Numbers

The U.S. banking system uses a variation of the UPC scheme that appends check digits to the numbers assigned to banks. Each bank has an eight-digit routing number $a_1a_2 \cdots a_8$ together with a check digit a_9 so that a_9 is the last digit of

$$7a_1 + 3a_2 + 9a_3 + 7a_4 + 3a_5 + 9a_6 + 7a_7 + 3a_8$$

In this formula, the weights are 7, 3, and 9. The weights were carefully chosen so that all single-digit errors and most transposition errors are detected.

EXAMPLE 2 ➔ Bank Routing Number

The check illustrated in Figure 16.2 has routing number 062000019. The check digit 9 is the last digit of

$$7 \cdot 0 + 3 \cdot 6 + 9 \cdot 2 + 7 \cdot 0 + 3 \cdot 0 + 9 \cdot 0 + 7 \cdot 0 + 3 \cdot 1 = 39$$

The first four digits of a nine-digit bank routing number identify the bank's Federal Reserve District, office, state, or special collection arrangement; the next four digits are the bank's identification number; the ninth digit is the check digit. The block of numbers 33 74489 shown in Figure 16.2 is the account number. The last block, 0134, is the check number.

Poportic/Shutterstock

Figure 16.2 A bank check with routing number 062000019. The 9 is the check digit.

You may wonder if there is any advantage to using three weights for the routing number as opposed to using two, as is the case for the UPC error-detection scheme. The answer is yes. While both the UPC scheme and the bank scheme detect 100% of single-position errors and most transposition errors involving adjacent digits, the bank scheme will detect most transposition errors of the form $\cdots abc \cdots \to \cdots cba \cdots$, whereas the UPC scheme does not detect such errors.

For example, say that we look at a number that begins with 241. In the UPC scheme, these digits contribute $3 \cdot 2 + 4 + 3 \cdot 1 = 13$ toward the total calculation, while the string 142 (the first and third digits are transposed) also contributes $3 \cdot 1 + 4 + 3 \cdot 2 = 13$ toward the total calculation. So the error is not detected. In contrast, using the bank scheme, 241 contributes $7 \cdot 2 + 3 \cdot 4 + 9 \cdot 1 = 35$ toward the total, while 142 contributes $7 \cdot 1 + 3 \cdot 4 + 9 \cdot 2 = 37$ toward the total. Because the total for the correct number ends with 0, the total for the number that had the transposition error would end with the digit 2. Thus, the error is detected.

Self Check 2

If a number of the form $71a_3a_4 \cdots a_9$ is a valid bank routing number, is $17a_3a_4 \cdots a_9$ a valid bank routing number? ■

Luhn Algorithm

One of the most efficient error-detection methods is the **Luhn algorithm,** created in 1953 by the IBM scientist Hans Peter Luhn, used by all major credit-card companies, as well as by many libraries, blood banks, and the South Dakota driver's license department. For a credit card number $a_1a_2a_3a_4a_5a_6a_7a_8a_9a_{10}a_{11}a_{12}a_{13}a_{14}a_{15}$, let T be the number of digits in positions 1, 3, 5, 7, 9, 11, 13, 15 that are 5 or greater. The check digit a_{16} is chosen so that $2a_1 + a_2 + 2a_3 + a_4 + 2a_5 + a_6 + 2a_7 + a_8 + 2a_9 + a_{10} + 2a_{11} + a_{12} + 2a_{13} + a_{14} + 2a_{15} + T + a_{16}$ is divisible by 10.

EXAMPLE 3 ➔ Luhn Algorithm Check Digit

To demonstrate how to calculate a check digit using the Luhn algorithm, suppose a bank intends to issue a credit card with the identification number 312560019643001. We calculate $2 \cdot 3 + 1 + 2 \cdot 2 + 5 + 2 \cdot 6 + 0 + 2 \cdot 0 + 1 + 2 \cdot 9 + 6 + 2 \cdot 4 + 3 + 2 \cdot 0 + 0 + 2 \cdot 1$ to get 66. Then we note that among the digits in the odd-numbered positions, only 6 and 9 are greater than or equal to 5. So we add 2 to 66 to get 68. The check digit is whatever is needed to bring the final tally to a number that ends with 0. Because 68 + 2 = 70, the check digit for our example is 2. This digit is appended to the end of the number the bank issues for identification purposes. Errors in input data are detected by applying the same algorithm to the input, including the check digit. If the correct number is entered into a computer, the result will end in 0. If the result doesn't end with 0, a mistake has been detected.

The credit card illustrated in Figure 16.3 has a check digit that is not valid because the algorithm yields

$$2 \cdot 4 + 1 + 2 \cdot 2 + 8 + 2 \cdot 0 + 0 + 2 \cdot 1 + 2 + 2 \cdot 3 + 4$$
$$+ 2 \cdot 5 + 6 + 2 \cdot 7 + 8 + 2 \cdot 9 + 3 + 0 = 94$$

which does not end in 0. This method allows computers to detect 100% of single-position errors and about 98% of other common errors. For credit cards such as American Express that use 14 digits plus a check digit, the weights are 1, 2, 1, 2, . . . , 1, 2 instead of 2, 1, 2, 1, . . . , 2, 1.

Figure 16.3 Credit card with an invalid number.

Besides detecting errors, a check digit offers partial protection against fraudulent numbers. A person who wanted to create a phony credit card, bank account number, or driver's license number would have to know the appropriate check-digit scheme for the number to go unchallenged by the computer.

Self Check 3

If the first three digits of a 16-digit credit card are 924, what do they contribute to the weighted sum total when calculating the check digit?

International Standard Book Numbers

Thus far, we have not discussed any schemes that detect 100% of single errors and 100% of transposition errors. As seen on the back of this book and most others published since 2007, there are two identification numbers—a 13-digit number called the *13-digit International Standard Book Number (ISBN-13)* and a 10-digit number, called the *10-digit ISBN (ISBN-10)*. The 10-digit ISBN detects 100% of single-digit errors and 100% of transposition errors.

A correctly coded 10-digit ISBN $a_1a_2 \cdots a_{10}$ has the property that $10a_1 + 9a_2 + 8a_3 + 7a_4 + 6a_5 + 5a_6 + 4a_7 + 3a_8 + 2a_9 + a_{10}$ is evenly divisible by 11. Consider the 10-digit ISBN of the book that you are now reading: 1-4641-2473-6 (1-4641-2488-4 for paperback version). The 1 at the beginning indicates that the book is published in an English-speaking country, while the next block of digits, 4641, identifies the publisher, W. H. Freeman and Company. The third block for the hardback edition, 2473, is assigned by the publisher and identifies this particular book. The last digit 6, for the hardback version, is the check digit. Let's verify that this number is a legitimate possibility. We must compute $10 \cdot 1 + 9 \cdot 4 + 8 \cdot 6 + 7 \cdot 4 + 6 \cdot 1 + 5 \cdot 2 + 4 \cdot 4 + 3 \cdot 7 + 2 \cdot 3 + 6 = 187$. Because $187 = 11 \cdot 17$, it is evenly divisible by 11, so no error has been detected.

Self Check 4

If the weighted sum of the digits of an ISBN excluding the check digit is 87, what is the check digit? ■

How can we be sure that this method detects 100% of the single-position errors? Let's say that a correct number is $a_1a_2a_3a_4a_5a_6a_7a_8a_9a_{10}$ and that a mistake is made in the second position. (The same argument applies equally well in every position.) We may write this incorrect number as $a_1a_2'a_3a_4a_5a_6a_7a_8a_9a_{10}$, where $a_2' \neq a_2$. For this error to go undetected, it must be the case that $10 \cdot a_1 + 9 \cdot a_2' + 8 \cdot a_3 + 7 \cdot a_4 + 6 \cdot a_5 + 5 \cdot a_6 + 4 \cdot a_7 + 3 \cdot a_8 + 2 \cdot a_9 + a_{10}$ is evenly divisible by 11. Then, because both $10a_1 + 9a_2 + 8a_3 + 7a_4 + 6a_5 + 5a_6 + 4a_7 + 3a_8 + 2a_9 + a_{10}$ and $10a_1 + 9a_2' + 8a_3 + 7a_4 + 6a_5 + 5a_6 + 4a_7 + 3a_8 + 2a_9 + a_{10}$ are divisible by 11, so is their difference:

$$(10 \cdot a_1 + 9 \cdot a_2 + 8 \cdot a_3 + \cdots + a_{10}) - (10 \cdot a_1 + 9 \cdot a_2' + 8 \cdot a_3 + \cdots + a_{10}) = 9 \cdot (a_2 - a_2')$$

Because a_2 and a_2' are distinct digits between 0 and 9, their difference must be one of $\pm 1, \ldots, \pm 9$. Thus, the only possibilities for the number $9 \cdot (a_2 - a_2')$ are ± 9, ± 18, ± 27, ± 36, ± 45, ± 54, ± 63, ± 72, ± 81—and none of these is divisible by 11. So a single-position error cannot go undetected.

EXAMPLE 4 ISBN-10 Single-Error Detection

To illustrate with a specific example why the ISBN-10 method detects single errors, let's say the valid ISBN 1-4292-0900-3 is mistaken as 1-2292-0900-3 (an error in position 2). Because the correct second digit in the weighted sum contributes $9 \cdot 4 = 36$ to the total of 176 and the incorrect second digit contributes $9 \cdot 2 = 18$, we see that the weighted sum of the incorrect number is $176 - 18 = 158$. But 158 is not evenly divisible by 11, so the error is detected. ■

To verify that the ISBN-10 method detects all adjacent transposition errors, let's suppose that the first two digits are transposed. (The same argument applies to all positions.) Say that the correct number is $a_1a_2a_3 \cdots a_{10}$. As before, for the incorrect number $a_2a_1a_3 \cdots a_{10}$ to go undetected, it must be the case that the difference of the correct number and the incorrect number is evenly divisible by 11. That is,

$$(10a_1 + 9a_2 + 8a_3 + \cdots + a_{10}) - (10a_2 + 9a_1 + 8a_3 + \cdots + a_{10})$$

is evenly divisible by 11. This reduces to the condition that $a_1 - a_2$ is divisible by 11. But the only possible differences of two numbers between 0 and 9 are plus or minus the numbers between 0 and 9, and of these only 0 is divisible by 11. Thus, $a_1 - a_2 = 0$. But then $a_1 = a_2$ and there is no error.

EXAMPLE 5 ISBN-10 Transposition-Error Detection

Let's trace through an example to see how the ISBN-10 method detects adjacent transposition errors. Say 1-4292-0900-3 is mistaken as 1-2492-0900-3 (digits in positions 2 and 3 are transposed). Because the second and third digits in the weighted sum of the correct number contribute $9 \cdot 4 + 8 \cdot 2 = 52$ to the total of 176 and the second and third digits of the incorrect number contribute $9 \cdot 2 + 8 \cdot 4 = 50$ to the weighted sum, we know that the weighted sum of the incorrect number is $176 - 2 = 174$, which is not evenly divisible by 11. ■

With a bit more work, we could prove that every transposition error is detected, not just the transpositions of adjacent digits. This is possible because 11 is prime.

Because the ISBN-10 method, in contrast to the other methods we have described, detects all single-position errors and all transposition errors, why is it not used more? Well, it does have a drawback. Say the next title published by W. H. Freeman is to have 0902 for the third block. (The 10-digit ISBN for all W. H. Freeman books begins with 1-4641.) What check digit should be assigned? Call it a. Then the weighted sum is $10 \cdot 1 + 9 \cdot 4 + 8 \cdot 6 + 7 \cdot 4 + 6 \cdot 1 + 5 \cdot 4 + 4 \cdot 5 + 3 \cdot 1 + 2 \cdot 3 + a = 177 + a$. Because the next integer after 177 that is divisible by 11 is 187, we see that $a = 10$. But appending 10 to the existing 9-digit number would result in an 11-digit number instead of a 10-digit one. This is the only flaw in the 10-digit ISBN scheme. To avoid this flaw, publishers use an X to represent the check digit 10. As a result, not all 10-digit ISBNs consist solely of digits; some end with X. Publishers could avoid this inconsistency by simply refraining from using numbers that require an X.

To expand the inventory of ISBNs and make them compatible with the UPC/EAN numbering scheme for other retail items worldwide, publishers began using a 13-digit ISBN in 2007. The 13-digit ISBN is the same as the 10-digit ISBN number except for a prefix of 978 or 979 and the check digit. The check digit for the 13-digit ISBN is calculated so that the weighted sum using the weights 1, 3, 1, 3, . . . , 1, 3, 1 ends with 0. Thus, the 13-digit ISBN and the 13-digit UPC/EAN numbers used for retail products employ the same check-digit method. During a phase-in period, publishers will use both the 10-digit and 13-digit numbers.

Although all the check-digit schemes that we have described in this chapter that use weighted sums detect 100% of single-digit errors, it is impossible to devise a check-digit scheme that employs a single check digit that detects 100% of multiple-digit errors. In the case of the UPC scheme, for instance, while one error might cause the weighted sum to be, say, 62 (and thereby detect the error), a second error may result in the sum 70 so that no errors are detected. Consider the single-digit error that we examined earlier for corn flakes. Recall the correct UPC number is 0 38000 00127 7 and its weighted sum is 60. For the incorrect corn flakes number 0 58000 00127 7 (with a single error in position 2), we have

$$3 \cdot 0 + 5 + 3 \cdot 8 + 0 + 3 \cdot 0 + 0 + 3 \cdot 0 + 0 + 3 \cdot 1 + 2 + 3 \cdot 7 + 7 = 62$$

But if we also make a second error of using 8 in position 4 instead of 0, so that the number becomes 0 58800 00127 7, the errors are not detected because its weighted sum is 70, which does end with 0. Nevertheless, the weight schemes detect most multiple errors because it is rare when a second or third error exactly cancels out previous errors.

A Multiplication by 13 Scheme

After single-digit errors and adjacent transposition errors, the third most common error is one of the form $\cdots abc \cdots \to \cdots cba \cdots$. In practice, these kinds of errors commonly occur in phone numbers that have matching digits separated by another digit such as 727 5856. A likely mistake when writing or dialing this number is to switch the 8 and the 6, resulting in the number 727 5658. Such an error is called a *jump transposition.* Remarkably, there is a simple way to encode identification numbers so that the three most common errors are detected 100% of the time without having to introduce an alphabetic character, as is done for the 10-digit ISBN numbers.

To illustrate the method, suppose that a math instructor wants to post student grades publicly without revealing any information about the students' ID numbers. Assuming the last four digits of each student ID number are different, she could assign each student a six-digit number by multiplying the last four digits of their identification numbers by 13 (adding leading 0s when necessary). For example, a student with an ID number that ends with 8912 is assigned 115856 = 8912 × 13. (To preserve confidentiality of the original four digits, students are not informed of the encoding method.) Of course, the instructor can recapture the original four-digit numbers by dividing the encoded numbers by 13. Since all encoded numbers are divisible by 13, the jump transposition error 115856 → 115658 is detected because 115658 is not divisible by 13.

The arguments for verifying that encoding identification numbers as multiples of 13 detect 100% of all single-digit errors, all transposition errors involving adjacent digits, and all jump transposition errors are similar to those we used to show that the 10-digit ISBN numbers detect errors. In particular, a single-digit error in the number $a_n a_{n-1} \cdots a_i \cdots a_0$ of the form $a_n a_{n-1} \cdots a_i' \cdots a_0$, where $a_i' \neq a_i$ is not detected if and only if $a_n a_{n-1} \cdots a_i' \cdots a_0$ is a multiple of 13. But if both $a_n a_{n-1} \cdots a_i \cdots a_0$ and $a_n a_{n-1} \cdots a_i' \cdots a_0$ are multiples of 13, then so is their difference $(a_n a_{n-1} \cdots a_i \cdots a_0) - (a_n a_{n-1} \cdots a_i' \cdots a_0) = (a_i - a_i')10^i$. But 13 does not divide the term on the right when $a_i \neq a_i'$. Similarly, the transposition of adjacent digits a_i and a_{i-1} is undetected if and only if $9(a_i - a_{i-1})10^{i-1}$ is divisible by 13, which happens only when $a_i = a_{i-1}$. And the jump transposition $\cdots a_i a_{i-1} a_{i-2} \cdots \to \cdots a_{i-2} a_{i-1} a_i \cdots$ is undetected if and only if $99(a_i - a_{i-2})10^{i-2}$ is divisible by 13, which happens only when $a_i = a_{i-2}$. Incidentally, the arguments just given reveal why we used multiplication by 13 rather than some smaller positive integer. For example, if multiplication by 11 were used to transform the identification numbers instead of 13, then all single-digit errors and all adjacent transposition errors are detected, but not all jump transpositions are, because 11 divides 99.

Euro banknotes, car rental companies, some state driver's license numbers, and the vehicle identification number (VIN) that identifies cars and trucks use alpha-numeric identification numbers. To calculate the check digit, the letters are assigned numerical values. See Spotlight 16.1.

The VIN System

SPOTLIGHT 16.1

Automobiles and trucks are given a VIN by the manufacturer. A typical VIN has 17 alphanumeric characters that code information such as the country where the vehicle was built, manufacturer, make, body style, engine type, plant where the vehicle was built, model year, model, type of restraint, a check digit, and a production sequence number. The check digit is calculated by converting the 26 consecutive letters of the alphabet to the numbers 1, 2, 3, 4, 5, 6, 7, 8, 9, 1, 2, 3, 4, 5, 6, 7, 8, 9, 2, 3, 4, 5, 6, 7, 8, 9 (note the skipped digit after the second 9) to obtain a 16-digit number $a_1a_2 \cdots a_{15}a_{16}$ that is weighted with 8, 7, 6, 5, 4, 3, 2, 10, 9, 8, 7, 6, 5, 4, 3, 2. The check digit is the remainder when the weighted sum $8 \cdot a_1 + 7 \cdot a_2 + \cdots + 3 \cdot a_{15} + 2 \cdot a_{16}$ is divided by 11 unless the remainder is 10, in which case an X is used instead. The check digit is inserted in position 9.

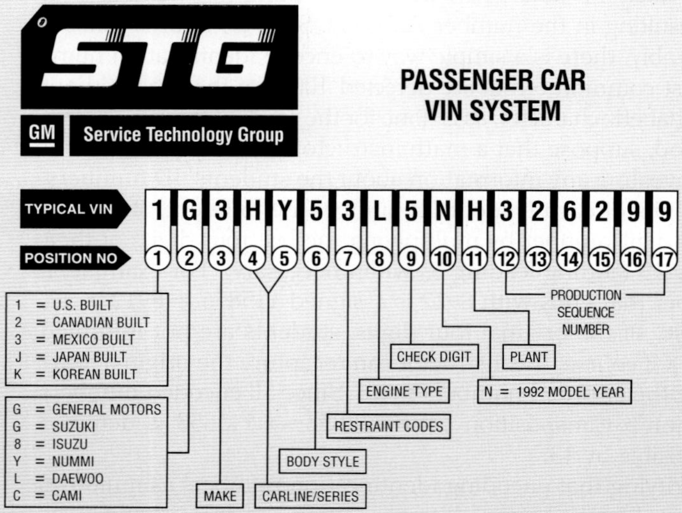

Summary of Error Detection Schemes

Postal money orders The last digit of a Postal Service money order number is the remainder obtained when the sum of the first 10 digits of the number is divided by 9. This method will detect any single mistake except replacing 0 with a 9 or vice versa.

American Express and VISA travelers cheques; Euro banknotes The last digit is the smallest nonnegative integer such that the sum of the digits, including the check digit, is divisible by 9. This method will detect any single mistake except replacing 0 with a 9 or vice versa.

Airlines; Avis and National rental cars The last digit is the remainder after division by 7 of the number itself, excluding the last digit. This method will detect any single error except the substitution of 0 for a 7, 1 for an 8, 2 for a 9, or vice versa.

It will detect transpositions of adjacent digits with the exceptions of the pairs 0, 7; 1, 8; and 2, 9.

UPCs For any UPC number $a_1a_2a_3a_4a_5a_6a_7a_8a_9a_{10}a_{11}a_{12}$, the last digit is chosen so that $3a_1 + a_2 + 3a_3 + a_4 + 3a_5 + a_6 + 3a_7 + a_8 + 3a_9 + a_{10} + 3a_{11} + a_{12}$ is divisible by 10. This method detects 100% of all single errors and 89% of most others.

EANs For any EAN $a_1a_2a_3a_4a_5a_6a_7a_8a_9a_{10}a_{11}a_{12}a_{13}$, the last digit is chosen so that $a_1 + 3a_2 + a_3 + 3a_4 + a_5 + 3a_6 + a_7 + 3a_8 + a_9 + 3a_{10} + a_{11} + 3a_{12} + a_{13}$ is evenly divisible by 10. This method detects 100% of all single-digit errors and 89% of most others.

Bank routing numbers For a bank routing number $a_1a_2a_3a_4a_5a_6a_7a_8$, the check digit a_9 is the last digit of $7a_1 + 3a_2 + 9a_3 + 7a_4 + 3a_5 + 9a_6 + 7a_7 + 3a_8$. This method detects 100% of all single errors and 89% of most others.

Luhn algorithm For a credit card number $a_1a_2a_3a_4a_5a_6a_7a_8a_9a_{10}a_{11}a_{12}a_{13}a_{14}a_{15}$, let T be the number of digits in positions 1, 3, 5, 7, 9, 11, 13, 15 that are greater than or equal to 5. The last digit a_{16} is chosen so that $2a_1 + a_2 + 2a_3 + a_4 + 2a_5 + a_6 + 2a_7 + a_8 + 2a_9 + a_{10} + 2a_{11} + a_{12} + 2a_{13} + a_{14} + 2a_{15} + T + a_{16}$ is divisible by 10. This method detects 100% of all single errors; all transposition errors involving adjacent digits except for the pair 0, 9; and 98% of most other errors.

ISBN-10s A correctly coded 10-digit ISBN $a_1a_2 \cdots a_{10}$ has the property that $10a_1 + 9a_2 + 8a_3 + 7a_4 + 6a_5 + 5a_6 + 4a_7 + 3a_8 + 2a_9 + a_{10}$ is evenly divisible by 11. This method detects 100% of all single errors and 100% of all transposition errors. If a_{10} is X, it is replaced with 10 in the weighted sum.

ISBN-13s This scheme is the same as the UPC scheme.

Multiplication by 13 This method detects all single-digit errors, all transposition errors involving adjacent digits, and all jump transposition errors.

16.2 The ZIP Code

Identification numbers sometimes encode geographic data. The ZIP code, social security numbers, and telephone numbers are the foremost examples. In 1963, the U.S. Postal Service numbered every American post office with a five-digit **ZIP code.** (ZIP is an acronym for Zone Improvement Plan.) The numbers begin with 0s at the points farthest east—00601 for Adjuntas, Puerto Rico—and work up to 9s at the points farthest west—99950 for Ketchikan, Alaska (see Figure 16.4).

Let's use 55812, one of the ZIP codes for Duluth, Minnesota, as an example.

5 The first digit represents one of 10 geographic areas, usually a group of states. The numbers begin at the points farthest east (0) and end at the points farthest west (9).

58 The second two digits, in combination with the first, identify a central mail-distribution point known as a sectional center. The location of a sectional center is based on geography, transportation facilities, and population density. Although just four centers serve the entire state of Utah, there are six of them to take care of New York City.

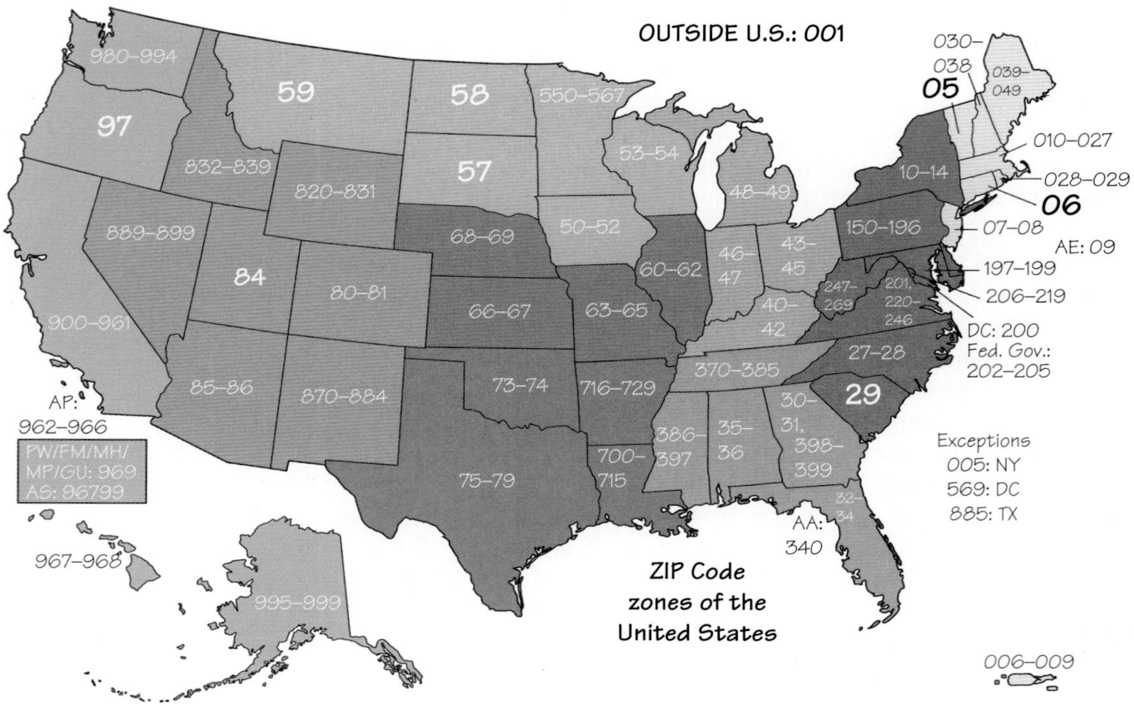

Figure 16.4 ZIP code scheme.

12 The last two digits indicate the town or local post office. In many cases, the largest city in a region will be given the digits 01 and surrounding towns assigned succeeding digits alphabetically.

In 1983, the U.S. Postal Service added four digits to the ZIP code. When four digits are added after a dash—for example, 68588-1234—the number is called the **ZIP + 4 code.** Mail with ZIP + 4 coding is eligible for cheaper bulk rates, being easier to sort with automated equipment. It's also helpful for businesses that wish to sort the recipients of their mailings by geographic location. The first two numbers of the four-digit suffix represent a delivery sector, which may be several blocks, a group of streets, several office buildings, or a small geographic area. The last two numbers narrow the area further. They might denote one floor of a large office building, a department in a large firm, or a group of post office boxes.

For businesses that receive an enormous volume of mail, the ZIP + 4 code permits automation of in-house mailroom sorting. For example, the first seven digits of all mail sent to the University of Minnesota Duluth, are 55812-24. The school has designated nine pairs of digits for the last two positions to direct the mail to the appropriate dormitory or apartment building.

16.3 Bar Codes

In modern applications, bar codes and identification numbers go hand in hand. **Bar coding** is a method for automated data collection. It is a way to transmit information rapidly, accurately, and efficiently to a computer.

Bar Code DEFINITION

A **bar code** is a series of dark bars and light spaces that represent characters.

To decode the information in a bar code, a beam of light is passed over the bars and spaces via a scanning device, such as a handheld wand or a fixed-beam device. The dark bars reflect very little back to the scanner, whereas the light spaces reflect much light. The differences in reflection intensities are detected by the scanner and converted to strings of 0s and 1s that represent specific numbers and letters. Such strings are called a **binary coding** of the numbers and letters.

Binary Code DEFINITION

Any system for representing data with only two symbols is a **binary code.**

ZIP Code Bar Coding

The simplest bar code is the **Postnet code** used by the U.S. Postal Service and commonly found on business reply forms (see Figure 16.5). For a ZIP + 4 code, there are 52 vertical bars of two possible lengths (long and short). The long bars at the beginning and end are called *guard bars* and together provide a frame for the

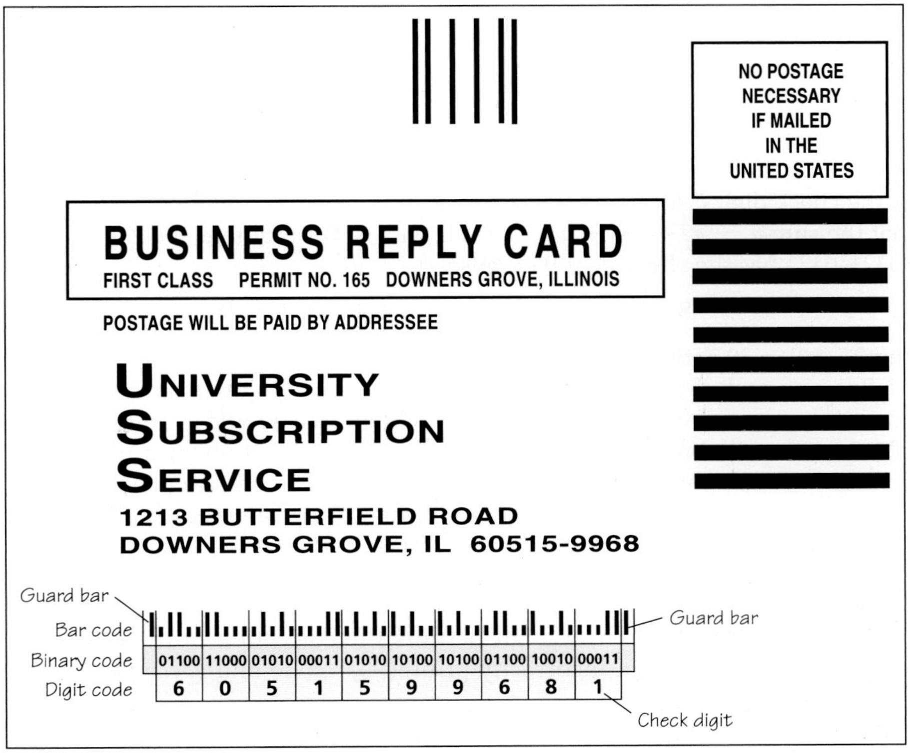

Figure 16.5 ZIP + 4 bar code.

remaining 50 bars. In blocks of 5, the 50 bars within the guard bars represent the ZIP + 4 code and a 10th digit for error correction. Each block of 5 is composed of exactly 2 long bars and 3 short bars, which a scanner converts to binary strings, according to the pattern shown below:

Decimal Digit	Bar Code	Binary Code
1		00011
2		00101
3		00110
4		01001
5		01010
6		01100
7		10001
8		10010
9		10100
0		11000

Handheld scanner reading the shipping bar code on a crate.

The 10th digit of a Postnet code number is a check digit chosen so that the sum of the 9 digits of the ZIP + 4 code and the 10th one is evenly divisible by 10. That is, the check digit C for the ZIP + 4 code $a_1a_2 \cdots a_9$ is the digit with the property that the sum $a_1 + a_2 + \cdots + a_9 + C$ ends with 0. For example, the ZIP + 4 code 80321-0421 has the check digit 9 because

$$8 + 0 + 3 + 2 + 1 + 0 + 4 + 2 + 1 = 21$$

and $21 + 9 = 30$ ends with 0.

Because each digit is represented by exactly two long bars and three short ones, any error in reading or printing a single bar would result in a block of five with only one long bar or three long bars. In either case, the error is detected. This is the reason behind the choice of five bars to code each digit rather than four bars. With five bars per digit, there are exactly 10 arrangements composed of two long bars and three short bars. Any misreading of a single bar in such a block is therefore recognizable because it does not match any of the other blocks for the 10 digits. And because the block location of the error is known, the check digit permits the correction of the error. Let's look at an example of an incorrectly printed bar code and see how the error is correctable.

EXAMPLE 6 Detecting and Correcting an Error

The scanner ignores the guard bars at the beginning and the end and reads the remaining bars in blocks of five, as shown below. (We have inserted dashed dividing lines for readability.)

3 0 7 2 2 ? 9 0 1 7

The sixth block is an incorrect one because it has only one long bar. To correct the error, the computer linked with the bar-code scanner sums the remaining 9 digits to obtain 31. Because the sum of all 10 digits ends with 0, the correct value for the sixth digit must be 9. ■

Self Check 5

Suppose the bar code for the first digit of the ZIP + 4 code has three long bars instead of the required two long bars. If the remaining bar code for the last 9 digits (including the check digit) represents 505582721, what is the correct first digit? ■

Beginning in 1993, large organizations and businesses that wanted to receive reduced rates for ZIP + 4 bar-coded mail were required to use a 12-digit bar code called the *delivery-point bar code.* This code permits machines to sort a letter into the order in which it will be delivered by the carrier. Mail for the first location on a mail route occurs first, mail for the second location on a route occurs second, and so on.

The 12-digit bar code uses the Postnet bar scheme to code the 12-digit string composed of the 9-digit ZIP + 4 number, followed by the last two digits of the street address or box number and a check digit chosen so that the sum of all 12 digits is evenly divisible by 10. For example, a letter addressed to 1738 Maple Street with ZIP + 4 code 55811-2742 would have the Postnet bar code for the digits 558112742384 (38 is from the street address and 4 is the check digit).

In May 2011, the Postnet code was replaced by a new bar code called the *Intelligent Mail Barcode.* The new bar code converts 31 digits of data into 65 vertical bars that encode the type of service, the mail owner, a unique serial number that enables the user to track the letter at every step from arrival at the post office to delivery, and the delivery-point ZIP code. The code uses bars of three lengths and multiple levels to create four states as shown below.

Source: Rolling Stone

Scott Camazine/Sue Trainor

Figure 16.6 Entomologist Stephen Buchmann developed a reliable, inexpensive way to track bees using the same technology that supermarkets use to speed up checkout lines and keep track of inventory. He glued barcode labels onto the backs of 100 bees and placed a laser scanner above the hive. In the past, researchers marked bees with paint or tags, but monitoring activity required the presence of a human observer.

Source: KMART

Figure 16.7 UPC identification number 4 41120 10640 9. The initial 4 indicates that the number is a savings coupon. The block 41120 identifies the retailer as Kmart. The block 10640 identifies the product. The last digit, 9, is a check digit.

The UPC Bar Code

The bar code that we encounter most often is the UPC that was first used on grocery items in 1974. The widespread use of the UPC code by manufacturers enabled businesses to cheaply and accurately manage much larger inventories than was previously possible. This, in turn, gave rise to big box retailers such as Target and Home Depot, which are now commonplace. As Figure 16.6 shows, it has other applications as well. The UPC bar code translates 12-digit UPC identification numbers discussed earlier into bars that can be read quickly and accurately by a laser scanner. The number has four components—two five-digit numbers sandwiched between two single digits—as shown in Figure 16.7.

For the UPC 0 38000 00127 7, here is what the four components represent:

0 The first digit identifies the kind of product. For example, a 0 represents general merchandise; a 2 signals variable-weight items, such as cheese and meat; a 3 means drug and certain other health-related products; a 4 means products marked for price reduction by the retailer (see Figure 16.7); a 5 signals cents-off coupons.

38000 The next five digits identify the company.

00127 The next five digits, assigned by the manufacturer to identify the product, can include size, color, or other important information (but not price).

7 The final digit is the check digit. This digit is often not printed, but it is always included in the bar code.

Each digit of the UPC code is represented by a space divided into seven modules of equal width, as illustrated in Figure 16.8. How these seven modules are filled depends on the digit being represented and whether the digit being represented is part of the manufacturer's number or the product number. In every case, there are two light spaces and two dark bars of various thicknesses that alternate. A UPC code has on each end two long bars of one-module thickness separated by a light space of one-module thickness. These two modules are called the *guard bar patterns* (Figure 16.9). The guard bar patterns define the thickness of a single module of each type. They are not part of the identification number. The manufacturer's number and the product number are separated by a center bar pattern consisting of the

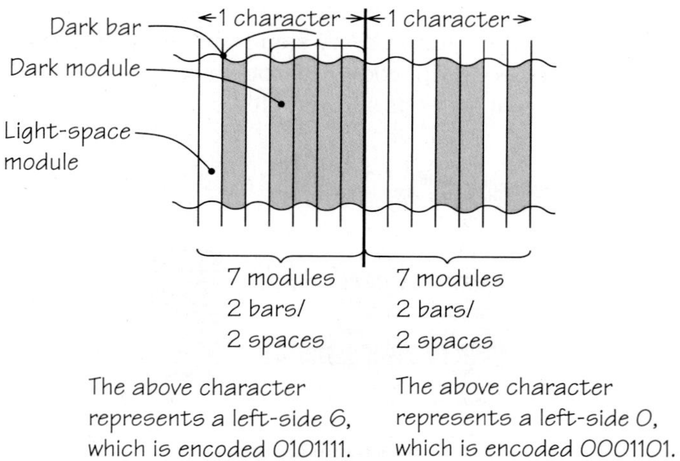

Figure 16.8 UPC bar coding for a left-side 6 and left-side 0, part of the manufacturer's number.

following five modules: a light space, a dark bar, a light space, a dark bar, and a light space (see Figure 16.9). The center bar pattern is not part of the identification number but merely serves to separate the manufacturer's number and product number. Figure 16.8 shows how the digits 6 and 0 in a manufacturer's number are coded.

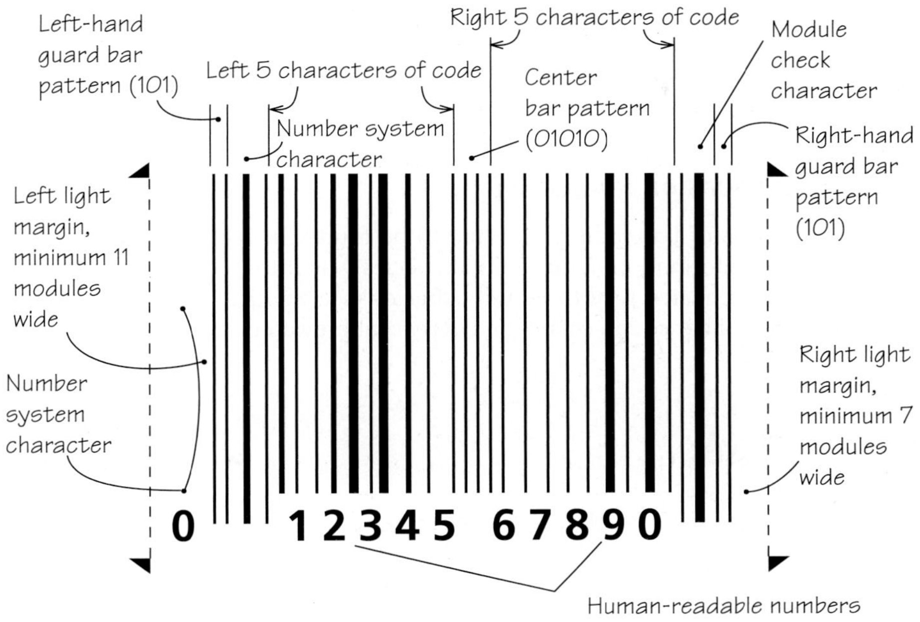

Figure 16.9 UPC bar-code format.

Observe the following pattern in Figure 16.8: a light space of one-module thickness, a dark bar of one-module thickness, a light space of one-module thickness, and a dark bar of four-module thickness. Symbolically, such a pattern of light spaces and dark bars is represented as 0101111. Here, each 0 means a one-module-thickness light space and each 1 means a one-module-thickness dark bar.

Table 16.1 shows the binary code for all digits. Notice that the code for the digits in the product number (the block of five digits on the right side) can be obtained

TABLE 16.1 Binary UPC Coding

Digit	Manufacturer's Number	Product Number
0	0001101	1110010
1	0011001	1100110
2	0010011	1101100
3	0111101	1000010
4	0100011	1011100
5	0110001	1001110
6	0101111	1010000
7	0111011	1000100
8	0110111	1001000
9	0001011	1110100

from the code for the digits in the manufacturer's number (the block of digits on the left side), and vice versa, by replacing each 0 by a 1 and each 1 by a 0. Thus, the code 0111011 for 7 in a manufacturer's number becomes 1000100 in the product number. Also notice that each manufacturer's number has an odd number of 1s, whereas each product number has an even number of 1s. This permits a computer linked with an optical scanner to determine whether the bar code was scanned left to right or right to left. (If the first block of digits has an even number of 1s for each digit, the scanning is being done right to left.) Thus, scanning can be done in either direction without ambiguity.

New Applications of Bar Coding

New applications of bar coding continue to be found. In 2003, a method of bar coding genetic information about animal species was introduced that provides a convenient, inexpensive way to identify species (see Spotlight 16.2). Recently, a new generation of bar codes, called the *QR code* (abbreviated from Quick Response

New Frontier: Bar Coding DNA

SPOTLIGHT 16.2

In 2003, Paul Hebert from the University of Guelph in Canada proposed the compilation of a public library of DNA bar codes for animal species. Rather than scanning an animal's entire genome, which is expensive and time-consuming, Hebert pinpointed a short piece of a section of a single gene that could be used to distinguish one animal species from another cheaply and quickly. For about $2 per sample, the genetic sequence of this tiny gene section can be converted to a four-color bar code that corresponds to the four nucleotides that make up the genetic code. The bar code identifies the species of its source in the same way that the UPC bar code identifies a retail item. By 2013, more than 300,000 species were bar-coded, and a new field of science was born. The technique has already resulted in improved food safety, disease prevention, and better environmental monitoring.

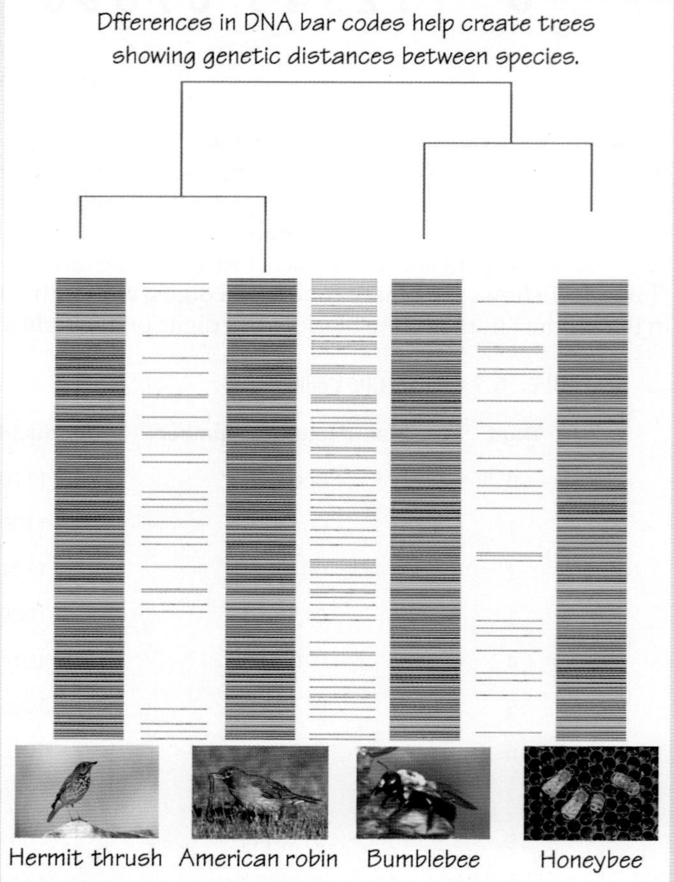

DNA bar codes for the hermit thrush *(Stubblefield Photography/Shutterstock)*, American robin *(Alan & Sandy Carey/Science Source)*, bumblebee *(Mark Stoeckle, The Rockefeller University)*, and honeybee *(Scott Camazine/Science Source)*.

History of Bar Codes

SPOTLIGHT 16.3

1948 Graduate students Norman Joseph Woodland and Bernard Silver at the Drexel Institute of Technology begin working on a bar code.
1952 Woodland and Silver receive a U.S. patent.
1967 The Association of American Railroads adopts an optical bar code.
1969 General Motors uses bar codes to keep track of inventory.
1971 The Uniform Code Council, originally called the Uniform Grocery Product Code Council, is formed to administer the UPC.
1972 U.S. Supermarket Ad Hoc Committee on a Uniform Grocery Product Code recommends the adoption of the 1972 UPC.
1973 An ad-hoc committee composed of grocery executives chooses the linear bar code with 11 digits and a 12th check digit.
1974 A 10-pack of Wrigley's Juicy Fruit chewing gum was the first product with a bar code; it was first scanned at a checkout counter in Troy, Ohio. Today, the pack of gum is on display at the Smithsonian Institution's National Museum of American History.
1974 Ninety-five percent of the railroad fleet is labeled with a bar code.
1977 European Article Numbering Association is formed in Belgium.
1981 U.S. Department of Defense adopts the use of bar codes for marking all products sold to the U.S. military.
1982 The U.S. Postal Service adopts the Postnet bar code.
1992 Norman Joseph Woodland is awarded the 1992 National Medal of Technology by President George H. W. Bush.
1994 The two-dimensional QR bar is introduced in Japan for use in the automotive industry.
2004 Consortium for the Barcode of Life is established.
2011 Intelligent Mail Barcode replaces the Postnet code for U.S. Postal Service mail.

Code), uses mosaics of black and white rectangles that encode much more information than traditional bar codes (see Figure 16.10). These bar codes can be read by specially equipped cell phones to display video, music, or text on the screen or to link the cell phone to a web page. A user can point his or her cell phone at the bar code in a magazine, on a billboard, or on the side of a building to receive information about a product or service. A bar code for a movie will allow the viewer to watch a trailer. Scanning the wrapper of a hamburger will provide nutrition information. See Spotlight 16.3 for a history of bar codes.

In June 2011, the Royal Dutch Mint issued the world's first official coin with a QR code. One study found that during the month of June 2011, 14 million mobile users scanned a QR code or a bar code.

Ko Sasaki/The New York Times/Redux

Figure 16.10 Bar code on a building in Japan that can be read by a properly equipped cell phone.

16.4 Encoding Personal Data

Consider the Social Security number 189-31-9431. The only information we can deduce about the holder of this number is that the person's mailing address when he or she applied for the number was in Pennsylvania (Social Security numbers are assigned based on the ZIP code of the mailing address; see Spotlight 16.4). Figure 16.11 shows an Illinois driver's license number: P142-4754-2173. What information about the holder can be deduced from this number? This time, we can determine the date of birth, sex, and much about the person's name.

These two examples illustrate the extremes in coding personal data. The Social Security number has no personal data encoded in the number. It is entirely determined by the place and time that it is issued, not the individual to whom it

Ten Fun Facts about Social Security Numbers

SPOTLIGHT 16.4

1. The first Social Security record was established for John David Sweeney on December 1, 1936, with a Social Security number (SSN) of 055-09-0001. Sweeney died at the age of 61 without receiving any Social Security benefits.
2. The lowest SSN, 001-01-0001, was given to Grace D. Owen, of Concord, New Hampshire, in 1936.
3. From 1937 until 1940, Social Security paid benefits in the form of a single, lump-sum payment. The average lump-sum payment during this period was $58.06. The first recipient of Social Security benefits was Ernest Ackerman, who retired one day after the program began. He paid in 5 cents and received a lump-sum payment of 17 cents.
4. Payment of monthly Social Security benefits began in 1940. Ida May Fuller was the first American to receive a monthly Social Security benefit check. She received the check, amounting to $22.54, on January 31, 1940. By the time of her death in 1975 at the age of 100, Fuller had collected $22,888.92 in monthly Social Security benefits, compared to her total contribution of $24.75 to the system.
5. Over 40,000 people have claimed to own the SSN 078-05-1120. In 1938, the wallet manufacturer E. H. Ferree Company included a sample Social Security card with the number 078-05-1120 in each wallet.
6. From the start of the program in 1936 until 2005, an estimated $8.9 trillion was paid out as Social Security benefits, while $10.7 trillion was collected as income.
7. Since 1936, over 420 million different Social Security numbers have been issued. Over 5.5 million new numbers are assigned every year. As of June 30, 2013, 18% of the U.S. population were receiving monthly Social Security benefits.
8. Prior to June 25, 2011, the first three digits of Social Security numbers were assigned geographically by state, with the lowest numbers in the Northeast and the highest in the Northwest. Since then, a randomized assignment methodology has been used.
9. Currently invalid Social Security numbers include numbers with sets of zeros (as in 000-xx-xxxx, xxx-00-xxxx, and xxx-xx-0000), numbers starting with 666, numbers from 987-65-4320 through 987-65-4329, and numbers with the starting three digits above 770.
10. Social Security numbers are not reassigned after people die.

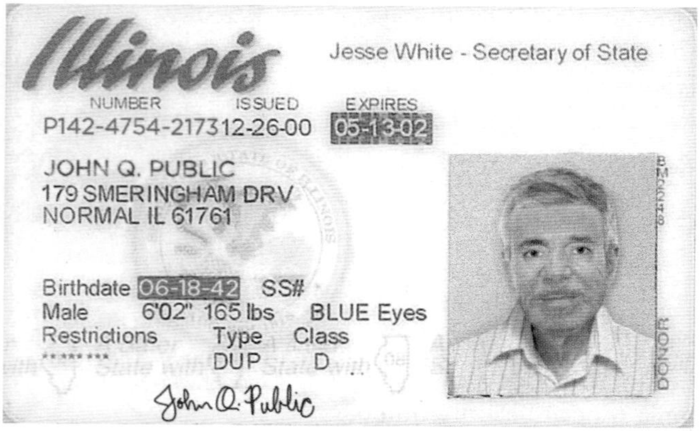

Figure 16.11 Illinois driver's license.

is assigned. In contrast, in some states, driver's license numbers are determined entirely by personal information about the holders. It is no coincidence that the unsophisticated Social Security numbering scheme predates computers. Agencies that have large databases that include personal information such as names, sex, and dates of birth find it convenient to encode these data into identification numbers. Examples of such agencies are the National Archives (where census records are kept), genealogical research centers, the Library of Congress, and state motor vehicle departments.

There are many methods in use to encode personal data such as name, sex, and date of birth. These methods are perhaps most widely used in assigning driver's license numbers in some states. Coding license numbers solely from personal data enables automobile insurers, government entities, and law enforcement agencies to determine the number from the personal data. Many states encode the surname, first name, middle initial, date of birth, and sex by very sophisticated schemes.

In one scheme that is based on sound, the first four characters of the license number are obtained by applying the **Soundex Coding System** to the surname as follows:

1. Delete all occurrences of *h* and *w*. (For example, *Schworer* becomes *Scorer* and *Hughgill* becomes *uggill*.)
2. Assign numbers to the remaining letters as follows:

$a, e, i, o, u, y \rightarrow 0$	$l \rightarrow 4$
$b, f, p, v \rightarrow 1$	$m, n \rightarrow 5$
$c, g, j, k, q, s, x, z \rightarrow 2$	$r \rightarrow 6$
$d, t \rightarrow 3$	

3. If two or more letters with the same numeric value are adjacent, omit all but the first. (For example, *Scorer* becomes *Sorer* and *uggill* becomes *ugil*.)
4. Delete the first character of the original name if still present (*Sorer* becomes *orer*).
5. Delete all occurrences of *a, e, i, o, u,* and *y*.
6. Retain only the first three digits corresponding to the remaining letters; append trailing 0s if fewer than three letters remain; precede the three digits obtained in Step 6 with the first letter of the surname.

Figure 16.12 shows three examples.

	Step 1		Step 2	
Schworer	→	Scorer	→	Scorer 220606

Step 3	Step 4	Step 5	Step 6
→ Sorer 20606	→ orer 0606	→ rr 66	→ S-660

	Step 1		Step 2	
Hughgill	→	uggill	→	uggill 022044

Step 3	Step 4	Step 5	Step 6
→ ugil 0204	→ ugil 0204	→ gl 24	→ H-240

	Step 1		Step 2	
Schmidlapper	→	Scmidlapper	→	Scmidlapper 22503401106

Step 3	Step 4	Step 5	Step 6
→ Smidlaper 250340106	→ midlaper 50340106	→ mdlpr 53416	→ S-534

Figure 16.12 The Soundex Coding System.

What is the advantage of this method? It is an error-correcting scheme. Indeed, it is designed so that likely misspellings of a name nevertheless result in the correct coding of the name. For example, frequent misspellings of the name *Erickson* are *Ericksen, Eriksen, Ericson,* and *Ericsen.* Observe that all of these yield the same coding as *Erickson.* If a law enforcement official, a genealogical researcher, or a librarian wanted to pull up the file from a data bank for someone whose name was pronounced "Erickson," the correct spelling isn't essential because the computer searches for records that are coded as E-625 for all spelling variations. The search feature of a website where many mathematicians post their research papers uses the Soundex Coding System. This system was designed for the U.S. Census Bureau when much census information was obtained orally (see Spotlight 16.5).

Census Records at the National Archives

SPOTLIGHT 16.5

One of the best places to look for information pertaining to family history is the old censuses that are kept up by the National Archives in Washington, D.C. By law, census records are open to the public 72 years after the census was taken. The data from 1880, 1900, 1910, and 1920 censuses (records from 1890 were destroyed by fire) were put on cards during the 1930s as a Works Progress Administration (WPA) project. This information was coded using the Soundex system so that names that sound alike regardless of how they are spelled are grouped together. On old documents, family names were so often misspelled—especially those that were not of British origin—that genealogists say several variations of a name may apply to one set of ancestors. To look for a surname on the index, the researchers must work out the Soundex code. This code, together with the state record, identifies a page number on microfilm where the data are located.

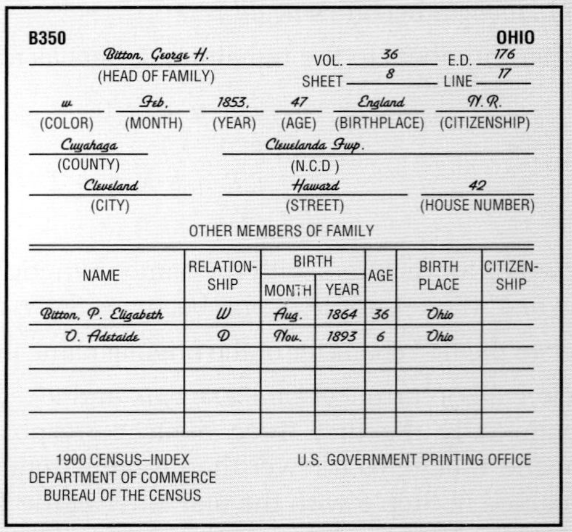
B350 OHIO
Bitton, George H. (HEAD OF FAMILY) VOL. 36 E.D. 176
SHEET 8 LINE 17
w (COLOR) Feb. (MONTH) 1853. (YEAR) 47 (AGE) England (BIRTHPLACE) N. R. (CITIZENSHIP)
Cuyahaga (COUNTY) Clevelanda Twp. (N.C.D)
Cleveland (CITY) Haward (STREET) 42 (HOUSE NUMBER)
OTHER MEMBERS OF FAMILY

NAME	RELATIONSHIP	BIRTH MONTH	BIRTH YEAR	AGE	BIRTH PLACE	CITIZENSHIP
Bitton, P. Eligabeth	W	Aug.	1864	36	Ohio	
V. Adetaide	D	Nov.	1893	6	Ohio	

1900 CENSUS–INDEX
DEPARTMENT OF COMMERCE
BUREAU OF THE CENSUS
U.S. GOVERNMENT PRINTING OFFICE

A typical census Soundex card is shown. Note the Soundex code in the upper-left corner (B350).

Self Check 6

What is the Soundex code for the surname *Jackman?* ■

There are many schemes for encoding the date of birth and the sex in driver's license numbers. For example, the last five digits of Illinois and Florida driver's license numbers capture the year and date of birth as well as the sex. In Illinois, each day of the year is assigned a three-digit number in sequence beginning with 001 for January 1. However, each month is assumed to have 31 days. Thus, March 1 is given the number 063 because both January and February are assumed to have 31 days. These numbers are then used to identify the month and day of birth of male drivers. For females, the scheme is identical except that 600 is added to the number. The last two digits of the year of birth, separated by a dash (probably to obscure the fact that they represent the year of birth), are listed in the fifth and fourth positions from the end of the driver's license number. Thus, a male born on October 13, 1940, would have the last five digits 4-0292 ($292 = 9 \cdot 31 + 13$), whereas a female born on the same day would have 4-0892.

The scheme to identify birth date and sex in Florida is the same as in Illinois except that each month is assumed to have 40 days and 500 is added for women. Moreover, a dash occurs between the two digits for the year and the three digits for the day. For example, the five digits 49-585 belong to a woman born on March 5, 1949.

In this chapter, we have investigated how mathematics is used to append a check digit to an identification number for error detection. In the next chapter, we will show how codes consisting of 0s and 1s can be devised so that errors can be corrected.

Review Vocabulary

Bar code A code that employs bars and spaces to represent information. (pp. 680, 681)

Binary code A coding scheme that uses two symbols, usually 0 and 1. (p. 681)

Check digit A digit included in an identification number for the purpose of error detection. (p. 670)

Error-detecting code A code in which certain types of errors can be detected. (p. 670)

European Article Number (EAN) A 13-digit identification number used on most retail items in Europe. It detects 100% of single-digit errors and most other errors. (p. 672)

International Standard Book Number (ISBN) An identification number used on books throughout the world that contains a check digit for error detection. (p. 669)

Luhn algorithm An error-detecting method used by credit cards, libraries, blood banks, the Canadian Social Insurance program, and others. (p. 673)

Postnet code The bar code used by the U.S. Postal Service for ZIP codes. (p. 681)

Soundex Coding System An encoding scheme for surnames based on sound. (p. 689)

Universal Product Code (UPC) A bar code and identification number that are used on most retail items. The UPC code detects 100% of all single-digit errors and most other types of errors. (p. 671)

Weights Numbers used in the calculation of check digits. (p. 672)

ZIP code A five-digit code used by the U.S. Postal Service to divide the country into geographic units to speed sorting of the mail. ZIP stands for Zone Improvement Plan. (p. 679)

ZIP + 4 code The nine-digit code used by the U.S. Postal Service to refine ZIP codes into smaller units. (p. 680)

Self Check Answers

1. 4

2. No. Since 71 contributes 2 to the last digit of the weighted sum of the digits of $71a_3a_4 \cdots a_8$ and 17 contributes 8 to the last digit of weighted sum of the digits of $17a_3a_4 \cdots a_8$, a_9 is not the last digit of the weighted sum of the digits $17a_2a_3 \cdots a_8$.

3. 29

4. 1

5. 5

6. J255

Skills Check

1. When a single incorrect digit is entered, an error-detecting code

(a) sometimes will detect the error.

(b) always will detect the error but may not be able to correct it.

(c) always will detect and correct the error.

2. If a U.S. Postal Service money order is numbered 1012065994?, where ? indicates that the last digit is obliterated, ? is ____________.

3. If the first five digits of a valid U.S. Postal Service money order are rearranged, the resulting number will ____________ have the same check digit as the original number.

(a) always

(b) sometimes

(c) never

4. If the number $19a_3a_4 \cdots a_{10}7$ is a valid Postal Service money order number and the number $7Xa_3a_4 \cdots a_{10}7$ is also a valid Postal Service money order number, the value of X is ____________.

5. If $a_1a_2 \cdots a_{11}$ is a valid U.S. Postal Service money order number, then $a_1a_2 \cdots a_{10} - a_{11}$ is evenly divisible by

(a) 9.

(b) 10.

(c) 11.

6. The sum of the digits of a correctly coded American Express Travelers Cheque identification number is evenly divisible by ________________.

7. Is the number 105408970012 a legitimate airline ticket number?

(a) Yes.

(b) No, but if the final digit is changed to a 5, the resulting number, 105408970015, is legitimate.

(c) No, but if the final digit is changed to a 3, the resulting number, 105408970013, is legitimate.

8. If the digits of a VISA travelers cheque, excluding the check digit, add up to 36, the check digit is ________.

9. If 8103955 is a valid Avis identification number, which of the three following numbers is *not* detected as invalid by the Avis check-digit scheme?

(a) 1103955

(b) 8173955

(c) 8703955

10. If an American Express Travelers Cheque is numbered *X*425036791, where *X* indicates that the first digit is obliterated, *X* is ____.

11. If the first two digits of a valid airline ticket identification number are transposed, the resulting number will ______________ be valid.

(a) always

(b) sometimes

(c) never

12. A correctly coded UPC number has a weighted sum that is evenly divisible by _____________.

13. Identify each binary UPC code below as the manufacturer's part of the number or as the product part of the number.

(a) 100111

(b) 0011001

(c) 0001011

14. The check digit that should be appended to the UPC code 0-14300-25433 is _____________.

15. The bank routing number error detection scheme detects

(a) all transportation and most single-digit errors.

(b) all single-digit errors and most transpositions.

(c) all single-digit errors and all transpositions.

16. The check digit that should be appended to the bank routing number 01500085 is _____________.

17. The bank routing number error detection scheme detects

(a) the same errors as the UPC scheme.

(b) fewer errors than the UPC scheme.

(c) more errors than the UPC scheme.

18. A correctly coded 10-digit ISBN has a weighted sum that is evenly divisible by _____________.

19. Suppose the 10-digit ISBN 0-1750-3549-0 is reported incorrectly as 0-1750-3540-1. Which of the following statements is true?

(a) This error will not be detected by the check digit.

(b) While this particular error will be detected, the check digit does not detect all 2-digit errors in ISBNs.

(c) All 2-digit errors in a 10-digit ISBN are detectable by the check digit.

20. If a 10-digit ISBN number has an *X* in the check-digit position, the remainder of the weighted sum divided by 11 is ________.

21. A valid 13-digit ISBN has a weighted sum evenly divisible by

(a) 9.

(b) 10.

(c) 11.

22. The ISBN-10 error detection scheme detects _________% of single-digit errors and __________% of transposition errors.

23. If an error in an identification number is made by transposing the first and third digits, the error

(a) usually is detected by the UPC scheme.

(b) always is detected by the bank scheme.

(c) always is detected by the ISBN-10 scheme.

24. If the weighted sum of the first 12 digits of a 13-digit ISBN is 56, the 13th character is _______.

25. If the ZIP code for a home begins with a 9, in which part of the United States is it located?

(a) The East

(b) The Midwest

(c) The West

26. The sum of the 10 digits of a Postnet code number is evenly divisible by _______.

27. If a scanner misreads exactly one bar of a Postnet code, the computer will

(a) not always detect the error.

(b) always detect the error but will not always be able to correct it.

(c) always detect the error and correct it.

28. Which of the codes listed below uses a check digit?

(a) Soundex

(b) Social Security

(c) VIN

29. Errors in ID numbers of the form

(a) $a_1a_2 \cdots a_i \cdots a_n \rightarrow a_1a_2 \cdots a_i' \cdots a_n$ are called ______________ errors.

(b) $a_1a_2 \cdots a_ia_{i+1} \cdots a_n \rightarrow a_1a_2 \cdots a_{i+1}a_i \cdots a_n$ are called ______________ errors.

(c) $a_1a_2 \cdots a_ia_{i+1}a_{i+2} \cdots a_n \rightarrow a_1a_2 \cdots a_{i+2}a_{i+1}a_i \cdots a_n$ are called ______________ errors.

30. The multiplication by 13 encoding scheme detects

(a) ______% of single-digit errors.

(b) ______% of adjacent transposition errors.

(c) ______% of jump transposition errors.

Chapter 16 Exercises

Challenge Discussion

16.1 Check Digits

1. Determine the check digit for a money order with identification number 3953981640.

2. Determine the value of ? so that 7?345417803 is a valid money order identification number.

3. Determine the check digit for the Avis identification number 873345672.

4. Determine two values for ? that will make 723459?0161 a valid Postal Service money order number.

5. Determine the check digit for the airline ticket number 30860422052.

6. Suppose a money order with the identification number and check digit 21720421168 is erroneously copied as 27750421168. Will the error be detected? Explain your reasoning.

7. Determine the check digit for the travelers cheque with identification number 661340874.

8. Determine the check digit for an Avis rental car with identification number 651421.

9. If a Postal Service money order with identification number $19a_3a_4 \cdots a_{10}$ has the check digit 5, what is the check digit for the Postal Service money order number $33a_3a_4 \cdots a_{10}$?

10. Which of the error detection schemes below will detect some errors of the form $abc \rightarrow cba$ (jump transposition)?

(a) UPC code

(b) Bank scheme

(c) 13-digit ISBN

11. If an Avis identification number $a_1a_2 \cdots a_{10}a_{11}8$ has the check digit 5, what is the check digit for the Avis identification number $a_1a_2 \cdots a_{10}a_{11}6$?

12. Find the check digit for the UPC number 03608072089.

13. Suppose that the packaging of a retail item were damaged in such a way that the first digit of a 12-digit UPC code was scratched off, but the remaining 11 digits were 88072303584; determine the first digit.

14. The fourth digit of the UPC number shown below is not discernible. Determine the correct value.

0 88 55 30143 3

15. If one randomly chooses a large number of 10-digit ISBNs, about what percent would have an X in the check digit position?

16. Determine the ISBN-10 check digit for the number 0-547-16509.

17. Determine the ISBN-13 check digit for the number 978-0-547-16509.

18. When the eighth edition of this textbook was in preparation, the publisher sent the author of this section the following ISBNs for the book: ISBN-10: 1-4292-0900-3; ISBN-13: 978-1-4292-0890-0. How did the author know the ISBN-13 was wrong, and how did he know how to correct it? (This really happened!)

19. When calculating the check digit for a 13-digit ISBN, why can you disregard the first two digits? (Try it for Exercise 16.)

20. Determine the check digit for the bank routing number 09100001.

21. Determine the check digit for an American Express Travelers Cheque with identification number 461212023.

22. Suppose that a check digit is assigned to a four-digit number by appending the remainder after division by 7. If the number 36806 has a single-digit error in the first position, determine the possibilities for the correct number.

23. Determine whether the MasterCard number 3541 0232 0033 2270 is valid.

24. Suppose that the digit indicated by a question mark in the MasterCard number 426452002177?337 is unreadable. What is the unreadable digit?

25. Suppose that a valid credit card number of the form $11a_3a_4 \cdots a_{16}$ is changed to $55a_3a_4 \cdots a_{16}$. Is the result a valid credit number?

26. Suppose that a valid credit card number of the form $22a_3a_4 \cdots a_{16}$ is changed to $55a_3a_4 \cdots a_{16}$. Is the result a valid credit number?

27. Suppose a correctly coded credit card number $a_1a_2 \cdots a_{16}$ is modified in such a way that the sum produced by the Luhn algorithm is increased by 10. Explain why the credit card error detection method does not detect the error. What happens if the sum is increased by 12?

28. If a number of the form $a_1a_2 \cdots a_{13}a_{14}34$ is a valid credit card number, what is the check digit for a credit number of the form $a_1a_2 \cdots a_{13}a_{14}5$?

29. If a credit card number of the form $83a_3 \cdots a_{15}4$ is a valid credit card number, what is the check digit for a credit number of the form $19a_3 \cdots a_{15}$?

30. Replace the question mark in the number JM1GD222?J1581570 with a digit that will result in a valid VIN. (See Spotlight 16.1 on page 678 for a description of the method to be used.)

31. Create a check digit for the UPC number 38137009213 using the weights 7, 1, 7, 1, 7, 1, . . ., 7, 1, instead of 3, 1, 3, 1, 3, 1, . . ., 3, 1. Test to see whether this check digit will detect single-digit errors by trying several examples.

32. Create a check digit for the UPC number 38137009213 using the weights 2, 1, 2, 1, 2, 1, . . ., 2, 1, instead of 3, 1, 3, 1, 3, 1, . . ., 3, 1. Is the error caused by replacing the 3 in the first position with an 8 detected? What about the error caused by replacing the 1 in the third position with a 6? Explain why or why not.

33. If the weights 5, 1, 5, 1, 5, 1, . . ., 5, 1 were used for the UPC code, which single-digit errors would go undetected?

34. Exercises 31–33 reveal that using the weights 1, 3, or 7 for a particular position detects all errors in that position, whereas using weights 2 or 5 in a position does not detect all errors. Using this observation, make a guess about error-detection capability using weights 9, 4, 6, or 8.

35. Enter the number 036000260809 in a Google search box. Is it a valid UPC number? Now change any one digit. Is the new number a valid UPC number? Try a different change.

36. Suppose that a valid credit card number $90a_3a_4 \cdots a_{16}$ is mistakenly entered into a credit card reader as $09a_3a_4 \cdots a_{16}$. Explain why the Luhn algorithm does not detect the error.

37. If a valid credit card number using the Luhn algorithm has the form $47a_3 \cdots a_{15}a_{16}$, determine the check digit of the number if the first two digits were transposed and the digits in positions 3 through 15 were unchanged.

38. If a credit card number of the form $83a_3 \cdots a_{15}a_{16}$ is a valid credit card number, what is the check digit for a credit card number of the form $18a_3 \cdots a_{15}a_{16}$?

39. Explain why encoding ID numbers by multiplying them by 15 would not detect the single-digit error of replacing 114180 by 114150.

40. Explain why encoding ID numbers by multiplying them by 15 would not detect the transposition error of replacing 125025 by 120525.

41. Explain why encoding ID numbers by multiplying them by 17 would detect all single-digit errors, adjacent transposition errors, and jump transposition errors.

42. State a general criterion for the detection of an error of the form $\cdots abc \cdots \rightarrow \cdots cba \cdots$ for the routing number of a checking account.

43. The 10-digit ISBN 0-669-03925-4 is the result of a transposition of two adjacent digits not involving the first or last digit. Determine the correct ISBN.

44. Explain why the bank scheme will detect the error $751 \cdots \rightarrow 157 \cdots$ but the UPC scheme will not.

45. Suppose that the check digit a_9 for the bank routing number was chosen to be the last digit of $3a_1 + 7a_2 + a_3 + 3a_4 + 7a_5 + a_6 + 3a_7 + 7a_8$ instead of using the method described in this chapter. How would this compare with the actual check digit?

46. Explain why an error caused by transposing the first two digits of a Postal Service money order is not detected by the check-digit scheme. Explain why the same is true for the second and third digits. What about the last two digits?

47. Suppose that a company assigns an extra digit to every employee Social Security number by appending a 0 if the sum of the digits is even and a 1 if the sum is odd. If a 2 were mistakenly read as a 7, would the error be detected? What if a 2 were mistakenly read as an 8? Try a few other experiments with single-digit errors. (For experiments, you can use three-digit numbers instead of nine-digit numbers.) Determine which errors are detected by this method. Explain your reasoning. Approximately what percentage of errors is detected by appending the extra digit?

48. Explain why the Postal Service money order check-digit scheme does not detect the mistake of substituting a 0 for a 9, or vice versa.

49. Which digit never appears as a check digit on a Postal Service money order?

50. Which digit never appears as a check digit on an American Express Travelers Cheque?

51. Which digits never appear as a check digit for an airline identification number?

52. Suppose that four-digit numbers $a_1a_2a_3a_4$ are assigned a check digit a_5 so that $a_1 + 2a_2 + a_3 + 2a_4 + a_5$ is evenly divisible by 10. Test the number 43216 created in this way to see whether the method detects adjacent transposition errors.

53. Starting with the 10-digit ISBN 0-7167-4782-0, create three new numbers by transposing any two different digits. (They need not be adjacent.) Are these errors detected by the scheme?

54. Suppose that in an Avis identification number, an 8 is mistaken for a 5. Is the error detected? What if a 9 is mistaken for a 2?

55. Give an argument to show that the 10-digit ISBN error-detection method will detect a transposition error involving the first and third digits. Does the same argument work for the fourth and sixth digits?

56. Suppose the check digit a_{10} of 10-digit ISBNs were chosen so that $a_1 + 2a_2 + 3a_3 + 4a_4 + 5a_5 + 6a_6 + 7a_7 + 8a_8 + 9a_9 + 10a_{10}$ is divisible by 11 instead of using the method described in the chapter. How would this compare with the actual check digit?

57. Consider a UPC number in which the digits 7 and 2 appear consecutively (i.e., the number has the form $\cdots 72 \cdots$). Will the error caused by transposing these digits (i.e., the number is taken as $\cdots 27 \cdots$) be detected? What if the digits 6 and 2 were transposed instead? State the general criterion for the detection of an error of the form $\cdots ab \cdots \rightarrow \cdots ba \cdots$ by the UPC scheme.

58. If the first three digits of a routing number for a checking account are 537 and the 5 and 3 are transposed, will the error be detected? If the first three numbers are 237 and the 2 and 7 are transposed, will the error be detected?

59. The Canadian province of Quebec assigns a check digit a_{12} to an 11-digit driver's license number $a_1a_2 \cdots a_{11}$ so that $12a_1 + 11a_2 + 10a_3 + 9a_4 + 8a_5 + 7a_6 + 6a_7 + 5a_8 + 4a_9 + 3a_{10} + 2a_{11} + a_{12}$ is divisible by 10. Criticize this method. Describe all single-digit errors that are undetected by this scheme.

60. Speculate on the reason why telephone numbers, Social Security numbers, and serial numbers on most currency do not have check digits.

61. Suppose that a company uses a check-digit scheme similar to the UPC scheme, except that instead of using the UPC weights 3, 1, 3, 1, . . . , it uses w, 1, w, 1, ... If two of the ID numbers used by the company are 73215674 and 73215661, determine w.

62. If a publishing company has headquarters in both the United States and Germany and publishes the same book in both countries, it is likely that the 10-digit ISBN for the book will be identical except for the first and last digits (because the first digit for U.S. publications is 0 and the first digit for German publications is 3). If the last digit of the U.S. edition is 1, what is the last digit for the German publication?

16.2 The ZIP Code

16.3 Bar Codes

63. Determine the ZIP + 4 code and check digit for each of the following Postnet bar codes.

(a)

(b)

(c)

64. In each of the following Postnet bar codes, exactly one mistake occurs (i.e., a long bar appears instead of a short one, or vice versa). Determine the correct ZIP code.

(a)

(b)

(c)

65. Below is a 12-digit delivery-point bar code. Determine the ZIP + 4 number, the last two digits of the street address, and the check digit.

66. If the check digit for the ZIP + 4 code for a house on 1738 Maple Street is 3, what would the check digit for the delivery-point code be?

67. Find the check digit for the ZIP + 4 code 50037-2452.

68. Explain why any two errors in a particular block of five bars in a Postnet code are always detectable. Explain why not all such errors can be corrected.

69. Form all possible strings consisting of exactly three *a*'s and two *b*'s and arrange the strings in alphabetical order. (For example, the first two possibilities are *aaabb* and *aabab*.) Do you see any relationship between your list and the Postnet code?

70. The back cover of recently published books includes a bar code that has the 10-digit ISBN above the bars and a 13-digit identification number below the bars (see below). Examine the bar code on several books. How does the number below the bar code differ from the UPC code?

ISBN-13: 978-1-4292-0900-7
ISBN-10: 1-4292-0900-3
EAN
9 781429 209007
90000

71. Suppose that the first block of a UPC bar code following the guard bar pattern that a scanner reads is 1000100. Is the scanner reading left to right or right to left?

16.4 Encoding Personal Data

72. If a Social Security number begins with 0, the number was issued in a state in the

(a) Northeast.

(b) Midwest.

(c) West.

73. The Canadian postal system has assigned each geographic region a six-character code composed of alternating letters (not including D, F, I, O, Q, and U) and digits, such as P7B5E1 and K7L3N6. Discuss the advantages that this scheme has over the five-digit ZIP code used in the United States.

74. Given that the letters D, F, I, O, Q, and U are not used in the Canadian postal code (see the previous exercise), determine the maximum possible number of postal codes.

75. Determine the Soundex code for the names *Hu*, *Lee*, and *Shaw*.

76. Determine the Soundex code for *Skow*, *Sachs*, *Lennon*, *Lloyd*, *Ehrheart*, and *Ollenburger*.

77. Determine the total number of codes possible using the Soundex Coding System.

78. Explain why none of the following blocks of digits can be a Soundex code for someone's last name: S-205, S-723, S-5513.

79. In Florida, the last three digits of the driver's license number of a female with birth month m and birth date b are $40(m - 1) + b + 500$. For both males and females, the fourth and fifth digits from the end give the year of birth. Determine the last five digits of a Florida driver's license number for a female born on July 18, 1942.

80. Thinking of the last three digits of a Florida driver's license number as a 3-digit integer, what is the largest possible integer it can be?

81. Suppose two females with Florida driver's licenses were born exactly one month apart. By how much would the 3-digit integers represented by the last three digits of their driver's license numbers differ?

82. Describe a situation in which the 3-digit integers represented by the last three digits of the Florida driver's licenses of identical twins could differ.

83. Explain why an Illinois driver's license number that ends with the last five digits 03217 is suspicious.

84. Determine the last five digits of an Illinois driver's license number for a male born on June 18, 1942.

85. In Illinois, one obtains the last three digits of the driver's license number for a female by adding 600 to the number for a male with the same birthday. In Florida, 500 is added to the number for a male. Why can't Florida use 600?

86. Explain why an Illinois driver's license number that ends with 77061 cannot be valid.

87. Determine the birth date of a person whose Illinois driver's license number ends with 58818.

88. Provide three names that share the same Soundex code as *Gallihan.*

89. The state of Washington encodes the last two digits of the year of birth into driver's license numbers (in positions 8 and 9) by subtracting the two-digit number from 100. For example, a person born in 1942 has 58 in positions 8 and 9, whereas a person born in 1971 has 29 in positions 8 and 9. Speculate on the reason for subtracting the birth year from 100.

90. Apply the Soundex code to common ways to misspell your name. Do they give the same code as your name does?

91. Why would the Soundex system of coding be a poor method for encoding names in China?

Chapter Review

92. Explain why an error caused by transposing any two digits of an American Express Travelers Cheque is not detected by the check-digit scheme.

93. Determine the Soundex code for *Smith, Schmid, Smyth,* and *Schmidt.*

94. Driver's license number-assignment schemes that use personal data sometimes produce the same number for different people. Speculate about circumstances under which this is more likely to occur.

95. When using the travelers cheque, credit card, or UPC number algorithms for detecting errors, does the computer have to know which digit is the check digit?

96. Change 173 into a Postnet code.

97. The following is an actual identification number and bar code from a roll of wallpaper. What appears to be wrong with them? Speculate on the reason for the apparent violation of the UPC format.

5 011419 194056

Building Regulations: 1985 Class 0
FINE ART WALLCOVERINGS LTD.
HOLMES CHAPEL, CHESHIRE
MADE IN ENGLAND
FABRIQUE EN ANGLETERRE

98. Explain why the last three digits of a valid Illinois driver's license cannot be 373.

Writing Projects

1. Prepare a two-page report on the driver's license coding schemes used by Michigan, Maryland, and Washington (Michigan and Maryland use the same method). J. Gallian's "Assigning Driver's License Numbers" has the information you will need (see the Suggested Readings below).

2. Prepare a report on the history of the bar code.

3. Prepare a report on the Barcode of Life project.

4. Prepare a report on smart card technology.

5. Prepare a report on the Soundex system.

Suggested Readings

GALLIAN, J. The mathematics of identification numbers, *College Mathematics Journal,* 22 (1991): 194–202. A survey of check-digit schemes associated with identification numbers. Available at www.d.umn.edu/~jgallian/ident.pdf.

GALLIAN, J. Assigning driver's license numbers, *Mathematics Magazine,* 64 (1992): 13–22. Discusses various methods used by the states to assign driver's license numbers. Several of these methods include check digits for error detection. Available at www.d.umn.edu/~jgallian/license.pdf.

GALLIAN, J. Error detection methods, *ACM Computing Surveys,* 28 (1996): 504–17. A detailed description of many error-detection methods. Available at www.d.umn.edu/~jgallian/detection.pdf.

GALLIAN, J., and S. WINTERS. Modular arithmetic in the marketplace, *American Mathematical Monthly*, 95 (1988): 548–51. A detailed analysis of the check-digit schemes presented in this chapter. In particular, the error-detection rates for the various schemes are given. Available at www.d.umn.edu/~jgallian/marketplace.pdf.

Suggested Website

www.d.umn.edu/~jgallian/fapp7 This website links to applets that generate the Soundex code for surnames, the driver's license numbers in various states based on personal information, VINs, and QR codes.

Encoding Information

17

Tomasz Wyszolmirski/iStock/360/Getty Images

What do these forms of electronic communication have in common: the Internet, smartphones, GPS trackers, satellite TV and radio, Facebook, Twitter, YouTube, and iTunes? One answer is that mathematics plays a hidden but central role in the way these media store, transmit, and secure digital data. In this chapter, we explain some of the basic ideas that underpin information science.

17.1 Binary Codes

Coded data made up of two states (or symbols) is called a **binary code.** Binary codes are the hidden language of computers. The Postnet code (short and long bars) and the Universal Product Code (UPC) bar code (white spaces and dark bars) explained in Chapter 16 are two examples of binary codes. Morse code (dots and dashes) and Braille (bumps and flat markings) are two more. The opinions of two critics rating movies with U for "thumbs up" and D for "thumbs down" could be conveyed by a binary code with the four messages UU, UD, DU, and DD. Fax machines, CDs, DVDs, high-definition television signals, smartphones, and space probes represent data as strings of 0s and 1s rather than the usual digits 0 through 9 and letters *A* through *Z*. In this section, we illustrate one way binary codes can be devised so that errors in the transmission of the code can be corrected.

The idea behind error-correction schemes is simple and one you often use. To illustrate, suppose that you are reading the employment section of a newspaper and you see the phrase "must have a minimum of bive years' experience." Instantly you detect an error because *bive* is not a word in the English language. Moreover, you are fairly confident that the intended word is *five.* Why is that? Because *five* is a word derived from *bive* by changing a single letter, and it makes the phrase understandable. In other phrases, words such as *bike* or *give* might be sensible alternatives to *bive.* Using the extra information provided by the context, we often are able to infer the intended meaning when errors occur.

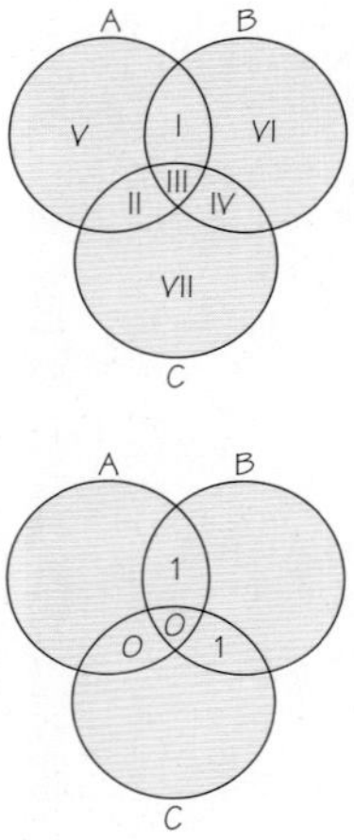

Figure 17.1 Diagram for message 1001.

To demonstrate the way binary error-correcting schemes work, suppose that the National Aeronautics Space Administration (NASA) sends a spacecraft to land at 1 of 16 possible landing sites on Mars. The spacecraft orbits Mars while surveying the sites for the most favorable landing conditions. NASA officials have coded the 16 landing sites with four-digit strings of 0s and 1s such as 0000, 0001, 0010, and 0100. (Recall when counting that if each of k events can occur in $n_1, n_2, \ldots, n_k$ ways, then the total number of ways all k events can occur is the product $n_1 n_2 \cdots n_k$. In this case, there are $2 \cdot 2 \cdot 2 \cdot 2 = 2^4 = 16$ strings.

Once the desired site has been selected, NASA informs the spacecraft where to land by sending the code for the site. However, signals sent through space are subject to interference called *noise.* The noise might cause the spacecraft to interpret the signal as 0001 when the signal actually sent was 1001. Fortunately, over the past 60 years, mathematicians, computer scientists, and engineers have devised highly sophisticated schemes to build extra information into messages composed of 0s and 1s that often permits the correct message to be inferred even though the message may have been received incorrectly (see Spotlight 17.1). The process of converting a message or other information into a code is called **encoding.**

The following example illustrates the basic ideas underlying binary error-correcting schemes.

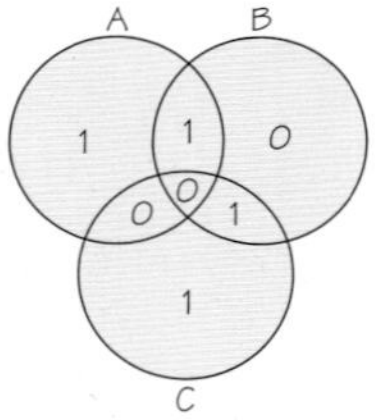

Figure 17.2 Diagram for encoded message 1001101.

EXAMPLE 1 An Error-Correcting Code

Assume that we want to send the message 1001. We append extra information to this message with the aid of the diagram in Figure 17.1. Begin by placing the four message digits in the four overlapping regions I, II, III, and IV, with the digit in the first position (starting at the left of the sequence) in region I, the digit in the second position in region II, and so on. For regions V, VI, and VII, assign 0 or 1 so that the total number of 1s in each circle is even (see Figure 17.2). The encoded message 1001101 is obtained by reading the entries in regions I through VII in order. ■

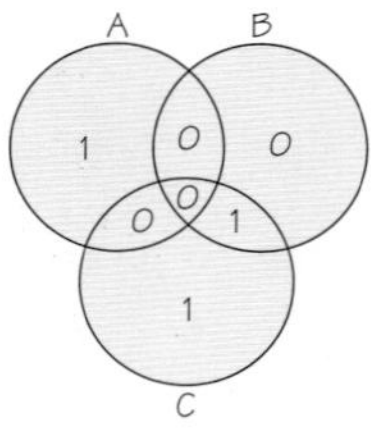

Figure 17.3 Diagram for received message 0001101.

Using the diagram, we have encoded our message 1001 as 1001101. Now suppose that this encoded message is received as 0001101 (an error in the first position). How would we know an error was made? We place each digit from the received message in its appropriate region, as in Figure 17.3.

Noting that there is an odd number of 1s in both circles *A* and *B,* we instantly realize that something is wrong because the intended message had an even

The Ubiquitous Reed–Solomon Codes

SPOTLIGHT 17.1

One of the mathematical ideas underlying current error-correcting techniques for everything from computer hard-disk drives to CD and DVD players was introduced in 1960 by Irving Reed and Gustave Solomon. Reed–Solomon codes made possible the stunning pictures of the outer planets sent back by the space probes *Voyager 1* and *Voyager 2.* They make it possible to scratch a CD or a DVD and still enjoy the content.

"When you talk about CD players, digital television, and various other digital systems—all of those Reed–Solomon [codes] are an integral part of the system," says Robert McEliece, a coding theorist at Caltech.

Why? Because digital information consists of 0s and 1s, and a physical device may confuse the two occasionally. *Voyager 2,* for example, was transmitting data at incredibly low power over billions of miles. Error-correcting codes are a kind of safety net, mathematical insurance against the vagaries of an imperfect material world.

In 1960, the theory of error-correcting codes was only about a decade old. Through the 1950s, a number of researchers began experimenting with a variety of error-correcting codes. But the Reed–Solomon paper, McEliece says, "hit the jackpot." "In hindsight, it seems obvious," Reed later said. However, he added, "Coding theory was not a subject when we published the paper." The two authors knew they had a nice result; they didn't know what impact the paper would have.

Rex Ridenoure

Irving Reed (left) and Gustave Solomon At the Jet Propulsion Laboratory in 1989 to monitor the encounter of *Voyager 2* with Neptune.

Five decades later, the impact is clear. The vast array of applications has settled the questions of the practicality and significance of Reed–Solomon codes. Billions of dollars in modern technology depend on ideas that stem from Reed and Solomon's original work.

Source: Information from the article "The Ubiquitous Reed–Solomon codes" by Barry Cipra from *SIAM News,* January 1993, p. 1.

number of 1s in each circle. How do we correct the error? The answer involves parity. (Parity refers to the oddness or evenness of a number; even integers have **even parity** and odd integers have **odd parity.**) Because circles A and B have the wrong parity and C does not, the error is located in the portion of the diagram in circles A and B, but not in circle C, that is, region I (shaded in Figure 17.4). Here we also see the advantage of using only 0s and 1s to encode data. If you have only two possibilities and one of them is incorrect, then the other one must be correct. Because the 0 in region I is incorrect, we know that 1 is correct.

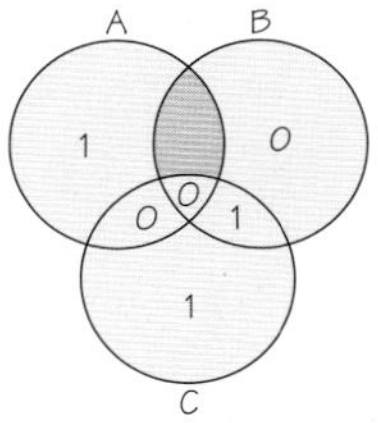

Figure 17.4 Circles *A* and *B*, but not *C*, have the wrong parity.

EXAMPLE 2 Error Correction

For practice, we will do another example. Consider the message 0111. Proceeding as before, we place 0 in region I and 1s in regions II, III, and IV. For regions V, VI, VII, we assign 0 or 1 so that the total number of 1s in each circle is even (see Figure 17.5). Then the message 0111 is encoded as 0111001. If this code word is received as, say, 0111011 (error in the sixth position), the diagram for the received word is shown in Figure 17.6. Then circle *B* has an odd number of 1s, and circles *A* and *C* have an even number. If the received message has only one error, then the error must be in circle *B* but not circles *A* and *C*. This tells us that the entry is region VI is incorrect. So, the error can be corrected. ■

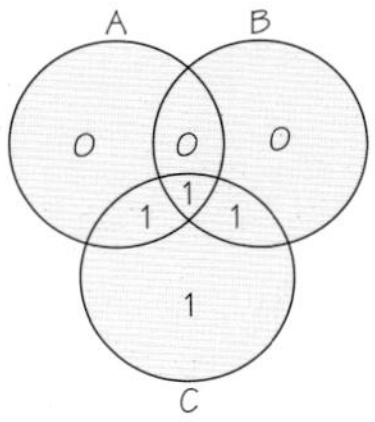

Figure 17.5 Diagram for encoded message 0111001.

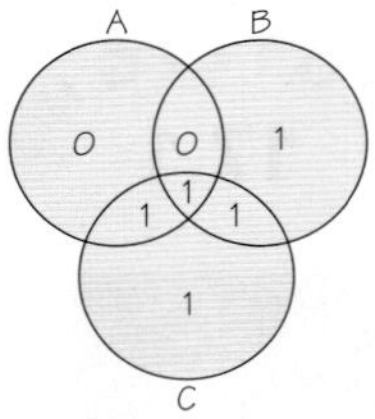

Figure 17.6 Diagram for encoded message 0111011.

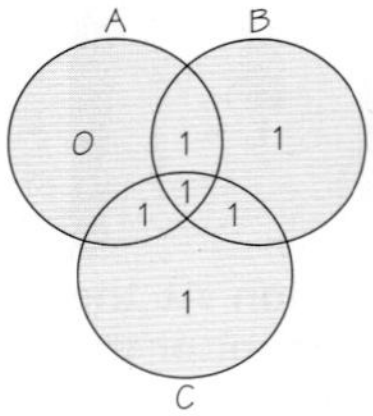

Figure 17.7 Diagram for encoded message 1111011.

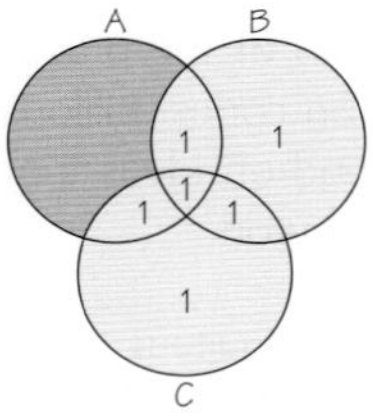

Figure 17.8 Circle *A*, but not *B* and *C*, has the wrong parity.

Self Check 1

Use the circle diagram shown in Figure 17.1 to encode the message 0101. ■

Because the circle diagram method was designed to correct a received message with exactly one error, if a received message has two or more errors, the diagram method will never yield the correct message. For example, if the encoded message 0111001 is received as 1111011 (errors in positions 1 and 6), then only circle *A* has an odd number of 1s (see Figure 17.7), so our decoding method assumes that there is an error in a region in circle *A* but not in regions in circles *B* and *C*. Thus, our method incorrectly assumes that a single error occurred in region V (shaded in Figure 17.8).

The diagram technique can be used to encode all 16 possible binary messages of length 4, as shown in Table 17.1. The encoded messages are called **code words.** The three digits appended to each string of length 4 provide the "extra information" that is sufficient to infer the intended four-digit message, so long as the received seven-digit message has at most one error.

TABLE 17.1 Binary Code

Message	→	Code Word	Message	→	Code Word
0000	→	0000000	0110	→	0110010
0001	→	0001011	0101	→	0101110
0010	→	0010111	0011	→	0011100
0100	→	0100101	1110	→	1110100
1000	→	1000110	1101	→	1101000
1100	→	1100011	1011	→	1011010
1010	→	1010001	0111	→	0111001
1001	→	1001101	1111	→	1111111

Because the diagram method can be used to encode only four-digit binary messages, it is not practical. We use it merely to illustrate a simple way to encode and decode messages. In the next section, we explain a method that can be used in more general settings.

17.2 Encoding with Parity-Check Sums

Strings of 0s and 1s with extra digits for error correction can be used to send full-text messages. A simple way to do this is to assign an empty space the string 00000, the letter *a* the string 00001, *b* the string 00010, *c* the string 00100, and so on. Because there are 32 possible binary strings of length 5, the five unassigned strings can be used for special purposes, such as indicating uppercase letters or numerals.

For example, we might use the string 11111 to indicate a "shift" from lowercase to uppercase when it precedes the code for a letter (i.e., 1111100010 represents *B*). This is analogous to the shift key on a keyboard. Similarly, we could use 11110 to indicate a "shift" from letters to numerals. Here, 11110 followed by the code for *a*

Neil Sloane

SPOTLIGHT 17.2

Neil Sloane is one of the world's leading researchers on the subject of sphere packing, a field that has become indispensable to modern communications. Without it, we might not have CDs, DVDs, or satellite photos of Neptune.

To exchange information rapidly and correctly, machines must code it. For example, imagine you want to transmit a child's drawing that uses every one of the 64 Crayola colors. For transmission, you could code each of those colors as a number—say, the integers from 1 to 64. Then you could divide the image into many pixels and assign 6-digit strings of 0s and 1s to each one, based on the color it contains. The transmission would then be a steady stream of those numbers, one for each pixel.

Because there are 64 possible combinations of 0s and 1s in a 6-digit string, you could handle the entire Crayola palette with 64 different 6-digit "code words." For instance, 000000 could represent the first color, 000001 the next color, 000010 the next, and so on.

But in a noisy signal, two different code words might look nearly identical. A bit of noise, for example, might shift a spike of current to the wrong place, so that 001000 looks like 000100. An efficient way to keep the colors straight in spite of noise is to add four extra digits to the 6-digit code words. The receiver, programmed to know the 64 permissible combinations, now could spot any other combination as an error introduced by noise; then it would correct the error automatically to the "nearest" permissible color.

Laine Whitcomb

Neil Sloane

In fact, says Sloane, "If any of those 10 digits were wrong, you could still figure out what the right crayon was."

Source: Information from the article "Math in a Million Dimensions" by David Berreby, *Discover*, October 1990.

represents the numeral 0, 11110 followed by the code for b represents the numeral 1, and so on up to 9. Punctuation marks could be handled in the same fashion. Rather than using the circle diagram method, messages are encoded by appending extra digits determined by the parity of various sums of certain portions of the messages. We illustrate this method for the 16 messages shown in the left-hand column of Table 17.1. (See also Spotlight 17.2.)

Our goal is to take any binary string $a_1a_2a_3a_4$ and append three check digits $c_1c_2c_3$ so that any single error in any of the seven positions can be corrected. We do this as follows. Choose

$$c_1 = \begin{cases} 0 \text{ if } a_1 + a_2 + a_3 \text{ is even} \\ 1 \text{ if } a_1 + a_2 + a_3 \text{ is odd} \end{cases}$$

$$c_2 = \begin{cases} 0 \text{ if } a_1 + a_3 + a_4 \text{ is even} \\ 1 \text{ if } a_1 + a_3 + a_4 \text{ is odd} \end{cases}$$

$$c_3 = \begin{cases} 0 \text{ if } a_2 + a_3 + a_4 \text{ is even} \\ 1 \text{ if } a_2 + a_3 + a_4 \text{ is odd} \end{cases}$$

The sums $a_1 + a_2 + a_3$, $a_1 + a_3 + a_4$, and $a_2 + a_3 + a_4$ are called **parity-check sums.** They are so named because their function is to guarantee that the sum of various components of the encoded message is even. Indeed, c_1 is defined so that $a_1 + a_2 + a_3 + c_1$ is even. (Recall that this is precisely how the value in region V in Figure 17.2 was defined.) Similarly, c_2 is defined so that $a_1 + a_3 + a_4 + c_2$ is even, and c_3 is defined so that $a_2 + a_3 + a_4 + c_3$ is even.

EXAMPLE 3 Encoding and Decoding a Message

Let's revisit the message 1001 that we considered in Figure 17.1. Then, $a_1a_2a_3a_4 = 1001$ and

$$c_1 = 1 \text{ because } 1 + 0 + 0 \text{ is odd}$$
$$c_2 = 0 \text{ because } 1 + 0 + 1 \text{ is even}$$
$$c_3 = 1 \text{ because } 0 + 0 + 1 \text{ is odd}$$

So, because $c_1c_2c_3 = 101$, we have $1001 \rightarrow 1001101$.

How is the intended message determined from a received encoded message? This process is called **decoding.** Say, for instance, that the message 1000, which has been encoded using parity-check sums as $u = 1000110$, is received as $v = 1010110$ (an error in the third position). We simply compare v with each of the 16 code words (that is, the possible correct messages) in Table 17.1 and decode it as the one that differs from v in the fewest positions. (Put another way, we decode v as the code word that agrees with v in the most positions.) This method works even if the error in the message is one of the check digits rather than one of the digits of the original message string. When there is more than one code word that differs from v in the fewest positions, we do not decode.

To carry out these comparisons, it is convenient to define the distance between two strings of equal length.

Distance Between Two Strings DEFINITION

The **distance between two strings** of equal length is the number of positions in which the strings differ.

For example, the distance between $v = 1010110$ and $u = 1000110$ is 1 because they differ in only one position (the third). In contrast, the distance between 1000110 and 0111001 is 7 because they differ in all seven positions. Thus, our decoding procedure is simply to decode any received message w as the code word w' that is "nearest" to it. More specifically, for any received word w, we determine the distance between w and all code words and decode w as the one for which the distance is a minimum. Table 17.2 shows the distance between $v = 1010110$ and all 16 code words. From this table, we see that v will be decoded

TABLE 17.2 Distances from 1010110 to Code Words

v	1010110	1010110	1010110	1010110	1010110	1010110	1010110	1010110
Code word	0000000	0001011	0010111	0100101	1000110	1100011	1010001	1001101
Distance	4	5	2	5	1	4	3	4
v	1010110	1010110	1010110	1010110	1010110	1010110	1010110	1010110
Code word	0110010	0101110	0011100	1110100	1101000	1011010	0111001	1111111
Distance	3	4	3	2	5	2	6	3

as 1000110 = u because it differs from u in only one position, whereas it differs from all others in the table in at least two positions. This method is called **nearest-neighbor decoding.**

Assuming that errors occur independently, the nearest-neighbor method decodes each received message as the one it most likely represents.

Nearest-Neighbor Decoding DEFINITION

The **nearest-neighbor decoding** method decodes a received message as the code word that agrees with the message in the most positions, provided that there is only one such code word.

Self Check 2

Find the distance between 10010111 and 11011110.

The scheme we have just described was first proposed in 1948 by Richard Hamming, a mathematician at Bell Laboratories. It is one of a family of codes that are called the *Hamming codes.* The collection of all possible strings obtained from messages of a given length of 0s and 1s by appending extra 0s and 1s using parity-check sums, as illustrated earlier, is called a **binary linear code.** The strings with the appended digits are called **code words.**

Binary Linear Code/Code Words DEFINITION

A **binary linear code** is a set of words composed of 0s and 1s obtained from all possible messages of a given length by using parity-check sums to append check digits to the messages. The resulting strings are called **code words.**

You should think of a binary linear code as a set of n-digit strings in which each string is composed of two parts: the message part, consisting of the original messages, and the remaining check-digit part.

The longer the messages are, the more check digits are required to correct errors. For example, binary messages consisting of six digits require four check digits to ensure that all messages with one error can be decoded correctly.

Given a binary linear code, how can we tell whether it will correct errors and how many errors it will detect? It is remarkably easy. We examine all the code words to find one that has the fewest number of 1s, excluding the *zero code word* consisting entirely of 0s. Call this minimum number of 1s in any nonzero code word the *weight* of the code.

Weight of a Binary Code DEFINITION

The **weight of a binary code** is the minimum number of 1s that occur among all nonzero code words of that code.

The test for the error-detecting and error-correcting capacities of a code is as follows.

Test for Error Detection and Correction Capacity PROCEDURE

1. Calculate the weight t of the code.
2. The code will detect any $t - 1$ or fewer errors.
3. If t is odd, the code will correct any $(t - 1)/2$ or fewer errors.
4. If t is even, the code will correct any $(t - 2)/2$ or fewer errors.

EXAMPLE 4 Error Detection and Correction Capacity

The binary code

$$C_1 = \{000, 001, 010, 100, 110, 101, 011, 111\}$$

has weight 1. It will not detect any error.

The binary code

$$C_2 = \{0000, 0101, 1010, 1111\}$$

has weight 2. It will detect any $2 - 1 = 1$ error but will not correct every 1 error.

The binary code

$$C_3 = \{00000, 10011, 01010, 11001, 00101, 10110, 01111, 11100\}$$

has weight 3. It will detect any $3 - 1 = 2$ or fewer errors or correct any $(4 - 2)/2 = 1$ error.

The binary code

$$C_4 = \{000000, 0010111, 0101011, 1001101, 1100110, 1011010, 0111100, 1110001\}$$

has weight 4. It will detect any $4 - 1 = 3$ or fewer errors or correct any $(4 - 2)/2 = 1$ error.

Applying this test to the code in Table 17.1, we see that the weight is 3, so it will correct any $(3 - 1)/2 = 1$ error or it will detect any $3 - 1 = 2$ or fewer errors. Be careful here. We must decide *in advance* whether we want to use our code to correct single errors or to detect any two or fewer errors. It can do whichever we choose, but not both. If we decide to detect errors, then we will not decode any message that was not on our original list of encoded messages (just as *bive* is not a word in the English language). Instead, we simply note that an error was made and, in most applications, request a retransmission. An example occurs when a bar-code reader at the supermarket detects an error and therefore does not emit a sound (in effect, requesting a rescanning). On the other hand, if we decide to correct errors, we will decode any received message as its nearest neighbor.

Here is an example of another binary linear code.

EXAMPLE 5 A Binary Linear Code

We let the set of messages be {000, 001, 010, 100, 110, 101, 011, 111} and append three check digits c_1, c_2, and c_3 using

$$c_1 = \begin{cases} 0 \text{ if } a_1 + a_2 + a_3 \text{ is even} \\ 1 \text{ if } a_1 + a_2 + a_3 \text{ is odd} \end{cases}$$

$$c_2 = \begin{cases} 0 \text{ if } a_1 + a_3 \text{ is even} \\ 1 \text{ if } a_1 + a_3 \text{ is odd} \end{cases}$$

$$c_3 = \begin{cases} 0 \text{ if } a_2 + a_3 \text{ is even} \\ 1 \text{ if } a_2 + a_3 \text{ is odd} \end{cases}$$

For instance, if we take $a_1a_2a_3$ as 101, we have

$$c_1 = 0 \text{ because } 1 + 0 + 1 \text{ is even}$$
$$c_2 = 0 \text{ because } 1 + 1 \text{ is even}$$
$$c_3 = 1 \text{ because } 0 + 1 \text{ is odd}$$

So we encode 101 by appending 001, that is, 101 → 101001. The entire code is shown in Table 17.3.

TABLE 17.3 Code Words

Message	→	Code Word	Message	→	Code Word
000	→	000000	110	→	110011
001	→	001111	101	→	101001
010	→	010101	011	→	011010
100	→	100110	111	→	111100

Because the minimum number of 1s of any nonzero code word is three, this code will either correct any single error or detect any two errors, whichever we choose.

It is natural to ask how the method of appending extra digits with parity-check sums enables us to detect or even correct errors. Error detection is obvious. Think of how a computer spell-checker works. If you type *bive* instead of *five*, the spell-checker detects the error because the string *bive* is not on its list of valid words. On the other hand, if you type *give* instead of *five*, the spell-checker will not detect the error because *give* is on its list of valid words.

Our error-detection scheme works the same way, except that if we add extra digits to ensure that our code words differ in many positions—say, t positions—then even as many as $t - 1$ mistakes will not convert one code word into another code word. And if every pair of code words differs from each other in at least three positions, we can correct any single error because the incorrect received word will differ from the correct code word in exactly one position, but it will differ from all others in two or more positions.

Thus, in this case, the correct word is the unique "nearest neighbor." So the role of the parity-check sums is to ensure that code words differ in many positions. For example, consider the code in Table 17.1. The messages 1000 and 1100 differ in only the second position. But the two parity-check sums $a_1 + a_2 + a_3$ and $a_2 + a_3 + a_4$ guarantee that encoded words for these messages have different values in positions 5 and 7 as well as in position 2. It is the job of mathematicians to discover the appropriate parity-check sums to correct several errors in long codes.

17.3 Data Compression

Binary linear codes are **fixed-length codes.** In a fixed-length code, each code word is represented by the same number of digits (or symbols). In contrast, Morse code (see Spotlight 17.3), designed for the telegraph, is a **variable-length code,** that is, a code in which the number of symbols for each code word may vary.

Morse Code

SPOTLIGHT 17.3

Morse code is a ternary code consisting of short marks, long marks, and spaces (see figure at right). It was invented in the early 1840s by Samuel Morse as an efficient way to transmit messages using electronic pulses through telegraph wires. The code enabled operators to send strings of short pulses, long pulses, and pauses representing characters transformed into indentations on paper tape that could be easily converted back to characters. Although Morse code was widely used up until the mid-twentieth century, it has gradually been supplanted by more machine-friendly codes. Because Morse code uses data compression, sending messages using Morse code is faster than text messaging. Many Nokia cell phones can convert text messages to Morse code.

A	·—	N	—·
B	—···	O	———
C	—·—·	P	·——·
D	—··	Q	——·—
E	·	R	·—·
F	··—·	S	···
G	——·	T	—
H	····	U	··—
I	··	V	···—
J	·———	W	·——
K	—·—	X	—··—
L	·—··	Y	—·——
M	——	Z	——··

Morse code

Notice that in Morse code, the letters that occur most frequently are represented by the fewest number of symbols, whereas letters that occur less frequently are coded with more symbols. By assigning the code in this manner, telegrams could be created more quickly and could convey more information per line than could fixed-length codes, such as the binary code for Postnet code (discussed in Section 16.3), where each digit from 0 through 9 is represented by a string of length 5. Morse code is an example of data compression.

Data Compression DEFINITION

Data compression is the process of encoding data so that data are represented by fewer symbols.

Figure 17.9 shows a typical frequency distribution for letters in English-language text material.

Data compression provides a means to reduce the costs of data storage and transmission. A **compression algorithm** converts data from an easy-to-use format to one designed for compactness. Conversely, an uncompression algorithm converts the compressed information back to its original (or approximately original) form. Downloaded files in the ZIP format are an example of a particular kind of data compression. When you "unzip" the file, you return the compressed data to its original state. In some applications, such as datasets that represent images, the original data need only be recaptured in approximate form. In these cases, there are algorithms that result in a great saving of space. Graphics Interchange Format (GIF) encoding returns compressed data to its exact original form, while Joint Photographic Experts Group (JPEG) encoding and Motion Picture Expert Group (MPEG) encoding return data only approximately to its original state.

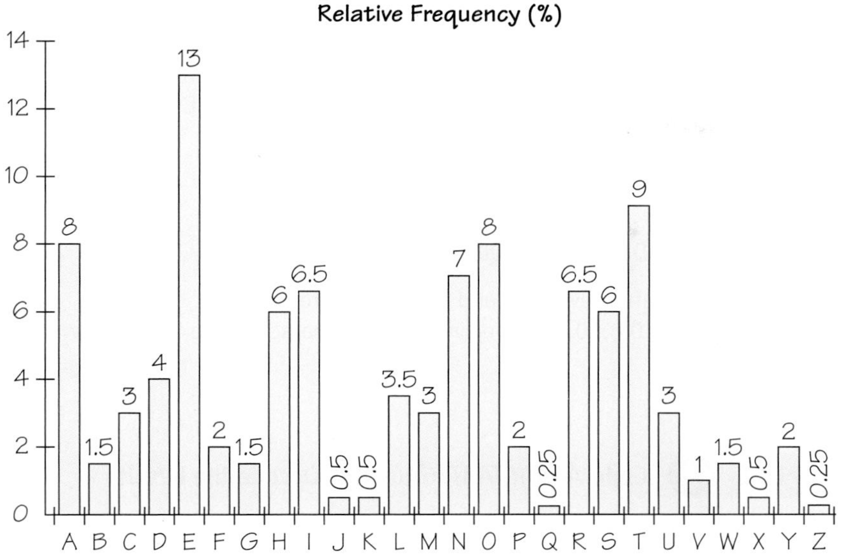

Figure 17.9 A widely used frequency table for letters in normal English usage.

EXAMPLE 6 Data Compression

Let's illustrate the principles of data compression with a simple example. Biologists are able to describe genes by specifying sequences composed of the four letters *A, T, G,* and *C,* which represent the four nucleotides adenine, thymine, guanine, and cytosine, respectively. One way to encode a sequence such as *AAACAGTAAC* in fixed-length binary form would be to encode the letters as

$$A \to 00 \qquad C \to 01 \qquad T \to 10 \qquad G \to 11$$

The corresponding binary code for the sequence *AAACAGTAAC* is then

00000001001110000001

On the other hand, if we knew from experience that *A* occurs most frequently, *C* second most frequently, and so on, and that *A* occurs much more frequently than *T* and *G* together, the most efficient binary encoding would be

$$A \to 0 \qquad C \to 10 \qquad T \to 110 \qquad G \to 111$$

For this encoding scheme, the sequence *AAACAGTAAC* is encoded as

0001001111100010

Notice that the encoded sequence in Example 6 has 20% fewer digits than our previous sequence, in which each letter was assigned a fixed length of 2 (16 digits versus 20 digits). However, to realize this savings, we have made decoding more difficult. For the binary sequence using the fixed length of two symbols per character, we decode the sequence by taking the digits two at a time in succession and converting them to the corresponding letters.

The following observation explains the algorithm for decoding our compressed sequences. Note that 0 occurs only at the end of a code word. Thus, each time you see a 0, it is the end of the code word. Also, because the code words 0, 10, and 110 end in a 0, the only circumstances under which there are three consecutive 1s is when the code word is 111. So, to quickly decode a compressed binary sequence using our coding scheme, insert a comma after every 0 and after every three consecutive 1s. The digits between the commas are code words. Example 7 illustrates the method.

EXAMPLE 7 Decode 0001001111100010

Inserting commas after every 0 and every occurrence of a block 111, we have 0,0,0,10,0,111,110,0,0,10. Converting these characters to letters, we obtain *AAACAGTAAC*.

EXAMPLE 8 Code *AGAACTAATTGACA* and Decode the Result

Recall: $A \to 0$, $C \to 10$, $T \to 110$, and $G \to 111$. So

$$AGAACTAATTGACA \to 0111001011000110110111 0100$$

To decode the encoded sequence, we insert commas after every 0 and after every occurrence of 111 and convert to letters:

0,	111,	0,	0,	10,	110,	0,	0,	110,	110,	111,	0,	10,	0
A,	*G*,	*A*,	*A*,	*C*,	*T*,	*A*,	*A*,	*T*,	*T*,	*G*,	*A*,	*C*,	*A*

Self Check 3

Using the coding scheme

$$A \to 0 \qquad C \to 10 \qquad T \to 110 \qquad G \to 111$$

convert the sequence *CCTGA* into binary code.

Given the encoding scheme

$$A \to 0 \qquad C \to 10 \qquad T \to 110 \qquad G \to 111$$

decode the binary string 0101110111010.

Delta Encoding

For datasets of numbers that vary little from one number to the next, the method of compression called *delta encoding* works well.

EXAMPLE 9 Delta Encoding

Consider the following closing prices (rounded to the nearest integer) of the Standard & Poor's (S&P) index of the stock prices of 500 companies in July 2013:

1615	1614	1615	1632	1640	1652	1653	1675	1680	1683	1676
1681	1689	1692	1696	1692	1686	1690	1692	1685	1686	1686

These numbers use 108 characters in all (counting spaces). To compress this dataset using the delta method, we start with the first number and continue by listing only the change from each entry to the next. So our list becomes

1615 −1 1 17 8 12 1 22 5 3 −7 5 8 3 4 −4 −6 4 2 −7 1 0

This time we have used only 52 characters, counting the minus signs, to represent the same data, a savings of almost 52%. ■

Self Check 4

Given that the following numbers were encoded using the delta function, determine the original numbers:

1020 15 12 −6 8 20 −4 ■

Huffman Coding

The coding methods we have shown so far are too simple for general use, but in 1951, a graduate student named David Huffman (see Spotlight 17.4) devised a scheme for data compression that became widely used. As in the first compression scheme we discussed, Huffman coding assigns short code words to the characters with high probabilities of occurring and long code words to those with low probabilities of occurring. A *Huffman code* is made by arranging the characters from top

David Huffman

SPOTLIGHT 17.4

In 1951, David Huffman and his classmates in an electrical engineering graduate course on information theory were given the choice of writing a term paper or taking a final exam. For the term paper, Huffman's professor had assigned what at first appeared to be a simple problem: Find the most efficient method of representing numbers, letters, or other symbols using binary code. Huffman worked on the problem for months, developing a number of approaches, but he couldn't prove that any of them were the most efficient.

Just as he was about to throw his notes in the garbage, the solution came to him. "It was the most singular moment of my life," Huffman says. "There was the absolute lightning of sudden realization. It was my luck to be there at the right time and also not have my professor discourage me by telling me that other good people had struggled with the problem," he says. When presented with his student's discovery, Huffman recalls, his professor exclaimed: "Is that all there is to it?!"

"The Huffman code is one of the fundamental ideas that people in computer science and data communications are using all the time," says Donald Knuth of Stanford University. Although others have used Huffman's encoding to help make millions of dollars, Huffman's main compensation was dispensation from the final exam. He never tried to patent an invention from his work and experienced only a twinge of regret at not having used his creation to make himself rich. "If I had the best of both worlds, I would have had recognition as a scientist, and I would have gotten monetary rewards," he says. "I guess I got one and not the other."

David Huffman died October 7, 1999.

Courtesy University of California, Santa Cruz

David Huffman

Source: Excerpted from the article "Profile: David Huffman" by Gary Stix, *Scientific American*, September 1991, pp. 54, 58.

to bottom in a so-called *code tree* according to increasing probability; it proceeds by combining, at each stage, the two least probable combinations and repeating this process until there is only one combination remaining.

EXAMPLE 10 Huffman Code

To illustrate the Huffman method, say we have six letters of a dataset that occur with the following probabilities:

A	0.125
B	0.051
C	0.215
D	0.173
E	0.210
F	0.226

Rearranging them in increasing order, we have:

B	0.051
A	0.125
D	0.173
E	0.210
C	0.215
F	0.226

Because *B* and *A* are the two letters least likely to occur, we begin our tree by merging them with the one with the smallest probability on the left (i.e., *BA* rather than *AB*), adding their probabilities, and rearranging the resulting items in increasing order:

D	0.173
BA	0.176
E	0.210
C	0.215
F	0.226

This time, *D* and *BA* are the two least likely remaining entries, so we merge them with *D* on the left (because it has smallest probability), add their probabilities, and re-sort from smallest to largest. This gives:

E	0.210
C	0.215
F	0.226
DBA	0.349

Next, we combine *E* and *C* with *E* on the left and re-sort to get:

F	0.226
DBA	0.349
EC	0.425

Then we combine *F* and *DBA* with *F* on the left and re-sort again:

EC	0.425
FDBA	0.575

And finally we combine *EC* and *FDBA* with *EC* on the left to obtain:

ECFDBA	1.000

To create the *Huffman tree,* start with a vertex at the top labeled with the final string *ECFDBA* 1.000, then reverse the steps used to merge letters to work your way down the tree by creating two branches corresponding to each merger of the two least likely entries at each stage; assign 0 to the branch with the lower probability and 1 to the other branch, as shown in Figure 17.10. To assign a binary code word to each letter, we follow the path from the end of the tree to each letter by assigning, at each merging juncture, 0 or 1 according to the label on the branch taken. Thus, we have

A	1111
B	1110
C	01
D	110
E	00
F	10

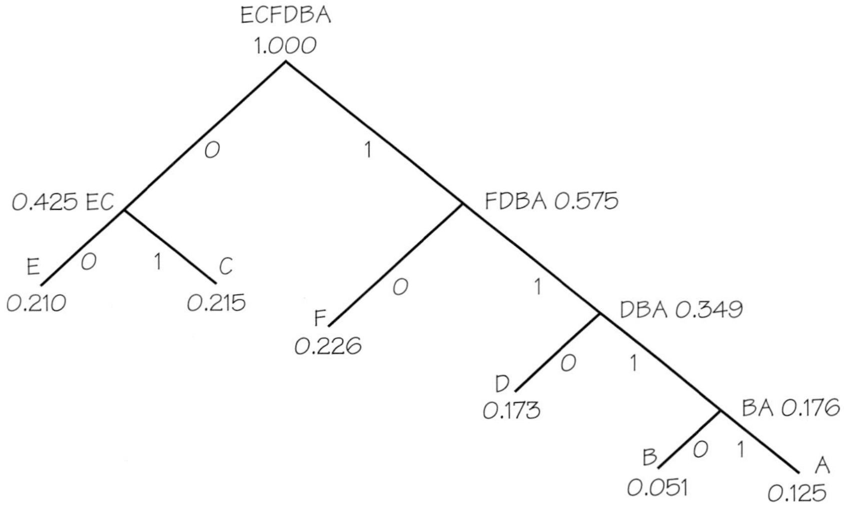

Figure 17.10 A Huffman tree.

Notice that the letters that occur least often have the longest codes and the letters that occur most often have the shortest codes.

Decoding codes created from a Huffman tree is possible because there is only one way a particular string could have occurred. Consider the Huffman code created using the code words given in the preceding display:

$$1110100001101001111010010$$

How can we determine the corresponding string of letters that has this Huffman code? The method is quite simple. Starting at the beginning of the string, every time we find a code word, we replace that string with its corresponding letter (*A, B, C, D, E,* or *F*) and continue. Since all of our code words have at least two digits, we look at the first two digits. If they form a code word, then replace them with the corresponding letter. If not, then look at the next digit. If the three digits form a code word, then replace them with the corresponding letter. If not, then these three digits and the next one are a four-digit code word.

EXAMPLE 11 ➔ Decoding a Huffman Code

Looking at our example 1110100001101001111010010, we see that neither 11 nor 111 is a code word but 1110 is the code word for *B*. So, replacing 1110 with *B*, we have *B*100001101001111010010. The next possibility is 10, which is the code word for *F*, so we have *BF*0001101001111010010. Next, we have 00, which is the code word for *E*, giving us *BFE*01101001111010010. Continuing in this way, we obtain *BFECFFCACEF*. Of course, in practice, coding and decoding are done by computers. ■

Self Check 5

Decode the binary string 001101111001001 that was encoded using the Huffman code given in Example 10. ■

17.4 Cryptography

Thus far, we have discussed ways in which data can be encoded to detect errors or correct errors in transmission. In many situations, there is also a desire for security against unauthorized interpretation of coded data (that is, a desire for secrecy). The process of disguising data is called *encryption.*

Cryptography DEFINITION

Cryptography is the study of methods to make and break secret codes. The process of coding information to prevent unauthorized use is called **encryption.** The process of making intelligible information from a deliberately distorted transmission is called **decryption.**

Historically, encryption was used primarily for military and diplomatic transmissions (see Spotlights 17.6 through 17.8, starting on page 724). Today, encryption is essential for securing electronic transactions of all kinds. Cryptography is what allows you to have a website safely receive your credit card number. Cryptographic schemes prevent hackers from charging calls to your cell phone. Cryptography is also used for authenticating electronic transactions. In 1998, history was made when President Bill Clinton and Ireland's Prime Minister Bertie Ahern used digital signatures to sign an intergovernmental document. Each leader had a unique signing code and a digital certificate that served as a "digital ID," thereby ensuring that the document truly was approved by them. Although modern encryption schemes are extremely complex, we will illustrate the fundamental concepts involved with a few simple examples.

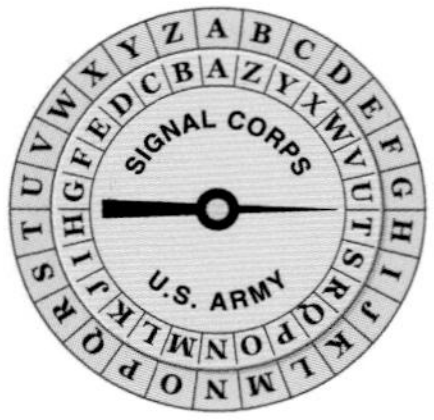

© Universal Images Group Limited/Alamy

Among the first known cryptosystems is the **Caesar cipher,** used by Julius Caesar to send messages to his troops. To encrypt a message with the method employed by Caesar, we use the following table to replace each letter in the top row with the letter below it:

A	B	C	D	E	F	G	H	I	J	K	L	M	N	O	P	Q	R	S	T	U	V	W	X	Y	Z
D	E	F	G	H	I	J	K	L	M	N	O	P	Q	R	S	T	U	V	W	X	Y	Z	A	B	C

For example, the message ATTACK AT DAWN is encrypted as DWWDFN DW GDZQ.

To decrypt the message, replace each letter with the letter above it in the table. Obviously, it would not require much effort for someone to "crack" this code, but it might be good enough to use on your friends who are not mathematically inclined.

Self Check 6

Use the Caesar cipher to encrypt the message SEND MONEY. ■

To describe more sophisticated schemes for transmitting messages secretly, it is convenient to introduce a special kind of arithmetic used in cryptography. Recall that if we divide a positive integer a by a positive integer n, there are unique integers q and r such that $a = nq + r$, where r is nonnegative and less than n. The number q is called the *quotient* and r is called the *remainder.* For any positive integers a and n, we define a mod n (read "a modulo n" or just "a mod n") to be the remainder when a is divided by n. For negative integers a, we can find a mod n by adding exactly enough multiples of n to a so that the sum is nonnegative (this works because modulo n, $n = 0$). Thus, $-84 \bmod 26 = 20$ because $-84 + 4 \cdot 26 = -84 + 104 = 20$.

EXAMPLE 12 ➜ ***a* mod *n***

$$3 \bmod 2 = 1 \text{ because } 3 = 1 \cdot 2 + 1$$
$$6 \bmod 2 = 0 \text{ because } 6 = 3 \cdot 2 + 0$$
$$5 \bmod 3 = 2 \text{ because } 5 = 1 \cdot 3 + 2$$
$$53 \bmod 9 = 8 \text{ because } 53 = 5 \cdot 9 + 8$$
$$37 \bmod 10 = 7 \text{ because } 37 = 3 \cdot 10 + 7$$
$$66 \bmod 11 = 0 \text{ because } 66 = 11 \cdot 6 + 0$$
$$105 \bmod 26 = 1 \text{ because } 105 = 4 \cdot 26 + 1$$
$$342 \bmod 85 = 2 \text{ because } 342 = 4 \cdot 85 + 2$$
$$62 \bmod 85 = 62 \text{ because } 62 = 0 \cdot 85 + 62$$
$$-22 \bmod 26 = 4 \text{ because } -22 + 26 = 4$$
$$-172 \bmod 25 = 3 \text{ because } -172 + 7 \cdot 25 = -172 + 175 = 3$$

To find the quotient and remainder for small integers, simply use long division. For larger numbers such as 2751 mod 13, use a calculator to divide 2751 by 13 to obtain 211.615385. Then the integer part 211 is the quotient q. To find the remainder r when a is divided by n, note that $r = a - nq$. So we have $r = 2751 - 13 \cdot 211 = 2751 - 2743 = 8$. An easy way to find 2751 mod 13 and −517 mod 13 is to enter "2751 mod 13" and "−517 mod 13" into a Google search box. ■

Arithmetic involving mod n is called **modular arithmetic.** Although this arithmetic may appear unfamiliar, you often unconsciously use it. For example, if it is now September, what month will it be 25 months from now? Of course, you answer "October," but the interesting fact is that you didn't arrive at the answer by starting with September and counting off 25 months. Instead, without even thinking about it, you simply observed that $25 = 2 \cdot 12 + 1$ so that $25 \bmod 12 = 1$, and you added one month to September.

Modeling the Genetic Code

SPOTLIGHT 17.5

The way that genetic material is composed can be conveniently modeled using modulo 4 arithmetic. A DNA molecule is made up of two long strands in the form of a double helix. Each strand is made up of strings of the four nitrogen bases adenine (A), thymine (T), guanine (G), and cytosine (C). Each base on one strand binds to a complementary base on the other strand. Adenine always binds to thymine, and guanine always binds to cytosine. To model this situation, we identify A with 0, T with 2, G with 1, and C with 3. Thus, the DNA segment ACGTAACAGGA and its complementary segment TGCATTGTCCT are identified by 03120030110 and 21302212332.

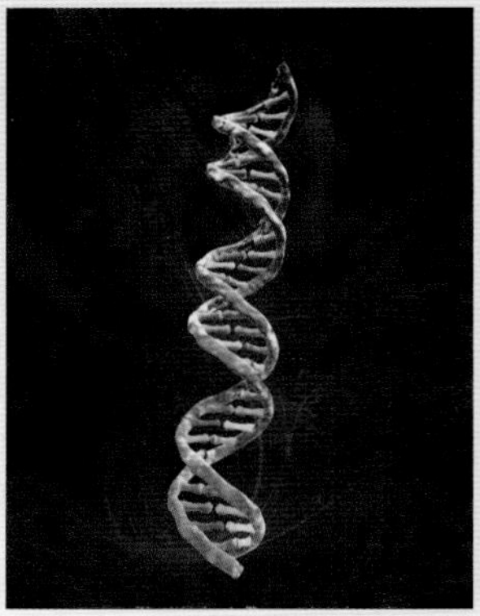

© Gunilla Elam/Science Source

Using modulo 4 arithmetic, $0 + 2 = 2$, $2 + 2 = 0$, $1 + 2 = 3$, and $3 + 2 = 1$, we see that adding 2 to any of the integers 0, 1, 2, or 3 interchanges 0 and 2 and 1 and 3. So, for any DNA segment $a_1a_2 \cdots a_n$ represented by strings of 0s, 1s, 2s, and 3s, we see that its complementary segment is represented by $a_1a_2 \cdots a_n + 22 \cdots 2$, where we add the integers in each component using modulo 4. In particular, $03120030110 + 22222222222 = 21302212332$.

Source: Information from *Discrete Mathematics* by S. Washburn, T. Marlow, and C. Ryan, Addison-Wesley, 1999.

Similarly, if it is now Wednesday, you know that in 23 days, it will be Friday. This time, you arrived at your answer by noting that $23 = 3 \cdot 7 + 2$ (that is, 23 mod 7 = 2), so you added 2 days to Wednesday instead of counting off 23 days. Because $51 = 2 \cdot 24 + 3$, if it is now 1 P.M., in 51 hours, it will be 4 P.M. (51 mod 24 = 3). If you travel east 390 degrees, you end up 30 degrees east of where you began (390 mod 360 = 30). Applications of modular arithmetic include the check-digit schemes described in Chapter 16, where arithmetic mod 7, 9, 10, and 11 were used. The parity-check sums (discussed in Section 17.2) use addition mod 2. An application of modular arithmetic to genetics is described in Spotlight 17.5.

With modular arithmetic, we can describe the Caesar cipher easily as follows. Begin by saying that the letter *A* is in position 0, *B* is in position 1, *C* is in position 2, and so on. Then the Caesar cipher replaces the letter in position i with the letter in position $(i + 3)$ mod 26. This formula expresses the fact that the Caesar cipher shifts each letter from *A* through *W* three positions to the right, while *X, Y,* and *Z* are replaced with *A, B,* and *C,* respectively. (Think of *X, Y,* and *Z* as "wrapping" around to the beginning of the alphabet.) It is customary to use the term *Caesar cipher* for any encryption scheme that shifts each letter a fixed number of positions. When the shift is anything other than three positions, we will specify the amount of the shift.

Decimation Cipher

Rather than encrypting a message by adding a fixed number to the position of every letter and using modular arithmetic, as is done in the Caesar cipher, another simple way to encrypt a message is to multiply each position by a fixed number and use modular arithmetic. This method of encryption is called the **decimation cipher.**

To begin, we assign the 26 letters the numbers 0 through 25, in order, and select any odd integer k between 3 and 25 except 13. (For the method to work, k and 26 must have no prime divisors in common.) Then, in the message, a letter with numerical value i is replaced with the letter with numerical value ki modulo 26. For example, if k is 5, then D is replaced with P because D has value 3 and $5 \times 3 = 15$, which is assigned to P. When ki exceeds 26, we use modulo 26 arithmetic to determine the replacement for the letter. Thus, J is replaced by T because J has the value 9 and $5 \times 9 = 45$ and 45 mod 26 = 19 and T has value 19. The value of k is called the **key.**

Algebra Review Appendix
Prime and Composite Numbers

To decode an encrypted message, we use the same method except that we multiply the numerical value for each encrypted letter by 21. The number 21 is used to decrypt the message because it has the property that $5 \times 21 = 105$ and 105 mod 26 = 1. As a consequence, for any integer x, we have $(x \times 5 \times 21)$ mod 26 equals $(x \times 1)$ mod 26 = x mod 26. Thus, multiplying an integer x by 5 and then the result by 21 gets us back to x when we use modulo 26. In general, if a number k is used to encrypt a message modulo 26, the value j used to decrypt the message has to be chosen so that kj mod 26 = 1. Given a particular value for k, we can find the corresponding j by trial and error. Table 17.4 shows the values corresponding to each choice of k.

TABLE 17.4 Decimation Cipher Decryption Values

Encryption Value	Decryption Value
3	9
5	21
7	15
9	3
11	19
15	7
17	23
19	11
21	5
23	17
25	25

EXAMPLE 13 Decimation Cipher

To illustrate the decimation cipher, let's encrypt the message ATTACK AT DAWN using the key 3.

Message	A	T	T	A	C	K	A	T	D	A	W	N
Position	0	19	19	0	2	10	0	19	3	0	22	13
Position × 3	0	57	57	0	6	30	0	57	9	0	66	39
New position	0	5	5	0	6	4	0	5	9	0	14	13
Encrypted message	A	F	F	A	G	E	A	F	J	A	O	N

Linear Cipher

We can combine the methods used in the Caesar cipher and the decimation cipher easily by replacing the letter represented by the integer x with the letter corresponding to the integer $(kx + s) \bmod 26$, where k is any allowable decimation key and s is the shift we want. For example, choosing $k = 11$ and $s = 6$, the letter D, which corresponds to 3, is replaced by M, which corresponds to 13, because $(11 \cdot 3 + 6) \bmod 26 = 13$. This method is called a **linear cipher.**

To decrypt a message that has been created using the linear cipher formula $(kx + s) \bmod 26$, we must undo shifting by s by shifting by $-s$, as well as undo the step of multiplying by the encryption value k by multiplying by the decryption value corresponding to k given in Table 17.4. In our example of encoding the letter with the numerical value x with the formula $(11x + 6) \bmod 26$, we can determine the starting value of x that results in 13 by solving the equation $(11x + 6) \bmod 26 = 13 \bmod 26$: We subtract 6 from both sides to get $11x \bmod 26 = 7 \bmod 26$, which reverses the shift of 6; then we multiply by 19, which we see from Table 17.4 reverses multiplying by 11, to get $19 \cdot 11x \bmod 26 = 19 \cdot 7 \bmod 26$. Because $19 \cdot 11 \bmod 26 = 1$ and $19 \cdot 7 \bmod 26 = 3$, this simplifies to $x = 3$, which corresponds to the letter D. In cases where working backward results in a negative value after the reverse shift such as $19 \cdot (-5) \bmod 26$, Google still gives the correct answer.

Undoing a message letter x encrypted by multiplying by k and shifting by s is tantamount to using the formula $k'(x - s)$, where k' is the decryption value for k given in Table 17.4 (notice that we undo the two steps in reverse order).

EXAMPLE 14 Linear Cipher

To illustrate a linear cipher, we encrypt the message ADVANCE using the formula $21x + 7$ and decrypt the message ANQANHQ that was encrypted using the formula $21x + 7$.

Message	A	D	V	A	N	C	E
Position	0	3	21	0	13	2	4
Position × 21 + 7	7	70	448	7	280	49	91
New position	7	18	6	7	20	23	13
Encrypted message	H	S	G	H	N	X	E

To decrypt ANQANHQ, we see from Table 17.4 that the decryption value for 21 is 5, so our decryption formula is $5(x - 7) = 5x - 35$. To make calculations easier, it is better to replace -35 by a positive integer by observing that when using modulo 26, we may add 26 to any integer without changing the value. Thus, $-35 = -35 + 26 = -9 = -9 + 26 = 17$, and our formula becomes $5x + 17$. (Or use Google to obtain $-35 \bmod 26 = 17$.)

Message	A	N	Q	A	N	H	Q
Position	0	13	16	0	13	7	16
Position × 5 + 17	17	82	97	17	82	52	97
New position	17	4	19	17	4	0	19
Decrypted message	R	E	T	R	E	A	T

Self Check 7

Use the decimation cipher with the key 5 to encrypt the message ATTACK.

Self Check 8

What is the decryption formula for messages encrypted using the linear cipher $7x + 3$? ■

In most cases, the encryption algorithm requires that a message be changed to a particular format that is still readable. Such messages are called **plaintext.** An encrypted version of a message is called **ciphertext.** Our next cipher provides an example of a plaintext format.

Permutation Cipher

Another easy-to-use cipher is the *permutation cipher.* For an integer n greater than 1, a **permutation** of the integers 1, 2, . . . , n is a reordering of those integers. To specify a reordering, we list the integers 1, 2, . . . , n in the top row of a 2-by-n array and the reordered integers in the bottom row. For $n = 3$, one possible permutation is

$$\alpha = \begin{bmatrix} 1 & 2 & 3 \\ 3 & 1 & 2 \end{bmatrix}$$

To use the permutation α to encrypt the message ATTACK AT DAWN, we first break up the message into blocks of three letters each, ignoring the spaces between the words to obtain the plaintext ATT ACK ATD AWN. (This has the added advantage of disguising the lengths of each word, which makes breaking the code more difficult.) We then use the permutation to reorder the three letters in each block in the same way we reordered the integers 1, 2, and 3. That is, the first letter is put in the third position, the second letter is put in the first position, and the third letter is put in the second position. Doing this for each block we have TTA CKA TDA WNA. Messages encrypted with permutations are popularly called *anagrams.*

To decrypt a message encrypted by any permutation α of the integers 1, 2, . . . , n, we use the permutation α^{-1} (called α *inverse*) that reverses the effect α of by reading α from bottom to top. That is, we create the 2-by-n permutation array as follows. The top row of α^{-1} is 1, 2, . . . , n. The first entry in the bottom row of α^{-1} is the integer in the top row of α that is above 1; the second entry in the bottom row of α^{-1} is the integer in the top row of α above 2; and continuing in this way, the nth entry of the bottom row of α^{-1} is the integer in the top row of α above n. For $\alpha = \begin{bmatrix} 1 & 2 & 3 \\ 3 & 1 & 2 \end{bmatrix}$, we have $\alpha^{-1} = \begin{bmatrix} 1 & 2 & 3 \\ 2 & 3 & 1 \end{bmatrix}$. Applying α^{-1} to the ciphertext

TTA CKA TDA WNA

yields the plaintext

ATT ACK ATD AWN

Notice that when α moves a letter in a block to a new position, α^{-1} moves that letter back to its original position. In effect, α^{-1} "undoes" α. Of course, the recipient of the encrypted message must disentangle the blocks to figure out the words involved by seeing which possibilities make sense.

EXAMPLE 15 Encryption Using a Permutation

We use the permutation $\alpha = \begin{bmatrix} 1 & 2 & 3 & 4 \\ 3 & 4 & 2 & 1 \end{bmatrix}$ to encrypt ATTACK AT DAWN. Since α involves four integers, our blocks are ATTA CKAT DAWN. Applying α to the plaintext, we obtain the ciphertext ATAT TACK NWDA.

In cases of messages encrypted with a permutation of 1, 2, . . . , n where there are not enough letters to fill the last block, we add extra nonsense letters like X, Q, and Z as filler. Thus, when the message is ATTACK AT DAWN and n is 5, the last block could be WNXQZ or WNQXX. The recipient of the message will recognize the nonsense letters as padding needed to complete the last block.

EXAMPLE 16 Decryption Using a Permutation

We decrypt the ciphertext

IEON LYFB TDAN YBWO XASE

that was encrypted with the permutation $\alpha = \begin{bmatrix} 1 & 2 & 3 & 4 \\ 3 & 4 & 2 & 1 \end{bmatrix}$. Applying $\alpha^{-1} = \begin{bmatrix} 1 & 2 & 3 & 4 \\ 4 & 3 & 1 & 2 \end{bmatrix}$ to the ciphertext, we obtain the plaintext

ONEI FBYL ANDT WOBY SEAX

Regrouping the letters to make sense and recognizing that the X at the end was filler, this converts to

ONE IF BY LAND TWO BY SEA

Although linear ciphers and permutation ciphers are simple to use, they have the serious flaw of preserving the frequency of the encrypted letters. That is, the frequency of each letter in the plaintext is the same as the frequency of its replacement in the ciphertext. When letter frequency is preserved, it is easy to deduce the method of encryption for long messages.

The next two ciphers we discuss are the type that encode the same letter in more than one way. For these ciphers, letters that occur frequently in the message may not correspond to letters that occur frequently in the ciphertext and vice versa.

Playfair Cipher

The **Playfair cipher** was invented in 1854 by the British scientist Charles Wheatstone; it was then popularized by Baron Lyon Playfair. The cipher was used in the Boer War during 1899–1902 and World War II (1939–1945). The Playfair cipher uses a 5-by-5 matrix and three rules for converting plaintext to ciphertext. Since the matrix accommodates only 25 letters, the letter Q is omitted in all plaintext and keywords (some authors retain Qs but replace all occurrences of the letter J in a plaintext or keyword with I). To encode a message, the sender groups successive letters into pairs. Because the method requires that each pair of letters be distinct letters, in situations where the same letter would occur twice, the letter X is inserted between them [if the duplicate pair is XX, a Z is inserted between them—the X (or Z) is ignored after the ciphertext has been decoded]. If this process results in a plaintext with an odd number of letters, an X is added at the end (or a Z if the last letter is X). Thus, the message CALL IT QUITS yields the plaintext pairs CA LX LI TU IT SX. To create the cipher, the sender selects a keyword and inserts it in the matrix starting in the first row and continuing as necessary with no letter used more than once. The keyword MATHEMATICS would yield the matrix entries

M	A	T	H	E
I	C	S		

The remaining slots of the matrix are filled with the unused letters in the alphabet in their usual order. In our case, the cipher matrix is

M	A	T	H	E
I	C	S	B	D
F	G	J	K	L
N	O	P	R	U
V	W	X	Y	Z

The rules for transforming each pair of plaintext letters to ciphertext letters are as follows.

Playfair Encryption Rules

1. If both letters are in the same row, each of the two letters is replaced by the letter to the immediate right of it: AH becomes TE. If a letter is at the end of a row, it is replaced by the letter at the beginning: KL becomes LF.
2. If both letters are in the same column, each of the two letters is replaced by the letter immediately below it: EU becomes DZ. If a letter is at the bottom of a column, it is replaced by the letter at the top: XS becomes TJ.
3. If the two letters in a pair are not in the same row or the same column, the two letters are at opposite corners of a rectangle. In this case, each letter is replaced by the letter in the corner of the rectangle in the same row (the letters on the left and right corners of the rectangle are swapped): DA becomes CE.

EXAMPLE 17 Encrypting with a Playfair Cipher

We encode the plaintext CALL IT QUITS using the key word MATHEMATICS.

Plaintext	CA	LX	LI	TU	IT	SX
Ciphertext	GC	JZ	FD	EP	SM	JT

Note that with the Playfair cipher, the same letter in plaintext can be encoded more than one way in ciphertext (in Example 17, the first L becomes J and the second L becomes F). This means that letters appearing with high frequency in English may not encode to letters with high frequency—which makes breaking the cipher more difficult. Of course, to decrypt a ciphertext we must reverse the encryption steps as demonstrated in the next example.

Self Check 9

Use the Playfair cipher with the keyword MATHEMATICS to encode the message GO FOR IT.

EXAMPLE 18 Decrypting a Playfair Ciphertext

To decode the ciphertext TB TZ PN OB AC HC EU WB that was encoded using the keyword BEATLES and the matrix

B	E	A	T	L
S	C	D	F	G
H	I	J	K	M
N	O	P	R	U
V	W	X	Y	Z

we use the same matrix but reverse the encryption rules 1 and 2.

Playfair Decryption Rules

1′. If both letters are in the same row, each of the two letters is replaced by the letter to the immediate left of it: ET becomes BA. If a letter is at the beginning of a row, it is replaced by the letter at the end: HM becomes MK.

2′. If both letters are in the same column, each of the two letters is replaced by the letter immediately above it: WO becomes OI. If a letter is at the top of a column, it is replaced by the letter at the bottom: LU becomes ZM.

Applying these rules gives us

Plaintext	TB	TZ	PN	OB	AC	HC	EU	WB
Ciphertext	AL	LY	OU	NE	ED	IS	LO	VE

Regrouping the letters to make sense, we get the message ALL YOU NEED IS LOVE.

Vigenère Cipher

Modular arithmetic also provides the basis for a more sophisticated cryptosystem called the **Vigenère cipher.** For this method, we first select a keyword, which can be any word. The letters of the keyword are then used to determine the amount of shifting for each letter of our message. As before, we say A is in position 0, B is in position 1, and so on.

EXAMPLE 19 Vigenère Cipher

We use the Vigenère cipher to encrypt the message ATTACK AT DAWN. Choosing the keyword MATH, we shift the first letter of the message by 12 because *M* is in position 12; the second letter of the message is shifted by 0 (unchanged) because *A* is in position 0; the third letter of the message is shifted by 19 because *T* is in position 19, and so on. A shift of j means that the letter in position i is replaced by the letter in position $(i + j) \bmod 26$. When we have used all the letters of the keyword, we start over at the beginning. To encrypt ATTACK AT DAWN using the keyword MATH, we first note that the letters in the keyword MATH are in positions 12, 0, 19, and 7, respectively. So the *A* in ATTACK is converted to *M* $(0 + 12 = 12)$, the first *T* in ATTACK is converted to *T* $(19 + 0 = 19)$, the second *T* in ATTACK is converted to *M* $((19 + 19) \bmod 26 = 12)$, and so on. The first two lines of Table 17.5 show the position numbers for the letters of the message and the keyword. The third line of the table is obtained from the first two by adding the values in the columns using mod 26 and converting the results back to letters.

TABLE 17.5 Encryption Using a Vigenère Cipher

ATTACK AT DAWN	0	19	19	0	2	10	0	19	3	0	22	13
MATHMA TH MATH	12	0	19	7	12	0	19	7	12	0	19	7
MTMHOK TA PAPU	12	19	12	7	14	10	19	0	15	0	15	20

The **Vigenère square** in Table 17.6 provides a quick way to use the Vigenère cipher to encrypt messages with any keyword. Use the left column for each letter of the keyword and the top row for the corresponding letter of the message to be encrypted. The letter at the intersection of that row and column is the shifted letter. For example, if the message letter is *Q* and the corresponding letter from the keyword is *M*, then the letter *C* at the intersection of the row that begins with *M* and the column that begins with *Q* is the encrypted letter for *Q*.

TABLE 17.6 Vigenère Square

	A	B	C	D	E	F	G	H	I	J	K	L	M	N	O	P	Q	R	S	T	U	V	W	X	Y	Z
A	A	B	C	D	E	F	G	H	I	J	K	L	M	N	O	P	Q	R	S	T	U	V	W	X	Y	Z
B	B	C	D	E	F	G	H	I	J	K	L	M	N	O	P	Q	R	S	T	U	V	W	X	Y	Z	A
C	C	D	E	F	G	H	I	J	K	L	M	N	O	P	Q	R	S	T	U	V	W	X	Y	Z	A	B
D	D	E	F	G	H	I	J	K	L	M	N	O	P	Q	R	S	T	U	V	W	X	Y	Z	A	B	C
E	E	F	G	H	I	J	K	L	M	N	O	P	Q	R	S	T	U	V	W	X	Y	Z	A	B	C	D
F	F	G	H	I	J	K	L	M	N	O	P	Q	R	S	T	U	V	W	X	Y	Z	A	B	C	D	E
G	G	H	I	J	K	L	M	N	O	P	Q	R	S	T	U	V	W	X	Y	Z	A	B	C	D	E	F
H	H	I	J	K	L	M	N	O	P	Q	R	S	T	U	V	W	X	Y	Z	A	B	C	D	E	F	G
I	I	J	K	L	M	N	O	P	Q	R	S	T	U	V	W	X	Y	Z	A	B	C	D	E	F	G	H
J	J	K	L	M	N	O	P	Q	R	S	T	U	V	W	X	Y	Z	A	B	C	D	E	F	G	H	I
K	K	L	M	N	O	P	Q	R	S	T	U	V	W	X	Y	Z	A	B	C	D	E	F	G	H	I	J
L	L	M	N	O	P	Q	R	S	T	U	V	W	X	Y	Z	A	B	C	D	E	F	G	H	I	J	K
M	M	N	O	P	Q	R	S	T	U	V	W	X	Y	Z	A	B	C	D	E	F	G	H	I	J	K	L
N	N	O	P	Q	R	S	T	U	V	W	X	Y	Z	A	B	C	D	E	F	G	H	I	J	K	L	M
O	O	P	Q	R	S	T	U	V	W	X	Y	Z	A	B	C	D	E	F	G	H	I	J	K	L	M	N
P	P	Q	R	S	T	U	V	W	X	Y	Z	A	B	C	D	E	F	G	H	I	J	K	L	M	N	O
Q	Q	R	S	T	U	V	W	X	Y	Z	A	B	C	D	E	F	G	H	I	J	K	L	M	N	O	P
R	R	S	T	U	V	W	X	Y	Z	A	B	C	D	E	F	G	H	I	J	K	L	M	N	O	P	Q
S	S	T	U	V	W	X	Y	Z	A	B	C	D	E	F	G	H	I	J	K	L	M	N	O	P	Q	R
T	T	U	V	W	X	Y	Z	A	B	C	D	E	F	G	H	I	J	K	L	M	N	O	P	Q	R	S
U	U	V	W	X	Y	Z	A	B	C	D	E	F	G	H	I	J	K	L	M	N	O	P	Q	R	S	T
V	V	W	X	Y	Z	A	B	C	D	E	F	G	H	I	J	K	L	M	N	O	P	Q	R	S	T	U
W	W	X	Y	Z	A	B	C	D	E	F	G	H	I	J	K	L	M	N	O	P	Q	R	S	T	U	V
X	X	Y	Z	A	B	C	D	E	F	G	H	I	J	K	L	M	N	O	P	Q	R	S	T	U	V	W
Y	Y	Z	A	B	C	D	E	F	G	H	I	J	K	L	M	N	O	P	Q	R	S	T	U	V	W	X
Z	Z	A	B	C	D	E	F	G	H	I	J	K	L	M	N	O	P	Q	R	S	T	U	V	W	X	Y

Enigma Machines

SPOTLIGHT 17.6

Enigma machines were cipher devices used by the Germans in World War II (1939–1945). An Enigma machine had three to five wheels that would scramble the letters of a message in a way similar to the Vigenère method. The machines were easy to use and offered a high degree of security when used properly. Although messages encoded with Enigma machines were difficult to decipher, operator negligence and the capture by the Allied forces of a number of Enigma machines and the tables of wheel settings allowed Polish and British cryptologists to break the code.

David Buchan/Getty Images for Great Britain Campaign

Jefferson Wheel Cipher

In 1795 Thomas Jefferson, who became the third president of the United States in 1801, invented an encryption device known as the Jefferson wheel cipher. Employing the idea underlying the Vigenère square, the Jefferson cipher consists of a set of 26 disks with a hole in the center, each with the 26 letters of the alphabet arranged around the edge (see the figure below), scrambled in a random way from the approximately 40,329,146,000,000,000,000,000,000 choices. The disks numbered from 1 to 26 are mounted on the axle in any order desired. Rather than using a keyword, the order in which the disks are arranged on the axle is the cipher key. Both sender and receiver must arrange the disks in the same order. The sender rotates each disk up and down to spell out the desired message on one row and then selects any row to determine the approprite substitutions. To decrypt the coded message, the receiver rotates the disks so that they spell out the encrypted message on one row and then looks for another row that make sense. In the very unlikely case of two sensible messages, the receiver can request the message to be re-encrypted.

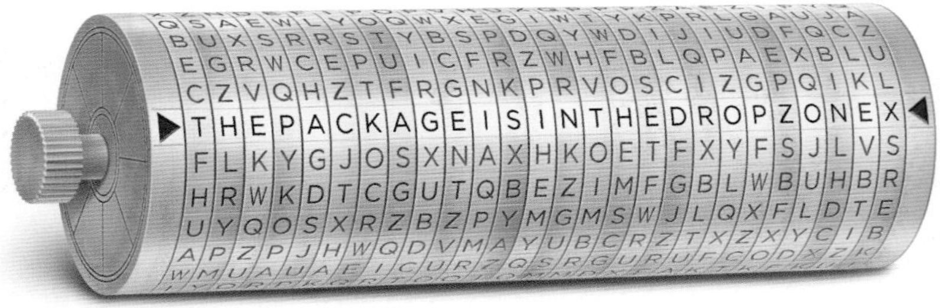

The same system was independently invented in 1891 by Etienne Bazeries, who used 20 disks. It was used by the United States Army from 1923 until 1942.

EXAMPLE 20 Jefferson Wheel Cipher

To illustrate the Jefferson cipher, we use a system with only six disks labeled as follows:

1: XWAZJGDLUBVIQHKYPNTCROSFEM
2: EBPLNKACZDTRXMJQOYHGSVUFIW
3: EQGYXPLOCKBDMAIZVRNSJUWFHT
4: UKTEBSXRPLNDVHGFCQYIZMJWAO
5: GJVDKCPMNZQWXYABEUOTSIHFRL
6: KGAMIHWPNYCJBFZDRUXLQOTVES

To send the message ATTACK, we select a key such as 4, 2, 5, 1, 6, 3 and arrange the disks in that order. We then rotate the six disks so that the word ATTACK appears as a row as shown below (only the first eight rows are provided).

4	2	5	1	6	3
A	T	T	A	C	K
O	R	S	Z	J	B
U	X	I	J	B	D
K	M	H	G	F	M
T	J	F	D	Z	A
E	Q	R	L	D	I
B	O	L	U	R	Z
S	Y	G	B	U	V
⋮	⋮	⋮	⋮	⋮	⋮

If we select as our substitution row the one immediately below the row with the word ATTACK, the encrypted message is ORSZJB. Had we selected as our substitution row the one that is three below our message, we would have obtained the encrypted message KMHGFM. Each of the 25 rows is equally effective for encryption purposes.

Conversely, had we received the ciphertext KMHGFM, we would have rotated disk 2 so that M appeared in the same row as K on disk 4, then rotated disk 5 so that H appeared in the same row as KM, then continued in the same way until the row became KMHGFM. Then we would have looked for another row that is a sensible word, such as ATTACK. ■

Mavis Batey

SPOTLIGHT 17.7

Mavis Batey was born Mavis Lever in south London on May 5, 1921. As a linguistics student at a London college during the early days of World War II, she was recruited by the British government to do intelligence work. One of her first assignments was to read the personal ads in the *Times* of London in search of secret German messages. Following that, she joined a team of code breakers at Bletchley Park, a Victorian estate north of London. Her team's efforts helped the Allies defeat the Italian navy in 1941. Their cracking of the Enigma codes employed by the German military intelligence service assisted the Allied forces' 1944 invasion of France to drive out the Germans. Britain's wartime prime minister, Winston Churchill, called the Bletchley Park code breakers "the geese that laid the golden eggs but never cackled."

Private Collection/Prismatic Pictures/Bridgeman Images

Batey was the model for the code breaker played by Kate Winslet in the 2001 film *Enigma*. She died in 2013 at age 92. Newspapers in the United States and the United Kingdom carried lengthy obituaries about her.

Alan Turing

SPOTLIGHT 17.8

On *Time* magazine's list of the 100 most influential people of the 20th century was Alan Turing, a British mathematician born in London in 1912. While a student at Cambridge, Turing imagined a machine that could execute logical processes. The device in his mind-experiment quickly acquired the name "Turing machine." At the time, no one knew that Turing's idea would provide a blueprint for what would eventually become the electronic digital computer. Now we know that everyone who opens a spreadsheet or a word-processing program is using an incarnation of a Turing machine.

After graduating from Cambridge, Turing became a crucial member of a team of cryptologists working for the British government who successfully broke the Enigma codes (see Spotlight 17.6 on page 724).

Turing's life took a tragic turn in 1952 when he admitted that he had engaged in homosexual acts in his home, a felony in Britain at that time. As punishment, he was chemically castrated and subjected to estrogen treatments. Made despondent by this treatment, he committed suicide two years later at the age of 41 by eating an apple laced with cyanide.

Today, Turing is widely honored for his fundamental contributions to computer science and his role in the defeat of Germany in World War II. Many rooms, lecture halls, and buildings at universities around the world have been named in honor of Turing. The annual award for contributions to the computing community, which is widely considered to be equivalent to a Nobel Prize, is called the "Turing Award." In December 2013, Queen Elizabeth II granted Turing a pardon and issued a statement saying that Turing's treatment was unjust: "Turing was an exceptional man with a brilliant mind who deserves to be remembered and recognized for his fantastic contribution to the war effort and his legacy to science."

famouspeople/Alamy

The 2014 film *Imitation Game,* loosely based on Turing's life, starred Benedict Cumberbatch as Turing and Keira Knightley as a member of the code-breaking team; it was nominated for eight Academy Awards.

Information from *Time*, Vol. 153, Issue 12 (March 29, 1999), p. 147.

Encrypting Credit Card Data on the Web

Suppose that you want to purchase a compact disc from Amazon.com. Should you be concerned that a hacker will intercept your credit card number during the transaction? As you might expect, your credit card number is sent to Amazon in encrypted form to protect the data.

To describe one way that this encryption can be done, we need to perform addition of binary strings. We add two binary strings $a_1a_2 \ldots a_n$ and $b_1b_2 \ldots b_n$ as follows:

$$\begin{array}{r} a_1a_2 \cdots a_n \\ +\ b_1b_2 \cdots b_n \\ \hline c_1c_2 \cdots c_n \end{array}$$

where $c_i = 0$ if $a_i = b_i$ and $c_i = 1$ if $a_i \neq b_i$. Equivalently, $c_i = (a_i + b_i) \bmod 2$. (Add a_i and b_i in the ordinary way, but replace 2 by 0.)

EXAMPLE 21 Sum of Binary Strings

$$\begin{array}{r} 11000111 \\ +\ 01110110 \\ \hline 10110001 \end{array} \qquad \begin{array}{r} 00111011 \\ +\ 01100101 \\ \hline 01011110 \end{array} \qquad \begin{array}{r} 10011100 \\ +\ 10011100 \\ \hline 00000000 \end{array}$$

Self Check 10

Find the sums

$$\begin{array}{r} 10110111 \\ +\ 11101001 \end{array} \qquad \begin{array}{r} 110110 \\ +\ 011101 \end{array}$$

■

We can now explain one way to send credit card numbers over the Web securely. When you place an order with Amazon, the company sends your computer a randomly generated string of 0s and 1s called a *key*. This key has the same length as the binary string corresponding to your credit card number, and the two strings are added (think of this process as "locking" the data). The resulting sum is then transmitted to Amazon. Amazon in turn adds the same key to the received string, which then produces the original string corresponding to your credit card number. (Adding the key a second time "unlocks" the data.)

To illustrate the idea, say you want to send an eight-digit binary string such as $s = 10101100$ to Amazon (actual credit card numbers have very long strings), and Amazon sends your computer the key $k = 00111101$. Your computer returns the string $s + k = 10101100 + 00111101 = 10010001$ to Amazon, and Amazon adds k to this string to get $10010001 + 00111101 = 10101100$, which is the string representing your credit card number. If someone intercepts the number $s + k = 10010001$ during transmission, it is of no value without knowing k. This method works because of the property of binary addition that $a_1a_2 \cdots a_n + b_1b_2 \cdots b_n = 00 \cdots 0$ if and only if the two strings are identical. Thus, $(s + k) + k = s + (k + k) = s + 00 \cdots 0 = s$. The method is secure because the key sent by Amazon is randomly generated and used only one time.

You can tell when you are using an encryption scheme on a web transaction by looking to see if the web address begins with "https" rather than the customary "http." You will also see a small padlock in the status bar at the bottom of the browser window.

Even with sophisticated encryption schemes and security precautions, however, computer system breaches are still common. See Spotlight 17.9 for a highly publicized instance.

Smart Cards

SPOTLIGHT 17.9

In December of 2013, hackers breached the computer system of Target, America's second leading general merchandise company, and stole the credit and debit card numbers of 70 million customers. Within days, the stolen credit and debit card numbers were selling on the black market for as little as $20 to a high of $100 per card. Such an incident was possible because most credit cards issued in the United States before 2014 employ a magnetic stripe and can easily be dupicated by criminals. In contrast, Europe, Mexico, Canada, and 80 other countries embed a microchip in their cards to encrypt the data. Unlike magnetic stripes, which always contain the same data, chip cards encrypt different data for each transaction and communicate with a card reader. This enables them to detect and react to tampering attempts, thereby helping to counter possible attacks.

Review Vocabulary

Binary code Coded data made up of two states (or symbols). (p. 699)

Binary linear code A code consisting of words composed of 0s and 1s obtained by using parity-check sums to append check digits to messages. (p. 705)

Caesar cipher A cryptosystem used by Julius Caesar whereby each letter is shifted the same amount. (p. 714)

Ciphertext An encrypted version of a message. (p. 719)

Code word A string of digits composed of a message and check digits. (p. 702)

Compression algorithm A procedure for converting data from one format to another one designed for compactness. (p. 708)

Cryptography The study of how to make and break secret codes. (p. 714)

Data compression The process of encoding data so that data are represented by fewer symbols. (p. 708)

Decimation cipher A cryptosystem that uses multiplication by a fixed value to shift each letter. (p. 716)

Decoding The process of converting code words into a message for information. (p. 704)

Decryption The process of making intelligible information from a deliberately distorted transmission. (p. 714)

Distance between two strings The distance between two strings of equal length is the number of positions in which they differ. (p. 704)

Encoding The process of converting a message or other information into code. (p. 700)

Encryption The process of encoding data to make it unreadable by unauthorized people. (p. 714)

Even parity Even integers are said to have even parity. (p. 701)

Fixed-length code A code in which the number of symbols for each code word is the same. (p. 707)

Key A string used to encode and decode data. (p. 717)

Linear cipher A cryptosystem that replaces a letter with numerical value x by the letter with numerical value $(kx + s)$ mod 26. (p. 718)

Modular arithmetic Addition and multiplication involving modulo n. (p. 715)

Nearest-neighbor decoding A method that decodes a received message as the code word that agrees with the message in the most positions. (p. 705)

Odd parity Odd integers are said to have odd parity. (p. 701)

Parity-check sums Sums of digits whose parities determine the check digits. (p. 703)

Permutation cipher A cryptosystem that reorders the letters of a message. (p. 719)

Plaintext A message in readable form. (p. 719)

Playfair cipher A cryptosystem that uses a 5-by-5 matrix to encrypt messages. (p. 720)

Variable-length code A code in which the number of symbols for each code word may vary. (p. 707)

Vigenère cipher A cryptosystem that uses a keyword to determine how much each letter is shifted. (p. 722)

Vigenère square A table that can be used to encrypt any message quickly using any keyword with the Vigenère cipher. (p. 723)

Weight of a binary code The minimum number of 1s that occur among all nonzero code words of a code. (p. 705)

Self Check Answers

1. 0101110

2. 3

3. 10101101110, ACGAGAC

4. 1020 1035 1047 1041 1049 1069 1065

5. EDAEFC

6. VHQG PRQHB

7. ARRAKY

8. $15(x - 3) = 15x - 45 = 15x + 7$

9. OW GN NB ST

10. 01011110; 101011

Skills Check

1. If three film critics listed their ratings of a film with thumbs up (U) or thumbs down (D)—for example, one outcome is UDU—and critic number 1's rating is first, critic number 2's rating is second, critic number 3's rating is third—the number of possible outcomes is

(a) 6.

(b) 8.

(c) 9.

2. A four-digit binary message was encoded using Table 17.1 and the message 1010010 was received. Using the nearest-neighbor method, the decoded four-digit message is ______________________.

3. Using the circle diagram in Figure 17.1 to encode the message 1011, the encoded message is

(a) 1011001.

(b) 1011010.

(c) 1010001.

4. The distance between received words 1011001 and 1000101 is ______________________.

5. Using the nearest-neighbor method and the code in Table 17.2, the word 1110011 decodes as

(a) 0110010.

(b) 1100011.

(c) 1010001.

6. The weight of the binary linear code {0000000, 0011111, 0101011, 0110100} is ______.

7. If the two messages 0 and 1 are encoded as 000 and 111, respectively, the number of errors that the code can correct is

(a) 0.

(b) 1.

(c) 2.

8. If a binary linear code has weight 4, the maximum number of errors that it will detect is ______________.

9. If every pair of code words differs in at least five positions, then nearest-neighbor decoding can decode words accurately that have

(a) two errors.

(b) three errors.

(c) four errors.

10. Using the encoding scheme A → 0, B → 10, C → 11, the string 010110 decodes as

(a) ABCB.

(b) ABCA.

(c) ABACA.

11. Use delta encoding to compress the following numbers: 1027 1023 1028 1060 1070.

12. Given that the following numbers were encoded using the delta function, determine the original numbers: 1221 10 −15 −3 12 8.

13. Find the following values: 57 mod 26; −17 mod 26; 82 mod 26.

14. Solve the equation $(3x + 5) \bmod 26 = 12$.

15. The Caesar cipher would encrypt GO HOME NOW as

(a) JR KRPH QRZ.

(b) DL ELJB KLT.

(c) Neither of these answers is correct.

16. The permissible values for the key of a decimation cipher are ________.

17. Use the decimation cipher with key 9 to encrypt the message RUN.

18. Describe the cipher obtained when the values $k = 1$ and $s = 0$ are used in the linear cipher formula $kx + s$.

19. If a message were encrypted using the decimation cipher with the key 9, what value would you use to decrypt it? (Answer this question without looking at Table 17.4.)

(a) 3

(b) 5

(c) 7

20. The name for the cipher obtained when the value $k = 1$ is used in the linear cipher formula $kx + s$ is

(a) Vigenère.

(b) Caesar.

(c) decimation.

21. Using the Vigenère cipher and the keyword ADAM to decrypt EIEIO, we obtain

(a) ELELR.

(b) EFEFL.

(c) EFEWO.

22. Counting the permutation $\alpha = \begin{bmatrix} 1 & 2 & 3 \\ 1 & 2 & 3 \end{bmatrix}$, how many permutations of 1, 2, 3 are there?

23. Given $\alpha = \begin{bmatrix} 1 & 2 & 3 & 4 \\ 2 & 1 & 4 & 3 \end{bmatrix}$, find α^{-1}.

24. If α is a permutation cipher involving 1, 2, 3, 4, 5, 6, what is the length of the plaintext blocks?

25. If a permutation of the integers 1, 2, 3, 4, 5 was to encode the message MEET THE BOSS AT NOON, what would the blocks be?

26. If the first word of a message to be encrypted using a Playfair cipher was FLEET, how would it be converted to pairs of letters?

27. If the first word of a message to be encrypted using a Playfair cipher was JET, how would it be converted to pairs of letters?

28. The sum of the binary strings 1011001 and 1001101 is ________.

29. If u is a binary string of length 8, what string is $u + u$?

30. Let $u = 1110011$. Find a string v such that $u + v = 1111111$.

Chapter 17 Exercises

Challenge Discussion

17.1 Binary Codes

1. Use the circle diagram method shown in Figures 17.1 and 17.2 (page 700) to verify the code words in Table 17.1 (page 702) for the messages 0101, 1011, and 1111.

2. Use the circle diagram method to decode the received messages 0111011 and 1000101.

3. Find the distance between each of the following pairs of words:

(a) 11011011 and 10100110

(b) 01110100 and 11101100

4. Referring to Table 17.1, use the nearest-neighbor method to decode the received words 0000110 and 1110100.

5. If the code word 0110010 is received as 1001101, how is it decoded using the circle diagram method?

6. Suppose that a received word has the circle diagram arrangement shown here:

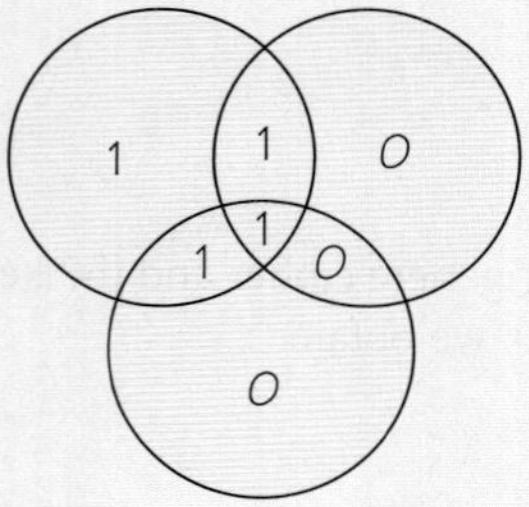

What can we conclude about the received word?

7. Suppose that a received word has the circle diagram arrangement shown here:

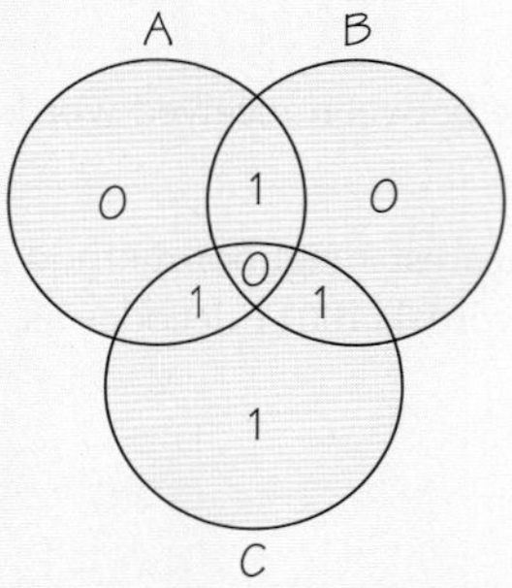

Assuming that exactly one error was made in transmission, what was the sent word?

8. If the code word 1010 in the code C_2 in Example 4 (page 706) is received as 1000, explain why the error will be detected but cannot be corrected.

9. For the code C_4 in Example 4, decode the following received messages:

(a) 1101011

(b) 1011011

(c) 1100001

10. If the code word 0010111 in the code C_4 in Example 4 is received as 1010110, will the error be detected? Can it be corrected?

17.2 Encoding with Parity-Check Sums

11. Determine the binary linear code that consists of all possible three-digit messages with three check digits appended using the parity-check sums $a_2 + a_3$, $a_1 + a_3$, and $a_1 + a_2$. (That is, $c_1 = 0$ if $a_2 + a_3$ is even, $c_1 = 1$ if $a_2 + a_3$ is odd, and similarly for c_2 and c_3.)

12. Let C be the code

{0000000, 0010111, 0101011, 1001101, 1100110, 1011010, 0111100, 1110001}

What is the error-correcting capability of C? What is the error-detecting capability of C?

13. Find all code words for binary messages of length 4 by appending three check digits using the parity-check sums $a_2 + a_3 + a_4$, $a_2 + a_4$, and $a_1 + a_2 + a_3$. Will this code correct any single error?

14. Consider the binary linear code

C = 000000, 100110, 010101, 001011, 110011, 101101, 011110, 111000

Use nearest-neighbor decoding to decode 101100 and 101011. If a received word is 001100, can you determine the intended code word? Explain your reasoning.

15. Construct a binary linear code using all eight possible binary messages of length 3 and appending three check digits using the parity-check sums $a_1 + a_2$, $a_2 + a_3$, and $a_1 + a_3$. Decode each of the received words below by the nearest-neighbor method:

001001, 011000, 000110, 100001

16. Extend the code words listed in Table 17.1 (page 702) to eight digits by appending a 0 to words of even weight and a 1 to words of odd weight. What are the error-detecting and error-correcting capabilities of the new code?

17. Extend the code words listed in Table 17.2 (page 704) to eight digits by appending a 0 to words of even weight and a 1 to words of odd weight. What are the error-detecting and error-correcting capabilities of the new code?

18. Suppose that the weight of a binary linear code is 6. How many errors can the code correct? How many errors can the code detect?

19. How many code words are there in a binary linear code that has all possible messages of length 5 with three check digits appended? How many possible received words are there with this code?

17.3 Data Compression

20. Suppose we code a four-symbol genetic set {A, C, T, G} into binary form as follows:

$A \to 0 \qquad C \to 10 \qquad T \to 110 \qquad G \to 111$

Convert the sequence *ACAAGTAAC* into binary code.

21. Use the code in Exercise 20 to determine the sequence of symbols represented by the binary code 001100001111000.

22. Suppose that we code a five-symbol set {A, B, C, D, E} into binary form as follows:

$A \to 0 \qquad B \to 10 \qquad C \to 110$
$D \to 1110 \qquad E \to 1111$

Convert the sequence of *AEAADBAABCB* into binary code. Determine the sequence of symbols represented by the binary code 010001101000111111110.

23. Use the code in Exercise 22 to convert the sequence *EABAADABB* into binary code. Determine the sequence of letters represented by the binary code 0010001100111101111010.

24. Devise a variable-length binary coding scheme for a six-symbol set {A, B, C, D, E, F}. Assume that A is the most frequently occurring symbol, B is the second most frequently occurring symbol, and so on.

25. Judging from the frequency table in Figure 17.9 (page 709), what are the three most frequently occurring consonants in English text material? What is the most frequently occurring vowel?

26. Explain why Morse code must include a space after each letter, but fixed-length codes do not.

27. Following are the closing values (rounded to the nearest integer) of the Dow Jones Industrial Average stock market values for a two-week period. Use the delta function method to compress these values. What percentage reduction in characters is there?

13403 13739 13816 13767 13820
13759 13779 13878 13913 13896

28. The following numbers were encoded using delta function encoding. Determine the original numbers.

1207 373 −57 −97 −234 −105 178
−73 275 79 −183 −146 −94 129

29. Assume that a dataset of four letters occurs with the following probabilities:

A	0.425
B	0.210
C	0.215
D	0.150

Use a Huffman tree to create a Huffman code for A, B, C, and D.

30. Given that the delta function was used to create values in the following list

13403 336 77 −49 53 −61 20 99 35 −17

recreate the original list.

31. Decode the binary string

11100100001001110110011010

which has been encoded using the Huffman code given on page 713.

32. Use a Huffman tree to assign a binary code to the letters that occur with the following probabilities:

A	0.025
B	0.150
C	0.015
D	0.170
E	0.200
F	0.225
G	0.215

17.4 Cryptography

33. Use hand computations to calculate the following.

(a) 80 mod 26

(b) −42 mod 26

(c) 269 mod 26

34. Solve for x by trial and error.

(a) $3x \bmod 26 = 1$

(b) $2x \bmod 31 = 1$

(c) $9x \bmod 40 = 1$

35. Use the Caesar cipher to encrypt the message RETREAT. Decrypt the message DGYDQFH, which was encrypted using the Caesar cipher.

36. Suppose a message encrypted with a Caesar shift can be decoded with the same shift. What is the shift?

37. Using 0, 1, 2, . . . , 25 to label the positions of the letters *A, B, C, . . . , Z*, suppose that we create a cipher by replacing the letter in position i with the letter in position $(i + 8) \bmod 26$. How many iterations of this cipher must be done before a message will return to its original state?

38. The message ADDAOS was encrypted using the decimation cipher with the key 7. Decrypt it.

39. Explain why 2 cannot be used as the key in a decimation cipher.

40. If you attempted to use the decimation cipher with the key 13, how would the word MESSAGE be encrypted?

41. Use the decimation cipher with the key 5 to encrypt RETREAT.

42. Suppose you wanted to break a code that employed a linear cipher by trying all possible linear ciphers. How many would you have to test to try every one?

43. Find a formula for a linear cipher that decrypts messages encrypted using the formula $11x + 20$.

44. List all possible permutations of 1, 2, 3, including 1, 2, 3 itself, in a systematic way. (The original ordering 1, 2, 3 is viewed as a permutation to simplify various theoretical arguments.)

45. Encrypt the message ADVANCE WHEN READY using the permutation $\alpha = \begin{bmatrix} 1 & 2 & 3 & 4 \\ 2 & 4 & 1 & 3 \end{bmatrix}$.

46. Encrypt the message ATTACK POSTPONED using the permutation $\alpha = \begin{bmatrix} 1 & 2 & 3 & 4 & 5 \\ 2 & 1 & 5 & 3 & 4 \end{bmatrix}$.

47. The message EMTENAOTXOQN was encrypted using the permutation $\alpha = \begin{bmatrix} 1 & 2 & 3 & 4 \\ 2 & 4 & 1 & 3 \end{bmatrix}$. Decrypt it.

48. The message TAACTPKSTOOPEDN was decrypted using the permutation $\alpha = \begin{bmatrix} 1 & 2 & 3 & 4 & 5 \\ 2 & 1 & 5 & 3 & 4 \end{bmatrix}$. Decrypt it.

49. How many permutations of 1, 2, 3, 4 are possible, including the original ordering 1, 2, 3, 4?

50. Use the Playfair cipher with the keyword IMAGINE to encrypt the plaintext CALL ME.

51. The ciphertext WR VI SE CW RY was encrypted with the Playfair cipher using the key WORLD WAR II. Decipher it.

52. Explain why the ciphertext BT TZ PM UO AW AC HK EU W cannot have been encoded using the Playfair cipher.

53. How many possible keys are there for a Jefferson wheel with six disks?

54. Referring to the Jefferson wheel with six disks and the key shown in Example 20 (page 725), what would the encrypted message be if we used the row that is just above the row ATTACK?

55. Referring to the Jefferson wheel with six disks and the key shown in Example 20, what row was used to create the encrypted message MCEMPL?

56. Referring to the Jefferson wheel with six disks shown in Example 20 and the key 5, 1, 4, 2, 6, 3, encrypt the message DEPLOY, using the row directly below the message.

57. Given that BEATLES was used as the keyword for the Vigenère cipher to encrypt SSLETRY TXOGPW, decrypt the message.

58. Use the Vigenère cipher with the keyword CLUE to encrypt the message THE WALRUS WAS PAUL.

59. Use the Vigenère cipher with the keyword HELP to encrypt the message PHONE HOME.

60. Use the linear cipher method with $k = 5$ and $s = 4$ to encrypt GOOD.

61. Given that the received message ZVW was encrypted using the linear cipher method with $k = 11$ and $s = 9$, decrypt the message.

62. Add the following pairs of binary strings.

(a) 10111011 and 01111011

(b) 11101000 and 01110001

63. For any two binary strings u and v of the same length, define $u + v$ to be the string obtained by summing the strings as explained in Example 21 (page 726). Find $u + v$ for each of the following cases.

(a) $u = 1100001$ and $v = 0011100$

(b) $u = 1011010$ and $v = 0111001$

64. Suppose that u and v are two binary words of length 7 whose distance apart is 7. What string is $u + v$?

65. Given the binary word $u = 1010110$, find all binary words of length 7 that are a distance of 1 from u.

66. Suppose that u and v are binary code words of the same length. Using the method of summing binary strings as shown in Example 21, explain why the distance from u to v is the same as the weight of $u + v$.

67. Suppose that u, v, and w are binary code words of the same length. Using the method of summing binary strings as shown in Example 21, explain why the distance from u to v is the same as the distance from $u + w$ to $v + w$.

68. All binary linear codes have the property that the sum of two code words is another code word. Use this fact to determine which of the following sets cannot be a binary linear code.

(a) {0000, 0011, 0111, 0110, 1001, 1010, 1100, 1111}

(b) {0000, 0010, 0111, 0001, 1000, 1010, 1101, 1111}

(c) {0000, 0110, 1011, 1101}

Chapter Review

69. Messages encrypted with the ________ cipher are the hardest to break.

(a) Vigenère

(b) decimation

(c) linear

70. Explain why for any choice of the key k shown in Table 17.4 (page 717) the number 13 is always encrypted as 13.

71. Suppose a Huffman tree has been used to create a binary code for the letters A through J, and the results include $B = 111110$, $J = 111111$, and $G = 11110$. If the code has only two code words of length 6 and one of length 5, what can you say about the probability of the occurrence of the letters B, J, and G?

72. What is the minimum weight of a binary code that can correct any 3 or fewer errors?

73. Explain why encoding ID numbers by multiplying them by 15 would not detect the single-digit error of replacing 1328 by 1326.

74. Explain why any error involving the code word 001 in the code C_1 in Example 4 (page 706) is not detected.

Suggested Websites and Videos

www.xarg.org/tools/caesar-cipher/ This site has software for encoding and decoding messages with many different ciphers.

mentalfloss.com/article/27921/how-cryptic-4-famous-unsolved-ciphers-and-codes This article describes four famous unsolved ciphers and codes.

www.youtube.com Search for the BBC TV film *The Strange Life and Death of Dr. Turing,* a 50-minute documentary film about the British code breaker and war hero.

www.youtube.com Search for the BBC TV film *Bletchley Park's Lost Heroes,* a one-hour documentary film revealing the secret story of how two men hacked into Hitler's personal super-code machine.

Part VI

On Size and Growth

Mathematics is the study of patterns and relationships. It can explain why there are no Godzillas in reality, point out unexpected similarities in ancient pottery, and suggest new and beautiful artistic designs. Mathematicians search for and classify numerical, geometric, and even abstract patterns. In these chapters, we follow some of those searches. We concentrate on geometric patterns but find that those lead to numerical considerations, too.

In Chapter 18, Growth and Form, we look at how the sizes of objects influence their forms. Godzilla, very tall trees, mile-high buildings, and stratospheric mountains: Are they possible? If not, why not? Seeing the underlying principles of scaling will help you to appreciate why objects in the world have the shapes and sizes that they do.

We start with a simple numerical pattern in Chapter 19, Symmetry and Patterns, which leads to questions such as: What proportions make a pattern esthetically pleasing? How important is bilateral symmetry? We expand our notion of symmetry but discover surprising limitations that even broader notions of symmetry face. We examine the ingredients of the beauty of fractal patterns, ones that resemble themselves at finer and finer scales, in nature and in traditional art from Africa and elsewhere.

Chapter 20, Tilings, asks how to arrange objects symmetrically on a surface. What shapes can we use? What patterns can arise if all the objects are the same? What if the objects themselves are symmetrical, or if we allow irregular shapes but demand that they all face the same way? Can you arrange shapes in a pattern that does not repeat but is nevertheless systematic?

Rob Koenen/Moment/Getty Images

Growth and Form

18

Armin Puschmann/Alamy

Films show us giant creatures, including King Kong, Godzilla, and the oliphants in *The Lord of the Rings.* Literature has the giant of "Jack and the Beanstalk," the Brobdingnagians of *Gulliver's Travels,* and the giants in the Battle of Hogwarts in *Harry Potter and the Deathly Hallows.* "Fakelore" gives us Paul Bunyan and his blue ox, Babe. Could such beings ever exist? (See Figure 18.1.)

Universal/Courtesy Everett Collection

Figure 18.1 Could King Kong exist?

Every species has to adapt and survive at different sizes from babyhood to mature adulthood. For example, the giant panda ranges from 1 pound (lb) at birth to 275 lb in adulthood. A baby panda could be crushed by its mother; an adult panda needs to eat a lot.

For contrast, consider the horse. A newborn foal that weighed as little as a newborn panda would be too small to keep up with the herd and could not survive. An adult horse weighs much more than a panda and has to consume far more food, but the horse can move much more quickly and cover greater distances to find sustenance.

There have been large land mammals (mammoths) and huge sea mammals (blue whales)—not to mention the dinosaurs. But the tallest humans have been only 9 to 10 feet (ft) tall, and even the biggest dinosaur, *Titanosaurus,* stood only 40 ft high.

What about supergiants and utterly huge monsters? That they have never existed suggests limits to size and shape. What does it mean for two objects to be similar in shape? What does enlarging an object do to its surface area and volume, and what would be the effect? How high could a mountain be, or a building, without collapsing? What would being a supergiant involve?

Using a few simple principles of geometry, we show that no objects or living beings could exist, unchanged in shape, on a vastly different scale, larger or smaller.

18.1 Geometric Similarity

We use the powerful mathematical concept of **geometric similarity.** You may have encountered this idea in geometry in connection with similar triangles (see Figure 18.2). Here we apply it to real objects.

Both: DeAgostini/Getty Images

Figure 18.2 A pair of similar triangles. Each side of the larger one is twice as long as the corresponding side in the smaller one.

Geometric Similarity DEFINITION

Two objects are **geometrically similar** if they have the same shape, regardless of the materials of which they are made; they do not have to be the same size.

Although similar objects need not be the same size, corresponding distances must be proportional, just as in similar triangles. For example, when a photo is enlarged, it is enlarged by the same factor in both the horizontal and vertical directions—in fact, in any direction (such as a diagonal). We call this magnification factor the **linear scaling factor** (or **length scaling factor**).

Linear (Length) Scaling Factor DEFINITION

The **linear (length) scaling factor** of two geometrically similar objects is the ratio of a length of any part of the second to the corresponding part of the first.

In Figure 18.3, the linear scaling factor is 3; the enlargement is 3 times as wide and 3 times as high as the original. In fact, every pair of points becomes 3 times as far apart.

Objects can be scaled down as well as up; for example, the smaller photograph in Figure 18.3 is geometrically similar to the larger one. The linear scaling factor to get the smaller from the larger is 1/3.

Both: Design Pics Inc./Alamy

Figure 18.3 Two geometrically similar photographs.

How Area and Volume Scale

The enlargement can be divided into $3 \times 3 = 9$ rectangles, each the size of the original. Hence, the enlargement has $3 \times 3 = 3^2 = 9$ times the area of the original. More generally, if the linear scaling factor is some general number L (not necessarily 3), the resulting enlargement has an area $L \times L = L^2$ ("L squared") times the area of the original.

Area Scaling Factor DEFINITION

The **area scaling factor** of two geometrically similar objects is the ratio of the area of any part of the second to the corresponding part of the first.

How Area Scales RULE

The *area* of a scaled-up object goes up in proportion to the square of the linear scaling factor.

We symbolize the relationship between the area A and the linear scaling factor L by

$$A \propto L^2$$

where the symbol $\propto$ is read as "is proportional to" or "scales as."

What about enlarging three-dimensional objects? If we take a cube and enlarge it by a linear scaling factor of 3, it becomes 3 times as long, 3 times as high, and 3 times as deep as the original (see Figure 18.4).

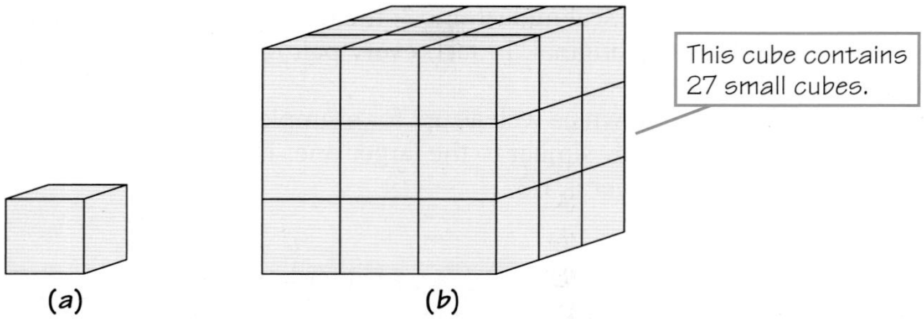
This cube contains 27 small cubes.
(a)
(b)

Figure 18.4 Cube (b) is made by enlarging cube (a) by a factor of 3.

What about volume? How much bigger is the volume of the enlarged cube? The enlarged cube has three layers, each with $3 \times 3 = 9$ little cubes, each the same size as the original. Thus, the total volume is $3 \times 3 \times 3 = 3^3 = 27$ times as much as the original cube. Denoting the volume by V, we can write

$$V \propto L^3$$

Thus, for an object enlarged or shrunken by a linear scaling factor L, the resulting volume is $L \times L \times L = L^3$ ("L cubed") times the volume of the original. Like the relationship between area and L^2, this relationship holds even for irregularly shaped objects, such as science-fiction monsters.

Volume Scaling Factor DEFINITION

The **volume scaling factor** of two geometrically similar objects is the ratio of the volume of any part of the second to the corresponding part of the first.

How Volume Scales RULE

The *volume* of a scaled-up object goes up in proportion to the cube (third power) of the linear scaling factor.

You can see, however, that the area of each face (side) of the enlarged cube is $3^2 = 9$ times as large as that of a face of the original cube, just as the area of the photo enlarged by a factor of 3 has 9 times the area of the original. The total surface area of the enlarged cube is 9 times the total surface area of the original cube.

More generally, for objects of any shape, the total *surface area* of a scaled-up object goes up with the *square* of the linear scaling factor. Thus, the surface area of an object scaled up by a linear scaling factor of L is L^2 times the surface area of the original. This feature holds true even for irregular shapes.

Before we discuss scaling real three-dimensional objects, you should understand some language pitfalls in describing increases and decreases.

The Language of Growth, Enlargement, and Decrease

Let's see if you can make sense of some incorrect and confusing claims you can find on the Internet:

1 out of 3 households earned 200 percent less than the poverty line in 2009.

—inthesetimes.com/working/entry/6793/since_great_recession_more_working_families_in_poverty

Well, 50% less than would be half as much, 100% less than would be zero, and 200% less than simply makes no sense. But here it's just that the word order is mixed up. What the writer meant to say is "earned less than 200% of the poverty line."

Americans . . . bike 25 times less than the Dutch.

—cleantechnica.com/2014/01/03/americans-drive-twice-much-brits-bike-25-times-less-dutch/

You no doubt guess that "25 times less" really means "one twenty-fifth" as much. But the latter doesn't sound as grand. Writing "25 times" does give the correct ratio, and "less" clues the reader about which is less. But mixing "times" with "less" or "more" offers potential for confusion, as you can see in the following example.

At a website giving information about federal student loans, you may find the following statement:

Under this plan, your monthly payments . . . won't be more than three times greater than any other payment.

Let's see. Suppose you have \$1. If I have \$1 greater than what you have, then I have \$2. If I have \$3 greater than what you have, then I have . . . \$4.

The terms *more, larger,* and *greater* refer to *adding* to the original amount. The terms *times* and *twice as much as* refer to *multiplication* of the original amount. For instance, if you have \$1 and I have "5 times as much," then I have \$5—which is \$4 more than you have, hence "4 times more than" what you have (the original plus 4 times as much).

What we have in the loan information is a mixing of additive and multiplicative terms, and it is not clear what is meant. Further investigation reveals that what was intended was "no more than 3 times your lowest payment." Using *times* together with *more* or *greater than* is, to use a football analogy, "piling on." It is often designed to impress the audience by attempted exaggeration.

The relationship of an original amount to a larger amount can be expressed either in multiplicative terms ("one-fifth as much" or "20% as much") or in subtractive terms ("four-fifths less than" or "80% less than")—but don't mix the two.

Texting while driving is the number one cause of death for American teenagers. What is wrong with the following claim, and how could you state it correctly?

Texting drivers looked at the road 400 percent less than drivers who did not text message while behind the wheel of a car.

—carlsbad.patch.com/groups/michael-bombergers-blog/p/bp–the-dangers-of-texting-while-driving

If we follow the same logic of the writer of the claim about American and Dutch cycling, we would conclude that texters look at the road one-fourth as much as non-texters. But we would be wrong; tracing this phrase to its original source (in Australia in 2006), we find:

Drivers spent up to 400 percent more time with their eyes off the road when text messaging, than when not text messaging. . . . The amount of time that drivers spent with their eyes off the road increased by up to 400% when retrieving and sending text messages.

—www.distraction.gov/downloads/pdfs/effects-of-text-Messaging.pdf

This phrasing would mean that texters spent 5 times as much time with their eyes off the road as non-texters; or put another way, non-texters had their eyes off the road one-fifth as often as non-texters. But alas! The details of the report reveal that non-texters spent 10% of the time with their eyes off the road and texters spent 40% (aargh!). So the phrasing by the report writers and the "interpretation" of it by the blogger are both misleading. We could correctly talk about "4 times as much" or "300% more" or "one-fourth as much" or "25% as much" or "75% less"—as long as we don't mix up additive and multiplicative terminology.

EXAMPLE 1 Generic Medicines

What is wrong with the following claim, and how would you state it correctly?

> *Most generic medications aren't 50% or 75% less expensive than their brand named equivalents,* ***they are 100 times cheaper!!***
>
> —David Belk, *The True Cost of Healthcare A View of Healthcare Costs from the Inside*, www.truecostofhealthcare.org

The claim was rephrased (by another pair of doctors) as the following:

> *Generic medications can cost 100 percent less than their brand-name equivalent.*
>
> —www.facebook.com/permalink.php?id=554443731254232\&story_fbid=619922344706370

What's wrong with that claim?

The first claim mixes "times" with a form of "more." The second claim would indicate that generic medicines cost nothing. Correct versions of the statements would be: "Generic medicines cost 1% as much as their brand-name equivalents" and "Brand-name medicines cost 100 times as much as their generic equivalents." ■

Self Check 1

What's wrong with the following claim, and how would you state it correctly?

> *Will the U.S. women's world champion soccer team really get paid 40 times less than the men?*
>
> —David Muir, *ABC World News Tonight*, 7 July 2015 ■

People often say "5 times more than" when they mean "5 times as much." All you can do is be aware of the potential confusion, try to figure out what was meant, and be careful in your own expression.

Correct Comparisons RULE

Don't use *times* with *more* or with *less*.

Finally, we need to distinguish *percent* from *percentage points:* If support for the president decreased from 60% to 30%, it dropped 30 *percentage points* but decreased 50% (because the drop of 30 percentage points is 50% of the original 60 percentage points).

18.2 How Much Is That in . . . ?

We are interested in the limits of size, so we need to compare objects of different sizes, such as a gorilla with King Kong. However, it is not easy to compare objects measured in different units—say, the gorilla is measured in inches and pounds, but King Kong is measured in centimeters and kilograms. Consequently, we explore how to convert units from one measurement system to another.

We introduce two systems of units in which physical quantities are commonly measured and give a table of *conversion factors* and examples of how to convert from one system to the other.

U.S. Customary System

Table 18.1 lists units of the U.S. *customary system* of measurement and their abbreviations. Please note in the table the systematic way of converting from one unit to another and the use of scientific notation. The symbol ≈ means "is approximately equal to."

TABLE 18.1 Units of the U.S. Customary System

Distance:	
1 mile (mi)	= 1760 yards (yd) = 5280 feet (ft)
1 yard (yd)	= 3 feet (ft) = 36 inches (in.)
1 foot (ft)	= 12 inches (in.)
Area:	
1 square mile	= 1 mi × 1 mi = 5280 ft × 5280 ft ≈ 2.8×10^7 ft^2
	= 640 acres
1 acre	= 43,560 ft^2
Volume:	
1 cubic mile	= 1 mi × 1 mi × 1 mi
	= 5280 ft × 5280 ft × 5280 ft ≈ 1.5×10^{11} ft^3
1 U.S. gallon (gal)	= 4 U.S. quarts (qt) = 231 $in.^3$
Mass:	
1 ton (t)	= 2000 pounds (lb)

Metric System

The world generally uses the metric system in science, industry, and commerce. It was first adopted in France in 1795. The fundamental unit of length, the *meter* (m), was originally 1/10,000,000 of the distance from the North Pole to the equator, along the meridian through Paris. The length is now defined as the distance that light travels in a vacuum in $\frac{1}{299{,}792{,}458}$ second. The *second* (s), in turn, is defined as the time that it takes an atom of the metal cesium to vibrate 9,192,631,770 times.

All other units of length, area, and volume are defined in terms of the meter. For example, a centimeter (cm) is one-hundredth of a meter.

Mass is the quantity of matter. The metric unit of mass, the *kilogram* (kg), is defined as the mass of a platinum-iridium standard kept in Paris. Since you can't determine the mass of a sack of potatoes by comparing it to that, we measure the mass indirectly by seeing how much force gravity exerts on it—that is, we weigh it on a scale calibrated in pounds or kilograms. However, a mass of 1 kg would "weigh" (register on the scale) only one-sixth as much on the moon, because of the weaker gravity there.

Table 18.2 lists the units of the metric system.

TABLE 18.2 Units of the Metric System

Distance:	
1 meter (m)	= 100 centimeters (cm)
1 kilometer (km)	= 1000 meters (m)
	= 100,000 centimeters (cm) = 1×10^5 cm
	= 1,000,000 millimeters (mm) = 1×10^6 mm
Area:	
1 square meter (m^2)	= 1 m × 1 m
	= 100 cm × 100 cm = 10,000 (cm^2) = 1×10^4 cm^2
1 hectare (ha)	= 10,000 m^2
Volume:	
1 liter (L)	= 1000 cm^3 = 0.001 m^3
1 cubic meter (m^3)	= 1 m × 1 m × 1 m
	= 100 cm × 100 cm × 100 cm
	= 1,000,000 cm^3 = 1×10^6 cm^3 (or cc)
Mass:	
1 kilogram (kg)	= 1000 grams (g) = 1×10^3 g
1 metric ton (tonne)	= 1000 kg

Converting Between Systems

The fundamental units of the U.S. customary system, the yard (for length) and the pound (for mass), are defined in terms of metric units, so that we have *exactly*

$$1 \text{ yd} = 0.9144 \text{ m}$$
$$1 \text{ lb} = 0.45359237 \text{ kg}$$

Table 18.3 illustrates the conversion factors.

Most Internet search engines offer conversions; try entering, for example, "180 cm in ft." You do not need to memorize conversion factors, but you will find it useful in life to memorize a few rough approximations:

$$1 \text{ cm} \approx 0.4 \text{ in.} \quad 1 \text{ m} \approx 3 \text{ ft} \quad 1 \text{ km} \approx 0.6 \text{ mi} \quad 1 \text{ mi} \approx 1.6 \text{ km}$$
$$1 \text{ m}^2 \approx 10 \text{ ft}^2 \quad 1 \text{ L} \approx 1 \text{ qt} \quad 1 \text{ gal} \approx 4 \text{ L} \quad 1 \text{ kg} \approx 0.5 \text{ lb}$$

In the following examples, we explain how to convert measurements between systems. For such conversions, you may also find it useful to be acquainted with various prefixes used to denote multiples of units. From your experience with computers, you are no doubt familiar with the prefixes kilo-, mega-, and giga- (as in kilobytes, megabytes, gigabytes), indicating, respectively, the numbers 10^3, 10^6, and 10^9. Perhaps you are also familiar with tera- (hard drives are now sold that

TABLE 18.3 Conversions Between the U.S. Customary System and the Metric System

Distance:	
1 in.	$= 2.54$ cm
1 ft	$= 12$ in. $= 12 \times 2.54$ cm $= 30.48$ cm $= 0.3048$ m ≈ 0.3 m
1 yd	$= 0.9144$ m ≈ 1 m
1 mi	$= 5280$ ft $= 5280 \times 30.48$ cm
	$= 160{,}934.4$ cm ≈ 1.609 km ≈ 1.6 km
1 cm	$\approx 1/2.54$ in. ≈ 0.4 in.
1 m	≈ 39.37 in. ≈ 3.281 ft ≈ 3 ft
1 km	≈ 0.621 mi ≈ 0.6 mi
Area:	
1 ft^2	≈ 1000 cm^2
1 m^2	≈ 10 ft^2
1 hectare (ha)	≈ 2.5 acres
Volume:	
1 ft^3	≈ 30 liters (L)
1 gallon	≈ 3.785 liters (L) ≈ 4 L
1 cubic meter (m^3)	$= 1000$ liters ≈ 264 U.S. gallons ≈ 35 ft^3
1 liter (L)	$= 1000$ cm^3 ≈ 1.06 U.S. quarts (qt) ≈ 0.26 U.S. gallons ≈ 1 qt
Mass:	
1 lb	$= 0.45359237$ kg ≈ 0.5 kg
1 kg	≈ 2.205 lb ≈ 2 lb

can contain several terabytes, where 1 terabyte = 1000 gigabytes = 10^{12} bytes of data). The corresponding scientific notation prefixes are k (thousand), M (million), G (billion), and T (trillion). What's beyond that? P (peta-), E (exa-), Z (zetta-), and Y (yotta-) for 10^{15}, 10^{18}, 10^{21}, and 10^{24}, respectively. (A hint on proper usage: "Gazillion" is not a scientific unit prefix!) There are also corresponding prefixes for ever-smaller quantities: m (milli-), μ (micro-), n (nano-), and p (pico-) for 10^{-3}, 10^{-6}, 10^{-9}, and 10^{-12}.

EXAMPLE 2 What's That in Feet?

An international student tells his American student friends that he is 180 cm tall. They ask, "How much is that in feet and inches?"

We approach this conversion by using the scaling factor 1 cm $= \frac{1}{2.54}$ in.:

$$\begin{aligned} 180 \text{ cm} &= 180 \times 1 \text{ cm} \\ &= 180 \times \frac{1}{2.54} \text{ in.} \\ &\approx 70.9 \text{ in.} = 70.9 \text{ in.} \times \frac{1 \text{ ft}}{12 \text{ in.}} = \frac{70.9}{12} \text{ ft} \approx 5.9 \text{ ft} \end{aligned}$$

However, because we normally give height in feet and a whole number of inches, the height is

$$70.9 \text{ in.} = 5 \times (12 \text{ in.}) + 10.9 \text{ in.}$$

or approximately 5 ft 11 in.

Another way to approach the problem is by means of a proportion:

$$\frac{\text{height in in.}}{\text{height in cm}} = \frac{\text{length of 1 in. in in.}}{\text{length of 1 in. in cm}} = \frac{1 \text{ in.}}{2.54 \text{ cm}}$$

so that

$$\text{height in in.} = \text{height in cm} \times \frac{1 \text{ in.}}{2.54 \text{ cm}} = 180 \text{ cm} \times \frac{1 \text{ in.}}{2.54 \text{ cm}} \approx 70.9 \text{ in.}$$

where the cm in the numerator cancels with the cm in the denominator.

A strategy for a calculation requiring two or more conversions is first to plan the calculation: Write out an equation that involves just the units involved in the proportions, leaving out the numbers. Then, after checking that the units cancel correctly, insert the corresponding numbers and do the resulting arithmetic. ■

Self Check 2

Suppose that you are 5 ft 3 in. tall. How much is that in cm? ■

EXAMPLE 3 Got Gas?

Although in the United States we have traditionally measured the efficiency of cars in miles per gallon (mpg), the rest of the world measures it in liters per 100 km. The conversion between these two measures is more complicated than other conversions because the U.S. measure has distance (mi) in the numerator and quantity of fuel (gal) in the denominator, whereas the other measure has quantity of fuel (L) in the numerator and distance (km) in the denominator. We need to take this difference into account when doing the conversion.

For example, according to the Environmental Protection Agency (EPA), the most efficient gasoline vehicle of all time was the two-passenger 2000 Honda Insight, at 61 mpg on the highway. (This model was discontinued in 2006 due to poor sales!) What is the equivalent in liters per 100 km?

$$\begin{aligned} 61 \text{ mpg} &= 61 \times \frac{1 \text{ mi}}{1 \text{ gal}} \\ &= 61 \times \frac{1.609 \text{ km}}{3.785 \text{ L}} \approx 25.93 \frac{\text{km}}{\text{L}} \\ &= \frac{25.93}{1} \times \frac{100 \text{ km}}{100 \text{ L}} = \frac{1}{\frac{1}{25.93}} \times \frac{100 \text{ km}}{100 \text{ L}} \\ &\approx \frac{100 \text{ km}}{3.9 \text{ L}} \end{aligned}$$

■

In other words, 100 km requires about 4 L. (Europeans would call such a car a "4-liter car.") The key steps in the solution are to multiply both units by 100 and then divide both the numerator and the denominator of the fraction by 25.93, so as to get exactly 100 km in the numerator of the result.

U.S. new-car labels, as shown in Figure 18.5, indicate fuel economy in terms of mpg but also—in much smaller print—gallons per 100 miles.

Self Check 3

The car indicated in Figure 18.5 uses 3.8 gallons per 100 miles. How many liters does it use for 100 km? ■

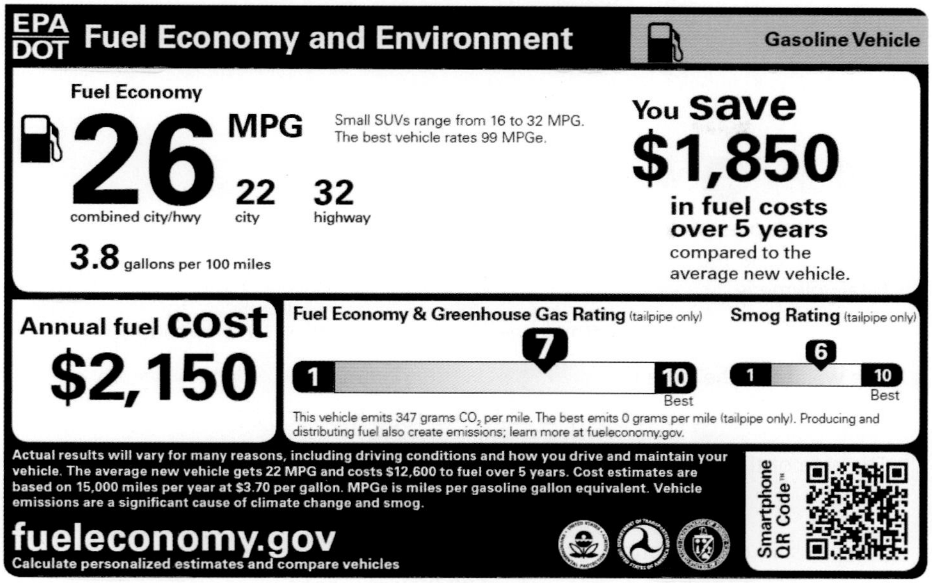

Source: Environmental Protection Agency

Figure 18.5 Fuel-economy label for a 2015-model U.S. gasoline-powered car.

18.3 Big Stuff

Gravity exerts an enormous effect on the size and shape of things around us. **Weight** (attractive force toward the center of the Earth) is the reading at sea level on a scale (such as your bathroom scale) calibrated in pounds or kilograms of mass.

Suppose that the two cubes in Figure 18.4 on page 740 are made of steel and that the first is 1 ft on a side and the second is 3 ft on a side. A cubic foot of steel weighs about 500 lb; we say that the **density** of steel is 500 lb per cubic foot, or 500 lb/ft^3. The cube that is 1 ft on a side weighs 1 ft^3 $\times$ 500 lb/ft^3 $=$ 500 lb. The weight W of an object of volume V and uniform density D is

$$W = DV$$

Each cube's bottom face supports the weight of the entire cube. **Pressure** is the force per unit area, so the pressure exerted on the bottom face by the weight of the cube is equal to the weight of the cube divided by the area of the bottom face, or

$$P = \frac{W}{A}$$

The first cube weighs 500 lb and has a bottom face with area of 1 ft^2, so the pressure exerted on this face is 500 lb/ft^2.

The second cube is 3 ft on a side. The area of the bottom face increases with the square of the linear scaling factor, so it is $3^2 \times 1$ ft^2 $=$ 9 ft^2. As we saw earlier, volume goes up with the cube of the linear scaling factor. So this larger cube has a volume of $3^3 \times 1$ ft^3 $=$ 27 ft^3. Because both cubes are made of the same steel, the larger cube has 27 times as much steel as the smaller one. Hence, it weighs 27 times as much as the smaller cube, or 27 $\times$ 500 lb $=$ 13,500 lb.

When we divide this weight by the area of the bottom face (9 ft^2), we find that the pressure exerted on the bottom face is 1500 lb/ft^2, or 3 times the pressure on the bottom face of the original cube. This makes sense because over each 1-ft^2 area stands 3 ft^3 of steel. In general, if the linear scaling factor for the cube is L, the pressure on the bottom face is L times as much. Using the notation of proportionality, where $\propto$ stands for "is proportional to," we have $A \propto L^2$ and $W \propto V \propto L^3$, so

$$P = \frac{W}{A} \propto \frac{L^3}{L^2} \propto L$$

EXAMPLE 4 What About a 10-Foot Steel Cube?

What is the pressure on the bottom of a steel cube 10 ft on a side?

If we scale the original cube of steel up to a cube that is 10 ft on a side, then the total volume is

$$V = \text{length} \times \text{width} \times \text{height} = 10\text{ ft} \times 10\text{ ft} \times 10\text{ ft} = 1000\text{ ft}^3$$

The weight of the cube is

$$W = D \times V = \frac{500\text{ lb}}{\text{ft}^3} \times 1000\text{ ft}^3 = 500{,}000\text{ lb}$$

The area of the bottom face is

$$A = \text{length} \times \text{width} = 10\text{ ft} \times 10\text{ ft} = 100\text{ ft}^2$$

The pressure on the bottom face is

$$P = \frac{W}{A} = \frac{500{,}000\text{ lb}}{100\text{ ft}^2} = \frac{5000\text{ lb}}{\text{ft}^2}$$

This is 10 *times*—not "10 times *more* than"—the pressure on the bottom face of the original 1-ft cube.

Self Check 4

What would be the pressure on the bottom face of a steel cube 100 ft on a side? Be sure to give not just a number but also the correct units for the answer.

EXAMPLE 5 What About Burj Khalifa?

Burj Khalifa in Dubai (see Figure 18.6), completed in 2010, is still the world's tallest skyscraper at 2684 ft (how much is that in meters?), not counting radio and television antennas (see Spotlight 18.1). What is the pressure at the bottom of its walls?

David James/© Paramount Pictures/ Courtesy Everett Collection

Figure 18.6 Actor Tom Cruise dangling from Burj Khalifa in the film *Mission Impossible—Ghost Protocol* (2012).

The building is made of reinforced concrete, which weighs 160 lb/ft^3. Although the building tapers toward the top, we are not far off if we model it as straight up and down—that is, as a rectangular solid. Consider one of its supporting walls. The volume of the wall is its height H times the area A of its base, or $V = HA$. The weight of the wall is $W = DV = DHA$. The pressure at the bottom from the wall's weight is

$$P = \frac{W}{A} = \frac{DV}{A} = \frac{DHA}{A} = DH = \frac{160\text{ lb}}{\text{ft}^3} \times 2684\text{ ft} = \frac{429{,}440\text{ lb}}{\text{ft}^2} \approx 430{,}000\text{ lb/ft}^2$$

where we have rounded off the calculated answer to agree with the degree of precision of the ingredient numbers.

Algebra Review Appendix
Significant Digits

That's not counting the contents of the tower, which also must be supported.

A Mile-High Building?

SPOTLIGHT 18.1

In 1956, the famous American architect Frank Lloyd Wright (1867–1959) proposed a mile-high "Illinois Sky-City" tower for the Chicago lakefront. Burj Khalifa, at 2684 ft, is just over half that high. In the text, we focus on the problem of holding up the weight of such a structure.

But there are other limits to the height of a building—for example, the bending of the building in the wind, which can go up dramatically with height. Bending can be controlled by making the building stiffer.

The terrorist destruction of the World Trade Center towers in 2001 resulted not directly from the aircraft impacts but from the subsequent fires and collapse of the towers' structure.

Even if designed to better resist fires and impacts, however, a mile-high building might not be practical. For example, the enormous number of people (perhaps 100,000) living, working, or visiting in such a building would create enormous traffic problems (pedestrian, parking, deliveries) for blocks around.

Cost per square foot of usable area is an important consideration. Even if the building did not taper, the space in the upper floors might not justify their additional expense. With increasing height, an increasingly larger proportion of the cross-sectional area of all floors must be devoted to services, such as elevators; everyone entering the building and going to any floor needs to start in an elevator on the ground floor. In an emergency evacuation, the people must *walk* downstairs! And that could take dangerously long.

Some architects, however, maintain that the main limit on the height of a building is human physiology. Differences in air pressure between the top and bottom of a building limit how fast elevators can rise or drop without discomfort to passengers, thereby enforcing long travel times for "vertical commuters." Human psychology also might present some limits.

Are skyscrapers now dinosaurs, except as exercises in one-upmanship—"vertical cities for the affluent," as Blaine Brownell of the University of Minnesota School of Architecture calls them? Telecommuting and outsourcing of work to decentralized locations may make it obsolete to bring office workers together at a single site in a dense, expensive downtown.

In 2012, renowned science fiction writer Neal Stephenson, together with structural engineer Dr. Keith Hjelmstad, designed a tower 20 km (over 60,000 ft) high. The base would be 10 mi wide, the building would require one billion tons of steel (half the annual output of the world), and the cost would be $1 trillion. Said Hjelmstad, it "would be the biggest project ever undertaken by humans." It would be a convenient launching pad for sending payloads into orbit and beyond.

Guess what they found to be the biggest hurdle? Not the challenge of building it, nor the fact that it would be in the way of airliners, but handling the jet stream—which at that height can gust to 310 mph. So the project is "on pause."

Meanwhile, Stephenson has written a science fiction story about an engineer trying to build such a building: "Atmosphera incognita," in *Hieroglyph: Stories and Visions for a Better Future,* edited by Ed Finn and Kathyrn Cramer (William Morrow, 2014), 1–37.

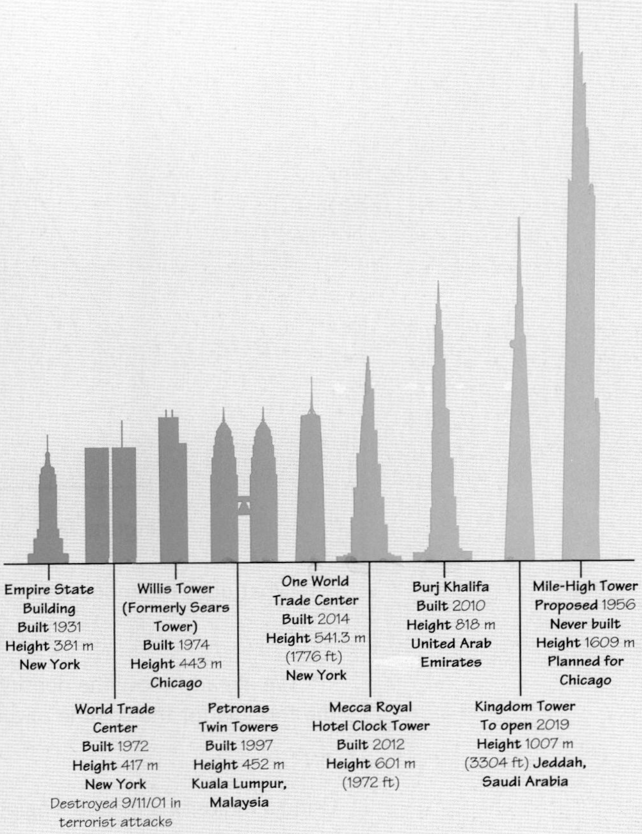

Self Check 5

Consider the possibility of a Super Burj Khalifa 10 times as high. How much weight would the bottom of its walls have to support per square foot? ■

EXAMPLE 6 How High Can a Mountain Be?

Just like skyscrapers, the height of mountains is also limited, by gravity, their composition, and their shape. How tall can a mountain be?

We build a simple mathematical model of a mountain. Suppose that it is made of granite, a common material, with uniform density. Granite weighs 165 lb/ft^3 and has a **crushing strength** of about 4 million lb/ft^2.

In the interests of both realism and simplicity, we assume that the mountain is a solid cone whose width at the base is the same as its height. Let's model Mount Everest, the tallest mountain on Earth, at about 6 mi high. The base, then, is a circle with a diameter (distance across) of 6 mi. The radius (half the diameter) is 3 mi (Figure 18.7). Because we took round numbers (6 mi) for the height and width, we record as significant only the first digit or two of the results of the calculations.

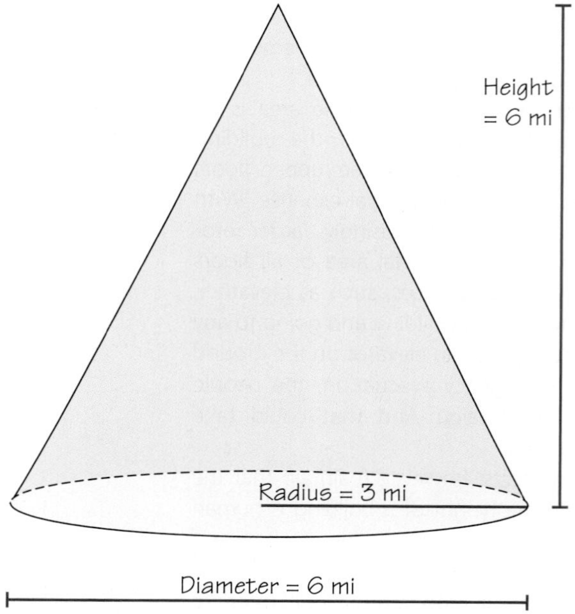

Figure 18.7 Model of Mount Everest as a cone of granite.

What does the model Everest weigh? The relevant formula is $W = DV$, or

$$\text{weight} = \text{density} \times \text{volume}$$

We already know the density of granite (165 lb/ft^3), so to find the weight, we need the formula for the volume of a cone of radius r and height h:

$$V = \text{volume} = \frac{1}{3}\pi r^2 h$$

Using a radius of 3 mi and a height of 6 mi, we find that the model Everest has a volume of about 57 mi^3.

To find the weight of 57 mi^3 of granite, we need to convert units because the density that we know is in pounds per cubic foot (lb/ft^3). Let's convert to units of feet as follows:

$$\begin{aligned} 1\text{ mi}^3 &= 1\text{ mi} \times 1\text{ mi} \times 1\text{ mi} \\ &= 5280\text{ ft} \times 5280\text{ ft} \times 5280\text{ ft} \\ &\approx 1.5 \times 10^{11}\text{ ft}^3 \end{aligned}$$

Thus,

$$57\text{ mi}^3 \approx 57 \times 1.5 \times 10^{11}\text{ ft}^3 \approx 8.6 \times 10^{12}\text{ ft}^3$$

So we have

$$\begin{aligned} W &= \text{weight of mountain} = \text{density} \times \text{volume} = D \times V \\ &= \frac{165\text{ lb}}{\text{ft}^3} \times 8.6 \times 10^{12}\text{ ft}^3 \\ &= 1.4 \times 10^{15}\text{ lb} \end{aligned}$$

Now that we know the mountain weighs 1.4 quadrillion pounds, we want to find the pressure on the base of the cone and compare it with the crushing strength of granite. (Everest is standing, so if our model is any good, that pressure will be below the crushing strength.)

Physics tells us that the weight of the mountain is spread evenly over the base of the cone (though we are oversimplifying the geology underlying mountains). Because

$$P = \frac{W}{A} \qquad \left(\text{pressure} = \frac{\text{weight}}{\text{area}}\right)$$

we need to calculate the area of the base of the cone. The shape is a circle, and the familiar formula

$$A = \text{area} = \pi r^2$$

gives for a radius $r = 3$ mi an area of 28 mi^2.

Once again, we need to convert units to express the pressure in pounds per square foot, the units in which the crushing strength is expressed. We get

$$A = 28\text{ mi}^2 = 28 \times 5280\text{ ft} \times 5280\text{ ft} \approx 7.8 \times 10^8\text{ ft}^2$$

Then

$$P = \frac{W}{A} = \frac{1.5 \times 10^{15}\text{ lb}}{7.8 \times 10^8\text{ ft}^2} \approx 1.8 \times 10^6\text{ lb/ft}^2$$

This number is about half the crushing strength of granite, 4×10^6 lb/ft^2. For a mountain to come close to the limitation of the crushing strength of granite, it would have to be only about 10 mi high, not quite twice as high as Everest. Other physical considerations suggest a maximum height of at most 15 mi. The fact that no current mountains are that high may be a consequence of the Earth's high amount of volcanic activity and the structural deformation of the Earth's crust. For a mountain that is too big, the rock below it would flow and the mountain would sink. ■

What about mountains made of other materials—glass, ice, wood, or old cars? They couldn't be nearly as high. The pressure would cause glass to flow, ice to melt, and old cars to compact. What about mountains on another planet, or on an asteroid? Their potential height depends on the gravity there. On the moon, whose gravity is one-sixth of Earth's, mountains could potentially be 6 times as high (not "6 times higher"). However, the tallest mountain on the moon is only a bit more than half as high as Mt. Everest.

Sorry, No King Kongs

Unfortunately, the resistance of bone to crushing is not nearly as great as that of steel or granite. This fact helps to explain why there couldn't be any King Kongs (unless they were made of steel or granite!). A King Kong scaled up by a factor of, say, 20 would weigh $20^3 = 8000$ times as much. Although the weight increases with the cube of the linear scaling factor, the ability to support the weight—as measured by the cross-sectional area of the bones, like the area of the bottom face of the cube in Figure 18.4—increases only with the square of the linear scaling factor.

These simple consequences of the geometry of scaling apply not only to supermonsters but also to other objects, such as trees.

Figure 18.8 Even giant sequoias can grow no taller than their form and materials allow.

EXAMPLE 7 How Tall Can a Tree Be?

Galileo suggested that no tree could grow taller than 300 ft. The world's tallest trees are giant sequoias (Figure 18.8), which grow only on the West Coast of the United States and hence were unknown to Galileo. The tallest known sequoia today is 380 ft in Redwood National Park in Northern California. (To protect it, the National Park Service does not disclose its exact location.)

What limits the height of a tree? If the roots do not adequately anchor it, a tall tree can blow over. (This happened in 1990 to the world's then-tallest tree, the Dyerville Giant in Humboldt Redwoods State Park in California.) The tree could buckle or snap under its own weight and the force of a strong wind. The wood at the bottom can be crushed if there is too much weight pressing upon it. Finally, there is a limit to how far the tree can lift water and minerals from the roots to the leaves.

Could a tree be a mile high? To make a rough estimate of the pressure at the base of the tree due to gravity, let's model the tree as a perfectly vertical cylinder. Over each square foot at the bottom, there are 5280 ft^3 of cells of wood, which weighs about half as much as water. A convenient scientific fact to know involving the metric system is that water weighs just about 1 gram (g) per cubic centimeter. So, to calculate the weight, we first translate 1 ft^3 into metric measurement:

$$1\ ft^3 = (12\ in.)^3 = (12 \times 2.54\ cm)^3 \approx 28{,}317\ cm^3$$

So, 1 ft^3 of water weighs about

$$28{,}317\ g = 28.317\ kg \approx 28.317 \times 2.205\ lb \approx 62\ lb$$

Consequently, 5280 ft^3 of water weighs about 5280 lb $\times$ 62 lb $\approx$ 330,000 lb. The weight of the same volume of wood is about half as much, or about 165,000 lb. Therefore, the pressure at the bottom of the tree would be about 165,000 lb/ft^2.

This is an overestimate because we assumed that the tree does not taper. A tree that tapers steadily looks like an elongated cone; using a more realistic cone model (as we did in the last section for a mountain), you would find that the pressure at the bottom of the tree would be one-third of 164,000 lb/ft^2, or about 55,000 lb/ft^2. A biological organism needs a safety factor of at least 2 to 4 times the absolute minimum physical limits, so a mile-high tree would need from 110,000 to 220,000 lb/ft^2 of upward pressure for water and minerals. Tension in the string of water molecules from root to leaf ranges from 80,000 to 3.2 million lb/ft^2, for different kinds and heights of trees, so this consideration does not rule out mile-high trees.

At more than about 500 $lb/in.^2 \approx 70{,}000\ lb/ft^2$, though, the bottom of the tree would begin to crush under this weight. On this basis, a mile-high tree is barely feasible, with little margin of safety. However, researchers who hauled themselves up to the top of the tallest trees in 2004 found a much lower limit, at least for giant sequoias. With increasing height, leaves are smaller, dryer, and less efficient at photosynthesis. The researchers estimated that trees can't top out higher than 400 to 427 ft. The tallest reliably measured tree was a North American Douglas fir, measured in 1902 at 413 ft.

There are other considerations. The taller the tree, the greater the area from which it must draw water and minerals, for which nearby trees also compete. Moreover, for a tree to grow very tall, it would have to live for a very long time. Evolution and time may select against extremely tall trees, or maybe, for no reason at all, they have just never evolved.

Why Would You Want to Be Big?

SPOTLIGHT 18.2

The size of some animals is astonishing to us. A blue whale can be 100 ft long and weigh 200 tons. Land-dwelling dinosaurs of 200 million years ago may have been almost as large.

Why so big? To be tall has advantages: A tall tree can gather more sunlight; a giraffe can harvest leaves that other herbivores can't reach; and a dog is more likely to be submissive to another dog or person who is taller. In the case of people, taller can be more attractive: Studies indicate that taller men earn more money, and women make themselves look taller by wearing high heels.

Sheer bulk and muscle can have its advantages, as when male bighorn sheep or red deer wrestle for dominance and the right to sire offspring. And a larger size can make it easier for an animal to defend territory.

But apart from intraspecies (within species) differences, what could be the advantage of being larger than other species?

A full-grown and healthy wildebeest is less likely to be eaten by predators; but a hippopotamus, a rhinoceros, or an elephant—which are all even bigger—are even less likely to become victims.

There are some gains in "efficiency" in being large. The metabolic rate goes down, which means that a species twice as large does not require quite twice as much food. A larger species can have a more complex digestive system, so it can eat a broader array of foods—or simply different foods—than smaller species.

A large animal needs a thick hide to hold in its bulk, and such a hide can also provide thermal insulation, allowing the animal to live in a colder climate.

Hallam creations/Shutterstock

Of course, there are also disadvantages to size. Really big birds, such as ostriches, emus, or even penguins, can't fly—but maybe they don't need to: Ostriches can run fast, and penguins can swim. A large animal can't easily take shelter under a small tree or hide by digging a burrow. Large animals tend to reproduce more slowly, with longer gestation periods (two years for an elephant) and smaller litters (with even twins rare). The biggest challenge is finding enough food; whales migrate to do so.

But why did some animals *become* large, even gigantic? They adapted and evolved to fit an open ecological niche, and becoming large aided their fitness for that niche. When that niche disappeared—the steamy plant jungles in the case of dinosaurs, the Great Plains of grasses in the case of American buffalo—the animals did, too.

18.4 Dimension Tension

A large change in scale forces a change in either materials or form. A major manifestation of this **problem of scale** is the tension between weight and the need to support it. For example, a real building or machine must differ from a scale model: The balsa wood or plastic of the model would never be strong enough for the real thing, which would need aluminum, steel, or reinforced concrete.

Another way to compensate is to redesign the object to distribute its weight better. Let's go back to the original cube. It supports all its weight on its bottom face. In the version scaled up by a factor of 3, each small cube of the bottom layer has a bottom face supporting that cube's weight plus the weight of the two cubes piled on top of it.

Let's redesign the scaled-up cube, concentrating for simplicity only on the front face, with its nine small cubes. We take the three cubes on top and move them to the bottom, alongside the three that are already there. We take the three cubes on the second level, cut each in half, and put a half-cube over each of the six ground-level cubes (see Figure 18.9).

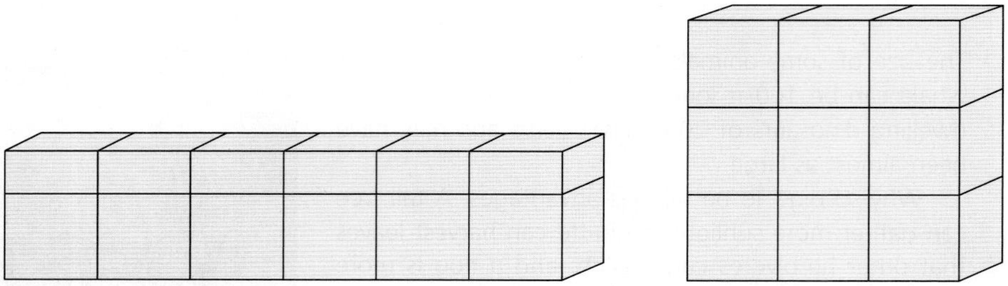

Figure 18.9 Nine small cubes rearranged to support greater weight.

We have the same volume and weight that we started with, but now there is less pressure on the bottom face of each small cube.

Of course, the new design is not geometrically similar to the object that we started with; it's no longer a cube. We have solved the scaling problem by changing the proportions.

We observe in nature both strategies for scaling: change of materials and change of form. Small animals (such as insects) do not have bony internal skeletons. Larger animals generally do. Animals made of similar materials but differing greatly in size, such as a mouse and an elephant, must differ in shape. If a mouse were scaled up to the size of an elephant, it would need the disproportionately thicker legs of the elephant to support its weight and the elephant's thick hide to contain its tissue.

Some dinosaurs, like *Supersaurus* (which weighed 30 tons), had special adaptations to lighten their weight, such as hollow bones, just as some birds have. Hollow bones are stronger: Of two bones of the same weight and length, the hollow one is wider across at its midpoint because of the air it contains, and the greater the width, the greater the resistance to fracture.

Falls, Jumps, and Flight

The need to support weight can be thought of as a tension between volume and area. As an object is scaled up, its volume and weight go up together, so long as the density remains constant (e.g., no air bubbles introduced into the steel to make it into a Swiss cheese). At the same time, the ability to support the weight goes up with the cross-sectional area, like the bottom face of the steel cube.

Area–Volume Tension DEFINITION

Area–volume tension is a result of the fact that as an object is scaled up, the volume increases faster than the surface area and faster than areas of cross sections.

Because volume V is proportional to the cube of the linear scaling factor L, we have $V \propto L^3$; taking each side to the one-third power gives $L \propto V^{1/3}$. The fact that surface area A is proportional to the square of the linear scaling factor is expressed in our notation as

$$A \propto L^2 \propto (V^{1/3})^2 = V^{2/3}$$

so that surface area scales as the two-thirds power of volume.

Algebra Review Appendix
Natural and Fractional Exponents

In any crowded city, you can observe "parking tension"—tension between length, area, and volume. Consider an apartment building that spans a city block. The area of parking spaces on the adjacent streets is proportional to the perimeter of (length around) the building. But the number of cars belonging to people in the building is proportional to the number of apartments, which is proportional to the volume of the building. So the higher the building, the greater the parking tension.

In some cities, zoning puts shops on the ground floor, which cuts out one floor of apartments. If the residents' cars are away during the day, customers and employees of the shops can park where the apartment dwellers do at night. A more common solution is an underground garage, with an area for cars proportional to the number of apartments, so in effect to the volume of the building. However, garages constructed with one car per apartment are inadequate now that families tend to have more than one car.

Other examples of the solutions to dimensional tension include the old-fashioned diner, with its serving counter in the form of an S-shape so as to expand its effective length, and your small intestine, which coils its 20-ft length to fit into your abdomen.

Though we can forget about humans "leaping tall buildings in a single bound," "soaring like an eagle," or "diving miles below the sea," area–volume tension has many practical consequences related to those fantastical ideas. Consider the following examples.

EXAMPLE 8 Falls

Area–volume tension affects how animals respond to falling, another of gravity's effects. How come a mouse may be unharmed by a 10-story fall, and a cat by a 2-story fall, but many humans are injured by falling while running, walking, or even just standing?

The explanation is that the energy acquired in falling is proportional to the weight of the falling object, and hence to its volume. This energy must be absorbed either by the object or by what it hits, or must be otherwise dissipated at impact—for example, as sound. The fall is absorbed over part of the surface area of the object, just as the weight of the cube was distributed over its base. With scaling up, volume—hence weight, hence falling energy—goes up much faster than area. As size increases, the hazards of falling from the same height increase. ■

EXAMPLE 9 Jumps

A flea can jump as high as 50 cm (20 in.) vertically, many times its own height. Some people believe that if a flea were as large as a person, it could jump 1000 ft into the air. Is that so?

Imagining—against our earlier arguments—that there could be so large a flea, we can deduce its limits: A scaled-up flea could jump about the same height as a small flea. The strength of a muscle is proportional to its cross-sectional *area* (see Spotlight 18.3). A jump involves suddenly contracting the muscle through its length, so it turns out that the ability to jump is proportional to the *volume* of muscle. But the volume of the flea and the volume of its leg muscles would go up in proportion as we enlarged the flea to a geometrically similar giant flea.

Let's say that a real flea's leg muscles account for 1% of its body. If we scale the flea up to the size of a person (without any change in its form), the enlarged flea's leg muscles will still make up 1% of its body. For either flea, each bit of muscle has the same power: In a jump, it propels 100 times its own weight, and it can do so to the same height. Both the weight of the flea and the power of its legs go up proportionately. In fact, the maximum heights that people, fleas, grasshoppers, and kangaroos can jump from standing are all within a factor of 3 of one another. ■

EXAMPLE 10 Flight

How does weight affect flight? After all, ostriches can't fly.

Wouldn't it be nice to be able to fly? Well, you have to be able to stay up. The power necessary for sustained flight is proportional to the **wing loading,** which is the weight supported divided by the area of the wings. We know that in scaling up, weight grows with the cube of the length of the bird or plane, and wing area with the square of the length. So the wing loading is proportional to the length of the flying object.

Quetzalcoatlus northropi

For example, if a bird or plane is scaled up proportionally by a linear scaling factor of 4, it will weigh $4^3 = 64$ times as much but will have only $4^2 = 16$ times as much wing area. So each square foot of wing must support 4 times as much weight. Once you're up, you have to keep moving. To stay level, an airborne object moving forward must fly fast enough to maintain the lift on the wings. It turns out that the minimum necessary speed is proportional to the square root of the wing loading. Combining this fact with the first consideration, we conclude that the minimum speed goes up with the square root of the length. A bird scaled up by a factor of 4 must fly $\sqrt{4} = 2$ times as fast. (Hovering helicopters, hummingbirds, and insects maintain lift by moving their wings directly rather than relying on aerodynamic lift from forward motion.)

Take, for instance, a sparrow, whose minimum speed is about 20 miles per hour (mph). An ostrich is about 16 times as long as a sparrow, so the minimum speed for an ostrich would be $\sqrt{16} \times 20 = 80$ mph. Have you seen any flying ostriches lately? Heavy birds have to fly fast or not at all!

Of course, ostriches are not just scaled-up sparrows, nor are eagles (nor are airplanes). Larger flying birds have disproportionately larger wings than a sparrow so as to keep the wing loading down.

The largest animal ever to take to the air was *Quetzalcoatlus northropi,* a flying reptile that lived 65 million years ago. It had a wingspan of 36 ft, weighed about 100 lb, and was as tall as a giraffe. Recent research suggests that it might have been able to "fly" 10,000 miles or more, nonstop, by taking advantage of air currents.

You have to stay up, you have to keep moving—and you have to get up there. Here, basic aerodynamics imposes further limits. Paleontologists originally thought that *Q. northropi* weighed 200 lb and had a 50-ft wingspan. Even though that works out to about the same wing loading as for 100 lb and a wingspan of 36 ft, other considerations from aerodynamics show that a reptile of the larger size couldn't have gotten off the ground.

Self Check 6

The largest and heaviest aircraft in service today is the An-225—and we indeed mean "the" because there is only one! It was used to bring humanitarian equipment from Japan for Haiti, as well as—in a single flight—216,000 meals for U.S. military personnel. The plane has a wing area of 905 m^2 and a maximum takeoff weight of 1.3 million lb. Fully loaded, what would be its wing loading, in lb/ft^2?

Scaled to Fit

SPOTLIGHT 18.3

Big isn't always beautiful when it comes to the U.S. military's physical fitness tests.

Paul Vanderburgh, of the Department of Health and Sport Science at the University of Dayton, has spent more than a dozen years researching how a person's body mass affects performance on such tests, which consist of distance runs, push-ups, sit-ups, and abdominal crunches. The Arnold Schwarzeneggers of the world actually tend to score lower.

Vanderburgh emphasizes that some larger people (like Schwarzenegger) have more muscle, not more fat. Nevertheless, he and fellow researcher Todd Crowder found that scores for larger and heavier (though muscular) men and women are 15% to 20% lower than for their smaller and lighter counterparts. "A person's strength doesn't increase as fast as their size," explains student Liz Trouten. "The extra muscle that big people have doesn't make up for their size."

Vanderburgh noticed at the U.S. Military Academy that, even at similar fitness levels, smaller cadets tend to score higher on physical fitness tests than larger cadets. "Fitness testing is a big part of cadets' grade point averages, and the stakes are pretty high. The test results affect class rank and even a cadet's first assignment, so it matters a lot how well a cadet does."

For example, a larger cadet with a fitness test score of 256, which Vanderburgh compares to a grade of C+, may not be eligible for certain awards and assignments. However, a smaller cadet with a perfect score of 300 "would get lots of attention." But that doesn't mean the C+ cadet isn't worthy. "In fact, if these two cadets were scale models of each other, these two performances would be biologically the same, and they should receive the same score."

Photo by Spc. Hannah Frenchick, 20th Public Affairs Detachment

Vanderburgh gives another example using the scale-model approach. Take a woman who is 5 feet 5 inches tall, weighs 130 pounds, and scores a perfect 300 on the fitness test. If she were 5 feet 8 inches tall and 30% heavier, she would score only 250.

To compensate for this body mass "penalty," Vanderburgh and Crowder developed a correction factor, which multiplies the score by a number based on weight, "to place everybody on an even playing field."

This formula is similar to the Flyer Handicap, developed by Vanderburgh and colleague Lloyd Laubach. The handicap adjusts a runner's race time based on age and body weight. "A higher body weight is definitely a handicap for performance, whether it be running a marathon or military physical fitness tests," Laubach said. (A web calculator for the Flyer Handicap is at http://academic.udayton.edu/PaulVanderburgh/weight_age_grading_calculator.htm.)

Source: Information from the article "Scaled to Fit," by Kristen Wicker in the *University of Dayton Quarterly* (Winter 2006–2007), 21–22.

Keeping Cool (and Warm)

Area–volume tension is also crucial to an animal's thermal equilibrium. Both warm-blooded and cold-blooded animals gain or lose heat from the environment in proportion to body surface area.

Warm-Blooded Animals

A warm-blooded animal's basal metabolism, or rate of food intake needed to maintain body heat, depends primarily on its surface area, the temperature of its environment, and the insulation provided by its coat or skin. Other factors being equal, a scaled-up mammal scales up its food consumption with *surface area* (proportional to the square of the linear scaling factor), *not with volume* (proportional to its cube). For example, a mouse eats about half of its weight in food every day, while a human consumes only about one-fiftieth of its own weight, because the mouse has more surface area per unit volume.

Thus, the metabolic rate should be proportional to the surface area. Using proportionality notation, we can find how the metabolic rate changes with the mass of the animal. We know that mass is proportional to volume, which in turn is proportional to the cube of length, or

$$M \propto V \propto L^3$$

Taking each side to the one-third power, we have

$$M^{1/3} \propto V^{1/3} \propto L, \quad \text{so} \quad L \propto M^{1/3}$$

Meanwhile, the metabolic rate (call it R) is proportional to surface area, so

$$R \propto A \propto L^2, \propto (M^{1/3})^2 = M^{2/3}$$

So, based on area–volume tension, we would expect metabolic rate to scale as the two-thirds power of body mass. But it doesn't—instead, it scales as the *three-quarters* power of body mass; that is, $R \propto M^{3/4}$. The least-squares line (see Section 6.4 on page 260) through the points in the "mouse-to-elephant" curve of Figure 18.10 has a slope of 0.74, very close to three-quarters. (The logarithmic coordinates used in this graph are explained in Section 18.5.)

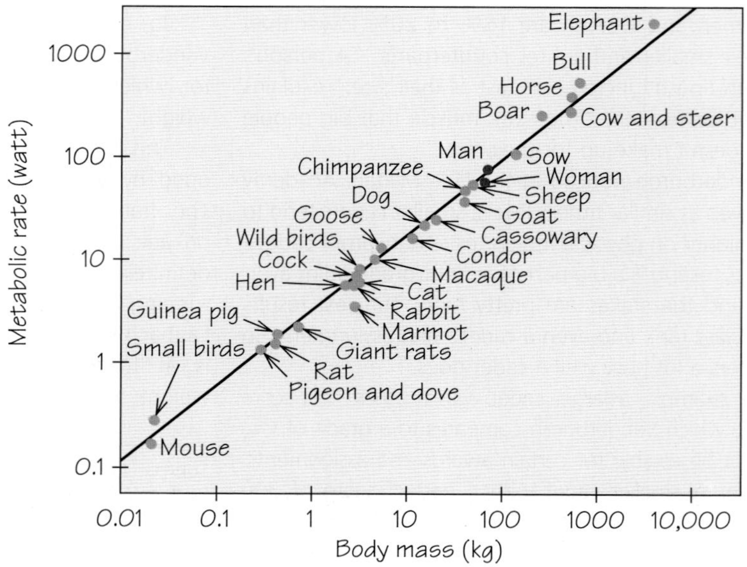

Figure 18.10 Metabolic rates for mammals and birds, when plotted against body mass on logarithmic coordinates, tend to fall along a single straight line. (Adapted from F. G. Benedict, *Vital Energetics: A Study in Comparative Basal Metabolism*, Carnegie Institute of Washington, Washington, DC, Publication No. 503, 1938.)

Why the difference from the two-thirds that area–volume tension would predict? And does the small difference between two-thirds and three-quarters matter? The answers lie in further considerations from geometry, physiology, and physics. A plant or animal needs a network of vessels (like the blood system) to transport resources to, and wastes away from, every part of the animal's tissues. The terminal branches (capillaries in the blood system) tend to be just about the same size in all species, for reasons of the physics involved.

To minimize the energy involved in transport, the network of vessels needs to be organized as a fractal-like tree, with smaller and smaller vessels branching off. (See Section 19.5 to learn about fractal patterns.) With same-size smallest branches at the ends, it turns out that minimization of energy demands that the metabolic rate scale as the three-quarters power of body mass. Fractal branching makes it possible for the circulatory system of a whale, with 10^7 times the mass of a mouse, to have only 70% more branches than the mouse has.

EXAMPLE 11 Dives

Sperm whales (and some other species) regularly hold their breath and stay underwater for an hour. Why can't we?

In part, it's because we aren't as large as whales. A mammal's breath-holding ability depends on how much air it can hold in its lungs, which is proportional to its mass. It also depends on how fast it uses up air—in other words, on its metabolic rate, which is proportional to the three-quarters power of its mass. Hence, the limit of duration of a dive should be proportional to the volume of the air in the lungs divided by the rate of using up the air, or

$$\frac{M}{M^{3/4}} = M^{1/4}$$

For a 90,000-lb sperm whale, this limit is proportional to $90{,}000^{1/4} \approx 17.3$, while the corresponding figure for a 150-lb human is $150^{1/4} \approx 3.5$. So the sperm whale should be able to hold its breath for about $17.3/3.5 \approx 5$ times as long. However, most humans cannot hold their breath for one-fifth of an hour (12 minutes)! (The record is 22 minutes; don't try it—the guy died the next time he attempted it.) This fact tells us the whale has special adaptations to make long dives possible. The stars of the 2005 film *March of the Penguins*, emperor penguins, weigh 80 to 90 lb but can dive for as long as 20 minutes. Their special adaptations are more blood per pound of body weight, an abundance of myoglobin (which can store oxygen) in their tissues, and the slowing of their heart rate during dives.

Self Check 7

Consider a dolphin weighing 450 lb. If a whale can hold its breath for a certain length of time, for what fraction of that length of time should a dolphin be able to hold its breath? What does your calculation assume?

Cold-Blooded Animals

Mammals and birds regulate their metabolism and maintain a constant internal body temperature. Cold-blooded animals, such as alligators and lizards, have a somewhat different issue. They absorb heat from the environment for energy, but they must also dissipate any excess heat to keep their temperatures below unsafe levels. The amount of heat that must be gained or lost is proportional to total volume because the entire animal must be warmed or cooled. But the heat is exchanged through the skin, so the rate is proportional to surface area.

Dimetrodon was a large, mammal-like reptile that roamed present-day Texas and Oklahoma 280 million years ago (see Figure 18.11a).

John Weinstein/Field Museum Library/Getty images

Picture Press/Alamy

Figure 18.11 (a) *Dimetrodon* may have evolved a sail to absorb and dissipate heat efficiently. (b) A toucan's bill helps it lose heat.

Dimetrodon had a great "sail," or fan, on its back. As an individual grew, and as the species evolved, the sail grew. But it did not grow according to geometric similarity, the kind of growth that is also called **proportional growth.**

Proportional Growth DEFINITION

Proportional growth is growth according to geometric similarity, where the length of every part of the organism enlarges by the same linear scaling factor.

Instead, the area of *Dimetrodon's* sail grew in proportion to the volume of the animal, a fact that strongly suggests to paleontologists that the sail was a temperature-regulating organ. Larger specimens of *Dimetrodon* didn't look like scaled-up smaller ones. We would say that the sail grew disproportionately compared with the rest of the animal. An individual twice as long would have 8 (2^3) times as much weight and volume and a sail with 8 times as much area. If it had grown according to geometric similarity, the sail would have been twice as high and twice as wide, and hence would have had only 4 times as much area.

Dimetrodon was a large animal, but heat regulation is even more important for small animals; like human babies, they can lose heat quickly because of their high ratio of surface area to volume. Paleontologists believe that birds evolved from dinosaurs and that feathers are modified reptilian scales. The wings of birds and insects may have evolved not for flight but as temperature-control devices. Birds that live in hot climates, such as the toco toucan (*Ramphastos toco*), tend to have large bills and use them to lose heat by increasing bloodflow to the bill (see Figure 18.11b).

Some scientists have speculated that African pygmies are small in part because a small body can better lose heat in the hot, humid climate of the Ituri Forest in the Congo, where pygmies live. The discovery announced in late 2004 of "hobbit-sized" people (1 m tall) who lived on the island of Flores in Indonesia 13,000 years ago suggests another explanation. Being marooned on the island with a self-limiting food supply (they hunted pygmy elephants) made large size—and a corresponding need for more calories—a disadvantage.

Other scientists have suggested that ancestors of human beings began walking on two legs in part to keep cool in a hot climate. Walking upright exposes less body area to the rays of the sun than walking on all fours and also reduces the amount of water needed by about one-half.

18.5 How We Grow

A large change of scale forces adaptive changes in materials or form. However, within narrow limits—in most cases, up to a factor of 2—creatures can grow according to geometric similarity. That is, they can grow proportionally, so that their shape is preserved. A striking example of such growth by a far greater factor is the chambered nautilus (*Nautilus pompilius*). Each new chamber that it adds to its shell is larger than,

Helping to Find Missing Children

SPOTLIGHT 18.4

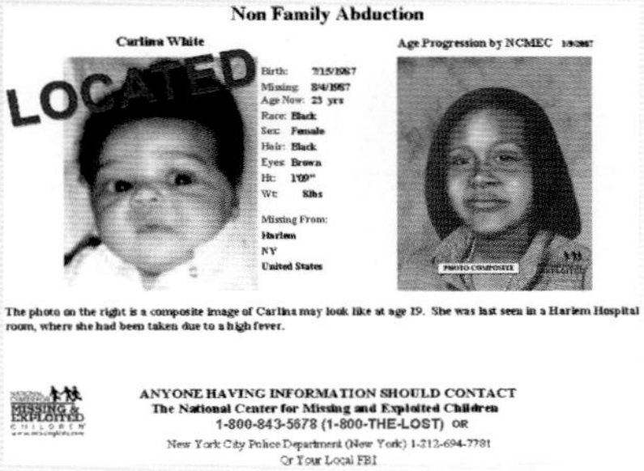

National Center for Missing and Exploited Children/EPA/Newscom

(a) Photograph at age 19 days of Carlina White, with age progression to age 19.

Julia Xanthos/NY Daily News via Getty Images

(b) Photograph of Carlina White at age 23 in January 2011.

What might a 19-day-old child look like 19 years later? The National Center for Missing and Exploited Children (NCMEC), in Arlington, Virginia, uses a computer and a more sophisticated version of the graph-paper technique to answer such questions. Computer age-progression specialists scan photographs of both the missing child and an older sibling or a biological parent at age 19. The face of the missing child is stretched, depending on age, to reflect craniofacial growth, and then merged with the image of the sibling or parent at 19 years old. The result is a rough idea of what the missing child may look like. As mathematicians and biologists refine their models of how faces change over time, this technique will improve. It may even become possible to gain an idea of how a child may look at age 40 or 65.

Carlina White, who was kidnapped at 19 days old, as a teenager began to realize that her "mother" was not her birth mother. Eventually, she contacted the NCMEC, which found three possible matches. In January 2011, at age 23, she was at last reunited with her birth mother, a happy resolution to the longest-known case of abduction by a stranger.

but geometrically similar to, the previous chamber and also similar to the shape of the shell as a whole—an *equiangular,* or *logarithmic,* spiral (see Figure 18.12).

Photodisc/Punchstock

Figure 18.12 A chambered nautilus shell.

Most living things grow over the course of their lives by a factor greater than 2. We've seen with *Dimetrodon* that a big specimen was not just a scaled-up small one. Nor is a human adult simply a scaled-up baby: A baby's head is relatively much larger than an adult's, and its arms are disproportionately shorter. In growth from baby to adult, the body does not scale up as a whole. Different parts of the body scale geometrically, each with a different linear scale factor. That is, a baby's eyes grow to perhaps twice their original size, while the arms grow by a factor of about 4.

Although the laws for growth can be much more complicated than for proportional growth (or even for the allometric growth that we discuss later), more sophisticated mathematics—for example, differential geometry, the geometry of curves and surfaces—permits analysis of complex and interlocking scalings. For a model of the process in which a baby's head changes shape to grow into an adult head, we can use graph paper: First, we put a picture of the baby's skull on graph paper. Then we determine how to deform the grid until the pattern matches an adult skull. (See Figure 18.13 and Spotlight 18.4.) The same idea lies at the heart of computerized "morphing," the process in which the face of one person can be changed smoothly into the face of another, with different scalings for different parts of the face.

Allometric Growth

If we measure the arm length or head size for humans of different ages and compare these measurements with body height, we observe that humans do not grow

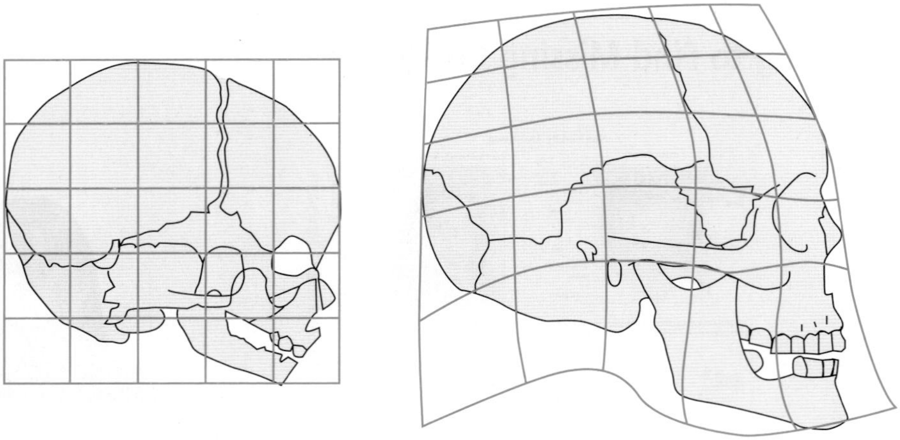

Figure 18.13 Modeling the changes in the shape of a human head from infancy to adulthood.

proportionally; that is, in a way that maintains geometric similarity. The head of a newborn baby may be one-third of the baby's length, but an adult's head is usually close to one-seventh of the individual's height. The arm, which at birth is one-third as long as the body, is by adulthood closer to two-fifths as long (see Figure 18.14a).

Graphing provides a way to test for differential growth. We plot body height on the horizontal axis and arm length on the vertical axis (see Figure 18.14b). A straight line would indicate proportional growth—that is, according to geometric similarity. We do get a straight line from 9 months (0.75 years) on up. But up to 9 months, we get a curve, which indicates that the ratio of arm length to height does not remain constant over the first year.

Algebra Review Appendix
Logarithms

Is there an orderly law by which we can relate arm length to height? Let's plot again, this time using a different scale (see Figure 18.14c). For this **base-10 logarithmic scale,** we mark off equal units, as usual. But instead of labeling the marked points with 0, 1, 2, 3, and so on, we label them with the corresponding powers of 10: $10^0 = 1$, $10^1 = 10$, $10^2 = 100$, $10^3 = 1000$, and so on, which are also called **orders of magnitude.** This is the procedure that we used to make Figure 18.10 on page 758. In that figure, the fact that the data cling close to a straight line indicates a power–law relationship between metabolic rate and body mass, which would not be obvious if we had graphed using ordinary scales instead of logarithmic scales.

Plotting a point on such a scale is not easy because the point midway between 1 and 10 on the logarithmic scale is not 5.5; rather, it is $10^{1/2} \approx 3$. Special graph paper marks smaller divisions and makes points easier to plot; paper marked with log scales on both axes is called **log-log paper,** while **semilog paper** has a logarithmic scale on just one axis. Many computer plotting packages can also produce logarithmic scales.

We could use a logarithmic scale for either height or arm length, or for both. When logarithmic scales are used for both, as in Figure 18.14c, the data plot closely to a straight line. Looking carefully at the figure, we can discern actually two different straight lines: a steeper one that fits early development (we will see shortly that it has slope 1.2), and a less steep one (with slope 1.0) that fits development after 9 months of age.

The change from one line to another after 9 months indicates a change in pattern of growth. The pattern after 9 months, characterized by the straight line with slope 1, is indeed proportional growth (sometimes called **isometric growth**). For the pattern before 9 months, the slope is 1.2. The fact that it is greater than 1 means that arm length is increasing relatively faster than height. This early growth also follows a definite pattern, called **allometric growth.**

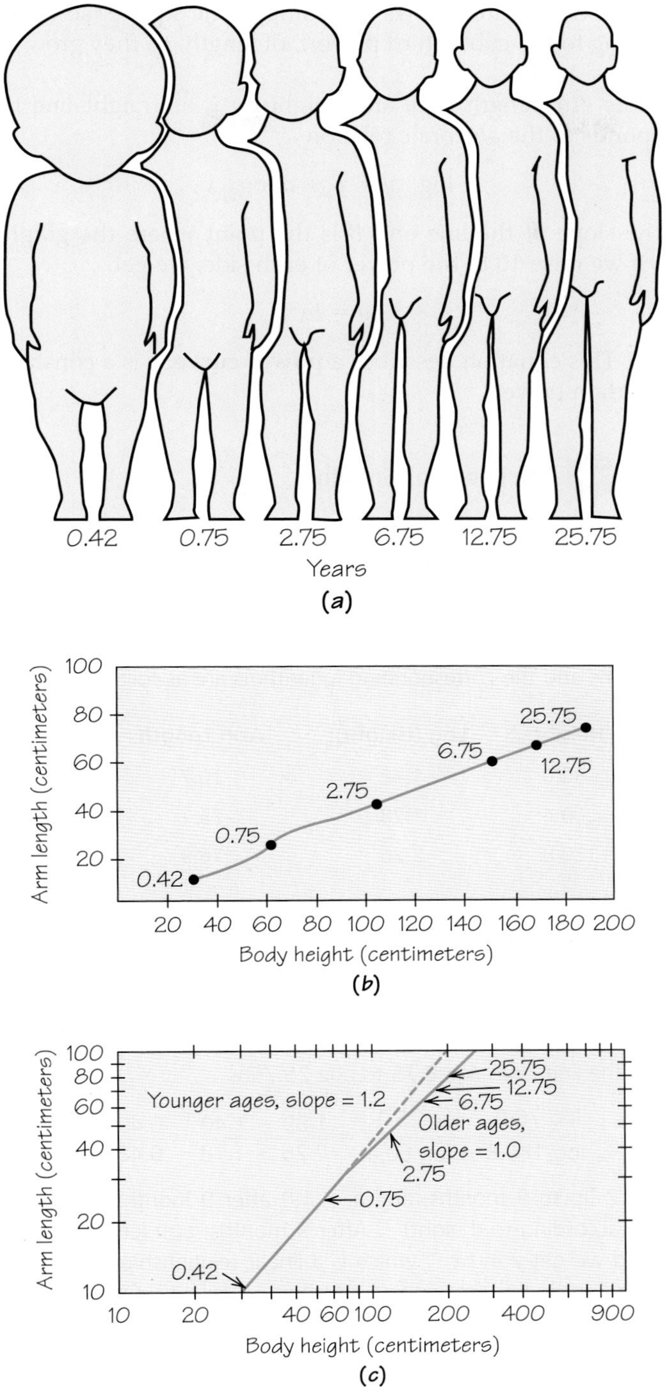

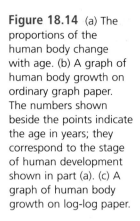

Figure 18.14 (a) The proportions of the human body change with age. (b) A graph of human body growth on ordinary graph paper. The numbers shown beside the points indicate the age in years; they correspond to the stage of human development shown in part (a). (c) A graph of human body growth on log-log paper.

Allometric Growth DEFINITION

Allometric growth is growth of the length of one feature at a rate proportional to a power of the length of another.

In geometric scaling, area grows according to the square (second power), and volume according to the cube (third power), of length, so they grow allometrically with length.

If we denote arm length by y and height by x, a straight-line fit on log-log paper corresponds to the algebraic relation

$$\log_{10} y = B + a \log_{10} x$$

where a is the slope of the line and B is the point where the graph crosses the vertical axis. If we raise 10 to the power of each side, we get

$$y = bx^a$$

where $b = 10^B$. This equation describes a **power curve:** y is a constant multiple of x raised to a certain power.

EXAMPLE 12 The Power of Growth

How do we arrive at those slopes of 1.0 and 1.2 in Figure 18.14?

You could use statistical software or your calculator to find the equation of the least-squares regression line (as discussed in Section 6.4 on page 260) through the points on the log-log plots. Here, we find approximate values for the slope a for each line from the coordinates of the points at the ends of the lines, for ages 0.42, 0.75, and 25.75. The observations and the corresponding logarithms are as follows:

Age	Height	Log (Height)	Arm Length	Log (Arm Length)
0.42	30.0	1.48	10.7	1.03
0.75	60.4	1.78	25.1	1.40
25.75	180.8	2.26	76.9	1.89

The slope for the line from age 0.42 to age 0.75 is the vertical change over the horizontal change in terms of log units:

$$\frac{\log 25.1 - \log 10.7}{\log 60.4 - \log 30.0} = \frac{1.40 - 1.03}{1.78 - 1.48} = \frac{0.37}{0.30} \approx 1.2$$

The slope for the line from age 0.75 to age 25.75 is

$$\frac{\log 76.9 - \log 25.1}{\log 180.8 - \log 60.4} = \frac{1.89 - 1.40}{2.26 - 1.78} = \frac{0.49}{0.48} \approx 1.0$$

So $a = 1.2$ up to 9 months, and $a = 1.0$ after 9 months. Up to 9 months, arm length grows according to $(\text{height})^{1.2}$. After 9 months, arm length grows according to $(\text{height})^{1.0}$, and we get $y = bx^{1.0}$, which is a linear relationship describing proportional growth—that is, growth according to geometric similarity. On ordinary graph paper, proportional growth appears as a straight line and allometric growth as a curve. On log-log paper, both patterns appear as straight lines.

Allometry was used by paleontologists to determine that all specimens (just six!) of the earliest bird, *Archaeopteryx,* are indeed of the same species, and that the puzzling minute fossil fish *Palaeospondylus* (found only in Scotland) was probably just the larval stage of a better-known fish. ■

In this chapter, we have explored the limitations on life imposed by dwelling in three dimensions. In Chapters 19 and 20, we will see that dimensionality also imposes surprising limits on artistic creativity in devising patterns.

Review Vocabulary

Allometric growth A pattern of growth in which the length of one feature grows at a rate proportional to a power of the length of another feature. (pp. 762, 763)

Area scaling factor The ratio of the area of any part of one figure to the corresponding part of another, equal to the square of the linear scaling factor. (p. 739)

Area–volume tension A result of the fact that as an object is scaled up, the volume increases faster than the surface area and faster than areas of cross sections. (p. 755)

Base-10 logarithmic scale A scale on which equal divisions correspond to powers of 10. (p. 762)

Crushing strength The maximum ability of a substance to withstand pressure without being crushed or deformed. (p. 750)

Density Mass per unit volume. (p. 747)

Geometric similarity Two objects are geometrically similar if they have the same shape, regardless of the materials of which they are made. They need not be the same size. Corresponding linear dimensions must have the same factor of proportionality. (p. 738)

Isometric growth Proportional growth. (p. 762)

Linear (length) scaling factor The number by which each linear dimension of an object is multiplied when it is scaled up or down; that is, the ratio of the length of any part of one of two geometrically similar objects to the length of the corresponding part of the second. (p. 738)

Log-log paper Graph paper on which both the vertical and the horizontal scales are logarithmic scales; that is, the scales are marked in orders of magnitude 1, 10, 100, 1000, . . . , instead of 1, 2, 3, 4, ... (p. 762)

Orders of magnitude Powers of 10. (p. 762)

Power curve A curve described by an equation $y = bx^a$, so that y is proportional to a power of x. (p. 764)

Pressure Force per unit area. (p. 747)

Problem of scale As an object or being is scaled up, its surface and cross-sectional areas increase at a rate different from its volume, forcing adaptations of materials or shape. (p. 754)

Proportional growth Growth according to geometric similarity, where the length of every part of the organism enlarges by the same linear scaling factor. (p. 760)

Semilog paper Graph paper on which only one of the scales is a logarithmic scale. (p. 762)

Volume scaling factor The ratio of a volume of any part of one figure to the corresponding part of another, equal to the cube of the linear scaling factor. (p. 740)

Weight Attractive force of a mass toward the center of the Earth, as measured at sea level. (p. 747)

Wing loading Weight supported divided by wing area. (p. 756)

Self-Check Answers

1. Nothing can be 40 times less than another quantity. Correct: The men are paid 40 times as much as the women; the women are paid one-fortieth as much as the men (in salaries from their pro teams).

2. 160 cm (rounded)

3. 8.9 L/100 km

4. 50,000 lb/ft^2

5. 4.3×10^6 lb/ft^2

6. 133 lb/ft^2

7. About one-fourth as long; the calculation assumes that a dolphin is a "scaled-down" whale.

Skills Check

1. A quarter and a nickel are

(a) geometrically similar because they both have the same shape (a squashed cylinder).

(b) not geometrically similar because they have different denominations.

(c) not geometrically similar because the quarter has a greater diameter while the nickel is thicker.

2. A scale model of a carillon stands 10 in. tall, and the actual carillon stands 100 ft tall. The linear scaling factor of the carillon compared with its model is ____________.

3. A popular scale for diecast toy model cars is 1 to 18. If the actual car has a length of 15 ft, what should be the length of the model?

(a) 10 in.

(b) 15 in.

(c) 18 in.

4. You want to enlarge a 2-in. by 3-in. photograph to a 5-in. by 7-in. copy. Assuming that the cost of photographic paper is proportional to its area and that 2-in. by 3-in. reprints cost 40 cents each, you would expect to pay ________________ for the large copy.

5. If a medium 10-in. pizza costs $10 and a similar 12-in. pizza costs $14, which costs less per square inch?

(a) The 10-in. pizza

(b) They are about the same price per square inch.

(c) The 12-in. pizza

6. An artist plans to melt 1000 pennies and re-form a larger penny proportional in all dimensions to an ordinary penny. The linear scaling factor of the large penny compared with the ordinary penny is ________.

7. The actor Elijah Wood, who plays Frodo Baggins in the movie version of J. R. R. Tolkien's *The Hobbit,* is 5 ft 6 in. tall, but his character is barely 4 ft tall. Put correctly, how much shorter is Frodo than Wood?

(a) 138% shorter

(b) 73% shorter

(c) 27% shorter

8. In December 2014, the telescope on NASA's Kepler spacecraft discovered a planet 180 light years away that has a diameter 2.5 times that of Earth. Assuming the same density as Earth, the newly discovered planet would have a mass about _______ times that of Earth.

(a) 2.5

(b) 6

(c) 15

9. Which of the following is precise language to express the comparison that Ahmed has $100 and Burke has $50?

(a) Ahmed has 100% more than Burke.

(b) Burke has 200% less than Ahmed.

(c) Burke has twice as less as Ahmed.

10. The Apple iPhone 6 Plus features a screen with 1920 physical pixels vertically and 1080 pixels horizontally, the same as "Full HDTV" high-definition video. For the iPhone screen to be geometrically similar to a 720p high-definition (just "HD") screen, which has 720 pixels horizontally, the high-definition screen would have to have _______ pixels vertically (which it does).

11. A kilometer is approximately equal in length to

(a) 5 mi.

(b) 3 mi.

(c) 3/5 mi.

12. The distance for the marathon race was established in 1921 as 42.195 km. Converted to the U.S. customary system, the distance is ______________________.

13. A 2-liter bottle contains approximately

(a) 2 quarts.

(b) 1 gallon.

(c) 10 pints.

14. A weight of 130 lb is approximately the same as ________ kg.

15. Coffee costs about $10 per pound in the United States. If a Canadian dollar (CAD$) exchanges for U.S. $0.86, what is the approximate equivalent cost in Canadian dollars of 500 g of coffee?

(a) CAD $9

(b) CAD $10

(c) CAD $13

16. A common speed limit in European neighborhoods is 30 km/h, which is about __________ mph.

17. Which of the following is a unit in the metric system?

(a) minimeter

(b) miniliter

(c) millimeter

18. A mile is approximately ______ km.

19. A speed of 60 mph is approximately

(a) 3.2×10^3 mm/min.

(b) 90 ft/s.

(c) 50 m/s.

20. A sculpture weighs 140 lb and is supported by three legs, each of which is 0.5 in. by 0.5 in. by 2 in. high. The legs exert a pressure of approximately _______ $lb/in.^2$ on the floor.

21. The mass of a scaled-up object goes up with the

(a) square root of the volume.

(b) square of the linear scaling factor.

(c) density.

22. In comparing flight speeds of birds, an analysis of wing loading leads to the conclusion that ___________ birds need to fly faster than ___________ birds.

23. Scaling analysis leads to the general conclusion that

(a) smaller animals can jump much higher than larger animals.

(b) larger animals can jump much higher than smaller animals.

(c) all animals can jump to about the same height.

24. If an object is scaled linearly so that its volume grows to 8 times its original volume, its surface area is scaled to ______ times its original surface area.

25. A large change in scale forces a change in

(a) form.

(b) form and materials.

(c) form or materials.

26. Assuming that a catfish maintains the same shape and proportions as its grows, and that a catfish 8 in. long weighs about 1 lb, a 2-lb catfish would be about _____ in. long.

27. Metabolic rate scales as the _______ power of mass.

(a) one-third

(b) two-thirds

(c) three-quarters

28. The base-10 logarithm of 1000 is ____.

29. The population of Mexico is expected to grow 1.2% per year from 2015 through 2018. The population will be growing

(a) proportionally.

(b) allometrically.

(c) by a constant amount each year.

30. Allometric growth is growth of the length of one feature at a rate ________ the length of another.

Chapter 18 Exercises

Challenge Discussion

Most of the exercises require scientific functions calculator.

18.1 Geometric Similarity

1. Your digital camera probably takes pictures with an aspect ratio of 4 to 3, meaning that the longer side is 4/3 times as long (in pixels) as the shorter side. For example, you probably can take a "small" picture with 640 pixels by 480 pixels, or perhaps a "large" picture with 2592 pixels by 1944 pixels (for a total of 2592 × 1944 pixels, or just a little more than 5 megapixels). Photographic prints from your digital camera are available in various sizes of paper, quoted in inches: 4 × 6, 5 × 7, and 8 × 10.

(a) Which of the paper sizes, if any, is geometrically similar to the original digital image?

(b) If a 4 × 6 print is made by scaling the shorter side of the digital image to be exactly 4 in., how long should the longer side of the image be on the print?

(c) If a 4 × 6 print is made by scaling the longer side of the digital image to be exactly 6 in., how long should the shorter side of the image be on the print? (*Hint:* The paper isn't wide enough!)

For Exercises 2–3, refer to the following: The area of a circle of radius r is πr^2. Expressed in terms of the diameter, $d = 2r$, the area is $\pi d^2/4$. If we apply a linear scaling factor L to the diameter, then the area of the scaled circle—just as is the case for a square—changes with L^2, the square of the linear scaling factor.

2. A natural application of this idea is to pizza. The prices at Vince's pizza restaurant in Beloit, Wisconsin, are \$7.57, \$8.71, \$9.49, \$11.13, and \$12.63, respectively, for small (10-in.), medium (12-in.), large (14-in.), extra-large (16-in.), and XX-large (18-in.) cheese pizzas.

(a) What is the linear scaling factor for an XX-large pizza compared with a small one?

(b) How many times as large in area is the extra-large pizza compared with the small one?

(c) How much pizza does each size give per dollar? What "hidden" assumptions are you making about how the pizzas are scaled up?

(d) The corresponding prices for a pizza with six toppings are \$12.35, \$15.49, \$18.42, \$21.60, and \$25.74. Is there any size of these for which you get more pizza per dollar than some size of the cheese pizzas? (Curiously, all the prices are, to the nearest cent, exactly \$0.10 higher than four years earlier! That is an example of arithmetic scaling.)

3. The *NBC Nightly News* on June 21, 2010, featured a story about food portions, stating that dinner plates were 9 in. in diameter in the 1960s but now they are 12 in. in diameter, "making room for one-third more food."

(a) What is the linear scaling factor for a 12-in. plate compared with a 9-in. plate?

(b) How many times as large in area is a 12-in. plate compared with a 9-in. plate?

(c) What percentage greater is the area of a 12-in. plate compared with a 9-in. plate?

4. Dollhouses and their furnishings are usually built to a scale of exactly 1 in. to 1 ft, meaning that an item 1 ft long in a real house is 1 in. long in a dollhouse.

(a) What is the linear scaling factor for a dollhouse?

(b) If a dollhouse were made of the same materials as a real house, how would their weights compare?

5. Shellac O'Lean, a freshman on your school's football team, is 6 ft tall and weighs 150 lb. Since he would rather play basketball, he plans to grow to 7 ft 1 in., the same height as Shaquille O'Neal, his idol. If Shellac scales geometrically as he grows, how much can he expect to weigh? (O'Neal weighs 325 lb.)

6. At our house, we have a 10-in. frying pan and a 12-in. frying pan; the 12-in. one weighs a lot more, cooks food more slowly, and never gets as hot. Suppose that a 10-in. frying pan weighs 1 lb, apart from its handle. How much would a geometrically similar 12-in. frying pan weigh? How much would it weigh if it had the same thickness of metal as the 10-in. pan?

7. The human figures in Lego sets are 4 cm tall (without hats or helmets).

(a) What is the linear scaling factor of a Lego figure if it represents a human who is 160 cm tall?

(b) How does the volume of a real human compare with the volume of a Lego figure?

(c) The car in one Lego set is 10 cm long. Using the linear scaling factor in part (a), how long would a real car be?

8. Recent dollar coins (Presidential, Susan B. Anthony, Sacagawea) have been largely rejected by the public, which finds them too small and too light. Suppose that you are appointed to the U.S. Citizens Coinage Advisory Committee and commissioned to design a new $5 coin. (Whom should it depict?) The sole requirement is that it be made of the same material as the quarter but weigh 4 times as much. A quarter can be described geometrically as a circular cylinder approximately 24.26 mm in diameter and 1.75 mm thick. Because your new dollar should weigh 4 times as much, it needs to have 4 times the volume of a quarter. [The formula for the volume of a cylinder is $h(d/2)^2$, where h is the height and d is the diameter.]

United States Mint

Sacagawea dollar.

United States Mint

Presidential series dollar. Coins shown actual size (26.50 mm in diameter, 2.00 mm thick).

(a) A member of your public advisory panel suggests just doubling the diameter and doubling the thickness. What do you tell this individual, in the most diplomatic terms?

(b) If you double the diameter, how thick does the coin need to be?

(c) Another member feels that the result of part (b) would be inconveniently large and proposes instead to scale up the quarter proportionally. (She studied a previous edition of this book.) What would the dimensions be for this coin?

9. Criticize the following statement and write a correct version.

Murders by firearms in Britain are 30 times fewer per capita than in the US.

—ragingbull.com/forum/topic/449338

10. Criticize the following statement and write a correct version.

Energizer Ultimate Lithium ... designed to last up to 9x longer, which means up to 9x less waste.

—Formerly stated on the Energizer website

11. Criticize the following statement and write a correct version.

Cadmium chloride is a nasty chemical ... used as a coating for thin-film solar cells because it increases the efficiency of converting sunlight to energy.... [In tests] magnesium chloride ... yielded comparable efficiency.... Magnesium chloride is also nontoxic, abundant and costs about 300 times less than cadmium chloride.

—*Scientific American* 311(3) (September 2014): 20

12. Criticize the following statement and write a correct version.

Antibiotics for children that are covered by private insurance cost five times more in the United States than in the United Kingdom.

—consumer.healthday.com/public-health-information-30/health-cost-news-348/briefs-emb-12-19-antibiotic-costs-pharmacotherapy-bumc-release-batch-1071-683273.html

13. Abuses of the language of comparison aren't hard to find. For example, the phrase "times less than" occurs more than 1 million times in documents on the Internet. Search on the Internet to find an abuse of "times" and "less than" together, figure out what the author meant to say, and write it in correct language.

14. The phrase "times more than" occurs suprisingly more frequently on the Internet than "times less than"—more than 29 million times. Search on the Internet to find an abuse of "times" and "more than" together, figure out what the author meant to say, and write it in correct language.

18.2 How Much Is That in . . . ?

15. In 1991, Edward N. Lorenz, a meteorologist who was an early researcher into chaos and dynamical systems (discussed in Chapter 23), received the Kyoto Prize in Basic Sciences, consisting of a gold medal and 45 million Japanese yen (¥). If US$1 = ¥125 at the time, what was the value of the cash award in 1991 U.S. dollars? (In Chapter 21, we show how to convert such an amount to its value in today's dollars.)

For Exercises 16–20, refer to the following. Energy can be thought of as power applied over time. Power can be measured in kilowatts and energy in kilowatt-hours (kWh), where 1 kilowatt-hour is 1000 watts of power exerted for 1 hour. For example, a 40-watt bulb burning for 24 hours consumes 960 watt-hours $\approx$ 1 kWh. Other forms of energy, such as food or gasoline, can be converted to an equivalent in kWh.

16. An average adult in the U.S. consumes about 2500 food calories per day; one food calorie has the same energy as 1.16×10^{-3} kWh. A gallon of gasoline contains 33.7 kWh of energy. Compare the energy consumption of a car that gets 30 mpg and is driven about 30 mi/day (say, 12,000 mi/yr) with the food energy consumption of its driver. What do you conclude?

17. The Toyota Prius recovers energy to its battery when the car brakes. In the "classic" Prius model (2002–2003), for each 50 watt-hours (Wh) of energy recovered, an icon appears on the dashboard screen. A gallon of gasoline contains 33.7 kWh of energy. Suppose gasoline costs $3.00/gal. How many icons would have to appear before the car saves $1 in gasoline cost? (We neglect consideration of the relative efficiency of the car in converting gasoline or recovered electrical energy into mechanical motion.)

18. About 435×10^3 km^2 of U.S. land could be devoted to wind energy ("without raising too many hackles," according to Englishman David J. C. Mackay, *Sustainable Energy—Without the Hot Air,* UIT Cambridge, 2009, p. 234). The windmills would generate on average 1.2 W/m^2. (The hardware required would be 100 times as much as is currently installed.)

(a) How much is that land area in square miles?

(b) How much electricity would be produced per year?

(c) Suppose instead that all that area were covered with solar cells producing electricity at an annual average of 1600 kWh/m^2. How much electricity would be produced per year?

(d) How much energy per day would that amount to per person in the United States, in kWh/d? (The United States has about 315 million people; over all forms of energy usage throughout society—manufacturing, transport, heat, electricity and its production—Americans average the equivalent of 250 kWh/d per person.)

For Exercises 19–20, refer to the following. How can an all-electric car get 99 mpg? The figure below shows the EPS fuel economy label for an all-electric car. The fuel economy is measured in miles per gallon gasoline equivalent (MPGe). A gallon of gasoline equivalent means the amount of alternative fuel (electricity, natural gas, or hydrogen) that contains the same energy as in a gallon of gasoline.

Source: Environmental Protection Agency

The fuel economy label cites how many kilowatt-hours (kWh) of energy the vehicle uses per 100 miles—namely, 34 kWh. That number in turn is converted to MPGe via the equation

$$\text{MPGe} = \frac{\text{kWh content of 1 gallon of gasoline}}{\text{kWh consumed per mile}} = \frac{33.705 \text{ kWh/gal}}{\text{number of kWh/mi}} = \frac{3370.5 \text{ kWh/gal}}{\text{number of kWh/100 mi}}$$

Thus, in the label shown, the number of kWh/100 mi is 34, so the MPGe = 3,370.5/34 ≈ 99 mi/gal.

19. The 2015 Kia Soul Electric vehicle uses a combined city/highway 32 kWh/100 mi.

(a) What does the electricity cost to drive it 15,000 mi/yr at a cost of $0.12/kWh?

(b) What is the car's MPGe?

(c) The fuel economy labels in Great Britain for electric cars give just the energy consumption in mi/kWh. For example, the 2012 Nissan Leaf is rated at 3.5 mi/kWh. What is the corresponding MPGe?

20. Light-duty vehicles (passenger cars and trucks) travel about 2.66 trillion miles per year. Suppose that all of them were converted to run on electric power at 34 kWh/100 mi. How much electricity would be required annually?

21. In mid-2015, the cost of mailing a letter from the United States to the rest of the world was $1.15. How much was that in euros (€), the currency of the European Union (EU) in mid-2015, when the exchange rate was €1 = $1.10? (For comparison, the cost then of a letter to the United States from the EU ranged from €0.51 to €1.70, depending on the country.)

22. The cost of mailing a letter from the United States to Canada in mid-2015 was USD $1.15. How much was that in Canadian dollars when the exchange rate was USD $1 = CAD $1.24? (The postage cost from Canada to the United States was CAD $1.10.)

23. In Germany, the fuel efficiency of cars is measured in liters of fuel per 100 km (L/100 km). A typical average in a compact station wagon is 7.3 L/100 km. What is that in miles per gallon (mpg)?

24. According to EPA ratings, the highest city gas mileage for non-electric 2015 cars was with the hybrid Toyota Prius *c*, at 53 mpg in the city. How many liters of gasoline does such a Prius use to travel 100 km in the city?

25. Consider a real locomotive that weighs 88 tons and an HO-gauge scale model of it, for which the linear scaling factor is 1/87.

(a) How much would an exact scale model weigh in tons?

(b) What assumptions are involved in your answer to part (a)?

(c) How much would an exact scale model weigh in pounds?

(d) How much would an exact scale model weigh in kilograms?

(e) How much would an exact scale model weigh in tonnes (1 tonne = 1 metric ton = 1000 kg)?

26. What's wrong in the following quotations?

(a) *President Bush visited California, where 12 forest fires have charred more than 700,000 square miles.*

—Steve Stadelman, WTVO television news, Channel 17, Rockford, Illinois, October 2007. (Curiously, exactly the same number, 700,000, also appeared in news reports for California fires in 2014, 2011, 2003, 2000, and 1987.)

(b) *The population of the USA has topped 300 million.... If current trends continue, it is expected to reach 400 billion by 2043. This makes it an acceleration of growth.*

—*Significance* 3(4) (December 2006): 146

27. Gasoline is sold in the United States by the U.S. gallon and in Europe by the liter (1 U.S. gal = 231 in.^{3}; 1 L= 1000 cm^3). What was the equivalent cost, in U.S. dollars per U.S. gallon, for gasoline in Germany priced in euros at €1.59 per liter, when €1 = $1.24, in December 2014?

28. Consider which uses more energy: your house—with heating, cooling, electric, and perhaps gas for cooking—or the cars in the household (say, 2). A common measure of energy is 1 BOE, one *barrel of oil equivalent,* the approximate energy released by burning one barrel of crude oil, for which we have

$$1\text{BOE} \approx 5800 \text{ ft}^3 \text{ natural gas} \approx 1.7 \text{ MWh electricity} \approx 40 \text{ gal gasoline}$$

(a) Suppose that in 1 year, your household uses 7000 kWh of electricity plus 700 therms of natural gas (1 therm ≈ 100 ft^3) for heating, and that the household has two cars that get an average of 25 mpg and are each driven 15,000 miles per year. Which uses more energy: your house or the cars?

(b) Recalculate your answer to part (a) to take into account that conversion of natural gas, oil, or coal into electricity is at best only 40% efficient, while natural gas for heating can be about 90% efficient.

29. Before 1933, the United States issued gold coins for general circulation, the highest denomination being \$20. (The U.S. mint still produces gold coins but for collecting and investment purposes.) Suppose that you want to make gold coins available again, with the value of the coin the same as the value of the gold in it. Such a coin would be round and have a thickness of 1 mm (that is half the thickness of current \$1 coins and thinner than a dime). Gold has a density of 19.3 g/cm^3; and the volume of a cylinder is $V = \pi r^2 h$, where r is the radius and h is the height (thickness). At a price of \$1200 per ounce, what would be the diameter of a \$20 gold coin?

18.3 Big Stuff

30. The weight of a 1-ft cube of steel is 500 lb. What is the pressure on the bottom face in

(a) pounds per square inch?

(b) atmospheres (1 atm = 14.7 lb/in.^{2})?

31. The express elevators to the observation deck of One World Trade Center in New York City travel a height of 1293 feet in 47 seconds, reaching a speed of 2000 ft/min.

(a) How much is 2000 ft/min in miles per hour?

(b) How much is 2000 ft/min in meters per second?

32. As is the case for One World Trade Center, elevators in others of the world's tallest buildings usually take less than 1 minute to get to the top. How fast, in meters per second, would the elevators in a mile-high building have to be to achieve that standard?

33. At birth, the average human female baby weighs 7 lb and is 19 in. in length (height). At 15 months of age, the average weight is 21 lb. What would you expect the height to be then? (Thanks for the idea to Shai Simonson, Stonehill College.)

For Exercises 34 and 35, refer to the following. A mature gorilla weighs 400 lb and stands 5 ft tall; its two feet combined have an area of about 1 ft^2.

34. (a) Give an estimate of the gorilla's weight when it was half as tall.

(b) What assumptions are involved in your estimate?

(c) When the gorilla is standing, what is the pressure on its feet in pounds per square inch?

35. Suppose that King Kong is a gorilla scaled up with a linear scaling factor of 10.

(a) How much does the King weigh?

(b) What is the pressure on the King's feet in pounds per square inch?

36. You may want a waterbed, but waterbeds are not allowed in your building. Apart from the danger of flood if the bed should puncture or leak, the weight is an issue.

(a) Suppose that a queen-size waterbed mattress is 80 in. long by 60 in. wide by 12 in. high, and water weighs 1kg/L. How much does the water in the mattress weigh in pounds?

(b) If the weight of the mattress and frame is carried by four legs, each 2 in. by 2 in., what is the pressure, in pounds per square inch, on each leg?

(c) How does the pressure on the legs of the waterbed compare with the pressure that a person exerts on his or her feet—for example, a 130-lb person with a total foot area of about one-quarter of a square foot in contact with the ground?

37. If you aren't allowed to have a waterbed, how about a spa (hot tub)? Find the weight of the water in a spa that is in the shape of a cylinder 6 ft in diameter and 3.5 ft deep. (*Hint:* The volume of a cylinder is $\pi r^2 h$, where r is the radius and h is the height.)

38. What does the largest giant sequoia tree (named "Hyperion") weigh? Model the tree as a (very elongated) cone. Assume that the tree is 379 ft high, has a circumference of 40 ft at the base, and that the density of the wood is 31 lb/ft^3. (The volume of a cone of height h and radius r is $\frac{1}{3}r^2 h$.)

39. A 6-ft-tall indoor holiday tree needs four strings of lights to decorate it. How many strings of lights are needed for an outdoor tree that is 30 ft high? (Thanks for the idea to Charlotte Chell, Carthage College, Kenosha, Wisconsin.)

For Exercises 40 and 41, refer to the following. An ancient measure of length, the cubit, was the distance from the elbow to the tip of the middle finger of a person's outstretched arm. So the length of a cubit depended on the person, though there was some attempt at standardization. Most estimates place the cubit between 17 and 22 in.

40. According to Genesis 6:15, Noah was to build an ark: "The length of the ark shall be three hundred cubits, the breadth of it fifty cubits, and the height of it thirty cubits." How does the volume of such an ark compare with that of a U.S. suburban house totaling 2000 ft^2, with 8-ft ceilings?

41. Goliath [of David and Goliath, as related in the Bible (I Samuel 17:4)] was "six cubits and a span." A "span" was originally the distance from the tip of the thumb to the tip of the little finger when the hand is fully extended, about 9 in. What range of heights would this indicate for Goliath in feet and inches? In centimeters?

For Exercises 42–44, refer to the following. The body mass index (BMI) is the basis for the National Heart, Lung, and Blood Institute's weight guidelines. BMI is body weight (in kilograms) divided by the square of height (in meters). A BMI of 25 through 29 is considered "overweight"; a BMI of 30 or over is considered "obese." Some 55% of American adults have a BMI of 25 or above. (*Note:* BMI is not a useful measure for young children, pregnant or breastfeeding women, the frail elderly, or very muscular people.) For practice with this concept, calculate your own BMI.

forestpath/Shutterstock

42. Calculate the BMI for a woman 160 cm tall who weighs 65 kg. Is she overweight according to the institute's guidelines?

43. Suppose that weight and height are measured instead in U.S. customary units of pounds and inches. We can still calculate body weight divided by the square of height using these units. What conversion factor is necessary to convert this number to the BMI?

44. Calculate the BMI of a man who is 6 ft 2 in. tall and weighs 217 lb.

18.4 Dimension Tension

45. (a) Refer to Example 10 (page 756). What would you expect an individual *Q. northropi* to weigh if it had half the wingspan of an adult?

(b) If an individual weighed half as much as an adult, what would you expect its wingspan to be?

46. In the children's story *Peter Pan,* Peter and Wendy can fly. We may suppose that they are 4 ft tall, so they are about 8 times as tall as a sparrow is long. What should their minimum flying speed be?

47. Icarus of Greek legend escaped from Crete with his father, Daedalus, on wings made by Daedalus and attached with wax. Against his father's advice, Icarus flew too close to the sun; as a result, the wax melted, the wings fell off, and he plunged into the sea and drowned. What must have been his minimum cruising speed? What assumptions does your answer involve?

48. A typical ultralight aircraft (with pilot on board) weighs about 400 lb, has a wingspan of 28 ft, and has a maximum speed of 65 mph. Also, solar and human-powered aircraft, such as the Gossamer Condor and other superlightweight planes, fly more slowly than Example 10 (page 756) predicts; for instance, the Gossamer Condor is far longer than an ostrich, but it flies at only 12 mph. How are these aircraft possible?

49. Justify the claim in Example 10 on page 756 that a *Q. northropi* weighing 200 lb with a wingspan of 50 ft would have had the same wing loading as one weighing 100 lb with a wingspan of 36 ft.

50. Jonathan Swift's Gulliver traveled to Lilliput, where the Lilliputians were human-shaped but only about 6 in. tall. In other words, they were geometrically similar in shape to ordinary human beings but only one-twelfth as tall. What would a Lilliputian weigh? Are Lilliputians ruled out by the size–shape and area–volume considerations in this chapter? If you think they are, what considerations do you find convincing? If not, why not?

51. The cult movie *Them* (1954) features enormous ants (8 m long by 3 m wide). We can investigate the feasibility of such a scaled-up insect by considering its oxygen consumption. A common ant, 1 cm long, needs 24 milliliters (mL) of oxygen per second for each cubic centimeter of its volume. Because an ant has no lungs, it absorbs oxygen through its "skin" at a rate of 6.2 mL per second per square centimeter. Suppose that the tissues of a scaled-up ant would have the same need for oxygen for each cubic centimeter, and that the ant's skin could absorb oxygen at the same rate as a common ant.

(a) Compared with a common ant, how many times as large is an enormous ant's length, surface area, and volume?

(b) What proportion of such an ant's oxygen need could its skin supply?

(c) What can you conclude about the existence of such insects? (Adapted from George Knill and George Fawcett, Animal form or keeping your cool, *Mathematics Teacher,* May 1982, 395–397.)

For Exercises 52–55, refer to the following. Maybe some trees could grow to a mile high, but they just don't live long enough to have the chance. In this problem, we try to determine how fast the height of a tree increases. We can measure indirectly how much mass the tree adds in a year by the area of the annual tree ring added. Here are two relevant facts:

- As you may have noticed from stumps, as a tree grows older, its annual rings get less wide. Although the width of the ring varies somewhat from year to year with the amount of rainfall and other factors, the total *area* of each annual ring is roughly the same over the years, meaning that *the tree adds roughly the same amount of mass each year.* Call that amount a; the mass M of the tree is at, where t is its age in years.
- Over a large range of tree sizes and tree species, the diameter d of a tree of a species is approximately proportional to the three-halves power of the height h of the tree. (Different species have different constants of proportionality.) Thus, $d \propto h^{3/2}$ (which is shown in Exercise 52).

Now, if we assume that the bulk of the mass of the tree is in the trunk, and if we model the trunk either as a long cylinder or as a thin cone, the mass is proportional to the volume, so $M \propto d^2h$. Then

$$at = M \propto d^2h \propto (h^{3/2})^2\, h = h^4$$

so $h \propto t^{1/4}$. In other words, *the tree grows in height as the fourth root of its age.*

52. Suppose that a tree grows to 10 m in 15 years. How tall will it be (if it lives long enough) when it is 60 years old?

53. How long would it take the tree in Exercise 52 to grow to be 40 m tall?

54. Giant sequoias can reach 100 m after about 1000 years. If it could keep on growing at the same rate of its addition of mass, how long would it take a giant sequoia 100 m tall to grow to 200 m?

55. The branching of trees is similar to the branching of circulatory systems in the bodies of animals. For similar reasons, the area of the cross section of the tree at its base scales as the three-fourths power of the tree's mass—that is, $A \propto M^{3/4}$. Assume that most of the mass is in the trunk and model the tree either as a tall cylinder ($V = \pi r^2 h$) or an elongated cone ($V = \pi r^2 h/3$). Show that the diameter d of a tree is approximately proportional to the three-halves power of the height—that is, $d \propto h^{3/2}$.

18.5 How We Grow

56. Listed below are the numbers of species of reptiles and amphibians on some Caribbean islands, together with the approximate areas of the islands. (Suggested by Florence Gordon of the New York Institute of Technology, with contributions from Kevin Mitchell and James Ryan of Hobart and William Smith Colleges, Geneva, New York. This table is adapted from Tables 15 and 16 in P. J. Darlington, *Zoogeography: The Geographic Distribution of Animals,* Wiley, New York, 1957, pp. 483–484.)

Island	Area (mi^2)	Species
Redonda	1	3
Saba	4.9	5
Montserrat	40	9
Trinidad	2000	80
Puerto Rico	3400	40
Jamaica	4500	39
Hispaniola	30,000	84
Cuba	40,000	76

(a) Plot the number of species versus area on ordinary graph paper and then on log-log graph paper. If you don't have log-log paper available, use a calculator or spreadsheet to take the logarithms ($\log_{10}$) of all the numbers and then graph logarithm of number of species versus logarithm of area on ordinary graph paper. (*Note:* Trinidad is an outlier from the general pattern; see Chapter 6.)

(b) Is the relationship that you graphed in part (a) proportional? Is it allometric?

(c) What would be the expected number of species on an island of 400 mi^2?

(d) For each 10-fold increase in the island's size, what happens to the number of species, approximately?

57. The accompanying table lists the weights and wingspans of some birds and of some fully loaded airplanes. (Idea and most of the data contributed by Florence Gordon, New York Institute of Technology.)

(a) Use a calculator or spreadsheet to take the logarithms ($\log_{10}$) of all the numbers and then graph logarithm of weight versus logarithm of wingspan on ordinary graph paper.

(b) For the birds, is the relationship that you graphed in part (a) proportional? Is it allometric? How about for the planes?

Birds	Weight (lb)	Wingspan (ft)
Crow	1	2.9
Harris hawk	2.6	3.2
Blue-footed booby	4	3
Red-tailed hawk	4	4
Horned owl	5	5
Turkey vulture	6.5	6
Eagle	12	7.5
Golden eagle	13	7.3
Whooping crane	16.1	7.5
Vulture	18.7	9.3
Condor	22	9.9
Quetzalcoatlus northropi	100	36
Planes		
Boeing 737	117,000	93
DC9	121,000	93.5
Boeing 727	209,500	108
Boeing 757	300,000	156.1
Boeing 707	330,000	145.7
DC8	350,000	148.5
Howard Hughes's "Spruce Goose"	400,000	320.9
DC10	572,000	165.4
Boeing 747	805,000	195.7
Boeing 747-400	895,000	212.6
Anton An-225	1,323,000	290.2

(c) Does the same relationship of wingspan to weight seem to hold for birds and planes?

(d) A 25-million-year-old fossil recently discovered in South Carolina is of a seabird with a wingspan of 24 ft, *Pelagornis sandersi.* (That wingspan is larger than that of the largest living bird that can fly, the wandering albatross, 11 ft.) From your graph in part (a), what would you estimate the weight of a *P. sandersi* to have been?

58. Each December, the Friends of the Rock River Philharmonic raises funds with a holiday sale of poinsettia plants. The choices available, with cost in 2014 and average number of flowers per plant, are shown in the accompanying table.

Diameter of Flowerpot	Average Number of Flowers	Cost
6 in.	6	\$11
7 in.	9	\$18.75
8 in.	18	\$30

(a) Plot on ordinary graph paper cost versus diameter of flowerpot. Do you observe linear scaling?

(b) Use a calculator or spreadsheet to take the base-10 logarithms ($\log_{10}$) of the cost and then graph the logarithm of cost versus diameter of flowerpot.

(c) Repeat parts (a) and (b) for cost versus average number of flowers.

(d) Use a model to estimate the number of flowers and the cost for a 9-in.-diameter flowerpot.

Chapter Review

59. In the photo below of a statue of Paul Bunyan in Akeley, Minn., he would appear to stand about 33 ft tall and be geometrically similar to an ordinary muscular man.

(a) What is the linear scaling factor of the Paul Bunyan statue compared to a man 6 ft tall?

(b) If a man 6 ft tall weighs 200 lb, how much would Paul Bunyan weigh?

Pierdelune/Shutterstock

60. Some humans, such as the Bushmen of the Kalahari Desert in Africa, live in desert environments, where it is important to be able to do without water for periods of time. Would you expect such an environment to favor short people or tall ones? (Adapted from A. Zherdev, Horseflies and flying horses, *Quantum,* May–June 1994, 32–37, 59–60.)

61. Smaller birds and mammals generally maintain higher body temperatures than do larger ones. Explain why you would expect this to be so. (Adapted from A. Zherdev, Horseflies and flying horses, *Quantum,* May–June 1994, 32–37, 59–60.)

62. According to *Time* (March 7, 2005), men's brains on average are 10% larger than women's, even though men on average are only 8% taller. (The article mainly discusses the many differences in brain structure that likely outweigh any size differences.) If the brain scales linearly with height, and men are 8% taller, what percentage larger would you expect their brains to be? (And this consideration has nothing to do with relative intelligence!)

63. A common measure of water in agriculture is the KAF, which is one thousand acre-feet. An acre-foot of water is the amount of water to cover a field of area one acre to a depth of one foot. A common measure of water usage in homes is the water billing unit, which is 748 gallons.

(a) How many gallons are there in 1 KAF?

(b) How much is one water billing unit in cubic feet? (Your water meter measures in cubic feet.)

(c) A typical two-person home in the Midwest uses about 100 water billing units per year. How much is that in KAF?

64. The largest animal ever to live is the blue whale. No blue whale has been weighed as a whole, but estimates are that the largest ones weigh 200 tons, are 30 m long, and have a lung volume of 5000 L. An adult human male 6 ft tall and weighing 200 lb has a lung capacity of about 6 L.

(a) If the human were scaled up to the height (length) of the whale, what would you project the lung volume to be?

(b) If the human were scaled up to the weight of the whale, what would you project the lung volume to be?

(c) What conclusions would you draw from the comparisons?

65. The U.S. Energy Security and Independence Act of 2007 required that 15 billion gallons of ethanol and other renewable fuels be blended with gasoline in 2015. In the United States, corn is the main source of ethanol. What percentage of total arable U.S. land (land that can be planted to crops) would be needed to raise that much corn ethanol? The contiguous United States (without Alaska or Hawaii) has 270×10^6 hectares of arable land, which could yield about 400 gal/acre of ethanol.

66. Unit conversion can matter, and the number of significant digits in the conversion factor can matter. New electronic components are designed using metric units, but accompanying circuit boards are often designed using U.S. customary units. Consider a connector with 100 pins on a side with design distance ("pitch") of 0.6 mm from one pin to the next. (Thanks for the idea to Mark Biegert, Calix, Inc., Plymouth, Minnesota.)

(a) How far, in mm, is the 100th pin from the first one?

(b) Use the conversion factor 1 mm $\approx$ 0.03937 in., which is accurate to five decimal places and four significant digits, to calculate the distance in inches of the 100th pin from the first one. Then convert that quantity back to millimeters by using the exact conversion 1 in. = 25.4 mm.

(c) Use the conversion factor 1 mm $\approx$ 0.039 in., which is accurate to three decimal places but only two significant digits, to calculate the distance of the 100th pin from the first one. Then convert that quantity back to millimeters by using the exact conversion 1 in. = 25.4 mm. Will that last pin fit into a slot that is the exact distance of 99×0.6 mm from the first pin?

Writing Projects

1. A human infant at birth usually weighs between 5 and 10 lb and has a height (length) between 1 and 2 ft, with the shorter babies having the lesser weight. Considering the weight and height of an adult human, write a paragraph arguing that human growth must not be just proportional growth.

2. That area scales with the square of length, and volume with the cube of length, has important consequences for the depiction and interpretation of data graphics. Suppose we wish to indicate in an artistic way that the disposable income (after taxes) of a typical U.S. worker is twice that of a typical worker in France, by showing that the U.S. worker's income can buy twice as expensive a car. We draw one car for the French worker and another one—fancier and "twice as large"—for the American.

What's the problem? Well, first, people tend to respond to graphics by comparing areas. Because the larger car is twice as high and twice as long as the smaller one, its image has 4 times the area. Second, we are used to interpreting depth and perspective in drawings in terms of three-dimensional objects. Because the larger car is also, by implication, twice as wide as the smaller, it has 8 times the volume. The graphic can leave the subconscious impression that the U.S. worker has 8 times as much disposable income, instead of just twice as much. (*Caution:* Psychological studies show that the area perceived by a viewer does not exactly match the mathematical area.)

With these ideas in mind, evaluate—in a paragraph each—the data depictions in parts (a) through (c).

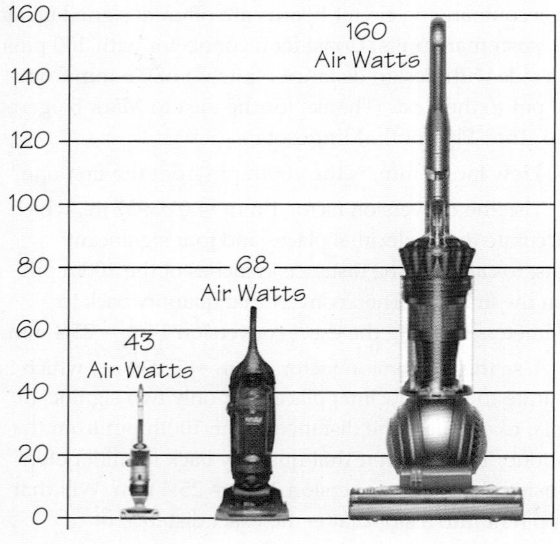

(a) "[A]s you can see, Dyson DC41 has twice the suction of any other vacuum." (www.dyson.com)

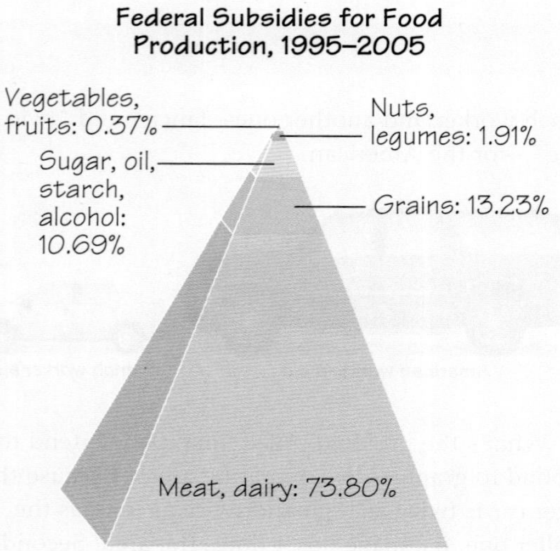

(b) Federal subsidies for food production, 1995–2005. (Why that salad costs more than a Big Mac, *Readers Digest,* October 2010, p. 72.)

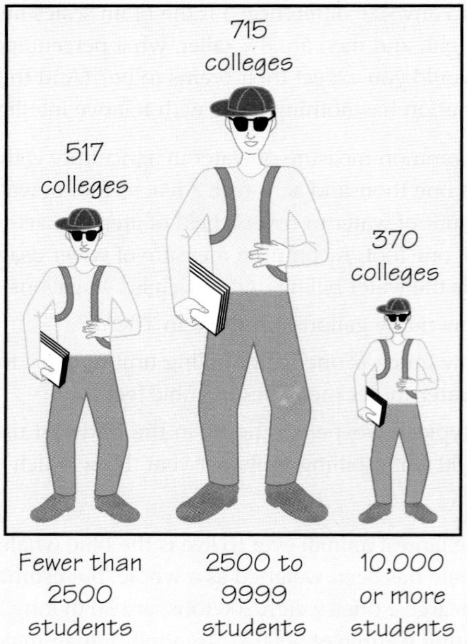

(c) U.S. colleges as classified by enrollment. (From David S. Moore, *Statistics: Concepts and Controversies,* 4th ed., W. H. Freeman, New York, 1997, p. 217.)

3. Evaluate in a paragraph each of the depictions in parts (a) through (c).

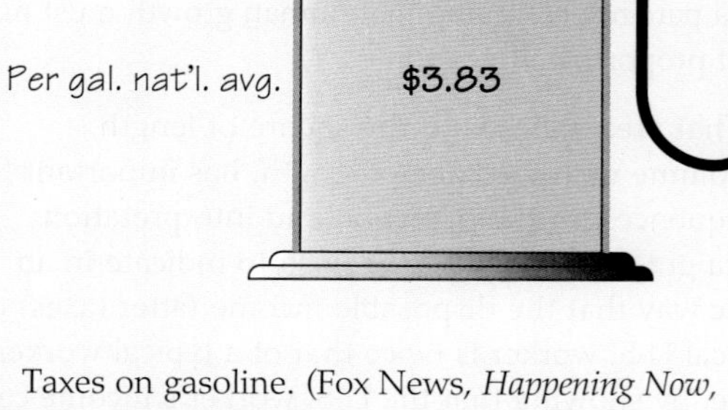

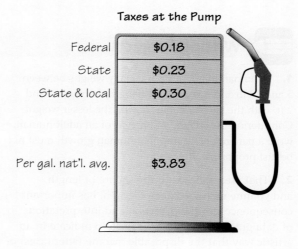

(a) Taxes on gasoline. (Fox News, *Happening Now,* 3/6/12 via *Media Matters.*)

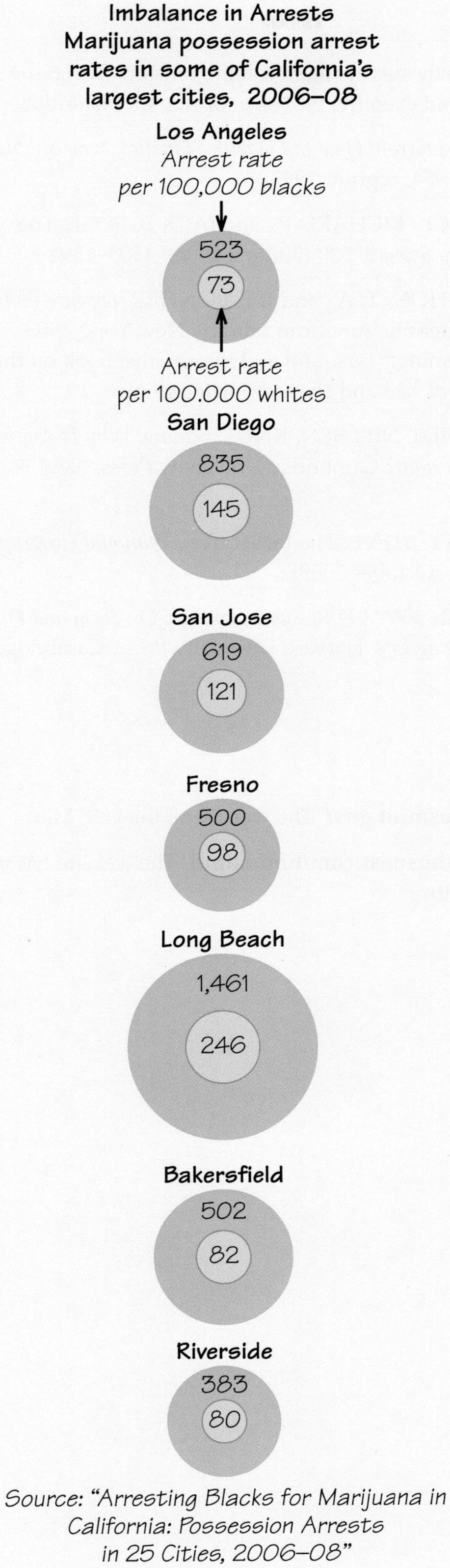

(b) Marijuana arrest rates in some California cities, 2006–2008. (*New York Times,* October 23, 2010, p. A19.)

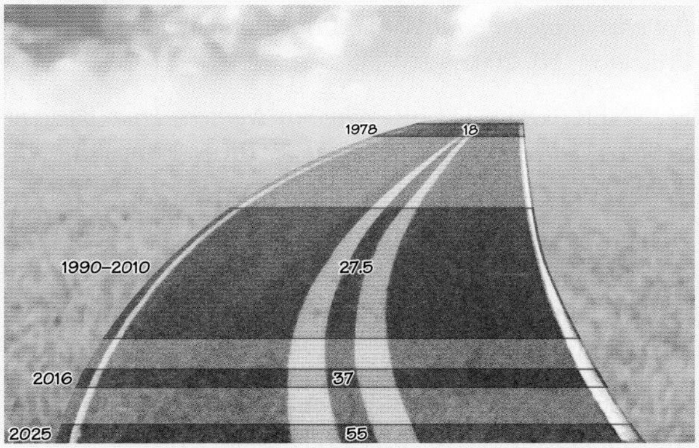

(c) U.S. fuel economy standards for autos (in miles per gallon) over the years.

4. With the ideas of Writing Projects 2 and 3 in mind, collect and evaluate similar depictions of data from magazines and newspapers.

5. Dolls and human figures are usually scaled to be geometrically similar to actual humans. But are dolls designed to represent babies or adult humans? Go to a toy store and measure the height, the vertical height of the head, and the arm length of some dolls and other figures. Scale your measurements to compare them with Figure 18.14 (page 763); from that comparison, try to estimate the ages of the humans that the figures resemble. Write up your procedure, data, calculations, and conclusions in a page or two.

6. (Refer to Exercises 42–44.) Because body weight is average density times body volume, BMI is average density times a quantity that has units of length. Discuss whether BMI makes sense as a measure of being overweight. Would dividing by a different power of height make for a better measure? (See Keith Devlin, "Top 10 Reasons Why the BMI Is Bogus," NPR, July 4, 2009, http://www.npr.org/templates/story/story.php?storyId5106268439.)

Suggested Readings

ADAM, JOHN A. *Mathematics in Nature: Modeling Patterns in the Natural World,* Princeton University Press, Princeton, NJ, 2003.

BONNER, JOHN TYLER. *Why Size Matters: From Bacteria to Blue Whales,* Princeton University Press, Princeton, NJ, 2006.

DUDLEY, BRIAN A. C. *Mathematical and Biological Interrelations,* Wiley, New York, 1977. An excellent and extended introduction to graphing, scale factors, and logarithmic plots.

GOULD, STEPHEN JAY. Size and shape. In *Ever Since Darwin,* Norton, New York, 1977, Chap. 21.

HALDANE, J. B. S. On being the right size. In *Possible Worlds and Other Papers,* Harper, New York, 1928. Reprinted in James R. Newman (ed.), *The World of Mathematics,* vol. 2, Simon & Schuster, New York, 1956, pp. 952–957. Also reprinted in John Maynard Smith (ed.), *On Being the Right Size and Other Essays* by J. B. S. Haldane, Oxford University Press, Oxford, 1985, pp. 1–8. Succinctly surveys area-volume tension, flying, the size of eyes, and even the best size for human institutions.

HUFF, Darrell. *How to Lie with Statistics,* Norton, New York, 1954; reprint, 1993.

LARRICK, RICHARD P., and JACK B. SOLL. The MPG illusion. *Science,* 320 (20 June 2008): 1593–1594.

McMAHON, T. A., and J. T. BONNER. *On Size and Life,* Scientific American Library, New York, 1983. Astonishingly beautiful and informative book on the effects of size and shape on living things.

SCHMIDT-NIELSEN, KNUT. *Scaling: Why Is Animal Size So Important?* Cambridge University Press, New York, 1984.

SILLETT, STEVE. The tallest trees. *National Geographic,* 216 (4) (October 2009).

WEIBEL, EWALD R. *Symmorphosis: On Form and Function in Shaping Life,* Harvard University Press, Cambridge, MA, 2000.

Suggested Websites

physics.nist.gov/cuu/Units/index.html A website with in-depth information on SI, the modern metric system.

www.missingkids.com The website for the National Center for Missing and Exploited Children.

www.usmint.gov/ The website of the U.S. Mint.

www.thusness.com/bmi.t.html The website has a BMI calculator.

Symmetry and Pattern

19

© Blaize Pascall/Alamy

What is symmetry? You may think symmetry refers just to a mirror-image reflection, such as between the sides of the human body. But you also sense symmetry in rotations, as in the two-dimensional spirals in the beauty of daisies and sunflowers (Figure 19.1) and even in three-dimensional form on pineapples, pinecones, sunflowers, and cacti (Figure 19.2a).

The chambered nautilus (Figure 19.2b) may stretch your notion of symmetry (as will other examples in this chapter). It has neither reflection nor rotational symmetry, but its successive sections are geometrically similar (in the sense of Chapter 18: same shape but different size), and the resulting spiral has

Figure 19.1 (a) This daisy has in its center 21 spirals clockwise and 34 spirals counterclockwise. (b) This heart of a sunflower has 34 spirals clockwise and 55 spirals counterclockwise.

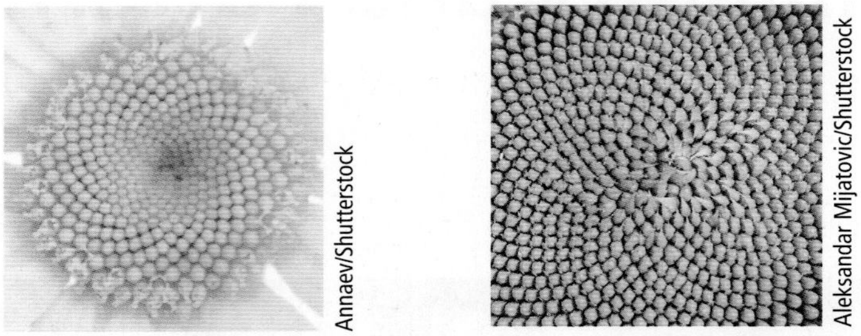

(a) (b)

Figure 19.2 (a) Spirals of scales on a pinecone: 8 right, 13 left. (b) A chambered nautilus shell.

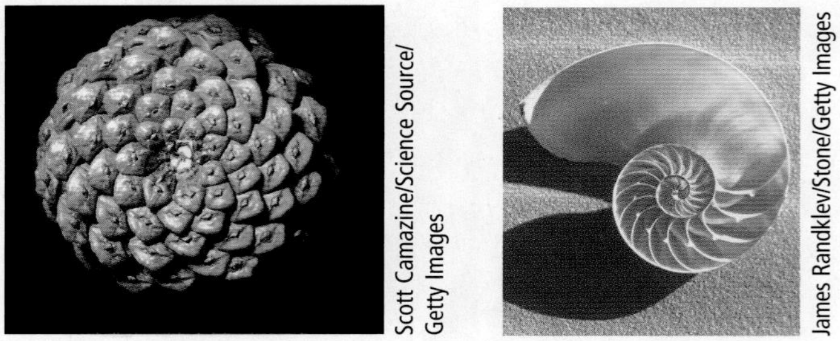

(a) (b)

the same shape at any size: A photographic enlargement superimposed on it would fit exactly.

In a broad sense, symmetry includes notions of *repetition* but also *balance* and *proportionality* (geometric similarity).

What is the connection of mathematics to the beauty and symmetries that we observe in nature? *Mathematics* is the *science of patterns;* it allows us to identify, classify, and appreciate more deeply different kinds of symmetry.

A simple number pattern is related both to the symmetric spirals in plants and to a proportion that offers a standard of beauty. The formula for calculating that proportion is also used in the Consumer Price Index (CPI) and was critical in formulating the standard for high-definition television (HDTV).

What does it mean for a pattern to be symmetrical, that is, have copies of a design element ("motif") arranged in a symmetrical way? Utterly surprising is that though there is an infinite variety of motifs, there are very few symmetrical patterns for arranging copies of one. We illustrate these patterns and provide (in Section 19.3) notation, decision trees, and examples.

Symmetries of a pattern can be thought of as geometric transformations that preserve a pattern, and they interact among themselves to generate further symmetries.

Finally, what happens if we scale a design element down (or up) in size as part of the notion of symmetry? Along with the chambered nautilus, we get *fractals,* with their own symmetry and beauty.

19.1 Fibonacci Numbers and the Golden Ratio

Fibonacci Numbers

Associated with the geometric symmetry in flowers and pinecones is a kind of *numeric symmetry,* with a "proportion" in the sense of a ratio of numbers. Strangely, the number of spirals in plants is not just any whole number but always comes from a particular sequence of numbers called **Fibonacci numbers** (see Spotlight 19.1).

Leonardo of Pisa ("Fibonacci")

SPOTLIGHT 19.1

Born in 1170, Leonardo of Pisa has been known as "Fibonacci" for the past century and a half. This nickname, which refers to his descent from an ancestor named Bonaccio, is modern, and there is no evidence that he went by it.

Leonardo was the greatest mathematician of the Middle Ages. He introduced calculation with Hindu-Arabic numerals into Italy to replace the Roman numerals and abacus then in use, and he also wrote about geometry, algebra, and number theory. We know little of Leonardo's life apart from a short autobiographical sketch:

> I joined my father after his assignment by his homeland Pisa as an officer in the customhouse located at Bugia [Algeria] for the Pisan merchants who were often there. He had me marvelously instructed in the Arabic-Hindu numerals and calculation. I enjoyed so much the instruction that I later continued to study mathematics while on business trips to Egypt, Syria, Greece, Sicily, and Provence and there enjoyed discussions and disputations with the scholars of those places.

(Translated by L. E. Sigler in Leonardo Pisano, *The Book of Squares: An Annotated Translation into Modern English,* Academic Press, New York, 1987, p. xvi.)

Leonardo of Pisa ("Fibonacci")
A portrait of unlikely authenticity.

The *Liber abbaci,* written in Latin, contains a famous problem about rabbits (see Exercise 4 for details), whose solution is now called the Fibonacci sequence.

The book mentions an earlier version written specifically for merchants. A copy of that book was finally found in 2003, making it the world's oldest-known arithmetic textbook. In addition to simpler instances of problems from the *Liber abbaci,* this earlier version includes material on geometry and on the calculation of interest and depreciation. The rabbits problem appears, too, but this time in terms of pigeons instead. (Maybe the outdoor merchants had had enough of pigeons by the time the second version of the book was to appear.)

Fibonacci Numbers (Fibonacci Sequence) DEFINITION

Fibonacci numbers occur in the sequence

$$1, 1, 2, 3, 5, 8, 13, 21, 34, 55, 89, 144, 233, 377, \ldots$$

This sequence begins with the numbers 1 and 1 again, and each next number is obtained by adding together the two preceding numbers.

Sometimes a sequence of numbers is specified by stating the value of the first term or first several terms and then giving an equation to calculate succeeding terms from preceding ones. This is called a *recursive rule,* and the sequence is said to be defined by **recursion.** Let's denote the nth Fibonacci number by F_n; then the Fibonacci sequence can be defined by the following.

Algebra Review Appendix
Sequences

Recursion for the Fibonacci Sequence DEFINITION

$$F_1 = 1, F_2 = 1, \quad \text{and} \quad F_{n+1} = F_n + F_{n-1} \quad \text{for} \quad n \geq 2$$

Self Check 1

What is the value of F_{12}?

The recursive rule just expresses in algebraic form that the next Fibonacci number is the sum of the previous two.

Look at the daisy in Figure 19.1a. There are 21 spirals streaming out in the clockwise direction and 34 harder-to-count ones in the counterclockwise direction—two consecutive Fibonacci numbers.

The common grocery pineapple (*Ananas comosus*) has three sets of spirals, one each along the three directions through each hexagonally shaped scale: 8 spirals to the right, 13 to the left, and 21 vertically—again, consecutive Fibonacci numbers.

Why are the numbers of spirals in plants the same numbers that appear next to each other in a purely mathematical sequence? There is no easy answer for this phenomenon, which is called **phyllotaxis.** There are several intricate theories about the dynamics of the plant's growth that offer explanations (e.g., optimizing total exposure of the leaves to sunlight). Spotlight 19.2 reveals what happens when phyllotaxis is applied to solar energy.

Fibonacci Solar Power

SPOTLIGHT 19.2

A concentrated solar plant (CSP) consists of an array of mirrors (*heliostats*) that track the sun to reflect and focus sunlight to a central tower, which heats up and converts the heat to steam and then to electricity. Such plants can be enormous, in terms of number of mirrors (hundreds of thousands), area they take up (thousands of acres), and power generated (hundreds of megawatts). The first such plant in the United States, and the largest in the world, is the Ivanpah Solar Power Facility, which has three collecting towers, cost $2.2 billion, and opened in 2014. The plant is in the Mojave Desert near the border between California and Nevada.

A key question in designing such a plant is how to arrange the mirrors efficiently, taking into account the area available, distance of mirrors from the tower (closer is better), and the passage of the sun overhead. But a crucial consideration in design is that at certain times of day a mirror can stand in the shadow of one or more other mirrors, or they can block its reflection to the tower.

Scientists at MIT and at Aachen University in Germany have proposed a new spiral *biomimetic* pattern for heliostat layout, based on mimicking phyllotaxis—as if the heliostats were leaves on a plant.

The photo below shows the PS20 and PS10 solar plants in Andalsuia, Spain. The accompanying figure shows the design of the PS10 CSP as constructed and also the scientists' recommended redesign.

In the redesign on the right, you can clearly see the spirals, just as you can see them on sunflowers and pineapples.

The redesign, with the same power output as the existing plant, would use 14% less area. In the case of a plant as large as Ivanpah, the savings in area would be closer to 20%. Alternatively, a plant occupying the same area could produce correspondingly more power.

Chris Satlberger/ Science Source

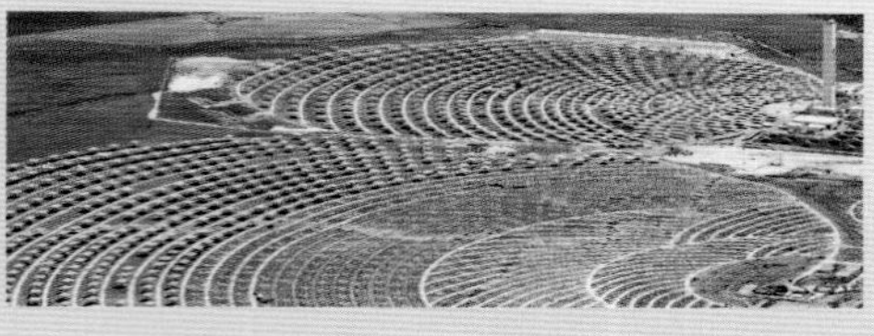

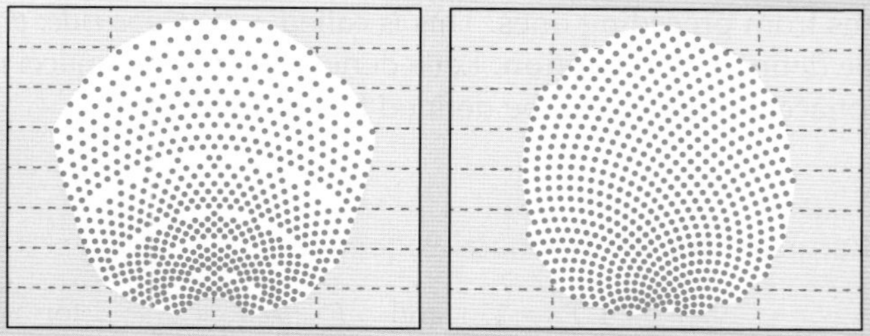

Drumming Up Mathematics

SPOTLIGHT 19.3

Manjul Bhargava, professor of mathematics at Princeton University, was awarded the Fields Medal a few days after his 40th birthday in 2014. There is no Nobel Prize in mathematics; the Fields Medal is considered on a par in honor (but unfortunately for mathematicians, involves much less prize money). The award was based on his work in number theory, in a field called quadratic forms.

His interest in the subject was prompted in part by rhythms he learned to play on the *tabla* (a pair of drums), and in part by the mathematical nature of rhythms in Sanskrit poetry. Bhargava has since taught courses on the mathematics involved in music, poetry, and magic.

Meter in Sanskrit poetry depends on sequencing combinations of long and short syllables. The long syllables are regarded as twice as long as the short ones; you may think of a short syllable as occupying a single beat and a long syllable as taking two beats. The question arose, 2000 years ago: For a fixed number n of beats, how many different rhythms of short and long syllables can there be?

Make yourself a table of the rhythms possible for each of $n = 1$ through 5. You should recognize, as the Sanskrit poets did, the pattern in the sequence! Can you explain how, for example, the number of 5-beat rhythms can be traced back to the numbers of 3-beat and 4-beat rhythms?

Other mathematical sequences of integers are inspired not by nature but by rhythms in music and poetry. Spotlight 19.3 tells how those aroused the interest of a mathematician who was recently honored for his achievements.

The Golden Ratio

During the last several centuries, an attractive myth arose suggesting that the ancient Greeks considered a specific numerical proportion essential to beauty and symmetry. This proportion has been known in modern times as the **golden ratio, golden mean,** or even **divine proportion** (these are great names, in terms of public relations!). This proportion had been investigated by Euclid in Book II of his *Elements,* but there is little evidence connecting this proportion to Greek aesthetics. We pursue the golden ratio briefly because of its intimate connection to the Fibonacci sequence and because it does have some appeal as a standard for beautiful proportion.

Golden Ratio DEFINITION

The value of the **golden ratio,** usually denoted by the Greek letter phi (ϕ), is

$$\phi = \frac{1 + \sqrt{5}}{2} \approx 1.618033$$

The basic aesthetic claim is that a **golden rectangle**—one whose height and width are in the ratio of 1 to 1.618—is the most pleasing of all rectangles. In Spotlight 19.4, we show how to construct a golden rectangle that is 1 unit by ϕ units.

How the Greeks Constructed a Golden Rectangle

SPOTLIGHT 19.4

In constructing a golden rectangle, the Greeks started from a 1-by-1 square (outlined in black in the figure), which they made by constructing perpendiculars at the two ends of a horizontal segment of unit length. To extend the square to a golden rectangle, they bisected the original segment, getting a new point that divides it into two pieces of length one-half each. Using this new point and a compass opening equal to the distance from it to a far corner of the square (shown by the diagonal line in the figure), they could add the blue length to the length one-half to get an interval (in red at bottom) with total length ϕ.

A golden rectangle has the "regenerative" property that if you cut a square-shaped piece off one end of it, the rectangle that remains is again a golden rectangle.

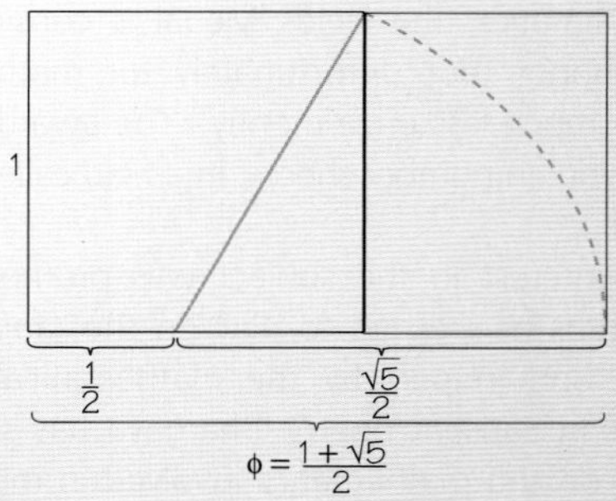

Why would anyone think that this is an attractive ratio? And where did it come from? The answer lies not in Fibonacci numbers but in the Greeks' pursuit of balance in their study of geometry.

Given two line lengths, one way to find a length that "strikes a balance" between the two is to average them. For lengths l (the larger) and w (the smaller), their average, or *arithmetic mean,* is $m = (l + w)/2$, and it satisfies

$$l - m = m - w$$

Thinking of the two sides of the equation as pans on a balance, the length m strikes a balance between l and w, in terms of a common *difference* from the two original lengths. More generally, the arithmetic mean of n numbers or lengths is their sum divided by n. (See Chapter 5 for its use in statistics.)

The Greeks, however, preferred a balance in terms of *ratios* rather than differences. They sought a length s, the **geometric mean,** that gives a common ratio

$$l \div s = s \div w \quad \text{or} \quad \frac{l}{s} = \frac{s}{w}$$

Multiplying both sides by sw, we obtain $lw = s^2$, which expresses the geometric fact that s is the side of a square whose area equals the area of an l-by-w rectangle (the Greeks thought in terms of geometric objects). In geometry, the geometric mean s is called the *mean proportional* between l and w (see Figure 19.3).

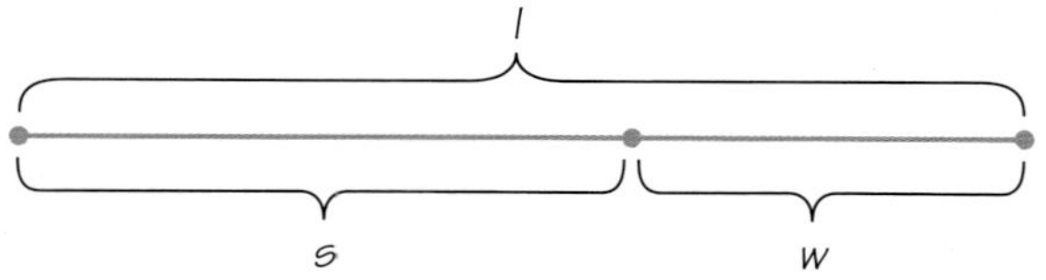

Figure 19.3 The line segment of length l is divided so that the length of s is the geometric mean between l and $w = l - s$. The dividing point divides the length l in the golden ratio.

Geometric Mean DEFINITION

The quantity $s = \sqrt{lw}$ is the **geometric mean** of l and w.

More generally, the geometric mean of n numbers is the nth root of the product of all n factors multiplied together: The geometric mean of $x_1, x_2, \ldots, x_n$ is $\sqrt[n]{x_1 \cdot x_2 \cdot \cdots \cdot x_n}$. For example, the geometric mean of 1, 2, 3, and 4 is $\sqrt[4]{1 \cdot 2 \cdot 3 \cdot 4} = \sqrt[4]{24} = 24^{1/4} \approx 2.213$.

Self Check 2

What is the geometric mean of 10, 100, and 1000?

The Greeks found symmetry and proportion in the geometric mean, but the geometric mean also has important practical applications (see Spotlight 19.5).

The Consumer Price Index: An Application of the Geometric Mean

SPOTLIGHT 19.5

The Bureau of Labor Statistics (BLS) uses the geometric mean—not the arithmetic mean—to calculate the Consumer Price Index (CPI), which tracks changes in the cost of the goods and services that people buy.

The geometric mean takes into account substitutions that consumers make when prices change. For example, if the price of beef goes up but the price of chicken doesn't, then consumers may buy less beef and substitute the cheaper chicken for some beef.

Suppose that, overall, U.S. families consume equal dollar values of beef and chicken. A typical family might consume weekly 5 lb of beef at \$4/lb and 10 lb of chicken at \$2/lb, for \$20 each and a total cost of \$40. We say that beef and chicken each have a *relative market share* of 0.5 (50% beef, 50% chicken, by dollar value).

What if beef goes up to \$6/lb but chicken stays at \$2/lb? The *relative price change* in beef is \$6/\$4 = 1.5 and the relative price change in chicken is \$2/\$2 = 1.00 (no change). If the average family continues to eat just as much beef and chicken as before, the cost is now \$50, an increase of 25%. Because \$30 goes for beef and \$20 for chicken, the relative market shares have changed to, by dollar value, 60% beef and 40% chicken. The relative price change for the family's meat is \$50/\$40 = 1.25, which is just the arithmetic mean of the two relative price changes (1.50 and 1.00). A more general formulation is

$$\begin{aligned}&\text{relative price change}\\&\quad = \text{old market share of beef} \times \frac{\text{new cost of beef}}{\text{old cost of beef}}\\&\qquad + \text{old market share of chicken} \times \frac{\text{new cost of chicken}}{\text{old cost of chicken}}\\&\quad = 0.5 \times \frac{6.00}{4.00} + 0.5 \times \frac{2.00}{2.00}\\&\quad = 0.75 + 0.50 = 1.25\end{aligned}$$

A family that eats no beef sees no increase. A family that eats only beef sees an increase of 50%. The CPI is an average over *all* families, weighted by the dollar value that each consumes.

If instead we use the geometric mean to calculate a price index, we get a relative price change of $\sqrt{1.50 \times 1.00} \approx 1.225$.

The more general formulation is

$$\begin{aligned}&\text{relative price change}\\&\quad = \left(\frac{\text{new cost of beef}}{\text{old cost of beef}}\right)^{(\text{old market share of beef})}\\&\qquad \times \left(\frac{\text{new cost of chicken}}{\text{old cost of chicken}}\right)^{(\text{old market share of chicken})}\\&\quad = \left(\frac{6.00}{4.00}\right)^{0.5} \times \left(\frac{2.00}{2.00}\right)^{0.5}\\&\quad = \sqrt{1.50 \times 1.00} \approx 1.225\end{aligned}$$

This relative price change, a 22.5% increase, is less than the 25% using the arithmetic mean.

The Consumer Price Index: An Application of the Geometric Mean *(continued)*

The intention of the CPI is to measure the change in the cost of goods and services that still yield the same level of satisfaction to consumers. Use of the arithmetic mean presumes that a family buys the same amount of beef and chicken (5 lb beef, 10 lb chicken) as before. Use of the geometric mean presumes that a family buys the same *relative dollar value* of each meat as before, hence \$24.50 (12.25 lb) of chicken and \$24.50 (4.08 lb) of beef, for a total of \$49 = 1.225 × \$40. Buying 2.25 lb more chicken and 0.92 lb less beef is supposed to yield the "same satisfaction" as before.

Because the geometric mean is always less than or equal to the arithmetic mean (see Exercise 19), the geometric mean gives a lower figure for inflation than using the arithmetic mean would produce.

Social Security payments, some wage increases, cost-of-living allowances (COLAs), and income tax rates are all automatically geared to the CPI, which we treat in detail in Chapter 21.

The Obama administration proposed federal budget for 2013–2014 called for changing to a "chained CPI" for calculating the federal payments and adjustments noted above. The chained CPI uses the geometric mean methodology in attempting to account for such substitutions as in our example. The term *chained* comes from the use of market baskets in adjacent months, thereby "chaining" together succeeding months. As in our example above, the chained CPI gives a lower estimate of inflation, on the order of 0.25% lower per year. Adoption of a chained CPI would result in the government saving money, with the cost borne by senior citizens and the bottom 90% of income-earners.

Glow Images, Inc./Getty Images

The Greeks were interested in cutting a single line segment of length l into lengths s and w, with $l = w + s$, so that s would be the mean proportional between w and l. Surprisingly, the ratio ϕ arises, as we show. Denote by x the common ratio

$$\frac{l}{s} = \frac{s}{w} = x$$

Substituting $l = s + w$, we get

$$x = \frac{l}{s} = \frac{s + w}{s} = \frac{s}{s} + \frac{w}{s} = 1 + \frac{w}{s}$$

But w/s is just $1/x$, so we have

$$x = 1 + \frac{1}{x}$$

Multiplying through by x gives

$$x^2 = x + 1 \quad \text{or} \quad x^2 - x - 1 = 0$$

This is a quadratic equation of the form

$$ax^2 + bx + c = 0$$

with $a = 1$, $b = -1$, and $c = -1$. We apply the famous quadratic formula,

$$x = \frac{-b \pm \sqrt{b^2 - 4ac}}{2a}$$

to get the two solutions

$$x = \frac{1 + \sqrt{5}}{2} \approx 1.618033 \quad \text{and} \quad x = \frac{1 - \sqrt{5}}{2} \approx -0.618033$$

The negative second solution does not correspond to a length. The first solution is the golden ratio ϕ. It occurs also in other contexts in geometry; for example, ϕ is the ratio of a diagonal to a side of a regular pentagon (see Figure 19.4).

The golden ratio has a great myth and mystique attached to it. Some claim that the Great Pyramid was designed with it as a fundamental principle. Others find the golden ratio in proportions of parts of the Greek Parthenon.

Did Leonardo da Vinci use it in his drawings of the human figure (Figure 19.5a)? In his *Mona Lisa,* the area from the top of the head of the woman to the top of her bodice appears to form a golden rectangle. It is certainly true that human bodies exhibit ratios close to the golden ratio, as you can see by comparing your overall height to the height of your navel. The Swiss-born architect Charles-Édouard Jeanneret-Gris (1887–1965) definitely used the golden ratio (including a navel-height feature) as the basis for his "Modulor" scale of proportions (Figure 19.5b). (Like many music stars today, he adopted a single name, "Le Corbusier"—a form of his grandfather's name—in his case as a declaration of reinventing himself.)

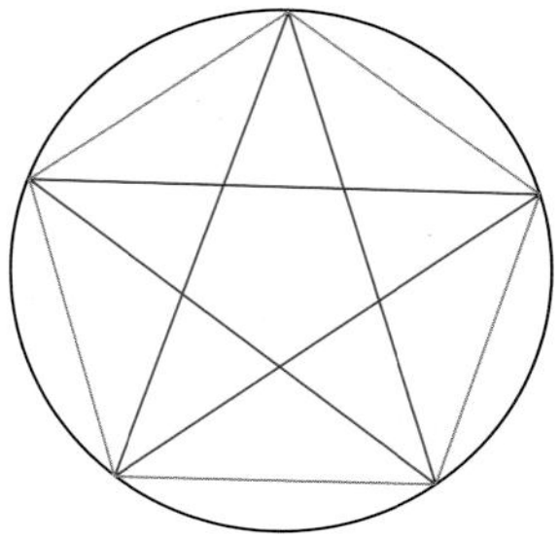

Figure 19.4 In a pentagon with equal sides, ϕ is the ratio of a diagonal to a side. The five-pointed star formed by the diagonals was the symbol of the followers of the ancient Greek mathematician Pythagoras.

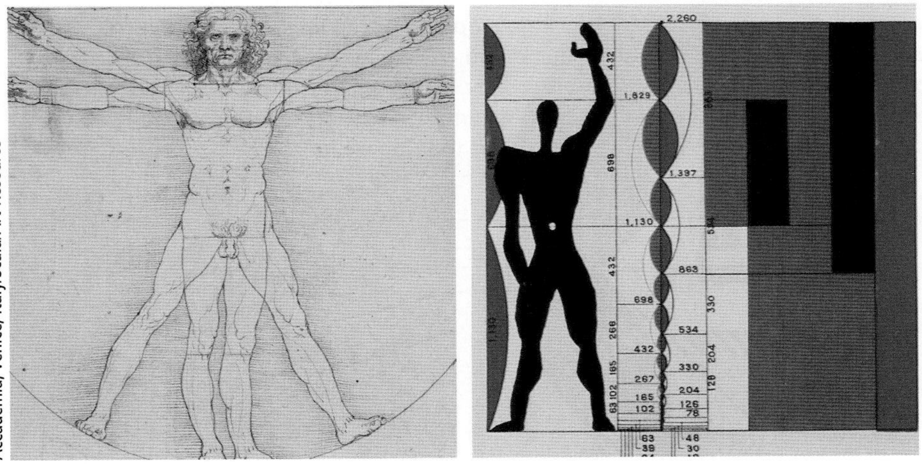

(*a*) (*b*)

Figure 19.5 (a) Leonardo da Vinci's *Vitruvian Man* (circa 1490), based on body proportions by Vitruvius (architect and engineer, 1st century B.C.). Despite claims on the Web and in the thriller *The Da Vinci Code,* neither Vitruvius nor Leonardo suggested using ϕ for human proportions or anything else. (b) Le Corbusier, however, did use ϕ in his "Modulor" scale of proportions.

More recently, paintings by the Impressionists Gustave Caillebotte (1848–1894) and Georges Seurat (1859–1891) exhibit rectangles in close proportion to the golden ratio. Wolfgang Amadeus Mozart (1756–1791), who was fascinated by mathematics as a student, may have constructed the lengths of parts of some of his piano sonatas with an eye to the golden ratio. Some modern and contemporary artists, such as István (Étienne) Beöthy (1897–1961) and Jo Niemeyer (1946–) have made explicit their inspiration and delight in using the golden ratio in their graphic art and sculpture.

However, in the cases of the Great Pyramid, the Parthenon, Leonardo, the Impressionists, Mozart, and others, we have no evidence of any intention on the part of the artists and architects to use the golden ratio consciously as an artistic organizing principle. The laudatory name *golden ratio* was not used in antiquity; it dates to the 19th century.

Moreover, experiments show that people's preferences for dimensions of rectangles are not focused on the golden ratio but cover a wide range, with golden rectangles not holding any special place. A ratio near 1.8 seems the most popular; this ratio is close to the aspect ratio of high-definition TV (see Spotlight 19.6).

Deliberate uses of the golden ratio and Fibonacci numbers include composition of a photograph (a modification of the "rule of thirds"), design of a tartan, and a (dubious) stock-trading strategy.

Are We Trying to Reclaim the "Glory That Was Greece"[1]?

SPOTLIGHT 19.6

The *aspect ratio* of an object, such as a computer screen, is the ratio of its width to its height. For a golden rectangle, this is phi, approximately 1.61. An aspect ratio is usually expressed as two numbers separated by a colon; for example, most digital cameras take pictures that have an aspect ratio of 4:3 (pronounced "4 to 3").

Through the ages, artists have produced canvases of varying sizes and aspect ratios. The introduction of photography on a mass-consumer scale involved standardizing sizes for photographic roll film and for photographic paper prints. A popular size was 36 mm × 24 mm (despite the 36, this was called "35 millimeter film," with an aspect ratio of 3:2 = 1.5).

Meanwhile, Thomas Edison had introduced movies with an aspect ratio of 4:3, but in 1932 the Academy of Motion Picture Arts and Sciences (which gives out the Academy Awards) settled on a standard film with an aspect ratio of 1.375.

Television came later, with an aspect ratio of 4:3, or 1.33; because that ratio was close to the movies' 1.375, movies could be shown on TV with very little cropping of the image. Until 1968, TV was in black and white only, while movies in theaters offered color. After TV transitioned to color broadcasting, filmmakers sought a different advantage by providing the "panoramic vision" of *widescreen.*

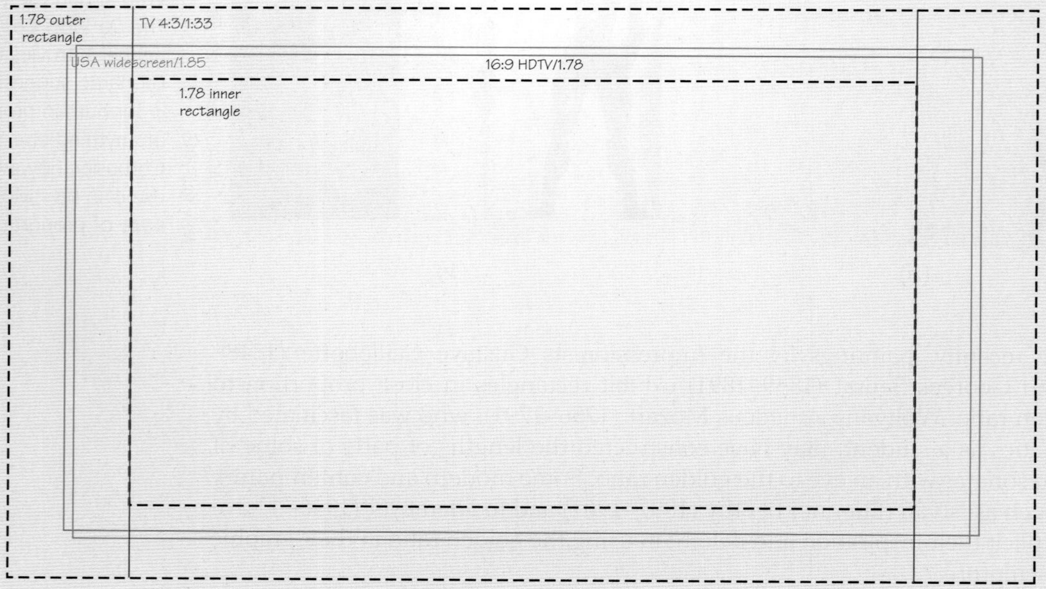

[1]From Edgar Allen Poe's poem "To Helen."

Are We Trying to Reclaim the "Glory That Was Greece"?

Movies in the United States evolved to an aspect ratio of 1.85, while those in some European countries evolved to 1.66, with some widescreen films at 2.35.

When high-definition TV (HDTV) was developed, a 16:9 ≈ 1.78 ratio was chosen as a compromise.

When HDTV is shown on a TV set with a 4:3 aspect ratio but with the original 16:9 aspect ratio preserved, either black bars appear above and below the image (*letterboxing*) or the image is cropped at the sides. When 4:3 video is presented on HDTV, the original material is often placed in the middle of the screen and black bars appear on sides of the image (*pillarbox* effect); the alternative is to enlarge the image but crop it at the top and bottom.

The standard HDTV resolutions of "720p" (1280 × 720 pixels) and "1080p" (1920 × 1080 pixels) both have exactly a 16:9 aspect.

Computers offer choices for the display resolution, such as 640 × 480 pixels (4:3) or 2560 × 1600 pixels (16:10, common on Android tablets). Those aspect ratios are not as high as 1.78, but there is often a "stretched" option that avoids letterboxing or pillarboxing. However, for the sharpest clarity, the display resolution of the computer video card should match the native resolution of the monitor, so that a pixel of an image file corresponds exactly to a screen pixel.

Smartphones are yet another display arena. The iPhone, for example, has an aspect ratio of 1.5. When you are watching a movie on a smartphone, either the image must be compressed horizontally or the edges of each frame are trimmed.

So over time, there has been an evolution of aspect ratios, from 1.5 for roll film and 1.33 for TV to an eventual 1.78 for HDTV—a compromise value closer to the Greeks' golden ratio.

Imageroller/Alamy

Chuck Eckert/Alamy

alexsi/iStock/360/Getty Images

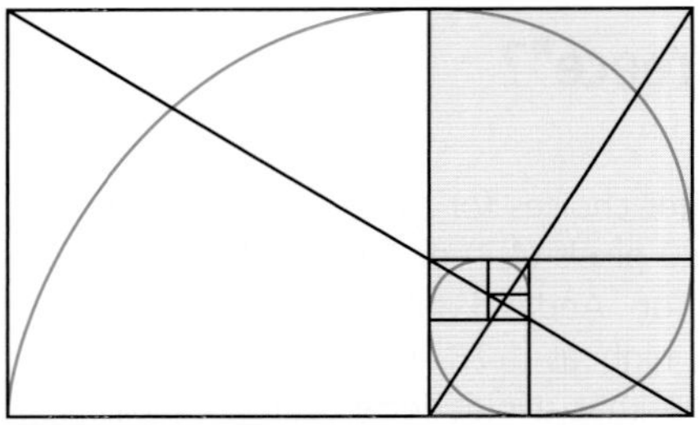

Figure 19.6 A logarithmic spiral determined by a sequence of golden rectangles and corresponding squares.

The spirals of the daisy are approximations to a special logarithmic (equiangular) spiral, the golden spiral (Figure 19.6). The mathematical reason for this connection is that the ratios of consecutive Fibonacci numbers

$\frac{1}{1}$	$\frac{2}{1}$	$\frac{3}{2}$	$\frac{5}{3}$	$\frac{8}{5}$	$\frac{13}{8}$	$\frac{21}{13}$
1.0	2.0	1.5	1.666 . . .	1.6	1.625	1.615. . .

provide alternately under- and overapproximations that narrow in on $\phi = 1.618033 \ldots$.

This fact is easy to show. We assume that there is a limit and show that it must be ϕ. For any n, we have

$$\frac{F_n}{F_{n-1}} = \frac{F_{n-1} + F_{n-2}}{F_{n-1}} = 1 + \frac{F_{n-2}}{F_{n-1}}$$

Suppose that F_n/F_{n-1} tends to a limit x; then F_{n-2}/F_{n-1} tends to $1/x$. Substituting, respectively, x and $1/x$ gives

$$x = 1 + \frac{1}{x}$$

which after multiplication by x and rearrangement becomes $x^2 - x - 1 = 0$, whose solutions we know to be ϕ and $1 - \phi$.

A golden spiral gets wider by a factor of ϕ for each quarter turn. The spiral of the chambered nautilus shell of Figure 19.2b on page 780 is also an equiangular spiral, but the widening factor is approximately 1.3 rather than the golden ratio.

For reasons that we do not understand, some ratios in the DNA molecule are close to the golden ratio; for example, the length of one full cycle of a strand in the double helix is about 1.62 times its width. Perhaps the most surprising appearance of the golden ratio is in connection with black holes, regions of space in which the gravitational field is so strong that nothing can escape (even light). A rotating black hole loses energy and, up to a point, heats up as it does so; after that point—when the mass of the hole equals its angular momentum times the square root of ϕ—the hole starts to cool down instead.

19.2 Rosette and Strip Patterns

The spiral distribution of the seeds in a sunflower head and the spiraling of leaves around a plant stem are instances of *similarity* and *repetition,* two key aspects of symmetry.

They also illustrate *balance,* which refers to regularity in *how* the repetitions are arranged. In considering patterns with repetition, we distinguish the individual element or figure of the design (the *motif*) from the *pattern* of the design—*how the copies of the motif are arranged.* See, for example, Figure 19.7, which shows two fundamentally different ways of arranging stars.

StockCube/Shutterstock

Anthony Berenyl/Shutterstock

Figure 19.7 When Alaska and Hawaii entered the Union, the U.S. flag was redesigned from a rectangular array of 48 stars to a different pattern for 50 stars. Here, we show both patterns.

We explore and classify the fundamentally different ways in which a flat design can be symmetrical. The ideas that we discuss were used by scientists to discover what crystalline forms are possible. Although there is a limitless number of chemical structures, and of motifs that people can make, what is quite surprising is that there is only a limited number of ways to arrange atoms in a structure, or motifs in a design, in a symmetrical way.

How can we identify all the ways that designs can be put together without counting the designs themselves? The key mathematical idea is to look at what you can do to the pattern without changing its appearance.

Rigid Motions

Mathematicians describe various kinds of symmetry by using the geometric notion of a **rigid motion,** also known as an **isometry** (which means "same size"). A transformation or "motion" of the plane is a mapping of the entire plane into itself. A rigid motion preserves the distances between every pair of points. You can imagine picking up the plane and moving it as a whole, perhaps rotating it, possibly flipping it over—but *without changing its size or shape.* For example, moving every point 1 unit in the same direction constitutes a rigid motion.

Rigid Motion DEFINITION

A **rigid motion** is a transformation of the entire plane that preserves the size and shape of figures. In particular, any pair of points is the same distance apart after the transformation as before.

Figure 19.8 shows the results of how various motions affect the rectangle and its interior in Figure 19.8a. In Figure 19.8b, each side is shrunk by 50%—which is not a rigid motion because the size of the rectangle changes.

For Figure 19.8c, we shear ("squash") the rectangle; again, this is not a rigid motion because the shape of the rectangle changes. In Figure 19.8d, we rotate the plane 90° (a quarter-turn) counterclockwise around the center of the rectangle: This is a rigid motion. Similarly, in Figure 19.8e, rotating the plane by 180° (a half-turn) is a rigid motion.

In Figure 19.8f, we reflect the rectangle along a vertical mirror down the middle. The right and left halves exchange places.

Figure 19.8g shows the result of reflecting across a diagonal of the rectangle. All reflections and all rotations are rigid motions. So are all **translations,** which move every point in the plane a certain distance in the same direction.

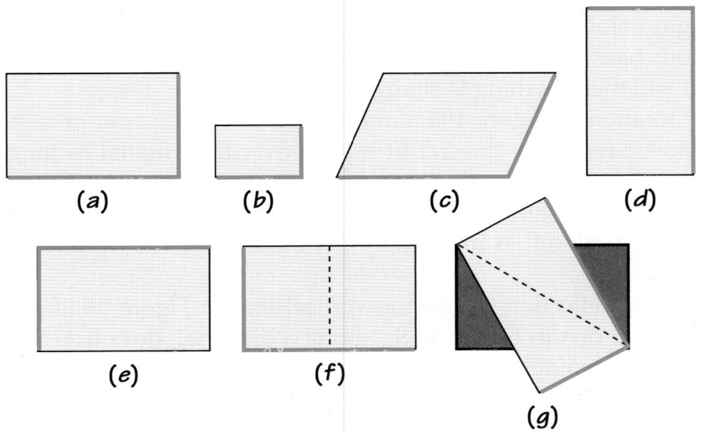

Figure 19.8 Results of various motions applied to a blue-edged rectangle and its interior. (a) the original rectangle and interior; (b) 50% reduction (not a rigid motion); (c) shearing (not a rigid motion); (d) quarter-turn; (e) half-turn; (f) reflection along the vertical line down the middle; (g) reflection along a diagonal line.

The only remaining kind of rigid motion of the plane is a combination of reflection and translation. Glide reflection is the kind of pattern that your footprints make as you walk: Each successive element of the design (footprint) is a reflection of the previous one (Figure 19.9). The motion combines translation (*glide*) with a reflection across a line parallel to the direction of the translation.

Glide Reflection DEFINITION

A **glide reflection** is a translation (= glide) combined with alternating reflection in a line parallel to the translation direction.

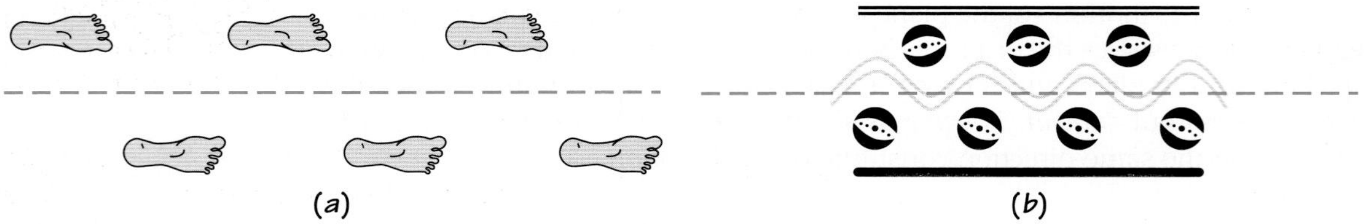

Figure 19.9 Glide reflection of (a) footprints; (b) design elements on a pot from San Ildefonso Pueblo, in New Mexico.

Any rigid motion of the plane must be one of the following:

- Reflection (across a line)
- Rotation (around a point)
- Translation (in a particular direction)
- Glide reflection (across a line)

Performing one rigid motion after another results in a rigid motion that (surprisingly) must be one of the four types that we have just explored.

Preserving the Pattern

In terms of symmetry, we are especially interested in rigid motions like those of Figures 19.8e and 19.8f that **preserve the pattern**—that is, motions for which the pattern looks exactly the same, *with all the parts appearing in the same relative places*, after the motion is applied.

You might enjoy thinking of applying these motions as "The Pattern Game": You turn your back, I apply a transformation, and then you turn back and see if you can tell whether anything is changed.

The 90° rotation of Figure 19.8a into Figure 19.8d does not preserve the pattern. The moved rectangle doesn't fit exactly over the original rectangle. On the other hand, the 180° rotation in Figure 19.8e does preserve the pattern. It's true that the top of the original rectangle is now on the bottom of the transformed version, but you can't tell. A rotation by any multiple of 180° also would preserve the pattern.

Similarly, the reflection across the vertical line in Figure 19.8f preserves the pattern, while the one in Figure 19.8g, along a diagonal, does not. Spotlight 19.7 discusses possible biological consequences of **reflection symmetry** or imperfections in it.

"Strive Then to Be Perfect"

SPOTLIGHT 19.7

Is there no such thing as objective and universal beauty, as claimed by American feminist Naomi Wolf in her book *The Beauty Myth?*

Stand in front of a mirror and look at yourself. Do your left and right sides look exactly symmetrical? What about the part in your hair, freckles on your face, evenness of your shoulders, bending of your ears?

Symmetry may be a proxy for fitness. Symmetrical racehorses tend to run faster; male lions with lopsided facial whisker-spot patterns die younger. The more symmetrical a flower is, the more nectar it produces, making it a better food source for pollinating insects; and correspondingly, insects prefer symmetrical flowers, giving such flowers a better chance of being pollinated.

Perhaps because of association with fitness, symmetry may affect mate selection among animals. Female zebra finches prefer males with symmetrical leg bands. Fruit flies and female barn swallows prefer males with symmetrical tails; a particular parasite can lead to an uneven tail.

Roger Antrobus/Getty Images

Michelangelo's David

What about people? Both male and female Britons, as well as Tanzanian hunter-gatherers, find facial symmetry more attractive than asymmetry.

Perfectly symmetrical female faces that are computer-generated from composites of individual photos appear more attractive to men than photos of actual women's faces.

The pattern of footsteps in Figure 19.9a is not preserved under reflection along the direction of walking; there is not a left footprint directly across from a right footprint. The pattern is preserved under a glide reflection along the direction of walking, as well as by a translation of two steps, or of four steps, and so on—but not by a translation of one step.

The rigid motions that preserve a pattern are the **symmetries of the pattern.**

You may think of a pattern as a recipe for repeating a figure (motif) indefinitely. Of course, any pattern in nature or art has only a finite number of copies of the figure. If the recipe for repetition is clear, we can imagine that we are looking at just a part of a pattern that extends indefinitely.

Patterns in the plane can be divided into those that have indefinitely many repetitions in one of the following ways:

- No direction—the **rosette patterns**
- Exactly one direction (and its reverse)—the **strip patterns** (sometimes called *band patterns* or *frieze patterns*)
- More than one direction (and their reverses)—the **wallpaper patterns**

Rosette Patterns

A rosette is an ornament or badge made to resemble a flower, and the term *rosette pattern* is used to describe the possible symmetries of such an ornament or a single flower. The repetition aspect of symmetry consists of the repetition of the petals around the stem. Translations and glide reflections do not come into play. The pattern is preserved under a rotation by certain angles corresponding to the number of petals. There may or may not be reflections that preserve the pattern, depending on whether the petal itself has reflection symmetry. Most flowers do (Figure 19.10a), but some do not.

(a)

(b)

Figure 19.10 (a) A flower; each petal has reflection symmetry. (b) A pinwheel with seven symmetrical "leaves," each asymmetric, hence pattern *c7*.

An everyday example of the rosette pattern—a human-made one—that does not have reflection symmetry is a pinwheel (Figure 19.10b). If there is no reflection symmetry, the motif of the pattern (the element that is repeated) is an entire petal. If there is reflection symmetry, the motif is just half a petal because the entire pattern can be generated by rotation and reflection of a half petal. The fact that these are the only possibilities is sometimes called *Leonardo's theorem,* after Leonardo da Vinci, who, in the course of planning the design of churches, needed to decide if chapels and niches could be added without destroying the symmetry of the central design.

Leonardo realized that there are two different classes of rosettes: those without reflection symmetry (*cyclic rosettes*) and those with it (*dihedral rosettes*) (see Figure 19.10). The respective notations for the patterns are *cn,* where the c stands for "cyclic," and *dn,* where the d stands for "dihedral," while *n* is the number of times that the rosette coincides with its original position in one complete turn around the center. A cyclic pattern has no lines of reflection symmetry, while the dihedral pattern *dn* has *n* different lines of reflection symmetry. The flower in Figure 19.10a with its 21 petals (a Fibonacci number!) has dihedral pattern *d21* (if we idealize it) because each petal has reflection symmetry, while the pinwheel in Figure 19.10b has pattern *c7.*

Strip Patterns

We illustrate the different kinds of strip patterns, and their "ingredient" symmetries, with patterns in the art of the Bakuba people of the Democratic Republic of the Congo, who are noted for their fascination with pattern and symmetry (see Spotlight 19.8).

All the strip patterns offer repetition and **translation symmetry** along the direction of the strip. For simplicity, we position patterns horizontally. For example, in Figure 19.11b, what you can imagine as eyes and a nose (if you rotate the page clockwise) is constantly shifted (translated) along the pattern by the same amount each time.

The pattern may have no other rigid motions that preserve it apart from translation, as in Figure 19.11a.

The other simplest rigid motion to check is **reflection symmetry** across a line. For a strip pattern, the horizontal center line of the strip may be a reflection line, as in Figure 19.11b; we say that the pattern has symmetry across a horizontal line.

Patterns Created by the Bakuba People

SPOTLIGHT 19.8

Among the Bakuba people of the Democratic Republic of the Congo (see the shaded area of the map), it is considered an achievement to invent a new pattern, and every Bakuba king had to create a new pattern at the outset of his reign. The pattern was displayed on the king's drum throughout his reign and, for some kings, on his dynastic statue.

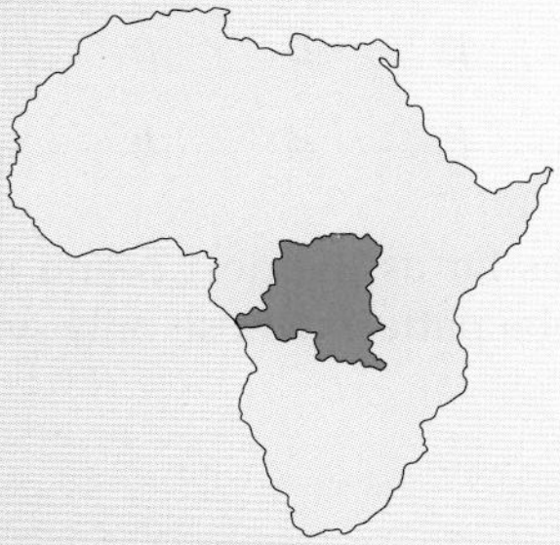

Two women with raffia cloths from the Bakuba village of Mbelo, July 1985. *Left:* Mpidi Muya with embroidered raffia (a kind of fiber) cloth. *Right:* Muema Kenye with plush and embroidered raffia cloth.

When missionaries first showed a motorcycle to a Bakuba king in the 1920s, he showed little interest in it. But the king was so enthralled by the novel pattern the tire tracks made in the sand that he had the pattern copied and gave it his name.

Source: Adapted from Jan Vansina, *The Children of Woot,* University of Wisconsin Press, Madison, 1978, p. 221.

The patterns made by tire tracks attracted the interest of the Bakuba people and its king.

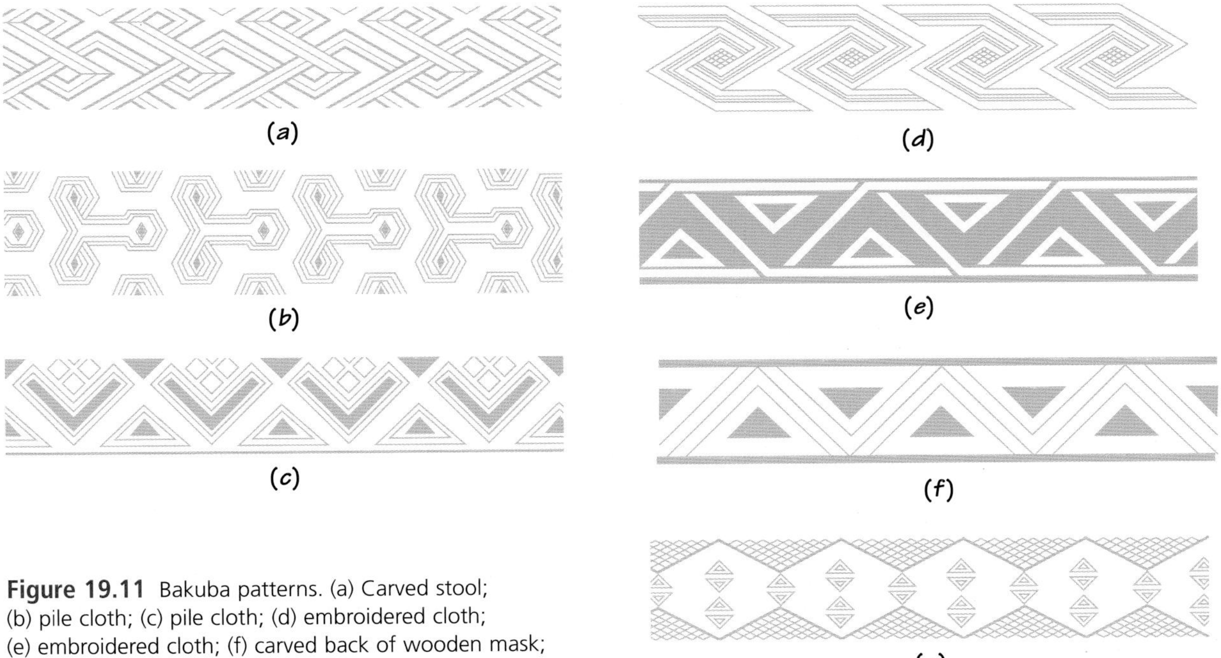

Figure 19.11 Bakuba patterns. (a) Carved stool; (b) pile cloth; (c) pile cloth; (d) embroidered cloth; (e) embroidered cloth; (f) carved back of wooden mask; (g) carved box.

There may instead be reflection across a *vertical* axis, such as the vertical lines through or between the *V*s in Figure 19.11c.

What kind of **rotational symmetry** can a strip pattern have? The only possibility for a strip pattern is a rotation by 180° (a half-turn), because any other angle won't even bring the strip back into itself. [We don't count rotations of 360° or integer multiples of 360° (full turns), because any pattern is preserved under these.] Figure 19.11d shows a strip pattern that is unchanged by a 180° rotation about any point at the center of the small crosshatched regions.

What about glide reflections? A row of alternating *p*'s and *b*'s has glide reflection, as shown in the illustration below, in which the dotted line through the middle of the pattern is the line in which reflection takes place.

Glide: p p p p p p p p p

Glide reflection: p b p b p b p b p

For glide reflection, a *p* is translated as far as the next *b* and is then reflected upside down. Figure 19.11e shows a Bakuba pattern whose only symmetry (except for translation) is glide reflection.

Self Check 3

What symmetries—translations, reflections, glide reflections, half-turns—does the pattern of Figure 19.11f have?

Having examined symmetries on strip patterns, we can ask: What *combinations* of the four are possible? It turns out that apart from the five kinds of patterns we have already seen, there are only two other possibilities: We can have the combination of vertical line reflection together with half-turns and glide reflection, either with horizontal line reflection (Figure 19.11g) or without (Figure 19.11f).

Mathematical analysis reveals the following.

There Are Only Seven Strip Patterns RULE

There are only seven ways to repeat a pattern along a strip.

That this number is so small is quite surprising because there are myriad different design elements (motifs). Two designs may look entirely different, yet share the same pattern of reproducing their design elements.

Wallpaper Patterns and Crystal Structures

So far, we have classified the patterns that have no translation repetition (the rosette patterns) and those with repetition in one direction (the strip patterns). What about repetitions in more than one direction—say, in two different directions across a plane? It turns out that there are exactly 17 ways to do so, called *wallpaper patterns.* We give illustrations, notation, and a flowchart later, in Spotlight 19.9.

We emphasize again that "pattern" does not refer to the basic design but to how its repetition is structured across the plane. There is an infinite variety of possible designs that artists can devise. You should imagine that the artist has created one copy of the design and is contemplating how to place equal-sized copies of it in other parts of the (infinite) plane, in a way that is symmetrical. There are very few—just 17—strategies for doing so.

Crystallographers (physicists and chemists interested in the ways that crystals can occur or be built) in the 19th century classified three-dimensional crystal structures in terms of combinations of symmetry elements. They proved—after several years of coming up with different totals!—that there are exactly 230 patterns for crystals. Mathematicians have refined the classification of patterns further to take into account colors that are repeated in a symmetrical way.

The 17 Wallpaper Patterns

SPOTLIGHT 19.9

We give an example of each of the 17 wallpaper patterns, together with a flowchart for identifying them. Crystallographers have standard notations and abbreviations for the patterns. The full notation consists of four symbols:

1. The first symbol is *c* (for "centered") if all rotation centers lie on the reflection lines, or *p* (for "primitive") otherwise.
2. The second symbol indicates rotational symmetry. It is either *1, 2, 3, 4,* or *6,* corresponding to the rotational symmetry of, respectively, 360°, 180°, 120°, 90°, and 60°. The symbol is the largest applicable number. For example, if symmetries of 360°, 120°, and 60° are present, the symbol is *6.*
3. The third symbol is either *m, g,* or *1,* corresponding to the presence of "mirror," "glide," or no reflection symmetry.
4. The fourth symbol (*m, g,* or *1*) is for describing symmetry relative to an axis at an angle to the symmetry axis of the third symbol.

(*Note:* The patterns *p31m* and *p3m1* are exceptions to this scheme.)

Below each pattern illustration, we give both the standard abbreviation (on top) and the full notation (below).

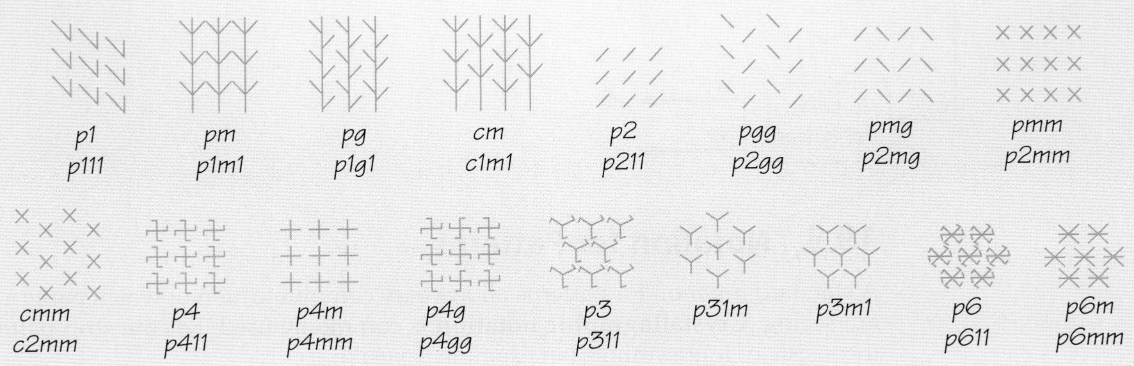

The 17 Wallpaper Patterns *(continued)*

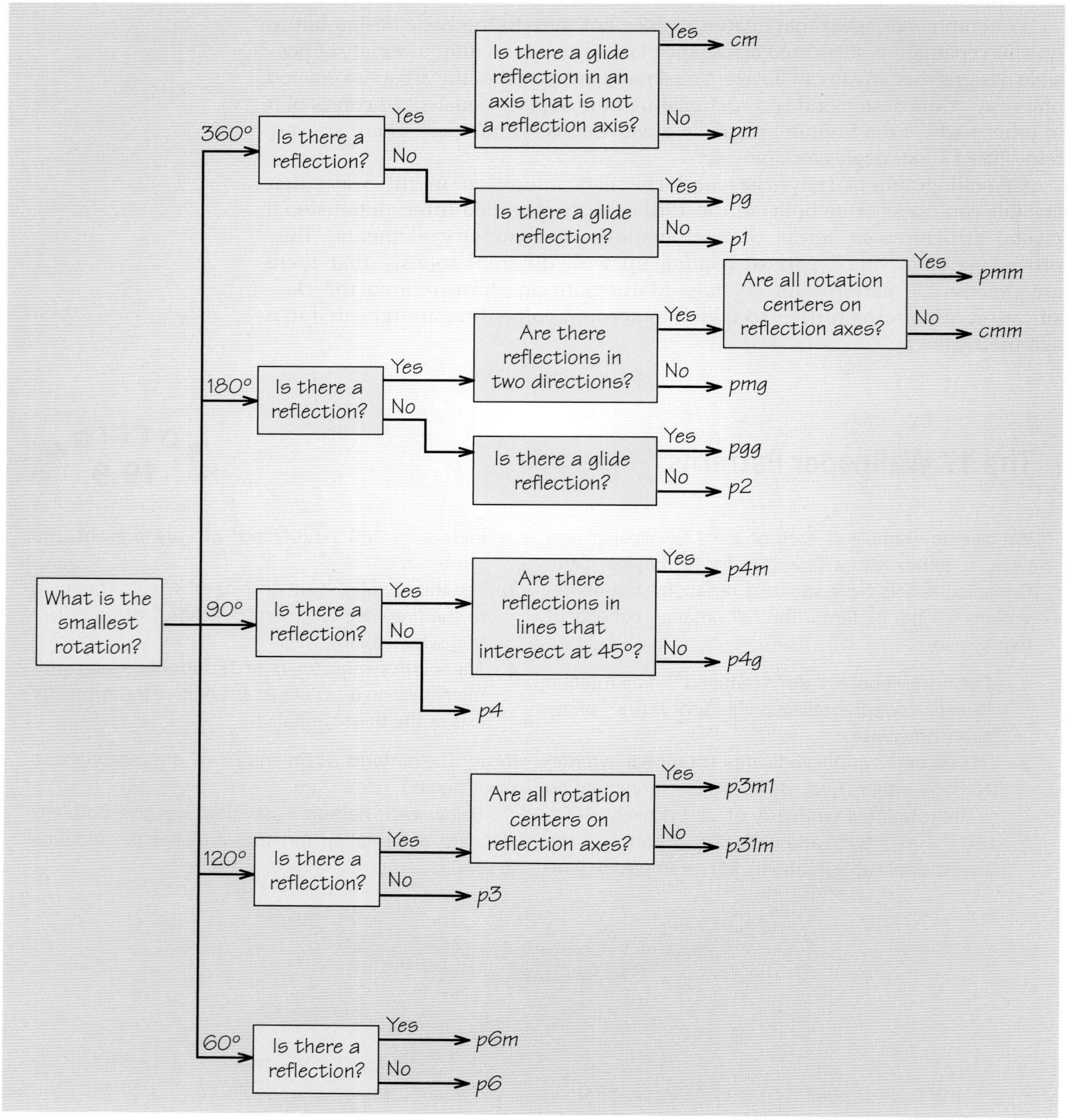

19.3 Notation for Patterns

A standard notation for patterns allows easy communication of which pattern you are seeing. **Crystallographic notation** is commonly used. For the strip patterns, it consists of four symbols (an example is *pma2*):

1. The first symbol is always a *p,* which indicates that the pattern repeats (is "periodic") in the horizontal direction.

2. The second symbol is *m* if there is a vertical line of reflection. Otherwise, it is *1*.
3. The third symbol is
   - *m* (for "mirror"), if there is a horizontal line of reflection (in which case, there is also glide reflection)
   - *a* (for "alternating"), if there is a glide reflection but no horizontal reflection
   - *1*, if there is no horizontal reflection or glide reflection
4. The fourth symbol is *2*, if there is half-turn rotational symmetry; otherwise, it is *1*.

A *1* always means that the pattern does *not* have the symmetry corresponding to that position. Figure 19.12 shows the notation schematically.

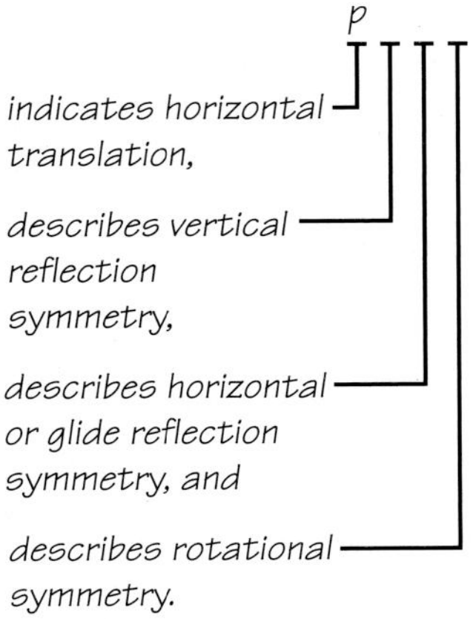

Figure 19.12 Scheme for strip pattern notation.

EXAMPLE 1 Bakuba Patterns

We use the flowchart of Figure 19.13 to analyze some of the Bakuba patterns of Figure 19.11.

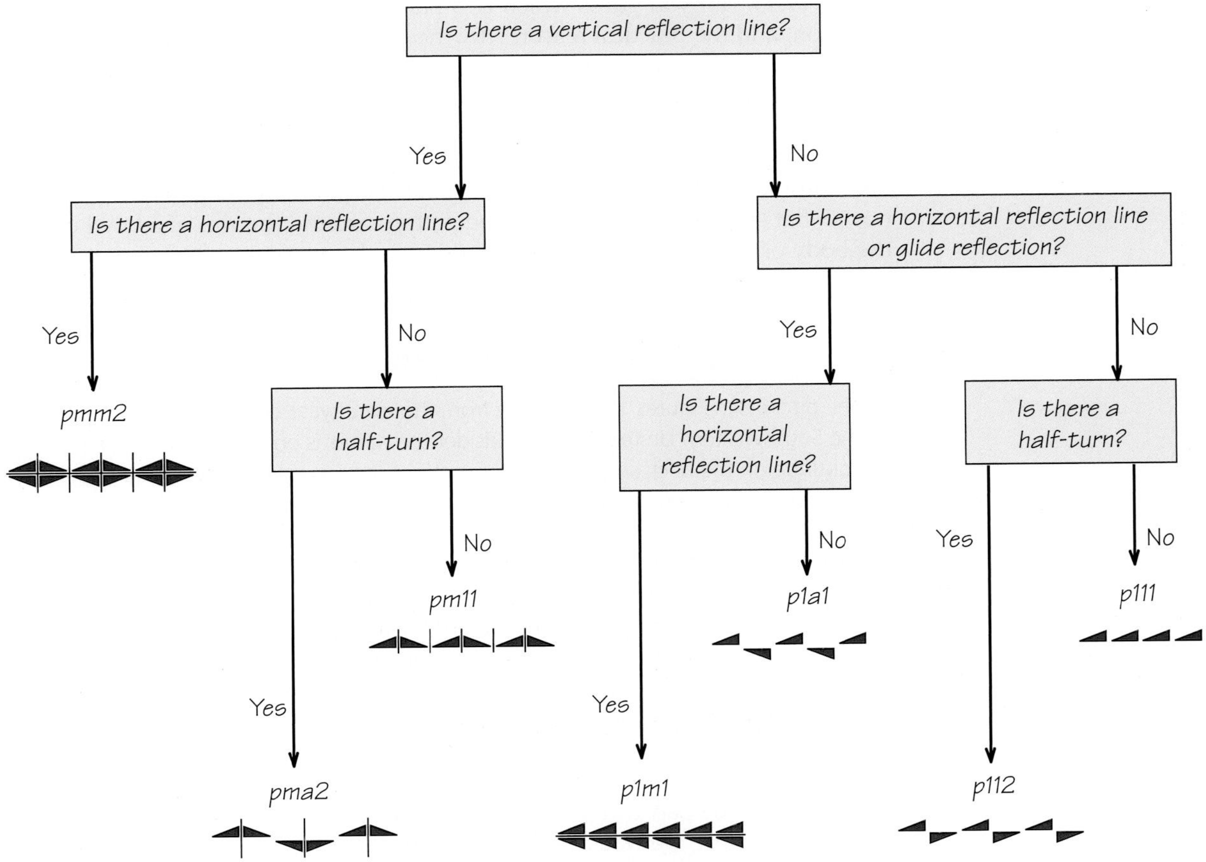

Figure 19.13 A flowchart for identifying the seven strip patterns and classifying them according to crystallographic notation.

Figure 19.11a does not have a vertical reflection, so we branch right, and the pattern notation begins to take shape as *p1_ _*. The figure has neither a horizontal reflection nor a glide reflection, so we branch right again, filling in the third position in the notation to get *p11_*. A half-turn preserves part of but not all the pattern, so we conclude that we have a *p111* pattern.

Figure 19.11b does not have vertical reflection, so we branch right to *p1_ _*. The figure does have horizontal reflection, so we branch left and left, concluding that the pattern is *p1m1*.

Figure 19.11f has vertical reflection, so we branch left to *pm_ _*. The figure does not have horizontal reflection, so we branch right but cannot yet fill in the third symbol. The figure does have a half-turn symmetry (and glide symmetry), with its center on the middle of the three lines between any pair of closest triangles. So the pattern is *pma2*.

Self Check 4

What is the pattern notation for the pattern in Figure 19.11c?

Remember the Bakuba people's fascination with tire treadmarks? Apart from aesthetic value, certain symmetries are important for practical purposes. The Museum of Transport in Glasgow, Scotland, includes all kinds of vehicle tires. However, only five of the seven strip patterns appear among treads of all the tires there. Examining Figure 19.13, can you guess which two patterns do not appear, what they have in common, and the practical reason why they are not used in tire treads?

Imperfect Patterns

In applying these classification schemes to patterns on real objects, we need to take into account that the pattern may not be perfectly rendered. Also, patterns not on flat surfaces—for example, the pattern around the rim of a bowl or around the body of a jar—require some latitude in interpretation.

EXAMPLE 2 ➔ Patterns on Pueblo Pottery

The pitchers in Figures 19.14abc are from thousand-year-old Pueblo sites in what is now the Southwestern United States. Consider the patterns on the bodies of the pots, which continue on the back sides. Let's suppose that they could be unwrapped and continued as strip patterns, but we'll disregard the patterns on the handles.

What are the patterns on the pots? We immediately come up against the question of the perfectness of the patterns. In Figure 19.14c, the rendering of the jagged pattern near the top of the pitcher is done rather crudely, and the pattern lower down is not done much better. Is this lack of pattern? It hardly seems so—more likely, there was just a lack of perfection in executing a pattern in the mind of the potmaker. Generosity compels us to opt for this latter interpretation for our analysis.

Similarly, in Figure 19.14d, if we regard the pot as having a single pattern going around the pot, we notice that the top part of the pattern is not matched exactly to the bottom part. Also, it's not clear if the vertical line in the middle of the pot face is supposed to be part of the pattern, or whether it is also reproduced on the unseen part. To be strict, we would decide at least that there are two separate patterns (upper and lower) separated by a white-line edge. But there also appears to be an attempted match-up between the two parts (however inexactly done) that we may not want to deny. We could attribute the variations (including the vertical line) to artistic license and consider the pot as having a single pattern around it.

We follow the flowchart in Figure 19.13 on page 799 and get the following results:

- Figure 19.14a: Is there a vertical reflection? *No.* Is there a horizontal reflection or glide reflection? *No.* Is there a half-turn? *No.* Hence, the pattern is *p111.* In fact, it looks a lot like the example at far right in Figure 19.13.
- Figure 19.14b: Is there a vertical reflection? *No.* Is there a horizontal reflection or glide reflection? *No.* Is there a half-turn? *Yes.* The pattern is *p112.* (The "jags" may represent zigzagging lightning bolts.)
- Figure 19.14c (top band): Is there a vertical reflection? *No* (notice the little "teeth" on the lower left and upper right edges only). Is there a horizontal reflection? *No.* Is there a half-turn? *Yes.* The pattern is *p112.*
- Figure 19.14c (lower band): Is there a vertical reflection? *No.* Is there a horizontal reflection? *No.* Is there a half-turn? *Yes.* This pattern too is *p112.*
- Figure 19.14d (lower band only): Is there a vertical reflection? *Yes.* Is there a horizontal reflection? *Yes.* The pattern is *pmm2.*

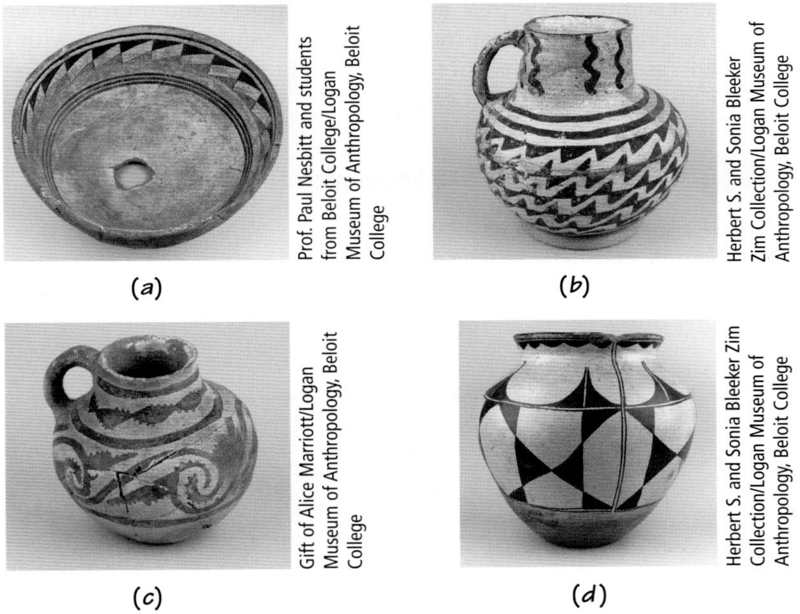

Prof. Paul Nesbitt and students from Beloit College/Logan Museum of Anthropology, Beloit College
(a)
Herbert S. and Sonia Bleeker Zim Collection/Logan Museum of Anthropology, Beloit College
(b)
Gift of Alice Marriott/Logan Museum of Anthropology, Beloit College
(c)
Herbert S. and Sonia Bleeker Zim Collection/Logan Museum of Anthropology, Beloit College
(d)

Figure 19.14 Pottery from the Late Pueblo II Phase (culture), in the Southwestern United States (A.D. 1050–1150), and a more recent piece. (All vessels are owned by Logan Museum of Anthropology, Beloit College, which graciously supplied the photos and granted permission for their reproduction.)

Women made the pots in the Pueblo culture, and they strongly preferred the symmetry of half-turns. Very few of their pots have any reflection symmetry, neither mirror or glide reflection. The avoidance of reflection symmetry was a consistent feature of ancient pottery of the indigenous peoples of the Western Hemisphere.

The pot in Figure 19.14d, as you might then surmise, is much more recent.

Self Check 5

What is the notation for the pattern of Figure 19.11f on page 795?

19.4 Symmetry Groups

We mentioned earlier that the key mathematical idea about detecting and analyzing symmetry is to look not at the motifs of a pattern but at its symmetries, the transformations that preserve the pattern.

The symmetries of a pattern have four notable properties:

- If we combine two symmetries by applying first one and then the other, we get another symmetry.
- There is an identity, or "null," symmetry that doesn't move anything but leaves every point of the pattern exactly where it is.
- Each symmetry has an inverse, or "opposite," that undoes it and also preserves the pattern. A rotation is undone by an equal rotation in the opposite direction, a reflection is its own inverse, and a translation or glide reflection is undone by another of the same distance in the opposite direction.
- In applying a number of symmetries one after the other, we may combine consecutive ones without affecting the result (*associativity*). For example, if we have symmetries A followed by B followed by C, we can do either of the following: first combine A with B, apply that symmetry, and then apply C; or first apply A and then follow that by applying the combination of B with C. That is, we can "associate" adjacent symmetries, but we must observe the overall order (A, B, C) in which they occur.

EXAMPLE 3 Symmetries of a Rectangle

What are the symmetries of a rectangle?

Consider the rectangle of Figure 19.15.

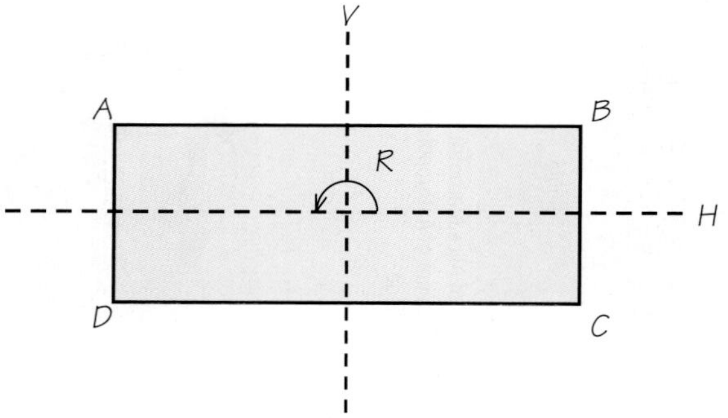

Figure 19.15 A rectangle, with reflection symmetries and 180° rotation symmetry marked.

Its symmetries, the rigid motions that bring it back to coincide with itself (even as they interchange the labeled corners), are as follows:

- The identity symmetry I, which leaves every point where it is
- A 180° (half-turn) rotation R around its center
- A reflection V in the vertical line through its center
- A reflection H in the horizontal line through its center

Now, convince yourself that the symmetries fulfill the four properties above:

- Combining any pair by applying first one and then the other is equivalent to one of the others. It's handy to have a notation for this combining; if we apply

V first and then *H,* we will write the result as *VH*—in other words, *we apply the sequence of actions from left to right.* You can check that the result is the same as applying *R;* that is, $VH = R$. Check this by following where the corner *A* goes under the symmetries. Practice combining symmetries by making yourself a "multiplication table" of them.

- The element *I* is an identity element.
- Each element is its own inverse. For example, rotating by 180°, then doing it again, gets the rectangle back to coincide with itself; in our notation, $RR = I$.
- Try some examples to verify that associativity holds. For instance, check that $RHV = (RH)V = R(HV)$. In other words, applying *R* then *H* then *V,* we get the same result if we combine the first two and then apply the third, or if we apply the first one and then apply the combination of the second two.

Self Check 6

What are the symmetries of a square?

The four properties of symmetries of an object are common to many kinds of mathematical objects. The properties characterize what mathematicians call a *group.* Various familiar collections of numbers, together with operations on them, form groups.

EXAMPLE 4 A Group of Numbers

Show that if we regard numbers as symmetries, with multiplication as the way of combining them, the positive real numbers have the four properties of symmetries noted above.

- Multiplying two positive real numbers yields another positive real number.
- The positive real number 1 is an identity element.
- Any positive real number *x* has an inverse ($1/x$) in the collection.
- In multiplying several numbers, it doesn't matter if we first multiply some adjacent pairs of numbers; that is, it doesn't matter how we group or parenthesize the multiplication. For instance, $2 \times 3 \times 4$ is equal to $2 \times (3 \times 4) = 2 \times 12$ and also to $(2 \times 3) \times 4 = 6 \times 4$.

Hence, we say that the positive real numbers form a group under multiplication.

Group — DEFINITION

A **group** is a collection of elements $\{A, B, \ldots\}$ and an operation $\circ$ between pairs of them such that the following properties hold:

- *Closure:* The result of one element operating on another is itself an element of the collection: $A \circ B$ is in the collection.
- *Identity element:* There is a special element *I,* called the *identity element,* such that the result of an operation involving the identity and any element is that same element: $I \circ A = A$ and $A \circ I = A$.
- *Inverses:* For any element *A,* there is another element, called its *inverse* and denoted A^{-1}, such that the result of an operation involving an element and its inverse is the identity element: $A \circ A^{-1} = I$ and $A^{-1} \circ A = I$.
- *Associativity:* The result of several consecutive operations is the same regardless of grouping or parenthesizing, provided that the consecutive order of operations is maintained: $A \circ B \circ C = A \circ (B \circ C) = (A \circ B) \circ C$.

EXAMPLE 5 A Group of Non-Numbers

With all your experience with arithmetic, numbers are concrete to you, even if thinking of them in terms of a group isn't yet familiar. Here, we look at a very simple "abstract" group. The group is a collection of just three elements {A, B, C}, and it is convenient to show how the operation behaves by giving a table of its results as follows:

▲	A	B	C
A	A	B	C
B	B	C	A
C	C	A	B

The table is organized so that, for example, we find the result of A ▲ B by looking in the row for A and the column for B, finding B. So A ▲ B = B. Similarly, C ▲ B = A. We confirm that indeed this set is a group under the operation.

Since all the entries in the table are from {A, B, C}, the set is closed under the operation. You should identify which element serves as an identity element. What is the inverse to A? To B? To C? To check associativity would require checking the results of all possible products X ▲ Y ▲ Z, where each of X, Y, and Z can be any of A, B, or C. We won't go to that (tedious) length, but you should check just one example. For instance, (A ▲ B) ▲ C = B ▲ C = A, while A ▲ (B ▲ C) = A ▲ A = A. Also, can you see why there are $3^3 = 27$ products to check?

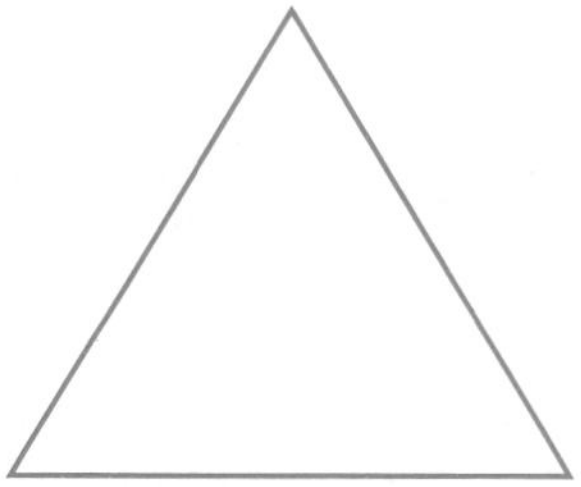

An equilateral triangle.

This particular abstract group can be interpreted concretely in several ways. One interpretation is in terms of an equilateral triangle.

Each of A, B, and C is a rotation of the triangle about its center. A is a rotation by 0°, B is a rotation clockwise by 120° (one-third of a complete turn), and C is a rotation counterclockwise by 120°. The operation ▲ just gives the result of doing one rotation followed by another. For example, B ▲ C is first to rotate the triangle 120° clockwise, then rotate it 120° counterclockwise—which leaves it as if it had not rotated at all; that is, as rotated by 0°. Hence, B ▲ C = A.

We have, in fact, explored here some of the symmetries of an equilateral triangle. (What other symmetries does an equilateral triangle have? Think of it as a rosette pattern.) ■

Symmetry Group of a Pattern DEFINITION

The symmetries that preserve a pattern form the **symmetry group of the pattern.**

EXAMPLE 6 Group Theory in Your Room, or More Than Once Upon a Mattress

Have you ever turned a mattress? The purpose of this usually seasonal practice is to even out lumps and sags and make the mattress last longer and be more comfortable.

There are various ways to "turn" a mattress, but they all require that the mattress fit back on the bed. What are all those ways?

For this example, make yourself a "mattress" from a (nonsquare) rectangular sheet of paper. (It's a lot easier to flip a piece of paper than a real mattress!) Label the initial position of the "mattress": Write "UP" in the middle of the top surface of the paper and label the edges in clockwise order "TOP," "RIGHT," "BOTTOM," and "LEFT." Then turn the sheet over so that the top edge is again at the top. Label this surface of the paper "BOTTOM" in the middle and label the edges in clockwise order "TOP," "RIGHT," "BOTTOM," and "LEFT." Now turn the sheet over again to the initial position.

We'll designate the turning over of the paper that you just did as follows:

- *Flip* (turn it over, left and right switch), denoted by *F*, amounts to a 180° rotation around an axis running from bottom to top of the mattress.
- *Rotation* (up side stays up, top and bottom switch), denoted by *R*, amounts to a 180° rotation of the mattress as seen from above, that is, around a vertical axis through the center of the mattress.
- *Toss* (turn it over and switch top and bottom), denoted by *T*, corresponds to a 180° rotation of the mattress around a horizontal axis across the middle of the short side of the mattress.

How many different positions are there for the mattress? Satisfy yourself that there are really only four positions for the mattress that fit it back on the bed: the results of the turns *F*, *R*, and *P*—together with the "un-turn," which we denote by *I*. These four turns together form a group {*I*, *F*, *R*, *P*}. The group operation is to perform one turn followed by another.

Performing one of the turns, followed by a subsequent one in the next season, actually gives the same result as a single turn; for example, doing *R* followed by *T* puts the mattress in the same position as doing just *F*. This fact is the *closure* property of the group.

The *identity element* of the group is the "un-turn" *I*. Each turn is its own *inverse:* Performing it twice in a row puts the mattress back in the initial position *I*.

Associativity of the turns is true, though it is tedious to check, but this necessary property of a group indeed holds.

The same structure of operations holds for turning your pillow. Any group with four elements, each of which is its own inverse, has basically the same abstract structure, called the *Klein 4-group* [after Felix Klein (1849–1925), who classified geometries by their symmetry groups].

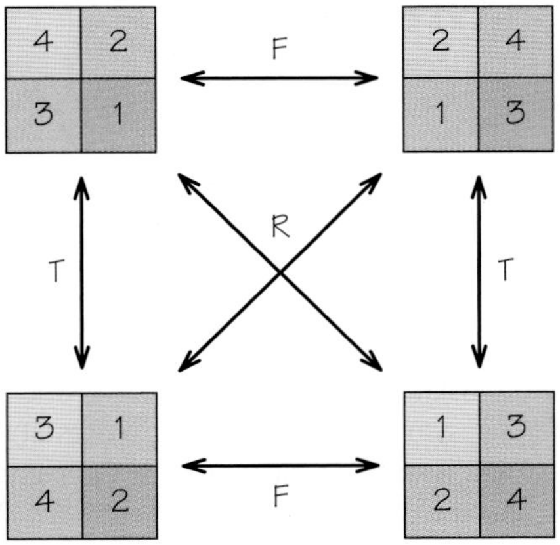

Klein 4-group

EXAMPLE 7 Symmetry Groups of Strip Patterns

Each of the strip patterns of Figure 19.11 on page 795 has a different group of symmetries. What do they have in common, and how do they differ?

The pattern of Figure 19.11a is preserved only by translations. If we let *T* denote the smallest translation to the right that preserves the pattern, then the pattern is also preserved by $T \circ T$ (which we write as T^2), by $T \circ T \circ T = T^3$, and so forth. Although the pattern looks the same after each of these translations by different distances, we can tell these translations apart if we number each copy of the motif and observe which other motif it is carried into under the symmetry.

For instance, T^2 takes each motif into another one that is located two motifs to the right. The symmetry T has an inverse T^{-1} among the symmetries of the pattern: the smallest translation to the *left* that preserves the pattern; and $T^{-1} \circ T^{-1}$ (which we write as T^{-2}), $T^{-1} \circ T^{-1} \circ T^{-1} = T^{-3}$, and so forth also are symmetries. The entire collection of symmetries of the pattern is

$$\{\ldots, T^{-3}, T^{-2}, T^{-1}, I, T, T^2, T^3, \ldots\}$$

From this listing, you see that it is natural to think of the identity I as being T^0. All of the strip patterns are preserved by translations, so the symmetry group of each includes the *subgroup* of all translations in this list. We say that the group is **generated** by T and we write the group as $\langle T \rangle$, where between the angle brackets we list symmetries that, in combination, produce all the group elements. The symmetry group of Figure 19.10e includes, in addition, a glide reflection G and all combinations of the glide reflection with the translations. Doing two glide reflections is equivalent to doing a translation, which we express as $G^2 = T$. The glide is only "half as far" as the shortest translation that preserves the pattern. Check that $G \circ T = T \circ G$. The symmetry group of the pattern is

$$\{\ldots, G^{-3}, G^{-2} = T^{-1}, G^{-1}, I, G, G^2 = T, G^3, \ldots\} = \langle G \rangle$$

The pattern of Figure 19.11c is preserved by vertical reflections at regular intervals. If we let V denote reflections at a fixed particular location, the other reflections can be obtained as combinations of V and T. To get a handle on what each of the symmetries does, it helps to make a "simplified" copy of the strip (we use Ws), number fixed positions on the page, and identify individual copies of the Ws with letters ("invisible," because the letters are not part of the pattern) as in the following:

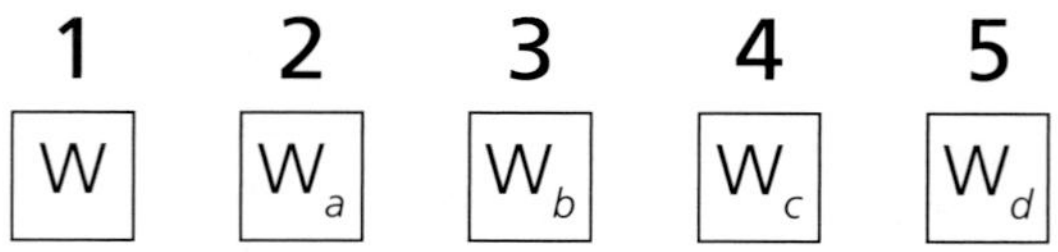

The symmetries move the Ws among the numbered positions. Let V be the reflection across the vertical line through the middle of position 3, and let T be the translation that moves each W one square to the right. To familiarize yourself with the symmetries, write out the result of each of V, T, and $TV = T \circ V$ (recall, this is V followed by T; for simplicity, we can omit the operation sign between symmetries). In each case, where does W_e end up? The symmetry group of the pattern, the list of all of the symmetries, is

$$\{\ldots, T^{-3}, T^{-2}, T^{-1}, I, T, T^2, T^3, \ldots; \\ \ldots, T^{-3}V, T^{-2}V, T^{-1}V, V, TV, T^2V, T^3V, \ldots\}$$

This group is notable because not all its elements satisfy the commutative property that $A \circ B = B \circ A$, which you are accustomed to for numerical operations ($ab = b + a$; $a \times b = b \times a$). In fact, we do not have $VT = TV$, but instead $VT = T^{-1}V$. We can express this group compactly as

$$\langle T, V \mid VT = T^{-1}V \rangle$$

where we list the symmetries that generate the group and indicate any relations that hold between them.

Self Check 7

Verify that $VT = T^{-1}V$ by working out the effect of each, using your simplified strip.

We have made a transition from thinking about patterns in geometrical terms to reasoning about them in algebraic notation—in effect, applying one branch of mathematics to another. This kind of cross-fertilization is characteristic of contemporary mathematics. The concept of a group is a fundamental one in the mathematical field of abstract algebra. The generality ("abstractness") is exactly why groups and other algebraic structures arise in so many applications, in areas ranging from crystallography, quantum physics, and cryptography, to error-correcting codes (see Chapters 16 and 17) and anthropology (describing kinship systems).

19.5 Fractal Patterns and Chaos

We noted earlier that similarity and repetition are key aspects of symmetry, as are balance and proportionality. In most of our examples, the repetitions of a motif have been at the same size. Exceptions were the chambers of the nautilus in Figure 19.2b (on page 780) and the varying sizes of leaves, seeds, and florets in plants that feature the spiral pattern of phyllotaxis (Figures 19.1 and 19.2a, on page 780). A further example in nature is a spiral galaxy. These phenomena exhibit a kind of "proportion," or numerical symmetry—symmetry with changes of scale.

Another example of similarity but at different sizes is the nested dolls (known as *matrioshka*) shown in Figure 19.16. They feature a linear scaling factor (see Section 18.1 on page 738) between one doll and another. Each part of one doll (face, arm, etc.) has the same proportion (scaling factor) to the corresponding part of a second doll.

C Squared Studios/Photodisc Green/Getty Images

Figure 19.16 Nested *matrioshka* dolls from Russia exhibit symmetry at different scales.

Fractals

Fractals are a kind of symmetry in which there is linear scaling of parts. The word **fractal** was invented in 1975 by Benoit Mandelbrot from the Latin word *fractus* meaning "broken into fragments" (of varied sizes), from which we get *fragment* and also *fracture* and *fraction* (see Spotlight 19.10). Mandelbrot defined a fractal in mathematical terms that we formulate more informally as follows.

The Father of Fractals

SPOTLIGHT 19.10

Benoit Mandelbrot (1924–2010) was born in Poland, spent part of his time growing up in France, and came to the United States to work for IBM. He found that many phenomena feature both repeating patterns and power curves (see Section 18.5 on page 764). The patterns are repeated at a change of scale, as Barnsley's fern in Figure 19.19 and the Sierpiński triangle in Figure 19.20 show, and can be described by very simple rules. In 1975, he coined the term *fractals*, and the systems of rules became known as iterated function systems. The website for the course that he pioneered at Yale University (see the Suggested Websites at the end of this chapter) lists 100 or so examples of fractal phenomena in nature and society. Mandelbrot's book *The (mis)Behavior of Markets: A Fractal View of Risk, Ruin, and Reward* (see the Suggested Readings at the end of this chapter) applies fractals to the movement of stock prices

Hank Morgan/Science Source/Getty Images

Benoit Mandelbrot

Fractal DEFINITION

A **fractal** is a pattern that exhibits similarity at ever finer scales.

What this means is that no matter how closely we zoom in, we still see the same pattern.

We show various fractals in Figure 19.17. Figure 19.17a looks to us like an orchid with pronounced "bee guides" to the pollen. With its vertical mirror line, the overall pattern has *d1* rosette symmetry. However, the basic motif of the lacy wings is repeated at an infinite number of scales. In Figure 19.17b, the "suckers" on the "tentacles" appear in smaller and smaller sizes as the tentacles wind their way toward the point at the center. The pattern in Figure 19.17c has overall rosette

(*a*) "Paradise." (*b*) "Purgatory." (*c*) "r-crest." (*d*) "Scarab 2."

All: Courtesy Noel Giffin/Spanky Fractal Database

Figure 19.17 Various fractals.

symmetry of type *c2*, but the "seahorse" motif, with two large seahorses foot-to-foot in the center, is repeated in diminishing sizes throughout. Figure 19.17d features (to our imagination) "spiky snowmen," with smaller ones growing out of the sides of larger ones. What do they look like to you? And does the overall pattern as a rosette have symmetry *c1, c2, d1,* or *d2?*

A famous example of a fractal pattern is M. C. Escher's print *Circle Limit IV* (Figure 19.18). As you examine the angels and devils closer and closer to the boundary of the circle, you notice that they are not necessarily geometrically similar to the ones at the center. However, if you imagine that the print is the image of a hemisphere, figures farther away from your viewpoint should indeed appear smaller.

Figure 19.18 M. C. Escher's fractal pattern *Circle Limit IV.*

Apart from their beauty and the opportunity that they offer as an art form (there is even "fractal music"!), fractals have become a basis for several major applications:

- Fractals with simple rules can mimic very well the structure of a leaf, a shell, a tree, or a mountain (see Figure 19.19).

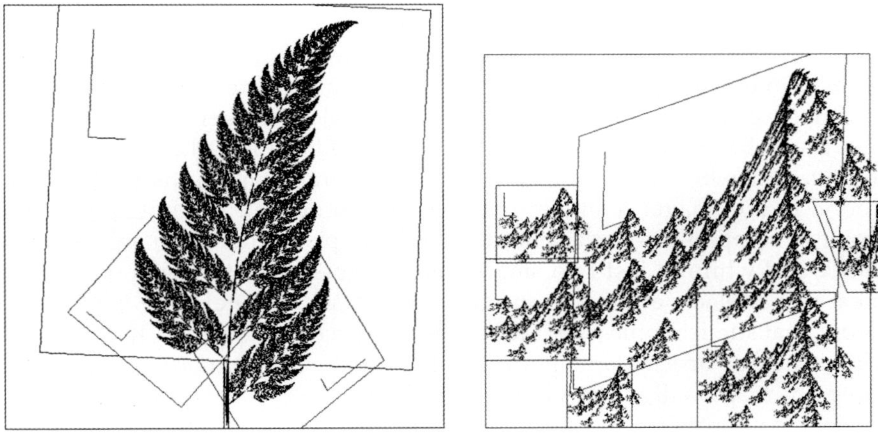

(*a*) (*b*)

Both: Frederick Solomon

Figure 19.19 (a) Barnsley's fractal fern and (b) a snowy mountain landscape, both with templates showing how they were formed using reflections and linear "distortions" in addition to linear scaling. Each leaf of the fern or each mountain peak is, in fact, just a smaller copy of the entire image.

This fact not only allows us to model natural phenomena using fractals but also suggests that they are produced by corresponding simple "rules of nature." Moreover, computer special effects in films can use fractals to mimic nature very closely, as in *Star Trek II: The Wrath of Khan,* for landscapes on the Genesis planet, and in *Return of the Jedi,* for the moons of Endor and the Death Star. In *The Perfect Storm,* fractals were used to add surface textures and even to scatter light inside water drops.

Figure 19.20 shows how a recursive rule, like the one for forming the Fibonacci numbers, can form a fractal. [This is an example of an *iterated function system* (IFS), a topic that is discussed in detail in Chapter 23 in connection with mathematical chaos.]

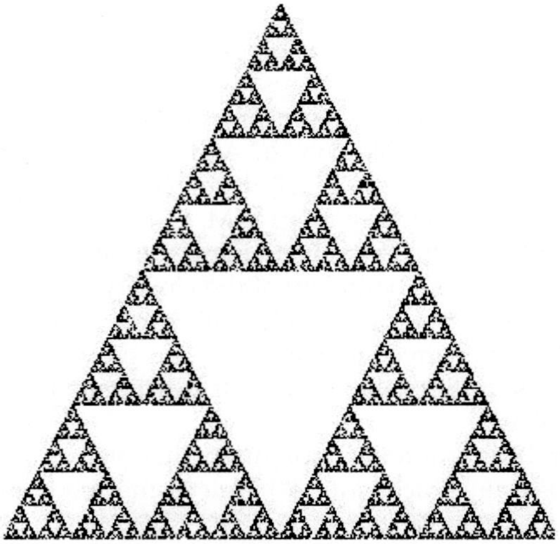

Figure 19.20 A Sierpiński "triangle." Start with the big triangle and remove its middle triangle. Then do the same for the three remaining smaller triangles. Recursively, do the same for each subsequent smaller triangle. Can you guess the area of the resulting figure? (*Hint:* It may be less than you think.)

- You are likely familiar with graphics in the JPEG compressed format. Similar in efficiency to the JPEG compression algorithm is another method of image compression that is based on fractals. The key idea is to store not the millions of bits that make up an image but instead a much smaller number (maybe thousands) of rules for generating patterns that can be found in the image. A simple example (which doesn't use fractals) is that you can compress a checkerboard of a million pixels that alternate between black and white to just two simple rules: If the current pixel is white, the next one is black, and if the current one is black, the next one is white. A more realistic example was Microsoft's *Encarta Encyclopedia.* It contained thousands of articles and photographs, plus color animations and hundreds of maps—all on one CD-ROM, thanks to fractal data compression. Fractal compression has been used for downloading satellite imagery. Although decoding a fractally compressed image is fast, the coding step is not, so fractal compression has not been applied with success to video compression.
- Antennas in a fractal shape are used in cellphone applications. Such a shape maximizes the effective length of the antenna for the space that it takes up; it also allows the antenna to operate well at a range of frequencies.

Symmetry in Chaos

While the patterns in rosettes, strips, wallpaper patterns, and some fractals can be produced from very simple rules for the symmetries involved, you may be surprised to learn that symmetry can also arise from apparently random behavior.

We think of symmetry as referring to order, and chaos to disorder and randomness. Scientists use the word *chaos* in a technical sense to describe systems whose behavior over time is inherently unpredictable. We explore chaotic systems in Chapter 23 (Section 23.5 on page 971). Here, we investigate how chaos can produce astonishingly beautiful designs on a computer screen.

One way to produce a graphic is to start with an initial pixel on the screen, apply a mathematical function (formula) to its coordinates to generate coordinates of a new pixel, light up the new pixel, then repeat the process with the new pixel—in other words, iterate recursively.

Iterate the process a large number of times—millions or even hundreds of millions of times. Since the screen has many fewer pixels than that, by what mathematicians call the *pigeonhole principle,* some pixels must be visited more than once—maybe even thousands of times.

The clue to producing art from this process is "color by number": Choose the color for each pixel according to how many times it is visited, and choose the colors with an eye toward beauty. Figure 19.21 shows an example with *d5* symmetry that was produced by 30 million iterations. The scale on the right in Figure 19.21 shows the colors for the number of times that pixels were hit; unhit pixels stay black. The order in which pixels are visited appears to be completely chaotic and is irrelevant to the final image.

- If you ignore the first thousand or so pixels visited, *it doesn't matter what pixel you start from—you get the same image!* But the pixels are visited in different orders.
- The formulas are variations on the *logistic map,* which we discuss in Chapter 23 in connection with biological populations.

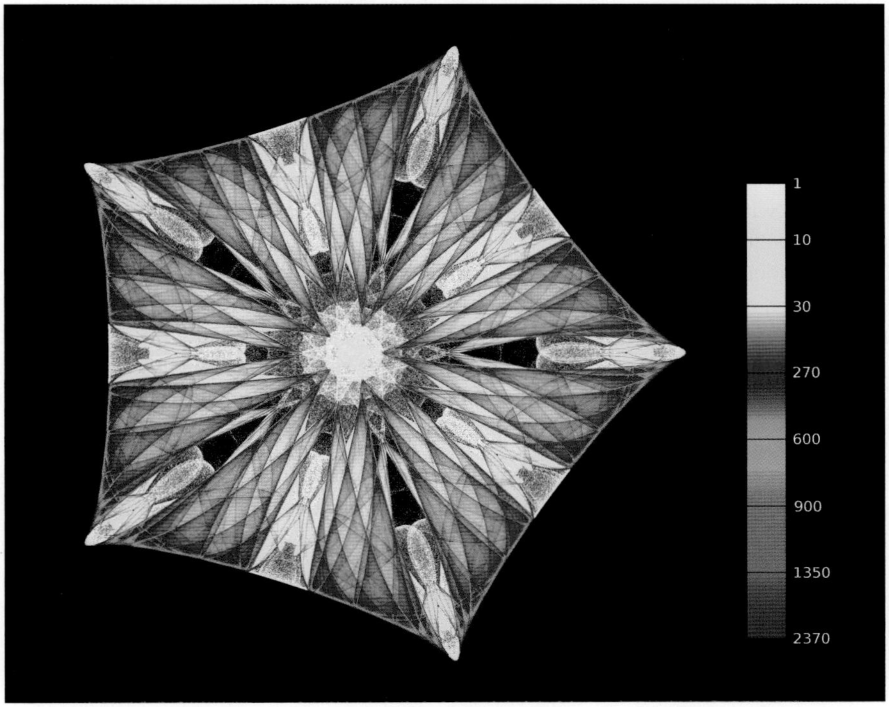

This image appears as Fig. 1.13, page 18, in Michael Field and Martin Golubitsky, *Symmetry in Chaos: A Search for Pattern in Mathematics, Art, and Nature,* 2nd Ed., Society for Industrial and Applied Mathematics (SIAM), 2009.

Figure 19.21 "Emperor's Cloak," with *d5* symmetry. This work of art was produced by iterating a chaos-producing function, starting at one point and successively generating new points according to a fixed rule. The color bar shows the coloring of pixels according to how often they are visited by the iterations.

Review Vocabulary

Crystallographic notation A four-symbol notation used by crystallographers (and mathematicians) to classify strip patterns and wallpaper patterns. (p. 798)

Divine proportion Another glorifying term for the golden ratio. (p. 783)

Fibonacci numbers The numbers in the sequence 1, 1, 2, 3, 5, 8, 13, 21, 34, ... Each number after the second one is obtained by adding the two preceding numbers. (p. 780)

Fractal A pattern that exhibits similarity at ever-finer scales. (pp. 807, 808)

Generated A group is generated by a particular set of elements if composing them and their inverses in combinations can produce all elements of the group. (p. 806)

Geometric mean The geometric mean of two numbers a and b is $\sqrt{ab}$. (pp. 784, 785)

Glide reflection A glide reflection is a translation (= glide) combined with alternating reflection in a line parallel to the translation direction. Example: pbpbpb. (p. 792)

Golden ratio, golden mean Inflated names for the number

$$\phi = \frac{1 + \sqrt{5}}{2} \approx 1.618033 \qquad \text{(p. 783)}$$

Golden rectangle A rectangle, the lengths of whose sides are in the golden ratio. (p. 783)

Group A group is a collection of elements with an operation on pairs of them such that the collection is closed under the operation, there is an identity for the operation, each element has an inverse, and the operation is associative. (p. 803)

Isometry Another word for rigid motion. Angles and distances, and consequently shape and size, remain unchanged by a rigid motion. For plane figures, there are only four possible isometries: reflection, rotation, translation, and glide reflection. (p. 791)

Phyllotaxis The spiral pattern of shoots, leaves, or seeds around the stem of a plant. (p. 782)

Preserves the pattern A transformation preserves a pattern if all parts of the pattern look exactly the same after the transformation has been performed. (p. 792)

Recursion A method of defining a sequence of numbers, in which the next number is given in terms of previous ones. (p. 781)

Reflection symmetry Mirror-image symmetry. (p. 792)

Rigid motion A transformation of the plane that preserves the size and shape of figures. In particular, any pair of points is the same distance apart after the transformation as before. (Also called *isometry*.) (p. 791)

Rosette pattern A pattern whose only symmetries are rotations about a single point and reflections through that point. (p. 793)

Rotational symmetry A figure has rotational symmetry if a rotation about its "center" leaves it looking the same, as occurs with the letter *S*. (p. 796)

Strip pattern A pattern that has indefinitely many repetitions in one direction. (p. 793)

Symmetry of the pattern A transformation of a pattern is a symmetry of the pattern if it preserves the pattern. (p. 793)

Symmetry group of the pattern The group of symmetries that preserve the pattern. (p. 804)

Translation A rigid motion that moves everything a certain distance in one direction. (p. 791)

Translation symmetry An infinite figure has translation symmetry if it can be translated (slid, without turning) along itself without appearing to have changed. Example: *AAA*. (p. 794)

Wallpaper pattern A pattern in the plane that has indefinitely many repetitions in more than one direction. (p. 793)

Self Check Answers

1. 144

2. 100

3. Translations, vertical reflections, half-turns, and glide reflections

4. *pm11*

5. *pma2*

6. $\{I, V, H, R, R^2, R^3, D_1, D_2\}$, where H and V are horizontal and vertical reflection (as in the rectangle case), R is a quarter turn, and D_1 and D_2 are reflections across the diagonals.

7. Both VT and $T^{-1}V$ produce $W_e\ W_d\ W_c\ W_b\ W_a \ldots$ with W_b in position 3 (the position of the axis of the vertical reflection V).

Skills Check

1. Symmetry includes notions of
(a) balance.
(b) similarity.
(c) repetition.
(d) all of the above.

2. Many people think that mathematics is just about numbers, but in fact mathematics includes the study of ________________.

3. Which of the following rectangles is closest to the aspect ratio of a golden rectangle?
(a) 16 by 9
(b) 13 by 6
(c) 11 by 8

4. The geometric mean of 4 and 36 is ____________.

5. Which artist claimed to use the golden ratio in his work?
(a) Leonardo da Vinci
(b) Wolfgang Amadeus Mozart
(c) Neither

6. In the Fibonacci sequence, ______ is the next element after 21.

7. Recursion means
(a) cursing over and over.
(b) giving a next number in terms of previous ones.
(c) giving previous numbers in terms of the current one.

8. Of pineapples, pine beetles, and pine bark, only ____________ exhibit Fibonacci numbers.

9. Phyllotaxis is
(a) the town in Pennsylvania where fractals were invented, near Fillodoughphia.
(b) more taxis than you need.
(c) a spiral pattern in plants.

10. The geometric mean of 4, 16, and 125 is _______.

11. A rigid motion always moves any pair of points
(a) in the same direction.
(b) to another pair of points the same distance apart.
(c) to their mirror images.

12. The capital letters ___, ___, ___, ___, ___, ___, and ___ each have a rotation isometry.

13. Assume that the following two patterns continue in both directions. Which of these patterns has a reflection isometry?

ZZZZZZZZZ
UUUUUUUUU

(a) ZZZZZZZZZ only
(b) UUUUUUUUU only
(c) Neither

14. This strip pattern

has ____________ and ____________ isometries.

15. What isometries does this wallpaper pattern have?

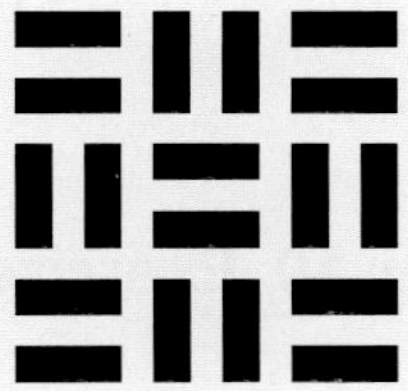

(a) Translation and reflection only
(b) Translation and rotation only
(c) Translation, rotation, and reflection

16. This wallpaper pattern

has ____________ and ____________ isometries.

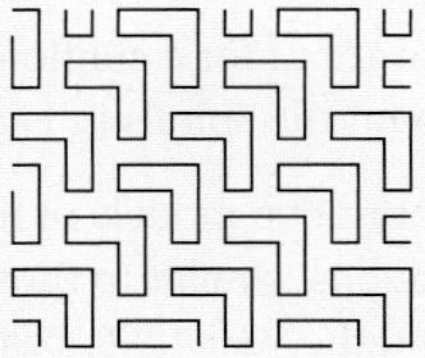

17. If a horizontal strip pattern has a glide reflection isometry, then
(a) it always has a horizontal reflection isometry.
(b) it may also have a horizontal reflection isometry.
(c) it cannot have a horizontal reflection isometry.

18. If a strip pattern has both vertical and horizontal reflection isometries, then it always has a ___________ isometry.

19. Consider the strip pattern in the raffia cloth held by the woman in the photo at right. What isometries does it have?

(a) Vertical reflection

(b) Horizontal reflection

(c) Glide reflection

Dorothy K. Washburn

20. There are ____ ways to repeat a pattern along a strip.

21. The key mathematical idea for detecting and analyzing patterns is

(a) cataloging possible motifs.

(b) figuring out the symmetries.

(c) looking for the golden ratio.

22. The symbol *p* indicates that a strip pattern has _____________ symmetry.

23. The symbol *2* indicates that a pattern has

(a) rotational symmetry.

(b) reflection symmetry.

(c) too much symmetry.

24. The symbol *m* indicates that a wallpaper pattern has ___________________ symmetry.

25. The symmetry group of a nonsquare rectangle has how many elements?

(a) 4

(b) 6

(c) 8

26. The symmetry group of the strip pattern *pmm2* has __________ elements.

27. The symmetry group of a square has how many elements?

(a) 2

(b) 4

(c) 8

28. In the mattress group, the result of a flip followed by a rotation is a _______________.

29. In mathematics, chaos refers to

(a) randomness.

(b) disorder.

(c) unpredictability.

30. A fractal is a pattern that exhibits ___________ at ever finer scales.

Chapter 19 Exercises

Challenge Discussion

19.1 Fibonacci Numbers and the Golden Ratio

1. Examine the "scales" on the surface of a pineapple, which are arranged in spirals around the fruit in three distinct directions. For each direction, how many spirals are there?

2. Repeat Exercise 1, but for a pinecone from your area.

3. Repeat Exercise 1, but for a sunflower. (If you can't obtain one, then count the spirals in Figure 19.1b.)

4. Here are two primitive models of natural increase of biological populations, similar to those that Fibonacci hypothesized around the year 1200, based on the following situation: A pair of newborn male and female rabbits is placed in an enclosure to breed.

(a) Suppose that the rabbits start to bear young one month after their own birth. At the end of each month, they have another male–female pair, which in turn mature and start to bear young one month later. Assuming that none of the rabbits die, how many pairs of rabbits will there be at the end of six months from the start (just before any births for that month)? (*Hint:* Draw a month-by-month chart of the situation at the end of the month, just before any births.)

(b) Repeat part (a), but assume instead that the rabbits start to bear young exactly two months after their own birth.

5. Put the golden ratio $\phi = (1 + \sqrt{5})/2$ into the memory of your calculator.

(a) Look at the value. Now square it (either use the x^2 button or multiply it by itself). What do you observe?

(b) Go back to ϕ. Now take its reciprocal (either use the $1/x$ button or divide it into 1). What do you observe?

(c) What formula explains what you saw in part (a)?

(d) What formula explains what you saw in part (b)?

6. The golden ratio ϕ satisfies the equation $x^2 = x + 1$. Show that $(1 - \phi)$ also satisfies the equation, so that $(1 - \phi)$ is the other solution to the equation.

7. The geometric mean has interpretations in both arithmetic and geometry.

(a) Find the geometric mean of 3 and 27.

(b) Find the length of a side of a square that has the same area as a rectangle that is 4 by 64.

(c) You are to make a golden rectangle with 6 inches of string. How long should it be, and how wide?

8. What is the geometric mean of 3, 9, 81, and 243?

9. What is the geometric mean of 2, 4, 8, 16, and 32? (Such a sequence, in which each successive number is the same constant times the previous one, is called a *geometric sequence.*)

10. Using its recursion rule, extend the Fibonacci sequence to the left. For example, the value of F_0 must satisfy $F_0 + F_1 = F_2$, or $F_0 + 1 = 1$, so $F_0 = 0$. Using the same idea, find $F_{-1}, F_{-2}, \ldots, F_{-6}$.

11. Another sequence closely related to the Fibonacci sequence is the *Lucas sequence,* which is formed using the same recursive rule but different starting numbers. The nth Lucas number L_n is given by

$$L_1 = 1,\ L_2 = 3, \quad \text{and} \quad L_{n+1} = L_n + L_{n-1} \quad \text{for} \quad n \geq 2$$

(a) Calculate L_3 through L_{10}.

(b) Calculate the ratio of successive terms of the Lucas sequence:

$$\frac{L_2}{L_1}, \frac{L_3}{L_2}, \ldots \frac{L_{10}}{L_9}$$

What do you notice?

12. Repeat Exercise 11, but start with the pair of numbers 1 and 4.

13. Repeat Exercise 11, but start with a pair of numbers of your choice. Based on your result and those in Exercises 11 and 12, what is your hunch?

14. For a sequence specified by a recursive rule, finding an explicit expression for the nth term is not easy, nor is the form necessarily simple. An exact expression for the nth term of the Fibonacci sequence is given by the Binet formula:

$$F_n = \frac{1}{\sqrt{5}}\left(\frac{1+\sqrt{5}}{2}\right)^n - \frac{1}{\sqrt{5}}\left(\frac{1-\sqrt{5}}{2}\right)^n$$

(a) Verify the formula for $n = 1$ and $n = 2$ by multiplying out, not by using a calculator.

(b) Use the Binet formula and your calculator to find F_5.

(c) In fact, the second term on the right of the equation gets closer and closer to 0 as n gets large. Because we know that the Fibonacci numbers are integers, we can just round off the result of calculating the first term. Find F_{13} by calculating the first term with your calculator and rounding.

15. For two positive numbers x and y, show that the arithmetic mean is always greater than or equal to the geometric mean. Try some values for x and y and convince yourself, then demonstrate algebraically that it is true in general. When does equality hold? [*Hint:* Suppose that the claim is false, so that $(x + y)/2 < \sqrt{xy}$. Square both sides of the inequality, bring all terms to one side, factor, and observe a contradiction.]

16. You may remember having to work problems like, "If Joe can dig a ditch in 3 days, and Sam can dig it in 4, how long will it take the two of them working together?" The answer is related to the harmonic mean of 3 and 4. The Environmental Protection Agency uses the harmonic mean to calculate the "average" fuel economy of the fleet of cars from a manufacturer. The formula for the harmonic mean of two numbers x and y is

$$\frac{2}{\frac{1}{x} + \frac{1}{y}} = \frac{2xy}{x + y}$$

(a) Calculate the answer for Joe and Sam, which is one-half of the harmonic mean of 3 and 4. Explain why this is the correct answer. (In terms of fuel economy, using the harmonic mean assumes that each car is driven the same number of miles; using the arithmetic mean would assume that each car uses the same amount of fuel.)

(b) Show that the harmonic mean of two positive numbers is always less than or equal to the geometric mean. (Thus, in light of Exercise 15, we have the general conclusion that $H \leq G \leq A$, where H stands for the harmonic mean, G for the geometric mean, and A for the arithmetic mean.) (*Hint:* Suppose that the claim is false. Simplify the fraction that is the harmonic mean, square both sides of the inequality, and proceed as in Exercise 15.)

(c) Show once more that the harmonic mean of two positive numbers is always less than the geometric mean, but this time do it with less work: Let $A = \frac{1}{x}$ and $B = \frac{1}{y}$, and discover one connection (equation) between the harmonic mean of x and y and the arithmetic mean of A and B, and a second connection between the geometric mean of x and y and the geometric mean of A and B. Then use Exercise 15 on A and B.

(d) What should be the formula for the harmonic mean of three numbers? Of n numbers?

17. New houses are to be built along one side of a street (Leonardo's Lane), divided into equal-sized lots. Each house is either a single-family detached house, taking up one lot, or a duplex, taking up two lots. Suppose that there are n lots on the street. How many different arrangements (orderings) of houses are there, for $n = 1, 2, 3, 4, 5$, and in general? (This exercise was inspired by a puzzle by Paul Dixon at the website by Ron Knott, cited in the Suggested Websites at the end of this chapter.)

18. When it snows in the winter, the local school district superintendent must decide by 5 A.M. whether to declare a snow day and cancel school. The 900 faculty and staff are now notified by a robocall broadcast, but formerly a binary "telephone tree" was used, in which the superintendent called two people and each person who received a call called two others. Suppose that each call takes exactly 1 minute.

(a) Draw the telephone tree of calls for, and determine how many calls take place in, the first 1 minute, 2 minutes, 3 minutes, 4 minutes, and 5 minutes.

(b) How many calls does it take to notify all the faculty and staff? How long does that take?

(This exercise was inspired by a puzzle at the website by Ron Knott.)

19. Here is a trick to "prove" that you can calculate faster than a person with a calculator. Turn your back and ask a friend to write down any two positive integers, then add them to get a third, then add the second and third to get a fourth, and so on, adding each time the last two integers until there are 10 numbers. Have your friend show you the list, whereupon you write down right away the total of all 10, while your friend begins to add them up on the calculator (to prove that you're right). The secret: The total is always 11 times the seventh number, and multiplying by 11 is fairly easy to do in your head—just add each pair of neighboring digits, carrying if necessary. Suppose that your friend writes down m and n as the first two numbers. Show that indeed the total of all 10 numbers is 11 times the seventh number. (Adapted from Martin Gardner, *Mathematical Circus*, Knopf, New York, 1979.)

20. The game of Fibonacci Nim begins with n counters. Two players take turns removing at least one counter, but no more than twice as many as the opponent just did. The winner is the player who takes the last counter. One other rule: The first player may not win immediately by taking all the counters on the first turn. (Adapted from Martin Gardner, *Mathematical Circus*, Knopf, New York, 1979.)

(a) Play this game taking turns with an opponent and starting with different numbers n of counters and try to come up with a strategy for one player or the other to win. (*Hint:* The key is that any positive integer can be represented uniquely as a sum of Fibonacci numbers.)

(b) Proceed as in part (a), but with the rule changes that the player who takes the last counter loses and the first player may not take all but one counter.

21. Shari Lynn Levine, as a high school student, published an article in *The Fibonacci Quarterly* that investigated the "Beta-nacci" sequence that results if, instead of bearing one pair of baby rabbits per month, mature rabbits bear two pairs every month, starting when they reach two months of age. Here, we ask you to rediscover some of Shari's results.

(a) How many rabbits will there be each month for the first 12 months?

(b) What is the recursive rule for the nth Beta-nacci number B_n?

(c) For the terms of the sequence in part (a), calculate the ratios of successive terms. (*Motivating hint:* It's not the golden ratio this time.)

(d) Suppose that the ratio of successive terms approaches a number x. We show how to find x exactly. For very large n, we have $B_{n+1} \approx xB_n \approx x^2B_{n-1}$. Inserting these values into the recursive rule for the sequence and dividing by B_{n-1} gives the equation $x^2 = x + 2$. Solve this equation for x (you can use the quadratic formula). Make a table of values of $3B_n$ versus 2^n. From the evidence, can you suggest a formula for B_n?

22. Generalize Exercise 21, parts (a) through (d), as follows:

(a) to the case of each pair of rabbits having three pairs of rabbits (the "Gamma-nacci" sequence).

(b) to the case of each pair of rabbits having q pairs of rabbits.

For Exercises 23 and 24, refer to the following. We have seen that the golden ratio is a positive root of the quadratic polynomial $x^2 - x - 1$. We can generalize this polynomial to $x^2 - mx - 1$ for $m = 1, 2, 3, \ldots$ and consider the positive roots of those polynomials as generalized means—the *metallic means family,* as they are sometimes known. In particular, for $m = 2, 3, 4$, and 5, we have, respectively, the silver, bronze, copper, and nickel means. It is surely surprising that these numbers arise both in connection with quasicrystals (investigated in Chapter 20) and in analyzing the behavior of some dynamical systems (investigated in Chapter 23) as the systems evolve into chaotic behavior.

23. Use the quadratic formula to find expressions in terms of square roots for the silver, bronze, copper, and nickel means, and approximate these to three decimal places. Find a general expression in terms of a square root for the *m*th metallic mean.

24. Just as the golden mean arises as the limiting ratio of consecutive terms of the Fibonacci sequence, each of the metallic means arises as the limiting ratio of consecutive terms of generalized Fibonacci sequences. A generalized Fibonacci sequence G can be defined by

$$G_1 = 1, G_2 = 1, \quad \text{and} \quad G_{n+1} = pG_n + qG_{n-1} \quad \text{for} \quad n \geq 2$$

where p and q are positive integers. The Fibonacci sequence itself is the case $p = q = 1$.

(a) Try various small values of p and q and determine which means they lead to.

(b) Divide the equation for G_{n+1} by G_n. Assume that G_{n+1}/G_n and G_n/G_{n-1} both tend toward the same number x as n gets large, replace those quantities by x, and simplify the resulting equation. What must be the value of x?

(c) What happens to the sequence and to the mean if we allow one or both of p and q to be negative integers?

19.2 Rosette and Strip Patterns

25. Determine whether each of the following statements is always or sometimes true. Drawing some sketches may be helpful.

(a) A line reflection preserves collinearity of points. That is, if the points *A, B,* and *C* are in a straight line (collinear), then their images reflected in some other line also lie in a straight line.

(b) A line reflection preserves betweenness. That is, if the collinear points *A, B,* and *C* (with *B* between *A* and *C*) are reflected about a line, then the image of *B* is between the images of *A* and *C*.

(c) The image of a line segment under a line reflection is a line segment of the same length.

(d) The image of an angle under a line reflection is an angle of the same measure.

(e) The image of a pair of parallel lines under a line reflection is a pair of parallel lines.

26. Determine whether each of the following statements is always or sometimes true. Drawing some sketches may be helpful.

(a) The image of a pair of perpendicular lines under a line reflection is a pair of perpendicular lines.

(b) The image of a square under a line reflection is a square.

(c) Label the vertices of a square *A, B, C,* and *D* in a clockwise direction. Then their images *A′, B′, C′,* and *D′* under a line reflection also follow a clockwise direction.

(d) The length of the perimeter of a geometric figure is equal to the length of the perimeter of its image under a line reflection.

(e) The image of a vertical line under a line reflection is always a vertical line.

27. Which of the capital letters of the alphabet, when drawn in the most symmetrical way, have the following symmetries? For example, assume that the upper and lower loops of B are the same size.

(a) A horizontal line of reflection symmetry

(b) A vertical line of reflection symmetry

(c) A rotational symmetry

28. Repeat Exercise 27 for the lowercase letters.

29. In *The Complete Walker III* (3d ed., Knopf, New York, 1984, p. 505), Colin Fletcher's answer to "What games should I take on a backpacking trip?" is the game he calls "Colinvert": "You strive to find words with meaningful mirror (or half-turn) images." Some of the words he found are as follows:

MOM WOW pod Mud bUM

(a) Which of his words reflect into themselves?

(b) Which of his words rotate into themselves?

(c) Find some more words or phrases of these various types—the longer, the better.

30. Repeat Exercise 29, but for words written vertically instead of horizontally.

31. For each of the following patterns, identify the rigid motions that preserve the pattern:

(a) CCCCCCCCCC

(b) GGGGGGGGGG

(c) HHHHHHHHHH

(d) MMMMMMMMMM

32. Repeat Exercise 31, but for the following patterns:

(a) SSSSSSSSSS

(b) bdbdbdbdbd

(c) dbpqdbpqdbpq

19.3 Notation for Patterns

33. What is the notation for the symmetry pattern of a regular pentagon (which has all five sides equal)—*d5, c5,* or something else?

34. What is the notation for the symmetry pattern of a snowflake with six symmetrical arms, with each arm having mirror symmetry?

35. Give the notation (such as *d4* or *c5*) for the symmetry patterns of the rosettes in hubcaps (a) through (c) below, disregarding the logos in the centers.

36. Repeat Exercise 35 for hubcaps (d) through (f).

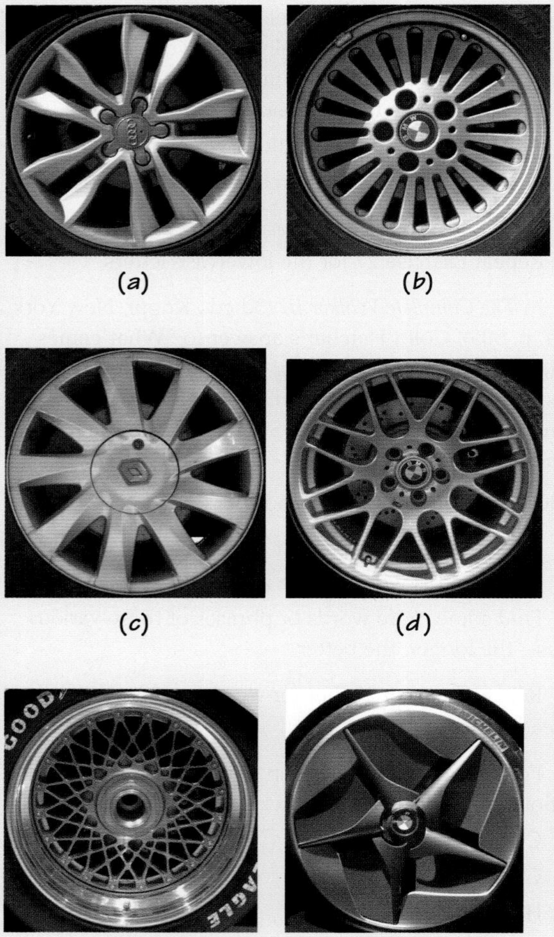

(a) (b) (c) (d) (e) (f)

All hubcap photos courtesy of Paul J. Campbell (Beloit College, Wisconsin), who thanks Joe Gallian (University of Minnesota, Duluth) for the idea.

37. Repeat Exercise 35 for corporate logos (a) through (c) below. (*Bonus:* Can you identify the corporations?)

(a) (b) (c)

38. Repeat Exercise 35 for automobile logos (d) through (f) below.

(d) (e) (f)

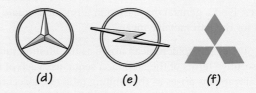

For Exercises 39–40, refer to the following. Step patterns are found in Celtic illuminated manuscripts, metalwork, and stone crosses. Square ones were constructed by first designing on a square lattice one-quarter of the pattern (say, the top right), using horizontal and diagonal lines to produce a prototype such as the following:

Then three copies were added, either by (1) rotating the original successively by 90° [as in accompanying illustration (a)], or else by (2) reflecting it across its right and bottom edges [as in illustration (b)]. (Based on research by Mark A. M. Lynch of Glasgow Caledonian University, Scotland.)

(a) (b)

39. Identify the rosette pattern for

(a) step pattern (a).

(b) step pattern (b).

40. Which rosette pattern would result if the prototype, unlike the one above, has reflection symmetry across its diagonal from the top left to the lower right and

(a) strategy (1) is used.

(b) strategy (2) is used.

41. Use the flowchart in Figure 19.13 (page 799) to identify the notation for the types of strip patterns from the pottery and basketry shown in the illustrations on the next page.

All: Photos courtesy of Logan Museum of Anthropology, Beloit College, Beloit, WI

(a)

(b)

(c)

(d)

(e)

(f)

(g)

Pottery and basketry from the Americas.
(a), (b) Mexico, modern.
(c) Lower Central America, pre-Columbian.
(d) Pomo people, California, early 20th century.
(e) Woodland Indians, central North America, early 20th century.
(f) Pomo people, California, mid-20th century; originally from the collection of Dr. Herbert Zim, editor of the Golden Guides series of nature books.
(g) Woodland Indians, central North America, early 20th century.

42. In each of the four accompanying examples, two adjacent triangles of an infinite strip are shown. (Contributed by Margaret A. Owens, California State University, Chico.)

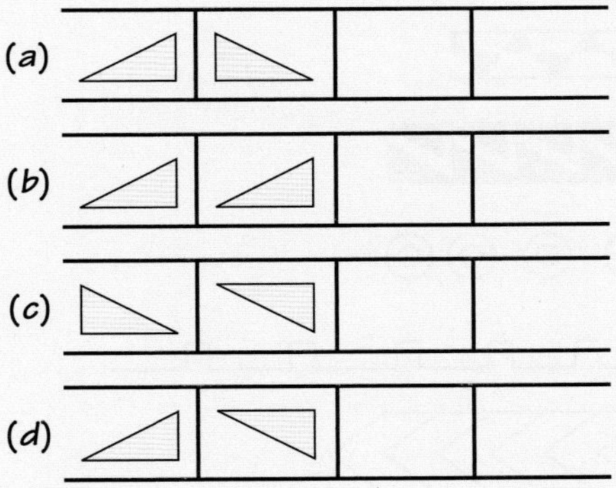

For each example:

(a) Determine a motion (translation, reflection, rotation, or glide reflection) that takes the first (left) triangle to the second (right) one.

(b) Draw the next four triangles of the infinite strip that would result if the second triangle is moved to the next space by another motion of the same kind, and so on.

(c) Identify (by notation) the resulting strip as one of the seven possible strip patterns.

43. Repeat Exercise 41 for the accompanying eight strip patterns, all of which appear on the brass straps for a single lamp from 19th-century Benin in West Africa. (From H. Ling Roth, in Great Benin.)

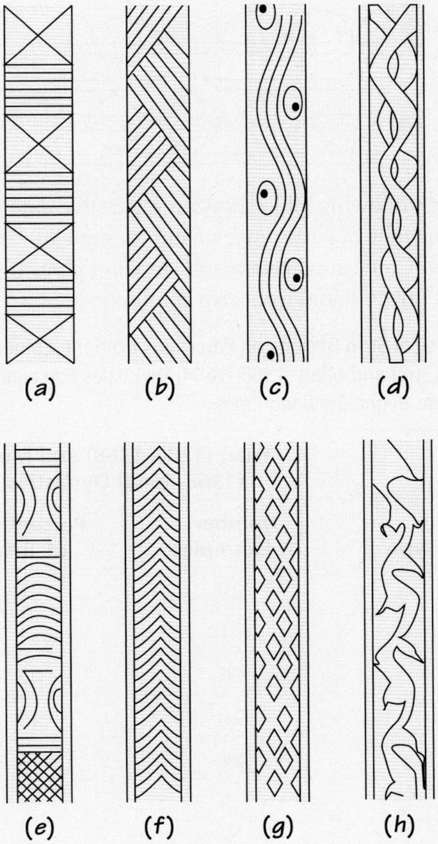

Note that the patterns are roughly carved, so you will need to discern the intent of the artist.

44. Repeat Exercise 41 for the accompanying patterns from San Ildefonso Pueblo, New Mexico.

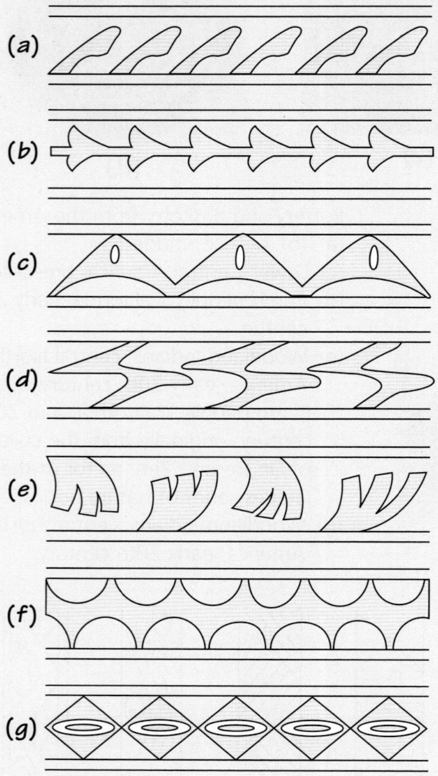

45. The following table shows comparative data about the frequency of occurrence of strip designs of various types on Chinese porcelain and smoking pipes (Begho, in what is now Ghana) from two continents.

Frequency of Strip Designs on Porcelains from the Chinese Yuan (1280–1368) and Ming (1368–1644) Dynasties Porcelains and on African Begho Smoking Pipes

	Yuan (1280–1368) and Ming (1368–1644) Dynasties	
Strip Type	**Number of Examples**	**Percentage of Total**
p111	29	18
p1m1	1	1
pm11	66	42
p112	20	13
p1a1	21	13
pma2	13	8
pmm2	9	6
TOTAL	**159**	

	Begho	
Strip Type	**Number of Examples**	**Percentage of Total**
p111	4	2
p1m1	9	4
pm11	22	10
p112	19	8
p1a1	2	1
pma2	9	4
pmm2	165	72
TOTAL	**230**	

(a) Which types of motions appear to be preferred for designs from each of the two localities?

(b) What other conclusions do you draw from the data of this table?

(c) On the evidence of the table alone, in which locality is each of the following strip patterns (but not necessarily the motif) most likely to have been found?

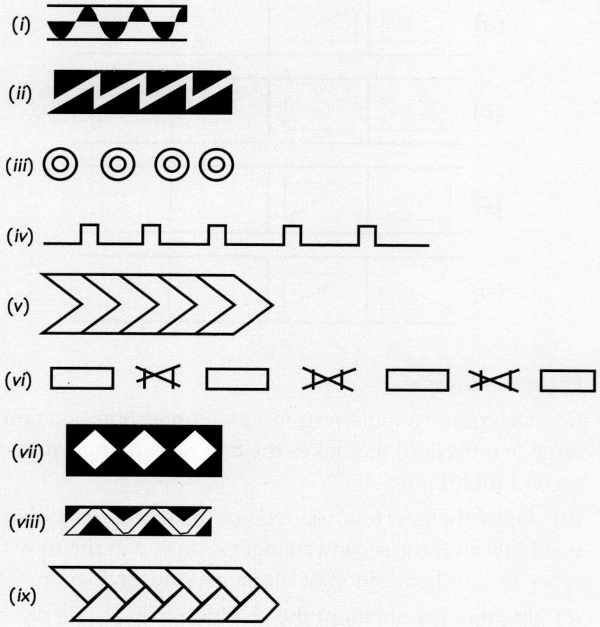

46. For the Nigerian Yoruba cloths (a) and (b) in the following illustration, use the flowchart in Spotlight 19.9 (page 797) to identify (by notation) the type of wallpaper pattern.

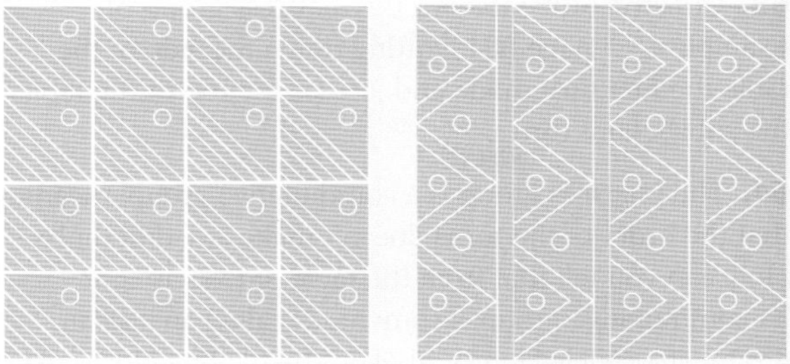

(a) (b)

Patterns on Yoruba (West Africa) adire cloth, made by starching a pattern onto white cloth, then dyeing the cloth before rinsing out the starch, so that the starched portion remains as a white design against a colored background.

47. For the Nigerian Yoruba cloths (c) and (d) in the accompanying illustration, use the flowchart in Spotlight 19.9 (page 797) to identify (by notation) the type of wallpaper pattern.

(c) (d)

48. For the floor tilings (a) and (b) below, use the flowchart in Spotlight 19.9 (page 797) to identify (by notation) the type of wallpaper pattern.

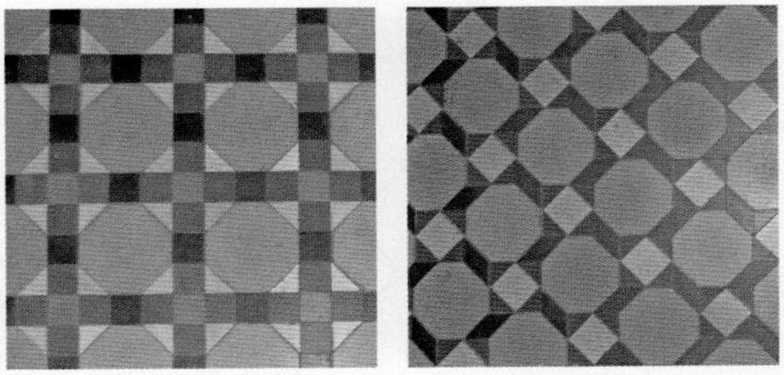

Both: Jaume de Dios Pont

(a) (b)

49. For the floor tilings (c) and (d) that follow, use the flowchart in Spotlight 19.9 (page 797) to identify (by notation) the type of wallpaper pattern.

Both: Jaume de Dios Pont

(c) (d)

50. The triangles in the grid at the top of the following figure show the beginning steps in forming instances of several of the wallpaper patterns by putting together a vertical motion and a horizontal motion.

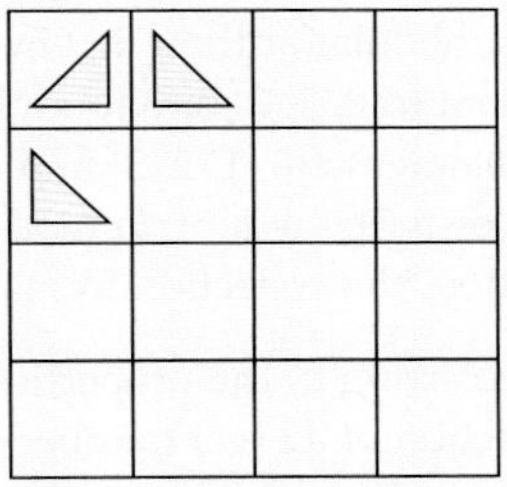

(a) Identify the horizontal motion.
(b) Identify the vertical motion.
(c) Fill in the remaining empty squares.
(d) Identify the wallpaper pattern.

51. Which of the 17 wallpaper patterns can be formed by the technique used in Exercise 50?

For Exercises 52–57, refer to the following. In Chapter 20, we study both repeating and nonrepeating plane patterns, from the point of view of their basic building blocks (tiles). Here, we ask you to analyze the repeating patterns from figures in that chapter according to wallpaper type, using the flowchart of Spotlight 19.9 (page 797). Identify all the symmetries and give the notational type for the wallpaper pattern of the illustrations that follow.

52. Figure 20.10 (page 837)

53. Figure 20.11 (page 838)

54. Figure 20.16 (page 841)

55. The figure in Spotlight 20.2 (page 836)

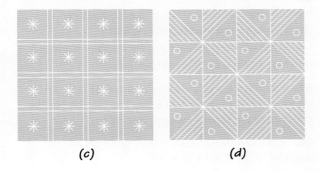

56. The hexagonal regular tiling at upper right in Figure 20.5 (page 831)

57. The convex hexagon tiling of type 3 in Figure 20.9 (page 835)

58. Visit the website escher.epfl.ch/escher/, which features an interactive Java program called Escher Web Sketch. Experiment with choosing wallpaper patterns using crystallographic notation. For each one, draw on the screen a colored design for the motif; the program will reproduce the motif using the pattern.

19.4 Symmetry Groups

59. For positive integers a and n, the expression a mod n means the remainder when a is divided by n. Thus, 23 mod 4 = 3, because $23 = 5 \times 4 + 3$, and we say that "23 is equivalent to 3 modulo 4." (See Chapter 17 for further details about this modular arithmetic.) Every positive integer is equivalent to 0, 1, 2, or 3 modulo 4. Consider the collection of elements {0, 1, 2, 3} and the operation on them defined by $a \oplus b = (a + b) \bmod 4$. Show that under this operation, the collection forms a group.

60. Explain, by referring to the properties of a group, whether the collection of all real numbers is a group under the operation of (a) addition and (b) multiplication.

61. Explain why the table for the operation * below shows that the elements indicated do not form a group under *.

*	A	B	C
A	B	B	B
B	B	C	A
C	C	A	B

62. Consider the table below for an operation #.

#	A	B	C	D	E	F
A	A	B	C	D	E	F
B	B	A	D	C	F	E
C	C	E	A	F	B	D
D	D	F	B	E	A	C
E	E	C	F	A	D	B
F	F	D	E	B	C	A

(a) Explain why the elements form a group under #. (Don't bother to check associativity.)

(b) What do you notice about $F \# E$ versus $E \# F$? (This is the smallest example of a group that is noncommutative.)

63. Construct the table for the mattress group of Example 6 on page 804, putting down the side of the table the first turn made and, across the top of the table, the subsequent turn.

64. The mattress group is a commutative group, meaning that the order of the turns doesn't make any difference. How can you tell that from the table of Exercise 63? [A commutative group is sometimes called an *abelian* group, after Niels Henrik Abel (1801–1829), a Norwegian mathematician who died of tuberculosis at a young age.]

65. A problem about mattress turning is that people usually don't remember the immediately previous position or immediately previous turn that they made months ago. So the next time, they could wind up just turning the mattress back to the position that it was in only two seasons ago. Show that the mnemonic "Spin in spring, flip in fall" (courtesy of Bill Sandidge of Atlanta, Georgia), in fact, cycles the mattress through its four positions.

66. A king mattress is officially 76 in. × 80 in., which is almost square. Make yourself a "square mattress" from a sheet of paper and investigate its group.

(a) What are the possible turns for a square mattress?

(b) Make a table for this mattress group.

67. Show that the collection of numbers {1, 3, 5, 7} under multiplication modulo 8 (see Exercise 59) has the structure of the Klein 4-group (page 805).

68. Show that the dihedral rosette group $d4$ has the structure of the Klein 4-group.

For Exercises 69–72, refer to the following. Like mattresses, car tires need to be rotated so as to promote even wear; wear on a tire varies with wheel position. The three main rotation schemes are shown below.

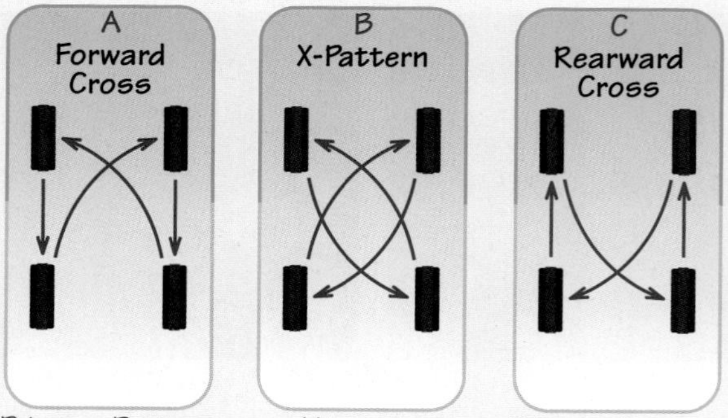

Primary Pattern for Front-Wheel Drive Vehicles — Alternate Pattern — Primary Pattern for Rear- and Four-Wheel Drive Vehicles

69. Show that successive tire rotations using scheme A form a group. Is it the Klein 4-group (page 805)?

70. Observe that repeating scheme B will never take the front right tire to the back right wheel. Is there any combination of schemes A and C that will produce the result of scheme B?

71. A tire rotation scheme is designed so that no tire will remain where it was. Such a rearrangement (permutation) of objects is called a *derangement.* Two more tire rotation schemes are shown below. Are there still more schemes that are derangements? One way to record a tire rotation is to label the original tire positions clockwise from the left front: 1 for left front, 2 for right front, 3 for right rear, and 4 for left rear. Record the results of a scheme by writing in turn where each tire goes; for example, scheme D below produces the derangement 4321 because tire 1 goes to position 4, tire 2 goes to position 3, and so forth.

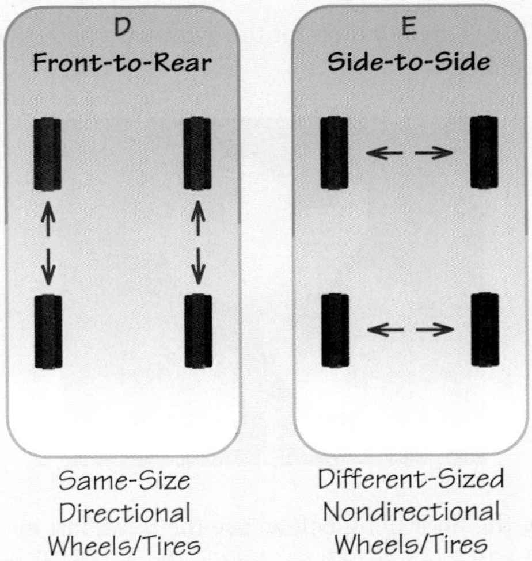

72. Do the five tire rotation schemes A, B, C, D, and E, plus the identity rotation and any others that you found in Exercise 69, form a group? Why or why not?

73. For the traditional North American beadwork shown below, answer the following questions.

(a) Which rosette pattern does it have?

(b) Specify two rigid motions that together generate the group of the pattern.

(c) List the elements of the group.

Courtesy Dr. Ron Eglash, RPI. www.rpi.edu/~eglash/csdt.html

74. Repeat Exercise 73 for the Plains Indian embroidery shown below.

Courtesy Dr. Ron Eglash, RPI. www.rpi.edu/~eglash/csdt.html

75. What are the elements for the group of symmetries of a square?

76. Using the notation of rosette patterns, describe the group of symmetries of

(a) an equilateral triangle (all three sides equal).

(b) an isosceles triangle (two equal sides) that is not equilateral.

(c) a scalene triangle (no pair of sides equal).

77. (a) Give a numerical example to show that the operation of subtraction on the integers is not associative.

(b) Repeat part (a), but for division on the positive real numbers.

78. What are the elements of the group of symmetries of

(a) Figure 19.11b?

(b) Figure 19.11f? (See page 795 for these images.)

79. What are the elements of the group of symmetries of

(a) Figure 19.11d?

(b) Figure 19.11g?

80. What are the elements of the group of symmetries of the dihedral pattern *d8*? (See the flower in Figure 19.10a on page 794.)

81. What are the elements of the group of symmetries of the cyclic pattern *c8*?

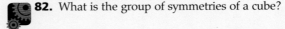

82. What is the group of symmetries of a cube?

83. What is the group of symmetries of a general rectangular solid (its length, width, and height are all unequal)?

19.5 Fractal Patterns and Chaos

84. Explore Sprott's Fractal Gallery at sprott.physics.wisc.edu/fractals.htm, which features a "Fractal of the Day" and accompanying fractal music. There are various "rooms" in the gallery—including "Iterated Function Systems," "Natural Fractals" (we particularly like "Broccoli" and "Trees"), and "Publication Quality Attractors" ("SMKBNZQA" is our favorite)—together with PC programs for generating such fractals. What are your favorites, and why?

85. Explain how the pattern of the following illustration is fractal.

Scott Camazine/Science Source

Exercises 86–88 use applets that require a computer with a web browser equipped with Java and Flash plug-ins. These plug-ins are available at links from the website csdt.rpi.edu/.

86. The website csdt.rpi.edu/African/MANG_DESIGN/culture/mang_homepage.html offers information about a fractal-patterned ivory hatpin from the Mangbetu culture in Africa. The site includes a tutorial on producing similar designs using reflection, rotation, translation, and scaling. Work your way through the tutorial and then create a Mangbetu-style artifact.

87. Cornrow hairstyles are fractal in nature. At the website csdt.rpi.edu/african/CORNROW_CURVES/, you can see how and why, including a tutorial on designing cornrow hairstyles using reflection, rotation, translation, and scaling. Work your way through the tutorial and then create a hairstyle. The website also includes instructions for actual braiding, with a short video.

88. Download fractal-creation software and accompanying documentation and use the software to create your own fractal. Links to software, most of it free, are available at www.Nahee.com/PNL/Fractal_Software.html and at fractalfoundation.org/resources/fractal-software/. Recommended software:

For Windows: Fractint, from www.Nahee.com/spanky/www/fractint/fractint.html

For Macintosh: XaoS, a fractal zoomer, from fractalfoundation.org/resources/fractal-software/.

Chapter Review

89. You have 42 in. of string that you want to use as the perimeter of a rectangle whose sides have lengths in inches that are Fibonacci numbers. What are the dimensions of your rectangle?

90. Identify the rigid motions that preserve the pattern of DDDDDDDD.

91. What is the notation for the symmetry pattern of DDDDDDDDD?

92. What is the notation for the symmetry pattern of the pinwheel below?

Ryan McVay/Photodisc/Getty Images

93. For the floor tiling below, use the flowchart in Spotlight 19.9 (page 797) to identify (by notation) the type of wallpaper pattern.

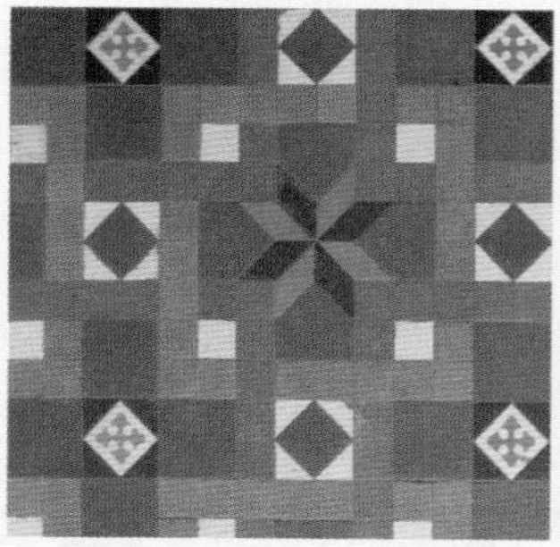

Jaume de Dios Pont

94. Do the real numbers form a group under division?

95. What are the elements of the group of symmetries of rosette pattern *d3*?

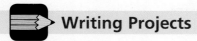

Writing Projects

1. Generations of children have enjoyed the popular toy Spirograph, which allows the user to trace out symmetric patterns. A pencil or pen is placed in a hole in one of several plastic circular disks with teeth on the outside rim. The disk is then meshed in the teeth of another plastic circle and rotated around its inside or outside. Each plastic piece is labeled with the number of teeth that it has on its circumference.

Obtain a copy of a Spirograph or closely related toy, or else visit www.wordsmith.org/~anu/java/spirograph.html, which offers an interactive Java application (which you can download) that mimics what the Spirograph toy does.

(a) Experiment to determine, from the numbers of teeth on the rotating circular disk and the fixed circle, what symmetry pattern the result will have.

(b) Choose a rotating circular disk and fixed circle for which the ratio of the number of teeth reduces to a whole number. For each of several "offsets" (holes to choose for the pencil or pen), trace overlapping designs. What symmetry pattern do you get for the design taken as a whole? Repeat this experiment for other pairs of pieces and try to reach a general conclusion.

(c) Write up, in a page or so, a description of your experiments and what conclusions you reached.

2. (Project for a team of two or three) Explore your campus looking for symmetrical patterns in decorative elements of the walls, floor, carpets, and ceilings. Find one example each of a rosette pattern, strip pattern, and wallpaper pattern. Take a digital photo of each and incorporate your photos into a document of three pages or so that explains to the reader where the pattern can be found, what symmetries (translation, rotation, reflection, or glide reflection) it has, and how you identified the notation for it.

3. (Project for a team of two or three) Visit a store that sells wallpaper and ask for a few old samples. Identify three that have different patterns according to the flowchart in Spotlight 19.9 (page 797). Write in a page or two your explanations of how you identified the patterns, and attach the wallpaper samples to your report.

Suggested Readings

BELCASTRO, SARAH-MARIE, and THOMAS C. HULL. Classifying frieze patterns without using groups, *College Mathematics Journal,* 33 (March 2002): 93–98. Elementary analysis of why there are only seven ways to repeat a pattern along a strip.

CARSON, JUDITHLYNNE. Fibonacci numbers and pineapple phyllotaxy, *Two-Year College Mathematics Journal,* 9 (3) (June 1978): 132–136.

CROWE, DONALD W. *Symmetry, Rigid Motions and Patterns, High School Mathematics and Its Applications (HiMAP)* Module 4, COMAP, Lexington, MA, 1987. Reprinted in smaller format in *The UMAP Journal,* 8 (3) (1987): 207–236. Instructional module on rigid motions of the plane, strip patterns, and wallpaper patterns, with worksheets.

HAYES, BRIAN. Group theory in the bedroom, *American Scientist,* 93 (5) (September–October 2005): 393ff, at www.americanscientist.org/issues/id.3465,y.2005,no.9,content.true,page.1,css.print/issue.aspx. Reprinted, with afterthoughts, in *Group Theory in the Bedroom, and Other Mathematical Diversions,* Hill and Wang, New York, 2008, pp. 219–237, 252–253.

LEE, KEVIN D. *KaleidoMania!: Interactive Symmetry,* Windows/Macintosh program, Key Curriculum Press, Emeryville, California, 1999. Lets the user construct rosette, strip, and wallpaper patterns.

LIVIO, MARIO. *The Golden Ratio: The Story of Phi, the World's Most Astonishing Number,* Broadway Books, New York, 2002. Reviewed in *Notices of the American Mathematical Society,* 52 (3) (March 2005): 344–347, at www.ams.org/notices/200503/rev-markowsky.pdf.

MANDELBROT, BENOIT B., and RICHARD L. HUDSON. *The (mis)Behavior of Markets: A Fractal View of Risk, Ruin, and Reward,* Basic Books, New York, 2004.

MARKOWSKY, GEORGE. Misconceptions about the golden ratio, *College Mathematics Journal,* 23 (1) (January 1992): 2–19.

POLSTER, BURKARD, and MARTY ROSS. Fibonacci or fairy tale? *Math Horizons* (February 2015): 16–17.

POSAMENTIER, ALFRED S., and INGMAR LEHMANN. *The (Fabulous) Fibonacci Numbers,* Prometheus Books, Amherst, NY, 2007.

POSAMENTIER, ALFRED S., and INGMAR LEHMANN. *The Glorious Golden Ratio,* Prometheus Books, Amherst, NY, 2011.

WALSER, HANS. *The Golden Section,* Mathematical Association of America, Washington, DC, 2001.

WASHBURN, DOROTHY K., and DONALD W. CROWE. *Symmetries of Culture: Theory and Practice of Plane Pattern Analysis,* University of Washington Press, Seattle, 1988. An introduction to the mathematics of symmetry, splendidly illustrated with photographs of patterns from cultures all over the world. Includes a complete analysis of patterns with two colors and proofs that there are only four rigid motions in the plane and exactly seven strip patterns.

Suggested Websites

www.geom.uiuc.edu/software/tilings/Tessellation resources. Lists programs for various platforms that allow the user to create designs featuring the rosette, strip, and wallpaper patterns.

escher.epfl.ch/escher/ Interactive Escher Web Sketch program that allows a user to design repeating patterns. Choose a wallpaper pattern using crystallographic notation and draw on the screen a colored design for the motif; the program then reproduces the motif using the pattern. The software (for Windows, Macintosh, and Unix) can also be downloaded.

www.geom.uiuc.edu/java/Kali/ Interactive Java Kali Web program that lets the user draw pictures under the action of rosette, strip, or wallpaper groups. Versions for various platforms can be downloaded.

www.wordsmith.org/~anu/java/spirograph.html Interactive Spirograph Java application (which you can download) that lets you do electronically what the Spirograph toy does.

csdt.rpi.edu/african/African_Fractals/An African fractals site.

classes.yale.edu/fractals/index.html Website for Benoit Mandelbrot's course in fractals at Yale. Features many applets for different kinds of IFS (e.g., incorporating randomness), including a fractal music composer.

www.math.smith.edu/phyllo//index.html Phyllotaxis: An interactive site for the study of plant pattern formation, created by Pau Atela and Christophe Golë.

www.maths.surrey.ac.uk/hosted-sites/R.Knott/Fibonacci/fib.html Fibonacci numbers and the golden section. Splendidly illustrated extensive web pages by Ron Knott about Fibonacci numbers and the golden ratio: their occurrences in nature, their applications, puzzles, and much more.

fractalfoundation.org/OFC/OFC-index.htm Fractal Foundation online course. Amazing photographs of fractals in nature, applications of fractals, and artwork of fractals of all kinds.

Tilings

20

Patrick McCabe/Alamy

Our ancestors were artistic; they covered floors and walls with patterns and mosaics, in Roman houses and Muslim mosques (see Figure 20.1). They featured patterns in other decorative arts, too—carpets, fabrics, baskets, and even wallpaper and linoleum.

Such patterns use repeated shapes ("tiles") to cover a surface, without gaps or overlaps. Apart from aesthetic appeal, repeated patterns can have practical applications. In manufacturing, for example, stamping components from a sheet of metal is most economical if the shapes of the components fit together without gaps—in other words, if the shapes form a **tiling.**

Figure 20.1 Glazed tiles of minarets and tiled dome, Friday Mosque, Yazd, Iran.

Mathematics is about discerning patterns—whether in physical objects (as in Chapter 18), numbers (as in Chapter 19), geometrical shapes (this chapter), or other structures—and crafting arguments to explain them.

In this chapter, we ask: What shapes can form a tiling that repeats in a regular way?

What if you have just one kind of tile—all the tiles are the same size and shape? For example, could a regular polygon work? Only a few do. In investigating tilings by polygons that don't all have equal sides, you will encounter significant contributions from one amateur mathematician—a housewife—and from another, despite a degree of autism.

Abandon polygons—let your imagination loose, and consider tiles in the shapes of horsemen, fish, or any other artistic shape you like. Which can tile? What if you let the tiles appear upside down as well as right side up?

What if you have two or more different shapes of tiles? Surprisingly, there is no way to decide if an arbitrary collection of shapes can tile. For some, we can exhibit tilings; for others, we can prove that they don't tile. But mathematicians have proved that there is no algorithm (mechanical step-by-step process) that can decide for every conceivable set of tile shapes whether they will tile or not. (See Chapter 9 for other examples of mathematically "unattainable ideals," in that case with regard to voting.)

Finally, what about "tiling" in three dimensions—filling space with solid shapes? There are more surprises in store, including quasicrystals, the discovery of which won a Nobel Prize, and their applications in new ultrastrong alloys and coatings for nonstick cookware.

20.1 Tilings with Regular Polygons

Tiling (Tessellation) DEFINITION

A **tiling (tessellation)** is a covering of the entire infinite plane by nonoverlapping regions called tiles.

The simplest tilings use only one size and shape of tile. They are known as **monohedral tilings.**

Monohedral Tiling DEFINITION

A **monohedral tiling** is a tiling that uses only one size and shape of tile.

The simplest tiles are **regular polygons,** all of whose sides have the same length and all of whose angles are equal. A square is a regular polygon with four equal sides and four equal interior angles; a triangle with all sides equal (an **equilateral triangle**) is also a regular polygon. A polygon with five sides is a *pentagon,* one with six sides is a *hexagon,* and one with n sides is an **n-gon.** Regular polygons have a high degree of symmetry. Each has the reflection and rotation symmetries of a dihedral rosette pattern (see Section 19.2 on page 790). In three dimensions, the corresponding highly symmetrical figures are called *regular polyhedra* (see Spotlight 20.1).

Regular Polyhedra and Buckyballs

SPOTLIGHT 20.1

The three-dimensional analogue of a regular polygon is a regular polyhedron, a convex solid whose faces are regular polygons that are all alike (same number of sides, same size), with each vertex surrounded by the same number of polygons. Although there are infinitely many regular polygons, there are only five regular polyhedra, a fact proved by Theaetetus (414–368 B.C.). They were called the *Platonic solids* by the ancient Greeks.

If the restriction that the same number of polygons meet at each vertex is relaxed, five additional convex polyhedra are obtained, all of whose faces are equilateral triangles. If we allow faces to be more than one kind of regular polygon, 13 further convex polyhedra are obtained, known as the semiregular polyhedra or Archimedean solids (although there is no documented evidence that Archimedes studied them—but in the early 1600s, Johannes Kepler catalogued them all). Once inflated, the truncated icosahedron—whose faces are pentagons and hexagons—is known throughout the world as a regulation soccer ball. Drawings of it appear in the work of Leonardo da Vinci. The truncated icosahedron is also the structure of C_{60}, a form of carbon known as "buckminsterfullerene" and, more familiarly, the *buckyball.* A total of 60 carbon atoms lie at the 60 vertices of this molecule, which was discovered in 1985. It is named after R. Buckminster Fuller (1895–1983), inventor and promoter of the geodesic dome. The molecule buckminsterfullerene resembles a dome.

The buckyball is part of a family of carbon molecules, the *fullerenes,* in which each carbon atom is joined to three others. Thirty years before the discovery of fullerenes, mathematicians had shown that a convex polyhedron having pentagons and hexagons as faces must have exactly 12 pentagon faces but may have any number of hexagon faces, from 0 on up, except for 1.

That there must be 12 pentagons follows from a famous equation due to Leonhard Euler (1707–1783). For any convex polyhedron, it must be true that $v - e + f = 2$, where v is the number of vertices, e is the number of edges, and f is the number of faces of the polyhedron.

In 2003, astronomers and mathematicians advanced a remarkable new theory about why the universe does not show as much historic fluctuation in temperature as other models predict. This lack of fluctuation could be explained by the universe being in the shape of a *dodecahedron* (the figure shown here with 12 pentagonal sides), with opposite faces coinciding. This theory harks back to Kepler, who had conceived of the universe in terms of the five regular polyhedra nested within one another.

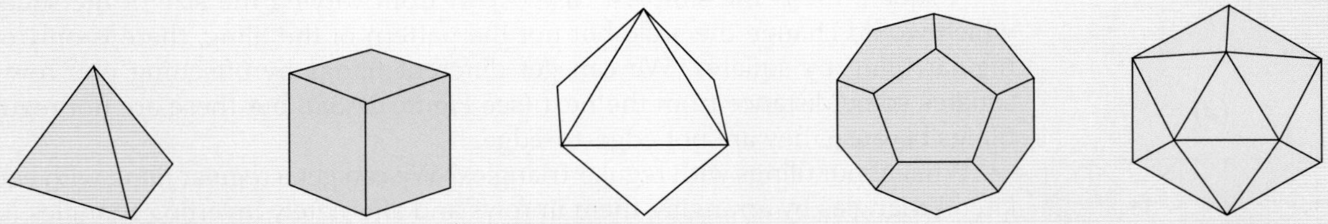

An **exterior angle** of a polygon is an angle formed by one side and the extension of an adjacent side (see Figure 20.2). At each vertex of the polygon, there are two exterior angles, depending on which side we extend; but we will consider only one of them. Let us agree to extend the sides consistently in turn as we proceed counterclockwise around the polygon, as in Figure 20.2, producing the set of exterior angles *A* through *E,* one at each vertex.

By a convention dating back to the ancient Babylonians, angles are measured in degrees (°), with the total measure of angles around a point being 360°. If we bring together a set of the exterior angles of a polygon at a point, we can see that they add up to 360° (see Figure 20.2). Hence, for a regular polygon with n sides, each exterior

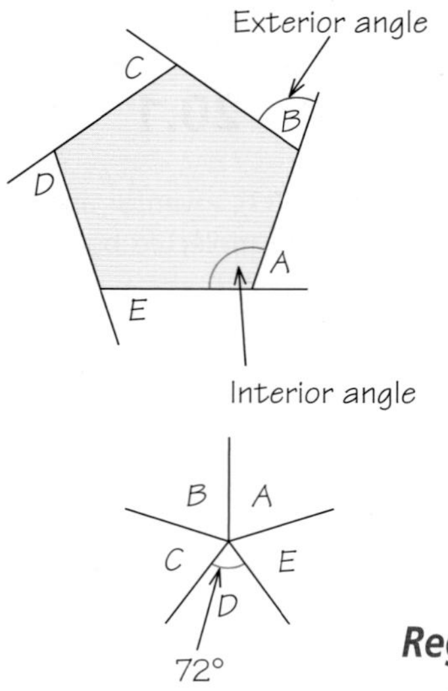

Figure 20.2 The exterior angles of a regular pentagon, like those of any regular polygon, add up to 360°. Each exterior angle measures 360°/5 = 72°.

angle must measure 360°/n. For example, a square, with $n = 4$ sides, has 4 exterior angles in a set, each measuring 90°; a regular pentagon, with $n = 5$ sides, has 5, each measuring 72°; a regular hexagon, with 6 sides, has 6, each measuring 60°.

Each exterior angle is paired with a corresponding **interior angle** (the angle inside the polygon formed by the two adjacent sides), and the measures of the pair of angles add up to a straight line, or 180°. For a regular polygon with more than six sides, each interior angle is between 120° and 180°. That's because if $n > 6$, each exterior angle measures 360°/n, which has to be less than 60°. Since the measures of the exterior angle and the adjacent interior angle add to 180°, the adjacent interior angle must measure more than 120° (but less than 180° if the polygon is not to bend back upon itself). This consideration will shortly prove crucial in determining how regular polygons can fit together to form tilings.

Regular Tilings

Edge-to-Edge DEFINITION

A tiling is **edge-to-edge** if for every tile, each edge coincides with the entire edge of a bordering tile.

Figure 20.3 shows one tiling that is not edge-to-edge and another that is.

Regular Tiling DEFINITION

A **regular tiling** is an edge-to-edge tiling that uses only one kind of regular polygon.

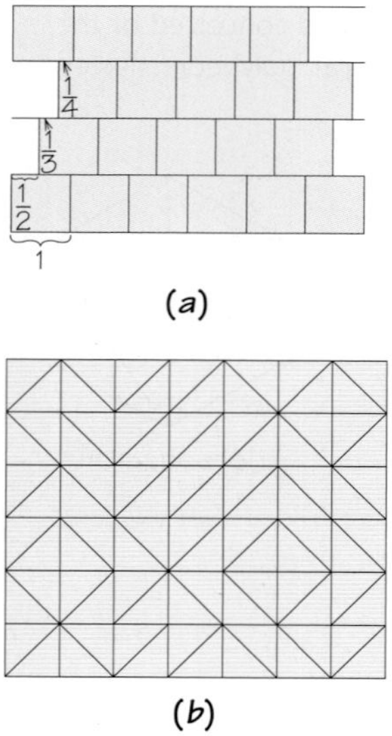

Figure 20.3 (a) A tiling that is not edge-to-edge. The horizontal edges of two adjoining squares do not exactly coincide. (b) A tiling by right triangles that is edge-to-edge.

A square tile is the simplest case. Apart from varying the size of the square, which would change the scale but not the pattern of the tiling, there is only one regular tiling by squares. We can get different tilings by offsetting one row of squares some distance from the next (see Figure 20.3a), but these are not regular tilings because they are not edge-to-edge.

What about tilings with regular triangles? We can get a regular tiling with equilateral triangles by arranging them in rows and alternately inverting triangles (see the upper-left pattern in Figure 20.4). As with squares, there is only one pattern of equilateral triangles that forms an edge-to-edge tiling. [Any tiling by squares can be refined to one by triangles by drawing a diagonal of each square (as in Figure 20.3b), but the resulting right triangles are not regular (equilateral).]

What about tiles with more than four sides? An edge-to-edge tiling with regular hexagons is easy to construct (see the third pattern in the first row of Figure 20.4).

However, if we look for a tiling with regular pentagons, we won't find one. How do we know whether we're just not being clever enough or there really isn't one to be found? This is the kind of question that mathematics is uniquely equipped to answer. In the other sciences, phenomena may exist even though we have not observed them; such was the case for bacteria before the invention of the microscope and the moons of Jupiter before the invention of the telescope. However, we can conclude with certainty that there is no edge-to-edge tiling with regular pentagons.

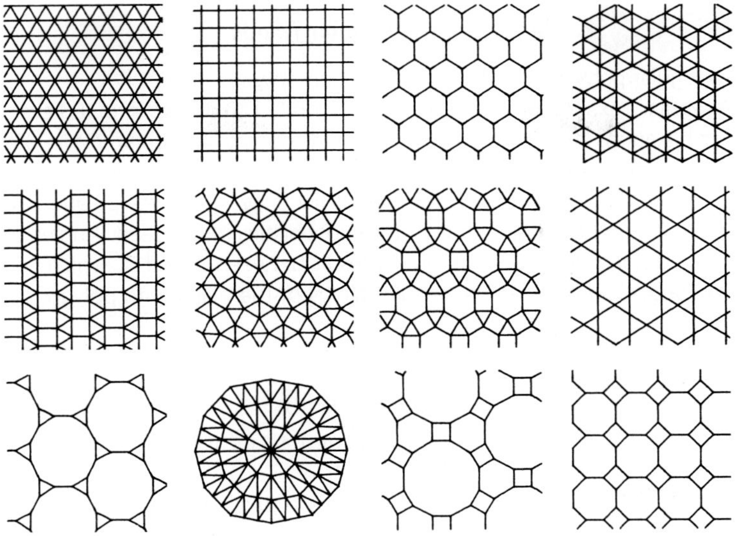

Figure 20.4 The three regular tilings . . . and the eight semiregular tilings, plus a "mystery" tiling that does not belong to either group. Can you identify it? (*Hint:* It uses just one shape of tile, which isn't regular.)

The proof is easy and numerical. As we calculated earlier, the exterior angles of a pentagon are each 72°; each corresponding interior angle is thus 108° (see Figure 20.5). How many pentagons can meet at a point? The total measure of all of the angles around a point must be 360°. As you can see in Figure 20.5, four pentagons at a point would be too many (their angles would add to $4 \times 108° = 432°$, so they'd have to overlap), and three would be too few (their angles would add to $3 \times 108° = 324°$, so some of the area wouldn't be covered). Because 108 does not divide evenly into 360, regular pentagons cannot tile the plane in an edge-to-edge manner.

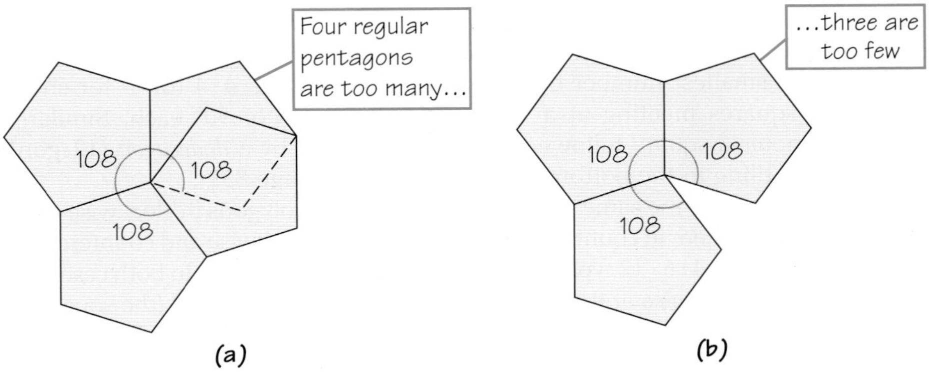

Figure 20.5 The adjacent interior angles of polygons that come together at a vertex in a tiling must add up to 360°—no more, no less.

With this argument, we can do something that is characteristic of mathematics—we can *generalize it* to a criterion for when a regular polygon can tile the plane; namely, when the measure of one of its interior angles divides 360 evenly, and not otherwise. We can apply this criterion to determine exactly which other regular polygons can tile the plane.

Only Three Regular Tilings THEOREM

The only regular tilings are the ones with equilateral triangles, with squares, and with regular hexagons.

EXAMPLE 1 Identifying the Regular Tilings

Which regular polygons can tile the plane, and how?

A regular hexagon has interior angles of 120°; 120 divides into 360 evenly, and three regular hexagons fit together exactly around a point. A regular 7-gon (heptagon)—or any regular polygon with more than 6 sides—has interior angles that are larger than 120° but smaller than 180°. Now, 360 divided by 120 gives 3, and 360 divided by 180 gives 2—and there aren't any other possibilities in between. Angles between 180° and 120° divided into 360° will give a result between 2 and 3, which consequently is not an integer. So there are no regular tilings of the plane with polygons of more than six sides. ■

Self Check 1

Explain why the plane cannot be tiled by regular decagons (which have 10 sides). ■

A natural follow-up question—which is typical of mathematical thinking—is to ask which *combinations* of regular polygons of different numbers of sides can tile the plane edge-to-edge.

To answer the question, we need to describe how such polygons meet around a point.

Vertex Type DEFINITION

In an edge-to-edge tiling by regular polygons, the **vertex type** of a vertex is the arrangement of the polygons around the vertex.

For a notation for a vertex type, we list the number of sides of polygons, separated by raised periods, in either clockwise or counterclockwise order, starting from the smallest number of sides. For example, $4 \cdot 4 \cdot 4 \cdot 4$ (or 4^4 for short) denotes 4 squares meeting at a vertex, with 4 angles of 90° each. Similarly, $4 \cdot 6 \cdot 12$ denotes a square followed by a hexagon then by a *dodecagon* (12-gon); see the third tiling in the bottom row of Figure 20.4 (page 831).

Two vertices have the same type even if one has the polygons in clockwise order and the other has them in counterclockwise order; both clockwise and counterclockwise versions of the $4 \cdot 6 \cdot 12$ type occur in that tiling in Figure 20.4. In both cases, the sum of the interior angles at the vertex is 360°. Note that 360° = 90° (the square) + 120° (the hexagon) + 150° (the dodecahedron—do you see why it is 150°?).

Semiregular Tiling DEFINITION

An edge-to-edge tiling that uses a mix of regular polygons with different numbers of sides but in which all vertex types are alike—that is, the same polygons in the same order, clockwise or counterclockwise—is called a **semiregular tiling** (see Figure 20.4 on page 831).

As before, the technique of adding up angles at a vertex—the sum must be 360°)—can eliminate some impossible combinations, such as "pentagon, pentagon, pentagon" (see Figure 20.5 on page 831).

Once we have found an arrangement that is numerically possible, we must check for the actual existence of such a tiling by trying to construct it—showing that it is also geometrically possible. For example, even though a numerically possible arrangement of regular polygons around a point is "triangle, square, square, hexagon" (because $60° + 90° + 90° + 120° = 360°$), it is nevertheless not possible to construct a tiling with that vertex figure occurring at every vertex.

The result of such investigations is that in a semiregular tiling, no polygon can have more than 12 sides. In fact, polygons with 5, 7, 9, 10, or 11 sides do not occur either. Figure 20.4 on page 831 exhibits all the semiregular tilings.

If we abandon the restriction that the vertex type be the same at every vertex, then there are infinitely many edge-to-edge tilings with regular polygons, even if we continue to insist that all polygons with the same number of sides have the same size.

20.2 Tilings with Irregular Polygons

What about edge-to-edge tilings with irregular polygons, which may have some sides longer than others or some interior angles larger than others? We will look just at monohedral tilings (in which all tiles have the same size and shape) and investigate, in turn, which triangles, **quadrilaterals** (four-sided polygons), hexagons, and so forth can tile the plane.

In the most general shape of a triangle, all sides are of different lengths and all interior angles of different sizes. Such a triangle is called a **scalene triangle,** from the Greek word for "uneven." We can always take two copies of any triangle and fit them together to form a **parallelogram,** a quadrilateral whose opposite sides are parallel (Figure 20.6a). It's easy to see that we then can use such parallelograms to tile the plane by making strips and then fitting together layers of strips edge-to-edge (Figure 20.6b).

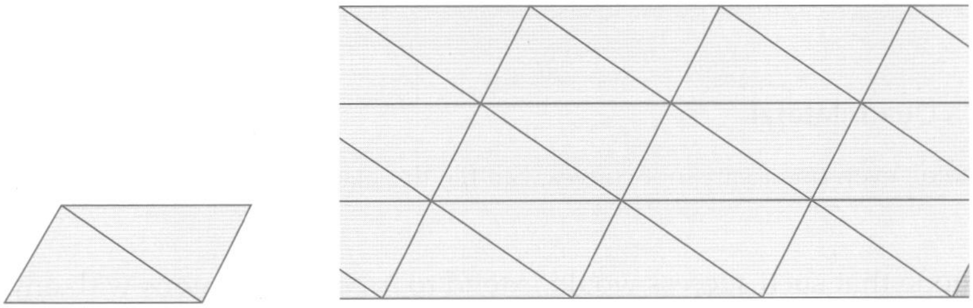

(*a*) (*b*)

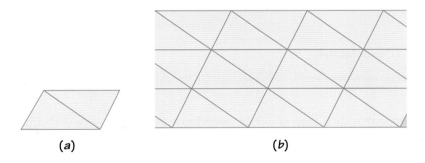

Figure 20.6 (a) Two triangles form a parallelogram. (b) Any triangle (even a scalene one) can tile the plane.

Tiling with a Triangle THEOREM

Any triangle whatsoever—whether it has three sides equal, two sides equal, or no sides equal—can tile the plane.

What about quadrilaterals? We have seen that squares can tile the plane, and rectangles certainly can, too. We have just noted that any parallelogram can tile. What about a quadrilateral with its opposite sides not parallel, as in Figure 20.7a? The same technique as for triangles will work. We fit together two copies of the quadrilateral, forming a hexagon whose opposite sides are parallel. Such hexagons fit next to each other to form a tiling, as in Figure 20.7b.

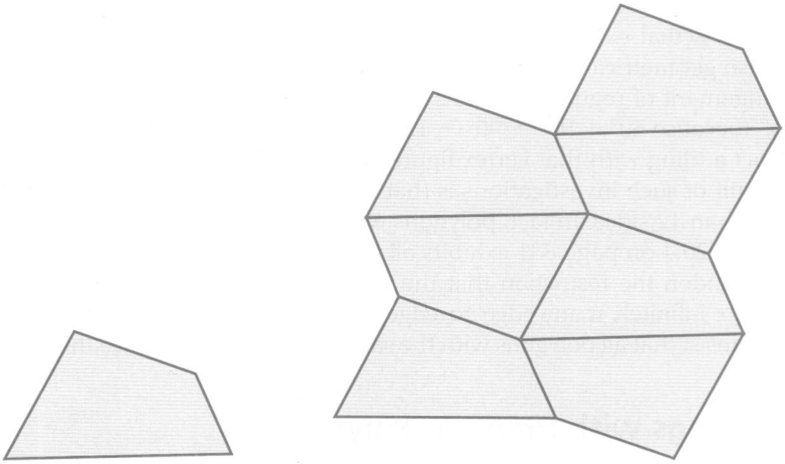

(a) (b)

Figure 20.7 (a) A general quadrilateral. (b) Any quadrilateral tiles the plane.

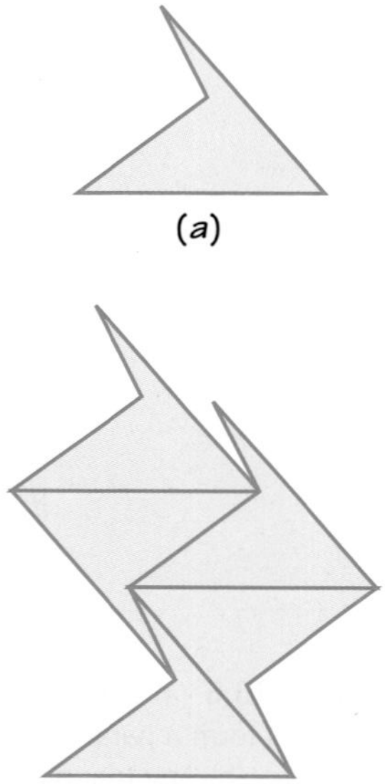

(a) (b)

Figure 20.8 (a) A general nonconvex quadrilateral. (b) Any quadrilateral, convex or not, tiles the plane.

The quadrilaterals shown in Figure 20.7 are all **convex,** meaning that if you take any two points on the tile (including the boundary), the line segment joining them lies entirely within the tile (again, including the boundary). The quadrilateral of Figure 20.8a is not convex, but the same technique works to form a tiling (Figure 20.8b).

Convex DEFINITION

A tile (including its boundary) is **convex** if for any two points on it, all the points on the line segment joining them also belong to the tile.

Tiling with a Quadrilateral THEOREM

Any quadrilateral, even one that is not convex, can tile the plane.

We could hope that such success would extend to irregular polygons with any numbers of sides, but it doesn't. Which convex hexagons can tile the plane was determined by Karl Reinhardt (1895–1941) in his 1918 doctoral thesis. He showed that for a convex hexagon to tile, it must belong to one of three classes. Examples of the three classes are shown in Figure 20.9. Tilings with a hexagon of type 2 use both ordinary and mirror-image versions of the hexagon.

Tiling with a Hexagon THEOREM

Exactly three classes of convex hexagons can tile the plane.

Reinhardt also explored tilings by convex pentagons and found five classes that tile. For example, any pentagon with two parallel sides can tile. Reinhardt did not complete the solution, as he did for hexagons, by proving conclusively that no other pentagons could tile. He claimed that it would be very tedious to finish the analysis. Still, he felt that he had found them all. In 1968, after 35 years of working

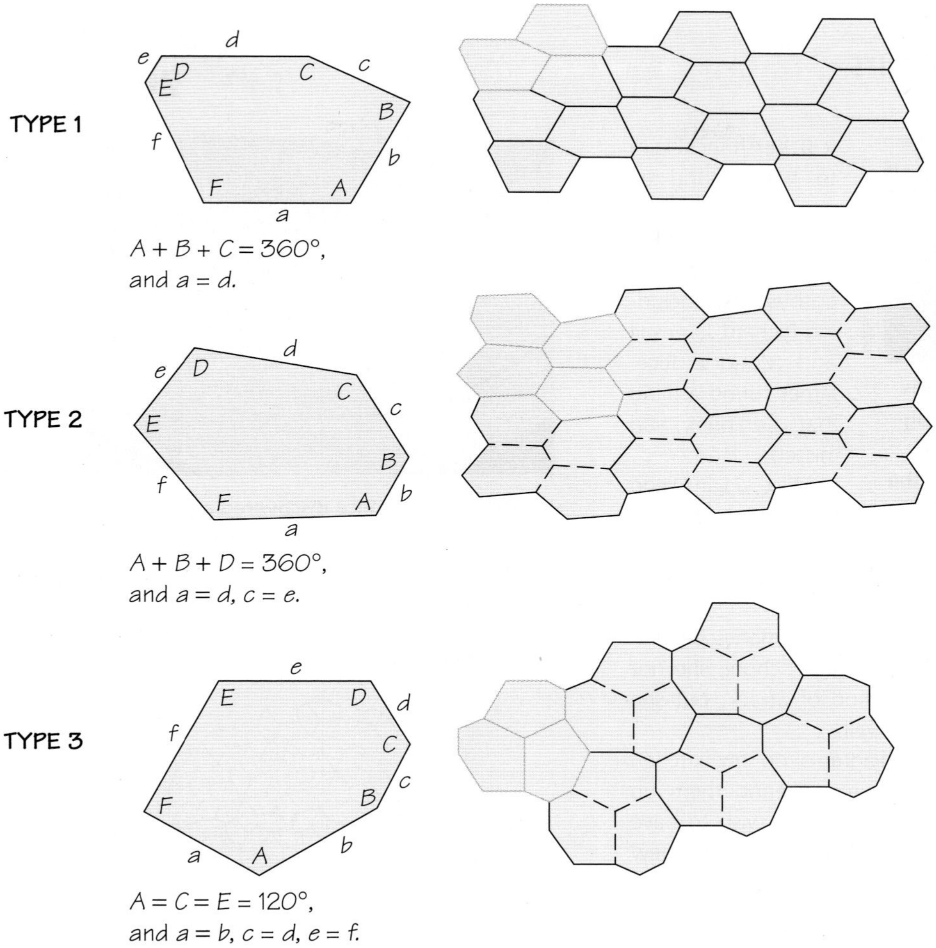

Figure 20.9 The three types of convex hexagon tiles.

on the problem on and off, R. B. Kershner, a physicist at Johns Hopkins University, discovered three more classes of pentagons that tile. Kershner, in turn, was sure that he had found all pentagons that tile; but, like Reinhardt, he did not offer a complete proof, which "would require a rather large book."

When an account of the alleged "complete" classification into 8 types appeared in the July 1975 issue of *Scientific American,* the article provoked an amateur mathematician to discover a 9th type! Marjorie Rice, a housewife with no formal education in mathematics beyond high school "general mathematics" taken 36 years earlier, devised her own mathematical notation and found four more types (see Spotlight 20.2). A 14th type was found by a mathematics graduate student in 1985. In July 2015, a 15th type was discovered by researchers at the University of Washington Bothell who used a computerized search. No one knows if that completes the classification.

With the situation so intricate for convex pentagons, you might think that it must be still worse for polygons with seven or even more sides. In fact, however, the situation is remarkably simple, as Reinhardt proved in 1927.

Tiling with a Polygon with More Sides THEOREM

A convex polygon with seven or more sides cannot tile.

In Praise of Amateurs

SPOTLIGHT 20.2

Courtesy of Doris Schattschneider

Marjorie Rice with Dr. Doris Schattschneider, who recognized Rice's achievements and brought her work to the attention of the world.

R. B. Kershner's claim to have found all convex pentagons that tile was read by many puzzle enthusiasts, including Richard James III and Marjorie Rice. James found a tiling that Kershner had missed.

Rice, a San Diego housewife and mother of five, read about James's new tile. "I thought I would see if I could find still another type. It was a delightful new puzzle to me."

With no formal education in mathematics beyond a general mathematics course in high school, she not only worked out her own method of attack, but she invented her own notation as well.

"I began drawing little diagrams on my kitchen counter when no one was there, covering them up quickly if someone came by, for I didn't wish to have to explain what I was doing. I was searching for a new type, and a few weeks later, I found it." Over the next two years, she found three additional new tilings.

Rice was born in 1923 in St. Petersburg, Florida, and went to a one-room country school.

"When I was in the sixth or seventh grade, our teacher pointed out to us the golden section in the proportions of a picture frame. This immediately caught my imagination and I never forgot it. I've . . . been especially interested in architecture and the ideas of architects and planners such as Buckminster Fuller. I've come across the golden section again in my reading and considered its use in painting and design." (The golden ratio is discussed in Chapter 19, on pages 783–790.)

After high school, Rice worked until her marriage in 1945. She was drawn back into mathematics by her children, finding solutions to their homework problems "by unorthodox means, since I did not know the correct procedures." She became especially interested in textile design and the works of M. C. Escher. As she pursued the pentagonal tilings, she produced some imaginative Escher-like patterns. (You can see some of them at one of the Suggested Websites for this chapter.)

Ivars Peterson

In 1995, Marjorie Rice discovered a pentagon tiling now installed in the entry foyer of the headquarters of the Mathematical Association of America in Washington, D.C. The angles of each pentagon tile are 60°, 120°, 90°, 120°, and 150°—all multiples of 30°.

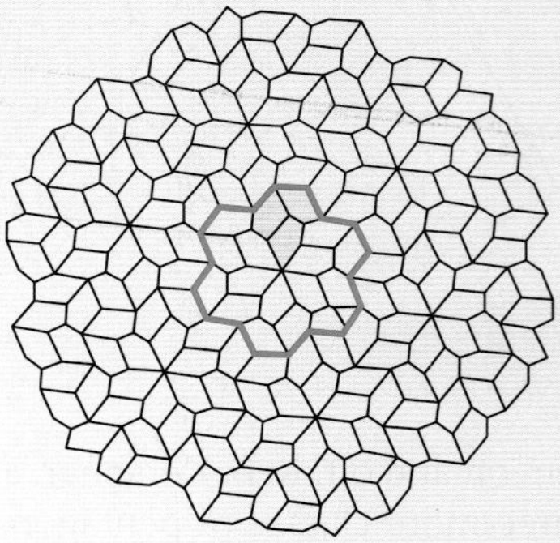

Diagram of the pentagon tiling at the Mathematical Association of America. The tiling is periodic, although not every pentagon is surrounded in the same way. Three pentagons form a fundamental block, and the outlined group of 18 pentagons tiles by translations.

What makes a person pursue a problem so patiently and persistently? Marjorie Rice was not trained for it, nor was she paid, but she gained great personal satisfaction from the pursuit.

Intense spirit of inquiry and keen perception are the forte of all such amateurs. No formal education provides these gifts. Lack of a mathematical degree separates these "amateurs" from the "professionals," yet their curiosity and ingenious methods make them true mathematicians.

Information from Doris Schattschneider, "In Praise of Amateurs," in David A. Klarner (ed.), *The Mathematical Gardner,* pp. 140–166, plus Plates I–VIII, Wadsworth, Belmont, CA, 1981.

M. C. Escher and Tilings

The Dutch artist M. C. Escher (1898–1972) was inspired by the great variety of decoration in tilings in the Alhambra, a 14th-century palace built during the last years of Islamic dominance in Spain. He devoted much of his career of making prints to creating tilings with tiles in the shapes of living beings. Those prints of interlocking animals and people have inspired awe and wonder all over the world. Figures 20.10 through 20.13 on pages 837–840 illustrate a few of his drawings and finished works. Like Marjorie Rice, he too developed his own mathematical notation for the different kinds of patterns for the tilings.

For more about Escher, see Spotlight 20.3.

Figure 20.10 Escher No. 128 (*Bird*), from Escher's 1941–1942 notebook.

Maurits Escher

SPOTLIGHT 20.3

Maurits Escher (1898–1972) was a Dutch graphic artist who specialized in lithograph and woodcut prints. Some of the images in his work were inspired by the landscapes and architecture that he observed while living and traveling in Italy early in his career.

Other works of his—particularly those involving tilings—were stimulated by a visit to the Alhambra, a 14th-century castle in Granada, Spain, built by Moors (Muslims from North Africa). The walls and floors of the Alhambra are tiled in various patterns,

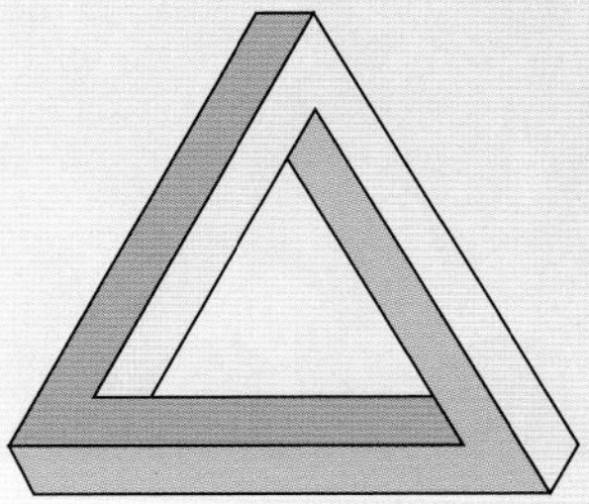

A Penrose triangle.

Maurits Escher (*continued*)

including strip patterns and wallpaper patterns. Escher took such patterns and changed their periodic geometric shapes into similarly periodic images of plants and animals.

His prints involving tilings came to the attention of mathematicians at the International Congress of Mathematicians in Amsterdam in 1954. Thereafter, he interacted with mathematicians, with a fruitful cross-fertilization of ideas.

Escher is also known for his prints of versions of impossible figures, including one (the Penrose triangle) inspired by the father of Sir Roger Penrose.

20.3 Using Only Translations

Following Escher's artistic genius, we investigate monohedral tilings in which the tile used may be nonconvex and have any curve whatever as a boundary (rather than one consisting of only straight edges). You may wonder just how much liberty can be taken in shaping a tile and how you could design an Escher-like tiling yourself.

[*Technical note:* We restrict ourselves to isohedral tilings, those in which any tile can be made to coincide with any other tile by one of the wallpaper groups of symmetries of Spotlight 19.9, page 797. In practical terms, if you copy the tiling onto a sheet of clear plastic, choose any tile on the paper original and any tile on the plastic, and make them coincide (perhaps by turning the plastic over), then all the tiles on the plastic coincide exactly with tiles on the original.]

In the simplest type of tiling, the tile is just moved along in two directions; that is, copies are laid edge-to-edge in rows, as in Figure 20.10.

Each tile must fit exactly into the ones next to it, including its neighbors above and below. We say that each tile is a **translation** of each other one; we can move them to coincide without doing any rotation or reflection.

When is it possible for a tile to cover the plane in this manner? The boundary of the tile must be divisible into matching pairs of opposing parts that will fit together. Figures 20.10 and 20.11 illustrate two basic ways that this can happen.

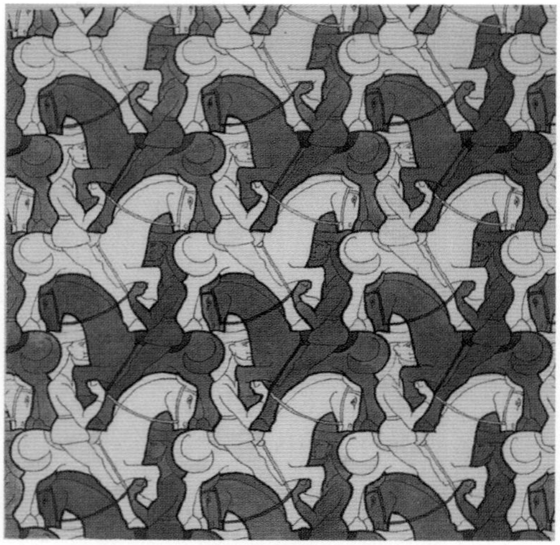

(*a*)

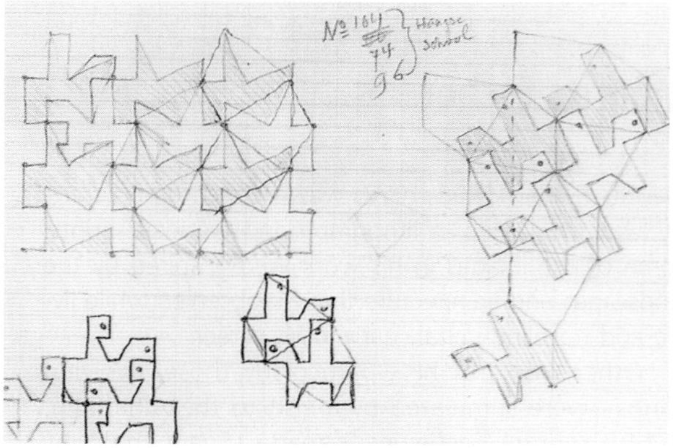

(*b*)

Figure 20.11 (a) Escher No. 67 (*Horseman*), from Escher's 1941–1942 notebook. (b) Sketch showing the tile design for the *Horseman* print.

Translation Criterion RULE

A tile can tile the plane by translations alone if either

1. There are four consecutive points *A, B, C,* and *D* on the boundary such that
   - **(a)** the boundary part from *A* to *B* is congruent by translation to the boundary part from *D* to *C,* and
   - **(b)** the boundary part from *B* to *C* is congruent by translation to the boundary part from *A* to *D* (see Figure 20.12a)

or

2. There are six consecutive points *A, B, C, D, E,* and *F* on the boundary such that the boundary parts *AB, BC,* and *CD* are congruent by translation, respectively, to the boundary parts *ED, FE,* and *AF* (see Figure 20.12b).

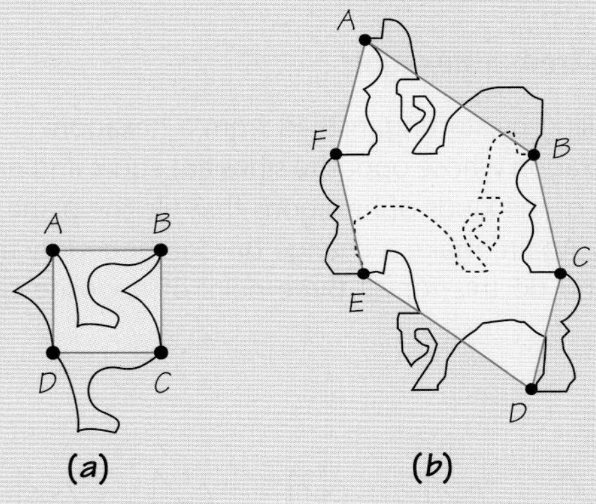

Figure 20.12 Individual tiles traced from the Escher prints of Figures 20.10 and 20.11, with points marked to show they fulfill the criteria for tiling by translations. The two horsemen form a block that tiles by translations, although a single horseman can tile by itself if we allow mirror-image reflections, too.

In the first, two opposite pairs of sides match; in the second, three opposite pairs of sides match.

The tiles for Figures 20.10 and 20.11 are shown in outline form in Figure 20.12, together with points marked to show how the tiles fulfill the criterion.

In fact, alternative 1 of the criterion is a special case of alternative 2 (see Exercises 21 and 22 on page 860). Moreover, alternative 2 completely "characterizes" tiles that can tile by translations. That is, not only *if alternative 2 is true, then the tile can tile by translations,* but also the criterion works "in reverse": *If a tile can tile by translations, then alternative 2 must be true* (for some choice of six consecutive points).

A nice feature of the Translation Criterion is that if you can find points as required for alternative 2, then you can join them in order, as in Figures 20.12a and b, to see how to do the tiling.

To create tilings, though, you can proceed exactly as Escher did. His notebooks show that he designed his patterns in just the way that we now describe.

EXAMPLE 2 ➡ Tiling Starting from a Parallelogram

How can you make an Escher-like tiling by starting from a parallelogram?

Applying the first alternative of the criterion, start from a parallelogram, make a change to the boundary on one side, and then copy that change to the opposite side. Similarly, change one of the other two sides and copy that change on the side opposite it (see Figure 20.13). Revise as necessary, always making the same change to opposite sides. You might find it useful (as Escher did) to make your designs on graph paper, or you can work by cutting and taping together pieces of heavy paper.

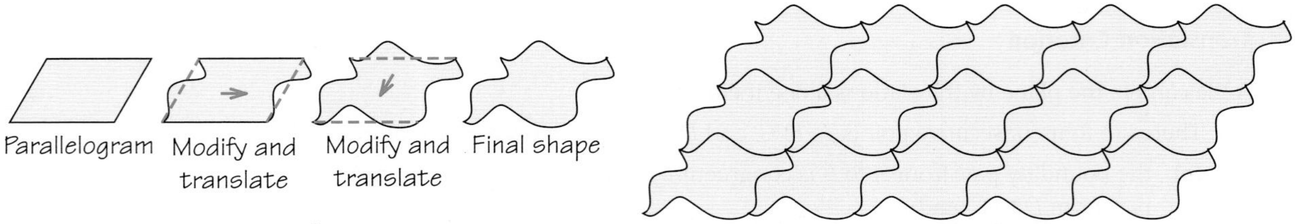

Figure 20.13 How to make an Escher-like tiling by translations from a parallelogram base.

Self Check 2

Start from a parallelogram and design your own Escher-like tiling.

EXAMPLE 3 Tiling Starting from a Hexagon

How can you make an Escher-like tiling by starting instead from a hexagon?

Start from a **par-hexagon,** a hexagon whose opposite sides are equal and parallel. A par-hexagon is a special case of one of the kinds of hexagons that tile the plane, Type 1, shown in Figure 20.9 on page 835. Again, make a change on one boundary and copy the change to the opposite side, and do this for all three pairs of opposite sides (see Figure 20.14).

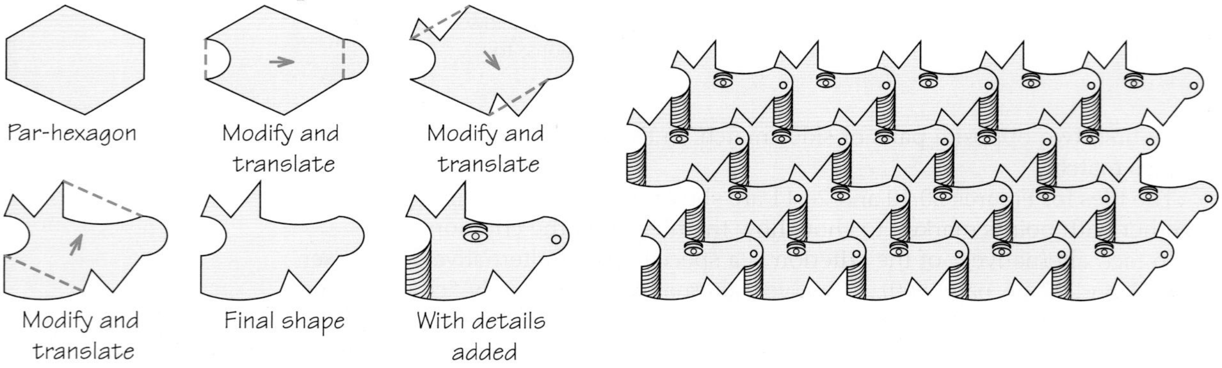

Figure 20.14 How to make an Escher-like tiling by translations, from a par-hexagon base.

Self Check 3

Start from a par-hexagon and design your own Escher-like tiling.

20.4 Using Translations Plus Half-Turns

If the tiling is to allow half-turns, then in order to match up the tiles, some of the figures will appear "upside down." The part of the boundary of a right-side-up figure has to match the corresponding part of itself in an upside-down position. For that to happen, that part of the boundary must be **centrosymmetric:** symmetric about the midpoint, that is, unaltered by a 180° rotation around that midpoint. The key to some of Escher's more sophisticated monohedral designs, and the fundamental principle behind some further easy recipes for making Escher-like tilings, is the **Conway Criterion,** formulated by John H. Conway (b. 1937) of Princeton University.

Conway Criterion RULE

A tile can tile the plane by translations plus half-turns (including possibly by translations alone) if there are six consecutive points on the boundary (some of which may coincide, but at least three of which are distinct)—call them *A, B, C, D, E,* and *F*—such that

- the boundary part from *A* to *B* is congruent by translation to the boundary part from *E* to *D,* and
- each of the boundary parts *BC, CD, EF,* and *FA* is centrosymmetric.

The first condition means that we can match up those two boundary parts exactly, curve for curve, angle for angle. The second condition means that each of the remaining boundary parts is brought back into itself by a half-turn around its center. The second condition is automatically fulfilled if the boundary part in question is a straight-line segment.

The tiles for Figures 20.15 and 20.16 are shown in outline form in Figure 20.17, together with points marked to show how the tiles fulfill the Conway Criterion.

Figure 20.15 shows that Escher sketched tiny circles exactly where we have red dots in Figure 20.17a.

Figure 20.15 Escher No. 6 (*Camel*), from Escher's 1941–1942 notebook.

Figure 20.16 Escher No. 88 (*Seahorse*).

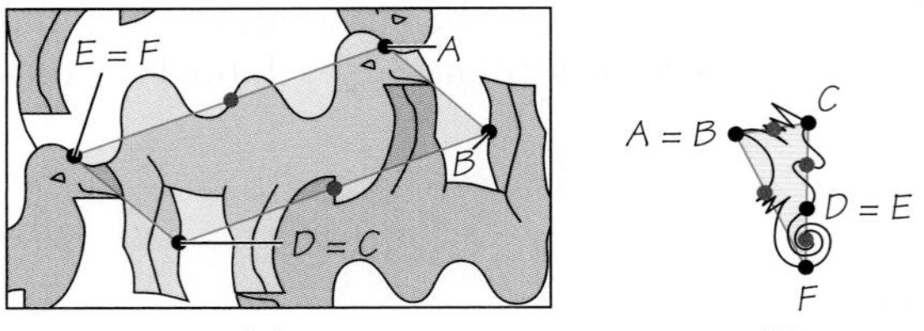

Figure 20.17 Individual tiles traced from the Escher prints of Figures 20.15 and 20.16, with points marked to show that they fulfill the Conway criterion for tiling by translations and half-turns *(around the red dots).*

Mathematicians do not know whether the Conway Criterion completely characterizes tiles that can tile by translations and half-turns. Tiles that fulfill the Conway Criterion can tile by translations and half-turns, but there might be tiles that tile that way but do not satisfy the criterion—however, nobody knows of any.

(The Conway Criterion does completely characterize tiles that produce the wallpaper pattern *p2* of Spotlight 19.9 on page 797: Any tile that satisfies the criterion can be used to make a *p2* pattern, and any tile that can produce that pattern must satisfy the criterion.)

We have considered tilings that tile by translations alone or by translations plus half-turns. You may have noticed differences among the figures, in the following ways:

- In Figures 20.10 and 20.14, all the birds and all the horseheads face the same way, and none are upside down. What if we wanted to allow some of them to be upside down?
- In Figures 20.15 and 20.16, although there are upside-down camels and seahorses, all the right-side-up ones face the same way and all the upside-down ones face the opposite way. What if we wanted to allow both kinds to face both ways?
- In Figure 20.11, there are no upside-down horsemen, but the dark horsemen face left, while the light horsemen face right. The two kinds of horsemen are mirror images of each other; but we avoided letting the tile occur in mirror-image reflections of itself by building the mirroring into the tile itself. However, we noted in Figure 20.12 that a single horseman could be a tile by itself if we allow mirror-image reflections.

In addition to the Translation Criterion and the Conway Criterion, there are several other sufficient criteria for a tile to produce a tiling, depending on what operations we allow on the tile: translations, mirror reflections, glide reflections (illustrated in Figure 19.9, page 792), or rotations by various angles (60°, 90°, 120°, or 180°). (We have already considered 180° rotations; they are half-turns in the Conway Criterion.) We can't consider them all here or show the extent to which Escher used them; but in the Suggested Readings and Suggested Websites sections at the end of the chapter, we point to relevant sources and explanations. There is a simple way to implement the Conway Criterion to make Escher-like tilings by starting from any triangle or any quadrilateral. We illustrate this creative process in the following examples.

EXAMPLE 4 ➔ Tiling Using a Triangle

How can you make an Escher-like tiling by starting from a triangle?

Modify half of one side of the triangle, then rotate that modified half side around the midpoint of the side to extend the modification to the rest of the side, thereby making the new side centrosymmetric. Then you can do the same to the second and third sides (Figure 20.18).

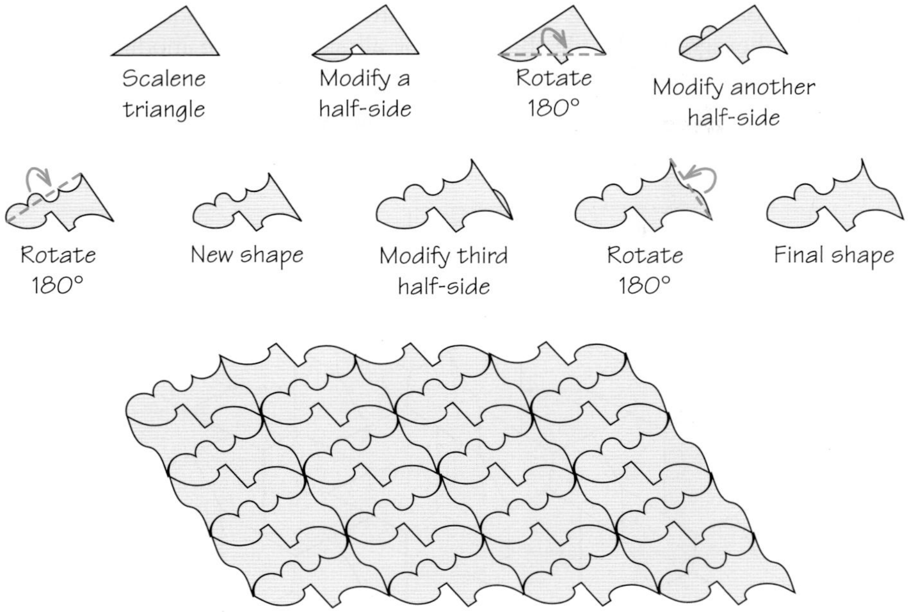

Figure 20.18 How to make an Escher-like tiling by translations and half-turns, using a scalene triangle base.

Self Check 4

Start from a triangle and design your own Escher-like tiling.

EXAMPLE 5 Tiling Using a Quadrilateral

How can you make an Escher-like tiling by starting from an arbitrary quadrilateral?

For the quadrilateral, use the same process as in Example 4, modifying some or all of the four sides—as many as you wish (Figure 20.19).

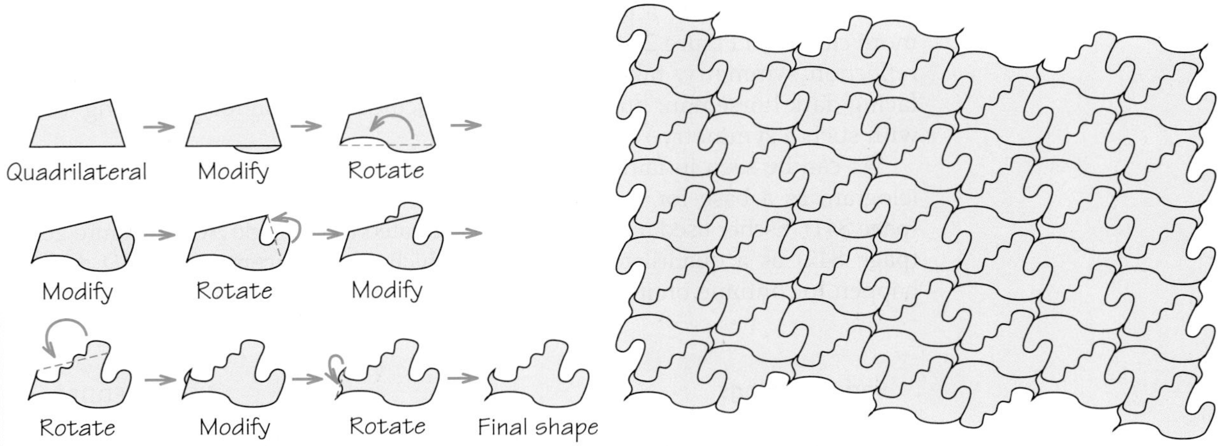

Figure 20.19 How to make an Escher-like tiling by translations and half-turns, using a quadrilateral base.

Self Check 5

Start from a quadrilateral and design your own Escher-like tiling.

The approach of Example 5 will work with some of the sides of some pentagons and hexagons that tile. Because not all sides can be modified, there is less freedom for designing tiles, so it is more difficult to make the resulting tiles resemble intended figures. Figure 20.20 shows the constraints for one kind of tiling by pentagons.

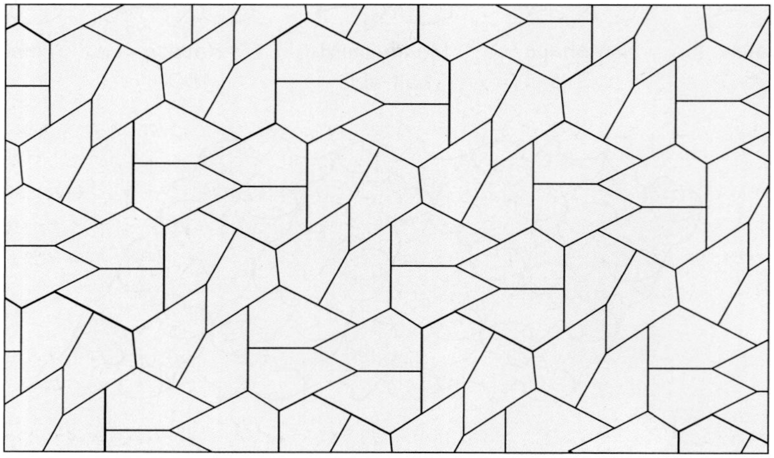

Figure 20.20 An example of type 12 of the 14 families of tilings by pentagons. The restrictions are that $\angle A = 90°$, $\angle C + \angle E = 180°$, $2\angle B + \angle C = 360°$, $2a = c + e = d$. *(This figure was adapted from a tiling produced by the "Pentagon Tilings" demonstration by Ed Pegg Jr. at http://demonstrations.wolfram.com/PentagonTilings/.)*

Check the Suggested Websites at the end of this chapter to access a demonstration of all 14 known pentagon tilings and see how much they can be varied in appearance. At another website listed there, you can look at art created by Marjorie Rice from her pentagon tilings.

Sketches in Escher's notebook indicate how he designed many of his prints. For the bird tiling of Figure 20.10, the single bird below the tiling shows that he modified the sides of a square. For the horsemen tiling of Figure 20.11a on page 838, the sketches in Figure 20.11b show that a pair of horsemen occupy the modified sides of a par-hexagon. We redrew the two fundamental figures more clearly in Figure 20.12 on page 839. (The horsemen tiling also has a glide-reflection symmetry, taking a leftward-facing light horseman to a rightward-facing dark horseman; but we have not discussed criteria for producing a tiling with such a symmetry.)

As can be seen in faint lines in Figure 20.15 (page 841), Escher used a parallelogram as a base for the camel tiling. For the seahorse tiling of Figure 20.16 (page 841), Escher used a triangle base. He treated the triangle *ACF* in Figure 20.17b (page 842) as a quadrilateral *ACDF* in which two adjacent sides (*CD* and *DF*) happen to continue on in a straight line.

Periodic Tilings

PROCEDURE

All the patterns that we have exhibited and discussed so far have been **periodic tilings.** If we transfer a periodic tiling to a transparency, it is possible to slide the transparency a certain distance in some direction, without rotating it, until the transparency exactly matches the tiling everywhere. We can also achieve the same result by moving the transparency in some second direction (possibly vertically) by a certain (possibly different) distance.

In a periodic tiling, you can identify a **fundamental region**—a tile, or a block of tiles—with which you can cover the plane by translations, rotations, mirror reflections, and/or glide reflections. For example, in Figure 20.10 (page 837), a single bird forms a fundamental region. In Figure 20.15 (page 841), two adjacent camels, one right side up and one upside down, form a fundamental region. In the terminology of Chapter 19, the periodic tilings are those that are preserved under translations in more than two directions.

20.5 Nonperiodic Tilings

In Figure 20.3a (page 830), the second row from the bottom is offset one-half of a unit to the right from the bottom row, the third row from the bottom is offset one-third of a unit farther, and so forth. Because whatever the value of n, the sum $\frac{1}{2} + \frac{1}{3} + \frac{1}{4} + \cdots + \frac{1}{n}$ never adds up to exactly a whole number, there is no direction (horizontal, vertical, or diagonal) in which we can move the entire tiling and have it coincide exactly with itself.

Nonperiodic Tiling DEFINITION

A **nonperiodic tiling** is a tiling that cannot be made to coincide with itself by any translation in any direction.

EXAMPLE 6 A Nonperiodic Tiling Through Randomness

How can you use chance to make a tiling?

Consider the usual edge-to-edge square tiling. For each square, flip a coin. Depending on the result, divide the square into two right triangles by adding either a rising or a falling diagonal (see Figure 20.3b on page 830). Because what happens in each individual square is unconnected to what happens in the rest of the tiling, this random tiling by right triangles has almost no chance of being periodic.

We are not much interested in the arbitrary "random" tilings that can result from a process such as the one described in Example 6. We are much more interested in tilings that are nonperiodic but nevertheless show some kind of order and symmetry.

Penrose Tilings

For all known cases, if a *single tile* can be used to make a nonperiodic tiling of the plane, then it also can be used to make a periodic tiling. It is still an open question whether this property is true for any shape of tile whatever. Curiously, the answer is known for the corresponding question in three dimensions: In 1993, Conway discovered a single convex polyhedron that tiles space nonperiodically but cannot be used to make a periodic tiling.

For a long time, mathematicians also tended to believe the more general assertion that if you can construct a nonperiodic tiling with a *set of one or more tiles,* you can construct a periodic tiling from the same tiles. In 1964, however, a set of tiles was found that permits only nonperiodic tiling. It contains 20,000 different shapes! Over the next several years, smaller sets were discovered with the

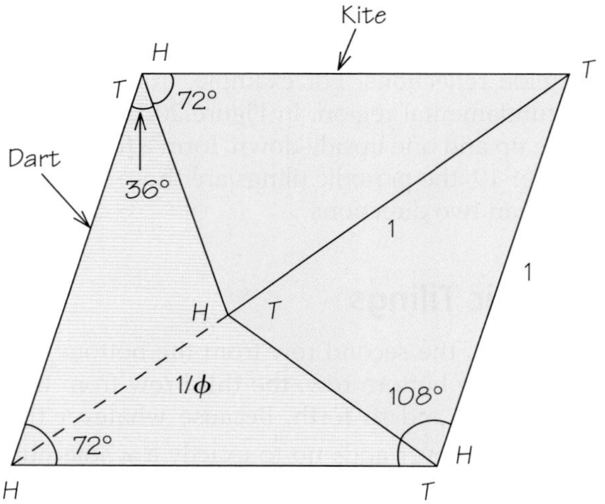

Figure 20.21 Construction of Penrose's "dart" (beige area) and "kite" (blue area). The length $1/\phi \approx 0.618$ is the reciprocal of the golden ratio ϕ.

same property, with as few as 100 shapes. But it was still amazing when Sir Roger Penrose (b. 1931), a mathematical physicist at Oxford, announced in 1975 that he had found a set that tiles only nonperiodically—consisting of just two tiles! (See Figure 20.21 and Spotlight 20.4.)

Sir Roger Penrose

SPOTLIGHT 20.4

Sir Roger Penrose, a professor at the University of Oxford, received a doctorate in mathematics but has been seriously interested in physics for many years. He demonstrated mathematically that the collapse of a black hole is "completely general and unstoppable." He discovered what are now called Penrose tiles in 1974, which were based on an earlier six-tile scheme of his. He has explored the relationship between artificial intelligence and human consciousness, and his more recent endeavor has to do with the development of conformal cyclic cosmology to explain the construction of the universe.

He was "totally fascinated" when he saw Maurits Escher's exhibit at the International Congress of Mathematicians in 1954. Subsequently, Penrose and his father produced a picture of an impossible staircase, while quite independently Escher did the same in his lithograph *Belvedere.* Escher followed up with other works of the same nature, as well as a water color *Ghosts* showing a non-isohedral tiling inspired by Penrose.

Courtesy Sir Roger Penrose

Roger Penrose standing on a Penrose tiling in the foyer of the Mitchell Institute Building at Texas A&M University.

These tiles, together known as *Penrose pieces,* were named individually by Conway as "darts" and "kites." One of each can be cut from a single rhombus. A **rhombus** is a quadrilateral with four equal sides and equal opposite interior angles; it's like a "leaning" square (see Figure 20.21). The particular rhombus from which the Penrose tiles are constructed has interior angles of 72° and 108°. If we cut the longer diagonal in two pieces so that the longer piece is the golden ratio, or $(1 + \sqrt{5})/2 \approx 1.618$ times as long as the shorter (see Chapter 19, page 783), and connect the dividing point to the remaining corners, we split the rhombus into a dart and a kite (Figure 20.21).

Label the front and back vertices of the dart with H (for head) and its two wing tips with T (for tail), and do the reverse for the kite. Then the rule for fitting the pieces together is that only vertices with the same letter may meet: Heads must go to heads, tails must go to tails. Thus, the rules don't allow for the pieces to fit together as a rhombus (which would permit them to tile periodically).

A nicer-looking method of enforcing the rules, proposed by Conway, is to draw circular arcs of different colors on the pieces and require that adjacent edges must join arcs of the same color. The result is the pretty patterns of Figure 20.22. In fact, Conway thinks of the darts as children, each with two hands. The rule for fitting the pieces together is that children are required to hold hands. Penrose patterns become dancing circles of children.

Penrose also devised another pair of pieces, a thick rhombus and a thin rhombus, with accompanying matching rules that also force them to tile only nonperiodically; Figure 20.23 shows them used in a tiling.

From Sir Roger Penrose

Figure 20.22 A Penrose tiling with specially marked tiles, forming what is known as the *cartwheel tiling.*

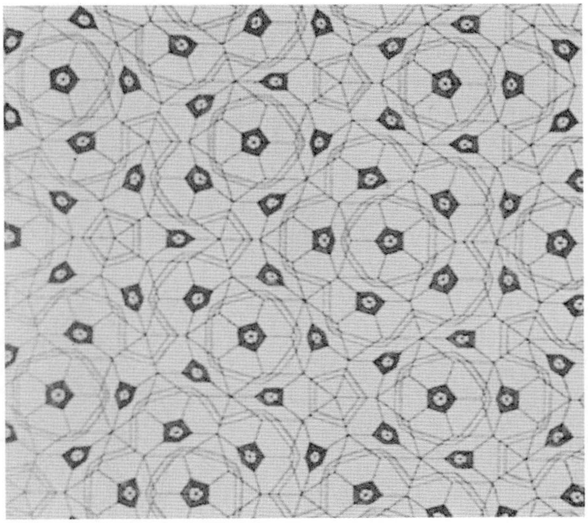

Tiling by Sir Roger Penrose

Figure 20.23 A Penrose nonperiodic tiling made with two rhombus shapes, one thin and one fatter. The fatter one has a yellow stripe across one end.

Figure 20.24 shows a modification of the Penrose pieces into two bird shapes. Figure 20.25 shows a coloring of a Penrose tiling in which no two adjacent pieces have the same color.

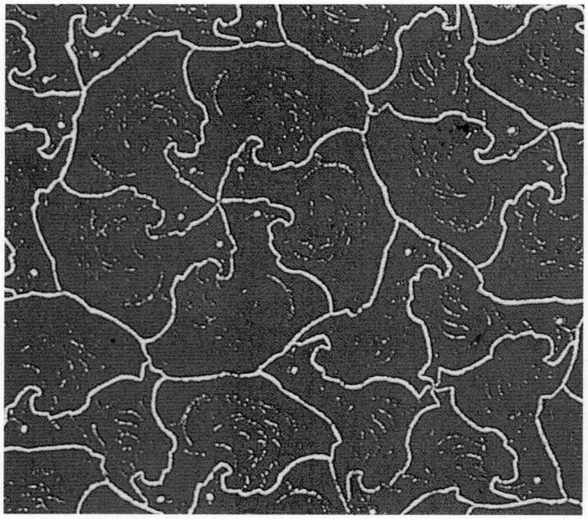

Figure 20.24 A modification of a Penrose tiling by refashioning the kites and darts into bird shapes.

Figure 20.25 A Penrose tiling by kites and darts, colored with five colors. A Penrose tiling can always be colored using just three colors in such a way that two tiles that share an edge have different colors.

Quasiperiodic Tilings

We show that tilings with Penrose's pieces (1) must be nonperiodic, but nonetheless (2) possess unexpected symmetry. We will say that they are *quasiperiodic,* a term that we explain carefully later.

Conway invented methods, *inflation* and *deflation,* to make bigger and smaller versions of Penrose patterns. Inflation turns any Penrose pattern into a different Penrose pattern with larger darts and kites, by systematically cutting each dart in half and attaching its halves instead either to a pair of kites, to make a new larger kite, or to a single kite, to make a new larger dart (see Figure 20.26).

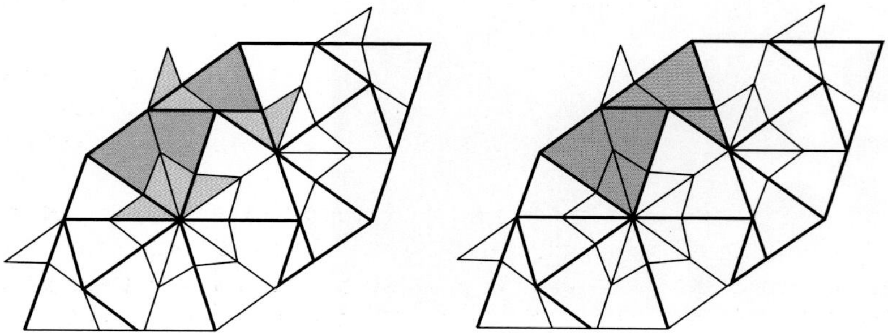

Figure 20.26 In inflation, each light blue dart is separated into two halves. Two such halves join two pink kites to become a larger orange kite, and two others attach to a single kite as wings of a larger dark blue dart.

Inflation is handy for showing that any Penrose pattern must be nonperiodic. We offer a proof by contradiction. Suppose (contrary to what we want to establish) that some Penrose pattern is periodic; that is, it has translation symmetry in some direction. Let d be the distance along the translation direction to the first repetition.

Performing inflation does the same thing to each repetition, so the inflated pattern must still have translation symmetry at the same distance d along the translation direction.

Keep performing inflation, time after time, until the darts and kites are so large that they are more than d across. The pattern, as we have just argued, must still have translation symmetry at a distance d—but it can't, because there's no repetition inside a single tile!

We reach a contradiction. So what's wrong? Our initial supposition, that the pattern was periodic in the first place, must be wrong. We conclude that all Penrose tilings are nonperiodic.

Penrose Tilings Are Nonperiodic THEOREM

Any tiling using Penrose's pieces (darts and kites) and his rules for fitting them together must be nonperiodic.

In Chapter 19, we explored our intuitions of symmetry in terms of balance, similarity, and repetition. Patterns made with the Penrose pieces certainly involve repetition, but it is the balance in the arrangement that we seek. What balance can there be in a nonperiodic pattern? It turns out that some Penrose patterns have a single line of reflection. But most surprising of all is that every Penrose pattern has a kind of fivefold rotational symmetry.

EXAMPLE 7 How Can a Penrose Pattern Have Fivefold Symmetry?

Fivefold symmetry is impossible for regular tilings: How could a Penrose pattern have such symmetry?

Look again at Figure 20.21 on page 846, which shows how to split a rhombus into the Penrose dart and kite pieces. Except in the recess of the dart and the matching part of the kite, each of the internal angles of the kite and of the dart is either 72° or 36°.

Now, 72° goes into 360° 5 times, and 36° goes into 360° 10 times. If we recall that it is the interior angles that matter in arranging polygons around a point, we see why it might be possible for a Penrose pattern to have fivefold or tenfold rotational symmetry.

A Penrose pattern with tenfold rotational symmetry is impossible, but there are exactly two Penrose patterns that tile the entire plane and have fivefold rotational symmetry about one particular point. We show finite parts of these patterns in Figure 20.27. For each pattern, the center of rotational symmetry is at the center of the figure, surrounded by either five darts or five kites.

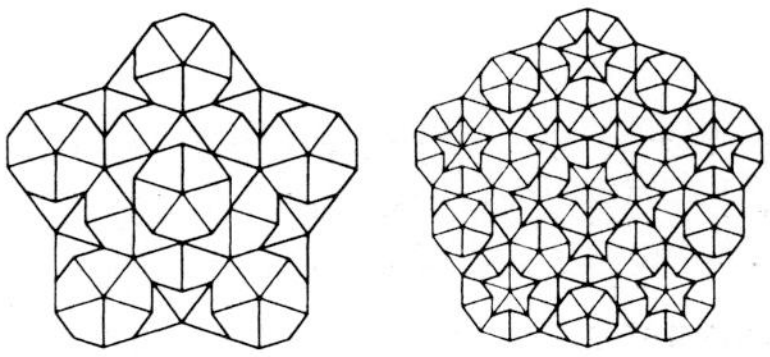

Figure 20.27 Successful deflation (i.e., the systematic cutting up of large tiles into smaller ones) of patches of tiles of a Penrose nonperiodic tiling.

In any other Penrose pattern, the pattern as a whole does not have fivefold rotational symmetry. However, what is surprising is that even so, the pattern must have arbitrarily large finite regions with fivefold rotational symmetry. You can see fivefold symmetry in the larger and larger regions of Figure 20.23 (page 847) that are enclosed by yellow lines. In Conway's metaphor, whenever a chain of children (the fatter rhombus shapes in the figure) closes, the region inside has fivefold symmetry.

Penrose patterns have another remarkable feature. Despite their being nonperiodic, all Penrose patterns are somewhat alike, in the following sense.

Penrose Inside of Penrose THEOREM

The subpattern of any finite region in one Penrose pattern is contained somewhere inside every other Penrose pattern. In fact, any subpattern occurs infinitely many times in every Penrose pattern.

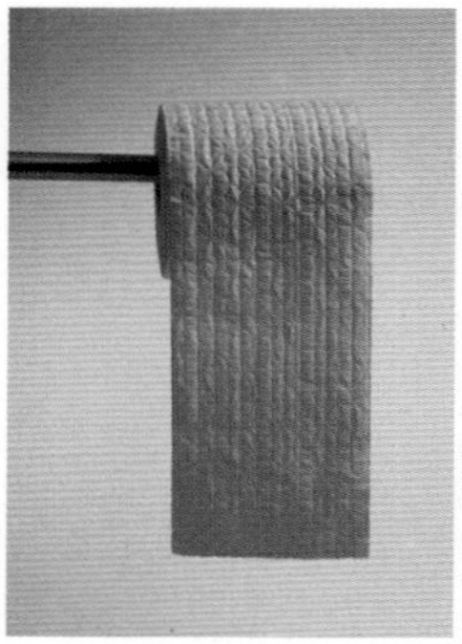
Mario Ruiz/The LIFE Images Collection/Getty Images

Figure 20.28 Penrose toilet paper.

The nonperiodicity of Penrose tilings found a surprising application in 1997—to bathroom tissue. Quilted bathroom tissue is embossed with a pattern to keep the layers together (Figure 20.28). If the pattern is regular, then the multiple layers on the roll can produce lumpy ridges and grooves. Using a nonrepeating Penrose pattern averts the lumpiness. However, the company used Penrose's copyrighted design without his permission, and Penrose sued for breach of copyright. They reached a settlement out of court.

Penrose tilings have found decorative use, not only in floor tilings as in Spotlight 20.2 on page 836, but also most notably on a grander scale in the new Transbay Transit Center in downtown San Francisco.

Penrose tilings are **quasiperiodic,** or somewhere between periodic and random.

Quasiperiodic Tiling DEFINITION

A tiling is **quasiperiodic** if it exhibits *local periodicity* under some transformations: You can translate or rotate it so that a finite number of tiles coincide perfectly, yet the entire tiling does not. We say that such a tiling is "locally" periodic but not "globally" periodic.

Musical Sequences

Looking at interpretations of a concept in other dimensions, or a problem in other sizes, is characteristic of the outlook of mathematicians in their search for patterns.

Although the quasiperiodic feature of Penrose tilings may seem apparent from examining them, we can make the idea clearer by considering tilings—in particular, quasiperiodic tilings—in just one dimension instead of in the plane of two dimensions. (Shortly we will also see what such tilings can be in three dimensions.)

The motivation for such a consideration came from Robert Ammann (1946–1994), who introduced lines onto the two rhombic Penrose pieces (used in Figure 20.23 on page 847) that are now known as *Ammann bars.* (For more about Robert Ammann, see Spotlight 20.5.)

In any Penrose tiling, these bars line up into five sets of parallel lines, each set rotated 72° from the next, forming a pentagonal grid (see Figure 20.29). The distance between two adjacent parallel bars is one of only two values, either A or B.

We can think of the pattern of As and Bs in the Ammann bars of Figure 20.29 as representing a tiling of a line by line segments (the tiles) of lengths A and B.

Exercise 43 (page 862) asks you to use your imagination to determine the monohedral tilings of the line, as well as periodic tilings with tiles of two different lengths. Exercises 49–62 (pages 863–864) define inflation and deflation operations on tilings of the line, show just what relationship between the lengths A and B is necessary for a quasiperiodic tiling, and make connections to the Fibonacci sequence and golden ratio ϕ of Chapter 19.

Mathematics and Autism

SPOTLIGHT 20.5

In 1975, Martin Gardner wrote in his "Mathematical Games" column in *Scientific American* about Penrose's discovery of a set of tiles that do not tile the plane periodically, but do tile it nonperiodically. Gardner did not show any tilings because Penrose was waiting for a patent.

In response to that column, Gardner received a letter from Robert Ammann, who wrote, "I am also interested in nonperiodic tiling, and have discovered both a set of two polygons which tile the plane only nonperiodically and a set of four solids which fill space only nonperiodically." Ammann included pictures: He had independently discovered Penrose tilings and generalized them to three dimensions.

Ammann went on to do the following:

- Find five more sets of planar tilings
- Discover the general organizing principle behind what are now known as Ammann bars (the one-dimensional analogue of Penrose tilings, which we discuss later and in the exercises)
- Make many other contributions to the theory of nonperiodic tiling (including being the first to find a three-dimensional analogue)

No one in the tiling community knew who Ammann was. He described himself as "an amateur doodler," and he declined all invitations to meetings and conferences.

In 1987, Marjorie Senechal, a geometer at Smith College, finally succeeded in tracking him down, arranging for him to meet Penrose, Donald Coxeter, and John Conway, and even getting Ammann to speak at two conferences.

After Ammann's early death from a heart attack in 1994, Senechal learned more about his life. At age three, he was precociously brilliant. At age four, he stopped talking and withdrew into his own world. He struggled through childhood and adolescence, "off the charts intellectually but impossible emotionally." During the years of his tiling discoveries, he was living in a motel and supporting himself by sorting mail in a post office. His mother said that the mathematical meetings were the high point of his life: "No one else reached out to him."

Perhaps we should draw the moral that it is important to encourage talent, mathematical or otherwise, wherever we find it, and to welcome into our community all able contributors.

The Mysterious Mr. Ammann, Media Highlights section of the *College Mathematics Journal* 36(2) (March 2005): 167–168.

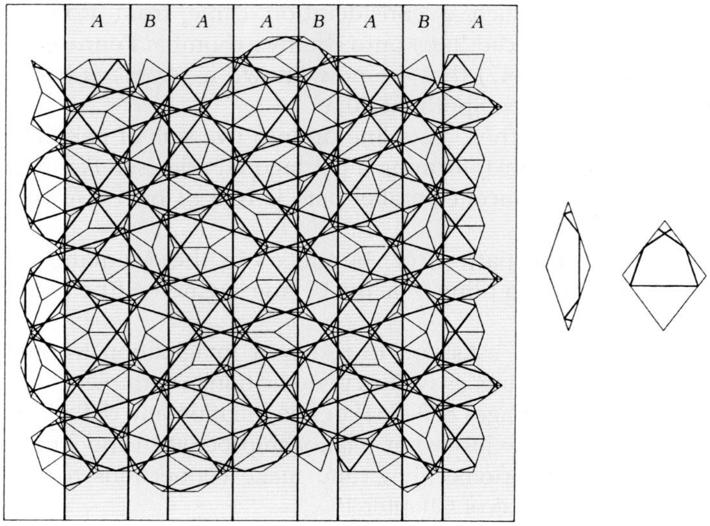

Figure 20.29 Penrose tilings with Ammann bars. Specially placed lines on the tiles produce five sets of parallel bars in different directions.

Here, we pursue mathematical thinking and ask what kind of patterns there can be in the Ammann bars, as sequences of *A*s and *B*s.

EXAMPLE 8 Musical Sequences

What about the order in which the *A*s and *B*s occur, as we move from left to right in Figure 20.29? Is there any pattern to that?

From the limited part of the pattern that we can observe, we see the sequence as

ABAABABA

You might think from the figure that the pattern continues repeating the group *ABAAB* indefinitely. But such is not the case. The sequence of intervals between Ammann bars is nonperiodic: It cannot be produced by repeating any finite group of symbols. We can think of it as a one-dimensional analogue of a Penrose tiling. The notation is reminiscent of the melody pattern of songs: Many popular songs follow the pattern *ABA*, with the first and the last sections having the same melody but the middle section being different. Consequently, a sequence of intervals between Ammann bars is known as a **musical sequence.**

There are some rules that musical sequences follow. Two *B*s can never be next to each other, nor can we have three *A*s in a row (see Exercises 54 and 55 on page 863). Just as any finite part of any Penrose tiling occurs infinitely often in any other Penrose tiling, any finite part of any musical sequence appears infinitely often in any other one. The order of the symbols is neither periodic nor random, but between the two—quasiperiodic.

Self Check 6

If we had looked at a smaller part of the pattern of Ammann bars, we would have seen *ABAABA* and possibly conjectured that the pattern continues by repeating *ABA* over and over. What are the possibilities for the next three bars if they are not to be *ABA* and there cannot be three *A*s in a row or two *B*s in a row?

The ratio of darts to kites in an infinite Penrose tiling, or of *A*s to *B*s in a musical sequence, is exactly the golden ratio, approximately 1.618, as you can show in Exercise 62 on page 864). So if you are going to play with sets of Penrose pieces to see what kinds of patterns you can create, you will need about 1.6 times as many darts as kites.

As pointed out by geometers Marjorie Senechal (Louise Wolff Kahn Professor Emerita in Mathematics and History of Science and Technology, Smith College) and Jean Taylor (Professor Emerita, Rutgers University; Visiting Scholar, the Courant Institute of Mathematical Sciences, New York University), Penrose tilings have three important properties:

- **Nonperiodic:** They are constructed according to rules that force nonperiodicity.
- **Self-similar:** They can be obtained from a substitution process (inflation and deflation) that features self-similarity at different scales (like the fractals in Chapter 19).
- **Quasiperiodic:** They are quasiperiodic, in the sense that we have defined.

These properties are somewhat independent, meaning that a tiling may have just one or two of them without having all three; however, all three are required for a Penrose tiling.

Quasicrystals and Barlow's Law

Although Penrose's discovery was a big hit among geometers and in recreational mathematics circles in the mid-1970s, few people thought that his work might have practical significance. In the early 1980s, some mathematicians even generalized Penrose tilings to three dimensions, using solid polyhedra to fill space nonperiodically. Like the two-dimensional Penrose patterns, these have orderly fivefold symmetry but are nonperiodic.

Yet in 1982, scientists at the U.S. National Bureau of Standards discovered unexpected fivefold symmetry while looking for new ultrastrong alloys of aluminum (mixtures of aluminum with other metals).

Manganese doesn't ordinarily alloy with aluminum, but the experimenters were able to produce small crystals of alloy by cooling mixtures of the two metals at a rate of millions of degrees per second. Following routine procedures, chemist Daniel Shechtman began a series of tests to determine the atomic structure of the special crystals. But there was nothing routine about what he found: The atomic structures of the manganese–aluminum crystals were so startling that it took Shechtman three years to convince his colleagues they were real.

Why did he encounter such resistance? His patterns—and the crystals that produced them—defied one of the fundamental laws of crystallography. Like our discovery that the plane cannot be tiled by regular pentagons, **Barlow's law,** also called the **crystallographic restriction,** says that since the atomic structure of a crystal must be periodic, it can have only rotational symmetries that are twofold, threefold, fourfold, or sixfold. If there were a center of fivefold symmetry, many such centers would have to exist. Barlow proved this impossible.

Peter Barlow (1776–1862) argued by contradiction, the same approach that we saw earlier in Conway's proof that Penrose patterns are not periodic. Suppose (contrary to what we intend to show) that there is more than one

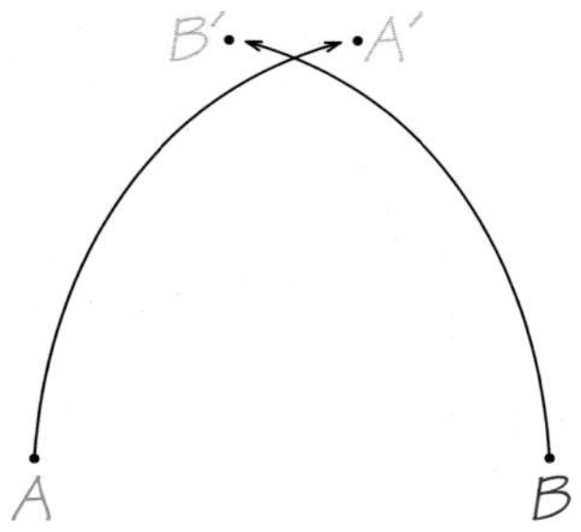

Figure 20.30 Barlow's proof that no pattern can have two centers of fivefold symmetry.

fivefold rotation center. Let *A* and *B* be two of these that are closest together (see Figure 20.30).

Rotate the pattern of Figure 20.29 by one-fifth of a turn clockwise around *B*, which carries *A* to some point *A*. Because the pattern has fivefold symmetry around *B*, the point *A*, which is the image of the fivefold center *A*, must itself be a fivefold center.

Now use *A* as a center and rotate the pattern by one-fifth of a turn clockwise, which carries *B* to some point *B*. As we just argued in the case of *A*, *B* must also be a fivefold center. But *A* and *B* are closer together than *A* and *B*, which is a contradiction. Hence, our original supposition must be false, and a pattern can have at most one fivefold rotation center (as the patterns in Figure 20.27 on page 849, in fact, do) and so cannot be periodic.

For chemists, crystals are modeled well by periodic three-dimensional tilings; an array of atoms with no symmetry whatsoever would not be considered a crystal. Since Barlow's law shows that fivefold symmetry is impossible in a periodic tiling, no one suspected until Penrose's discovery that there could be symmetric nonperiodic tilings, or, until Shechtman's alloys, that real atoms could arrange themselves in such a way.

Shechtman's alloys, since they are not periodic, are not crystals, though in other respects they do resemble crystals. It is scientifically more fruitful to extend the concept of crystals to include them than to rule them out, so the term *crystal* now officially encompasses both traditional periodic crystals as well as atomic structures that are not periodic but ordered in some way. The generalized kind of crystals are now known as *quasicrystals* (see Spotlight 20.6).

Once again, as so often happens in history, pure mathematical research anticipated scientific applications. Penrose's discovery, once just a delightful piece of recreational mathematics, has prompted a major reexamination of the theory of crystals. Barlow's law is not refuted, since it applies only to periodic crystals, not to quasicrystals.

Quasicrystals

SPOTLIGHT 20.6

In 1984, working at the University of Pennsylvania, Paul Steinhardt and Don Levine did a computer simulation of what a three-dimensional Penrose pattern would be like. They decided to call such structures *quasicrystals.* Later that fall, their chemist colleague Daniel Shechtman showed that quasicrystals really exist by synthesizing some. He produced images of an alloy of aluminum and manganese that were amazingly similar to images from the computer simulations. In short order, sevenfold, ninefold, and other symmetries were also shown to occur in real materials.

In 1991, Sergei Burkov showed that quasiperiodic tilings can be made using only a single kind of 10-sided tile, provided the tiles are allowed to overlap. With overlaps, the resulting patterns are no longer tilings. They are called *coverages.* In late 1998, scientists presented electron microscope photos that demonstrated that atoms really can form such coverages.

The current theory is that quasicrystals are packings of copies of a single type of atom cluster, with each cluster sharing atoms with its neighbors; that is, each cluster overlaps nearby clusters. The clusters

Quasicrystals

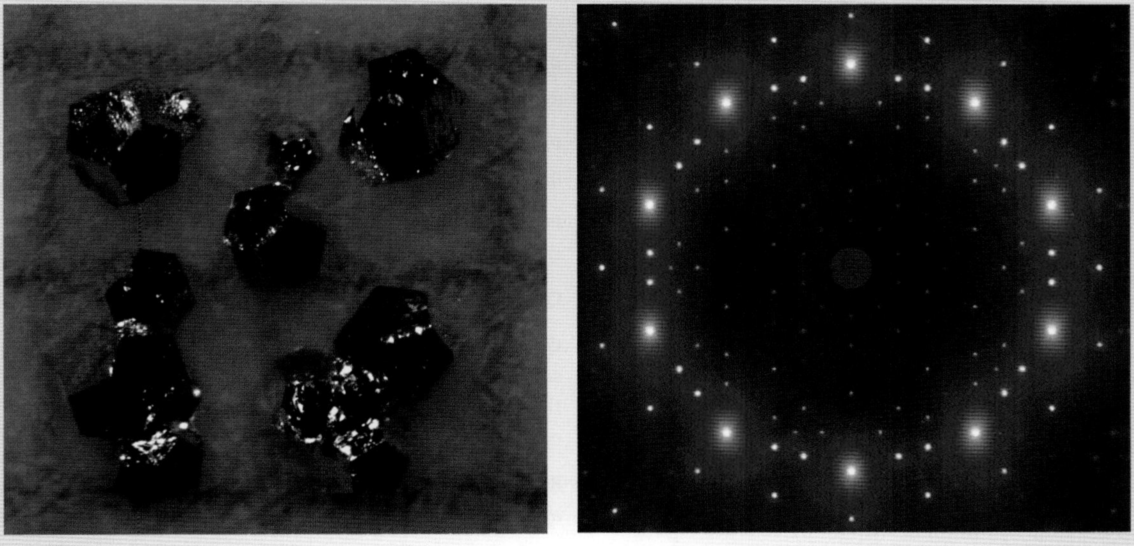

Both: Courtesy of the U.S. Department of Energy's Ames Laboratory

(*a*) (*b*)

(a) High-energy x-ray diffraction pattern for a single grain of i-Gd-Cd, made up of gadolinium, a rare earth, and cadmium. The pattern shows a new kind of quasicrystal discovered in 2013, an icosahedral binary quasicrystal; note the 10-fold symmetry. (An icosahedron is a regular polyedron; see Spotlight 20.1 on page 829.) The investigators, who used mathematics to pinpoint the existence of such crystals, were exploring their magnetic properties. Rare earths are important for their magnetic properties; high-performance magnets made from them are used in windmill turbines, hybrid cars, cellphones, and headphones. (b) Photo of actual grain of i-Gd-Cd, showing its (imperfect) icosahedral shape.

form a quasiperiodic pattern that maximizes their density, thereby minimizing the energy of the atoms involved.

The irregular atomic structure of quasicrystals results in the following:

- Low surface energy, making them suitable as coatings for nonstick cookware (a French firm has produced a line of such pots)
- Low friction, desirable in shavers and scalpels
- Quick absorption of heat, useful as an ingredient in the plastic powder used in 3D printers

In 2007, Steinhardt and Peter J. Lu announced the possible discovery of what appeared, to them, to be fragments of decagonal and Penrose tilings in medieval Islamic architecture in Iran; but other scientists have disputed their conclusion.

In 2009, naturally occurring quasicrystals were discovered for the first time in a remote area of Russia, in what appears to be the remains of a meteorite that was formed at the beginning of the solar system.

In 2011, Daniel Shechtman was awarded the Nobel Prize in Chemistry for his discovery of quasicrystals.

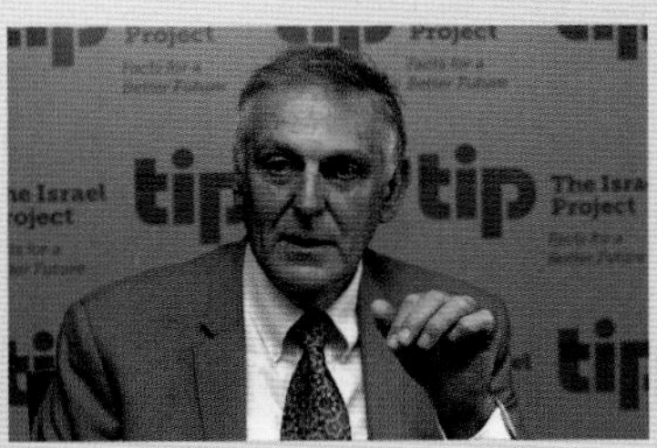

Gali Tibbon/AFP/Getty Images

Daniel Shechtman. Note the Penrose pattern on his tie!

Review Vocabulary

Barlow's law (crystallographic restriction) A law of crystallography stating that a crystal can have only rotational symmetries that are twofold, threefold, fourfold, or sixfold. (p. 853)

Centrosymmetric Symmetric by 180° of rotation around its center. (p. 840)

Convex A tile (including its boundary) is convex if for any two points on it, all the points on the line segment joining them also belong to the tile. (p. 834)

Conway Criterion A criterion for determining whether a shape can tile by means of translations and half-turns. (p. 841)

Edge-to-edge tiling A tiling is edge-to-edge if for every tile, each edge coincides with the entire edge of a bordering tile. (p. 830)

Equilateral triangle A triangle with all three sides equal. (p. 828)

Exterior angle The angle outside a polygon formed by one side and the extension of an adjacent side. (p. 829)

Fundamental region A tile or group of adjacent tiles that can tile by translations, rotations, mirror reflections, and/or glide reflections. (p. 845)

Interior angle The angle inside a polygon formed by two adjacent sides. (p. 830)

Monohedral tiling A tiling with only one size and shape of tile. (The tile is allowed to occur also in "turned-over," or mirror-image, form.) (p. 828)

Musical sequence A sequence of intervals between Ammann bars. (p. 852)

n-gon A polygon with n sides. (p. 828)

Nonperiodic tiling A tiling that cannot be made to coincide with itself by any translation. (p. 845)

Parallelogram A convex quadrilateral whose opposite sides are equal and parallel. (p. 833)

Par-hexagon A hexagon whose opposite sides are equal and parallel. (p. 840)

Periodic tiling A tiling that repeats by translations in two different directions, possibly horizontal and vertical. (p. 844)

Quadrilateral A polygon with four sides. (p. 833)

Quasiperiodic A tiling that exhibits local periodicity under some transformations: It can be translated or rotated so that a finite number of tiles coincide perfectly, yet the entire tiling will not. (p. 850)

Regular polygon A polygon whose sides and angles are all equal. (p. 828)

Regular tiling An edge-to-edge tiling that uses only one kind of regular polygon. (p. 830)

Rhombus A parallelogram whose sides are all equal—four equal sides and equal opposite interior angles. (p. 847)

Scalene triangle A triangle with no sides equal. (p. 833)

Semiregular tiling An edge-to-edge tiling that uses a mix of regular polygons with different numbers of sides but in which all vertex types are alike—the same polygons in the same order, clockwise or counterclockwise. (p. 832)

Tiling (tessellation) A covering of the entire infinite plane by nonoverlapping regions, called tiles. (p. 828)

Translation A rigid motion that moves everything a certain distance in one direction. (p. 838)

Vertex type The pattern of polygons surrounding a vertex in a tiling. (p. 832)

Self Check Answers

1. The measure of an exterior angle of a regular decagon is 36°, so each interior angle measures 144°. But 144 does not divide 360 evenly, so we cannot fit an integral number of decagons around a point (two is too few, and three is too many).

2. Answers will vary.

3. Answers will vary.

4. Answers will vary.

5. Answers will vary.

6. The eight possibilities for three symbols in a row are *AAA, AAB, ABA, ABB, BAA, BAB, BBA,* and *BBB.* (Notice how they can all be listed systematically in terms of alphabetical order.) *AAA* is not possible (three *A*s in a row), and neither are *ABB, BBA,* or *BBB* (two *B*s in a row). *AAB* is also eliminated because it cannot follow *ABA* (three *A*s in a row). That leaves *ABA, BAA,* and *BAB;* in a musical sequence, it is *BAA* that, in fact, occurs after *ABAABA.*

Skills Check

1. In a tiling of the plane, the tiles

(a) must all be the same size.

(b) must all be the same shape.

(c) may be of different shapes and sizes.

2. The measure of an exterior angle of a regular octagon is ___________.

3. In a tiling of the plane by polygons, the tiles

(a) must all have the same number of sides.

(b) need not be regular polygons.

(c) can have different numbers of sides.

4. A regular tiling can be constructed using polygons with ___________, ___________, or ___________ sides.

5. Regular octagons and squares can form a semiregular tiling of the plane with

(a) two octagons and one square at each vertex.

(b) two octagons and two squares at each vertex.

(c) varying configurations at the vertices.

6. A semiregular tiling has a square, a regular dodecagon (12-gon), and another regular polygon at each vertex. This other polygon has ___________ sides.

7. A tessellation

(a) allows overlapping pieces.

(b) is not the same as a tiling.

(c) covers the entire infinite plane.

8. There are ___________ regular polyhedra.

9. How many semiregular tilings are there?

(a) 5

(b) 8

(c) Infinitely many

10. The smallest number of sides that a polygon can have and not be able to tile the plane is ___________.

11. A tiling of the plane can be formed using as a tile

(a) any convex quadrilateral but no nonconvex quadrilateral.

(b) any nonconvex quadrilateral but no convex quadrilateral.

(c) any quadrilateral.

12. How many regular tilings are there?

(a) None

(b) 3

(c) Infinitely many

13. A tiling of the plane can be formed using which of the following as a tile?

(a) Some but not all pentagons

(b) Any pentagon with at least two right angles

(c) Any pentagon with at least three right angles

14. Any quadrilateral can tile the plane using which operations?

(a) Only translations

(b) Translations plus half-turns

(c) Only half-turns

15. Regular pentagons

(a) can't tile the plane.

(b) can tile the plane, but only if you are very careful.

(c) don't occur in any tilings.

16. An artist famous for works based on tilings is ___________.

17. A convex irregular polygon

(a) can never tile the plane.

(b) can always tile the plane.

(c) cannot tile the plane if it has more than six sides.

18. The tile below can be used to tile the plane using which operations?

(a) Only translations

(b) Translations plus half-turns

(c) Only half-turns

19. Which of the following statements is true?

(a) If a polygon fulfills the Conway Criterion, it can tile the plane by translations.

(b) If a polygon fulfills the Conway Criterion, it can tile the plane by translations and half-turns.

(c) If a polygon doesn't fulfill the Conway Criterion, it can't tile the plane at all.

20. The ___________ Criterion says that the tile below can be used to create a tiling of the plane using ___________ and ___________.

21. In a nonperiodic tiling of the plane,

(a) the pattern never repeats.

(b) the pattern is not repeated by any translation.

(c) there must be at least three kinds of tiles.

22. Penrose tilings are ___________-periodic.

23. A Penrose dart has the property that

(a) opposite angles are congruent.

(b) it is nonconvex.

(c) the edges are all of different lengths.

24. A rhombus always has the property that ___________.

25. Ammann bars are

(a) an Arab delicacy that comes from Jordan.

(b) jazz venues where musical sequences are played.

(c) sets of parallel bars in a Penrose pattern.

26. Barlow's law prohibits the existence of crystals with ___________ symmetry.

27. Quasicrystals

(a) do not exist in nature.

(b) are not regular enough to be used in New Age ceremonies.

(c) are symmetric nonperiodic tilings.

28. In a Penrose tiling, the proportion of darts to kites is ___________.

29. A nonperiodic tiling

(a) is an impossibility.

(b) requires at least two kinds of tiles.

(c) does not have translation symmetry.

30. The process that takes a Penrose pattern into a different Penrose pattern with larger darts and kites is called ___________.

Chapter 20 Exercises

Challenge Discussion

Hint: For the exercises about determining whether a shape can tile the plane, you should make copies of the shape and experiment with placing them. One easy way to make copies is to trace the shape onto a piece of paper, staple half a dozen other blank sheets behind that sheet, and use scissors to cut through all the sheets along the edges of the traced shape on the top sheet.

20.1 Tilings with Regular Polygons

1. Determine the measure of an exterior angle and of an interior angle of a regular dodecagon (12 sides).

2. Determine the measure of an exterior angle and of an interior angle of a regular decagon (10 sides).

3. Specify a formula for the measure of an interior angle of a regular n-gon.

4. Using the formula from Exercise 3 and a calculator, make a table of the interior-angle measures of regular polygons with 3, 4, …, 12 sides.

5. Use the table of interior-angle measures from Exercise 4 to determine all the possible vertex types of regular polygons (with at most 12 sides) surrounding a point.

6. Which of the vertex types of Exercise 5 do not occur in a semiregular tiling?

7. In addition to the vertex types of Exercise 5, exactly five others are possible, each involving one polygon with more than 12 sides. None of these vertex types leads to a semiregular tiling. The five many-sided polygons involved in these five vertex types have 15, 18, 20, 24, and 42 sides. Determine the other polygons in each of these five vertex types.

For Exercises 8 and 9, refer to the second tiling in the bottom row of Figure 20.4 on page 831, which shows a tiling by isosceles triangles.

8. Use the center of the tiling to determine the measures of the angles of the isosceles triangle tile.

9. Every vertex except the center vertex has the same vertex type, in terms of the measures of the angles surrounding the vertex. What is that vertex type?

20.2 Tilings with Irregular Polygons

10. For each tile below, find a tiling of the plane. (Adapted from Branko Grünbaum and G. C. Shephard, *Tilings and Patterns,* W. H. Freeman, New York, 1987, p. 25.)

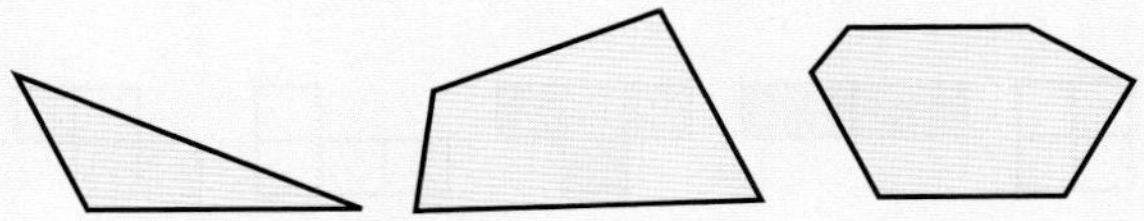

(a) (b) (c)

11. You know that a regular pentagon cannot tile the plane. Suppose that you cut one in half into two mirror-symmetric pieces. Can this new shape tile the plane? (See Figure 19.4 on page 787 for a regular pentagon that you can trace.)

12. Design a nonconvex hexagon that can tile the plane.

13. If you "unbulge" an ordinary soccer ball so that each of its sewn pieces is flat, you get a polyhedron, but it is not a regular polyhedron. It is a truncated icosahedron, one of the semiregular polyhedra. Some of the faces are regular hexagons, and some are regular pentagons. Explain why not all of its faces can be regular hexagons. All the vertices have the same vertex type. What is it? How many pentagons are there and how many hexagons? How many vertices?

milos luzanin/Alamy

14. A soccer ball should be as round as possible, which suggests using many polygons, each of which is short across. Of course, making a ball with a large number of small pieces requires more sewing. A ball with 92 faces was used for some European soccer competitions: 12 pentagons, 20 hexagons, and 60 triangles (see the accompanying figure), all of them regular polygons. What are the possible vertex types for a polyhedron made up of pentagons, hexagons, and triangles? Explain why the polyhedron underlying this soccer ball is not semiregular.

20.3 Using Only Translations

Refer to tiles (a) through (g) below in doing Exercises 15 and 16.

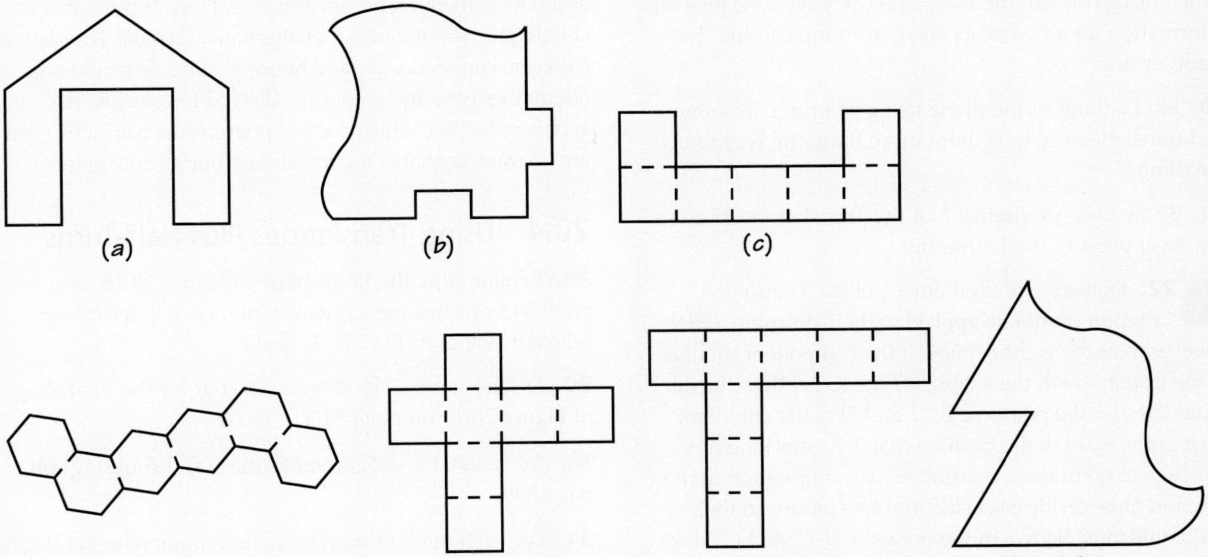

(a) (b) (c)

(d) (e) (f) (g)

15. For each tile (a) through (c), determine whether it can be used to tile the plane by translations. (From Frederick Barber et al., *Tiling the Plane,* COMAP, Lexington, MA, 1989, pp. 1, 8, 9.)

16. Repeat Exercise 15, but for tiles (d) through (g).

17. Start from a par-hexagon of your choice and modify it to tile the plane by translations. (You will probably find it useful to do your work on graph paper. If you choose a regular hexagon, special graph paper is available, ruled into regular hexagons, that would be particularly good to use.) Can you draw a design on the tile so as to make an Escher-like pattern?

18. Start from a parallelogram of your choice and modify it to tile the plane by translations. (You will probably find it useful to do your work on graph paper.) Can you draw a design on the tile so as to make an Escher-like pattern?

Refer to the following information in doing Exercises 19–22. A particularly simple kind of polygon, called a *polyomino,* is made of squares joined edge-to-edge. The name is a generalization of "domino"; indeed, there is only one kind of domino (two squares joined at an edge to form a rectangle). There are just two *trominos* (short for "triominos"), the straight tromino and the L-tromino. The straight tromino has the shape of a rectangle, so it can tile the plane by translations; the L-tromino has the shape of a hexagon.

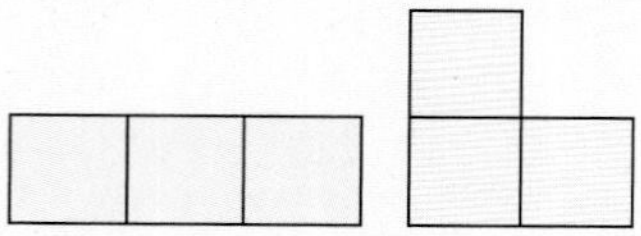

19. Is the L-tromino convex? Does the result about which hexagons can tile the plane (on page 834) give any information about whether the L-tromino can tile the plane or not?

20. Find a tiling of the plane using just the L-tromino and translations of it. Is there more than one way to do the tiling?

21. Show how alternative 2 of the Translation Criterion can be applied to the L-tromino.

22. Explain why alternative 1 of the Translation Criterion cannot be applied to the L-tromino. (*Hint:* Label each of the eight corners of the component squares of the tromino with the letters *S, T,* ... *Z.* Let these be our candidates for the points *A, B, C,* and *D* of the criterion.) Each of the sides of the tromino that is 2 units long has nowhere to go under a translation. Any application of the criterion must divide each side into two pieces, so their midpoints must be two of the points *A, B, C,* and *D.* Make a similar argument about two corners of the tromino. Thus, we have four points, which can be labeled consecutively *A, B, C,* and *D,* starting at any one of them. Show that none of the four possibilities "works." (This argument can be generalized to show that trying *A, B, C,* and *D* at points other than the corners of the squares won't work either.)

23. Explain why there are no other tetrominos (each made of four squares joined at edges) than the five shown below—plus differing mirror images of two of them (which ones?). In the order shown, they are called the *square, straight,* T-, L-, and *skew* tetrominos.

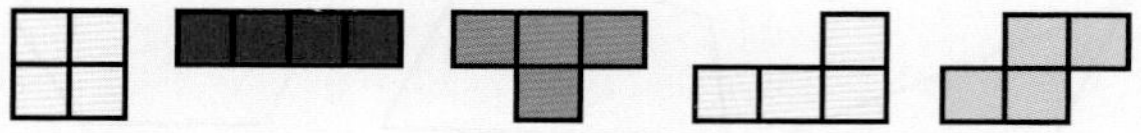

For Exercises 24–28, you will find it useful to make yourself several copies of each of the polyominos (e.g., by cutting them out of graph paper).

24. Show how alternative 2 of the Translation Criterion applies to the T-tetromino.

25. Apply alternative 1 of the Translation Criterion to the skew-tetromino and show how it can tile.

26. Show how alternative 2 of the Translation Criterion applies to the L-tetromino.

27. Show how alternative 2 of the Translation Criterion can be applied to the skew-tetromino and show how it can tile.

28. In Exercises 25 and 27, we indulged in what would appear to be "overkill," demonstrating the same result in two different ways. Doing so can be useful in mathematics for giving greater understanding about why and how something occurs or is true. Here, those exercises should give you the idea that alternative 2 of the Translation Criterion can reduce to (and hence is more general than) alternative 1 if some points are allowed to coincide. For such a reduction, which pairs of points must coincide? (You are allowed to relabel the remaining four distinct points.)

20.4 Using Translations Plus Half-Turns

29. Explain how the final shape of Figure 20.18 on page 843 satisfies the Conway Criterion by identifying relevant points *A, B, C, D, E,* and *F.*

30. Do the same as in Exercise 29, but for the final shape of Figure 20.19 on page 843.

For Exercises 31 and 32, refer to tiles (a) through (g) on page 859.

31. For each tile (a) through (c), determine whether it can be used to tile the plane by translations and half-turns.

32. Repeat Exercise 31 for tiles (d) through (g).

33. Show how an arbitrary pentagon with two parallel sides, such as the one shown below, can tile the plane.

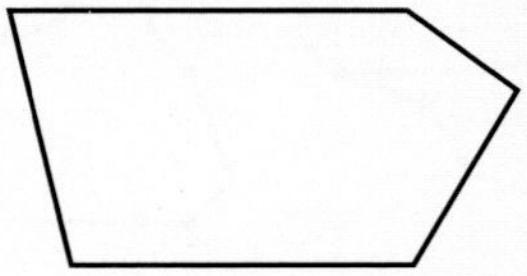

34. The pentagon below is a pentagonal tile of type 13, which was discovered by Marjorie Rice. Show how it can tile the plane. (*Hint:* Carefully trace and cut out a dozen or so copies and try fitting them together.)

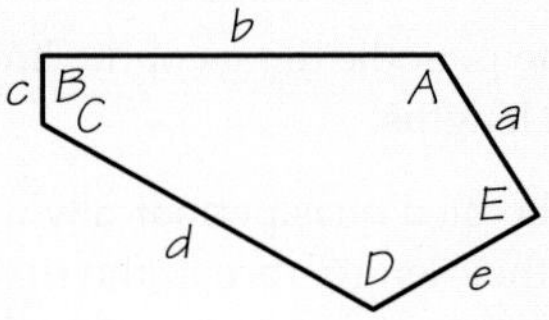

The parts of this pentagon satisfy the following relations:

$$A = C = D = 120^\circ$$
$$B = E = 90^\circ$$
$$a = e$$
$$a + e = d$$

[Adapted from Doris Schattschneider, "In Praise of Amateurs," in David A. Klarner (ed.), *The Mathematical Gardner,* Wadsworth, Belmont, CA, 1981, p. 162.]

35. Start from a triangle of your choice and modify it to tile the plane by translations and half-turns. (You will probably find it useful to do your work on graph paper.) Can you draw a design on the tile so as to make an Escher-like pattern of a plant, animal, or other object?

36. Start from a quadrilateral of your choice and modify it to tile the plane by translations and half-turns. (You will probably find it useful to do your work on graph paper.) Can you draw a design on the tile so as to make an Escher-like pattern of a plant, animal, or other object?

For Exercises 37–40, refer to the information about polyominos preceding Exercise 19. We saw earlier that all the dominos, trominos, and tetrominos tile the plane by translations. Here, we investigate the 12 pentominos, shown below with a letter notation for each. (If you allow mirror images to count as different pentominos, there are 18.) It will be useful for you to make several copies of each of them.

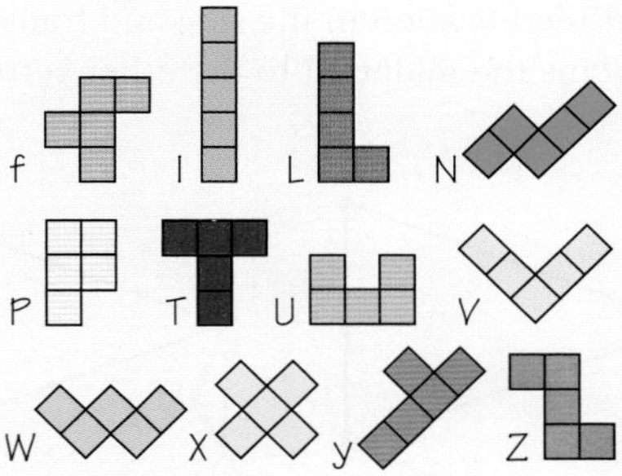

37. By experimenting, determine which of the pentominos can tile the plane by translations. (*Hint:* There are nine.)

38. Apply the Conway Criterion to the f-pentomino and show how it can tile by translations and half-turns.

39. Apply the Conway Criterion to the U-pentomino and show how it can tile by translations and half-turns.

40. Apply the Conway Criterion to the T-pentomino and show how it can tile by translations and half-turns.

41. (Thanks to Doris Schattschneider, Moravian College.) In the text, we discuss some criteria and methods for generating Escher-like patterns that involve just translations or translations and half-turns. A slight variation on one such method allows construction of tilings that feature a tile and its mirror image. We modify a parallelogram so that two reflected tiles are joined and the joined pair tiles by translation. You get to design the shape of the tile and put whatever art you like on it.

Begin with an isosceles triangle and mark the midpoint of each of its sides (it's best to use graph paper or a dynamic geometry program). Half-turn the triangle about the midpoint of one of its two equal sides to produce a parallelogram.

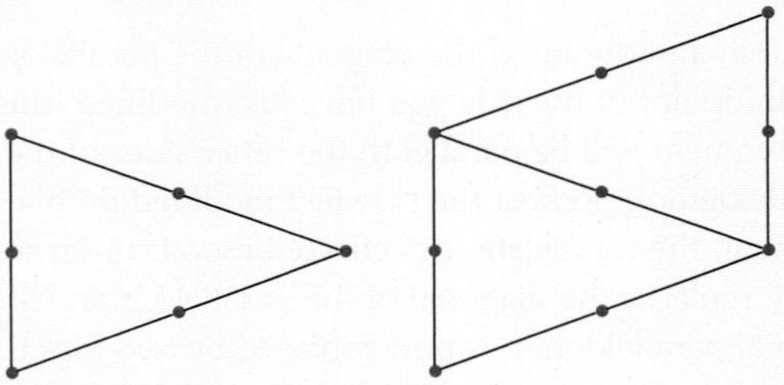

Modify half of the (odd) third side in any manner—here is where you get to be creative!—joining a vertex to the midpoint of that side. Then reflect that modification in the side and translate it so that it joins the midpoint to the other vertex of that side.

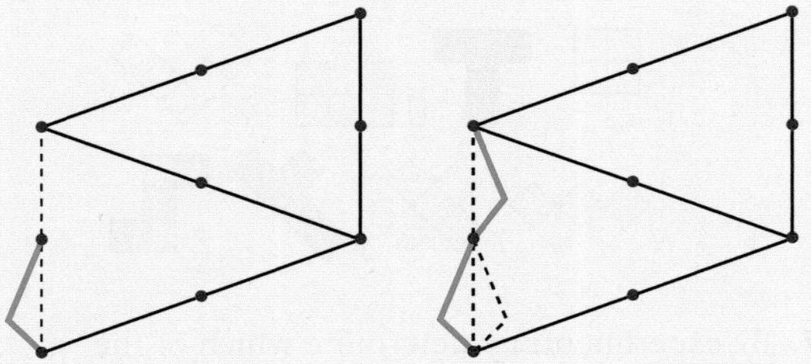

Now translate the lower half of the modified side to the upper half of the opposite side of the parallelogram, and translate the upper half of the modified side to the lower half of the opposite side of the parallelogram.

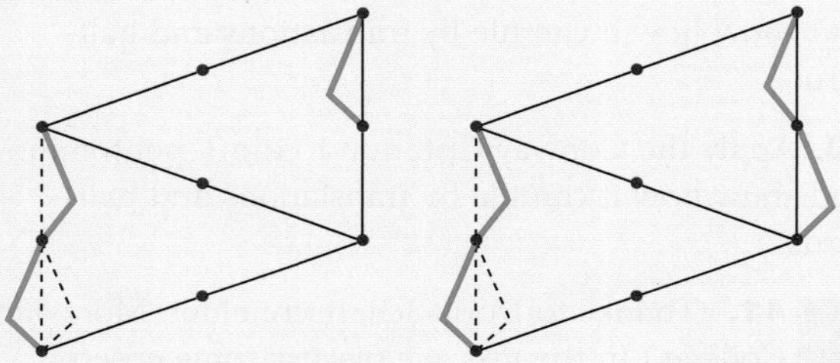

Now modify one of the remaining sides of the parallelogram in any way you wish, and then translate that modification to the opposite side of the parallelogram.

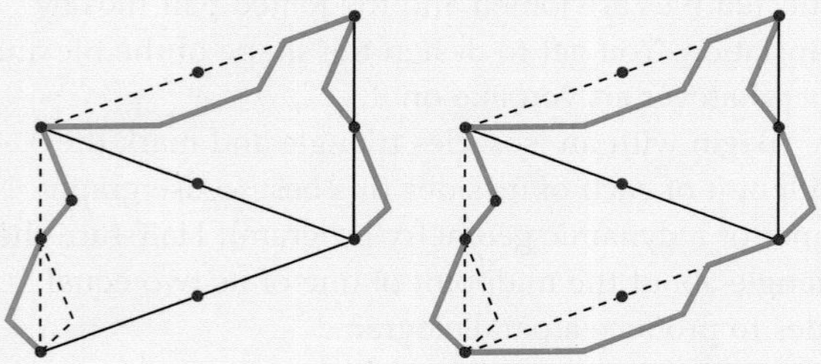

Join a midpoint of the diagonal of the parallelogram to a midpoint of the side you have just modified (this line segment will be parallel to the other sides of the parallelogram). Reflect the side just modified in this segment, then translate the reflected modification so that it replaces the diagonal of the parallelogram. Your original parallelogram is now replaced by two tiles that are reflected images of each other. The outline of the joined pair of tiles is a modified par-hexagon that tiles by translations.

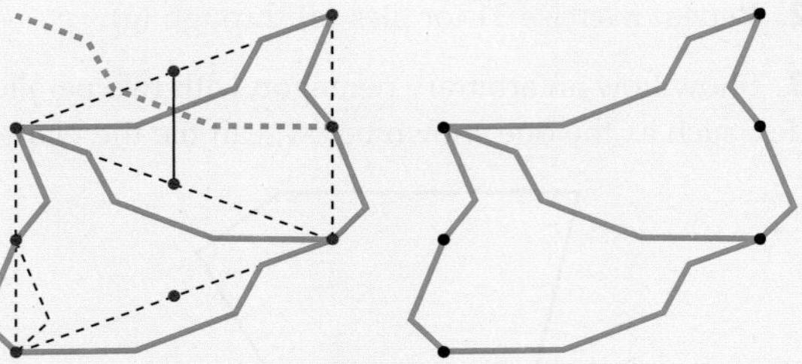

42. Does the Conway Criterion apply to the tilings created by the method of Exercise 41?

20.5 Nonperiodic Tilings

43. (a) Describe the monohedral tilings of the line.

(b) Describe the periodic tilings of the line that use two tiles of different lengths.

44. A line can be tiled quasiperiodically with a pair of tiles but only if their lengths are in the right proportion, so that the pattern can be scaled up. We determine what that proportion must be. Let the two lengths be S (for "short") and L (for "long"), with $L > S$ and $L = cS > 1$, where the scaling factor is $c > 1$. The key requirement for the scaling factor c is that scaling up a tile of length L produces a total length of $S + L$, that is, $cL = S + L$.

(a) Using substitution, reduce the two equations $L = cS$ and $cL = S + L$ to a single equation, eliminating L and leaving just S and c involved.

(b) Use the quadratic formula to solve for two possible values of c. Since we want $c > 1$, we choose the larger value.

Exercises 45–48 illustrate the connection between *nonperiodic* tilings of the line and the Fibonacci numbers and golden ratio of Chapter 19 (but these exercises do not require any information from that chapter). (Thanks for this idea to David J. Wright, Oklahoma State University.) The rabbit problem in Chapter 19 (Exercise 4, page 814) leads us directly into quasiperiodic patterns (and even to musical sequences, as later exercises show). Let A (for "adult") denote an adult pair of rabbits and B (for "baby") denote a baby pair. We record the population at the end of each month, just before any births, in a systematic way, as a string of As and Bs. At the end of their second month of life, a rabbit pair will be considered to be adult and give birth to a baby pair.

At the end of the first month, the sequence is just A, and the same is true at the end of the second month. When an adult pair A has a baby pair B, we write the new B immediately to the right of the A. So at the end of the third month, the sequence is *AB;* at the end of the fourth, it is *ABA* because the first baby pair is now adult; at the end of the fifth month, we have *ABAAB.* (In this model, rabbits breed like clockwork, month after month, and never die!)

45. What is the sequence at the end of the sixth month?

46. Explain why we can never have two *B*s next to each other.

47. Explain why we can never have three *A*s in a row.

48. Show that from the fourth month on, the sequence for the current month consists of the sequence for last month followed by the sequence for two months ago.

For Exercises 49–51, we define inflation and deflation of a sequence of "tiles" consisting of *S*s ("shorts") and *L*s ("longs").

- Inflation: Replace each *S* by *L* and each *L* by *LS.* For example, the inflation of (*L*)(*S*)(*L*) would be (*LS*)(*L*)(*LS*) = *LSLLS,* where we have inserted parentheses for clarity.
- Deflation: Replace each *LS* by *L* and each lone *L* by *S.* For example, the deflation of (*LS*)(*LS*)(*L*) would be (*L*)(*L*)(*S*) = *LLS,* where we have inserted parentheses for clarity.

49. (a) Start with just a single *S* and repeat the inflation process, showing the stages, until you reach a stage with 21 tiles.

(b) How many tiles are there at each stage that you reached? (If you continue this process forever, you tile a half-line to the right; you could tile the entire line by reflecting this right half-line over to cover the left half-line. The result is called a *Fibonacci tiling* of the line because of the appearance of the Fibonacci sequence in the numbers of tiles.)

(c) If a line segment contains m copies of the *L* tile and n copies of the *S* tile, how many tiles will the inflation of the segment contain?

50. (a) Apply deflation repeatedly to your answer to Exercise 49a, which has 21 letters, until you can deflate no further. What do you end up with?

(b) Apply deflation repeatedly instead to the periodic sequence *LSLSLSLS*.... What do you end up with?

(c) Devise your own periodic sequence of *L* and *S* tiles. Apply the deflation process to it repeatedly. What do you end up with? What do you conjecture?

(d) Explain how the Fibonacci tiling is quasiperiodic.

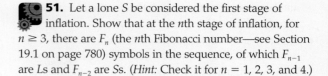

51. Let a lone *S* be considered the first stage of inflation. Show that at the nth stage of inflation, for $n \geq 3$, there are F_n (the nth Fibonacci number—see Section 19.1 on page 780) symbols in the sequence, of which F_{n-1} are *L*s and F_{n-2} are *S*s. (*Hint:* Check it for $n = 1, 2, 3,$ and 4.)

For Exercises 52–62, refer to the following. Call a sequence *musical* if repeatedly applying deflation to it eventually results in a single *S.*

52. Explain why inflation and deflation preserve musicality: If we inflate or deflate a musical sequence, we get another musical sequence. (Another way to think of this relationship is that a musical sequence is self-similar under inflation and deflation.)

53. Explain why no musical sequence contains *SS.*

54. Explain why no musical sequence contains *LLL.*

55. Explain why no musical sequence ends in *LL.*

56. Explain why no musical sequence ends in *LSLS.*

57. Show that apart from the lone sequence *S,* every musical sequence is an initial subsequence of all the musical sequences that are successive inflations of it.

58. Exercises 53–56 give necessary conditions for a sequence to be musical. Are those conditions sufficient? That is, if a sequence does not contain *LLL* or *SS,* and it does not end in *LL* or *LSLS,* must it be a musical sequence? (This is the kind of question that mathematicians ask in an effort to pin down the characteristics of a musical sequence.)

59. We show how to check whether a finite block of *A*s and *B*s can be a part (subsequence) of some musical sequence or else is never a part of any musical sequence. We do so by deflating the block: If the deflation arrives at a single symbol, the block is part of a musical sequence; if the deflation cannot arrive at a single symbol, the block is not part of any musical sequence. However, we use a slightly modified form of deflation:

- If at any stage of the deflation of the block we arrive at *S* as the starting symbol, we can tack on an initial *L* in front of it (because as an internal part of a musical sequence, the deflated block would have to have an *L* preceding the *S*—it can't have two *S*s in a row and be musical).
- We add the rule that if at any stage of the deflation of the block we arrive at *LL* as the ending symbols, we replace the *LL* with *SL* (because as an internal part of a musical sequence, the deflated block would have to be followed by an *S*—it can't have three *L*s in a row

and be musical; at the next deflation, the *LLS* would become *SL*.)

- If, at any stage of this modified deflation, we arrive at two or more *S*s in a row or three or more *L*s in a row, then the original block could not be part of a musical sequence. Otherwise, the original block will deflate eventually to a single symbol, at which point we conclude that the original block is a part of a musical sequence.

Check the two blocks below. Is either a part of a musical sequence?

(a) *LSLLSLSLLS*

(b) *LSLLSLSLSL*

60. (For students who have covered Chapter 19.) From Exercise 53, we know that each application of inflation to a musical sequence simply extends it. By successive inflation, then, we can build an infinite sequence. Argue that as we approach this limiting sequence, the ratio of *L*s to *S*s tends toward the golden ratio ϕ of page 780.

61. (For students who have covered Chapter 19.) Conclude from Exercise 56 that the sequence cannot be periodic or settle into a periodic repetition after a finite "burn-in" period. Thus, the sequence is nonperiodic. (*Hint:* ϕ is not a rational number; that is, it cannot be represented as a ratio m/n of whole numbers m and n.)

62. (For students who have covered Chapter 19.) After applying deflation a large number of times, a large area of a Penrose pattern will contain some number k kites and d darts. At each deflation, a kite is broken down into two kites and two halves of darts, and a dart is broken down into a single kite and two halves of darts. Thus, each of the k kites gives rise to $2k$ kites, and each of the d darts gives rise to d kites, for a total of $2k + d$ new kites. Similarly, there will be $d + k$ new darts. The ratio $x = k/d$ of kites to darts will become

$$\frac{2k + d}{k + d} = \frac{2\frac{k}{d} + 1}{\frac{k}{d} + 1} = \frac{2x + 1}{x + 1}$$

Assume that the ratio $x = k/d$ eventually stabilizes as the number of darts and kites gets larger and larger. What value does it tend to? *Hint:* Set $x = (2x + 1)/(x + 1)$ and solve for x.

Chapter Review

63. Specify a formula for the measure of an exterior angle of a regular n-gon.

64. Give a numerical reason why a semiregular tiling could not include both regular polygons with 12 sides and regular polygons with 8 sides (with or without any regular polygons with other numbers of sides).

65. The Translation Criterion for tiling of the plane by translations alone offers two possibilities that guarantee such a tiling. State the details of one of these possibilities.

66. State the Conway Criterion in full detail: What kind of tilings does it refer to? What is the condition for such a tiling?

67. What is a quasicrystal? What is the connection of quasicrystals to Penrose tilings? What was so revolutionary about Penrose tilings?

Writing Projects

1. Use computer software to make some tilings of your own (see the Suggested Websites). Print out your tilings and describe, in a sentence or two each, how you made them.

2. You can build a model of a buckyball by weaving strips of paper in a hexagonal pattern, much as peoples in Africa and elsewhere weave baskets and balls. Background material and instructions on doing this are available at www.ccd.rpi.edu/Eglash/csdt/african/hex/intro.html. Make such a buckyball, preferably with colored strips. Explain why the result is a buckyball. The construction depends on the fact that when a strand is wrapped around a rim of the same width, a 60° angle results. Explain why this is so, and experiment to try to determine what angles are formed when the strand and rim are not the same width (e.g., when a the strand is twice as wide as the rim or half as wide).

Suggested Readings

BEYER, JINNY. *Designing Tessellations: The Secrets of Interlocking Patterns,* McGraw-Hill, New York, 1999.

FATHAUER, ROBERT. *Designing and Drawing Tessellations,* Tessellations Publishing, Phoenix, Ariz., 2010.

KAPLAN, CRAIG. Escherization. www.cgl.uwaterloo.ca/~csk/projects/escherization/. Gives examples of tilings, using recognizable motifs, that were produced by the author's algorithm, including ones based on Penrose tilings.

PENROSE, ROGER. Pentaplexity, *Eureka* 39 (1978): 16–22; reprinted under the title Pentaplexity: A class of non-periodic tilings of the plane, *Mathematical Intelligencer* 2(1) (March 1979): 32–37; reprinted with striking color figures, *Eureka* 62 (2012): 10–15. http://world.mathigon.org/resources/Polygons_and_Polyhedra/Pentaplexity.pdf

RANUCCI, ERNEST, and JOSEPH TEETERS. *Creating Escher-Type Patterns,* Creative Publications, Oak Lawn, IL, 1977.

SCHATTSCHNEIDER, DORIS. *M. C. Escher: Visions of Symmetry,* 2d ed., Harry N. Abrams, New York, 2004.

SCHATTSCHNEIDER, DORIS. Will it tile? Try the Conway criterion!, *Mathematics Magazine* 53(4) (September 1980): 224–233.

SENECHAL, MARJORIE. The mysterious Mr. Ammann, *Mathematical Intelligencer* 26(4) (Fall 2004): 10–21.

SERRA, MICHAEL. *Discovering Geometry: An Investigative Approach,* 4th ed., Kendall Hunt, Dubuque, IA, 2007. Chapter 7, Transformations and Tessellations, treats regular and semiregular tessellations, tessellations with irregular polygons, and the Conway criterion.

SEYMOUR, DALE, and JILL BRITTON. *Introduction to Tessellations,* Dale Seymour Publications, Palo Alto, CA, 1990. An excellent introduction to tessellations, including how to make Escher-like tessellations.

STEPHENS, PAM, and JIM McNEILL. *Tessellations: The History and Making of Symmetrical Designs,* Crystal Productions, Glenview, IL, 2001.

TEETERS, JOSEPH L. How to draw tessellations of the Escher type, *Mathematics Teacher* 67 (1974): 307–310.

WILLSON, JOHN. *Mosaic and Tessellated Patterns: How to Create Them,* Dover, Mineola, NY, 1983.

Suggested Websites

http://mathcs.slu.edu/escher/index.php/Math_and_the_Art_of_M._C._Escher *Math and the Art of M. C. Escher,* by Anneke Bart and Bryan Clair. Textbook (with explorations and exercises) for a course about symmetry as exhibited in the works of Escher.

www.geometrygames.org/KaleidoTile/index.html Interactive Windows and Macintosh program that lets the user design tilings on the plane, the sphere, and the hyperbolic plane. Spherical tilings can be realized as polyhedra.

demonstrations.wolfram.com/ComplementTiling/ Interactive Web program to make Escher-like tilings, with links to other demonstrations.

demonstrations.wolfram.com/TilingConstructor/ Free software for creating tilings, with a great many options; author Karl Scherer offers several related demonstrations, available from this page.

demonstrations.wolfram.com/PentagonTilings/ Free demonstration of all 14 known families of tilings by pentagons; the user can adjust angles, sides, and directions of repetition.

www.cgl.uwaterloo.ca/~csk/software/penrose/ Applet (online or downloadable) that lets you "warp" Penrose's darts and kites to make tilings with other shapes of tiles.

www.tilingsearch.org Database of high-quality images of tilings, which allows searching by shapes and by numerous other criteria.

www.tess-elation.co.uk/ David Bailey's World of Escherlike Tessellations, including essays, artwork, and explanations of Penrose tiles and other tessellations.

www.dbsullivanstudio.com/downloads/tutorials/tessellation_tutorial_part_1.pdf A tutorial on how to draw a tessellation with tiles of varying shapes. (There appears to be no Part 2 available.)

www.tesselmaniac.com/tess/TesselManiac.html Macintosh and Windows tessellation software (payware).

tessellation.info/en/info/artists/17/Marjorie_Rice/ Site with some of Marjorie Rice's artistic works created from pentagon tilings.

www.eschertiles.com/links.html Links to sites with Escher-inspired tilings and also to the official M. C. Escher site.

nlvm.usu.edu/en/nav/vlibrary.html Applets for tessellations and fractals.

Part VII

HOUSE FOR SALE
ANOTHER SUCCESS LET US HELP YOU
BUY/SELL YOUR NEXT HOME
SOLD
LICENSED AGENT
NO REASONABLE OFFER REFUSED

Your Money and Resources

This part of the text concentrates on numerical patterns of growth and decline in the realms of finance, resources, and biology. The unifying concept is a *population,* whether of dollars, barrels of oil, or tons of fish.

How much interest will your savings account earn in the next year? How much will the monthly payment be on your credit card, your car loan, or a home mortgage? How much would you need to save to pay for a child's college education or for your retirement? What will inflation do to your savings? How much should you pay for a stock?

These are problems of daily life for which mathematics provides custom-tailored models. In Chapter 21, "Savings Models," and Chapter 22, "Borrowing Models," you become familiar with the mathematics and terminology of situations that you will face repeatedly in everyday life.

The financial models of these chapters apply broadly to important problems in other areas of life. Growth of a biological population is like growth of money at interest. Decay of a "population" of a radioactive substance is like depreciation of an asset or inflation of a currency. Determining how long it will be before a "population" of a nonrenewable resource, such as oil or coal, may be exhausted is like calculating how long a retirement "nest egg" will last. Managing a renewable biological resource, such as a forest or a fishery, presents problems similar to the management of a trust fund, such as the endowment of a college. In Chapter 23, "The Economics of Resources," we explore these similarities, together with the profound effect that economic conditions can have on natural resources. Finally, you will see the surprisingly large and puzzling consequences that very small changes can produce in a physical system or biological population as a result of behavior that mathematicians call chaos.

SnapshotPhotos/Shutterstock

[illegible]

Borrowing Models

[illegible]

The Economics of Resources

This part of the text [illegible]

[illegible]

[illegible] loan [illegible] What [illegible]

[illegible]

[illegible] Chapter [illegible]

[illegible] mathematics and economics of [illegible]

The financial models [illegible] problems in other areas [illegible]

[illegible]

Savings Models

21

JGI/Tom Grill/Getty Images

How much interest will your savings account earn in the next year? What difference does it make how the interest is calculated? How much should you save for a comfortable retirement in the face of inflation? In this chapter, we consider such questions and show how the underlying mathematical models also help explain the recent mortgage and bank crisis and the effect of interest-rate changes on stock prices.

We look at two ways that money can grow at interest: *simple interest,* which is *arithmetic* (or *linear*) growth, and *compound interest,* which is *geometric* (or *exponential*) growth. A compound interest rate can be converted to an equivalent simple interest rate, called the *effective rate,* which for a period of a year is known as the annual percentage yield (APY). That is the rate that must be advertised by banks that try to attract your savings. Interest can be compounded any number of times in a year; but there is a limit to how much can be realized, even with compounding infinitely often (known as continuous compounding).

For regular deposits to a savings account—to save up for a car, the down payment on a house, or other purchases—there is a formula that relates the size of the regular deposit, the interest rate, the amount to be accumulated, and the time to realize it. We derive this formula and show you how to use it to find any of the quantities if you know the others.

However, dollars saved may lose value due to inflation, so we show how to take inflation into account in your savings plans. Inflation, a "decay" of dollars, is usually expressed in terms of the Consumer Price Index (CPI). You learn how to convert from the CPI to the rate of inflation and how to determine the real rate of growth of an investment under inflation.

This chapter and the next encompass the world of savings and loans that you will experience in your lifetime, ranging from your parents (and you) saving for your college education to making time payments on a car or home loan to saving—and withdrawing savings—for retirement.

21.1 Simple Interest and Arithmetic Growth

When you open a savings account, your primary concerns are the safety and the growth of the "population" of your savings. We assume the safety and concentrate on measuring the growth.

EXAMPLE 1 Simple Interest

Interest is the money earned on a savings account or other fund. Suppose that you deposit \$1000 in an account that "pays interest at a rate of 10% annually." (This is an unrealistic rate, particularly in this era of very low interest rates. We use it solely because it makes the calculations easy.) Assuming that you make no other deposits or withdrawals, how much is in the account after 5 years?

The \$1000 is the **principal,** the initial balance of the account. At the end of one year, interest is added. The amount of interest is 10% of the principal, or

$$10\% \times \$1000 = 0.10 \times \$1000 = \$100$$

So the balance at the beginning of the second year is \$1000 + \$100 = \$1100. We express an interest rate either as a percentage or as a decimal fraction. "Percent" means "per 100," so you can think of the symbol "%" as standing for "1 per 100" or $\frac{1}{100} = 0.01$. So to convert from a percentage to a decimal fraction, divide the percentage by 100 by moving the decimal point two places to the left. An interest rate of 10% is 10/100 or 0.10; an interest rate r as a decimal number (such as 0.10) is $100r\%$ (10%). (*Caution:* A common error in using the formulas in this chapter is to forget to express the percentage as a decimal; for example, for $r = 5\%$, don't substitute 5 for the r in the formula, but instead substitute 0.05.)

With simple interest, interest is paid only on the original balance, no matter how much interest has accumulated. At the end of the first year, the account will contain \$1100. But at the end of the second year, you again receive only \$100; so at the beginning of the third year, the account contains \$1200. In fact, at the end of each year, you receive just \$100 in interest, amounting to a final balance of \$1500 at the end of the fifth year. ■

Self Check 1

What will be the balance after 10 years?

Simple Interest DEFINITION

Simple interest is interest that is paid on the original principal only, not on any accumulated interest.

Algebra Review Appendix
Fractions, Percents, and Percentages

The formulas for simple interest are themselves simple.

Simple Interest RULE

For a principal P and an annual rate of interest r, the interest earned in t years is

$$I = Prt$$

and the total amount A accumulated in the account is

$$A = P + I = P + Prt = P(1 + rt)$$

You may find this method for interest rather strange if you are used to a different system—compound interest, which we will consider shortly. However, simple interest is often used for the following transactions:

- Private loans between individuals, because it is easy to calculate
- Commercial loans for less than one year—not just because it is easy to calculate, but also because for low interest rates over short terms, simple interest differs negligibly from compound interest
- Financing of corporations and the government through bonds. A **bond** is a loan with repayment at the end of a fixed term and usually simple interest paid annually or semiannually
- Some student loans

EXAMPLE 2 Simple Interest on a Student Loan

Let's suppose that you have exhausted the amount that you can borrow under federal loan programs and need a private direct student loan for \$10,000. The lowest fixed interest rate from PNC Bank, Pittsburgh, Pennsylvania, in July 2015 was 6.49%. There is an interest-only repayment option, under which you make monthly interest payments while you are in school and pay toward the principal only after graduation. Under this plan, PNC earns simple interest from you while you are in school.

How much monthly interest would you pay for such a \$10,000 loan? The principal is $P = \$10{,}000$, the annual interest rate is $r = 6.49\% = 0.0649$ per year, and the number of years is $t = \frac{1}{12}$ year. The interest for one month would be

$$I = Prt = \$10{,}000 \times 0.0649 \times \tfrac{1}{12} \approx \$54.08$$

(Actually, the interest rate might not be 6.49% but could be as much as 12.99%, since it would depend on the creditworthiness of you and any cosigner.)

Self Check 2

If you opt not to make the monthly interest payments, whe... hen you are supposed to begin to repay the loan, the accumulated simple intere... italized, that is, added to the principal of the loan. If you are supposed to begi... ent after 51 months, what would be the principal of the loan at that time, ass... the interest rate stays the same throughout?

We frequently observe the kind of growth corresponding to simple in... called **arithmetic growth** or **linear growth,** in other contexts.

Arithmetic Growth DEFINITION

Arithmetic growth (pronounced with accent on the "met" syllable) (also called **linear growth**) is growth by a constant amount in each time period.

For example, the population of active medical doctors in the United States grows arithmetically because the medical schools graduate the same number of doctors each year and the numbers of doctors dying and retiring are also fairly constant but smaller. The concept of linear growth has appeared already in this book in the discussions of linear programming (Chapter 4) and linear regression (Chapter 6).

21.2 Compound Interest and Geometric Growth

What you probably expected to happen to the savings account discussed in the last section is that during the second year, the account would earn interest of 10%, not on just the initial balance of \$1000 (as with simple interest) but on the new balance of \$1100. Then, at the end of the second year, 10% of \$1100, or \$110, would be added to the account.

Thus, during the second year, you would earn interest on both the principal of \$1000 and on the \$100 interest added. With this method, you receive more interest during the second year than during the first; that is, the account grows more during the second year. At the beginning of the third year, the account contains \$1210, so at the end of the third year, you receive \$121 in interest. Again, this is more than at the end of the preceding year.

Compound Interest DEFINITION

Compound interest is interest that is paid on both the original principal and accumulated interest.

Savings institutions usually compound interest and credit it to accounts more often than once a year—for example, quarterly (four times per year). With an interest rate of 10% per year and quarterly compounding, you get one-fourth of the rate, or 2.5%, paid in interest each quarter year. The "quarter" (three months) is the **compounding period,** or the time elapsing before interest is paid.

Compounding Period DEFINITION

The **compounding period** is the fundamental interval on which compounding is based, within which no compounding is done.

EXAMPLE 3 Interest Compounded Quarterly

Suppose again, as in Example 1, that you have $1000 deposited at 10% annual interest, but this time with interest compounded quarterly. How much is in the account at the end of one year?

At the end of the first quarter, you have the original balance plus $25 interest, so the balance at the beginning of the second quarter is $1025. During the second quarter, you receive interest equal to 2.5% of $1025, or $25.625, which is rounded up in posting to your account (since the fraction is half a cent or more) to $25.63. Continuing in this manner, the balance at the end of the first year is $1103.82; see Table 21.1. (You should confirm all calculations in this chapter on your calculator.)

Even though the account was advertised as paying 10% interest, the interest for the year is $103.82, which is 10.382% of the original principal of $1000.

Self Check 3

How much would be in the account at the end of one year at 5% annual interest compounded semiannually (twice a year)?

TABLE 21.1 Compound Interest on $1000, at an Interest Rate of 10% Compounded Quarterly

Date	Beginning Balance	Interest on Principal	Interest on Interest	Total Interest Added	Ending Balance
January 1	1000.00				
March 31	1000.00	25.00	0.00	25.00	1025.00
June 30	1025.00	25.00	0.63	25.63	1050.63
September 30	1050.63	25.00	1.27	26.27	1076.90
December 31	1076.90	25.00	1.92	26.92	1103.82

Practical note: Without rounding the interest for each quarter, the interest for the year would have been not $1103.82 but $1103.81 (as shown in Table 21.2). Table 21.1 shows the results with rounding done only at the end of the year, while savings institutions must round at each posting and credit the rounded amount to your account. A spreadsheet program could duplicate the results of their computer programs; but in this table and in later calculations, we take the simpler route of rounding only at the final answer. Any differences will be very small; and if your answers differ by just a few cents, that will be OK.

If interest is compounded monthly (12 times per year) or daily (365 times per year), the resulting balance is even larger, as shown in Table 21.2. We will show you shortly the formula for these calculations. (The table also shows the results of continuous compounding, which we discuss later.)

TABLE 21.2 Comparing Compound Interest: The Value of $1000, at 10% Annual Interest, for Different Compounding Periods*

Years	Compounded Yearly	Compounded Quarterly	Compounded Monthly	Compounded Daily	Compounded Continuously
1	1100.00	1103.81	1104.71	1105.16	1105.17
5	1610.51	1638.62	1645.31	1648.61	1648.72
10	2593.74	2685.06	2707.04	2717.91	2718.28

*Without rounding at posting of interest and neglecting leap years; the difference in most cases is no more than 1 cent.

Interest Rates

Two accounts at the same annual rate of interest can produce different *yields* (amounts of interest), depending on how the compounding is done. To help prevent confusion for consumers, the Truth in Savings Act establishes terminology and calculation methods for interest.

A **nominal rate** is any stated rate of interest for a specified length of time, such as a 3% annual interest rate on a savings account or a 1.5% monthly rate on a credit-card balance. But by itself, a nominal rate *does not indicate nor take into account whether or how often interest is compounded.*

To keep interest rates straight, we use the following:

- r for a nominal annual rate, which may or may not be compounded.
- i for the rate during a compounding period, which can be a day, a month, or a year. There is no compounding done in a shorter interval than the compounding period. We call i the *periodic rate.*
- m for the number of compounding periods in a year.

We use r only for an annual rate, and t for a number of years. To avoid confusion, we don't use the terminology *annual percentage rate* because that term has a special legal meaning just for loans (see Section 22.2, page 911).

Rate Per Compounding Period RULE

For a nominal annual rate r compounded m times per year, the rate per compounding period is

$$\text{periodic rate} = \boxed{i = \frac{r}{m} = \frac{\text{nominal annual interest rate}}{\text{number of compounding periods per year}}}$$

For that \$1000 in savings at 10% compounded quarterly, we have $r = 10\%$ and $m = 4$, so $i = 2.5\%$ per quarter.

Geometric Growth

We look for the underlying mathematical pattern of compounding. We continue to use the values from our previous example—namely, an initial balance of \$1000, an annual interest rate $r = 10\%$, quarterly compounding (so $m = 4$), and hence quarterly interest rate $i = 2.5\%$. For quarterly compounding, you have at the end of the first quarter,

$$\text{initial balance} + \text{interest} = \$1000 + \$1000(0.025) = \$1000(1 + 0.025)$$

and at the end of the second quarter,

$$\begin{aligned}\text{initial balance} + \text{interest} &= \$1000(1 + 0.025) + [\$1000(1 + 0.025)](0.025)\\ &= [\$1000(1 + 0.025)] \times (1 + 0.025)\\ &= \$1000(1 + 0.025)^2\end{aligned}$$

The pattern continues in this way, so that you have $\$1000(1 + 0.025)^4$ at the end of the fourth quarter. You use the calculator button marked $\boxed{y^x}$ to evaluate expressions such as $(1.025)^4$; on a spreadsheet, use the caret key $\boxed{\wedge}$ (Shift-6), as in 1.025 $\boxed{\wedge}$ 4.

More generally, with initial principal P and interest rate i (= $100i\%$) per compounding period, you have at the end of the first compounding period,

$$P + Pi = P(1 + i)$$

This amount can be viewed as a new starting balance. Hence, in the next compounding period, the amount $P(1 + i)$ grows to

$$P(1 + i) + P(1 + i)i = P(1 + i)(1 + i) = P(1 + i)^2$$

The pattern continues, and we reach the following **compound interest formula.**

Compound Interest Formula RULE

An initial principal P in an account that pays interest at a periodic interest rate i per compounding period grows after n compounding periods to

$$A = P(1 + i)^n$$

For convenience, we convert the general interest formula into one that is specific for years and annual rates. An annual rate of interest r with m compounding periods per year gives a rate $i = r/m$ per compounding period, and t years contains $n = mt$ compounding periods.

Compound Interest Formula for Several Years RULE

An initial principal P in an account that pays interest at a nominal annual rate r, compounded m times per year, grows after t years to

$$A = P\left(1 + \frac{r}{m}\right)^{mt}$$

Notation for Savings DEFINITION

A	amount accumulated, sometimes denoted FV for "future value"
P	initial principal, sometimes denoted PV for "present value"
r	nominal annual rate of interest
t	number of years
m	number of compounding periods per year
$n = mt$	total number of compounding periods
$i = r/m$	periodic rate, the interest rate per compounding period

The amount added each compounding period is proportional to the amount present at the time of compounding; we are adding Pi to the amount P. This type of growth is called **geometric growth.**

Geometric Growth (Exponential Growth) DEFINITION

Geometric growth (also called **exponential growth**) is growth proportional to the amount present.

EXAMPLE 4 Compound Interest for Several Years

Suppose that you have a principal of $P = \$1000$ invested at 10% nominal interest per year. Using the compound interest formula $A = P(1 + i)^n$, we determine the amount in the account after 10 years, for annual, quarterly, and monthly compounding.

- *Annual compounding.* The annual rate of 10% gives $i = 0.10$, and after 10 years, the account has

$$\$1000(1 + 0.10)^{10} = \$1000(1.10)^{10} \approx \$2593.74$$

- *Quarterly compounding.* Then $i = r/m = 0.10/4 = 0.025$, and after 10 years ($mt = 4 \times 10 = 40$ quarters) the account contains

$$\$1000\left(1 + \frac{0.10}{4}\right)^{4\times10} = \$1000(1.025)^{40} \approx \$2685.06$$

- *Monthly compounding.* Then $i = r/m = 0.10/12 \approx 0.008333$. The amount in the account after 10 years ($mt = 12 \times 10 = 120$ months) is

$$\$1000\left(1 + \frac{0.10}{12}\right)^{12\times10} \approx \$2707.04$$

These entries are found in the last row of Table 21.2 on page 873.

Self Check 4

How much would be in the account after one year (not a leap year) with daily compounding?

Algebra Review Appendix
Natural and Fractional Exponents

In doing the calculations, *be sure to enter the interest rate as a decimal,* and use as many decimal places as your calculator or spreadsheet carries (don't round off until the final result). We show intermediate results with enough decimal places to give the final result to the nearest cent.

Simple Interest Versus Compound Interest

The amounts in accounts paying interest at 10% per year with compound and simple interest are shown in Table 21.3 and in the graph in Figure 21.1. They dramatically illustrate exponential growth at compound interest (the red curve above) compared with linear growth at simple interest (the blue straight line below).

TABLE 21.3 The Growth of \$1000: Compound Interest Versus Simple Interest

Years	Amount in Account with Compounded Interest	Amount with Simple Interest
1	1100.00	1100.00
2	1210.00	1200.00
3	1331.00	1300.00
4	1464.10	1400.00
5	1610.51	1500.00
10	2593.74	2000.00
20	6727.50	3000.00
50	117,390.85	6000.00
100	13,780,612.34	11,000.00

In some situations, the contrast between linear and exponential growth is not so immediately dramatic at first glance. In fact, for low rates of interest, or over a small number of years, the two are hard to distinguish. The much-overused phrase "growing exponentially" is often misused to mean "growing rapidly," but in fact exponential growth need not be rapid.

The concepts of linear growth and exponential growth are realized in "populations" other than the dollars in banking, as we note below.

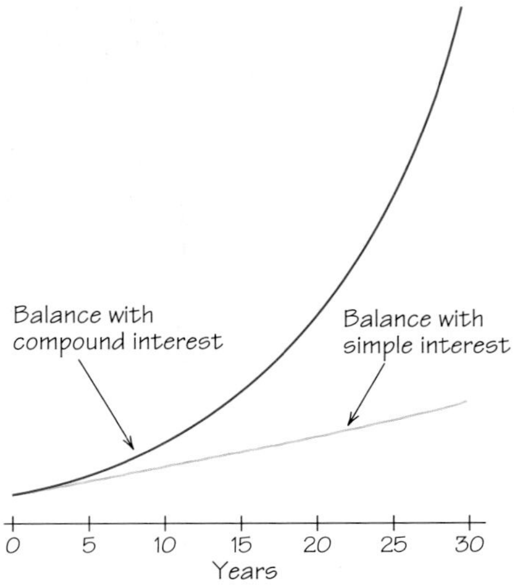

Figure 21.1 The growth of $1000: compound interest and simple interest. The straight line explains why growth at simple interest is also known as *linear growth*.

Algebra Review Appendix
Graphs of Exponential Equations

- **Medical doctors:** We noted earlier (page 872) that the population of U.S. medical doctors grows as if it were at simple interest (arithmetic growth) because the same number of doctors are added each year.
- **Populations and food:** On the other hand, general human populations tend to grow as if at compound interest (geometric growth), because the number of children born—the "interest"—increases as the population—the "balance"—increases. We examine models for population growth and for consumption of resources in Chapter 23. The distinction between arithmetic growth and geometric growth is fundamental to the major theory of demographer and economist Thomas Robert Malthus (1766–1834). He claimed that human populations grow geometrically but food supplies grow arithmetically, so that populations tend to outstrip their ability to feed themselves (see Spotlight 21.1).
- **Global warming:** The amount of carbon dioxide in the atmosphere, which contributes to global warming, has been growing since 1750 as a result of burning fuels. The amount is growing *super*exponentially—that is, faster than exponentially. The growth rate, or "interest rate," itself increases every year. The current growth rate is about 0.57% per year. That seems like a very low rate of interest, but it is "interest" on what is already a large "principal" of carbon dioxide. The international Kyoto Protocol that went into effect in early 2005 (without U.S. participation) aims to lower worldwide emissions. Even if we limited emissions to a constant amount per year, the "principal" of carbon dioxide atoms in the atmosphere would still go up—and global warming would intensify—but just arithmetically, instead of superexponentially. We are in effect "saving" carbon dioxide into the atmosphere, at an unknown future cost.
- **Nuclear waste:** The situation of radioactive waste generated and stored at a nuclear power plant is more complicated. The absolute volume of waste added each year depends on the size and output of the power plant, not on the growing amount of waste in storage. Hence, the volume of waste grows arithmetically. What about the total amount of radioactive material in the storage dump? The radioactive ingredients decay very slowly into nonradioactive ones. While the radioactivity of waste already in storage is decreasing, new amounts of radioactive material are added each year. The situation requires a hybrid model that incorporates positive arithmetic growth (adding to the dump) accompanied by (much smaller) negative geometric growth (radioactive decay). The situation is like turning on the faucet in the bathtub while leaving the drain hole open a little. The faucet determines how fast water runs in, the height of the water determines how fast it runs out, and the two rates combined determine what happens to the volume of water in the tub.

Thomas Robert Malthus

SPOTLIGHT 21.1

Thomas Robert Malthus (1766–1834), a 19th-century English demographer and economist, based a well-known prediction on his perception of the different patterns of growth of the human population and growth of the "population" of food supplies.

Although he believed that agricultural productivity would not be able to keep up with geometric growth in the human population, he concluded that over the long run, population growth could not remain geometric. It would be limited by war, disease, and starvation. This perspective was hardly an optimistic forecast and doubtless was responsible for a dreary image associated with his views.

Some observers suggest that the genocide in Rwanda in 1994 was indirectly a result of overpopulation compared with available food resources.

The Granger Collection, New York

Thomas Robert Malthus

21.3 Effective Rate and APY

Table 21.2 on page 873 shows that at an annual interest rate of 10% (a nominal rate) compounded daily for one year, \$1000 yields \$105.16 in interest, which is 10.516% of the principal. If instead you had \$1000 just at simple interest of 10.516% for one year, you would earn exactly the same amount. Thus, 10% compounded daily is effectively the same as 10.516% simple interest, and we say that 10.516% is the **effective rate.**

Since there is no compounding done inside a compounding period, the effective rate for a compounding period is the nominal rate.

Effective Rate and APY DEFINITION

The **effective rate** is the rate of simple interest that would realize exactly the same amount of interest over the same length of time. For a year, the effective rate is called the **annual percentage yield (APY).**

For an interest rate i per compounding period, a principal of \$1 grows to $\$(1 + i)^n$ in n periods, so the interest earned on that \$1 is the final principal $\$(1 + i)^n$ minus the original principal of \$1, or $\$(1 + i)^n - \$1$. Hence, we have the following formula for the effective rate:

Formula for Effective Rate RULE

$$\text{effective rate} = (1 + i)^n - 1$$

Mostly, we will be interested in the effective rate on an annual basis. For a nominal annual interest rate r compounded m times per year, the interest rate per compounding period is $i = r/m$, and an amount of \$1 grows in one year to

$$\$\left(1 + \frac{r}{m}\right)^m$$

The effective *annual* rate of interest (the APY) is the amount of interest earned

$$\$\left(1 + \frac{r}{m}\right)^m - \$1$$

divided by the original principal. Since that principal is \$1, we have the following.

Formula for APY RULE

$$\text{APY} = \left(1 + \frac{r}{m}\right)^m - 1$$

where

APY = annual percentage yield (effective annual rate)
r = nominal annual interest rate
m = number of compounding periods per year

The only difference between the effective rate and the APY is that the term APY is used only for an annual rate (the "A" in APY). The effective rate could be over any length of time.

EXAMPLE 5 Finding the APY

For a nominal annual rate of 10% compounded monthly, what is the APY?

$$\text{APY} = \left(1 + \frac{0.10}{12}\right)^{12} - 1 \approx 0.10471 \approx 10.47\%$$

Self Check 5

What is the APY for an account that earns 5% per year, compounded quarterly?

In some cases, you know the principal, the current balance, and the interval of time, and you want to learn the interest rate. For example, money market funds typically report earnings to investors each month, based on interest rates that vary from day to day, but often do not report the average interest rate. We can find the equivalent average effective *daily* rate, from which we can calculate the APY.

The compound interest formula gives the end-of-month balance as $A = P(1 + i)^n$, where P is the balance at the beginning of the month, i is the average daily interest rate, and n is the number of days that the statement covers. So we have

$$(1 + i)^n = \frac{A}{P}$$

Taking the nth root (the $\frac{1}{n}$th power) gives

$$1 + i = \left(\frac{A}{P}\right)^{1/n} \qquad i = \left(\frac{A}{P}\right)^{1/n} - 1$$

EXAMPLE 6 Daily Interest Rate on a Money Market Account

Suppose that the monthly statement from the fund reports a beginning balance (P) of \$7373.93 and a closing balance (A) of \$7382.59 for 28 days (n). What is the effective daily rate?

We have

$$i = \left(\frac{7382.59}{7373.93}\right)^{1/28} - 1 \approx 0.0000419194$$

Thus, the average effective daily rate is 0.00419194%. Compounding daily for a (non-leap) year, we would have $(1 + 0.0000419194)^{365} = 1.01542$, for an APY of 1.54%.

Self Check 6

For an account that is compounded daily and earns 10% APY, what is the effective daily rate? *Hint:* For any initial principal P in such an account, at the end of 365 days we have $A/P = 1.10$.

21.4 A Limit to Compounding

The rows in Table 21.2 (page 873) show a trend: More frequent compounding yields more interest. But as the frequency of compounding increases, the interest tends to a limiting amount, shown in the far-right column.

Why is this so? Basically, because the extra interest added to the principal from more frequent compounding is not on deposit for the entire year, interest is credited to the account (and begins earning interest) at the end of each quarter. For example, in the case of quarterly compounding, the interest earned in the first quarter is then on deposit for only three of the four quarters of the year. In the first row of Table 21.1, in the case of quarterly compounding, \$25 in interest is posted at the end of the first quarter and is part of the principal for only the remaining three quarters of the year, earning

$$\$25\left[\left(1 + \frac{0.10}{4}\right) - 1\right] \approx \$25(1.07689 - 1) \approx \$1.92$$

As compounding is done more and more often, smaller and smaller amounts of interest on interest are added.

Let's see what happens with the crazy interest rate of 100% per year compounded m times per year. For an initial balance of \$1, the amount at the end of one year—from the compound interest formula, with $P = \$1$ and $i = 100\%$—is

$$A = \$1 \times \left(1 + \frac{100\%}{m}\right)^m = \$\left(1 + \frac{1}{m}\right)^m$$

As m increases, this amount gets closer and closer to a special number called $e = 2.71828\ldots$ (see Spotlight 21.2). This is illustrated in Table 21.4, where the dots (ellipses) indicate that more decimal places follow. Try those values of m on your calculator.

TABLE 21.4 Yield of \$1 at 100% Interest, Compounded m Times Per Year

m	$\left(1 + \frac{1}{m}\right)^m$
1	2.0000000 . . .
10	2.5937424 . . .
100	2.7048138 . . .
1,000	2.7169239 . . .
1,000,000	2.7182804 . . .

The Number e

SPOTLIGHT 21.2

The number **e** is similar to the familiar number π in several respects. Both arise naturally, π in finding the area and circumference of circles, and e in compounding interest continuously (e is also the base for the system of "natural" logarithms). In addition, neither number is rational (expressible as the ratio of two integers, such as 7/2) or even algebraic (the solution of a polynomial equation with integer coefficients, such as $x^2 = 2$); we say that they are *transcendental* numbers. Finally, no pattern has ever been found in the digits of the decimal expansion of either number.

In addition to its fundamental importance in banking and growth of populations, the number e occurs naturally in several other common contexts as well.

A custom in some families is for each member to buy a holiday gift for just one other member (colloquially, "guy"). This practice allows everyone to take part in giving and receiving without having to buy a gift for every family member. In advance of the holiday, all the members' names are put into a hat and each member draws out a name at random. If anyone draws his or her own name, the drawing is annulled and is redone. What is the probability that the first drawing is successful? (This problem is often called the *hat-check problem*, after a whimsical imaginary 19th-century situation in which men who check their hats at a theater checkroom get hats back at random.)

The answer tends toward $1/e \approx 0.37$ as the family size increases. In other words, we can expect such a drawing to be successful only about 37% of the time. For families of sizes 2, 3, and 4, the chances are 50%, 33%, and 38%. So $1/e$ is a good approximation even for small families.

The situation is usually complicated further by the additional restriction that the drawing is also annulled if any husband or wife draws the other's name (you're supposed to give a gift to your spouse, regardless!). In this case, for a large family in which all members are paired off, the probability of a successful drawing turns out to be approximately $1/e^2 \approx 14\%$, with the chances for 2, 3, and 4 couples being 17%, 11%, and 14%.

The same results hold if, instead of the husband–wife restriction, no one can have the same "guy" as the previous year. But if that rule is imposed in addition to the husband–wife rule, then the chance of a successful drawing goes down to $1/e^3 \approx 5\%$—too small for the drawing ceremony to be fun anymore!

For a general interest rate r, as m becomes larger and larger, the limiting amount is e^r, and the interest method is called **continuous compounding.** The APY is $(e^r - 1)$. (You can calculate powers of e using the [e^x] or [exp] button on your calculator. On some calculators, this button is the [2nd] function of the button marked [LN] or [ln x]. For example, to calculate $e^{0.10}$, press [2nd], press [ln x], and enter 0.10. You get 1.105170918.)

Continuous Compounding DEFINITION

Continuous compounding is the method of calculating interest that yields what compound interest tends toward with more and more frequent compounding per period.

EXAMPLE 7 Continuous Compounding

For \$1000 at an annual rate of 10%, compounded m times in the course of a single year, what is the balance at the end of the year?

This quantity gets closer and closer to $\$1000e^{0.10} = \$1105.17\ldots$ as the number m of compoundings increases. No matter how frequently interest is compounded—daily, hourly, every second, infinitely often ("continuously")—the original \$1000 at the end of one year cannot grow beyond \$1105.17. The values after 5 and 10 years are shown in the lower rows of Table 21.2 (page 873).

Self Check 7

Suppose that \$1,000,000 earns 2% annual interest compounded one million times per year. How much interest is earned at the end of the year?

Continuous Interest Formula RULE

For a principal P deposited in an account at a nominal annual rate r, compounded continuously, the balance after t years is

$$A = Pe^{rt}$$

We illustrate with \$1000 at 10%. For one year, we have $t = 1$ and

$$A = \$1000e^{0.10} \approx \$1105.17$$

To find the amount in the account after 5 years, we have $t = 5$:

$$A = \$1000e^{(0.10)(5)} = \$1000e^{0.5} \approx \$1648.72$$

exactly as shown in the rightmost column of Table 21.2 (page 873).

It makes virtually no difference whether compounding is done daily or continuously over the course of a year. Most banks apply a daily periodic rate (based on compounding continuously) to the balance in the account each day and post interest daily (rounded to the nearest cent). The daily nominal rate (for a non-leap year) is $r/365$, so each day the balance of the account is multiplied by $e^{r/365}$, the daily effective rate. Except for the rounding in posting interest, the effect is the same as continuous compounding throughout the year, because the compound interest formula gives $A = P\,(e^{r/365})^{365}$, which simplifies to the formula $A = Pe^{r}$ from the continuous interest formula.

For example, for a principal of \$1000 and an interest rate of 5%, interest compounded daily over a year yields an amount

$$\$1000\left(1 + \frac{0.05}{365}\right)^{365} \approx \$1051.2675$$

while continuous compounding yields $\$1000e^{0.05} \approx \$1051.2711$. Both round to the same \$1051.27.

PhotoAlto/Eric Audras/Getty Images

21.5 A Model for Saving

The compound interest formula tells the fate over time of a single deposited amount, but another common question that arises in finance is: What size deposit do you need to make regularly in an account with a fixed rate of interest, to have a specified amount at a particular time in the future?

This question is important in planning for a major purchase in the future or accumulating a retirement nest egg. In Chapter 22, we apply the same concepts and formula to paying off a mortgage and making installment payments on a car.

EXAMPLE 8 A Savings Plan

A graduate at her first job saves \$100 per month, deposited directly into her credit union account on payday, the last day of the month. The account earns 1.8% per year, compounded monthly. How much will she have at the end of five years, assuming that the credit union continues to pay the same interest rate?

She makes the first deposit at the end of the first month and the last deposit at the end of the 60th month. The monthly interest rate is $i = r/12 = 0.018/12 = 0.0015$.

It's easier to look at the deposits in reverse time order. The last deposit is made on the last day of the five years, so it earns no interest and contributes just \$100 to the total.

The second-last deposit earns interest for one month, contributing $\$100(1 + i)$. Similarly, the third-last contribution is on deposit for two months, contributing $\$100(1 + i)^2$.

Continuing in the same way, we find that the first deposit earns interest for 59 months and contributes $\$100(1 + i)^{59}$. The total of all of the contributions is

$$\$100 + \$100(1 + i)^1 + \$100(1 + i)^2 + \cdots + \$100(1 + i)^{59}$$
$$= \$100[1 + (1 + i)^1 + (1 + i)^2 + \cdots + (1 + i)^{59}]$$

We will return to this example after developing a formula needed for the solution.

The expression in Example 8 for the total of all contributions is known as a **geometric series** because the successive terms have geometric growth: Each succeeding term is a constant—in this case, $(1 + i)$—times the preceding term. For the sum of such a series with general ratio x, we have the following rule.

Sum of a Geometric Series RULE

The sum S of the geometric series with first term a and common ratio x

$$a + a_0x + ax^2 + ax^3 + \cdots ax^{n-2} + ax^{n-1}$$

is

$$S = a\left[\frac{x^n - 1}{x - 1}\right]$$

provided $x \neq 1$.

The formula is easy to derive. Multiply the sum S by x, getting

$$xS = ax + ax^2 + ax^3 + \cdots + ax^{n-1} + ax^n$$

and then subtract that quantity from S itself:

$$S - xS = a + ax + ax^2 + ax^3 + \cdots + ax^{n-2} + ax^{n-1}$$
$$- ax - ax^2 - ax^3 - \cdots - ax^{n-1} - ax^n$$

With most terms canceling, we are left with

$$S - xS = a - ax^n,\ S(1 - x) = a\,(1 - x^n),\ S = a\left[\frac{x^n - 1}{x - 1}\right] = a\left[\frac{1 - x^n}{1 - x}\right]$$

with the last step of dividing by $(1 - x)$ allowed only if $x \neq 1$. (If $x = 1$, what is each of the terms of the series?)

That this formula works can be confirmed by multiplying both sides by $(x - 1)$ and watching terms on the left cancel. (You should do this confirmation for $n = 4$.) Why doesn't this approach confirm that the formula should work for $x = 1$?

In the case of periodic deposits at periodic interest rate i, we have $x = 1 + i$, and the formula becomes

$$1 + (1 + i)^1 + (1 + i)^2 + \cdots + (1 + i)^{n-1} = \frac{(1 + i)^n - 1}{i}$$

We return to the graduate at her first job. Saving \$100 per month to an account earns 1.8% per year, compounded monthly, for five years. We have $a = \$100$, $n = 60$ months, and monthly interest rate $i = \frac{0.018}{12} = 0.0015$, so $x = 1 + i = \left(1 + \frac{0.018}{12}\right) = 1.0015$. The total accumulation after five years is

$$A = \$100\left[\frac{(1 + 0.0015)^{60} - 1}{1.0015 - 1}\right] \approx \$6273.37$$

Self Check 8

Suppose that the annual interest rate is 1.2% per year, compounded monthly, and our graduate saves for 4 years. How much will she have? ■

We can generalize the result of the example to a periodic deposit of d per compounding period (deposited at the end of the period) and an interest rate i per period. The amount A accumulated after n compounding periods is given by the following **savings formula.**

Savings Formula RULE

$$A = d\left[\frac{(1 + i)^n - 1}{i}\right] = d\left[\frac{(1 + \frac{r}{m})^{mt} - 1}{\frac{r}{m}}\right]$$

where

$A =$	amount accumulated
$d =$	regular deposit of payment at the end of each period
$n = mt$	number of periods
$r =$	nominal annual interest rate
$m =$	number of compounding periods per year
$t =$	number of years
$i = r/m$	periodic rate, the interest rate per compounding period

Algebra Review Appendix
Arithmetic and Geometric Sequences and Series

The expression on the right gives the amount accumulated in terms of the nominal annual interest rate r, the number m of compounding periods per year, and the number t of years, using the relations $i = r/m$ and $n = mt$.

The savings formula involves four quantities: A, d, i, and n. If any three are known, the fourth can be found. A common situation is for A, i, and n to be known, with d (the regular payment) to be found.

Since we often want to find d, we solve the savings formula algebraically once and for all for d to get the following **payment formula.**

Payment Formula RULE

$$d = A\left[\frac{i}{(1 + i)^n - 1}\right] = A\left[\frac{\frac{r}{m}}{(1 + \frac{r}{m})^{mt} - 1}\right]$$

Sometimes the purpose of saving is to accumulate a fixed sum by a particular date. Such a savings plan is called a **sinking fund** because you sink money into it. The opposite of a sinking fund is an **annuity,** where an amount is paid out regularly.

Sinking Fund DEFINITION

A **sinking fund** is a savings plan to accumulate a fixed sum by a particular date, usually through equal periodic deposits.

Annuity DEFINITION

An **annuity** is a sequence of periodic payments.

EXAMPLE 9 Sinking Fund

Suppose that your parents had started saving for your college education when you were born. How much would they have had to save each month to accumulate \$15,000 (to pay for tuition, room, and board for just your first year of college!) over 18 years, with an account earning a steady 5% per year, compounded monthly?

Applying the payment formula with $A = \$15{,}000$, monthly rate $i = r/m = 0.05/12$, and $n = mt = 12 \times 18 = 216$, we get

$$d = \$15{,}000\left[\frac{\frac{0.05}{12}}{\left(1 + \frac{0.05}{12}\right)^{12 \times 18} - 1}\right] \approx \$42.96$$

Self Check 9

Suppose that you start saving for your child's college education when your child is born, but you can earn only a steady 3% interest per year (about what 30-year U.S. Treasury bonds are currently earning). How much would you have to save each month to accumulate \$15,000 when the child turns 18?

Under \$50 sounds like a manageable amount to contribute, but it doesn't take into account inflation, costs beyond the first year, the higher cost of a private college, or putting a younger brother or sister through college, too. In the next section, we investigate how to take inflation into account.

Saving for Retirement (Why It's Never Too Early to Start)

Financial advisers stress the importance of beginning early to save for retirement. Many firms offer a 401(k) plan (named after a section of law regulating pensions), which allows an employee to make monthly contributions to a retirement account. Often, the employer also contributes. The plan has the advantage that income tax on the contributions is deferred until the employee withdraws the money during retirement. Meanwhile, the account accumulates earnings on what would have been taxed.

That means, for example, that an employee making a \$100 monthly contribution may see a reduction in take-home pay of only \$75 or less, since taxes are not withheld on the contribution.

Sometimes a company's pension plan consists of just contributing company stock to the employee's individual 401(k) account. In 2002, the bankruptcy of Enron Corporation resulted in thousands of its employees losing almost their entire retirement savings. Those savings consisted largely of Enron stock contributed by Enron, which fell from \$90 per share to under \$1 per share in just a couple of months. The Enron bankruptcy illustrated how unwise it is for most of an employee's retirement fund to consist of stock in just one company, particularly if—as was the case for Enron—the employee is not free to sell the stock. Even more people lost retirement savings and jobs when the stock of WorldCom declined more than 99% in 2002, after news of financial fraud by its management.

EXAMPLE 10 **Retirement Fund Annuity Savings**

Suppose that you start a 401(k) plan when you turn 23 and contribute \$50 at the end of each month until you turn 65 and retire. Suppose that you put your contributions into a very safe long-term investment that returns a steady 5% annual interest compounded monthly. How much will be in your fund at retirement?

Apply the savings formula with $d = \$50$, $i = 0.05/12$, $n = mt = 12 \times (65 - 23) = 504$. We get

$$d = \$50\left[\frac{\left(1 + \frac{0.05}{12}\right)^{504} - 1}{\frac{0.05}{12}}\right] \approx \$85{,}567.43$$

At first glance, that may seem like a lot of money, but it is not so much if that's all you have to live off for the rest of your life. (Of course, there may also be Social Security payments.) In the exercises later in this chapter, we explore the effects of saving more each month, getting a higher interest rate, saving on taxes, and—especially—having inflation erode the value of your savings.

Self Check 10

Suppose that you wait until you are 45 to start saving. How much would you have?

Annuities are a common way for retirees to receive funds saved up for retirement. We examine an example in Section 22.4 (page 929), where we turn the savings formula around to get a formula (the amortization formula) to relate the amount of savings to a regular payout. In that section, you can find out how much monthly income for a fixed period—or for life—\$86,000 could buy.

The winner of a grand prize in a lottery often has a choice between an immediate cash payment and an annuity of a fixed number of annual payments. Lottery winners almost elect the immediate cash payment, but the annual payments are an attractive alternative for a few.

For most major U.S. lotteries today, the annuity's annual payments increase by a certain percentage each year. The lottery uses the immediate cash amount to purchase a contract with an insurance company to make the payments. The starting payment amount is determined so that the *present value* of the entire stream of payments is the same as the immediate cash value. So we next look into the meaning of present value.

Present Value

Suppose that you want to make a one-time deposit of amount P that will grow to a specific amount A in n compounding periods from now by earning interest at a rate i per period. The quantities A, P, i, and n are related through the compound interest formula, $A = P(1 + i)^n$. The quantity P is called the **present value** of the amount A that is to be paid n compounding periods in the future.

Present Value DEFINITION

The **present value** of an amount to be paid at a specific time in the future is what that future payment would be worth today, as determined from a given interest rate and compounding period.

Present Value RULE

The present value of an amount A to be paid t years in the future, earning in the mean time a nominal rate of interest r compounded m times per year—that is, after $n = mt$ compounding periods at a rate $i = r/m$ per compounding period—is

$$P = \frac{A}{(1+i)^n} = A\left(1 + \tfrac{r}{m}\right)^{-mt}$$

If the interest is instead compounded continuously, then the present value is

$$P = \frac{A}{e^{rt}} = Ae^{-rt}$$

The present value takes into account that the amount P would grow under compound interest, using the compound interest formula (page 875), to the amount A at the end of the t years.

EXAMPLE 11 Certificates of Deposit

A certificate of deposit (CD) pays a fixed rate of interest for a term specified in advance, which may range from 1 month to 10 years. The best local rate in June 2015 was for a 60-month CD at 0.747%, compounded daily. How much would need to have been set aside in such a CD to have \$12,000 in 60 months?

We find the present value of receiving \$12,000 60 months from now, with $r = 0.00747$, $m = 365$, and $t = 60/12 = 5$, so that $mt = 365 \times 5 = 1825$. (Fine point: We use $mt = 365 \times 5$ despite the fact that one or two leap days will occur during the 5 years. For a leap year, the bank may apply a daily rate of either 1/366 or 1/365 of the interest rate for the year. We neglect those complications, which would make little difference.) The present value formula gives

$$P = \frac{A}{\left(1+\frac{r}{m}\right)^{mt}} = \frac{\$12{,}000}{\left(1+\frac{0.00747}{365}\right)^{365\times 5}} = \frac{\$12{,}000}{(1.000020465753425)^{1825}} \approx \$11{,}560.07$$

Self Check 11

An alternative in June 2015 was a 24-month CD at 0.300%, also compounded daily. How much would need to have been set aside in such a CD to have \$12,000 at the end of the 24-month term? (Neglect leap days and use a 365-day year.)

Spotlight 21.3 shows how to use a spreadsheet to do various kinds of financial calculations.

Using a Spreadsheet for Financial Calculations

SPOTLIGHT 21.3

Both commercial (e.g., Microsoft Excel) and open-source (e.g., OpenOffice) spreadsheets have commands that use the formulas developed in this chapter in situations of compound interest:

PMT (*rate, nper, pv, fv, type*)
PV (*rate, nper, pmt, fv, type*)
FV (*rate, nper, pmt, pv, type*)
RATE (*nper, pmt, pv, fv, type, guess*)

where

rate = interest rate per payment (compounding) period
nper = total number of payments
pv = principal, or present value
pmt = payment made each (compounding) period
fv = future value, or cash balance after last payment
type = 0 if payment made at end of period, 1 if at beginning
guess = guess at interest rate per payment period

Both *pv* and *pmt* will be negative if they correspond to payment by you. The guess can be omitted. Also, for simplicity, all problems in this chapter have payments at the end of the period, so *type* = 0; since that is the default for the spreadsheet, it too can be omitted. (*Note:* OpenOffice uses semicolons for separators instead of commas.)

We show how some of our previous examples can be solved by using a spreadsheet:

Example 3 (compound interest, page 873), but with monthly compounding: We want the amount accumulated in an account of \$1000 after 10 years at 10% interest compounded monthly. We have *rate* = 10%/12, *nper* = 10 × 12 = 120, *pmt* = 0 (since we make no payments after depositing the principal), and *pv* = −1000. Put into a cell in the spreadsheet = FV(10%/12, 10*12, 0, −1000) and see \$2707.04 emerge, as in Table 21.2, page 873.

Example 8 (savings plan, page 882): We want to determine how much a regular \$100 per month deposit will amount to after 5 years at 1.8% per year, compounded monthly. We have *rate* = 1.8%/12, *nper* = 5 × 10 = 60, *pmt* = −100 (remember, payments are outlays), and *pv* = 0. Put into a cell = FV(1.8%/12, 5*12, −100, 0) and see \$6273.37 emerge.

Example 9 (sinking fund, page 885): Your parents want to save \$15,000 over 18 years at 5% compounded monthly. Put into a cell = PMT(5%/12, 18*12, 0, 15000) and see (\$42.96)—it's in red because it is \$−42.96, meaning that this amount must be deposited monthly.

Example 11 (certificate of deposit, page 887): We want to deposit an amount that will accumulate \$12,000 in 60 months = 1825 days at 0.747% interest compounded daily. Put in = PV(0.747%/365, 1825, 0, 12000) and see (\$11,560.07), which is in red because it is a deposit.

21.6 Inflation

In times of economic **inflation,** prices increase. When the rate of inflation is constant, the compound interest formula (page 875) can be used to project prices.

Inflation — DEFINITION

Inflation is a rise in prices from a set base year.

Annual Rate of Inflation — DEFINITION

The **annual rate of inflation** a (= $100a$%) is the additional proportionate cost of goods one year later. Goods that cost \$1 in the base year will cost \$$(1 + a)$ one year later.

For a simple model of inflation, we can assume that inflation is uniform throughout the year and constant over a period of years. Later we will look at varying inflation rates.

EXAMPLE 12 Inflation

Suppose that there is constant 2% annual inflation from mid-2015 through mid-2020. What will be the projected price in mid-2020 of an item that costs \$100 in mid-2015?

The compound interest formula applies with $P = \$100$, $a = r = 2\%$, $m = 1$, and $t = 5$. The projected price is $A = P(1 + r)^t = \$100(1 + 0.02)^5 \approx \$110.41$.

Self Check 12

Suppose instead that inflation is 3% from mid-2015 to mid-2020. What will be the projected price in mid-2020 of an item that costs \$100 in mid-2017?

Present Value

During constant-rate inflation, prices grow geometrically (exponentially) and the value of the dollar "decays" geometrically (also exponentially).

Exponential Decay DEFINITION

Exponential decay is geometric growth with a negative rate of growth.

Let a (for "additional") represent the annual rate of inflation; what costs \$1 now will cost $\$(1 + a)$ this time next year. For example, if the annual inflation rate is $a = 25\%$ throughout the coming year, then what costs \$1 now would cost \$1.25 this time next year. A dollar next year would buy only $1/1.25 = 0.8$ times as much as a dollar buys today. In other words, a dollar next year would be worth only \$0.80 in today's dollars; by next year, a dollar would have lost 20% of its **purchasing power.** Notice that although the inflation rate is 25%, the loss in purchasing power is 20%. This may seem peculiar; why aren't they the same percentage?

The situation may be clearer if you consider inflation at a rate of 100%. Then a dollar next year is worth only 50% of a dollar today; what costs \$1 today will cost \$2 next year. The reason for the difference in the percentages is that the percentage of inflation (100%) uses today's price level (\$1) as a base (prices rise from \$1 to \$2, hence they rise 100%), while the percentage loss in purchasing power uses as a base the larger price level next year (\$2—so loss in purchasing power is 50% of \$2).

Purchasing Power RULE

The purchasing power of a dollar a year from now, relative to today, with annual inflation rate a is

$$\frac{\$1}{1 + a} = \$1 - \frac{\$a}{1 + a}$$

For an annual inflation rate a, the loss in purchasing power is the fraction $a/(1 + a)$. (You should calculate what this expression is for $a = 25\%$.) The quantity $i = -a/(1 + a)$ behaves like a negative interest rate. We can use the compound interest formula to find the relative purchasing power of P dollars t years from now as

$$A = P(1 + i)^t = P\left(1 + \frac{-a}{1 + a}\right)^t$$

The actual posted price of an item, at any point in time, is said to be in **current dollars.** That price can be compared with prices at other times by adjusting for inflation, which means converting all prices to **constant dollars,** dollars of a particular year.

EXAMPLE 13 Deflated Dollars

Suppose that there is 25% annual inflation from mid-2015 through mid-2019. What would be the value of a dollar in mid-2019 in constant mid-2015 dollars? The inflation figure is unrealistic (we hope!), but it makes the calculations easy so that you can focus on the ideas.

We have $a = 0.25$, so $i = -a/(1 + a) = -0.25/1.25 = -0.20$. This—not 25%—is the negative interest rate, the rate at which the dollar is losing purchasing power. We have $t = 4$ years, so the purchasing power of \$1 four years from mid-2015, in terms of 2015 dollars, would be

$$\$1(1 + i)^4 = \$(1 - 0.20)^4 = \$(0.80)^4 \approx \$0.41$$

For a more realistic rate of 3% annual inflation, we would have $a = 0.03$ and negative interest rate $i = -as/(1 + a) = -0.03/1.03 = -0.0291262$. In 2019, the purchasing power of \$1 would be

$$\$1(1 + i)^4 = \$(1 - 0.0291262)^4 = \$(0.970874)^4 \approx \$0.89$$

in 2015 dollars. Notice that "losing" 3% to inflation each year for 4 years amounts to losing almost—but not quite as much as—a total of $4 \times 3\% = 12\%$. (This is just as with the previous inflation rate, where losing 25% per year for 4 years doesn't completely reduce the value to 0.)

Self Check 13

Suppose that there is 2% inflation from mid-2015 through mid-2020. What would be the value of a dollar in mid-2020 in constant mid-2015 dollars?

In Example 13, we may think of the value of the dollar as "depreciating" each year. Depreciation of the value of equipment or a building is similar.

EXAMPLE 14 Depreciation of a Car

If you bought a car at the beginning of 2015 for \$12,000 and its value in current dollars depreciates at a rate of 15% per year, what will be its value at the beginning of 2018 in current dollars?

We have $P = \$12{,}000$, $i = -0.15$, and $n = 3$. The compound interest formula gives

$$A = P(1 + i)^n = \$12{,}000(1 - 0.15)^3 \approx \$7369.50$$

Self Check 14

Suppose that the car cost $21,000 in mid-2015 and its value in current dollars depreciates 17% per year. What will be its value in current dollars in mid-2019, when you want to trade it in? ■

The Consumer Price Index

In our preceding model, we supposed that inflation stayed constant over a period of time. That is not generally the case. However, based on measures of inflation, we can determine the equivalent today of a price in an earlier year or how much a dollar in that year would be worth today in purchasing power.

The official measure of inflation is the Consumer Price Index (CPI), determined by the Bureau of Labor Statistics (BLS). Here, we describe and use the CPI-U, the index for all urban consumers, which covers about 80% of the U.S. population and is the index of inflation that is usually referred to in newspaper and magazine articles. Each month, the BLS determines the average cost of a "market basket" of goods, including food, housing, transportation, clothing, and other items. It compares this cost with the cost of the same (or comparable) goods in a base period. The base period used to construct the CPI-U is 1982–1984. The index for 1982–1984 is set to 100, and the CPI-U for other years is calculated by using the proportion

$$\frac{\text{CPI for other year}}{100} = \frac{\text{cost of market basket in other year}}{\text{cost of market basket in base period}}$$

For example, the cost of the market basket in 2013 (in 2013 dollars) was 2.330 times the cost in 1982–1984 (in 1982–1984 dollars), so the CPI for 2013 was $\$100 \times 2.330$, or 233.0.

Table 21.5 shows the average CPI for each year from 1913 through 2013, with estimates for 2014 and 2015. This table can be used to convert the cost of an item in dollars for one year to what it would cost in dollars in a different year, using the proportion cost in CPI for year A:

$$\frac{\text{cost in year A}}{\text{cost in year B}} = \frac{\text{CPI for year A}}{\text{CPI for year B}}$$

EXAMPLE 15 ➔ The Price of Our House and the Value of a Dollar

Where my family and I live, housing is relatively inexpensive. We bought our house in mid-1992 for $133,000 (close to the median price of U.S. housing then). What would be the equivalent cost in mid-2015 dollars?

We see from Table 21.5 that the CPI for 1992 is 140.3 and the CPI for 2015 is estimated to be 242.3. The table gives the average value for each year, which is very close to the value at mid-year. Month-by-month values are available at the Bureau of Labor Statistics Web site (data.bls.gov/pdq/).

Using the proportion, we have

$$\frac{\text{cost in 2015}}{\text{cost in 1992}} = \frac{\text{CPI for 2015}}{\text{CPI for 1992}}$$

or

$$\frac{\text{cost in 2015}}{\$133{,}000} = \frac{241.4}{140.3}$$

so that

$$\text{cost in 2015} = \$133{,}000 \times \frac{241.4}{140.3} \approx \$229{,}000$$

That's what our house would sell for if its price exactly matched inflation. The ratio 241.4/140.3 ≈ 1.721 is the *scaling factor* for converting 1992 dollars to 2015 dollars. What we are observing is a proportion, or *numerical similarity,* between 1992 dollars and 2015 dollars, analogous to the geometric similarity discussed in Chapter 18 (page 738). To convert from 2015 dollars to 1992 dollars, we would multiply by 1/1.721 ≈ 0.581.

TABLE 21.5 U.S. Consumer Price Index (1982–1984 = 100)

—	—	1931	15.2	1951	26.0	1971	40.5	1991	136.2
—	—	1932	13.7	1952	26.6	1972	41.8	1992	140.3
1913	9.9	1933	13.0	1953	26.7	1973	44.4	1993	144.5
1914	10	1934	13.4	1954	26.9	1974	49.3	1994	148.2
1915	10.1	1935	13.7	1955	26.8	1975	53.8	1995	152.4
1916	10.9	1936	13.9	1956	27.2	1976	56.9	1996	156.9
1917	12.8	1937	14.4	1957	28.1	1977	60.6	1997	160.5
1918	15.1	1938	14.1	1958	28.9	1978	65.2	1998	163.0
1919	17.3	1939	13.9	1959	29.1	1979	72.6	1999	166.6
1920	20.0	1940	14	1960	29.6	1980	82.4	2000	172.2
1921	17.9	1941	14.7	1961	29.9	1981	90.9	2001	177.1
1922	16.8	1942	16.3	1962	30.2	1982	96.5	2002	179.9
1923	17.1	1943	17.3	1963	30.6	1983	99.6	2003	184.0
1924	17.1	1944	17.6	1964	31.0	1984	103.9	2004	188.9
1925	17.5	1945	18.0	1965	31.5	1985	107.6	2005	195.3
1926	17.7	1946	19.5	1966	32.4	1986	109.6	2006	201.6
1927	17.4	1947	22.3	1967	33.4	1987	113.6	2007	207.3
1928	17.1	1948	24.1	1968	34.8	1988	118.3	2008	215.3
1929	17.1	1949	23.8	1969	36.7	1989	124.0	2009	214.5
1930	16.7	1950	24.1	1970	38.8	1990	130.7	2010	218.1
								2011	224.9
								2012	229.6
								2013	233.0
								2014	236.7
								2015 (est.)	241.4

Note: This is the CPI-U index, which covers all urban consumers, about 80% of the U.S. population. Each index is an average for all cities for the year. The basis for the index is the period 1982–1984, for which the index was set equal to 100. For each year, the figure is the average during the year, which is usually close to the value at mid-year.

Source: data.bls.gov/cgi-bin/surveymost?cu

Self Check 15

What is the scaling factor for converting 2000 dollars to 2015 dollars?

Patti McConville/Getty Images

Spotlight 19.5 (page 785) describes how the CPI is calculated, using the geometric mean (introduced on page 785). That spotlight also discusses the *chained CPI,* a different method of calculating inflation that produces a lower CPI value. Spotlight 21.4 details the various ways in which people can invest their money.

What Is a Financial Derivative?

SPOTLIGHT 21.4

Ways to save or invest offer different levels of risk:

- Savings accounts, CDs, bonds, and annuities can offer guaranteed rates of interest.
- Money market accounts and inflation-protected savings bonds offer rates that vary with interest rates in the general economy.
- Stocks, mutual funds, and **derivatives** are not savings but investments, on which you could lose money.

Thanks to the Great Depression, when many banks failed and depositors lost everything, accounts in most banks and credit unions are now insured by the federal government for up to $250,000. If the bank or credit union fails, the government takes it over and depositors get their money back.

A bond is a loan at simple interest with repayment at the end of a fixed term. By buying bonds, investors loan to states, municipalities, and corporations for civic construction projects, company expansions, and other purposes. Bonds do not offer any guarantee: The bond issuer (the borrower) could default, meaning that the bond holder (the investor) would no longer receive interest payments and might lose the original investment.

Bonds usually can be bought and sold after their initial purchase. The selling price can vary with sentiment about creditworthiness of the issuer and with changes in interest rates.

- If prevailing interest rates go above the rate that the bond is paying, the bond becomes less valuable; investors can get a better deal elsewhere, so the price of the bond may fall. Consequently, the bond's "yield" (effective interest rate relative to the current price) may rise toward the prevailing interest rate.
- If interest rates instead fall, the bond may become more valuable because it is paying a higher rate than otherwise available; its price may rise but then its yield would decline toward the prevailing interest rate.

Some bonds can be "called," meaning that the issuer pays back the bond holder before the end of the bond term. Bonds tend to be called when interest rates fall: The issuer calls bonds previously issued at a high interest rate and then offers new bonds at a lower rate, thus saving on the cost of borrowing.

Stocks (also called *securities* or *equities*) are another way for a company to raise capital, and an opportunity for investors to share in the financial future of the company by owning shares in it. Usually, a part of the company's profits is distributed each year to shareholders as dividends; if not, either because the company lost money, made little profit, or reinvested all of its profits, then the price of its shares may fall. The price of a company's stock depends on "fundamentals" (measures of the "health" of the company), on the general state of the economy as a whole, and on current events.

Derivative DEFINITION

A **derivative** is a financial instrument whose value "derives" from the value of an underlying asset such as a stock, bond, commodity, mortgage, option, and so on.

What Is a Financial Derivative? *(continued)*

An example of a derivative is shares in a *mutual fund.* A mutual fund holds an array of other companies' stocks and bonds, perhaps concentrating on stocks of a particular kind ("green" companies, foreign companies, companies focused on growth, etc.). Investors in the mutual fund do not individually own any shares of the stocks that it invests in; the value of the mutual fund's own shares derives from the investments that it holds. Hedge funds are basically mutual funds that are subject to less regulation than other mutual funds.

A popular kind of mutual fund is an *index fund.* An index fund selects investments to try to mirror the aggregate ups and downs of a particular stock exchange, as expressed by some "index" (summary) of it, and thus captures proportionate gains (but will also tend to suffer proportionate losses). Well-known indexes include the Dow Jones Industrial Average and the Standard and Poor's (S&P) 500 Index. At another level removed, there are index funds of index funds.

Traditional derivatives include agricultural commodity futures (corn, soybeans, pork bellies), used by farmers and food processors to hedge against the risk of poor crops, as well as currency futures, used by companies to lock in a fixed exchange rate for future purchases of foreign goods and raw materials. There are even weather derivatives, whose price depends on the number of sunny days or amount of rainfall in a particular region! Derivatives arise also in the context of loans. A credit derivative bases its value not on the underlying loan, such as a mortgage, but on the risk that the loan will not be repaid. The buyer of the credit derivative does not own the mortgage loan itself, hence is not entitled to be paid by the homeowner.

A common form of credit derivative is a *credit default swap,* which currently amount to $26 trillion worldwide (almost $4,000 for every person on the globe). One party buys protection from a second against default on a debt owed by a third. The "protection buyer" pays a fee to the "protection seller" in return for the seller making good on the debt if the debtor defaults. The buyer thus "swaps" risk to the seller. But you aren't going to want to believe this: The debtor can owe the debt to someone else entirely unrelated to the protection buyer or seller! In such a case, the protection buyer and seller are gambling together on whether the debtor will default. This has happened in connection with the "sovereign debt" of countries such as Greece and Ireland.

Where do *you* come into all this? A handful of investors made billions in the past few years through credit default swaps on "bundles" of home mortgages. Too many turned out to be *subprime* mortgages, that is, loans to people with poor credit ratings, whose mortgages consequently had higher interest rates. These homeowners defaulted in large numbers on their mortgage loans, in part because of the following:

- The loans had high interest rates, because the people were judged poor credit risks.
- The homes were no longer worth as much as the remaining amount of the loan principal to be paid off (the homes were "under water").
- The changing economy forced the people out of their jobs.

The banks then foreclosed on the defaulted loans and forced the people out of their homes. (Spotlight 22.3, on page 922, looks in more detail at the mortgage crisis.)

The credit default swaps mentioned involved not the original mortgages, nor even the bundles of them—which themselves were derivatives of the underlying assets of the homes involved. The owners of the bundles had issued securities backed by the bundles—not by the homes—and the credit default swaps were on those securities. But the decline in housing prices reduced the value of the underlying asset of homes, hence of the bundles, hence of the derivatives based on the bundles. We are talking about derivatives of derivatives! The securities became "toxic assets": No one was willing to buy them at anywhere near the original price, and banks have continued to be unwilling to sell them at the big losses that would be involved.

Attempts by holders of the derivatives to foreclose on the homes have been to some degree thwarted by the fact that the derivative owners are not the direct owners of the mortgage loans. In some cases, there has been so much trading of loans and of derivatives on them, without sufficient documentation, that it is impossible to establish to whom a loan is owed.

Warren Buffett, one of the world's richest men, was prescient in 2002 when he called financial derivatives "time bombs" and "financial weapons of mass destruction."

Financial derivatives, combined with greed and exploitation, indeed destroyed the prosperity and optimism that the world enjoyed in the early years of the 21st century.

It's natural to think that if your investment is growing at 6% per year and inflation is at 3% per year, then the real growth in the value (purchasing power) of your investment is 6%−3% = 3%. That is a handy approximation, especially for low interest rates, but it is not exactly right. Let's see why.

Suppose that you invest \$500 for a year at 6% and inflation is 3%. At the beginning of the year, you have \$500, which at \$5 per pound could buy 100 pounds of steak. At the end of the year, you have \$500(1 + 0.06) = \$530. But the price of steak has gone up with inflation, so steak now costs \$5(1 + 0.03) = \$5.15 per pound. How much steak would your \$530 buy? Just \$530/\$5.15/lb = 102.91 lb. In other words, in terms of purchasing power, or real gain, your investment has grown only 2.91%. This is not a great deal different from 3%, but it is different, and the difference is greater for higher rates of interest and inflation.

More generally, consider an investment principal P and a market basket of goods with value m. Let the annual yield (rate of interest) of the investment be r and the rate of inflation be a. We calculate the rate of real growth g of the investment as follows:

At the beginning of the year, the investment would buy quantity

$$q_{\text{old}} = P/m$$

of the market basket. At the end of the year, the investment would buy quantity

$$q_{\text{new}} = \frac{P(1 + r)}{m(1 + a)}$$

of the market basket. For the example above, put in $P = \$100$, $r = 0.06$, $m = \$5$, and $a = 0.03$. Notice that the gain of r in the investment multiplies the principal by $(1 + r)$, while the erosion due to inflation divides the principal by $(1 + a)$. The two influences on the investment have directly opposite effects.

The growth of the investment, relative to how many market baskets it could have bought originally, is

$$g = \frac{q_{\text{new}} - q_{\text{old}}}{q_{\text{old}}} = \frac{\frac{P(1 + r)}{m(1 + a)} - \frac{P}{m}}{\frac{P}{m}} = \frac{1 + r}{1 + a} - 1 = \frac{r - a}{1 + a}$$

In the last expression, the numerator is the difference of the two rates (6% − 3% in our example), which is divided by a quantity greater than 1 if there is inflation. One way to understand why this is the correct formula is to realize that the gain itself is not in original dollars but in deflated dollars.

You should confirm that this formula gives 2.91% for $r = 6\%$ and $a = 3\%$. You can see that if the rate of inflation a is very small—only a small percentage—then $1 + a \approx 1$ and so $g \approx r - a$.

The relationship between interest rate, inflation rate, and rate of real growth is called *Fisher's effect,* after the American economist Irving Fisher (1867–1947).

Real Rate of Growth RULE

The real (effective) annual rate of growth of an investment at annual interest rate r with annual inflation rate a is

$$g = \frac{r - a}{1 + a}$$

Review Vocabulary

Annual percentage yield (APY) The effective interest rate per year. (p. 878)

Annual rate of inflation The additional proportionate cost of goods one year later. (p. 888)

Annuity A sequence of periodic payments. (pp. 884, 885)

Arithmetic growth Growth by a constant amount in each time period. (p. 872)

Bond A loan with repayment at the end of a fixed term and usually simple interest paid annually or semiannually. (p. 871)

Compound interest Interest that is paid on both the original principal and the accumulated interest. (p. 872)

Compound interest formula The formula for the amount in an account that pays compound interest periodically. For an initial principal P and an effective rate i per compounding period, the amount after n compounding periods is $A = P(1 + i)^n$. (p. 875)

Compounding period The fundamental interval on which compounding is based, within which no compounding is done. Also called simply *period.* (p. 872)

Constant dollars Costs are expressed in constant dollars if inflation or deflation has been taken into account by converting the costs to their equivalent in dollars of a particular year. (p. 890)

Continuous compounding Payment of interest in an amount toward which compound interest tends with more and more frequent compounding. (p. 881)

Current dollars The actual cost of an item at a point in time; inflation or deflation before or since then has not been taken into account. (p. 890)

Derivative A financial instrument whose value "derives" from the value of an underlying asset. (p. 893)

e The base for continuous compounding, geometric (exponential) growth, and natural logarithms; $e = 2.71828\ldots$ (p. 880)

Effective rate The rate of simple interest that would realize exactly as much interest over the same period of time. (p. 878)

Exponential decay Geometric "growth" at a negative rate. (p. 889)

Exponential growth Geometric growth. (p. 875)

Geometric growth Growth proportional to the amount present. (p. 875)

Geometric series A sum of terms, each of which is the same constant times the previous term; that is, the terms undergo geometric growth. (p. 883)

Inflation A rise in prices from a set base year. (p. 888)

Interest Money earned on a savings account or other fund. (p. 870)

Linear growth Arithmetic growth. (p. 872)

Nominal rate A stated rate of interest for a specified length of time; a nominal rate does not take into account any compounding. (p. 874)

Payment formula The formula for the amount of a regular payment d to pay off an amount A in n compounding periods, with interest rate i per compounding period and m compounding periods per year.

$$d = A\left[\frac{i}{(1 + i)^n - 1}\right] = A\left[\frac{\frac{r}{m}}{(1 + \frac{r}{m})^{mt} - 1}\right] \qquad \text{(p. 884)}$$

Present value The value today of an amount to be paid or received at a specific time in the future, as determined from a given interest rate and compounding period. (p. 887)

Principal Initial balance. (p. 870)

Purchasing power The purchasing power of a dollar a year from now, relative to today, with annual inflation rate a is

$$\frac{\$1}{1 + a} = \$1 - \frac{\$a}{1 + a} \qquad \text{(p. 889)}$$

Savings formula The formula for the amount A accumulated after $n = mt$ periods, with a uniform deposit of d at the end of each compounding period and interest rate $i = r/m$ per period:

$$A = d\left[\frac{(1 + i)^n - 1}{i}\right] = d\left[\frac{(1 + \frac{r}{m})^{mt} - 1}{\frac{r}{m}}\right] \qquad \text{(p. 884)}$$

Simple interest Interest that is paid on the original principal only, not on any accumulated interest. (p. 871)

Sinking fund A savings plan to accumulate a fixed sum by a particular date, usually through equal periodic deposits. (pp. 884, 885)

Self Check Answers

1. $2000
2. $\$10{,}000 + 51 \times \dfrac{\$10{,}000 \times 0.0649}{12} \approx \$12{,}758.25$
3. $1050.62
4. $1105.16
5. 5.09%
6. $(1.10)^{1/365} - 1 \approx 0.000261158 = 0.0261158\%$
7. $20,201.34
8. $4914.55
9. $52.46
10. $20,551.68
11. $11,928.22
12. $115.93
13. $0.91
14. $9966.25
15. 1.407

Skills Check

1. Simple interest is an example of
(a) linear growth.
(b) arithmetic growth.
(c) constant growth.

2. If a savings account pays 3% simple annual interest, a deposit of $250 will earn ____________ in 2 years.

3. An 18% annual rate on a credit-card balance is an example of
(a) an effective rate.
(b) a nominal rate.
(c) an adjusted rate.

4. If you deposit $1000 at 6.2% simple interest, the balance after three years is ____________.

5. Suppose that you deposit $180 at the beginning of each year into a savings account that pays 2% simple interest per year at the end of the year. After two years, the amount in the account is
(a) $360.00.
(b) $370.80.
(c) $374.40.

6. If $800 is invested for one year at 6% compounded quarterly, the amount of interest earned is ____________.

7. If a single deposit is made into a compound interest CD, the account earns
(a) interest only for the first period.
(b) the same amount of interest each period.
(c) more interest in each subsequent period.

8. If you deposit $1000 at 6.2% interest compounded quarterly, the balance after three years is ____________.

9. Compound interest is an example of
(a) geometric growth.
(b) exponential growth.
(c) humongous growth.

10. Suppose that you deposit $15 at the end of each month into a savings account that pays 2% interest compounded monthly. After a year, ____________ is in the account.

11. Compound interest is paid on
(a) just the initial principal.
(b) the current balance.
(c) just the accumulated interest.

12. Suppose that you deposit $10 at the end of each day into a savings account that pays 2% interest compounded daily. After a year, ____________ is in the account.

13. Suppose that you open an account that pays 1% interest compounded monthly with a deposit of $1000. At the end of one year, you will have approximately
(a) $1010.
(b) $1120.
(c) $1200.

14. The APY for 3% compounded monthly is ____________.

15. Suppose that you invest $250 in an account that pays 4.5% interest compounded quarterly. After 30 months, how much is in your account?

(a) $279.08

(b) $279.59

(c) $279.71

16. The APY for 3% compounded daily is ___________.

17. Which of the following pays more interest?

(a) 6% compounded annually

(b) 6% compounded monthly

(c) 6% compounded continuously

18. If you deposit $1000 at 6.2% interest compounded continuously, the balance after three years is ___________.

19. If an account with $100 pays 6% interest compounded continuously, at the end of one year, you will

(a) be fabulously wealthy.

(b) be getting 6% APY.

(c) have earned only pennies more interest than 6% simple interest.

20. The APY for 3% compounded continuously is ___________.

21. Which of the following is the most generous interest rate for a one-year CD?

(a) 6% simple interest

(b) 5.9% compounded annually

(c) 5.9% compounded continuously

22. The value of *e* is approximately ___________.

23. The number *e* is

(a) irrational.

(b) irrelevant.

(c) irrotational.

24. Depositing $100 on a child's annual birth date at ages 1 through 18 years is an example of a(n) ___________.

25. The sum of the geometric infinite series $1 + \frac{1}{2} + \frac{1}{4} + \frac{1}{8} + \dots$ is

(a) infinite.

(b) 2.

(c) *e*.

26. If your investments earned 4.5% last year but inflation was 1.0%, the real rate of growth of your investments was ___________%.

27. An example of exponential decay is

(a) the depreciation of factory equipment.

(b) a retirement annuity.

(c) the CPI.

28. If a new car costs $18,000 and loses value at a rate of 20% per year, its value after 3 years is ___________.

29. If your investment is growing but at a rate less than the rate of inflation

(a) you have a positive real growth in your investment.

(b) you do not have a positive real growth in your investment.

(c) you do not have enough information to determine whether the real growth is positive or negative.

30. According to the CPI, a market basket of goods that cost $100 in mid-2000 would cost ___________ in mid-2015.

Chapter 21 Exercises

Challenge   Discussion

The exercises require either a scientific calculator with buttons for powers exponential and natural logarithm or else a computer spreadsheet program.

21.1 Simple Interest and Arithmetic Growth

1. In Example 2 on page 871, we considered a private student loan at 6.49%, but the rate for such a loan could have been as high as 12.99%. Under the same circumstances as in Example 1 ($10,000 principal, interest-only repayment), and at the higher rate, how much interest would you pay over 51 months of deferred payments on the principal?

2. Suppose that you need $30,000 for your last year of college. You could go to a private lending institution and apply for a signature student loan;

rates range from 7% to 14%. However, your Aunt Sally is willing to lend you the money from her retirement savings, with no repayment until after graduation. All she asks is that in the meantime, you pay her each month the amount of interest that she would otherwise get on her savings (since she needs that to live on), which is 4%. What is your monthly payment to her, and how much interest will you pay her over the academic year (9 months)? (Aunt Sally will be glad to hear from you every month, anyway!)

3. In April 2014, you could buy a 10-year U.S. Treasury note ("T-note," a kind of bond) for $10,000 that pays 2.726% simple interest every year through April 4, 2024. How much total interest would it earn by then?

4. In April 2014, you could buy a 30-year U.S. Treasury bond for $10,000 that pays 3.59% simple interest every year through April 2044. How much total interest would it earn by then?

5. In 2015, the concentration of the greenhouse gas carbon dioxide in Earth's atmosphere reached 400 parts per million (ppm) by volume.

(a) The concentration in recent years has been increasing at about 2 ppm per year. If that trend continues, what will the concentration be at about the time that you retire, in, let us say, 2060?

(b) Explain how this situation is like simple interest on a bank account.

21.2 Compound Interest and Geometric Growth

6. You deposit $100,000 at 0.70% per year in a money market account. What is the interest at the end of one year if the interest paid is

(a) simple interest?

(b) compounded annually?

(c) compounded quarterly?

(d) compounded daily?

7. Repeat Exercise 6, but for $100,000 at 1.2% per year.

8. In 2015, the concentration of the greenhouse gas carbon dioxide in the Earth's atmosphere reached 400 parts per million (ppm) by volume. Although the concentration has been increasing at about 2 ppm per year in recent years, that annual increase itself has been steadily increasing, due to increased burning of fuels. The rate of increase of carbon dioxide in 2015 was about 0.57% per year. If that trend continues from the 2015 level of 400 ppm, what will the concentration be when you retire, in, say, 2060?

9. Your Uncle Blake wants to help you get a start in life after college by setting aside $10,000 now—but you get it only after you finish college, three years from now. Since you are taking this course (and have an interest in what happens with the money in the meantime!), he asks for your advice about three fairly safe investment opportunities:

- Investment A pays 3.75% annual interest, compounded twice a year.
- Investment B pays 4% annual interest, but compounded only once a year.
- Investment C pays nothing until the end, when it pays 12%.

Which investment do you recommend to your uncle?

10. Suppose that you inherit $100,000 and invest it until you retire 40 years from now, in a fund that averages 6.5% annual earnings, which you reinvest.

(a) How much will you have at the end of the 40 years?

(b) Managers of investment funds usually charge a percentage each year of the amount invested (even if the investments lose money!). Suppose that the investment fund management charges you 0.25%, so you earn 6.25% instead of 6.5%. How much will you have at the end of the 40 years?

11. The *rule of 72* is a handy rule of thumb for estimating how long it takes money to double with annual compounding: If r is the annual interest rate, *expressed as a decimal,* then the doubling time is approximately $72/(100r)$ years. If you express the interest rate as $R\%$, then the doubling time is approximately $72/R$ years.

(a) Calculate the balance at the end of the predicted doubling time for each $1000, with annual compounding, for the small growth rates of 3%, 4%, and 6%.

(b) Repeat part (a) for the intermediate interest rates of 8% and 9%.

(c) Repeat part (a) for the larger interest rates of 12%, 24%, and 36%.

(d) What do you conclude about the rule of 72?

12. If carbon dioxide in the atmosphere continues to increase at 0.57% per year, how long will it take for the concentration to double from 400 ppm in 2015?

13. An urban legend is that "the amount of information in the world doubles every three days." Presumably the claim refers to the amount in bits, the smallest unit of storage in a computer. Show that the claim is absolutely preposterous by doing a little arithmetic and comparing your result with 10^{70} (the number of particles in the universe).

(a) Start with one bit of data and double the number of bits every third day. How long does it take to get past 10^{70}? (*Hint:* Don't just keep multiplying by 2 over and over. Convince yourself that since the amount of data increases by a factor of 2 every 3 days, then it increases by a factor of $2^2 = 4$ every 6 days, a factor of $4^2 = 16$ every 12 days, a factor of $16^2 = 256$ every 24 days, and so forth.)

(b) Part (a) involves a lot of multiplying by 2, even if you do it efficiently. Another approach is to use the fact that $2^{10} = 1024 \approx 1000$. Thus, the amount of data increases by a factor of more than 1000 every $3 \times 10 = 30$ days, or every month (except February, but the 31-day months make up for it). By when will the total surely be past 10^{70}?

14. In a *FoxTrot* cartoon by Bill Amend (9/10/2006), shown below, Paige confronts a math problem in which "a math teacher assigns one second of homework the first week of school, two seconds the second week, four seconds the third, and so on." She is asked whether she would agree to this weekly homework doubling for the duration of the 36-week school year. How much homework (in hours) would this plan require in week 36?

15. (Requires a spreadsheet) Write a spreadsheet program that reproduces Table 21.1 (page 873).

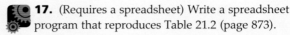

16. (Requires a spreadsheet) Write a spreadsheet program that reproduces Table 21.1 but rounds the interest added to the nearest cent.

17. (Requires a spreadsheet) Write a spreadsheet program that reproduces Table 21.2 (page 873).

18. (Requires a spreadsheet) Write a spreadsheet program that reproduces Table 21.2 but rounds to the nearest cent the interest added for yearly and quarterly compounding.

21.3 Effective Rate and APY

19. For the deposit in Exercise 6c, what is the APY?

20. For the deposit in Exercise 7b, what is the APY?

21. I had a CD with National City Bank through 2010 that paid 4.69% interest compounded daily. What was the APY for this rate?

22. First Community Credit Union of Beloit, Wisconsin, currently pays dividends at 0.10% per year, compounded monthly. What is the APY for such a rate?

23. U.S. savings bonds are sometimes used as patriotic presents or awards, since they help the government by loaning it money. The interest is exempt from state and local income taxes and from federal income tax if used to pay for college tuition and fees. In December 2013, the Rodel Exemplary Teacher Initiative, which addresses the shortage of effective teachers in Arizona's neediest schools and encourages excellent teachers to stay in the profession, chose 11 teachers each to receive a U.S. savings bond. These bonds cost $5000 each and earn

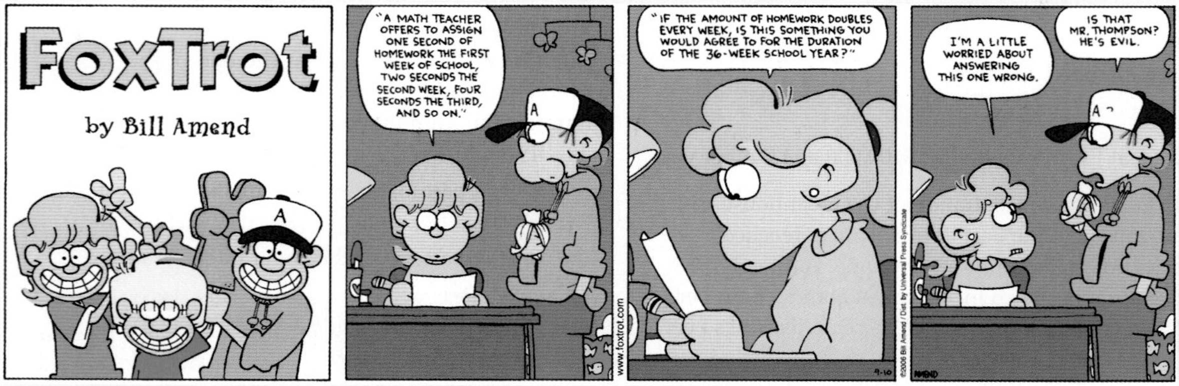

0.10% annual interest for up to 30 years. Unlike many other bonds, the interest on these bonds is compounded semiannually. What will be the value of such a bond after 30 years?

24. A Series I savings bond earns a variable rate of interest that can change every six months, with compounding done semiannually. The initial rate was 1.48% in early 2015. If that rate continues unchanged for the 30 years of the bond's duration, what will be the final accumulation on a $10,000 bond? What is the APY?

25. Suppose that on the statement for a money market account, the initial balance was $7744.70, the statement was for 34 days, the final balance was $7770.84, and there were no deposits or withdrawals. Calculate the APY.

26. Repeat Exercise 25, but for the preceding month, which had initial balance $7722.54, a period of 27 days, final balance $7744.70, and no deposits or withdrawals. Calculate the APY.

21.4 A Limit to Compounding

27. Use your calculator to evaluate for n = 1, 10, 100, 1000, and 1,000,000:

$$\left(1 + \frac{1}{m}\right)^m$$

As m gets larger, what number do these powers seem to be approaching?

28. Repeat Exercise 27 but for $\left(1 + \frac{2}{m}\right)^m$. *Hint:*

$$\left(1 + \frac{1}{m}\right)^m \times \left(1 + \frac{1}{m}\right)^m \approx \left(1 + \frac{2}{m}\right)^m$$

29. (Contributed by John Oprea of Cleveland State University.) Use your calculator to evaluate for m = 1, 10, 100, 1000, and 1,000,000:

$$\left(1 - \frac{1}{m}\right)^m$$

As m gets larger, what number do these powers seem to be approaching? *Hint:*

$$\left(1 + \frac{1}{m}\right)^m \times \left(1 - \frac{1}{m}\right)^m \approx (1)^m$$

30. Repeat Exercise 27 but for $\left(1 - \frac{2}{m}\right)^m$. *Hint:* See the hints for Exercises 28 and 29.

31. Suppose that you have $1000 on deposit at your bank at an annual rate of 1.2%. How much interest do you receive after one year if the bank compounds

(a) continuously?

(b) daily, using 365 days in a year?

32. Suppose that you have a bank account with a balance of $4532.10 at the beginning of the year and $4632.10 at the end of the year. Your bank advertises "continuous compounding," but in fact, it compounds continuously over each 24-hour day and posts interest to accounts daily.

(a) What effective rate did you receive?

(b) What nominal rate is the calculation based on?

(c) What difference is there between what the bank is doing and true continuous compounding?

33. Suppose that you have an investment that earns 0% in the first year, but 10% in the second year.

(a) What rate of interest, compounded annually, would yield the same return after two years?

(b) What rate of interest, compounded continuously, would yield the same return after two years?

34. Suppose that you have an investment that earns 10% in the first year, 20% in the second year, and 30% in the third year.

(a) What rate of interest, compounded annually, would yield the same return after three years? (The answer here is related to the geometric mean discussed in Chapters 14 and 19, but you do not need to use that to solve the problem.)

(b) What rate of interest, compounded continuously, would yield the same return after three years? (Thanks for the idea to Yi Cheng, Indiana University South Bend.)

35. Suppose that your house went down in value a total of 25% over the last three years but then will go up a total of 25% over the next three years. Will you end up with more than, less than, or the same as the original value?

36. Suppose that your friend's house went up in value a total of 25% over three years but then went down a total of 25% over the succeeding three years. Did your friend's house end up at more than, less than, or the same as the original value?

Refer to the following in doing Exercises 37 and 38. For small interest rates, there is little difference between compounding annually, quarterly, monthly, daily, or continuously. Investigating doubling times with continuous compounding leads to understanding why the rule of 72 in Exercise 11 (page 899) works. Recall that for continuous compounding of an initial principal of P at annual rate r, the balance A at the end of t years is Pe^{rt}. For the initial principal to double, we have $2P = A = Pe^{rt}$, so $e^{rt} = 2$. Taking the natural logarithm of both sides yields $rt = \ln 2$, where ln stands for the natural logarithm, represented on a calculator by a button marked either $\boxed{\text{ln}}$ or $\boxed{\text{LN}}$—*not* $\boxed{\text{log}}$ or $\boxed{\text{log}_{10}}$ (those stand for a different kind of logarithm). Using the button gives $\ln 2 \approx 0.693$. So we have $rt \approx 0.693$, from which we can determine t if we know r.

37. Calculate the doubling times for continuous compounding at 2%, 3%, and 4%, and compare them with those predicted by the rule of 72.

38. Calculate the doubling times for continuous compounding at 6%, 8%, and 9%, and compare them with those predicted by the rule of 72. What do you conclude? Why do you think people prefer the rule of 72 over a rule of 69.3?

21.5 A Model for Saving

39. Suppose that you want to save up \$2000 for a semester abroad two years from now. How much do you have to put away each month in a savings account that earns 2% interest compounded monthly?

40. Repeat Exercise 39, except that you have found a better deal: 3% interest compounded monthly.

41. Parents may struggle for the first few years after a child is born but often are finally able to start saving toward the child's college education when the child starts school at age 6 (because the parents stop paying for daycare). If they save \$400 per month in an account paying 2.5% interest compounded monthly, how much will they have for college expenses 12 years later?

42. Suppose that you save for retirement by contributing the same amount each month from your 23rd birthday until your 65th birthday, in a retirement account that pays a steady 4% annual interest compounded monthly.

(a) How much will be in your fund at age 65 if you save \$100 a month?

(b) How much will be in your fund if you get a steady return of 7.5% compounded monthly?

(c) How much will be in your fund if you get a steady return of 10% compounded monthly? (This is approximately the average annual return on the New York Stock Exchange from 1950 to 2000.)

43. A colleague feels that he will need \$1 million in savings to afford to retire at age 65 and still maintain his current standard of living. A younger colleague, age 30, decides to begin saving for retirement based on that advice. How much does the younger colleague need to save per month to have \$1 million at retirement if the fund earns a steady 3% annual interest compounded monthly?

44. The younger colleague of Exercise 43 is not satisfied with a 3% return, which he could get with U.S. Treasury bonds or perhaps long-term CDs. Instead, he wants to take the riskier route of investing in the stock market, which has over its history returned an APY of about 10% per year (although between October 2007 and February 2009, it lost almost half its value). Assuming that over the 35 years until his retirement the stock market behaves just that way (a big assumption!), how much would he need to invest each month to achieve his goal of \$2 million by age 65?

45. Many young people do not start saving right away for retirement, although by the time that they do, they may be earning more and thus could afford to save more each month. How much will be in your fund at age 65 if you don't start saving until age 35 (by which time you hope interest rates will have risen) and at that age, start saving \$100 per month in an account paying a steady 6% annual interest compounded monthly?

46. Suppose that you have children young, pay for their college expenses, and finally start saving for retirement at age 50. How much do you have to save per month, with a steady return of 6% compounded monthly, to accumulate \$250,000 by age 65?

47. Suppose that when you turn 25, you are single and are in a 25% bracket for federal income tax and a 7% bracket for state and local income taxes. (In 2015 this corresponded to an income, beyond exemptions and deductions, of between \$37,000 and \$90,000.) This means that you pay an income tax rate ("marginal rate") of 32% on part of your income (but a lower rate on the rest). Suppose that you commit to saving \$100 per month toward retirement.

(a) How much will be in your fund at age 65 if you can get a steady return of 6% compounded monthly?

(b) How much will you pay in income tax on that \$100?

48. We continue the circumstances of Exercise 47. Instead of saving \$100 per month—money on which you have already paid taxes, "after-tax" dollars—you have the alternative option offered in the tax code of participating in a tax-deferred retirement account (TDA), either through payroll deduction at work [e.g., as part of a 401(k) plan] or through an independent retirement account (traditional IRA). The money that goes into such a fund consists of "pre-tax" dollars: You do not pay tax on the money until you withdraw it (usually at retirement). Since you don't pay income tax on the money as you put it in, you can actually put in more than \$100 per month while reducing your take-home pay by only \$100.

(a) How much can you put into the retirement fund each month if you reduce your take-home pay by exactly \$100?

(b) How much will be in your fund at age 65 if you can get a steady return of 6% compounded monthly?

(c) Suppose that when you turn 65, you withdraw the entire amount in your account and pay the deferred taxes that are owed on it, say a total of 32% (federal, state, and local combined). How much do you net?

49. There is yet another alternative to the two options for saving toward retirement in Exercises 47 and 48. Instead of saving after-tax dollars or contributing to a tax-deferred plan, you can take the money as income, pay income tax on it, and make a deposit into a Roth IRA. For this special kind of retirement account, the interest earned over the years is not taxed. (There are further advantages and disadvantages.)

Suppose that you put \$100 per month in after-tax dollars into a Roth IRA account. Assuming the same savings account or safe investment as in Exercises 47 and 48 that pays a steady return of 6% compounded monthly, how much will be in your account, tax-free, at age 65? How does that compare with the answers to Exercises 47a and 48c?

50. We continue the theme of Exercises 47–49 by comparing in algebraic terms three kinds of investments for retirement: an ordinary after-tax investment, a tax-deferred investment (such as a tax-deferred annuity or an IRA), and a Roth IRA. Let an investment earn interest at a steady annual yield r and let your income (in whatever year you receive it) be taxed at rate τ.

(a) Ordinary after-tax investment: Explain why if you earn \$$E$, pay taxes on it, let what remains earn interest, and pay tax each year on that year's interest, the \$$E$ grows after n years to \$$E(1-\tau)\times[1+r(1-\tau)]^n$.

(b) Ordinary IRA: Explain why if you earn \$$E$, defer taxes on it, let it earn interest, and defer taxes on all the interest, then the \$$E$ grows after n years to \$$E(1+r)^n(1-\tau)$.

(c) Roth IRA: Explain why if you earn \$$E$, pay taxes on it, let what remains earn interest, and pay no taxes on any of the interest, the \$$E$ grows after n years to \$$E(1-\tau)(1+r)^n$.

(d) Which investment gives the best return after n years?

51. Apart from CDs, returns on investments are rarely the same from year to year, since they vary with prevailing interest rates. How should you calculate an "average" rate of return over several years? Consider a mutual fund that delivers 100% return one year and loses 50% the next year. Calculate just the ordinary average (the arithmetic mean) of the percentages to get [100% + (−50%)]/2 = 25%. That sounds good, but check what happens to a \$1000 investment: It grows to \$2000, then halves back to \$1000—for a 0% gain. Because the average of the percentages is not useful, the customary way used in finance to calculate the "average" return is to use the geometric mean. If the initial value of the portfolio was P, and its value after n years is A, then the average annual rate of return is the value of r that solves $(1+r)^n = A/P$, or $r = (A/P)^{1/n} - 1$.

(a) Use this formula to determine the average annual rate of return for a portfolio with returns of 10%, −25%, and 25% in three consecutive years.

(b) Is the average rate that the formula finds a nominal rate or an effective rate?

52. Repeat Exercise 51a, but for consecutive returns over four years of 10%, −20%, −10%, and 7%.

53. (Adapted from Terence Blows, Northern Arizona University.) Classify the following growth and decay scenarios as linear (arithmetic), exponential (geometric), or neither.

(a) The amount of caffeine in the bloodstream decreases by 10% every hour.

(b) The amount of trash in a landfill increases by 350 tons per week.

(c) The amount of alcohol in the bloodstream decreases by 10 grams (the amount in a standard drink) per hour.

(d) Your age increases every day.

54. (Adapted from Terence Blows, Northern Arizona University.) Classify as in Exercise 53, but for the following circumstances.

(a) The mean concentration of carbon dioxide in the atmosphere increases by 2 ppm per year.

(b) The mean concentration of carbon dioxide in the atmosphere increases 0.5% per year.

(c) Your knowledge of mathematics and its applications increases with each section of this book that you study.

(d) The number of people in the world increases by 1.2% per year.

55. What is the present value of \$10,000, 4 years from now, at an APY of 5%?

56. What is the present value of \$150,000, 10 years from now, at an APY of 3%?

57. As you will see in Chapter 22, if you had a 30-year \$200,000 mortgage at 8% on a house or apartment, three-quarters of the way through the mortgage—after 22.5 years of payments—you would still owe half the amount, or about \$100,000! You also would have paid about \$300,000 in interest. (Current interest rates are much lower, but even at 4.75%, you would still owe \$80,000 after 22.5 years.) What is the present value of \$100,000, 22.5 years from now, at an interest rate of 8%? (If you put this much into a down payment, but made the same-size payments as for the 30-year mortgage on \$200,000, you would own the house free and clear after 22.5 years.)

58. If you have a 30-year \$200,000 mortgage at 6.48% on a house or apartment, after 10 years of payments you will still owe about \$170,000. What is the present value of \$170,000, 10 years from now, at an interest rate of 6.48%?

21.6 Inflation

59. Suppose that inflation proceeds at a constant rate of 2% per year from mid-2015 through mid-2018.

(a) Find the cost in mid-2018 of a basket of goods that cost \$1 in mid-2015.

(b) What will be the value of a dollar in mid-2018 in constant mid-2015 dollars?

60. The Ford Mustang automobile celebrated its 50th anniversary in 2015. The original model cost about \$2400 new in mid-1965. How much would that be in mid-2015 dollars? (The sticker price of the 2015 model year Mustang was \$23,335.)

Scott Eells/Bloomberg via Getty Images

61. A first-semester college mathematics book cost \$10.75 in 1962. What would the equivalent price be in 2015 dollars? How does that compare with what you paid to buy or rent this book? (That book had black-and-white text and figures and no photographs, color or otherwise.)

62. In 1970, before the oil embargo by the Organization of the Petroleum Exporting Countries (OPEC), gasoline cost about 25¢ per gallon. In 1974, after the embargo, it cost about 70¢ per gallon. What would the equivalent prices be in 2015 dollars? How do they compare with the price of gasoline today?

Refer to the following in doing Exercises 63 and 64. From Table 21.5 (page 892), you can determine the average rate of inflation from one year to another. For example, you find the inflation from 1990 to 2000 by subtracting the two index numbers and dividing by the earlier one: $(172.2 - 130.7)/130.7 = 31.752\%$. However, the average rate of inflation is not this number divided by the number of years (10). We must take into account compounding of the rate of inflation. We set $(1 + a)^{10} = 1.31752$ and find $a = (1.31752)^{1/10} - 1 \approx 2.80\%$.

63. Find the average rate of inflation from 2005 to 2015. Is 3% a good approximation?

64. If inflation had been 3% each year from 2005 to 2015, what would the CPI have been in 2015?

65. (Based on a problem in Ed Barbeau's "More Fallacies, Flaws, & Flimflam"; Washington, D.C., Mathematical Association of America, 2013, p. 2.) Suppose that you get a year-end pay raise of 5%, but over the year, there has been inflation of 10%—so in effect you have suffered a pay decrease in terms of what your salary will buy. What is the percentage decrease?

66. Suppose that you get a year-end pay raise of 5% but over the year there has been inflation of 2%. What is the percentage increase in your purchasing power?

67. According to state law introduced by Governor Scott Walker in 2011, teachers and state employees in Wisconsin can negotiate only salary increases and only up to the level of inflation. A typical elementary school teacher who began teaching in Wisconsin in 2015 at age 22 earned a salary of about \$35,000. What would be the purchasing power in today's dollars of her salary when she retires in 2055 at age 62 if each year her salary increase matches inflation?

Refer to the following in doing Exercises 68–70. A typical new assistant professor at a liberal arts college starts at age 30 with a salary of \$45,000, while colleagues retiring now at age 65 may make about twice that. One college gives annual pay raises of inflation plus 1%, plus a promotion raise (to associate professor) of \$1500 after (usually) 6 years and another promotion raise (to full professor) of \$1500 after (usually) another 6 years.

68. (Spreadsheet helpful) Can a new assistant professor who starts now expect to make the equivalent of \$90,000 in today's dollars when she retires 35 years from now if inflation holds steady at 1.5%?

69. Repeat Exercise 68 but with inflation steady at 2.5%.

70. (Spreadsheet helpful) Suppose that you are the vice president for academic affairs at the liberal arts college. Suggest a salary policy that would result in the new assistant professor, when she retires in 35 years, making the equivalent of

(a) \$90,000 in today's dollars.

(b) \$135,000 in today's dollars. (The candidate for your faculty position would prefer that—hence she would be more likely to accept an offer to come work at your college.)

71. (Spreadsheet helpful) Your roommate (a business major) has already planned her retirement and started funding it in 2015. She plans to retire in 2050 at age 57 on \$100,000 per year in 2050 dollars, living on just the interest on her investments. Assume that she realizes a steady 7.2% interest rate and assume a steady 3% annual inflation. (Suggested by Terence Blows, Northern Arizona University.)

(a) What must the size of her nest egg be, and what should her monthly investment be over the 35 years, to achieve this goal?

(b) What will be the value in 2015 dollars of her 2050 income of \$100,000?

(c) What will be the value in 2015 dollars of her income of \$100,000 in 2078 (when she is 85)?

72. (Spreadsheet helpful) You think what your roommate means in Exercise 71 is that she wants to retire in 2050 with a steady income of \$100,000 a year in 2015 dollars. You also feel that she should plan to receive that same value of income for 43 years in case she lives to 100 (2% of your classmates will). What is the present value in 2015 of the planned stream of 43 years of retirement income?

Refer to the following in doing Exercises 73–75. In the savings formula, the interest rate i appears twice. The particular ways in which i is involved make it impossible to solve it algebraically to get an explicit formula for i. However, with the help of a spreadsheet, you can find i approximately when the other quantities are given.

73. (Spreadsheet helpful) Suppose that you decide to lease a new car. (Leasing is cheaper than buying, because over the lease period you pay only about half the cost of buying the car.) At the end of a 48-month lease period, you either return the car or else make a lump-sum payment of \$10,000 if you want to keep the car. You decide to save up, just in case you decide to keep the car; if you don't keep this car, you will still have a down payment for a new leased or purchased car. You feel comfortable with saving \$140/month (over and above your lease payments). How high an annual nominal interest rate on savings do you need to accumulate \$10,000 in 48 months, with interest compounded monthly?

74. (Spreadsheet helpful) A 1990 advertisement read, "If you had put \$100 per month into this fund starting in 1980, you'd have \$37,747 today." Assume that deposits were made on the last day of the month, starting in January 1980, through December 1989, and that interest was paid monthly on the last day of the month (120 months).

(a) How much money was deposited during this period?

(b) What annual rate of interest, compounded monthly, would lead to the result described in the advertisement? What is the APY?

75. (Spreadsheet helpful) Suppose that your parents are willing to lend you \$20,000 for part of the cost of your college education and living expenses. They want you to repay them the \$20,000, without any interest, in a lump sum 15 years after you graduate, when they plan to retire and move. Meanwhile, you will be busy repaying federally guaranteed loans for the first 10 years after graduation. But you realize that you won't be able to repay the lump sum without saving up. So you decide that you will put aside money in an interest-bearing account every month for the five years before the payment is due. You feel comfortable with putting aside \$275 a month (the amount of the payment on your college loans, which will be paid off after 10 years).

How high an annual nominal interest rate on savings do you need to accumulate the \$20,000 in 60 months, if interest is compounded monthly? Enter into a spreadsheet the values $d = 275$, $r = 0.05$ (annual rate), and $n = 60$, and the savings formula with r replaced by $r/12$ (the monthly interest rate). You will find that the amount accumulated is not enough. Change r to 0.09; it's more than enough. Try other values until you determine r to two decimal places.

Chapter Review

76. Tuition, room, and board cost $3400 per year at a certain Minnesota liberal arts college in 1970. What would be the equivalent cost in 2015 dollars?

77. Suppose that a 25-year-old who makes $40,000 a year gets a raise of 1% above the inflation rate each year, through age 65. What would be the purchasing power in today's dollars of the final year's salary?

78. What is the APY for an interest rate of 5% compounded daily?

79. Postage for a first-class letter cost $0.34 in 2001. If the cost had just kept up with inflation since then, what would it have been in 2015?

80. We noted that exponential growth is not necessarily fast growth. Confirm this fact for yourself by comparing the growth of $100,000 at 1% interest over 10 years for the two situations of

(a) simple interest (which corresponds to linear growth).

(b) interest compounded continuously (which corresponds to exponential growth).

Applet Exercise

To do this exercise, go to www.macmillanhighered.com/fapp10e.

1. How important is it to begin a retirement fund at an early stage of one's career? In the *Saving for Retirement* applet, you will discover that early funding of a retirement plan can make a huge difference in the ultimate amount that will be available when a person retires.

Writing Projects

1. Plan your retirement. Decide on a retirement age and desired income (in today's dollars). Estimate yield on investments, inflation rate, and Social Security benefits. (*Note:* By the time of your retirement, a woman retiring in her mid-60s will likely live an average of 20 years more, a man 18 years more. Social Security income goes up with inflation. Annuities are available whereby income grows to keep up with inflation. Various other financial products, such as a life annuity, can make sure that you don't outlive your retirement income. We discuss life annuities in Chapter 22. Also, do not neglect consideration of taxes and tax deferral of income, as considered in Exercises 47–50.) Write up your assumptions, justifications for them, calculations, and conclusions in three to four pages. Be sure to note any additional factors that you think should be taken into account but which your analysis does not include; don't be afraid to consider possibilities with financial impacts you can't calculate exactly.

2. Exercises 47–50 (pages 902–903) ask you to look at various forms of tax-deferred and ordinary savings and compare them on the basis of amount of tax-free income accumulation at age 65.

Ordinary savings have the important advantage that, at any time, you can do anything you want with the money accumulated so far (buy a car, put down money on a house, and so on). A second advantage is that the money is free and clear, in that taxes have already been paid on it.

A tax-deferred [e.g., IRA or 401(k)] retirement fund has the advantage of postponing taxation of the funds, but the disadvantage that if you withdraw funds before age $59\frac{1}{2}$, you must pay income tax in the year of withdrawal and in addition a 10% penalty for "early withdrawal." (These plans were given the advantage of tax deferral to encourage individuals to save for retirement—hence the penalty for withdrawing money earlier.)

A third option, the Roth IRA, has some of the advantages and disadvantages of each of the other plans.

Look into the details of the rules for 401(k) plans and Roth IRAs, compare your answers in Exercises 47–50, and devise and describe your own plan for how you will save for retirement. Your report should run two to three pages.

3. Should you invest in "forever" stamps? In 2007 the U.S. Postal Service began selling such stamps, which will suffice for first-class postage at any time in the future, even when rates rise. These stamps cost $0.49 in early 2015. Would it be a good investment to buy a "lifetime supply" of them? To render your judgment, compare the historic increases of postal rates (see www.vaughns-1-pagers.com/economics/postal-rates.htm) with the CPI (page 891). For example, you could convert postal rates at each of the dates of change to the cost in today's dollars, and then see if there appears to be a recent trend that you can project into the future.

4. Based on ticket sales and current dollars, the top-grossing domestic film of all time was *Avatar,* which earned $761 million since release in 2009. But adjusted for inflation of ticket prices, *Gone with the Wind* (1939, plus re-releases) comes out on top at $1,608 million in 2014 dollars (see boxofficemojo.com/alltime/adjusted.htm).

(a) Use a spreadsheet to analyze how movie ticket prices have risen since 1913 compared with the CPI. The CPI data from Table 21.5 (page 892) can be downloaded from data.bls.gov/cgi-bin/surveymost?cu. (Check the first box, click "Retrieve data," then in the next screen adjust the year "From" to 1913.) Find the data on movie ticket prices at boxofficemojo.com/about/adjuster.htm. Prepare a short report with a graph. (Thanks to Martin Campbell for the idea.)

(b) The comparison of ticket sales adjusts to the estimated average ticket price in 2014 ($7.96); and since it considers proceeds rather than audience share, it does not take into account the larger population today compared with the past. Examine the full details of the adjustment at boxofficemojo.com/about/adjuster.htm. What do you think of the methodology used? How would you do the adjustment?

5. The federal minimum wage has been $7.25 per hour since July 24, 2009. Has it kept pace with inflation over the years? Use a spreadsheet to analyze how it has risen since 1938 compared with the CPI. The CPI data from Table 21.5 can be downloaded from data.bls.gov/cgi-bin/surveymost?cu. (Check the first box, click "Retrieve data," then in the next screen adjust the year "From" to 1938.) Find the data on federal minimum wage rates at www.dol.gov/whd/minwage/chart.htm. Prepare a short report with a graph.

6. The assumptions in many of the exercises of this chapter of a constant interest rate, rate of return on an investment, or tax rates (as in Exercises 47–50) holding over a long period of time are simplifications that simply won't be true. Interest rates fluctuate (though you can lock in a long-term constant rate by buying a long-term bond or certificate of deposit), and the tax rate may change (with your income, your state of residence, and changes in tax laws). If your marginal tax rate (the rate you pay on one additional dollar of income) is lower in one year than the tax rate you expect to pay in retirement, what kind of retirement investment is better for you that year? If you have a windfall one year and your marginal tax rate is higher that year than the tax rate that you expect to pay in retirement, what kind of retirement investment is better for you that year? How should you take the various factors into account as you make investment and savings decisions early in life versus later in life, near retirement?

Suggested Readings

LINDSTROM, PETER A. Nominal vs. Effective Rates of Interest. UMAP Modules in Undergraduate Mathematics and Its Applications: Module 474. COMAP, Inc., Arlington, MA, 1988. Reprinted in Paul J. Campbell (ed.), UMAP Modules: Tools for Teaching 1988, COMAP, Inc., Arlington, MA, 1989, pp. 21–53. A learning module, requiring no more background than this chapter, that teaches about nominal and effective rates of interest and how to calculate them. It gives real examples of banks using different options for calculating interest.

MILLER, CHARLES D., VERN E. HEEREN, and JOHN HORNSBY. Consumer mathematics. In *Mathematical Ideas,* 11th ed., Pearson Education/Addison Wesley Longman, Boston, MA, 2007.

VEST, FLOYD, and REYNOLDS GRIFFITH. The mathematics of bond pricing and interest rate risk. *Consortium* (COMAP), 59 (Fall 1996): HiMAP Pullout Section 1–6.

Suggested Websites

bls.gov/cpi/ Home page for the inflation tables prepared by the BLS.

data.bls.gov/cgi-bin/cpicalc.pl CPI Inflation Calculator. Converts dollar value from any year to its equivalent buying power in any other year.

www.westegg.com/inflation/ Inflation calculator for the United States (1800–2014).

Borrowing Models

22

John Lamb/Getty Images

How much interest are you paying on your student loans? When will they eventually be paid off? How much will they eventually cost you?

In the preceding chapter, we looked at consumer financial models for saving and formulas for calculating the amount accumulated. Savings or investments would not earn interest unless they could be loaned to someone to make productive use of the money.

In this chapter, we examine the other side of consumer finance: borrowing. You may have a student loan, you will probably need to borrow (or have already borrowed) to buy a car, you will almost certainly borrow if you buy a house or apartment, and you are borrowing if you use a credit card. For any such loan, you pay "finance charges," which include interest and perhaps other "fees" as well. We investigate and compare some common kinds of loans.

We begin by (re)acquainting you briefly with simple interest and compound interest, in the contexts of student loans and credit cards. "Conventional" loans, such as the mortgage on a house, use the savings formula from Chapter 21 to calculate the monthly payment. Finally, we consider annuities, a way to provide income security in retirement.

If you have a grasp of the ideas behind compound interest and can use the compound interest formula (page 911) and the savings formula (page 915), you can proceed with this chapter without first reading Chapter 21.

Dennis MacDonald/Alamy

22.1 Simple Interest

The amount of **interest** charged on a loan is determined by the **principal** (the amount borrowed) and by the method used to calculate the interest. With **simple interest,** the borrower pays a fixed amount of interest for each period of the loan. The interest rate is usually quoted as an annual rate.

For a principal P and an annual rate of interest r, the interest owed after t years is

$$I = Prt$$

and the total amount A due on the loan is

$$A = P(1 + rt)$$

EXAMPLE 1 Simple Interest on a Federal Direct Student Loan

The U.S. Department of Education offers direct loans to students to use for tuition, fees, housing, and textbooks (such as this one), with repayment deferred until after graduation. Any eligible student can get such a loan, regardless of financial need or credit history.

For 2014–2015, the interest rate was 4.29% for undergraduates. Interest is charged from when you receive the loan until repayment is scheduled to start, usually six months after you leave school. This interest is on the original principal only (including an origination fee of 1.073%), not on accrued (accumulated) interest, hence it is simple interest on the principal.

The interest is actually calculated according to a method called the Simplified Daily Interest Formula: The interest rate is divided by the number of days in a year, taken to be 365.25, and that daily rate (called the **interest rate factor**) is multiplied by the number of days in the billing period. For example, for the interest rate of 4.29%, the interest rate factor is 0.0429/365.25 ≈ 0.000117454, or 0.0117454% per day. *However, for simplicity, we calculate as if the year were evenly divisible into 12 months and into 4 quarters. We also regard the origination fee as included in the original loan principal.*

One option for repayment is to pay the interest as it comes due, billed on a quarterly basis, but defer starting to pay back the principal until six months after leaving school (the "grace period").

Suppose things played out as follows:

- You took out such a loan for \$5500 (the maximum amount for a dependent first-year student) on September 1, 2014, at the start of your freshman year.
- You pay the interest as you go, on a quarterly basis as billed.
- Paying back the loan is to begin on December 1, 2019, after you graduate on June 1, 2019 (so you will have had the loan for 4 years + 3 months = 51 months).
- We disregard the origination fee, which is deducted from what you get.

How much is the monthly interest, how much total interest will you have paid over the 51 months, and how much will you owe when you start to pay back the loan?

We have $P = \$5500$ and $r = 4.29\% = 0.0429$, and for one quarter, we have $t = \frac{3}{12} = 0.25$ years. So the interest for one quarter is $I = Prt = \$5500 \times 0.0429 \times 0.25 \approx \$58.99$. Over the 51 months (17 quarters), you will have paid $17 \times \$58.99 = \$1002.83$. At the time when repayment begins, you would still owe (just) the original principal of \$5500.

If instead you do not pay the interest as you go, you would owe the principal of \$5500 plus the accumulated simple interest of \$1002.83, for a total of \$6502.83.

Self Check 1

Suppose that you borrow \$3000 at the start of your junior year. What is your quarterly interest payment?

Caution: The misleading phrase "simple interest loan" is now often used in the loan trade to refer to a loan in which interest is charged each day based on the balance owed that day, as opposed to some other kinds of loans, such as add-on loans (which we consider later). Such a "simple interest loan" is usually a loan at compound interest with a compounding period of one day. We consider compound interest loans next.

22.2 Compound Interest

Compounding is the calculation of interest on interest. A common example is the balance on a credit card. So long as there is an outstanding balance, the interest owed is calculated on the entire balance, including any part of it that was previously calculated as interest and added to the balance in earlier months.

We will be using two formulas from Chapter 21: the compound interest formula (page 875) and the savings formula (page 884), phrasing them for loans. Here is the **compound interest formula,** followed by an example.

Compound Interest Formula RULE

If a principal P is loaned at interest rate i per compounding period, then after n compounding periods (with no repayment), the amount owed is

$$A = P(1 + i)^n$$

To make the connection to multiple compoundings per year, we give the formula in a slightly more elaborate and detailed version below.

General Compound Interest Formula RULE

For a principal P loaned

- at a nominal annual rate of interest rate r
- with m compounding periods per year (so the interest rate is $i = r/m$ per compounding period)

the amount owed after t years (hence after $n = mt$ compounding periods), with no payment meanwhile of interest or principal, is

$$A = P(1 + i)^n = P(1 + \tfrac{r}{m})^{mt}$$

EXAMPLE 2 What Happens If You Don't Make the Payments on the Principal?

After you begin to repay your federal direct student loan, what happens? The loan is capitalized, meaning that any unpaid interest accrued while you were in school and during the grace period is added to the principal and becomes the new balance on which interest is to be paid. That balance will decline as you **amortize** (pay off) the loan. You will be responsible for fixed monthly payments. The interest is calculated monthly, and the proportion of your payment going to interest will decline as you pay off the loan.

What if you do not make the payments due for the repayment? Well, the interest accrues and is *capitalized* every quarter into the principal: The interest is added to the loan balance (the amount owed), and interest for the next payment is calculated on the new loan balance. In other words, the compounding period is one quarter.

Suppose that six months after you graduate, when the grace period expires and you have to begin making payments, you owe \$10,000 at 4.29% interest per year. However, you haven't been able to find a job, so you fail to make any payments for the next year. How much would you owe then?

The principal P is \$10,000. The quarterly interest rate is $i = 4.29\%/4 = 1.0725\%$ and there are $n = 4$ compounding periods. The compound interest formula gives the new amount owed as $A = \$10{,}000(1 + 0.010725)^4 = \$10{,}436.00$.

(It would be very foolish to just ignore the payments due, since after 270 days of nonpayment, the loan would be in default and all kinds of bad things would happen! Instead, you should contact the loan server about possible reduction or postponement of payments. Also, by executive order of President Obama, after December 2015 student loan repayments will be capped at 10% of the borrower's monthly income.) ■

Self Check 2

Suppose that your federal student loans total \$20,000 and you fail to make the first six monthly payments. How much does the loan grow to? ■

Terminology for Loan Rates

The interest on a loan depends on whether compounding is done and how the interest is calculated.

A **nominal rate** is any stated rate of interest for a specified length of time. For instance, a nominal rate could be a 1.5% monthly rate on a credit card balance. By itself, such a rate does not indicate or take into account whether or how often interest is compounded.

The **effective rate** *takes into account compounding*. It is the rate of simple interest that would realize exactly as much interest over the same period of time.

As we saw, a student loan of \$10,000 at a yearly interest rate of 4.29% (a nominal rate), calculated as 1.0725% per quarter compounded quarterly, yields \$436.00 in interest owed at the end of the year, which is 4.36% of the original principal. Hence, the effective annual rate is 4.36%. In other words, a \$10,000 loan at *simple* interest of 4.36% for one year would owe exactly the same interest.

When stated per year ("annualized"), the effective rate is called the **effective annual rate (EAR).** (In connection with savings, the effective annual rate is the annual percentage yield discussed in Section 21.3.) Thus, for the student loan example above, the EAR (rounded) is 4.36%.

To keep the rates straight, we use i for a nominal rate for the specified **compounding period**—such as a day, a month, or a year—*within which no compounding is done*; this rate is the effective rate for that length of time. For a nominal rate compounded m times per year, we have $i = r/m$. For that \$10,000 student loan at 4.29% compounded quarterly, we have $r = 4.29\%$ and $m = 4$, so $i = r/m = 1.0725\%$ per quarter.

Just as the Truth in Savings Act (mentioned in Chapter 21, page 874) does for savings, the Truth in Lending Act establishes terminology and calculation methods for interest for loans. The Truth in Lending Act introduced the term **annual percentage rate (APR).**

Annual Percentage Rate (APR) DEFINITION

The **annual percentage rate (APR)** equals the number of compounding periods per year times the rate of interest per compounding period:

$$\text{APR} = m \times i$$

For the student loan in repayment, the interest is compounded quarterly, or $m = 4$ times per year, and the interest rate for the compounding period is $i = 1.0725\%$; so the APR is $4 \times 1.0725\% = 4.29\%$. The APR is the rate that the Truth in Lending Act requires the lender to disclose to the borrower. *The APR is usually smaller than the effective annual rate (EAR).* In the case of the student loan of Example 2, the APR is 4.29%, while we calculated subsequently that the EAR is 4.36%. Spotlight 22.1 explains further.

What's the Real Rate?

SPOTLIGHT 22.1

Financial experts agree that the real, "true" rate of interest for savings or loans is the effective annual rate.

The 1991 Federal Truth in *Savings* Act requires that savers be told the annual percentage yield (APY) (discussed on page 878 in Chapter 21), which is just the effective annual rate.

The 1968 Federal Truth in *Lending* Act, however, requires that borrowers be told the APR, which is *not* the same as the effective annual rate. The APR is the rate of interest per compounding period times the number of compounding periods per year. Thus, a credit card rate of 1.5% per month translates

What's the Real Rate? *(continued)*

to an APR of 18%. The APR does not take into account compounding. Hence, it is not equivalent to—indeed, it understates—the true cost of borrowing; that is, the effective annual rate. For the credit card loan, with monthly compounding, the effective annual rate is

$$(1 + 0.015)^{12} - 1 \approx 19.562\%$$

The APR also ignores costs that are sometimes involved in borrowing, such as a flat charge for making the loan in the first place (called a "loan-processing fee"), charges for late payments, and charges for failing to make a minimum payment.

In 2015–2016, Federal Direct Loans had a 1.073% origination fee, and one-half of the loan is disbursed at the start of each semester; both these factors raise the effective interest rate. For Federal Direct PLUS Loans to parents of students, the origination fee is 4%. The fees are intended to cover in part the cost of loans that default.

For home mortgage loans, however, the Truth in Lending Act requires that lenders include in the APR some of the upfront costs referred to as "closing costs": any "loan origination" fee, "loan-processing" fee, and "points" (additional charges to get a reduced interest rate). The APR does not include title insurance, appraisal, credit-report fees, or transaction taxes.

Closing costs are paid at the closing of the sale, while interest is paid over the life of the loan. However, the APR treats the closing costs included in it as if they will be amortized over the term of the mortgage, despite the fact that they were paid beforehand. Here, too, the APR understates the true costs.

However, very few people hold a mortgage to its maturity. The median life of a 30-year mortgage is only about 5 years; that is, half of all mortgage holders pay off their mortgage before 5 years are up, usually because they sell their homes and move elsewhere. Thus, for almost all home loans, the APR also includes interest that will never be paid.

Also, we must take into account inflation. One advantage of buying a home with a fixed-rate mortgage is that your payment stays the same, but your earnings and the value of your home are likely to go up with inflation. You are thus paying back the loan with dollars of lesser value. For any loan in a time of inflation, *Fisher's effect* comes into play: If your loan has an effective annual rate of 7% but inflation is running at 3.5% per year, the true cost to you of the loan is not exactly 7% − 3.5% = 3.5%. Instead, for an effective annual rate of r and an inflation rate of a, the cost of the loan at the beginning of the first year is indeed $r - a$ (3.5% in our example), but at the end of the first year, it is

$$g = \frac{r - a}{1 + a}$$

For $r = 7\%$ and $a = 3.5\%$, we get $g = 3.38\%$. The reason that this is less than the expected 3.5% is that at the end of the first year, you are paying back the loan with dollars that have been inflated for a year. As inflation mounts over the term of a mortgage, the cost g goes down steadily each year. For example, at the end of 5 years of steady inflation at 3.5%, the total inflation has been $a = (1 + 0.035)^5 - 1 \approx 18.8\%$, and we have $g = 2.95\%$.

A final—and major—consideration is that interest paid on your home mortgage is deductible from taxable income on federal, state, and some local income tax returns. Thus, your home ownership is subsidized by other taxpayers (just as you help subsidize other home buyers), and the cost to you of the loan is reduced further.

22.3 Conventional Loans

A common situation that you are likely to encounter is a loan—for a house, a car, or college expenses—that is to be paid back in equal periodic installments. Your payments are said to amortize (pay back) the loan. In these so-called **conventional loans,** each payment pays the current interest and also repays part of the principal. As the principal is reduced, there is less interest owed, so less of each payment goes to the interest and more toward paying off the principal.

We remind you of the **savings formula** from Chapter 21 (page 884).

Savings Formula RULE

The amount A that is accumulated

- at a nominal annual rate of interest rate r
- with m compounding periods per year (so interest rate $i = r/m$ per compounding period)
- after t years (hence $n = mt$ compounding periods)
- by a uniform deposit d at the end of each compounding period

is

$$A = d\left[\frac{(1+i)^n - 1}{i}\right] = d\left[\frac{(1+\frac{r}{m})^{mt} - 1}{\frac{r}{m}}\right]$$

For the loan situation, the "uniform deposit" becomes the monthly payment:

d = the monthly payment, made at the end of each month

EXAMPLE 3 Buying a House

Suppose that you buy a house with a \$100,000 loan to be paid off over 30 years in equal monthly installments. Suppose that the interest rate for the loan is 6.00%. How much is your monthly payment?

Imagine changing the setup slightly so that instead of making monthly payments, you are supposed to pay off the entire principal and interest at the end. Meanwhile, you make payments to a savings fund that you're building up to pay off the loan, and the savings fund earns the same rate of interest that the loan costs. The interest rate of 6.00% on the loan is compounded monthly, so the monthly rate is 0.5%. At the end of 30 years, the principal and interest on the loan would (by the compound interest formula) amount to

$$\$100{,}000 \times (1 + 0.005)^{12\times 30} \approx \$602{,}257.52$$

On the other hand, saving d each month for 30 years at 6.00% interest compounded monthly, we know from the savings formula that you will accumulate

$$d\left[\frac{(1 + 0.005)^{360} - 1}{0.005}\right]$$

To make d just the right amount to pay off the loan exactly, we need to solve the equation

$$d\left[\frac{(1 + 0.005)^{360} - 1}{0.005}\right] = \$100{,}000\,(1 + 0.005)^{12\times 30} \approx \$602{,}257.52$$

for d, getting $d \approx \$599.55$ as the monthly payment.

Now, \$600,000+ is not what you will be paying over the course of the loan of \$100,000! It is only an intermediate amount that we come across in calculating your payment. The total of your loan payments will be ("only") 360 × \$599.55 = \$215,838.00—on a loan of just \$100,000. (Usually, the bank will round up the regular monthly payment to the next nearest cent, with the consequence that the very last payment will be slightly less than the usual monthly payment.)

It is useful to note that for a 30-year mortgage at 6%, the monthly payment is almost exactly 0.6% of the amount of the loan—here, the payment is \$599.55 and 0.6% of \$100,000 is \$600. Hence, for lower interest rates, the monthly payment will be a lower percentage of the amount of the loan, though the relationship is not proportional. ■

Self Check 3

What is the monthly payment on a 30-year mortgage for $100,000 if the interest rate is 3.5% per year? ■

Algebra Review Appendix
Natural and Fractional Exponents

We put the idea behind the calculation in Example 3 into a more general equivalence: *Paying off a conventional loan is like saving.* You can think of paying off the loan as making payments to a savings account that earns interest at the same rate as the loan. At the end of the loan term, the savings balance will exactly equal the principal and interest on the loan. Let the loan amount be P, the effective interest rate per compounding period be i, the number of compounding periods be n, and the loan payment be d. We equate the principal and interest on the loan (from the compound interest formula) with the savings balance (from the savings formula):

$$P(1+i)^n = d\left[\frac{(1+i)^n - 1}{i}\right]$$

The quantity P is sometimes called the *present value of an annuity* of n payments of d, each at the end of a compounding period with interest i per period. This terminology is used in the financial mode of some calculators, such as the TI-83.

Algebra Review Appendix
Solving for One Variable in Terms of Another

Solving the above equation for d requires a little algebra. To make things simpler, let $b = (1+i)^n$, so

$$Pb = d\left[\frac{b-1}{i}\right]$$

Then

$$d = P\left[\frac{b}{\frac{b-1}{i}}\right] = P\left[\frac{bi}{b-1}\right]$$

Now divide numerator and denominator by b, getting

$$d = P\left[\frac{i}{1-b^{-1}}\right]$$

Substituting $(1+i)^n$ back for b, we get the usual form of the **amortization payment formula.**

Amortization Payment Formula RULE

A conventional loan amount P

- at a nominal annual rate of interest rate r
- with m compounding periods per year (so interest rate $i = r/m$ per compounding period)
- for t years (hence $n = mt$ compounding periods)

can be paid off by uniform payments at the end of each compounding period in the amount

$$d = P\left[\frac{i}{1-(1+i)^{-n}}\right] = P\left[\frac{\frac{r}{m}}{1-\left(1+\frac{r}{m}\right)^{-mt}}\right]$$

The spreadsheet function PMT can be used to check your calculations of the payment amount; see Spotlight 21.3 (page 888) for details and examples.

EXAMPLE 4 Repaying Your Student Loan

The standard repayment option for federal direct student loans is repayment over 10 years with a minimum monthly payment of $50 at the end of each month, beginning six months after you graduate, and a minimum monthly payment of $50. For the $5500 student loan of Example 1 (page 910), if you didn't pay the interest over the 51 months, what will your monthly payments be on the $6502.83 that you will owe at the start of repayment?

With the amortization payment formula, it's easy to figure out your monthly payment. We have $P = \$6502.83$, monthly interest rate $i = \frac{r}{m} = \frac{0.0429}{12} \approx 0.003575$, and $n = mt = 12 \times 10 = 120$ months for the payback. We find the payment d as

$$d = P\left[\frac{\frac{r}{m}}{1 - \left(1 + \frac{r}{m}\right)^{-mt}}\right] = \$6502.83\left[\frac{\frac{0.0429}{12}}{1 - \left(1 + \frac{0.0429}{12}\right)^{-12\times 10}}\right] \approx \$66.74$$

So your monthly payment will be $66.74. (That is for this loan; you may owe still more per month for loans for your other years in college.) Hence, over the lifetime of the loan, you will pay almost 120 × $66.74 = $8008.80. We say "almost," because your last payment will differ by a few cents. Of the total, $8008.80 − $6502.83 = $1505.97 is interest during the 10-year payback period, plus the $1002.83 interest accrued during the deferment and grace periods (calculated in Example 1), for a total of $2508.80 in interest, on an original loan principal of $5500.

In fact, the amount you receive is reduced by an origination fee of 1.073% of the loan amount, or 0.0173 × $5500 = $95.15; so altogether, you will pay almost 50% as much in interest and fees as the original principal. If you were to stretch your payments over more years (permissible in some circumstances), an even greater proportion would be interest.

You can check these amounts, and those for your own loans, at studentloans.gov/myDirectLoan/mobile/repayment/repaymentEstimator.action and (with more options and details) at www.finaid.org/calculators/loandiscountanalyzer.phtml.

Self Check 4

Suppose that you graduate with the average amount of student loans to repay, $25,000. What would your monthly payment be if the interest rate on all the loans is 4.29% and the repayment term is the standard 10 years?

EXAMPLE 5 Buying a Car

You decide to buy a new Wheelmobile car. After a down payment, you need to finance (borrow) an additional $12,000. After you compare interest rates offered by the car dealership, local banks, and your credit union, the lowest monthly payment you can find is for an 84-month loan at 5.9% from the credit union, compounded monthly. What is your monthly payment?

We have $P = \$12{,}000$, monthly interest rate $i = 0.059/12 \approx 0.00491667$, and $n = 84$. Using the amortization payment formula, we have

$$d = P\left[\frac{\frac{r}{m}}{1 - \left(1 + \frac{r}{m}\right)^{-mt}}\right] = \$12{,}000\left[\frac{0.00491667}{1 - (1 + 0.00491667)^{-84}}\right] \approx \$174.73$$

How much interest do you pay? You make payments totaling 84 × $174.73 = $14,677.32, so the interest is $14,677.32 − $12,000 = $2,677.32.

If you had bought a Plushmobile instead, with $24,000 to finance, you would have borrowed twice as much, and your monthly payment would have been twice as much.

Self Check 5

What would be your monthly payment on the $12,000 loan, and the total interest paid over the term of the loan, if you got a 60-month loan instead, at the advertised rate of 2.99%? ■

Suppose that you want to buy a car, and you know how much you want to afford for a monthly payment and how long a loan term you would agree to. How much car can you afford? We turn the the amortization payment formula around to solve for P instead of for d:

$$P = d\left[\frac{1-(1+i)^{-n}}{i}\right]$$

EXAMPLE 6 How Much Car Can You Afford?

Suppose that you feel comfortable with a monthly car payment of $200, you don't want to pay on the car for more than four years, and you can get financing at 1.99%. How large a loan can you get?

$$P = d\left[\frac{1-(1+i)^{-n}}{i}\right] = \$200\left[\frac{1-\left(1+\frac{.0199}{12}\right)^{-48}}{\frac{.0199}{12}}\right] \approx \$9221$$

We have not taken into account the costs of insurance, registration, parking, gasoline, or tolls. ■

Self Check 6

Suppose that you can afford $250 per month for the car payment (apart from insurance, gas, etc.), you can get a 0% car loan, and you are willing to pay over 7 years. How much car can you afford? (Careful! The amortization payment formula does not apply. Why not?) ■

A car loan is often for 48 or 60 months; but when you buy a home, you usually borrow a great deal more money and pay it off over a much longer period. The usual term for a home mortgage is 30 years.

EXAMPLE 7 A 30-Year Mortgage on a Median-Priced Home

Suppose that you are a family with the U.S. median household income of about $53,000 and you want to buy a median-priced home for $205,000, with a 30-year fixed-rate, federally insured mortgage at 3.25% (the data are for the beginning of 2015). Recall that the median (discussed in Section 5.4, pages 196–200) means that half are below and half are above. Suppose that you can make a down payment of only $7000 (just above the minimum of 3.5% required), plus pay closing costs of about $4000. Can you afford such a home?

Mortgage lenders have "affordability" guidelines that suggest that a family cannot afford to spend more than 31% of its monthly income on housing. Thus, by these guidelines, you can afford $0.31 \times \$53{,}000/12 \approx \$1370$ per month.

What is the monthly payment on the loan? The principal is $P = \$205{,}000 - \$7000 = \$198{,}000$, the monthly interest rate is $i = 0.0325/12 \approx 0.00270833$, and $n = 360$ months. The amortization payment formula gives a monthly payment of

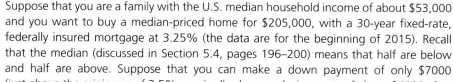

$$d = P\left[\frac{\frac{r}{m}}{1-\left(1+\frac{r}{m}\right)^{-mt}}\right] = \$198{,}000\left[\frac{\frac{0.0325}{12}}{1-\left(1+\frac{0.0325}{12}\right)^{-12\times 30}}\right] \approx \$862$$

Well, that sounds good. Unfortunately, there is more to the mortgage than just the amount needed to amortize the loan. Your payment will also have to cover real estate taxes, mortgage insurance, and homeowner's insurance on the property. On a \$198,000 home, these may add \$500 or more to the monthly payment, which will then total about \$1362.

It's really close! So, the median household may just barely be able to afford the median-priced home (depending on the area of the country), even while interest rates remain at record lows.

A payment on an amortized loan includes both the current interest and a portion toward repaying the principal. You are "building equity" in a house as you pay off the mortgage.

Equity DEFINITION

Equity is the amount of principal of a loan that has been repaid.

The equity is the sum of the payments made minus the part that went to interest.

EXAMPLE 8 Home Equity

My wife's parents sold their house in rural Minnesota to move to the town where we live. They had bought their house in 1980 with a 30-year mortgage for \$100,000 at an 8% interest rate. After 22 years, how much equity did they have in the house—that is, how much of the principal had been repaid? And how much did they still owe on the house?

Michael Siluk/The Image Works

What may shock you (and disappointed them) is that when they sold their house in May 2002—after 269 months of payments, almost exactly three-quarters of the 30 years of the mortgage—they had only \$50,000 in equity (hence, they still owed \$50,000 on the house) but had already paid \$147,000 in interest. *Three-quarters of their payments had gone to interest.*

We can use the amortization payment formula to determine just how much equity they had after 269 months of payments, but first we need to determine their monthly payment. We see $P = \$100{,}000$, $n = 360$ months, and $i = \frac{0.08}{12}$ monthly interest, getting $d = \$733.76$.

Now we use the formula again, this time "in reverse." Knowing i and d, we find out how much of the loan would have been paid off by the remaining $360 - 269 = 91$ payments of \$733.76:

$$P = d\left[\frac{1-(1+i)^{-n}}{i}\right] = \$733.76\left[\frac{1-\left(1+\frac{0.08}{12}\right)^{-91}}{\frac{0.08}{12}}\right] \approx \$49{,}940$$

This is how much my parents-in-law had yet to pay, so their equity was $\$100{,}000 - \$49{,}940 = \$50{,}060$. (The above formula would not apply if they had made larger or additional payments.)

Self Check 7

Is it true for any mortgage that after three-quarters of the term of the loan, only half of the loan will be paid off?

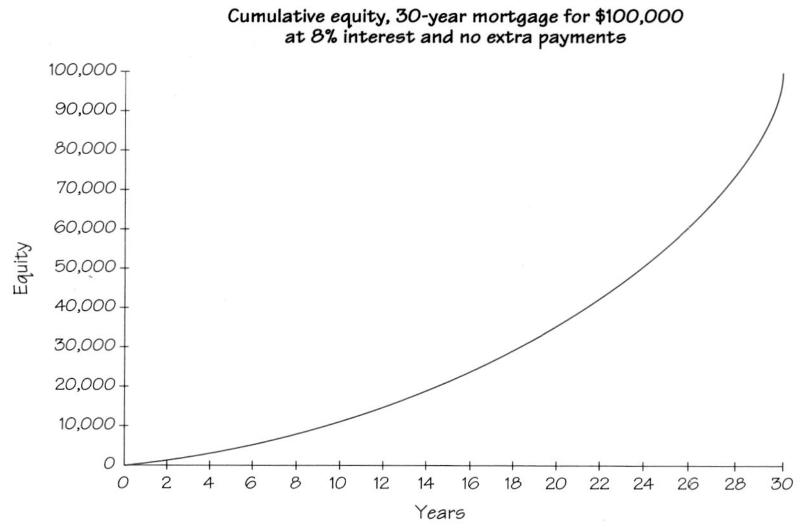

Figure 22.1 Equity grows almost exponentially, especially in the later years of a mortgage.

Figure 22.1 and Table 22.1 show that equity builds up very slowly at first but rapidly later. (The values shown do not take into account any increase or decrease in the value of the house itself, the effect of inflation, or the effect of making higher monthly payments or other extra payments.) In fact, the amount of principal in a payment grows by a factor of $(1 + i)$ from one payment to the next, so the equity at any point is the sum of a geometric series (discussed in Section 21.5, pages 882–887) whose common ratio is $(1 + i)$.

TABLE 22.1 Equity in a 30-Year Mortgage for $200,000 at 5% Interest

End of Year	1	2	3	4	5	10	15	20	25	30
Equity ($\$ \times 10^3$)	3	6	9	13	16	37	64	99	143	200

When you buy a home, you have several options:

- A conventional 30-year mortgage
- A conventional 15-year mortgage
- An adjustable-rate mortgage (ARM) for either length of time but with an interest rate that can vary

You might expect the payment on a 15-year mortgage to be double that of a 30-year mortgage. On the contrary, the payment is only 55% more (for a 4% mortgage) to 26% more (for a 9% mortgage). This range includes the prevailing mortgage rates over the past 20 years. Moreover, over the course of a $200,000 mortgage at 5%, you would pay $187,000 in interest over 30 years but only $85,000 over 15 years. At 9%, the interest totals are $380,000 versus $166,000.

(Some financial counselors advise taking a 30-year mortgage and making extra payments when you can afford them, rather than incurring the higher payment obligation of a 15-year loan, on which, if you encounter tight personal financial circumstances, you might not be able to make the payments.)

In Spotlight 22.2, we discuss what we did in our own circumstances and mention other options.

What We Did with Our House, and What Else You Could Do

SPOTLIGHT 22.2

We bought our house in 1992. We were offered a choice between a 30-year fixed-rate mortgage at 8.375% and a 30-year ARM at 6.875% whose rate could be raised (or lowered) by up to 2% every year. When we asked, we were also quoted slightly lower rates for corresponding 15-year mortgages.

We were planning to stay in the house much longer than the median of 5 years, and we were concerned that inflation might force the ARM considerably higher. Also, we did not want the obligation of the higher payments of a 15-year mortgage, in case our circumstances changed (such as through job loss or death). Some loans provide for penalties for paying off the loan early, but in our case (thanks to state law), there was no penalty for making extra payments (if we could afford them).

We chose the 8.375% fixed-rate 30-year mortgage (and made some extra payments). People in other circumstances, or with a different tolerance for risk, would no doubt have decided otherwise. Had we been sure then that interest rates would not go higher in the 1990s, we would have gone for the ARM. But hindsight is always better than foresight. Homeowners with mortgage interest rates such as ours later refinanced at much lower prevailing rates, near 5% for a fixed-rate 30-year mortgage.

Currently, about one-third of borrowers take ARMs rather than fixed-rate mortgages. Newer mortgage "products" include interest-only mortgages and shared appreciation mortgages (SAMs). With an interest-only ARM, payments are (just slightly) lower than for a conventional 30-year mortgage, but you accumulate no equity (at least, not by paying off the loan; the market value of the house may rise). After five to seven years, you start also paying off the principal—which means that your payments go up then. In some such loans, the interest rate—and your payments—fluctuate as frequently as every month.

In a SAM, interest payments are lower or nonexistent, but the lender receives a portion of any appreciation (rise in value) when the house is sold. In a nationally reported instance in 2003, a single mother received a no-interest SAM loan to finance the $30,000 down payment on a $223,000 house in Pleasanton, California, through the city's affordable housing program. Four years later, she sold the house for $385,000, and the "affordable housing" lenders got 60% of the $162,000 appreciation, or $97,000. She herself realized $65,000 (minus the cost of the sale) but complained bitterly, saying that she would have been better off to have put the loan on her credit card! Critics have termed SAMs an urban form of sharecropping.

Because very few mortgages are held for the full term, it is useful to compare the status of mortgages after 5 years, the median length of time that Americans remain in a home. Table 22.2 shows the equity after 5 years for a variety of interest rates. For a 30-year mortgage, the equity after 5 years may be less than the cost of selling the home through a realtor, which is normally 6% to 7% of the sale price. Of course, the resale value of the home also may be higher after 5 years.

A mortgage with an interest rate that can vary is called an **adjustable-rate mortgage (ARM).** Usually such mortgages have a substantially lower interest rate (hence, a lower payment) than a fixed-rate mortgage. The ARM's interest rate may go up or down with interest rates in the economy. Normally, the rate can be raised or lowered only every year or two, and then by a limited percentage. An ARM may

TABLE 22.2 Equity (in thousands of dollars) on a $200,000 Mortgage After 5 Years

Term (years)	Interest Rate					
	4%	5%	6%	7%	8%	9%
15	54	51	48	45	42	40
30	19	16	14	12	10	8

be attractive if you plan to pay off the mortgage after only a few years, or because it allows for lower payments or buying a more expensive home, or because you do not plan to keep the home long (hence, you would be selling before the interest rate could rise substantially).

Does it pay to buy a house or apartment? Apart from the joys of ownership, you need to take into account the up-front expenses of closing costs (perhaps $3000), plus the back-end expense of selling the house (usually 6% to 7% if through a realtor, so say $12,000). Consulting the table, and imagining that you get a loan at 5%, you might think that you would finally be in the black on your house as an investment after 6 years (and you have had the pleasure of living in it rent-free!).

However, we have not yet taken into account the ongoing expenses of maintenance, repairs, insurance, and real estate taxes (perhaps $5000 to $10,000 per year). Of course, if your house is rising in value by, say, 5% ($10,000) per year or more, it's a different story. The growth of home ownership, which rose to 73% before the bursting of the housing bubble in 2007, depended on just such a steady rise in value. Renting may be more attractive if you anticipate moving in just a few years; each year, one-sixth of Americans move.

Spotlight 22.3 details the role of mortgage loans in the Great Recession of 2008 and subsequent years.

The Mortgage Crisis

SPOTLIGHT 22.3

Late 2007 saw the development and widening consequences of what has become known as the "mortgage crisis." To understand what that was, why it took place, and how it will have widespread effects for some time, you need to know what happens when you get a mortgage to buy a house compared with what used to happen.

In the "good" old days, you would go to a local bank (or savings and loan, or credit union). If you proved "creditworthy"—meaning that after careful consideration, the personnel felt that you could repay the loan—the bank would lend you its money, raised from its depositors. The interest rate depended on your credit rating and down payment. You paid back the loan at a fixed rate of interest, usually over 30 years. If interest rates went down, you could refinance the loan at a lower rate by taking out a new loan to pay off the old one; if rates went up, your payments would not rise.

Meanwhile, the value of your house usually went up 5% to 10% per year, your income went up, and your payments stayed the same. (In fact, if you have only 10% equity in your house and it goes up in value 10% in one year, you have made 100% on your investment! This kind of "leveraging" can make real estate investments very profitable—so long as prices keep rising.)

What changed? Efforts to extend home ownership to a wider proportion of the population resulted in banks making more "subprime" loans (loans to people with poorer credit histories who are less likely or able to repay them), with lower down payments but higher rates of interest (because of the greater risk). Some of those loans were "predatory"—lending at high rates of interest to people who could not possibly make the payments.

Certainly, real estate speculation played a role, as did greed. House prices doubled from 2000 to 2006 (the "housing bubble"), but median income adjusted for inflation remained stagnant—and then fell with increasing unemployment, making it increasingly hard to afford houses. Banks countered with ARMs and interest-only mortgages. They also maximized their income with upfront charges ("origination fee" plus "points" paid by the buyer to lower the interest rate); and at the same time, they minimized their risk by immediately "flipping" the mortgages (selling them to bigger banks or other investors, to whom buyers would then make their payments). All these factors led to banks making more loans that were riskier.

The Mortgage Crisis

What went wrong? Interest rates rose, and people with ARMs saw their payments rise beyond their ability to pay. At the same time, the "pyramid" of housing prices could not continue with the higher mortgage rates, so housing prices fell. When your house becomes "underwater"—worth less than the balance remaining on the mortgage—you might be better off just walking away from it (especially because you build up almost no equity in the first few years of the mortgage). Mortgage defaults lowered the value of investments in bundles of mortgages.

Houses are worth perhaps hundreds of billions of dollars less than a few years ago, and big investment banks still have mortgages on their hands that are worth hundreds of billions of dollars less than they paid for them. That means that for some time to come, banks have less money to lend and can (indeed, must) demand more creditworthy clients and higher rates of interest.

How does all this affect you? Financial institutions have less money to lend for any purpose—buying a home, buying a car, starting a business, etc.—hence the "credit crunch" following the mortgage crisis. We can only hope that the worst will be over before you are in the market for a house, car, or business loan.

Ian Waldie/Bloomberg via Getty Images

Where you won't see a direct effect is in the Consumer Price Index (CPI) (discussed in Section 19.5, on pages 785–786, and Section 21.6, on pages 891–893), which is based on the rental value of houses, not their prices. If the doubling of housing prices in 2000–2006 had been taken into account, the CPI would have risen by 5% per year rather than 3%.

22.4 Other Loans

Discounted Loans

A federal student loan is in fact a hybrid loan. It is a combination of a conventional loan, in which the principal is amortized through monthly payments, and a discounted loan, in which in effect some of the interest is paid in advance as a "loan origination fee."

A similar circumstance exists with mortgage loans for which at the time of purchase the borrower can pay "points" (a percentage of the house price) to obtain a lower interest rate.

Consequently, we explore **discounted loans,** as well as the related add-on loans.

Discounted Loan DEFINITION

A **discounted loan** is a loan in which the borrower receives the principal minus the interest but must pay back the entire principal, usually with equal payments.

For a discounted loan, the interest is computed as simple interest. Instead of getting the principal P, the borrower gets the *proceeds,* the principal minus the interest: $P - I$. However, the borrower must pay back the entire principal P over the term of the loan; if there are N payments, usually the amount of each payment is P/N. (*What is discounted is not the cost of the loan but how much the borrower gets!*) The interest is based on the entire P; the borrower never has the use of that much, yet pays—up front—interest on the entire amount P.

We first consider a simple imaginary situation to show how a discounted loan works. Suppose you need to borrow \$7200 to buy a car. So you go to Joe's Friendly Loan Service at a nearby off-campus location, and Joe offers you a discounted loan, at an interest rate of 5% per year (but he's *really* friendly), with all of the interest to be paid up front and repayment of the principal to be in monthly installments over 3 years. The interest up front comes to $\$7200 \times 0.05 \times 3 = \$1080$. How much is your monthly payment? It's just $\$7200/36 = \$200$ per month.

There's just one (big) problem: With a discounted loan, Joe gives you the proceeds of the loan, which is \$7200 minus the interest of \$1080, or \$6120. That leaves you \$1080 short of the cost of the car!

Joe suggests that maybe you want to borrow a principal P large enough so that the proceeds comes to \$7200. How much should the amount of the loan be then? With the 15% interest subtracted, the proceeds are 85% of the principal P. You need the proceeds to be $\$7200 = P(0.85)$, or $P = \$7200/0.85 \approx \$8470.59$. Then you get \$7200, Joe gets a slightly bigger cut up front ($0.15 \times \$8470.59 = \$1270.59$), and your payment is $\$8470.59/36 \approx \$235.29$.

In general, for a principal P, annual interest rate r, and a term of t years, the proceeds are $P/(1 - rt)$.

Discounted Loan Formula RULE

For a principal P loaned at annual interest rate r for t years, the monthly payment is

$$d = \frac{\frac{P}{1-rt}}{12t} = \frac{P}{12t(1 - rt)}$$

Being a law-abiding (mostly) loan shark, Joe knows that he is supposed to tell you the APR for your loan. To do that presents a problem: The APR is supposed to be the number of compounding periods per year times the rate of interest per compounding period. But Joe isn't doing any compounding on your loan—except for the interest paid up front, all of your payments go toward paying back the principal.

Joe is confronting an instance of a larger problem: How do you compare different loans if up-front fees are involved?

We know how to compare some kinds of different loans for the same principal, if there are no additional fees involved:

- If the loans are over the same term, the loan with a lower interest rate is better for the borrower.
- If the loans are over different terms, as for a 15-year mortgage versus a 30-year mortgage, or a 48-month car loan versus a 72-month car loan, the borrower has to balance higher monthly payments for a shorter loan against paying more interest in the long run on a longer loan.

But for two loans with the same term, which of the following is less expensive?

- A loan with a 5% interest rate but 4% in fees up front
- A loan with a 4% interest rate but 5% in fees up front

The answer depends in part on the time value of money—how much more it is worth to you to pay a dollar at some time in the future compared with paying it today. That in turn may depend on the rate of inflation (see page 914) and on what you could gain by investing the money instead. In other words, it depends on the *net present value* of the stream of payments, given an investment interest rate and a rate for inflation. We cannot go into the details of such considerations here; that would be taken up in a course in mathematics of finance.

In fact, people tend to compare loans on the basis of the monthly payments rather than the interest rates. What we can do is take a loan with up-front fees and calculate the interest rate for an equivalent no-fee loan.

EXAMPLE 9 Equivalencing to a No-Fee Conventional Loan

In the case of Joe financing your car purchase, we can imagine that you are not borrowing \$8470.59 and making an immediate payment of \$1270.59, but simply borrowing \$7200 and then repaying \$7200 as a conventional loan at \$235.29 per month over 36 months. The question is, what APR does that amortization correspond to?

We know d, P, and n and need to solve the amortization payment formula for the annual interest rate r, which is the APR:

$$d = P\left[\frac{\frac{r}{12}}{1-\left(1+\frac{r}{12}\right)^{-n}}\right]$$

That doesn't look easy to do! And it isn't—it cannot be done; it's algebraically impossible to separate the r out by itself. However, you can find the value of r by putting the formula into a spreadsheet and trying successive approximations, or else by entering =12*RATE (36, 235.53, −7200) into the spreadsheet. (See Spotlight 21.3, page 888, for details about using a spreadsheet for financial calculations.) You find $r \approx 10.87\%$. Just don't miss any payments. ■

Self Check 8

Joe's competitor Sam has a counteroffer: a discounted loan at 5%, with proceeds of \$7200, but repaid with monthly payments over 4 years (instead of 3 years). How much interest do you pay, what is the amount of your monthly payment, and what is the APR of the corresponding conventional loan? ■

EXAMPLE 10 Your Federal Student Loan

We can do a similar analysis for a federal student loan. For such a loan, in effect you pay 1.073% interest in advance as a "loan origination fee" and then have a conventional loan for the principal at a set interest rate of 4.29%. For simplicity, we assume that you make the monthly interest payments until your repayment starts.

So suppose you need \$10,000. You need to borrow \$10,000/(1 − 0.01073) = \$10,108.46. To pay off the \$10,108.46 loan over 120 months, your payment, either from the amortization payment formula or from =PMT (0.0429/12,120,−10108.46), is \$103.74.

Although you are paying back as if the loan were for \$10,108.46, you got only \$10,000. We can use a spreadsheet to calculate the APR for a corresponding no-fee loan as in Example 9 above via =12*RATE (120, 103.74, −10000) and arrive at 4.521%.

The effective annual rate (EAR) of the corresponding no-fee loan is

$$\left(1 + \frac{r}{12}\right)^{12} - 1 = \left(1 + \frac{0.048923327}{12}\right)^{12} - 1 \approx 4.62\%$$

Self Check 9

Does the EAR depend on how much you borrowed?

The situation with mortgages is similar. Banks often offer choices of mortgages with various combinations of interest rates and "points." A **point** is 1% of the mortgage amount. You may "pay points" to "buy down" the interest rate for the mortgage to a lower rate. The cost of the points is paid up front to the bank at the closing of the house sale, so it is like interest on a discounted loan.

EXAMPLE 11 Paying Points

Suppose that you are offered a 30-year mortgage at 3.5% if you are willing to pay 2 points. What is the APR of the corresponding conventional loan, and what is its EAR?

The amount of the mortgage is not specified; in fact, to answer the question, we don't need to know that. But to make the example concrete, suppose that the mortgage is for \$100,000.

To have the full \$100,000, you need to borrow \$100,000/(1 − 0.02) ≈ \$102,041. To pay off that amount over 360 months at 3.5% annual interest, the payment can be found to be \$458.21 from =PMT (0.035/12,360, −102041). We apply this payment to determine what interest rate it would correspond to in repaying a loan of only \$100,000 by finding =12*RATE (360, 458.21, −100000) and we arrive at an APR of 3.663%. The corresponding EAR is

$$\left(1 + \frac{r}{12}\right)^{12} - 1 = \left(1 + \frac{0.036633054}{12}\right)^{12} - 1 \approx 3.725\%$$

Self Check 10

Here is an extreme example: Suppose that you are offered a 30-year mortgage for \$360,000 at 0% if you are willing to pay 50 points. You don't have the 50 points (= \$180,000), so you need to borrow a total of \$540,000. What is the APR of the corresponding conventional loan, and what is its EAR?

In equivalencing a discounted loan to a no-fee loan, we are spreading some up-front fees over the term of the repayment. What is being equated between the two loans is the total amount of interest and fees that would be paid over the entire term of the loan. However, since the point of making payments over time is to make possible a large purchase for which you do not have enough money now, it does not make sense (as in Self Check 10) to pay a large amount of interest up front.

That is especially true if there is some likelihood of your paying the loan off early. For example, it does not pay to "pay points" if you expect to move in a few years and hence pay off the mortgage early.

Figure 22.2 on the next page shows parts of a model mortgage loan estimate disclosure form created by the government's Consumer Financial Protection Bureau, as directed by Congress.

The form includes new disclosure requirements effective from August 2015. Although the Bureau believes that including almost all of the up-front costs that the consumer must pay to get the loan as part of the "finance charge" (and the APR) would be more meaningful and beneficial for the consumer, it did not change policy to do so because of potential cost to the mortgage industry. Such a change would discourage lenders from shifting costs to fees that do not have to be declared as part of the finance charge. So the policy for the foreseeable future is "some fees in, some fees out."

Add-on Loans

Another type of consumer loan is the **add-on loan,** which is often used for payday loans and sometimes for other purchases, such as furniture or a car. Such a loan has the great convenience of easy calculation of the interest and payments.

Add-on Loan DEFINITION

An **add-on loan** is a loan in which the borrower receives the principal, the interest is calculated in advance as simple interest, and the borrower pays back the principal plus interest in equal installments.

You borrow an amount P to be repaid in t years; the interest is simple interest at an annual rate r (= $100r$%), for a total of $I = Prt$. You must pay $P + I = P(1 + rt)$ in installments; with n payments, each payment is

$$d = \frac{P(1 + rt)}{n}$$

In effect, you pay $\frac{1}{n}$th of the principal and $\frac{1}{n}$th of the total interest with each payment. With monthly installments, we have $n = 12t$ and the formula becomes

$$d = \frac{P(1 + rt)}{12t}$$

Add-on Loan Formula RULE

For a principal P loaned at annual interest rate r for t years, the monthly payment is

$$d = \frac{P(1 + rt)}{12t}$$

Figure 22.2 Portions of a model loan estimate form for a mortgage for $162,000 on a house with sale price $180,000. The borrower had already put down a deposit of $10,000 toward a 10% down payment of $18,000. Note under the second "tab" the additional monthly payments for mortgage insurance and for real estate taxes ("Estimated Escrow") and under J the amount of Total Closing Costs. The APR appears to be based on the monthly payment of $761.78 on a no-fee mortgage for $154,444—that is, the $162,000 principal minus $8054 closing costs except for $85 taxes and $413 initial escrow.

Loan Terms

		Can this amount increase after closing?
Loan Amount	$162,000	NO
Interest Rate	3.875%	NO
Monthly Principal & Interest *See Projected Payments below for your Estimated Total Monthly Payment*	$761.78	NO
		Does the loan have these features?
Prepayment Penalty		NO
Balloon Payment		NO

Projected Payments

Payment Calculation	Years 1–7	Years 8–30
Principal & Interest	$761.78	$761.78
Mortgage Insurance	+ 82	+ —
Estimated Escrow *Amount can increase over time*	+ 206	+ 206
Estimated Total Monthly Payment	$1,050	$968

		This estimate includes	In escrow?
Estimated Taxes, Insurance & Assessments *Amount can increase over time*	$206 a month	☒ Property Taxes	YES
		☒ Homeowner's Insurance	YES
		☐ Other:	

Closing Cost Details

Loan Costs

A. Origination Charges	**$1,802**
.25% of Loan Amount (Points)	$405
Application Fee	$300
Underwriting Fee	$1,097
B. Services You Cannot Shop For	**$672**
Appraisal Fee	$405
Credit Report Fee	$30
Flood Determination Fee	$20
Flood Monitoring Fee	$32
Tax Monitoring Fee	$75
Tax Status Research Fee	$110
C. Services You Can Shop For	**$3,198**
Pest Inspection Fee	$135
Survey Fee	$65
Title – Insurance Binder	$700
Title – Lender's Title Policy	$535
Title – Settlement Agent Fee	$502
Title – Title Search	$1,261
D. TOTAL LOAN COSTS (A + B + C)	**$5,672**

Other Costs

E. Taxes and Other Government Fees	**$85**
Recording Fees and Other Taxes	$85
Transfer Taxes	
F. Prepaids	**$867**
Homeowner's Insurance Premium (6 months)	$605
Mortgage Insurance Premium (months)	
Prepaid Interest ($17.44 per day for 15 days @ 3.875%)	$262
Property Taxes (months)	
G. Initial Escrow Payment at Closing	**$413**
Homeowner's Insurance $100.83 per month for 2 mo.	$202
Mortgage Insurance per month for mo.	
Property Taxes $105.30 per month for 2 mo.	$211
H. Other	**$1,017**
Title – Owner's Title Policy (optional)	$1,017
I. TOTAL OTHER COSTS (E + F + G + H)	**$2,382**
J. TOTAL CLOSING COSTS	**$8,054**
D + I	$8,054
Lender Credits	

Calculating Cash to Close

Total Closing Costs (J)	$8,054
Closing Costs Financed (Paid from your Loan Amount)	$0
Down Payment/Funds from Borrower	$18,000
Deposit	– $10,000
Funds for Borrower	$0
Seller Credits	$0
Adjustments and Other Credits	$0
Estimated Cash to Close	$16,054

Comparisons

	Use these measures to compare this loan with other loans.
In 5 Years	$56,582 Total you will have paid in principal, interest, mortgage insurance, and loan costs. $15,773 Principal you will have paid off.
Annual Percentage Rate (APR)	4.274% Your costs over the loan term expressed as a rate. This is not your interest rate.
Total Interest Percentage (TIP)	69.45% The total amount of interest that you will pay over the loan term as a percentage of your loan amount.

EXAMPLE 12 **Add-on Loan**

Suppose that you need to borrow $8000 to buy a used car. The dealer offers you a 5% add-on loan to be repaid in monthly installments over 4 years. This sounds like a much better deal than the 8% conventional loan that you can get at the credit union. How much is your payment *d* on the dealer's add-on loan?

The total amount to be paid over the 4 years (48 months) is

$$P + I = P(1 + iy) = \$8000 \times (1 + .05 \times 4) = \$8000 \times 1.2 = \$9600$$

for a monthly payment of an even $200.00.

Self Check 11

What would be the monthly payment on the loan from the credit union at 8% over 48 months?

With an add-on loan, everything sounds simple and straightforward (especially the calculation). The interest is calculated on the entire principal; however, while you are slowly paying back the principal, you do not have the use of the whole amount for the entire loan period. (In fact, you have the use of the full principal for just one period!) You do indeed have the use of the car! But the net value that you have at any point is the value of the car minus the amount of principal already repaid (we neglect depreciation of the car). It is on this net value, and the amount of interest, that a "true" interest rate could be calculated.

22.5 Annuities

We encountered in Section 21.5 (page 885) the concept of an **annuity.**

Annuity DEFINITION

An **annuity** is a sequence of periodic payments.

An annuity is much like a mortgage but from the opposite point of view: Instead of making a payment every month, you receive a check every month! In purchasing an annuity, you make a loan to a company in exchange for periodic "payback" payments.

We restrict our discussion to annuities for which payments are made at the end of each period and for which the compounding period is the same as the payment period.

The basis for the annuity payments can be a lump sum or even the winnings from a lottery. For example, winners of lotteries are often offered the choice of receiving either the jackpot amount paid as an annuity over a number of years or else a smaller lump sum to be paid immediately. The cost to the lottery administration is the same. If the winner wants an annuity, the administration uses the lump sum to buy either government bonds that pay at the specified rate or else an equivalent annuity from an insurance company.

You can think of the government or the insurance company as borrowing the lump sum and paying it back via the annuity payments. In effect, the government or the insurance company amortizes the lump sum over the term of the annuity.

Annuities can vary in the amount of the payments (level or graduated) and the term (how long payments continue).

EXAMPLE 13 Winning a Lottery

Tim Boyle/Getty Images

The world record lottery jackpot with a single winner was $370.9 million in May 2013. The winner had the option of receiving the "annuity value" of the prize, $590.5 million, in 30 graduated annual installments, the first payment being right away. To take into account inflation, each installment was to be 4% larger than the previous one. However, the winner chose instead the instant lump sum of $370.9 million. Well, she was 84 years old at the time ... (but had she chosen the annuity, after her death any remaining payments would have been made to her estate). What was the interest rate of the annuity?

We don't offer a solution here. In the past, lotteries used to offer annuities with level payments over 25 to 30 years. For such an annuity, we could use the amortization formula and software to determine the effective interest rate on which the payments are based. Now the annuities offered by lotteries usually feature graduated payments (4% or 5% per year), a situation that offers more complications than we want to go into here. Such calculations are commonly handled by actuaries (see Spotlight 22.4).

What Is an Actuary?

SPOTLIGHT 22.4

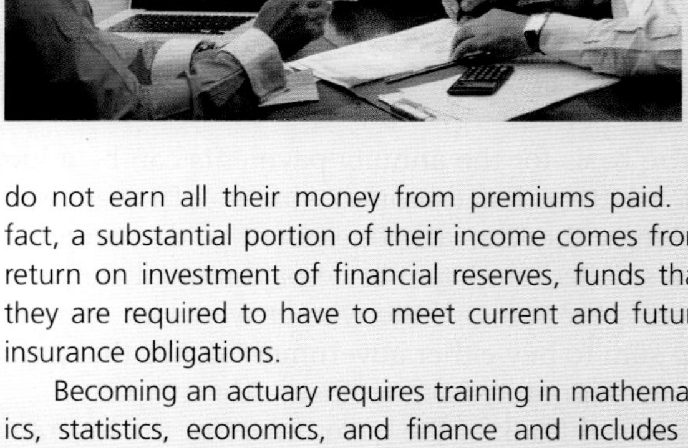

Getty Images/Images Bazaar

The Truth in Savings Act and the Truth in Lending Act specify that the APY for savings and the APR for loans must be calculated "according to the actuarial method."

Actuaries are financial experts who manage risks. They assess the costs and likelihood of risks such as tornadoes, floods, auto accidents, and deaths. Actuaries are crucially involved in setting the premiums for insurance against risks. Their calculations take into account historical rates—such as the percentage of female 85-year-olds who live to be 86, or the percentage of unmarried male drivers under age 25 who have auto accidents—and project those rates and the accompanying costs into the future.

Other actuaries concentrate on setting up and evaluating healthcare plans or pension and fringe benefit plans. For example, the city of Beloit, Wisconsin, hired a consulting actuary to estimate the current and future costs of free lifetime medical benefits to families of police and firefighters.

Another major activity of actuaries is managing return on investment. Contrary to popular belief, insurance companies (particularly life insurance companies) do not earn all their money from premiums paid. In fact, a substantial portion of their income comes from return on investment of financial reserves, funds that they are required to have to meet current and future insurance obligations.

Becoming an actuary requires training in mathematics, statistics, economics, and finance and includes a sequence of professional exams taken over several years.

Self Check 12

Why would lotteries offer graduated payments?

A more common situation than winning a jackpot is saving toward retirement and then at retirement purchasing an annuity. One possibility is an annuity for a fixed number of years (perhaps as long as you expect to live), called an **annuity-certain.**

EXAMPLE 14 How Much Do You Need to Retire?

Suppose that your father is ready to retire at 65 and wants to purchase an annuity that pays $5000 per month for 25 years. The insurance company offering the annuity is willing to assume that the long-range steady interest rate will be 3% per year compounded monthly (and that rate also takes into account its costs and profit). What should be the cost of the annuity—that is, how much should your father expect to pay for such a stream of income?

We apply the amortization formula "in reverse." We know the amount of the monthly payment *d* and we need to find the principal *P*.

We apply the amortization formula with $d = \$5000$, $r = 0.03$, $m = 12$, and $t = 25$, to find the amount *P*:

$$P = d\left[\frac{1-\left(1+\frac{r}{m}\right)^{-mt}}{\frac{r}{m}}\right] = \$5000\left[\frac{1-\left(1+\frac{0.03}{12}\right)^{-12\times 25}}{\frac{0.03}{12}}\right] \approx \$1{,}054{,}382.27$$

So such an annuity would cost about a million dollars in retirement savings.

Self Check 13

How much would such an annuity cost if the insurance company expects the long-range steady interest rate to be 4%?

However, if your father retires at 65 and purchases a 25-year annuity, he might be in trouble if he lives longer than the term of the annuity (past 90), because the payments would stop and he would have no further income from the annuity. (About 2% of U.S. children born since 2000 can expect to live to age 100.) Similarly, if your father were to die sooner, his designated beneficiaries would still get the payments due after his death, but that money wouldn't have helped him meet living expenses while he was alive.

Another drawback to such an annuity is that the purchasing power of the level payments will decline with inflation.

An approach that avoids some of these disadvantages is a **life annuity:** You receive payments for as long as you live. How much you receive is based on the life expectancy of people your age, as determined from population data. For example, Social Security is in effect a nonlevel life annuity:

- You get payments until you die.
- How much you get depends on your age when you choose to start taking payments (as well as on your pre-retirement earnings).
- The amount of payments in a given year is determined by the consumer price index (CPI) (see Chapter 21, pages 888–893).

There are many variations on life annuities, such as payments that increase with anticipated cost-of-living increases, or payments that last until both you and your life partner die (see Spotlight 22.4). We focus on a simple one-life annuity.

The insurance company that sells you the annuity makes money on your policy if you die younger than average and loses money if you die older than average. As in any kind of insurance, over a large number of people, the company expects gains to balance losses (actually, overbalance losses, to account for expenses and profit). This is a manifestation of the law of large numbers (discussed in Section 8.5 on page 378). The company invests the annuity funds, and its profits depend on the rate that its investments earn compared with the rate that it pays on the annuity.

A drawback of an annuity-certain is that you may outlive your annuity (and then have no income). A drawback of a life annuity is that you may die before receiving many (or even any) payments, thus in effect "losing" your retirement savings. On the other hand, if you live a long time, a life annuity may turn out to have been a better deal.

The amount of the payments for each annuity variation depends on prevailing and predicted rates of interest and (in the case of life annuities) on your age and sex. How much the annuitant (the purchaser of the annuity) receives per month depends on gender. Because women on average live longer than men, the monthly payment to a woman may be lower.

EXAMPLE 15 Life Income Annuity

Suppose that your 65-year-old father retires and purchases for $250,000 a life income annuity. According to the table from one particular insurance company, he would receive $6.3448 per month for every $1000, so his monthly income would be $1586. According to the Social Security Administration actuarial life table, his life expectancy at age 65 is about 17 years = 204 months. If he lived exactly that long, he would receive a total of 204 × $1586 = $323,544.

However, simple algebra cannot be used to find the rate of interest that the annuity would need to earn to last that long. We use the RATE function in a spreadsheet (for more details, see Spotlight 21.3 (page 888); entering =RATE(204, 1586, −250000) gives a monthly rate of 0.2636%, for an effective annual rate of $(1 + 0.002636)^{12} - 1 \approx 3.21\%$.

Now let's consider instead the case of your mother retiring now, also at age 65 and also with a $250,000 life income annuity; she would receive $5.9010 per month for every $1000, or $1475 per month. Her life expectancy would be about 19.72 years = 237 months. If she lived exactly that long, she would receive a total of $237 × $1475 = $349,575. The rate of interest that her annuity would need to earn to last that long can be calculated from the amortization formula; using =RATE(237, 1475, −250000) gives a monthly rate of 0.2997%, for an effective annual rate of 3.66%. The difference between this rate and the one for your father probably reflects the fact that the company uses life expectancies (which vary by region of the country) that differ from those for the nation as a whole.

Notice that a man and a woman who save the same amount receive different monthly incomes at retirement: The woman receives less per month but for longer—93% as much for 16% longer. Yet their living expenses are likely to be the same. That consideration has resulted in some companies offering "merged gender" rate schedules for annuity payments, so that the individual receives the same monthly payment regardless of gender. ■

Review Vocabulary

Add-on loan A loan in which the borrower receives the principal, the interest is calculated in advance as simple interest, and the borrower pays back the principal plus interest in equal installments. (p. 927)

Adjustable-rate mortgage (ARM) A loan whose interest rate can vary during the course of the loan. (p. 921)

Amortization payment formula The formula for installment loans that relates the principal P, the interest rate i per compounding period, the payment d at the end of each period, and the number n of compounding periods needed to pay off the loan:

$$d = P\left[\frac{i}{1-(1+i)^{-n}}\right],\ P = d\left[\frac{1-(1+i)^{-n}}{i}\right]$$

(p. 916)

Amortize To repay in regular installments. (p. 912)

Annual percentage rate (APR) The number of compounding periods per year times the rate of interest per compounding period. (p. 913)

Annuity A sequence of periodic payments. (p. 929)

Annuity-certain A specified number of equal periodic payments. (p. 931)

Compound interest formula The formula for the amount in an account that pays compound interest periodically. For an initial principal A and effective rate i per compounding period, the amount after n compounding periods is $A = P(1 + i)^n$. (p. 911)

Compounding period The fundamental interval for compounding, within which no compounding is done. Also called simply *period*. (p. 913)

Conventional loan A loan in which each payment pays all the current interest and also repays part of the principal. (p. 914)

Discounted loan A loan in which the borrower receives the principal minus the interest but must pay back the entire principal, usually with equal payments. (p. 923)

Effective annual rate (EAR) The effective rate per year. (p. 913)

Effective rate The actual percentage rate, taking into account compounding. (p. 913)

Equity The amount of principal of a loan that has been repaid. (p. 919)

Interest Money charged for a loan. (p. 910)

Interest rate factor The annual interest rate divided by the number of days in a year, taken to be 365.25. (p. 910)

Life annuity An annuity with regular payments for as long as you live. (p. 931)

Nominal rate A stated rate of interest for a specified length of time; a nominal rate does not take into account any compounding. (p. 912)

Points (on a mortgage) Payments up front to secure a lower interest rate; one point equals 1% of the principal. (p. 926)

Principal Amount borrowed. (p. 910)

Savings formula The formula for the amount in an account to which a regular deposit is made (equal for each period) and interest is credited, both at the end of each period. For a regular deposit of d and an interest rate i per compounding period, the amount A accumulated is

$$A = d\left[\frac{(1+i)^n - 1}{i}\right]$$

(p. 915)

Simple interest The method of paying interest on only the initial balance in an account and not on any accrued interest. For a principal P, an interest rate r per year, and t years, the interest I is $I = Prt$. (p. 910)

Self Check Answers

1. $3000 × 0.0429 × 0.25 = $32.18, rounded up

2. $20,000 × (1.010725)² ≈ $20,431.30

3. $449.04

4. $261.03

5. $215.57 payment; $934.20 interest

6. $21,000

7. No. For example, if the loan is at 0%, then three-quarters of it will be paid off after three-quarters of the term.

8. $1800 interest; $187.50 payment; 11.4% APR

9. No

10. 2.91% APR; 2.95% EAR

11. $195.30

12. One reason is that because of inflation, the value (purchasing power) of a level payment would decline over the 25 or 30 years of the term of the annuity.

13. $947,262.41

Skills Check

1. The interest charged on federal direct student loans while you are in school is
(a) simple interest.
(b) compound interest, compounded daily.
(c) compound interest, compounded quarterly.

2. If you borrow $1000 at 5% interest per year, compounded quarterly, and pay back the principal and interest after four years, the amount that you pay back is ______________.

3. A nominal rate of interest
(a) takes into account any compounding involved.
(b) is always stated as an annual rate.
(c) Neither of the above is correct.

4. If at the end of each month, you put $100 in an account that pays 6% annual interest compounded monthly, at the end of the two years, you will have ______________.

5. An effective interest rate
(a) always takes inflation into account.
(b) is the same as the nominal rate.
(c) takes compounding into account.

6. APR stands for ______________.

7. The nominal rate of interest for a loan is
(a) the same as the effective rate.
(b) less than the effective rate.
(c) never greater than the effective rate.

8. If a store credit account charges 1.5% interest each month, the effective annual rate is ______________.

9. Credit card interest is
(a) computed using compound interest.
(b) computed using simple interest.
(c) included in the late fees.

10. If a store credit account charges 1.5% interest each month, the APR is ______________.

11. The APR calculated for a loan takes into account the amount of
(a) the loan.
(b) the loan plus any loan-processing fee.
(c) the loan plus any loan-origination fee and any other closing costs.

12. The median length of time that Americans hold a mortgage is ______________.

13. The Truth in Lending Act requires that borrowers be told the ______________ of the loan.
(a) APY
(b) APR
(c) EAR

14. Your credit union offers to finance a $6000 conventional loan at 4% to be repaid in four years of monthly payments. Your monthly payment is ______________.

15. After 15 years of minimum payments on a 30-year mortgage, the balance remaining is about ______________ of the original balance.
(a) one-third
(b) one-half
(c) two-thirds

16. If you finance $15,000 for 3 years at 6% compounded monthly, the monthly payments will be ______________.

17. Equity in a 30-year conventional mortgage grows
(a) linearly.
(b) logarithmically.
(c) exponentially, but slowly.

18. Monthly payments for a 15-year, 6% mortgage are about ______________ times the payments for a 30-year mortgage of the same amount and the same interest rate.

19. An ARM
(a) has variable interest rates but maintains a fixed payment amount.

(b) has variable payment amounts.

(c) is always a better alternative to a fixed-rate mortgage.

20. In a 30-year mortgage, most of the amount of the first several payments goes toward ______________.

21. With a conventional loan,

(a) each payment pays current interest and repays part of the principal.

(b) you pay exactly the same current interest rate as everyone else.

(c) you have to buy a conventional house—no condos, triple-deckers, or yurts allowed.

22. A convenient rule of thumb is that for a 30-year mortgage at 6%, the monthly payment is about 0.6% of the loan. So, on a $100,000 mortgage, the monthly payment is about $600. About ______________ of the first payment goes toward interest.

23. Most people

(a) never pay off their home mortgage loan.

(b) pay off their home mortgage loan early.

(c) make late payments on their home mortgage loan.

24. A high rate of inflation is likely to mean ______________ payments on your ARM.

25. In this chapter, equity refers to

(a) fair lending practices.

(b) principal paid off on a loan.

(c) a union that represents actors and stage managers for live theatrical performances.

26. ARM stands for ______________.

27. Paying off a conventional loan is like

(a) saving.

(b) slaving.

(c) shaving.

28. If you just won a lottery jackpot paid in 25 equal annual installments of $1 million each at 6% annual effective interest, the present value of the jackpot is ______________.

29. A life income annuity is designed to pay a fixed amount each period until

(a) the annuity runs out of money.

(b) you die.

(c) you reach your life expectancy.

30. A professional who assesses the costs of risks for life insurance, auto insurance, health insurance, or pensions is called a(n) ______________.

Chapter 22 Exercises

Challenge Discussion

22.1 Simple Interest

1. Suppose that you take out a federal direct loan on September 1 before your senior year for $7500 (the maximum allowed for a dependent student) and plan to begin paying it back on December 1 after graduation (so you will have had the loan for 15 months, including the six-month grace period after leaving school). The interest rate is 4.29% and you pay the interest every quarter until that December 1. How much will you owe on that December 1, and how much of that will be interest?

2. Suppose that you borrow $7500 on September 1 before your junior year. You will graduate 21 months later, on June 1, and you will have 6 months of grace period. So you plan to begin paying the loan back on December 1, 27 months after you took it out. But you do not pay any of the interest as it accumulates. How much will you owe on that December 1, and how much of that will be interest?

3. Suppose that you borrow $5500 for your first year and $6500 for your second year (the maximum amounts for a dependent student), as federal direct student loans at a 4.29% interest rate. Suppose that each loan begins on September 1 of its year, that you finish college in four years, that you do not pay the accruing interest in the meantime, and that you begin repayment on December 1 after graduation. What is your total debt on that December 1, and how much of that is interest?

4. Assume the same situation as in Exercise 3, but you also borrow $7500 for each of your third and fourth years (again, the maximum amounts), again on September 1, all at a 4.29% interest rate. You finish college in four years, and you begin repayment on December 1 after graduation. What is your total debt on that December 1, and how much of that is interest?

22.2 Compound Interest

5. If you borrowed $15,000 to buy a new car at 4.9% interest per year, compounded annually, and paid back all the principal and interest at the end of 5 years, how much would you pay back?

6. Assume the same situation as in Exercise 5, but the interest is compounded monthly. How much would you pay back?

7. If you borrowed $200,000 to buy a house at 6% interest per year, compounded annually, and paid back the principal and interest at the end of 30 years, how much would you pay back?

8. Assume the same situation as in Exercise 7, but the interest is compounded monthly (this is the usual case). How much would you pay back?

9. A credit card bill showed an APR of 17.24%.

(a) What is the corresponding daily interest rate (the bank uses a 365-day year for this purpose)?

(b) What is the effective annual rate (EAR)?

10. You receive an offer for a credit card with 0% fixed APR for the first 12 months, after which the card would have one of several rates depending on credit history. The highest rate was a 22.74% APR (and the company reserves the right to change the APR "at any time for any reason").

(a) What is the corresponding daily interest rate for the 22.74% APR?

(b) What is the effective annual rate (EAR)?

22.3 Conventional Loans

11. Suppose that your federal direct student loans plus accumulated interest total $20,000 at the time that you start repayment, and the interest rate is 4.29%.

(a) What is your monthly payment?

(b) How much will you pay in interest over 10 years?

12. Suppose that your federal direct student loans plus accumulated interest total $31,811 at the time that you start repayment, and the interest rate is 4.29%.

(a) If you elect the standard repayment plan of a fixed amount each month for 10 years, what is your monthly payment?

(b) How much will you pay in interest?

However, because your accumulated outstanding federal loans total more than $30,000, you can elect to repay over 25 years instead. If you do that:

(c) What is your monthly payment?

(d) How much in total will you pay in interest?

Refer to the following for Exercises 13 and 14. Your parents (if their credit rating qualifies) can take out a federal Direct PLUS loan to pay for the total remaining cost of your undergraduate education, after any other financial aid (such as a a federal direct student loan). The simple interest rate was 6.84% for 2015–2016. (There was also a loan origination fee of 4.292%, which we disregard in these exercises.) The standard repayment plan is fixed monthly payments over 10 years, and your parents can elect to defer the start of repayment until six months after your graduation.

13. Suppose that your parents take out a PLUS loan on your behalf on September 1 before your senior year for $10,000, at the rates mentioned above, and begin paying it back six months after you graduate on June 1. How much is their monthly payment?

14. If your parents instead take out a PLUS loan for $10,000 on September 1 before each of your four years of college, how much is their monthly payment if they begin paying it back six months after your graduation?

Ekasit Wangprasert/ Alamy

15. In late November 2014, a car dealership in southern Wisconsin was offering a new 2014 Toyota Corolla LE sedan for $18,299 (not including sales tax, registration, license plates, title, and $125 "dealer documentation fee") at 0% annual interest over 36 months. What would be the monthly payment?

16. Repeat Exercise 15, but opt instead for 1.9% over 60 months. What would be the monthly payment?

17. Repeat Exercise 15, but opt instead for a further $500 manufacturer rebate and pay 2.9% interest over 60 months. What would be the monthly payment?

18. In December 2010, Kevin Lauterbach, 29, of Coral Springs, Florida, who had "mildly damaged credit," bought a 2008 Jeep Liberty with no money down and a 72-month loan for $19,000 with a 4.75% rate. (*New York Times*, February 28, 2011, p. A3).

(a) What would the monthly payment have been?

(b) How much interest would he have paid over the course of the 72 months?

(c) In fact, he was instead required to make a payment every two weeks. How much was that payment?

19. Suppose that you have good credit and can get a 30-year mortgage for $100,000 at 5%. What is your monthly payment?

20. Assume the same situation as in Exercise 19, except that your credit is not as good and the rate that you are offered is 7.125%. What is your monthly payment?

21. Assume the same situation as in Exercise 19, but you inquire about a 15-year loan instead. You are offered 3.75%. What is your monthly payment?

22. Assume the same situation as in Exercise 21, but your credit is not as good, and you are offered 6.75%. What is your monthly payment?

23. For the mortgage in Exercise 19, how much equity would you have after 5 years?

24. For the mortgage in Exercise 20, how much equity would you have after 5 years?

25. For the mortgage in Exercise 21, how much equity would you have after 5 years?

26. For the mortgage in Exercise 22, how much equity would you have after 5 years?

Refer to the following for Exercises 27 and 28. When interest rates drop, it may become attractive to refinance your home. Refinancing means that you acquire a new mortgage to borrow the current principal due on your home and use the proceeds to pay off your old mortgage. You then begin a new 15- or 30-year mortgage at the new, lower interest rate. A second factor that reduces your monthly payment is that the equity you accumulated under the old mortgage reduces the amount that you have to borrow under the new mortgage. Suppose that you have been paying for 5 years on a 30-year mortgage for $200,000 with a fixed rate of 6%. Your monthly payment is $1199.10, and you have $13,890.81 in equity, so $200,000 − $13,891.20 = $186,108.80 remains to be paid. We consider two refinancing offers.

27. The first offer is from a local bank for a 30-year, fixed-rate mortgage at 4.25% with closing costs of $2639.07. (You must pay the closing costs right away—you cannot include them in the mortgage.)

(a) What is the new monthly payment?

(b) How much less is that per month than the old payment?

(c) How many months will it take for the savings in payments to make up for the closing costs?

28. The second offer is on the Internet (from a company you have never heard of) for a 30-year, fixed-rate mortgage at 3.99% with closing costs of $5000.

(a) What is the new monthly payment?

(b) How much less is that per month than the old payment?

(c) How many months will it take for the savings in payments to make up for the additional closing costs?

29. In a 2/28 "hybrid" adjustable-rate mortgage (ARM), the initial interest rate is fixed for 2 years and then is adjusted every 6 months. (You usually pay "points" up front at closing in exchange for the "rate lock" for the first 2 years.) Suppose you buy a house with a $200,000 mortgage, with a 2/28 ARM with initial rate of 3%; and suppose that 2 years later, the interest rate goes up to 5%.

(a) What was your payment originally, at 3%?

(b) What is your new payment? (*Hint:* The amount of the loan is no longer $200,000, and you have only 28 years to pay it off.)

30. In a 5/1 "hybrid" adjustable-rate mortgage (ARM), the initial interest rate is fixed for 5 years and then is adjusted annually. (You usually pay "points" up front at closing in exchange for the "rate lock" for the first 5 years.) Suppose that you buy a house with a $200,000 mortgage with a 5/1 ARM with initial rate of 4%; suppose that 5 years later, the interest rate goes up to 6%.

(a) What was your payment originally, at 4%?

(b) What is your new payment? (*Hint:* The amount of the loan is no longer $200,000, and you have only 25 years to pay it off.)

Refer to the following for Exercises 31–35, about credit card payments. Many credit cards use a similar formula for the minimum payment, which is the new balance (if less than $25), or else the greatest of $25 or 1% of the new balance (excluding interest and late fees), plus the interest billed, rounded down to the nearest dollar. Any late fees are then added on to this calculated amount. Moreover, when any interest is due, there is a minimum charge of $1.50.

31. (Requires a spreadsheet) Suppose that your credit card has an APR of 18% interest rate, corresponding to approximately 1.5% per month. (The actual interest applied is daily interest, at a daily rate of 18%/365; but for simplicity we use a uniform approximate monthly rate. Also, the amount of interest owed depends on exactly when in the month your payment is received.)

(a) Why do we say "approximately" 1.5% per month for an APR of 18%?

(b) How many months will it take to pay off a new balance of $3117.83 by making the minimum payment each month?

(c) How much will you have paid altogether? How much of that is interest?

32. (Requires a spreadsheet) Repeat Exercise 31, but this time you miss the first payment, incurring a $35 late fee and an increase to a penalty APR of 30%, corresponding to approximately 2.5% per month.

(a) How many months—of paying your bill on time!—will it take to pay off the balance of $2500 by making the minimum payment each month? (You must pay the $35 late fee the first month, over and above the minimum payment on the $2500 and one month's interest.)

(b) How much will you pay altogether?

33. The purpose of such a complicated formula for the minimum payment on a credit card is to avoid the situation of a customer who makes just the minimum payment but nevertheless falls farther and farther behind. For example, formerly some banks set the minimum payment at balance due (if < $10) or else the larger of $10 or 2% of the total new balance (including interest). However, for a high enough interest rate, paying 2% of the balance due will not cover the interest, so the balance actually would increase (this is called negative amortization). How high would the APR have to be to make this happen? (*Hint:* It's not just 12 × 2%.)

34. (Requires a spreadsheet) A well-known national credit card calculates minimum payment due as the new balance (if less than $35), or else the greatest of the following:

- $35
- 2% of the new balance (excluding new late fees)
- Interest charged on the statement plus 1% of the new balance (excluding late fees and new interest charged on the statement), not to exceed 4% of the new balance

Then any late fees are added and the total is rounded to the nearest whole dollar.

(a) With a monthly interest rate of 1.5%, how many months will it take to pay off a new balance of $5000 by making the minimum payment each month?

(b) How much will you pay altogether?

35. (Requires a spreadsheet) Repeat Exercise 34, but for a monthly interest rate of 2.5%.

(a) How many months will it take to pay off a new balance of $5000 by making the minimum payment each month?

(b) How much will you pay altogether?

36. (Requires a spreadsheet) A bank or credit union may offer to let you agree in advance to skip a payment (e.g., on a car loan but usually not on a mortgage or a credit card)—in exchange for a processing fee (such as $35) to be added to the principal. If you skip the payment, interest continues to accrue for that month on the remaining principal plus the added processing fee. You continue regular payments as usual in the same amount as before, except that the last payment is a larger "balloon" payment to pay off the loan. Suppose that you borrowed $11,158.05 from your credit union for a 60-month home improvement loan at 9%. Verify that your monthly payment is $231.62 and that after 12 months of payments you still owe $9307.74. You receive an offer to skip the 13th payment, for $35 added to the principal, and you do so. How much will the balloon payment be?

37. (Requires a spreadsheet) Ads for purchasing cars often cite the monthly payment per $1000 borrowed. For example, a recent ad quoted $17.48 per $1000 borrowed for a 60-month loan.

(a) What is the corresponding APR? (*Hint:* Use the RATE function in your spreadsheet.)

(b) What is the corresponding EAR?

38. (Requires a spreadsheet) An ad I saw on TV quoted 0.9% interest and $17.05 monthly payment per $1000 for an auto loan—but the ad went by too fast for me to see the term of the loan. For how many months would it be? (*Hint:* Use the NPER function in your spreadsheet.)

39. As we noted, the Consumer Financial Protection Bureau feels that almost all of the up-front costs that the consumer must pay to get a loan should be included as part of the "finance charge" and consequently factor into the quoted APR. Citicorp (a mortgage lender) argues that "the APR calculation includes interest that you will never pay and spreads the closing costs over too many years." Consider the closing costs noted in Figure 22.2 (page 928). Which of these costs do you feel should be included as part of the fee for getting the loan and hence should be entered into the APR?

40. Should "truth in lending" require disclosure of the EAR rather than the APR? Why or why not?

22.4 Other Loans

41. You need to buy a car and finance $5000 of the cost. The dealer offers you a 5.9% add-on loan to be repaid in monthly installments over four years. How much is your monthly payment?

42. You have to make some home improvements—well, really maintenance that you can't defer any longer!—and need to borrow $3000 to pay for them. You can get an 8.5% add-on loan from a savings and loan association to be repaid in installments over 2 years. How much is your monthly payment?

43. Repeat Exercise 42, except that you can get a 9% discounted loan from a loan company to be repaid in monthly installments over 4 years. What is the monthly payment on this loan?

44. Repeat Exercise 42, except that you can get an 8.5% discounted loan from a loan company to be repaid in monthly installments over 5 years. What is the monthly payment on this loan?

45. Suppose you can get either an add-on loan or a discounted loan, both for the same proceeds (principal), at the same interest rate, and for the same period. Show in general that the add-on loan has a lower monthly payment.

For Exercises 46 and 47, refer to the following. Payday lenders provide small loans until the borrower's next payday. The borrower receives the desired cash in exchange for a postdated check in the amount of the loan plus a fee, which is usually a percentage of the loan amount (often 15% to 20% for a two-week loan). In many states, there are now more payday loan offices than McDonald's fast-food outlets. The average loan amount is \$300. You can think of such a loan as an add-on loan with a single payment at the end of the loan term.

46. For one payday lender, the fee for a \$100 loan for up to two weeks is \$15. What is the APR if the loan is for the full two weeks?

47. Another payday lender charges \$26.10 for a \$100 loan for 7 to 14 days. What is the APR if the loan is for 7 days?

48. The leading British payday lender Wonga made profits of \$1.9 billion on loans in 2012. Wonga charges a daily rate of 1%, compounded daily.

(a) What is the corresponding APR?

(b) What is the corresponding EAR?

49. All Credit Lenders (with storefronts in Illinois, Wisconsin, and South Carolina) offers "line of credit" loans. With such a loan, you receive a cash advance, much as you might from using a credit card. The interest, calculated on a daily basis, comes to the currently advertised APR of 24% (below the Illinois usury cap of 36%). But if you have not paid back the loan by the end of the month, you are charged a "required account protection fee," usually \$15 per \$50 borrowed, whose alleged purpose is to protect the borrower in case the borrower becomes unemployed and is unable to make payments. In March 2012, Loralty Harden (who is retired and disabled) borrowed \$100 under such an arrangement at the Machesney Park, Illinois, office, with an interest rate of 18%. During the subsequent year, she paid \$360 in protection fees and \$18 in interest. She still owed \$100. Using her case, the attorney general of Illinois sued parent company CMK Investments for "unfair and deceptive business practices."

(a) If the "account protection fee" were considered interest, what would be the APR of Ms. Harden's loan?

(b) If the "account protection fee" were considered interest, what would be the EAR on her loan?

50. Should your state cap the interest rate on short-term loans, such as payday loans? According to one source, the average payday loan is "flipped" eight times, so the loan system is trapping borrowers in a "cycle of debt." Lenders counter that they are doing the borrowers a favor because some borrowers have no other alternative (except theft or robbery!), and that small loans and high rates of nonrepayment make high rates of interest essential. One representative said that capping interest rates at a proposed 36% in Wisconsin would "eliminate the industry." Another objected to figuring the interest rate on an annual basis, claiming that doing so is like calculating the cost of staying in a hotel for a year even though you stay only a couple of nights.

For Exercises 51 and 52, refer to the following. Some loans, for cars or leases, penalize early repayment by using the Rule of 78s (also known as the "sum of the digits rule") to calculate interest paid, because it loads the interest toward the early months of the loan.

Suppose you have an add-on loan for \$1122 for 12 months with monthly payments of \$100 and total interest for the year of \$78. According to the Rule of 78s, 12/78 of the year's interest, or \$12, is considered as having been paid in the first month, \$11 in the second, and so forth, up to \$1 in the twelfth month. It is called the Rule of 78s because $12 + 11 + \cdots + 1 = 78$. So, if you pay off the remainder of the loan after just 6 months, you will have made payments totalling \$600, of which \$12 + \$11 + \$10 + \$9 + \$8 + \$7 = \$57 is considered interest. So, according to the rule, you will have paid off only \$600 − \$57 = \$543 of the principal. Thus, you still owe \$1122 − \$543 = \$579. That is less than the \$600 you would have paid in the succeeding 6 months—but you had the use of the money for only 6 months instead of the full year.

51. For the 12-month loan:

(a) What is the APR of the original add-on loan?

(b) What is the APR for the paid-back-six-months-early option?

52. For a 24-month loan, the Rule of 78s uses 300 instead of 78: $24 + 23 + \cdots + 1 = 300$. (The general formula for the sum of the first n integers is $n(n + 1)/2$.) Suppose that you take out an add-on loan for a car that costs \$10,000 at a 3.0% interest rate, with monthly payments over a term of 24 months and a contract that specifies use of the Rule of 78s for early repayment. How much will you have to pay if you pay off the balance on the loan after 12 months?

53. Put off by the high monthly car payments of Exercises 15–17, you might be attracted to leasing a car instead of buying one. With the "college grad discount," in May 2014 in Rhode Island you could lease a 2014 Toyota Corolla LE A4 for \$65 per month for 24 months (plus tax, tag, title, registration, and "dealer documentation fee" of \$200). You would also get no-cost maintenance and 24-hour roadside assistance during the 24 months. Because this was a lease, not a loan, the dealer does not have to disclose anything about interest rates. However, the "capitalized cost" was the buy-it price of \$15,490 minus the required down payment of \$1900, so \$13,590. You would have the option at the end of the lease to purchase the car for \$13,542. So in effect, your monthly payments would have paid for the difference in value of the car of $\$15{,}490 - \$13{,}542 = \$1948$. However, those payments would have totaled only $\$65 \times 24 = \$1560$.

The purchase price, if you had bought instead of leased, would have been \$15,490, with (after the down payment) financing needed for \$13,590.

(a) What would be the monthly payment over 48 months, at 1.9% interest?

(b) What would be the monthly payment over 60 months, at 2.9% interest?

54. The manufacturer's suggested retail price (MSRP) on the car in Exercise 53 was \$19,335. How do the manufacturer and the dealer make money on leasing? Why are the payments for leasing a car so much lower than for purchasing?

22.5 Annuities

55. (Requires a spreadsheet) Jack Whittaker of West Virginia, on Christmas Day in 2002, won the jackpot worth \$314.9 million in a lottery with "annuity value." (His subsequent life has been far from a fairy tale, as a Google search will reveal.) Instead of receiving \$314.9 million in 30 equal annual payments, including one immediately, he chose a lump sum, which came to \$170 million. What was the corresponding interest rate of the annuity?

56. (Requires a spreadsheet) Winners of the Powerball lottery can elect either an immediate lump sum (almost all do) or an annuity. In the latter case, the advertised jackpot amount is paid in 30 annual payments, including one immediate payment. To keep up with inflation, each payment is 4% more than the previous year's; such an annuity is called a graduated annuity. On October 10, 2007, Eugene and Stanislawa Markiewicz took their prize of \$20 million in the form of a graduated annuity.

(a) What was the amount of their first payment, and how much will they receive in their last payment in October 2036?

(b) The winners could have chosen a lump sum of \$9,402,914.90 instead. What was the corresponding interest rate of the annuity?

57. Suppose a man retires at age 65, and in addition to Social Security, he needs \$2000 per month in income. Based on an expected lifetime of 204 more months, how much would he have to invest in a life income annuity earning 3% APR to pay that much per year?

Amble Design/Shutterstock

58. Repeat Exercise 57, but for a woman at age 65, whose expected lifetime is 237 more months.

59. (Requires a spreadsheet) Long ago, Darryl Strawberry played for the New York Mets baseball team. Part of his compensation was an annuity of \$8,891.82 to be paid out over the 30 years after the end of his career. But to pay off his tax debts, in January 2015 the government auctioned off the right to receive the 12 years of remaining payments. The IRS set a minimum bid of \$550,000. What interest rate would that correspond to?

60. Some annuity companies offer "merged gender" rate schedules, so that for the same purchase price, a man or a woman at the same age receives the same monthly annuity payment. (This is in effect what happens with the "life annuity" of Social Security payments.) Since a woman can expect to live longer, is that fair?

Chapter Review

61. You get a 30-year mortgage for $150,000 at 4.5%. What is your monthly payment?

62. The simple interest rate on a federal PLUS loan was 6.84% for 2015–2016. There was also a loan origination fee of 4.292%, so you receive as proceeds only (100 − 4.292)% = 95.708% of the loan amount. Assume that your parents make interest payments until repayment starts and then repay with fixed monthly payments over 10 years. For the equivalent no-fee loan:

(a) What is the APR?

(b) What is the EAR?

63. A payday lender charges $25 for a $100 loan for 7 to 14 days. What is the APR if the loan is for 14 days?

64. Suppose that you bought your house with a 30-year adjustable-rate mortgage (ARM) at 4.5% for $150,000. After 5 years, the rate was raised to 6.5%. What was your new payment?

65. In July 2015, you could buy a 2015 Toyota Camry LE for $22,790 at 0% interest for 60 months, or you could get a $1000 Toyota rebate but then pay 2.9% over 60 months. Which deal offers a lower monthly payment?

Applet Exercise

To do this exercise, go to www.macmillanhighered.com/fapp10e.

1. There are two ways to buy a car: save up and pay cash or borrow the money. In the *Buying a Car: Cash vs. Loan* applet, you can explore just how much more expensive it is to borrow the money.

Writing Projects

1. Locate current advertised incentives for a car that you would consider buying and compare them in an essay of two to three pages. For each option, give the price, the interest rate, the term, and how much interest you would pay over the course of the loan.

2. A substantial proportion of new cars today are not sold but leased. Contact a local car dealer about a car that you are interested in and find out the details on leasing. Compare the cost of the lease and associated expenses with the cost of purchasing and owning the car. Include estimated maintenance, repair, and insurance costs for each option. Which seems like a better deal, and why? Consult the Suggested Websites at the end of this chapter. Write two to three pages describing and comparing the two options.

3. Suppose that you have a choice between a mortgage at 6% with 2 points (2%) and a mortgage at 8% with no points. Which would you choose, and why? Does it make a difference how long you are planning to own the home, or how expensive the home is? Write a page justifying your decision.

4. One of the advantages of buying a home with a fixed-rate mortgage is that your payment stays the same but your earnings and the value of your home are likely to go up, if only because of inflation. You will be paying back the loan with dollars of lesser value. Suppose you buy a "starter" two-bedroom home for $105,000 under a special program for first-time home buyers that requires a down payment of only $5000. You have a 30-year, fixed-rate mortgage for $100,000 at 7%, on which the monthly payment is $665.30. You also have a $2000 one-time expense in closing costs and annual costs of $200 for insurance and $2000 for property taxes. You live in the home for 5 years and spend $10,000 on maintenance, upkeep, and improvements. You then sell the home for $125,000, pay a realtor $9000 to sell it, and pay closing costs of $500 (for title insurance and other costs). Finally, it costs $3000 to move.

(a) Create a balance sheet of revenue and expenses. How did you make out on owning the home?

(b) Remember that you also got to live in the home without paying rent. Translate the cost of owning the home into an equivalent monthly rent.

5. (Team exercise) Explore actual costs of homes in your area, mortgages with local banks (including closing costs), and property taxes and insurance. Come up with data on a particular mortgage, as well as the costs and benefits of refinancing, and create a corresponding balance sheet for five-year ownership.

Suggested Readings

MILLER, CHARLES D., VERN E. HEEREN, and JOHN HORNSBY. Consumer mathematics. In *Mathematical Ideas*, 11th ed., Pearson Education/Addison Wesley, Reading, MA, 2008.

YAREMA, CONNIE H., and JOHN H. SAMPSON. Just say "Charge it!" *Mathematics Teacher* 94 (7) (October 2001), 558–564. Shows how to apply the savings formula and the amortization formula and graph the results on the TI-83 calculator. Notes that the 78% of undergraduates in the United States who have credit cards carry an average debt of more than \$2700, with 10% owing more than \$7000.

Suggested Websites

www.leaseguide.com/ A guide to how car leasing works, including what "money factor" means and how leasing cost is determined.

www.edmunds.com/calculators/ A commercial site offering a calculator to compare rebates versus interest-rate offers for car purchases. (*Note:* Edmunds is a loan broker; the mention here of calculators at its website does not imply endorsement of its other services by this book's authors, editors, or publisher.)

nytimes.com/studentloancalculator An interactive student debt and payment calculator, including information about average student debt amounts at many colleges and universities.

www.edpubs.gov/document/en1214p.pdf A two-page leaflet listing kinds of federal student aid.

www.edpubs.gov/document/en1168p.pdf A 32-page booklet with details about federal student loans.

The Economics of Resources

23

Hans-Peter Merten/The Image Bank/Getty Images

How many people will there be in the world when you retire? Will there be any oil left by then? Will there be any fish left?

In this chapter, we explore the growth and decline of populations, such as populations of people, barrels of oil, and fish. How long can a nonrenewable resource last us?

In Chapters 21 and 22, we explored mathematical models for saving, accumulating, and borrowing—the building up of resources. In this chapter, we model processes in the other direction—the use, decay, depletion, and spending down of resources, even resources that tend to replenish themselves regularly.

The Formulas We Use in This Chapter

SPOTLIGHT 23.1

We use here only three formulas from Chapter 21, "Savings Models," specialized to an annual interest rate r and number of years n (formerly t). We change our perspective from a t that could be any amount of time to a number n of years, because populations of resources and animals are usually tallied on an annual basis:

Compound interest formula (page 875): If a principal P is deposited into an account that pays interest at rate r per year, then after n years the account contains the amount

$$A_n = P(1 + r)^n$$

Savings formula (page 884): For a uniform deposit of d per year (deposited at the end of the year) and an interest rate r per year, the amount A accumulated after n years is

$$A_n = d\left[\frac{(1 + r)^n - 1}{r}\right]$$

Continuous interest formula (page 882): For a principal P in an account at a nominal annual rate r, compounded continuously, the balance after n years is

$$A_n = Pe^{rn}$$

If you understand and can use these formulas, you can proceed in this chapter without first reading Chapter 21 or Chapter 22.

We begin with a geometric growth model for human populations, applying it to the current U.S. population. Limits to growth of any population lead us to the logistic model, which takes into account the carrying capacity of the environment. Surprisingly, such a model applies also to the spread of a technology or new product.

We then calculate how long a nonrenewable resource, such as oil, can last at a current rate of use and also at a constantly increasing rate of use; and we show how the logistic model fits well the history of oil consumption.

In the case of renewable resources, such as forests and fisheries, we examine how the growth curve determines potential equilibrium population sizes and what harvesting policies can produce yields that are sustainable year after year.

There are some resources, such as nuclear waste, that decay and we *want* to become exhausted. How long will it take until Fukushima, Japan, or Chernobyl, Ukraine, is inhabitable again?

Economic factors influence harvesting efforts and consequently the population of the resource. But even biologically and economically sound policies are subject to the unpredictability of weather and other factors. What's worse, as we will see, even in the absence of chance effects, the dynamics of a population can vary in an apparently chaotic fashion—even to extinction.

Spotlight 23.1 reviews the formulas that will be used in this chapter.

23.1 Growth Models for Biological Populations

We encountered geometric growth models for savings accounts in Chapter 21. Growth is proportional to the amount present, and such growth is expressed in terms of compound interest and its formula.

We now use a geometric growth model to estimate the sizes of human populations. In addition to the **rate of natural increase**—the annual birth rate minus the annual death rate—we must take into account net immigration. The sum of the natural rate of increase and net immigration, in the terminology of financial models, is the effective rate of growth.

Birth, death, and migration rates rarely remain constant for long, so projections must be made with care. In the short run, however, predictions based on the model may be useful. Let's apply this model to two questions about the population of the United States.

EXAMPLE 1 Predicting the U.S. Population

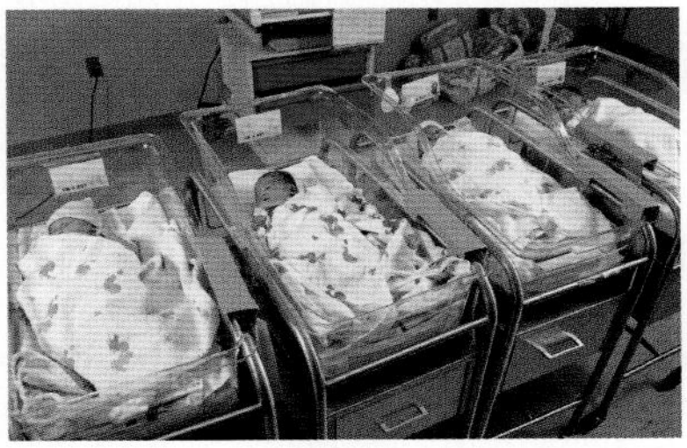
Diane Macdonald/Getty Images

The U.S. population increased at an average effective growth rate of 0.77% per year (including immigration) to 321 million at mid-2015. What would be the anticipated population at mid-2020? What would it be if the effective rate of growth changes to 1% per year, or to 0.5% per year?

We apply the compound interest formula with initial population size ("principal") 321 million. Using one year as the compounding period and the compound interest formula $A = P(1 + r)^n$, where $n = 5$, the projected population size in mid-2020 for a rate $r = 0.0077$ is

$$\begin{aligned}\text{population in 2020} &= (\text{population in 2015}) \times (1 + \text{growth rate})^5\\ &= 321{,}000{,}000(1 + 0.0077)^5\\ &= 321{,}000{,}000(1.03909)\\ &\approx 334{,}000{,}000\end{aligned}$$

Because of the limited accuracy of the estimates of population and growth rate, we round off the final answer. The result of a calculation can't be more precise than the ingredients; we round back to millions because that was the precision of our original data.

Using the same formula, a growth rate of 1% per year leads to a population of 337 million, while a growth rate of 0.5% per year yields 329 million. ■

Self Check 1

Why is a formula from banking relevant here? ■

So an uncertainty of about one-fourth of one percentage point (0.23–0.27%) in the growth rate has major implications, even over fairly short time horizons. The presence or absence of 3 to 5 million people would have a significant impact on our social and economic systems, in terms of need (or lack thereof) of daycare centers, schools, and products for babies and children. (About one-third of population growth in the United States is projected to come from net immigration.)

At the other end of the age distribution, much of the concern over long-range funding of the Social Security and Medicare programs results from uncertainties

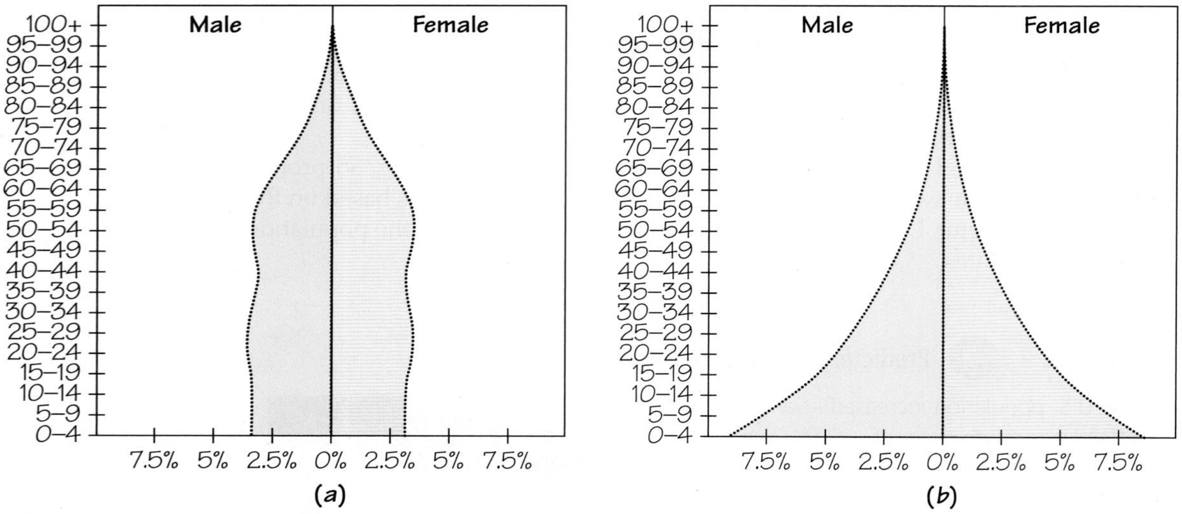

Figure 23.1 Graphs of populations in 2015, grouped by age and gender, for (a) the United States and (b) Nigeria. The horizontal scales are in percentages, not in absolute numbers, so in both graphs the total area for the population is the same (100%). However, the U.S. population was 321 million, while that of Nigeria was 184 million. (Data from populationpyramid.net/)

over birth and immigration rates. Figure 23.1a gives a graph of the U.S. population in 2015, structured by age and sex, and Figure 23.1b does the same for Nigeria.

Rates of increase in most developing nations are much higher than in industrialized nations. At a rate of 2.8%, the population of Nigeria, Africa's most populous country, will grow from 184 million in mid-2015 to 240 million in mid-2025, a shocking increase of one-third—and 60 million people!—in only 10 years. Such projections raise concern over providing sufficient food and resources for all people.

It is not just the number of people that is crucial, but also the **population structure.** In poorer countries, the proportion of the population over 60 years of age will be 20% by 2050, compared with 8% now; in Japan, where the overall population is expected to decline by one-sixth by then, it will be more than 40%.

Limitations on Growth

A population that keeps adding a fixed percentage each year, like a bank account accumulating compound interest, would eventually grow to astronomical numbers. But no biological population can continue to increase without limit (see Spotlight 23.2). Its growth is eventually constrained by the availability of resources such as food, shelter, and psychological and social "space." There may be a maximum population size that can be supported by the available resources, the **carrying capacity** of the environment.

Carrying Capacity DEFINITION

The **carrying capacity** of an environment is the maximum population size that it can support indefinitely with the available stream of resources.

As the population increases toward the carrying capacity—which we will denote by K—the growth rate decreases.

12 Billion by 2050—or Only 9 Billion?

SPOTLIGHT 23.2

How many people can the world hold? Are developing countries heading for a population disaster? Will falling fertility play havoc with Social Security in the United States? Will aging result in 50% of Japanese being over 60 in 2100?

Potential answers come from mathematical modeling of the future, using predictions of trends. The best analyses suggest a probability distribution over a range of estimates. They project separately by age, gender, education, and other characteristics. They try to factor in improvements in agriculture, spread of diseases (e.g., HIV), changes in urbanization, increases in economic aspirations, and the potential for climate change (e.g., from global warming).

A basic concept is *total fertility rate* (TFR), the average number of births per woman. Absent catastrophes (such as war or epidemic), a rate of 2.1 continues a population at the same size. A model that assumes a value above 2.1 will predict an increasing population; one that assumes a value below 2.1 will predict a dwindling one. Most of the world's population growth will occur in the lesser-developed countries, whose TFR values are well above 2.1 (e.g., it is 4.8 for Africa as a whole). However, all countries in Europe have TFR values below 2.1; without immigration, they will lose population and struggle with fewer workers to provide social benefits to the elderly. The value for the United States is 1.86, but the population is increasing nevertheless because of immigration.

China's situation illustrates demographic momentum. Its fertility rate has been below the replacement level for 20 years and is now about 1.5. To some extent, this reduction may be due to the government's policy of one child per family, though economic gains and widespread education of women may be more responsible. However, the number of women in the childbearing years was (and still is) so large that China's population will continue to grow until 2020. If present trends continue, it will then decline by 100 million by 2040 and by 250 million more by 2060. Such a swift and large decline would have disastrous economic and social consequences.

Sophisticated models try to assess how the TFR will vary with changing social circumstances. The most important single factor impacting fertility is education of women. There is a strong negative association between level of education and TFR, and the effect can be very large: In India and China, women with some college education have, on average, only half as many children as women with no education. Hence, a country's policies about education, and their success, may directly impact future population levels.

As we have revised this book for successive editions, we have seen population projections change. The estimates have decreased, because fertility rates have declined. The key questions are how to model such declines, whether they will continue, and how they will adapt to other world changes. In Spotlight 21.1 (page 878), we saw Malthus's oversimplified prediction that arithmetic growth in food supplies would limit the geometric increase of human populations. Some demographers now think that population growth will remain a serious problem in some parts of the world but that global population may stabilize or even decline after 2050.

How many people the world will have depends on how well we as a world conserve the environment, distribute food, provide jobs, produce and consume energy, and make other critical decisions about our money and resources. The key concern is the quality of life of all people. Political and economic events in far corners of the world, and even natural disasters—such as the tsunami in the Indian Ocean at the end of 2004 and the earthquakes in Haiti in 2010 and Fukushima, Japan, in 2011—impact us all. Neglecting problems faced by increasing numbers of poor people provides no security, peace, or moral refuge for anyone.

Mira/Alamy

The carrying capacity is the long-range capacity to support the population, so the population could exceed it for short periods of time. This could happen either because the population grows very rapidly and surges above the carrying capacity or because the food supply suddenly decreases, thus temporarily lowering the carrying capacity, as happens to deer and other animals in winter. The discrete **logistic model** is a simple model that provides excellent predictions for some populations.

Logistic Model DEFINITION

The discrete **logistic model** for population growth takes carrying capacity into account by reducing the population by a term that measures how close the population size P is to the carrying capacity K. The population grows from a population of P_n in year n to P_{n+1} in year $(n + 1)$ according to

$$\begin{aligned} P_{n+1} &= P_n + rP_n\left(1 - \frac{P_n}{K}\right) \\ &= (1 + r)P_n - rP_n\frac{P_n}{K} \\ &= (1 + r)P_n - \frac{r}{K}P_n^2 \end{aligned}$$

where r is the natural rate of increase when the population is small.

When the population P_n is close to 0, then P_n/K is also close to 0, the factor $\left(1 - \frac{P_n}{K}\right)$ is close to 1, and we have $P_{n+1} \approx (1 + r)P_n$, which is exponential growth. However, when the population P_n is close to K, then P_n/K is close to 1 and provides a "brake" on further population growth.

Starting from small numbers, a population following the logistic model will at first grow exponentially. Then its rate of increase will slow to a linear pace before tapering off as the population approaches the carrying capacity. Figure 23.2 shows such growth, with the population measured in terms of a fraction of its carrying capacity.

In fact, the fastest growth of the population—corresponding to the steepest slope of the population curve—occurs when the population is at the halfway point $K/2$ toward the carrying capacity K.

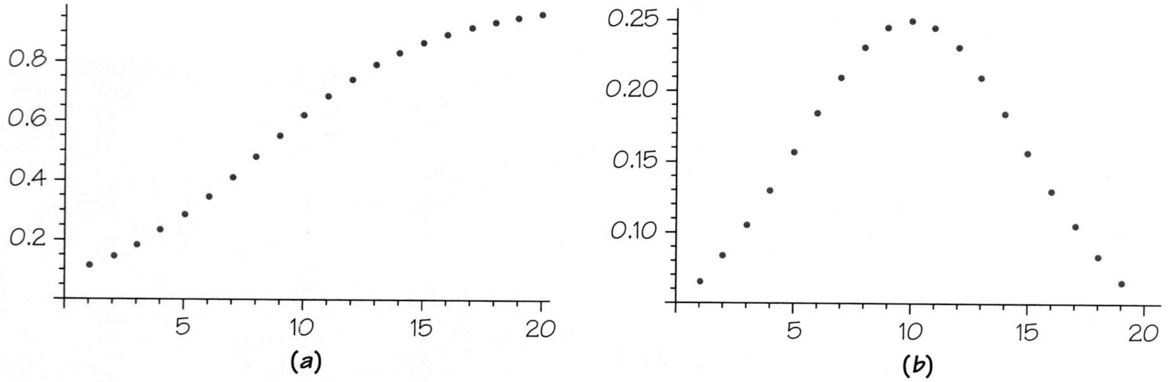

Figure 23.2 (a) Cumulative logistic growth of a population. (b) Yearly growth of a logistic population.

EXAMPLE 2 Logistic Model for the U.S. Population

How well does the historical U.S. population fit a logistic model?

The U.S. population from 1790 to 1950 closely followed a logistic model with $r = 0.031 = 3.1\%$, $P =$ population in 1790 $= 3{,}900{,}000$, and $K = 201$ million. In the first decades after 1790, the population was a small fraction of this carrying capacity, and it grew at close to the rate $r = 3.1\%$ per year (a rate higher than in many developing nations today). By 1920, the U.S. population had reached 106 million, and the growth rate had slowed by about one-half, to 1.5% per year (see Figure 23.3).

The 2015 U.S. population of 321 million far exceeds the carrying capacity of 201 million that the model suggested. Why? What was wrong with the model? The structure of the U.S. population changed, from a large proportion of people making their living on family farms to a highly urbanized society. The average number of children per family shrank. As the structure changed, the model based on assumptions of the prior structure gradually became invalid. A logistic model to the data through 2010 suggests a carrying capacity in excess of 400 million, which leaves room for growth toward the 420 million projected by the Census Bureau in 2050.

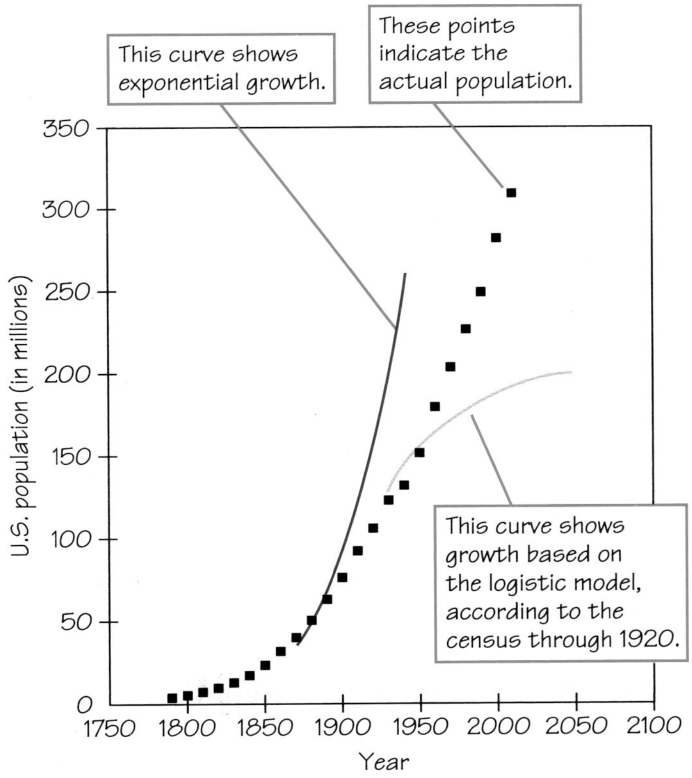

Figure 23.3 U.S. population by census year, showing actual growth, exponential (geometric) growth, and logistic growth.

Self Check 2

One data point in Figure 23.3 is noticeably lower than you might otherwise expect from the indicated trend. Why do you think that is so?

What about the world population? Figure 23.4 shows the historical trend—which does not show any sign of tapering off yet toward a carrying capacity—and three projections for the future. By the time you may retire, say, in the 2050s, you will know which came true.

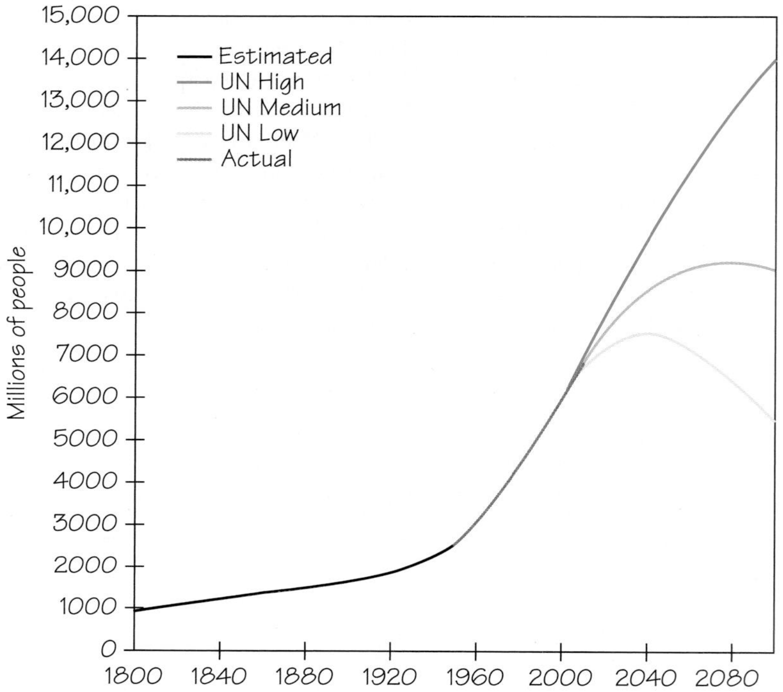

Figure 23.4 World population from 1800, with high, medium, and low projections by the United Nations to 2100. (Courtesy Loren Cobb, GFDL and Creative Commons license)

The logistic model applies not only to population growth limited by carrying capacity but also to modeling the spread of a technology or product, such as smartphones or flat-screen plasma TVs. Consider again Figure 23.2 (page 948) and think of the population as a population of smartphone sales. Initially, sales are slow. Then they begin to climb rapidly. Finally, as the market gets saturated, sales slow. We will see in the next section that the logistic model can apply also to exhaustion of nonrenewable resources.

23.2 How Long Can a Nonrenewable Resource Last?

People use resources. Some resources are renewable, but others are not. In this section, we model depletion of **nonrenewable resources.** In the next, we treat renewable resources.

Nonrenewable Resource DEFINITION

A **nonrenewable resource** is one that does not tend to replenish itself.

Gasoline, coal, and natural gas are important examples, while lottery winnings and inheritances could be examples from personal affairs. There is no practical way to recover or reconstitute these resources after use. Some substances, such as aluminum or the sand used to make glass, are potentially recyclable, but to the extent that we do not recycle them, they, too, are nonrenewable.

Neil Lee Sharp/Alamy

For a nonrenewable resource, only a fixed supply S is available. Even without human population increases, we face dwindling nonrenewable resources. We are interested in this question: How long will the supply of a resource last?

As long as the rate of use of the resource remains constant, the answer is easy. If we are using U units per year and continue using U units per year, then the supply will last S/U years.

The expression S/U is called the **static reserve.**

Static Reserve DEFINITION

The **static reserve** is how long the supply will last at a particular constant annual rate of use.

This kind of calculation is the basis for claims that at the current rate of consumption, U.S. coal reserves will last 250 years, or that the U.S. strategic reserve of gasoline (stored in underground salt domes in the South) will last 60 days.

However, the rate of use of resources tends to increase with population and with a higher standard of living. For example, projections for use of electric power often assume that use will increase by a fixed percentage each year—that is, geometrically or exponentially. This is the simplest situation (apart from constant usage) and one that we can easily model.

Suppose that $U_1 = U$ is the rate of use of the resource in the first year (this year), and that usage increases $r = 0.05 = 5\%$ each year. Then the usage in the second year is

$$U_2 = U_1 + 0.05U_1 = 1.05U$$

and usage in the third year is

$$U_3 = U_2 + 0.05U_2 = 1.05U_2 = 1.05(1.05U) = (1.05)^2U$$

This is just like compound interest. Generalizing, we see that usage in year i will be $(1.05)^{i-1}U$. Total usage over the next five years, for example, will be

$$U + (1.05)^1U + (1.05)^2U + (1.05)^3U + (1.05)^4U$$

This situation should remind you of the accumulation of regular deposits plus interest (see Chapter 21). Here, the usage U corresponds to a deposit, and the increasing rate of use r corresponds to the annual interest rate. We may think of the situation as making regular withdrawals (with interest) from a fixed supply of the nonrenewable resource. The savings formula gives

$$A = d\left[\frac{(1 + r)^n - 1}{r}\right]$$

In the resource situation, A is the accumulated amount of the resource that has been used up at the end of n years, and $U = d$ is the initial rate of use.

To find out how long the supply S will last, we set the supply S equal to the cumulative use A over n years and then determine what n has to be. We have

$$S = U\left[\frac{(1 + r)^n - 1}{r}\right]$$

We perform some algebra to isolate the term involving n, to get

$$(1 + r)^n = 1 + \frac{S}{U}r$$

At this point, to isolate n, we have to take the natural logarithm of both sides:

$$\ln[(1 + r)^n] = n \ln(1 + r) = \ln\left(1 + \frac{S}{U}r\right)$$

Doing this gives the final expression

$$n = \frac{\ln\left(1 + \frac{S}{U}r\right)}{\ln(1 + r)}$$

Algebra Review Appendix
Using Logarithms to Solve Equations

which may look complicated but is quite easy to evaluate on a calculator. The expression for n is called the **exponential reserve.**

Exponential Reserve DEFINITION

The **exponential reserve** is how long the supply S will last at an initial rate of use U that is increasing by a proportion r each year, namely

$$n = \frac{\ln\left(1 + \frac{S}{U}r\right)}{\ln(1 + r)}$$

EXAMPLE 3 U.S. Coal Reserves

Coal accounts for 30% of U.S. energy use, including 40% of U.S. electricity. Proved reserves of U.S. coal would last about 250 years at the current rate of use, so the static reserve is 250 years. How long would the supply last if the rate of use increases 0.77% per year indefinitely, the current rate of growth of the U.S. population?

The corresponding exponential reserve is

$$n = \frac{\ln[(1 + (250)(0.0077)]}{\ln(1.0077)} = \frac{\ln 2.925}{\ln 1.0077} \approx 140 \text{ years}$$

That's quite a difference!

Self Check 3

Actually, we will never run out of coal. Why not?

We must not take such projections as exact predictions. Estimates of supplies of a resource may underestimate how much is available, and previously unknown sources may be discovered or the technology improved to extract previously unavailable supplies. In addition, as supplies dwindle, the economic considerations of supply, demand, and price come into play. We will never completely run

What Are Reserves? Or, How Much Is Out There, Really?

SPOTLIGHT 23.3

There are different categories for existing amounts of nonrenewable resources, and the standard terminology reflects how certain we are about the estimates of the amounts.

The amounts are classified as either *identified* or *undiscovered.* For identified resources, we know the location, the quality, and the quantity, from geologic and engineering evidence. Undiscovered resources, however, may be hypothetical or based on speculative reasoning; and experts may disagree greatly about such amounts.

Of identified resources, the most important component is *proved reserves,* measured amounts that we know from geologic and engineering data and that can be recovered with current technology under current economic conditions. Probable reserves are further amounts, arrived at from sampling and projection from the data. Finally, there are inferred reserves, amounts about which we are less certain, since their estimates may or may not be based on samples or evidence.

out of any resource. It will always be available "at a price." In fact, climate scientists suggest that if we are to avoid severe global warming, we will have to leave two-thirds of Earth's oil and gas resources in the ground.

We must not take such projections lightly, either, because we are discussing resources that, once used, are gone forever. In any projection, it is very important to examine the assumptions, because small differences in the rate of increase of use can make big differences in the exponential reserve (see Spotlight 23.3).

EXAMPLE 4 Using Up Retirement Savings

Suppose that you begin retirement with $1 million in savings, and you don't trust banks or the stock market, so you keep it all under your mattress. Suppose it costs you $50,000 per year to live at your accustomed standard of living and there is no inflation. How long will your retirement nest egg last? How long will it last if inflation is constant at 2.35%, the average annual inflation over the past 20 years from 1995 to 2015?

The static reserve is $1,000,000/($50,000/year) = 20 years. With inflation, it will cost you increasingly more per year to live, so you should realize that your savings will last only for the length of the exponential reserve, which is

$$n = \frac{\ln[(1 + (20)(0.0235)]}{\ln(1.0235)} = \frac{\ln 1.47}{\ln 1.0235} \approx 16.6 \text{ years}$$

You have a fine strategy if you expect to live just about 17 more years and plan to die broke! ■

Self Check 4

Under those circumstances, how long would $2 million last? ■

In our examples so far, we have assumed that the resource is just sitting there, waiting to be used up. For many natural resources, however, we have to find and develop new sources. As doing that becomes more difficult and more costly, at some point the exponentially increasing demand outstrips the ability to meet that demand.

Such a situation is modeled well by the logistic model famously applied to oil by M. King Hubbert, director of Shell Oil Company's research laboratory. Figure 23.5 shows data for cumulative U.S. oil production through 2013, compared with the logistic curve for the ultimate production of $K = 240$ gigabarrels (240 billion barrels).

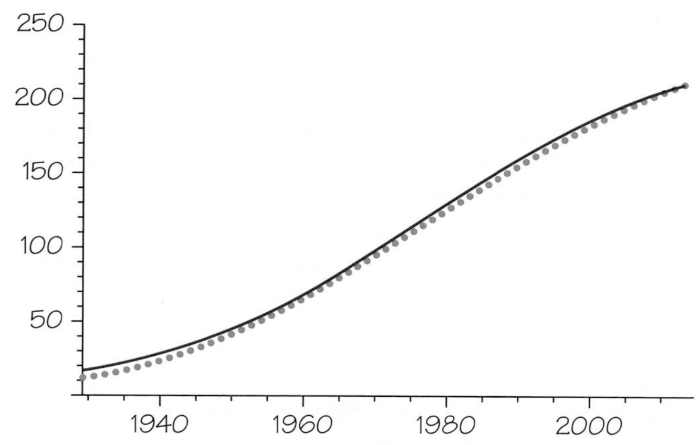

Figure 23.5 Cumulative U.S. crude oil production (blue dots) in billions of barrels, with a logistic model (solid red curve) that assumes ultimate production of 240 billion barrels.

In 1956, Hubbert predicted that U.S. production would peak in the early 1970s (it did) and decline steadily thereafter. It has done that, too, except for a blip from the Prudhoe Bay oilfield in Alaska in the 1980s and now a big bump from *fracking* (fracturing rock to release oil and natural gas) in the past few years. See Figure 23.6 (the curve is similar to but "heavier in the tails" than the normal distribution curve of Chapter 5).

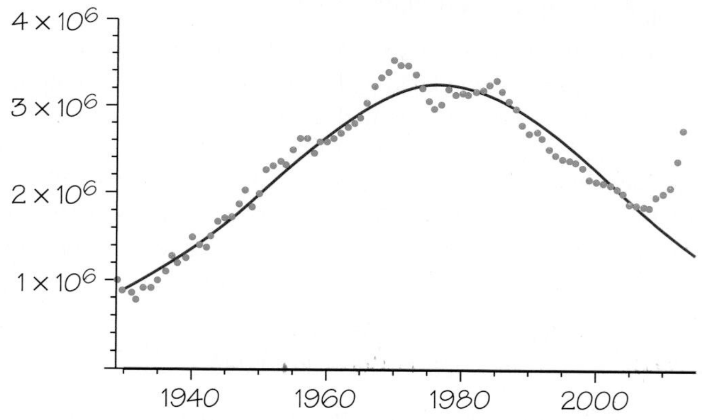

Figure 23.6 Annual U.S. crude oil production (blue dots) in billions of barrels versus year, with production as predicted by a logistic model (solid red curve) that assumes ultimate production of 240 billion barrels.

If we rearrange the logistic equation on page 948 by dividing both sides by P and doing a little algebra, we get

$$\frac{P_{n+1} - P_n}{P_n} = r - \frac{r}{K}P_n$$

Algebra Review Appendix
Graphing a Line in Slope-Intercept Form

You can recognize this as the equation of a straight line $y = a + bx$, where

- $(P_{n+1} - P_n)/P_n$ takes the role of y
- r is a
- $-r/K$ is b
- P_n takes the role of x

In other words, for a logistic model, if we graph $(P_{n+1} - P_n)/P_n$ against P_n, we get a straight line. Figure 23.7 shows that the data fit the Hubbert model fairly well until the fracking era. Whether the "bump" from fracking will be just a blip or will add substantially to ultimate production remains to be seen. In any case, we cannot count on being able to find and exploit more and more large quantities of oil.

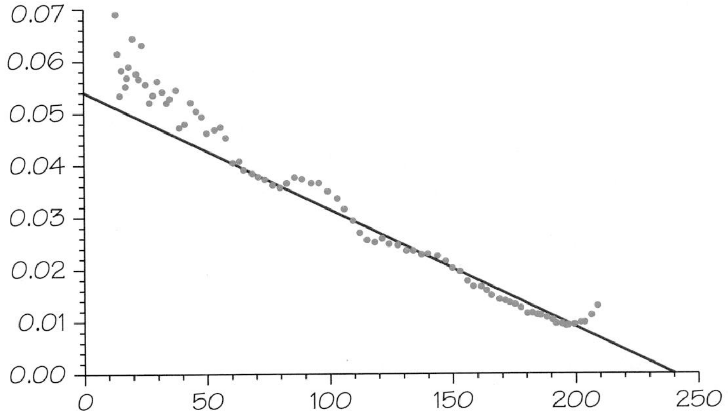

Figure 23.7 Data (blue dots) for U.S. oil production with line (solid red line) based on Hubbert's logistic model, which depicts the equation $y = 0.054 - 0.000225x$.

The world's oil and gas are running out far faster than most people realize or than their governments are willing to acknowledge (see Exercises 11 and 12). The need for "affordable" fuels will likely soon dominate the political agendas of the entire world, particularly since experts have long predicted that world oil output would peak around 2006. In fact, production leveled off in 2005–2009 at about 85 million barrels per day, or 31 billion barrels per year.

23.3 Radioactive Decay

Some populations grow (Section 23.1), others are eaten away through use (Section 23.2), and some just decay away on their own. In fact, there are some "resources" that we *want* to become exhausted, such as nuclear waste.

"Decaying" resources can have enormous economic consequences, which could involve the following:

- Unintended release of radioactive materials
- Safe disposal of nuclear waste
- Building and securing nuclear power plants
- Assuring reliable supplies of radioactive materials for medical purposes

Radioactive materials are characterized by exponential decay, which we encountered in Section 21.5 (page 890) as geometric growth with a negative rate of growth, in connection with depreciation and inflation. Here, we see it in a different light, with a different way of measuring it. Unlike the purchasing power of the dollar, which declines at an irregular rate as inflation varies, a radioactive substance emits particles and decreases in quantity at a predictable continuous rate. The negative growth rate r is usually written instead as $-\lambda$, where the (positive) quantity λ is called the **decay constant.**

Decay Constant DEFINITION

The **decay constant** for a substance decaying exponentially is the fraction of the substance that decays per unit of time.

The amount of radioactive substance remaining is given by the continuous interest formula of Section 21.4 (page 882): For an original amount P, the amount A remaining after t time units is

$$A = Pe^{-\lambda t}$$

However, the rate of decay of the substance is often described instead in terms of the **half-life.** A substance decaying geometrically never completely vanishes, even after millions of years. Since there is no time until it is all gone, we settle for measuring how long it takes until half of it is gone.

Half-life DEFINITION

The **half-life** of a substance decaying exponentially is the time that it takes for one-half of a quantity of the substance to decay.

Most chemical elements can occur in several **isotopes** (versions with the same atomic number but different atomic weights), some of which may be radioactive.

Isotope DEFINITION

An **isotope** of a chemical element is a form whose atomic nucleus contains the same number of protons (the atomic number for the element) as other forms but a different number of neutrons (giving it a different atomic weight).

For instance, iodine with atomic weight 131, iodine-131, is produced in nuclear reactors and nuclear explosions; it is a *radioisotope* (also called *radionuclide*), meaning that it is radioactive. There was great concern about the release of iodine-131 across Japan and the Pacific after the earthquake, tsunami, and subsequent nuclear reactor disasters in Japan in early 2011. Iodine in the atmosphere falls to the ground and gets into water supplies, milk, and other foods. Iodine is absorbed into the human body from food and concentrated in the thyroid gland; radioactive iodine can cause thyroid cancer.

Iodine-131 from the earlier reactor disaster at Chernobyl in Ukraine in 1986 has caused thousands of cancers over the years. Iodine-131 was also released from 1945 to 1963 as a result of atmospheric testing of nuclear weapons by the United States, the Soviet Union, and Great Britain, by France until 1974, and by China until 1980. (Humans need "good" iodine in their diet, iodine-127; but iodine-131 can replace it and cause cancer.) The half-life of iodine-131 is 8 days. This means that of 1 gram (g) of iodine-131 now, in 8 days only 0.5 g will remain (the rest will decay into nonradioactive xenon gas). Further, in $2 \times 8 = 16$ days, only 0.25 g will remain; and in $3 \times 8 = 24$ days, only 0.125 g (one-eighth of the original amount) will remain.

We can determine the half-life from the decay constant. Let H be the half-life, which means that the amount A remaining of a quantity P after time $t = H$ is $\frac{1}{2}P$. Hence, we have

$$\frac{P}{2} = A = Pe^{-\lambda H} \quad \text{or} \quad \frac{1}{2} = e^{-\lambda H}$$

To analyze this further and solve for H, we need to take the natural logarithm of each side:

$$\ln\left(\frac{1}{2}\right) = -\lambda H, \qquad H = \frac{\ln\frac{1}{2}}{-\lambda} = \frac{-\ln 2}{-\lambda} = \frac{\ln 2}{\lambda} \approx \frac{0.693}{\lambda}$$

where we have made use of the general fact that $\ln\frac{1}{C} = \ln 1 - \ln C = -\ln C$. You can check on your calculator that $\ln 2 \approx 0.693$.

Thus, we have the folllowing fundamental relationship.

Decay Constant and Half-Life Relationship RULE

$$\lambda H = \ln 2 \approx 0.693$$

EXAMPLE 5 Radioactivity in Your Home

The major radioactivity exposure for most Americans is from radon, which is formed from the decay of uranium. Radon-222 is a gas that enters many U.S. homes from the underlying soil and rock. It causes 10% to 15% of lung cancers in the United States (the rest are due to smoking). The half-life of radon-222 is 3.82 days. How long does it take for a quantity of radon-222 to decay to one-thousandth (0.1%) of the original amount?

For radon-222, we have $H = 3.82$ days and want to determine the t when $A = 0.001P$:

$$0.001P = P\left(\frac{1}{2}\right)^{t/3.82}$$

Dividing both sides by P gives

$$0.001 = \left(\frac{1}{2}\right)^{t/3.82}$$

and taking the natural logarithm of each side gives

$$\ln 0.001 \approx -6.91 = \frac{t}{3.82}\ln\frac{1}{2} = \frac{t}{3.82}(-0.693)$$

where the values $\ln 0.001 \approx -6.91$ and $\ln\frac{1}{2} \approx -0.693$ you get from your calculator. Solving for t, we find

$$t = \frac{6.91 \times 3.82}{0.693} \approx 38 \text{ days}$$

Self Check 5

Would it take only one-tenth as long for the radon to decay to only 1% of the original amount?

Carbon-14 Dating

Carbon-14 dating is a method of determining the age of organic materials, including mummies, charcoal from ancient fires, parchment, and cloth.

The element carbon, which is present in the food that we eat and in all living things, always has small traces of a radioactive form, called carbon-14. Plants and animals continually absorb carbon-14 during their lives, from the air (for plants) and from food (for animals), so that the concentration in their bodies stays the same while they are alive. Once they die, however, no new carbon-14 gets absorbed, and the carbon-14 already present decays.

Because we know the concentration of carbon-14 in living things, and we know how long it takes carbon-14 to decay, we can calculate from a sample how long ago a plant or animal was living.

The half-life of carbon-14 is 5730 years; its decay constant is

$$\lambda = \frac{\ln 2}{H} \approx \frac{0.693}{5730 \text{ yr}} \approx 1.209 \times 10^{-4}/\text{yr} = 0.0001209 \text{ yr}^{-1}$$

In other words, about 12 in 100,000 carbon-14 atoms decay each year. In each gram of carbon, approximately 814 carbon-14 atoms decay each hour. An approximate age of a sample can be determined by working backwards by half-lives. Suppose that a sample is decaying at 26 atoms per hour per gram of carbon. Table 23.1 shows that the 814 atoms per hour per gram of carbon would decrease to approximately 26 atoms per hour per gram of carbon in approximately 29,000 years, so that is the approximate age of the sample. (An age of 0 for the sample denotes the time of death of the living body.)

TABLE 23.1 Estimating the Age of a Carbon Sample

Age in Half-Lives	Age in Years	Decays per Hour per Gram of Carbon
0	0	$\left(\frac{1}{2}\right)^0 (814) = 814$
1	5,730	$\left(\frac{1}{2}\right)^1 (814) = 407$
2	11,460	$\left(\frac{1}{2}\right)^2 (814) = 203.5$
3	17,190	$\left(\frac{1}{2}\right)^3 (814) \approx 101.8$
4	22,920	$\left(\frac{1}{2}\right)^4 (814) \approx 50.9$
5	28,650	$\left(\frac{1}{2}\right)^5 (814) \approx 25.5$

EXAMPLE 6 Carbon-14 Dating

How much of the original carbon-14 would be left after 50,000 years? (This is roughly the practical age limit for carbon-14 dating of the typically small samples available.) How many atoms would be decaying per hour per gram of carbon?

We have

$$A = Pe^{-\lambda t} \approx Pe^{-0.0001209 \times 50,000} \approx Pe^{-6.05} \approx 0.0024P$$

so only about 0.24% of the original amount remains. This remaining amount would be decaying at a rate of $0.0024 \times 814 \approx 2$ atoms per hour per gram of carbon. An accurate estimate for the age of a sample of 1 milligram (mg) of carbon might take weeks or months of tallying counts on a decay counter. (The analysis is now often done by much faster atomic mass spectrometry rather than by counting decays, and calibration adjustments must be made for varying amounts of atmospheric carbon-14 over the years. Our calculations do not take such calibration, much of which is done from tree rings, into account.)

Self Check 6

For a 1-mg sample of carbon 50,000 years old, how many decays would you expect over a one-month period?

EXAMPLE 7 How Old?

We turn the question around to determine the age from the observed number of decays: How old is a sample that is decaying at a rate of 105 grams per hour per gram of carbon?

The formula that relates t, the age of the sample in years, and N, the number of carbon-14 atoms disintegrating per gram per hour, is

$$\left(\frac{1}{2}\right)^{t/5730} = \frac{N}{814}$$

Solving for N gives

$$N = 814 \times \left(\frac{1}{2}\right)^{t/5730}$$

Using natural logarithms, we can solve for t as

$$\ln N = \ln 814 + \frac{t}{5730} \ln \frac{1}{2}, \qquad t = 55{,}403 - 8267 \ln N$$

Thus, a sample decaying at a rate of 105 atoms per hour per gram of carbon would be $t = 55{,}403 - 8267 \times \ln 105 \approx 16{,}930 \approx 17{,}000$ years old.

Self Check 7

How much difference would it make if the decay rate was instead 110 atoms per hour per gram of carbon?

Although we are interested in the exhaustion and decay of nuclear waste, we are also interested in maintaining supplies of some radioactive materials:

- Even before the Japanese disaster, there was concern about nuclear power—not just in terms of safety, but also whether enough uranium could be mined for a "nuclear future" that would involve a great many more reactors to meet the world's demand for electricity without adding to its output of carbon dioxide.
- Some radioactive isotopes are used for medical imaging and as tracers in the body, such as technetium-99m (the "m" stands for "metastable"), which is used in 20 million diagnostic procedures each year. In 2010, two of the five reactors that produce molybdenum-99 (which decays into technetium-99m) closed for repairs, reducing capacity by two-thirds. The short half-life of molybdenum-99 (66 hr) and the even shorter half-life of technetium-99m (6 hr) mean that weekly supplies of "fresh" molybdenum-99 are essential.

23.4 Harvesting Renewable Resources

A **renewable natural resource** is a resource that tends to replenish itself, such as fish, wildlife, and forests. How much can we harvest and still allow for the resource to replenish itself?

Renewable Resource DEFINITION

A **renewable resource** is one that tends to replenish itself.

Other renewable resources are biological populations. We concentrate on the subpopulation with commercial value. For a forest, this might be trees of a commercially useful species and appropriate size. We measure the population size in terms of its **biomass,** the physical mass of the population. For example, we measure a fish population in pounds rather than in number of fish, and a forest not by counting the trees but by estimating the number of board feet of usable timber.

Reproduction Curves

Our models for growth include many simplifications. Complicated factors that can affect populations, such as climatic or economic change, may mean that the only way to understand a population is to plot a graph of its size over time. Either from data, a model, or both, we can construct a **reproduction curve** to predict next year's likely population size (biomass) based on this year's size. Although the precise shape of the curve varies from one population to another, the shape in Figure 23.8 is typical. For all possible sizes, it shows the size this year (on the horizontal axis) and the size next year (on the vertical axis), taking into account the

Algebra Review Appendix
Function Notation

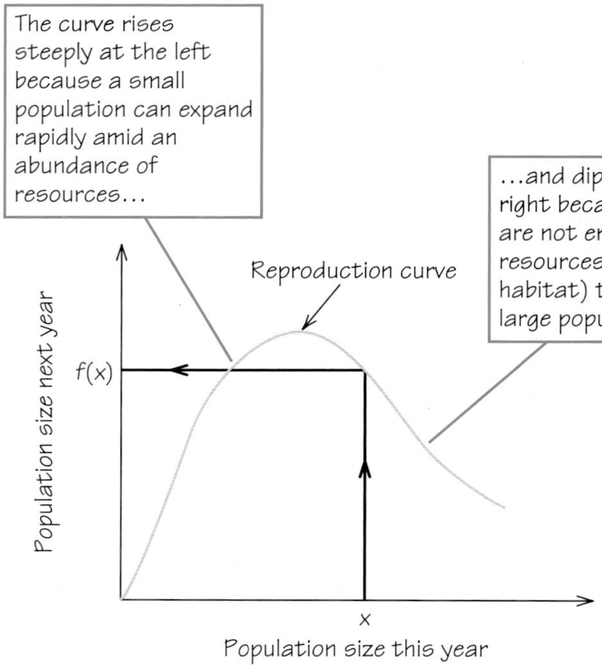
The curve rises steeply at the left because a small population can expand rapidly amid an abundance of resources...
...and dips on the right because there are not enough resources (food, good habitat) to support a large population.
Reproduction curve
Population size next year
f(x)
x
Population size this year

Figure 23.8 A typical reproduction curve.

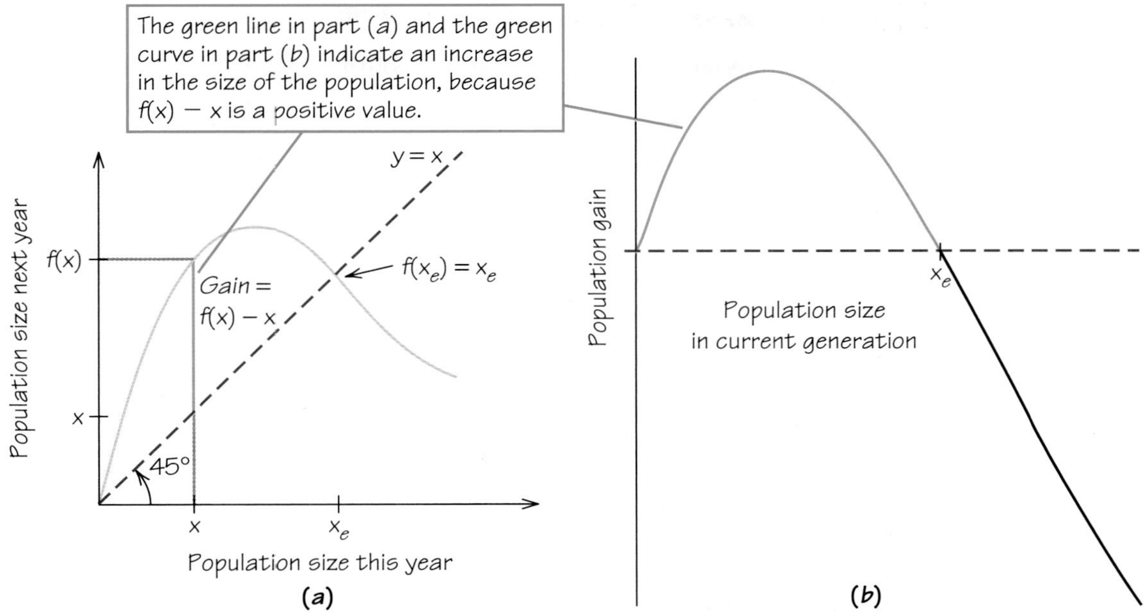

Figure 23.9 Depiction of the natural increase (gain) in population from one year to the next. The population size x_e is the equilibrium population size, for which the population one year later is the same, or $f(x_e) = x_e$.

growth in size of continuing members and addition of new members, minus losses due to death and other factors.

Let x on the horizontal axis be a typical size of the population in the current year. The size next year is given by the height of the curve above the point marked x. This value is denoted by $f(x)$. (You can think of f as standing for "function of," or even as "forthcoming.") Figure 23.9a shows the reproduction curve, plus the *line of exact replacement:* the broken line $f(x) = x$ (which makes a 45° angle with the horizontal axis) that corresponds to the population next year being exactly the same as this year's. You can trace what happens for various choices for x. For an x for which the curve is above the broken line, next year's size, $f(x)$, is larger than this year's, x.

In Figure 23.9b, the **natural increase,** or gain in population size, is shown as the length of the green vertical line from the broken line to the curve, which in algebraic terms is $f(x) - x$. For an x for which the curve is below the broken line, next year's size is smaller than this year's and $f(x) - x$ is negative. For the size labeled x_e, for which the curve crosses the broken line, the size is the same next year as it was this year. This is the **equilibrium population size.** (Zero is also an equilibrium size for the population, but not one that we are interested in attaining!)

Equilibrium Population Size DEFINITION

An **equilibrium population size** is one that does not change from one year to the next.

Sustainable-Yield Harvesting

For fishing, the **yield** (harvest) depends on both the population of fish and the amount of fishing effort (number of boats, hours of fishing).

We model a fishery whose goal is to harvest the same yield every year on a sustainable basis. A more sophisticated model could take into account fishing effort (number of boats, number of days out) and adjust the harvest accordingly. However, the simpler model will bring out the main features about sustainable yield. Also, a policy of a fixed harvest limit is easier for authorities to enforce than trying to estimate how many fish are out there.

Before we consider the harvest, in Figure 23.10 we take another look at the reproduction curve and the line of exact replacement, this time with numbers involved. The green area corresponds to situations for which there will be more fish next year than this year, and the red area denotes fewer fish next year. The point between the two areas is the natural equilibrium: 10 million tons of fish this year reproduce to give 10 million tons next year.

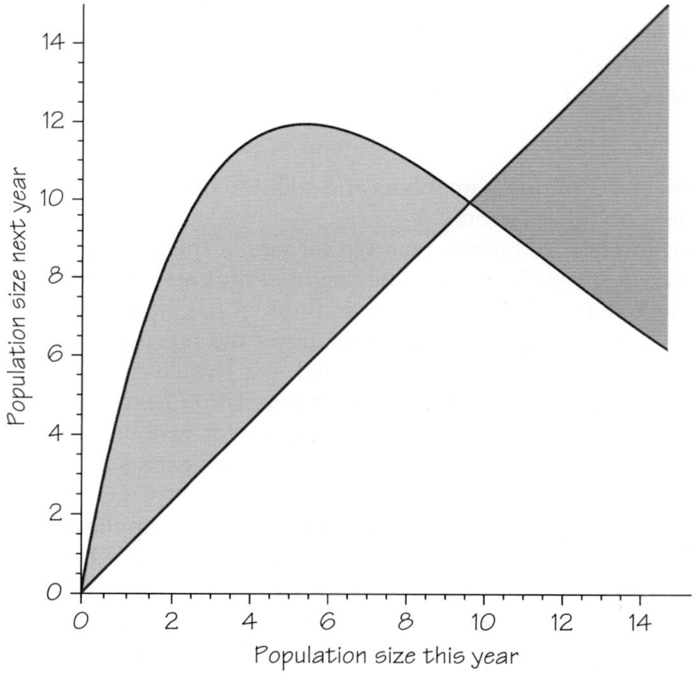

Figure 23.10 Population gain (green) and loss (red) from one year to the next, for an after-harvest population this year.

How much fish could we harvest on a sustainable basis? For a sustainable harvest, we harvest all of the gain and bring the population back to the same level as before the current season's growth, from which the population should generate the same gain next year. For a **sustainable yield,** the same amount is harvested every year and the population remaining after each year's harvest is the same. To achieve this stability, the harvest must exactly equal the natural increase each year, the length of the green vertical line in Figure 23.9a.

Sustainable-Yield Harvesting Policy DEFINITION

A **sustainable-yield harvesting policy** is a policy that would maintain the same yield if continued indefinitely.

Let's try out a few possible fish harvests.

EXAMPLE 8 Harvest Strategies

What are the consequences of some harvest strategies?

- Suppose that we harvest 4 million tons of fish out of the 10 available this year and leave 6 million tons. That means we leave 6 million as the now remaining population this year to reproduce for next year. We follow up from 6 on the horizontal axis to the blue curve and find that next year the population will be 12 million. Next year we could then harvest 6 million [the distance between the reproduction curve (blue) and the line of exact replacement (red)] and again leave 6 million. We could then continue that pattern indefinitely, for a sustainable catch of 6 million tons.
- Suppose that we are greedier and take 8 million tons of fish this year out of the 10 million available, leaving 2 million tons. We are lucky—those 2 million will reproduce to 8 million. But we can't take 8 million tons next year, or we would wipe out the population. However, if next year we take 6 million, we would again be leaving 2 million. We could then continue that pattern indefinitely, for a sustainable catch of 6 million tons.

Self Check 8

Describe what happens in the next several years if we constantly harvest 6 million tons of fish.

Both of the scenarios in the example yield (after the first year) a sustainable 6 million tons. Which is better for the stock of fish? Surely, we would be better off to leave 6 million in the sea rather than 2 million. After all, our simplistic model is not allowing for chance events (bad weather, changed ocean currents, etc.) that might not allow as much reproduction as our reproduction curve estimates. And there would always be the temptation to take still more of the 6 million.

Is 6 million tons of fish the best we could do for a sustainable harvest? Figure 23.11 on the next page shows the population gain for each level of population left this year to reproduce. From that, you can see that a **maximum sustainable yield (harvest)** of a little over 7 million tons is possible, together with each year leaving 4 million tons to reproduce.

How much fish can be harvested "safely" and on a sustainable basis depends on knowing where we are on the population gain curve. But that curve is difficult to determine, as is estimating how many fish are out there. Moreover, the

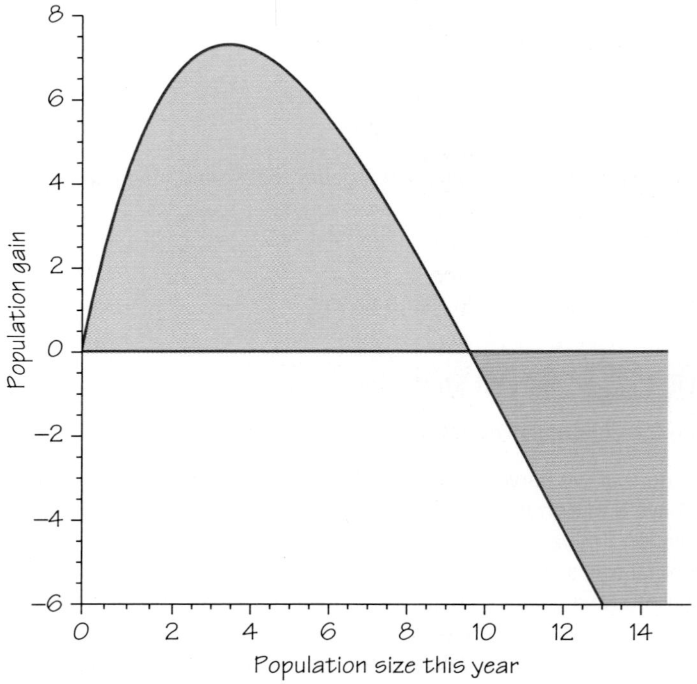

Figure 23.11 Population gain (green) and loss (red) from one year to the next, for an after-harvest population of fish this year.

curve is an ideal that varies with changes in the weather, the environment, and other fish populations. Finally, we will see that not just environmental factors but other ones—plus the possible involvement of mathematical chaos—also come into play.

In reality, a successful fishery attracts more fishers and more boats. A result may be more fish, at greater expenditure of overall effort, but then a lower price for fish. As a result, fishers have to fish more hours to maintain incomes, meaning more pressure on the fishery. Either cooperation among the fishers (as in Alaska's Bristol Bay) or government regulation (as for lobsters off the Maine coast) may be needed to avoid overexploitation and possible extinction of the resource (see Spotlight 23.4).

Dynamics of a Population over Time

The line $y = x$ provides a convenient way to trace the evolution of the population over several years (see Figure 23.12a), by alternating steps vertically to the curve and horizontally to the line $y = x$. Begin with the first year's population on the horizontal axis and go up vertically to the curve. The height is the population in the second year. Proceed horizontally from the curve over to the line $y = x$. Proceeding vertically from there to the curve yields a height that is the population in the third year. The result is a **cobweb diagram,** so-called because it resembles a spiderweb. A cobweb diagram gives a convenient visual representation of the evolution of the population over time.

Fishing: Out to Sea?

SPOTLIGHT 23.4

Patrick Ford/LatitudeStock/Alamy

Fishing in Malawi's Senga Bay for a catch of young kapenta fish.

Natural resources are the heritage of all humans, and societies should take steps to maintain that heritage for future generations. Major tools for doing so in connection with hunting and fishing of public resources are limits on harvests: seasonal (when), catch (how much), and size/length (what).

The goal of such limits is to regulate the harvesting in a sustainable way, leaving enough of the resource unharvested to allow for reproduction. Size/length limits, in particular, are aimed at avoiding harvesting of immature members of the population, so that they can grow to maturity and reproduce (and be harvested later when they are larger). Such limits are imposed in both commercial fishing [e.g., Maine lobsters, which must be at least 3.25 in. (83 mm) from eye socket to end of carapace] and sport fishing [e.g., muskellunge fish ("muskies") in Wisconsin must be at least 40 in. long]. In the ocean, trawlers must have nets that allow smaller fish to escape.

Such selective harvesting has major effects:

- As you might expect, the average size and the average age of fish in the sea both decrease. This is called an *age truncation effect.*
- The average size of fish caught also decreases. The "big ones" that didn't get away just aren't as big as they used to be, as attested by photos of trophy fish caught in Florida in the 1950s compared to those caught today. This result, too, is expected, both because of increased fishing over time and because the fish tend to be harvested as soon as they reach minimum size rather than having the opportunity to grow larger.
- The age of maturation of the fish decreases, which in turn has the paradoxical effect of increasing the growth rate of the population. Scientists have observed this phenomenon also in natural predator–prey relationships:
  - With little fishing in the Mediterranean during World War I, comparison of data for fish harvests before, during, and after the war shows that catching equal percentages of both sharks and their prey at a constant rate raises the population of the prey.
  - In his video *Earth: A New Wild: Oceans* (2015), conservation scientist M. Sanjayan is shown swimming amidst absolutely immense numbers of sharks at the Palmyra Atoll in the Pacific—so many that "all the little fish in the ocean around here have to be reproducing fast in order to feed all these hungry mouths. . . . So everything is forced to reproduce younger."
- Though an increase in the reproduction rate is good for the population, less obvious is that age truncation also increases variability and instability in the dynamics of the fish population. The dynamics are nonlinear, meaning that an effect can be out of proportion to its cause, so there is potential for chaos (see Section 23.5). Age truncation creates greater potential for booms and busts, and such wild fluctuations can result in systematic and permanent decline in fish levels.

Nevertheless, it is not selective harvesting but overharvesting that has endangered (and even eliminated) stocks of wild fish and animals. Economic circumstances, including increases in population, can bring pressure to bear on conservation decisions. In particular, lifting a fishing ban when the total stock of fish has recovered—but the age stratification has not—carries risk to the resource and to the livelihoods of those who depend on it.

Figure 23.12 shows several traces for the same reproduction curve, each starting from a different initial population on the horizontal axis. The resulting variation is surprising and can be "chaotic" in a very specific mathematical sense, showing how apparently random behavior can result from strict rules.

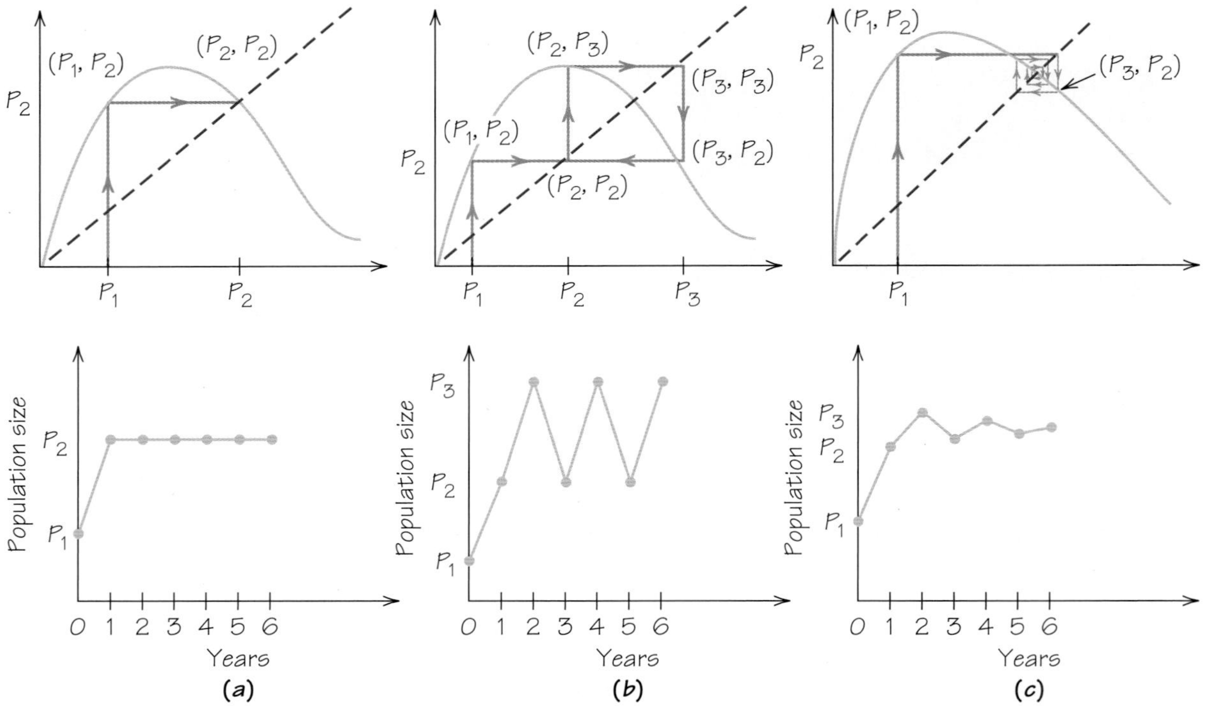

Figure 23.12 Examples of the dynamics, over time, for the same reproduction curve but different starting populations. (a) The population goes in one year to the equilibrium population and stays there year after year. (b) After initial adjustment, the population cycles between values over and under the equilibrium population. (c) The population spirals in toward the equilibrium population.

We consider two models: one for a cattle ranch and one that can apply to either a fishing boat or a tree farm.

We assume that the price p received is the same for each harvested unit and does not depend on the size of our harvest. In effect, we assume that our operation is a small part of the total market, not substantially affecting overall supply and hence price.

We want to stay in business, so we do not extinguish the resource for quick profits. For any given population size, we harvest just the natural increase.

EXAMPLE 9 Cattle Ranching

We assume that the cost of raising and taking a steer to market is the same for every steer and does not depend on how many steers we take to market. What should our sustainable-yield harvesting policy be?

Because the cost does not depend on the population size, the cost curve is a horizontal line (Figure 23.13). As long as the selling price per unit is higher than the harvest cost per unit, we make a profit. The points of view of economics and biology agree, because the maximum profit occurs for the maximum sustainable yield.

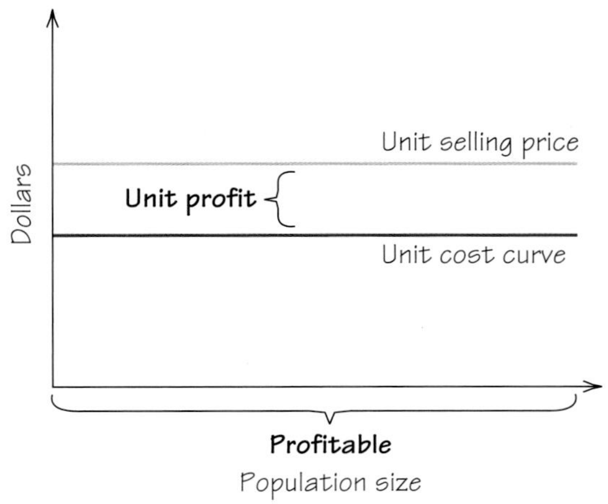

Figure 23.13 The unit cost, unit revenue, and unit profit of harvesting 1 unit, as a function of population size, for the cattle ranch.

EXAMPLE 10 Fishing and Logging

In this model, we assume that the cost of harvesting a unit of the population decreases as the size of the population increases. This is the familiar principle of **economy of scale.** For example, the same fishing effort yields more fish when fish are more abundant. Similarly, a logger's harvest costs per tree are less when the trees are clumped together. This is the logger's motivation to clear-cut large stands. What would a sustainable-yield harvesting policy be?

The cost curve slopes downward and to the right, as in Figure 23.14. The size of the population from which 1 unit is harvested is shown on the horizontal axis. The (average) cost of harvesting a single unit is measured on the vertical axis.

An optimal sustainable-yield harvesting policy depends on the relation between price and costs. There are two cases, as shown in Figure 23.15 on the next page, for a fixed price level.

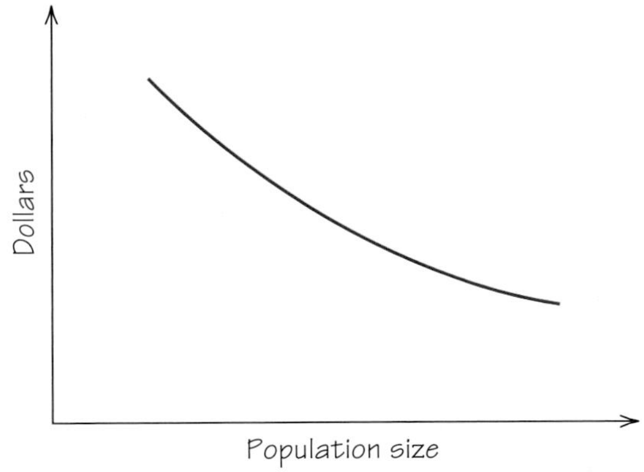

Figure 23.14 The unit cost, as a function of population size, for fishing or logging.

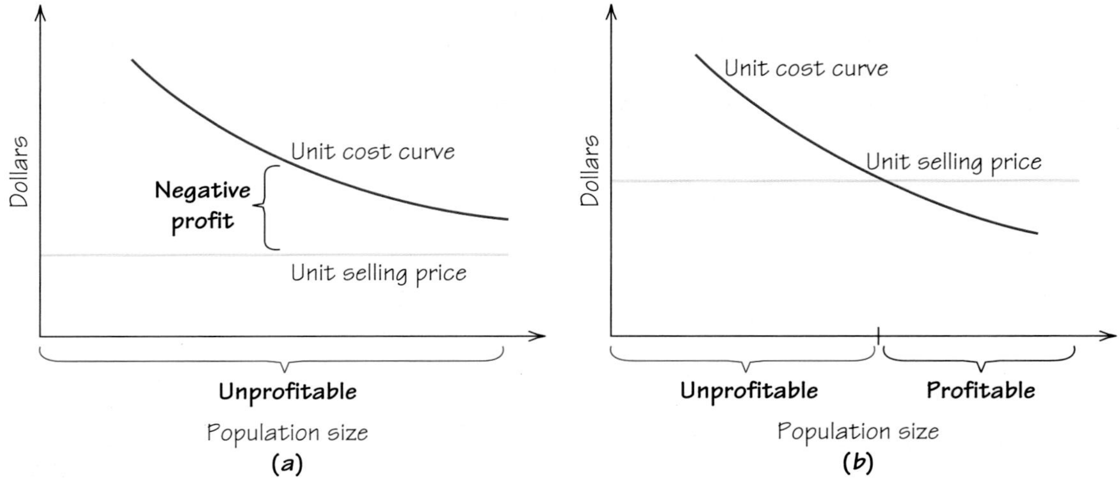

Figure 23.15 The unit cost, unit revenue, and unit profit of harvesting 1 unit, as a function of population size, for fishing or logging. (a) The market price is below the harvesting cost for all population sizes. (b) The operation is profitable for populations above a certain minimum size.

- The unit cost curve lies entirely above the unit price line (Figure 23.15a). The revenue received for a harvested unit is less than the cost of harvesting it, no matter how large the population. It is impossible to make a profit.
- The unit cost curve intersects the unit price line (Figure 23.15b). Above a certain population size, the price for a harvested unit is more than the cost of harvesting it, so profit is possible. Some population size, the *economically optimal harvest,* gives a maximum net profit. That harvest is lower than the maximum sustainable yield but leaves a greater number x_Q to reproduce for next year.

Using calculus, it can be shown that x_Q is actually larger than x_M, the population that gives the maximum sustainable yield (see Figure 23.16).

Compared to harvesting the maximum sustainable yield, economic considerations result in harvesting less, but also in maintaining a larger stock of the population.

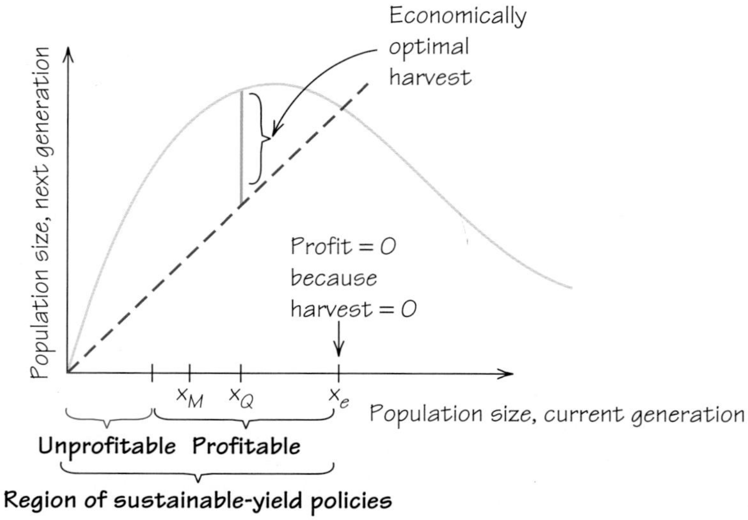

Figure 23.16 Reproduction curve showing regions of profitability for sustainable-yield policy, with the economically optimal population size x_Q marked.

This consequence may seem paradoxical, but the simple reason is that it is cheaper—it takes less effort—to catch fish when there are more of them out there.

Why Eliminate a Renewable Resource?

Our simple models fail to take into account a very critical feature of a modern economy: the time value of money, as measured by the interest that capital can earn. We now investigate how the time value of money makes biological populations susceptible to overexploitation and even extinction.

Some species, such as the passenger pigeon, were harvested to extinction. In other cases, an entire ecosystem has been destroyed (see Spotlight 23.5). Why would anyone eliminate a renewable resource? Our approach helps explain why.

The Tragedy of Easter Island

SPOTLIGHT 23.5

Lisa Strachan/Shutterstock

Easter Island is famous for its isolation—it is 1400 miles from the nearest island—and for its hundreds of huge stone statues. For 30,000 years before the arrival of people in about 400, Easter Island maintained a lush forest, with several species of land birds. By the time of the first visit by Europeans in 1722, the island was barren—denuded of all trees and bushes over 10 feet high, and with no native animals larger than an insect. The 2000 or so islanders had only three or four leaky canoes made of small pieces of wood.

What happened? Careful analysis of pollen in soil samples tells the sad story. The settlers and their descendants cut wood to plant gardens, build canoes, make sledges and rollers to move the huge statues, and burn for cooking and warmth in the winter. In addition to crops that they raised and chickens that they brought to the island and cultivated, they ate palm fruit, fish, shellfish, the meat and eggs of birds, and the meat of porpoises that they hunted from seagoing canoes. The population of the island grew to 7000 (or perhaps even 20,000).

By 1500, the forest was gone. Most tree species, all land birds, half the seabirds, and all large- and medium-sized shellfish had been extinguished. There was no firewood, no wood for sledges and rollers to transport hundreds of statues at various stages of completion, and no wood for seaworthy canoes. Without canoes, fishing declined and porpoises could not be taken. Stripping the trees exposed the soil, which eroded, so crop yields fell. The people continued raising chickens, but warfare and cannibalism ensued. By 1700, the population had shrunk to 10% to 25% of its former size.

Why didn't the people notice earlier what was happening, imagine the consequences of keeping on as they had been, and act to avert catastrophe? After all, the trees did not disappear overnight.

From one year to the next, changes may not have been very noticeable. The forests may have been regarded as communal property, with no one charged with limiting exploitation or ensuring new growth. There was no quantitative assessment of the resources available and the need for conservation versus the long-term needs of the "public works" program of erecting statues. Moreover, the religion of the people, the prestige of the chiefs, and the livelihood of hundreds depended on the statue industry. There was no perceived need to limit the population and no technology for birth control.

Once the large trees were gone, there was no means for excess population to emigrate.

Information from Jared Diamond, "Easter's end," *Discover*, 16(8) (August 1995): 63–69.

Sustainable-yield policies involve revenues that will be received, year after year, in the future. The value of these revenues is discounted to reflect the lost investment income that we could earn if instead we had the revenues today. For funds invested at a return of $100r\%$ per year, compounded annually, the present value P of an amount A to be received in n years in the future is related to A by the compound interest formula $A = P(1 + r)^n$.

The economic goal is to maximize the sum of the present values of all future receipts from harvesting. The optimal harvesting policy thus must depend on the expected rate of return r. We don't delve into the details of the calculations here, but instead just give the results of the analysis.

Again, there are several cases to consider:

1. The unit cost of harvesting exceeds the unit price received for all population sizes. Then it is impossible to make a profit.
2. For small r: For some population size x, the unit cost of harvesting equals the unit price received. Then there is a size between x and x_e (the equilibrium population size) for which the present value of the total return is maximized and the population and its yield are sustained. This size may be larger or smaller than x_M, the population that gives the maximum sustainable yield; but for large enough r, it will be smaller.
3. For larger r, *the economically optimal policy may be to harvest the entire population, immediately extinguish the resource, and invest the proceeds in something else.* The unit price exceeds the unit cost for all population sizes.

Let's put this in the simplest and starkest terms. Suppose that you own a resource, such as a forest, whose cost of harvesting is small relative to its value. If the rate at which the trees grow is greater than what you can earn on other investments, it pays to let the forest keep on growing.

On the other hand, if the forest grows more slowly than the rate of return on other investments, the economically optimal harvesting policy is to cut down all the trees now and invest the money in some other asset. For example, you could then start raising cattle on the land—and right there you have the scenario that is resulting in deforestation all over the world.

The sobering fact is that *very few economically significant renewable resources can sustain annual growth rates over 10%.* Many, like whales and most forests, have growth rates in the 4% to 5% range. These values—even a growth rate of 10%—are far below the return that many investors expect on their investment. For example, over the long run, the U.S. stock market has yielded an average 10% annual return, but venture capital firms expect to exceed a 25% annual profit.

The concept of maximum sustainable yield is an attractive ideal if the expectations of investors are low enough. However, there are still difficult problems:

- One problem is the "tragedy of the commons," discussed by ecologist Garrett Hardin. Several hundred years ago, English shepherds would graze their flocks together on common land. The grass of the commons could support only a fixed number of sheep. Each shepherd could reasonably think that adding just one or two more sheep to his flock would not overtax the commons. Yet if each did so, there could be disaster, with all the sheep starving. Many natural-products industries, such as fisheries, are a form of commons. Small overexploitation by each harvester can produce disastrous results for all. Global warming may be a tragedy of a worldwide commons.
- How, in the presence of human needs or greed, can we anticipate and prevent overexploitation and possible extinction of a resource? By and large, it has been

politically impossible to force a harvesting industry to reduce current harvests to ensure stability in the future.

- In some industries, such as a fishery, growth of the population may be abundant one year but meager another. Moreover, varied harvesting pressure and selectivity can magnify fluctuations in the age structure and the abundance of the resource (see Spotlight 23.4, page 965). So a steady yield can be neither guaranteed nor sustained without some risk of damaging the resource. For example, a few good years in a row may provoke increased investment in fishing capacity; then attempting to harvest at the same levels in succeeding normal or below-normal years results in overfishing. This exact scenario destroyed the California sardine fishery in the 1930s, the Peruvian anchovy fishery in 1972, and much of the North Atlantic fishery in the 1980s. The ocean off northwest Africa was nearly picked clean during the past decade.

Were the fishing fleets and regulators mentioned above at fault for extinguishing these fisheries by overexploiting a dependable resource? Or were the extinctions due to chance variations of the fish stocks? In the next section, we examine a third possibility—that the fish stocks followed simple rules that nevertheless produced "chaotic" behavior of stocks, that is, wide variation from one year to the next. When we do not see the pattern, we interpret such behavior as randomness, much as the moves in a chess game may appear random and inexplicable to someone who does not know the rules of the game.

23.5 Dynamical Systems and Chaos

In this and the two preceding chapters, we have considered systems that change over time: bank accounts, the amount due on a loan, and the size of a population. Other examples are a dripping faucet, a playground swing, a pinball play, the solar system, the business cycle, epidemics, the passage of a drug through the human body, and the weather. Some of these systems are very predictable (interest on a bank account), while others are notoriously unpredictable (the weather). Some involve no outside influences (the amount due on a loan, assuming that you don't get behind on payments!), while others are the result of many contributing factors (the business cycle).

In some systems (such as the population of a country), the state of the system may depend largely on its states at previous times (e.g., last year's population), while in other systems (such as an epidemic) chance may play a large role (e.g., in who and how many become infected). We are interested in modeling systems, such as a fishery, as they operate without influence from outside or from chance. The applicable mathematical tool is a **dynamical system.**

Dynamical System DEFINITION

A **dynamical system** is a mathematical model for a system whose state evolves with time and whose future states depend deterministically on its present and past states.

To make this definition meaningful, we need to be explicit about what is meant by **deterministic.**

Deterministic DEFINITION

A system is **deterministic** if its changes through time depend only on natural and mathematical laws and are not substantially affected by what we consider to be chance or free will.

An example of a deterministic system is the path of a golf putt, which is governed by gravity, terrain, wind, and the force imparted by the golfer. A nonexample is the outcome of a vigorous toss of a coin or a random number generator; although the result, like the golf putt, is determined by physical laws, we consider the result to be random. Another nonexample is the outcome of an election, which involves choices by humans.

Mathematical Chaos

We think of **chaos** as referring to general confusion, unpredictability, and apparent randomness. Mathematicians and other scientists use the word to describe systems whose behavior over time is inherently unpredictable.

Chaos DEFINITION

A dynamical system exhibits **chaos** if it is:

1. *Orbitally dense*—any state is near one that eventually will recur
2. *Transitive*—from any state you can eventually get close to any other
3. *Sensitive*—a small change in the initial state can produce widely diverging results later

EXAMPLE 11 Chaos in Manhattan

You may already know from experience that getting around Manhattan can be a chaotic experience, in the ordinary sense of the word. How is Manhattan's subway system also chaotic in the mathematical sense?

1. Orbitally dense: Subway trains "orbit," periodically visiting subway stops, and everybody lives near a subway stop.
2. Transitive: You can get close to anywhere else in Manhattan by taking the subway.
3. Sensitive: If you get on the wrong train, you could wind up miles from where you want to be.

Because the system covers the island of Manhattan, #1 is actually a consequence of #2. Also, anyone who has gotten on the wrong subway train or bus realizes that #3 is an inevitable consequence of #1 and #2. So, in fact, #1 and #3 both follow from #2—a conclusion that is true not just of this Manhattan example but of a large class of dynamical systems.

Self Check 9

Why does #1 follow from #2 for the Manhattan subway system?

The most noticeable property of a chaotic system is sensitivity—that a small change now can make a big difference later.

This feature is sometimes known as the **butterfly effect,** from the title of a 1979 talk by meteorologist E. N. Lorenz (1917–2008): "Predictability: Does the Flap of a Butterfly's Wings in Brazil Set Off a Tornado in Texas?" (The phrase probably traces to a 1953 science fiction story by Ray Bradbury, "A Sound of Thunder," in which history is changed by a time-traveler who steps on and kills a prehistoric butterfly.)

We can get a feel for chaotic systems by playing with some examples of an **iterated function system (IFS).** The fancy name just means that we take an initial value, apply a function to it, then repeat over and over. This is exactly what we did earlier, geometrically, with reproduction curves for populations. (See Section 19.5, pages 807–811, for more about IFS and their connection to fractals.)

Iterated function system (IFS) DEFINITION

An **iterated function system (IFS)** is a sequence of numbers or geometric objects in which each next element is produced from the previous one by applying a consistent function (rule) to the previous element.

EXAMPLE 12 Doubling on a "Stone Age" Calculator

Imagine that you have a calculator that keeps only the last two digits of a number. It has a special key marked [DBL] that doubles the number in the display and keeps only the last two digits. For example, [DBL] applied to 52 gives 04 (not 104). Let's start with two numbers that are as close together as can be on this calculator, such as 37 and 38. As we push the [DBL] key over and over again, will the result stay close?

We get 37, 74, 48, 96, 92, 84, 68, 36, 72, 44, 88, 76, 52, 04, . . . 38, 76, 52, 04, 08, 16, 32, 64, 28, 56, 12, 24, 48, 96, . . . Already, by the fourth iteration, the two sequences are far apart. The function used in this iterated function system is

$$f(x) = 2x \bmod 100$$

where the mod notation of modular arithmetic (introduced in Chapter 17 on page 715) means to take the remainder when $2x$ is divided by 100.

Self Check 10

What happens if you start with 01?

EXAMPLE 13 The Solar System

The American moon landings in 1969 and later, as well as all other space missions, were possible because of the predictability, or determinism, of the solar system. The moon and planets follow their orbits like clockwork. So how could the solar system be chaotic?

Over tens of millions of years, the orbit of each planet is chaotic, meaning that the slightest change in its position or velocity—due to, say, a comet passing nearby—could produce a huge difference later. In mid-2015, we learned that at least two of Pluto's five moons (Nix and Hydra, discovered only in 2005) rotate and tumble chaotically.

More down-to-earth examples of physical systems that can exhibit chaos include the fluttering of a falling autumn leaf, heart arrhythmias, and the Tilt-A-Whirl amusement park ride.

Chaos in Biological Populations

If we measure this year's population as a fraction x of the carrying capacity, and do the same for next year's population as a fraction $f(x)$, the logistic population model takes the form

$$x_{n+1} = f(x_n) = \lambda x_n(1 - x_n)$$

where the Greek letter lambda $\lambda = 1 + r$ is the amount by which the population is multiplied when the population x is near 0. When expanded, the equation has the familiar form of a quadratic in x:

$$x_{n+1} = -\lambda x_n^2 + \lambda x_n$$

Values of λ slighlty larger than 1 correspond to natural growth rates r near 0, and the behavior we noted earlier, of the population gradually creeping toward the carrying capacity, takes place. But for larger values of λ, corresponding to high growth rates of the population, there are other outcomes, as the next example illustrates.

The population can more quickly expand and (especially) contract (as it gets close to the carrying capacity).

EXAMPLE 14 The Logistic Population Model

What behaviors can occur in the logistic model?

For different values of the parameter λ and different starting values for the population fraction, each of the behaviors of Figure 23.12 on page 966 can occur:

- $\lambda = 2.8$ and starting population fraction $x = 0.357$ produces the values 0.357, 0.643, 0.643, . . . , the pattern shown in the lower graph in Figure 23.12a.
- $\lambda = 3.1$ and starting population fraction $x = 0.235$ produces the values 0.235, 0.557, 0.765, 0.558, 0.765, 0.558, . . . , the pattern shown in the lower graph in Figure 23.12b.
- $\lambda = 2.5$ and starting population fraction $x = 0.550$ produces the values 0.550, 0.619, 0.590, 0.605, 0.598, 0.601, 0.599, . . . , the pattern shown in the lower graph in Figure 23.12c.

In other words, for population growth rates (values of λ) that are fairly close together (2.5, 2.8, 3.1), the population evolves very differently. This is a surprising and nonintuitive conclusion.

But there is more. For $\lambda = 4$ and any starting population fraction, the population does not settle down into any of the patterns of Figure 23.12; year after year, it wanders "unpredictably" all over the place (Figure 23.17). This is chaotic behavior: It is deterministic, complex, but—in the long run—unpredictable. In the short run, the behavior is completely predictable. For example, from this year's population fraction, the equation tells us exactly what next year's will be. Repeating the use of the equation, we can determine what it will be the following year. But as the years pass, any sense of pattern gets lost in the complexity.

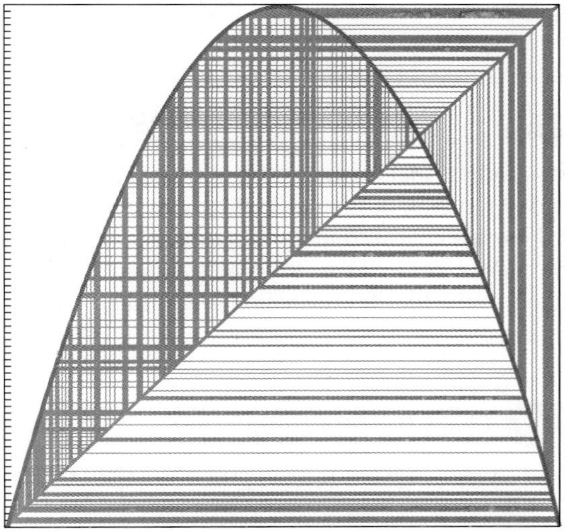

Figure 23.17 Cobweb diagram of chaotic behavior of a population.

Self Check 11

Which pattern do you arrive at if you start with $\lambda = 2.7$ and starting population fraction $x = 0.5$?

The potentially chaotic behavior of a biological population is bad news for those who manage a ranch or any biological population in the wild or in captivity. In recent years, the lobster catch in Maine has been much higher than in previous years, reaching record levels, for no discernible reason. On the other hand, in the late 1950s, the annual harvest of Dungeness crabs off the central California coast declined from 12 million pounds to less than 1 million pounds without any evidence of disease, heightened predation, or increased crabbing effort. Researchers who modeled that population in 1994 found that booms and busts are the rule. The population can remain nearly level for generations before suddenly exploding or crashing without warning.

Searching for an environmental cause for these fluctuations could be futile, because there may not be one. Moreover, observing the population over a few generations provides no help in predicting future behavior.

EXAMPLE 15 Childhood Disease Epidemics

The incidence of childhood diseases such as chickenpox and measles varies greatly from year to year. Why?

There are four plausible explanations for the fluctuations:

1. There is an underlying regular cycle that is perturbed and occasionally overwhelmed by random events ("noise").
2. There is no pattern, because the fluctuations are due solely to chance.
3. There is no discernible pattern, because such fluctuations are inherent in the epidemiology of the disease, a chaotic system.
4. Parents refuse to vaccinate their children against the disease, or the vaccine becomes less effective, thereby increasing the numbers of children susceptible.

- The first explanation, a perturbed cycle, fits chickenpox, with a cycle of one year.
- For measles, either the second or the third explanation may be correct, depending on the size of the community. For small communities, chance is an adequate explanation. For large communities, historical data from before the era of mass immunization suggest that measles cases were chaotic. That doesn't mean that they occurred at random, but rather that they were unpredictable. Research also shows that there is a critical community size above which a disease will not die out solely by chance. For measles, this size is about 250,000.
- The two factors in the fourth explanation seems to account for recent flare-ups in whooping cough.

EXAMPLE 16 Chaos in Your Laptop

Does a computer always do what it is supposed to? How could your laptop computer be chaotic?

The Intel i5 chip common in many computers has 731 million transistors. They are subject to physical laws and environmental conditions, and they don't all always respond in the same amount of time to do their commanded tasks. Researchers have found that running the same program (of several billion instructions) repeatedly under identical conditions can take greatly varying amounts of time. However, for some programs, the pattern of times displays clear evidence of mathematical chaos.

What you need to understand about chaos is that behavior that appears to be random can be produced even with very simple systems that are completely governed by deterministic rules. Just because the behavior appears chaotic does not mean a lack of underlying order and structure, though discovering that structure may be difficult.

Even if we discover the structure, prediction may elude us because of chaos. If we had an absolutely correct model of how weather behaves and measurements at every location on and above Earth, we might still not be able to forecast the weather accurately a week ahead.

What about the fishery extinctions? Perhaps the fishers and the fish were victims not of greed or chance, but of the chaotic nature of the reproduction curve for the fish.

Review Vocabulary

Biomass A measure of a population in common units of equal value. (p. 960)

Butterfly effect A small change in initial conditions of a system making an enormous difference later on. (p. 973)

Carrying capacity The maximum population size that can be supported indefinitely by the available resources. (p. 946)

Chaos Complex but deterministic behavior that is unpredictable in the long run. (p. 972)

Cobweb diagram A kind of graphical portrayal of the evolution of a dynamical system, such as a population. (p. 964)

Compound interest formula Formula for the amount in an account that pays compound interest periodically.

For an initial principal P and effective rate r per year, the amount after n years is $A = P(1 + r)^n$. (p. 944)

Continuous interest formula For a principal P in an account at a nominal annual rate r, compounded continuously, the balance after n years is $A_n = Pe^{rn}$. (p. 944)

Decay constant For a substance decaying exponentially, the fraction that decays per unit time. (pp. 955, 956)

Deterministic A system is deterministic if its future behavior is completely determined by its present state, past history, and known laws. (p. 972)

Dynamical system A system whose state depends only on its states at previous times. (p. 971)

Economy of scale Costs per unit decrease with increasing volume. (p. 967)

Equilibrium population size A population size that does not change from year to year. (p. 961)

Exponential reserve How long a fixed amount of a resource will last at a constantly increasing rate of use. A supply S, at an initial rate of use U that is increasing by a proportion r each year, will last

$$n = \frac{\ln\left(1 + \frac{S}{U}r\right)}{\ln(1 + r)}$$

years. (p. 952)

Half-life For a substance decaying exponentially, the time that it takes for one-half of a quantity to decay. (p. 956)

Isotope A form of a chemical element whose atomic nucleus contains the same number of protons as other forms (the atomic number of the element) but a different number of neutrons (giving it a different atomic weight). (p. 956)

Iterated function system (IFS) A sequence of elements (numbers or geometric objects) in which each next element is produced from the previous one by applying a consistent function (rule) to the previous element. (p. 973)

Logistic model A particular population model that begins with near-geometric growth but then tapers off toward a limiting population (the carrying capacity). (p. 948)

Maximum sustainable yield (harvest) The largest harvest that can be repeated indefinitely. (p. 963)

Natural increase The growth of a population that is not harvested. (p. 961)

Nonrenewable resource A resource that does not tend to replenish itself. (p. 950)

Population structure The division of a population into subgroups. (p. 946)

Rate of natural increase Birthrate minus death rate; the annual rate of population growth without taking into account net migration. (p. 945)

Renewable natural resource A resource that tends to replenish itself; examples are fish, forests, wildlife. (p. 960)

Reproduction curve A curve that shows population size in the next year plotted against population size in the current year. (p. 960)

Savings formula Formula for the amount in an account to which a regular deposit is made (equal for each period) and interest is credited, both at the end of each period. For a regular deposit of d and an effective interest rate r per year, the amount A accumulated after n years is

$$A = d\left[\frac{(1 + r)^n - 1}{r}\right] \qquad \text{(p. 944)}$$

Static reserve How long a fixed amount of a resource will last at a constant rate of use; a supply S used at an annual rate U will last S/U years. (p. 951)

Sustainable yield A harvest that can be continued at the same level indefinitely. (p. 962)

Sustainable-yield harvesting policy A harvesting policy that could be continued indefinitely to maintain the same yield. (p. 963)

Yield The amount harvested. (p. 962)

Self Check Answers

1. The compound interest formula is not just for banking but for any "population" that increases geometrically (exponentially).

2. The data point for 1940 reflects a low birthrate during the Great Depression in the 1930s.

3. We could import it, a rising price would reduce use, or we would find substitutes.

4. 28.5 years

5. Reducing to 0.1% $= \frac{1}{1000}$ takes about 10 half-lives, since $\left(\frac{1}{2}\right)^{10} = \frac{1}{1024}$. Since $\left(\frac{1}{2}\right)^{7} = \frac{1}{124}$, reducing to 1% $= \frac{1}{100}$

takes not quite 7 half-lives, hence a little less than $7 \times 3.82 \approx 27$ days.

6. $\dfrac{2 \text{ decays}}{\text{hour-gram}} \times \dfrac{1 \text{ gram}}{1000 \text{ milligrams}} \times \dfrac{31 \times 24 \text{ hours}}{\text{month}} \approx 1.5 \dfrac{\text{decays}}{\text{month}}$

7. The age would then be about 16,500 years.

8. The populations over the next years are 10 (this year), minus harvest of 6, leaving 4; 7.5, minus harvest of 6, leaving 1.5; 5, minus harvest of 6, resulting in extinction.

9. Just go to a subway stop and wait for a train that (via connections) goes to where you want to go.

10. After 02, you wind up eventually in the same loop as when starting with 37 or 38: 04, 08, 16, 32, 64, 28, 56, 12, 24, 48, 96, 92, 84, 68, 36, 72, 44, 88, 76, 52, 04, ... It can be shown that the length of the loop (20) must divide the modulus (100).

11. The sequence goes 0.500, 0.675, 0.592, 0.652, 0.613, 0.641, ..., and centers down on $1.7/2.7 \approx 0.630$.

Skills Check

1. The carrying capacity of a population is the
(a) largest recorded population.
(b) largest supportable population.
(c) change in population.

2. The logistic population model can model not only human or animal populations but also ________.

3. A population with a total fertility rate of 2 will
(a) double in each generation.
(b) remain exactly the same size from generation to generation.
(c) decrease.

4. China's population exhibits demographic ________.

5. The logistic curve is a model for a population that is growing
(a) linearly.
(b) exponentially.
(c) with a ceiling.

6. U.S. oil production has grown according to a(n) ________ model.

7. Management of a nonrenewable resource can be modeled by
(a) an annuity.
(b) a savings account.
(c) an add-on loan.

8. If we have enough of a resource to last 200 years at the current rate of use, but the rate of use increases by 10% per year, the supply will last ________ years.

9. The static reserve is always ________ the exponential reserve.
(a) greater than
(b) equal to
(c) less than

10. If we have enough of a resource to last 1000 years at the current rate of use, but the rate of use increases by 1% per year, the supply will last ________ years.

11. If a sample contains 20 grams of carbon-14, whose half-life is 5730 years, in how many years will only 5 grams of carbon-14 remain in the sample?
(a) 4 years
(b) 11,460 years
(c) 22,920 years

12. Plutonium-239 (produced in breeder reactors and used in atomic bombs) has a half-life of approximately 24,200 years. Of 4000 grams of plutonium-239 now (the approximate content of one atomic bomb), in how many years will only 500 grams remain?

13. The equilibrium population size is the same as the
(a) carrying capacity.
(b) intersection point of the reproduction curve and the diagonal.
(c) natural increase of the population.

14. The shape of the reproduction curve reflects that a small population has abundant resources and can grow quickly by the curve ________ steeply at the left.

15. A sustainable-yield harvest is
(a) possible at exactly one size of a population.
(b) not always possible.
(c) possible at various sizes of a population.

16. A cobweb diagram shows the ________ of a population.

17. If the starting population for a reproduction curve is changed, the subsequent population pattern will
(a) eventually return to the same pattern.
(b) always change to a different pattern.
(c) sometimes change to a different pattern.

18. Harvesting at an economically optimal level maintains a __________ stock of the population than harvesting the maximum sustainable yield.

19. Economic considerations __________ the conservation of a resource.

(a) always work against

(b) never interfere with

(c) always affect in some way

20. The harvest yield that leads to the maximum net profit under sustainable-yield harvesting is always __________ than the maximum sustainable yield.

21. A system whose current state depends solely on its previous states is called

(a) a dynamical system.

(b) a stable system.

(c) an optimal system.

22. For the logistic model $f(x) = 4x(1 - x)$, a starting population fraction that will immediately lead to 0 population is __________.

23. Chaotic behavior appears to be random and is __________ due to chance.

(a) actually not

(b) actually always

(c) sometimes

24. For the logistic model $x_{n+1} = f(x) = 3x_n(1 - x_n)$, if the starting population fraction is $x_0 = 0.5$, the next population fraction is __________.

25. Pressing a digit and then repeatedly pressing the [SIN] key is a model of

(a) an iterated function system.

(b) chaos.

(c) randomness.

26. The butterfly effect refers to a feature of chaos called __________.

27. Mathematical chaos occurs in

(a) heart arrhytmias.

(b) the solar system.

(c) amusement park rides.

28. Biological populations can be at risk of extinction from any of __________, __________, and/or __________.

29. Mathematical chaos

(a) happens only in mathematics.

(b) is just another word for random events.

(c) can happen in deterministic systems.

30. The three characteristics of mathematical chaos in a dynamical system are __________, __________, and __________.

Chapter 23 Exercises

Challenge   Discussion

23.1 Growth Models for Biological Populations

1. (Spreadsheet helpful) For many years, China has been the world's most populous country. However, India has been catching up, with 1282 million in mid-2015 and growing at 1.5% per year, versus China in mid-2015 with 1402 million and growing at only 0.5% per year. If these rates continue, when would India have more people than China?

2. The population of the less-developed countries (excluding China) in mid-2015 was 4.7 billion and expected to grow at 1.7% per year (this is an annual yield, so you may think of it as compounded annually). If this growth rate were to continue until mid-2025, what would the size of the population be then?

3. If the growth rate of the less developed countries of Exercise 2 is actually 1.6% rather than 1.7%, what would the size of the population be in mid-2025?

4. (Spreadsheet helpful) If the growth rate of the less developed countries of Exercise 2 can be reduced from 1.7% by 0.04% per year from 2015 through 2025, beginning in mid-2015, what would the size of the population be in mid-2025?

5. An advertisement for Paul Kennedy's book *Preparing for the Twenty-First Century* (Random House, 1993) asked: "By 2025, Africa's population will be: 50%, 150% or 300% greater than Europe's?" The population of 740 million in Europe in mid-2015 was expected to increase 1%—not per year, just 1%—by 2025. The population of 1166 million in Africa in mid-2015 was expected to increase at about 2.6% per year. Which answer would you give to the question?

6. Similar to the rule of 72 used in banking and explained in Exercise 11 in Chapter 21, the rule of 69 says that if a country's population continues to grow at a constant rate of r% per year, then it will double in size every $69/r$ years. Apply the rule of 69 to estimate the doubling times for the following populations (figures are for mid-2015).

(a) Africa, 1166 million, 2.6%

(b) United States, 321 million, 0.77%

7. Do the calculations as in Exercise 6 for:

(a) China, 1.402 billion, 0.5%

(b) The world as a whole, 7.309 billion, 1.2%

8. Wisconsin's electricity demand increased 3.03% per year over the 35 years from 1970 to 2005. If that trend continued, when would Wisconsin need to have twice as much generating capacity as it did in 2005?

NASA Earth Observatory image by Robert Simmon, using Suomi NPP VIIRS data provided courtesy of Chris Elvidge (NOAA National Geophysical Data Center)

Is Warren Sanderson right about world population growth slowing down (see Spotlight 23.2, page 947)? How much difference does it make in projections if we look at the world as a whole or break it down by countries or regions? In Exercises 9 and 10, we investigate this question, first projecting as a whole and then projecting by regions and adding.

9. The world population of 7.309 billion in mid-2015 was expected to increase 1.2% per year.

(a) Project the population to mid-2035 (by then you may have finished having children, if you have any) and to mid-2050 (by then you may be thinking about retiring).

(b) What are the assumptions involved in your projections?

10. Divide the countries of the world into three groups with differing rates of increase (see the accompanying table). (Why is this useful?)

(a) Redo the projections in Exercise 9a for the years 2035 and 2050 by projecting each group separately and adding the totals. Is there a major difference from the results in Exercise 9a?

Group	Population Mid-2015 (billions)	Rate of Growth (%)
More developed countries	1.25	0.1
Less developed countries (excluding China)	4.70	1.7
China	1.40	0.5

(b) Will the world be able to support the numbers of people that you project? What problems could these greater numbers of people cause? What could be done to avert those problems? Do you think that anything will be done before there is some kind of worldwide crisis?

(In mid-1995, the world population was 5.7 billion and growing at 1.5% per year. Those figures led to projections, as in Exercise 9, of 8.3 billion for 2020 and 11.1 billion for 2040. The corresponding growth rates for the groups of Exercise 10 were 0.2%, 2.2%, and 1.1%, which led to projections of 8.9 billion for 2020 and 12.7 billion for 2040.)

23.2 How Long Can a Nonrenewable Resource Last?

11. At the start of 2015, world oil proved reserves (including oil sands suitable for fracking) totaled 1326 billion barrels, while daily consumption was 93 million barrels. The U.S. Energy Information Administration (EIA) projected in 2010 that world consumption would increase 1.0% per year through 2035.

(a) What was the static reserve for oil at the start of 2015?

(b) What was the exponential reserve for oil at the start of 2015?

(c) What considerations may affect the answers to parts (a) and (b) over time?

12. At the start of 2010, world natural gas proved reserves totaled 6609 trillion cubic feet, while annual consumption was 119 trillion cubic feet in 2011. The EIA projected in 2010 that world consumption would increase 1.0% per year through 2035.

(a) What was the static reserve for natural gas at the start of 2010?

(b) What was the exponential reserve for natural gas at the start of 2010?

(c) What considerations may affect the answers to parts (a) and (b) over time?

13. In his State of the Union address in January 2012, President Barack Obama said, "We have a supply of natural gas that can last America nearly one hundred years." The U.S. Energy Information Administration had stated in October 2011 that the United States has 2543 trillion cubic feet (Tcf) of "potential" natural gas resources (including both identified and undiscovered resources). (One-third of that total was gas from shale, extractable by fracking.) In 2010, U.S. consumption of natural gas was 23.8 Tcf and projected to increase at 0.6% per year.

(a) What was the static reserve for natural gas in 2011?

(b) What was the exponential reserve then?

14. In an update to its 2011 report noted in Exercise 13, in 2014 the Energy Information Administration estimated the United States had 1932 Tcf of "technically recoverable dry natural gas resources," consumption in 2012 was 25.6 Tcf, and growth in consumption was projected to be 0.8% per year from 2012 to 2040. (Actual reserves amounted to just 308 Tcf.)

(a) What was the static reserve for natural gas "resources" in 2014?

(b) What was the exponential reserve then?

15. Not just the chrome trim on your car, but stainless steel and superalloys also require the metal chromium. In *The Limits to Growth* in 1972, author Donella H. Meadows and colleagues noted that the world reserves of chromium were 775 million tons, with consumption of 1.85 million tons and consumption growing at 2.6% per year.

(a) What was the static reserve for chromium in 1972?

(b) What was the exponential reserve then?

16. In 2013, world reserves of chromium were about 880 million tons, with consumption of 26 million tons and consumption growing at 4% per year.

(a) What was the static reserve for chromium in 2013?

(b) What was the exponential reserve then?

17. Can our energy problems be solved by increasing the supply? [Thanks for the idea to Evar D. Nering of Arizona State University, The mirage of a growing fuel supply, *New York Times* (June 4, 2001), op-ed page.]

(a) Suppose that we have a 100-year "supply" of a resource (such as oil, for which known world reserves will last less than 100 years at the current world rate of use). That is, the resource would last 100 years at the current rate of consumption. Suppose that the resource is consumed at a rate that increases 2.5% per year (this is the average increase in consumption for oil in the United States since 1973). How long would the resource last?

(b) Suppose that we greatly underestimated the supply and actually have a 1000-year supply at the current rate of use. How long would that last if consumption increases 2.5% per year?

(c) Let's think big and suppose that there is 100 times as much of the resource as we thought—a 10,000-year static reserve. How long would that last if consumption increases 2.5% per year?

18. We explore the consequences of reducing the rate of growth of oil use. Suppose that we halve the growth rate from the 2.5% per year given in Exercise 17 to 1.25% per year. [Thanks for the idea to Evar D. Nering of Arizona State University, The mirage of a growing fuel supply, *New York Times* (June 4, 2001), op-ed page.]

(a) How long would the 100-year static reserve last?

(b) How long would the 1000-year static reserve last?

(c) How long would the 10,000-year supply last?

19. We continue the ideas of Exercises 17 and 18, but with a more radical hypothesis.

(a) How long would the 100-year supply last if we reduced our consumption by just 0.5% per year—that is, if we used 0.5% less each year instead of 2.5% more?

(b) How long would it last if we used 1% less each year?

20. By the time there is concern about using up a nonrenewable resource, it may be too late. Suppose that a resource has a static reserve of 10,000 years, but consumption is growing at 3.5% per year.

(a) How long would the resource last?

(b) How long before half the resource would be gone?

(c) How long to use up the second half?

(d) How much longer would the resource last if after half of it is gone, consumption is stabilized at the then-current level?

(e) What implications do you see to your answers?

21. Do a calculation to criticize the claim in the following quotation: "The United States holds 437 billion tons of known (coal) reserves, enough energy to keep 100 million large electric generating plants going for the next 800 years or so." [*Forbes* (December 15, 1975), p. 28; thanks to Albert A. Bartlett.]

For Exercises 22–24, refer to the following. The formula for the average growth rate over a period of time is even simpler than the one for the exponential reserve of a resource. If usage at the beginning is N_0 and at the end of an interval of t years it is N, then the average annual rate of growth is

$$\frac{1}{t}\ln\left(\frac{N}{N_0}\right)$$

22. The U.S. population was 63 million in 1890 and 321 million in 2015 (at roughly the same times of the year). What was the average annual rate of growth, to two decimal places?

23. The U.S. population was 3.9 million in 1790 and 63 million in 1890. What was the average annual rate of growth, to two decimal places?

24. The average annual increase in oil consumption in the United States during 1993–2004 was nearly 2%. In fact, consumption in 1993 was 6.291 billion barrels and consumption in 2004 was 7.588 billion barrels. What is a more accurate (two-decimal-place) estimate of the average annual percentage increase in oil consumption? (From 2004 through 2007, growth was zero, due to much higher prices for oil; and growth was again zero from 2008 to 2011, due to the world recession.)

23.3 Radioactive Decay

25. The carbon-14 in a carbon sample is decaying at approximately 6.5 atoms per hour per gram of carbon. Determine the approximate age of the sample.

26. The "Ice Man" is the popular name for the body of a man that was found in 1991 preserved in a glacier in the Tyrolean Alps. Carbon-14 dating revealed that he had died about 5200 years ago. About how many atoms of carbon-14 are breaking down today per hour per gram of carbon in his tissue?

27. (Contributed by John Oprea of Cleveland State University.) The Nuclear Test-Ban Treaty of 1963 brought an end to atmospheric testing of nuclear weapons (except by France and China). Testing had released into the atmosphere the radioactive isotope strontium-90. Just as with iodine-131, it settled out of the air onto grass in fields, was eaten by cows, and wound up in children's milk. In the body, strontium-90 mimics calcium and is absorbed into the bones, where its radiation can cause cancer; its half-life is 28.8 years. Assuming that none of the strontium-90 absorbed into the bones of children in 1962 has been otherwise excreted, approximately what proportion will still remain 58 years later, in 2020?

28. After one week, what percentage of a supply of molybdenum-99 (with a half-life of 66 hr) remains?

For Exercises 29–36, refer to the following. The physical half-life of a radioactive substance is unaffected by the biochemistry inside the human body. However, such a substance may not remain in the body for good but instead be excreted at some rate, thereby reducing the radiation exposure. The rate of excretion can be expressed in the form of a biological half-life. Let the physical half-life be H_p and the biological half-life be H_b. Then a combined effective half-life H_e can be calculated from

$$\frac{1}{H_e} = \frac{1}{H_p} + \frac{1}{H_b}$$

The two paths for eliminating the radioisotope act in parallel. In mathematical terms, H_e is one-half the harmonic mean of the two half-lives (Chapter 19, Exercise 16). Sometimes physical decay is the dominant influence (Exercises 29–31) and sometimes biological clearing is (Exercise 32).

29. Iodine-131 actually has benign uses; it is used in treating thyroid cancer, hyperthyroidism, and non-Hodgkins lymphoma. Its physical half-life is 8 days; its biological half-life is 138 days. What is its effective half-life in the body?

30. Apart from uranium itself, the four major radioactive isotopes present in reactors and their waste are iodine-131, strontium-90, strontium-89, and cesium-137. Of the four, the isotope with the longest physical half-life is cesium-137, at 30 years. Found in high levels in water near the ruined Japanese Fukushima reactors in 2011, it is absorbed by ocean plants and moves up the food chain with increasing concentrations. Its biological half-life is 70 days. What is its effective half-life in the body?

31. Technetium-99m has a physical half-life of 0.25 days and a biological half-life of 1 day. What is its effective half-life in the body?

32. Phosphorus-32 is used to treat excess production of red blood cells in bone marrow. It has a physical half-life of 14.3 days and a biological half-life of 1155 days. What is its effective half-life in the body?

33. The physical half-life of carbon-14 is 5730 years, while its biological half-life is 40 days. The relatively short biological half-life explains why over a lifetime the proportion of carbon-14 in a body equilibrates (becomes equal) to the level in the atmosphere. What is the effective half-life of carbon-14 in the body?

34. In Exercise 27, we asked what proportion of strontium-90 remains in the body after 58 years, assuming no biological clearing and asking you to apply the physical half-life of 28.8 years. However, strontium-90 is indeed cleared from the body, with a biological half-life of about 50 years. Based on the effective half-life, give a different answer to what proportion of strontium-90 remains in the body after 58 years.

35. Reinterpret the formula involving H_e in terms of decay constants and explain why the formula makes sense.

36. Joe can split a cord of wood in 1 hour and Sam can split a cord in 2 hours. Explain why the amount of time that it would take both of them working together to split a cord is one-half the harmonic mean of 1 and 2. *Hint:* Consider first the case where it takes each of them separately 1 hour (Chapter 19, Exercise 16).

23.4 Harvesting Renewable Resources

For Exercises 37–41, refer to the following. We suppose that a population has the reproduction curve shown in the accompanying figure, with units of thousands of tons of biomass. The mathematical description is that the population in the following year, x_{n+1}, depends on the population x_n in the current year (after any harvest) according to

$$x_{n+1} = f(x_n) = 4x_n(1 - 0.05x_n)$$

for x_n between 0 and 10, in units of millions of pounds. We start with a population x_0 this year whose value we can vary. (For these exercises, a spreadsheet or a programmable calculator is useful.)

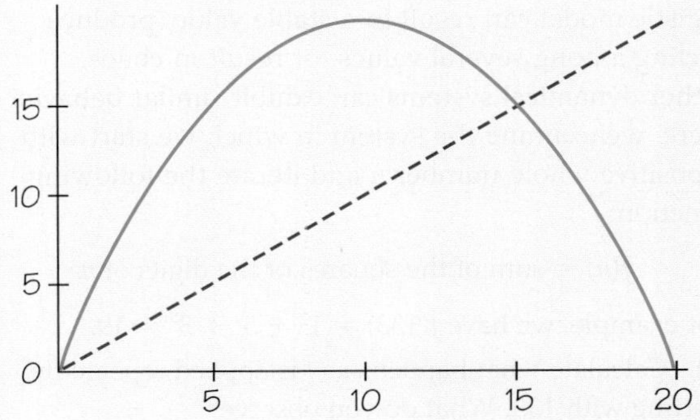

37. Start with $x_0 = 5$. Calculate numerically the population in the first few years, draw a cobweb diagram, and briefly describe the qualitative behavior of the population.

38. Repeat Exercise 37 with the starting value $x_0 = 10$.

39. Repeat Exercise 37 with the starting value $x_0 = 7$, going at least as far as x_9.

40. Try to find a starting value (besides 0, 5, 10, and 15) that leads to a stable population over time.

41. What is the equilibrium population size?

For Exercises 42–52, refer to the following. We suppose that the population has the reproduction curve shown in the following figure, with units of thousands of tons of biomass, whose mathematical description is

$$x_{n+1} = f(x_n) = 3x_n(1 - 0.05x_n)$$

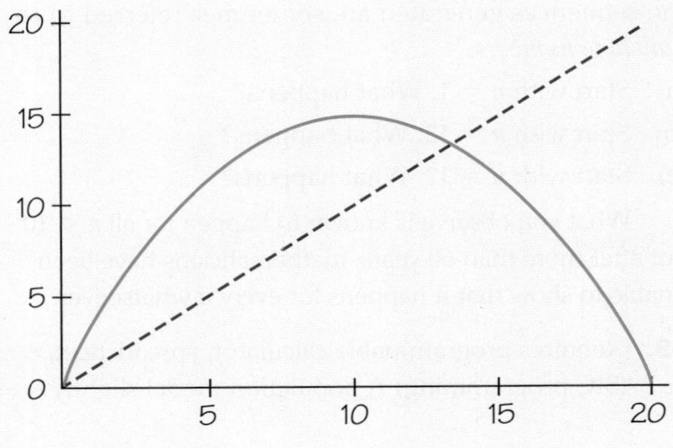

42. Start with $x_0 = 5$. Calculate numerically a population in the first 10 years, draw a cobweb diagram, and briefly describe the qualitative behavior of the population.

43. Repeat Exercise 42 with the starting value $x_0 = 10$.

44. Try to find a starting value (besides 0 and 20) that leads to extinction of the population.

45. What is the equilibrium population size?

46. Which of the reproduction curves—the one for Exercises 37–41 or the one for Exercises 42–45—seems to you more realistic as a model of a biological population, and why?

47. What is the significance of the red dashed line in the preceding figures and its intersection with the blue curve?

48. The following figure shows the annual population gain in the absence of any harvesting. Determine the maximum sustainable yield to one decimal point.

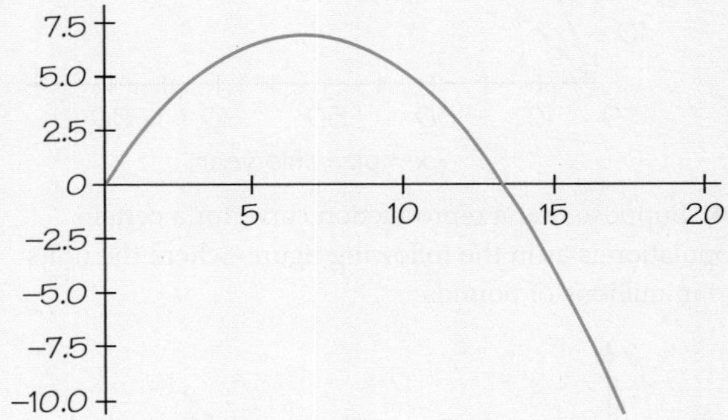

49. Suppose that the population of Exercises 42–48 starts with $x_0 = 11$ and each year we harvest half of the population. For example, in year 1, we harvest 5.5 million pounds, leaving the remaining 5.5 million pounds to reproduce (according to the reproduction curve) for the next year. Calculate numerically the population in the first 10 years, draw a cobweb diagram, and briefly describe the qualitative behavior of the population.

50. Repeat Exercise 49 for a starting population $x_0 = 5$.

51. Harvesting a set proportion of a population is unrealistic for some situations, such as fishing, in which we can't know the size of the population or when we have harvested half of it. A more realistic situation for fishing is that increasing harvests attract increasing fishing effort (e.g., more boats). Repeat Exercise 49 with $x_0 = 11$ and a harvesting strategy that harvests 1 million tons the first year and every year harvests an extra 1 million tons (over the harvest of the previous year).

52. Suppose that the population of Exercises 42–51 has been overharvested to the point that only 1 million tons remain at the end of a particular year. If there is no harvesting at all until a year after a year with a population of 11 million tons, when can harvesting resume?

53. A reproduction curve for a population is shown in the following figure. Estimate the equilibrium population size and the maximum sustainable yield. (The units are in millions of pounds.)

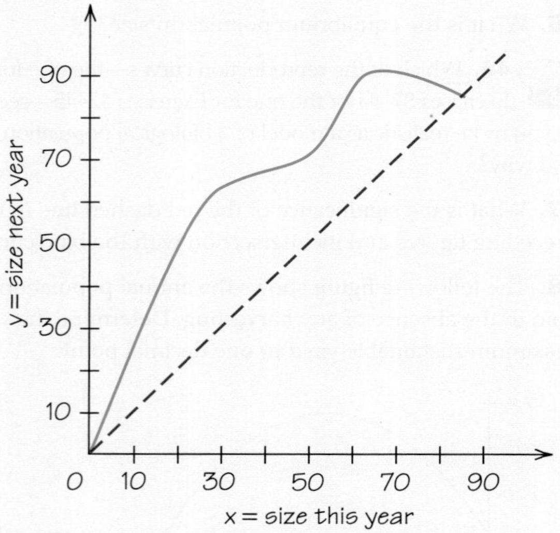

54. Suppose that a reproduction curve for a certain population is as in the following figure, where the units are in millions of pounds.

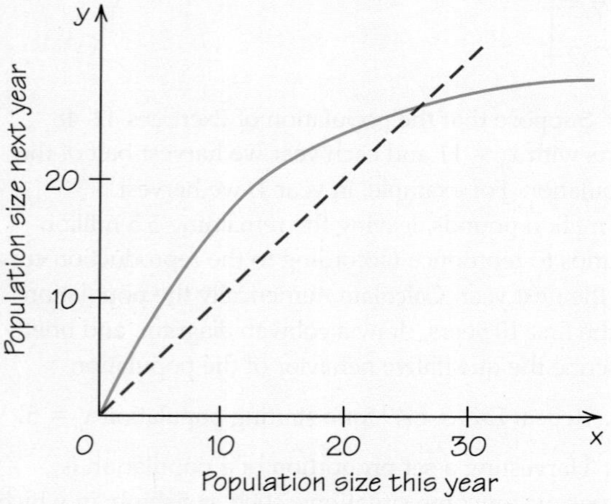

(a) Estimate the sustainable yields corresponding to a population of size 10 remaining after the harvest.

(b) Estimate the maximum sustainable yield.

23.5 Dynamical Systems and Chaos

55. In doubling on a "Stone Age" calculator (see Example 12), both sequences and also the sequence in the Self Check question all end up in the same *orbit of attraction,* or loop.

(a) What happens if we start with 39?

(b) Can you find a starting value that is not "attracted" to the indicated orbit?

(c) Explain what you observe in parts (a) and (b) and give a general argument about why every starting value must eventually lead to a repeating orbit.

56. What happens if you successively multiply by 3 on a "Stone Age" calculator? What are the orbits of attraction?

57. You saw in Figure 23.12 on page 966 that a logistic model can result in a stable value, produce cycling among several values, or result in chaos. Other dynamical systems can exhibit similar behavior. Here, we examine the system in which we start with a positive whole number n and iterate the following function:

$$f(n) = \text{sum of the squares of the digits of } n$$

For example, we have $f(133) = 1^2 + 3^2 + 3^2 = 19$.

(a) Calculate what happens as f is applied repeatedly, starting with 133. What do you observe?

(b) Pick a number different from 133 and different from 1, and iterate f repeatedly. What do you observe?

(c) Why did we exclude 1 in part (b)?

(d) Try some other values. Can you offer a general conclusion? Can you offer an argument why your conclusion is correct?

58. The behavior of some very simple dynamical systems is still not completely known. Consider the system that starts with a positive whole number n and gives as the next number

$$f(n) = \begin{cases} 3n + 1 & \text{if } n \text{ is odd} \\ n/2 & \text{if } n \text{ is even} \end{cases}$$

This iterative function system was devised by Lothar O. Collatz, later of the University of Hamburg, during his student days before World War II. It is sometimes called the "$3n + 1$ problem" or the "Syracuse problem" (because it became popular in the Mathematics Department of Syracuse University), and the sequences generated are sometimes referred to as *hailstone numbers.*

(a) Start with $n = 1$. What happens?

(b) Start with $n = 13$. What happens?

(c) Start with $n = 12$. What happens?

What you observe is known to happen for all $n < 10^{40}$, but after more than 60 years, mathematicians have been unable to show that it happens for every n whatsoever.

59. (Requires programmable calculator, spreadsheet, or BASIC programming) A population model slightly different from the logistic model is given by the iterative function system

$$g(x) = x + rx(1 - x)$$

where x is a fraction of the limiting population and r is a growth rate.

(a) Set $r = 3$, start with $x = x_0 = 0.01$, and calculate the first 20 values $x_0, \ldots, x_{19}$.

(b) In part (a), you should have found $x_9 = 0.722914$. Replace this value with the rounded-up value $x_9 = 0.723$ and continue on to calculate x_{19}.

(c) Now replace x_9 with the rounded-down value $x_9 = 0.722$ and continue on to calculate x_{19}.

60. A dynamical system expressed as an iterated functional system $f(x)$ has an equilibrium point at a value x^* if, once the system reaches x^*, it always stays at that value. In terms of an equation, an equilibrium point exists at x^* if $f(x^*) = x^*$. For the dynamical system of Exercise 59, find all equilibrium points.

61. (Requires programmable calculator, spreadsheet, BASIC programming, or preferably the use of software listed in the Suggested Websites section) The behavior of the logistic population model $f(x) = \lambda x(1 - x)$ depends on the value of the positive parameter λ. As λ increases from 0 to 4, the system changes from one behavior to another through the following possible states:

- The population simply dies out.
- The population tends toward a nonzero equilibrium point.
- The population oscillates between 2 points.
- The population oscillates between 4 points, then 8 points, then 16 points, and so on.
- The population oscillates between numbers of points that are not powers of 2, until at last the population oscillates between three values.
- The population behaves chaotically.

Explore what happens for various values of λ between 0 and 4, trying to identify where the shifts in the system's behavior take place.

Chapter Review

62. The population of the U.S. in mid-2015 was 321 million, increasing at 0.77% per year. If that growth rate continues, what will the population be in 2050?

63. (Spreadsheet helpful) In 2015, the United States had the third largest population, after China and India. The population of Nigeria was 184 million, increasing at 2.8% per year; and the population of the U.S. was 321 million, increasing at 0.77% per year. If those rates continue, in what year would the population of Nigeria

(a) reach the level of the 2015 U.S. population?

(b) surpass the population of the United States?

64. Indium is an element (atomic number 49) used in solar cells and in the screens of tablet computers and cell phones; refined indium costs about $600 per kilogram. The current world reserves are about 11,000 metric tonnes, with annual world consumption of about 770 metric tonnes.

(a) What is the static reserve of indium?

(b) World demand increased from 570 tonnes to 770 tonnes from 2008 to 2013. What was the average annual compounded rate of increase?

(c) Suppose that demand for indium increases at 6% per year for the foreseeable future. What is the corresponding exponential reserve?

65. You are radioactive! So are bananas. Your body and bananas both contain potassium. (But you are fine because your body excretes it.) The half-life of potassium-40 is 1.25 billion years. (Fortunately, the biological half-life in you is only about two weeks.)

(a) What is the decay constant for potassium-40, in proportion per year? This is the tiny proportion that decays per year.

(b) What is the decay constant for potassium-40, in proportion per second? This is the very tiny proportion that decays per second.

(c) A typical banana contains about 450 mg of potassium, of which 0.0117% is radioactive potassium-40. Every 40 grams of potassium-40 is composed of 6.02×10^{23} atoms (i.e., Avogadro's number of atoms) of potassium-40. How many decays per second of potassium-40 does a banana experience? [The dose of radiation from eating a banana is known as a *banana equivalent dose (BED)*. The average daily exposure to radiation from all sources is about 100 BED, while a chest CT scan delivers 70,000 BED.]

66. Suppose that a population has a reproduction curve with the mathematical description

$$x_{n+1} = f(x_n) = 1.5x_n(1 - 0.05x_n)$$

Draw a cobweb diagram starting with $x_0 = 5$ and describe what happens to the population in the long run. What happens if you start with a different value?

67. The world population of whooping cranes (*Grus americana*) was 15 in 1942. Thanks to intensive efforts to save the species, there were 600 in 2014. What was the average rate of growth in population over that period? (See Exercises 22–24.)

Writing Projects

1. Based on the calculations you did in Exercises 9 and 10, write a one- to two-page guest editorial for a newspaper. Describe your projections and how you arrived at them, how serious a problem you think population growth is, what problems it is likely to cause, what you think needs to be done, and what the implications are for your own life.

2. Identify a particular regional, national, or world nonrenewable primary resource (such as coal) or a secondary resource (one, such as electric power, that is produced from primary resources). Research how much of it is available now and what the current rate of consumption is. Determine the static reserve. Estimate the growth rate in consumption, taking into account human population increase, and determine the exponential reserve. What social and technological factors contribute to the increasing rate of consumption? Brainstorm how those factors could be changed. Write an essay of three to five pages.

3. Identify a particular regional, national, or world renewable resource (such as timber or clean drinking water). Research how much of it is produced now, how much is harvested now, and what the current rate of consumption is. Estimate the growth rate in consumption, taking into account human population increase. For how long can this resource continue to meet the demand? What social and technological factors contribute to the increasing rate of consumption? Brainstorm how those factors could be changed. Write an essay of three to five pages.

Suggested Readings

BARTLETT, ALBERT A. *The Essential Exponential! For the Future of Our Planet,* Center for Science, Mathematics, & Computer Education, University of Nebraska–Lincoln (126 Morrill Hall, Lincoln, NE 68588-0350), 2004.

CLOVER, CHARLES. *The End of the Line: How Overfishing Is Changing the World and What We Eat,* The New Press, New York, 2006.

COHEN, JOEL. *How Many People Can the Earth Support?* Norton, New York, 1995.

GLEICK, JAMES. *Chaos: Making a New Science,* Viking, New York, 1987.

KALMAN, DAN. Chapter 13: Logistic Growth, and Chapter 14: Chaos in Logistic Models, in *Elementary Mathematical Models: Order Aplenty and a Glimpse of Chaos,* Mathematical Association of America, Washington, D.C., 1997.

MEYER, MICHAEL. How much is left? *Scientific American* 303(3) (September 2010): 74–81.

PETERSON, IVARS. *Newton's Clock: Chaos in the Solar System,* W. H. Freeman, New York, 1993.

RANDERS, JORGEN. *2052: A Global Forecast for the Next Forty Years,* Chelsea Green, White River Junction, Vt., 2012.

SCHWARTZ, RICHARD H. *Mathematics and Global Survival,* 4th ed., Ginn Press, Needham Heights, Mass., 1998.

Suggested Websites

http://www.census.gov/population/international/data/ U.S. Census Bureau population statistics and projections for all countries, with the option to display population pyramids.

www.prb.org/ Population Reference Bureau population statistics and rates of growth by regions.

www.clickrepair.net/chaos/ Web-based chaos applet.

populationpyramid.net/ Population pyramids for virtually all countries from 1950 to 2050.

math.bu.edu/DYSYS/applets/ Downloadable Java applets designed to accompany Robert L. Devaney's *A Toolkit of Dynamics Activities,* but can be used independently.

forio.com/simulate/simulation/pontifexconsult/how-much-oil-is-left/ Computer "lab" that allows the user to make various assumptions about reserves, consumption, productivity, and cost of production and to then simulate the future reserves and consumption of oil.

demonstrations.wolfram.com/DiscreteLogisticEquation/ Interactive Internet program that allows you to set the parameter λ and two separate initial population values and to observe the resulting population fluctuations.

APPENDIX | Algebra Review

CONTENTS

Algebra Review I: Handling Operations

A. Order of Operations

When expressions have more than one operation, we need to know the order in which we perform the operations. Here are the rules for the order of operations:

1. Perform operations within parentheses (or brackets), working from the innermost out.
2. Evaluate exponents or roots.
3. Do any multiplication or division, going from left to right.
4. Do any addition or subtraction, going from left to right.

Most calculators (scientific calculators and graphing calculators) follow the order of operations given above.

Example 1. Evaluate the expression $2 \times 6 - 36 \div (7 - 4)^2$.

$2 \times 6 - 36 \div \underbrace{(7 - 4)}_{\text{Start here.}}{}^2$	Perform the operation within the parentheses.
$= 2 \times 6 - 36 \div (3)^2$	Perform the square.
$= 2 \times 6 - 36 \div 9$	Perform the multiplication and division from left to right.
$= 12 - 4$	Finally, perform the subtraction.
$= 8$	

When a fraction is involved, we treat the numerator and denominator as each being enclosed in parentheses. Perform the calculation in the numerator and then perform the calculation in the denominator. Finish the problem by performing the division.

Example 2. Use order of operations to simplify

$$\frac{128 - 4(2 + 14)}{7 + 5^2}.$$

$\frac{128 - 4(2+14)}{7+5^2}$	Working on the numerator, perform the calculation in the parentheses.
$= \frac{128 - 4(16)}{7+5^2}$	Continuing with the numerator, perform the multiplication.
$= \frac{128 - 64}{7+5^2}$	Finishing with the numerator, perform the subtraction.
$= \frac{64}{7+5^2}$	Switching to the denominator, square 5 to get 25.
$= \frac{64}{7+25}$	Finishing with the denominator, compute the sum.
$= \frac{64}{32}$	Perform the division.
$= 2$	

Note: When evaluating expressions such as -2^4, treat the expression as $-1 \cdot 2^4$. In this case, you would perform the power first and then the multiplication:

$$-2^4 = -1 \cdot 2^4 = -1 \cdot 16 = -16$$

Practice Exercises

Use order of operations to simplify each of the following by hand. Leave any fractions as fractions, without computing a decimal equivalent. If you have a scientific or graphing calculator, check your answers using your calculator.

1. $(5)(6) - 10 \div 2 + 1$
2. $5 + 44 - (6 + 8 \div 2) + (2)(3)$
3. $-3^2 + 2 \cdot 3 + 10$
4. $5 - (2 + 3)^2 + (8)(2) \div 4$
5. $\frac{5 - 4}{7 - (-2)}$
6. $\frac{5 - (-3)}{-2 - (-7)}$
7. $\frac{(-5)(3) + 4^2}{6 + 3^2}$
8. $\frac{15 \div 5 \cdot 4 \div 6 - 8}{-6 - (-5) - 8 \div 2}$

B. Distributive Law

In expressions involving multiplication and addition/subtraction, you can distribute the multiplication over the addition/subtraction. Here is a statement of the distributive law:

$$a(b + c) = ab + ac \quad \text{or} \quad a(b - c) = ab - ac$$

Example 1. Use the distributive law to remove the parentheses in the expression $3(2x + 5)$.

In the expression $3(2x + 5)$, $a = 3$, $b = 2x$, and $c = 5$. Applying the distributive property gives:

$$3(2x + 5) = (3)(2x) + (3)(5) = 6x + 15$$

Example 2. Apply the distributive law: $-2(x - 3)$.

$$-2(x - 3) = (-2)(x) - (-2)(3) = -2x + 6$$

Factoring an expression means writing it as a product of factors, or "unmultiplying" it. One way to factor an expression is to check to see whether the terms have a factor in common. If so, use the distributive law (in reverse) as follows:

$$ab + ac = a(b + c) \quad \text{or} \quad ab - ac = a(b - c)$$

Example 3. Factor $2x + 4$.

Both terms have a common factor of 2. So we factor out the 2 as follows:

$$4x + 2 = (2)(2x) + (2)(1) = 2(2x + 1)$$

Example 4. Factor $6x^2 - 9x$.

Both terms have a common factor of $3x$. So we factor out the $3x$ as follows:

$$6x^2 - 9x = (3x)(2x) - (3x)(3) = 3x(2x - 3)$$

Practice Exercises

In Practice Exercises 1–5, apply the distributive law to remove the parentheses.

1. $2(x + 1)$
2. $7(2w - 3)$
3. $(x + 4)(-3)$
4. $3.1(2p + 1.5)$
5. $5(2x + y - 3)$

In Practice Exercises 6–8, use the distributive law to factor.

6. $12x - 6$
7. $24x^2 + 10x$
8. $2 + 4x + 8x^2$

C. Operations with Rational Numbers (Fractions)

Rational numbers (fractions) are numbers of the form $\frac{m}{n}$ where m and n ($n \neq 0$) are integers.

Adding and Subtracting Fractions

To add or subtract fractions, convert all the individual fractions to equivalent fractions that share the same denominator. Then add or subtract numerators and place the result over the shared denominator. The lowest common denominator (or LCD) is the simplest denominator that fractions have in common. Using the lowest common denominator usually makes the work easier.

Example 1. Subtract $\frac{1}{6}$ from $\frac{3}{4}$.

The denominators are $6 = 2 \cdot 3$ and $4 = 2 \cdot 2$. The lowest common denominator is $2 \cdot 2 \cdot 3 = 12$.

$\frac{3}{4} - \frac{5}{6} = \frac{3 \cdot 3}{4 \cdot 3} - \frac{5 \cdot 2}{6 \cdot 2}$	Replace each fraction with an equivalent fraction with a denominator of 12.
$\frac{3 \cdot 3}{4 \cdot 3} - \frac{5 \cdot 2}{6 \cdot 2} = \frac{9}{12} - \frac{10}{12} = \frac{9 - 10}{12} = \frac{-1}{12}$	Since both fractions have the same denominator, subtract the numerators.
$\frac{-1}{12} = -\frac{1}{12}$	It doesn't matter if the "−" is in the numerator or out in front of the entire fraction.

Multiplying Fractions

To multiply two fractions, multiply the numerators and multiply the denominators. You want to cancel any common factors before carrying out the multiplication, so write the numerators and denominators in factored form.

Example 2. Multiply $\frac{15}{49}$ by $\frac{14}{25}$.

$$\frac{15}{49} \times \frac{14}{25} = \frac{15 \times 14}{49 \times 25} = \frac{(\cancel{5} \cdot 3) \times (2 \cdot \cancel{7})}{(7 \cdot \cancel{7}) \times (\cancel{5} \cdot 5)} = \frac{6}{35}$$

Dividing Fractions

To divide any expression by a fraction, multiply the expression by the reciprocal of the fraction.

Example 3. Divide 5 by $\frac{1}{6}$.

$$\frac{5}{\frac{1}{6}} = 5 \cdot 6 = 30$$

Example 4. Determine $\frac{1}{2} \div \frac{3}{4}$.

$$\frac{1}{2} \div \frac{3}{4} = \frac{\frac{1}{2}}{\frac{3}{4}} = \frac{1}{2} \cdot \frac{4}{3} = \frac{1 \cdot 4}{2 \cdot 3} = \frac{\cancel{2} \cdot 2}{\cancel{2} \cdot 3} = \frac{2}{3}$$

Practice Exercises

Determine the values of the following. Express your answers in reduced form.

1. $\frac{6}{35} - \frac{3}{20}$
2. $\left(\frac{1}{9}\right)\left(\frac{3}{4}\right)$
3. $\frac{\frac{4}{9}}{5}$
4. $\frac{\frac{4}{5}}{\frac{2}{3}}$
5. $\frac{\frac{2}{5} + \frac{1}{10}}{2}$
6. $\frac{1}{3}\left(\frac{6}{5} - \frac{3}{10}\right)$
7. $\frac{2}{15} + \frac{6}{25}$
8. $\frac{\frac{2}{3}}{\frac{3}{4} + \frac{4}{5}}$

Algebra Review II: Representing Numbers

A. Fractions, Percents, and Percentages

Fractions are a way to represent a portion of a whole. When working with fractions, we often are required to convert them to decimals or express them as percentages. **Percent** means *per hundred,* and it is accompanied by a number and the % symbol (or the word *percent*).

Example 1. Explain the meaning of 15% (or 15 percent).

15% (or 15 percent) means 15 hundredths or $\frac{15}{100} = 0.15$.

A **percentage** is a portion of a whole expressed in hundredths.

Example 2. Suppose 15 out of 20 students in a class are female. What percentage of the class is female?

The portion consists of 15 female students and the whole is the class of 20 students. Obtain the percentage as follows:

- Divide the portion by the whole.

  $\frac{15}{20} = 0.75$
- Multiply by 100 and attach the % symbol.

  $(0.75 \times 100)\% = 75\%$

The words *percent* and *percentage* are related to each other, but they do not have the same meaning. However, the use of percent for percentage is becoming more and more common, so don't worry about the difference.

Some fraction-to-percent conversions require rounding, thereby changing the value slightly. Example 3 demonstrates how to round a percent.

Example 3. Express $\frac{3}{13}$ as a percent and round to the nearest hundredth of a percent.

- First, convert the fraction to a decimal. A calculator with 10 places of accuracy will yield 0.2307692308.
- Second, multiply this number by 100 to obtain 23.07692308.
- Third, round to two decimal places (for the nearest hundredth of a percent) and then affix the percent symbol: 23.08%. (See Algebra Review II, item F, Rounding Numbers, page AR-5.)

Practice Exercises

1. Explain the meaning of 25%.
2. Suppose 10 out of 25 students in a class are male. What percentage of the class is male?
3. In a survey of 500 students, 347 circled "Yes" in answer to a question. What percentage of those surveyed responded "Yes"?
4. Express $\frac{6}{18}$ as a percent and round to the nearest hundredth of a percent.

B. Remainders

To find the remainder when an integer *a* is divided by an integer *b,* we divide *a* by *b* using long division.

Example 1. Divide 95 by 7. What is the remainder?

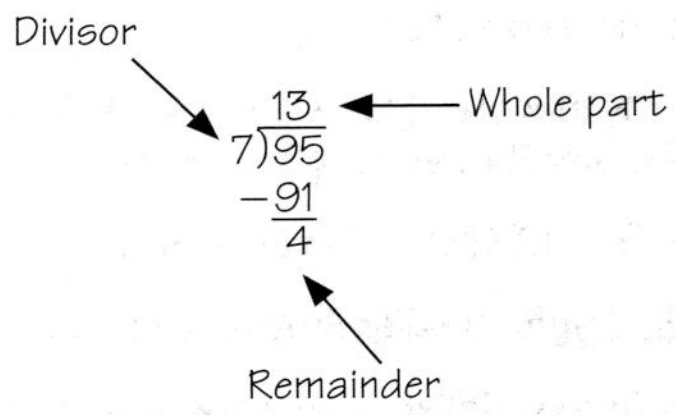

An important feature of a remainder is that it is always less than the divisor, and it can be 0.

Calculator Note: To use a calculator to find the remainder when a is divided by b, perform the following steps:

STEP 1. Enter $a \div b$.

STEP 2. Enter $a -$ (whole part from Step 1) $\times b$. The result will be the remainder.

Example 2. Use a calculator to find the remainder when 95 is divided by 7.

STEP 1. Enter $95 \div 7$. The result is 13.57142857.

STEP 2. Enter $95 - (13 \times 7)$. The result gives a remainder of 4. (See calculator screenshot.)

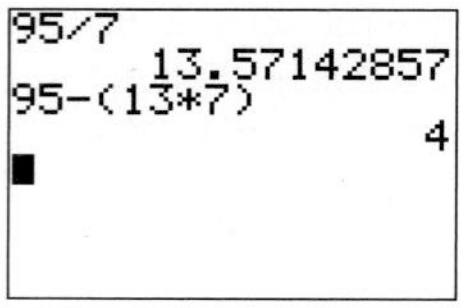

Practice Exercises

1. Use long division to divide 108 by 8. What is the whole part? What is the remainder?
2. Use long division to divide 227 by 6. What is the whole part? What is the remainder?
3. Use long division to divide 141 by 3. What is the whole part? What is the remainder?
4. A number divided by 5 results in a whole part of 6 and a remainder of 2. What is the number?
5. A number divided by 9 results in a whole part of 14 and a remainder of 7. What is the number?
6. Use a calculator to find the whole part and remainder when 76 is divided by 14.
7. Use a calculator to find the whole part and the remainder when 141 is divided by 6.
8. When two numbers are divided, can the remainder ever be larger than the whole part? Can the remainder ever be larger than the divisor?

C. Prime and Composite Numbers

An integer greater than 1 is prime if the only positive integers that divide it with a remainder equal to 0 are the number itself and 1. There are infinitely many prime numbers. The first ten are 2, 3, 5, 7, 11, 13, 17, 19, 23, and 29. A list of the first 1000 prime numbers can be found by searching Google for "1000 primes."

A composite number is a positive integer that is greater than 1 and is not prime. One can use a list of primes and a calculator to determine whether any positive integer N is prime by dividing N, *one by one,* by all the primes less than or equal to $\sqrt{N}$. If one of the quotients is an integer, then N is *not* prime.

Example 1. Determine whether 147 is a prime or composite number.

Since $\sqrt{147} \approx 12.12$, we need only divide 147 by 2, 3, 5, 7, and 11. In this case, $147 \div 3 = 49$, so 147 is not prime.

Example 2. Determine whether 151 is a prime or composite number.

Since $\sqrt{151} \approx 12.28$, we need only divide 151 by 2, 3, 5, 7, and 11. When 151 is divided by 2, 3, 5, 7, and 11, the quotient is never an integer, so 151 is prime.

The following are divisibility tests that can quickly identify that a positive integer is a composite number.

- Integers whose last digit is an even number are divisible by 2.
- Integers whose last digit is 0 or 5 are divisible by 5.
- Integers for which the sum of their digits is divisible by 3 are themselves divisible by 3.

Example 3. Determine whether 4785 and 2301 are prime or composite numbers.

Both are composite numbers. Using the divisibility tests:

- 4785 is divisible by 5, since its last digit is 5.
- 2301 is divisible by 3, since $2 + 3 + 0 + 1 = 6$, which is divisible by 3.

Practice Exercises

Determine whether the following numbers are prime or composite numbers.

1. 187
2. 229
3. 3342
4. 899
5. 1295
6. 4221
7. 421
8. 3,476,210

D. Significant Digits

All measurements introduce some degree of uncertainty due to the deficiencies of the measuring device or the person doing the measuring. It is this uncertainty that we take into consideration when dealing with significant digits. The rules for determining the number of significant digits are as follows:

- Any non-zero number is significant.
- Any 0 that lies both to the right of the decimal point and to the right of a non-zero digit is significant.
- Any 0 between significant digits is significant.

Example 1. Determine the number of significant digits for 10,134 and 2300.00.

The number 10,134 has five significant digits and 2300.00 has six significant digits.

Example 2. Determine the number of significant digits for 0.200 and 0.000034.

Because the leading zero is not considered a significant digit, 0.200 has three significant digits, while 0.000034 has two.

In the context of significant digits, when multiplying (or dividing) numbers, we determine which of the original numbers has the smallest number of significant digits and round our product (or quotient) to match this smallest accuracy. (See Algebra Review II, item F, Rounding Numbers below.)

Example 3. Suppose that you had a box and measured to find a height of 4.8 inches, length of 5.4 inches, and width of 9.4 inches. Find the volume of the box.

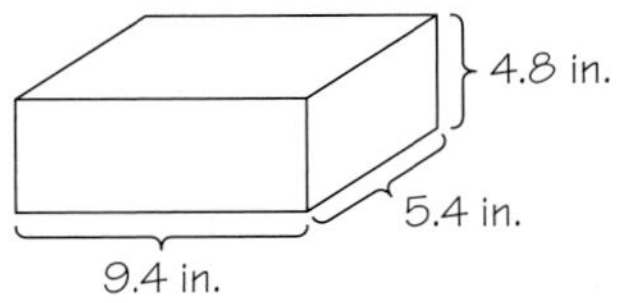

Each of these measurements has two significant digits. The volume of the box is found by multiplying length, width, and height. When this is done, we get a result of 243.648 cubic inches. However, because we had only two significant digits in our measurements, we must round this answer to 240 cubic inches.

When adding or subtracting, first perform the calculation with all the digits, and then round the sum or difference to have the same decimal places as the number with the least number of digits following the decimal point.

Example 4. Find the product and sum of the measurements 0.012, 2.30, and 0.001, and then round appropriately.

The product of these numbers is 0.0000276. Because 0.001 has only one significant digit, we round the answer to 0.00003, which has one significant digit. The sum of the numbers is 2.313. Because 2.30 has the smallest accuracy in terms of decimal places, we round this sum to 2.31.

Some calculations involve numbers that are considered exact. Numbers such as exact conversion factors do not figure into determining the number of significant digits in a calculation.

Practice Exercises

For Practice Exercises 1–4, determine the number of significant digits.

1. 1002
2. 12,000
3. 0.0035
4. 0.0210

For Practice Exercises 5–8, use the box in the diagram that follows. Round your answers according to the conventions for significant figures and number of decimal places discussed above.

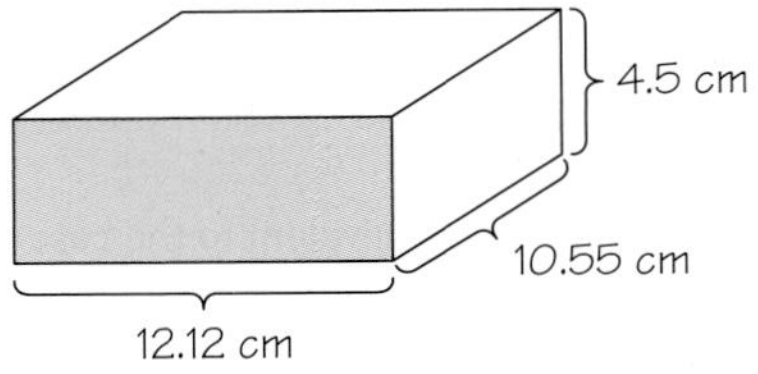

5. Determine the volume of the box. Round appropriately.
6. Determine the perimeter of the front face of the box shaded in blue.
7. Determine the area of the front face of the box shaded in blue.
8. Determine the area of the top of the box.

E. Scientific Notation

A number written in the form $a \times 10^n$, where $1 \le a < 10$ and n is an integer, is written in scientific notation. Scientific notation is a useful way to express very large or very small numbers.

Example 1. Write 568,000 in scientific notation.

$$568{,}000 = 5.68 \times 10^5$$

Example 2. Write 0.0028 in scientific notation.

$$0.0028 = 2.8 \times 10^{-3}$$

(See Algebra Review VI, item B, Natural and Fractional Exponents, page AR-23.)

Most calculators and many computer programs use a shorthand version of scientific notation. Instead of indicating a multiplication by 10^n, the calculator/computer program will write "E" followed by the power n. The base is understood to be 10.

Example 3. Write the numbers in Examples 1 and 2 as they might appear on a calculator.

568,000 might appear as 5.68 E5.

0.0028 might appear as 2.8 E–3.

Warning: This is entirely different from $(5.65)^5$ or $(2.8)^{-3}$.

Practice Exercises

1. Write 495,000,000 in scientific notation.
2. Write 0.00000072 in scientific notation.
3. Write your answer to Practice Exercise 1 as it might appear on a calculator.
4. Write your answer to Practice Exercise 2 as it might appear on a calculator.
5. Use your calculator to find the value of 52^5. Write the answer both with and without using scientific notation.
6. Use your calculator to divide 467 by 200,000,000. Write the answer without using scientific notation.

F. Rounding Numbers

In various exercises, you will be asked to round numbers to a certain place, such as the nearest thousands place or to two decimal places.

Case 1: Rounding to whole number places, such as ones, tens, hundreds, thousands, and so forth.

To round a number to a particular place, look at the digit in that place—we'll call this the target digit—and then look at the digit to its right.

- If the digit to the right of the target digit is a 0, 1, 2, 3, or 4, then leave the target digit as it is and replace any digits to its right with 0s. Do not include a decimal point or any 0s to the right of a decimal point.
- If the digit to the right of the target digit is a 5, 6, 7, 8, or 9, then add 1 to the target digit and replace any digits to its right with 0s. Do not include a decimal point or any 0s to the right of a decimal point.

Case 2: Rounding to decimal places, such as tenths, hundredths, thousandths, and so forth.

To round a number to a particular place, look at the digit in that place—the target digit—and then look at the digit to its right.

- If the digit to the right of the target digit is a 0, 1, 2, 3, or 4, then leave the target digit as it is and omit any digits to its right.
- If the digit to the right of the target digit is a 5, 6, 7, 8, or 9, then add 1 to the target digit and omit any digits to its right.

Example 1. Round 67,853 (a) to the nearest hundreds and (b) to the nearest tens.

(a) Locate the target digit, the number in the hundreds place—in this case, 8. Look at the digit one place to its right, which is 5. Since this digit is 5, add one to the target digit ($8 + 1 = 9$) and replace the digits to its right with 0s. The rounded number is 67,900.

(b) Locate the target digit, the number in the tens place—in this case, 5. Look at the digit one place to its right, which is 3. Since this digit is less than 5, leave the target digit as it is and replace the digit to its right with 0. The rounded number is 67,850.

Example 2. Round 4.64779 (a) to one decimal place (the nearest tenths place) and (b) to two decimal places (the nearest hundredths place).

(a) Locate the target digit, the digit in the tenths place—in this case, 6. Look at the digit one place to its right, which is 4. Since this digit is less than 5, leave the target digit as it is and omit any digits to its right. The rounded number is 4.6.

(b) Locate the target digit, the digit in the hundredths place—in this case, 4. Look at the digit one place to its right, which is 7. Since this digit is greater than 5, add 1 to the target digit ($4 + 1 = 5$) and omit any numbers to its right. The rounded number is 4.65.

Example 3. Round 7.4997 to two decimal places.

Locate the target digit, the digit in the hundredths place—in this case, 9. Look at the digit one place to its right, which is also 9. Since this digit is greater than 5, add 1 to the target digit ($9 + 1 = 10$). Notice that this increases the digit in the tenths place. Omit any digits to the right of the target digit. The rounded number is 7.50.

Practice Exercises

1. Round 48,749 to the nearest thousands place.
2. Round 48,749 to the nearest hundreds place.
3. Round 2.267 to one decimal place.
4. Round 2.263 to two decimal places.
5. Round 5.1997 to two decimal places.

G. Counting in Binary

The numeration system that we use every day is a base-10 system, which involves 10 digits, 0 through 9. In order to exceed 9, we add a place and write the number 10. A number such as 307 can be written in expanded form as $307 = 3 \times 100 + 0 \times 10 + 7 \times 1$. Using exponential notation, we could also write the following:

$$307 = 3 \times 10^2 + 0 \times 10^1 + 7 \times 10^0$$

Example 1. Write 1027 in expanded form using exponential notation.

$$1027 = 1 \times 10^3 + 0 \times 10^2 + 2 \times 10^1 + 7 \times 10^0$$

Notice in the exponential form shown in Example 1 that the base is 10. Binary is a base-2 system and has only two digits, namely 0 and 1. If we were to convert a binary number to its base-10 (decimal) equivalent, we would use a base of 2 in the same way we used the base of 10 in Example 1.

Example 2. Convert the binary number 1011 (read as "one zero one one") to its base-10 equivalent.

The binary number 1011 is equivalent to the base-10 number:

$$1 \times 2^3 + 0 \times 2^2 + 1 \times 2^1 + 1 \times 2^0 = 8 + 0 + 2 + 1 = 11$$

After the number 1, a binary number will have at least two places. Starting with 0, the string of binary numbers in order is

0, 1, 10, 11, 100, 101, 110, 111, 1000, 1001,
1010, 1011, 1100, 1101, 1110, 1111 . . .

When writing these numbers, we do not put commas in, as we do with decimal numbers. For example, the binary number 11011 would *not* be written as 11,011. Also, when binary numbers are used and the context is not clear whether the number is binary or base-10, a subscript of 2 is often included. Thus, 11_2 represents 11 base 2 (which is 3 in base-10) while 11 is the usual number "eleven."

Practice Exercises

For Practice Exercises 1–5, convert the base-2 number to its equivalent in base-10.

1. 111100_2
2. 1011011_2
3. 10001100_2
4. 11111_2
5. 1001001_2
6. If you were counting in base-2, what number comes after 11111_2?
7. What base-2 number is equivalent to the base-10 number 35?
8. What base-2 number is equivalent to the base-10 number 65?

Algebra Review III: Working with Formulas

A. Using Formulas

A formula is a statement that claims that two expressions are equal. When dealing with formulas, it is important to know what each variable represents. Generally, formulas are case-sensitive, which means that an uppercase X may represent a different variable than a lowercase x. If more than one variable from the same type of variables is needed, then subscripts are often used.

Here are three examples of formulas.

- The formula for finding the mean, $\bar{x}$, of n numbers $x_1, x_2, \ldots, x_n$ is

  $$\bar{x} = \frac{x_1 + x_2 + \cdots + x_n}{n}.$$

- The slope of a line is determined by two points often expressed as (x_1, y_1) and (x_2, y_2). The formula that relates these two points to the slope of the line that passes through them, commonly denoted as m, is

  $$m = \frac{y_2 - y_1}{x_2 - x_1}$$

- For a principal P invested at an annual rate of interest r, the total amount A accumulated in the account after t years is

  $$A = P(1 + rt)$$

In most formulas, the value of the variable on the left is found by substituting values for the variable(s) on the right.

The following three steps occur when evaluating a formula to determine the numerical value of one of the variables:

1. Understand what the variables in the formula represent.
2. Carefully substitute the values that are given into the formula.
3. Apply order of operations to simplify the expression.

Next, we consider some examples involving finding the mean, slope, and total amount.

Example 1. Find the mean of the numbers 2, −3, and 5.

- First, assign a meaning to the numbers.

  $n = 3$ and $x_1 = 2$, $x_2 = -3$, and $x_3 = 5$

- Second, substitute the constants for the variables.

  $$\bar{x} = \frac{2 + (-3) + 5}{3}$$

- Third, simplify by applying order of operations.

  $$\bar{x} = \frac{2 + (-3) + 5}{3} = \frac{4}{3}$$

Example 2. Calculate the slope of the line that passes through (−3, 4) and (2, 5).

- First, assign a meaning to the numbers.

  $(x_1, y_1) = (-3, 4)$ and $(x_2, y_2) = (2, 5)$

- Second, substitute the constants for the variables.

  $$m = \frac{y_2 - y_1}{x_2 - x_1} = \frac{5 - 4}{2-(-3)}$$

- Third, simplify by applying order of operations.

  $$\frac{5 - 4}{2-(-3)} = \frac{5 - 4}{2 + 3} = \frac{1}{5}$$

(Notice that subtracting a negative number gives the same result as adding its opposite: $2 - (-3) = 2 + 3 = 5$.)

Example 3. Suppose that you invest \$1000 in an account that pays 4% interest on the initial amount invested. How much money is in your account after 5 years?

- First, assign a meaning to the numbers: The principal P is the amount invested, $P = 1000$. The 4% interest rate must be converted to its decimal equivalent, $r = 0.04$. The principal is invested for 5 years, so $t = 5$. We want to find A, the amount in the account.
- Second, substitute the constants for the variables.

  $$A = P(1 + rt) = 1000(1 + (0.04)(5))$$

- Third, simplify by applying order of operations.

$1000(1 + (0.04)(5))$	Order of operations tells us to compute the value inside the parentheses first. We begin by computing the multiplication that appears in the expression inside the parentheses.
$= 1000(1 + 0.2)$	Next, complete the addition that appears in the expression inside the parentheses.
$= 1000(1.02)$	Perform the multiplication.
$= 1020$	

The amount of money in the account after 5 years is $A = \$1020$.

Sometimes when working with formulas, the variable for which you want to solve is not isolated on the left-hand side of the equation. In that case, additional algebraic manipulation is required to solve for the unknown variable. The following two common algebraic manipulations will often do the trick:

- Multiply or divide both sides of the equation by the same nonzero number.
- Add or subtract the same number from both sides of the equation.

Example 4. Suppose you are told that the mean of four numbers is 10, and you know that the first three numbers are 8, 12, and 6. What is the value of the fourth number?

- First, assign a meaning to the numbers.

  $\bar{x} = 10$, $x_1 = 8$, $x_2 = 12$, and $x_3 = 6$. We need to find x_4.

- Second, substitute the constants for the variables.

  $$\bar{x} = \frac{x_1 + x_2 + x_3 + x_4}{n}$$

  $$10 = \frac{8 + 12 + 6 + x_4}{4}$$

- Third, simplify by applying order of operations.

$$10 = \frac{26 + x_4}{4}$$

- Fourth, apply further algebraic manipulation to solve for the unknown variable.

$(4)(10) = (4)\left(\frac{26 + x_4}{4}\right)$ Multiply both sides of the equation by 4 and then simplify.

$40 = 26 + x_4$ Subtract 26 from both sides of the equation.

$40 - 26 = 26 + x_4 - 26$ Simplify.

$14 = x_4$

We have found that the fourth number must be 14.

Example 5. The slope of a line through two points is ½. The first point has coordinates (1, 1). The y-coordinate of the second point is 3. What is the x-coordinate of the second point?

- First, assign a meaning to the numbers.

$m = 1/2$, $(x_1, y_1) = (1, 1)$, $(x_2, y_2) = (x_2, 3)$. We need to find x_2.

- Second, substitute the constants for the variables.

$$m = \frac{y_2 - y_1}{x_2 - x_1}$$

$$\frac{1}{2} = \frac{3 - 1}{x_2 - 1}$$

- Third, simplify by applying order of operations.

$$\frac{1}{2} = \frac{2}{x_2 - 1}$$

- Fourth, apply further algebraic manipulation to solve for the unknown variable.

$\frac{1}{2} = \frac{2}{x_2 - 1}$ To clear the fractions, multiply both sides of the equation by $2(x_2 - 1)$, which is valid as long as $2(x_2 - 1) \neq 0$ or $x_2 \neq 1$.

$\cancel{2}(x_2 - 1)\frac{1}{\cancel{2}} = 2\cancel{(x_2 - 1)}\frac{2}{\cancel{x_2 - 1}}$ Simplify.

$x_2 - 1 = 4$ Add +1 to both sides of the equation.

$x_2 - 1 + 1 = 4 + 1$ Simplify.

$x_2 = 5$

Calculator note: When calculating the value of a fraction, such as when you find the slope of a line given two points on the line, you must separately enclose both the numerator and denominator in parentheses.

Example 6. Using a calculator, find the slope of a line that passes through the points (5, 9) and (17, 6).

The keystrokes for a TI-84 graphing calculator would be:

[(] [6] [−] [9] [)] [÷] [(] [1] [7] [−] [5] [)] [ENTER],

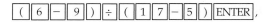

which yields –.25.

This is how this calculation should appear on your calculator screen.

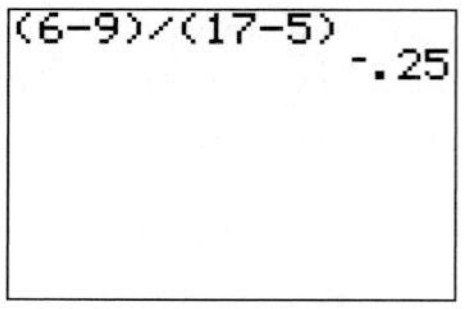

Practice Exercises

Use the formulas above and the three-step method to find the answers to Practice Exercises 1–3.

1. Find the mean of the numbers −4, 8, 3, and −2.
2. Calculate the slope of a line passing through (3, −6) and (5, −4).
3. A principal of $5000 is invested in an account that pays an annual rate of 3%. How much money is in the account after four years?
4. The mean of five numbers is 4.4. Four of the numbers are 4, 2, 3, and 6. What is the fifth number?
5. The slope of a line through two points, (2, 3) and $(x, 7)$, is $\frac{3}{2}$. Find x.
6. The slope of a line through two points, (1, 7) and $(3, y)$, is −5. Find y.
7. A principal P was invested in an account paying an annual interest rate of 5%. After 10 years, the total amount in the account was $3750. Find P.
8. The formula for finding the area of a rectangle is $A = l \times w$, where A is the area, l is the length, and w is the width. If a rectangle's length is 7 cm and its area is 84 cm^2, what is its width?

B. Solving for One Variable in Terms of Another

Sometimes when working with formulas, the variable of interest is not isolated on the left-hand side of the equation. In that case, additional algebraic manipulation is required to solve for the variable of interest in terms of the remaining variables. The following two common algebraic manipulations will often do the trick:

- Multiply or divide both sides of the equation by the same nonzero number.
- Add or subtract the same number from both sides of the equation.

For example, $C = \pi d$ is the formula for finding the circumference of a circle, C, given its diameter, d. However, in the context of forestry, a forester would measure a tree's circumference in order to calculate its diameter. So, the forester would prefer having a formula that gives d in terms of C. Dividing both sides of this formula by π gives

$$d = \frac{C}{\pi}$$

Example 1. Solve $F = \frac{9}{5}C + 32$ for C.

$F = \frac{9}{5}C + 32$	The original formula gives F in terms of C. Subtract 32 from both sides of the equation.
$F - 32 = \frac{9}{5}C$	Multiply both sides by $\frac{5}{9}$. (Notice that $\frac{5}{9} \cdot \frac{9}{5} = 1$.)
$\frac{5}{9}(F - 32) = C$	Switch the left and right sides of the equation for ease of reading C in terms of F.
$C = \frac{5}{9}(F - 32)$	

If the equation involves a single radical expression involving the variable of interest, a different approach is needed. In that case, first isolate the radical on one side of the equation. Raise both sides of the equation to the power required to remove the radical. Finally, use the algebraic manipulations in the two bullets above to solve for the variable of interest in terms of the other variables. [For information on radicals and powers, see Algebra Review VI, item A, Powers and Roots (page AR-22), and item D, Rules for Exponents and Roots (page AR-25).]

Example 2. Solve $m = 2\sqrt{\frac{0.25}{n}}$ for n.

$m = 2\sqrt{\frac{0.25}{n}}$	The original formula gives m in terms of n. Isolate the radical expression on one side of the equation by dividing both sides by 2.
$\frac{m}{2} = \sqrt{\frac{0.25}{n}}$	Square both sides of the equation to remove the radical.
$\left(\frac{m}{2}\right)^2 = \frac{0.25}{n}$	Remove the parentheses from the left-hand side by squaring.
$\frac{m^2}{4} = \frac{0.25}{n}$	Multiply both sides of the equation by $4n$ to clear the fractions.
$n \cdot m^2 = 4(0.25) = 1$	Divide both sides by m^2 to solve for n.
$n = \frac{1}{m^2}$	We now have a formula where n is given in terms of m.

You can use the same technique used in Example 2 if the variable of interest is raised to a power.

Example 3. Solve $C = \pi r^2$ for r.

$C = \pi r^2$	The original formula gives C in terms of r. Isolate the expression raised to a power on one side of the equation by dividing both sides by π.
$\frac{C}{\pi} = r^2$	Raise both sides to the one-half power.
$\left(\frac{C}{\pi}\right)^{\frac{1}{2}} = (r^2)^{\frac{1}{2}} = r$	Switch the left and right sides of the equation for ease of reading r in terms of C.
$r = \left(\frac{C}{\pi}\right)^{\frac{1}{2}} = \sqrt{\frac{C}{\pi}}$	We now have a formula in which r is given in terms of C.

Practice Exercises

Use the simple interest formula (from Chapter 21, page 871), $A = P(1 + rt)$, for Practice Exercises 1 and 2.

1. Solve the simple interest formula for P.
2. Solve the simple interest formula for t.

Use the compound interest formula (from Chapter 21, page 875), $A = P(1 + i)^n$, for Practice Exercises 3 and 4.

3. Solve the compound interest formula for P.
4. Solve the compound interest formula for i.

C. Formulas Related to Geometric Shapes

In Chapters 18 and 20, you will encounter a variety of two-dimensional and three-dimensional geometric shapes. Formulas for perimeter and area are useful when dealing with two-dimensional shapes. Formulas for surface area and volume are useful when dealing with three-dimensional shapes. This section focuses on a few of the most commonly used two- and three-dimensional geometric shapes.

Two-Dimensional Shapes

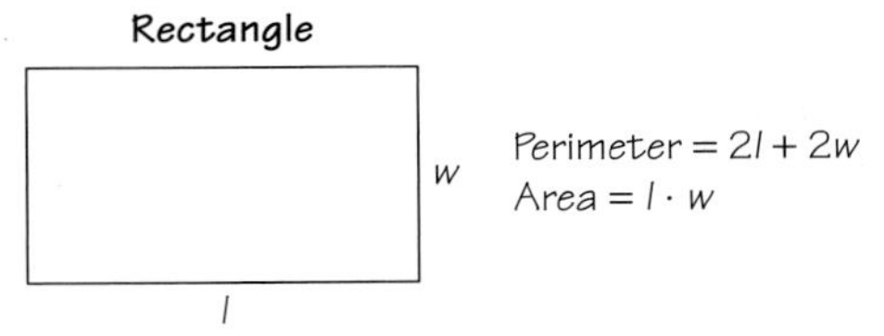

Note: Squares are special cases of rectangles, where $l = w$.

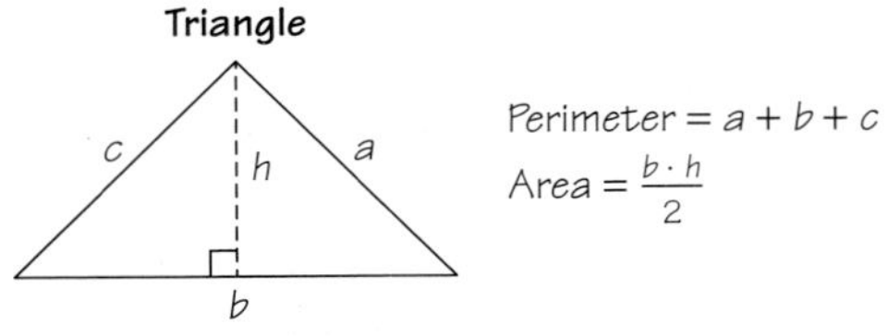

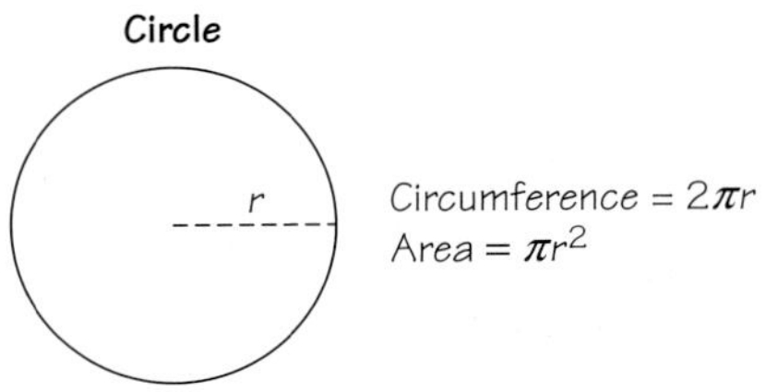

Note: For circles, the length around the circle is called the circumference rather than the perimeter.

Example 1. Jordon plans to fence in a 12-ft-by-8-ft rectangular enclosure for her dog. Determine the length of fencing needed to make the enclosure and the area that will be enclosed.

The perimeter gives the length of fencing that is needed: perimeter = 2(12 ft) + 2(8 ft) = 40 ft. The enclosed area is (12 ft)(8 ft) = 96 ft^2.

Three-Dimensional Shapes

Rectangular Solid

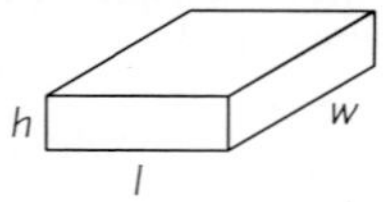

Surface area = $2(w \cdot h + l \cdot h + w \cdot l)$
Area = $l \cdot w \cdot h$

Cylinder

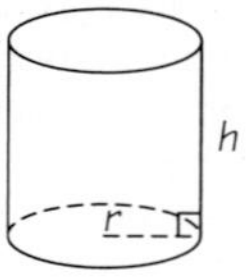

Surface area = $2\pi r^2 + 2\pi r \cdot h$
Volume = $\pi r^2 \cdot h$

Sphere

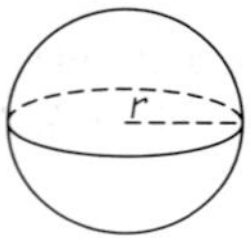

Surface area = $4\pi r^2$
Volume = $\frac{4}{3}\pi r^3$

Cone

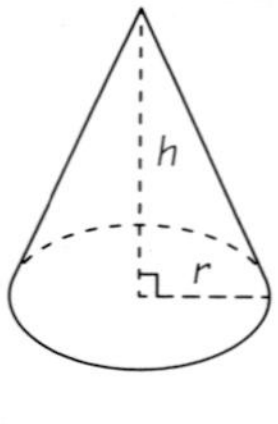

Surface area = $\pi r^2 + \pi r\sqrt{r^2 + h^2}$
Volume = $\frac{1}{3}\pi r^2 \cdot h$

Example 2. For a party, Cara made paper cones to hold candy and popcorn (see diagram).

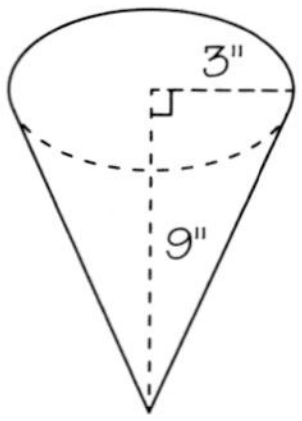

(a) After placing candy in some of the cones, Cara added a top to contain the candy. What is the surface area of each cone? Round your answer to two decimal places and include the units of measurement.

surface area = $\pi(3)^2 + \pi(3)\sqrt{(3)^2 + (9)^2} \approx 117.7 \text{ in.}^2$

(b) Cara filled the remaining open cones with popcorn. What is the surface area of one of the open cones?

For an open cone, we need to remove the area of the circular top. The area of a circle with a 3-in. radius is $\pi(3)^2 \approx 28.3$. The surface area is approximately 117.7 − 28.3, or 89.4 in.^2

(c) What is the volume of one of Cara's paper cones?

Volume = $\frac{1}{3}\pi(3)^2(9) \approx 84.8 \text{ in.}^3$

Practice Exercises

1. A bicycle tire has a diameter of 26 in. How many inches does the bicycle travel each time the wheel makes one revolution?
2. Suppose a square has an area of 144 cm^2. What is the side length of the square?
3. A landscaper is creating a garden in the shape of a triangle (see diagram). He plans to put a metal edge around the garden plot. Find the length of metal edging needed.

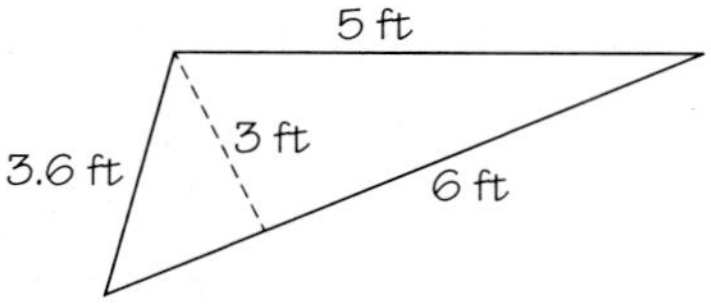

4. In order to get the correct amount of mulch for the garden in Practice Exercise 3, the landscaper needs to know the area of the plot. What is the area?
5. A cylindrical grain silo has a radius of 15 ft and height of 110 ft. What is the volume of grain that can be stored in the silo? What is the surface area of the silo?
6. The diameter of a sphere is 12 in. What is the volume and surface area of the sphere?
7. The height of a cone is 25 cm. The diameter of its circular base is 10 cm. Find the volume and surface area of the cone.

Algebra Review IV: Linear Equations and Inequalities

A. Linear Equations in One Variable

A linear equation in one variable is an equation that involves only the first power of the variable (typically represented by *x*) and no division by any expression involving the variable. Here are some examples of linear and nonlinear equations.

Linear Equations	Nonlinear Equations
$2x = 6$	$x^2 + 1 = 6$
$0.3(w - 5) + 4 = 6.1 + 0.1w$	$\frac{3}{x} + 4 = 5$
$\frac{1}{2}(x + 2) = \frac{3}{7}$	$\sqrt{\frac{p(1-p)}{25}} = 0.1$
$(2x - 1) = 4$	

A linear equation can be solved by performing the following steps:

1. Simplify both sides of the equation. This includes removing all parentheses and combining all like terms. If the equation has fractions, clearing these will generally simplify the remaining steps. To clear the fractions, multiply both sides of the equation by a number divisible by all denominators of the fractions.
2. Place all terms involving the variable on one side of the equation; place everything else on the other side of the equation.
3. Combine like terms or factor the variable out of all terms.
4. Divide both sides by the coefficient of the variable.

Not every equation will require all four steps.

Example 1. Solve $\frac{1}{2}(x + 2) + \frac{3}{7}(2x - 1) = 4$ for x.

STEP 1. Simplify both sides of the equation.

$\frac{1}{2}(x + 2) + \frac{3}{7}(2x - 1) = 4$	Clear the fractions: Multiply both sides of the equation by 14, the smallest number divisible by both 2 and 7.
$14\left(\frac{1}{2}(x + 2) + \frac{3}{7}(2x - 1)\right) = 14(4)$	Multiply each term on the left-hand side by 14. (Use the distributive law.)
$14\left(\frac{1}{2}(x + 2)\right) + 14\left(\frac{3}{7}(2x - 1)\right) = 14[4]$	Perform the multiplications by 14.
$7(x + 2) + 6(2x - 1) = 56$	Apply the distributive law to remove the parentheses.
$7x + 14 + 12x - 6 = 56$	Simplify by combining terms involving x and constant terms.
$19x + 8 = 56$	

STEP 2. There is only one term involving x, which is on the left-hand side. Place all terms not involving x to the right-hand side of the equation.

$19x + 8 = 56$	Subtract 8 from both sides of the equation.
$19x + 8 - 8 = 56 - 8$	

STEP 3. Combine like terms.

$19x = 48$

STEP 4. Solve for x.

$\frac{19x}{19} = \frac{48}{19}$	Divide both sides of the equation by 19, the coefficient of x.
$x = \frac{48}{19}$	

Example 2. Solve $0.3(w - 5) + 4 = 6.1 + 0.1w$ for w.

STEP 1. Simplify both sides of the equation.

$0.3(w - 5) + 4 = 6.1 + 0.1w$	Remove parentheses on the left-hand side by multiplication.
$0.3w - 1.5 + 4 = 6.1 + 0.1w$	Combine the constant terms on the left-hand side.
$0.3w + 2.5 = 6.1 + 0.1w$	

STEP 2. Place all terms with w on the left side of the equation and all other terms on the right side.

$-0.3w + 2.5 = 6.1 + 0.1w$	Add $-0.1w - 2.5$ to both sides of the equation.
$0.3w + 2.5 - 0.1w - 2.5 = 6.1 + 0.1w - 0.1w - 2.5$	

STEP 3. Combine like terms.

$0.2w = 3.6$

STEP 4. Solve for w.

$\frac{0.2w}{0.2} = \frac{3.6}{0.2}$	Divide both sides of the equation by the coefficient of w.
$w = 18$	

Practice Exercises

Solve the following equations for x.

1. $2x = 6$
2. $40 - 2x = 19 + 5x$
3. $2.8 - 2(x - 1.3) = 1.6x$
4. $\frac{2}{3}x - 6 = 4 - \frac{1}{2}x$
5. $4(1 - 5x) = 2(x - 1) + 6x$
6. $1 - 4(2 - x) = 6 - (x + 6)$
7. $\frac{1}{5}(2 - x) + 1 = \frac{1}{10}x$
8. $\frac{1}{4}x + \frac{1}{2}(x - 4) = 3 - \frac{1}{6}(x - 2)$

B. Plotting Points in the Plane

To form a coordinate plane, place two number lines at right angles, so they cross at 0. The number lines are called axes. Usually, the horizontal axis is labeled as the x-axis and the vertical axis is labeled as the y-axis. A point in the coordinate plane is specified by an ordered pair (x, y). To plot an ordered pair, such as (3, 2), draw a vertical line through the x-axis at $x = 3$ and a horizontal line through the y-axis at $y = 2$. The point of intersection of these two lines is the point (3, 2).

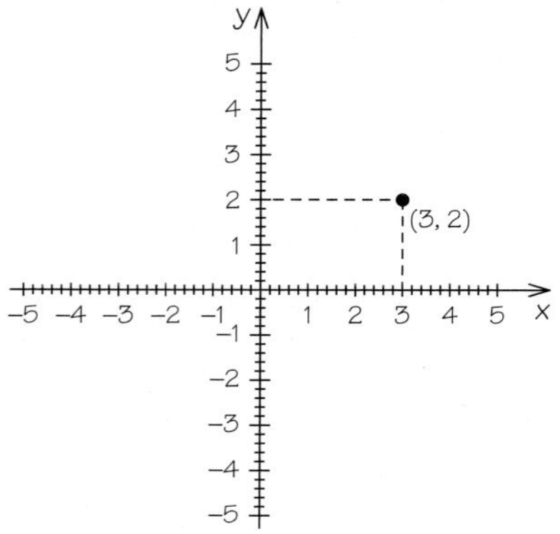

Example 1. Plot the points (2, 3), (–1, 4), (4, –2), and (–3, –1).

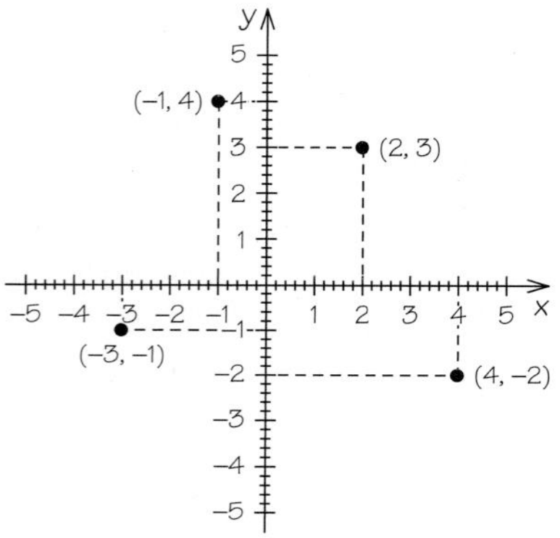

Practice Exercises

1. Plot the points (1, 3), (1, –2), (1, 1), and (1, –4). What pattern do you notice about these points?
2. Plot the points (2, –2), (4, –2), (–2, –2), and (1, –2). What pattern do you notice about these points?
3. Plot the points (–4, –4), (–2, –2), (0, 0), (1, 1), and (3, 3). What pattern do you notice about these points?

C. Distance and Midpoint between Two Points in the Plane

Given two points in the plane, sometimes we will want to calculate the distance between them; other times, we will want to place a point midway between the two points. Given two points in the plane, (x_1, y_1) and (x_2, y_2), we use the following formula to find the distance between them.

Distance between (x_1, y_1) and (x_2, y_2):

$$D = \sqrt{(x_2 - x_1)^2 + (y_2 - y_1)^2}$$

Example 1. Find the distance between the points (1, 2) and (6, 5).

$$D = \sqrt{(6 - 1)^2 + (5 - 2)^2} = \sqrt{5^2 + 3^2} = \sqrt{25 + 9}$$
$$= \sqrt{34} \approx 5.83$$

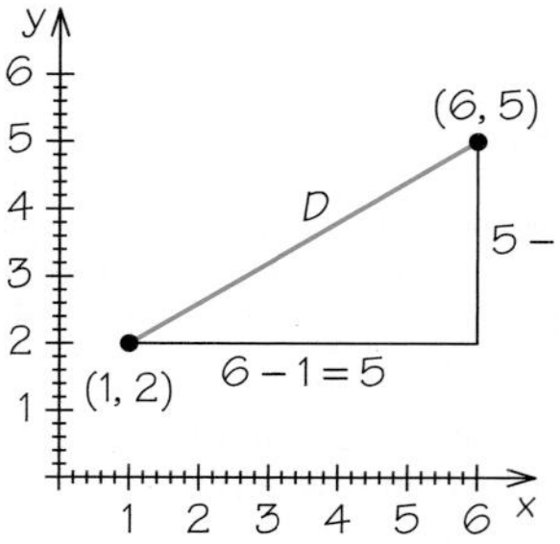

To find the midpoint between two numbers on a number line, simply average the two numbers.

Midpoint between x_1 and x_2: Midpoint $= \frac{x_1 + x_2}{2}$

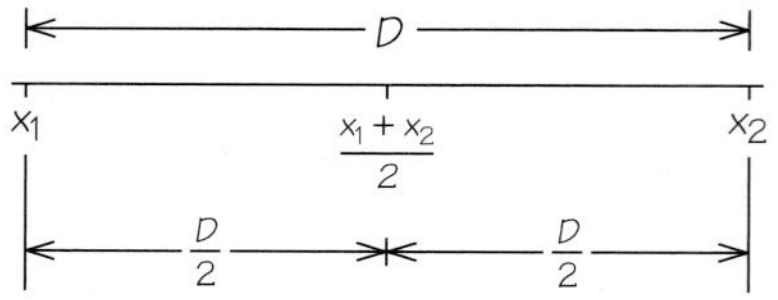

Example 2. Determine the midpoint between 2 and 12.

The distance between 2 and 12 is 10. So the midpoint should be $\frac{10}{2}$, or 5, units from both 2 and 12. Using the midpoint formula we get:

$$\text{Midpoint} = \frac{2 + 12}{2} = \frac{14}{2} = 7$$

Notice that 7 is 5 units above 2 and 5 units below 12.

We can also find the midpoint of a line segment between two points (x_1, y_1) and (x_2, y_2). The x- and y-coordinates of the midpoint will be the average of the x- and y-coordinates of the two points, respectively.

Midpoint between (x_1, y_1) and (x_2, y_2) is

$$\left(\frac{x_1 + x_2}{2}, \frac{y_1 + y_2}{2}\right)$$

Example 3. Find the midpoint between the points (1, 2) and (6, 5). Then find the distance between (1, 2) and the midpoint and the distance between the midpoint and (6, 5). What is true of these two distances?

$$\text{Midpoint} = \left(\frac{1 + 6}{2}, \frac{2 + 5}{2}\right) = \left(\frac{7}{2}, \frac{7}{2}\right)$$

The distance between (1, 2) and $\left(\frac{7}{2}, \frac{7}{2}\right)$ is

$$D = \sqrt{\left(\frac{7}{2} - 1\right)^2 + \left(\frac{7}{2} - 2\right)^2} = \sqrt{\left(\frac{5}{2}\right)^2 + \left(\frac{3}{2}\right)^2}$$
$$= \sqrt{\frac{34}{4}} = \frac{\sqrt{34}}{2}$$

The distance between $\left(\frac{7}{2}, \frac{7}{2}\right)$ and (6, 5) is

$$D = \sqrt{\left(6 - \frac{7}{2}\right)^2 + \left(5 - \frac{7}{2}\right)^2} = \sqrt{\left(\frac{5}{2}\right)^2 + \left(\frac{3}{2}\right)^2}$$
$$= \sqrt{\frac{34}{4}} = \frac{\sqrt{34}}{2}$$

These two distances are equal, half the distance between (1, 2) and (6, 5).

Practice Exercises

1. Find the distance between (2, 1) and (7, 10).
2. (a) Find the midpoint of the line segment between (2, 1) and (7, 10).
   (b) Use the distance formula to find the distance between (2, 1) and the midpoint from part (a), as well as the distance between the midpoint and (7, 10). What is true of these two distances?
3. (a) Find the distance between (–2, –7) and (3, 4).
   (b) Determine the midpoint of the line segment between (–2, –7) and (3, 4).
4. (a) Find the distance between (–6, –7) and (1, –3).
   (b) Determine the midpoint of the line segment between (–6, –7) and (1, –3).

D. Linear Equations in Two Variables

A linear equation in two variables (typically x and y) can be written in the form $ax + by = c$. A solution to such an equation is an ordered pair, (x, y), that satisfies the equation.

Example 1. Verify that $(4, -3)$ is a solution to $3x + 2y = 6$.

Substitute 4 for x and -3 for y, and then simplify to show that the expression on the left side of the equation is equal to 6, the number on the right side.

$$3(4) + 2(-3) \stackrel{?}{=} 6$$
$$12 + (-6) \stackrel{?}{=} 6$$
$$6 = 6 \text{ True}$$

Example 2. Show that $(1, 1)$ is not a solution to $3x + 2y = 6$.

$$3(1) + 2(1) \stackrel{?}{=} 6$$
$$3 + 2 \stackrel{?}{=} 6$$
$$5 = 6 \text{ False}$$

A line is a visual representation of the linear equation's solution set and can be graphed by drawing a line through two distinct points in the solution set. When the linear equation is in the form $ax + by = c$, an efficient method of graphing is to graph by intercepts. The y-intercept can be found by substituting $x = 0$ and then solving for y. The x-intercept can be found by substituting $y = 0$ and then solving for x. To graph $ax + by = c$, draw a line through the x- and y-intercepts.

Example 3. Determine the x and y intercepts for $3x + 2y = 6$. Draw a line that represents the solution set.

To find the y-intercept, substitute $x = 0$ and solve for y:

$$3(0) + 2y = 6$$
$$0 + 2y = 6$$
$$2y = 6$$
$$y = 3$$

The y-intercept is the ordered pair (0, 3).

To find the x-intercept, substitute $y = 0$ and solve for x:

$$3x + 2(0) = 6$$
$$3x + 0 = 6$$
$$3x = 6$$
$$x = 2$$

The x-intercept is the ordered pair (2, 0).

To graph $3x + 2y = 6$, plot the two intercepts and then draw a line through them as shown below. Notice that $(4, -3)$ (see Example 1) is on the line, but (1, 1) (see Example 2) is not.

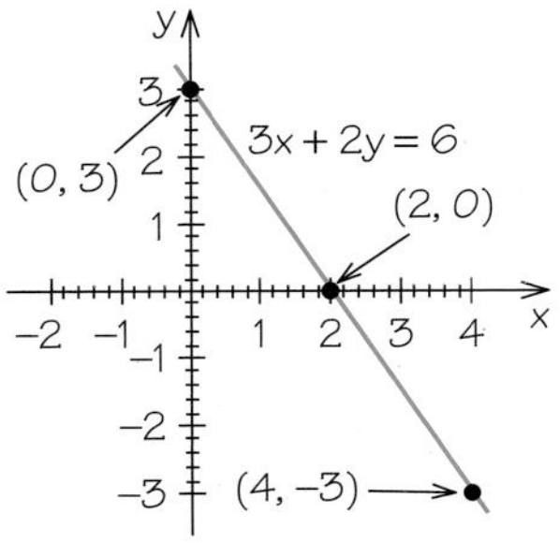

Two special kinds of lines of interest are horizontal and vertical lines. A horizontal line can be written in the form $y = b$ (b is where the line crosses the y-axis). A vertical line can be written in the form $x = a$ (a is where the line crosses the x-axis).

Example 4. Graph the solution set of $4x + 6y = 4(3 + x)$.

Start by simplifying the equation and getting all variables to one side of the equation.

$4x + 6y = 4(3 + x)$	Remove the parentheses by multiplying. (Use the distributive law.)
$4x + 6y = 12 + 4x$	Subtract $4x$ from both sides of the equation.
$6y = 12$	Divide both sides of the equation by 6, the coefficient of y.
$y = 2$	

The graph of the solution set is a horizontal line with y-intercept at (0, 2).

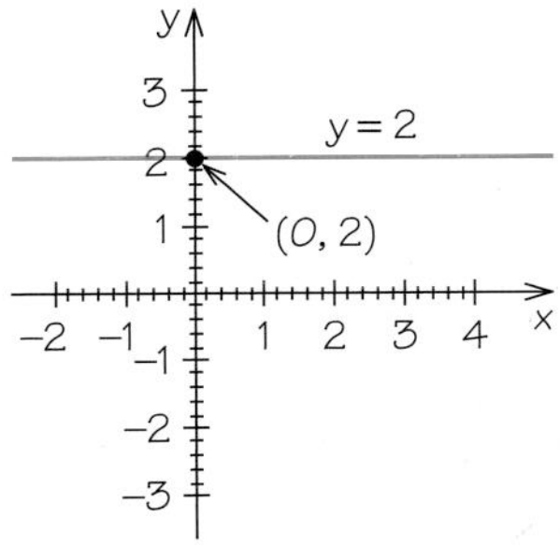

Example 5. Graph the line $x = 1$.

The graph is a vertical line with x-intercept (1, 0).

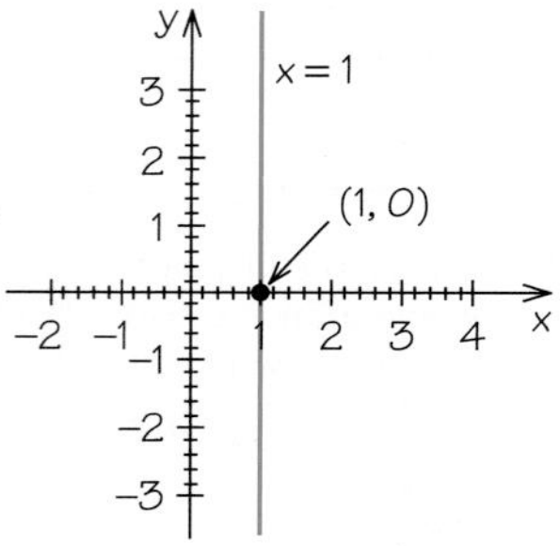

Practice Exercises

1. Is (3, 4) a solution to $4x + 3y = 10$?
2. Is (1, 2) a solution to $4x + 3y = 10$?
3. Find the x- and y-intercepts for $4x + 3y = 10$. Draw a line that represents the solution set.
4. Express $4x - 5y = 2(3 + x)$ in the form $ax + by = c$.
5. Is $\left(2, -\frac{2}{5}\right)$ a solution to the linear equation in Practice Exercise 4?
6. Find the x- and y-intercepts for the linear equation in Practice Exercise 4. Draw a line that represents the solution set.
7. Graph the solution set of $6x + 4y = 4(2 + y)$.
8. Graph the solution set of $14x + 2y = 2(5x - 3) + 4x$.

E. Slope of a Line

The slope of a line is defined to be the ratio of the vertical change to the horizontal change. If we think of slope as $m = \frac{\text{change in } y}{\text{change in } x}$, the slope of the line below would be $m = \frac{1}{3}$.

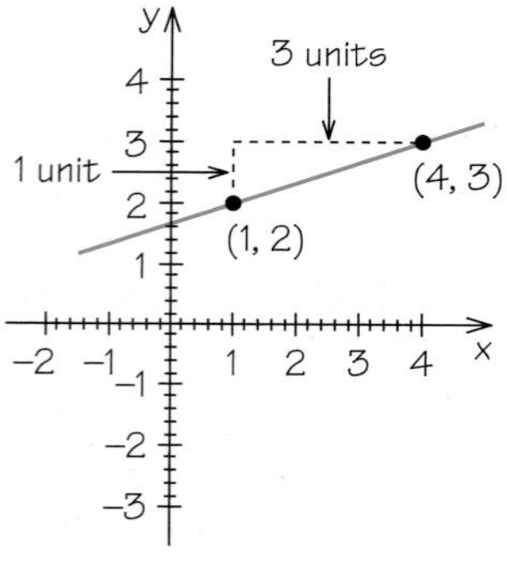

Given two points (x_1, y_1) and (x_2, y_2), the formula to calculate slope is $m = \frac{y_2 - y_1}{x_2 - x_1}$.

Example 1. Calculate the slope of a line that passes through the points (1, 3) and (4, 0).

By letting $(x_1, y_1) = (1, 3)$ and $(x_2, y_2) = (4, 0)$, the slope of the line below would be $m = \frac{y_2 - y_1}{x_2 - x_1} = \frac{0 - 3}{4 - 1} = \frac{-3}{3} = -1$

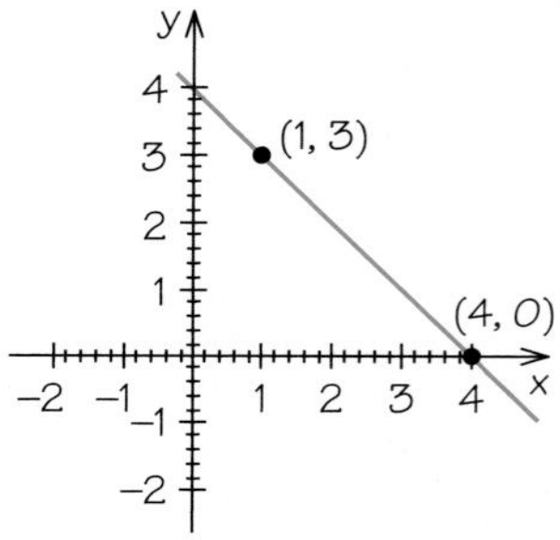

Given a graph of a line, we can determine visually whether the slope is positive or negative. If the y-values increase as the x-values increase, as is the case in the graph that appears before Example 1, then the slope is positive. On the other hand, if the y-values decrease as the x-values increase, as is the case for the graph in Example 1, then the slope is negative. Horizontal lines have a slope of 0. Vertical lines have undefined slope.

Example 2. Graph the line that passes through the points (1, 3) and (4, 3). Determine its slope.

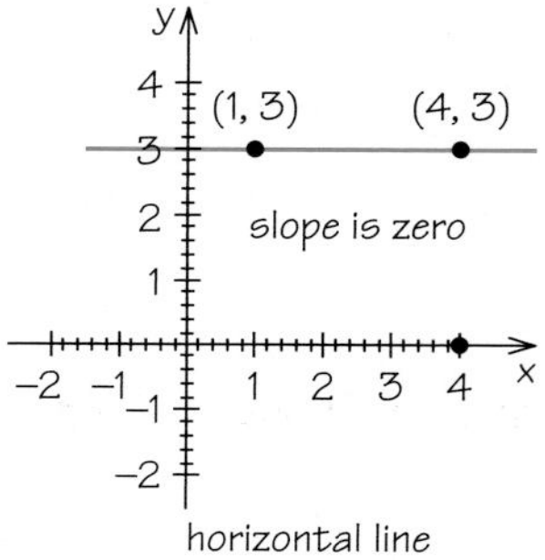

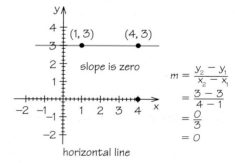

$$m = \frac{y_2 - y_1}{x_2 - x_1} = \frac{3 - 3}{4 - 1} = \frac{0}{3} = 0$$

Example 3. Graph the line that passes through the points (1, 3) and (1, 1). Determine its slope.

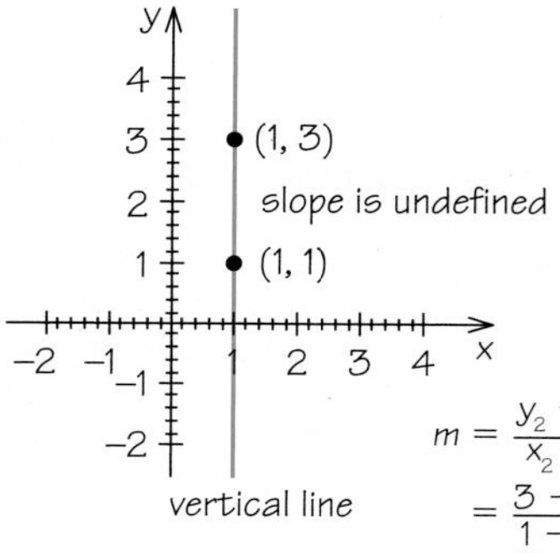

$$m = \frac{y_2 - y_1}{x_2 - x_1} = \frac{3 - 1}{1 - 1} = \frac{2}{0}, \text{ undefined}$$

(For additional help in computing slopes, see Example 2 in Algebra Review III, item A, Using Formulas, page AR-7.)

Practice Exercises

For Practice Exercises 1–4, determine the slope of the line that passes through the given points.

1. (5, 1) and (20, 6)
2. (4, −2) and (8, −6)

3. (5, −2) and (5, 2)
4. (−3, 8) and (−5, 8)

5–8. Determine whether the slope of each line in the graph below is positive, negative, zero, or undefined.

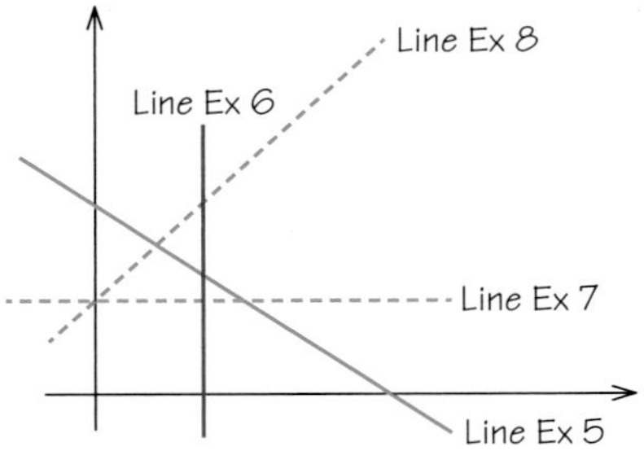

9. Determine the slope of the line given the information in the graph below.

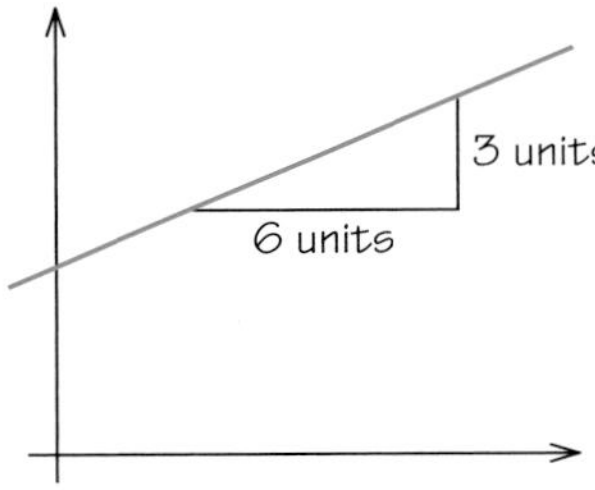

F. Graphing a Line in Slope-Intercept Form

When a linear equation is written in the form $y = mx + b$, it is said to be in slope-intercept form. In this form, b is the location where its graph intersects the vertical axis, which is generally referred to as the y-intercept. This intercept represents one point on the graph of the line. In order to graph a line, two points are required. A second point can be obtained by using the slope.

Example 1. Graph the line $y = \frac{2}{3}x - 1$ by first plotting the y-intercept and then using the slope to find a second point on the line.

In the graph below, first plot the y-intercept, (0, −1). Using the slope of $\frac{2}{3}$, find a second point on the line by starting at (0, −1) moving up 2 units and to the right 3 units, which gives the point (3, 1). Draw a line through (0, −1) and (3, 1).

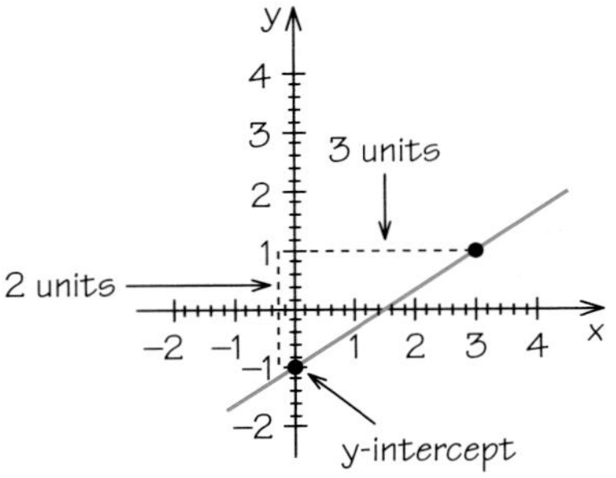

Example 2. Graph the line $y = -0.5x + 2$.

First, plot the y-intercept, (0, 2). Instead of using the slope to obtain a second point, it may be easier to evaluate the formula with a value for x other than $x = 0$. For example, if we substitute $x = 4$, we find that $y = -0.5(4) + 2 = -2 + 2 = 0$. This corresponds to the point (4, 0). By drawing a line through the y-intercept and the point (4, 0), we have the following graph:

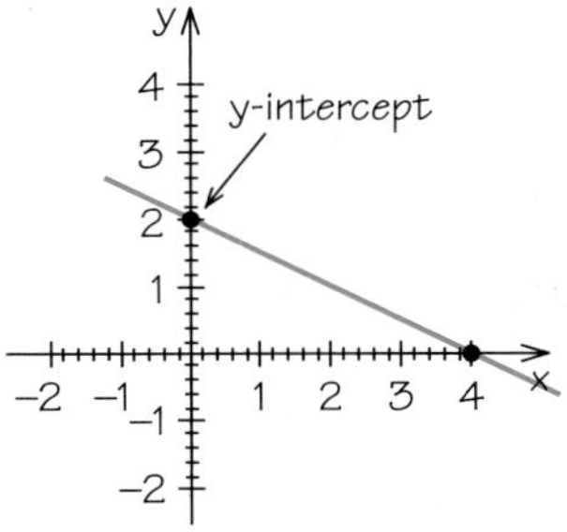

Calculator Note: You can use a graphing calculator to graph a line in the form $y = mx + b$. Here's how:

- Press [Y=]. Clear out any stored functions.
- Enter the function, $mx + b$, as Y1 and then press [ZOOM] [6] to graph the line in the standard viewing window (Xmin = −10, Xmax = 10, Ymin = −10, Ymax = 10]).
- If needed, press [WINDOW] and adjust the scaling.

Example 3. Using a graphing calculator, graph $y = 2x + 5$ in the standard viewing window.

Following the note above, enter $2x + 5$ opposite Y1. Your graph should be similar to the screenshot shown below.

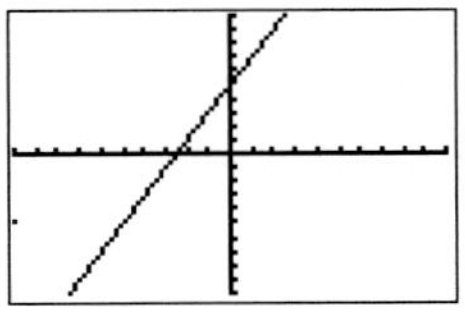

Practice Exercises

1. A line crosses the y-axis at 4 and has slope $\frac{4}{5}$. Write an equation that describes the line.
2. A line has y-intercept of (0, −3) and slope of 2. Identify a second point on the line.
3. A line is described by the equation $y = \frac{5}{4}x + 2$. Starting at the y-intercept, use the slope to plot a second point that lies on the line that has an x-coordinate equal to 4. Draw a line through the y-intercept and second point.
4. Plot two points on the line described by $y = -0.25x + 5$. Draw a graph of the line. (You may have to extend your horizontal axis in order to see where your graph crosses the x-axis.)
5. (a) The points (2, 5) and (4, 9) lie on a line. What is the slope of the line?

   (b) Substitute the slope into the equation $y = mx + b$. Substitute one of the points and solve for b.

(c) Finally, write an equation that describes the line through the points (2, 5) and (4, 9).

6. Repeat Practice Exercise 5 using the points (1, 7) and (3, 2).

For Practice Exercises 7 and 8, use a graphing calculator to graph the equations.

7. $y = 2.8x + 3$

8. $y = -\frac{3}{2}x + \frac{1}{2}$. (Adjust the window settings: Xmin $= -5$, Xmax $= 5$, Ymin$= -5$, and Ymax $= 5$.)

G. Linear Inequalities in Two Variables

To graph linear inequalities, we first graph the line, determine whether the line should be drawn solid or dashed, and then determine which side of the line to shade.

- We draw a solid line in cases where the points on the line are included in the solution set (the inequality is inclusive, $\le$ or $\ge$) and a dashed line when the points on the line are not part of the solution set (the inequality is not inclusive, $<$ or $>$).
- To determine which side of the line to shade, we substitute a test point. A test point can be any point in the plane that is not on the line. In most cases, choosing the origin as a test point allows for easy calculation. If substituting the test point results in an inequality that is true, shade the region on the side of the line containing the test point; otherwise, shade the region on the opposite side of the line to the test point.

Example 1. Graph $3x + 2y \le 12$ in the first quadrant.

This line has a y-intercept of (0, 6) and an x-intercept of (4,0). We draw a solid line because of the inclusive inequality symbol $\le$, which indicates that the points on the line are part of the solution. Next, we test to see if (0, 0) is in the solution set: $3(0) + 2(0) \le 12$, or $0 \le 12$, is a true statement. Therefore, we shade the region on the side of the line containing our test point.

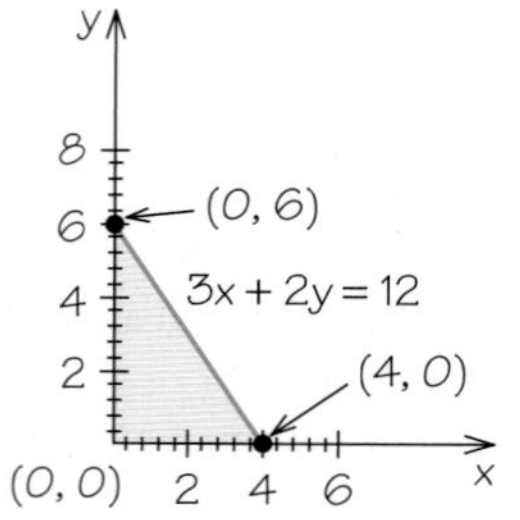

Calculator Note: To graph a linear inequality, you will first need to solve the inequality for y. In other words, express the inequality in the form $y \ge mx + b$, where the inequality sign could be any of the four possibilities: $\ge$, $\le$, $>$, or $<$.

The following algebraic manipulations do not change the solution set of an inequality:

- Adding or subtracting the same quantity to both sides of the inequality.
- Multiplying or dividing both sides of the inequality by the same positive quantity.
- Multiplying or dividing both sides of the inequality by the same negative number and *reversing the direction of the inequality sign.*

Example 2. Use a graphing calculator to graph $3x + 2y \le 12$.

STEP 1. Solve the inequality for y.

$3x + 2y \le 12$	Subtract $3x$ from both sides of the inequality.
$2y \le -3x + 12$	Divide both sides of the inequality by 2, the coefficient of y.
$y \le -\dfrac{3}{2}x + 6$	

STEP 2. Using the test point (0, 0), we find that $0 \le -\frac{3}{2}(0) + 6$ or $0 \le 6$, which is a true statement. The solution set is the region on the side of the line containing (0, 0).

STEP 3. Enter the linear function $y = -\frac{3}{2}x + 6$ into your calculator's function list.

- Press [Y=] and clear any stored functions.
- Enter the function as Y1. Press [(] [(−)] [3] [÷] [2] [)] [X,T,θ,n] [+] [6] and then [ZOOM] [6] to graph the line in the standard viewing window.
- The test point (0, 0) lies below the line. So next we shade that region. Press [Y =]. The cursor should be to the right of Y1. Use the left arrow key to move the cursor to the left of Y1. Press [ENTER] repeatedly until the "shade below" symbol is displayed. Press [GRAPH]. The result appears below.

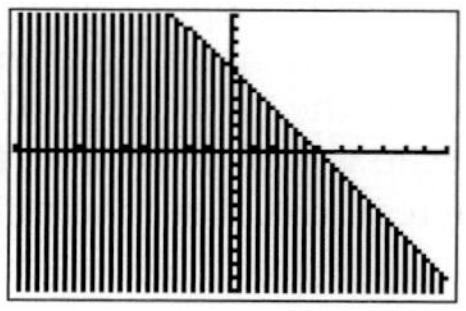

Unlike in Example 1, for Example 2 we did not restrict the solution set to the first quadrant.

Note: You will need to determine whether to draw a solid line or a dotted line. In this case, a solid line should be drawn because of the inclusive inequality $\le$.

Example 3. Use algebraic manipulation to solve the following inequality for y: $5x - 10y \le 5$. Use a graphing calculator to graph the solution set.

STEP 1. Solve the inequality for y.

$5x - 10y \le 5$	Subtract $5x$ from both sides of the inequality.
$-10y \le -5x + 5$	Divide both sides of the inequality by -10 and reverse the direction of the inequality.
$y \ge \dfrac{-5}{-10}x + \dfrac{5}{-10}$	Simplify.
$y \ge \frac{1}{2}x - \frac{1}{2}$	

STEP 2. Using the test point (0, 0), we find $0 \geq \frac{1}{2}(0) - \frac{1}{2} = -\frac{1}{2}$, which is true. So, we will shade the side of the line that contains the test point.

STEP 3. Graph $y = \frac{1}{2}x - \frac{1}{2}$, determine which side of the line to shade and then graph the solution set. The result appears below. The viewing window is set so that the values for x and y are between -5 and 5. The region above the line is shaded because the test point (0, 0) lies above the line.

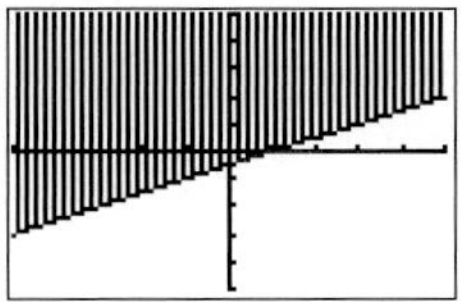

Practice Exercises

1. Does (1, 3) belong to the solution set of $4x - 2y > 1$?
2. Does (4, 2) belong to the solution set of $2x - 4y \leq 0$?
3. Graph the solution set of $5y - 10x > 20$ as follows:
   (a) Graph $5y - 10x = 20$. (Should you draw this line solid or dashed?)
   (b) Select a test point. Does your test point make the inequality true or false?
   (c) Shade the region corresponding to the solution set.
4. Graph the solution set of $4x - 2y > 1$. Use the test point from Practice Exercise 1 to determine which region should be shaded. (If this test point does not help, select a different test point.)
5. Graph the solution set of $2x - 4y \leq 0$. Use the test point from Practice Exercise 2 to determine which region should be shaded. (If this test point does not help, select a different test point.)
6. Solve $3x - 4y \geq 6$ for y. [In other words, express in the form y ($\geq$ or $\leq$) $mx + b$.]
7. Use a graphing calculator to graph the solution set of $y \leq -3x + 6$. Make a sketch of the graph.
8. Use a graphing calculator to graph the solution set of $3x - 4y \geq 6$. Make a sketch of the graph.

H. Systems of Linear Equations and Inequalities

Systems of Linear Equations

Two or more linear equations that are to hold true at the same time are called a *system* of linear equations. Systems of linear equations can be represented by graphing the linear equations and determining the point of intersection, if one exists. The point of intersection is the solution to the system of linear equations.

Here is a general strategy for solving a system of two equations in two variables:

STEP 1. Solve one of the equations for one of the variables. Substitute this expression into the second equation to obtain one equation in one variable.

STEP 2. Solve this new equation using the techniques from Part IV, item A, Linear Equations in One Variable (page AR-10).

STEP 3. Use the solution from Step 2 and either of the original equations to compute the corresponding value for the second variable.

Example 1. Solve the system:

$x + y = 5 \quad (1)$

$x + 2y = 7 \quad (2)$

STEP 1. Solve equation (1) for y and then substitute the expression for y into equation (2).

$x + y = 5$	Add $-x$ to both sides of the equation.
$y = -x + 5$	Substitute $-x + 5$ for y in equation (2).
$x + 2(-x + 5) = 7$	

STEP 2. Solve the equation from Step 1 for x.

$x + 2(-x + 5) = 7$	Use the distributive law to remove the parentheses.
$x - 2x + 10 = 7$	Subtract 10 from both sides of the equation.
$x - 2x + 10 - 10 = 7 - 10$	Simplify by combining terms.
$-x = -3$	Multiply both sides of the equation by -1.
$x = 3$	

STEP 3. Find the corresponding value for y by substituting 3 for x in either equation (1) or (2).

$x + y = 5$	Substitute 3 for x in equation (1).
$3 + y = 5$	Subtract 3 from both sides of the equation.
$y = 5 - 3 = 2$	

The solution is (3, 2).

Finally, we check that our solution is correct:

Check (3, 2) in	Check (3, 2) in
$x + y = 5$	$x + 2y = 7$
$3 + 2 \stackrel{?}{=} 5$	$3 + 2(2) \stackrel{?}{=} 7$
$5 = 5$ True	$7 = 7$ True

Example 2. Represent graphically the system of equations from Example 1.

We can use the graph-by-intercept approach (see Algebra Review IV, item D, Linear Equations in Two Variables, page AR-13) to graph the two linear equations. Below are graphs of equations (1) and (2), along with their corresponding x- and y-intercepts. The solution to the system of equations is the point of intersection of the two lines, which appears to be (3, 2).

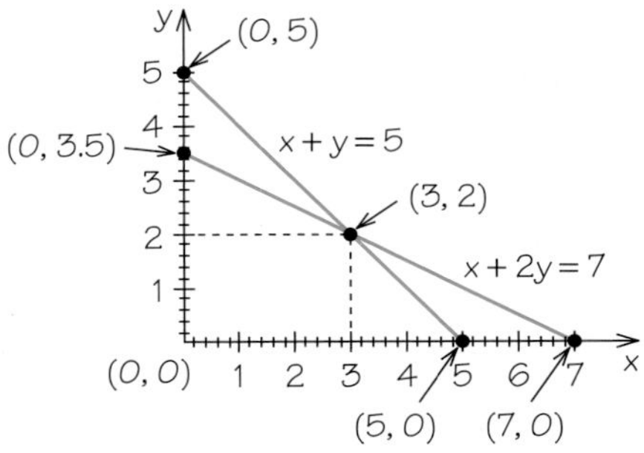

Calculator Note: Next, we find the solution to the system of linear equations in Example 1 using a TI-84 graphing calculator.

First we solve equations (1) and (2) for y, which results in the following equations:

$y = -x + 5$ (1)′

$y = -\frac{1}{2}x + \frac{7}{2}$ (2)′

Next, use a TI-84 graphing calculator to graph equations (1)′ and (2)′:

- Press [Y=], clear any previously stored functions, and enter the expressions for (1)′ and (2)′ as functions Y1 and Y2, respectively.

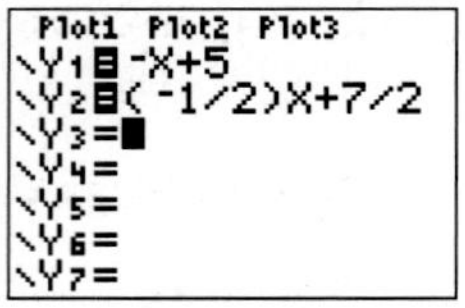

- Press [WINDOW] and adjust the window settings to Xmin = −1, Xmax = 8, Xscl = 1; Ymin = −1, Ymax = 6, YScl = 1. Press [GRAPH].

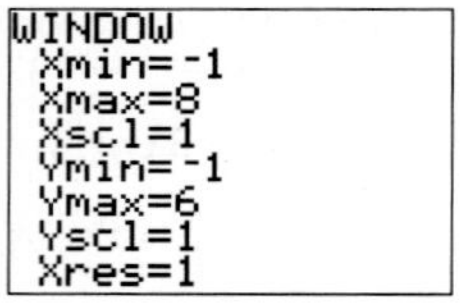

Now you need to find the point of intersection. Here's how:

- Press [2nd] [TRACE] (for CALC) [5] (for intersect).
- Press [ENTER] twice in response to "First curve?" and "Second curve?".
- Use the arrow keys to move the cursor close to the point of intersection and press [ENTER].
- Read off the solution $x = 3$ and $y = 2$ or (3, 2) as shown in the screenshot.

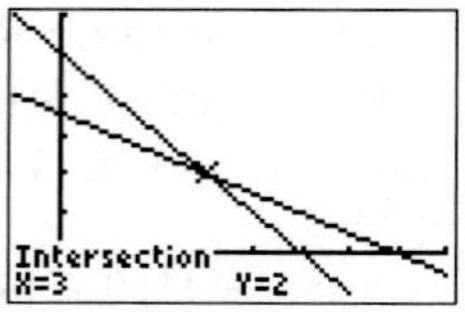

Systems of Linear Inequalities

If you are given more than one inequality, then you have a system of linear inequalities. In graphing a system of linear inequalities, you look for the region that is common to the graphs of all the individual inequalities.

Example 3. Graph the solution set of $x \geq 1$, $y \geq 2$, and $3x + 2y \leq 12$.

Begin by graphing the equations $x = 1$, $y = 2$, and $3x + 2y = 12$. We have drawn solid lines in the graph that follows because each of the inequalities is inclusive ($\leq$ or $\geq$).

We can see that the shaded region satisfies all three inequalities. In addition, the three points of intersection are shown. These were found by examining the three lines two at a time.

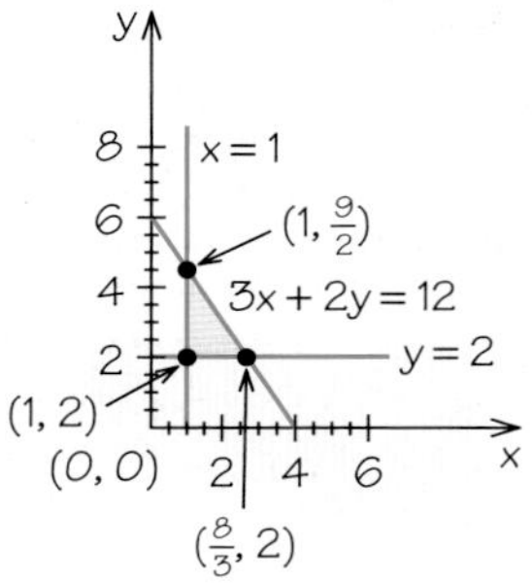

Calculator Note: Since $x = 1$ is not a function (it fails the vertical line test), you cannot graph it using the [Y=] menu on a TI-84 graphing calculator. Graph the other functions first and approximate the graph of $x = 1$ by pressing [2nd] [PRGM] (for DRAW) and then [4] (for Vertical), followed by [ENTER]. Use the left or right arrow keys to move the vertical line as close as possible to its desired location. (You won't be able to get it exact.)

Practice Exercises

1. Check to see if (4, 1) is a solution to the system of equations

   $2y - x = 2$

   $y + 2x = 9$

For Practice Exercises 2 and 3, use algebraic procedures to solve the system of equations.

2. $y + 3x = 6$

   $2y - x = 4$

3. $2y + 3x = 12$

   $4y + 7x = 12$

4. (a) Use algebraic procedures to solve the following system of equations:

   $3y + 2x = 7$

   $6y + 4x = 10$

   (b) What problem did you encounter?

   (c) Graph the system of equations, and explain why this system has no solution.

5. Use a graphing calculator to solve the following system of equations:

   $y = 3x + 9$

   $y = -2x + 1$

6. Check whether (1, 3) is in the solution set to $x \geq 0$, $y + 2x \leq 4$, and $y \geq 3x$.
7. Graph the solution set of $x \leq 6$, $y \geq 4$, and $y \leq 2x + 3$.
8. Graph the solution set of $x > 1$, $2y - x \geq 2$, $y + 2x \leq 9$.

Algebra Review V: Summation Notation, Sequences, and Series

A. Summation Notation

A sequence of numbers is an ordered list of numbers. If there are k numbers in a list, the sequence can be written as $a_1, a_2, \ldots, a_k$. A *summation* is the addition of the numbers in such a sequence. Summation notation has the following parts:

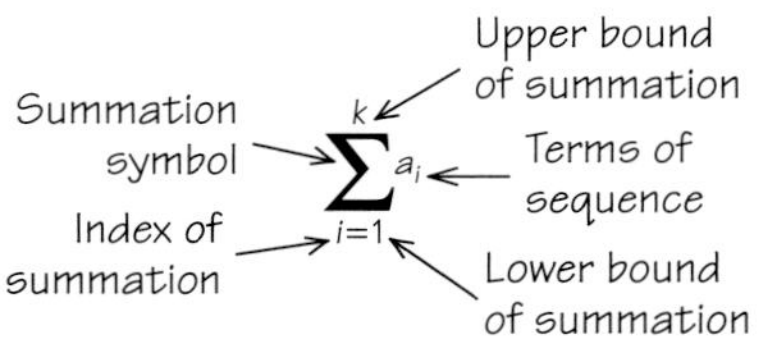

Example 1. Find the value of $\sum_{i=1}^{3} i^2$.

The index i indicates that the summation starts at 1 and that successive values of i can be found by adding 1 until the upper bound of 3 is reached. Thus, we get

$$\sum_{i=1}^{3} i^2 = 1^2 + 2^2 + 3^2 = 14$$

Example 2. Find the value of $\sum_{i=1}^{5} 3i$.

We find $\sum_{i=1}^{5} 3i$ as follows:

$$\sum_{i=1}^{5} 3i = 3(1) + 3(2) + 3(3) + 3(4) + 3(5)$$
$$= 3 + 6 + 9 + 12 + 15 = 45$$

If each term in a sum contains the same factor, we can use the distributive law to pull the factor out in front of the sum:

$$\sum_{i=1}^{n} ka_i = k\sum_{i=1}^{n} a_i$$

Example 3. Return to Example 2. This time, pull the constant out in front of the summation.

$$\sum_{i=1}^{5} 3i = 3\sum_{i=1}^{5} i = 3(1 + 2 + 3 + 4 + 5)$$
$$= 3(15) = 45$$

We can also use summation notation to shorten formulas. In the next two examples, we revisit formulas for calculating the mean and standard deviation of a sample, this time using summation notation.

Example 4. In Chapter 5 (page 197), the formula for the mean of n numbers is given as $\bar{x} = \frac{x_1 + x_2 + \cdots x_n}{n}$. Express this formula using summation notation.

We express the formula for $\bar{x}$ using summation notation in the numerator: $\bar{x} = \frac{\sum_{i=1}^{n} x_i}{n}$.

Example 5. In Chapter 5 (page 204), the formula for the standard deviation of n data values is given as $s = \sqrt{\frac{(x_1 - \bar{x})^2 + (x_2 - \bar{x})^2 + \cdots + (x_n - \bar{x})^2}{n - 1}}$. Express this formula using summation notation.

We express the formula for s using summation notation in the numerator of the fraction:

$$s = \sqrt{\frac{\sum_{i=1}^{n} (x_i - \bar{x})^2}{n - 1}}$$

Up to this point, we have dealt with a single list of numbers. However, sometimes we combine corresponding elements of two ordered lists of numbers. When there are two lists of numbers, make sure to pair the values from both lists correctly.

Example 6. Determine $\sum_{i=1}^{5} a_i b_i$ from the two lists of numbers in the following table.

i	1	2	3	4	5
a at *i*th position	2	−5	0	3	−1
b at *i*th position	3	2	−3	1	4

$$\sum_{i=1}^{5} a_i b_i = (2)(3) + (-5)(2) + (0)(-3) + (3)(1) +$$
$$(-1)(4) = 6 + (-10) + 0 + 3 + (-4) = -5$$

Practice Exercises

Find the value of the sums in Practice Exercises 1–3.

1. $\sum_{i=1}^{3} i^3$
2. $\sum_{i=1}^{5} 2i$
3. $\sum_{i=1}^{4} (2i - 1)$
4. Suppose that $x_1 = 3$, $x_2 = 5$, and $x_3 = 4$.
   (a) Use the formula in Example 4 to find $\bar{x}$.
   (b) Use your answer to part (a) and the formula in Example 5 to find s.
5. The table below contains two lists of numbers.

i	1	2	3	4
a at *i*th position	2	5	7	9
b at *i*th position	8	6	4	2

   (a) Determine $\sum_{i=1}^{4} a_i b_i$.
   (b) Determine $\sum_{i=1}^{4} (a_i - b_i)$.

B. Sequences

A sequence is a listing of objects, generally numbers, called *terms.* Terms in a sequence are listed in a specific order and are separated by commas.

Example 1. Write a sequence of positive multiples of 3.

3, 6, 9, 12 . . .

The sequence in Example 1 goes on forever, and is therefore called an *infinite* sequence. Sequences that have a definite end are called *finite.*

Subscripts are used in sequences to describe where in the ordered list a term is located. For example, t_1 could represent the first term of the sequence in Example 1. In this case, $t_1 = 3$. (Sometimes the initial term in a sequence is represented by t_0 instead of t_1.) Because the pattern in Example 1 continues, we know that $t_5 = 15$.

Example 2. For the sequence in Example 1, given that t_1 represents the first term, what is t_8?

We can extend the sequence out to eight terms: 3, 6, 9, 12, 15, 18, 21, 24, ... Thus, $t_8 = 24$.

In general, t_n represents the nth term, t_{n-1} is the term that precedes t_n, and t_{n+1} is the term that follows t_n.

There are two types of rules that govern how terms in a sequence are calculated: *explicit* and *recursive.* An explicit rule defines how to find the nth term by direct calculation. A recursive rule defines the first or first few terms and then makes a general statement about how to obtain other terms from previous terms.

Example 3. Find an explicit rule for expressing the sequence of positive multiples of three. Use the explicit rule to determine t_{35}.

We begin by examining the list below to look for an *explicit* pattern.

Term Number	Term	Comment
1	3	$t_1 = 3 \cdot 1$
2	6	$t_2 = 3 \cdot 2$
3	9	$t_3 = 3 \cdot 3$
4	12	$t_4 = 3 \cdot 4$

We can summarize the pattern in the last column of the table above with the explicit rule $t_n = 3 \cdot n$. Using this rule, we find $t_{35} = 3 \cdot 35 = 105$.

Example 4. Find a recursive rule for describing the sequence of positive multiples of three. Use the recursive rule to determine t_9 from Example 2's answer.

We begin by examining the following list to look for a *recursive* pattern, a pattern that describes how to obtain the next term in a sequence from the previous term (or terms).

Term Number	Term	Comment
1	3	$t_1 = 3$
2	6	$t_2 = t_1 + 3$
3	9	$t_3 = t_2 + 3$
4	12	$t_4 = t_3 + 3$

A recursive rule about this sequence is $t_1 = 3$ and $t_n = t_{n-1} + 3$ for $n \geq 2$ (or $t_{n+1} = t_n + 3$ for $n \geq 1$). To find t_9, we need the value of the previous term, $t_8 = 24$, which we get from Example 2. Next, we use the recursive rule to find t_9: $t_9 = t_8 + 3 = 24 + 3 = 27$.

Calculator Note: Calculators that store an answer for use with the next calculation can be used to easily generate a recursively defined sequence. Example 5 shows how to use this feature on a TI-84 graphing calculator. Example 6 shows how to use the ENTRY feature on a TI-84 to quickly generate terms of an explicitly defined sequence.

Example 5. Use a calculator to generate the first four terms of the following recursively defined sequence: $t_1 = 5$ and $t_n = 4 \cdot t_{n-1} + 1$.

Using a TI-84, here's one way to generate terms in this sequence:

- Enter the value of the first term: press [5] [ENTER].
- Press [×] [4] [+] [1] [ENTER] to obtain the second term in the sequence.
- Press [ENTER] to obtain the third term. Continue pressing [ENTER] to generate more terms.

Below is a calculator screen showing the first four terms of this sequence.

```
                5
Ans*4+1
               21
Ans*4+1
               85
Ans*4+1
              341
█
```

Example 6. Use a calculator to generate the first six terms of the following explicitly defined sequence: $t_n = n^2 - 2$.

Using a TI-84, here's one way to generate terms in this sequence:

- Enter $n = 1$ into the formula to get the first term, t_1: press [1] [x^2] [−] [2] [ENTER].
- To get the second term, press [2nd] [ENTER] (for ENTRY), use the left arrow key to move the cursor over 1, and then press [2] (to replace 1 with 2) and [ENTER].
- To get the third term, press [2nd] [ENTER], use the left arrow key to move the cursor over 2, and then press [3] (to replace 2 with 3) and [ENTER].
- Repeat the process above to replace the value of n by 4, 5, and 6.

Below is a calculator screen that shows the first three terms of this sequence and entry of the fourth term prior to pressing ENTER. The first six terms of this sequence are −1, 2, 7, 14, 23, and 34.

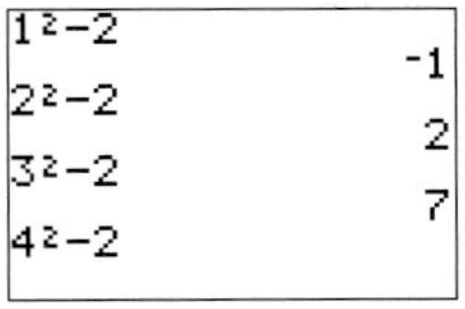

Practice Exercises

1. Write a sequence of positive multiples of 5.
2. Write an explicit rule to generate positive multiples of 5. Use this rule to find the tenth term in the sequence.
3. Write a recursive rule to generate positive multiples of 5.

For Practice Exercises 4–8, use the given rules to generate the first six terms in the sequence.

4. $a_n = \frac{1}{2}n$
5. $b_1 = 3$ and $b_n = b_{n-1} + 2$
6. $c_1 = 1$ and $c_{n+1} = 3 \cdot c_n - 1$
7. $d_n = 3n - 2$
8. $t_1 = 1, t_2 = 2$, and $t_n = t_{n-1} + t_{n-2}$

C. Arithmetic and Geometric Sequences and Series

An arithmetic sequence occurs when you add the same amount (common difference) to a term to obtain the next term. This relationship between term and term number is linear. A geometric sequence occurs when you multiply the same amount to a term (common ratio) to obtain the next term. This relationship between term and term number is exponential.

Suppose that each sequence has a first term *a*. For the arithmetic sequence, we add *d* to obtain the next term, and for the geometric sequence we multiply by *r*. We can find the *n*th term of each sequence by looking for a pattern.

Term Number	Arithmetic	Geometric
1	a	a
2	$a + d$	ar
3	$a + 2d$	ar^2
4	$a + 3d$	ar^3
⋮	⋮	⋮
n	$a + (n - 1)d$	ar^{n-1}

The *n*th term of an arithmetic sequence is $a + (n - 1)d$, and the *n*th term of a geometric sequence is ar^{n-1}.

Example 1. Consider two specific sequences where the first term is 3. The first sequence, an arithmetic sequence, has a common difference of $d = 2$. The second sequence, a geometric sequence, has a common ratio of $r = 2$. Write the first eight terms of each sequence.

Arithmetic sequence: 3, 5, 7, 9, 11, 13, 15, 17

Geometric sequence: 3, 6, 12, 24, 48, 96, 192, 384

A series is made from a sequence of terms that are summed.

Example 2. Find the arithmetic series formed from the eight terms of the arithmetic sequence in Example 1.

The series is the sum of the terms in the sequence. You could find the sum by adding up all eight terms: $3 + 5 + 7 + 9 + 11 + 13 + 15 + 17$. However, we would prefer to find the sum using a method that can be generalized when more terms are included. Notice the following pattern: The sum of the first and last term is 20 and that sum appears three more times.

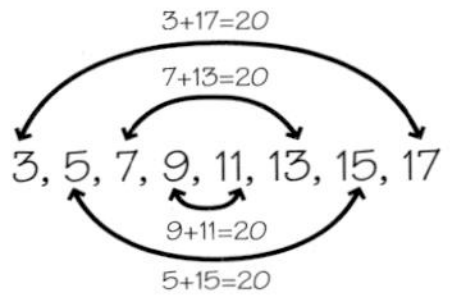

The series (the sum of the eight terms in the sequence) is $4 \cdot 20 = 80$.

In general, we can extend the pattern noted in Example 2 to find the sum of the first *n* terms of an arithmetic sequence. Here's the result:

$$\left(\frac{\text{number of terms, } n}{2}\right) \times (\text{first term} + \text{last term})$$

We can express the series formed from the first *n* terms in an arithmetic sequence a_1, a_2, a_3, using summation notation:

$$S_n = \sum_{i=1}^{n} a_i = \frac{n(a_1 + a_n)}{2}$$

Example 3. Given an arithmetic sequence with terms specified by $a_n = 3 + 2(n - 1)$, find the value of the series S_{10}, which is the sum of the first 10 terms of the sequence.

Start by determining the first and tenth term in the sequence: $a_1 = 3 + 2(1 - 1) = 3$ and $a_{10} = 3 + 2\,(10 - 1) = 21$. Next, use the formula above for finding the sum of terms in an arithmetic sequence: $S_{10} = \frac{10(3 + 21)}{2} = \frac{240}{2} = 120$.

A geometric series is made by summing the terms of a geometric sequence. The sum of terms in a geometric sequence can be computed from the geometric series rule, which is

$$S_n = a\left(\frac{r^n - 1}{r - 1}\right)$$

where *a* is the first term and *r* is the common ratio.

Example 4. Find the geometric series formed from the sum of the eight terms in the geometric sequence in Example 1.

Substituting 3 for *a*, 2 for *r*, and 8 for *n* into the geometric series rule, we have

$$S_8 = 3\left(\frac{2^8 - 1}{2 - 1}\right) = 3\left(\frac{256 - 1}{1}\right) = 3(255) = 765$$

Practice Exercises

For Practice Exercises 1–4, determine whether the sequence is arithmetic, geometric, or neither, and then write the next two terms in the sequence.

1. 5, 10, 20, 40, 80
2. 2, −2, 4, −4, 6, −6
3. 4, 7, 10, 13, 16
4. 200, 100, 50, 25
5. An arithmetic sequence has first term −12 and common difference $d = 5$. What are the first five terms of this sequence? What is the sum of the first five terms, S_5?
6. A geometric sequence has first term 5 and common ratio $r = 4$. What are the first five terms of this sequence? What is the sum of the first five terms, S_5?
7. Given an arithmetic sequence with terms specified by $a_n = 1 + 4(n - 1)$, find S_{15}.
8. Given a geometric sequence with first term $a = 2$ and common ratio $r = 3$, find S_{15}.

Algebra Review VI: Exponents, Roots, and Logarithms

A. Powers and Roots

Powers

A positive integer exponent or power is a convenient shorthand for indicating repeated multiplication of a number a called the *base*. For example, the expression a^2, read as "a squared," means a multiplied by a. When either a positive or negative number is squared, the outcome will always be positive. When 0 is squared, the outcome is 0.

Example 1. Simplify $(-8)^2$.

$$(-8)^2 = (-8)(-8) = 64$$

The expression a^3 is read as "a cubed" and a^n is read as "a to the nth power," or simply "a to the n." For positive integer powers of n, it means that a is multiplied times itself n times.

$$a^n = \underbrace{a \cdot a \cdot a \cdots a}_{n \text{ times}}$$

Example 2. Simplify 2^5, $(-2)^3$, and $(-3)^4$.

$$2^5 = (2)(2)(2)(2)(2) = 32$$
$$(-2)^3 = (-2)(-2)(-2) = -8$$
$$(-3)^4 = (-3)(-3)(-3)(-3) = 81$$

Notice that when a negative number is raised to an odd power, the result is negative, but when a negative number is raised to an even power, the result is positive.

Calculator Note: When raising negative numbers to powers, always be sure to enclose the negative number in parentheses before hitting the power key.

Example 3. Use a calculator to determine the square of −4.

On a TI-84, use the keystrokes [(] [(−)] [4] [)] [^] [2] [ENTER]. (Other calculators should be similar.) The result is $(-4)^2 = 16$ (see the calculator screenshot below on the left).

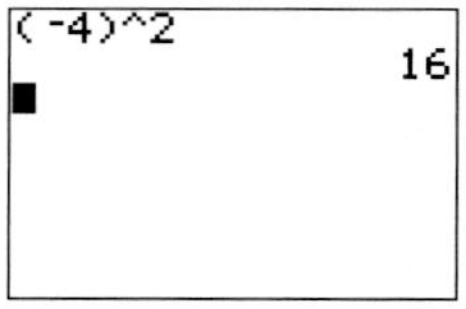

Calculation of $(-4)^2$

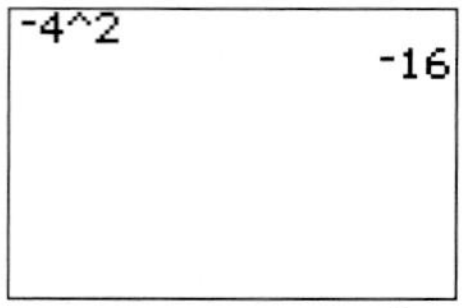

Calculation of -4^2

Example 4. Repeat Example 3, but don't enclose −4 in parentheses.

On a TI-84, use the keystrokes [(−)] [4] [^] [2] [ENTER], which gives the result $-4^2 = -16$ (see the calculator screenshot above on the right). In this case (based on order of operations), 4 was squared first, resulting in 16, and then its opposite was taken, resulting in a final answer of −16.

Roots

The expression $\sqrt{a}$ indicates a number that we square to get a. The expression $\sqrt{a}$ is called the "principal square root of a," and the outcome is never negative. The expression $\sqrt[n]{a}$, read as the "nth root of a," is the number that we raise to the nth power to get a.

$$\sqrt[n]{a} = b \text{ means } b^n = a$$

Example 5. Determine $\sqrt{144}$ and $\sqrt[5]{32}$.

$$\sqrt{144} = 12 \text{ because } 12^2 = 144$$
$$\sqrt[5]{32} = 2 \text{ because } 2^5 = 32$$

When a square root (or an nth root) involves operations, it may be easier to simplify some of the calculations before taking the square root (or the nth root).

Example 6. Approximate $\sqrt{\frac{(3.1)(2.7)}{5}}$ to one decimal place.

$$\sqrt{\frac{(3.1)(2.7)}{5}} = \sqrt{\frac{8.37}{5}} = \sqrt{1.674} \approx 1.3$$

Example 7. The standard deviation of a set of numbers (introduced in Chapter 5 on page 204) involves a square root. Find the standard deviation of 2, −3, and 5. Give a simplified exact answer and an approximation to two decimal places.

The formula for computing the standard deviation of a set of numbers is

$$s = \sqrt{\frac{(x_1 - \bar{x})^2 + (x_2 - \bar{x})^2 + \cdots + (x_n - \bar{x})^2}{n - 1}}$$

We follow the three steps from Algebra Review III, item A, Using Formulas (page AR-7).

- First, assign a meaning to the numbers.
  $n = 3$ and, $x_1 = 2$, $x_2 = -3$, and $x_3 = 5$. From Example 1 in item A, Using Formulas, we calculated the mean of 2, −3, and 5 and found that $\bar{x} = \frac{4}{3}$.

- Second, substitute the constants for the variables.

$$s = \sqrt{\frac{(x_1 - \bar{x})^2 + (x_2 - \bar{x})^2 + (x_3 - \bar{x})^2}{n - 1}}$$

$$= \sqrt{\frac{\left(2 - \frac{4}{3}\right)^2 + \left(-3 - \frac{4}{3}\right)^2 + \left(5 - \frac{4}{3}\right)^2}{3 - 1}}$$

- Third, simplify by applying order of operations.

$$s = \sqrt{\frac{\left(2 - \frac{4}{3}\right)^2 + \left(-3 - \frac{4}{3}\right)^2 + \left(5 - \frac{4}{3}\right)^2}{3 - 1}}$$

$$= \sqrt{\frac{\left(\frac{2}{3}\right)^2 + \left(-\frac{13}{3}\right)^2 + \left(\frac{11}{3}\right)^2}{2}} = \sqrt{\frac{\frac{4}{9} + \frac{169}{9} + \frac{121}{9}}{2}}$$

$$= \sqrt{\frac{\frac{294}{9}}{2}} = \sqrt{\frac{294}{18}} = \sqrt{\frac{49}{3}} = \frac{7}{\sqrt{3}} \approx 4.04$$

Calculator Note: You can find a decimal approximation for the problem in Example 7 by entering the entire expression into your calculator before pressing the [ENTER] key. The keystrokes for using a TI-84 to calculate the standard deviation of 2, −3, and 5 are given below.

- Press [2nd] [x^2] (for √( )
- [(] [(] [2] [−] [4] [÷] [3] [)] [^] [2] [+]
- [(] [(−)] [3] [−] [4] [÷] [3] [)] [^] [2] [+]
- [(] [5] [−] [4] [÷] [3] [)] [^] [2] [)] [÷] [(]
  [3] [−] [1] [)] [)]
- Finally, press [ENTER].

After entering these keystrokes, your calculator screen should match the one below.

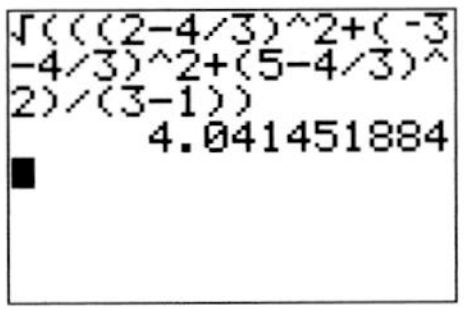

Warning: When entering expressions involving parentheses into a calculator, make sure every left parenthesis has a matching right parenthesis. Also, notice that some commands, such as the square root command, may automatically include a left parenthesis that you will need to pair with a right parenthesis.

Calculator Note: The TI-84 has a square root key but no nth root key. Here's how to compute the nth root of a number a, $\sqrt[n]{a}$, when $n \geq 3$:

- Enter the value for n.
- Press [MATH] [5] (for $\sqrt[x]{\ }$).
- Enter the number a and press [ENTER].

Practice Exercises

1. Evaluate the following by hand. (Feel free to check your results using a calculator.)
   (a) $(-3)^4$
   (b) $(-2)^5$
   (c) 4^3
2. Evaluate the following by hand. (Feel free to check your results using a calculator.)
   (a) $\sqrt[4]{10{,}000}$
   (b) $\sqrt[3]{0.008}$
   (c) $\sqrt{2500}$
3. Approximate $\sqrt{\frac{(2.7)^2 + 3(4.6)}{4}}$ to one decimal place.
4. The dataset 1, 3, 5, 7 has mean $\bar{x} = 4$. Determine the standard deviation (round your answer to two decimal places).
5. Use your calculator to approximate $(1 + 0.05)^6$ to two decimal places.

B. Natural and Fractional Exponents

We can extend the definition of exponent to 0 and negative integers as follows:

$$a^0 = 1,\ a \neq 0$$
$$a^{-n} = \frac{1}{a^n},\ a \neq 0$$

Example 1. Determine the values of 3^0, 2^{-4}, and -3^{-2}.

$3^0 = 1$

$2^{-4} = \frac{1}{2^4} = \frac{1}{16}$

$-3^{-2} = -\left(\frac{1}{3^2}\right)$
$= -\frac{1}{9}$

Order of operations tells us that powers are performed before multiplication. Taking the opposite of a number is equivalent to multiplication by −1.

Next, we extend the definition of exponent to fractions, in other words, exponents in the form $\frac{m}{n}$ with m and n integers and $n > 0$.

$$a^{\frac{1}{n}} = \sqrt[n]{a},\ a \geq 0 \text{ if } n \text{ is even}$$
$$a^{\frac{m}{n}} = \sqrt[n]{a^m} = (\sqrt[n]{a})^m$$

In defining $a^{\frac{m}{n}}$ (with $a \geq 0$ if n is even), it doesn't matter if you raise a to the mth power first or take the nth root of a first. You can do whichever is easier.

Example 2. Determine $36^{\frac{1}{2}}$ and $8^{\frac{2}{3}}$.

$36^{\frac{1}{2}} = \sqrt{36} = 6$

$8^{\frac{2}{3}} = (\sqrt[3]{8})^2 = (2)^2 = 4$ or $8^{\frac{2}{3}} = \sqrt[3]{8^2} = \sqrt[3]{64} = 4$

In the computation of $8^{\frac{2}{3}}$, notice that it does not matter whether we took the cube root of 8 first and then squared the result or squared 8 first and then took the cube root.

Example 3. Write the following radical expressions using exponents: $\sqrt[5]{x}$ and $(\sqrt[3]{x})^2$.

$$\sqrt[5]{x} = x^{\frac{1}{5}}$$

$$(\sqrt[3]{x})^2 = x^{\frac{2}{3}}$$

Calculator Note: When using a calculator, it may be easier to compute a constant raised to an exponent rather than the equivalent radical expression. See Example 4.

Example 4. Convert the following radical expressions to expressions involving exponents, and then use your calculator to approximate the values to two decimal places: $\sqrt[7]{89}$ and $(\sqrt{45})^3$.

$$\sqrt[7]{89} = 89^{\frac{1}{7}} \approx 1.90$$

$$(\sqrt{45})^3 = 45^{\frac{3}{2}} \approx 301.87$$

Here's how these computations should appear on a TI-84 graphing calculator. Notice that the fractional exponents are enclosed in parentheses.

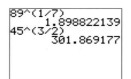

```
89^(1/7)
          1.898822139
45^(3/2)
           301.869177
```

Practice Exercises

1. Evaluate the following by hand.
   (a) 42^0
   (b) 5^{-2}
2. Evaluate the following by hand.
   (a) $(-27)^{\frac{1}{3}}$
   (b) $(-27)^{\frac{2}{3}}$
3. Write the following radical expressions using exponents.
   (a) $\sqrt[10]{x}$
   (b) $\sqrt[5]{x^3}$
4. Convert the following radical expressions to expressions involving exponents, and then use your calculator to approximate the values to two decimal places.
   (a) $\sqrt{0.0026}$
   (b) $(\sqrt[5]{256})^2$
   (c) $\dfrac{1}{\sqrt{40}}$

C. Graphs of Exponential Equations

Equations of the form $y = c \cdot a^x$ are called exponential equations (or exponential functions). In such equations, the independent variable, x, is an exponent. To understand the behavior of such a relation, we can sketch a graph by plotting points and then drawing a smooth curve through the plotted points.

Example 1. Determine points on the graph of $y = 10^x$, plot these points, and then draw a graph $y = 10^x$. Compare the graph of this exponential function with the graph of $y = x$.

We begin by determining points on the graph of $y = 10^x$.

x	$y = 10^x$
2	$10^2 = 100$
1	$10^1 = 10$
1/2	$10^{\frac{1}{2}} = \sqrt{10} \approx 3.16$
0	$10^0 = 1$
−1/2	$10^{-\frac{1}{2}} = \dfrac{1}{\sqrt{10}} \approx 0.32$
−1	$10^{-1} = \dfrac{1}{10} = 0.1$
−2	$10^{-2} = \dfrac{1}{10^2} = \dfrac{1}{100} = 0.01$

After plotting these points, we draw a smooth curve through them to obtain a graph of $y = 10^x$. Then we add a graph $y = x$. The graphs of these two equations appear below.

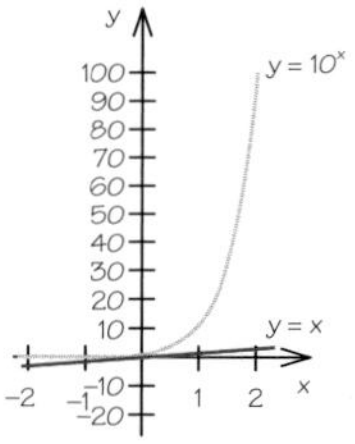

By comparing the graphs of $y = 10^x$ and $y = x$, we can see that as the values of x increase, the y-values of the exponential equation increase faster than those of the linear equation. This will be true for exponential equations of the form $y = a^x$, where $a > 1$.

Practice Exercises

1. Without the aid of a calculator, complete the table below to get coordinates of some points on the graph of $y = 4^x$.

x	−2.0	−1.5	−1.0	−0.5	0.0	0.5	1.0	1.5	2.0
$y = 4^x$									

2. Sketch by hand the graph of $y = 4^x$. (Use an x-interval from −2 to 2.)

3. Graph $y = e^x$ and $y = 2^x$ on the same set of axes. Use an x-interval from 0 to 2. You will need a graphing calculator. [$e \approx 2.718$ and is called the natural base. On a TI-84 graphing calculator, press [2nd] [LN] (for e^x).]

D. Rules for Exponents and Roots

Rules for Exponents

When multiplying two exponential expressions with the same base, the rule is to add the exponents.

$$a^m a^n = a^{m+n}$$

The statement above is easily understood when expanding the multiplication in a problem such as the following:

$$2^3 \cdot 2^4 = \underbrace{\underbrace{2 \cdot 2 \cdot 2}_{\text{3 times}} \cdot \underbrace{2 \cdot 2 \cdot 2 \cdot 2}_{\text{4 times}}}_{\text{7 times}} = 2^{3+4} = 2^7$$

Example 1. Simplify $3^5 \cdot 3^4$ and $7^{\frac{4}{3}} \cdot 7^{\frac{1}{3}}$.

$$3^5 \cdot 3^4 = 3^{5+4} = 3^9$$
$$7^{\frac{4}{3}} \cdot 7^{\frac{1}{3}} = 7^{\frac{4}{3}+\frac{1}{3}} = 7^{\frac{5}{3}}$$

When dividing two exponential expressions with the same base, subtract the exponents.

$$\frac{a^m}{a^n} = a^{m-n},\ a \neq 0$$

This last statement is easily understood when simplifying a problem, such as $\frac{2^5}{2^3}$, by cancelling as follows:

- Using cancellation: $\frac{2^5}{2^3} = \frac{2 \cdot 2 \cdot \cancel{2 \cdot 2 \cdot 2}}{\cancel{2 \cdot 2 \cdot 2}} = 2^2$
- Using the rule for quotients: $\frac{2^5}{2^3} = 2^{5-3} = 2^2$

Example 2. Determine the value of $\frac{4^3}{4^3}$ and $\frac{4^3}{4^6}$ by cancellation and by using the rule for quotients.

$$\frac{4^3}{4^3} = \frac{\cancel{4 \cdot 4 \cdot 4}}{\cancel{4 \cdot 4 \cdot 4}} = 1 \text{ and } \frac{4^3}{4^3} = 4^{3-3} = 4^0 = 1$$

$$\frac{4^3}{4^6} = \frac{\cancel{4 \cdot 4 \cdot 4}}{\cancel{4 \cdot 4 \cdot 4} \cdot 4 \cdot 4 \cdot 4} = \frac{1}{4^3} \text{ and } \frac{4^3}{4^6} = 4^{3-6} = 4^{-3} = \frac{1}{4^3}$$

When raising an exponential expression to a power, multiply powers.

$$(a^m)^n = a^{m \times n}$$

To see how this rule works, we expand $(2^3)^2$.

$$(2^3)^2 = \underbrace{2^3 \cdot 2^3}_{\text{2 times}} = \underbrace{\underbrace{2 \cdot 2 \cdot 2}_{\text{3 times}} \cdot \underbrace{2 \cdot 2 \cdot 2}_{\text{3 times}}}_{2\times3=6 \text{ times}} = 2^{2\times3} = 2^6$$

Example 3. Simplify $(4^5)^3$ both by expanding and by using the rule for raising to powers.

- Using expansion:

$$(4^5)^3 = \underbrace{4^5 \times 4^5 \times 4^5}_{\text{3 times}}$$
$$= \underbrace{\underbrace{4 \cdot 4 \cdot 4 \cdot 4 \cdot 4}_{\text{5 times}} \times \underbrace{4 \cdot 4 \cdot 4 \cdot 4 \cdot 4}_{\text{5 times}} \times \underbrace{4 \cdot 4 \cdot 4 \cdot 4 \cdot 4}_{\text{5 times}}}_{3\times5=15 \text{ times}} = 4^{15}$$

- Using the rule for raising to powers: $(4^5)^3 = 4^{5\times3} = 4^{15}$

Warning: Don't touch this! There is no "distributive law for exponents." That is, in general,

$$(a + b)^n \neq a^n + b^n$$

Example 4. A student wrote the following equation on a test: $(x + 2)^2 = x^2 + 4$. What is wrong with the student's work? Correct the error.

The student tried to distribute the square power over the addition in $x + 2$. However, there is no distributive law for exponents. Instead the student should use the distributive law to expand the square.

$(x + 2)^2 = (x + 2)(x + 2)$	Expand the square.
$= (x + 2)(x) + (x + 2)(2)$	Use the distributive law.
$= (x)(x) + (2)(x) + (x)(2) + (2)(2)$	Use the distributive law a second time.
$= x^2 + 4x + 4$	Combine like terms and simplify.

Rules for Roots

$$\sqrt[n]{ab} = \sqrt[n]{a} \cdot \sqrt[n]{b}$$

$$\sqrt[n]{\frac{a}{b}} = \frac{\sqrt[n]{a}}{\sqrt[n]{b}}$$

The rules for roots may help us calculate a root by hand, as shown in Example 5.

Example 5. Evaluate $\sqrt{490{,}000}$ and $\sqrt{0.0036}$ by hand.

$$\sqrt{490{,}000} = \sqrt{49 \cdot 100 \cdot 100} = \sqrt{49} \cdot \sqrt{100} \cdot \sqrt{100} = 7 \cdot 10 \cdot 10$$
$$= 700$$

$$\sqrt{0.0036} = \sqrt{\frac{36}{10{,}000}} = \frac{\sqrt{36}}{\sqrt{10{,}000}} = \frac{\sqrt{6 \cdot 6}}{\sqrt{100 \cdot 100}}$$
$$= \frac{\sqrt{6^2}}{\sqrt{100} \cdot \sqrt{100}} = \frac{6}{10 \cdot 10} = 0.06$$

The rules of roots can sometimes be used to simplify an expression containing a radical, as shown in Example 6.

Example 6. Simplify $\sqrt{9x^3}$.

$$\sqrt{9x^3} = \sqrt{9x^2 \cdot x} = \sqrt{9x^2} \cdot \sqrt{x} = 3x\sqrt{x}$$

Warning: Don't touch this! There is no rule that is the "distributive law for radicals." That is, in general,

$$\sqrt[n]{a + b} \neq \sqrt[n]{a} + \sqrt[n]{b}$$

Example 7. A student wrote the following on a test: $\sqrt{16 + 9} = \sqrt{16} + \sqrt{9} = 7$. What is wrong with the student's work? Show the correct calculations.

There is no rule for addition of radicals. Instead, simplify the number under the radical sign first and then take the square root: $\sqrt{16 + 9} = \sqrt{25} = 5$.

Practice Exercises

In Practice Exercises 1–10, use the rules of exponents to simplify.

1. $4^3 \cdot 4^7$
2. $\frac{10^8}{10^4}$

3. $(4^2)^5$
4. $6^{\frac{3}{2}} \cdot 6^{-\frac{1}{2}}$
5. $3^5 \cdot 3$
6. $(5^3)^{-2}$
7. $\dfrac{7^{\frac{5}{2}}}{7^{\frac{1}{2}}}$
8. $\left(6^{\frac{1}{2}}\right)^4$
9. $\dfrac{3^3}{3^{10}}$
10. $5^{\frac{1}{6}} \cdot 5^{\frac{1}{3}}$

In Practice Exercises 11 and 12, use the rules of roots (as shown in Example 5) to evaluate the root by hand.

11. $\sqrt{6400}$
12. $\sqrt{0.0049}$
13. A student wrote the following on a test: $\sqrt{64 + 4} = \sqrt{64} + \sqrt{4} = 8 + 2$. What is wrong with the student's work? Use your calculator to approximate the correct answer to two decimal places.

E. Logarithms

A logarithm is an exponent, which represents the power that a number must be raised to in order to get another number: $\log_b x$ is the power that b must be raised to in order to get x. We read the expression $\log_b x$ as "the base-b logarithm of x."

Example 1. Determine $\log_3 9$ and $\log_3 81$.

$\log_3 9 = 2$ because 3 must be raised to the second power in order to get 9.

$\log_3 81 = 4$ because 3 must be raised to the fourth power in order to get 81.

Notice that a logarithmic equation has a corresponding exponential equation:

$\log_3 9 = 2$ because $3^2 = 9$

$\log_3 81 = 4$ because $3^4 = 81$

The subscript (or base) of the logarithmic equation is also the base of the corresponding exponential equation.

Because our number system is base 10, we often use a base-10 logarithm, which is also called a *common logarithm.* For common logarithms, the base is often omitted. On most calculators the LOG key computes the base-10 logarithm of a number.

$$\log x = \log_{10} x$$

Example 2. Determine log(1000), log(10), and log(1).

- $\log(1000) = \log_{10} 1000$ = the power to which 10 must be raised in order to get 1000. Since $10^3 = 1000$, we find that $\log(1000) = 3$.
- $\log(10) = \log_{10} 10 = 1$ because $10^1 = 10$.
- $\log(1) = \log_{10} 1 = 0$ because $10^0 = 1$.

The following two properties are true for logarithms when $x > 0$:

1. $a^{\log_a x} = x$, and for base 10, $10^{\log x} = x$
2. $\log_a x^r = r \log_a x$, and for base 10, $\log x^r = r \log x$

Example 3. Simplify $10^{2+3\log x}$ for $x > 0$.

$10^{2+3\log x}$	Apply logarithm property 2.
$10^{2+3\log x} = 10^{2+\log x^3}$	Use the rule of exponents: $a^{m+n} = a^m a^n$.
$= 10^2 10^{\log x^3}$ $= 100x^3$	Apply logarithm property 1.

Another important base is e, which is used in continuously compounding interest problems. Like the more familiar number π, e is an irrational number—that is, a nonrepeating, nonterminating decimal. Its value is approximately 2.718. The base-e logarithm could be written as $\log_e x$. Mathematics, however, uses a special notation for base-e logarithms, $\ln x$, and a special title, the *natural logarithm.*

Practice Exercises

For Practice Exercises 1–8, determine the value of the logarithm.

1. $\log_2 16$
2. $\log_5 1$
3. $\log_5 25$
4. log(10,000)
5. $\log\left(\frac{1}{10}\right)$
6. $\log\left(\frac{1}{100}\right)$
7. ln(1)
8. $\ln(e)$

Simplify the expressions in Practice Exercises 9 and 10.

9. $10^{3+2\log x}$
10. $10^{1-4\log x}$

F. Using Logarithms to Solve Equations

When an equation contains the variable of interest as an exponent, using logarithms can help solve the equation. In this section, we use the natural logarithm, the logarithm with a base of e, $\ln x = \log_e x$. (Another approach would be to use the common logarithm, $\log x = \log_{10} x$.)

Example 1. Solve $2^x = 10$.

Because $2^3 = 8$ and $2^4 = 16$, we can estimate that the value of x is between 3 and 4. Although we know the value of x is closer to 3, we can't state its exact value without logarithms. By taking the natural logarithm of both sides of the equation, we can get an exact solution. Here are the steps:

$2^x = 10$	Take the natural logarithm of both sides of the equation.

$\ln(2^x) = \ln(10)$ — Use logarithm property 2 [see Algebra Review VI, item E, Logarithms (page AR-26)] to move the x in front of the logarithm.

$x\ln(2) = \ln(10)$ — Divide both sides of the equation by $\ln(2)$.

$$x = \frac{\ln(10)}{\ln(2)}$$

Using a calculator and rounding to two decimal places, we find that the approximate power is $x = 3.32$.

In the equation $A = P(1 + i)^n$, when the variables P, i, and n are stated, A can be found by direct calculation. Solving for P or i is also straightforward (see Practice Exercises 3 and 4 in Algebra Review III, item B, Solving for One Variable in Terms of Another, page AR-9). However, if we want to solve $A = P(1 + i)^n$ for the exponent n, we again turn to the technique of taking the natural logarithm of both sides of the equation.

Example 2. Solve $A = P(1 + i)^n$ for n.

$P(1 + i)^n = A$ — We switched the right and left sides of the equation. Isolate the expression with n by dividing both sides of the equation by P.

$(1 + i)^n = \dfrac{A}{P}$ — Take the natural logarithm of both sides.

$\ln(1 + i)^n = \ln\left(\dfrac{A}{P}\right)$ — Use logarithm property 2 to move the exponent in front of the logarithm.

$n\ln(1 + i) = \ln\left(\dfrac{A}{P}\right)$ — Divide both sides by $\ln(1 + i)$.

$$n = \frac{\ln\left(\frac{A}{P}\right)}{\ln(1 + i)}$$

Practice Exercises

In Practice Exercises 1–3, find an exact solution, and then use your calculator to determine an approximate solution to two decimal places.

1. $3^x = 15$
2. $2 \cdot 5^x = 28$
3. $80(1.03)^t = 100$
4. Given $A = P(1 + i)^n$, where $A = 2000$, $P = 1000$, and $i = 0.04$, find the exact value for n. Then find an approximation to two decimal places.

Algebra Review VII: Functions

A. Determining If a Relation Is a Function

A function is a relationship between two sets. Those two sets are called the domain and range. The domain can be thought of as the set of allowable input values and is generally associated with the variable x. The range can be thought of as the set of output values and is generally associated with the variable y.

A relationship defined between the two sets is a function if every value of x is assigned to only one value of y. Given a graph, there is an easy way to tell if it represents a function—it's called the *vertical line test.*

Vertical Line Test
If you can draw a vertical line that cuts through the graph in more than one place, the graph cannot represent a function; otherwise, it can.

Example 1. Given the graphs below, which ones represent functions?

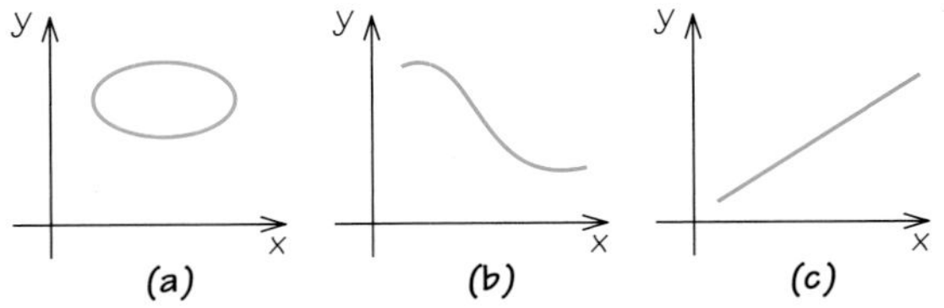

Graph (a) does not represent a function. It is possible to draw a vertical line that cuts the graph in two places. Graphs (b) and (c) do represent functions. For either of these graphs, any vertical line drawn will intersect the graph in at most one place.

Practice Exercises

Use the vertical line test to determine whether the graphs in Practice Exercises 1–4 represent functions.

1.

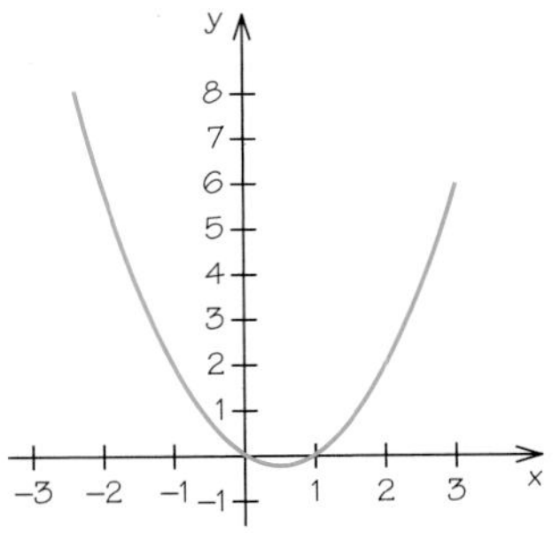

2.

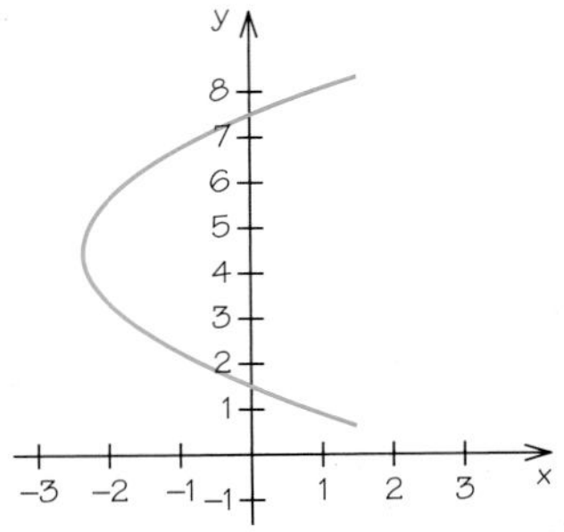

3.

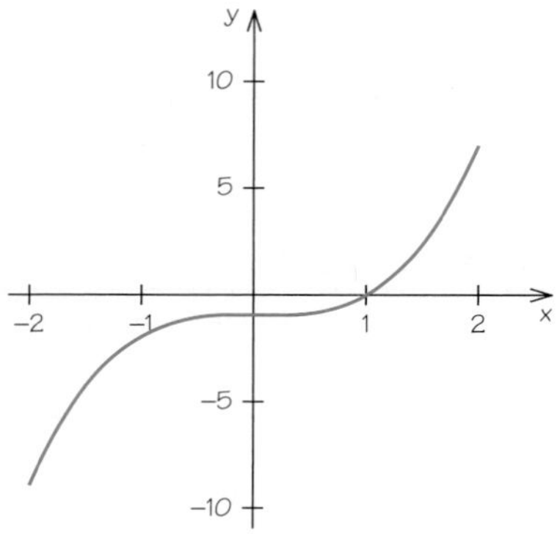

4.

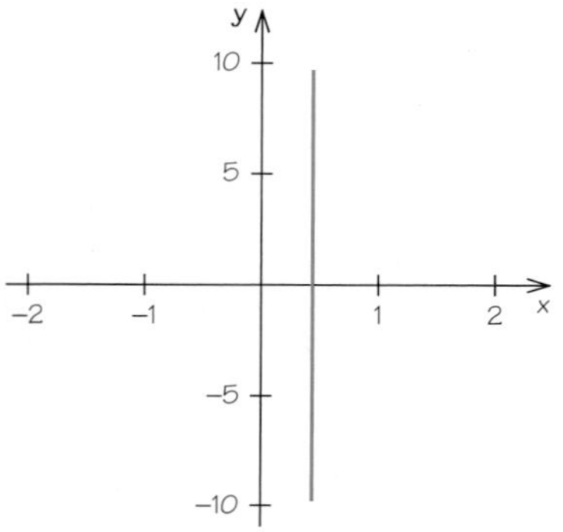

B. Function Notation

The most common notation used when identifying a function is $f(x)$.

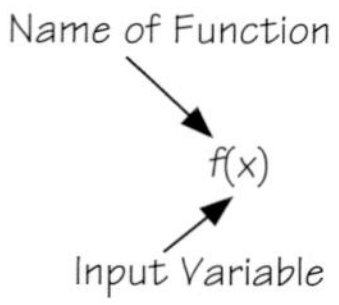

This notation is read as "f evaluated at x" or "f of x." *Note:* In function notation, multiplication is *not* implied by $f(x)$, as it would be in $(a)(b)$ or ab.

As an example, let's look at the squaring function, which takes a number and squares it. The squaring function can be written as $f(x) = x^2$. In this function, x would be the input and the square of x would be the output, so if 2 is in our domain, then 4 would be in our range.

Example 1. Given the squaring function $f(x) = x^2$, evaluate $f(-3)$ and $f(3)$.

$$f(-3) = (-3)^2 = 9 \text{ and } f(3) = 3^2 = 9$$

For a function, each input can be assigned to at most one output. However, from Example 1 we see that it is possible for more than one input to be assigned to the same output.

When we graph a function $f(x)$, our points are of the form $(x, f(x))$. For the squaring function, the points are of the form (x, x^2).

Example 2. Specify four points on the graph of $f(x) = x^2$ and sketch the graph.

From Example 1, we know that $(-3, 9)$ and $(3, 9)$ are points on the graph. Two more points are $(0, 0^2) = (0, 0)$ and $(2, 2^2) = (2, 4)$. A graph of the function is shown below.

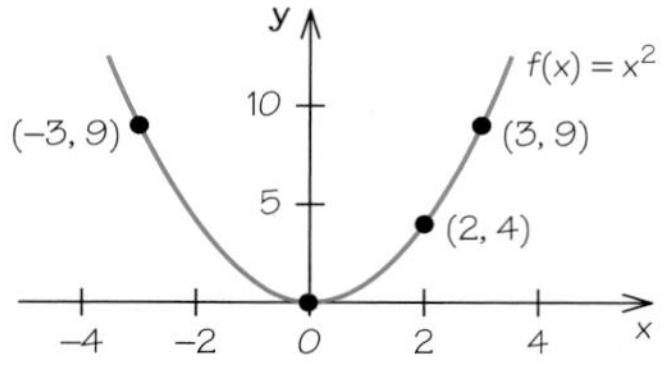

Another important function is the identity function: $f(x) = x$. In the identify function, a number is assigned to itself. So, with an input of a, the output is also a. Some points on the graph of the identity function are (0, 0), (1, 1), and (2, 2).

Example 3. Graph the identity function.

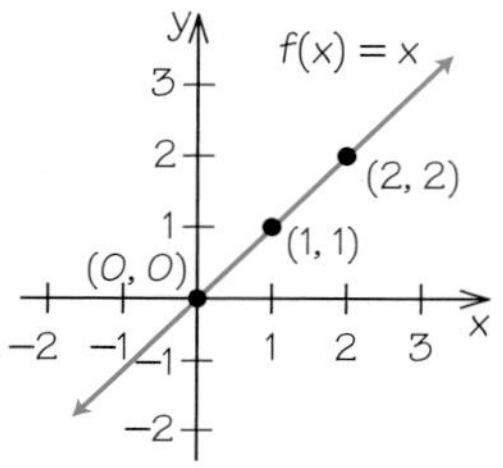

The identity function is an example of a linear function, a function of the form $f(x) = mx + b$. (For the identity function, $m = 1$ and $b = 0$.) In this book, you will also find exponential functions, which have the form $f(x) = a \cdot b^x$.

Calculator Note: After entering a function into a TI-84 graphing calculator, you can graph the function and also evaluate the function at specific input values.

Here's how to graph a function such as $f(x) = x^2$ using a TI-84 graphing calculator:

- Press [Y=] and clear any stored functions.
- Enter the function opposite Y1 = as follows: Press [X,T,θ,n] [x^2].
- Press [ZOOM] [6] to graph the function in the standard viewing window. Your graph should match the one below.

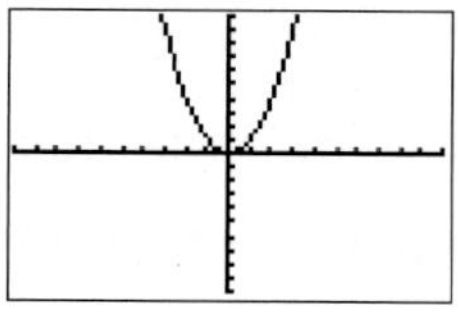

You can also use your calculator to evaluate a function for a specific input value. We continue with the function $f(x) = x^2$, which we have already stored in a graphing calculator as Y1. Next, we use the calculator to evaluate $f(-4)$. Keep in mind that the calculator knows this function as Y1 instead of as *f*. So, for the calculator you want to determine Y1(–4):

- Press [VARS], highlight Y-VARS, and then press [1] (for function).
- Press [1] (for Y1, which is where we stored our function).
- Press [(] [(−)] [4] [)] [ENTER]. You should get 16.

Practice Exercises

1. The cubing function can be expressed as $f(x) = x^3$. Evaluate $f(-2)$ and $f(2)$.
2. What is the form of the points for the cubing function?
3. Sketch the graph of the function $f(x) = x^3$ on the x-interval from -3 to 3.
4. $f(x) = x^2 + x$ is an example of a quadratic function. Evaluate $f(-2)$ and $f(2)$.
5. What is the form of the points for $f(x) = x^2 + x$?
6. Sketch the graph of $f(x) = x^2 + x$.
7. Given $f(x) = 3 \cdot 2^x$, evaluate $f(2)$ and $f(3)$.
8. Given $f(x) = 3x + 7$, evaluate $f(1)$ and $f(4)$.

Algebra Review VIII: Permutations and Combinations

A. Permutations

A permutation is an arrangement or order of distinct elements in a set. The number of permutations or arrangements of n distinct objects is $n!$ (pronounced "n factorial").

$$n! = (n)(n-1)(n-2)\cdots(2)(1)$$

Example 1. How many ways can you arrange five books on a shelf?

There are 5! possible ways to arrange five books on a shelf.

$$5! = (5)(4)(3)(2)(1) = 120$$

Example 2. Use the answer in Example 1 to help determine the number of ways to arrange seven books on a shelf.

There are 7! possible ways to arrange seven books on a shelf.

$$7! = 7 \cdot 6 \cdot \overbrace{5 \cdot 4 \cdot 3 \cdot 2 \cdot 1}^{5!} = 7 \cdot 6 \cdot 5! = 7 \cdot 6 \cdot 120 = 5040$$

Suppose that instead of choosing all five of the books and arranging them as in Example 1, you only had space on your shelf for three books. In this case, you want "the number of permutations of five books taken three at a time," which we express as ${}_5P_3$. More generally, the number of permutations of n distinct objects taken k at a time, expressed as ${}_nP_k$, can be computed as follows:

$${}_nP_k = \frac{n!}{(n-k)!} = (n)(n-1)\cdots(n-k+1)$$

Example 3. How many ways can you select three books from five books and arrange them on a shelf?

$${}_5P_3 = \frac{5!}{(5-3)!} = \frac{5 \cdot 4 \cdot 3 \cdot \cancel{2 \cdot 1}}{\cancel{2 \cdot 1}} = 60$$

Example 4. Show the calculations for the number of permutations of 25 distinct objects taken 4 at a time.

$${}_{25}P_4 = \frac{25!}{(25-4)!} = \frac{25!}{21!} = \frac{25 \cdot 24 \cdot 23 \cdot 22 \cdot \cancel{21!}}{\cancel{21!}}$$
$$= 303{,}600$$

Calculator Note: TI-84 graphing calculators will compute both factorials and permutations. Here are the steps for using a TI-84 to compute the answers to Examples 1 and 4:

To compute 5!:

- Press [5].
- Press [MATH], highlight PRB, and press [4] (for !).
- Press [ENTER] to complete the calculation.

To compute ${}_{25}P_4$:

- Press [2] [5].
- Press [MATH], highlight PRB, and press [2] (for ${}_nP_r$).
- Press [4] followed by [ENTER].

The output below shows the screen from the TI-84 after performing these two calculations.

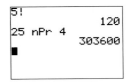

```
5!
                 120
25 nPr 4
              303600
█
```

Practice Exercises

Calculate the following showing the mathematical steps needed for the calculation. (In other words, do not use the built-in calculator functions for factorials or permutations.)

1. 8!
2. ${}_{15}P_5$
3. ${}_{30}P_2$
4. ${}_{27}P_3$
5. Use your calculator's built-in function for ${}_nP_k$ to calculate ${}_{50}P_{10}$.

B. Combinations

A combination is a selection of a subset of a specific size, say k, from a set of n distinct elements where the order of the selection does not matter. We can express the number of combinations as ${}_nC_k$, which is often pronounced as "n choose k."

Example 1. Suppose you have five books and want to choose three of the five to give to a friend as a gift. How many possible ways can this gift be selected?

In Example 3 in Algebra Review VIII, item A, Permutations (above), we calculated the number of ways that we could select three books out of five and arrange them: ${}_5P_3 = 60$. But here, the order of the selection does not matter—the selected books will all go in the same gift box. So Book A, Book B, and Book C is the same gift as Book C,

Book B, and Book A. There are 3! arrangements of the three selected books. So our answer should be $_5P_3/3! = \frac{60}{6} = 10$.

The number of ways to choose k distinct objects from n distinct objects (where the order of the selection does not matter) is

$$_nC_k = \frac{_nP_k}{k!} = \frac{n!}{(n-k)!k!}$$

Example 2. Find the number of possible ways to choose a three-person committee from a club with 50 members.

$$_{50}C_3 = \frac{50!}{(50-3)!3!} = \frac{50\cdot 49\cdot 48\cdot \cancel{47!}}{\cancel{47!}3!}$$

$$= \frac{\overset{25}{\cancel{50}}\cdot 49\cdot \overset{16}{\cancel{48}}}{\cancel{3}\cdot\cancel{2}\cdot 1} = 25\cdot 49\cdot 16 = 19{,}600$$

Example 3. Show the calculations for $_{25}C_4$. Compare its value with $_{25}P_4$ in Example 4 from Algebra Review VIII, item A, Permutations (page AR-29).

$$_{25}C_4 = \frac{25!}{(25-4)!4!} = \frac{25\cdot 24\cdot 23\cdot 22\cdot \cancel{21!}}{\cancel{21!}4!}$$

$$= \frac{25\cdot \overset{6}{\cancel{24}}\cdot 23\cdot 22}{\cancel{4}\cdot\cancel{3}\cdot\cancel{2}\cdot 1} = 25\cdot 23\cdot 22 = 12{,}650$$

$_{25}P_4 = 303{,}600$ which is $4! \times {_{25}C_4}$ or $24 \times {_{25}C_4}$

Calculator Note: TI-84 graphing calculators will compute combinations. Here are the steps for using a TI-84 to compute $_{25}C_4$ from Example 3:

- Press [2] [5].
- Press [MATH], highlight PRB, and press [3] (for $_nC_r$).
- Press [4] followed by [ENTER].

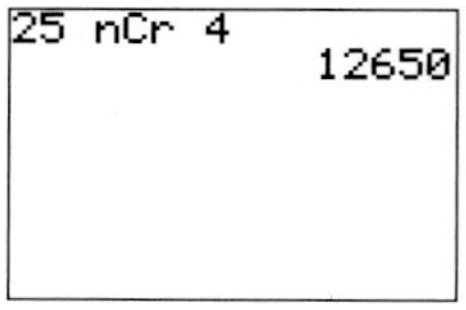

Practice Exercises

Calculate the following showing the mathematical steps needed for the calculation. (In other words, do not use the built-in calculator function for combinations.)

1. $_{15}C_5$
2. $_{30}C_2$
3. $_{27}C_3$
4. $_{50}C_{10}$
5. Use your calculator's built-in function for $_nC_k$ to calculate $_{50}C_{20}$.

Answers to Algebra Review Exercises

Algebra Review I

A. *Order of Operations*

1. 26
2. 45
3. 7
4. -16
5. $\frac{1}{9}$
6. $\frac{8}{5}$
7. $\frac{1}{15}$
8. $\frac{6}{5}$

B. *Distributive Law*

1. $2x + 2$
2. $14w - 21$
3. $-3x - 12$
4. $6.2p + 4.65$
5. $10x + 5y - 15$
6. $6(2x - 1)$
7. $2x(12x + 5)$
8. $2(1 + 2x + 4x^2)$

C. *Operations with Rational Numbers (Fractions)*

1. $\frac{3}{140}$
2. $\frac{1}{12}$
3. $\frac{4}{45}$
4. $\frac{6}{5}$
5. $\frac{1}{4}$
6. $\frac{3}{10}$
7. $\frac{28}{75}$
8. $\frac{40}{93}$

Algebra Review II

A. *Fractions, Percents, and Percentages*

1. 25% means 25 hundredths or $\frac{25}{100} = 0.25$.
2. $\left(\frac{10}{25} \times 100\right)\% = 40\%$
3. $\left(\frac{347}{500} \times 100\right)\% = 69.4\%$
4. $\left(\frac{6}{18} \times 100\right)\% = 33.33\%$

B. *Remainders*

1. Whole part = 13; remainder = 4
2. Whole part = 37; remainder = 5
3. Whole part = 47; remainder = 0
4. 32
5. 133
6. Whole part = 5; remainder = 6
7. Whole part = 23; remainder = 3
8. The remainder can be larger than the whole part. For example, 23 divided by 6 results in a whole part of 3 with a remainder of 5. The remainder is always smaller than the divisor.

C. *Prime and Composite Numbers*

1. $187 \div 11 = 17$, so 187 is a composite number.
2. 229 is prime.
3. 3342 is a composite number. It is divisible by 2 since its last digit is an even number.
4. $899 \div 29 = 31$, so 899 is a composite number.
5. 1295 is a composite number. It is divisible by 5 since its last digit is 5.
6. 4221 is a composite number; $4 + 2 + 2 + 1 = 9$, which is divisible by 3, so 4221 is divisible by 3.
7. 421 is prime.
8. 3,476,210 is a composite number. It is divisible by 5 since its last digit is 0.

D. *Significant Digits*

1. 4
2. 2
3. 2
4. 3
5. Volume = (12.12)(10.55)(4.5) = 575.397. Because one of the numbers in the product had only two significant digits, we round this answer to 580 cubic inches.
6. Perimeter = 12.12 + 12.12 + 4.5 + 4.5 = 33.24. Because one of the numbers in the sum had only one decimal place, we round this answer to 33.2 inches.
7. Area = (12.12)(4.5) = 54.54. Because one of the numbers in the product had only two significant digits, we round this answer to 55 square inches.
8. Area = (12.12)(10.55) = 127.866. Because both of the numbers in the product had four significant digits, we round this answer to 127.9 square inches.

E. Scientific Notation

1. 4.95×10^8
2. 7.2×10^{-7}
3. 4.95 E8
4. 7.2 E–7
5. 3.80204032 E8 or 380,204,032
6. 2.335 E–6 or 0.000002335

F. Rounding Numbers

1. 49,000
2. 48,700
3. 2.3
4. 2.26
5. 5.20

G. Counting in Binary

1. $32 + 16 + 8 + 4 = 60$
2. $64 + 16 + 8 + 2 + 1 = 91$
3. $128 + 8 + 4 = 140$
4. $16 + 8 + 4 + 2 + 1 = 31$
5. $64 + 8 + 1 = 73$
6. 100000
7. $35 = 32 + 2 + 1$ is equivalent to 100011_2.
8. $65 = 64 + 1$ is equivalent to 1000001_2.

Algebra Review III

A. Using Formulas

1. 1.25
2. 1
3. $5600
4. $x_5 = 7$
5. $x = \frac{14}{3}$
6. $y = -3$
7. $2500
8. 12 cm

B. Solving for One Variable in Terms of Another

1. $P = \frac{A}{1 + rt}$
2. $t = \frac{A - P}{P \cdot r}$
3. $P = \frac{A}{(1 + i)^n} = A(1 + i)^{-n}$
4. $i = \left(\frac{A}{P}\right)^{\frac{1}{n}} - 1 = \sqrt[n]{\frac{A}{P}} - 1$

C. Formulas Related to Geometric Shapes

1. $r = \frac{26}{2} = 13$; $C = 2\pi(13) \approx 81.7$ in.
2. Area = (side length)^2 = 144 cm^2; side length = 12 cm
3. Perimeter = $3.6' + 5' + 6' = 14.6'$
4. Area = $\frac{(6)(3)}{2} = 9$ ft^2
5. Volume = $\pi(15)^2(110) \approx 77{,}754.4$ ft^3; surface area = $2\pi(15)^2 + 2\pi(15)(110) \approx 11{,}781.0$ ft^2
6. First, determine the radius: $r = \frac{12}{2} = 6$.
   Volume = $\frac{4}{3}\pi(6)^3 \approx 904.8$ in.^{3}; surface area = $4\pi(6)^2 \approx 452.4$ in.^{2}
7. First, determine the radius: $r = \frac{10}{2} = 5$.
   Volume = $\frac{1}{3}\pi(5)^2(25) \approx 654.5$ cm^3;
   surface area = $\pi(5)^2 + \pi(5)\left(\sqrt{5^2 + 25^2}\right) \approx 479.0$ cm^2.

Algebra Review IV

A. Linear Equations in One Variable

1. 3
2. 3
3. 1.5
4. $\frac{60}{7}$
5. $\frac{3}{14}$
6. $\frac{7}{5}$
7. $\frac{14}{3}$
8. $\frac{64}{11}$

B. Plotting Points in the Plane

1. The points appear to fall on a vertical line. That's because all these points have the same x-coordinate.

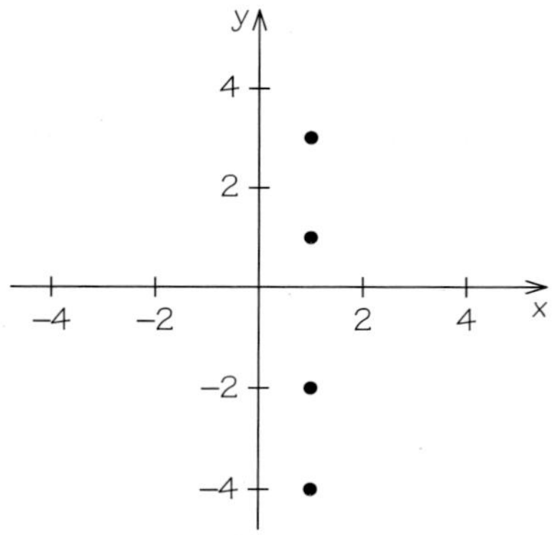

2. The points appear to fall on a horizontal line. That's because all these points have the same y-coordinate.

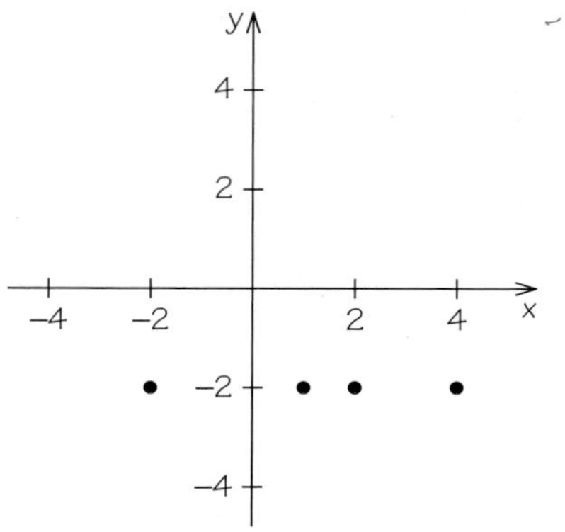

points all appear to fall on a diagonal line.

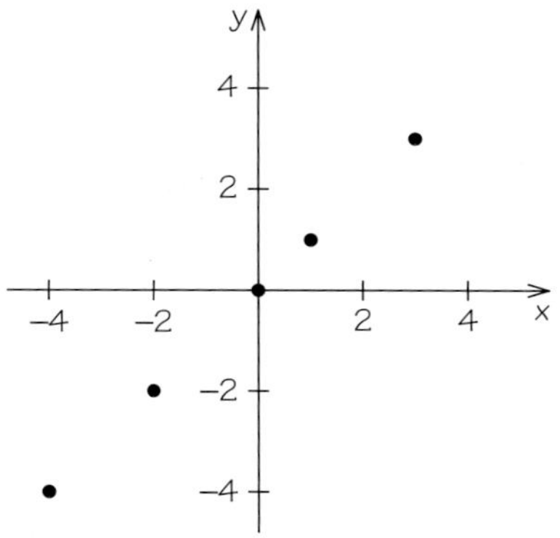

C. Distance and Midpoint Between Two Points in the Plane

1. $D = \sqrt{(7-2)^2 + (10-1)^2} = \sqrt{5^2 + 9^2} = \sqrt{106} \approx 10.3$

2. (a) Midpoint $= \left(\frac{7+2}{2}, \frac{10+1}{2}\right) = \left(\frac{9}{2}, \frac{11}{2}\right)$

   (b) Distance between (2, 1) and midpoint:

$$D = \sqrt{\left(\frac{9}{2} - 2\right)^2 + \left(\frac{11}{2} - 1\right)^2} = \sqrt{\left(\frac{5}{2}\right)^2 + \left(\frac{9}{2}\right)^2}$$

$$= \sqrt{\frac{106}{4}} = \frac{\sqrt{106}}{2}$$

   Distance between midpoint and (7, 10):

$$D = \sqrt{\left(7 - \frac{9}{2}\right)^2 + \left(10 - \frac{11}{2}\right)^2} = \sqrt{\left(\frac{5}{2}\right)^2 + \left(\frac{9}{2}\right)^2}$$

$$= \sqrt{\frac{106}{4}} = \frac{\sqrt{106}}{2}$$

These two distances are equal, half of the distance between (2, 1) and (7, 10).

3. (a) $D = \sqrt{(3-(-2))^2 + (4-(-7))^2}$

$$= \sqrt{(5)^2 + (11)^2} = \sqrt{146} \approx 12.1$$

   (b) Midpoint $= \left(\frac{-2+3}{2}, \frac{-7+4}{2}\right) = \left(\frac{1}{2}, -\frac{3}{2}\right)$

4. (a) $D = \sqrt{(1-(-6))^2 + (-3-(-7))^2}$

$$= \sqrt{(7)^2 + (4)^2} = \sqrt{65} \approx 8.1$$

   (b) Midpoint $= \left(\frac{-6+1}{2}, \frac{-7+(-3)}{2}\right) = \left(\frac{-5}{2}, -5\right)$

D. Linear Equations in Two Variables

1. No
2. Yes
3. Intercepts: $\left(\frac{5}{2}, 0\right), \left(0, \frac{10}{3}\right)$

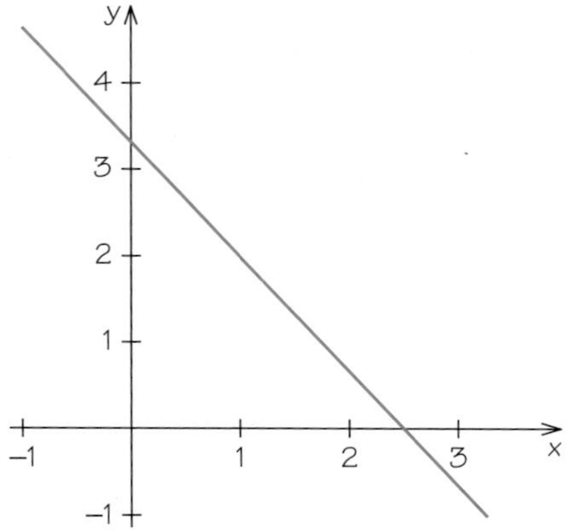

4. $2x - 5y = 6$
5. Yes
6. Intercepts: $(3, 0), \left(0, -\frac{6}{5}\right)$

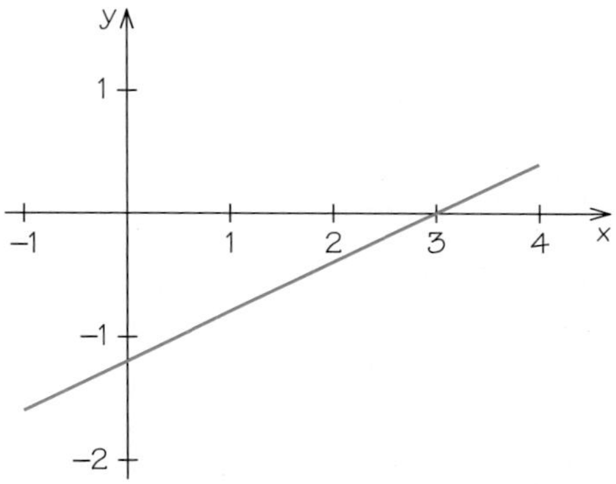

7. 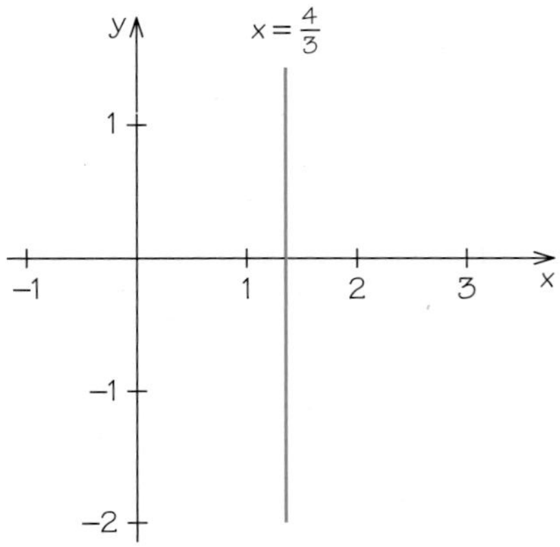

8.

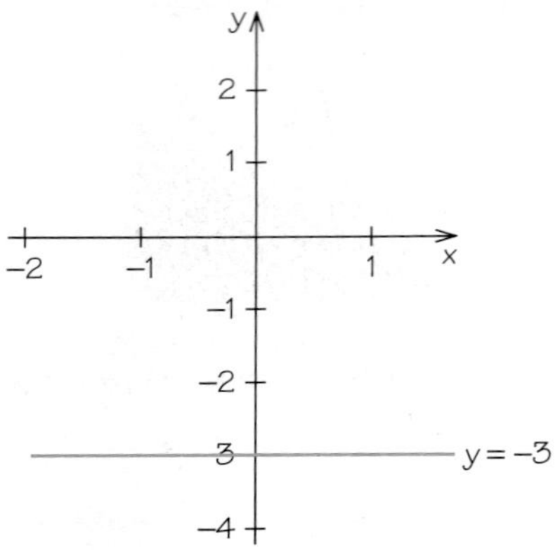

E. *Slope of a Line*

1. $m = \dfrac{6 - 1}{20 - 5} = \dfrac{1}{3}$
2. $m = \dfrac{-6 - (-2)}{8 - 4} = -1$
3. It is a vertical line; the slope is undefined.
4. It is a horizontal line; the slope is 0.
5. The slope is negative.
6. The slope is undefined.
7. The slope is 0.
8. The slope is positive.
9. $m = \frac{3}{6} = \frac{1}{2}$

F. *Graphing a Line in Slope-Intercept Form*

1. $y = \frac{4}{5}x + 4$
2. Sample answer: $(1, -1)$
3. Sample answer: Starting at $(0, 2)$, increasing the x-value by 4 will result in an increase of $\left(\frac{5}{4}\right)(4) = 5$ in the y-value. This gives a second point of $(4, 2 + 5) = (4, 7)$.

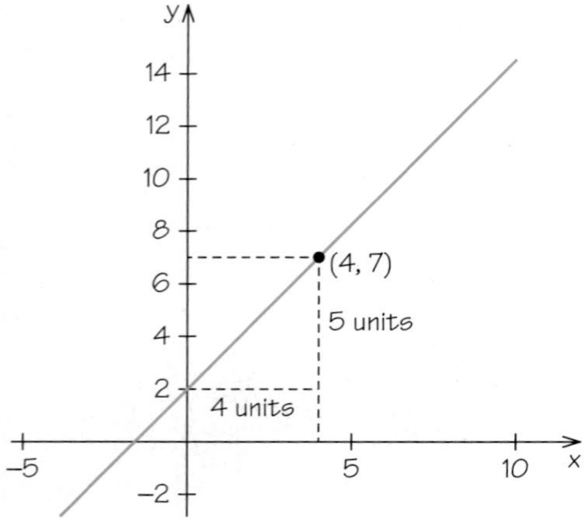

4.

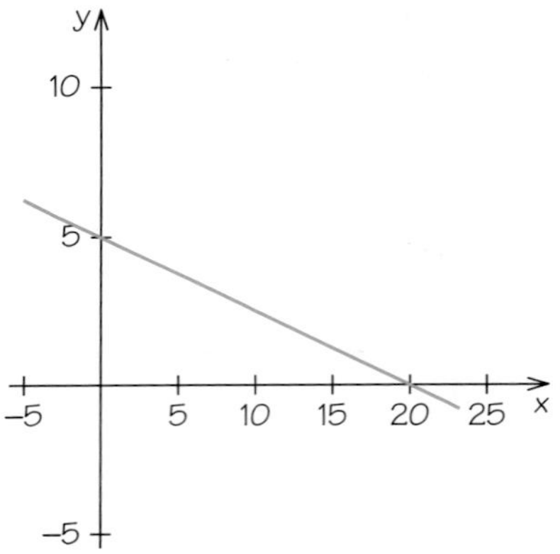

5. (a) $m = \dfrac{9 - 5}{4 - 2} = 2$

   (b) $y = 2x + b$; $5 = 2\,(2) + b$ and so $b = 1$

   (c) $y = 2x + 1$

6. (a) $m = \dfrac{2 - 7}{3 - 1} = \dfrac{-5}{2}$

   (b) $y = -\frac{5}{2}x + b$; $7 = -\frac{5}{2}(1) + b$ and so $b = 7 + \frac{5}{2} = \frac{19}{2}$ $= 9.5$

   (c) $y = -\frac{5}{2}x + \frac{19}{2}$ or $y = -2.5x + 9.5$

7. Viewing window $[-10, 10] \times [-10, 10]$

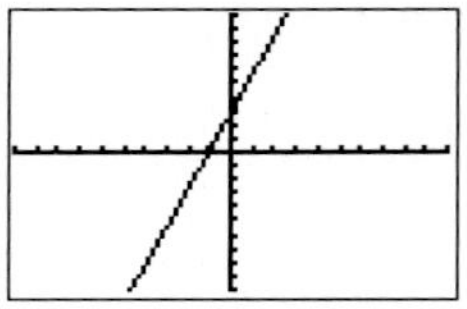

8. Viewing window $[-5, 5] \times [-5, 5]$

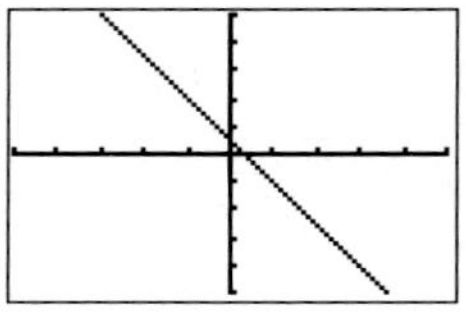

G. *Linear Inequalities in Two Variables*

1. No; $4(1) - 2(3) > 1$ is false.
2. Yes; $2(4) - 4(2) \leq 0$ is true.
3. (a) See the line in part (c).

   (b) The test point $(0, 0)$ makes the inequality false. Hence, the side of the line opposite the test point has been shaded.

(c)

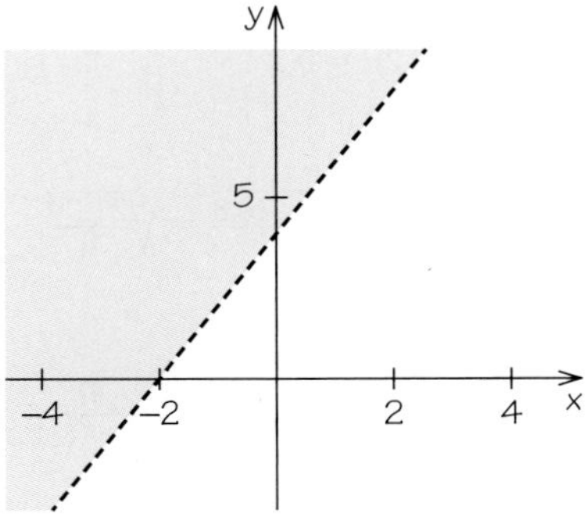

4. The test point (1, 3), which lies above the line, makes the inequality false. Hence, the solution set is shaded below the line.

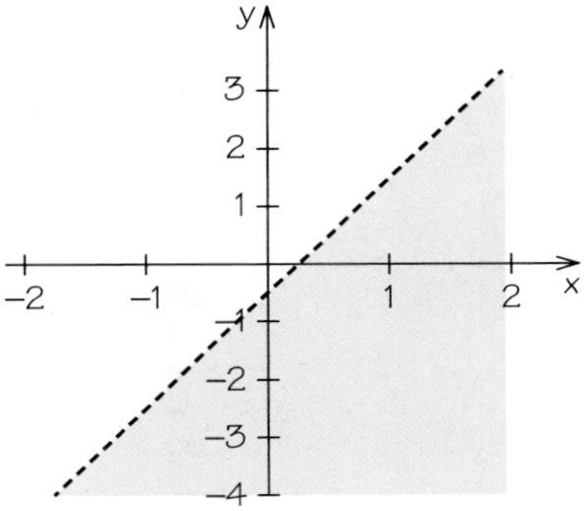

5. The test point (4, 2) makes the inequality true. But the test point lies on the line $2x - 4y = 0$. So we need to choose a different test point. Test point (1, 1), which lies above the line, makes the inequality true. Therefore, we shade above the line.

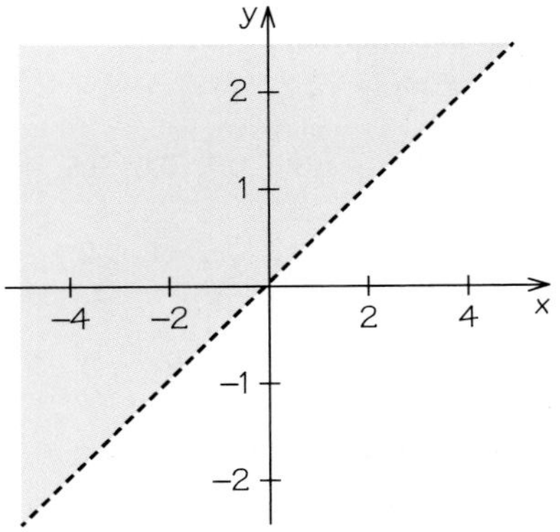

6. $y \le \frac{3}{4}x - \frac{3}{2}$

7. Window $[-5, 5] \times [-10, 10]$

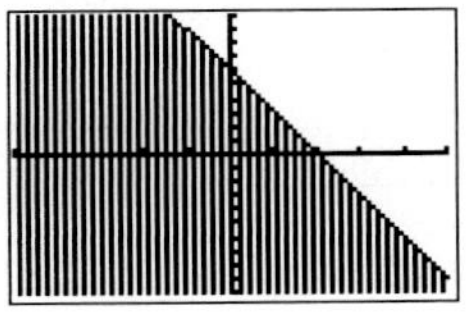

8. Window $[-10, 10] \times [-10, 10]$; graph $y \le \frac{3}{4}x - \frac{3}{2}$.

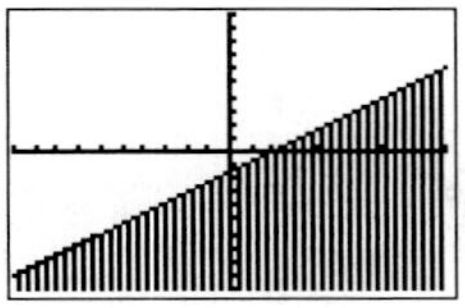

H. Systems of Linear Equations and Inequalities

1. This is not a solution.
2. $\left(\frac{8}{7}, \frac{18}{7}\right) \approx (1.14, 2.57)$
3. $(-12, 24)$
4. (a) There is no solution.
   (b) After substitution for y, the x's cancelled out.
   (c) The graphs of the two linear equations turn out to be parallel lines, which do not intersect.
5. $(-1.6, 4.2)$

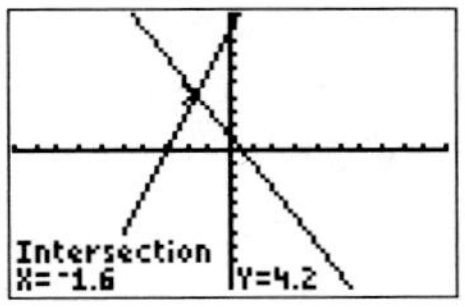

6. It is not in the solution set of the system of inequalities; (1, 3) does not satisfy the second inequality.
7.

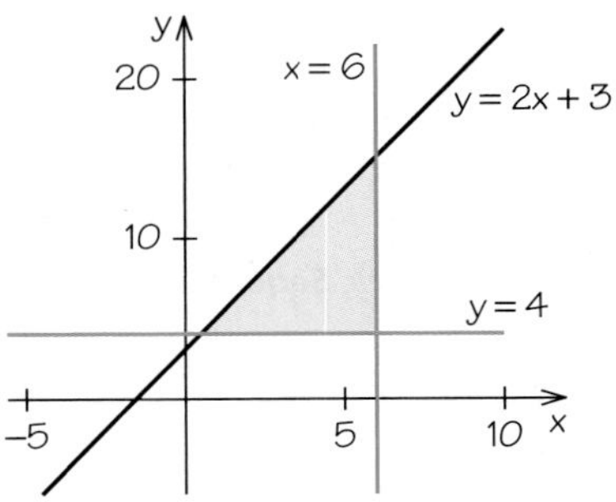

8.

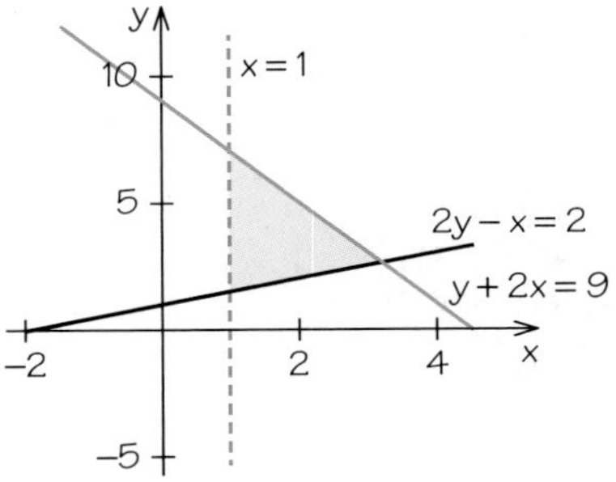

Algebra Review V

A. Summation Notation

1. $\sum_{i=1}^{3} i^3 = 1^3 + 2^3 + 3^3 = 1 + 8 + 27 = 36$

2. $\sum_{i=1}^{5} 2i = 2 \cdot 1 + 2 \cdot 2 + 2 \cdot 3 + 2 \cdot 4 + 2 \cdot 5 = 30$

3. $\sum_{i=1}^{4} (2i - 1) = 1 + 3 + 5 + 7 = 16$

4. (a) $\bar{x} = \frac{3 + 5 + 4}{3} = 4$

   (b) $s = \sqrt{\frac{(3-4)^2 + (5-4)^2 + (4-4)^2}{3-1}}$

   $= \sqrt{\frac{2}{2}} = 1$

5. (a) $\sum_{i=1}^{4} a_i b_i = (2)(8) + (5)(6) + (7)(4) + (9)(2) = 16 + 30 + 28 + 18 = 92$

   (b) $\sum_{i=1}^{4} (a_i - b_i) = (2 - 8) + (5 - 6) + (7 - 4) + (9 - 2) = (-6) + (-1) + 3 + 7 = 3$

B. Sequences

1. 5, 10, 15, 20, 25, . . .
2. $t_n = 5n$; $t_{10} = 50$
3. $t_1 = 5$ and $t_n = t_{n-1} + 5$ (or $t_{n+1} = t_n + 5$)
4. $\frac{1}{2}, 1, \frac{3}{2}, 2, \frac{5}{2}, 3$
5. 3, 5, 7, 9, 11, 13
6. 1, 2, 5, 14, 41, 122
7. 1, 4, 7, 10, 13, 16
8. 1, 2, 3, 5, 8, 13

C. Arithmetic and Geometric Sequences and Series

1. Geometric; 160, 320
2. Neither; 8, −8
3. Arithmetic; 19, 22
4. Geometric; 12.5, 6.25
5. −12, −7, −2, 3, 8; $S_5 = -10$
6. 5, 20, 80, 320, 1280; $S_5 = 1705$
7. $a_1 = 1, a_{15} = 57; S_{15} = \frac{15(57 + 1)}{2} = 435$
8. $S_{15} = 2\left(\frac{3^{15} - 1}{3 - 1}\right) = 14{,}348{,}906$

Algebra Review VI

A. Powers and Roots

1. (a) $(-3)^4 = 81$
   (b) $(-2)^5 = -32$
   (c) $4^3 = 64$

2. (a) $\sqrt[4]{10{,}000} = \sqrt[4]{10^4} = 10$
   (b) $\sqrt[3]{0.008} = \sqrt[3]{(0.2)^3} = 0.2$
   (c) $\sqrt{2500} = \sqrt{50^2} = 50$

3. $\sqrt{\frac{(2.7)^2 + 3(4.6)}{4}} = \sqrt{\frac{7.29 + 13.8}{4}} = \sqrt{\frac{21.09}{4}}$

   $= \sqrt{5.2725} \approx 2.3$

4. $\sqrt{\frac{(1-4)^2 + (3-4)^2 + (5-4)^2 + (7-4)^2}{4-1}}$

   $= \sqrt{\frac{(-3)^2 + (-1)^2 + 1^2 + 3^2}{3}} = \sqrt{\frac{20}{3}} \approx 2.58$

5. $(1 + 0.05)^6 \approx 1.34$

B. Natural and Fractional Exponents

1. (a) $42^0 = 1$
   (b) $5^{-2} = \frac{1}{5^2} = \frac{1}{25}$ or 0.04

2. (a) $(-27)^{\frac{1}{3}} = \sqrt[3]{-27} = -3$
   (b) $(-27)^{\frac{2}{3}} = (\sqrt[3]{-27})^2 = (-3)^2 = 9$

3. (a) $\sqrt[10]{x} = x^{\frac{1}{10}}$
   (b) $\sqrt[5]{x^3} = x^{\frac{3}{5}}$

4. (a) $\sqrt{0.0026} = (0.0026)^{\frac{1}{2}} \approx 0.05$
   (b) $(\sqrt[5]{256})^2 = (256)^{\frac{2}{5}} = 9.19$
   (c) $\frac{1}{\sqrt{40}} = (40)^{-\frac{1}{2}} \approx 0.16$

C. Graphs of Exponential Equations

1.

x	−2.0	−1.5	−1.0	−0.5	0.0	0.5	1.0	1.5	2.0
$y = 4^x$	$\frac{1}{16}$	$\frac{1}{8}$	$\frac{1}{4}$	$\frac{1}{2}$	1	2	4	8	16

2.

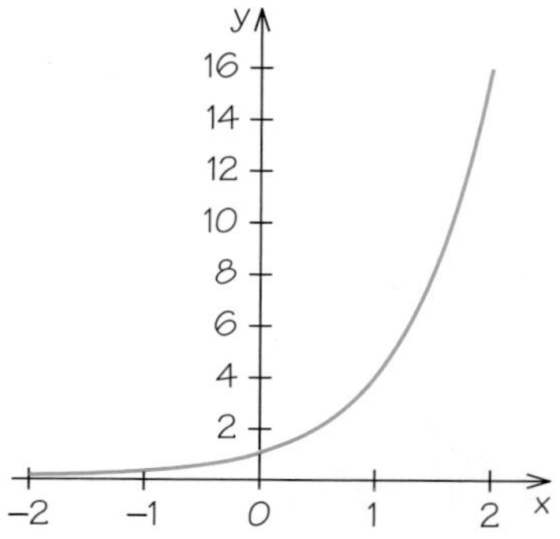

3.

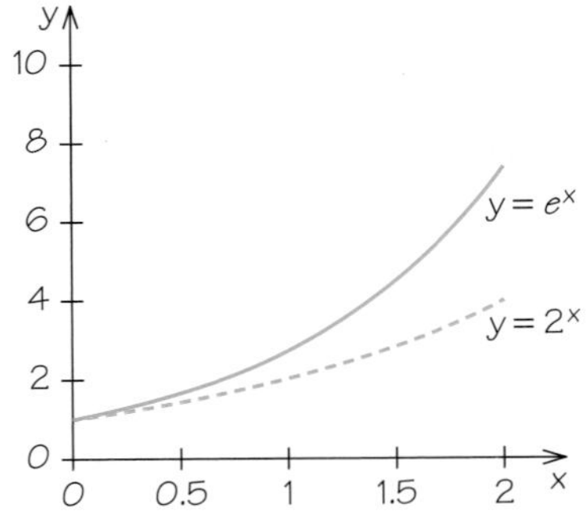

D. Rules for Exponents and Roots

1. $4^3 \cdot 4^7 = 4^{10}$
2. $\frac{10^8}{10^4} = 10^{8-4} = 10^4$
3. $(4^2)^5 = 4^{2\times 5} = 4^{10}$
4. $6^{\frac{3}{2}} \cdot 6^{-\frac{1}{2}} = 6^{\frac{3}{2}+\left(-\frac{1}{2}\right)} = 6$
5. $3^5 \cdot 3 = 3^{5+1} = 3^6$
6. $(5^3)^{-2} = 5^{3\times(-2)} = 5^{-6} = \frac{1}{5^6}$
7. $\frac{7^{\frac{5}{2}}}{7^{\frac{1}{2}}} = 7^{\frac{5}{2}-\frac{1}{2}} = 7^2$
8. $\left(6^{\frac{1}{2}}\right)^4 = 6^{\frac{1}{2}\times 4} = 6^2$
9. $\frac{3^3}{3^{10}} = 3^{3-10} = 3^{-7} = \frac{1}{3^7}$
10. $5^{\frac{1}{6}} \cdot 5^{\frac{1}{3}} = 5^{\frac{1}{6}+\frac{1}{3}} = 5^{\frac{3}{6}} = 5^{\frac{1}{2}}$
11. $\sqrt{6400} = \sqrt{64 \cdot 100} = \sqrt{64} \cdot \sqrt{100} = 8 \cdot 10 = 80$
12. $\sqrt{0.0049} = \sqrt{\frac{49}{10{,}000}} = \frac{\sqrt{49}}{\sqrt{10{,}000}} = \frac{7}{100} = 0.07$
13. The student should not have tried to distribute a square root over addition. $\sqrt{64 + 4} = \sqrt{68} \approx 8.25$

E. Logarithms

1. $\log_2 16 = 4$
2. $\log_5 1 = 0$
3. $\log_5 25 = 2$
4. $\log (10{,}000) = 4$
5. $\log\left(\frac{1}{10}\right) = -1$
6. $\log\left(\frac{1}{100}\right) = -2$
7. $\ln(1) = 0$
8. $\ln(e) = 1$
9. $1000x^2$
10. $10x^{-4} = \frac{10}{x^4}$

F. Using Logarithms to Solve Equations

1. $x = \frac{\ln(15)}{\ln(3)} \approx 2.46$
2. $x = \frac{\ln(14)}{\ln(5)} \approx 1.64$
3. $t = \frac{\ln(1.25)}{\ln(1.03)} \approx 7.55$
4. $n = \frac{\ln\left(\frac{A}{P}\right)}{\ln(1 + i)} = \frac{\ln\left(\frac{2000}{1000}\right)}{\ln(1.04)} = \frac{\ln(2)}{\ln(1.04)} \approx 17.67$

Algebra Review VII

A. Determining If a Relation Is a Function

1. Function
2. Not a function; it is possible to draw a vertical line that cuts through the graph in two places.
3. Function
4. Not a function; the vertical line itself cuts through its own graph in an infinite number of places.

B. Function Notation

1. $f(-2) = (-2)^3 = -8$ and $f(2) = (2)^3 = 8$
2. (x, x^3)
3.

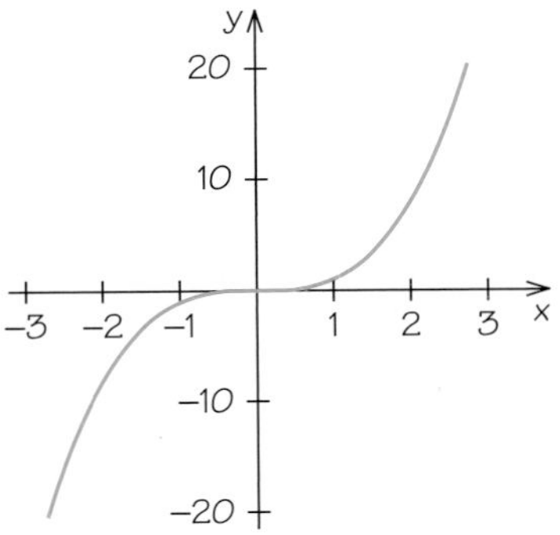

4. $f(-2) = (-2)^2 + (-2) = 4 - 2 = 2$ and $f(2) = 2^2 + 2 = 6$
5. $(x, x^2 + x)$
6.

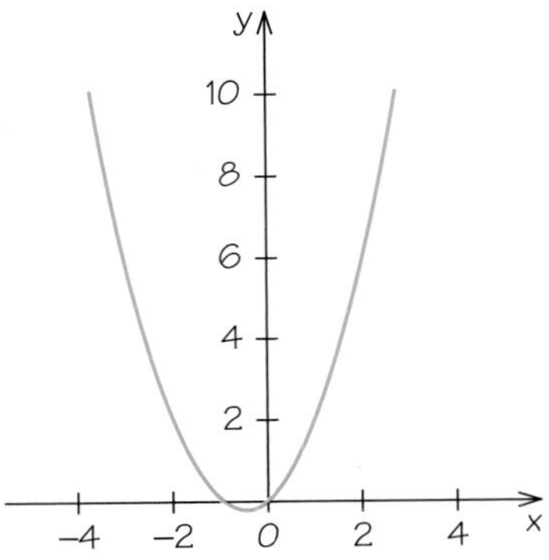

7. $f(2) = 3 \cdot 2^2 = 3(4) = 12; f(3) = 3 \cdot 2^3 = 3(8) = 24$

8. $f(1) = 3(1) + 7 = 10; f(4) = 3(4) + 7 = 19$

Algebra Review VIII

A. Permutations

1. $8! = 8 \cdot \overbrace{7 \cdot 6 \cdot 5 \cdot 4 \cdot 3 \cdot 2 \cdot 1}^{7!} = 8 \cdot 5040 = 40{,}320$

2. ${}_{15}P_5 = \dfrac{15!}{(15-5)!} = \dfrac{15!}{10!} = 15 \cdot 14 \cdot 13 \cdot 12 \cdot 11 = 360{,}360$

3. ${}_{30}P_2 = \dfrac{30!}{28!} = 30 \cdot 29 = 870$

4. ${}_{27}P_3 = 27 \cdot 26 \cdot 25 = 17{,}550$

5. 3.727604302E16

B. Combinations

1. ${}_{15}C_5 = \dfrac{15!}{(15-5)!5!} = \dfrac{15 \cdot 14 \cdot 13 \cdot 12 \cdot 11 \cdot \cancel{10!}}{(\cancel{10!})(5!)}$

$= \dfrac{\cancel{15} \cdot \overset{7}{\cancel{14}} \cdot 13 \cdot \overset{3}{\cancel{12}} \cdot 11}{\cancel{5} \cdot \cancel{4} \cdot \cancel{3} \cdot \cancel{2} \cdot 1} = 7 \cdot 13 \cdot 3 \cdot 11 = 3003$

2. ${}_{30}C_2 = \dfrac{30!}{28!2!} = \dfrac{30 \cdot 29}{2} = 15 \cdot 29 = 435$

3. ${}_{27}C_3 = \dfrac{27!}{24!3!} = \dfrac{27 \cdot 26 \cdot 25}{3 \cdot 2 \cdot 1} = 2925$

4. ${}_{50}C_{10} = \dfrac{50!}{40!10!} = \dfrac{\cancel{50} \cdot 49 \cdot \cancel{48} \cdot 47 \cdot 46 \cdot \overset{5}{\cancel{45}} \cdot \overset{11}{\cancel{44}} \cdot 43 \cdot \cancel{42} \cdot 41}{\cancel{10} \cdot \cancel{9} \cdot \cancel{8} \cdot \cancel{7} \cdot \cancel{6} \cdot \cancel{5} \cdot \cancel{4} \cdot \cancel{3} \cdot \cancel{2} \cdot 1}$

$= 49 \cdot 47 \cdot 46 \cdot 5 \cdot 11 \cdot 43 \cdot 41 = 10{,}272{,}278{,}170$

5. ${}_{50}C_{20} = 4.712921224\text{E}13$

Answers to Skills Check Exercises

Chapter 1

1. c
2. 7; 9
3. b
4. four
5. a
6. 20
7. a
8. *B*; *E*
9. c
10. 8; 7
11. b
12. 1; 2; 4; 5; 10
13. c
14. 10
15. a
16. 9; 23
17. c
18. 8
19. c
20. a
21. 4
22. 12
23. a
24. 4
25. a
26. digraph; digraph; graph
27. b
28. 9; 15
29. a
30. 13

Chapter 2

1. c
2. 90
3. a
4. 0
5. b
6. 27
7. c
8. 26
9. c
10. 32
11. c
12. *V*
13. b
14. 94
15. c
16. 72
17. b
18. 2340
19. a
20. 234
21. b
22. 14; 3
23. a
24. 7; 8; 9
25. 14
26. b
27. b
28. 18
29. c
30. 16

Chapter 3

1. c
2. 20
3. c
4. b
5. 14 min
6. 4; 2; 3
7. 2 min
8. b
9. b
10. T_6; T_4
11. c
12. 4; 2; 3
13. a
14. 10 min
15. c
16. 14
17. b
18. 17
19. b
20. 3
21. b
22. 1; 4
23. a
24. 1
25. a
26. c
27. 2
28. 2
29. b
30. 3

Chapter 4

1. a
2. −2; −3
3. a
4. 3; 3
5. c
6. 14; 4
7. b
8. $x + 2y = 8$
9. 25; 6
10. 92
11. a
12. 3; 5
13. b
14. b
15. b
16. $15
17. a
18. convex
19. b
20. c
21. 4; 4
22. c
23. 3
24. 3; 2
25. 3; 1
26. a
27. a
28. 59
29. a
30. −3

Chapter 5

1. The students (identified by Student ID)
2. 5; gender and residency are qualitative; scholarship award, high school percentile, and first-year GPA are quantitative.
3. 8
4. 8/64 = 0.125 or 12.5%
5. 37.5%
6. left
7. c
8. 32
9. c
10. leaf
11. b
12. $\bar{x} \approx 27.41$ cm
13. 27.2 cm
14. 28.0 cm
15. left
16. 50%
17. b
18. less
19. b
20. 2
21. c
22. 25.0, 26.2, 27.2, 28.0, 32.8
23. b
24. grams
25. a
26. standard deviation
27. c
28. b
29. c
30. 95%

Chapter 6

1. a
2. response
3. a
4. *x* or horizontal
5. b
6. positive
7. c
8. b
9. c
10. −5
11. b
12. $100x + 500$
13. b
14. 727.6
15. a
16. $0.02x + 3$
17. c
18. the same: $r = 0.86$
19. b
20. $r = 0.96$
21. $m \approx (0.96)\left(\frac{0.884}{0.163}\right) \approx 5.2$
22. a
23. b
24. predicted (or estimated or fitted)
25. c
26. above
27. b
28. extrapolation
29. c
30. cause
31. It's a positive association; as number of days attended increases, GPA tends to increase.
32. Strong
33. $m = (0.905)\left(\frac{0.913}{19.87}\right) \approx 0.0416$; $b = 2.989 - 0.0416 \times 156 \approx -3.501$; $\hat{y} = 0.0416x - 3.501$
34. $2.20 - (0.0416(156) - 3.501) \approx -0.789$
35. Predicted GPA = $0.0416(60) - 3.501 = -1.005$. It does not make sense to have a negative GPA. This example shows the danger of extrapolation.

Chapter 7

1. b
2. whole
3. c
4. more
5. c
6. voluntary response
7. b
8. 25
9. a
10. 6694
11. c
12. 10
13. b
14. 52%
15. c
16. 40%
17. a
18. cause
19. b
20. an observational study
21. b
22. placebo
23. c
24. 0.35
25. b
26. 0.015
27. a
28. 0.065 (or 6.5%)
29. c
30. 0.06 (or 6%)

Chapter 8

1. a
2. 24
3. b
4. 0.85
5. a
6. c
7. 0.3025
8. 2/36 = 1/18
9. c
10. 1
11. b

12. 0.3
13. b
14. 24
15. b
16. $\frac{39}{52} \times \frac{38}{51} \times \frac{37}{50} \times \frac{36}{49} \approx 0.304$
17. c
18. 0.4
19. b
20. 1
21. b
22. a
23. the population mean μ
24. b
25. $50
26. c
27. $\sqrt{1.1779} \approx 1.09$
28. 2
29. 516
30. 11.6
31. c

Chapter 9

1. b
2. if any two voters exchange ballots, the election outcome is unchanged
3. b
4. a switch in a ballot from being a vote for the loser to being a vote for the winner doesn't change the election outcome
5. c
6. majority rule
7. b
8. defeats every other candidate in a one-on-one contest
9. d
10. monotonicity
11. a
12. sometimes produces no winner at all
13. c
14. receives the most first-place votes
15. a
16. has the highest Borda score
17. a
18. reverses the order in which this nonwinner and the winner were ranked
19. c
20. the Hare system
21. c
22. one-on-one contests take place according to an ordering of the candidates called an "agenda"
23. b
24. approval voting
25. c
26. it is to a voter's advantage to submit a ballet that misrepresents his or her preferences
27. a
28. they fail to satisfy monotonicity
29. d
30. satisfies the Condorcet winner criterion and independence of irrelevant alternatives, and always produces at least one winner in every election

Chapter 10

1. c
2. Borda
3. c
4. an insincere (or disingenuous) ballot
5. c
6. *A* over *D* over *C* over *B*
7. a
8. two
9. c
10. one in which the winner has the fewest first-place votes
11. a
12. monotonicity
13. b
14. treats both candidates equally and all voters equally
15. b
16. there are only three candidates
17. c
18. placing the additional *j* candidates at the bottom of each ballot (in any order whatsoever)
19. c
20. agenda manipulation
21. a
22. group manipulation
23. d
24. at least as good as, and sometimes better than, the other
25. c
26. Gibbard–Satterthwaite theorem
27. d
28. manipulable
29. a
30. the chair has the most power, but fares the worst

Chapter 11

1. b
2. 101
3. c
4. 3
5. c
6. *C*
7. d
8. $\frac{1}{12}$, or 8.33%
9. c
10. the last juror
11. b
12. 4 and 3
13. b
14. *A*, *B*, and *C*
15. a
16. veto
17. a
18. 2^{n-1}
19. c
20. $\{A, B, D\}$
21. b
22. 20
23. c
24. $_4C_1 = 4$
25. d
26. $_3C_2 + {_3C_3} = 4$
27. d
28. $\{A, B, C\}$, $\{A, B, D\}$, and $\{A, C, D\}$
29. a
30. *B*

Chapter 12

1. Candidates *A*, *B*, and *C* receive 49.11%, 41.96%, and 8.93% of the vote, respectively. Under the Democratic Delegate Selection Rules, Candidate *C* would lose her support because she received less than 15% of the popular vote.
2. Candidates *A* and *B* receive 53.92% and 46.08% of the popular vote after *C*'s votes are dropped from the total.
3. Candidate *A*'s quota is (14,056/23,903)12 = 7.057.
4. Candidates *A*, *B*, *C*, and *D* receive 5, 4, 3, and 2 delegates, respectively.
5. Candidates *A*, *B*, and *C* receive 10, 8, and 2 delegates, respectively.
6. Because Candidate *C* receives less than 15% of the vote, Candidate *C* receives no delegates. Candidates *A* and *B* receive 11 and 9 delegates, respectively.
7. The median is 5.
8. The extended median is the interval [5, 6].
9. Answers will vary. One such answer is 1, 2, 2, 3.
10. No. The median has to be a single value.
11. If the extended median is a single value, then there is a single equilibrium. If the extended median is not a single value, then there is more than one equilibrium.
12. b
13. *B*
14. c
15. c
16. *B*
17. a
18. just to the left or just to the right of *M*
19. c
20. a
21. *C*
22. Yes; *C* ultimately decides which candidate wins.
23. The median is 5. It is the same.
24. True
25. 28
26. Awarding electoral votes based on votes in electoral districts would be the same process as awarding a primary by using a winner-take-all rule.
27. c
28. the electoral votes of the states that passed the National Popular Vote law would award their electoral votes to the candidate with the most votes

Chapter 13

1. a
2. reflects the relative worth of each issue to that party
3. a
4. the transfer of items (or parts thereof) from one party to the other until points are equalized
5. d
6. point ratio
7. d
8. equitable
9. c
10. the boat, car, and part of the land
11. c
12. the Knaster inheritance procedure
13. b
14. cash only
15. b
16. never willingly choose his or her least-preferred item, and avoid wasting a choice on an item that he or she knows will remain available and can be chosen later
17. b
18. no other player received more than he or she did
19. a
20. each nondivider receives a portion that he or she has approved
21. b
22. leaves the game
23. a
24. they use the divide-and-choose procedure
25. c
26. the first player
27. d
28. the amount of the second-highest bid
29. c
30. Brønislaw Knaster

Chapter 14

1. a
2. 380; 47.5; 4.2; 2.1; 1.7
3. c
4. 15 + 12 + 18 + 16 + 19 = 80
5. b
6. 300
7. a
8. 15 + 12 + 18 + 16 + 19 = 80
9. c
10. population
11. c
12. 0 + 0 + 100 = 100
13. b
14. 1 + 1 + 98 = 100
15. b
16. Hamilton
17. a
18. 0.5 seats
19. a
20. 2; 3
21. a
22. Hill–Huntington
23. c
24. Jefferson
25. d
26. *D*; *A*; *B*; *C*
27. c
28. Webster
29. b
30. 5; quota

Chapter 15

1. c
2. third
3. a
4. second
5. Yes; no
6. Yes, the last game has a saddlepoint.
7. a
8. Three
9. c
10. 3
11. c
12. its payoff
13. a
14. prevent a player from being exploited by always choosing a pure strategy
15. c
16. 0
17. c
18. 0.4
19. 0.5
20. No
21. (a) Guess fastball
    (b) Guess curveball
    (c) Pitch curveball
    (d) Pitch fastball
22. 0.8
23. pitch fastballs and curveballs each with $\frac{1}{2}$ probability
24. (a) Guess fastball
    (b) Guess curveball
    (c) Pitch fastball
    (d) Pitch fastball
25. pitch fastball
26. guess fastball
27. selecting *B* is optimal if the other player selects *B*
28. b
29. a
30. Player I prefers Choice *B* to Choice *A*
31. b
32. The equilibrium outcome is for Player I to Not Swerve and for Player II to Swerve.

Chapter 16

1. a
2. 1
3. a
4. 3
5. a
6. 9
7. b
8. 0
9. a; b
10. 8
11. b
12. 10
13. (a) Product
    (b) Manufacturer
    (c) Manufacturer
14. 9
15. b
16. 9
17. c
18. 11
19. b
20. 10
21. b
22. 100%; 100%
23. c
24. 4
25. c
26. 10
27. c
28. c
29. (a) single-digit
    (b) adjacent transposition
    (c) jump transposition
30. (a) 100%
    (b) 100%
    (c) 100%

Chapter 17

1. b
2. 1011
3. b
4. 3
5. b
6. 3
7. b
8. 3
9. a
10. b
11. 1027 −4 5 32 10
12. 1221 1231 1216 1213 1225 1233
13. 5; 9; 4
14. 11
15. a
16. 3, 5, 7, 9, 11, 15, 17, 19, 21, 23, 25
17. XYN
18. Every letter is unchanged.
19. a
20. b
21. c
22. $3! = 6$
23. $\alpha^{-1} = \begin{bmatrix} 1 & 2 & 3 & 4 \\ 2 & 1 & 4 & 3 \end{bmatrix}$
24. 6
25. MEETT HEBOS SATNO ONXQZ
26. FL EX ET
27. IE TX
28. 0010100
29. 00000000
30. 0001100

Chapter 18

1. c
2. 120
3. a
4. \$2.33
5. b
6. 10
7. c
8. c
9. a
10. 1280
11. c
12. 26.2187575 mi (officially, 26 mi and 385 yd)
13. a
14. 60
15. c
16. 19
17. c
18. 1.6
19. b
20. 200
21. c
22. heavy (or big); light (or small)
23. c
24. 4
25. c
26. 10
27. c
28. 3
29. a (or b)
30. proportional to a power of

Chapter 19

1. d
2. patterns
3. a
4. 12
5. c
6. 34
7. b
8. pineapples
9. c
10. 20
11. b
12. H; I; N; O; S; X; Z
13. b
14. translation; rotation
15. c
16. translation; reflection (along diagonal lines)
17. b
18. half-turn rotation
19. c
20. 7
21. b
22. translation
23. a
24. reflection
25. a
26. infinitely many

27. c
28. toss
29. c
30. self-similarity

Chapter 20

1. c
2. 45°
3. b or c
4. 3; 4; 6
5. a
6. 6
7. c
8. 5
9. b
10. 5
11. c
12. b
13. a
14. b
15. a
16. M. C. Escher
17. c
18. b
19. b
20. Conway; translations; half-turns
21. b
22. quasi
23. b
24. opposite sides are equal in length (or parallel, or opposite angles are congruent)
25. c
26. fivefold
27. c
28. the golden ratio
29. c
30. inflation

Chapter 21

1. a, b, or c
2. $15
3. b
4. $1186.00
5. b
6. $49.09
7. c
8. $1202.71
9. a (or b)
10. $181.66
11. b
12. $3686.64 ($3696.74 in a leap year)
13. a
14. 3.0416%
15. b
16. 3.04533%
17. c
18. $1204.42
19. c
20. 3.04545%
21. c
22. 2.718
23. a
24. annuity (or sinking fund)
25. b
26. 3.46535%
27. c (or a)
28. $9216
29. b
30. $140.71

Chapter 22

1. a
2. $1219.89
3. c
4. $2543.20
5. c
6. annual percentage rate
7. c
8. 19.56%
9. a
10. 18%
11. a
12. 5 years
13. b
14. $135.47
15. c
16. $456.33
17. c
18. 1.25 to 1.5
19. b
20. interest
21. a
22. $500
23. b
24. higher
25. b
26. adjustable rate mortgage
27. a
28. $13.55 million if you get the first million right away and $12.78 million if you have to wait a year for the first million
29. b
30. actuary

Chapter 23

1. b
2. the spread of a technology or of a product
3. c
4. momentum
5. c
6. logistic
7. b
8. 32
9. a
10. 241
11. b
12. 72,600
13. b
14. rising
15. b
16. dynamics (or growth, or evolution, or variation)
17. c
18. larger
19. c
20. smaller
21. a
22. 0 (or 1)
23. a
24. 0.75
25. a
26. sensitivity to initial conditions
27. a, b, and c
28. greed; chance; chaotic variation
29. c
30. near-periodic; transitive; sensitive (to initial conditions)

Answers to Odd-Numbered Exercises

Chapter 1

1. (a) Six vertices
   (b) Nine edges
3. (a)

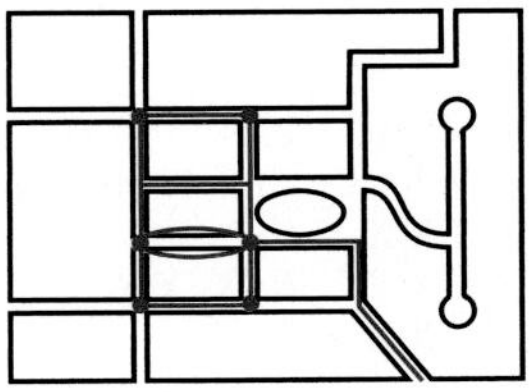

   (b) *ADGIJHKLMLKHGHEDEBA* (other answers possible)
5. (a) No
   (b) *EC*, *AD*, *BD*, and *AC*
   (c) Five vertices; six edges
7. (a) Eight vertices
   (b) 13 edges
   (c) *A*: 4; *B*: 2; *C*: 3; *D*: 3; *E*: 4; *F*: 4; *G*: 3; *H*: 3
   (d) *A*, *D*, and *F*
   (e) *E*, *G*, and *H*
9. *E*: 0; *A*: 1; *H*, *D*, and *G*: 2; *B* and *F*: 3; *C*: 5. *E* might be on an island and have no road access to the other cities.
11. (a) 5; 5
    (b) 7; 6
    (c) 10; 14
13. (a) *CGDBC* (Answers can vary.)
    (b) (i) *BD*; *BFD*
    (ii) *CBF*; *CGDF*; *CGDBF*
    (iii) *GDBCG* (Answers can vary.)
15. Drawings can vary. Possible renderings for parts (a) through (c) include the following:
    (a)

(b)

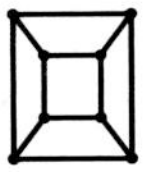

(c)

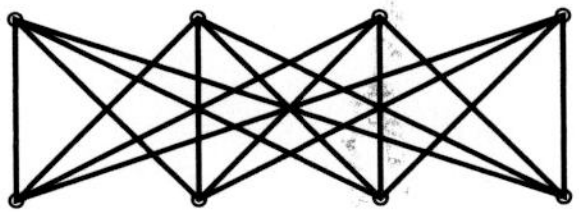

(d) Yes
17. 2
19. (a)

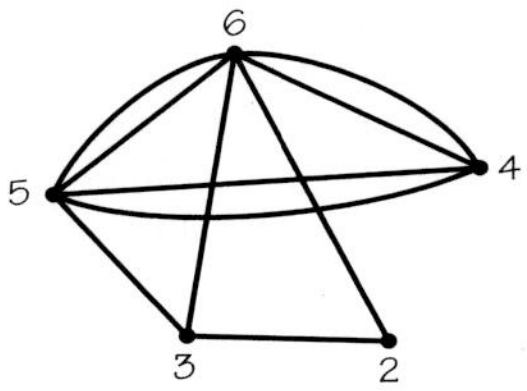

(b) No. To have an Euler circuit a graph cannot have odd-valent vertices.
21. Drawings can vary. Possible renderings include the following:

23. Drawings can vary. Possible renderings include the following:

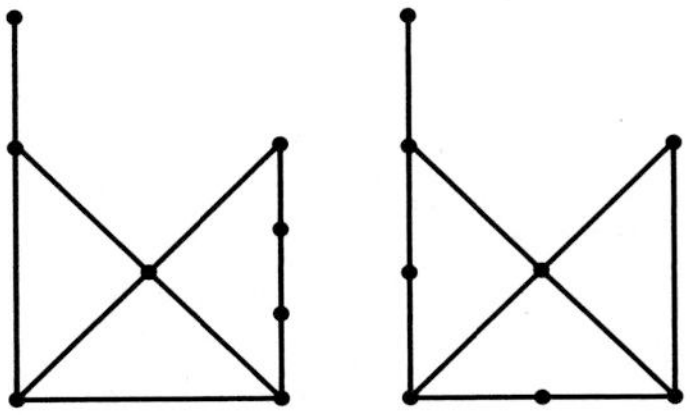

(*a*) (*b*)

25. With part (a), not all edges are traveled by the worker. With part (b), the end of the route is not the same as the beginning of the route. The description is not realistic because there is no Euler circuit in the graph.
27. Since this graph is connected and even-valent, it has an Euler circuit. Any Euler circuit will solve the problem efficiently.
29.

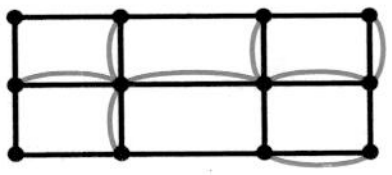

31. (a) 3
    (b) Answers will vary.
    (c) Answers will vary.
33. (a) No
    (b) 4
    (c) Answers will vary. One solution would involve duplicating edges *CD*, *DE*, *FG*, and *AB* in the accompanying graph.
35. Do not choose edge 2, but edges 1 or 10 could be chosen.
37. Answers will vary.
39. (a) *A*, *C*, *E*, and *H* are odd-valent.
    (b) 2
41. Answers will vary; no.
43. Answers will vary. Possible answers include *AECDABDCBEA*.
45. Drawings can vary. Possible renderings for parts (a), (b), and (c) include the following:

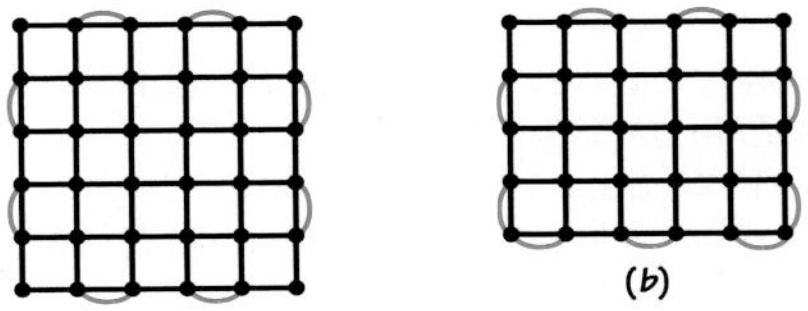

(*a*) (*b*)

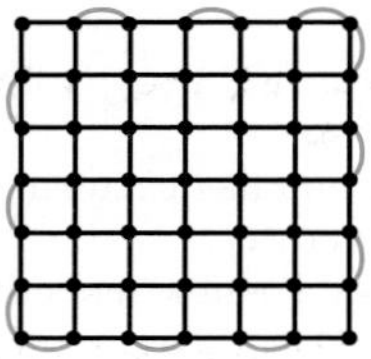

(c)

(d) Yes; no

47. (a) 2
(b) Yes
(c) 2
(d) No

49. (a) *ADEADCAFCFBA*
(b) 95 minutes

51.

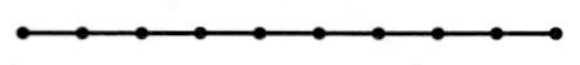

53. There are many circuits that achieve a minimum length of 44,000 feet.

55. Graphs (a) and (c); additional answers will vary.

57. For each edge *e* of *G*, add another edge joining the vertices that are endpoints of *e* to obtain *H*. If *G* is connected, *H* does have an Euler circuit because whatever the valences of *G*, graph *H* has valences that are doubles of these and remains connected. Hence, *H* is even-valent and connected.

59. (a) Drawings can vary. Possible renderings include the following:

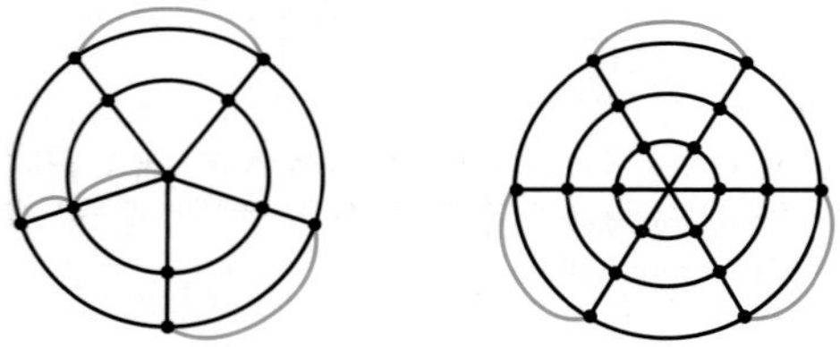

(b) The best eulerization for the four-circle, four-ray case adds two edges.
(c) Answers will vary.

61. (a) Yes
(b) 15

63. Answers will vary.

65. 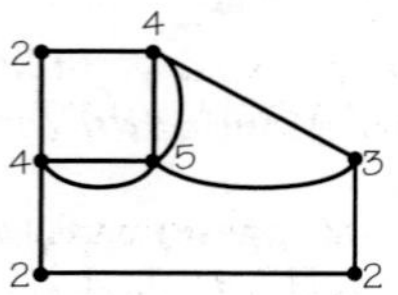

; connected

67. Answers will vary.

69. Answers will vary. A possible answer is *ABDEFBEBFEDBACDCBCA*.

71. (a) 10 vertices; 19 edges
(b) *A* has valence 4; *F* has valence 4; *D* has valence 4.
(c) There are Euler circuits. Answers will vary.

Chapter 2

1. (a) $X_2X_6X_7X_3X_4X_8X_{12}X_{11}X_{10}X_9X_5X_1X_2$
(b) $X_2X_5X_8X_3X_4X_7X_6X_1X_2$ is one possibility, as is $X_2X_1X_4X_3X_8X_5X_6X_7X_2$.
(c) $X_2X_8X_1X_{10}X_7X_6X_9X_5X_4X_3X_2$ is one possibility, as is $X_2X_1X_8X_9X_{10}X_7X_6X_5X_4X_3X_2$.

3. (a) Not possible
(b) A possible answer is $X_3X_2X_{11}X_{12}X_6X_{13}X_{20}X_{16}X_{17}X_{14}X_{10}X_1X_4X_9X_{15}X_{18}X_{19}X_8X_7X_5X_3$
(c) $X_3X_4X_2X_5X_6X_1X_3$

5. Possible answers include:
(a) Answers will vary. One Hamiltonian circuit is $X_3X_1X_2X_4X_5X_6X_3$.
(b) $X_3X_2X_1X_6X_7X_8X_9X_{10}X_{11}X_{12}X_5X_4X_3$
(c) $X_3X_1X_2X_7X_6X_9X_8X_5X_4X_3$

7. (a) Yes
(b) Yes
(c) Yes

9. (a) No for (a); yes for (b); no for (c)
(b) No longer possible to send messages between these two sites

11. Answers will vary.

13. (a) Yes for both.
(b) Answers will vary.
(c) Add edges X_2X_8, X_8X_6, X_6X_4, and X_4X_2.

15. Drawings can vary. Possible renderings include the following:

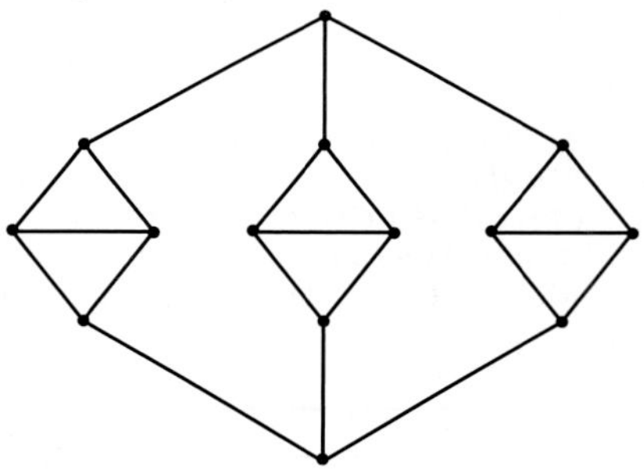

17. (a) No
(b) No

19. (a) Yes
(b) No
(c) No

21. (a) Yes
(b) No
(c) Answers will vary.

23. (a) Hamiltonian circuit: yes; Euler circuit: yes; additional answers will vary.
(b) Hamiltonian circuit: yes; Euler circuit: yes
(c) Hamiltonian circuit: yes; Euler circuit: no
(d) Hamiltonian circuit: no; Euler circuit: yes; additional answers will vary.

25. (a) Hamiltonian circuit: yes; Euler circuit: no
(b) Hamiltonian circuit: yes; Euler circuit: no
(c) Hamiltonian circuit: yes; Euler circuit: no
(d) Hamiltonian circuit: no; Euler circuit: no

27. (a) Drawings will vary.
(b) Drawings will vary.
(c) Answers will vary.

29. (a) 2520
(b) 16,807
(c) 16,800

31. (a) 15,120
(b) 17,576

33. Yes; 172

35. (a) 17,558,424
(b) Answers will vary.

37. 10,000,000; 900

39. Drawings will vary.; 6, 10, and 15 edges, respectively; $\frac{n(n-1)}{2}$ edges; 3, 12, and 60, respectively

41. (a) Possible drawings include the following:

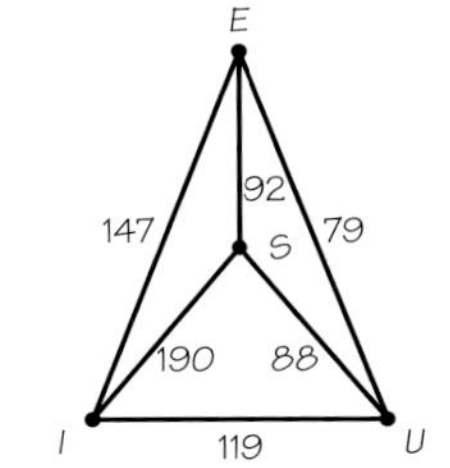

(b) Tour 1: *UISEU*: 480
Tour 2: *USIEU*: 504
Tour 3: *UIESU*: 446
(c) Tour 3
(d) No
(e) Tour 1; yes; yes; no
(f) Tour 2; no

43. *FCMRF*

45. (a) *HBCAH*
(b) *HBCAH*
(c) The cheapest route costs 132 and coincides with the nearest-neighbor solution from *H* and the sorted-edges solution.

47. *MACBM*

49. A traveling salesman problem

51. Yes; Hamiltonian circuit; Chinese postman problem; answers will vary and require at least 9 reuses of edges.

53. Answers will vary.

55. The optimal tour is the same but its cost is now 4700.

57. Diagram (a): (a) There is a circuit and wiggled edges do not include all vertices. (b) The circuit does not include all the vertices of the graph.
Diagram (b): (a) The tree does not include all vertices of the graph. (b) Not a circuit
Diagram (c): (a) Not a tree (b) Not a circuit
Diagram (d): (a) Not a tree (b) Not a circuit

59. (a) 1, 2, 3, 4, 5, 8; cost is 23.
(b) 1, 1, 1, 2, 2, 3, 3, 4, 5, 6, 6; cost is 34.
(c) 1, 1, 1, 2, 2, 2, 2, 2, 3, 3, 3, 4, 4, 4, 5, 5, 6, 7; cost is 60.
(d) 1, 2, 2, 3, 3, 3, 4, 5, 5, 5, 6, 6; cost is 45.

61. *H*'s spanning tree has 24 vertices; *H* has 24 vertices. *H* has at least 23 edges and, if there are no multiple edges, at most (24)(23)/2 edges.

63. Yes

65. Yes; additional answers will vary.

67. Yes; yes

69. There are three different trees with the same cost.

71. (a) True
(b) False (unless all the edges of the graph have the same weight)
(c) True
(d) False
(e) False

73. (a) Answers will vary for each edge.
(b) 5; one less than the number of vertices in the graph
(c) No (*CD* and *AG* must be included in any spanning tree.)

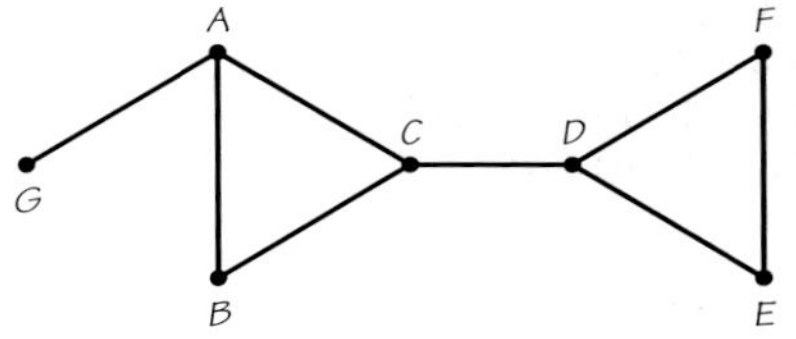

75.

	A	B	C	D
A	0	16	13	5
B	16	0	19	11
C	13	19	0	8
D	5	11	8	0

77. (a) 22; $T_3T_2T_5$
(b) 30; $T_3T_5T_7$

79. T_1, T_5, and T_7 if shortened would reduce the earliest completion time, while shortening the other tasks will not; 29; $T_1T_4T_7$.

81. Answers will vary.

83. Drawings can vary. Possible renderings include the following:

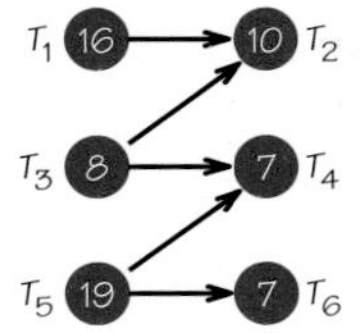

85. The number of passwords can be found by performing the following multiplication: (26)(26)(26)(25)(3)(9)(8)(7).

87. The edges are added to the minimum-cost spanning tree in this order: *GF, JD, IH, KL, HK, IC, JL, FI, LE, AI, AB*.

Chapter 3

1. Answers will vary.
3. Answers will vary.
5. The tasks would be scheduled so that T_1 gets done from 0 to 1, T_2 from 1 to 6, T_4 from 6 to 12, T_3

from 12 to 14, T_5 from 14 to 17, and T_6 from 17 to 21. Even though there is only one machine, the tasks don't appear in the schedule in the order they appear in the list because of restrictions in task order due to the order-requirement digraph.

7. (a) Processor 1: T_1, T_2, T_3, T_5, T_7; Processor 2: Idle 0 to 2, T_4, T_6, idle 4 to 5
   (b) Processor 1: T_1, T_2, T_3, T_6, T_7; Processor 2: Idle 0 to 2, T_4, T_5, idle 4 to 5
   (c) Yes
   (d) No
   (e) T_3 and T_5
9. (a) 2
   (b) 1
11. The two vertices with no incoming edges correspond to the only tasks that are ready at time 0. These tasks are assigned to different processors at time 0, forcing the third processor to be idle for some period of time.
13. (a) (i) Processor 1: T_1 from 0 to 13, T_3 from 13 to 25, T_6 from 25 to 45; Processor 2: T_2 from 0 to 18, T_4 from 18 to 27, T_5 from 27 to 35, idle from 35 to 45
   (ii) Processor 1: T_1 from 0 to 13, T_3 from 13 to 25, T_4 from 25 to 34, T_5 from 34 to 42; Processor 2: T_2 from 0 to 18, T_6 from 18 to 38, idle from 38 to 42
   (b) Yes
   (c) T_2, T_6, and 38; sum of the task times divided by 2 is 40.
15. (a) Yes (b) No
17. Answers will vary.
19. (a) Processor 1: T_1, T_6, idle 15 to 21, T_7, idle 27 to 31; Processor 2: T_2, T_5, T_8; Processor 3: T_3, T_4, idle from 13 to 31
   (b) Processor 1: T_1, T_6, idle 15 to 21, T_7, idle 27 to 31; Processor 2: T_3, T_4, idle from 13 to 21, T_8; Processor 3: T_2, T_5, idle from 21 to 31
   (c) Processor 1: T_4, idle 10 to 11, T_6, idle 18 to 21, T_8; Processor 2: T_2, T_5, T_7, idle 27 to 31; Processor 3: T_1, T_3, idle 11 to 31
21. Yes
23. (a) 15
   (b) 19
   (c) Use the list T_1, T_4, T_3, T_2, T_5, T_6, T_7 to obtain the following schedule:

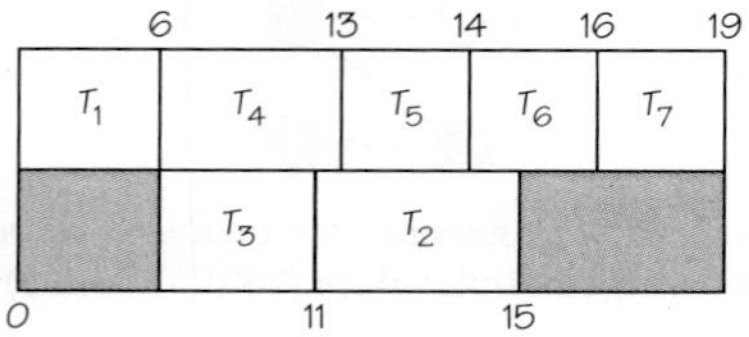

25. (a) No
   (b) If T_5 could be started on Machine 3 and Machine 2 was also free at the same time, proper use of the list-processing algorithm should have assigned this task to Machine 2.
   (c) Use the digraph with no edges and the list T_2, T_5, T_4, T_3, T_1.
27. (a) T_1, T_2, T_3, and T_6
   (b) No tasks require that T_1 and T_6 be done before these other tasks can begin.
   (c) T_6
   (d) Processor 1: T_1, T_6; Processor 2: T_2, T_4, idle from 18 to 30; Processor 3: T_3, T_5, idle from 12 to 30
   (e) No
   (f) Processor 1: T_6, idle from 20 to 22; Processor 2: T_3, T_5, T_1; Processor 3: T_2, T_4, idle from 18 to 22
   (g) Yes
   (h) Yes
29. (a) 120
   (b) No; T_1 must be assigned to the first machine at time 0.
   (c) No; when 2 divides 31, there is a remainder of 1.
   (d) No
31. Yes
33. Answers will vary.
35. No
37. (a) Task times: $T_1 = 3$, $T_2 = 3$, $T_3 = 2$, $T_4 = 3$, $T_5 = 3$, $T_6 = 4$, $T_7 = 5$, $T_8 = 3$, $T_9 = 2$, $T_{10} = 1$, $T_{11} = 1$, and $T_{12} = 3$. This schedule would be produced from the list T_1, T_3, T_2, T_5, T_4, T_6, T_7, T_8, T_{11}, T_{12}, T_9, T_{10}.
   (b) Task times: $T_1 = 3$, $T_2 = 3$, $T_3 = 3$, $T_4 = 2$, $T_5 = 2$, $T_6 = 4$, $T_7 = 3$, $T_8 = 5$, $T_9 = 8$, $T_{10} = 4$, $T_{11} = 7$, $T_{12} = 9$, and $T_{13} = 3$. This schedule would be produced from the list T_1, T_5, T_7, T_4, T_3, T_6, T_{11}, T_8, T_{12}, T_9, T_2, T_{10}, T_{13}.
39. (a) (i) Processor 1: T_1, T_3, T_5, T_7, idle from 16 to 20; Processor 2: T_2, T_4, T_6, T_8
   (ii) Processor 1: T_8, T_5, T_4, T_1; Processor 2: T_7, T_6, T_3, T_2
   (b) Yes
41. Answers will vary.
43. In part (a), 33 is not exactly divisible by 4; in part (b), 56 is not exactly divisible by 5.
45. (a) (i) Machine 1: 12, 9, 15, idle from 36 to 50; Machine 2: 7, 10, 13, 20
   (ii) Machine 1: 12, 13, 20; Machine 2: 7, 9, 15, 10, idle from 41 to 45
   (iii) Machine 1: 20, 12, 9, idle from 41 to 45; Machine 2: 15, 13, 10, 7
   (b) None of the schedules found is optimal
   (c) The critical path list is T_6, T_5, T_4, T_1, T_7, T_2, T_3. This list does not lead to an optimal schedule, but there is a schedule where the completion time is 43, achievable by scheduling the tasks of length 20, 13, and 10 on Machine 1 and tasks of length 24 and 19 on Machine 2.
47. (a) Machine 1: 129; Machine 2: 129
   (b) Machine 1: 123; Machine 2: 123
   (c) Yes
49. (a) Processor 1: 12, 13, 45, 34, 63, 43, 16, idle 226 to 298; Processor 2: 23, 24, 23, 53, 25, 74, 76; Processor 3: 32, 23, 14, 21, 18, 47, 23, 43, 16, idle 237 to 298
   (b) Processor 1: 12, 24, 14, 34, 25, 23, 16, 16, 76; Processor 2: 23, 23, 21, 63, 43, idle 173 to 240; Processor 3: 32, 23, 53, 74, idle 182 to 240; Processor 4: 13, 45, 18, 47, 43, idle 166 to 240
   (c) *Three machines:* Processor 1: 76, 45, 43, 24, 23, 18, 16, 13; Processor 2: 74, 47, 34, 32, 23, 21, 14, 12, idle 257 to 258; Processor 3: 63, 53, 43, 25, 23, 23, 16, idle 246 to 248

Four machines: Processor 1: 76, 43, 24, 23, 16, idle 182 to 194; Processor 2: 74, 43, 25, 23, 16, 13; Processor 3: 63, 45, 32, 23, 18, 12, idle 193 to 194; Processor 4: 53, 47, 34, 23, 21, 14, idle 192 to 194

(d) Processor 1: 84, 45, 43, 25, 23, 23, 16, 12; Processor 2: 82, 55, 34, 32, 23, 18, 14, 13; Processor 3: 71, 61, 43, 24, 23, 21, 16, idle 259 to 271

51. Answers will vary.
53. Each task heads a path of length equal to the time to do that task.
55. 9; number of bins would not change, but the placement of the items in the bins would differ.
57. (a) 17
(b) 16
(c) 16
(d) 13
59. Yes, both are acceptable.
61. (a) Answers will vary.
(b) It is possible.
63. No; yes
65. Answers will vary.
67. Answers will vary.
69. (a) 152 min; 124 min
(b) 155 min; 120 min
(c) Yes; five-processor decreasing-time schedule
(d) 11
(e) NFD: 13; WFD: 11
(f) An optimal packing with 10 bins exists.
71. (a) Answers will vary.
(b) Answers will vary.
(c) Packing rectangles of width 1 in an $m \times 1$ rectangle is a special case of the two-dimensional problem.
(d) Answers will vary.
73. Answers will vary.
75.

77. (a) Graph in: (a) Yes (b) No (c) Yes (d) Yes (e) No (f) No
(b) Graph in: (a) Yes (b) No (c) Yes (d) Yes (e) Yes (f) Yes
(c) Graph in: (a) 3 (b) 5 (c) 3 (d) 2 (e) 4 (f) 4
79. (a) Drawings can vary. Possible renderings include the following:

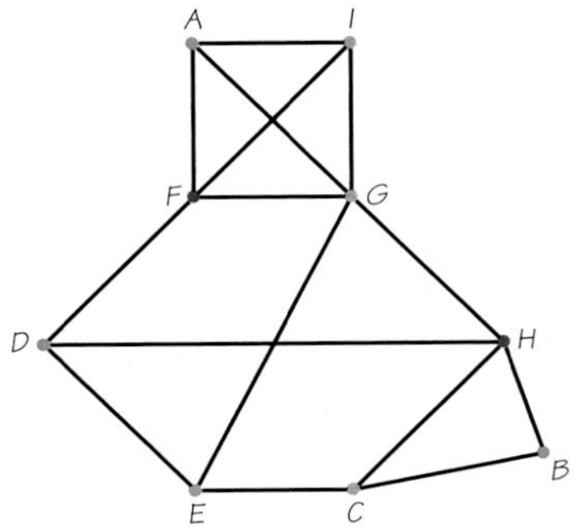

(b) 4
(c) The coloring in part (a) indicates one possible arrangement.
81. (a) Drawings can vary. Possible renderings include the following:

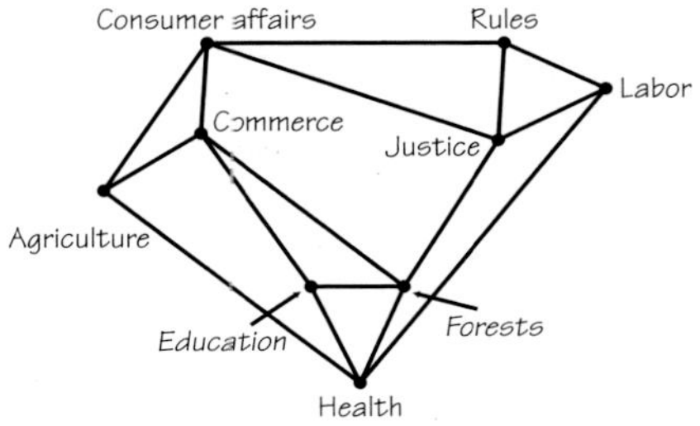

(b) 3
(c) 3; additional answers will vary.
83. (a) Drawings can vary. Possible renderings include the following:

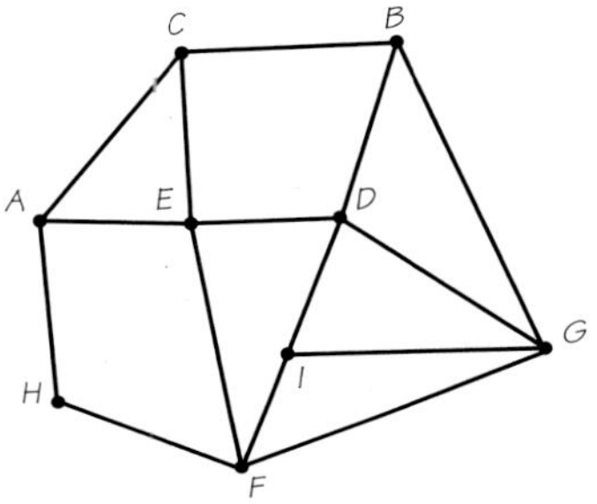

(b) 3
(c) 3
85. (a)

	F	E	H	P	T	A	D	C
F		×		×	×			
E	×					×	×	
H						×	×	×
P	×					×		×
T	×						×	×
A		×	×	×				
D		×	×		×			
C			×	×	×			

(b) 2
(c) 3 (Two rooms will be used for three committee meetings and one room for two committee meetings.)
87. Answers will vary.
89. Graph in: (a) 8 (b) 5 (c) 6 (d) 4 (e) 3 (f) 4; the minimum is either the largest valence of any vertex or 1 more than the largest valence.
91. (a) Graph (a) 2 (b) 4 (c) 4 (d) 4 (e) 3 (f) 3
(b) Answers will vary.
93. 3
95. Three play groups will ensure no conflict. Four equal-size conflict-free groups can be formed. These groups would have size 2.
97. Yes
99. (a) Contents of bins listed as items from the bottom to the top: *First fit:* Bin 1: 3, 6, 1; Bin 2: 9; Bin 3: 5, 3; Bin 4: 7; Bin 5: 4, 5; Bin 6: 6. *Next fit:* Bin 1: 3; Bin 2: 9; Bin 3: 6; Bin 4: 5, 3, 1; Bin 5: 7; Bin 6: 4, 5; Bin 7: 6. *First-fit decreasing:* Bin 1: 9, 1; Bin 2: 3, 7; Bin 3: 6, 4; Bin 4: 6, 3; Bin 5: 5, 5.
(b) 19
(c) 17
(d) Yes

Chapter 4

1. The intersection points are (1, 7), (1, 5), and (3, 5).
3. The lines $x = 3$ and $y = 4$ meet at the point (3, 4); the lines $x = 3$ and $x + y = 10$ meet at the point (3, 7); the lines $y = 4$ and $x + y + 10$ meet at the point (6, 4).

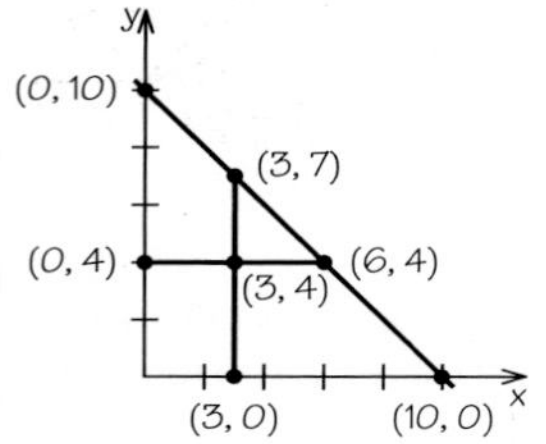

5. (a)

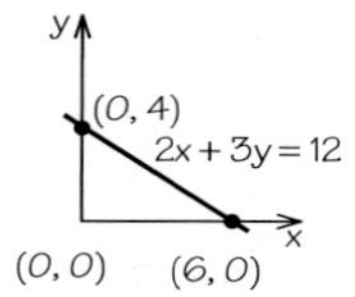

(b)

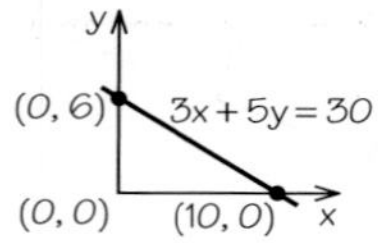

(c)

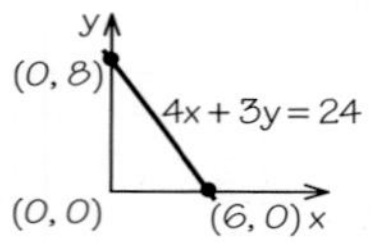

7. (a)

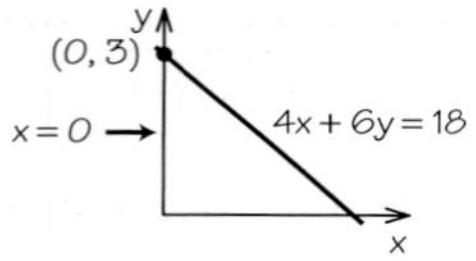

(b)

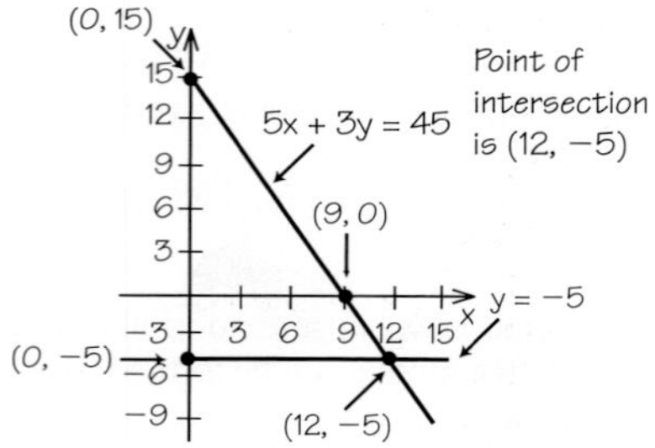

(c)

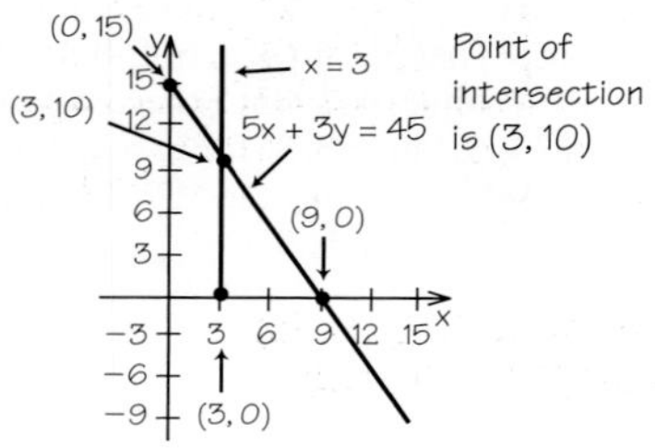

Note: For Exercises 9 and 11, first quadrant only is shown. Point of intersection is labeled.

9. (a)

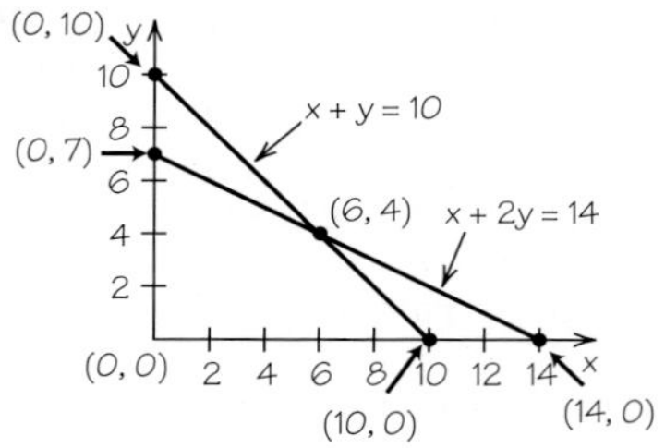

(b)

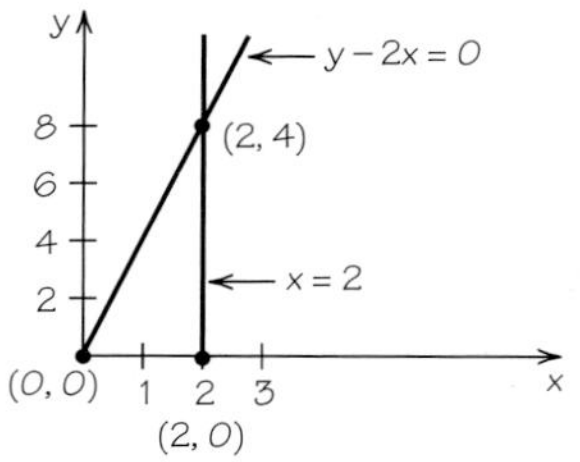

11. (a)

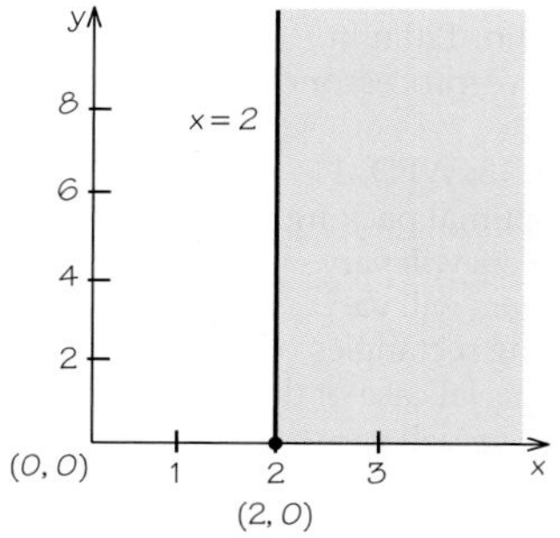

(b)

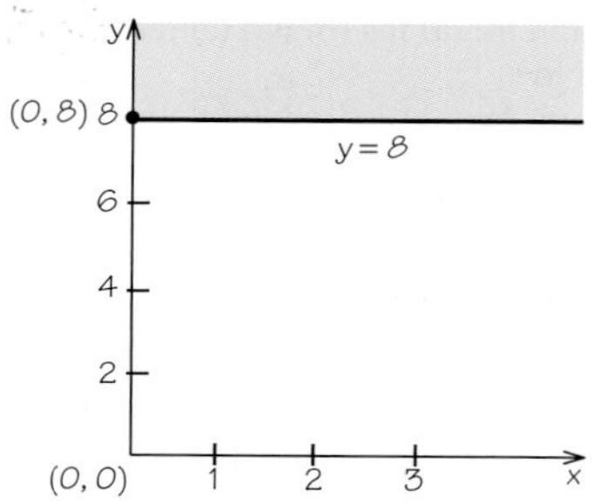

(c)

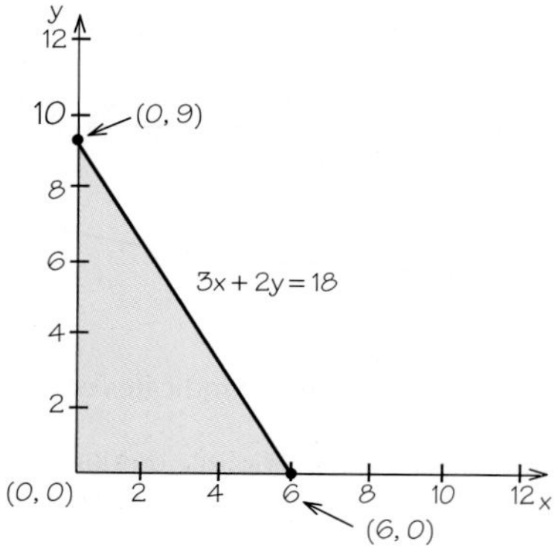

(d)

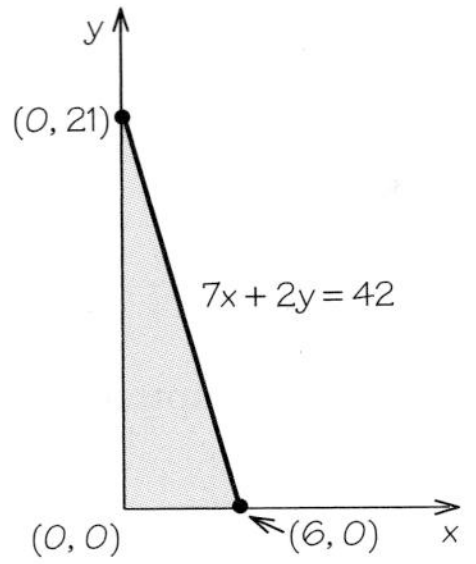

13. (a) $6x + 4y \leq 300$
(b) $30x + 72y \leq 420$

15.

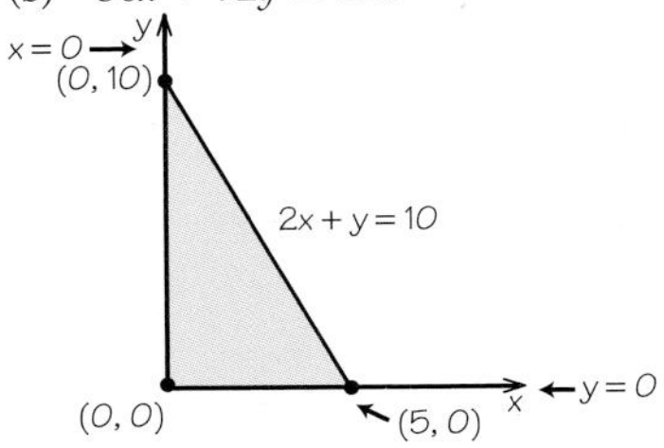

17.

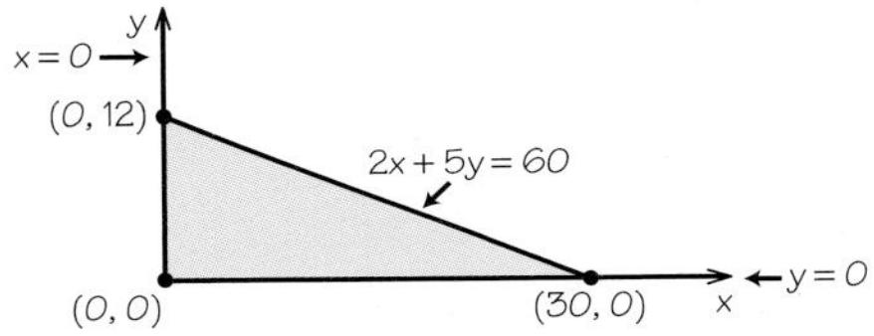

19.

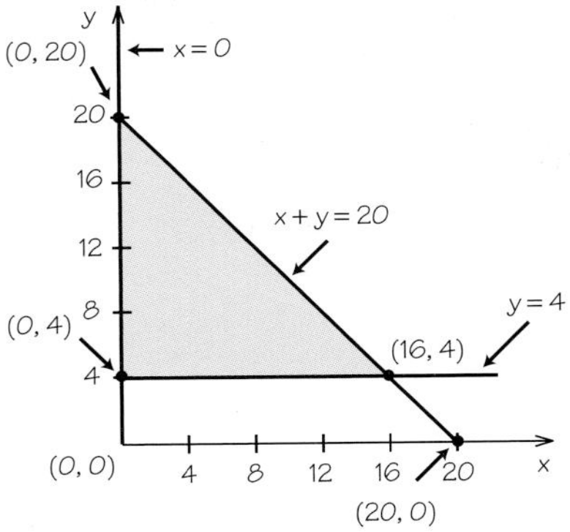

21. Exercise 15: (2, 4): yes; (10, 6): no; Exercise 17: (2, 4): yes; (10, 6): yes; Exercise 19: (2, 4): yes; (10, 6): yes
23. Make 0 skateboards and 30 dolls for a profit of $111.
25. *Note:* These situations are shown only for the first quadrant.

(a)

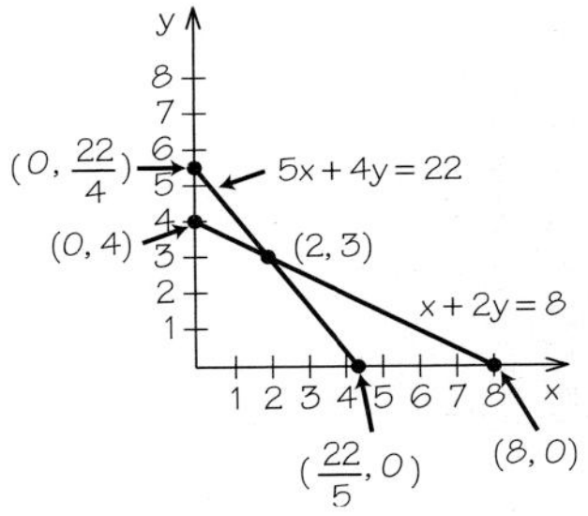

(b)

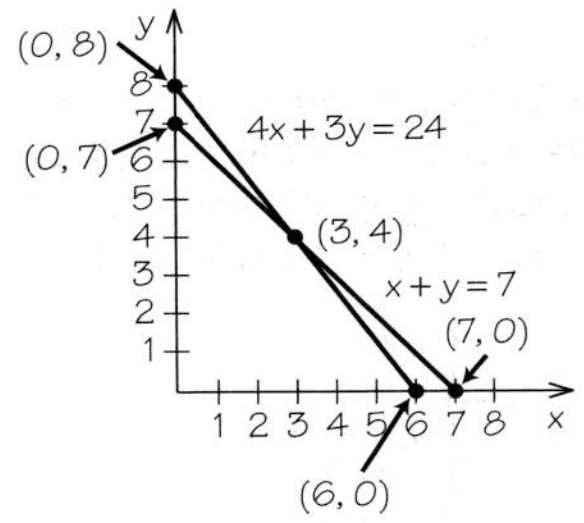

27.

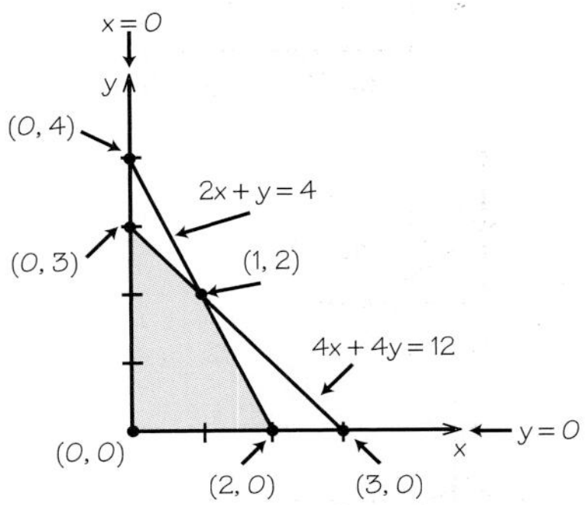

29.

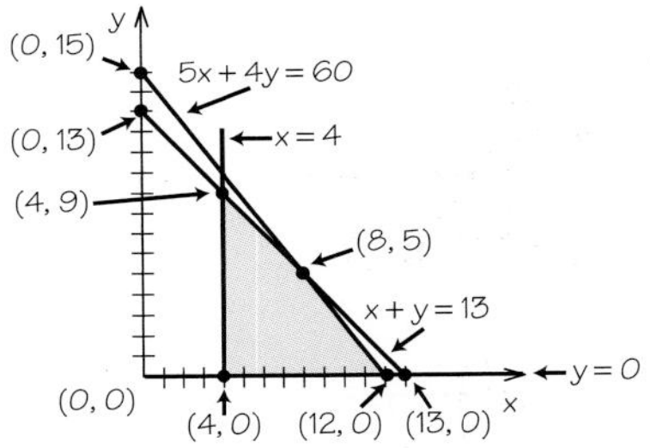

31. The maximum value of P is 31 and occurs at the point (7, 5).
33. (a) $x = 0$ and $x = 4$ are vertical constraints, while $y = 0$ is a horizontal constraint.
(b)

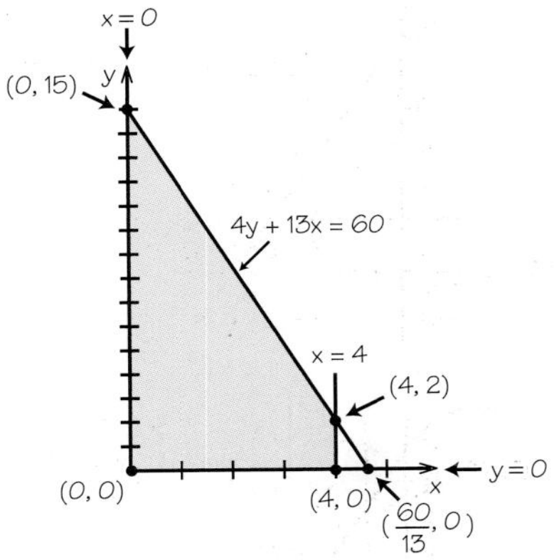

(c) (0, 0); (4, 0); (4, 2); (0, 15)
(d) (i) 16; (ii) (0, 15); (4, 2)

35. 28
37. 38
39. (a) (2, 0)
 (b) They do not.
 (c) Profit at (2, 0) is greater than the profit at $(\frac{13}{7}, 0)$.
 (d) Yes; profit at R is less than the profit at Q.
 (e) Answers will vary.
41. Schedule 400 oil changes and no tune-ups; schedule 300 oil changes and 20 tune-ups.
43. Schedule 360 routine visits and no comprehensive visits; schedule 210 routine visits and 30 comprehensive visits.
45. Take four math courses and no other courses; take two math courses and two other courses.
47. Make two grade A and five grade B batches in both cases.
49. Make 3000 cartons of regular and 2000 cartons of diet in both cases.
51. Make no desk lamps and 1200 floor lamps; make 150 desk lamps and 1080 floor lamps.
53. Make 50 chairs, 10 tables, and no beds each month.
55. Make 470 pounds of Excellent, none of Southern, 2400 pounds of World, and 320 pounds of Special.
57. 43
59. Make 60 business and no charity calls; make 45 business and 10 charity calls.
61. Make 3 bikes and 2 wagons in both cases.
63. (a)

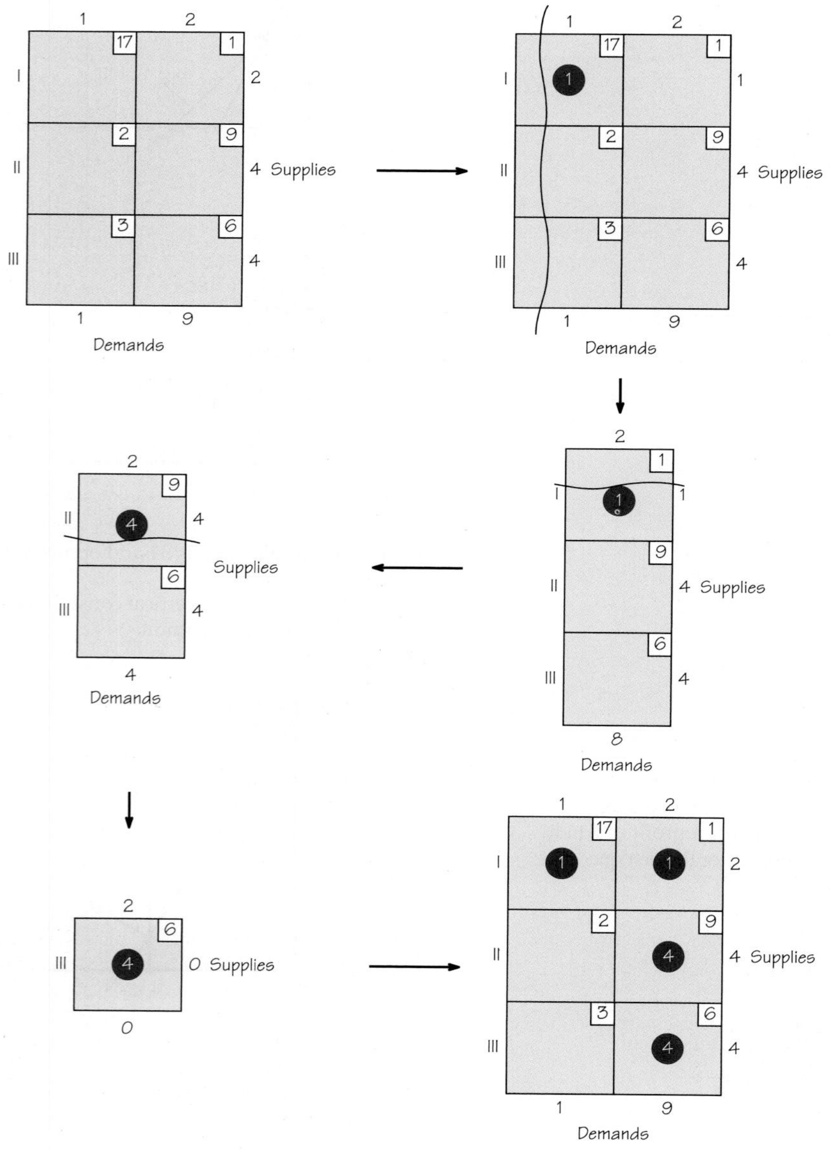

(b)

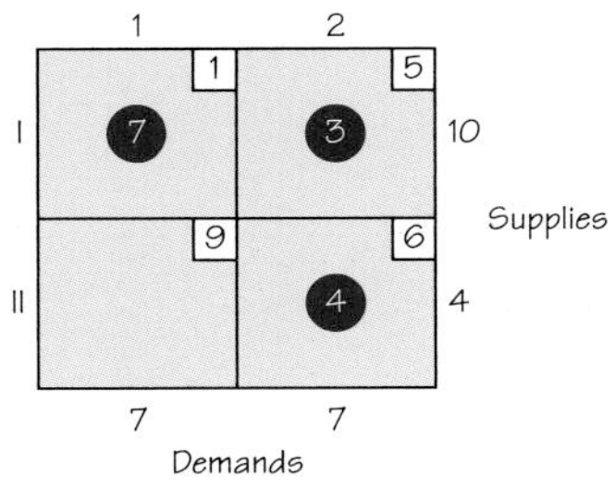

(c) This tableau could represent how to meet the needs of two organic groceries with organic fruit from two suppliers.
(d) The cost associated with this way to ship is 46.

65. (a)

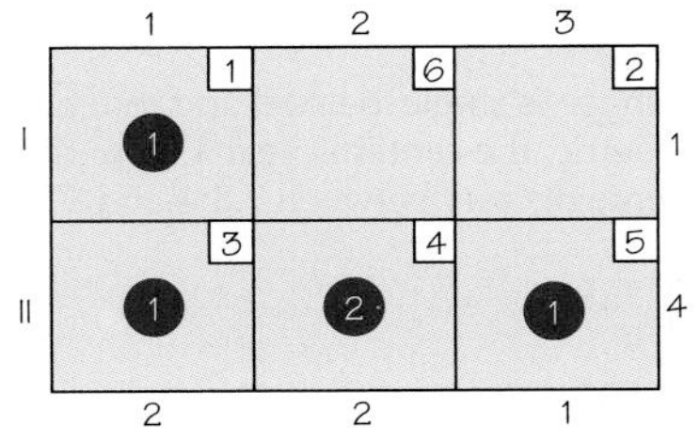

(b) 17
(c) (I, 2): 4; (I, 3): −1

67. (a) Connected and has no circuit
(b) Add edge joining vertex I to vertex 2; add edge from vertex I to vertex 3.
(c) Circuit 2, I, 1, II, 2 corresponds to the circuit of cells, (I, 2), (I, 1), (II, 1), (II, 2), (I, 2). Circuit 3, I, 1, II, 3 corresponds to the circuit of cells, (I, 3), (I, 1), (II, 1), (II, 3), (I, 3).

69. (a) (i)

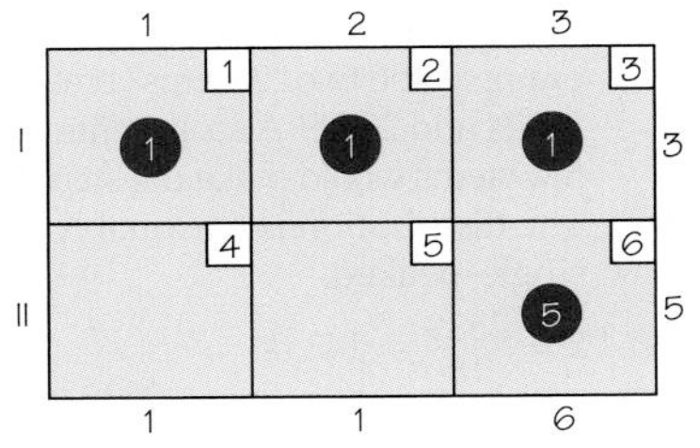

(ii)

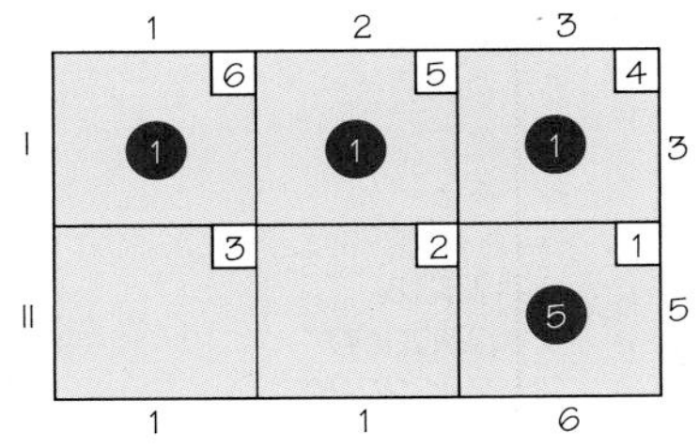

(iii)

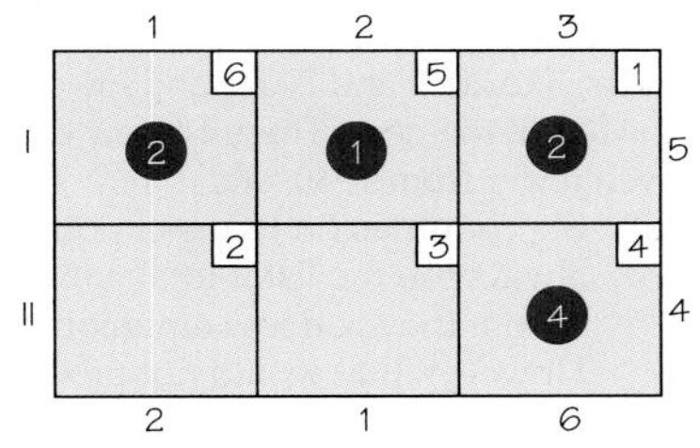

(b) For both (i) and (ii) the tableaux shown are optimal. However, there are also other optimal tableaux. For (iii) the tableau shown is not optimal. Using the stepping stone algorithm, the cost can be reduced to 16.

71. (a) Yes
(b) No
(c) Yes

73. (a) The cost associated with the feasible solution shown is 72.

(b)

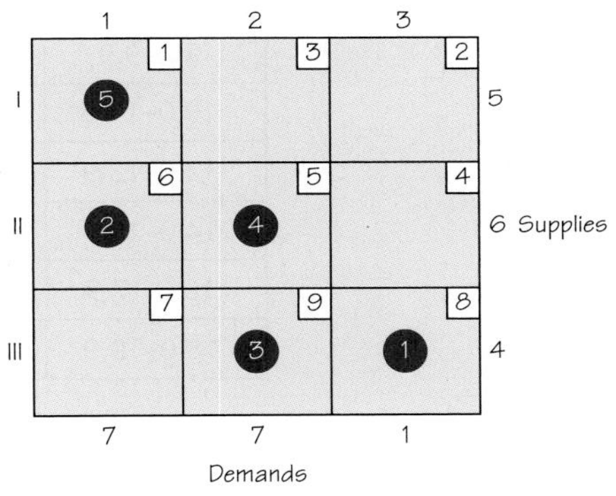

The cost associated with the feasible solution obtained is the same as for the Northwest Corner Rule. However, this will not always be true.
(c) Answers will vary.

75. 49

Chapter 5

1. (a) Vehicle makes and models (i.e., the four cars)
(b) Vehicle type, transmission type, number of cylinders, city mpg, and highway mpg
(c) Cylinders (maybe) and city mpg and highway mpg (certainly)

3. (a) The interval between 0.1 μg/g and 0.2 μg/g; around 28 data values fell within this interval, which means that $(28/83 \times 100)$% or approximately 33.7% of the fish had mercury concentrations that fell in this class interval.
(b) Approximately 56 of the fish had mercury levels below 0.30 μg/g.
(c) Approximately 27 of the fish from the sample had mercury levels at or above 0.30 μg/g. Hence, around 32.5% of the fish in the sample had levels of mercury concentration above the USEPA guidelines.

5. Most coins in circulation were minted in recent years, so we would expect a peak at the right (highest-numbered years, like 2012 and 2014) and lower bars trailing out to the left of the peak. There are few coins from 1990 and even fewer from 1980, etc.

7. (a) Big countries (in terms of population) would always top the list if total emissions were used, even if they had low emissions for their size. However, that would not provide a measure of the energy consumption per person.

(b) Using class widths of 2 metric tons per person, we have the following:

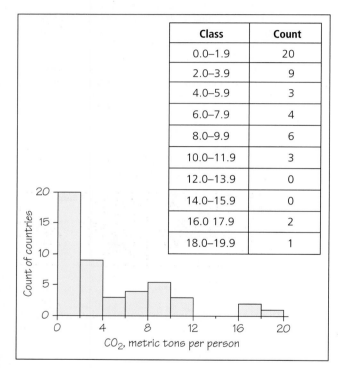

Class	Count
0.0–1.9	20
2.0–3.9	9
4.0–5.9	3
6.0–7.9	4
8.0–9.9	6
10.0–11.9	3
12.0–13.9	0
14.0–15.9	0
16.0 17.9	2
18.0–19.9	1

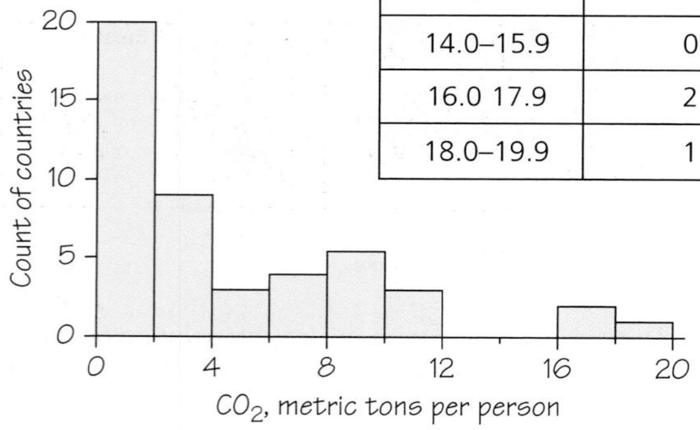

The distribution is skewed to the right. There appear to be three high outliers: Canada, Australia, and the United States.

9. (a)

Class Interval	Frequency
0 ≤ career home runs < 100	45
100 ≤ career home runs < 200	30
200 ≤ career home runs < 300	7
300 ≤ career home runs < 400	10
400 ≤ career home runs < 500	3
500 ≤ career home runs < 600	4
600 ≤ career home runs < 700	0
700 ≤ career home runs < 800	1

(b)

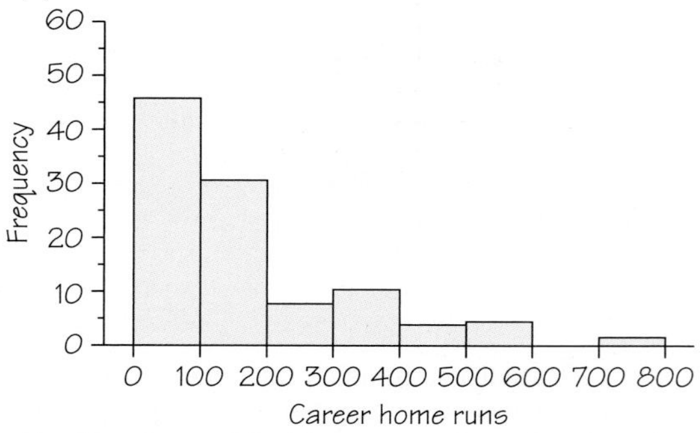

(c) The shape of the histogram is skewed to the right. There is a gap in the data between 600 and 700 and one potential outlier between 700 and 800 (Babe Ruth's 714 career home runs).

11. (a) 17.4%

(b) The shape is single-peaked and roughly symmetric; the center is near 13.5%; the percentages vary between 7.8% and 17.4%.

13. (a)

```
10 | 139
11 | 5
12 | 669
13 | 77
14 | 08
15 | 244
16 | 55
17 | 8
18 |
19 |
20 | 0
```

There is one high outlier: 200.

(b) The center of the 17 observations other than the outlier is 137 (9th of 17). Ignoring the outlier, there are values between 101 and 178.

15. (a) The repeated stems break up the intervals further. For example, the two "2 stems" break the twenties into 20–24 and 25–29. Also, if stems were not repeated, too few stems would make the stemplot less informative.

(b) The distribution is reasonably symmetric and single-peaked.

17. (a) $\bar{x} = \frac{2539}{18} \approx 141.06$

(b) Without the outlier, $\bar{x} = \frac{2539 - 200}{17} = \frac{2339}{17} \approx 137.6.$

(c) The high outlier pulls the mean up.

19. (a)

```
5 | 66
5 |
6 | 0
6 | 2
6 | 4
6 |
6 | 8888
7 | 000000
7 | 2222222
7 | 44444
7 | 66
7 | 88
```

(b) Mean ≈ 70.1 beats/min; median = 72 beats/min; mode = 72 beats/min

(c) The median or mode of 72 beats/min best describes a "typical" pulse rate for this man. There are a few days when the man's pulse rate was very low. These low values tend to pull the mean down.

21. The mean household income will decrease. Even though separately the two divorced parties have the same combined income, the total number of households has increased. By thinking about the formula for $\bar{x}$, the numerator will remain the same, but the denominator will increase by the number of divorces that establish new households.

23. Examples will vary. One possible answer is 1, 2, 2, 2, 3, 3, 4, 17. The third quartile is 3.5; $\bar{x} = 4.25$, which is above Q_3.

25. The five-number summary is 7.8, 12.4, 13.5, 14.3, 17.4.

27. (a) Minimum = 7519, Q_1 = 29,407, M = 35,473.5, Q_3 = 43,718, maximum = 49,812. (If using software, the results for Q_1 and Q_3 may differ from the hand calculations above.)

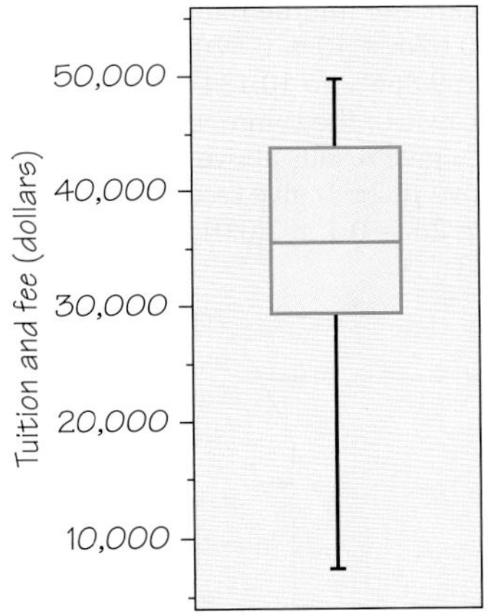

(b) The boxplot does not show the two distinctive clusters of values corresponding to the public and private colleges and universities.

29. The five-number summary is 0.0, 0.75, 3.2, 7.8, 19.9. The third quartile and maximum are much farther from the median than the first quartile and minimum, showing that the right side of the distribution has more variability than the left side.

31. The income distribution for bachelor's degree holders is generally higher than for high school graduates: The median for bachelor's is greater than Q_3 for high school. The bachelor's distribution has much more variability, especially at the high-income end but also between the quartiles.

33. (a) The histogram below shows the distribution to be unimodal and right-skewed. There are some potential outliers. (Histograms can vary depending on the choice of class width.)

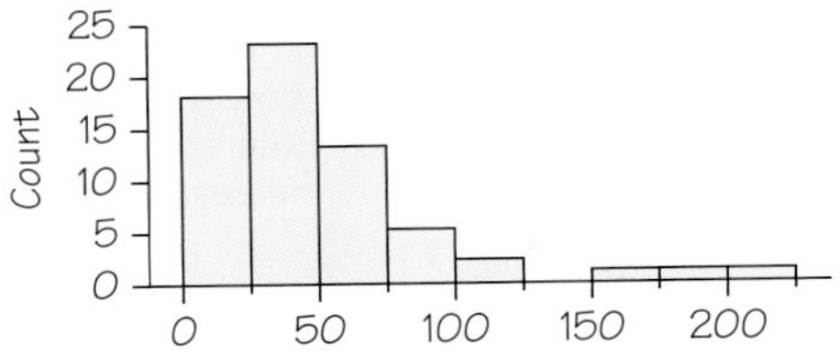

(b) $\bar{x}$ = 48.25, M = 37.8; the long right tail inflates the mean.

(c) The five-number summary is 2.0, 21.5, 37.8, 60.1, 204.9. (*Note:* Results for Q_1 and Q_3 may differ if calculated using computer software.) Q_3 and the maximum are much farther above the median than Q_1 and the minimum are below it, showing that the right side of the distribution has more variability than the left side.

35. (a) The five-number summary is 1.0, 3.5, 6.85, 10.2, 42.3.

(b) IQR = Q3 − Q1 = 10.2 − 3.5 = 6.7; $Q_1 - 1.5 \times$ IQR = −6.55; $Q_3 + 1.5 \times$ IQR = 20.25

(c) There are six data values above 20.25: 21.1, 22.3, 25.0, 33.1, 33.6, and 42.3. These values correspond to the states Florida, Nevada, Arizona, California, Texas, and New Mexico, respectively.

37. The standard deviation of the tuition and fees of Massachusetts's public colleges will be smaller than the standard deviation of the private colleges. The tuition and fees for the public colleges is spread over two class intervals, whereas the data for the private colleges is spread over six class intervals.

39. (a) $\bar{x}$ = 5.448, s = 0.221

(b) M = 5.46; yes

41. (a) Using the TI-83, we have the following for datasets A and B, respectively:

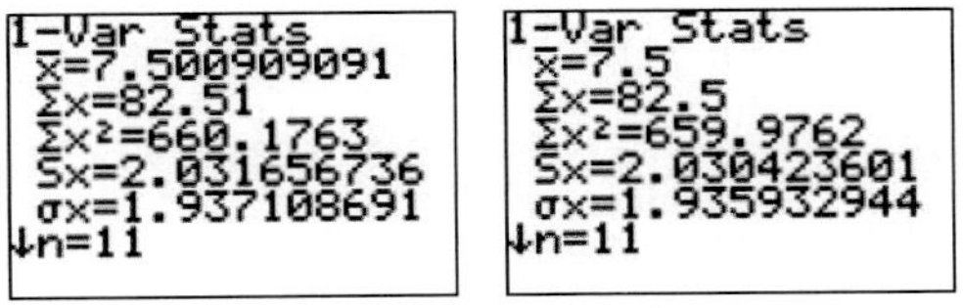

Thus, for each data set we have $\bar{x} \approx 7.50$ and $s \approx 2.03$.

(b) From the stemplots below, we observe that dataset A has two low potential outliers and dataset B has one high potential outlier. (For these stemplots, the data have been rounded.)

Dataset A:

3	1
4	7
5	
6	1
7	3
8	1178
9	113

Dataset B:

5	47
6	148
7	158
8	28
9	
10	
11	
12	7

43. (a) One possible answer is 1, 1, 1, 1.

(b) 0, 0, 10, 10

(c) Yes. *Any* set of four equal numbers yields the smallest possible value for *s*: 0.
(d) No. Within the 0 to 10 constraint, numbers can't deviate any further from the mean.

45. (a) Figure 5.33d
(b) Figure 5.33b
(c) Figure 5.33c
(d) Figure 5.33a

47. Approximately 68% of the students will receive a grade of C. Approximately $\left(\frac{95-68}{2}\right)\% = \frac{27}{2}\% = 13.5\%$ of students will receive a grade of B. Thus, $0.135(200) = 27$ students will receive a grade of B.

49. The distribution is left-skewed, so the mean is pulled toward the long tail. Therefore, *A* is the mean and *B* is the median, as shown in the diagram.

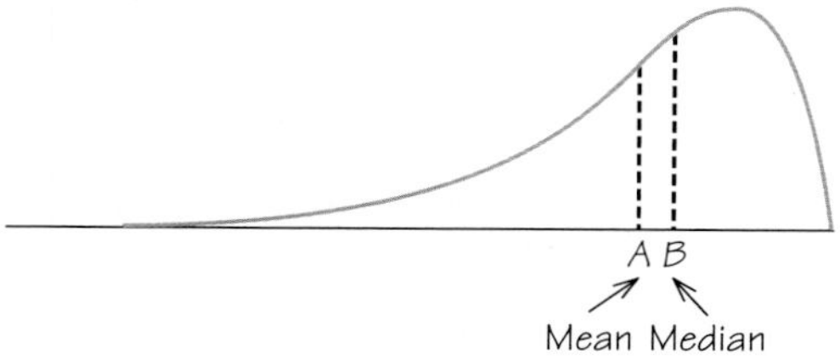

51. 24.1%; 36.25 (about 1 standard deviation above the mean); −11.95 (about 1 standard deviation below the mean)

53. (a) $\mu \pm 3\sigma = 336 \pm 3(3) = 336 \pm 9$, or 327 to 345 days
(b) 16% lie above 339.

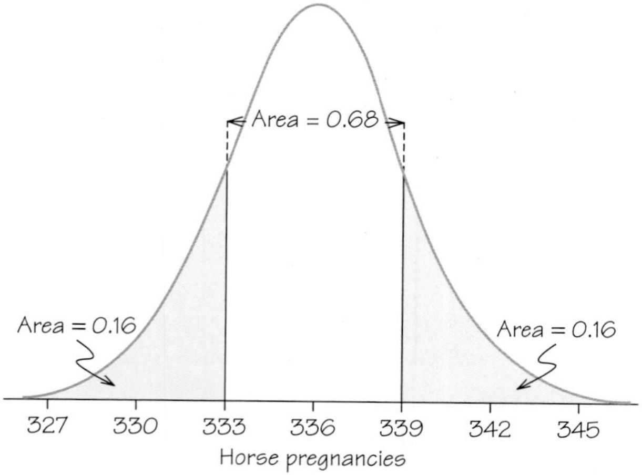

55. The quartiles are $\mu \pm 0.67\sigma = 516 \pm 0.67(116) \approx 516 \pm 78$, or $Q_1 = 438$ and $Q_3 = 594$.

57. (a) $\mu \pm 2\sigma = 10.98 \pm 2(17.46) = 10.98 \pm 34.92$, or −23.94% to 45.90% (see diagram)
(b) A loss of at least 23.94%

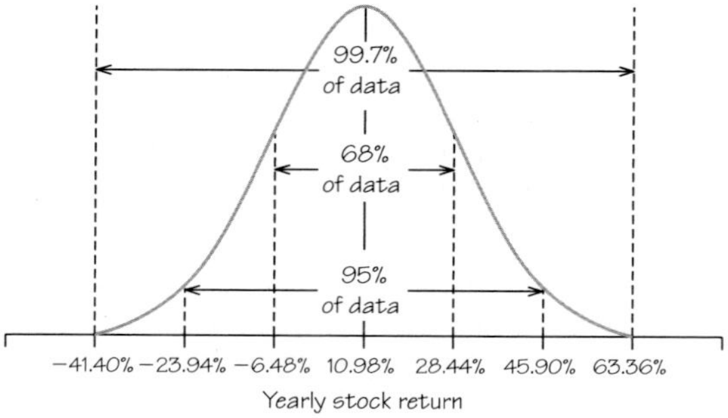

59. (a) Normal curves are symmetric, so median = mean = 10%.
(b) Because 95% of values lie within 2σ of μ, $\mu \pm 2\sigma = 10 \pm 2(0.2) = 10 \pm 0.4$ implies that 9.6% to 10.4% is the interval of concentrations that cover the middle 95% of all the capsules.
(c) The interval between the two quartiles covers the middle half of all capsules. Thus, $\mu \pm 0.67\sigma = 10 \pm 0.67(0.2) = 10 \pm 0.134$ implies that 9.866% to 10.134% is the desired range.

61. (a) Because of the symmetry of the normal curves, 50% give results above 0.4; because 0.43 is 2σ above μ, 2.5% give results above 0.43.
(b) $\mu \pm 2\sigma = 0.4 \pm 2(0.015) = 0.4 \pm 0.03$, or 0.37 to 0.43

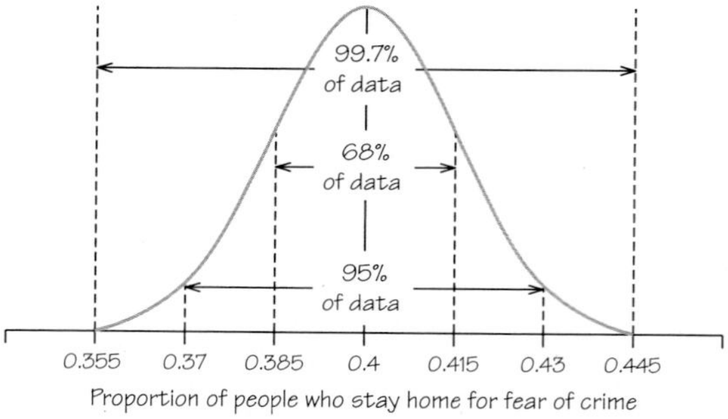

63. (a) *z*-score = (70 − 63.8)/4.2 = 1.48. A woman's height must be 1.48 standard deviations above the mean height for women in order to join the Boston Beanstalks.
(b) *z*-score = (74 − 69.4)/4.7 = 0.98. A man's height must be 0.98 standard deviations above the mean height for men in order to join the Boston Beanstalks.
(c) The height requirements for women are more stringent than for men. Women need to be a half standard deviation further from the mean than their male counterparts.

65. As can be seen from the following stemplots, lengths of red flowers are somewhat right skewed with no outliers; lengths of yellow flowers are reasonably symmetric, also with no outliers. For the stemplots, values are rounded to the nearest tenth.

Red

37	489
38	00112289
39	268
40	67
41	5799
43	02
43	1

Yellow

34	66
35	257
36	0015788
37	01
38	1

67. (a) Red: $\bar{x} = 39.71$, $s = 1.799$; yellow: $x = 36.18$, $s = 0.975$
(b) The mean and standard deviation are better suited to the symmetrical yellow distribution.

69. The top 2.5% of the distribution lies above

$$36.18 + 2(0.975) = 38.13 \text{ millimeters}$$

The top 16% of the distribution lies above

$$36.18 + 1(0.975) = 37.155 \text{ millimeters}$$

The top 25% of the distribution lies above

$$36.18 + 0.67(0.975) = 36.83 \text{ millimeters}$$

The value 37.4 is between 37.155 and 38.13, so between 2.5% and 16% of yellow flowers are longer that 37.4 millimeters.

71. If every number in a dataset is increased by 10, then the mode, mean, and median will each increase by 10. The range and standard deviation, however, will not change.

73. (a) Since the uniform density curve forms a rectangle, the area is found by multiplying the length of the rectangle by its height: Area = (1)(1) = 1.
(b) $\mu = 0.5$—that's the balance point for the region under the density function.
(c) Area under the density curve over this interval is $(0.8 - 0.2)(1) = 0.6$. (Since it is rectangular in shape, just multiply width times height.)
(d) 50%

Chapter 6

1. (a) Latter case; study time
(b) Relationship only
(c) Latter case; rainfall
(d) Relationship only

3. (a) Life expectancy increases with GDP in a curved pattern. The increase is very rapid at first, but it levels off for GDP above roughly $5000 per person.
(b) Sample response: As countries go from very poor to moderate economies, improvements in quality/quantity of food, housing, and medical care have a great impact on increasing life expectancy. As countries move from moderate economies to rich economies, such improvements still improve life expectancy but at a lower rate.

5. The scatterplot below shows a fairly strong, positive, straight-line relationship.

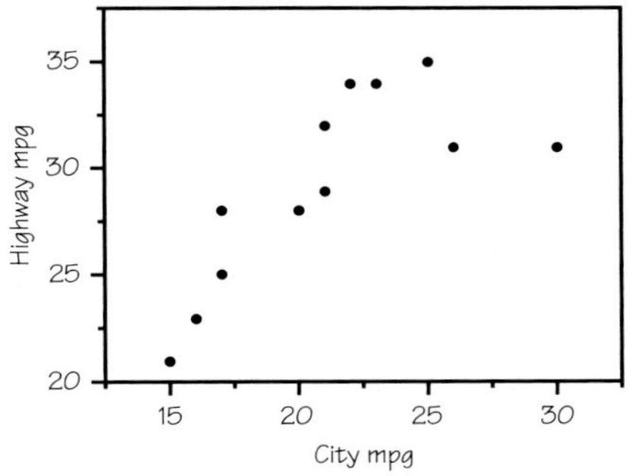

7. (a) Speed
(b)

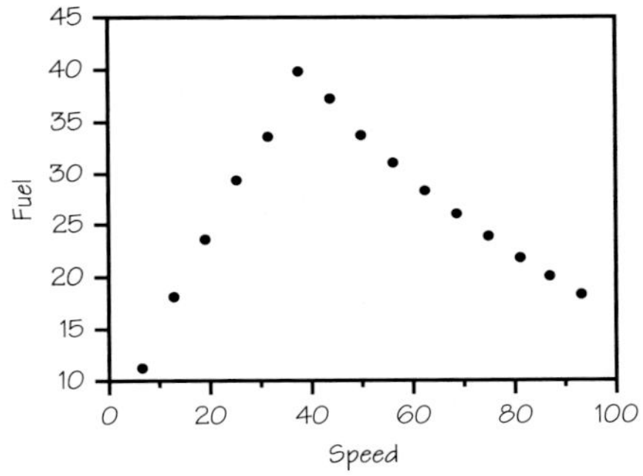

Sample response: Initially, the form is linear, but then it switches direction and bends upward.
(c) Relationship is positive for speeds under 40 mph; relationship is negative for speeds above 40 mph.
(d) Quite strong; little scatter about the piecewise linear then curved pattern

9. (a) Ground temperature
(b) Positive association

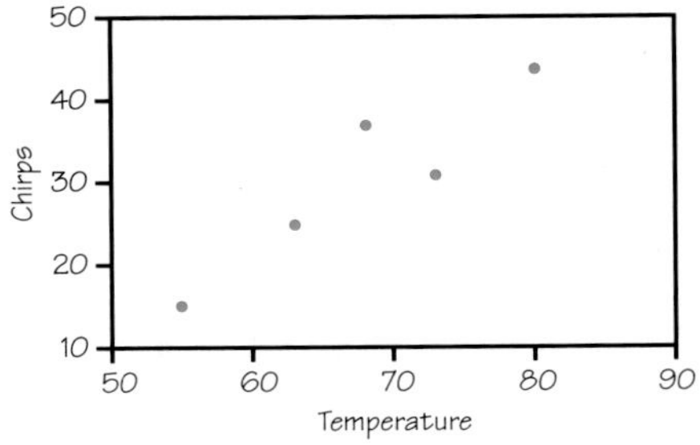

(c) Strong, straight-line pattern
(d) Yes

11. (a) Fish length is the explanatory variable; mercury level is the response variable.

(b)

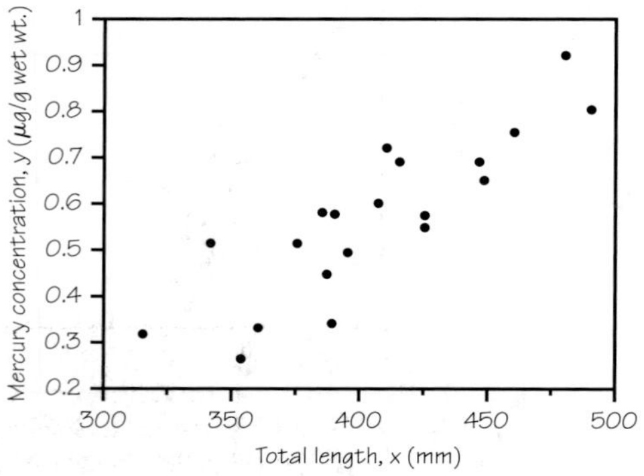

(c) Increase; positive association
(d) Linear; the dots appear to form a pattern about a straight line.
(e) The relationship between mercury in fish tissue may be different for fish that are below the edible/legal size. In terms of health, only the fish that are of edible size pose a health risk if their mercury levels are too high.

13. 2.31

15. (a) The pH decreases as the number of weeks increases.

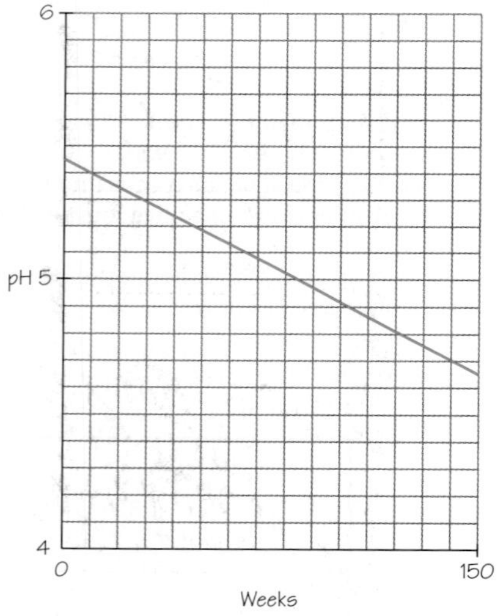

(b) 5.42; 4.64 (both rounded to two decimal places)
(c) −0.0053; on average, pH declined by 0.0053 per week during the study period.

17. (a) predicted BAC = 0.15 − 0.015 × number of hours after drinking stopped
(b) $4\frac{2}{3} \approx 4.67$ hours
(c) 10

19. $r = 0.7514$, reflecting the moderate, positive, straight-line pattern in the Exercise 5 scatterplot.

21. $r = 0.9262$; correlation has increased; the point is an outlier that pulls the line toward it.

23. $r = -0.0435$; the relationship is strong because the points form a pattern that shows little scatter, but the pattern is curved, not linear.

25. $r = 1$; the data would fall exactly on a straight line.

27. (a) Negative
(b) Negative
(c) Positive
(d) Small

29. (a) Dividend Growth: 0.98; Small Cap Stock: 0.81; Emerging Markets: 0.35
(b) No, just that they moved in the same direction

31. (a) Correlation is positive because the association between fish length and mercury concentration in fish tissue is positive.
(b) $r \approx 0.852$; there is a strong linear relationship between these two variables.

33. (a) One way to graph the least-squares regression line is to mark its y-intercept and then use the rise/run from the slope, $m = 0.75 = \frac{3}{4}$, to mark a second point and draw the line.

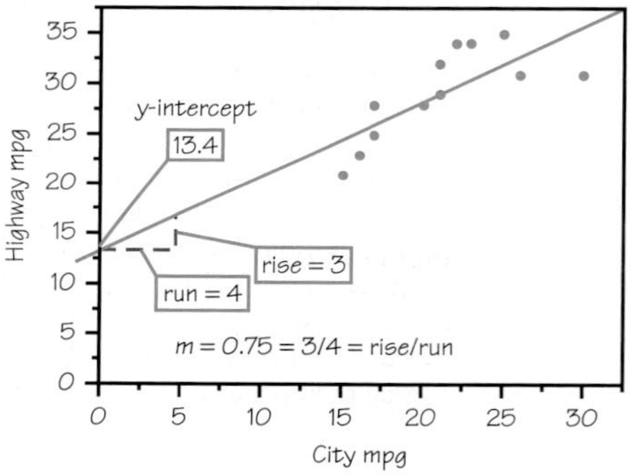

(b) Approximately 27 mpg, as shown below

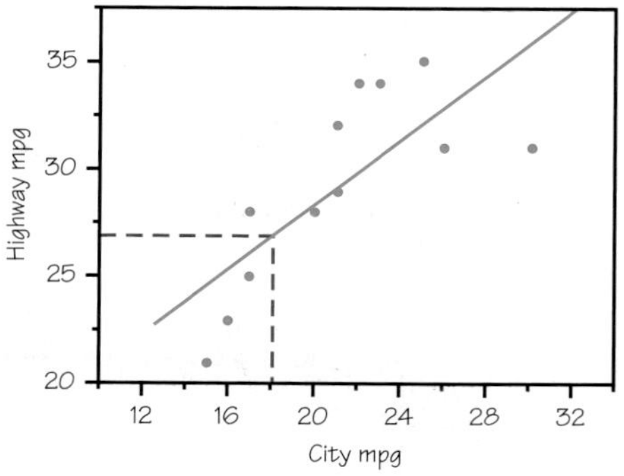

(c) 26.9 mpg
(d) Sample response: If we look at the two data points associated with 17 City mpg (the closest in City mpg to what we are trying to predict), we see that one car gets 25 City mpg and another gets 28. The corresponding residuals are −1.18 and 1.82,

respectively. So, in our prediction, we might expect to be off by as much as 1.82 mpg.

35. (a) 11.2, 37.3, and 18.3, respectively
(b) 26.96, 26.49, and 25.87, respectively
(c)

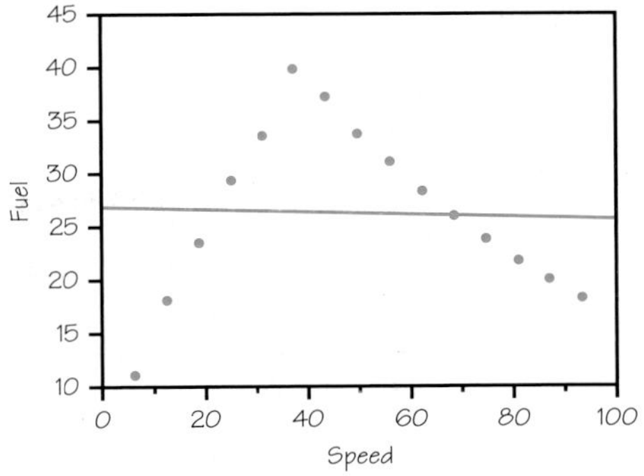

(d) No; the least-squares line gives the best straight-line fit and these data do not show a straight-line pattern.

37. (a) Predicted height = 21.48 + 0.3265 × 422 = 159.263; residual = 160 − 159.263 = 0.737. Since the residual is positive, Person 2's data point lies above the regression line.
(b) Predicted height = 21.48 + 0.3265 × 459 = 171.3435; residual = 170 − 171.3435 = −1.3435. Since the residual is negative, Person 4's data point lies below the regression line.

39. (a) Predicted height of husband = 33.67 + 0.54 × (height of woman)
(b) 69.85 in.

41. The predicted y corresponding to $x = \bar{x}$ is $\hat{y} = m\bar{x} + b$. Now substitute $b = \bar{y} - \left(r\frac{s_y}{s_x}\right)\bar{x}$ and $m = r\frac{s_y}{s_x}$ into this equation to get $\hat{y} = \bar{y}$.

43. Answers will vary. Below are plots that would be useful for the analyses.

One-variable analysis:

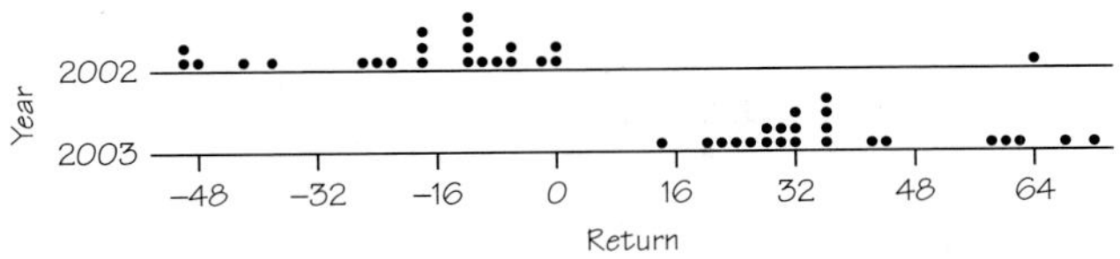

If the outlier for the 2002 returns is removed, the mean and standard deviation become −19.68 and 16.06, respectively.

Two-variable analysis:

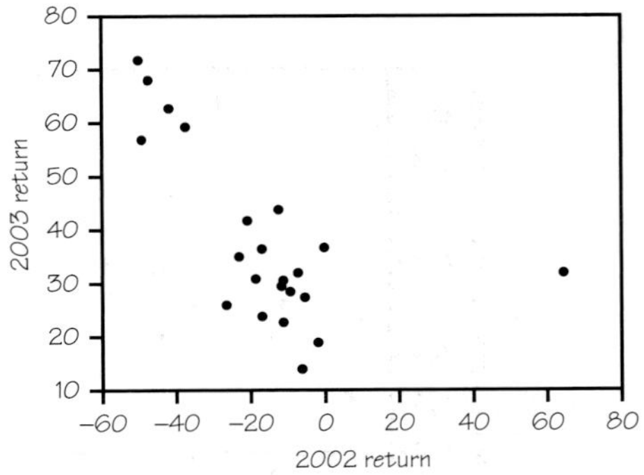

The equation of the least-squares regression line is predicted 2003 return = 31.12 − 0.413 × 2002 return. After removing the outlier, the equation becomes predicted 2003 return = 21.46 − 0.840 × 2002 return. This regression line appears to do a better job of summarizing the pattern in the data.

45. (a) All four have $r \approx 0.816$ and $\hat{y} = 3.0 + 0.5x$; $y = 8.0$.
(b)

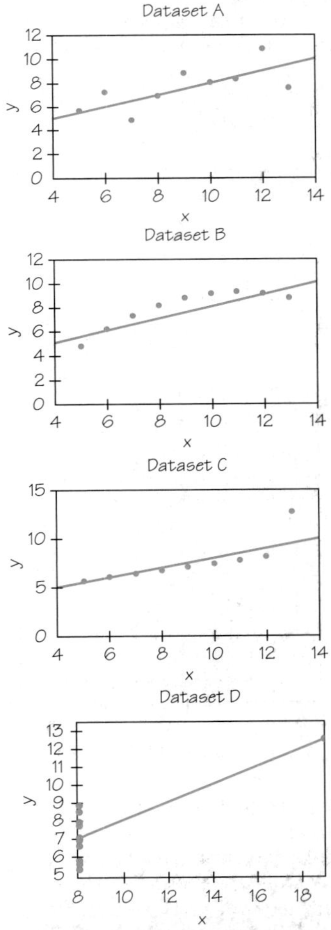

(c) Dataset A; additional answers will vary.

Variable Return	Year	Mean	StDev	Minimum	Q_1	Median	Q_3	Maximum
	2002	−16.03	23.51	−50.50	−26.90	−12.80	−6.70	64.30
	2003	37.74	15.78	14.10	27.50	32.30	43.90	71.90

47. Sample response: People who consume products with artificial sweeteners get hooked on the taste of sweet foods and thus end up consuming more sweet foods than those who do not consume products with artificial sweeteners.

49. Sample response: Higher income means that more money can be spent on health care, which leads to overall better health. On the other hand, better health means workers can be more productive (because they feel good) and will take fewer days off due to illness. The higher productivity and higher number of work days will lead to a higher national income.

51. (a) $\bar{x} = 6, \bar{y} = 9.5$

$$s_x = \sqrt{\frac{(4-6)^2 + (8-6)^2}{2-1}} = \sqrt{8} \approx 2.828,$$

$$s_y = \sqrt{\frac{(7-9.5)^2 + (12-9.5)^2}{2-1}} = \sqrt{12.5} \approx 3.536$$

(b) $$r = \frac{1}{2-1}\left[\left(\frac{4-6}{2.828}\right)\left(\frac{7-9.5}{3.536}\right) + \left(\frac{8-6}{2.828}\right)\left(\frac{12-9.5}{3.536}\right)\right] \approx 1.00$$

(c) $m = r \cdot \frac{s_y}{s_x} = 1 \cdot \frac{3.536}{2.828} \approx 1.25, b = \bar{y} - 1.25 \cdot \bar{x} = 9.5 - 1.25 \cdot 6 = 2$, equation $y = 1.25x + 2$

(d) No. You can find the equation using basic algebra. Using the slope formula, we get $m = \frac{12-7}{8-4} = \frac{5}{4} = 1.25$. The equation of the line will have the form $y = 1.25x + b$. Substitute one of the two points into this equation and solve for b: $7 = 1.25 \times 4 + b$, which gives $b = 2$. The equation is $y = 1.25x + 2$.

53. (a) Lead level is the explanatory variable; reading score is the response variable.

(b) Negative; sample response: Due to lead being toxic, expect that increases in lead levels will affect children's brains and impede reading; yes, the scatterplot supports this answer.

55. (a) Positive; not close to ± 1

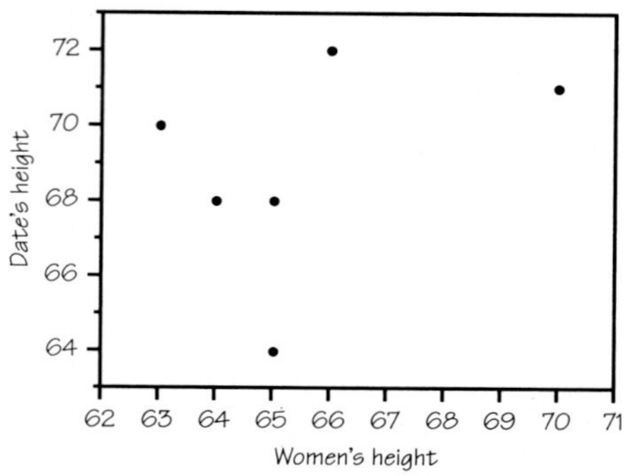

(b) $r = 0.3602$

57. (a) 0.42

(b) For each additional inch of a woman's height, the height of the next person dated goes up by 0.42 in., on average.

(c) 69.22 in.

59. (a) The 100 points lie on a line with slope $m = 1$. Because the points lie on a line with positive slope, $r = 1$.

(b) Here is a scatterplot of the data with the additional data point. (*Note:* Each dot in the lower right represents 25 data points.) Guesses for the slope and correlation will vary. However, the outlier will pull the line toward it, possibly turning the slope from positive to negative. If that happens, then the correlation will also be negative.

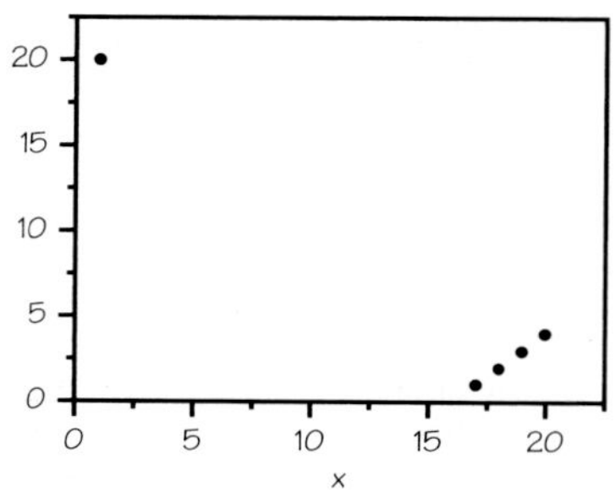

To find the value of r, we first find s_x and s_y. (See Chapter 5, page 204, for the standard deviation formula.) First, we will need $\bar{x}$ and $\bar{y}$:

$$\bar{x} = \frac{1851}{101} \text{ and } \bar{y} = \frac{270}{101}$$

Next, we find the squared deviations from the mean and then the standard deviations.

Times observed	Observations x_i	Deviations $x_i - \bar{x}$	Squared deviations $(x_i - \bar{x})^2$
25	17	$17 - \frac{1851}{101} = -\frac{134}{101}$	$\frac{17{,}956}{101^2}$
25	18	$18 - \frac{1851}{101} = -\frac{33}{101}$	$\frac{1089}{101^2}$
25	19	$19 - \frac{1851}{101} = \frac{68}{101}$	$\frac{4624}{101^2}$
25	20	$20 - \frac{1851}{101} = \frac{169}{101}$	$\frac{28{,}561}{101^2}$
1	1	$1 - \frac{1851}{101} = -\frac{1750}{101}$	$\frac{3{,}062{,}506}{101^2}$

$$s_x^2 = \frac{\frac{17{,}956}{101^2} \cdot 25 + \frac{1089}{101^2} \cdot 25 + \frac{4624}{101^2} \cdot 25 + \frac{28{,}561}{101^2} \cdot 25 + \frac{3{,}062{,}506}{101^2}}{101 - 1}$$

$$= \frac{\frac{4{,}368{,}250}{101^2}}{100} = \frac{4{,}368{,}250}{101^2 \cdot 100}$$

$$s_x = \sqrt{\frac{4{,}368{,}250}{101^2 \cdot 100}} = \frac{\sqrt{4{,}368{,}250}}{101 \cdot 10} = \frac{5\sqrt{174{,}730}}{101 \cdot 10}$$

$$= \frac{\sqrt{174{,}730}}{101 \cdot 2} \approx 2.07$$

Repeating the process for s_y, we find that $s_y = s_x$.

Since $r = \frac{1}{n-1}\left[\left(\frac{x_1-\bar{x}}{s_x}\right)\left(\frac{y_1-\bar{y}}{s_y}\right)+\left(\frac{x_2-\bar{x}}{s_x}\right)\left(\frac{y_2-\bar{y}}{s_y}\right)\right.$

$\left.+\cdots+\left(\frac{x_n-\bar{x}}{s_x}\right)\left(\frac{y_n-\bar{y}}{s_y}\right)\right]$, we have the following:

$$r = \frac{1}{101-1}\cdot\frac{1}{s_xs_y}\left[(x_1-\bar{x})(y_1-\bar{y})+(x_2-\bar{x})(y_2-\bar{y})\right.$$
$$\left.+\cdots+(x_n-\bar{x})(y_n-\bar{y})\right]$$

$$r = \frac{1}{100}\cdot\frac{1}{\sqrt{\frac{174{,}730}{101\cdot 2}}\cdot\sqrt{\frac{174{,}730}{101\cdot 2}}}\left[(x_1-\bar{x})(y_1-\bar{y})\right.$$
$$\left.+(x_2-\bar{x})(y_2-\bar{y})+\cdots+(x_n-\bar{x})(y_n-\bar{y})\right]$$

$$=\frac{101^2}{174{,}730\cdot 25}\left[-\frac{134}{101}\cdot\left(-\frac{169}{101}\right)\cdot 25+-\frac{33}{101}\cdot\left(-\frac{68}{101}\right)\cdot 25\right.$$
$$\left.+\frac{68}{101}\cdot\frac{33}{101}\cdot 25+\frac{169}{101}\cdot\frac{134}{101}\cdot 25+\left(-\frac{1750}{101}\right)\cdot\frac{1750}{101}\right]$$

$$=\frac{101^2}{174{,}730\cdot 25\cdot 101^2}\left[-134\cdot -169\cdot 25+-33\cdot -68\cdot 25\right.$$
$$\left.+68\cdot 33\cdot 25+169\cdot 134\cdot 25+(-1750)(1750)\right]$$

$$=\frac{1}{174{,}730\cdot 25}\left[-1{,}818{,}000\right]=-\frac{72}{173}\approx 0.4162$$

Next, we determine the slope using the formula $m = r\frac{s_y}{s_x}$.

However, since $s_x = s_y$ we know $m = r$. Thus, $m = -\frac{72}{173} \approx -0.4162$.

Chapter 7

1. (a) U.S. adults
   (b) The 1021 who responded
3. (a) The population would be all the Hudson Valley Patch Facebook readers or it could be all residents of the Hudson Valley region in New York State. [If the latter, the sample in part (b) will miss all of the non-Facebook readers in Hudson Valley.]
   (b) The sample would be the readers who went to the Facebook page and voted for the worst Valentine's Day gift.
   (c) No. First, not all Hudson Valley residents are on Facebook with the Hudson Valley Patch. In particular, the votes do not represent the opinions of non-Facebook users.
5. Population: "constituents," probably voters living in her district. Sample: the 361 who wrote letters. Those who wrote probably feel strongly about gun control and may not represent all constituents (voluntary response).
7. Sample response: (a) Print a coupon in the campus newspaper asking students to check their opinion, cut out the coupon, and mail it in.
   (b) Ask all the students in a large sociology course to record their opinion before giving an exam in the course. (This is a convenience sample.)
9. Selected labels: 13, 15, 05, and 09. These labels correspond to the names Sophia, Grace, Hannah, and Elizabeth.
11. (a) 001 to 371
    (b) Area codes labeled 214, 235, 119
13. If you always start at the same point in the table, your sample is predictable in advance. Repeated samples of the same size from the same population will always be the same—that's not random.
15. Sample response for TI-84: First, assign labels to the names. The first column on names (Max through Cooper) are assigned labels 1–5; the second column, labels 6–10; the third column, labels 11–15; the fourth column, labels 16–20. Then use a TI-84 calculator and the randIntNoRep command to generate five randomly chosen numbers from the integers 1 to 20 with no repeats. The randomly chosen labels were 8, 7, 16, 15, 17. So, the puppies' names will be Jack, Bear, Teddy, Milo, and Jax.
17. (a) Because $\frac{200}{5} = 40$, we divide the list into 5 groups of 40. (By the way, if the list has 204 rooms, we divide it into 5 groups of 40 and a final group of 4. A sample contains a room from the final group only when the first room chosen is among the first 4 in the list.) Label the first 40 rooms 01 to 40. Line 120 chooses room 35. The sample consists of rooms 35, 75, 115, 155, and 195.
    (b) Each of the first 40 rooms has chance 1 in 40 of being chosen. Each later room is chosen exactly when the corresponding room in the first 40 is chosen. Thus, every room has an equal chance: 1 in 40.
    (c) The only possible samples consist of 5 rooms spaced 40 apart in the list. An SRS gives *all* samples of 5 rooms an equal chance to be chosen.
19. (a) All people aged 18 and over living in the United States.
    (b) Of the 1334 called, 403 did not respond. The rate is $\frac{403}{1334} \approx 0.30$, or about 30%.
    (c) It is hard to remember exactly how many movies you saw in the past 12 months.
21. Sample response: (a) Yes. The only way a student can answer "no" is if the coin landed on tails, after which the student would have to answer the question honestly.
    (b) Even though a student has *not* cheated, if the result of the coin flip is heads, then the student must answer "yes." So the statement "It is true that all students who have not cheated on an exam in high school or in college answered 'no'" would be a false statement.
    (c) About 16 students have not cheated on an exam in high school or in college.
    (d) $\frac{34}{50} = 0.68$, or 68%
    (e) No. There is no way of telling which of the "yes" answers are true.
23. Sample response: People are more reluctant to "change" the Constitution than to "add to" it. So the wording "adding to" will produce a higher percentage in favor.

25. (a) Journal response type is the explanatory variable.
(b) Personal well-being is the response variable.
(c) We have three groups, where each was given a random assignment.
(d) Gratitude causes well-being.

27. The design resembles Figure 7.3 (page 307). Be sure to show randomization, two groups and their treatments, and the response variable (change in obesity).

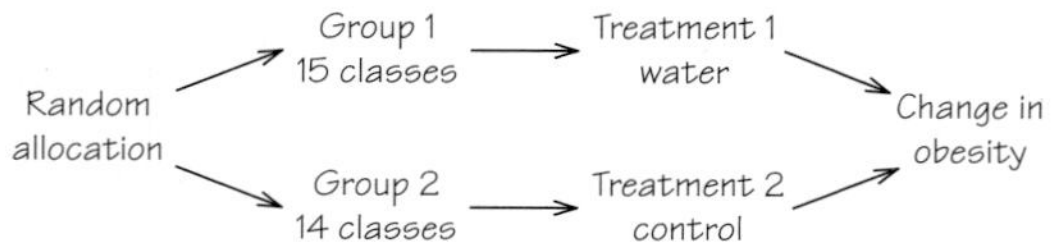

If we label the 29 classes 01 to 29 and choose 15 for the treatment group, this group contains classes 17, 09, 22, 13, 07, 02, 27, 01, 18, 25, 29, 19, 14, 15, and 08. We used lines 103 to 106 of Table 7.1 (page 298), skipping any duplicate pairs of digits. In order, these classes are 1, 2, 7, 8, 9, 13, 14, 15, 17, 18, 19, 22, 25, 27, and 29. The remaining 14 classes make up the control group.

29. (a) Sample response: The type of instruction that each teacher would use was determined not by chance, but rather by the fact that only one of the teachers had computers in her classroom. Furthermore, this was the classroom of the more experienced teacher. Confounding variables are teacher experience, class size, classroom setup (computers versus no computers), and the fact that only one of the teachers drafted the test (which could be biased toward the animated science lessons).
(b) The design resembles Figure 7.3.

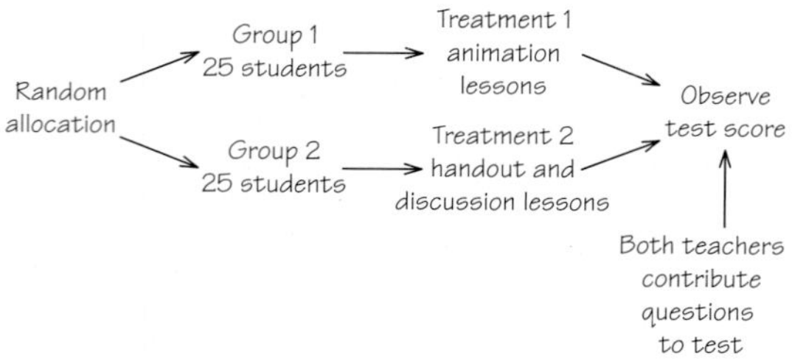

To choose the students, each is given a label, 1–50. Using either Excel and Rand() or the TI-84 calculator and randIntNoRep, a random sample of size 25 is chosen. Here is one possible sample with labels written in order from smallest to largest: 1, 3, 6, 9, 11, 12, 13, 14, 16, 17, 18, 19, 24, 30, 32, 33, 34, 35, 36, 38, 40, 41, 42, 43, 48.

31. This is a randomized comparative experiment with four branches, similar to Figure 7.3 but with four groups. The "flowchart" outline must show random assignment of subjects to groups, the four treatments, and the response variable (healthcare spending).

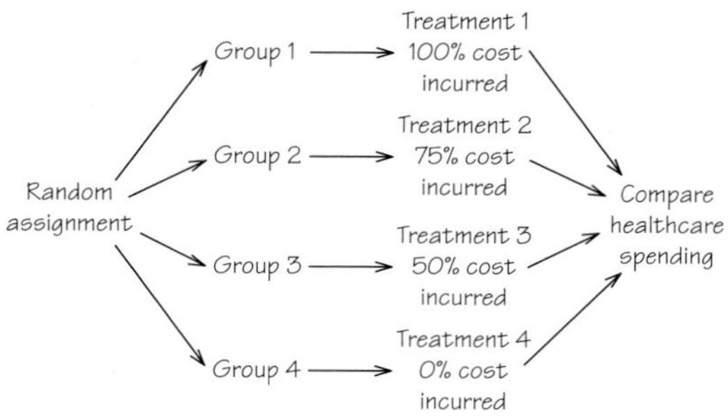

We can't show the group sizes because we don't know how many people or households are available to participate.

33. (a) There are six treatments, each combination of a level of discount and fraction on sale. In table form, the treatments are as follows:

	Discount Level		
	20%	**40%**	**60%**
50% on sale	1	2	3
100% on sale	4	5	6

The six treatments are as follows (enumerated for ease of reading):

Treatment 1: 20% off shoes; 50% of shoes on sale
Treatment 2: 40% off shoes; 50% of shoes on sale
Treatment 3: 60% off shoes; 50% of shoes on sale
Treatment 4: 20% off shoes; 100% of shoes on sale
Treatment 5: 40% off shoes; 100% of shoes on sale
Treatment 6: 60% off shoes; 100% of shoes on sale

(b) The outline randomly assigns 10 students to each of the 6 treatment groups, then compares the attractiveness ratings. It resembles Figure 7.3, but with 6 branches.

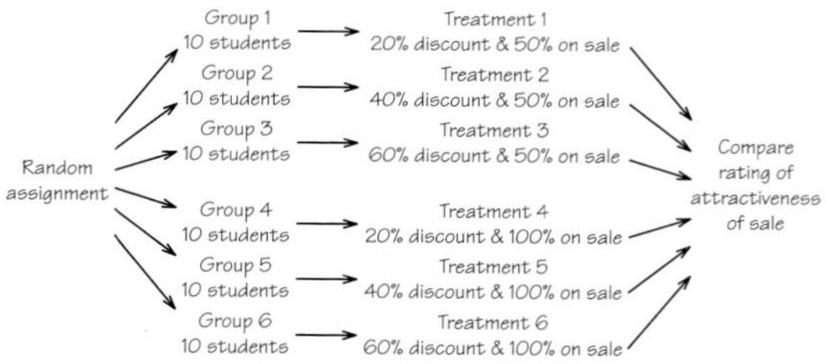

Label the subjects 01 to 60 and read line 123 of Table 7.1 (page 298). The first group contains subjects labeled 54, 58, 08, 15, 07, 27, 10, 25, 60, and 55. In order, these are 7, 8, 10, 15, 25, 27, 54, 55, 58, and 60.

35. (a) The design resembles Figure 7.3.

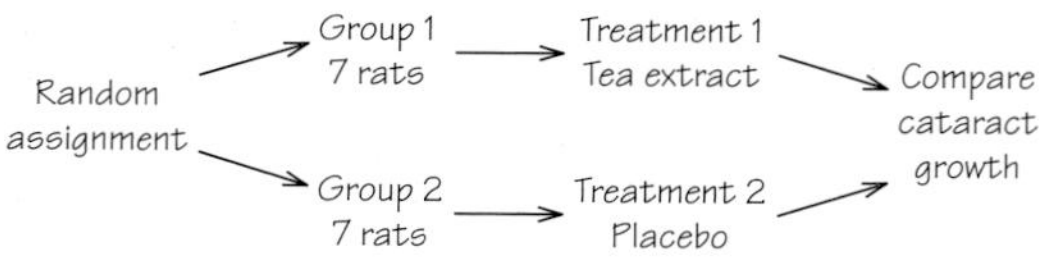

(b) Label the rats 01 to 14. The tea group contains 07, 09, 06, 08, 12, 04, and 11. In order, these are 4, 6, 7, 8, 9, 11, and 12.

37. (a) This is an observational study—a prospective study. It takes a group of people, both smokers and nonsmokers, and observes them over a nine-year period. The response variable is whether or not the subject develops diabetes. The purpose of the study is to describe the response variable (diabetic/not diabetic) for those who were smokers versus nonsmokers at the start of the study, as well as those who were smokers and later quit smoking.
(b) You cannot conclude that quitting smoking causes diabetes. Most people who quit smoking also gain weight. Weight gain is also associated with diabetes. So, it would be impossible to tell whether the diabetes was caused by the cessation of smoking or the gain in weight.

39. No treatment was imposed on the subjects. This observational study collected detailed information about the subjects but made no attempt to influence them. This is a retrospective study since the children were broken into two groups at the start of the study, those with leukemia and those who did not have leukemia, and then it looked for possible links between magnetic fields in the children's environment and the illness.

41. Sample response: (a) It is an observational study that gathers information (e.g., through interviews) without imposing any treatment.
(b) *Significant* means "unlikely to be due simply to chance."
(c) Nondrinkers might be more elderly or in poorer health than moderate drinkers.

43. (a) This is a randomized comparative experiment with four branches. The "flowchart" outline must show random assignment of subjects to groups, the group sizes and treatments, and the response variable (colon cancer). It is best to use groups of equal size, 216 people in each group.

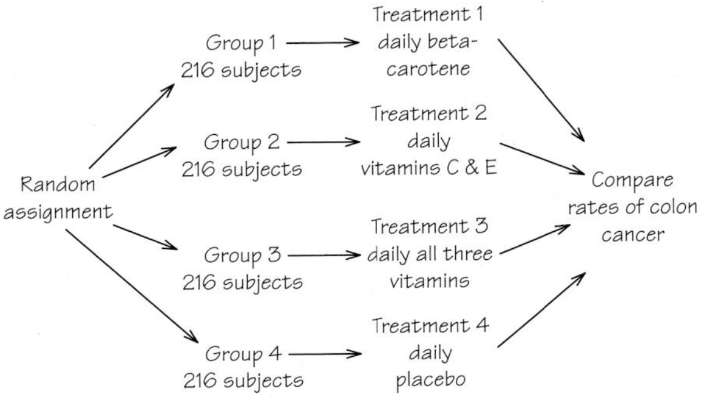

(b) With labels 001 to 864, the first five chosen are 731, 253, 304, 470, and 296. In order, these are 253, 296, 304, 470, and 731.
(c) Neither those working with the subjects nor the subjects know the contents of the pill each subject took daily.
(d) The differences in colon cancer cases in the four groups were so small that they could easily be due to the chance assignment of subjects to groups.
(e) People who eat lots of fruits and vegetables may eat less meat or more cereals than other people. They may drink less alcohol or exercise more.

45. This is an observational study. There was a measurement of information, but no attempt was made to influence the response.

47. Both are statistics because both describe the sample (the subjects who took part in the study).

49. (a) The distribution is approximately normal with mean of $p = 0.14$ and a standard deviation as follows:

$$\sqrt{\frac{p(1-p)}{n}} = \sqrt{\frac{0.14\,(1-0.14)}{500}}$$
$$= \sqrt{\frac{0.14(0.86)}{500}} \approx 0.0155$$

(b) $0.14 \pm (2)(0.0155) = 0.14 \pm 0.031$

$0.14 - 0.031 = \underline{0.109}$ to $0.14 + 0.031 = \underline{0.171}$

51. (a) Each digit in the table has 1 chance in 10 to be any of the 10 possible digits 0, 1, 2, 3, 4, 5, 6, 7, 8, 9. So in the long run, 60% of the digits we encounter will be 0, 1, 2, 3, 4, or 5, and 40% will be 6, 7, 8, or 9.
(b) Line 101 contains 29 digits 0 to 5. This stands for a sample with $\frac{29}{40} = 0.725 = 72.5\%$ "yes" responses. If we use lines 101 to 110 to simulate 10 samples, the counts of "yes" responses are 29, 24, 23, 23, 20, 24, 23, 19, 24, and 18. Thus, three samples are exactly correct ($\frac{24}{40} = 0.60 = 60\%$), one overestimates, and six underestimate.

53. (a) The sample proportion who claim to have attended is $\hat{p} = \frac{750}{1785} \approx 0.420$.
(b) The approximate 95% confidence interval is calculated as follows:

$$\hat{p} \pm 2\sqrt{\frac{\hat{p}(1-\hat{p})}{n}} = 0.420 \pm 2\sqrt{\frac{0.420(1-0.420)}{1785}}$$
$$= 0.420 \pm 2\sqrt{\frac{0.420(0.580)}{1785}} \approx 0.420 \pm 0.023$$

$0.420 - 0.023 = 0.397$ to $0.420 + 0.023 = 0.443$

The interval would be (0.397, 0.443).

55. (a) The sample proportion who admit running a red light is $\hat{p} = \frac{171}{880} \approx 0.194$. The approximate 95% confidence interval is calculated as follows:

$$\hat{p} \pm 2\sqrt{\frac{\hat{p}(1-\hat{p})}{n}} = 0.194 \pm 2\sqrt{\frac{0.194(1-0.194)}{880}}$$
$$= 0.194 \pm 2\sqrt{\frac{0.194(0.806)}{880}} \approx 0.194 \pm 0.027$$

$0.194 - 0.027 = 0.167$ to $0.194 + 0.027 = 0.221$

The interval would be (0.167, 0.221).

(b) It is likely that more than 171 ran a red light, because some people are reluctant to admit illegal or antisocial acts.

57. (a) $\hat{p} = 0.5$

(b) $2\sqrt{\frac{\hat{p}(1-\hat{p})}{n}} = 2\sqrt{\frac{0.5(1-0.5)}{n}} = 2\sqrt{\frac{0.5(0.5)}{n}}$

$= 2(0.5)\sqrt{\frac{1}{n}} = \sqrt{\frac{1}{n}} = \frac{1}{\sqrt{n}}$

59. (a) $\frac{1468}{13{,}000} \approx 0.113 = 11.3\%$

(b) No. The response rate is so low that it is likely that those who responded differ from the population as a whole. That is, there is a bias that the margin of error does not include.

61. (a) No. The number of e-filed returns in all states is much larger than the sample size. When this is true, the margin of error depends only on the size of the sample, not on the size of the population.

(b) Yes. The sample sizes vary from 970 to 49,000, so the margins of error will also vary.

63. The margin of error for 90% confidence comes from the central 90% of a normal sampling distribution. We need not go as far out to cover 90% of the distribution as to cover 95%. So the margin of error for 90% confidence is smaller than for 95% confidence.

65. The sample proportion of successes is $\hat{p} = \frac{7}{97} \approx 0.072$. That is, there were 7.2% successes in the sample. The approximate 95% confidence interval is calculated as follows:

$$\hat{p} \pm 2\sqrt{\frac{\hat{p}(1-\hat{p})}{n}} = 0.072 \pm 2\sqrt{\frac{0.072(1-0.072)}{97}}$$

$$= 0.072 \pm 2\sqrt{\frac{0.072(0.928)}{97}} \approx 0.072 \pm 0.052$$

$$0.072 - 0.052 = 0.020 \text{ to } 0.072 + 0.052 = 0.124$$

We are 95% confident that the true proportion of articles that discuss the success of blinding is between 0.020 and 0.124 (that is, 2.0% to 12.4%).

67. The distribution of the sample proportion $\hat{p}$ is approximately normal with mean $p = 0.1$ (i.e., 10%) and standard deviation

$$\sqrt{\frac{p(1-p)}{n}} = \sqrt{\frac{0.1(1-0.1)}{97}} = \sqrt{\frac{0.1(0.9)}{97}} \approx 0.030$$

or 3%. Notice that 7% is 1 standard deviation below the mean. By the 68 part of the 68–95–99.7 rule, 68% of all samples will have between 7% and 13% that discuss blinding. Half of the remaining 32% of all samples lie on either side. So 16% of samples will have fewer than 7% articles that discuss blinding. That is, the probability is about 0.16.

69. (a) This is an observational study because there was no manipulation of environment or assignment of treatment.

(b) This is a prospective study.

(c) This study could not have been done as an experiment because one cannot manipulate or control the environment throughout the 21+ years.

71. (a) $0.63 \pm 2\sqrt{\frac{(0.63)(1-0.63)}{1000}} \approx 0.63 \pm 0.0305$; from approximately 0.60 to 0.66, or from 60% to 66%. The margin of error, to two decimals, is 0.03 or ±3%. (It was 3.05%, which we rounded to 3%.)

(b) They match if rounded to the nearest whole percent.

(c) Corresponding to 50%: margin of error is $2\sqrt{\frac{(0.5)(0.5)}{1000}} \approx 0.0316$; 3.1%. So, the margin of error would be 3% only if we round to the nearest whole percent.

Corresponding to 80%, margin of error is $2\sqrt{\frac{(0.8)(0.2)}{1000}} \approx 0.0253$, or approximately 2.5%.

(d) Solve $2\sqrt{\frac{(0.5)(0.5)}{n}} = 0.03$ for n; $n = \left(\frac{0.5}{0.015}\right)^2 \approx$ 1111.11. In order to guarantee that the margin of error was less than 3%, a sample size of at least 1112 should be used.

Chapter 8

1. Sample response: The time it takes to get from the dorm to this class each day; the outside temperature when I wake up in the morning; the number of chocolate chips in a cookie from the student center.

3. (a) Results will vary, but the probability of a head is usually greater than 0.5 when spinning pennies. One possible explanation is the "bottle cap effect." The rim on a penny is slightly wider on the head side, so just as spinning bottle caps almost always fall with the open side up, pennies fall more often with the head side up. Additional results will vary.

(b) Sample response: In 54 of the 100 tosses, the tack landed point up. Hence, the estimated probability is 0.54.

5. The first five lines contain 200 digits, of which 21 are 0s. The proportion of 0s is $\frac{21}{200} = 0.105$ (see the table on next page).

7. (a) We will use H for "hit" and M for "miss" in this exercise.

S = {HHHH, HHHM, HHMH, HMHH, MHHH, HHMM, HMMH, MMHH, HMHM, MHHM, MHMH, HMMM, MMMH, MHMM, MMHM, MMMM}

(b) $S = \{0, 1, 2, 3, 4\}$

9. (a) It is usually easier to add further branching to a tree than further dimensions to a table.

Table of Random Digits

101	19223	95034	05756	28713	96409	12531	42544	82853
102	73676	47150	99400	01927	27754	42648	82425	36290
103	45467	71709	77558	00095	32863	29485	82226	90056
104	52711	38889	93074	60227	40011	85848	48767	52573
105	95592	94007	69971	91481	60779	53791	17297	59335

(b)

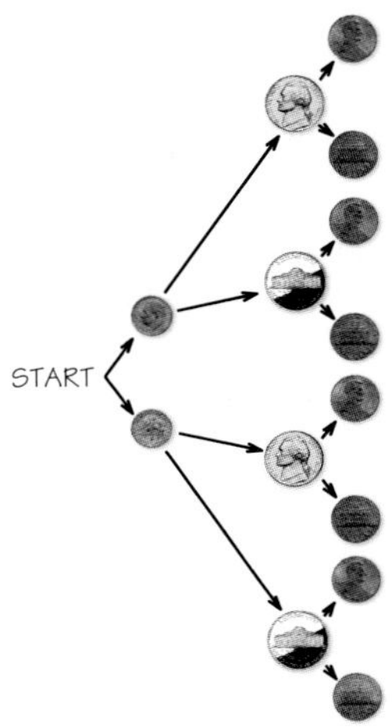

$S =$ {HHH, HHT, HTH, HTT, THH, THT, TTH, TTT}

(c) There are seven out of the eight outcomes that yield at least one of the three coins landing on heads. Thus, the probability of at least one of the three coins landing on heads is $\frac{7}{8}$.

11. (a) Probability 0
(b) Probability 1
(c) Probability 0.01, once per 100 trials on average in the long run
(d) Probability 0.6

13. If the next flip is heads, Player A wins; if tails, the two players will be tied and will have to flip the coin again. Thus, the probability of Player A winning is $\frac{1}{2} + \frac{1}{4} = \frac{3}{4}$. The probability of Player B winning is $\frac{1}{4}$. To be fair, Player A should receive $\frac{3}{4}$ of the jackpot money and Player B should receive $\frac{1}{4}$ of the jackpot money.

15. We know that $P(A^c)/P(A) = 3/2$ and $P(A^c) = 1 - P(A)$. Now, we need to solve the following equation for $P(A)$: $\frac{1 - P(A)}{P(A)} = \frac{3}{2}$. This gives $P(A) = \frac{2}{5} = 0.4$.

17. Mutually exclusive events with positive probabilities cannot be independent. Sample explanation: Suppose A and B are mutually exclusive with $P(A) > 0$ and $P(B) > 0$. Since A and B are mutually exclusive, $P(A \text{ and } B) = 0$; $P(A|B) = P(A \text{ and } B)/P(B) = 0$. But then, $0 = P(A|B) \neq P(A) > 0$, which means that A and B are dependent events.

19. The three events of tossing a pair of dice are independent events. The probability of rolling an even sum is 1/2 (there are 18 out of 36 ways to roll an even sum). Because there are three independent events, the probability that she will roll an even sum three times is $(1/2)(1/2)(1/2) = 1/8 = 0.125$.

21. Let G = Worker 1 lives in a household where number of vehicles exceeds workers in the household, and H = Worker 2 lives in a household where number of vehicles exceeds workers in the household.
(a) $P(G \text{ and } H) = P(G)P(H) = (0.445)(0.445) \approx 0.198$
(b) $P(G^C \text{ and } H^C) = P(G^C)P(H^C) = (1 - 0.445)(1 - 0.445) \approx 0.308$

23. $P(\text{South}) = 0.320$; $P(\text{"\$100,000+" and South}) = 0.079$; $P(\text{\$100,000+} \mid \text{South}) = P(\text{"\$100,000+" and South})/P(\text{South}) = 0.079/0.320 \approx 0.247$. Approximately 24.7% of Southern households earn \$100,000 or more per year.

25. (a) $P(\text{male}) = 0.498$; $P(\text{female}) = 0.502$; $P(\text{above average} \mid \text{female}) = 0.603$; $P(\text{above average} \mid \text{male}) = 0.686$
(b) $P(\text{female and above average}) = P(\text{above average} \mid \text{female})P(\text{female}) = (0.603)(0.502) \approx 0.303$
(c) $P(\text{male and above average}) = P(\text{above average} \mid \text{male})P(\text{male}) = (0.686)(0.498) \approx 0.342$
(d) $P(\text{above average}) = P(\text{female and above average}) + P(\text{male and above average}) \approx 0.303 + 0.342 = 0.645$
(e) $P(\text{female} \mid \text{above average}) = P(\text{female and above average})/P(\text{above average}) = 0.303/0.645 \approx 0.470$
(f) The probability that a randomly selected student is female is 0.502. On the other hand, if you know that the student rated his/her intelligence as above average, then the probability that the student is female decreases to 0.470.

27. (a) The probabilities were all between 0 and 1 inclusive, and the sum of the probabilities is 1. Hence, the probability model is legitimate.
(b) The model is discrete because there are only five possible categories for outcomes.
(c) $0.174 + 0.218 = 0.392$
(d) $1 - 0.283 = 0.717$
(e) $(0.717)(0.717) \approx 0.514$

29. (a) Here is the probability histogram:

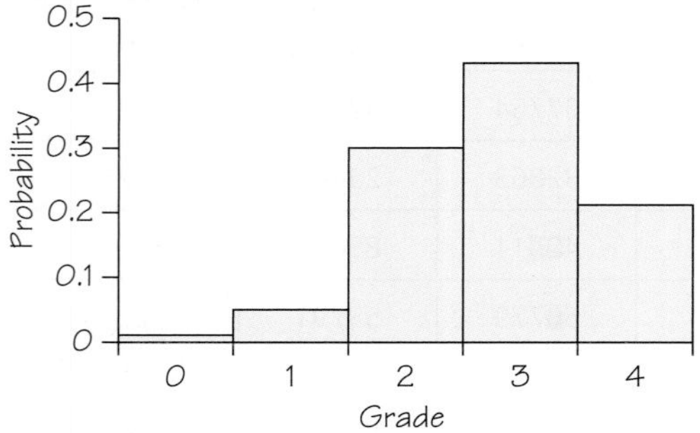

The histogram does not look to be normally distributed.

(b) $0.43 + 0.21 = 0.64$

31. (a) Not legitimate: The probabilities are between 0 and 1, but the sum is greater than 1. Rule 2 is violated.

$$0.56 + 0.24 + 0.44 + 0.17 = 1.41$$

(b) Legitimate: The probabilities are between 0 and 1, inclusively, and have sum 1.

$$0.39 + 0.28 + 0 + 0.33 = 1$$

One may find it surprising that no student surveyed was born in fall, but that outcome is possible.

33. Like tossing a pair of standard dice, the sample space contains sums between 2 and 12 from the 36 possible outcomes of rolling this pair of "weird dice." (See bottom of this page.) The probability for each sum is the same as the pair of standard dice.

Outcome	2	3	4	5	6	7	8	9	10	11	12
Probability	$\frac{1}{36}$	$\frac{1}{18}$	$\frac{1}{12}$	$\frac{1}{9}$	$\frac{5}{36}$	$\frac{1}{6}$	$\frac{5}{36}$	$\frac{1}{9}$	$\frac{1}{12}$	$\frac{1}{18}$	$\frac{1}{36}$

35. All 90 guests are equally likely to get the prize, so $P(\text{woman}) = 48/90 = 8/15$.

37. (a) ${}_{10}P_5 = \dfrac{10!}{5!} = 10 \cdot 9 \cdot 8 \cdot 7 \cdot 6 = 30{,}240$

(b) ${}_{10}C_5 = \dfrac{10!}{5!5!} = \dfrac{10 \cdot 9 \cdot 8 \cdot 7 \cdot 6}{5 \cdot 4 \cdot 3 \cdot 2} = 252$

39. (a) $2 \times 2 \times 2 \times 2 \times 2 \times 2 \times 2 \times 2 \times 2 \times 2 = 2^{10} = 1024$

(b) $\dfrac{2}{1024} = \dfrac{1}{512}$

41. (a) There are $36 \times 36 \times 36 = 36^3 = 46{,}656$ different codes. The probability of no x is as follows:

$$\frac{35 \times 35 \times 35}{46{,}656} = \frac{42{,}875}{46{,}656} \approx 0.919$$

(b) The probability of no digits is as follows:

$$\frac{26 \times 26 \times 26}{46{,}656} = \frac{17{,}576}{46{,}656} = \frac{2197}{5832} \approx 0.377$$

43. (a) The possibilities are *aps, asp, pas, psa, sap, spa.*

(b) *Asp, pas, sap,* and *spa* are English words.

(c) The probability is $\frac{4}{6} = \frac{2}{3} \approx 0.667$. The answer can also be expressed as exactly $66\frac{2}{3}\%$ or approximately 66.7%.

45. (a) There are four possible royal flush hands.

(b) There are ${}_{52}C_5 = \frac{52!}{5!(52-5)!} = \frac{52!}{5!47!} = \frac{52 \cdot 51 \cdot 50 \cdot 49 \cdot 48}{5 \cdot 4 \cdot 3 \cdot 2 \cdot 1} = 2{,}598{,}960$ possible five-card hands.

(c) The probability would be $\frac{4}{2{,}598{,}960} = \frac{1}{649{,}740} \approx 0.00000154$.

47. There are $({}_{20}C_4 \times {}_{10}C_3) = 4{,}845 \times 120 = 581{,}400$ ways to select the 4 inkjet and 3 laser printers. Therefore, $P(A) = 581/400/\ {}_{30}C_7 = 581{,}400/2{,}035{,}800 \approx 0.286$.

49. (a) $\frac{1}{2} \times \text{base} \times \text{height} = \frac{1}{2}(2)(1) = 1$

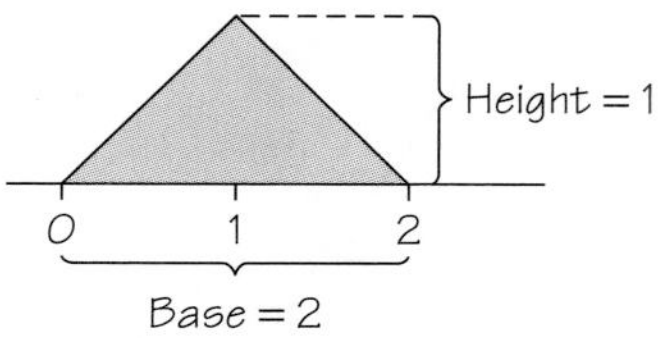

(b) The probability is 1/2 by symmetry or finding the area:

$$\frac{1}{2} \times \text{base} \times \text{height} = \frac{1}{2}(1)(1) = \frac{1}{2}$$

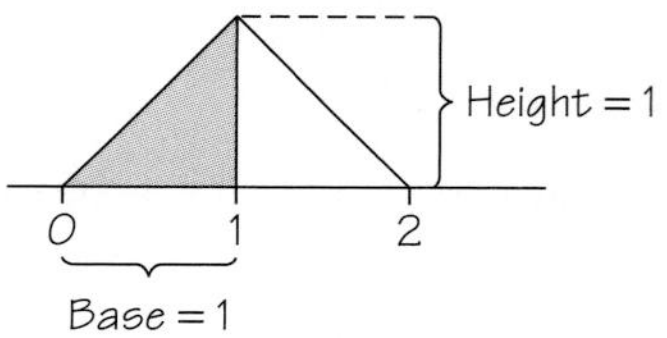

(c) The area representing this event is

$$\left(\frac{1}{2}\right)(0.5)(0.5) = 0.125$$

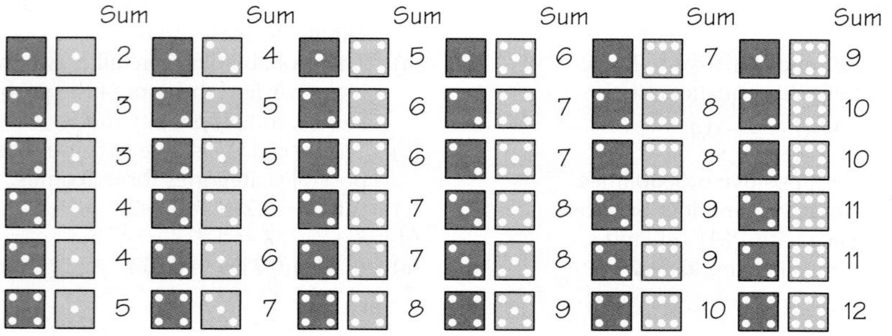

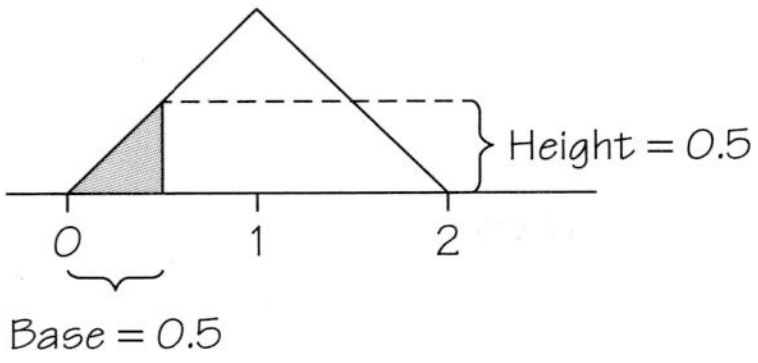

51. The probability would be $\frac{150}{600} = \frac{1}{4} = 0.25$.

53. The mean is as follows:

$$\mu = (0)(0.01) + (1)(0.05) + (2)(0.30) + (3)(0.43) + (4)(0.21) = 2.78$$

The variance is calculated as follows:

$$\sigma^2 = (0-2.78)^2(0.01)+(1-2.78)^2(0.05)+(2-2.78)^2 \times (0.30)+(3-2.78)^2(0.43)+(4-2.78)^2(0.21) = 0.7516;\ \sigma = \sqrt{0.7516} \approx 0.8669.$$

55. The mean for owner-occupied units is

$$\mu = (1)(0.000) + (2)(0.001) + \cdots + (10)(0.047) = 6.248$$

The mean for rented units is

$$\mu = (1)(0.011) + (2)(0.027) + \cdots + (10)(0.005) = 4.321$$

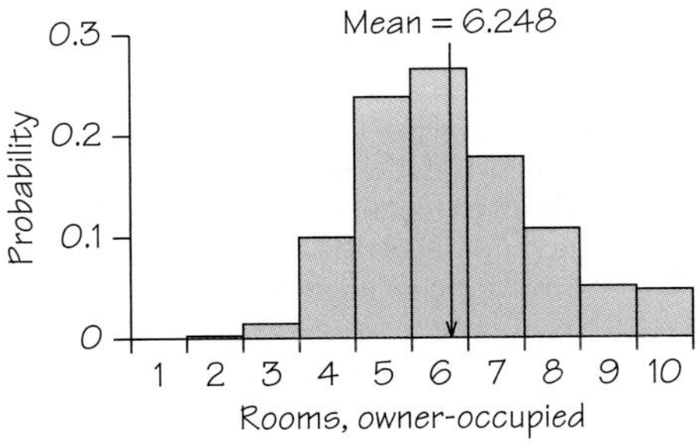

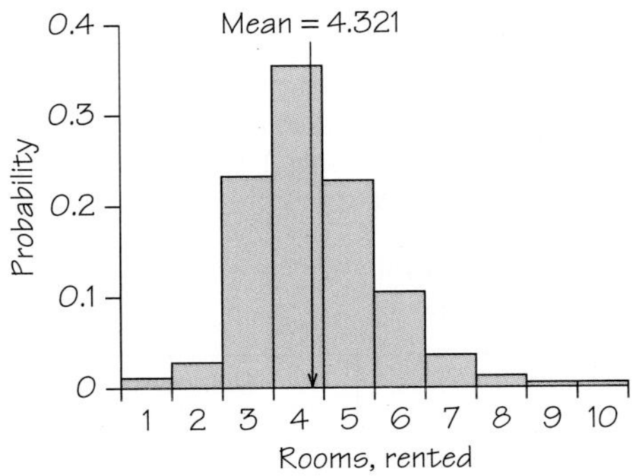

57. (a) Model has mean 1, because density curve is symmetric about 1.
(b) Model has mean 0, because density curve is symmetric about 0.

59. Since $\mu = (0)(0.85) + (200)(0.15) = 0 + 30 = 30$, the breakeven price would be \$30.

61. A net prize would be what you have after expenses (cost to play, taxes, other fees, etc.). In this exercise, prizes are given as "net" prizes, so it is understood that they include the cost to play the game. The probability of receiving a prize is as follows:

$$\frac{1}{10{,}000{,}000} + \frac{1}{10{,}000} + \frac{1}{1{,}000} + \frac{3}{100}$$
$$= \frac{1}{10{,}000{,}000} + \frac{1000}{10{,}000{,}000} + \frac{10{,}000}{10{,}000{,}000} + \frac{300{,}000}{10{,}000{,}000} = \frac{311{,}001}{10{,}000{,}000}$$

Thus, the probability of not receiving a prize is

$$1 - \frac{311{,}001}{10{,}000{,}000} = \frac{9{,}688{,}999}{10{,}000{,}000}$$

We now have

$$\mu = (1{,}000{,}000)\left(\frac{1}{10{,}000{,}000}\right) + (1000)\frac{1}{10{,}000} + (100)\frac{1}{1000} + (4)\frac{3}{100} + (-1)\frac{9{,}688{,}999}{10{,}000{,}000} \approx -0.55$$

Approximately \$0.45 "comes back" in prizes on the \$1 ticket.

63. (a) The expected value on a single question is $\left(\frac{1}{5}\right)(1) + \left(\frac{4}{5}\right)\left(-\frac{1}{4}\right) = \left(\frac{1}{5}\right) - \left(\frac{1}{5}\right) = 0$. Thus, random guessing makes no difference.
(b) The expected value on a single question is now $\left(\frac{1}{4}\right)(1) + \left(\frac{3}{4}\right)\left(-\frac{1}{4}\right) = \frac{1}{4} - \frac{3}{16} = \frac{4}{16} - \frac{3}{16} = \frac{1}{16}$. Thus, guessing will increase your score on average.

65. 95% of all samples have an $\bar{x}$ between $0.15 - 2(0.02) = 0.15 - 0.04 = 0.11$ and $0.15 + 2(0.02) = 0.15 + 0.04 = 0.19$. The interval is (0.11, 0.19).

67. (a) The standard deviation of the average measurement is $\frac{\sigma}{\sqrt{n}} = \frac{10}{\sqrt{3}} \approx 5.77$ mg
(b) To cut the standard deviation in half (from 10 mg to 5 mg), we need $n = 4$ measurements because $\frac{\sigma}{\sqrt{n}}$ is then $\frac{\sigma}{\sqrt{4}} = \frac{\sigma}{2}$. Averages of several measurements are less variable than individual measurements, so an average is more likely to give about the same result each time.

69. (a)

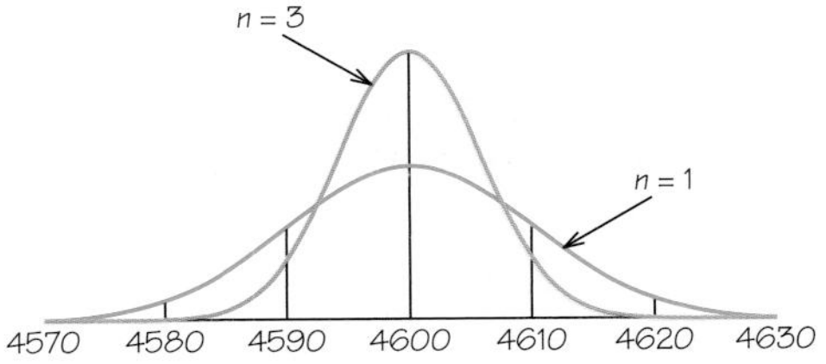

(b) Use the 95 part of the 68–95–99.7 rule with $\sigma = 10$: $4600 \pm 2(10)$ results in the interval (4580, 4620).
(c) Now the standard deviation is 5.77, so we have the following: $4600 \pm 2(5.77)$ results in the interval (4588.46, 4611.54).

71. (a) Because 26.2 is 1 standard deviation above the mean, the probability is about 0.16.
(b) The mean remains $\mu = 21.0$. The standard deviation is $\frac{\sigma}{\sqrt{9}} = \frac{5.2}{3} = \frac{26}{15} \approx 1.7$.

(c) Because $26.2 = 21.0 + 5.2 = 21.0 + 3\left(\frac{26}{15}\right)$ is 3 standard deviations above the mean, the probability is about 0.0015. (This is half of the 0.003 probability for outcomes more than 3 standard deviations from the mean, using the 99.7 part of the 68–95–99.7 rule.)

73. high; since the standard deviation for a sampling distribution is $\frac{\sigma}{\sqrt{n}}$, smaller schools will have a smaller value for n. This will result in more variable means and therefore will be overrepresented in the list of very highest and very lowest performers.

75. (a) There are $26 \times 10 \times 10 \times 26 \times 26 \times 26 =$ 45,697,600 different license plates of this form.
(b) There are $26 \times 10 \times 10 = 2600$ plates ending in AAA because that leaves only the first three characters free.
(c) The probability is $\frac{2600}{45{,}697{,}600} \approx 0.0000569$.

77. (a) The probability is $0.10 + 0.08 = 0.18$.
(b) The complement to the event of working out at least one day is working out no days. Thus, using the complement rule, the desired probability is $1 - 0.61 = 0.39$.

79. (a) The variance is calculated as follows:

$$\sigma^2 = (0-0.77)^2\,(0.61) + (1-0.77)^2\,(0.17) + (2-0.77)^2\,(0.10) + (3-0.77)^2\,(0.08) + (4-0.77)^2\,(0.04) = 1.3371$$

Thus, the standard deviation is $\sigma = \sqrt{1.3371} \approx$ 1.156 days.
(b) The sampling distribution of $\bar{x}$ is approximately normal with $\mu = 0.77$ and standard deviation $\frac{\sigma}{\sqrt{n}} = \frac{1.156}{\sqrt{100}} \approx 0.1156$. Applying the 68–95–99.7 rule: with probability 0.95, values of $\bar{x}$ lie in the interval (0.54, 1.00).

81. (a) Because there are only 7 days in the week and 10 people are chosen at random, the probability that there is a match is 100% (certainty).
(b) Assuming 31 days in a month, the probability we seek is

$$1-\left(\frac{30}{31}\right)\times\left(\frac{29}{31}\right)\times\cdots\times\left(\frac{22}{31}\right) = 1-\frac{{}_{31}P_{10}}{31^{10}}$$
$$\approx 1-0.196 = 0.804$$

or approximately 80%.
(c) Assuming 365 days in a year, the probability we seek is

$$1-\left(\frac{364}{365}\right)\times\left(\frac{363}{365}\right)\times\cdots\times\left(\frac{355}{365}\right) = 1-\frac{{}_{365}P_{10}}{365^{10}}$$
$$\approx 1-0.883 = 0.117$$

or approximately 12%.

Chapter 9

1. Answers will vary.
3. Answers will vary.
5. Answers will vary.
7. With n voters (or n equal-sized groups of voters) and n alternatives $A_1, \ldots, A_n$ it is possible to have $\frac{n-1}{n}$ prefer A_1 to A_2, $\frac{n-1}{n}$ prefer A_2 to A_3, …, $\frac{n-1}{n}$ prefer A_{n-1} to A_n, and $\frac{n-1}{n}$ prefer A_n to A_1.
9. Beth would get the offer.
11. *A* is the winner.
13. (a) *A*
(b) *B*
(c) *C*
(d) *D*
15. (a) Five-way tie
(b) *C*
(c) Five-way tie
(d) *E*
17. (a) *C*
(b) *E*
(c) *E*
(d) *E*
19. (a) If everyone prefers *B* to *D*, for example, then *D* has no first-place votes.
(b) Moving up a winning candidate one spot on some list neither decreases the number of first-place votes for the winning candidate nor increases the number of first-place votes for any other candidate.
21. (a) A Condorcet winner always wins this kind of one-on-one contest.
(b) Moving up a candidate on some list only improves that candidate's chances in one-on-one contests.
23. In the plurality runoff method, having one candidate ranked consistently higher than another would imply that only one candidate is being considered. This candidate would have received all first-place votes and is therefore the winner. Thus, none of the other candidates is being considered and cannot be the winner. The Pareto condition is therefore satisfied.
25. Answers will vary.
27. Answers will vary.
31. (a) A three-way tie with both methods
(b) In both the Hare procedure and plurality runoff, the candidate with 2 or more first-place votes will become the sole winner.
(c) No, either the situation in part (a) or the situation in part (b) must occur.
33. (a) Because *D* has the most votes, *D* is chosen for the board.
(b) The top four are *A*, *B*, *D*, and *F*.
(c) Candidates *B*, *D*, and *F* have at least 80% (8 out of 10) approval.
(d) Candidates *A*, *B*, *D*, *F*, and *G* have at least 60% (6 out of 10) approval. But because at most four candidates can be elected, only *A*, *B*, *D*, and *F* are considered.
35. Answers will vary.
37. (a) *E*
(b) Answers will vary.
39. One possibility is:

	Number of voters (7)		
Rank	**3**	**2**	**2**
First	C	A	B
Second	A	B	A
Third	B	C	C

41. *P* is the voting system in which the winner (or winners) is the candidate with the fewest last-place votes.

Chapter 10

1. One example of two such elections is the following:

Election 1			
Rank	**Number of voters (3)**		
First	*A*	*A*	*B*
Second	*B*	*B*	*A*

Election 2			
Rank	**Number of voters (3)**		
First	*B*	*A*	*B*
Second	*A*	*B*	*A*

3. One example of two such elections is the following:

Election 1			
Rank	**Number of voters (3)**		
First	*A*	*B*	*B*
Second	*B*	*A*	*A*

Election 2			
Rank	**Number of voters (3)**		
First	*B*	*B*	*B*
Second	*A*	*A*	*A*

5. (a) The voting system does not treat all *voters* the same.
   (b) A dictatorship in which Voter 1 is the dictator.
   (c) A dictatorship in which Voter 2 is the dictator and a dictatorship in which Voter 3 is the dictator.
7. One way is to alter the elections in Example 4 of the text by adding *F* and *G* to the bottom of each of the six ballots in both elections, and then adding the two rightmost columns as shown.

Election 1								
	Number of voters (8)							
Rank	**1**	**1**	**1**	**1**	**1**	**1**	**1**	**1**
First	*A*	*B*	*A*	*E*	*A*	*E*	*A*	*G*
Second	*B*	*C*	*B*	*D*	*B*	*D*	*B*	*F*
Third	*C*	*A*	*C*	*C*	*C*	*C*	*C*	*E*
Fourth	*D*	*D*	*D*	*B*	*D*	*B*	*D*	*D*
Fifth	*E*	*E*	*E*	*A*	*E*	*A*	*E*	*C*
Sixth	*F*	*F*	*F*	*F*	*F*	*F*	*F*	*B*
Seventh	*G*	*G*	*G*	*G*	*G*	*G*	*G*	*A*

B has the highest Borda score and is the winner.

The voter on the far left prefers *A* to *B*. By casting a disingenuous ballot (still preferring *A* to *B*, though), the outcome of the election is altered.

Election 2								
	Number of voters (8)							
Rank	**1**	**1**	**1**	**1**	**1**	**1**	**1**	**1**
First	*A*	*B*	*A*	*E*	*A*	*E*	*A*	*G*
Second	*D*	*C*	*B*	*D*	*B*	*D*	*B*	*F*
Third	*C*	*A*	*C*	*C*	*C*	*C*	*C*	*E*
Fourth	*B*	*D*	*D*	*B*	*D*	*B*	*D*	*D*
Fifth	*E*	*E*	*E*	*A*	*E*	*A*	*E*	*C*
Sixth	*F*	*F*	*F*	*F*	*F*	*F*	*F*	*B*
Seventh	*G*	*G*	*G*	*G*	*G*	*G*	*G*	*A*

Now *A* has the highest Borda score and is the winner.

9. The desired ballots (obtained as suggested in the statement of the exercise) are as follows:

Election 1					
	Number of voters (5)				
Rank	**1**	**1**	**1**	**1**	**1**
First	*A*	*B*	*B*	*A*	*D*
Second	*B*	*A*	*A*	*B*	*C*
Third	*C*	*C*	*C*	*C*	*B*
Fourth	*D*	*D*	*D*	*D*	*A*

B has the highest Borda score and is the winner.

The voter on the far left prefers *A* to *B*. By casting a disingenuous ballot (still preferring *A* to *B*, though), the outcome of the election is altered.

Election 2					
	Number of voters (5)				
Rank	**1**	**1**	**1**	**1**	**1**
First	*A*	*B*	*B*	*A*	*D*
Second	*D*	*A*	*A*	*B*	*C*
Third	*C*	*C*	*C*	*C*	*B*
Fourth	*B*	*D*	*D*	*D*	*A*

Now, A has the highest Borda score and is the winner.

11.

Election 1				
	Number of voters (4)			
Rank	**1**	**1**	**1**	**1**
First	*B*	*D*	*C*	*B*
Second	*C*	*C*	*A*	*A*
Third	*D*	*A*	*B*	*C*
Fourth	*A*	*B*	*D*	*D*

C has the highest Borda score and is the winner. But the winner becomes *B* if the leftmost voter changes his or her ballot to *B*, *D*, *A*, *C*. Thus, *B* has the highest Borda score and is declared the winner.

13.

Election 1					
	Number of voters (5)				
Rank	1	1	1	1	1
First	*A*	*B*	*C*	*C*	*D*
Second	*B*	*A*	*B*	*B*	*B*
Third	*C*	*C*	*A*	*A*	*C*
Fourth	*D*	*D*	*D*	*D*	*A*

A, *B*, and *D* have the fewest first-place votes and are thus eliminated, leaving *C* as the winner using the Hare system. But the winner becomes *B* if the leftmost voter changes his or her ballot to *B*, *A*, *C*, *D*.

15.

Election 1			
	Number of voters (3)		
Rank	1	1	1
First	*A*	*B*	*C*
Second	*B*	*C*	*A*
Third	*C*	*A*	*B*

Thus, *C* is the winner with sequential pairwise voting and the agenda *A*, *B*, *C*. But the winner becomes *B* if the leftmost voter changes his or her ballot to *B*, *A*, *C*.

17. A winner with the Hare system must be ranked at the top of at least one voter's ballot, or else it would be eliminated in the first round. For such a voter, there is no outcome preferred to his or her top choice being the single winner.

19. Answers will vary.

21.

Election 1					
22%	23%	15%	29%	7%	4%
D	*D*	*H*	*H*	*J*	*J*
H	*J*	*D*	*J*	*H*	*D*
J	*H*	*J*	*D*	*D*	*H*

Since *D* has the most first-place votes, Alfonse D'Amato (*D*) is the winner by plurality voting. The plurality rule is group manipulable if the voters in the 7% group all change their ballots.

Election 2					
22%	23%	15%	29%	7%	4%
D	*D*	*H*	*H*	*H*	*J*
H	*J*	*D*	*J*	*J*	*D*
J	*H*	*J*	*D*	*D*	*H*

Since *H* has the most first-place votes, Elizabeth Holtzman (*H*) is the winner by plurality voting.

23. If the first voter changes his preference so that *C* is in first place and *D* is still in third, fourth, or fifth place, then Candidate *C* will be the winner since the candidate has 2 first- or second-place votes and has a first-place vote, whereas Candidate *D* does not.

Example of Election 2			
Rank	Number of voters (3)		
First	*C*	*C*	*E*
Second	*B*	*D*	*D*
Third	*A*	*A*	*A*
Fourth	*D*	*B*	*B*
Fifth	*E*	*E*	*C*

25. Properties 1, 2, and 3

27. Properties 1, 2, and 4

29. If the faculty votes for Terms, then the students will get their first choice—Terms—by voting for Terms, but they will get their third choice—the J-plan—by voting for Semesters.

31. Consider what happens if the leftmost voter changes his or her ballot to *B*, *A*, *D*, *C*.

Election 1				
	Number of voters (4)			
Rank	1	1	1	1
First	*A*	*C*	*B*	*D*
Second	*B*	*A*	*D*	*C*
Third	*C*	*B*	*C*	*A*
Fourth	*D*	*D*	*A*	*B*

Because *C* is the only candidate in the first election that either wins or ties each other candidate in a head-to-head match-up, by the weak Condorcet method, *C* wins outright in the first election. However, the winner becomes *B* if the voter on the left changes his or her ballot as follows:

Election 2				
	Number of voters (4)			
Rank	1	1	1	1
First	*B*	*C*	*B*	*D*
Second	*A*	*A*	*D*	*C*
Third	*D*	*B*	*C*	*A*
Fourth	*C*	*D*	*A*	*B*

B wins, thus showing that the weak Condorcet method is manipulable.

33. A winner in plurality voting must be ranked at the top of at least one voter's ballot. For such a voter, there is no outcome preferred to his or her top choice being the single winner.

Chapter 11

1. Any two of the other three voters have enough votes to carry a motion, but the weight-12 voter cannot combine his vote with any one voter to make a winning coalition. Therefore, his vote can never affect a committee decision.
3. The voters with weights 5 and 4 each have veto power, and the weight-3 voter is a dummy.
5. (a) In 1958, *B*, *G*, and *L* were dummies. In 1964, *N*, *G*, and *L* were dummies. There were no dummies in 1970 and after.
   (b) In 1958 and 1964, the Hempstead supervisors would be dictators; *N*, *B*, *G*, and *L* would be dummies. Following 1970 there would have been no dummies until 1982, when *G* was a dummy.
7. (a) *A* is pivotal in these permutations: *BACD*, *BADC*, *BCAD*, *BDAC*, *CABD*, *CADB*, *CBAD*, *CDAB*, *DABC*, *DACB*, *DBAC*, and *DCAB*.
   (b) *B* is pivotal in these permutations: *ABCD*, *ABDC*, *CDBA*, and *DCBA*.
   (c) The Shapley–Shubik index of *A*, which is pivot in 12 permutations, is $\frac{12}{4!} = \frac{1}{2}$. Each of the other voters has four pivots and a Shapley–Shubik power index of $\frac{1}{6}$.
9. (a) Voter *A* is pivotal in the third or fourth position of any permutation.
   (b) $2 \times 5! = 240$
   (c) Voter *A* has a Shapley–Shubik power index of $\frac{2\times5!}{6!} = \frac{1}{3}$. Each of the five weight-2 voters has a Shapley–Shubik power index equal to $\frac{1}{5}$ of the remaining $\frac{2}{3}$ of the power; that is, $\frac{2}{15}$.
11. No. Before *D* left, he was pivot in no permutations. The Shapley–Shubik power indices of the other participants were their shares of the permutations in which they were pivot. With *D* removed, the shares would remain the same.
13. (a) 3
    (b) {*A*, *B*, *C*}, {*A*, *B*, *D*}, and {*A*, *C*, *D*}
15. In listing winning coalitions, the critical voters are marked with an asterisk.
    (a) *A* is dictator; the winning coalitions are {*A**} and {*A*,* *B*} and *A* is the only critical voter in each. Hence *A* has a Banzhaf power index of 2, and *B* has a Banzhaf power index of 0.
    (b) This is a majority-rule system: The winning coalitions are all with two voters, and the grand coalition. The grand coalition has no critical voters, and each member of a two-voter coalition is critical. For example, *A* is critical in {*A*,* *B**} and {*A*,* *C**}, so the Banzhaf power index of *A* is 2. *B* and *C* also have Banzhaf power indices of 2.
    (c) The winning coalitions are {*A*,* *B**}, {*A*,* *C**}, and {*A*,* *B*, *C*}. In the first two, both voters are critical, but in the grand coalition, only *A* is critical. Hence the Banzhaf power index of *A* is 3, while the Banzhaf power indices of *B* and *C* are both equal to 1.
    (d) Any two of *A*, *B*, and *C* can form a winning coalition. *D* cannot form a winning coalition by joining with a single other voter, and so is a dummy. The winning coalitions are: {*A*,* *B**}, {*A*,* *C**}, {*B*,* *C**}, {*A*,* *B*,* *D*}, {*A*,* *C*,* *D*}, {*B*,* *C*,* *D*}, and {*A*, *B*, *C*, *D*}. Thus, *A*, *B*, and *C* have Banzhaf power indices equal to 4, while the Banzhaf power index of *D* is 0.
    (e) *A* can form a winning coalition with any one of the other three voters. Without *A* it takes all three of the other voters to form a winning coalition. Thus, the winning coalitions are {*A*,* *B**}, {*A*,* *C**}, {*A*,* *D**}, {*A*,* *B*, *C*}, {*A*,* *B*, *D*}, {*A*,* *C*, *D*}, {*B*,* *C*,* *D**}, and {*A*, *B*, *C*, *D*}. The Banzhaf power indices are: *A*, 6; *B*, 2:, *C*, 2; and *D*, 2.
17. For each quota, the winning coalitions, with the number of extra votes, are listed. Asterisks identify the critical voters. The notation for Banzhaf power indices is such that (5,3,3,1) designates 5 for *A*, 3 for *B* and *C*, and 1 for *D*. The abbreviation "e.v." means "extra votes."
    (a) $q = 52$ {*A*,* *B**}, 3 e.v.; {*A*,* *C**}, 2 e.v.; {*B*,* *C*,* *D**}, 18 e.v.; {*A*,* *B*, *C*}, 27 e.v.; {*A*,* *B*,* *D*}, 24 e.v.; {*A*,* *C*,* *D*}, 23 e.v.; {*A*, *B*, *C*, *D*}, 48 e.v.
    Banzhaf power index: (5,3,3,1)
    (b) $q = 55$ {*A*,* *B**}, 0 e.v.; {*B*,* *C*,* *D**}, 15 e.v.; {*A*,* *B*,* *C*}, 24 e.v.; {*A*,* *B*,* *D*}, 21 e.v.; {*A*,* *C*,* *D**}, 20 e.v.; {*A*, *B*, *C*, *D*}, 45 e.v.
    Banzhaf power index: (4,4,2,2)
    (c) $q = 56$ {*B*,* *C*,* *D**}, 14 e.v.; {*A*,* *B*,* *C**}, 23 e.v.; {*A*,* *B*,* *D**}, 20 e.v.; {*A*,* *C*,* *D**}, 19 e.v.; {*A*, *B*, *C*, *D*}, 44 e.v.
    Banzhaf power index: (3,3,3,3)
    (d) $q = 57$ {*B*,* *C*,* *D**}, 13 e.v.; {*A*,* *B*,* *C**}, 22 e.v.; {*A*,* *B*,* *D**}, 19 e.v.; {*A*,* *C*,* *D**}, 18 e.v.; {*A*, *B*, *C*, *D*}, 43 e.v.
    Banzhaf power index: (3,3,3,3)
    (e) $q = 71$ {*A*,* *B*,* *C**}, 8 e.v.; {*A*,* *B*,* *D**}, 5 e.v.; {*A*,* *C*,* *D**}, 4 e.v.; {*A*,* *B*, *C*, *D*}, 29 e.v.
    Banzhaf power index: (4,2,2,2)
    (f) $q = 76$ {*A*,* *B*,* *C**}, 3 e.v.; {*A*,* *B*,* *D**}, 0 e.v.; {*A*,* *B*,* *C*, *D*}, 24 e.v.
    Banzhaf power index: (3,3,1,1)
    (g) $q = 80$ {*A*,* *B*,* *C*,* *D**}, 20 e.v.
    Banzhaf power index: (1,1,1,1)
19. (a) 20 (b) 4950 (c) 4950 (d) 126
21. (a) [7: 4, 4, 1, 1, 1, 1, 1]
    (b) Three voters must precede Gerry in the permutation. One must be Essie or Franklin, and two must be chosen from the four weight-1 voters other than Gerry, who already has been placed in the middle of the permutation. The number of ways to select these voters is ${}_2C_1 \times {}_4C_2 = 12$. Once selected, there are $3! = 6$ ways to order them. The remaining three voters, who will be after Gerry in the permutation, can also be ordered in 6 ways, so the number of permutations in which Gerry is pivotal is $12 \times 6 \times 6 = 432$.
    (c) Gerry's Shapley–Shubik index is $\frac{432}{7!} = \frac{3}{35}$, and each of the other four weight-1 voters has the same

index. The sum of the indices of the weight-1 voters is $5 \times \frac{3}{35} = \frac{3}{7}$. The remaining $\frac{4}{7}$ of the power is shared equally by Essie and Franklin, so the Shapley–Shubik power index of each is $\frac{2}{7}$.

(d) Such a coalition would include either Essie or Franklin and two weight-1 voters in addition to Gerry. There are ${}_2C_1 \times {}_4C_2 = 12$ such coalitions. That is Gerry's Banzhaf power index.

(e) To be a winning coalition, at least 3 votes are needed in addition to Franklin's. For Franklin to be a critical voter, the coalition must not include more than 6 votes in addition to Franklin's. There are ${}_5C_3 = 10$ ways to assemble three weight-1 voters to get 3 more votes to join Franklin. To get 4 more votes, Franklin could be joined by Essie alone or by four weight-1 voters; there are $1 + {}_5C_4 = 6$ of these coalitions. To get 5 more votes, one needs either all of the weight-1 voters or Essie and four weight-1 voters. There are 6 ways to assemble a coalition of this type. To get 6 more votes, we would need Essie and two weight-1 voters; there are ${}_5C_2$ such coalitions.

(f) Franklin's Banzhaf power index, which is the same as Essie's, is $10+6+6+10 = 32$, and we have noted that Gerry, and each other weight-1 voter, has an index of 12.

23. A juror is critical in a coalition with four other like-minded jurors. There are ${}_5C_2 = 5$ four-member coalitions of the five other jurors, so the Banzhaf power index of this juror is 5. The probability is this index, divided by $2^5 = 32$, or $\frac{5}{32}$.

25. There would be no significant difference in Nebraska's share of the voting power in the Electoral College.

27. No. For example, consider a jury whose vote must be unanimous to pass a motion. A single juror would be a minimal blocking coalition, and two such minimal blocking coalitions would not intersect.

29. (a) [4: 2, 1, 1, 1]
(b) [9: 3, 3, 1, 1, 1, 1, 1]

31. A faculty member F will be a critical voter in a coalition that includes, in addition to F, exactly two of the other three faculty members and at least two administrators. There are ${}_3C_2 = 3$ ways to assemble the other faculty members and ${}_3C_2 + {}_3C_3 = 4$ ways to assemble the administrators. Thus, each faculty member has a Banzhaf power index of $3 \times 4 = 12$.

An administrator A will be critical in a winning coalition that includes at least three faculty members and exactly one of the other two administrators. There are ${}_4C_3 + {}_4C_4 = 5$ ways to assemble the faculty members and ${}_2C_1 = 2$ ways to get the other administrator. An administrator, with a Banzhaf power index of $5 \times 2 = 10$, is a bit less powerful on this committee than a faculty member, according to the Banzhaf model.

33. If there is only one minimal winning coalition, then every voter who belongs to that coalition has veto power, and every voter who doesn't belong is a dummy. If there are just two minimal winning coalitions, then they must intersect. Those voters in the intersection will have veto power.

35. (a) In the following table, all coalitions that include E are omitted.

Winning Coalition	Extra Votes	Losing Coalition	Vote Deficit
{A, B}	15	{ }	51
{A, C}	13	{A}	11
{A, B, C}	39	{B}	25
{A, B, D}	25	{C}	27
{A, B, C, D}	49	{D}	41
{A, C, D}	23	{A, D}	1
{B, C, D}	9	{B, C}	1
		{B, D}	15
		{C, D}	17

(b) A can sell zero shares, because if B gains just one share, the losing coalition $\{B, C\}$ will become a winning coalition.

(c) A can sell 13 shares. All winning coalitions that include A have enough votes to support that, and the losing coalitions that include D would still be losing after the sale. ($\{A, D\}$ is one share short of meeting the quota, but any transfer of shares between A and D would not affect its total.)

(d) A can sell zero shares. If E had just one share, the losing coalition $\{B, C, E\}$ would become a winning coalition.

37. (a) [8 : 6, 1, 1, 1, 1, 1, 1, 1, 1]
(b) The Banzhaf power index of the chairperson is 246, and each ordinary member's index is 8. There are $246 + 8 \times 8 = 310$ critical votes in all. According to the Banzhaf model, the chair has $\frac{246}{310} = 79.35\%$ of the power, and each of the other members has 2.90% of the power.
(c) The chair is pivot in a permutation when located in positions 3 through 8 of a permutation; that is, in 6 of the 9 locations of a permutation. Hence the Shapley–Shubik power index of the chair is $\frac{6}{9} = 66.67\%$. The other members share the remaining $\frac{1}{3}$ of the power, so the Shapley–Shubik power index of each is $\frac{1}{8} \times \frac{1}{3} = 4.17\%$.
(d) The two models do not agree closely in this case.

39. The borough presidents, taken together, have a total voting weight of 36, and the voting weight of each borough president is more than 1. Thus, the minimal winning coalitions consist of one of the following:

- All three city officials (34 extra votes)
- Two city officials and one borough president (at least 0.8 and not more than 10.3 extra votes)
- One city official and all five borough presidents (0 extra votes)

Therefore, the weighted voting system [9 : 4, 4, 4, 1, 1, 1, 1, 1] is equivalent to the system with minimal winning coalitions (a) through (c).

The Supreme Court would reject this system since—like the system that it was intended to replace—all of the borough presidents wield the same amount of voting power. It is interesting to note that the system where the city officials had 2 votes and the quota was 6, the Banzhaf and Shapley–Shubik models both gave each city official about 19% of the power. With the proposed change, the city officials each had about 29% of the power with the Banzhaf model (27% with the Shapley–Shubik model).

41. The minimal winning coalitions would be formed by either all three senior members or two senior members and one junior member. The winning coalitions that include A as a critical voter would be $\{A, B, C\}$; $\{A, S, J\}$, with S being either B or C and J being one of D, E, F (there are $2 \times 3 = 6$ of these); $\{A, S, J_1, J_2\}$, with S as before and J_1, J_2 representing two of D, E, F (there are 6 of these coalitions, too); or $\{A, S, D, E, F\}$ (two coalitions). Hence the Banzhaf power index of each senior member is $1 + 6 + 6 + 2 = 15$.

A junior member can only be critical in a coalition with two senior members and no other junior members. There are 3 such coalitions—that is the Banzhaf power index of a junior member.

43. Use S_1, S_2 to denote a senior member (either B or C) and J_1, J_2, J_3, J_4 to represent junior members. In the committee with three junior members, A is pivotal in the following permutations:

- $S_1S_2AJ_1J_2J_3$ (12 permutations)
- $J_1S_1AS_2J_2J_3$ (${}_3C_1 \times {}_2C_1 \times 12 = 72$ permutations)
- $J_1J_2S_1AS_2J_3$ (also 72 permutations)
- $J_1J_2J_3S_1AS_2$ (${}_2C_1 \times 4! = 48$ permutations)

Thus, the Shapley–Shubik power index of A is $(12 + 2 \times 72 + 48) \div 6! = 28.33\%$. By the Shapley–Shubik model, the three senior members hold between them $3 \times 28.33\% = 85\%$ of the power. The remaining 15% is divided between the junior members, so each has a Shapley–Shubik power index of 5%.

With four junior members, A would be pivotal in these permutations:

- $S_1S_2AJ_1J_2J_3J_4$ (48 permutations)
- $J_1S_1AS_2J_2J_3J_4$ (${}_4C_1 \times {}_2C_1 \times 48 = 384$ permutations)
- $J_1J_2S_1AS_2J_3J_4$ (${}_4C_2 \times {}_2C_1 \times 36 = 432$ permutations)
- $J_1J_2J_3S_1AS_2J_4$ (384 permutations)
- $J_1J_2J_3J_4AS_1S_2$ ($4! \times 2! = 48$ permutations)

This comes to a total of 1296 permutations. Dividing by $7! = 5040$, we obtain the Shapley–Shubik power index of A, 25.714%. The three senior members have 77.142% of the power in the Shapley–Shubik model. The remaining 22.858% belongs to the junior members, so each has 5.714% of the voting power in this model. Thus, by including another junior member, the power of *each* junior member has slightly increased.

45. (a) The winning coalitions must have a total of at least 8 members, including at least 2 rich alumni. The *minimal* winning coalitions will have exactly 8 members: either 3 rich alumni and 5 recent graduates or 2 rich alumni and 6 recent graduates.
(b) The total weight of a coalition including 3 rich alumni and 5 recent graduates would be $3r + 5$, and the total weight of a coalition with 2 rich alumni and 6 recent graduates would be $2r + 6$.
(c) This losing coalition has total weight of $r + 12$. Its total weight is less than the total weight of the winning coalition with 2 rich almuni and 6 recent graduates. Therefore, $2r + 6 > r + 12$. Subtract $r + 6$ from both sides to get $r > 6$.
(d) The largest losing coalition with 3 rich alumni would include 4 recent graduates, and its total weight would be $3r + 4$; thus $2r + 6 > 3r + 4$. Subtract $2r + 4$ from both sides to get $2 > r$; in other words, r is less than 2. But the inequality from part (c) says r is greater than 6. No number can be both more than 6 and less than 2, so the two inequalities are inconsistent.
(e) No

47. We will name the voters A, B, C.
(a) Minimal winning coalitions: $\{A, B\}$, $\{A, C\}$. This system is known as the "chair veto."
(b) Minimal winning coalitions: $\{A, B\}$, $\{A, C\}$, $\{B, C\}$. This is the "majority" system.
(c) Minimal winning coalition: $\{A\}$. This is the "dictator" system.
(d) Minimal winning coalition: $\{A, B\}$. This is the "clique" system.
(e) Minimal winning coalition: $\{A, B, C\}$. This is the "consensus" system.

49. Call the voters A, B, C, D, and E, in decreasing order by weight. The total weight of this system is 27, so the quota must be at least 14. If A is not to have veto power, then the remaining voters, whose total weight is 15, must be able to pass a motion. Thus, the only possible quotas that prevent any voter from having veto power are 14 and 15. If the quota is 14, E will be a dummy voter. The winning coalitions that include E are $\{B, C, D, E\}$, which has total weight 15, and $\{A, E\}$ combined with some nonempty subset of $\{B, C, D\}$, which would have total weight at least equal to 16, which is the weight of $\{A, D, E\}$. Voter E would not be critical in any of these coalitions, since they all have at least 1 extra vote.

If the quota is 15, then $\{B, C, D, E\}$ has 0 extra votes, which makes it a minimal winning coalition. The weight-1 voter is not a dummy. Therefore, 15 is the only quota that meets the specifications. The other minimal winning coalitions are $\{A, B\}$, $\{A, C\}$, and $\{A, D\}$.

51. The minimal winning coalitions consist of the following:

- The president, the vice president, 50 senators, and 218 members of the House of Representatives
- The president, 51 senators, and 218 members of the House of Representatives
- 67 senators and 290 members of the House of Representatives

53. Focus on the weight-3 voter, who is singular. This voter is pivot if and only if placed in the third, fourth, or fifth position in a permutation. The permutation has length 7, so the Shapley–Shubik power index for this voter is $\frac{3}{7}$. The six weight-1 voters share the remaining $\frac{4}{7}$ of the power; thus, each has Shapley–Shubik power index equal to $\frac{4}{7} \div 6 = \frac{2}{21}$.

Chapter 12

1. Ann, Bob, and Carl receive 12, 5, and 2 delegates, respectively.

3.

Candidates	Popular Vote	Percentage of Votes	Percentage after drops	Quota	Initial Delegates	Final Delegates
Clark	14,010	12.29				
Dean	33,493	29.37	43.87	3.071	3	3
Edwards	13,396	11.75				
Kerry	42,847	37.58	56.13	3.929	3	4
Lieberman	1,784	1.56				
Others	8,507	7.46				

5. This can happen with a minimum number of 7 candidates because 1/7 < 15%, while 1/6 > 15%.
7. Anchor, Barber, and Coachman, respectively, receive 4, 3, and 1 delegates in District 1; 5, 2, and 1 delegates in District 2; and 8, 6, and 2 delegates if the districts are combined.
9. No, it is not always the case. Answers will vary for the last part. One such example is when *A, B,* and *C* receive 2700, 1000, and 1000 votes, respectively, and 8 delegates are to be awarded. Candidates *A, B,* and *C* receive 4, 2, and 2 delegates, respectively, under Hamilton's method. If candidates *B* and *C* are combined, then *A* receives 5 delegates and the combined *B*/*C* receives 3.
11. Yes. With no cutoff, Candidate *B* receives 1 delegate. With a cutoff, Candidate *B* receives no delegates.
13. Candidate *A*'s best position is 9; she would receive 17 votes.
15. Twelve voters would have to be introduced so that *A* wins 18 to 17.
17. Because there is no median position, there must be an even number of voters. List the ideal positions of the voters in left-to-right/numerical order. Assume that there are $2n$ voters. Because there is no single median, the nth and $(n + 1)$st ideal positions are not the same. Any position in between these two ideal points has an equal number of voters' ideal points to both the left and the right. These two ideal points are the endpoints of the extended median.
19. Answers will vary. However, a business wants to be closer to more customers than its rival business.
21. Yes, because shoppers to the left of 5 would go to the store at 2.5, and shoppers to the right of 5 would go to the store at 7.5. The shoppers at 5 are equidistant between the stores; 50 could go to one and 50 to the other.
23. No, *B* cannot do better. However, *B* could still get two of the voters to vote for him by announcing a policy position closer to two of the voters. For example, by announcing $(\frac{1}{2}, \frac{1}{2})$, voters at (1, 0) and (0, 1) would vote for *B* and the other voters for *A*.
25. *B* wins the election if she announces a position between 1 and 3, but loses if she announces positions between 0 and 1 or between 3 and 4. The median is 2. If *B* announces the median position, then voters between 0 and 2.5 would vote for *B*.
27. If *B*'s position is not the median, then *A* can do no better than to just settle next to *B* on the side of the median. *A*'s position should be close enough to *B*'s position that *A*'s position is closer to the nearest voter to *B*'s position that is in between *B*'s position and the median. This way, *A* maximizes her vote, getting all voters on the side with the median to vote for her. If *B*'s policy position is at the median, then *A* can do no better than to also select the median (by the median-voter theorem).
29. The extended median is [2, 3]. The greatest value that *B* can announce and still be in equilibrium with *A* is 3. The least value that *B* can announce and still be in equilibrium with *A* is 2.
31. Yes, there is more than one position that *C* can announce to maximize his or her vote total. For example, the positions 4 and 4.1 both yield 5 votes. *C* cannot win the election but can tie in the election.
33. If, say, *A* takes a position at *M*, and *B* takes a position to the right of *M*, *C* should take a position just to the left of *M* that is closer to *M* than *B*'s position, giving *C* essentially half the votes and enabling him or her to win the election. If neither *A* nor *B* takes a position at *M*, *C* should take a position next to the candidate closer to *M*. The position that *C* takes to maximize his or her vote may be either closer to *M* (if the candidates are far apart) or farther from *M* (if the candidates are closer together), but this position may not be winning.
35. Yes; by announcing a policy position of 4, *C* could win 9 of the votes, while *A* and *B* receive 8 and 7 votes, respectively.
37. Following the hint, *C* will obtain $\frac{1}{3}$ of the vote by taking a position at *M*, as will *A* and *B*, so there will be a three-way tie among the candidates. Because a non-unimodal distribution can be bimodal, with the two modes close to *M*, *C* can win if he or she picks up most of the vote near the two modes, enabling *C* to win with more than $\frac{1}{3}$ of the vote. In this case, Candidate *C* has taken advantage of a $\frac{2}{3}$-separation opportunity.
39. *B* should enter just to the right of $\frac{3}{4}$, making it advantageous for *C* to enter just to the left of *A*, giving *C* essentially $\frac{1}{4}$ of the vote.
41. No, Gore should not set off this chain of events. Gore would receive 1% percent of the electorate ($\frac{1}{3}$ of the 3% of voters who prefer Nader). Bush would receive all

of 2% of the electorate who vote for Buchanan. Gore and Bush would be tied, each receiving 49% of the electorate.

43. Answers will vary. However, such vote trading does not solve problems created by the Electoral College.
45. No, because state C is so large, its median position will always be the median of the three states.
Large states have more influence on the outcome of an election because winning a large state's electoral votes brings a candidate closer to the majority.
47. Answers will vary.
49. Answers will vary.
51. A wins the primaries in states 1 and 2, while B wins the primaries in states 3, 4, and 5 and the primary election. A now wins the primaries in states 1, 2, and 3, as well as the primary election. For questions about money, answers will vary.
53. Yes, the point of the National Popular Vote law is to elect the direct popular-vote winner.

Chapter 13

1. Donald receives the Trump Tower triplex and about 87% ownership of the Palm Beach mansion. Ivana gets the rest.
3. Phil gets his way on the stereo level issue, the smoking rights issue, the phone time issue, the visitor policy issue, and about 87% of his way on the alcohol issue. Mike gets his way on the rest.
5. (a) Mary will handle salary recommendations and external reviews.
(b) Fred will handle class schedules, department meetings, and calculus placement.
(c) They share hiring. Since Fred retained $\frac{12}{17}$, Mary will be taking on more of the burden.
7. Answers will vary.
9. Mary gets the car and places $\frac{32,100}{2} = \$16,050$ in a kitty. John takes out $\frac{28,225}{2} = \$14,112.50$. The remaining $\$16,050 - \$14,112.50 = \$1937.50$ is split equally. The net effect of this is that Mary receives the car and pays John \$15,081.25.
11. Answers will vary.
13. See table below.
15. Bob first chooses the investments, and the final allocation has him also receiving the car and the MP3 player.
17. Mark first chooses the tractor, and the final allocation has him also receiving the truck and the tools.
19. Donald first chooses the Palm Beach mansion, and the final allocation has him also receiving the Trump Tower triplex and the cash and jewelry.
21. One way is to have Bob divide the cake into four pieces and to let Carol choose any three. Another is to have Bob divide the cake into two pieces and then let Carol choose one. They can do divide-and-choose on the piece that Carol did not choose.
23. Bob should be the divider. That way, he can get 12 units of value instead of 9 units of value.
25. (a) See figures below.

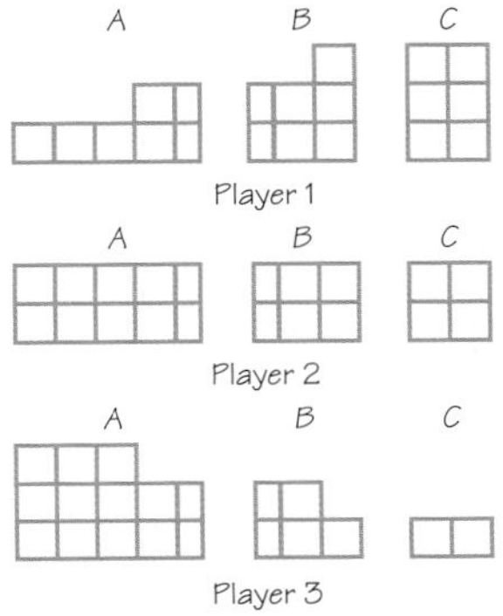

(b) Player 2 finds A acceptable (9 square units), but not B (5 square units) or C (4 square units). Player 3 finds A acceptable (12 square units), but not B (4 square units) or C (2 square units).
(c) Players 2 and 3 both find B and C unacceptable. (C is on the right.)
(d) (i) Player 1 gets a piece he thinks is 6 square units. Player 2 gets a piece he thinks is 7 square units, and Player 3 gets a piece he thinks is 10 square units.
(ii) Player 1 gets a piece she thinks is 6 square units, Player 2 gets a piece she thinks is $8\frac{2}{3}$ square units, and Player 3 gets a piece she thinks is 8 square units.
27. (a) Ted thinks he is getting at least one-third of the piece that Bob initially received and at least one-third of the piece that Carol initially received. Thus, Ted thinks he is getting at least one-third of part of

Potential Recipient	Months Waiting, Position in Terms of Waiting, and Points Awarded	Antigens Matched and Points Awarded	Percent Sensitized and Points Awarded	Total Points
A	9 (1) $10 \times \left(\frac{4}{4}\right) = 10$ points	2 $2 \times 2 = 4$ points	20% 2 points	16.0
B	6 (2) $10 \times \left(\frac{3}{4}\right) = 7.5$ points	3 $2 \times 3 = 6$ points	0% 0 points	13.5
C	5 (3) $10 \times \left(\frac{2}{4}\right) = 5$ points	4 $2 \times 4 = 8$ points	40% 4 points	17.0
D	2 (4) $10 \times \left(\frac{1}{4}\right) = 2.5$ points	6 $2 \times 6 = 12$ points	60% 6 points	20.5

the cake (Bob's piece) plus one-third of the rest of the cake (Carol's piece).

(b) Bob gets to keep exactly two-thirds (in his own view) of the piece that he initially received and thought was at least of size one-half. Two-thirds times one-half equals one-third. The same argument applies to Carol.

(c) If, for example, Ted thinks the half Carol initially gets is worthless, then Ted may wind up thinking that he (Ted) has only slightly more than one-third of the cake, while Bob has (in Ted's view) almost two-thirds of the cake. In such a case, Ted will envy Bob.

29. (a) See figures below.

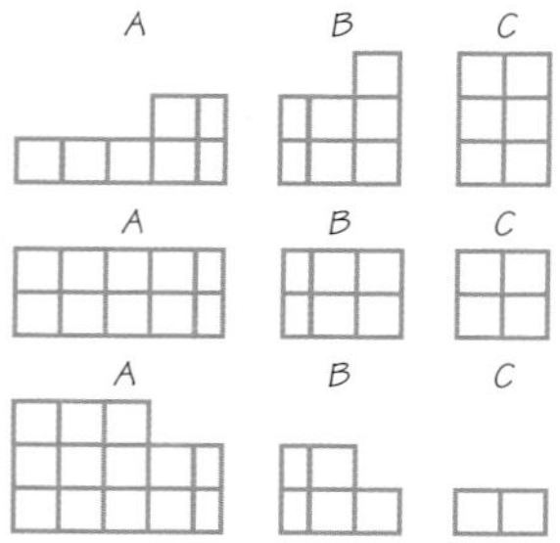

(b) See figures below.

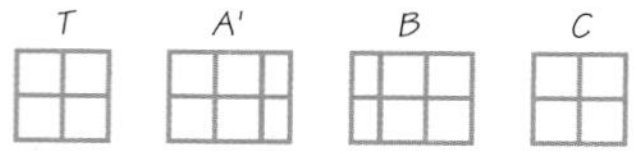

(c) Player 3 will choose *A* (which he thinks is of size 6 square units). Player 2 will choose *B* (which he thinks is of size 5 square units). Player 1 will receive *C* (which he thinks is of size 6 square units). The proviso does not come into play (since Player 3 took the trimmed piece).

31. (a) This is because Player 1 views all three pieces as being of equal size or value.

(b) Player 2 created a two-way tie for largest or most valuable piece. Player 3 may take one of these, but at least one will still be available for Player 2 to choose.

(c) This is because Player 3 gets to choose first.

(d) Player 1 will not envy Player 3 because Player 3 received the trimmed piece, and so Player 1's piece is as large in his eyes as Player 3's piece plus all of the trimmings. Player 1 will not envy Player 2 because he is choosing before Player 2.

(e) This is because Player 2 views all three pieces of the trimmings as being of equal size or value.

(f) This is because Player 3 gets to choose first.

33. If I bid, say, $110, then Bob wins the lamp but pays $10 more than he thinks it's worth. This is worse for him than bidding $100 and losing the auction.

35. For the final resolution, *E* receives the Duesenberg and Cord and pays $8500, *F* receives the Bentley and Pierce-Arrow and pays $7500, and *G* receives the Ferrari plus $16,000.

37. Fred will do the calculus placement, department meetings, and class schedules. Mary will do external reviews, hiring, and salary recommendations.

39. It is better to be the chooser. As divider, I'd get exactly 50% (or risk getting less). As chooser, I have a guarantee of getting at least 50% and the possibility (depending on the division) of getting more than 50%.

Chapter 14

1.

Jane's Expenses		
Category	**Percentage**	**Rounded**
Rent	43.66%	44%
Food	22.54%	23%
Transportation	9.86%	10%
Gym	16.90%	17%
Miscellaneous	7.04%	7%
Totals	100.00%	101%

The sum of the rounded percentages is not 100%.

3. No, although if the quotas are rounded in the usual way, there will be 21 sections. One must reduce the apportionment of one course, and a logical way to do so will lead to the original apportionment.

5. If you round the numbers to the nearest whole number, the sum will be 62. One approach to solving this problem would be to round the two numbers whose fractional parts are less than 0.6 down: round 14.53 to 14, and round 16.55 to 16. The rounded sum would be

$$12 + 10 + 8 + 14 + 16 = 60$$

7. $0 + 2 + 1 + 2 + 3 + 2 = 10$

9. The sum of the lower quotas is 100, leaving two seats to be given to the parties with the largest fractions. The fractional parts of the quotas are 0.8, 0.6, and 0.6. Therefore the Anti-UFO and Who Cares parties are tied for the 102nd seat.

11. Geometry, 3 sections; algebra, 1 section; calculus, 1 section

13. (a) Abe, 36 coins; Beth, 19 coins; Charles, 23 coins; David, 20 coins; Esther, 2 coins

(b) Charles must give Esther a coin.

(c) Blame the population paradox.

15. As the following table shows, the Alabama paradox occurred. Party *E* lost a seat when the house size increased from 89 to 90 and did not recover the seat when the house size increased to 91. (If you like, you can see what happens when the house size is 92.)

Party	Population	Seats		
A	5,576,330	27	28	28
B	1,387,342	7	7	7
C	3,334,241	16	17	17
D	7,512,860	37	37	38
E	310,968	2	1	1
Totals	18,121,741	89	90	91

17. Geometry gets 4 sections, calculus gets 1 section, and there is no algebra class.
19. The Jefferson and Webster methods give identical apportionments for 36 or 37 pearls: With 36 pearls, Abe gets 14, Beth gets 19, and Charles gets 3. The 37th pearl is awarded to Abe. Although the Jefferson and Webster methods yield the same result, that does not mean that they are right. If there is a principle on which to choose a method, it would probably be to choose the method by which the cost per pearl is as close as possible to the same for each of the friends. The cost per pearl is the district size, so they should use the Dean method (see Exercise 48 and Writing Project 2), which minimizes absolute differences in district size. Charles might want to study up on it, because it allocates the 37th pearl to him!
21. See table below.
23. See table on next page
25. The 34th seat
27. Let d be the divisor used to apportion before the new state is added. Let p^* be the population of the new state. The new state should be apportioned $\left[\frac{p^*}{d}\right]$, where square brackets indicate the method of rounding that is specified for the particular divisor method in use. This is the formula that would be used for each of the states before and after the new state joined; therefore, none of the original apportionments would change.
29. **Hamilton:** $92 + 2 + 2 + 2 + 1 + 1 = 100$. The quota condition is satisfied.
Jefferson: $95 + 1 + 1 + 1 + 1 + 1 = 100$. Because 92.15 is rounded up to 95, the quota condition is violated.
Webster: $90 + 2 + 2 + 2 + 2 + 2 = 100$. Again, the quota condition is violated: This time 92.15 is rounded down to 90.
31.

	Before Tax	After Tax
Abe	36	36
Beth	19	19
Charles	22	23
David	20	20
Esther	3	2
Total	100	100

Now it is Esther who must give a coin to Charles!
33. (a) One quota will be rounded up and the other down to obtain the Webster apportionment. The quota that is rounded up will have a fractional part greater than 0.5 and will be greater than the fractional part of the quota that is rounded down. The Hamilton method will give the party whose quota has the larger fractional part an additional seat. Thus, the apportionments will be identical.
(b) These paradoxes never occur with the Webster method, which gives the same apportionment in this case.
(c) The Hamilton method, which always satisfies the quota condition, gives the same apportionment.
(d) No, each of these methods is capable of producing

Seats	Demopublicans		Repocrats		Greenocrats		Greenicans		Independents	
	Priority	Seat	Priority	Seat	Priority	Seat	Priority	Seat	Priority	Seat
1	~~44,856~~	1	~~34,944~~	2	~~20,004~~	4	~~19,002~~	5	~~9,804~~	11
2	~~22,428~~	3	~~17,472~~	6	~~10,002~~	10	~~9,501~~	12	4,902	
3	~~14,952~~	7	~~11,648~~	8	~~6,668~~	17	~~6,334~~	19	3,268	
4	~~11,214~~	9	~~8,736~~	14	5,001		4,751		2,451	
5	~~8,971~~	13	~~6,989~~	16	4,001		3,800		1,961	
6	~~7,476~~	15	~~5,824~~	20	3,334		3,167		1,634	
7	~~6,408~~	18	4,992		2,858		2,715		1,401	
8	5,607		4,368		2,501		2,375		1,226	

The Demopublicans get 7 seats, the Repocrats get 6, the Greenocrats and the Greenicans each get 3, and the Independents get 1 seat.

Seats	Whigs		Tories		Liberals		Centrists	
	Priority	Seat #	Priority	Seat #	Priority	Seat #	Priority	Seat #
1	~~5,525,381~~	1	~~3,470,152~~	3	~~3,864,226~~	2	201,203	
2	~~1,841,794~~	4	~~1,156,717~~	6	~~1,288,075~~	5	67,068	
3	~~1,105,076~~	7	~~694,030~~	10	~~772,845~~	9	40,241	
4	~~789,340~~	8	~~495,736~~	14	~~552,032~~	12	28,743	
5	~~613,931~~	11	~~385,572~~	17	~~429,358~~	15	22,356	
6	~~502,307~~	13	~~315,468~~	21	~~351,293~~	19	18,291	
7	~~425,029~~	16	~~266,935~~	24	~~297,248~~	22	15,477	
8	~~368,359~~	18	~~231,343~~	28	~~257,615~~	26	13,414	
9	~~325,022~~	20	204,127		~~227,307~~	29	11,835	
10	~~290,810~~	23	182,640		203,380		10,590	
11	~~263,113~~	25	165,245		184,011		9,581	
12	~~240,234~~	27	150,876		168,010		8,748	
13	~~221,015~~	30	138,806		154,569		8,048	
14	204,644		128,524		143,119		7,452	
15	190,530		119,660		133,249		6,938	

The Whigs get 13 of the first 30 seats, the Liberals get 9, the Tories get 8, and the Centrists get none of the first 30 seats.

a different apportionment. For example, suppose that one party receives 99.9% of the vote and the other receives 0.1%. If the house size is 100, the Jefferson method would apportion all 100 seats to the dominant party. (So would the Hamilton and Webster methods.) The Hill–Huntington method would apportion the dominant party 99 seats, and 1 seat to the other party. If the dominant party received 99.4% and the other party received 0.6%, the Jefferson method would still apportion all of the seats to the dominant party, but the Hamilton and Webster methods would give 1 seat to the other party.

35. 10.77%

37. The standard divisor is 18.2, but with this divisor the rounded quotas (rounding by Hill–Huntington) are 3, 2, and 1, so 6 sections are apportioned. We will choose a larger divisor; 20 is convenient. The apportionment quotients are 2.8 (which rounds to 3), 1.4, and 0.35. Remember that all numbers between 0 and $\sqrt{2}$ are rounded to 1; therefore, there will be 3 sections of algebra and 1 section each for geometry and calculus.

39. (a) The rounding point between 12 and 13 is $\sqrt{156}$. Dividing North Carolina's apportionment population by $\sqrt{156}$, we obtain $d^* = 645{,}930.8$. With a divisor greater than d^*, North Carolina will receive fewer than 13 seats; with a divisor less than d^*, North Carolina's apportionment will be at least 13.

(b) With a population greater than $\sqrt{12}d^* = 2{,}237{,}570$, Utah would be apportioned 4 seats.

(c) 856

41. The quotas for the three districts are $1\frac{3}{7}$, $8\frac{4}{7}$, and 10, respectively. The Hamilton method simply rounds the first two districts to 1 and 9, respectively, and the Webster and Jefferson methods give the same apportionment. Therefore, these methods do not lead to ties. The Hill–Huntington rounding points are $\sqrt{2} < 1\frac{3}{7}$ and $\sqrt{72} < 8\frac{4}{7}$. (It is important to notice that $\sqrt{72} = \sqrt{6^2 \times 2} = 6\sqrt{2}$.) The Hill–Huntington roundings of the quotas are therefore 2, 9, and 10, respectively. The apportionment quotient for the first district will be less than $\sqrt{2}$ (so that it will receive 1 seat), if the divisor is greater than $\frac{100{,}000}{\sqrt{2}}$. Using this divisor, the apportionment quotients would be $\sqrt{2}$ for the first district and $6\sqrt{2}$, the rounding point between 8 and 9, for the second district. Any divisor large enough to cause the first district's quota to be rounded down will also cause the second district's quota to be rounded down. The result would be 1 seat for the first district, 8 seats for the second district, and 10 seats for the third district—a total of 19 seats. The first and second districts are tied for the last seat.

43. (a) Lowndes favors small states.

(b) Yes

(c) Yes

(d)

State	Quota	Lower Quota	Priority	Apportionment
DE	1.843	1	84.30%	2
VT	2.839	2	41.95%	3
NJ	5.959	5	19.18%	6
NH	4.707	4	17.68%	5
GA	2.351	2	17.55%	3
SC	6.844	6	14.07%	7
KY	2.280	2	14.00%	3
RI	2.271	2	13.55%	3
CT	7.860	7	12.29%	8
NC	11.732	11	6.65%	11
MA	15.774	15	5.16%	15
VA	20.926	20	4.63%	20
MD	9.243	9	2.70%	9
PA	14.366	14	2.61%	14
NY	11.004	11	0.04%	11
Totals	120	111		120

45. (a) No
(b) The divisor will have to be greater than the standard divisor.
(c) It favors small states.
(d) No, because an apportionment quotient q between 0 and 1 must be rounded up to get $\lceil q \rceil$.
(e) The Adams method always uses a divisor that is greater than the standard divisor. Therefore the apportionment quotients are less than the quotas that are obtained by dividing populations by the standard divisor. In other words, if q is the quota for some state, and q^* is the apportionment quotient for the same state, then $q^* < q$. It follows that $\lceil q^* \rceil \leq \lceil q \rceil$. But $\lceil q^* \rceil$ is the Adams apportionment and $\lceil q \rceil$ is the upper quota; hence the Adams method cannot exceed the upper quota for any state.

47. Consider an apportionment problem where we will compare two methods, the Hamilton method and the X method. (We do not specify what the X method is, and we leave the apportionment problem also unspecified.) Our objective is to show that the maximum absolute deviation (MAD) for the X apportionment is at least as large as the MAD for the Hamilton apportionment.

Since the Hamilton method satisfies the quota condition, we will assume that the X method has given each state either its upper or lower quota; otherwise, its MAD would be greater than 1, and thus worse than Hamilton's. If the methods give different apportionments, they must differ for more than one state; otherwise, they would not have the same house size.

Let A be the state that has the largest absolute deviation with the Hamilton method, and let q_A be its quota. The apportionment for A is either the lower quota, $\lfloor q_A \rfloor$, or the upper quota, $\lceil q_A \rceil$, so the absolute deviation for A (and the MAD for the Hamilton apportionment) is either $q_A - \lfloor q_A \rfloor$ or $\lceil q_A \rceil - q_A$.

If the X apportionment for the state A is the same as the Hamilton apportionment for A, then the MAD for the X apportionment is at least equal to the absolute deviation for A, so it cannot be less than the MAD for the Hamilton apportionment.

Suppose that the X apportionment for A differs from the Hamilton apportionment: Say Hamilton assigns A its lower quota and X assigns A its upper quota. There must be another state B to which Hamilton assigns the upper quota and X assigns the lower quota. Since the Hamilton method rounds q_A down and q_B up, the fractional part of q_B is greater than or equal to the fractional part of q_A. The absolute deviation in the X apportionment for state B is the fractional part of q_B, so the MAD for the X apportionment, which must be at least the absolute deviation for state B, is greater than or equal to the MAD for the Hamilton method in this apportionment problem.

If the Hamilton method awards state A its upper quota, we can reason as before and reach the same conclusion—that the X apportionment has a MAD no less than that of the Hamilton method.

49. (a) Divide each township's population by 3000 and round the quotient down to get the apportionment.

Township	Population	Seats	Quota
Alpha	16,210	5	5.062
Beta	40,052	13	12.508
Gamma	8,284	2	2.587
Delta	48,018	16	14.996
Epsilon	2,711	0	0.847

The house size is the total number of seats apportioned: 36.
(b) Epsilon Township is unrepresented on the Board, but if a divisor less than 2711 had been chosen, it would have a seat.
(c) The standard divisor is the county's population, 115,275, divided by the house size, 36. This is $3202\frac{1}{12}$. The quotas, obtained by dividing the populations by the standard divisor, are shown in the table displayed in the answer for part (a).
(d) No. The upper quota for Delta Township is 15 seats, but it is awarded 16 seats.

51. (a) The total population of this county is 301,500, and the house size is 301. The standard divisor is therefore $301{,}500 \div 301 = 1001.66$. The quotas, determined by dividing the township populations by the house size, are shown in the following table. The sum of the rounded quotas (using the Webster rounding) is equal to the house size. Thus, each township's Webster apportionment is its rounded quota.

Township	Population	Quota	Rounded Quota	District Population
A	109,050	108.87	109	1000.46
B	55,920	55.83	56	998.57
C	67,770	67.66	68	996.62
D	61,260	61.16	61	1004.28
E	7,500	7.49	7	1071.43

(b) The district populations are shown in the table displayed in part (a). Township *E* is disadvantaged, because its district population is greater than the district population of each of the others by approximately 70.
(c) The following table shows the absolute difference between the district populations of Township *E* and each of the other townships, as well as the difference if a seat were transferred from the other township to Township *E*.

Township	Abs Diff Before Transfer	Abs Diff After Transfer
A	70.97	72.22
B	72.86	79.23
C	74.81	73.99
D	67.17	83.50

The table shows that if a seat is taken from Township *C* and given to Township *E*, the absolute difference in district population will be less.

Chapter 15

1. (a) This game has a saddlepoint.
(b) The maximin strategy is row 1; the minimax strategy is column 2; the value is 5.
(c) Row 2 is dominated; column 1 is dominated.
3. (a) This game does not have a saddlepoint.
(b) The maximin strategies are rows 1 and 2, as the row minima are equal; the minimax strategy is column 1.
(c) There are no dominated strategies.
5. (a) This game has a saddlepoint.
(b) The maximin strategy is row 3; the minimax strategy is column 3; the value is –20.
(c) None of Player I's row strategies are dominated; Player II's column 3 dominates both columns 1 and 2.
7. Answers will vary. For example, payoffs of 4, 3, 3, and 4 for (Top, Left), (Top, Right), (Middle, Left), and (Middle, Right) give the desired property.
9. The value of the game is $\frac{123}{250}$, or 0.492. This is achieved by the kicker kicking the ball to the goalie's left with probability $\frac{11}{25}$ and to the goalie's right with probability $\frac{14}{25}$, and by the goalie diving to the left with probability $\frac{11}{25}$ and to the right with probability $\frac{14}{25}$.
11. The batter's batting average at equilibrium is 0.240.
13. The value of the game is 0.525.
15. (a)

	Officer Does Not Patrol	Officer Patrols
You park in street	0	−$40
You park in lot	−$32	−$16

(b) Your optimal mixed strategy: $\left(\frac{2}{7}, \frac{5}{7}\right)$; officer's optimal mixed strategy: $\left(\frac{3}{7}, \frac{4}{7}\right)$; value: −$22.86
(c) It is unlikely that the officer's payoffs are the opposite of yours.
(d) Answers will vary.
17. (a) Answers will vary.
(b) An optimal strategy will guarantee a tie.
19. She should always play row 1. Her expected value under row 1 is $\left(\frac{1}{2}\right)5 + \left(\frac{1}{2}\right)(-3) = 1$, while her expected value under row 2 is $\left(\frac{1}{2}\right)(-3) + \left(\frac{1}{2}\right)(1) = -1$.
21. (a) "Bet, then call" should be avoided by Player I.
(b) Player I: $\left(\frac{1}{3}, \frac{2}{3}, 0\right)$; Player II: $\left(\frac{2}{3}, 0, \frac{1}{3}\right)$; value $-\frac{1}{12}$
(c) Player II
23. When *A* succeeds in inducing *B* to think that the threat is real and, as a consequence, *B* defers to the threatener—without the threat being carried out.
25. (a) 50% chance of rain: leave umbrella; 75% chance of rain: carry umbrella
(b) Carry umbrella
(c) Saddlepoint at "carry umbrella" and "rain," giving value −2
(d) Leave umbrella

27. There are two equilibria: (Boxing, Boxing) and (Ballet, Ballet). These are sensible equilibrium outcomes, but they require some coordination between the two players.

29. The outcome (row 1, column 1) is an equilibrium. This is sensible because row 1 dominates row 2 and column 1 dominates column 2.

31. Your choice will depend on whether you put more value on obtaining a payoff of 4 while avoiding a payoff of 1 by choosing your first strategy, or "playing it safe" by never doing worse than a payoff of 2, and sometimes obtaining a payoff of 3, by choosing your second strategy.

33. ($250, $50, $50) is an equilibrium. Anneliese receives the stamp and pays $50. She cannot pay less for the stamp. If Binh or Charlie were to receive the stamp, then either would have to pay Anneliese's bid of $250, which exceeds each of their values of the stamp. ($150, $100, $50) is not an equilibrium. If Anneliese's and Charlie's bids are held at their respective values, then Binh would prefer to bid $151; he would receive the stamp and pay $150. ($400, $299, $1) is an equilibrium. Anneliese receives the stamp and pays $299; she cannot pay less. If Binh or Charlie were to receive the stamp, then either would have to pay Anneliese's bid of $400, which exceeds each of their values of the stamp.

35. The associated matrix is

		Even Player		
		2	**4**	**6**
	1	(2, 1)	(2, 1)	(2, 1)
Odd Player	**3**	(2, 4)	(6, 3)	(6, 3)
	5	(2, 4)	(4, 8)	(10, 5)

Even always does better playing 4 instead of 6. So, 6 is weakly dominated by 4. Once 6 is eliminated, Odd always prefers 3 to 1 and 3 to 5. So 1 and 5 are eliminated. After these eliminations, Even prefers playing 2 to 4. This yields the outcome of Odd playing 3 and Even player 2 as an equilibrium with payoffs of (2, 4). Another equilibrium has Odd playing 1 and Even playing 2. This game is similar to the Prisoners' Dilemma because both players can do better by playing nonequilibrium strategies (e.g., when Odd plays 5 and Even plays 6). It is different from the Prisoners' Dilemma because, although an outcome with the lowest payoff is an equilibrium, there is another equilibrium with a better payoff for Even.

37. In the sequential duel, the first player will shoot in the air because if the first player shoots another player, then he or she becomes the lone target for the remaining player. This choice is not optimal if the second player shoots in the air simultaneously. It would be better for Player 1 to shoot Player 3 instead so that Player 3 cannot shoot Player 1.

39. The following zero-sum game has no weakly dominated strategies.

$$\begin{bmatrix} 4 & 1 & -1 \\ 3 & 2 & 3 \\ 5 & 0 & 1 \end{bmatrix}$$

Player 1's maximin strategy is to play row 2, while Player 2's minimax strategy is to play column 2. However, no row or column can be eliminated by the successive deletion of dominated strategies.

41. The zero-sum payoff matrix from Example 1 has been transformed below by using ordinal payoffs.

(2, 7)	(3, 6)	(1, 8)
(6, 3)	(5, 4)	(7, 2)
(8, 1)	(2, 7)	(4, 5)

The (Suburban, 8 A.M.–4 P.M. shift) outcome is still an equilibrium. Mark has no incentive to change the hospital location (the row) because payoffs of 2 and 3 are the other two payoffs in column 2. Lisa has no incentive to switch the shift schedule because she would receive payoffs of 2 or 3, instead of the 4 for the 8 A.M. to 4 P.M. shift.

43. No, the analysis does not depend on which country goes first. This is the case because *A* is a dominant strategy.

Chapter 16

1. 3
3. 3
5. 5
7. 6
9. 1
11. 3
13. 8
15. 10%
17. 7
19. The lead digits 97 contribute 30 to the weighted sum. Thus, the digit needed to make the sum evenly divisible by 10 is the same if you leave the 9 and 7 out of the calculation. (For example, if the sum were 162, including the 97 digits, then the sum would be 132 if it didn't include them. In either case, the check digit is 8.)
21. 6
23. Yes, it is valid.
25. No. Because 11 contributes 3 to the running total, whereas 55 contributes 16, the running total for the changed number will end with 3 instead of 0.
27. The Luhn algorithm for a valid credit card results in a running total that ends with 0, so increasing the sum by 10 would still give a total that ends with 0. Thus, the new number is valid. Increasing the sum by 12 changes the last digit of the running total to 2, which is not a valid credit number.
29. 3
31. The check digit is 7. This check digit method detects all single-digit errors.
33. In the odd-numbered positions, an error caused by replacing an odd digit by an odd digit or an even digit replaced by an even digit is not detected.
35. It is the UPC code for Kleenex. Changing exactly one digit will not result in a valid UPC number.
37. The new check digit is $x - 4$ or $x + 6$, depending on whether x is greater than or equal to 4 or less than 4.
39. Because the difference, $114180 - 114150 = 30$, is a multiple of 15.

41. The same arguments in the text for multiplying by 13 apply when multiplying by 17.
43. 0-669-09325-4
45. If c_1 is the check digit for the weights 7, 3, 9, 7, 3, 9, 7, 3 and c_2 is the check digit for the weights 3, 7, 1, 3, 7, 1, 3, 7, then $c_2 = 0$ when $c_1 = 0$. Otherwise, $c_2 = 10 - c_1$.
47. The mistake of reading a 2 as a 7 is detected because the sum of the digits of the incorrect number would be odd. The mistake of reading a 2 as an 8 is not detected because the sum of the digits of the incorrect number would remain even. An error is detected when an odd digit is misread as an even one (or vice versa) because the sum of the digits changes from even to odd (or vice versa). Approximately 50% of errors are detected.
49. Because the remainder upon dividing by 9 is less than 9, 9 cannot be a remainder.
51. Because the remainder after dividing by 7 is less than 7, the digits 7, 8, and 9 cannot be a check digit.
53. Yes. The ISBN-10 scheme detects all transposition errors.
55. For the transposition to go undetected, it must be the case that the difference of the correct number and the incorrect number is evenly divisible by 11. That is, $(10a_1 + 9a_2 + 8a_3 + \cdots + a_{10}) - (10a_3 + 9a_2 + 8a_1 + \cdots + a_{10})$ is divisible by 11. This reduces to $2a_1 - 2a_3 = 2(a_1 - a_3)$ is divisible by 11. But $2(a_1 - a_3)$ is divisible by 11 only when $a_1 - a_3$ is divisible by 11 and this happens only when $a_1 - a_3 = 0$. In this case, there is no error. The same argument works for the fourth and sixth digits.
57. The combination 72 contributes $3 \cdot 7 + 2 = 23$ or $7 + 3 \cdot 2 = 13$ (depending on the location of the combination) toward the total sum, while the combination 27 contributes $3 \cdot 2 + 7 = 13$ or $2 + 7 \cdot 3 = 23$. So, the total sum resulting from the number with the transposition is still divisible by 10. Therefore, the error is not detected. Similarly, the combination 26 contributes $3 \cdot 2 + 6 = 12$ toward the total sum, whereas the combination 62 contributes $3 \cdot 6 + 2 = 20$ toward the total sum; so the new sum will not be divisible by 10. Similarly, when the combination 26 contributes $2 + 3 \cdot 6 = 20$ to the total, the combination 62 contributes $6 + 3 \cdot 2 = 12$ to the total. So, the total for the number resulting from the transposition will not be divisible by 10 and the error is detected. In general, an error that occurs by transposing *ab* to *ba* is undetected if and only if $a - b$ is 5 or -5.
59. Many single-digit errors go undetected. Substitution of *b* for *a* where *a* is 5 or -5 in positions 1, 5, 7, 9, and 11 is undetected; all errors in position 3 are undetected; substitution of *b* for *a* where $b - a$ is even in position 8 is undetected.
61. Because both numbers are valid, the difference of the weighted sums is divisible by 10. That is, $(7w + 3 + 2w + 1 + 5w + 6 + 7w + 4) - (7w + 3 + 2w + 1 + 5w + 6 + 6w + 1)$ is divisible by 10. The difference simplifies to $w + 3$. So, $w = 7$.
63. (a) 51593-2067; 2
    (b) 50347-0055; 1
    (c) 44138-9901; 1
65. 55811-2742; 22; 1
67. 2
69. If you replace each short bar in the bar code table by an *a* and replace each long bar in the bar code table by a *b*, the resulting strings are listed in alphabetical order.
71. Right to left
73. The Canadian scheme detects any transposition error involving adjacent characters. Also, because the Canadian code has 6 characters and there are more than 10 choices for the alphabetic characters, there are many more possible Canadian codes than U.S codes. Hence, the Canadian scheme can target a location more precisely.
75. H-000; L-000; S-000
77. $26 \cdot 7^3 = 8918$
79. 42758
81. By 40 if they are the same gender; by 540 if they are different genders.
83. The digits 03 indicate that the person was born in 1903 or 2003.
85. For a woman born in November or December, the formula $40(m - 1) + b + 600$ gives a number requiring four digits.
87. August 1, 1958
89. Since many people do not like to make their age public, this method is used to make it less likely that people would notice that the license number encodes year of birth.
91. Because of the short names and large population, there would be a significant percentage of people whose names would be coded the same.
93. S 530 for each name
95. The computer need not know which digit is the check digit because it merely checks to see if the weighted sum is divisible by 9 for traveler's checks and divisible by 10 for the other two.
97. The manufacturer's number and the product number are 6 digits long instead of 5. This bar code is the European article number bar code.

Chapter 17

1. See Table 17.1 for answer.
3. (a) 6
   (b) 3
5. 1001101
7. 1101000
9. (a) 0101011
   (b) 1011010
   (c) 1110001
11. 000000, 100011, 010101, 001110, 110110, 1[illegible]1101, 011011, 111000
13. 0000000, 1000001, 0100111, 0010101, 00[illegible]1110, 1100110, [illegible]11011, 1110011, 1010100, 1001111, 0110010, 0101001, 0[illegible] 1101000, 1011010, 0111100, 1111101. No, because 1000001 has weight 2.
15. 000000, 100101, 010110, 001011, 1100[illegible]11, 101110, [illegible] 011101, 111000. 001001 is decoded as 001011; 011000 is decoded as 111000; 000110 is deco[illegible]ded as 010110; 100001 is decoded as 100101.
17. 00000000, 00010111, 00101110, 0100[illegible]1011, [illegible]1010, 01100101, 10001101,11000110, 10100011, 1001[illegible]

01011100, 00111001, 11101000, 11010001, 10110100, 01110010, 11111111. The code will detect any three errors or correct any single error.

19. $2^5 = 32$; $2^8 = 256$
21. *AATAAAGCAA*
23. 111101000111001010; *AABAACAEADB*
25. T, N, and R; E
27. 13403 336 77 −49 53 −61 20 99 35 −17; we go from 59 characters to 36 characters, a reduction of almost 39%.
29. *A* is 0; *B* is 111; *C* is 10; *D* is 110.
31. *BCEEFCDDCFF*
33. (a) 2
    (b) 10
    (c) 9
35. UHWUHDW; ADVANCE
37. 13
39. There is no integer j such that $2j = 1$ modulo 26.
41. HURHUAR
43. $19x + 10$
45. VAADENWCNHREDEYA
47. MEET AT NOON
49. $4! = 4 \cdot 3 \cdot 2 \cdot 1 = 24$
51. DO NOT CALL
53. $6! = 6 \cdot 5 \cdot 4 \cdot 3 \cdot 2 \cdot 1 = 720$
55. Third above the row ATTACK
57. ROLLING STONES
59. WLZCL LZBL
61. SUN
63. (a) 1111101
    (b) 1100011
65. 0010110, 1110110, 1000110, 1011110, 1010010, 1010100, 1010111
67. Recall that the distance between two strings is the number of positions by which they differ. Observe that everywhere u and v agree, so do $u + w$ and $v + w$. Conversely, in every position that u and v disagree, so do $u + w$ and $v + w$.
69. a
71. *B* is the least likley, *J* is the second least likely, and *G* is the third least likley.
73. The difference $1328 \cdot 15 - 1326 \cdot 15 = 19{,}920 - 19{,}890 = 30$ is a multiple of 15.

Chapter 18

1. (a) None
   (b) 4 in. $\times \frac{4}{3} = 5\frac{1}{3}$ in.
   (c) 6 in. $\times \frac{3}{4} = 4\frac{1}{2}$ in.
3. (a) $1\frac{1}{3} \approx 1.33$
   (b) 1.78
   (c) 78%
5. 247 lb
7. (a) $\frac{1}{40} = 0.025$
   (b) 64,000 tines as large
   (c) 400 cm = 4.00 m = 13.1 ft
9. The writer usesboth multiplicative and subtractive language togetler. Better: "…one-thirtieth as common per capita.…"
11. The writer uses both multiplicative and subtractive language togethe. Better: "…one three-hundredth as much as.…"
13. Answers will vary.
15. $360,000
17. 225 (So it's mainly a "feel-good" icon.)
19. (a) $576
    (b) 105 MPGe
    (c) 118 MPGe
21. €1.05
23. 32 mpg
25. (a) 0.00013364 tons
    (b) That all parts of the scale model are made of the same materials as the real locomotive
    (c) 0.267 lb
    (d) 0.121 kg
    (e) 0.000121 metric tonnes
27. $4.85
29. 0.28 cm = 2.9 mm. You would barely be able to see it!
31. (a) 23 mph
    (b) 10 m/s
33. Assuming geometric scaling, $\sqrt[3]{21/7} \times 19 \approx 27$ in.
35. (a) 400,000 lb
    (b) 28 lb/in^2.
37. 6200 lb, or 2800 kg
39. The lights are strung around the outside of the tree branches, so in effect they cover the outside "area" of the tree (thought of as a cone). Hence, the number of strings needed grows in proportion to the square of the height: a 30-ft tree will need $5^2 = 25$ times as many strings as a 6-ft tree. However, you could also argue that a 30-ft tree is meant to be viewed from farther away, so that stringing the lights farther apart on the 30-ft tree would produce the same effect as with the shorter tree.
41. 9 ft 3 in. to 11 ft 9 in. (in modern times, there have been men over 9 ft tall); 282 cm to 358 cm.
43. You must multiply the value calculated from pounds and inches by 703.
45. Assuming geometric growth:
    (a) $\left(\frac{1}{2}\right)^3$ = one-eighth as much, or about 12 lb
    (b) $\frac{1}{2^3}$ as much, or about 29 ft
47. Sample response: Icarus would have been about as tall as an adult female ostrich; hence, if shaped like an ostrich, he would have had to fly about 100 mph.
49. The square of wingspan is proportional to wing area, so the wing loading is proportional to weight divided by square of wingspan. For the 200-pounder, that ratio is $200/(50^2) = 0.080$, while for the 100-pounder it is $100/(36^2) \approx 0.077$—close enough.
51. (a) Length = 800; surface area = 640,000; volume = 512,000,000
    (b) One eight-hundredth
    (c) There couldn't be any such giant ants.
53. $4^4 = 256$ times as long, or $256 \times 15 = 3840$ years, so $3840 - 15 = 3825$ more years.
55. $A \propto d^2$ and $A \propto M^{3/4} \propto (d^2h)^{3/4} = d^{3/2}h^{3/4}$, so $d^2 \propto d^{3/2}h^{3/4}$, hence $d^{1/2} \propto h^{3/4}$ and $d \propto h^{3/2}$.
57. (a) The accompanying graph shows birds along the lower line (in green) and planes along the upper line (in red).
    (b) Both relationships are allometric, since the results are good fits to straight lines whose slopes are not 1.

(c) The slope for birds is less steep than for planes.
(d) 64 lb

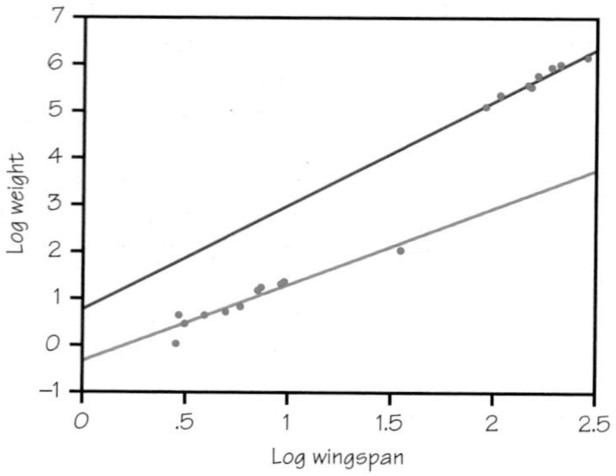

59. (a) 5.5
(b) About 33,000 lb
61. A small warm-blooded animal has a large surface-area-to-volume ratio. Pound for pound, it loses heat more rapidly than a larger animal and hence must produce more heat per pound, resulting in a higher body temperature.
63. (a) 326×10^6 gal
(b) 100 ft^3 (that is the actual definition of a water billing unit)
(c) 2.3×10^{-4} KAF
65. The arable land totals 667 million acres, which could produce 2.67×10^{11} gal; the 15 billion gal required would be $(15 \times 10^9)/(267 \times 10^9) \approx 6\%$ of the total. In 2014, actual acreage planted to field corn was 92 million acres; at 400 gal/acre, that would supply 37 billion gal, so the ethanol would require about 40% of the corn crop.

Chapter 19

1. 8, 13, and 21 for the usual pineapples in U.S. stores; 5, 8, and 13 or 6, 9, and 15 for other species.
3. Answers will vary but will be Fibonacci numbers.
5. (a), (b) The digits after the decimal point do not change.
(c) $\phi^2 = \phi + 1$
(d) $1/\phi = \phi - 1$
7. (a) 9
(b) 16
(c) Length $\ell = 6/(2 + 2\phi) \approx 1.15$, width $\phi\ell \approx 1.85$; or vice versa.
9. 8
11. (a) 4, 7, 11, 18, 29, 47, 76, 123
(b) 3, 1.333, 1.75, 1.571, 1.636, 1.611, 1.621, 1.617, 1.618. The ratios approach ϕ.
13. Answers will vary, but the ratio will always approach ϕ by alternating under- and overapproximations.
15. Answers will vary. Equality holds exactly when $x = y$.
17. 1, 2, 3, 5, 8, . . . , F_{n+1}
19. The seventh number is $5m + 8n$, and the total is $55m + 88n$.
21. (a) 1, 1, 3, 5, 11, 21, 43, 85, 171, 341, 683, 1,365
(b) $B_n = B_{n-1} + 2B_{n-2}$
(c) 1, 3, 1.667, 2.2, 1.909, 2.048, 1.977, 2.012, 1.994, 2.003, 1.999
(d) $x = 2, -1$; we discard the -1 root. $B_n = [2^n - (-1)^n]/3$
23. Silver mean: $1 + \sqrt{2} \approx 2.414$; bronze mean: $\frac{1}{2}(3 + \sqrt{13}) \approx 3.303$; copper mean: $2 + \sqrt{5} \approx 4.236$; nickel mean: $\frac{1}{2}(5 + \sqrt{29}) \approx 5.193$. General expression: $\frac{1}{2}(m + \sqrt{m^2 + 4})$
25. All are always true.
27. (a) B, C, D, E, H, I, K, O, X
(b) A, H, I, M, O, T, U, V, W, X, Y
(c) H, I, N, O, S, X, Z
29. (a) MOM, WOW; MUd and bUM reflect into each other, as do MOM and WOW.
(b) pod rotates into itself; MOM and WOW rotate into each other.
(c) Here are some possibilities: NOW NO; SWIMS; CHECK BOOK BOX; OX HIDE.
31. For all parts, translations.
(a) Reflection in the horizontal midline
(b) None other than translations
(c) Reflection in the horizontal midline, reflections in vertical lines through the centers of the Hs or between them, 180° rotation around the centers of the Hs or the midpoints between them, glide reflections.
(d) Reflections in vertical lines through the centers of the Ms or between them.
33. *d5*
35. (a) *d5* (Audi)
(b) *d20* without considering bolts, *d5* with bolts (BMW)
(c) *c9* (Renault)
37. (a) *c6*
(b) (CBS) *d2*
(c) (Dodge Ram) *d1*
39. (a) *c4*
(b) *d2*
41. (a) *p1a1*
(b) *p1m1*
(c) *p111*
(d) *p112*
(e) *pm11*
(f) *pma2*
(g) *pmm2*
43. (a) *pmm2*
(b) *p1a1*
(c) *pma2*
(d) *p112*
(e) *pmm2* (perhaps)
(f) *p1m1*
(g) *pma2*
(h) *p111*
45. (a) Patterns with vertical reflections are preferred on the Chinese pieces, while patterns with both horizontal and vertical reflections are strongly preferred on the Beghó pipes
(b) Neither culture completely excludes any strip type.

(c) (i) *pm11* or *pma2*: China. (ii) *p112*: China. (iii) *pmm2*: Begho. (iv) *pm11*: China. (v) *p1m1*: Begho. (vi) *pmm2*: Begho. (vii) *pmm2*: Begho. (viii) *pma2*: China. (ix) *p1a1*: China.

47. (c) Smallest rotation is 90°, there are reflections, there are reflections in lines that intersect at 45°: *p4m*.
(d) Smallest rotation is 90°, there are no reflections: *p4*.

49. (c) *p4m*
(d) Regarding the red/brown shapes as squares: *p4m*, if color is disregarded; otherwise *cmm*.

51. None of the five patterns with hexagonal symmetry can be realized, nor any of *p4g*, *p4*, *cm*, and *cmm*. The remaining eight can all be formed by the technique.

53. *pg*, if color is disregarded; otherwise, *p1*.

55. *p6*

57. *p3*

59. Answers will vary.

61. There is no identity element.

63.

	I	**F**	**R**	**T**
I	I	F	R	T
F	F	I	T	R
R	R	T	I	F
T	T	R	F	I

65. The notation of Example 6 uses *F* for flip, *R* (rotation) for spin. Start in summer in the upper left position of the figure for Example 6 and follow the arrows: The *F* in the fall brings the mattress to the position in the upper right, following that by *R* takes it to the position in the lower left, doing another *F* brings it to the position in the lower right, and yet another *R* brings it back to the original position at upper left.

67. Answers will vary.

69. Answers will vary. No, it is the cyclic group of order 4—e.g., the tire at position 1 goes to 3, then to 2, then to 4, and then back to 1.

71. There are four more derangements: CW = 2341, CCW = 4123, CWcross = 2413, and CCWcross = 3142, where CW stands for clockwise and CCW for counterclockwise.

73. (a) *d2*
(b) Any two of: *R* (180° rotation around the center), *V* (reflection in vertical line through its center), *H* (reflection in horizontal line through its center).
(c) $\{I, R, V, H\}$

75. There are four rotational symmetries (including the identity), two reflection symmetries, and two reflections across diagonal lines: $\{I, R, R^2, R^3, H, V, RH = VR, RV = HR\}$.

77. Answers will vary.

79. As in Example 6, number fixed positions, label with letters copies of the pattern elements in the positions, and pick a fixed position about which to make a half-turn *R*.
(a) $< T, R \mid R^2 = I, T \circ R = R \circ T^{-1} > = \{\ldots, T^{-1}, I, T^1, \ldots; \ldots, R \circ T^{-1}, R, R \circ T^1, \ldots\}$.
(b) $< T, R, H \mid R^2 = H^2 = I, T \circ H = H \circ T, R \circ H = H \circ R, (R \circ T)^2 = I > = \{\ldots, T^{-1}, I, T, \ldots; \ldots, R \circ T^{-1}, R, R \circ T, \ldots; \ldots, H \circ T^{-1}, H, H \circ T, \ldots; \ldots, R \circ H \circ T^{-1}, R \circ H, R \circ H \circ T, \ldots\}$.

81. $< R \mid R^8 = I > = \{I, R, R^2, R^3, R^4, R^5, R^6, R^7\}$, where *R* is a rotation by 45°.

83. There are four rotational symmetries (including the identity), three reflection symmetries, and an inversion through the center that swaps diagonally opposite corners.

85. The green branches have, in turn, smaller branches extending from them, and even smaller ones branch from those.

87. Answers will vary.

89. 8×13

91. *p1m1*

93. *p4*

95. The pattern *d3* is the pattern of symmetries of an equilateral triangle. Let *R* be a clockwise rotation of 120° about the center and let *V* be a reflection about the vertical line of symmetry. Then the elements of the group are $\{I, R, R^2, V, VR = R^2V, VR^2 = RV\}$. To convince yourself, label the vertices *A*, *B*, and *C*, and apply the symmetries.

Chapter 20

1. Exterior: 30°; Interior: 150°

3. $180° - \frac{360°}{n}$

5. The usual notation for a vertex type is to denote a regular *n*-gon by *n*, separate the sizes of polygons by periods, and list the polygons in clockwise order starting from the smallest number of sides, so that, for example, 3.3.3.3.3.3 denotes six equilateral triangles meeting at a vertex. The possible vertex types are 3.3.3.3.3.3, 3.3.3.3.6, 3.3.3.4.4, 3.3.4.3.4, 3.3.4.12, 3.4.3.12, 3.3.6.6, 3.6.3.6, 3.4.4.6, 3.4.6.4, 3.12.12, 4.4.4.4, 4.6.12, 4.8.8, 5.5.10, and 6.6.6.

7. 3.7.42, 3.9.18, 3.8.24, 3.10.15, and 4.5.20

9. At each of the vertices except the center one, six triangles meet, with angles (in clockwise order) of 75°, 75°, 30°, 30°, 75°, and 75°.

11. Yes

13. Three hexagonal faces meeting at at point would form a solid angle of $3 \times 120° = 360°$, hence would form a flat surface. The vertex type on an ordinary soccer ball is 5.6.6. The ball has 32 faces (of which 12 are pentagons and 20 are hexagons), 60 vertices, and 92 edges.

15. (a) No
(b) No
(c) No

17. Answers will vary.

19. No; no

21. The only way to tile by translations is to fit the outer "elbow" of one tile into the inner "elbow" of another. Labeling the corners as follows works: the corners on the top *A* and *B*, those on the rightmost side *C* and *D*, the middle of the bottom *E*, and the middle of the leftmost side *F*.

23. Answers will vary.
25. Place the skew-tetromino on a coordinate system with unit length for the side of a square and with the lower left corner at (0, 0). Then $A = (1, 2)$, $B = (3, 2)$, $C = (2, 0)$, and $D = (0, 0)$ works.
27. Place the skew-tetromino on a coordinate system with unit length for the side of a square and with the lower left corner at (0, 0). Then $A = (0, 1)$, $B = (2, 2)$, $C = (3, 2)$, $D = (3, 1)$, $E = (1, 0)$, and $F = (0, 0)$ works.
29. Answers will vary.
31. (a) Yes
(b) No
(c) No
33. See figure below.

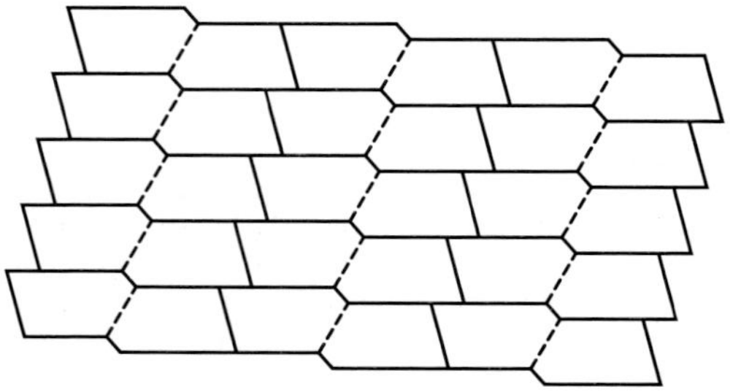

35. Answers will vary.
37. N, Z, W, P, y, I, L, V, X. See www.srcf.ucam.org/~jsm28/tiling/5-omino-trans.ps.gz.
39. Place the U on a coordinate system with unit length for the side of a square and with the lower left corner at (0, 0). Then $A = (2, 2)$, $B = (3, 2)$, $C = (3, 0)$, $D = (1, 0)$, $E = (0, 0)$, and $F = (0, 1)$ works.
41. Answers will vary.
43. (a) Consecutive segments, all of the same length; each length gives a different tiling.
(b) Let the lengths be S and L. Then the tilings are consecutive repetitions of any finite sequence of Ss and Ls.
45. *ABAABABA*
47. The two leftmost As would have had to come from two Bs in a row in the preceding month.
49. (a) *S; L; LS; LSL; LSLLS; LSLLSLSL; LSLLSLSLLSLLS; LSLLSLSLLSLLSLSLLSLSL*
(b) F_n segments at stage n
(c) $2m + n$
51. Let L_n and S_n be the number of Ls and the number of Ss at the nth stage. Each S at the nth stage (for $n \geq 3$) must have come from one L in the previous stage, so $S_n = L_{n-1}$. Similarly, we get an L at the nth stage from each S and each L in the previous stage, so $L_n = S_{n-1} + L_{n-1}$. Using both these facts together, we have $L_n = L_{n-1} + L_{n-2}$. We note that $L_1 = 0$, $L_2 = 1$, $L_3 = 1$, $L_4 = 2, \ldots$. The L_n sequence obeys the same recurrence rule as the Fibonacci sequence and starts with the same values one step later; in fact, it is always just one step behind the Fibonacci sequence: $L_n = F_{n-1}$. Consequently, $S_n = L_{n-1} = F_{n-2}$, and the total number of symbols is $S_n + L_n = F_{n-1} + F_{n-2} = F_n$.
53. In an inflated sequence, the only way an S enters is as an S preceded by an L, so two Ss in a row cannot occur in a musical sequence.
55. If a sequence ends in *LL*, its deflation ends in *SS*, which Exercise 53 shows is impossible for a musical sequence.
57. It suffices to show that every musical sequence is an initial subsequence of its inflation. Note that for any musical sequence except S, its inflation is longer; and for any except L, its deflation is shorter. Let M be the *shortest* musical sequence that is not an initial subsequence of its inflation I. Now by hypothesis M is not S; nor can M be L, because the inflation of L is LS. Since M is neither S nor L, it can be deflated and the deflation Q is shorter. Because Q is shorter than M, Q is an initial subsequence of M. But under inflation of M to I, the initial part of M that is Q inflates to all of M, followed by inflation of the rest of M. Hence, M is an initial subsequence of I, contrary to the supposition that it isn't.
59. (a) Yes
(b) No
61. If the sequence were periodic, the limiting ratio of Ls to Ss would be the same as the ratio in the repeating part, which would be a rational number, contrary to the result of Exercise 56.
63. $360°/n$
65. See page 839.
67. See Spotlight 20.6 (page 854).

Chapter 21

1. $51 \times \$10{,}000 \times \frac{.1299}{12} = \$5520.75$
3. $(2024 - 2014) \times \$10{,}000 \times .02726 = \$2726$
5. (a) 490 ppm
(b) Answers will vary.
7. (a) \$1200
(b) \$1200
(c) \$1205.41
(d) \$1207.20
9. Investment B
11. (a) \$2,032.79; \$2,025.82; \$2,012.20
(b) \$1,999.00; \$1,992.56
(c) \$1,973.82; \$1,906.62; \$1,849.60
(d) For small and intermediate interest rates, the rule of 72 gives good approximations to the doubling time.
13. (a) 698 days
(b) After 24 months
15. Answers will vary.
17. Answers will vary.
19. 0.702%
21. 4.8014%
23. \$5152.23
25. 3.68%
27. e
29. e^{-1}
31. In both cases, \$12.07, not taking into account any rounding to the nearest cent of the daily posted interest.
33. (a) We seek r for which $(1 + r)^2 = (1.00)(1.10)$, so $r = (1.1)^{1/2} - 1 = 4.88\%$.
(b) We solve $e^{2r} = 1.1$, getting $r = \frac{1}{2} \ln 1.1 = 4.77\%$.
35. It will be worth $(1 - .25) \times (1 + .25) \approx 94\%$ of the original value.

37. 34.7, 23.1, and 17.3 years; all close to the predictions of 36, 24, and 18 years.
39. \$81.75
41. \$67,092.02
43. \$1348.50
45. \$100,451.50
47. (a) \$199,149.07
    (b) \$32—but you must also pay tax on that \$32 (so $0.32 \times \$32 = \$10.24$), and tax on that \$10.24, and so forth. All in all, you pay $\$100(0.32 + 0.32^2 + 0.32^3 + \cdots) = \$100 \frac{.32}{1-0.32} = \$47.06$.
49. Using the Roth IRA, the entire \$199,149.07 calculated in Exercise 47a is yours tax-free at age 65. However, for the situation of Exercise 47, you will still owe tax on the interest earned: $32\% \times (\$199{,}149.07 - \$100 \times 12 \times 40) = 0.32 \times \$151{,}149.07 = \$48{,}367.70$, so your tax-free net is $\$199{,}149.07 - \$48{,}367.70 = \$150{,}781.37$. For the situation of Exercise 48, per the answer to part (c), you have \$199,150.67 tax-free.
51. (a) $A = P(1.10)(0.75)(1.25) = 1.03125P$, so $r = (1.03125)^{1/3} - 1 \approx 1.031\%$.
    (b) It is the effective rate.
53. (a) Exponential (decay)
    (b) Linear
    (c) Linear
    (d) Linear
55. $\$10{,}000/(1.05)^4 \approx \$8227.02$
57. $\$100{,}000/(1.08)^{22.5} \approx \$17{,}699.68$
59. (a) $\$(1.02)^3 \approx \$1.06$
    (b) $\$1/1.06121 \approx \$0.94$
61. $\frac{242.3}{30.2} \times \$10.75 \approx \$86.25$; answers about price of student's textbook will vary.
63. 2.14%, so 2% would be a better approximation than 3%.
65. $(0.05 - 0.10)/(1 + 0.10) \approx -4.55\%$
67. No more than \$35,000
69. Nowhere close. The equivalent in today's dollars would be about \$64,000.
71. (a) Nest egg: $\$100{,}000/0.072 \approx \$1{,}388{,}888.89$

    Monthly deposit: $\dfrac{(\$1{,}388{,}888.89)\frac{0.072}{12}}{\left(1 + \frac{0.072}{12}\right)^{420} - 1} \approx \$735.16$

    (b) $\$100{,}000/(1.03)^{35} \approx \$35{,}538.34$
    (c) $\$100{,}000/(1.03)^{63} \approx \$15{,}532.98$
73. 1.60% per month, or 19.2% annual rate
75. 0.634% per month, or 7.61% annual rate; you can get the same result in a spreadsheet with = 12*RATE(60,−275,0,20000,0,0.05/12). Since there is no exact formula for the interest rate, the spreadsheet uses a similar but more efficient method of successive approximation.
77. $\$40{,}000 \times (1.01)^{40} \approx \$60{,}000$
79. About \$0.47

Chapter 22

1. You will owe just \$7500 because you paid the interest as it accumulated.
3. The first loan accumulates interest of $\$5500 \times \frac{0.0429}{12} \times 51 \approx \$1002.79$, and the second loan accumulates interest of $\$6500 \times \frac{0.0429}{12} \times 39 = \$906.26$. Your total debt is $\$5500 + \$1002.79 + \$6500 + \$906.26 = \$13{,}909.05$, including a total of \$1909.05 in interest.
5. $\$15{,}000(1.049)^5 \approx \$19{,}053.23$
7. $\$200{,}000(1.06)^{30} \approx \$1{,}148{,}698.24$
9. (a) $0.1724/365 \approx 0.0472329\%$
   (b) $(1.000472329)^{365} - 1 \approx 18.81\%$
11. (a) \$205.258 (which would be rounded up to \$205.26, with the last payment less)
    (b) Ignoring the rounding up: $(120 \times \$205.258 - \$20{,}000) \approx \$4630.96$
13. The interest for the 15 months until start of repayment is $15 \times \$10{,}000 \times \frac{.0684}{12} = \$855.00$, so the starting principal when repayment begins is \$10,855.00. The monthly payment will be \$125.15.
15. \$508.31
17. \$319.03
19. \$536.82
21. \$727.22
23. Using the amortization formula "in reverse": \$8169.83; rounding interest at each payment: \$8171.48.
25. Using the amortization formula "in reverse": \$27,321.51; rounding interest at each payment: \$27,322.42.
27. (a) \$915.54
    (b) \$283.56
    (c) 10 months
29. (a) \$843.21
    (b) \$1062.22, for an initial balance of \$191,521.75
31. (a) Months, and billing periods, differ in their numbers of days. Also, the daily interest rate is $18\%/365 \approx 0.0493151\%$; compounded for a 30-day month, the monthly rate is then $1.000493151^{30} - 1 \approx 1.4901\%$.
    (b) 179 months, or almost 15 years. The first payment is \$31 with no interest due, the second payment is \$77. *Hint:* Put the principal in column A and the interest due in column B. For the interest rounded to the nearest penny, use =ROUND(A2*.015, 2); for the payment rounded down to the nearest dollar, use =MIN(A2 + B2, MAX(25, FLOOR(0.01*A2 + B2, 1))). Then adjust the last few months' interest charges by hand to be the minimum \$1.50.
    (c) \$6873.25; \$3755.42
33. Let r be the APR, with OB the old balance (after the preceding payment) and NB the new balance (after addition of interest for this period), and let the bill be for 30 days. The daily interest rate is $i = r/365$, and we have $\text{NB} = (1 + r/365)^{30}\text{OB}$. The interest is $[(1 + r/365)^{30} - 1]\text{OB}$. If the interest is greater than 0.02NB, a payment of 2% of NB will not keep up with the interest due. Solving

$$[(1 + r/365)^{30} - 1]\text{OB} > 0.02\text{NB} = 0.02(1 + r/365)^{30}\text{OB}$$

gives first $[(1 + r/365)^{30} - 1] > 0.02(1 + r/365)^{30}$, then $0.98[(1 + r/365)^{30}] > 1$, and $(1 + r/365)^{30} > 1/0.98$, so that $1 + r/365 > (1/0.98)^{1/30} \approx 1.00067$, yielding finally $r > 24.59\%$. Using a 31-day month gives $r > 23.79\%$.
35. (a) 210 months, or 17.5 years. The first payment is \$100 with no interest due, the second payment is

$172. *Hint:* With no late charges, we can neglect the provision about 4%. Put the principal in column A and the interest due in column B. For the interest rounded to the nearest penny, use =ROUND(A2*.025, 2); for the payment rounded to the nearest dollar, use =MIN(A2 + B2, ROUND(MAX(35, 0.02*A2, B2 + 0.01*A2), 0)) .
(b) $15,524.70

37. (a) 1.89% (The rate is actually 1.9%, with the $17.48 being rounded down from $17.484.)
(b) 1.91%

39. Answers will vary.

41. $128.75

43. $97.66

45. The payment on the add-on loan is always less than on the discounted loan:

$$\frac{P(1 + rt)}{12t} < \frac{P}{12t(1 - rt)}$$

if and only if

$$1 + rt < \frac{1}{1 - rt}$$

in turn (since $1 - rt > 0$) if and only if $(1 + rt)(1 - rt) = 1 - r^2t^2 < 1$, which is always true for $r > 0$.

47. 1361% if calculated as 7 days of 365 (1357% if calculated as 1 week of 52)

49. (a) 378%
(b) 379.72%: Only the 18% is compounded; mathematically, the "protection fee" is simple interest.

51. (a) $78/1122 \approx 6.95\%$
(b) $2 \times 57/1122 \approx 10.16\%$

53. (a) $294.24
(b) $243.59

55. 4.41%

57. $319,297.70

59. 16.8%

61. $760.03

63. 652% if calculated as 14 days of 365 (650% if calculated as 1 week of 52)

65. 0% interest: $379.84; 2.9% with rebate: $390.57

Chapter 23

1. Late 2024

3. 5.51 billion

5. 84% greater than Europe's population

7. (a) 138 years
(b) 58 years

9. (a) 2035: 9.3 billion; 2050: 11.1 billion
(b) No change in growth rate, no change in death rates, no global catastrophes, etc.

11. (a) 14 years
(b) 13 years
(c) Answers will vary.

13. (a) 107 years
(b) 83 years

15. (a) 419 years
(b) 96 years

17. (a) 51 years
(b) 131 years
(c) 224 years

19. (a) 139 years
(b) Forever!

21. That's about 5.5 tons/plant/year ≈ 30 lb/plant/day, which is unreasonable.

23. 2.78%.

25. 6.5 atoms per hour is one-fourth of 26 atoms per hour, so the sample is two half-lives (of 5730 years each) older than the sample in the text example and Table 23.1 (page 958): $29,000 + 2 \times 5730 \approx 40,000$ years. Using the formula $t = 55,403 - 8267 \ln N$ gives 40,253 years, which again—given the imprecision of the data of 6.5 atoms per hour—should be rounded to 40,000 years.

27. Two half-lives have passed, so about one-fourth of the strontium-90 remains.

29. 7.6 days

31. 0.2 days = 5 hr

33. 40 days

35. Answers will vary.

37. After the first year, the population stays at 15.

39. We calculate with all digits but report the numbers rounded to 1 decimal place: 7, 18.2, 6.6, 17.6, 8.4, 19.5, 2.0, 7.3, 18.6, 5.3.

41. We must have $f(x_n) = x_n$, or $4x_n(1 - 0.05x_n) = x_n$. The only solutions are $x_n = 0$ and $4(1 - 0.05x_n) = 1$, or $x_n = 15$.

43. We calculate with all digits but report the numbers rounded to 1 decimal place: 10, 15.0, 11.3, 14.8, 11.6, 14.6, 11.8, 14.5, 11.9, 14.4. The population is oscillating but slowly converging to $40/3 \approx 13.3$.

45. $x_n = 0, 40/3 \approx 13.3$

47. The red dashed line indicates the same size population next year as this year; where it intersects the blue curve is the equilibrium population size.

49. The population sizes are 11, 12.0, 12.6, 12.9, 13.1, 13.2, 13.3, 13.3, 13.3, 13.3.

51. The population sizes are 11, 15.0, 13.7, 14.9, 14.9, 15.0, 14.8, 14.3, 13.0, 9.5—and the following year the population is wiped out.

53. Equilibrium population is about 85 million pounds. Maximum sustainable yield (greatest difference between line and curve) is about 35 million pounds, leaving a population of 25 million pounds to reproduce for next year.

55. (a) The last entry shown for the first sequence is the fourth entry of the second sequence, so the first "joins" the second and they then both end up going through the same cycle (loop) of numbers over and over.
(b) 39, 78, 56, and we have "joined" the second sequence. However, an initial 00 stays 00 forever; and any other initial number ending in 0 "joins" the loop sequence 20, 40, 80, 60, 20, ...
(c) Regardless of the original number, after the second push of the key, we have a number divisible by 4 and all subsequent numbers are divisible by 4. There are 25 such numbers between 00 and 99. You can verify that an initial number either joins the self-loop 00 (the only such numbers are 00, 50, and 25); joins the loop 20, 40, 80, 60, 20, ... (the only such numbers are the multiples of 5 other than 00, 50, and 25); or joins the big loop of the other 20 multiples of 4.

57. (a) 133, 19, 82, 68, 100, 1, 1, ... The sequence stabilizes at 1.
(b) Answers will vary.
(c) That would trivialize the exercise!
(d) For simplicity, limit consideration to 3-digit numbers. Then the largest value of f for any 3-digit number is $9^2 + 9^2 + 9^2 = 243$. For numbers between 1 and 243, the largest value of f is $1^2 + 9^2 + 9^2 = 163$. Thus, if we iterate f over and over—say, 164 times—starting with any number between 1 and 163, we must eventually repeat a number, since there are only 163 potentially different results. And once a number repeats, we have a cycle. Thus, applying f to any 3-digit number eventually produces a cycle. How many different cycles are there? That we leave you to work out. *Hints:* (1) There aren't very many cycles. (2) There is symmetry in the problem, in that some pairs of numbers give the same result; for example, $f(68) = f(86)$.

59. (a) 0.0397, 0.15407173, 0.545072626, 1.288978, 0.171519142, 0.59782012, 1.31911379, 0.0562715776, 0.215586839, **0.722914301**, 1.32384194, 0.0376952973, 0.146518383, 0.521670621, 1.27026177, 0.240352173, 0.78810119, 1.2890943, 0.171084847, **0.596529312**
(b) 0.723, 1.323813, 0.0378094231, 0.146949035, 0.523014083, 1.27142514, 0.236134903, 0.777260536, 1.29664032, 0.142732915, **0.509813606**
(c) 0.722, 1.324148, 0.0364882223, 0.141958718, 0.507378039, 1.25721473, 0.287092278, 0.901103183, 1.16845189, **0.577968093**

61. Period 2 begins at $\lambda = 3$, period 4 at $1 + \sqrt{6} \approx 3.449$, period 8 at 3.544, period 3 at $1 + 2\sqrt{2} \approx 3.828$, and chaotic behavior onsets at about 3.57.

63. (a) 2035
(b) 2043

65. (a) 5.55×10^{-10}/yr
(b) 1.76×10^{-17}/sec
(c) 14/sec

67. 5%

Index

Note: Page numbers followed by f indicate figures; those followed by t indicate tables.